TABLES OF SPECTRAL LINES OF NEUTRAL AND IONIZED ATOMS

TABLES OF SPECTRAL LINES OF NEUTRAL AND IONIZED ATOMS

By A.R. Striganov and N. S. Sventitskii
Atomic Spectroscopy Laboratory
I. V. Kurchatov Institute of
Atomic Energy

Translated from Russian

Springer Science+Business Media, LLC 1968

Born in the Kalinin region in 1907 and a graduate of Moscow State University, Prof. Striganov is presently head of the Atomic Spectroscopy Laboratory of the I. V. Kurchatov Institute of Atomic Energy and a member of the Committee on Spectroscopy of the Academy of Sciences of the USSR. His published work deals with the development of methods of atomic emission spectrum analysis, the study of optical spectra, the hyperfine structure of the spectra of heavy elements, and development of the spectroscopic diagnosis of plasmas.

ISBN 978-1-4757-6612-7 ISBN 978-1-4757-6610-3 (eBook)
DOI 10.1007/978-1-4757-6610-3

The original Russian text was published by Atomizdat in Moscow in 1966 and has been corrected and updated by the authors for this edition. In order to make this volume as useful and convenient as possible, last-minute additions and corrections have been incorporated into the tables in their proper locations at the expense of such typographic niceties as page length and spacing. This procedure also accounts for the lettered—occasionally only partly filled—pages interpolated at various points.

А. Р. СТРИГАНОВ, Н. С. СВЕНТИЦКИЙ

ТАБЛИЦЫ СПЕКТРАЛЬНЫХ ЛИНИЙ
НЕЙТРАЛЬНЫХ И ИОНИЗОВАННЫХ АТОМОВ

TABLITSY SPEKTRAL'NYKH LINII
NEITRAL'NYKH I IONIZOVANNYKH ATOMOV

Library of Congress Catalog Card Number 68-28091

Originally published by Plenum US in 1968.
Softcover reprint of the hardcover 1st edition 1968

FOREWORD

Tables of Spectral Lines of Neutral and Ionized Atoms was first published in Moscow in 1966. All misprints and errors that have come to our attention have been corrected, and additions based on journal articles have been made for the Plenum Press edition. In particular, additions have been made in the tables for Li [4], C I [1], N I [1], N IV [12], and N V [14].

Such highly important spectra as those of N IV, N V, O IV, O V, and O VI in the visible and partially in the ultraviolet regions have, until recently, received almost no attention in the laboratory. The tables of these spectra include astrophysical data from B. Edlen (Z. Astrophys., 7:378, 1933) and C. E. Moore (A Multiplet Table of Astrophysical Interest, Part I, N.B.S., 1945) with rather rough estimates of the wavelengths of the spectral lines. But as the spectra of highly ionized atoms have been studied in the laboratory, these values have been determined more precisely, and we have striven to incorporate them in the American edition of the book. For the spectra of N IV and N V, we have employed the recent, comprehensive papers of R. Hallin (Arkiv for Fysik, 32:201, 1966; 31:511, 1966), in which the system of energy levels was refined and expanded, and many classified lines in the visible, ordinary ultraviolet, and vacuum ultraviolet regions are cited. In addition, our tables retain certain lines in the spectra of N IV, N V, O IV, and O V from the above-mentioned papers of Edlen and Moore, for which we calculated wavelengths according to the energy levels of Hallin. These lines are easily detected, since there are no intensity estimates in the second column of the tables.

We shall be very happy if the data in this book on the spectra of 22 elements are of use to physicists and engineers.

A. R. Striganov

April 10, 1968

Commission on Spectroscopy of the Academy of Sciences of the USSR, Moscow

PREFACE

Optical atomic spectroscopy finds ever-widening application in various fields of physics and technology. Spectroscopic methods have acquired great importance for the determination of the most important parameters of plasma in discharge tubes, in plasma apparatus, and in ion sources. The development of quantum generators is closely linked with the utilization of the results of optical spectroscopy, and it is therefore very important for investigators to have detailed data on the spectra of atoms and their ions. However, to date no books have been published, either here or abroad, containing sufficiently complete tabulations of the spectra of neutral and ionized atoms in all regions of the optical spectrum, and it has therefore been necessary to refer to individual journal articles, which has made work on the interpretation of spectra difficult.

Harrison's tables of spectral lines, although fairly complete, only cover the spectra of neutral and singly ionized atoms. The tables of A. N. Zaidel, V. K. Prokof'ev, S. M. Raiskii, and E. Ya. Shreider are mainly designed for spectral analysis, and only isolated very intense lines are given in the spectra of higher degrees of ionization. In Charlotte Moore's tables of multiplets the classification of the most intense lines of almost all the elements for many degrees of ionization is given, but these tables are not complete enough, and their arrangement according to multiplets makes work on the analysis of spectra difficult. Very complete tables on the far-UV region of the spectrum have recently been published by Kelly, but the classification of the lines is not given.

This book contains the emission spectra of 22 of the elements with which we most frequently have to deal in the investigations of plasmas in all their forms and modifications. For each spectral line the wavelength, intensity, and classification are given. The book contains the spectra of all the gaseous elements except radon, of all the alkali metals except rubidium, which is rarely met in nature, and the radioactive francium, and of carbon, magnesium, aluminum, silicon, calcium, titanium, iron, and copper, which sometimes are present in plasma as impurities. The choice of elements is to some extent arbitrary, and it might have been possible to include a few more if space had permitted.

For each element, in addition to the spectrum of the neutral atom, the spectra of several of its ions are given. In all, the tables contain over 30,000 lines in the IR, visible, near-UV and far-UV regions of the spectrum. In the compilation of the tables there is no demarcation of the region of the optical spectrum at either end.

These tables surpass all previously published tables in completeness for most of the elements (the exceptions are the spectra of Ca I and II, Ti I and II, and Fe I and II), and in particular the spectra of the gases have been greatly extended. Moreover, we give all necessary characteristics of the lines – their classification. In the examination of original papers we have come across many inaccuracies and errors in the classification of lines, and all the errors detected have been corrected.

In compilation of the tables for certain elements assistance was given by A. I. Odintsov, G. A. Odintsova, Z. I. Shlepkova, P. P. Gavrin, Yu. P. Dontsov, V. V. Eliseev, L. N. Kaporskii, and V. P. Kachalov. The general editing and checking of all the tables were done by Prof. A. R. Striganov.

CONTENTS

TABLES OF SPECTRAL LINES OF NEUTRAL AND IONIZED ATOMS

INTRODUCTION

The book consists of four main sections, in which spectral lines are assembled in the order of diminishing wavelength. In the first section the most intense lines are given for each element, and these are arranged according to the degrees of ionization of the atoms. The first column contains the wavelength in angstroms, the second the intensity of the line, and the third the energy of the upper level in electron-volts. The second section gives a summary table of the most intense lines arranged according to wavelength. Here the first column contains the wavelength in angstroms, the second the element to which it belongs and its degree of ionization, and the third the intensity of the line. The tables of the most intense lines in the first and second sections of the book are intended to facilitate rapid orientation to the spectrum under investigation and the detection of the main components and impurities in the plasma.

The third section gives complete tables of the spectral lines and their classification. As in the first section, the lines of elements are arranged in order of increasing atomic number, and in the table for each particular element, in order of increasing degree of ionization. In these tables the first column lists the wavelength in angstroms according to the most reliable literature sources, the second the intensity according to the same source, and the third and fourth the energies of the lower and upper levels in electron-volts. The relation 1 eV = 8066.1 cm^{-1} was used for conversion from cm^{-1} to eV. The fifth column lists the transition expressed by the usual spectroscopic symbols for the lower and upper levels. The sixth column gives the corresponding internal quantum numbers of the total angular momentum of the electron. The tables of this section contain not only classified lines, but also lines which have been shown to belong to the given degree of ionization of the atom but whose classification has not yet been established. For some elements there are fairly intense lines which cannot yet be definitely assigned to the neutral atom or one or another ion. Such lines are collected in supplementary tables of unidentified lines.

The accuracies with which wavelengths have been measured by different authors are different. Lines are found in the tables for which the wavelength stated has a possible error of more than 1 Å. For some elements lines are given whose wavelengths were measured accurately to the fourth decimal place by the interferometric method. Such lines can be used as secondary standards for the accurate measurement of wavelengths of lines in the spectra under investigation. All wavelengths of more than 2000 Å were measured in air; shorter wavelengths are the values in a vacuum.

Intensities of lines are taken directly from the original papers, so that in the spectra of different atoms, and also in different regions of the spectrum of a given atom, different intensity scales are met. In most of the spectra the 10-grade intensity scale is used with some extension beyond 10 for very bright lines and with the use of the symbols 0 and 00 for very weak lines. This scale is used in almost all the spectra in the far-UV region, and also in the spectra of highly ionized atoms. For neutral and singly ionized atoms a 1000-grade scale is used most frequently with some extension for bright lines. In a number of spectra the 100-grade scale is used. Only in a few cases, when different authors used different scales for the estimation of the intensities of the lines which they measured, the intensities of a small part of the lines have been adjusted to a common scale. This was done in the spectra of Ne I, Ar I, Kr I, and Xe I, as a result of which the intensities of lines in the IR, visible, and near-UV regions are all expressed on the 1000-grade scale. In the IR and visible regions of the spectrum of He I [2] we used a 10,000-grade scale. For convenience of comparison with lines in the UV region these intensities are halved. In the spectrum of F I [1] a scale is used which reaches 500,000 units for the very brightest lines. In this case, in all regions of the spectrum except the far-UV the intensities of the lines are diminished tenfold.

From this it will be seen that the comparison of the intensities of lines in different regions of the spectrum, and especially of spectra of different degrees of ionization and of different elements, must be approached with great caution. This is not entirely because some authors use different scales and often estimate intensities visually without allowing for the spectral sensitivity of the photographic plate. It is necessary to remember also that the relative intensities of lines depend greatly on the source in which the spectrum is excited and on its parameters. Lines of different degrees of ionization are particularly sensitive in this respect. It was not possible to describe the source and conditions of excitement for each of the spectra presented in this book. It may be stated, however, that for metals and carbon the spectra of neutral atoms were excited with the aid of an electric arc or a discharge tube with a hollow cathode, while the spectra of ionized atoms were obtained with a condensed spark. In the case of gases the spectra of neutral and singly ionized atoms were excited in a Geissler tube. The spectra of higher degrees of ionization were obtained with the aid of a powerful spark in a vacuum. More detailed information on this question will be found in the literature, which is given in the book for each element.

The classification of the lines is derived from the original papers and also from Moore's tables [5, 6]. All old notations of levels have been replaced by modern ones in accordance with Moore's book "Atomic Energy Levels," since the publication of which many new levels have been established and the classification of a number of spectra has been greatly extended (C I, C II, N II, O I, F I, Na I, Mg I, Mg II, AC I, Si I, Si II, Si III, Si N, Cl I, Ar II, K I, Ca III). All this has been taken into account in the present book.

In the spectral tables in Section III for H, He, Li, C, N, O, Ne, Na, Ar, K, Kr, Xe, and Cs all classified spectral lines are included, but weak unclassified

TABLE 1

Element	Degree of ionization								Unidentified lines	Total
	I	II	III	IV	V	VI	VII	VIII		
Hydrogen	150	–	–	–	–	–	–	–	–	150
Helium	153	62	–	–	–	–	–	–	–	215
Lithium	41	45	12	–	–	–	–	–	–	98
Carbon	482	500	327	81	13	3	–	–	–	1406
Nitrogen	576	565	304	112	74	4	–	–	–	1635
Oxygen	263	527	453	326	172	44	9	–	–	1799
Neon	1002	329	115	133	67	33	19	13	99	1810
Sodium	89	123	216	72	102	134	–	–	12	748
Argon	1012	1751	284	105	36	39	25	25	–	3277
Potassium	107	171	99	85	142	119	–	–	50	773
Krypton	645	1005	510	61	2	8	2	2	–	2235
Xenon	589	953	421	46	1	3	5	4	23	2045
Cesium	127	372	17	–	–	–	–	–	476	992

lines with intensities of I = 1, 0, or 00 on the 10-grade scale for which the transitions are unknown and for which it is unknown whether the spectrum belongs to the neutral atom or an ion are excluded. In the case of the spectrum of Ar II [11, 12], in view of its very large size classified lines with intensities I of 0.5 and 0 are also excluded. In the spectra of high degrees of ionization there is no abridgment. In Table 1 we give the numbers of lines in Section III for each of the elements listed and in some cases the number of unidentified lines which have not yet been assigned to the neutral atom or an ion. The table also gives the total number of lines for each element.

TABLE 2

Element	Degree of ionization						Unidentified lines	Total
	I	II	III	IV	V	VI		
Fluorine	438	255	281	208	113	61	–	1356
Magnesium	210	145	87	79	114	124	12	771
Aluminum	136	451	70	66	69	85	13	790
Silicon	438	311	323	118	13	63	–	1206
Chlorine	602	581	296	84	45	35	–	1643
Calcium	222	85	206	63	121	113	13	823
Titanium	1006	396	90	31	4	12	–	1539
Iron	1159	844	427	73	145	101	–	2749
Copper	619	972	291	–	–	–	151	2033

From the spectra of neutral and singly ionized atoms of F, Mg, Al, Si, Cl, Ca, Ti, and Cu weak classified lines with I = 1 or 0 on the 10-grade scale and unclassified lines with I = 2, 1, or 0 are excluded. From the spectrum of Ti I classified lines with I = 2 and unclassified lines with I = 4 or 3 are also excluded. From the spectra of Fe I, II, and III all lines with I = 4, 3, 2, 1, or 0 are excluded. Information on the numbers of lines given for these elements in this book is given in Table 2.

The fourth section of the book consists of a summary table of all lines arranged according to wavelength. For each line it is stated whether it belongs to the neutral atom or an ion, and its intensity is given.

The fifth section lists forbidden lines belonging to quadrupole and magnetic dipole transitions. Many of these lines are observed under ordinary conditions, and some are found in the spectra of the sun and the stars. In the same section the edges of the bands of the more important molecules are given, and there is a table of the wavelengths of lines in the molecular spectrum of hydrogen H_2.

Below, for each element we list the literature used in the compilation of the data of the tables.

Literature

LITERATURE

TABLES OF SPECTRAL LINES AND ENERGY LEVELS OF ATOMS

1. G.R. Harrison, Wavelength Tables, M.I.T. Press, New York, 1939.
2. W. F. Meggers, C. H. Corliss, and B. F. Scribner, Tables of Spectral-Line Intensities, Parts I, II, N.B.S., 1961.
3. A. N. Zaidel et al., Tables of Spectrum Lines, Pergamon Press, New York, 1961.
4. R. L. Kelly, A Table of Emission Lines in the Vacuum Ultraviolet for All Elements, California, 1961.
5. C. E. Moore, A Multiplet Table of Astrophysical Interest, Parts I, II, N.B.S., Washington, 1945.
6. C.E. Moore, An Ultraviolet Multiplet Table, Circular 488, N.B.S., Washington, Sect. 1, 1950 (H-V); Sect. 2, 1952 (Cr-Nb); Sect. 3, 1962 (Mo-Rd, without rare-earth elements); summary tables of spectral lines; Sect. 4, 1962 (H-Nb); Sect. 5, 1962 (Mo-La, Hf-Rd).
7. Landolt-Börnstein, Zahlenwerte und Funktion aus Physik, Chemie, Astronomie, Geophysik und Technik. Vol. I, Part I, Atome und Ionen, Sixth edition, Berlin, 1950.
8. C.E. Moore, Atomic Energy Levels, Circular 467, N.B.S., Washington; Vol. I, 1949 (H-V); Vol. II, 1952 (Cr-Nb); Vol. III, 1958 (Mo-La).
9. R. W. B. Pearse and A. G. Gaydon, Identification of Molecular Spectra [Russian translation], Moscow, Izd. inostr. lit., 1949.

SPECTRA OF INDIVIDUAL ELEMENTS

Hydrogen

1. C. E. Moore, A Multiplet Table of Astrophysical Interest, Parts I, II, N.B.S., Washington, 1945; An Ultraviolet Multiplet Table, Circular 488, Section I, N.B.S., Washington, 1950.
2. C. J. Humphreys, J. Res. Natl. Bur. Standards 50:1 (1953); classification (cl.) of H, IR region.

Helium

1. C. E. Moore, A Multiplet Table of Astrophysical Interest, Parts I, II, N.B.S., Washington, 1945; An Ultraviolet Multiplet Table, Circular 488, Section I, N.B.S., Washington, 1950.
2. W. C. Martin, J. Res. Natl. Bur. Standards 64A:19 (1960); cl. of He I, secondary standards, region 21132-320 Å.
3. G. Herzberg, Proc. Roy. Soc. A 248:309 (1958); cl. of He I, secondary standards, region 9210-537 Å.

Lithium

1. H. Kayser and H. Konen, Handb. Spektroskopie 7(3):979 (1934); cl. of Li I, Li II, region 74,500-170 Å.
2. R. W. France, Proc. Roy. Soc. A 129:354 (1930); cl. Li I, main series in absorption to n = 32.
3. K. W. Meissner, L. G. Mundie, and P. H. Stelson, Phys. Rev. 74:932 (1948); 75:891 (1949); interferometer measurements of λ.
4. I. Johansson, Arkiv. fys. 15(2):169 (1959); cl. of Li I, regions 26,878-7135 Å and 8126-3915 Å.
5. Y. G. Torresson and B. Edlen, Arkiv. fys. 23:117 (1963); lines of Li II.
6. A. Ericson and B. Edlen, Nature 125:233 (1930); Z. Phys. 59:656 (1930); cl. of Li II, far-UV region.
7. H. G. Gale and J. B. Hoag, Phys. Rev. 37:1703 (1931); cl. of Li III, region 730-100 Å.
8. E. Freytag, Naturwissenschaften 46:314 (1959); cl. of Li II, Li III, far-UV region.

Carbon

1. I. Johansson, Arkiv fys. 29(2):175 (1965); cl. of C I; region 25,843-11,619 Å; 31(3):201 (1966); cl. of C I, regions 11,331-2478 Å and 1994-945 Å.
2. S. Clad, Arkiv fys. 7(1–2):7 (1954); cl. of C II, region 8800-440 Å.
3. B. Edlen, Nova acta Reg. soc. sci. upsaliensis, 9:6, 31-153 (1934); cl. of C I–V, far-UV region.
4. K. Bockasten, Arkiv fys. 9(5):457 (1955); cl. of C III, region 9950-343 Å.
5. K. Bockasten, Arkiv fys. 10(6):567 (1956); cl. of C IV, region 9950-384 Å.
6. F. Z. Tyren, Phys. 98:768 (1935); cl. of C V and VI, region 40-30 Å.

Nitrogen

1. K. B. S. Eriksson, Arkiv fys. 19(2–3):235 (1961); cl. of N I, region 18,751-11,998 Å.
2. K. B. S. Eriksson, Arkiv fys. 13(5):429 (1958); cl. of N I, region 11,651-7424 Å.
3. O. S. Duffendack and R. A. Wolfe, Phys. Rev. 34:409 (1929); cl. of N I, region 6982-3437 Å.
4. E. Ekefors, Z. Phys. 63:437 (1930); cl. of N I, region 1900-1009 Å.
5. G. Herzberg, Proc. Roy. Soc. A, 248:309 (1958); cl. of N I, regions 11,323-6623 Å and 965-910 Å.
6. K. B. S. Eriksson, Arkiv fys. 13(4):303 (1958); cl. of N II, region 10,547-475 Å.
7. L. J. Freeman, Proc. Roy. Soc. A 121:318 (1928); cl. of N III, region 6488-374 Å.
8. B. Edlen, Z. Phys. 98:561 (1936); cl. of N III, region 2192-1906 Å.
9. B. Edlen, Nova acta Reg. soc. sci. upsaliensis, 9:6, 78, 62, 41 (1934); cl. of N III and IV, region 4500-180 Å, cl. of N V, far-UV region.
10. L. J. Freeman, Proc. Roy. Soc. A 127:330 (1930); cl. of N IV, region 7127-922 Å.
11. B. Edlen, Z. Astrophys. 7:378 (1933); cl. of N IV and V, region 7600-3100 Å.
12. R. Hallin, Arkiv fys. 32(2-3):201 (1966); cl. of N IV, region 7703-948 Å.
13. S. C. Tilford, J. Opt. Soc. Am. 53:1051 (1963); cl. of N V, region 300-100 Å.
14. R. Hallin, Arkiv fys. 31(6):511 (1966); cl. of N V, region 7618-425 Å.
15. F. Tyren, Nova acta Reg. soc. sci. upsaliensis (4) 12:1, 24 (1940); cl. of N VI, far-UV region.

Oxygen

1. K. B. S. Eriksson and H. B. S. Isberg, Arkiv fys. 24(6):549 (1963); cl. of O I, region 18,214-4654 Å.
2. B. Edlen, Kungliga Svenska Vetenskapsakademiens Handlingar, 20(10):31 (1943); cl. of O I, regions 13,163-2876 and 1358-448 Å.
3. R. Frerichs, Phys. Rev. 34:1239 (1929); 36:398 (1930); cl. of O I, region 8820-2884 Å.
4. C. Mihul, Ann. Phys. 9:294 (1928); cl. of O II, III, and IV, region 6910-2045 Å.

5. B. Edlen, Nova acta Reg. soc. sci. upsaliensis 9:6, 136, 115, 87, 62, 44 (1934); cl. of 0 II-VI, region 6606-104 Å.
6. B. Edlen, Z. Phys., 93:726 (1935); cl. of O II, III, and IV, regions 4489-740 Å, 3384-554 Å, and 3492-153 Å.
7. A. Fowler, Proc. Roy. Soc. A 117:317 (1928); cl. of O III, region 5592-1903 Å.
8. I.S. Bowen, Phys. Rev. 29:231 (1927); cl. of O II, region 4925-430 Å, cl. of O III and IV, far-UV region.
9. L. J. Freeman, Proc. Roy. Soc. A 127:330 (1930); cl. of O IV, region 4813-2449 Å.
10. F. Tyren, Nova acta Reg. soc. sci. upsaliensis, 12:1, 24 (1940); cl. of O VI and VII, 20 Å region.
11. B. Edlen, Arkiv fys. 4:441 (1952); cl. of O VII, 120 Å region.

Fluorine

1. K. Liden, Arkiv fys. 1(3):229 (1949); cl. and new levels of F I, regions 11,550-2970 and 980-680 Å.
2. H. Dingle, Proc. Roy. Soc. A 128:600 (1930); cl. of F II, regions 6000-2500 and 1750 470 Å.
3. Edlen, B., Z. Phys. 93:433 (1935); cl. of F II, regions 6000-2400 Å and 740-250 Å; cl. of F III, region 3350-2400 Å.
4. I. S. Bowen, Phys. Rev. 45 (II):82 (1934); cl. of F II and III, regions 608-353 and 3437-214 Å.
5. H. Dingle, Proc. Roy. Soc. A 122:144 (1929); cl. of F III, region 3270-530 Å.
6. B. Edlen, Z. Phys. 92:19 (1934); cl. of F IV, regions 3170-2170 and 680-140 Å.
7. B. Edlen, Z. Phys. 89:597 (1934); cl. of F V, region 2700-120 Å.
8. B. Edlen, Z. Phys. 94:47 (1935); cl. of F V and VI, region 2700-120 Å.
9. A. S. Kaufman, T. P. Hughes, and R. V. Williams, Proc. Phys. Soc. A 76:17 (1960); cl. of F VI, three lines in 2320 Å region.
10. B. Edlen, Z. Phys. 89:179 (1934); cl. of F VI and VII, region 1140-86 Å.
11. H. Flemberg, Arkiv mat. astr. fys. 28A:18 (1942); cl. of F VIII, far-UV region.
12. B. Edlen, Arkiv fys. 4(28)441 (1952); cl. of F VIII, far-UV region.
13. F. Tyren, Nova acta Reg. soc. sci. upsaliensis 12:1, 25 (1940); cl. of F VIII, far-UV region.

Neon

1. G. Hepner, Compt. rend. Acad. sci. colon. 248(8):1142 (1959); cl. of Ne I, region 25,000-19,570 Å.
2. C. J. Humphreys and H. J. Kostkowski, J. Res. Natl. Bur. Standards 49:73 (1952); cl. of Ne I, region 18,624-18,035 Å.
3. W.F. Meggers, J. Res. Natl. Bur. Standards 14:487 (1935); cl. of Ne I, region 12,690-10,562 Å.
4. W. F. Meggers and C. J. Humphreys, J. Res. Natl. Bur. Standards 10:427 (1933); cl. of Ne I, region 18,549-7724 Å.
5. F. Paschen, Ann. Physik 60:405 (1919); cl. of Ne I, region 9840-2550 Å.
6. B. Petersson, Arkiv fys. 27(4):317 (1965); Ne I, far- and near-UV regions.
7. L. Bloch, E. Bloch, and C. Dejardin, J. Phys. et radium 7:129 (1926); lines of Ne I-III, visible and UV regions.
8. W. F. Meggers and C. J. Humphreys, J. Res. Natl. Bur. Standards 13:293 (1934); cl. and secondary standards of Ne I, region 9665-4334 Å.
9. C. J. Humphreys, J. Res. Natl. Bur. Standards 20:17 (1938); cl. and secondary standards of Ne, region 3754-3369 Å.
10. K. Burns, K. B. Adams and J. Longwell, J. Opt. Soc. Am. 40:339 (1950); cl. and secondary standards of Ne I, region 8919-3126 Å.

11. S.A. Sullivan, J. Opt. Soc. Am. 45:1031 (1955); secondary standards of Ne I, region 8865-7059 Å.
12. P.G. Wilkinson and K.L. Andrew, J. Opt. Soc. Am. 53(6):710 (1963); secondary Standards of Ne I and II, far-UV region.
13. T. L. de Bruin and C.J. Bakker, Z. Phys. 69:19 (1931); cl. of Ne II, region 4922-353 Å.
14. T. L. de Bruin, Z. Phys. 77:505 (1932); cl. of Ne III, region 2825-2086 Å.
15. J.C. Boyce, Phys. Rev. 46:378 (1934); cl. of Ne I-IV, far-UV region.
16. J.C. Boyce and H.A. Robinson, J. Opt. Soc. Am., 26:133 (1933); lines of Ne II, far-UV region.
17. S. Frish, Z. Phys. 64:499 (1930); lines of Ne II, far-UV region.
18. F.W. Paul and H.D. Polster, Phys. Rev. 59:424 (1941); cl. of Ne IV-VI, far-UV region.
19. A.S. Kaufman, T.P. Hughes, and R.V. Williams, Proc. Roy. Soc. A76:17 (1960); cl. and levels of Ne IV-VI, region 2300-2200 Å.
20. B.C. Fawcett, B.B. Jones, and R. Wilson, Proc. Phys. Soc. A 78:1223 (1961); cl. of Ne VI-VIII, far-UV region.
21. K. Bockasten, R. Hallin, and T.P. Hughes, Proc. Phys. Soc. A 81:522 (1963); cl. and levels of Ne IV-VII, region 2300-430 Å.
22. S. Goldsmith and A.S. Kaufman, Proc. Phys. Soc. A 81:544 (1963); cl. and levels of Ne IV-VI, region 2400-2200 Å.
23. L.L. House and G.A. Sawyer, Astrophys. J. 139:775 (1964); cl. of Ne VII and VIII, far-UV region.

Sodium

1. P. Risberg, Arkiv fys. 10:583 (1956); cl. of Na I, region 23,380-2543 Å.
2. S.E. Frich, Z. Phys. 70:498 (1931); cl. of Na II, region 4455-2315 Å.
3. J.S. Bowen, Phys. Rev. 31:967 (1928); cl. of Na II, region 3711-2506 Å.
4. B.B. Vance, Phys. Rev. 41:480 (1932); cl. of Na II, region 302-269 Å.
5. D.H. Tombonlian, Phys. Rev. 54:347 (1938); cl. of Na III, region 2564-1100 Å.
6. J. Söderqvist, Nova acta Reg. soc. sci. upsaliensis, 9:39, 51 (1934); cl. of Na IV, region 412-129 Å.
7. J.C. Boyce and H.A. Robinson, J. Opt. Soc. Am. 26:133 (1936); lines of Na III, far-UV region.
8. J. Söderqvist, Arkiv mat. astr. fys., 32A(19):1 (1946); lines of Na V and VI, far-UV region.
9. J. Söderqvist, Arkiv mat. astr. fys., 30A(11):1 (1944); cl. of Na VII and VIII, far-UV region.

Magnesium

1. G. Risberg, Arkiv. fys. 28(5):381 (1965); cl. of Mg I, region 26,393-2025 Å.
2. R.A. Fisher and F.E. Eshbach, J. Opt. Soc. Am. 43(11):1030 (1953); cl. of Mg I, region 17,109-11,828 Å.
3. E.W. Selwyn, Proc. Phys. Soc. A 41:392 (1929); cl. of Mg I, region 12,083-1668 Å.
4. P. Risberg, Arkiv fys. 9(5):483 (1955); cl. of Mg II, region 11,620-824 Å.
5. J. Söderqvist, Nova a'cta Reg. soc. sci. upsaliensis 9:7, 22, 39 (1934); cl. of Mg III, region 2529-158 Å; cl. of Mg IV, region 1956-112 Å.
6. J. Söderqvist, Arkiv mat. astr. fys. 32A:4 (1946); cl. of Mg V and VI, region 403-72 Å.

Aluminum

1. K.B.S. Eriksson and H.B.S. Isberg, Arkiv fys. 23:527 (1963); cl. of Al I, region 21,164-1762 Å.
2. F. Paschen and R. Ritschl, Ann. Physik. 18:867 (1933); cl. of Al II, region 10,122-5971 Å.
3. R.A. Sawyer and F. Paschen, Ann. Physik 84:1 (1927); cl. of Al II, region 7471-954 A.

4. F. Paschen, Ann. Physik 12:509 (1932); cl. of Al II, region 1989-1834 Å.
5. E. Ekefors, Z. Phys. 51:471 (1928); cl. of Al III, region 1862-486 Å.
6. F. Paschen, Ann. Physik 71:142 (1923); cl. of Al III, region 5722-1352 Å.
7. J. Söderqvist, Nova acta Reg. soc. sci. upsaliensis, 9:34, 39, 51 (1934); cl. of Al IV-VI, far-UV region.
8. E. Ferner, Arkiv mat. astr. fys. 36A(1):48, 57 (1948); cl. of Al V and VI, far-UV region.

Silicon

1. U. Litzen, Arkiv. fys. 28(3):239 (1965); cl. of Si I, region 25,854-10,289 Å.
2. L. J. Radziemski and K. L. Andrew, J. Opt. Soc. Am. 55(5):474 (1965); cl. of Si I, region 12,270-1255 Å.
3. C.C. Kiess, J. Res. Natl. Bur. Standards 21:185 (1938); cl. of Si I, region 12,270-1565 Å.
4. A. G. Shenstone, Proc. Roy. Soc. A 261(1305):153 (1961); cl. of Si II, region 9412-843 Å.
5. C. E. Moore, Selected Tables of Atomic Spectra Si II, Si III, Si IV, N. B. S., Washington, 1965.
6. Y. G. Torresson, Arkiv fys. 18:389 (1961); cl. of Si III, region 9800-466 Å.
7. Y. G. Torresson, Arkiv fys. 17:179 (1960); cl. of Si IV, region 9018-457 Å.
8. E. Ferner, Arkiv mat. astr. fys. 28A:4 (1941); cl. of Si V and VI, region 249-65 Å.

Chlorine

1. C. J. Humphreys and E. Paul, J. Opt. Soc. Am. 49(12):1180 (1959); cl. of Cl I, region 25,324-6920 Å, new levels.
2. L. Minnhagen, J. Opt. Soc. Am. 51(3):298 (1961); cl. of Cl I, region 16,284-10,280 Å, new levels.
3. C. C. Kiess, J. Res. Natl. Bur. Standards 10:827 (1933); cl. of Cl I, region 10,428-3945 Å.
4. J. B. Green and J. T. Lynn, Phys. Rev. 69:165 (1946); cl. of Cl I, region 9192-3692 Å.
5. S. Avellen, Arkiv fys. 8:211 (1954); cl. of Cl I, region 1396-1189 Å.
6. C. C. Kiess and T. L. de Bruin, J. Res. Natl. Bur. Standards 23:443 (1939); cl. of Cl II, region 9483-558 Å.
7. I. S. Bowen, Phys. Rev. 45:401 (1934); cl. of Cl III, region 4972-406 Å; cl. of Cl IV, region 3168-319 Å, cl. of Cl V, region 555-286 Å.
8. I. S. Bowen, Phys. Rev. 31:34 (1928); cl. of Cl III, region 3851-573 Å; cl. of Cl IV, region 986-463 Å; cl. of Cl V, region 895-390 Å.
9. J. S. Bowen, Phys. Rev. 46:377(1934); cl. of Cl IV, region 757-332 Å.
10. L.W. Phillips and W.L. Parker; Phys. Rev. 60:301 (1941); cl. of Cl V and VI, region 580:236 Å.
11. W. L. Parker and L. W. Phillips, Phys. Rev. 57:140 (1940); cl. of Cl VI, region 737-195 Å.

Argon

1. M. C. de Hepner. Compt. rend. Acad. sci. colon 248(8):1142 (1959); cl. of Ar I, region 25,660-18,430 Å.
2. E. Paul and C. J. Humphreys, J. Opt. Soc. Am. 49(12):1186 (1959); cl. of Ar I, region 25,125-13,870 Å.
3. W. R. Sittner and E. R. Peck, J. Opt. Soc. Am. 39:474 (1949); cl. of Ar I, region 18,000-11,500 Å.
4. C. J. Humphreys and H. J. Kostkowski, J. Res. Natl. Bur. Standards 49:73 (1952); cl. of Ar I, region 17,000-12,000 Å.
5. W. F. Meggers, J. Res. Natl. Bur. Standards 14:487 (1935); cl. of Ar I, region 13,000-10,500 Å.

6. W. F. Meggers and C. J. Humphreys, J. Res. Natl. Bur. Standards 10:427 (1933); cl. of Ar I, region 12,000-7000 Å.
7. K.W. Meissner, Z. Phys. 39:172 (1926); 40:839 (1927); cl. of Ar I, regions 10,000-3056 and 1066-797 Å.
8. K. Burns and K.A. Adams, J. Opt. Soc. Am. 43(11):1020 (1953); cl. and secondary standards of Ar I, region 10,000-3170 Å.
9. T.A. Littlefield and D.T. Turnbull, Proc. Roy. Soc. A 218:577 (1953); cl. and secondary standards of Ar I, region 6538-3554 Å.
10. J. C. Boyce, Phys. Rev. 48(II):396 (1935); cl. of Ar I-V, region 2000-325 Å.
11. L. Minnhagen, Arkiv fys. 25(19):203 (1964); cl. of Ar II, region 11,253-2000 Å, new levels.
12. L. Minnhagen, Arkiv fys. 14(6):483 (1959); cl. of Ar II, region 1989-489 Å, new levels.
13. L. Minnhagen, Arkiv fys. 18(2):97 (1960); ionization potential of Ar II, 222,848, 2 cm^{-1}, new levels.
14. T.L. de Bruin, Proc. Koninkl. nederl. acad. wet. am. (II)36(7):724 (1933); (I)40(4):340 (1937); cl. of Ar II and III, region 4200-210 Å.
15. T. L. de Bruin, Zeeman Verhandelingen 413 (1935); cl. of Ar III, region 2750-2100 Å.
16. J. C. Boyce, Phys. Rev. 49:351 (1936); cl. of Ar III, far-UV region.
17. A. B. Rao, Indian J. Phys. 12:399 (1938); unclassified lines of Ar IV, region 5838-2300 Å.
18. T. L. de Bruin, Physica 3(8):809 (1936); cl. of Ar IV, region 3134-2300 Å.
19. L. W. Phillips and W. L. Parker, Phys. Rev. 60:301 (1941); cl. of ArV-IX, region 840-48 Å.
20. B. C. Fawcett, B. B. Jones, and R. Wilson, Proc. Phys. Soc. A 78:1223 (1961); cl. of Ar VI and VIII, far-UV region.

Potassium

1. P. Risberg, Arkiv fys. 10:583 (1956); cl. of K I, region 15,168-2992 Å.
2. T. L. de Bruin, Z. Phys. 38:94 (1926); cl. of K II, region 7699-1725 Å.
3. I. S. Bowen, Phys. Rev. 31:499 (1928); lines of K II, III, and IV, far-UV region.
4. T. L. de Bruin, Z. Phys. 53:658 (1929); cl. of K III, region 3515-2550 Å.
5. E. Ekerfors, Z. Phys. 71:53 (1931); lines of K II-VI, far-UV region.
6. B. Edlen, Z. Phys. 104:410 (1937); Phys. Rev. 62:434 (1942); cl. of K III and IV, far-UV region.
7. I.S. Bowen, Phys. Rev. 46:791 (1934); cl. of K IV and V, far-UV region.
8. H. A. Robinson, Phys. Rev. 52:725 (1937); cl. of K VI, far-UV region.
9. W.-Z. Tsien, Chinese J. Phys. 3:117 (1939); cl. of K III, IV, and V, far-UV region.
10. M. Ram, Indian J. Phys. 8:151 (1933); cl. of K III, IV, and VI, far-UV region.
11. A. E. Whitford, Phys. Rev. 46:793 (1934); cl. of K VI, VII, and VIII, far-UV region.
12. L. W. Phillips, Phys. Rev. 55:708 (1939); cl. of K VII, far-UV region.
13. W. L. Parker and L. W. Phillips, Phys. Rev. 57:140 (1940); cl. of K VIII, region 573-156 Å.

Calcium

1. C.J. Humphreys, J. Res. Natl. Bur. Standards 47:262 (1951); cl. of Ca I, region 22,651-12,815 Å.
2. C. E. Moore, A Multiplet Table of Astrophysical Interest, Parts I, II, N. B. S., Washington, 1945; An Ultraviolet Multiplet Table, Circular 488, Section 2, N. B. S., Washington, 1952.
3. B. Edlen and P. Risberg, Arkiv fys. 10(6):553 (1956); cl. of Ca II, region 21,428-1341 Å.
4. I. S. Bowen, Phys. Rev. 31(4):497 (1928); cl. of Ca III, region 4081-1271 Å.
5. E. Ekefors, Z. Phys. 71:53 (1931); lines of Ca III-VI, region 1030-135 Å.
6. W.-Z. Tsien, Chinese J. Phys. 3:117 (1939); cl. of Ca IV and V, region 766-228 Å.

7. I.S. Bowen, Phys. Rev. 46:791 (1934); cl. of Ca IV-VI, region 373-184 Å.

Titanium

1. H.N. Russell, Astrophys. J. 65:317 (1927); cl. of Ti I, region 11,970-2117 Å; cl. of Ti II; region 9252-1906 Å.
2. A.K. Wardcikee, J. Opt. Soc. Am. 45(5):354 (1955); cl. of Ti I, region 5929-5618 Å.
3. C.M. Wilson and M.P. Thekaekara, J. Opt. Soc. Am. 51(3):289 (1961); cl. of Ti I, region 3072-2117 Å, new levels.
4. H.N. Russell and R.J. Lang, Astrophys. J. 66:12 (1927); cl. of Ti III, region 4215-1002 Å; cl. of Ti IV, region 5492-423 A.
5. P.G. Kruger, S.G. Weissberg, and L.W. Phillips, Phys. Rev. 51:1090 (1937); cl. of Ti V, region 228-163 Å.
6. G. Racah, Phys. Rev. 61:537 (1942); levels of Ti V.
7. S.G. Weissberg and P. G. Kruger, Phys. Rev. 49:872 (1936); cl. of Ti VI, region 524-182 Å.

Iron

1. C.E. Moore, A. Multiplet Table of Astrophysical Interest, Parts I, II, N.B.S., Washington, 1945; An Ultraviolet Multiplet Table, Circular 488, Section 2, N.B.S., Washington, 1952.
2. B. Edlen and P. Swings, Astrophys. J. 95:532 (1942); cl. of Fe III, region 6500-500 Å.
3. S. Glad, Arkiv fys. 10:291 (1956); cl. of Fe III, regions 8600-1966 and 1611-1465 Å.
4. H.T. Gilroy, Phys. Rev. 38:2217 (1931); cl. of Fe IV, region 1825-574 Å.
5. P.G. Kruger and H.T. Gilroy, Phys. Rev. 48:720 (1935); cl. of Fe IV, region 587-525 Å.
6. I.S. Bowen, Phys. Rev. 52:1153 (1937); cl. of Fe V, region 1554-364 Å.
7. I.S. Bowen, Phys. Rev. 47:924 (1935); cl. of Fe VI, region 318-276 Å.

Copper

1. A.G. Shenstone, Philos. Trans. Roy. Soc. London A 241(832):297 (1948); cl. of Cu I, region 9717-1523 Å.
2. A.G. Shenstone, Philos. Trans. Roy. Soc. London A 235:195 (1936); cl. of Cu II, region 10,166-675 Å.
3. A.G. Shenstone and L. Wilets, Phys. Rev. 83(1):104 (1951); cl. of Cu III, region 2822-672 Å.

Krypton

1. E. Paul and C.J. Humphreys, J. Opt. Soc. Am. 49(12):1186 (1959); cl. of Kr I, region 25,200-20,200 Å.
2. C.J. Humphreys and H.J. Kostkowski, J. Res. Natl. Bur. Standards 49(2):73 (1952); cl. of Kr I, region 18,800-11,800 Å.
3. W.R. Sittner and E.R. Peck, J. Opt. Soc. Am. 39:474 (1949); cl. of Kr I, region 21,900-12,100 Å.
4. W.F. Meggers, J. Res. Natl. Bur. Standards 14:487 (1935); cl. of Kr I, region 21,900-12,100 Å.
5. W.F. Meggers and C.J. Humphreys, J. Res. Natl. Bur. Standards 10:427 (1933); cl. of Kr I, region 12,100-7600 Å.
6. W.F. Meggers, T.L. de Bruin, and C.J. Humphreys, J. Res. Natl. Bur. Standards 7:643 (1931); cl. of Kr I, region 9850-3180 Å.
7. W.Z. Gremmer, Phys. 73:620 (1932); cl. of Kr I, region 9850-5050 Å.
8. E.Z. Rasmussen, Phys 73:779 (1932); cl. of Kr I, region 9740-5430 Å.
9. W.F. Meggers and C. J. Humphreys, J. Res. Natl. Bur. Standards 13:293 (1934); cl. and secondary standards of Kr I, region 8930-4270 Å.

10. C.J. Humphreys, J. Res. Natl. Bur. Standards 20:29 (1938); cl. and secondary standards of Kr I, region 4800-3420 Å.
11. T.A. Littlefield, Proc. Roy. Soc. A 187:220 (1946); cl. and secondary standards of Kr I, region 4464-4274 Å.
12. J.C. Boyce, Phys. Rev. 47:718 (1935); cl. of Kr I-IV, far-UV region.
13. T.L. de Bruin, C.J. Humphreys, and W.F. Meggers, J. Res. Natl. Bur. Standards 11:409 (1933); cl. of Kr II, regions 10,659-2080 and 965-576 Å.
14. C.J. Humphreys, Phys. Rev. 47:712 (1935); cl. of Kr III, region 7353-2116 Å.
15. T.L. de Bruin, Zeeman Verhandelingen, p. 413 (1935); cl. of Kr III, region 7057-2100 Å.
16. A.B. Rao and S.G. Krishnamutry, Proc. Phys. Soc. A 51:772 (1939); cl. of Kr IV, region 3934-2237 Å.
17. B.C. Fawcett, B.B. Jones, and R. Wilson, Proc. Phys. Soc. A 78:1223 (1961); cl. of Kr V-VIII, far-UV region.

Xenon

1. B. Petersson, Arkiv fys. 27(4):317 (1965); cl. of Xe I, region 1469-1170 Å.
2. M.G. Hepner, Compt. rend. Acad. sci. colon. 242:1430 (1956); cl. of Xe I, region 21,000-10,300 Å.
3. W.R. Sittner and E.R. Peck, J. Opt. Soc. Am. 39:474 (1949); cl. of Xe I, region 22,000-12,000 Å.
4. C.J. Humphreys and H.J. Kostkowski, J. Res. Natl. Bur. Standards 49:73 (1952); cl. of Xe I, region 16,730-11,740 Å.
5. W.F. Meggers, J. Res. Natl. Bur. Standards 14:487 (1935); cl. of Xe I, region 12,620-10,550 Å.
6. C.J. Humphreys and W.F. Meggers, J. Res. Natl. Bur. Standards 10:139 (1933); cl. of Xe I, secondary standards, region 11,140-3340 Å.
7. W. Gremmer, Z. Phys. 59:154 (1930); cl. of Xe I, region 9920-3420 Å.
8. E. Rasmussen, Z. Phys. 73:779 (1932); cl. of Xe I, region 10,250-5530 Å.
9. W.F. Meggers and C.J. Humphreys, J. Res. Natl. Bur. Standards 13:293 (1934); cl. of Xe I, secondary standards, region 9920-3950 Å.
10. J.C. Boyce, Phys. Rev. 49:730 (1936); cl. of Xe I-III, region 1978-627 Å.
11. J.H. Ablink and H.B. Dorgelo, Z. Phys. 47:221 (1928); lines of Xe I and II, far-UV region.
12. C.J. Humphreys, J. Res. Natl. Bur. Standards 22:19 (1939); cl. of Xe II, region 10,220-2230 Å.
13. C.J. Humphreys, J. Res. Natl. Bur. Standards 16:639 (1936); cl. of Xe III, region 7653-2235 Å.
14. C.J. Humphreys, W.F. Meggers, and T. L. de Bruin, J. Res. Natl. Bur. Standards 23:683 (1939); cl. of Xe III, visible and UV region.
15. B.C. Fawcett, B.B. Jones, and R. Wilson, Proc. Phys. Soc. A 78:1223 (1961); cl. of Xe V-IX, far-UV region.

Cesium

1. F. Paschen, Ann. Physik. 33:717 (1910); cl. of Cs I, region 74,000-13,000 Å.
2. J. Seguier, Compt. rend. Acad. sci. colon. 255:489 (1962); cl. of Cs I, region 34,900-10,900 Å.
3. I. Johannsson, Arkiv fys. 20:135 (1961); cl. of Cs I, region 30,100-8520 Å.
4. H. Kleiman, J. Opt. Soc. Am. 52:441 (1962); cl. of Cs I, region 10,120-3880 Å.
5. K.W. Meissner, Ann. Physik. 65:378 (1921); lines of Cs I, region 8080-6010 Å.
6. K.W. Meissner and W. Weinmann, Ann. Physik. 29:758 (1937); cl. of Cs I, secondary standards.
7. H.R. Kratz, Phys. Rev. 75:1844 (1949); cl. of Cs I, main series in absorption.
8. A. Fowler, Report on Series in Line Spectra, p. 106, London, 1922, cl. of Cs I.

9. J. Olthoff and R. A. Sawyer, Phys. Rev. 42:766 (1932); cl. of Cs II, region 6960-610 Å.
10. M. A. Wheatley and R. A. Sawyer, Phys. Rev. 61:591 (1942); cl. of Cs II, region 5950-560 Å.
11. L. A. Sommer, Ann. Physik. 75:163 (1924); lines of Cs II, visible and UV regions.
12. G. J. Balasse, Phys. et radium 8:311 (1927); lines of Cs II, visible and UV regions.
13. R. Ricard, Compt. rend. Acad. sci. colon. 206:905 (1938); cl. of Cs II, region 5410-2150 Å.
14. R. Ricard, M. Givord, F. George, Compt. rend. Acad. sci. colon. 205:1229 (1937); cl. of Cs II, region 3750-1180 Å.
15. M. A. Fitzgerald and R. A. Sawyer, Phys. Rev. 46:576 (1934); cl. of Cs III.
16. W. W. Shaver, Trans. Roy. Soc. Canada 18:23 (1924); lines of Cs in far-UV region.
17. W. Finkelnburg and W. Humbach, Naturwissenschaften 42:35 (1955); ionization potential of Cs III.

Forbidden Lines

1. I. S. Bowen, Astrophys. J. 121:306 (1955); classification of lines of N, O, F, Ne, S, Cl, Ar, K, Ca, Mn, Fe.
2. C. E. Moore, A Multiplet Table of Astrophysical Interest, Parts I, II, N. B. S., Washington, 1945, cl. of lines of Be, C, N, O, F, Ne, Na, Mg, Al, Si, P, S, Cl, Ar, K, Ca, Se, Ti, V, Cr, Mn, Fe, Co, Ni, Cu, Kr, Sr, Y, Zr, Xe, La, Eu.

Section I

Most Intense Lines for Each Element Arranged According to Degree of Ionization

λ, Å	I	E_B, eV

HYDROGEN, Z = 1

Ionization potential 13,597 eV

λ, Å	I	E_B, eV
40511,4	120	13,06
18751,1	700	12,75
12818,05	140	13,06
H_α 6562,793	3000	12,09
H_β 4861,332	500	12,75
H_γ 4340,468	200	13,06
H_δ 4101,737	100	13,22
1215,670	3000	10,20
1025,722	1000	12,09
972,537	400	12,75
949,743	200	13,06

DEUTERIUM, Z = 1

Ionization potential 13,601 eV

λ, Å	I	E_B, eV
18746,0	700	12,75
12814,56	140	13,06
D_α 6561,032	3000	12,09
D_β 4860,029	500	12,75
D_γ 4339,287	200	13,06
D_δ 4100,621	100	13,22
1215,340	3000	10,20
1025,443	1000	12,09
972,272	400	12,75
949,485	200	13,06

TRITIUM, Z = 1

Ionization potential 13,603 eV

λ, Å	I	E_B, eV
18744,3	700	12,75
12813,40	140	13,06
T_α 6560,435	3000	12,09
T_β 4859,595	500	12,75
T_γ 4338,893	200	13,06
T_δ 4100,249	100	13,22
1215,229	3000	10,20
1025,350	1000	12,09
972,184	400	12,75
945,401	200	13,06

HELIUM, Z = 2

He I, ionization potential 24,586 eV

λ, Å	I	E_B, eV
20581,30	10000	21,22
18696,94	1500	23,74
18685,96	3600	23,74
17002,38	1800	23,73
10830,337	25000	20,96
10830,248	15000	20,96
10829,088	5000	20,96
7065,190	2500	22,72
6678,151	1000	23,07
5875,966	1000	23,07
5875,621	7500	23,07
4471,479	1000	23,73
3888,648	5000	23,01
3187,745	200	23,71
2945,106	100	24,03
584,334	500	21,22
537,030	200	23,09
522,213	80	23,74

He II, ionization potential 54,414 eV

λ, Å	I	E_B, eV
6560,099	100	52,90
5411,524	50	53,30
4685,682	300	51,01
3203,104	200	52,24
2733,32	100	52,90
2511,22	50	53,30
1640,474	10	48,37
1640,332	5	48,37
1215,171	5	51,01
303,783	500	40,81
256,317	150	48,37
243,027	70	51,01

LITHIUM, Z = 3

Li I, ionization potential 5,391 eV

λ, Å	I	E_B, eV
26877,82	8	3,83
24464,66	6	4,34
18703,09	7	4,54
17546,05	7	4,54
6707,84	1000	1,85
6103,64	500	3,88
4602,86	100	4,54
3232,66	50	3,83
2741,20	10	4,52

Li II, ionization potential 75,635 eV

λ, Å	I	E_B, eV
5484,7	10	61,28
5037,8	6	72,10
4788,8	8	72,23
4677,7	8	72,23
3249,8	5	73,46
3199,43	7	73,46
2767,0	4	74,12
2730,7	5	74,12
1756,0	5	69,28

λ, Å	I	E_B, eV
1682,4	4	69,59
1653,9	8	68,78
1493,7	6	69,58
1198,6	7	69,37

Li III, ionization potential 122,446 eV

λ, Å	I	E_B, eV
729,1	—	108,84
135,02	—	91,94
113,93	—	108,84

CARBON, Z = 6

C I, ionization potential 11.259 eV

λ, Å	I	E_B, eV
14542,50	179	8,54
14420,12	61	9.71
11754,76	144	9,70
11753,32	142	9,70
11748,22	82	9,69
9405,73	16	9,00
9094,83	12	8,85
8335,15	13	9,17
7115,19	9	10,38
7113,18	9	10,38
6014,85	9	10,70
6013,22	10	10,70
6006,03	9	10,71
5380,34	10	9,99
5052,17	8	10,14
4932,05	8	10,20
4771,75	8	10,08
2967,244	5	4,18
2582,901	5	7,48
2478,556	16	7,68
1930,905	10	7,68
1657,008	10	7,49
1561,435	20	7,95
1560,691	15	7,95
1277,551	10	9,71
1277,282	9	9,71
1193,252	10	10,40

C II, ionization potential 24,381 eV

λ, Å	I	E_B, eV
7236,42	20	18,05
7231,32	18	18,04
6582,88	15	16,33
6578,05	18	16,33
5891,59	12	20,15
5889,77	15	20,15
5151,09	13	20,71
5145,16	15	23,12
5143,49	12	23,11
5133,28	12	23,12
5132,94	12	23,11
4267,258	20	20,95
4267,003	18	20,95
3920,693	18	19,49
3918,978	15	19,49
2992,618	18	22,19
2837,603	18	16,33
2836,710	20	16,33
1335,684	14	9,29
1334,515	13	9,29
1037,017	13	11,96
1036,330	12	11,96
1010,369	10	17,61
1010,074	10	17,61
904,468	10	13,71
904,134	12	13,72
903,950	11	13,71
903,609	10	13,72
687,355	11	18,05
687,059	10	18,04

C III, ionization potential 47,881 eV

λ, Å	I	E_B, eV
8500,32	10	32,10
8332,99	7	41,33
5695,92	12	34,28
4651,47	11	32,19
4650,25	13	32,19
4647,42	14	32,19
4325,560	8	41,30
4186,900	9	42,97
4070,261	9	42,96
4068,912	9	42,96
4067,940	8	42,96
2982,106	8	38,43
2725,90	7	44,46
2725,30	7	44,46
2724,85	6	44,46
2697,75	7	43,99
2296,870	16	18,09
2162,944	9	40,01
977,026	18	12,69
574,279	12	34,28
538,312	13	29,53
538,150	12	29,53
538,075	11	29,53
535,288	10	41,25
511,527	10	42,32
459,633	15	33,47
459,521	14	33,47
459,462	13	33,47
386,203	14	32,10

C IV, ionization potential 64,489 eV

λ, Å	I	E_B, eV
5811,98	9	39,68
5801,33	10	39,68
4658,30	9	58,44
2906,29	5	60,05
2530,6	6	55,78
2529,98	11	55,78
2524,41	9	55,78
2405,10	6	55,78
2404,44	5	55,78
1550,771	19	7,99

λ, Å	I	E_B, eV
1548,185	20	8,01
419,714	14	37,55
419,525	13	37,55
384,178	17	40,28
384,032	16	40,28
312,455	14	39,68
312,418	15	39,68
289,143	9	50,87
244,907	10	50,62

C V, ionization potential 392,067 eV

λ, Å	I	E_B, eV
248,744	0	354,24
248,668	0	354,24
40,731	—	304,38
40,270	—	307,87
34,973	—	354,49
33,426	—	370,90

C VI, ionization potential 489,946 eV

λ, Å	I	E_B, eV
33,734	—	367,44
28,464	—	435,51
26,988	—	459,33

NITROGEN, Z = 7

N I, ionization potential 14,548 eV

λ, Å	I	E_B, eV
13581,33	1200	11,60
13429,61	670	11,60
12469,62	1350	13,00
12461,25	680	12,99
10114,644	13	12,99
10112,483	12	12,99
10108,893	11	13,00
10105,130	10	12,86
8711,708	15	11,75
8683,400	16	11,76
8680,270	17	11,76
8629,238	16	12,12
8594,005	15	12,12
8216,317	15	11,84
7915,419	7	13,92
7898,985	8	13,92
7468,309	16	11,99
7442.299	15	11,99
7423,639	14	11,99
6723,12	9	13,69
6484,88	9	13,67
6482,74	9	13,68
6008,48	10	13,66
5564,37	9	13,99
5560,37	9	14,00
4935,03	10	13,20
4151,46	12	13,32
4109,959	12	13,70
4099,951	9	13,70
3830,39	9	13,92
3822,07	6	13,92
1494,668	60	10,68
1492,817	30	10,69
1492,624	80	10,69
1411,939	30	12,36
1319,684	30	12,97
1200,711	30	10,33
1199,549	50	10,34
1101,293	40	13,64
1100,362	30	13,65
1098,264	40	13,67
1098,103	40	13,68
1097,245	50	13,69
1096,749	35	13,69
1096,322	35	13,69
1095,940	35	13,70
1069,984	30	13,94
1068,476	35	13,99
1067,607	35	14,00
965,042	10	12,85
953,658	15	13,00
953,415	15	13,00
906,426	15	13,68
906,202	10	13,68

N II, ionization potential 29,611 eV

λ, Å	I	E_B, eV
10065,15	7	27,44
10035,45	7	27,45
10023,27	8	27,45
9969,34	7	27,44
9891,09	7	27,42
6610,565	13	23,47
6482,053	13	20,41
5941,653	12	23,24
5931,779	11	23,24
5710,766	10	20,64
5686,213	10	20,64
5679,562	14	20,66
5676,019	11	20,64
5666,627	12	20,66
5010,620	10	20,94
5007,325	11	23,41
5005,149	14	23,14
5001,477	12	23,13
5001,136	11	23,12
4994,363	10	23,42
4643,085	11	21,15
4630,543	14	21,15
4621,394	10	21,15
4607,157	10	21,15
4601,480	11	21,16
4447,033	12	23,19
4041,311	11	26,21
3994,998	15	21,60
3955,851	10	21,60
3918,999	9	23,57
3838,374	8	24,39
3437,147	9	22,10
3328,730	7	24,39

λ, Å	I	E_B, eV
3006 ,830	7	24 ,53
2522 ,227	7	26 ,07
2520 ,791	6	26 ,06
2317 ,046	8	26 ,01
2316 ,690	6	25 ,99
2316 ,493	7	26 ,00
1844 ,4	10	25 ,19
1085 ,701	12	11 ,43
1085 ,542	9	11 ,44
1084 ,572	11	11 ,44
1083 ,990	10	11 ,44
916 ,700	12	13 ,54
916 ,004	11	13 ,54
915 ,955	10	13 ,54
915 ,603	10	13 ,54
775 ,957	12	17 ,88
746 ,976	7	18 ,50
671 ,391	8	18 ,47
660 ,280	9	20 ,67
645 ,167	10	19 ,23
644 ,825	9	19 ,23
644 ,621	8	19 ,23

N III, ionization potential 47,436 eV

λ, Å	I	E_B, eV
4641 ,90	7	33 ,13
4640 ,64	10	33 ,13
4634 ,16	8	33 ,13
4514 ,89	7	38 ,40
4379 ,09	10	42 ,54
4103 ,37	9	30 ,46
4097 ,31	10	30 ,46
3771 ,08	7	38 ,96
3754 ,62	6	38 ,96
3374 ,06	6	39 ,34
3367 ,36	7	39 ,35
2983 ,58	6	42 ,49
2862 ,26	6	44 ,04
2063 ,99	10	48 ,13
2063 ,50	10	48 ,12
991 ,579	17	12 ,52
991 ,514	14	12 ,52
989 ,790	16	12 ,52
772 ,385	12	23 ,16
771 ,901	11	23 ,16
771 ,544	10	23 ,16
764 ,357	15	16 ,24
763 ,340	14	16 ,24
686 ,335	14	18 ,08
685 ,816	16	18 ,10
685 ,513	15	18 ,08
684 ,996	14	18 ,10
452 ,226	11	27 ,44
451 ,869	10	27 ,44
374 ,441	12	33 ,13
374 ,204	11	33 ,13

N IV, ionization potential 77,466 eV

λ, Å	I	E_B, eV
6380,77	8	50,15
3484,96	13	50,33
3482,99	14	50,33
3478,71	15	50,34
2646,956	12	68,73
2646,176	11	68,73
2645,654	10	68,73
1718,551	20	23,42
955,335	20	29,18
924,283	14	21,76
923,675	14	21,75
923,220	16	21,78
923,057	14	21,76
922,519	14	21,76
921,992	14	21,78
765,148	15	16,20
335,050	11	53,20
283,579	12	52,07
283,470	11	52,07
283,420	10	52,07

N V, ionization potential 97,881 eV

λ, Å	I	E_B, eV
4944,56	9	90,94
4619,98	10	59,23
4603,73	12	59,24
2981,31	10	88,44
2980,78	8	88,44
1242,804	19	9,97
1238,821	20	10,00
266,378	9	56,55
266,197	8	56,55
247,709	7	60,06
247,564	6	60,05

N VI, ionization potential 552,04 eV

λ, Å	I	E_B, eV
29 ,084	—	426 ,27
28 ,787	—	430 ,67
24 ,898	—	497 ,94
23 ,771	—	521 ,55

OXYGEN, Z = 8

O I, ionization potential 13,617 eV

λ, Å	I	E_B, eV
13165 ,11	24	11 ,93
13164 ,85	26	11 ,93
13163 ,89	25	11 ,93
8446 ,758	29	10 ,99
8446 ,359	30	10 ,99
8446 ,250	27	10 ,99
7775 ,388	26	10 ,74
7774 ,166	27	10 ,74
7771 ,943	28	10 ,74
7254 ,529	17	12 ,70
7254 ,447	20	12 ,70
7254 ,154	19	12 ,70
7002 ,228	17	12 ,76
7001 ,915	15	12 ,76
6455 ,975	19	12 ,66
6454 ,445	18	12 ,66
6453 ,602	17	12 ,66

λ, Å	I	E_B, eV
6158,183	21	12,75
6158,183	21	12,75
6156,765	20	12,75
6155,975	19	12,75
3954,387	10	14,12
3947,489	7	12,28
3947,301	10	12,28
3823,469	10	15,78
3692,440	7	12,88
1306,025	25	9,52
1304,866	30	9,52
1302,173	30	9,52
1040,941	15	11,93
1039,233	20	11,93
1027,433	20	12,09
988,776	15	12,54
O II, ionization potential 35,146 eV		
4705,355	8	28,88
4676,234	8	25,65
4661,635	9	25,64
4649,139	10	25,66
4641,811	9	25,65
4596,174	8	28,36
4590,971	9	28,36
4416,972	8	26,22
4414,909	10	26,25
4349,426	8	25,85
4319,631	8	25,85
4317,139	8	25,84
4189,788	10	31,32
4185,456	8	31,32
4119,221	8	28,86
4075,868	10	28,71
4072,164	8	28,69
3973,263	10	26,55
3911,960	10	28,83
3749,49	9	26,30
3727,33	8	26,30
3712,75	7	26,30
3470,81	8	29,25
3407,38	7	32,15
3390,25	8	28,94
3377,20	7	28,95
3138,44	8	29,60
3134,82	10	29,62
2575,300	10	29,06
2571,476	8	29,06
2530,30	8	29,06
2445,55	10	28,51
2433,538	9	28,91
2300,35	8	28,83
2293,32	6	28,82
932,046	10	—
919,78	15	—
834,462	15	26,30
833,326	15	26,30
832,754	14	26,30
796,661	10	20,58
718,562	16	20,58
718,484	17	20,58
672,948	8	23,44
644,148	12	28,12
539,547	8	22,98
539,086	8	23,00
538,256	10	26,36
537,830	9	28,38
O III, ionization potential 54,934 eV		
5592,37	6	36,07
5268,06	2	41,26
3961,59	8	41,14
3759,87	9	36,48
3754,67	7	36,45
3265,46	10	40,27
3260,98	8	40,25
3047,13	8	37,25
2983,78	9	38,01
2686,14	10	46,62
2874,57	8	54,01
2665,69	7	54,01
2597,69	8	45,62
2390,44	8	41,26
2382,32	7	45,47
835,292	16	14,88
835,096	14	14,88
833,742	16	14,88
832,927	14	14,88
703,850	18	17,65
702,899	17	17,65
702,822	16	17,65
599,598	18	23,19
597,818	15	26,09
525,795	18	26,09
508,182	18	24,43
507,683	17	24,43
434,975	10	33,86
374,075	10	33,18
345,309	10	41,26
328,448	10	40,26
320,979	12	41,14
305,769	10	40,58
305,656	9	40,58
303,799	9	40,87
O IV, ionization potential 77,394 eV		
5305,3	15	62,18
4799,2	10	—
4798,25	5	61,94
4786,4	20	—
4783,43	4	61,94
3071,66	5	48,37
3063,46	6	48,38
2836,25	6	58,79
2517,40	7	59,35
2509,23	8	59,36
2507,77	7	59,33
2501,80	8	—
2499,29	6	59,33

λ, Å	I	E_B, eV
2493,75	10	59,36
2493,40	7	59,35
2450,06	10	—
2449,36	8	—
790,103	13	15,74
787,710	15	15,74
779,905	10	31,63
625,852	14	28,67
625,130	14	28,67
624,617	13	28,67
609,829	15	20,38
608,395	14	20,38
555,262	16	22,38
554,514	18	22,41
554,074	17	22,38
553,328	16	22,41
279,937	11	44,34
279,633	10	44,34
260,389	10	63,35
238,573	15	52,01
238,361	14	52,01

O V, ionization potential 113,873 eV

λ, Å	I	E_B, eV
6830	8	87,34
4158,8	0	87,79
4123,9	2	84,04
2789,86	3	72,28
2787,03	4	72,28
2781,04	5	72,29
1371,287	10	28,73
762,001	10	26,51
761,130	10	26,49
760,445	12	26,54
760,229	10	26,51
759,440	10	26,51
758,677	10	26,54
629,732	15	19,69
220,352	13	75,95
207,794	10	86,12
207,794	10	88,39
192,906	14	74,50
192,800	13	74,50
192,751	12	74,50
172,163	12	72,01

O VI, ionization potential 138,080 eV

λ, Å	I	E_B, eV
3834,24	1	82,58
3811,35	2	82,60
1037,613	9	11,95
1031,912	10	12,01
184,117	9	79,35
183,937	8	79,35
173,082	13	83,64
172,935	12	83,64
150,124	9	82,58
150,088	10	82,60
129,872	6	107,48
129,786	5	107,47

O VII, ionization potential 739,114 eV

λ, Å	I	E_B, eV
128,500	0	665,14
128,412	00	665,14
120,331	00	664,07
21,804	—	568,59
21,602	—	573,91

FLUORINE, Z = 9

F I, ionization potential 17,422 eV

λ, Å	I	E_B, eV
8900,92	1000	15,90
8298,581	2000	15,88
8274,615	1500	15,88
8230,773	3000	15,88
8214,726	2500	15,88
8040,931	1000	14,53
7800,212	15000	14,61
7754,696	18000	14,58
7398,688	10000	14,37
7311,019	15000	14,68
7202,360	15000	14,75
7127,890	30000	14,76
7037,469	45000	14,75
6902,475	15000	14,53
6870,215	8000	14,55
6856,030	50000	14,50
6834,264	9000	14,54
6773,984	7000	14,53
6413,651	8000	14,68
6348,508	10000	14,68
6239,651	13000	14,68
977,745	100	12,73
976,217	100	12,75
973,895	350	12,73
958,524	500	12,98
955,545	750	13,02
954,825	1000	12,98
951,871	500	13,02
809,599	125	15,36
806,964	150	15,36

F II, ionization potential 34,985 eV

λ, Å	I	E_B, eV
4447,18	12	31,56
4446,71	10	31,56
4299,177	10	29,55
4246,16	15	31,58
4103,525	15	28,77
4103,085	10	28,77
4025,495	15	25,75
4025,010	10	25,75
4024,727	20	25,75
3851,667	10	25,12
3849,987	15	25,12
3847,086	20	25,12
3505,614	15	28,66
3503,095	12	28,66
3501,416	10	28,66

λ, Å	I	E_B, eV
3264,16	7	33,01
3202,740	10	30,53
3153,492	6	32,10
3059,960	8	29,17
3058,141	7	29,17
3057,083	6	29,17
608,065	7	20,46
607,472	6	20,47
606,805	8	20,46
606,284	6	20,49
605,668	7	20,47
546,846	6	22,67
514,945	6	26,66
484,600	8	28,17
471,990	6	26,27
457,177	6	29,71

F III, ionization potential 62,659 eV

λ, Å	I	E_B, eV
3264,164	9	—
3174,725	10	44,13
3174,125	12	44,18
3121,515	12	43,30
3115,669	10	43,27
3042,808	10	46,72
2916,335	10	43,58
2860,308	9	49,02
2811,422	10	44,69
2788,093	20	44,67
2759,589	10	47,14
2484,360	9	44,32
658,337	12	18,83
656,878	11	18,87
656,125	10	18,90
567,737	9	26,06
567,676	10	26,06
508,384	10	30,78
465,113	10	33,05
464,284	9	33,09
430,154	11	33,05
429,511	10	33,09

F IV, ionization potential 87,157 eV

λ, Å	I	E_B, eV
2826,13	5	56,10
2820,74	4	56,05
2456,92	5	61,15
2451,58	4	61,10
2298,29	5	57,11
2171,44	4	58,23
679,217	16	18,33
679,003	13	18,33
677,224	15	18,33
677,154	13	18,34
676,130	14	18,34
572,637	16	21,73
571,384	15	21,73
571,302	14	21,73
570,636	14	21,73
490,997	16	28,38
490,566	13	31,91
430,758	15	31,91
420,727	16	29,54
420,041	15	29,54
419,644	14	29,54

F V, ionization potential 114.237 eV

λ, Å	I	E_B, eV
2707,17	2	81,68
2450,63	2	70,11
2252,72	2	82,60
466,995	5	26,64
465,978	7	26,70
465,374	6	26,64
464,370	5	26,70
190,839	7	65,06
190,571	6	65,06
186,842	5	77,10
178,434	5	88,43
166,177	10	74,70
165,983	9	74,69
163,558	5	86,54
148,002	5	83,86
134,539	5	92,24

F VI, ionization potential 157,151 eV

λ, Å	I	E_B, eV
2327,28	5	97,97
2323,35	7	97,98
2315,39	9	98,00
535,204	10	23,16
156,247	6	102,51
139,900	7	100,70
139,800	6	100,69
139,758	5	100,69
126,923	5	97,67

NEON, Z = 10

Ne I, ionization potential 21,564 eV

λ, Å	I	E_B, eV
23636,3	205	20,19
18390,10	180	20,81
18385,17	160	20,81
18282,58	200	20,71
18276,59	260	20,71
11177,59	300	19,66
11143,09	300	19,69
10844,54	200	19,78
10562,43	200	20,14
9665,424	1000	19,66
9534,167	500	20,02
9486,680	500	19,69
9425,38	500	20,02
9326,52	600	20,04
9300,85	600	20,04
9201,76	600	20,04
9148,68	600	20,05
8865,7562	500	20,03
8853,8669	700	20,04

λ, Å	I	E_B, eV
8783,7539	1000	20,14
8780,6223	1200	20,05
8681,9216	500	20,04
8679,4898	500	20,14
8654,3837	1500	20,14
8634,6472	600	20,05
8495,3591	500	20,03
8377,6062	800	20,03
8300,3248	600	20,05
7245,1665	1000	18,38
7173,9380	1000	18,57
7032,4128	1000	18,38
6929,4672	1000	18,63
6598,9529	1000	18,72
6402,2460	2000	18,55
6382,9914	1000	18,61
6334,4279	1000	18,57
6266,4950	1000	18,69
6217,2813	1000	18,61
6163,5939	1000	18,72
6143,0623	1000	18,63
6074,3377	1000	18,71
6029,9971	1000	18,72
5975,5340	600	18,69
5881,8950	1000	18,72
5852,4878	2000	18,96
5764,4188	700	20,70
5400,5616	2000	18,96
5343,2834	600	20,70
5341,0938	1000	20,70
5330,7775	600	20,71
4957,0335	1000	21,11
4884,9170	1000	21,11
4827,3444	1000	21,95
4788,9270	1000	21,14
4715,3466	1500	21,18
4712,0660	1000	21,18
4710,0669	1000	21,01
4708,8619	1200	21,01
4704,3949	1500	21,02
4537,7545	1000	21,11
3593,5263	500	20,30
3520,4714	1000	20,37
3472,5706	500	20,19
3417,9031	500	20,30
3369,9069	700	20,30
3369,8076	500	20,30
3057,388	300	20,90
2982,663	300	20,77
2974,714	300	20,78
2675,64	200	21,30
2675,24	200	21,30
2647,42	200	21,30
743,721	12	16,67
735,892	30	16,85
629,729	6	19,69
626,819	6	19,78
618,668	5	20,04

λ, Å	I	E_B, eV
615,623	5	20,14

Ne II, ionization potential 41,079 eV

λ, Å	I	E_B, eV
4428,54	6	37,55
4409,30	7	37,65
4397,94	6	37,55
4391,94	7	37,62
4379,50	6	37,63
4290,40	6	37,62
4219,76	6	37,54
3777,16	8	30,55
3766,29	8	30,52
3734,94	7	30,55
3727,08	9	31,18
3713,084	10	31,12
3709,64	7	30,57
3694,197	10	30,52
3664,112	9	30,55
3542,90	7	34,86
3355,05	7	30,93
3334,87	10	30,88
3323,75	7	31,51
3297,74	7	30,93
3218,21	8	34,74
2955,73	7	31,36
1938,827	8	34,25
1930,033	8	34,28
1916,081	10	34,25
1907,494	8	34,28
462,388	14	26,91
460,725	15	26,91
447,813	8	27,78
446,252	8	27,78
407,136	8	30,55
405,852	9	30,55

Ne III. ionization potential 63,742 eV

λ, Å	I	E_B, eV
2678,64	25	44,23
2677,90	30	44,23
2613,41	12	48,53
2610,03	15	48,53
2595,68	20	43,72
2593,60	30	43,72
2590,04	40	47,73
2412,94	12	49,37
2412,73	15	49,37
2263,21	12	54,01
2216,07	15	54,12
2213,76	12	54,13
2163,77	15	49,46
2095,54	20	54,16
2092,44	12	54,16
2089,43	15	54,17
491,050	9	25,33
490,310	7	25,40
489,501	10	25,33
488,868	7	25,44
488,103	8	25,40
379,308	7	35,89

λ, Å	I	E_B, eV
Ne IV, ionization potential 97,044 eV		
2384,95	7	64,66
2373,21	9	—
2372,16	7	64,62
2357,96	10	64,72
2352,52	8	64,66
2285,79	9	68,86
543,891	150	22,79
542,073	100	22,87
469,865	200	31,47
469,817	200	31,47
421,609	150	37,11
388,218	100	39,64
387,141	125	39,73
358,721	200	39,64
212,556	150	63,44
208,734	100	59,40
208,485	100	59,47
Ne V, ionization potential 126,287 eV		
2265,71	6	79,55
2263,39	3	79,40
2259,57	3	79,45
2245,48	3	92,05
2232,41	4	92,15
2227,42	3	92,09
572,336	80	21,80
569,830	30	21,81
482,987	50	25,81
416,198	80	33,54
365,594	100	37,67
359,385	50	34,63
358,472	50	34,63
Ne VI, ionization potential 157,94 eV		
2253,22	3	109,20
2055,93	3	95,62
2042,382	3	95,66
562,805	15	22,19
403,262	10	30,91
401,939	25	31,01
401,138	15	30,91
122,686	10	101,22
122,520	20	101,22
Ne VII		
1992,060	3	127,51
1981,974	6	127,54
465,21	10	26,65
106,2	7	—
106,1	7	—
Ne VIII		
780,324	4	15,87
770,409	8	16,09
98,2	9	—
98,1	9	—
88,1	9	—

SODIUM, Z = 11

λ, Å	I	E_B, eV
Na I, ionization potential 5,139 eV		
22056,44	300	3,75
14767,48	1155	4,59
11403,78	12	3,19
11381,45	11	3,19
10834,87	8	4,76
10749,29	9	4,34
10746,44	10	4,34
9961,281	7	4,86
8649,922	7	4,62
8194,8237	9	3,62
5895,9236	16	2,10
5889,9504	32	2,10
5688,2046	9	4,28
4545,186	8	4,83
4541,633	7	4,83
4497,658	11	4,86
4494,177	10	4,86
4423,246	7	4,91
4393,340	9	4,93
4390,029	8	4,93
4324,615	7	4,97
3302,979	18	3,75
3302,369	19	3,75
2853,013	15	4,34
2852,811	16	4,34
Na II, ionization potential 47,30 eV		
3631,266	8	36,35
3533,043	10	36,35
3285,603	8	37,09
3129,368	6	36,90
3092,729	10	36,85
3053,664	6	41,27
2984,183	7	41,12
2951,231	8	41,05
2904,914	7	37,21
2893,946	6	41,25
2881,140	6	37,24
2841,721	7	37,21
2671,829	6	40,99
2660,996	7	41,01
2611,815	7	41,10
2531,548	6	41,25
376,375	3	32,94
372,069	6	33,32
Na III, ionization potential 71,66 eV		
2563,32	25	50,40
2553,61	25	50,36
2497,05	50	50,36
2474,69	40	50,40
2468,86	30	51,47

λ, Å	I	E_B, eV
2459,40	45	51,36
2309,96	30	51,69
2297,14	25	51,10
2296,64	25	51,87
2285,72	35	51,87
2278,48	40	51,89
2251,44	45	51,10
2246,66	40	51,02
2239,43	45	51,10
2232,17	40	51,87
2230,30	50	50,95
2225,90	45	51,07
2202,78	40	51,02
2011,88	30	57,52
2005,24	30	57,65
1985,58	30	51,75
1965,04	18	57,26
1960,76	20	57,40
1951,21	40	51,75
1946,43	20	60,98
1933,87	30	57,43
1926,27	45	57,46
1856,73	20	57,08
1850,39	18	57,10
1850,24	20	61,32
1849,58	35	57,06
1844,36	20	57,08

Na IV, ionization potential 98,902 eV

λ, Å	I	E_B, eV
412,240	8	30,21
411,333	7	30,34
410,540	6	30,34
410,371	10	30,21
409,615	8	30,40
408,682	8	30,34
319,638	10	42,64
190,835	8	65,10
190,440	10	65,10
181,758	8	68,21
168,409	8	73,75
168,084	10	73,76
162,445	8	80,18
156,536	8	79,21

Na V, ionization potential 138,627 eV

λ, Å	I	E_B, eV
463,263	12	26,76
461,051	10	26,89
459,897	7	26,96
445,190	7	36,84
400,722	10	36,84
360,367	8	43,39
360,319	8	43,39
333,910	9	46,12
332,550	8	46,26
308,264	10	46,12
307,152	8	46,26

Na VI, ionization potential 172,36 eV

λ, Å	I	E_B, eV
491,340	6	25,32
489,580	5	25,32
417,595	6	29,92
415,505	4	29,92
366,110	4	59,17
362,444	4	43,41
361,250	8	38,59
317,641	6	43,41
313,748	5	39,75
311,921	4	39,75
127,837	4	101,36
124,153	4	100,10
124,059	4	100,10
123,929	5	100,27
109,896	5	117,20
107,683	5	115,36
107,608	4	115,30
107,288	4	115,78

MAGNESIUM, Z = 12

Mg I, ionization potential 7,645 eV

λ, Å	I	E_B, eV
15024,99	35	5,93
11828,18	45	5,39
10811,085	35	7,09
5711,0880	30	6,52
5528,4047	40	6,59
5183,6042	45	5,11
5172,6843	44	5,11
5167,3216	42	5,11
4702,9909	30	6,98
3838,293	40	5,95
3832,302	38	5,95
3829,355	36	5,95
2852,127	50	4,35
1827,97	8	6,78
1747,81	5	7,09

Mg II, ionization potential 15,034 eV

λ, Å	I	E_B, eV
10951,78		10,00
10914,23	10	10,00
10092,16	14	12,86
9632,435	11	12,86
9631,888	12	12,86
9340,544	10	14,18
9327,545	10	14,18
9244,266	13	10,00
9218,248	14	10,00
8835,082	11	13,49
8824,323	10	13,49
8745,657	11	13,50
8734,990	10	13,50
8234,639	11	11,50
7896,368	13	11,57
7877,051	12	11,57
6545,973	11	13,52
6346,737	10	13,52
4481,327	13	11,63
4481,130	14	11,63

λ, Å	I	E_B, eV
4390,564	10	12,82
2936,509	10	8,65
2802,704	12	4,42
2797,998	10	8,86
2795,528	13	4,43
2660,817	8	13,52
2660,755	8	13,52
1753,474	60	11,50
1750,664	50	11,50

Mg III, ionization potential 80,134 eV

λ, Å	I	E_B, eV
2092,64	4	58,85
2065,54	5	58,77
1749,02	5	65,94
1738,91	6	65,90
234,258	12	52,92
231,730	14	53,50
187,194	8	65,83
186,510	9	66,47
170,802	5	72,59

Mg IV, ionization potential 109,318 eV

λ, Å	I	E_B, eV
323,310	18	38,62
320,999	20	38,62
181,345	8	68,64
180,796	9	68,85
180,617	10	68,64
180,070	8	68,85
172,306	7	72,23
171,653	8	72,23

Mg V, ionization potential 141.262 eV

λ, Å	I	E_B, eV
355,326	12	35,12
354,223	10	35,31
353,300	9	35,31
353,094	14	35,12
352,202	10	35,42
351,089	12	35,31
312,311	10	49,33
276,581	16	49,33
137,414	8	90,22

Mg VI, ionization potential 186.898 eV

λ, Å	I	E_B, eV
403,315	8	30,74
400,676	7	30,94
399,289	6	31,05
349,155	10	42,22
270,394	12	52,56
268,986	10	52,80

ALUMINUM, Z = 13

Al I, ionization potential 5, 985 eV

λ, Å	I	E_B, eV
21163,75	13	4,67
21093,04	12	4,67
16750,56	12	4,83
16718,96	11	4,83
13150,76	14	4,09
13123,41	15	4,09
11254,881	15	5,12
11253,190	14	5,12
8773,896	14	5,43
8772,866	13	5,43
7836,134	12	5,60
7835,309	11	5,60
6698,673	11	4,99
6696,023	13	4,99
5557,063	10	5,37
3961,5200	26	3,14
3944,0058	24	3,14
3092,8386	20	4,02
3092,7099	26	4,02
3082,1529	24	4,02
2660,386	12	4,67
2652,475	12	4,67
2575,095	10	4,83
2567,983	10	4,83
2372,070	10	8,84
2369,304	10	8,83
1769,140	4	7,02
1766,385	4	7,03
1765,636	4	7,02

Al II, ionization potential 18,827 eV

λ, Å	I	E_B, eV
8640,7	8	13,26
8363,52	8	16,54
8359,57	9	16,54
8354,35	10	16,54
6201,70	9	17,31
6201,52	10	17,31
6183,42	10	17,30
6182,28	8	17,30
5593,23	10	15,41
5316,07	7	17,92
5283,77	8	17,93
4666,8	11	18,26
4663,054	10	13,20
3900,68	10	10,59
3655,00	8	16,47
3587,057	8	15,30
3586,546	9	15,30
2868,52	9	17,97
2816,179	20	11,82
2669,166	10	4,63
1862,34	15	11,32
1858,05	10	11,32
1828,61	10	18,10
1764,01	10	11,66
1725,01	15	11,85
1721,31	10	11,85
1670,81	15	7,42
1539,74	10	15,41

Al III, ionization potential 28,447 eV

λ, Å	I	E_B, eV
5722,65	6	17,80
5696,47	8	17,81
5163,90	7	25,94
5150,86	6	25,94
4701,65	6	23,41

λ, Å	I	E_B, eV
4529,176	6	20,55
3601,623	6	17,81
2907,05	10	25,04
2762,815	9	25,03
1935,83	10	20,78
1862,899	10	6,65
1854,67	10	6,65
1611,90	8	14,37
1605,70	8	14,37
893,905	5	20,55
856,768	5	21,15
695,817	5	17,81
560,390	7	22,15

Al IV, ionization potential 119,983 eV

λ, Å	I	E_B, eV
161,686	14	76,68
160,073	16	77,45
130,403	11	95,10
129,729	12	95,57
124,034	8	99,55
116,459	7	106,46

Al V, ionization potential 153.806 eV

λ, Å	I	E_B, eV
281,397	14	44,48
278,699	16	44,48
131,441	20	94,75
131,003	20	95,06
130,848	20	94,75
130,413	20	95,06
126,065	15	98,77
125,525	15	98,77
107,945	20	–

Al VI, ionization potential 190,466 eV

λ, Å	I	E_B, eV
243,760	12	56.02
109,514	20	113,21
107,620	14	120,35
104,344	16	119,15
104,047	20	119,15
92,626	15	133,85
90,858	12	141,61
90,200	20	142,60
88,376	15	140,61
88,273	15	145,61
88,170	20	140,61
87,655	13	141,44
85,515	15	144,98

SILICON, Z = 14

Si I, ionization potential 8,151 eV

λ, Å	I	E_B, eV
10869,5408	130	6,22
10827,091	140	7,33
8556,7803	120	7,32
7944,0011	140	7,54
7932,3490	120	7,53
7423,4969	425	7,29
7415,9462	275	7,29
7405,774	375	7,29
7289,1730	400	7,32
2881,5792	1000	5,08
2528,5086	450	4,93
2524,1079	425	4,92
2519,2023	350	4,93
2516,1125	500	4,95
2514,3161	375	4,93
2506,8973	425	4,95
2435,1545	300	5,87
1901,331	1000	7,30
1874,838	500	7,39
1850,668	500	6,73
1847,468	400	6,72
1845,510	300	6,72
1814,068	500	7,62

Si II, ionization potential 16,342 eV

λ, Å	I	E_B, eV
9412,72	100	14,15
7849,72	500	14,10
7848,80	400	14,10
6371,359	1000	10,07
6347,103	1000	10,07
5978,929	500	12,15
5957,561	500	12,15
5669.562	1000	16,38
5466,868	500	14,79
5466,432	500	14,79
5202,413	500	18,73
5055,981	1000	12,52
5041,026	1000	12,52
4130,893	500	12,84
3856,017	500	10,07
3339,819	500	13,78
2905,692	500	14,01
2904,283	300	14,10
1816,921	200	6,86
1808,003	150	6,86
1533,445	1000	8,12
1526,719	500	8,12
1350,057	150	14,53
1309,274	200	9,50
1265,023	200	9,84
1264,730	2000	9,80
1260,418	1000	9,84
1251,164	200	15,25
1229,388	200	15,43
1194,496	250	10,41
1193,284	200	10,39
992,675	200	12,52
891,999	200	13,93

Si III, ionization potential 33.466 eV

λ, Å	I	E_B, eV
8103,448	11	30,08
7612,356	12	28,22
5739,733	20	21,88
4828,968	18	28,55
4574,759	20	21,72
4567,823	25	21,73
4552,616	30	21,74

λ, Å	I	E_B, eV
3924,468	20	28,55
3806,544	30	24,99
3796,114	25	24,99
3791,41	20	24,99
3590,465	20	25,33
3241,622	15	25,56
3233,954	14	25,56
3210,534	15	28,55
3196,504	14	28,55
3186,022	13	28,55
3185,125	16	25,77
3096,826	16	21,72
3093,424	20	21,73
3086,236	25	21,74
2655,512	14	30,06
2559,210	14	25,39
2541,818	25	15,15
1303,320	16	16,10
1301,146	14	16,08
1298,960	18	16,13
1298,891	15	16,10
1296,726	14	16,10
1294,543	17	16,13
1206,533	30	20,55
1206,510	30	10,27
1113,228	18	17,72
1109,965	16	17,72
1108,368	14	17,72
997,389	16	19,01
994,787	13	19,01
Si IV, ionization potential 45,140 eV		
4654,323	10	39,09
4631,241	9	39,09
4212,407	7	39,08
4116,097	9	27,06
4088,854	10	27,08
3762,435	8	34,29
3165,710	9	31,00
3149,561	7	31,00
1673,374	150	—
1402,770	12	8,34
1393,755	15	8,90
1291,969	30	—
1280,336	20	—
1128,340	10	19,88
1128,325	10	19,88
1056,582	12	—
1051,596	70	—
818,128	8	24,05
815,053	7	24,05
Si V, ionization potential 166,762 eV		
118,968	20	104,21
117,860	20	105,19
97,143	10	127,62
96,439	15	128,56
85,175	10	145,56
Si VI, ionization potential 205,157 eV		
99,460	15	124,65
84,082	12	148,07
83,128	15	149,14
80,577	12	153,86

CHLORINE, Z = 17

λ, Å	I	E_B, eV
Cl I, ionization potential 13,017 eV		
20199,36	227	10,92
19766,78	185	10,93
19755,28	717	10,90
19370,30	227	10,92
16198,47	259	11,39
15970,49	283	11,28
15959,97	735	11,25
15928,92	342	11,43
15883,34	277	11,35
15869,66	2780	11,18
15730,06	1487	11,22
15520,29	1094	11,29
15465,07	381	11,39
15108,04	269	11,25
14931,70	294	11,32
13821,72	525	11,30
13346,76	550	11,49
13296,01	310	12,11
13243,83	350	11,37
11436,34	1000	11,36
11409,68	269	11,37
11122,97	300	11,39
10392,51	331	10,47
8585,96	100	10,43
8428,25	100	10,50
8375,95	150	10,40
8333,27	5000	10,47
8212,00	100	10,43
7878,22	75	10,49
7744,04	125	10,63
7717,57	100	10,59
7547,06	100	10,63
7414,10	90	10,59
7256,63	125	10,63
4601,00	20	11,97
4526,20	30	11,94
4438,48	20	11,71
4389,76	25	11,74
4379,90	20	11,82
4363,30	20	11,83
4323,35	20	11,85
1379,529	11	8,98
1363,449	10	9,20
1351,657	10	9,28
1347,238	12	9,20
1201,358	11	10,43
1188,768	12	10,43
Cl II, ionization potential 23,80 eV		
6094,65	100	18,03
5443,42	100	15,95

λ, Å	I	E_B, eV
5423,52	100	15,96
5423,25	150	15,96
5392,12	100	18,30
5217,93	150	16,34
5103,04	125	18,14
5099,30	100	18,14
5078,25	150	18,16
4917,72	125	18,23
4904,76	135	18,24
4896,77	200	18,25
4819,46	200	15,95
4810,06	225	15,95
4794,54	250	15,96
4768,68	150	19,68
4740,40	150	20,00
4572,13	100	19,05
4343,62	100	18,57
4132,48	200	19,00
3860,98	100	19,17
3860,80	150	19,17
3850,97	100	19,17
3843,26	100	20,50
3833,40	200	21,48
3827,62	150	21,48
3820,25	100	21,48
3353,39	125	18,03
3329,12	150	20,06
3315,44	100	18,59
2688,04	150	18,57
2676,95	100	18,59
2658,74	100	19,00
1079,08	15	11,58
1071,76	10	11,65
1071,05	20	11,58
1063,83	10	11,65
961,49	10	14,34
834,67	10	14,85

Cl III, ionization potential 39,912 eV

λ, Å	I	E_B, eV
4608,21	5	27,01
4596,22	4	26,87
4591,10	4	27,02
4523,33	4	25,01
4370,91	4	25,09
4059,07	6	25,31
4018,50	6	25,36
3748,81	8	25,42
3720,45	8	25,53
3612,85	8	25,01
3602,10	9	25,09
3560,68	8	26,84
3530,03	9	26,87
3393,45	8	27,01
3392,89	8	27,02
3340,42	9	25,29
3329,06	8	25,31
3191,45	9	25,53
3139,34	8	25,53
2710,37	7	30,10
2416,42	7	30,42
2283,93	7	30,52
2253,07	7	30,52
1822,50	6	25,09
1015,023	7	12,21
1008,777	6	12,29
561,738	7	24,32
561,680	7	24,32
561,530	7	24,32
557,118	7	22,25
556,605	7	22,27
556,232	6	22,29

Cl IV, ionization potential 53,462 eV

λ, Å	I	E_B, eV
3076,68	6	30,87
3063,13	5	30,75
2782,47	7	31,29
2751,23	5	31,21
2724,03	5	31,21
984,952	7	12,75
554,619	7	22,52
552,017	7	22,52
537,606	9	23,23
535,666	7	23,21
534,727	8	23,18
486,172	8	27,21

Cl V, ionization potential 67,81 eV

λ, Å	I	E_B, eV
547,630	10	33,47
546,329	6	33,52
545,114	10	33,47
542,297	6	23,05
542,229	8	33,52
538,032	5	23,04
392,433	5	31,78

Cl VI, ionization potential 96,70 eV

λ, Å	I	E_B, eV
555,485	20	34,70
551,992	10	34,70
325,161	25	50,51
323,936	20	50,51
323,356	15	50,51
243,854	12	63,22

ARGON, Z = 18

Ar I, ionization potential 15,759 eV

λ, Å	I	E_B, eV
20986,10	155	13,86
20616,21	356	13,90
13718,77	1000	13,98
13622,38	500	14,06
13503,99	850	14,01
13367,38	800	14,10
13313,39	600	14,21
13273,05	750	14,24

λ, Å	I	E_B, eV
12487,63	700	14,07
12439,19	500	13,90
10470,051	500	12,91
9784,5010	1000	13,09
9657,7841	1500	12,91
9224,4955	1000	13,17
9122,9660	500	12,91
8521,4428	2000	13,28
8424,6473	2500	13,09
8408,2094	3000	13,30
8264,5221	1500	13,33
8115,3108	5000	13,08
8103,6920	2000	13,15
8014,7853	800	13,09
8006,1566	600	13,17
7635,1056	500	13,17
7503,8685	700	13,48
5650,7054	1500	15,10
5495,8760	1000	15,33
5187,7507	800	15,30
4702,3155	1200	14,46
4628,4409	1000	14,51
4596,0964	1000	14,52
4522,3238	800	14,46
4510,7335	1000	14,58
4345,167	1000	14,66
4335,3381	800	14,69
4333,5612	1000	14,69
4300,1011	1200	14,51
4272,1690	1200	14,52
4266,2868	1200	14,53
4259,3617	1200	14,74
4251,1850	800	14,46
4200,6746	1200	14,50
4198,3176	1200	14,58
4191,0288	1200	14,66
4190,7138	600	14,51
4181,8837	1000	14,69
4164,1795	1000	14,52
4158,5906	1200	14,53
4044,4185	1200	14,69
3948,9785	2000	14,69
3947,5048	1000	14,69
3834,6788	800	15,06
3770,3698	400	15,01
3690,8960	300	14,91
3675,2367	300	15,20
3649,8330	800	15,22
3634,4605	300	15,03
3632,6837	300	15,03
3606,5224	1000	15,06
3572,2960	300	15,30
3567,6562	300	15,02
3554,3056	300	15,03
3461,0785	300	15,20
3373,4823	300	15,30
3319,3446	300	15,28
1066,660	15	11,62
1048,218	25	11,83
835,003	6	14,85
834,397	6	14,86

Ar II, ionization potential 27,628 eV

λ, Å	I	E_B, eV
10812,901	12	19,76
10764,378	8	24,31
10683,050	12	19,49
10638,121	8	24,65
10519,510	9	24,19
10467,173	20	19,68
10111,595	8	21,50
9967,045	12	24,19
9854,065	8	25,45
9849,460	10	25,45
8771,855	15	19,87
6886,618	20	19,49
6863,535	20	19,55
6756,548	20	19,61
6684,307	50	19,55
6643,716	100	19,49
6639,743	30	19,64
6638,226	50	19,61
6483,076	20	19,97
6243,125	25	19,68
6172,290	40	21,13
6114,929	50	21,14
5145,319	25	19,55
5141,790	20	21,14
5062,036	30	19,26
5017,160	20	21,13
5009,334	30	19,22
4965,073	25	19,76
4933,206	25	19,26
4879,860	30	19,68
4847,815	25	19,30
4806,017	35	19,22
4764,862	25	19,87
4735,905	25	19,26
4726,859	25	19,76
4657,893	25	19,80
4609,560	25	21,14
4589,896	25	21,13
4579,346	25	19,97
4545,045	25	19,87
4430,192	20	19,61
4426,005	25	19,55
4400,988	20	19,22
4379,667	20	19,64
4371,329	20	19,26
4348,063	50	19,49
4331,199	25	19,61
4277,524	20	21,35
4266,528	25	19,55
4228,162	20	19,68
4103,913	20	22,51
4072,006	25	21,50

λ, Å	I	E_B, eV	λ, Å	I	E_B, eV
4013,858	25	19,49	883,179	9	14,23
3968,360	20	19,55	879,622	8	14,23
3928,629	25	19,97	878,728	12	14,11
3868,524	20	23,17	875,534	9	14,30
3850,578	30	19,97	871,099	10	14,23
3780,841	25	22,77	769,152	12	17,86
3765,269	20	22,51	643,256	9	19,47
3729,310	30	19,97	641,808	12	19,46
3588,448	30	22,95	637,282	20	19,46
3582,362	20	23,07	553,470	9	22,40
3581,608	18	23,10	529,900	9	23,40
3576,611	25	23,01	508,434	9	24,38
3561,031	20	24,62	**Ar IV, ionization potential 59,806 eV**		
3559,508	25	23,16	2926,33	11	35,99
3545,842	18	24,62	2913,00	12	36,16
3545,597	18	23,26	2809,44	16	35,65
3535,319	18	22,81	2788,96	14	35,55
3514,388	20	22,79	2784,47	12	37,70
3491,538	25	22,77	2757,92	14	37,74
3491,243	20	22,81	2640,34	15	35,93
3476,749	20	22,79	2624,92	12	37,97
3249,801	15	23,12	2621,36	12	37,98
3243,689	14	23,08	2615,68	12	35,85
3169,667	15	23,17	2599,47	12	36,67
2979,051	15	21,43	2562,17	12	35,86
2942,892	20	21,35	2513,28	12	36,17
2891,612	18	21,43	850,602	25	14,58
1889,029	6	24,19	843,772	20	14,69
1873,140	6	24,31	840,029	15	14,76
1600,694	6	24,15	801,409	10	18,10
1574,992	6	21,35	801,086	10	18,09
932,0528	10	13,48	689,007	12	20,62
919,7815	10	13,48	683,278	10	20,76
745,323	7	16,81	**Ar V, ionization potential 75,02 eV**		
744,925	8	16,64	827,055	5	15,09
740,270	10	16,75	709,195	5	17,58
679,400	6	18,43	558,481	5	24,22
676,241	6	18,33	527,693	6	23,75
666,010	6	18,62	524,189	5	23,75
Ar III, ionization potential 40,908 eV			463,938	7	26,97
3795,37	20	29,79	461,227	6	26,97
3503,58	15	27,91	449,065	18	27,86
3480,55	20	27,94	446,949	8	27,83
3391,85	15	30,18	445,997	5	27,80
3358,49	15	28,06	337,998	6	36,93
3344,72	20	28,08	**Ar VI, ionization potential 91,32 eV**		
3336,13	25	28,10	551,371	8	22,76
3311,25	15	25,36	462,007	25	27,11
3301,88	20	25,37	461,227	6	39,28
3285,85	25	25,39	459,320	10	39,64
2317,47	15	31,28	457,475	20	27,10
2302,17	15	31,31	294,052	6	42,44
2192,06	15	33,75	282,423	6	56,55
2177,22	25	31,08	**Ar VII, ionization potential 124,03 eV**		
2170,23	20	31,08	585,754	15	21,17
2166,19	15	31,08	479,379	12	40,19
2133,87	15	33,75			
887,404	10	14,11			

λ, Å	I	E_B, eV
475,656	8	40,19
250,940	7	63,73
192,635	7	78,69
176,566	10	70,22

Ar VIII, ionization potential 148,49 eV

λ, Å	I	E_B, eV
700,24	10	17,70
230,875	7	71,40
180,254	15	86,48
179,400	10	86,47
158,923	8	78,01

POTASSIUM, Z = 19

K I, ionization potential 4,340 eV

λ, Å	I	E_B, eV
11772,83	17	2,67
11769,62	16	2,67
11690,21	17	2,67
11022,67	16	3,80
11019,87	17	3,80
7698,959	24	1,61
7664,899	25	1,62
6938,767	20	3,40
6911,084	19	3,40
5831,887	17	3,74
5801,752	17	3,75
5782,384	16	3,75
4047,206	17	3,06
4044,136	18	3,07

K II, ionization potential 31,817 eV

λ, Å	I	E_B, eV
5056,27	7	22,71
5005,60	8	22,71
4829,23	9	22,71
4608,45	8	23,33
4505,33	6	23,14
4388,16	7	23,46
4309,10	7	23,51
4305,00	7	23,33
4263,40	7	23,14
4225,67	7	23,33
4222,97	7	23,57
4186,24	8	23,11
4149,19	7	23,46
4134,72	7	23,14
4114,99	6	23,25
4001,24	7	23,57
3972,58	6	23,51
3966,72	6	23,57
3897,92	8	23,33
3783,19	6	23,51
3767,36	6	23,53
3681,54	6	23,51
3618,49	6	23,57
3530,75	7	24,15
612,621	4	—
607,931	5	—
600,765	6	—
495,144	6	—
485,084	5	—

K III, ionization potential 45,747 eV

λ, Å	I	E_B, eV
497,104	15	24,94
471,569	15	26,56
470,089	20	26,37
466,793	15	26,56
448,595	15	27,91
444,344	15	27,90
440,429	15	—
434,722	15	27,90

K IV, ionization potential 60,909 eV

λ, Å	I	E_B, eV
745,264	10	16,64
741,950	10	16,92
737,144	10	16,82
646,188	15	21,22
393,142	10	31,74

K V, ionization potential 82,6 eV

λ, Å	I	E_B, eV
724,420	8	17,11
603,429	8	25,51
586,322	8	24,15
580,319	7	24,34
425,588	7	32,14
372,148	10	36,29
300,252	7	44,26

K VI, ionization potential 99,741 eV

λ, Å	I	E_B, eV
968,518	6	—
724,420	8	17,48
623,016	8	20,26
616,136	6	20,26
488,120	10	27,75
464,270	10	27,07
460,438	8	27,07
458,048	7	27,07

CALCIUM, Z = 20

Ca I, ionization potential 6,113 eV

λ, Å	I	E_B, eV
19861,70	500	2,52
19852,96	250	4,53
19776,67	2000	2,53
19505,62	500	2,52
19452,82	1500	2,52
19309,43	500	2,52
13134,96	400	5,39
13033,41	300	5,39
12909,07	200	5,39
12818,69	400	4,88
10343,85	500	4,13
7326,146	400	4,62
7202,194	200	4,43

λ, Å	I	E_B, eV
7148,147	500	4,44
6717,685	500	4,55
6462,566	125	4,44
6439,073	150	4,45
6162,172	150	3,91
6122,219	100	3,91
5857,454	100	5,05
5594,468	60	4,74
5588,757	80	4,74
5270,270	60	4,88
4454,781	80	4,68
4434,960	60	4,68
4226,728	500	2,93

Ca II, ionization potential 11.870 eV

λ, Å	I	E_B, eV
9890,63	11	9,69
8927,36	11	8,44
8912,07	10	8,44
8662,140	16	3,12
8542,089	17	3,15
8498,018	13	3,15
8248,797	11	9,02
8201,720	10	9,02
3968,468	22	3,12
3933,663	23	3,15
3736,901	18	6,47
3706,026	17	6,47
3181,275	15	7,05
3179,332	18	7,05
3158,869	17	7,05
1840,061	8	8,44
1838,008	7	8,44

Ca III, ionization potential 51.218 eV

λ, Å	I	E_B, eV
3537,75	7	33,75
3372,68	8	33,75
3119,66	8	34,68
2988,61	7	34,39
2924,33	8	34,95
2899,78	9	34,34
2881,80	7	34,54
2869,95	7	34,39
2866,57	7	35,03
2813,88	7	34,85
2687,78	8	34,68
528,286	8	—
409,971	18	30,24
403,732	20	30,71
357,973	8	—

Ca IV, ionization potential 67.196 eV

λ, Å	I	E_B, eV
669,725	10	18,90
656,038	15	18,90
450,565	10	27,52
443,821	15	28,32
434,570	12	28,53
336,555	15	37,22
335,374	25	36,97
321,593	10	38,94
318,093	15	38,98

Ca V, ionization potential 84,39 eV

λ, Å	I	E_B, eV
646,570	8	19,17
558,602	10	—
542,290	10	—
425,000	15	—
352,915	9	37,46
322,166	10	—
286,965	9	48,63
267,772	8	48,63

Ca VI, ionization potential 109 eV

λ, Å	I	E_B, eV
505,199	8	—
373,997	7	36,55
370,022	7	36,85
340,528	8	39,81
239,535	7	55,16
229,734	7	53,97
228,628	7	54,23

TITANIUM, Z = 22

Ti I, ionization potential 6,836 eV

λ, Å	I	E_B, eV
9705,64	80	2,10
9675,55	90	2,12
9638,28	100	2,13
8692,34	100	2,47
8682,99	125	2,48
8675,38	150	2,50
8548,07	100	3,32
8518,37	100	3,33
8468,46	100	3,35
8435,68	300	2,31
8434,98	300	2,32
8426,50	200	2,30
8412,36	150	2,29
8396,93	90	2,29
8382,82	90	2,29
8382,54	100	2,30
8377,90	100	2,31
4981,732	60	3,34
4534,782	60	3,57
4533,238	80	3,58
4305,910	60	3,73
3998,635	100	3,15
3989,758	80	3,13
3981,761	70	3,11
3958,206	80	3,18
3956,336	60	3,15
3948,670	60	3,14
3752,860	80	3,35
3741,059	60	3,33
3653,497	100	3,44
3642,675	80	3,42
3635,462	80	3,41
3371,447	80	3,72

λ, Å	I	E_B, eV
3354,634	60	3,72
3199,915	100	3,92
3191,994	80	3,90
3186,451	60	3,89
2956,18	70	4,24
2948,38	60	4,22
2941,963	60	4,21

Ti II, ionization potential 13,637 eV

λ, Å	I	E_B, eV
4549,622	60	4,31
4395,031	60	3,90
4300,052	60	4,06
3913,464	60	4,28
3900,546	70	4,31
3761,320	200	3,87
3759,291	200	3,90
3685,192	250	3,97
3641,330	100	4,64
3624,826	70	4,64
3504,890	80	5,43
3383,761	125	3,60
3372,800	100	3,69
3361,213	125	3,72
3349,399	125	3,75
3349,035	75	4,31
3341,875	100	4,28
3329,455	70	3,86
3322,936	75	3,88
3236,573	70	3,86
3234,517	75	3,88
3088,027	75	4,06
2884,099	70	5,43

Ti III, ionization potential 28,143 eV

λ, Å	I	E_B, eV
2516,01	20	9,69
1498,65	30	9,32
1455,22	40	10,30
1422,41	25	10,05
1298,95	40	9,57
1298,67	50	9,60
1295,91	30	9,57
1294,67	50	9,60
1293,26	30	9,64
1289,32	30	9,64
1286,38	40	9,69

Ti IV, ionization potential 43,245 eV

λ, Å	I	E_B, eV
2546,85	12	29,28
2068,16	15	15,96
1467,25	30	24,41
1451,75	30	24,40
781,78	20	15,86
779,14	20	15,96

Ti V, ionization potential 99,8 eV

λ, Å	I	E_B, eV
228,898	75	54,16
225,337	100	55,02

Ti VI, ionization potential 119,762 eV

λ, Å	I	E_B, eV
199,759	6	62,79
198,974	8	62,31
194,900	7	64,32
192,747	8	64,34

IRON, Z = 26

Fe I, ionization potential 7,897 eV

λ, Å	I	E_B, eV
8688,632	1500	3,60
8661,907	600	3,65
8387,780	1200	3,65
8327,063	1200	3,68
8046,073	600	4,42
7998,972	700	5,92
7937,166	700	5,87
7511,045	800	5,82
7207,406	500	5,87
7187,341	800	5,82
6677,994	600	4,55
6494,985	1000	4,31
6400,013	800	5,54
3878,5745	100	3,28
3859,9132	300	3,21
3825,8834	200	4,16
3820,4274	250	4,11
3749,4875	200	4,22
3737,1333	150	3,36
3734,8659	300	4,18
3719,9367	250	3,33
3581,195	250	4,32
3440,6069	150	3,60
3021,0743	150	4,16
3020,6405	200	4,11
3020,4918	150	4,19
2983,5714	125	4,16
2166,773	100	5,72
2084,117	50	5,94

Fe II, ionization potential 6,182 eV

λ, Å	I	E_B, eV
7320,70	40	5,58
7307,957	50	5,58
6456,376	200	5,82
6247,562	80	5,87
6147,735	30	5,90
5962,4	30	—
5427,832	30	—
5136,795	35	5,25
2453,935	25	10,27
2445,114	40	10,27
2255,691	50	10,28
2251,831	80	10,27
2249,063	30	10,32
2247,692	35	10,30
2245,505	45	10,28
2228,761	30	10,37
2218,289	30	10,38
2208,419	30	10,38
2093,683	35	9,69
2000,368	30	8,71

λ, Å	I	E_B, eV
1787,997	35	9,82
1786,738	40	9,83
1785,262	40	9,81
1639,403	30	7,68
1636,334	30	7,68
1635,389	35	8,57
1631,124	30	7,68
1629,155	30	7,69
1621,685	30	7,69
1608,446	35	7,71
1144,946	35	10,82
1112,086	35	11,26
1096,886	30	11,29
1071,596	30	11,64
932,687	30	13,37
932,244	30	13,40
930,558	30	13,40
930,165	30	13,37
930,030	30	13,41
929,612	30	13,42
929,538	30	13,38
928,107	30	13,40
927,176	30	13,42
923,884	30	13,42

Fe III, ionization potential 30,647 eV

λ, Å	I	E_B, eV
5929,69	18	20,60
4372,81	20	25,74
4164,73	20	23,61
3288,81	15	14,09
3276,08	15	14,09
3266,88	20	14,10
3120,847	20	15,12
3013,167	20	14,42
3007,275	20	22,91
2295,859	15	16,30
2174,658	15	14,84
2151,776	15	16,66
2097,480	15	14,67
2078,989	14	11,04
1994,073	13	14,09
1987,503	15	14,10
1960,318	13	16,22
1953,322	13	15,11
1943,481	14	14,25
1937,345	14	14,27
1931,507	14	15,06
1930,387	15	14,29
1926,304	18	10,16
1922,789	15	14,32
1915,083	15	14,34
1914,056	19	10,21
1895,456	20	10,27
1890,669	13	14,42

Fe IV, ionization potential 57,1 eV

λ, Å	I	E_B, eV
1647,05	45	21,00
1640,03	65	21,10
1630,99	75	21,21
1479,65	39	23,16
1475,67	28	23,05
1472,13	35	23,10
1464,81	40	28,29
1459,92	40	28,22
1259,54	30	30,90
576,8	40	21,49
574,5	50	21,58
526,28	75	21,32
525,68	100	21,58

Fe V, ionization potential 78 eV

λ, Å	I	E_B, eV
1464,73	6	32,87
1448,91	6	31,76
1440,59	7	31,88
1430,61	8	32,02
1409,51	7	32,15
1409,19	6	32,00
1406,78	7	33,22
1402,45	6	33,13
1376,45	6	32,16
1373,68	6	32,30
422,287	6	33,05
418,033	6	32,78
417,382	6	32,87
392,907	6	34,72
384,957	6	32,36
365,858	6	33,98
365,440	6	34,08

Fe VI, ionization potential 102 eV

λ, Å	I	E_B, eV
312,263	7	43,26
311,702	7	43,39
304,551	7	43,26
304,221	7	43,39
297,568	8	45,28
297,308	7	45,26
296,988	6	44,10
294,520	7	42,09
294,265	7	42,19
293,966	8	42,32
293,745	8	42,45
292,736	7	42,60
291,229	6	46,13
291,184	6	42,82

COPPER, Z = 29

Cu I, ionization potential 7,726 eV

λ, Å	I	E_B, eV
8092,634	2000	5,35
7933,130	1500	5,35
5782,132	1500	3,79
5700,240	1500	3,82
5292,517	1650	7,74

λ, Å	I	E_B, eV
5218,202	2500	6,19
5153,235	2000	6,19
5105,541	1500	3,82
4651,124	2000	7,74
4062,641	2000	6,87
3530,383	2000	5,15
3307,948	2500	8,82
3279,815	2000	5,42
3273,957	10000	3,79
3247,540	10000	3,82
3063,411	2500	5,69
3036,101	2500	5,72
3010,838	2000	5,51
2997,364	2000	5,78
2961,165	2500	5,57
2766,371	2500	6,12
2618,366	2500	6,12
2492,146	2000	4,97
2406,665	1500	6,79
2392,627	2500	6,82
2293,842	2500	6,79
2263,079	2200	7,12
2244,265	2300	5,52
2230,084	2500	6,95
2225,697	2100	5,57
2214,581	1600	6,98
2199,752	1300	7,28
2199,583	1700	7,02
2181,720	1700	5,68
2178,944	1600	5,69
2165,093	1300	5,72
1825,348	100	6,79
1774,820	200	6,98
1741,574	50	7,12
1725,664	50	7,18
1713,364	50	7,24
1703,843	30	7,28
1691,076	30	8,72
1688,093	30	8,73
1655,318	30	8,88

Cu II, ionization potential 20,291 eV

λ, Å	I	E_B, eV
8283,21	60	16,56
7988,17	60	16,56
7807,66	75	16,56
7664,70	75	16,58
7404,34	100	16,56
6273,330	60	16,95
6216,910	60	16,96
5051,778	60	16,88
4931,653	100	16,85
4909,726	100	16,85
4043,502	75	11,85
2884,196	60	13,39
2544,806	100	13,39
2403,338	100	13,39
2369,890	100	8,49
2246,995	75	8,23
2210,259	60	8,86
2192,268	75	8,49
2179,399	60	8,66
2148,974	60	8,49
2135,976	75	8,52
2054,969	50	8,86
2043,791	60	8,78
1621,426	60	16,56
1593,556	60	16,56
1541,703	75	16,56
1519,837	60	16,82
1488,637	75	16,56
1063,003	60	14,64
1060,630	60	14,52
1059,096	60	14,96
1056,955	60	14,99
1054,690	60	15,01
1044,743	80	15,12
1044,516	80	14,59
1039,569	60	14,76
1039,345	60	14,65
1036,470	60	15,22
945,524	60	16,09
943,328	60	15,97
935,892	60	15,97
935,074	60	16,23
932,940	60	16,01
914,209	80	16,28
896,753	60	16,66
893,674	80	16,59
890,567	60	16,75
886,946	60	16,70

Cu III, ionization potential 36,834 eV

λ, Å	I	E_B, eV
1750,391	500	15,63
1741,378	500	15,43
1722,379	1000	14,74
1709,036	700	14,95
1702,994	500	15,11
1687,134	600	15,04
1684,642	500	15,19
1674,602	500	15,95
1671,886	500	15,72
1670,140	500	15,73
1642,208	2000	15,09
1600,194	500	15,44
1593,758	1000	15,32
1543,438	500	19,07

KRYPTON, Z = 36

Kr I, ionization potential 13,999 eV

λ, Å	I	E_B, eV
21900,51	2250	12,11
18167,12	1500	12,12
16935,71	800	12,26
16896,58	700	12,04
16890,40	1000	12,18

λ, Å	I	E_B, eV
16784,65	950	12,28
15335,29	850	12,11
15239,85	900	12,26
14734,46	900	12,28
14426,93	1100	12,38
13634,22	1700	12,35
13622,28	800	12,35
13177,38	850	12,38
12204,39	700	13,14
11819,43	2000	12,35
9751,759	2000	11,30
8928,6920	2000	11,30
8776,7490	6000	11,44
8508,8700	3000	12,10
8298,1077	5000	11,53
8281,0495	1500	12,14
8263,2398	3000	12,14
8190,0543	3000	11,55
8112,900	6000	11,44
8104,3642	4000	11,44
8059,5038	1500	12,10
7854,8215	800	12,14
7694,5393	500	11,53
7685,2460	400	12,26
7601,5443	2000	11,55
7587,4130	1000	11,67
5870,9153	3000	12,14
5570,2890	2000	12,14
5562,2254	500	12,14
4502,3546	600	12,78
4463,6901	800	12,81
4453,9177	600	12,82
4362,6424	500	12,76
4319,5798	1000	12,78
4318,5523	400	12,78
4273,9700	1000	12,82
3837,81	30	13,14
3812,2155	20	13,28
3800,5437	30	13,29
3796,8839	20	13,29
3773,4241	50	13,32
3679,58	100	13,28
3665,3259	80	13,29
3615,4755	20	13,46
3502,5537	20	13,45
3431,7217	20	13,53
3424,9433	15	13,53
1235,839	13	10,03
1164,868	4	10,64

Kr II, ionization potential 24.570 eV

λ, Å	I	E_B, eV
10221,46	1000	16,83
9803,14	500	—
9619,61	400	—
9605,80	500	—
9577,52	500	17,37
9361,95	300	18,49
9293,82	500	22,17
9238,48	500	18,21
7407,02	400	16,60
7289,78	400	16,60
5681,89	400	16,87
5333,41	500	20,86
5208,32	500	16,65
5125,73	400	19,57
4846,60	700	17,25
4832,07	800	16,83
4765,74	1000	16,87
4739,00	3000	16,60
4680,41	500	17,65
4658,87	2000	16,65
4633,88	800	18,49
4619,15	1000	17,37
4615,28	500	17,37
4577,20	800	18,56
4523,14	400	19,57
4489,88	400	21,32
4475,00	800	18,62
4436,81	600	17,37
4431,67	500	17,38
4355,47	3000	16,83
4317,81	500	19,47
4292,92	600	17,16
4088,33	500	18,88
4065,11	300	18,87
4057,01	300	18,87
3920,14	200	20,00
3906,25	150	22,06
3875,44	150	20,77
3783,13	500	20,11
3778,09	500	20,15
3744,80	150	20,47
3741,69	200	21,81
3721,35	150	20,71
3718,63	200	20,71
3718,02	300	21,89
3680,37	100	20,02
3653,97	250	20,00
3631,87	200	20,02
3607,88	100	21,04
2833,00	100	20,86
2464,77	100	20,89
964,962	30	13,51
911,384	25	14,27
886,302	30	13,99
884,144	30	14,69
868,869	25	14,69
844,058	25	14,69
818,147	25	15,82
782,084	25	15,85
752,051	30	16,48
722,036	50	17,17

Kr III, ionization potential 36,947 eV

λ, Å	I	E_B, eV
6651,75	10	22,33
6602,90	10	24,56
6310,22	10	24,56
6078,38	10	22,33
6037,17	10	24,65

λ, Å	I	E_B, eV
5501,43	10	24,17
5016,45	20	25,15
4988,52	10	25,86
4710,48	10	24,26
4536,46	10	26,32
4443,28	15	23,89
4294,83	10	24,56
4226,58	25	24,03
4225,92	20	24,56
4171,79	15	21,76
4154,46	40	24,17
4131,33	40	21,79
4067,37	50	24,23
3564,23	100	22,27
3507,42	200	22,33
3488,59	100	24,07
3439,46	100	23,89
3351,93	100	21,76
3325,75	200	21,79
3268,48	100	24,03
3264,81	150	24,26
3245,69	300	21,88
3191,21	80	24,17
3189,11	100	22,33
3124,39	100	25,15
3024,45	80	24,56
2892,18	100	24,26
897,801	40	14,37
876,674	22	14,80
870,825	20	14,80
862,578	35	14,37
854,733	25	15,07
837,666	22	14,80
785,968	25	17,59
722,036	50	17,17
698,037	20	18,32
686,254	20	18,07
680,119	22	18,79
676,564	25	18,32
672,330	25	18,44
663,039	20	20,51
659,716	22	18,79
646,417	20	19,18
	Kr IV	
3224,99	6	25,89
2774,70	6	25,84
2748,18	8	25,89
2621,11	7	25,32
2615,3	8	25,72
2609,5	10	26,13
2519,38	6	25,90
2459,74	6	25,89
2291,26	6	26,39
842,035	22	14,72
816,822	18	15,18
805,763	7	15,39
	Kr V	
708,85	8	—
472,16	3	—
	Kr VI	
742,83	8	—
705,84	8	—
569,13	5	—
554,52	5	—
544,03	5	—
465,27	6	—
	Kr VII	
618,67	1	—
585,37	8	—
	Kr VIII	
695,91	8	—
651,57	10	—

XENON, Z = 54

Xe I, ionization potential 12,129 eV

15418,01	110	10,60
14732,38	200	10,56
13656,48	150	10,60
12623,32	300	10,56
11085,25	250	11,27
10895,32	200	11,05
10838,34	1000	9,58
9923,192	3000	89,69
9799,699	2000	9,58
9162,654	500	9,79
9045,446	400	9,69
8987,57	200	11,06
8952,254	1000	9,82
8930,83	200	10,96
8908,73	200	10,97
8862,32	300	10,98
8819,412	5000	9,72
8739,39	300	11,00
8409,190	2000	9,79
8346,823	2000	11,05
8280,1163	7000	9,93
8266,519	500	11,07
8231,6348	10000	9,82
8206,341	700	10,96
7967,341	500	11,00
7887,395	300	11,14
7642,025	500	11,07
7119,598	500	11,46
6882,155	300	11,49
6469,705	300	11,50
6318,062	500	11,68
6182,420	300	11,69
5823,890	300	11,57
4923,1522	500	10,95
4916,508	500	10,96

λ, Å	I	E_B, eV
4843,294	300	10,99
4829,709	400	11,00
4807,019	500	11,01
4734,1524	600	11,05
4697,020	300	10,95
4671,226	2000	10,97
4624,2757	1000	10,99
4582,7474	300	11,14
4524,6805	400	11,05
4500,9772	500	11,07
1469,610	5	8,44
1295,587	8	9,57
1100,46	15	—

Xe II, ionization potential 21,208 eV

λ, Å	I	E_B, eV
9698,68	50	14,48
9591,35	50	13,89
8804,61	30	—
8716,19	50	14,48
8347,24	100	14,07
8297,55	100	—
8151,80	100	—
8038,26	100	—
8031,64	100	—
8008,45	300	—
7164,83	800	15,98
6990,88	2000	14,10
6942,11	1000	—
6805,74	1000	14,07
6595,01	800	—
6356,35	500	17,36
6097,59	1000	13,86
6051,15	1000	13,89
6036,20	500	13,89
5976,46	1000	13,86
5419,15	2000	14,07
5339,38	1000	13,86
5313,87	800	16,43
5292,22	1000	13,89
5080,62	600	16,51
4921,48	800	15,26
4883,53	600	15,88
4876,50	500	16,12
4862,54	800	16,43
4844,33	2000	14,10
4603,03	600	14,48
4585,48	500	16,80
4480,86	500	17,24
4462,19	1000	—
4448,13	500	—
4395,77	500	17,05
4393,20	500	17,90
4330,52	1000	16,93
4310,51	500	18,14
4296,40	500	16,74
4245,38	500	16,80
4238,25	500	16,80
4193,15	500	—
4180,10	1000	16,82
3907,91	100	17,24
3461,26	100	17,38
3366,72	300	17,78
3121,87	250	17,36
3104,40	70	17,38
3017,43	100	—
2979,32	300	17,36
2907,18	80	17,40
2895,22	150	18,51
2864,73	150	17,38
2854,53	60	17,40
1100,432	10	11,27
1074,476	15	11,54
1051,920	10	11,79
1048,272	8	11,83
1041,306	9	11,91

Xe III, ionization potential 32,121 eV

λ, Å	I	E_B, eV
6238,24	60	20,39
6221,66	25	20,39
5524,39	40	21,32
5401,04	50	20,69
5367,06	30	19,92
5238,95	60	20,69
4869,47	40	20,16
4723,57	30	18,20
4683,53	60	18,22
4673,66	30	20,39
4537,33	30	19,92
4434,16	50	19,92
4285,89	30	20,63
4145,73	100	20,12
4109,07	100	19,71
4060,43	60	22,81
4050,05	200	18,63
3950,56	300	18,20
3922,53	500	18,22
3880,46	60	19,71
3877,80	200	20,39
3841,52	100	19,92
3780,98	300	18,85
3624,05	600	18,48
3583,64	80	20,65
3579,69	100	20,16
3522,83	80	18,63
3444,23	60	20,12
3268,96	80	18,85
3242,86	100	18,85
3091,06	50	19,92
3023,80	100	20,62
2948,06	40	20,12
2947,53	40	20,62
2945,25	60	24,37
2940,22	40	24,33
2906,56	50	20,65
1232,074	25	12,18
1130,344	30	12,18
1017,680	35	12,18

λ, Å	I	E_B, eV
1003,370	35	13,57
896,003	20	13,84
893,989	20	13,87
852,950	25	19,17
824,881	30	15,03
823,210	25	15,06
779,126	25	15,91
698,541	20	17,75

Xe IV

Experimental data have not been published. On p. 605 we give wavelengths calculated from levels given in tables.

Xe V

λ, Å	I	E_B, eV
682,56	3	—

Xe VI

λ, Å	I	E_B, eV
880,04	2	—
800,84	2	—
599,84	3	—

Xe VII

λ, Å	I	E_B, eV
995,50	3	—
723,71	3	—
698,02	10	—
566,04	2	—

Xe VIII

λ, Å	I	E_B, eV
858,59	3	—
740,44	7	—
562,55	2	—
517,00	2	—

CESIUM, Z = 55

Cs I, ionization potential 3,894 eV

λ, Å	I	E_B, eV
14694,93	1000	2,30
10123,6025	1200	3,03
10024,3595	1000	3,03
9172,3217	1000	2,80
8943,483	2000	1,38
8761,415	500	2,80
8521,149	4000	1,45
8079,0332	1000	3,34
7943,8820	800	3,01
7608,9032	500	3,01
7279,9570	500	3,51
7228,5356	500	3,51
6973,2966	500	3,23
6723,2943	500	3,23
6586,5096	500	3,34
4593,172	1000	2,70
4555,280	2000	2,72
3888,610	150	3,19
3876,146	300	3,20
3611,459	200	3,43
3476,814	100	3,56

Cs II, ionization potential 25,076 eV

λ, Å	I	E_B, eV
5925,651	60	16,01
5831,159	60	15,88
5563,019	125	16,21
5419,687	60	18,50
5370,979	80	15,68
5358,53	500	20,23
5249,373	80	16,12
5227,002	200	15,68
5043,800	80	16,21
4603,755	60	16,01
4363,275	50	19,05
4277,100	50	16,21
3965,187	25	19,00
3959,495	20	16,51
3925,583	25	19,04
3906,933	20	19,05
3805,096	25	18,94
3785,424	20	19,39
3368,555	30	17,78
3271,626	20	17,77
3267,135	30	17,78
3265,924	30	17,55
2940,953	20	17,55
2931,09	20	19,56
2816,943	20	23,40
2573,03	30	23,87
2273,83	20	24,50
2267,61	20	24,50
1191,55	8	24,50
1178,65	10	24,50
926,75	20	13,38
901,34	20	13,75
813,85	20	15,23
808,77	20	15,33
668,43	12	18,55
639,42	12	19,39

Cs III, ionization potential 34,6 eV

λ, Å	I	E_B, eV
877,9	7	15,84
782,6	3	15,84
722,2	2	17,16
645,0	4	19,22
595,7	2	20,81
550,2	2	22,53
547,8	2	22,63

Section II

Summary Table of Most Intense Lines Arranged According to Wavelength

λ, Å	Symbol	I	λ, Å	Symbol	I
40511,4	H	120	14426,93	Kr I	1100
26877,82	Li I	8	13821,72	Cl I	525
24464,66	Li I	6	13718,77	Ar I	1000
23636,3	Ne I	205	13656,48	Xe I	150
22056,44	Na I	300	13634,22	Kr I	1700
21900,51	Kr I	2250	13622,38	Ar I	500
21163,75	Al I	13	13622,28	Kr I	800
21093,04	Al I	12	13581,33	N I	1200
20986,10	Ar I	155	13503,99	Ar I	850
20616,21	Ar I	356	13429,61	N I	670
20581,30	He I	10000	13367,38	Ar I	800
20199,36	Cl I	227	13346,76	Cl I	550
19861,70	Ca I	500	13313,39	Ar I	600
19852,96	Ca I	250	13296,01	Cl I	310
19776,67	Ca I	2000	13273,05	Ar I	750
19766,78	Cl I	185	13243,83	Cl I	350
19755,28	Cl I	717	13177,38	Kr I	850
19505,62	Ca I	500	13165,11	O I	24
19452,82	Ca I	1500	13164,85	O I	26
19370,30	Cl I	227	13163,89	O I	25
19309,43	Ca I	500	13150,76	Al I	14
18751,1	H	700	13134,96	Ca I	400
18746,0	D	700	13123,41	Al I	15
18744,3	T	700	13033,41	Ca I	300
18703,09	Li I	7	12909,07	Ca I	200
18696,94	He I	1500	12818,69	Ca I	400
18685,96	He I	3600	12818,05	H	140
18390,10	Ne I	180	12814,56	D	140
18385,17	Ne I	160	12813,40	T	140
18282,58	Ne I	200	12623,32	Xe I	300
18276,59	Ne I	260	12614,8	C I	200
18167,12	Kr I	1500	12487,63	Ar I	700
17546,05	Li I	7	12469,62	N I	1350
17002,38	He I	1800	12461,25	N I	680
16935,71	Kr I	800	12439,19	Ar I	500
16896,58	Kr I	700	12204,39	Kr I	700
16890,40	Kr I	1000	11894,9	C I	200
16784,65	Kr I	950	11828,18	Mg I	45
16750,56	Al I	12	11819,43	Kr I	2000
16718,96	Al I	11	11772,83	K I	17
16198,47	Cl I	259	11769,62	K I	16
15970,49	Cl I	283	11754,0	C I	600
15959,97	Cl I	735	11747,5	C I	300
15928,92	Cl I	342	11690,21	K I	17
15883,34	Cl I	277	11656,0	C I	200
15869,66	Cl I	2780	11436,34	Cl I	1000
15730,06	Cl I	1487	11409,68	Cl I	269
15520,29	Cl I	1094	11403,78	Na I	12
15465,07	Cl I	381	11381,45	Na I	11
15418,01	Xe I	110	11254,881	Al I	15
15335,29	Kr I	850	11253,190	Al I	14
15239,85	Kr I	900	11177,59	Ne I	300
15108,04	Cl I	269	11143,09	Ne I	300
15024,99	Mg I	35	11122,97	Cl I	300
14931,70	Cl I	294	11085,25	Xe I	250
14767,48	Na I	1155	11022,67	K I	16
14734,46	Kr I	900	11019,87	K I	17
14732,38	Xe I	200	10951,78	Mg II	10
14694,93	Cs I	1000	10914,23	Mg II	10

λ, Å	Symbol	I	λ, Å	Symbol	I
10895,32	Xe I	200	9534,167	Ne I	500
10869,5408	Si I	130	9486,680	Ne I	500
10844,54	Ne I	200	9425,38	Ne I	500
10838,34	Xe I	1000	9412,72	Si II	100
10834,87	Na I	8	9405,73	C I	16
10830,337	He I	25000	9361,95	Kr II	300
10830,248	He I	15000	9340,544	Mg II	10
10829,088	He I	5000	9327,545	Mg II	10
10827,091	Si I	140	9326,52	Ne I	600
10812,901	Ar II	12	9300,85	Ne I	600
10811,085	Mg I	35	9293,82	Kr II	500
10764,378	Ar II	8	9244,266	Mg II	13
10749,29	Na I	9	9238,48	Kr II	500
10746,44	Na I	10	9224,4955	Ar I	1000
10683,050	Ar II	12	9218,248	Mg II	14
10638,121	Ar II	8	9201,76	Ne I	600
10562,43	Ne I	200	9172,3217	Cs I	1000
10519,510	Ar II	9	9162,654	Xe I	500
10470,051	Ar I	500	9148,68	Ne I	600
10467,173	Ar II	20	9122,9660	Ar I	500
10392,51	Cl I	331	9094,83	C I	12
10343,85	Ca I	500	9045,446	Xe I	400
10221,46	Kr II	1000	8987,57	Xe I	200
10123,6025	Cs I	1200	8952,254	Xe I	1000
10114,644	N I	13	8943,483	Cs I	2000
10112,483	N I	12	8930,83	Xe I	200
10111,595	Ar II	8	8928,6920	Kr I	2000
10108,893	N I	11	8927,36	Ca II	11
10105,130	N I	10	8912,07	Ca II	10
10092,16	Mg II	14	8908,73	Xe I	200
10065,15	N II	7	8900,92	F I	1000
10035,45	N II	7	8865,7562	Ne I	500
10024,3595	Cs I	1000	8862,32	Xe I	300
10023,27	N II	8	8853,8669	Ne I	700
9969,34	N II	7	8835,082	Mg II	11
9967,045	Ar II	12	8824,323	Mg II	10
9961,281	Na I	7	8819,412	Xe I	5000
9923,192	Xe I	3000	8804,61	Xe II	30
9891,09	N II	7	8783,7539	Ne I	1000
9890,63	Ca II	11	8780,6223	Ne I	1200
9854,065	Ar II	8	8776,7490	Kr I	6000
9849,460	Ar II	10	8773,896	Al I	14
9803,14	Kr II	500	8772,866	Al I	13
9799,699	Xe I	2000	8771,855	Ar II	15
9784,5010	Ar I	1000	8761,415	Cs I	500
9751,759	Kr I	2000	8745,657	Mg II	11
9705,64	Ti I	80	8739,39	Xe I	300
9698,68	Xe II	50	8734,990	Mg II	10
9675,55	Ti I	90	8716,19	Xe II	50
9665,424	Ne I	1000	8711,708	N I	15
9657,7841	Ar I	1500	8692,34	Ti I	100
9638,28	Ti I	100	8688,632	Fe I	1500
9632,435	Mg II	11	8683,400	N I	16
9631,888	Mg II	12	8682,99	Ti I	125
9619,61	Kr II	400	8681,9216	Ne I	500
9605,80	Kr II	500	8680,270	N I	17
9691,35	Xe II	50	8679,4898	Ne I	500
9577,52	Kr II	500	8675,38	Ti I	150

λ, Å	Symbol	I	λ, Å	Symbol	I
8662,140	Ca II	16	8234,639	Mg II	11
8661,907	Fe I	600	8231,6348	Xe I	10000
8654,3837	Ne I	1500	8230,773	F I	3000
8649,922	Na I	7	8216,317	N I	15
8640,7	Al II	8	8214,726	F I	2500
8634,6472	Ne I	600	8212,00	Cl I	100
8629,238	N I	16	8206,341	Xe I	700
8594,005	N I	15	8201,720	Ca II	10
8585,96	Cl I	100	8194,8237	Na I	9
8556,7803	Si I	120	8190,0543	Kr I	3000
8548,07	Ti I	100	8151,80	Xe II	100
8542,089	Ca II	17	8115,3108	Ar I	5000
8521,4428	Ar I	2000	8112,900	Kr I	6000
8521,149	Cs I	4000	8104,3642	Kr I	4000
8518,37	Ti I	100	8103,6920	Ar I	2000
8508,8700	Kr I	3000	8103,448	Si III	11
8500,32	C III	10	8092,634	Cu I	2000
8498,018	Ca II	13	8079,0332	Cs I	1000
8495,3591	Ne I	500	8059,5038	Kr I	1500
8468,46	Ti I	100	8046,073	Fe I	600
8446,758	O I	29	8040,931	F I	1000
8446,359	O I	30	8038,26	Xe II	100
8446,250	O I	27	8031,64	Xe II	100
8435,68	Ti I	300	8014,7853	Ar I	800
8434,98	Ti I	300	8008,45	Xe II	300
8428,25	Cl I	100	8006,1566	Ar I	600
8426,50	Ti I	200	7998,972	Fe I	700
8424,6473	Ar I	2500	7988,17	Cu II	60
8412,36	Ti I	150	7967,341	Xe I	500
8409,190	Xe I	2000	7944,0011	Si I	140
8408,2094	Ar I	3000	7943,8820	Cs I	800
8396,93	Ti I	90	7937,166	Fe I	700
8387,780	Fe I	1200	7933,130	Cu I	1500
8382,82	Ti I	90	7932,3490	Si I	120
8382,54	Ti I	100	7915,419	N I	7
8377,90	Ti I	100	7898,985	N I	8
8377,6062	Ne I	800	7896,368	Mg II	13
8375,95	Cl I	150	7887,395	Xe I	300
8363,52	Al II	8	7878,22	Cl I	75
8359,57	Al II	9	7877,051	Mg II	12
8354,35	Al II	10	7854,8215	Kr I	800
8347,24	Xe II	100	7849,72	Si II	500
8346,823	Xe I	2000	7848,80	Si II	400
8335,15	C I	13	7836,134	Al I	12
8333,27	Cl I	5000	7835,309	Al I	11
8332,99	C III	7	7807,66	Cu II	75
8327,063	Fe I	1200	7800,212	F I	15000
8300,3248	Ne I	600	7775,388	O I	26
8298,581	F I	2000	7774,166	O I	27
8298,1077	Kr I	5000	7771,943	O I	28
8297,55	Xe II	100	7754,696	F I	18000
8283,21	Cu II	60	7744,94	Cl I	125
8281,0495	Kr I	1500	7717,57	Cl I	100
8280,1163	Xe I	7000	7698,959	K I	24
8274,615	F I	1500	7694,5393	Kr I	500
8266,519	Xe I	500	7685,2460	Kr I	400
8264,5221	Ar I	1500	7664,899	K I	25
8263,2398	Kr I	3000	7664,70	Cu II	75
8248,797	Ca II	11	7642,025	Xe I	500

λ, Å	Symbol	I
7635,1056	Ar I	500
7612,356	Si III	12
7608,9032	Cs I	500
7601,5443	Kr I	2000
7587,4130	Kr I	1000
7547,06	Cl I	100
7511,045	Fe I	800
7503,8685	Ar I	700
7468,309	N I	16
7442,259	N I	15
7423,635	N I	14
7423,4969	Si I	425
7415,9462	Si I	275
7414,10	Cl I	90
7407,02	Kr II	400
7405,774	Si I	374
7404,34	Cu II	100
7398,688	F I	10000
7326,146	Ca I	400
7320,70	Fe II	40
7311,019	F I	15000
7307,957	Fe II	50
7289,78	Kr II	400
7289,1730	Si I	400
7279,9570	Cs I	500
7256,63	Cl I	125
7254,529	O I	17
7254,447	O I	20
7254,154	O I	19
7245,1665	Ne I	1000
7236,42	C II	20
7231,32	C II	18
7228,5256	Cs I	500
7207,406	Fe I	500
7202,360	F I	15000
7202,194	Ca I	200
7187,341	Fe I	800
7173,9380	Ne I	1000
7164,83	Xe II	800
7148,147	Ca I	500
7127,890	F I	30000
7119,598	Xe I	500
7115,19	C I	9
7113,18	C I	9
7065,190	He I	2500
7037,469	F I	45000
7032,4128	Ne I	1000
7002,228	O I	17
7001,915	O I	15
6990,88	Xe II	2000
6973,2966	Cs I	500
6942,11	Xe II	1000
6938,767	K I	20
6929,4672	Ne I	1000
6911,084	K I	19
6902,475	F I	15000
6886,618	Ar II	20
6882,155	Xe I	300
6870,215	F I	8000
6863,535	Ar II	20
6856,030	F I	50000
6834,264	F I	9000
6830	O V	8
6805,74	Xe II	1000
6773,984	F I	7000
6756,548	Ar II	20
6723,2943	Cs I	500
6723,12	N I	9
6717,685	Ca I	500
6707,84	Li I	1000
6698,673	Al I	11
6696,023	Al I	13
6684,307	Ar II	50
6678,151	He I	1000
6677,994	Fe I	600
6651,75	Kr III	10
6643,716	Ar II	100
6639,743	Ar II	30
6638,226	Ar II	50
6610,565	N II	13
6602,90	Kr III	10
6598,9529	Ne I	1000
6595,01	Xe II	800
6586,5096	Cs I	500
6582,88	C II	15
6578,05	C II	18
6562,793	H	3000
6561,032	D	3000
6560,435	T	3000
6560,099	He II	100
6545,973	Mg II	11
6494,985	Fe I	1000
6484,88	N I	9
6483,076	Ar II	20
6482,74	N I	9
6482,053	N II	13
6469,705	Xe I	300
6462,566	Ca I	125
6456,376	Fe II	200
6455,975	O I	19
6454,445	O I	18
6453,602	O I	17
6439,073	Ca I	150
6413,651	F I	8000
6402,2460	Ne I	2000
6400,013	Fe I	800
6382,9914	Ne I	1000
6380,77	N IV	8
6371,359	Si II	1000
6356,35	Xe II	500
6348,508	F I	10000
6347,103	Si II	1000
6346,737	Mg II	10
6334,4279	Ne I	1000
6318,062	Xe I	500
6310,22	Kr III	10
6273,330	Cu II	60
6266,4950	Ne I	1000

λ, Å	Symbol	I	λ, Å	Symbol	I
6247,562	Fe II	80	5801,33	C IV	10
6243,125	Ar II	25	5782,384	K I	16
6239,651	F I	13000	5782,132	Cu I	1500
6238,24	Xe III	60	5764,4188	Ne I	700
6221,66	Xe III	25	5739,733	Si III	20
6217,2813	Ne I	1000	5722,65	Al III	6
6216,910	Cu II	60	5711,0880	Mg I	30
6201,70	Al II	9	5710,766	N II	10
6201,52	Al II	10	5700,240	Cu I	1500
6183,42	Al II	10	5696,47	Al III	8
6182,420	Xe I	300	5695,92	C III	12
6182,28	Al II	8	5688,2046	Na I	9
6172,290	Ar II	40	5686,213	N II	10
6163,5939	Ne I	1000	5681,89	Kr II	400
6162,172	Ca I	150	5679,562	N II	14
6158,183	O I	21	5676,019	N II	11
6156,765	O I	20	5669,562	Si II	1000
6155,975	O I	19	5666,627	N II	12
6147,735	Fe II	30	5650,7054	Ar I	1500
6143,0623	Ne I	1000	5594,468	Ca I	60
6122,219	Ca I	100	5593,23	Al II	10
6114,929	Ar II	50	5592,37	O III	6
6103,64	Li I	500	5588,757	Ca I	80
6097,59	Xe II	1000	5570,2890	Kr I	2000
6094,65	Cl II	100	5564,37	N I	9
6078,38	Kr III	10	5563,019	Cs II	125
6074,3377	Ne I	1000	5562,2254	Kr I	500
6051,15	Xe II	1000	5560,37	N I	9
6037,17	Kr III	10	5557,063	Al I	10
6036,20	Xe II	500	5528,4047	Mg I	40
6029,9971	Ne I	1000	5524,39	Xe III	40
6014,85	C I	9	5501,43	Kr III	10
6013,22	C I	10	5495,876	Ar I	1000
6008,48	N I	10	5484,7	Li II	10
6006,03	C I	9	5466,868	Si II	500
5978,929	Si II	500	5466,432	Si II	500
5976,46	Xe II	1000	5443,42	Cl II	100
5975,5340	Ne I	600	5427,832	Fe II	30
5962,4	Fe II	30	5423,52	Cl II	100
5957,561	Si II	500	5423,25	Cl II	150
5941,653	N II	12	5419,687	Cs II	60
5931,779	N II	11	5419,15	Xe II	2000
5929,69	Fe III	18	5411,524	He II	50
5925,651	Cs II	60	5401,04	Xe III	50
5895,9236	Na I	16	5400,5616	Ne I	2000
5891,59	C II	12	5392,12	Cl II	100
5889,9504	Na I	32	5380,34	C I	10
5889,77	C II	15	5370,979	Cs II	80
5881,8950	Ne I	1000	5367,06	Xe III	30
5875,966	He I	1000	5358,53	Cs II	500
5875,621	He I	7500	5343,2834	Ne I	600
5870,9153	Kr I	3000	5341,0938	Ne I	1000
5857,454	Ca I	100	5339,38	Xe II	1000
5852,4878	Ne I	2000	5333,41	Kr II	500
5831,887	K I	17	5330,7775	Ne I	600
5831,159	Cs II	60	5316,07	Al II	7
5823,890	Xe I	300	5313,87	Xe II	800
5811,98	C IV	9	5305,3	O IV	15
5801,752	K I	17	5292,517	Cu I	1650

λ, Å	Symbol	I	λ, Å	Symbol	I
5292,22	Xe II	1000	4923,1522	Xe I	500
5283,77	Al II	8	4921,48	Xe II	800
5270,270	Ca I	60	4917,72	Cl II	125
5268,06	O III	2	4916,508	Xe I	500
5249,373	Cs II	80	4909,726	Cu II	100
5238,95	Xe III	60	4904,76	Cl II	135
5227,002	Cs II	200	4896,77	Cl II	200
5218,202	Cu I	2500	4884,9170	Ne I	1000
5217,93	Cl II	150	4883,53	Xe II	600
5208,32	Kr II	500	4879,860	Ar II	30
5202,413	Si II	500	4876,50	Xe II	500
5187,7507	Ar I	800	4869,47	Xe III	40
5183,6042	Mg I	45	4862,54	Xe II	800
5172,6843	Mg I	44	4861,332	H	500
5167,3216	Mg I	42	4860,029	D	500
5163,90	Al III	7	4859,595	T	500
5153,235	Cu I	2000	4847,815	Ar II	25
5151,09	C II	13	4846,60	Kr II	700
5150,86	Al III	6	4844,33	Xe II	2000
5145,319	Ar II	25	4843,294	Xe I	300
5145,16	C II	15	4832,07	Kr II	800
5143,49	C II	12	4829,709	Xe I	400
5141,790	Ar II	20	4829,23	K II	9
5136,795	Fe II	35	4828,968	Si III	18
5133,28	C II	12	4827,3444	Ne I	1000
5132,94	C II	12	4819,46	Cl II	200
5125,73	Kr II	400	4810,06	Cl II	225
5105,541	Cu I	1500	4807,019	Xe I	500
5103,04	Cl II	125	4806,017	Ar II	35
5099,30	Cl II	100	4799,2	O IV	10
5080,62	Xe II	600	4798,25	O IV	5
5078,25	Cl II	125	4794,54	Cl II	250
5062,036	Ar II	30	4788,9270	Ne I	1000
5056,27	K II	7	4788,8	Li II	8
5055,981	Si II	1000	4786,4	O IV	20
5052,17	C I	8	4783,43	O IV	4
5051,778	Cu II	60	4771,75	C I	8
5043,800	Cs II	80	4768,68	Cl II	150
5041,026	Si II	1000	4765,74	Kr II	1000
5037,8	Li II	6	4764,862	Ar II	25
5017,160	Ar II	20	4740,40	Cl II	150
5016,45	Kr III	20	4739,00	Kr II	3000
5010,620	N II	10	4735,905	Ar II	25
5009,334	Ar II	30	4734,1524	Xe I	600
5007,325	N II	11	4726,859	Ar II	25
5005,60	K II	8	4723,57	Xe III	30
5005,149	N II	14	4715,3466	Ne I	1500
5001,477	N II	12	4712,066	Ne I	1000
5001,136	N II	11	4710,48	Kr III	10
4994,363	N II	10	4710,0669	Ne I	1000
4988,52	Kr III	10	4708,8619	Ne I	1200
4981,732	Ti I	60	4705,355	O II	8
4965,073	Ar II	25	4704,3949	Ne I	1500
4957,0335	Ne I	1000	4702,9909	Mg I	30
4944,56	N V	9	4702,3155	Ar I	1200
4935,03	N I	10	4701,65	Al III	6
4933,206	Ar II	25	4697,020	Xe I	300
4932,05	C I	8	4685,682	He II	300
4931,653	Cu II	100	4683,53	Xe III	60

λ, Å	Symbol	I	λ, Å	Symbol	I
4680,41	Kr II	500	4537,7545	Ne I	1000
4677,7	Li II	8	4537,33	Xe III	30
4676,234	O II	8	4536,46	Kr III	10
4673,66	Xe III	30	4534,782	Ti I	60
4671,226	Xe I	2000	4533,238	Ti I	80
4666,8	Al II	11	4529,176	Al III	6
4663,054	Al II	10	4526,20	Cl I	30
4661,635	O II	9	4524,6805	Xe I	400
4658,87	Kr II	2000	4523,33	Cl III	4
4658,30	C IV	9	4523,14	Kr II	400
4657,893	Ar II	25	4522,3238	Ar I	800
4654,323	Si IV	10	4514,89	N III	7
4651,47	C III	11	4510,7335	Ar I	1000
4651,124	Cu I	2000	4505,33	K II	6
4650,25	C III	13	4502,3546	Kr I	600
4649,139	O II	10	4500,9772	Xe I	500
4647,42	C III	14	4497,658	Na I	11
4643,085	N II	11	4494,177	Na I	10
4641,90	N III	7	4489,88	Kr II	400
4641,811	O II	9	4481,327	Mg II	13
4640,64	N III	10	4481,130	Mg II	14
4634,16	N III	8	4480,86	Xe II	500
4633,88	Kr II	800	4475,00	Kr II	800
4631,241	Si IV	9	4471,479	He I	1000
4630,543	N II	14	4463,6901	Kr I	800
4628,4409	Ar I	1000	4462,19	Xe II	1000
4624,2757	Xe I	1000	4454,781	Ca I	80
4621,394	N II	10	4453,9177	Kr I	600
4619,98	N V	10	4448,13	Xe II	500
4619,15	Kr II	1000	4447,18	F II	12
4615,28	Kr II	500			
4609,560	Ar II	25	4447,033	N II	12
4608,45	K II	8	4446,71	F II	10
4608,21	Cl III	5	4443,28	Kr III	15
4607,157	N II	10	4438,48	Cl I	20
4603,755	Cs II	60	4436,81	Kr II	600
4603,73	N V	12	4434,960	Ca I	60
4603,03	Xe II	600	4434,16	Xe III	50
4602,86	Li I	100	4431,67	Kr II	500
4601,480	N II	11	4430,192	Ar II	20
4601,00	Cl I	20	4428,54	Ne II	6
4596,22	Cl III	4			
4596,174	O II	8	4426,005	Ar II	25
4596,0964	Ar I	1000	4423,246	Na I	7
4593,172	Cs I	1000	4416,972	O II	8
4591,10	Cl III	4	4414,909	O II	10
4590,971	O II	9	4409,30	Ne II	7
4589,896	Ar II	25	4400,988	Ar II	20
4585,48	Xe II	500	4397,94	Ne II	6
4582,7474	Xe I	300	4395,77	Xe II	500
4579,346	Ar II	25	4395,031	Ti II	60
4577,20	Kr II	800	4393,340	Na I	9
4574,759	Si III	20	4393,20	Xe II	500
4572,13	Cl II	100	4391,94	Ne II	7
4567,823	Si III	25	4390,564	Mg II	10
4555,280	Cs I	2000	4390,029	Na I	8
4552,616	Si III	30	4389,76	Cl I	25
4549,622	Ti II	60	4388,16	K II	7
4545,186	Na I	8	4379,90	Cl I	20
4545,045	Ar II	25	4379,667	Ar II	20
4541,633	Na I	7	4379,50	Ne II	6

λ, Å	Symbol	I
4379,09	N III	10
4372,81	Fe III	20
4371,329	Ar II	20
4370,91	Cl III	4
4363,30	Cl I	20
4363,275	Cs II	50
4362,6424	Kr I	500
4355,47	Kr II	3000
4349,426	O II	8
4348,063	Ar II	50
4345,167	Ar I	1000
4343,62	Cl II	100
4340,468	H	200
4339,287	D	200
4338,893	T	200
4335,3381	Ar I	800
4333,5612	Ar I	1000
4331,199	Ar II	25
4330,52	Xe II	1000
4325,560	C III	8
4324,615	Na I	7
4323,35	Cl I	20
4319,631	O II	8
4319,5798	Kr I	1000
4318,5523	Kr I	400
4317,81	Kr II	500
4317,139	O II	8
4310,51	Xe II	500
4309,10	K II	7
4305,910	Ti I	60
4305,00	K II	7
4300,1011	Ar I	1200
4300,052	Ti II	60
4299,177	F II	10
4296,40	Xe II	500
4294,83	Kr III	10
4292,92	Kr II	600
4290,40	Ne II	6
4285,89	Xe III	30
4277,524	Ar II	20
4277,100	Cs II	50
4273,9700	Kr I	1000
4272,1690	Ar I	1200
4267,258	C II	20
4267,003	C II	18
4266,528	Ar II	25
4266,2868	Ar I	1200
4263,40	K II	7
4259,3617	Ar I	1200
4251,1850	Ar I	800
4246,16	F II	15
4245,38	Xe II	500
4238,25	Xe II	500
4228,162	Ar II	20
4226,728	Ca I	500
4226,58	Kr III	25
4225,92	Kr III	20
4225,67	K II	7
4222,97	K II	7
4219,76	Ne II	6
4212,407	Si IV	7
4200,6746	Ar I	1200
4198,3176	Ar I	1200
4193,15	Xe II	500
4191,0288	Ar I	1200
4190,7138	Ar I	600
4189,788	O II	10
4186,900	C III	9
4186,24	K II	8
4185,456	O II	8
4181,8837	Ar I	1000
4180,10	Xe II	1000
4171,79	Kr III	15
4164,73	Fe III	20
4164,1795	Ar I	1000
4158,8	O V	0
4158,5906	Ar I	1200
4154,46	Kr III	40
4151,46	N I	12
4149,19	K II	7
4145,73	Xe III	100
4134,72	K II	7
4132,48	Cl II	200
4131,33	Kr III	40
4130,893	Si II	500
4123,9	O V	2
4119,221	O II	8
4116,097	Si IV	9
4114,99	K II	6
4109,959	N I	12
4109,07	Xe III	100
4103,913	Ar II	20
4103,525	F II	15
4103,37	N III	9
4103,085	F II	10
4101,737	H	100
4100,621	D	100
4100,249	T	100
4099,951	N I	9
4097,31	N III	10
4088,854	Si IV	10
4088,33	Kr II	500
4075,868	O II	10
4072,164	O II	8
4072,006	Ar II	25
4070,261	C III	9
4068,912	C III	9
4067,940	C III	8
4067,37	Kr III	50
4065,11	Kr II	300
4062,641	Cu I	2000
4060,43	Xe III	60
4059,07	Cl III	6
4057,01	Kr II	300
4050,05	Xe III	200
4047,206	K I	17
4044,4185	Ar I	1200

λ, Å	Symbol	I	λ, Å	Symbol	I
4044,136	K I	18	3860,80	Cl II	150
4043,502	Cu II	75	3859,9132	Fe I	300
4041,311	N II	11	3856,017	Si II	500
4025,495	F II	15	3851,667	F II	10
4025,010	F II	10	3850,97	Cl II	100
4024,727	F II	20	3850,578	Ar II	30
4018,50	Cl III	6	3849,987	F II	15
4013,858	Ar II	25	3847,086	F II	20
4001,24	K II	7	3843,26	Cl II	100
3998,635	Ti I	100	3841,52	Xe III	100
3994,998	N II	15	3838,374	N II	8
3989,758	Ti I	80	3838,293	Mg I	40
3981,761	Ti I	70	3837,81	Kr I	30
3973,263	O II	10	3834,6788	Ar I	800
3972,58	K II	6	3834,24	O VI	1
3968,468	Ca II	22	3833,40	Cl II	200
3968,360	Ar II	20	3832,302	Mg I	38
3966,72	K II	6	3830,39	N I	9
3965,187	Cs II	25	3829,355	Mg I	36
3961,59	O III	8	3827,62	Cl II	150
3961,5200	Al I	26	3825,8834	Fe I	200
3959,495	Cs II	20	3823,469	O I	10
3958,206	Ti I	80	3822,07	N I	6
3956,336	Ti I	60	3820,4274	Fe I	250
3955,851	N II	10	3820,25	Cl II	100
3954,387	O I	10	3812,2155	Kr I	20
3950,56	Xe III	300	3811,35	O VI	2
3948,9785	Ar I	2000	3806,544	Si III	30
3948,670	Ti I	60	3805,096	Cs II	25
3947,5048	Ar I	1000	3800,5437	Kr I	30
3947,489	O I	7	3796,8839	Kr I	20
3947,301	O I	10	3796,114	Si III	25
3944,0058	Al I	24	3795,37	Ar III	20
3933,663	Ca II	23	3791,41	Si III	20
3928,629	Ar II	25	3785,424	Cs II	20
3925,583	Cs II	25	3783,19	K II	6
3924,468	Si III	20	3783,13	Kr II	500
3922,53	Xe III	500	3780,98	Xe III	300
3920,693	C II	18	3780,841	Ar II	25
3920,14	Kr II	200	3778,09	Kr II	500
3918,999	N II	9	3777,16	Ne II	8
3918,978	C II	15	3773,4241	Kr I	50
3913,464	Ti II	60	3771,08	N III	7
3911,960	O II	10	3770,3698	Ar I	400
3907,91	Xe II	100	3767,36	K II	6
3906,933	Cs II	20	3766,29	Ne II	8
3906,25	Kr II	150	3765,269	Ar II	20
3900,68	Al II	10	3762,435	Si IV	8
3900,546	Ti II	70	3761,320	Ti II	200
3897,92	K II	8	3759,87	O III	9
3888,648	He I	5000	3759,291	Ti II	200
3888,610	Cs I	150	3754,67	O III	7
3880,46	Xe III	60	3754,62	N III	6
3878,5745	Fe I	100	3752,860	Ti I	80
3877,80	Xe III	200	3749,49	O II	9
3876,146	Cs I	300	3749,4875	Fe I	200
3875,44	Kr II	150	3748,81	Cl III	8
3868,524	Ar II	20	3744,80	Kr II	150
3860,98	Cl II	100	3741,69	Kr II	200

λ, Å	Symbol	I	λ, Å	Symbol	I
3741,059	Ti I	60	3572,2960	Ar I	300
3737,1333	Fe I	150	3567,6562	Ar I	300
3736,901	Ca II	18	3564,23	Kr III	100
3734,94	Ne II	7	3561,031	Ar II	20
3734,8659	Fe I	300	3560,68	Cl III	8
3729,310	Ar II	30	3559,508	Ar II	25
3727,33	O II	8	3554,3056	Ar I	300
3727,08	Ne II	9	3545,842	Ar II	18
3721,35	Kr II	150	3545,597	Ar II	18
3720,45	Cl III	8	3542,90	Ne II	7
3719,9367	Fe I	250	3537,75	Ca III	7
3718,63	Kr II	200	3535,319	Ar II	18
3718,02	Kr II	300	3533,043	Na II	10
3713,084	Ne II	10	3530,75	K II	7
3712,75	O II	7	3530,383	Cu I	2000
3709,64	Ne II	7	3530,03	Cl III	9
3706,026	Ca II	17	3522,83	Xe III	80
3694,197	Ne II	10	3520,4714	Ne I	1000
3692,44	O I	7	3514,388	Ar II	20
3690,8960	Ar I	300	3507,42	Kr III	200
3685,192	Ti II	250	3505,614	F II	15
3681,54	K II	6	3504,890	Ti II	80
3680,37	Kr II	100	3503,58	Ar III	15
3679,58	Kr I	100	3503,095	F II	12
3675,2367	Ar I	300	3502,5537	Kr I	20
3665,3259	Kr I	80	3501,416	F II	10
3664,112	Ne II	9	3491,538	Ar II	25
3655,00	Al II	8	3491,243	Ar II	20
3653,97	Kr II	250	3488,59	Kr III	100
3653,497	Ti I	100	3484,96	N IV	13
			3482,99	N IV	14
3649,8330	Ar I	800	3480,55	Ar III	20
3642,675	Ti I	80	3478,71	N IV	15
3641,330	Ti II	100	3476,814	Cs I	100
3635,462	Ti I	80	3476,749	Ar II	20
3634,4605	Ar I	300	3472,5706	Ne I	500
3632,6837	Ar I	300	3470,81	O II	8
3631,87	Kr II	200	3461,26	Xe II	100
3631,266	Na II	8	3461,0785	Ar I	300
3624,826	Ti II	70	3444,23	Xe III	60
3624,05	Xe III	600	3440,6069	Fe I	150
3618,49	K II	6	3439,46	Kr III	100
3615,4755	Kr I	20	3437,147	N II	9
3612,85	Cl III	8	3431,7217	Kr I	20
3611,459	Cs I	200	3424,9433	Kr I	15
3607,88	Kr II	100	3417,9031	Ne I	500
3606,5224	Ar I	1000	3407,38	O II	7
3602,10	Cl III	9	3393,45	Cl III	8
3601,623	Al III	6	3392,89	Cl III	8
3593,5263	Ne I	500	3391,85	Ar III	15
3590,465	Si III	20	3390,25	O II	8
3588,448	Ar II	30	3383,761	Ti II	125
3587,057	Al II	8	3377,20	O II	7
3586,546	Al II	9	3374,06	N III	6
3583,64	Xe III	80	3373,4823	Ar I	300
3582,362	Ar II	20	3372,800	Ti II	100
3581,608	Ar II	18	3372,68	Ca III	8
3581,195	Fe I	250	3371,447	Ti I	80
3579,69	Xe III	100	3369,9069	Ne I	700
3576,611	Ar II	25	3369,8076	Ne I	500

λ, Å	Symbol	I	λ, Å	Symbol	I
3368,555	Cs II	30	3233,954	Si III	14
3367,36	N III	7	3232,66	Li I	50
3366,72	Xe II	300	3224,99	Kr IV	6
3361,213	Ti II	125	3218,21	Ne II	8
3358,49	Ar III	15	3210,534	Si III	15
3355,05	Ne II	7	3203,104	He II	200
3354,634	Ti I	60	3202,740	F II	10
3353,39	Cl II	125	3199,915	Ti I	100
3351,93	Kr III	100	3199,43	Li II	7
3349,399	Ti II	125	3196,504	Si III	14
3349,035	Ti II	75	3191,994	Ti I	80
3344,72	Ar III	20	3191,45	Cl III	9
3341,875	Ti II	100	3191,21	Kr III	80
3340,42	Cl III	9	3189,11	Kr III	100
3339,819	Si II	500	3187,745	He I	200
3336,13	Ar III	25	3186,451	Ti I	60
3334,87	Ne II	10	3186,022	Si III	13
3333,139	Si II	300	3185,125	Si III	16
3329,455	Ti II	70	3181,275	Ca II	15
3329,12	Cl II	150	3179,332	Ca II	18
3329,06	Cl III	8	3174,725	F III	10
3328,730	N II	7	3174,125	F III	12
3325,75	Kr III	200	3169,667	Ar II	15
3323,75	Ne II	7	3165,710	Si IV	9
3322,936	Ti II	75	3158,869	Ca II	17
3319,3446	Ar I	300	3153,492	F II	6
3315,44	Cl II	100	3149,561	Si IV	7
3311,25	Ar III	15	3139,34	Cl III	8
3307,948	Cu I	2500	3138,44	O II	8
3302,979	Na I	18	3134,82	O II	10
3302,369	Na I	19	3129,368	Na II	6
3301,88	Ar III	20	3124,39	Kr III	100
3297,74	Ne II	7	3121,87	Xe II	250
3288,81	Fe III	15	3121,515	F III	12
3285,85	Ar III	25	3120,847	Fe III	20
3285,603	Na II	8	3119,66	Ca III	8
3279,815	Cu I	2000	3115,669	F III	10
3276,08	Fe III	15	3104,40	Xe II	70
3273,957	Cu I	10000	3096,826	Si III	16
3271,626	Cs II	20	3093,424	Si III	20
3268,96	Xe III	80	3092,8386	Al I	20
3268,48	Kr III	100	3092,729	Na II	10
3267,135	Cs II	30	3092,7099	Al I	26
3266,88	Fe III	20	3091,06	Xe III	50
3265,924	Cs II	30	3088,027	Ti II	75
3265,46	O III	10	3086,236	Si III	25
3264,81	Kr III	150	3082,1529	Al I	24
3264,164	F III	9	3076,68	Cl IV	6
3264,16	F II	7	3071,66	O IV	5
3260,98	O III	8	3063,46	O IV	6
3249,801	Ar II	15	3063,411	Cu I	2500
3249,8	Li II	5	3063,13	Cl IV	5
3247,540	Cu I	10000	3059,960	F II	8
3245,69	Kr III	300	3058,141	F II	7
3243,689	Ar II	14	3057,388	Ne I	300
3242,86	Xe III	100	3057,083	F II	6
3241,622	Si III	15	3053,664	Na II	6
3236,573	Ti II	70	3047,13	O III	8
3234,517	Ti II	75	3042,808	F III	10

λ, Å	Symbol	I	λ, Å	Symbol	I
3036,101	Cu I	2500	2881,140	Na II	6
3024,45	Kr III	80	2869,95	Ca III	7
3023,80	Xe III	100	2868,52	Al II	9
3021,0743	Fe I	150	2866,57	Ca III	7
3020,6405	Fe I	200	2864,73	Xe II	150
3020,4918	Fe I	150	2862,26	N III	6
3017,43	Xe II	100	2860,308	F III	9
3013,167	Fe III	20	2854,53	Xe II	60
3010,838	Cu I	2000	2853,013	Na I	15
3007,275	Fe III	20	2852,811	Na I	16
3006,830	N II	7	2852,127	Mg I	50
2997,364	Cu I	2000	2841,721	Na II	7
2992,618	C II	18	2837,603	C II	18
2988,61	Ca III	7	2836,710	C II	20
2984,183	Na II	7	2836,25	O IV	6
2983,78	O III	9	2833,00	Kr II	100
2983,58	N III	6	2826,13	F IV	5
2983,5714	Fe I	125	2820,74	F IV	4
2982,663	Ne I	300	2816,943	Cs II	20
2982,106	C III	8	2816,179	Al II	20
2981,31	N V	10	2813,88	Ca III	7
2980,78	N V	8	2811,422	F III	10
2979,32	Xe II	300	2809,44	Ar IV	16
2979,051	Ar II	15	2802,704	Mg II	12
2974,714	Ne I	300	2797,998	Mg II	10
2967,244	C I	5	2795,528	Mg II	13
2961,165	Cu I	2500	2789,86	O V	3
2956,18	Ti I	70	2788,96	Ar IV	14
2955,73	Ne II	7	2788,093	F III	20
2951,231	Na II	8	2787,03	O V	4
2948,38	Ti I	60	2784,47	Ar IV	12
2948,06	Xe III	40	2782,47	Cl IV	7
2947,53	Xe III	40	2781,04	O V	5
2945,25	Xe III	60	2774,70	Kr IV	6
2945,106	He I	100	2767,0	Li II	4
2942,892	Ar II	20	2766,371	Cu I	2500
2941,963	Ti I	60	2762,815	Al III	9
2940,953	Cs II	20	2759,589	F III	10
2940,22	Xe III	40	2757,92	Ar IV	14
2936,509	Mg II	10	2751,23	Cl IV	5
2931,09	Cs II	20	2748,18	Kr IV	8
2926,33	Ar IV	11	2741,20	Li I	10
2924,33	Ca III	8	2733,32	He II	100
2916,335	F III	10	2730,7	Li II	5
2913,00	Ar IV	12	2725,90	C III	7
2907,18	Xe II	80	2725,30	C III	7
2907,05	Al III	10	2724,85	C III	6
2906,56	Xe III	50	2724,03	Cl III	5
2906,29	C IV	5	2710,37	Cl III	7
2905,692	Si II	500	2707,17	F V	2
2904,914	Na II	7	2697,75	C III	7
2904,283	Si II	300	2688,04	Cl II	150
2899,78	Ca III	9	2687,78	Ca III	8
2895,22	Xe II	150	2686,14	O III	10
2893,946	Na II	6	2678,64	Ne III	25
2892,18	Kr III	100	2677,90	Ne III	30
2891,612	Ar II	18	2676,95	Cl II	100
2884,196	Cu II	60	2675,64	Ne I	200
2884,099	Ti II	70	2675,24	Ne I	200
2881,80	Ca III	7			
2881,5792	Si I	1000			

λ, Å	Symbol	I
2674,57	O III	—
2671,829	Na II	6
2669,166	Al II	10
2665,69	O III	7
2660,996	Na II	7
2660,817	Mg II	8
2660,755	Mg II	8
2660,386	Al I	12
2658,74	Cl II	100
2655,512	Si III	14
2655,475	Al I	12
2647,42	Ne I	200
2646,956	N IV	12
2646,176	N IV	11
2645,654	N IV	10
2640,34	Ar IV	15
2624,92	Ar IV	12
2621,36	Ar IV	12
2621,11	Kr IV	7
2618,366	Cu I	2500
2615,68	Ar IV	12
2615,3	Kr IV	8
2613,41	Ne III	12
2611,815	Na II	7
2610,03	Ne III	15
2609,5	Kr IV	10
2599,47	Ar IV	12
2597,69	O III	8
2595,68	Ne III	20
2593,60	Ne III	30
2590,04	Ne III	40
2582,901	C I	5
2575,300	O II	10
2575,095	Al I	10
2573,03	Cs II	30
2571,476	O II	8
2567,983	Al I	10
2563,32	Na III	25
2562,17	Ar IV	12
2559,210	Si III	14
2553,61	Na III	25
2546,85	Ti IV	12
2544,806	Cu II	100
2541,818	Si III	25
2531,548	Na II	6
2530,6	C IV	6
2530,30	O II	8
2529,98	C IV	11
2528,5086	Si I	450
2524,41	C IV	9
2524,1079	Si I	425
2522,227	N II	7
2520,791	N II	6
2519,38	Kr IV	6
2519,2023	Si I	350
2517,40	O IV	7
2516,1125	Si I	500
2516,01	Ti III	20
2514,3161	Si I	375
2513,28	Ar IV	12
2511,22	He II	50
2509,23	O IV	8
2507,77	O IV	7
2506,8973	Si I	425
2501,80	O IV	8
2499,29	O IV	6
2497,05	Na III	50
2493,75	O IV	10
2493,40	O IV	7
2492,146	Cu I	2000
2484,360	F III	9
2478,556	C I	10
2474,69	Na III	40
2468,86	Na III	30
2464,77	Kr II	100
2459,74	Kr IV	6
2459,40	Na III	45
2456,92	F IV	5
2453,935	Fe II	25
2451,58	F IV	4
2450,63	F V	2
2450,06	O IV	10
2449,36	O IV	8
2445,55	O II	10
2445,114	Fe II	40
2435,1545	Si I	300
2433,538	O II	9
2416,42	Cl III	7
2412,94	Ne III	12
2412,73	Ne III	15
2406,665	Cu I	1500
2405,10	C IV	6
2404,44	C IV	5
2403,338	Cu II	100
2392,627	Cu I	2500
2390,44	O III	8
2384,95	Ne IV	7
2382,32	O III	7
2373,21	Ne IV	9
2372,16	Ne IV	7
2372,070	Al I	10
2369,890	Cu II	100
2369,304	Al I	10
2357,96	Ne IV	10
2352,52	Ne IV	8
2327,28	F VI	5
2323,35	F VI	7
2317,47	Ar III	15
2317,046	N II	8
2316,690	N II	6
2316,493	N II	7
2315,39	F VI	9
2309,96	Na III	30
2302,17	Ar III	15
2300,35	O II	8
2298,29	F IV	5
2297,14	Na III	25
2296,870	C III	16

λ, Å	Symbol	I	λ, Å	Symbol	I
2296,64	Na III	25	2163,77	Ne III	15
2295,859	Fe III	15	2162,944	C III	9
2293,842	Cu I	2500	2151,776	Fe III	15
2293,32	O II	6	2148,974	Cu II	60
2291,26	Kr IV	6	2135,976	Cu II	75
2285,79	Ne IV	9	2133,87	Ar III	15
2285,72	Na III	35	2097,480	Fe III	15
2283,93	Cl III	7	2095,54	Ne III	20
2278,48	Na III	40	2093,683	Fe II	35
2273,83	Cs II	20	2092,64	Mg III	4
2267,61	Cs II	20	2092,44	Ne III	12
2265,71	Ne V	6	2089,43	Ne III	15
2263,39	Ne V	3	2084,117	Fe I	50
2263,21	Ne III	12	2078,989	Fe III	14
2263,079	Cu I	2200	2068,16	Ti IV	15
2259,57	Ne V	3	2065,54	Mg III	5
2255,691	Fe II	50	2063,99	N III	10
2253,22	Ne VI	3	2063,50	N III	10
2253,07	Cl III	7	2055,93	Ne VI	3
2252,72	F V	2	2054,969	Cu II	50
2251,831	Fe II	80	2043,791	Cu II	60
2251,44	Na III	45	2042,382	Ne VI	3
2249,063	Fe II	30	2011,88	Na III	30
2247,692	Fe II	35	2005,98	Na III	6
2246,995	Cu II	75	2000,368	Fe II	30
2246,66	Na III	40	1994,073	Fe III	13
2245,505	Fe II	45	1992,060	Ne VII	3
2245,48	Ne V	3	1987,503	Fe III	15
2244,265	Cu I	2300	1985,58	Na III	30
2239,43	Na III	45	1981,974	Ne VII	6
2232,41	Ne V	4	1965,04	Na III	18
2232,17	Na III	40	1960,76	Na III	20
2230,30	Na III	50	1960,318	Fe III	13
2230,084	Cu I	2500	1953,322	Fe III	13
2228,761	Fe II	30	1951,21	Na III	40
2227,42	Ne V	3	1946,43	Na III	20
2225,90	Na III	45	1943,481	Fe III	14
2225,697	Cu I	2100	1938,827	Ne II	8
2218,289	Fe II	30	1937,345	Fe III	14
2216,07	Ne III	15	1935,83	Al III	10
2214,581	Cu I	1600	1933,87	Na III	30
2213,76	Ne III	12	1931,507	Fe III	14
2210,259	Cu II	60	1930,930	C I	10
2208,419	Fe II	30	1930,387	Fe III	15
2202,78	Na III	40	1930,033	Ne II	8
2199,752	Cu I	1300	1926,304	Fe III	18
2199,583	Cu I	1700	1926,27	Na III	45
2192,268	Cu II	75	1922,789	Fe III	15
2192,06	Ar III	15	1916,081	Ne II	10
2181,720	Cu I	1700	1915,083	Fe III	15
2179,399	Cu II	60	1914,056	Fe III	19
2178,944	Cu I	1600	1907,494	Ne II	8
2177,22	Ar III	25	1901,331	Si I	1000
2174,658	Fe III	15	1895,456	Fe III	20
2171,44	F IV	4	1890,669	Fe III	13
2170,23	Ar III	20	1889,029	Ar II	6
2166,773	Fe I	100	1874,838	Si I	500
2166,19	Ar III	15	1873,140	Ar II	6
2165,093	Cu I	1300	1862,899	Al III	10

λ, Å	Symbol	I	λ, Å	Symbol	I
1862,34	Al II	15	1647,05	Fe IV	45
1858,05	Al II	10	1642,208	Cu III	2000
1856,73	Na III	20	1640,474	He II	10
1854,67	Al III	10	1640,332	He II	5
1850,668	Si I	500	1640,03	Fe IV	65
1850,39	Na III	18	1639,403	Fe II	30
1850,24	Na III	20	1636,334	Fe II	30
1849,58	Na III	35	1635,389	Fe II	35
1847,468	Si I	400	1631,124	Fe II	30
1845,510	Si I	300	1630,99	Fe IV	75
1844,4	N II	10	1629,155	Fe II	30
1844,36	Na III	20	1621,685	Fe II	30
1840,061	Ca II	8	1621,426	Cu II	60
1838,008	Ca II	7	1611,90	Al III	8
1828,61	Al II	10	1608,446	Fe II	35
1827,97	Mg I	8	1605,70	Al III	8
1825,348	Cu I	100	1600,694	Ar II	6
1822,50	Cl III	6	1600,194	Cu III	500
1816,921	Si II	200	1593,738	Cu III	1000
1814,068	Si I	500	1593,556	Cu II	60
1808,003	Si II	150	1574,992	Ar II	6
1787,997	Fe II	35	1561,40	C I	20
1786,738	Fe II	40	1560,702	C I	15
1785,262	Fe II	40	1550,771	C IV	19
1774,820	Cu I	200	1548,185	C IV	20
1769,140	Al I	4	1543,438	Cu III	500
1766,385	Al I	4	1541,703	Cu II	75
1765,636	Al I	4	1539,74	Al II	10
1764,01	Al II	10	1533,445	Si II	1000
1756,0	Li II	5	1526,719	Si II	500
1753,474	Mg II	60	1519,837	Cu II	60
1750,664	Mg II	50	1498,65	Ti III	30
1750,391	Cu III	500	1494,668	N I	60
1749,02	Mg III	5	1493,7	Li II	6
1747,81	Mg I	5	1492,817	N I	30
1741,574	Cu I	50	1492,624	N I	80
1741,378	Cu III	500	1488,637	Cu II	75
1738,91	Mg III	6	1479,65	Fe IV	38
1725,664	Cu I	50	1475,67	Fe IV	28
1725,01	Al II	15	1472,13	Fe IV	35
1722,379	Cu III	1000	1469,610	Xe I	5
1721,31	Al II	10	1467,25	Ti IV	30
1718,551	N IV	20	1464,81	Fe IV	40
1713,364	Cu I	50	1464,73	Fe V	6
1709,036	Cu III	700	1459,92	Fe IV	40
1703,843	Cu I	30	1455,22	Ti III	40
1702,994	Cu III	500	1451,75	Ti IV	30
1691,076	Cu I	30	1448,91	Fe V	6
1688,093	Cu I	30	1440,59	Fe V	7
1687,134	Cu III	600	1432,538	C I	10
1684,642	Cu III	500	1432,115	C I	15
1682,4	Li II	4	1431,595	C I	20
1674,602	Cu III	500	1430,61	Fe V	8
1673,374	Si IV	150	1422,41	Ti III	25
1671,886	Cu III	500	1411,939	N I	30
1670,81	Al II	15	1409,51	Fe V	7
1670,140	Cu III	500	1409,19	Fe V	6
1656,998	C I	10	1406,78	Fe V	7
1655,318	Cu I	30	1402,770	Si IV	12
1653,9	Li II	8			

λ, Å	Symbol	I	λ, Å	Symbol	I
1402,45	Fe V	6	1188,768	Cl I	12
1393,755	Si IV	15	1178,65	Cs II	10
1379,529	Cl I	11	1164,868	Kr I	4
1376,45	Fe V	6	1144,946	Fe II	35
1373,68	Fe V	6	1130,344	Xe III	30
1371,287	O V	10	1128,340	Si IV	10
1363,449	Cl I	10	1128,325	Si IV	10
1351,657	Cl I	10	1113,228	Si III	18
1350,057	Si II	150	1112,086	Fe II	35
1347,238	Cl I	12	1109,965	Si III	16
1335,684	C II	14	1108,368	Si III	14
1334,515	C II	13	1101,293	N I	40
1319,684	N I	30	1100,46	Xe I	15
1309,274	Si II	200	1100,432	Xe II	10
1306,025	O I	25	1100,362	N I	30
1304,866	O I	30	1098,264	N I	40
1303,320	Si III	16	1098,103	N I	40
1302,173	O I	30	1097,245	N I	50
1301,146	Si III	14	1096,886	Fe II	30
1298,95	Ti III	40	1096,749	N I	35
1298,960	Si III	18	1096,322	N I	35
1298,891	Si III	15	1095,940	N I	35
1298,67	Ti III	50	1085,701	N II	12
1296,726	Si III	14	1085,542	N II	9
1295,91	Ti III	30	1084,572	N II	11
1295,587	Xe I	8	1083,990	N II	10
1294,67	Ti III	50	1079,08	Cl II	15
1294,543	Si III	17	1074,476	Xe II	15
1293,26	Ti III	30	1071,76	Cl II	10
1291,969	Si IV	30	1071,596	Fe II	30
1289,32	Ti III	30	1071,05	Cl II	20
1286,38	Ti III	40	1069,984	N I	30
1280,336	Si IV	20	1068,476	N I	35
1277,617	C I	10	1067,607	N I	35
1277,274	C I	9	1066,660	Ar I	15
1265,023	Si II	200	1063,83	Cl II	10
1264,730	Si II	2000	1063,003	Cu II	60
1260,418	Si II	1000	1060,630	Cu II	60
1259,54	Fe IV	30	1059,096	Cu II	60
1251,164	Si II	200	1056,955	Cu II	60
1242,804	N V	19	1056,582	Si IV	12
1238,821	N V	20	1054,690	Cu II	60
1235,839	Kr I	13	1051,920	Xe II	10
1232,074	Xe III	25	1051,596	Si IV	70
1229,388	Si II	200	1048,272	Xe II	8
1215,670	H	3000	1048,218	Ar I	25
1215,340	D	3000	1044,743	Cu II	80
1215,229	T	3000	1044,516	Cu II	80
1215,171	He II	5	1041,306	Xe II	9
1206,533	Si III	30	1040,941	O I	15
1206,510	Si III	30	1039,569	Cu II	60
1201,358	Cl I	11	1039,345	Cu II	60
1200,711	N I	30	1039,233	O I	20
1199,549	N I	50	1037,613	O VI	9
1198,6	Li II	7	1037,017	C II	13
1194,496	Si II	250	1036,470	Cu II	60
1193,284	Si II	200	1036,330	C II	12
1193,284	C I	10	1031,912	O VI	10
1191,55	Cs II	8	1027,433	O I	20

λ, Å	Symbol	I	λ, Å	Symbol	I
1025,722	H	1000	923,220	N IV	16
1025,443	D	1000	923,057	N IV	14
1025,350	T	1000	922,519	N IV	14
1017,680	Xe III	35	921,992	N IV	14
1015,023	Cl III	7	919,7815	Ar II	10
1010,369	C II	10	919,78	O II	15
1010,074	C II	10	916,700	N II	12
1008,777	Cl III	6	916,004	N II	11
1003,370	Xe III	35	915,955	N II	10
997,389	Si III	16	915,603	N II	10
995,50	Xe VII	3	914,209	Cu II	80
994,787	Si III	13	911,384	Kr II	25
992,675	Si II	200	906,426	N I	15
991,579	N III	17	906,202	N I	10
991,514	N III	14	904,468	C II	10
989,790	N III	16	904,134	C II	12
988,776	O I	15	903,950	C II	11
984,952	Cl IV	7	903,609	C II	10
977,745	F I	100	901,34	Cs II	20
977,026	C III	18	897,801	Kr III	40
976,217	F I	100	896,753	Cu II	60
973,895	F I	350	896,003	Xe III	20
972,537	H	400	893,989	Xe III	20
972,272	D	400	893,905	Al III	5
972,184	T	400	893,674	Cu II	80
968,518	K VI	6	891,999	Si II	200
965,042	N I	10	890,567	Cu II	60
964,962	Kr II	30	887,404	Ar III	10
961,49	Cl II	10	886,946	Cu II	60
958,524	F I	500	886,302	Kr II	30
955,545	F I	750	884,144	Kr II	30
955,335	N IV	20	883,179	Ar III	9
954,825	F I	1000	880,04	Xe VI	2
953,658	N I	15	879,622	Ar III	8
953,415	N I	15	878,728	Ar III	12
951,871	F I	500	877,9	Cs III	7
949,743	H	200	876,674	Kr III	22
949,485	D	200	875,534	Ar III	9
949,401	T	200	871,099	Ar III	10
945,524	Cu II	60	870,825	Kr III	20
943,328	Cu II	60	868,869	Kr II	25
935,892	Cu II	60	862,578	Kr III	35
935,074	Cu II	60	858,59	Xe VIII	3
932,940	Cu II	60	856,768	Al III	5
932,687	Fe II	30	854,733	Kr III	25
932,244	Fe II	30	852,950	Xe III	25
932,0528	Ar II	10	850,602	Ar IV	25
932,046	O II	10	844,058	Kr II	25
930,558	Fe II	30	843,772	Ar IV	20
930,165	Fe II	30	842,035	Kr IV	22
930,030	Fe II	30	840,029	Ar IV	15
929,612	Fe II	30	837,666	Kr III	22
929,538	Fe II	30	835,292	O III	16
928,107	Fe II	30	835,096	O III	14
927,176	Fe II	30	835,003	Ar I	6
926,75	Cs II	20	834,67	Cl II	10
924,283	N IV	14	834,462	O II	15
923,884	Fe II	30	834,397	Ar I	6
923,675	N IV	14	833,742	O III	16

λ, Å	Symbol	I	λ, Å	Symbol	I
833,326	O II	15	723,71	Xe VII	3
832,927	O III	14	722,2	Cs III	2
832,754	O II	14	722,036	Kr III	50
827,055	Ar V	5	722,036	Kr II	50
824,881	Xe III	30	718,562	O II	16
823,210	Xe III	25	718,484	O II	17
818,147	Kr II	25	709,195	Ar V	5
818,128	Si IV	8	708,85	Kr V	8
816,822	Kr IV	18	705,84	Kr VI	8
815,053	Si IV	7	703,850	O III	18
813,85	Cs II	20	702,899	O III	17
809,599	F I	125	702,822	O III	16
808,77	Cs II	20	700,24	Ar VIII	10
806,964	F I	150	698,541	Xe III	20
805,763	Kr IV	7	698,037	Kr III	20
801,409	Ar IV	10	698,02	Xe VII	10
801,086	Ar IV	10	695,91	Kr VIII	8
800,84	Xe VI	2	695,817	Al III	5
796,661	O II	10	689,007	Ar IV	12
790,103	O IV	13	687,355	C II	11
787,710	O IV	15	687,059	C II	10
785,968	Kr III	25	686,335	N III	14
782,6	Cs III	3	686,254	Kr III	20
782,084	Kr II	25	685,816	N III	16
781,78	Ti IV	20	685,513	N III	15
780,324	Ne VIII	4	684,996	N III	14
779,905	O IV	10	683,278	Ar IV	10
779,126	Xe III	25	682,56	Xe V	3
779,14	Ti IV	20	680,119	Kr III	22
775,957	N II	12	679,400	Ar II	6
772,385	N III	12	679,217	F IV	16
771,901	N III	11	679,003	F IV	13
771,544	N III	10	677,224	F IV	15
770,409	Ne VIII	8	677,154	F IV	13
769,152	Ar III	12	676,564	Kr III	25
765,148	N IV	15	676,241	Ar III	6
764,357	N III	15	676,130	F IV	14
763,340	N III	14	672,948	O II	8
762,001	O V	10	672,330	Kr III	25
761,130	O V	10	671,391	N II	8
760,445	O V	12	669,725	Ca IV	10
760,229	O V	10	668,43	Cs II	12
759,440	O V	10	666,010	Ar II	6
758,677	O V	10	663,039	Kr III	20
752,051	Kr II	30	660,280	N II	9
746,976	N II	7	659,716	Kr III	22
745,323	Ar II	7	658,337	F III	12
745,264	K IV	10	656,878	F III	11
744,925	Ar II	8	656,125	F III	10
743,721	Ne I	12	656,038	Ca IV	15
742,83	Kr VI	8	651,57	Kr VIII	10
741,950	K IV	10	646,570	Ca V	8
740,44	Xe VIII	7	646,417	Kr III	20
740,270	Ar II	10	646,188	K IV	15
737,144	K IV	10	645,167	N II	10
735,892	Ne I	30	645,0	Cs III	4
729,1	Li III	—	644,825	N II	9
724,420	K VI	8	644,621	N II	8
724,420	K V	8	644,148	O II	12

λ, Å	Symbol	I	λ, Å	Symbol	I
643,256	Ar III	9	555,485	Cl VI	20
641,808	Ar III	12	555,262	O IV	16
639,42	Cs II	12	554,619	Cl IV	7
637,282	Ar III	20	554,52	Kr VI	5
629,732	O V	15	554,514	O IV	18
629,729	Ne I	6	554,074	O IV	17
626,819	Ne I	6	553,470	Ar III	9
625,852	O IV	14	553,328	O IV	16
625,130	O IV	14	552,017	Cl IV	7
624,617	O IV	13	551,992	Cl VI	10
623,016	K VI	8	551,371	Ar VI	8
618,67	Kr VII	1	550,2	Cs III	2
618,668	Ne I	5	547,8	Cs III	2
616,136	K VI	6	547,630	Cl V	10
615,623	Ne I	5	546,846	F II	6
612,621	K II	4	546,329	Cl V	6
609,829	O IV	15	545,114	Cl V	10
608,395	O IV	14	544,03	Kr VI	5
608,065	F II	7	543,891	Ne IV	150
607,931	K II	5	542,297	Cl V	6
607,472	F II	6	542,290	Ca V	10
606,805	F II	8	542,229	Cl V	8
606,284	F II	6	542,073	Ne IV	100
605,668	F II	7	539,547	O II	8
603,429	K V	8	539,086	O II	8
600,765	K II	6	538,3120	C III	13
599,84	Xe VI	3	538,256	O II	10
599,598	O III	18	538,150	C III	12
597,818	O III	15	538,075	C III	11
595,7	Cs III	2	538,032	Cl V	5
586,322	K V	8	537,830	O II	9
585,754	Ar VII	15	537,606	Cl IV	9
585,37	Kr VII	8	537,030	He I	200
584,334	He I	500	535,666	Cl IV	7
580,319	K V	7	535,288	C III	10
576,8	Fe IV	40	535,204	F VI	10
574,5	Fe IV	50	534,727	Cl IV	8
574,279	C III	12	529,900	Ar III	9
572,637	F IV	16	528,286	Ca III	8
572,336	Ne V	80	527,693	Ar V	6
571,384	F IV	15	526,28	Fe IV	75
571,302	F IV	14	525,795	O III	18
570,636	F IV	14	525,68	Fe IV	100
569,830	Ne V	50	524,189	Ar V	5
569,13	Kr VI	5	522,213	He I	80
567,737	F III	9	517,00	Xe VIII	2
567,676	F III	10	514,945	F II	6
566,04	Xe VII	2	511,527	C III	10
562,805	Ne VI	15	508,434	Ar III	9
562,55	Xe VIII	2	508,384	F III	10
561,738	Cl III	7	508,182	O III	18
561,680	Cl III	7	507,683	O III	17
561,530	Cl III	7	505,199	Ca VI	8
560,390	Al III	7	497,104	K III	15
558,602	Ca V	10	495,144	K II	6
558,481	Ar V	5	491,340	Na VI	6
557,118	Cl III	7	491,050	Ne III	9
556,605	Cl III	7	490,997	F IV	16
556,232	Cl III	6	490,566	F IV	13

λ, Å	Symbol	I	λ, Å	Symbol	I
490,310	Ne III	7	434,722	K III	15
489,580	Na VI	5	434,570	Ca IV	12
489,501	Ne III	10	430,758	F IV	15
488,868	Ne III	7	430,154	F III	11
488,120	K VI	10	429,511	F III	10
488,103	Ne III	8	425,588	K V	7
486,172	Cl IV	8	425,000	Ca V	15
485,084	K II	5	422,287	Fe V	6
484,600	F II	8	421,609	Ne IV	150
482,987	Ne V	50	420,727	F IV	16
479,379	Ar VII	12	420,041	F IV	15
475,656	Ar VII	8	419,714	C IV	14
472,16	Kr V	3	419,644	F IV	14
471,990	F II	6	419,525	C IV	13
471,569	K III	15	418,033	Fe V	6
470,089	K III	20	417,595	Na VI	6
469,865	Ne IV	200	417,382	Fe V	6
469,817	Ne IV	200	416,198	Ne V	80
466,995	F V	5	415,505	Na VI	4
466,793	K III	15	412,240	Na IV	8
465,978	F V	7	411,333	Na IV	7
465,374	F V	6	410,540	Na IV	6
465,27	Kr VI	6	410,371	Na IV	10
465,21	Ne VII	10	409,971	Ca III	18
465,113	F III	10	409,615	Na IV	8
464,370	F V	5	408,682	Na IV	8
464,284	F III	9	407,136	Ne II	8
464,270	K VI	10	405,852	Ne II	9
463,938	Ar V	7	403,732	Ca III	20
463,263	Na V	12	403,315	Mg VI	8
462,388	Ne II	14	403,262	Ne VI	10
462,007	Ar VI	25	401,939	Ne VI	25
461,227	Ar VI	6	401,138	Ne VI	15
461,227	Ar V	6	400,722	Na V	10
461,051	Na V	10	400,676	Mg VI	7
460,725	Ne II	15	399,289	Mg VI	6
460,438	K VI	8	393,142	K IV	10
459,897	Na V	7	392,907	Fe V	6
459,633	C III	15	392,433	Cl V	5
459,521	C III	14	388,218	Ne IV	100
459,462	C III	13	387,141	Ne IV	125
459,320	Ar VI	10	386,203	C III	14
458,048	K VI	7	384,957	Fe V	6
457,475	Ar VI	20	384,178	C IV	17
457,177	F II	6	384,032	C IV	16
452,226	N III	11	379,308	Ne III	7
451,869	N III	10	376,375	Na II	3
450,565	Ca IV	10	374,441	N III	12
449,065	Ar V	18	374,204	N III	11
448,595	K III	15	374,075	O III	10
447,813	Ne II	8	373,997	Ca VI	7
446,949	Ar V	8	372,148	K V	10
446,252	Ne II	8	372,069	Na II	6
445,997	Ar V	5	370,022	Ca VI	7
445,190	Na V	7	366,110	Na VI	4
444,344	K III	15	365,858	Fe V	6
443,821	Ca IV	15	365,594	Ne V	100
440,429	K III	15	365,440	Fe V	6
434,975	O III	10	362,444	Na VI	4

λ, Å	Symbol	I
361,250	Na VI	8
360,367	Na V	8
360,319	Na V	8
359,385	Ne V	50
358,721	Ne IV	200
358,472	Ne V	50
357,973	Ca III	8
355,326	Mg V	12
354,223	Mg V	10
353,300	Mg V	9
353,094	Mg V	14
352,915	Ca V	9
352,202	Mg V	10
351,089	Mg V	12
349,155	Mg VI	10
345,309	O III	10
340,528	Ca VI	8
337,998	Ar V	6
336,555	Ca IV	15
335,374	Ca IV	25
335,050	N IV	11
333,910	Na V	9
332,550	Na V	8
328,448	O III	10
325,161	Cl VI	25
323,936	Cl VI	20
323,356	Cl VI	15
323,310	Mg IV	18
322,166	Ca V	10
321,593	Ca IV	10
320,999	Mg IV	20
320,979	O III	12
319,638	Na IV	10
318,093	Ca IV	15
317,641	Na VI	6
313,748	Na VI	5
312,455	C IV	14
312,418	C IV	15
312,311	Mg V	10
312,263	Fe VI	7
311,921	Na VI	4
311,702	Fe VI	7
308,264	Na V	10
307,152	Na V	8
305,769	O III	10
305,656	O III	9
304,551	Fe VI	7
304,221	Fe VI	7
303,799	O III	9
303,783	He II	500
300,252	K V	7
297,568	Fe VI	8
297,308	Fe VI	7
296,988	Fe VI	6
294,520	Fe VI	7
294,265	Fe VI	7
294,052	Ar VI	6
293,966	Fe VI	8
293,745	Fe VI	8

λ, Å	Symbol	I
292,736	Fe VI	7
291,229	Fe VI	6
291,184	Fe VI	6
289,143	C IV	9
286,965	Ca V	9
283,579	N IV	12
283,470	N IV	11
283,420	N IV	10
282,423	Ar VI	6
281,397	Al V	14
279,937	O IV	11
279,633	O IV	10
278,699	Al V	16
276,581	Mg V	16
270,394	Mg VI	12
268,986	Mg VI	10
267,772	Ca V	8
266,378	N V	9
266,197	N V	8
260,389	O IV	10
256,317	He I	150
250,940	Ar VII	7
248,744	C V	0
248,668	C V	0
247,709	N V	7
247,564	N V	6
244,907	C IV	10
243,854	Cl VI	12
243,760	Al VI	12
243,027	He II	70
239,535	Ca VI	7
238,573	O IV	15
238,361	O IV	14
234,258	Mg III	12
231,730	Mg III	14
230,875	Ar VIII	7
229,734	Ca VI	7
228,898	Ti V	75
228,628	Ca VI	7
225,337	Ti V	100
220,352	O V	13
212,556	Ne IV	150
208,734	Ne IV	100
208,485	Ne IV	100
207,794	O V	10
199,759	Ti VI	6
198,974	Ti VI	8
194,900	Ti VI	7
192,906	O V	14
192,800	O V	13
192,751	O V	12
192,747	Ti VI	8
192,635	Ar VII	7
190,839	F V	7
190,835	Na IV	8
190,571	F V	6
190,440	Na IV	10

λ, A	Symbol	I	λ, A	Symbol	I
187,194	Mg III	8	125,525	Al V	15
186,842	F V	5	124,153	Na VI	4
186,510	Mg III	9	124,059	Na VI	4
184,117	O VI	9	124,034	Al IV	8
183,937	O VI	8	123,929	Na VI	5
181,758	Na IV	8	122,686	Ne VI	10
181,345	Mg IV	8	122,520	Ne VI	20
180,796	Mg IV	9	120,331	O VII	0
180,617	Mg IV	10	118,968	Si V	20
180,254	Ar VIII	15	117,860	Si V	20
			116,459	Al IV	7
180,070	Mg IV	8	113,93	Li III	—
179,400	Ar VIII	10	109,896	Na VI	5
178,434	F V	5	109,514	Al VI	20
176,566	Ar VII	10	107,945	Al V	20
173,082	O VI	13	107,683	Na VI	5
			107,620	Al VI	14
172,935	O VI	12	107,608	Na VI	4
172,306	Mg IV	7	107,288	Na VI	4
172,163	O V	12	106,2	Ne VII	7
171,653	Mg IV	8	106,1	Ne VII	7
170,802	Mg III	5			
			104,344	Al VI	16
168,409	Na IV	8	104,047	Al VI	20
168,084	Na IV	10	99,460	Si VI	15
166,177	F V	10	98,2	Ne VIII	9
165,983	F V	9	98,1	Ne VIII	9
163,558	F V	5			
			97,143	Si V	10
162,445	Na IV	8	96,439	Si V	15
161,686	Al IV	14	92,626	Al VI	15
160,073	Al IV	16	88,469	Al VI	5
158,923	Ar VIII	8	88,1	Ne VIII	9
156,536	Na IV	8	85,764	Al VI	8
156,247	F VI	6	85,724	Al VI	6
150,124	O VI	9	85,622	Al VI	6
150,088	O VI	10	85,175	Si V	10
148,002	F V	5	84,082	Si VI	12
139,900	F VI	7	83,128	Si VI	15
139,800	F VI	6	80,577	Si VI	12
139,758	F VI	5	77,945	Al VI	10
137,414	Mg V	8	74,656	Al VI	5
135,02	Li III	—	74,444	Al VI	6
134,539	F V	5	72,810	Al VI	5
131,441	Al V	20	40,731	C V	—
131,003	Al V	20	40,270	C V	—
130,848	Al V	20	34,973	C V	—
130,413	Al V	20	33,734	C VI	—
130,403	Al IV	11	33,426	C V	—
129,872	O VI	6	29,084	N VI	—
129,786	O VI	5	28,787	N VI	—
129,729	Al IV	12	28,464	C VI	—
128,500	O VII	0	26,988	C VI	—
128,412	O VII	0	24,898	N VI	—
127,837	Na VI	4	23,771	N VI	—
126,923	F VI	5	21,804	O VII	—
126,065	Al V	15	21,602	O VII	—

Section III

Complete Tables of Spectral Lines with Their Classification Arranged by Element and Degree of Ionization

HYDROGEN, Z = 1

H, ground state $1s\ ^2S_{1/2}$

Ionization potential 109678,758 cm^{-1}; 13,597 eV

λ, Å	I	E_H, eV	E_B, eV	Transition	J
190569	—	13,32	13,39	$7i\ ^2I-8k\ ^2K^\circ$ etc.	$^{13}/_2,\ ^{11}/_2-^{15}/_2,\ ^{13}/_2$
123684	3	13,22	13,32	$6h\ ^2H^\circ-7i\ ^2I$ etc.	$^{11}/_2,\ ^{9}/_2-^{13}/_2,\ ^{11}/_2$
113057	—	13,32	13,43	$7i\ ^2I-9k\ ^2K^\circ$ etc.	$^{13}/_2,\ ^{11}/_2-^{15}/_2,\ ^{13}/_2$
75004,5	—	13,22	13,39	$6h\ ^2H^\circ-8i\ ^3I$ etc.	$^{11}/_2,\ ^{9}/_2-^{13}/_2,\ ^{11}/_2$
74577,6	20	13,06	13,22	$5g\ ^2G-6h\ ^2H^\circ$ etc.	$^{9}/_2,\ ^{7}/_2-^{11}/_2,\ ^{9}/_2$
46524,7	3	13,06	13,32	$5g\ ^2G-7h\ ^2H^\circ$ etc.	
40511,4	120	12,75	13,06	$4f\ ^2F^\circ-5g\ ^3G$ etc.	$^{7}/_2,\ ^{5}/_2-^{9}/_2,\ ^{7}/_2$
26251,3	40	12,75	13,22	$4f\ ^2F^\circ-6g\ ^3G$ etc.	
21655,2	13	12,75	13,32	$4f\ ^2F^\circ-7g\ ^2G$ etc.	
18751,1	700	12,09	12,75	$3d\ ^2D-4f\ ^2F^\circ$ etc.	For all the transitions indicated of this series $^{5}/_2,\ ^{3}/_2-^{7}/_2,\ ^{5}/_2$
12818,05	140	12,09	13,06	$3d\ ^2D-5f\ ^2F^\circ$ etc.	
10938,09	28	12,09	13,22	$3d\ ^2D-6f\ ^2F^\circ$ etc.	
10049,38	6	12,09	13,32	$3d\ ^2D-7f\ ^2F^\circ$ etc.	
9545,974	5	12,09	13,39	$3d\ ^2D-8f\ ^2F^\circ$ etc.	
9229,017	4	12,09	13,43	$3d\ ^2D-9f\ ^2F^\circ$ etc.	
9014,911	3	12,09	13,46	$3d\ ^2D-10f\ ^2F^\circ$ etc.	
8862,787	2	12,09	13,48	$3d\ ^2D-11f\ ^2F^\circ$ etc.	
8750,475	—	12,09	13,50	$3d\ ^2D-12f\ ^2F^\circ$ etc.	
8665,021	—	12,09	13,52	$3d\ ^2D-13f\ ^2F^\circ$ etc.	
8598,394	—	12,09	13,53	$3d\ ^2D-14f\ ^2F^\circ$ etc.	
8545,384	—	12,09	13,54	$3d\ ^2D-15f\ ^2F^\circ$ etc.	
8502,487	—	12,09	13,54	$3d\ ^2D-16f\ ^2F^\circ$ etc.	
8467,256	—	12,09	13,55	$3d\ ^2D-17f\ ^2F^\circ$ etc.	
8437,958	—	12,09	13,56	$3d\ ^2D-18f\ ^2F^\circ$ etc.	
8413,321	—	12,09	13,56	$3d\ ^2D-19f\ ^2F^\circ$ etc.	
8392,400	—	12,09	13,56	$3d\ ^2D-20f\ ^2F^\circ$ etc.	
8374,478	—	12,09	13,57	$3d\ ^2D-21f\ ^2F^\circ$ etc.	
8359,006	—	12,09	13,57	$3d\ ^2D-22f\ ^2F^\circ$ etc.	
8345,553	—	12,09	13,57	$3d\ ^2D-23f\ ^2F^\circ$ etc.	
8333,785	—	12,09	13,57	$3d\ ^2D-24f\ ^2F^\circ$ etc.	
8323,428	—	12,09	13,58	$3d\ ^2D-25f\ ^2F^\circ$ etc.	
8314,262	—	12,09	13,58	$3d\ ^2D-26f\ ^2F^\circ$ etc.	
8306,115	—	12,09	13,58	$3d\ ^2D-27f\ ^2F^\circ$ etc.	
8298,837	—	12,09	13,58	$3d\ ^2D-28f\ ^2F^\circ$ etc.	
8292,309	—	12,09	13,58	$3d\ ^2D-29f\ ^2F^\circ$ etc.	
8286,434	—	12,09	13,58	$3d\ ^2D-30f\ ^2F^\circ$ etc.	
8281,125	—	12,09	13,58	$3d\ ^2D-31f\ ^2F^\circ$ etc.	
8276,310	—	12,09	13,58	$3d\ ^2D-32f\ ^2F^\circ$ etc.	
8203,572	—	12,09	13,60	Limit of series	
H_α 6562,849	2000	10,20	12,09	$2p\ ^2P^\circ-3d\ ^2D$ etc.	$^{3}/_2-^{5}/_2,\ ^{3}/_2$
H_α 6562,725	1000	10,20	12,09	$2p\ ^2P^\circ-3d\ ^2D$ etc.	$^{1}/_2-^{3}/_2$
H_β 4861,332	500	10,20	12,75	$2p\ ^2P^\circ-4d\ ^2D$ etc.	For all the transitions indicated of this series $^{3}/_2,\ ^{1}/_2-^{5}/_2,\ ^{3}/_2$
H_γ 4340,468	200	10,20	13,06	$2p\ ^2P^\circ-5d\ ^2D$ etc.	
H_δ 4101,737	100	10,20	13,22	$2p\ ^2P^\circ-6d\ ^2D$ etc.	
H_ε 3970,074	80	10,20	13,32	$2p\ ^2P^\circ-7d\ ^2D$ etc.	
3889,051	60	10,20	13,39	$2p\ ^2P^\circ-8d\ ^2D$ etc.	
3835,386	40	10,20	13,43	$2p\ ^2P^\circ-9d\ ^2D$ etc.	
3797,900	20	10,20	13,46	$2p\ ^2P^\circ-10d\ ^2D$ etc.	
3770,632	15	10,20	13,48	$2p\ ^2P^\circ-11d\ ^2D$ etc.	
3750,154	10	10,20	13,50	$2p\ ^2P^\circ-12d\ ^2D$ etc.	
3734,370	8	10,20	13,52	$2p\ ^2P^\circ-13d\ ^2D$ etc.	
3721,940	6	10,20	13,53	$2p\ ^2P^\circ-14d\ ^2D$ etc.	

λ, Å	I	E_H, eV	E_B, eV	Transition	J
3711,974	5	10,20	13,54	$2p\ ^2P^\circ$—$15d\ ^2D$ etc.	
3703,855	4	10,20	13,54	$2p\ ^2P^\circ$—$16d\ ^2D$ etc.	
3697,154	3	10,20	13,55	$2p\ ^2P^\circ$—$17d\ ^2D$ etc.	
3691,557	2	10,20	13,56	$2p\ ^2P^\circ$—$18d\ ^2D$ etc.	
3686,833	—	10,20	13,56	$2p\ ^2P^\circ$—$19d\ ^2D$ etc.	
3682,810	—	10,20	13,56	$2p\ ^2P^\circ$—$20d\ ^2D$ etc.	
3679,355	—	10,20	13,57	$2p\ ^2P^\circ$—$21d\ ^2D$ etc.	
3676,365	—	10,20	13,57	$2p\ ^2P^\circ$—$22d\ ^2D$ etc.	
3673,761	—	10,20	13,57	$2p\ ^2P^\circ$—$23d\ ^2D$ etc.	
3671,478	—	10,20	13,57	$2p\ ^2P^\circ$—$24d\ ^2D$ etc.	
3669,466	—	10,20	13,58	$2p\ ^2P^\circ$—$25d\ ^2D$ etc.	
3667,684	—	10,20	13,58	$2p\ ^2P^\circ$—$26d\ ^2D$ etc.	
3666,097	—	10,20	13,58	$2p\ ^2P^\circ$—$27d\ ^2D$ etc.	
3664,679	—	10,20	13,58	$2p\ ^2P^\circ$—$28d\ ^2D$ etc.	
3663,406	—	10,20	13,58	$2p\ ^2P^\circ$—$29d\ ^2D$ etc.	
3662,258	—	10,20	13,58	$2p\ ^2P^\circ$—$30d\ ^2D$ etc.	
3661,221	—	10,20	13,58	$2p\ ^2P^\circ$—$31d\ ^2D$ etc.	
3660,279	—	10,20	13,58	$2p\ ^2P^\circ$—$32d\ ^2D$ etc.	
3645,981	—	10,20	13,60	Limit of series	
1215,670	3000	0,00	10,20	$1s\ ^2S$—$2p\ ^2P^\circ$	For all the transitions of this series $^1/_2$—$^3/_2$, $^1/_2$
1025,722	1000	0,00	12,09	$1s\ ^2S$—$3p\ ^2P^\circ$	
972,537	400	0,00	12,75	$1s\ ^2S$—$4p\ ^2P^\circ$	
949,743	200	0,00	13,06	$1s\ ^2S$—$5p\ ^2P^\circ$	
937,804	120	0,00	13,22	$1s\ ^2S$—$6p\ ^2P^\circ$	
930,748	80	0,00	13,32	$1s\ ^2S$—$7p\ ^2P^\circ$	
926,226	50	0,00	13,39	$1s\ ^2S$—$8p\ ^2P^\circ$	
923,150	40	0,00	13,42	$1s\ ^2S$—$9p\ ^2P^\circ$	
920,963	30	0,00	13,45	$1s\ ^2S$—$10p\ ^2P^\circ$	
919,351	20	0,00	13,48	$1s\ ^2S$—$11p\ ^2P^\circ$	
918,129	16	0,00	13,50	$1s\ ^2S$—$12p\ ^2P^\circ$	
917,181	12	0,00	13,52	$1s\ ^2S$—$13p\ ^2P^\circ$	
916,429	10	0,00	13,53	$1s\ ^2S$—$14p\ ^2P^\circ$	
915,824	8	0,00	13,54	$1s\ ^2S$—$15p\ ^2P^\circ$	
915,329	7	0,00	13,54	$1s\ ^2S$—$16p\ ^2P^\circ$	
914,919	6	0,00	13,55	$1s\ ^2S$—$17p\ ^2P^\circ$	
914,576	5	0,00	13,56	$1s\ ^2S$—$18p\ ^2P^\circ$	
914,286	—	0,00	13,56	$1s\ ^2S$—$19p\ ^2P^\circ$	
911,753	—	0,00	13,60	Limit of series	

DEUTERIUM, Z = 1

D, ground state $1s\ ^2S_{1/2}$

Ionization potential 109708,596 **cm^{-1}; 13,601 eV**

λ, Å	I	E_H, eV	E_B, eV	Transition	J
18746,0	700	12,09	12,75	$3d\ ^2D—4f\ ^2F^\circ$ etc.	For all the transitions indicated of this series $^5/_2$, $^3/_2—^7/_2$, $^5/_2$
12814,56	140	12,09	13,06	$3d\ ^2D—5f\ ^2F^\circ$ etc.	
10935,11	28	12,09	13,22	$3d\ ^2D—6f\ ^2F^\circ$ etc.	
10046,64	6	12,09	13,32	$3d\ ^2D—7f\ ^2F^\circ$ etc.	
9543,376	5	12,09	13,39	$3d\ ^2D—8f\ ^2F^\circ$ etc.	
9226,505	4	12,09	13,43	$3d\ ^2D—9f\ ^2F^\circ$ etc.	
9012,457	3	12,09	13,46	$3d\ ^2D—10f\ ^2F^\circ$ etc.	
8860,374	2	12,09	13,48	$3d\ ^2D—11f\ ^2F^\circ$ etc.	
8748,093	—	12,09	13,50	$3d\ ^2D—12f\ ^2F^\circ$ etc.	
8201,334	—	12,09	13,60	Limit of series	
D_α 6561,032	3000	10,20	12,09	$2p\ ^2P^\circ—3d\ ^2D$ etc.	For all the transitions indicated of this series $^3/_2$, $^1/_2—^5/_2$, $^3/_2$
D_β 4860,029	500	10,20	12,75	$2p\ ^2P^\circ—4d\ ^2D$ etc.	
D_γ 4339,287	200	10,20	13,06	$2p\ ^2P^\circ—5d\ ^2D$ etc.	
D_δ 4100,621	100	10,20	13,22	$2p\ ^2P^\circ—6d\ ^2D$ etc.	
D_ε 3968,995	80	10,20	13,32	$2p\ ^2P^\circ—7d\ ^2D$ etc.	
3887,993	60	10,20	13,39	$2p\ ^2P^\circ—8d\ ^2D$ etc.	
3834,342	40	10,20	13,43	$2p\ ^2P^\circ—9d\ ^2D$ etc.	
3796,866	20	10,20	13,46	$2p\ ^2P^\circ—10d\ ^2D$ etc.	
3769,606	15	10,20	13,48	$2p\ ^2P^\circ—11d\ ^2D$ etc.	
3644,989	—	10,20	13,60	Limit of series	
1215,340	3000	0,00	10,20	$1s\ ^2S—2p\ ^2P^\circ$	For all the transitions of this series $^1/_2—^3/_2$, $^1/_2$
1025,443	1000	0,00	12,09	$1s\ ^2S—3p\ ^2P^\circ$	
972,272	400	0,00	12,75	$1s\ ^2S—4p\ ^2P^\circ$	
949,485	200	0,00	13,06	$1s\ ^2S—5p\ ^2P^\circ$	
937,548	120	0,00	13,22	$1s\ ^2S—6p\ ^2P^\circ$	
930,495	80	0,00	13,32	$1s\ ^2S—7p\ ^2P^\circ$	
925,974	50	0,00	13,39	$1s\ ^2S—8p\ ^2P^\circ$	
922,899	40	0,00	13,43	$1s\ ^2S—9p\ ^2P^\circ$	
920,713	30	0,00	13,46	$1s\ ^2S—10p\ ^2P^\circ$	
911,505	—	0,00	13,60	Limit of series	

TRITIUM, Z = 1

T, o ground state $1s\ ^2S_{1/2}$

Ionization potential 109718,526 cm⁻¹; 13,603 eV

λ, Å	I	E_H, eV	E_B, eV	Transition	J
18744,3	700	12,09	12,75	$3d\ ^2D$—$4f\ ^2F^\circ$ etc.	For all the transitions indicated of this series $^5/_2, ^3/_2$—$^7/_2, ^5/_2$
12813,40	140	12,09	13,06	$3d\ ^2D$—$5f\ ^2F^\circ$ etc.	
10934,12	28	12,09	13,22	$3d\ ^2D$—$6f\ ^2F^\circ$ etc.	
10045,73	6	12,09	13,32	$3d\ ^2D$—$7f\ ^2F^\circ$ etc.	
9542,509	5	12,09	13,39	$3d\ ^2D$—$8f\ ^2F^\circ$ etc.	
9225,667	4	12,09	13,43	$3d\ ^2D$—$9f\ ^2F^\circ$ etc.	
9011,639	3	12,09	13,46	$3d\ ^2D$—$10f\ ^2F^\circ$ etc.	
8859,570	2	12,09	13,48	$3d\ ^2D$—$11f\ ^2F^\circ$ etc.	
8747,298	—	12,09	13,50	$3d\ ^2D$—$12f\ ^2F^\circ$ etc.	
8200,594	—	12,09	13,60	Limit of series	
T_α 6560,435	3000	10,20	12,09	$2p\ ^2P^\circ$—$3d\ ^2D^\circ$ etc.	For all the transitions indicated of this series $^3/_2, ^1/_2$—$^5/_2, ^3/_2$
T_β 4859,595	500	10,20	12,75	$2p\ ^2P^\circ$—$4d\ ^2D^\circ$ etc.	
T_γ 4338,893	200	10,20	13,06	$2p\ ^2P^\circ$—$5d\ ^2D^\circ$ etc.	
T_δ 4100,249	100	10,20	13,22	$2p\ ^2P^\circ$—$6d\ ^2D$ etc.	
T_ε 3968,637	80	10,20	13,32	$2p\ ^2P^\circ$—$7d\ ^2D$ etc.	
3887,640	60	10,20	13,39	$2p\ ^2P^\circ$—$8d\ ^2D$ etc.	
3833,994	40	10,20	13,43	$2p\ ^2P^\circ$—$9d\ ^2D$ etc.	
3796,522	20	10,20	13,46	$2p\ ^2P^\circ$—$10d\ ^2D$ etc.	
3769,264	15	10,20	13,48	$2p\ ^2P^\circ$—$11d\ ^2D$ etc.	
3644,656	—	10,20	13,60	Limit of series	
1215,229	3000	0,00	10,20	$1s\ ^2S$—$2p\ ^2P^\circ$	For all the transitions of this series $^1/_2$—$^3/_2, ^1/_2$
1025,350	1000	0,00	12,09	$1s\ ^2S$—$3p\ ^2P^\circ$	
972,184	400	0,00	12,75	$1s\ ^2S$—$4p\ ^2P^\circ$	
945,401	200	0,00	13,06	$1s\ ^2S$—$5p\ ^2P^\circ$	
937,464	120	0,00	13,22	$1s\ ^2S$—$6p\ ^2P^\circ$	
930,410	80	0,00	13,32	$1s\ ^2S$—$7p\ ^2P^\circ$	
925,890	50	0,00	13,39	$1s\ ^2S$—$8p\ ^2P^\circ$	
922,815	40	0,00	13,43	$1s\ ^2S$—$9p\ ^2P^\circ$	
920,629	30	0,00	13,46	$1s\ ^2S$—$10p\ ^2P^\circ$	
911,422	—	0,00	13,60	Limit of series	

HELIUM, Z = 2

He I, ground state $1s^2\ ^1S_0$

Ionization potential 198310,8 cm^{-1}; 24,586 **eV**

λ, Å	I	E_H, eV	E_B, eV	Transition	J
21132,04	40	23,09	23,67	$3p\ ^1P^\circ\ -4s\ ^1S$	1—0
21121,31	150	23,01	23,59	$3p\ ^3P^\circ\ -4s\ ^3S$	0—1
21120,04	150	23,01	23,59	$3p\ ^3P^\circ\ -4s\ ^3S$	2, 1—1
20581,30	10 000	20,61	21,22	$2s\ ^1S\ -2p\ ^1P^\circ$	0—1
19543,13	65	23,07	23,71	$3d\ ^3D\ -4p\ ^3P^\circ$	3, 2, 1—2, 1, 0
19089,37	550	23,09	23,74	$3p\ ^1P^\circ\ -4d\ ^1D$	1—2
18696,94	1500	23,07	23,74	$3d\ ^1D\ -4f\ ^1F^\circ$	2—3
18685,96	3600	23,07	23,74	$3d\ ^3D\ -4f\ ^3F^\circ$	3, 2, 1—4, 3, 2
18555,55	6	23,07	23,74	$3d\ ^1D\ -4p\ ^1P^\circ$	2—1
17003,15	200	23,01	23,73	$3p\ ^3P^\circ\ -4d\ ^3D$	0—1
17002,38	1800	23,01	23,73	$3p\ ^3P^\circ\ -4d\ ^3D$	2, 1—3, 2, 1
15083,66	60	22,92	23,74	$3s\ ^1S\ -4p\ ^1P^\circ$	0—1
12968,44	50	23,09	24,04	$3p\ ^1P^\circ\ -5d\ ^1D$	1—2
12845,95	30	23,01	23,97	$3p\ ^3P^\circ\ -5s\ ^3S$	2, 1, 0—1
12790,27	125	23,07	24,04	$3d\ ^1D\ -5f\ ^1F^\circ$	2—3
12784,79	400	23,07	24,04	$3d\ ^3D\ -5f\ ^3F^\circ$	3, 2, 1—4, 3, 2
12527,51	100	22,72	23,71	$3s\ ^3S\ -4p\ ^3P^\circ$	1—2, 1, 0
11969,48	—	23,01	24,04	$3p\ ^3P^\circ\ -3d\ ^3D$	0—1
11969,07	220	23,01	24,04	$3p\ ^3P^\circ\ -5d\ ^3D$	2, 1—3, 2, 1
11225,90	—	23,09	24,19	$3p\ ^1P^\circ\ -6s\ ^1S$	1—0
11045,00	8	23,09	24,21	$3p\ ^1P^\circ\ -6d\ ^1D$	1—2
11013,07	8	22,92	24,04	$3s\ ^1S\ -5p\ ^1P^\circ$	0—1
10996,56	3	23,07	24,20	$3d\ ^3D\ -6p\ ^3P^\circ$	3, 2, 1—2, 1, 0
10916,98	25	23,07	24,21	$3d\ ^1D\ -6f\ ^1F^\circ$	2—3
10912,92	60	23,07	24,21	$3d\ ^3D\ -6f\ ^3F^\circ$	3, 2, 1—4, 3, 2
10902,16	1	23,07	24,21	$3d\ ^1D\ -6p\ ^1P^\circ$	2—1
10830,337	25 000	19,82	20,96	$2s\ ^3S\ -2p\ ^3P^\circ$	1—2
10830,248	15 000	19,82	20,96	$2s\ ^3S\ -2p\ ^3P^\circ$	1—1
10829,088	5000	19,82	20,96	$2s\ ^3S\ -2p\ ^3P^\circ$	1—0
10667,65	15	23,01	24,17	$3p\ ^3P^\circ\ -6s\ ^3S$	2, 1, 0—1
10311,54	7	23,01	24,21	$3p\ ^3P^\circ\ -6d\ ^3D$	0—1
10311,23	50	23,01	24,21	$3p\ ^3P^\circ\ -6d\ ^3D$	2, 1—3, 2, 1
10233,06	2	23,09	24,30	$3p\ ^1P^\circ\ -7s\ ^1S$	1—0
10138,50	5	23,09	24,31	$3p\ ^1P^\circ\ -7d\ ^1D$	1—2
10072,04	3	23,07	24,30	$3d\ ^3D\ -7p\ ^3P^\circ$	3, 2, 1—2, 1, 0
10031,16	10	23,07	24,31	$3d\ ^1D\ -7f\ ^1F^\circ$	2—3
10027,73	30	23,07	24,31	$3d\ ^3D\ -7f\ ^3F^\circ$	3, 2, 1—4, 3, 2
9702,60	15	23,01	24,28	$3p\ ^3P^\circ\ -7s\ ^3S$	2, 1, 0—1
9682,19	1	23,09	24,37	$3p\ ^1P^\circ\ -8s\ ^1S$	1—0
9625,64	3	23,09	24,37	$3p\ ^1P^\circ\ -8d\ ^1D$	1—2
9603,42	5	22,92	24,21	$3s\ ^1S\ -6p\ ^1P^\circ$	0—1
9552,89	2	23,07	24,37	$3d\ ^3D\ -8p\ ^3P^\circ$	3, 2, 1—2, 1, 0
9529,27	5	23,07	24,37	$3d\ ^1D\ -8f\ ^1F^\circ$	2—3
9526,17	15	23,07	24,37	$3d\ ^3D\ -8f\ ^3F^\circ$	3, 2, 1—4, 3, 2
9516,87	3	23,01	24,31	$3p\ ^3P^\circ\ -7d\ ^3D$	0—1
9516,60	20	23,01	24,31	$3p\ ^3P^\circ\ -7d\ ^3D$	2, 1—3, 2, 1
9463,61	50	22,72	24,03	$3s\ ^3S\ -5p\ ^3P^\circ$	1—2, 1, 0
9210,337	10	23,07	24,42	$3d\ ^3D\ -9f\ ^3F^\circ$	3, 2, 1—4, 3, 2
9174,52	2	23,01	24,37	$3p\ ^3P^\circ\ -8s\ ^3S$	2, 1, 0—1
9063,27	6	23,01	24,37	$3p\ ^3P^\circ\ -8d\ ^3D$	2, 1, 0—3, 2, 1
8996,978	2	23,07	24,45	$3d\ ^3D\ -10f\ ^3F^\circ$	3, 2, 1—4, 3, 2
8914,74	2	22,92	24,31	$3s\ ^1S\ -7p\ ^1P^\circ$	0—1

λ, Å	I	E_H, eV	E_B, eV	Transition	J
8776,74	2	23,01	24,42	$3p\ ^3P^\circ$—$9d\ ^3D$	2, 1, 0—3, 2, 1
8361,69	10	22,72	24,20	$3s\ ^3S$—$6p\ ^3P^\circ$	1—2, 1, 0
7816,15	5	22,72	24,30	$3s\ ^3S$—$7p\ ^3P^\circ$	1—2, 1, 0
7281,349	500	21,22	22,92	$2p\ ^1P^\circ$—$3s\ ^1S$	1—0
7065,707	300	20,96	22,72	$2p\ ^3P^\circ$—$3s\ ^3S$	0—1
7065,190	2500	20,96	22,72	$2p\ ^3P^\circ$—$3s\ ^3S$	2, 1—1
6678,151	1000	21,22	23,07	$2p\ ^1P^\circ$—$3d\ ^1D$	1—2
5875,966	1000	20,96	23,07	$2p\ ^3P^\circ$—$3d\ ^3D$	0—1
5875,621	7500	20,96	23,07	$2p\ ^3P^\circ$—$3d\ ^3D$	2, 1—3, 2, 1
5047,738	50	21,22	23,67	$2p\ ^1P^\circ$—$4s\ ^1S$	1—0
5015,6779	500	20,61	23,09	$2s\ ^1S$—$3p\ ^1P^\circ$	0—1
4921,9310	100	21,22	23,74	$2p\ ^1P^\circ$—$4d\ ^1D$	1—2
4713,376	20	20,96	23,59	$2p\ ^3P^\circ$—$4s\ ^3S$	0—1
4713,1455	150	20,96	23,59	$2p\ ^3P^\circ$—$4s\ ^3S$	2, 1—1
4471,682	120	20,96	23,73	$2p\ ^3P^\circ$—$4d\ ^3D$	0—1
4471,479	1000	20,96	23,73	$2p\ ^3P^\circ$—$4d\ ^3D$	2, 1—3, 2, 1
4437,551	15	21,22	24,01	$2p\ ^1P^\circ$—$5s\ ^1S$	1—0
4387,9294	50	21,22	24,04	$2p\ ^1P^\circ$—$5d\ ^1D$	1—2
4168,967	3	21,22	24,19	$2p\ ^1P^\circ$—$6s\ ^1S$	1—0
4143,761	10	21,22	24,21	$2p\ ^1P^\circ$—$6d\ ^1D$	1—2
4120,992	7	20,96	23,97	$2p\ ^3P^\circ$—$5s\ ^3S$	0—1
4120,815	60	20,96	23,97	$2p\ ^3P^\circ$—$5s\ ^3S$	2, 1—1
4026,359	25	20,96	24,04	$2p\ ^3P^\circ$—$5d\ ^3D$	0—1
4026,1912	250	20,96	24,04	$2p\ ^3P^\circ$—$5d\ ^3D$	2, 1—3, 2, 1
4023,973	2	21,22	24,30	$2p\ ^1P^\circ$—$7s\ ^1S$	1—0
4009,268	5	21,22	24,31	$2p\ ^1P^\circ$—$7d\ ^1D$	1—2
3964,7289	100	20,61	23,74	$2s\ ^1S$—$4p\ ^1P^\circ$	0—1
3935,912	2	21,22	24,37	$2p\ ^1P^\circ$—$8s\ ^1S$	1—0
3926,534	7	21,22	24,37	$2p\ ^1P^\circ$—$8d\ ^1D$	1—2
3888,648	5000	19,82	23,01	$2s\ ^3S$—$3p\ ^3P^\circ$	1—2, 1, 0
3878,181	3	21,22	24,41	$2p\ ^1P^\circ$—$9s\ ^1S$	1—0
3871,791	5	21,22	24,42	$2p\ ^1P^\circ$—$9d\ ^1D$	1—2
3867,630	5	20,96	24,17	$2p\ ^3P^\circ$—$6s\ ^3S$	0—1
3867,475	30	20,96	24,17	$2p\ ^3P^\circ$—$6s\ ^3S$	2, 1—1
3838,100	2	21,22	24,45	$2p\ ^1P^\circ$—$10s\ ^1S$	1—0
3833,554	4	21,22	24,45	$2p\ ^1P^\circ$—$10d\ ^1D$	1—2
3819,758	10	20,96	24,21	$2p\ ^3P^\circ$—$6d\ ^3D$	0—1
3819,6072	100	20,96	24,21	$2p\ ^3P^\circ$—$6d\ ^3D$	2, 1—3, 2, 1
3805,740	3	21,22	24,47	$2p\ ^1P^\circ$—$11d\ ^1D$	1—2
3784,862	2	21,22	24,49	$2p\ ^1P^\circ$—$12d\ ^1D$	1—2
3768,784	2	21,22	24,51	$2p\ ^1P^\circ$—$13d\ ^1D$	1—2
3756,107	1	21,22	24,52	$2p\ ^1P^\circ$—$14d\ ^1D$	1—2
3733,010	3	20,96	24,28	$2p\ ^3P^\circ$—$7s\ ^3S$	0—1
3732,865	10	20,96	24,28	$2p\ ^3P^\circ$—$7s\ ^3S$	2, 1—1
3705,148	3	20,96	24,31	$2p\ ^3P^\circ$—$7d\ ^3D$	0—1
3705,005	30	20,96	24,31	$2p\ ^3P^\circ$—$7d\ ^3D$	2, 1—3, 2, 1
3652,130	2	20,96	24,36	$2p\ ^3P^\circ$—$8s\ ^3S$	0—1
3651,990	7	20,96	24,36	$2p\ ^3P^\circ$—$8s\ ^3S$	2, 1—1
3634,369	2	20,96	24,37	$2p\ ^3P^\circ$—$8d\ ^3D$	0—1
3634,232	15	20,96	24,37	$2p\ ^3P^\circ$—$8d\ ^3D$	2, 1—3, 2, 1
3613,643	30	20,61	24,04	$2s\ ^1S$—$5p\ ^1P^\circ$	0—1
3599,448	2	20,96	24,41	$2p\ ^3P^\circ$—$9s\ ^3S$	0—1
3599,314	5	20,96	24,41	$2p\ ^3P^\circ$—$9s\ ^3S$	2, 1—1
3587,405	2	20,96	24,42	$2p\ ^3P^\circ$—$9d\ ^3D$	0—1
3587,270	10	20,96	24,42	$2p\ ^3P^\circ$—$9d\ ^3D$	2, 1—3, 2, 1
3562,979	4	20,96	24,44	$2p\ ^3P^\circ$—$10s\ ^3S$	2, 1, 0—1
3554,547	1	20,96	24,45	$2p\ ^3P^\circ$—$10d\ ^3D$	0—1
3554,415	7	20,96	24,45	$2p\ ^3P^\circ$—$10d\ ^3D$	2, 1—3, 2, 1

λ, Å	I	E_H, eV	E_B, eV	Transition	J
3536,809	3	20,96	24,47	$2p\,^3P^\circ-11d\,^3S$	2, 1, 0—1
3530,491	5	20,96	24,47	$2p\,^3P^\circ-11d\,^3D$	2, 1, 0—3, 2, 1
3517,317	2	20,96	24,49	$2p\,^3P^\circ-12d\,^3S$	2, 1, 0—1
3512,512	4	20,96	24,49	$2p\,^3P^\circ-12d\,^3D$	2, 1, 0—3, 2, 1
3502,379	2	20,96	24,50	$2p\,^3P^\circ-13s\,^3S$	2, 1, 0—1
3498,645	3	20,96	24,51	$2p\,^3P^\circ-13d\,^3D$	2, 1, 0—3, 2, 1
3490,685	2	20,96	24,51	$2p\,^3P^\circ-14s\,^3S$	2, 1, 0—1
3487,723	2	20,96	24,52	$2p\,^3P^\circ-14d\,^3D$	2, 1, 0—3, 2, 1
3478,957	2	20,96	24,53	$2p\,^3P^\circ-15d\,^3D$	2, 1, 0—3, 2, 1
3471,818	1	20,96	24,53	$2p\,^3P^\circ-16d\,^3D$	2, 1, 0—3, 2, 1
3447,586	15	20,61	24,21	$2s\,^1S-6p\,^1P^\circ$	0—1
3354,550	10	20,61	24,31	$2s\,^1S-7p\,^1P^\circ$	0—1
3296,773	7	20,61	24,37	$2s\,^1S-8p\,^1P^\circ$	0—1
3258,275	5	20,61	24,42	$2s\,^1S-9p\,^1P^\circ$	0—1
3231,266	3	20,61	24,45	$2s\,^1S-10p\,^1P^\circ$	0—1
3211,568	2	20,61	24,47	$2s\,^1S-11p\,^1P^\circ$	0—1
3196,742	2	20,61	24,49	$2s\,^1S-12p\,^1P^\circ$	0—1
3187,745	200	19,82	23,71	$2s\,^3S-4p\,^3P^\circ$	1—2, 1, 0
2945,106	100	19,82	24,03	$2s\,^3S-5p\,^3P^\circ$	1—2, 1, 0
2829,076	40	19,82	24,20	$2s\,^3S-6p\,^3P^\circ$	1—2, 1, 0
2763,804	20	19,82	24,30	$2s\,^3S-7p\,^3P^\circ$	1—2, 1, 0
2723,191	10	19,82	24,37	$2s\,^3S-8p\,^3P^\circ$	1—2, 1, 0
2696,119	7	19,82	24,42	$2s\,^3S-9p\,^3P^\circ$	1—2, 1, 0
2677,135	5	19,82	24,45	$2s\,^3S-10p\,^3P^\circ$	1—2, 1, 0
2663,271	4	19,82	24,47	$2s\,^3S-11p\,^3P^\circ$	1—2, 1, 0
2652,848	3	19,82	24,49	$2s\,^3S-12p\,^3P^\circ$	1—2, 1, 0
2644,802	2	19,82	24,50	$2s\,^3S-13p\,^3P^\circ$	1—2, 1, 0
601,4041	5	0,00	20,61	$1s^2\,^1S-2s\,^1S$	0—0
591,4117	20	0,00	20,96	$1s^2\,^1S-2p\,^3P^\circ$	0—1
584,3340	500	0,00	21,22	$1s^3\,^1S-2p\,^1P^\circ$	0—1
537,0296	200	0,00	23,09	$1s^2\,^1S-3p\,^1P^\circ$	0—1
522,2128	80	0,00	23,74	$1s^2\,^1S-4p\,^1P^\circ$	0—1
515,6165	50	0,00	24,04	$1s^2\,^1S-5p\,^1P^\circ$	0—1
512,0982	35	0,00	24,21	$1s^2\,^1S-6p\,^1P^\circ$	0—1
509,9979	25	0,00	24,31	$1s^2\,^1S-7p\,^1P^\circ$	0—1
508,6431	20	0,00	24,37	$1s^2\,^1S-8p\,^1P^\circ$	0—1
507,7178	15	0,00	24,42	$1s^2\,^1S-9p\,^1P^\circ$	0—1
507,0576	10	0,00	24,45	$1s^2\,^1S-10p\,^1P^\circ$	0—1
506,5702	7	0,00	24,47	$1s^2\,^1S-11p\,^1P^\circ$	0—1
506,2000	5	0,00	24,49	$1s^2\,^1S-12p\,^1P^\circ$	0—1
505,9122	4	0,00	24,51	$1s^2\,^1S-13p\,^1P^\circ$	0—1
505,6840	3	0,00	24,52	$1s^3\,^1S-14p\,^1P^\circ$	0—1
320,392	10	20,96	59,66	$1s2p\,^3P^\circ-2p^2\,^3P$	2, 1, 0—2, 1, 0

He II, ground state $1s\ ^2S_{1/2}$

Ionization potential 438908,670 cm^{-1}; 54,414 eV

λ, Å	I	E_H, eV	E_B, eV	Transition	J
11626,40	—	52,24	53,30	$5g\,^2G-7h\,^2H^\circ$ etc.	9/2, 7/2—11/2, 9/2
10123,61	—	51,01	52,24	$4f\,^2F^\circ-5g\,^2G$ etc.	7/2, 5/2—9/20, 7/2
9344,93	—	52,24	53,56	$5g\,^2G-8h\,^2H^\circ$ etc.	9/2, 7/2—11/2, 9/2
8236,77	—	52,24	53,74	$5g\,^2G-9h\,^2H^\circ$ etc.	9/2, 7/2—11/2, 9/2
7592,74	—	52,24	53,87	$5g\,^2G-10h\,^2H^\circ$ etc.	9/2, 7/2—11/2, 9/2

λ, A	I	E_H, eV	E_B, eV	Transition	J
7177,50	—	52,24	53,96	$5g\ ^2G$—$11h\ ^2H^\circ$ etc.	9/2, 7/2—11/2, 9/2
6890,88	—	52,24	54,04	$5g\ ^2G$—$12h\ ^2H^\circ$ etc.	9/2, 7/2—11/2, 9/2
6683,26	—	52,24	54,09	$5g\ ^2G$—$13h\ ^2H^\circ$ etc.	9/2, 7/2—11/2, 9/2
6560,099	100	51,01	52,90	$4f\ ^2F^\circ$—$6g\ ^2G$ etc.	7/2, 5/2—9/2, 7/2
6527,16	—	52,24	54,14	$5g\ ^2G$—$14h\ ^2H^\circ$ etc.	9/2, 7/2—11/2, 9/2
6406,44	—	52,24	54,17	$5g\ ^2G$—$15h\ ^2H^\circ$ etc.	9/2, 7/2—11/2, 9/2
6310,8	—	52,24	54,20	$5g\ ^2G$—$16h\ ^2H^\circ$ etc.	9/2, 7/2—11/2, 9/2
6233,8	—	52,24	54,23	$5g\ ^2G$—$17h\ ^2H^\circ$ etc.	9/2, 7/2—11/2, 9/2
6170,6	—	52,24	54,25	$5g\ ^2G$—$18h\ ^2H^\circ$ etc.	9/2, 7/2—11/2, 9/2
5694,46	—	52,24	54,41	Limit of series	
5411,524	50	51,01	53,30	$4f\ ^2F^\circ$—$7g\ ^2G$ etc.	7/2, 5/2—9/2, 7/2
4859,323	7	51,01	53,56	$4f\ ^2F^\circ$—$8g\ ^2G$ etc.	7/2, 5/2—9/2, 7/2
4685,682	300	48,37	51,01	$3d\ ^2D$—$4f\ ^2F^\circ$ etc.	5/2, 3/2—7/2, 5/2
4541,59	5	51,01	53,74	$4f\ ^2F^\circ$—$9g\ ^2G$ etc.	For all the transitions
4338,67	3	51,01	53,87	$4f\ ^3F^\circ$—$10g\ ^2G$ etc.	indicated of this series
4199,87	2	51,01	53,96	$4f\ ^2F^\circ$—$11g\ ^2G$ etc.	7/2, 5/2—9/2, 7/2
4100,04	2	51,01	54,04	$4f\ ^2F^\circ$—$12g\ ^2G$ etc.	
4025,60	—	51,01	54,09	$4f\ ^2F^\circ$—$13g\ ^2G$ etc.	
3968,43	—	51,01	54,14	$4f\ ^2F^\circ$—$14g\ ^2G$ etc.	
3644,47	—	51,01	54,41	Limit of series	
3203,104	200	48,37	52,24	$3d\ ^2D$—$5f\ ^2F^\circ$ etc.	For all the transitions
2733,32	100	48,37	52,90	$3d\ ^2D$—$6f\ ^2F^\circ$ etc.	indicated of this series
2511,22	50	48,37	53,30	$3d\ ^2D$—$7f\ ^2F^\circ$ etc.	5/2, 3/2—7/2, 5/2
2385,42	30	48,37	53,56	$3d\ ^2D$—$8f\ ^2F^\circ$ etc.	
2306,22	20	48,37	53,74	$3d\ ^2D$—$9f\ ^2F^\circ$ etc.	
2252,71	10	48,37	53,87	$3d\ ^2D$—$10f\ ^2F^\circ$ etc.	
2214,67	6	48,37	53,96	$3d\ ^2D$—$11f\ ^2F^\circ$ etc.	
2186,61	4	48,37	54,04	$3d\ ^2D$—$12f\ ^2F^\circ$ etc.	
2165,24	2	48,37	54,09	$3d\ ^2D$—$13f\ ^2F^\circ$ etc.	
2049,94	—	48,37	54,41	Limit of series	
1640,490	1	40,81	48,37	$2p\ ^3P^\circ$—$3d\ ^2D$ etc.	3/2—3/2
1640,474	10	40,81	48,37	$2p\ ^2P^\circ$—$3d\ ^2D$ etc.	3/2—5/2
1640,332	5	40,81	48,37	$2p\ ^2P^\circ$—$3d\ ^2D$ etc.	1/2—3/2
1215,171	5	40,81	51,01	$2p\ ^2P^\circ$—$4d\ ^2D$ etc.	3/2—5/2, 3/2
1215,088	2	40,81	51,01	$2p\ ^2P^\circ$—$4d\ ^2D$ etc.	1/2—3/2
1084,951	3	40,81	52,24	$2p\ ^2P^\circ$—$5d\ ^2D$ etc.	For all the transitions
1025,280	2	40,81	52,90	$2p\ ^2P^\circ$—$6d\ ^2D$ etc.	indicated of this series
992,370	1	40,81	53,30	$2p\ ^2P^\circ$—$7d\ ^2D$ etc.	3/2, 1/2—5/2, 3/2
972,118	0,7	40,81	53,56	$2p\ ^2P^\circ$—$8d\ ^2D$ etc.	
958,705	0,5	40,81	53,74	$2p\ ^2P^\circ$—$9d\ ^2D$ etc.	
949,335	0,3	40,81	53,87	$2p\ ^2P^\circ$—$10d\ ^2D$ etc.	
942,52	0,2	40,81	53,96	$2p\ ^2P^\circ$—$11d\ ^2D$ etc.	
937,40	—	40,81	54,04	$2p\ ^2P^\circ$—$12d\ ^2D$ etc.	
933,46	—	40,81	54,09	$2p\ ^2P^\circ$—$13d\ ^2D$ etc.	
911,37	—	40,81	54,41	Limit of series	
303,783	500	0,00	40,81	$1s\ ^2S$—$2p\ ^2P^\circ$	For all the transitions
256,317	150	0,00	48,37	$1s\ ^2S$—$3p\ ^2P^\circ$	of this series 1/2—3/2, 1/2
243,027	70	0,00	51,01	$1s\ ^2S$—$4p\ ^2P^\circ$	
237,331	35	0,00	52,24	$1s\ ^2S$—$5p\ ^2P^\circ$	
234,347	20	0,00	52,90	$1s\ ^2S$—$6p\ ^2P^\circ$	
232,584	13	0,00	53,30	$1s\ ^2S$—$7p\ ^2P^\circ$	
231,454	8	0,00	53,56	$1s\ ^2S$—$8p\ ^2P^\circ$	
230,686	5	0,00	53,74	$1s\ ^2S$—$9p\ ^2P^\circ$	
230,139	4	0,00	53,87	$1s\ ^2S$—$10p\ ^2P^\circ$	
229,736	3	0,00	53,96	$1s\ ^2S$—$11p\ ^2P^\circ$	
229,431	2	0,00	54,04	$1s\ ^2S$—$12p\ ^2P^\circ$	
227,83	—	0,00	54,41	Limit of series	

LITHIUM, Z = 3

Li I, ground state $1s^2\,2s\,^2S_{1/2}$

Ionization potential 43487,19 cm^{-1}; 5,391 eV

λ, Å	I	E_H, eV	E_B, eV	Transition	J
26877,82	8	3,37	3,83	$3s\,^2S - 3p\,^2P^\circ$	1/2 – 3/2, 1/2
24464,66	6	3,83	4,34	$3p\,^2P^\circ - 4s\,^2S$	3/2, 1/2 – 1/2
19274,78	4	3,88	4,52	$3d\,^2D - 4p\,^2P^\circ$	5/2, 3/2 – 1/2
18703,09	7	3,88	4,54	$3d\,^2D - 4f\,^2F^\circ$	5/2, 3/2 – 7/2, 5/2
17546,05	7	3,83	4,54	$3p\,^2P^\circ - 4d\,^2D$	3/2, 1/2 – 5/2, 3/2
13557,75	4	3,83	4,75	$3p\,^2P^\circ - 5s\,^2S$	3/2, 1/2 – 1/2
12793,31	5	3,88	4,85	$3d\,^2D - 5f\,^2F^\circ$	5/2, 3/2 – 7/2, 5/2
12237,67	4	3,83	4,85	$3p\,^2P^\circ - 5d\,^2D$	3/2, 1/2 – 5/2, 3/2
11032,09	1	3,83	4,96	$3p\,^2P^\circ - 6s\,^2S$	3/2, 1/2 – 1/2
10976,06	0	3,88	5,01	$3d\,^2D - 6p\,^2P^\circ$	5/2, 3/2 – 3/2, 1/2
10919,07	3	3,88	5,01	$3d\,^2D - 6f\,^2F^\circ$	5/2, 3/2 – 7/2, 5/2
10510,60	3	3,83	5,01	$3p\,^2P^\circ - 6d\,^2D$	3/2, 1/2 – 5/2, 3/2
10032,81	2	3,88	5,11	$3d\,^2D - 7f\,^2F^\circ$	5/2, 3/2 – 7/2, 5/2
9955,09	2	3,83	5,08	$3p\,^2P^\circ - 7s\,^2S$	3/2, 1/2 – 1/2
9686,37	2	3,83	5,11	$3p\,^2P^\circ - 7d\,^2D$	3/2, 1/2 – 5/2, 3/2
9530,73	1	3,88	5,18	$3d\,^2D - 8f\,^2F^\circ$	5/2, 3/2 – 7/2, 5/2
9376,71	1	3,83	5,16	$3p\,^2P^\circ - 8s\,^2S$	3/2, 1/2 – 1/2
9217,32	2	3,83	5,18	$3p\,^2P^\circ - 8d\,^2D$	3/2, 1/2 – 5/2, 3/2
9214,61	1	3,88	5,22	$3d\,^2D - 9f\,^2F^\circ$	5/2, 3/2 – 7/2, 5/2
8921,14	0	3.83	5,22	$3p\,^2P^\circ - 9d\,^2D$	3/2, 1/2 – 5/2, 3/2
8465,352	4	3,37	4,84	$3s\,^2S - 5p\,^2P^\circ$	1/2 – 3/2, 1/2
8126,378	300	1,85	3,37	$2p\,^2P^\circ - 3s\,^2S$	3/2, 1/2 – 1/2
7582,169	3	3,37	5,01	$3s\,^2S - 6p\,^2P^\circ$	1/2 – 3/2, 1/2
7135,040	1	3,37	5,11	$3s\,^2S - 7p\,^2P^\circ$	1/2 – 3/2, 1/2
6707,807	1000	0,00	1,85	$2s\,^2S - 2p\,^2P^\circ$	1/2—3/2, 1/2
6240,4	2	1,85	3,83	$2p\,^2P^\circ - 3p\,^2P^\circ$	3/2, 1/2—3/2, 1/2
6103,61	500	1,85	3,88	$2p\,^2P^\circ - 3d\,^2D$	3/2, 1/2—5/2, 3/2
4971,720	50	1,85	4,34	$2p\,^2P^\circ - 4s\,^2S$	3/2, 1/2—1/2
4636,0	1	1,85	4,52	$2p\,^2P^\circ - 4p\,^2P^\circ$	3/2, 1/2—3/2, 1/2
4602,871	100	1,85	4,54	$2p\,^2P^\circ - 4d\,^2D$	3/2, 1/2—5/2, 3/2
4602,02	1	1,85	4,54	$2p\,^3P^\circ - 4f\,^2F$	3/2, 1/2—5/2, 3/2
4273,107	10	1,85	4,75	$2p\,^2P^\circ - 5s\,^2S$	3/2, 1/2—1/2
4148,4	—	1,85	4,84	$2p\,^2P^\circ - 5p\,^2P^\circ$	3/2, 1/2—3/2, 1/2
4132,598	50	1,85	4,85	$2p\,^2P^\circ - 5d\,^2D$	3/2, 1/2—5/2, 3/2
3985,520	5	1,85	4,96	$2p\,^2P^\circ - 6s\,^2S$	3/2, 1/2—1/2
3921,6	—	1,85	5,01	$2p\,^2P^\circ - 6p\,^2P^\circ$	3/2, 1/2—3/2, 1/2
3915,329	25	1,85	5,01	$2p\,^2P^\circ - 6d\,^2D$	3/2, 1/2—5/2, 3/2
3838,15	3	1,85	5,08	$2p\,^2P^\circ - 7s\,^2S$	3/2, 1/2—1/2
3794,72	10	1,85	5,11	$2p\,^2P^\circ - 7d\,^2D$	3/2, 1/2—5/2, 3/2
3718,7	5	1,85	5,18	$2p\,^2P^\circ - 8d\,^2D$	3/2, 1/2—5/2, 3/2
3670,4	3	1,85	5,22	$2p\,^2P^\circ - 9d\,^2D$	3/2, 1/2—5/2, 3/2
3232,634	50	0,00	3,83	$2s\,^2S - 3p\,^2P^\circ$	1/2—3/2, 1/2
3195,6	3	0,00	3,88	$2s\,^2S - 3d\,^2D$	1/2—3/2
2741,186	10	0,00	4,52	$2s\,^2S - 4p\,^2P^\circ$	1/2—3/2, 1/2
2732,3	2	0,00	4,54	$2s\,^2S - 4d\,^2D$	1/2—3/2
2562,305	5	0,00	4,84	$2s\,^2S - 5p\,^2P^\circ$	1/2—3/2, 1/2
2557,4	—	0,00	4,85	$2s\,^2S - 5d\,^2D$	1/2—3/2
2475,057	4	0,00	5,01	$2s\,^2S - 6p\,^2P^\circ$	1/2—3/2, 1/2
2425,414	3	0,00	5,11	$2s\,^2S - 7p\,^2P^\circ$	1/2—3/2, 1/2
2394,355	2	0,00	5,18	$2s\,^2S - 8p\,^2P^\circ$	1/2—3/2, 1/2
2373,548	1	0,00	5,22	$2s\,^2S - 9p\,^2P^\circ$	1/2—3/2, 1/2
2358,917	1	0,00	5,25	$2s\,^2S - 10p\,^2P^\circ$	1/2—3/2, 1/2

Li II, ground state $1s^2\ ^1S_0$

Ionization potential 610079, cm^{-1}; 75,635 eV

λ, Å	I	E_H, eV	E_B, eV	Transition	J
9581,42	—	60,92	62,21	$2s\ ^1S-2p\ ^1P^\circ$	0—1
5484,7	10	59,02	61,28	$2s\ ^3S-2p\ ^3P^\circ$	1—2, 1, 0
5037,8	6	69,64	72,10	$3p\ ^1P^\circ-4s\ ^1S$	1—0
4881,3	3	69,37	71,90	$3p\ ^3P^\circ-4s\ ^3S$	2, 1, 0—1
4788,8	8	69,64	72,23	$3p\ ^1P^\circ-4d\ ^1D$	1—2
4677,7	8	69,59	72,23	$3d\ ^1D-4f\ ^1F^\circ$	2—3
4671,8	4	69,58	72,23	$3d\ ^3D-4f\ ^3F^\circ$	3, 2, 1—4, 3, 2
4325,7	3	69,37	72,23	$3p\ ^3P^\circ-4d\ ^3D$	2, 1, 0—3, 2, 1
4156,3	1	69,28	72,26	$3s\ ^1S-4p\ ^1P^\circ$	0—1
3684,1	3	68,78	72,14	$3s\ ^3S-4p\ ^3P^\circ$	1—2, 1, 0
3305,2	4	69,64	73,39	$3p\ ^1P^\circ-5s\ ^1S$	1—0
3249,8	5	69,64	73,46	$3p\ ^1P^\circ-5d\ ^1D$	1—2
3199,43	7	69,59	73,46	$3d\ ^1D-5f\ ^1F^\circ$	2—3
3195,8	3	69,58	73,46	$3d\ ^3D-5f\ ^3F^\circ$	3, 2, 1—4, 3, 2
3155,4	2	69,37	73,29	$3p\ ^3P^\circ-5s\ ^3S$	2, 1, 0—1
3029,1	2,5	69,37	73,46	$3p\ ^3P^\circ-5d\ ^3D$	2, 1, 0—3, 2, 1
2952,7	0,5	69,28	73,47	$3s\ ^1S-5p\ ^1P^\circ$	0—1
2790,39	2	69,64	74,09	$3p\ ^1P^\circ-6s\ ^1S$	1—0
2767,0	4	69,64	74,12	$3p\ ^1P^\circ-6d\ ^1D$	1—2
2730,7	5	69,59	74,12	$3d\ ^1D-6f\ ^1F^\circ$	2—3
2728,4	2	69,58	74,12	$3d\ ^3D-6f\ ^3F^\circ$	3, 2, 1—4, 3, 2
2674,4	2	68,78	73,41	$3s\ ^3S-5p\ ^3P^\circ$	1—2, 1, 0
2657,3	1,5	69,37	74,03	$3p\ ^3P^\circ-6s\ ^3S$	2, 1, 0—1
2605,1	1,5	69,37	74,12	$3p\ ^3P^\circ-6d\ ^3D$	2, 1, 0—3, 2, 1
2551,7	1	69,64	74,50	$3p\ ^1P^\circ-7s\ ^1S$	1—0
2539,4	2	69,64	74,52	$3p\ ^1P^\circ-7d\ ^1D$	1—2
2508,83	3	69,59	74,53	$3d\ ^1D-7f\ ^1F^\circ$	2—3
2430,0	1	69,37	74,47	$3p\ ^3P^\circ-7s\ ^3S$	2, 1, 0—1
2410,85	1	69,64	74,78	$3p\ ^1P^\circ-8d\ ^1D$	1—2
2402,3	1	69,37	74,52	$3p\ ^3P^\circ-7d\ ^3D$	2, 1, 0—3, 2, 1
2383,26	2	69,59	74,79	$3d\ ^1D-8f\ ^1F^\circ$	2—3
2330,0	1	68,78	74,10	$3s\ ^3S-6p\ ^3P^\circ$	1—2, 1, 0
1756,0	5	62,21	69,28	$2p\ ^1P^\circ-3s\ ^1S$	1—0
1682,4	4	62,21	69,59	$2p\ ^1P^\circ-3d\ ^1D$	1—2
1653,9	8	61,28	68,78	$2p\ ^3P^\circ-3s\ ^3S$	2, 1, 0—1
1493,7	6	61,28	69,58	$2p\ ^3P^\circ-3d\ ^3D$	2, 1, 0—3, 2, 1
1420,89	1	60,92	69,64	$2s\ ^1S-3p\ ^1P^\circ$	0—1
1254,6	1	62,21	72,10	$2p\ ^1P^\circ-4s\ ^1S$	1—0
1198,6	7	59,02	69,37	$2s\ ^3S-3p\ ^3P^\circ$	1—2, 1, 0
1167,0	2	61,28	71,90	$2p\ ^3P^\circ-4s\ ^3S$	2, 1, 0—1
1132,8	1	61,28	72,23	$2p\ ^3P^\circ-4d\ ^3D$	2, 1, 0—3, 2, 1
199,282	3	0,00	62,21	$1s^2\ ^1S-2p\ ^1P^\circ$	0—1
178,015	1	0,00	69,64	$1s^2\ ^1S-3p\ ^1P^\circ$	0—1
171,582	1	0,00	72,26	$1s^2\ ^1S-4p\ ^1P^\circ$	0—1
168,741	—	0,00	73,47	$1s^2\ ^1S-5p\ ^1P^\circ$	0—1

Li III, ground state $1s\ ^2S_{1/2}$

Ionization potential 987657,8 cm^{-1}; 122,446 eV

λ, Å	I	E_H, eV	E_B, eV	Transition	J
729,1	—	91,94	108,84	$2p\ ^2P^\circ-3d\ ^2D$	$^3/_2$, $^1/_2$—$^5/_2$, $^3/_2$
540,0	—	91,94	114,79	$2p\ ^2P^\circ-3d\ ^2D$	$^3/_2$, $^1/_2$—$^5/_2$, $^3/_2$

λ, Å	I	E_H, eV	E_B, eV	Transition	J
482,1	—	91,94	117,55	$2p\ ^2P^\circ-5d\ ^2D$	$^3/_2,\ ^1/_2-^5/_2,\ ^3/_2$
455,6	—	91,94	119,05	$2p\ ^2P^\circ-6d\ ^2D$	$^3/_2,\ ^1/_2-^5/_2,\ ^3/_2$
441,0	—	91,94	119,95	$2p\ ^2P^\circ-7d\ ^2D$	$^3/_2,\ ^1/_2-^5/_2,\ ^3/_2$
135,02	—	0,00	91,94	$1s\ ^2S-2p\ ^2P^\circ$	$^1/_2-^3/_2,\ ^1/_2$
113,93	—	0,00	108,84	$1s\ ^2S-3p\ ^2P^\circ$	$^1/_2-^3/_2,\ ^1/_2$
108,01	—	0,00	114,79	$1s\ ^2S-4p\ ^3P^\circ$	$^1/_2-^3/_2,\ ^1/_2$
105,49	—	0,00	117,55	$1s\ ^2S-5p\ ^2P^\circ$	$^1/_2-^3/_2,\ ^1/_2$
104,17	—	0,00	119,05	$1s\ ^2S-6p\ ^2P^\circ$	$^1/_2-^3/_2,\ ^1/_2$
103,40	—	0,00	119,95	$1s\ ^2S-7p\ ^2P^\circ$	$^1/_2-^3/_2,\ ^1/_2$
102,86	—	0,00	120,53	$1s\ ^2S-8p\ ^2P^\circ$	$^1/_2-^3/_2,\ ^1/_2$

CARBON, Z = 6

C I, ground state $1s^2\ 2s^2\ 2p^2\ ^3P_0$

Ionization potential 90820,42 cm^{-1}; 11,269 eV

λ, Å	I	E_H, eV	E_B, eV	Transition	J
25842,20	1	8,85	9,33	$3p\ ^3P - 2p^3\ ^3P^o$	2 – 1
25833,66	1	8,85	9,33	$3p\ ^3P - 2p^3\ ^3P^o$	2 – 2
25706,03	1	8,85	9,33	$3p\ ^3P - 2p^3\ ^3P^o$	1 – 1
25697,56	1	8,85	9,33	$3p\ ^3P - 2p^3\ ^3P^o$	1 – 2
22906,56	7	9,17	9,71	$3p\ ^1S - 4s\ ^1P^o$	0 – 1
21295,27	1	9,83	10,42	$3d\ ^3P^o - 4f\,D(2\frac{1}{2})$	1 – 2
21259,89	8	9,83	10,42	$3d\ ^3P^o - 4f\,D(2\frac{1}{2})$	2 – 3, 2
21211,55	2	9,83	10,42	$3d\ ^3P^o - 4f\,D(1\frac{1}{2})$	0 – 1
21191,41	4	9,83	10,42	$3d\ ^3P^o - 4f\,D(1\frac{1}{2})$	1 – 2, 1
21023,13	8	9,17	9,76	$3p\ ^1S - 3d\ ^1P^o$	0 – 1
19721,99	23	9,00	9,63	$3p\ ^1D - 3d\ ^1D^o$	1 – 2
18926,54	3	9,76	10,42	$3d\ ^1P^o - 4f\,D(2\frac{1}{2})$	1 – 2
18844,42	2	9,76	10,42	$3d\ ^1P^o - 4f\,D(1\frac{1}{2})$	1 – 2, 1
18320,67	8	9,74	10,42	$3d\ ^1F^o - 4f\,G(3\frac{1}{2})$	3 – 4, 3
18221,12	8	9,74	10,42	$3d\ ^1F^o - 4f\,G(4\frac{1}{2})$	3 – 4
18139,80	13	8,65	9,33	$3p\ ^3D - 2p^3\ ^3P^o$	3 – 2
18034,86	5	8,64	9,33	$3p\ ^3D - 2p^3\ ^3P^o$	2 – 1
18030,47	2	8,64	9,33	$3p\ ^3D - 2p^3\ ^3P^o$	2 – 2
17966,12	2	8,64	9,33	$3p\ ^3D - 2p^3\ ^3P^o$	1 – 1
17959,24	3	8,64	9,33	$3p\ ^3D - 2p^3\ ^3P^o$	1 – 0
17918,38	4	9,33	10,02	$2p^3\ ^3P^o - 4p\ ^3D$	2 – 3
17826,33	4	9,71	10,41	$3d\ ^3D^o - 4f\,F(3\frac{1}{2})$	3 – 4, 3
17814,03	3	9,71	10,41	$3d\ ^3D^o - 4f\,F(2\frac{1}{2})$	2 – 3, 2
17768,94	3	9,71	10,41	$3d\ ^3D^o - 4f\,F(2\frac{1}{2})$	1 – 2
17637,38	3	9,71	10,42	$3d\ ^3D^o - 4f\,G(3\frac{1}{2})$	3 – 4, 3
17505,64	3	9,70	10,41	$3d\ ^3F^o - 4f\,G(3\frac{1}{2})$	3 – 4, 3
17455,97	2	9,70	10,41	$3d\ ^3F^o - 4f\,F(3\frac{1}{2})$	2 – 3
17448,60	11	9,00	9,71	$3p\ ^1D - 4s\ ^1P^o$	2 – 1
17338,56	10	9,70	10,42	$3d\ ^3F^o - 4f\,G(4\frac{1}{2})$	4 – 5, 4
17323,51	2	9,70	10,42	$3d\ ^3F^o - 4f\,G(3\frac{1}{2})$	3 – 4, 3
17274,99	3	9,70	10,42	$3d\ ^3F^o - 4f\,G(3\frac{1}{2})$	2 – 3
17234,48	2	9,70	10,42	$3p\ ^3F^o - 4f\,G(4\frac{1}{2})$	3 – 4
16890,38	50	9,00	9,74	$3d\ ^1D - 3d\ ^1F^o$	2 – 3
16021,64	3	9,63	10,41	$3d\ ^1D^o - 4f\,F(2\frac{1}{2})$	2 – 3, 2
16004,81	2	9,63	10,41	$3d\ ^1D^o - 4f\,F(3\frac{1}{2})$	2 – 3
14782,98	4	8,85	9,69	$3p\ ^3P - 4s\ ^3P^o$	2 – 2
14637,03	2	8,85	9,70	$3p\ ^3P - 3d\ ^3F^o$	2 – 3
14542,50	179	7,68	8,54	$3s\ ^1P^o - 3p\ ^1P$	1 – 1
14442,24	13	8,85	9,71	$3p\ ^3P - 3d\ ^3D^o$	2 – 2
14429,03	12	8,85	9,71	$3p\ ^3P - 3d\ ^3D^o$	1 – 1
14420,12	61	8,85	9,71	$3p\ ^3P - 3d\ ^3D^o$	2 – 3
14403,25	16	8,85	9,71	$3p\ ^3P - 3d\ ^3D^o$	0 – 1
14399,65	38	8,85	9,71	$3p\ ^3P - 3d\ ^3D^o$	1 – 2
13765,29	1	7,95	8,85	$2p^3\ ^3D^o - 3p\ ^3P$	1 – 0
13743,93	3	7,95	8,85	$2p^3\ ^3D^o - 3p\ ^3P$	2 – 1
13741,86	1	7,95	8,85	$2p^3\ ^3D^o - 3p\ ^3P$	1 – 1
13705,41	1	7,95	8,85	$2p^3\ ^3D^o - 3p\ ^3P$	2 – 2
13697,81	6	7,95	8,85	$2p^3\ ^3D^o - 3p\ ^3P$	3 – 2
13581,35	5	8,77	9,68	$3p\ ^3S - 4s\ ^3P^o$	1 – 0
13559,66	12	8,77	9,68	$3p\ ^3S - 4s\ ^3P^o$	1 – 1

λ, Å	I	E_H, eV	E_B, eV	Transition	J
13502,27	20	8,77	9,69	$3p\,^3S-4s\,^3P^\circ$	1 – 2
12614,10	26	8,65	9,83	$3p\,^3P-3d\,^3P^\circ$	2 – 2
12601,48	8	8,85	9,83	$3p\,^3P-3d\,^3P^\circ$	2 – 1
12581,59	6	8,85	9,83	$3p\,^3P-3d\,^3P^\circ$	1 – 2
12569,04	5	8,85	9,83	$3p\,^3P-3d\,^3P^\circ$	1 – 1
12562,12	6	8,85	9,83	$3p\,^3P-3d\,^3P^\circ$	1 – 0
12549,48	5	8,85	9,83	$3p\,^3P-3d\,^3P^\circ$	0 – 1
11895,75	30	8,65	9,69	$3p\,^3D-4s\,^3P^\circ$	3 – 2
11892,91	17	8,64	9,68	$3p\,^3D-4s\,^3P^\circ$	2 – 1
11879,59	8	8,64	9,68	$3p\,^3D-4s\,^3P^\circ$	1 – 0
11862,99	5	8,64	9,68	$3p\,^3D-4s\,^3P^\circ$	1 – 1
11848,73	6	8,64	9,69	$3p\,^3D-4s\,^3P^\circ$	1 – 2
11801,08	7	8,65	9,70	$3p\,^3D-3d\,^3F^\circ$	3 – 3
11777,54	11	8,64	9,69	$3p\,^3D-3d\,^3F^\circ$	2 – 2
11754,76	114	8,64	9,70	$3p\,^3D-3d\,^3F^\circ$	2 – 3
11753,32	142	8,65	9,70	$3p\,^3D-3d\,^3F^\circ$	3 – 4
11748,22	82	8,64	9,69	$3p\,^3D-3d\,^3F^\circ$	1 – 2
11674,14	7	8,65	9,71	$3p\,^3D-3d\,^3D^\circ$	3 – 2
11669,63	24	8,77	9,83	$3p\,^3S-3d\,^3P^\circ$	1 – 2
11659,68	47	8,65	9,71	$3p\,^3D-3d\,^3D^\circ$	3 – 3
11658,85	13	8,77	9,83	$3p\,^3S-3d\,^3P^\circ$	1 – 1
11652,91	5	8,77	9,83	$3p\,^3S-3d\,^3P^\circ$	1 – 0
11647,99	5	8,64	9,71	$3p\,^3D-3d\,^3D^\circ$	2 – 1
11628,83	23	8,64	9,71	$3p\,^3D-3d\,^3D^\circ$	2 – 2
11619,29	12	8,64	9,71	$3p\,^3D-3d\,^3D^\circ$	1 – 1
11330,285	6	8,54	9,63	$3p\,^1P-3d\,^1D^\circ$	1 – 2
10753,985	2	7,49	8,64	$3s\,^3P^\circ-3p\,^3D$	2 – 1
10729,533	6	7,49	8,64	$3s\,^3P^\circ-3p\,^3D$	2 – 2
10707,333	6	7,48	8,64	$3s\,^3P^\circ-3p\,^3D$	1 – 1
10691,250	10	7,49	8,65	$3s\,^3P^\circ-3p\,^3D$	2 – 3
10685,345	6	7,48	8,64	$3s\,^3P^\circ-3p\,^3D$	0 – 1
10683,082	8	7,48	8,64	$3s\,^3P^\circ-3p\,^3D$	1 – 2
10541,226	4	8,54	9,71	$3p\,^1P-4s\,^1P^\circ$	1 – 1
10123,871	6	8,54	9,76	$3p\,^1P-3d\,^1P^\circ$	1 – 1
9658,44	10	7,49	8,77	$3s\,^3P^\circ—3p\,^3S$	2—1
9620,80	9	7,48	8,77	$3s\,^3P^\circ—3p\,^3S$	1—1
9603,03	7	7,48	8,77	$3s\,^3P^\circ—3p\,^3S$	0—1
9405,73	16	7,68	9,00	$3s\,^1P^\circ—3p\,^1D$	1—2
9182,83	4	9,00	10,35	$3p\,^1D—4d\,^1D^\circ$	2—2
9111,80	10	7,49	8,85	$3s\,^3P^\circ—3p\,^3P$	2—1
9094,83	12	7,49	8,85	$3s\,^3P^\circ—3p\,^3P$	2—2
9088,51	9	7,48	8,85	$3s\,^3P^\circ—3p\,^3P$	1—0
9078,28	8	7,48	8,85	$3s\,^3P^\circ—3p\,^3P$	1—1
9062,47	8	7,48	8,85	$3s\,^3P^\circ—3p\,^3P$	0—1
9061,43	9	7,48	8,85	$3s\,^3P^\circ—3p\,^3P$	1—2
8960,75	2	9,33	10,71	$2p^3\,^3P^\circ—5f\,F\,(2^1/_2)$	2—3
8904,34	2	9,33	10,72	$2p^3\,^3P^\circ—5f\,D\,(2^1/_2)$	2—3
8903,20	1	9,33	10,72	$2p^3\,^3P^\circ—5f\,D\,(2^1/_2)$	1—2
8890,67	2	9,33	10,72	$2p^3\,^3P^\circ—5f\,D\,(1^1/_2)$	1—2
8873,39	3	9,00	10,40	$3p\,^1D—5s\,^1P^\circ$	2—1
8753,08	3	9,00	10,42	$3p\,^1D—4d\,^1P^\circ$	2—1
8536,26	1	9,33	10,78	$2p^3\,^3P^\circ—6p\,^3D$	1—2
3510,45	1	9,33	10,79	$2p^3\,^3P^\circ—6p\,^3D$	2—3
8430,88	1	9,33	10,80	$2p^3\,^3P^\circ—6p\,^3P$	2—2
8335,15	13	7,68	9,17	$3s\,^1P^\circ—3p\,^1S$	1—0
8083,80	5	8,85	10,38	$3p\,^3P—5s\,^3P^\circ$	2—1

λ, Å	I	E_H, eV	E_B, eV	Transition	J
8078,48	4	8,85	10,38	$3p\ ^3P-5s\ ^3P^\circ$	1—0
8070,42	3	8,85	10,38	$3p\ ^3P-5s\ ^3P^\circ$	1—1
8062,36	3	8,85	10,38	$3p\ ^3P-5s\ ^3P^\circ$	0—1
8058,62	8	8,85	10,39	$3p\ ^3P-5s\ ^3P^\circ$	2—2
8045,33	4	8,85	10,39	$3p\ ^3P-5s\ ^3P^\circ$	1—2
8021,26	3	8,85	10,40	$3p\ ^3P-4d\ ^3D^\circ$	2—3
8018,56	1	8,85	10,39	$3p\ ^3P-4d\ ^3D^\circ$	0—1
7993,42	3	9,33	10,88	$2p^3\ ^3P^\circ-6f\ F\ (2^1/_2)$	2—3, 2
7992,53	0	9,33	10,88	$2p^3\ ^3P^\circ-6f\ F\ (2^1/_2)$	1—2
7987,89	2	9,17	10,72	$3p\ ^1S-5d\ ^1P^\circ$	0—1
7952,19	3	9,33	10,89	$2p^3\ ^3P^\circ-6f\ D\ (2^1/_2)$	2—3, 2
7951,35	1	9,33	10,89	$2p^3\ ^3P^\circ-6f\ D\ (2^1/_2)$	1—2
7944,60	3	9,33	10,89	$2p^3\ ^3P^\circ-6f\ D\ (1^1/_2)$	1—2, 1
7860,89	8	8,85	10,43	$3p\ ^3P-4d\ ^3P^\circ$	2—2
7852,86	4	8,85	10,43	$3p\ ^3P-4d\ ^3P^\circ$	2—1
7848,25	4	8,85	10,43	$3p\ ^3P-4d\ ^3P^\circ$	1—2
7840,28	2	8,85	10,43	$3p\ ^3P-4d\ ^3P^\circ$	1—1
7837,11	3	8,85	10,43	$3p\ ^3P-4d\ ^3P^\circ$	1—0
7832,63	3	8,85	10,43	$3p\ ^3P-4d\ ^3P^\circ$	0—1
7692,50	2	8,77	10,38	$3p\ ^3S-5s\ ^3P^\circ$	1—0
7685,20	4	8,77	10,38	$3p\ ^3S-5s\ ^3P^\circ$	1—1
7662,43	5	8,77	10,39	$3p\ ^3S-5s\ ^3P^\circ$	1—2
7505,67	1	9,33	10,98	$2p^3\ ^3P^\circ-7f\ F\ (2^1/_2)$	2—3, 2
7483,44	3	8,77	10,43	$3p\ ^3S-4d\ ^3P^\circ$	1—2
7476,18	2	8,77	10,43	$3p\ ^3S-4d\ ^3P^\circ$	1—1
7473,30	1	8,77	10,43	$3p\ ^3S-4d\ ^3P^\circ$	1—0
7470,09	1	9,33	10,99	$2p^3\ ^3P^\circ-7f\ D\ (2^1/_2)$	2—3, 2
7465,45	1	9,33	10,99	$2p^3\ ^3P^\circ-7f\ D\ (1^1/_2)$	1—2, 1
7364,73	3	9,00	10,69	$3p\ ^1D-5d\ ^1D^\circ$	2—2
7286,11	0	9,00	10,70	$3p\ ^1D-6s\ ^3P^\circ$	2—1
7241,32	2	9,00	10,71	$3p\ ^1D-6s\ ^1P^\circ$	2—1
7224,24	1	9,00	10,72	$3p\ ^1D-5d\ ^1F^\circ$	2—3
7216,03	0	9,17	10,89	$3p\ ^1S-6d\ ^1P^\circ$	0—1
7202,26	2	9,00	10,72	$3p\ ^1D-5d\ ^1P^\circ$	2—1
7132,11	1	8,65	10,38	$3p\ ^3D-4d\ ^3F^\circ$	3—3
7122,20	1	8,64	10,38	$3p\ ^3D-4d\ ^3F^\circ$	2—2
7119,67	7	8,64	10,38	$3p\ ^3D-5s\ ^3P^\circ$	2—1
7116,99	8	8,65	10,39	$3p\ ^3D-5s\ ^3P^\circ$	3—2
7115,19	9	8,64	10,38	$3p\ ^3D-5s\ ^3P^\circ$	1—0
		8,64	10,38	$3p\ ^3D-4d\ ^3F^\circ$	2—3
7113,18	9	8,65	10,38	$3p\ ^3D-4d\ ^3F^\circ$	3—4
7111,48	7	8,64	10,38	$3p\ ^3D-4d\ ^3F^\circ$	1—2
7108,94	3	8,64	10,38	$3p\ ^3D-5s\ ^3P^\circ$	1—1
7100,12	5	8,64	10,39	$3p\ ^3D-5s\ ^3P^\circ$	2—2
7093,25	3	8,65	10,39	$3p\ ^3D-4d\ ^3D^\circ$	3—2
7087,83	4	8,65	10,40	$3p\ ^3D-4d\ ^3D^\circ$	3—3
7085,51	0	8,64	10,39	$3p\ ^3D-4d\ ^3D^\circ$	2—1
7076,48	2	8,64	10,39	$3p\ ^3D-4d\ ^3D^\circ$	2—2
7074,86	1	8,64	10,39	$3p\ ^3D-4d\ ^3D^\circ$	1—1
7056,87	0	8,64	10,38	$3p\ ^3D-5s\ ^1p^\circ$	2—1
6962,31	0	8,65	10,43	$3p\ ^3D-4d\ ^3P^\circ$	3—2
6828,12	6	8,54	10,35	$3p\ ^1P-4d\ ^1D^\circ$	1—2
6711,29	1	8,54	10,38	$3p\ ^1P-5s\ ^3P^\circ$	1—1
6688,79	4	8,85	10,70	$3p\ ^3P-6s\ ^3P^\circ$	2—1
6683,95	4	8,85	10,70	$3p\ ^3P-6s\ ^3P^\circ$	1—0
6674,11	4	8,85	10,70	$3p\ ^3P-6s\ ^3P^\circ$	0—1

λ, Å	I	E_H, eV	E_B, eV	Transition	J
6671,84	5	8,85	10,71	$3p\ ^3P-6s\ ^3P^\circ$	2—2
6663,04	4	8,85	10,71	$3p\ ^3P-5d\ ^3D^\circ$	2—2
6662,73	3	8,85	10,71	$3p\ ^3P-6s\ ^3P^\circ$	1—2
6655,51	6	8,54	10,40	$3p\ ^1P-5s\ ^1P^\circ$	1—1
6654,61	3	9,00	10,86	$3p\ ^1D-6d\ ^1D^\circ$	2—2
6653,95	1	8,85	10,71	$3p\ ^3P-5d\ ^3D^\circ$	1—2
6617,23	0	9,00	10,87	$3p\ ^1D-7s\ ^3P^\circ$	2—1
6611,35	4	8,85	10,72	$3p\ ^3P-5d\ ^3P^\circ$	2—2
6605,79	1	8,85	10,73	$3p\ ^3P-5d\ ^3P^\circ$	2—1
6602,42	2	8,85	10,72	$3p\ ^3P-5d\ ^3P^\circ$	1—2
6596,85	1	8,85	10,73	$3p\ ^3P-5d\ ^3P^\circ$	1—1
6595,24	1	8,85	10,73	$3p\ ^3P-5d\ ^3P^\circ$	1—0
6591,45	1	8,85	10,73	$3p\ ^3P-5d\ ^3P^\circ$	0—1
6587,61	8	8,54	10,42	$3p\ ^1P-4d\ ^1P^\circ$	1—1
6586,27	2	9,00	10,88	$3p\ ^1D-7s\ ^1P^\circ$	2—1
6578,77	2	9,00	10,89	$3p\ ^1D-6d\ ^1F^\circ$	2—3
6568,71	2	9,00	10,89	$3p\ ^1D-6d\ ^1P^\circ$	2—1
6417,54	2	8,77	10,70	$3p\ ^3S-6s\ ^3P^\circ$	1—0
6413,55	3	8,77	10,70	$3p\ ^3S-6s\ ^3P^\circ$	1—1
6397,98	5	8,77	10,71	$3p\ ^3S-6s\ ^3P^\circ$	1—2
6389,87	2	8,77	10,71	$3p\ ^3S-5d\ ^3D^\circ$	1—2
6378,79	0	8,77	10,71	$3p\ ^3S-6s\ ^1P^\circ$	1—1
6342,32	2	8,77	10,72	$3p\ ^3S-5d\ ^3P^\circ$	1—2
6337,20	1	8,77	10,73	$3p\ ^3S-5d\ ^3P^\circ$	1—1
6335,70	0	8,77	10,73	$3p\ ^3S-5d\ ^3P^\circ$	1—0
6292,37	2	9,00	10,97	$3p\ ^1D-7d\ ^1D^\circ$	2—2
6242,70	1	9,00	10,99	$3p\ ^1D-7d\ ^1F^\circ$	2—3
6237,27	1	9,00	10,99	$3p\ ^1D-7d\ ^1P^\circ$	2—1
6120,82	2	8,85	10,87	$3p\ ^3P-7s\ ^3P^\circ$	2—1
6115,85	2	8,85	10,87	$3p\ ^3P-7s\ ^3P^\circ$	1—0
6113,15	1	8,85	10,87	$3p\ ^3P-7s\ ^3P^\circ$	1—1
6108,53	2	8,85	10,87	$3p\ ^3P-7s\ ^3P^\circ$	0—1
6107,65	1	8,85	10,88	$3p\ ^3P-7s\ ^3P^\circ$	2—2
6100,46	4	8,85	10,88	$3p\ ^3P-6d\ ^3D$	2—2
6100,03	2	8,85	10,88	$3p\ ^3P-7s\ ^3P^\circ$	1—2
6098,92	1	8,85	10,88	$3p\ ^3P-6d\ ^3D^\circ$	2—3
6094,30	0	8,85	10,88	$3p\ ^3P-7s\ ^1P^\circ$	2—1
6092,84	1	8,85	10,88	$3p\ ^3P-6d\ ^3D^\circ$	1—2
6086,69	0	8,85	10,88	$3p\ ^3P-7s\ ^1P^\circ$	1—1
6079,77	1	9,00	11,04	$3p\ ^1D-8d\ ^1D^\circ$	2—2
6078,40	2	8,85	10,89	$3p\ ^3P-6d\ ^3P^\circ$	2—2
6070,83	1	8,85	10,89	$3p\ ^3P-6d\ ^3P^\circ$	1—2
6068,25	0	8,64	10,69	$3p\ ^3D-5d\ ^1D^\circ$	2—2
6062,09	0	8,85	10,89	$3p\ ^3P-6d\ ^3P^\circ$	0—1
6044,79	0	9,00	11,05	$3p\ ^1D-9s\ ^1P^\circ$	2—1
6042,46	1	9,00	11,05	$3p\ ^1D-8d\ ^1F^\circ$	2—3
6039,17	0	9,00	11,05	$3p\ ^1D-8d\ ^1P^\circ$	2—1
6019,87	0	8,64	10,70	$3p\ ^3D-5d\ ^3F^\circ$	2—2
6016,45	6	8,64	10,70	$3p\ ^3D-5d\ ^3F^\circ$	2—3
6014,85	9	8,64	10,70	$3p\ ^3D-6s\ ^3P^\circ$	2—1
6013,22	10	8,65	10,71	$3p\ ^3D-6s\ ^3P^\circ$	3—2
		8,65	10,70	$3p\ ^3D-5d\ ^3F^\circ$	3—4
6012,24	5	8,64	10,70	$3p\ ^3D-5d\ ^3F^\circ$	1—2
6010,68	7	8,64	10,70	$3p\ ^3D-6s\ ^3P^\circ$	1—0
6007,18	6	8,64	10,70	$3p\ ^3D-6s\ ^3P^\circ$	1—1
6006,03	9	8,65	10,71	$3p\ ^3D-5d\ ^3D^\circ$	3—2
6003,67	1	8,64	10,71	$3p\ ^3D-5d\ ^3D^\circ$	2—1
6002,98	4	8,65	10,71	$3p\ ^3D-5d\ ^3D^\circ$	3—3

λ, Å	I	E_H, eV	E_B, eV	Transition	J
6001,13	8	8,64	10,71	$3p\ {}^3D-6s\ {}^3P^\circ$	2—2
5996,06	2	8,64	10,71	$3p\ {}^3D-5d\ {}^3D^\circ$	1—1
5989,40	1	7,95	10,02	$2p^3\ {}^3D^\circ-{}^4p\ {}^3D$	2—1
5989,03	2	7,95	10,02	$2p^3\ {}^3D^\circ-4p\ {}^3D$	1—1
5984,26	3	8,64	10,71	$3p\ {}^3D-6s\ {}^1P^\circ$	2—1
5982,67	2	7,95	10.02	$2p^3\ {}^3D^\circ-4p\ {}^3D$	2—2
5982,27	0	7,95	10,02	$2p^3\ {}^3D^\circ-4p\ {}^3D$	1—2
5981,22	1	7,95	10,02	$2p^3\ {}^3D^\circ-4p\ {}^3D$	3—2
5972,59	0	8,64	10,72	$3p\ {}^3D-5d\ {}^1F^\circ$	2—3
5970,73	0	7,95	10,02	$2p^3\ {}^3D^\circ-4p\ {}^3D$	2—3
5969,33	4	7,95	10,02	$2p^3\ {}^3D^\circ-4p\ {}^3D$	3—3
5963,99	4	8,65	10,72	$3p\ {}^3D-5d\ {}^3P^\circ$	3—2
5952,13	2	8,64	10,72	$3p\ {}^3D-5d\ {}^3P^\circ$	2—2
5950,04	1	8,64	10,72	$3p\ {}^3D-5d\ {}^1P^\circ$	2—1
5947,61	1	8,64	10,73	$3p\ {}^3D-5d\ {}^3P^\circ$	2—1
5943,39	0	9,00	11,09	$3p\ {}^1D-9d\ {}^1D^\circ$	2—2
5940,10	0	8,64	10,73	$3p\ {}^3D-5d\ {}^3P^\circ$	1—1
5912,58	0	9,00	11,10	$3p\ {}^1D-9d\ {}^1F^\circ$	2—3
5892,00	1	8,77	10,87	$3p\ {}^3S-7s\ {}^3P^\circ$	1—0
5889,52	2	8,77	10,87	$3p\ {}^3S-7s\ {}^3P^\circ$	1—1
5877,31	2	8,77	10,88	$3p\ {}^3S-7s\ {}^3P^\circ$	1—2
5870,66	3	8,77	10,88	$3p\ {}^3S-6d\ {}^3D^\circ$	1—2
5864,95	0	8,77	10,88	$3p\ {}^3S-7s\ {}^1P^\circ$	1—1
5850,25	0	8.77	10,89	$3p\ {}^3P-7d\ {}^3D^\circ$	1—2
5846,35	0	8,77	10,89	$3p\ {}^3S-6d\ {}^3P^\circ$	1—1
5824,64	1	8,85	10,98	$3p\ {}^3P-8s\ {}^3P^\circ$	2—1
5819,50	1	8,85	10,98	$3p\ {}^3P-8s\ {}^3P^\circ$	1—0
5817,70	0	8,85	10,98	$3p\ {}^3P-8s\ {}^3P^\circ$	1—1
5813,51	1	8,85	10,98	$3p\ {}^3P-8s\ {}^3P^\circ$	0—1
5805,80	3	8,85	10,98	$3p\ {}^3P-7d\ {}^3D^\circ$	2—2
		8,85	10,98	$3p\ {}^3P-8s\ {}^3P^\circ$	1—2
5805,19	4	7,95	10,08	$2p^3\ {}^3D^\circ-4p\ {}^3P$	1—0
5800,59	6	7,95	10,08	$2p^3\ {}^3D^\circ-4p\ {}^3P$	2—1
5800,23	3	7,95	10,08	$2p^3\ {}^3D^\circ-4p\ {}^3P$	1—1
5798,90	0	8,85	10,98	$3p\ {}^3P-7d\ {}^3D^\circ$	1—2
5794,46	3	7,95	10,08	$2p^3\ {}^3D^\circ-4p\ {}^3P$	2—2
5793,12	7	7,95	10,08	$2p^3\ {}^3D^\circ-4p\ {}^3P$	3—2
5720,78	2	8,54	10,70	$3p\ {}^1P-6s\ {}^3P^\circ$	1—1
5693,11	3	8,54	10,71	$3p\ {}^1P-6s\ {}^1P^\circ$	1—1
5668,96	7	8,54	10,72	$3p\ {}^1P-5d\ {}^1P^\circ$	1—1
5629,93	1	8.85	11,05	$3p\ {}^3P-8d\ {}^3D^\circ$	2—2
5676,49	0	8,77	10.98	$3p\ {}^3S-8s\ {}^3P^\circ$	1—0
5674,81	0	8,77	10,98	$3p\ {}^3S-8s\ {}^3P^\circ$	1—1
5603,73	0	8,77	10,98	$3p\ {}^3S-8s\ {}^3P^\circ$	1—2
5597,30	1	8,77	10,98	$3p\ {}^3S-7d\ {}^3D^\circ$	1—2
5553,17	1	8,64	10,87	$3p\ {}^3D-6d\ {}^3F^\circ$	2—3
5551,59	5	8,64	10,87	$3p\ {}^3D-7s\ {}^3P^\circ$	2—1
5551,03	2	8,65	10,88	$3p\ {}^3D-7s\ {}^3P^\circ$	3—2
5548,90	1	8,65	10,87	$3p\ {}^3D-6d\ {}^3F^\circ$	3—4
5548,24	0	8,64	10,87	$3p\ {}^3D-6d\ {}^3F^\circ$	1—2
5547,27	3	8,64	10,87	$3p\ {}^3D-7s\ {}^3P^\circ$	1—0
5545,07	6	8,64	10,87	$3p\ {}^3D-7s\ {}^3P^\circ$	1—1
		8,65	10,88	$3p\ {}^3D-6d\ {}^3D^\circ$	3—2
5543,82	0	8,65	10,88	$3p\ {}^3D-6d\ {}^3D^\circ$	3—3
5540,76	2	8,64	10,88	$3p\ {}^3D-7s\ {}^3P^\circ$	2—2
5534,81	1	8,64	10,88	$3p\ {}^3D-6d\ {}^3D^\circ$	2—2
5529,78	1	8,64	10,88	$3p\ {}^3D-7s\ {}^1P^\circ$	2—1
5526,84	2	8,65	10,89	$3p\ {}^3D-6d\ {}^3P^\circ$	3—2
5516,64	0	8,64	10,89	$3p\ {}^3D-6d\ {}^3P^\circ$	2—2
5515,69	0	8,85	11,10	$3p\ {}^3P-9d\ {}^3D^\circ$	2—2
5380,34	10	7,68	9,99	$3s\ {}^1P^\circ-4p\ {}^1P$	1—1
5317,46	1	7,68	10,02	$3s\ {}^1P^\circ-4p\ {}^3D$	1—1
5306,84	2	8,64	10,98	$3p\ {}^3D-8s\ {}^3P^\circ$	2—1

λ, Å	I	E_H, eV	E_B, eV	Transition	J
5306,32	0	8,65	10,98	$3p\ ^3D-8s\ ^3P^\circ$	3–2
5302,35	1	8,64	10,98	$3p\ ^3D-8s\ ^3P^\circ$	1–0
5300,84	1	8,64	10,98	$3p\ ^3D-8s\ ^3P^\circ$	1–1
5300,55	3	8,65	10,99	$3p\ ^3D-7d\ ^3D^\circ$	3–2
5300,12	1	8,54	10,87	$3p\ ^1P-7s\ ^3P^\circ$	1–1
5296,93	0	8,64	10,98	$3p\ ^3D-8s\ ^3P^\circ$	2–2
5291,22	1	8,64	10,99	$3p\ ^3D-7d\ ^3D^\circ$	2–2
5288,32	0	8,64	10,98	$3p\ ^3D-8s\ ^1P^\circ$	2–1
5280,24	2	8,54	10,88	$3p\ ^1P-7s\ ^1P^\circ$	1–1
5268,96	4	8,54	10,89	$3p\ ^1P-6d\ ^1P^\circ$	1–1
5159,92	1	8,64	11,04	$3p\ ^3D-9s\ ^3P^\circ$	2–1
5155,29	1	8,64	11,04	$3p\ ^3D-9s\ ^3P^\circ$	1–0
5153,57	2	8,65	11,05	$3p\ ^3D-8d\ ^3D^\circ$	3–2
5089,63	0	8,54	10,97	$3p\ ^1P-7d\ ^1D^\circ$	1–2
5076,59	1	8,54	10,98	$3p\ ^1P-8s\ ^3P^\circ$	1–1
5064,15	0	8,64	11,09	$3p\ ^3D-10s\ ^3P^\circ$	2–1
5059,66	0	8,54	10,98	$3p\ ^1P-8s\ ^1P^\circ$	1–1
5057,68	0	8,65	11,10	$3p\ ^3D-9d\ ^3D^\circ$	3–2
5053,52	2	8,54	10,99	$3p\ ^1P-7d\ ^1P^\circ$	1–1
5052,17	8	7,68	10,14	$3s\ ^1P^\circ-4p\ ^1D$	1–2
5041,80	6	7,95	10,41	$2p^3\ ^3D^\circ-4f\ F\ (2^1/_2)$	2–3, 2
5041,48	6	7,95	10,41	$2p^3\ ^3D^\circ-4f\ F\ (2^1/_2)$	1–2
5040,13	4	7,95	10,41	$2p^3\ ^3D^\circ-4f\ F\ (3^1/_2)$	2–3
5039,07	7	7,95	10,41	$2p^3\ ^3D^\circ-4f\ F\ (3^1/_2)$	3–4, 3
5024,92	3	7,95	10,42	$2p^3\ ^3D^\circ-4f\ G\ (3^1/_2)$	2–3
5023,85	7	7,95	10,42	$2p^3\ ^3D^\circ-4f\ G\ (3^1/_2)$	3–4, 3
5018,06	2	7,95	10,42	$2p^3\ ^3D^\circ-4f\ D\ (2^1/_2)$	2–3, 2
5017,76	1	7,95	10,42	$2p^3\ ^3D^\circ-4f\ D\ (2^1/_2)$	1–2
5017,09	3	7,95	10,42	$2p^3\ ^3D^\circ-4f\ D\ (2^1/_2)$	3–3, 2
5012,28	2	7,95	10,42	$2p^3\ ^3D^\circ-4f\ D\ (1^1/_2)$	2–2, 1
5012,00	2	7,95	10,42	$2p^3\ ^3D^\circ-4f\ D\ (1^1/_2)$	1–2, 1
4991,41	0	8,65	11,13	$3p\ ^3D-10d\ ^3D^\circ$	3–2
4943,58	0	8,65	11,15	$3p\ ^3D-11d\ ^3D^\circ$	3–2
4942,02	0	8,54	11,04	$3p\ ^1P-9s\ ^3P^\circ$	1–1
4932,05	8	7,68	10,20	$3s\ ^1P^\circ-4p\ ^1S$	1–0
4926,40	0	8,54	11,05	$3p\ ^1P-9s\ ^1P^\circ$	1–1
4922,68	1	8,54	11,05	$3n\ ^1P-8d\ ^1P^\circ$	1–1
4898,63	1	7,49	10.02	$3s\ ^3P^\circ-4p\ ^3D$	2–2
4893,43	0	7,48	10.02	$3s\ ^3P^\circ-4p\ ^3D$	1–1
4890,65	2	7,49	10,02	$3s\ ^3P^\circ-4p\ ^3D$	2–3
4888,91	1	7,48	10,02	$3s\ ^3P^\circ-4p\ ^3D$	1–2; 0–1
4836,76	0	8,54	11,10	$3p\ ^1P-9d\ ^1P^\circ$	1–1
4826,80	3	7,49	10,06	$3s\ ^3P^\circ-4p\ ^3S$	2–1
4817,37	4	7,48	10,06	$3s\ ^3P^\circ-4p\ ^3S$	1–1
4812,92	2	7,48	10,06	$3s\ ^3P^\circ-4p\ ^3S$	0–1
4796,08	0	7,95	10,53	$2p^3\ ^3D^\circ-5p\ ^3D$	2–1
4795,88	0	7,95	10,53	$2p^3\ ^3D^\circ-5p\ ^3D$	1–1
4792,65	0	7,95	10,53	$2p^3\ ^3D^\circ-5p\ ^3D$	2–2
4791,71	0	7,95	10,53	$2p^3\ ^3D^\circ-5p\ ^3D$	3–2
4783,80	1	7,95	10,54	$2p^3\ ^3D^\circ-5p\ ^3D$	3–3
4775,91	6	7,49	10,08	$3s\ ^3P^\circ-4p\ ^3P$	2–1
4771,75	8	7,49	10,08	$3s\ ^3P^\circ-4p\ ^3P$	2–2
4770,03	5	7,48	10,08	$3s\ ^3P^\circ-4p\ ^3P$	1–0
4766,68	4	7,48	10,08	$3s\ ^3P^\circ-4p\ ^3P$	1–1
4762,54	5	7,48	10,08	$3s\ ^3P^\circ-4p\ ^3P$	1–2
4762,31	5	7,48	10,08	$3s\ ^3P^\circ-4p\ ^3P$	0–1
4742,57	2	7,95	10,56	$2p^3\ ^3D^\circ-5p\ ^3P$	1–0
4738,47	3	7,95	10,57	$2p^3\ ^3D^\circ-5p\ ^3P$	2–1
4738,21	1	7,95	10,57	$2p^3\ ^3D^\circ-5p\ ^3P$	1–1
4735,17	2	7,95	10,57	$2p^3\ ^3D^\circ-5p\ ^3P$	2–2
4734,26	5	7,95	10,57	$2p^3\ ^3D^\circ-5p\ ^3P$	3–2

λ, Å	I	E_H, eV	E_B, eV	Transition	J
4478,83	4	7,95	10,72	$2p^3\ {}^3D^\circ - 5f\ F\ (2^1/_2)$	2−3, 2
4478,59	4	7,95	10,72	$2p^3\ {}^3D^\circ - 5f\ F\ (2^1/_2)$	1—2
4478,32	2	7,95	10,72	$2p^3\ {}^3D^\circ - 5f\ F\ (3^1/_2)$	2—3
4477,47	4	7,95	10,72	$2p^3\ {}^3D^\circ - 5f\ F\ (3^1/_2)$	3−4, 3
4467,31	2	7,95	10,72	$2p^3\ {}^3D^\circ - 5f\ G\ (3^1/_2)$	2—3
4466,48	5	7,95	10,72	$2p^3\ {}^3D^\circ - 5f\ G\ (3^1/_2)$	3−4, 3
4464,68	2	7,95	10,72	$2p^3\ {}^3D^\circ - 5f\ D\ (2^1/_2)$	2−3, 2
4464,45	1	7,95	10,72	$2p^3\ {}^3D^\circ - 5f\ D\ (2^1/_2)$	1—2
4463,89	2	7,95	10,72	$2p^3\ {}^3D^\circ - 5f\ D\ (2^1/_2)$	3−3, 2
4461,50	1	7,95	10,72	$2p^3\ {}^3D^\circ - 5f\ D\ (1^1/_2)$	2−2, 1
4461,30	1	7,95	10,72	$2p^3\ {}^3D^\circ - 5f\ D\ (1^1/_2)$	1−2, 1
4371,37	6	7,68	10,52	$3s\ {}^1P^\circ - 5p\ {}^1P$	1—1
4355,41	1	7,68	10,53	$3s\ {}^1P^\circ - 5p\ {}^3D$	1—1
4348,97	1	7,95	10,80	$2p^3\ {}^3D^\circ - 6p\ {}^3P$	1—0
4344,31	1	7,95	10,80	$2p^3\ {}^3D^\circ - 6p\ {}^3P$	2—1
4344,11	0	7,95	10,80	$2p^3\ {}^3D^\circ - 6p\ {}^3P$	1—1
4342,40	0	7,95	10,80	$2p^3\ {}^3D^\circ - 6p\ {}^3P$	2—2
4341,64	2	7,95	10,80	$2p^3\ {}^3D^\circ - 6p\ {}^3P$	3—2
4269,02	6	7,68	10,59	$3s\ {}^1P^\circ - 5p\ {}^1D$	1—2
4228,33	5	7,68	10,62	$3s\ {}^1P^\circ - 5p\ {}^1S$	1—0
4223,36	3	7,95	10,88	$2p^3\ {}^3D^\circ - 6f\ F\ (2^1/_2)$	2−3, 2
4223,16	4	7,95 7,95	10,88 10,88	$2p^3\ {}^3D^\circ - 6f\ F\ (2^1/_2)$ $2p^3\ {}^3D^\circ - 6f\ F\ (3^1/_2)$	1—2 2—3
4222,47	3	7,95	10,88	$2p^3\ {}^3D^\circ - 6f\ F\ (3^1/_2)$	3−4, 3
4213,07	2	7,95	10,89	$2p^3\ {}^3D^\circ - 6f\ G\ (3^1/_2)$	2—3
4212,33	4	7,95	10,89	$2p^3\ {}^3D^\circ - 6f\ G\ (3^1/_2)$	3−4, 3
4211,82	2	7,95	10,89	$2p^3\ {}^3D^\circ - 6f\ D\ (2^1/_2)$	2−3, 2
4211,61	1	7,95	10,89	$2p^3\ {}^3D^\circ - 6f\ D\ (2^1/_2)$	1—2
4211,12	2	7,95	10,89	$2p^3\ {}^3D^\circ - 6f\ D\ (2^1/_2)$	3−3, 2
4209,91	0	7,95	10,89	$2p^3\ {}^3D^\circ - 6f\ D\ (1^1/_2)$	2−2, 1
4209,71	0	7,95	10,89	$2p^3\ {}^3D^\circ - 6f\ D\ (1^1/_2)$	1−2, 1
4153,37	0	7,95	10,93	$2p^3\ {}^3D^\circ - 7p\ {}^3P$	1—0
4147,98	1	7,95	10,93	$2p^3\ {}^3D^\circ - 7p\ {}^3P$	2—1
4146,97	0	7,95	10,93	$2p^3\ {}^3D^\circ - 7p\ {}^3P$	2—2
4146,26	2	7,95	10,93	$2p^3\ {}^3D^\circ - 7p\ {}^3P$	3—2
4083,16	1	7,95	10,98	$2p^3\ {}^3D^\circ - 7f\ F\ (2^1/_2)$	2−3, 2
4082,98	1	7,95	10,98	$2p^3\ {}^3D^\circ - 7f\ F\ (2^1/_2)$	1—2
4082,40	1	7,95	10,98	$2p^3\ {}^3D^\circ - 7f\ F\ (3^1/_2)$	3−4, 3
4073,33	1	7,95	10,99	$2p^3\ {}^3D^\circ - 7f\ G\ (3^1/_2)$	2—3
4072,64	3	7,95	10,99	$2p^3\ {}^3D^\circ - 7f\ G\ (3^1/_2)$	3—4
4070,97	2	7,49	10,53	$3s\ {}^3P^\circ - 5p\ {}^3D$	2—2
4066,75	2	7,48	10,53	$3s\ {}^3P^\circ - 5p\ {}^3D$	1—1
4065,25	4	7,49	10,54	$3s\ {}^3P^\circ - 5p\ {}^3D$	2—3
4064,27	3	7,48	10,53	$3s\ {}^3P^\circ - 5p\ {}^3D$	1—2
4063,58	2	7,48	10,53	$3s\ {}^3P^\circ - 5p\ {}^3D$	0—1
4033,23	0	7,95	11,02	$2p^3\ {}^3D^\circ - 8p\ {}^3P$	3—2
4031,80	3	7,49	10,57	$3s\ {}^3P^\circ - 5p\ {}^3P$	2—1
4029,41	4	7,49	10,57	$3s\ {}^3P^\circ - 5p\ {}^3P$	2—2
4028,36	2	7,48	10,56	$3s\ {}^3P^\circ - 5p\ {}^3P$	1—0
4025,22	1	7,48	10,57	$3s\ {}^3P^\circ - 5p\ {}^3P$	1—1
4022,84	3	7,48	10,57	$3s\ {}^3P^\circ - 5p\ {}^3P$	1—2
4022,12	2	7,48	10,57	$3s\ {}^3P^\circ - 5p\ {}^3P$	0—1
4009,93	4	7,68	10,77	$3s\ {}^1P^\circ - 6p\ {}^1P$	1—1
4002,98	2	7,68	10,78	$3s\ {}^1P^\circ - 6p\ {}^3D$	1—1
3997,14	1	7,95	11,05	$2p^3\ {}^3D^\circ - 8f\ F\ (2^1/_2)$	2−3, 2
3996,97	0	7,95	11,05	$2p^3\ {}^3D^\circ - 8f\ F\ (2^1/_2)$	1—2
3996,49	0	7,95	11,05	$2p^3\ {}^3D^\circ - 8f\ F\ (3^1/_2)$	3−4, 3
3986,88	1	7,95	11,06	$2p^3\ {}^3D^\circ - 8f\ G\ (3^1/_2)$	3−4, 3

λ, Å	I	E_H, eV	E_B, eV	Transition	J
3961,40	3	7,68	10,81	$3s\ ^1P^\circ$—$6p\ ^1D$	1—2
3942,22	3	7,68	10,82	$3s\ ^1P^\circ$—$6p\ ^1S$	1—0
3833,35	3	7,68	10,91	$3s\ ^1P^\circ$—$7p\ ^1P$	1—1
3828,85	2	7,68	10,92	$3s\ ^1P^\circ$—$7p\ ^3D$	1—1
3804,31	2	7,68	10,94	$3s\ ^1P^\circ$—$7p\ ^1D$	1—2
3793,68	2	7,68	10,95	$3s\ ^1P^\circ$—$7p\ ^1S$	1—0
3763,96	0	7,48	10,77	$3s\ ^3P^\circ$—$6p\ ^1P$	1—1
3762,25	2	7,49	10,78	$3s\ ^3P^\circ$—$6p\ ^3D$	2—2
3757,84	1	7,48	10,78	$3s\ ^3P^\circ$—$6p\ ^3D$	1—1
3757,05	3	7,49	10,79	$3s\ ^3P^\circ$—$6p\ ^3D$	2—3
3756,52	2	7,48	10,78	$3s\ ^3P^\circ$—$6p\ ^3D$	1—2
3755,12	1	7,48	10,78	$3s\ ^3P^\circ$—$6p\ ^3D$	0—1
3742,85	1	7,49	10,80	$3s\ ^3P^\circ$—$6p\ ^3P$	2—1
3741,44	2	7,49	10,80	$3s\ ^3P^\circ$—$6p\ ^3P$	2—2
3740,79	0	7,48	10,80	$3s\ ^3P^\circ$—$6p\ ^3P$	1—0
3737,19	0	7,48	10,80	$3s\ ^3P^\circ$—$6p\ ^3P$	1—1
3735,78	1	7,48	10,80	$3s\ ^3P^\circ$—$6p\ ^3P$	1—2
3734,51	0	7,48	10,80	$3s\ ^3P^\circ$—$6p\ ^3P$	0—1
3732,35	2	7,68	11,00	$3s\ ^1P^\circ$—$8p\ ^1P$	1—1
3729,03	1	7,68	11,01	$3s\ ^1P^\circ$—$8p\ ^3D$	1—1
3712,04	1	7,68	11,02	$3s\ ^1P^\circ$—$8p\ ^1D$	1—2
3705,56	1	7,68	11,02	$3s\ ^1P^\circ$—$8p\ ^1S$	1—0
3668,60	1	7,68	11,06	$3s\ ^1P^\circ$—$9p\ ^1P$	1—1
3652,82	0	7,68	11,08	$3s\ ^1P^\circ$—$9p\ ^1D$	1—2
3648,62	0	7,68	11,08	$3s\ ^1P^\circ$—$9p\ ^1S$	1—0
3625,61	0	7,68	11,10	$3s\ ^1P^\circ$—$10p\ ^1P$	1—1
3609,56	0	7,68	11,12	$3s\ ^1P^\circ$—$10p\ ^1s$	1—0
3608,70	1	7,49	10,92	$3s\ ^3P^\circ$—$7S\ ^3D$	2—2
3607,94	0	7,48	10,91	$3s\ ^3P^\circ$—$7p\ ^1P$	1—1
3603,95	0	7,48	10,92	$3s\ ^3P^\circ$—$7p\ ^3D$	1—1
3603,53	2	7,49	10,93	$3s\ ^3P^\circ$—$7p\ ^3D$	2—3
3603,44	1	7,48	10,92	$3s\ ^3P^\circ$—$7p\ ^3D$	1—2
3601,47	0	7,48	10,92	$3s\ ^3P^\circ$—$7p\ ^3D$	0—1
3595,46	0	7,49	10,93	$3s\ ^3P^\circ$—$7p\ ^3P$	2—2
3595,14	0	7,68	11,13	$3s\ ^1P^\circ$—$11p\ ^1P$	1—1
3518,31	0	7,48	11,00	$3s\ ^3P^\circ$—$8p\ ^1P$	1—1
3514,80	2	7,49	11,02	$3s\ ^3P^\circ$—$8p\ ^3D$	2—3
3458,50	1	7,49	11,07	$3s\ ^3P^\circ$—$9p\ ^3D$	2—3
3420,41	0	7,49	11,11	$3s\ ^3P^\circ$—$10p\ ^3D$	2—3
2967,244	5	0,00	4,18	$2p^2\ ^3P$—$2p^3\ ^5S^\circ$	2—2
2964,846	2	0,00	4,18	$2p^2\ ^3P$—$2p^3\ ^5S^\circ$	1—2
2582,901	5	2,68	7,48	$2p^2\ ^1S$—$3s\ ^3P^\circ$	0—1
2478,556	16	2,68	7,68	$2p^2\ ^1S$—$3s\ ^1P^\circ$	0—1
1993,627	2	1,26	7,48	$2p^2\ ^1D$—$3s\ ^3P^\circ$	2—1
1930,905	10	1,26	7,68	$2p^2\ ^1D$—$3s\ ^1P^\circ$	2—1
1765,366	1	2,68	9,71	$2p^2\ ^1S$—$3d\ ^3D^\circ$	0—1
1763,909	2	2,68	9,71	$2p^2\ ^1S$—$4s\ ^1P^\circ$	0—1
1751,827	8	2,68	9,76	$2p^2\ ^1S$—$3d\ ^1P^\circ$	0—1
1658,121	5	0,00	7,48	$2p^2\ ^3P$—$3s\ ^3P^\circ$	2—1
1657,907	4	0,00	7,48	$2p^2\ ^3P$—$3s\ ^3P^\circ$	1—0
1657,379	2	0,00	7,48	$2p^2\ ^3P$—$3s\ ^3P^\circ$	1—1
1657,008	10	0,00	7,49	$2p^2\ ^3P$—$3s\ ^3P^\circ$	2—2
1656,930	3	0,00	7,48	$2p^2\ ^3P$—$3s\ ^3P^\circ$	0—1
1656,268	5	0,00	7,49	$2p^2\ ^3P$—$3s\ ^3P^\circ$	1—2
1602,971	5	2,68	10,42	$2p^2\ ^1S$—$4d\ ^1P^\circ$	0—1
1561,435	20	0,00	7,95	$2p^2\ ^3P$—$2p^3\ ^3D^\circ$	2—3
1561,337	2	0,00	7,95	$2p^2\ ^3P$—$2p^3\ ^3D^\circ$	2—2, 1
1560,691	15	0,00	7,95	$2p^2\ ^3P$—$2p^3\ ^3D^\circ$	1—2, 1
1560,306	8	0,00	7,95	$2p^2\ ^3P$—$2p^3\ ^3D^\circ$	0—1
1542,177	2	2,68	10,72	$2p^2\ ^1S$—$5d\ ^1P^\circ$	0—1

λ, Å	I	E_H, eV	E_B, eV	Transition	J
1510,924	1	2,68	10,88	$2p^2\,{}^1S - 7s\,{}^1P^\circ$	0—1
1481,762	7	1,26	9,63	$2p^2\,{}^1D - 3d\,{}^1D^\circ$	2—2
1472,231	0	1,26	9,68	$2p^2\,{}^1D - 4s\,{}^3P^\circ$	2—1
1470,082	1	1,26	9,70	$2p^2\,{}^1D - 3d\,{}^3F^\circ$	2—3
1468,410	15	1,26	9,71	$2p^2\,{}^1D - 3d\,{}^3D^\circ$	2-1
1467,405	3	1,26	9,71	$2p^2\,{}^1D - 4s\,{}^1P^\circ$	2—1
1463,336	6	1,26	9,74	$2p^2\,{}^1D - 3d\,{}^1F^\circ$	2—3
1459,032	2	1,26	9,76	$2p^2\,{}^1D - 3d\,{}^1P^\circ$	2—1
1432,530	1	4,18	12,84	$2p^3\,{}^5S^\circ - 3s\,{}^5P$	2—1
1432,105	2	4,18	12,84	$2p^3\,{}^5S^\circ - 3s\,{}^5P$	2—2
1431,597	2	4,18	12,84	$2p^3\,{}^5S^\circ - 3s\,{}^5P$	2—3
1364,165	6	1,26	10,35	$2p^2\,{}^1D - 4d\,{}^1D^\circ$	2—2
1359,329	2	1,26	10,38	$2p^2\,{}^1D - 4d\,{}^3F^\circ$	2—3
1357,140	3	1,26	10,40	$2p^2\,{}^1D - 5s\,{}^1P^\circ$	2—1
1355,825	6	1,26	10,41	$2p^2\,{}^1D - 4d\,{}^1F^\circ$	2—3
1354,292	5	1,26	10,42	$2p^2\,{}^1D - 4d\,{}^1P^\circ$	2—1
1329,588	6	0,00	9,33	$2p^2\,{}^3P - 2p^3\,{}^3P^\circ$	2—2, 1
1329,103	5	0,00	9,33	$2p^2\,{}^3P - 2p^3\,{}^3P^\circ$	1—2, 1, 0
1328,834	3	0,00	9,33	$2p^2\,{}^3P - 2p^3\,{}^3P^\circ$	0—1
1315,903	4	1,26	10,69	$2p^2\,{}^1D - 5d\,{}^1D^\circ$	2—2
1313,471	6	1,26	10,70	$2p^2\,{}^1D - 5d\,{}^3F^\circ$	2—3
1312,261	2	1,26	10,71	$2p^2\,{}^1D - 5d\,{}^3D^\circ$	2—3
1311,985	2	1,26	10,71	$2p^2\,{}^1D - 6s\,{}^1P^\circ$	2—1
1311,365	8	1,26	10,72	$2p^2\,{}^1D - 5d\,{}^1F^\circ$	2—3
1310,646	4	1,26	10,72	$2p^2\,{}^1D - 5d\,{}^1P^\circ$	2—1
1291,380	1	1,26	10,86	$2p^2\,{}^1D - 6d\,{}^1D^\circ$	2—2
1289,983	3	1,26	10,87	$2p^2\,{}^1D - 6d\,{}^3F^\circ$	2—3
1288,633	2	1,26	10,88	$2p^2\,{}^1D - 7s\,{}^1P^\circ$	2—1
1288,445	5	1,26	10,89	$2p^2\,{}^1D - 6d\,{}^1F^\circ$	2—3
1288,055	1	1,26	10,89	$2p^2\,{}^1D - 6d\,{}^1P^\circ$	2—1
1280,852	4	0,00	9,68	$2p^2\,{}^3P - 4s\,{}^3P^\circ$	2—1
1280,604	2	0,00	9,68	$2p^2\,{}^3P - 4s\,{}^3P^\circ$	1—0
1280,403	2	0,00	9,69	$2p^2\,{}^3P - 4s\,{}^3P^\circ$	1-1
1280,340	6	0,00	9,68	$2p^2\,{}^3P - 4s\,{}^3P^\circ$	2-2
1280,140	2	0,00	9,68	$2p^2\,{}^3P - 4s\,{}^3P^\circ$	0—1
1279,897	5	0,00	9,69	$2p^2\,{}^3P - 4s\,{}^3P^\circ$	1—2
1279,230	6	0,00	9,70	$2p^2\,{}^3P - 3d\,{}^3F^\circ$	2—3
1277,727	3	0,00	9,71	$2p^2\,{}^3P - 3d\,{}^3D^\circ$	2—2
1277,551	10	0,00	9,71	$2p^2\,{}^3P\ \ 3d\,{}^3D^\circ$	2-3; 1-1
1277,282	9	0,00	9,71	$2p^2\,{}^3P - 3d\,{}^3D^\circ$	1-2; 0-1
1277,154	2	0,00	9,71	$2p^2\,{}^3P - 4s\,{}^1P^\circ$	0—1
1276,754	4	0,00	9,71	$2p^2\,{}^3P - 4s\,{}^1P^\circ$	1-1
1275,021	5	1,26	10,99	$2p^2\,{}^1D - 7d\,{}^1F^\circ$	2—3
1274,880	2	1,26	10,99	$2p^2\,{}^1D - 7d\,{}^1P^\circ$	2—1
1274,131	5	0,00	9,74	$2p^2\,{}^3P - 3d\,{}^1F^\circ$	2—3
1267,633	1	1,26	11,04	$2p^2\,{}^1D - 8d\,{}^3F^\circ$	2—3, 2
1266,449	3	1,26	11,05	$2p^2\,{}^1D - 8d\,{}^1F^\circ$	2—3
1261,560	8	0,00	9,83	$2p^2\,{}^3P - 3d\,{}^3P^\circ$	2—2
1261,430	5	0,00	9,83	$2p^2\,{}^3P - 3d\,{}^3P^\circ$	2-1
1261,128	7	0,00	9,83	$2p^2\,{}^3P - 3d\,{}^3P^\circ$	1—2
1261,000	3	0,00	9,83	$2p^2\,{}^3P - 3d\,{}^3P^\circ$	1—1
1260,930	4	0,00	9,83	$2p^2\,{}^3P - 3d\,{}^3P^\circ$	1-0
1260,738	4	0,00	9,83	$2p^2\,{}^3P - 3d\,{}^3P^\circ$	0—1
1260,670	2	1,26	11,10	$2p^2\,{}^1D - 9d\,{}^1F^\circ$	2—3
1253,538	0,5	1,26	11,15	$2p^2\,{}^1D - 11d\,{}^1F^\circ$	2-3
1197,812	0,5	0,00	10,35	$2p^2\,{}^3P - 4d\,{}^1D$	1-2
1194,656	5	0,00	10,38	$2p^2\,{}^3P - 5s\,{}^3P^\circ$	2—1
1194,494	7	0,00	10,38	$2p^2\,{}^3P - 4d\,{}^3F^\circ$	2—3
1194,291	2	0,00	10,38	$2p^2\,{}^3P - 4d\,{}^3F^\circ$ $2p^2\,{}^3P - 5s\,{}^3P^\circ$	1—2 1—1
1194,060	5	0,00	10,39	$2p^2\,{}^3P - 5s\,{}^3P^\circ$	2—2

λ, Å	I	E_H, eV	E_B, eV	Transition	J
1194,027	3	0,00	10,38	$2p^2\,{}^3P-5s\,{}^3P^\circ$	0—1
1193,674	4	0,00	10,39	$2p^2\,{}^3P-5s\,{}^3P^\circ$	1—2
1193,388	3	0,00	10,39	$2p^2\,{}^3P-4d\,{}^3D^\circ$	2—2
1193,252	10	0,00	10,40	$2p^2\,{}^3P-4d\,{}^3D^\circ$	2−3; 1−1
1193,013	8	0,00	10,39	$2p^2\,{}^3P-4d\,{}^3D^\circ$	0−1; 1−2
1192,923	2	0,00	10,40	$2p^2\,{}^3P-5s\,{}^1P$	2−1
1192,480	2	0,00	10,40	$2p^2\,{}^3P-5s\,{}^1P^\circ$	1—1
1191,855	1	0,00	10,41	$2p^2\,{}^3P-4d\,{}^1F^\circ$	2—3
1189,628	6	0,00	10,43	$2p^2\,{}^3P-4d\,{}^3P^\circ$	2—2
1189,556	4	0,00	10,43	$2p^2\,{}^3P-4d\,{}^3P^\circ$	2—1
1189,244	4	0,00	10,43	$2p^2\,{}^3P-4d\,{}^3P^\circ$	1−2
1189,074	3	0,00	10,43	$2p^2\,{}^3P-4d\,{}^3P^\circ$	1−1, 0
1188,935	1	0,00	10,43	$2p^2\,{}^3P-4d\,{}^3P^\circ$	0—1
1159,004	5	0,00	10,70	$2p^2\,{}^3P-5d\,{}^3F^\circ$	2—3
1158,729	3	0,00	10,70	$2p^2\,{}^3P-5d\,{}^3F^\circ$	1—2
1158,398	2	0,00	10,71	$2p^2\,{}^3P-6s\,{}^3P^\circ$	2—2.
1158,138	8	0,00	10,71	$2p^2\,{}^3P-5d\,{}^3D^\circ$	1−1; 2−3
1158,030	7	0,00	10,71	$2p^2\,{}^3P-6s\,{}^3P^\circ$ $2p^2\,{}^3P-5d\,{}^3D^\circ$	1—2 2—3
1157,825	3	0,00	10,71	$2p^2\,{}^3P-6s\,{}^1P^\circ$	2—1
1157,391	2	0,00	10,71	$2p^2\,{}^3P-6s\,{}^1P^\circ$	1—1
1157,333	1	0,00	10,72	$2p^2\,{}^3P-5d\,{}^1F^\circ$	2—3
1156,619	5	0,00	10,72	$2p^2\,{}^3P-5d\,{}^3P^\circ$	2—2
1156,502	1	0,00	10,73	$2p^2\,{}^3P-5d\,{}^3P^\circ$ $2p^2\,{}^3P-5d\,{}^1P^\circ$	2—1 1−1
1156,059	2	0,00	10,73	$2p^2\,{}^3P-5d\,{}^3P^\circ$	1—1
1155,839	1	0,00	10,73	$2p^2\,{}^3P-5d\,{}^3P^\circ$	0—1
1141,705	1	0,00	10,86	$2p^2\,{}^3P-6d\,{}^1D^\circ$	2—2
1140,688	3	0,00	10,87	$2p^2\,{}^3P-6d\,{}^3F^\circ$	2—3
1140,391	1	0,00	10,87	$2p^2\,{}^3P-6d\,{}^3F^\circ$ $2p^2\,{}^3P-6d\,{}^3D^\circ$	1—2 2—1
1140,070	1	0,00	10,88	$2p^2\,{}^3P-6d\,{}^3D^\circ$	2—2
1139,894	7	0,00	10,88	$2p^2\,{}^3P-6d\,{}^3D^\circ$	2−3, 2
1139,794	6	0,00	10,88	$2p^2\,{}^3P-6d\,{}^3D^\circ$	0−1
1139,142	2	0,00	10,89	$2p^2\,{}^3P-6d\,{}^3P^\circ$	2—2
1139,037	1	0,00	10,89	$2p^2\,{}^3P-6d\,{}^3P^\circ$	2—1
1138,625	1	0,00	10,89	$2p^2\,{}^3P-6d\,{}^3P^\circ$	1−1, 2
1129,927	1	0,00	10,98	$2p^2\,{}^3P-7d\,{}^3F^\circ$	2—3
1129,626	1	0,00	10,98	$2p^2\,{}^3P-7d\,{}^3F^\circ$	1—2
1129,161	6	0,00	10,98	$2p^2\,{}^3P-7d\,{}^3D^\circ$	0−1; 2−3, 2
1128,748	1	0,00	10,99	$2p^2\,{}^3P-7d\,{}^3P^\circ$ $2p^2\,{}^3P-7d\,{}^1P^\circ$	2—2 2—1
1122,325	4	0,00	11,05	$2p^2\,{}^3P-8d\,{}^3D^\circ$	2−3, 2
1122,179	1	0,00	11,05	$2p^2\,{}^3P-8d\,{}^3P^\circ$	2—2
1117,706	3	0,00	11,10	$2p^2\,{}^3P-9d\,{}^3D^\circ$	2−3, 2
1114,414	2	0,00	11,13	$2p^2\,{}^3P-10d\,{}^3D^\circ$	2−3, 2
945,566	3	0,00	13,12	$2p^2\,{}^3P-2p^3\,{}^3S^\circ$	2—1
945,336	2	0,00	13,12	$2p^2\,{}^3P-2p^3\,{}^3S^\circ$	1—1
945,193	1	0,00	13,12	$2p^2\,{}^3P-2p^3\,{}^3S^\circ$	0—1

C II, ground state $1s^2\,2s^2\,2p\;{}^2P^0{}_{1/2}$

Ionization potential 196659,0 cm⁻¹; 24,381 eV

λ, Å	I	E_H, eV	E_B, eV	Transition	J
8799,9	0	23,38	24,79	$3p\,{}^2D-3d\,{}^2F^\circ$	$^3/_2$—$^5/_2$
8793,8	1	23,38	24,79	$3p\,{}^2D-3d\,{}^2F^\circ$	$^5/_2$—$^7/_2$

λ, Å	I	E_H, eV	E_B, eV	Transition	J
8696,71	5	19,49	20,92	$4s\,^2S-2p^3\,^2P^\circ$	1/2—1/2
8682,56	8	19,49	20,92	$4s\,^2S-2p^3\,^2P^\circ$	1/2—3/2
8414,49	1	27,22 27,22	28,70 28,70	$4d\,^4F^\circ-5f\,^4G$ $4d\,^4F^\circ-5f\,^4G$	5/2—7/2 3/2—5/2
8413,42	2	27,23 27,22	28,70 28,70	$4d\,^4F^\circ-5f\,^4G$ $4d\,^4F^\circ-5f\,^4G$	9/2—11/2 7/2—9/2
8076,64	8	23,12	24,65	$3p\,^4P-3d\,^4P^\circ$	5/2—5/2
8062,78	6	23,12	24,65	$3p\,^4P-3d\,^4P^\circ$	5/2—3/2
8062,12	5	23,11	24,65	$3p\,^4P-3d\,^4P^\circ$	3/2—5/2
8048,32	3	23,11	24,65	$3p\,^4P-3d\,^4P^\circ$	3/2—3/2
8039,39	6	23,11	24,66	$3p\,^4P-3d\,^4P^\circ$	3/2—1/2
8037,76	5	23,11	24,65	$3p\,^4P-3d\,^4P^\circ$	1/2—3/2
8028,86	2	23,11	24,66	$3p\,^4P-3d\,^4P^\circ$	1/2—1/2
7530,60	2	20,92	22,57	$2p^3\,^2P^\circ-3p\,^2P$	3/2—1/2
7519,86	4	21,73 20,92	23,38 22,57	$5p\,^2P^\circ-3p\,^2D$ $2p^3\,^2P^\circ-3p\,^2P$	1/2—3/2 1/2—1/2
7519,50	7	20,92	22,57	$2p^3\,^2P^\circ-3p\,^2P$	3/2—3/2
7508,90	3	20,92	22,57	$2p^3\,^2P^\circ-3p\,^2P$	1/2—3/2
7505,31	2	21,73	23,38	$5p\,^2P^\circ-3p\,^2D$	3/2—5/2
7237,17	7	16,33	18,04	$3p\,^2P^\circ-3d\,^2D$	3/2—3/2
7236,42	20	16,33	18,05	$3p\,^2P^\circ-3d\,^2D$	3/2—5/2
7231,32	18	16,33	18,04	$3p\,^2P^\circ-3d\,^2D$	1/2—3/2
7144,19	1	22,54	24,27	$3p\,^4D-3d\,^4F^\circ$	7/2—5/2
7134,11	6	22,54	24,27	$3p\,^4D-3d\,^4F^\circ$	7/2—7/2
7132,45	1	22,53	24,27	$3p\,^4D-3d\,^4F^\circ$	5/2—3/2
7125,73	7	22,53	24,27	$3p\,^4D-3d\,^4F^\circ$	5/2—5/2
7119,90	12	22,54 22,53	24,28 24,27	$3p\,^4D-3d\,^4F^\circ$ $3p\,^4D-3d\,^4F^\circ$	7/2—9/2 3/2—3/2
7115,63	10	22,53	24,27	$3p\,^4D-3d\,^4F^\circ$	5/2—7/2
7113,04	7	22,53	24,27	$3p\,^4D-3d\,^4F^\circ$	3/2—5/2
7112,48	6	22,53	24,27	$3p\,^4D-3d\,^4F^\circ$	1/2—3/2
7063,70	8	22,90	24,65	$3p\,^4S-3d\,^4P^\circ$	3/2—5/2
7053,09	6	22,90	24,65	$3p\,^4S-3d\,^4P^\circ$	3/2—3/2
7046,26	4	22,90	24,66	$3p\,^4S-3d\,^4P^\circ$	3/2—1/2
6812,29	3	20,71	22,53	$3s\,^4P^\circ-3p\,^4D$	5/2—3/2
6800,68	7	20,71	22,53	$3s\,^4P^\circ-3p\,^4D$	5/2—5/2
6798,11	3	20,70	22,53	$3s\,^4P^\circ-3p\,^4D$	3/2—1/2
6791,47	7	20,70	22,53	$3s\,^4P^\circ-3p\,^4D$	3/2—3/2
6787,22	6	20,70	22,53	$3s\,^4P^\circ-3p\,^4D$	1/2—1/2
6783,90	10	20,71	22,54	$3s\,^4P^\circ-3p\,^4D$	5/2—7/2
6780,61	5	20,70	22,53	$3s\,^4P^\circ-3p\,^4D$	1/2—3/2
6779,93	8	20,70	22,53	$3s\,^4P^\circ-3p\,^4D$	3/2—5/2
6755,16	3	22,54	24,37	$3p\,^4D-3d\,^4D^\circ$	7/2—5/2
6750,55	8	22,54	24,37	$3p\,^4D-3d\,^4D^\circ$	7/2—7/2
6742,43	3	22,53	24,37	$3p\,^4D-3d\,^4D^\circ$	5/2—3/2
6738,62	6	22,53	24,37	$3p\,^4D-3d\,^4D^\circ$	5/2—5/2
6734,00	2	22,53	24,37	$3p\,^4D-3d\,^4D^\circ$	5/2—7/2
6733,58	2	22,53	24,37	$3p\,^4D-3d\,^4D^\circ$	3/2—1/2
6731,07	5	22,53	24,37	$3p\,^4D-3d\,^4D^\circ$	3/2—3/2
6727,19	4	22,53 22,53	24,37 24,37	$3p\,^4D-3d\,^4D^\circ$ $3p\,^4D-3d\,^4D^\circ$	3/2—5/2 1/2—1/2
6724,56	2	22,53	24,37	$3p\,^4D-3d\,^4D^\circ$	1/2—3/2
6723,65	1	20,84	22,68	$4d\,^2D-6p\,^2P^\circ$	5/2—3/2
6622,05	1	20,95	22,82	$4f\,^2F^\circ-6d\,^2D$	5/2, 7/2—3/2, 5/2
6582,88	15	14,45	16,33	$3s\,^2S-3p\,^2P^\circ$	1/2—1/2
6578,05	18	14,45	16,33	$3s\,^2S-3p\,^2P^\circ$	1/2—3/2
6461,95	5	20,95	22,87	$4f\,^2F^\circ-6g\,^2G$	5/2, 7/2—7/2, 9/2
6454,77	1	20,95	22,87	$4f\,^2F^\circ-6h\,^2H$	7/2—9/2

λ, Å	I	E_H, eV	E_B, eV	Transition	J
6385,72	1	26,63	28,58	$4p\ ^4D-5d\ ^4F^{\circ}$	7/2—9/2
6295,20	0	24,66	26,63	$3d\ ^4P^{\circ}-4p\ ^4D$	1/2—1/2
6290,01	1	24,66	26,63	$3d\ ^4P^{\circ}-4p\ ^4D$	1/2—3/2
6284,56	0	24,65	26,63	$3d\ ^4P^{\circ}-4p\ ^4D$	3/2—3/2
6275,79	1	24,65	26,63	$3d\ ^4P^{\circ}-4p\ ^4D$	3/2—5/2
6259,59	4	20,15	22,13	$4p\ ^2P^{\circ}-5d\ ^2D$	3/2—3/2, 5/2
6257,18	2	20,15	22,13	$4p\ ^2P^{\circ}-5d\ ^2D$	1/2—3/2
6256,54	2	24,60	26,58	$3d\ ^2D^{\circ}-4p\ ^2P$	3/2—1/2
6253,84	2	24,65	26,63	$3d\ ^4P^{\circ}-4p\ ^4D$	5/2—7/2
6250,74	4	24,60	26,58	$3d\ ^2D^{\circ}-4p\ ^2P$	5/2—3/2
6246,57	1	24,60	26,58	$3d\ ^2D^{\circ}-4p\ ^2P$	3/2—3/2
6151,43	4	20,84	22,86	$4d\ ^2D-6f\ ^2F^{\circ}$	3/2, 5/2—5/2, 7/2
6102,56	4	22,57	24,60	$3p\ ^2P-3d\ ^2D^{\circ}$	3/2—3/2
6098,51	9	22,57	24,60	$3p\ ^2P-3d\ ^2D^{\circ}$	3/2—5/2
6095,29	7	22,57	24,60	$3p\ ^2P-3d\ ^2D^{\circ}$	1/2—3/2
6037,96	0	27,29	29,35	$4d\ ^4D^{\circ}-6f\ ^4F$	7/2—9/2
5919,45	3	24,66	26,75	$3d\ ^4P^{\circ}-4p\ ^4S$	1/2—3/2
5914,64	4	24,65	26,75	$3d\ ^4P^{\circ}-4p\ ^4S$	3/2—3/2
5907,21	6	24,65	26,75	$3d\ ^4P^{\circ}-4p\ ^4S$	5/2—3/2
5891,59	12	18,04	20,15	$3d\ ^2D-4p\ ^2P^{\circ}$	3/2—1/2
5889,77	15	18,05	20,15	$3d\ ^2D-4p\ ^2P^{\circ}$	5/2—3/2
5889,27	6	18,04	20,15	$3d\ ^2D-4p\ ^2P^{\circ}$	3/2—3/2
5856,04	5	22,54	24,65	$3p\ ^4D-3d\ ^4P^{\circ}$	7/2—5/2
5843,61	2	22,53	24,65	$3p\ ^4D-3d\ ^4P^{\circ}$	5/2—5/2
5836,35	4	22,53	24,65	$3p\ ^4D-3d\ ^4P^{\circ}$	5/2—3/2
5827,85	2	22,53	24,65	$3p\ ^4D-3d\ ^4P^{\circ}$	3/2—3/2
5823,14	2	22,53	24,66	$3p\ ^4D-3d\ ^4P^{\circ}$	3/2—1/2
		22,53	24,65	$3p\ ^4D-3d\ ^4P^{\circ}$	1/2—3/2
5818,30	2	22,53	24,66	$3p\ ^4D-3d\ ^4P^{\circ}$	1/2—1/2
5791,77	1	27,23	29,37	$4d\ ^4F^{\circ}-6f\ ^4G$	9/2—11/2
5713,56	0	24,65	26,83	$3d\ ^4P^{\circ}-4p\ ^4P$	3/2—1/2
5712,51	1	24,66	26,83	$3d\ ^4P^{\circ}-4p\ ^4P$	1/2—3/2
5708,03	0	24,65	26,83	$3d\ ^4P^{\circ}-4p\ ^4P$	3/2—3/2
5701,16	2	24,65	26,83	$3d\ ^4P^{\circ}-4p\ ^4P$	3/2—5/2
		24,65	26,83	$3d\ ^4P^{\circ}-4p\ ^4P$	5/2—3/2
5694,30	2	24,65	26,83	$3d\ ^4P^{\circ}-4p\ ^4P$	5/2—5/2
5662,47	12	20,71	22,90	$3s\ ^4P^{\circ}-3p\ ^4S$	5/2—3/2
5648,07	10	20,71	22,90	$3s\ ^4P^{\circ}-3p\ ^4S$	3/2—3/2
5640,55	8	20,70	22,90	$3s\ ^4P^{\circ}-3p\ ^4S$	1/2—3/2
5537,61	3	19,49	21,73	$4s\ ^2S-5p\ ^2P^{\circ}$	1/2—1/2
5535,35	5	19,49	21,73	$4s\ ^2S-5p\ ^2P^{\circ}$	1/2—3/2
5490,16	1	24,37	26,63	$3d\ ^4D^{\circ}-4p\ ^4D$	3/2—3/2
5488,95	1	24,37	26,63	$3d\ ^4D^{\circ}-4p\ ^4D$	7/2—5/2
5485,90	2	24,37	26,63	$3d\ ^4D^{\circ}-4p\ ^4D$	5/2—5/2
5483,35	1	24,37	26,63	$3d\ ^4D^{\circ}-4p\ ^4D$	3/2—5/2
5478,59	4	24,37	26,63	$3d\ ^4D^{\circ}-4p\ ^4D$	7/2—7/2
5368,58	1	20,84	23,15	$4d\ ^2D-7p\ ^2P^{\circ}$	3/2, 5/2—1/2, 3/2
5367,67	1	25,99	28,30	$4s\ ^4P^{\circ}-5p\ ^4D$	5/2—7/2
5342,40	2	20,95	23,27	$4f\ ^2F^{\circ}-7g\ ^2G$	5/2, 7/2—7/2, 9/2
5339,85	1	20,95	23,27	$4f\ ^2F^{\circ}-7h\ ^2H$	7/2—9/2
5334,79	6	20,15	22,47	$4p\ ^2P^{\circ}-6s\ ^2S$	3/2—1/2
5332,89	4	20,15	22,47	$4p\ ^2P^{\circ}-6s\ ^2S$	1/2—1/2
5290,09	0	26,63	28,97	$4p\ ^4D-6s\ ^4P^{\circ}$	5/2—3/2
		26,63	28,97	$4p\ ^4D-6s\ ^4P^{\circ}$	3/2—1/2
5286,47	1	26,63	28,98	$4p\ ^4D-6s\ ^4P^{\circ}$	7/2—5/2
5259,71	5	24,27	26,63	$3d\ ^4F^{\circ}-4p\ ^4D$	5/2—3/2
		24,27	26,63	$3d\ ^4F^{\circ}-4p\ ^4D$	3/2—1/2

λ, Å	I	E_H, eV	E_B, eV	Transition	J
5259,06	5	24,27	26,63	$3d\,^4F^\circ-4p\,^4D$	$7/2-5/2$
5257,24	7	24,28	26,63	$3d\,^4F^\circ-4p\,^4D$	$9/2-7/2$
5256,09	2	24,27	26,63	$3d\,^4F^\circ-4p\,^4D$	$3/2-3/2$
5253,57	4	24,27	26,63	$3d\,^4F^\circ-4p\,^4D$	$5/2-5/2$
5249,51	2	24,27	26,63	$3d\,^4F^\circ-4p\,^4D$	$7/2-7/2$
5151,09	13	20,71	23,11	$3s\,^4P^\circ-3p\,^4P$	$5/2-3/2$
5145,16	15	20,71	23,12	$3s\,^4P^\circ-3p\,^4P$	$5/2-5/2$
5143,49	12	20,70	23,11	$3s\,^4P^\circ-3p\,^4P$	$3/2-1/2$
5139,17	9	20,70	23,11	$3s\,^4P^\circ-3p\,^4P$	$3/2-3/2$
5137,26	7	20,70	23,11	$3s\,^4P^\circ-3p\,^4P$	$1/2-1/2$
5133,28	12	20,70	23,12	$3s\,^4P^\circ-3p\,^4P$	$3/2-5/2$
5132,94	12	20,70	23,11	$3s\,^4P^\circ-3p\,^4P$	$1/2-3/2$
5126,93	2	20,15	22,57	$4p\,^2P^\circ-3p\,^2P$	$3/2-1/2$
5125,20	4	20,15	22,57	$4p\,^2P^\circ-3p\,^2P$	$1/2-1/2$
5122,15	2	20,84	23,26	$4d\,^2D-7f\,^2F^\circ$	$3/2,\ 5/2-5/2,\ 7/2$
5121,82	5	20,15	22,57	$4p\,^2P^\circ-3p\,^2P$	$3/2-3/2$
5120,10	3	25,07	27,49	$3d\,^2P^\circ-4f\,^4D$	$1/2-3/2$
		20,15	22,57	$4p\,^2P^\circ-3p\,^2P$	$1/2-3/2$
5119,45	4	25,07	27,49	$3d\,^2P^\circ-4f\,^2D$	$3/2-5/2$
5116,75	2	25,07	27,49	$3d\,^2P^\circ-4f\,^4D$	$1/2-1/2$
5114,26	4	25,07	27,49	$3d\,^2P^\circ-4f\,^2D$	$1/2-3/2$
5113,69	4	25,07	27,49	$3d\,^2P^\circ-4f\,^4D$	$3/2-3/2,\ 5/2$
5107,91	1	25,07	27,49	$3d\,^2P^\circ-4f\,^2D$	$3/2-3/2$
5049,24	2	24,37	26,83	$3d\,^4D^\circ-4p\,^4P$	$3/2-1/2$
5047,11	3	24,37	26,83	$3d\,^4D^\circ-4p\,^4P$	$5/2-3/2$
5044,98	1	24,37	26,83	$3d\,^4D^\circ-4p\,^4P$	$3/2-3/2$
5044,35	5	24,37	26,83	$3d\,^4D^\circ-4p\,^4P$	$7/2-5/2$
5041,76	2	24,37	26,83	$3d\,^4D^\circ-4p\,^4P$	$5/2-5/2$
5040,74	2	20,92	23,38	$2p^3\,^2P^\circ-3p\,^2D$	$3/2-3/2$
5035,91	5	20,92	23,38	$2p^3\,^2P^\circ-3p\,^2D$	$1/2-3/2$
5032,07	7	20,92	23,38	$2p^3\,^2P^\circ-3p\,^2D$	$3/2-5/2$
4964,73	4	22,57	25,07	$3p\,^2P-3d\,^2P^\circ$	$3/2-3/2$
4959,92	1	22,57	25,07	$3p\,^2P-3d\,^2P^\circ$	$1/2-3/2$
4958,67	1	22,57	25,07	$3p\,^2P-3d\,^2P^\circ$	$3/2-1/2$
4953,85	3	22,57	25,07	$3p\,^2P-3d\,^2P^\circ$	$1/2-1/2$
4867,07	2	19,49	22,09	$4s\,^2S-3s\,^2P^\circ$	$1/2-1/2$
4862,57	4	19,49	22,10	$4s\,^2S-3s\,^2P^\circ$	$1/2-3/2$
4802,70	1	20,95	23,53	$4f\,^2F^\circ-8g\,^2G$	$5/2,\ 7/2-7/2,\ 9/2$
4747,28	2	13,72	16,33	$2p^2\,^2P-3p\,^2P^\circ$	$3/2-1/2$
4744,77	5	13,72	16,33	$2p^2\,^2P-3p\,^2P^\circ$	$3/2-3/2$
4737,97	3	13,71	16,33	$2p^2\,^2P-3p\,^2P^\circ$	$1/2-1/2$
4735,46	2	13,71	16,33	$2p^2\,^2P-3p\,^2P^\circ$	$1/2-3/2$
4734,60	2	24,79	27,41	$3d\,^2F^\circ-4f\,^2F$	$7/2-7/2$
4727,41	2	24,79	27,41	$3d\,^2F^\circ-4f\,^2F$	$5/2-5/2$
4638,91	2	20,15	22,82	$4p\,^2P^\circ-6d\,^2D$	$3/2-5/2$
4637,63	1	20,15	22,82	$4p\,^2P^\circ-6d\,^2D$	$1/2-3/2$
4629,98	2	24,79	27,47	$3d\,^2F^\circ-4f\,^4G$	$7/2-9/2$
4627,44	1	24,79	27,47	$3d\,^2F^\circ-4f\,^2G$	$7/2-7/2$
4625,56	3	24;79	27,47	$3d\,^2F^\circ-4f\,^4G$	$5/2-7/2$
4619,23	8	24,79	27,48	$3d\,^2F^\circ-4f\,^2G$	$7/2-9/2$
4618,40	6	24,79	27,47	$3d\,^2F^\circ-4f\,^2G$	$5/2-7/2$
4413,255	1	24,60	27,41	$3d\,^2D^\circ-4f\,^2F$	$5/2-5/2$
4411,506	7	24,60	27,41	$3d\,^2D^\circ-4f\,^2F$	$5/2-7/2$
4411,163	6	24,60	27,41	$3d\,^2D^\circ-4f\,^2F$	$3/2-5/2$
4409,979	5	24,60	27,41	$3d\,^2D^\circ-4f\,^4F$	$5/2-7/2$
4409,161	2	24,60	27,41	$3d\,^2D^\circ-4f\,^4F$	$3/2-5/2$

λ, Å	I	E_H, eV	E_B, eV	Transition	J
4376,562	5	24,65	27,49	$3d\ ^4P^\circ - 4f\ ^2D$	3/2—5/2
4375,009	4	24,66	27,49	$3d\ ^4P^\circ - 4f\ ^4D$	1/2—3/2
4374,272	9	24,65	27,49	$3d\ ^4P^\circ - 4f\ ^4D$	5/2—7/2
4372,487	7	24,65	27,49	$3d\ ^4P^\circ - 4f\ ^2D$	5/2—5/2
		24,66	27,49	$3d\ ^4P^\circ - 4f\ ^4D$	1/2—1/2
4372,350	6	24,65	27,49	$3d\ ^4P^\circ - 4f\ ^4D$	3/2—3/2, 5/2
4370,661	1	24,66	27,49	$3d\ ^4P^\circ - 4f\ ^2D$	1/2—3/2
4369,857	2	24,65	27,49	$3d\ ^4P^\circ - 4f\ ^4D$	3/2—1/2
4368,263	4	24,65	27,49	$3d\ ^4P^\circ - 4f\ ^4D$	5/2—3/2, 5/2
4368,047	1	24,65	27,49	$3d\ ^4P^\circ - 4f\ ^2D$	3/2—3/2
4326,156	5	23,11	25,98	$3p\ ^4P - 4s\ ^4P^\circ$	3/2—1/2
4325,827	4	23,12	25,98	$3p\ ^4P - 4s\ ^4P^\circ$	5/2—3/2
4323,102	3	23,11	25,98	$3p\ ^4P - 4s\ ^4P^\circ$	1/2—1/2
4321,647	3	23,11	25,98	$3p\ ^4P - 4s\ ^4P^\circ$	3/2—3/2
4318,600	5	23,11	25,98	$3p\ ^4P - 4s\ ^4P^\circ$	1/2—3/2
4317,260	8	23,12	25,98	$3p\ ^4P - 4s\ ^4P^\circ$	5/2—5/2
4313,100	6	23,11	25,98	$3p\ ^4P - 4s\ ^4P^\circ$	3/2—5/2
4307,59	2	20,15	23,03	$4p\ ^2P^\circ - 7s\ ^2S$	3/2—1/2
4306,33	1	20,15	23,03	$4p\ ^2P^\circ - 7s\ ^2S$	1/2—1/2
4295,920	4	24,60	27,49	$3d\ ^2D^\circ - 4f\ ^2D$	5/2—5/2
4291,819	3	24,60	27,49	$3d\ ^2D^\circ - 4f\ ^4D$	5/2—3/2, 5/2
4289,876	2	24,60	27,49	$3d\ ^2D^\circ - 4f\ ^4D$	3/2—3/2, 5/2
4285,704	3	24,60	27,49	$3d\ ^2D^\circ - 4f\ ^2D$	3/2—3/2
4267,258	20	18,05	20,95	$3d\ ^2D - 4f\ ^2F^\circ$	5/2—7/2
4267,003	18	18,04	20,95	$3d\ ^2D - 4f\ ^2F^\circ$	3/2—5/2
4077,778	4	24,37	27,41	$3d\ ^4D^\circ - 4f\ ^2F$	7/2—7/2
4077,625	2	24,37	27,41	$3d\ ^4D^\circ - 4f\ ^2F$	5/2—5/2
		24,37	27,41	$3d\ ^4D^\circ - 4f\ ^4F$	7/2—5/2
4076,83	0	24,37	27,41	$3d\ ^4D^\circ - 4f\ ^4F$	5/2—3/2
4076,526	4	24,37	27,41	$3d\ ^4D^\circ - 4f\ ^4F$	7/2—7/2
4076,251	3	24,37	27,41	$3d\ ^4D^\circ - 4f\ ^2F$	3/2—5/2
4076,142	5	24,37	27,41	$3d\ ^4D^\circ - 4f\ ^2F$	5/2—7/2
4075,851	12	24,37	27,41	$3d\ ^4D^\circ - 4f\ ^4F$	5/2—5/2
		24,37	27,41	$3d\ ^4D^\circ - 4f\ ^4F$	7/2—9/2
4075,395	4	24,37	27,41	$3d\ ^4D^\circ - 4f\ ^4F$	3/2—3/2
4074,845	8	24,37	27,41	$3d\ ^4D^\circ - 4f\ ^4F$	5/2—7/2
4074,518	10	24,37	27,41	$3d\ ^4D^\circ - 4f\ ^4F$	3/2—5/2
		24,37	27,41	$3d\ ^4D^\circ - 4f\ ^4F$	1/2—3/2
4021,167	3	22,90	25,98	$3p\ ^4S - 4s\ ^4P^\circ$	3/2—1/2
4017,278	5	22,90	25,98	$3p\ ^4S - 4s\ ^4P^\circ$	3/2—3/2
4009,884	7	22,90	25,99	$3p\ ^4S - 4s\ ^4P^\circ$	3/2—5/2
3980,323	8	24,37	27,49	$3d\ ^4D^\circ - 4f\ ^4D$	7/2—7/2
3978,759	4	24,37	27,49	$3d\ ^4D^\circ - 4f\ ^2D$	7/2—5/2
		24,37	27,49	$3d\ ^4D^\circ - 4f\ ^4D$	5/2—7/2
3977,269	5	24,37	27,49	$3d\ ^4D^\circ - 4f\ ^2D$	5/2—5/2
3975,953	1	24,37	27,49	$3d\ ^4D^\circ - 4f\ ^2D$	3/2—5/2
3975,341	2	24,37	27,49	$3d\ ^4D^\circ - 4f\ ^4D$	7/2—5/2
3973,760	7	24,37	27,49	$3d\ ^4D^\circ - 4f\ ^4D$	5/2—3/2, 5/2
3972,439	6	24,37	27,49	$3d\ ^4D - 4f\ ^4D$	3/2—3/2, 5/2
3971,574	2	24,37	27,49	$3d\ ^4D^\circ - 4f\ ^4D$	1/2—3/2
3970,386	4	24,37	27,49	$3d\ ^4D^\circ - 4f\ ^4D$	3/2—1/2
		24,37	27,49	$3d\ ^4D^\circ - 4f\ ^2D$	5/2—3/2
3969,520	3	24,37	27,49	$3d\ ^4D^\circ - 4f\ ^4D$	1/2—1/2
3968,92	0	24,37	27,49	$3d\ ^4D^\circ - 4f\ ^2D$	3/2—3/2
3953,95	0	24,28	27,41	$3d\ ^4F^\circ - 4f\ ^2F$	9/2—7/2
3952,679	1	24,28	27,41	$3d\ ^4F^\circ - 4f\ ^4F$	9/2—7/2

λ, Å	I	E_H, eV	E_B, eV	Transition	J
3952,058	9	24,28	27,41	$3d\ ^4F^\circ-4f\ ^4F$	9/2—9/2
3949,530	4	24,27	27,41	$3d\ ^4F^\circ-4f\ ^2F$	7/2—7/2
3949,373	1	24,27	27,41	$3d\ ^4F^\circ-4f\ ^4F$	7/2—5/2
3948,333	6	24,27	27,41	$3d\ ^4F^\circ-4f\ ^4F$	7/2—7/2
3947,715	6	24,27	27,41	$3d\ ^4F^\circ-4f\ ^2F$	5/2—5/2
		24,27	27,41	$3d\ ^4F^\circ-4f\ ^4F$	7/2—9/2
3947,079	2	24,27	27,41	$3d\ ^4F^\circ-4f\ ^4F$	5/2—3/2
3946,429	1	24,27	27,41	$3d\ ^4F^\circ-4f\ ^2F$	5/2—7/2
3946,278	5	24,27	27,41	$3d\ ^4F^\circ-4f\ ^4F$	5/2—5/2
3945,197	4	24,27	27,41	$3d\ ^4F^\circ-4f\ ^4F$	5/2—7/2
3945,003	5	24,27	27,41	$3d\ ^4F^\circ-4f\ ^4F$	3/2—3/2
3944,193	3	24,27	27,41	$3d\ ^4F^\circ-4f\ ^4F$	3/2—5/2
3920,693	18	16,33	19,49	$3p\ ^2P^\circ-4s\ ^2S$	3/2—1/2
3918,978	15	16,33	19,49	$3p\ ^2P^\circ-4s\ ^2S$	1/2—1/2
3883,824	1	24,28	27,47	$3d\ ^4F^\circ-4f\ ^4G$	9/2—7/2
3880,588	7	24,28	27,47	$3d\ ^4F^\circ-4f\ ^4G$	9/2—9/2
3879,640	7	24,27	27,47	$3d\ ^4F^\circ-4f\ ^4G$	7/2—7/2
3878,028	7	24,27	27,47	$3d\ ^4F^\circ-4f\ ^4G$	5/2—5/2
3876,664	12	24,27	27,47	$3d\ ^4F^\circ-4f\ ^4G$	5/2—7/2
3876,408	12	24,27	27,47	$3d\ ^4F^\circ-4f\ ^4G$	7/2—9/2
3876,187	12	24,28	27,47	$3d\ ^4F^\circ-4f\ ^4G$	9/2—11/2
3876,055	9	24,27	27,47	$3d\ ^4F^\circ-4f\ ^4G$	3/2—5/2
3874,666	2	24,27	27,47	$3d\ ^4F^\circ-4f\ ^2G$	7/2—7/2
3873,067	0	24,28	27,48	$3d\ ^4F^\circ-4f\ ^2G$	9/2—9/2
3871,669	7	24,27	27,47	$3d\ ^4F^\circ-4f\ ^2G$	5/2—7/2
3868,874	6	24,27	27,48	$3d\ ^4F^\circ-4f\ ^2G$	7/2—9/2
3862,181	2	24,28	27,49	$3d\ ^4F^\circ-4f\ ^4D$	9/2—7/2
3856,62	0	24,27	27,49	$3d\ ^4F^\circ-4f\ ^2D$	7/2—5/2
3836,683	2	20,15	23,38	$4p\ ^2P^\circ-3p\ ^2D$	3/2—3/2
3835,730	6	20,15	23,38	$4p\ ^2P^\circ-3p\ ^2D$	1/2—3/2
3831,743	8	20,15	23,38	$4p\ ^2P^\circ-3p\ ^2D$	3/2—5/2
3590,862	8	22,53	25,98	$3p\ ^4D-4s\ ^4P^\circ$	5/2—3/2
		22,53	25,98	$3p\ ^4D-4s\ ^4P^\circ$	3/2—1/2
3589,657	9	22,54	25,99	$3p\ ^4D-4s\ ^4P^\circ$	7/2—5/2
3588,915	5	22,53	25,98	$3p\ ^4D-4s\ ^4P^\circ$	1/2—1/2
3587,657	6	22,53	25,98	$3p\ ^4D-4s\ ^4P^\circ$	3/2—3/2
3585,809	3	22,53	25,98	$3p\ ^4D-4s\ ^4P^\circ$	1/2—3/2
3584,977	7	22,53	25,99	$3p\ ^4D-4s\ ^4P^\circ$	5/2—5/2
3581,763	3	22,53	25,99	$3p\ ^4D-4s\ ^4P^\circ$	3/2—5/2
3406,361	2	25,07	28,69	$3d\ ^2P^\circ-5f\ ^2D$	3/2—5/2
3404,33	1	25,07	28,70	$3d\ ^2P^\circ-5f\ ^2D$	1/2—3/2
3403,66	1	25,07	28,70	$3d\ ^2P^\circ-5f\ ^4D$	3/2—5/2, 3/2
3401,53	0	25,07	28,70	$3d\ ^2P^\circ-5f\ ^2D$	3/2—3/2
3393,946	1	24,60	28,25	$3d\ ^2D^\circ-5p\ ^2P$	3/2—1/2
3392,146	2	24,60	28,25	$3d\ ^2D^\circ-5p\ ^2P$	5/2—3/2
3361,721	6	18,04	21,73	$3d\ ^2D-5p\ ^2P^\circ$	3/2—1/2
3361,051	8	18,05	21,73	$3d\ ^2D-5p\ ^2P^\circ$	5/2—3/2
3360,891	3	18,04	21,73	$3d\ ^2D-5p\ ^2P^\circ$	3/2—3/2
3357,19	0	24,66	28,35	$3d\ ^4P^\circ-5p\ ^4S$	1/2—3/2
3355,690	1	24,65	28,35	$3d\ ^4P^\circ-5p\ ^4S$	3/2—3/2
3353,302	2	24,65	28,35	$3d\ ^4P^\circ-5p\ ^4S$	5/2—3/2
3174,58	0	24,79	28,70	$3d\ ^2F^\circ-5f\ ^4G$	7/2—9/2
3172,62	1	24,79	28,70	$3d\ ^2F^\circ-5f\ ^4G$	5/2—7/2
3170,03	2	24,79	28,70	$3d\ ^2F^\circ-5f\ ^2G$	7/2—9/2
3169,66	1	24,79	28,70	$3d\ ^2F^\circ-5f\ ^2G$	5/2—7/2
3167,931	8	18,65	22,57	$2p^3\ ^2D^\circ-3p\ ^2P$	3/2—1/2
3165,974	4	18,65	22,57	$2p^3\ ^2D^\circ-3p\ ^2P$	3/2—3/2
3165,467	9	18,65	22,57	$2p^3\ ^2D^\circ-3p\ ^2P$	5/2—3/2
3157,13	0	24,37	28,30	$3d\ ^4D^\circ-5p\ ^4D$	7/2—7/2

λ, Å	I	$E_{\text{н}}$, eV	$E_{\text{в}}$, eV	Transition	J
3142,04	0	20,84	24,79	$4d\ ^2D-3d\ ^2F^\circ$	$3/2-5/2$
3137,92	1	20,84	24,79	$4d\ ^2D-3d\ ^2F^\circ$	$5/2-7/2$
3124,133	2	23,38	27,35	$3p\ ^2D-4d\ ^2D^\circ$	$5/2-5/2$
3122,086	1	23,38	27,35	$3p\ ^2D-4d\ ^2D^\circ$	$3/2-3/2$
3100,570	2	18,05	22,10	$3d\ ^2D-3s\ ^2P^\circ$	$5/2-3/2$
3087,90	0	24,37	28,38	$3d\ ^4D^\circ-5p\ ^4P$	$5/2-3/2$
3086,903	1	24,37	28,38	$3d\ ^4D^\circ-5p\ ^4P$	$7/2-5/2$
3083,052	2	24,27	28,29	$3d\ ^4F^\circ-5p\ ^4D$	$7/2-5/2$
		24,27	28,29	$3d\ ^4F^\circ-5p\ ^4D$	$5/2-3/2$
		24,27	28,29	$3d\ ^4F^\circ-5p\ ^4D$	$3/2-1/2$
3082,381	2	24,28	28,30	$3d\ ^4F^\circ-5p\ ^4D$	$9/2-7/2$
3060,64	1	24,65	28,69	$3d\ ^4P^\circ-5f\ ^2D$	$3/2-5/2$
3059,83	0	24,66	28,70	$3d\ ^4P^\circ-5f\ ^4D$	$1/2-3/2$
3059,091	3	24,65	28,70	$3d\ ^4P^\circ-5f\ ^4D$	$5/2-7/2$
3058,45	2	24,65	28,70	$3d\ ^4P^\circ-5f\ ^4D$	$3/2-3/2,\ 5/2$
		24,66	28,70	$3d\ ^4P^\circ-5f\ ^4D$	$1/2-1/2$
3056,85	1	24,66	28,70	$3d\ ^4P^\circ-5f\ ^2D$	$1/2-3/2$
		24,65	28,70	$3d\ ^4P^\circ-5f\ ^4D$	$5/2-3/2,\ 5/2$
3049,671	3	24,60	28,67	$3d\ ^2D^\circ-5f\ ^2F$	$5/2-7/2$
3049,398	3	24,60	28,67	$3d\ ^2D^\circ-5f\ ^2F$	$3/2-5/2$
3048,933	2	24,60	28,67	$3d\ ^2D^\circ-5f\ ^4F$	$5/2-7/2$
3048,69	0	24,60	28,67	$3d\ ^2D^\circ-5f\ ^4F$	$3/2-5/2$
3040,512	2	23,38	27,46	$3p\ ^2D-4d\ ^2F^\circ$	$3/2-5/2$
3039,714	3	23,38	27,46	$3p\ ^2D-4d\ ^2F^\circ$	$5/2-7/2$
2992,618	18	18,04	22,19	$3d\ ^2D-5f\ ^2F^\circ$	$3/2,\ 5/2-5/2,\ 7/2$
2969,59	0	23,12	27,29	$3p\ ^4P-4d\ ^4D^\circ$	$5/2-3/2$
2968,836	2	23,12	27,29	$3p\ ^4P-4d\ ^4D^\circ$	$5/2-5/2$
2967,868	7	23,12	27,29	$3p\ ^4P-4d\ ^4D^\circ$	$5/2-7/2$
2967,629	3	23,11	27,29	$3p\ ^4P-4d\ ^4D^\circ$	$3/2-3/2$
2966,871	5	23,11	27,29	$3p\ ^4P-4d\ ^4D^\circ$	$3/2-5/2$
2966,655	3	23,11	27,29	$3p\ ^4P-4d\ ^4D^\circ$	$1/2-1/2$
2966,187	3	23,11	27,29	$3p\ ^4P-4d\ ^4D^\circ$	$1/2-3/2$
2910,729	3	23,12	27,38	$3p\ ^4P-4d\ ^4P^\circ$	$5/2-5/2$
2908,957	2	23,12	27,38	$3p\ ^4P-4d\ ^4P^\circ$	$5/2-3/2$
		23,11	27,38	$3p\ ^4P-4d\ ^4P^\circ$	$3/2-5/2$
2907,09	1	23,11	27,38	$3p\ ^4P-4d\ ^4P^\circ$	$3/2-3/2$
2906,011	2	23,11	27,38	$3p\ ^4P-4d\ ^4P^\circ$	$3/2-1/2$
2905,715	2	23,11	27,38	$3p\ ^4P-4d\ ^4P^\circ$	$1/2-3/2$
2885,496	6	24,37	28,67	$3d\ ^4D^\circ-5f\ ^2F$	$5/2-7/2$
		24,37	28,67	$3d\ ^4D^\circ-5f\ ^2F$	$3/2-5/2$
		24,37	28,67	$3d\ ^4D^\circ-5f\ ^4F$	$7/2-9/2$
2884,808	4	24,37	28,67	$3d\ ^4D^\circ-5f\ ^4F$	$3/2-5/2$
		24,37	28,67	$3d\ ^4D^\circ-5f\ ^4F$	$5/2-7/2$
		24,37	28,67	$3d\ ^4D^\circ-5f\ ^4F$	$1/2-3/2$
2861,060	2	24,37	28,70	$3d\ ^4D^\circ-5f\ ^4D$	$7/2-7/2$
2858,00	1	24,37	28,70	$3d\ ^4D^\circ-5f\ ^4D$	$5/2-5/2$
2837,603	18	11,96	16,33	$2p^2\ ^2S-3p\ ^2P^\circ$	$1/2-1/2$
2836,710	20	11,96	16,33	$2p^2\ ^2S-3p\ ^2P^\circ$	$1/2-3/2$
2822,812	2	24,28	28,67	$3d\ ^4F^\circ-5f\ ^4F$	$9/2-9/2$
2821,54	1	24,27	28,67	$3d\ ^4F^\circ-5f\ ^4F$	$7/2-5/2$
		24,27	28,67	$3d\ ^4F^\circ-5f\ ^2F$	$7/2-7/2$
2820,70	1	24,27	28,67	$3d\ ^4F^\circ-5f\ ^4F$	$7/2-7/2$
		24,27	28,67	$3d\ ^4F^\circ-5f\ ^2F$	$5/2-5/2$
2820,00	1	24,27	28,67	$3d\ ^4F^\circ-5f\ ^4F$	$5/2-5/2$
2819,13	1	24,27	28,67	$3d\ ^4F^\circ-5f\ ^4F$	$5/2-7/2$
		24,27	28,67	$3d\ ^4F^\circ-5f\ ^4F$	$3/2-3/2,\ 5/2$
2803,45	0	24,28	28,70	$3d\ ^4F^\circ-5f\ ^4G$	$9/2-9/2$
2802,95	0	24,27	28,70	$3d\ ^4F^\circ-5f\ ^4G$	$7/2-7/2$
2802,39	0	24,27	28,70	$3d\ ^4F^\circ-5f\ ^4G$	$5/2-5/2$

λ, Å	I	E_H, eV	E_B, eV	Transition	J
2801,43	5	24,27	28,70	$3d\ ^4F^\circ$—$5f\ ^4G$	$3/2$—$5/2$
		24,27	28,70	$3d\ ^4F^\circ$—$5f\ ^4G$	$5/2$—$7/2$
2801,21	5	24,27	28,70	$3d\ ^4F^\circ$—$5f\ ^4G$	$7/2$—$9/2$
		24,28	28,70	$3d\ ^4F^\circ$—$5f\ ^4G$	$9/2$—$11/2$
2799,15	1	24,27	28,70	$3d\ ^4F^\circ$—$5f\ ^2G$	$5/2$—$7/2$
2797,70	1	24,27	28,70	$3d\ ^4F^\circ$—$5f\ ^2G$	$7/2$—$9/2$
2767,673	3	22,90	27,38	$3p\ ^4S$—$4d\ ^4P^\circ$	$3/2$—$5/2$
2766,118	2	22,90	27,38	$3p\ ^4S$—$4d\ ^4P^\circ$	$3/2$—$3/2$
2765,120	1	22,90	27,38	$3p\ ^4S$—$4d\ ^4P^\circ$	$3/2$—$1/2$
2747,282	12	16,33	20,84	$3p\ ^2P^\circ$—$4d\ ^2D$	$3/2$—$3/2$, $5/2$
2746,488	10	16,33	20,84	$3p\ ^2P^\circ$—$4d\ ^2D$	$1/2$—$1/2$
2730,61	1	22,10	26,58	$3s\ ^2P^\circ$—$4p\ ^2P$	$3/2$—$1/2$
2729,213	2	22,09	26,58	$3s\ ^2P^\circ$—$4p\ ^2P$	$1/2$—$1/2$
2728,707	4	22,10	26,58	$3s\ ^2P^\circ$—$4p\ ^2P$	$3/2$—$3/2$
2727,36	2	22,09	26,58	$3s\ ^2P^\circ$—$4p\ ^2P$	$1/2$—$3/2$
2712,32	0	24,79	29,36	$3d\ ^2F^\circ$—$6f\ ^4G$	$7/2$—$9/2$
2710,59	1	24,79	29,36	$3d\ ^2F^\circ$—$6f\ ^4G$	$5/2$—$7/2$
2708,4	1	24,79	29,37	$3d\ ^2F^\circ$—$6f\ ^2G$	$7/2$—$9/2$
		24,79	29,37	$3d\ ^2F^\circ$—$6f\ ^2G$	$5/2$—$7/2$
2669,960	3	18,05	22,68	$3d\ ^2D$—$6p\ ^2P^\circ$	$5/2$—$3/2$
2643,427	3	22,54	27,22	$3p\ ^4D$—$4d\ ^4F^\circ$	$7/2$—$7/2$
2642,331	3	22,53	27,22	$3p\ ^4D$—$4d\ ^4F^\circ$	$5/2$—$5/2$
2641,425	8	22,53	27,22	$3p\ ^4D$—$4d\ ^4F^\circ$	$3/2$—$3/2$
		22,54	27,23	$3p\ ^4D$—$4d\ ^4F^\circ$	$7/2$—$9/2$
2640,894	5	22,53	27,22	$3p\ ^4D$—$4d\ ^4F^\circ$	$5/2$—$7/2$
2640,560	6	22,53	27,22	$3p\ ^4D$—$4d\ ^4F^\circ$	$3/2$—$5/2$
		22,53	27,22	$3p\ ^4D$—$4d\ ^4F^\circ$	$1/2$—$3/2$
2628,46	1	24,65	29,38	$3d\ ^4P^\circ$—$6f\ ^4D$	$5/2$—$7/2$
2622,90	2	18,65	23,38	$2p^3\ ^2D^\circ$—$3p\ ^2D$	$3/2$—$3/2$
2620,20	3	18,65	23,38	$2p^3\ ^2D^\circ$—$3p\ ^2D$	$5/2$—$5/2$
2612,45	2	24,60	29,35	$3d\ ^2D^\circ$—$6f\ ^2F$	$5/2$—$7/2$
		24,60	29,35	$3d\ ^2D^\circ$—$6f\ ^2F$	$3/2$—$5/2$
2605,62	1	22,54	27,29	$3p\ ^4D$—$4d\ ^4D^\circ$	$7/2$—$5/2$
2604,863	4	22,54	27,29	$3p\ ^4D$—$4d\ ^4D^\circ$	$7/2$—$7/2$
2603,72	1	22,53	27,29	$3p\ ^4D$—$4d\ ^4D^\circ$	$5/2$—$3/2$
2603,161	3	22,53	27,29	$3p\ ^4D$—$4d\ ^4D^\circ$	$5/2$—$5/2$
2602,39	2	22,53	27,29	$3p\ ^4D$—$4d\ ^4D^\circ$	$5/2$—$7/2$
		22,53	27,29	$3p\ ^4D$—$4d\ ^4D^\circ$	$3/2$—$1/2$
2602,02	2	22,53	27,29	$3p\ ^4D$—$4d\ ^4D^\circ$	$3/2$—$3/2$
2601,42	2	22,53	27,29	$3p\ ^4D$—$4d\ ^4D^\circ$	$3/2$—$5/2$
		22,53	27,29	$3p\ ^4D$—$4d\ ^4D^\circ$	$1/2$—$1/2$
2601,05	1	22,53	27,29	$3p\ ^4D$—$4d\ ^4D^\circ$	$1/2$—$3/2$
2592,71	1	22,57	27,35	$3p\ ^2P$—$4d\ ^2D^\circ$	$3/2$—$3/2$
2591,845	4	22,57	27,35	$3p\ ^2P$—$4d\ ^2D^\circ$	$3/2$—$5/2$
2591,410	2	22,57	27,35	$3p\ ^2P$—$4d\ ^2D^\circ$	$1/2$—$3/2$
2574,826	10	18,04	22,86	$3d\ ^2D$—$6f\ ^2F^\circ$	$3/2$, $5/2$—$5/2$, $7/2$
2571,76	1	24,37	29,19	$3d\ ^4D^\circ$—$6p\ ^4P$	$7/2$—$5/2$
2570,57	2				
2556,12	0	21,73	26,58	$5p\ ^2P^\circ$—$4p\ ^2P$	$3/2$—$1/2$
2555,66	1	21,73	26,58	$5p\ ^2P^\circ$—$4p\ ^2P$	$1/2$—$1/2$
2554,478	3	21,73	26,58	$5p\ ^2P^\circ$—$4p\ ^2P$	$3/2$—$3/2$
2547,35	1	24,27	29,13	$3d\ ^4F^\circ$—$6p\ ^4D$	$7/2$—$5/2$
		24,27	29,13	$3d\ ^4F^\circ$—$6p\ ^4D$	$5/2$—$3/2$
		24,27	29,13	$3d\ ^4F^\circ$—$6p\ ^4D$	$3/2$—$1/2$
2546,81	2	24,28	29,14	$3d\ ^4F^\circ$—$6p\ ^4D$	$9/2$—$7/2$
2543,45	2	23,11	28,00	$3p\ ^4P$—$5s\ ^4P^\circ$	$3/2$—$1/2$
		23,12	28,00	$3p\ ^4P$—$5s\ ^4P^\circ$	$5/2$—$3/2$
2540,88	1	23,11	28,00	$3p\ ^4P$—$5s\ ^4P^\circ$	$1/2$—$3/2$
2540,39	3	23,12	28,00	$3p\ ^4P$—$5s\ ^4P^\circ$	$5/2$—$5/2$

λ, Å	I	E_H, eV	E_B, eV	Transition	J
2538,98	2	23,11	28,00	$3p\ ^4P-5s\ ^4P^\circ$	$3/2-5/2$
2512,065	12	13,72	18,65	$2p^2\ ^2P-2p^3\ ^2D^\circ$	$3/2-5/2$
2511,734	5	13,72	18,65	$2p^2\ ^2P-2p^3\ ^2D^\circ$	$3/2-3/2$
2509,121	10	13,71	18,65	$2p^2\ ^2P-2p^3\ ^2D^\circ$	$1/2-3/2$
2491,37	2	24,37	29,36	$3d\ ^4D^\circ-6f\ ^4F$	$7/2-9/2$
2490,87	2	24,37	29,35	$3d\ ^4D^\circ-6f\ ^4F$	$3/2-5/2$
		24,37	29,35	$3d\ ^4D^\circ-6f\ ^4F$	$5/2-7/2$
		24,37	29,35	$3d\ ^4D^\circ-6f\ ^4F$	$1/2-3/2$
2434,81	1	24,27	29,36	$3d\ ^4F^\circ-6f\ ^4G$	$7/2-9/2$
		24,27	29,36	$3d\ ^4F^\circ-6f\ ^4G$	$5/2-7/2$
		24,27	29,36	$3d\ ^4F^\circ-6f\ ^4G$	$3/2-5/2$
2434,24	2	24,28	29,37	$3d\ ^4F^\circ-6f\ ^4G$	$9/2-11/2$
2433,49	0	22,90	28,00	$3p\ ^4S-5s\ ^4P^\circ$	$3/2-3/2$
2432,90	0	24,27	29,37	$3d\ ^4F^\circ-6f\ ^2G$	$5/2-7/2$
2432,12	0	24,27	29,37	$3d\ ^4F^\circ-6f\ ^2G$	$7/2-9/2$
2430,78	1	22,90	28,00	$3p\ ^4S-5s\ ^4P^\circ$	$3/2-5/2$
2426,70	2	18,04	23,15	$3d\ ^2D-7p\ ^2P^\circ$	$3/2,\ 5/2-1/2,\ 3/2$
2402,402	7	16,33	21,49	$3p\ ^2P^\circ-5s\ ^2S$	$3/2-1/2$
2401,761	5	16,33	21,49	$3p\ ^2P^\circ-5s\ ^2S$	$1/2-1/2$
2375,08	4	18,04	23,26	$3d\ ^2D-7f\ ^2F^\circ$	$3/2,\ 5/2-5/2,\ 7/2$
2270,20	2	22,53	28,00	$3p\ ^4D-5s\ ^4P^\circ$	$5/2-3/2$
		22,53	28,00	$3p\ ^4D-5s\ ^4P^\circ$	$3/2-1/2$
2269,70	2	22,54	28,00	$3p\ ^4D-5s\ ^4P^\circ$	$7/2-5/2$
2269,36	0	22,53	28,00	$3p\ ^4D-5s\ ^4P^\circ$	$1/2-1/2$
2268,91	1	22,53	28,00	$3p\ ^4D-5s\ ^4P^\circ$	$3/2-3/2$
2267,77	0	22,53	28,00	$3p\ ^4D-5s\ ^4P^\circ$	$5/2-5/2$
2256,79	0	23,12	28,61	$3p\ ^4P-5d\ ^4D^\circ$	$5/2-5/2$
2256,19	2	23,12	28,61	$3p\ ^4P-5d\ ^4D^\circ$	$5/2-7/2$
		23,11	28,61	$3p\ ^4P-5d\ ^4D^\circ$	$3/2-3/2$
2255,68	1	23,11	28,61	$3p\ ^4P-5d\ ^4D^\circ$	$3/2-5/2$
2255,23	0	23,11	28,61	$3p\ ^4P-5d\ ^4D^\circ$	$1/2-3/2$
2242,10	1	23,12	28,64	$3p\ ^4P-5d\ ^4P^\circ$	$5/2-5/2$
2241,05	1	23,12	28,64	$3p\ ^4P-5d\ ^4P^\circ$	$5/2-3/2$
		23,11	28,64	$3p\ ^4P-5d\ ^4P^\circ$	$3/2-5/2$
2189,62	1	20,92	26,58	$2p^3\ ^2P^\circ-4p\ ^2P$	$3/2-1/2$
2188,72	1	20,92	26,58	$2p^3\ ^2P^\circ-4p\ ^2P$	$1/2-1/2$
2188,39	2	20,92	26,58	$2p^3\ ^2P^\circ-4p\ ^2P$	$3/2-3/2$
2187,48	1	20,92	26,58	$2p^3\ ^2P^\circ-4p\ ^2P$	$1/2-3/2$
2174,168	3	14,45	20,15	$3s\ ^2S-4p\ ^2P^\circ$	$1/2-1/2$
2173,848	5	14,45	20,15	$3s\ ^2S-4p\ ^2P^\circ$	$1/2-3/2$
2156,28	1	22,90	28,64	$3p\ ^4S-5d\ ^4P^\circ$	$3/2-5/2$
2155,39	0	22,90	28,64	$3p\ ^4S-5d\ ^4P^\circ$	$3/2-3/2$
2154,70	0	22,90	28,64	$3p\ ^4S-5d\ ^4P^\circ$	$3/2-1/2$
2137,897	5	16,33	22,13	$3p\ ^2P^\circ-5d\ ^2D$	$3/2-3/2,\ 5/2$
2137,417	3	16,33	22,13	$3p\ ^2P^\circ-5d\ ^2D$	$1/2-3/2$
2114,72	0	23,12	28,98	$3p\ ^4P-6s\ ^4P^\circ$	$5/2-5/2$
2093,13	1	20,71	26,63	$3s\ ^4P^\circ-4p\ ^4D$	$5/2-5/2$
2091,63	2	20,70	26,63	$3s\ ^4P^\circ-4p\ ^4D$	$1/2-1/2$
		20,71	26,63	$3s\ ^4P^\circ-4p\ ^4D$	$5/2-7/2$
2091,17	2	20,70	26,63	$3s\ ^4P^\circ-4p\ ^4D$	$3/2-5/2$
		20,70	26,63	$3s\ ^4P^\circ-4p\ ^4D$	$1/2-3/2$
2052,16	2	22,54	28,58	$3p\ ^4D-5d\ ^4F^\circ$	$7/2-9/2$
2051,79	2	22,53	28,58	$3p\ ^4D-5d\ ^4F^\circ$	$5/2-7/2$
		22,53	28,58	$3p\ ^4D-5d\ ^4F^\circ$	$3/2-5/2$
		22,53	28,58	$3p\ ^4D-5d\ ^4F^\circ$	$1/2-3/2$
2018,38	2	16,33	22,47	$3p\ ^2P^\circ-6s\ ^2S$	$3/2-1/2$
2017,94	1	16,33	22,47	$3p\ ^2P^\circ-6s\ ^2S$	$1/2-1/2$
1988,51	1	16,33	22,57	$3p\ ^2P^\circ-3p\ ^2P$	$3/2-1/2$
1988,09	2	16,33	22,57	$3p\ ^2P^\circ-3p\ ^2P$	$1/2-1/2$

λ, Å	I	E_H, eV	E_B, eV	Transition	J
1987,76	3	16,33	22,57	$3p\,^2P^\circ-3p\,^2P$	$3/2-3/2$
1987,33	1	16,33	22,57	$3p\,^2P^\circ-3p\,^2P$	$1/2-3/2$
1760,81	3	9,29	16,33	$2p^2\,^2D-3p\,^2P^\circ$	$3/2-1/2$
1760,40	4	9,29	16,33	$2p^2\,^2D-3p\,^2P^\circ$	$5/2-3/2$
1722,21	0	13,72	20,92	$2p^2\,^2P-2p^3\,^2P^\circ$	$3/2-1/2$
1721,66	2	13,72	20,92	$2p^2\,^2P-2p^3\,^2P^0$	$3/2-3/2$
1720,99	1	13,71	20,92	$2p^2\,^2P-2p^3\,^2P^\circ$	$1/2-1/2$
1720,44	0	13,71	20,92	$2p^2\,^2P-2p^3\,^2P^\circ$	$1/2-3/2$
1335,684	14	0,01	9,29	$2p\,^2P^\circ-2p^2\,^2D$	$3/2-5/2$
1334,515	13	0,00	9,29	$2p\,^2P^\circ-2p^2\,^2D$	$1/2-3/2$
1323,916	8	9,29	18,65	$2p^2\,^2D-2p^3\,^2D^\circ$	$3/2,\ 5/2-3/2,\ 5/2$
1141,746	2	9,29	20,15	$2p^2\,^2D-4p\,^2P^\circ$	$3/2-1/2$
1141,630	3	9,29	20,15	$2p^2\,^2D-4p\,^2P^\circ$	$5/2-3/2$
1139,49	0	13,72	24,60	$2p^2\,^2P-3d\,^2D^\circ$	$3/2-3/2$
1139,330	3	13,72	24,60	$2p^2\,^2P-3d\,^2D^\circ$	$3/2-5/2$
1138,936	2	13,71	24,60	$2p^2\,^2P-3d\,^2D^\circ$	$1/2-3/2$
1092,740	2	13,72	25,07	$2p^2\,^2P-3d\,^2P^\circ$	$3/2-3/2$
1092,422	0	13,72	25,07	$2p^2\,^2P-3d\,^2P^\circ$	$3/2-1/2$
1092,240	0	13,71	25,07	$2p^2\,^2P-3d\,^2P^\circ$	$1/2-3/2$
1091,930	1	13,71	25,07	$2p^2\,^2P-3d\,^2P^\circ$	$1/2-1/2$
1066,121	6	9,29	20,92	$2p^2\,^2D-2p^3\,^2P^\circ$	$3/2-1/2$
1065,883	7	9,29	20,92	$2p^2\,^2D-2p^3\,^2P^\circ$	$5/2-3/2$
1063,30	0	9,29	20,95	$2p^2\,^2D-4f\,^2F^\circ$	$3/2-5/2$
		9,29	20,95	$2p^2\,^2D-4f\,^2F^\circ$	$5/2-7/2,\ 5/2$
1037,017	13	0,01	11,96	$2p^2\,P^\circ-2p^2\,^2S$	$3/2-1/2$
1036,330	12	0,00	11,96	$2p\,^2P^\circ-2p^2\,^2S$	$1/2-1/2$
1010,369	10	5,34	17,61	$2p\,^4P-2p^3\,^4S^\circ$	$5/2-3/2$
1010,074	10	5,33	17,61	$2p^2\,^4P-2p^3\,^4S^\circ$	$3/2-3/2$
1009,854	9	5,33	17,61	$2p^2\,^4P-2p^3\,^4S^\circ$	$1/2-3/2$
946,208	2	11,96	25,07	$2p^2\,^2S-3d\,^2P^\circ$	$1/2-3/2$
945,981	1	11,96	25,07	$2p^2\,^2S-3d\,^2P^\circ$	$1/2-1/2$
904,468	10	0,01	13,71	$2p\,^2P^\circ-2p^2\,^2P$	$3/2-1/2$
904,134	12	0,01	13,72	$2p\,^2P^\circ-2p^2\,^2P$	$3/2-3/2$
903,950	11	0,00	13,71	$2p\,^2P^\circ-2p^2\,^2P$	$1/2-1/2$
903,609	10	0,00	13,72	$2p\,^2P^\circ-2p^2\,^2P$	$1/2-3/2$
858,561	9	0,01	14,45	$2p\,^2P^\circ-3s\,^2S$	$3/2-1/2$
858,094	8	0,00	14,45	$2p\,^2P^\circ-3s\,^2S$	$1/2-1/2$
809,770	3	9,29	24,60	$2p^2\,^2D-3d\,^2D^\circ$	$3/2,\ 5/2-3/2$
809,682	4	9,29	24,60	$2p^2\,^2D-3d\,^2D^\circ$	$3/2,\ 5/2-5/2$
806,846	6	5,34	20,70	$2p^2\,^4P-3s\,^4P^\circ$	$5/2-3/2$
		5,33	20,70	$2p^2\,^4P-3s\,^4P^\circ$	$3/2-1/2$
806,684	4	5,33	20,70	$2p^2\,^4P-3s\,^4P^\circ$	$1/2-1/2$
		5,33	20,70	$2p^2\,^4P-3s\,^4P^\circ$	$3/2-3/2$
806,555	7	5,34	20,71	$2p^2\,^4P-3s\,^4P^\circ$	$5/2-5/2$
		5,33	20,70	$2p^2\,^4P-3s\,^4P^\circ$	$1/2-3/2$
806,384	5	5,33	20,71	$2p^2\,^4P-3s\,^4P^\circ$	$3/2-5/2$
799,947	4	9,29	24,79	$2p^2\,^2D-3d\,^2F^\circ$	$3/2-5/2$
799,660	5	9,29	24,79	$2p^2\,^2D-3d\,^2F^\circ$	$5/2-7/2$
687,355	11	0,01	18,05	$2p\,^2P^\circ-3d\,^2D$	$3/2-5/2$
687,059	10	0,00	18,04	$2p\,^2P^\circ-3d\,^2D$	$1/2-3/2$
686,480	2	9,29	27,35	$2p^2\,^2D-4d\,^2D^\circ$	$3/2-3/2$
		9,29	27,35	$2p^2\,^2D-4d\,^2D^\circ$	$5/2-5/2$
651,342	8	5,34	24,37	$2p^2\,^4P-3d\,^4D^\circ$	$5/2-7/2$
651,269	7	5,33	24,37	$2p^2\,^4P-3d\,^4D^\circ$	$3/2-5/2$
651,216	7	5,33	24,37	$2p^2\,^4P-3d\,^4D$	$1/2-1/2,\ 3/2$
641,875	6	5,34	24,65	$2p^2\,^4P-3d\,^4P^\circ$	$5/2-5/2$
641,772	6	5,34	24,65	$2p^2\,^4P-3d\,^4P^\circ$	$5/2-3/2$
		5,33	24,65	$2p^2\,^4P-3d\,^4P^\circ$	$3/2-5/2$

λ, Å	I	E_H, eV	E_B, eV	Transition	J
641,591	6	5,33	24,66	$2p^2\ {}^4P-3d\ {}^4P^\circ$	3/2—1/2
		5,33	24,65	$2p^2\ {}^4P-3d\ {}^4P^\circ$	1/2—3/2
636,247	4	0,01	19,49	$2p\ {}^2P^\circ-4s\ {}^2S$	3/2—1/2
635,988	3	0,00	19,49	$2p\ {}^2P^\circ-4s\ {}^2S$	1/2—1/2
600,532	2	5,33	25,98	$2p^2\ {}^4P-4s\ {}^4P^\circ$	5/2—3/2
		5,32	25,98	$2p^2\ {}^4P-4s\ {}^4P^\circ$	3/2—1/2
600,369	3	5,33	25,99	$2p^2\ {}^4P-4s\ {}^4P^\circ$	5/2—5/2
600,265	1	5,32	25,99	$2p^2\ {}^4P-4s\ {}^4P^\circ$	3/2—5/2
595,032	7	0,01	20,84	$2p\ {}^2P^\circ-4d\ {}^2D$	3/2—5/2
594,808	6	0,00	20,84	$2p\ {}^2P^\circ-4d\ {}^2D$	1/2—3/2
577,108	2	0,01	21,49	$2p\ {}^2P^\circ-5s\ {}^2S$	3/2—1/2
576,900	1	0,00	21,49	$2p\ {}^2P^\circ-5s\ {}^2S$	1/2—1/2
564,645	5	5,34	27,29	$2p^2\ {}^4P-4d\ {}^4D^\circ$	5/2—5/2, 7/2
		5,33	27,29	$2p^2\ {}^4P-4d\ {}^4D^\circ$	3/2—3/2, 5/2
		5,33	27,29	$2p^2\ {}^4P-4d\ {}^4D^\circ$	1/2—1/2, 3/2
562,577	3	5,34	27,38	$2p^2\ {}^4P-4d\ {}^4P^\circ$	5/2—5/2
562,498	3	5,34	27,38	$2p^2\ {}^4P-4d\ {}^4P^\circ$	5/2—3/2
		5,33	27,38	$2p^2\ {}^4P-4d\ {}^4P^\circ$	3/2—5/2
562,355	3	5,33	27,38	$2p^2\ {}^4P-4d\ {}^4P^\circ$	3/2—1/2
		5,33	27,38	$2p^2\ {}^4P-4d\ {}^4P^\circ$	1/2—3/2
560,443	5	0,01	22,13	$2p\ {}^2P^\circ-5d\ {}^2D$	3/2—5/2
560,244	4	0,00	22,13	$2p\ {}^2P^\circ-5d\ {}^2D$	1/2—3/2
551,894	0	0,01	22,47	$2p\ {}^2P^\circ-6s\ {}^2S$	3/2—1/2
549,568	3	0,01	22,57	$2p\ {}^2P^\circ-3p\ {}^2P$	3/2—1/2
549,507	5	0,01	22,57	$2p\ {}^2P^\circ-3p\ {}^2P$	3/2—3/2
549,375	4	0,00	22,57	$2p\ {}^2P^\circ-3p\ {}^2P$	1/2—1/2
549,317	3	0,00	22,57	$2p\ {}^2P^\circ-3p\ {}^2P$	1/2—3/2
547,288	0	5,33	28,00	$2p^2\ {}^4P-5s\ {}^4P^\circ$	5/2—3/2
		5,32	28,00	$2p^2\ {}^4P-5s\ {}^4P^\circ$	3/2—1/2
547,169	0	5,33	28,00	$2p^2\ {}^4P-5s\ {}^4P^\circ$	5/2—5/2
543,475	3	0,01	22,82	$2p\ {}^2P^\circ-6d\ {}^2D$	3/2—5/2
543,291	2	0,00	22,82	$2p\ {}^2P^\circ-6d\ {}^2D$	1/2—3/2
532,716	3	5,34	28,61	$2p^2\ {}^4P-5d\ {}^4D^\circ$	5/2—7/2
		5,33	28,61	$2p^2\ {}^4P-5d\ {}^4D^\circ$	3/2—5/2
531,917	1	5,34	28,64	$2p^2\ {}^4P-5d\ {}^4P^\circ$	5/2—5/2
531,775	0	5,33	28,64	$2p^2\ {}^4P-5d\ {}^4P^\circ$	3/2—1/2
		5,33	28,64	$2p^2\ {}^4P-5d\ {}^4P^\circ$	1/2—3/2
530,386	4	0,01	23,38	$2p\ {}^2P^\circ-3p\ {}^2D$	3/2—5/2
530,290	3	0,00	23,38	$2p\ {}^2P^\circ-3p\ {}^2D$	1/2—3/2
466,536	0	0,01	26,58	$2p\ {}^2P^\circ-4p\ {}^2P$	3/2—1/2
466,492	2	0,01	26,58	$2p\ {}^2P^\circ-4p\ {}^2P$	3/2—3/2
466,404	1	0,00	26,58	$2p\ {}^2P^\circ-4p\ {}^2P$	1/2—1/2
466,358	0	0,00	26,58	$2p\ {}^2P^\circ-4p\ {}^2P$	1/2—3/2
438,910	1	0,01	28,25	$2p\ {}^2P^\circ-5p\ {}^2P$	3/2—3/2

C III, ground state $1s^2\,2s^2\ {}^1S_0$
Ionization potential 386213,9 cm^{-1}; 47,881 eV

λ, Å	I	E_H, eV	E_B, eV	Transition	J
9717,73	2	32,19	33,47	$3p\ {}^3P^\circ-3d\ {}^3D$	2—2
9715,11	5	32,19	33,47	$3p\ {}^3P^\circ-3d\ {}^3D$	2—3
9706,44	2	32,19	33,47	$3p\ {}^3P^\circ-3d\ {}^3D$	1—1
9705,39	3	32,19	33,47	$3p\ {}^3P^\circ-3d\ {}^3D$	1—2
9701,12	2	32,19	33,47	$3p\ {}^3P^\circ-3d\ {}^3D$	0—1

λ, Å	I	E_H, eV	E_B, eV	Transition	J
9358,37	1	38,65	39,97	$4s\ ^1S - 4p\ ^1P^\circ$	0—1
8665,22	3	43,03	44,46	$5f\ ^3F^\circ - 6g\ ^3G$	4—5
8663,65	2	43,03	44,46	$5f\ ^3F^\circ - 6g\ ^3G$	3—4
8652,6	1	42,83	44,26	$5d\ ^3D - 6p\ ^3P^\circ$	3—2
8500,32	10	30,64	32,10	$3s\ ^1S - 3p\ ^1P^\circ$	0—1
8358,72	2	39,84	41,33	$4d\ ^3D - 3d\ ^3F^\circ$	2—2
8357,86	2	39,85	41,33	$4d\ ^3D - 3d\ ^3F^\circ$	3—3
8347,94	5	39,84	41,33	$4d\ ^3D - 3d\ ^3F^\circ$	1—2
8341,59	6	39,84	41,33	$4d\ ^3D - 3d\ ^3F^\circ$	2—3
8332,99	7	39,85	41,33	$4d\ ^3D - 3d\ ^3F^\circ$	3—4
8296,51	1	41,33	42,83	$3d\ ^3F^\circ - 5d\ ^3D$	4—3
8272,26	1	41,33	42,83	$3d\ ^3F^\circ - 5d\ ^3D$	3—2
8255,62	1	41,33	42,83	$3d\ ^3F^\circ - 5d\ ^3D$	2—1
8196,48	10	—	—	—	—
8021,14	1	42,98	44,52	$5d\ ^1D - 6f\ ^1F^\circ$	2—3
7796,00	4	40,57	42,16	$3p\ ^3S - 3d\ ^3P^\circ$	1—2
7780,42	3	40,57	42,16	$3p\ ^3S - 3d\ ^3P^\circ$	1—1
7707,43	6	39,64	41,25	$3p\ ^1P - 3d\ ^1D^\circ$	1—2
7625,94	2	38,22	39,84	$3s\ ^3P^\circ - 4d\ ^3D$	2—2
7612,65	7	41,33	42,96	$3d\ ^3F^\circ - 5g\ ^3G$	4—5
		38,22	39,85	$3s\ ^3P^\circ - 4d\ ^3D$	2—3
7595,29	2	38,21	39,84	$3s\ ^3P^\circ - 4d\ ^3D$	1—1
7592,28	5	41,33	42,96	$3d\ ^3F^\circ - 5g\ ^3G$	3—4
7586,40	4	38,21	39,84	$3s\ ^3P^\circ - 4d\ ^3D$	1—2
7578,16	4	41,33	42,96	$3d\ ^3F^\circ - 5g\ ^3G$	2—3
7576,68	2	38,21	39,84	$3s\ ^3P^\circ - 4d\ ^3D$	0—1
7486,52	3	42,83	44,48	$5d\ ^3D - 6f\ ^3F$	1—2
		42,83	44,48	$5d\ ^3D - 6f\ ^3F$	3—4
		42,83	44,48	$5d\ ^3D - 6f\ ^3F$	2—3
7353,96	0	41,30	42,98	$3p\ ^1D - 3d\ ^1P^\circ$	2—1
7212,29	1	42,67	44,39	$5p\ ^3P^\circ - 6d\ ^3D$	1—2
7210,52	2	42,67	44,39	$5p\ ^3P^\circ - 6d\ ^3D$	2—3
7037,25	7	38,43	40,19	$3s\ ^1P^\circ - 4d\ ^1D$	1—2
6899,64	1	40,87	42,67	$3p\ ^3P - 5p\ ^3P^\circ$	2—2
6881,09	1	40,06	41,86	$3p\ ^3D - 3d\ ^3D^\circ$	3—2
6872,05	4	40,06	41,86	$3p\ ^3D - 3d\ ^3D^\circ$	3—3
6868,80	1	40,05	41,86	$3p\ ^3D - 3d\ ^3D^\circ$	2—1
6862,71	3	40,05	41,86	$3p\ ^3D - 3d\ ^3D^\circ$	2—2
6857,27	2	40,05	41,86	$3p\ ^3D - 3d\ ^3D^\circ$	1—1
6853,70	1	40,05	41,86	$3p\ ^3D - 3d\ ^3D^\circ$	2—3
6851,20	1	40,05	41,86	$3p\ ^3D - 3d\ ^3D^\circ$	1—2
6774,93	0	42,16	43,99	$3d\ ^3P^\circ - 6s\ ^3S$	2—1
6773,37	1	38,22	40,05	$3s\ ^3P^\circ - 3p\ ^3D$	2—1
6762,17	4	38,22	40,05	$3s\ ^3P^\circ - 3p\ ^3D$	2—2
6744,38	7	38,22	40,06	$3s\ ^3P^\circ - 3p\ ^3D$	2—3
6742,24	5	38,21	40,05	$3s\ ^3P^\circ - 3p\ ^3D$	1—1
6731,04	6	38,21	40,05	$3s\ ^3P^\circ - 3p\ ^3D$	1—2
6727,39	6	38,21	40,05	$3s\ ^3P^\circ - 3p\ ^3D$	0—1
6460,33	0	42,56	44,47	$5p\ ^1P^\circ - 6d\ ^1D$	1—2
6350,76	2	41,30	43,25	$3p\ ^1D - 5f\ ^1F^\circ$	2—3
6205,56	5	39,97	41,97	$4p\ ^1P^\circ - 5s\ ^1S$	1—0
6163,96	0	39,85	41,86	$4d\ ^3D - 3d\ ^3D^\circ$	3—2
6159,97	0	39,84	41,86	$4d\ ^3D - 3d\ ^3D^\circ$	2—1
6156,68	3	39,85	41,86	$4d\ ^3D - 3d\ ^3D^\circ$	3—3
6155,09	2	39,84	41,86	$4d\ ^3D - 3d\ ^3D^\circ$	2—2
6154,13	1	39,84	41,86	$4d\ ^3D - 3d\ ^3D^\circ$	1—1
6149,23	0	39,84	41,86	$4d\ ^3D - 3d\ ^3D^\circ$	1—2
6147,81	0	39,84	41,86	$4d\ ^3D - 3d\ ^3D^\circ$	2—3

λ, Å	I	E_H, eV	E_B, eV	Transition	J
5894,07	3	40,06	42,16	$3p\ ^3D—3d\ ^3P°$	3—2
5880,54	1	40,05	42,16	$3p\ ^3D—3d\ ^3P°$	2—2
5871,69	2	40,05	42,16	$3p\ ^3D—3d\ ^3P°$	2—1
5863,24	1	40,05	42,16	$3p\ ^3D—3d\ ^3P°$	1—1
5858,35	1	40,05	42,16	$3p\ ^3D—3d\ ^3P°$	1—0
5826,42	1	40,19	42,32	$4d\ ^1D—3d\ ^1F°$	2—3
5771,66	2	42,32	44,47	$3d\ ^1F°—6g\ ^1G$	3—4
5695,92	12	32,10	34,28	$3p\ ^1P°—3d\ ^1D$	1—2
5359,95	2	39,85	42,16	$4d\ ^3D—3d\ ^3P°$	3—2
5353,12	0	39,84	42,16	$4d\ ^3D—3d\ ^3P°$	2—2
5345,84	1	39,84	42,16	$4d\ ^3D—3d\ ^3P°$	2—1
5341,46	0	39,84	42,16	$4d\ ^3D—3d\ ^3P°$	1—1
5337,42	0	39,84	42,16	$4d\ ^3D—3d\ ^3P°$	1—0
5305,10	2	43,03	45,37	$5f\ ^3F°—7g\ ^3G$	—
5272,53	6	38,22	40,57	$3s\ ^3P°—3p\ ^3S$	2—1
5253,58	5	38,21	40,57	$3s\ ^3P°—3p\ ^3S$	1—1
5249,11	4	40,19	42,56	$4d\ ^1D—5p\ ^1P°$	2—1
5244,67	3	38,21	40,57	$3s\ ^3P°—3p\ ^3S$	0—1
4859,6	0	42,83	45,38	$5d\ ^3D—7f\ ^3F°$	3—4
4793,66	2	40,06	42,67	$3p\ ^3D—5p\ ^3P°$	3—2
4730,16	1	40,05	42,67	$3p\ ^3D—5p\ ^3P°$	2—1
4724,33	1	40,05	42,67	$3p\ ^3D—5p\ ^3P°$	1—0
4673,95	6	38,22	40,87	$3s\ ^3P°—3p\ ^3P$	2—1
4665,86	8	38,22	40,87	$3s\ ^3P°—3p\ ^3P$	2—2
4663,64	6	38,21	40,87	$3s\ ^3P°—3p\ ^3P$	1—0
4659,06	5	38,21	40,87	$3s\ ^3P°—3p\ ^3P$	1—1
4652,06	5	38,21	40,87	$3s\ ^3P°—3p\ ^3P$	0—1
4651,47	11	29,53	32,19	$3s\ ^3S—3p\ ^3P°$	1—0
4651,01	5	38,21	40,87	$3s\ ^3P°—3p\ ^3P$	1—2
4650,25	13	29,53	32,19	$3s\ ^3S—3p\ ^3P°$	1—1
4647,42	14	29,53	32,19	$3s\ ^3S—3p\ ^3P°$	1—2
4593,3	1	48,31	51,01	$4f\ ^3D—5g\ ^3F°$	3—4
4587,6	0	48,31	51,01	$4f\ ^3D—5g\ ^3F°$	—
4516,77	6	39,39	42,14	$4p\ ^3P°—5s\ ^3S$	2—1
4515,78	5	39,39	42,14	$4p\ ^3P°—5s\ ^3S$	1—1
4515,33	3	39,39	42,14	$4p\ ^3P°—5s\ ^3S$	0—1
4443,08	2	40,19	42,98	$4d\ ^1D—3d\ ^1P°$	2—1
4388,016	4	39,85	42,67	$4d\ ^3D—5p\ ^3P°$	3—2
4383,544	2	39,84	42,67	$4d\ ^3D—5p\ ^3P°$	2—2
4382,898	3	39,84	42,67	$4d\ ^3D—5p\ ^3P°$	2—1
4379,952	2	39,84	42,67	$4d\ ^3D—5p\ ^3P°$	1—1
4379,481	2	39,84	42,67	$4d\ ^3D—5p\ ^3P°$	1—0
4367,50	3	48,13	50,97	$4f\ ^3F°—5g\ ^3G$	4—5
4361,87	4	48,13	50,97	$4f\ ^3F°—5g\ ^3G$	—
4358,90	2	48,14	50,96	$4f\ ^3G—5g\ ^3H°$	5—6
4325,560	8	38,43	41,30	$3s\ ^1P°—3p\ ^1D$	1—2
4315,44	3	48,06	50,93	$4d\ ^3D°—5f\ ^3F$	3—4
4257,894	2	39,92	42,83	$4f\ ^3F°—5d\ ^3D$	4—3
4256,455	1	39,92	42,83	$4f\ ^3F°—5d\ ^3D$	3—2
4255,42	1	39,91	42,83	$4f\ ^3F°—5d\ ^3D$	2—1
4247,308	4	39,64	42,56	$3p\ ^1P—5p\ ^1P°$	1—1
4186,900	9	40,01	42,97	$4f\ ^1F°—5g\ ^1G$	3—4
4173,089	2	40,01	42,98	$4f\ ^1F°—5d\ ^1D$	3—2
4166,95	1	41,30	44,27	$3p\ ^1D—6p\ ^1P°$	2—1
4163,26	2	40,06	43,03	$3p\ ^3D—5f\ ^3F°$	3—3
4162,86	7	40,06	43,03	$3p\ ^3D—5f\ ^3F°$	3—4
4156,76	2	40,05	43,03	$3p\ ^3D—5f\ ^3F°$	2—2
4156,49	6	40,05	43,03	$3p\ ^3D—5f\ ^3F°$	2—3

λ, Å	I	E_H, eV	E_B, eV	Transition	J
4152,512	5	40,05	43,03	$3p\ ^3D—5f\ ^3F°$	1—2
4121,843	5	39,97	42,98	$4p\ ^1P°—5d\ ^1D$	1—2
4070,261	9	39,92	42,96	$4f\ ^3F°—5g\ ^3G$	4—5
4068,912	9	39,92	42,96	$4f\ ^3F°—5g\ ^3G$	3—4
4067,940	8	39,91	42,96	$4f\ ^3F°—5g\ ^3G$	2—3
4056,062	7	40,19	43,25	$4d\ ^1D—5f\ ^1F°$	2—3
4001,56	0	47,83	50,93	$4d\ ^3F°—5f\ ^3G$	—
3999,92	0	47,83	50,93	$4d\ ^3F°—5f\ ^3G$	4—5
3889,475	1	39,85	43,03	$4d\ ^3D—5f\ ^3F°$	3—3
3889,144	6	39,85	43,03	$4d\ ^3D—5f\ ^3F°$	3—4
3885,941	5	39,84	43,03	$4d\ ^3D—5f\ ^3F°$	2—3
3883,816	4	39,84	43,03	$4d\ ^3D—5f\ ^3F°$	1—2
3703,71	4	39,64	42,98	$3p\ ^1P—3d\ ^1P°$	1—1
3609,625	6	39,39	42,83	$4p\ ^3P°—5d\ ^3D$	2—2, 3
3609,063	5	39,39	42,83	$4p\ ^3P°—5d\ ^3D$	1—1, 2
3608,81	3	39,39	42,83	$4p\ ^3P°—5d\ ^3D$	0—1
3262,272	3	38,36	42,16	$4s\ ^3S—3d\ ^3P°$	1—2
3259,541	2	38,36	42,16	$4s\ ^3S—3d\ ^3P°$	1—1
3258,00	1	38,36	42,16	$4s\ ^3S—3d\ ^3P°$	1—0
3170,016	4	38,65	42,56	$4s\ ^1S—5p\ ^1P°$	0—1
3161,92	2	38,22	42,14	$3s\ ^3P°—5s\ ^3S$	2—1
3155,09	1	38,21	42,14	$3s\ ^3P°—5s\ ^3S$	1—1
3151,85	0	38,21	42,14	$3s\ ^3P°—5s\ ^3S$	0—1
3038,91	1	40,19	44,27	$4d\ ^1D—6p\ ^1P°$	2—1
2982,106	8	34,28	38,43	$3d\ ^1D—3s\ ^1P°$	2—1
2874,722	3	38,36	42,67	$4s\ ^3S—5p\ ^3P°$	1—2
2874,43	2	38,36	42,67	$4s\ ^3S—5p\ ^3P°$	1—1
2874,24	0	38,36	42,67	$4s\ ^3S—5p\ ^3P°$	1—0
2863,712	4	40,19	44,52	$4d\ ^1D—6f\ ^1F°$	2—3
2857,013	1	38,65	42,98	$4s\ ^1S—3d\ ^1P°$	0—1
2854,13	0	48,13	52,47	$4f\ ^3F°—6g\ ^3G$	—
2853,13	0	48,13	52,47	$4f\ ^3F°—6g\ ^3G$	4—5
2849,050	5	38,43	42,78	$3s\ ^1P°—3p\ ^1S$	1—0
2844,117	2	12,69	17,04	$2p\ ^1P°—2p^2\ ^3P$	1—2
2808,07	1	39,85	44,26	$4d\ ^3D—6p\ ^3P$	3—2
2806,31	1	39,84	44,26	$4d\ ^3D—6p\ ^3P$	2—1
2805,13	0	39,84	44,26	$4d\ ^3D—6p\ ^3P°$	1—0, 1
2799,47	4	40,06	44,48	$3p\ ^3D—6f\ ^3F°$	3—4
2796,46	3	40,05	44,48	$3p\ ^3D—6f\ ^3F°$	2—3
2794,56	2	40,05	44,48	$3p\ ^3D—6f\ ^3F°$	1—2
2777,714	5	40,01	44,47	$4f\ ^1F°—6g\ ^1G$	3—4
2751,828	3	39,97	44,47	$4p\ ^1P°—6d\ ^1D$	1—2
2725,90	7	39,92	44,46	$4f\ ^3F°—6g\ ^3G$	4—5
2725,30	7	39,92	44,46	$4f\ ^3F°—6g\ ^3G$	3—4
2724,85	6	39,91	44,46	$4f\ ^3F°—6g\ ^3G$	2—3
2697,75	7	39,39	43,99	$4p\ ^3P°—6s\ ^3S$	2—1
2697,42	3	39,39	43,99	$4p\ ^3P°—6s\ ^3S$	1—1
2672,959	5	39,85	44,48	$4d\ ^3D—6f\ ^3F°$	3—4
2671,318	4	39,84	44,48	$4d\ ^3D—6f\ ^3F°$	2—3
2670,240	3	39,84	44,48	$4d\ ^3D—6f\ ^3F°$	1—2
2616,627	4	33,47	38,21	$3d\ ^3D—3s\ ^3P°$	1—0
2614,478	5	33,47	38,21	$3d\ ^3D—3s\ ^3P°$	1, 2—1
2610,020	6	33,47	38,22	$3d\ ^3D—3s\ ^3P°$	3—2
2609,83	1	33,47	38,22	$3d\ ^3D—3s\ ^3P°$	2—2
2480,861	4	39,39	44,39	$4p\ ^3P°—6d\ ^3D$	2—3
2480,502	4	39,39	44,39	$4p\ ^3P°—6d\ ^3D$	1, 0—2, 1
2296,870	16	12,69	18,09	$2p\ ^1P°—2p^2\ ^1D$	1—2

λ, Å	I	E_H, eV	E_B, eV	Transition	J
2202,54	1	38,65	44,27	$4s\ ^1S—6p\ ^1P^\circ$	0—1
2176,963	4	34,28	39,97	$3d\ ^1D—4p\ ^1P^\circ$	2—1
2162,944	9	34,28	40,01	$3d\ ^1D—4f\ ^1F^\circ$	2—3
2145,58	0	40,87	46,69	$3p\ ^3P—4s\ ^3P^\circ$	2, 1—1, 0
2142,49	1	40,87	46,69	$3p\ ^3P—4s\ ^3P^\circ$	2—2
2140,92	1	41,86	47,65	$3d\ ^3D^\circ—4p\ ^3P$	1, 2—0, 1
2139,86	1	41,86	47,65	$3d\ ^3D^\circ—4p\ ^3P$	3—2
2100,46	0	38,36	44,26	$4s\ ^3S—6p\ ^3P^\circ$	1—2
		33,47	39,39	$3d\ ^3D—4p\ ^3P^\circ$	1—0, 1
2091,999	6	33,47	39,39	$3d\ ^3D—4p\ ^3P^\circ$	2—1, 2
		33,47	39,39	$3d\ ^3D—4p\ ^3P^\circ$	3—2
2016,84	1	42,16	48,31	$3d\ ^3P^\circ—4f\ ^3D$	2—3
2015,7	0	42,16	48,31	$3d\ ^3P^\circ—4f\ ^3D$	—
2010,094	5	32,19	38,36	$3p\ ^3P^\circ—4s\ ^3S$	2—1
2009,570	4	32,19	38,36	$3p\ ^3P^\circ—4s\ ^3S$	1—1
2009,327	2	32,19	38,36	$3p\ ^3P^\circ—4s\ ^3S$	0—1
1979,62	1	41,86	48,13	$3d\ ^3D^\circ—4f\ ^3F$	—
1979,16	2	41,86	48,13	$3d\ ^3D^\circ—4f\ ^3F$	3—4
1931,027	4	—	—	—	—
1923,31	2	33,47	39,91	$3d\ ^3D—4f\ ^3F^\circ$	1, 2—2
		33,47	39,92	$3d\ ^3D—4f\ ^3F^\circ$	3—3
1923,14	4	33,47	39,92	$3d\ ^3D—4f\ ^3F^\circ$	2—3
1922,93	5	33,47	39,92	$3d\ ^3D—4f\ ^3F^\circ$	3—4
1894,49	2	—	—	—	—
1645,06	1	—	—	—	—
1620,68	1	—	—	—	—
1620,33	2	—	—	—	—
1620,05	3	—	—	—	—
1591,48	2	—	—	—	—
1577,89	2	—	—	—	—
1577,32	2	—	—	—	—
1576,40	3	—	—	—	—
1531,85	2	—	—	—	—
1478,30	1	—	—	—	—
1478,05	2	—	—	—	—
1477,68	3	—	—	—	—
1428,95	1	—	—	—	—
1428,53	2	—	—	—	—
1428,17	2	—	—	—	—
1427,85	3	—	—	—	—
1426,78	1	—	—	—	—
1426,45	4	—	—	—	—
1308,73	2	—	—	—	—
1296,30	2	—	—	—	—
1256,52	1	—	—	—	—
1247,383	3	12,69	22,63	$2p\ ^1P^\circ—2p^2\ ^1S$	1—0
1176,370	3	6,49	17,03	$2p\ ^3P^\circ—2p^2\ ^3P$	2—1
1175,987	3	6,49	17,03	$2p\ ^3P^\circ—2p^2\ ^3P$	1—0
1175,711	5	6,49	17,04	$2p\ ^3P^\circ—2p^2\ ^3P$	2—2
1175,590	2	6,49	17,03	$2p\ ^3P^\circ—2p^2\ ^3P$	1—1
1175,263	3	6,49	17,03	$2p\ ^3P^\circ—2p^2\ ^3P$	0—1
1174,933	3	6,49	17,04	$2p\ ^3P^\circ—2p^2\ ^3P$	1—2
1165,87	1	—	—	—	—
977,026	18	0,00	12,69	$2s^2\ ^1S—2p\ ^1P^\circ$	0—1
884,516	8	—	—	—	—
784,393	3	—	—	—	—
714,879	1	—	—	—	—
690,526	7	12,69	30,64	$2p\ ^1P^\circ—3s\ ^1S$	1—0

λ, Å	I	E_H, eV	E_B, eV	Transition	J
622,144	2	—	—	—	—
609,275	6	—	—	—	—
609,025	4	—	—	—	—
585,666	6	—	—	—	—
585,608	6	—	—	—	—
585,496	5	—	—	—	—
585,417	8	—	—	—	—
585,261	6	—	—	—	—
574,279	12	12,69	34,28	$2p\ ^1P^\circ$—$3d\ ^1D$	1—2
566,490	4	—	—	—	—
565,530	7	18,09	40,01	$2p^2\ ^1D$—$4f\ ^1F^\circ$	2—3
554,655	2	—	—	—	—
538,312	13	6,49	29,53	$2p\ ^3P^\circ$—$3s\ ^3S$	2—1
538,150	12	6,49	29,53	$2p\ ^3P^\circ$—$3s\ ^3S$	1—1
538,075	11	6,49	29,53	$2p\ ^3P^\circ$—$3s\ ^3S$	0—1
535,288	10	18,09	41,25	$2p^2\ ^1D$—$3d\ ^1D^\circ$	2—2
511,527	10	18,09	42,32	$2p^2\ ^1D$—$3d\ ^1F^\circ$	2—3
499,583	7	—	—	—	—
499,530	9	—	—	—	—
499,462	8	—	—	—	—
499,425	7	—	—	—	—
497,910	1	—	—	—	—
493,587	7	—	—	—	—
493,519	5	—	—	—	—
493,464	5	—	—	—	—
493,396	5	—	—	—	—
493,364	5	—	—	—	—
493,341	5	—	—	—	—
492,649	7	18,09	43,25	$2p^2\ ^1D$—$5f\ ^1F^\circ$	2—3
483,733	5	—	—	—	—
483,618	4	—	—	—	—
483,567	3	—	—	—	—
477,625	3	12,69	38,65	$2p\ ^1P^\circ$—$4s\ ^1S$	1—0
460,050	8	12,69	39,64	$2p\ ^1P^\circ$—$3p\ ^1P$	1—1
459,633	15	6,49	33,47	$2p\ ^3P^\circ$—$3d\ ^3D$	2—1
459,521	14	6,49	33,47	$2p\ ^3P^\circ$—$3d\ ^3D$	1—1
459,462	13	6,49	33,47	$2p\ ^3P^\circ$—$3d\ ^3D$	0—1
450,732	9	12,69	40,19	$2p\ ^1P^\circ$—$4d\ ^1D$	1—2
433,337	8	12,69	41,30	$2p\ ^1P^\circ$—$3p\ ^1D$	1—2
418,609	2	—	—	—	—
416,769	5	—	—	—	—
411,958	3	12,69	42,78	$2p\ ^1P^\circ$—$3p\ ^1S$	1—0
409,325	6	—	—	—	—
399,688	6	—	—	—	—
399,637	6	—	—	—	—
398,551	1	—	—	—	—
398,168	1	—	—	—	—
390,055	3	—	—	—	—
389,090	7	6,49	38,36	$2p\ ^3P^\circ$—$4s\ ^3S$	2—1
389,005	6	6,49	38,36	$2p\ ^3P^\circ$—$4s\ ^3S$	1—1
388,965	5	6,49	38,36	$2p\ ^3P^\circ$—$4s\ ^3S$	0—1
386,203	14	0,00	32,10	$2s^2\ ^1S$—$3p\ ^1P^\circ$	0—1
379,065	1	—	—	—	—
371,784	8	6,49	39,84	$2p\ ^3P^\circ$—$4d\ ^3D$	2—2
371,747	10	6,49	39,84	$2p\ ^3P^\circ$—$4d\ ^3D$	2—3
371,694	10	{ 6,49 6,49	39,84 39,84	$2p\ ^3P^\circ$—$4d\ ^3D$ $2p\ ^3P^\circ$—$5d\ ^3D$	1—2 0—1
369,472	2	—	—	—	—
369,415	5	—	—	—	—

λ, Å	I	E_H, eV	E_B, eV	Transition	J
366,169	4	—	—	—	—
365,778	1	—	—	—	—
363,864	6	6,49	40,57	$2p\ ^3P^\circ—3p\ ^3S$	2—1
363,790	5	6,49	40,57	$2p\ ^3P^\circ—3p\ ^3S$	1—1
363,761	4	6,49	40,57	$2p\ ^3P^\circ—3p\ ^3S$	0—1
360,675	5	—	—	—	—
360,623	7	—	—	—	—
360,557	6	—	—	—	—
358,740	4	—	—	—	—
353,000	3	—	—	—	—
350,330	2	—	—	—	—
347,854	3	—	—	—	—
347,777	3	—	—	—	—
341,242	7	—	—	—	—
341,179	6	—	—	—	—
341,143	5	—	—	—	—
339,773	1	—	—	—	—
330,687	1	—	—	—	—
330,637	1	—	—	—	—
327,784	1	—	—	—	—
327,176	4	—	—	—	—
327,112	4	—	—	—	—
325,570	1	—	—	—	—
322,575	8	0,00	38,43	$2s^2\ ^1S—3s\ ^1P^\circ$	0—1
319,266	3	—	—	—	—
314,395	1	—	—	—	—
310,171	7	0,00	39,97	$2s^2\ ^1S—4p\ ^1P^\circ$	0—1
303,468	1	—	—	—	—
303,432	4	—	—	—	—
301,279	1	—	—	—	—
301,243	3	—	—	—	—
301,206	2	—	—	—	—
291,330	5	0,00	42,56	$2s^2\ ^1S—5p\ ^1P^\circ$	0—1
281,390	2	—	—	—	—
280,522	2	—	—	—	—
280,043	3	—	—	—	—
274,051	2	—	—	—	—
271,014	1	—	—	—	—
270,583	1	—	—	—	—

C IV, ground state $1s^2 2s\ ^2S_{1/2}$

Ionization potential 520177,8 cm^{-1}; 64,490 eV

λ, Å	I	E_H, eV	E_B, eV	Transition	J
7726,2	6	58,44	60,05	$6f\ ^2F^\circ—7g\ ^2G$ etc.	—
5811,98	9	37,55	39,68	$3s\ ^2S—3p\ ^2P^\circ$	$^1/_2—^1/_2$
5801,33	10	37,55	39,68	$3s\ ^2S—3p\ ^2P^\circ$	$^1/_2—^3/_2$
5018,39	2	55,65	58,12	$5p\ ^2P^\circ—6s\ ^2S$	$^3/_2—^1/_2$
5016,58	1	55,65	58,12	$5p\ ^2P^\circ—6s\ ^2S$	$^1/_2—^1/_2$
4786,7	0	55,78	58,36	$5d\ ^2D—6p\ ^2P^\circ$	$^3/_2—^1/_2$
4785,88	1	55,78	58,36	$5d\ ^2D—6p\ ^2P^\circ$	$^5/_2—^3/_2$
4685,4	1	58,44	61,09	$6f\ ^2F^\circ—8g\ ^2G$ etc.	—
4658,30	9	55,78	58,44	$5f\ ^2F^\circ—6g\ ^2G$ etc.	—
4646	coincident with C III	55,78	58,44	$5d\ ^2D—6f\ ^2F^\circ$	$^5/_2—^5/_2$, $^7/_2$

λ, Å	I	E_H, eV	E_B, eV	Transition	J
4441,49	3	55,65	58,44	$5p\ ^2P^\circ - 6d\ ^2D$	3/2—5/2
4440,34	2	55,65	58,44	$5p\ ^2P^\circ - 6g\ ^2D$	1/2—5/2
3934,89	1	55,22	58,36	$5s\ ^2S - 6p\ ^2P^\circ$	1/2—1/2
3934,29	2	55,22	58,36	$5s\ ^2S - 6p\ ^2P^\circ$	1/2—3/2
3689,6	2	58,44	61,80	$6f\ ^2F^\circ - 9g\ ^2G$ etc.	—
2953,95	1	55,65	59,84	$5p\ ^2P^\circ - 7s\ ^2S$	3/2—1/2
2953,40	0	55,65	59,84	$5p\ ^2P^\circ - 7s\ ^2S$	1/2—1/2
2935,12	1	55,78	60,00	$5d\ ^2D - 7p\ ^2P^\circ$	5/2, 3/2—3/2, 1/2
2906,29	5	55,77	60,05	$5f\ ^2F^\circ - 7g\ ^2G$ etc.	—
2901,60	2	55,78	60,05	$5d\ ^2D - 7g\ ^2G$	5/2—7/2
2819,24	1	55,65	60,04	$5p\ ^2P^\circ - 7d\ ^2D$	3/2, 1/2—5/2, 3/2
2698,67	4	50,62	55,22	$4p\ ^2P^\circ - 5s\ ^2S$	3/2—1/2
2697,75	4	50,62	55,22	$4p\ ^2P^\circ - 5s\ ^2S$	1/2—1/2
2595,295	3	50,87	55,65	$4d\ ^2D - 5p\ ^2P^\circ$	3/2—1/2
2595,089	4	50,87	55,65	$4d\ ^2D - 5p\ ^2P^\circ$	5/2—3/2
2533,77	2	50,88	55,78	$4f\ ^2F^\circ - 5d\ ^2D$	7/2, 5/2—5/2, 3/2
2530,6	6	50,88	55,78	$4f\ ^2F^\circ - 5f\ ^2F^\circ$	7/2—5/2, 7/2
2529,98	11	50,88	55,78	$4f\ ^2F^\circ - 5g\ ^2G$	7/2—7/2, 9/2
2527,7	1	50,87	55,78	$4d\ ^2D - 5d\ ^2D$	5/2, 3/2—5/2, 3/2
2524,41	9	50,87	55,78	$4d\ ^2D - 5f\ ^2F^\circ$	5/2, 3/2—5/2
2523,7	4	50,87	55,78	$4d\ ^2D - 5g\ ^2G$	5/2—7/2
2405,10	6	50,62	55,78	$4p\ ^2P^\circ - 5d\ ^2D$	3/2—5/2
2404,44	5	50,62	55,78	$4p\ ^2P^\circ - 5d\ ^2D$	1/2—3/2
2335,90	2	55,77	61,09	$5f\ ^2F^\circ - 8g\ ^2G$ etc.	—
2104,24	1	49,76	55,65	$4s\ ^2S - 5p\ ^2P^\circ$	1/2—1/2
2103,94	2	49,76	55,65	$4s\ ^2S - 5p\ ^2P^\circ$	1/2—3/2
1550,771	19	0,00	7,99	$2s\ ^2S - 2p\ ^2P^\circ$	1/2—1/2
1548,185	20	0,00	8,01	$2s\ ^2S - 2p\ ^2P^\circ$	1/2—3/2
1230,511	3	39,68	49,76	$3p\ ^2P^\circ - 4s\ ^2S$	3/2—1/2
1230,046	2	39,68	49,76	$3p\ ^2P^\circ - 4s\ ^2S$	1/2—1/2
1198,58	1	—	—	—	—
1168,990	4	40,28	50,88	$3d\ ^2D - 4f\ ^2F^\circ$	5/2—7/2
1168,873	3	40,28	50,88	$3d\ ^2D - 4f\ ^2F^\circ$	5/2—5/2
1107,933	2	39,68	50,87	$3p\ ^2P^\circ - 4d\ ^2D$	3/2—5/2
1107,600	1	39,68	50,87	$3p\ ^2P^\circ - 4d\ ^2D$	1/2—3/2
948,214	1	37,55	50,62	$3s\ ^2S - 4p\ ^2P^\circ$	1/2—1/2
948,098	2	37,55	50,62	$3s\ ^2S - 4p\ ^2P^\circ$	1/2—3/2
770,379	0	—	—	—	—
419,714	14	8,01	37,55	$2p\ ^2P^\circ - 3s\ ^2S$	3/2—1/2
419,525	13	7,99	37,55	$2p\ ^2P^\circ - 3s\ ^2S$	1/2—1/2
384,178	17	8,01	40,28	$2p\ ^2P^\circ - 3d\ ^2D$	3/2—5/2
384,032	16	7,99	40,28	$2p\ ^2P^\circ - 3d\ ^2D$	1/2—3/2
312,455	14	0,00	39,68	$2s\ ^2S - 3p\ ^2P^\circ$	1/2—1/2
312,418	15	0,00	39,68	$2s\ ^2S - 3p\ ^2P^\circ$	1/2—3/2
307,806	1	0,00	40,28	$2s\ ^2S - 3d\ ^2D$	1/2—3/2
296,951	7	7,99	49,76	$2p\ ^2P^\circ - 4s\ ^2S$	3/2—1/2
296,857	6	8,01	49,76	$2p\ ^2P^\circ - 4s\ ^2S$	1/2—1/2
289,143	9	7,99	50,87	$2p\ ^2P^\circ - 4d\ ^2D$	3/2—3/2, 5/2
289,048	3	8,01	50,87	$2p\ ^2P^\circ - 4d\ ^2D$	1/2—3/2
262,627	4	7,99	55,22	$2p\ ^2P^\circ - 5s\ ^2S$	3/2—1/2
262,553	3	8,01	55,22	$2p\ ^2P^\circ - 5s\ ^2S$	1/2—1/2
259,542	7	7,99	55,78	$2p\ ^2P^\circ - 5d\ ^2D$	3/2—3/2
259,471	6	8,01	55,78	$2p\ ^2P^\circ - 5d\ ^2D$	1/2—5/2
247,415	2	7,99	55,22	$2p\ ^2P^\circ - 6s\ ^2S$	3/2—1/2
247,357	1	8,01	55,22	$2p\ ^2P^\circ - 6s\ ^2S$	1/2—1/2
245,830	5	7,99	58,44	$2p\ ^2P^\circ - 6d\ ^2D$	3/2—3/2, 5/2
245,775	4	8,01	58,44	$2p\ ^2P^\circ - 6d\ ^2D$	1/2—3/2
244,907	10	0,00	50,62	$2s\ ^2S - 4p\ ^2P^\circ$	1/2—3/2, 1/2

λ, Å	I	E_H, eV	E_B, eV	Transition	J
239,196	1	8,01	59,84	$2p\,^2P^\circ - 7s\,^2S$	$3/2,\ 1/2 - 1/2$
238,250	3	7,99	60,04	$2p\,^2P^\circ - 7d\,^2D$	$3/2 - 5/2,\ 3/2$
238,200	2	8,01	60,04	$2p\,^2P^\circ - 7d\,^2D$	$1/2 - 3/2$
233,530	3	8,01	61,09	$2p\,^2P^\circ - 8d\,^2D$	$3/2,\ 1/2 - 5/2,\ 3/2$
230,43	2	8,01	61,80	$2p\,^2P^\circ - 9d\,^2D$	$3/2,\ 1/2 - 5/2,\ 3/2$
228,27	1	8,01	61,31	$2p\,^2P^\circ - 10d\,^2D$	$3/2,\ 1/2 - 5/2,\ 5/2$
222,791	7	0,00	55,65	$2s\,^2S - 5p\,^2P^\circ$	$1/2 - 3/2,\ 1/2$
212,421	5	0,00	58,36	$2s\,^2S - 6p\,^2P^\circ$	$1/2 - 3/2,\ 1/2$
206,641	3	0,00	60,00	$2s\,^2S - 7p\,^2P^\circ$	$1/2 - 3/2,\ 1/2$
203,057	2	0,00	61,06	$2s\,^2S - 8p\,^2P^\circ$	$1/2 - 3/2,\ 1/2$
200,68	1	0,00	61,78	$2s\,^2S - 9p\,^2P^\circ$	$1/2 - 3/2,\ 1/2$
199,04	1	0,00	62,29	$2s\,^2S - 10p\,^2P^\circ$	$1/2 - 3/2,\ 1/2$
197,82	1	0,00	62,95	$2s\,^2S - 11p\,^2P^\circ$	$1/2 - 3/2,\ 1/2$

C V, ground state $1s^2\,^1S_0$

Ionization potential 3162450 cm^{-1}; 392,067 eV

λ, Å	I	E_H, eV	E_B, eV	Transition	J
2277,92	2	298,94	304,38	$2s\,^3S - 2p\,^3P^\circ$	1—1
2277,25	1	298,94	304,38	$2s\,^3S - 2p\,^3P^\circ$	1—0
2270,91	3	298,94	304,91	$2s\,^3S - 2p\,^3P^\circ$	1—2
248,744	0	304,40	354,24	$2p\,^3P^\circ - 3d\,^3D$	2—1, 2, 3
248,668	0	304,38	354,24	$2p\,^3P^\circ - 3d\,^3D$	1—1, 2, 3
40,731	—	0,00	304,38	$1s^2\,^1S - 2p\,^3P^\circ$	0—1
40,270	—	0,00	307,87	$1s^2\,^1S - 2p\,^1P^\circ$	0—1
34,973	—	0,00	354,49	$1s^2\,^1S - 3p\,^1P^\circ$	0—1
33,426	—	0,00	370,90	$1s^2\,^1S - 4p\,^1P^\circ$	0—1
32,754	—	0,00	378,51	$1s^2\,^1S - 5p\,^1P^\circ$	0—1
32,400	—	0,00	382,65	$1s^2\,^1S - 6p\,^1P^\circ$	0—1
32,188	—	0,00	385,17	$1s^2\,^1S - 7p\,^1P^\circ$	0—1
32,064	—	0,00	386,66	$1s^2\,^1S - 8p\,^1P^\circ$	0—1

C VI, ground state $1s\,^2S_{1/2}$

Ionization potential 3951950 cm^{-1}; 489,946 eV

λ, Å	I	E_H, eV	E_B, eV	Transition	J
33,734	—	0,00	367,44	$1s\,^2S - 2p\,^2P^\circ$	$1/2 - 1/2$
28,464	—	0,00	435,51	$1s\,^2S - 3p\,^2P^\circ$	$1/2 - 1/2$
26,988	—	0,00	459,33	$1s\,^2S - 4p\,^2P^\circ$	$1/2 - 1/2$

NITROGEN, Z = 7

N I, ground state $1s^2\,2s^2\,2p^3\,{}^4S^{\circ}_{3/2}$

Ionization potential 117345 cm^{-1}; 14,548 eV

λ, Å	I	E_H, eV	E_B, eV	Transition	J
18751,01	2	13,02	13,68	$3d\,^4D-4f\,D(3)^\circ$	$7/2-7/2,\ 5/2$
18670,00	4	13,02	13,68	$3d\,^4D-4f\,D(3)^\circ$	$5/2-7/2,\ 5/2$
18658,16	32	13,04	13,70	$3d\,^2D-4f\,F(4)^\circ$	$5/2-7/2$
18630,19	13	13,03	13,70	$3d\,^2D-4f\,F(2)^\circ$	$3/2-5/2,\ 3/2$
18587,24	13	13,03	13,70	$3d\,^2D-4f\,F(3)^\circ$	$3/2-5/2$
18566,75	4	13,02	13,69	$3d\,^4D-4f\,G(4)^\circ$	$7/2-9/2,\ 7/2$
18251,58	11	13,02	13,70	$3d\,^4D-4f\,F(2)^\circ$	$5/2-5/2,\ 3/2$
18240,54	13	13,02	13,70	$3d\,^4D-4f\,F(3)^\circ$	$7/2-7/2,\ 5/2$
18229,66	60	13,02	13,70	$3d\,^4D-4f\,F(4)^\circ$	$7/2-9/2,\ 7/2$
18210,56	32	13,02	13,70	$3d\,^4D-4f\,F(3)^\circ$ $3d\,^4D-4f\,F(2)^\circ$	$5/2-7/2,\ 5/2$ $3/2-5/2,\ 3/2$
18199,13	8	13,02	13,70	$3d\,^4D-4f\,F(4)^\circ$	$5/2-7/2$
18171,60	13	13,01	13,70	$3d\,^4D-4f\,F(2)^\circ$	$1/2-3/2$
18169,74	13	13,02	13,70	$3d\,^4D-4f\,F(3)^\circ$	$3/2-5/2$
18116,27	6	13,00	13,69	$3d\,^4P-4f\,D(2)^\circ$	$1/2-3/2$
18108,61	12	13,00	13,69	$3d\,^2F-4f\,G(4)^\circ$	$7/2-9/2,\ 7/2$
18097,71	10	13,00	13,68	$3d\,^4P-4f\,D(3)^\circ$	$5/2-7/2,\ 5/2$
18049,56	33	12,99	13,68	$3d\,^2F-4f\,D(3)^\circ$	$5/2-7/2,\ 5/2$
18029,95	80	13,00	13,69	$3d\,^4P-4f\,D(2)^\circ$	$3/2-5/2,\ 3/2$
17979,89	51	13,00	13,69	$3d\,^4P-4f\,G(3)^\circ$	$5/2-7/2,\ 5/2$
17936,55	17	13,00	13,69	$3d\,^4P-4f\,D(1)^\circ$	$1/2-$
17925,70	8	13,00	13,69	$3d\,^4P-4f\,G(4)^\circ$	$5/2-7/2$
17918,06	7	13,00	13,69	$3d\,^4P-4f\,D(2)^\circ$	$5/2-5/2,\ 3/2$
17878,26	0	13,00 12,99	13,70 13,69	$3d\,^2F-4f\,G(5)^\circ$ $3d\,^2F-4f\,G(4)^\circ$	$7/2-9/2$ $5/2-7/2$
17852,09	10	13,00	13,69	$3d\,^4P-4f\,D(1)^\circ$	$3/2-$
17787,27	8	13,00	13,70	$3d\,^2F-4f\,F(4)^\circ$	$7/2-9/2,\ 7/2$
17643,98	42	12,98	13,68	$3d\,^4F-4f\,D(3)^\circ$	$5/2-7/2,\ 5/2$
17636,83	8	12,98	13,69	$3d\,^4F-4f\,G(3)^\circ$	$7/2-7/2,\ 5/2$
17584,86	100	12,98 12,98	13,69 13,68	$3d\,^4F-4f\,G(4)^\circ$ $3d\,^4F-4f\,D(3)^\circ$	$7/2-9/2,\ 7/2$ $3/2-5/2$
17531,99	18	12,98	13,69	$3d\,^4F-4f\,G(3)^\circ$	$5/2-7/2,\ 5/2$
17516,58	125	12,99	13,70	$3d\,^4F-4f\,G(5)^\circ$	$9/2-11/2,\ 9/2$
17480,41	27	12,98	13,69	$3d\,^4F-4f\,G(4)^\circ$	$5/2-7/2$
17474,16	32	12,98	13,69	$3d\,^4F-4f\,G(3)^\circ$	$3/2-5/2$
17436,22	24	12,97	13,68	$3d\,^2P-4f\,D(3)^\circ$	$3/2-5/2$
17429,23	16	12,99	13,70	$3d\,^4F-4f\,F(4)^\circ$	$9/2-9/2,\ 7/2$
17385,13	12	12,97	13,69	$3d\,^2P-4f\,D(2)^\circ$	$1/2-3/2$
17367,55	23	12,98	13,70	$3d\,^4F-4f\,G(5)^\circ$	$7/2-9/2$
17326,86	16	12,97	13,69	$3d\,^2P-4f\,G(3)^\circ$	$3/2-5/2$
17291,81	6	12,98	13,70	$3d\,^4F-4f\,F(3)^\circ$	$7/2-7/2,\ 5/2$
17282,04	4	12,98	13,70	$3d\,^4F-4f\,F(4)^\circ$	$7/2-9/2,\ 7/2$
17269,17	11	12,97	13,69	$3d\,^2P-4f\,D(2)^\circ$	$3/2-5/2,\ 3/2$
17219,55	10	12,97	13,69	$3d\,^2P-4f\,D(1)^\circ$	$1/2-$
15771,10	22	12,12	12,91	$3p\,^2P^\circ-4s\,^2P$	$3/2-1/2$
15682,86	54	12,12	12,91	$3p\,^2P^\circ-4s\,^2P$	$1/2-1/2$
15582,27	200	12,12	12,92	$3p\,^2P^\circ-4s\,^2P$	$3/2-3/2$
15496,13	34	12,12	12,92	$3p\,^2P^\circ-4s\,^2P$	$1/2-3/2$
15146,66	75	10,93	11,75	$2p^4\,^4P-3p\,^4D^\circ$	$1/2-1/2$
15102,29	26	10,93	11,75	$2p^4\,^4P-3p\,^4D^\circ$	$3/2-1/2$

λ, Å	I	E_H, eV	E_B, eV	Transition	J
15094,96	75	10,93	11,75	$2p^4\,{}^4P - 3p\,{}^4D^\circ$	1/2 – 3/2
15050,88	80	10,93	11,75	$2p^4\,{}^4P - 3p\,{}^4D^\circ$	3/2 – 3/2
14966,60	180	10,93	11,76	$2p^4\,{}^4P - 3p\,{}^4D^\circ$	3/2 – 5/2
14952,07	15	10,92	11,75	$2p^4\,{}^4P - 3p\,{}^4D^\circ$	5/2 – 3/2
14868,87	100	10,92	11,76	$2p^4\,{}^4P - 3p\,{}^4D^\circ$	5/2 – 5/2
14757,07	300	10,92	11,76	$2p^4\,{}^4P - 3p\,{}^4D^\circ$	5/2 – 7/2
14681,04	55	12,12	12,97	$3p\,{}^2P^\circ - 3d\,{}^2P$	3/2 – 3/2
14604,64	27	12,12	12,97	$3p\,{}^2P^\circ - 3d\,{}^2P$	1/2 – 3/2
14598,42	17	12,12	12,97	$3p\,{}^2P^\circ - 3d\,{}^2P$	3/2 – 1/2
14548,55	20	11,99	12,85	$3p\,{}^4S^\circ - 4s\,{}^4P$	3/2 – 1/2
14522,81	36	12,12	12,97	$3p\,{}^2P^\circ - 3d\,{}^2P$	1/2 – 1/2
14454,62	29	11,99	12,85	$3p\,{}^4S^\circ - 4s\,{}^4P$	3/2 – 3/2
14313,21	80	11,99	12,86	$3p\,{}^4S^\circ - 4s\,{}^4P$	3/2 – 5/2
13686,03	14	10,93	11,84	$2p^4\,{}^4P - 3p\,{}^4P^\circ$	1/2 – 1/2
13668,60	65	12,12	13,03	$3p\,{}^2P^\circ - 3d\,{}^2D$	3/2 – 3/2
13651,63	60	10,93	11,84	$2p^4\,{}^4P - 3p\,{}^4P^\circ$	1/2 – 3/2
13649,74	58	10,93	11,84	$2p^4\,{}^4P - 3p\,{}^4P^\circ$	3/2 – 1/2
13624,18	350	12,12	13,04	$3p\,{}^2P^\circ - 3d\,{}^2D$	3/2 – 5/2
13615,56	35	10,93	11,84	$2p^4\,{}^4P - 3p\,{}^4P^\circ$	3/2 – 3/2
13602,27	190	12,12	13,03	$3p\,{}^2P^\circ - 3d\,{}^2D$	1/2 – 3/2
13588,55	115	12,00	12,91	$3p\,{}^2D^\circ - 4s\,{}^2P$	3/2 – 1/2
13587,73	200	12,01	12,92	$3p\,{}^2D^\circ - 4s\,{}^2P$	5/2 – 3/2
13581,33	1200	10,69	11,60	$3s\,{}^2P - 3p\,{}^2S^\circ$	3/2 – 1/2
13544,61	65	10,93	11,84	$2p^4\,{}^4P - 3p\,{}^4P^\circ$	3/2 – 5/2
13534,64	60	10,92	11,84	$2p^4\,{}^4P - 3p\,{}^4P^\circ$	5/2 – 3/2
13464,53	185	10,92	11,84	$2p^4\,{}^4P - 3p\,{}^4P^\circ$	5/2 – 5/2
13448,12	21	12,00	12,92	$3p\,{}^2D^\circ - 4s\,{}^2P$	3/2 – 3/2
13429,61	670	10,68	11,60	$3s\,{}^2P - 3p\,{}^2S^\circ$	1/2 – 1/2
12897,32	51	12,01	12,97	$3p\,{}^2D^\circ - 3d\,{}^2P$	5/2 – 3/2
12778,5	5	13,04	14,00	$3d\,{}^2D - 5f\,F(4)^\circ$	5/2 – 7/2
12771,51	15	12,00	12,97	$3p\,{}^2D^\circ - 3d\,{}^2P$	3/2 – 3/2
12730,68	35	12,01	12,98	$3p\,{}^2D^\circ - 3d\,{}^4P$	5/2 – 7/2
12708,89	30	12,00	12,97	$3p\,{}^2D^\circ - 3d\,{}^2P$	3/2 – 1/2
12662,16	27	12,00	12,98	$3p\,{}^2D^\circ - 3d\,{}^4F$	3/2 – 5/2
12581,00	27	12,01	12,99	$3p\,{}^2D^\circ - 3d\,{}^2F$	5/2 – 5/2
12578,8	3	13,02	14,00	$3d\,{}^4D - 5f\,F(3)^\circ$	7/2 – 7/2
12575,99	8	13,02	14,00	$3d\,{}^4D - 5f\,F(4)^\circ$	7/2 – 9/2
12564,4	4	13,02	14,00	$3d\,{}^4D - 5f\,F(3)^\circ$	5/2 – 7/2
12557,66	14	12,01	13,00	$3p\,{}^2D^\circ - 3d\,{}^4P$	5/2 – 5/2
		13,02	14,00	$3d\,{}^4D - 5f\,F(2)^\circ$	3/2 –
12469,62	1350	12,99	14,00	$3d\,{}^2F - 5f\,D(3)^\circ$	5/2 –
		12,01	13,00	$3p\,{}^2D^\circ - 3d\,{}^2F$	5/2 – 7/2
12464,2	5	13,00	14,00	$3d\,{}^4P - 5f\,D(2)^\circ$	3/2 – 5/2
12461,25	680	12,00	12,99	$3p\,{}^2D^\circ - 3d\,{}^2F$	3/2 – 5/2
12438,40	195	12,00	13,00	$3p\,{}^2D^\circ - 3d\,{}^4P$	3/2 – 5/2
12428,81	6	13,00	13,99	$3d\,{}^4P - 5f\,G(3)^\circ$	5/2 – 7/2
12404,27	98	11,99	12,99	$3p\,{}^4S^\circ - 3d\,{}^2F$	3/2 – 5/2
12391,9	5	12,99	13,99	$3d\,{}^2F - 5f\,G(4)^\circ$	5/2 – 7/2
12384,83	12	13,00	14,00	$3d\,{}^2F - 5f\,G(5)^\circ$	7/2 – 9/2
		12,00	13,00	$3p\,{}^2D^\circ - 3d\,{}^4P$	3/2 – 3/2
12381,65	375	11,99	13,00	$3p\,{}^4S^\circ - 3d\,{}^4P$	3/2 – 5/2
12328,76	350	11,99	13,00	$3p\,{}^4S^\circ - 3d\,{}^4P$	3/2 – 3/2
12298,55	120	11,84	12,85	$3p\,{}^4P^\circ - 4s\,{}^4P$	3/2 – 1/2

λ, Å	I	E_H, eV	E_B, eV	Transition	J
12288,97	260	11,84	12,85	$3p\,^4P^\circ - 4s\,^4P$	$5/2 - 3/2$
		11,99	13,00	$3p\,^4S^\circ - 3d\,^4P$	$3/2 - 1/2$
12270,80	20	11,84	12,85	$3p\,^4P^\circ - 4s\,^4P$	$1/2 - 1/2$
12261,28	27	12,01	13,02	$3p\,^2D^\circ - 3d\,^4D$	$5/2 - 7/2$
12250,11	11	12,98	13,99	$3d\,^4F - 5f\,G(4)^\circ$	$7/2 - 9/2$
12231,32	75	11,84	12,85	$3p\,^4P^\circ - 4s\,^4P$	$3/2 - 3/2$
12210,17	12	12,99	14,00	$3d\,^4F - 5f\,G(5)^\circ$	$9/2 - 11/2$
12203,93	150	11,84	12,85	$3p\,^4P^\circ - 4s\,^4P$	$1/2 - 3/2$
12186,32	480	11,84	12,86	$3p\,^4P^\circ - 4s\,^4P$	$5/2 - 5/2$
12142,16	12	11,99	13,01	$3p\,^4S^\circ - 3d\,^4D$	$3/2 - 1/2$
12129,97	170	11,84	12,86	$3p\,^4P^\circ - 4s\,^4P$	$3/2 - 5/2$
12124,60	35	11,99	13,02	$3p\,^4S^\circ - 3d\,^4D$	$3/2 - 3/2$
12109,30	25	12,01	13,03	$3p\,^2D^\circ - 3d\,^2D$	$5/2 - 3/2$
12106,59	45	11,99	13,02	$3p\,^4S^\circ - 3d\,^4D$	$3/2 - 5/2$
12074,51	230	12,01	13,04	$3p\,^2D^\circ - 3d\,^2D$	$5/2 - 5/2$
11998,36	110	12,00	13,03	$3p\,^2D^\circ - 3d\,^2D$	$3/2 - 3/2$
11651,45	2	10,93	11,99	$2p^4\,^4P - 3p\,^4S^\circ$	$1/2 - 3/2$
11625,173	3	10,93	11,99	$2p^4\,^4P - 3p\,^4S^\circ$	$3/2 - 3/2$
11566,114	4	10,92	11,99	$2p^4\,^4P - 3p\,^4S^\circ$	$5/2 - 3/2$
11323,169	3	11,75	12,85	$3p\,^4D^\circ - 4s\,^4P$	$3/2 - 1/2$
11313,891	4	11,76	12,85	$3p\,^4D^\circ - 4s\,^4P$	$5/2 - 3/2$
11294,238	2	11,75	12,85	$3p\,^4D^\circ - 4s\,^4P$	$1/2 - 1/2$
11291,657	5	11,76	12,86	$3p\,^4D^\circ - 4s\,^4P$	$7/2 - 5/2$
11266,198	3	11,75	12,85	$3p\,^4D^\circ - 4s\,^4P$	$3/2 - 3/2$
11237,582	2	11,75	12,85	$3p\,^4D^\circ - 4s\,^4P$	$1/2 - 3/2$
11227,076	3	11,76	12,86	$3p\,^4D^\circ - 4s\,^4P$	$5/2 - 5/2$
11180,114	1	11,75	12,86	$3p\,^4D^\circ - 4s\,^4P$	$3/2 - 5/2$
10884,60	2	11,84	12,98	$3p\,^4P^\circ - 3d\,^4F$	$5/2 - 7/2$
10879,19	1	11,84	12,98	$3p\,^4P^\circ - 3d\,^4F$	$3/2 - 5/2$
10774,993	3	11,84	12,99	$3p\,^4P^\circ - 3d\,^2F$	$5/2 - 5/2$
10757,888	7	11,84	13,01	$3p\,^4P^\circ - 3d\,^4P$	$5/2 - 5/2$
10730,510	4	11,84	12,99	$3p\,^4P^\circ - 3d\,^2F$	$3/2 - 5/2$
10717,954	6	11,84	13,00	$3p\,^4P^\circ - 3d\,^4P$	$5/2 - 3/2$
10713,550	8	11,84	13,01	$3p\,^4P^\circ - 3d\,^4P$	$3/2 - 5/2$
10693,167	3	11,85	13,00	$3p\,^4P^\circ - 3d\,^2F$	$5/2 - 7/2$
10653,034	8	11,84	13,00	$3p\,^4P^\circ - 3d\,^4P$	$1/2 - 3/2$
10643,981	6	11,84	13,00	$3p\,^4P^\circ - 3d\,^4P$	$3/2 - 1/2$
10623,177	5	11,84	13,00	$3p\,^4P^\circ - 3d\,^4P$	$1/2 - 1/2$
10596,958	6	13,72	14,98	$3p'\,^2F - 3d'\,^2G$	$7/2 - 9/2$
10591,905	5	13,72	14,98	$3p'\,^2F - 3d'\,^2G$	$5/2 - 7/2$
10563,339	5	11,84	13,02	$3p\,^4P^\circ - 3d\,^4D$	$5/2 - 3/2$
10549,635	8	11,84	13,02	$3p\,^4P^\circ - 3d\,^4D$	$5/2 - 5/2$
10539,554	10	11,84	13,02	$3p\,^4P^\circ - 3d\,^4D$	$5/2 - 7/2$
10533,775	5	11,84	13,01	$3p\,^4P^\circ - 3d\,^4D$	$3/2 - 1/2$
10520,574	8	11,84	13,02	$3p\,^4P^\circ - 3d\,^4D$	$3/2 - 3/2$
10513,403	7	11,84	13,02	$3p\,^4P^\circ - 3d\,^4D$	$1/2 - 1/2$
10506,998	8	11,84	13,02	$3p\,^4P^\circ - 3d\,^4D$	$3/2 - 5/2$
10500,266	6	11,84	13,02	$3p\,^4P^\circ - 3d\,^4D$	$1/2 - 3/2$
10199,98	2	11,76	12,98	$3p\,^4D^\circ - 3d\,^4F$	$7/2 - 5/2$
10166,79	3	11,76	12,98	$3p\,^4D^\circ - 3d\,^4F$	$5/2 - 3/2$
10164,849	7	11,76	12,98	$3p\,^4D^\circ - 3d\,^4F$	$7/2 - 7/2$
10147,274	8	11,76	12,98	$3p\,^4D^\circ - 3d\,^4F$	$5/2 - 5/2$
10128,285	7	11,75	12,98	$3p\,^4D^\circ - 3d\,^4F$	$3/2 - 3/2$
10114,644	13	11,76	12,99	$3p\,^4D^\circ - 3d\,^4F$	$7/2 - 9/2$
10112,484	12	11,76	12,98	$3p\,^4D^\circ - 3d\,^4F$	$5/2 - 7/2$
10108,895	11	11,75	12,98	$3p\,^4D^\circ - 3d\,^4F$	$3/2 - 5/2$
10105,147	10	11,75	12,98	$3p\,^4D^\circ - 3d\,^4F$	$1/2 - 3/2$
10054,259	4	11,76	13,01	$3p\,^4D^\circ - 3d\,^4P$	$7/2 - 5/2$
10017,822	5	11,76	12,99	$3p\,^4D^\circ - 3d\,^2F$	$5/2 - 5/2$

λ, Å	I	E_H, eV	E_B, eV	Transition	J
10003,055	5	11,76	13,01	$3p\ ^4D°—3d\ ^4P$	$^5/_2—^5/_2$
9997,750	4	11,76	13,00	$3p\ ^4D°—3d\ ^2F$	$^7/_2—^7/_2$
9980,424	3	11,75	12,99	$3p\ ^4D°—3d\ ^2F$	$^3/_2—^5/_2$
9965,736	3	11,75	13,01	$3p\ ^4D°—3d\ ^4P$	$^3/_2—^5/_2$
9947,066	4	11,76	13,00	$3p\ ^4D°—3d\ ^2F$	$^5/_2—^7/_2$
9931,474	5	11,75	13,00	$3p\ ^4D°—3d\ ^4P$	$^3/_2—^3/_2$
9909,220	2	11,75	13,00	$3p\ ^4D°—3d\ ^4P$	$^1/_2—^3/_2$
9905,54	0	11,75	13,00	$3p\ ^4D°—3d\ ^4P$	$^3/_2—^1/_2$
9883,369	3	11,75	13,00	$3p\ ^4D°—3d\ ^4P$	$^1/_2—^1/_2$
9872,159	6	11,76	13,02	$3p\ ^4D°—3d\ ^4D$	$^7/_2—^5/_2$
9863,332	9	11,76	13,02	$3p\ ^4D°—3d\ ^4D$	$^7/_2—^7/_2$
9834,623	6	11,76	13,02	$3p\ ^4D°—3d\ ^4D$	$^5/_2—^3/_2$
9822,754	7	11,76	13,02	$3p\ ^4D°—3d\ ^4D$	$^5/_2—^5/_2$
9814,026	4	11,76	13,02	$3p\ ^4D°—3d\ ^4D$	$^5/_2—^7/_2$
9810,018	5	11,75	13,01	$3p\ ^4D°—3d\ ^4D$	$^3/_2—^1/_2$
9798,565	5	11,75	13,02	$3p\ ^4D°—3d\ ^4D$	$^3/_2—^3/_2$
9788,298	4	11,75	13,01	$3p\ ^4D°—3d\ ^4D$	$^1/_2—^1/_2$
9786,788	4	11,75	13,02	$3p\ ^4D°—3d\ ^4D$	$^3/_2—^5/_2$
9776,904	4	11,75	13,02	$3p\ ^4D°—3d\ ^4D$	$^1/_2—^3/_2$
9694,01	1	11,76	13,04	$3p\ ^4D°—3d\ ^2D$	$^5/_2—^5/_2$
9464,23	1	11,60	12,91	$3p\ ^2S°—4s\ ^2P$	$^1/_2—^1/_2$
9460,676	10	10,69	12,00	$3s\ ^2P—3p\ ^2D°$	$^3/_2—^3/_2$
9392,789	15	10,69	12,01	$3s\ ^2P—3p\ ^2D°$	$^3/_2—^5/_2$
9386,805	14	10,68	12,00	$3s\ ^2P—3p\ ^2D°$	$^1/_2—^3/_2$
9208,001	8	12,36	13,70	$3s'\ ^2D—3p'\ ^2D°$	$^3/_2—^3/_2$
9207,59	3	12,36	13,70	$3s'\ ^2D—3p'\ ^2D°$	$^5/_2—^3/_2$
9187,84	3	12,36	13,70	$3s'\ ^2D—3p'\ ^2D°$	$^3/_2—^5/_2$
9187,449	9	12,36	13,70	$3s'\ ^2D—3p'\ ^2D°$	$^5/_2—^5/_2$
9060,472	10	11,60	12,97	$3p\ ^2S°—3d\ ^2P$	$^1/_2—^3/_2$
9049,890	12	12,36	13,72	$3s'\ ^2D—3p'\ ^2F°$	$^3/_2—^5/_2$

λ, Å	I	E_H, eV	E_B, eV	Transition	J
9049,47	5	12,36	13,72	$3s'\,{}^2D$—$3p'\,{}^2F^\circ$	$^5/_2$—$^5/_2$
9045,878	13	12,36	13,72	$3s'\,{}^2D$—$3p'\,{}^2F^\circ$	$^5/_2$—$^7/_2$
9028,918	9	11,60	12,98	$3p\,{}^2S^\circ$—$3d\,{}^2P$	$^1/_2$—$^1/_2$
8747,357	9	10,34	11,75	$3s\,{}^4P$—$3p\,{}^4D^\circ$	$^5/_2$—$^3/_2$
8728,909	10	10,33	11,75	$3s\,{}^4P$—$3p\,{}^4D^\circ$	$^3/_2$—$^1/_2$
8718,841	14	10,34	11,76	$3s\,{}^4P$—$3p\,{}^4D^\circ$	$^5/_2$—$^5/_2$
8711,708	15	10,33	11,75	$3s\,{}^4P$—$3p\,{}^4D^\circ$	$^3/_2$—$^3/_2$
8703,255	14	10,33	11,75	$3s\,{}^4P$—$3p\,{}^4D^\circ$	$^1/_2$—$^1/_2$
8686,161	14	10,33	11,75	$3s\,{}^4P$—$3p\,{}^4D^\circ$	$^1/_2$—$^3/_2$
8683,400	16	10,33	11,76	$3s\,{}^4P$—$3p\,{}^4D^\circ$	$^3/_2$—$^5/_2$
8680,270	17	10,34	11,76	$3s\,{}^4P$—$3p\,{}^4D^\circ$	$^5/_2$—$^7/_2$
8655,869	14	10,69	12,12	$3s\,{}^2P$—$3p\,{}^2P^\circ$	$^3/_2$—$^1/_2$
8629,238	16	10,69	12,12	$3s\,{}^2P$—$3p\,{}^2P^\circ$	$^3/_2$—$^3/_2$
8594,005	15	10,68	12,12	$3s\,{}^2P$—$3p\,{}^2P^\circ$	$^1/_2$—$^1/_2$
8567,735	14	10,68	12,12	$3s\,{}^2P$—$3p\,{}^2P^\circ$	$^1/_2$—$^3/_2$
8242,374	13	10,34	11,84	$3s\,{}^4P$—$3p\,{}^4P^\circ$	$^5/_2$—$^3/_2$
8223,121	13	10,33	11,84	$3s\,{}^4P$—$3p\,{}^4P^\circ$	$^3/_2$—$^1/_2$
8216,317	15	10,34	11,84	$3s\,{}^4P$—$3p\,{}^4P^\circ$	$^5/_2$—$^5/_2$
8210,708	11	10,33	11,84	$3s\,{}^4P$—$3p\,{}^4P^\circ$	$^3/_2$—$^3/_2$
8201,766	7	12,36	13,87	$3s'\,{}^2D$—$5p\,{}^2D^\circ$	$^3/_2$—$^3/_2$
8201,43	2	12,36	13,87	$3s'\,{}^2D$—$5p\,{}^2D^\circ$	$^5/_2$—$^3/_2$
8200,357	10	10,33	11,84	$3s\,{}^4P$—$3p\,{}^4P^\circ$	$^1/_2$—$^1/_2$
8188,005	13	10,33	11,84	$3s\,{}^4P$—$3p\,{}^4P^\circ$	$^1/_2$—$^3/_2$
8184,852	13	10,33	11,84	$3s\,{}^4P$—$3p\,{}^4P^\circ$	$^3/_2$—$^5/_2$
8174,50	1	12,12	13,64	$3p\,{}^2P^\circ$—$5s\,{}^2P$	$^3/_2$—$^1/_2$
8166,51	2	12,36	13,87	$3s'\,{}^2D$—$5p\,{}^2D^\circ$	$^3/_2$—$^5/_2$
8166,235	8	12,36	13,87	$3s'\,{}^2D$—$5p\,{}^2D^\circ$	$^5/_2$—$^5/_2$
8150,66	1	12,12	13,64	$3p\,{}^2P^\circ$—$5s\,{}^2P$	$^1/_2$—$^1/_2$
8129,170	3	12,12	13,65	$3p\,{}^2P^\circ$—$5s\,{}^2P$	$^3/_2$—$^3/_2$
8105,631	2	12,12	13,65	$3p\,{}^2P^\circ$—$5s\,{}^2P$	$^1/_2$—$^3/_2$
7915,419	7	12,36	13,92	$3s'\,{}^2D$—$3p'\,{}^2P^\circ$	$^3/_2$—$^1/_2$
7899,27	3	12,36	13,92	$3s'\,{}^2D$—$3p'\,{}^2P^\circ$	$^3/_2$—$^3/_2$
7898,985	8	12,36	13,92	$3s'\,{}^2D$—$3p'\,{}^2P^\circ$	$^5/_2$—$^3/_2$
7468,309	16	10,34	11,99	$3s\,{}^4P$—$3p\,{}^4S^\circ$	$^5/_2$—$^3/_2$
7442,299	15	10,33	11,99	$3s\,{}^4P$—$3p\,{}^4S^\circ$	$^3/_2$—$^3/_2$
7423,639	14	10,33	11,99	$3s\,{}^4P$—$3p\,{}^4S^\circ$	$^1/_2$—$^3/_2$
6982,02	00	11,84	13,61	$3p\,{}^4P^\circ$—$5s\,{}^4P$	$^3/_2$—$^1/_2$
6979,10	1	11,84	13,62	$3p\,{}^4P^\circ$—$5s\,{}^4P$	$^5/_2$—$^3/_2$
6951,50	1	11,84	13,62	$3p\,{}^4P^\circ$—$5s\,{}^4P$	$^1/_2$—$^3/_2$
6945,22	4	11,84	13,63	$3p\,{}^4P^\circ$—$5s\,{}^4P$	$^5/_2$—$^5/_2$
6926,90	1	11,84	13,63	$3p\,{}^4P^\circ$—$5s\,{}^4P$	$^3/_2$—$^5/_2$
6874,30	1	—	—	—	—
6793,82	00	11,84	13,67	$3p\,{}^4P^\circ$—$4d\,{}^4F$	$^5/_2$—$^7/_2$
6758,60	4	11,84	13,67	$3p\,{}^4P^\circ$—$4d\,{}^4D$	$^3/_2$—$^5/_2$
6752,40	4	11,84	13,68	$3p\,{}^4P^\circ$—$4d\,{}^4D$	$^5/_2$—$^7/_2$
6741,29	3	11,84	13,68	$3p\,{}^4P^\circ$—$4d\,{}^4P$	$^3/_2$—$^1/_2$
6733,48	6	11,84	13,68	$3p\,{}^4P^\circ$—$4d\,{}^4P$	$^1/_2$—$^1/_2$
6723,12	9	11,84	13,69	$3p\,{}^4P^\circ$—$4d\,{}^4P$	$^5/_2$—$^5/_2$
6713,12	1	—	—	—	—
6708,81	4	—	—	—	—
6706,20	4	11,84	13,69	$3p\,{}^4P^\circ$—$4d\,{}^4P$	$^3/_2$—$^5/_2$
6666,75	0	12,12	13,99	$3p\,{}^2P^\circ$—$5d\,{}^2P$	$^3/_2$—$^3/_2$
6656,510	1	11,75	13,61	$3p\,{}^4D^\circ$—$5s\,{}^4P$	$^3/_2$—$^1/_2$
6653,458	5	11,76	13,62	$3p\,{}^4D^\circ$—$5s\,{}^4P$	$^5/_2$—$^3/_2$
6646,510	2	11,75	13,61	$3p\,{}^4D^\circ$—$5s\,{}^4P$	$^1/_2$—$^1/_2$
6644,963	9	11,76	13,63	$3p\,{}^4D^\circ$—$5s\,{}^4P$	$^7/_2$—$^5/_2$
6636,938	4	11,75	13,62	$3p\,{}^4D^\circ$—$5s\,{}^4P$	$/_2$—$^3/_2$

λ, Å	I	E_H, eV	E_B, eV	Transition	J
6627,02	0	11,75	13,62	3p $^4D^\circ$—5s ^4P	$1/2$—$3/2$
6622,543	3	11,76	13,63	3p $^4D^\circ$—5s ^4P	$5/2$—$5/2$
6606,77	00	11,75	13,63	3p $^4D^\circ$—5s ^4P	$3/2$—$5/2$
6506,45	0	11,76	13,67	3p $^4D^\circ$—4d ^4F	$7/2$—$7/2$
6499,52	3	11,76	13,66	3p $^4D^\circ$—4d ^4F	$5/2$—$5/2$
6491,28	3	11,75	13,66	3p $^4D^\circ$—4d ^4F	$3/2$—$3/2$
6484,88	9	11,76	13,67	3p $^4D^\circ$—4d ^4F	$5/2$—$7/2$
6483,75	3	11,75	13,66	3p $^4D^\circ$—4d ^4F	$3/2$—$5/2$
6482,74	9	11,76	13,68	3p $^4D^\circ$—4d ^4F	$7/2$—$9/2$
6481,73	2	11,75	13,66	3p $^4D^\circ$—4d ^4F	$1/2$—$3/2$
6480,50	0	11,75	13,66	3p $^4D^\circ$—4d ^4D	$3/2$—$1/2$
6471,03	1	11,75	13,66	3p $^4D^\circ$—4d ^4D	$1/2$—$1/2$
6468,32	4	11,76	13,67	{ 3p $^4D^\circ$—4d ^4D 3p $^4D^\circ$—4d ^4D	$7/2$—$7/2$ $5/2$—$5/2$
6457,93	3	11,75	13,67	3p $^4D^\circ$—4d ^4D	$3/2$—$3/2$
6452,75	1	11,75	13,67	3p $^4D^\circ$—4d ^4D	$3/2$—$5/2$
6448,49	0	11,75	13,67	3p $^4D^\circ$—4d ^4D	$1/2$—$1/2$
6441,70	5	11,76	13,69	3p $^4D^\circ$—4d ^4P	$7/2$—$5/2$
6440,95	3	—	—	—	—
6437,01	4	11,75	13,68	3p $^4D^\circ$—4d ^4P	$3/2$—$1/2$
6428,05	00	11,75	13,68	3p $^4D^\circ$—4d ^4P	$1/2$—$1/2$
6422,93	3	—	—	—	—
6420,47	3	11,76	13,69	3p $^4D^\circ$—4d ^4P	$5/2$—$5/2$
6417,05	2	—	—	—	—
6321,70	00	11,99	13,93	3p $^4S^\circ$—6s ^4P	$3/2$—$1/2$
6303,68	0	11,99	13,93	3p $^4S^\circ$—6s ^4P	$3/2$—$3/2$
6285,78	1	—	—	—	—
6275,43	1	11,99	13,94	3p $^4S^\circ$—6s ^4P	$3/2$—$5/2$
6272,83	1	12,01	13,99	3p $^2D^\circ$—5d ^2P	$5/2$—$3/2$
6075,83	3	—	—	—	—
6017,70	2	—	—	—	—
6015,40	1	—	—	—	—
6008,48	10	11,60	13,66	3p $^2S^\circ$—4d ^2P	$1/2$—$3/2$
5999,47	6	11,60	13,67	3p $^2S^\circ$—4d ^2P	$1/2$—$1/2$
5856,23	1	11,84	13,93	3p $^4P^\circ$—6s ^4P	$3/2$—$1/2$
5854,16	2	11,84	13,93	3p $^4P^\circ$—6s ^4P	$5/2$—$3/2$
5841,01	2	11,84	13,93	3p $^4P^\circ$—6s ^4P	$3/2$—$3/2$
5834,71	1	11,84	13,93	3p $^4P^\circ$—6s ^4P	$1/2$—$3/2$
5829,53	6	11,84	13,94	3p $^4P^\circ$—6s ^4P	$5/2$—$5/2$
5816,48	2	11,84	13,94	3p ^4P —6s ^4P	$3/2$—$5/2$
5793,51	1	—	—	—	—
5752,64	4	11,84	14,00	3p $^4P^\circ$—5d ^4P	$5/2$—$5/2$
5747,36	2	11,99	14,15	3p $^4S^\circ$—7s ^4P	$3/2$—$5/2$
5740,65	2	—	—	—	—
5735,63	1	—	—	—	—
5667,04	1				
5625,43	2	11,75	13,93	3p $^4D^\circ$—6s ^4P	$3/2$—$1/2$
5623,20	4	11,76	13,93	3p $^4D^\circ$—6s ^4P	$5/2$—$3/2$
5618,18	1	11,75	13,93	3p $^4D^\circ$—6s ^4P	$1/2$—$1/2$
5616,54	5	11,76	13,94	3p $^4D^\circ$—6s ^4P	$7/2$—$5/2$
5611,36	1	11,75	13,93	3p $^4D^\circ$—6s ^4P	$3/2$—$3/2$
5604,28	0	11,75	13,93	3p $^4D^\circ$—6s ^4P	$1/2$—$3/2$
5600,54	0	11,76	13,94	3p $^4D^\circ$—6s ^4P	$5/2$—$5/2$
5567,63	1	—	—	—	—
5564,37	9	11,76	13,99	3p $^4D^\circ$—5d ^4F	$5/2$—$7/2$
5563,84	3	—	—	—	—
5560,37	9	11,76	14,00	3p $^4D^\circ$—5d ^4F	$7/2$—$9/2$
5557,44	2	—	—	—	—
5545,11	3	11,76	14,00	3p $^4D^\circ$—5d ^4P	$7/2$—$5/2$

λ, Å	I	$E_н$, eV	$E_в$, eV	Transition	J
5540,36	1	—	—	—	—
5535,37	1	—	—	—	—
5411,881	5	12,12	14,42	$3p\ ^2P^\circ$–$3s\ ^2S$	$3/2 - 1/2$
5401,450	4	12,12	14,42	$3p\ ^2P^\circ$–$3s\ ^2S$	$1/2 - 1/2$
5378,45	0	10,93	13,23	$2p^4\ ^4P$—$4p\ ^4D^\circ$	$^1/_2$—$^1/_2$
5372,66	3	10,93	13,24	$2p^4\ ^4P$—$4p\ ^4D^\circ$	$^1/_2$—$^3/_2$
5371,10	1	11,84	14,15	$3p\ ^4P$—$7s\ ^4P$	$^5/_2$—$^5/_2$
5367,27	1	10,93	13,24	$2p^4\ ^4P$—$4p\ ^4D^\circ$	$^3/_2$—$^3/_2$
5356,77	5	10,93	13,24	$2p^4\ ^4P$—$4p\ ^4D^\circ$	$^3/_2$—$^5/_2$
5344,23	00	10,92	13,24	$2p^4\ ^4P$—$4p\ ^4D^\circ$	$^5/_2$—$^5/_2$
5334,42	1	11,84	14,17	$3p\ ^4P^\circ$—$6d\ ^4P$	$^5/_2$—$^5/_2$
5328,70	5	10,92	13,25	$2p^4\ ^4P$—$4p\ ^4D^\circ$	$^5/_2$—$^7/_2$
5310,52	1	10,93	13,26	$2p^4\ ^4P$—$4p\ ^4P^\circ$	$^1/_2$—$^3/_2$
5309,48	1	10,93	13,26	$2p^4\ ^4P$—$4p\ ^4P^\circ$	$^3/_2$—$^1/_2$
5304,9	1	10,93	13,26	$2p^4\ ^4P$—$4p\ ^4P^\circ$	$^3/_2$—$^3/_2$
5292,75	0	10,94 10,93	13,26 13,27	$2p^4\ ^4P$—$4p\ ^4P^\circ$ $2p^4\ ^4P$—$4p\ ^4P^\circ$	$^5/_2$—$^3/_2$ $^3/_2$—$^5/_2$
5281,18	3	10,92	13,27	$2p^4\ ^4P$—$4p\ ^4P^\circ$	$^5/_2$—$^5/_2$
5201,71	2	11,60	13,99	$3p\ ^2S^\circ$—$5d\ ^2P$	$^1/_2$—$^3/_2$
5189,51	1	11,76	14,15	$3p\ ^4D$—$7s\ ^4P$?	$^7/_2$—$^5/_2$
5187,1	1	10,93	13,32	$2p^4\ ^4P$—$4p\ ^4S^\circ$	$^1/_2$—$^3/_2$
5181,47	0	10,93	13,32	$2p^4\ ^4P$—$4p\ ^4S^\circ$	$^3/_2$—$^3/_2$
5169,45	1	10,92	13,32	$2p^4\ ^4P$—$4p\ ^4S^\circ$	$^5/_2$—$^3/_2$
5162,78	1	—	—	—	—
4935,03	10	10,69	13,20	$3s\ ^2P$—$4p\ ^2S^\circ$	$^3/_2$—$^1/_2$
4914,90	5	10,68	13,20	$3s\ ^2P$—$4p\ ^2S^\circ$	$^1/_2$—$^1/_2$
4886,30	2	—	—	—	—
4881,79	1	—	—	—	—
4847,38	2	—	—	—	—
4837,93	1	—	—	—	—
4831,16	1	—	—	—	—
4753,13	2	—	—	—	—
4750,26	2	—	—	—	—
4744,04	3	—	—	—	—
4742,90	2	—	—	—	—
4731,22	1	—	—	—	—
4685,74	3	—	—	—	—
4669,77	3	—	—	—	—
4660,05	2	—	—	—	—
4657,72	1	—	—	—	—
4656,65	1	—	—	—	—
4651,08	1	—	—	—	—
4625,61	1	—	—	—	—
4554,21	1	—	—	—	—
4553,38	1	—	—	—	—
4503,53	1	—	—	—	—
4502,27	2	—	—	—	—
4497,45	1	—	—	—	—
4494,67	5	—	—	—	—
4492,40	7	—	—	—	—
4358,27	10	—	—	—	—
4343,41	1	—	—	—	—
4336,48	5	—	—	—	—
4321,99	1	—	—	—	—
4317,70	5	—	—	—	—
4313,11	4	—	—	—	—
4305,46	6	—	—	—	—
4284,92	2	—	—	—	—
4282,20	1	—	—	—	—
4281,39	2	—	—	—	—

λ, Å	I	E_H, eV	E_B, eV	Transition	J
4254,7	4	10,33	13,24	$3s\,^4P$—$4p\,^4D^\circ$	$^3/_2$—$^5/_2$
		10,33	13,24	$3s\,^4P$—$4p\,^4D^\circ$	$^1/_2$—$^3/_2$
4253,28	4	10,34	13,24	$3s\,^4P$—$4p\,^4D^\circ$	$^5/_2$—$^7/_2$
4230,35	4	10,34	13,26	$3s\,^4P$—$4p\,^4P^\circ$	$^5/_2$—$^3/_2$
4229,59	2	—	—	—	—
4224,74	4	10,33	13,26	$3s\,^4P$—$4p\,^4P^\circ$	$^3/_2$—$^1/_2$
4223,04	5	10,34	13,27	$3s\,^4P$—$4p\,^4P^\circ$	$^5/_2$—$^5/_2$
4220,79	2	—	—	—	—
4215,92	2	10,33	13,26	$3s\,^4P$—$4p\,^4P^\circ$	$^1/_2$—$^3/_2$
4214,73	5	10,33	13,27	$3s\,^4P$—$4p\,^4P^\circ$	$^3/_2$—$^5/_2$
4209,05	1	—	—	—	—
4206,29	1	—	—	—	—
4205,65	2	—	—	—	—
4193,49	3	—	—	—	—
4187,06	1	—	—	—	—
4166,64	1	—	—	—	—
4151,46	12	10,34	13,32	$3s\,^4P$—$4p\,^4S^\circ$	$^5/_2$—$^3/_2$
4145,78	2	—	—	—	—
4143,4	—	10,33	13,32	$3s\,^4P$—$4p\,^4S^\circ$	$^3/_2$—$^3/_2$
4137,63	7	10,33	13,32	$3s\,^4P$—$4p\,^4S^\circ$	$^1/_2$—$^3/_2$
4129,16	1	—	—	—	—
4113,972	6	10,69	13,70	$3s\,^2P$—$3p'\,^2D^\circ$	$^3/_2$—$^3/_2$
4109,959	12	10,69	13,70	$3s\,^2P$—$3p'\,^2D^\circ$	$^3/_2$—$^5/_2$
4102,18	2	—	—	—	—
4099,951	9	10,68	13,70	$3s\,^2P$—$3p'\,^2D^\circ$	$^1/_2$—$^3/_2$
4037,35	1	—	—	—	—
4033,64	1	—	—	—	—
4010,99	2	—	—	—	—
4001,65	1	—	—	—	—
3999,98	4	—	—		—
3994,86	3	—	—	—	—
3969,95	1	—	—	—	—
3957,20	3	—	—	—	—
3952,21	3	—	—	—	—
3869,10	4	—	—	—	—
3834,84	2	—	—	—	—
3834,24	4	10,69	13,92	$3s\,^2P$—$3p'\,^2P^\circ$	$^3/_2$—$^1/_2$
3830,39	9	10,69	13,92	$3s\,^2P$—$3p'\,^2P^\circ$	$^3/_2$—$^3/_2$
3822,07	6	10,68	13,92	$3s\,^2P$—$3p'\,^2P^\circ$	$^1/_2$—$^1/_2$
3818,27	2	10,68	13,92	$3s\,^2P$—$3p'\,^2P^\circ$	$^1/_2$—$^3/_2$
3687,88	2	—	—	—	—
3681,10	3	—	—	—	—
3650,19	5	—	—	—	—
3545,62	2	—	—	—	—
3532,65	4	—	—	—	—
3437,14	4	—	—	—	—
1889,056	2	—	—	—	—
1873,217	1	—	—	—	—
1846,399	6	—	—	—	—
1836,739	4	3,57	10,33	$2p^3\,^2P^\circ$—$3s\,^4P$	$^3/_2$, $^1/_2$—$^1/_2$
1757,223	1	—	—	—	—
1750,079	2	—	—	—	—
1745,249	30	3,57	10,68	$2p^3\,^2P^\circ$—$3s\,^2P$	$^3/_2$, $^1/_2$—$^1/_2$
1742,724	10	3,57	10,69	$2p^3\,^2P^\circ$—$3s\,^2P$	$^3/_2$, $^1/_2$—$^3/_2$
1729,481	1	—	—	—	—
1728,170	3	—	—	—	—
1721,746	3	—	—	—	—
1677,906	4	—	—	—	—
1671,020	1	—	—	—	—

λ, Å	I	E_H, eV	E_B, eV	Transition	J
1647,493	3	—	—	—	—
1640,437	2	—	—	—	—
1602,973	1	—	—	—	—
1597,181	2	—	—	—	—
1592,867	3	—	—	—	—
1577,131	1	—	—	—	—
1542,499	1	—	—	—	—
1494,668	60	2,38	10,68	$2p^3\ ^2D^\circ—3s\ ^2P$	$^3/_2—^1/_2$
1492,817	30	2,38	10,69	$2p^3\ ^2D^\circ—3s\ ^2P$	$^3/_2—^3/_2$
1492,624	80	2,38	10,69	$2p^3\ ^2D^\circ—3s\ ^2P$	$^5/_2—^3/_2$
1481,750	4	—	—	—	—
1467,384	3	—	—	—	—
1466,723	5	—	—	—	—
1462,822	2	—	—	—	—
1411,939	30	3,57	12,36	$2p^3\ ^2P^\circ—3s'\ ^2D$	$^1/_2,\ ^3/_2—^3/_2,\ ^5/_2$
1411,510	1	—	—	—	—
1360,566	1	—	—	—	—
1355,887	6	—	—	—	—
1327,927	10	3,57	12,91	$2p^3\ ^2P^\circ—4s\ ^2P$	$^3/_2,\ ^1/_2—^1/_2$
1326,572	15	3,57	12,92	$2p^3\ ^2P^\circ—4s\ ^2P$	$^3/_2,\ ^1/_2—^3/_2$
1321,60	4	—	—	—	—
1320,83	3	—	—	—	—
1319,684	30	3,57	12,97	$2p^3\ ^2P^\circ—3d\ ^2P$	$^3/_2,\ ^1/_2—^3/_2$
1319,003	20	3,57	12,98	$2p^3\ ^2P^\circ—3d\ ^2P$	$^3/_2,\ ^1/_2—^1/_2$
1318,13	3	—	—	—	—
1317,41	2	—	—	—	—
1316,287	1	3,57	12,99	$2p^3\ ^2P^\circ—3d\ ^2F$	$^3/_2—^5/_2$
1315,484	1	3,57	13,00	$2p^3\ ^2P^\circ—3d\ ^4P$	$^3/_2,\ ^1/_2—^1/_2$
1315,23	2	—	—	—	—
1313,47	3	—	—	—	—
1313,20	3	3,57	13,00	$2p^3\ ^2P^\circ—3d\ ^4P$	$^3/_2,\ ^1/_2—^3/_2$
1312,86	3	3,57	13,01	$2p^3\ ^2P^\circ—3d\ ^4P$	$^3/_2—^5/_2$
1312,44	3	—	—	—	—
1310,952	25	3,57	13,04	$2p^3\ ^2P^\circ—3d\ ^2D$	$^3/_2—^3/_2$
1310,548	25	3,57	13,04	$2p^3\ ^2P^\circ—3d\ ^2D$	$^3/_2—^5/_2$
1310,057	1	—	—	—	—
1309,30	3	—	—	—	—
1308,86	3	—	—	—	—
1280,362	1	—	—	—	—
1279,995	1	—	—	—	—
1243,309	15	2,38	12,36	$2p^3\ ^2D^\circ—3s'\ ^2D$	$^5/_2,\ ^3/_2—^5/_2$
1243,179	20	2,38	12,36	$2p^3\ ^2D^\circ—3s'\ ^2D$	$^5/_2,\ ^3/_2—^3/_2$
1233,20	2	3,57	13,63	$2p^3\ ^2P^\circ—5s\ ^4P$	$^3/_2—^5/_2$
1231,588	1	3,57	13,65	$2p^3\ ^2P^\circ—5s\ ^2P$	$^3/_2,\ ^1/_2—^1/_2$
1230,288	0	3,57	13,65	$2p^3\ ^2P^\circ—5s\ ^2P$	$^3/_2,\ ^1/_2—^3/_2$
1229,40	2	—	—	—	—
1229,172	1	—	—	—	—
1228,790	10	3,57	13,66	$\{2p^3\ ^2P^\circ—4d\ ^2P$ $2p^3\ ^2P^\circ—4d\ ^4D$	$^3/_2,\ ^1/_2—^3/_2$ $^3/_2,\ ^1/_2—^1/_2$
1228,410	5	3,57	13,67	$2p^3\ ^2P^\circ—4d\ ^2P$	$^3/_2,\ ^1/_2—^1/_2$
1227,788	2	3,57	13,67	$2p^3\ ^2P^\circ—4d\ ^4D$	$^3/_2—^5/_2$
1227,226	1	3,57	13,68	$2p^3\ ^2P^\circ—4d\ ^4P$	$^3/_2,\ ^1/_2—^1/_2$
1227,00	3	3,57	13,68	$2p^3\ ^2P^\circ—4d\ ^4D$	$^3/_2—^7/_2$
1226,831	1	3,57	13,68	$2p^3\ ^2P^\circ—4d\ ^4P$	$^3/_2,\ ^1/_2—^3/_2$
1225,85	2	3,57	13,69	$2p^3\ ^2P^\circ—4d\ ^4P$	$^3/_2—^5/_2$
1225,372	10	3,57	13,69	$2p^3\ ^2P^\circ—4d\ ^2D$	$^3/_2,\ ^1/_2—^3/_2$
1225,028	15	3,57	13,70	$2p^3\ ^2P^\circ—4d\ ^2D$	$^3/_2—^5/_2$
1223,80	2	—	—	—	—

λ, Å	I	E_H, eV	E_B, eV	Transition	J
1223,20	3	—	—	—	—
1200,711	30	0,00	10,33	$2p^3\ {}^4S^\circ$—$3s\ {}^4P$	3/2—1/2
1200,224	0	0,00	10,33	$2p^3\ {}^4S^\circ$—$3s\ {}^4P$	3/2—3/2
1199,718	2	—	—	—	—
1199,549	50	0,00	10,34	$2p^3\ {}^4S^\circ$—$3s\ {}^4P$	3/2—5/2
1192,55	2	3,57	13,97	$2p^3\ {}^2P^\circ$—$6s\ {}^4P$	3/2—5/2
1191,99	4	3,57	13,98	$2p^3\ {}^2P^\circ$—$6s\ {}^2P$	3/2, 1/2—1/2
1191,03	5	3,57	13,99	$2p^3\ {}^2P^\circ$—$5d\ {}^2P$	3/2, 1/2—3/2
1190,84	5	3,57	13,99	$2p^3\ {}^2P^\circ$—$5d\ {}^2P$ $2p^3\ {}^2P^\circ$—$5d\ {}^4D$	3/2, 1/2—1/2 3/2—5/2
1190,52	3	—	—	—	—
1189,628	5	—	—	—	—
1189,244	3	3,57	14,00	$2p^3\ {}^2P^\circ$—$5d\ {}^2D$	5/2, 3/2—3/2
1188,972	5	3,57	14,00	$2p^3\ {}^2P^\circ$—$5d\ {}^2D$	5/2, 3/2—5/2
1183,998	3	—	—	—	—
1177,694	15	2,38	12,91	$2p^3\ {}^2D^\circ$—$4s\ {}^2P$	3/2—1/2
1176,626	3	—	—	—	—
1176,508	15	2,38	12,92	$2p^3\ {}^2D^\circ$—$4s\ {}^2P$	5/2, 3/2—3/2
1174,84	3	—	—	—	—
1172,55	3	3,57	14,15	$2p^3\ {}^2P^\circ$—$7s\ {}^2P$	3/2, 1/2—1/2
1172,02	2	3,57	14,15	$2p^3\ {}^2P^\circ$—$7s\ {}^2P$	3/2, 1/2—3/2
1171,60	2	3,57	14,16	$2p^3\ {}^2P^\circ$—$6d\ {}^2P$	3/2, 1/2—3/2
1171,39	2	3,57	14,16	$2p^3\ {}^2P^\circ$—$6d\ {}^2P$	3/2, 1/2—1/2
1171,067	0	2,38 3,57	12,98 14,16	$2p^3\ {}^2D^\circ$—$3d\ {}^2P$ $2p^3\ {}^2P^\circ$—$6d\ {}^2D$	5/2, 3/2—3/2 3/2—3/2
1170,276	1	2,38	12,98	$2p^3\ {}^2D^\circ$—$3d\ {}^4F$	3/2—5/2
1169,692	1	2,38	12,98	$2p^3\ {}^2D^\circ$—$3d\ {}^4F$	5/2—7/2
1168,537	20	2,38	12,99	$2p^3\ {}^2D^\circ$—$3d\ {}^2F$	3/2—5/2
1168,334	8	2,38	13,01	$2p^3\ {}^2D^\circ$—$3d\ {}^4P$	3/2—5/2
1167,450	25	2,38	13,00	$2p^3\ {}^2D^\circ$—$3d\ {}^2F$	5/2—7/2
1165,566	2	2,38	13,02	$2p^3\ {}^2D^\circ$—$3d\ {}^4D$	5/2—7/2
1164,322	8	2,38	13,03	$2p^3\ {}^2D$ —$3d\ {}^2D$	5/2—3/2
1163,884	12	2,38	13,04	$2p^3\ {}^2D^\circ$—$3d\ {}^2D$	5/2—5/2
1161,26	2	3,57	14,25	$2p^3\ {}^2P^\circ$—$8s\ {}^2P$	3/2, 1/2—3/2, 1/2
1160,67	2	3,57	14,26	$2p^3\ {}^2P^\circ$—$7d\ {}^2P$	3/2, 1/2—3/2
1160,45	3	3,57	14,26	$2p^3\ {}^2P^\circ$—$7d\ {}^2P$	3/2, 1/2—1/2
1159,858	1	3,57	14,26	$2p^3\ {}^2P^\circ$—$7d\ {}^2D$	3/2, 1/2—3/2
1159,285	1	3,57	14,27	$2p^3\ {}^2P^\circ$—$7d\ {}^2D$	3/2—5/2
1158,051	2	—	—	—	—
1154,23	3	3,57	14,32	$2p^3\ {}^2P^\circ$—$9s\ {}^2P$	3/2, 1/2—1/2
1153,52	4	3,57	14,32	$2p^3\ {}^2P^\circ$—$8d\ {}^2P$	3/2, 1/2—3/2
1152,75	1	3,57	14,33	$2p^3\ {}^2P^\circ$—$8d\ {}^2D$	3/2, 1/2—3/2
1152,35	4	3,57	14,33	$2p^3\ {}^2P^\circ$—$8d\ {}^2D$	3/2—5/2
1149,47	2	3,57	14,36	$2p^3\ {}^2P^\circ$—$10s\ {}^2P$	3/2, 1/2—1/2
1148,76	4	3,57	14,37	$2p^3\ {}^2P^\circ$—$9d\ {}^2P$	3/2, 1/2—3/2
1147,69	4	3,57	14,38	$2p^3\ {}^2P^\circ$—$9d\ {}^2D$	3/2—5/2
1145,92	1	3,57	14,39	$2p^3\ {}^2P^\circ$—$11s\ {}^2P$	3/2, 1/2—1/2
1145,28	2	3,57	14,40	$2p^3\ {}^2P^\circ$—$10d\ {}^2P$	3/2, 1/2—3/2
1144,24	2	3,57	14,41	$2p^3\ {}^2P^\circ$—$10d\ {}^2D$	3/2—5/2
1143,649	5	3,57	14,41	$2p^3\ {}^2P^\circ$—$3s''\ {}^2S$	3/2, 1/2—1/2
1143,32	1	3,57	14,42	$2p^3\ {}^2P^\circ$—$12s\ {}^2P$	3/2, 1/2—1/2
1142,73	2	3,57	14,42	$2p^3\ {}^2P^\circ$—$11d\ {}^2P$	3/2, 1/2—3/2
1141,70	2	3,57	14,43	$2p^3\ {}^2P^\circ$—$11d\ {}^2D$	3/2—5/2
1141,20	1	3,57	14,44	$2p^3\ {}^2P^\circ$—$13s\ {}^2P$	3/2, 1/2—1/2
1140,76	2	3,57	14,44	$2p^3\ {}^2P^\circ$—$12d\ {}^2P$	3/2, 1/2—3/2
1139,818	1	—	—	—	—
1139,15	1	3,57	14,46	$2p^3\ {}^2P^\circ$—$12d\ {}^2D$	3/2—5/2
1134,981	25	0,00	10,92	$2p^3\ {}^4S^\circ$—$2p^4\ {}^4P$	3/2—5/2
1134,417	25	0,00	10,93	$2p^3\ {}^4S^\circ$—$2p^4\ {}^4P$	3/2—3/2

λ, Å	I	E_H, eV	E_B, eV	Transition	J
1134,168	20	0,00	10,93	$2p^3\,^4S^\circ$—$2p^4\,^4P$	3/2—1/2
1101,293	40	2,38	13,64	$2p^3\,^2D^\circ$—$5s\,^2P$	3/2—1/2
1100,362	30	2,38	13,65	$2p^3\,^2D^\circ$—$5s\,^2P$	5/2, 3/2—3/2
1099,80	2	—	—	—	—
1099,153	25	2,38	13,66	$2p^3\,^2D^\circ$—$4d\,^4F$	5/2, 3/2—5/2
1098,98	2	2,38	13,67	$2p^3\,^2D^\circ$—$4d\,^2P$	5/2, 3/2—3/2
1098,78	1	—	—	—	—
1098,63	2	2,38	13,67	$2p^3\,^2D^\circ$—$4d\,^4F$	5/2—7/2
1098,264	40	2,38	13,67	$2p^3\,^2D^\circ$—$4d\,^4D$	5/2, 3/2—3/2
1098,103	40	2,38	13,68	$2p^3\,^2D^\circ$—$4d\,^2F$	5/2, 3/2—5/2
1097,990	25	—	—	—	—
1097,245	50	2,38	13,69	$2p^3\,^2D^\circ$—$4d\,^2F$	5/2—7/2
1096,749	35	2,38	13,69	$2p^3\,^2D^\circ$—$4d\,^4P$	5/2, 3/2—5/2
1096,322	35	2,38	13,69	$2p^3\,^2D^\circ$—$4d\,^2D$	5/2, 3/2—3/2
1095,940	35	2,38	13,70	$2p^3\,^2D^\circ$—$4d\,^2D$	5/2, 3/2—5/2
1095,279	4	—	—	—	—
1071,656	1	—	—	—	—
1070,821	0	2,38	13,93	$2p^3\,^2D^\circ$—$6s\,^4P$	5/2, 3/2—3/2
1069,984	30	2,38	13,94	$2p^3\,^2D^\circ$—$6s\,^4P$	5/2, 3/2—5/2
1069,198	2	2,38 2,38	13,99 13,98	$2p^3\,^2D^\circ$—$5d\,^4D$ $2p^3\,^2D^\circ$—$6s\,^2P$	5/2, 3/2—5/2 3/2—1/2
1068,66	4	2,38	13,99	$2p^3\,^2D^\circ$—$5d\,^2F$	5/2, 3/2—7/2
1068,476	35	2,38	13,99	$2p^3\,^2D^\circ$—$5d\,^4F$	5/2—5/2
1067,607	35	2,38	14,00	$2p^3\,^2D^\circ$—$5d\,^2F$	5/2—7/2
1067,37	4	2,38	14,00	$2p^3\,^2D^\circ$—$5d\,^2D$	5/2, 3/2—3/2
1066,97	4	2,38	14,00	$2p^3\,^2D^\circ$—$5d\,^2D$	5/2, 3/2—5/2
1066,56	3	—	—	—	—
1066,126	1	—	—	—	—
1063,351	1	—	—	—	—
1053,90	3	2,38	14,15	$2p^3\,^2D^\circ$—$7s\,^4P$	5/2, 3/2—5/2
1053,65	3	2,38	14,15	$2p^3\,^2D^\circ$—$7s\,^2P$	3/2—1/2
1053,38	5	2,38	14,15	$2p^3\,^2D^\circ$—$6d\,^4F$	5/2—7/2
1053,03	3	2,38	14,15	$2p^3\,^2D^\circ$—$7s\,^3p$	5/2, 3/2—5/2
1052,72	2	—	—	—	—
1052,16	3	—	—	—	—
1052,07	3	2,38	14,17	$2p^3\,^2D^\circ$—$6d\,^2F$	5/2—7/2
1051,89	2	2,38	14,17	$2p^3\,^2D^\circ$—$6d\,^2D$	5/2, 3/2—5/2
1044,60	4	2,38	14,25	$2p^3\,^2D^\circ$—$8s\,^2P$	3/2—1/2
1044,13	5	2,38	14,26	$2p^3\,^2D^\circ$—$7d\,^2F$	5/2—7/2
1043,58	2	2,38	14,26	$2p^3\,^2D^\circ$—$7d\,^2D$	5/2, 3/2—3/2
1043,12	5	2,38	14,27	$2p^3\,^2D^\circ$—$7d\,^2D$	5/2, 3/2—5/2
1038,90	1	2,38	14,32	$2p^3\,^2D^\circ$—$9s\,^2P$	3/2—1/2
1038,76	1	—	—	—	—
1038,34	3	2,38	14,32	$2p^3\,^2D^\circ$—$8d\,^2F$	5/2—7/2
1037,64	1	2,38	14,33	$2p^3\,^2D^\circ$—$8d\,^2D$	5/2, 3/2—3/2
1037,38	4	2,38	14,33	$2p^3\,^2D^\circ$—$8d\,^2D$	5/2, 3/2—5/2
1034,96	1	3,57	14,36	$2p^3\,^2P^\circ$—$10s\,^2P$	3/2, 1/2—1/2
1034,39	2	2,38	14,37	$2p^3\,^2D^\circ$—$9d\,^2F$	5/2—7/2
1033,65	0	2,38	14,38	$2p^3\,^2D^\circ$—$9d\,^2D$	5/2, 3/2—3/2
1033,48	3	2,38	14,38	$2p^3\,^2D^\circ$—$9d\,^2D$	5/2, 3/2—5/2
1032,958	2	—	—	—	—
1032,19	1	2,38	14,39	$2p^3\,^2D^\circ$—$11s\,^2P$	3/2—1/2
1031,65	2	2,38	14,40	$2p^3\,^2D^\circ$—$10d\,^2F$	5/2—7/2
1030,72	2	2,38	14,41	$2p^3\,^2D^\circ$—$10d\,^2D$	5/2, 3/2—5/2
1029,53	1	2,38	14,43	$2p^3\,^2D^\circ$—$11d\,^2F$	5/2—7/2
1028,64	2	2,38	14,44	$2p^3\,^2D^\circ$—$11d\,^2D$	5/2, 3/2—5/2
1008,875	1	—	—	—	—
965,042	10	0,00	12,85	$2p^3\,^4S^\circ$—$4s\,^4P$	3/2—1/2
964,626	5	0,00	12,85	$2p^3\,^4S^\circ$—$4s\,^4P$	3/2—3/2

λ, Å	I	E_H, eV	E_B, eV	Transition	J
963,991	5	0,00	12,86	$2p^3\ ^4S^\circ - 4s\ ^4P$	$^3/_2 - ^5/_2$
959,54	3	0,00	12,92	$2p^3\ ^4S^\circ - 4s\ ^2P$	$^3/_2 - ^3/_2$
955,91	3	0,00	12,97	$2p^3\ ^4S^\circ - 3d\ ^2P$	$^3/_2 - ^3/_2$
955,438	—	0,00	12,98	$2p^3\ ^4S^\circ - 3d\ ^4F$	$^3/_2 - ^3/_2$
955,265	—	0,00	12,98	$2p^3\ ^4S^\circ - 3d\ ^4F$	$^3/_2 - ^5/_2$
954,11	3	0,00	12,99	$2p^3\ ^4S^\circ - 3d\ ^2F$	$^3/_2 - ^5/_2$
953,98	6	—	—	—	—
953,658	15	0,00	13,00	$2p^3\ ^4S^\circ - 3d\ ^4P$	$^3/_2 - ^1/_2$
953,415	15	0,00	13,00	$2p^3\ ^4S^\circ - 3d\ ^4P$	$^3/_2 - ^3/_2$
953,399	6	—	—	—	—
952,789	3	0,00	13,01	$2p^3\ ^4S^\circ - 3d\ ^4P$	$^3/_2 - ^5/_2$
952,522	4	0,00	13,01	$2p^3\ ^4S^\circ - 3d\ ^4D$	$^3/_2 - ^1/_2$
952,414	—	0,00	13,02	$2p^3\ ^4S^\circ - 3d\ ^4D$	$^3/_2 - ^3/_2$
952,304	8	0,00	13,02	$2p^3\ ^4S^\circ - 3d\ ^4D$	$^3/_2 - ^5/_2$
951,35	1	0,00	13,03	$2p^3\ ^4S^\circ - 3d\ ^2D$	$^3/_2 - ^3/_2$
951,08	3	0,00	13,04	$2p^3\ ^4S^\circ - 3d\ ^2D$	$^3/_2 - ^5/_2$
910,6456	0	0,00	13,61	$2p^3\ ^4S^\circ - 5s\ ^4P$	$^3/_2 - ^1/_2$
910,2785	0	0,00	13,62	$2p^3\ ^4S^\circ - 5s\ ^4P$	$^3/_2 - ^3/_2$
909,6976	0	0,00	13,63	$2p^3\ ^4S^\circ - 5s\ ^4P$	$^3/_2 - ^5/_2$
908,83	2	0,00	13,64	$2p^3\ ^4S^\circ - 5s\ ^2P$	$^3/_2 - ^1/_2$
908,23	3	0,00	13,65	$2p^3\ ^4S^\circ - 5s\ ^2P$	$^3/_2 - ^3/_2$
907,278	4	0,00	13,66	$2p^3\ ^4S^\circ - 4d\ ^4D$	$^3/_2 - ^1/_2$
906,833	2	0,00	13,67	$2p^3\ ^4S^\circ - 4d\ ^4D$	$^3/_2 - ^3/_2$
906,722	1	0,00	13,67	$2p^2\ ^4S^\circ - 4d\ ^4D$	$^3/_2 - ^5/_2$
906,63	4	—	—	—	—
906,426	15	0,00	13,68	$2p^3\ ^4S^\circ - 4d\ ^4P$	$^3/_2 - ^1/_2$
906,202	10	0,00	13,68	$2p^3\ ^4S^\circ - 4d\ ^4P$	$^3/_2 - ^3/_2$
905,829	5	0,00	13,69	$2p^3\ ^4S^\circ - 4d\ ^4P$	$^3/_2 - ^5/_2$
905,53	2	0,00	13,69	$2p^3\ ^4S^\circ - 4d\ ^2D$	$^3/_2 - ^3/_2$
905,23	4	0,00	13,70	$2p^3\ ^4S^\circ - 4d\ ^2D$	$^3/_2 - ^5/_2$
888,363	0	0,00	13,93	$2p^3\ ^4S^\circ - 6s\ ^4P$	$^3/_2 - ^1/_2$
888,019	0	0,00	13,93	$2p^3\ ^4S^\circ - 6s\ ^4P$	$^3/_2 - ^3/_2$
887,41	4	0,00	13,94	$2p^3\ ^4S^\circ - 6s\ ^4P$	$^3/_2 - ^5/_2$
886,95	3	0,00	13,98	$2p^3\ ^4S^\circ - 5d\ ^4F$	$^3/_2 - ^3/_2$
886,80	3	0,00	13,98	$2p^3\ ^4S^\circ - 5d\ ^4F$	$^3/_2 - ^5/_2$
886,33	6	0,00	13,99	$2p^3\ ^4S^\circ - 5d\ ^4D$	$^3/_2 - ^5/_2$
885,93	3	0,00	14,00	$2p^3\ ^4S^\circ - 5d\ ^4P$	$^3/_2 - ^3/_2$
885,67	5	0,00	14,00	$2p^3\ ^4S^\circ - 5d\ ^4P$	$^3/_2 - ^5/_2$
885,36	3	0,00	14,00	$2p^3\ ^4S^\circ - 5d\ ^2D$	$^3/_2 - ^5/_2$
877,11	2	0,00	14,13	$2p^3\ ^4S^\circ - 7s\ ^4P$	$^3/_2 - ^1/_2$
876,79	2	0,00	14,14	$2p^3\ ^4S^\circ - 7s\ ^4P$	$^3/_2 - ^3/_2$
876,32	4	0,00	14,15	$2p^3\ ^4S^\circ - 7s\ ^4P$	$^3/_2 - ^5/_2$
875,764	0	0,00	14,16	$2p^3\ ^4S^\circ - 6d\ ^4D$	$^3/_2 - ^5/_2$
875,25	5	0,00	14,17	$2p^3\ ^4S^\circ - 6d\ ^4P$	$^3/_2 - ^3/_2$
875,092	5	0,00	14,17	$2p^3\ ^4S^\circ - 6d\ ^4P$	$^3/_2 - ^5/_2$
871,01	1	0,00	14,23	$2p^3\ ^4S^\circ - 8s\ ^4P$	$^3/_2 - ^1/_2$
870,40	3	0,00	14,24	$2p^3\ ^4S^\circ - 8s\ ^4P$	$^3/_2 - ^3/_2$
870,00	3	0,00	14,25	$2p^3\ ^4S^\circ - 8s\ ^4P$	$^3/_2 - ^5/_2$
869,66	5	0,00	14,26	$2p^3\ ^4S^\circ - 7d\ ^4D$	$^3/_2 - ^5/_2$
868,98	5	0,00	14,27	$2p^3\ ^4S^\circ - 7d\ ^4P$	$^3/_2 - ^5/_2$
865,93	3	0,00	14,32	$2p^3\ ^4S^\circ - 9s\ ^4P$	$^3/_2 - ^5/_2$
865,63	5	0,00	14,32	$2p^3\ ^4S^\circ - 8d\ ^4D$	$^3/_2 - ^5/_2$
864,93	5	0,00	14,33	$2p^3\ ^4S^\circ - 8d\ ^4P$	$^3/_2 - ^5/_2$
863,15	3	0,00	14,36	$2p^3\ ^4S^\circ - 10s\ ^4P$	$^3/_2 - ^5/_2$
862,90	5	0,00	14,37	$2p^3\ ^4S^\circ - 9d\ ^4D$	$^3/_2 - ^5/_2$
862,15	5	0,00	14,38	$2p^3\ ^4S^\circ - 9d\ ^4P$	$^3/_2 - ^5/_2$
861,15	1	0,00	14,39	$2p^3\ ^4S^\circ - 11s\ ^4P$	$^3/_2 - ^5/_2$
860,85	4	0,00	14,40	$2p^3\ ^4S^\circ - 10d\ ^4D$	$^3/_2 - ^5/_2$

λ, Å	I	E_H, eV	E_B, eV	Transition	J
860,15	4	—	—	—	—
859,75	2	0,00	14,42	$2p^3\ ^4S^\circ$—$12s\ ^4P$	$^3/_2$—$^5/_2$
859,35	3	0,00	14,43	$2p^3\ ^4S^\circ$—$11d\ ^4D$	$^3/_2$—$^5/_2$
858,80	2	0,00	14,44	$2p^3\ ^4S^\circ$—$11d\ ^4P$	$^3/_2$—$^5/_2$
857,76	2	0,00	14,45	$2p^3\ ^4S^\circ$—$12d\ ^4P$	$^3/_2$—$^5/_2$
856,80	2	—	—	—	—
856,24	2	—	—	—	—
855,70	2	—	—	—	—

N II, ground state $1s^2\,2s^2\,2p^2\ ^3P_0$
Ionization potential 238846,7 cm⁻¹; 29,611 eV

λ, Å	I	E_H, eV	E_B, eV	Transition	J
10546,76	4	25,46	26,63	$4p\ ^1D$—$5s\ ^1P^\circ$	2—1
10126,27	5	26,19	27,42	$4f\ G\ (3^1/_2)$—$5g\ G\ (4^1/_2)^\circ$	4—4, 5
10118,49	4	26,19	27,42	$4f\ G\ (3^1/_2)$—$5g\ G\ (4^1/_2)^\circ$	3—4
10070,12	6	26,21	27,44	$4f\ D\ (2^1/_2)$—$5g\ F\ (3^1/_2)^\circ$	2—3
10065,15	7	26,21	27,44	$4f\ D\ (2^1/_2)$—$5g\ F\ (3^1/_2)^\circ$	3—3, 4
10035,45	7	26,21	27,45	$4f\ G\ (4^1/_2)$—$5g\ H\ (5^1/_2)^\circ$	4—5
10023,27	8	26,21	27,45	$4f\ G\ (4^1/_2)$—$5g\ H\ (5^1/_2)^\circ$	5—5, 6
9969,34	7	26,19	27,44	$4f\ G\ (3^1/_2)$—$5g\ H\ (4^1/_2)^\circ$	4—5
9961,86	6	26,19	27,44	$4f\ G\ (3^1/_2)$—$5g\ H\ (4^1/_2)^\circ$	3—4
9891,09	7	26,17	27,42	$4f\ F\ (3^1/_2)$—$5g\ G\ (4^1/_2)^\circ$	4—4, 5
		26,17	27,42	$4f\ F\ (3^1/_2)$—$5g\ G\ (3^1/_2)^\circ$	3—4, 3
9887,39	6	26,17	27,42	$4f\ F\ (3^1/_2)$—$5g\ G\ (4^1/_2)^\circ$	3—4
9868,21	5	26,16	27,42	$4f\ F\ (2^1/_2)$—$5g\ G\ (3^1/_2)^\circ$	2—3
9865,41	6	26,17	27,42	$4f\ F\ (2^1/_2)$—$5g\ G\ (3^1/_2)^\circ$	3—4, 3
9794,01	3	26,17	27,43	$4d\ ^1F^\circ$—$5f\ G\ (4^1/_2)^\circ$	3—4
9741,43	4	26,17	27,44	$4f\ F\ (3^1/_2)$—$5g\ H\ (4^1/_2)^\circ$	4—4, 5
9737,75	4	26,17	27,44	$4f\ F\ (3^1/_2)$—$5g\ H\ (4^1/_2)^\circ$	3—4
9722,36	1	26,17	27,44	$4f\ F\ (3^1/_2)$—$5g\ F\ (3^1/_2)^\circ$	4—3, 4
9718,66	1	26,17	27,44	$4f\ F\ (3^1/_2)$—$5g\ F\ (3^1/_2)^\circ$	3—3, 4
9696,77	1	26,16	27,44	$4f\ F\ (2^1/_2)$—$5g\ F\ (3^1/_2)^\circ$	2—3
9694,0	1	26,17	27,44	$4f\ F\ (2^1/_2)$—$5g\ F\ (3^1/_2)^\circ$	3—3, 4
9480,73	1	26,13	27,44	$4d\ ^3P^\circ$—$5f\ D\ (2^1/_2)$	1—2
9453,50	1	26,13	27,45	$4d\ ^3P^\circ$—$5f\ D\ (1^1/_2)$	0—1
9442,82	3	26,12	27,44	$4d\ ^3P^\circ$—$5f\ D\ (2^1/_2)$	2—3
9439,40	1	26,12	27,44	$4d\ ^3P^\circ$—$5f\ D\ (2^1/_2)$	2—2
9431,20	1	26,13	27,44	$4d\ ^3P^\circ$—$5f\ D\ (1^1/_2)$	1—1
9325,84	0	25,23	26,56	$4p\ ^3S$—$5s\ ^3P^\circ$	1—1
9281,06	3	26,07	27,41	$4d\ ^3D^\circ$—$5f\ F\ (3^1/_2)$	3—4
9266,61	1	26,06	27,41	$4d\ ^3D^\circ$—$5f\ F\ (2^1/_2)$	2—3
9253,98	1	26,06	27,41	$4d\ ^3D^\circ$—$5f\ F\ (3^1/_2)$	2—3
9242,02	2	26,06	27,41	$4d\ ^3D^\circ$—$5f\ F\ (2^1/_2)$	1—2
9217,10	2	25,23	26,58	$4p\ ^3S$—$5s\ ^3P^\circ$	1—2
9146,40	2	26,07	27,43	$4d\ ^3D^\circ$—$5f\ G\ (3^1/_2)$	3—4
9121,00	1	26,06	27,43	$4d\ ^3D^\circ$—$5f\ G\ (3^1/_2)$	2—3
9096,17	1	26,07	27,44	$4d\ ^3D^\circ$—$5f\ D\ (2^1/_2)$	3—3
9092,93	0	26,07	27,44	$4d\ ^3D^\circ$—$5f\ D\ (2^1/_2)$	3—2
9089,45	1	25,20	26,56	$4p\ ^3P$—$5s\ ^3P^\circ$	2—1
9069,51	1	25,19	26,56	$4p\ ^3P$—$5s\ ^3P^\circ$	1—0
9063,78	0	26,06	27,44	$4d\ ^3D^\circ$—$5f\ D\ (2^1/_2)$	2—2
9032,04	1	25,19	26,56	$4p\ ^3P$—$5s\ ^3P^\circ$	1—1
9010,39	1	25,19	26,56	$4p\ ^3P$—$5s\ ^3P^\circ$	0—1

λ, Å	I	E_H, eV	E_B, eV	Transition	J
8986,15	4	25,20	26,58	$4p\ ^3P-5s\ ^3P^\circ$	2—2
8983,28	3	26,03	27,41	$4d\ ^1D^\circ-5f\ F\ (2^1/_2)$	2—3
8971,36	1	26,03	27,41	$4d\ ^1D^\circ-5f\ F\ (3^1/_2)$	2—3
8930,04	1	25,19	26,58	$4p\ ^3P-5s\ ^3P^\circ$	1—2
8893,32	1	26,25	27,65	$3s\ ^3P-3p\ ^3S^\circ$	2—1
8855,40	0	26,25	27,65	$3s\ ^3P-3p\ ^3S^\circ$	1—1
8846,46	1	26,03	27,43	$4d\ ^1D^\circ-5f\ G\ (3^1/_2)$	2—3
8819,56	2	26,00	27,41	$4d\ ^3F^\circ-5f\ F\ (3^1/_2)$	3—4
8772,95	3	25,99	27,41	$4d\ ^3F^\circ-5f\ F\ (3^1/_2)$	2—3
8763,39	1	26,01	27,43	$4d\ ^3F^\circ-5f\ G\ (3^1/_2)$	4—4
8710,54	6	26,01	27,44	$4d\ ^3F^\circ-5f\ G\ (4^1/_2)$	4—5
8699,002	5	25,14	26,56	$4p\ ^3D-5s\ ^3P^\circ$	2—1
8697,79	3	26,00	27,43	$4d\ ^3F^\circ-5f\ G\ (3^1/_2)$	3—4
8694,900	4	25,13	26,56	$4p\ ^3D-5s\ ^3P^\circ$	1—0
8687,430	5	20,67	22,10	$2p^3\ ^1P^\circ-3p\ ^1S$	1—0
8676,076	7	25,15	26,58	$4p\ ^3D-5s\ ^3P^\circ$	3—2
8660,52	3	25,13	26,56	$4p\ ^3D-5s\ ^3P^\circ$	1—1
8653,38	3	25,99	27,43	$4d\ ^3F^\circ-5f\ G\ (3^1/_2)$	2—3
8638,31	3	26,00	27,44	$4d\ ^3F^\circ-5f\ G\ (4^1/_2)$	3—4
8604,32	3	25,14	26,58	$4p\ ^3D-5s\ ^3P^\circ$	2—2
8438,742	11	22,10	23,57	$3p\ ^1S-3d\ ^1P^\circ$	0—1
8296,205	4	23,57	25,06	$3d\ ^1P^\circ-4p\ ^1P$	1—1
789,62	4	25,06	26,63	$3p\ ^1P-5s\ ^1P^\circ$	1—1
776,237	10	21,60	23,19	$3p\ ^1D-3d\ ^1D^\circ$	2—2
7256,53	2	23,42	25,13	$3d\ ^3P^\circ-4p\ ^3D$	0—1
7215,06	3	23,42	25,14	$3d\ ^3P^\circ-4p\ ^3D$	1—2
7188,20	2	23,41	25,14	$3d\ ^3P^\circ-4p\ ^3D$	2—2
7138,87	4	23,41	25,15	$3d\ ^3P^\circ-4p\ ^3D$	2—3
7014,73	2	23,42	25,19	$3d\ ^3P^\circ-4p\ ^3P$	0—1
7013,98	2	23,42	25,19	$3d\ ^3P^\circ-4p\ ^3P$	1—0
6975,64	4	23,41	25,19	$3d\ ^3P^\circ-4p\ ^3P$	2—1
6966,81	3	23,42	25,20	$3d\ ^3P^\circ-4p\ ^3P$	1—2
6941,752	5	23,41	25,20	$3d\ ^3P^\circ-4p\ ^3P$	2—2
6887,834	5	28,49	30,29	$3p\ ^5S^\circ-3d\ ^5P$	2—3
6869,580	4	28,49	30,29	$3p\ ^5S^\circ-3d\ ^5P$	2—2
6857,030	3	28,49	30,30	$3p\ ^5S^\circ-3d\ ^5P$	2—1
6847,237	4	23,42	25,23	$3d\ ^3P^\circ-4p\ ^3S$	0—1
6834,094	6	23,42	25,23	$3d\ ^3P^\circ-4p\ ^3S$	1—1
6809,989	7	23,41	25,23	$3d\ ^3P^\circ-4p\ ^3S$	2—1
6801,31	1	24,39	26,21	$4s\ ^3P^\circ-4f\ D\ (2^1/_2)$	2—3
6634,789	3	24,39	26,25	$4s\ ^3P^\circ-3s\ ^3P$	2—1
6629,795	7	23,19	25,06	$3d\ ^1D^\circ-4p\ ^1P$	2—1
6613,622	5	24,39	26,26	$4s\ ^3P^\circ-3s\ ^3P$	2—2
6610,565	13	21,60	23,47	$3p\ ^1D-3d\ ^1F^\circ$	2—3
6595,666	3	24,37	26,25	$4s\ ^3P^\circ-3s\ ^3P$	1—0
6564,20	3	23,57	25,46	$3d\ ^1P^\circ-4p\ ^1D$	1—2
6561,78	3	24,37	26,26	$4s\ ^3P^\circ-3s\ ^1P^\circ$	1—2
6560,203	3	24,37	26,25	$4s\ ^3P^\circ-3s\ ^1P^\circ$	0—1
6554,47	3	23,24	25,13	$3d\ ^3D^\circ-4p\ ^3D$	2—1
6545,530	3	23,24	25,14	$3d\ ^3D^\circ-4p\ ^3D$	3—2
6544,162	4	23,24	25,13	$3d\ ^3D^\circ-4p\ ^3D$	1—1
6532,550	5	23,24	25,14	$3d\ ^3D^\circ-4p\ ^3D$	2—2
6522,39	2	23,24	25,14	$3d\ ^3D^\circ-4p\ ^3D$	1—2
6504,608	6	23,24	25,15	$3d\ ^3D^\circ-4p\ ^3D$	3—3
6491,79	2	23,24	25,15	$3d\ ^3D^\circ-4p\ ^3D$	2—3
6482,053	13	18,50	20,41	$3s\ ^1P^\circ-3p\ ^1P$	1—1
6457,69	0	19,23	21,15	$2p^3\ ^3S^\circ-3p\ ^3P$	1—1
6433,45	1	19,23	21,16	$2p^3\ ^3S^\circ-3p\ ^3P$	1—2

λ, Å	I	E_H, eV	E_B, eV	Transition	J
6399,16	2	23,19	25,13	$3d\ ^1D^\circ$—$4p\ ^3D$	2—1
6384,31	2	23,12	25,06	$3d\ ^3F^\circ$—$4p\ ^1P$	2—1
6379,615	9	18,46	20,41	$3s\ ^3P^\circ$—$3p\ ^1P$	1—1
6357,569	5	23,24	25,19	$3d\ ^3D^\circ$—$4p\ ^3P$	1—0
6356,545	6	23,24	25,19	$3d\ ^3D^\circ$—$4p\ ^3P$	2—1
6346,86	5	23,24	25,19	$3d\ ^3D^\circ$—$4p\ ^3P$	1—1
6340,569	7	23,24	25,20	$3d\ ^3D^\circ$—$4p\ ^3P$	3—2
6328,39	5	23,24	25,20	$3d\ ^3D^\circ$—$4p\ ^3P$	2—2
6318,80	1	23,24	25,20	$3d\ ^3D^\circ$—$4p\ ^3P$	1—2
6285,70	2	21,16	23,13	$3p\ ^3P$—$3d\ ^3F^\circ$	2—3
6284,322	6	21,60	23,57	$3p\ ^1D$—$3d\ ^1P^\circ$	2—1
6242,412	7	23,47	25,46	$3d\ ^1F^\circ$—$4p\ ^1D$	3—2
6218,67	0	23,24	25,23	$3d\ ^3D^\circ$—$4p\ ^3S$	2—1
6183,68	0	28,36	30,36	$3p\ ^3D^\circ$—$3d\ ^5D$	3—3
6173,313	7	23,13	25,14	$3d\ ^3F^\circ$—$4p\ ^3D$	3—2
6170,166	6	23,12	25,13	$3d\ ^3F^\circ$—$4p\ ^3D$	2—1
6167,755	8	23,14	25,15	$3d\ ^3F^\circ$—$4p\ ^3D$	4—3
6150,755	4	23,12	25,14	$3d\ ^3F^\circ$—$4p\ ^3D$	2—2
6136,894	4	23,13	25,15	$3d\ ^3F^\circ$—$4p\ ^3D$	3—3
6065,00	3	21,15	23,19	$3p\ ^3P$—$3d\ ^1D^\circ$	1—2
5960,901	4	21,16	23,24	$3p\ ^3P$—$3d\ ^3D^\circ$	2—1
5954,276	5	23,57	25,58	$3d\ ^1P^\circ$—$4p\ ^1S$	1—0
5952,388	8	21,16	23,24	$3p\ ^3P$—$3d\ ^3D^\circ$	2—2
5941,653	12	21,16	23,24	$3p\ ^3P$—$3d\ ^3D^\circ$	2—3
5940,240	8	21,15	23,24	$3p\ ^3P$—$3d\ ^3D^\circ$	1—1
5931,779	11	21,15	23,24	$3p\ ^3P$—$3d\ ^3D^\circ$	1—2
5927,811	9	21,15	23,24	$3p\ ^3P$—$3d\ ^3D^\circ$	0—1
5899,83	1	26,25	28,35	$3s\ ^3P$—$3p\ ^3D^\circ$	0—1
5897,25	2	26,26	28,35	$3s\ ^3P$—$3p\ ^3D^\circ$	1—2
5893,15	3	26,26	28,36	$3s\ ^3P$—$3p\ ^3D^\circ$	2—3
5767,440	7	18,50	20,64	$3s\ ^1P^\circ$—$3p\ ^3D$	1—1
5747,296	8	18,50	20,64	$3s\ ^1P^\circ$—$3p\ ^3D$	1—2
5730,65	5	18,47	20,64	$3s\ ^3P^\circ$—$3p\ ^3D$	2—1
5710,766	10	18,47	20,64	$3s\ ^3P^\circ$—$3p\ ^3D$	2—2
5686,213	10	18,46	20,64	$3s\ ^3P^\circ$—$3p\ ^3D$	1—1
5679,562	14	18,47	20,66	$3s\ ^3P^\circ$—$3p\ ^3D$	2—3
5676,019	11	18,46	20,64	$3s\ ^3P^\circ$—$3p\ ^3D$	0—1
5666,627	12	18,46	20,66	$3s\ ^3P^\circ$—$3p\ ^3D$	1—2
5631,72	1	25,46	27,66	$4p\ ^1D$—$6s\ ^1P^\circ$	2—1
5565,25	3	25,51	27,72	$3s\ ^5P$—$3p\ ^5D^\circ$	3—2
5552,67	4	25,50	27,72	$3s\ ^5P$—$3p\ ^5D^\circ$	2—1
5551,922	5	25,51	27,73	$3s\ ^5P$—$3p\ ^5D^\circ$	3—3
5543,471	5	25,50	27,72	$3s\ ^5P$—$3p\ ^5D^\circ$	2—2
5540,059	4	25,49	27,72	$3s\ ^5P$—$3p\ ^5D^\circ$	1—2
5535,363	8	25,49 25,51	27,72 27,74	$3s\ ^5P$—$3p\ ^5D^\circ$ $3s\ ^5P$—$3p\ ^5D^\circ$	1—1 3—4
5530,244	7	25,50	27,73	$3s\ ^5P$—$3p\ ^5D^\circ$	2—3
5526,239	5	25,49	27,72	$3s\ ^5P$—$3p\ ^5D^\circ$	1—2
5495,666	10	21,16	23,41	$3p\ ^3P$—$3d\ ^3P^\circ$	2—2
5493,22	1	20,94	23,19	$3p\ ^3S$—$3d\ ^1D^\circ$	1—2
5480,062	7	21,16	23,42	$3p\ ^3P$—$3d\ ^3P^\circ$	2—1
5478,096	7	21,15	23,42	$3p\ ^3P$—$3d\ ^3P^\circ$	1—2
5475,29	4	23,19	25,46	$3d\ ^1D^\circ$—$4p\ ^1D$	2—2
5462,592	7	21,15	23,42	$3p\ ^3P$—$3d\ ^3P^\circ$	1—1
5454,221	7	21,15	23,42	$3p\ ^3P$—$3d\ ^3P^\circ$	1—0
5452,083	7	21,15	23,42	$3p\ ^3P$—$3d\ ^3P^\circ$	0—1
5390,68	1	20,94	23,24	$3p\ ^3S$—$3d\ ^3D^\circ$	1—1
5383,71	2	20,94	23,24	$3p\ ^3S$—$3d\ ^3D^\circ$	1—2

λ, Å	I	E_H, eV	E_B, eV	Transition	J
5351,220	4	27,98	30,29	$3p\ {}^5P^\circ - 3d\ {}^5P$	3—3
5340,213	3	27,98	30,29	$3p\ {}^5P^\circ - 3d\ {}^5P$	3—2
5338,732	4	27,97	30,29	$3p\ {}^5P^\circ - 3d\ {}^5P$	2—3
5327,76	1	27,97	30,29	$3p\ {}^5P^\circ - 3d\ {}^5P$	2—2
5320,953	4	27,97	30,29	$3p\ {}^5P^\circ - 3d\ {}^5P$	1—2
5320,203	3	27,97	30,30	$3p\ {}^5P^\circ - 3d\ {}^5P$	2—1
5313,419	2	27,97	30,30	$3p\ {}^5P^\circ - 3d\ {}^5P$	1—1
5260,57	2	26,26	28,62	$3s\ {}^3P - 3p\ {}^3P^\circ$?	2—2
5205,11	0	25,23	27,62	$4p\ {}^3S - 6s\ {}^3P^\circ$	1—1
5199,48	1	27,74	30,12	$3p\ {}^5D^\circ - 3d\ {}^5F$	4—3
5191,97	2	27,73	30,12	$3p\ {}^5D^\circ - 3d\ {}^5F$	3—2
5190,380	4	27,74	30,13	$3p\ {}^5D^\circ - 3d\ {}^5F$	4—4
5186,200	2	27,98	30,36	$3p\ {}^5P^\circ - 3d\ {}^5D$	3—2
5184,964	4	27,72	30,12	$3p\ {}^5P^\circ - 3d\ {}^5F$	2—1
		27,73	30,12	$3p\ {}^5D^\circ - 3d\ {}^5F$	3—3
5183,200	4	27,98	30,36	$3p\ {}^5P^\circ - 3d\ {}^5D$	3—3
5180,352	4	27,72	30,12	$3p\ {}^5D^\circ - 3d\ {}^5F$	2—2
5179,52	7	27,74	30,13	$3p\ {}^5D^\circ - 3d\ {}^5F$	4—5
5179,35	7	27,98	30,36	$3p\ {}^5P^\circ - 3d\ {}^5D$	3—4
5177,060	4	27,72	30,12	$3p\ {}^5D^\circ - 3d\ {}^5F$	1—1
5176,563	2	27,97	30,36	$3p\ {}^5P^\circ - 3d\ {}^5D$	2—1
5175,891	6	27,73	30,13	$3p\ {}^5D^\circ - 3d\ {}^5F$	3—4
5174,463	4	27,97	30,36	$3p\ {}^5P^\circ - 3d\ {}^5D$	2—2
5173,386	5	27,72	30,12	$3p\ {}^5D^\circ - 3d\ {}^5F$	2—3
5172,970	3	27,72	30,12	$3p\ {}^5D^\circ - 3d\ {}^5F$	0—1
5172,346	4	27,72	30,12	$3p\ {}^5D^\circ - 3d\ {}^5F$	1—2
5171,45	4	27,97	30,36	$3p\ {}^5P^\circ - 3d\ {}^5D$	2—3
5171,30	2	27,97	30,36	$3p\ {}^5P^\circ - 3d\ {}^5D$	1—0
5170,168	4	27,97	30,36	$3p\ {}^5P^\circ - 3d\ {}^5D$	1—1
5168,99	1	25,23	27,64	$4p\ {}^3S - 6s\ {}^3P^\circ$	1—2
5168,056	4	27,97	30,36	$3p\ {}^5P^\circ - 3d\ {}^5D$	1—2
5104,437	5	22,10	24,53	$3p\ {}^1S - 4s\ {}^1P^\circ$	0—1
5095,58	1	25,20	27,64	$4p\ {}^3P - 6s\ {}^3P^\circ$	2—2
5073,590	5	18,50	20,94	$3s\ {}^1P^\circ - 3p\ {}^3S$	1—1
5046,51	2	25,20	27,65	$4p\ {}^3P - 3p\ {}^3S^\circ$	2—1
5045,100	11	18,47	20,94	$3s\ {}^3P^\circ - 3p\ {}^3S$	2—1
5040,72	3	20,66	23,12	$3p\ {}^3D - 3d\ {}^3F^\circ$	3—2
5028,81	1	25,19	27,65	$4p\ {}^3P - 3p\ {}^3S^\circ$	1—1
5025,662	9	20,66	23,13	$3p\ {}^3D - 3d\ {}^3F^\circ$	3—3
5023,048	5	25,51	27,97	$3s\ {}^5P - 3p\ {}^5P^\circ$	3—2
5022,06	0	25,19	27,65	$4p\ {}^3P - 3p\ {}^3S^\circ$	0—1
5016,387	9	20,65	23,12	$3p\ {}^3D - 3d\ {}^3F^\circ$	2—2
5012,029	6	25,51	27,98	$3s\ {}^5P - 3p\ {}^5P^\circ$	3—3
5011,30	5	25,50	27,97	$3s\ {}^5P - 3p\ {}^5P^\circ$	2—1
5010,620	10	18,46	20,94	$3s\ {}^3P^\circ - 3p\ {}^3S$	1—1
5007,325	11	20,94	23,41	$3p\ {}^3S - 3d\ {}^3P^\circ$	1—2
5005,149	14	25,50	27,97	$3s\ {}^5P - 3p\ {}^5P^\circ$	2—2
		20,66	23,14	$3p\ {}^3D - 3d\ {}^3F^\circ$	3—4
5003,88	0	25,14	27,62	$4p\ {}^3D - 6s\ {}^3P^\circ$	2—1
5002,703	9	18,46	20,94	$3s\ {}^3P^\circ - 3p\ {}^3S$	0—1
5001,477	12	20,64	23,13	$3p\ {}^3D - 3d\ {}^3F^\circ$	2—3
5001,136	11	20,64	23,12	$3p\ {}^3D - 3d\ {}^3F^\circ$	1—2
4997,227	4	25,49	27,97	$3s\ {}^5P - 3p\ {}^5P^\circ$	1—1
4994,363	10	20,94	23,42	$3p\ {}^3S - 3d\ {}^3P^\circ$	1—1
		25,50	27,98	$3s\ {}^5P - 3p\ {}^5P^\circ$	2—3
		25,15	27,64	$4p\ {}^3D - 6s\ {}^3P^\circ$	3—2
4991,240	5	25,49	27,97	$3s\ {}^5P - 3p\ {}^5P^\circ$	1—2
4987,367	8	20,94	23,42	$3p\ {}^3S - 3d\ {}^3P^\circ$	1—0
4895,111	8	17,88	20,41	$2p^3\ {}^1D^\circ - 3p\ {}^1P$	2—1

λ, Å	I	E_H, eV	E_B, eV	Transition	J
4860,170	4	20,64	23,19	$3p\,^3D-3d\,^1D^\circ$	1—2
4810,306	4	20,66	23,24	$3p\,^3D-3d\,^3D^\circ$	3—2
4803,289	10	20,66	23,24	$3p\,^3D-3d\,^3D^\circ$	3—3
4793,650	4	20,64	23,24	$3p\,^3D-3d\,^3D^\circ$	2—1
4788,131	8	20,64	23,24	$3p\,^3D-3d\,^3D^\circ$	2—2
4781,190	4	20,64	23,24	$3p\,^3D-3d\,^3D^\circ$	2—3
4779,722	7	20,64	23,24	$3p\,^3D-3d\,^3D^\circ$	1—1
4774,241	4	20,64	23,24	$3p\,^3D-3d\,^3D^\circ$	1—2
4721,57	2	27,74	30,36	$3p\,^5D^\circ-3d\,^5D$	4—3
4718,38	4	27,74	30,36	$3p\,^5D^\circ-3d\,^5D$	4—4
4712,07	2	27,73	30,36	$3p\,^5D^\circ-3d\,^5D$	3—2
4709,59	2	27,73	30,36	$3p\,^5D^\circ-3d\,^5D$	3—3
4706,40	2	27,73	30,36	$3p\,^5D^\circ-3d\,^5D$	3—4
4704,24	2	27,72	30,36	$3p\,^5D^\circ-3d\,^5D$	2—1
4702,51	2	27,72	30,36	$3p\,^5D^\circ-3d\,^5D$	2—2
4700,04	2	27,72	30,36	$3p\,^5D^\circ-3d\,^5D$	2—3
4698,55	1	27,72	30,36	$3p\,^5D^\circ-3d\,^5D$	1—0
4695,89	2	27,72	30,36	$3p\,^5D^\circ-3d\,^5D$	1—2
4694,637	6	23,57	26,21	$3d\,^1P^\circ-4f\,D\,(2^1/_2)$	1—2
4678,14	6	23,57	26,22	$3d\,^1P^\circ-4f\,D\,(1^1/_2)$	1—2
4674,909	5	18,50	21,15	$3s\,^1P^\circ-3p\,^3P$	1—0
4667,206	5	18,50	21,15	$3s\,^1P^\circ-3p\,^3P$	1—1
4654,532	5	18,50	21,16	$3s\,^1P^\circ-3p\,^3P$	1—2
4643,085	11	18,47	21,15	$3s\,^3P^\circ-3p\,^3P$	2—1
4630,543	14	18,47	21,15	$3s\,^3P^\circ-3p\,^3P$	2—2
4621,394	10	18,46	21,15	$3s\,^3P^\circ-3p\,^3P$	1—0
4613,866	9	18,46	21,15	$3s\,^3P^\circ-3p\,^3P$	1—1
4608,085	3	23,47	26,16	$3d\,^1F^\circ-4f\,F\,(2^1/_2)$	3—3
4607,157	10	18,46	21,15	$3s\,^3P^\circ-3p\,^3P$	0—1
4602,53	3	23,47	26,17	$3d\,^1F^\circ-4f\,F\,(3^1/_2)$	3—3
4601,480	11	18,46	21,16	$3s\,^3P^\circ-3p\,^3P$	1—2
4564,764	3	20,41	23,12	$3p\,^1P-3d\,^3F^\circ$	1—2
4552,527	7	23,47	26,19	$3d\,^1F^\circ-4f\,G\,(3^1/_2)$	3—4
4530,410	9	23,47	26,21	$3d\,^1F^\circ-4f\,G\,(4^1/_2)$	3—4
4508,77	2	23,41	26,16	$3d\,^3P^\circ-4f\,F\,(2^1/_2)$	2—3
4507,557	6	20,66	23,41	$3p\,^3D-3d\,^3P^\circ$	3—2
4488,12	2	20,64	23,41	$3p\,^3D-3d\,^3P^\circ$	2—2
4477,691	4	20,64	23,42	$3p\,^3D-3d\,^3P^\circ$	2—1
4465,527	2	20,64	23,42	$3p\,^3D-3d\,^3P^\circ$	1—1
4459,933	3	20,64	23,41	$3p\,^3D-3d\,^3D^\circ$	1—0
4447,033	12	20,41	23,19	$3p\,^3D-3d\,^3D^\circ$	1—2
4442,018	6	23,42	26,21	$3d\,^3P^\circ-4f\,D\,(2^1/_2)$	1—2
4433,475	5	23,42	26,22	$3d\,^3P^\circ-4f\,D\,(1^1/_2)$	0—1
4432,735	8	23,41	26,21	$3d\,^3P^\circ-4f\,D\,(2^1/_2)$	2—3
4431,816	3	23,41	26,21	$3d\,^3P^\circ-4f\,D\,(2^1/_2)$	2—2
4427,964	4	23,42	26,22	$3d\,^3P^\circ-4f\,D\,(1^1/_2)$	1—1
4427,236	5	23,42	26,22	$3d\,^3P^\circ-4f\,D\,(1^1/_2)$	1—2
4417,82	1	23,41	26,21	$3d\,^3P^\circ-4f\,D\,(1^1/_2)$	2—1
4417,07	4	23,41	26,21	$3d\,^3P^\circ-4f\,D\,(1^1/_2)$	2—2
4374,98	2	20,41	23,24	$3p\,^1P-3d\,^3D^\circ$	1—2
4247,31	1	20,64	23,57	$3p\,^3D-3d\,^1P^\circ$	2—1
4247,20	1	23,24	26,16	$3d\,^3D^\circ-4f\,F\,(2^1/_2)$	3—3
4242,489	3	23,24	26,17	$3d\,^3D^\circ-4f\,F\,(3^1/_2)$	3—3
4241,784	10	23,24	26,17	$3d\,^3D^\circ-4f\,F\,(3^1/_2)$	3—4
		23,24	26,16	$3d\,^3D^\circ-4f\,F\,(2^1/_2)$	2—3
4241,240	3	23,24	26,16	$3d\,^3D^\circ-4f\,F\,(2^1/_2)$	2—2
4237,05	7	23,24	26,17	$3d\,^3D^\circ-4f\,F\,(3^1/_2)$	2—3
4236,91	8	23,24	26,16	$3d\,^3D^\circ-4f\,F\,(2^1/_2)$	1—2

λ, Å	I	E_H, eV	E_B, eV	Transition	J
4227,743	8	21,60	24,53	$3p\ ^1D—4s\ ^1P^\circ$	2—1
4209,09	0	30,36	33,32	$3d\ ^5D—4f\ ^5F^\circ$	4—4
4207,50	3	30,36	33,32	$3d\ ^5D—4f\ ^5F^\circ$	4—5
4206,51	2	30,36	33,32	$3d\ ^5D—4f\ ^5F^\circ$	3—4
4206,11	1	30,36	33,32	$3d\ ^5D—4f\ ^5F^\circ$	2—3
4201,35	1	23,24	26,19	$3d\ ^3D^\circ—4f\ G\ (3^1/_2)$	3—3
4199,980	5	23,24	26,19	$3d\ ^3D^\circ—4f\ G\ (3^1/_2)$	3—4
4195,974	3	23,24	26,19	$3d\ ^3D^\circ—4f\ G\ (3^1/_2)$	2—3
4181,10	2	23,24	26,21	$3d\ ^3D^\circ—4f\ G\ (4^1/_2)$	3—4
4179,674	5	23,24	26,21	$3d\ ^3D^\circ—4f\ D\ (2^1/_2)$	3—3
4178,86	0	23,24	26,21	$3d\ ^3D^\circ—4f\ D\ (2^1/_2)$	3—2
4176,161	8	23,19	26,16	$3d\ ^1D^\circ—4f\ F\ (2^1/_2)$	2—3
4173,572	3	23,24	26,21	$3d\ ^3D^\circ—4f\ D\ (2^1/_2)$	2—2
4171,607	6	23,19	26,17	$3d\ ^1D^\circ—4f\ F\ (3^1/_2)$	2—3
4169,38	1	23,24	26,21	$3d\ ^3D^\circ—4f\ D\ (2^1/_2)$	1—2
4161,14	1	23,24	26,22	$3d\ ^3D^\circ—4f\ D\ (1^1/_2)$	2—1
4160,50	2	23,24	26,22	$3d\ ^3D^\circ—4f\ D\ (1^1/_2)$	2—2
4157,01	3	23,24	26,22	$3d\ ^3D^\circ—4f\ D\ (1^1/_2)$	1—1
4156,39	1	23,24	26,22	$3d\ ^3D^\circ—4f\ D\ (1^1/_2)$	1—2
4154,77	2	30,29	33,28	$3d\ ^5P—4f\ ^5D^\circ$	3—4
4145,776	6	25,51	28,49	$3s\ ^5P—3p\ ^5S^\circ$	3—2
4133,672	5	25,50	28,49	$3s\ ^5P—3p\ ^5S^\circ$	2—2
4131,782	4	23,19	26,19	$3d\ ^1D^\circ—4f\ G\ (3^1/_2)$	2—3
4124,078	4	25,49	28,49	$3s\ ^5P—3p\ ^5S^\circ$	1—2
4114,36	0	20,41	23,42	$3p\ ^1P—3d\ ^3P^\circ$	1—1
4110,83	2	23,19	26,21	$3d\ ^1D^\circ—4f\ D\ (2^1/_2)$	2—3
4110,04	3	23,19	26,21	$3d\ ^1D^\circ—4f\ D\ (2^1/_2)$	2—2
4097,3 coinc.N III		23,19	26,22	$3d\ ^1D^\circ—4f\ D\ (1^1/_2)$	2—2
4096,58	0	23,14	26,17	$3d\ ^3F^\circ—4f\ F\ (3^1/_2)$	4—3
4095,904	4	23,14	26,17	$3d\ ^3F^\circ—4f\ F\ (3^1/_2)$	4—4
4087,303	3	23,13	26,16	$3d\ ^3F^\circ—4f\ F\ (2^1/_2)$	3—3
4082,89	1	23,13	26,17	$3d\ ^3F^\circ—4f\ F\ (3^1/_2)$	3—3
4082,270	5	23,13	26,17	$3d\ ^3F^\circ—4f\ F\ (3^1/_2)$	3—4
4076,908	3	23,12	26,16	$3d\ ^3F^\circ—4f\ F\ (2^1/_2)$	2—2
4073,042	6	23,12	26,17	$3d\ ^3F^\circ—4f\ F\ (3^1/_2)$	2—3
4056,90	4	23,14	26,19	$3d\ ^3F^\circ—4f\ G\ (3^1/_2)$	4—4
4044,777	4	23,13	26,19	$3d\ ^3F^\circ—4f\ G\ (3^1/_2)$	3—3
4043,529	9	23,13	26,19	$3d\ ^3F^\circ—4f\ G\ (3^1/_2)$	3—4
4041,311	11	23,14	26,21	$3d\ ^3F^\circ—4f\ G\ (4^1/_2)$	4—5
4039,345	2	23,14	26,21	$3d\ ^3F^\circ—4f\ G\ (4^1/_2)$	4—4
4037,96	1	23,14	26,21	$3d\ ^3F^\circ—4f\ D\ (2^1/_2)$	4—3
4035,080	9	23,12	26,19	$3d\ ^3F^\circ—4f\ G\ (3^1/_2)$	2—3
4026,075	7	23,13	26,21	$3d\ ^3F^\circ—4f\ G\ (4^1/_2)$	3—4
3994,998	15	18,50	21,60	$3s\ ^1P^\circ—3p\ ^1D$	1—2
3955,851	10	18,46	21,60	$3s\ ^3P^\circ—3p\ ^1D$	1—2
3941,23	1	30,12	33,28	$3d\ ^5F—4f\ ^5G^\circ$	3—4
3940,66	2	30,13	33,28	$3d\ ^5F—4f\ ^5G^\circ$	4—5
3939,57	4	30,13	33,29	$3d\ ^5F—4f\ ^5G^\circ$	5—6
3918,999	9	20,41	23,57	$3p\ ^1P—3d\ ^1P^\circ$	1—1
3856,057	6	21,16	24,38	$3p\ ^3P—4s\ ^3P^\circ$	2—1
3855,100	5	21,15	24,37	$3p\ ^3P—4s\ ^3P^\circ$	1—0
3847,409	5	21,15	24,38	$3p\ ^3P—4s\ ^3P^\circ$	1—1
3842,183	5	21,15	24,38	$3p\ ^3P—4s\ ^3P^\circ$	0—1
3838,374	8	21,16	24,39	$3p\ ^3P—4s\ ^3P^\circ$	2—2
3829,793	6	21,15	24,39	$3p\ ^3P—4s\ ^3P^\circ$	1—2
3615,858	2	20,94	24,37	$3p\ ^3S—4s\ ^3P^\circ$	1—0
3609,097	4	20,94	24,38	$3p\ ^3S—4s\ ^3P^\circ$	1—1
3593,597	5	20,94	24,39	$3p\ ^3S—4s\ ^3P^\circ$	1—2

λ, Å	I	E_H, eV	E_B, eV	Transition	J
3437,147	9	18,50	22,10	$3s\,^1P^\circ - 3p\,^1S$	1—0
3408,127	5	18,46	22,10	$3s\,^3P^\circ - 3p\,^1S$	1—0
3331,310	6	20,64	24,38	$3p\,^3D - 4s\,^3P^\circ$	2—1
3330,314	5	20,61	24,37	$3p\,^3D - 4s\,^3P^\circ$	1—0
3329,704	5	17,88	21,60	$2p^3\,^1D^\circ - 3p\,^1D$	2—2
3328,730	7	20,66	24,39	$3p\,^3D - 4s\,^3P^\circ$	3—2
3324,573	5	20,64	24,37	$3p\,^3D - 4s\,^3P^\circ$	1—1
3318,098	5	20,65	24,39	$3p\,^3D - 4s\,^3P^\circ$	2—2
3206,709	2	23,57	27,44	$3d\,^1P^\circ - 5f\,D\,(2^1/_2)$	1—2
3200,685	2	23,57	27,45	$3d\,^1P^\circ - 5f\,D\,(1^1/_2)$	1—2
3126,40	3	23,47	27,44	$3d\,^1F^\circ - 5f\,G\,(4^1/_2)$	3—4
3086,78	2	23,42	27,44	$3d\,^3P^\circ - 5f\,D\,(2^1/_2)$	1—2
3084,155	2	23,42	27,45	$3d\,^3P^\circ - 5f\,D\,(1^1/_2)$	0—1
3082,191	4	23,41	27,44	$3d\,^3P^\circ - 5f\,D\,(2^1/_2)$	2—3
3081,485	2	23,42	27,45	$3d\,^3P^\circ - 5f\,D\,(1^1/_2)$	1—1
3081,222	2	23,42	27,45	$3d\,^3P^\circ - 5f\,D\,(1^1/_2)$	1—2
3023,668	4	22,10	26,20	$3p\,^1S - 4d\,^1P^\circ$	0—1
3006,830	7	20,41	24,53	$3p\,^1P - 4s\,^1P^\circ$	1—1
2976,971	4	23,24	27,41	$3d\,^3D^\circ - 5f\,F\,(3^1/_2)$	3—4
2974,65	2	23,24	27,41	$3d\,^3D^\circ - 5f\,F\,(3^1/_2)$	2—3
2973,601	3	23,24	27,41	$3d\,^3D^\circ - 5f\,F\,(2^1/_2)$	1—2
2962,953	4	23,24	27,43	$3d\,^3D^\circ - 5f\,G\,(3^1/_2)$	3—4
2943,495	4	23,19	27,41	$3d\,^1D^\circ - 5f\,F\,(2^1/_2)$	2—3
2942,17	3	23,19	27,41	$3d\,^1D^\circ - 5f\,F\,(3^1/_2)$	2—3
2928,655	3	23,19	27,43	$3d\,^1D^\circ - 5f\,G\,(3^1/_2)$	2—3
2923,050	1	23,19	27,44	$3d\,^1D^\circ - 5f\,D\,(2^1/_2)$	2—3
2922,76	1	23,19	27,44	$3d\,^1D^\circ - 5f\,D\,(2^1/_2)$	2—2
2917,734	1	23,19	27,45	$3d\,^1D^\circ - 5f\,D\,(1^1/_2)$	2—2
2904,357	1	23,14	27,41	$3d\,^3F^\circ - 5f\,F\,(3^1/_2)$	4—4
2899,086	1	23,13	27,41	$3d\,^3F^\circ - 5f\,F\,(2^1/_2)$	3—3
2897,503	4	23,13	27,41	$3d\,^3F^\circ - 5f\,F\,(3^1/_2)$	3—4
2893,889	1	23,12	27,41	$3d\,^3F^\circ - 5f\,F\,(2^1/_2)$	2—2
2892,868	4	23,12	27,41	$3d\,^3F^\circ - 5f\,F\,(3^1/_2)$	2—3
2891,046	3	23,14	27,43	$3d\,^3F^\circ - 5f\,G\,(3^1/_2)$	4—4
2885,273	6	23,14	27,44	$3d\,^3F^\circ - 5f\,G\,(4^1/_2)$	4—5
2884,685	2	23,13	27,43	$3d\,^3F^\circ - 5f\,G\,(3^1/_2)$	3—3
2884,246	4	23,13	27,43	$3d\,^3F^\circ - 5f\,G\,(3^1/_2)$	3—4
2879,751	4	23,12	27,43	$3d\,^3F^\circ - 5f\,G\,(3^1/_2)$	2—3
2877,681	4	23,13	27,44	$3d\,^3F^\circ - 5f\,G\,(4^1/_2)$	3—4
2830,36	0	30,13	34,52	$3d\,^5F - 5f\,^5G^\circ$	4—5
2829,358	1	30,13	34,52	$3d\,^5F - 5f\,^5G^\circ$	5—6
2823,635	5	20,67	25,06	$2p^3\,^1P^\circ - 4p\,^1P$	1—1
2799,216	5	21,60	26,03	$3p\,^1D - 4d\,^1D^\circ$	2—2
2734,702	2	22,10 23,57	26,63 28,10	$3p\,^1S - 5s\,^1P^\circ$ $3d\,^1P^\circ - 6f\,D\,(2^1/_2)$	0—1 1—2
2731,37	1	23,57	28,10	$3d\,^1P^\circ - 6f\,D\,(1^1/_2)$	1—2
2709,837	6	21,60	26,17	$3p\,^1D - 4d\,^1F^\circ$	2—3
2690,49	1	23,47	28,08	$3d\,^1F^\circ - 6f\,F\,(3^1/_2)$	3—4
2679,60	1	23,47	28,09	$3d\,^1F^\circ - 6f\,G\,(3^1/_2)$	3—4
2675,78	2	23,47	28,10	$3d\,^1F^\circ - 6f\,G\,(4^1/_2)$	3—4
2646,87	0	23,42	28,10	$3d\,^3P^\circ - 6f\,D\,(2^1/_2)$	1—2
2646,02	0	23,42	28,10	$3d\,^3P^\circ - 6f\,D\,(1^1/_2)$	0—1
2643,93	1	23,42	28,10	$3d\,^3P^\circ - 6f\,D\,(1^1/_2)$	1—2
2643,413	2	23,41	28,10	$3d\,^3P^\circ - 6f\,D\,(2^1/_2)$	2—3
2590,938	5	20,67	25,46	$2p^3\,^1P^\circ - 4p\,^1D$	1—2
2563,319	3	23,24	28,08	$3d\,^3D^\circ - 6f\,F\,(3^1/_2)$	3—4
2561,943	2	23,24	28,08	$3d\,^3D^\circ - 6f\,F\,(2^1/_2)$	2—3
2561,545	1	23,24	28,08	$3d\,^3D^\circ - 6f\,F\,(3^1/_2)$	2—3

λ, Å	I	E_H, eV	E_B, eV	Transition	J
2560,243	3	23,24	28,08	$3d\ ^3D^\circ-6f\ F\ (2^1/_2)$	1—2
2558,62	0	21,16	26,00	$3p\ ^3P-4d\ ^3F^\circ$	2—3
2553,422	4	23,24	28,09	$3d\ ^3D^\circ-6f\ G\ (3^1/_2)$	3—4
2551,64	2	23,24	28,09	$3d\ ^3D^\circ-6f\ G\ (3^1/_2)$	2—3
2537,873	3	23,19	28,08	$3d\ ^1D^\circ-6f\ F\ (2^1/_2)$	2—3
2537,49	0	23,19	28,08	$3d\ ^1D^\circ-6f\ F\ (3^1/_2)$	2—3
2527,762	2	23,19	28,09	$3d\ ^1D^\circ-6f\ G\ (3^1/_2)$	2—3
2526,17	0	21,16	26,06	$3p\ ^3P-4d\ ^3D^\circ$	2—1
2525,48	0	23,19	28,10	$3d\ ^1D^\circ-6f\ D\ (2^1/_2)$	2—3
2524,488	4	21,16	26,06	$3p\ ^3P-4d\ ^3D^\circ$	2—2
2522,458	4	21,15	26,06	$3p\ ^3P-4d\ ^3D^\circ$	1—1
2522,227	7	21,16	26,07	$3p\ ^3P-4d\ ^3D^\circ$	2—3
2520,791	6	21,15	26,06	$3p\ ^3P-4d\ ^3D^\circ$	1—2
2520,222	5	21,15	26,06	$3p\ ^3P-4d\ ^3D^\circ$	0—1
2504,188	4	23,13	28,08	$3d\ ^3F^\circ-6f\ F\ (3^1/_2)$	3—4
2500,672	4	23,12	28,08	$3d\ ^3F^\circ-6f\ F\ (3^1/_2)$	2—3
2499,825	2	23,14	28,09	$3d\ ^3F^\circ-6f\ G\ (4^1/_2)$	4—4
2496,97	4	23,14	28,10	$3d\ ^3F^\circ-6f\ G\ (4^1/_2)$	4—5
2496,83	5	21,16	26,12	$3p\ ^3P-4d\ ^3P^\circ$	2—2
2494,92	0	23,13	28,09	$3d\ ^3F^\circ-6f\ G\ (3^1/_2)$	3—3
2494,71	3	23,13	28,09	$3d\ ^3F^\circ-6f\ G\ (3^1/_2)$	3—4
2493,940	3	21,16	26,13	$3p\ ^3P-4d\ ^3P^\circ$	2—1
2493,16	2	21,15	26,12	$3p\ ^3P-4d\ ^3P^\circ$	1—2
2491,46	3	23,13	28,10	$3d\ ^3F^\circ-6f\ G\ (4^1/_2)$	3—4
2491,21	3	23,12	28,09	$3d\ ^3F^\circ-6f\ G\ (3^1/_2)$	2—3
2490,281	4	21,15	26,13	$3p\ ^3P-4d\ ^3P^\circ$	1—1
		20,67	25,58	$2p^3\ ^1P^\circ-4p\ ^1S$	1—0
2488,746	3	21,15	26,13	$3p\ ^3P-4d\ ^3P^\circ$	1—0
2488,120	2	21,15	26,13	$3p\ ^3P-4d\ ^3P^\circ$	0—1
2461,83	0	23,47	28,51	$3d\ ^1F^\circ-7f\ G\ (4^1/_2)$	3—4
2461,270	6	21,60	26,63	$3p\ ^1D-5s\ ^1P^\circ$	2—1
2390,866	4	20,94	26,12	$3p\ ^3S-4d\ ^3P^\circ$	1—2
2388,230	3	20,94	26,13	$3p\ ^3S-4d\ ^3P^\circ$	1—1
2386,78	1	20,94	26,13	$3p\ ^3S-4d\ ^3P^\circ$	1—0
2364,04	0	23,24	28,48	$3d\ ^3D^\circ-7f\ F\ (3^1/_2)$	2—3
2356,90	0	23,24	28,50	$3d\ ^3D^\circ-7f\ G\ (3^1/_2)$	3—4
2330,855	2	—	—	—	—
2326,340	3	22,10	27,43	$3p\ ^1S-5d\ ^1P^\circ$	0—1
2325,16	0	20,66	25,99	$3p\ ^3D-4d\ ^3F^\circ$	3—2
2321,650	4	20,66	26,00	$3p\ ^3D-4d\ ^3F^\circ$	3—3
2319,941	4	20,65	25,99	$3p\ ^3D-4d\ ^3F^\circ$	2—2
2317,046	8	20,66	26,01	$3p\ ^3D-4d\ ^3F^\circ$	3—4
2316,690	6	20,64	25,99	$3p\ ^3D-4d\ ^3F^\circ$	1—2
2316,493	7	20,65	26,00	$3p\ ^3D-4d\ ^3F^\circ$	2—3
2315,25	0	23,13	28,48	$3d\ ^3F^\circ-7f\ F\ (3^1/_2)$	3—4
2312,13	0	23,12	28,48	$3d\ ^3F^\circ-7f\ F\ (3^1/_2)$	2—3
2309,53	1	23,14	28,51	$3d\ ^3F^\circ-7f\ G\ (4^1/_2)$	4—5
2293,318	4	21,16	26,56	$3p\ ^3P-5s\ ^3P^\circ$	2—1
2292,652	3	21,15	26,56	$3p\ ^3P-5s\ ^3P^\circ$	1—0
2291,652	4	20,66	26,07	$3p\ ^3D-4d\ ^3D^\circ$	3—3
2290,259	3	21,15	26,56	$3p\ ^3P-5s\ ^3P^\circ$	1—1
2289,84	0	20,65	26,06	$3p\ ^3D-4d\ ^3D^\circ$	2—1
2288,444	5	20,65	26,06	$3p\ ^3D-4d\ ^3D^\circ$	2—2
		21,15	26,56	$3p\ ^3P-5s\ ^3P^\circ$	0—1
2286,689	6	21,16	26,58	$3p\ ^3P-5s\ ^3P^\circ$	2—2
		20,64	26,06	$3p\ ^3D-4d\ ^3D^\circ$	1—1
2283,652	4	21,15	26,58	$3p\ ^3P-5s\ ^3P^\circ$	1—2
2238,974	4	20,67	26,21	$2p^3\ ^1P^\circ-4f\ D\ (2^1/_2)$	1—2
2235,208	4	20,67	26,22	$2p^3\ ^1P^\circ-4f\ D\ (1^1/_2)$	1—2

λ, Å	I	E_H, eV	E_B, eV	Transition	J
2218,41	0	20,41	25,99	$3p\,^1P-4d\,^3F^\circ$	1—2
2206,088	6	20,41	26,03	$3p\,^1P-4d\,^1D^\circ$	1—2
2203,633	3	20,94	26,56	$3p\,^3S-5s\,^3P^\circ$	1—1
2197,506	4	20,94	26,58	$3p\,^3S-5s\,^3P^\circ$	1—2
2189,78	2	—	—	—	—
2159,927	3	21,60	27,33	$3p\,^1D-5d\,^1D^\circ$	2—2
2142,775	6	0,02	5,80	$2p^2\,^3P-2p^3\,^5S^\circ$	2—2
2139,007	4	0,01	5,80	$2p^2\,^3P-2p^3\,^5S^\circ$	1—2
2130,179	5	21,60	27,42	$3p\,^1D-5d\,^1F^\circ$	2—3
2096,856	5	20,65	26,56	$3p\,^3D-5s\,^3P^\circ$	2—1
2096,192	4	20,64	26,56	$3p\,^3D-5s\,^3P^\circ$	1—0
2095,532	6	20,66	26,58	$3p\,^3D-5s\,^3P^\circ$	3—2
2094,183	3	20,64	26,56	$3p\,^3D-5s\,^3P^\circ$	1—1
2091,316	3	20,65	26,58	$3p\,^3D-5s\,^3P^\circ$	2—2
2079,968	3	19,23	25,19	$2p^3\,^3S^\circ-4p\,^3P$	1—1
2076,944	4	19,23	25,20	$2p^3\,^3S^\circ-4p\,^3P$	1—2
1991,64	—	20,41	26,63	$3p\,^1P-5s\,^1P^\circ$	1—1
1887,45	4	18,50	25,06	$3s\,^1P^\circ-4p\,^1P$	1—1
1878,60	2	18,46	25,06	$3s\,^3P^\circ-4p\,^1P$	1—1
1868,21	0	18,50	25,13	$3s\,^1P^\circ-4p\,^3D$	1—1
1862,57	2	18,47	25,14	$3s\,^3P^\circ-4p\,^3D$	2—2
1859,22	5	18,47	25,15	$3s\,^3P^\circ-4p\,^3D$	2—3
1858,47	2	18,46	25,13	$3s\,^3P^\circ-4p\,^3D$	0—1
1857,77	3	18,46	25,14	$3s\,^3P^\circ-4p\,^3D$	1—2
1849,41	1	18,50	25,20	$3s\,^1P^\circ-4p\,^3P$	1—2
1848,1	0	18,47	25,19	$3s\,^3P^\circ-4p\,^3P$	2—1
1845,7	—	18,47	25,20	$3s\,^3P^\circ-4p\,^3P$	2—2
1844,	10	18,46	25,19	$3s\,^3P^\circ-4p\,^3P$	1—0
1843,4	0	18,46	25,19	$3s\,^3P^\circ-4p\,^3P$	1—1
1842,5	1	18,46	25,19	$3s\,^3P^\circ-4p\,^3P$	0—1
1841,14	—	18,46	25,20	$3s\,^3P^\circ-4p\,^3P$	1—2
1836,3 6	1	18,47	25,23	$3s\,^3P^\circ-4p\,^3S$	2—1
1831,78	5	18,46	25,23	$3s\,^3P^\circ-4p\,^3S$	1—1
1830,458	4	18,46	25,23	$3s\,^3P^\circ-4p\,^3S$	0—1
1766,08	1	19,23	26,25	$2p^3\,^3S^\circ-3s\,^3P$	1—0
1765,13	1	19,23	26,25	$2p^3\,^3S^\circ-3s\,^3P$	1—1
1763,63	2	19,23	26,26	$2p^3\,^3S^\circ-3s\,^3P$	1—2
1743,197	3	13,54	20,64	$2p^3\,^3P^\circ-3p\,^3D$	1, 2—2
1740,309	4	13,54	20,66	$2p^3\,^3P^\circ-3p\,^3D$	2—3
1675,920	1	13,54	20,94	$2p^3\,^3P^\circ-3p\,^3S$	0—1
1675,744	4	13,54	20,94	$2p^3\,^3P^\circ-3p\,^3S$	1, 2—1
1629,830	4	13,54	21,15	$2p^3\,^3P^\circ-3p\,^3P$	1—0
1629,02	1	13,54	21,15	$2p^3\,^3P^\circ-3p\,^3P$	0, 1, 2—1
1627,42	1	13,54	21,16	$2p^3\,^3P^\circ-3p\,^3P$	1, 2—2
1616,06	1	—	—	—	—
1590,25	2	—	—	—	—
1573,21	1	—	—	—	—
1346,44	0	11,44	20,64	$2p^3\,^3D^\circ-3p\,^3D$	1, 2—1
1345,330	1	11,44	20,64	$2p^3\,^3D^\circ-3p\,^3D$	1, 2—2
1343,338	2	11,43	20,66	$2p^3\,^3D^\circ-3p\,^3D$	3—3
1330,816	3	—	—	—	—
1276,800	2	11,44	21,15	$2p^3\,^3D^\circ-3p\,^3P$	1—0
1276,206	3	11,44	21,15	$2p^3\,^3D^\circ-3p\,^3P$	1, 2—1
1275,247	1	11,44	21,16	$2p^3\,^3D^\circ-3p\,^3P$	1, 2—2
1275,038	4	11,43	21,16	$2p^3\,^3D^\circ-3p\,^3P$	3—2
1258,75	3	—	—	—	—
1085,701	12	0,02	11,43	$2p^2\,^3P^\circ-2p^3\,^3D^\circ$	2—3

λ, Å	I	$E_н$, eV	$E_в$, eV	Transition	J
1085,542	9	0,02	11,44	$2p^2\,^3P—2p^3\,^3D^\circ$	2—2, 1
1084,572	11	0,01	11,44	$2p^2\,^3P—2p^3\,^3D^\circ$	1—1, 2
1083,990	10	0,00	11,44	$2p^2\,^3P—2p^3\,^3D^\circ$	0—1
916,700	12	0,02	13,54	$2p^2\,^3P—2p^3\,^3P^\circ$	2—1, 2
916,004	11	0,01	13,54	$2p^2\,^3P—2p^3\,^3P^\circ$	1—1, 2
915,955	10	0,01	13,54	$2p^2\,^3P—2p^3\,^3P^\circ$	1—0
915,603	10	0,00	13,54	$2p^2\,^3P—2p^3\,^3P^\circ$	0—1
860,205	0	4,05	18,46	$2p^2\,^1S—3s\,^3P^\circ$	0—1
858,374	1	4,05	18,50	$2p^2\,^1S—3s\,^1P^\circ$	0—1
836,837	1	11,44	26,25	$2p^3\,^3D^\circ—3s\,^3P$	1, 2—0
836,618	3	11,44	26,25	$2p^3\,^3D^\circ—3s\,^3P$	1, 2—1
836,281	0	11,44	26,26	$2p^3\,^3D^\circ—3s\,^3P$	1, 2—2
836,184	3	11,43	26,26	$2p^3\,^3D^\circ—3s\,^3P$	3—2
775,957	12	1,90	17,88	$2p^2\,^1D—2p^3\,^1D^\circ$	2—2
748,364	5	1,90	18,46	$2p^2\,^1D—3s\,^3P^\circ$	2—1
746,976	7	1,90	18,50	$2p^2\,^1D—3s\,^1P^\circ$	2—1
745,836	6	4,05	20,67	$2p^2\,^1S—2p^3\,^1P^\circ$	0—1
671,999	6	0,02	18,46	$2p^2\,^3P—3s\,^3P^\circ$	2—1
671,770	6	0,01	18,46	$2p^2\,^3P—3s\,^3P^\circ$	1—0
671,629	6	0,01	18,46	$2p^2\,^3P—3s\,^3P^\circ$	1—1
671,391	8	0,01	18,47	$2p^2\,^3P—3s\,^3P^\circ$	{ 2—2 { 0—1
671,014	6	0,01	18,47	$2p^2\,^3P—3s\,^3P^\circ$	1—2
670,881	1	0,02	18,50	$2p^2\,^3P—3s\,^1P^\circ$	2—1
670,508	1	0,01	18,50	$2p^2\,^3P—3s\,^1P^\circ$	1—1
670,289	2	0,00	18,50	$2p^2\,^3P—3s\,^1P^\circ$	0—1
660,280	9	1,90	20,67	$2p^2\,^1D—2p^3\,^1P^\circ$	2—1
657,327	1	—	—	—	—
645,167	10	0,02	19,23	$2p^2\,^3P—2p^3\,^3S^\circ$	2—1
644,825	9	0,01	19,23	$2p^2\,^3P—2p^3\,^3S^\circ$	1—1
644,621	8	0,00	19,23	$2p^2\,^3P—2p^3\,^3S^\circ$	0—1
635,180	5	4,05	23,57	$2p^2\,^1S—3d\,^1P^\circ$	0—1
629,434	2	5,80	25,50	$2p^3\,^5S^\circ—3s\,^5P$	2—2
629,161	3	5,80	25,51	$2p^3\,^5S^\circ—3s\,^5P$	2—3
582,150	5	1,90	23,19	$2p^2\,^1D—3d\,^1D^\circ$	2—2
574,650	6	1,90	23,47	$2p^2\,^1D—3d\,^1F^\circ$	2—3
559,760	0	4,05	26,20	$2p^2\,^1S—4d\,^1P^\circ$	0—1
547,813	0	1,90	24,53	$2p^2\,^1D—4s\,^1P^\circ$	2—1
533,809	4	0,02	23,24	$2p^2\,^3P—3d\,^3D^\circ$	2—2
533,726	6	0,02	23,24	$2p^2\,^3P—3d\,^3D^\circ$	2—3
533,644	4	0,01	23,24	$2p^2\,^3P—3d\,^3D^\circ$	1—1
533,577	5	0,01	23,24	$2p^2\,^3P—3d\,^3D^\circ$	1—2
533,504	4	0,00	23,24	$2p^2\,^3P—3d\,^3D^\circ$	0—1
529,860	5	0,02	23,41	$2p^2\,^3P—3d\,^3P^\circ$	2—2
529,713	3	0,02	23,41	$2p^2\,^3P—3d\,^3P^\circ$	2—1
529,627	3	0,01	23,41	$2p^2\,^3P—3d\,^3P^\circ$	1—2
529,481	3	0,01	23,42	$2p^2\,^3P—3d\,^3P^\circ$	1—1
529,405	3	0,01	23,42	$2p^2\,^3P—3d\,^3P^\circ$	1—0
529,343	3	0,00	23,42	$2p^2\,^3P—3d\,^3P^\circ$	0—1
513,845	2	1,90	26,03	$2p^2\,^1D—4d\,^1D^\circ$	2—2
510,757	3	1,90	26,17	$2p^2\,^1D—4d\,^1F^\circ$	2—3
509,018	0	0,02	24,37	$2p^2\,^3P—4s\,^3P^\circ$	2—1
508,903	0	0,01	24,37	$2p^2\,^3P—4s\,^3P^\circ$	1—0
508,700	2	0,02	24,39	$2p^2\,^3P—4s\,^3P^\circ$	2—2
508,459	0	0,01	24,39	$2p^2\,^3P—4s\,^3P^\circ$	1—2
506,160	3	5,80	30,29	$2p^3\,^5S^\circ—3d\,^5P$	2—3
506,057	2	5,80	30,29	$2p^3\,^5S^\circ—3d\,^5P$	2—2
505,985	1	5,80	30,30	$2p^3\,^5S^\circ—3d\,^5P$	2—1
485,857	0	1,90	27,42	$2p^2\,^1D—5d\,^1F^\circ$	2—3

λ, Å	I	E_H, eV	E_B, eV	Transition	J
475,876	1	0,02	26,06	$2p^2\ {}^3P-4d\ {}^3D^\circ$	2—2
475,800	3	0,02	26,07	$2p^2\ {}^3P-4d\ {}^3D^\circ$	2—3
475,697	2	0,01	26,06	$2p^2\ {}^3P-4d\ {}^3D^\circ$	1—2
475,638	1	0,00	26,06	$2p^2\ {}^3P-4d\ {}^3D^\circ$	0—1
474,883	2	0,02	26,13	$2p^2\ {}^3P-4d\ {}^3P^\circ$	2—2
474,774	0	0,02	26,13	$2p^2\ {}^3P-4d\ {}^3P^\circ$	2—1
474,698	0	0,01	26,13	$2p^2\ {}^3P-4d\ {}^3P^\circ$	1—2
474,601	0	0,01	26,13	$2p^2\ {}^3P-4d\ {}^3P^\circ$	1—1
474,540	0	0,01	26,13	$2p^2\ {}^3P-4d\ {}^3P^\circ$	1—0
474,493	0	0,00	26,13	$2p^2\ {}^3P-4d\ {}^3P^\circ$	0—1
453,340	1	0,02	27,36	$2p^2\ {}^3P-5d\ {}^3D^\circ$	2—3
453,257	0	0,01	27,36	$2p^2\ {}^3P-5d\ {}^3D^\circ$	1—2

N III, ground state $1s^2 2s^2 2p\ {}^2P^0_{1/2}$

Ionization potential 382625,5 cm⁻¹; 47,436 eV

λ, Å	I	E_H, eV	E_B, eV	Transition	J
6487,55	0	39,35	41,26	$3p\ {}^4P-3d\ {}^4D^\circ$	5/2—3/2
6478,69	2	39,35	41,26	$3p\ {}^4P-3d\ {}^4D^\circ$	5/2—5/2
6468,77	00	39,34	41,26	$3p\ {}^4P-3d\ {}^4D^\circ$	3/2—1/2
6466,86	4	39,35	41,27	$3p\ {}^4P-3d\ {}^4D^\circ$	5/2—7/2
6463,03	2	39,34	41,26	$3p\ {}^4P-3d\ {}^4D^\circ$	3/2—3/2
6453,95	3	39,34	41,26	$3p\ {}^4P-3d\ {}^4D^\circ$	3/2—5/2
6450,78	2	39,34	41,26	$3p\ {}^4P-3d\ {}^4D^\circ$	1/2—1/2
6445,05	2	39,34	41,26	$3p\ {}^4P-3d\ {}^4D^\circ$	1/2—3/2
5314,45	2	39,35	41,68	$3p\ {}^4P-3d\ {}^4P^\circ$	5/2—5/2
5298,93	1	39,35	41,69	$3p\ {}^4P-3d\ {}^4P^\circ$	5/2—3/2
5297,86	1	39,34	41,68	$3p\ {}^4P-3d\ {}^4P^\circ$	3/2—5/2
5282,52	00	39,34	41,69	$3p\ {}^4P-3d\ {}^4P^\circ$	3/2—3/2
5272,60	1	39,34	41,69	$3p\ {}^4P-3d\ {}^4P^\circ$	3/2—1/2
5270,59	1	39,34	41,69	$3p\ {}^4P-3d\ {}^4P^\circ$	1/2—3/2
5260,91	1	39,34	41,69	$3p\ {}^4P-3d\ {}^4P^\circ$	1/2—1/2
4896,71	0	38,40	40,94	$3p\ {}^4D-3d\ {}^4F^\circ$	7/2—5/2
4884,14	1	38,40	40,95	$3p\ {}^4D-3d\ {}^4F^\circ$	7/2—7/2
4881,81	0	38,40	40,94	$3p\ {}^4D-3d\ {}^4F^\circ$	5/2—3/2
4873,58	2	38,40	40,94	$3p\ {}^4D-3d\ {}^4F^\circ$	5/2—5/2
4867,18	5	38,40	40,96	$3p\ {}^4D-3d\ {}^4F^\circ$	7/2—5/2
		38,39	40,94	$3p\ {}^4D-3d\ {}^4F^\circ$	3/2—3/2
4861,33	4	38,40	40,95	$3p\ {}^4D-3d\ {}^4F^\circ$	5/2—7/2
4858,88	3	38,39	40,94	$3p\ {}^4D-3d\ {}^4F^\circ$	3/2—5/2
4858,74	2	38,39	40,94	$3p\ {}^4D-3d\ {}^4F^\circ$	1/2—3/2
4641,90	7	30,46	33,13	$3p\ {}^2P^\circ-3d\ {}^2D$	3/2—3/2
4640,64	10	30,46	33,13	$3p\ {}^2P^\circ-3d\ {}^2D$	3/2—5/2
4634,16	8	30,46	33,13	$3p\ {}^2P^\circ-3d\ {}^2D$	1/2—3/2
4547,34	0	35,67	38,39	$3s\ {}^4P^\circ-3p\ {}^4D$	5/2—3/2
4546,36	3	38,96	41,68	$3p\ {}^4S-3d\ {}^4P^\circ$	3/2—5/2
4544,80	0	38,64	41,37	$4p\ {}^2P^\circ-5s\ {}^2S$	3/2—1/2
4535,11	2	38,96	41,69	$3p\ {}^4S-3d\ {}^4P^\circ$	3/2—3/2
4534,57	3	35,67	38,40	$3s\ {}^4P^\circ-3p\ {}^4D$	5/2—5/2
4530,84	1	35,66	38,39	$3s\ {}^4P^\circ-3p\ {}^4D$	3/2—1/2
4527,86	0	38,96	41,69	$3p\ {}^4S-3d\ {}^4P^\circ$	3/2—1/2
4523,60	4	35,66	38,39	$3s\ {}^4P^\circ-3p\ {}^4D$	2—3/2
4518,18	3	35,65	38,39	$3s\ {}^4P^\circ-3p\ {}^4D$	1/2—1/2

λ, Å	I	E_H, eV	E_B, eV	Transition	J
4514,89	7	35,67	38,40	$3s\ ^4P^\circ$—$3p\ ^4D$	$5/2$—$7/2$
4510,92	6	35,66	38,40	$3s\ ^4P^\circ$—$3p\ ^4D$	$3/2$—$5/2$
		35,65	38,39	$3s\ ^4P^\circ$—$3p\ ^4D$	$1/2$—$3/2$
4379,09	10	39,71	42,54	$4f\ ^2F^\circ$—$5g\ ^2G$	$5/2$, $7/2$—$7/2$, $9/2$
4353,66	2	38,40	41,26	$3p\ ^4D$—$3d\ ^4D^\circ$	$7/2$—$5/2$
4348,36	5	38,40	41,27	$3p\ ^4D$—$3d\ ^4D^\circ$	$7/2$—$7/2$
4339,52	3	38,40	41,26	$3p\ ^4D$—$3d\ ^4D^\circ$	$5/2$—$3/2$
4335,53	4	38,40	41,26	$3p\ ^4D$—$3d\ ^4D^\circ$	$5/2$—$5/2$
4330,44	2	38,39	41,26	$3p\ ^4D$—$3d\ ^4D^\circ$	$3/2$—$1/2$
4330,14	2	38,40	41,27	$3p\ ^4D$—$3d\ ^4D^\circ$	$5/2$—$7/2$
4328,15	3	38,39	41,26	$3p\ ^4D$—$3d\ ^4D^\circ$	$3/2$—$3/2$
4323,93	2	38,39	41,26	$3p\ ^4D$—$3d\ ^4D^\circ$	$1/2$—$1/2$ $3/2$—$5/2$
4321,37	1	38,39	41,26	$3p\ ^4D$—$3d\ ^4D^\circ$	$1/2$—$3/2$
4294,76	0	—	—	—	—
4290,80	3	—	—	—	—
4290,55	1	—	—	—	—
4288,72	1	—	—	—	—
4288,21	0	—	—	—	—
4284,51	1	—	—	—	—
4215,69	3	36,85	39,79	$3s\ ^2P^\circ$—$3p\ ^2D$	$3/2$—$3/2$
4200,02	6	36,85	39,80	$3s\ ^2P^\circ$—$3p\ ^2D$	$3/2$—$5/2$
4195,70	5	36,84	39,79	$3s\ ^2P^\circ$—$3p\ ^2D$	$1/2$—$3/2$
4103,37	9	27,44	30,46	$3s\ ^2S$—$3p\ ^2P^\circ$	$1/2$—$1/2$
4097,31	10	27,44	30,46	$3s\ ^2S$—$3p\ ^2P^\circ$	$1/2$—$3/2$
4003,64	4	39,40	42,50	$4d\ ^2D$—$5f\ ^2F^\circ$	$5/2$—$7/2$
3998,69	3	39,39	42,50	$4d\ ^2D$—$5f\ ^2F^\circ$	$3/2$—$5/2$
3942,78	1	38,33	41,47	$3p\ ^2P$—$3d\ ^2D^\circ$	$3/2$—$3/2$
3938,52	4	38,33	41,48	$3p\ ^2P$—$3d\ ^2D^\circ$	$3/2$—$5/2$
3934,41	3	38,32	41,48	$3p\ ^2P$—$3d\ ^2D^\circ$	$1/2$—$3/2$
3792,87	1	38,40	41,68	$3p\ ^4D$—$3d\ ^4P^\circ$	$7/2$—$5/2$
3779,23	—	38,40	41,68	$3p\ ^4D$—$3d\ ^4P^\circ$	$5/2$—$5/2$
3771,45	2	38,40	41,69	$3p\ ^4D$—$3d\ ^4P^\circ$	$5/2$—$3/2$
3771,08	7	35,67	38,96	$3s\ ^4P^\circ$—$3p\ ^4S$	$5/2$—$3/2$
3770,37	—	38,39	41,68	$3p\ ^4D$—$3d\ ^4P^\circ$	$3/2$—$5/2$
3762,62	—	38,39	41,69	$3p\ ^4D$—$3d\ ^4P^\circ$	$3/2$—$3/2$
3757,66	—	38,39	41,69	$3p\ ^4D$—$3d\ ^4P^\circ$	$3/2$—$1/2$
3757,60	—	38,39	41,69	$3p\ ^4D$—$3d\ ^4P^\circ$	$1/2$—$3/2$
3754,62	6	35,66	38,96	$3s\ ^4P^\circ$—$3p\ ^4S$	$3/2$—$3/2$
3752,65	3	38,39	41,69	$3p\ ^4D$—$3d\ ^4P^\circ$	$1/2$—$1/2$
3745,83	4	35,65	38,96	$3s\ ^4P^\circ$—$3p\ ^4S$	$1/2$—$3/2$
3374,06	6	35,67	39,34	$3s\ ^4P^\circ$—$3p\ ^4P$	$5/2$—$3/2$
3367,36	7	35,67	39,35	$3s\ ^4P^\circ$—$3p\ ^4P$	$5/2$—$5/2$
3365,79	3	35,66	39,34	$3s\ ^4P^\circ$—$3p\ ^4P$	$3/2$—$1/2$
3361,90	2	35,66	39,34	$3s\ ^4P^\circ$—$3p\ ^4P$	$3/2$—$3/2$
3358,72	1	35,65	39,34	$3s\ ^4P^\circ$—$3p\ ^4P$	$1/2$—$1/2$
3355,47	2	36,85	40,55	$3s\ ^2P^\circ$—$3p\ ^2S$	$3/2$—$1/2$
3354,29	4	35,66	39,35	$3s\ ^4P^\circ$—$3p\ ^4P$	$3/2$—$5/2$
3353,78	4	35,65	39,34	$3s\ ^4P^\circ$—$3p\ ^4P$	$1/2$—$3/2$
3342,77	1	36,84	40,55	$3s\ ^2P^\circ$—$3p\ ^2S$	$1/2$—$1/2$
3172,97	2	—	—	—	—
3171,14	1	—	—	—	—
2983,58	6	38,33	42,49	$3p\ ^2P$—$3d\ ^2P^\circ$	$3/2$—$3/2$
2982,07	1	—	—	—	—
2978,87	3	38,33	42,49	$3p\ ^2P$—$3d\ ^2P^\circ$	$1/2$—$3/2$
2977,32	3	38,33	42,50	$3p\ ^2P$—$3d\ ^2P^\circ$	$3/2$—$1/2$
2972,60	4	38,32	42,50	$3p\ ^2P$—$3d\ ^2P^\circ$	$1/2$—$1/2$
2862,26	6	39,71	44,04	$4f\ ^2F^\circ$—$6g\ ^2G$	$7/2$, $9/2$—$7/2$, $9/2$
2714,35	1	28,56	33,13	$2p^3\ ^2P^\circ$—$3d\ ^2D$	$3/2$—$3/2$

λ, Å	I	E_H, eV	E_B, eV	Transition	J
2714,08	3	28,56	33,13	$2p^3\ ^2P^\circ - 3d\ ^2D$	1/2—3/2
2713,95	5	28,56	33,13	$2p^3\ ^2P^\circ - 3d\ ^2D$	3/2—5/2
2696,64	1	—	—	—	—
2696,19	2	—	—	—	—
2689,26	4	39,40	44,01	$4d\ ^2D - 6f\ ^2F^\circ$	5/2—7/2
2687,01	3	39,39	44,01	$4d\ ^2D - 6f\ ^2F^\circ$	3/2—5/2
2622,85	2	42,13	46,86	$3d\ ^2F^\circ - 4p\ ^2D$	7/2—5/2
2621,19	1	42,12	46,85	$3d\ ^2F^\circ - 4p\ ^2D$	5/2—3/2
2486,49	3	41,47	46,46	$3d\ ^2D^\circ - 4p\ ^2P$	3/2—1/2
2484,56	4	41,48	46,47	$3d\ ^2D^\circ - 4p\ ^2P$	5/2—3/2
2482,85	1	41,47	46,47	$3d\ ^2D^\circ - 4p\ ^2P$	3/2—3/2
2471,24	00	41,69	46,71	$3d\ ^4P^\circ - 4p\ ^4D$	1/2—1/2
2468,36	0	41,69	46,71	$3d\ ^4P^\circ - 4p\ ^4D$	1/2—3/2
2466,24	1	41,69	46,71	$3d\ ^4P^\circ - 4p\ ^4D$	3/2—3/2
2463,04	00	41,68	46,71	$3d\ ^4P^\circ - 4p\ ^4D$	5/2—3/2
2462,56	1	41,69	46,72	$3d\ ^4P^\circ - 4p\ ^4D$	3/2—5/2
2459,26	0	41,68	46,72	$3d\ ^4P^\circ - 4p\ ^4D$	5/2—5/2
2453,85	4	41,68	46,73	$3d\ ^4P^\circ - 4p\ ^4D$	5/2—7/2
2372,46	2	41,69	46,92	$3d\ ^4P^\circ - 4p\ ^4S$	1/2—3/2
2370,49	3	41,69	46,92	$3d\ ^4P^\circ - 4p\ ^4S$	3/2—3/2
2367,43	4	41,68	46,92	$3d\ ^4P^\circ - 4p\ ^4S$	5/2—3/2
2322,81	1	41,69	47,02	$3d\ ^4P^\circ - 4p\ ^4P$	3/2—1/2
2322,23	0	41,69	47,03	$3d\ ^4P^\circ - 4p\ ^4P$	1/2—3/2
2320,33	00	41,69	47,03	$3d\ ^4P^\circ - 4p\ ^4P$	3/2—3/2
2317,35	0	41,68 41,69	47,03 47,04	$3d\ ^4P^\circ - 4p\ ^4P$ $3d\ ^4P^\circ - 4p\ ^4P$	5/2—3/2 3/2—5/2
2314,56	1	41,69	47,02	$3d\ ^4P^\circ - 4p\ ^4P$	1/2—1/2
2274,12	0	41,26	46,71	$3d\ ^4D^\circ - 4p\ ^4D$	1/2—1/2
2273,51	1	41,26	46,71	$3d\ ^4D^\circ - 4p\ ^4D$	5/2—3/2
2272,42	0	41,26	46,71	$3d\ ^4D^\circ - 4p\ ^4D$	3/2—3/2
2271,79	0	41,27 41,26	46,72 46,71	$3d\ ^4D^\circ - 4p\ ^4D$ $3d\ ^4D^\circ - 4p\ ^4D$	7/2—5/2 1/2—3/2
2270,43	2	41,26	46,72	$3d\ ^4D^\circ - 4p\ ^4D$	5/2—5/2
2269,30	0	41,26	46,72	$3d\ ^4D^\circ - 4p\ ^4D$	3/2—5/2
2267,28	3	41,27	46,73	$3d\ ^4D^\circ - 4p\ ^4D$	7/2—7/2
2265,87	0	41,26	46,73	$3d\ ^4D^\circ - 4p\ ^4D$	5/2—7/2
2248,88	5	33,13	38,64	$3d\ ^2D - 4p\ ^2P^\circ$	3/2—1/2
2247,92	6	33,13	38,64	$3d\ ^2D - 4p\ ^2P^\circ$	5/2—3/2
2247,65	2	33,13	38,64	$3d\ ^2D - 4p\ ^2P^\circ$	3/2—3/2
2237,21	—	36,85	42,39	$3s\ ^2P^\circ - 5d\ ^2D$	3/2—5/2
2231,65	—	36,84	42,39	$3s\ ^3P^\circ - 5d\ ^2D$	1/2—3/2
2192,52	1	42,50	48,15	$3d\ ^2P^\circ - 4f\ ^4D$	1/2—3/2
2191,39	3	42,49	48,14	$3d\ ^2P^\circ - 4f\ ^4D$	3/2—5/2
2188,52	3	42,50	48,16	$3d\ ^2P^\circ - 4f\ ^2D$	1/2—3/2
2188,27	5	42,49	48,15	$3d\ ^2P^\circ - 4f\ ^2D$	3/2—5/2
2185,13	1	42,49	48,16	$3d\ ^2P^\circ - 4f\ ^2D$	3/2—3/2
2151,61	0	—	—	—	—
2149,96	0	41,26	47,02	$3d\ ^4D^\circ - 4p\ ^4P$	3/2—1/2
2148,99	1	40,94 41,26	46,71 47,03	$3d\ ^4F^\circ - 4p\ ^4D$ $3d\ ^4D^\circ - 4p\ ^4P$	3/2—1/2 3/2—3/2
2148,47	3	40,94	46,71	$3d\ ^4F^\circ - 4p\ ^4D$	5/2—3/2
2148,09	3	40,95	46,72	$3d\ ^4F^\circ - 4p\ ^4D$	7/2—5/2
2147,79	2	41,27	47,04	$3d\ ^4D^\circ - 4p\ ^4P$	7/2—5/2
2147,27	4	40,96	46,73	$3d\ ^4F^\circ - 4p\ ^4D$	9/2—7/2
2146,59	00	41,26	47,04	$3d\ ^4D^\circ - 4p\ ^4P$	5/2—3/2
2145,74	1	40,94	46,72	$3d\ ^4F^\circ - 4p\ ^4D$	5/2—5/2
2143,96	0	40,95	46,73	$3d\ ^4F^\circ - 4p\ ^4D$	7/2—7/2
2142,67	0	—	—	—	—

λ, Å	I	E_H, eV	E_B, eV	Transition	J
2079,86	6	—	—	—	—
2074,74	2	42,13	48,11	$3d\ ^2F^\circ-4f\ ^4G$	$7/2-5/2$
2072,86	1	42,13	48,11	$3d\ ^2F^\circ-4f\ ^4G$	$7/2-7/2$
2071,79	2	—	—	—	—
2070,63	5	42,13	48,12	$3d\ ^2F^\circ-4f\ ^4G$	$7/2-9/2$
2068,25	6	42,13	48,12	$3d\ ^2F^\circ-4f\ ^2G$	$7/2-7/2$
2063,99	10	42,13	48,13	$3d\ ^2F^\circ-4f\ ^2G$	$7/2-9/2$
2063,50	10	42,12	48,12	$3d\ ^2F^\circ-4f\ ^2G$	$5/2-7/2$
2035,62	2	—	—	—	—
2035,02	3	—	—	—	—
1953,80	3	39,34	45,69	$3p\ ^4P-4s\ ^4P^\circ$	$3/2-1/2$
1953,66	3	39,35	45,70	$3p\ ^4P-4s\ ^4P^\circ$	$5/2-3/2$
1952,20	1	39,34	45,69	$3p\ ^4P-4s\ ^4P^\circ$	$1/2-1/2$
1951,43	2	39,34	45,70	$3p\ ^4P-4s\ ^4P^\circ$	$3/2-3/2$
1949,81	4	39,34	45,70	$3p\ ^4P-4s\ ^4P^\circ$	$1/2-3/2$
1949,22	6	39,35	45,71	$3p\ ^4P-4s\ ^4P^\circ$	$5/2-5/2$
1946,99	5	39,34	45,71	$3p\ ^4P-4s\ ^4P^\circ$	$3/2-5/2$
1923,86	2	—	—	—	—
1923,11	2	—	—	—	—
1921,49	4	41,69	48,14	$3d\ ^4P^\circ-4f\ ^4D$	$3/2-5/2$
1920,86	8	41,69 41,68	48,15 48,14	$3d\ ^4P^\circ-4f\ ^4D$ $3d\ ^4P^\circ-4f\ ^4D$	$1/2-3/2$ $5/2-7/2$
1919,99	2	41,69	48,15	$3d\ ^4P^\circ-4f\ ^4D$	$1/2-1/2$
1919,71	2	41,69	48,15	$3d\ ^4P^\circ-4f\ ^4D$	$3/2-3/2$
1919,44	1	41,68	48,14	$3d\ ^4P^\circ-4f\ ^4D$	$5/2-5/2$
1919,06	0	41,69	48,14	$3d\ ^4P^\circ-4f\ ^4D$	$3/2-5/2$
1918,69	0	41,69	48,15	$3d\ ^4P^\circ-4f\ ^4D$	$3/2-1/2$
1908,96	1	41,48	47,97	$3d\ ^2D^\circ-4f\ ^2F$	$5/2-5/2$
1908,11	7	41,47	47,98	$3d\ ^2D^\circ-4f\ ^2F$	$5/2-7/2$ $3/2-5/2$
1907,28	4	41,28	47,98	$3d\ ^2D^\circ-4f\ ^4F$	$5/2-5/2$
1906,89	1	41,48	47,98	$3d\ ^2D^\circ-4f\ ^4F$	$5/2-7/2$
1906,22	1	41,47	47,98	$3d\ ^2D^\circ-4f\ ^4F$	$3/2-5/2$
1885,25	10	33,13	39,71	$3d\ ^2D-4f\ ^2F^\circ$	$3/2$, $5/2-5/2$, $7/2$
1845,80	4	41,26	47,98	$3d\ ^4D^\circ-4f\ ^4F$	$5/2-7/2$
1845,64	5	41,27	47,98	$3d\ ^4D^\circ-4f\ ^4F$	$7/2-9/2$
1841,68	1	38,96	45,69	$3p\ ^4S-4s\ ^4P^\circ$	$3/2-1/2$
1839,59	2	38,96	45,70	$3p\ ^4S-4s\ ^4P^\circ$	$3/2-3/2$
1835,587	6	38,96	45,71	$3p\ ^4S-4s\ ^4P^\circ$	$3/2-5/2$
1805,5	7	30,46	37,33	$3p\ ^2P^\circ-4s\ ^2S$	$3/2-1/2$
1804,3	6	30,46	37,33	$3p\ ^2P^\circ-4s\ ^2S$	$1/2-1/2$
1751,75	10	18,10	25,18	$2p^2\ ^2P-2p^3\ ^2D^\circ$	$3/2-5/2$
1751,24	6	18,10	25,18	$2p^2\ ^2P-2p^3\ ^2D^\circ$	$3/2-3/2$
1747,86	9	18,08	25,18	$2p^2\ ^2P-2p^3\ ^2D^\circ$	$1/2-3/2$
1730,04	8	40,95	48,12	$3d\ ^4F^\circ-4f\ ^4G$	$7/2-9/2$
1699,95	4	38,40	45,69	$3p\ ^4D-4s\ ^4P^\circ$	$5/2-3/2$ $3/2-1/2$
1699,32	5	38,40	45,71	$3p\ ^4D-4s\ ^4P^\circ$	$7/2-5/2$
1699,00	2	38,39	45,69	$3p\ ^4D-4s\ ^4P^\circ$	$1/2-1/2$
1698,16	2	38,39	45,70	$3p\ ^4D-4s\ ^4P^\circ$	$3/2-3/2$
1697,19	0	38,39	45,70	$3p\ ^4D-4s\ ^4P^\circ$	$1/2-3/2$
1696,54	3	38,40	45,71	$3p\ ^4D-4s\ ^4P^\circ$	$5/2-5/2$
1694,79	0	38,39	45,71	$3p\ ^4D-4s\ ^4P^\circ$	$3/2-5/2$
1471,69	2	39,35	47,77	$3p\ ^4P-4d\ ^4D^\circ$	$5/2-7/2$
1471,02	1	39,34	47,77	$3p\ ^4P-4d\ ^4D^\circ$	$3/2-5/2$
1470,68	0	39,34	47,77	$3p\ ^4P-4d\ ^4D^\circ$	$1/2-3/2$
1387,31	4	30,46 30,46	39,39 39,39	$3p\ ^2P^\circ-4d\ ^2D$ $3p\ ^2P^\circ-4d\ ^2D$	$1/2-3/2$ $3/2-5/2$
1347,56	0	38,40	47,61	$3p\ ^4D-4d\ ^4F^\circ$	$7/2-7/2$

λ, Å	I	E_H, eV	E_B, eV	Transition	J
1346,27	4	38,40	47,62	$3p\,^4D-4d\,^4F^\circ$	7/2—9/2
1345,69	4	38,39	47,61	$3p\,^4D-4d\,^4F^\circ$	5/2, 3/2—7/2, 5/2
1324,40	3	38,40	47,75	$3p\,^4D-4d\,^2D^\circ$	7/2—7/2
1184,544	8	18,10	28,56	$2p^2\,^2P-2p^3\,^2P^\circ$	3/2—1/2, 3/2
1183,030	7	18,08	28,56	$2p^2\,^2P-2p^3\,^2P^\circ$	1/2—1/2, 3/2
1006,015	6	16,24	28,56	$2p^2\,^2S-2p^3\,^2P^\circ$	1/2—1/2, 3/2
991,579	17	0,02	12,52	$2p\,^2P^\circ-2p^2\,^2D$	3/2—5/2
991,514	14	0,02	12,52	$2p\,^2P^\circ-2p^2\,^2D$	3/2—3/2
989,790	16	0,00	12,52	$2p\,^2P^\circ-2p^2\,^2D$	1/2—3/2
979,919	9	12,52	25,18	$2p^2\,^2D-2p^3\,^2D^\circ$	5/2—5/2
979,842	8	12,52	25,18	$2p^2\,^2D-2p^3\,^2D^\circ$	3/2—3/2
871,850	0	16,24	30,46	$2p^2\,^2S-3p\,^2P^\circ$	1/2—3/2
772,975	8	12,52	28,56	$2p^2\,^2D-2p^3\,^2P^\circ$	3/2—1/2
772,891	9	12,52	28,56	$2p^2\,^2D-2p^3\,^2P^\circ$	5/2—3/2
772,385	12	7,11	23,16	$2p^2\,^4P-2p^3\,^4S^\circ$	5/2—3/2
771,901	11	7,10	23,16	$2p^2\,^4P-2p^3\,^4S^\circ$	3/2—3/2
771,544	10	7,09	23,16	$2p^2\,^4P-2p^3\,^4S^\circ$	1/2—3/2
764,357	15	0,02	16,24	$2p\,^2P-2p^2\,^2S^\circ$	3/2—1/2
763,340	14	0,00	16,24	$2p\,^2P-2p^2\,^2S^\circ$	1/2—1/2
691,388	1	12,52	30,46	$2p^2\,^2D-3p\,^2P^\circ$	3/2—1/2
691,187	2	12,52	30,46	$2p^2\,^2D-3p\,^2P^\circ$	5/2—3/2
686,335	14	0,02	18,08	$2p\,^2P^\circ-2p^2\,^2P$	3/2—1/2
685,816	16	0,02	18,10	$2p\,^2P^\circ-2p^2\,^2P$	3/2—3/2
685,513	15	0,00	18,08	$2p\,^2P^\circ-2p^2\,^2P$	1/2—1/2
684,996	14	0,00	18,10	$2p\,^2P^\circ-2p^2\,^2P$	1/2—3/2
601,878	0	16,24	36,84	$2p^2\,^2S-3s\,^2P^\circ$	1/2—1/2
601,468	1	16,24	36,85	$2p^2\,^2S-3s\,^2P^\circ$	1/2—3/2
530,268	3	18,10	41,48	$2p^2\,^2P-3d\,^2D^\circ$	3/2—5/2
530,037	2	18,08	41,47	$2p^2\,^2P-3d\,^2D^\circ$	1/2—3/2
509,897	4	12,52	36,84	$2p^2\,^2D-3s\,^2P^\circ$	3/2—1/2
509,586	5	12,52	36,85	$2p^2\,^2D-3s\,^2P^\circ$	5/2—3/2
472,392	5	16,24	42,49	$2p^2\,^2S-3d\,^2P^\circ$	1/2—3/2
472,232	4	16,24	42,50	$2p^2\,^2S-3d\,^2P^\circ$	1/2—1/2
456,078	1	12,52	39,71	$2p^2\,^2D-4f\,^2F^\circ$	5/2—7/2
452,226	11	0,02	27,44	$2p\,^2P^\circ-3s\,^2S$	3/2—1/2
451,869	10	0,00	27,44	$2p\,^2P^\circ-3s\,^2S$	1/2—1/2
434,280	6	7,11	35,66	$2p^2\,^4P-3s\,^4P^\circ$	5/2—3/2
434,246	6	7,10	35,65	$2p^2\,^4P-3s\,^4P^\circ$	3/2—1/2
434,129	5	7,10	35,66	$2p^2\,^4P-3s\,^4P^\circ$	{ 3/2—3/2 { 1/2—1/2
434,066	7	7,11	35,67	$2p^2\,^4P-3s\,^4P^\circ$	5/2—5/2
434,014	6	7,09	35,66	$2p^2\,^4P-3s\,^4P^\circ$	1/2—3/2
433,911	6	7,10	35,67	$2p^2\,^4P-3s\,^4P^\circ$	3/2—5/2
428,244	5	12,52	41,47	$2p^2\,^2D-3d\,^2D^\circ$	3/2—3/2
428,180	6	12,52	41,48	$2p^2\,^2D-3d\,^2D^\circ$	5/2—5/2
418,910	6	12,52	42,12	$2p^2\,^2D-3d\,^2F^\circ$	3/2—5/2
418,705	7	12,52	42,13	$2p^2\,^2D-3d\,^2F^\circ$	5/2—7/2
413,797	0	12,52	42,49	$2p^2\,^2D-3d\,^2P^\circ$	5/2—3/2
413,681	0	12,52	42,50	$2p^2\,^2D-3d\,^2P^\circ$	3/2—1/2
399,084	1	18,10	49,17	$2p^2\,^2P-3d'\,^2D^\circ$	3/2—3/2
399,045	4	18,10	49,17	$2p^2\,^2P-3d'\,^2D^\circ$	3/2—5/2
398,885	3	18,08	49,17	$2p^2\,^2P-3d'\,^2D^\circ$	1/2—3/2
374,441	12	0,02	33,13	$2p\,^2P^\circ-3d\,^2D$	3/2—5/2
374,31	1	0,02	33,13	$2p\,^2P^\circ-3d\,^2D$	3/2—3/2
374,204	11	0,00	33,13	$2p\,^2P^\circ-3d\,^2D$	1/2—3/2
362,985	6	7,11	41,26	$2p^2\,^4P-3d\,^4D^\circ$	5/2—5/2
362,946	8	7,11	41,27	$2p^2\,^4P-3d\,^4D^\circ$	5/2—7/2
362,881	8	7,10	41,26	$2p^2\,^4P-3d\,^4D^\circ$	3/2—3/2, 5/2
362,833	7	7,09	41,26	$2p^2\,^4P-3d\,^4D^\circ$	1/2—1/2, 3/2

λ, Å	I	E_H, eV	E_B, eV	Transition	J
358,578	6	7,11	41,68	$2p^2\ ^4P—3d\ ^4P^\circ$	5/2—5/2
358,509	5	7,11	41,69	$2p^2\ ^4P—3d\ ^4P^\circ$	5/2—3/2
358,469	5	7,10	41,68	$2p^2\ ^4P—3d\ ^4P^\circ$	3/2—5/2
358,401	3	7,10	41,69	$2p^2\ ^4P—3d\ ^4P^\circ$	3/2—3/2
358,356	5	7,10	41,69	$2p^2\ ^4P—3d\ ^4P^\circ$	3/2—1/2
358,327	5	7,09	41,69	$2p^2\ ^4P—3d\ ^4P^\circ$	1/2—3/2
358,278	3	7,09	41,69	$2p^2\ ^4P—3d\ ^4P^\circ$	1/2—1/2
351,979	1	12,52	47,75	$2p^2\ ^2D—4d\ ^2D^\circ$	3/2—3/2
348,800	0	12,52	48,07	$2p^2\ ^2D—4d\ ^2D^\circ$	3/2—5/2
348,690	0	12,52	48,08	$2p^2\ ^2D—4d\ ^2D^\circ$	5/2—7/2
338,345	2	12,52	49,17	$2p^2\ ^2D—4d\ ^2D^\circ$	5/2—5/2 3/2—3/2
332,327	3	0,02	37,33	$2p\ ^2P^\circ—4s\ ^2S$	3/2—1/2
332,133	2	0,00	37,33	$2p\ ^2P^\circ—4s\ ^2S$	1/2—1/2
323,671	4	0,02	38,32	$2p\ ^2P^\circ—3p\ ^2P$	3/2—1/2
323,615	6	0,02	38,33	$2p\ ^2P^\circ—3p\ ^2P$	3/2—3/2
323,488	5	0,00	38,32	$2p\ ^2P^\circ—3p\ ^2P$	1/2—1/2
323,431	4	0,00	38,33	$2p\ ^2P^\circ—3p\ ^2P$	1/2—3/2
321,270	1	7,11	45,69	$2p^2\ ^4P—4s\ ^4P^\circ$	3/2—1/2 5/2—3/2
321,161	2	7,11	45,71	$2p^2\ ^4P—4s\ ^4P^\circ$	5/2—5/2
321,071	1	7,10	45,71	$2p^2\ ^4P—4s\ ^4P^\circ$	3/2—5/2
314,877	6	0,02	39,39	$2p\ ^2P^\circ—4d\ ^2D$	3/2—3/2
314,850	9	0,02	39,40	$2p\ ^2P^\circ—4d\ ^2D$	3/2—5/2
314,715	8	0,00	39,39	$2p\ ^2P^\circ—4d\ ^2D$	1/2—3/2
311,628	3	0,02	39,80	$2p\ ^2P^\circ—3p\ ^2D$	3/2—5/2
311,539	2	0,00	39,79	$2p\ ^2P^\circ—3p\ ^2D$	1/2—3/2
305,918	1	0,02	40,55	$2p\ ^2P^\circ—3p\ ^2S$	3/2—1/2
304,912	3	7,11	47,77	$2p^2\ ^4P—4d\ ^4D^\circ$	5/2—5/2
304,874	4	7,11	47,77	$2p^2\ ^4P—4d\ ^4D^\circ$	5/2—7/2
304,818	4	7,10	47,77	$2p^2\ ^4P—4d\ ^4D^\circ$	3/2—5/2
304,032	2	7,11	47,88	$2p^2\ ^4P—4d\ ^4P^\circ$	5/2—5/2
303,981	2	7,11 7,10	47,88 47,88	$2p^2\ ^4P—4d\ ^4P^\circ$ $2p^2\ ^4P—4d\ ^4P^\circ$	3/2—5/2 5/2—3/2
303,891	2	7,10 7,09	47,88 47,88	$2p^2\ ^4P—4d\ ^4P^\circ$ $2p^2\ ^4P—4d\ ^4P^\circ$	3/2—1/2 1/2—3/2
299,820	1	0,02	41,37	$2p\ ^2P^\circ—5s\ ^2S$	3/2—1/2
299,670	0	0,00	41,37	$2p\ ^2P^\circ—5s\ ^2S$	1/2—1/2
292,595	4	0,02	42,39	$2p\ ^2P^\circ—5d\ ^2D$	3/2—5/2
292,447	3	0,00	42,39	$2p\ ^2P^\circ—5d\ ^2D$	1/2—3/2
284,346	2	7,11	50,71	$2p^2\ ^4P—5d\ ^4D^\circ$	5/2—7/2
284,296	1	7,10 7,09	50,71 50,71	$2p^2\ ^4P—5d\ ^4D^\circ$ $2p^2\ ^4P—5d\ ^4D^\circ$	1/2—3/2 3/2—5/2
282,213	2	0,02	43,95	$2p\ ^2P^\circ—6d\ ^2D$	3/2—5/2
282,093	1	0,00	43,95	$2p\ ^2P^\circ—6d\ ^2D$	1/2—3/2
267,952	1	0,02	46,28	$2p\ ^2P^\circ—3p'\ ^2D$	3/2—5/2
267,851	0	0,00	46,28	$2p\ ^2P^\circ—3p'\ ^2D$	1/2—3/2
264,948	2	0,02	46,81	$2p\ ^2P^\circ—3p'\ ^2P$	3/2—3/2
264,837	1	0,00	46,81	$2p\ ^2P^\circ—3p'\ ^2P$	1/2—1/2

N IV, ground state $1s^2 2s^2\ ^1S^0$
Ionization potential 624864,7 cm⁻¹; 77,468 eV

λ, Å	I	E_H, eV	E_B, eV	Transition	J
7702,96	4	71,42	73,03	$2s6hH^\circ—2s7iI$	—
7582,40	2	71,39	73,02	$2s6g\ ^3G—2s7hH^\circ$	—

λ, Å	I	E_H, eV	E_B, eV	Transition	J
7129,18	–	50,34	52,08	$2s3p\ ^3P^\circ - 2s3d\ ^3D$	2 – 1
7127,27	1	50,34	52,08	$2s3p\ ^3P^\circ - 2s3d\ ^3D$	2 – 2
7122,98	5	50,34	52,08	$2s3p\ ^3P^\circ - 2s3d\ ^3D$	2 – 3
7111,30	1	50,33	52,08	$2s3p\ ^3P^\circ - 2s3d\ ^3D$	1 – 1
7109,40	3	50,33	52,08	$2s3p\ ^3P^\circ - 2s3d\ ^3D$	1 – 2
7103,28	2	50,33	52,08	$2s3p\ ^3P^\circ - 2s3d\ ^3D$	0 – 1
6380,77	8	48,21	50,15	$2s3s\ ^1S - 2s3p\ ^1P^\circ$	0 – 1
6219,89	4	60,45	62,44	$2p3p\ ^3S - 2s4p\ ^3P^\circ$	1 – 2
6215,43	3	60,45	62,45	$2p3p\ ^3S - 2s4p\ ^3P^\circ$	1 – 1
6212,41	1	60,45	62,45	$2p3p\ ^3S - 2s4p\ ^3P^\circ$	1 – 0
6119,23	–	61,78	63,80	$2p3d\ ^1D^\circ - 2s4d\ ^1D$	2 – 2
5843,84	–	61,29	63,42	$2s3p\ ^3P - 2s3d\ ^3P^\circ$	2 – 2
5826,44	–	61,29	63,42	$2s3p\ ^3P - 2s3d\ ^3P^\circ$	2 – 1
5812,30	–	61,28	63,42	$2s3p\ ^3P - 2s3d\ ^3P^\circ$	1 – 2
5795,08	–	61,28	63,42	$2s3p\ ^3P - 2s3d\ ^3P^\circ$	1 – 1
5736,94	4	59,62	61,78	$2p3p\ ^1P - 2p3d\ ^1D^\circ$	1 – 2
5245,61	3	57,72	60,08	$2p3s\ ^3P^\circ - 2p3p\ ^3D$	2 – 2
5226,69	3	57,70	60,07	$2p3s\ ^3P^\circ - 2p3p\ ^3D$	1 – 1
5205,15	3	57,69	60,07	$2p3s\ ^3P^\circ - 2p3p\ ^3D$	0 – 1
5204,29	5	57,72	60,10	$2p3s\ ^3P^\circ - 2p3p\ ^3D$	2 – 3
5200,40	4	57,70	60,08	$2p3s\ ^3P^\circ - 2p3p\ ^3D$	1 – 2
5071,62	–	61,95	64,40	$2p3p\ ^1D - 2p3d\ ^1P^\circ$	2 – 1
4762,10	–	60,10	62,68	$2p3p\ ^3D - 2p3d\ ^3D^\circ$	2 – 2
4752,50	–	60,10	62,69	$2p3p\ ^3D - 2p3d\ ^3D^\circ$	2 – 3
4740,26	–	60,08	62,69	$2p3p\ ^3D - 2p3d\ ^3D^\circ$	1 – 2
4707,31	4	68,75	71,39	$2s5f\ ^3F^\circ - 2s6g\ ^3G$	–
4606,33	6	68,73	71,42	$2s5g\ ^3G - 2s6h\ H^\circ$	–
4529,61	–	57,72	60,45	$2p3s\ ^3P^\circ - 2p3p\ ^3S$	2 – 1
4495,86	–	57,70	60,45	$2p3s\ ^3P^\circ - 2p3p\ ^3S$	1 – 1
4479,92	–	57,69	60,45	$2p3s\ ^3P^\circ - 2p3p\ ^3S$	0 – 1
4182,60	–	60,45	63,42	$2p3p\ ^3S - 2p3d\ ^3P^\circ$	1 – 2
4173,67	–	60,45	63,42	$2p3p\ ^3S - 2p3d\ ^3P^\circ$	1 – 1
4057,759	8	50,15	53,21	$2s3p\ ^1P^\circ - 2s3d\ ^1D$	1 – 2
3823,95	0	59,62	62,86	$2p3p\ ^1P - 2s4p\ ^1P^\circ$	1 – 1
3747,54	6	58,64	61,95	$2p3s\ ^1P^\circ - 2p3p\ ^1D$	1 – 2
3714,43	–	60,10	63,42	$2p3p\ ^3D - 2p3d\ ^3P^\circ$	2 – 2
3694,15	–	60,08	63,42	$2p3p\ ^3D - 2p3d\ ^3P^\circ$	1 – 1
3689,95	–	60,08	63,42	$2p3p\ ^3D - 2p3d\ ^3P^\circ$	1 – 0
3484,96	13	46,77	50,33	$2s3s\ ^3S - 2s3p\ ^3P^\circ$	1 – 0
3482,99	14	46,77	50,33	$2s3s\ ^3S - 2s3p\ ^3P^\circ$	1 – 1
3478,71	15	46,77	50,34	$2s3s\ ^3S - 2s3p\ ^3P^\circ$	1 – 2
3474,55	3	57,72	61,28	$2p3s\ ^3P^\circ - 2p3p\ ^3P$	2 – 1
3463,37	6	57,72	61,29	$2p3s\ ^3P^\circ - 2p3p\ ^3P$	2 – 2
3461,36	2	57,70	61,28	$2p3s\ ^3P^\circ - 2p3p\ ^3P$	1 – 0
3454,70	2	57,70	61,28	$2p3s\ ^3P^\circ - 2p3p\ ^3P$	1 – 1
3445,20	2	57,69	61,28	$2p3s\ ^3P^\circ - 2p3p\ ^3P$	0 – 1
3443,59	3	57,70	61,29	$2p3s\ ^3P^\circ - 2p3p\ ^3P$	1 – 2
3141,16	3	60,10	64,04	$2p3p\ ^3D - 2s4f\ ^3F^\circ$	3 – 4
3127,41	2	60,08	64,04	$2p3p\ ^3D - 2s4f\ ^3F^\circ$	2 – 3
3118,79	1	60,07	64,04	$2p3p\ ^3D - 2s4f\ ^3F^\circ$	1 – 2
3078,25	6	64,70	68,73	$2s4f\ ^1F^\circ - 2s5g\ ^1G$	3 – 4
2884,77	4	68,73	73,02	$2s5g\ ^3G - 2s7h H^\circ$	–
2809,35	2	63,80	68,22	$2s4d\ ^1D - 2s5p\ ^1P^\circ$	2 – 1

λ, Å	I	E_H, eV	E_B, eV	Transition	J
2646,956	12	64,04	68,73	$2s4f\,^3F^\circ - 2s5g\,^3G$	4 – 5
2646,176	11	64,04	68,73	$2s4f\,^3F^\circ - 2s5g\,^3G$	3 – 4
2645,654	10	64,04	68,73	$2s4f\,^3F^\circ - 2s5g\,^3G$	2 – 3
2594,34	2	59,62	64,40	$2p3p\,^1P - 2p3d\,^1P^\circ$	1 – 1
2477,69	8	63,80	68,81	$2s4d\,^1D - 2s5f\,^1F^\circ$	2 – 3
2431,55	0	62,45	67,54	$2s4p\,^3P^\circ - 2s5s\,^3S$	0 – 1
2431,07	2	62,45	67,54	$2s4p\,^3P^\circ - 2s5s\,^3S$	1 – 1
2430,41	3	62,44	67,54	$2s4p\,^3P^\circ - 2s5s\,^3S$	2 – 1
2426,54	1	63,43	68,53	$2p3d\,^3P^\circ - 2s5d\,^3D$	0 – 1
2424,73	2	63,42	68,53	$2p3d\,^3P^\circ - 2s5d\,^3D$	1 – 2
2421,65	3	63,42	68,53	$2p3d\,^3P^\circ - 2s5d\,^3D$	2 – 3
2402,05	5	58,64	63,80	$2p3s\,^1P^\circ - 2s4d\,^1D$	1 – 2
2318,09	6	63,41	68,75	$2s4d\,^3D - 2s5f\,^3F^\circ$	–
2080,34	6	62,77	68,73	$2p3d\,^1F^\circ - 2s5g\,^1G$	3 – 4
2036,42	1	62,45	68,53	$2s4p\,^3P^\circ - 2s5d\,^3D$	0 – 1
2036,10	4	62,45	68,53	$2s4p\,^3P^\circ - 2s5d\,^3D$	1 – 2
2035,57	5	62,44	68,53	$2s4p\,^3P^\circ - 2s5d\,^3D$	2 – 3
1718,551	20	16,20	23,42	$2s2p\,^1P^\circ - 2p^2\,^1D$	1 – 2
1702,006	5	61,44	68,73	$2p3d\,^3F^\circ - 2s5g\,^3G$	4 – 5
1699,03	4	61,43	68,73	$2p3d\,^3F^\circ - 2s5g\,^3G$	3 – 4
1696,86	3	61,42	68,73	$2p3d\,^3F^\circ - 2s5g\,^3G$	2 – 3
1688,11	3	64,04	71,39	$2s4f\,^3F^\circ - 2s6g\,^3G$	4 –
1687,82	2	64,04	71,39	$2s4f\,^3F^\circ - 2s6g\,^3G$	3 –
1687,60	1	64,04	71,39	$2s4f\,^3F^\circ - 2s6g\,^3G$	2 –
1486,496	2	0,00	8,34	$2s^2\,^1S - 2s2p\,^3P^\circ$	0 – 1
1446,114	5	53,21	61,78	$2s3d\,^1D - 2p3d\,^1D^\circ$	2 – 2
1438,37	3	62,77	71,39	$2p3d\,^1F^\circ - 2s6g\,^1G$	3 – 4
1326,964	0	52,08	61,42	$2s3d\,^3D - 2p3d\,^3F^\circ$	1 – 2
1325,685	1	52,08	61,43	$2s3d\,^3D - 2p3d\,^3F^\circ$	2 – 3
1323,98	2	52,08	61,44	$2s3d\,^3D - 2p3d\,^3F^\circ$	3 – 4
1309,557	4	50,15	59,62	$2s3p\,^1P^\circ - 2p3p\,^1P$	1 – 1
1296,600	5	53,21	62,77	$2s3d\,^1D - 2p3d\,^1F^\circ$	2 – 3
1284,218	3	53,21	62,86	$2s3d\,^1D - 2s4p\,^1P^\circ$	2 – 1
1273,716	2	50,33	60,07	$2s3p\,^3P^\circ - 2p3p\,^3D$	1 – 1
1273,47	3	50,33	60,07	$2s3p\,^3P^\circ - 2p3p\,^3D$	0 – 1
1272,74	2	50,34	60,08	$2s3p\,^3P^\circ - 2p3p\,^3D$	2 – 2
1272,160	4	50,33	60,08	$2s3p\,^3P^\circ - 2p3p\,^3D$	1 – 2
1270,28	5	50,34	60,10	$2s3p\,^3P^\circ - 2p3p\,^3D$	2 – 3
1246,51	2	61,44	71,39	$2p3d\,^3F^\circ - 2s6g\,^3G$	4 –
1244,92	1	61,43	71,39	$2p3d\,^3F^\circ - 2s6g\,^3G$	3 –
1243,73	coincides with N V	61,42	71,39	$2p3d\,^3F^\circ - 2s6g\,^2G$	2 –
1225,719	4	50,34	60,45	$2s3p\,^3P^\circ - 2p3p\,^3S$	2 – 1
1225,192	3	50,33	60,45	$2s3p\,^3P^\circ - 2p3p\,^3S$	1 – 1
1224,960	1	50,33	60,45	$2s3p\,^3P^\circ - 2p3p\,^3S$	0 – 1
1188,006	6	48,21	58,64	$2s3s\,^1S - 2p3s\,^1P^\circ$	0 – 1
1169,478	1	52,08	62,68	$2s3d\,^3D - 2p3d\,^3D^\circ$	1 – 1
1169,063	2	52,08	62,68	$2s3d\,^3D - 2p3d\,^3D^\circ$	2 – 2
1168,599	3	52,08	62,69	$2s3d\,^3D - 2p3d\,^3D^\circ$	3 – 3
1136,241	2	46,77	57,69	$2s3s\,^3S - 2p3s\,^3P^\circ$	1 – 0
1135,244	3	46,77	57,70	$2s3s\,^3S - 2p3s\,^3P^\circ$	1 – 1
1133,117	4	46,77	57,72	$2s3s\,^3S - 2p3s\,^3P^\circ$	1 – 2
1086,691	2	50,34	61,75	$2s3p\,^3P^\circ - 2s4s\,^3S$	2 – 1
1086,269	1	50,33	61,75	$2s3p\,^3P^\circ - 2s4s\,^3S$	1 – 1
1086,084	coincides with N I	50,33	61,75	$2s3p\,^3P^\circ - 2s4s\,^3S$	0 – 1

λ, Å	I	E_H, eV	E_B, eV	Transition	J
1078,708	6	53,21	64,70	$2s3d\,^1D - 2s4f\,^1F^{\circ}$	2–3
1036,16	8	52,08	64,04	$2s3d\,^3D - 2s4f\,^3F^{\circ}$	–
955,335	20	16,20	29,18	$2s2p\,^1P^{\circ} - 2p^2\,^1S$	1–0
948,540	5	50,34	63,41	$2s3p\,^3P^{\circ} - 2s4d\,^3D$	2–3
948,244	4	50,33	63,41	$2s3p\,^3P^{\circ} - 2s4d\,^3D$	1–2
948,155	2	50,33	63,41	$2s3p\,^3P^{\circ} - 2s4d\,^3D$	0–1
924,274	10	8,36	21,77	$2s2p\,^3P^{\circ} - 2p^2\,^3P$	2–1
923,669	10	8,34	21,76	$2s2p\,^3P^{\circ} - 2p^2\,^3P$	1–0
923,211	12	8,36	21,79	$2s2p\,^3P^{\circ} - 2p^2\,^3P$	2–2
923,045	10	8,34	21,77	$2s2p\,^3P^{\circ} - 2p^2\,^3P$	1–1
922,507	10	8,33	21,77	$2s2p\,^3P^{\circ} - 2p^2\,^3P$	0–1
921,982	10	8,34	21,79	$2s2p\,^3P^{\circ} - 2p^2\,^3P$	1–2
823,273	2	8,36	23,42	$2s2p\,^3P^{\circ} - 2p^2\,^1D$	2–2
765,140	17	0,00	16,20	$2s^2\,^1S - 2s2p\,^1P^{\circ}$	0–1
463,743	3	23,42	50,15	$2p^2\,^1D - 2s3p\,^1P^{\circ}$	2–1
420,758	1	29,18	58,64	$2p^2\,^1S - 2p3s\,^1P^{\circ}$	0–1
387,353	4	16,20	48,21	$2s2p\,^1P^{\circ} - 2s3s\,^1S$	1–0
352,058	4	29,18	64,40	$2p^2\,^1S - 2p3d\,^1P^{\circ}$	0–1
351,931	5	23,42	58,64	$2p^2\,^1D - 2p3s\,^1P^{\circ}$	2–1
345,201	3	21,77	57,69	$2p^2\,^3P - 2p3s\,^3P^{\circ}$	1–0
345,107	3	21,77	57,70	$2p^2\,^3P - 2p3s\,^3P^{\circ}$	1–1
345,063	5	21,79	57,72	$2p^2\,^3P - 2p3s\,^3P^{\circ}$	2–2
345,023	3	21,76	57,70	$2p^2\,^3P - 2p3s\,^3P^{\circ}$	0–1
344,915	3	21,77	57,72	$2p^2\,^3P - 2p3s\,^3p^{\circ}$	1–2
335,050	11	16,20	53,21	$2s2p\,^1P^{\circ} - 2s3d\,^1D$	1–2
323,175	7	23,42	61,78	$2p^2\,^1D - 2p3d\,^1D^{\circ}$	2–2
322,724	9	8,36	46,77	$2s2p\,^3P^{\circ} - 2s3s\,^3S$	2–1
322,570	8	8,34	46,77	$2s2p\,^3P^{\circ} - 2s3s\,^3S$	1–1
322,503	7	8,33	46,77	$2s2p\,^3P^{\circ} - 2s3s\,^3S$	0–1
315,053	8	23,42	62,77	$2p^2\,^1D - 2p3d\,^1F^{\circ}$	2–3
303,163	4	21,79	62,68	$2p^2\,^3P - 2p3d\,^3D^{\circ}$	2–2
303,123	6	21,79	62,69	$2p^2\,^3P - 2p3d\,^3D^{\circ}$	2–3
303,079	4	21,77	62,68	$2p^2\,^3P - 2p3d\,^3D^{\circ}$	1–1
303,048	5	21,77	62,68	$2p^2\,^3P - 2p3d\,^3D^{\circ}$	1–2
303,009	4	21,76	62,68	$2p^2\,^3P - 2p3d\,^3D^{\circ}$	0–1
300,316	3	23,42	64,70	$2p^2\,^1D - 2s4f\,^1F^{\circ}$	2–3
297,815	5	21,79	63,42	$2p^2\,^3P - 2p3d\,^3P^{\circ}$	2–2
297,768	3	21,79	63,42	$2p^2\,^3P - 2p3d\,^3P^{\circ}$	2–1
297,712	3	21,77	63,42	$2p^2\,^3P - 2p3d\,^3P^{\circ}$	1–2
297,644	4	21,77	63,42	$2p^2\,^3P - 2p3d\,^3P^{\circ}$	1–0,1
297,595	3	21,76	63,42	$2p^2\,^3P - 2p3d\,^3P^{\circ}$	0–1
285,563	5	16,20	59,62	$2s2p\,^1P^{\circ} - 2p3p\,^1P$	1–1
283,579	12	8,36	52,08	$2s2p\,^3P^{\circ} - 2s3d\,^3D$	2–3
283,470	11	8,34	52,08	$2s2p\,^3P^{\circ} - 2s3d\,^3D$	1–2
283,420	10	8,33	52,08	$2s2p\,^3P^{\circ} - 2s3d\,^3D$	0–1
270,995	6	16,20	61,95	$2s2p\,^1P^{\circ} - 2p3p\,^1D$	1–2
260,455	2	16,20	63,80	$2s2p\,^1P^{\circ} - 2s4d\,^1D$	1–2
247,205	10	0,00	50,15	$2s^2\,^1S - 2s3p\,^1P^{\circ}$	0–1
239,693	1	8,36	60,08	$2s2p\,^3P^{\circ} - 2p3p\,^3D$	2–2
239,618	4	8,36	60,10	$2s2p\,^3P^{\circ} - 2p3p\,^3D$	2–3
		8,34	60,08	$2s2p\,^3P^{\circ} - 2p3p\,^3D$	1–2
		8,33	60,07	$2s2p\,^3P^{\circ} - 2p3p\,^3D$	0–1
239,210	2	21,79	73,61	$2p^2\,^3P - 2p4d\,^3D^{\circ}$	2–3
239,161	1	21,77	73,61	$2p^2\,^3P - 2p4d\,^3D^{\circ}$	0–1; 1–2
237,983	3	8,36	60,45	$2s2p\,^3P^{\circ} - 2p3p\,^3S$	2–1
237,903	2	8,34	60,45	$2s2p\,^3P^{\circ} - 2p3p\,^3S$	1–1
237,860	1	8,33	60,45	$2s2p\,^3P^{\circ} - 2p3p\,^3S$	0–1

λ, Å	I	E_H, eV	E_B, eV	Transition	J
234,249	2	8,36	61,28	$2s2p\,^3P^\circ - 2p3p\,^3P$	2—1
234,195	4	8,36	61,29	$2s2p\,^3P^\circ - 2p3p\,^3P$	2—2
234,124	3	{ 8,33	61,28	$2s2p\,^3P^\circ - 2p3p\,^3P$	0—1
		8,34	61,29	$2s2p\,^3P^\circ - 2p3p\,^3P$	1—2
225,205	5	8,36	63,41	$2s2p\,^3P^\circ - 2s4d\,^3D$	2—3
225,136	4	8,34	63,41	$2s2p\,^3P^\circ - 2s4d\,^3D$	1—2
225,098	3	8,33	63,41	$2s2p\,^3P^\circ - 2s4d\,^3D$	0—1
218,085	1	21,77	78,63	$2p^2\,^3P - 2p5d\,^3D^\circ$	2—3
217,227	1	16,20	73,28	$2s2p\,^1P^\circ - 2p4p\,^1D$	1—2
211,396	0	0,00	58,64	$2s^2\,^1S - 2p3s\,^1P^\circ$	0—1
206,021	2	8,36	68,53	$2s2p\,^3P^\circ - 2s5d\,^3D$	2—3
205,960	2	8,33	68,53	$2s2p\,^3P^\circ - 2s5d\,^3D$	0—1; 1—2
197,230	3	0,00	62,86	$2s^2\,^1S - 2s4p\,^1P^\circ$	0—1
197,007	1	8,36	71,29	$2s2p\,^3P^\circ - 2s6d\,^3D$	2—3
196,954	0	8,33	71 29	$2s2p\,^3P^\circ - 2s6d\,^3D$	{ 0−1; 1−2
181,746	1	0,00	68.22	$2s^2\,^1S - 2s5p\,^1P^\circ$	0—1

N V, ground state $1s^2 2s\ ^2S_{1/2}$

Ionization potential 789537,2 cm^{-1}; 97,883 eV

λ, Å	I	E_H, eV	E_B, eV	Transition	J
7628,0	—	90,94	92,57	$7f\,^2F^\circ - 8d\,^2D$	—
7618,46	5	90,94	92,57	$7ghi - 8hik$	—
7615,2	—	90,94	92,57	$7fgh - 8ghi$	—
7600,2	—	90,94	92,57	$7d\,^2D - 8f\,^2F$	—
7327,8	—	90,87	92,57	$7p\,^2P^\circ - 8d\,^2D$	—
6717,2	—	90,68	92,53	$7s\,^2S - 8p\,^2P^\circ$	—
5273,49	—	88,33	90,68	$6p\,^2P^\circ - 7s\,^2S$	—
5067,0	—	88,43	90,87	$6d\,^2D - 7p\,^2P^\circ$	—
4951,27	—	88,44	90,94	$6f\,^2F^\circ - 7d\,^2D$	—
4944,56	9	88,44	90,94	$6fgh - 7ghi$	—
4933,8	—	88,43	90,94	$6d\,^2D - 7f\,^2F^\circ$	—
4749,73	—	88,33	90,94	$6d\,^2P^\circ - 7d\,^2D$	—
4619,98	10	56,55	59,23	$3s\,^2S - 3p\,^2P^\circ$	½ − ½
4603,73	12	56,55	59,24	$3s\,^2S - 3p\,^2P^\circ$	½ − 3/2
4334,0	—	88,02	90,87	$6s\,^2S - 7p\,^2P^\circ$	—
3161,68	3	84,10	88,02	$5p\,^2P^\circ - 6s\,^2S$	3/2 − ½
3159,75	2	84,09	88,02	$5p\,^2P^\circ - 6s\,^2S$	½ − ½
2998,43	5	88,44	92,57	$6fgh - 8ghi$	—
2981,31	10	84,28	88,44	$5g\,^2G - 6h\,^3H^\circ$	—
2980,78	8	84,28	88,44	$5f\,^2F^\circ - 6g\,^2G$	—
2974,52	6	84,27	88,44	$5d\,^2D - 6f\,^2F^\circ$	—
2859,16	5	84,10	88,43	$5p\,^2P^\circ - 6d\,^2D$	3/2 − 5/2
2858,03	4	84,09	88,43	$5p\,^2P^\circ - 6d\,^2D$	½ − 3/2
2591,44	1	83,55	88,33	$5s\,^2S - 6p\,^2P^\circ$	½ − ½
2590,81	2	83,55	88,33	$5s\,^2S - 6p\,^2P^\circ$	½ − 3/2
1882,92	1	84,10	90,68	$5p\,^2P^\circ - 7s\,^2S$	3/2 − ½
1882,36	0	84,09	90,68	$5p\,^2P^\circ - 7s\,^2S$	½ − ½
1860,37	6	84,28	90,94	$5f\,^2F^\circ - 7y\,^2G$	—
1857,88	3	84,27	90,94	$5d\,^2D - 7f\,^2F^\circ$	5/2 − 7/2, 5/2
1857,69	3	84,27	90,94	$5d\,^2D - 7f\,^2F^\circ$	3/2 − 5/2
1811,62	1	84,10	90,94	$5p\,^2P^\circ - 7d\,^2D$	3/2 − 5/2, 3/2

λ, Å	I	E_H, eV	E_B, eV	Transition	J
1811,08	0	84,09	90,94	$5p\,^2P^\circ-7d\,^2D$	$1/2-3/2$
1703,218	4	76,27	83,55	$4p\,^2P^\circ-5s\,^2S$	$3/2-1/2$
1702,25	3	76,26	83,55	$4p\,^2P^\circ-4s\,^2S$	$1/2-1/2$ coincides with NIV
1655,922	2	76,61	84,10	$4d\,^2D-5p\,^2P^\circ$	$3/2-1/2$; $5/2-3/2$
1621,966	1	76,63	84,27	$4f\,^2F^\circ-5d\,^2D$	$5/2-3/2$; $7/2-5/2$
1619,688	12	76,63	84,28	$4f\,^2F^\circ-5g\,^2G$	–
1616,328	9	76,61	84,28	$4d\,^2D-5f\,^2F^\circ$	–
1549,336	6	76,27	84,27	$4p\,^2P^\circ-5d\,^2D$	$3/2-5/2$
1548.	–	76,26	84,27	$4p\,^2P^\circ-5d\,^2D$	$1/2-3/2$ coincides with NIV
1495,5	2	84,28	92,57	$5f\,^2F^\circ-8g\,^2G$	–
1389,822	2	75,17	84,09	$4s\,^2S-5p\,^2P^\circ$	$1/2-1/2$
1389,514	3	75,17	84,10	$4s\,^2S-5p\,^2P^\circ$	$1/2-3/2$
1242,804	19	0,00	9,98	$2s\,^2S-2p\,^2P^\circ$	$1/2-1/2$
1238,821	20	0,00	10,01	$2s\,^2S-2p\,^2P^\circ$	$1/2-3/2$
1049,65	3	76,63	88,44	$4f\,^2F^\circ-6g\,^2G$	–
1048,20	2	76,61	88,44	$4d\,^2D-6f\,^2F^\circ$	–
778,172	2	59,24	75,17	$3p\,^2P^\circ-4s\,^2S$	$3/2-1/2$
777,712	1	59,23	75,17	$3p\,^2P^\circ-4s\,^2S$	$1/2-1/2$
748,291	9	60,06	76,63	$3d\,^2D-4f\,^2F^\circ$	$5/2-7/2$
748,195	8	60,06	76,63	$3d\,^2D-4f\,^2F^\circ$	$3/2-5/2$
713,860	8	59,24	76,61	$3p\,^2P^\circ-4d\,^2D$	$3/2-5/2$
713,518	6	59,23	76,61	$3p\,^2P^\circ-4d\,^2D$	$1/2-3/2$
628,874	3	56,55	76,26	$3s\,^2S^\circ-4p\,^2P^\circ$	$1/2-1/2$
628,744	5	56,55	76,27	$2s\,^2S-4p\,^2P^\circ$	$1/2-3/2$
511,834	5	60,06	84,28	$2d\,^2D-5f\,^2F^\circ$	–
450,08	3	56,55	84,10	$3s\,^2S-5p\,^2P^\circ$	$1/2-3/2$, $1/2$
436,85	4	60,06	88,44	$3d\,^2D-6g\,^2F^\circ$	–
424,75	2	59,24	88,43	$3p\,^2P^\circ-6d\,^2D$	$3/2-5/2$
424,61	1	59,23	88,43	$3p\,^2P^\circ-6d\,^2D$	$1/2-3/2$
266,378	84	10,01	56,55	$2p\,^2P^\circ-3s\,^2S$	$3/2-1/2$
266,197	80	9,98	56,55	$2p\,^2P^\circ-3s\,^2S$	$1/2-1/2$
247,709	100	10,01	60,06	$2p\,^2P^\circ-3d\,^2D$	$3/2-5/2$
247,564	85	9,98	60,05	$2p\,^2P^\circ-3d\,^2D$	$1/2-3/2$
209,306	80	0,00	59,23	$2s\,^2S-3p\,^2P^\circ$	$1/2-1/2$
209,274	80	0,00	59,24	$2s\,^2S-3p\,^2P^\circ$	$1/2-3/2$
190,250	32	10,01	75,17	$2p\,^2P^\circ-4s\,^2S$	$3/2-1/2$
190,158	20	9,98	75,17	$2p\,^2P^\circ-4s\,^2S$	$1/2-1/2$
186,153	62	10,01	76,61	$2p\,^2P^\circ-4d\,^2D$	$3/2-5/2$
186,069	52	9,98	76,61	$2p\,^2P^\circ-4d\,^2D$	$1/2-3/2$
168,590	12	10,01	83,54	$2p\,^2P^\circ-5s\,^2S$	$3/2-1/2$
168,517	5	9,98	83,57	$2p\,^2P^\circ-5s\,^2S$	$1/2-1/2$
166,947	52	10,01	84,27	$2p\,^2P^\circ-5d\,^2D$	$3/2-5/2$
166,881	44	9,98	84,27	$2p\,^2P^\circ-5d\,^2D$	$1/2-3/$
162,565	48	0,00	76,26	$2s\,^2S-4p\,^2P^\circ$	$1/2-3/$
158,933	7	10,01	88,02	$2p\,^2P^\circ-6s\,^2S$	$3/2-1/2$
158,867	4	9,98	88,02	$2p\,^2P^\circ-6s\,^2S$	$1/2-1/2$
158,090	36	10,01	88,43	$2p\,^2P^\circ-6d\,^2D$	$3/2-5/2$
158,030	24	9,98	88,43	$2p\,^2P^\circ-6d\,^2D$	$1/2-3/2$
153,683	6	10,01	90,68	$2p\,^2P^\circ-7s\,^2S$	$3/2-1/2$
153,624	3	9,98	90,68	$2p\,^2P^\circ-7s\,^2S$	$1/2-1/2$
153,192	28	10,01	90,93	$2p\,^2P^\circ-7s\,^2S$	$3/2-5/2$
153,136	18	9,98	90,93	$2p\,^2P^\circ-7d\,^2D$	$1/2-3/2$
150,488	5	10,01	92,39	$2p\,^2P^\circ-8s\,^2S$	$3/2-1/2$

λ, Å	I	E_H, eV	E_B, eV	Transition	J
150,429	2	9,98	92,39	$2p\ ^2P^\circ$—$8s\ ^2S$	$^1/_2$—$^1/_2$
150,171	14	10,01	92,57	$2p\ ^2P^\circ$—$8d\ ^2D$	$^3/_2$—$^5/_2$
150,116	7	9,98	92,57	$2p\ ^2P^\circ$—$8d\ ^2D$	$^1/_2$—$^3/_2$
148,387	4	10,01	93,55	$2p\ ^2P^\circ$—$9s\ ^2S$	$^3/_2$—$^1/_2$
148,328	1	9,98	93,55	$2p\ ^2P^\circ$—$9s\ ^2S$	$^1/_2$—$^1/_2$
148,168	7	10,01	93,68	$2p\ ^2P^\circ$—$9d\ ^2D$	$^3/_2$—$^5/_2$
148,116	4	9,98	93,68	$2p\ ^2P^\circ$—$9d\ ^2D$	$^1/_2$—$^3/_2$
147,433	24	0,00	84,09	$2s\ ^2S$—$5p\ ^2P^\circ$	$^1/_2$—$^3/_2$
146,921	3	10,01	94,39	$2p\ ^2P^\circ$—$10s\ ^2S$	$^3/_2$—$^1/_2$
146,767	6	10,01	94,48	$2p\ ^2P^\circ$—$10d\ ^2D$	$^3/_2$—$^3/_2$
146,716	3	9,98	94,48	$2p\ ^2P^\circ$—$10d\ ^2D$	$^1/_2$—$^3/_2$
145,742	5	10,01	95,07	$2p\ ^2P^\circ$—$11d\ ^2D$	$^3/_2$—$^5/_2$
144,978	4	10,01	95,51	$2p\ ^2P^\circ$—$12d\ ^2D$	$^3/_2$—$^5/_2$
144,392	3	10,01	95,85	$2p\ ^2P^\circ$—$13d\ ^2D$	$^3/_2$—$^5/_2$
143,914	2	10,01	96,15	$2p\ ^2P^\circ$—$14d\ ^2D$	$^3/_2$—$^5/_2$
143,520	1	10,01	96,39	$2p\ ^2P^\circ$—$15d\ ^2D$	$^3/_2$—$^5/_2$
143,241	1	10,01	96,55	$2p\ ^2P^\circ$—$16d\ ^2D$	$^3/_2$—$^5/_2$
142,981	0	10,01	96,72	$2p\ ^2P^\circ$—$17d\ ^2D$	$^3/_2$—$^5/_2$
142,797	0	10,01	96,83	$2p\ ^2P^\circ$—$18d\ ^2D$	$^3/_2$—$^5/_2$
140,364	16	0,00	88,33	$2s\ ^2S$—$6p\ ^2P^\circ$	$^1/_2$—$^3/_2$
136,429	8	0,00	90,87	$2s\ ^2S$—$7p\ ^2P^\circ$	$^1/_2$—$^3/_2$
133,994	7	0,00	92,52	$2s\ ^2S$—$8p\ ^2P^\circ$	$^1/_2$—$^3/_2$
132,383	6	0,00	93,65	$2s\ ^2S$—$9p\ ^2P^\circ$	$^1/_2$—$^3/_2$
131,254	5	0,00	94,46	$2s\ ^2S$—$10p\ ^2P^\circ$	$^1/_2$—$^3/_2$
130,431	4	0,00	95,05	$2s\ ^2S$—$11p\ ^2P^\circ$	$^1/_2$—$^3/_2$
129,811	3	0,00	95,50	$2s\ ^2S$—$12p\ ^2P^\circ$	$^1/_2$—$^3/_2$
129,337	2	0,00	95,84	$2s\ ^2S$—$13p\ ^2P^\circ$	$^1/_2$—$^3/_2$
128,954	1	0,00	96,14	$2s\ ^2S$—$14p\ ^2P^\circ$	$^1/_2$—$^3/_2$
128,662	1	0,00	96,36	$2s\ ^2S$—$15p\ ^2P^\circ$	$^1/_2$—$^3/_2$
128,430	0	0,00	96,53	$2s\ ^2S$—$16p\ ^2P^\circ$	$^1/_2$—$^3/_2$
128,229	0	0,00	96,68	$2s\ ^2S$—$17p\ ^2P^\circ$	$^1/_2$—$^3/_2$

N VI, ground state $1s^2\ ^1S_0$

Ionization potential 4452800 cm^{-1}; 552,04 eV

λ, Å	I	E_H, eV	E_B, eV	Transition	J
29,084	—	0,00	426,27	$1s^2\ ^1S$—$2p\ ^3P^\circ$	0—1
28,787	—	0,00	430,67	$1s^2\ ^1S$—$2p\ ^1P^\circ$	0—1
24,898	—	0,00	497,94	$1s^2\ ^1S$—$3p\ ^1P^\circ$	0—1
23,771	—	0,00	521,55	$1s^2\ ^1S$—$4p\ ^1P^\circ$	0—1

OXYGEN, Z = 8

O I, ground state $1s^2 2s^2 2p^4\ ^3P_2$
Ionization potential 109 837,03 cm^{-1}; 13,617 eV

λ, Å	I	E_H, eV	E_B, eV	Transition	J
18243,63	22	12,08	12,76	$3d\ ^3D^\circ$—$4f\ ^3F$	3, 2, 1—4, 3, 2
18021,21	23	12,08	12,76	$3d\ ^5D^\circ$—$4f\ ^5F$	—
13165,11	24	10,99	11,93	$3p\ ^3P$—$4s\ ^3S^\circ$	0—1
13164,85	26	10,99	11,93	$3p\ ^3P$—$4s\ ^3S^\circ$	2—1
13163,89	25	10,99	11,93	$3p\ ^3P$—$4s\ ^3S^\circ$	1—1
12570,04	20	12,09	13,07	$3d\ ^3D^\circ$—$5f\ ^3F$	3, 2, 1—4, 3, 2
12464,02	21	12,08	13,07	$3d\ ^5D^\circ$—$5f\ ^5F$	—
11302,376	23	10,74	11,84	$3p\ ^5P$—$4s\ ^5S^\circ$	3—2
11297,682	22	10,74	11,84	$3p\ ^5P$—$4s\ ^5S^\circ$	2—2
11295,104	21	10,74	11,84	$3p\ ^5P$—$4s\ ^5S^\circ$	1—2
11287,318	21	10,99	12,09	$3p\ ^3P$—$3d\ ^3D^\circ$	0—1
11287,022	21	10,99	12,09	$3p\ ^3P$—$3d\ ^3D^\circ$	2—2, 1
11286,914	24	10,99	12,09	$3p\ ^3P$—$3d\ ^3D^\circ$	2—3
11286,344	23	10,99	12,09	$3p\ ^3P$—$3d\ ^3D^\circ$	1—2, 1
10753,530	17	12,09	13,24	$3d\ ^3D^\circ$—$6f\ ^3F$	3, 2, 1—4, 3, 2
10675,940	16	12,08	13,24	$3d\ ^5D^\circ$—$6f\ ^5F$	—
10675,725	17	12,08	13,24	$3d\ ^5D^\circ$—$6f\ ^5F$	—
10167,252	10	9,52	10,74	$3s\ ^3S^\circ$—$3p\ ^5P$	1—2
9891,743	13	12,08	13,34	$3d\ ^3D^\circ$—$7f\ ^3F$	3, 2, 1—4, 3, 2
9826,002	12	12,08	13,34	$3d\ ^5D^\circ$—$7f\ ^5F$	2, 1, 0—3, 2
9825,847	13	12,08	13,34	$3d\ ^5D^\circ$—$7f\ ^5F$	4, 3—4, 3, 2
9760,65	5	14,13	15,40	$3p'\ ^1F$—$3d'\ ^1G^\circ$	3—4
		14,13	15,40	$3p'\ ^1F$—$3d'\ ^3G^\circ$	3—4
9741,49	4	14,13	15,40	$3p'\ ^1F$—$3d'\ ^3G^\circ$	3—4
9677,41	1	14,13	15,41	$3p'\ ^1F$—$3d'\ ^1F^\circ$	3—3
9522,01	4	14,10	15,40	$3p'\ ^3F$—$3d'\ ^3F^\circ$	4—4
9505,67	5	14,10	15,40	$3p'\ ^3F$—$3d'\ ^1G^\circ$	3—4
9499,39	0	14,10	15,40	$3p'\ ^3F$—$3d'\ ^1G^\circ$	4—4
9498,04	8	14,10	15,40	$3p'\ ^3F$—$3d'\ ^3G^\circ$	4—5
		14,10	15,40	$3p'\ ^3F$—$3d'\ ^3G^\circ$	2—3
9492,76	1	14,10	15,40	$3p'\ ^3F$—$3d'\ ^3G^\circ$	3—3
9487,49	6	14,10	15,40	$3p'\ ^3F$—$3d'\ ^3G^\circ$	3—4
9399,24	1	12,73	14,05	$3s'\ ^1D^\circ$—$3p'\ ^3D$	2—1
9266,006	24	10,74	12,08	$3p\ ^5P$—$3d\ ^5D^\circ$	3—4
9265,938	21	10,74	12,08	$3p\ ^5P$—$3d\ ^5D^\circ$	3—3
9262,774	23	10,74	12,08	$3p\ ^5P$—$3d\ ^5D^\circ$	2—3
9262,671	22	10,74	12,08	$3p\ ^5P$—$3d\ ^5D^\circ$	2—2
9262,584	19	10,74	12,08	$3p\ ^5P$—$3d\ ^5D^\circ$	2—1
9260,935	20	10,74	12,08	$3p\ ^5P$—$3d\ ^5D^\circ$	1—2
9260,845	21	10,74	12,08	$3p\ ^5P$—$3d\ ^5D^\circ$	1—1
9260,806	20	10,74	12,08	$3p\ ^5P$—$3d\ ^5D^\circ$	1—0
9156,02	4	14,05	15,40	$3p'\ ^3D$—$3d'\ ^3F^\circ$	3—4
8820,45	15	12,73	14,13	$3s'\ ^1D^\circ$—$3p'\ ^1F$	2—3
8819,60	5	—	—	—	—
8586,00	2	—	—	—	—
8508,66	2	14,37	15,83	$3s''\ ^1P^\circ$—$3p''\ ^1P$	1—1
8446,758	29	9,52	10,99	$3s\ ^3S^\circ$—$3p\ ^3P$	1—1
8446,359	30	9,52	10,99	$3s\ ^3S^\circ$—$3p\ ^3P$	1—2
8446,250	27	9,52	10,99	$3s\ ^3S^\circ$—$3p\ ^3P$	1—0
8429,128	1	14,12	15,59	$3s''\ ^3P^\circ$—$4p'\ ^3D$	0—1
8428,342	2	14,12	15,59	$3s''\ ^3P^\circ$—$4p'\ ^3D$	1—2
8426,326	4	14,12	15,59	$3s''\ ^3P^\circ$—$4p'\ ^3D$	2—3
8424,780	1	14,12	15,59	$3s''\ ^3P^\circ$—$4p'\ ^3D$	1—1
8420,968	1	14,12	15,59	$3s''\ ^3P^\circ$—$4p'\ ^3D$	2—2

λ, Å	I	E_H, eV	E_B, eV	Transition	J
8235,408	5	12,54	14,05	$3s'\,^3D^\circ$—$3p'\,^3D$	1—2
8233,085	13	12,54	14,05	$3s'\,^3D^\circ$—$3p'\,^3D$	1—1
8230,016	10	12,54	14,05	$3s'\,^3D^\circ$—$3p'\,^3D$	2—2, 3
8227,680	10	12,54	14,05	$3s'\,^3D^\circ$—$3p'\,^3D$	2—1
8221,84	15	12,54	14,05	$3s'\,^3D^\circ$—$3p'\,^3D$	3—3
8221,829	20	12,54	14,05	$3s'\,^3D^\circ$—$3p'\,^3D$	3—2
7995,074	15	10,99	12,54	$3p\,^3P$—$3s'\,^3D^\circ$	2—3
7987,333	11	10,99	12,54	$3p\,^3P$—$3s'\,^3D^\circ$	2—2
7986,977	13	10,99	12,54	$3p\,^3P$—$3s'\,^3D^\circ$	1—2
7982,398	11	10,99	12,54	$3p\,^3P$—$3s'\,^3D^\circ$	0—1
7981,941	10	10,99	12,54	$3p\,^3P$—$3s'\,^3D^\circ$	1—1
7952,182	9	12,54	14,10	$3s'\,^3D^\circ$—$3p'\,^3F$	1—2
7950,824	10	12,54	14,10	$3s'\,^3D^\circ$—$3p'\,^3F$	2—3
7947,566	10	12,54	14,10	$3s'\,^3D^\circ$—$3p'\,^3F$	3—4
7947,204	3	12,54	14,10	$3s'\,^3D^\circ$—$3p'\,^3F$	2—2
7943,178	6	12,54	14,10	$3s'\,^3D^\circ$—$3p'\,^3F$	3—3
7939,49	1	12,54	14,10	$3s'\,^3D^\circ$—$3p'\,^3F$	3—2
7886,31	4	14,37	15,94	$3s''\,^1P^\circ$—$3p'\,^1D$	1—2
7775,388	26	9,14	10,74	$3s\,^5S^\circ$—$3p\,^5P$	2—1
7774,166	27	9,14	10,74	$3s\,^5S^\circ$—$3p\,^5P$	2—2
7771,943	28	9,14	10,74	$3s\,^5S^\circ$—$3p\,^5P$	2—3
7706,77	5	14,05	15,65	$3p'\,^3D$—$2p^5\,^3P^\circ$	2, 3—2
7665,48	1	14,05	15,66	$3p'\,^3D$—$2p^5\,^3P^\circ$	1—1
7663,45	3	14,05	15,66	$3p'\,^3D$—$2p^5\,^3P^\circ$	2—1
7639,99	1	14,05	15,67	$3p'\,^3D$—$2p^5\,^3P^\circ$	1—0
7480,652	8	14,12	15,78	$3s''\,^3P^\circ$—$3p''\,^3D$	0—1
7479,148	8	14,12	15,78	$3s''\,^3P^\circ$—$3p''\,^3D$	1—2
7477,264	7	14,12	15,78	$3s''\,^3P^\circ$—$3p''\,^3D$	1—1
7476,473	12	14,12	15,78	$3s''\,^3P^\circ$—$3p''\,^3D$	2—3
7473,226	5	14,12	15,78	$3s''\,^3P^\circ$—$3p''\,^3D$	2—2
7471,374	2	14,12	15,78	$3s''\,^3P^\circ$—$3p''\,^3D$	2—1
7254,529	17	10,99	12,70	$3p\,^3P$—$5s\,^3S^\circ$	0—1
7254,447	20	10,99	12,70	$3p\,^3P$—$5s\,^3S^\circ$	2—1
7254,154	19	10,99	12,70	$3p\,^3P$—$5s\,^3S^\circ$	1—1
7157,360	7	—	—	—	—
7156,80	12	12,73	14,46	$3s'\,^1D^\circ$—$3p'\,^1D$	2—2
7025,52	3	12,36	14,12	$4p\,^3P$—$3s''\,^3P^\circ$	2, 1—2
7002,228	17	10,99	12,76	$3p\,^3P$—$4d\,^3D^\circ$	2, 0—3, 2, 1
7001,915	15	10,99	12,76	$3p\,^3P$—$4d\,^3D^\circ$	1—2, 1
6726,538	6	9,14	10,99	$3s\,^5S^\circ$—$3p\,^3P$	2—1
6726,283	9	9,14	10,99	$3s\,^5S^\circ$—$3p\,^3P$	2—2
6653,78	5	14,37	16,23	$3s''\,^1P^\circ$—$3p''\,^1S$	1—0
6455,975	19	10,74	12,66	$3p\,^5P$—$5s\,^5S^\circ$	3—2
6454,445	18	10,74	12,66	$3p\,^5P$—$5s\,^5S^\circ$	2—2
6453,602	17	10,74	12,66	$3p\,^5P$—$5s\,^5S^\circ$	1—2
6374,292	4	14,13	16,08	$3p'\,^1F$—$4d'\,^1G^\circ$	3—4
6366,282	3	14,13	16,08	$3p'\,^1F$—$4d'\,^3G^\circ$	3—4
6324,682	3	12,09	14,05	$3d\,^3D^\circ$—$3p'\,^3D$	3, 2, 1—2, 3
6323,283	1	12,09	14,05	$3d\,^3D^\circ$—$3p'\,^3D$	2, 1—1
6266,89	3	14,10	16,08	$3p'\,^3F$—$4d'\,^3F^\circ$	4—4
6264,346	3	14,10	16,08	$3p'\,^3F$—$4d'\,^1G^\circ$	3—4
6261,55	6	14,10	16,08	$3p'\,^3F$—$4d'\,^3G^\circ$	2—3
		14,10	16,08	$3p'\,^3F$—$4d'\,^3G^\circ$	4—5
6259,22	0	14,10	16,08	$3p'\,^3F$—$4d'\,^3G^\circ$	3—3
6256,84	4	14,10	16,08	$3p'\,^3F$—$4d'\,^3G^\circ$	3—4
6158,183	21	10,74	12,75	$3p\,^5P$—$4d\,^5D^\circ$	3—4, 3, 2
6156,765	20	10,74	12,75	$3p\,^5P$—$4d\,^5D^\circ$	2—3, 2, 1
6155,975	19	10,74	12,75	$3p\,^5P$—$4d\,^5D^\circ$	1—2, 1, 0
6106,398	4	14,05	16,08	$3p'\,^3D$—$4d'\,^3F^\circ$	3—4

λ, Å	I	E_H, eV	E_B, eV	Transition	J
6046,494	10	10,99	13,04	$3p\ {}^3P-6s\ {}^3S^\circ$	0—1
6046,438	13	10,99	13,04	$3p\ {}^3P-6s\ {}^3S^\circ$	2—1
6046,232	12	10,99	13,04	$3p\ {}^3P-6s\ {}^3S^\circ$	1—1
5995,28	3	14,05	16,11	$3p'\ {}^3D-4d'\ {}^3P^\circ$	2, 3—2
5993,18	1	14,05	16,11	$3p'\ {}^3D-4d'\ {}^3P^\circ$	1—1
5991,93	2	14,05	16,11	$3p'\ {}^3D-4d'\ {}^3P^\circ$	2—1
5991,34	1	14,05	16,11	$3p'\ {}^3D-4d'\ {}^3P^\circ$	1—0
5958,583	13	10,99	13,07	$3p\ {}^3P-5d\ {}^3D^\circ$	0—1
		10,99	13,07	$3p\ {}^3P-5d\ {}^3D^\circ$	2—3, 2, 1
5958,386	12	10,99	13,07	$3p\ {}^3P-5d\ {}^3D^\circ$	1—2, 1
5750,424	5	13,13	15,29	$6p\ {}^3P-3d'\ {}^3P^\circ$	2, 1—2
5731,103	3	13,13	15,29	$6p\ {}^3P-3d'\ {}^3P^\circ$	2, 1, 0—1
5720,613	1	13,13	15,30	$6p\ {}^3P-3d'\ {}^3P^\circ$	1—0
5555,003	9	10,99	13,22	$3p\ {}^3P-7s\ {}^3S^\circ$	2—1
5554,832	8	10,99	13,22	$3p\ {}^3P-7s\ {}^3S^\circ$	1—1
5512,770	8	10,99	13,24	$3p\ {}^3P-6d\ {}^3D^\circ$	2—3, 2, 1
5512,603	7	10,99	13,24	$3p\ {}^3P-6d\ {}^3D^\circ$	1—2, 1
5492,8	3	14,13	16,39	$3p'\ {}^1F-5d'\ {}^1G^\circ$	3—4
5486,6	3	14,13	16,39	$3p'\ {}^1F-5d'\ {}^3G^\circ$	3—4
5436,861	11	10,74	13,02	$3p\ {}^5P-6s\ {}^5S^\circ$	3—2
5435,775	10	10,74	13,02	$3p\ {}^5P-6s\ {}^5S^\circ$	2—2
5435,176	9	10,74	13,02	$3p\ {}^5P-6s\ {}^5S^\circ$	1—2
5410,76	4	14,10	16,39	$3p'\ {}^3F-5d'\ {}^1G^\circ$	3—4
		14,10	16,39	$3p'\ {}^3F-5d'\ {}^3F^\circ$	4—4
5408,87	3	14,10	16,39	$3p'\ {}^3F-5d'\ {}^3G^\circ$	3—4
5408,59	4	14,10	16,39	$3p'\ {}^3F-5d'\ {}^3G^\circ$	4—5
5404,87	3	—	—	—	—
5330,739	13	10,74	13,06	$3p\ {}^5P-5d\ {}^5D^\circ$	3—4, 3, 2
5329,685	12	10,74	13,06	$3p\ {}^5P-5d\ {}^5D^\circ$	2—3, 2, 1
5329,101	11	10,74	13,06	$3p\ {}^5P-5d\ {}^5D^\circ$	1—2, 1, 0
5299,044	5	10,99	13,33	$3p\ {}^5P-8s\ {}^3S^\circ$	2—1
5298,887	4	10,99	13,33	$3p\ {}^5P-8s\ {}^3S^\circ$	1—1
5275,121	4	10,99	13,34	$3p\ {}^5P-7d\ {}^3D^\circ$	2—3, 2, 1
5274,968	2	10,99	13,34	$3p\ {}^5P-7d\ {}^3D^\circ$	1—2, 1
5146,06	5	10,99	13,40	$3p\ {}^3P-9s\ {}^3S^\circ$	2—1
		12,87	15,29	$5p\ {}^3P-3d'\ {}^3P^\circ$	2, 1—2
5130,53	3	12,87	15,29	$5p\ {}^3P-3d'\ {}^3P^\circ$	2, 1, 0—1
		10,99	13,40	$3p\ {}^3P-8d\ {}^3D^\circ$	2—1, 2, 3
5047,70	5	—	—	—	—
5037,16	15	—	—	—	—
5020,217	7	10,74	13,21	$3p\ {}^5P-7s\ {}^5S^\circ$	3—2
5019,291	6	10,74	13,21	$3p\ {}^5P-7s\ {}^5S^\circ$	2—2
5018,783	5	10,74	13,21	$3p\ {}^5P-7s\ {}^5S^\circ$	1—2
4968,793	8	10,74	13,23	$3p\ {}^5P-6d\ {}^5D^\circ$	3—4, 3, 2
4967,882	7	10,74	13,23	$3p\ {}^5P-6d\ {}^5D^\circ$	2—3, 2, 1
4967,378	6	10,74	13,23	$3p\ {}^5P-6d\ {}^5D^\circ$	1—2, 1, 0
4802,981	4	10,74	13,32	$3p\ {}^5P-8s\ {}^5S^\circ$	3—2
4802,132	3	10,74	13,32	$3p\ {}^5P-8s\ {}^5S^\circ$	2—2
4801,80	2	10,74	13,32	$3p\ {}^5P-8s\ {}^5S^\circ$	1—2
4773,752	5	10,74	13,37	$3p\ {}^5P-7d\ {}^5D^\circ$	3—4, 3, 2
4772,913	4	10,74	13,37	$3p\ {}^5P-7d\ {}^5D^\circ$	2—3, 2, 1
4772,448	3	10,74	13,37	$3p\ {}^5P-7d\ {}^5D^\circ$	1—2, 1, 0
4673,70	3	10,74	13,39	$3p\ {}^5P-9s\ {}^5S^\circ$	3—2
4672,75	3	10,74	13,39	$3p\ {}^5P-9s\ {}^5S^\circ$	1, 2—2
4655,359	3	10,74	13,40	$3p\ {}^5P-8d\ {}^5D^\circ$	3—4, 3, 2
4654,558	2	10,74	13,40	$3p\ {}^5P-8d\ {}^5D^\circ$	2—3, 2, 1
4654,118	1	10,74	13,40	$3p\ {}^5P-8d\ {}^5D^\circ$	1—2, 1, 0
4589,89	3	—	—	—	—
4588,98	2	—	—	—	—

λ, Å	I	E_H, eV	E_B, eV	Transition	J
4577,66	3	—	—	—	—
4576,79	2	—	—	—	—
4368,30	10	9,52	12,36	$3s\,^3S^\circ$—$4p\,^3P$	1—2, 1, 0
4233,32	7	12,36	15,29	$4p\,^3P$—$3d'\,^3P^\circ$	2, 1—2
4222,78	5	12,36	15,29	$4p\,^3P$—$3d'\,^3P^\circ$	2, 1, 0—1
4217,09	4	12,36	15,30	$4p\,^3P$—$3d'\,^3P^\circ$	1—0
3954,596	5	10,99	14,12	$3p\,^3P$—$3s''\,^3P^\circ$	1—2
3954,687	10	10,99	14,12	$3p\,^3P$—$3s''\,^3P^\circ$	2—2
3953,056	2	10,99	14,12	$3p\,^3P$—$3s''\,^3P^\circ$	2—1
		10,99	14,12	$3p\,^3P$—$3s''\,^3P^\circ$	0—1
3952,982	1	10,99	14,12	$3p\,^3P$—$3s''\,^3P^\circ$	1—1
3951,987	3	10,99	14,12	$3p\,^3P$—$3s''\,^3P^\circ$	1—0
3947,594	4	9,14	12,28	$3p\,^5S^\circ$—$4p\,^5P$	2—1
3947,489	7	9,14	12,28	$3p\,^5S^\circ$—$4p\,^5P$	2—2
3947,301	10	9,14	12,28	$3p\,^5S^\circ$—$4p\,^5P$	2—3
3830,26	—	12,88	16,11	$5p\,^3P$—$4d'\,^3P^\circ$	2, 1—2
3825,530	1	12,54	15,78	$3s'\,^3D^\circ$—$3p''\,^3D$	1—2
3825,249	4	12,54	15,78	$3s'\,^3D^\circ$—$3p''\,^3D$	2—3
3825,090	3	12,54	15,78	$3s'\,^3D^\circ$—$3p''\,^3D$	1—1
3824,425	3	12,54	15,78	$3s'\,^3D^\circ$—$3p''\,^3D$	2—2
		12,54	15,78	$3s'\,^3D^\circ$—$3p''\,^3D$	2—1
3823,469	10	12,54	15,78	$3s'\,^3D^\circ$—$3p''\,^3D$	3—3
3822,63	—	12,54	15,78	$3s'\,^3D^\circ$—$3p''\,^3D$	3—2
3692,44	7	9,52	12,88	$3s\,^3S^\circ$—$5p\,^3P$	1—2, 1, 0
2883,78	3	10,99	15,29	$3p\,^3P$—$3d'\,^3P^\circ$	2, 1—2
2878,95	2	10,99	15,29	$3p\,^3P$—$3d'\,^3P^\circ$	2, 1, 0—1
2876,30	1	10,99	15,30	$3p\,^3P$—$3d'\,^3P^\circ$	1—0
1358,524	5	0,02	9,14	$2p^4\,^3P$—$3s\,^5S^\circ$	1—2
1355,605	2	0,00	9,14	$2p^4\,^3P$—$3s\,^5S^\circ$	2—2
1306,025	25	0,03	9,52	$2p^4\,^3P$—$3s\,^3S^\circ$	0—1
1304,866	30	0,02	9,52	$2p^4\,^3P$—$3s\,^3S^\circ$	1—1
1302,173	30	0,00	9,52	$2p^4\,^3P$—$3s\,^3S^\circ$	2—1
1217,643	2	4,19	14,37	$2p^4\,^1S$—$3s''\,^1P^\circ$	0—1
1152,149	2	1,97	12,73	$2p^4\,^1D$—$3s'\,^1D^\circ$	2—2
1066,660	9	—	—		—
1048,218	8	—	—	—	—
1041,688	1	0,03	11,93	$2p^4\,^3P$—$4s\,^3S^\circ$	0—1
1040,941	15	0,02	11,93	$2p^4\,^3P$—$4s\,^3S^\circ$	1—1
1039,233	20	0,00	11,93	$2p^4\,^3P$—$4s\,^3S^\circ$	2—1
1028,162	8	0,03	12,09	$2p^4\,^3P$—$3d\,^3D^\circ$	0—1
1027,433	20	0,02	12,09	$2p^4\,^3P$—$3d\,^3D^\circ$	1—2, 1
1025,766	9	0,00	12,09	$2p^4\,^3P$—$3d\,^3D^\circ$	2—3, 2, 1
999,493	2	1,97	14,37	$2p^4\,^1D$—$3s''\,^1P^\circ$	2—1
990,805	2	0,03	12,54	$2p^4\,^3P$—$3s'\,^3D^\circ$	0—1
990,210	8	0,02	12,54	$2p^4\,^3P$—$3s'\,^3D^\circ$	1—1, 2
990,132	1	—	—	—	—
988,776	15	0,00	12,54	$2p^4\,^3P$—$3s'\,^3D^\circ$	2—3
988,661	2	0,00	12,54	$2p^4\,^3P$—$3s'\,^3D^\circ$	2—2
988,581	3	0,00	12,54	$2p^4\,^3P$—$3s'\,^3D^\circ$	2—1
978,616	4	0,03	12,70	$2p^4\,^3P$—$5s\,^3S^\circ$	0—1
977,967	1	0,02	12,70	$2p^4\,^3P$—$5s\,^3S^\circ$	1—1
976,452	5	0,00	12,70	$2p^4\,^3P$—$5s\,^3S^\circ$	2—1
973,884	4	0,03	12,76	$2p^4\,^3P$—$4d\,^3D^\circ$	0—1
973,240	5	0,02	12,76	$2p^4\,^3P$—$4d\,^3D^\circ$	1—2, 1
971,741	8	0,00	12,76	$2p^4\,^3P$—$4d\,^3D^\circ$	2—3, 2, 1
952,940	4	0,03	13,04	$2p^4\,^3P$—$6s\,^3S^\circ$	0—1
952,414	8	0,02	13,04	$2p^4\,^3P$—$6s\,^3S^\circ$	1—1
950,888	4	0,00	13,04	$2p^4\,^3P$—$6s\,^3S^\circ$	2—1
950,732	—	0,03	13,07	$2p^4\,^3P$—$5d\,^3D^\circ$	0—1

λ, Å	I	E_H, eV	E_B, eV	Transition	J
950,114	0	0,02	13,07	$2p^4\ ^3P—5d\ ^3D°$	1—2, 1
948,689	4	0,00	13,07	$2p^4\ ^3P—5d\ ^3D°$	2—3, 2, 1
939,837	—	0,03	13,22	$2p^4\ ^3P—7s\ ^3S°$	0—1
939,237	—	0,02	13,22	$2p^4\ ^3P—7s\ ^3S°$	1—1
938,621	—	0,03	13,24	$2p^4\ ^3P—6d\ ^3D°$	0—1
938,022	—	0,02	13,24	$2p^4\ ^3P—6d\ ^3D°$	1—2, 1
937,841	3	0,00	13,22	$2p^4\ ^3P—7s\ ^3S°$	2—1
936,630	3	0,00	13,24	$2p^4\ ^3P—6d\ ^3D°$	2—3, 2, 1
935,183	4	1,97	15,22	$2p^4\ ^1D—4s'\ ^1D°$	2—2
931,479	—	0,03	13,34	$2p^4\ ^3P—7d\ ^3D°$	0—1
930,889	—	0,02	13,34	$2p^4\ ^3P—7d\ ^3D°$	1—2, 1
929,517	—	0,00	13,34	$2p^4\ ^3P—7d\ ^3D°$	2—3, 2, 1
922,011	—	1,97	15,41	$2p^4\ ^1D—3d'\ ^1F°$	2—3
882,88	—	1,94	16,01	$2p^4\ ^1D—5s'\ ^1D°$	2—2
879,553	1	0,03	14,12	$2p^4\ ^3P—3s''\ ^3P°$	0—1
879,108	1	0,02	14,12	$2p^4\ ^3P—3s''\ ^3P°$	1—1
879,079	1	0,02	14,12	$2p^4\ ^3P—3s''\ ^3P°$	1—2
879,027	1	—	—	—	—
878,979	1	0,02	14,12	$2p^4\ ^3P—3s''\ ^3P°$	1—0
877,885	2	0,00	14,12	$2p^4\ ^3P—3s''\ ^3P°$	2—2
877,804	2	0,00	14,12	$2p^4\ ^3P—3s''\ ^3P°$	2—1
861,63	—	1,97	16,36	$2p^4\ ^1D—6s'\ ^1D°$	2—2
850,74	—	1,97	16,54	$2p^4\ ^1D—7s'\ ^1D°$	2—2
812,158	—	0,03	15,29	$2p^4\ ^3P—3d'\ ^3P°$	0—1
812,096	3	0,02	15,29	$2p^4\ ^3P—3d'\ ^3P°$	1—2
811,710	1	0,02	15,29	$2p^4\ ^3P—3d'\ ^3P°$	1—1
811,501	1	0,02	15,30	$2p^4\ ^3P—3d'\ ^3P°$	1—0
811,052	4	0,00	15,29	$2p^4\ ^3P—3d'\ ^3P°$	2—2
810,667	1	0,00	15,29	$2p^4\ ^3P—3d'\ ^3P°$	2—1
792,971	—	0,02	15,65	$2p^4\ ^3P—2p^5\ ^3P°$	1—2
792,937	3	0,03	15,66	$2p^4\ ^3P—2p^5\ ^3P°$	0—1
792,510	1	0,02	15,66	$2p^4\ ^3P—2p^5\ ^3P°$	1—1
792,237	1	0,02	15,67	$2p^4\ ^3P—2p^5\ ^3P°$	1—0
791,976	3	0,00	15,65	$2p^4\ ^3P—2p^5\ ^3P°$	2—2
791,516	1	0,00	15,66	$2p^4\ ^3P—2p^5\ ^3P°$	2—1
770,698	0	0,03	16,11	$2p^4\ ^3P—4d'\ ^3P°$	0—1
770,350	—	0,02	16,11	$2p^4\ ^3P—4d'\ ^3P°$	1—2
770,294	1	0,02	16,11	$2p^4\ ^3P—4d'\ ^3P°$	1—1
770,264	1	0,02	16,11	$2p^4\ ^3P—4d'\ ^3P°$	1—0
769,411	1	0,00	16,11	$2p^4\ P—4d'\ ^3P°$	2—2
769,355	1	0,00	16,11	$2p^4\ ^3P—4d'\ ^3P°$	2—1
756,7	1	0,02	16,40	$2p^4\ ^3P—5d'\ ^3P°$	1—2, 1, 0
755,8	2	0,00	16,40	$2p^4\ ^3P—5d'\ ^3P°$	2—2, 1
749,3	1	0,02	16,56	$2p^4\ ^3P—6d'\ ^3P°$	1—2, 1
748,4	1	0,00	16,56	$2p^4\ ^3P—6d'\ ^3P°$	2—2, 1

O II, **ground state** $1s^2 2s^2 2p^3\ ^4S^0_{3/2}$

Ionization potential 283550,9 cm^{-1}; 35,146 eV

λ, Å	I	E_H, eV	E_B, eV	Transition	J
6910,75	3	28,68	30,48	$3d\ ^4F—4p\ ^4D°$	$^5/_2—^3/_2$
6908,11	2	28,67	30,47	$3d\ ^4F—4p\ ^4D°$	$^3/_2—^1/_2$
6906,54	4	28,69	30,49	$3d\ ^4F—4p\ ^4D°$	$^7/_2—^5/_2$

λ, Å	I	E_H, eV	E_B, eV	Transition	J
6895,29	5	28,70	30,50	$3d\,{}^4F-4p\,{}^4D^\circ$	$^9/_2-{}^7/_2$
6885,07	1	28,67	30,48	$3d\,{}^4F-4p\,{}^4D^\circ$	$^3/_2-{}^3/_2$
6869,74	1	28,68	30,49	$3d\,{}^4F-4p\,{}^4D^\circ$	$^5/_2-{}^5/_2$
6846,97	1	28,69	30,50	$3d\,{}^4F-4p\,{}^4D^\circ$	$^7/_2-{}^7/_2$
6721,35	5	23,44	25,28	$3s\,{}^2P-3p\,{}^2S^\circ$	$^3/_2-{}^1/_2$
6678,19	0	28,95	30,81	$3d\,{}^2P-4p\,{}^2P^\circ$	$^1/_2-{}^3/_2$
6666,94	1	28,94	30,80	$3d\,{}^2P-4p\,{}^2P^\circ$	$^3/_2-{}^1/_2$
6640,90	4	23,42	25,28	$3s\,{}^2P-3p\,{}^2S^\circ$	$^1/_2-{}^1/_2$
6627,62	3	28,94	30,81	$3d\,{}^2P-4p\,{}^2P^\circ$	$^3/_2-{}^3/_2$
5206,73	5	26,56	28,94	$3p\,{}^2P^\circ-3d\,{}^2P$	$^3/_2-{}^3/_2$
5190,56	3	26,55	28,94	$3p\,{}^2P^\circ-3d\,{}^2P$	$^1/_2-{}^3/_2$
5176,00	2	26,56	28,95	$3p\,{}^2P^\circ-3d\,{}^2P$	$^3/_2-{}^1/_2$
5160,02	4	26,55	28,95	$3p\,{}^2P^\circ-3d\,{}^2P$	$^1/_2-{}^1/_2$
4955,78	3	26,56	29,06	$3p\,{}^2P^\circ-3d\,{}^2D$	$^3/_2-{}^3/_2$
4943,06	7	26,56	29,07	$3p\,{}^2P^\circ-3d\,{}^2D$	$^3/_2-{}^5/_2$
4941,12	5	26,55	29,06	$3p\,{}^2P^\circ-3d\,{}^2D$	$^1/_2-{}^3/_2$
4924,60	6	26,30	28,82	$3p\,{}^4S^\circ-3d\,{}^4P$	$^3/_2-{}^5/_2$
4906,88	5	26,30	28,83	$3p\,{}^4S^\circ-3d\,{}^4P$	$^3/_2-{}^3/_2$
4890,93	4	26,30	28,84	$3p\,{}^4S^\circ-3d\,{}^4P$	$^3/_2-{}^1/_2$
4871,58	5	28,83	31,37	$3p'\,{}^2P^\circ-3d'\,{}^2D$	$^3/_2-{}^5/_2$
4864,95	3	26,30	28,85	$3p\,{}^4S^\circ-3d\,{}^4D$	$^3/_2-{}^1/_2$
4861,03	3	28,82	31,37	$3p'\,{}^2P^\circ-3d'\,{}^2D$	$^1/_2-{}^3/_2$
4856,76	2	26,30	28,85	$3p\,{}^4S^\circ-3d\,{}^4D$	$^3/_2-{}^3/_2$
4856,49	2	26,30	28,85	$3p\,{}^4S^\circ-3d\,{}^4D$	$^3/_2-{}^5/_2$
4845,01	1	26,30	28,86	$3p\,{}^4S^\circ-3d\,{}^2F$	$^3/_2-{}^5/_2$
4843,45	0	31,69	34,25	$3d'\,{}^2S-4f'\,{}^2P^\circ$	$^1/_2-{}^1/_2,\ {}^3/_2$
4843,26	1	31,69	34,25	$3d'\,{}^2S-4f'\,{}^2P^\circ$	$^3/_2-{}^1/_2,\ {}^3/_2$
4752,70	2	26,25	28,85	$3p\,{}^2D^\circ-3d\,{}^4D$	$^5/_2-{}^5/_2$
4751,34	4	26,25	28,86	$3p\,{}^2D^\circ-3d\,{}^4D$	$^5/_2-{}^7/_2$
4741,71	3	26,25	28,86	$3p\,{}^2D^\circ-3d\,{}^2F$	$^5/_2-{}^5/_2$
4710,04	5	26,22	28,85	$3p\,{}^2D^\circ-3d\,{}^4D$	$^3/_2-{}^5/_2$
4707,80	0	29,07	31,70	$3d\,{}^2D-4f\,{}^2D^\circ$	$^5/_2-{}^5/_2$
4705,355	8	26,25	28,88	$3p\,{}^2D^\circ-3d\,{}^2F$	$^5/_2-{}^7/_2$
4703,18	3	28,51	31,15	$3p'\,{}^2D^\circ-3d'\,{}^2F$	$^3/_2-{}^5/_2$
4701,76	0	28,83	31,46	$3p'\,{}^2P^\circ-3d'\,{}^2P$	$^3/_2-{}^1/_2$
4701,23	2	28,83	31,46	$3p'\,{}^2P^\circ-3d'\,{}^2P$	$^3/_2-{}^3/_2$
4699,21	7	26,22	28,86	$3p\,{}^2D^\circ-3d\,{}^2F$	$^3/_2-{}^5/_2$
		28,51	31,15	$3p'\,{}^2D^\circ-3d'\,{}^2F$	$^5/_2-{}^7/_2$
4698,99	3	28,51	31,15	$3p'\,{}^2D^\circ-3d'\,{}^2F$	$^5/_2-{}^3/_2$
4698,48	1	28,51	31,15	$3p'\,{}^2D^\circ-3d'\,{}^2F$	$^5/_2-{}^5/_2$
4696,36	2	23,00	25,64	$3s\,{}^4P-3p\,{}^4D^\circ$	$^5/_2-{}^3/_2$
4691,47	1	28,82	31,46	$3p'\,{}^2P^\circ-3d'\,{}^2P$	$^1/_2-{}^1/_2$
4690,97	0	28,82	31,46	$3p'\,{}^2P^\circ-3d'\,{}^2P$	$^1/_2-{}^3/_2$
4677,00	0	29,07	31,72	$3d\,{}^2D-4f\,{}^2G^\circ$	$^5/_2-{}^7/_2$
4676,234	8	23,00	25,65	$3s\,{}^4P-3p\,{}^4D^\circ$	$^5/_2-{}^5/_2$
4673,75	4	22,98	25,63	$3s\,{}^4P-3p\,{}^4D^\circ$	$^3/_2-{}^1/_2$
4669,53	0	29,06	31,71	$3d\,{}^2D-4f\,{}^2D^\circ$	$^3/_2-{}^3/_2$
4669,33	0	29,06	31,71	$3d\,{}^2D-4f\,{}^4D^\circ$	$^3/_2-{}^5/_2$
4661,635	9	22,98	25,64	$3s\,{}^4P-3p\,{}^4D^\circ$	$^3/_2-{}^3/_2$
4650,841	6	22,96	25,63	$3s\,{}^4P-3p\,{}^4D^\circ$	$^1/_2-{}^1/_2$
4649,139	10	23,00	25,66	$3s\,{}^4P-3p\,{}^4D^\circ$	$^5/_2-{}^7/_2$
4641,811	9	22,98	25,65	$3s\,{}^4P-3p\,{}^4D^\circ$	$^3/_2-{}^5/_2$
4638,854	6	22,96	25,64	$3s\,{}^4P-3p\,{}^4D^\circ$	$^1/_2-{}^3/_2$
4621,28	0	29,07	31,75	$3d\,{}^2D-4f\,{}^4F^\circ$	$^5/_2-{}^5/_2$
4613,67	1	29,07	31,75	$3d\,{}^2D-4f\,{}^4F^\circ$	$^5/_2-{}^7/_2$
4613,11	0	29,07	31,75	$3d\,{}^2D-4f\,{}^2F^\circ$	$^5/_2-{}^5/_2$
4610,14	3	29,06	31,75	$3d\,{}^2D-4f\,{}^4F^\circ$	$^3/_2-{}^5/_2$
4609,42	4	29,07	31,76	$3d\,{}^2D-4f\,{}^2F^\circ$	$^5/_2-{}^7/_2$

λ, Å	I	E_H, eV	E_B, eV	Transition	J
4602,11	2	29,06	31,75	$3d\ ^2D-4f\ ^2F^\circ$	3/2–5/2
4596,174	8	25,66	28,36	$3s'\ ^2D-3p'\ ^2F^\circ$	3/2–5/2
4590,971	9	25,66	28,36	$3s'\ ^2D-3p'\ ^2F^\circ$	5/2–7/2
4506,50	2	—	—	—	
4491,25	3	28,94	31,70	$3d\ ^2P-4f\ ^2D^\circ$	3/2–5/2
4489,48	1	28,95	31,71	$3d\ ^2P-4f\ ^2D^\circ$	1/2–3/2
4488,17	2	31,46	34,23	$3d'\ ^2P-4f'\ ^2D^\circ$	3/2–3/2, 5/2
4488,09	2	31,46	34,23	$3d'\ ^2P-4f'\ ^2D^\circ$	3/2–3/2, 5/2
4487,72	0	31,46	34,23	$3d'\ ^2P-4f'\ ^2D^\circ$	1/2–3/2
4477,88	2	28,94	31,77	$3d\ ^2P-4f\ ^4G^\circ$	3/2–5/2
4476,08	0	28,95	31,73	$3d\ ^2P-4f\ ^4D^\circ$	1/2–3/2
4469,41	4	30,42	33,20	$3s'''\ ^6S^\circ-3p'''\ ^6P$	5/2–3/2
4469,32	3	28,83	31,60	$3p'\ ^2P^\circ-4d\ ^4D$	3/2–3/2, 5/2
4467,88	4	30,42	33,20	$3s'''\ ^6S^\circ-3p'''\ ^6P$	5/2–5/2
4466,32	2	28,94	31,71	$3d\ ^2P-4f\ ^4D^\circ$	3/2–5/2
4466,28	4	28,94	31,73	$3d\ ^2P-4f\ ^4D^\circ$	3/2–3/2
4465,40	4	30,42	33,20	$3s'''\ ^6S^\circ-3p'''\ ^6P$	5/2–7/2
4452,377	6	23,44	26,22	$3s\ P-3p\ ^2D^\circ$	3/2–3/2
4448,21	6	28,36	31,15	$3p'\ ^2F^\circ-3d'\ ^2F$	7/2–7/2
4443,05	5	28,36	31,15	$3p'\ ^2F^\circ-3d'\ ^2F$	5/2–5/2
4416,972	8	23,42	26,22	$3s\ ^2P-3p\ ^2D^\circ$	1/2–3/2
4414,909	10	23,44	26,25	$3s\ ^2P-3p\ ^2D^\circ$	3/2–5/2
4414,37	1	28,83	31,64	$3p'\ ^2P^\circ-4d\ ^2P$	3/2–3/2
4406,02	1	26,25	29,06	$3p\ ^2D^\circ-3d\ ^2D$	5/2–3/2
4395,95	7	26,25	29,07	$3p\ ^2D^\circ-3d\ ^2D$	5/2–5/2
4379,55	3	—	—	—	—
4378,41	0	31,37	34,20	$3d'\ ^2D-4f'\ ^2F^\circ$	5/2–7/2
4378,01	0	31,37	34,20	$3d'\ ^2D-4f'\ ^2F^\circ$	3/2–5/2
4371,65	2	28,88	31,72	$3d\ ^2F-4f\ ^4G^\circ$	7/2–9/2
4369,28	4	26,22	29,06	$3p\ ^2D^\circ-3d\ ^2D$	3/2–3/2
4366,896	7	23,00	25,84	$3s\ ^4P-3p\ ^4P^\circ$	5/2–3/2
4359,38	1	26,23	29,07	$3p\ ^2D^\circ-3d\ ^2D$	3/2–5/2
4358,40	0	28,86	31,70	$3d\ ^4D-4f\ ^4D^\circ$	7/2–7/2
4357,25	0	28,85	31,70	$3d\ ^4D-4f\ ^4D^\circ$	5/2–7/2
		28,85	31,71	$3d\ ^4D-4f\ ^4D^\circ$	3/2–5/2
		25,85	28,69	$3p\ ^4P^\circ-3d\ ^4F$	5/2–7/2
4353,60	1	28,86	31,71	$3d\ ^2F-4f\ ^4G^\circ$	5/2–7/2
4351,269	6	25,66	28,51	$3s'\ ^2D-3p'\ ^2D^\circ$	5/2–5/2
4349,426	8	23,00	25,85	$3s\ ^4P-3p\ ^4P^\circ$	5/2–5/2
4347,425	6	25,66	28,51	$3s'\ ^2D-3p'\ ^2D^\circ$	3/2–3/2
4345,562	7	22,98	25,83	$3s\ ^4P-3p\ ^4P^\circ$	3/2–1/2
4344,42	0	28,85	31,71	$3d\ ^4D-4f\ ^4G^\circ$	5/2–7/2
4343,36	0	28,86	31,71	$3d\ ^2F-4f\ ^4D^\circ$	5/2–5/2
		31,37	34,23	$3d'\ ^2D-4f'\ ^2D^\circ$	5/2–3/2, 5/2
4342,83	1	31,37	34,23	$3d'\ ^2D-4f'\ ^2D^\circ$	3/2–3/2, 5/2
4342,00	4	28,88	31,74	$3d\ ^2F-4f\ ^2G^\circ$	7/2–9/2
4340,36	2	28,86	31,72	$3d\ ^2F-4f\ ^2G^\circ$	5/2–7/2
4336,865	6	22,98	25,84	$3s\ ^4P-3p\ ^4P^\circ$	3/2–3/2
4334,29	0	28,85	31,71	$3d\ ^4D-4f\ ^2D^\circ$	5/2–3/2
		28,86	31,71	$3d\ ^4D-4f\ ^2D^\circ$	3/2–3/2
		28,85	31,71	$3d\ ^4D-4f\ ^4D^\circ$	5/2–5/2
4332,76	1	28,86	31,72	$3d\ ^4D-4f\ ^4G^\circ$	7/2–9/2
4331,89	2	28,51	31,37	$3p'\ ^2D^\circ-3d\ ^2D$	3/2–3/2
4331,47	0	28,51	31,37	$3p'\ ^2D^\circ-3d'\ ^2D$	3/2–5/2
4331,13		28,85	31,72	$3d\ ^4D-4f\ ^4G^\circ$	5/2–7/2
		28,86	31,73	$3d\ ^2F-4f\ ^4D^\circ$	5/2–3/2
4328,62	2	28,83	31,69	$3p'\ ^2P^\circ-3d'\ ^2S$	3/2–1/2
4327,89	0	28,51	31,37	$3p'\ ^2D^\circ-3d'\ ^2D$	5/2–3/2
4327,48	3	28,51	31,37	$3p'\ ^2D^\circ-3d'\ ^2D$	5/2–5/2

λ, Å	I	E_H, eV	E_B, eV	Transition	J
4325,77	3	22,96	25,83	$3s\ ^4P-3p\ ^4P^\circ$	1/2—1/2
4319,93	1	28,82	31,69	$3p'\ ^2P^\circ-3d'\ ^2S$	1/2—1/2
4319,631	8	22,98	25,85	$3s\ ^4P-3p\ ^4P^\circ$	3/2—5/2
4317,65	0	28,83	31,70	$3d\ ^4P-4f\ ^2D^\circ$	3/2—5/2
4317,139	8	22,96	25,84	$3s\ ^4P-3p\ ^4P^\circ$	1/2—3/2
4315,80	00	22,88	31,75	$3d\ ^2F-4f\ ^4F^\circ$	7/2—7/2
4315,35	0	28,88	31,75	$3d\ ^4D-4f\ ^4F^\circ$	7/2—5/2
		28,85	31,73	$3d\ ^4D-4f\ ^4D^\circ$	3/2—1/2
		28,85	31,73	$3d\ ^4D-4f\ ^4D^\circ$	1/2—3/2
4313,43	1	28,88	31,75	$3d\ ^2F-4f\ ^4F^\circ$	7/2—9/2
4312,10	0	28,88	31,76	$3d\ ^2F-4f\ ^2F^\circ$	7/2—7/2
4308,96	1	28,85	31,73	$3d\ ^4D-4f\ ^4D^\circ$	1/2—1/2
4307,31	1	28,84	31,71	$3d\ ^4P-4f\ ^2D^\circ$	1/2—3/2
4305,53	0	28,83	31,71	$3d\ ^4P-4f\ ^4G^\circ$	3/2—5/2
4303,82	5	28,82	31,70	$3d\ ^4P-4f\ ^4D^\circ$	5/2—7/2
4303,06	0	31,32	34,20	$3d'\ ^2G-4f'\ ^2G^\circ$	7/2—7/2, 9/2
4302,81	0	31,32	34,20	$3d'\ ^2G-4f'\ ^2G^\circ$	9/2—7/2, 9/2
4294,82	3	28,83	31,71	$3d\ ^4P-4f\ ^4D^\circ$	3/2—5/2
4292,23	0	28,86	31,75	$3d\ ^2F-4f\ ^4F^\circ$	5/2—5/2
4291,25	1	28,82	31,71	$3d\ ^4P-4f\ ^4G^\circ$	5/2—7/2
4288,83	1	28,84	31,73	$3d\ ^4P-4f\ ^4D^\circ$	1/2—1/2
4285,70	3	28,86	31,75	$3d\ ^2F-4f\ ^4F^\circ$	5/2—7/2
4283,75	0	28,86	31,75	$3d\ ^4D-4f\ ^4F^\circ$	3/2—3/2
4283,13	0	28,86	31,75	$3d\ ^4D-4f\ ^4F^\circ$	5/2—5/2
4282,96	1	28,86	31,75	$3d\ ^4D-4f\ ^4F^\circ$	3/2—5/2
4282,82	0	28,83	31,73	$3d\ ^4P-4f\ ^4D^\circ$	3/2—3/2
4281,40	0	28,82	31,71	$3d\ ^4P-4f\ ^4D^\circ$	5/2—5/2
4281,25	0	28,82	31,73	$3d\ ^4P-4f\ ^4D^\circ$	5/2—3/2
4277,90	1	28,66	31,75	$3d\ ^4D-4f\ ^4F^\circ$	7/2—7/2
4277,40	1	28,86	31,75	$3d\ ^4D-4f\ ^2F^\circ$	7/2—5/2
		28,85	31,75	$3d\ ^4D-4f\ ^4F^\circ$	1/2—3/2
4276,71	1	28,85	31,75	$3d\ ^4D-4f\ ^4F^\circ$	5/2—7/2
		28,83	31,73	$3d\ ^4P-4f\ ^4D^\circ$	3/2—1/2
4276,64	3	28,86	31,75	$3d\ ^4D-4f\ ^4F^\circ$	5/2—5/2
4276,21	0	28,85	31,75	$3d\ ^4D-4f\ ^2F^\circ$	5/2—5/2
4275,90	0	28,86	31,75	$3d\ ^4D-4f\ ^2F^\circ$	3/2—5/2
4275,52	4	28,86	31,76	$3d\ ^4D-4f\ ^4F^\circ$	7/2—9/2
4274,13	00	28,86	31,75	$3d\ ^4D-4f\ ^2F^\circ$	7/2—7/2
4273,17	0	28,86	31,75	$3d\ ^4D-4f\ ^2F^\circ$	5/2—7/2
4253,98	4	31,32	34,23	$3d'\ ^2G-4f'\ ^2H^\circ$	7/2—9/2
4253,74	4	31,32	34,23	$3d'\ ^2G-4f'\ ^2H^\circ$	9/2—9/2, 11/2
4196,72	1	28,51	31,46	$3p'\ ^2D^\circ-3d'\ ^2P$	3/2—1/2
4196,26	00	28,51	31,46	$3p'\ ^2D^\circ-3d'\ ^2P$	3/2—3/2
4192,50	2	28,51	31,46	$3p'\ ^2D^\circ-3d'\ ^2P$	5/2—3/2
4189,788	10	28,36	31,32	$3p'\ ^2F^\circ-3d'\ ^2G$	7/2—9/2
4185,456	8	28,36	31,32	$3p'\ ^2F^\circ-3d'\ ^2G$	5/2—7/2
4169,230	4	25,85	28,82	$3p\ ^4P^\circ-3d\ ^4P$	5/2—5/2
4156,54	3	25,85	28,83	$3p\ ^4P^\circ-3d\ ^4P$	5/2—3/2
4153,302	7	25,84	28,82	$3p\ ^4P^\circ-3d\ ^4P$	3/2—5/2
4146,09	3	33,20	36,19	$3p'''\ ^6P-3d'''\ ^6D^\circ$	7/2—9/2
4145,90	0	33,20	36,19	$3p'''\ ^6P-3d'''\ ^6D^\circ$	7/2—7/2
4143,77	2	33,20	36,19	$3p'''\ ^6P-3d'''\ ^6D^\circ$	5/2—7/2
4143,52	1	33,20	36,19	$3p'''\ ^6P-3d'''\ ^6D^\circ$	5/2—5/2
4142,24	0	33,20	36,19	$3p'''\ ^6P-3d'''\ ^6D^\circ$	3/2—5/2
4142,08	1	33,20	36,19	$3p'''\ ^6P-3d'''\ ^6D^\circ$	3/2—1/2
4141,96	1	33,20	36,19	$3p'''\ ^6P-3d'''\ ^6D^\circ$	3/2—3/2
4140,76	0	25,83	28,83	$3p\ ^4P^\circ-3d\ ^4P$	3/2—3/2
4132,806	7	25,83	28,83	$3p\ ^4P^\circ-3d\ ^4P$	1/2—3/2
4129,34	2	25,84	28,84	$3p\ ^4P^\circ-3d\ ^4P$	3/2—1/2

λ, Å	I	E_H, eV	E_B, eV	Transition	J
4121,48	4	25,83	28,84	$3p\ {}^4P^\circ-3d\ {}^4P$	$1/2-1/2$
4120,554	2	25,85	28,86	$3p\ {}^4P^\circ-3d\ {}^4D$	$5/2-3/2$
4120,279	3	25,85	28,86	$3p\ {}^4P^\circ-3d\ {}^4D$	$5/2-5/2$
4119,221	8	25,85	28,86	$3p\ {}^4P^\circ-3d\ {}^4D$	$5/2-7/2$
4113,82	1	25,36	31,37	$3p'\ {}^2F^\circ-3d'\ {}^2D$	$7/2-5/2$
4112,029	4	25,85	28,86	$3p\ {}^4P^\circ-3d\ {}^2F$	$5/2-5/2$
4110,795	3	25,84	28,85	$3p\ {}^4P^\circ-3d\ {}^4D$	$3/2-1/2$
4110,20	1	28,36	31,37	$3p'\ {}^2F^\circ-3d'\ {}^2D$	$5/2-3/2$
4108,75	0	28,69	31,71	$3d\ {}^4F-4f\ {}^4G^\circ$	$7/2-7/2$
4107,07	1	28,68	31,70	$3d\ {}^4F-4f\ {}^4D^\circ$	$5/2-7/2$
4106,03	0	25,66	28,68	$3p\ {}^4D^\circ-3d\ {}^4F$	$7/2-5/2$
4105,000	7	25,84	28,86	$3p\ {}^4P^\circ-3d\ {}^4D$	$3/2-3/2$
4104,743	5	25,84	28,86	$3p\ {}^4P^\circ-3d\ {}^4D$	$3/2-5/2$
4103,017	5	25,83	28,85	$3p\ {}^4P^\circ-3d\ {}^4D$	$1/2-1/2$
4098,27	0	28,67	31,70	$3d\ {}^4F-4f\ {}^2D^\circ$	$3/2-5/2$
4097,260	4	28,69	31,72	$3d\ {}^4F-4f\ {}^4G^\circ$	$7/2-9/2$
		25,83	28,86	$3p\ {}^4P^\circ-3d\ {}^4D$	$1/2-3/2$
4096,543	3	25,84	28,86	$3p\ {}^4P^\circ-3d\ {}^2F$	$3/2-5/2$
4096,18	0	28,68	31,71	$3d\ {}^4F-4f\ {}^4G^\circ$	$5/2-5/2$
4095,63	0	28,68	31,71	$3d\ {}^4F-4f\ {}^4G^\circ$	$5/2-7/2$
4094,18	0	25,65	28,68	$3p\ {}^4D^\circ-3d\ {}^4F$	$5/2-3/2$
4092,940	8	25,66	28,69	$3p\ {}^4D^\circ-3d\ {}^4F$	$7/2-7/2$
4089,295	4	28,70	31,73	$3d\ {}^4F-4f\ {}^4G^\circ$	$9/2-11/2$
4087,16	2	28,68	31,71	$3d\ {}^4F-4f\ {}^4G^\circ$	$3/2-5/2$
4085,124	3	25,65	28,68	$3p\ {}^4D^\circ-3d\ {}^4F$	$5/2-5/2$
4084,66	1	25,85	28,88	$3p\ {}^4P^\circ-3d\ {}^2F$	$5/2-7/2$
4083,907	2	28,68	31,72	$3d\ {}^4F-4f\ {}^2G^\circ$	$5/2-7/2$
4078,862	4	25,64	28,68	$3p\ {}^4D^\circ-3d\ {}^4F$	$3/2-3/2$
4075,868	10	25,66	28,70	$3p\ {}^4D^\circ-3d\ {}^4F$	$7/2-9/2$
4072,164	8	25,65	28,69	$3p\ {}^4D^\circ-3d\ {}^4F$	$5/2-7/2$
4071,20	0	28,69	31,74	$3d\ {}^4F-4f\ {}^2G^\circ$	$7/2-9/2$
4069,897	6	25,64	28,68	$3p\ {}^4D^\circ-3d\ {}^4F$	$3/2-5/2$
4069,634	4	25,63	28,68	$3p\ {}^4D^\circ-3d\ {}^4F$	$1/2-3/2$
4062,90	1	28,70	31,75	$3d\ {}^4F-4f\ {}^4F^\circ$	$9/2-9/2$
4061,00	2	31,15	34,20	$3d'\ {}^2F-4f'\ {}^2G^\circ$	$5/2-7/2$
4060,98	2	31,15	34,20	$3d'\ {}^2F-4f'\ {}^2G^\circ$	$5/2-7/2$
4060,58	3	31,15	34,20	$3d'\ {}^2F-4f'\ {}^2G^\circ$	$7/2-9/2$
4054,55	00	31,15	34,20	$3d'\ {}^2F-4f'\ {}^2F^\circ$	$5/2-5/2$
4054,10	0	31,15	34,20	$3d'\ {}^2F-4f'\ {}^2F^\circ$	$7/2-7/2$
		28,69	31,75	$3d\ {}^4F-4f\ {}^4F^\circ$	$7/2-5/2$
4048,22	1	28,70	31,76	$3d\ {}^4F-4f\ {}^4F^\circ$	$7/2-7/2$
4046,15	00	28,69	31,76	$3d\ {}^4F-4f\ {}^4F^\circ$	$7/2-9/2$
4044,96	0	28,69	31,75	$3d\ {}^4F-4f\ {}^2F^\circ$	$7/2-7/2$
4041,31	0	28,68	31,75	$3d\ {}^4F-4f\ {}^4F^\circ$	$5/2-5/2$
4035,09	0	28,68	31,75	$3d\ {}^4F-4f\ {}^2F^\circ$	$5/2-5/2$
4033,18	0	28,67	31,75	$3d\ {}^4F-4f\ {}^4F^\circ$	$3/2-3/2$
4026,40	0	28,67	31,75	$3d\ {}^4F-4f\ {}^2F^\circ$	$3/2-5/2$
4024,04	1	31,15	34,23	$3d'\ {}^2F-4f'\ {}^2D^\circ$	$3/2-5/2$
3985,46	0	25,83	28,94	$3p\ {}^4P^\circ-3d\ {}^2P$	$1/2-3/2$
3982,719	5	23,44	26,55	$3s\ {}^2P-3p\ {}^2P^\circ$	$3/2-1/2$
3973,263	10	23,44	26,56	$3s\ {}^2P-3p\ {}^2P^\circ$	$3/2-3/2$
3967,441	1	25,83	28,95	$3p\ {}^4P^\circ-3d\ {}^2P$	$1/2-1/2$
3963,13	0	28,51	31,64	$3p'\ {}^2D^\circ-4d\ {}^2P$	$5/2-3/2$
3954,372	7	23,42	26,55	$3s\ {}^2P-3p\ {}^2P^\circ$	$1/2-1/2$
3945,048	5	23,42	26,56	$3s\ {}^2P-3p\ {}^2P^\circ$	$1/2-3/2$
3926,58	—	25,66	28,82	$3p\ {}^4D^\circ-3d\ {}^4P$	$7/2-5/2$
3919,287	6	25,66	28,82	$3s'\ {}^2D-3p'\ {}^2P^\circ$	$3/2-1/2$
3912,088	2	25,66	28,83	$3s'\ {}^2D-3p'\ {}^2P^\circ$	$3/2-3/2$
3911,960	10	25,66	28,83	$3s'\ {}^2D-3p'\ {}^2P^\circ$	$5/2-3/2$

λ, Å	I	E_H, eV	E_B, eV	Transition	J
3907,45	4	25,65	28,83	$3p\ ^4D^\circ—3d\ ^4P$	$^5/_2—^5/_2$
3896,30	1	25,65	28,83	$3p\ ^4D^\circ—3d\ ^4P$	$^5/_2—^3/_2$
3893,53	2	25,64	28,82	$3p\ ^4D^\circ—3d\ ^4P$	$^3/_2—^5/_2$
3883,15	3	25,66	28,85	$3p\ ^4D^\circ—3d\ ^4D$	$^7/_2—^5/_2$
3882,45	1	25,64	28,83	$3p\ ^4D^\circ—3d\ ^4P$	$^3/_2—^3/_2$
3882,197	7	25,66	28,86	$3p\ ^4D^\circ—3d\ ^4D$	$^7/_2—^7/_2$
3875,82	4	25,66	28,86	$3p\ ^4D^\circ—3d\ ^2F$	$^7/_2—^5/_2$
3874,10	2	25,63	28,83	$3p\ ^4D^\circ—3d\ ^4P$	$^1/_2—^3/_2$
3872,45	1	25,64	28,84	$3p\ ^4D^\circ—3d\ ^4P$	$^3/_2—^1/_2$
3864,68	1	25,65	28,85	$3p\ ^4D^\circ—3d\ ^4P$	$^5/_2—^3/_2$
3864,45	1	25,65	28,85	$3p\ ^4D^\circ—3d\ ^4D$	$^5/_2—^5/_2$
3864,13	7	25,63	28,84	$3p\ ^4D^\circ—3d\ ^4P$	$^1/_2—^1/_2$
3863,50	2	25,65	28,86	$3p\ ^4D^\circ—3d\ ^4D$	$^5/_2—^7/_2$
3857,18	4	25,65	28,86	$3p\ ^4D^\circ—3d\ ^2F$	$^5/_2—^5/_2$
3856,16	5	25,64	28,85	$3p\ ^4D^\circ—3d\ ^4D$	$^3/_2—^1/_2$
3851,47	0	25,66	28,88	$3p\ ^4D^\circ—3d\ ^2F$	$^7/_2—^7/_2$
3851,04	3	25,64	28,86	$3p\ ^4D^\circ—3d\ ^4D$	$^3/_2—^3/_2$
3850,81	2	25,64	28,86	$3p\ ^4D^\circ—3d\ ^4D$	$^3/_2—^5/_2$
3847,89	3	25,63	28,85	$3p\ ^4D^\circ—3d\ ^4D$	$^1/_2—^1/_2$
3843,58	3	25,64	28,86	$3p\ ^4D^\circ—3d\ ^2F$	$^3/_2—^5/_2$
3842,82	3	25,63	28,86	$3p\ ^4D^\circ—3d\ ^4D$	$^1/_2—^3/_2$
3833,10	3	25,65	28,88	$3p\ ^4D^\circ—3d\ ^2F$	$^5/_2—^7/_2$
3830,45	4	26,56	29,80	$3p\ ^2P^\circ—4s\ ^2P$	$^3/_2—^1/_2$
3821,68	4	26,55	29,80	$3p\ ^2P^\circ—4s\ ^2P$	$^1/_2—^1/_2$
3803,14	6	26,56	29,82	$3p\ ^2P^\circ—4s\ ^2P$	$^3/_2—^3/_2$
3794,48	3	26,55	29,82	$3p\ ^2P^\circ—4s\ ^2P$	$^1/_2—^3/_2$
3785,01	0	30,81	34,08	$4p\ ^2P^\circ—4d'\ ^2D$	$^3/_2—^5/_2$
3777,60	4	26,30	29,58	$3p\ ^4S^\circ—4s\ ^4P$	$^3/_2—^1/_2$
3762,63	5	26,30	29,60	$3p\ ^4S^\circ—4s\ ^4P$	$^3/_2—^3/_2$
3749,49	9	23,00	26,30	$3s\ ^4P—3p\ ^4S^\circ$	$^5/_2—^3/_2$
3741,69	0	28,36	31,67	$3p'\ ^2F^\circ—4d\ ^2F$	$^7/_2—^7/_2$
3739,92	6	26,30	29,62	$3p\ ^4S^\circ—4s\ ^4P$	$^3/_2—^5/_2$
3735,94	3	28,83	32,15	$3p'\ ^2P^\circ—4s'\ ^2D$	$^3/_2—^5/_2$
3729,34	2	28,82	32,15	$3p'\ ^2P^\circ—4s'\ ^2D$	$^1/_2—^3/_2$
3727,33	8	22,98	26,30	$3s\ ^4P—3p\ ^4S^\circ$	$^3/_2—^3/_2$
3712,75	7	22,96	26,30	$3s\ ^4P—3p\ ^4S^\circ$	$^1/_2—^3/_2$
3533,97	00	28,86	32,36	$3d\ ^4D—5p\ ^4D^\circ$	$^7/_2—^5/_2$
3516,92	0	28,86	32,38	$3d\ ^4D—5p\ ^4D^\circ$	$^7/_2—^7/_2$
3506,02	0	28,85	32,39	$3d\ ^4D—5p\ ^4P^\circ$	$^5/_2—^3/_2$
3501,67	00	28,85	32,39	$3d\ ^4D—5p\ ^4P^\circ$	$^1/_2—^3/_2$
3500,5	00	28,86	32,40	$3d\ ^2F—5p\ ^4P^\circ$	$^5/_2—^5/_2$
3496,27	1	25,28	28,83	$3p\ ^2S^\circ—3d\ ^4P$	$^1/_2—^3/_2$
3495,44	0	28,86	32,40	$3d\ ^4D—5p\ ^4P^\circ$	$^7/_2—^5/_2$
3494,66	00	28,85	32,40	$3d\ ^4D—5p\ ^4P^\circ$	$^5/_2—^5/_2$
3488,18	0	25,28	28,84	$3p\ ^2S^\circ—3d\ ^4P$	$^1/_2—^1/_2$
3474,94	1	25,28	28,85	$3p\ ^2S^\circ—3d\ ^4D$	$^1/_2—^1/_2$
3470,81	8	26,25	29,82	$3p\ ^2D^\circ—4s\ ^2P$	$^5/_2—^3/_2$
3470,42	5	26,22	29,79	$3p\ ^2D^\circ—4s\ ^2P$	$^3/_2—^1/_2$
3459,07	0	28,86	32,44	$3d\ ^2F—5p\ ^2D^\circ$	$^5/_2—^3/_2$
3457,99	1	28,88	32,47	$3d\ ^2F—5p\ ^2D^\circ$	$^7/_2—^5/_2$
3453,31	0	28,85	32,44	$3d\ ^4D—5p\ ^2D^\circ$	$^5/_2—^3/_2$
3447,98	1	26,22	29,82	$3p\ ^2D^\circ—4s\ ^2P$	$^3/_2—^3/_2$
3420,61	3	—	—	—	—
3419,87	2	—	—	—	—
3409,84	6	28,51	32,15	$3p'\ ^2D^\circ—4s'\ ^2D$	$^3/_2—^3/_2$
3407,38	7	28,51	32,15	$3p'\ ^2D^\circ—4s'\ ^2D$	$^5/_2—^3/_2$
3390,25	8	25,28	28,94	$3p\ ^{2\circ}S—3d\ ^2P$	$^1/_2—^3/_2$

λ, Å	I	E_H, eV	E_B, eV	Transition	J
3377,20	7	25,28	28,95	$3p\ ^2S^\circ—3d\ ^2P$	$^1/_2—^1/_2$
3375,77	0	28,69	32,36	$3d\ ^4F—5p\ ^4D^\circ$	$^7/_2—^5/_2$
3374,77	00	30,81	34,48	$4p\ ^2P^\circ—5s'\ ^2D$	$^3/_2—^5/_2$
3371,85	2	28,70	32,38	$3d\ ^4F—5p\ ^4D^\circ$	$^9/_2—^7/_2$
3370,23	00	28,67	32,35	$3d\ ^4F—5p\ ^4D^\circ$	$^3/_2—^3/_2$
3367,00	00	28,68	32,36	$3d\ ^4F—5p\ ^4D^\circ$	$^5/_2—^5/_2$
3360,15	00	28,69	32,38	$3d\ ^4F—5p\ ^4D^\circ$	$^7/_2—^7/_2$
3306,60	6	25,84	29,58	$3p\ ^4P^\circ—4s\ ^4P$	$^3/_2—^1/_2$
3305,15	6	25,85	29,58	$3p\ ^4P^\circ—4s\ ^4P$	$^5/_2—^3/_2$
3301,56	3	25,83	29,59	$3p\ ^4P^\circ—4s\ ^4P$	$^1/_2—^1/_2$
3295,13	4	25,84	29,58	$3p\ ^4P^\circ—4s\ ^4P$	$^3/_2—^3/_2$
3290,13	5	25,83	29,60	$3p\ ^4P^\circ—4s\ ^4P$	$^1/_2—^3/_2$
3287,59	9	25,85	29,62	$3p\ ^4P^\circ—4s\ ^4P$	$^5/_2—^5/_2$
3277,69	7	25,84	29,62	$3p\ ^4P^\circ—4s\ ^4P$	$^3/_2—^5/_2$
3273,52	7	28,36	32,15	$3p'\ ^2F^\circ—4s'\ ^2D$	$^7/_2—^5/_2$
3270,98	7	28,36	32,15	$3p'\ ^2F^\circ—4s'\ ^2D$	$^5/_2—^3/_2$
3218,10	2	33,20	37,05	$3p'''\ ^6P—4s'''\ ^6S^\circ$	$^7/_2—^5/_2$
3216,76	1	33,20	37,05	$3p'''\ ^6P—4s'''\ ^6S^\circ$	$^5/_2—^5/_2$
3216,08	0	33,20	37,05	$3p'''\ ^6P—4s'''\ ^6S^\circ$	$^3/_2—^5/_2$
3169,2	1	29,07	32,98	$3d\ ^2D—5f\ ^2F^\circ$	$^5/_2—^7/_2$
3165,1	1	29,06	32,98	$3d\ ^2D—5f\ ^2F^\circ$	$^3/_2—^5/_2$
3139,77	4	25,64	29,58	$3p\ ^4D^\circ—4s\ ^4P$	$^3/_2—^1/_2$
3138,44	8	25,65	29,60	$3p\ ^4D^\circ—4s\ ^4P$	$^5/_2—^3/_2$
3134,82	10	25,66	29,62	$3p\ ^4D^\circ—4s\ ^4P$	$^7/_2—^5/_2$
3134,32	3	25,63	29,58	$3p\ ^4D^\circ—4s\ ^4P$	$^1/_2—^1/_2$
3129,44	6	25,64	29,60	$3p\ ^4D^\circ—4s\ ^4P$	$^3/_2—^3/_2$
3124,02	2	25,63	29,60	$3p\ ^4D^\circ—4s\ ^4P$	$^1/_2—^3/_2$
3122,62	6	25,65	29,62	$3p\ ^4D^\circ—4s\ ^4P$	$^5/_2—^5/_2$
3113,71	1	25,64	29,62	$3p\ ^4D^\circ—4s\ ^4P$	$^3/_2—^5/_2$
3097,52	0	28,94	32,94	$3d\ ^2P—5f\ ^4D^\circ$	$^3/_2—^5/_2$
3081,46	2	—	—	—	—
3047,9	0	28,88	32,95	$3d\ ^2F—5f\ ^4G^\circ$	$^7/_2—^9/_2$
3039,76	1	28,86	32,93	$3d\ ^4D—5f\ ^4D^\circ$	$^7/_2—^7/_2$
3039,51	1	28,85	32,93	$3d\ ^4D—5f\ ^4D^\circ$	$^5/_2—^7/_2$
3032,50	1	28,86	32,95	$3d\ ^2F—5f\ ^2G^\circ$	$^5/_2—^7/_2$
3032,08	2	28,88	32,97	$3d\ ^2F—5f\ ^2G^\circ$	$^7/_2—^9/_2$
3028,82	1	28,86	32,95	$3d\ ^4D—5f\ ^4G^\circ$	$^7/_2—^9/_2$
3025,75	1	28,88	32,98	$3d\ ^2F—5f\ ^4F^\circ$	$^7/_2—^9/_2$
3016,14	1	28,85	32,96	$3d\ ^4D—5f\ ^4D^\circ$	$^1/_2—^1/_2$
3014,50	1	28,86	32,97	$3d\ ^2F—5f\ ^4F^\circ$	$^5/_2—^5/_2$
3013,37	3	28,82	32,93	$3d\ ^4P—5f\ ^4D^\circ$	$^5/_2—^7/_2$
3012,83	1	28,85	32,97	$3d\ ^4D—5f\ ^4F^\circ$	$^5/_2—^5/_2$
3009,83	1	28,85	32,97	$3d\ ^4D—5f\ ^4F^\circ$	$^3/_2—^5/_2$
3009,62	1	28,85	32,97	$3d\ ^4D—5f\ ^4F^\circ$	$^1/_2—^3/_2$
3008,83	3	28,83	32,95	$3d\ ^4P—5f\ ^4D^\circ$	$^3/_2—^3/_2$
3008,28	1	28,86	32,98	$3d\ ^4D—5f\ ^4F^\circ$	$^7/_2—^7/_2$
3007,74	3	28,85	32,98	$3d\ ^4D—5f\ ^4F^\circ$	$^5/_2—^7/_2$
3007,08	3	28,86	32,98	$3d\ ^4D—5f\ ^4F^\circ$	$^7/_2—^9/_2$
3006,82	3	—	—	—	—
3006,01	2	28,84	32,96	$3d\ ^4P—5f\ ^4D^\circ$	$^1/_2—^1/_2$
3005,62	2	—	—	—	—
3002,93	1	—	—	—	—
2997,74	2	—	—	—	—
2995,94	1	—	—	—	—
2915,65	1	28,68	32,93	$3d\ ^4F—5f\ ^4D^\circ$	$^5/_2—^7/_2$
2911,85	2	28,69	32,95	$3d\ ^4F—5f\ ^4G^\circ$	$^7/_2—^9/_2$

λ, Å	I	E_H, eV	E_B, eV	Transition	J
2911,20	2	—	—	—	—
2908,74	1	28,67	32,94	3d ^4F—5f $^4G^\circ$	3/2—5/2
2906,62	3	28,70	32,97	3d ^4F—5f $^4G^\circ$	9/2—11/2
2905,00	2	28,68	32,95	3d ^4F—5f $^2G^\circ$	5/2—7/2
2904,29	2	—	—	—	—
2892,47	2	—	—	—	—
2891,88	1	28,69	32,98	3d ^4F—5f $^4F^\circ$	7/2—9/2
2887,91	3	—	—	—	—
2885,90	1	—	—	—	—
2883,96	4	—	—	—	—
2879,04	3	—	—	—	—
2836,35	2	26,38	30,75	2p^4 ^2P—4p $^2D^\circ$	1/2—3/2
2808,84	2	26,36	30,77	2p^4 ^2P—4p $^2D^\circ$	3/2—5/2
2803,11	1	26,38	30,80	2p^4 ^2P—4p $^2P^\circ$	1/2—1/2
2783,15	2	26,36	30,81	2p^4 ^2P—4p $^2P^\circ$	3/2—3/2
2747,46	6	25,28	29,79	3p $^2S^\circ$—4s ^2P	1/2—1/2
2733,34	10	25,28	29,82	2p $^2S^\circ$—4s ^2P	1/2—3/2
2718,90	3	—	—	—	—
2715,45	5	—	—	—	—
2575,433	1	26,56	31,37	3p $^2P^\circ$—3d′ ^2D	3/2—3/2
2575,300	10	26,56	31,37	3p $^2P^\circ$—3d′ ^2D	3/2—5/2
2571,476	8	26,55	31,37	2p $^2P^\circ$—3d′ ^2D	1/2—3/2
2530,30	8	26,25	31,15	3p $^2D^\circ$—3d′ ^2F	5/2—7/2
2527,03	1	26,56	31,46	3p $^2P^\circ$—3d′ ^2P	3/2—1/2
2526,91	7	26,56	31,46	3p $^2P^\circ$—3d′ ^2P	3/2—3/2
2523,20	1	26,55	31,46	3p $^2P^\circ$—3d′ ^2P	1/2—1/2
2523,09	1	26,55	31,46	3p $^2P^\circ$—3d′ ^2P	1/2—3/2
2517,97	6	26,23	31,15	3p $^2D^\circ$—3d′ ^2F	3/2—5/2
2445,55	10	23,44	28,51	3s ^2P—3p′ $^2D^\circ$	3/2—5/2
2444,26	5	23,44	28,51	3s ^2P—3p′ $^2D^\circ$	3/2—3/2
2441,67	2	26,56	31,64	3p $^2P^\circ$—4d ^2P	3/2—3/2
2438,09	1	26,55	31,64	3p $^2P^\circ$—4d ^2P	1/2—3/2
2436,06	5	25,66	30,75	3s′ ^2D—4p $^2D^\circ$	3/2—3/2
2433,538	9	23,42	28,51	3s ^2P—3p′ $^2D^\circ$	1/2—3/2
2431,66	0	26,55	31,65	3p $^2P^\circ$—4d ^2P	1/2—1/2
2425,55	2	25,66	30,77	3s′ ^2D—4p $^2D^\circ$	5/2—5/2
2418,60	1	26,25	31,37	3p $^2D^\circ$—3d′ ^2D	5/2—3/2
2418,46	7	26,25	31,37	3p $^2D^\circ$—3d′ ^2D	5/2—5/2
2415,13	4	26,56	31,69	3p $^2P^\circ$—3d′ ^2S	3/2—1/2
2411,60	6	26,55	31,69	3p $^2P^\circ$—3d′ ^2S	1/2—1/2
2407,49	6	26,23	31,37	3p $^2D^\circ$—3d′ ^2D	3/2—3/2
2407,37	1	26,23	31,37	3p $^2D^\circ$—3d′ ^2D	3/2—5/2
2406,41	6	25,66	30,81	3s′ ^2D—4p $^2P^\circ$	5/2—3/2
2375,73	4	26,25	31,46	3p $^2D^\circ$—3d′ ^2P	5/2—3/2
2365,15	3	26,22	31,46	3p $^2D^\circ$—3d′ ^2P	3/2—1/2
2365,03	1	26,22	31,46	3p $^2D^\circ$—3d′ ^2P	3/2—3/2
2339,31	3	26,30	31,60	3p $^4S^\circ$—4d ^4D	3/2—3/2, 5/2
2331,16	0	28,86	34,20	3d ^2F—4f′ $^2G^\circ$	7/2—7/2
2327,97	2	26,30	31,63	3p $^4S^\circ$—4d ^4P	3/2—3/2
2324,83	0	26,30	31,63	3p $^4S^\circ$—4d ^4P	3/2—1/2
2322,15	3	26,38	31,73	2p^4 ^2P—4f $^4D^\circ$	1/2—3/2
2319,68	4	26,36	31,71	2p^4 ^2P—4f $^2D^\circ$	3/2—5/2
2316,79	3	26,38	31,73	2p^4 ^2P—4f $^4D^\circ$	1/2—1/2
2316,12	3	26,36	31,71	2p^4 ^2P—4f $^4D^\circ$	3/2—5/2
2313,05	3	26,36	31,73	2p^4 ^2P—4f $^4D^\circ$	3/2—3/2
2307,72	1	26,36	31,73	2p^4 ^2P—4f $^4D^\circ$	3/2—1/2
2302,83	5	23,44	28,82	3s ^2P—3p′ $^3P^\circ$	3/2—1/2
2300,35	8	23,44	28,83	3s ^2P—3p′ $^2P^\circ$	3/2—3/2

λ, Å	I	$E_н$, eV	$E_в$, eV	Transition	J
2293,32	6	23,42	28,82	$3s\ ^2P-3p'\ ^2P^\circ$	$1/2-1/2$
2290,88	6	23,42	28,83	$3s\ ^2P-3p'\ ^2P^\circ$	$1/2-3/2$
2284,89	3	26,25	31,67	$3p\ ^2D^\circ-4d\ ^2F$	$5/2-7/2$
2283,42	3	26,22	31,65	$3p\ ^2D^\circ-4d\ ^2F$	$3/2-5/2$
2262,80	1	26,56	32,04	$3p\ ^2P^\circ-5s\ ^2P$	$3/2-1/2$
2259,66	2	26,55	32,04	$3p\ ^2P^\circ-5s\ ^2P$	$1/2-1/2$
2252,90	3	26,56	32,06	$3p\ ^2P^\circ-5s\ ^2P$	$3/2-3/2$
2250,00	1	26,55	32,06	$3p\ ^2P^\circ-5s\ ^2P$	$1/2-3/2$
2239,89	0	28,51	34,06	$3p'\ ^2D^\circ-4d'\ ^2F$	$3/2-5/2$
2229,66	0	28,51	34,07	$3p'\ ^2D^\circ-4d'\ ^2F$	$5/2-7/2$
2218,70	2	26,56	32,15	$3p\ ^2P^\circ-4s'\ ^2D$	$3/2-5/2$
2215,67	1	26,55	32,15	$3p\ ^2P^\circ-4s'\ ^2D$	$1/2-3/2$
2195,43	2	26,30	31,95	$3p\ ^4S^\circ-5s\ ^4P$	$3/2-1/2$
2191,44	2	28,83	34,48	$3p'\ ^2P^\circ-5s'\ ^2D$	$3/2-5/2$
2190,42	2	26,30	31,96	$3p\ ^4S^\circ-5s\ ^4P$	$3/2-3/2$
2189,51	1	—	—	—	—
2189,20	2	28,82	34,48	$3p'\ ^2P^\circ-5s'\ ^2D$	$1/2-3/2$
2182,64	4	26,30	31,98	$3p\ ^4S^\circ-5s\ ^4P$	$3/2-5/2$
2161,65	0	26,25	31,98	$3p\ ^2D^\circ-5s\ ^4P$	$5/2-5/2$
2160,52	0	26,22	31,96	$3p\ ^2D^\circ-5s\ ^4P$	$3/2-3/2$
2149,47	0	25,84	31,60	$3p\ ^4P^\circ-4d\ ^4D$	$3/2-3/2,\ 5/2$
2148,23	0	25,85	31,60	$3p\ ^4P^\circ-4d\ ^4D$	$5/2-3/2,\ 5/2$
2131,99	4	26,22	32,04	$3p\ ^2D^\circ-5s\ ^2P$	$3/2-1/2$
2131,76	5	26,25	32,06	$3p\ ^2D^\circ-5s\ ^2P$	$5/2-3/2$
2123,39	0	26,22	32,06	$3p\ ^2D^\circ-5s\ ^2P$	$3/2-3/2$
2101,29	4	26,25	32,15	$3p\ ^2D^\circ-4s'\ ^2D$	$5/2-5/2$
2100,69	1	25,66	31,56	$3p\ ^4D^\circ-4d\ ^4F$	$7/2-9/2$
2099,91	1	25,65	31,55	$3p\ ^4D^\circ-4d\ ^4F$	$5/2-7/2$
2092,90	1	26,22	32,15	$3p\ ^2D^\circ-4s'\ ^2D$	$3/2-3/2$
2074,17	1	28,51	34,48	$3p'\ ^2D^\circ-5s'\ ^2D$	$3/2-5/2$
2022,83	1	28,36	34,48	$3p'\ ^2F^\circ-5s'\ ^2D$	$5/2-3/2$
2021,45	1	25,83	31,96	$3p\ ^4P^\circ-5s\ ^4P$	$1/2-3/2$
2020,44	2	25,85	31,98	$3p\ ^4P^\circ-5s\ ^4P$	$5/2-5/2$
2016,60	2	25,84	31,98	$3p\ ^4P^\circ-5s\ ^4P$	$3/2-5/2$
1964,25	0	25,64	31,95	$3p\ ^4D^\circ-5s\ ^4P$	$3/2-1/2$
1963,84	2	25,65	31,96	$3p\ ^4D^\circ-5s\ ^4P$	$5/2-3/2$
1962,24	3	25,66	31,98	$3p\ ^4D^\circ-5s\ ^4P$	$7/2-5/2$
1960,34	1	25,64	31,96	$3p\ ^4D^\circ-5s\ ^4P$	$3/2-3/2$
1957,42	0	25,65	31,98	$3p\ ^4D^\circ-5s\ ^4P$	$5/2-5/2$
932,046	10	—	—	—	—
919,78	15	—	—	—	—
834,462	15	0,00	14,86	$2p^3\ ^4S^\circ-2p^4\ ^4P$	$3/2-5/2$
833,326	15	0,00	14,88	$2p^3\ ^4S^\circ-2p^4\ ^4P$	$3/2-3/2$
832,754	14	0,00	14,89	$2p^3\ ^4S^\circ-2p^4\ ^4P$	$3/2-1/2$
796,661	10	5,02	20,58	$2p^3\ ^2P^\circ-2p^4\ ^2D$	$1/2,\ 3/2-5/2,\ 3/2$
750,226	2	—	—	—	—
749,662	1	—	—	—	—
741,293	00	14,89	31,61	$3p^4\ ^4P-3s'''\ ^4S^\circ$	$1/2-3/2$
740,838	0	14,88	31,61	$2p^4\ ^4P-3s'''\ ^4S^\circ$	$3/2-3/2$
739,949	1	14,86	31,61	$2p^4\ ^4P-3s'''\ ^4S^\circ$	$5/2-3/2$
718,562	16	3,33	20,58	$2p^3\ ^2D^\circ-2p^4\ ^2D$	$3/2-3/2,\ 5/2$
718,484	17	3,32	20,58	$2p^3\ ^2D^\circ-2p^4\ ^2D$	$5/2-3/2,\ 5/2$
673,768	7	5,02	23,42	$2p^3\ ^2P^\circ-3s\ ^2P$	$3/2,\ 1/2-1/2$
672,948	8	5,02	23,44	$2p^3\ ^2P^\circ-3s\ ^2P$	$3/2,\ 1/2-1/2$
644,148	12	5,02	24,26	$2p^3\ ^2P^\circ-2p^4$	$3/2,\ 1/2-1/2$
617,051	6	3,33	23,42	$2p^3\ ^2D^\circ-3s\ ^2P$	$3/2-1/2$
616,363	4	3,33	23,44	$2p^3\ ^2D^\circ-3s\ ^2P$	$3/2-3/2$

λ, Å	I	E_H, eV	E_B, eV	Transition	J
616,291	7	3,32	23,44	$2p^3\ ^2D^\circ-3s\ ^2P$	5/2—3/2
600,585	6	5,02	25,66	$2p^3\ ^2P^\circ-3s'\ ^2D$	1/2, 3/2—3/2, 5/2
580,967	7	5,02	26,36	$2p^3\ ^2P^\circ-2p^4\ ^2P^\circ$	1/2, 3/2—3/2
580,400	6	5,02	26,38	$2p^3\ ^2P^\circ-2p^4\ ^2P^\circ$	1/2, 3/2—1/2
555,121	5	3,33	25,66	$2p^3\ ^2D^\circ-3s'\ ^2D$	3/2—5/2, 3/2
555,056	5	3,32	25,66	$2p^3\ ^2D^\circ-3s'\ ^2D$	5/2—5/2, 3/2
539,853	7	0,00	22,96	$2p^3\ ^4S^\circ-3s^4\ P$	3/2—1/2
539,547	8	0,00	22,98	$2p^3\ ^4S^\circ-3s\ ^4P$	3/2—3/2
539,086	8	0,00	23,00	$2p^3\ ^4S^\circ-3s\ ^4P$	3/2—5/2
538,318	7	3,33	26,36	$2p^3\ ^2D^\circ-2p^4\ ^2P$	3/2—3/2
538,256	10	3,32	26,36	$2p^3\ ^2D^\circ-2p^4\ ^2P$	5/2—3/2
537,830	9	3,33	26,38	$2p^3\ ^2D^\circ-2p^4\ ^2P$	3/2—1/2
518,242	5	5,02	28,94	$2p^3\ ^2P^\circ-3d\ ^2P$	3/2, 1/2—3/2
517,937	4	5,02	28,95	$2p^3\ ^2P^\circ-3d\ ^2P$	3/2, 1/2—1/2
515,640	4	5,02	29,06	$2p^3\ ^2P^\circ-3d\ ^2D$	3/2, 1/2—3/2
515,498	5	5,02	29,07	$2p^3\ ^2P^\circ-3d\ ^2D$	3/2—5/2
500,343	1	5,02	29,80	$2p^3\ ^2P^\circ-4s\ ^2P$	3/2, 1/2—1/2
499,871	2	5,02	29,82	$2p^3\ ^2P^\circ-4s\ ^2P$	3/2, 1/2—3/2
485,631	4	3,33	28,86	$2p^3\ ^2D^\circ-3d\ ^4D$	3/2—5/2
485,572	1	3,32	28,86	$2p^2\ ^2D^\circ-3d\ ^4D$	5/2—5/2, 7/2
485,515	5	3,33	28,86	$2p^3\ ^2D^\circ-3d\ ^2F$	3/2—5/2
485,465	0	3,32	28,86	$2p^3\ ^2D^\circ-3d\ ^2F$	5/2—5/2
485,086	6	3,32	28,88	$2p^3\ ^2D^\circ-3d\ ^2F$	5/2—7/2
484,025	2	3,33	28,94	$2p^3\ ^2D^\circ-3d\ ^2P$	3/2—3/2
483,976	5	3,32	28,94	$2p^3\ ^2D^\circ-3d\ ^2P$	5/2—3/2
483,752	4	3,33	28,95	$2p^3\ ^2D^\circ-3d\ ^2P$	3/2—1/2
481,755	3	3,33	29,06	$2p^3\ ^2D^\circ-3d\ ^2D$	3/2—3/2
481,704	1	3,32	29,06	$2p^3\ ^2D^\circ-3d\ ^2D$	5/2—3/2
481,635	0	3,33	29,07	$2p^3\ ^2D^\circ-3d\ ^2D$	3/2—5/2
481,587	4	3,32	29,07	$2p^3\ ^2D^\circ-3d\ ^2D$	5/2—5/2
470,408	4	5,02	31,37	$2p^3\ ^2P^\circ-3d'\ ^2D$	3/2—3/2, 5/2
468,766	2	5,02	31,46	$2p^3\ ^2P^\circ-3d'\ ^2P$	3/2—3/2, 5/2
467,926	0	3,32	29,82	$2p^3\ ^2D^\circ-4s\ ^2P$	5/2—3/2
465,760	2	5,02	31,64	$2p^3\ ^2P^\circ-4d\ ^2P$	1/2, 3/2—3/2
465,529	1	5,02	31,65	$2p^3\ ^2P^\circ-4d\ ^2P$	1/2, 3/2—1/2
464,785	3	5,02	31,69	$2p^3\ ^2P^\circ-3d'\ ^2S$	1/2, 3/2—1/2
464,310	1	5,02	31,72	$2p^3\ ^2P^\circ-4d\ ^2D$	1/2—3/2
464,194	2	5,02	31,73	$2p^3\ ^2P^\circ-4d\ ^2D$	3/2—5/2
458,422	0	5,02	32,06	$2p^3\ ^2P^\circ-5s\ ^2P$	1/2, 3/2—3/2
456,997	1	5,02	32,15	$2p^3\ ^2P^\circ-4s'\ ^2P$	1/2, 3/2—3/2, 5/2
445,638	4	3,33	31,15	$2p^3\ ^2D^\circ-3d'\ ^2F$	3/2—5/2
445,601	4	3,32	31,15	$2p^3\ ^2D^\circ-3d'\ ^2F$	5/2—7/2
443,681	0	5,02	32,96	$2p^3\ ^2P^\circ-5d\ ^2D$	3/2—5/2
442,048	4	3,33	31,37	$2p^3\ ^2D^\circ-3d'\ ^2D$	3/2—3/2
442,001	4	3,32	31,37	$2p^3\ ^2D^\circ-3d'\ ^2D$	5/2—5/2
440,598	2	3,33	31,46	$2p^3\ ^2D^\circ-3d'\ ^2P$	3/2—1/2
440,552	3	3,32	31,46	$2p^3\ ^2D^\circ-3d'\ ^2P$	5/2—3/2
437,683	3	3,33	31,65	$2p^3\ ^2D^\circ-4d\ ^2F$	3/2—5/2
437,332	3	3,32	31,67	$2p^3\ ^2D^\circ-4d\ ^2F$	5/2—7/2
436,649	0	3,33	31,72	$2p^3\ ^2D^\circ-4d\ ^2D$	3/2—3/2
436,510	1	3,32	31,73	$2p^3\ ^2D^\circ-4d\ ^2D$	5/2—5/2
430,177	6	0,00	28,82	$2p^3\ ^4S^\circ-3d\ ^4P$	3/2—5/2
430,041	6	0,00	28,83	$2p^3\ ^4S^\circ-3d\ ^4P$	3/2—3/2
429,918	5	0,00	28,84	$2p^3\ ^4S^\circ-3d\ ^4P$	3/2—1/2
429,716	4	0,00	28,85	$2p^3\ ^4S^\circ-3d\ ^4D$	3/2—1/2
429,647	5	0,00	28,86	$2p^3\ ^4S^\circ-3d\ ^4D$	3/2—3/2, 5/2
429,557	2	0,00	28,86	$2p^3\ ^4S^\circ-3d\ ^2F$	3/2—5/2
426,526	1	5,02	34,08	$2p^3\ ^2P^\circ-4d'\ ^2D$	1/2, 3/2—1/2, 3/2

λ, Å	I	E_H, eV	E_B, eV	Transition	J
425,273	0	5,02	34,17	$2p^3\ ^2P°—4d'\ ^2P$	$^1/_2,\ ^3/_2—^1/_2,\ ^3/_2$
424,577	0	5,02	34,22	$2p^3\ ^2P°—4d'\ ^2S$	$^1/_2,\ ^3/_2—^1/_2$
418,812	0	3,32	32,93	$2p^3\ ^2D°—5d\ ^2F$	$^5/_2—^7/_2$
418,598	1	0,00	29,62	$2p^3\ ^4S°—4s\ ^4P$	$^3/_2—^5/_2$
403,372	0	3,33	34,06	$2p^3\ ^2D°—4d'\ ^2F$	$^3/_2—^5/_2$
403,273	0	3,32	34,07	$2p^3\ ^2D°—4d'\ ^2F$	$^5/_2—^7/_2$
403,087	0	3,33	34,08	$2p^3\ ^2D°—4d'\ ^2D$	$^3/_2—^3/_2,\ ^5/_2$
403,035	0	3,32	34,08	$2p^3\ ^2D°—4d'\ ^2D$	$^5/_2—^3/_2,\ ^5/_2$
392,322	3	0,00	31,60	$2p^3\ ^4S°—4d\ ^4D$	$^1/_2—^1/_2,\ ^3/_2$
392,002	3	0,00	31,63	$2p^3\ ^4S°—4d\ ^4P$	$^3/_2—^5/_2$
391,943	2	0,00	31,63	$2p^3\ ^4S°—4d\ ^4P$	$^3/_2—^3/_2$
391,912	1	0,00	31,63	$2p^3\ ^4S°—4d\ ^4P$	$^3/_2—^1/_2$
377,045	0	0,00	32,88	$2p^3\ ^4S°—5d\ ^4D$	$^3/_2—^3/_2,\ ^5/_2$
376,745	0	0,00	32,91	$2p^3\ ^4S°—5d\ ^4P$	$^3/_2—^5/_2$
376,693	0	0,00	32,91	$2p^3\ ^4S°—5d\ ^4P$	$^3/_2—^3/_2,\ ^1/_2$

O III, ground state $1s^2 2s^2 2p^2\ ^3P_0$
Ionization potential 443193,5 cm⁻¹; 54,934 eV

λ, Å	I	E_H, eV	E_B, eV	Transition	J
5592,37	6	33,86	36,07	$3s\ ^1P°—3p\ ^1P$	1—1
5592,01	—	38,01	40,23	$3p\ ^1D—3d\ ^3F°$	2—2
5508,11	1	38,01	40,26	$3p\ ^1D—3d\ ^1D°$	2—2
5268,06	2	38,90	41,26	$3p\ ^1S—3d\ ^1P°$	0—1
4569,50	1	52,86	55,56	$3p'\ ^1D°—3d'\ ^1F$	2—3
4555,30	0	46,91	49,63	$3p\ ^3P°—3d\ ^3P$	2—2
4529,7	00	46,62	49,36	$3p\ ^5S°—3d\ ^5D$	2—3
4474,95	1	52,86	55,60	$3p'\ ^1D°—3d'\ ^1D$	2—2
4461,56	1	46,62	49,36	$3p\ ^5S°—3d\ ^5P$	2—3
4447,82	0	46,62	49,41	$3p\ ^5S°—3d\ ^5P$	2—2
4440,1	0	46,62	49,41	$3p\ ^5S°—3d\ ^5P$	2—1
4434,43	2	—	—	—	—
4239,5	00	33,15	36,07	$3s\ ^3P°—3p\ ^1P$	1—1
4081,10	1	43,43	46,47	$3s\ ^3P—3p\ ^3D°$	2—3
4073,90	0	43,41	46,45	$3s\ ^3P—3p\ ^3D°$	1—2
3961,59	8	38,01	41,14	$3p\ ^1D—3d\ ^1F°$	2—3
3935,0	2	—	—	—	—
3816,75	1	38,01	41,26	$3p\ ^1D—3d\ ^1P°$	2—1
3810,99	2	33,18	36,43	$3s\ ^3P°—3p\ ^3D$	2—1
3791,26	6	33,18	36,45	$3s\ ^3P°—3p\ ^3D$	2—2
3774,00	6	33,15	36,43	$3s\ ^3P°—3p\ ^3D$	1—1
3759,87	9	33,18	36,48	$3s\ ^3P°—3p\ ^3D$	2—3
3757,21	5	33,13	36,43	$3s\ ^3P°—3p\ ^3D$	0—1
3754,67	7	33,15	36,45	$3s\ ^3P°—3p\ ^3D$	1—2
3734,80	1	42,01	45,33	$3s\ ^5P—3p\ ^5D°$	3—2
3732,13	1	37,25	40,57	$3p\ ^3P—3d\ ^3D°$	2—1
3729,70	1	46,44	49,76	$3p\ ^3D°—3d\ ^3F$	1—2
3728,82	1	46,47	49,79	$3p\ ^3D°—3d\ ^3F$	3—4
3728,49	0	46,45	49,77	$3p\ ^3D°—3d\ ^3F$	2—3
3725,30	3	37,25	40,57	$3p\ ^3P—3d\ ^3D°$	2—2
3721,95	1	41,99	45,32	$3s\ ^5P—3p\ ^5D°$	2—1
3720,86	3	42,01	45,34	$3s\ ^5P—3p\ ^5D°$	3—3
3715,08	6	37,25	40,58	$3p\ ^3P—3d\ ^3D°$	2—3

λ, Å	I	E_H, eV	E_B, eV	Transition	J
3714,03	2	37,23	40,57	$3p\ ^3P—3d\ ^3D°$	1—1
3712,48	2	41,99	45,33	$3s\ ^5P—3p\ ^5D°$	2—2
3709,52	2	41,97	45,31	$3s\ ^5P—3p\ ^5D°$	1—0
3707,24	6	37,23	40,57	$3p\ ^3P—3d\ ^3D°$	1—2
3704,73	3	41,97	45,32	$3s\ ^5P—3p\ ^5D°$	1—1
3703,37	5	42,01	45,36	$3s\ ^5P—3p\ ^5D°$	3—4
3702,75	5	37,22	40,57	$3p\ ^3P—3d\ ^3D°$	0—1
3698,70	5	41,99	45,34	$3s\ ^5P—3p\ ^5D°$	2—3
3695,37	4	41,97	45,33	$3s\ ^5P—3p\ ^5D°$	1—2
3653,00	1	46,92	50,31	$3p\ ^3P°—3d\ ^3D$	0—1
3650,70	0	46,91	50,31	$3p\ ^3P°—3d\ ^3D$	1—1
3649,20	00	46,91	50,31	$3p\ ^3P°—3d\ ^3D$	2—1
3646,84	2	46,91	50,31	$3p\ ^3P°—3d\ ^3D$	1—2
3645,20	1	46,91	50,31	$3p\ ^3P°—3d\ ^3D$	2—2
3638,70	3	46,91	50,32	$3p\ ^3P°—3d\ ^3D$	2—3
3556,90	1	43,43	46,91	$3s\ ^3P—3p\ ^3P°$	2—2
3475,26	—	45,36	48,92	$3p\ ^5D°—3d\ ^5F$	4—3
3466,90	0	45,34	48,92	$3p\ ^5D°—3d\ ^5F$	3—2
3466,15	2	45,36	48,93	$3p\ ^5D°—3d\ ^5F$	4—4
3459,98	2	45,34	48,92	$3p\ ^5D°—3d\ ^5F$	3—3
3459,52	0	45,33	48,91	$3p\ ^5D°—3d\ ^5F$	2—1
3455,12	5	45,36	48,94	$3p\ ^5D°—3d\ ^5F$	4—5
3454,90	2	45,33	48,92	$3p\ ^5D°—3d\ ^5F$	2—2
3451,33	1	45,32	48,91	$3p\ ^5D°—3d\ ^5F$	1—1
3450,94	4	45,34	48,93	$3p\ ^5D°—3d\ ^5F$	3—4
3448,05	0	45,33	48,92	$3p\ ^5D°—3d\ ^5F$	2—3
3447,22	1	45,32	48,91	$3p\ ^5D°—3d\ ^5F$	0—1
3446,73	2	45,32	48,92	$3p\ ^5D°—3d\ ^5F$	1—2
3444,10	5	37,25	40,85	$3p\ ^3P—3d\ ^3P°$	2—2
3440,39	4	36,98	40,58	$2p^4\ ^1D—3d\ ^3D°$	2—3
3430,60	4	37,25	40,86	$3p\ ^3P—3d\ ^3P°$	2—1
3428,67	3	37,23	40,86	$3p\ ^3P—3d\ ^3P°$	1—2
3427,42	3	—	—	—	—
3415,29	3	37,23	40,86	$3p\ ^3P—3d\ ^3P°$	1—1
3408,13	1	37,23	40,87	$3p\ ^3P—3d\ ^3P°$	1—0
3405,74	2	37,22	40,86	$3p\ ^3P—3d\ ^3P°$	0—1
3399,71	2	—	—	—	—
3394,26	1	45,67	49,36	$3p\ ^5P°—3d\ ^5D$	3—3
3384,95	4	45,71	49,37	$3p\ ^5P°—3d\ ^5D$	3—4
3383,85	2	45,70	49,36	$3p\ ^5P°—3d\ ^5D$	2—2
3382,69	3	45,70	49,36	$3p\ ^5P°—3d\ ^5D$	2—3
3376,82	1	45,69	49,36	$3p\ ^5P°—3d\ ^5D$	1—1
3376,66	2	—	—	—	—
3369,40	00	36,89	40,57	$3p\ ^3S—3d\ ^3D°$	1—1
3363,83	1	36,89	40,57	$3p\ ^3S—3d\ ^3D°$	1—2
3362,38	4	42,01	45,70	$3s\ ^5P—3p\ ^5P°$	3—2
3355,92	3	45,71	49,40	$3p\ ^5P°—3d\ ^5P$	3—3
3350,99	4	42,01	45,71	$3s\ ^5P—3p\ ^5P°$	3—3
3350,68	3	41,99	45,69	$3s\ ^5P—3p\ ^5P°$	2—1
3348,05	2	45,71	49,41	$3p\ ^5P°—3d\ ^5P$	3—2
3344,26	2	45,70	49,40	$3p\ ^5P°—3d\ ^5P$	2—3
		41,99	45,70	$3s\ ^5P—3p\ ^5P°$	2—2
3340,74	6	33,18	36,89	$3s\ ^3P°—3p\ ^3S$	2—1
3336,78	3	45,70	49,41	$3p\ ^5P°—3d\ ^5P$	2—2
		41,97	45,69	$3s\ ^5P—3p\ ^5P°$	1—1
3333,00	4	41,99	45,71	$3s\ ^5P—3p\ ^5P°$	2—3
3332,49	1	45,70	49,41	$3p\ ^5P°—3d\ ^5P$	2—1
3330,40	4	45,69	49,41	$3p\ ^5P°—3d\ ^5P$	1—2
		41,97	45,70	$3s\ ^5P—3p\ ^5P°$	1—2

λ, Å	I	E_H, eV	E_B, eV	Transition	J
3326,16	0	45,69	49,41	$3p\ ^5P^\circ—3d\ ^5P$	1—1
3312,30	5	33,15	36,89	$3s\ ^3P^\circ—3p\ ^3S$	1—1
3305,77	0	36,48	40,23	$3p\ ^3D—3d\ ^3F^\circ$	3—2
3299,36	3	33,13	36,89	$3s\ ^3P^\circ—3p\ ^3S$	0—1
3284,57	4	36,48	40,25	$3p\ ^3D—3d\ ^3F^\circ$	3—3
3281,94	3	36,45	40,23	$3p\ ^3D—3d\ ^3F^\circ$	2—2
3279,97	1	46,25	50,03	$4p\ ^1S—5d\ ^1P^\circ$	0—1
3267,31	5	36,43	40,23	$3p\ ^3D—3d\ ^3F^\circ$	1—2
3265,46	10	36,48	40,27	$3p\ ^3D—3d\ ^3F^\circ$	3—4
3260,98	8	36,45	40,25	$3p\ ^3D—3d\ ^3F^\circ$	2—3
3252,94	2	36,45	40,26	$3p\ ^3D—3d\ ^1D^\circ$	2—2
3238,57	5	36,43	40,26	$3p\ ^3D—3d\ ^1D^\circ$	1—2
3215,97	1	46,47	50,32	$3p\ ^3D^\circ—3d\ ^3D$	3—3
3207,12	1	46,45	50,31	$3p\ ^3D^\circ—3d\ ^3D$	2—2
3200,95	1	46,44	50,31	$3p\ ^3D^\circ—3d\ ^3D$	1—1
3132,86	6	36,89	40,85	$3p\ ^3S—3d\ ^3P^\circ$	1—2
3121,71	5	36,89	40,86	$3p\ ^3S—3d\ ^3P^\circ$	1—1
3115,73	4	36,89	40,87	$3p\ ^3S—3d\ ^3P^\circ$	1—0
3095,81	00	45,36	49,36	$3p\ ^5D^\circ—3d\ ^5D$	4—3
3088,04	2	45,36	49,37	$3p\ ^5D^\circ—3d\ ^5D$	4—4
3084,63	0	45,34	49,36	$3p\ ^5D^\circ—3d\ ^5D$	3—2
3083,65	1	45,34	49,36	$3p\ ^5D^\circ—3d\ ^5D$	3—3
3075,95	0	45,34	49,37	$3p\ ^5D^\circ—3d\ ^5D$	3—4
3075,19	0	45,33	49,36	$3p\ ^5D^\circ—3d\ ^5D$	2—2
3074,68	00	45,33	49,36	$3p\ ^5D^\circ—3d\ ^5D$	2—1
3074,15	0	45,33	49,36	$3p\ ^5D^\circ—3d\ ^5D$	2—3
3068,68	0	45,32	49,36	$3p\ ^5D^\circ—3d\ ^5D$	1—2
3068,06	00	45,32	49,36	$3p\ ^5D^\circ—3d\ ^5D$	1—0
3065,01	00	45,32	49,36	$3p\ ^5D^\circ—3d\ ^5D$	0—1
3059,30	6	33,18	37,23	$3s\ ^3P^\circ—3p\ ^3P$	2—1
3047,13	8	33,18	37,25	$3s\ ^3P^\circ—3p\ ^3P$	2—2
3043,02	5	33,15	37,22	$3s\ ^3P^\circ—3p\ ^3P$	1—0
3035,43	4	33,15	37,23	$3s\ ^3P^\circ—3p\ ^3P$	1—1
3034,32	0	41,26	45,34	$3d\ ^1P^\circ—4p\ ^1P$	1—1
3024,57	4	33,13	37,23	$3s\ ^3P^\circ—3p\ ^3P$	0—1
3024,36	1	36,48	40,57	$3p\ ^3D—3d\ ^3D^\circ$	3—2
3023,45	5	33,15	37,25	$3s\ ^3P^\circ—3p\ ^3P$	1—2
3017,63	5	36,48	40,58	$3p\ ^3D—3d\ ^3D^\circ$	3—3
3008,79	3	36,45	40,57	$3p\ ^3D—3d\ ^3P^\circ$	2—1
3004,35	4	36,45	40,57	$3p\ ^3D—3d\ ^3D^\circ$	2—2
2997,71	2	36,45	40,58	$3p\ ^3D—3d\ ^3D^\circ$	2—3
2996,51	3	36,43	40,57	$3p\ ^3D—3d\ ^3D^\circ$	1—1
2992,11	2	36,43	40,57	$3p\ ^3D—3d\ ^3D^\circ$	1—2
2983,78	9	33,86	38,01	$3s\ ^1P^\circ—3p\ ^1D$	1—2
2983,66	1	36,07	40,23	$3p\ ^1P—3d\ ^3F^\circ$	1—2
2959,74	5	36,07	40,26	$3p\ ^1P—3d\ ^1D^\circ$	1—2
2926,33	2	—	—	—	—
2916,40	4	—	—	—	—
2912,98	2	—	—	—	—
2903,30	4	—	—	—	—
2898,48	1	36,98	41,26	$2p^4\ ^1D—3d\ ^1P^\circ$	2—1
2895,45	2	—	—	—	—
2893,70	3	—	—	—	—
2885,36	3	—	—	—	—
2881,70	4	—	—	—	—
2881,28	2	—	—	—	—
2879,80	0	45,47	49,78	$4p\ ^3D—5d\ ^3F^\circ$	3—2
2873,82	2	—	—	—	—

λ, Å	I	E_H, eV	E_B, eV	Transition	J
2866,73	2	—	—	—	—
2863,57	3	—	—	—	—
2862,52	3	—	—	—	—
2862,26	3	45,45	49,78	$4p\ ^3D—5d\ ^3F°$	2—2
2861,38	3	—	—	—	—
2857,89	3	—	—	—	—
2856,78	3	—	—	—	—
2853,78	1	45,44	49,78	$4p\ ^3D—5d\ ^3F°$	1—2
2848,91	6	—	—	—	—
2845,84	3	—	—	—	—
2844,39	2	—	—	—	—
2836,34	4	36,48	40,85	$3p\ ^3D—3d\ ^3P°$	3—2
2818,68	1	36,45	40,85	$3p\ ^3D—3d\ ^3P°$	2—2
2809,63	3	36,45	40,86	$3p\ ^3D—3d\ ^3P°$	2—1
2799,01	2	36,43	40,86	$3p\ ^3D—3d\ ^3P°$	1—1
2797,97	1	45,47	49,90	$4p\ ^3D—5d\ ^3D°$	3—3
2794,09	5	36,43	40,87	$3p\ ^3D—3d\ ^3P°$	1—0
2789,89	3	—	—	—	—
2784,48	2	—	—	—	—
2772,04	2	45,34	49,81	$4p\ ^1P—5d\ ^1D°$	1—2
2770,15	2	—	—	—	—
2766,50	2	—	—	—	—
2761,30	3	—	—	—	—
2756,22	1	36,07	40,57	$3p\ ^1P—3d\ ^3D°$	2—1
2752,47	0	36,07	40,57	$3p\ ^1P—3d\ ^3D°$	1—2
2745,00	2	—	—	—	—
2739,15	0	49,65	54,18	$3d\ ^3P—4p\ ^3S°$	0—1
2735,14	1	49,65	54,18	$3d\ ^3P—4p\ ^3S°$	1—1
2726,95	1	49,63	54,18	$3d\ ^3P—4p\ ^3S°$	2—1
2713,40	2	40,87	45,44	$3p\ ^3P°—4p\ ^3D$	0—1
2708,87	1	40,86	45,44	$3d\ ^3P°—4p\ ^3D$	1—1
2701,05	3	40,86	45,45	$3d\ ^3P°—4p\ ^3D$	1—2
2696,11	2	—	—	—	—
2696,00	3	—	—	—	—
2695,49	6	45,04	49,63	$3p\ ^3S°—3d\ ^3P$	1—2
2692,74	1	40,85	45,45	$3d\ ^3P°—4p\ ^3D$	2—2
2687,53	5	45,04	49,65	$3p\ ^3S°—3d\ ^3P$	1—1
2686,14	10	42,01	46,62	$3s\ ^5P—3p\ ^5S°$	3—2
2683,65	4	45,04	49,65	$3p\ ^3S°—3d\ ^3P$	1—0
2681,42	2	—	—	—	—
2677,81	3	40,85	45,47	$3p\ ^3P°—4p\ ^3D$	2—3
2674,57	8	41,99	46,62	$3s\ ^5P—3p\ ^5S°$	2—2
2670,10	2	—	—	—	—
2665,69	7	41,97	46,62	$3s\ ^5P—3p\ ^5S°$	1—2
2641,53	3	—	—	—	—
2640,98	2	—	—	—	—
2640,68	2	—	—	—	—
2639,04	2	—	—	—	—
2628,53	2	—	—	—	—
2623,69	3	—	—	—	—
2622,41	3	—	—	—	—
2622,32	2	41,26	45,98	$3d\ ^1P°—4p\ ^1D$	1—2
2609,59	4	40,87	45,62	$3d\ ^3P°—4p\ ^3S$	0—1
2605,41	6	40,86	45,62	$3d\ ^3P°—4p\ ^3S$	1—1
2597,69	8	40,85	45,62	$3d\ ^3P°—4p\ ^3S$	2—1
2596,79	00	40,57	45,34	$3d\ ^3D°—4p\ ^1P$	1—1
2588,23	0	36,07	40,85	$3p\ ^1P—3d\ ^3P°$	1—1
2582,99	3	—	—	—	—

λ, Å	I	$E_н$, eV	$E_в$, eV	Transition	J
2578,27	00	36,45	41,26	$3p\ ^3D—3d\ ^1P°$	2—1
2572,30	2	—	—	—	—
2569,21	2	—	—	—	—
2569,11	1	36,43	41,26	$3p\ ^3D—3d\ ^1P°$	1—1
2558,06	8	41,14	45,98	$3d\ ^1F°—4p\ ^1D$	3—2
2549,62	2	40,57	45,44	$3p\ ^3D°—4p\ ^3D$	2—1
2547,45	2	40,58	45,45	$3p\ ^3D°—4p\ ^3D$	3—2
2546,43	4	40,57	45,44	$3d\ ^3D°—4p\ ^3D$	1—1
2542,68	5	40,57	45,45	$3d\ ^3D°—4p\ ^3D$	2—2
2539,50	2	40,57	45,45	$3d\ ^3D°—4p\ ^3D$	1—2
2534,08	6	40,58	45,47	$3d\ ^3D°—4p\ ^3D$	3—3
2529,36	1	40,57	45,47	$3p\ ^3D°—4p\ ^3D$	2—3
2492,2	00	49,36	54,34	$3d\ ^5D—4p\ ^5D°$	2—1
2489,0	00	49,36	54,34	$3d\ ^5D—4p\ ^5D°$	3—2
2488,3	00	49,37	54,35	$3d\ ^5D—4p\ ^5D°$	4—3
2485,27	00	49,36	54,34	$3d\ ^5D—4p\ ^5D°$	2—2
2483,24	0	49,36	54,35	$3d\ ^5D—4p\ ^5D°$	3—3
2482,60	0	49,36	54,35	$3d\ ^5D—4p\ ^5D°$	2—3
2480,73	1	49,37	54,37	$3d\ ^5D—4p\ ^5D°$	4—4
2475,73	00	49,36	54,37	$3d\ ^5D—4p\ ^5D°$	3—4
2457,8	00	40,57	45,62	$3d\ ^3D°—4p\ ^3S$	2—1
2454,99	8	33,86	38,90	$3s\ ^1P°—3p\ ^1S$	1—0
2454,21	0	49,41	54,46	$3d\ ^5P—4p\ ^5P°$	2—1
2450,06	2	—	—	—	—
2449,38	2	—	—	—	—
2446,92	00	49,40	54,47	$3d\ ^5P—4p\ ^5P°$	3—2
2441,06	2	49,40	54,48	$3d\ ^5P—4p\ ^5P°$	3—3
2438,83	5	40,26	45,34	$3d\ ^1D°—4p\ ^1P$	2—1
2434,96	2	40,85	45,94	$3d\ ^3P°—4p\ ^3P$	2—2
2431,69	0	49,79	54,89	$3d\ ^3F—4p\ ^3D°$	4—3
2429,65	0	49,36	54,46	$3d\ ^5D—4p\ ^5P°$	1—1
2429,35	1	49,36	54,46	$3d\ ^5D—4p\ ^5P°$	2—1
2426,94	2	49,36	54,47	$3d\ ^5D—4p\ ^5P°$	3—2
2426,35	0	49,36	54,46	$3d\ ^5D—4p\ ^5P°$	2—1, 2
2425,93	2	49,37	54,48	$3d\ ^5D—4p\ ^5P°$	4—3
2422,84	5	40,23	45,34	$3d\ ^3F°—4p\ ^1P$	2—1
2421,2	00	49,36	54,48	$3d\ ^5D—4p\ ^5P°$	3—3
2394,33	4	40,26	45,44	$3d\ ^1D°—4p\ ^3D$	2—1
2390,44	8	36,07	41,26	$3p\ ^1P—3d\ ^1P°$	1—1
2388,20	1	40,26	45,45	$3d\ ^1D°—4p\ ^3D$	2—2
2383,92	6	40,25	45,45	$3d\ ^3F°—4p\ ^3D$	3—2
2382,32	7	40,27	45,47	$3d\ ^3F°—4p\ ^3D$	4—3
2378,90	4	40,23	45,44	$3d\ ^3F°—4p\ ^3D$	2—1
2372,82	2	40,23	45,45	$3d\ ^3F°—4p\ ^3D$	2—2
2372,21	3	40,25	45,47	$3d\ ^3F°—4p\ ^3D$	3—3
2319,52	2	40,57	45,91	$3p\ ^3D°—4p\ ^3P$	1—0
2317,37	3	40,57	45,92	$3p\ ^3D°—4p\ ^3P$	2—1
2315,52	4	40,58	45,94	$3p\ ^3D°—4p\ ^3P$	3—2
2314,76	2	40,57	45,92	$3p\ ^3D°—4p\ ^3P$	1—1
2311,58	2	40,57	45,94	$3p\ ^3D°—4p\ ^3P$	2—2
2308,70	1	35,21	40,57	$2p^4\ ^3P—3d\ ^3D°$	1—2
2288,36	00	48,92	54,33	$3d\ ^5F—4p\ ^5D°$	2—1
2288,20	2	—	—	—	—
2288,12	00	48,92	54,34	$3d\ ^5F—4p\ ^5D°$	3—2
2287,21	1	48,93	54,35	$3d\ ^5F—4p\ ^5D°$	4—3
2286,40	0	48,91	54,33	$3d\ ^5F—4p\ ^5D°$	1—1
2285,66	2	48,94	54,37	$3d\ ^5F—4p\ ^5D°$	5—4
2285,07	00	48,92	54,34	$3d\ ^5F—4p\ ^5D°$	2—2

λ, Å	I	E_H, eV	E_B, eV	Transition	J
2239,10	2	—	—	—	—
2228,15	3	38,90	44,47	$3p\,^1S—4s\,^1P^\circ$	0—1
2186,97	1	35,18	40,85	$3p^4\,^3P—3d\,^3P^\circ$	2—2
2181,66	1	35,18	40,86	$2p^4\,^3P—3d\,^3P^\circ$	2—1
2165,32	3	40,26	45,98	$3d\,^1D^\circ—4p\,^1D$	2—2
2162,88	5	—	—	—	—
2107,13	5	—	—	—	—
2106,07	5	—	—	—	—
2089,27	2	—	—	—	—
2052,53	4	—	—	—	—
2045,41	5	—	—	—	—
1916,48	2	—	—	—	—
1907,06	1	—	—	—	—
1902,89	1	—	—	—	—
1153,773	3	24,43	35,18	$2p^3\,^3S^\circ—2p^4\,^3P$	1—2
1150,882	2	24,43	35,21	$2p^3\,^3S^\circ—2p^4\,^3P$	1—1
1149,603	1	24,43	35,22	$2p^3\,^3S^\circ—2p^4\,^3P$	1—0
1138,545	2	26,09	36,98	$2p^3\,^1P^\circ—2p^4\,^1D$	1—2
898,957	8	23,19	36,98	$2p^3\,^1D^\circ—2p^4\,^1D$	2—2
887,404	10	—	—	—	—
878,728	11	—	—	—	—
875,534	9	—	—	—	—
871,099	10	—	—	—	—
835,292	16	0,04	14,88	$2p^2\,^3P—2p^3\,^3D^\circ$	2—3
835,096	14	0,04	14,88	$2p^2\,^3P—2p^3\,^3D^\circ$	2—1, 2
833,742	16	0,01	14,88	$2p^2\,^3P—2p^3\,^3D^\circ$	1—1, 2
832,927	14	0,00	14,88	$2p\,^3P—2p^3\,^3D^\circ$	0—1
752,762	4	26,09	42,56	$2p^3\,^1P^\circ—2p^4\,^1S$	1—0
707,315	4	17,65	35,18	$2p^3\,^3P^\circ—2p^4\,^3P$	1, 2—2
706,298	2	17,65	35,21	$2p^3\,^3P^\circ—2p^4\,^3P$	0—1
706,224	3	17,65	35,21	$2p^3\,^3P^\circ—2p^4\,^3P$	1, 2—1
705,762	2	17,65	35,22	$2p^3\,^3P^\circ—2p^4\,^3P$	1—0
703,850	18	0,04	17,65	$2p^2\,^3P—2p^3\,^3P^\circ$	2—1, 2
702,899	17	0,01	17,65	$2p^2\,^3P—2p^3\,^3P^\circ$	1—1, 2
702,822	16	0,01	17,65	$2p^2\,^3P—2p^3\,^3P^\circ$	1—0
702,332	16	0,00	17,65	$2p^2\,^3P—2p^3\,^3P^\circ$	0—1
659,538	0	17,65	36,45	$2p^3\,^3P^\circ—3p\,^3D$	1—2
658,758	1	17,65	36,48	$2p^3\,^3P^\circ—3p\,^3D$	2—3
644,159	6	17,65	36,89	$2p^3\,^3P^\circ—3p\,^3S$	1, 2—1
610,850	6	14,88	35,18	$2p^3\,^3D^\circ—2p^4\,^3P$	2—2
610,746	8	14,88	35,18	$2p^3\,^3D^\circ—2p^4\,^3P$	3—2
610,043	7	14,88	35,21	$2p^3\,^3D^\circ—2p^4\,^3P$	2, 1—1
609,705	6	14,88	35,22	$2p^3\,^3D^\circ—2p^4\,^3P$	1—0
599,598	18	2,51	23,19	$2p^2\,^1D—2p^3\,^1D^\circ$	2—2
597,818	15	5,35	26,09	$2p^2\,^1S—2p^3\,^1P^\circ$	0—1
574,065	00	14,88	36,48	$2p^3\,^3D^\circ—3p\,^3D$	3—3
554,275	0	14,88	37,25	$2p^3\,^3D^\circ—3p\,^3P$	3—2
525,795	18	2,51	26,09	$2p^2\,^1D—2p^3\,^1P^\circ$	2—1
508,182	18	0,01	24,43	$2p^2\,^3P—2p^3\,^3S^\circ$	2—1
507,683	17	0,01	24,43	$2p^2\,^3P—2p^3\,^3S^\circ$	1—1
507,391	16	0,00	24,43	$2p^2\,^3P—2p^3\,^3S^\circ$	0—1
491,980	1	24,43	49,63	$2p^3\,^3S^\circ—3d\,^3P$	1—2
491,714	0	24,43	49,65	$2p^3\,^3S^\circ—3d\,^3P$	1—1
481,587	4	17,65	43,39	$2p^3\,^3P^\circ—3s\,^3P$	1—0
481,381	2	17,65	43,41	$2p^3\,^3P^\circ—3s\,^3P$	0—1
481,354	3	17,65	43,21	$2p^3\,^3P^\circ—3s\,^3P$	1, 2—1
480,955	4	17,65	43,43	$2p^3\,^3P^\circ—3s\,^3P$	1, 2—2

λ, Å	I	E_H, eV	E_B, eV	Transition	J
459,896	1	—	—	—	—
434,975	10	5,35	33,86	$2p^2\ ^1S—3s\ ^1P^\circ$	0—1
434,840	2	14,88	43,39	$2p^3\ ^3D^\circ—3s\ ^3P$	1—0
434,646	3	14,88	43,41	$2p^3\ ^3D^\circ—3s\ ^3P$	2—1
434,256	4	14,88	43,43	$2p^3\ ^3D^\circ—3s\ ^3P$	3—2
397,310	0	17,65	48,86	$2p^3\ ^3P^\circ—3s'\ ^3D$	0—1
397,231	1	17,65	48,86	$2p^3\ ^3P^\circ—3s'\ ^3D$	1—2
397,120	2	17,65	48,87	$2p^3\ ^3P^\circ—3s'\ ^3D$	2—3
395,558	2	2,51	33,86	$2p^2\ ^1D—3s\ ^1P^\circ$	2—1
387,639	4	17,65	49,63	$2p^3\ ^3P^\circ—3d\ ^3P$	1,2—2
387,482	3	17,65	49,65	$2p^3\ ^3P^\circ—3d\ ^3P$	0, 1, 2—1
387,398	2	17,65	49,65	$2p^3\ ^3P^\circ—3d\ ^3P$	1—0
382,903	1	23,19	55,55	$2p^3\ ^1D^\circ—3d'\ ^1F^\circ$	2—3
382,214	1	23,19	55,55	$2p^3\ ^1D^\circ—3d'\ ^1D^\circ$	2—2
379,631	2	17,65	50,31	$2p^3\ ^3P^\circ—3d\ ^3D$	0, 1—1
379,575	3	17,65	50,31	$2p^3\ ^3P^\circ—3d\ ^3D$	1, 2—2
379,505	4	17,65	50,32	$2p^3\ ^3P^\circ—3d\ ^3D$	2—3
374,436	8	0,04	33,15	$2p^2\ ^3P—3s\ ^3P^\circ$	2—1
374,331	8	0,01	33,13	$2p^2\ ^3P—3s\ ^3P^\circ$	1—0
374,165	8	0,01	33,15	$2p^2\ ^3P—3s\ ^3P^\circ$	1—1
374,075	10	0,04	33,18	$2p^2\ ^3P—3s\ ^3P^\circ$	2—2
374,005	8	0,00	33,15	$2p^2\ ^3P—3s\ ^3P^\circ$	0—1
373,805	8	0,01	33,18	$2p^2\ ^3P—3s\ ^3P^\circ$	1—2
364,940	1	14,88	48,86	$2p^3\ ^3D^\circ—3s'\ ^3D$	1—1
364,867	2	14,88	48,86	$2p^3\ ^3D^\circ—3s'\ ^3D$	2—2
364,739	3	14,88	48,87	$2p^3\ ^3D^\circ—3s'\ ^3D$	3—3
359,616	1	—	—	—	—
359,415	2	7,48	41,97	$2p^3\ ^5S^\circ—3s\ ^5P$	2—1
359,384	7	7,48	41,97	$2p^3\ ^5S^\circ—3s\ ^5P$	2—1
359,223	8	7,48	41,99	$2p^3\ ^5S^\circ—3s\ ^5P$	2—2
359,016	8	7,48	42,01	$2p^3\ ^5S^\circ—3s\ ^5P$	2—3
356,768	0	14,88	49,63	$2p^3\ ^3D^\circ—3d\ ^3P$	2—2
356,725	2	14,88	49,63	$2p^3\ ^3D^\circ—3d\ ^3P$	3—2
356,625	1	14,88	49,65	$2p^3\ ^3D^\circ—3d\ ^3P$	2—1
356,558	0	14,88	49,65	$2p^3\ ^3D^\circ—3d\ ^3P$	1—0
355,469	5	14,88	49,76	$2p^3\ ^3D^\circ—3d\ ^3F$	1—2
355,333	5	14,88	49,77	$2p^3\ ^3D^\circ—3d\ ^3F$	2—3
355,293	3	14,88	49,77	$2p^3\ ^3D^\circ—3d\ ^3F$	3—3
355,137	6	14,88	49,79	$2p^3\ ^3D^\circ—3d\ ^3F$	3—4
350,703	8	—	—	—	—
349,961	1	14,88	50,31	$2p^3\ ^3D^\circ—3d\ ^3D$	1—1
349,918	2	14,88	50,31	$2p^3\ ^3D^\circ—3d\ ^3D$	2—2
349,825	3	14,88	50,32	$2p^3\ ^3D^\circ—3d\ ^3D$	3—3
345,309	10	5,35	41,26	$2p^2\ ^1S—3d\ ^1P^\circ$	0—1
328,742	9	2,51	40,23	$2p^2\ ^1D—3d\ ^3F^\circ$	2—2
328,448	10	2,51	40,26	$2p^2\ ^1D—3d\ ^1D^\circ$	2—2
320,979	12	2,51	41,14	$2p^2\ ^1D—3d\ ^1F^\circ$	2—3
320,720	2	17,65	56,74	$2p^3\ ^3P^\circ—3d'\ ^3P$	—
319,996	3	2,51	41,26	$2p^2\ ^1D—3d\ ^1P^\circ$	2—1
317,265	1	17,65	56,74	$2p^3\ ^3P^\circ—3d'\ 3P$	—
316,967	3	5,35	44,47	$2p^2\ ^1S—4s\ ^1P^\circ$	0—1
308,306	2	0,04	40,25	$2p^2\ ^3P—3d\ ^3F^\circ$	2—3
308,051	1	0,01	40,26	$2p^2\ ^3P—3d\ ^1D^\circ$	1—2
305,879	4	0,04	40,57	$2p^2\ ^3P—3d\ ^3D^\circ$	2—1
305,836	8	0,04	40,57	$2p^2\ ^3P—3d\ ^3D^\circ$	2—2
305,769	10	0,04	40,58	$2p^2\ ^3P—3d\ ^3D^\circ$	2—3
305,703	8	0,01	40,57	$2p^2\ ^3P—3d\ ^3D^\circ$	1—1
305,656	9	0,01	40,58	$2p^2\ ^3P—3d\ ^3D^\circ$	1—2

λ, Å	I	E_H, eV	E_B, eV	Transition	J
305,596	8	0,00	40,57	$2p^2\ ^3P—3d\ ^3D^\circ$	0—1
303,799	9	0,04	40,85	$2p^2\ ^3P—3d\ ^3P^\circ$	2—2
303,693	7	0,04	40,86	$2p^2\ ^3P—3d\ ^3P^\circ$	2—1
303,621	7	0,01	40,85	$2p^2\ ^3P—3d\ ^3P^\circ$	1—2
303,515	7	0,01	40,86	$2p^2\ ^3P—3d\ ^3P^\circ$	1—1
303,460	7	0,01	40,87	$2p^2\ ^3P—3d\ ^3P^\circ$	1—0
303,411	7	0,00	40,86	$2p^2\ ^3P—3d\ ^3P^\circ$	0—1
300,455	3	14,88	56,14	$2p^3\ ^3D^\circ—3d'\ ^3F$	1, 2, 3—2, 3, 4
299,275	2	14,88	56,31	$2p^3\ ^3D^\circ—3d'\ ^3D$	1, 2, 3—1, 2, 3
296,270	1	14,88	56,74	$2p^3\ ^3D^\circ—3d'\ ^3P$	1, 2, 3—0, 2, 1
296,012	4	7,48	49,36	$2p^3\ ^5S^\circ—3d\ ^5D$	2—3
295,944	3	5,35	47,25	$2p^2\ ^1S—4d\ ^1P^\circ$	0—1
295,716	6	7,48	49,40	$2p^3\ ^5S^\circ—3d\ ^5P$	2—3
295,657	6	7,48	49,41	$2p^3\ ^5S^\circ—3d\ ^5P$	2—2
295,619	5	7,48	49,41	$2p^3\ ^5S^\circ—3d\ ^5P$	2—1
295,511	3	2,51	44,47	$2p^2\ ^1D—4s\ ^1P^\circ$	2—1
286,038	0	5,35	48,69	$2p^2\ ^1S—5s\ ^1P^\circ$	0—1
280,483	1	0,04	44,24	$2p^2\ ^3P—4s\ ^3P^\circ$	2—1
280,412	1	0,01	44,23	$2p^2\ ^3P—4s\ ^3P^\circ$	1—0
280,328	1	0,01	44,24	$2p^2\ ^3P—4s\ ^3P^\circ$	1—1
280,265	3	0,04	44,27	$2p^2\ ^3P—4s\ ^3P^\circ$	2—2
280,234	1	0,00	44,24	$2p^2\ ^3P—4s\ ^3P^\circ$	0—1
280,116	1	0,01	44,27	$2p^2\ ^3P—4s\ ^3P^\circ$	1—2
280,030	2	2,51	46,79	$2p^2\ ^1D—4d\ ^3F^\circ$	2—2
279,787	3	2,51	46,82	$2p^2\ ^1D—4d\ ^1D^\circ$	2—2
277,514	1	5,35	50,03	$2p^2\ ^1S—5d\ ^1P^\circ$	0—1
277,385	7	2,51	47,21	$2p^2\ ^1D—4d\ ^1F^\circ$	2—3
275,513	4	0,04	45,04	$2p^2\ ^3P—3p\ ^3S^\circ$	2—1
275,366	3	0,01	45,04	$2p^2\ ^3P—3p\ ^3S^\circ$	1—1
275,281	2	0,00	45,04	$2p^2\ ^3P—3p\ ^3S^\circ$	0—1
271,611	0	7,48	53,12	$2p^3\ ^5S^\circ—4s\ ^5P$	2—1
271,523	1	7,48	53,14	$2p^3\ ^5S^\circ—4s\ ^5P$	2—2
271,403	1	7,48	53,16	$2p^3\ ^5S^\circ—4s\ ^5P$	2—3
268,451	1	2,51	48,69	$2p^2\ ^1D—5s\ ^1P^\circ$	2—1
267,121	4	0,04	46,45	$2p^2\ ^3P—3p\ ^3D^\circ$	2—2
267,050	3	0,01	46,44	$2p^2\ ^3P—3p\ ^3D^\circ$	1—1
267,030	7	0,04	46,47	$2p^2\ ^3P—3p\ ^3D^\circ$	2—3
266,985	7	0,01	46,45	$2p^2\ ^3P—3p\ ^3D^\circ$	1—2
266,967	6	0,00	46,44	$2p^2\ ^3P—3p\ ^3D^\circ$	0—1
264,480	6	0,04	46,91	$2p^2\ ^3P—3p\ ^3P^\circ$	2—2, 1
264,338	5	0,01	46,91	$2p^2\ ^3P—3p\ ^3P^\circ$	1—2, 1, 0
264,257	4	0,00	46,91	$2p^2\ ^3P—3p\ ^3P^\circ$	0—1
263,903	0	0,04	47,02	$2p^2\ ^3P—4d\ ^3D^\circ$	2—1
263,861	3	0,04	47,02	$2p^2\ ^3P—4d\ ^3D^\circ$	2—2
263,818	5	0,04	47,03	$2p^2\ ^3P—4d\ ^3D^\circ$	2—3
263,768	3	0,01	47,02	$2p^2\ ^3P—4d\ ^3D^\circ$	1—1
263,728	4	0,01	47,02	$2p^2\ ^3P—4d\ ^3D^\circ$	1—2
263,692	3	0,00	47,02	$2p^2\ ^3P—4d\ ^3D^\circ$	0—1
262,882	1	0,04	47,20	$2p^2\ ^3P—4d\ ^3P^\circ$	2—2
262,729	0	0,01	47,20	$2p^2\ ^3P—4d\ ^3P^\circ$	2—2
262,289	0	2,51	49,78	$2p^2\ ^1D—5d\ ^3F^\circ$	2—2
262,113	2	2,51	49,81	$2p^2\ ^1D—5d\ ^1D^\circ$	2—2
261,027	4	2,51	50,01	$2p^2\ ^1D—5d\ ^1F^\circ$	2—3
256,506	3	7,48	55,81	$2p^3\ ^5S^\circ—4d\ ^5P$	2—3
256,460	3	7,48	55,82	$2p^3\ ^5S^\circ—4d\ ^5P$	2—2
256,425	2	7,48	55,83	$2p^3\ ^5S^\circ—4d\ ^5P$	2—1
255,302	0	0,04	48,62	$2p^2\ ^3P—5s\ ^3P^\circ$	2—1

λ, Å	I	E_H, eV	E_B, eV	Transition	J
255,158	1	0,04	48,62	$2p^2\ {}^3P—5s\ {}^3P^\circ$	2—2
255,044	0	0,01	48,62	$2p^2\ {}^3P—5s\ {}^3P^\circ$	1—2
253,548	0	2,51	51,41	$2p^2\ {}^1D—6d\ {}^1D^\circ$	2—2
248,618	2	0,04	49,90	$2p^2\ {}^3P—5d\ {}^3D^\circ$	2—3
248,574	1	0,01	49,90	$2p^2\ {}^3P—5d\ {}^3D^\circ$	1—2
248,538	1	0,00	49,90	$2p^2\ {}^3P—5d\ {}^3D^\circ$	0—1
248,320	1	2,51	52,44	$2p^2\ {}^1D—7d\ {}^1F^\circ$	2—3
247,080	1	2,51	52,69	$2p^2\ {}^1D—3p'\ {}^1F^\circ$	2—3
246,265	3	2,51	52,86	$2p^2\ {}^1D—3p'\ {}^1D^\circ$	2—2
244,049	2	2,51	53,31	$2p^2\ {}^1D—3p'\ {}^1P^\circ$	2—1
241,875	1	7,48	58,73	$2p^3\ {}^5S^\circ—5d\ {}^5P$	2—3
241,819	1	7,48	58,73	$2p^3\ {}^5S^\circ—5d\ {}^5P$	2—2, 1
241,037	2	0,04	51,47	$2p^2\ {}^3P—6d\ {}^3D^\circ$	2—3
240,979	2	0,04	51,47	$2p^2\ {}^3P—6d\ {}^3D^\circ$	2—3
236,710	1	—	—	—	—
228,988	0	0,04	54,18	$2p^2\ {}^3P—4p\ {}^3S^\circ$	2—1
228,893	0	0,01	54,18	$2p^2\ {}^3P—4p\ {}^3S^\circ$	1—1
226,038	1	0,04	54,89	$2p^2\ {}^3P—4p'\ {}^3D^\circ$	2—3

O IV, ground state $1s^2 2s^2 2p\ {}^2P^0_{1/2}$

Ionization potential 624396,5 cm⁻¹; 77,394 eV

λ, Å	I	E_H, eV	E_B, eV	Transition	J
7004, 06	—	56,17	57,94	$3s\ {}^2P^\circ—3p\ {}^2P$	$^3/_2—^3/_2$
6931, 39	—	56,14	57,93	$3s\ {}^2P^\circ—3p\ {}^2P$	$^1/_2—^1/_2$
5452, 5	—	71,33	73,61	$4p\ {}^2P—4d\ {}^2D^\circ$	$^3/_2—^5/_2$
5426, 5	—	71,31	73,60	$4p\ {}^2P—4d\ {}^2D^\circ$	$^1/_2—^3/_2$
5362,42	—	59,87	62,18	$3p\ {}^2D—3d\ {}^2D^\circ$	$^5/_2—^5/_2$
5305,32	15	59,84	62,18	$3p\ {}^2D—3d\ {}^2D^\circ$	$^3/_2—^3/_2$
4823,93	—	59,36	61,93	$3p\ {}^4P—3d\ {}^4D^\circ$	$^5/_2—^3/_2$
4813,07	1	59,36	61,94	$3p\ {}^4P—3d\ {}^4D^\circ$	$^5/_2—^5/_2$
4800,77	—	59,35	61,93	$3p\ {}^4P—3d\ {}^4D^\circ$	$^3/_2—^1/_2$
4799,2	10	—	—	—	
4798,25	5	59,36	61,94	$3p\ {}^4P—3d\ {}^4D^\circ$	$^5/_2—^7/_2$
4794,22	2	59,35	61,93	$3p\ {}^4P—3d\ {}^4D^\circ$	$^3/_2—^3/_2$
4787,7	3	—	—	—	—
4786,4	20	—	—	—	—
4783,43	4	59,35	61,94	$3p\ {}^4P—3d\ {}^4D^\circ$	$^3/_2—^5/_2$
4779,09	2	59,33	61,93	$3p\ {}^4P—3d\ {}^4D^\circ$	$^1/_2—^1/_2$
4772,57	2	59,33	61,93	$3p\ {}^4P—3d\ {}^4D^\circ$	$^1/_2—^3/_2$
4568, 5	—	68,50	71,21	$5f\ {}^2F^\circ—6d\ {}^2D$	$^5/_2$, $^7/_2—^3/_2$, $^5/_2$
4472, 4	—	71,33	74,10	$4p\ {}^2P—4d\ {}^2P^\circ$	$^3/_2—^3/_2$
4262, 3	—	67,86	70,76	$3p'\ {}^2D—3d'\ {}^2F^\circ$	$^5/_2—^7/_2$
3995,17	2	59,36	62,46	$3p\ {}^4P—3d\ {}^4P^\circ$	$^5/_2—^5/_2$
3977,10	1	59,36	62,48	$3p\ {}^4P—3d\ {}^4P^\circ$	$^5/_2—^3/_2$
3974,66	—	59,35	62,46	$3p\ {}^4P—3d\ {}^4D^\circ$	$^3/_2—^5/_2$
3956,82	—	59,35	62,48	$3p\ {}^4P—3d\ {}^4P^\circ$	$^3/_2—^3/_2$
3945,29	—	59,35	62,49	$3p\ {}^4P—3d\ {}^4P^\circ$	$^3/_2—^1/_2$
3942,14	—	59,33	62,48	$3p\ {}^4P—3d\ {}^4P^\circ$	$^1/_2—^3/_2$
3930,63	—	59,33	62,49	$3p\ {}^4P—3d\ {}^4P^\circ$	$^1/_2—^1/_2$
3774,38	—	58,08	61,37	$3p\ {}^4D—3d\ {}^4F^\circ$	$^7/_2—^5/_2$
3758,45	0	58,08	61,38	$3p\ {}^4D—3d\ {}^4F^\circ$	$^7/_2—^7/_2$
3755,82	—	58,06	61,36	$3p\ {}^4D—3d\ {}^4F^\circ$	$^5/_2—^3/_2$

λ, Å	I	E_H, eV	E_B, eV	Transition	J
3744,73	0	58,06	61,37	3p ^4D—3d $^4F^\circ$	$^5/_2$—$^5/_2$
3736,78	4	{ 58,08	61,40	3p ^4D—3d $^4F^\circ$	$^7/_2$—$^9/_2$
		58,04	61,36	3p ^4D—3d $^4F^\circ$	$^3/_2$—$^3/_2$
3734	3	—	—	—	—
3729,03	3	58,06	61,38	3p ^4D—3d $^4F^\circ$	$^5/_2$—$^7/_2$
3725,81	2	{ 58,04	61,37	3p ^4D—3d $^4F^\circ$	$^3/_2$—$^5/_2$
		58,03	61,36	3p ^4D—3d $^4F^\circ$	$^1/_2$—$^3/_2$
3722	2	—	—	—	—
3567	2	—	—	—	—
3563,36	2	59,87	63,35	3p^2 D—3d $^2F^\circ$	$^5/_2$—$^7/_2$
3560,42	1	59,84	63,32	3p ^2D—3d $^2F^\circ$	$^3/_2$—$^5/_2$
3520,9	—	60,23	63,75	4s ^2S—3d $^2P^\circ$	$^1/_2$—$^3/_2$
3502,2	—	60,23	63,77	4s ^2S—3d $^2P^\circ$	$^1/_2$—$^1/_2$
3492,24	0	64,30	67,85	3s′ $^2P^\circ$—3p′ ^2D	$^1/_2$—$^3/_2$
3489,84	1	64,31	67,86	3s′ $^2P^\circ$—3p′ ^2D	$^3/_2$—$^5/_2$
3425,57	0	54,42	58,04	3s $^4P^\circ$—3p ^4D	$^5/_2$—$^3/_2$
3413,71	1	48,38	52,01	3p $^2P^\circ$—3d ^2D	$^3/_2$—$^3/_2$
3411,76	4	48,38	52,01	3p $^2P^\circ$—3d ^2D	$^3/_2$—$^5/_2$
3409,75	2	54,42	58,06	3s $^4P^\circ$—3p ^4D	$^5/_2$—$^5/_2$
3405,97	—	54,39	58,03	3s $^4P^\circ$—3p ^4D	$^3/_2$—$^1/_2$
3403,58	3	48,37	52,01	3p $^2P^\circ$—3d ^2D	$^1/_2$—$^3/_2$
3396,83	2	54,39	58,04	3s $^4P^\circ$—3p ^4D	$^3/_2$—$^3/_2$
3390,37	—	54,37	58,03	3s $^4P^\circ$—3p ^4D	$^1/_2$—$^1/_2$
3385,55	6	54,42	58,08	3s $^4P^\circ$—3p ^4D	$^5/_2$—$^7/_2$
3381,33	—	54,37	58,04	3s $^4P^\circ$—3p ^4D	$^1/_2$—$^3/_2$
3381,28	4	54,39	58,06	3s $^4P^\circ$—3p ^4D	$^3/_2$—$^5/_2$
3378,09	0	56,17	59,84	3s $^2P^\circ$—3p^2 D	$^3/_2$—$^3/_2$
3375,50	3	58,79	62,46	3p ^4S—3d $^4P^\circ$	$^3/_2$—$^5/_2$
3371,38	4	—	—	—	—
3363,46	6	—	—	—	—
3362,63	—	58,79	62,46	3p ^4S—3d $^4P^\circ$	$^3/_2$—$^3/_2$
3354,31	—	58,79	62,49	3p ^4S—3d $^4P^\circ$	$^3/_2$—$^1/_2$
3349,11	3	56,17	59,87	3s $^2P^\circ$—3p ^2D	$^3/_2$—$^5/_2$
3348,08	2	56,14	59,84	3s $^2P^\circ$—3p ^2D	$^1/_2$—$^3/_2$
3216,31	—	58,08	61,94	3p ^4D—3d $^4D^\circ$	$^7/_2$—$^5/_2$
3209,64	3	58,08	61,94	3p ^4D—3d $^4D^\circ$	$^7/_2$—$^7/_2$
3199,53	1	58,06	61,93	3p ^4D—3d $^4D^\circ$	$^5/_2$—$^3/_2$
3194,75	1	58,06	61,94	3p ^4D—3d $^4D^\circ$	$^5/_2$—$^5/_2$
3185,72	0	58,04	61,93	3p ^4D—3d $^4D^\circ$	$^3/_2$—$^3/_2$
3180,98	—	58,04	61,94	3p ^4D—3d $^4D^\circ$	$^3/_2$—$^5/_2$
3180,72	0	58,04	61,94	3p ^4D—3d $^4D^\circ$	$^1/_2$—$^1/_2$
3177,80	0	58,03	61,93	3p ^4D—3d $^4D^\circ$	$^1/_2$—$^3/_2$
3071,66	5	44,37	48,37	3s ^2S—3p $^2P^\circ$	$^1/_2$—$^1/_2$
3063,46	6	44,37	48,38	3s ^2S—3p $^2P^\circ$	$^1/_2$—$^3/_2$
3052,54	1	56,17	60,23	3s $^2P^\circ$—4s ^2S	$^3/_2$—$^1/_2$
3028,04	0	56,14	60,23	3s $^2P^\circ$—4s ^2S	$^1/_2$—$^1/_2$
2926,14	1	57,94	62,18	3p ^2P—3d $^2D^\circ$	$^3/_2$—$^3/_2$
2921,43	3	57,94	62,18	3p ^2P—3d $^2D^\circ$	$^3/_2$—$^5/_2$
2916,29	2	57,93	62,18	3p ^2P—3d $^2D^\circ$	$^1/_2$—$^3/_2$
2836,25	6	54,42	58,79	3s $^4P^\circ$—3p ^4S	$^5/_2$—$^3/_2$
2829,18	2	58,08	62,46	3p ^4D—3d $^4P^\circ$	$^7/_2$—$^5/_2$
2816,53	4	58,06	62,46	3p ^4D—3d $^4P^\circ$	$^5/_2$—$^5/_2$
2803,60	2	58,06	62,48	3p ^4D—3d $^4P^\circ$	$^5/_2$—$^3/_2$
2787,04	8	{ 58,03	62,48	3p ^4D—3d $^4P^\circ$	$^1/_2$—$^3/_2$
		58,03	62,49	3p ^4D—3d $^4P^\circ$	$^3/_2$—$^1/_2$
2782,46	3	—	—	—	—
2781,05	7	—	—	—	—
2724,01	2	—	—	—	—

λ, Å	I	E_H, eV	E_B, eV	Transition	J
2682,41	2	—	—	—	—
2677,09	2	—	—	—	—
2675,34	2	—	—	—	—
2663,22	2	—	—	—	—
2662,29	2	—	—	—	—
2632,78	4	—	—	—	—
2620,04	3	—	—	—	—
2578,24	4	—	—	—	—
2553,61	2	—	—	—	—
2538,94	3	—	—	—	—
2531,76	2	—	—	—	—
2529,92	3	—	—	—	—
2528,08	2	—	—	—	—
2517,40	7	54,42	59,35	$3s\,^4P^\circ-3p\,^4P$	$5/2-3/2$
2509,23	8	54,42	59,36	$3s\,^4P^\circ-3p\,^4P$	$5/2-5/2$
2507,77	7	54,39	59,33	$3s\,^4P^\circ-3p\,^4P$	$3/2-1/2$
2506,56	3	—	—	—	—
2504,70	5	—	—	—	—
2504,20	2	—	—	—	—
2501,84	4	54,39	59,35	$3s\,^4P^\circ-3p\,^4P$	$3/2-3/2$
2499,29	6	54,37	59,33	$3s\,^4P^\circ-3p\,^4P$	$1/2-1/2$
2497,10	3	—	—	—	—
2493,75	10	54,39	59,36	$3s\,^4P^\circ-3p\,^4P$	$3/2-5/2$
2493,40	7	54,37	59,35	$3s\,^4P^\circ-3p\,^4P$	$1/2-3/2$
2459,45	3	—	—	—	—
2450,06	10	—	—	—	—
2449,36	8	—	—	—	—
2384,67	1	63,30	68,50	$4d\,^2D-5f\,^2F^\circ$	$3/2,\ 5/2-5/2,\ 7/2$
1343,507	7	22,41	31,63	$3p^2\,^2P-2p^3\,^2D^\circ$	$3/2-5/2$
1342,995	4	22,41	31,64	$2p^2\,^2P-2p^3\,^2D^\circ$	$3/2-3/2$
1338,603	6	22,38	31,64	$2p^2\,^2P-2p^3\,^2D^\circ$	$1/2-3/2$
923,433	4	22,41	35,83	$2p^2\,^2P-2p^3\,^2P^\circ$	$3/2-1/2$
923,353	6	22,41	35,83	$2p^2\,^2P-2p^3\,^2P^\circ$	$3/2-3/2$
921,364	5	22,38	35,83	$2p^2\,^2P-2p^3\,^2P^\circ$	$1/2-1/2$
921,301	4	22,38	35,83	$2p^2\,^2P-2p^3\,^2P^\circ$	$1/2-3/2$
802,250	5	20,38	35,83	$2p^2\,^2S-2p^3\,^2P^\circ$	$1/2-1/2$
802,198	6	20,38	35,83	$2p^2\,^2S-2p^3\,^2P^\circ$	$1/2-3/2$
790,203	16	0,05	15,74	$2p\,^2P^\circ-2p^2\,^2D$	$3/2-5/2$
790,103	13	0,05	15,74	$2p\,^2P^\circ-2p^2\,^2D$	$3/2-3/2$
787,710	15	0,00	15,74	$2p\,^2P^\circ-2p^2\,^2D$	$1/2-3/2$
779,997	6	15,74	31,63	$2p^2\,^2P-2p^3\,^2P^\circ$	$3/2-5/2$
779,905	10	15,74	31,63	$2p^2\,^2D-2p^3\,^2D^\circ$	$5/2-5/2$
779,821	9	15,74	31,64	$2p^2\,^2D-2p^3\,^2D^\circ$	$3/2-3/2$
779,734	6	15,74	31,64	$2p^2\,^2P-2p^3\,^2P^\circ$	$5/2-3/2$
625,852	14	8,86	28,67	$2p^2\,^4P-2p^3\,^4S^\circ$	$5/2-3/2$
625,130	14	8,84	28,67	$2p^2\,^4P-2p^3\,^4S^\circ$	$3/2-3/2$
624,617	13	8,82	28,67	$2p^2\,^4P-2p^3\,^4S^\circ$	$1/2-3/2$
617,033	7	15,74	35,83	$2p^2\,^2D-2p^3\,^2P^\circ$	$3/2-1/2$
616,933	8	15,74	35,83	$2p^2\,^2D-2p^3\,^2P^\circ$	$5/2-3/2$
609,829	15	0,05	20,38	$2p\,^2P^\circ-2p^2\,^2S$	$3/2-1/2$
608,395	14	0,05	20,38	$2p\,^2P^\circ-2p^2\,^2S$	$1/2-1/2$
555,262	16	0,05	22,38	$2p^2\,P^\circ-2p^2\,^2P$	$3/2-1/2$
554,514	18	0,00	22,41	$2p\,^2P^\circ-2p^2\,^2P$	$3/2-3/2$
554,074	17	0,00	22,38	$2p\,^2P^\circ-2p^2\,^2P$	$1/2-1/2$
553,328	16	0,00	22,41	$2p\,^2P^\circ-2p^2\,^2P$	$1/2-3/2$
471,603	0	31,64	57,93	$2p^3\,^2D^\circ-3p^2P$	$3/2-1/2$
471,273	1	31,63	57,94	$2p^3\,^2D^\circ-3p\,^2P$	$5/2-3/2$
442,873	0	20,38	48,37	$2p^2\,^2S-3p\,^2P^\circ$	$1/2-1/2$
442,705	1	20,38	48,38	$2p^2\,^2S-3p\,^2P^\circ$	$1/2-3/2$

λ, Å	I	E_H, eV	E_B, eV	Transition	J
399,85	4	—	—	—	—
399,71	2	—	—	—	—
399,62	2	—	—	—	—
399,50	3	—	—	—	—
379,919	3	15,74	48,37	$2p^2\ ^2D—3p^2\ P^\circ$	$3/2—1/2$
379,775	4	15,74	48,38	$2p^2\ ^2D—3p\ ^2P^\circ$	$5/2—3/2$
367,192	2	22,38	56,14	$2p^2\ ^2P—3s\ ^2P^\circ$	$1/2—1/2$
		22,41	56,17	$2p^2\ ^2P—3s\ ^2P^\circ$	$3/2—3/2$
346,688	3	20,38	56,14	$2p^2\ ^2S—3s\ ^2P^\circ$	$1/2—1/2$
346,372	4	20,38	56,17	$2p^2\ ^2S—3s\ ^2P^\circ$	$1/2—3/2$
339,436	0	31,64	68,16	$2p^3\ ^2D^\circ—3p'\ ^2P$	$3/2—1/2$
339,330	1	31,63	68,17	$2p^3\ ^2D^\circ—3p'\ ^2P$	$5/2—3/2$
327,519	0	35,83	73,68	$2p^3\ ^2P^\circ—4f\ ^2D$	$1/2—3/2$
327,320	1	35,83	73,71	$2p^3\ ^2P^\circ—4f\ ^2D$	$3/2—5/2$
321,457	1	35,83	74,40	$2p^3\ ^2P^\circ—3s'''\ ^2D$	$1/2,\ 3/2—3/2—5/2$
311,726	3	22,41	62,18	$2p^2\ ^2P—3d\ ^2D^\circ$	$3/2—3/2$
311,679	6	22,41	62,18	$2p^2\ ^2P—3d\ ^2D^\circ$	$3/2—5/2$
311,490	5	22,38	62,18	$2p^2\ ^2P—3d\ ^2D^\circ$	$1/2—3/2$
306,882	7	15,74	56,14	$2p^2\ ^2D—3s\ ^2P^\circ$	$3/2—1/2$
306,621	8	15,74	56,17	$2p^2\ ^2D—3s^2\ P^\circ$	$5/2—3/2$
299,850	4	22,41	63,75	$2p^2\ ^2P—3d\ ^2P^\circ$	$3/2—3/2$
299,710	2	22,41	63,77	$2p^2\ ^2P—3d\ ^2P^\circ$	$3/2—1/2$
299,620	2	22,38	63,75	$2p^2\ ^2P—3d\ ^2P^\circ$	$1/2—3/2$
299,495	3	22,38	63,77	$2p^3\ ^2P—3d\ ^2P^\circ$	$1/2—1/2$
295,874	2	22,41	64,31	$2p^2\ ^2P—3s'\ ^2P^\circ$	$3/2—3/2$
295,140	1	31,64	73,64	$2p^3\ ^2D^\circ—4f\ ^2F$	$3/2—5/2$
295,051	1	31,63	73,65	$2p^3\ ^2D^\circ—4f\ ^2F$	$5/2—7/2$
294,853	1	31,64	73,68	$2p^3\ ^2D^\circ—4f\ ^2D$	$3/2—3/2$
294,650	1	31,63	73,71	$2p^3\ ^2D^\circ—4f\ ^2D$	$5/2—5/2$
291,203	1	35,83	78,41	$2p^3\ ^2P^\circ—3d''\ ^2D$	$3/2—5/2$
291,054	1	35,83	78,43	$2p^3\ ^2P^\circ—3d''\ ^2D$	$1/2—3/2$
289,933	1	31,64	74,40	$3p^3\ ^2D^\circ—3s'''\ ^2D$	$3/2—1/2$
289,898	2	31,63	74,40	$2p^3\ ^2D^\circ—3s'''\ ^2D$	$5/2—3/2$
289,590	1	28,67	71,48	$2p^3\ ^4S^\circ—3s''\ ^4P$	$3/2—1/2$
289,469	2	28,67	71,50	$2p^3\ ^4S^\circ—3s''\ ^4P$	$3/2—3/2$
289,292	3	28,67	71,53	$2p^3\ ^4S^\circ—3s''\ ^4P$	$3/2—5/2$
285,838	7	20,38	63,75	$2p^2\ ^2S—3d\ ^2P^\circ$	$1/2—3/2$
285,714	6	20,38	63,77	$2p^2\ ^2S—3d\ ^2P^\circ$	$1/2—1/2$
282,213	1	20,38	64,31	$2p^2\ ^2S—3s'\ ^2P^\circ$	$1/2—3/2$
279,937	11	0,05	44,34	$2p\ ^2P^\circ—3s\ ^2S$	$3/2—1/2$
279,633	10	0,00	44,34	$2p\ ^2P^\circ—3s\ ^2S$	$1/2—1/2$
279,456	2	35,83	80,20	$2p^3\ ^2P^\circ—3d'''\ ^2D$	$1/2,\ 3/2—3/2,\ 5/2$
272,311	6	8,86	54,39	$2p^2\ ^4P—3s\ ^4P^\circ$	$5/2—3/2$
272,270	6	8,84	54,37	$2p^2\ ^4P—3s\ ^4P^\circ$	$3/2—1/2$
272,174	7	8,84	54,39	$2p^2\ ^4P—3s\ ^4P^\circ$	$3/2—3/2$
272,125	7	8,86	54,42	$2p^2\ ^4P—3s\ ^4P^\circ$	$5/2—5/2$
272,076	6	8,82	54,39	$2p^2\ ^4P—3s\ ^4P^\circ$	$1/2—3/2$
271,989	6	8,84	54,42	$2p^2\ ^4P—3s\ ^4P^\circ$	$3/2—5/2$
269,559	1	35,83	81,82	$2p^3\ ^2P^\circ—3d'''\ ^2S$	$1/2,\ 3/2—1/2$
266,967	5	15,74	62,18	$2p^2\ ^2D—3d\ ^2D^\circ$	$3/2—3/2$
266,932	6	15,74	62,18	$2p^2\ ^2D—3d\ ^2D^\circ$	$5/2—5/2$
266,729	0	31,64	78,12	$2p^3\ ^2D^\circ—3d''\ ^2F$	$3/2—5/2$
266,690	0	31,63	78,12	$2p^3\ ^2D^\circ—3d'''\ ^2F$	$5/2—7/2$
265,062	0	31,64	78,42	$2p^3\ ^2D^\circ—3d''\ ^2D$	$3/2,\ 5/2—3/2,\ 5/2$
260,556	9	15,74	63,32	$2p^2\ ^2D—3d\ ^2F^\circ$	$3/2—5/2$
260,389	10	15,74	63,35	$2p^2\ ^2D—3d\ ^2F^\circ$	$5/2—7/2$
258,207	3	15,74	63,75	$2p^2\ ^2D—3d\ ^2P^\circ$	$5/2—3/2$
258,116	2	15,74	63,77	$2p^2\ ^2D—3d\ ^2P^\circ$	$3/2—1/2$

λ, Å	I	E_H, eV	E_B, eV	Transition	J
255,252	5	15,74	64,31	$2p^2\ ^2D—3s'\ ^2P^{\circ}$	$^5/_2—^3/_2$
253,082	7	22,41	71,39	$2p^2\ ^2P—3d'\ ^2D^{\circ}$	$^3/_2—^5/_2$
252,948	6	22,38	72,12	$2p^2\ ^2P—3d'\ ^2P^{\circ}$	$^1/_2—^3/_2$
252,564	6	{ 31,64	80,72	$2p^3\ ^2D^{\circ}—3d'''\ ^2F$	$^3/_2—^5/_2$
		{ 31,63	80,72	$2p^3\ ^2D^{\circ}—3d'''\ ^2F$	$^5/_2—^7/_2$
251,148	1	31,63	81,00	$2p^3\ ^2D^{\circ}—3d'''\ ^2P$	$^5/_2—^3/_2$
251,114	1	31,64	81,01	$2p^3\ ^2D^{\circ}—3d'''\ ^2P$	$^3/_2—^1/_2$
249,365	4	22,41	72,12	$2p^2\ ^2P—3d'\ ^2P^{\circ}$	$^3/_2—^3/_2$
249,223	3	22,38	72,12	$2p^2\ ^2P—3d'\ ^2P^{\circ}$	$^1/_2—^1/_2$
246,563	4	28,67	78,95	$3p^3\ ^4S^{\circ}—3d''\ ^4P$	$^3/_2—^5/_2$
246,503	3	28,67	78,97	$2p^3\ ^4S^{\circ}—3d''\ ^4P$	$^3/_2—^3/_2$
246,465	2	28,67	78,97	$2p^3\ ^4S^{\circ}—3d''\ ^4P$	$^3/_2—^1/_2$
245,720	1	—	—	—	—
242,183	0	22,41	73,60	$2p^2\ ^2P—4d\ ^2D^{\circ}$	$^3/_2—^3/_2$
242,140	3	22,41	73,61	$2p^2\ ^2P—4d\ ^2D^{\circ}$	$^3/_2—^5/_2$
242,045	2	22,38	73,60	$2p^2\ ^2P—4d\ ^2D^{\circ}$	$^1/_2—^3/_2$
240,079	1	22,41	74,05	$2p^2\ ^2P—3p''\ ^2S^{\circ}$	$^3/_2—^1/_2$
239,935	0	22,38	74,05	$2p^2\ ^2P—3p''\ ^2S^{\circ}$	$^1/_2—^1/_2$
239,592	3	20,38	72,12	$2p^2\ ^2S—3d'\ ^2P^{\circ}$	$^1/_2—^1/_2,\ ^3/_2$
238,573	15	0,05	52,01	$2p\ ^2P^{\circ}—3d\ ^2D$	$^3/_2—^5/_2$
238,361	14	0,00	52,01	$2p\ ^2P^{\circ}—3d\ ^2D$	$^1/_2—^3/_2$
236,071	1	8,86	61,38	$2p^2\ ^4P—3d\ ^4F^{\circ}$	$^5/_2—^7/_2$
234,988	3	15,74	68,50	$2p^2\ ^2D—5f\ ^2F^{\circ}$	$^5/_2,\ ^7/_2—^5/_2$
233,596	6	8,86	61,94	$2p^2\ ^4P—3d\ ^4D^{\circ}$	$^5/_2—^5/_2$
233,561	8	8,86	61,94	$2p^2\ ^4P—3d\ ^4D^{\circ}$	$^5/_2—^7/_2$
233,521	6	8,84	61,93	$2p^2\ ^4P—3d\ ^4D^{\circ}$	$^3/_2—^3/_2$
233,495	7	8,84	61,94	$2p^2\ ^4P—3d\ ^4D^{\circ}$	$^3/_2—^5/_2$
233,457	7	8,82	61,93	$2p^2\ ^4P—3d\ ^4D^{\circ}$	$^1/_2—^3/_2,\ ^1/_2$
231,302	7	8,86	62,46	$2p^2\ ^4P—3d\ ^4P^{\circ}$	$^5/_2—^5/_2$
231,240	6	8,86	62,48	$2p^2\ ^4P—3d\ ^4P^{\circ}$	$^5/_2—^3/_2$
231,200	6	8,84	62,46	$2p^2\ ^4P—3d\ ^4P^{\circ}$	$^3/_2—^5/_2$
231,144	4	8,84	62,48	$2p^2\ ^4P—3d\ ^4P^{\circ}$	$^3/_2—^3/_2$
231,101	6	8,84	62,49	$2p^2\ ^4P—3d\ ^4P^{\circ}$	$^3/_2—^1/_2$
231,070	7	8,82	62,48	$2p^2\ ^4P—3d\ ^4P^{\circ}$	$^1/_2—^3/_2$
231,031	3	8,82	62,49	$2p^2\ ^4P—3d\ ^4P^{\circ}$	$^1/_2—^1/_2$
230,755	2	20,38	74,10	$2p^2\ ^2S—4d\ ^2P^{\circ}$	$^1/_2—^3/_2$
230,682	1	20,38	74,12	$2p^2\ ^2S—4d\ ^2P^{\circ}$	$^1/_2—^1/_2$
230,040	0	22,41	76,30	$2p^2\ ^2P—3p''\ ^2D^{\circ}$	$^3/_2—^5/_2$
229,896	0	22,38	76,30	$2p^2\ ^2P—3p''\ ^2D^{\circ}$	$^1/_2—^3/_2$
225,299	5	15,74	70,76	$2p^2\ ^2D—3d'\ ^2F^{\circ}$	$^3/_2,\ ^5/_2—^5/_2,\ ^7/_2$
223,841	0	15,74	71,12	$2p^2\ ^2D—4s\ ^2P^{\circ}$	$^3/_2—^1/_2$
223,728	0	15,74	71,15	$2p^2\ ^2D—4s\ ^2P^{\circ}$	$^5/_2—^3/_2$
222,777	4	15,74	71,39	$2p^2\ ^2D—3d'\ ^2D^{\circ}$	$^3/_2—^3/_2$
222,763	5	15,74	71,39	$2p^2\ ^2D—3d'\ ^2D^{\circ}$	$^5/_2—^5/_2$
221,648	4	15,74	71,67	$2p^2\ ^2D—6f\ ^2F^{\circ}$	$^5/_2,\ ^3/_2—^5/_2,\ ^7/_2$
221,515	0	20,38	76,34	$2p^2\ ^2S—3p''\ ^2P^{\circ}$	$^1/_2—^3/_2,\ ^1/_2$
216,960	0	15,74	72,88	$2p^2\ ^2D—7f\ ^2F^{\circ}$	$^5/_2—^5/_2,\ ^7/_2$
214,290	1	15,74	73,60	$2p^2\ ^2D—4d\ ^2D^{\circ}$	$^3/_2—^3/_2$
214,249	1	15,74	73,61	$2p^2\ ^2D—4d\ ^2D^{\circ}$	$^5/_2—^5/_2$
214,209	4	0,05	57,93	$4p\ ^2P^{\circ}—3p\ ^2P$	$^3/_2—^1/_2$
214,155	6	0,05	57,94	$2p^2\ P^{\circ}—3p\ ^2P$	$^3/_2—^3/_2$
214,032	5	0,00	57,93	$2p^2\ P^{\circ}\ 3p\ ^2P$	$^1/_2—^1/_2$
213,978	4	0,00	57,94	$2p^2\ P^{\circ}—3p^2\ P$	$^1/_2—^3/_2$
213,061	3	15,74	73,93	$2p^2\ ^2D—4d\ ^2F^{\circ}$	$^3/_2—^5/_2$
212,974	3	15,74	73,95	$2p^2\ ^2D—4d\ ^2F^{\circ}$	$^5/_2—^7/_2$
212,578	2	15,74	74,06	$2p^2\ ^2D—8f\ ^2F^{\circ}$	$^5/_2—^5/_2,\ ^7/_2$
212,808	0	20,38	78,91	$2p^2\ ^2S—5d\ ^2P^{\circ}$	$^1/_2—^1/_2,\ ^3/_2$
207,348	4	0,05	59,84	$2p\ ^2P^{\circ}—3p\ ^2D$	$^3/_2—^3/_2$
207,239	7	0,05	59,87	$2p\ ^2P^{\circ}—3p\ ^2D$	$^3/_2—^5/_2$

λ, Å	I	E_H, eV	E_B, eV	Transition	J
207,183	6	0,00	59,84	$2p\ ^2P^\circ - 3p\ ^2D$	$1/2-3/2$
206,002	1	0,05	60,23	$2p\ ^2P^\circ - 4s\ ^2S$	$3/2-1/2$
205,842	0	0,00	60,23	$2p\ ^2P^\circ - 4s\ ^2S$	$1/2-1/2$
204,996	0	22,41	82,88	$2p^2\ ^2P - 4d'\ ^2D^\circ$	$3/2-5/2$
204,905	0	22,38	82,88	$2p^2\ ^2P - 4d'\ ^2D^\circ$	$1/2-3/2$
204,708	0	15,74	76,30	$2p^2\ ^2D - 3p''\ ^2D^\circ$	$5/2,\ 3/2-5/2,\ 3/2$
203,048	5	0,05	61,11	$2p\ ^2P^\circ - 3p\ ^2S$	$3/2-1/2$
202,891	4	0,00	61,11	$2p\ ^2P^\circ - 3p\ ^2S$	$1/2-1/2$
201,098	1	8,86	70,51	$2p^2\ ^4P - 4s\ ^4P^\circ$	$5/2-3/2$
201,073	1	8,84	70,50	$2p^2\ ^4P - 4s\ ^4P^\circ$	$3/2-1/2$
201,022	0	—	—		—
200,995	2	8,86	70,54	$2p^2\ ^4P - 4s\ ^4P^\circ$	$5/2-5/2$
200,966	1	8,82	70,51	$2p^2\ ^4P - 4s\ ^4P^\circ$	$1/2-3/2$
200,915	1	8,84	70,54	$2p^2\ ^4P - 4s\ ^4P^\circ$	$3/2-5/2$
200,827	1	15,74	77,47	$2p^2\ ^2D - 3p'''\ ^2F^\circ$	—
196,435	0	15,74	78,85	$2p^2\ ^2D - 5d\ ^2F^\circ$	$3/2-5/2$
196,348	0	15,74	78,88	$2p^2\ ^2D - 5d\ ^2F^\circ$	$5/2-7/2$
196,009	8	0,05	63,30	$2p\ ^2P^\circ - 4d\ ^2D$	$3/2-5/2$
195,863	7	0,00	63,30	$2p\ ^2P^\circ - 4d\ ^2D$	$1/2-3/2$
192,244	3	8,86	73,37	$2p^2\ ^4P - 4d\ ^4D^\circ$	$5/2-5/2$
192,206	5	8,86	73,37	$2p^2\ ^4P - 4d\ ^4D^\circ$	$5/2-7/2$
192,169	4	8,84	73,37	$2p^2\ ^4P - 4d\ ^4D^\circ$	$3/2-5/2$
192,139	4	8,82	73,37	$2p^2\ ^4P - 4d\ ^4D^\circ$	$1/2-3/2$
191,752	3	8,86	73,52	$2p^2\ ^4P - 4d\ ^4P^\circ$	$5/2-5/2$
191,695	2	8,84	73,52	$2p^2\ ^4P - 4d\ ^4P^\circ$	$3/2-5/2$
191,640	0	8,84	73,52	$2p^2\ ^4P - 4d\ ^4P^\circ$	$3/2-3/2$
191,609	2	8,82	73,52	$2p^2\ ^4P - 4d\ ^4P^\circ$	$1/2-5/2$
188,190	0	8,86	74,75	$2p^2\ ^4P - 3p''\ ^4D^\circ$	$5/2-5/2$
188,152	2	8,86	74,75	$2p^2\ ^4P - 3p''\ ^4D^\circ$	$5/2-7/2$
186,982	0	8,86	75,18	$2p^2\ ^4P - 3p''\ ^4P^\circ$	$5/2-3/2$
186,936	2	8,86	75,18	$2p^2\ ^4P - 3p''\ ^4P^\circ$	$5/2-5/2$
186,872	1	8,84	75,18	$2p^2\ ^4P - 3p''\ ^4P^\circ$	$3/2-5/2$
185,544	1	—	—	—	—
185,384	0	—	—	—	—
183,454	1	8,86	76,44	$2p^2\ ^4P - 3p''\ ^4S^\circ$	$5/2-5/2$
183,395	1	8,84	76,44	$2p^2\ ^4P - 3p''\ ^4S^\circ$	$3/2-3/2$
183,353	0	8,82	76,44	$2p^2\ ^4P - 3p''\ ^4S^\circ$	$1/2-3/2$
183,139	0	0,05	66,87	$2p\ ^2P^\circ - 5s\ ^2S$	$3/2-1/2$
182,832	4	0,05	67,86	$2p\ ^2P^\circ - 3p'\ ^2D$	$3/2-5/2$
182,711	3	0,00	67,85	$2p\ ^2P^\circ - 3p'\ ^2D$	$1/2-3/2$
181,995	4	0,05	68,17	$2p\ ^2P^\circ - 3p'\ ^2P$	$3/2-3/2$
181,876	3	0,00	68,17	$2p\ ^2P^\circ - 3p'\ ^2P$	$1/2-1/2,\ 3/2$
181,275	5	0,05	68,44	$2p\ ^2P^\circ - 5d\ ^2D$	$3/2-5/2$
181,150	4	0,00	68,44	$2p\ ^2P^\circ - 5d\ ^2D$	$1/2-3/2$
180,481	2	0,05	68,74	$2p\ ^2P^\circ - 3p'\ ^2S$	$3/2-1/2$
180,351	1	0,00	68,74	$2p\ ^2P^\circ - 3p'\ ^2S$	$1/2-1/2$
177,808	2	8,86	78,59	$2p^2\ ^4P - 5d\ ^4D^\circ$	$5/2-7/2$
177,761	2	8,84	78,59	$2p^2\ ^4P - 5d\ ^4D^\circ$	$3/2-7/2,\ 5/2$
177,698	1	8,86	78,63	$2p^2\ ^4P - 5d\ ^4P^\circ$	$5/2-5/2$
177,659	0	8,84	78,63	$2p^2\ ^4P - 5d\ ^4P^\circ$	$3/2-5/2$
177,598	0	8,82	78,63	$2p^2\ ^4P - 5d\ ^4P^\circ$	$1/2-5/2$
174,220	3	0,05	71,21	$2p\ ^2P^\circ - 6d\ ^2D$	$3/2-5/2$
174,105	2	0,00	71,21	$2p\ ^2P^\circ - 6d\ ^2D$	$1/2-5/2$
173,968	0	0,05	71,31	$2p\ ^2P^\circ - 4p\ ^2P$	$3/2-1/2$
173,917	2	0,05	71,33	$2p\ ^2P^\circ - 4p\ ^2P$	$3/2-3/2$
173,851	1	0,00	71,31	$2p\ ^2P^\circ - 4p\ ^2P$	$1/2-1/2$
173,803	0	0,00	71,33	$2p\ ^2P^\circ - 4p\ ^2P$	$1/2-3/2$

λ, Å	I	E_H, eV	E_B, eV	Transition	J
171,191	0	0,05	72,47	$2p\ ^2P^\circ—4p\ ^2D$	$^3/_2—^3/_2$
171,121	2	0,05	72,50	$2p\ ^2P^\circ—4p\ ^2D$	$^3/_2—^5/_2$
171,071	2	0,00	72,47	$2p\ ^2P^\circ—4p\ ^2D$	$^1/_2—^3/_2$
170,988	0	8,86	81,37	$2p^2\ ^4P—6d\ ^4D^\circ$	$^5/_2—^7/_2$
170,940	0	8,84	81,37	$2p^2\ ^4P—6d\ ^4D^\circ$	$^3/_2—^7/_2$
167,145	0	8,86	83,03	$2p^2\ ^4P—7d\ ^4D^\circ$	$^5/_2—^7/_2$
160,141	0	0,05	77,92	$2p\ ^2P^\circ—5p\ ^2P$	$^3/_2—^3/_2$
158,606	1	0,05	78,21	$2p\ ^2P^\circ—5p\ ^2D$	$^3/_2—^5/_2$
158,553	0	0,00	78,19	$2p\ ^2P^\circ—5p\ ^2D$	$^1/_2—^3/_2$
153,162	1	0,00	81,01	$2p\ ^2P^\circ—3d'''\ ^2P$	—
152,355	0	0,05	81,42	$2p\ ^2P^\circ—4p'\ ^2D$	$^3/_2—^5/_2$
152,264	0	0,00	81,42	$2p\ ^2P^\circ—4p'\ ^2D$	$^1/_2—^3/_2$

O V, ground state $1s^2 2s^2\ ^1S_0$

Ionization potential 918702 cm^{-1}; 113,873 eV

λ, Å	I	E_H, eV	E_B, eV	Transition	J
7437,3	—	89,61	91,26	$4s\ ^3S—4p\ ^3P^\circ$	1—1
6909,0	—	85,53	87,32	$3p\ ^3P—3d\ ^3D^\circ$	2—1
6878,5	—	85,53	87,33	$3p\ ^3P—3d\ ^3D^\circ$	2—2
6830,1	8	85,53	87,34	$3p\ ^3P—3d\ ^3D^\circ$	2—3
6819,4	—	85,51	87,32	$3p\ ^3P—3d\ ^3D^\circ$	1—1
6789,8	—	85,51	87,33	$3p\ ^3P—3d\ ^3D^\circ$	1—2
6766,8	—	85,49	87,32	$3p\ ^3P—3d\ ^3D^\circ$	0—1
6328,6	—	86,43	88,39	$3p\ ^1D—3d\ ^1F^\circ$	2—3
5607,51	—	72,29	74,50	$3p\ ^3P^\circ—3d\ ^3D$	2—1
5604,11	—	72,29	74,50	$3p\ ^3P^\circ—3d\ ^3D$	2—2
5597,90	—	72,29	74,50	$3p\ ^3P^\circ—3d\ ^3D$	2—3
5583,29	—	72,29	74,50	$3p\ ^3P^\circ—3d\ ^3D$	1—1
5579,93	—	72,28	74,50	$3p\ ^3P^\circ—3d\ ^3D$	1—2
5572,00	—	72,28	74,50	$3p\ ^3P^\circ—3d\ ^3D$	0—1
5473,7	—	85,53	87,79	$3p\ ^3P—3d\ ^3P^\circ$	2—2
5431,5	—	85,53	87,81	$3p\ ^3P—3d\ ^3P^\circ$	2—1
5417,4	—	85,51	87,79	$3p\ ^3P—3d\ ^2P^\circ$	1—2
5376,0	—	85,51	87,81	$3p\ ^3P—3d\ ^3P^\circ$	1—1
5352,1	—	85,51	87,82	$3p\ ^3P—3d\ ^3P^\circ$	1—0
5343,3	—	85,49	87,81	$3p\ ^3P—3d\ ^3P^\circ$	0—1
5114,2	—	69,59	72,01	$3s\ ^1S—3p\ ^1P^\circ$	0—1
4554,28	0	83,40	86,12	$3p\ ^1P—3d\ ^1D^\circ$	1—2
4522,2	—	86,43	89,17	$3p\ ^1D—3d\ ^1P^\circ$	2—1
4213,2	—	81,03	83,97	$3s\ ^3P^\circ—3p\ ^3D$	2—1
4178,2	—	81,03	83,99	$3s\ ^3P^\circ—3p\ ^3D$	2—2
4158,76	0	84,82	87,79	$3p\ ^3S—3d\ ^3P^\circ$	1—2
4153,2	—	80,99	83,97	$3s\ ^3P^\circ—3p\ ^3D$	1—1
4135,9	—	84,82	87,81	$3p\ ^3S—3d\ ^3P^\circ$	1—1
4125,4	—	80,97	83,97	$3s\ ^3P^\circ—3p\ ^3D$	0—1
4123,90	2	81,03	84,04	$3s\ ^3P^\circ—3p\ ^3D$	2—3
4121,7	—	84,82	87,82	$3p\ ^3S—3d\ ^3P^\circ$	1—0
4119,2	—	80,99	83,99	$3s\ ^3P^\circ—3p\ ^3D$	1—2
3761,6	—	84,04	87,33	$3p\ ^3D—3d\ ^3D^\circ$	3—2
3747,1	—	84,04	87,34	$3p\ ^3D—3d\ ^3D^\circ$	3—3
3726,4	—	83,99	87,32	$3p\ ^3D—3d\ ^3D^\circ$	2—1

λ, Å	I	E_H, eV	E_B, eV	Transition	J
3717,5	—	83,99	87,33	$3p\,{}^3D$—$3d\,{}^3D^\circ$	2—2
3703,3	—	83,99	87,34	$3p\,{}^3D$—$3d\,{}^3D^\circ$	2—3
3702,2	—	89,17	92,52	$3d\,{}^1P^\circ$—$4d\,{}^1D$	1—2
3699,0	—	83,97	87,32	$3p\,{}^3D$—$3d\,{}^3D^\circ$	1—1
3690,2	—	83,97	87,32	$3p\,{}^3D$—$3d\,{}^3D^\circ$	1—2
3298,7	—	84,04	87,79	$3p\,{}^3D$—$3d\,{}^3P^\circ$	3—2
3275,67	0	81,03	84,82	$3s\,{}^3P^\circ$—$3p\,{}^3S$	2—1
3264,7	—	83,99	87,79	$3p\,{}^3D$—$3d\,{}^3P^\circ$	2—2
3249,7	—	83,99	87,81	$3p\,{}^3D$—$3d\,{}^3P^\circ$	2—1
3243,6	—	83,97	87,79	$3p\,{}^3D$—$3d\,{}^3P^\circ$	1—2
3239,3	—	80,99	84,82	$3s\,{}^3P^\circ$—$3p\,{}^3S$	1—1
3228,8	—	83,97	87,81	$3p\,{}^3D$—$3d\,{}^3P^\circ$	1—1
3222,3	—	80,97	84,82	$3s\,{}^3P^\circ$—$3p\,{}^3S$	0—1
3220,1	—	83,97	87,82	$3p\,{}^3D$—$3d\,{}^3P^\circ$	1—0
3144,68	1	72,01	75,95	$3p\,{}^1P^\circ$—$3d\,{}^1D$	1—2
3058,68	0	82,38	86,43	$3s\,{}^1P^\circ$—$3p\,{}^1D$	1—2
2789,86	3	67,83	72,28	$3s\,{}^3S$—$3p\,{}^3P^\circ$	1—0
2787,03	4	67,83	72,28	$3s\,{}^3S$—$3p\,{}^3P^\circ$	1—1
2781,04	5	67,83	72,29	$3s\,{}^3S$—$3p\,{}^3P^\circ$	1—2
2769,69	1	81,03	85,51	$3s\,{}^3P^\circ$—$3p\,{}^3P$	2—1
2755,13	2	81,03	85,53	$3s\,{}^3P^\circ$—$3p\,{}^3P$	2—2
2752,24	0	80,99	85,49	$3s\,{}^3P^\circ$—$3p\,{}^3P$	1—0
2743,62	0	80,99	85,51	$3s\,{}^3P^\circ$—$3p\,{}^3P$	1—1
2731,44	0	80,97	85,51	$3s\,{}^3P^\circ$—$3p\,{}^3P$	0—1
2729,35	1	80,99	85,53	$3s\,{}^3P^\circ$—$3p\,{}^3P$	1—2
1371,287	10	19,69	28,73	$2p\,{}^1P^\circ$—$2p^2\,{}^1D$	1—2
774,522	7	19,69	35,69	$2p\,{}^1P^\circ$—$2p^2\,{}^1S$	1—0
762,001	10	10,24	26,51	$2p\,{}^3P^\circ$—$2p^2\,{}^3P$	2—1
761,130	10	10,20	26,49	$2p\,{}^3P^\circ$—$2p^2\,{}^3P$	1—0
760,445	12	10,24	26,54	$2p\,{}^3P^\circ$—$2p^2\,{}^3P$	2—2
760,229	10	10,20	26,51	$2p\,{}^3P^\circ$—$2p^2\,{}^3P$	1—1
759,440	10	10,18	26,51	$2p\,{}^3P^\circ$—$2p^2\,{}^3P$	0—1
758,677	10	10,20	26,54	$2p\,{}^3P^\circ$—$2p^2\,{}^3P$	1—2
712,668	3	10,20	26,54	$2p^3\,P^\circ$—$2p^2\,{}^3P$	1—2
629,732	15	0,00	19,69	$2s^2\,{}^1S$—$2p\,{}^1P^\circ$	0—1
341,391	0	35,69	72,01	$2p^2\,{}^1S$—$3p\,{}^1P^\circ$	0—1
286,448	6	28,73	72,01	$2p^2\,{}^1D$—$3p\,{}^1P^\circ$	2—1
270,982	0	26,54	72,29	$2p^2\,{}^3P$—$3p\,{}^3P^\circ$	2—2
265,550	4	35,69	82,38	$2p^2\,{}^1S$—$3s\,{}^1P^\circ$	0—1
248,459	6	19,69	69,59	$2p\,{}^1P^\circ$—$3s\,{}^1S$	1—0
231,823	7	35,69	89,17	$2p^2\,{}^1S$—$3d\,{}^1P^\circ$	0—1
227,688	5	28,73	82,38	$2p^2\,{}^1D$—$3s\,{}^1P^\circ$	2—1
227,662	4	26,54	80,99	$2p^2\,{}^3P$—$3s\,{}^3P^\circ$	2—1
227,636	5	26,51	80,97	$2p^2\,{}^3P$—$3s\,{}^3P^\circ$	1—0
227,549	5	26,51	80,99	$2p^2\,{}^3P$—$3s\,{}^3P^\circ$	1—1
227,510	7	26,54	81,03	$2p^2\,{}^3P$—$3s\,{}^3P^\circ$	2—2
227,468	5	26,49	80,98	$2p^2\,{}^3P$—$3s\,{}^3P^\circ$	0—1
227,374	5	26,51	81,03	$2p^2\,{}^3P$—$3s\,{}^3P^\circ$	1—2
222,235	3	35,69	91,48	$2p^2\,{}^1S$—$4p\,{}^1P^\circ$	0—1
220,352	13	19,69	75,95	$2p\,{}^1P^\circ$—$3d\,{}^1D$	1—2
216,018	8	28,73	75,95	$2p^2\,{}^1D$—$3d\,{}^1D^\circ$	2—2
215,245	9	10,24	67,83	$2p\,{}^3P^\circ$—$3s\,{}^3S$	2—1
215,104	8	10,20	67,83	$2p\,{}^3P^\circ$—$3s\,{}^3S$	1—1
215,034	7	10,18	67,83	$2p\,{}^3P^\circ$—$3s\,{}^3S$	0—1
207,794	10	28,73	88,39	$2p^2\,{}^1D$—$3d\,{}^1F^\circ$	2—3
205,102	3	28,73	89,17	$2p^2\,{}^1D$—$3d\,{}^1P^\circ$	2—1
203,935	6	26,54	87,33	$2p^2\,{}^3P$—$3d\,{}^3D^\circ$	2—2
203,890	8	26,54	87,34	$2p^2\,{}^3P$—$3d\,{}^3D^\circ$	2—3
203,851	6	26,51	87,32	$2p^2\,{}^3P$—$3d\,{}^3D^\circ$	1—1

λ, Å	I	E_H, eV	E_B, eV	Transition	J
203,821	7	26,51	87,33	$2p^2\ {}^3P—3d\ {}^3D^\circ$	1—2
203,783	6	26,49	87,32	$2p^2\ {}^3P—3d\ {}^3D^\circ$	0—1
202,393	7	26,54	87,79	$2p^2\ {}^3P—3d\ {}^3P^\circ$	2—2
202,335	5	26,54	87,81	$2p^2\ {}^3P—3d\ {}^3P^\circ$	2—1
202,282	5	26,51	87,79	$2p^2\ {}^3P—3d\ {}^3P^\circ$	1—2
202,226	5	26,51	87,81	$2p^2\ {}^3P—3d\ {}^3P^\circ$	1—1
202,191	5	26,51	87,82	$2p^2\ {}^3P—3d\ {}^3P^\circ$	1—0
202,158	5	26,49	87,81	$2p^2\ {}^3P—3d\ {}^3P^\circ$	0—1
198,031	3	—	—	—	—
194,593	8	19,69	83,40	$2p\ {}^1P^\circ—3p\ {}^1P$	1—1
192,906	14	10,24	74,50	$2p\ {}^3P^\circ—3d\ {}^3D$	2—3
192,800	13	10,20	74,50	$2p\ {}^3P^\circ—3d\ {}^3D$	1—2
192,751	12	10,18	74,50	$2p\ {}^3P^\circ—3d\ {}^3D$	0—1
191,556	2	26,54	91,26	$2p^2\ {}^3P—4p\ {}^3P^\circ$	2—1, 2
191,458	1	26,51	91,26	$2p^2\ {}^3P—4p\ {}^3P^\circ$	1—0, 1, 2
191,397	0	26,49	91,26	$2p^2\ {}^3P—4p\ {}^3P^\circ$	0—1
185,747	9	19,69	86,43	$2p\ {}^1P^\circ—3p\ {}^1D$	1—2
185,455	2	35,69	102,19	$2p^2\ {}^1S—4s\ {}^1P^\circ$	0—1
182,205	2	19,69	87,73	$2p\ {}^1P^\circ—3p\ {}^1S$	1—0
178,713	2	35,69	105,07	$2p^2\ {}^1S—4d\ {}^1P^\circ$	0—1
174,558	2	19,69	90,71	$2p\ {}^1P^\circ—4s\ {}^1S$	1—0
172,168	12	0,00	72,01	$2s^2\ {}^1S—3p\ {}^1P^\circ$	0—1
170,218	5	19,69	92,52	$2p\ {}^1P^\circ—4d\ {}^1D$	1—2
169,586	0	—	—	—	—
169,478	0	—	—	—	—
168,077	4	10,24	83,99	$2p\ {}^3P^\circ—3p\ {}^3D$	2—2
168,042	4	10,20	83,97	$2p\ {}^3P^\circ—3p\ {}^3D$	1—1
		10,20	83,99	$2p\ {}^3P^\circ—3p\ {}^3D$	1—2
167,991	8	10,24	84,04	$2p\ {}^3P^\circ—3p\ {}^3D$	2—3
		10,18	83,97	$2p\ {}^3P^\circ—3p\ {}^3D$	0—1
166,234	5	10,24	84,82	$2p\ {}^3P^\circ—3p\ {}^3S$	2—1
166,152	4	10,20	84,82	$2p\ {}^3P^\circ—3p\ {}^3S$	2—1
166,113	3	10,18	84,82	$2p\ {}^3P^\circ—3p\ {}^3S$	0—1
164,986	2	28,73	103,87	$2p^2\ {}^1D—4d\ {}^1D^\circ$	2—2
164,710	4	10,24	85,51	$2p\ {}^3P^\circ—3p\ {}^3P$	2—1
164,656	6	10,24	85,53	$2p\ {}^3P^\circ—3p\ {}^3P$	2—2
164,628	4	10,20	85,51	$2p\ {}^3P^\circ—3p\ {}^3P$	1—1
164,578	5	10,20	85,53	$2p\ {}^3P^\circ—3p\ {}^3P$	1—2
164,178	2	28,73	104,24	$2p^2\ {}^1D—6f\ {}^1F^\circ$	2—3
162,494	4	28,73	105,02	$2p^2\ {}^1D—4d\ {}^1F^\circ$	2—3
159,380	4	26,54	104,33	$2p^2\ {}^3P—4d\ {}^3D^\circ$	2—3
159,343	4	26,50	104,31	$2p^2\ {}^3P—4d\ {}^3D^\circ$	1—2
158,926	2	26,54	104,55	$2p^2\ {}^3P—4d\ {}^3P^\circ$	2—2
158,813	1	26,51	104,56	$2p^2\ {}^3P—4d\ {}^3P^\circ$	1—1
156,225	3	10,24	89,59	$2p^3\ P^\circ—4s\ {}^3S$	2—1
156,158	2	10,20	89,59	$2p\ {}^3P^\circ—4s\ {}^3S$	1—1
156,126	1	10,18	89,59	$2p\ {}^3P^\circ—4s\ {}^3S$	0—1
153,948	3	19,69	100,22	$2p\ {}^1P^\circ—5d\ {}^1D$	1—2
151,548	6	10,24	92,04	$2p^3\ P^\circ—4d\ {}^3D$	2—3
151,481	5	10,20	92,04	$2p^3\ P^\circ—4d\ {}^3D$	1—3, 2, 1
151,449	4	10,18	92,04	$2p\ {}^3P^\circ—4d\ {}^3D$	0—1
149,078	2	19,69	102,85	$2p\ {}^1P^\circ—4p\ {}^1P$	1—1
149,034	0	28,73	111,90	$2p^2\ {}^1D—5d\ {}^1D^\circ$	2—2
147,261	3	19,69	103,87	$2p\ {}^1P^\circ—4p\ {}^1D$	1—2
146,345	1	19,69	104,40	$2p\ {}^1P^\circ—6d\ {}^1D$	1—2
144,837	1	26,54	112,14	$2p^2\ {}^3P—5d\ {}^3D^\circ$	2—3
144,802	1	26,50	112,14	$2p^2\ {}^3P—5d\ {}^3D^\circ$	1, 0—2, 1
142,119	0	19,69	106,92	$2p\ {}^1P^\circ—7d\ {}^1D$	1—2
140,109	0	10,24	98,72	$2p\ {}^3P^\circ—5s\ {}^3S$	2—1

λ, Å	I	E_H, eV	E_B, eV	Transition	J
140,045	0	10,18	98,72	$2p\ ^3P^\circ$—$5s\ ^3S$	1, 0—1
139,492	0	19,69	—	$2p\ ^1P^\circ$—$8d\ ^1D$	1—2
139,025	5	0,00	89,17	$2s^2\ ^1S$—$3d\ ^1P^\circ$	0—1
138,108	4	10,24	100,00	$2p\ ^3P^\circ$—$5d\ ^3D$	2—3
138,054	3	10,20	—	$2p\ ^3P^\circ$—$5d\ ^3D$	1—2
138,030	2	10,18	—	$2p\ ^3P^\circ$—$5d\ ^3D$	0—1
135,523	5	0,00	91,48	$2s^2\ ^1S$—$4p\ ^1P^\circ$	0—1
133,521	3	10,24	103,09	$2p\ ^3P^\circ$—$4p\ ^3D$	2—3
133,395	0	10,24	103,18	$2p\ ^3P^\circ$—$4p\ ^3S$	2—1
133,328	0	10,18	103,18	$2p\ ^3P^\circ$—$4p\ ^3S$	1, 0—1
132,885	0	10,24	103,54	$2p\ ^3P^\circ$—$4p\ ^3P$	2—1
132,851	2	10,24	103,56	$2p\ ^3P^\circ$—$4p\ ^3P$	2—2
132,800	1	10,18	103,54	$2p\ ^3P^\circ$—$4p\ ^3P$	0, 1—1, 2
131,807	1	10,24	104,29	$2p\ ^3P^\circ$—$6d\ ^3D$	2—1, 2, 3
131,750	1	10,20	104,29	$2p\ ^3P^\circ$—$6d\ ^3D$	0, 1, 2—1, 2, 3
128,297	0	10,24	106,86	$2p\ ^3P^\circ$—$7d\ ^3D$	2—3
128,235	0	10,24	106,87	$2p\ ^3P^\circ$—$7d\ ^3D$	0, 1, 2—1, 2, 3
124,598	3	0,00	99,45	$2s^2\ ^1S$—$5p\ ^1P^\circ$	0—1
122,372	0	10,24	111,54	$2p\ ^3P^\circ$—$5p\ ^3D$	2—3
122,128	0	10,24	111,75	$2p\ ^3P^\circ$—$5p\ ^3P$	2—2

O VI, ground state $1s^2 2s\ ^2S_{1/2}$
Ionization potential 1113999,5 cm^{-1}; 138,080 eV

λ, Å	I	E_H, eV	E_B, eV	Transition	J
5602	—	128,02	130,24	$7p\ ^2P^\circ$—$8s\ ^2S$	1/2, 3/2—1/2
5410	—	128,11	130,39	$7d\ ^2D$—$8p\ ^2P^\circ$	3/2, 5/2—1/2, 3/2
5298	—	128,12	130,46	$7f\ ^2F^\circ$—$8d\ ^2D$	5/2, 7/2—3/2, 5/2
5292	—	127,57	129,90	$7g\ ^2G$—$8f\ ^2F^\circ$	—
5291	—	128,12	130,46	$7g\ ^2G$ etc. —$8h\ ^2H^\circ$ etc.	7/2, 9/2—9/2, 11/2
5289	—	128,12	130,46	$7f\ ^2F^\circ$—$8g\ ^2G$ etc.	5/2, 7/2—7/2, 9/2
5279	—	128,11	130,46	$7d\ ^2D$—$8f\ ^2F^\circ$	3/2, 5/2—5/2, 7/2
5112	—	128,02	130,46	$7p\ ^2P^\circ$—$8d\ ^2D$	1/2, 3/2—3/2, 5/2
4751	—	127,79	130,39	$7s\ ^2S$—$8p\ ^2P^\circ$	1/2—1/2, 3/2
3834,24	1	79,35	82,58	$3s\ ^2S$—$3p\ ^2P^\circ$	1/2—1/2
3811,35	2	79,35	82,60	$3s\ ^2S$—$3p\ ^2P^\circ$	1/2—3/2
3622	—	124,36	127,79	$6p\ ^2P^\circ$—$7s\ ^2S$	1/2, 3/2—1/2
3509	—	124,49	128,02	$6d\ ^2D$—$7p\ ^2P^\circ$	3/2, 5/2—1/2, 3/2
3438	—	124,51	128,11	$6f\ ^2F^\circ$—$7d\ ^2D$	5/2, 7/2—3/2, 5/2
3434	—	124,51	128,12	$6g\ ^2G$—$7h\ ^2H^\circ$ etc.	7/2 to 11/2— 7/2 to 13/2
3433	—	124,51	128,12	$6f\ ^2F^\circ$—$7g\ ^2G$	5/2, 7/2—7/2, 9/2
3426	—	124,49	128,12	$6d\ ^2D$—$7f\ ^2F^\circ$	3/2, 5/2—5/2, 7/2
3314	—	124,36	128,11	$6p\ ^2P^\circ$—$7d\ ^2D$	1/2, 3/2—3/2, 5/2
3068	—	123,99	128,02	$6s\ ^2S$—$7p\ ^2P^\circ$	1/2—1/2, 3/2
1037,613	9	0,00	11,95	$2s\ ^2S$—$2p\ ^2P^\circ$	1/2—1/2
1031,912	10	0,00	12,01	$2s\ ^3S$—$2p\ ^2P^\circ$	1/2—3/2
519,723	2	83,64	107,50	$3d\ ^2D$—$4f\ ^2F^\circ$	5/2—7/2
519,610	2	83,64	107,50	$3d\ ^2D$—$4f\ ^2F^\circ$	3/2—5/2
498,431	1	82,60	107,48	$3p\ ^2P^\circ$—$4d\ ^2D$	3/2—5/2
498,090	0	82,58	107,47	$3p\ ^2P^\circ$—$4d\ ^2D$	1/2—3/2
447,840	0	79,35	107,03	$3s\ ^2S$—$4p\ ^2P^\circ$	1/2—1/2
447,712	0	79,35	107,04	$3s\ ^2S$—$4p\ ^2P^\circ$	1/2—3/2
184,117	9	12,01	79,35	$2p\ ^2P^\circ$—$3s\ ^2S$	3/2—1/2

λ, Å	I	E_H, eV	E_B, eV	Transition	J
183,937	8	11,95	79,35	$2p\ ^2P^\circ—3s^2\ S$	$^1/_2—^1/_2$
173,082	13	12,01	83,64	$2p\ ^2P^\circ—3d\ ^2D$	$^3/_2—^5/_2$
172,935	12	11,95	83,64	$2p\ ^2P^\circ—3d\ ^2D$	$^1/_2—^3/_2$
150,124	9	0,00	82,58	$2s\ ^2S—3p\ ^2P^\circ$	$^1/_2—^1/_2$
150,088	10	0,00	82,60	$2s\ ^2S—3p\ ^2P^\circ$	$^1/_2—^3/_2$
132,312	2	12,01	105,71	$2p\ ^2P^\circ—4s\ ^2S$	$^3/_2—^1/_2$
132,219	1	11,95	105,71	$2p\ ^2P^\circ—4s\ ^2S$	$^1/_2—^1/_2$
129,872	6	12,01	107,48	$2p\ ^2P^\circ—4d\ ^2D$	$^3/_2—^5/_2$
129,786	5	11,95	107,47	$2p\ ^2P^\circ—4d\ ^2D$	$^1/_2—^3/_2$
116,419	2	12,01	118,50	$2p\ ^2P^\circ—5d\ ^2D$	$^3/_2—^5/_2$
116,347	1	11,95	118,50	$2p\ ^2P^\circ—5d\ ^2D$	$^1/_2—^3/_2$
115,824	4	0,00	107,04	$2s\ ^2S—4p\ ^2P^\circ$	$^1/_2—^1/_2$, $^3/_2$
110,824	1	12,01	124,49	$2p\ ^2P^\circ—6d\ ^2D$	$^3/_2—^5/_2$
110,220	0	11,95	124,49	$2p\ ^2P^\circ—6d\ ^2D$	$^1/_2—^3/_2$
110,148	0	—	—	—	—
104,811	2	0,00	118,28	$2s\ ^2S—5p\ ^2P^\circ$	$^1/_2—^1/_2$, $^3/_2$

O VII, ground state $1s^2\ ^1S_0$

Ionization potential 5963000 cm⁻¹; 739,114 eV

λ, Å	I	E_H, eV	E_B, eV	Transition	J
128,500	0	568,66	665,14	$2p\ ^3P^\circ—3d\ ^3D$	2—3
128,412	00	568,58	665,14	$2p\ ^3P^\circ—3d\ ^3D$	1, 0—2, 1
120,331	00	561,04	664,07	$2s\ ^3S—3p\ ^3P^\circ$	1—2, 1, 0
21,804	—	0,00	568,59	$1s^2\ ^1S—2p\ ^3P^\circ$	0—1
21,602	—	0,00	573,91	$1s^2\ ^1S—2p\ ^1P^\circ$	0—1
18,627	—	0,00	665,58	$1s^2\ ^1S—3p\ ^1P^\circ$	0—1
17,768	—	0,00	697,76	$1s^2\ ^1S—4p\ ^1P^\circ$	0—1
17,396	—	0,00	712,68	$1s^2\ ^1S—5p\ ^1P^\circ$	0—1
17,200	—	0,00	720,80	$1s^2\ ^1S—6p\ ^1P^\circ$	0—1

FLUORINE, Z = 9

F I, ground state $1s^22s^22p^5\ ^2P^0_{3/2}$

Ionization potential 140524,5 cm^{-1}; 17,422 eV

λ, Å	I	E_H, eV	E_B, eV	Transition	J
11557,17	5	14,50	15,58	3p $^4D^\circ$—4s ^4P	7/2—5/2
11544,65	2	14,58	15,66	3p $^2D^\circ$—4s ^2P	5/2—3/2
11480,22	1	14,61	15,69	3p $^2D^\circ$—4s ^2P	3/2—1/2
11473,70	3	14,53	15,61	3p $^4D^\circ$—4s ^4P	5/2—3/2
11414,20	1,5	14,54	15,63	3p $^4D^\circ$—4s ^4P	3/2—1/2
10940,37	4	14,76	15,90	3p $^2P^\circ$—3d ^2D	1/2—3/2
10924,81	2,5	14,75	15,88	3p $^2P^\circ$—3d ^4D	3/2—5/2
10883,28	2,5	14,75	15,88	3p $^2P^\circ$—3d ^4D	3/2—3/2
10862,31	20	14,75	15,88	3p $^2P^\circ$—3d ^2D	3/2—5/2
10769,43	4	14,75	15,90	3p $^2P^\circ$—3d ^2D	3/2—3/2
10592,28	2	14,76	15,93	3p $^2P^\circ$—3d ^2P	1/2—1/2
10588,71	5	14,76	15,93	3p $^2P^\circ$—3d ^4P	1/2—3/2
10490,21	1	14,76	15,94	3p $^2P^\circ$—3d ^4F	1/2—3/2
10431,92	1	14,75	15,93	3p $^2P^\circ$—3d ^2P	3/2—1/2
10426,29	6	14,75	15,93	3p $^2P^\circ$—3d ^4F	3/2—5/2
10417,29	7	14,39	15,58	3p $^2P^\circ$—3d ^4F	3/2—5/2
10380,84	7	14,76	15,96	3p $^2P^\circ$—3d ^2P	1/2—3/2
10332,95	2,5	14,75	15,94	3p $^2P^\circ$—3d ^4F	3/2—3/2
10293,01	3,5	14,68	15,88	3p $^4S^\circ$—3d ^2D	3/2—5/2
10287,96	1,5	14,68	15,88	3p $^2S^\circ$—3d ^4D	1/2—3/2
10285,45	15	14,37	15,58	3p $^4P^\circ$—4s ^4P	5/2—5/2
10270,75	4	14,40	15,61	3p $^4P^\circ$—4s ^4P	1/2—3/2
10241,98	4	14,75	15,96	3p $^2P^\circ$—3d ^2F	3/2—5/2
10226,82	3	14,75	15,96	3p $^2P^\circ$—3d ^2P	3/2—3/2
10222,50	2	14,68	15,89	3s $^2S^\circ$—3d ^4D	1/2—1/2
10209,57	4	14,68	15,90	3p $^4S^\circ$—3d ^2D	3/2—3/2
10186,15	5	14,68	15,90	3p $^2S^\circ$—3d ^2D	1/2—3/2
10163,50	3	14,39	15,61	3p $^4P^\circ$—4s ^4P	3/2—3/2
10087,13	6	14,68	15,91	3p $^4S^\circ$—3d ^4P	3/2—1/2
10074,17	1	14,40	15,63	3p $^4P^\circ$—4s ^4P	1/2—1/2
10064,25	4	14,68	15,91	3p $^2S^\circ$—3d ^4P	1/2—1/2
10038,03	4	14,37	15,61	3p $^4P^\circ$—4s ^4P	5/2—3/2
9970,92	4	14,39	15,63	3p $^4P^\circ$—4s ^4P	3/2—1/2
9905,65	1,5	14,68	15,93	3p $^4S^\circ$—3d ^2P	3/2—1/2
9902,65	12	14,68	15,93	3p $^4S^\circ$—3d ^4P	3/2—3/2
9883,58	8	14,68	15,93	3p $^2S^\circ$—3d ^2P	1/2—1/2
9822,11	15	14,68	15,94	3p $^4S^\circ$—3d ^4P	3/2—5/2
9794,80	6	14,68	15,94	3p $^2S^\circ$—3d ^4F	1/2—3/2
9753,57	1,5	14,61	15,88	3p $^2D^\circ$—3d ^4D	3/2—3/2
9736,70	9	14,61	15,88	3p $^2D^\circ$—3d ^2D	3/2—5/2
9734,34	25	14,68	15,96	3p $^4S^\circ$—3d ^2F	3/2—5/2
9720,57	1	14,68	15,96	3p $^4S^\circ$—3d ^2P	3/2—3/2
9699,40	7	14,68	15,96	3p $^2S^\circ$—3d ^2P	1/2—3/2
9662,04	12	14,61	15,90	3p $^2D^\circ$—3d ^2D	3/2—3/2
9574,80	3	14,58	15,88	3p $^2D^\circ$—3d ^4D	5/2—7/2
9552,99	0,7	14,58	15,88	3p $^2D^\circ$—3d ^4D	5/2—5/2
9552,30	0,7	14,61	15,91	3p $^2D^\circ$—3d ^4P	3/2—1/2
9505,30	25	14,58	15,88	3p $^2D^\circ$—3d ^2D	5/2—5/2
9433,67	200	14,58	15,90	3p $^2D^\circ$—3d ^2F	5/2—7/2
9389,47	0,8	14,61	15,93	3p $^2D^\circ$—3d ^2P	3/2—1/2
9386,75	2,5	14,61	15,93	3p $^2D^\circ$—3d ^4P	3/2—3/2
9384,96	40	14,61	15,93	3p $^2D^\circ$—3d ^4F	3/2—5/2
9314,34	60	14,61	15,94	3p $^2D^\circ$—3d ^4P	3/2—5/2

λ, Å	I	E_H, eV	E_B, eV	Transition	J
9262,69	8	14,55	15,89	$3p\ ^4D^\circ$—$3d\ ^4D$	$^1/_2$—$^1/_2$
9244,57	15	14,54	15,88	$3p\ ^4D^\circ$—$3d\ ^4D$	$^3/_2$—$^3/_2$
9235,38	50	14,61	15,96	$3p\ ^2D^\circ$—$3d\ ^2F$	$^3/_2$—$^5/_2$
9232,85	6	14,55	15,90	$3p\ ^4D^\circ$—$3d\ ^2D$	$^1/_2$—$^3/_2$
9229,40	2,5	14,54	15,88	$3p\ ^4D^\circ$—$3d\ ^2D$	$^3/_2$—$^5/_2$
9223,05	6	14,61	15,96	$3p\ ^2D^\circ$—$3d\ ^2P$	$^3/_2$—$^3/_2$
9191,65	10	14,54	15,89	$3p\ ^4D^\circ$—$3d\ ^4D$	$^3/_2$—$^1/_2$
9178,68	350	14,58	15,93	$3p\ ^2D^\circ$—$3d\ ^4F$	$^5/_2$—$^7/_2$
9169,76	3	14,58	15,93	$3p\ ^2D^\circ$—$3d\ ^4F$	$^5/_2$—$^5/_2$
9162,33	2	14,54	15,90	$3p\ ^4D^\circ$—$3d\ ^2D$	$^3/_2$—$^3/_2$
9151,78	180	14,53	15,88	$3p\ ^4D^\circ$—$3d\ ^4D$	$^5/_2$—$^5/_2$
9132,53	1,4	14,55	15,91	$3p\ ^4D^\circ$—$3d\ ^4P$	$^1/_2$—$^1/_2$
9122,63	40	14,53	15,88	$3p\ ^4D^\circ$—$3d\ ^4D$	$^5/_2$—$^3/_2$
9107,87	10	14,53	15,88	$3p\ ^4D^\circ$—$3d\ ^2D$	$^5/_2$—$^5/_2$
9102,33	50	14,58	15,94	$3p\ ^2D^\circ$—$3d\ ^4P$	$^5/_2$—$^5/_2$
9097,49	3,5	14,58	15,94	$3p\ ^2D^\circ$—$3d\ ^4F$	$^5/_2$—$^3/_2$
9042,10	400	14,53	15,90	$3p\ ^4D^\circ$—$3d\ ^2F$	$^5/_2$—$^7/_2$
9025,49	350	14,50	15,88	$3p\ ^4D^\circ$—$3d\ ^4D$	$^7/_2$—$^7/_2$
9015,19	0,7	14,58	15,96	$3p\ ^2D^\circ$—$3d\ ^2P$	$^5/_2$—$^3/_2$
9006,19	50	14,50	15,88	$3p\ ^4D^\circ$—$3d\ ^4D$	$^7/_2$—$^5/_2$
8983,65	3,5	14,55	15,93	$3p\ ^4D^\circ$—$3d\ ^2P$	$^1/_2$—$^1/_2$
8981,18	12	14,55	15,93	$3p\ ^4D^\circ$—$3d\ ^4P$	$^1/_2$—$^3/_2$
8963,66	4	14,50	15,88	$3p\ ^4D^\circ$—$3d\ ^2D$	$^7/_2$—$^5/_2$
8936,61	0,7	12,98	14,37	$3s\ ^2P$—$3p\ ^4P^\circ$	$^3/_2$—$^5/_2$
8916,89	2	14,54	15,93	$3p\ ^4D^\circ$—$3d\ ^2P$	$^3/_2$—$^1/_2$
8914,43	7	14,54	15,93	$3p\ ^4D^\circ$—$3d\ ^4P$	$^3/_2$—$^3/_2$
8912,78	300	14,54	15,93	$3p\ ^4D^\circ$—$3d\ ^4F$	$^3/_2$—$^5/_2$
8910,27	140	14,55	15,94	$3p\ ^4D^\circ$—$3d\ ^4F$	$^1/_2$—$^3/_2$
8900,92	1000	14,50	15,90	$3p\ ^4D^\circ$—$3d\ ^4F$	$^7/_2$—$^9/_2$
8899,92	60	14,50	15,90	$3p\ ^4D^\circ$—$3d\ ^2F$	$^7/_2$—$^7/_2$
8849,06	70	14,54	15,94	$3p\ ^4D^\circ$—$3d\ ^4P$	$^3/_2$—$^5/_2$
8844,502	120	14,54	15,94	$3p\ ^4D^\circ$—$3d\ ^4F$	$^3/_2$—$^3/_2$
8831,232	100	14,55	15,96	$3p\ ^4D^\circ$—$3d\ ^2P$	$^1/_2$—$^3/_2$
8807,582	900	14,53	15,93	$3p\ ^4D^\circ$—$3d\ ^4F$	$^5/_2$—$^7/_2$
8799,36	70	14,53	15,93	$3p\ ^4D^\circ$—$3d\ ^4F$	$^5/_2$—$^5/_2$
8792,50	35	17,06	18,47	$3p'\ ^2F^\circ$—$3d'\ ^2G$	$^7/_2$—$^9/_2$
8785,63	14	17,06	18,47	$3p'\ ^2F^\circ$—$3d'\ ^2G$	$^5/_2$—$^7/_2$
8777,73	120	14,54	15,96	$3p\ ^4D^\circ$—$3d\ ^2F$	$^3/_2$—$^5/_2$
8766,61	10	14,54	15,96	$3p\ ^4D^\circ$—$3d\ ^2P$	$^3/_2$—$^3/_2$
8737,270	140	14,53	15,94	$3p\ ^4D^\circ$—$3d\ ^4P$	$^5/_2$—$^5/_2$
8732,80	,6	14,53	15,94	$3p\ ^4D^\circ$—$3d\ ^4F$	$^5/_2$—$^3/_2$
8672,62	35	14,50	15,93	$3p\ ^4D^\circ$—$3d\ ^4F$	$^7/_2$—$^7/_2$
8667,71	1	14,53	15,96	$3p\ ^4D^\circ$—$3d\ ^2F$	$^5/_2$—$^5/_2$
8664,63	6	14,50	15,93	$3p\ ^4D^\circ$—$3d\ ^4F$	$^7/_2$—$^5/_2$
8656,93	0,8	14,53	15,96	$3p\ ^4D^\circ$—$3d\ ^2P$	$^5/_2$—$^3/_2$
8612,58	6	—	—	—	—
8606,06	6	—	—	—	—
8604,47	1,4	14,50	15,94	$3p\ ^4D^\circ$—$3d\ ^4P$	$^7/_2$—$^5/_2$
8537,04	7	14,50	15,96	$3p\ ^4D^\circ$—$3d\ ^2F$	$^7/_2$—$^5/_2$
8345,556	120	14,40	15,88	$3p\ ^4P^\circ$—$3d\ ^4D$	$^1/_2$—$^3/_2$
8302,40	600	14,40	15,89	$3p\ ^4P^\circ$—$3d\ ^4D$	$^1/_2$—$^1/_2$
8298,581	2000	14,39	15,88	$3p\ ^4P^\circ$—$3d\ ^4D$	$^3/_2$—$^5/_2$
8278,44	2	14,40	15,90	$3p\ ^4P^\circ$—$3d\ ^2D$	$^1/_2$—$^3/_2$
8274,615	1500	14,39	15,88	$3p\ ^4P^\circ$—$3d\ ^4D$	$^3/_2$—$^3/_2$
8262,49	12	14,39	15,88	$3p\ ^4P^\circ$—$3d\ ^2D$	$^3/_2$—$^5/_2$
8232,19	500	14,39	15,89	$3p\ ^4P^\circ$—$3d\ ^4D$	$^3/_2$—$^1/_2$
8230,773	3000	14,37	15,88	$3p\ ^4P^\circ$—$3d\ ^4D$	$^5/_2$—$^7/_2$
8214,726	2500	14,37	15,88	$3p\ ^4P^\circ$—$3d\ ^4D$	$^5/_2$—$^5/_2$

λ, Å	I	E_H, eV	E_B, eV	Transition	J
8208,634	350	14,39	15,90	$3p\ ^4P^\circ - 3d\ ^2D$	3/2—3/2
8197,734	60	14,40	15,91	$3p\ ^4P^\circ - 3d\ ^4P$	1/2—1/2
8191,241	300	14,37	15,88	$3p\ ^4P^\circ - 3d\ ^4D$	5/2—3/2
8179,339	600	14,37	15,88	$3p\ ^4P^\circ - 3d\ ^2D$	5/2—5/2
8159,51	300	13,02	14,54	$3s\ ^2P - 3p\ ^4D^\circ$	1/2—3/2
8129,26	600	14,39	15,89	$3p\ ^4P^\circ - 3d\ ^4P$	3/2—1/2
8126,56	350	14,37	15,90	$3p\ ^4P^\circ - 3d\ ^2D$	5/2—3/2
8077,521	350	14,40	15,93	$3p\ ^4P^\circ - 3d\ ^2P$	1/2—1/2
8075,519	900	14,40	15,93	$3p\ ^4P^\circ - 3d\ ^4P$	1/2—3/2
8040,931	1000	12,98	14,53	$3s\ ^2P - 3p\ ^4D^\circ$	3/2—5/2
8018,12	8	14,40	15,94	$3p\ ^4P^\circ - 3d\ ^4F$	1/2—3/2
8011,05	5	14,39	15,93	$3p\ ^4P^\circ - 3d\ ^2P$	3/2—1/2
8009,05	1	14,39	15,93	$3p\ ^4P^\circ - 3d\ ^4P$	3/2—3/2
8007,79	15	14,39	15,93	$3p\ ^4P^\circ - 3d\ ^4F$	3/2—5/2
7978,96	5	—	—	—	—
7956,32	300	14,39	15,94	$3p\ ^4P^\circ - 3d\ ^4P$	3/2—5/2
7954,09	60	14,40	15,96	$3p\ ^4P^\circ - 3d\ ^2P$	1/2—3/2
7952,66	2	14,39	15,94	$3p\ ^4P^\circ - 3d\ ^4F$	3/2—3/2
7948,52	40	12,98	14,54	$3s\ ^2P^\circ - 3p\ ^4D^\circ$	3/2—3/2
7936,314	350	14,37	15,93	$3p\ ^4P^\circ - 3d\ ^4F$	5/2—7/2
7930,93	220	14,37	15,93	$3p\ ^4P^\circ - 3d\ ^4P$	5/2—3/2
7929,65	4	14,37	15,93	$3p\ ^4P^\circ - 3d\ ^4F$	5/2—5/2
7898,558	500	14,39	15,96	$3p\ ^4P^\circ - 3d\ ^2F$	3/2—5/2
7889,62	8	14,39	15,96	$3p\ ^4P^\circ - 3d\ ^2P$	3/2—3/2
7879,18	300	14,37	15,94	$3p\ ^4P^\circ - 3d\ ^4P$	5/2—5/2
7875,56	8	14,37	15,94	$3p\ ^4P^\circ - 3d\ ^4F$	5/2—3/2
7822,59	80	14,37	15,96	$3p\ ^4P^\circ - 3d\ ^2F$	5/2—5/2
7820,79	6	—	—	—	—
7813,76	10	14,37	15,96	$3p\ ^4P^\circ - 3d\ ^2P$	5/2—3/2
7800,212	15000	13,02	14,61	$3s\ ^2P - 3p\ ^2D^\circ$	1/2—3/2
7754,696	18000	12,98	14,58	$3s\ ^2P - 3p\ ^2D^\circ$	3/2—5/2
7607,170	7000	12,98	14,61	$3s\ ^2P - 3p\ ^2D^\circ$	3/2—3/2
7573,384	5000	12,75	14,39	$3s\ ^4P - 3p\ ^4P^\circ$	1/2—3/2
7552,235	5000	12,73	14,37	$3s\ ^4P - 3p\ ^4P^\circ$	3/2—5/2
7514,919	900	12,75	14,40	$3s\ ^4P - 3p\ ^4P^\circ$	1/2—1/2
7489,155	2500	13,02	14,68	$3s\ ^2P - 3p\ ^2S^\circ$	1/2—1/2
7482,723	2200	12,73	14,39	$3s\ ^4P - 3p\ ^4P^\circ$	3/2—3/2
7476,54	70	13,02	14,68	$3s\ ^2P - 3p\ ^4S^\circ$	1/2—3/2
7465,645	4000	12,73	14,40	$3s\ ^4P - 3p\ ^4P^\circ$	3/2—1/2
7398,688	10000	12,70	14,37	$3s\ ^4P - 3p\ ^4P^\circ$	5/2—5/2
7331,957	5000	12,70	14,39	$3s\ ^4P - 3p\ ^4P^\circ$	5/2—3/2
7319,33	6	14,75	16,44	$3p\ ^2P^\circ - 5s\ ^4P$	3/2—5/2
7314,303	700	15,36	17,06	$3s'\ ^2D - 3p'\ ^2F^\circ$	3/2—5/2
7313,77	40	15,36	17,06	$3s'\ ^2D - 3p'\ ^2F^\circ$	5/2—5/2
7311,019	15000	12,98	14,68	$3s\ ^2P - 3p\ ^2S^\circ$	3/2—1/2
7309,033	1000	15,36	17,06	$3s'\ ^2D - 3p'\ ^2F^\circ$	5/2—7/2
7298,98	150	12,98	14,68	$3s\ ^2P - 3p\ ^4S^\circ$	3/2—3/2
7240,12	2	14,75	16,46	$3p\ ^2P^\circ - 5s\ ^4P$	3/2—3/2
7202,360	15000	13,02	14,75	$3s\ ^2P - 3p\ ^2P^\circ$	1/2—3/2
7127,890	30000	13,02	14,76	$3s\ ^2P - 3p\ ^2P^\circ$	1/2—1/2
7109,61	3	14,75	16,49	$3p\ ^2P^\circ - 5s\ ^2P$	3/2—3/2
7037,469	45000	12,98	14,75	$3s\ ^2P - 3p\ ^2P^\circ$	3/2—3/2
6982,69	1	14,68	16,46	$3p\ ^4S^\circ - 5s\ ^4P$	3/2—3/2
6971,71	1	14,68	16,46	$3p\ ^2S^\circ - 5s\ ^4P$	1/2—3/2
6966,349	4000	12,98	14,76	$3s\ ^2P - 3p\ ^2P^\circ$	3/2—1/2
6909,816	6000	12,75	14,54	$3s\ ^4P - 3p\ ^4D^\circ$	1/2—3/2
6902,475	15000	12,73	14,53	$3s\ ^4P - 3p\ ^4D^\circ$	3/2—5/2
6870,215	8000	12,75	14,55	$3s\ ^4P - 3p\ ^4D^\circ$	1/2—1/2
6856,030	50000	12,70	14,50	$3s\ ^4P - 3p\ ^4D^\circ$	5/2—7/2

λ, Å	I	E_H, eV	E_B, eV	Transition	J
6837,74	5	14,75	16,56	$3p\,{}^2P^\circ - 4d\,{}^4D$	$^3/_2 - {}^5/_2$
6834,264	9000	12,73	14,54	$3s\,{}^4P - 3p\,{}^4D^\circ$	$^3/_2 - {}^3/_2$
6825,56	15	14,75	16,56	$3p\,{}^2P^\circ - 4d\,{}^2D$	$^3/_2 - {}^5/_2$
6811,67	1	14,75	16,56	$3p\,{}^2P^\circ - 4d\,{}^4D$	$^3/_2 - {}^1/_2$
6806,85	10	14,75	16,57	$3p\,{}^2P^\circ - 4d\,{}^2D$	$^3/_2 - {}^3/_2$
6795,528	1500	12,73	14,55	$3s\,{}^4P - 3p\,{}^4D^\circ$	$^3/_2 - {}^1/_2$
6773,984	7000	12,70	14,53	$3s\,{}^4P - 3p\,{}^4D^\circ$	$^5/_2 - {}^5/_2$
6767,007	50	15,36	17,20	$3s'\,{}^2D - 3p'\,{}^2D^\circ$	$^3/_2 - {}^3/_2$
6766,54	5	15,36	17,20	$3s'\,{}^2D - 3p'\,{}^2D^\circ$	$^5/_2 - {}^3/_2$
6763,325	5	15,36	17,20	$3s'\,{}^2D - 3p'\,{}^2D^\circ$	$^3/_2 - {}^5/_2$
6762,934	70	15,36	17,20	$3s'\,{}^2D - 3p'\,{}^2D^\circ$	$^5/_2 - {}^5/_2$
6741,90	1	14,76	16,60	$3p\,{}^2P^\circ - 4d\,{}^2P$	$^1/_2 - {}^1/_2$
6737,76	6	14,76	16,60	$3p\,{}^2P^\circ - 4d\,{}^4P$	$^1/_2 - {}^3/_2$
6708,282	400	12,70	14,54	$3s\,{}^4P - 3p\,{}^4D^\circ$	$^5/_2 - {}^3/_2$
6690,481	1800	12,73	14,58	$3s\,{}^4P - 3p\,{}^2D^\circ$	$^3/_2 - {}^5/_2$
6667,00	7	14,75	16,60	$3p\,{}^2P^\circ - 4d\,{}^4F$	$^3/_2 - {}^5/_2$
6660,62	12	14,76	16,62	$3p\,{}^2P^\circ - 4d\,{}^2P$	$^1/_2 - {}^3/_2$
6651,97	0,5	14,75	16,61	$3p\,{}^2P^\circ - 4d\,{}^4P$	$^3/_2 - {}^5/_2$
6650,405	400	12,75	14,61	$3s\,{}^4P - 3p\,{}^2D^\circ$	$^1/_2 - {}^3/_2$
6649,51	6	14,75	16,61	$3p\,{}^2P^\circ - 4d\,{}^4F$	$^3/_2 - {}^3/_2$
6611,04	5	14,58	16,46	$3p\,{}^2D^\circ - 5s\,{}^4P$	$^5/_2 - {}^3/_2$
6609,55	2,5	14,61	16,49	$3p\,{}^2D^\circ - 5s\,{}^2P$	$^3/_2 - {}^3/_2$
6607,73	2,5	14,68	16,56	$3p\,{}^4S^\circ - 4d\,{}^4D$	$^3/_2 - {}^5/_2$
6604,86	0,6	14,61	16,49	$3p\,{}^2D^\circ - 5s\,{}^4P$	$^3/_2 - {}^1/_2$
6599,725	6	14,75	16,62	$3p\,{}^2P^\circ - 4d\,{}^2F$	$^3/_2 - {}^5/_2$
6596,90	0,8	14,75	16,62	$3p\,{}^2P^\circ - 4d\,{}^2P$	$^3/_2 - {}^3/_2$
6596,35	8	14,68	16,56	$3p\,{}^4S^\circ - 4d\,{}^2D$	$^3/_2 - {}^5/_2$
6589,21	4	14,68	16,56	$3p\,{}^2S^\circ - 4d\,{}^4D$	$^1/_2 - {}^3/_2$
6583,36	2	14,68	16,56	$3p\,{}^4S^\circ - 4d\,{}^4D$	$^3/_2 - {}^1/_2$
6580,389	300	12,73	14,61	$3s\,{}^4P - 3p\,{}^2D^\circ$	$^3/_2 - {}^3/_2$
6578,871	12	14,68	16,57	$3p\,{}^4S^\circ - 4d\,{}^2D$	$^3/_2 - {}^3/_2$
6573,61	5	14,68	16,56	$3p\,{}^2S^\circ - 4d\,{}^4D$	$^1/_2 - {}^1/_2$
6569,694	450	12,70	14,58	$3s\,{}^4P - 3p\,{}^2D^\circ$	$^5/_2 - {}^5/_2$
6569,14	7	14,68	16,57	$3p\,{}^2S^\circ - 4d\,{}^2D$	$^1/_2 - {}^3/_2$
6563,59	7	14,68	16,57	$3p\,{}^4S^\circ - 4d\,{}^4P$	$^3/_2 - {}^1/_2$
6553,93	6	14,68	16,57	$3p\,{}^2S^\circ - 4d\,{}^4P$	$^1/_2 - {}^1/_2$
6512,71	12	14,61	16,52	$3p\,{}^2D^\circ - 5s\,{}^2P$	$^3/_2 - {}^1/_2$
6502,08	18	14,58	16,49	$3p\,{}^2D^\circ - 5s\,{}^2P$	$^5/_2 - {}^3/_2$
6478,45	8	14,53	16,44	$3p\,{}^4D^\circ - 5s\,{}^4P$	$^5/_2 - {}^5/_2$
6476,39	7	14,54	16,46	$3p\,{}^4D^\circ - 5s\,{}^4P$	$^3/_2 - {}^3/_2$
6463,50	70	12,70	14,61	$3s\,{}^4P - 3p\,{}^2D^\circ$	$^5/_2 - {}^3/_2$
6457,06	0,8	14,68	16,60	$3p\,{}^4S^\circ - 4d\,{}^2P$	$^3/_2 - {}^1/_2$
6453,32	10	14,68	16,60	$3p\,{}^4S^\circ - 4d\,{}^4P$	$^3/_2 - {}^3/_2$
6448,14	0,6	14,68	16,60	$3p\,{}^4S^\circ - 4d\,{}^4F$	$^3/_2 - {}^5/_2$
6447,69	6	14,68	16,60	$3p\,{}^2S^\circ - 4d\,{}^2P$	$^1/_2 - {}^1/_2$
6434,11	15	14,68	16,61	$3p\,{}^4S^\circ - 4d\,{}^4P$	$^3/_2 - {}^5/_2$
6422,87	2,5	12,75	14,68	$3s\,{}^4P - 3p\,{}^2S^\circ$	$^1/_2 - {}^1/_2$
6422,43	7	14,68	16,61	$3p\,{}^2S^\circ - 4d\,{}^4F$	$^1/_2 - {}^3/_2$
6416,31	18	14,53	16,46	$3p\,{}^4D^\circ - 5s\,{}^4P$	$^5/_2 - {}^3/_2$
6413,651	8000	12,75	14,68	$3s\,{}^4P - 3p\,{}^4S^\circ$	$^1/_2 - {}^3/_2$
6405,171	60	14,50	16,44	$3p\,{}^4D^\circ - 5s\,{}^4P$	$^7/_2 - {}^5/_2$
6385,17	10	14,68	16,62	$3p\,{}^4S^\circ - 4d\,{}^2F$	$^3/_2 - {}^5/_2$
6373,33	2	14,68	16,62	$3p\,{}^2S^\circ - 4d\,{}^2P$	$^1/_2 - {}^3/_2$
6371,77	2	14,54	16,49	$3p\,{}^4D^\circ - 5s\,{}^2P$	$^3/_2 - {}^3/_2$
6367,43	10	14,54	16,49	$3p\,{}^4P^\circ - 5s\,{}^4P$	$^3/_2 - {}^1/_2$
6365,84	0,8	14,61	16,56	$3p\,{}^2D^\circ - 4d\,{}^4D$	$^3/_2 - {}^3/_2$
6363,34	8	14,61	16,56	$3p\,{}^2D^\circ - 4d\,{}^2D$	$^3/_2 - {}^5/_2$
6348,508	10000	12,73	14,68	$3s\,{}^4P - 3p\,{}^4S^\circ$	$^3/_2 - {}^3/_2$

λ, Å	I	E_H, eV	E_B, eV	Transition	J
6314,74	0,8	14,55	16,52	$3p\ ^4D^\circ—5s\ ^2P$	$^1/_2—^1/_2$
6313,61	3	14,53	16,49	$3p\ ^4D^\circ—5s\ ^2P$	$^5/_2—^3/_2$
6279,028	9	14,58	16,56	$3p\ ^2D^\circ—4d\ ^4D$	$^5/_2—^7/_2$
6263,696	18	14,58	16,56	$3p\ ^2D^\circ—4d\ ^2D$	$^5/_2—^5/_2$
6254,690	80	14,58	16,56	$3p\ ^2D^\circ—4d\ ^2F$	$^5/_2—^7/_2$
6239,651	13000	12,70	14,68	$3s\ ^4P—3p\ ^4S^\circ$	$^5/_2—^3/_2$
6230,11	3	14,61	16,60	$3p\ ^2D^\circ—4d\ ^4P$	$^3/_2—^3/_2$
6225,356	18	14,61	16,60	$3p\ ^2D^\circ—4d\ ^4F$	$^3/_2—^5/_2$
6212,249	18	14,61	16,61	$3p\ ^2D^\circ—4d\ ^4P$	$^3/_2—^5/_2$
6210,87	400	12,75	14,75	$3s\ ^4P—3p\ ^2P^\circ$	$^1/_2—^3/_2$
6166,628	25	14,61	16,62	$3p\ ^2D^\circ—4d\ ^2F$	$^3/_2—^5/_2$
6164,136	1,5	14,61	16,62	$3p\ ^2D^\circ—4d\ ^2P$	$^3/_2—^3/_2$
6155,359	1	12,75	14,76	$3s\ ^4P—3p\ ^2P^\circ$	$^1/_2—^1/_2$
6152,55	2	14,54	16,56	$3p\ ^4D^\circ—4d\ ^4D$	$^3/_2—^5/_2$
6149,76	800	12,73	14,75	$3s\ ^4P—3p\ ^2P^\circ$	$^3/_2—^3/_2$
6145,029	8	14,54	16,56	$3p\ ^4D^\circ—4d\ ^4D$	$^3/_2—^3/_2$
6133,220	70	14,58	16,60	$3p\ ^2D^\circ—4d\ ^4F$	$^5/_2—^7/_2$
6131,43	1	14,54	16,56	$3p\ ^4D^\circ—4d\ ^4D$	$^3/_2—^1/_2$
6129,93	1	14,58	16,60	$3p\ ^2D^\circ—4d\ ^4F$	$^5/_2—^5/_2$
6127,49	0,6	14,54	16,57	$3p\ ^4D^\circ—4d\ ^2D$	$^3/_2—^3/_2$
6117,222	10	14,58	16,61	$3p\ ^2D^\circ—4d\ ^4P$	$^5/_2—^5/_2$
6098,34	25	14,53	16,56	$3p\ ^4D^\circ—4d\ ^4D$	$^5/_2—^5/_2$
6090,902	6	14,53	16,56	$3p\ ^4D^\circ—4d\ ^4D$	$^5/_2—^3/_2$
6088,61	2,5	14,53	16,56	$3p\ ^4D^\circ—4d\ ^2D$	$^5/_2—^5/_2$
6080,113	100	14,53	16,56	$3p\ ^4D^\circ—4d\ ^2F$	$^5/_2—^7/_2$
6052,19	1	14,55	16,60	$3p\ ^4D^\circ—4d\ ^2P$	$^1/_2—^1/_2$
6048,80	1	14,55	16,60	$3p\ ^4D^\circ—4d\ ^4P$	$^1/_2—^3/_2$
6047,54	900	12,70	14,75	$3s\ ^4P—3p\ ^2P^\circ$	$^5/_2—^3/_2$
6038,944	18	14,39	16,44	$3p\ ^4P^\circ—5s\ ^4P$	$^3/_2—^5/_2$
6038,04	80	14,50	16,56	$3p\ ^4D^\circ—4d\ ^4D$	$^7/_2—^7/_2$
6033,34	8	14,50	16,56	$3p\ ^4D^\circ—4d\ ^4D$	$^7/_2—^5/_2$
6029,95	20	14,55	16,61	$3p\ ^4D^\circ—4d\ ^4F$	$^1/_2—^3/_2$
6021,91	10	14,40	16,46	$3p\ ^4P^\circ—5s\ ^4P$	$^1/_2—^3/_2$
6018,47	1	14,54	16,60	$3p\ ^4D^\circ—4d\ ^4P$	$^3/_2—^3/_2$
6015,828	150	14,50	16,56	$3p\ ^4D^\circ—4d\ ^4F$	$^7/_2—^9/_2$
6014,03	40	14,54	16,60	$3p\ ^4D^\circ—4d\ ^4F$	$^3/_2—^5/_2$
6001,78	8	14,54	16,61	$3p\ ^4D^\circ—4d\ ^4P$	$^3/_2—^5/_2$
5999,753	15	14,54	16,61	$3p\ ^4D^\circ—4d\ ^4F$	$^3/_2—^3/_2$
5994,425	50	14,37	16,44	$3p\ ^4P^\circ—5s\ ^4P$	$^5/_2—^5/_2$
5986,635	30	14,55	16,62	$3p\ ^4D^\circ—4d\ ^2P$	$^1/_2—^3/_2$
5984,94	1,5	14,39	16,46	$3p\ ^4P^\circ—5s\ ^4P$	$^3/_2—^3/_2$
5969,056	1	14,75	16,82	$3p\ ^2P^\circ—6s\ ^4P$	$^3/_2—^3/_2$
5965,28	70	14,53	16,60	$3p\ ^4D^\circ—4d\ ^4F$	$^5/_2—^7/_2$
5962,166	3	14,53	16,60	$3p\ ^4D^\circ—4d\ ^4F$	$^5/_2—^5/_2$
5959,187	25	14,54	16,62	$3p\ ^4D^\circ—4d\ ^2F$	$^3/_2—^5/_2$
5956,87	2	14,54	16,62	$3p\ ^4D^\circ—4d\ ^2P$	$^3/_2—^3/_2$
5950,147	12	14,53	16,61	$3p\ ^4D^\circ—4d\ ^4P$	$^5/_2—^5/_2$
5941,179	5	14,37	16,46	$3p\ ^4P^\circ—5s\ ^4P$	$^5/_2—^3/_2$
5940,697	5	17,06	19,15	$3p'\ ^2F^\circ—4d'\ ^2G$	$^7/_2—^9/_2$
5937,56	2,5	17,06	19,15	$3p'\ ^2F^\circ—4d'\ ^2G$	$^5/_2—^7/_2$
5931,39	6	14,40	16,49	$3p\ ^4P^\circ—5s\ ^2P$	$^1/_2—^3/_2$
5927,60	1	14,40	16,49	$3p\ ^4P^\circ—5s\ ^4P$	$^1/_2—^1/_2$
5903,06	0,8	14,50	16,60	$3p\ ^4D^\circ—4d\ ^4F$	$^7/_2—^7/_2$
5891,74	8	14,39	16,49	$3p\ ^4P^\circ—5s\ ^4P$	$^3/_2—^1/_2$
5853,06	2,5	14,37	16,49	$3p\ ^4P^\circ—5s\ ^2P$	$^5/_2—^3/_2$
5827,48	0,6	14,75	16,87	$3p\ ^2P^\circ—5d\ ^2D$	$^3/_2—^5/_2$
5785,45	0,5	14,68	16,82	$3p\ ^2S^\circ—6s\ ^4P$	$^1/_2—^3/_2$
5734,39	3	14,40	16,56	$3p\ ^4P^\circ—4d\ ^4D$	$^1/_2—^3/_2$

λ, Å	I	E_H, eV	E_B, eV	Transition	J
5722,59	6	14,40	16,56	$3p\ ^4P^\circ—4d\ ^4D$	$^1/_2—^1/_2$
5719,16	1,5	14,40	16,57	$3p\ ^4P^\circ—4d\ ^2D$	$^1/_2—^3/_2$
5707,62	2	14,40	16,57	$3p\ ^4P^\circ—4d\ ^4P$	$^1/_2—^1/_2$
5707,31	25	14,39	16,56	$3p\ ^4P^\circ—4d\ ^4D$	$^3/_2—^5/_2$
5700,82	25	14,39	16,56	$3p\ ^4P^\circ—4d\ ^4D$	$^3/_2—^3/_2$
5689,14	18	14,39	16,56	$3p\ ^4P^\circ—4d\ ^4D$	$^3/_2—^1/_2$
5685,74	8	14,39	16,57	$3p\ ^4P^\circ—4d\ ^2D$	$^3/_2—^3/_2$
5674,39	8	14,39	16,57	$3p\ ^4P^\circ—4d\ ^4P$	$^3/_2—^1/_2$
5671,668	90	14,37	16,56	$3p\ ^4P^\circ—4d\ ^4D$	$^5/_2—^7/_2$
5667,532	40	14,37	16,56	$3p\ ^4P^\circ—4d\ ^4D$	$^5/_2—^5/_2$
5661,106	7	14,37	16,56	$3p\ ^4P^\circ—4d\ ^4D$	$^5/_2—^3/_2$
5659,15	15	14,37	16,56	$3p\ ^4P^\circ—4d\ ^2D$	$^5/_2—^5/_2$
5652,601	0,5	14,68	16,87	$3p\ ^4S^\circ—5d\ ^2D$	$^3/_2—^3/_2$
5646,254	8	14,37	16,57	$3p\ ^4P^\circ—4d\ ^2D$	$^5/_2—^3/_2$
5626,93	12	14,40	16,60	$3p\ ^4P^\circ—4d\ ^2P$	$^1/_2—^1/_2$
5624,06	20	14,40	16,60	$3p\ ^4P^\circ—4d\ ^4P$	$^1/_2—^3/_2$
5607,66	0,5	14,40	16,61	$3p\ ^4P^\circ—4d\ ^4F$	$^1/_2—^3/_2$
5591,734	1	14,39	16,60	$3p\ ^4P^\circ—4d\ ^4P$	$^3/_2—^3/_2$
5587,888	1	14,39	16,60	$3p\ ^4P^\circ—4d\ ^4F$	$^3/_2—^5/_2$
5577,33	10	14,39	16,61	$3p\ ^4P^\circ—4d\ ^4P$	$^3/_2—^5/_2$
5570,216	2,5	14,40	16,62	$3p\ ^4P^\circ—4d\ ^2P$	$^1/_2—^3/_2$
5553,53	0,7	14,37	16,60	$3p\ ^4P^\circ—4d\ ^4P$	$^5/_2—^3/_2$
5552,43	12	14,37	16,60	$3p\ ^4P^\circ—4d\ ^4F$	$^5/_2—^7/_2$
5546,74	0,7	14,68	16,92	$3p\ ^4S^\circ—5d\ ^4P$	$^3/_2—^5/_2$
5540,52	18	14,39	16,62	$3p\ ^4P^\circ—4d\ ^2F$	$^3/_2—^5/_2$
5539,33	6	14,37	16,61	$3p\ ^4P^\circ—4d\ ^4P$	$^5/_2—^5/_2$
5538,61	0,5	14,68	16,92	$3p\ ^2S^\circ—5d\ ^4F$	$^1/_2—^3/_2$
		14,39	16,62	$3p\ ^4P—4d\ ^2P$	$^3/_2—^3/_2$
5534,862	2	14,58	16,82	$3p\ ^2D^\circ—6s\ ^4P$	$^5/_2—^3/_2$
5450,05	1	14,58	16,86	$3p\ ^2D^\circ—6s\ ^2P$	$^5/_2—^3/_2$
5412,900	0,6	14,58	16,87	$3p\ ^2D^\circ—5d\ ^2D$	$^5/_2—^5/_2$
5410,15	4	14,58	16,87	$3p\ ^2D^\circ—5d\ ^2F$	$^5/_2—^7/_2$
5397,718	1	14,53	16,82	$3p\ ^4D^\circ—6s\ ^4P$	$^5/_2—^3/_2$
5385,88	0,6	14,61	16,91	$3p\ ^2D^\circ—5d\ ^4F$	$^3/_2—^5/_2$
5381,03	0,8	14,61	16,92	$3p\ ^2D^\circ—5d\ ^4P$	$^3/_2—^5/_2$
5370,10	5	14,50	16,81	$3p\ ^4D^\circ—6s\ ^4P$	$^7/_2—^5/_2$
5342,80	1	14,61	16,93	$3p\ ^2D^\circ—5d\ ^2F$	$^3/_2—^5/_2$
5316,98	0,5	14,53	16,86	$3p\ ^4D^\circ—6s\ ^2P$	$^5/_2—^3/_2$
5315,97	3	14,58	16,91	$3p\ ^2D^\circ—5d\ ^4F$	$^5/_2—^7/_2$
5285,57	0,8	14,53	16,87	$3p\ ^4D^\circ—5d\ ^4D$	$^5/_2—^5/_2$
5279,01	12	14,53	16,87	$3p\ ^4D^\circ—5d\ ^2F$	$^5/_2—^7/_2$
5244,28	0,8	14,55	16,92	$3p\ ^4D^\circ—5d\ ^4F$	$^1/_2—^3/_2$
5238,69	2	14,50	16,87	$3p\ ^4D^\circ—5d\ ^4D$	$^7/_2—^7/_2$
5230,41	15	14,50	16,87	$3p\ ^4D^\circ—5d\ ^4F$	$^7/_2—^9/_2$
5226,96	3	14,54	16,91	$3p\ ^4D^\circ—5d\ ^4F$	$^3/_2—^5/_2$
5221,42	0,5	14,54	16,92	$3p\ ^4D^\circ—5d\ ^4F$	$^3/_2—^3/_2$
5207,96	1	14,55	16,93	$3p\ ^4D^\circ—5d\ ^2P$	$^1/_2—^3/_2$
5189,27	4	14,53	16,91	$3p\ ^4D^\circ—5d\ ^4F$	$^5/_2—^7/_2$
5186,41	0,8	14,54	16,93	$3p\ ^4D^\circ—5d\ ^2F$	$^3/_2—^5/_2$
5110,269	1	14,39	16,81	$3p\ ^4P^\circ—6s\ ^4P$	$^3/_2—^5/_2$
5078,352	3	14,37	16,81	$3p\ ^4P^\circ—6s\ ^4P$	$^5/_2—^5/_2$
5040,69	0,6	14,58	17,04	$3p\ ^2D^\circ—6d\ ^2F$	$^5/_2—^7/_2$
4989,31	2	14,39	16,87	$3p\ ^4P^\circ—5d\ ^4D$	$^3/_2—^5/_2$
4986,43	2,5	14,39	16,87	$3p\ ^4P^\circ—5d\ ^4D$	$^3/_2—^3/_2$
4981,54	1,5	14,61	17,10	$3p\ ^2D^\circ—6d\ ^2F$	$^3/_2—^5/_2$
		14,39	16,87	$3p\ ^4P^\circ—5d\ ^4D$	$^3/_2—^1/_2$
4980,45	0,8	14,39	16,87	$3p\ ^4P^\circ—5d\ ^2D$	$^3/_2—^3/_2$
4960,65	6	14,37	16,87	$3p\ ^4P^\circ—5d\ ^4D$	$^5/_2—^7/_2$

λ, Å	I	E_H, eV	E_B, eV	Transition	J
4958,85	3	14,37	16,87	$3p\ ^4P^\circ$—$5d\ ^4D$	5/2—5/2
4956,01	1	14,37	16,87	$3p\ ^4P^\circ$—$5d\ ^4D$	5/2—3/2
4955,41	1,5	14,37	16,87	$3p\ ^4P^\circ$—$5d\ ^2D$	5/2—5/2
4952,20	2,5	14,50	17,00	$3p\ ^4D^\circ$—$7s\ ^4P$	7/2—5/2
4950,16	1	14,37	16,87	$3p\ ^4P^\circ$—$5d\ ^2D$	5/2—3/2
4930,45	1	14,40	16,91	$3p\ ^4P^\circ$—$5d\ ^2P$	1/2—1/2
4928,83	2	14,40	16,91	$3p\ ^4P^\circ$—$5d\ ^4P$	1/2—3/2
4926,67	0,8	14,53	17,04	$3p\ ^4D^\circ$—$6d\ ^2F$	5/2—7/2
4898,06	1	14,39	16,92	$3p\ ^4P^\circ$—$5d\ ^4P$	3/2—5/2
4888,37	0,8	14,50	17,04	$3p\ ^4D^\circ$—$6d\ ^4D$	7/2—7/2
4884,25	2,5	14,50	17,04	$3p\ ^4D^\circ$—$6d\ ^4F$	7/2—9/2
4880,46	0,6	14,54	17,08	$3p\ ^4D^\circ$—$6d\ ^4F$	3/2—5/2
4874,02	1,5	14,37	16,91	$3p\ ^4P^\circ$—$5d\ ^4F$	5/2—7/2
4868,69	0,5	14,37	16,92	$3p\ ^4P^\circ$—$5d\ ^4P$	5/2—5/2
4866,33	2	14,39	16,93	$3p\ ^4P^\circ$—$5d\ ^2F$	3/2—5/2
4847,13	0,5	14,53	17,08	$3p\ ^4D^\circ$—$6d\ ^4F$	5/2—7/2
4703,06	1	14,37	17,00	$3p\ ^4P^\circ$—$7s\ ^4P$	5/2—5/2
4697,00	0,6	14,50	17,14	$3p\ ^4D^\circ$—$7d\ ^4F$	7/2—9/2
4645,42	2	14,37	17,04	$3p\ ^4P^\circ$—$6d\ ^4D$	5/2—7/2
4644,43	1	14,37	17,04	$3p\ ^4P^\circ$—$6d\ ^4D$	5/2—5/2
4564,83	0,6	14,39	17,10	$3p\ ^4P^\circ$—$6d\ ^2F$	3/2—5/2
3992,210	2	12,98	16,09	$3s\ ^2P$—$4p\ ^4D^\circ$	3/2—5/2
3966,14	1,5	12,98	16,11	$3s\ ^2P$—$4p\ ^4D^\circ$	3/2—3/2
3953,799	2	12,98	16,12	$3s\ ^2P$—$4p\ ^2D^\circ$	3/2—5/2
3948,563	5	13,02	16,16	$3s\ ^2P$—$4p\ ^2P^\circ$	1/2—3/2
3934,262	5	12,98	16,13	$3s\ ^2P$—$4p\ ^2D^\circ$	3/2—3/2
3933,11	3	12,98	16,13	$3s\ ^2P$—$4p\ ^2S^\circ$	3/2—1/2
3930,689	8	13,02	16,18	$3s\ ^2P$—$4p\ ^2P^\circ$	1/2—1/2
3898,478	5	12,98	16,16	$3s\ ^2P$—$4p\ ^2P^\circ$	3/2—3/2
3712,733	0,8	12,73	16,07	$3s\ ^4P$—$4p\ ^4P^\circ$	3/2—1/2
3705,93	1,5	12,70	16,04	$3s\ ^4P$—$4p\ ^4P^\circ$	5/2—5/2
3691,859	2	12,70	16,05	$3s\ ^4P$—$4p\ ^4P^\circ$	5/2—3/2
3690,018	4	12,73	16,09	$3s\ ^4P$—$4p\ ^4D^\circ$	3/2—5/2
3689,40	1	12,75	16,11	$3s\ ^4P$—$4p\ ^4D^\circ$	1/2—3/2
3672,85	2,5	12,75	16,12	$3s\ ^4P$—$4p\ ^4D^\circ$	1/2—1/2
3668,174	12	12,70	16,07	$3s\ ^4P$—$4p\ ^4D^\circ$	5/2—7/2
3667,757	4	12,73	16,11	$3s\ ^4P$—$4p\ ^4D^\circ$	3/2—3/2
3661,793	3	12,75	16,13	$3s\ ^4P$—$4p\ ^2D^\circ$	1/2—3/2
3657,187	3	12,73	16,12	$3s\ ^4P$—$4p\ ^2D^\circ$	3/2—5/2
3652,982	1,5	12,70	16,09	$3s\ ^4P$—$4p\ ^4D^\circ$	5/2—5/2
3651,174	2	12,75	16,14	$3s\ ^4P$—$4p\ ^4S^\circ$	1/2—3/2
3630,776	1,5	12,75	16,16	$3s\ ^4P$—$4p\ ^2P^\circ$	1/2—3/2
3629,963	4	12,73	16,14	$3s\ ^4P$—$4p\ ^4S^\circ$	3/2—3/2
3620,789	1	12,70	16,12	$3s\ ^4P$—$4p\ ^2D^\circ$	5/2—5/2
3609,808	1,5	12,73	16,16	$3s\ ^4P$—$4p\ ^2P^\circ$	3/2—3/2
3604,401	2	12,70	16,13	$3s\ ^4P$—$4p\ ^2D^\circ$	5/2—3/2
3594,103	6	12,70	16,14	$3s\ ^4P$—$4p\ ^4S^\circ$	5/2—3/2
3574,346	1,5	12,70	16,16	$3s\ ^4P$—$4p\ ^2P^\circ$	5/2—3/2
977,745	100	0,05	12,73	$2p^5\ ^2P^\circ$—$3s\ ^4P$	1/2—3/2
976,505	40	0,00	12,70	$2p^5\ ^2P^\circ$—$3s\ ^4P$	3/2—5/2
976,217	100	0,05	12,75	$2p^5\ ^2P^\circ$—$3s\ ^4P$	1/2—1/2
973,895	350	0,00	12,73	$2p^5\ ^2P^\circ$—$3s\ ^4P$	3/2—3/2
972,401	20	0,00	12,75	$2p^5\ ^2P^\circ$—$3s\ ^4P$	3/2—1/2
958,524	500	0,05	12,98	$2p^5\ ^2P^\circ$—$3s\ ^2P$	1/2—3/2
955,545	750	0,05	13,02	$2p^5\ ^2P^\circ$—$3s\ ^2P$	1/2—1/2
954,825	1000	0,00	12,98	$2p^5\ ^2P^\circ$—$3s\ ^2P$	3/2—3/2
951,871	500	0,00	13,02	$2p^5\ ^2P^\circ$—$3s\ ^2P$	3/2—1/2
809,599	125	0,05	15,36	$2p^5\ ^2P^\circ$—$3s'\ ^2D$	1/2—3/2

λ, Å	I	E_H, eV	E_B, eV	Transition	J
806,964	150	0,00	15,36	$2p^5\,^2P^\circ$—$3s'\,^2D$	3/2—5/2
796,982	3	0,05	15,61	$2p^5\,^2P^\circ$—$4s\,^4P$	1/2—3/2
795,774	2	0,05	15,63	$2p^5\,^2P^\circ$—$4s\,^4P$	1/2—1/2
794,417	10	{ 0,05 0,00	15,66 15,61	$2p^5\,^2P^\circ$—$4s\,^2P$ $2p^5\,^2P^\circ$—$4s\,^4P$	1/2—3/2 3/2—3/2
792,536	10	0,05	15,69	$2p^5\,^2P^\circ$—$4s\,^2P$	1/2—1/2
791,875	12	0,00	15,66	$2p^5\,^2P^\circ$—$4s\,^2P$	3/2—3/2
790,006	7	0,00	15,69	$2p^5\,^2P^\circ$—$4s\,^2P$	3/2—1/2
782,976	5	0,05	15,88	$2p^5\,^2P^\circ$—$3d\,^4D$	1/2—3/2
782,575	2	0,05	15,89	$2p^5\,^2P^\circ$—$3d\,^4D$	1/2—1/2
782,378	10	0,05	15,90	$2p^5\,^2P^\circ$—$3d\,^2D$	1/2—3/2
781,654	3	0,05	15,91	$2p^5\,^2P^\circ$—$3d\,^4P$	1/2—1/2
780,713	5	0,00	15,88	$2p^5\,^2P^\circ$—$3d\,^4D$	3/2—5/2
780,519	10	{ 0,05 0,05	15,93 15,93	$2p^5\,^2P^\circ$—$3d\,^2P$ $2p^5\,^2P^\circ$—$3d\,^4P$	1/2—1/2 1/2—3/2
780,390	15	0,00	15,87	$2p^5\,^2P^\circ$—$3d\,^2D$	3/2—5/2
779,972	2	0,05	15,94	$2p^5\,^2P^\circ$—$3d\,^4F$	1/2—3/2
779,910	5	0,00	15,90	$2p^5\,^2P^\circ$—$3d\,^2D$	3/2—3/2
779,365	6	0,05	15,96	$2p^5\,^2P^\circ$—$3d\,^2P$	1/2—3/2
779,192	2	0,00	15,91	$2p^5\,^2P^\circ$—$3d\,^4P$	3/2—1/2
778,059	6	{ 0,00 0,00	15,93 15,93	$2p^5\,^2P^\circ$—$3d\,^2P$ $2p^5\,^2P^\circ$—$3d\,^4P$	3/2—1/2 3/2—3/2
777,531	4	0,00	15,94	$2p^5\,^2P^\circ$—$3d\,^4F$	3/2—3/2
777,010	5	0,00	15,96	$2p^5\,^2P^\circ$—$3d\,^2F$	3/2—5/2
776,926	4	0,00	15,96	$2p^5\,^2P^\circ$—$3d\,^2P$	3/2—3/2
755,603	2	0,05	16,46	$2p^5\,^2P^\circ$—$5s\,^4P$	1/2—3/2
754,148	2	{ 0,00 0,05	16,44 16,49	$2p^5\,^2P^\circ$—$5s\,^4P$ $2p^5\,^2P^\circ$—$5s\,^2P$	3/2—5/2 1/2—3/2
753,303	4	0,00	16,46	$2p^5\,^2P^\circ$—$5s\,^4P$	3/2—3/2
752,884	4	0,05	16,52	$2p^5\,^2P^\circ$—$5s\,^2P$	1/2—1/2
751,861	4	0,00	16,49	$2p^5\,^2P^\circ$—$5s\,^2P$	3/2—3/2
750,610	3	0,00	16,52	$2p^5\,^2P^\circ$—$5s\,^2P$	3/2—1/2
748,946	3	0,05	16,60	$2p^5\,^2P^\circ$—$4d\,^4P$	1/2—3/2
748,709	2	0,05	16,61	$2p^5\,^2P^\circ$—$4d\,^4F$	1/2—3/2
748,580	4	0,00	16,56	$2p^5\,^2P^\circ$—$4d\,^2D$	3/2—5/2
748,338	2	0,00	16,57	$2p^5\,^2P^\circ$—$4d\,^2D$	3/2—3/2
747,999	2	0,05	16,62	$2p^5\,^2P^\circ$—$4d\,^2P$	1/2—3/2
746,627	3	0,00	16,60	$2p^5\,^2P^\circ$—$4d\,^4F$	3/2—5/2
745,767	2	0,00	16,61	$2p^5\,^2P^\circ$—$4d\,^4F$	3/2—3/2
736,987	2	0,00	16,82	$2p^5\,^2P^\circ$—$6s\,^4P$	3/2—3/2
682,581	2	0,05	18,21	$2p^5\,^2P^\circ$—$4s'\,^2D$	1/2—3/2
680,709	2	0,00	18,21	$2p^5\,^2P^\circ$—$4s'\,^2D$	3/2—5/2

F II, ground state $1s^2 2s^2 2p^4\,^3P_2$

Ionization potential 282190,2 cm^{-1}; 34,985 eV

λ, Å	I	E_H, eV	E_B, eV	Transition	J
5589,31	0	33,02	35,24	$3p''\,^1S$—$3d''\,^1P^\circ$	0—1
5173,16	2	29,71	32,10	$2p^5\,^1P^\circ$—$3p''\,^1D$	1—2
5001,98	3	30,53	33,01	$3p'\,^1D$—$3d'\,^1D^\circ$	2—2
4933,25	5	30,53	33,05	$3p'\,^1D$—$3d'\,^1F^\circ$	2—3
4859,37	7	26,66	29,21	$3s'\,^1D^\circ$—$3p'\,^1P$	2—1
4738,0	0	26,66	29,28	$3s'\,^1D^\circ$—$3p'\,^3D$	2—1
4734,37	2	30,53	33,15	$3p'\,^1D$—$3d'\,^1P^\circ$	2—1

λ, Å	I	E_H, eV	E_B, eV	Transition	J
4576,1	0	25,75	28,46	$3p\ ^3P—3s''\ ^3P^\circ$	2—2
4447,18	12	28,77	31,56	$3d\ ^3D^\circ—4f\ ^3F$	3—4, 3, 2
4446,71	10	28,77	31,56	$3d\ ^3D^\circ—4f\ ^3F$	2—3, 2
4446,51	6	28,77	31,56	$3d\ ^3D^\circ—4f\ ^3F$	1—2
4417,3	3	—	—	—	—
4299,177	10	26,66	29,55	$3s'\ ^1D^\circ—3p'\ ^1F$	2—3
4278,89	4	—	—	—	—
4277,51	6	—	—	—	—
4275,21	8	—	—	—	—
4246,16	15	28,66	31,58	$3d\ ^5D^\circ—4f\ ^5F$	—
4225,12	4	—	—	—	—
4207,87	2	28,46	31,41	$3s''\ ^3P^\circ—3p''\ ^3S$	0—1
4207,442	5	28,46	31,41	$3s''\ ^3P^\circ—3p''\ ^3S$	1—1
4207,162	7	28,46	31,41	$3s''\ ^3P^\circ—3p''\ ^3S$	2—1
4192,62	2	32,10	35,06	$3p''\ ^1D—3d''\ ^1D^\circ$	2—2
4126,96	2	30,53	33,54	$3p'\ ^1D—4s'\ ^1D^\circ$	2—2
4119,219	7	26,27	29,28	$3s'\ ^3D—3p'\ ^3D^\circ$	1—1
4118,756	3	26,27	29,28	$3s'\ ^3D—3p'\ ^3D^\circ$	1—2
4117,008	5	26,27	29,28	$3s'\ ^3D—3p'\ ^3D^\circ$	2—1
4116,547	7	26,27	29,28	$3s'\ ^3D—3p'\ ^3D^\circ$	2—2
4112,975	5	26,27	29,28	$3s'\ ^3D—3p'\ ^3D^\circ$	3—2
4112,734	4	26,27	29,28	$3s'\ ^3D—3p'\ ^3D^\circ$	2—3
4109,173	8	26,27	29,28	$3s'\ ^3D—3p'\ ^3D^\circ$	3—3
4103,871	7	25,75	28,77	$3p\ ^3P—3d\ ^3D^\circ$	2—2
4103,724	7	25,75	28,77	$3p\ ^3P—3d\ ^3D^\circ$	0—1
4103,525	15	25,75	28,77	$3p\ ^3P—3d\ ^3D^\circ$	2—3
4103,217	5	25,75	28,77	$3p\ ^3P—3d\ ^3D^\circ$	1—1
4103,085	10	25,75	28,77	$3p\ ^3P—3d\ ^3D^\circ$	1—2
4083,919	6	32,10	35,14	$3p''\ ^1D—3d''\ ^1F^\circ$	2—3
4025,495	15	22,67	25,75	$3s\ ^3S^\circ—3p\ ^3P$	1—1
4025,010	10	22,67	25,75	$3s\ ^3S^\circ—3p\ ^3P$	1—0
4024,727	20	22,67	25,75	$3s\ ^3S^\circ—3p\ ^3P$	1—2
3974,791	6	28,46	31,58	$3s''\ ^3P^\circ—3p''\ ^3D$	2—3
3972,670	4	28,46	31,58	$3s''\ ^3P^\circ—3p''\ ^3D$	1—2
3972,411	2	28,46	31,58	$3s''\ ^3P^\circ—3p''\ ^3D$	2—2
3972,047	6	28,46	31,58	$3s''\ ^3P^\circ—3p''\ ^3D$	0—1
3971,626	3	28,46	31,58	$3s''\ ^3P^\circ—3p''\ ^3D$	1—1
3952,26	2	32,10	35,24	$3p''\ ^1D—3d''\ ^1P^\circ$	2—1
3945,65	4	29,78	32,92	$3p'\ ^3P—3d'\ ^3D^\circ$	0—1
3944,33	6	29,77	32,92	$3p'\ ^3P—3d'\ ^3D^\circ$	1—2
3941,52	3	29,77	32,92	$3p'\ ^3P—3d'\ ^3D^\circ$	1—1
3939,03	7	29,77	32,91	$3p'\ ^3P—3d'\ ^3D^\circ$	2—3
3935,00	3	29,77	32,92	$3p'\ ^3P—3d'\ ^3D^\circ$	1—2
3903,819	4	26,27	29,45	$3s'\ ^3D^\circ—3p'\ ^3F$	1—2
3901,955	5	26,27	29,45	$3s'\ ^3D^\circ—3p'\ ^3F$	2—3
3901,852	2	26,27	29,45	$3s'\ ^3D^\circ—3p'\ ^3F$	2—2
3898,833	6	26,27	29,45	$3s'\ ^3D^\circ—3p'\ ^3F$	3—4
3898,725	2	26,27	20,45	$3s'\ ^3D^\circ—3p'\ ^3F$	3—3
3896,66	3	31,90	35,08	$3p''\ ^3P—3d''\ ^3P^\circ$	2—2
3896,12	1	31,90	35,08	$3p''\ ^3P—3d''\ ^3P^\circ$	1—1
3893,04	2	31,90	35,08	$3p''\ ^3P—3d''\ ^3P^\circ$	1—2
3851,667	10	21,90	25,12	$3s\ ^5S^\circ—3p\ ^5P$	2—1
3849,987	15	21,90	25,12	$3s\ ^5S^\circ—3p\ ^5P$	2—2
3847,086	20	21,90	25,12	$3s\ ^5S^\circ—3p\ ^5P$	2—3
3837,68	0,5	28,46	31,69	$3s''\ ^3P^\circ—3p''\ ^1P$	0—1
3827,68	0,5	29,77	33,01	$3p'\ ^3P—3d'\ ^1D^\circ$	1—2
3818,52	2	29,78	33,02	$3p'\ ^3P—3d'\ ^3S^\circ$	0—1
3814,65	4	29,77	33,02	$3p'\ ^3P—3d'\ ^3S^\circ$	1—1

λ, Å	I	E_H, eV	E_B, eV	Transition	J
3805,90	5	29,77	33,02	$3p'\,^3P-3d'\,^3S^\circ$	2—1
3801,09	3	29,77	33,03	$3p'\,^3P-3d'\,^3P^\circ$	1—2
3798,46	3	29,78	33,04	$3p'\,^3P-3d'\,^3P^\circ$	0—1
3794,60	2	29,77	33,04	$3p'\,^3P-3d'\,^3P^\circ$	1—1
3792,40	4	29,77	33,03	$3p'\,^3P-3d'\,^3P^\circ$	2—2
3792,12	1	29,77	33,04	$3p'\,^3P-3d'\,^3P^\circ$	1—0
3785,97	3	29,77	33,04	$3p'\,^3P-3d'\,^3P^\circ$	2—1
3781,63	2	26,27	29,55	$3s'\,^3D^\circ-3p'\,^1F$	2—3
3753,3	0	29,55	32,85	$3p'\,^1F-3d'\,^3F^\circ$	3—3
3752,36	2	29,55	32,85	$3p'\,^1F-3d'\,^3F^\circ$	3—4
3739,60	4	29,71	33,02	$2p^5\,^1P^\circ-3p''\,^1S$	1—0
3710,365	4	29,55	32,89	$3p'\,^1F-3d'\,^3G^\circ$	3—4
3706,63	0,5	31,89	35,24	$3p''\,^3P-3d''\,^1P^\circ$	0—1
3704,51	8	29,55	32,89	$3p'\,^1F-3d'\,^1G^\circ$	3—4
3679,67	5	31,69	35,06	$3p''\,^1P-3d''\,^1D^\circ$	1—2
3668,9	0	29,77	33,15	$3p'\,^3P-3d'\,^1P^\circ$	1—1
3656,50	0,5	31,69	35,08	$3p''\,^1P-3d''\,^3P^\circ$	1—2
3642,798	7	29,45	32,85	$3p'\,^3F-3d'\,^3F^\circ$	3, 2—2
3641,985	8	29,45	32,85	$3p'\,^3F-3d'\,^3F^\circ$	3—3
3641,011	3	29,45	32,85	$3p'\,^3F-3d'\,^3F^\circ$	3—4
3640,891	9	29,45	32,85	$3p'\,^3F-3d'\,^3F^\circ$	4—4
3608,89	3	28,46	31,89	$3s''\,^3P^\circ-3p''\,^3P$	1—0
3607,32	3	28,46	31,90	$3s''\,^3P^\circ-3p''\,^3P$	0—1
3606,80	4	28,46	31,90	$3s''\,^3P^\circ-3p''\,^3P$	2, 1—1
3603,72	6	28,46	31,90	$3s''\,^3P^\circ-3p''\,^3P$	2, 1—2
3602,85	8	29,45	32,89	$3p'\,^3F-3d'\,^3G^\circ$	4—5
3601,403	7	29,45	32,89	$3p'\,^3F-3d'\,^3G^\circ$	3—4
3598,704	7	29,45	32,89	$3p'\,^3F-3d'\,^3G^\circ$	2—3
3595,917	5	29,45	32,89	$3p'\,^3F-3d'\,^1G^\circ$	3—4
3590,63	7	31,58	35,03	$3p''\,^3D-3d''\,^3F^\circ$	3—4
3589,345	6	31,58	35,03	$3p''\,^3D-3d''\,^3F^\circ$	2—3
3587,980	5	31,58	35,03	$3p''\,^3D-3d''\,^3F^\circ$	1—2
3587,42	3	31,58	35,03	$3p''\,^3D-3d''\,^3F^\circ$	3—3
3587,13	3	31,58	35,03	$3p''\,^3D-3d''\,^3F^\circ$	2—2
3577,23	2	29,55	33,01	$3p'\,^1F-3d'\,^1D^\circ$	3—2
3574,92	3	29,45	32,91	$3p'\,^1F-3d'\,^1D^\circ$	4, 3, 2—3
3571,68	3	29,45	32,92	$3p'\,^1F-3d'\,^1D^\circ$	3, 2—2
3569,47	2	29,45	32,92	$3p'\,^1F-3d'\,^1D^\circ$	2—1
3563,87	0	31,58	35,06	$3p''\,^3D-3d''\,^1D^\circ$	1—2
3548,5	0,5	31,58	35,07	$3p''\,^3D-3d''\,^3P^\circ$	1—0
3546,6	0	31,58	35,07	$3p''\,^3D-3d''\,^3P^\circ$	1—1
3546,06	1	31,58	35,08	$3p''\,^3D-3d''\,^3P^\circ$	1—2
3545,5	0,5	31,58	35,07	$3p''\,^3D-3d''\,^3P^\circ$	2—1
3544,392	3	26,27	29,77	$3s'\,^3D^\circ-3p'\,^3P$	2—2
3541,937	8	29,55	33,05	$3p'\,^1F-3d'\,^1F^\circ$	3—3
3541,765	9	26,77	29,77	$3s'\,^3D^\circ-3p'\,^3P$	3—2
3539,45	1	31,58	35,08	$3p''\,^3D-3d''\,^3P^\circ$	3—2
3538,474	3	26,27	29,77	$3s'\,^3D^\circ-3p'\,^3P$	1—1
3536,838	7	26,27	29,77	$3s'\,^3D^\circ-3p'\,^3P$	2—1
3535,162	4	26,27	29,78	$3s'\,^3D^\circ-3p'\,^3P$	1—0
3522,883	6	28,17	31,69	$3s''\,^1P^\circ-3p''\,^1P$	1—1
3505,763	4	—	—	—	—
3505,614	15	25,12	28,66	$3p\,^5P-3d\,^5D^\circ$	3—4
3505,508	6	25,12	28,66	$3p\,^5P-3d\,^5D^\circ$	3—3
3503,095	12	25,12	28,66	$3p\,^5P-3d\,^5D^\circ$	2—3
3502,954	8	25,12	28,66	$3p\,^5P-3d\,^5D^\circ$	2—2
3502,859	4	25,12	28,66	$3p\,^5P-3d\,^5D^\circ$	2—1
3501,562	5	25,12	28,66	$3p\,^5P-3d\,^5D^\circ$	1—2

λ, Å	I	E_H, eV	E_B, eV	Transition	J
3501,487	3	25,12	28,66	$3p\ ^5P$—$3d\ ^5D^\circ$	1—1
3501,416	10	25,12	28,66	$3p\ ^5P$—$3d\ ^5D^\circ$	1—0
3493,215	5	31,69	35,24	$3p''\ ^1P$—$3d''\ ^1P^\circ$	1—1
3476,2	0	29,28	32,85	$3p'\ ^3D$—$3d'\ ^3F^\circ$	3—2
3475,68	2	29,28	32,85	$3p'\ ^3D$—$3d'\ ^3F^\circ$	3—3
3474,800	7	29,28	32,85	$3p'\ ^3D$—$3d'\ ^3F^\circ$	3—4
3473,621	2	29,28	32,85	$3p'\ ^3D$—$3d'\ ^3F^\circ$	2—2
3473,314	5	29,28	32,85	$3p'\ ^3D$—$3d'\ ^3F^\circ$	1—2
3472,964	6	29,28	32,85	$3p'\ ^3D$—$3d'\ ^3F^\circ$	2—3
3453,8	0	29,45	33,04	$3p'\ ^3F$—$3d'\ ^3P^\circ$	3—2
3442,5	0	29,45	33,05	$3p'\ ^3F$—$3d'\ ^1F^\circ$	3—3
3436,57	4	—	—	—	—
3433,69	2	29,28	32,89	$3p'\ ^3D$—$3d'\ ^1G^\circ$	3—4
3417,21	4	—	—	—	—
3417,02	6	25,75	29,38	$3p\ ^3P$—$4s\ ^3S^\circ$	2—1
3416,80	1	25,75	29,38	$3p\ ^3P$—$4s\ ^3S^\circ$	0—1
3416,58	4	—	—	—	—
3416,45	4	25,75	29,38	$3p\ ^3P$—$4s\ ^3S^\circ$	1—1
3414,663	5	29,28	32,91	$3p'\ ^3D$—$3d'\ ^3D^\circ$	3—3
3412,04	2	29,28	32,91	$3p'\ ^3D$—$3d'\ ^3D^\circ$	2—3
3411,66	3	29,28	32,92	$3p'\ ^3D$—$3d'\ ^3D^\circ$	3—2
3410,82	1,5	29,21	32,85	$3p'\ ^1P$—$3d'\ ^3F^\circ$	1—2
3409,02	3	29,28	32,92	$3p'\ ^3D$—$3d'\ ^3D^\circ$	2—2
3408,68	1,5	29,28	32,92	$3p'\ ^3D$—$3d'\ ^3D^\circ$	1—2
3406,83	1	29,28	32,92	$3p'\ ^3D$—$3d'\ ^3D^\circ$	2—1
3406,56	2	29,28	32,92	$3p'\ ^3D$—$3d'\ ^3D^\circ$	1—1
3405,980	4	—	—	—	—
3399,29	3	31,90	35,54	$3p''\ ^3P$—$4s''\ ^3P^\circ$	2—2
3398,78	1	31,90	35,55	$3p''\ ^3P$—$4s''\ ^3P^\circ$	2—1
3396,63	1	31,90	35,54	$3p''\ ^3P$—$4s''\ ^3P^\circ$	1—2
3395,77	1	31,90	35,55	$3p''\ ^3P$—$4s''\ ^3P^\circ$	1—0
3394,22	2	31,89	35,54	$3p''\ ^3P$—$4s''\ ^3P^\circ$	0—1
3393,40	1	31,41	35,06	$3p''\ ^3S$—$3d''\ ^1D^\circ$	1—2
3379,29	2	31,41	35,07	$3p''\ ^3S$—$3d''\ ^3P^\circ$	1—0
3377,44	4	31,41	35,07	$3p''\ ^3S$—$3d''\ ^3P^\circ$	1—1
3373,49	5	31,41	35,08	$3p''\ ^3S$—$3d''\ ^3P^\circ$	1—2
3348,43	0,5	29,21	32,92	$3p'\ ^1P$—$3d'\ ^3D^\circ$	1—2
3346,41	0,5	29,21	32,92	$3p'\ ^1P$—$3d'\ ^3D^\circ$	1—1
3324,13	0,5	29,28	33,01	$3p'\ ^3D$—$3d'\ ^1D^\circ$	3—2
3321,30	2	29,28	33,01	$3p'\ ^3D$—$3d'\ ^1D^\circ$	1—2
3311,63	1	29,28	33,02	$3p'\ ^3D$—$3d'\ ^3S^\circ$	2—1
3303,89	6	29,28	33,03	$3p'\ ^3D$—$3d'\ ^3P^\circ$	3—2
3301,41	3	29,28	33,03	$3p'\ ^3D$—$3d'\ ^3P^\circ$	2—2
3296,56	5	29,28	33,04	$3p'\ ^3D$—$3d'\ ^3P^\circ$	2—1
3296,19	2	29,28	33,04	$3p'\ ^3D$—$3d'\ ^3P^\circ$	1—1
3294,37	4	29,28	33,04	$3p'\ ^3D$—$3d'\ ^3P^\circ$	1—0
3264,16	7	29,21	33,01	$3p'\ ^1P$—$3d'\ ^1D^\circ$	1—2
3239,91	0,5	29,21	33,04	$3p'\ ^1P$—$3d'\ ^3P^\circ$	1—1
3238,10	0	29,21	33,04	$3p'\ ^1P$—$3d'\ ^3P^\circ$	1—0
3215,10	0,5	31,69	35,54	$3p''\ ^1P$—$4s''\ ^3P^\circ$	1—2
3214,67	0,5	31,69	35,55	$3p''\ ^1P$—$4s''\ ^3P^\circ$	1—1
3202,740	10	26,66	30,53	$3s'\ ^1D^\circ$—$3p'\ ^1D$	2—2
3201,17	1	29,28	33,15	$3p'\ ^3D$—$3d'\ ^1P^\circ$	1—1
3162,42	1	26,66	30,58	$3s'\ ^1D^\circ$—$4p\ ^4P$	2—2
3153,492	6	28,17	32,10	$3s''\ ^1P^\circ$—$3p''\ ^1D$	1—2
3147,965	5	29,21	33,15	$3p'\ ^1P$—$3d'\ ^1P^\circ$	1—1
3125,73	1	31,58	35,55	$3p''\ ^3D$—$4s''\ ^3P^\circ$	1—1, 0

λ, Å	I	E_H, eV	E_B, eV	Transition	J
3125,59	0,5	31,58	35,54	$3p''\ {}^3D$—$4s''\ {}^3P^\circ$	2—2
3125,15	1	31,58	35,55	$3p''\ {}^3D$—$4s''\ {}^3P^\circ$	2—1
3124,19	2	31,58	35,54	$3p''\ {}^3D$—$4s''\ {}^3P^\circ$	3—2
3120,12	1	29,45	33,42	$3p'\ {}^3F$—$4s'\ {}^3D^\circ$	4—3
3118,69	0,5	29,45	33,42	$3p'\ {}^3F$—$4s'\ {}^3D^\circ$	3—2
3117,75	0	29,45	33,42	$3p'\ {}^3F$—$4s'\ {}^3D^\circ$	2—1
3106,16	4	29,55	33,54	$3p'\ {}^1F$—$4s'\ {}^1D^\circ$	3—2
3059,960	8	25,12	29,17	$3p\ {}^5P$—$4s\ {}^5S^\circ$	3—2
3058,141	7	25,12	29,17	$3p\ {}^5P$—$4s\ {}^5S^\circ$	2—2
3057,083	6	25,12	29,17	$3p\ {}^5P$—$4s\ {}^5S^\circ$	1—2
2988,45	3	28,66	32,81	$3d\ {}^5D^\circ$—$5f\ {}^5F$	—
2904,61	0	26,27	30,53	$3s'\ {}^3D^\circ$—$3p'\ {}^1D$	3—2
2876,49	3	26,27	30,58	$3s'\ {}^3D^\circ$—$4p\ {}^3P$	1—0
2875,88	2	26,27	30,58	$3s'\ {}^3D^\circ$—$4p\ {}^3P$	1—1
2874,80	4	26,27	30,58	$3s'\ {}^3D^\circ$—$4p\ {}^3P$	2—1
2874,22	1	26,27	30,58	$3s'\ {}^3D^\circ$—$4p\ {}^3P$	1—2
2873,13	2	26,27	30,58	$3s'\ {}^3D^\circ$—$4p\ {}^3P$	2—2
2871,40	5	26,27	30,58	$3s'\ {}^3D^\circ$—$4p\ {}^3P$	3—2
2867,30	3	29,21	33,54	$3p'\ {}^1P$—$4s'\ {}^1D^\circ$	1—2
2739,63	0	30,53	35,06	$3p'\ {}^1D$—$3d''\ {}^1D^\circ$	2—2
2692,790	5	30,53	35,14	$3p'\ {}^1D$—$3d''\ {}^1F^\circ$	2—3
2556,10	4	28,17	33,02	$3s''\ {}^1P^\circ$—$3p''\ {}^1S$	1—0
2522,5	0	26,66	31,58	$3s'\ {}^1D^\circ$—$3p''\ {}^3D$	2—3
2497,72	2	30,58	35,55	$4p\ {}^3P$—$4s''\ {}^3P^\circ$	2—1, 0
2496,79	1	30,58	35,54	$4p\ {}^3P$—$4s''\ {}^3P^\circ$	1—2
2466,162	4	26,66	31,69	$3s'\ {}^1D^\circ$—$3p''\ {}^1P$	2—1
2334,99	0,5	26,27	31,58	$3s'\ {}^3D^\circ$—$3p''\ {}^3D$	2—3
2334,12	1	26,27	31,58	$3s'\ {}^3D^\circ$—$3p''\ {}^3D$	2—2
2333,78	2	26,27	31,58	$3s'\ {}^3D^\circ$—$3p''\ {}^3D$	3—3
2329,93	0,5	20,46	25,75	$2p^5\ {}^3P^\circ$—$3p\ {}^3P$	2—2
2279,36	3	26,66	32,10	$3s'\ {}^1D^\circ$—$3p''\ {}^1D$	2—2
2217,34	5	29,55	35,14	$3p'\ {}^1F$—$3d''\ {}^1F^\circ$	3—3
1747,40	3	22,67	29,77	$3s\ {}^3S^\circ$—$3p'\ {}^3P$	1—2
1745,57	2	22,67	29,77	$3s\ {}^3S^\circ$—$3p'\ {}^3P$	1—1
1744,86	1	22,67	29,78	$3s\ {}^3S^\circ$—$3p'\ {}^3P$	1—0
1704,90	0,5	25,75	33,02	$3p\ {}^3P$—$3d'\ {}^3S^\circ$	2—1
608,065	7	0,04	20,46	$2p^4\ {}^3P$—$2p^5\ {}^3P^\circ$	1—2
607,472	6	0,06	20,47	$2p^4\ {}^3P$—$2p^5\ {}^3P^\circ$	0—1
606,925	5	0,04	20,47	$2p^4\ {}^3P$—$2p^5\ {}^3P^\circ$	1—1
606,805	8	0,00	20,46	$2p^4\ {}^3P$—$2p^5\ {}^3P^\circ$	2—2
606,284	6	0,04	20,49	$2p^4\ {}^3P$—$2p^5\ {}^3P^\circ$	1—0
605,668	7	0,00	20,47	$2p^4\ {}^3P$—$2p^5\ {}^3P^\circ$	2—1
548,517	2	5,59	28,17	$2p^4\ {}^1S$—$3s''\ {}^1P^\circ$	0—1
548,324	3	0,06	22,67	$2p^4\ {}^3P$—$3s\ {}^3S^\circ$	0—1
547,873	4	0,04	22,67	$2p^4\ {}^3P$—$3s\ {}^3S^\circ$	1—1
546,846	6	0,00	22,67	$2p^4\ {}^3P$—$3s\ {}^3S^\circ$	2—1
514,945	6	2,59	26,66	$2p^4\ {}^1D$—$3s'\ {}^1D^\circ$	2—2
513,649	4	5,59	29,71	$2p^4\ {}^1S$—$2p^5\ {}^1P^\circ$	0—1
484,600	8	2,59	28,17	$2p^4\ {}^1D$—$3s''\ {}^1P^\circ$	2—1
473,021	3	0,06	26,27	$2p^4\ {}^3P$—$3s'\ {}^3D^\circ$	0—1
472,710	5	0,04	26,27	$2p^4\ {}^3P$—$3s'\ {}^3D^\circ$	1—2
471,990	6	0,00	26,27	$2p^4\ {}^3P$—$3s'\ {}^3D^\circ$	2—3
471,949	3	—	—	—	—
457,177	6	2,59	29,71	$2p^4\ {}^1D$—$3p^5\ {}^1P^\circ$	2—1
436,563	1	0,06	28,46	$2p^4\ {}^3P$—$3s''\ {}^3P^\circ$	0—1
436,279	2	0,04	28,46	$2p^4\ {}^3P$—$3s''\ {}^3P^\circ$	1—2, 1, 0
435,634	3	0,00	28,46	$2p^4\ {}^3P$—$3s''\ {}^3P^\circ$	2—2, 1
431,826	2	0,06	28,77	$2p^4\ {}^3P$—$3d\ {}^3D^\circ$	0—1

λ, Å	I	E_H, eV	E_B, eV	Transition	J
431,545	3	0,04	28,77	$2p^4\,{}^3P - 3d\,{}^3D^\circ$	1—2, 1
430,909	4	0,00	28,77	$2p^4\,{}^3P - 3d\,{}^3D^\circ$	2—3, 2
422,012	1	0,00	29,38	$2p^4\,{}^3P - 3s\,{}^3S^\circ$	2—1
417,874	1	5,59	35,24	$2p^4\,{}^1S - 3d''\,{}^1P^\circ$	0—1
407,511	4	2,59	33,01	$2p^4\,{}^1D - 3d'\,{}^1D^\circ$	2—2
407,053	5	2,59	33,05	$2p^4\,{}^1D - 3d'\,{}^1F^\circ$	2—3
405,644	4	2,59	33,15	$2p^4\,{}^1D - 3d'\,{}^1P^\circ$	2—1
400,579	1	2,59	33,54	$2p^4\,{}^1D - 3s'\,{}^1D^\circ$	2—2
393,676	1	0,00	31,49	$2p^4\,{}^3P - 4d\,{}^3D^\circ$	2—3
380,902	2	2,59	35,14	$2p^4\,{}^1D - 3d''\,{}^1F^\circ$	2—2
376,686	1	0,00	32,91	$2p^4\,{}^3P - 3d'\,{}^3D^\circ$	2—3
375,928	1	{ 0,04	33,02	$2p^4\,{}^3P - 3d'\,{}^3P^\circ$	1—0
		0,06	33,04	$2p^4\,{}^3P - 3d'\,{}^3P^\circ$	0—1
375,793	1	0,04	33,03	$2p^4\,{}^3P - 3d'\,{}^3P^\circ$	1—2
375,718	1	0,04	33,04	$2p^4\,{}^3P - 3d'\,{}^3P^\circ$	1—1, 0
375,434	1	0,00	33,02	$2p^4\,{}^3P - 3d'\,{}^3S^\circ$	2—1
375,300	2	0,00	33,03	$2p^4\,{}^3P - 3d'\,{}^3P^\circ$	2—2
353,421	0	0,00	35,08	$2p^4\,{}^3P - 3d''\,{}^3P^\circ$	2—2

F III, ground state $1s^2 2s^2 2p^3\,{}^4S^0_{3/2}$

Ionization potential 505410 cm^{-1}; 62,659 eV

λ, Å	I	E_H, eV	E_B, eV	Transition	J
3436,57	4	44,69	48,29	$3p\,{}^2P^\circ - 3d\,{}^2P$	$^3/_2-{}^3/_2$
3426,34	2	44,67	48,29	$3p\,{}^2P^\circ - 3d\,{}^2P$	$^1/_2-{}^3/_2$
3411,66	3	44,69	48,32	$3p\,{}^2P^\circ - 3d\,{}^2P$	$^3/_2-{}^1/_2$
3401,62	2	44,67	48,32	$3p\,{}^2P^\circ - 3d\,{}^2P$	$^1/_2-{}^1/_2$
3372,24	1	44,69	48,36	$3p\,{}^2P^\circ - 3d\,{}^4D$	$^3/_2-{}^3/_2$
3367,65	1	44,69	48,37	$3p\,{}^2P^\circ - 3d\,{}^4D$	$^3/_2-{}^1/_2$
3358,32	4	50,18	53,87	$3s'''\,{}^4S^\circ - 3p'''\,{}^4P$	$^3/_2-{}^5/_2$
3357,82	1	44,67	48,37	$3p\,{}^2P^\circ - 3d\,{}^4D$	$^1/_2-{}^1/_2$
3355,98	3	50,18	53,88	$3s'''\,{}^4S^\circ - 3p'''\,{}^4P$	$^3/_2-{}^3/_2$
3354,34	2	50,18	53,88	$3s'''\,{}^4S^\circ - 3p'''\,{}^4P$	$^3/_2-{}^1/_2$
3267,202	4	—	—	—	—
3264,164	9	—	—	—	—
3253,43	2,5	—	—	—	—
3213,972	6	40,28	44,13	$3s\,{}^2P - 3p\,{}^2D^\circ$	$^3/_2-{}^3/_2$
3174,725	10	40,23	44,13	$3s\,{}^2P - 3p\,{}^2D^\circ$	$^1/_2-{}^3/_2$
3174,125	12	40,28	44,18	$3s\,{}^2P - 3p\,{}^2D^\circ$	$^3/_2-{}^5/_2$
3156,11	0	47,65	51,59	$3p'\,{}^2P^\circ - 3d'\,{}^2D$	$^3/_2-{}^3/_2$
3154,387	4	47,67	51,60	$3p'\,{}^2P^\circ - 3d'\,{}^2P$	$^3/_2-{}^5/_2$
3146,962	8	39,33	43,27	$3s\,{}^4P - 3p\,{}^4D^\circ$	$^5/_2-{}^5/_2$
3145,536	4	39,29	43,23	$3s\,{}^4P - 3p\,{}^4D^\circ$	$^3/_2-{}^1/_2$
3142,777	3	47,65	51,59	$3p'\,{}^2P^\circ - 3d'\,{}^2D$	$^1/_2-{}^3/_2$
3134,208	8	39,29	43,25	$3s\,{}^4P - 3p\,{}^4D^\circ$	$^3/_2-{}^3/_2$
3124,762	8	39,26	43,23	$3s\,{}^4P - 3p\,{}^4D^\circ$	$^1/_2-{}^1/_2$
3124,18	3	—	—	—	—
3121,515	12	39,33	43,30	$3s\,{}^4P - 3p\,{}^4D^\circ$	$^5/_2-{}^7/_2$
3115,669	10	39,29	43,27	$3s\,{}^4P - 3p\,{}^4D^\circ$	$^3/_2-{}^5/_2$
3113,579	8	39,26	43,25	$3s\,{}^4P - 3p\,{}^4D^\circ$	$^1/_2-{}^3/_2$
3066,71	2	44,32	48,36	$3p\,{}^4S^\circ - 3d\,{}^4D$	$^3/_2-{}^5/_2$
3049,139	8	42,65	46,72	$3s'\,{}^2D - 3p'\,{}^2P^\circ$	$^3/_2-{}^5/_2$
3048,80	2	42,65	46,72	$3s'\,{}^2D - 3p'\,{}^2F^\circ$	$^5/_2-{}^5/_2$

λ, Å	I	E_H, eV	E_B, eV	Transition	J
3047,077	4	—	—	—	—
3042,808	10	42,65	46,72	$3s'\,^2D—3p'\,^2F^\circ$	5/3—7/2
3039,746	6	47,15	51,23	$3p'\,^2D^\circ—3d'\,^2F$	3/2—5/2
3039,254	7	47,14	51,22	$3p'\,^2D^\circ—3d'\,^2F$	5/2—7/2
3034,54	1,5	47,14	51,23	$3p'\,^2D^\circ—3d'\,^2F$	5/2—5/2
3003,15	2	53,88	58,01	$3p'''\,^4P—3d'''\,^4D^\circ$	1/2—3/2, 1/2
3001,920	3	53,88	58,01	$3p'''\,^4P—3d'''\,^4D^\circ$	3/2—1/2, 3/2, 5/2
3000,097	4	53,87	58,01	$3p'''\,^4P—3d'''\,^4D^\circ$	5/2—7/2
2999,465	6	48,59	52,72	$3s'''\,^6S^\circ—3p'''\,^6P$	5/2—3/2
2997,513	6	48,59	52,72	$3s'''\,^6S^\circ—3p'''\,^6P$	5/2—5/2
2997,168	6	44,32	48,45	$3p\,^4S^\circ—3d\,^4P$	3/2—5/2
2994,273	8	48,59	52,73	$3s'''\,^6S^\circ—3p'''\,^6P$	5/2—7/2
2988,45	4	—	—	—	—
2984,479	5	44,32	48,47	$3p\,^4S^\circ—3d\,^4P$	3/2—3/2
2983,765	4	—	—	—	—
2978,145	4	44,32	48,48	$3p\,^4S^\circ—3d\,^4P$	3/2—1/2
2966,89	1	47,67	51,84	$3p'\,^2P^\circ—3d'\,^2P$	3/2—1/2
2961,596	5	47,67	51,85	$3p'\,^2P^\circ—3d'\,^2P$	3/2—3/2
2959,666	2	44,32	48,51	$3p\,^4S^\circ—3d\,^2P$	3/2—5/2
2955,13	2	47,65	51,84	$3p'\,^2P^\circ—3d'\,^2P$	1/2—1/2
2954,37	0,5	49,00	53,20	$3d\,^2D—4p\,^2D^\circ$	3/2—3/2
2949,91	1,5	47,65	51,85	$3p'\,^2P^\circ—3d'\,^2P$	1/2—3/2
2932,479	8	39,33	43,56	$3s\,^4P—3p\,^4P^\circ$	5/2—3/2
2920,887	4	46,20	50,45	$3s''\,^2S—3p''\,^2P^\circ$	1/2—1/2
2920,538	6	46,20	50,45	$3s''\,^2S—3p''\,^2P^\circ$	1/2—3/2
2916,335	10	39,33	43,58	$3s\,^4P—3p\,^4P^\circ$	5/2—5/2
2913,279	8	39,29	43,55	$3s\,^4P—3p\,^4P^\circ$	3/2—1/2
2905,301	6	39,29	43,56	$3s\,^4P—3p\,^4P^\circ$	3/2—3/2
2895,458	4	39,26	43,55	$3s\,^4P—3p\,^4P^\circ$	1/2—1/2
2889,447	8	39,29	43,58	$3s\,^4P—3p\,^4P^\circ$	3/2—5/2
2887,559	8	39,26	43,56	$3s\,^4P—3p\,^4P^\circ$	1/2—3/2
2876,49	3	—	—	—	—
2875,87	3	—	—	—	—
2874,81	5	—	—	—	—
2873,12	4	—	—	—	—
2871,40	8	—	—	—	—
2869,993	3	44,69	49,00	$3p\,^2P^\circ—3d\,^2D$	3/2—3/2
2867,30	5	—	—	—	—
2865,670	4	44,18	48,51	$3p\,^2D^\circ—3d\,^2F$	5/2—5/2
2862,866	6	44,67	49,00	$3p\,^2P^\circ—3d\,^2D$	1/2—3/2
2860,308	9	44,69	49,02	$3p\,^2P^\circ—3d\,^2D$	3/2—5/2
2841,7	3	—	—	—	—
2835,606	9	44,18	48,55	$3p\,^2D^\circ—3d\,^2F$	5/2—7/2
2833,962	8	44,13	48,51	$3p\,^2D^\circ—3d\,^2F$	3/2—5/2
2826,081	5	—	—	—	—
2823,77	3	—	—	—	—
2820,695	4	—	—	—	—
2818,302	5	40,28	44,67	$3s\,^2P^\circ—3p\,^2P$	3/2—1/2
2811,422	10	40,28	44,69	$3s\,^2P^\circ—3p\,^2P$	3/2—3/2
2803,97	0,5	48,32	52,74	$3d\,^2P—4p\,^2S^\circ$	1/2—1/2
2795,500	4	—	—	—	—
2794,2	3	—	—	—	—
2793,18	2	49,00	53,44	$3d\,^2D—4p\,^2P^\circ$	3/2—1/2
2789,352	3	49,02	53,46	$3d\,^2D—4p\,^2P^\circ$	5/2—3/2
2788,093	20	40,23	44,67	$3s\,^2P^\circ—3p\,^2P$	1/2—1/2
2787,72	2	47,15	51,59	$3p'\,^2D^\circ—3d'\,^2D$	3/2—3/2
2787,38	0,5	48,29	52,74	$3d\,^2P—4p\,^2S^\circ$	3/2—1/2
2783,30	0	47,14	51,59	$3p'\,^2D^\circ—3d'\,^2D$	5/2—3/2

λ, Å	I	E_H, eV	E_B, eV	Transition	J
2781,956	4	47,14	51,60	$3p'\,^2D^\circ$—$3d'\,^2D$	$^5/_2$—$^5/_2$
2781,350	5	40,23	44,69	$3s\,^2P^\circ$—$3p\,^2P$	$^1/_2$—$^3/_2$
2759,81	5	42,65	47,14	$3s'\,^2D$—$3p'\,^2D^\circ$	$^3/_2$—$^5/_2$
2759,589	10	42,65	47,14	$3s'\,^2D$—$3p'\,^2D^\circ$	$^5/_2$—$^5/_2$
2756,664	5	46,72	51,22	$3p'\,^2F^\circ$—$3d'\,^2F$	$^7/_2$—$^7/_2$
2755,556	7	42,65	47,15	$3s'\,^2D$—$3p'\,^2D^\circ$	$^3/_2$—$^3/_2$
2755,307	4	42,65	47,15	$3s'\,^3D$—$3p'\,^2D^\circ$	$^5/_2$—$^3/_2$
2752,8	1	46,72	51,23	$3p'\,^2F^\circ$—$3d'\,^2F$	$^7/_2$—$^5/_2$
2751,8	1	46,72	51,22	$3p'\,^2F^\circ$—$3d'\,^2F$	$^5/_2$—$^7/_2$
2747,870	5	46,72	51,23	$3p'\,^2F^\circ$—$3d'\,^2F$	$^5/_2$—$^5/_2$
2740,31	1	48,47	52,99	$3d\,^4P$—$4p\,^4P^\circ$	$^3/_2$—$^1/_2$
2739,11	1	48,48	53,00	$3d\,^4P$—$4p\,^4P^\circ$	$^1/_2$—$^3/_2$
2737,954	4	47,67	52,19	$3p'\,^2P^\circ$—$3d'\,^2S$	$^3/_2$—$^1/_2$
2733,8	0,5	48,47	53,00	$3d\,^4P^\circ$—$4p\,^4P^\circ$	$^3/_2$—$^3/_2$
2727,93	2	47,65	52,19	$3p'\,^2P^\circ$—$3d'\,^2S$	$^1/_2$—$^1/_2$
2727,47	0	48,36	52,90	$3d\,^4D$—$4p\,^4D^\circ$	$^5/_2$—$^5/_2$
2723,25	1	48,45	53,00	$3d\,^4P$—$4p\,^4D^\circ$	$^5/_2$—$^3/_2$
2719,89	1,5	48,47	53,03	$3d\,^4P$—$4p\,^4P^\circ$	$^3/_2$—$^5/_2$
2718,14	1	48,38	52,94	$3d\,^4D$—$4p\,^4D^\circ$	$^7/_2$—$^7/_2$
2709,408	3	48,45	53,03	$3d\,^4P$—$4p\,^4P^\circ$	$^5/_2$—$^5/_2$
2695,45	4	—	—	—	—
2677,42	0	48,37	52,99	$3d\,^4D$—$4p\,^4P^\circ$	$^1/_2$—$^1/_2$
2674,54	1	48,36	52,99	$3d\,^4D$—$4p\,^4P^\circ$	$^3/_2$—$^1/_2$
2668,25	2	48,36	53,00	$3d\,^4D$—$4p\,^4P^\circ$	$^5/_2$—$^3/_2$
2664,390	3	48,38	53,03	$3d\,^4D$—$4p\,^4P^\circ$	$^7/_2$—$^5/_2$
2656,475	6	52,73	57,39	$3p'''\,^6P$—$3d'''\,^6D^\circ$	$^7/_2$—$^9/_2$
2656,294	3	52,73	57,39	$3p'''\,^6P$—$3d'''\,^6D^\circ$	$^7/_2$—$^7/_2$
2653,757	5	52,72	57,39	$3p'''\,^6P$—$3d'''\,^6D^\circ$	$^5/_2$—$^7/_2$
2653,491	4	52,72	57,39	$3p'''\,^6P$—$3d'''\,^6D^\circ$	$^5/_2$—$^5/_2$
2653,252	2	52,72	57,39	$3p'''\,^6P$—$3d'''\,^6D^\circ$	$^5/_2$—$^3/_2$
2651,958	3	52,72	57,39	$3p'''\,^6P$—$3d'''\,^6D^\circ$	$^3/_2$—$^5/_2$
2651,723	3	52,72	57,39	$3p'''\,^6P$—$3d'''\,^6D^\circ$	$^3/_2$—$^3/_2$
2651,550	3	52,72	57,39	$3p'''\,^6P$—$3d'''\,^6D^\circ$	$^3/_2$—$^1/_2$
2645,5	0	44,32	49,00	$3p\,^4S^\circ$—$3d\,^2D$	$^3/_2$—$^3/_2$
2641,24	2	48,51	53,20	$3d\,^2F$—$4p\,^2D^\circ$	$^5/_2$—$^3/_2$
2639,47	3	48,55	53,25	$3d\,^2F$—$4p\,^2D^\circ$	$^7/_2$—$^5/_2$
2639,05	4	47,15	51,84	$3p'\,^2D$—$3d'\,^2P$	$^3/_2$—$^1/_2$
2634,8	0,5	47,15	51,85	$3p'\,^2D^\circ$—$3d'\,^2P$	$^3/_2$—$^3/_2$
2630,93	1	47,14	51,85	$3p'\,^2D^\circ$—$3d'\,^2P$	$^5/_2$—$^3/_2$
2629,686	8	46,72	51,44	$3p'\,^2F^\circ$—$3d'\,^2G$	$^7/_2$—$^9/_2$
2625,000	7	46,72	51,44	$3p'\,^2F^\circ$—$3d'\,^2G$	$^5/_2$—$^7/_2$
2617,3	1	43,56	48,29	$3p\,^4P^\circ$—$3d\,^2P$	$^3/_2$—$^3/_2$
2613,083	5	43,30	48,04	$3p\,^4D^\circ$—$3d\,^4F$	$^7/_2$—$^7/_2$
2610,8	0,5	43,55	48,29	$3p\,^4P^\circ$—$3d\,^2P$	$^1/_2$—$^3/_2$
2606,045	5	43,27	48,02	$3p\,^4D^\circ$—$3d\,^4F$	$^5/_2$—$^5/_2$
2602,9	0	43,56	48,32	$3p\,^4P^\circ$—$3d\,^2P$	$^3/_2$—$^1/_2$
2600,551	4	43,25	48,01	$3p\,^4D^\circ$—$3d\,^4F$	$^3/_2$—$^3/_2$
2599,230	8	43,30	48,07	$3p\,^4D^\circ$—$3d\,^4F$	$^7/_2$—$^9/_2$
2596,5	1	43,55	48,32	$3p\,^4P^\circ$—$3d\,^2P$	$^1/_2$—$^1/_2$
2595,488	7	43,27	48,04	$3p\,^4D^\circ$—$3d\,^4F$	$^5/_2$—$^7/_2$
2593,195	6	43,25	48,02	$3p\,^4D^\circ$—$3d\,^4F$	$^3/_2$—$^5/_2$
2592,804	6	43,23	48,01	$3p\,^4D^\circ$—$3d\,^4F$	$^1/_2$—$^3/_2$
2592,65	1	43,58	48,36	$3p\,^4P^\circ$—$3d\,^4D$	$^5/_2$—$^5/_2$
2592,43	1	43,58	48,36	$3p\,^4P^\circ$—$3d\,^4D$	$^5/_2$—$^3/_2$
2583,760	7	43,58	48,38	$3p\,^4P^\circ$—$3d\,^4D$	$^5/_2$—$^7/_2$
2580,031	6	43,56	48,36	$3p\,^4P^\circ$—$3d\,^4D$	$^3/_2$—$^5/_2$
2579,846	3	43,56	48,36	$3p\,^4P^\circ$—$3d\,^4D$	$^3/_2$—$^3/_2$

λ, Å	I	E_H, eV	E_B, eV	Transition	J
2577,19	2	43,56	48,37	$3p\,^4P^\circ - 3d\,^4D$	$3/2 - 1/2$
2573,596	4	43,55	48,36	$3p\,^4P^\circ - 3d\,^4D$	$1/2 - 3/2$
2570,917	4	43,55	48,37	$3p\,^4P^\circ - 3d\,^4D$	$1/2 - 1/2$
2562,413	4	44,18	49,02	$3p\,^2D^\circ - 3d\,^2D$	$5/2 - 5/2$
2552,29	0	48,01	52,87	$3d\,^4F - 4p\,^4D^\circ$	$3/2 - 1/2$
2550,89	1	48,02	52,88	$3d\,^4F - 4p\,^4D^\circ$	$5/2 - 3/2$
2549,68	2	48,04	52,90	$3d\,^4F - 4p\,^4D^\circ$	$7/2 - 5/2$
2546,23	3	48,07	52,94	$3d\,^4F - 4p\,^4D^\circ$	$9/2 - 7/2$
2544,63	4	44,13	49,00	$3p\,^2D^\circ - 3d\,^2D$	$3/2 - 3/2$
2543,4	1	46,72	51,60	$3p'\,^2F^\circ - 3d'\,^2D$	$7/2 - 5/2$
2542,767	6	43,58	48,45	$3p\,^4P^\circ - 3d\,^4P$	$5/2 - 5/2$
2541,03	1	46,72	51,59	$3p'\,^2F^\circ - 3d'\,^2D$	$5/2 - 3/2$
2539,58	00	48,02	52,90	$3d\,^4F - 4p\,^4D^\circ$	$5/2 - 5/2$
2533,644	4	43,58	48,47	$3p\,^4P^\circ - 3d\,^4P$	$5/2 - 3/2$
2530,66	2	43,56	48,45	$3p\,^4P^\circ - 3d\,^4P$	$3/2 - 5/2$
2521,590	4	43,56	48,47	$3p\,^4P^\circ - 3d\,^4P$	$3/2 - 3/2$
2517,07	3	43,56	48,48	$3p\,^4P^\circ - 3d\,^4P$	$3/2 - 1/2$
2515,62	4	43,55	48,47	$3p\,^4P^\circ - 3d\,^4P$	$1/2 - 3/2$
2511,16	1	43,55	48,48	$3p\,^4P^\circ - 3d\,^4P$	$1/2 - 1/2$
2492,58	0	43,58	48,55	$3p\,^4P^\circ - 3d\,^2F$	$5/2 - 7/2$
2484,360	9	39,33	44,32	$3s\,^4P - 3p\,^4S^\circ$	$5/2 - 3/2$
2478,709	6	42,65	47,65	$3s'\,^2D - 3p'\,^2P^\circ$	$3/2 - 1/2$
2470,48	3	42,65	47,67	$3s'\,^2D - 3p'\,^2P^\circ$	$3/2 - 3/2$
2470,279	7	42,65	47,67	$3s'\,^2D - 3p'\,^2P^\circ$	$5/2 - 3/2$
2466,162	4	—	—	—	—
2464,834	8	39,29	44,32	$3s\,^4P - 3p\,^4S^\circ$	$3/2 - 3/2$
2455,81	0,5	43,25	48,29	$3p\,^4D^\circ - 3d\,^2P$	$3/2 - 3/2$
2452,070	7	39,29	44,32	$3s\,^4P - 3p\,^4S^\circ$	$1/2 - 3/2$
2451,56	4	—	—	—	—
2449,5	5	43,30	48,36	$3p\,^4D^\circ - 3d\,^4D$	$7/2 - 5/2$
2441,622	8	43,30	48,38	$3p\,^4D^\circ - 3d\,^4D$	$7/2 - 7/2$
2434,13	6	43,27	48,36	$3p\,^4D^\circ - 3d\,^4D$	$5/2 - 5/2$
2433,95	3	43,27	48,36	$3p\,^4D^\circ - 3d\,^4D$	$5/2 - 3/2$
2426,280	4	43,27	48,38	$3p\,^4D^\circ - 3d\,^4D$	$5/2 - 7/2$
2422,91	3	43,25	48,36	$3p\,^4D^\circ - 3d\,^4D$	$3/2 - 5/2$
2422,78	4	43,25	48,36	$3p\,^4D^\circ - 3d\,^4D$	$3/2 - 3/2$
2420,44	3	43,25	48,37	$3p\,^4D^\circ - 3d\,^4D$	$3/2 - 1/2$
2419,36	0,5	48,32	53,44	$3d\,^2P - 4p\,^2P^\circ$	$1/2 - 1/2$
2416,05	4	43,23	48,36	$3p\,^4D^\circ - 3d\,^4D$	$1/2 - 3/2$
2413,69	3	43,23	48,37	$3p\,^4D^\circ - 3d\,^4D$	$1/2 - 1/2$
2405,01	4	43,30	48,45	$3p\,^4D^\circ - 3d\,^4P$	$7/2 - 5/2$
2397,29	1	48,29	53,46	$3d\,^2P - 4p\,^2P^\circ$	$3/2 - 3/2$
2381,99	2	43,27	48,47	$3p\,^4D^\circ - 3d\,^4P$	$5/2 - 3/2$
2367,2	1	43,25	48,48	$3p\,^4D^\circ - 3d\,^4P$	$3/2 - 1/2$
2298,31	3				
2217,34	5	42,70	48,29	$3p\,^2S^\circ - 3d\,^2P$	$1/2 - 3/2$
2206,94	3	42,70	48,32	$3p\,^2S^\circ - 3d\,^2P$	$1/2 - 1/2$
658,337	12	0,00	18,83	$2p^3\,^4S^\circ - 2p^4\,^4P$	$3/2 - 5/2$
656,878	11	0,00	18,87	$2p^3\,^4S^\circ - 2p^4\,^4P$	$3/2 - 3/2$
656,125	10	0,00	18,90	$2p^3\,^4S^\circ - 2p^4\,^4P$	$3/2 - 1/2$
630,194	7	6,39	26,06	$2p^3\,^2P^\circ - 2p^4\,^2D$	$3/2 - 5/2$
630,131	6	6,39	26,06	$2p^3\,^2P^\circ - 2p^4\,^2D$	$3/2,\ 1/2 - 3/2$
567,794	6	4,23	26,06	$2p^3\,^2D^\circ - 2p^4\,^2D$	$3/2 - 5/2$
567,737	9	4,23	26,06	$2p^3\,^2D^\circ - 2p^4\,^2D$	$3/2 - 3/2$
567,676	10	4,23	26,06	$2p^3\,^2D^\circ - 2p^4\,^2D$	$5/2 - 5/2$
567,629	6	4,23	26,06	$2p^3\,^2D^\circ - 2p^4\,^2D$	$5/2 - 3/2$
523,661	4	26,06	49,74	$2p^4\,^2D - 2p^5\,^2P^\circ$	$5/2 - 3/2$
522,288	3	26,06	49,80	$2p^4\,^2D - 2p^5\,^2P^\circ$	$3/2 - 1/2$

λ, Å	I	E_H, eV	E_B, eV	Transition	J
508,384	10	6,39	30,78	$2p^3\,{}^2P^\circ - 2p^4\,{}^2S$	$^3/_2,\ ^1/_2 - {}^1/_2$
465,113	10	6,39	33,05	$2p^3\,{}^2P^\circ - 2p^4\,{}^2P$	$^3/_2,\ ^1/_2 - {}^3/_2$
464,284	9	6,39	33,09	$2p^3\,{}^2P^\circ - 2p^4\,{}^2P$	$^3/_2,\ ^1/_2 - {}^1/_2$
430,218	8	4,23	33,05	$2p^3\,{}^2D^\circ - 2p^4\,{}^2P$	$^3/_2 - {}^3/_2$
430,154	11	4,23	33,05	$2p^3\,{}^2D^\circ - 2p^4\,{}^3P$	$^5/_2 - {}^3/_2$
429,511	10	4,23	33,09	$2p^3\,{}^2D^\circ - 2p^4\,{}^3P$	$^3/_2 - {}^1/_2$
396,247	1	18,90	50,18	$2p^4\,{}^4P - 3s'''\,{}^4S^\circ$	$^1/_2 - {}^3/_2$
395,968	2	18,87	50,18	$2p^4\,{}^4P - 3s'''\,{}^4S^\circ$	$^3/_2 - {}^3/_2$
395,442	3	18,83	50,18	$2p^4\,{}^4P - 3s'''\,{}^4S^\circ$	$^5/_2 - {}^3/_2$
378,603	2	26,06	58,81	$2p^4\,{}^2D - 3s^{IV}\,{}^2D^\circ$	$^5/_2 - {}^5/_2$
378,563	1	26,06	58,82	$2p^4\,{}^2D - 3s^{IV}\,{}^2D^\circ$	$^3/_2 - {}^3/_2$
366,391	6	6,39	40,23	$2p^3\,{}^2P^\circ - 3s\,{}^2P$	$^3/_2,\ ^1/_2 - {}^1/_2$
365,874	7	6,39	40,28	$2p^3\,{}^2P^\circ - 3s\,{}^2P$	$^3/_2,\ ^1/_2 - {}^3/_2$
344,388	6	4,23	40,23	$2p^3\,{}^2D^\circ - 3s\,{}^2P$	$^3/_2 - {}^1/_2$
343,931	4	4,23	40,28	$2p^3\,{}^3D^\circ - 3s\,{}^2P$	$^3/_2 - {}^3/_2$
343,892	7	4,23	40,28	$2p^3\,{}^3D^\circ - 3s\,{}^2P$	$^5/_2 - {}^3/_2$
341,924	7	6,39	42,65	$2p^3\,{}^2P^\circ - 3s'\,{}^2D$	$^3/_2,\ ^1/_2 - {}^5/_2,\ ^3/_2$
322,685	7	4,23	42,65	$2p^3\,{}^2D^\circ - 3s'\,{}^2D$	$^3/_2 - {}^5/_2,\ ^3/_2$
322,650	8	4,23	42,65	$2p^3\,{}^2D^\circ - 3s'\,{}^2D$	$^5/_2 - {}^5/_2,\ ^3/_2$
316,998	2	18,90	58,01	$2p^4\,{}^4P - 3d'''\,{}^4D^\circ$	$^1/_2 - {}^3/_2,\ ^1/_2$
316,823	3	18,87	58,01	$2p^4\,{}^4P - 3d'''\,{}^4D^\circ$	$^3/_2 - {}^5/_2$
316,488	4	18,83	58,01	$2p^4\,{}^4P - 3d'''\,{}^4D^\circ$	$^5/_2 - {}^7/_2$
315,748	6	0,00	39,26	$2p^3\,{}^4S^\circ - 3s\,{}^4P$	$^3/_2 - {}^1/_2$
315,539	7	0,00	39,29	$2p^3\,{}^4S^\circ - 3s\,{}^4P$	$^3/_2 - {}^3/_2$
315,221	8	0,00	39,33	$2p^3\,{}^4S^\circ - 3s\,{}^4P$	$^3/_2 - {}^5/_2$
311,415	4	6,39	46,20	$2p^3\,{}^2P^\circ - 3s''\,{}^2S$	$^3/_2,\ ^1/_2 - {}^1/_2$
295,886	6	6,39	48,29	$2p^3\,{}^2P^\circ - 3d\,{}^2P$	$^3/_2,\ ^1/_2 - {}^3/_2$
295,710	5	6,39	48,32	$2p^3\,{}^2P^\circ - 3d\,{}^2P$	$^3/_2,\ ^1/_2 - {}^1/_2$
295,405	1	6,39	48,36	$2p^3\,{}^2P^\circ - 3d\,{}^4D$	$^3/_2,\ ^1/_2 - {}^5/_2,\ ^3/_2$
295,365	2	6,39	48,37	$2p^3\,{}^2P^\circ - 3d\,{}^4D$	$^3/_2,\ ^1/_2 - {}^1/_2$
290,947	5	6,39	49,00	$2p^3\,{}^2P^\circ - 3d\,{}^2D$	$^3/_2,\ ^1/_2 - {}^3/_2$
290,848	6	6,39	49,02	$2p^3\,{}^2P^\circ - 3d\,{}^2D$	$^3/_2 - {}^5/_2$
281,350	4	4,23	48,29	$2p^3\,{}^2D^\circ - 3d\,{}^2P$	$^5/_2 - {}^3/_2$
281,207	3	4,23	48,32	$2p^3\,{}^2D^\circ - 3d\,{}^2P$	$^3/_2 - {}^1/_2$
280,905	1	4,23	48,36	$2p^3\,{}^2D^\circ - 3d\,{}^4D$	$^5/_2,\ ^3/_2 - {}^5/_2,\ ^3/_2,\ ^1/_2$
280,010	6	4,23	48,51	$2p^3\,{}^2D^\circ - 3d\,{}^2F$	$^3/_2 - {}^5/_2$
279,692	7	4,23	48,55	$2p^3\,{}^2D^\circ - 3d\,{}^2F$	$^5/_2 - {}^7/_2$
276,895	4	4,23	49,00	$2p^3\,{}^2D^\circ - 3d\,{}^2D$	$^3/_2 - {}^3/_2$
276,786	5	4,23	49,02	$2p^3\,{}^2D^\circ - 3d\,{}^2D$	$^5/_2 - {}^5/_2$
274,260	6	6,39	51,60	$2p^3\,{}^2P^\circ - 3d'\,{}^2D$	$^3/_2,\ ^1/_2 - {}^5/_2,\ ^3/_2$
273,207	2	6,39	51,77	$2p^3\,{}^2P^\circ - 4s\,{}^2P$	$^3/_2,\ ^1/_2 - {}^1/_2$
272,915	3	6,39	51,82	$2p^3\,{}^2P^\circ - 4s\,{}^2P$	$^3/_2,\ ^1/_2 - {}^3/_2$
272,758	3	6,39	51,84	$2p^3\,{}^2P^\circ - 3d'\,{}^2P$	$^3/_2,\ ^1/_2 - {}^1/_2$
272,710	4	6,39	51,85	$2p^3\,{}^2P^\circ - 3d'\,{}^2P$	$^3/_2,\ ^1/_2 - {}^3/_2$
270,675	4	6,39	52,19	$2p^3\,{}^2P^\circ - 3d'\,{}^2S$	$^3/_2,\ ^1/_2 - {}^1/_2$
263,807	8	4,23	51,23	$2p^3\,{}^2D^\circ - 3d'\,{}^2F$	$^5/_2,\ ^3/_2 - {}^5/_2,\ ^7/_2$
261,751	6	4,23	51,59	$2p^3\,{}^2D^\circ - 3d'\,{}^2D$	$^3/_2 - {}^3/_2$
261,716	7	4,23	51,60	$2p^3\,{}^2D^\circ - 3d'\,{}^2D$	$^5/_2 - {}^5/_2$
260,782	1	4,23	51,77	$2p^3\,{}^2D^\circ - 4s\,{}^2P$	$^3/_2 - {}^1/_2$
260,498	3	4,23	51,82	$2p^3\,{}^2D^\circ - 4s\,{}^2P$	$^5/_2 - {}^3/_2$
260,375	3	4,23	51,84	$2p^3\,{}^2D^\circ - 3d'\,{}^2P$	$^3/_2 - {}^1/_2$
260,313	4	4,23	51,85	$2p^3\,{}^2D^\circ - 3d'\,{}^2P$	$^5/_2 - {}^3/_2$
256,890	2	6,39	54,65	$2p^3\,{}^2P^\circ - 4s'\,{}^2D$	$^3/_2,\ ^1/_2 - {}^5/_2,\ ^3/_2$
256,673	1	6,39	54,69	$2p^3\,{}^2P^\circ - 4d\,{}^2P$	$^3/_2,\ ^1/_2 - {}^3/_2$
256,525	1	6,39	54,72	$2p^3\,{}^2P^\circ - 4d\,{}^2P$	$^3/_2,\ ^1/_2 - {}^1/_2$
256,360	5	0,00	48,36	$2p^3\,{}^4S^\circ - 3d\,{}^4D$	$^3/_2 - {}^5/_2,\ ^3/_2$
255,865	7	0,00	48,45	$2p^3\,{}^4S^\circ - 3d\,{}^4P$	$^3/_2 - {}^5/_2$

λ, Å	I	E_H, eV	E_B, eV	Transition	J
255,772	6	0,00	48,47	$2p^3\,^4S^\circ - 3d\,^4P$	$3/2 - 3/2$
255,725	5	0,00	48,48	$2p^3\,^4S^\circ - 3d\,^4P$	$3/2 - 1/2$
255,624	1	6,39	54,89	$2p^3\,^2P^\circ - 3d''\,^2D$	$3/2,\ 1/2 - 5/2,\ 3/2$
254,193	3	6,39	55,16	$2p^3\,^2P^\circ - 4d\,^2D$	$3/2,\ 1/2 - 3/2$
254,162	4	6,39	55,17	$2p^3\,^2P^\circ - 4d\,^2D$	$3/2 - 5/2$
245,860	1	4,23	54,65	$2p^3\,^2D^\circ - 4s'\,^2D$	$5/2,\ 3/2 - 5/2,\ 3/2$
245,002	3	4,23	54,83	$2p^3\,^2D^\circ - 4d\,^2F$	$3/2 - 5/2$
244,768	4	4,23	54,88	$2p^3\,^2D^\circ - 4d\,^2F$	$5/2 - 7/2$
244,698	2	4,23	54,89	$2p^3\,^2D^\circ - 3d''\,^2D$	$5/2 - 5/2$
243,364	1	4,23	55,17	$2p^3\,^2D^\circ - 4d\,^2D$	$5/2,\ 3/2 - 5/2,\ 3/2$
240,730	2	6,39	57,92	$2p^3\,^2P^\circ - 4d'\,^2D$	$3/2,\ 1/2 - 5/2,\ 3/2$
240,550	1	0,00	51,54	$2p^3\,^4S^\circ - 4s\,^4P$	$3/2 - 5/2$
240,233	1	6,39	58,00	$2p^3\,^2P^\circ - 4d'\,^2P$	$3/2,\ 1/2 - 3/2,\ 1/2$
231,100	3	4,23	57,87	$2p^3\,^2D^\circ - 4d'\,^2F$	$5/2,\ 3/2 - 7/2,\ 5/2$
231,015	2	4,23	57,92	$2p^3\,^2D^\circ - 4d'\,^2D$	$5/2,\ 3/2 - 5/2,\ 3/2$
230,117	5	0,00	53,88	$2p^3\,^4S^\circ - 3p'''\,^4P$	$3/2 - 5/2,\ 3/2,\ 1/2$
226,166	4	0,00	54,82	$2p^3\,^4S^\circ - 4d\,^4P$	$3/2 - 5/2$
226,091	3	0,00	54,84	$2p^3\,^4S^\circ - 4d\,^4P$	$3/2 - 3/2$
226,051	2	0,00	54,84	$2p^3\,^4S^\circ - 4d\,^4P$	$3/2 - 1/2$
214,865	1	0,00	57,70	$2p^3\,^4S^\circ - 5d\,^4P$	$3/2 - 5/2$
214,804	1	0,00	57,72	$2p^3\,^4S^\circ - 5d\,^4P$	$3/2 - 3/2,\ 1/2$

F IV, **ground state** $1s^2\,2s^2\,2p^2\,^3P_0$

Ionization potential 703020 cm^{-1}; 87,157 eV

λ, Å	I	E_H, eV	E_B, eV	Transition	J
3176,08	2	64,45	68,36	$3s\,^3P - 3p\,^3D^\circ$	2—3
3167,74	1	64,41	68,32	$3s\,^3P - 3p\,^3D^\circ$	1—2
2882,99	0	51,72	56,02	$3s\,^3P^\circ - 3p\,^3D$	2—1
2861,40	2	51,72	56,05	$3s\,^3P^\circ - 3p\,^3D$	2—2
2841,72	2	51,65	56,02	$3s\,^3P^\circ - 3p\,^3D$	1—1
2826,13	5	51,72	56,10	$3s\,^3P^\circ - 3p\,^3D$	2—3
2823,80	3	51,63	56,02	$3s\,^3P^\circ - 3p\,^3D$	0—1
2820,74	4	51,65	56,05	$3s\,^3P^\circ - 3p\,^3D$	1—2
2807,46	0	62,36	66,77	$3s\,^5P - 3p\,^5D^\circ$	2—1
2806,00	1	62,40	66,81	$3s\,^5P - 3p\,^5D^\circ$	3—3
2796,80	2	62,36	66,79	$3s\,^5P - 3p\,^5D^\circ$	2—2
2794,26	3	58,23	62,66	$3p\,^1D - 3d\,^1F^\circ$	2—3
2788,56	1	62,33	66,77	$3s\,^5P - 3p\,^5D^\circ$	1—1
2785,96	3	62,40	66,84	$3s\,^5P - 3p\,^5D^\circ$	3—4
2781,18	2	62,36	66,81	$3s\,^5P - 3p\,^5D^\circ$	2—3
2778,03	1	62,33	66,79	$3s\,^5P - 3p\,^5D^\circ$	1—2
2764,60	0	64,45	68,94	$3s\,^3P - 3p\,^3P^\circ$	2—1
2744,51	1	64,45	68,97	$3s\,^3P - 3p\,^3P^\circ$	2—2
2723,25	0	64,39	68,94	$3s\,^3P - 3p\,^3P^\circ$	0—1
2718,34	0	64,41	68,97	$3s\,^3P - 3p\,^3P^\circ$	1—2
2713,54	0	57,11	61,68	$3p\,^3P - 3d\,^3D^\circ$	2—1
2706,66	1	57,11	61,69	$3p\,^3P - 3d\,^3D^\circ$	2—2
2695,45	3	57,11	61,71	$3p\,^3P - 3d\,^3D^\circ$	2—3
2695,03	1	57,08	61,68	$3p\,^3P - 3d\,^3D^\circ$	1—1
2688,11	2	57,08	61,69	$3p\,^3P - 3d\,^3D^\circ$	1—2
2682,60	1	57,06	61,68	$3p\,^3P - 3d\,^3D^\circ$	0—1
2648,18	0	66,84	71,52	$3p\,^5D^\circ - 3d\,^5F$	4—4
2640,63	1	66,81	71,51	$3p\,^5D^\circ - 3d\,^5F$	3—3

λ, Å	I	E_H, eV	E_B, eV	Transition	J
2635,37	3	66,84	71,55	$3p\,^5D^\circ - 3d\,^5F$	4—5
2634,49	0	66,79	71,49	$3p\,^5D^\circ - 3d\,^5F$	2—2
2630,28	3	66,81	71,52	$3p\,^5D^\circ - 3d\,^5F$	3—4
2626,74	2	66,79	71,51	$3p\,^5D^\circ - 3d\,^5F$	2—3
2625,51	0	{66,76	71,48	$3p\,^5D^\circ - 3d\,^5F$	0—1
		66,77	71,49	$3p\,^5D^\circ - 3d\,^5F$	1—2
2557,93	2	67,31	72,15	$3p\,^5P^\circ - 3d\,^5D$	3—4
2555,59	00	67,28	72,13	$3p\,^5P^\circ - 3d\,^5D$	2—1
2554,47	1	67,28	72,13	$3p\,^5P^\circ - 3d\,^5D$	2—2
2551,61	1	67,28	72,14	$3p\,^5P^\circ - 3d\,^5D$	2—3
2548,45	00	67,27	72,13	$3p\,^5P^\circ - 3d\,^5D$	1—0
2548,10	0	67,27	72,13	$3p\,^5P^\circ - 3d\,^5D$	1—1
2546,98	0	67,27	72,13	$3p\,^5P^\circ - 3d\,^5D$	1—2
2536,62	1	62,40	67,28	$3s\,^5P - 3p\,^5P^\circ$	3—2
		{62,40	67,31	$3s\,^5P - 3p\,^5P^\circ$	3—3
2523,67	3	62,36	67,27	$3s\,^5P - 3p\,^5P^\circ$	2—1
2516,27	00	62,36	67,28	$3s\,^5P - 3p\,^5P^\circ$	2—2
2515,57	2	51,72	56,64	$3s\,^3P^\circ - 3p\,^3S$	2—1
2515,01	2	57,11	62,04	$3p\,^3P - 3d\,^3P^\circ$	2—2
2508,31	0	62,33	67,27	$3s\,^5P - 3p\,^5P^\circ$	1—1
2503,57	1	62,36	67,31	$3s\,^5P - 3p\,^5P^\circ$	2—3
2501,66	0	57,11	62,06	$3p\,^3P - 3d\,^3P^\circ$	2—1
2501,10	1	62,33	67,28	$3s\,^5P - 3p\,^5P^\circ$	1—2
2498,95	0	57,08	62,04	$3p\,^3P - 3d\,^3P^\circ$	1—2
2485,79	0	57,08	62,06	$3p\,^3P - 3d\,^3P^\circ$	1—1
2484,06	1	51,65	56,64	$3s\,^3P^\circ - 3p\,^3S$	1—1
2479,77	0	56,05	61,05	$3p\,^3D - 3d\,^3F^\circ$	2—2
2478,05	1	56,10	61,10	$3p\,^3D - 3d\,^3F^\circ$	3—3
2475,31	0	57,06	62,06	$3p\,^3P - 3d\,^3P^\circ$	0—1
2463,79	2	56,02	61,05	$3p\,^3D - 3d\,^3F^\circ$	1—2
2456,92	5	56,10	61,15	$3p\,^3D - 3d\,^3F^\circ$	3—4
2451,58	4	56,05	61,10	$3p\,^3D - 3d\,^3F^\circ$	2—3
2435,62	2	56,02	61,10	$3p\,^3D - 3d\,^1D^\circ$	1—2
2311,83	3	51,72	57,08	$3s\,^3P^\circ - 3p\,^3P$	2—1
2298,29	5	51,72	57,11	$3s\,^3P^\circ - 3p\,^3P$	2—2
2297,82	2	56,64	62,04	$3p\,^3S - 3d\,^3P^\circ$	1—2
2294,17	2	51,65	57,06	$3s\,^3P^\circ - 3p\,^3P$	1—0
2286,69	1	56,64	62,06	$3p\,^3S - 3d\,^3P^\circ$	1—1
2285,22	2	51,65	57,08	$3s\,^3P^\circ - 3p\,^3P$	1—1
2280,72	0	56,64	62,08	$3p\,^3S - 3d\,^3P^\circ$	1—0
2273,65	2	51,63	57,08	$3s\,^3P^\circ - 3p\,^3P$	0—1
2271,97	3	51,65	57,11	$3s\,^3P^\circ - 3p\,^3P$	1—2
2211,07	1	56,10	61,71	$3p\,^3D - 3d\,^3D^\circ$	3—3
2197,36	0	56,05	61,69	$3p\,^3D - 3d\,^3D^\circ$	2—2
2171,44	4	52,52	58,23	$3s\,^1P^\circ - 3p\,^1D$	1—2
679,217	16	0,08	18,33	$2p^2\,^3P - 2p^3\,^3D^\circ$	2—3
679,003	13	0,08	18,33	$2p^2\,^3P - 2p^3\,^3D^\circ$	2—2
677,224	15	0,03	18,33	$2p^2\,^3P - 2p^3\,^3D^\circ$	1—2
677,154	13	0,03	18,34	$2p^2\,^3P - 2p^3\,^3D^\circ$	1—1
676,130	14	0,00	18,34	$2p^2\,^3P - 2p^3\,^3D^\circ$	0—1
577,737	4	21,73	43,18	$2p^3\,^3P^\circ - 2p^4\,^3P$	2, 1—2
576,349	2	21,73	43,24	$2p^3\,^3P^\circ - 2p^4\,^3P$	0—1
576,266	3	21,73	43,24	$2p^3\,^3P^\circ - 2p^4\,^3P$	2, 1—1
575,633	2	21,73	43,26	$2p^3\,^3P^\circ - 2p^4\,^3P$	1—0
572,637	16	0,08	21,73	$2p^2\,^3P - 2p^3\,^3P^\circ$	2—2, 1
571,384	15	0,03	21,73	$2p^2\,^3P - 2p^3\,^3P^\circ$	1—2, 1
571,302	14	0,03	21,73	$2p^3\,^3P - 2p^3\,^3P^\circ$	1—0
570,636	14	0,00	21,73	$2p^2\,^3P - 2p^3\,^3P^\circ$	0—1
498,911	4	18,33	43,18	$2p^3\,^3D^\circ - 2p^4\,^3P$	2—2

λ, Å	I	E_H, eV	E_B, eV	Transition	J
498,790	7	18,33	43,18	$2p^3\ ^3D^\circ-2p^4\ ^3P$	3—2
497,842	4	18,34	43,24	$2p^3\ ^3D^\circ-2p^4\ ^3P$	1—1
497,802	6	18,33	43,24	$2p^3\ ^3D^\circ-2p^4\ ^3P$	2—1
497,363	5	18,34	43,26	$2p^3\ ^3D^\circ-2p^4\ ^3P$	1—0
490,997	16	3,13	28,38	$2p^2\ ^1D-2p^3\ ^1D^\circ$	2—2
490,566	13	6,64	31,91	$2p^2\ ^1S-2p^3\ ^1P^\circ$	0—1
430,758	15	3,13	31,91	$2p^2\ ^1D-2p^3\ ^1P^\circ$	2—1
420,727	16	0,08	29,54	$2p^2\ ^3P-2p^3\ ^3S^\circ$	2—1
420,041	15	0,03	29,54	$2p^2\ ^3P-2p^2\ ^3S^\circ$	1—1
419,644	14	0,00	29,54	$2p^2\ ^3P-2p^3\ ^3S^\circ$	0—1
360,635	1	21,73	56,10	$2p^3\ ^3P^\circ-3p\ ^3D$	2—3
355,045	2	21,73	56,64	$2p^3\ ^3P^\circ-3p^3\ S$	2, 1—1
320,192	1	18,34	57,06	$2p^3\ ^3D^\circ-3p\ ^3P$	1—0
320,004	2	18,33	57,08	$2p^3\ ^3D^\circ-3p\ ^3P$	2, 1—1
319,695	3	18,33	57,11	$2p^3\ ^3D^\circ-3p\ ^3P$	3—2
290,608	2	21,73	64,39	$2p^3\ ^3P^\circ-3s\ ^3P$	1—0
290,461	2	21,73	64,41	$2p^3\ ^3P^\circ-3s\ ^3P$	0—1
290,440	3	21,73	64,41	$2p^3\ ^3P^\circ-3s\ ^3P$	2, 1—1
290,147	4	21,73	64,45	$2p^3\ ^3P^\circ-3s\ ^3P$	2, 1—2
288,267	1	29,54	72,55	$2p\ ^3S^\circ-3d\ ^3P$	1—2
279,834	3	28,38	72,68	$2p^3\ ^1D^\circ-3s'\ ^1D$	2—2
270,225	6	6,64	52,52	$2p^2\ ^1S-3s\ ^1P^\circ$	0—1
269,225	2	18,34	64,39	$2p^3\ ^3D^\circ-3s\ ^3P$	1—0
269,076	3	18,33	64,41	$2p^3\ ^3D^\circ-3s\ ^3P$	2, 1—1
268,817	1	18,33	64,45	$2p^3\ ^3D^\circ-3s\ ^3P$	2—2
268,785	4	18,33	64,45	$2p^3\ ^3D^\circ-3s\ ^3P$	3—2
254,595	1	21,73	70,42	$2p^3\ ^3P^\circ-3s'\ ^3D$	2, 1—2
254,491	2	21,73	70,44	$2p^3\ ^3P^\circ-3s'\ ^3D$	2—3
251,026	10	3,13	52,52	$2p^3\ ^1D-3s\ ^1P^\circ$	2—1
249,228	1	31,91	81,65	$2p^3\ ^1P^\circ-3d'\ ^1P$	1—1
243,922	4	21,73	72,55	$2p^3\ ^3P^\circ-3d\ ^3P$	2, 1—2
243,796	3	21,73	72,58	$2p^3\ ^3P^\circ-3d\ ^3P$	2, 1, 0—1
243,736	2	21,73	72,59	$2p^3\ ^3P^\circ-3d\ ^3P$	1—0
240,371	7	0,08	51,65	$2p^2\ ^3P-3s^3\ P^\circ$	2—1
240,275	7	0,03	51,63	$2p^2\ ^3P-3s\ ^3P^\circ$	1—0
240,146	7	0,03	51,65	$2p^2\ ^3P-3s\ ^3P^\circ$	1—1
240,079	9	0,08	51,72	$2p^2\ ^3P-3s\ ^3P^\circ$	2—2
240,017	7	0,00	51,65	$2p^2\ ^3P-3s\ ^3P^\circ$	0—1
239,856	7	0,03	51,72	$2p^2\ ^3P-3s\ ^3P^\circ$	1—2
238,099	1	18,33	70,41	$2p^3\ ^3D^\circ-3s'\ ^3D$	2, 1—1
238,042	2	21,73	73,81	$2p^3\ ^3P^\circ-3d\ ^3D$	1, 0—1
238,012	3	21,73	73,82	$2p^3\ ^3P^\circ-3d\ ^3D$	2, 1—2
		18,33	70,42	$2p^3\ ^3D^\circ-3s'\ ^3D$	2—2
237,955	4	21,73	73,88	$2p^3\ ^3P^\circ-3d\ ^3D$	2—3
237,913	3	18,33	70,44	$2p^3\ ^3D^\circ-3s'\ ^3D$	3—3
233,526	4	9,24	62,33	$2p^3\ ^5S^\circ-3s\ ^5P$	2—1
233,393	5	9,24	62,36	$2p^3\ ^5S^\circ-3s\ ^5P$	2—2
233,297	2	28,38	81,52	$2p^3\ ^1D^\circ-3d'\ ^1F$	2—3
233,222	6	9,24	62,40	$2p^3\ ^5S^\circ-3s\ ^5P$	2—3
233,159	2	28,38	81,55	$2p^3\ ^1D^\circ-3d'\ ^1D$	2—2
229,261	2	—	—	—	—
228,645	1	18,33	72,55	$2p^3\ ^3D^\circ-3d\ ^3P$	3—2
227,211	4	18,34	72,90	$2p^3\ ^3D^\circ-3d\ ^3F$	1—2
227,101	5	18,33	72,93	$2p^3\ ^3D^\circ-3d\ ^3F$	2—3
227,079	3	18,33	72,93	$2p^3\ ^3D^\circ-3d\ ^3F$	3—3
226,944	6	18,33	72,96	$2p^3\ ^3D^\circ-3d\ ^3F$	3—4
223,497	1	18,34	73,81	$2p^3\ ^3D^\circ-3d\ ^3D$	2, 1—1
223,456	2	18,33	73,82	$2p^3\ ^3D^\circ-3d\ ^3D$	2—2
223,394	3	18,33	73,83	$2p^3\ ^3D^\circ-3d\ ^3D$	3—3

λ, Å	I	E_H, eV	E_B, eV	Transition	J
220,765	7	6,64	62,80	$2p^2\,{}^1S$—$3d\,{}^1P^\circ$	0—1
214,062	7	3,13	61,05	$2p^2\,{}^1D$—$3d\,{}^3F^\circ$	2—2
213,848	7	3,13	61,10	$2p^2\,{}^1D$—$3d\,{}^1D^\circ$	2—2
210,547	1	21,73	80,61	$2p^3\,{}^3P^\circ$—$3d\,{}^3D$	1, 0—2, 1
210,480	1	21,73	80,63	$2p^3\,{}^3P^\circ$—$3d'\,{}^3D$	2—2
208,549	2	21,73	81,17	$2p^3\,{}^3P^\circ$—$3d'\,{}^3S$	2, 1, 0—1
208,254	9	3,13	62,66	$2p^2\,{}^1D^3$—$3d\,{}^1F^\circ$	2—3
203,152	1	0,08	61,10	$2p^2\,{}^3P$—$3d\,{}^3F^\circ$	2—3
201,465	4	18,33	79,87	$2p^3\,{}^3D^\circ$—$3d'\,{}^3F$	3, 2, 1—4, 3, 2
201,222	6	0,08	61,69	$2p^2\,{}^3P$—$3d\,{}^3D^\circ$	2—2
201,160	8	0,08	61,71	$2p^2\,{}^3P$—$3d\,{}^3D^\circ$	2—3
201,101	6	0,03	61,68	$2p^2\,{}^3P$—$3d\,{}^3D^\circ$	1—1
201,063	7	0,03	61,69	$2p^2\,{}^3P$—$3d\,{}^3D^\circ$	1—2
201,011	6	0,00	61,68	$2p^2\,{}^3P$—$3d\,{}^3D^\circ$	0—1
200,089	7	0,08	62,04	$2p^2\,{}^3P$—$3d\,{}^3P^\circ$	2—2
200,001	5	0,08	62,06	$2p^2\,{}^3P$—$3d\,{}^3P^\circ$	2—1
199,934	5	0,03	62,04	$2p^2\,{}^3P$—$3d\,{}^3P^\circ$	1—2
199,849	5	0,03	62,06	$2p^2\,{}^3P$—$3d\,{}^3P^\circ$	1—1
199,804	5	0,03	62,08	$2p^2\,{}^3P$—$3d\,{}^3P^\circ$	1—0
199,761	5	0,00	62,06	$2p^2\,{}^3P$—$3d\,{}^3P^\circ$	0—1
199,607	1	18,33	80,44	$2p^3\,{}^3D^\circ$—$3d'\,{}^3P$	3—2
199,086	3	18,31	80,61	$2p^3\,{}^3D^\circ$—$3d'\,{}^3D$	2, 1—2, 1
199,004	3	18,33	80,63	$2p^3\,{}^3D^\circ$—$3d'\,{}^3D$	3—3
197,108	2	9,24	72,14	$2p^3\,{}^5S^\circ$—$3d\,{}^5D$	2—3, 2
196,968	1	6,64	69,58	$2p^2\,{}^1S$—$4s\,{}^1P^\circ$	0—1
196,448	6	9,24	72,35	$2p^3\,{}^5S^\circ$—$3d\,{}^5P$	2—3
196,390	5	9,24	72,37	$2p^3\,{}^5S^\circ$—$3d\,{}^5P$	2—2
196,351	4	9,24	72,38	$2p^3\,{}^5S^\circ$—$3d\,{}^5P$	2—1
188,758	1	18,33	84,04	$2p^3\,{}^3D^\circ$—$4d\,{}^3F$	2—3
188,656	2	18,33	84,00	$2p^3\,{}^3D^\circ$—$4d\,{}^3F$	3—4
187,916	1	18,33	84,32	$2p^3\,{}^3D^\circ$—$4d\,{}^3D$	3—3
187,240	3	0,08	66,29	$2p^2\,{}^3P$—$3p\,{}^3S^\circ$	2—1
187,105	2	0,03	66,29	$2p^2\,{}^3P$—$3p\,{}^3S^\circ$	1—1
186,558	1	3,13	69,58	$2p^2\,{}^1D$—$4s\,{}^1P^\circ$	2—1
185,484	3	6,64	73,48	$2p^2\,{}^1S$—$4d\,{}^1P^\circ$	0—1
181,655	2	0,08	68,32	$2p^2\,{}^3P$—$3p\,{}^3D^\circ$	2—2
181,571	4	0,08	68,36	$2p^2\,{}^3P$—$3p\,{}^3D^\circ$	2—3
181,521	4	{0,03	68,32	$2p^2\,{}^3P$—$3p\,{}^3D^\circ$	1—2
		0,00	68,30	$2p^2\,{}^3P$—$3p\,{}^3D^\circ$	0—1
179,943	2	0,08	68,97	$2p^2\,{}^3P$—$3p\,{}^3P^\circ$	2—2
179,827	1	{0,00	68,94	$2p^2\,{}^3P$—$3p\,{}^3P^\circ$	0—1
		0,03	68,97	$2p^2\,{}^3P$—$3p\,{}^3P^\circ$	1—2
178,805	1	0,08	69,41	$2p^2\,{}^3P$—$4s\,{}^3P^\circ$	2—1
178,724	1	0,03	69,40	$2p^2\,{}^3P$—$4s\,{}^3P^\circ$	1—0
178,670	3	{0,08	69,46	$2p^2\,{}^3P$—$4s\,{}^3P^\circ$	2—2
		0,03	69,41	$2p^2\,{}^3P$—$4s\,{}^3P^\circ$	1—1
178,540	1	0,03	69,46	$2p^2\,{}^3P$—$4s\,{}^3P^\circ$	1—2
178,126	1	3,13	72,73	$2p^2\,{}^3D$—$4d\,{}^3F^\circ$	2—2
177,971	2	3,13	72,79	$2p^2\,{}^1D$—$4d\,{}^1D^\circ$	2—2
176,367	4	3,13	73,42	$2p^2\,{}^1D$—$4d\,{}^1F^\circ$	2—3
171,066	3	3,13	75,60	$2p^2\,{}^1D$—$3p'\,{}^1F^\circ$	2—3
170,187	2	3,13	75,98	$2p^2\,{}^1D$—$3p'\,{}^1D^\circ$	2—2
169,839	3	0,08	73,07	$2p^2\,{}^3P$—$4d\,{}^3D^\circ$	2—3
169,790	3	0,03	73,06	$2p^2\,{}^3P$—$4d\,{}^3D^\circ$	1—2
169,748	2	0,00	73,04	$2p^2\,{}^3P$—$4d\,{}^3D^\circ$	0—1
169,661	2	0,08	73,15	$2p^2\,{}^3P$—$4d\,{}^3P^\circ$	2—2
169,610	1	0,08	73,17	$2p^2\,{}^3P$—$4d\,{}^3P^\circ$	2—1
169,502	2	0,03	73,17	$2p^2\,{}^3P$—$4d\,{}^3P^\circ$	1—1

λ, Å	I	E_H, eV	E_B, eV	Transition	J
169,481	1	0,03	73,18	$2p^2\ ^3P—4d\ ^3P^\circ$	1—0
169,166	2	18,33	91,62	$2p^3\ ^3D^\circ—4d'\ ^3F$	3, 2, 1—4, 3, 2
168,450	2	3,13	76,73	$2p^2\ ^1D—3p'\ ^1P^\circ$	2—1
166,499	2	9,24	83,70	$2p^2\ ^5S^\circ—4d\ ^5P$	2—3
166,444	2	9,24	83,72	$2p^2\ ^5S^\circ—4d\ ^5P$	2—2, 1
164,612	2	3,13	78,44	$2p^2\ ^1D—5d\ ^1F^\circ$	2—3
158,925	1	3,13	81,14	$2p^2\ ^1D—6d\ ^1F^\circ$	2—3
158,601	1	0,08	78,25	$2p^2\ ^3P—5d\ ^3D^\circ$	2—3
151,005	1	0,08	82,18	$2p^2\ ^3P—4p\ ^3D^\circ$	2—3
150,977	1	—	—	—	—

F V, ground state $1s^2\ 2s^2\ 2p\ ^2P^0_{1/2}$
Ionization potential 921 450 cm^{-1}; 114,237 eV

λ, Å	I	E_H, eV	E_B, eV	Transition	J
2736,91	0	77,10	81,63	$3s\ ^4P^\circ—3p\ ^4D$	$^5/_2—^5/_2$
2721,06	0	77,04	81,59	$3s\ ^4P^\circ—3p\ ^4D$	$^3/_2—^3/_2$
2712,88	0	77,01	81,58	$3s\ ^4P^\circ—3p\ ^4D$	$^1/_2—^1/_2$
2707,17	2	77,10	81,68	$3s\ ^4P^\circ—3p\ ^4D$	$^5/_2—^7/_2$
2703,96	1	70,11	74,70	$3p\ ^2P^\circ—3d\ ^2D$	$^3/_2—^5/_2$
2702,30	1	{77,04 77,01	81,63 81,59	$3s\ ^4P^\circ—3p\ ^4D$ $3s\ ^4P^\circ—3p\ ^4D$	$^3/_2—^5/_2$ $^1/_2—^3/_2$
2693,98	1	70,09	74,69	$3p\ ^2P^\circ—3d\ ^2D$	$^1/_2—^3/_2$
2461,33	1	65,06	70,09	$3s\ ^2S—3p\ ^2P^\circ$	$^1/_2—^1/_2$
2450,63	2	65,06	70,11	$3s\ ^2S—3p\ ^2P^\circ$	$^1/_2—^3/_2$
2252,72	2	77,10	82,60	$3s\ ^4P^\circ—3p\ ^4S$	$^5/_2—^3/_2$
2229,18	1	77,04	82,60	$3s\ ^4P^\circ—3p\ ^4S$	$^3/_2—^3/_2$
1088,41	0	26,70	38,09	$2p^2\ ^2P—2p^3\ ^2D^\circ$	$^3/_2—^5/_2$
757,08	4	26,70	43,07	$2p^2\ ^2P—2p^3\ ^2P^\circ$	$^3/_2—^3/_2$
657,335	4	0,09	18,95	$2p\ ^2P^\circ—2p^2\ ^2D$	$^3/_2—^5/_2$
657,220	1	0,09	18,96	$2p\ ^2P^\circ—2p^2\ ^2D$	$^3/_2—^3/_2$
654,034	3	0,00	18,95	$2p\ ^2P^\circ—2p^2\ ^2D$	$^1/_2—^3/_2$
647,879	1	18,95	38,09	$2p^2\ ^2D—2p^3\ ^2D^\circ$	$^5/_2—^5/_2$
526,298	3	10,74	34,30	$2p^2\ ^4P—2p^3\ ^4S^\circ$	$^5/_2—^3/_2$
525,292	3	10,70	34,30	$2p^2\ ^4P—2p^3\ ^4S^\circ$	$^3/_2—^3/_2$
524,597	2	10,67	34,30	$2p^2\ ^4P—2p^3\ ^4S^\circ$	$^1/_2—^3/_2$
514,087	1	18,96	43,07	$2p^2\ ^2D—2p^3\ ^2P^\circ$	$^3/_2—^1/_2$
513,975	2	18,95	43,07	$2p^2\ ^2D—2p^3\ ^2P^\circ$	$^5/_2—^3/_2$
508,079	4	0,09	24,49	$2p\ ^2P^\circ—2p^2\ ^2S$	$^3/_2—^1/_2$
506,163	3	0,00	24,49	$2p\ ^2P^\circ—2p^2\ ^2S$	$^1/_2—^1/_2$
466,995	5	0,09	26,64	$2p\ ^2P^\circ—2p^2\ ^2P$	$^3/_2—^1/_2$
465,978	7	0,09	26,70	$2p\ ^2P^\circ—2p^2\ ^2P$	$^3/_2—^3/_2$
465,374	6	0,00	26,64	$2p\ ^2P^\circ—2p^2\ ^2P$	$^1/_2—^1/_2$
464,370	5	0,00	26,70	$2p\ ^2P^\circ—2p^2\ ^2P$	$^1/_2—^3/_2$
242,439	2	18,95	70,09	$2p^2\ ^2D—3p\ ^2P^\circ$	$^3/_2—^1/_2$
242,324	3	18,95	70,11	$2p^2\ ^2D—3p\ ^2P^\circ$	$^5/_2—^3/_2$
235,840	1	{26,64 26,70	79,20 79,27	$2p^2\ ^2P—3s\ ^2P^\circ$ $2p^2\ ^2P—3s\ ^2P^\circ$	$^1/_2—^1/_2$ $^3/_2—^3/_2$
226,608	2	24,49	79,20	$2p^2\ ^2S—3s\ ^2P^\circ$	$^1/_2—^1/_2$
226,341	2	24,49	79,27	$2p^2\ ^2S—3s\ ^2P^\circ$	$^1/_2—^3/_2$
206,594	3	26,70	86,71	$2p^2\ ^2P—3d\ ^2D^\circ$	$^3/_2—^5/_2$
206,430	2	26,64	86,70	$2p^2\ ^2P—3d\ ^2D^\circ$	$^1/_2—^3/_2$
205,778	3	18,95	79,20	$2p^2\ ^2D—3s\ ^2P^\circ$	$^3/_2—^1/_2$
205,552	4	18,95	79,27	$2p^2\ ^2D—3s\ ^2P^\circ$	$^5/_2—^3/_2$

λ, Å	I	E_H, eV	E_B, eV	Transition	J
200,861	1	26,64	88,37	$2p^2\ ^2P—3s'\ ^2P^\circ$	1/2—3/2, 1/2
198,476	1	38,09	100,55	$2p^3\ ^2D^\circ—3s'''\ ^2D$	5/2, 3/2—5/2, 3/2
196,870	1	34,30	97,27	$2p^3\ ^4S^\circ—3s''\ ^4P$	3/2—3/2
196,713	2	34,30	97,32	$2p^3\ ^4S^\circ—3s''\ ^4P$	3/2—5/2
194,108	3	24,49	88,37	$2p^2\ ^2S—3s'\ ^2P^\circ$	1/2—3/2, 1/2
191,973	4	24,49	89,07	$2p^2\ ^2S—3d\ ^2P^\circ$	1/2—3/2
191,892	3	24,49	89,10	$2p^2\ ^2S—3d\ ^2P^\circ$	1/2—1/2
190,839	7	0,09	65,06	$2p\ ^2P^\circ—3s\ ^2S$	3/2—1/2
190,571	6	0,00	65,06	$2p\ ^2P^\circ—3s\ ^2S$	1/2—1/2
189,943	2	43,07	108,34	$2p^3\ ^2P^\circ—3d'''\ ^2D$	3/2, 1/2—5/2, 3/2
187,008	4	10,74	77,04	$2p^2\ ^4P—3s\ ^4P^\circ$	5/2—3/2
186,968	4	10,70	77,01	$2p^2\ ^4P—3s\ ^4P^\circ$	3/2—1/2
186,879	3	10,67	77,01	$2p^2\ ^4P—3s\ ^4P^\circ$	1/2—1/2
		10,70	77,04	$2p^2\ ^4P—3s\ ^4P^\circ$	3/2—3/2
186,842	5	10,74	77,10	$2p^2\ ^4P—3s\ ^4P^\circ$	5/2—5/2
186,788	4	10,67	77,04	$2p^2\ ^4P—3s\ ^4P^\circ$	1/2—3/2
186,715	4	10,70	77,10	$2p^2\ ^4P—3s\ ^4P^\circ$	3/2—5/2
183,016	3	18,95	86,70	$2p^2\ ^2D—3d\ ^2D^\circ$	3/2—3/2
182,979	4	18,95	86,71	$2p^2\ ^2D—3d\ ^2D^\circ$	5/2—5/2
178,612	3	18,95	88,37	$2p^2\ ^2D—3s'\ ^2P^\circ$	5/2, 3/2—3/2, 1/2
178,590	4	18,95	88,38	$2p^2\ ^2D—3d\ ^2F^\circ$	3/2—5/2
178,434	5	18,95	88,43	$2p^2\ ^2D—3d\ ^2F^\circ$	5/2—7/2
174,698	4	26,70	97,66	$2p^2\ ^2P—3d'\ ^2D^\circ$	3/2—5/2
174,568	3	26,64	97,66	$2p^2\ ^2P—3d'\ ^2D^\circ$	1/2—3/2
174,513	3	38,09	109,14	$2p^3\ ^2D^\circ—3d'''\ ^2F$	3/2—5/2
174,490	3	38,09	109,14	$2p^3\ ^2D^\circ—3d'''\ ^2F$	5/2—7/2
173,020	1	26,70	98,35	$2p^2\ ^2P—3d'\ ^2P^\circ$	3/2—3/2, 1/2
171,302	2	34,30	106,67	$2p^3\ ^4S^\circ—3d''\ ^4P$	3/2—5/2
171,241	1	34,30	106,70	$2p^3\ ^4S^\circ—3d''\ ^4P$	3/2—3/2
167,858	1	24,49	98,35	$2p^2\ ^2S—3d'\ ^2P^\circ$	1/2—3/2, 1/2
166,177	10	0,09	74,70	$2p^2\ ^2P^\circ—3d\ ^2D$	3/2—5/2
165,983	9	0,00	74,69	$2p^2\ ^2P^\circ—3d\ ^2D$	1/2—3/2
163,596	2	10,74	86,53	$2p^2\ ^4P—3d\ ^4D^\circ$	5/2—5/2
163,558	5	10,74	86,54	$2p^2\ ^4P—3d\ ^4D^\circ$	5/2—7/2
163,501	4	10,70	86,53	$2p^2\ ^4P—3d\ ^4D^\circ$	3/2—5/2
163,456	3	10,67	86,51	$2p^2\ ^4P—3d\ ^4D^\circ$	1/2—3/2, 1/2
162,270	4	10,74	87,14	$2p^2\ ^4P—3d\ ^4P^\circ$	5/2—5/2
162,215	3	10,74	87,17	$2p^2\ ^4P—3d\ ^4P^\circ$	5/2—3/2
162,172	3	10,70	87,14	$2p^2\ ^4P—3d\ ^4P^\circ$	3/2—5/2
162,121	2	10,70	87,17	$2p^2\ ^4P—3d\ ^4P^\circ$	3/2—3/2
162,082	3	10,70	87,19	$2p^2\ ^4P—3d\ ^4P^\circ$	3/2—1/2
162,053	3	10,67	87,17	$2p^2\ ^4P—3d\ ^4P^\circ$	1/2—3/2
162,013	2	10,67	87,19	$2p^2\ ^4P—3d\ ^4P^\circ$	1/2—1/2
158,537	4	18,95	97,15	$2p^2\ ^2D—3d'\ ^2F^\circ$	5/2, 3/2—7/2, 5/2
157,515	1	18,95	97,66	$2p^2\ ^2D—3d'\ ^2D^\circ$	5/2, 3/2—5/2, 3/2
152,563	2	0,09	81,35	$2p\ ^2P^\circ—3p\ ^2P$	3/2—1/2
152,511	4	0,09	81,38	$2p\ ^2P^\circ—3p\ ^2P$	3/2—3/2
152,391	3	0,00	81,35	$2p\ ^2P^\circ—3p\ ^2P$	1/2—1/2
152,339	2	0,00	81,38	$2p\ ^3P^\circ—3p\ ^2P$	1/2—3/2
148,108	1	0,09	83,80	$2p\ ^2P^\circ—3p\ ^2D$	3/2—3/2
148,002	5	0,09	83,86	$2p\ ^2P^\circ—3p\ ^2D$	3/2—5/2
147,946	4	0,00	83,80	$2p\ ^2P^\circ—3p\ ^2D$	1/2—3/2
145,547	3	0,09	85,27	$2p\ ^2P^\circ—3p\ ^2S$	3/2—1/2
145,392	2	0,00	85,27	$2p\ ^2P^\circ—3p\ ^2S$	1/2—1/2
145,177	1	18,95	104,35	$2p^2\ ^2D—4d\ ^2D^\circ$	5/2—5/2
144,673	1	18,95	104,65	$2p^2\ ^2D—3p'''\ ^2F^\circ$	3/2—5/2
144,637	1	18,95	104,67	$2p^2\ ^2D—3p'''\ ^2F^\circ$	5/2—7/2
143,897	1	18,95	105,11	$2p^2\ ^2D—4d\ ^2F^\circ$	5/2—7/2
140,414	1	0,09	88,39	$2p\ ^2P^\circ—4s\ ^2S$	3/2—1/2

λ, Å	I	E_H, eV	E_B, eV	Transition	J
138,181	1	10,74	100,46	$2p^2\,{}^4P$—$4s\,{}^4P^\circ$	5/2—5/2
136,902	3	10,74	101,30	$2p^2\,{}^4P$—$3p''\,{}^4D^\circ$	5/2, 3/2—7/2, 5/2
134,539	5	0,09	92,24	$2p\,{}^2P^\circ$—$4d\,{}^2D$	3/2—5/2
134,407	4	0,00	92,24	$2p\,{}^2P^\circ$—$4d\,{}^2D$	1/2—3/2
133,662	1	10,74	103,49	$2p^2\,{}^4P$—$3p''\,{}^4S^\circ$	5/2—3/2
133,208	1	0,09	93,16	$2p\,{}^2P^\circ$—$3p'\,{}^2D$	3/2—5/2
132,819	2	0,09	93,44	$2p\,{}^2P^\circ$—$3p'\,{}^2P$	3/2—3/2
132,699	1	0,00	93,44	$2p\,{}^2P^\circ$—$3p'\,{}^2P$	1/2—3/2, 1/2
132,511	3	10,74	104,30	$2p^2\,{}^4P$—$4d\,{}^4D^\circ$	5/2—7/2
132,484	3	10,70	104,28	$2p^2\,{}^4P$—$4d\,{}^4D^\circ$	3/2—5/2, 3/2
132,453	2	10,67	104,27	$2p^2\,{}^4P$—$4d\,{}^4D^\circ$	1/2—3/2, 1/2
132,310	1	10,74	104,44	$2p^2\,{}^4P$—$4d\,{}^4P^\circ$	5/2—5/2
131,638	0	0,09	94,27	$2p\,{}^2P^\circ$—$3p'\,{}^2S$	3/2—1/2
131,516	0	0,00	94,27	$2p\,{}^2P^\circ$—$3p'\,{}^2S$	1/2—1/2
123,774	1	0,09	100,26	$2p\,{}^2P^\circ$—$5d\,{}^2D$	3/2—5/2
123,665	0	0,00	100,26	$2p\,{}^2P^\circ$—$5d\,{}^2D$	1/2—3/2
120,032	0	0,09	103,39	$2p\,{}^2P^\circ$—$4p\,{}^2D$	3/2—5/2
119,986	0	0,00	103,34	$2p\,{}^2P^\circ$—$4p\,{}^2D$	1/2—3/2

F VI, ground state $1s^2\,2s^2\,{}^1S_0$
Ionization potential 1 267 581 cm^{-1}; 157,151 eV

λ, Å	I	E_H, eV	E_B, eV	Transition	J
4592,85	1	98,00	100,70	$3p\,{}^3P^\circ$—$3d\,{}^3D$	2—3
2327,28	5	92,65	97,97	$3s\,{}^3S$—$3p\,{}^3P^\circ$	1—0
2323,35	7	92,65	97,98	$3s\,{}^3S$—$3p\,{}^3P^\circ$	1—1
2315,39	9	92,65	98,00	$3s\,{}^3S$—$3p\,{}^3P^\circ$	1—2
1139,60	2	23,16	34,04	$2p\,{}^1P^\circ$—$2p^2\,{}^1D$	1—2
648,50	1	12,08	31,20	$2p\,{}^3P^\circ$—$2p^2\,{}^3P$	2—1
647,31	1	12,01	31,16	$2p\,{}^3P^\circ$—$2p^2\,{}^3P$	1—0
646,36	3	12,08	31,26	$2p\,{}^3P^\circ$—$2p^2\,{}^3P$	2—2
646,10	1	12,01	31,20	$2p\,{}^3P^\circ$—$2p^2\,{}^3P$	1—1
645,02	1	11,98	31,20	$2p\,{}^3P^\circ$—$2p^2\,{}^3P$	0—1
643,98	1	12,01	31,26	$2p\,{}^3P^\circ$—$2p^2\,{}^3P$	1—2
535,204	10	0,00	23,16	$2s^2\,{}^1S$—$2p\,{}^1P^\circ$	0—1
194,840	1	34,04	97,67	$2p^2\,{}^1D$—$3p\,{}^1P^\circ$	2—1
173,145	1	23,16	94,77	$2p\,{}^1P^\circ$—$3s\,{}^1S$	1—0
164,015	1	34,04	109,63	$2p^2\,{}^1D$—$3s\,{}^1P^\circ$	2—1
163,138	2	42,20	118,20	$2p^2\,{}^1S$—$3d\,{}^1P^\circ$	0—1
161,477	1	31,26	108,04	$2p^2\,{}^3P$—$3s\,{}^3P^\circ$	2—1
161,414	1	31,20	108,00	$2p^2\,{}^3P$—$3s\,{}^3P^\circ$	1—0
161,341	1	31,20	108,04	$2p^2\,{}^3P$—$3s\,{}^3P^\circ$	1—1
161,308	3	31,26	108,12	$2p^2\,{}^3P$—$3s\,{}^3P^\circ$	2—2
161,257	1	31,16	108,04	$2p^2\,{}^3P$—$3s\,{}^3P^\circ$	0—1
161,174	1	31,20	108,12	$2p^2\,{}^3P$—$3s\,{}^3P^\circ$	1—2
156,247	6	23,16	102,51	$2p\,{}^1P^\circ$—$3d\,{}^1D$	1—2
154,506	3	34,04	114,28	$2p^2\,{}^1D$—$3d\,{}^1D^\circ$	2—2
153,880	4	12,08	92,65	$2p\,{}^3P^\circ$—$3s\,{}^3S$	2—1
153,741	3	12,01	92,65	$2p\,{}^3P^\circ$—$3s\,{}^3S$	1—1
153,678	2	11,98	92,65	$2p\,{}^3P^\circ$—$3s\,{}^3S$	0—1
148,653	4	34,04	117,44	$2p^2\,{}^1D$—$3d\,{}^1F^\circ$	2—3
146,718	2	31,26	115,76	$2p^2\,{}^3P$—$3d\,{}^3D^\circ$	2—2
146,676	4	31,26	115,78	$2p^2\,{}^3P$—$3d\,{}^3D^\circ$	2—3

λ, Å	I	E_H, eV	E_B, eV	Transition	J
146,613	3	31,20	115,76	$2p^2\ ^3P—3d\ ^3D°$	1—2
146,576	2	31,16	115,74	$2p^2\ ^3P—3d\ ^3D°$	0—1
145,691	3	31,26	116,36	$2p^2\ ^3P—3d\ ^3P°$	2—2
145,630	1	31,26	116,39	$2p^2\ ^3P—3d\ ^3P°$	2—1
145,585	1	31,20	116,36	$2p^2\ ^3P—3d\ ^3P°$	1—2
145,489	1	31,20	116,41	$2p^2\ ^3P—3d\ ^3P°$	1—0
145,462	1	31,16	116,39	$2p^2\ ^3P—3d\ ^3P°$	0—1
141,154	2	23,16	111,00	$2p\ ^1P°—3p\ ^1P$	1—1
139,900	7	12,08	100,70	$2p\ ^3P°—3d\ ^3D$	2—3
139,800	6	12,01	100,69	$2p\ ^3P°—3d\ ^3D$	1—2, 1
139,758	5	11,98	100,69	$2p\ ^3P°—3d\ ^3D$	0—1
135,397	3	23,16	114,73	$2p\ ^1P°—3d\ ^1D$	1—2
126,923	5	0,00	97,67	$2s^2\ ^1S—3p\ ^1P°$	0—1
124,474	0	12,08	111,68	$2p\ ^3P°—3p\ ^3D$	2—2
124,400	0	12,01	111,63	$2p\ ^3P°—3p\ ^3D$	1—1
124,387	3	12,01	111,68	$2p\ ^3P°—3p\ ^3D$	2, 1, 0—3, 2, 1
123,175	1	12,08	112,73	$2p\ ^3P°—3p\ ^3S$	2—1
123,091	0	12,01	112,73	$2p\ ^3P°—3p\ ^3S$	1—1
123,051	0	11,98	112,73	$2p\ ^3P°—3p\ ^3S$	0—1
122,251	0	12,08	113,49	$2p\ ^3P°—3p\ ^3P$	2—1
122,200	2	12,08	113,53	$2p\ ^3P°—3p\ ^3P$	1—0; 2—2,
122,169	0	12,01	113,49	$2p\ ^3P°—3p\ ^3P$	1—1
122,122	1	12,01	113,53	$2p\ ^3P°—3p\ ^3P$	0—1; 1—2
120,116	1	23,16	126,38	$2p\ ^1P°—4d\ ^1D$	1—2
116,094	0	34,04	140,83	$2p^2\ ^1D—4d\ ^1F°$	2—3
113,840	0	31,26	140,14	$2p^2\ ^3P—4d\ ^3D°$	2—3
109,040	1	12,08	125,77	$2p\ ^3P°—4d\ ^3D$	2—3
108,975	1	12,01	—	$2p\ ^3P°—4d\ ^3D$	1, 0—2, 1
99,203	0	0,00	124,95	$2s^2\ ^1S—4p\ ^1P°$	0—1
99,105	0	12,08	137,17	$2p\ ^3P°—5d\ ^3D$	2—3
99,044	0	12,01	—	$2p\ ^3P°—5d\ ^3D$	1, 0—2, 1

NEON, Z = 10

Ne I, ground state $1s^2\,2s^2\,2p^6\ {}'S_0$
Ionization potential 173 931,7 cm^{-1}; 21,564 eV

λ, Å	I	E_H, eV	E_B, eV	Transition	J
24935,6	7	20,21	20,71	$4p\,[1^1/_2]-4d\,[2^1/_2]^\circ$	2—2
24777,7	15	20,21	20,71	$4p\,[1^1/_2]-4d\,[2^1/_2]^\circ$	1—2
24458,7	36	20,30	29,80	$4p'\,[1^1/_2]-4d'\,[2^1/_2]^\circ$	2—3
24448,5	20	20,30	20,80	$4p'\,[^1/_2]-4d'\,[1^1/_2]^\circ$	1—2
24366,4	95	19,69	20,20	$4s\,[1^1/_2]^\circ-4p\,[2^1/_2]$	1—2
24248,9	32	19,78	20,29	$4s'\,[^1/_2]^\circ-4p'\,[1^1/_2]$	1—1
24161,5	25	20,29	20,80	$4p'\,[1^1/_2]-4d'\,[2^1/_2]^\circ$	1—2
24097,8	11	20,20	20,71	$4p\,[2^1/_2]-4d\,[2^1/_2]^\circ$	2—2
23978,4	68	20,19	20,70	$4p\,[2^1/_2]-4d\,[3^1/_2]^\circ$	3—4
23956,2	47	19,78	20,30	$4s'\,[^1/_2]^\circ-4p'\,[^1/_2]$	1—1
23951,3	119	19,78	20,30	$4s'\,[^1/_2]^\circ-4p'\,[1^1/_2]$	1—2
23709,4	62	19,69	20,21	$4s\,[1^1/_2]^\circ-4p\,[1^1/_2]$	1—1
23636,3	205	19,66	20,19	$4s\,[1^1/_2]^\circ-4p\,[2^1/_2]$	2—3
23565,6	40	19,69	20,21	$4s\,[1^1/_2]^\circ-4p\,[1^1/_2]$	1—2
23372,1	62	19,76	20,29	$4s'\,[^1/_2]^\circ-4p'\,[1^1/_2]$	0—1
23260,7	45	19,66	20,20	$4s\,[1^1/_2]^\circ-4p\,[2^1/_2]$	2—2
23101,0	62	19,76	20,30	$4s'\,[^1/_2]^\circ-4p'\,[^1/_2]$	0—1
22662,5	15	19,66	20,21	$4s\,[1^1/_2]^\circ-4p\,[1^1/_2]$	2—1
22529,7	105	19,66	20,21	$4s\,[1^1/_2]^\circ-4p\,[1^1/_2]$	2—2
22468,4	8	20,15	20,70	$4p\,[^1/_2]-4d\,[^1/_2]^\circ$	1—0
22428,2	15	20,15	20,71	$4p\,[^1/_2]-4d\,[1^1/_2]^\circ$	1—1
22245,3	12	20,15	20,71	$4p\,[^1/_2]-4d\,[1^1/_2]^\circ$	1—2
21707,4	25	19,69	20,26	$4s\,[1^1/_2]^\circ-4p\,[^1/_2]$	1—0
21040,9	27	19,78	20,37	$4s'\,[^1/_2]^\circ-4p'\,[^1/_2]$	1—0
20350,6	10	19,69	20,30	$4s\,[1^1/_2]^\circ-4p'\,[1^1/_2]$	1—2
19574,0	10	19,69	20,30	$4s\,[1^1/_2]^\circ-4p'\,[1^1/_2]$	2—2
18624,94	20	20,05	20,71	$3d\,[2^1/_2]^\circ-4f\,[2^1/_2]$	3—2, 3
18618,69	15	20,05	20,71	$3d\,[2^1/_2]^\circ-4f\,[2^1/_2]$	2—2, 3
18597,30	120	20,05	20,71	$3d\,[2^1/_2]^\circ-4f\,[3^1/_2]$	3—3, 4
18591,12	25	20,05	20,71	$3d\,[2^1/_2]^\circ-4f\,[3^1/_2]$	2—3
18475,79	3	20,14	20,81	$3d'\,[1^1/_2]^\circ-4f'\,[2^1/_3]$	1—2
18458,58	10	20,04	20,71	$3d\,[1^1/_2]^\circ-4f\,[1^1/_2]$	1—1, 2
18422,43	110	20,14	20,81	$3d'\,[1^1/_2]^\circ-4f'\,[2^1/_2]$	2—2, 3
18403,16	60	20,04	20,71	$3d\,[1^1/_2]^\circ-4f\,[2^1/_2]$	1—2
18390,10	180	20,14	20,81	$3d'\,[2^1/_2]^\circ-4f'\,[3^1/_2]$	3—3, 4
18385,17	160	20,14	20,81	$3d'\,[2^1/_2]^\circ-4f'\,[3^1/_2]$	2—3
18359,21	6	20,04	20,71	$3d\,[1^1/_2]^\circ-4f\,[1^1/_2]$	2—1, 2
18304,00	140	20,04	20,71	$3d\,[1^1/_2]^\circ-4f\,[2^1/_2]$	2—2, 3
18282,58	200	20,03	20,71	$3d\,[3^1/_2]^\circ-4f\,[4^1/_2]$	3—4
18276,59	260	20,03	20,71	$3d\,[3^1/_2]^\circ-4f\,[4^1/_2]$	4—4, 5
18226,57	10	20,03	20,71	$3d\,[3^1/_2]^\circ-4f\,[3^1/_2]$	3—3, 4
18220,76	15	20,03	20,71	$3d\,[3^1/_2]^\circ-4f\,[3^1/_2]$	4—3, 4
18082,71	130	20,02	20,71	$3d\,[\ ^1/_2]^\circ-4f\,[1^1/_2]$	1—1, 2
18035,49	20	20,02	20,71	$3d\,[^1/_2]^\circ-4f\,[1/^1_2]$	0—1
15234,4	2	18,96	19,78	$3p'\,[^1/_2]-4s'\,[^1/_2]^\circ$	0—1
12912,4	2	18,70	19,66	$3p'\,[1^1/_2]-4s\,[1^1/_2]^\circ$	2—2
12690,1	2	18,71	19,69	$3p\,[^1/_2]-4s\,[1^1/_2]^\circ$	0—1
12594,8	1	18,70	19,69	$3p'\,[1^1/_2]-4s\,[1^1/_2]^\circ$	2—1
12459,49	2	18,69	19,69	$3p'\,[1^1/_2]-4s\,[1^1/_2]^\circ$	1—1
12066,38	15	18,63	19,66	$3p\,[1^1/_2]-4s\,[1^1/_2]^\circ$	2—2
11984,99	10	18,72	19,76	$3p'\,[^1/_2]-4s'\,[^1/_2]^\circ$	1—0
11789,93	10	18,61	19,66	$3p\,[1^1/_2]-4s\,[1^1/_2]^\circ$	1—2
11789,11	50	18,63	19,69	$3p\,[1^1/_2]-4s\,[1^1/_2]^\circ$	2—1

λ, Å	I	E_H, eV	E_B, eV	Transition	J
11766,87	60	18,72	19,78	$3p'\ [1^1/_2]$—$4s'\ [^1/_2]^\circ$	1—1
11688,08	10	18,96	20,02	$3p'\ [^1/_2]$—$3d\ [^1/_2]^\circ$	0—1
11614,18	80	18,69	19,76	$3p'\ [1^1/_2]$—$4s'\ [^1/_2]^\circ$	1—0
11601,62	25	18,71	19,78	$3p\ [^1/_2]$—$4s'\ [^1/_2]^\circ$	0—1
11536,41	50	18,96	20,04	$3p'\ [^1/_2]$—$4d\ [1^1/_2]^\circ$	0—1
11525,11	90	18,61	19,69	$3p\ [1^1/_2]$—$4s\ [1^1/_2]^\circ$	1—1
11522,82	150	18,70	19,78	$3p'\ [1^1/_2]$—$4s'\ [^1/_2]^\circ$	2—1
11409,24	100	18,69	19,78	$3p'\ [1^1/_2]$—$4s'\ [^1/_2]^\circ$	1—1
11390,53	110	18,57	19,66	$3p\ [2^1/_2]$—$4s\ [1^1/_2]^\circ$	2—2
11366,80	3	19,69	20,78	$4s\ [1^1/_2]^\circ$—$5p\ [2^1/_2]$	1—2
11333,60	3	19,78	20,87	$4s'\ [^1/_2]^\circ$—$5p'\ [1^1/_2]$	1—1
11329,56	1	19,78	20,87	$4s'\ [^1/_2]^\circ$—$5p'\ [^1/_2]$	1—1
11304,47	2	19,69	20,78	$4s\ [1^1/_2]^\circ$—$5p\ [1^1/_2]$	1—1
11303,96	5	19,78	20,88	$4s'\ [^1/_2]^\circ$—$5p'\ [1^1/_2]$	1—2
11298,45	1	19,66	20,76	$4s\ [1^1/_2]^\circ$—$5p\ [^1/_2]$	2—1
11293,00	2	19,69	20,78	$4s\ [1^1/_2]^\circ$—$5p\ [1^1/_2]$	1—2
11177,59	300	18,55	19,66	$3p\ [2^1/_2]$—$4s\ [1^1/_2]^\circ$	3—2
11160,29	10	19,66	20,77	$4s\ [1^1/_2]^\circ$—$5p\ [2^1/_2]$	2—3
11143,09	300	18,57	19,69	$3p\ [2^1/_2]$—$4s\ [1^1/_2]^\circ$	2—1
11138,55	4	19,76	20,87	$4s'\ [^1/_2]^\circ$—$5p'\ [1^1/_2]$	0—1
11134,62	4	19,76	20,87	$4s'\ [^1/_2]^\circ$—$5p'\ [^1/_2]$	0—1
11120,37	5	19,66	20,78	$4s\ [1^1/_2]^\circ$—$5p\ [2^1/_2]$	2—2
11060,88	2	19,66	20,78	$4s\ [1^1/_2]^\circ$—$5p\ [1^1/_2]$	2—1
11049,80	20	19,66	20,78	$4s\ [1^1/_2]^\circ$—$5p\ [1^1/_2]$	2—2
11044,06	15	19,78	20,90	$4s'\ [^1/_2]^\circ$—$5p'\ [^1/_2]$	1—0
11020,93	10	19,69	20,81	$4s\ [1^1/_2]^\circ$—$5p\ [^1/_2]$	1—0
10888,53	8	20,05	21,18	$3d\ [2^1/_2]$—$6f\ [3^1/_2]$	3—3, 4
10886,35	5	20,05	21,18	$3d\ [2^1/_2]^\circ$—$6f\ [3^1/_2]$	2—3
10844,54	200	18,63	19,78	$3p\ [1^1/_2]$—$4s'\ [^1/_2]^\circ$	2—1
10838,30	3	20,04	21,18	$3d\ [1^1/_2]^\circ$—170863	1— —
10830,33	4	—	—	—	—
10819,95	5	20,14	21,28	$3d'\ [1^1/_2]^\circ$—$6f'\ [2^1/_2]$	2—2, 3
10814,83	4	20,04	21,18	$3d\ [1^1/_2]^\circ$—$6f\ [2^1/_2]$	1—2
10808,22	7	20,14	21,28	$3d'\ [2^1/_2]^\circ$—$6f'\ [3^1/_2]$	3—3, 4
10806,43	5	20,14	21,28	$3d'\ [2^1/_2]^\circ$—$6f'\ [3^1/_2]$	2—3
10798,12	150	18,61	19,76	$3p\ [1^1/_2]$—$4s'\ [^1/_2]^\circ$	0—1
10789,37	2	20,03	21,18	$3d\ [3^1/_2]^\circ$—170863	3——
10780,57	6	20,04	21,18	$3d\ [1^1/_2]^\circ$—$6f\ [2^1/_2]$	2—2, 3
10766,15	10	20,05	21,20	$3d\ [2^1/_2]^\circ$—170990	3—2
10764,09	12	20,05	21,20	$3d\ [2^1/_2]^\circ$—170990	2—2
10760,34	1	20,03	21,18	$3d\ [3^1/_2]^\circ$—$6f\ [3^1/_2]$	3—3, 4
10758,28	2	20,03	21,18	$3d\ [3^1/_2]$—$6f\ [3^1/_2]$	4—3, 4
10690,48	6	20,02	21,18	$3d\ [^1/_2]^\circ$—$6f\ [1^1/_2]$	1—1, 2
10673,80	2	20,02	21,18	$3d\ [^1/_2]^\circ$—$6f\ [1^1/_2]$	0—1
10620,63	40	18,61	19,78	$3p\ [1^1/_2]$—$4s'\ [^1/_2]^\circ$	1—1
10562,43	200	18,96	20,14	$3p'\ [^1/_2]$—$3d\ [1^1/_2]^\circ$	0—1
10432,53	3	19,69	20,88	$4s\ [1^1/_2]^\circ$—$5p'\ [1^1{}_2]$	1—2
10295,40	80	18,57	19,78	$3p\ [2^1/_2]$—$4s'\ [^1/_2]^\circ$	2—1
10245,70	7	19,66	20,87	$4s\ [1/^1{}_2]^\circ$—$5p'\ [^1/_2]$	2—1
10224,6	2	19,66	20,88	$4s\ [1^1/_2]^\circ$—$5p'\ [1^1/_2]$	2—2
10210,73	2	19,69	20,90	$4s\ [1^1/_2]^\circ$—$5p'\ [^1/_2]$	1—0
10091,53	3	20,14	21,36	$3d'\ [2^1/_2]^\circ$—172319	3—2
10038,9	2	20,05	21,28	$3d\ [2^1/_2]^\circ$—$6f'\ [3^1/_2]$	3—3, 4
10037,1	2	20,05	21,28	$3d\ [2^1/_2]^\circ$—$6f'\ [3^1/_2]$	2—3
10008,55	4	20,05	21,29	$3d\ [2^1/_2]^\circ$—$7f\ [2^1/_2]$	3—2, 3
10007,31	30	20,05	21,29	$3d\ [2^1/_2]^\circ$—$7f\ [3^1/_2]$	3—3, 4
10005,54	20	20,05	21,29	$3d\ [2^1/_2]^\circ$—$7f\ [3^1/_2]$	2—3

λ, Å	I	E_H, eV	E_B, eV	Transition	J
9974,2	2	20,04	21,28	$3d\ [1^1/_2]^\circ-6f'\ [2^1/_2]$	1—2
9963,55	6	20,14	21,38	$3d'\ [1^1/_2]^\circ-7f'\ [2^1/_2]$	1—2
9947,94	15	20,14	21,38	$3d'\ [1^1/_2]^\circ-7f'\ [2^1/_2]$	2—2, 3
9944,9	2	20,04	21,28	$3d\ [1^1/_2]^\circ-6f'\ [2^1/_2]$	2—2, 3
9944,1	7	20,04	21,29	$3d\ [1^1/_2]^\circ-7f\ [2^1/_2]$	1—2
9938,35	15	20,14	21,38	$3d'\ [2^1/_2]^\circ-7f'\ [2^1/_2]$	3—2, 3
9936,83	10	20,14	21,38	$3d'\ [2^1/_2]^\circ-7f'\ [2^1/_2]$	2—2, 3
9918,52	4	20,04	21,29	$3d\ [1^1/_2]^\circ-7f\ [1^1/_2]$	2—1, 2
9915,13	20	20,04	21,29	$3d\ [1^1/_2]^\circ-7f\ [2^1/_2]$	2—2, 3
9902,31	30	20,05	21,30	$3d\ [2^1/_2]^\circ-171799$	3—2
9900,58	40	20,05	21,30	$3d\ [2^1/_2]^\circ-171799$	2—2
9899,06	2	20,03	21,29	$3d\ [3^1/_2]^\circ-4f\ [3^1/_2]$	3—3, 4
9897,30	3	20,03	21,29	$3d\ [3^1/_2]^\circ-4f\ [3^1/_2]$	4—3, 4
9875,90	2	20,05	21,30	$3d\ [2^1/_2]^\circ-7p'\ [1^1/_2]$	2—1
9837,47	20	20,02	21,29	$3d\ [^1/_2]^\circ-7f\ [1^1/_2]$	1—1, 2
9823,42	5	20,04	21,31	$3d\ [1^1/_2]^\circ-7p'\ [^1/_2]$	1—0
9788,1	2	19,78	21,05	$4s'\ [^1/_2]^\circ-6p\ [^1/_2]$	1—1
9741,3	1	20,02	21,29	$3d\ [^1/_2]^\circ-8p\ [^1/_2]$	0—1
9728,2	1	20,14	21,41	$3d'\ [1^1/_2]^\circ-172698$	2——
9724,8	1	19,78	21,05	$4s'\ [1^1/_2]^\circ-6p\ [2^1/_2]$	1—2
9702,40	3	20,02	21,30	$3d\ [^1/_2]^\circ-7p'\ [1^1/_2]$	1—2
9665,424	1000	18,38	19,66	$3p\ [^1/_2]-4s\ [1^1/_2]^\circ$	1—2
9642,2	1	19,76	21,05	$4s'\ [^1/_2]^\circ-6p\ [^1/_2]$	0—1
9592,19	5	—	—	—	—
9584,79	3	—	—	—	—
9573,99	2	19,78	21,07	$4s'\ [^1/_2]^\circ-6p\ [^1/_2]$	1—0
9547,40	300	18,72	20,02	$3p'\ [^1/_2]-3d\ [^1/_2]^\circ$	1—0
9534,167	500	18,72	20,02	$3p'\ [^1/_2]-3d\ [^1/_2]^\circ$	1—1
9508,4	5	20,05	21,35	$3d\ [2^1/_2]^\circ-8f\ [3^1/_2]$	2—3
9506,59	3	—	—	—	—
9486,680	500	18,38	19,69	$3p\ [^1/_2]-4s\ [1^1/_2]^\circ$	1—1
9467,81	2	20,14	21,45	$3d\ [1^1/_2]^\circ-8f'\ [2^1/_2]$	1—2
9459,21	300	18,72	20,04	$3p'\ [^1/_2]-3d\ [1^1/_2]^\circ$	1—2
9454,0	1	20,14	21,45	$3d'\ [1^1/_2]^\circ-8f'\ [2^1/_2]$	2—2, 3
9452,08	10	20,04	21,35	$3d\ [1^1/_2]^\circ-8f\ [1^1/_2]$	1—1, 2
9445,26	3	20,14	21,45	$3d'\ [2^1/_2]^\circ-8f'\ [2^1/_2]$	3—2, 3
9443,8	2	20,14	21,45	$3d'\ [2^1/_2]^\circ-8f'\ [2^1/_2]$	2—2, 3
9432,94	40	18,72	20,04	$3p'\ [^1/_2]-3d\ [1^1/_2]^\circ$	1—2
9425,38	500	18,71	20,02	$3p\ [^1/_2]-3d\ [^1/_2]^\circ$	0—1
9412,32	4	20,03	21,35	$3d\ [3^1/_2]^\circ-8f\ [3^1/_2]$	3—3, 4
9410,75	6	20,03	21,35	$3d\ [3^1/_2]^\circ-8f\ [3^1/_2]$	4—3, 4
9405,75	8	20,03	21,35	$3d\ [3^1/_2]^\circ-8f\ [2^1/_2]$	3—2, 3
9377,2	5	18,72	20,05	$3p'\ [^1/_2]-3d\ [2^1/_2]^\circ$	1—2
9373,28	200	18,70	20,02	$3p'\ [1^1/_2]-3d\ [^1/_2]^\circ$	2—1
9353,3	3	20,02	21,35	$3d\ [^1/_2]^\circ-8f\ [^1/_2]$	1—1, 2
9340,5	2	20,02	21,35	$3d\ [^1/_2]^\circ-8f\ [^1/_2]$	0—1
9326,52	600	18,71	20,04	$3p\ [^1/_2]-3d\ [1^1/_2]^\circ$	0—1
9313,98	300	18,70	20,03	$3p'\ [1^1/_2]-3d\ [3^1/_2]^\circ$	2—3
9310,58	150	18,70	20,02	$3p'\ [1^1/_2]-3d\ [^1/_2]^\circ$	1—0
9300,85	600	18,70	20,04	$3p'\ [1^1/_2]-3d\ [1^1/_2]^\circ$	2—1
9275,53	100	18,70	20,04	$3p'\ [1^1/_2]-3d\ [1^1/_2]^\circ$	2—1
9226,67	200	18,70	20,04	$3p'\ [1^1/_2]-3d\ [1^1/_2]^\circ$	1—2
9221,88	150	18,70	20,04	$3p'\ [1^1/_2]-3d\ [1^1/_2]^\circ$	1—1
9221,59	200	18,70	20,05	$3p'\ [1^1/_2]-3d\ [2^1/_2]^\circ$	2—2
9220,05	400	18,70	20,05	$3p'\ [1^1/_2]-3d\ [2^1/_2]^\circ$	2—3
9212,9	2	20,05	21,39	$3d\ [2^1/_2]^\circ-172553$	2—3
9201,76	600	18,69	20,04	$3p'\ [1^1/_2]-3d\ [1^1/_2]^\circ$	1—1
9191,8	3	—	—	—	—

λ, Å	*I*	E_H, eV	E_B, eV	Transition	*J*
9148,68	600	18,69	20,05	3*p*′ [1½]—3*d* [2½]°	1—2
9121,14	20	20,03	21,39	3*d* [3½]°—172553	4—3
9103,53	3	20,03	21,40	3*d* [3½]°—9*f* [3½]	3—3, 4
9102,1	1	20,03	21,40	3*d* [3½]°—9*f* [3½]	4—3, 4
9073,04	8	19,69	21,05	4*s* [1½]°—6*p* [2½]	1—2
9052,54	6	19,69	21,06	4*s* [1½]°—6*p* [1½]	1—1
9049,06	3	19,69	21,06	4*s* [1½]°—6*p* [1½]	1—2
9046,8	1	20,02	21,40	3*d* [½]°—9*f* [2½]	1—2
9039,0	3	20,04	21,41	3*d* [1½]°—172698	1——
9036,98	6	19,78	21,15	4*s*′ [½]°—6*p*′ [1½]	1—2
8988,58	200	18,38	19,76	3*p* [½]°—4*s*′ [½]°	1—0
8968,6	2	19,66	21,05	4*s* [1½]°—6*p* [½]	2—1
8962,34	3	19,78	21,16	4*s*′ [½]°—6*p*′ [½]	1—0
8948,12	7	20,02	21,41	3*d* [½]°—172698	1——
8941,47	6	19,69	21,07	4*s* [1½]°—6*p* [½]	1—0
8929,24	10	19,66	21,05	4*s* [1½]°—6*p* [2½]	2—3
8927,4	2	19,76	21,15	4*s*′ [½]°—6*p*′ [½]	0—1
8919,4987	300	18,63	20,02	3*p* [1½]—3*d* [½]°	2—1
8915,44	3	19,66	21,05	4*s* [1½]°—6*p* [2½]	2—2
8913,0	3	—	—	—	—
8895,6	2	19,66	21,06	4*s* [1½]°—6*p* [1½]	2—1
8892,22	10	19,66	21,06	4*s* [1½]°—6*p* [1½]	2—2
8865,7562	500	18,63	20,03	3*p* [1½]—3*d* [3½]°	2—3
8865,3057	100	18,38	19,78	3*p* [½]—4*s*′ [½]°	1—1
8853,8669	700	18,63	20,04	3*p* [1½]—3*d* [1½]°	2—2
8830,9078	50	18,63	20,04	3*p* [1½]—3*d* [1½]°	2—1
8820,36	6	—	—	—	—
8792,51	30	18,72	20,14	3*p*′ [½]—3*d*′ [2½]°	1—2
8783,7539	1000	18,72	20,14	3*p*′ [½]—3*d*′ [1½]°	1—2
8782,014	50	18,63	20,05	3*p* [1½]—3*d* [2½]°	2—2
8780,6223	1200	18,63	20,05	3*p* [1½]—3*d* [2½]°	2—3
8778,75	150	18,61	20,02	3*p* [1½]—3*d* [½]°	1—0
8771,6592	400	18,72	20,14	3*p*′ [½]—3*d*′ [1½]°	1—1
8767,55	15	18,61	20,02	3*p* [1½]—3*d* [½]°	1—1
8714,52	5	19,78	21,20	4*s*′ [½]°—7*p* [½]	1—1
8704,1132	200	18,61	20,04	3*p* [1½]—3*d* [1½]°	1—2
8681,9216	500	18,61	20,04	3*p* [1½]—3*d* [1½]°	1—1
8679,4898	500	18,71	20,14	3*p* [½]—3*d*′ [1½]°	0—1
8655,5206	400	18,70	20,14	3*p*′ [1½]—3*d*′ [2½]°	2—2
8654,3837	1500	18,70	20,14	3*p*′ [1½]—3*d*′ [2½]°	2—3
8647,0400	300	18,70	20,14	3*p*′ [1½]—3*d*′ [1½]°	2—2
8635,31	50	18,70	20,14	3*p*′ [1½]—3*d*′ [1½]°	2—1
8634,6472	600	18,61	20,05	3*p* [1½]—3*d* [2½]°	1—2
8591,2583	400	18,69	20,14	3*p*′ [1½]—3*d*′ [2½]°	1—2
8582,91	60	18,69	20,14	3*p*′ [1½]—3*d*′ [1½]°	1—2
8571,3535	100	18,69	20,14	3*p*′ [1½]—3*d*′ [1½]°	1—1
8544,6952	60	18,57	20,02	3*p* [2½]—3*d* [½]°	2—1
8495,3591	500	18,57	20,03	3*p* [2½]—3*d* [3½]°	2—3
8484,4424	80	18,57	20,04	3*p* [2½]—3*d* [1½]°	2—2
8470,72	5	19,69	21,15	4*s* [1½]°—6*p*′ [1½]	1—2
8463,3569	150	18,57	20,04	3*p* [2½]—3*d* [1½]°	2—2
8418,4265	400	18,57	20,05	3*p* [2½]—3*d* [2½]°	2—2
8417,161	100	18,57	20,05	3*p* [2½]—3*d* [2½]°	2—3
8377,6062	800	18,55	20,03	3*p* [2½]—3*d* [3½]°	3—4
8376,41	200	18,55	20,03	3*p* [2½]—3*d* [3½]°	3—3
8365,7464	150	18,55	20,04	3*p* [2½]—3*d* [1½]°	3—2
8301,54	150	18,55	20,05	3*p* [2½]—3*d* [2½]°	3—2

λ, Å	I	E_H, eV	E_B, eV	Transition	J
8300,3248	600	18,55	20,05	$3p\ [2^1/_2]—3d\ [2^1/_2]^\circ$	3—3
8267,117	80	18,63	20,14	$3p\ [1^1/_2]—3d'\ [2^1/_2]^\circ$	2—2
8266,0788	200	18,63	20,14	$3p\ [1^1/_2]—3d'\ [2^1/_2]^\circ$	2—3
8259,3795	150	18,63	20,14	$3p\ [1^1/_2]—3d'\ [1^1/_2]^\circ$	2—2
8248,6812	30	18,63	20,14	$3p\ [1^1/_2]—3d'\ [1^1/_2]^\circ$	2—1
8136,4061	300	18,61	20,14	$3p\ [1^1/_2]—3d'\ [2^1/_2]^\circ$	1—2
8128,908	60	18,61	20,14	$3p\ [1^1/_2]—3d'\ [1^1/_2]^\circ$	1—2
8118,5495	100	18,61	20,14	$3p\ [1^1/_2]—3d'\ [1^1/_2]^\circ$	1—1
8093,08	2	19,69	21,22	$4s\ [1^1/_2]^\circ—7p\ [^1/_2]$	1—0
8082,4576	200	16,85	18,38	$3s'\ [^1/_2]^\circ—3p\ [^1/_2]$	1—1
8076,06	1	19,78	21,31	$4s'\ [^1/_2]^\circ—7p'\ [^1/_2]$	1—0
8041,79	2	19,66	21,20	$4s\ [1^1/_2]^\circ—7p\ [2^1/_2]$	2—3
8024,11	2	19,66	21,21	$4s\ [1^1/_2]^\circ—7p\ [1^1/_2]$	2—2
7944,16	20	18,57	20,14	$3p\ [2^1/_2]—3d'\ [2^1/_2]^\circ$	2—2
7943,1805	200	18,57	20,14	$3p\ [2^1/_2]—3d'\ [2^1/_2]^\circ$	2—3
7936,9946	70	18,57	20,14	$3p\ [2^1/_2]—3d'\ [1^1/_2]^\circ$	2—2
7927,1172	40	18,57	20,14	$3p\ [2^1/_2]—3d'\ [1^1/_2]^\circ$	2—1
7840,04	1	18,55	20,14	$3p\ [2^1/_2]—3d'\ [2^1/_2]^\circ$	3—2
7839,0550	30	18,55	20,14	$3p\ [2^1/_2]—3d'\ [2^1/_2]^\circ$	3—3
7833,06	7	18,55	20,14	$3p\ [2^1/_2]—3d'\ [1^1/_2]^\circ$	3—2
7724,6281	10	18,96	20,57	$3p'\ [^1/_2]—5s\ [1^1/_2]^\circ$	0—1
7670,85	5	19,69	21,30	$4s\ [1^1/_2]^\circ—7p'\ [1^1/_2]$	1—2
7621,33	5	19,69	21,30	$4s\ [1^1/_2]^\circ—8p\ [^1/_2]$	1—0
7572,06	5	—	—	—	—
7544,0439	100	13,38	20,02	$3p\ [^1/_2]—3d\ [^1/_2]^\circ$	1—0
7535,7739	300	18,38	20,02	$3p\ [^1/_2]—3d\ [^1/_2]^\circ$	1—1
7488,8712	500	18,38	20,04	$3p\ [^1/_2]—3d\ [1^1/_2]^\circ$	1—2
7472,4383	50	18,38	20,04	$3p\ [^1/_2]—3d\ [1^1/_2]^\circ$	1—1
7438,8981	300	16,71	18,38	$3s'\ [^1/_2]^\circ—3p\ [^1/_2]$	0—1
7325,57	15	19,78	21,46	$4s'\ [^1/_2]^\circ—9p'\ [^1/_2]$	1—1,0
7307,93	15	19,66	21,36	$4s\ [1^1/_2]^\circ—9p\ [2^1/_2]$	2—3
7304,82	30	18,96	20,66	$3p'\ [^1/_2]—5s'\ [^1/_2]^\circ$	0—1
7245,1665	1000	16,67	18,38	$3s\ [1^1/_2]^\circ—3p\ [^1/_2]$	1—1
7173,9380	1000	16,85	18,57	$3s'\ [^1/_2]^\circ—3p\ [2^1/_2]$	1—1
7138,70	30	—	—	—	—
7112,2	10	18,96	20,71	$3p'\ [^1/_2]—3d\ [1^1/_2]^\circ$	0—1
7064,42	2	18,38	20,14	$3p\ [^1/_2]—3d'\ [2^1/_2]^\circ$	1—2
7059,1079	200	18,38	20,14	$3p\ [^1/_2]—3d'\ [1^1/_2]^\circ$	1—2
7051,2937	70	18,38	20,14	$3p\ [^1/_2]—3d'\ [1^1/_2]^\circ$	1—1
7032,4128	1000	16,62	18,38	$3s\ [1^1/_2]^\circ—3p\ [^1/_2]$	2—1
7024,0500	500	16,85	18,61	$3s'\ [^1/_2]^\circ—3p\ [1^1/_2]$	1—1
6929,4672	1000	16,85	18,63	$3s'\ [^1/_2]^\circ—3p\ [1^1/_2]$	1—2
6759,586	15	18,72	20,56	$3p'\ [^1/_2]—5s\ [1^1/_2]^\circ$	1—2
6738,058	70	18,96	20,80	$3p'\ [^1/_2]—4d'\ [1^1/_2]^\circ$	0—1
6717,0428	70	16,85	18,69	$3s'\ [^1/_2]^\circ—3p'\ [1^1/_2]$	1—1
6678,2764	500	16,85	18,70	$3s'\ [^1/_2]^\circ—3p'\ [1^1/_2]$	1—2
6666,8967	100	18,71	20,57	$3p\ [^1/_2]—5s\ [1^1/_2]^\circ$	0—1
6652,0925	150	16,85	18,71	$3s'\ [^1/_2]^\circ—3p\ [^1/_2]$	1—0
6640,80	5	18,70	20,57	$3p'\ [1^1/_2]—5s\ [1^1/_2]^\circ$	2—1
6640,012	10	18,69	20,56	$3p'\ [1^1/_2]—5s\ [1^1/_2]^\circ$	1—2
6602,907	100	18,69	20,57	$3p'\ [1^1/_2]—5s\ [1^1/_2]^\circ$	1—1
6598,9529	1000	16,85	18,72	$3s'\ [^1/_2]^\circ—3p'\ [^1/_2]$	1—1
6532,8824	100	16,71	18,61	$3s'\ [^1/_2]^\circ—3p'\ [1^1/_2]$	0—1
6506,5279	100	16,67	18,57	$3s\ [1^1/_2]^\circ—3p\ [2^1/_2]$	1—2
6444,7118	150	18,63	20,56	$3p\ [1^1/_2]—5s\ [1^1/_2]^\circ$	2—2
6421,7108	100	18,72	20,66	$3p'\ [^1/_2]—5s'\ [^1/_2]^\circ$	1—0
6409,753	150	18,63	20,57	$3p\ [1^1/_2]—5s\ [1^1/_2]^\circ$	2—1
6402,2460	2000	16,62	18,55	$3s\ [1^1/_2]^\circ—3p\ [2^1/_2]$	2—3

λ, Å	I	E_H, eV	E_B, eV	Transition	J
6401,076	100	18,72	20,66	$3p'\ [^1/_2]-5s'\ [^1/_2]^{\circ}$	1—1
6382,9914	1000	16,67	18,61	$3s\ [1^1/_2]^{\circ}-3p\ [1^1/_2]$	1—1
6365,013	100	18,61	20,56	$3p\ [1^1/_2]-5s\ [1^1/_2]^{\circ}$	1—1
6351,8618	100	18,71	20,66	$3p\ [^1/_2]-5s'\ [^1/_2]^{\circ}$	0—1
6334,4279	1000	16,62	18,57	$3s\ [1^1/_2]^{\circ}-3p\ [2^1/_2]$	2—2
6330,901	150	18,61	20,57	$3p\ [1^1/_2]-3s\ [1^1/_2]^{\circ}$	1—1
6328,1646	300	18,70	20,66	$3p'\ [1^1/_2]-5s'\ [^1/_2]^{\circ}$	2—1
6313,6921	150	18,69	20,66	$3p'\ [1^1/_2]-5s'\ [^1/_2]^{\circ}$	1—0
6304,7892	100	16,67	18,63	$3s\ [1^1/_2]^{\circ}-3p\ [1^1/_2]$	1—2
6293,7447	100	18,69	20,66	$3p'\ [1^1/_2]-5s'\ [^1/_2]^{\circ}$	1—1
6276,039	50	18,72	20,70	$3p'\ [^1/_2]-4d\ [^1/_2]^{\circ}$	1—0
6273,018	70	18,72	20,70	$3p'\ [^1/_2]-4d\ [^1/_2]^{\circ}$	1—1
6266,4950	1000	16,71	18,69	$3s'\ [^1/_2]^{\circ}-3p'\ [1^1/_2]$	0—1
6258,796	100	18,72	20,71	$3p'\ [^1/_2]-4d\ [1^1/_2]^{\circ}$	1—2
6252,732	2	18,72	20,71	$3p'\ [^1/_2]-4d\ [1^1/_2]^{\circ}$	1—1
6249,593	5	18,96	20,95	$3p'\ [^1/_2]-6s\ [1^1/_2]^{\circ}$	0—1
6246,7294	100	18,57	20,56	$3p\ [2^1/_2]-5s\ [1^1/_2]^{\circ}$	2—2
6225,742	50	18,71	20,70	$3p\ [^1/_2]-4d\ [^1/_2]^{\circ}$	0—1
6217,2813	1000	16,62	18,61	$3s\ [1^1/_2]^{\circ}-3p\ [1^1/_2]$	2—1
6213,8758	150	18,57	20,57	$3p\ [2^1/_2]-5s\ [1^1/_2]^{\circ}$	2—1
6205,7775	100	18,71	20,71	$3p\ [^1/_2]-4d\ [1^1/_2]^{\circ}$	0—1
6202,981	15	18,70	20,70	$3p'\ [1^1/_2]-4d\ [^1/_2]^{\circ}$	2—1
6193,0663	50	18,70	20,70	$3p'\ [1^1/_2]-4d\ [3^1/_2]^{\circ}$	2—3
6189,0649	70	18,70	20,71	$3p'\ [1^1/_2]-4d\ [1^1/_2]^{\circ}$	2—2
6183,169	5	18,70	20,71	$3p'\ [1^1/_2]-4d\ [1^1/_2]^{\circ}$	2—1
6182,1460	150	18,55	20,56	$3p\ [2^1/_2]-5s\ [1^1/_2]^{\circ}$	3—2
6175,291	50	18,70	20,71	$3p'\ [1^1/_2]-4d\ [2^1/_2]^{\circ}$	2—2
6174,8829	70	18,70	20,71	$3p'\ [1^1/_2]-4d\ [2^1/_2]^{\circ}$	2—3
6172,821	15	18,69	20,70	$3p'\ [1^1/_2]-4d\ [^1/_2]^{\circ}$	1—0
6163,5939	1000	16,71	18,72	$3s'\ [^1/_2]^{\circ}-3p'\ [^1/_2]$	0—1
6156,145	50	18,69	20,71	$3p'\ [1^1/_2]-4d\ [1^1/_2]^{\circ}$	1—2
6150,303	100	18,69	20,71	$3p'\ [1^1/_2]-4d\ [1^1/_2]^{\circ}$	1—1
6143,0623	1000	16,62	18,63	$3s\ [1^1/_2]^{\circ}-3p\ [1^1/_2]$	2—2
6142,508	100	18,69	20,71	$3p\ [1^1/_2]-4d\ [2^1/_2]^{\circ}$	1—2
6128,4598	100	16,67	18,69	$3s\ [1^1/_2]^{\circ}-3p'\ [1^1/_2]$	1—1
6118,027	15	18,63	20,66	$3p\ [1^1/_2]-5s'\ [^1/_2]^{\circ}$	2—1
6096,1630	300	16,67	18,70	$3s\ [1^1/_2]^{\circ}-3p'\ [1^1/_2]$	1—2
6074,3377	1000	16,67	18,71	$3s\ [1^1/_2]^{\circ}-3p\ [^1/_2]$	1—0
6064,5359	50	18,61	20,66	$3p\ [1^1/_2]-5s'\ [^1/_2]^{\circ}$	1—0
6046,1348	50	18,61	20,66	$3p\ [1^1/_2]-5s'\ [^1/_2]^{\circ}$	1—1
6042,013	15	18,96	21,02	$3p'\ [^1/_2]-5d\ [1^1/_2]^{\circ}$	0—1
6029,9971	1000	16,67	18,72	$3s\ [1^1/_2]^{\circ}-3p'\ [^1/_2]$	1—1
6000,9275	100	18,63	20,70	$3p\ [1^1/_2]-4d\ [^1/_2]^{\circ}$	2—1
5991,6532	75	18,63	20,70	$3p\ [1^1/_2]-4d\ [3^1/_2]^{\circ}$	2—3
5987,9074	150	18,63	20,71	$3p\ [1^1/_2]-4d\ [1^1/_2]^{\circ}$	2—2
5982,401	8	18,63	20,71	$3p\ [1^1/_2]-4d\ [1^1/_2]^{\circ}$	2—1
5975,5340	600	16,62	18,69	$3s\ [1^1/_2]^{\circ}-3p'\ [1^1/_2]^{\circ}$	2—1
5974,6273	500	18,63	20,71	$3p\ [1^1/_2]-4d\ [2^1/_2]^{\circ}$	2—3
5966,171	35	18,72	20,80	$3p'\ [^1/_2]-4d'\ [2^1/_2]^{\circ}$	1—2
5965,4710	500	18,72	20,80	$3p'\ [^1/_2]-4d'\ [1^1/_2]^{\circ}$	1—2
5961,6228	70	18,72	20,80	$3p'\ [^1/_2]-4d'\ [1^1/_2]^{\circ}$	1—1
5944,8342	500	16,62	18,70	$3s\ [1^1/_2]^{\circ}-3p'\ [1^1/_2]$	2—2
5939,319	50	18,57	20,66	$3p\ [2^1/_2]-5s'\ [^1/_2]^{\circ}$	2—1
5934,458	75	18,61	20,70	$3p\ [1^1/_2]-4d\ [1^1/_2]^{\circ}$	1—0
5933,958	8	—	—	—	—
5922,709	25	18,61	20,70	$3p\ [1^1/_2]-4d\ [3^1/_2]^{\circ}$	1—3
5919,037	8	18,61	20,71	$3p\ [1^1/_2]-4d\ [1^1/_2]^{\circ}$	1—2

λ, Å	I	E_H, eV	E_B, eV	Transition	J
5918,9068	250	18,71	20,80	$3p\ [^1/_2]-4d'\ [1^1/_2]^\circ$	0—1
5913,6327	250	18,61	20,71	$3p\ [1^1/_2]-4d\ [1^1/_2]^\circ$	1—1
5906,4294	50	18,61	20,71	$3p\ [1^1/_2]-4d\ [2^1/_2]^\circ$	1—2
5902,783	5	18,70	20,80	$3p'\ [1^1/_2]-4d'\ [2^1/_2]^\circ$	2—2
5902,4623	50	18,70	20,80	$3p'\ [1^1/_2]-4d'\ [2^1/_2]^\circ$	2—3
5902,097	3	18,70	20,80	$3p'\ [1^1/_2]-4d'\ [1^1/_2]^\circ$	2—2
5898,406	20	18,70	20,80	$3p'\ [1^1/_2]-4d'\ [1^1/_2]^\circ$	2—1
5881,8950	1000	16,62	18,72	$3s\ [1^1/_2]^\circ-3p'\ [^1/_2]$	2—1
5872,8275	500	18,69	20,80	$3p'\ [1^1/_2]-4d'\ [2^1/_2]^\circ$	1—2
5872,149	75	18,69	20,80	$3p'\ [1^1/_2]-4d'\ [1^1/_2]^\circ$	1—2
5870,971	3	—	—	—	—
5868,4183	75	18,69	20,80	$3p'\ [1^1/_2]-4d'\ [1^1/_2]^\circ$	1—1
5852,4878	2000	16,85	18,96	$3s'\ [^1/_2]^\circ-3p'\ [^1/_2]$	1—0
5828,910	75	18,57	20,70	$3p\ [2^1/_2]-4d\ [^1/_2]^\circ$	2—1
5820,1558	500	18,57	20,70	$3p\ [2^1/_2]-4d\ [3^1/_2]^\circ$	2—3
5816,645	50	18,57	20,71	$3p\ [2^1/_2]-4d\ [1^1/_2]^\circ$	2—2
5812,400	15	—	—	—	—
5811,4066	300	18,57	20,71	$3p\ [2^1/_2]-4d\ [1^1/_2]^\circ$	2—1
5804,4496	500	18,57	20,71	$3p\ [2^1/_2]-4d\ [2^1/_2]^\circ$	2—2
5804,098	75	18,57	20,71	$3p\ [2^1/_2]-4d\ [2^1/_2]^\circ$	2—3
5770,307	50	18,96	21,11	$3p'\ [^1/_2]-5d'\ [1^1/_2]^\circ$	0—1
5764,4188	700	18,55	20,70	$3p\ [2^1/_2]-4d\ [3^1/_2]^\circ$	3—4
5764,063	3	18,55	20,70	$3p\ [2^1/_2]-4d\ [3^1/_2]^\circ$	3—3
5760,5885	70	18,55	20,71	$3p\ [2^1/_2]-4d\ [1^1/_2]^\circ$	3—2
5748,650	70	18,55	20,71	$3p\ [2^1/_2]-4d\ [2^1/_2]^\circ$	3—2
5748,2985	500	18,55	20,71	$3p\ [2^1/_2]-4d\ [2^1/_2]^\circ$	3—3
5719,532	75	18,63	20,80	$3p\ [1^1/_2]-4d'\ [2^1/_2]^\circ$	2—2
5719,2248	500	18,63	20,80	$3p\ [1^1/_2]-4d'\ [2^1/_2]^\circ$	2—3
5718,899	150	18,63	20,80	$3p\ [1^1/_2]-4d'\ [1^1/_2]^\circ$	2—2
5715,339	35	18,63	20,80	$3p\ [1^1/_2]-4d'\ [1^1/_2]^\circ$	2—1
5689,8163	150	18,38	20,56	$3p\ [^1/_2]-5s\ [1^1/_2]^\circ$	1—2
5684,647	25	18,96	21,14	$3p'\ [^1/_2]-7s\ [1^1/_2]^\circ$	0—1
5662,5489	75	18,38	20,56	$3p\ [^1/_2]-5s\ [1^1/_2]^\circ$	1—1
5656,6588	500	18,61	20,80	$3p\ [1^1/_2]-4d'\ [2^1/_2]^\circ$	1—2
5656,030	75	18,61	20,80	$3p\ [1^1/_2]-4d'\ [1^1/_2]^\circ$	1—2
5652,5664	75	18,61	20,80	$3p\ [1^1/_2]-4d'\ [1^1/_2]^\circ$	1—1
5591,15	8	18,96	21,18	$3p'\ [^1/_2]-6d\ [^1/_2]^\circ$	0—1
5589,378	50	18,72	20,94	$3p'\ [^1/_2]-6s\ [1^1/_2]^\circ$	1—2
5585,905	5	18,96	21,18	$3p'\ [^1/_2]-6d\ [1^1/_2]^\circ$	0—1
5576,049	35	18,72	20,95	$3p'\ [^1/_2]-6s\ [1^1/_2]^\circ$	1—1
5563,047	75	18,57	20,80	$3p\ [2^1/_2]-4d'\ [2^1/_2]^\circ$	2—2
5562,7662	500	18,57	20,80	$3p\ [2^1/_2]-4d'\ [2^1/_2]^\circ$	2—3
5562,441	150	18,57	20,80	$3p\ [2^1/_2]-4d'\ [1^1/_2]^\circ$	2—2
5559,087	35	18,57	20,80	$3p\ [2^1/_2]-4d'\ [1^1/_2]^\circ$	2—1
5538,651	50	18,71	20,95	$3p\ [^1/_2]-6s\ [1^1/_2]^\circ$	0—1
5533,6788	75	18,70	20,94	$3p'\ [1^1/_2]-6s\ [1^1/_2]^\circ$	2—2
5520,63	3	18,70	20,95	$3p'\ [1^1/_2]-6s\ [1^1/_2]^\circ$	2—1
5511,485	15	18,55	20,80	$3p\ [2^1/_2]-4d'\ [2^1/_2]^\circ$	3—3
5511,176	3	18,55	20,80	$3p\ [2^1/_2]-4d'\ [1^1/_2]^\circ$	3—2
5507,339	25	18,69	20,94	$3p'\ [1^1/_2]-6s\ [1^1/_2]^\circ$	1—2
5494,4158	50	18,69	20,95	$3p'\ [1^1/_2]-6s\ [1^1/_2]^\circ$	1—1
5448,5091	150	18,38	20,66	$3p\ [^1/_2]-5s'\ [^1/_2]^\circ$	1—0
5447,120	8	18,96	21,24	$3p'\ [^1/_2]-7s'\ [^1/_2]^\circ$	0—1
5433,6513	250	18,38	20,66	$3p\ [^1/_2]-5s'\ [^1/_2]^\circ$	1—1
5420,155	50	18,72	21,01	$3p'\ [^1/_2]-5d\ [^1/_2]^\circ$	1—0
5418,5584	150	18,72	21,01	$3p'\ [^1/_2]-5d\ [^1/_2]^\circ$	1—1
5412,649	250	18,72	21,02	$3p'\ [^1/_2]-5d\ [1^1/_2]^\circ$	1—2
5410,12	5	18,72	21,02	$3p'\ [^1/_2]-5d\ [1^1/_2]^\circ$	1—1

λ, Å	I	E_H, eV	E_B, eV	Transition	J
5400,5616	2000	16,67	18,96	$3s\,[1{}^1/_2]^\circ - 3p'\,[{}^1/_2]$	1—0
5383,250	25	18,71	21,01	$3p\,[{}^1/_2] - 5d\,[{}^1/_2]^\circ$	0—1
5374,9774	50	18,71	21,02	$3p\,[{}^1/_2] - 5d\,[1{}^1/_2]^\circ$	0—1
5372,3110	75	18,63	20,94	$3p\,[1{}^1/_2] - 6s\,[1{}^1/_2]^\circ$	2—2
5366,222	25	18,70	21,01	$3p'\,[1{}^1/_2] - 5d\,[{}^1/_2]^\circ$	2—1
5362,248	25	18,70	21,01	$3p'\,[1{}^1/_2] - 5d\,[3{}^1/_2]^\circ$	2—3
5360,442	35	18,70	21,02	$3p'\,[1{}^1/_2] - 5d\,[1{}^1/_2]^\circ$	2—2
5360,0121	150	18,63	20,95	$3p\,[1{}^1/_2] - 6s\,[1{}^1/_2]^\circ$	2—1
5358,020	10	18,70	21,02	$3p'\,[1{}^1/_2] - 5d\,[1{}^1/_2]^\circ$	2—1
5355,422	150	18,70	21,02	$3p'\,[1{}^1/_2] - 5d\,[2{}^1/_2]^\circ$	2—2
5355,176	150	18,70	21,02	$3p'\,[1{}^1/_2] - 5d\,[2{}^1/_2]^\circ$	2—3
5353,513	5	18,96	21,28	$3p'\,[{}^1/_2] - 6d'\,[1{}^1/_2]^\circ$	0—1
5349,204	150	18,72	21,04	$3p'\,[{}^1/_2] - 6s'\,[{}^1/_2]^\circ$	1—1
5343,2834	600	18,38	20,70	$3p\,[{}^1/_2] - 4d\,[{}^1/_2]^\circ$	1—0
5342,700	1	18,96	21,28	$3p'\,[{}^1/_2] - 7d\,[1{}^1/_2]^\circ$	1—1
5341,0938	1000	18,38	20,70	$3p\,[{}^1/_2] - 4d\,[{}^1/_2]^\circ$	1—1
5335,710	10	18,69	21,02	$3p'\,[1{}^1/_2] - 5d\,[1{}^1/_2]^\circ$	1—2
5333,323	50	18,69	21,02	$3p'\,[1{}^1/_2] - 5d\,[1{}^1/_2]^\circ$	1—1
5330,7775	600	18,38	20,71	$3p\,[{}^1/_2] - 4d\,[1{}^1/_2]^\circ$	1—2
5326,3968	75	18,38	20,71	$3p\,[{}^1/_2] - 4d\,[1{}^1/_2]^\circ$	1—1
5320,550	2	18,38	20,71	$3p\,[{}^1/_2] - 4d\,[2{}^1/_2]^\circ$	1—2
5316,806	25	18,61	20,94	$3p\,[1{}^1/_2] - 6s\,[1{}^1/_2]^\circ$	1—2
5314,781	30	18,71	21,04	$3p\,[{}^1/_2] - 6s'\,[{}^1/_2]^\circ$	0—1
5304,7580	70	18,61	20,95	$3p\,[1{}^1/_2] - 6s\,[1{}^1/_2]^\circ$	1—1
5298,1891	150	18,70	21,04	$3p'\,[1{}^1/_2] - 6s'\,[{}^1/_2]^\circ$	2—1
5280,0853	50	18,69	21,04	$3p'\,[1{}^1/_2] - 6s'\,[{}^1/_2]^\circ$	1—0
5274,0393	40	18,69	21,04	$3p'\,[1{}^1/_2] - 6s'\,[{}^1/_2]^\circ$	1—1
5234,0271	50	18,57	20,94	$3p\,[2{}^1/_2] - 6s\,[1{}^1/_2]^\circ$	2—2
5222,3517	50	18,57	20,95	$3p\,[2{}^1/_2] - 6s\,[1{}^1/_2]^\circ$	2—1
5214,339	35	18,63	21,01	$3p\,[1{}^1/_2] - 5d\,[{}^1/_2]^\circ$	2—1
5210,5672	50	18,63	21,01	$3p\,[1{}^1/_2] - 5d\,[3{}^1/_2]^\circ$	2—3
5208,8648	70	18,63	21,02	$3p\,[1{}^1/_2] - 5d\,[1{}^1/_2]^\circ$	2—2
5206,565	3	18,63	21,02	$3p\,[1{}^1/_2] - 5d\,[1{}^1/_2]^\circ$	2—1
5203,8962	150	18,63	21,02	$3p\,[1{}^1/_2] - 5d\,[2{}^1/_2]^\circ$	2—3
5193,2227	150	18,72	21,11	$3p'\,[{}^1/_2] - 5d'\,[2{}^1/_2]^\circ$	1—2
5193,1302	150	18,72	21,11	$3p'\,[{}^1/_2] - 5d'\,[1{}^1/_2]^\circ$	1—2
5191,322	35	18,72	21,11	$3p'\,[{}^1/_2] - 5d'\,[1{}^1/_2]^\circ$	3—2
5188,6122	150	18,55	20,94	$3p\,[2{}^1/_2] - 6s\,[1{}^1/_2]^\circ$	3—2
5182,320	2	18,96	21,36	$3p'\,[{}^1/_2] - 8s'\,[{}^1/_2]^\circ$	0—1
5163,474	10	18,61	20,01	$3p\,[1{}^1/_2] - 5d\,[{}^1/_2]^\circ$	1—0
5158,902	50	18,71	21,11	$3p\,[{}^1/_2] - 5d'\,[1{}^1/_2]^\circ$	0—1
5156,667	50	18,61	21,02	$3p\,[1{}^1/_2] - 5d\,[1{}^1/_2]^\circ$	1—2
5154,4271	50	18,61	21,02	$3p\,[1{}^1/_2] - 5d\,[1{}^1/_2]^\circ$	1—1
5151,9610	75	18,61	21,02	$3p\,[1{}^1/_2] - 5d\,[2{}^1/_2]^\circ$	1—2
5150,077	35	18,63	20,04	$3p\,[1{}^1/_2] - 6s'\,[{}^1/_2]^\circ$	2—1
5145,122	35	18,70	21,11	$3p'\,[1{}^1/_2] - 5d'\,[2{}^1/_2]^\circ$	2—2
5145,011	500	18,70	21,11	$3p'\,[1{}^1/_2] - 5d'\,[1{}^1/_2]^\circ$	2—2
5144,9384	500	18,70	21,11	$3p'\,[1{}^1/_2] - 5d'\,[2{}^1/_2]^\circ$	2—3
5143,2665	5	18,70	21,11	$3p'\,[1{}^1/_2] - 5d'\,[1{}^1/_2]^\circ$	2—1
5128,280	2	18,72	21,14	$3p'\,[{}^1/_2] - 7s\,[1{}^1/_2]^\circ$	1—2
5122,337	150	18,69	21,11	$3p'\,[1{}^1/_2] - 5d'\,[2{}^1/_2]^\circ$	1—2
5122,257	150	18,69	21,11	$3p'\,[1{}^1/_2] - 5d'\,[1{}^1/_2]^\circ$	1—2
5121,866	2	18,72	21,14	$3p'\,[{}^1/_2] - 7s\,[1{}^1/_2]^\circ$	1—1
5120,506	25	18,69	21,11	$3p'\,[1{}^1/_2] - 5d'\,[1{}^1/_2]^\circ$	1—1
5117,011	35	18,38	20,80	$3p\,[{}^1/_2] - 4d'\,[2{}^1/_2]^\circ$	1—2
5116,5032	150	18,38	20,80	$3p\,[{}^1/_2] - 4d'\,[1{}^1/_2]^\circ$	1—2
5113,6724	75	18,38	20,80	$3p\,[{}^1/_2] - 4d'\,[1{}^1/_2]^\circ$	1—1

λ, Å	I	E_H, eV	E_B, eV	Transition	J
5104,701	35	18,61	21,04	$3p\ [1^1/_2]—6s'\ [^1/_2]°$	1—0
5099,042	25	18,61	21,04	$3p\ [1^1/_2]—6s'\ [^1/_2]°$	1—1
5090,321	8	18,71	21,14	$3p\ [^1/_2]—7s\ [1^1/_2]°$	0—1
5083,968	25	18,57	21,01	$3p\ [2^1/_2]—5d\ [^1/_2]°$	2—1
5081,360	15	18,70	21,14	$3p'\ [1^1/_2]—7s\ [1^1/_2]°$	2—2
5080,3852	150	18,57	21,01	$3p\ [2^1/_2]—5d\ [3^1/_2]°$	2—3
5078,762	15	18,57	21,01	$3p\ [2^1/_2]—5d\ [1^1/_2]°$	2—2
5076,581	35	18,57	21,01	$3p\ [2^1/_2]—5d\ [1^1/_2]°$	2—1
5074,201	35	18,57	21,02	$3p\ [2^1/_2]—5d\ [2^1/_2]°$	2—2
5074,062	3	18,57	21,02	$3p\ [2^1/_2]—5d\ [2^1/_2]°$	2—3
5059,150	2	18,69	21,14	$3p'\ [1^1/_2]—7s\ [1^1/_2]°$	1—2
5052,930	25	18,69	21,14	$3p'\ [1^1/_2]—7s\ [1^1/_2]°$	1—1
5046,608	3	18,72	21,18	$3p'\ [^1/_2]—6d\ [^1/_2]°$	1—0
5045,816	15	18,72	21,18	$3p'\ [^1/_2]—6d\ [^1/_2]°$	1—1
5042,853	15	18,72	21,18	$3p'\ [^1/_2]—6d\ [1^1/_2]°$	1—2
5041,598	1	18,72	21,18	$3p'\ [^1/_2]—6d\ [1^1/_2]°$	1—1
5037,7512	500	18,55	21,01	$3p\ [2^1/_2]—5d\ [3^1/_2]°$	3—4
5037,577	3	18,55	21,01	$3p\ [2^1/_2]—5d\ [3^1/_2]°$	3—3
5035,989	35	18,55	21,02	$3p\ [2^1/_2]—5d\ [1^1/_2]°$	3—2
5031,483	2	18,55	21,02	$3p\ [2^1/_2]—5d\ [2^1/_2]°$	3—2
5031,3504	250	18,55	21,02	$3p\ [2^1/_2]—5d\ [2^1/_2]°$	3—3
5022,870	25	18,57	21,04	$3p\ [2^1/_2]—6s'\ [^1/_2]°$	2—1
5015,187	5	18,71	21,18	$3p\ [^1/_2]—6d\ [^1/_2]°$	0—1
5011,003	25	18,71	21,18	$3p\ [^1/_2]—6d\ [1^1/_2]°$	0—1
5005,333	50	18,63	21,11	$3p\ [1^1/_2]—5d'\ [2^1/_2]°$	2—2
5005,1587	500	18,63	21,11	$3p\ [1^1/_2]—5d'\ [2^1/_2]°$	2—3
5003,561	2	18,63	21,11	$3p\ [1^1/_2]—5d'\ [1^1/_2]°$	2—1
5000,395	3	18,70	21,18	$3p'\ [1^1/_2]—6d\ [^1/_2]°$	2—1
4998,502	10	18,70	21,18	$3p'\ [1^1/_2]—6d\ [3^1/_2]°$	2—3
4997,482	15	18,70	21,18	$3p'\ [1^1/_2]—6d\ [1^1/_2]°$	2—2
4996,209	2	18,70	21,18	$3p'\ [1^1/_2]—6d\ [1^1/_2]°$	2—1
4994,930	150	18,70	21,18	$3p'\ [1^1/_2]—6d\ [2^1/_2]°$	2—3
4979,625	5	18,69	21,18	$3p'\ [1^1/_2]—6d\ [^1/_2]°$	1—0
4975,961	10	18,69	21,18	$3p'\ [1^1/_2]—6d\ [1^1/_2]°$	1—2
4974,760	50	18,69	21,18	$3p'\ [1^1/_2]—6d\ [1^1/_2]°$	1—1
4973,538	100	18,69	21,18	$3p'\ [1^1/_2]—6d\ [2^1/_2]°$	1—2
4957,122	150	18,61	21,11	$3p\ [1^1/_2]—5d'\ [2^1/_2]°$	1—2
4957,0335	1000	18,61	21,11	$3p\ [1^1/_2]—5d'\ [1^1/_2]°$	1—2
4955,382	150	18,61	21,11	$3p\ [1^1/_2]—5d'\ [1^1/_2]°$	1—1
4944,9899	100	18,63	21,14	$3p\ [1^1/_2]—7s\ [1^1/_2]°$	2—2
4939,0457	100	18,63	21,14	$3p\ [1^1/_2]—7s\ [1^1/_2]°$	2—1
4930,944	50	18,72	21,24	$3p'\ [^1/_2]—7s'\ [^1/_2]°$	1—0
4928,235	70	18,72	21,24	$3p'\ [^1/_2]—7s'\ [^1/_2]°$	1—1
4899,013	50	18,71	21,24	$3p\ [^1/_2]—7s'\ [^1/_2]°$	0—1
4897,924	70	18,61	21,14	$3p\ [1^1/_2]—7s\ [1^1/_2]°$	1—2
4892,228	10	18,72	21,26	$3p'\ [^1/_2]—8s\ [1^1/_2]°$	1—2
4892,1007	500	18,61	21,14	$3p\ [1^1/_2]—7s\ [1^1/_2]°$	1—1
4888,365	5	18,72	21,26	$3p'\ [^1/_2]—8s\ [1^1/_2]°$	1—1
4885,084	100	18,57	21,11	$3p\ [2^1/_2]—5d'\ [2^1/_2]°$	2—2
4884,9170	1000	{ 18,70 18,57	21,24 21,11	$3p'\ [1^1/_2]—7s'\ [^1/_2]°$ $3p\ [2^1/_2]—5d'\ [2^1/_2]°$	2—1 2—3
4883,403	15	18,57	21,11	$3p\ [2^1/_2]—5d'\ [1^1/_2]°$	2—1
4868,268	70	18,63	21,18	$3p\ [1^1/_2]—6d\ [^1/_2]°$	2—1
4867,010	70	18,69	21,24	$3p'\ [1^1/_2]—7s'\ [^1/_2]°$	1—0
4866,476	80	18,63	21,18	$3p\ [1^1/_2]—6d\ [3^1/_2]°$	2—3
4865,501	100	18,63	21,18	$3p\ [1^1/_2]—6d\ [1^1/_2]°$	2—2
4864,351	30	18,69	21,24	$3p'\ [1^1/_2]—7s'\ [^1/_2]°$	1—1
4863,0800	100	18,63	21,18	$3p\ [1^1/_2]—6d\ [2^1/_2]°$	2—3

λ, Å	I	E_H, eV	E_B, eV	Transition	J
4859,604	15	18,71	21,26	$3p\ [^1/_2]-8s\ [1^1/_2]^\circ$	0—1
4852,6571	100	18,72	21,28	$3p'\ [^1/_2]-6d'\ [2^1/_2]^\circ$	1—2
4851,501	60	18,72	21,28	$3p'\ [^1/_2]-6d\ [1^1/_2]^\circ$	1—1
4849,530	30	18,70	21,26	$3p'\ [1^1/_2]-8s\ [1^1/_2]^\circ$	2—2
4845,767	5	18,72	21,28	$3p'\ [^1/_2]-7d\ [^1/_2]^\circ$	1—0
4845,145	15	18,72	21,28	$3p'\ [^1/_2]-7d\ [^1/_2]^\circ$	1—1
4842,941	50	18,72	21,28	$3p'\ [^1/_2]-7d\ [1^1/_2]^\circ$	1—2
4842,566	10	18,72	21,28	$3p'\ [^1/_2]-7d\ [1^1/_2]^\circ$	1—1
4837,3139	500	18,38	20,94	$3p\ [^1/_2]-6s\ [1^1/_2]^\circ$	1—2
4829,288	5	18,69	21,26	$3p'\ [1^1/_2]-8s\ [1^1/_2]^\circ$	1—2
4827,587	300	18,57	21,14	$3p\ [2^1/_2]-7s\ [1^1/_2]^\circ$	2—2
4827,3444	1000	18,38	21,95	$3p\ [^1/_2]-6s\ [1^1/_2]^\circ$	1—1
4825,529	50	18,69	21,26	$3p'\ [1^1/_2]-8s\ [1^1/_2]^\circ$	1—1
4823,370	50	18,61	21,18	$3p\ [1^1/_2]-6d\ [^1/_2]^\circ$	1—0
4823,174	100	18,71	21,28	$3p\ [^1/_2]-6d'\ [1^1/_2]^\circ$	0—1
4821,9236	300	18,57	21,14	$3p\ [2^1/_2]-7s\ [1^1/_2]^\circ$	2—1
4819,937	70	18,61	21,18	$3p\ [1^1/_2]-6d\ [1^1/_2]^\circ$	1—2
4818,789	150	18,61	21,18	$3p\ [1^1/_2]-6d\ [1^1/_2]^\circ$	1—1
4817,6386	300	18,61	21,18	$3p\ [1^1/_2]-6d\ [2^1/_2]^\circ$	1—2
4816,900	1	18,71	21,28	$3p\ [^1/_2]-7d\ [^1/_2]^\circ$	0—1
4814,338	50	18,71	21,28	$3p\ [^1/_2]-7d\ [1^1/_2]^\circ$	0—1
4810,634	100	18,70	21,28	$3p'\ [1^1/_2]-6d'\ [2^1/_2]^\circ$	2—2
4810,0640	150	18,70	21,28	$3p'\ [1^1/_2]-6d'\ [2^1/_2]^\circ$	2—3
4809,500	10	18,70	21,28	$3p'\ [1^1/_2]-6d'\ [1^1/_2]^\circ$	2—1
4803,225	1	18,70	21,28	$3p'\ [1^1/_2]-7d\ [^1/_2]^\circ$	2—1
4802,363	10	18,70	21,28	$3p'\ [1^1/_2]-7d\ [3^1/_2]^\circ$	2—3
4801,076	2	18,70	21,28	$3p'\ [1^1/_2]-7d\ [1^1/_2]^\circ$	2—2
4800,111	15	18,70	21,29	$3p'\ [1^1/_2]-7d\ [2^1/_2]^\circ$	2—3
4790,728	30	18,69	21,28	$3p'\ [1^1/_2]-6d'\ [2^1/_2]^\circ$	1—2
4790,218	500	18,69	21,28	$3p'\ [1^1/_2]-6d'\ [1^1/_2]^\circ$	1—2
4789,600	100	18,69	21,28	$3p'\ [1^1/_2]-6d'\ [1^1/_2]^\circ$	1—1
4788,9270	1000	18,55	21,14	$3p\ [2^1/_2]-7s\ [1^1/_2]^\circ$	3—2
4784,022	2	18,69	21,28	$3p'\ [1^1/_2]-7d\ [^1/_2]^\circ$	1—0
4781,239	2	18,69	21,28	$3p'\ [1^1/_2]-7d\ [1^1/_2]^\circ$	1—2
4780,884	30	18,69	21,28	$3p'\ [1^1/_2]-7d\ [1^1/_2]^\circ$	1—1
4780,338	300	18,69	21,29	$3p'\ [1^1/_2]-7d\ [2^1/_2]^\circ$	1—2
4758,728	150	18,63	21,24	$3p\ [1^1/_2]-7s'\ [^1/_2]^\circ$	2—1
4754,440	100	18,57	21,18	$3p\ [2^1/_2]-6d\ [^1/_2]^\circ$	2—1
4753,123	1	18,72	21,33	$3p'\ [^1/_2]-9s\ [1^1/_2]^\circ$	1—2
4752,7320	500	18,57	21,18	$3p'\ [^1/_2]-6d\ [3^1/_2]^\circ$	2—3
4751,802	30	18,57	21,18	$3p\ [2^1/_2]-6d\ [1^1/_2]^\circ$	2—2
4750,686	30	18,57	21,18	$3p\ [2^1/_2]-6d\ [1^1/_2]^\circ$	2—1
4749,5754	300	18,57	21,18	$3p\ [2^1/_2]-6d\ [2^1/_2]^\circ$	2—2
4725,145	70	18,63	21,26	$3p\ [1^1/_2]-8s\ [1^1/_2]^\circ$	2—2
4724,162	5	18,72	21,35	$3p'\ [^1/_2]-8d\ [^1/_2]^\circ$	1—0
4723,810	70	18,71	21,33	$3p\ [^1/_2]-9s\ [1^1/_2]^\circ$	0—1
		18,72	21,35	$3p'\ [^1/_2]-8d\ [^1/_2]^\circ$	1—1
4722,714	15	18,72	21,35	$3p'\ [^1/_2]-8d\ [1^1/_2]^\circ$	1—2
4722,150	5	18,72	21,35	$3p'\ [^1/_2]-8d\ [1^1/_2]^\circ$	1—1
4721,536	70	18,63	21,26	$3p\ [1^1/_2]-8s\ [1^1/_2]^\circ$	2—1
4717,608	70	18,61	21,24	$3p\ [1^1/_2]-7s'\ [^1/_2]^\circ$	1—0
4715,3466	1500	18,55	21,18	$3p\ [2^1/_2]-6d\ [3^1/_2]^\circ$	3—4
4715,246	30	18,55	21,18	$3p\ [2^1/_2]-6d\ [3^1/_2]^\circ$	3—3
4715,132	30	18,61	21,24	$3p\ [1^1/_2]-7s'\ [^1/_2]^\circ$	1—1
4714,336	70	18,55	21,18	$3p\ [2^1/_2]-6d\ [1^1/_2]^\circ$	3—2
4712,800	10	18,70	21,33	$3p\ [1^1/_2]-9s\ [1^1/_2]^\circ$	2—2
4712,135	15	18,72	21,36	$3p'\ [^1/_2]-8s'\ [^1/_2]^\circ$	1—0
4712,066	1000	18,55	21,18	$3p\ [2^1/_2]-6d\ [2^1/_2]^\circ$	3—3
4710,478	30	18,72	21,36	$3p'\ [^1/_2]-8s'\ [^1/_2]^\circ$	1—1

λ, Å	I	E_H, eV	E_B, eV	Transition	J
4710,0669	1000	18,38	21,01	3p [1/2]—5d [1/2]°	1—0
4708,8619	1200	18,38	21,01	3p [1/2]—5d [1/2]°	1—1
4704,3949	1500	18,38	21,02	3p [1/2]—5d [1 1/2]°	1—2
4702,526	150	18,38	21,02	3p [1/2]—5d [1 1/2]°	1—1
4700,469	5	18,38	21,02	3p [1/2]—5d [2 1/2]°	1—2
4696,943	5	18,71	21,35	3p [1/2]—8d [1/2]°	0—1
4695,363	20	18,71	21,35	3p [1/2]—8d [1 1/2]°	0—1
4691,580	15	18,69	21,33	3p′ [1 1/2]—9s [1 1/2]°	1—1
4688,191	2	18,63	21,28	3p [1 1/2]—6d′ [2 1/2]°	2—2
4687,6724	100	18,63	21,28	3p [1 1/2]—6d′ [2 1/2]°	2—3
4683,764	30	18,71	21,36	3p′ [1/2]—8s′ [1/2]°	0—1
4683,238	5	18,70	21,35	3p′ [1 1/2]—8d [3 1/2]°	2—3
4682,910	10	18,70	21,35	3p′ [1 1/2]—8d [1 1/2]°	2—2
4682,146	20	18,61	21,26	3p [1 1/2]—8s [1 1/2]°	1—2
4681,930	20	18,70	21,35	3p′ [1 1/2]—8d [2 1/2]°	2—3
4681,200	50	18,63	21,28	3p [1 1/2]—7d [1/2]°	2—1
4680,363	100	18,63	21,28	3p [1 1/2]—7d [3 1/2]°	2—3
4679,135	150	18,63	21,28	3p [1 1/2]—7d [1 1/2]°	2—2
4678,604	50	18,61	21,26	3p [1 1/2]—8s [1 1/2]°	1—1
4678,218	300	18,63	21,29	3p [1 1/2]—7d [2 1/2]°	2—3
4670,884	70	18,70	21,36	3p′ [1 1/2]—8s [1/2]°	2—1
4667,356	100	18,72	21,38	3p′ [1/2]—7d′ [2 1/2]°	1—2
4666,654	50	18,72	21,38	3p′ [1/2]—7d′ [1 1/2]°	1—1
4663,518	20	18,72 18,69	21,35 21,35	3p′ [1/2]—10s [1 1/2]° 3p′ [1 1/2]—8d [1 1/2]°	1—2 1—1
4663,092	40	18,69	21,35	3p′ [1 1/2]—8d [2 1/2]°	1—2
4661,1054	150	18,38	21,04	3p [1/2]—6s′ [1/2]°	1—0
4656,3936	300	18,38	21,04	3p [1/2]—6s′ [1/2]°	1—1
4653,699	50	18,69	21,36	3p′ [1 1/2]—8s′ [1/2]°	1—0
4652,101	30	18,69	21,36	3p′ [1 1/2]—8s′ [1/2]°	1—1
4649,904	70	18,57	21,24	3p [2 1/2]—7s′ [1/2]°	2—1
4645,885	1	18,61	21,28	3p [1 1/2]—6d′ [1 1/2]°	1—2
4645,4180	300	18,61	21,28	3p [1 1/2]—6d′ [2 1/2]°	1—2
4644,833	40	18,61	21,28	3p [1/2]—6d′ [1 1/2]°	1—1
4643,931	2	18,72	21,39	3p′ [1/2]—9d [1/2]°	1—1
4643,182	5	18,72	21,39	3p′ [1/2]—9d [1 1/2]°	1—2
4640,443	70	18,71	21,38	3p [1/2]—7d′ [1 1/2]°	0—1
4639,591	30	18,61	21,28	3p [1 1/2] 7d [1/2]°	1—0
4636,974	50	18,61	21,28	3p [1 1/2]—7d [1 1/2]°	1—2
4636,634	70	18,61	21,28	3p [1 1/2]—7d [1 1/2]°	1—1
4636,125	70	18,61	21,29	3p [1 1/2]—7d [2 1/2]°	1—2
4628,460	30	18,70	21,38	3p′ [1 1/2]—7d′ [1 1/2]°	2—2
4628,3113	150	18,70	21,38	3p′ [1 1/2]—7d′ [2 1/2]°	2—2, 3
4627,799	2	18,70	21,38	3p′ [1 1/2]—7d′ [1 1/2]°	2—1
4617,837	70	18,57	21,26	3p [2 1/2]—8s [1 1/2]°	2—2
4616,911	5	18,71	21,39	3p [1/2]—9d [1 1/2]°	0—1
4614,391	100	18,57	21,26	3p [2 1/2]—8s [1 1/2]°	2—1
4609,910	150	18,69	21,38	3p′ [1 1/2]—7d′ [1 1/2]°	1—2
4609,365	30	18,69	21,38	3p [1 1/2]—7d′ [1 1/2]°	1—1
4604,938	5	18,69	21,38	3p′ [1 1/2]—10s [1 1/2]°	1—1
4604,680	1	18,70	21,39	3p′ [1 1/2]—9d [1 1/2]°	2—2
4604,095	15	18,70	21,40	3p′ [1 1/2]—9d [2 1/2]°	2—3
4595,249	50	18,63	21,33	3p [1 1/2]—9s [1 1/2]°	2—2
4593,243	50	18,63	21,33	3p [1 1/2]—9s [1 1/2]°	2—1
4586,145	2	18,69	21,39	3p′ [1 1/2]—9d [1 1/2]°	1—1
4585,876	10	18,69	21,40	3p′ [1 1/2]—9d [2 1/2]°	1—2
4582,980	2	18,72	21,43	3p′ [1/2]—9s′ [1/2]°	1—0
4582,556	15	18,57	21,28	3p [2 1/2]—6d′ [2 1/2]°	2—2

λ, Å	I	E_H, eV	E_B, eV	Transition	J
4582,4521	150	18,55	21,26	$3p\ [2^1/_2]-8s\ [1^1/_2]^\circ$	3—2
4582,105	15	18,57	21,28	$3p\ [2^1/_2]-6d'\ [1^1/_2]^\circ$	2—2
4582,035	150	18,57	21,28	$3p\ [2^1/_2]-6d'\ [2^1/_2]^\circ$	2—3
4575,858	20	18,57	21,28	$3p\ [2^1/_2]-7d\ [^1/_2]^\circ$	2—1
4575,0620	300	18,57	21,28	$3p\ [2^1/_2]-7d\ [3^1/_2]^\circ$	2—3
4573,759	30	18,57	21,28	$3p\ [2^1/_2]-7d\ [1^1/_2]^\circ$	2—2
4573,557	50	18,57	21,28	$3p\ [2^1/_2]-7d\ [1^1/_2]^\circ$	2—1
4573,066	5	18,57	21,29	$3p\ [2^1/_2]-7d\ [2^1/_2]^\circ$	2—2
4567,845	10	18,63	21,35	$3p\ [1^1/_2]-8d\ [^1/_2]^\circ$	2—1
4567,139	15	18,63	21,35	$3p\ [1^1/_2]-8d\ [3^1/_2]^\circ$	2—3
4566,830	40	18,63	21,35	$3p\ [1^1/_2]-8d\ [1^1/_2]^\circ$	2—2
4565,888	60	18,63	21,35	$3p\ [1^1/_2]-8d\ [2^1/_2]^\circ$	2—3
4556,698	2	18,71	21,43	$3p\ [^1/_2]-9s'\ [^1/_2]^\circ$	0—1
4555,392	30	18,63	21,36	$3p\ [1^1/_2]-8s'\ [^1/_2]^\circ$	2—1
4554,824	40	18,72	21,45	$3p'\ [^1/_2]-8d'\ [2^1/_2]^\circ$	1—2
4554,561	5	18,61	21,33	$3p\ [1^1/_2]-9s\ [1^1/_2]^\circ$	1—2
4554,415	10	18,72	21,45	$3p'\ [^1/_2]-8d'\ [1^1/_2]^\circ$	1—1
4552,598	30	18,61	21,33	$3p\ [1^1/_2]-9s\ [1^1/_2]^\circ$	1—1
4550,640	1	18,70	21,43	$3p'\ [1^1/_2]-10d\ [3^1/_2]^\circ$	2—3
4547,728	15	18,55	21,28	$3p\ [2^1/_2]-6d\ [2^1/_2]^\circ$	3—2
4547,218	10	18,55	21,28	$3p\ [2^1/_2]-6d\ [2^1/_2]^\circ$	3—3
4545,729	1	18,69	21,42	$3p'\ [1^1/_2]-11s\ [1^1/_2]^\circ$	1—1
4544,502	50	18,70	21,43	$3p'\ [1^1/_2]-9s'\ [^1/_2]^\circ$	2—1
4540,380	50	18,55	21,28	$3p\ [2^1/_2]-7d\ [3^1/_2]^\circ$	3—4
4539,168	50	18,55	21,28	$3p\ [2^1/_2]-7d\ [1^1/_2]^\circ$	3—2
4538,293	300	18,55	21,29	$3p\ [2^1/_2]-7d\ [2^1/_2]^\circ$	3—3
4537,7545	1000	18,38	21,11	$3p\ [^1/_2]-5d'\ [2^1/_2]^\circ$	1—2
4537,683	300	18,38	21,11	$3p\ [^1/_2]-5d'\ [1^1/_2]^\circ$	1—2
4536,312	150	18,38	21,11	$3p\ [^1/_2]-5d'\ [1^1/_2]^\circ$	1—1
4532,395	1	18,69	21,43	$3p'\ [1^1/_2]-10d\ [1^1/_2]^\circ$	1—1
4529,476	30	18,71	21,45	$3p\ [^1/_2]-8d'\ [1^1/_2]^\circ$	0—1
4527,973	1	18,61	21,35	$3p\ [1^1/_2]-8d\ [^1/_2]^\circ$	1—0
4527,725	15	18,69	21,43	$3p'\ [1^1/_2]-9s'\ [^1/_2]^\circ$	1—0
4526,685	15	18,69	21,43	$3p'\ [1^1/_2]-9s'\ [^1/_2]^\circ$	1—1
		18,61	21,35	$3p\ [1^1/_2]-8d\ [1^1/_2]^\circ$	1—2
4526,177	50	18,61	21,35	$3p\ [1^1/_2]-8d\ [1^1/_2]^\circ$	1—1
4525,764	70	18,61	21,35	$3p\ [1^1/_2]-8d\ [2^1/_2]^\circ$	1—2
4517,736	100	18,70	21,45	$3p'\ [1^1/_2]-8d'\ [2^1/_2]^\circ$	2—3
4516,936	50	18,61	21,36	$3p\ [1^1/_2]-8s'\ [^1/_2]^\circ$	1—0
4515,411	30	18,61	21,36	$3p\ [1^1/_2]-8s'\ [^1/_2]^\circ$	1—1
4515,022	2	18,63	21,38	$3p\ [1^1/_2]-7d'\ [2^1/_2]^\circ$	2—2
4514,891	70	18,63	21,38	$3p\ [1^1/_2]-7d'\ [2^1/_2]^\circ$	2—3
4511,509	20	18,63	21,38	$3p\ [1^1/_2]-10s\ [1^1/_2]^\circ$	2—2
4510,170	15	18,63	21,38	$3p\ [1^1/_2]-10s\ [1^1/_2]^\circ$	2—1
4500,182	50	18,69	21,45	$3p'\ [1^1/_2]-8d'\ [1^1/_2]^\circ$	1—2
4499,843	5	18,69	21,45	$3p'\ [1^1/_2]-8d'\ [1^1/_2]^\circ$	1—1
		18,72	21,48	$3p'\ [^1/_2]-10s'\ [^1/_2]^\circ$	1—0
4499,000	2	18,72	21,48	$3p'\ [^1/_2]-10s'\ [^1/_2]^\circ$	1—1
4493,699	50	18,57	21,33	$3p\ [2^1/_2]-9s\ [1^1/_2]^\circ$	2—2
4493,108	5	18,63	21,39	$3p\ [1^1/_2]-5d\ [^1/_2]^\circ$	2—1
4492,689	15	18,63	21,39	$3p\ [1^1/_2]-5d\ [3^1/_2]^\circ$	2—3
4492,412	30	18,63	21,39	$3p\ [1^1/_2]-5d\ [1^1/_2]^\circ$	2—2
4492,132	5	18,63	21,39	$3p\ [1^1/_2]-5d\ [1^1/_2]^\circ$	2—1
4491,838	50	18,63	21,40	$3p\ [1^1/_2]-5d\ [2^1/_2]^\circ$	2—2, 3
4491,771	80	18,57	21,33	$3p\ [2^1/_2]-9s\ [1^1/_2]^\circ$	2—1
4488,0926	300	18,38	21,14	$3p\ [^1/_2]-7s\ [1^1/_2]^\circ$	1—2
4483,190	150	18,38	21,14	$3p\ [^1/_2]-7s\ [1^1/_2]^\circ$	1—1
4480,823	15	18,72	21,49	$3p'\ [^1/_2]-9d'\ [2^1/_2]^\circ$	1—2

λ, Å	I	E_H, eV	E_B, eV	Transition	J
4475,656	100	18,61	21,38	$3p\ [1^1/_2]—7d'\ [1^1/_2]^\circ$	1—2
4475,131	5	18,61	21,38	$3p\ [1^1/_2]—7d\ [1^1/_2]^\circ$	1—1
4472,246	1	18,61	21,38	$3p\ [1^1/_2]—10s\ [1^1/_2]^\circ$	1—2
4470,971	5	18,61	21,38	$3p\ [1^1/_2]—10s\ [1^1/_2]^\circ$	1—1
4467,491	1	18,57	21,35	$3p\ [2^1/_2]—8d\ [^1/_2]^\circ$	2—1
4466,8120	70	18,57	21,35	$3p\ [2^1/_2]—8d\ [3^1/_2]^\circ$	2—3
4466,503	2	18,57	21,35	$3p\ [2^1/_2]—8d\ [1^1/_2]^\circ$	2—2
4466,045	5	18,57	21,35	$3p\ [2^1/_2]—8d\ [1^1/_2]^\circ$	2—1
4465,651	50	18,57	21,35	$3p\ [2^1/_2]—8d\ [2^1/_2]^\circ$	2—2
4462,856	2	18,70	21,48	$3p'\ [1^1/_2]—10s'\ [^1/_2]^\circ$	2—1
4460,175	100	18,55	21,33	$3p\ [2^1/_2]—9s\ [1^1/_2]^\circ$	3—2
4456,380	1	18,71	21,49	$3p\ [^1/_2]—9d'\ [1^1/_2]^\circ$	0—1
4455,564	15	18,57	21,36	$3p\ [2^1/_2]—8s'\ [^1/_2]^\circ$	2—1
4454,285	5	18,63	21,42	$3p\ [1^1/_2]—11s\ [1^1/_2]^\circ$	2—2
4453,528	1	18,61	21,39	$3p\ [1^1/_2]—9d\ [1^1/_2]^\circ$	1—2
4453,324	2	18,63	21,42	$3p\ [1^1/_2]—11s\ [1^1/_2]^\circ$	2—1
4453,253	5	18,61	21,39	$3p\ [1^1/_2]—9d\ [1^1/_2]^\circ$	1—1
4452,983	15	18,61	21,40	$3p\ [1^1/_2]—9d\ [2^1/_2]^\circ$	1—2
4446,538	1	18,69	21,48	$3p'\ [1^1/_2]—10s'\ [^1/_2]^\circ$	1—0
4445,671	1	18,69	21,48	$3p'\ [1^1/_2]—10s'\ [^1/_2]^\circ$	1—1
4444,978	30	18,70	21,49	$3p'\ [1^1/_2]—9d'\ [2^1/_2]^\circ$	2—2
4440,890	1	18,63	21,43	$3p\ [1^1/_2]—10d\ [3^1/_2]^\circ$	2—3
4440,812	2	18,63	21,43	$3p\ [1^1/_2]—10d\ [1^1/_2]^\circ$	2—2
4440,363	15	18,63	21,43	$3p\ [1^1/_2]—10d\ [2^1/_2]^\circ$	2—3
4435,094	5	18,63	21,43	$3p\ [1^1/_2]—9s'\ [^1/_2]^\circ$	2—1
4433,7239	70	18,55	21,35	$3p\ [2^1/_2]—8d\ [3^1/_2]^\circ$	3—4
4433,398	10	18,55	21,35	$3p\ [2^1/_2]—8d\ [1^1/_2]^\circ$	3—2
4432,526	20	18,55	21,35	$3p\ [2^1/_2]—8d\ [2^1/_2]^\circ$	3—3
4429,410	1	18,72	21,52	$3p'\ [^1/_2]—10d'\ [2^1/_2]^\circ$	1—2
4427,981	15	18,69	21,49	$3p'\ [1^1/_2]—9d'\ [2^1/_2]^\circ$	1—2
4427,755	30	18,69	21,49	$3p'\ [1^1/_2]—9d'\ [1^1/_2]^\circ$	1—1
4425,400	150	18,38	21,18	$3p\ [^1/_2]—6d\ [^1/_2]^\circ$	1—0
4424,8096	300	18,38	21,18	$3p\ [^1/_2]—6d\ [^1/_2]^\circ$	1—1
4422,5205	300	18,38	21,18	$3p\ [^1/_2]—6d\ [1^1/_2]^\circ$	1—2
4421,559	50	18,38	21,18	$3p\ [^1/_2]—6d\ [1^1/_2]^\circ$	1—1
4420,558	1	18,38	21,18	$3p\ [^1/_2]—6d\ [2^1/_2]^\circ$	1—2
4416,817	50	18,57	21,38	$3p\ [2^1/_2]—7d'\ [2^1/_2]^\circ$	2—3
4415,141	5	18,61	21,42	$3p\ [1^1/_2]—11s\ [1^1/_2]^\circ$	1—1
4413,561	15	18,57	21,38	$3p\ [2^1/_2]—10s\ [1^1/_2]^\circ$	2—2
4412,285	20	18,57	21,38	$3p\ [2^1/_2]—10s\ [1^1/_2]^\circ$	2—1
4409,620	20	18,63	21,45	$3p\ [1^1/_2]—8d'\ [2^1/_2]^\circ$	2—3
4405,582	2	18,71	21,52	$3p\ [^1/_2]—10d'\ [1^1/_2]^\circ$	0—1
4402,985	1	18,63	21,45	$3p\ [1^1/_2]—11d\ [2^1/_2]^\circ$	2—3
4402,580	1	18,61	21,43	$3p\ [1^1/_2]—10d\ [1^1/_2]^\circ$	1—1
4402,374	2	18,61	21,43	$3p\ [1^1/_2]—10d\ [2^1/_2]^\circ$	1—2
4398,136	5	18,61	21,43	$3p\ [1^1/_2]—9s'\ [^1/_2]^\circ$	1—0
4397,175	1	18,61	21,43	$3p\ [1^1/_2]—9s'\ [^1/_2]^\circ$	1—1
4395,969	1	18,57	21,39	$3p\ [2^1/_2]—9d\ [^1/_2]^\circ$	2—1
4395,556	50	18,57	21,39	$3p\ [2^1/_2]—9d\ [3^1/_2]^\circ$	2—3
4395,306	1	18,57	21,39	$3p\ [2^1/_2]—9d\ [1^1/_2]^\circ$	2—2
4395,008	1	18,57	21,39	$3p\ [2^1/_2]—9d\ [1^1/_2]^\circ$	2—1
4394,773	15	18,57	21,40	$3p\ [2^1/_2]—9d\ [2^1/_2]^\circ$	2—2
4394,370	15	18,70	21,52	$3p\ [1^1/_2]—10d'\ [2^1/_2]^\circ$	2—3, 2
4381,220	30	18,55	21,38	$3p\ [2^1/_2]—10s\ [1^1/_2]^\circ$	3—2
4377,754	2	18,69	21,52	$3p'\ [1^1/_2]—10d'\ [2^1/_2]^\circ$	1—2
4374,997	2	18,63 18,61	21,47 21,44	$3p\ [1^1/_2]—12d\ [2^1/_2]^\circ$ $3p\ [1^1/_2]—12s\ [1^1/_2]^\circ$	2—3 1—1
4372,157	30	18,61	21,45	$3p\ [1^1/_2]—8d'\ [1^1/_2]^\circ$	1—2

λ, Å	I	E_H, eV	E_B, eV	Transition	J
4371,796	2	18,61	21,45	$3p\ [1^1/_2]—8d'\ [1^1/_2]^\circ$	1—1
4365,705	1	18,61	21,45	$3p\ [1^1/_2]—11d\ [2^1/_2]^\circ$	1—2
4363,524	70	18,55	21,39	$3p\ [2^1/_2]—9d\ [3^1/_2]^\circ$	3—4
4363,228	2	18,55	21,39	$3p\ [2^1/_2]—9d\ [1^1/_2]^\circ$	3—2
4362,690	30	18,55	21,40	$3p\ [2^1/_2]—9d\ [2^1/_2]^\circ$	3—3
4358,816	2	18,57	21,42	$3p\ [2^1/_2]—11s\ [1^1/_2]^\circ$	2—2
4357,918	5	18,57	21,42	$3p\ [2^1/_2]—11s\ [1^1/_2]^\circ$	2—1
4357,298	2	18,63	21,48	$3p\ [1^1/_2]—10s'\ [^1/_2]^\circ$	2—1
4346,036	15	18,57	21,43	$3p\ [2^1/_2]—10d\ [3^1/_2]^\circ$	2—3
4345,762	1	18,61	21,46	$3p\ [1^1/_2]—13s\ [1^1/_2]^\circ$	1—2
4345,479	2	18,57	21,43	$3p\ [2^1/_2]—10d\ [2^1/_2]^\circ$	2—2
4340,420	2	18,57	21,43	$3p\ [2^1/_2]—9s'\ [^1/_2]^\circ$	2—1
4340,256	2	18,63	21,49	$3p\ [1^1/_2]—9d'\ [2^1/_2]^\circ$	2—2
4338,200	2	18,61	21,47	$3p\ [1^1/_2]—12d\ [2^1/_2]^\circ$	1—2
4336,221	50	18,38	21,24	$3p\ [^1/_2]—7s'\ [^1/_2]^\circ$	1—0
4334,1267	70	18,38	21,24	$3p\ [^1/_2]—7s'\ [^1/_2]^\circ$	1—1
4327,265	10	18,55	21,42	$3p\ [2^1/_2]—11s\ [1^1/_2]^\circ$	3—2
4321,492	2	18,61	21,48	$3p\ [1^1/_2]—10s'\ [^1/_2]^\circ$	1—0
4319,511	1	18,57	21,44	$3p\ [2^1/_2]—12s\ [1^1/_2]^\circ$	2—2
4318,834	5	18,57	21,44	$3p\ [2^1/_2]—12s\ [1^1/_2]^\circ$	2—1
4316,008	15	18,57	21,45	$3p\ [2^1/_2]—8d'\ [2^1/_2]^\circ$	2—3, 2
4314,695	30	18,55	21,43	$3p\ [2^1/_2]—10d\ [3^1/_2]^\circ$	3—4, 3
4314,110	1	18,55	21,43	$3p\ [2^1/_2]—10d\ [2^1/_2]^\circ$	3—3
4310,130	2	18,57	21,45	$3p\ [2^1/_2]—11d\ [3^1/_2]^\circ$	2—3
4306,2625	70	18,38	21,26	$3p\ [^1/_2]—8s\ [1^1/_2]^\circ$	1—2
4303,955	5	18,61	21,49	$3p\ [1^1/_2]—9d'\ [2^1/_2]^\circ$	1—2
4303,695	1	18,61	21,49	$3p\ [1^1/_2]—9d'\ [1^1/_2]^\circ$	1—1
4303,248	30	18,38	21,26	$3p\ [^1/_2]—8s\ [1^1/_2]^\circ$	1—1
4291,976	2	18,63	21,52	$3p\ [1^1/_2]—10d'\ [2^1/_2]^\circ$	2—2, 3
4289,799	2	18,57	21,46	$3p\ [2^1/_2]—13s\ [1^1/_2]^\circ$	2—1
4288,541	5	18,55	21,44	$3p\ [2^1/_2]—12s\ [1^1/_2]^\circ$	3—2
4283,242	10	18,57	21,47	$3p\ [2^1/_2]—12d\ [3^1/_2]^\circ$	2—3
4279,279	15	18,55	21,45	$3p\ [2^1/_2]—11d\ [3^1/_2]^\circ$	3—4, 3
4278,850	5	18,55	21,45	$3p\ [2^1/_2]—11d\ [2^1/_2]^\circ$	3—3, 2
4275,5598	70	18,38	21,28	$3p\ [^1/_2]—6d'\ [1^1/_2]^\circ$	1—2
4275,167	1	18,38	21,28	$3p\ [^1/_2]—6d'\ [2^1/_2]^\circ$	1—2
4274,656	50	18,38	21,28	$3p\ [^1/_2]—6d'\ [1^1/_2]^\circ$	1—1
4270,267	50	18,38	21,28	$3p\ [^1/_2]—7d\ [^1/_2]^\circ$	1—0
4269,724	70	18,38	21,28	$3p\ [^1/_2]—7d\ [^1/_2]^\circ$	1—1
4268,009	70	18,38	21,28	$3p\ [^1/_2]—7d\ [1^1/_2]^\circ$	1—2
4267,724	5	18,38	21,28	$3p\ [^1/_2]—7d\ [1^1/_2]^\circ$	1—1
4267,286	1	18,38	21,29	$3p\ [^1/_2]—7d\ [2^1/_2]^\circ$	1—2
4262,479	2	18,57	21,48	$3p\ [2^1/_2]—13d\ [3^1/_2]^\circ$	2—3
4259,739	1	18,55	21,46	$3p\ [2^1/_2]—13s\ [1^1/_2]^\circ$	3—2
4256,498	2	18,61	21,52	$3p\ [1^1/_2]—10d'\ [2^1/_2]^\circ$	1—2
4252,775	2	18,55	21,47	$3p\ [2^1/_2]—12d\ [3^1/_2]^\circ$	3—4
4252,418	2	18,55	21,47	$3p\ [2^1/_2]—12d\ [2^1/_2]^\circ$	3—3
4249,538	2	18,57	21,49	$3p\ [2^1/_2]—9d'\ [2^1/_2]^\circ$	2—2
4232,323	1	18,55	21,48	$3p\ [2^1/_2]—13d\ [3^1/_2]^\circ$	3—3
4221,554	1	18,55	21,48	$3p\ [2^1/_2]—13d\ [2^1/_2]^\circ$	3—3
4203,270	2	18,57	21,52	$3p\ [2^1/_2]—10d'\ [2^1/_2]^\circ$	2—3
4198,099	70	18,38	21,33	$3p\ [^1/_2]—9s\ [1^1/_2]^\circ$	1—2
4196,415	15	18,38	21,33	$3p\ [^1/_2]—9s\ [1^1/_2]^\circ$	1—1
4175,488	40	18,38	21,35	$3p\ [^1/_2]—8d\ [^1/_2]^\circ$	1—0
4175,223	60	18,38	21,35	$3p\ [^1/_2]—8d\ [^1/_2]^\circ$	1—1
4174,369	70	18,38	21,35	$3p\ [^1/_2]—8d\ [1^1/_2]^\circ$	1—2
4173,966	2	18,38	21,35	$3p\ [^1/_2]—8d\ [1^1/_2]^\circ$	1—1
4166,091	30	18,38	21,36	$3p\ [^1/_2]—8s'\ [^1/_2]^\circ$	1—0

λ, Å	I	E_H eV	E_B, eV	Transition	J
4164,802	50	18,38	21,36	$3p\ [^1/_2]—8s'\ [^1/_2]^\circ$	1—1
4131,054	70	18,38	21,38	$3p\ [^1/_2]—7d'\ [2^1/_2]^\circ$	1—2
4130,512	20	18,38	21,38	$3p\ [^1/_2]—7d'\ [1^1/_2]^\circ$	1—1
4128,072	30	18,38	21,38	$3p\ [^1/_2]—10s\ [1^1/_2]^\circ$	1—2
4126,941	2	18,38	21,38	$3p\ [^1/_2]—10s\ [1^1/_2]^\circ$	1—1
4112,885	10	18,38	21,39	$3p\ [^1/_2]—9d\ [^1/_2]^\circ$	1—0
4112,694	20	18,38	21,39	$3p\ [^1/_2]—9d\ [^1/_2]^\circ$	1—1
4112,100	15	18,38	21,39	$3p\ [^1/_2]—9d\ [1^1/_2]^\circ$	1—2
4111,882	1	18,38	21,39	$3p\ [^1/_2]—9d\ [1^1/_2]^\circ$	1—1
4080,148	50	18,38	21,42	$3p\ [^1/_2]—11s\ [1^1/_2]^\circ$	1—2
4079,359	2	18,38	21,42	$3p\ [^1/_2]—11s\ [1^1/_2]^\circ$	1—1
4069,389	5	18,38	21,43	$3p\ [^1/_2]—10d\ [^1/_2]^\circ$	1—0
4069,243	30	18,38	21,43	$3p\ [^1/_2]—10d\ [^1/_2]^\circ$	1—1
4068,835	30	18,38	21,43	$3p\ [^1/_2]—10d\ [1^1/_2]^\circ$	1—2
4064,829	15	18,38	21,43	$3p\ [^1/_2]—9s'\ [^1/_2]^\circ$	1—0
4064,036	50	18,38	21,43	$3p\ [^1/_2]—9s'\ [^1/_2]^\circ$	1—1
4045,662	2	18,38	21,44	$3p\ [^1/_2]—12s\ [1^1/_2]^\circ$	1—2
4042,642	50	18,38	21,45	$3p\ [^1/_2]—8d'\ [2^1/_2]^\circ$	1—2
4042,327	10	18,38	21,45	$3p\ [^1/_2]—8d'\ [1^1/_2]^\circ$	1—1
4037,696	5	18,38	21,45	$3p\ [^1/_2]—11d\ [^1/_2]^\circ$	1—0
4037,615	15	18,38	21,45	$3p\ [^1/_2]—11d\ [^1/_2]^\circ$	1—1
4037,262	5	18,38	21,45	$3p\ [^1/_2]—11d\ [1^1/_2]^\circ$	1—2, 1
4020,015	2	18,38	21,46	$3p\ [^1/_2]—13s\ [1^1/_2]^\circ$	1—2
4013,995	2	18,38	21,47	$3p\ [^1/_2]—12s\ [^1/_2]^\circ$	1—1
4013,752	1	18,38	21,47	$3p\ [^1/_2]—12d\ [1^1/_2]^\circ$	1—2
3999,263	1	18,38	21,48	$3p\ [^1/_2]—10s'\ [^1/_2]^\circ$	1—0
3998,594	1	18,38	21,48	$3p\ [^1/_2]—10s'\ [^1/_2]^\circ$	1—1
3995,721	1	18,38	21,48	$3p\ [^1/_2]—13d\ [^1/_2]^\circ$	1—1
3984,253	7	18,38	21,49	$3p\ [^1/_2]—9d'\ [2^1/_2]^\circ$	1—2
3984,065	2	18,38	21,49	$3p\ [^1/_2]—9d'\ [1^1/_2]^\circ$	1—1
3943,540	2	18,38	21,52	$3p\ [^1/_2]—10d'\ [2^1/_2]^\circ$	1—2
3899,723	2	16,85	20,02	$3s'\ [^1/_2]^\circ—3d\ [^1/_2]^\circ$	1—1
3889,427	5	16,85	20,03	$3s'\ [^1/_2]^\circ—3d\ [3^1/_2]^\circ$	1—3
3887,134	1	16,85	20,04	$3s'\ [^1/_2]^\circ—3d\ [1^1/_2]^\circ$	1—2
3882,698	2	16,85	20,04	$3s'\ [^1/_2]^\circ—3d\ [1^1/_2]^\circ$	1—1
3769,654	5	16,85	20,14	$3s'\ [^1/_2]^\circ—3d'\ [2^1/_2]^\circ$	1—2
3769,449	7	16,85	20,14	$3s'\ [^1/_2]^\circ—3d'\ [2^1/_2]^\circ$	1—3
3768,047	5	16,85	20,14	$3s'\ [^1/_2]^\circ—3d'\ [1^1/_2]^\circ$	1—2
3765,819	5	16,85	20,14	$3s'\ [^1/_2]^\circ—3d'\ [1^1/_2]^\circ$	1—1
3754,2148	50	16,85	20,15	$3s'\ [^1/_2]^\circ—4p\ [^1/_2]$	1—1
3701,2247	40	16,85	20,20	$3s'\ [^1/_2]^\circ—4p\ [2^1/_2]$	1—2
3685,7351	100	16,85	20,21	$3s'\ [^1/_2]^\circ—4p\ [1^1/_2]$	1—1
3682,2421	100	16,85	20,21	$3s'\ [^1/_2]^\circ—4p\ [1^1/_2]$	1—2
3633,6643	100	16,85	20,26	$3s'\ [^1/_2]^\circ—4p\ [^1/_2]$	1—0
3609,1787	50	16,71	20,15	$3s'\ [^1/_2]^\circ—4p\ [^1/_2]$	0—1
3600,1694	100	16,85	20,29	$3s'\ [^1/_2]^\circ—4p'\ [1^1/_2]$	1—1
3593,640	300	16,85	20,30	$3s'\ [^1/_2]^\circ—4p'\ [^1/_2]$	1—1
3593,5263	500	16,85	20,30	$3s'\ [^1/_2]^\circ—4p'\ [1^1/_2]$	1—2
3562,9551	15	16,67	20,15	$3s\ [1^1/_2]^\circ—4p\ [^1/_2]$	1—1
3520,4714	1000	16,85	20,37	$3s'\ [^1/_2]^\circ—4p'\ [^1/_2]$	1—0
3515,1900	200	16,67	20,20	$3s\ [1^1/_2]^\circ—4p\ [2^1/_2]$	1—2
3510,7207	50	16,62	20,15	$3s\ [1^1/_2]^\circ—4p\ [^1/_2]$	2—1
3501,2154	200	16,67	20,21	$3s\ [1^1/_2]^\circ—4p\ [1^1/_2]$	1—1
3498,0632	100	16,67	20,21	$3s\ [1^1/_2]^\circ—4p\ [1^1/_2]$	1—2
3472,5706	500	16,62	20,19	$3s\ [1^1/_2]^\circ—4p\ [2^1/_2]$	2—3
3466,5781	200	16,71	20,29	$3s'\ [^1/_2]^\circ—4p'\ [1^1/_2]$	0—1
3464,3385	100	16,62	20,20	$3s\ [1^1/_2]^\circ—4p\ [2^1/_2]$	2—2

λ, Å	I	E_H, eV	E_B, eV	Transition	J
3460,5235	100	16,71	20,30	$3s' [^1/_2]^\circ - 4p' [^1/_2]$	0—1
3454,1942	100	16,67	20,26	$3s [1^1/_2]^\circ - 4p [^1/_2]$	1—0
3450,7641	50	16,62	20,21	$3s [1^1/_2]^\circ - 4p [1^1/_2]$	2—1
3447,7022	200	16,62	20,21	$3s [1^1/_2]^\circ - 4p [1^1/_2]$	2—2
3423,9120	50	16,67	20,29	$3s [1^1/_2]^\circ - 4p' [1^1/_2]$	1—1
3418,007	50	16,67	20,30	$3s [1^1/_2]^\circ - 4p' [^1/_2]$	1—1
3417,9031	500	16,67	20,30	$3s [1^1/_2]^\circ - 4p' [1^1/_2]$	1—2
3375,6489	50	16,62	20,29	$3s [1^1/_2]^\circ - 4p' [1^1/_2]$	2—1
3369,9069	700	16,62	20,30	$3s [1^1/_2]^\circ - 4p' [^1/_2]$	2—1
3369,8076	500	16,62	20,30	$3s [1^1/_2]^\circ - 4p' [1^1/_2]$	2—2
3351,744	25	16,67	20,37	$3s [1^1/_2]^\circ - 4p' [^1/_2]$	1—0
3167,5762	50	16,85	20,76	$3s' [^1/_2]^\circ - 5p [^1/_2]$	1—1
3153,4107	100	16,85	20,78	$3s' [^1/_2]^\circ - 5p [2^1/_2]$	1—2
3148,6107	100	16,85	20,78	$3s' [^1/_2]^\circ - 5p [1^1/_2]$	1—1
3147,701	25	16,85	20,78	$3s' [^1/_2]^\circ - 5p [1^1/_2]$	1—2
3126,1986	200	16,85	20,81	$3s' [^1/_2]^\circ - 5p [^1/_2]$	1—0
3079,175	100	16,85	20,87	$3s' [^1/_2]^\circ - 5p' [1^1/_2]$	1—1
3078,875	100	16,85	20,87	$3s' [^1/_2]^\circ - 5p' [^1/_2]$	1—1
3076,971	200	16,85	20,88	$3s' [^1/_2]^\circ - 5p' [1^1/_2]$	1—2
3063,695	200	16,71	20,76	$3s' [^1/_2]^\circ - 5p [^1/_2]$	0—1
3057,388	300	16,85	20,90	$3s' [^1/_2]^\circ - 5p' [^1/_2]$	1—0
3045,949	7	16,71	20,78	$3s' [^1/_2]^\circ - 5p [1^1/_2]$	0—1
3030,313	50	16,67	20,76	$3s [1^1/_2]^\circ - 5p [^1/_2]$	1—1
3017,348	50	16,67	20,78	$3s [1^1/_2]^\circ - 5p [2^1/_2]$	1—2
3012,955	50	16,67	20,78	$3s [1^1/_2]^\circ - 5p [1^1/_2]$	1—1
3012,129	50	16,67	20,78	$3s [1^1/_2]^\circ - 5p [1^1/_2]$	1—2
2994,250	3	16,67	20,81	$3s [1^1/_2]^\circ - 4f' [2^1/_2]$	1—2
2992,438	200	16,62	20,76	$3s [1^1/_2]^\circ - 5p [^1/_2]$	2—1
2992,420	200	16,67	20,81	$3s [1^1/_2]^\circ - 5p [^1/_2]$	1—0
2982,663	300	16,62	20,77	$3s [1^1/_2]^\circ - 5p [2^1/_2]$	2—3
2980,922	50	16,71	20,87	$3s' [^1/_2]^\circ - 5p' [1^1/_2]$	0—1
2980,642	40	16,71	20,87	$3s' [^1/_2]^\circ - 5p' [^1/_2]$	0—1
2979,806	50	16,62	20,78	$3s [1^1/_2]^\circ - 5p [2^1/_2]$	2—2
2975,518	35	16,62	20,78	$3s [1^1/_2]^\circ - 5p [1^1/_2]$	2—1
2974,714	300	16, 2	20,78	$3s [1^1/_2]^\circ - 5p [1^1/_2]$	2—2
2957,293	8	16,62	20,81	$3s [1^1/_2]^\circ - 4f' [2^1/_2]$	2—2, 3
2952,527	5	16,85	21,05	$3s' [^1/_2]^\circ - 5p [^1/_2]$	1—1
2949,316	15	16,67	20,87	$3s [1^1/_2]^\circ - 5p' [1^1/_2]$	1—1
2949,043	10	16,67	20,87	$3s [1^1/_2]^\circ - 5p' [^1/_2]$	1—1
2947,297	200	16,67	20,88	$3s [1^1/_2]^\circ - 5p' [1^1/_2]$	1—2
2946,732	2	16,85	21,05	$3s' [^1/_2]^\circ - 6p [2^1/_2]$	1—2
2944,575	2	16,85	21,06	$3s' [^1/_2]^\circ - 6p [1^1/_2]$	1—1
2932,721	100	16,85	21,07	$3s' [^1/_2]^\circ - 6p [^1/_2]$	1—0
2929,312	15	16,67	20,90	$3s [1^1/_2]^\circ - 5p' [^1/_2]$	1—0
2913,417	2	16,62	20,87	$3s [1^1/_2]^\circ - 5p' [1^1/_2]$	2—1
2913,168	200	16,62	20,87	$3s [1^1/_2]^\circ - 5p' [^1/_2]$	2—1
2911,461	25	16,62	20,88	$3s [1^1/_2]^\circ - 5p' [^1/_2]$	2—2
2881,852	2	16,85	21,15	$3s' [^1/_2]^\circ - 6p' [^1/_2]$	1—1
2880,290	3	16,85	21,15	$3s' [^1/_2]^\circ - 6p' [1^1/_2]$	1—2
2872,663	35	16,85	21,16	$3s' [^1/_2]^\circ - 6p' [^1/_2]$	1—0
2862,070	8	16,71	21,05	$3s' [^1/_2]^\circ - 6p [^1/_2]$	0—1
2846,490	2	16,85	21,20	$3s' [^1/_2]^\circ - 7p [^1/_2]$	1—1
2842,632	15	16,85	21,21	$3s' [^1/_2]^\circ - 7p [1^1/_2]$	1—1
2835,233	15	16,85	21,22	$3s' [^1/_2]^\circ - 7p [^1/_2]$	1—0
2832,921	8	16,67	21,05	$3s [1^1/_2]^\circ - 6p [^1/_2]$	1—1
2827,584	3	16,67	21,05	$3s [1^1/_2]^\circ - 6p [2^1/_2]$	1—2
2825,609	8	16,67	21,06	$3s [1^1/_2]^\circ - 6p [1^1/_2]$	1—1
2825,259	10	16,67	21,06	$3s [1^1/_2]^\circ - 6p [1^1/_2]$	1—2

λ, Å	I	E_H, eV	E_B, eV	Transition	J
2814,685	20	16,67	21,07	$3s\ [1^1/_2]°—6p\ [^1/_2]$	1—0
2799,80	2	16,62	21,05	$3s\ [1^1/_2]°—6p\ [^1/_2]$	2—1
2795,963	8	16,62	21,05	$3s\ [1^1/_2]°—6p\ [2^1/_2]$	2—3
2795,613	1	16,71	21,15	$3s'\ [^1/_2]°—6p'\ [^1/_2]$	0—1
2795,101	35	16,71	21,15	$3s'\ [^1/_2]°—6p\ [1^1/_2]$	0—1
2794,592	5	16,62	21,05	$3s\ [1^1/_2]°—6p\ [2^1/_2]$	2—2
2792,660	3	16,62	21,06	$3s\ [1^1/_2]°—6p\ [1^1/_2]$	2—1
2792,318	20	16,62	21,06	$3s\ [1^1/_2]°—6p\ [1^1/_2]$	2—2
2782,07	2	16,85	21,30	$3s'\ [^1/_2]°—7p'\ [1^1/_2]$	1—1
2781,68	3	16,85	21,30	$3s'\ [^1/_2]°—7p'\ [1^1/_2]$	1—2
2775,049	5	16,85	21,31	$3s'\ [^1/_2]°—7p'\ [1^1/_2]$	1—0
2767,77	2	16,67	21,15	$3s\ [1^1/_2]°—6p'\ [^1/_2]$	1—1
2767,28	3	16,67	21,15	$3s\ [1^1/_2]°—6p'\ [1^1/_2]$	1—1
2766,364	3	16,67	21,15	$3s\ [1^1/_2]°—6p'\ [1^1/_2]$	1—2
2762,324	3	16,71	21,20	$3s'\ [^1/_2]°—7p\ [^1/_2]$	0—1
2759,323	2	16,67	21,16	$3s\ [1^1/_2]°—6p'\ [^1/_2]$	1—0
2758,64	3	16,71	21,21	$3s'\ [^1/_2]°—7p\ [1^1/_2]$	0—1
2755,82	15	16,62	21,12	$3s\ [1^1/_2]°—5f'\ [2^1/_2]$	2—2, 3
2743,53	15	16,85	21,36	$3s'\ [^1/_2]°—9p\ [^1/_2]$	1—0
2736,177	5	16,62	21,15	$3s\ [^1/_2]°—6p'\ [^1/_2]$	2—1
2735,69	8	16,62	21,15	$3s\ [1^1/_2]°—6p'\ [1^1/_2]$	2—1
2735,168	3	16,67	21,20	$3s\ [1^1/_2]°—7p\ [^1/_2]$	1—1
2734,755	2	16,62	21,15	$3s\ [1^1/_2]°—6p'\ [1^1/_2]$	2—2
2732,61	1	16,67	21,21	$3s\ [1^1/_2]°—7p\ [2^1/_2]$	1—2
2731,528	3	16,67	21,21	$3s\ [1^1/_2]°—7p\ [1^1/_2]$	1—1
2731,358	3	16,67	21,21	$3s\ [1^1/_2]°—7p\ [1^1/_2]$	1—2
2724,772	3	16,67	21,22	$3s\ [1^1/_2]°—7p\ [^1/_2]$	1—0
2704,32	2	16,62	21,20	$3s\ [1^1/_2]°—7p\ [^1/_2]$	2—1
2702,554	3	16,62	21,20	$3s\ [1^1/_2]°—7p\ [2^1/_2]$	2—3
2701,766	2	16,62	21,21	$3s\ [1^1/_2]°—7p\ [2^1/_2]$	2—2
2701,653	2	16,71	21,30	$3s'\ [^1/_2]°—7p'\ [1^1/_2]$	0—1
2700,681	2	16,62	21,21	$3s\ [1^1/_2]°—7p\ [1^1/_2]$	2—1
2700,555	8	16,62	21,21	$3s\ [1^1/_2]°—7p\ [1^1/_2]$	2—2
2680,685	1	16,85	21,46	$3s'\ [^1/_2]°—9p\ [^1/_2]$	1—0
2679,19	2	16,67	21,29	$3s\ [1^1/_2]°—8p\ [^1/_2]$	1—1
2677,87	2	16,67	21,30	$3s\ [1^1/_2]°—8p\ [2^1/_2]$	1—2
2677,020	1	16,67	21,30	$3s\ [1^1/_2]°—8p\ [1^1/_2]$	1—1
2675,64	200	16,67	21,30	$3s\ [1^1/_2]°—7p'\ [1^1/_2]$	1—1
2675,24	200	16,67	21,30	$3s\ [1^1/_2]°—7p'\ [1^1/_2]$	1—2
2669,13	3	{ 16,67	21,30	$3s\ [1^1/_2]°—8p\ [^1/_2]$	1—0
		16,67	21,31	$3s\ [1^1/_2]°—7p'\ [^1/_2]$	1—0
2657,52	15	16,85	21,51	$3s'\ [^1/_2]°—10p'\ [^1/_2]$	1—0
2648,56	25	16,62	21,30	$3s\ [1^1/_2]°—8p\ [2^1/_2]$	2—3
2648,21	15	16,62	21,30	$3s\ [1^1/_2]°—8p\ [2^1/_2]$	2—2
2647,76	8	16,71	21,39	$3s'\ [^1/_2]°—8p'\ [1^1/_2]$	0—1
2647,42	200	16,62	21,30	$3s\ [1^1/_2]°—8p\ [1^1/_2]$	2—2, 1
2646,19	15	16,62	21,30	$3s\ [1^1/_2]°—7p'\ [1^1/_2]$	2—1
2645,645	35	16,62	21,30	$3s\ [1^1/_2]°—7p'\ [1^1/_2]$	2—2
2644,16	5	16,85	21,53	$3s'\ [^1/_2]°—11p'\ [^1/_2]$	1—0
2642,47	8	16,67	21,36	$3s\ [1^1/_2]°—9p\ [1^1/_2]$	1—2, 1
2639,97	15	16,67	21,36	$3s\ [1^1/_2]°—9p\ [^1/_2]$	1—0
2636,070	25	16,85	21,55	$3s'\ [^1/_2]°—12p'\ [^1/_2]$	1—0
2622,90	15	16,67	21,40	$3s\ [1^1/_2]°—8p'\ [1^1/_2]$	1—2, 1
2616,62	25	16,67	21,40	$3s\ [1^1/_2]°—8p'\ [1^1/_2]$	1—0
2614,26	5	16,62	21,36	$3s\ [1^1/_2]°—9p\ [2^1/_2]$	2—3
2613,94	2	16,62	21,36	$3s\ [1^1/_2]°—9p\ [2^1/_2]$	2—2
2613,59	30	16,62	21,36	$3s\ [1^1/_2]°—9p\ [1^1/_2]$	2—2, 1
2594,56	2	16,62	21,39	$3s\ [1^1/_2]°—8p'\ [1^1/_2]$	2—2, 1

λ, Å	I	E_H, eV	E_B, eV	Transition	J
2591,15	3	16,62	21,40	$3s\,[1^1/_2]^\circ-10p\,[2^1/_2]$	2—3
2590,67	10	16,62	21,40	$3s\,[1^1/_2]^\circ-10p\,[1^1/_2]$	2—2, 1
2589,48	2	16,67	21,46	$3s\,[1^1/_2]^\circ-9p'\,[1^1/_2]$	1—2, 1
2574,55	8	16,62	21,43	$3s\,[1^1/_2]^\circ-11p\,[1^1/_2]$	2—2, 1
2561,79	8	16,62	21,46	$3s\,[1^1/_2]^\circ-9p'\,[1^1/_2]$	2—2, 1
958,86	1	—	—	—	—
743,721	12	0,00	16,67	$2p^6\,{}^1S-3s\,[1^1/_2]^\circ$	0—1
735,892	30	0,00	16,85	$2p^6\,{}^1S-3s'\,[^1/_2]^\circ$	0—1
660,04	2	—	—	—	—
629,729	6	0,00	19,69	$2p^6\,{}^1S-4s\,[1^1/_2]^\circ$	0—1
626,819	6	0,00	19,78	$2p^6\,{}^1S-4s'\,[^1/_2]^\circ$	0—1
619,092	4	0,00	20,02	$2p^6\,{}^1S-3d\,[^1/_2]^\circ$	0—1
618,668	5	0,00	20,04	$2p^6\,{}^1S-3d\,[1^1/_2]^\circ$	0—1
615,623	5	0,00	20,14	$2p^6\,{}^1S-3d'\,[1^1/_2]^\circ$	0—1
602,712	4	0,00	20,56	$2p^6\,{}^1S-5s\,[1^1/_2]^\circ$	0—1
600,04	2	0,00	20,66	$2p^6\,{}^1S-5s'\,[^1/_2]^\circ$	0—1
598,86	1	0,00	20,70	$2p^6\,{}^1S-4d\,[^1/_2]^\circ$	0—1
598,698	2	0,00	20,71	$2p^6\,{}^1S-4d\,[1^1/_2]^\circ$	0—1
595,911	3	0,00	20,80	$2p^6\,{}^1S-4d'\,[1^1/_2]^\circ$	0—1
591,82	2	0,00	20,95	$2p^6\,{}^1S-6s\,[1^1/_2]^\circ$	0—1
589,92	1	0,00	21,02	$2p^6\,{}^1S-5d\,[1^1/_2]^\circ$	0—1
589,16	1	0,00	21,04	$2p^6\,{}^1S-6s'\,[^1/_2]^\circ$	0—1
587,20	1	0,00	21,11	$2p^6\,{}^1S-5d\,[1^1/_2]^\circ$	0—1
586,30	—	—	—	—	—
585,25	—	—	—	—	—
582,46	—	—	—	—	—
581,14	—	—	—	—	—
580,64	—	—	—	—	—
580,50	—	—	—	—	—
579,75	—	—	—	—	—
579,40	—	—	—	—	—
578,82	—	—	—	—	—

Ne II, **ground state** $1s^2\,2s^2\,2p^5\,{}^2P^0_{3/2}$

Ionization potential 331350 cm^{-1}; 41,079 eV

λ, Å	I	E_H, eV	E_B, eV	Transition	J
4922,3	0	35,05	37,56	$4s\,{}^4P-4f\,{}^4D^\circ$	$^1/_2-^1/_2$
4869,8	0	35,01	37,55	$4s\,{}^4P-4f\,{}^4D^\circ$	$^3/_2-^3/_2$
4849,4	0	35,01	37,56	$4s\,{}^4P-4f\,{}^4D^\circ$	$^3/_2-^1/_2$
4795,62	2	34,96	37,55	$4s\,{}^4P-4f\,{}^4D^\circ$	$^5/_2-^5/_2$
4781,95	1	34,96	37,55	$4s\,{}^4P-4f\,{}^4D^\circ$	$^5/_2-^3/_2$
4732,53	1	34,93	37,55	$3d\,{}^2P-4f\,{}^4D^\circ$	$^3/_2-^5/_2$
4730,24	0,5	35,01	37,63	$4s\,{}^4P-4f\,{}^4F^\circ$	$^3/_2-^3/_2$
4719,37	1,5	34,93	37,55	$3d\,{}^2P-4f\,{}^4D^\circ$	$^3/_2-^3/_2$
4710,04	2	35,01	37,64	$4s\,{}^4P-4f\,{}^4G^\circ$	$^3/_2-^5/_2$
4701,2	0	—	—	—	—
4700,1	0	34,93	37,56	$3d\,{}^2P-4f\,{}^4D^\circ$	$^3/_2-^1/_2$
4647,34	0,5	34,96	37,63	$4s\,{}^4P-4f\,{}^4F^\circ$	$^5/_2-^3/_2$
4634,73	2	34,88	37,55	$3d\,{}^2P-4f\,{}^4D^\circ$	$^1/_2-^3/_2$
4627,85	3	34,96	37,64	$4s\,{}^4P-4f\,{}^4G^\circ$	$^5/_2-^5/_2$
4615,98	4	34,88	37,56	$3d\,{}^2P-4f\,{}^4D^\circ$	$^1/_2-^1/_2$
		34,86	37,54	$3d\,{}^4P-4f\,{}^4D^\circ$	$^5/_2-^7/_2$

λ, Å	I	E_H, eV	E_B, eV	Transition	J
4612,89	1	34,86	37,55	$3d\ ^4P-4f\ ^4D^\circ$	5/2—5/2
4600,11	1	34,86	37,55	$3d\ ^4P-4f\ ^4D^\circ$	5/2—3/2
4588,13	3	34,93	37,63	$3d\ ^2P-4f\ ^4F^\circ$	3/2—3/2
4580,35	3	34,96	37,67	$4s\ ^4P-4f\ ^4F^\circ$	5/2—5/2
4574,49	1	34,84	37,55	$3d\ ^4P-4f\ ^4D^\circ$	3/2—5/2
4569,01	5	34,93	37,64	$3d\ ^2P-4f\ ^4G^\circ$	3/2—5/2
4565,49	1	34,83	37,55	$3d\ ^4F-4f\ ^4D^\circ$	3/2—5/2
4562,05	1	34,84	37,55	$3d\ ^4P-4f\ ^4D^\circ$	3/2—3/2
4553,16	4	34,83	37,55	$3d\ ^4F-4f\ ^4D^\circ$	3/2—3/2
4544,11	1	34,84	37,56	$3d\ ^4P-4f\ ^4D^\circ$	3/2—1/2
4535,47	3	34,83	37,56	$3d\ ^4F-4f\ ^4D^\circ$	3/2—1/2
4534,66	2	34,81	37,55	$3d\ ^2F-4f\ ^4D^\circ$	5/2—5/2
4522,66	4	34,93	37,67	$3d\ ^2P-4f\ ^4F^\circ$	3/2—5/2
4517,79	2	34,80	37,54	$3d\ ^4F-4f\ ^4D^\circ$	7/2—7/2
4514,80	2	34,80	37,55	$3d\ ^4F-4f\ ^4D^\circ$	7/2—5/2
4511,37	4	34,93	37,67	$3d\ ^2P-4f\ ^2D^\circ$	3/2—5/2
4511,29	2	34,88	37,62	$3d\ ^2P-4f\ ^2D^\circ$	1/2—3/2
4508,21	3	34,88	37,63	$3d\ ^2P-4f\ ^4F^\circ$	1/2—3/2
4502,52	0,5	34,80	37,55	$3d\ ^4F-4f\ ^4F^\circ$	7/2—9/2
4498,94	5	34,81	37,56	$3d\ ^4P-4f\ ^4D^\circ$	1/2—1/2
4475,22	1	34,86	37,63	$3d\ ^4P-4f\ ^4F^\circ$	5/2—3/2
4471,52	3	34,86	37,63	$3d\ ^4P-4f\ ^4F^\circ$	5/2—7/2
4468,91	5	34,77	37,54	$3d\ ^2D-4f\ ^4D^\circ$	3/2—5/2
4456,95	5	34,77	37,55	$3d\ ^2D-4f\ ^4D^\circ$	3/2—3/2
4452,55	1	—	—	—	—
4446,46	3	34,84	37,63	$3d\ ^4F-4f\ ^4F^\circ$	5/2—3/2
4442,67	3	34,84	37,63	$3d\ ^4F-4f\ ^4F^\circ$	5/2—7/2
4439,95	2	34,77	37,56	$3d\ ^2D-4f\ ^4D^\circ$	3/2—1/2
4439,30	3	34,84	37,63	$3d\ ^4P-4f\ ^4F^\circ$	3/2—3/2
4432,26	1	37,86	40,66	$3p''\ ^2P^\circ-3d''\ ^2D$	3/2—5/2
4431,67	1	37,86	40,66	$3p''\ ^2P^\circ-3d''\ ^2D$	1/2—3/2
4430,90	4	34,83	37,63	$3d\ ^4F-4f\ ^4F^\circ$	3/2—3/2
4429,60	2	37,86	40,66	$3p''\ ^2P^\circ-3d''\ ^2D$	3/2—3/2
4428,54	6	34,84	37,64	$3d\ ^4F-4f\ ^4G^\circ$	5/2—5/2
		34,75	37,55	$3d\ ^2D-4f\ ^4D^\circ$	5/2—5/2
4421,38	3	34,84	37,64	$3d\ ^4P-4f\ ^4G^\circ$	3/2—5/2
4416,77	2	34,75	37,55	$3d\ ^2D-4f\ ^4D^\circ$	5/2—3/2
4413,20	4	34,83	37,64	$3d\ ^4F-4f\ ^4G^\circ$	3/2—5/2
		34,86	37,67	$3d\ ^4P-4f\ ^4F^\circ$	5/2—5/2
4412,54	2	34,74	37,54	$3d\ ^4F-4f\ ^4D^\circ$	9/2—7/2
4409,30	7	34,84	37,65	$3d\ ^4F-4f\ ^4F^\circ$	5/2—7/2
4397,94	6	34,74	37,55	$3d\ ^4F-4f\ ^4F^\circ$	9/2—9/2
4391,94	7	34,80	37,62	$3d\ ^4F-4f\ ^4F^\circ$	7/2—9/2
4385,00	2	34,84	37,67	$3d\ ^4F-4f\ ^4F^\circ$	5/2—5/2
4384,08	1	34,81	37,64	$3d\ ^2F-4f\ ^4G^\circ$	5/2—5/2
4379,50	6	34,80	37,63	$3d\ ^4F-4f\ ^4F^\circ$	7/2—7/2
4377,95	2	34,84	37,67	$3d\ ^4P-4f\ ^4F^\circ$	3/2—5/2
4369,77	5	34,83	37,67	$3d\ ^4F-4f\ ^4F^\circ$	3/2—5/2
4365,72	2	34,80	37,64	$3d\ ^4F-4f\ ^4G^\circ$	7/2—5/2
4346,12	1	—	—	—	—
4341,42	2	34,81	37,67	$3d\ ^2F-4f\ ^4F^\circ$	5/2—5/2
4339,78	1	34,77	37,63	$3d\ ^2D-4f\ ^4F^\circ$	3/2—3/2
4325,15	1	—	—	—	—
4322,66	1	34,77	37,64	$3d\ ^2D-4f\ ^4G^\circ$	3/2—5/2
4322,26	2	—	—	—	—
4290,40	6	34,74	37,62	$3d\ ^4F-4f\ ^4G^\circ$	9/2—11/2
4257,82	3	34,64	37,55	$3d\ ^4D-4f\ ^4D^\circ$	1/2—3/2
4257,25	1	—	—	—	—
4250,68	4	34,63	37,55	$3d\ ^4D-4f\ ^4D^\circ$	3/2—5/2

λ, Å	I	E_H, eV	E_B, eV	Transition	J
4244,17	0	34,75	37,67	$3d\,^2D—4f\,^4F^\circ$	$^5/_2—^5/_2$
4242,20	1	34,64	37,56	$3d\,^4D—4f\,^4D^\circ$	$^1/_2—^1/_2$
4239,95	2	34,63	37,55	$3d\,^4D—4f\,^4D^\circ$	$^3/_2—^3/_2$
4233,86	3	—	—	—	—
4231,60	4	34,62	37,55	$3d\,^4D—4f\,^4D^\circ$	$^5/_2—^5/_2$
4224,57	1	34,63	37,56	$3d\,^4D—4f\,^4D^\circ$	$^3/_2—^1/_2$
4220,92	2	34,62	37,55	$3d\,^4D—4f\,^4D^\circ$	$^5/_2—^3/_2$
4219,76	6	34,61	37,54	$3d\,^4D—4f\,^4D^\circ$	$^7/_2—^7/_2$
4217,15	3	34,61	37,55	$3d\,^4D—4f\,^4D^\circ$	$^7/_2—^5/_2$
4206,43	2	34,61	37,55	$3d\,^4D—4f\,^4F^\circ$	$^7/_2—^9/_2$
4150,67	3	34,64	37,63	$3d\,^4D—4f\,^4F^\circ$	$^1/_2—^3/_2$
4133,65	3	34,63	37,63	$3d\,^4D—4f\,^4F^\circ$	$^3/_2—^3/_2$
4118,10	0	34,63	37,64	$3d\,^4D—4f\,^4G^\circ$	$^3/_2—^5/_2$
4100,30	1	34,62	37,64	$3d\,^4D—4f\,^4G^\circ$	$^5/_2—^5/_2$
4098,77	4	34,61	37,63	$3d\,^4D—4f\,^4F^\circ$	$^7/_2—^7/_2$
4086,69	1	34,61	37,64	$3d\,^4D—4f\,^4G^\circ$	$^7/_2—^5/_2$
4080,48	2	34,63	37,67	$3d\,^4D—4f\,^4F^\circ$	$^3/_2—^5/_2$
4062,90	3	34,62	37,67	$3d\,^4D—4f\,^4F^\circ$	$^5/_2—^5/_2$
3999,86	1	—	—	—	—
3942,19	3	—	—	—	—
3840,48	1	—	—	—	—
3829,77	7	31,51	34,75	$3d\,^2P^\circ—3d\,^2D$	$^3/_2—^5/_2$
3823,19	1	—	—	—	—
3818,44	6	31,53	34,77	$3p\,^2P^\circ—3d\,^2D$	$^1/_2—^3/_2$
3806,30	2	31,36	34,62	$3p\,^4S^\circ—3d\,^4D$	$^3/_2—^5/_2$
3800,02	5	31,51	34,77	$3p\,^2P^\circ—3d\,^2D$	$^3/_2—^3/_2$
3790,96	3	31,36	34,63	$3p\,^4S^\circ—3d\,^4D$	$^3/_2—^3/_2$
3777,16	8	27,27	30,55	$3s\,^4P—3p\,^4P^\circ$	$^1/_2—^3/_2$
3766,29	8	27,23	30,52	$3s\,^4P—3p\,^4P^\circ$	$^3/_2—^5/_2$
3753,83	5	31,51	34,81	$3p\,^2P^\circ—3d\,^2F$	$^3/_2—^5/_2$
3751,26	5	27,27	30,57	$3s\,^4P—3p\,^4P^\circ$	$^1/_2—^1/_2$
3744,66	4	31,53	34,84	$3p\,^4P^\circ—3d\,^4P$	$^1/_2—^3/_2$
3734,94	7	27,23	30,55	$3s\,^4P—3p\,^4P^\circ$	$^3/_2—^3/_2$
3727,08	9	27,86	31,18	$3s\,^2P—3p\,^2D^\circ$	$^1/_2—^3/_2$
3721,86	2	31,51	34,84	$3p\,^2P^\circ—3d\,^4F$	$^3/_2—^5/_2$
3713,084	10	27,78	31,12	$3s\,^2P—3p\,^2D^\circ$	$^3/_2—^5/_2$
3709,64	7	27,23	30,57	$3s\,^4P—3p\,^4P^\circ$	$^3/_2—^1/_2$
3701,81	4	31,51	34,86	$3p\,^2P^\circ—3d\,^4P$	$^3/_2—^5/_2$
3697,09	2	31,53	34,88	$3p\,^2P^\circ—3d\,^2P$	$^1/_2—^1/_2$
3694,197	10	27,17	30,52	$3s\,^4P—3p\,^4P^\circ$	$^5/_2—^5/_2$
3679,80	2	31,51	34,88	$3p\,^2P^\circ—3d\,^2P$	$^3/_2—^1/_2$
3664,112	9	27,17	30,55	$3s\,^4P—3p\,^4P^\circ$	$^5/_2—^3/_2$
3659,93	3	31,36	34,75	$3p\,^4S^\circ—3d\,^2D$	$^3/_2—^5/_2$
3644,86	4	31,53	34,93	$3p\,^2P^\circ—3d\,^2P$	$^1/_2—^3/_2$
3643,89	5	27,78	31,18	$3s\,^2P—3p\,^2D^\circ$	$^3/_2—^3/_2$
3632,75	2	31,36	34,77	$3p\,^4S^\circ—3d\,^2D$	$^3/_2—^3/_2$
3628,06	4	31,51	34,93	$3p\,^2P^\circ—3d\,^2P$	$^3/_2—^3/_2$
3612,35	3	31,34	34,77	$3p\,^2S^\circ—3d\,^2D$	$^1/_2—^3/_2$
3594,18	4	31,36	34,81	$3p\,^4S^\circ—3d\,^4P$	$^3/_2—^1/_2$
3590,47	2	31,36	34,81	$3p\,^4S^\circ—3d\,^2F$	$^3/_2—^5/_2$
3574,64	5	30,55	34,02	$3s'\,^2D—3p'\,^2F^\circ$	$^3/_2—^5/_2$
3574,23	0	30,55	34,02	$3s'\,^2D—3p'\,^2F^\circ$	$^5/_2—^5/_2$
3571,26	4	31,36	34,83	$3p\,^4S^\circ—3d\,^4F$	$^3/_2—^5/_2$
3568,53	6	30,55	34,02	$3s'\,^2D—3p'\,^2F^\circ$	$^5/_2—^7/_2$
3565,84	4	31,36	34,84	$3p\,^4S^\circ—3d\,^4P$	$^3/_2—^3/_2$
3561,23	4	31,36	34,84	$3p\,^4S^\circ—3d\,^4F$	$^3/_2—^5/_2$
3557,84	4	27,86	31,34	$3s\,^2P—3p\,^2S^\circ$	$^1/_2—^1/_2$
3554,39	1	31,12	34,61	$3p\,^2D^\circ—3d\,^4D$	$^5/_2—^7/_2$

λ, Å	I	E_H, eV	E_B, eV	Transition	J
3551,52	1	31,34	34,83	$3p\ ^2S^\circ - 3d\ ^4F$	$^1/_2 - ^3/_2$
3546,22	1	31,34	34,84	$3p\ ^2S^\circ - 3d\ ^4P$	$^1/_2 - ^3/_2$
3542,90	7	34,36	34,86	$3p\ ^4S^\circ - 3d\ ^4P$	$^3/_2 - ^5/_2$
3542,28	2	34,38	37,88	$3p'\ ^2D^\circ - 3d'\ ^2P$	$^5/_2 - ^3/_2$
3539,94	0,5	34,38	37,88	$3p'\ ^2D^\circ - 3d'\ ^2P$	$^3/_2 - ^3/_2$
3537,99	3	37,38	37,89	$3p'\ ^2D^\circ - 3d'\ ^2P$	$^3/_2 - ^1/_2$
3522,72	1	31,36	34,88	$3p\ ^4S^\circ - 3d\ ^2P$	$^3/_2 - ^1/_2$
3503,61	5	31,34	34,88	$3p\ ^2S^\circ - 3d\ ^2P$	$^1/_2 - ^1/_2$
3481,96	6	27,78	31,34	$3s\ ^2P - 3p\ ^2S^\circ$	$^3/_2 - ^1/_2$
3480,75	2	34,30	37,86	$3s''\ ^2S - 3p''\ ^2P^\circ$	$^1/_2 - ^3/_2$
3479,53	1	34,30	37,86	$3s''\ ^2S - 3p''\ ^2P^\circ$	$^1/_2 - ^1/_2$
3477,69	3	31,18	34,75	$3p\ ^2D^\circ - 3d\ ^2D$	$^3/_2 - ^5/_2$
3475,25	1	31,36	34,93	$3p\ ^4S^\circ - 3d\ ^2P$	$^3/_2 - ^3/_2$
3459,38	2	34,38	37,97	$3p'\ ^2D^\circ - 3d'\ ^2D$	$^5/_2 - ^3/_2$
3457,16	4	34,38	37,97	$3p'\ ^2D^\circ - 3d'\ ^2D$	$^3/_2 - ^3/_2$
3456,68	4	31,34	34,93	$3p\ ^2S^\circ - 3d\ ^2P$	$^1/_2 - ^3/_2$
3453,10	3	31,18	34,77	$3p\ ^2D^\circ - 3d\ ^2D$	$^3/_2 - ^3/_2$
3443,70	2	31,53	35,13	$3p\ ^2P^\circ - 4s\ ^2P$	$^1/_2 - ^3/_2$
3442,12	1	31,36	34,96	$3p\ ^4S^\circ - 4s\ ^4P$	$^3/_2 - ^5/_2$
3440,80	1	34,28	37,89	$3p'\ ^2P^\circ - 3d'\ ^2P$	$^1/_2 - ^1/_2$
3438,97	2	34,28	37,88	$3p'\ ^2P - 3d'\ ^2P$	$^1/_2 - ^3/_2$
3428,76	5	31,51	35,13	$3p\ ^2P^\circ - 4s\ ^2P$	$^3/_2 - ^3/_2$
3417,71	5	31,12	34,75	$3p\ ^2D^\circ - 3d\ ^2F$	$^5/_2 - ^7/_2$
3416,87	4	31,12	34,75	$3p\ ^2D^\circ - 3d\ ^2D$	$^5/_2 - ^5/_2$
3414,82	2	31,18	34,81	$3p\ ^2D^\circ - 3d\ ^2F$	$^3/_2 - ^5/_2$
3413,13	3	34,25	37,89	$3p'\ ^2P^\circ - 3d'\ ^2P$	$^3/_2 - ^1/_2$
3411,38	1	34,25	37,88	$3p'\ ^2P^\circ - 3d'\ ^2P$	$^3/_2 - ^3/_2$
3406,88	5	34,38	38,02	$3p'\ ^2D^\circ - 3d'\ ^2D$	$^5/_2 - ^5/_2$
3404,77	4	34,38	38,02	$3p'\ ^2D^\circ - 3d'\ ^2D$	$^3/_2 - ^5/_2$
3397,90	1	31,36	35,01	$3p\ ^4S^\circ - 4s\ ^4P$	$^3/_2 - ^3/_2$
3392,78	5	27,86	31,51	$3s\ ^2P - 3p\ ^2P^\circ$	$^1/_2 - ^3/_2$
3390,56	2	30,97	34,63	$3p\ ^4D^\circ - 3d\ ^4D$	$^1/_2 - ^3/_2$
3388,46	6	31,18	34,84	$3p\ ^2D^\circ - 3d\ ^4F$	$^3/_2 - ^5/_2$
3386,24	2	30,96	34,62	$3p\ ^4D^\circ - 3d\ ^4D$	$^3/_2 - ^5/_2$
3379,39	1	30,97	34,64	$3p\ ^4D^\circ - 3d\ ^4D$	$^1/_2 - ^1/_2$
3378,28	5	27,86	31,53	$3s\ ^2P - 3p\ ^2P^\circ$	$^1/_2 - ^1/_2$
3377,23	1	31,53	35,20	$3p\ ^2P^\circ - 4s\ ^2P$	$^1/_2 - ^1/_2$
3374,10	3	30,96	34,63	$3p\ ^4D^\circ - 3d\ ^4D$	$^3/_2 - ^3/_2$
3371,87	4	31,18	34,86	$3p\ ^2D^\circ - 3d\ ^4P$	$^3/_2 - ^5/_2$
3367,20	6	31,12	34,80	$3p\ ^2D^\circ - 3d\ ^4F$	$^5/_2 - ^7/_2$
3362,89	2	30,96	34,64	$3p\ ^4D^\circ - 3d\ ^4D$	$^3/_2 - ^1/_2$
3360,63	5	27,27	30,96	$3s\ ^4P - 3p\ ^4D^\circ$	$^1/_2 - ^3/_2$
3357,90	3	30,93	34,62	$3p\ ^4D^\circ - 3d\ ^4D$	$^5/_2 - ^5/_2$
3356,35	2	31,12	34,81	$3p\ ^2D^\circ - 3d\ ^2F$	$^5/_2 - ^5/_2$
3355,05	7	27,23	30,93	$3s\ ^4P - 3p\ ^4D^\circ$	$^3/_2 - ^5/_2$
3353,63	2	31,18	34,88	$3p\ ^2D^\circ - 3d\ ^2P$	$^3/_2 - ^1/_2$
3345,88	1	30,93	34,63	$3p\ ^4D^\circ - 3d\ ^4D$	$^5/_2 - ^3/_2$
		30,55	34,25	$3s'\ ^2D - 3p'\ ^2P^\circ$	$^3/_2 - ^3/_2$
3345,49	3	30,55	34,25	$3s'\ ^2D - 3p'\ ^2P^\circ$	$^5/_2 - ^3/_2$
3344,43	5	27,27	30,97	$3s\ ^4P - 3p\ ^4D^\circ$	$^1/_2 - ^1/_2$
3336,12	2	34,25	37,97	$3p'\ ^2P^\circ - 3d'\ ^2D$	$^3/_2 - ^3/_2$
3334,87	10	27,17	30,88	$3s\ ^4P - 3p\ ^4D^\circ$	$^5/_2 - ^7/_2$
3330,78	2	31,12	34,84	$3p\ ^2D^\circ - 3d\ ^4F$	$^5/_2 - ^5/_2$
3329,20	4	30,88	34,61	$3p\ ^4D^\circ - 3d\ ^4D$	$^7/_2 - ^7/_2$
3327,16	5	27,23	30,96	$3s\ ^4P - 3p\ ^4D^\circ$	$^3/_2 - ^3/_2$
3323,75	7	27,78	31,51	$3s\ ^2P - 3p\ ^2P^\circ$	$^3/_2 - ^3/_2$
3320,29	2	30,88	34,62	$3p\ ^4D^\circ - 3d\ ^4D$	$^7/_2 - ^5/_2$
3319,75	3	30,55	34,28	$3s'\ ^2D - 3p'\ ^2P^\circ$	$^3/_2 - ^1/_2$
3314,60	1	31,12	34,86	$3p\ ^2D^\circ - 3d\ ^4P$	$^5/_2 - ^5/_2$

λ, Å	I	E_H, eV	E_B, eV	Transition	J
3311,30	3	27,23	30,97	$3s\,^4P—3p\,^4D°$	$^3/_2—^1/_2$
3310,55	1	31,18	34,93	$3p\,^2D°—3d\,^2P$	$^3/_2—^3/_2$
3309,78	3	27,78	31,53	$3s\,^2P—3p\,^2P°$	$^3/_2—^1/_2$
3297,74	7	27,17	30,93	$3s\,^4P—3p\,^4D°$	$^5/_2—^5/_2$
3275,20	2	31,34	35,13	$3p\,^2S°—4s\,^2P$	$^1/_2—^3/_2$
3270,79	2	27,17	30,96	$3s\,^4P—3p\,^4D°$	$^5/_2—^3/_2$
3269,86	3	30,96	34,75	$3p\,^4D°—3d\,^2D$	$^3/_2—^5/_2$
3263,43	3	30,97	34,77	$3p\,^4D°—3d\,^2D$	$^1/_2—^3/_2$
3255,39	2	31,12	34,93	$3p\,^2D°—3d\,^2P$	$^5/_2—^3/_2$
3248,15	3	30,96	34,77	$3p\,^4D°—3d\,^2D$	$^3/_2—^3/_2$
3244,15	5	30,93	34,75	$3p\,^4D°—3d\,^2F$	$^5/_2—^7/_2$
3243,34	2	30,93	34,75	$3p\,^4D°—3d\,^2D$	$^5/_2—^5/_2$
3232,38	3	30,55	34,38	$3s'\,^2D—3p'\,^2D°$	$^3/_2—^3/_2$
3231,97	0	30,55	34,38	$3s'\,^2D—3p'\,^2D°$	$^5/_2—^3/_2$
3230,16	5	30,55	34,38	$3s'\,^2D—3p'\,^2D°$	$^5/_2—^5/_2$
3229,50	3	34,02	37,86	$3p'\,^2F°—3d'\,^2G$	$^7/_2—^9/_2$
3224,82	4	34,02	37,86	$3p'\,^2F°—3d'\,^2G$	$^5/_2—^7/_2$
3218,21	8	30,88	34,74	$3p\,^4D°—3d\,^4F$	$^7/_2—^9/_2$
3214,38	5	30,96	34,81	$3p\,^4D°—3d\,^2F$	$^3/_2—^5/_2$
3213,70	3	30,97	34,83	$3p\,^4D°—3d\,^4F$	$^1/_2—^3/_2$
3209,38	3	30,97	34,84	$3p\,^4D°—3d\,^4P$	$^1/_2—^3/_2$
3208,99	2	30,88	34,75	$3p\,^4D°—3d\,^2F$	$^7/_2—^7/_2$
3198,62	5	30,93	34,80	$3p\,^4D°—3d\,^4F$	$^5/_2—^7/_2$
3194,61	4	30,96	34,84	$3p\,^4D°—3d\,^4P$	$^3/_2—^3/_2$
3190,86	2	30,96	34,84	$3p\,^4D°—3d\,^4F$	$^3/_2—^5/_2$
3188,74	3	30,93	34,81	$3p\,^4D°—3d\,^2F$	$^5/_2—^5/_2$
3187,60	2	27,23	31,12	$3s\,^4P—3p\,^2D°$	$^3/_2—^5/_2$
3176,16	3	30,96	34,86	$3p\,^4D°—3d\,^4P$	$^3/_2—^5/_2$
3173,58	3	30,93	34,83	$3p\,^4D°—3d\,^4F$	$^5/_2—^3/_2$
3169,30	0	30,93	34,84	$3p\,^4D°—3d\,^4P$	$^5/_2—^3/_2$
3165,70	4	30,93	34,84	$3p\,^4D°—3d\,^4F$	$^5/_2—^5/_2$
3164,46	3	30,88	34,80	$3p\,^4D°—3d\,^4F$	$^7/_2—^7/_2$
3154,82	1	30,88	34,81	$3p\,^4D°—3d\,^2F$	$^7/_2—^5/_2$
3151,16	2	30,93	34,86	$3p\,^4D°—3d\,^4P$	$^5/_2—^5/_2$
3143,74	2	31,18	35,13	$3p\,^2D°—4s\,^2P$	$^3/_2—^3/_2$
3141,35	3	34,25	38,20	$3p'\,^2P°—3d'\,^2F$	$^3/_2—^5/_2$
3135,82	1	27,17	31,12	$3s\,^4P—3p\,^2D°$	$^5/_2—^5/_2$
3132,22	2	30,88	34,84	$3p\,^4D°—3d\,^4F$	$^7/_2—^5/_2$
3118,02	4	30,88	34,86	$3p\,^4D°—3d\,^4P$	$^7/_2—^5/_2$
3097,15	3	34,02	38,02	$3p'\,^2F°—3d'\,^2D$	$^7/_2—^5/_2$
3094,08	4	31,12	35,13	$3p\,^2D°—4s\,^2P$	$^5/_2—^3/_2$
3092,91	2	34,02	38,02	$3p'\,^2F°—3d'\,^2D$	$^5/_2—^5/_2$
3088,23	3	31,18	35,20	$3p\,^2D°—4s\,^2P$	$^3/_2—^1/_2$
3072,68	1	{ 30,97	35,01	$3p\,^4D°—4s\,^4P$	$^1/_2—^3/_2$
		34,28	38,32	$3p'\,^2P°—3d'\,^2S$	$^1/_2—^1/_2$
3071,08	2	30,93	34,96	$3p\,^4D°—4s\,^4P$	$^5/_2—^5/_2$
3059,16	3	30,96	35,01	$3p\,^4D°—4s\,^4P$	$^3/_2—^3/_2$
3054,69	5	30,57	34,63	$3p\,^4P°—3d\,^4D$	$^1/_2—^3/_2$
3050,57	1	34,25	38,32	$3p'\,^2P°—3d'\,^2S$	$^3/_2—^1/_2$
3047,57	6	30,55	34,62	$3p\,^4P°—3d\,^4D$	$^3/_2—^5/_2$
3045,58	4	30,57	34,64	$3p\,^4P°—3d\,^4D$	$^1/_2—^1/_2$
3044,16	2	30,97	35,05	$3p\,^4D°—4s\,^4P$	$^1/_2—^1/_2$
3039,65	3	30,88	34,96	$3p\,^4D°—4s\,^4P$	$^7/_2—^5/_2$
3037,73	4	30,55	34,63	$3p\,^4P°—3d\,^4D$	$^3/_2—^3/_2$
3035,98	3	30,93	35,01	$3p\,^4D°—4s\,^4P$	$^5/_2—^3/_2$
3034,48	5	30,52	34,61	$3p\,^4P°—3d\,^4D$	$^5/_2—^7/_2$
3030,792	2	30,96	35,05	$3p\,^4D°—4s\,^4P$	$^3/_2—^1/_2$
3028,860	4	27,27	31,36	$3s\,^4P—3p\,^4S°$	$^1/_2—^3/_2$

λ, Å	I	E_H, eV	E_B, eV	Transition	J
3027,011	4	30,52	34,62	$3p\ ^4P^\circ—3d\ ^4D$	$^5/_2—^5/_2$
3017,348	3	30,52	34,63	$3p\ ^4P^\circ—3d\ ^4D$	$^5/_2—^3/_2$
3001,663	6	27,23	31,36	$3s\ ^4P—3p\ ^4S^\circ$	$^3/_2—^3/_2$
2973,07	1	34,02	38,18	$3p'\ ^2F^\circ—3d'\ ^2F$	$^5/_2—^7/_2$
2967,181	3	34,02	38,20	$3p'\ ^2F^\circ—3d'\ ^2F$	$^7/_2—^5/_2$
2963,235	2	34,02	38,20	$3p'\ ^2F^\circ—3d'\ ^2F$	$^5/_2—^5/_2$
2955,73	7	27,17	31,36	$3s\ ^4P—3p\ ^4S^\circ$	$^5/_2—^3/_2$
2953,10	0	30,55	34,75	$3p\ ^4P^\circ—3d\ ^2D$	$^3/_2—^5/_2$
2951,10	2	30,57	34,77	$3p\ ^4P^\circ—3d\ ^2D$	$^1/_2—^3/_2$
2935,30	1	30,55	34,77	$3p\ ^4P^\circ—3d\ ^2D$	$^3/_2—^3/_2$
2933,70	2	30,52	34,75	$3p\ ^4P^\circ—3d\ ^2D$	$^5/_2—^5/_2$
2925,623	3	30,57	34,81	$3p\ ^4P^\circ—3d\ ^4P$	$^1/_2—^1/_2$
2916,16	1	30,52	34,77	$3p\ ^4P^\circ—3d\ ^2D$	$^5/_2—^3/_2$
2910,059	5	30,55	34,81	$3p\ ^4P^\circ—3d\ ^4P$	$^3/_2—^1/_2$
2906,815	3	30,57	34,84	$3p\ ^4P^\circ—3d\ ^4P$	$^1/_2—^3/_2$
2897,03	2	30,52	34,80	$3p\ ^4P^\circ—3d\ ^4F$	$^5/_2—^7/_2$
2891,36	0,5	30,55	34,84	$3p\ ^4P^\circ—3d\ ^4P$	$^3/_2—^3/_2$
2888,43	1	30,55	34,84	$3p\ ^4P^\circ—3d\ ^4F$	$^3/_2—^5/_2$
2878,13	0,5	30,57	34,88	$3p\ ^4P^\circ—3d\ ^2P$	$^1/_2—^1/_2$
2876,43	4	30,52	34,83	$3p\ ^4P^\circ—3d\ ^4F$	$^5/_2—^3/_2$
2873,00	3	30,52	34,84	$3p\ ^4P^\circ—3d\ ^4P$	$^5/_2—^3/_2$
2869,95	2	30,52	34,84	$3p\ ^4P^\circ—3d\ ^4F$	$^5/_2—^5/_2$
2858,01	2	30,52	34,86	$3p\ ^4P^\circ—3d\ ^4P$	$^5/_2—^5/_2$
2809,50	4	30,55	34,96	$3p\ ^4P^\circ—4s\ ^4P$	$^3/_2—^5/_2$
2794,220	3	30,57	35,01	$3p\ ^4P^\circ—4s\ ^4P$	$^1/_2—^3/_2$
2792,015	5	30,52	34,96	$3p\ ^4P^\circ—4s\ ^4P$	$^5/_2—^5/_2$
2780,023	2	30,55	35,01	$3p\ ^4P^\circ—4s\ ^4P$	$^3/_2—^3/_2$
2770,06	1	30,57	35,05	$3p\ ^4P^\circ—4s\ ^4P$	$^1/_2—^1/_2$
2762,922	3	30,52	35,01	$3p\ ^4P^\circ—4s\ ^4P$	$^5/_2—^3/_2$
2756,716	3	30,55	35,05	$3p\ ^4P^\circ—4s\ ^4P$	$^3/_2—^1/_2$
1938,827	8	27,86	34,25	$3s\ ^2P—3p'\ ^2P^\circ$	$^1/_2—^3/_2$
1930,033	8	27,86	34,28	$3s\ ^2P—3p'\ ^2P^\circ$	$^1/_2—^1/_2$
1928,787	1	—	—	—	—
1916,082	10	27,78	34,25	$3s\ ^2P—3p'\ ^2P^\circ$	$^3/_2—^3/_2$
1907,494	8	27,78	34,28	$3s\ ^2P—3p'\ ^2P^\circ$	$^3/_2—^1/_2$
1889,714	1	—	—	—	—
1888,110	1	—	—	—	—
1883,799	1	—	—	—	—
1880,21	3	—	—	—	—
1854,11	1	—	—	—	—
1853,22	1	—	—	—	—
1732,69	1	—	—	—	—
1688,356	4	26,91	34,25	$2p^6\ ^2S—3p'\ ^2P^\circ$	$^1/_2—^3/_2$
1681,683	3	26,91	34,28	$2p^6\ ^2S—3p'\ ^2P^\circ$	$^1/_2—^1/_2$
462,388	14	0,10	26,91	$2p^5\ ^2P^\circ—2p^6\ ^2S$	$^1/_2—^1/_2$
460,725	15	0,00	26,91	$2p^5\ ^2P^\circ—2p^6\ ^2S$	$^3/_2—^1/_2$
456,895	5	0,10	27,23	$2p^5\ ^2P^\circ—3s\ ^4P$	$^1/_2—^3/_2$
456,344	4	0,00	27,17	$2p^5\ ^2P^\circ—3s\ ^4P$	$^3/_2—^5/_2$
455,270	7	0,00	27,23	$2p^5\ ^2P^\circ—3s\ ^4P$	$^3/_2—^3/_2$
454,648	5	0,00	27,27	$2p^5\ ^2P^\circ—3s\ ^4P$	$^3/_2—^1/_2$
447,813	8	0,10	27,78	$2p^5\ ^2P^\circ—3s\ ^2P$	$^1/_2—^3/_2$
446,591	7	0,10	27,86	$2p^5\ ^2P^\circ—3s\ ^2P$	$^1/_2—^1/_2$
446,252	8	0,00	27,78	$2p^5\ ^2P^\circ—3s\ ^2P$	$^3/_2—^3/_2$
445,032	7	0,00	27,86	$2p^5\ ^2P^\circ—3s\ ^2P$	$^3/_2—^1/_2$
407,136	8	0,10	30,55	$2p^5\ ^2P^\circ—3s'\ ^2D$	$^1/_2—^3/_2$
405,852	9	0,00	30,55	$2p^5\ ^2P^\circ—3s'\ ^2D$	$^3/_2—^5/_2$, $^3/_2$
362,456	4	0,10	34,30	$2p^5\ ^2P^\circ—3s''\ ^2S$	$^1/_2—^1/_2$
361,427	5	0,00	34,30	$2p^5\ ^2P^\circ—3s''\ ^2S$	$^3/_2—^1/_2$

λ, Å	I	E_H, eV	E_B, eV	Transition	J
357,534	5	0,10	34,77	$2p^5\ ^2P^\circ—3d\ ^2D$	$^1/_2—^3/_2$
356,795	5	0,00	34,75	$2p^5\ ^2P^\circ—3d\ ^2D$	$^3/_2—^5/_2$
356,534	3	0,00	34,77	$2p^5\ ^2P^\circ—3d\ ^2D$	$^3/_2—^3/_2$
356,436	2	0,10	34,88	$2p^5\ ^2P^\circ—3d\ ^2P$	$^1/_2—^1/_2$
356,131	4	0,00	34,81	$2p^5\ ^2P^\circ—3d\ ^2F$	$^3/_2—^5/_2$
355,946	2	0,10	34,93	$2p^5\ ^2P^\circ—3d\ ^2P$	$^1/_2—^3/_2$
355,848	1	0,00	34,84	$2p^5\ ^2P^\circ—3d\ ^4F$	$^3/_2—^5/_2$
355,647	3	0,00	34,86	$2p^5\ ^2P^\circ—3d\ ^4P$	$^3/_2—^5/_2$
355,450	2	0,00	34,88	$2p^5\ ^2P^\circ—3d\ ^2P$	$^3/_2—^1/_2$
354,954	4	0,00	34,93	$2p^5\ ^2P^\circ—3d\ ^2P$	$^3/_2—^3/_2$
353,922	2	0,10	35,13	$2p^5\ ^2P^\circ—4s\ ^2P$	$^1/_2—^3/_2$
353,206	3	0,10	35,20	$2p^5\ ^2P^\circ—4s\ ^2P$	$^1/_2—^1/_2$
352,946	4	0,00	35,13	$2p^5\ ^2P^\circ—4s\ ^2P$	$^3/_2—^3/_2$
352,237	2	0,00	35,20	$2p^5\ ^2P^\circ—4s\ ^2P$	$^3/_2—^1/_2$
331,50	2	0,10	37,50	$2p^5\ ^2P^\circ—4d\ ^2D$	$^1/_2—^3/_2$
331,06	1	0,10	37,55	$2p^5\ ^2P^\circ—4d\ ^2P$	$^1/_2—^3/_2,\ ^1/_2$
330,77	3	0,00	37,48	$2p^5\ ^2P^\circ—4d\ ^2D$	$^3/_2—^5/_2$
330,62	2	0,00	37,50	$2p^5\ ^2P^\circ—4d\ ^2D$	$^3/_2—^3/_2$
330,20	2	0,00	37,55	$2p^5\ ^2P^\circ—4d\ ^2P$	$^3/_2—^3/_2,\ ^1/_2$
328,08	2	0,10	37,89	$2p^5\ ^2P^\circ—3d'\ ^2P$	$^1/_2—^3/_2,\ ^1/_2$
327,63	2	0,10	37,94	$2p^5\ ^2P^\circ—4s'\ ^2D$	$^1/_2—^3/_2$
327,33	3	0,10	37,97	$2p^5\ ^2P^\circ—3d'\ ^2D$	$^1/_2—^3/_2$
327,25	2	0,00	37,89	$2p^5\ ^2P^\circ—3d'\ ^2P$	$^3/_2—^3/_2,\ ^1/_2$
326,77	3	0,00	37,94	$2p^5\ ^2P^\circ—4s'\ ^2D$	$^3/_2—^5/_2,\ ^3/_2$
326,54	5	0,00	37,97	$2p^5\ ^2P^\circ—3d'\ ^2D$	$^3/_2—^3/_2$
324,56	2	0,00	38,20	$2p^5\ ^2P^\circ—3d'\ ^2F$	$^3/_2—^5/_2$

Ne III, **ground state** $1s^2\,2s^2\,2p^4\ ^3P_2$

Ionization potential 514148 cm⁻¹; 63,742 eV

λ, Å	I	E_H, eV	E_B, eV	Transition	J
2825,82	5	46,43	50,81	$3s''\ ^3P^\circ—3p''\ ^3D$	0—1
2825,28	4	46,42	50,81	$3s''\ ^3P^\circ—3p''\ ^3D$	1—2
2824,47	3	46,42	50,82	$3s''\ ^3P^\circ—3p''\ ^3D$	1—1
2822,95	7	46,42	50,81	$3s''\ ^3P^\circ—3p''\ ^3D$	2—3
2802,34	2	46,42	50,85	$3s''\ ^3P^\circ—3p''\ ^3S$	1—1
2800,24	3	46,42	50,85	$3s''\ ^3P^\circ—3p''\ ^3S$	2—1
2787,73	4	43,79	48,23	$3s'\ ^3D^\circ—3p'\ ^3D$	1—1
2786,89	3	43,79	48,23	$3s'\ ^3D^\circ—3p'\ ^3D$	1—2
2786,17	2	43,79	48,23	$3s'\ ^3D^\circ—3p'\ ^3D$	2—1
2785,29	5	43,79	48,23	$3s'\ ^3D^\circ—3p'\ ^3D$	2—2
2783,03	2	43,78	48,24	$3s'\ ^3D^\circ—3p'\ ^3D$	3—2
2777,65	7	43,78	48,24	$3s'\ ^3D^\circ—3p'\ ^3D$	3—3
2678,64	25	39,60	44,23	$3s\ ^3S^\circ—3p\ ^3P$	1—1
2677,90	30	39,60	44,23	$3s\ ^3S^\circ—3p\ ^3P$	1—2, 0
2642,42	3	46,42	51,12	$3s''\ ^3P^\circ—3p''\ ^3P$	1—0
2642,25	2	46,42	51,12	$3s''\ ^3P^\circ—3p''\ ^3P$	0—1
2641,07	10	46,42	51,12	$3s''\ ^3P^\circ—3p''\ ^3P$	1—1
2640,56	6	46,42	51,12	$3s''\ ^3P^\circ—3p''\ ^3P$	1—2
2639,18	5	46,42	51,12	$3s''\ ^3P^\circ—3p''\ ^3P$	2—1
2638,70	10	46,42	51,12	$3s''\ ^3P^\circ—3p''\ ^3P$	2—2
2615,87	10	43,79	48,53	$3s'\ ^3D^\circ—3p'\ ^3F$	1—2
2614,51	4	43,79	48,53	$3s'\ ^3D^\circ—3p'\ ^3F$	2—2

λ, Å	I	E_H, eV	E_B, eV	Transition	J
2613,41	12	43,79	48,53	$3s'\ {}^3D^\circ - 3p'\ {}^3F$	2—3
2611,42	4	43,78	48,53	$3s'\ {}^3D^\circ - 3p'\ {}^3F$	3—3
2610,03	15	43,78	48,53	$3s'\ {}^3D^\circ - 3p'\ {}^3F$	3—4
2595,68	20	38,95	43,72	$3s\ {}^5S^\circ - 3p\ {}^5P$	2—1
2593,60	30	38,95	43,72	$3s\ {}^5S^\circ - 3p\ {}^5P$	2—2
2590,04	40	38,95	47,73	$3s\ {}^5S^\circ - 3p\ {}^5P$	2—3
2473,40	10	—	—	—	—
2468,20	4	49,48	54,50	$3p'\ {}^3P - 3d'\ {}^3P^\circ$	1—2
2463,38	2	49,48	54,51	$3p'\ {}^3P - 3d'\ {}^3P^\circ$	0—1
2462,35	6	49,47	54,50	$3p'\ {}^3P - 3d'\ {}^3P^\circ$	2—2
2460,84	1	49,48	54,51	$3p'\ {}^3P - 3d'\ {}^3P^\circ$	1—1
2457,55	2	49,48	54,52	$3p'\ {}^3P - 3d'\ {}^3P^\circ$	1—0
2454,98	5	49,47	54,51	$3p'\ {}^3P - 3d'\ {}^3P^\circ$	2—1
2441,90	2	49,48	54,56	$3p'\ {}^3P - 3d'\ {}^3S^\circ$	0—1
2439,34	5	49,48	54,56	$3p'\ {}^3P - 3d'\ {}^3S^\circ$	1—1
2433,62	8	49,47	54,56	$3p'\ {}^3P - 3d'\ {}^3S^\circ$	2—1
2413,78	10	44,23	49,37	$3p\ {}^3P - 3d\ {}^3D^\circ$	2, 0—1
2413,54	6	44,23	49,37	$3p\ {}^3P - 3d\ {}^3D^\circ$	2—2
2413,18	8	44,23	49,37	$3p\ {}^3P - 3d\ {}^3D^\circ$	1—1
2412,94	12	44,23	49,37	$3p\ {}^3P - 3d\ {}^3D^\circ$	1—2
2412,73	15	44,23	49,37	$3p\ {}^3P - 3d\ {}^3D^\circ$	2—3
2266,98	5	48,53	54,00	$3p'\ {}^3F - 3d'\ {}^3F^\circ$	3—2
2266,16	8	48,53	54,00	$3p'\ {}^3F - 3d'\ {}^3F^\circ$	2—2
2264,91	10	48,53	54,00	$3p'\ {}^3F - 3d'\ {}^3F^\circ$	3—3
2264,11	3	48,53	54,00	$3p'\ {}^3F - 3d'\ {}^3F^\circ$	2—3
2263,21	12	48,53	54,01	$3p'\ {}^3F - 3d'\ {}^3F^\circ$	4—4
2262,16	2	48,53	54,01	$3p'\ {}^3F - 3d'\ {}^3F^\circ$	3—4
2216,07	15	48,53	54,12	$3p'\ {}^3F - 3d'\ {}^3G^\circ$	4—5
2214,77	4	48,53	54,13	$3p'\ {}^3F - 3d'\ {}^3G^\circ$	4—4
2213,76	12	48,53	54,13	$3p'\ {}^3F - 3d'\ {}^3G^\circ$	3—4
2212,63	5	48,53	54,13	$3p'\ {}^3F - 3d'\ {}^3G^\circ$	3—3
2211,85	10	48,53	54,13	$3p'\ {}^3F - 3d'\ {}^3G^\circ$	2—3
2209,35	10	—	—	—	—
2208,04	4	—	—	—	—
2207,29	8	—	—	—	—
2205,95	5	—	—	—	—
2204,98	7	—	—	—	—
2202,22	7	48,53	54,16	$3p'\ {}^3F - 3d'\ {}^3D^\circ$	4—3
2201,23	4	48,53	54,16	$3p'\ {}^3F - 3d'\ {}^3D^\circ$	3—3
2197,86	7	48,53	54,17	$3p'\ {}^3F - 3d'\ {}^3D^\circ$	3—2
2197,10	3	48,53	54,17	$3p'\ {}^3F - 3d'\ {}^3D^\circ$	2—2
2194,92	5	48,53	54,17	$3p'\ {}^3F - 3d'\ {}^3D^\circ$	2—1
2190,29	7	—	—	—	—
2183,24	2	43,79	49,47	$3s'\ {}^3D^\circ - 3p'\ {}^3P$	1—2
2182,28	3	43,79	49,47	$3s'\ {}^3D^\circ - 3p'\ {}^3P$	2—2
2180,89	10	43,78	49,47	$3s'\ {}^3D^\circ - 3p'\ {}^3P$	3—2
2178,69	4	43,79	49,48	$3s'\ {}^3D^\circ - 3p'\ {}^3P$	1—1
2177,73	8	43,79	49,48	$3s'\ {}^3D^\circ - 3p'\ {}^3P$	2—1
2176,67	5	43,79	49,48	$3s'\ {}^3D^\circ - 3p'\ {}^3P$	1—0
2163,77	15	43,73	49,46	$3p\ {}^5P - 3d\ {}^5D^\circ$	3—4
2161,22	10	43,72	49,46	$3p\ {}^5P - 3d\ {}^5D^\circ$	2—3
2161,04	6	43,72	49,46	$3p\ {}^5P - 3d\ {}^5D^\circ$	2—2
2160,88	2	43,72	49,46	$3p\ {}^5P - 3d\ {}^5D^\circ$	2—1
2159,60	4	43,72	49,46	$3p\ {}^5P - 3d\ {}^5D^\circ$	1—2
2159,44	5	43,72	49,46	$3p\ {}^5P - 3d\ {}^5D^\circ$	1—1
2153,15	2	48,24	54,00	$3p'\ {}^3D - 3d'\ {}^3F^\circ$	3—3
2151,78	3	48,23	54,00	$3p'\ {}^3D - 3d'\ {}^3F^\circ$	2—2
2151,26	5	48,23	54,00	$3p'\ {}^3D - 3d'\ {}^3F^\circ$	1—2

λ, Å	I	E_H, eV	E_B, eV	Transition	J
2150,70	8	48,24	54,01	$3p'\,^3D-3d'\,^3F^\circ$	3—4
2149,92	6	48,23	54,00	$3p'\,^3D-3d'\,^3F^\circ$	2—3
2095,54	20	48,24	54,16	$3p'\,^3D-3d'\,^3D^\circ$	3—3
2092,44	12	48,23	54,16	$3p'\,^3D-3d'\,^3D^\circ$	2—3
		48,24	54,17	$3p'\,^3D-3d'\,^3D^\circ$	3—2
2089,43	15	48,23	54,17	$3p'\,^3D-3d'\,^3D^\circ$	2—2
2088,92	5	48,23	54,17	$3p'\,^3D-3d'\,^3D^\circ$	1—2
2087,44	7	48,23	54,17	$3p'\,^3D-3d'\,^3D^\circ$	2—1
2086,96	10	48,24	54,17	$3p'\,^3D-3d'\,^3D^\circ$	1—1
1257,190	6	39,60	49,47	$3s\,^3S^\circ-3p'\,^3P$	1—2
1255,685	5	39,60	49,48	$3s\,^3S^\circ-3p'\,^3P$	1—1
1255,026	2	39,60	45,48	$3s\,^3S^\circ-3p'\,^3P$	1—0
491,050	9	0,08	25,33	$2p^4\,^3P-2p^5\,^3P^\circ$	1—2
490,310	7	0,11	25,40	$2p^4\,^3P-2p^5\,^3P^\circ$	0—1
489,641	4	0,08	25,40	$2p^4\,^3P-2p^5\,^3P^\circ$	1—1
489,501	10	0,00	25,33	$2p^4\,^3P-2p^5\,^3P^\circ$	2—2
488,868	7	0,08	25,44	$2p^4\,^3P-2p^5\,^3P^\circ$	1—0
488,103	8	0,00	25,40	$2p^4\,^3P-2p^5\,^3P^\circ$	2—1
427,840	3	6,91	35,89	$2p^4\,^1S-2p^5\,^1P^\circ$	0—1
379,308	7	3,20	35,89	$2p^4\,^1D-2p^5\,^1P^\circ$	2—1
313,92	1	0,11	39,60	$2p^4\,^3P-3s\,^3S^\circ$	0—1
313,677	3	0,08	39,60	$2p^4\,^3P-3s\,^3S^\circ$	1—1
313,048	4	0,00	39,60	$2p^4\,^3P-3s\,^3S^\circ$	2—1
308,559	1	6,91	47,09	$2p^4\,^1S-3s''\,^1P^\circ$	0—1
301,124	4	3,20	44,38	$2p^4\,^1D-3s'\,^1D^\circ$	2—2
283,894	3	0,11	43,79	$2p^4\,^3P-3s'\,^3D^\circ$	0—1
283,690	5	0,08	43,79	$2p^4\,^3P-3s'\,^3D^\circ$	1—2, 1
283,206	6	0,00	43,78	$2p^4\,^3P-3s'\,^3D^\circ$	2—3
283,178	5	0,00	43,79	$2p^4\,^3P-3s'\,^3D^\circ$	2—2, 1
282,50	0	3,20	47,09	$2p^4\,^1D-3s''\,^1P$	2—1
267,709	2	0,11	46,42	$2p^4\,^3P-3s''\,^3P^\circ$	0—1
267,516	3	0,08	46,42	$2p^4\,^3P-3s''\,^3P^\circ$	1—2, 1, 0
267,059	3	0,00	46,42	$2p^4\,^3P-3s''\,^3P^\circ$	2—2, 1
251,726	2	0,11	49,37	$2p^4\,^3P-3d\,^3D^\circ$	0—1
251,558	2	0,08	49,37	$2p^4\,^3P-3d\,^3D^\circ$	1—2, 1
251,145	2	0,00	49,37	$2p^4\,^3P-3d\,^3D^\circ$	2—3, 2, 1

Ne IV, ground state $1s^2\,2s^2\,2p^3\,^4S^0_{3/2}$

Ionization potential 782768 cm^{-1}; 97,044 eV

λ, Å	I	E_H, eV	E_B, eV	Transition	J
2405,19	1	59,47	64,62	$3s\,^4P-3p\,^4D^\circ$	$^5/_2$—$^3/_2$
2404,28	0	60,64	65,80	$3s\,^2P-3p\,^2D^\circ$	$^3/_2$—$^3/_2$
2384,95	7	59,47	64,66	$3s\,^4P-3p\,^4D^\circ$	$^5/_2$—$^5/_2$
2384,20	3	59,40	64,59	$3s\,^4P-3p\,^4D^\circ$	$^3/_2$—$^1/_2$
2373,21	9	—	—	—	—
2372,16	7	59,40	64,62	$3s\,^4P-3p\,^4D^\circ$	$^3/_2$—$^3/_2$
2365,49	4	60,56	65,80	$3s\,^2P-3p\,^2D^\circ$	$^1/_2$—$^3/_2$
2363,28	6	60,64	65,89	$3s\,^2P-3p\,^2D^\circ$	$^3/_2$—$^5/_2$
2362,68	6	59,35	64,59	$3s\,^4P-3p\,^4D^\circ$	$^1/_2$—$^1/_2$
2357,96	10	59,47	64,72	$3s\,^4P-3p\,^4D^\circ$	$^5/_2$—$^7/_2$
2352,52	8	59,40	64,66	$3s\,^4P-3p\,^4D^\circ$	$^3/_2$—$^5/_2$
2350,84	6	59,35	64,62	$3s\,^4P-3p\,^4D^\circ$	$^1/_2$—$^3/_2$
2293,49	6	63,44	68,84	$3s'\,^2D-3p'\,^2F^\circ$	$^3/_2$—$^5/_2$

λ, Å	I	E_H, eV	E_B, eV	Transition	J
2293,14	2	63,44	68,84	$3s'\ ^2D—3p'\ ^2F^\circ$	$^5/_2—^5/_2$
2285,79	9	63,44	68,86	$3s'\ ^2D—3p'\ ^2F^\circ$	$^5/_2—^7/_2$
2264,54	4	66,76	72,23	$3s'\ ^6S^\circ—3p'\ ^6P$	$^5/_2—^3/_2$
2262,08	5	66,76	72,24	$3s'\ ^6S^\circ—3p'\ ^6P$	$^5/_2—^5/_2$
2258,02	6	66,76	72,25	$3s'\ ^6S^\circ—3p'\ ^6P$	$^5/_2—^7/_2$
2220,81	1	59,47	65,05	$3s\ ^4P—3p\ ^4P^\circ$	$^5/_2—^3/_2$
2203,88	2	59,47	65,09	$3s\ ^4P—3p\ ^4P^\circ$	$^5/_2—^5/_2$
2022,192	4	63,44	69,57	$3s'\ ^2D—3p'\ ^2D^\circ$	$^5/_2—^5/_2$
2018,441	3	63,44	69,58	$3s'\ ^2D—3p'\ ^2D^\circ$	$^3/_2—^3/_2$
786,141	1	60,20	75,97	$2p^5\ ^2P^\circ—3d'\ ^2P$	$^1/_2—^3/_2$
780,250	3	60,08	75,97	$2p^5\ ^2P^\circ—3d'\ ^2P$	$^3/_2—^3/_2$
758,317	3	60,08	76,43	$2p^5\ ^2P^\circ—3d'\ ^2S$	$^3/_2—^1/_2$
609,168	1	39,73	60,08	$2p^4\ ^2P—2p^5\ ^2P^\circ$	$^1/_2—^3/_2$
606,527	5	39,64	60,08	$2p^4\ ^2P—2p^5\ ^2P^\circ$	$^3/_2—^3/_2$
605,595	2	39,73	60,20	$2p^4\ ^2P—2p^5\ ^2P^\circ$	$^1/_2—^1/_2$
602,999	2	39,64	60,08	$2p^4\ ^2P—2p^5\ ^2P^\circ$	$^3/_2—^1/_2$
543,891	150	0,00	22,79	$2p^3\ ^4S^\circ—2p^4\ ^4P$	$^3/_2—^5/_2$
542,073	100	0,00	22,87	$2p^3\ ^4S^\circ—2p^4\ ^4P$	$^3/_2—^3/_2$
541,127	80	0,00	22,91	$2p^3\ ^4S^\circ—2p^4\ ^4P$	$^3/_2—^1/_2$
539,731	3	37,11	60,08	$2p^4\ ^2S—2p^5\ ^2P^\circ$	$^1/_2—^3/_2$
536,965	1	37,11	60,20	$2p^4\ ^2S—2p^5\ ^2P^\circ$	$^1/_2—^1/_2$
521,813	25	7,71	31,47	$2p^3\ ^2P^\circ—2p^4\ ^2D$	$^3/_2—^5/_2$
521,742	25	7,71	31,47	$2p^3\ ^2P^\circ—2p^4\ ^2D$	$^1/_2—^3/_2$
469,865	200	5,08	31,47	$2p^3\ ^2D^\circ—2p^4\ ^2D$	$^3/_2—^5/_2$
469,817	200	5,08	31,47	$2p^3\ ^2D^\circ—2p^4\ ^2D$	$^5/_2—^5/_2$ $^3/_2—^3/_2$
433,237	50	31,47	60,08	$2p^4\ ^2D—2p^5\ ^2P^\circ$	$^5/_2$, $^3/_2—^3/_2$
431,472	25	31,47	60,20	$2p^4\ ^2D—2p^5\ ^2P^\circ$	$^3/_2—^1/_2$
421,609	150	7,71	37,11	$2p^3\ ^2P^\circ—2p^4\ ^2S$	$^3/_2$, $^1/_2—^1/_2$
388,218	100	7,71	39,64	$2p^3\ ^2P^\circ—2p^4\ ^2P$	$^3/_2$, $^1/_2—^3/_2$
387,141	125	7,71	39,73	$2p^3\ ^2P^\circ—2p^4\ ^2P$	$^3/_2$, $^1/_2—^1/_2$
358,721	200	5,08	39,64	$2p^3\ ^2D^\circ—2p^4\ ^2P$	$^5/_2$, $^3/_2—^3/_2$
357,831	50	5,08	39,73	$2p^3\ ^2D^\circ—2p^4\ ^2P$	$^3/_2—^1/_2$
294,390	3	22,91	65,01	$2p^4\ ^4P—3p\ ^4P^\circ$	$^1/_2—^1/_2$
294,100	3	22,87	65,01	$2p^4\ ^4P—3p\ ^4P^\circ$	$^3/_2—^1/_2$
293,947	1	22,87	65,05	$2p^4\ ^4P—3p\ ^4P^\circ$	$^3/_2—^3/_2$,
293,649	5	22,87	65,09	$2p^4\ ^4P—3p\ ^4P^\circ$	$^3/_2—^5/_2$
293,429	10	22,79	65,05	$2p^4\ ^4P—3p\ ^4P^\circ$	$^5/_2—^3/_2$
293,123	15	22,79	65,09	$2p^4\ ^4P—3p\ ^4P^\circ$	$^5/_2—^5/_2$
287,206	10	22,91	66,08	$2p^4\ ^4P—3p\ ^4S^\circ$	$^1/_2—^3/_2$
286,934	15	22,87	66,08	$2p^4\ ^4P—3p\ ^4S^\circ$	$^3/_2—^3/_2$
286,688	15	22,79	66,08	$2p^4\ ^4P—3p\ ^4S^\circ$	$^5/_2—^3/_2$
248,004	8	22,91	72,9	$2p^4\ ^4P—3s'''\ ^4S^\circ$	$^1/_2—^3/_2$
247,807	8	22,87	72,90	$2p^4\ ^4P—3s'''\ ^4S^\circ$	$^3/_2—^3/_2$
247,422	10	22,79	72,90	$2p^4\ ^4P—3s'''\ ^4S^\circ$	$^5/_2—^3/_2$
234,701	25	7,71	60,56	$2p^3\ ^2P^\circ—3s\ ^2P$	$^3/_2$, $^1/_2—^1/_2$
234,316	25	7,71	60,64	$2p^3\ ^2P^\circ—3s\ ^2P$	$^3/_2$, $^1/_2—^3/_2$
223,605	25	5,08	60,56	$2p^3\ ^2D^\circ—3s\ ^2P$	$^3/_2—^1/_2$
223,241	25	5,08	60,64	$2p^3\ ^2D^\circ—3s\ ^2P$	$^5/_2—^3/_2$
222,600	40	7,71	63,44	$2p^3\ ^2P^\circ—3s'\ ^2D$	$^3/_2$, $^1/_2—^5/_2$, $^3/_2$
218,766	5	22,91	79,58	$2p^4\ ^4P—4p\ ^4D^\circ$	$^1/_2—^1/_2$
218,643	25	22,87 22,91	79,58 79,62	$2p^4\ ^4P—4p\ ^4D^\circ$ $2p^4\ ^4P—4p\ ^4D^\circ$	$^3/_2—^1/_2$ $^1/_2—^3/_2$
218,483	20	22,87	79,62	$2p^4\ ^4P—4p\ ^4D^\circ$	$^3/_2—^3/_2$
218,343	15	22,87	79,65	$2p^4\ ^4P—4p\ ^4D^\circ$	$^3/_2—^5/_2$
218,184	10	22,79	79,62	$2p^4\ ^4P—4p\ ^4D^\circ$	$^5/_2—^3/_2$
218,131	20	22,91	79,75	$2p^4\ ^4P—4p\ ^4P^\circ$	$^1/_2—^1/_2$
217,830	25	22,79	79,71	$2p^4\ ^4P—4p\ ^4D^\circ$	$^5/_2—^7/_2$
217,777	15	22,87	79,80	$2p^4\ ^4P—4p\ ^4P^\circ$	$^3/_2—^3/_2$

λ, Å	I	E_H, eV	E_B, eV	Transition	J
217,640	15	22,87	79,84	$2p^4\,{}^4P$—$4p\,{}^4P^\circ$	$^3/_2$—$^5/_2$
217,337	15	22,79	79,84	$2p^4\,{}^4P$—$4p\,{}^4P^\circ$	$^5/_2$—$^5/_2$
215,843	15	22,91	80,34	$2p^4\,{}^4P$—$4p\,{}^4S^\circ$	$^1/_2$—$^3/_2$
215,711	3	22,87	80,34	$2p^4\,{}^4P$—$4p\,{}^4S^\circ$	$^3/_2$—$^3/_2$
215,396	3	22,79	80,34	$2p^4\,{}^4P$—$4p\,{}^4S^\circ$	$^5/_2$—$^3/_2$
212,556	150	5,08	63,44	$2p^3\,{}^2D^\circ$—$3s'\,{}^2D$	$^5/_2$, $^3/_2$—$^5/_2$, $^3/_2$
208,899	80	0,00	59,35	$2p^3\,{}^4S^\circ$—$3s\,{}^4P$	$^3/_2$—$^1/_2$
208,734	100	0,00	59,40	$2p^3\,{}^4S^\circ$—$3s\,{}^4P$	$^3/_2$—$^3/_2$
208,485	100	0,00	59,47	$2p^3\,{}^4S^\circ$—$3s\,{}^4P$	$^3/_2$—$^5/_2$
204,908	5	22,91	83,41	$2p^4\,{}^4P$—$3d'''\,{}^4D^\circ$	$^1/_2$—$^5/_2$, $^3/_2$, $^1/_2$
204,786	15	22,87	83,41	$2p^4\,{}^4P$—$3d'''\,{}^4D^\circ$	$^3/_2$—$^5/_2$, $^3/_2$, $^1/_2$
204,531	25	22,79	83,41	$2p^4\,{}^4P$—$3d'''\,{}^4D^\circ$	$^5/_2$—$^7/_2$, $^5/_2$, $^3/_2$
204,270	15	7,71	68,40	$2p^3\,{}^2P^\circ$—$3s''\,{}^2S$	$^3/_2$, $^1/_2$—$^1/_2$
194,623	50	7,71	71,41	$2p^3\,{}^2P^\circ$—$3d\,{}^2P$	$^3/_2$, $^1/_2$—$^3/_2$
194,477	40	7,71	71,45	$2p^3\,{}^2P^\circ$—$3d\,{}^2P$	$^3/_2$, $^1/_2$—$^1/_2$
194,276	100	7,71	71,52	$2p^3\,{}^2P^\circ$—$3d''\,{}^2D$	$^3/_2$, $^1/_2$—$^5/_2$, $^3/_2$
190,645	15	7,71	72,74	$2p^3\,{}^2P^\circ$—$3d\,{}^2D$	$^3/_2$, $^1/_2$—$^3/_2$
190,565	25	7,71	72,76	$2p^3\,{}^2P^\circ$—$3d\,{}^2D$	$^3/_2$, $^1/_2$—$^5/_2$
186,915	15	5,08	71,41	$2p^3\,{}^2D^\circ$—$3d\,{}^2P$	$^5/_2$, $^3/_2$—$^3/_2$
186,787	5	5,08	71,45	$2p^3\,{}^2D^\circ$—$3d\,{}^2P$	$^3/_2$—$^1/_2$
186,575	150	5,08	71,52	$2p^3\,{}^2D^\circ$—$3d''\,{}^2D$	$^5/_2$, $^3/_2$—$^5/_2$, $^3/_2$
185,479	20	5,08	71,92	$2p^3\,{}^2D^\circ$—$3d\,{}^2F$	$^5/_2$—$^7/_2$
183,247	12	5,08	72,74	$2p^3\,{}^2D^\circ$—$3d\,{}^2D$	$^5/_2$, $^3/_2$—$^3/_2$
183,165	15	5,08	72,76	$2p^3\,{}^2D^\circ$—$3d\,{}^2D$	$^5/_2$, $^3/_2$—$^5/_2$
181,691	20	7,71	75,96	$2p^3\,{}^2P^\circ$—$3d'\,{}^2P$	$^3/_2$, $^1/_2$—$^1/_2$
181,614	20	7,71	75,97	$2p^3\,{}^2P^\circ$—$3d'\,{}^2P$	$^3/_2$, $^1/_2$—$^3/_2$
180,402	15	7,71	76,43	$2p^3\,{}^2P^\circ$—$3d'\,{}^2S$	$^3/_2$, $^1/_2$—$^1/_2$
177,161	80	5,08	75,06	$2p^3\,{}^2D^\circ$—$3d'\,{}^2F$	$^5/_2$, $^3/_2$—$^5/_2$
176,007	50	5,08	75,52	$2p^3\,{}^2D^\circ$—$3d'\,{}^2D$	$^5/_2$, $^3/_2$—$^5/_2$, $^3/_2$
174,920	8	5,08	75,96	$2p^3\,{}^2D^\circ$—$3d'\,{}^2P$	$^3/_2$—$^1/_2$
174,880	10	5,08	75,97	$2p^3\,{}^2D^\circ$—$3d'\,{}^2P$	$^5/_2$, $^3/_2$—$^3/_2$
174,303	3	7,71	78,83	$2p^3\,{}^2P^\circ$—$4s\,{}^2P$	$^3/_2$, $^1/_2$—$^1/_2$
172,620	80	0,00	71,82	$2p^3\,{}^4S^\circ$—$3d\,{}^4P$	$^3/_2$—$^5/_2$
172,525	50	0,00	71,86	$2p^3\,{}^4S^\circ$—$3d\,{}^4P$	$^3/_2$—$^3/_2$
172,492	40	0,00	71,87	$2p^3\,{}^4S^\circ$—$3d\,{}^4P$	$^3/_2$—$^1/_2$
168,101	2	5,08	78,83	$2p^3\,{}^2D^\circ$—$4s\,{}^2P$	$^3/_2$—$^1/_2$
167,921	5	5,08	78,91	$2p^3\,{}^2D^\circ$—$4s\,{}^2P$	$^5/_2$, $^3/_2$—$^3/_2$
163,602	2	7,71	83,49	$2p^3\,{}^2P^\circ$—$4d\,{}^2D$	$^3/_2$, $^1/_2$—$^3/_2$
163,562	12	7,71	83,51	$2p^3\,{}^2P^\circ$—$4d\,{}^2D$	$^3/_2$—$^5/_2$
160,471	10	5,08	82,34	$2p^3\,{}^2D^\circ$—$4s'\,{}^2D$	$^5/_2$, $^3/_2$—$^5/_2$, $^3/_2$
158,822	15	5,08	83,14	$2p^3\,{}^2D^\circ$—$4d\,{}^2F$	$^5/_2$, $^3/_2$—$^5/_2$
158,646	15	5,08	83,22	$2p^3\,{}^2D^\circ$—$4d\,{}^2F$	$^5/_2$—$^7/_2$
158,105	2	5,08	83,49	$2p^3\,{}^2D^\circ$—$4d\,{}^2D$	$^5/_2$, $^3/_2$—$^3/_2$
158,063	5	5,08	83,51	$2p^3\,{}^2D$—$4d\,{}^2D$	$^5/_2$, $^3/_2$—$^5/_2$
157,862	2	0,00	78,54	$2p^3\,{}^4S^\circ$—$4s\,{}^4P$	$^3/_2$—$^1/_2$
157,781	3	0,00	78,58	$2p^3\,{}^4S^\circ$—$4s\,{}^4P$	$^3/_2$—$^3/_2$
157,626	5	0,00	78,65	$2p^3\,{}^4S^\circ$—$4s\,{}^4P$	$^3/_2$—$^5/_2$
156,873	3	7,71	86,74	$2p^3\,{}^2P^\circ$—$4d'\,{}^2D$	$^3/_2$, $^1/_2$—$^5/_2$, $^3/_2$
156,480	5	7,71	86,94	$2p^3\,{}^2P^\circ$—$4d'\,{}^2P$	$^3/_2$, $^1/_2$—$^3/_2$, $^1/_2$
154,488	5	7,71	87,96	$2p^3\,{}^2P^\circ$—$4d''\,{}^2D$	$^3/_2$, $^1/_2$—$^5/_2$, $^3/_2$
152,231	15	5,08	86,52	$2p^3\,{}^2D^\circ$—$4d'\,{}^2F$	$^5/_2$, $^3/_2$—$^5/_2$, $^7/_2$
151,817	15	5,08	86,74	$2p^3\,{}^2D^\circ$—$4d'\,{}^2D$	$^5/_2$, $^3/_2$—$^5/_2$, $^3/_2$
150,931	1	7,71	89,85	$2p^3\,{}^2P^\circ$—$5s'\,{}^2D$	$^3/_2$, $^1/_2$—$^5/_2$, $^3/_2$
149,589	2	5,08	87,96	$2p^3\,{}^2D^\circ$—$4d''\,{}^2D$	$^5/_2$, $^3/_2$—$^5/_2$, $^3/_2$
148,942	4	0,00	83,24	$2p^3\,{}^4S^\circ$—$4d\,{}^4P$	$^3/_2$—$^5/_2$
148,787	3	0,00	83,33	$2p^3\,{}^4S^\circ$—$4d\,{}^4P$	$^3/_2$—$^3/_2$
148,660	1	0,00	83,40	$2p^3\,{}^4S^\circ$—$4d\,{}^4P$	$^3/_2$—$^1/_2$
146,262	2	5,08	89,85	$2p^3\,{}^2D^\circ$—$5s'\,{}^2D$	$^5/_2$, $^3/_2$—$^5/_2$, $^3/_2$

λ, Å	I	E_H, eV	E_B, eV	Transition	J
144,288	1	0,00	85,93	$2p^3\ ^4S^\circ - 5s\ ^4P$	$^3/_2 - ^1/_2$
144,151	2	0,00	86,01	$2p^3\ ^4S^\circ - 5s\ ^4P$	$^3/_2 - ^3/_2$
144,019	2	0,00	86,08	$2p^3\ ^4S^\circ - 5s\ ^4P$	$^3/_2 - ^5/_2$
142,929	3	5,08	91,82	$2p^3\ ^2D^\circ - 5d'\ ^2F$	$^5/_2,\ ^3/_2 - ^7/_2,\ ^5/_2$
140,127	3	5,08	93,55	$2p^3\ ^2D^\circ - 6s'\ ^2D$	$^5/_2,\ ^3/_2 - ^5/_2,\ ^3/_2$

Ne V, ground state $1s^2\,2s^2\,2p^2\ ^3P_0$
Ionization potential 1 018 634 cm^{-1}; 126,287 eV

λ, Å	I	E_H, eV	E_B, eV	Transition	J
2306,31	2	74,08	79,45	$3s\ ^3P^\circ - 3p\ ^3D$	2—2
2282,61	1	73,97	79,40	$3s\ ^3P^\circ - 3p\ ^3D$	1—1
2274,54	0	86,60	92,05	$3s\ ^5P - 3p\ ^5D^\circ$	3—2
2265,71	6	74,08	79,55	$3s\ ^3P^\circ - 3p\ ^3D$	2—3
2263,39	3	73,92	79,40	$3s\ ^3P^\circ - 3p\ ^3D$	0—1
2259,57	3	73,97	79,45	$3s\ ^3P^\circ - 3p\ ^3D$	1—2
2256,05	1	86,60	92,09	$3s\ ^5P - 3p\ ^5D^\circ$	3—3
2245,48	3	86,53	92,05	$3s\ ^5P - 3p\ ^5D^\circ$	2—2
2236,29	2	86,47	92,02	$3s\ ^5P - 3p\ ^5D^\circ$	1—1
2232,41	4	86,60	92,15	$3s\ ^5p - 3p\ ^5D^\circ$	3—4
2227,42	3	86,53	92,09	$3s\ ^5p - 3p\ ^5D^\circ$	2—3
2224,12	1	86,47	92,05	$3s\ ^5P - 3p\ ^5D^\circ$	1—2
572,336	80	0,14	21,80	$2p^2\ ^3P - 2p^3\ ^3D^\circ$	2—3
572,106	25	0,14	21,81	$2p^2\ ^3P - 2p^3\ ^3D^\circ$	2—2
569,830	50	0,05	21,81	$2p^2\ ^3P - 2p^3\ ^3D^\circ$	1—2
569,759	25	0,05	21,81	$2p^2\ ^3P - 2p^3\ ^3D^\circ$	1—1
568,418	40	0,00	21,81	$2p^2\ ^3P - 2p^3\ ^3D^\circ$	0—1
487,070	3	25,81	51,26	$2p^3\ ^3P^\circ - 2p^4\ ^3P$	2, 1—1
482,987	50	0,14	25,81	$2p^2\ ^3P - 2p^3\ ^3P^\circ$	2—2, 1
481,361	25	0,05	25,81	$2p^2\ ^3P - 2p^3\ ^3P^\circ$	1—2, 1
481,281	15	0,05	25,81	$2p^2\ ^3P - 2p^3\ ^3P^\circ$	1—0
480,406	25	0,00	25,81	$2p^2\ ^3P - 2p^3\ ^3P^\circ$	0—1
422,347	5	21,81	51,16	$2p^3\ ^3D^\circ - 2p^4\ ^3P$	2, 1—2
422,214	15	21,80	51,16	$2p^3\ ^3D^\circ - 2p^4\ ^3P$	3—2
420,951	15	21,81	51,26	$2p^3\ ^3D^\circ - 2p^4\ ^3P$	2, 1—1
420,386	10	21,81	51,30	$2p^3\ ^3D^\circ - 2p^4\ ^3P$	1—0
416,834	25	7,92	37,67	$2p^2\ ^1S - 2p^3\ ^1P^\circ$	0—1
416,198	80	3,76	33,54	$2p^2\ ^1D - 2p^3\ ^1D^\circ$	2—2
365,594	100	3,76	37,67	$2p^2\ ^1D - 2p^3\ ^1P^\circ$	2—1
359,385	50	0,14	34,63	$2p^2\ ^3P - 2p^3\ ^3S^\circ$	2—1
358,472	50	0,05	34,63	$2p^2\ ^3P - 2p^3\ ^3S^\circ$	1—1
357,955	40	0,00	34,63	$2p^2\ ^3P - 2p^3\ ^3S^\circ$	0—1
195,621	2	25,81	89,18	$2p^3\ ^3P^\circ - 3s\ ^3P$	1—0
195,553	3	25,81	89,20	$2p^3\ ^3P^\circ - 3s\ ^3P$	2, 1—1
195,368	5	25,81	89,26	$2p^3\ ^3P^\circ - 3s\ ^3P$	2, 1—2
184,730	10	7,92	75,03	$2p^2\ ^1S - 3s\ ^1P^\circ$	0—1
173,932	50	3,76	75,03	$2p^2\ ^1D - 3s\ ^1P^\circ$	2—1
167,921	5	0,14	73,97	$2p^2\ ^3P - 3s\ ^3P^\circ$	2—1
167,837	5	0,05	73,92	$2p^2\ ^3P - 3s\ ^3P^\circ$	1—0
167,670	25	0,14	74,08	$2p^2\ ^3P - 3s\ ^3P^\circ$	2—2
		0,05	73,97	$2p^3\ ^3P - 3s\ ^3P^\circ$	1—1
167,610	3	0,00	73,97	$2p^2\ ^3P - 3s\ ^3P^\circ$	0—1
167,483	15	0,05	74,08	$2p^2\ ^3P - 3s\ ^3P^\circ$	1—2
164,294	8	11,01	86,47	$2p^3\ ^5S^\circ - 3s\ ^5P$	2—1

λ, Å	I	E_H, eV	E_B, eV	Transition	J
164,145	10	11,01	86,53	$2p^3\ ^5S^\circ$—$3s\ ^5P$	2—2
164,023	10	11,01	86,60	$2p^3\ ^5S^\circ$—$3s\ ^5P$	2—3
156,610	2	7,92	87,08	$2p^2\ ^1S$—$3d\ ^1P^\circ$	0—1
151,424	12	3,76	85,63	$2p^2\ ^1D$—$3d\ ^1D^\circ$	2—2
148,787	3	3,76	87,08	$2p^2\ ^1D$—$3d\ ^1P^\circ$	2—1
147,132	15	3,76	88,02	$2p^2\ ^1D$—$3d\ ^1F^\circ$	2—3
143,344	15	0,14	86,63	$2p^2\ ^3P$—$3d\ ^3D^\circ$	2—3
143,273	10	0,05	86,58	$2p^2\ ^3P$—$3d\ ^3D^\circ$	1—2
143,219	5	0,00	86,56	$2p^2\ ^3P$—$3d\ ^3D^\circ$	0—1
142,724	15	0,14	87,00	$2p^2\ ^3P$—$3d\ ^3P^\circ$	2—2
142,661	4	0,14	87,04	$2p^2\ ^3P$—$3d\ ^3P^\circ$	2—1
142,503	10	0,05	87,04	$2p^2\ ^3P$—$3d\ ^3P^\circ$	1, 0—2 ,1
142,441	10	0,05	87,09	$2p^2\ ^3P$—$3d\ ^3P^\circ$	1—0
140,791	15	11,01	99,07	$2p^3\ ^5S^\circ$—$3d\ ^5P$	2—3
140,757	15	11,01	99,09	$2p^3\ ^5S^\circ$—$3d\ ^5P$	2—2
140,716	5	11,01	99,12	$2p^3\ ^5S^\circ$—$3d\ ^5P$	2—1
136,215	2	11,01	102,03	$2p^3\ ^5S^\circ$—$4s^5\ P$	2—3, 2, 1
129,034	5	3,76	99,84	$2p^2\ ^1D$—$4s\ ^1P^\circ$	2—1
128,793	1	11,01	107,28	$2p^3\ ^5S^\circ$—$4d\ ^5P$	2—3, 2, 1
125,830	2	0,05	98,60	$2p^2\ ^3P$—$4s\ ^3P^\circ$	2, 1, 0—2, 1, 0
123,712	3	3,76	103,97	$2p^2\ ^1D$—$4d\ ^1D^\circ$	2—2
122,520	20	3,76	105,03	$2p^2\ ^1D$—$4d\ ^1F$	2—3
118,841	1	0,05	104,39	$2p^2\ ^3P$—$4d\ ^3D^\circ$	2, 1, 0—3, 2, 1
118,715	5	0,05	104,50	$2p^2\ ^3P$—$4d\ ^3P^\circ$	2, 1, 0—2, 1, 0

Ne VI, ground state $1s^2 2s^2 2p\ ^2P^0_{1/2}$
Ionization potential 1 274 000 cm⁻¹; 157,94 eV

λ, Å	I	E_H, eV	E_B, eV	Transition	J
2289,36	1	103,70	109,11	$3s\ ^4P^\circ$—$3p\ ^4D$	5/2—5/2
2253,22	3	103,70	109,20	$3s\ ^4P^\circ$—$3p\ ^4D$	5/2—7/2
2247,76	1	103,60	109,11	$3s\ ^4P^\circ$—$3p\ ^4D$	3/2—5/2
2055,93	3	89,59	95,62	$3s\ ^2S$—$3p\ ^2P^\circ$	1/2—1/2
2042,382	3	89,59	95,66	$3s\ ^2S$—$3p\ ^2P^\circ$	1/2—3/2
562,805	15	0,16	22,19	$2p\ ^2P^\circ$—$2p^2\ ^2D$	3/2—5/2
562,735	1	0,16	22,19	$2p\ ^2P^\circ$—$2p^2\ ^2D$	3/2—3/2
558,595	5	0,00	22,19	$2p\ ^2P^\circ$—$2p^2\ ^2D$	1/2—3/2
454,072	3	12,60	39,90	$2p^2\ ^4P$—$2p^3\ ^4S^\circ$	5/2—3/2
452,745	3	12,52	39,90	$2p^2\ ^4P$—$2p^3\ ^4S^\circ$	3/2—3/2
451,843	2	12,46	39,90	$2p^2\ ^4P$—$2p^3\ ^4S^\circ$	1/2—3/2
440,60	0	22,19	50,33	$2p^2\ ^2D$—$2p^3\ ^2P^\circ$	3/2—1/2
440,404	1	22,19	50,34	$2p^2\ ^2D$—$2p^3\ ^2P^\circ$	5/2, 3/2—3/2
435,649	4	0,16	28,62	$2p\ ^2P^\circ$—$2p^2\ ^2S$	3/2—1/2
433,176	4	0,00	28,62	$2p\ ^2P^\circ$—$2p^2\ ^2S$	1/2—1/2
403,262	10	0,16	30,91	$2p\ ^2P^\circ$—$2p^2\ ^2P$	3/2—1/2
401,939	25	0,16	31,01	$2p\ ^2P^\circ$—$2p^2\ ^2P$	3/2—3/2
401,138	15	0,00	30,91	$2p\ ^2P^\circ$—$2p^2\ ^2P$	1/2—1/2
399,820	5	0,00	31,01	$2p\ ^2P^\circ$—$2p^2\ ^2P$	1/2—3/2
194,936	2	31,01	95,62	$2p^2\ ^2P$—$3p\ ^2P^\circ$	3/2—1/2
194,839	2	31,01	95,66	$2p^2\ ^2P$—$3p\ ^2P^\circ$	3/2—3/2
188,424	3	28,62	95,66	$2p^2\ ^2S$—$3p\ ^2P^\circ$	1/2—3/2
171,212	2	22,19	95,62	$2p^2\ ^2D$—$3p^2\ P^\circ$	3/2—1/2
171,114	5	22,19	95,66	$2p^2\ ^2D$—$3p\ ^2P^\circ$	5/2, 3/2—3/2
138,630	3	0,16	89,59	$2p\ ^2P^\circ$—$3s\ ^2S$	3/2—1/2

λ, Å	I	E_H, eV	E_B, eV	Transition	J
138,397	3	0,00	89,59	$2p\ ^2P^\circ$—$3s\ ^2S$	$^1/_2$—$^1/_2$
136,089	4	12,60	103,70	$2p^2\ ^4P$—$3s\ ^4P^\circ$	$^5/_2$—$^5/_2$, $^3/_2$
122,686	10	0,16	101,22	$2p\ ^2P^\circ$—$3d\ ^2D$	$^3/_2$—$^5/_2$, $^3/_2$
122,520	20	0,00	101,22	$2p\ ^2P^\circ$—$3d\ ^2D$	$^1/_2$—$^3/_2$
121,140	5	12,60	114,65	$2p^2\ ^4P$—$3d\ ^4D^\circ$	$^5/_2$—$^7/_2$, $^5/_2$, $^3/_2$
113,870	5	0,00	108,96	$2p\ ^2P^\circ$—$3p\ ^2P$	$^3/_2$, $^1/_2$—$^3/_2$, $^1/_2$
111,142	1	0,00	111,63	$2p\ ^2P^\circ$—$3p\ ^2S$	$^3/_2$, $^1/_2$—$^1/_2$
110,410	2	0,00	112,65	$2p\ ^2P^\circ$—$3p\ ^2D$	$^3/_2$, $^1/_2$—$^3/_2$, $^5/_2$

Ne VII, ground state $1s^2\ 2s^2\ ^1S_0$

λ, Å	I	E_H, eV	E_B, eV	Transition	J
1997,345	1	121,27	127,49	$3s\ ^3S$—$3p\ ^3P^\circ$	1—0
1992,060	3	121,27	127,51	$3s\ ^3S$—$3p\ ^3P^\circ$	1—1
1981,974	6	121,27	127,54	$3s\ ^3S$—$3p\ ^3P^\circ$	1—2
564,529	2	13,97	35,93	$2p\ ^3P^\circ$—$2p^2\ ^3P$	2—1
562,992	2	13,85	35,87	$2p\ ^3P^\circ$—$2p^2\ ^3P$	1—0
561,728	4	13,97	36,04	$2p\ ^3P^\circ$—$2p^2\ ^3P$	2—2
561,378	2	13,85	35,93	$2p\ ^3P^\circ$—$2p^2\ ^3P$	1—1
559,947	3	13,79	35,93	$2p\ ^3P^\circ$—$2p^2\ ^3P$	0—1
558,61	4	13,85	36,04	$2p\ ^3P^\circ$—$2p^2\ ^3P$	1—2
465,21	10	0,00	26,65	$2s^2\ ^1S$—$2p\ ^1P^\circ$	0—1
127,7	2	—	—	$2p\ ^1P^\circ$—$3s\ ^1S$	—
116,7	5	—	—	$2p\ ^1P^\circ$—$3d\ ^1D$	—
115,5	3	—	—	$2p\ ^3P^\circ$—$3s\ ^3S$	—
115,4	3	—	—	$2p\ ^3P^\circ$—$3s\ ^3S$	—
106,2	7	—	—	$2p\ ^3P^\circ$—$3d\ ^3D$	—
106,1	7	—	—	$2p\ ^3P^\circ$—$3d\ ^3D$	—
97,5	6	—	—	$2s^2\ ^1S$—$3p\ ^1P^\circ$	—
89,4	3	—	—	$2p\ ^1P^\circ$—$4d\ ^1D$	—
82,3	5	—	—	$2p\ ^3P^\circ$—$4d\ ^3D$	—

Ne VIII, ground state $1s^2\ 2s\ ^2S_{1/2}$

λ, Å	I	E_H, eV	E_B, eV	Transition	J
780,324	4	0,00	15,87	$2s\ ^2S$—$2p\ ^2P^\circ$	$^1/_2$—$^1/_2$
770,409	8	0,00	16,09	$2s\ ^2S$—$2p\ ^2P^\circ$	$^1/_2$—$^3/_2$
103,1	6	—	—	$2p\ ^2P^\circ$—$3s\ ^2S$	—
102,9	5	—	—	$2p\ ^2P^\circ$—$3s\ ^2S$	—
98,2	9	—	—	$2p\ ^2P^\circ$—$3d\ ^2D$	—
98,1	9	—	—	$2p\ ^2P^\circ$—$3d\ ^2D$	—
88,1	9	—	—	$2s\ ^2S$—$3p\ ^2P^\circ$	—
74,7	4	—	—	$2p\ ^2P^\circ$—$4s\ ^2S$	—
73,6	8	—	—	$2p\ ^2P^\circ$—$4d\ ^2D$	—
67,3	8	—	—	$2s\ ^2S$—$4p\ ^2P^\circ$	—
65,9	6	—	—	$2p\ ^2P^\circ$—$5d\ ^2D$	—
62,4	3	—	—	$2p\ ^2P^\circ$—$6d\ ^2D$	—
60,7	3	—	—	$2s\ ^2S$—$5p\ ^2P^\circ$	—

Unclassified Lines of Neon [4, 5, 6, 12, 20]

λ, Å	I	Expected assignment	λ, Å	I	Expected assignment
9760,57	2	Ne I	2706,74	1	Ne I
9497,9	2	Ne I	2686,75	3	Ne I
9393,8	2	Ne I	2685,58	1	Ne I
9368,02	2	Ne I	2685,33	1	Ne I
9326,66	2	Ne I	2677,36	3	Ne I
9069,7	2	Ne I	2669,36	1,5	Ne I
9045,4	2	Ne I	2663,38	2	Ne I
8842,1	2	Ne I	2651,01	6	Ne I
7125,37	3	Ne I	2645,51	6	Ne I
4517,29	2	Ne II	2621,10	2	Ne I
3786,29	2	Ne III or Ne IV	2619,77	1	Ne I
3783,92	1	Ne III or Ne IV	2619,02	1,5	Ne I
3782,31	1	Ne III or Ne IV	2595,21	6	Ne I
3771,64	1	Ne III or Ne IV	2507,08	1	Ne III or Ne IV
3740,60	1	Ne II	2367,02	5	Ne III or Ne IV
3454,83	1	Ne II	2365,77	6	Ne III or Ne IV
3267,22	1	Ne III or Ne IV	2362,85	5	Ne III or Ne IV
3265,37	3	Ne III or Ne IV	2307,27	2	Ne III or Ne IV
3260,87	3	Ne III or Ne IV	2306,61	6	Ne III or Ne IV
3250,34	1	Ne II	2305,50	2	Ne III or Ne IV
3238,47	1	Ne III or Ne IV	2304,87	4	Ne III or Ne IV
3207,906	2,5	Ne I	2303,94	3	Ne III or Ne IV
3101,407	2	Ne I	2300,38	2	Ne III or Ne IV
3081,45	1	Ne II	2298,96	1	Ne III or Ne IV
3067,214	4	Ne I	2278,98	10	Ne III or Ne IV
3065,668	1,5	Ne I	2273,58	20	Ne III or Ne IV
3062,58	2	Ne II	2204,16	2	Ne III or Ne IV
3043,02	2	Ne III or Ne IV	2203,89	6	Ne III or Ne IV
3028,424	1	Ne I	2200,82	5	Ne III or Ne IV
3026,913	3	Ne I	2192,74	7	Ne III or Ne IV
3024,63	2	Ne III or Ne IV	2191,45	1	Ne III or Ne IV
3007,82	1	Ne II	2191,16	4	Ne III or Ne IV
2983,82	5	Ne III or Ne IV	2186,62	3	Ne III or Ne IV
2974,527	1	Ne I	2129,54	6	Ne III or Ne IV
2910,44	2	Ne II	2124,27	7	Ne III or Ne IV
2905,85	4	Ne III or Ne IV	2102,33	2	Ne III or Ne IV
2895,05	1	Ne III or Ne IV	2099,59	4	Ne III or Ne IV
2893,11	1	Ne III or Ne IV	2099,34	10	Ne III or Ne IV
2866,65	5	Ne III or Ne IV	2098,00	1	Ne III or Ne IV
2827,98	0	Ne III or Ne IV	2097,43	2	Ne III or Ne IV
2821,68	0	Ne III or Ne IV	2096,23	12	Ne III or Ne IV
2820,44	0	Ne III or Ne IV	2094,15	2	Ne III or Ne IV
2818,88	1	Ne III or Ne IV	2093,64	3	Ne III or Ne IV
2798,96	0	Ne III or Ne IV	2091,90	4	Ne III or Ne IV
2781,42	1	Ne I	2089,20	2	Ne III or Ne IV
2767,00	7	Ne II or Ne III	2085,56	5	Ne III or Ne IV
2766,07	6	Ne III	2078,95	15	Ne III or Ne IV
2764,70	2	Ne III or Ne IV	2065,18	20	Ne III or Ne IV
2764,38	1	Ne III or Ne IV	2062,62	2	Ne III or Ne IV
2713,76	1	Ne I			

SODIUM, Z = 11

Na I, ground state $1s^2\,2s^2\,2p^6\,3s\ ^2S_{1/2}$

Ionization potential 41 449,65 cm^{-1}; 5,139 eV

λ, Å	I	E_H, eV	E_B, eV	Transition	J
90850	40	—	—	—	—
90480	30	—	—	—	—
74430	10	—	—	—	—
40449	80	—	—	—	—
23379,13	240	3,75	4,28	$4p\ ^2P^\circ - 4d\ ^2D$	3/2—5/2
23348,41	237	3,75	4,28	$4p\ ^2P^\circ - 4d\ ^2D$	1/2—3/2
22083,67	276	3,19	3,75	$4s\ ^2S - 4p\ ^2P^\circ$	1/2—1/2
22056,44	300	3,19	3,75	$4s\ ^2S - 4p\ ^2P^\circ$	1/2—3/2
18465,25	2	3,62	4,29	$3d\ ^2D - 4f\ ^2F^\circ$	3/2, 5/2—5/2, 7/2
16388,85	27	3,75	4,51	$4p\ ^2P^\circ - 6s\ ^2S$	3/2—1/2
16373,85	30	3,75	4,51	$4p\ ^2P^\circ - 6s\ ^2S$	1/2—1/2
14779,73	36	3,75	4,59	$4p\ ^2P^\circ - 5d\ ^2D$	3/2—5/2
14767,48	1155	3,75	4,59	$4p\ ^2P^\circ - 5d\ ^2D$	1/2—3/2
12679,17	83	3,62	4,59	$3d\ ^2D - 5f\ ^2F^\circ$	3/2, 5/2—5/2, 7/2
11403,78	12	2,10	3,19	$3p\ ^2P^\circ - 4s\ ^2S$	3/2—1/2
11381,45	11	2,10	3,19	$3p\ ^2P^\circ - 4s\ ^2S$	1/2—1/2
11197,21	2	3,75	4,86	$4p\ ^2P^\circ - 7d\ ^2D$	3/2—5/2
11190,19	1	3,75	4,86	$4p\ ^2P^\circ - 7d\ ^2D$	1/2—3/2
10834,87	8	3,62	4,76	$3d\ ^2D - 6f\ ^2F^\circ$	3/2, 5/2—5/2, 7/2
10749,29	9	3,19	4,34	$4s\ ^2S - 5p\ ^2P^\circ$	1/2—1/2
10746,44	10	3,19	4,34	$4s\ ^2S - 5p\ ^2P^\circ$	1/2—3/2
10572,28	3	3,75	4,93	$4p\ ^2P^\circ - 8d\ ^2D$	3/2—5/2
10566,00	1	3,75	4,93	$4p\ ^2P^\circ - 8d\ ^2D$	1/2—3/2
9961,281	7	3,62	4,86	$3d\ ^2D - 7f\ ^2F^\circ$	3/2, 5/2—5/2, 7/2
9465,938	6	3,62	4,93	$3d\ ^2D - 8f\ ^2F^\circ$	3/2, 5/2—5/2, 7/2
9153,878	4	3,62	4,97	$3d\ ^2D - 9f\ ^2F^\circ$	3/2, 5/2—5/2, 7/2
8942,962	2	3,62	5,00	$3d\ ^2D - 10f\ ^2F^\circ$	3/2, 5/2—5/2, 7/2
8650,889	6	3,19	4,62	$4s\ ^2S - 6p\ ^2P^\circ$	1/2—1/2
8649,922	7	3,19	4,62	$4s\ ^2S - 6p\ ^2P^\circ$	1/2—3/2
8194,8237	9	2,10	3,62	$3p\ ^2P^\circ - 3d\ ^2D$	3/2—5/2
8194,7905	1	2,10	3,62	$3p\ ^2P^\circ - 3d\ ^2D$	3/2—3/2
8183,2556	5	2,10	3,62	$3p\ ^2P^\circ - 3d\ ^2D$	1/2—3/2
7810,237	3	3,19	4,78	$4s\ ^2S - 7p\ ^2P^\circ$	1/2—1/2
7809,781	4	3,19	4,78	$4s\ ^2S - 7p\ ^2P^\circ$	1/2—3/2
7373,491	1	3,19	4,87	$4s\ ^2S - 8p\ ^2P^\circ$	1/2—1/2
7373,229	2	3,19	4,87	$4s\ ^2S - 8p\ ^2P^\circ$	1/2—3/2
6160,7470	2	2,10	4,12	$3p\ ^2P^\circ - 5s\ ^2S$	3/2—1/2
6154,2253	1	2,10	4,12	$3p\ ^2P^\circ - 5s\ ^2S$	1/2—1/2
5895,9236	16	0,00	2,10	$3s\ ^2S - 3p\ ^2P^\circ$	1/2—1/2
5889,9504	32	0,00	2,10	$3s\ ^2S - 3p\ ^2P^\circ$	1/2—3/2
5688,2046	9	2,10	4,28	$3p\ ^2P^\circ - 4d\ ^2D$	3/2—5/2
5688,1934	1	2,10	4,28	$3p\ ^2P^\circ - 4d\ ^2D$	3/2—3/2
5682,6333	5	2,10	4,28	$3p\ ^2P^\circ - 4d\ ^2D$	1/2—3/2
5669,8	3	2,10	4,29	$3p\ ^2P^\circ - 4f\ ^2F^\circ$	3/2— —
5532,0	2	—	—	—	—
5153,4024	2	2,10	4,51	$3p\ ^2P^\circ - 6s\ ^2S$	3/2—1/2
5148,8381	1	2,10	4,51	$3p\ ^2P^\circ - 6s\ ^2S$	1/2—1/2
4982,8134	2	2,10	4,59	$3p\ ^2P^\circ - 5d\ ^2D$	3/2—5/2
4978,5414	1	2,10	4,59	$3p\ ^2P^\circ - 5d\ ^2D$	1/2—3/2
4751,8218	2	2,10	4,71	$3p\ ^2P^\circ - 7s\ ^2S$	3/2—1/2
4747,9410	1	2,10	4,71	$3p\ ^2P^\circ - 7s\ ^2S$	1/2—1/2
4668,5595	2	2,10	4,76	$3p\ ^2P^\circ - 6d\ ^2D$	3/2—5/2
4664,8107	1	2,10	4,76	$3p\ ^2P^\circ - 6d\ ^2D$	1/2—3/2

λ, Å	I	E_H, eV	E_B, eV	Transition	J
4545,186	8	2,10	4,83	$3p\ ^2P^\circ - 8s\ ^2S$	$^3/_2 - ^1/_2$
4541,633	7	2,10	4,83	$3p\ ^2P^\circ - 8s\ ^2S$	$^1/_2 - ^1/_2$
4497,658	11	2,10	4,86	$3p\ ^2P^\circ - 7d\ ^2D$	$^3/_2 - ^5/_2$
4494,177	10	2,10	4,86	$3p\ ^2P^\circ - 7d\ ^2D$	$^1/_2 - ^3/_2$
4423,246	7	2,10	4,91	$3p\ ^2P^\circ - 9s\ ^2S$	$^3/_2 - ^1/_2$
4419,885	6	2,10	4,91	$3p\ ^2P^\circ - 9s\ ^2S$	$^1/_2 - ^1/_2$
4393,340	9	2,10	4,93	$3p\ ^2P^\circ - 8d\ ^2D$	$^3/_2 - ^5/_2$
4390,029	8	2,10	4,93	$3p\ ^2P^\circ - 8d\ ^2D$	$^1/_2 - ^3/_2$
4344,736	5	2,10	4,96	$3p\ ^2P^\circ - 10s\ ^2S$	$^3/_2 - ^1/_2$
4341,489	4	2,10	4,96	$3p\ ^2P^\circ - 10s\ ^2S$	$^1/_2 - ^1/_2$
4324,615	7	2,10	4,97	$3p\ ^2P^\circ - 9d\ ^2D$	$^3/_2 - ^5/_2$
4321,400	6	2,10	4,97	$3p\ ^2P^\circ - 9d\ ^2D$	$^1/_2 - ^3/_2$
4291,006	3	2,10	4,99	$3p\ ^2P^\circ - 11s\ ^2S$	$^3/_2 - ^1/_2$
4287,838	2	2,10	4,99	$3p\ ^2P^\circ - 11s\ ^2S$	$^1/_2 - ^1/_2$
4276,787	5	2,10	5,00	$3p\ ^2P^\circ - 10d\ ^2D$	$^3/_2 - ^5/_2$
4273,642	4	2,10	5,00	$3p\ ^2P^\circ - 10d\ ^2D$	$^1/_2 - ^3/_2$
4252,520	2	2,10	5,02	$3p\ ^2P^\circ - 12s\ ^2S$	$^3/_2 - ^1/_2$
4249,410	1	2,10	5,02	$3p\ ^2P^\circ - 12s\ ^2S$	$^1/_2 - ^1/_2$
4242,082	3	2,10	5,03	$3p\ ^2P^\circ - 11d\ ^2D$	$^3/_2 - ^5/_2$
4238,987	2	2,10	5,03	$3p\ ^2P^\circ - 11d\ ^2D$	$^1/_2 - ^3/_2$
4198,3	10	—	—	—	—
3426,862	6	0,00	3,62	$3s\ ^2S - 3d\ ^2D$	$^1/_2 - ^3/_2$
3302,979	18	0,00	3,75	$3s\ ^2S - 4p\ ^2P^\circ$	$^1/_2 - ^1/_2$
3302,369	19	0,00	3,75	$3s\ ^2S - 4p\ ^2P^\circ$	$^1/_2 - ^3/_2$
2893,618	1	0,00	4,28	$3s\ ^2S - 4d\ ^2D$	$^1/_2 - ^3/_2$
2853,013	15	0,00	4,34	$3s\ ^2S - 5p\ ^2P^\circ$	$^1/_2 - ^1/_2$
2852,811	16	0,00	4,34	$3s\ ^2S - 5p\ ^2P^\circ$	$^1/_2 - ^3/_2$
2680,433	7	0,00	4,62	$3s\ ^2S - 6p\ ^2P^\circ$	$^1/_2 - ^1/_2$
2680,340	8	0,00	4,62	$3s\ ^2S - 6p\ ^2P^\circ$	$^1/_2 - ^3/_2$
2593,919	2	0,00	4,78	$3s\ ^2S - 7p\ ^2P^\circ$	$^1/_2 - ^1/_2$
2593,869	3	0,00	4,78	$3s\ ^2S - 7p\ ^2P^\circ$	$^1/_2 - ^3/_2$
2543,872	2	0,00	4,87	$3s\ ^2S - 8p\ ^2P^\circ$	$^1/_2 - ^1/_2$
2543,841	1	0,00	4,87	$3s\ ^2S - 8p\ ^2P^\circ$	$^1/_2 - ^3/_2$
2512,210	4	0,00	4,93	$3s\ ^2S - 9p\ ^2P^\circ$	$^1/_2 - ^3/_2$
2512,128	2	0,00	4,93	$3s\ ^2S - 9p\ ^2P^\circ$	$^1/_2 - ^1/_2$
2490,733	3	0,00	4,98	$3s\ ^2S - 10p\ ^2P^\circ$	$^1/_2 - ^1/_2,\ ^3/_2$

Na II, ground state $1s^2\ 2s^2\ 2p^6\ ^1S_0$

Ionization potential 381 528 cm^{-1}; 47,30 eV

λ, Å	I	E_H, eV	E_B, eV	Transition	J
4123,069	3	38,29	41,30	$3p'\ [^1/_2] - 4s'\ [^1/_2]^\circ$	0—1
4114,95	3	38,29	41,31	$3p'\ [^1/_2] - 3d'\ [1^1/_2]^\circ$	0—1
4087,60	0	33,32	36,35	$3s'\ [^1/_2]^\circ - 3p\ [^1/_2]$	1—1
3711,074	6	33,01	36,35	$3s'\ [^1/_2]^\circ - 3p\ [^1/_2]$	0—1
3631,266	8	32,94	36,35	$3s\ [1^1/_2]^\circ - 3p\ [^1/_2]$	1—1
3533,043	10	32,84	36,35	$3s\ [1^1/_2]^\circ - 3p\ [^1/_2]$	2—1
3462,494	3	33,32	36,90	$3s'\ [^1/_2]^\circ - 3p\ [2^1/_2]$	1—2
3400,110	2	33,32	36,97	$3s'\ [^1/_2]^\circ - 3p\ [1^1/_2]$	1—1
3327,685	4	37,26	40,98	$3p'\ [^1/_2] - 3d\ [^1/_2]^\circ$	1—0
3318,032	4	37,26	40,99	$3p'\ [^1/_2] - 3d\ [^1/_2]^\circ$	1—1
3304,950	0	37,24	40,99	$3p\ [^1/_2] - 3d\ [^1/_2]^\circ$	0—1
3301,346	2	37,26	41,01	$3p'\ [^1/_2] - 3d\ [1^1/_2]^\circ$	1—2
3285,603	8	33,32	37,09	$3s'\ [^1/_2]^\circ - 3p\ [1^1/_2]$	1—2

λ, Å	I	E_H, eV	E_B, eV	Transition	J
3274,220	5	37,21	40,99	$3p'\,[1^1/_2]-3d\,[^1/_2]^\circ$	2—1
3260,218	3	37,17	40,98	$3p'\,[1^1/_2]-3d\,[^1/_2]^\circ$	1—0
3257,965	6	37,21	41,01	$3p'\,[1^1/_2]-3d\,[1^1/_2]^\circ$	2—2
3250,949	3	37,17	40,99	$3p'\,[1^1/_2]-3d\,[^1/_2]^\circ$	1—1
3234,926	4	37,17	41,01	$3p'\,[1^1/_2]-3d\,[1^1/_2]^\circ$	1—2
3225,976	4	37,26	41,10	$3p'\,[^1/_2]-4s\,[1^1/_2]^\circ$	1—2
3216,284	2	37,21	41,06	$3p'\,[1^1/_2]-3d\,[3^1/_2]^\circ$	2—3
3212,186	6	33,32	37,17	$3s'\,[^1/_2]^\circ-3p'\,[1^1/_2]$	1—1
3189,783	6	37,32	37,21	$3s'\,[^1/_2]^\circ-3p'\,[1^1/_2]$	1—2
3179,055	5	37,09	40,99	$3p\,[1^1/_2]-3d\,[^1/_2]^\circ$	2—1
3175,008	3	37,24	41,15	$3p\,[^1/_2]-4s\,[1^1/_2]^\circ$	0—1
3167,487	2	37,21	41,12	$3p'\,[1^1/_2]-3d\,[2^1/_2]^\circ$	2—2
3163,731	6	37,09	41,01	$3p\,[1^1/_2]-3d\,[1^1/_2]^\circ$	2—2
3163,247	1	37,21	41,12	$3p'\,[1^1/_2]-3d\,[2^1/_2]^\circ$	2—3
3161,16	0	33,32	37,24	$3s'\,[^1/_2]^\circ-3p\,[^1/_2]$	1—0
3159,53	0	37,21	41,13	$3p'\,[1^1/_2]-3d\,[1^1/_2]^\circ$	2—1
3149,266	5	33,32	37,26	$3s'\,[^1/_2]^\circ-3p'\,[^1/_2]$	1—1
3145,697	3	37,17	41,12	$3p'\,[1^1/_2]-3d\,[2^1/_2]^\circ$	1—2
3137,852	3	37,17	41,13	$3p'\,[1^1/_2]-3d\,[1^1/_2]^\circ$	1—1
3135,483	5	33,01	36,97	$3s'\,[^1/_2]^\circ-3p\,[1^1/_2]$	0—1
3129,368	6	32,94	36,90	$3s\,[1^1/_2]^\circ-3p\,[2^1/_2]$	1—2
3125,208	1	37,17	41,15	$3p'\,[1^1/_2]-4s\,[1^1/_2]^\circ$	1—1
3124,414	3	37,09	41,06	$3p\,[1^1/_2]-3d\,[3^1/_2]^\circ$	2—3
3104,396	4	37,26	41,25	$3p'\,[^1/_2]-4s'\,[^1/_2]^\circ$	1—0
3092,729	10	32,84	36,85	$3s\,[1^1/_2]^\circ-3p\,[2^1/_2]$	2—3
3087,047	2	36,97	40,98	$3p\,[1^1/_2]-3d\,[^1/_2]^\circ$	1—0
3080,250	3	37,26	41,28	$3p'\,[^1/_2]-3d'\,[1^1/_2]^\circ$	1—2
3078,733	1	36,97	40,99	$3p\,[1^1/_2]-3d\,[^1/_2]^\circ$	1—1
3078,315	6	32,94	36,97	$3s\,[1^1/_2]^\circ-3p\,[1^1/_2]$	1—1
		37,09	41,12	$3p\,[1^1/_2]-3d\,[2^1/_2]^\circ$	2—2
3074,334	6	37,09	41,12	$3p\,[1^1/_2]-3d\,[2^1/_2]^\circ$	2—3
3070,84	0	37,09	41,13	$3p\,[1^1/_2]-3d\,[1^1/_2]^\circ$	2—1
3066,536	4	37,26	41,30	$3p'\,[^1/_2]-4s'\,[^1/_2]^\circ$	1—1
3066,238	1	—	—	—	—
3064,372	4	36,97	41,01	$3p\,[1^1/_2]-3d\,[1^1/_2]^\circ$	1—2
3061,333	1	37,26	41,31	$3p'\,[^1/_2]-3d'\,[1^1/_2]^\circ$	1—1
3060,260	1	—	—	—	—
3058,727	1	37,09	41,15	$3p\,[1^1/_2]-4s\,[1^1/_2]^\circ$	2—1
3056,157	6	32,84	36,90	$3s\,[1^1/_2]^\circ-3p\,[2^1/_2]$	2—2
3055,346	1	37,24	41,30	$3p\,[^1/_2]-4s'\,[^1/_2]^\circ$	0—1
3053,664	6	37,21	41,27	$3p'\,[1^1/_2]-3d'\,[2^1/_2]^\circ$	2—3
3050,211	1	—	—	—	—
3045,593	5	37,17	41,25	$3p'\,[1^1/_2]-4s'\,[^1/_2]^\circ$	1—0
3037,071	5	37,17	41,26	$3p'\,[1^1/_2]-3d'\,[2^1/_2]^\circ$	1—2
3029,068	6	37,21	41,30	$3p'\,[1^1/_2]-4s'\,[^1/_2]^\circ$	2—1
		36,90	40,99	$3p\,[2^1/_2]-3d\,[^1/_2]^\circ$	2—1
3015,400	6	36,90	41,01	$3p\,[2^1/_2]-3d\,[1^1/_2]^\circ$	2—2
3009,138	4	37,17	41,30	$3p'\,[1^1/_2]-4s'\,[^1/_2]^\circ$	1—1
3007,442	4	32,85	36,97	$3s\,[1^1/_2]^\circ-3p\,[1^1/_2]$	2—1
2984,183	7	32,94	37,09	$3s\,[1^1/_2]^\circ-3p\,[1^1/_2]$	1—2
		36,97	41,12	$3p\,[1^1/_2]-3d\,[2^1/_2]^\circ$	1—2
2980,622	3	36,85	41,01	$3p\,[2^1/_2]-3d\,[2^1/_2]^\circ$	3—2
2979,662	5	36,90	41,06	$3p\,[2^1/_2]-3d\,[3^1/_2]^\circ$	2—3
2979,050	2	—	—	—	—
2977,132	3	36,97	41,13	$3p\,[1^1/_2]-3d\,[1^1/_2]^\circ$	1—1
2974,991	6	33,01	37,17	$3s'\,[^1/_2]^\circ-3p'\,[1^1/_2]$	0—1
2974,236	2	37,09	41,26	$3p\,[1^1/_2]-3d'\,[2^1/_2]^\circ$	2—1
2970,725	1	37,09	41,27	$3p\,[1^1/_2]-3d'\,[2^1/_2]^\circ$	2—3
2965,750	2	36,97	41,15	$3p\,[1^1/_2]-4s\,[1^1/_2]^\circ$	1—1
2960,112	1	37,09	41,28	$3p\,[1^1/_2]-3d'\,[1^1/_2]^\circ$	2—2

λ, Å	I	E_H, eV	E_B, eV	Transition	J
2952,395	3	36,90	41,10	$3p\,[2^1/_2]-4s\,[1^1/_2]^\circ$	2—2
2951,231	8	36,85	41,05	$3p\,[2^1/_2]-3d\,[3^1/_2]^\circ$	3—4
2947,441	5	37,09	41,30	$3p\,[1^1/_2]-4s'\,[^1/_2]^\circ$	2—1
2945,695	4	36,85	41,06	$3p\,[2^1/_2]-3d\,[3^1/_2]^\circ$	3—3
2937,725	5	36,90	41,12	$3p\,[2^1/_2]-3d\,[2^1/_2]^\circ$	2—2
2934,065	2	36,90	41,12	$3p\,[2^1/_2]-3d\,[2^1/_2]^\circ$	2—3
2930,883	1	36,90	41,13	$3p\,[2^1/_2]-3d\,[1^1/_2]^\circ$	2—1
2923,474	3	32,94	37,17	$3s\,[1^1/_2]^\circ-3p'\,[1^1/_2]$	1—1
2920,940	4	33,01	37,26	$3s'\,[^1/_2]^\circ-3p'\,[^1/_2]$	0—1
2919,845	2	36,90	41,15	$3p\,[2^1/_2]-4s\,[1^1/_2]^\circ$	2—1
2919,048	5	36,85	41,10	$3p\,[2^1/_2]-4s\,[1^1/_2]^\circ$	3—2
2917,516	5	32,84	37,09	$3s\,[1^1/_2]^\circ-3p\,[1^1/_2]$	2—2
2904,914	7	32,94	37,21	$3s\,[1^1/_2]^\circ-3p'\,[1^1/_2]$	1—2
2901,136	4	36,85	41,12	$3p\,[2^1/_2]-3d\,[2^1/_2]^\circ$	3—3
2893,946	6	36,97	41,25	$3p\,[1^1/_2]-4s'\,[^1/_2]^\circ$	1—0
2886,250	4	36,97	41,26	$3p\,[1^1/_2]-3d'\,[2^1/_2]^\circ$	1—2
2881,140	6	32,94	37,24	$3s\,[1^1/_2]^\circ-3p\,[^1/_2]$	1—0
2872,95	0	36,97	41,28	$3p\,[1^1/_2]-3d'\,[1^1/_2]^\circ$	1—2
2871,270	5	32,94	37,26	$3s\,[1^1/_2]^\circ-3p'\,[^1/_2]$	1—1
2861,011	1	36,97	41,30	$3p\,[1^1/_2]-3d'\,[1^1/_2]^\circ$	1—1
2859,481	5	32,84	37,18	$3s\,[1^1/_2]^\circ-3p'\,[1^1/_2]$	2—1
2841,721	7	32,84	37,21	$3s\,[1^1/_2]^\circ-3p'\,[1^1/_2]$	2—2
2839,555	4	36,90	41,27	$3p\,[2^1/_2]-3d'\,[2^1/_2]^\circ$	2—3
2829,854	2	36,90	41,28	$3p\,[2^1/_2]-3d'\,[1^1/_2]^\circ$	2—2
2818,271	2	36,90	41,30	$3p\,[2^1/_2]-3d'\,[1^1/_2]^\circ$	2—1
2809,514	5	32,84	37,26	$3s\,[1^1/_2]^\circ-3p'\,[^1/_2]$	2—1
2808,685	1	36,85	41,27	$3p\,[2^1/_2]-3d'\,[2^1/_2]^\circ$	3—3
2678,086	5	36,35	40,98	$3p\,[^1/_2]-3d\,[^1/_2]^\circ$	1—0
2671,829	6	36,35	40,99	$3p\,[^1/_2]-3d\,[^1/_2]^\circ$	1—1
2660,996	7	36,35	41,01	$3p\,[^1/_2]-3d\,[1^1/_2]^\circ$	1—2
2611,815	7	36,35	41,10	$3p\,[^1/_2]-4s\,[1^1/_2]^\circ$	1—2
2594,965	1	36,35	41,13	$3p\,[^1/_2]-3d\,[1^1/_2]^\circ$	1—1
2586,312	2	36,35	41,15	$3p\,[^1/_2]-4s\,[1^1/_2]^\circ$	1—1
2531,548	6	36,35	41,25	$3p\,[^1/_2]-4s'\,[^1/_2]^\circ$	1—0
2515,460	2	36,35	41,28	$3p\,[^1/_2]-3d'\,[1^1/_2]^\circ$	1—2
2506,295	2	36,35	41,30	$3p\,[^1/_2]-4s'\,[^1/_2]^\circ$	1—1
2493,153	5	33,32	38,29	$3s'\,[^1/_2]^\circ-3p'\,[^1/_2]$	1—0
2394,051	2	—	—	—	—
2387,026	2	—	—	—	—
2315,65	0	32,94	38,29	$3s\,[1^1/_2]^\circ-3p'\,[^1/_2]$	1—0
376,375	3	0,00	32,94	$2p^6\,{}^1S-3s\,[1^1/_2]^\circ$	0—1
372,069	6	0,00	33,32	$2p^6\,{}^1S-3s'\,[^1/_2]^\circ$	0—1
302,28	0	0,00	40,99	$2p^6\,{}^1S-3d\,[^1/_2]^\circ$	0—1
301,432	1	0,00	41,13	$2p^6\,{}^1S-3d\,[1^1/_2]^\circ$	0—1
301,311	0	0,00	41,15	$2p^6\,{}^1S-4s\,[1^1/_2]^\circ$	0—1
300,151	1	0,00	41,31	$2p^6\,{}^1S-3d'\,[1^1/_2]^\circ$	0—1
282,96	1	0,00	43,80	$2p^6\,{}^1S-5s\,[1^1/_2]^\circ$	0—1
282,827	0	0,00	43,83	$2p^6\,{}^1S-4d\,[1^1/_2]^\circ$	0—1
281,81	1	0,00	43,98	$2p^6\,{}^1S-5s'\,[^1/_2]^\circ$	0—1
275,10	0	0,00	45,07	$2p^6\,{}^1S-6s\,[1^1/_2]^\circ$	0—1
273,99	0	0,00	45,25	$2p^6\,{}^1S-6s'\,[^1/_2]^\circ$	0—1
271,01	0	0,00	45,75	$2p^6\,{}^1S-7s\,[1^1/_2]^\circ$	0—1
269,98	0	0,00	45,91	$2p^6\,{}^1S-7s'\,[^1/_2]^\circ$	0—1

Na III, ground state $1s^2\,2s^2\,2p^5\;^2P^0_{3/2}$

Ionization potential 578 033 cm^{-1}; 71,662 eV

λ, Å	I	E_H, eV	E_B, eV	Transition	J
2563,32	25	45,57	50,40	$3s\,^4P-3p\,^4P^\circ$	1/2—3/2
2553,61	25	45,51	50,36	$3s\,^4P-3p\,^4P^\circ$	3/2—5/2
2542,89	10	45,57	50,44	$3s\,^4P-3p\,^4P^\circ$	1/2—1/2
2530,21	15	45,51	50,40	$3s\,^4P-3p\,^4P^\circ$	3/2—3/2
2510,37	20	45,51	50,44	$3s\,^4P-3p\,^4P^\circ$	3/2—1/2
2497,05	50	45,40	50,36	$3s\,^4P-3p\,^4P^\circ$	5/2—5/2
2474,69	40	45,40	50,40	$3s\,^4P-3p\,^4P^\circ$	5/2—3/2
2468,86	30	46,45	51,47	$3s\,^2P-3p\,^2D^\circ$	1/2—3/2
2459,40	45	46,32	51,36	$3s\,^2P-3p\,^2D^\circ$	3/2—5/2
2421,00	8	49,49	54,61	$3s'\,^2D-3p'\,^2F^\circ$	5/2—5/2
2406,58	18	46,32	51,47	$3s\,^2P-3p\,^2D^\circ$	3/2—3/2
2367,33	4	46,45	51,69	$3s\,^2P-3p\,^2S^\circ$	1/2—1/2
2314,66	3	51,75	57,10	$3p\,^4S^\circ-3d\,^4D$	3/2—3/2
2309,96	30	46,32	51,69	$3s\,^2P-3p\,^2S^\circ$	3/2—1/2
2297,14	25	—	—	—	—
2296,64	25	49,49	54,87	$3s'\,^2D-3p'\,^2P^\circ$	3/2—3/2
2285,72	35	46,45	51,87	$3s\,^2P-3p\,^2P^\circ$	1/2—3/2
2279,85	3	51,69	57,12	$3p\,^2S^\circ-3d\,^4D$	1/2—1/2
2278,48	40	46,45	51,89	$3s\,^2P-3p\,^2P^\circ$	1/2—1/2
2267,96	8	49,49	54,95	$3s'\,^2D-3p'\,^2P^\circ$	3/2—1/2
2251,44	45	45,57	51,07	$3s\,^4P-3p\,^4D^\circ$	1/2—3/2
2251,17	20	51,89	57,40	$3p\,^2P^\circ-3d\,^4P$	1/2—3/2
2246,66	40	45,51	51,02	$3s\,^4P-3p\,^4D^\circ$	3/2—5/2
2244,17	3	51,87	57,40	$3p\,^2P^\circ-3d\,^4P$	3/2—3/2
2239,43	45	45,57	51,10	$3s\,^4P-3p\,^4D^\circ$	1/2—1/2
2232,17	40	46,32	51,87	$3s\,^2P-3p\,^2P^\circ$	3/2—3/2
2230,30	50	45,39	50,95	$3s\,^4P-3p\,^4D^\circ$	5/2—7/2
2229,56	15	51,87	57,43	$3p\,^2P^\circ-3d\,^4P$	3/2—5/2
2226,19	8	51,89	57,46	$3p\,^2P^\circ-3d\,^4F$	1/2—3/2
2225,90	45	45,51	51,07	$3s\,^4P-3p\,^4D^\circ$	3/2—3/2
2225,29	12	46,32	51,89	$3s\,^2P-3p\,^2P^\circ$	3/2—1/2
2222,79	0	51,75	57,33	$3p\,^4S^\circ-3d\,^4P$	3/2—1/2
2214,17	25	45,51	51,10	$3s\,^4P-3p\,^4D^\circ$	3/2—1/2
2211,16	1	51,87	57,48	$3p\,^2P^\circ-3d\,^4F$	3/2—5/2
2202,78	40	45,39	51,02	$3s\,^4P-3p\,^4D^\circ$	5/2—5/2
2200,37	2	51,47	57,10	$3p\,^2D^\circ-3d\,^4D$	3/2—3/2
2198,01	2	51,69	57,33	$3p\,^2S^\circ-3d\,^4P$	1/2—1/2
2194,85	1	51,75	57,40	$3p\,^4S^\circ-3d\,^4P$	3/2—3/2
2193,08	1	51,47	57,12	$3p\,^2D^\circ-3d\,^4D$	3/2—1/2
2190,18	5	49,49	55,15	$3s'\,^2D-3p'\,^2D^\circ$	3/2—5/2
2182,74	15	45,39	51,07	$3s\,^4P-3p\,^4D^\circ$	5/2—3/2
2180,63	0	51,75	57,43	$3p\,^4S^\circ-3d\,^4P$	3/2—5/2
2174,43	3	51,87	57,57	$3p\,^2P^\circ-3d\,^2D$	3/2—5/2
2173,84	1	51,36	57,06	$3p\,^2D^\circ-3d\,^4D$	5/2—7/2
2170,63	0	51,69	57,40	$3p\,^2S^\circ-3d\,^4P$	1/2—3/2
2166,67	3	51,36	57,08	$3p\,^2D^\circ-3d\,^4D$	5/2—5/2
2163,32	1	51,75	57,48	$3p\,^4S^\circ-3d\,^4F$	3/2—5/2
2147,37	1	51,69	57,46	$3p\,^2S^\circ-3d\,^4F$	1/2—3/2
2117,08	0	51,47	57,33	$3p\,^2D^\circ-3d\,^4P$	3/2—1/2
2116,70	8	45,51	51,36	$3s\,^4P-3p\,^2D^\circ$	3/2—5/2
2107,62	1	51,89	57,77	$3p\,^2P^\circ-3d\,^2P$	1/2—1/2
2099,60	1	45,57	51,47	$3s\,^4P-3p\,^2D^\circ$	1/2—3/2
		51,75	57,65	$3p\,^4S^\circ-3d\,^2D$	3/2—3/2
2079,06	0	51,47	57,43	$3p\,^2D^\circ-3d\,^4P$	3/2—5/2
2073,38	10	51,89	57,87	$3p\,^2P^\circ-3d\,^2P$	1/2—3/2
2067,41	18	51,87	57,87	$3p\,^2P^\circ-3d\,^2P$	3/2—3/2

λ, Å	I	E_H, eV	E_B, eV	Transition	J
2065,23	10	51,10	57,10	$3p\ ^4D°—3d\ ^4D$	$^1/_2—^3/_2$
2062,97	8	51,07	57,08	$3p\ ^4D°—3d\ ^4D$	$^3/_2—^5/_2$
2058,72	8	51,10	57,12	$3p\ ^4D°—3d\ ^4D$	$^1/_2—^1/_2$
2055,17	10	51,07	57,10	$3p\ ^4D°—3d\ ^4D$	$^3/_2—^3/_2$
2051,90	8	51,02	57,06	$3p\ ^4D°—3d\ ^4D$	$^5/_2—^7/_2$
2048,67	8	51,07	57,12	$3p\ ^4D°—3d\ ^4D$	$^3/_2—^1/_2$
2045,41	18	51,02	57,08	$3p\ ^4D°—3d\ ^4D$	$^5/_2—^5/_2$
2041,09	0	51,36	57,43	$3p\ ^2D°—3d\ ^4P$	$^5/_2—^5/_2$
2037,70	10	51,02	57,10	$3p\ ^4D°—3d\ ^4D$	$^5/_2—^3/_2$
2035,84	3	55,14	61,23	$3p'\ ^2D°—3d'\ ^2D$	$^3/_2—^3/_2$
2032,65	0	51,36	57,46	$3p\ ^2D°—3d\ ^4F$	$^5/_2—^3/_2$
2031,10	10	51,47	57,57	$3p\ ^2D°—3d\ ^2D$	$^3/_2—^5/_2$
2028,55	25	50,95	57,06	$3p\ ^4D°—3d\ ^4D$	$^7/_2—^7/_2$
2025,44	8	51,75	57,87	$3p\ ^4S°—3d\ ^2P$	$^3/_2—^3/_2$
2022,27	10	{ 51,87	57,99	$3p\ ^2P°—4s\ ^4P$	$^3/_2—^5/_2$
		50,95	57,08	$3p\ ^4D°—3d\ ^4D$	$^7/_2—^5/_2$
2011,88	30	51,36	57,52	$3p\ ^2D°—3d\ ^2F$	$^5/_2—^7/_2$
2008,43	8	55,15	61,32	$3p'\ ^2D°—3d'\ ^2D$	$^5/_2—^5/_2$
2005,33	6	55,14	61,32	$3p'\ ^2D°—3d'\ ^2D$	$^3/_2—^5/_2$
2005,24	30	{ 45,57	51,75	$3s\ ^4P—3p\ ^4S°$	$^1/_2—^3/_2$
		51,47	57,65	$3p\ ^2D°—3d\ ^2D$	$^3/_2—^3/_2$
2004,80	0	51,69	57,87	$3p\ ^2S°—3d\ ^2P$	$^1/_2—^3/_2$
1995,62	3	{ 51,36	57,57	$3p\ ^2D°—3d\ ^2D$	$^5/_2—^5/_2$
		51,87	58,08	$3p\ ^2P°—4s\ ^4P$	$^3/_2—^3/_2$
1985,58	30	45,51	51,75	$3s\ ^4P—3p\ ^4S°$	$^3/_2—^3/_2$
1980,95	0	54,89	61,14	$3p'\ ^2P°—3d'\ ^2P$	$^3/_2—^1/_2$
1977,14	1	54,89	61,16	$3p'\ ^2P°—3d'\ ^2P$	$^3/_2—^3/_2$
1976,62	1	54,95	61,23	$3p'\ ^2P°—3d'\ ^2D$	$^1/_2—^3/_2$
1975,58	0	55,15	61,42	$3p'\ ^2D°—3d'\ ^2F$	$^5/_2—^7/_2$
1965,04	18	50,95	57,26	$3p\ ^4D°—3d\ ^4F$	$^7/_2—^9/_2$
1960,76	20	51,07	57,40	$3p\ ^4D°—3d\ ^4P$	$^3/_2—^3/_2$
1956,48	0	51,75	58,08	$3p\ ^4S°—4s\ ^4P$	$^3/_2—^3/_2$
1955,31	8	54,89	61,23	$3p'\ ^2P°—3d'\ ^2D$	$^3/_2—^3/_2$
1951,21	40	45,40	51,75	$3s\ ^4P—3p\ ^4S°$	$^5/_2—^3/_2$
1950,79	15	51,10	57,46	$3p\ ^4D°—3d\ ^4F$	$^1/_2—^3/_2$
1946,70	0	45,51	51,87	$3s\ ^4P—3p\ ^2P°$	$^3/_2—^3/_2$
1946,43	20	54,62	60,98	$3p'\ ^2F°—3d'\ ^2G$	$^7/_2—^7/_2$
1944,99	3	51,02	57,40	$3p\ ^4D°—3d\ ^4P$	$^5/_2—^3/_2$
1943,40	6	54,61	60,98	$3p'\ ^2F°—3d'\ ^2G$	$^5/_2—^7/_2$
1942,19	6	51,36	57,74	$3p\ ^2D°—3d\ ^2F$	$^5/_2—^5/_2$
1941,77	0	51,07	57,46	$3p\ ^4D°—3d\ ^4F$	$^3/_2—^3/_2$
1941,61	0	45,51	51,89	$3s\ ^4P—3p\ ^2P°$	$^3/_2—^1/_2$
1939,32	0	51,02	57,42	$3p\ ^4D°—3d\ ^4F$	$^5/_2—^7/_2$
1935,54	0	51,07	57,48	$3p\ ^4D°—3d\ ^4F$	$^3/_2—^5/_2$
1933,87	30	51,02	57,43	$3p\ ^4D°—3d\ ^4P$	$^5/_2—^5/_2$
1927,21	15	54,89	61,32	$3p'\ ^2P°—3d'\ ^2D$	$^3/_2—^5/_2$
1926,27	45	51,02	57,46	$3p\ ^4D°—3d\ ^4F$	$^5/_2—^3/_2$
1920,12	6	51,02	57,48	$3p\ ^4D°—3d\ ^4F$	$^5/_2—^5/_2$
1918,46	6	50,95	57,42	$3p\ ^4D°—3d\ ^4F$	$^7/_2—^7/_2$
1913,17	8	50,95	57,43	$3p\ ^4D°—3d\ ^4P$	$^7/_2—^5/_2$
1899,70	3	50,95	57,48	$3p\ ^4D°—3d\ ^4F$	$^7/_2—^5/_2$
1890,75	12	51,89	58,45	$3p\ ^2P°—4s\ ^2P$	$^1/_2—^3/_2$
1887,48	15	50,95	57,52	$3p\ ^4D°—3d\ ^2F$	$^7/_2—^7/_2$
1885,75	4	51,87	58,45	$3p\ ^2P°—4s\ ^2P$	$^3/_2—^3/_2$
1874,22	0	51,47	58,08	$3p\ ^2D°—4s\ ^4P$	$^3/_2—^3/_2$
1873,32	4	54,61	61,23	$3p'\ ^2F°—3d'\ ^2D$	$^5/_2—^3/_2$
1872,45	1	50,95	57,57	$3p\ ^4D°—3d\ ^2D$	$^7/_2—^5/_2$
1869,43	1	51,36	57,99	$3p\ ^2D°—4s\ ^4P$	$^5/_2—^5/_2$

λ, Å	I	E_H, eV	E_B, eV	Transition	J
1862,40	6	51,89	58,55	$3p\,{}^2P^\circ-4s\,{}^2P$	$^1/_2-{}^1/_2$
1861,19	15	50,44	57,10	$3p\,{}^4P^\circ-3d\,{}^4D$	$^1/_2-{}^3/_2$
1859,61	0	51,47	58,14	$3p\,{}^2D^\circ-4s\,{}^4P$	$^3/_2-{}^1/_2$
1859,20	0	51,10	57,77	$3p\,{}^4D^\circ-3d\,{}^2P$	$^1/_2-{}^1/_2$
1857,57	5	51,87	58,55	$3p\,{}^2P^\circ-4s\,{}^2P$	$^3/_2-{}^1/_2$
1856,73	20	50,40	57,08	$3p\,{}^4P^\circ-3d\,{}^4D$	$^3/_2-{}^5/_2$
1855,91	15	50,44	57,12	$3p\,{}^4P^\circ-3d\,{}^4D$	$^1/_2-{}^1/_2$
1850,39	18	50,40	57,10	$3p\,{}^4P^\circ-3d\,{}^4D$	$^3/_2-{}^3/_2$
1850,24	20	54,62	61,32	$3p'\,{}^2F^\circ-3d'\,{}^2D$	$^7/_2-{}^5/_2$
1849,58	35	50,36	57,06	$3p\,{}^4P^\circ-3d\,{}^4D$	$^5/_2-{}^7/_2$
1847,54	10	54,61	61,32	$3p'\,{}^2F^\circ-3d'\,{}^2D$	$^5/_2-{}^5/_2$
1845,10	12	50,40	57,12	$3p\,{}^4P^\circ-3d\,{}^4D$	$^3/_2-{}^1/_2$
1844,36	20	50,36	57,08	$3p\,{}^4P^\circ-3d\,{}^4D$	$^3/_2-{}^5/_2$
		51,02	57,74	$3p\,{}^4D^\circ-3d\,{}^2F$	$^5/_2-{}^5/_2$
1843,43	2	51,36	58,09	$3p\,{}^2D^\circ-4s\,{}^4P$	$^5/_2-{}^3/_2$
1838,11	6	50,36	57,10	$3p\,{}^4P^\circ-3d\,{}^4D$	$^5/_2-{}^3/_2$
1835,22	15	54,95	61,71	$3p'\,{}^2P^\circ-3d'\,{}^2S$	$^1/_2-{}^1/_2$
1825,44	10	50,95	57,74	$3p\,{}^4D^\circ-3d\,{}^2F$	$^7/_2-{}^5/_2$
1824,52	0	51,07	57,87	$3p\,{}^4D^\circ-3d\,{}^2P$	$^3/_2-{}^3/_2$
1821,68	12	54,62	61,42	$3p'\,{}^2F^\circ-3d'\,{}^2F$	$^7/_2-{}^7/_2$
1819,01	2	54,61	61,42	$3p'\,{}^2F^\circ-3d'\,{}^2F$	$^5/_2-{}^7/_2$
1816,83	2	54,89	61,71	$3p'\,{}^2P^\circ-3d'\,{}^2S$	$^3/_2-{}^1/_2$
1814,35	3	54,62	61,45	$3p'\,{}^2F^\circ-3d'\,{}^2F$	$^7/_2-{}^5/_2$
1811,70	5	54,61	61,45	$3p'\,{}^2F^\circ-3d'\,{}^2F$	$^5/_2-{}^5/_2$
1810,74	4	51,02	57,87	$3p\,{}^4D^\circ-3d\,{}^2P$	$^5/_2-{}^3/_2$
1801,27	7	50,44	57,33	$3p\,{}^4P^\circ-3d\,{}^4P$	$^1/_2-{}^1/_2$
1791,80	8	51,07	57,99	$3p\,{}^4D^\circ-4s\,{}^4P$	$^3/_2-{}^5/_2$
1791,23	10	50,40	57,33	$3p\,{}^4P^\circ-3d\,{}^4P$	$^3/_2-{}^1/_2$
1782,92	12	50,44	57,40	$3p\,{}^4P^\circ-3d\,{}^4P$	$^1/_2-{}^3/_2$
1775,32	0	51,10	58,08	$3p\,{}^4D^\circ-4s\,{}^4P$	$^1/_2-{}^3/_2$
1773,00	0	50,40	57,40	$3p\,{}^4P^\circ-3d\,{}^4P$	$^3/_2-{}^3/_2$
1767,21	0	50,44	57,46	$3p\,{}^4P^\circ-3d\,{}^4F$	$^1/_2-{}^3/_2$
1763,84	3	50,40	57,43	$3p\,{}^4P^\circ-3d\,{}^4P$	$^3/_2-{}^5/_2$
1762,13	0	51,10	58,14	$3p\,{}^4D^\circ-4s\,{}^4P$	$^1/_2-{}^1/_2$
1761,05	1	50,95	57,99	$3p\,{}^4D^\circ-4s\,{}^4P$	$^7/_2-{}^5/_2$
1754,97	0	51,02	58,08	$3p\,{}^4D^\circ-4s\,{}^4P$	$^5/_2-{}^3/_2$
1752,65	3	50,36	57,43	$3p\,{}^4P^\circ-3d\,{}^4P$	$^5/_2-{}^5/_2$
1752,06	1	51,47	58,55	$3p\,{}^2D^\circ-4s\,{}^2P$	$^3/_2-{}^1/_2$
1746,39	0	50,36	57,46	$3p\,{}^4P^\circ-3d\,{}^4F$	$^5/_2-{}^3/_2$
1741,33	1	50,36	57,48	$3p\,{}^4P^\circ-3d\,{}^4F$	$^5/_2-{}^5/_2$
1731,08	0	50,36	57,52	$3p\,{}^4P^\circ-3d\,{}^2F$	$^5/_2-{}^7/_2$
1719,60	0	50,44	57,65	$3p\,{}^4P^\circ-3d\,{}^2D$	$^1/_2-{}^3/_2$
1718,48	0	50,36	57,57	$3p\,{}^4P^\circ-3d\,{}^2D$	$^5/_2-{}^5/_2$
1691,70	1	50,44	57,77	$3p\,{}^4P^\circ-3d\,{}^2P$	$^1/_2-{}^1/_2$
1678,74	1	50,36	57,74	$3p\,{}^4P^\circ-3d\,{}^2F$	$^5/_2-{}^5/_2$
1669,52	3	51,02	58,45	$3p\,{}^4D^\circ-4s\,{}^2P$	$^5/_2-{}^3/_2$
		50,44	57,87	$3p\,{}^4P^\circ-3d\,{}^2P$	$^1/_2-{}^3/_2$
1658,71	2	51,07	58,55	$3p\,{}^4D^\circ-4s\,{}^2P$	$^3/_2-{}^1/_2$
1650,91	1	50,36	57,87	$3p\,{}^4P^\circ-3d\,{}^2P$	$^5/_2-{}^3/_2$
1633,64	4	50,40	57,99	$3p\,{}^4P^\circ-4s\,{}^4P$	$^3/_2-{}^5/_2$
1624,07	12	50,36	57,99	$3p\,{}^4P^\circ-4s\,{}^4P$	$^5/_2-{}^5/_2$
1621,94	5	50,44	58,08	$3p\,{}^4P^\circ-4s\,{}^4P$	$^1/_2-{}^3/_2$
1613,77	8	50,40	58,08	$3p\,{}^4P^\circ-4s\,{}^4P$	$^3/_2-{}^3/_2$
1610,97	4	50,44	58,14	$3p\,{}^4P^\circ-4s\,{}^4P$	$^1/_2-{}^1/_2$
1604,47	6	50,36	58,08	$3p\,{}^4P^\circ-4s\,{}^4P$	$^5/_2-{}^3/_2$
1602,91	5	50,40	58,14	$3p\,{}^4P^\circ-4s\,{}^4P$	$^3/_2-{}^1/_2$
1548,68	8	50,44	58,45	$3p\,{}^4P^\circ-4s\,{}^2P$	$^1/_2-{}^3/_2$
1541,19	1	50,40	58,45	$3p\,{}^4P^\circ-4s\,{}^2P$	$^3/_2-{}^3/_2$
1529,67	1	50,44	58,55	$3p\,{}^4P^\circ-4s\,{}^2P$	$^1/_2-{}^1/_2$

λ, Å	I	E_H, eV	E_B, eV	Transition	J
1458,15	3	46,45	54,95	$3s\,^2P - 3p'\,^2P^{\circ}$	$1/2-1/2$
1436,21	12	46,32	54,95	$3s\,^2P - 3p'\,^2P^{\circ}$	$3/2-1/2$
1427,27	0	46,45	55,14	$3s\,^2P - 3p'\,^2D^{\circ}$	$1/2-3/2$
1337,39	6	51,87	61,14	$3p\,^2P^{\circ} - 3d'\,^2P$	$3/2-1/2$
1265,66	2	51,36	61,16	$3p\,^2D^{\circ} - 3d'\,^2P$	$5/2-3/2$
1256,68	1	51,36	61,23	$3p\,^2D^{\circ} - 3d'\,^2D$	$5/2-3/2$
1235,40	4	50,95	60,98	$3p\,^4D^{\circ} - 3d'\,^2G$	$7/2-7/2$
1224,73	4	51,10	61,23	$3p\,^4D^{\circ} - 3d'\,^2D$	$1/2-3/2$
1223,44	4	51,02	61,16	$3p\,^4D^{\circ} - 3d'\,^2P$	$5/2-3/2$
1221,12	5	51,07	61,23	$3p\,^4D^{\circ} - 3d'\,^2D$	$3/2-3/2$
1180,40	8	55,12	65,64	$3p'\,^2D^{\circ} - 3d''\,^2D$	$3/2-5/2$
1153,04	2	50,40	61,16	$3p\,^4P^{\circ} - 3d'\,^2P$	$3/2-3/2$
1122,30	0	50,40	61,45	$3p\,^4P^{\circ} - 3d'\,^2F$	$3/2-5/2$
1100,49	5	50,44	61,71	$3p\,^4P^{\circ} - 3d'\,^2S$	$1/2-1/2$
380,107	8	0,17	32,78	$2p^5\,^2P^{\circ} - 2p^6\,^2S$	$1/2-1/2$
378,143	10	0,00	32,78	$2p^5\,^2P^{\circ} - 2p^6\,^2S$	$3/2-1/2$
272,441	0	0,00	45,51	$2p^5\,^2P^{\circ} - 3s\,^4P$	$3/2-3/2$
268,623	5	0,17	46,32	$2p^5\,^2P^{\circ} - 3s\,^2P$	$1/2-3/2$
267,868	6	0,17	46,45	$2p^5\,^2P^{\circ} - 3s\,^2P$	$1/2-1/2$
267,642	8	0,00	46,32	$2p^5\,^2P^{\circ} - 3s\,^2P$	$3/2-3/2$
266,893	5	0,00	46,45	$2p^5\,^2P^{\circ} - 3s\,^2P$	$3/2-1/2$
251,371	6	0,17	49,49	$2p^5\,^2P^{\circ} - 3s'\,^2D$	$1/2-3/2$
250,515	8	0,00	49,49	$2p^5\,^2P^{\circ} - 3s'\,^2D$	$3/2-3/2,\ 5/2$
230,593	2	0,17	53,93	$2p^5\,^2P^{\circ} - 3s''\,^2S$	$1/2-1/2$
229,868	3	0,00	53,93	$2p^5\,^2P^{\circ} - 3s''\,^2S$	$3/2-1/2$
216,120	1	—	—	—	—
215,671	4	0,17	57,65	$2p^5\,^2P^{\circ} - 3d\,^2D$	$1/2-3/2$
215,340	4	0,00	57,57	$2p^5\,^2P^{\circ} - 3d\,^2D$	$3/2-5/2$
215,230	4	0,17	57,77	$2p^5\,^2P^{\circ} - 3d\,^2P$	$1/2-1/2$
215,042	2	0,00	57,67	$2p^5\,^2P^{\circ} - 3d\,^2D$	$3/2-3/2$
214,868	4	0,17	57,87	$2p^5\,^2P^{\circ} - 3d\,^2P$	$1/2-3/2$
214,596	2	0,00	57,77	$2p^5\,^2P^{\circ} - 3d\,^2P$	$3/2-1/2$
214,235	4	0,00	57,87	$2p^5\,^2P^{\circ} - 3d\,^2P$	$3/2-3/2$
203,324	2	0,17	61,14	$2p^5\,^2P^{\circ} - 3d'\,^2P$	$1/2-1/2$
203,282	2	0,17	61,16	$2p^5\,^2P^{\circ} - 3d'\,^2P$	$1/2-3/2$
203,050	3	0,17	61,23	$2p^5\,^2P^{\circ} - 3d'\,^2D$	$1/2-3/2$
202,760	3	0,00	61,14	$2p^5\,^2P^{\circ} - 3d'\,^2P$	$3/2-1/2$
202,720	3	0,00	61,16	$2p^5\,^2P^{\circ} - 3d'\,^2P$	$3/2-3/2$
202,490	2	0,00	61,23	$2p^5\,^2P^{\circ} - 3d'\,^2D$	$3/2-3/2$
202,184	4	0,00	61,32	$2p^5\,^2P^{\circ} - 3d'\,^2D$	$3/2-5/2$
195,538	0	0,00	63,40	$2p^5\,^2P^{\circ} - 4s'\,^2D$	$3/2-5/2$
194,306	1	0,00	63,81	$2p^5\,^2P^{\circ} - 4d\,^2D$	$3/2-5/2$
194,166	0	0,00	63,85	$2p^5\,^2P^{\circ} - 4d\,^2P$	$3/2-1/2$
194,032	1	0,00	63,90	$2p^5\,^2P^{\circ} - 4d\,^2P$	$3/2-3/2$
189,346	1	0,17	65,65	$2p^5\,^2P^{\circ} - 3d''\,^2D$	$1/2-3/2$
188,870	2	0,00	65,64	$2p^5\,^2P^{\circ} - 3d''\,^2D$	$3/2-3/2,\ 5/2$
184,218	0	0,17	67,47	$2p^5\,^2P^{\circ} - 4d'\,^2P$	$1/2-1/2,\ 3/2$
183,747	0	0,00	67,47	$2p^5\,^2P^{\circ} - 4d'\,^2P$	$3/2-1/2,\ 3/2$
183,575	0	0,00	67,53	$2p^5\,^2P^{\circ} - 4d'\,^2D$	$3/2-3/2,\ 5/2$

Na IV, **ground state** $1s^2\,2s^2\,2p^4\,^3P_2$

Ionization potential $797\,741\ \text{cm}^{-1}$; 98,902 eV

λ, Å	I	E_H, eV	E_B, eV	Transition	J
412,240	8	0,14	30,21	$2p^4\,^3P - 2p^5\,^3P^{\circ}$	1—2
411,333	7	0,20	30,34	$2p^4\,^3P - 2p^5\,^3P^{\circ}$	0—1

λ, Å	I	E_H, eV	E_B, eV	Transition	J
410,540	6	0,14	30,34	$2p^4\,^3P—2p^5\,^3P^\circ$	1—1
410,371	10	0,00	30,21	$2p^4\,^3P—2p^5\,^3P^\circ$	2—2
409,615	8	0,14	30,40	$2p^4\,^3P—2p^5\,^3P^\circ$	1—0
408,682	8	0,00	30,34	$2p^4\,^3P—2p^5\,^3P^\circ$	2—1
360,761	6	8,28	42,64	$2p^4\,^1S—2p^5\,^1P^\circ$	0—1
319,638	10	3,86	42,64	$2p^4\,^1D—2p^5\,^1P^\circ$	2—1
206,155	3	0,20	60,33	$2p^4\,^3P—3s\,^3S^\circ$	0—1
205,956	4	0,14	60,33	$2p^4\,^3P—3s\,^3S^\circ$	1—1
205,487	6	0,00	60,33	$2p^4\,^3P—3s\,^3S^\circ$	2—1
203,959	2	8,28	69,07	$2p^4\,^1S—3s''\,^1P^\circ$	0—1
199,769	6	3,86	65,92	$2p^4\,^1D—3s'\,^1D^\circ$	2—2
191,000	6	0,20	65,10	$2p^4\,^3P—3s'\,^3D^\circ$	0—1
190,835	8	0,14	65,10	$2p^4\,^3P—3s'\,^3D^\circ$	1—2
190,440	10	0,00	65,10	$2p^4\,^3P—3s'\,^3D^\circ$	2—3
190,126	6	3,86	69,07	$2p^4\,^1D—3s''\,^1P^\circ$	2—1
182,282	4	0,20	68,21	$2p^4\,^3P—3s''\,^3P^\circ$	0—1
182,128	6	0,14	68,21	$2p^4\,^3P—3s''\,^3P^\circ$	1—0, 1, 2
181,758	8	0,00	68,21	$2p^4\,^3P—3s''\,^3P^\circ$	2—1, 2
174,008	0	8,28	79,53	$2p^4\,^1S—3d'\,^1P^\circ$	0—1
168,544	5	0,20	73,75	$2p^4\,^3P—3d\,^3D^\circ$	0—1
168,409	8	0,14	73,75	$2p^4\,^3P—3d\,^3D^\circ$	1—1, 2
168,084	10	0,00	73,76	$2p^4\,^3P—3d\,^3D^\circ$	2—3
164,841	4	—	—	—	—
163,840	4	3,86	79,53	$2p^4\,^1D—3d'\,^1P^\circ$	2—1
163,187	6	3,86	79,83	$2p^4\,^1D—3d'\,^1D^\circ$	2—2
162,445	8	3,86	80,18	$2p^4\,^1D—3d'\,^1F^\circ$	2—3
157,782	3	3,86	82,43	$2p^4\,^1D—3d''\,^1D^\circ$	2—2
157,599	1	3,86	82,52	$2p^4\,^1D—3d''\,^1P^\circ$	2—1
157,090	4	3,86	82,78	$2p^4\,^1D—3d''\,^1F^\circ$	2—3
156,887	3	0,20	79,21	$2p^4\,^3P—3d'\,^3D^\circ$	0—1
156,780	5	0,14	79,21	$2p^4\,^3P—3d'\,^3D^\circ$	1—2
156,536	8	0,00	79,21	$2p^4\,^3P—3d'\,^3D^\circ$	2—3
155,832	0	0,20	79,76	$2p^4\,^3P—3d'\,^3P^\circ$	0—1
155,781	1	0,14	79,72	$2p^4\,^3P—3d'\,^3P^\circ$	1—2
155,693	2	0,14	79,77	$2p^4\,^3P—3d'\,^3P^\circ$	1—0, 1
155,622	0	0,20	79,86	$2p^4\,^3P—3d'\,^3S^\circ$	0—1
155,515	4	{ 0,00	79,72	$2p^4\,^3P—3d'\,^3P^\circ$	2—2
		{ 0,14	79,86	$2p^4\,^3P—3d'\,^3S^\circ$	1—1
155,445	3	0,00	79,76	$2p^4\,^3P—3d'\,^3P^\circ$	2—1
155,354	0	0,14	79,94	$2p^4\,^3P—4s\,^3S^\circ$	1—1
155,248	2	0,00	79,86	$2p^4\,^3P—3d'\,^3S^\circ$	2—1
155,090	1	0,00	79,94	$2p^4\,^3P—4s\,^3S^\circ$	2—1
151,303	1	3,86	85,80	$2p^4\,^1D—4s'\,^1D^\circ$	2—2
151,048	0	0,20	82,27	$2p^4\,^3P—3d''\,^3P^\circ$	0—1
150,968	2	0,14	82,27	$2p^4\,^3P—3d''\,^3P^\circ$	1—0, 1, 2
150,695	2	0,00	82,27	$2p^4\,^3P—3d''\,^3P^\circ$	2—1, 2
150,647	2	0,20	82,49	$2p^4\,^3P—3d''\,^3D^\circ$	0—1
150,545	3	0,14	82,49	$2p^4\,^3P—3d''\,^3D^\circ$	1—1, 2
150,297	4	0,00	82,49	$2p^4\,^3P—3d''\,^3D^\circ$	2—1, 2, 3
146,397	0	0,20	84,88	$2p^4\,^3P—4d\,^3D^\circ$	0—1
146,297	1	0,14	84,88	$2p^4\,^3P—4d\,^3D^\circ$	1—1, 2
146,060	3	0,00	84,88	$2p^4\,^3P—4d\,^3D^\circ$	2—1, 2, 3
145,846	0	3,86	88,86	$2p^4\,^1D—4s''\,^1P^\circ$	2—2
144,979	0	0,00	85,51	$2p^4\,^3P—4s'\,^3D^\circ$	2—1, 2, 3
142,688	0	3,86	90,74	$2p^4\,^1D—4d'\,^1P^\circ$	2—1
142,363	1	3,86	90,94	$2p^4\,^1D—4d'\,^1D^\circ$	2—2
142,232	2	3,86	91,02	$2p^4\,^1D—4d'\,^1F^\circ$	2—3
139,869	0	0,00	88,64	$2p^4\,^3P—4s''\,^3P^\circ$	2—1, 2
137,945	0	3,86	93,73	$2p^4\,^1D—4d''\,^1D^\circ$	2—2

λ, Å	I	E_H, eV	E_B, eV	Transition	J
137,714	0	3,86	93,88	$2p^4\,{}^1D-4d''\,{}^1F^\circ$	2—3
137,144	0	0,20	90,60	$2p^4\,{}^3P-4d'\,{}^3D^\circ$	0—1
137,062	0	0,14	90,60	$2p^4\,{}^3P-4d'\,{}^3D^\circ$	1—1, 2
136,854	1	0,00	90,60	$2p^4\,{}^3P-4d'\,{}^3D^\circ$	2—1, 2, 3
136,748	0	0,14	90,80	$2p^4\,{}^3P-4d'\,{}^3P^\circ$	1—0, 1, 2
136,645	0	0,14	90,87	$2p^4\,{}^3P-4d'\,{}^3S^\circ$	1—1
136,550	1	0,00	90,80	$2p^4\,{}^3P-4d'\,{}^3P^\circ$	2—1, 2
136,435	0	0,00	90,87	$2p^4\,{}^3P-4d'\,{}^3S^\circ$	2—1
132,740	0	0,00	93,40	$2p^4\,{}^3P-5s'\,{}^3D^\circ$	2—1, 2, 3
132,465	0	0,14	93,77	$2p^4\,{}^3P-4d''\,{}^3D^\circ$	1—1, 2
132,211	0	0,00	93,77	$2p^4\,{}^3P-4d''\,{}^3D^\circ$	2—1, 2, 3
129,464	0	0,00	95,76	$2p^4\,{}^3P-5d'\,{}^3D^\circ$	2—1, 2, 3

Na V, ground state $1s^2\,2s^2\,2p^3\,{}^4S^\circ_{3/2}$

Ionization potential 1 116 312 cm^{-1}; 138,840 eV

λ, Å	I	E_H, eV	E_B, eV	Transition	J
514,350	0	46,26	70,37	$2p^4\,{}^2P-2p^5\,{}^2P^\circ$	1/2—3/2
511,193	1	46,12	70,37	$2p^4\,{}^2P-2p^5\,{}^2P^\circ$	3/2—3/2
510,102	0	46,26	70,57	$2p^4\,{}^2P-2p^5\,{}^2P^\circ$	1/2—1/2
463,263	12	0,00	26,76	$2p^3\,{}^4S^\circ-2p^4\,{}^4P$	3/2—5/2
461,051	10	0,00	26,89	$2p^3\,{}^4S^\circ-2p^4\,{}^4P$	3/2—3/2
459,897	7	0,00	26,96	$2p^3\,{}^4S^\circ-2p^4\,{}^4P$	3/2—1/2
445,190	7	8,99	36,84	$2p^3\,{}^2P^\circ-2p^4\,{}^2D$	3/2—5/2
445,046	6	8,98	36,84	$2p^3\,{}^2P^\circ-2p^4\,{}^2D$	1/2—3/2
400,722	10	5,90	36,84	$2p^3\,{}^2D^\circ-2p^4\,{}^2D$	3/2, 5/2—3/2, 5/2
369,743	3	36,84	70,37	$2p^4\,{}^2D-2p^5\,{}^2P^\circ$	5/2—3/2
367,557	2	36,84	70,57	$2p^4\,{}^2D-2p^5\,{}^2P^\circ$	3/2—1/2
360,367	8	8,99	43,39	$2p^3\,{}^2P^\circ-2p^4\,{}^2S$	3/2—1/2
360,319	8	8,98	43,39	$2p^3\,{}^2P^\circ-2p^4\,{}^2S$	1/2—1/2
333,910	9	8,99	46,12	$2p^3\,{}^2P^\circ-2p^4\,{}^2P$	3/2—3/2
332,550	8	8,98	46,26	$2p^3\,{}^2P^\circ-2p^4\,{}^2P$	1/2—1/2
330,718	0	5,90	43,39	$2p^3\,{}^2D^\circ-2p^4\,{}^2S$	3/2—1/2
308,264	10	5,90	46,12	$2p^3\,{}^2D^\circ-2p^4\,{}^2P$	5/2—3/2
307,152	8	5,90	46,26	$2p^3\,{}^2D^\circ-2p^4\,{}^2P$	3/2—1/2
171,076	1	26,96	99,42	$2p^4\,{}^4P-3s'''\,{}^4S^\circ$	1/2—3/2
170,923	1	26,89	99,42	$2p^4\,{}^4P-3s'''\,{}^4S^\circ$	3/2—3/2
170,631	1	26,76	99,42	$2p^4\,{}^4P-3s'''\,{}^4S^\circ$	5/2—3/2
167,510	1	36,84	110,84	$2p^4\,{}^2D-3s^{IV}\,{}^2D^\circ$	3/2, 5/2—3/2, 5/2
163,930	2	8,98	84,61	$2p^3\,{}^2P^\circ-3s\,{}^2P$	1/2—1/2
163,616	3	8,98	84,76	$2p^3\,{}^2P^\circ-3s\,{}^2P$	1/2—3/2
157,511	2	5,90	84,61	$2p^3\,{}^2D^\circ-3s\,{}^2P$	3/2—1/2
157,209	3	5,90	84,76	$2p^3\,{}^2D^\circ-3s\,{}^2P$	3/2, 5/2—3/2
157,036	2	8,98	87,93	$2p^3\,{}^2P^\circ-3s'\,{}^2D$	3/2, 5/2—3/2, 5/2
151,303	1	26,96	108,89	$2p^4\,{}^4P-3s^{IV}\,{}^4D^\circ$	1/2—1/2, 3/2
151,188	1	26,89	108,89	$2p^4\,{}^4P-3s^{IV}\,{}^4D^\circ$	3/2—1/2, 3/2, 5/2
151,127	4	5,90	87,93	$2p^3\,{}^2D^\circ-3s'\,{}^2D$	3/2, 5/2—3/2, 5/2
150,968	2	26,76	108,89	$2p^4\,{}^4P-3s^{IV}\,{}^4D^\circ$	5/2—3/2, 5/2, 7/2
149,001	2	0,00	83,21	$2p^3\,{}^4S^\circ-3s\,{}^4P$	3/2—1/2
148,856	3	0,00	83,29	$2p^3\,{}^4S^\circ-3s\,{}^4P$	3/2—3/2
148,642	4	0,00	83,41	$2p^3\,{}^4S^\circ-3s\,{}^4P$	3/2—5/2
147,897	2	8,98	92,81	$2p^3\,{}^2P^\circ-3s''\,{}^2S$	1/, 3/2—1/2

λ, Å	I	E_H, eV	E_B, eV	Transition	J
144,661	1	26,96	112,66	$2p^4\,{}^4P-3d'''\,{}^4D^\circ$	1/2—1/2, 3/2
144,546	2	26,89	112,66	$2p^4\,{}^4P-3d'''\,{}^4D^\circ$	3/2—1/2, 3/2, 5/2
144,330	2	26,76	112,66	$2p^4\,{}^4P-3d'''\,{}^4D^\circ$	5/2—3/2, 5/2, 7/2
142,415	0	26,89	113,94	$2p^4\,{}^4P-3s^{V}\,{}^4P^\circ$	3/2—1/2, 3/2, 5/2
142,232	2	26,76	113,94	$2p^4\,{}^4P-3s^{V}\,{}^4P^\circ$	5/2—3/2, 5/2
140,258	0	36,84	125,23	$2p^4\,{}^2D-3d^{IV}\,{}^2F^\circ$	5/2—7/2
140,171	0	36,84	125,23	$2p^4\,{}^2D-3d^{IV}\,{}^2F^\circ$	3/2—5/2
138,917	3	8,98	98,23	$2p^3\,{}^2P^\circ-3d\,{}^2P$	1/2, 3/2—3/2
138,812	2	8,98	98,29	$2p^3\,{}^2P^\circ-3d\,{}^2P$	1/2, 3/2—1/2
135,854	3	8,98	100,24	$2p^3\,{}^2P^\circ-3d\,{}^2D$	1/2—3/2
135,791	3	8,99	100,28	$2p^3\,{}^2P^\circ-3d\,{}^2D$	3/2—5/2
134,272	2	5,90	98,23	$2p^3\,{}^2D^\circ-3d\,{}^2P$	5/2—3/2
134,183	0	5,90	98,29	$2p^3\,{}^2D^\circ-3d\,{}^2P$	3/2—1/2
133,388	4	5,90	98,84	$2p^3\,{}^2D^\circ-3d\,{}^2F$	3/2—5/2
133,162	5	5,90	99,00	$2p^3\,{}^2D^\circ-3d\,{}^2F$	5/2—7/2
131,635	3	8,98	103,16	$2p^3\,{}^2P^\circ-3d'\,{}^2D$	1/2, 3/2—3/2
131,413	2	5,90	100,24	$2p^3\,{}^2D^\circ-3d\,{}^2D$	3/2—3/2
131,345	3	5,90	100,28	$2p^3\,{}^2D^\circ-3d\,{}^2D$	5/2—5/2
130,723	1	8,98	103,82	$2p^3\,{}^2P^\circ-3d'\,{}^2P$	1/2, 3/2—1/2
130,680	2	8,98	103,86	$2p^3\,{}^2P^\circ-3d'\,{}^2P$	1/2, 3/2—3/2
129,942	1	8,98	104,40	$2p^3\,{}^2P^\circ-3d'\,{}^2S$	1/2, 3/2—1/2
128,051	4	5,90	102,72	$3p^3\,{}^2D^\circ-3d'\,{}^2F$	5/2—7/2
128,025	4	5,90	102,74	$2p^3\,{}^2D^\circ-3d'\,{}^2F$	3/2—5/2
127,473	4	5,90	103,16	$2p^3\,{}^2D^\circ-3d'\,{}^2D$	3/2—3/2
127,444	4	5,90	103,18	$2p^3\,{}^2D^\circ-3d'\,{}^2D$	5/2—5/2
127,036	0	26,96	124,55	$2p^4\,{}^4P-3d^{IV}\,{}^4P^\circ$	1/2—3/2
126,985	0	26,89	124,52	$2p^4\,{}^4P-3d^{IV}\,{}^4P^\circ$	3/2—5/2
126,920	0	26,89	124,57	$2p^4\,{}^4P-3d^{IV}\,{}^4P^\circ$	3/2—1/2
126,814	1	26,76	124,52	$2p^4\,{}^4P-3d^{IV}\,{}^4P^\circ$	5/2—5/2
126,779	0	26,76	124,55	$2p^4\,{}^4P-3d^{IV}\,{}^4P^\circ$	5/2—3/2
126,608	1	5,90	103,82	$2p^3\,{}^2D^\circ-3d'\,{}^2P$	3/2—1/2
126,557	2	5,90	103,86	$2p^3\,{}^2D^\circ-3d'\,{}^2P$	5/2—3/2
126,458	0	26,96	124,99	$2p^4\,{}^4P-3d^{IV}\,{}^4D^\circ$	1/2—1/2, 3/2
126,368	0	26,89	124,99	$2p^4\,{}^4P-3d^{IV}\,{}^4D^\circ$	3/2—1/2, 3/2, 5/2
126,210	1	26,76	124,99	$2p^4\,{}^4P-3d^{IV}\,{}^4D^\circ$	5/2—3/2, 5/2, 7/2
126,090	0	26,76	125,08	$2p^4\,{}^4P-3d^{IV}\,{}^4S^\circ$	5/2—3/2
125,899	2	8,98	107,46	$2p^3\,{}^2P^\circ-3d''\,{}^2D$	1/2, 3/2—3/2, 5/2
125,461	3	0,00	98,82	$2p^3\,{}^4S^\circ-3d\,{}^4D$	3/2—3/2, 5/2
125,428	3	0,00	98,84	$2p^3\,{}^4S^\circ-3d\,{}^4D$	3/2—1/2
125,286	5	0,00	98,96	$2p^3\,{}^4S^\circ-3d\,{}^4P$	3/2—5/2
125,216	4	0,00	99,01	$2p^3\,{}^4S^\circ-3d\,{}^4P$	3/2—3/2
125,178	4	0,00	99,04	$2p^3\,{}^4S^\circ-3d\,{}^4P$	3/2—1/2
121,263	0	8,99	111,23	$2p^2P-4s\,{}^2P$	3/2—3/2
117,990	4	0,00	105,08	$2p^3\,{}^4S^\circ-3p'''\,{}^4P$	3/2—1/2, 3/2, 5/2
117,876	0	5,90	111,08	$2p^3\,{}^2D^\circ-4s\,{}^2P$	3/2—1/2
114,738	1	8,98	117,04	$2p^3\,{}^2P^\circ-4d\,{}^2D$	1/2—3/2
114,700	1	8,99	117,07	$2p^3\,{}^2P^\circ-4d\,{}^2D$	3/2—5/2
113,574	0	5,90	115,06	$2p^3\,{}^2D^\circ-4s'\,{}^2D$	3/2, 5/2—3/2, 5/2
112,347	0	5,90	116,25	$2p^3\,{}^2D^\circ-4d\,{}^2P$	5/2—3/2
112,186	0	5,90	116,58	$2p^3\,{}^2D^\circ-4d\,{}^2F$	3/2—5/2
112,077	0	0,00	110,62	$2p^3\,{}^4S^\circ-4s\,{}^4P$	3/2—1/2
112,009	3	5,90	116,58	$2p^3\,{}^2D^\circ-4d\,{}^2F$	3/2—5/2
111,879	0	5,90	116,71	$2p^3\,{}^2D^\circ-4d\,{}^2F$	5/2—7/2
		0,00	110,81	$2p^3\,{}^4S^\circ-4s\,{}^4P$	3/2—5/2
111,552	0	5,90	117,04	$2p^3\,{}^2D^\circ-4d\,{}^2D$	3/2—3/2
111,512	1	5,90	117,07	$2p^3\,{}^2D^\circ-4d\,{}^2D$	5/2—5/2

λ, Å	I	E_H, eV	E_B, eV	Transition	J
110,921	0	8,98	120,76	$2p^3\ {}^2P^\circ - 4d'\ {}^2D$	1/2, 3/2—3/2, 5/2
110,878	2	5,90	117,71	$2p^3\ {}^2D^\circ - 3p^{IV}\ {}^2F$	5/2—7/2
110,817	2	5,90	117,78	$2p^3\ {}^2D^\circ - 3p^{IV}\ {}^2F$	3/2—5/2
108,017	2	5,90	120,67	$2p^3\ {}^2D^\circ - 4d'\ {}^2F$	3/2, 5/2—5/2, 7/2
107,934	2	5,90	120,76	$2p^3\ {}^2D^\circ - 4d'\ {}^2D$	3/2, 5/2—3/2, 5/2
106,490	1	0,00	116,42	$2p^3\ {}^4S^\circ - 4d\ {}^4D$	3/2—3/2, 5/2
106,399	1	0,00	116,52	$2p^3\ {}^4S^\circ - 4d\ {}^4D$	3/2—1/2
106,302	1	0,00	116,62	$2p^3\ {}^4S^\circ - 4d\ {}^4P$	3/2—5/2
106,278	1	0,00	116,65	$2p^3\ {}^4S^\circ - 4d\ {}^4P$	3/2—3/2
103,482	0	8,98	128,80	$2p^3\ {}^2P^\circ - 5d'\ {}^2D$	1/2, 3/2—3/2, 5/2
100,945	0	5,90	128,71	$2p^3\ {}^2D^\circ - 5d'\ {}^2F$	3/2, 5/2—5/2, 7/2
100,851	0	5,90	128,80	$2p^3\ {}^2D^\circ - 5d'\ {}^2D$	3/2, 5/2—3/2, 5/2

Na VI, ground state $1s^2\ 2s^2\ 2p^2\ {}^3P_0$

Ionization potential 1388419 cm⁻¹; 172,130 eV

λ, Å	I	E_H, eV	E_B, eV	Transition	J
638,21	0	39,75	59,17	$2p^3\ {}^3S^\circ - 2p^4\ {}^3P$	1—2
632,90	0	39,75	59,34	$2p^3\ {}^3S^\circ - 2p^4\ {}^3P$	1—1
528,730	0	43,41	66,86	$2p^3\ {}^1P^\circ - 2p^4\ {}^1D$	1—2
494,382	7	0,23	25,31	$2p^2\ {}^3P - 2p^3\ {}^3D^\circ$	2—3
494,160	3	0,23	25,32	$2p^2\ {}^3P - 2p^3\ {}^3D^\circ$	2—2
491,340	6	0,08	25,32	$2p^2\ {}^3P - 2p^3\ {}^3D^\circ$	1—2
491,240	3	0,08	25,32	$2p^2\ {}^3P - 2p^3\ {}^3D^\circ$	1—1
489,580	5	0,00	25,32	$2p^2\ {}^3P - 2p^3\ {}^3D^\circ$	0—1
440,266	3	38,59	66,86	$2p^3\ {}^1D^\circ - 2p^4\ {}^1D$	2—2
423,821	2	29,92	59,17	$2p^3\ {}^3P^\circ - 2p^4\ {}^3P$	2, 1—2
421,465	1	29,92	59,34	$2p^3\ {}^3P^\circ - 2p^4\ {}^3P$	0, 1, 2—1
417,595	6	0,23	29,92	$2p^2\ {}^3P - 2p^3\ {}^3P^\circ$	2—2, 1
415,505	4	0,08	29,92	$2p^2\ {}^3P - 2p^3\ {}^3P^\circ$	1—0, 1, 2
414,370	2	0,00	29,92	$2p^2\ {}^3P - 2p^3\ {}^3P^\circ$	0—1
366,240	0	25,32	59,17	$2p^3\ {}^3D^\circ - 2p^4\ {}^3P$	2—2
366,110	4	25,31	59,17	$2p^3\ {}^3D^\circ - 2p^4\ {}^3P$	3—2
364,477	3	25,32	59,34	$2p^3\ {}^3D^\circ - 2p^4\ {}^3P$	2—1
363,774	2	25,32	59,40	$2p^3\ {}^3D^\circ - 2p^4\ {}^3P$	1—0
362,444	4	9,21	43,41	$2p^2\ {}^1S - 2p^3\ {}^1P^\circ$	0—1
361,250	8	4,38	38,59	$2p^2\ {}^1D - 2p^3\ {}^1D^\circ$	2—2
317,641	6	4,38	43,41	$2p^2\ {}^1D - 2p^3\ {}^1P^\circ$	2—1
313,748	5	0,23	39,75	$2p^2\ {}^3P - 2p^3\ {}^3S^\circ$	2—1
312,608	3	0,08	39,75	$2p^2\ {}^3P - 2p^3\ {}^3S^\circ$	1—1
311,921	4	0,00	39,75	$2p^2\ {}^3P - 2p^3\ {}^3S^\circ$	0—1
149,621	0	25,32	108,18	$2p^3\ {}^3D^\circ - 3p\ {}^3P$	2—1
149,442	0	25,31	108,27	$2p^3\ {}^3D^\circ - 3p\ {}^3P$	3—2
146,398	0	43,41	128,10	$2p^3\ {}^1P^\circ - 3s'\ {}^1D$	1—2
141,040	0	29,92	117,82	$2p^3\ {}^3P^\circ - 3s\ {}^3P$	0, 1, 2—1
140,833	2	29,92	117,95	$2p^3\ {}^3P^\circ - 3s\ {}^3P$	1, 2—2
138,693	2	38,59	128,10	$2p^3\ {}^1D^\circ - 3s'\ {}^1D$	2—2
137,589	0	39,75	132,40	$2p^3\ {}^3S^\circ - 3d\ {}^3D$	1—2
134,532	3	9,21	101,36	$2p^2\ {}^1S - 3s\ {}^1P^\circ$	0—1
134,135	0	25,32	117,75	$2p^3\ {}^3D^\circ - 3s\ {}^3P$	1—0
134,021	1	25,32	117,82	$2p^3\ {}^3D^\circ - 3s\ {}^3P$	1, 2—1
133,825	2	25,31	117,95	$2p^3\ {}^3D^\circ - 3s\ {}^3P$	3—2

λ, Å	I	E_H, eV	E_B, eV	Transition	J
129,040	2	29,92	125,99	$2p^3\,{}^3P^\circ$—$3s'\,{}^3D$	0, 1, 2—1, 2, 3
127,837	4	4,38	101,36	$2p^2\,{}^1D$—$3s\,{}^1P^\circ$	2—1
125,383	0	43,41	142,29	$2p^3\,{}^1P^\circ$—$3d'\,{}^1D$	1—2
124,850	0	43,41	142,71	$2p^3\,{}^1P^\circ$—$3d'\,{}^1P$	1—1
124,153	4	0,23	100,10	$2p^2\,{}^3P$—$3s\,{}^3P^\circ$	2—1
124,059	4	0,08 29,92	100,10 129,85	$2p^2\,{}^3P$—$3s\,{}^3P^\circ$ $2p^3\,{}^3P^\circ$—$3d\,{}^3P$	1—0 2—2
123,970	2	0,08	100,10	$2p^2\,{}^3P$—$3s\,{}^3P^\circ$	1—1
123,929	5	0,23	100,27	$2p^2\,{}^3P$—$3s\,{}^3P^\circ$	2—2
123,868	3	0,00	100,10	$2p^2\,{}^3P$—$3s\,{}^3P^\circ$	0—1
123,744	4	0,08	100,27	$2p^2\,{}^3P$—$3s\,{}^3P^\circ$	1—2
123,134	4	25,32	125,99	$2p^3\,{}^3D^\circ$—$3s'\,{}^3D$	1, 2, 3—1, 2, 3
122,018	3	12,83	114,44	$2p^3\,{}^5S^\circ$—$3s\,{}^5P$	2—1
121,913	3	12,83	114,53	$2p^3\,{}^5S^\circ$—$3s\,{}^5P$	2—2
121,773	4	12,83	114,64	$2p^3\,{}^5S^\circ$—$3s\,{}^5P$	2—3
121,004	1	29,92	132,38	$2p^3\,{}^3P^\circ$—$3d\,{}^3D$	0—1
120,973	2	29,92	132,40	$2p^3\,{}^3P^\circ$—$3d\,{}^3D$	1—2
120,931	3	29,92	132,44	$2p^3\,{}^3P^\circ$—$3d\,{}^3D$	2—3
119,684	3	38,59	142,29	$2p^3\,{}^1D^\circ$—$3d'\,{}^1D$ $2p^3\,{}^1D^\circ$—$3d'\,{}^1F$	2—2 2—3
119,204	0	38,59	142,71	$2p^3\,{}^1D^\circ$—$3d'\,{}^1P$	2—1
118,585	0	25,31	129,85	$2p^3\,{}^3D^\circ$—$3d\,{}^3P$	3—2
118,500	0	25,32	129,94	$2p^3\,{}^3D^\circ$—$3d\,{}^3P$	2—1
117,699	3	25,32	130,66	$2p^3\,{}^3D^\circ$—$3d\,{}^3F$	1—2
117,609	3	25,32	130,73	$2p^3\,{}^3D^\circ$—$3d\,{}^3F$	2—3
117,491	4	25,31	130,83	$2p^3\,{}^3D^\circ$—$3d\,{}^3F$	3—4
115,780	0	25,32	132,40	$2p^3\,{}^3D^\circ$—$3d\,{}^3D$	2—2
115,729	2	25,31	132,44	$2p^3\,{}^3D^\circ$—$3d\,{}^3D$	3—3
114,666	4	9,21	117,33	$2p^2\,{}^1S$—$3d\,{}^1P^\circ$	0—1
113,125	4	4,38	113,99	$2p^2\,{}^1D$—$3d\,{}^3F^\circ$	2—2
112,950	4	4,38	114,15	$2p^2\,{}^1D$—$3d\,{}^1D^\circ$	2—2
112,448	3	29,92	140,17	$2p^3\,{}^3P^\circ$—$3d'\,{}^3P$	0, 1, 2—0, 1, 2
112,009	3	29,92	140,60	$2p^3\,{}^3P^\circ$—$3d'\,{}^3D$	2—3
111,793	1	4,38	115,30	$2p^2\,{}^1D$—$3d\,{}^3D^\circ$	2—2
111,725	1	4,38	115,36	$2p^2\,{}^1D$—$3d\,{}^3D^\circ$	2—3
110,750	2	29,92	141,86	$2p^3\,{}^3P^\circ$—$3d'\,{}^3S$	0, 1, 2—1
109,896	5	4,38	117,20	$2p^2\,{}^1D$—$3d\,{}^1F^\circ$	2—3
109,763	0	4,38	117,33	$2p^2\,{}^1D$—$3d\,{}^1P^\circ$	2—1
108,678	0	0,08	114,15	$2p^2\,{}^3P$—$3d\,{}^1D^\circ$	1—2
108,555	4	25,32	139,51	$2p^3\,{}^3D^\circ$—$3d'\,{}^3F$	1, 2, 3—2, 3, 4
107,934	2	25,32	140,17	$2p^3\,{}^3D^\circ$—$3d'\,{}^3P$	1, 2, 3—0, 1, 2
107,742	2	0,23	115,30	$2p^2\,{}^3P$—$3d\,{}^3D^\circ$	2—2
107,683	5	0,23	115,36	$2p^2\,{}^3P$—$3d\,{}^3D^\circ$	2—3
107,608	4	0,08	115,30	$2p^2\,{}^3P$—$3d\,{}^3D^\circ$	1—2
107,553	3	0,00	115,27	$2p^2\,{}^3P$—$3d\,{}^3D^\circ$	0—1
107,535	3	25,32	140,60	$2p^3\,{}^3D^\circ$—$3d'\,{}^3D$	2, 3—1, 2, 3
107,288	4	0,23	115,78	$2p^2\,{}^3P$—$3d\,{}^3P^\circ$	2—2
107,227	3	0,23	115,85	$2p^2\,{}^3P$—$3d\,{}^3P^\circ$	2—1
107,158	1	0,08	115,78	$2p^2\,{}^3P$—$3d\,{}^3P^\circ$	1—2
107,093	3	0,08	115,85	$2p^2\,{}^3P$—$3d\,{}^3P^\circ$	1—1
107,061	3	0,08	115,89	$2p^2\,{}^3P$—$3d\,{}^3P^\circ$	1—0
107,014	2	0,00	115,85	$2p^2\,{}^3P$—$3d\,{}^3P^\circ$	0—1
106,580	0	12,83	129,16	$2p^3\,{}^5S^\circ$—$3d\,{}^5D$	2—2, 3
106,125	4	12,83	129,65	$2p^3\,{}^5S^\circ$—$3d\,{}^5P$	2—3
106,077	3	12,83	129,71	$2p^3\,{}^5S^\circ$—$3d\,{}^5P$	2—2
106,040	3	12,83	129,75	$2p^3\,{}^5S^\circ$—$3d\,{}^5P$	2—1
103,210	2	0,23	120,36	$2p^2\,{}^3P$—$3p\,{}^3S^\circ$	2—1
103,078	1	0,08	120,36	$2p^2\,{}^3P$—$3p\,{}^3S^\circ$	1—1
103,002	0	0,00	120,36	$2p^2\,{}^3P$—$3p\,{}^3S^\circ$	0—1

λ, Å	I	E_H, eV	E_B, eV	Transition	J
100,590	1	0,23	123,48	$2p^2\,^3P$—$3p\,^3D^\circ$	2—2
100,519	3	0,23	123,57	$2p^2\,^3P$—$3p\,^3D^\circ$	2—3
100,469	2	0,08	123,48	$2p^2\,^3P$—$3p\,^3D^\circ$	0, 1—1, 2
99,680	1	0,23	124,61	$2p^2\,^3P$—$3p\,^3P^\circ$	2—1
99,617	1	0,23	124,68	$2p^2\,^3P$—$3p\,^3P^\circ$	2—2
99,500	0	0,08	124,68	$2p^2\,^3P$—$3p\,^3P^\circ$	1—2
99,004	0	25,31	150,53	$2p^3\,^3D^\circ$—$4s\,^3P$	3—2
98,309	0	29,92	156,04	$2p^3\,^3P^\circ$—$4d\,^3D$	2—1, 2, 3
97,636	0	29,92	156,90	$2p^3\,^3P^\circ$—$3p^{IV}\,^3P$	0, 1, 2—0, 1, 2
96,475	3	4,38	132,89	$2p^2\,^1D$—$3p'\,^1F^\circ$	2—3
96,307	1	0,23	128,96	$2p^2\,^3P$—$3p'\,^3D^\circ$	2—1, 2, 3
96,196	1	0,08	128,96	$2p^2\,^3P$—$3p'\,^3D^\circ$	1—1, 2
96,124	0	0,00	128,96	$2p^2\,^3P$—$3p'\,^3D^\circ$	0—1
95,933	3	4,38	133,62	$2p^2\,^1D$—$3p'\,^1D^\circ$	2—2
95,319	0	25,32	155,39	$2p^3\,^3D^\circ$—$4d\,^3F$	1—2
95,263	1	25,32	155,46	$2p^3\,^3D^\circ$—$4d\,^3F$	2—3
95,182	1	25,31	155,56	$2p^3\,^3D^\circ$—$4d\,^3F$	3—4
94,827	0	25,32	156,04	$2p^3\,^3D^\circ$—$4d\,^3D$	1, 2, 3—1, 2, 3
94,208	1	25,31	156,90	$2p^3\,^3D^\circ$—$3p^{IV}\,^3P$	3—2
91,836	0	0,23	135,23	$2p^2\,^3P$—$4s\,^3P^\circ$	2—2
91,737	0	0,08	135,23	$2p^2\,^3P$—$4s\,^3P^\circ$	1—2
91,475	0	4,38	139,93	$2p^2\,^1D$—$4d\,^3F^\circ$	2—2
91,414	0	29,92	165,57	$2p^3\,^3P^\circ$—$4d'\,^3P$	0, 1, 2—0, 1, 2
91,268	1	4,38	140,22	$2p^2\,^1D$—$4d\,^1D^\circ$	2—2
90,746	0	12,83	149,45	$2p^3\,^5S^\circ$—$4s\,^5P$	2—3
90,468	3	4,38	141,42	$2p^2\,^1D$—$4d\,^1F^\circ$	2—3
88,460	1	25,32	165,46	$2p^3\,^3D^\circ$—$4d'\,^3F$	1, 2, 3—2, 3, 4
88,387	0	25,32	165,57	$2p^3\,^3D^\circ$—$4d'\,^3P$	1, 2, 3—0, 1, 2
88,340	1	0,23	140,57	$2p^2\,^3P$—$4d\,^3D^\circ$	2—2
88,270	3	{ 0,23	140,68	$2p^2\,^3P$—$4d\,^3D^\circ$	2—3
		{ 25,32	165,76	$2p^3\,^3D^\circ$—$4d'\,^3D$	1, 2, 3—1, 2, 3
88,246	2	0,08	140,57	$2p^2\,^3P$—$4d\,^3D^\circ$	1—2
88,223	1	0,00	140,53	$2p^2\,^3P$—$4d\,^3D^\circ$	0—1
88,143	2	0,23	140,88	$2p^2\,^3P$—$4d\,^3P^\circ$	2—2
88,038	1	0,08	140,88	$2p^2\,^3P$—$4d\,^3P^\circ$	1—1, 2
87,211	7	12,83	154,99	$2p^3\,^5S^\circ$—$4d\,^5P$	2—3
87,141	1	12,83	154,99	$2p^3\,^5S^\circ$—$4d\,^5P$	2—2
83,639	1	4,38	152,61	$2p^2\,^1D$—$5d\,^1F^\circ$	2—3
81,584	1	{ 25,32	177,27	$2p^3\,^3D^\circ$—$5d'\,^3F$	1, 2, 3—2, 3, 4
		{ 0,23	152,27	$2p^2\,^3P$—$5d\,^3D^\circ$	2—2
81,543	1	0,23	152,27	$2p^2\,^3P$—$5d\,^3D^\circ$	2—3
81,498	2	{ 0,08	152,27	$2p^2\,^3P$—$5d\,^3D^\circ$	1—2
		{ 0,23	152,35	$2p^2\,^3P$—$5d\,^3P^\circ$	2—2
80,645	0	12,83	166,56	$2p^3\,^5S^\circ$—$5d\,^5P$	2—3
80,345	0	4,38	158,69	$2p^2\,^1D$—$6d\,^1F^\circ$	2—3

Unclassified Lines of Sodium

λ, Å	I	Expected assignment	λ, Å	I	Expected assignment
1787,4	4	—	1658,7	40	—
1770,8	6	—	138,628	2	—
1749,3	8	—	128,112	3	—
1698,9	10	—	110,085	2	—
1669,3	40	—	95,551	2	—
1668,7	40	—	87,524	2	—

MAGNESIUM, Z = 12

Mg I, ground state $1s^2\,2s^2\,2p^6\,3s^2\,^1S_0$

Ionization potential 61671,02 cm^{-1}; 7,646 eV

λ, Å	I	E_H, eV	E_B, eV	Transition	J
26392,9	5	6,12	6,59	$4p\,^1P^\circ-4d\,^1D$	1—2
17108,66	30	5,39	6,12	$4s\,^1S-4p\,^1P^\circ$	0—1
15765,84	10	5,93	6,72	$4p\,^3P^\circ-4d\,^3D$	2—3
15748,99	8	5,93	6,72	$4p\,^3P^\circ-4d\,^3D$	1—2
15740,71	6	5,93	6,72	$4p\,^3P^\circ-4d\,^3D$	0—1
15047,70	25	5,11	5,93	$4s\,^3S-4p\,^3P^\circ$	1—0
15040,24	30	5,11	5,93	$4s\,^3S-4p\,^3P^\circ$	1—1
15024,99	35	5,11	5,93	$4s\,^3S-4p\,^3P^\circ$	1—2
14877,62	28	5,95	6,78	$3d\,^3D-4f\,^3F^\circ$	—
12083,66	30	5,75	6,78	$3d\,^1D-4f\,^1F^\circ$	2—3
11828,18	45	4,35	5,39	$3p\,^1P^\circ-4s\,^1S$	1—0
11033,661	14	5,95	7,07	$3d\,^3D-6p\,^3P^\circ$	2—1
11032,103	15	5,95	7,07	$3d\,^3D-6p\,^3P^\circ$	3—2
10965,450	28	5,93	7,06	$4p\,^3P^\circ-5d\,^3D$	2—3
10957,304	27	5,93	7,06	$4p\,^3P^\circ-5d\,^3D$	1—2
10953,320	25	5,93	7,06	$4p\,^3P^\circ-5d\,^3D$	0—1
10811,085	35	5,95	7,09	$3d\,^3D-5f\,^3F^\circ$	—
9993,209	18	5,93	7,17	$4p\,^3P^\circ-7s\,^3S$	2—1
9986,475	17	5,93	7,17	$4p\,^3P^\circ-7s\,^3S$	1—1
9983,20	15	5,93	7,17	$4p\,^3P^\circ-7s\,^3S$	0—1
9505,433	5	5,95	7,25	$3d\,^3D-7p\,^3P^\circ$	1—0
9503,108	7	5,95	7,25	$3d\,^3D-7p\,^3P^\circ$	2—1
9502,454	8	5,95	7,25	$3d\,^3D-7p\,^3P^\circ$	3—2
9438,783	20	5,93	7,25	$4p\,^3P^\circ-6d\,^3D$	2—3
9432,764	19	5,93	7,25	$4p\,^3P^\circ-6d\,^3D$	1—2
9429,814	17	5,93	7,25	$4p\,^3P^\circ-6d\,^3D$	0—1
9414,964	25	5,95	7,26	$3d\,^3D-6f\,^3F^\circ$	—
9255,778	30	5,75	7,09	$3d\,^1D-5f\,^1F^\circ$	2—3
9246,499	12	5,75	7,09	$3d\,^1D-6p\,^1P^\circ$	2—1
8997,156	10	5,93	7,31	$4p\,^3P^\circ-8s\,^3S$	2—1
8991,692	9	5,93	7,31	$4p\,^3P^\circ-8s\,^3S$	1—1
8989,026	7	5,93	7,31	$4p\,^3P^\circ-8s\,^3S$	0—1
8923,569	20	5,39	6,78	$4s\,^1S-5p\,^1P^\circ$	0—1
8806,757	50	4,35	5,75	$3p\,^1P^\circ-3d\,^1D$	1—2
8736,021	17	5,95	7,36	$3d\,^3D-7f\,^3F^\circ$	—
8717,825	13	5,93	7,36	$4p\,^3P^\circ-7d\,^3D$	2—3
8712,689	12	5,93	7,36	$4p\,^3P^\circ-7d\,^3D$	1—2
8710,175	10	5,93	7,36	$4p\,^3P^\circ-7d\,^3D$	0—1
8473,694	7	5,93	7,40	$4p\,^3P^\circ-9s\,^3S$	2—1
8468,845	5	5,93	7,40	$4p\,^3P^\circ-9s\,^3S$	1—1
8466,483	2	5,93	7,40	$4p\,^3P^\circ-9s\,^3S$	0—1
8346,120	15	5,95	7,43	$3d\,^3D-8f\,^3F^\circ$	—
8310,264	10	5,93	7,42	$4p\,^3P^\circ-8d\,^3D$	2—3
8305,596	9	5,93	7,42	$4p\,^3P^\circ-8d\,^3D$	1—2
8303,313	7	5,93	7,42	$4p\,^3P^\circ-8d\,^3D$	0—1
8213,034	20	5,75	7,26	$3d\,^1D-6f\,^1F^\circ$	2—3
8209,839	10	5,75	7,26	$3d\,^1D-7p\,^1P^\circ$	2—1
8159,132	2	5,93	7,45	$4p\,^3P^\circ-10s\,^3S$	2—1
8154,644	1	5,93	7,45	$4p\,^3P^\circ-10s\,^3S$	1—1
8098,724	10	5,95	7,48	$3d\,^3D-9f\,^3F^\circ$	—
8054,232	7	5,93	7,47	$4p\,^3P^\circ-9d\,^3D$	2—1, 2, 3
8049,854	5	5,93	7,47	$4p\,^3P^\circ-9d\,^3D$	1—1, 2
8047,73	3	5,93	7,47	$4p\,^3P^\circ-9d\,^3D$	0—1

λ, Å	I	$E_{\text{н}}$, eV	$E_{\text{в}}$, eV	Transition	J
7930,806	7	5,95	7,51	$3d\ ^3D-10f\ ^3F^\circ$	—
7881,667	2	5,93	7,50	$4p\ ^3P^\circ-10d\ ^3D$	2— 1, 2, 3
7811,135	3	5,95	7,53	$3d\ ^3D-11f\ ^3F^\circ$	—
7759,297	1	5,93	7,53	$4p\ ^3P^\circ-11d\ ^3D$	2—1, 2, 3
7746,343	1	4,35	5,95	$3p\ ^1P^\circ-3d\ ^3D$	1—2
7722,614	1	5,95	7,55	$3d\ ^3D-12f\ ^3F^\circ$	—
7691,550	15	5,75	7,36	$3d\ ^1D-7f\ ^1F^\circ$	2—3
7690,165	8	5,75	7,36	$3d\ ^1D-8p\ ^1P^\circ$	2—1
7659,902	17	5,11	6,73	$4s\ ^3S-5p\ ^3P^\circ$	1—0
7659,152	19	5,11	6,73	$4s\ ^3S-5p\ ^3P^\circ$	1—1
7657,603	20	5,11	6,73	$4s\ ^3S-5p\ ^3P^\circ$	1—2
7486,225	5	—	—	—	—
7387,685	12	5,75	7,43	$3d\ ^1D-8f\ ^1F^\circ$	2—3
7387,004	5	5,75	7,43	$3d\ ^1D-9p\ ^1P^\circ$	2—1
7291,060	10	5,39	7,09	$4s\ ^1S-6p\ ^1P^\circ$	0—1
7193,172	10	5,75	7,48	$3d\ ^1D-9f\ ^1F^\circ$	2—3
7060,409	8	5,75	7,51	$3d\ ^1D-10f\ ^1F^\circ$	2—3
6972,674		—	—	—	—
6965,404	6	5,75	7,53	$3d\ ^1D-11f\ ^1F^\circ$	2—3
6894,898	4	5,75	7,55	$3d\ ^1D-12f\ ^1F^\circ$	2—3
6630,834	2	5,39	7,26	$4s\ ^1S-7p\ ^1P$	0—1
6406,619	6	—	—	—	—
6319,493	7	5,11	7,07	$4s\ ^3S-6p\ ^3P^\circ$	1—0
6319,236	9	5,11	7,07	$4s\ ^3S-6p\ ^3P^\circ$	1—1
6318,716	10	5,11	7,07	$4s\ ^3S-6p\ ^3P^\circ$	1—2
6256,750	7	—	—	—	—
6208,440	3	—	—	—	—
5839,820	3	—	—	—	—
5785,560	4	5,11	7,25	$4s\ ^3S-7p\ ^3P^\circ$	1—1
5785,312	5	5,11	7,25	$4s\ ^3S-7p\ ^3P^\circ$	1—2
5711,0880	30	4,35	6,52	$3p\ ^1P^\circ-5s\ ^1S$	1—0
5528,4047	40	4,35	6,59	$3p\ ^1P^\circ-4d\ ^1D$	1—2
5509,597	2	5,11	7,36	$4s\ ^3S-8p\ ^3P^\circ$	1—0, 1, 2
5345,977	1	5,11	7,43	$4s\ ^3S-9p\ ^3P^\circ$	1—0, 1, 2
5183,6042	45	2,72	5,11	$3p\ ^3P^\circ-4s\ ^3S$	2—1
5172,6843	44	2,71	5,11	$3p\ ^3P^\circ-4s\ ^3S$	1—1
5167,3216	42	2,71	5,11	$3p\ ^3P^\circ-4s\ ^3S$	0—1
4730,0285	10	4,35	6,97	$3p\ ^1P^\circ-6s\ ^1S$	1—0
4702,9909	30	4,35	6,98	$3p\ ^1P^\circ-5d\ ^1D$	1—2
4621,299	3	2,71	5,39	$3p\ ^3P^\circ-4s\ ^1S$	1—0
4571,0956	28	0,00	2,71	$3s^2\ ^1S-3p\ ^3P^\circ$	0—1
4409,84	1	7,19	10,00	$6d\ ^1D-3d\ ^1D^\circ$	2—2
4380,375	6	4,35	7,18	$3p\ ^1P^\circ-3p^2\ ^3P$	1—2
4354,529	6	4,35	7,19	$3p\ ^1P^\circ-7s\ ^1S$	1—0
4351,9056	20	4,35	7,19	$3p\ ^1P^\circ-6d\ ^1D$	1—2
4212,497	2	—	—	—	—
4205,096	2	—	—	—	—
4177,109	2	—	—	—	—
4167,2712	15	4,35	7,31	$3p\ ^1P^\circ-7d\ ^1D$	1—2
4165,101	4	4,35	7,31	$3p\ ^1P^\circ-8s\ ^1S$	1—0
4099,77	2	6,98	10,00	$3d\ ^1D^\circ-3d\ ^1D^\circ$	2—2
4081,833	2	2,72	5,75	$3p\ ^3P^\circ-3d\ ^1D$	2—2
4075,059	3	2,71	5,75	$3p\ ^3P^\circ-3d\ ^1D$	1—2
4057,5052	10	4,35	7,40	$3p\ ^1P^\circ-8d\ ^1D$	1—2
4054,689	2	4,35	7,40	$3p\ ^1P^\circ-9s\ ^1S$	1—0
3986,7533	8	4,35	7,45	$3p\ ^1P^\circ-9d\ ^1D$	1—2
3984,212	1	4,35	7,46	$3p\ ^1P^\circ-10s\ ^1S$	1—0
3938,400	6	4,35	7,49	$3p\ ^1P^\circ-10d\ ^1D$	1—2

λ, Å	I	E_H, eV	E_B, eV	Transition	J
3903,859	4	4,35	7,52	$3p\ ^1P^\circ$—$11d\ ^1D$	1—2
3899,542	1	7,18	10,35	$3p^2\ ^3P$—$3d\ ^3D^\circ$	2—1
3898,120	4	7,18	10,35	$3p^2\ ^3P$—$3d\ ^3D^\circ$	2—2
3895,572	3	7,18	10,36	$3p^2\ ^3P$—$3d\ ^3D^\circ$	2—3
3893,304	2	7,17	10,35	$3p^2\ ^3P$—$3d\ ^3D^\circ$	1—1
3891,906	2	7,17	10,35	$3p^2\ ^3P$—$3d\ ^3D^\circ$	1—2
3890,241	3	7,17	10,35	$3p^2\ ^3P$—$3d\ ^3D^\circ$	0—1
3878,306	3	4,35	7,54	$3p\ ^1P^\circ$—$12d\ ^1D$	1—2
3858,860	2	4,35	7,56	$3p\ ^1P^\circ$—$13d\ ^1D$	1—2
3854,965	1	2,72	5,93	$3p\ ^3P^\circ$—$4p\ ^3P^\circ$	2—1
3853,960	2	2,72	5,93	$3p\ ^3P^\circ$—$4p\ ^3P^\circ$	2—2
3848,914	1	2,71	5,93	$3p\ ^3P^\circ$—$4p\ ^3P^\circ$	1—1
3838,2943	20	2,72	5,95	$3p\ ^3P^\circ$—$3d\ ^3D$	2—2
3838,2918	20	2,72	5,95	$3p\ ^3P^\circ$—$3d\ ^3D$	2—3
3832,3037	20	2,71	5,95	$3p\ ^3P^\circ$—$3d\ ^3D$	1—2
3832,2996	18	2,71	5,95	$3p\ ^3P^\circ$—$3d\ ^3D$	1—1
3829,3549	36	2,71	5,95	$3p\ ^3P^\circ$—$3d\ ^3D$	0—1
3627,63	4	6,59	10,00	$4d\ ^1D$—$3d\ ^1D^\circ$	2—3
3515,602	4	—	—	—	—
3453,616	3	—	—	—	—
3444,409	2	—	—	—	—
3336,674	20	2,72	6,43	$3p\ ^3P^\circ$—$5s\ ^3S$	2—1
3332,146	19	2,71	6,43	$3p\ ^3P^\circ$—$5s\ ^3S$	1—1
3329,919	17	2,71	6,43	$3p\ ^3P^\circ$—$5s\ ^3S$	0—1
3299,050	4	—	—	—	—
3237,266	3	—	—	—	—
3201,796	3	2,72	6,59	$3p\ ^3P^\circ$—$4d\ ^1D$	2—2
3197,625	2	2,71	6,59	$3p\ ^3P^\circ$—$4d\ ^1D$	1—2
3126,380	2	—	—	—	—
3096,890	24	2,72	6,72	$3p\ ^3P^\circ$—$4d\ ^3D$	2—2
3092,984	22	2,71	6,72	$3p\ ^3P^\circ$—$4d\ ^3D$	1—2
3091,065	20	2,71	6,72	$3p\ ^3P^\circ$—$4d\ ^3D$	0—1
2941,995	13	2,72	6,93	$3p\ ^3P^\circ$—$6s\ ^3S$	2—1
2938,473	12	2,71	6,93	$3p\ ^3P^\circ$—$6s\ ^3S$	1—1
2936,739	10	2,71	6,93	$3p\ ^3P^\circ$—$6s\ ^3S$	0—1
2915,453	3	5,75	10,00	$3d\ ^1D$—$3d\ ^1D^\circ$	2—2
2906,360	4	2,72	6,98	$3p\ ^3P^\circ$—$5d\ ^1D$	2—2
2902,923	2	2,71	6,98	$3p\ ^3P^\circ$—$5d\ ^1D$	1—2
2852,127	50	0,00	4,35	$3s^2\ ^1S$—$3p\ ^1P^\circ$	0—1
2851,660	16	2,72	7,06	$3p\ ^3P^\circ$—$5d\ ^3D$	2—3
2848,342	14	2,71	7,06	$3p\ ^3P^\circ$—$5d\ ^3D$	1—2
2846,716	12	2,71	7,06	$3p\ ^3P^\circ$—$5d\ ^3D$	0—1
2842,647	6	—	—	—	—
2811,781	1	5,95	10,35	$3d\ ^3D$—$3d\ ^3D^\circ$	1, 2—1
2811,112	2	5,95	10,35	$3d\ ^3D$—$3d^3\ D^\circ$	1, 2, 3—2
2809,761	3	5,95	10,36	$3d\ ^3D$—$3d\ ^3D^\circ$	2, 3—3
2782,972	18	2,72	7,17	$3p\ ^3P^\circ$—$3p^2\ ^3P$	2—1
2781,416	18	2,71	7,17	$3p\ ^3P^\circ$—$3p^2\ ^3P$	1—0
2781,228	8	2,72	7,17	$3p\ ^3P^\circ$—$7s\ ^3S$	2—1
2779,831	20	2,72	7,18	$3p\ ^3P^\circ$—$3p^2\ ^3P$	2—2
		2,71	7,17	$3p\ ^3P^\circ$—$3p^2\ ^3P$	1—1
2778,270	18	2,71	7,17	$3p\ ^3P^\circ$—$3p^2\ ^3P$	0—1
		2,71	7,17	$3p\ ^3P^\circ$—$7s\ ^3S$	1—1
2776,690	18	2,71	7,18	$3p\ ^3P^\circ$—$3p^2\ ^3P$	1—2
		2,71	7,17	$3p\ ^3P^\circ$—$7s\ ^3S$	0—1
2768,339	7	2,72	7,19	$3p\ ^3P^\circ$—$6d\ ^1D$	2—2
2765,222	5	2,71	7,19	$3p\ ^3P^\circ$—$6d\ ^1D$	1—2
2736,542	12	2,72	7,25	$3p\ ^3P^\circ$—$6d\ ^3D$	2—3
2733,493	10	2,71	7,25	$3p\ ^3P^\circ$—$6d\ ^3D$	1—1, 2

λ, Å	I	E_H, eV	E_B, eV	Transition	J
2731,993	8	2,71	7,25	$3p\ ^3P^\circ-6d\ ^3D$	0—1
2698,145	6	2,72	7,31	$3p\ ^3P^\circ-8s\ ^3S$	2—1
2695,181	5	2,71	7,31	$3p\ ^3P^\circ-8s\ ^3S$	1—1
2693,723	3	2,71	7,31	$3p\ ^3P^\circ-8s\ ^3S$	0—1
2692,45	2	2,72	7,32	$3p\ ^3P^\circ-7d\ ^1D$	2—2
2689,49	1	2,71	7,32	$3p\ ^3P^\circ-7d\ ^1D$	1—2
2672,460	10	2,72	7,36	$3p\ ^3P^\circ-7d\ ^3D$	2—3
2669,553	8	2,71	7,36	$3p\ ^3P^\circ-7d\ ^3D$	1—1, 2
2668,124	6	2,71	7,36	$3p\ ^3P^\circ-7d\ ^3D$	0—1
2649,062	4	2,72	7,40	$3p\ ^3P^\circ-9s\ ^3S$	2—1
2646,206	3	2,71	7,40	$3p\ ^3P^\circ-9s\ ^3S$	1—1
2644,801	2	2,71	7,40	$3p\ ^3P^\circ-9s\ ^3S$	0—1
2632,873	8	2,72	7,42	$3p\ ^3P^\circ-8d\ ^3D$	2—3
2630,053	6	2,71	7,42	$3p\ ^3P^\circ-8d\ ^3D$	1—1, 2
2628,664	3	2,71	7,42	$3p\ ^3P^\circ-8d\ ^3D$	0—1
2617,513	3	2,72	7,45	$3p\ ^3P^\circ-10s\ ^3S$	2—1
2614,726	2	2,71	7,45	$3p\ ^3P^\circ-10s\ ^3S$	1—1
2613,357	1	2,71	7,45	$3p\ ^3P^\circ-10s\ ^3S$	0—1
2606,621	5	2,72	7,47	$3p\ ^3P-9d\ ^3D$	2—1, 2, 3
2603,854	4	2,71	7,47	$3p\ ^3P^\circ-9d\ ^3D$	1—1, 2
2602,495	2	2,71	7,47	$3p\ ^3P^\circ-9d\ ^3D$	0—1
2595,973	2	2,72	7,49	$3p\ ^3P^\circ-11s\ ^3S$	2—1
2593,231	1	2,71	7,49	$3p\ ^3P^\circ-11s\ ^3S$	1—1
2591,891	0	2,71	7,49	$3p\ ^3P^\circ-11s\ ^3S$	0—1
2588,285	3	2,72	7,50	$3p\ ^3P^\circ-10d\ ^3D$	2—1, 2, 3
2585,558	2	2,71	7,50	$3p\ ^3P^\circ-10d\ ^3D$	1—1, 2
2584,216	1	2,71	7,50	$3p\ ^3P^\circ-10d\ ^3D$	0—1
2580,587	1	2,72	7,52	$3p\ ^3P^\circ-12s\ ^3S$	2—1
2577,888	0	2,71	7,52	$3p\ ^3P^\circ-12s\ ^3S$	1—1
2574,945	2	2,72	7,53	$3p\ ^3P^\circ-11d\ ^3D$	2—1, 2, 3
2572,248	1	2,71	7,53	$3p\ ^3P^\circ-11d\ ^3D$	1—1, 2
2570,908	0	2,71	7,53	$3p\ ^3P^\circ-11d\ ^3D$	0—1
2564,937	1	2,72	7,55	$3p\ ^3P^\circ-12d\ ^3D$	2—1, 2, 3
2562,259	1	2,71	7,55	$3p\ ^3P^\circ-12d\ ^3D$	1—1, 2
2560,941	0	2,71	7,55	$3p\ ^3P^\circ-12d\ ^3D$	0—1
2557,226	0	2,72	7,56	$3p\ ^3P^\circ-13d\ ^3D$	2—1, 2, 3
2395,150	4	—	—	—	—
2025,824	9	0,00	6,12	$3s^2\ ^1S-4p\ ^1P^\circ$	0—1
1827,97	8	0,00	6,78	$3s^2\ ^1S-5p\ ^1P^\circ$	0—1
1747,81	5	0,00	7,09	$3s^2\ ^1S-6p\ ^1P^\circ$	0—1
1707,10	3	0,00	7,26	$3s^2\ ^1S-7p\ ^1P^\circ$	0—1
1683,51	1	0,00	7,36	$3s^2\ ^1S-8p\ ^1P^\circ$	0—1
1668,57	0	0,00	7,43	$3s^2\ ^1S-9p\ ^1P^\circ$	0—1

Mg II, ground state $1s^2\ 2s^2\ 2p^6\ 3s\ ^2S_{1/2}$
Ionization potential 121267,41 cm^{-1}; 15,034 eV

λ, Å	I	E_H, eV	E_B, eV	Transition	J
21432,11	5	11,50	12,08	$5s\ ^2S-5p\ ^2P^\circ$	$^1/_2-^1/_2$
21368,91	7	11,50	12,08	$5s\ ^2S-5p\ ^2P^\circ$	$^1/_2-^3/_2$
18622,68	25	12,86	13,52	$5g\ ^2G-6h\ ^2H^\circ$	—
18574,80	20	12,86	13,52	$5f\ ^2F^\circ-6g\ ^2G$	—
17717,72	15	12,82	13,52	$5d\ ^2D-6f\ ^2F^\circ$	—
11620,14	3	12,86	13,92	$5g\ ^2G-7h\ ^2H^\circ$	—
11600,56	3	12,86	13,92	$5f\ ^2F^\circ-7g\ ^2G$	—

λ, Å	I	$E_н$, eV	$E_в$, eV	Transition	J
11256,35	4	12,82	13,92	$5d\ ^2D-7f\ ^2F°$	$^3/_2-^5/_2$
11255,93	5	12,82	13,92	$5d\ ^2D-7f\ ^2F°$	$^5/_2-^7/_2$
10951,78	10	8,86	10,00	$3d\ ^2D-4p\ ^2P°$	$^3/_2-^1/_2$
10915,27	7	8,86	10,00	$3d\ ^2D-4p\ ^2P°$	$^3/_2-^3/_2$
10914,23	10	8,86	10,00	$3d\ ^2D-4p\ ^2P°$	$^5/_2-^3/_2$
10392,23	6	11,63	12,82	$4f\ ^2F°-5d\ ^2D$	$^7/_2-^5/_2$
10391,76	5	11,63	12,82	$4f\ ^2F°-5d\ ^2D$	$^5/_2-^5/_2$
10092,16	14	11,63	12,86	$4f\ ^2F°-5g\ ^2G$	—
9632,435	11	11,57	12,86	$4d\ ^2D-5f\ ^2F°$	$^3/_2-^5/_2$
9631,888	12	11,57	12,86	$4d\ ^2D-5f\ ^2F°$	$^5/_2-^7/_2$
9340,544	10	12,86	14,18	$5g\ ^2G-8h\ ^2H°$	—
9327,545	10	12,86	14,18	$5f\ ^2F°-8g\ ^2G$	—
9244,266	13	8,65	10,00	$4s\ ^2S-4p\ ^2P°$	$^1/_2-^1/_2$
9218,248	14	8,65	10,00	$4s\ ^2S-4p\ ^2P°$	$^1/_2-^3/_2$
8835,082	11	12,08	13,49	$5p\ ^2P°-7s\ ^2S$	$^3/_2-^1/_2$
8824,323	10	12,08	13,49	$5p\ ^2P°-7s\ ^2S$	$^1/_2-^1/_2$
8745,657	11	12,08	13,50	$5p\ ^2P°-6d\ ^2D$	$^3/_2-^5/_2$
8734,990	10	12,08	13,50	$5p\ ^2P°-6d\ ^2D$	$^1/_2-^3/_2$
8234,639	11	10,00	11,50	$4p\ ^2P°-5s\ ^2S$	$^3/_2-^1/_2$
8233,194	7	12,86	14,36	$5g\ ^2G-9h\ ^2H°$	—
8222,924	7	12,86	14,36	$5f\ ^2F°-9g\ ^2G$	—
8213,989	10	10,00	11,50	$4p\ ^2P°-5s\ ^2S$	$^1/_2-^1/_2$
8120,434	8	11,57	13,09	$4d\ ^2D-6p\ ^2P°$	$^3/_2-^1/_2$
8115,220	9	11,57	13,10	$4d\ ^2D-6p\ ^2P°$	$^5/_2-^3/_2$
7896,368	13	10,00	11,57	$4p\ ^2P°-4d\ ^2D$	$^3/_2-^5/_2$
7877,051	12	10,00	11,57	$4p\ ^2P°-4d\ ^2D$	$^1/_2-^3/_2$
7790,978	4	11,50	13,09	$5s\ ^2S-6p\ ^2P°$	$^1/_2-^1/_2$
7786,500	5	11,50	13,10	$5s\ ^2S-6p\ ^2P°$	$^1/_2-^3/_2$
7589,558	3	12,86	14,49	$5g\ ^2G-10h\ ^2H°$	—
7580,764	4	12,86	14,49	$5f\ ^2F°-10g\ ^2G$	—
7166,676	2	12,86	14,58	$5f\ ^2F°-11g\ ^2G$	—
6819,270	8	12,08	13,90	$5p\ ^2P°-8s\ ^2S$	$^3/_2-^1/_2$
6812,860	7	12,08	13,90	$5p\ ^2P°-8s\ ^2S$	$^1/_2-^1/_2$
6787,851	8	12,08	13,91	$5p\ ^2P°-7d\ ^2D$	$^3/_2-^5/_2$
6781,451	7	12,08	13,91	$5p\ ^2P°-7d\ ^2D$	$^1/_2-^3/_2$
6620,569	6	11,63	13,50	$4f\ ^2F°-6d\ ^2D$	$^7/_2-^5/_2$
6620,440	5	11,63	13,50	$4f\ ^2F°-6d\ ^2D$	$^5/_2-^3/_2$
6545,973	11	11,63	13,52	$4f\ ^2F°-6g\ ^2G$	—
6346,962	9	11,57	13,52	$4d\ ^2D-6f\ ^2F°$	$^3/_2-^5/_2$
6346,737	10	11,57	13,52	$4d\ ^2D-6f\ ^2F°$	$^5/_2-^7/_2$
5943,499	4	12,08	14,17	$5p\ ^2P°-9s\ ^2S$	$^3/_2-^1/_2$
5938,629	3	12,08	14,17	$5p\ ^2P°-9s\ ^2S$	$^1/_2-^1/_2$
5928,233	4	12,08	14,18	$5p\ ^2P°-8d\ ^2D$	$^3/_2-^5/_2$
5923,366	3	12,08	14,18	$5p\ ^2P°-8d\ ^2D$	$^1/_2-^3/_2$
5918,158	6	11,57	13,66	$4d\ ^2D-7p\ ^2P°$	$^3/_2-^1/_2$
5916,429	7	11,57	13,66	$4d\ ^2D-7p\ ^2P°$	$^5/_2-^3/_2$
5464,136	2	12,08	14,35	$5p\ ^2P°-10s\ ^2S$	$^3/_2-^1/_2$
5460,019	1	12,08	14,35	$5p\ ^2P°-10s\ ^2S$	$^1/_2-^1/_2$
5451,259	1	12,08	14,35	$5p\ ^2P°-9d\ ^2D$	$^1/_2-^3/_2$
5434,039	4	11,63	13,91	$4f\ ^2F°-7d\ ^2D$	—
5401,543	9	11,63	13,92	$4f\ ^2F°-7g\ ^2G$	—
5264,368	7	11,57	13,92	$4d\ ^2D-7f\ ^2F°$	$^5/_2-^5/_2$
5264,215	8	11,57	13,92	$4d\ ^2D-7f\ ^2F°$	$^5/_2-^7/_2$
5069,802	3	11,57	14,01	$4d\ ^2D-8p\ ^2P°$	$^3/_2-^1/_2$
5068,937	4	11,57	14,01	$4d\ ^2D-8p\ ^2P°$	$^5/_2-^3/_2$
4868,845	2	11,63	14,18	$4f\ ^2F°-8d\ ^2D$	—
4851,082	7	11,63	14,18	$4f\ ^2F°-8g\ ^2G$	—
4739,712	5	11,57	14,18	$4d\ ^2D-8f\ ^2F°$	$^3/_2-^5/_2$

λ, Å	I	E_H, eV	E_B, eV	Transition	J
4739,588	6	11,57	14,18	$4d\ ^2D-8f\ ^2F^\circ$	$^5/_2-^7/_2$
4631,405	1	11,57	14,24	$4d\ ^2D-9p\ ^2P^\circ$	$^3/_2-^1/_2$
4630,878	2	11,57	14,24	$4d\ ^2D-9p\ ^2P^\circ$	$^5/_2-^3/_2$
4534,291	6	11,63	14,36	$4f\ ^2F^\circ-9g\ ^2G$	—
4481,327	13	8,86	11,63	$3d\ ^2D-4f\ ^2F^\circ$	$^3/_2-^5/_2$
4481,130	14	8,86	11,63	$3d\ ^2D-4f\ ^2F^\circ$	$^5/_2-^7/_2$
4436,598	4	11,57	14,36	$4d\ ^2D-9f\ ^2F^\circ$	$^3/_2-^5/_2$
4436,486	5	11,57	14,36	$4d\ ^2D-9f\ ^2F^\circ$	$^5/_2-^7/_2$
4433,990	9	10,00	12,79	$4p\ ^2P^\circ-6s\ ^2S$	$^3/_2-^1/_2$
4427,994	8	10,00	12,79	$4p\ ^2P^\circ-6s\ ^2S$	$^1/_2-^1/_2$
4390,564	10	10,00	12,82	$4p\ ^2P^\circ-5d\ ^2D$	$^3/_2-^5/_2$
4384,637	9	10,00	12,82	$4p\ ^2P^\circ-5d\ ^2D$	$^1/_2-^3/_2$
4331,945	3	11,63	14,49	$4f\ ^2F^\circ-10g\ ^2G$	—
4242,543	2	11,57	14,49	$4d\ ^2D-10f\ ^2F^\circ$	$^3/_2-^5/_2$
4242,445	3	11,57	14,49	$4d\ ^2D-10f\ ^2F^\circ$	$^5/_2-^7/_2$
4193,482	2	11,63	14,58	$4f\ ^2F^\circ-11g\ ^2G$	—
4109,54	3	11,57	14,58	$4d\ ^2D-11f\ ^2F^\circ$	—
4093,90	1	11,63	14,66	$4f\ ^2F^\circ-12g\ ^2G$	—
4013,80	2	11,57	14,66	$4d\ ^2D-12f\ ^2F^\circ$	—
3850,385	7	8,86	12,08	$3d\ ^2D-5p\ ^2P^\circ$	$^3/_2-^1/_2$
3848,209	8	8,86	12,08	$3d\ ^2D-5p\ ^2P^\circ$	$^5/_2-^3/_2$
3615,583	3	8,65	12,08	$4s\ ^2S-5p\ ^2P^\circ$	$^1/_2-^1/_2$
3613,781	4	8,65	12,08	$4s\ ^2S-5p\ ^2P^\circ$	$^1/_2-^3/_2$
3553,366	8	10,00	13,49	$4p\ ^2P^\circ-7s\ ^2S$	$^3/_2-^1/_2$
3549,516	7	10,00	13,49	$4p\ ^2P^\circ-7s\ ^2S$	$^1/_2-^1/_2$
3538,813	8	10,00	13,50	$4p\ ^2P^\circ-6d\ ^2D$	$^3/_2-^5/_2$
3534,972	7	10,00	13,50	$4p\ ^2P^\circ-6d\ ^2D$	$^1/_2-^3/_2$
3175,783	7	10,00	13,90	$4p\ ^2P^\circ-8s\ ^2S$	$^3/_2-^1/_2$
3172,706	6	10,00	13,90	$4p\ ^2P^\circ-8s\ ^2S$	$^1/_2-^1/_2$
3168,951	6	10,00	13,91	$4p\ ^2P^\circ-7d\ ^2D$	$^3/_2-^5/_2$
3165,878	2	10,00	13,91	$4p\ ^2P^\circ-7d\ ^2D$	$^1/_2-^3/_2$
3104,809	8	8,86	12,86	$3d\ ^2D-5f\ ^2F^\circ$	$^3/_2-^5/_2$
3104,722	9	8,86	12,86	$3d\ ^2D-5f\ ^2F^\circ$	$^5/_2-^7/_2$
2971,839	1	10,00	14,17	$4p\ ^2P^\circ-9s\ ^2S$	$^3/_2-^1/_2$
2969,145	0	10,00	14,17	$4p\ ^2P^\circ-9s\ ^2S$	$^1/_2-^1/_2$
2968,020	2	10,00	14,18	$4p\ ^2P^\circ-8d\ ^2D$	$^3/_2-^3/_2$
2967,87	1	10,00	14,18	$4p\ ^2P^\circ-8d\ ^2D$	$^3/_2-^5/_2$
2965,19	0	10,00	14,18	$4p\ ^2P^\circ-8d\ ^2D$	$^1/_2-^3/_2$
2936,509	10	4,43	8,65	$3p\ ^2P^\circ-4s\ ^2S$	$^3/_2-^1/_2$
2928,634	9	4,42	8,65	$3p\ ^2P^\circ-4s\ ^2S$	$^1/_2-^1/_2$
2802,704	12	0,00	4,42	$3s\ ^2S-3p\ ^2P^\circ$	$^1/_2-^1/_2$
2797,998	10	4,43	8,86	$3p\ ^2P^\circ-3d\ ^2D$	$^3/_2-^5/_2$
2795,528	13	0,00	4,43	$3s\ ^2S-3p\ ^2P^\circ$	$^1/_2-^3/_2$
2790,776	9	4,42	8,86	$3p\ ^2P^\circ-3d\ ^2D$	$^1/_2-^3/_2$
2660,817	8	8,86	13,52	$3d\ ^2D-6f\ ^2F^\circ$	$^3/_2-^5/_2$
2660,755	8	8,86	13,52	$3d\ ^2D-6f\ ^2F^\circ$	$^5/_2-^5/_2,\ ^7/_2$
2449,590	6	8,86	13,92	$3d\ ^2D-7f\ ^2F^\circ$	—
2329,578	3	8,86	14,18	$3d\ ^2D-8f\ ^2F^\circ$	—
1753,474	60	4,43	11,50	$3p\ ^2P^\circ-5s\ ^2S$	$^3/_2-^1/_2$
1750,664	50	4,42	11,50	$3p\ ^2P^\circ-5s\ ^2S$	$^1/_2-^1/_2$
1737,628	10	4,43	11,57	$3p\ ^2P^\circ-4d\ ^2D$	$^3/_2-^5/_2$
1737,612	10	4,43	11,57	$3p\ ^2P^\circ-4d\ ^2D$	$^3/_2-^3/_2$
1734,852	10	4,42	11,57	$3p\ ^2P^\circ-4d\ ^2D$	$^1/_2-^3/_2$
1482,890	—	4,43	12,79	$3p\ ^2P^\circ-6s\ ^2S$	$^3/_2-^1/_2$
1480,880	—	4,42	12,79	$3p\ ^2P^\circ-6s\ ^2S$	$^1/_2-^1/_2$
1478,004	—	4,43	12,82	$3p\ ^2P^\circ-5d\ ^2D$	$^3/_2-^5/_2$
1477,997	—	4,43	12,82	$3p\ ^2P^\circ-5d\ ^2D$	$^3/_2-^3/_2$
1476,000	—	4,42	12,82	$3p\ ^2P^\circ-5d\ ^2D$	$^1/_2-^3/_2$

λ, Å	I	E_H, eV	E_B, eV	Transition	J
1369,423	—	4,43	13,49	$3p\ ^2P^\circ$—$7s\ ^2S$	$^3/_2$—$^1/_2$
1367,708	—	4,42	13,49	$3p\ ^2P^\circ$—$7s\ ^2S$	$^1/_2$—$^1/_2$
1367,256	—	4,43	13,50	$3p\ ^2P^\circ$—$6d\ ^2D$	$^3/_2$—$^3/_2$, $^5/_2$
1365,544	—	4,42	13,50	$3p\ ^2P^\circ$—$6d\ ^2D$	$^1/_2$—$^3/_2$
1309,443	—	4,43	13,90	$3p\ ^2P^\circ$—$8s\ ^2S$	$^3/_2$—$^1/_2$
1308,280	—	4,43	13,91	$3p\ ^2P^\circ$—$7d\ ^2D$	$^3/_2$—$^3/_2$, $^5/_2$
1307,875	—	4,42	13,90	$3p\ ^2P^\circ$—$8s\ ^2S$	$^1/_2$—$^1/_2$
1306,714	—	4,42	13,91	$3p\ ^2P^\circ$—$7d\ ^2D$	$^1/_2$—$^3/_2$
1273,423	—	4,43	14,17	$3p\ ^2P^\circ$—$9s\ ^2S$	$^3/_2$—$^1/_2$
1272,720	—	4,43	14,18	$3p\ ^2P^\circ$—$8d\ ^2D$	$^3/_2$—$^3/_2$, $^5/_2$
1271,940	—	4,42	14,17	$3p\ ^2P^\circ$—$9s\ ^2S$	$^1/_2$—$^1/_2$
1271,239	—	4,42	14,18	$3p\ ^2P^\circ$—$8d\ ^2D$	$^1/_2$—$^3/_2$
1240,395	—	0,00	10,00	$3s\ ^2S$—$4p\ ^2P^\circ$	$^1/_2$—$^1/_2$
1239,925	—	0,00	10,00	$3s\ ^2S$—$4p\ ^2P^\circ$	$^1/_2$—$^3/_2$
1026,113	—	0,00	12,08	$3s\ ^2S$—$5p\ ^2P^\circ$	$^1/_2$—$^1/_2$
1025,968	—	0,00	12,08	$3s\ ^2S$—$5p\ ^2P^\circ$	$^1/_2$—$^3/_2$
946,769	—	0,00	13,09	$3s\ ^2S$—$6p\ ^2P^\circ$	$^1/_2$—$^1/_2$
946,703	—	0,00	13,10	$3s\ ^2S$—$6p\ ^2P^\circ$	$^1/_2$—$^3/_2$
907,412	—	0,00	13,66	$3s\ ^2S$—$7p\ ^2P^\circ$	$^1/_2$—$^1/_2$
907,375	—	0,00	13,66	$3s\ ^2S$—$7p\ ^2P^\circ$	$^1/_2$—$^3/_2$
884,719	—	0,00	14,01	$3s\ ^2S$—$8p\ ^2P^\circ$	$^1/_2$—$^1/_2$
884,697	—	0,00	14,01	$3s\ ^2S$—$8p\ ^2P^\circ$	$^1/_2$—$^3/_2$

Mg III, ground state $1s^2\ 2s^2\ 2p^6\ ^1S_0$

Ionization potential 646 364 cm⁻¹; 80,134 eV

λ, Å	I	E_H, eV	E_B, eV	Transition	J
2529,97	2	53,05	57,95	$3s'\ [^1/_2]^\circ$—$3p\ [^1/_2]$	0—1
2468,50	3	52,92	57,95	$3s\ [1^1/_2]^\circ$—$3p\ [^1/_2]$	1—1
2396,04	3	52,77	57,95	$3s\ [1^1/_2]^\circ$—$3p\ [^1/_2]$	2—1
2318,83	1	53,50	58,85	$3s'\ [^1/_2]^\circ$—$3p\ [2^1/_2]$	1—2
2178,37	3	53,50	59,19	$3s'\ [^1/_2]^\circ$—$3p\ [1^1/_2]$	1—2
2134,72	3	53,50	59,31	$3s'\ [^1/_2]^\circ$—$3p'\ [1^1/_2]$	1—1
2113,45	2	53,50	59,37	$3s'\ [^1/_2]^\circ$—$3p'\ [1^1/_2]$	1—2
2098,62	2	53,05	58,95	$3s'\ [^1/_2]^\circ$—$3p\ [1^1/_2]$	0—1
2092,64	4	52,92	58,85	$3s\ [1^1/_2]^\circ$—$3p\ [2^1/_2]$	1—2
2086,55	2	53,50	59,44	$3s'\ [^1/_2]^\circ$—$3p'\ [^1/_2]$	1—1
2065,54	5	52,77	58,77	$3s\ [1^1/_2]^\circ$—$3p\ [2^1/_2]$	2—3
2056,13	3	52,92	58,95	$3s\ [1^1/_2]^\circ$—$3p\ [1^1/_2]$	1—1
2040,23	3	52,77	58,85	$3s\ [1^1/_2]^\circ$—$3p\ [2^1/_2]$	2—2
2005,55	0	52,77	58,95	$3s\ [1^1/_2]^\circ$—$3p\ [1^1/_2]$	2—1
1979,31	1	53,05	59,31	$3s'\ [^1/_2]^\circ$—$3p'\ [1^1/_2]$	0—1
1977,56	1	52,92	59,19	$3s\ [1^1/_2]^\circ$—$3p\ [1^1/_2]$	1—2
1971,57	0	59,44	65,73	$3p'\ [^1/_2]$—$3d\ [^1/_2]^\circ$	1—0
1962,18	0	59,44	65,76	$3p'\ [^1/_2]$—$3d\ [^1/_2]^\circ$	1—1
1954,87	0	59,42	65,76	$3p\ [^1/_2]$—$3d\ [^1/_2]^\circ$	0—1
1941,50	0	52,92	59,31	$3s\ [1^1/_2]^\circ$—$3p'\ [1^1/_2]$	1—1
		59,44	65,83	$3p'\ [^1/_2]$—$3d\ [1^1/_2]^\circ$	1—2
1938,95	0	59,37	65,76	$3p'\ [1^1/_2]$—$3d\ [^1/_2]^\circ$	2—1
1937,80	2	53,05	59,44	$3s'\ [^1/_2]^\circ$—$3p'\ [^1/_2]$	0—1
1933,59	0	60,06	66,47	$3p'\ [^1/_2]$—$3d'\ [1^1/_2]^\circ$	0—1
1930,64	3	52,77	59,19	$3s\ [1^1/_2]^\circ$—$3p\ [1^1/_2]$	2—2
1923,87	3	52,92	59,37	$3s\ [1^1/_2]^\circ$—$3p'\ [1^1/_2]$	1—2

λ, Å	I	E_H, eV	E_B, eV	Transition	J
1918,76	1	59,37	65,83	$3p'\ [1^1/_2]$—$3d\ [1^1/_2]^\circ$	2—2
1908,46	3	52,92	59,42	$3s\ [1^1/_2]^\circ$—$3p\ [^1/_2]$	1—0
1901,55	1	52,92	59,44	$3s\ [1^1/_2]^\circ$—$3p\ [^1/_2]$	1—1
1896,26	0	52,77	59,31	$3s\ [1^1/_2]^\circ$—$3p'\ [1^1/_2]$	2—1
1890,35	2	53,50	60,06	$3s'\ [^1/_2]^\circ$—$3p'\ [^1/_2]$	1—0
1887,31	0	59,19	65,76	$3p\ [1^1/_2]$—$3d\ [^1/_2]^\circ$	2—1
1879,46	4	52,77	59,37	$3s\ [1^1/_2]^\circ$—$3p'\ [1^1/_2]$	2—2
1868,23	1	59,19	65,83	$3p\ [1^1/_2]$—$3d\ [1^1/_2]^\circ$	2—2
1858,19	2	52,77	59,44	$3s\ [1^1/_2]^\circ$—$3p'\ [^1/_2]$	2—1
1838,32	1	59,19	65,94	$3p\ [1^1/_2]$—$3d\ [3^1/_2]^\circ$	2—3
1820,47	1	59,42	66,23	$3p[^1/_2]$—$3d\ [1^1/_2]^\circ$	0—1
1800,75	4	59,19	66,08	$3p\ [1^1/_2]$—$3d\ [2^1/_2]^\circ$	2—3
1794,68	3	59,44	66,35	$3p'\ [^1/_2]$—$3d'\ [1^1/_2]^\circ$	1—2
1791,50	1	59,31	66,23	$3p'\ [1^1/_2]$—$3d\ [1^1/_2]^\circ$	1—1
1788,05	1	59,37	66,30	$3p'\ [1^1/_2]$—$3d'\ [2^1/_2]^\circ$	2—2
1783,36	4	59,37	66,32	$3p'\ [1^1/_2]$—$3d'\ [2^1/_2]^\circ$	2—3
1773,09	3	59,31	66,30	$3p'\ [1^1/_2]$—$3d'\ [2^1/_2]^\circ$	1—2
1763,93	2	59,44	66,47	$3p'\ [^1/_2]$—$3d'\ [1^1/_2]^\circ$	1—1
1758,01	1	59,42	66,47	$3p\ [^1/_2]$—$3d'\ [1^1/_2]^\circ$	0—1
1749,02	5	58,85	65,94	$3p\ [2^1/_2]$—$3d\ [3^1/_2]^\circ$	2—3
1747,64	4	58,95	66,05	$3p\ [1^1/_2]$—$3d\ [2^1/_2]^\circ$	1—2
1744,08	0	59,19	66,30	$3p\ [1^1/_2]$—$3d'\ [2^1/_2]^\circ$	2—2
1739,56	0	59,19	66,32	$3p\ [1^1/_2]$—$3d'\ [2^1/_2]^\circ$	2—3
1738,91	6	58,77	65,90	$3p\ [2^1/_2]$—$3d\ [3^1/_2]^\circ$	3—4
1731,88	1	59,19	66,35	$3p\ [1^1/_2]$—$3d'\ [1^1/_2]^\circ$	2—2
1730,81	1	59,31	66,47	$3p'[1^1/_2]$—$3d'\ [1^1/_2]^\circ$	1—1
		58,77	65,94	$3p\ [2^1/_2]$—$3d\ [3^1/_2]^\circ$	3—3
1722,10	2	58,85	66,05	$3p\ [2^1/_2]$—$3d\ [2^1/_2]^\circ$	2—2
1714,85	0	58,85	66,08	$3p\ [2^1/_2]$—$3d\ [2^1/_2]^\circ$	2—3
1703,78	1	58,95	66,23	$3p\ [1^1/_2]$—$3d\ [1^1/_2]^\circ$	1—1
1703,43	0	59,19	66,47	$3p\ [1^1/_2]$—$3d'\ [1^1/_2]^\circ$	2—1
1697,32	2	58,77	66,08	$3p\ [2^1/_2]$—$3d\ [2^1/_2]^\circ$	3—3
1679,56	0	58,85	66,23	$3p\ [2^1/_2]$—$3d\ [1^1/_2]^\circ$	2—1
1675,76	0	58,95	66,35	$3p\ [1^1/_2]$—$3d'\ [1^1/_2]^\circ$	1—2
1659,28	0	58,85	66,32	$3p\ [2^1/_2]$—$3d'\ [2^1/_2]^\circ$	2—3
1652,26	0	58,85	66,35	$3p\ [2^1/_2]$—$3d'\ [1^1/_2]^\circ$	2—2
1648,88	0	58,95	66,47	$3p\ [1^1/_2]$—$3d'\ [1^1/_2]^\circ$	1—1
1642,86	1	58,77	66,32	$3p\ [2^1/_2]$—$3d'\ [2^1/_2]^\circ$	3—3
1592,39	1	57,95	65,73	$3p\ [^1/_2]$—$3d\ [^1/_2]^\circ$	1—0
1586,26	3	57,95	65,76	$3p\ [^1/_2]$—$3d\ [^1/_2]^\circ$	1—1
1572,72	4	57,95	65,83	$3p\ [^1/_2]$—$3d\ [1^1/_2]^\circ$	1—2
234,258	12	0,00	52,92	$2p^6\ {}^1S$—$3s\ [1^1/_2]^\circ$	0—1
231,730	14	0,00	53,50	$2p^6\ {}^1S$—$3s'\ [^1/_2]^\circ$	0—1
188,526	3	0,00	65,76	$2p^6\ {}^1S$—$3d\ [^1/_2]^\circ$	0—1
187,194	8	0,00	65,83	$2p^6\ {}^1S$—$3d\ [1^1/_2]^\circ$	0—1
186,510	9	0,00	66,47	$2p^6\ {}^1S$—$3d'\ [1^1/_2]^\circ$	0—1
182,973	2	0,00	67,76	$2p^6\ {}^1S$—$4s\ [1^1/_2]^\circ$	0—1
182,240	3	0,00	68,03	$2p^6\ {}^1S$—$4s'\ [^1/_2]^\circ$	0—1
171,896	0	0,00	72,12	$2p^6\ {}^1S$—$4d\ [^1/_2]^\circ$	0—1
171,395	4	0,00	72,33	$2p^6\ {}^1S$—$4d\ [1^1/_2]^\circ$	0—1
170,802	5	0,00	72,59	$2p^6\ {}^1S$—$4d'\ [1^1/_2]^\circ$	0—1
169,746	1	0,00	73,04	$2p^6\ {}^1S$—$5s\ [1^1/_2]^\circ$	0—1
169,150	1	0,00	73,29	$2p^6\ {}^1S$—$5s'\ [^1/_2]^\circ$	0—1
165,195	0	0,00	75,05	$2p^6\ {}^1S$—$5d\ [^1/_2]^\circ$	0—1
164,954	2	0,00	75,16	$2p^6\ {}^1S$—$5d\ [1^1/_2]^\circ$	0—1
164,384	2	0,00	75,42	$2p^6\ {}^1S$—$5d'\ [1^1/_2]^\circ$	0—1
164,159	0	0,00	75,52	$2p^6\ {}^1S$—$6s\ [1^1/_2]^\circ$	0—1
163,586	0	0,00	75,79	$2p^6\ {}^1S$—$6s'\ [^1/_2]^\circ$	0—1

λ, Å	I	E_H, eV	E_B, eV	Transition	J
161,683	0	0,00	76,68	$2p^6\,^1S$—$6d$ $[1^1/_2]^\circ$	0—1
161,135	0	0,00	76,94	$2p^6\,^1S$—$6d'$ $[1^1/_2]^\circ$	0—1
159,755	0	0,00	77,60	$2p^6\,^1S$—$7d$ $[1^1/_2]^\circ$	0—1
159,209	0	0,00	77,87	$2p^6\,^1S$—$7d'$ $[1^1/_2]^\circ$	0—1
158,530	0	0,00	78.20	$2p^6\,^1S$—$8d$ $[1^1/_2]^\circ$	0—1

Mg IV, ground state $1s^2\,2s^2\,2p^5\,^2P^\circ_{3/2}$
Ionization potential 881 759 cm⁻¹; 109,318 eV

λ, Å	I	E_H, eV	E_B, eV	Transition	J
1956,58	0	67,69	74,02	$3s\,^4P$—$3p\,^4P^\circ$	$^1/_2$—$^3/_2$
1946,20	0	67,59	73,96	$3s\,^4P$—$3p\,^4P^\circ$	$^3/_2$—$^5/_2$
1925,99	0	67,59	74,02	$3s\,^4P$—$3p\,^4P^\circ$	$^3/_2$—$^3/_2$
1906,71	0	67,59	74,09	$3s\,^4P$—$3p\,^4P^\circ$	$^3/_2$—$^1/_2$
1893,87	1	67,41	73,96	$3s\,^4P$—$3p\,^4P^\circ$	$^5/_2$—$^5/_2$
1874,59	0	67,41	74,02	$3s\,^4P$—$3p\,^4P^\circ$	$^5/_2$—$^3/_2$
1698,83	2	67,59	74,88	$3s\,^4P$—$3p\,^4D^\circ$	$^3/_2$—$^5/_2$
1683,04	3	67,41	74,78	$3s\,^4P$—$3p\,^4D^\circ$	$^5/_2$—$^7/_2$
1680,02	0	67,59	74,96	$3s\,^4P$—$3p\,^4D^\circ$	$^3/_2$—$^3/_2$
1658,92	0	67,41	74,88	$3s\,^4P$—$3p\,^4D^\circ$	$^5/_2$—$^5/_2$
1508,82	0	67,69	75,90	$3s\,^4P$—$3p\,^4S^\circ$	$^1/_2$—$^3/_2$
1490,41	0	67,59	75,90	$3s\,^4P$—$3p\,^4S^\circ$	$^3/_2$—$^3/_2$
1459,52	1	67,41	75,90	$3s\,^4P$—$3p\,^4S^\circ$	$^5/_2$—$^3/_2$
323,310	18	0,28	38,62	$2p^5\,^2P^\circ$—$2p^6\,^2S$	$^1/_2$—$^1/_2$
320,999	20	0,00	38,62	$2p^5\,^2P^\circ$—$2p^6\,^2S$	$^3/_2$—$^1/_2$
184,189	0	0,28	67,59	$2p^5\,^2P^\circ$—$3s\,^4P$	$^1/_2$—$^3/_2$
183,915	1	0,00	67,41	$2p^5\,^2P^\circ$—$3s\,^4P$	$^3/_2$—$^5/_2$
183,439	4	0,00	67,59	$2p^5\,^2P^\circ$—$3s\,^4P$	$^3/_2$—$^3/_2$
181,345	8	0,28	68,64	$2p^5\,^2P^\circ$—$3s\,^2P$	$^1/_2$—$^3/_2$
180,796	9	0,28	68,85	$2p^5\,^2P^\circ$—$3s\,^2P$	$^1/_2$—$^1/_2$
180,617	10	0,00	68,64	$2p^5\,^2P^\circ$—$3s\,^2P$	$^3/_2$—$^3/_2$
180,070	8	0,00	68,85	$2p^5\,^2P^\circ$—$3s\,^2P$	$^3/_2$—$^1/_2$
172,306	7	0,28	72,23	$2p^5\,^2P^\circ$—$3s'\,^2D$	$^1/_2$—$^3/_2$
171,653	8	0,00	72,23	$2p^5\,^2P^\circ$—$3s'\,^2D$	$^3/_2$—$^3/_2$, $^5/_2$
160,804	4	0,28	77,37	$2p^5\,^2P^\circ$—$3s''\,^2S$	$^1/_2$—$^1/_2$
160,230	6	0,00	77,37	$2p^5\,^2P^\circ$—$3s''\,^2S$	$^3/_2$—$^1/_2$
148,121	2	0,28	84,03	$2p^5\,^2P^\circ$—$3d\,^4P$	$^1/_2$—$^3/_2$
147,887	1	—	—	—	—
147,746	4	0,00	83,91	$2p^5\,^2P^\circ$—$3d\,^4P$	$^3/_2$—$^5/_2$
147,632	0	0,00	84,03	$2p^5\,^2P^\circ$—$3d\,^4P$	$^3/_2$—$^3/_2$
147,535	5	0,28	84,31	$2p^5\,^2P^\circ$—$3d\,^2D$	$^1/_2$—$^3/_2$
147,405	5	0,00	84,11	$2p^5\,^2P^\circ$—$3d\,^2D$	$^3/_2$—$^5/_2$
147,321	4	0,28	84,43	$2p^5\,^2P^\circ$—$3d\,^2P$	$^1/_2$—$^1/_2$
147,252	3	—	—	—	—
147,052	4	0,00	84,31	$2p^5\,^2P^\circ$—$3d\,^2D$	$^3/_2$—$^3/_2$
147,006	4	0,28	84,61	$2p^5\,^2P^\circ$—$3d\,^2P$	$^1/_2$—$^3/_2$
146,949	4	—	—	—	—
146,836	3	0,00	84,43	$2p^5\,^2P^\circ$—$3d\,^2P$	$^3/_2$—$^1/_2$
146,526	4	0,00	84,61	$2p^5\,^2P^\circ$—$3d\,^2P$	$^3/_2$—$^3/_2$
140,966	4	0,28	88,22	$2p^5\,^2P^\circ$—$3d'\,^2P$	$^1/_2$—$^1/_2$
140,918	2	0,28	88,26	$2p^5\,^2P^\circ$—$3d'\,^2P$	$^1/_2$—$^3/_2$
140,867	4	0,28	88,29	$2p^5\,^2P^\circ$—$3d'\,^2D$	$^1/_2$—$^3/_2$
140,564	2	0,28	88,56	$2p^5\,^2P^\circ$—$3d'\,^2S$	$^1/_2$—$^1/_2$

λ, Å	I	E_H, eV	E_B, eV	Transition	J
140,523	2	0,00	88,22	$2p^5\ ^2P^\circ—3d'\ ^2P$	3/2—1/2
140,475	2	0,00	88,26	$2p^5\ ^2P^\circ—3d'\ ^2P$	3/2—3/2
140,425	2	0,00	88,29	$2p^5\ ^2P^\circ—3d'\ ^2D$	3/2—3/2
140,176	4	0,00	88,44	$2p^5\ ^2P^\circ—3d'\ ^2D$	3/2—5/2
140,120	4	0,00	88,56	$2p^5\ ^2P^\circ—3d'\ ^2S$	3/2—1/2
139,995	1	—	—	—	—
138,693	1	0,28	89,67	$2p^5\ ^2P^\circ—4s\ ^2P$	1/2—3/2
138,394	2	0,28	89,86	$2p^5\ ^2P^\circ—4s\ ^2P$	1/2—1/2
138,262	3	0,00	89,67	$2p^5\ ^2P^\circ—4s\ ^2P$	3/2—3/2
137,966	1	0,00	89,86	$2p^5\ ^2P^\circ—4s\ ^2P$	3/2—1/2
133,202	3	0,28	93,35	$2p^5\ ^2P^\circ—3d''\ ^2D$	1/2—3/2
132,815	3	0,00	93,35	$2p^5\ ^2P^\circ—3d''\ ^2D$	3/2—3/2, 5/2
130,630	1	0,28	95,19	$2p^5\ ^2P^\circ—4d\ ^4P$	1/2—3/2
130,350	3	0,28	95,39	$2p^5\ ^2P^\circ—4d\ ^2P$	1/2—1/2
130,294	2	0,00	94,77	$2p^5\ ^2P^\circ—4d\ ^2D$	3/2—5/2
130,243	1	0,28	95,47	$2p^5\ ^2P^\circ—4d\ ^2P$	1/2—3/2
130,118	2	0,28	95,56	$2p^5\ ^2P^\circ—4d\ ^2D$	1/2—3/2
130,085	2	0,00	95,30	$2p^5\ ^2P^\circ—4d\ ^4P$	3/2—5/2
129,969	3	0,00	95,39	$2p^5\ ^2P^\circ—4d\ ^2P$	3/2—1/2
129,855	4	0,00	95,47	$2p^5\ ^2P^\circ—4d\ ^2P$	3/2—3/2
129,710	2	0,00	95,56	$2p^5\ ^2P^\circ—4d\ ^2D$	3/2—3/2
125,811	1	0,28	98,82	$2p^5\ ^2P^\circ—4s''\ ^2S$	1/2—1/2
125,459	0	0,00	98,82	$2p^5\ ^2P^\circ—4s''\ ^2S$	3/2—1/2
124,990	2	0,28	99,46	$2p^5\ ^2P^\circ—4d'\ ^2P$	1/2—1/2, 3/2
124,870	2	0,28	99,56	$2p^5\ ^2P^\circ—4d\ ^2D$	1/2—3/2
124,759	2	0,28	99,65	$2p^5\ ^2P^\circ—4d'\ ^2S$	1/2—1/2
124,649	3	0,00	99,46	$2p^5\ ^2P^\circ—4d'\ ^2P$	3/2—1/2, 3/2
124,538	2	0,00	99,56	$2p^5\ ^2P^\circ—4d'\ ^2D$	3/2—5/2
124,414	2	0,00	99,65	$2p^5\ ^2P^\circ—4d'\ ^2S$	3/2—1/2
123,722	0	0,28	100,49	$2p^5\ ^2P^\circ—5d\ ^2P$	1/2—1/2, 3/2
123,588	0	0,28	100,59	$2p^5\ ^2P^\circ—5d\ ^2D$	1/2—3/2
123,500	3	0,00	100,38	$2p^5\ ^2P^\circ—5d\ ^2D$	3/2—5/2
123,377	1	0,00	100,49	$2p^5\ ^2P^\circ—5d\ ^2P$	3/2—1/2, 3/2
123,273	2	0,00	100,59	$2p^5\ ^2P^\circ—5d\ ^2D$	3/2—3/2
120,283	1	—	—	—	—
118,603	1	—	—	—	—

Mg V, ground state $1s^22s^22p^4\ ^3P_2$

Ionization potential 1 139 421 cm^{-1}; 141,262 eV

λ, Å	I	E_H, eV	E_B, eV	Transition	J
355,326	12	0,22	35,12	$2p^4\ ^3P—2p^5\ ^3P^\circ$	1—2
354,223	10	0,31	35,31	$2p^4\ ^3P—2p^5\ ^3P^\circ$	0—1
353,300	9	0,22	35,31	$2p^4\ ^3P—2p^5\ ^3P^\circ$	1—1
353,094	14	0,00	35,12	$2p^4\ ^3P—2p^5\ ^3P^\circ$	2—2
352,202	10	0,22	35,42	$2p^4\ ^3P—2p^5\ ^3P^\circ$	1—0
351,089	12	0,00	35,31	$2p^4\ ^3P—2p^5\ ^3P^\circ$	2—1
312,311	10	9,63	49,33	$2p^4\ ^1S—2p^5\ ^1P^\circ$	0—1
276,581	16	4,51	49,33	$2p^4\ ^1D—2p^5\ ^1P^\circ$	2—1
152,591	0	35,42	116,67	$2p^5\ ^3P^\circ—3s'''\ ^3P$	0—1
152,527	1	35,31	116,59	$2p^5\ ^3P^\circ—3s'''\ ^3P$	1—2
152,384	1	35,31	116,67	$2p^5\ ^3P^\circ—3s'''\ ^3P$	1—1
152,149	3	35,12	116,59	$2p^5\ ^3P^\circ—3s'''\ ^3P$	2—2
152,019	0	35,12	116,67	$2p^5\ ^3P^\circ—3s'''\ ^3P$	2—1

λ, Å	I	E_H, eV	E_B, eV	Transition	J
146,621	4	0,31	84,87	$2p^4\ ^3P—3s\ ^3S^\circ$	0—1
146,464	5	0,22	84,87	$2p^4\ ^3P—3s\ ^3S^\circ$	1—1
146,083	6	0,00	84,87	$2p^4\ ^3P—3s\ ^3S^\circ$	2—1
145,485	5	9,63	94,85	$2p^4\ ^1S—3s''\ ^1P^\circ$	0—1
142,933	6	4,51	91,24	$2p^4\ ^1D—3s'\ ^1D^\circ$	2—2
137,880	6	0,31	90,23	$2p^4\ ^3P—3s'\ ^3D^\circ$	0—1
137,748	7	0,22	90,23	$2p^4\ ^3P—3s'\ ^3D^\circ$	1—2
137,414	8	0,00	90,22	$2p^4\ ^3P—3s'\ ^3D^\circ$	2—3
137,234	6	4,51	94,85	$2p^4\ ^1D—3s''\ ^1P^\circ$	2—1
136,128	0	35,42	126,50	$2p^5\ ^3P^\circ—3s^{IV}\ ^3D$	0—1
135,953	1	35,31	126,50	$2p^5\ ^3P^\circ—3s^{IV}\ ^3D$	1—2
135,638	2	35,12	126,50	$2p^5\ ^3P^\circ—3s^{IV}\ ^3D$	2—3
132,623	3	0,31	93,79	$2p^4\ ^3P—3s''\ ^3P^\circ$	0—1
132,485	5	0,22	93,80	$2p^4\ ^3P—3s''\ ^3P^\circ$	1—1, 2
132,171	6	0,00	93,80	$2p^4\ ^3P—3s''\ ^3P^\circ$	2—2
126,677	0	35,42	133,29	$2p^5\ ^3P^\circ—3d'''\ ^3D$	0—1
126,544	2	35,31	133,29	$2p^5\ ^3P^\circ—3d'''\ ^3D$	1—1, 2
126,280	4	35,12	133,29	$2p^5\ ^3P^\circ—3d'''\ ^3D$	2—1, 2, 3
125,600	4	9,63	108,34	$2p^4\ ^1S—3d'\ ^1P^\circ$	0—1
122,034	4	0,31	101,90	$2p^4\ ^3P—3d\ ^3D^\circ$	0—1
121,922	5	0,22	101,91	$2p^4\ ^3P—3d\ ^3D^\circ$	1—2
121,644	6	0,00	101,92	$2p^4\ ^3P—3d\ ^3D^\circ$	2—3
119,447	4	—	—	—	—
119,401	4	4,51	108,34	$2p^4\ ^1D—3d'\ ^1P^\circ$	2—1
118,810	5	4,51	108,86	$2p^4\ ^1D—3d'\ ^1D^\circ$	2—2
118,083	5	4,51	109,50	$2p^4\ ^1D—3d'\ ^1F^\circ$	2—3
115,537	4	4,51	111,81	$2p^4\ ^1D—3d''\ ^1D^\circ$	2—2
115,399	4	4,51	111,94	$2p^4\ ^1D—3d''\ ^1P^\circ$	2—1
115,093	4	4,51	112,23	$2p^4\ ^1D—3d''\ ^1F^\circ$	2—3
115,013	6	0,22	108,01	$2p^4\ ^3P—3d'\ ^3D^\circ$	1—1, 2
114,785	6	0,00	108,01	$2p^4\ ^3P—3d'\ ^3D^\circ$	2—1, 2, 3
114,324	3	0,31	108,76	$2p^4\ ^3P—3d'\ ^3P^\circ$	0—1
114,285	3	0,22	108,70	$2p^4\ ^3P—3d'\ ^3P^\circ$	1—2
114,226	3	0,22	108,76	$2p^4\ ^3P—3d'\ ^3P^\circ$	1—1
114,199	3	0,22	108,78	$2p^4\ ^3P—3d'\ ^3P^\circ$	1—0
114,059	4	0,00	108,70	$2p^4\ ^3P—3d'\ ^3P^\circ$	2—2
114,029	2	0,31	109,04	$2p^4\ ^3P—3d'\ ^3S^\circ$	0—1
113,990	3	0,00	108,76	$2p^4\ ^3P—3d'\ ^3P^\circ$	2—1
113,934	3	0,22	109,04	$2p^4\ ^3P—3d'\ ^3S^\circ$	1—1
113,823	1	35,12	144,03	$2p^5\ ^3P^\circ—4s'''\ ^3P$	2—2
113,703	4	0,00	109,04	$2p^4\ ^3P—3d'\ ^3S^\circ$	2—1
113,518	1	35,42	144,64	$2p^5\ ^3P^\circ—3d^{IV}\ ^3D$	0—1
113,414	2	35,31	144,63	$2p^5\ ^3P^\circ—3d^{IV}\ ^3D$	1—2
113,217	2	35,12	144,62	$2p^5\ ^3P^\circ—3d^{IV}\ ^3D$	2—3
111,496	2	0,22	111,41	$2p^4\ ^3P—3d''\ ^3P^\circ$	1—0
111,467	2	0,22	111,44	$2p^4\ ^3P—3d''\ ^3P^\circ$	1—1
111,419	2	0,22	111,49	$2p^4\ ^3P—3d''\ ^3P^\circ$	1—2
111,247	2	0,00	111,44	$2p^4\ ^3P—3d''\ ^3P^\circ$	2—1
111,199	4	0,00	111,49	$2p^4\ ^3P—3d''\ ^3P^\circ$	2—2
111,091	3	0,31	111,91	$2p^4\ ^3P—3d''\ ^3D^\circ$	0—1
111,031	3	0,22	111,88	$2p^4\ ^3P—3d''\ ^3D^\circ$	1—2
110,939	2	0,22	111,44	$2p^4\ ^3P—3d''\ ^3P^\circ$	1—1
110,859	4	0,00	111,83	$2p^4\ ^3P—3d''\ ^3D^\circ$	2—3
110,809	2	0,00	111,88	$2p^4\ ^3P—3d''\ ^3D^\circ$	2—2
110,121	0	0,31	112,89	$2p^4\ ^3P—4s\ ^3S^\circ$	0—1
110,029	1	0,22	112,89	$2p^4\ ^3P—4s\ ^3S^\circ$	1—1
109,812	2	0,00	112,89	$2p^4\ ^3P—4s\ ^3S^\circ$	2—1

λ, Å	I	E_H, eV	E_B, eV	Transition	J
109,174	0	9,63	123,21	$2p^4\,{}^1S$—$4s''\,{}^1P^\circ$	0—1
107,661	2	4,51	119,66	$2p^4\,{}^1D$—$4s'\,{}^1D^\circ$	2—2
104,432	2	4,51	123,21	$2p^4\,{}^1D$—$4s''\,{}^1P^\circ$	2—1
104,214	1	0,31	119,27	$2p^4\,{}^3P$—$4s'\,{}^3D^\circ$	0—1
104,182	1	0,31	119,31	$2p^4\,{}^3P$—$4d\,{}^3D^\circ$	0—1
104,140	2	0,22	119,27	$2p^4\,{}^3P$—$4s'\,{}^3D^\circ$	1—1, 2
104,100	2	0,22	119,32	$2p^4\,{}^3P$—$4d\,{}^3D^\circ$	1—2
103,947	3	0,00	119,27	$2p^4\,{}^3P$—$4s'\,{}^3D^\circ$	2—1, 2, 3
103,904	4	0,00	119,32	$2p^4\,{}^3P$—$4d\,{}^3D^\circ$	2—3
103,333	0	35,12	155,10	$2p^5\,{}^3P^\circ$—$5s'''\,{}^3P$	2—2
102,079	2	4,51	125,96	$2p^4\,{}^1D$—$4d'\,{}^1P^\circ$	2—1
101,782	3	4,51	126,31	$2p^4\,{}^1D$—$4d'\,{}^1D^\circ$	2—2
101,671	3	4,51	126,45	$2p^4\,{}^1D$—$4d'\,{}^1F^\circ$	2—3
100,949	0	0,00	122,81	$2p^4\,{}^3P$—$4s''\,{}^3P^\circ$	2—1, 2
99,788	1	0,00	124,24	$2p^4\,{}^3P$—$5s\,{}^3S^\circ$	2—1
99,067	2	4,51	129,65	$2p^4\,{}^1D$—$4d''\,{}^1D^\circ$	2—2
99,025	2	4,51	129,71	$2p^4\,{}^1D$—$4d''\,{}^1P^\circ$	2—1
98,983	1	4,51	129,76	$2p^4\,{}^1D$—$4d''\,{}^1F^\circ$	2—3
98,872	1	0,31	125,70	$2p^4\,{}^3P$—$4d'\,{}^3D^\circ$	0—1
98,805	2	0,22	125,70	$2p^4\,{}^3P$—$4d'\,{}^3D^\circ$	1—1, 2
98,636	2	0,00	125,70	$2p^4\,{}^3P$—$4d'\,{}^3D^\circ$	2—1, 2, 3
98,444	1	0,22	126,16	$2p^4\,{}^3P$—$4d'\,{}^3P^\circ$	1—2
98,406	1	0,22	126,20	$2p^4\,{}^3P$—$4d'\,{}^3P^\circ$	1—0, 1
98,271	2	0,00	126,16	$2p^4\,{}^3P$—$4d'\,{}^3P^\circ$	2—2
98,235	1	0,00	126,20	$2p^4\,{}^3P$—$4d'\,{}^3P^\circ$	2—1
97,686	1	0,31	127,24	$2p^4\,{}^3P$—$3p'''\,{}^3D^\circ$	0—1
97,606	2	0,22	127,24	$2p^4\,{}^3P$—$3p'''\,{}^3D^\circ$	1—1, 2
97,563	1	0,22	127,30	$2p^4\,{}^3P$—$5d\,{}^3D^\circ$	1—1, 2
97,439	2	0,00	127,24	$2p^4\,{}^3P$—$3p'''\,{}^3D^\circ$	2—1, 2, 3
97,391	1	0,00	127,30	$2p^4\,{}^3P$—$5d\,{}^3D^\circ$	2—1, 2, 3
96,159	0	0,31	129,24	$2p^4\,{}^3P$—$4d''\,{}^3P^\circ$	0—1
96,085	1	0,22	129,24	$2p^4\,{}^3P$—$4d''\,{}^3P^\circ$	1—1
96,019	2	0,31	129,41	$2p^4\,{}^3P$—$4d''\,{}^3D^\circ$	0—1
95,965	1	0,22	129,41	$2p^4\,{}^3P$—$4d''\,{}^3D^\circ$	1—1, 2, 3
95,909	1	0,00	129,27	$2p^4\,{}^3P$—$4d''\,{}^3P^\circ$	2—2
95,803	2	0,00	129,41	$2p^4\,{}^3P$—$4d''\,{}^3D^\circ$	2—1, 2, 3
95,592	0	4,51	134,20	$2p^4\,{}^1D$—$5d'\,{}^1D^\circ$	2—2
95,556	1	4,51	134,25	$2p^4\,{}^1D$—$5d'\,{}^1F^\circ$	2—3
94,793	0	0,00	130,79	$2p^4\,{}^3P$—$5s'\,{}^3D^\circ$	2—1, 2, 3
93,109	1	4,51	137,66	$2p^4\,{}^1D$—$5d''\,{}^1D^\circ$	2—2
92,641	0	0,00	133,82	$2p^4\,{}^3P$—$5d'\,{}^3D^\circ$	2—1, 2, 3
92,588	0	0,22	134,13	$2p^4\,{}^3P$—$5d'\,{}^3P^\circ$	1—2
92,428	0	0,00	134,13	$2p^4\,{}^3P$—$5d'\,{}^3P^\circ$	2—2
92,409	0	0,00	134,16	$2p^4\,{}^3P$—$5d'\,{}^3P^\circ$	2—1

Mg VI, ground state $1s^2\,2s^22p^3\,{}^4S^0_{3/2}$

Ionization potential 1 507 520 cm⁻¹; 186,898 eV

λ, Å₀	I	E_H, eV	E_B, eV	Transition	J
403,315	8	0,00	30,74	$2p^3\,{}^4S^\circ$—$2p^4\,{}^4P$	$^3/_2$—$^5/_2$
400,676	7	0,00	30,94	$2p^3\,{}^4S^\circ$—$2p^4\,{}^4P$	$^3/_2$—$^3/_2$
399,289	6	0,00	31,05	$2p^3\,{}^4S^\circ$—$2p^4\,{}^4P$	$^3/_2$—$^1/_2$
388,020	3	10,27	42,22	$2p^3\,{}^2P^\circ$—$2p^4\,{}^2D$	$^3/_2$—$^5/_2$
387,787	2	10,25	42,22	$2p^3\,{}^2P^\circ$—$2p^4\,{}^2D$	$^1/_2$—$^3/_2$

λ, $Å_0$	I	E_H, eV	E_B, eV	Transition	J
349,155	10	6,72	42,22	$2p^3\ ^2D^\circ - 2p^4\ ^2D$	$3/2 - 3/2,\ 5/2$
314,676	4	10,27	49,67	$2p^3\ ^2P^\circ - 2p^4\ ^2S$	$3/2 - 1/2$
314,554	3	10,25	49,67	$2p^3\ ^2P^\circ - 2p^4\ ^2S$	$1/2 - 1/2$
293,124	4	10,27	52,56	$2p^3\ ^2P^\circ - 2p^4\ ^2P$	$3/2 - 3/2$
293,026	2	10,25	52,56	$2p^3\ ^2P^\circ - 2p^4\ ^2P$	$1/2 - 3/2$
291,458	2	10,27	52,80	$2p^3\ ^2P^\circ - 2p^4\ ^2P$	$3/2 - 1/2$
291,348	3	10,25	52,80	$2p^3\ ^2P^\circ - 2p^4\ ^2P$	$1/2 - 1/2$
288,652	0	6,72	49,67	$2p^3\ ^2D^\circ - 2p^4\ ^2S$	$3/2 - 1/2$
270,394	12	6,71	52,56	$2p^3\ ^2D^\circ - 2p^4\ ^2P$	$5/2 - 3/2$
268,986	10	6,72	52,80	$2p^3\ ^2D^\circ - 2p^4\ ^2P$	$3/2 - 1/2$
137,814	0	52,56	142,53	$2p^4\ ^2P - 3s^{IV}\ ^2D^\circ$	$3/2 - 3/2,\ 5/2$
130,701	0	52,80	147,67	$2p^4\ ^2P - 3s^{V}\ ^2P^\circ$	$1/2 - 1/2$
130,630	1	52,80	147,71	$2p^4\ ^2P - 3s^{V}\ ^2P^\circ$	$1/2 - 3/2$
130,294	2	52,56	147,71	$2p^4\ ^2P - 3s^{V}\ ^2P^\circ$	$3/2 - 3/2$
126,488	1	49,67	147,67	$2p^4\ ^2S - 3s^{V}\ ^2P^\circ$	$1/2 - 1/2$
126,450	1	49,67	147,71	$2p^4\ ^2S - 3s^{V}\ ^2P^\circ$	$1/2 - 3/2$
125,600	4	31,05	129,76	$2p^4\ ^4P - 3s'''\ ^4S^\circ$	$1/2 - 3/2$
125,459	3	30,94	129,76	$2p^4\ ^4P - 3s'''\ ^4S^\circ$	$3/2 - 3/2$
125,206	3	30,74	129,76	$2p^4\ ^4P - 3s'''\ ^4S^\circ$	$5/2 - 3/2$
123,590	1	42,22	142,53	$2p^4\ ^2D - 3s^{IV}\ ^2D^\circ$	—
121,290	3	10,27	112,47	$2p^3\ ^2P^\circ - 3s\ ^2P$	$1/2,\ 3/2 - 1/2$
121,025	5	10,27	112,71	$2p^3\ ^2P^\circ - 3s\ ^2P$	$1/2,\ 3/2 - 3/2$
117,527	1	42,22	147,71	$2p^4\ ^2D - 3s^{V}\ ^2P^\circ$	$5/2 - 3/2$
117,226	3	6,72	112,47	$2p^3\ ^2D^\circ - 3s\ ^2P$	$3/2 - 1/2$
116,968	5	{ 6,71	112,71	$2p^3\ ^2D^\circ - 3s\ ^2P$	$5/2 - 3/2$
		10,27	116,24	$2p^3\ ^2P^\circ - 3s'\ ^2D$	—
114,725	0	31,05	139,11	$2p^4\ ^4P - 3s^{IV}\ ^4D^\circ$	$1/2 - 1/2,\ 3/2$
114,624	0	30,94	139,11	$2p^4\ ^4P - 3s^{IV}\ ^4D^\circ$	$3/2 - 1/2,\ 3/2,\ 5/2$
114,412	0	30,74	139,11	$2p^4\ ^4P - 3s^{IV}\ ^4D^\circ$	$5/2 - 3/2,\ 5/2,\ 7/2$
113,189	5	6,72	116,24	$2p^3\ ^2D^\circ - 3s'\ ^2D$	—
111,864	4	0,00	110,83	$2p^3\ ^4S^\circ - 3s\ ^4P$	$3/2 - 1/2$
111,746	4	0,00	110,95	$2p^3\ ^4S^\circ - 3s\ ^4P$	$3/2 - 3/2$
111,552	5	0,00	111,14	$2p^3\ ^4S^\circ - 3s\ ^4P$	$3/2 - 5/2$
111,199	4	10,27	121,77	$2p^3\ ^2P^\circ - 3s''\ ^2S$	$3/2 - 1/2$
111,160	3	10,25	121,77	$2p^3\ ^2P^\circ - 3s''\ ^2S$	$1/2 - 1/2$
108,441	0	31,05	145,38	$2p^4\ ^4P - 3s^{V}\ ^4P^\circ$	$1/2 - 1/2,\ 3/2$
108,338	1	30,94	145,38	$2p^4\ ^4P - 3s^{V}\ ^4P^\circ$	$3/2 - 1/2,\ 3/2,\ 5/2$
108,148	1	30,74	145,38	$2p^4\ ^4P - 3s^{V}\ ^4P^\circ$	$5/2 - 3/2,\ 5/2$
108,114	2	31,05	145,72	$2p^4\ ^4P - 3d'''\ ^4D^\circ$	$1/2 - 3/2$
108,015	3	30,94	145,72	$2p^4\ ^4P - 3d'''\ ^4D^\circ$	$3/2 - 5/2$
107,820	4	30,74	145,72	$2p^4\ ^4P - 3d'''\ ^4D^\circ$	$5/2 - 7/2$
105,502	3	42,22	159,73	$2p^4\ ^2D - 3d^{IV}\ ^2F^\circ$	$5/2 - 7/2$
105,410	2	42,22	159,84	$2p^4\ ^2D - 3d^{IV}\ ^2F^\circ$	$3/2 - 5/2$
104,597	5	10,27	128,79	$2p^3\ ^2P^\circ - 3d\ ^2P$	$1/2,\ 3/2 - 3/2$
104,519	3	10,27	128,87	$2p^3\ ^2P^\circ - 3d\ ^2P$	$1/2,\ 3/2 - 1/2$
102,239	5	10,25	131,52	$2p^3\ ^2P^\circ - 3d\ ^2D$	$1/2 - 3/2$
102,189	5	10,27	131,59	$2p^3\ ^2P^\circ - 3d\ ^2D$	$3/2 - 5/2$
101,556	3	6,72	128,79	$2p^3\ ^2D^\circ - 3d\ ^2P$	$3/2,\ 5/2 - 3/2$
101,508	2	6,72	128,87	$2p^3\ ^2D^\circ - 3d\ ^2P$	$3/2 - 1/2$
100,904	4	6,72	129,58	$2p^3\ ^2D^\circ - 3d\ ^2F$	$3/2 - 5/2$
100,702	5	6,71	129,83	$2p^3\ ^2D^\circ - 3d\ ^2F$	$5/2 - 7/2$
99,738	3	10,25	134,56	$2p^3\ ^2P^\circ - 3d'\ ^2D$	$1/2 - 3/2$
99,713	3	10,27	134,60	$2p^3\ ^2P^\circ - 3d'\ ^2D$	$3/2 - 5/2$
99,333	4	6,72	131,52	$2p^3\ ^2D^\circ - 3d\ ^2D$	$3/2 - 3/2$
99,279	4	6,71	131,59	$2p^3\ ^2D^\circ - 3d\ ^2D$	$5/2 - 5/2$
99,025	2	10,27	135,45	$2p^3\ ^2P^\circ - 3d'\ ^2P$	$1/2,\ 3/2 - 1/2$

λ, $Å_0$	I	E_H, eV	E_B, eV	Transition	J
98,983	4	10,27	135,51	$2p^3\,{}^2P^\circ—3d'\,{}^2P$	$^1/_2$, $^3/_2—^3/_2$
98,508	3	10,27	136,12	$2p^3\,{}^2P^\circ—3d'\,{}^2S$	$^1/_2$, $^3/_2—^1/_2$
97,278	5	6,71	134,16	$2p^3\,{}^2D^\circ—3d'\,{}^2F$	$^5/_2—^7/_2$
97,251	5	6,72	134,20	$2p^3\,{}^2D^\circ—3d'\,{}^2F$	$^3/_2—^5/_2$
96,973	4	6,72	134,56	$2p^3\,{}^2D^\circ—3d'\,{}^2D$	$^3/_2—^3/_2$
96,939	4	6,71	134,60	$2p^3\,{}^2D^\circ—3d'\,{}^2D$	$^5/_2—^5/_2$
96,903	0	31,05	158,99	$2p^4\,{}^4P—3d^{IV}\,{}^4P^\circ$	$^1/_2—^3/_2$
96,857	1	30,94	158,94	$2p^4\,{}^4P—3s^{IV}\,{}^4P^\circ$	$^3/_2—^5/_2$
96,797	1	30,94	159,02	$2p^4\,{}^4P—3d^{IV}\,{}^4P^\circ$	$^3/_2—^1/_2$
96,704	2	30,74	158,94	$2p^4\,{}^4P—3d^{IV}\,{}^4P^\circ$	$^5/_2—^5/_2$
96,670	2	30,74	158,99	$2p^4\,{}^4P—3d^{IV}\,{}^4P^\circ$	$^5/_2—^3/_2$
96,467	0	31,05	159,56	$2p^4\,{}^4P—3d^{IV}\,{}^4D^\circ$	$^1/_2—^1/_2$, $^3/_2$
96,388	1	30,94	159,56	$2p^4\,{}^4P—3d^{IV}\,{}^4D^\circ$	$^3/_2—^1/_2$, $^3/_2$, $^5/_2$
96,303	2	6,72	135,45	$2p^3\,{}^2D^\circ—3d'\,{}^2P$	$^3/_2—^1/_2$
96,256	2	6,71	135,51	$2p^3\,{}^2D^\circ—3d'\,{}^2P$	$^5/_2—^3/_2$
96,240	1	30,74	159,56	$2p^4\,{}^4P—3d^{IV}\,{}^4D^\circ$	$^5/_2—^3/_2$, $^5/_2$, $^7/_2$
96,159	1	30,74	159,67	$2p^4\,{}^4P—3d^{IV}\,{}^4S^\circ$	$^5/_2—^3/_2$
96,085	1	10,27	139,31	$2p^3\,{}^2P^\circ—3d''\,{}^2D$	—
95,803	2	6,72	136,12	$2p^3\,{}^2D^\circ—3d'\,{}^2S$	$^3/_2—^1/_2$
95,675	3	0,00	129,58	$2p^3\,{}^4S^\circ—3d\,{}^4D$	$^3/_2—^3/_2$, $^5/_2$
95,637	3	0,00	129,63	$2p^3\,{}^4S^\circ—3d\,{}^4D$	$^3/_2—^1/_2$
95,483	5	0,00	129,84	$2p^3\,{}^4S^\circ—3d\,{}^4P$	$^3/_2—^5/_2$
95,421	4	0,00	129,93	$2p^3\,{}^4S^\circ—3d\,{}^4P$	$^3/_2—^3/_2$
95,385	4	0,00	129,98	$2p^3\,{}^4S^\circ—3d\,{}^4P$	$^3/_2—^1/_2$
93,493	3	6,72	139,31	$2p^3\,{}^2D^\circ—3d''\,{}^2D$	—
93,109	1	30,94	164,10	$2p^4\,{}^4P—4s'''\,{}^4S^\circ$	$^3/_2—^3/_2$
92,964	1	30,74	164,10	$2p^4\,{}^4P—4s'''\,{}^4S^\circ$	$^5/_2—^3/_2$
90,897	6	0,00	136,39	$2p^3\,{}^4S^\circ—3p'''\,{}^4P$	$^3/_2—^1/_2$, $^3/_2$, $^5/_2$
89,649	0	10,27	148,56	$2p^3\,{}^2P^\circ—4s\,{}^2P$	$^3/_2—^3/_2$
89,021	0	31,05	170,31	$2p^4\,{}^4P—4d'''\,{}^4D^\circ$	$^1/_2—^3/_2$
88,952	2	30,94	170,31	$2p^4\,{}^4P—4d'''\,{}^4D^\circ$	$^3/_2—^5/_2$
88,827	2	30,74	170,31	$2p^4\,{}^4P—4d'''\,{}^4D^\circ$	$^5/_2—^7/_2$
87,406	0	6,71	148,56	$2p^3\,{}^2D^\circ—4s\,{}^2P$	$^5/_2—^3/_2$
86,807	2	10,27	153,05	$2p^3\,{}^2P^\circ—4s'\,{}^2D$	—
85,622	3	6,71	151,51	$2p^3\,{}^2D^\circ—3p^{IV}\,{}^2F$	$^5/_2—^7/_2$
85,577	2	6,72	151,59	$2p^3\,{}^2D^\circ—3p^{IV}\,{}^2F$	$^3/_2—^5/_2$
85,153	0	10,27	155,86	$2p^3\,{}^2P^\circ—4d\,{}^2D$	$^3/_2—^5/_2$
84,745	2	6,72	153,05	$2p^3\,{}^2D^\circ—4s'\,{}^2D$	—
83,560	2	0,00	148,37	$2p^3\,{}^4S^\circ—4s\,{}^4P$	$^3/_2—^5/_2$
83,519	3	6,72	155,16	$2p^3\,{}^2D^\circ—4d\,{}^2F$	$^3/_2—^5/_2$
83,403	4	6,71	155,36	$2p^3\,{}^2D^\circ—4d\,{}^2F$	$^5/_2—^7/_2$
82,853	1	10,27	159,90	$2p^3\,{}^2P^\circ—4d'\,{}^2D$	—
82,475	1	10,27	160,59	$2p^3\,{}^2P^\circ—4d'\,{}^2S$	$^1/_2$, $^3/_2—^1/_2$
82,238	2	30,74	181,49	$2p^4\,{}^4P—5d'''\,{}^4D^\circ$	$^5/_2—^7/_2$
81,106	3	6,72	159,57	$2p^3\,{}^2D^\circ—4d'\,{}^2F$	—
80,930	2	6,72	159,90	$2p^3\,{}^2D^\circ—4d'\,{}^2D$	—
80,724	0	6,72	160,29	$2p^3\,{}^2D^\circ—4d'\,{}^2P$	—
80,075	2	0,00	154,83	$2p^3\,{}^4S^\circ—4d\,{}^4D$	$^3/_2—^3/_2$, $^5/_2$
80,032	2	0,00	154,91	$2p^3\,{}^4S^\circ—4d\,{}^4D$	$^3/_2—^1/_2$
79,857	4	0,00	155,25	$2p^3\,{}^4S^\circ—4d\,{}^4P$	$^3/_2—^5/_2$
79,830	4	0,00	155,30	$2p^3\,{}^4S^\circ—4d\,{}^4P$	$^3/_2—^3/_2$
79,817	2	0,00	155,33	$2p^3\,{}^4S^\circ—4d\,{}^4P$	$^3/_2—^1/_2$
78,239	0	6,72	165,17	$2p^3\,{}^2D^\circ—4d''\,{}^2D$	—
77,511	1	6,72	166,66	$2p^3\,{}^2D^\circ—5d\,{}^2F$	$^3/_2—^5/_2$
77,405	2	6,71	166,88	$2p^3\,{}^2D^\circ—5d\,{}^2F$	$^5/_2—^7/_2$

λ, Å₀	I	E_H, eV	E_B, eV	Transition	J
76,908	0	10,27	171,47	$2p^3\,{}^2P^\circ$—$5d'\,{}^2D$	—
75,890	0	0,00	163,36	$2p^3\,{}^4S^\circ$—$5s\,{}^4P$	$^3/_2$—$^3/_2$
75,834	2	0,00	163,48	$2p^3\,{}^4S^\circ$—$5s\,{}^4P$	$^3/_2$—$^5/_2$
75,334	1	6,72	171,28	$2p^3\,{}^2D^\circ$—$5d'\,{}^2F$	$^5/_2$—$^5/_2$
75,248	1	6,72	171,47	$2p^3\,{}^2D^\circ$—$5d'\,{}^2D$	—
74,574	2	0,00	166,25	$2p^3\,{}^4S^\circ$—$4p'''\,{}^4P$	$^3/_2$—$^1/_2$, $^3/_2$, $^5/_2$
74,461	0	0,00	166,50	$2p^3\,{}^4S^\circ$—$5d\,{}^4D$	$^3/_2$—$^3/_2$, $^5/_2$
74,319	3	0,00	166,82	$2p^3\,{}^4S^\circ$—$5d\,{}^4P$	$^3/_2$—$^5/_2$
72,430	1	0,00	171,17	$2p^3\,{}^4S^\circ$—$6s\,{}^4P$	$^3/_2$—$^5/_2$

Unclassified Lines of Magnesium

λ, Å	I	Expected assignment	λ, Å	I	Expected assignment
106,453	2	—	93,650	2	—
105,066	5	—	88,016	2	—
102,906	3	—	86,440	2	—
100,597	2	—	86,417	2	—
100,545	2	—	79,880	2	—
99,610	2	—	75,666	3	—

ALUMINUM, Z = 13

Al I, ground state $1s^2\,2s^2\,2p^6\,3s^2\,3p\;{}^2P^0_{1/2}$

Ionization potential 48 279,16 cm^{-1}; 5,985 eV

λ, Å	I	E_H, eV	E_B, eV	Transition	J
21163,75	13	4,09	4,67	$4p\ {}^2P^\circ$—$5s\ {}^2S$	$^3/_2$—$^1/_2$
21093,04	12	4,09	4,67	$4p\ {}^2P^\circ$—$5s\ {}^2S$	$^1/_2$—$^1/_2$
19727,33	18	4,99	5,62	$5p\ {}^2P^\circ$—$7d\ {}^2D$	$^3/_2$—$^5/_2$
17699,09	13	4,67	5,37	$5s\ {}^2S$—$6p\ {}^2P^\circ$	$^1/_2$—$^3/_2$
16763,36	9	4,09	4,83	$4p\ {}^2P^\circ$—$4d\ {}^2D$	$^3/_2$—$^3/_2$
16750,56	12	4,09	4,83	$4p\ {}^2P^\circ$—$4d\ {}^2D$	$^3/_2$—$^5/_2$
16718,96	11	4,09	4,83	$4p\ {}^2P^\circ$—$4d\ {}^2D$	$^1/_2$—$^3/_2$
13150,76	14	3,14	4,09	$4s\ {}^2S$—$4p\ {}^2P^\circ$	$^1/_2$—$^1/_2$
13123,41	15	3,14	4,09	$4s\ {}^2S$—$4p\ {}^2P^\circ$	$^1/_2$—$^3/_2$
12757,26	4	4,02	4,99	$3d\ {}^2D$—$5p\ {}^2P^\circ$	$^3/_2$—$^1/_2$
12749,83	21	4,02	4,99	$3d\ {}^2D$—$5p\ {}^2P^\circ$	$^5/_2$—$^3/_2$
12747,65	2	4,02	4,99	$3d\ {}^2D$—$5p\ {}^2P^\circ$	$^3/_2$—$^3/_2$
11254,881	15	4,02	5,12	$3d\ {}^2D$—$4f\ {}^2F^\circ$	$^5/_2$—$^7/_2$
11253,190	14	4,02	5,12	$3d\ {}^2D$—$4f\ {}^2F^\circ$	$^3/_2$—$^5/_2$
10891,733	11	4,09	5,22	$4p\ {}^2P^\circ$—$6s\ {}^2S$	$^3/_2$—$^1/_2$
10872,975	10	4,09	5,22	$4p\ {}^2P^\circ$—$6s\ {}^2S$	$^1/_2$—$^1/_2$
10786,770	4	4,09	5,24	$4p\ {}^2P^\circ$—$5d\ {}^2D$	$^3/_2$—$^3/_2$
10782,045	9	4,09	5,24	$4p\ {}^2P^\circ$—$5d\ {}^2D$	$^3/_2$—$^5/_2$
10768,364	8	4,09	5,24	$4p\ {}^2P^\circ$—$5d\ {}^2D$	$^1/_2$—$^3/_2$
9172,14	4	4,02	5,37	$3d\ {}^2D$—$6p\ {}^2P^\circ$	$^3/_2$—$^1/_2$
9170,86	5	4,02	5,37	$3d\ {}^2D$—$6p\ {}^2P^\circ$	$^5/_2$—$^3/_2$
9163,261	2	7,03	8,39	$3p^2\ {}^2P$—$3d\ {}^2D^\circ$	$^3/_2$—$^3/_2$
9139,950	6	7,03	8,39	$3p^2\ {}^2P$—$3d\ {}^2D^\circ$	$^3/_2$—$^5/_2$
9089,906	5	7,02	8,39	$3p^2\ {}^2P$—$3d\ {}^2D^\circ$	$^1/_2$—$^3/_2$
8925,504	4	4,09	5,48	$4p\ {}^2P^\circ$—$6d\ {}^2D$	$^3/_2$—$^3/_2$
8923,555	9	4,09	5,48	$4p\ {}^2P^\circ$—$6d\ {}^2D$	$^3/_2$—$^5/_2$
8912,900	7	4,09	5,48	$4p\ {}^2P^\circ$—$6d\ {}^2D$	$^1/_2$—$^3/_2$
8841,277	10	4,09	5,49	$4p\ {}^2P^\circ$—$7s\ {}^2S$	$^3/_2$—$^1/_2$
8828,909	8	4,09	5,49	$4p\ {}^2P^\circ$—$7s\ {}^2S$	$^1/_2$—$^1/_2$
8773,896	14	4,02	5,43	$3d\ {}^2D$—$5f\ {}^2F^\circ$	$^5/_2$—$^7/_2$
8772,866	13	4,02	5,43	$3d\ {}^2D$—$5f\ {}^2F^\circ$	$^3/_2$—$^5/_2$
8076,298	2	4,09	5,62	$4p\ {}^2P^\circ$—$7d\ {}^2D$	$^3/_2$—$^3/_2$
8075,353	8	4,09	5,62	$4p\ {}^2P^\circ$—$7d\ {}^2D$	$^3/_2$—$^5/_2$
8065,968	6	4,09	5,62	$4p\ {}^2P^\circ$—$7d\ {}^2D$	$^1/_2$—$^3/_2$
8003,186	7	4,09	5,64	$4p\ {}^2P^\circ$—$8s\ {}^2S$	$^3/_2$—$^1/_2$
7993,048	5	4,09	5,64	$4p\ {}^2P^\circ$—$8s\ {}^2S$	$^1/_2$—$^1/_2$
7836,134	12	4,02	5,60	$3d\ {}^2D$—$6f\ {}^2F^\circ$	$^5/_2$—$^5/_2$, $^7/_2$
7835,309	11	4,02	5,60	$3d\ {}^2D$—$6f\ {}^2F^\circ$	$^3/_2$—$^5/_2$, $^7/_2$
7615,339	1	4,09	5,71	$4p\ {}^2P^\circ$—$8d\ {}^2D$	$^3/_2$—$^3/_2$
7614,820	7	4,09	5,71	$4p\ {}^2P^\circ$—$8d\ {}^2D$	$^3/_2$—$^5/_2$
7606,159	5	4,09	5,71	$4p\ {}^2P^\circ$—$8d\ {}^2D$	$^1/_2$—$^3/_2$
7563,214	3	4,09	5,72	$4p\ {}^2P^\circ$—$9s\ {}^2S$	$^3/_2$—$^1/_2$
7554,162	1	4,09	5,72	$4p\ {}^2P^\circ$—$9s\ {}^2S$	$^1/_2$—$^1/_2$
7362,297	9	4,02	5,70	$3d\ {}^2D$—$7f\ {}^2F^\circ$	$^5/_2$—$^5/_2$
7361,568	8	4,02	5,70	$3d\ {}^2D$—$7f\ {}^2F^\circ$	$^3/_2$—$^5/_2$
7084,644	6	4,02	5,77	$3d\ {}^2D$—$8f\ {}^2F^\circ$	$^5/_2$—$^5/_2$
7083,968	5	4,02	5,77	$3d\ {}^2D$—$8f\ {}^2F^\circ$	$^3/_2$—$^5/_2$
6698,673	11	3,14	4,99	$4s\ {}^2S$—$5p\ {}^2P^\circ$	$^1/_2$—$^1/_2$
6696,023	13	3,14	4,99	$4s\ {}^2S$—$5p\ {}^2P^\circ$	$^1/_2$—$^3/_2$
5557,948	8	3,14	5,37	$4s\ {}^2S$—$6p\ {}^2P^\circ$	$^1/_2$—$^1/_2$
5557,063	10	3,14	5,37	$4s\ {}^2S$—$6p\ {}^2P^\circ$	$^1/_2$—$^3/_2$
5107,943	4	3,14	5,57	$4s\ {}^2S$—$7p\ {}^2P^\circ$	$^1/_2$—$^1/_2$
5107,520	6	3,14	5,57	$4s\ {}^2S$—$7p\ {}^2P^\circ$	$^1/_2$—$^3/_2$

λ, Å	I	E_H, eV	E_B, eV	Transition	J
3961,5200	26	0,01	3,14	$3p\ ^2P^\circ$—$4s\ ^2S$	$^3/_2$—$^1/_2$
3944,0058	24	0,00	3,14	$3p\ ^2P^\circ$—$4s\ ^2S$	$^1/_2$—$^1/_2$
3935,677	4	5,24	8,39	$5d\ ^2D$—$3d\ ^2D^\circ$	$^3/_2$—$^3/_2$
3931,996	5	5,24	8,39	$5d\ ^2D$—$3d\ ^2D^\circ$	$^5/_2$—$^5/_2$
3482,628	5	4,83	8,39	$4d\ ^2D$—$3d\ ^2D^\circ$	$^3/_2$—$^3/_2$
3479,806	5	4,83	8,39	$4d\ ^2D$—$3d\ ^2D^\circ$	$^5/_2$—$^5/_2$
3458,216	6	0,01	3,60	$3p\ ^2P^\circ$—$3p^2\ ^4P$	$^3/_2$—$^1/_2$
3452,657	2	0,01	3,60	$3p\ ^2P^\circ$—$3p^2\ ^4P$	$^3/_2$—$^3/_2$
3444,865	6	0,00	3,60	$3p\ ^2P^\circ$—$3p^2\ ^4P$	$^1/_2$—$^1/_2$
3443,640	9	0,01	3,61	$3p\ ^2P^\circ$—$3p^2\ ^4P$	$^3/_2$—$^5/_2$
3439,347	6	0,00	3,60	$3p\ ^2P^\circ$—$3p^2\ ^4P$	$^1/_2$—$^3/_2$
3092,8386	20	0,01	4,02	$3p\ ^2P^\circ$—$3d\ ^2D$	$^3/_2$—$^3/_2$
3092,7099	26	0,01	4,02	$3p\ ^2P^\circ$—$3d\ ^2D$	$^3/_2$—$^5/_2$
3082,1529	24	0,00	4,02	$3p\ ^2P^\circ$—$3d\ ^2D$	$^1/_2$—$^3/_2$
3066,145	5	3,61	7,66	$3p^2\ ^4P$—$4s\ ^4P^\circ$	$^5/_2$—$^3/_2$
3064,290	7	3,60	7,65	$3p^2\ ^4P$—$4s\ ^4P^\circ$	$^3/_2$—$^1/_2$
3059,924	4	3,60	7,65	$3p^2\ ^4P$—$4s\ ^4P^\circ$	$^1/_2$—$^1/_2$
3059,029	4	3,60	7,66	$3p^2\ ^4P$—$4s\ ^4P^\circ$	$^3/_2$—$^3/_2$
3057,144	14	3,61	7,67	$3p^2\ ^4P$—$4s\ ^4P^\circ$	$^5/_2$—$^5/_2$
3054,679	5	3,60	7,66	$3p^2\ ^4P$—$4s\ ^4P^\circ$	$^1/_2$—$^3/_2$
3050,073	13	3,60	7,67	$3p^2\ ^4P$—$4s\ ^4P^\circ$	$^3/_2$—$^5/_2$
2913,267	3	5,24	9,49	$5d\ ^2D$—$4d\ ^2P^\circ$	$^5/_2$—$^5/_2$
2902,258	2	5,48	9,75	$6d\ ^2D$—$6s\ ^2P^\circ$	$^3/_2$—$^5/_2$
2894,228	3	5,48	9,76	$6d\ ^2D$—$6s\ ^2P^\circ$	$^5/_2$—$^7/_2$
2840,205	2	4,02	8,39	$3d\ ^2D$—$3d\ ^2D^\circ$	$^5/_2$—$^3/_2$
2840,099	7	4,02	8,39	$3d\ ^2D$—$3d\ ^2D^\circ$	$^3/_2$—$^3/_2$
2837,963	7	4,02	8,39	$3d\ ^2D$—$3d\ ^2D^\circ$	$^5/_2$—$^5/_2$
2837,856	2	4,02	8,39	$3d\ ^2D$—$3d\ ^2D^\circ$	$^3/_2$—$^5/_2$
2748,065	3	5,24	9,75	$5d\ ^2D$—$6s\ ^2P^\circ$	$^3/_2$—$^5/_2$
2740,980	4	5,24	9,76	$5d\ ^2D$—$6s\ ^2P^\circ$	$^5/_2$—$^7/_2$
2660,386	12	0,01	4,67	$3p\ ^2P^\circ$—$5s\ ^2S$	$^3/_2$—$^1/_2$
2657,406	3	4,83	9,49	$4d\ ^2D$—$4d\ ^2P^\circ$	$^5/_2$—$^5/_2$
2652,475	12	0,00	4,67	$3p\ ^2P^\circ$—$5s\ ^2S$	$^1/_2$—$^1/_2$
2575,397	8	0,01	4,83	$3p\ ^2P^\circ$—$4d\ ^2D$	$^3/_2$—$^3/_2$
2575,095	10	0,01	4,83	$3p\ ^2P^\circ$—$4d\ ^2D$	$^3/_2$—$^5/_2$
2567,983	10	0,00	4,83	$3p\ ^2P^\circ$—$4d\ ^2D$	$^1/_2$—$^3/_2$
2519,514	1	4,83	9,75	$4d\ ^2D$—$6s\ ^2P^\circ$	$^5/_2$—$^5/_2$
2519,222	4	4,83	9,75	$4d\ ^2D$—$6s\ ^2P^\circ$	$^3/_2$—$^5/_2$
2513,305	5	4,83	9,76	$4d\ ^2D$—$6s\ ^2P^\circ$	$^5/_2$—$^7/_2$
2378,395	7	0,01	5,22	$3p\ ^2P^\circ$—$6s\ ^2S$	$^3/_2$—$^1/_2$
2374,496	4	3,61	8,83	$3p^2\ ^4P$—$3d\ ^4D^\circ$	$^5/_2$—$^3/_2$
2373,571	8	3,61	8,83	$3p^2\ ^4P$—$3d\ ^4D^\circ$	$^5/_2$—$^5/_2$
2373,351	15	0,01	5,24	$3p\ ^2P^\circ$—$5d\ ^2D$	$^3/_2$—$^3/_2$
2373,122	7	0,01	5,24	$3p\ ^2P^\circ$—$5d\ ^2D$	$^3/_2$—$^5/_2$
2372,070	5	0,00	5,22	$3p\ ^2P^\circ$—$6s\ ^2S$	$^1/_2$—$^1/_2$
2372,070	10	3,61	8,84	$3p^2\ ^4P$—$3d\ ^4D^\circ$	$^5/_2$—$^7/_2$
2370,726	6	3,60	8,83	$3p^2\ ^4P$—$3d\ ^4D^\circ$	$^3/_2$—$^1/_2$
2370,225	9	3,60	8,83	$3p^2\ ^4P$—$3d\ ^4D^\circ$	$^3/_2$—$^3/_2$
2369,304	10	3,60	8,83	$3p^2\ ^4P$—$3d\ ^4D^\circ$	$^3/_2$—$^5/_2$
2368,112	8	3,60	8,83	$3p^2\ ^4P$—$3d\ ^4D^\circ$	$^1/_2$—$^1/_2$
2367,611	8	3,60	8,83	$3p^2\ ^4P$—$3d\ ^4D^\circ$	$^1/_2$—$^3/_2$
2367,053	6	0,00	5,24	$3p\ ^2P^\circ$—$5d\ ^2D$	$^1/_2$—$^3/_2$
2321,562	9	3,61	8,95	$3p^2\ ^4P$—$3d\ ^4P^\circ$	$^5/_2$—$^5/_2$
2319,057	5	3,61	8,96	$3p^2\ ^4P$—$3d\ ^4P^\circ$	$^5/_2$—$^3/_2$
2317,482	7	3,60	8,95	$3p^2\ ^4P$—$3d\ ^4P^\circ$	$^3/_2$—$^5/_2$
2314,983	4	3,60	8,96	$3p^2\ ^4P$—$3d\ ^4P^\circ$	$^3/_2$—$^3/_2$
2313,526	6	3,60	8,96	$3p^2\ ^4P$—$3d\ ^4P^\circ$	$^3/_2$—$^1/_2$
2312,491	5	3,60	8,96	$3p^2\ ^4P$—$3d\ ^4P^\circ$	$^1/_2$—$^3/_2$
2311,035	4	3,60	8,96	$3p^2\ ^4P$—$3d\ ^4P^\circ$	$^1/_2$—$^1/_2$

λ, Å	I	E_H, eV	E_B, eV	Transition	J
2269,222	1	0,01	5,48	$3p\ ^2P^\circ$—$6d\ ^2D$	3/2—3/2
2269,096	7	0,01	5,48	$3p\ ^2P^\circ$—$6d\ ^2D$	3/2—5/2
2266,014	3	4,02	9,49	$3d\ ^2D$—$4d\ ^2P^\circ$	5/2—5/2
2263,738	1	0,01	5,49	$3p\ ^2P^\circ$—$7s\ ^2S$	3/2—1/2
2263,463	7	0,00	5,48	$3p\ ^2P^\circ$—$6d\ ^2D$	1/2—3/2
2258,008	1	0,00	5,49	$3p\ ^2P^\circ$—$7s\ ^2S$	1/2—1/2
2210,060	2	0,01	5,62	$3p\ ^2P^\circ$—$7d\ ^2D$	3/2—3/2, 5/2
2204,668	1	0,00	5,62	$3p\ ^2P^\circ$—$7d\ ^2D$	1/2—3/2
2204,619	1	0,01	5,64	$3p\ ^2P^\circ$—$8s\ ^2S$	3/2—1/2
2199,183	1	0,00	5,64	$3p\ ^2P^\circ$—$8s\ ^2S$	1/2—1/2
2180,996	8	3,14	8,82	$4s\ ^2S$—$3d\ ^2P^\circ$	1/2—3/2
2177,396	6	3,14	8,83	$4s\ ^2S$—$3d\ ^2P^\circ$	1/2—1/2
2174,071	2	0,01	5,71	$3p\ ^2P^\circ$—$8d\ ^2D$	3/2—3/2, 5/2
2168,826	2	0,00	5,71	$3p\ ^2P^\circ$—$8d\ ^2D$	1/2—3/2
2164,915	2	4,02	9,75	$3d\ ^2D$—$6s\ ^2P^\circ$	3/2—1/2
2160,388	3	4,02	9,76	$3d\ ^2D$—$6s\ ^2P^\circ$	5/2—3/2
2150,699	5	0,01	5,78	$3p\ ^2P^\circ$—$9d\ ^2D$	3/2—5/2
2145,555	3	0,00	5,78	$3p\ ^2P^\circ$—$9d\ ^2D$	1/2—3/2
2134,733	2	0,01	5,82	$3p\ ^2P^\circ$—$10d\ ^2D$	3/2—5/2
2129,663	1	0,00	5,82	$3p\ ^2P^\circ$—$10d\ ^2D$	1/2—3/2
2123,362	1	0,01	5,85	$3p\ ^2P^\circ$—$11d\ ^2D$	3/2—5/2
1769,140	4	0,01	7,02	$3p\ ^2P^\circ$—$3p^2\ ^2P$	3/2—1/2
1766,385	4	0,01	7,03	$3p\ ^2P^\circ$—$3p^2\ ^2P$	3/2—3/2
1765,636	4	0,00	7,02	$3p\ ^2P^\circ$—$3p^2\ ^2P$	1/2—1/2
1762,899	2	0,00	7,03	$3p\ ^2P^\circ$—$3p^2\ ^2P$	1/2—3/2

Al II, ground state $1s^2\ 2s^2\ 2p^6\ 3s^2\ ^1S_0$

Ionization potential 151 860,4 cm^{-1}; 18,827 eV

λ, Å	I	E_H, eV	E_B, eV	Transition	J
10107,19	4	11,85	13,07	$3d\ ^3D$—$4p\ ^3P^\circ$	2—1
10077,32	1	11,85	13,07	$3d\ ^3D$—$4p\ ^3P^\circ$	2—2
10076,29	6	11,85	13,07	$3d\ ^3D$—$4p\ ^3P^\circ$	3—2
9331,979	2 }	15,31	16,64	$4f\ ^1F^\circ$—$5g\ ^1G$	3—4
9331,546	3 }				
9290,747	5	15,30	16,64	$4f\ ^3F^\circ$—$5g\ ^3G$	4—3, 4, 5
9290,649	6	15,30	16,64	$4f\ ^3F^\circ$—$5g\ ^3G$	4—3, 4
9288,550	2	15,30	16,64	$4f\ ^3F^\circ$—$5g\ ^3G$	3—3, 4
9288,145	3	15,30	16,64	$4f\ ^3F^\circ$—$5g\ ^3G$	3—3, 4
9286,794	2 }	15,30	16,64	$4f\ ^3F^\circ$—$5g\ ^3G$	2—3
9286,578	1 }				
9249,41	1	16,64	17,98	$5g\ ^1G$—$8h\ ^1H^\circ$	4—5
8858,39	1	16,57	17,97	$5f\ ^1F^\circ$—$8g\ ^1G$	3—4
8680,31	3	16,54	17,97	$5f\ ^3F^\circ$—$8g\ ^3G$	4—3, 4, 5
8675,28	1	16,54	17,97	$5f\ ^3F^\circ$—$8g\ ^3G$	3—3, 4
8674,92	2	16,54	17,97	$5f\ ^3F^\circ$—$8g\ ^3G$	3—3, 4
8671,28	1	16,54	17,97	$5f\ ^3F^\circ$—$8g\ ^3G$	2—3
8640,7	8	11,82	13,26	$4s\ ^1S$—$4p\ ^1P^\circ$	0—1
8363,52	8	15,06	16,54	$4d\ ^3D$—$5f\ ^3F^\circ$	1—2
8363,30	1	15,06	16,54	$4d\ ^3D$—$5f\ ^3F^\circ$	2—2
8359,57	9	15,06	16,54	$4d\ ^3D$—$5f\ ^3F^\circ$	2—3
8359,23	1	15,06	16,54	$4d\ ^3D$—$5f\ ^3F^\circ$	3—3
8354,35	10	15,06	16,54	$4d\ ^3D$—$5f\ ^3F^\circ$	3—4
8160,15	3	16,64	18,15	$5g\ ^1G$—$9h\ ^1H^\circ$	4—5
8119,72	1,5	16,47	17,99	$5d\ ^3D$—$8f\ ^3F^\circ$	3—4

λ, Å	I	E_H, eV	E_B, eV	Transition	J
7823,72	2	15,58	17,17	$5p\ ^3P^\circ—7s\ ^3S$	2—1
7815,83	1	15,58	17,17	$5p\ ^3P^\circ—7s\ ^3S$	1—1
7635,33	2	15,58	17,21	$5p\ ^3P^\circ—6d\ ^3D$	2—1, 2, 3
7627,85	1	15,58	17,21	$5p\ ^3P^\circ—6d\ ^3D$	1—1, 2
7471,37	1	13,65	15,31	$3d\ ^1D—4f\ ^1F^\circ$	2—3
7449,12	5	15,60	17,27	$5p\ ^1P^\circ—6d\ ^1D$	1—2
7063,624	3	11,32	13,07	$4s\ ^3S—4p\ ^3P^\circ$	1—0
7056,56	4	11,32	13,07	$4s\ ^3S—4p\ ^3P^\circ$	1—1
7042,056	5	11,32	13,08	$4s\ ^3S—4p\ ^3P^\circ$	1—2
6919,96	1	13,26	15,05	$4p\ ^1P^\circ—5s\ ^1S$	1—0
6917,93	1	15,47	17,26	$4d\ ^1D—6f\ ^1F^\circ$	2—3
6837,094	3	13,07	14,89	$4p\ ^3P^\circ—5s\ ^3S$	2—1
6823,382	2	13,07	14,89	$4p\ ^3P^\circ—5s\ ^3S$	1—1
6816,827	1	13,07	14,89	$4p\ ^3P^\circ—5s\ ^3S$	0—1
6335,70	5	13,65	15,60	$3d\ ^1D—5p\ ^1P^\circ$	2—1
6243,347	10	13,07	15,06	$4p\ ^3P^\circ—4d\ ^3D$	2—3
6231,759	7	13,07	15,06	$4p\ ^3P^\circ—4d\ ^3D$	1—2
6226,193	5	13,07	15,06	$4p\ ^3P^\circ—4d\ ^3D$	0—1
6201,70 6201,52	9 10 }	15,31	17,31	$4f\ ^1F^\circ—6g\ ^1G$	3—4
6183,42	10	15,30	17,30	$4f\ ^3F^\circ—6g\ ^3G$	4—3, 4, 5
6182,45	7	15,30	17,30	$4f\ ^3F^\circ—6g\ ^3G$	3—3, 4
6182,28	8	15,30	17,30	$4f\ ^3F^\circ—6g\ ^3G$	3—3, 4
6181,68	6	15,30	17,30	$4f\ ^3F^\circ—6g\ ^3G$	2—3
6181,57	5	15,30	17,30	$4f\ ^3F^\circ—6g\ ^3G$	2—3
6073,17	3	15,59	17,63	$5p\ ^3P^\circ—8s\ ^3S$	2—1
6068,37	1	15,58	17,63	$5p\ ^3P^\circ—8s\ ^3S$	1—1
6066,40	2	15,58	17,63	$5p\ ^3P^\circ—8s\ ^3S$	0—1
6061,06	6	15,60	17,65	$5p\ ^1P^\circ—8s\ ^1S$	1—0
6006,38	10	15,58	17,65	$5p\ ^3P^\circ—7d\ ^3D$	2—1, 2, 3
6001,81	4	15,58	17,65	$5p\ ^3P^\circ—7d\ ^3D$	1—1, 2
6001,18	1	16,46	18,56	$6s\ ^1S—15p\ ^1P^\circ$	0—1
5999,83	2	15,58	17,65	$5p\ ^3P^\circ—7d\ ^3D$	0—1
5972,05	5	15,60	17,68	$5p\ ^1P^\circ—7d\ ^1D$	1—2
5867,81	3	15,06	17,17	$4d\ ^3D—6f\ ^3F^\circ$	1—2
5861,53	7	15,06	17,18	$4d\ ^3D—6f\ ^3F^\circ$	2—3
5853,62	5	15,06	17,19	$4d\ ^3D—6f\ ^3F^\circ$	3—4
5613,19	3	15,47	17,68	$4d\ ^1D—7f\ ^1F^\circ$	2—3
5593,23	10	13,26	15,47	$4p\ ^1P^\circ—4d\ ^1D$	1—2
5502,88	3	15,47	17,72	$4d\ ^1D—8p\ ^1P^\circ$	2—1
5388,48	1	15,05	17,35	$5s\ ^1S—7p\ ^1P^\circ$	0—1
5371,84	1	15,06	17,37	$4d\ ^3D—7p\ ^3P^\circ$	2, 3—1, 2
5324,61	4	15,60	17,93	$5p\ ^1P^\circ—9s\ ^1S$	1—0
5316,07	7	15,58	17,92	$5p\ ^3P^\circ—9s\ ^3S$	2—1
5312,32	5	15,58	17,92	$5p\ ^3P^\circ—9s\ ^3S$	1—1
5310,76	2	15,58	17,92	$5p\ ^3P^\circ—9s\ ^3S$	0—1
5285,85	6	15,60	17,95	$5p\ ^1P^\circ—8d\ ^1D$	1—2
5283,77	8	15,58	17,93	$5p\ ^3P^\circ—8d\ ^3D$	2—1, 2, 3
5280,21	6	15,58	17,93	$5p\ ^3P^\circ—8d\ ^3D$	1—1, 2
5278,62	3	15,58	17,93	$5p\ ^3P^\circ—8d\ ^3D$	0—1
5277,68	2	15,30	17,65	$4f\ ^3F^\circ—7d\ ^3D$	4—3
5276,81	2	15,30	17,65	$4f\ ^3F^\circ—7d\ ^3D$	3—2, 3
5276,42	2	15,30	17,65	$4f\ ^3F^\circ—7d\ ^3D$	2—1, 2, 3
5158,187	1	15,31	17,71	$4f\ ^1F^\circ—7g\ ^1G$	3—4
5145,654	1,5	15,30	17,71	$4f\ ^3F^\circ—7g\ ^3G$	4—3, 4, 5
5144,998	1	15,30	17,71	$4f\ ^3F^\circ—7g\ ^3G$	3—3, 4
5100,34	1	15,06	17,49	$4d\ ^3D—3d\ ^3F^\circ$	1—2
5093,65	2	15,06	17,49	$4d\ ^3D—3d\ ^3F^\circ$	2—3

λ, Å	I	E_H, eV	E_B, eV	Transition	J
5085,02	4	15,06	17,50	$4d\,^3D - 3d\,^3F^\circ$	3—4
5000,97	3	15,47	17,95	$4d\,^1D - 8f\,^1F^\circ$	2—3
4962,10	3	15,47	17,97	$4d\,^1D - 9p\,^1P^\circ$	2—1
4918,98	3	15,60	18,12	$5p\,^1P^\circ - 10s\,^1S$	1—0
4902,77	5	15,58	18,11	$5p\,^3P^\circ - 10s\,^3S$	2—1
4899,64	3	15,58	18,11	$5p\,^3P^\circ - 10s\,^3S$	1—1
4898,76	5	15,60	18,13	$5p\,^1P^\circ - 9d\,^1D$	1—2
4898,52	2	15,58	18,11	$5p\,^3P^\circ - 10s\,^3S$	0—1
4666,8	11	15,60	18,26	$5p\,^1P^\circ - 11s\,^1S$	1—0
4663,054	10	10,60	13,26	$3p^2\,^1D - 4p\,^1P^\circ$	2—1
4650,646	1,5	15,31	17,97	$4f\,^1F^\circ - 8g\,^1G$	3—4
4650,544	2				
4648,62	1	15,47	18,14	$4d\,^1D - 10p\,^1P^\circ$	2—1
4640,384	3,5	15,30	17,97	$4f\,^3F^\circ - 8g\,^3G$	4—3, 4, 5
4640,362	4				
4639,833	1,5	15,30	17,97	$4f\,^3F^\circ - 8g\,^3G$	3—3, 4
4639,725	2				
4639,384	1	15,30	17,97	$4f\,^3F^\circ - 8g\,^3G$	2—3
4635,7	1	15,58	18,26	$5p\,^3P^\circ - 10d\,^3D$	2—1, 2, 3
4629,7	1	15,05	17,72	$5s\,^1S - 8p\,^1P^\circ$	0—1
4609,7	1	15,06	17,75	$4d\,^3D - 8p\,^3P^\circ$	1, 2, 3—0, 1, 2
4589,75	4	15,06	17,76	$4d\,^3D - 7f\,^3F^\circ$	1—2
4589,689	1	15,06	17,76	$4d\,^3D - 7f\,^3F^\circ$	2—2
4588,194	5	15,06	17,76	$4d\,^3D - 7f\,^3F^\circ$	2—3
4585,82	6	15,06	17,76	$4d\,^3D - 7f\,^3F^\circ$	3—4
4447,8	3	15,47	18,26	$4d\,^1D - 11p\,^1P^\circ$	2—1
4356,807	1,5	15,31	18,15	$4f\,^1F^\circ - 9g\,^1G$	3—4
4356,711	2				
4347,802	3,5	15,30	18,15	$4f\,^3F^\circ - 9g\,^3G$	4—3, 4, 5
4347,785	4				
4347,316	1,5	15,30	18,15	$4f\,^3F^\circ - 9g\,^3G$	3—3, 4
4347,223	2				
4346,918	1	15,30	18,15	$4f\,^3F^\circ - 9g\,^3G$	2—3
4307,20	3	15,47	18,35	$4d\,^1D - 4s\,^1P^\circ$	2—1
4240,75	3	15,05	17,97	$5s\,^1S - 9p\,^1P^\circ$	0—1
4227,982	4	15,06	17,99	$4d\,^3D - 8f\,^3F^\circ$	1—2
4227,923	1,5	15,06	17,99	$4d\,^3D - 8f\,^3F^\circ$	2—2
4227,493	5	15,06	17,99	$4d\,^3D - 8f\,^3F^\circ$	2—3
4227,406	2	15,06	17,99	$4d\,^3D - 8f\,^3F^\circ$	3—3
4226,812	6	15,06	17,99	$4d\,^3D - 8f\,^3F^\circ$	3—4
4202,40	2	15,47	18,42	$4d\,^1D - 12p\,^1P^\circ$	2—1
4168,424	1	15,31	18,28	$4f\,^1F^\circ - 10g\,^1G$	3—4
4160,263	3	15,30	18,28	$4f\,^3F^\circ - 10g\,^3G$	4—3, 4, 5
4160,239	2,5	15,30	18,28	$4f\,^3F^\circ - 10g\,^3G$	4—3, 4, 5
4159,809	1	15,30	18,28	$4f\,^3F^\circ - 10g\,^3G$	3—3, 4
4159,725	1,5	15,30	18,28	$4f\,^3F^\circ - 10g\,^3G$	3—3, 4
4159,450	1	15,30	18,28	$4f\,^3F^\circ - 10g\,^3G$	2—3
4026,5	5	13,65	16,73	$3d\,^1D - 6p\,^1P^\circ$	2—1
4009,58	1	15,05	18,14	$5s\,^1S - 10p\,^1P^\circ$	0—1
3996,381	3	15,06	18,16	$4d\,^3D - 9f\,^3F^\circ$	1—2
3996,159	4	15,06	18,16	$4d\,^3D - 9f\,^3F^\circ$	2—3
3996,075	1	15,06	18,16	$4d\,^3D - 9f\,^3F^\circ$	3—3
3995,86	5	15,06	18,16	$4d\,^3D - 9f\,^3F^\circ$	3—4
3900,680	10	7,42	10,60	$3p\,^1P^\circ - 3p^2\,^1D$	1—2
3866,160	2	13,26	16,46	$4p\,^1P^\circ - 6s\,^1S$	1—0
3859,33	3	15,05	18,26	$5s\,^1S - 11p\,^1P^\circ$	0—1
3842,317	1	15,06	18,29	$4d\,^3D - 10f\,^3F^\circ$	1—2

λ, Å	I	E_H, eV	E_B, eV	Transition	J
3842,213	2	15,06	18,29	$4d\ ^3D—10f\ ^3F^\circ$	2—3
3842,037	3	15,06	18,29	$4d\ ^3D—10f\ ^3F^\circ$	3—4
3753,10	1	15,05	18,35	$5s\ ^1S—4s\ ^1P^\circ$	0—1
3738,003	3	13,08	16,39	$4p\ ^3P^\circ—6s\ ^3S$	2—1
3734,567	1	15,06	18,37	$4d\ ^3D—11f\ ^3F^\circ$	3—4
3733,910	2	13,07	16,39	$4p\ ^3P^\circ—6s\ ^3S$	1—1
3731,950	1	13,07	16,39	$4p\ ^3P^\circ—6s\ ^3S$	0—1
3703,217	4	13,26	16,60	$4p\ ^1P^\circ—5d\ ^1D$	1—2
3655,00	8	13,08	16,47	$4p\ ^3P^\circ—5d\ ^3D$	2—3
3654,979	4	13,08	16,47	$4p\ ^3P^\circ—5d\ ^3D$	2—1, 2
3651,090	4	13,07	16,47	$4p\ ^3P^\circ—5d\ ^3D$	1—1, 2
3651,064	6	13,07	16,47	$4p\ ^3P^\circ—5d\ ^3D$	1—1, 2
3649,221 3649,182	1 1,5	} 13,07	16,47	$4p\ ^3P^\circ—5d\ ^3D$	0—1
3597,50	2	15,06	18,51	$4d\ ^3D—13f\ ^3F^\circ$	3—2, 3, 4
3587,441	7	11,85	15,30	$3d\ ^3D—4f\ ^3F^\circ$	1—2
3587,327	2	11,85	15,30	$3d\ ^3D—4f\ ^3F^\circ$	2—2
3587,176	1	11,85	15,30	$3d\ ^3D—4f\ ^3F^\circ$	3—2
3587,057	8	11,85	15,30	$3d\ ^3D—4f\ ^3F^\circ$	2—3
3586,908	3,5	11,85	15,30	$3d\ ^3D—4f\ ^3F^\circ$	3—3
3586,692	2	11,85	15,30	$3d\ ^3D—4f\ ^3F^\circ$	2—4
3586,546	9	11,85	15,30	$3d\ ^3D—4f\ ^3F^\circ$	3—4
3552,00	1	15,06	18,55	$4d\ ^3D—14f\ ^3F^\circ$	3—2, 3, 4
3428,916	6	13,65	17,26	$3d\ ^1D—6f\ ^1F^\circ$	2—3
3351,456	3	13,65	17,35	$3d\ ^1D—7p\ ^1P^\circ$	2—1
3315,614	1	11,85	15,58	$3d\ ^3D—5p\ ^3P^\circ$	1—0
3314,889	2	11,85	15,58	$3d\ ^3D—5p\ ^3P^\circ$	2—1
3313,351	3	11,85	15,58	$3d\ ^3D—5p\ ^3P^\circ$	3—2
3275,776	4	11,82	15,60	$4s\ ^1S—5p\ ^1P^\circ$	0—1
3135,875	5	13,26	17,21	$4p\ ^1P^\circ—7s\ ^1S$	1—0
3088,523	3	13,26	17,27	$4p\ ^1P^\circ—6d\ ^1D$	1—2
3074,665	6	13,65	17,68	$3d\ ^1D—7f\ ^1F^\circ$	2—3
3041,278	6	13,65	17,72	$3d\ ^1D—8p\ ^1P^\circ$	2—1
3026,762	1,5	13,08	17,17	$4p\ ^3P^\circ—7s\ ^3S$	2—1
3024,074	1	13,07	17,17	$4p\ ^3P^\circ—7s\ ^3S$	1—1
3001,82	3	—	—	—	—
2998,174	2	13,08	17,21	$4p\ ^3P^\circ—6d\ ^3D$	2—1, 2, 3
2995,524	1,5	13,07	17,21	$4p\ ^3P^\circ—6d\ ^3D$	1—1, 2
2994,280	1	13,07	17,21	$4p\ ^3P^\circ—6d\ ^3D$	0—1
2924,52	3	—	—	—	—
2903,19	1	11,32	15,58	$4s\ ^3S—5p\ ^3P^\circ$	1—1
2902,08	2	11,32	15,58	$4s\ ^3S—5p\ ^3P^\circ$	1—2
2884,20	4	—	—	—	—
2881,463	4	13,65	17,95	$3d\ ^1D—8f\ ^1F^\circ$	2—3
2868,52	9	13,65	17,97	$3d\ ^1D—9p\ ^1P^\circ$	2—1
2820,632	1	13,26	17,65	$4p\ ^1P^\circ—8s\ ^1S$	1—0
2816,179	20	7,42	11,82	$3p\ ^1P^\circ—4s\ ^1S$	1—0
2805,65	4	—	—	—	—
2762,460	2	13,65	18,13	$3d\ ^1D—9f\ ^1F^\circ$	2—3
2760,852	1	13,65	18,14	$3d\ ^1D—10p\ ^1P^\circ$	2—1
2723,091	2	13,08	17,63	$4p\ ^3P^\circ—8s\ ^3S$	2—1
2720,918	1	13,07	17,63	$4p\ ^3P^\circ—8s\ ^3S$	1—1
2709,582	1,5	13,08	17,65	$4p\ ^3P^\circ—7d\ ^3D$	2—1, 2, 3
2688,728	2	13,65	18,26	$3d\ ^1D—11p\ ^1P^\circ$	2—1
2683,280	3	13,65	18,27	$3d\ ^1D—10f\ ^1F^\circ$	2—3
2669,166	10	0,00	4,64	$3s^2\ ^1S—3p\ ^3P^\circ$	0—1
2650,10	4	13,26	17,93	$4p\ ^1P^\circ—9s\ ^1S$	1—0
2640,36	3	13,26	17,95	$4p\ ^1P^\circ—8d\ ^1D$	1—2

λ, Å	I	E_H, eV	E_B, eV	Transition	J
2638,695	3	11,85	16,54	$3d\ ^3D—5f\ ^3F^\circ$	1—2
2638,263	4	11,85	16,54	$3d\ ^3D—5f\ ^3F^\circ$	2—3
2637,696	5	11,85	16,54	$3d\ ^3D—5f\ ^3F^\circ$	3—4
2636,725	6	13,65	18,35	$3d\ ^1D—4s\ ^1P^\circ$	2—1
2635,17	1	10,60	15,30	$3p^2\ ^1D—4f\ ^3F^\circ$	2—2
2635,03	3	10,60	15,30	$3p^2\ ^1D—4f\ ^3F^\circ$	2—3
2631,553	7	10,60	15,31	$3p^2\ ^1D—4f\ ^1F^\circ$	2—3
2627,68	7	13,65	18,37	$3d\ ^1D—11f\ ^1F^\circ$	2—3
2597,18	6	13,65	18,42	$3d\ ^1D—12p\ ^1P^\circ$	2—1
2586,95	6	13,65	18,44	$3d\ ^1D—12f\ ^1F^\circ$	2—3
2565,68	4	13,65	18,48	$3d\ ^1D—13p\ ^1P^\circ$	2—1
2559,614	3	13,08	17,92	$4p\ ^3P^\circ—9s\ ^3S$	2—1
2557,71	5	13,07	17,92	$4p\ ^3P^\circ—9s\ ^3S$	1—1
2556,78	3	13,07	17,92	$4p\ ^3P^\circ—9s\ ^3S$	0—1
2556,01	4	13,65	18,50	$4d\ ^1D—13f\ ^1F^\circ$	2—3
2552,12	5	13,08	17,93	$4p\ ^3P^\circ—8d\ ^3D$	2—1, 2, 3
2550,23	3	13,07	17,93	$4p\ ^3P^\circ—8d\ ^3D$	1—1, 2
2549,30	1,5	13,07	17,93	$4p\ ^3P^\circ—8d\ ^3D$	0—1
2545,60	6	13,26	18,12	$4p\ ^1P^\circ—10s\ ^1S$	1—0
2540,70	1	13,65	18,53	$3d\ ^1D—14p\ ^1P^\circ$	2—1
2540,12	4	13,26	18,13	$4p\ ^1P^\circ—9d\ ^1D$	1—2
2533,16	1	11,85	16,74	$3d^3\ D—6p\ ^3P^\circ$	2—1
2532,655	2	11,85	16,74	$3d\ ^3D—6p\ ^3P^\circ$	3—2
2532,10	3	13,65	18,54	$3d\ ^1D—14f\ ^1F^\circ$	2—3
2526,477	1	11,82	16,73	$4s\ ^1S—6p\ ^1P^\circ$	0—1
2520,64	2	13,65	18,57	$3d\ ^1D—15p\ ^1P^\circ$	2—1
2513,15	1	13,65	18,58	$3d\ ^1D—15f\ ^1F^\circ$	2—3
2504,25	1	13,65	18,60	$3d\ ^1D—16p\ ^1P^\circ$	2—1
2497,85	2	13,65	18,61	$3d\ ^1D—16f\ ^1F^\circ$	2—3
2485,35	1	13,65	18,63	$3d\ ^1D—17f\ ^1F^\circ$	2—3
2476,30	4	13,26	18,26	$4p\ ^1P^\circ—11s\ ^1S$	1—0
2475,260	4	10,60	15,60	$3p^2\ ^1D—5p\ ^1P^\circ$	2—1
2472,95	1	13,26	18,27	$4p\ ^1P^\circ—10d\ ^1D$	1—2
2459,82	4	13,08	18,11	$4p\ ^3P^\circ—10s\ ^3S$	2—1
2458,05	2	13,07	18,11	$4p\ ^3P^\circ—10s\ ^3S$	1—1
2457,20	1	13,07	18,11	$4p\ ^3P^\circ—10s\ ^3S$	0—1
2455,22	2	13,08	18,12	$4p\ ^3P^\circ—9d\ ^3D$	2—1 ,2, 3
2453,47	1	13,07	18,12	$4p\ ^3P^\circ—9d\ ^3D$	1—1, 2
2427,70	3	13,26	18,36	$4p\ ^1P^\circ—12s\ ^1S$	1—0
2393,835	2	13,08	18,25	$4p\ ^3P^\circ—11s\ ^3S$	2—1
2392,15	4	13,07	18,25	$4p\ ^3P^\circ—11s\ ^3S$	1—1
2391,35	1	13,07	18,25	$4p\ ^3P^\circ—11s\ ^3S$	0—1
2390,755	2	13,07	18,26	$4p\ ^3P^\circ—10d\ ^3D$	2—1, 2, 3
2389,08	1	13,07	18,26	$4p\ ^3P^\circ—10d\ ^3D$	1—1, 2
2365,49	1,5	13,26	18,49	$4p\ ^1P^\circ—14s\ ^1S$	1—0
2350,20	4	—	—	—	—
2347,54	2,5	13,08	18,36	$4p\ ^3P^\circ—12s\ ^3S$	2—1
2345,92	1,5	13,07	18,36	$4p\ ^3P^\circ—12s\ ^3S$	1—1
2344,69	1	13,26	18,54	$4p\ ^1P^\circ—15s\ ^1S$	1—0
2326,498	2	11,85	17,17	$3d\ ^3D—6f\ ^3F^\circ$	1—2
2325,497	3	11,85	17,18	$3d\ ^3D—6f\ ^3F^\circ$	2—3
2324,20	4	11,85	17,18	$3d\ ^3D—6f\ ^3F^\circ$	3—4
2313,77	1	13,08	18,43	$4p\ ^3P^\circ—13s\ ^3S$	2—1
2285,52	2	11,32	16,74	$4s\ ^3S—6p\ ^3P^\circ$	1—1
2285,17	3	11,32	16,74	$4s\ ^3S—6p\ ^3P^\circ$	1—2
2243,05	4	11,82	17,35	$4s\ ^1S—7p\ ^1P^\circ$	0—1
2194,251	1	11,85	17,49	$3d\ ^3D—3d\ ^3F^\circ$	2—3

λ, Å	I	E_H, eV	E_B, eV	Transition	J
2192,607	1,5	11,85	17,50	$3d\,^3D—3d\,^3F^\circ$	3—4
2099,68	5	11,82	17,72	$4s\,^1S—8p\,^1P^\circ$	0—1
2095,2	5	11,85	17,76	$3d\,^3D—7f\,^3F^\circ$	1—2
2094,8	5,5	11,85	17,76	$3d\,^3D—7f\,^3F^\circ$	2—3
2094,3	6	11,85	17,76	$3d\,^3D—7f\,^3F^\circ$	3—4
2087,0	5	4,66	10,60	$3p\,^3P^\circ—3p^2\,^1D$	2—2
2081,5	2	4,64	10,60	$3p\,^3P^\circ—3p^2\,^1D$	1—2
2073,8	3	10,60	16,57	$3p^2\,^1D—5f\,^1F^\circ$	2—3
2047,72	1	11,32	17,37	$4s\,^3S—7p\,^3P^\circ$	1—0, 1
2039,93	3	—	—	—	—
2022,14	2	10,60	16,73	$3p^2\,^1D—6p\,^1P^\circ$	2—1
2016,09	1	11,85	17,99	$3d\,^3D—8f\,^3F^\circ$	1, 2, 3—2, 3, 4
1990,53	7	7,42	13,65	$3p\,^1P^\circ—3d\,^1D$	1—2
1965,23	4	11,69	18,00	$3p^2\,^3P—3d\,^3D^\circ$	2—3
1962,67	7	11,35	18,16	$3d\,^3D—9f\,^3F^\circ$	1, 2, 3—2, 3, 4
		11,82	18,14	$4s\,^1S—10p\,^1P^\circ$	0—1
1960,70	3	11,67	18,00	$3p^2\,^3P—3d\,^3D^\circ$	1—1, 2
1958,29	1	11,66	18,00	$3p^2\,^3P—3d\,^3D^\circ$	0—1
1945,35	5	—	—	—	—
1939,30	5	11,69	18,08	$3p^2\,^3P—4s\,^3P^\circ$	2—1
1936,96	4	11,67	18,07	$3p^2\,^3P—4s\,^3P^\circ$	1—0
1934,75	10	11,67	18,08	$3p^2\,^3P—4s\,^3P^\circ$	1—1
1934,54	10	11,69	18,10	$3p^2\,^3P—4s\,^3P^\circ$	2—2
1932,43	5	11,66	18,08	$3p^2\,^3P—4s\,^3P^\circ$	0—1
1930,03	5	11,67	18,10	$3p^2\,^3P—4s\,^3P^\circ$	1—2
1926,99	1	11,32	17,75	$4s\,^3S—8p\,^3P^\circ$	1—0, 1, 2
1925,99	2	11,82	18,26	$4s\,^1S—11p\,^1P^\circ$	0—1
1924,81	4	11,85	18,29	$3d\,^3D—10f\,^3F^\circ$	1, 2, 3—2, 3, 4
1910,91	5	11,69	18,17	$3p^2\,^3P—3d\,^3P^\circ$	2—2
1906,57	4	11,67	18,17	$3p^2\,^3P—3d\,^3P^\circ$	1—1
1904,38	2	11,66	18,17	$3p^2\,^3P—3d\,^3P^\circ$	0—1
1899,17	4	11,82	18,35	$4s\,^1S—4s\,^1P^\circ$	0—1
1897,49	2	11,85	18,38	$3d\,^3D—11f\,^3F^\circ$	1, 2, 3—2, 3, 4
1878,48	3	11,82	18,42	$4s\,^1S—12p\,^1P^\circ$	0—1
1877,13	1	11,85	18,45	$3d\,^3D—12f\,^3F^\circ$	1, 2, 3—2, 3, 4
1862,34	15	4,66	11,32	$3p\,^3P^\circ—4s\,^3S$	2—1
1859,99	3	10,60	17,26	$3p^2\,^1D—6f\,^1F^\circ$	2—3
1858,05	10	4,64	11,32	$3p\,^3P^\circ—4s\,^3S$	1—1
1855,95	8	4,64	11,32	$3p\,^3P^\circ—4s\,^3S$	0—1
		11,32	18,00	$4s\,^3S—3d\,^3D^\circ$	1—1, 2
1854,76	3	11,32	18,00	$4s\,^3S—9p\,^3P^\circ$	1—0, 1, 2
1848,90	2	11,85	18,55	$3d\,^3D—14f\,^3F^\circ$	1, 2, 3—2, 3, 4
		11,82	18,53	$4s\,^1S—14p\,^1P^\circ$	0—1
1839,64	2	11,85	18,58	$3d\,^3D—15f\,^3F^\circ$	1, 2, 3—2, 3, 4
1838,27	1	11,82	18,56	$4s\,^1S—15p\,^1P^\circ$	0—1
1836,97	1	10,60	17,35	$3p^2\,^1D—7p\,^1P^\circ$	2—1
1834,82	6	11,32	18,05	$4s\,^3S—4s\,^3P^\circ$	1—0
1832,87	8	11,32	18,07	$4s\,^3S—4s\,^3P^\circ$	1—1
1828,61	10	11,32	18,10	$4s\,^3S—4s\,^3P^\circ$	1—2
1807,40	4	—	—	—	—
1767,76	7	4,66	11,67	$3p\,^3P^\circ—3p^2\,^3P$	2—1
1765,82	4	4,64	11,66	$3p\,^3P^\circ—3p^2\,^3P$	1—0
1764,01	10	4,66	11,69	$3p\,^3P^\circ—3p^2\,^3P$	2—2
1763,85	8	4,64	11,67	$3p\,^3P^\circ—3p^2\,^3P$	1—1
1762,00	5	4,64	11,67	$3p\,^3P^\circ—3p^2\,^3P$	0—1
1760,15	7	4,64	11,69	$3p\,^3P^\circ—3p^2\,^3P$	1—2
1750,56	6	10,60	17,68	$3p^2\,^1D—7f\,^1F^\circ$	2—3
1739,64	5	10,60	17,72	$3p^2\,^1D—8p\,^1P^\circ$	2—1

λ, Å	I	E_H, eV	E_B, eV	Transition	J
1725,01	15	4,66	11,85	$3p\,^3P^\circ-3d\,^3D$	2—1, 2, 3
1721,31	10	4,64	11,85	$3p\,^3P^\circ-3d\,^3D$	1—1, 2
1719,43	8	4,64	11,85	$3p\,^3P^\circ-3d\,^3D$	0—1
1686,19	5	10,60	17,95	$3p^2\,^1D-8f\,^1F^\circ$	2—3
1681,78	5	10,60	17,97	$3p^2\,^1D-9p\,^1P^\circ$	2—1
1670,81	15	0,00	7,42	$3s^2\,^1S-3p\,^1P^\circ$	0—1
1644,78	5	10,60	18,13	$3p^2\,^1D-9f\,^1F^\circ$	2—3
1644,15	5	10,60	18,14	$3p^2\,^1D-10p\,^1P^\circ$	2—1
1625,60	3	7,42	15,05	$3p\,^1P^\circ-5s\,^1S$	1—0
1618,38	4	10,60	18,26	$3p^2\,^1D-11p\,^1P^\circ$	2—1
1616,41	4	10,60	18,27	$3p^2\,^1D-10f\,^1F^\circ$	2—3
1599,44	3	10,60	18,35	$3p^2\,^1D-4s\,^1P^\circ$	2—1
1596,02	3	10,60	18,36	$3p^2\,^1D-11f\,^1F^\circ$	2—3
1584,77	2	10,60	18,42	$3p^2\,^1D-12p\,^1P^\circ$	2—1
1580,93	2	10,60	18,44	$3p^2\,^1D-12f\,^1F^\circ$	2—3
1572,97	1	10,60	18,48	$3p^2\,^1D-13p\,^1P^\circ$	2—1
1569,35	1	10,60	18,50	$3p^2\,^1D-13f\,^1F^\circ$	2—3
1563,56	1	10,60	18,52	$3p^2\,^1D-14p\,^1P^\circ$	2—1
1560,35	1	10,60	18,54	$3p^2\,^1D-14f\,^1F^\circ$	2—3
1539,74	10	7,42	15,47	$3p\,^1P^\circ-4d\,^1D$	1—2
1371,26	2	7,42	16,46	$3p\,^1P^\circ-6s\,^1S$	1—0
1350,15	6	7,42	16,60.	$3p\,^1P^\circ-5d\,^1D$	1—2
1258,88	4	7,42	17,27	$3p\,^1P^\circ-6d\,^1D$	1—2
1211,93	3	4,66	14,89	$3p\,^3P^\circ-5s\,^3S$	2—1
1210,15	2	{ 7,42	17,65	$3p\,^1P^\circ-8s\,^1S$	1—0
		4,64	14,89	$3p\,^3p^\circ-5s\,^3s$	1—1
1209,19	1	4,64	14,89	$3p\,^3P^\circ-5s\,^3S$	0—1
1208,35	3	7,42	17,68	$3p\,^1P^\circ-7d\,^1D$	1—2
1191,86	5	4,66	15,06	$3p\,^3P^\circ-4d\,^3D$	2—1, 2, 3
1190,07	4	4,64	15,06	$3p\,^3P^\circ-4d\,^3D$	1—1, 2
1189,21	2	4,64	15,06	$3p\,^3P^\circ-4d\,^3D$	0—1
1179,38	1	7,42	17,93	$3p\,^1P^\circ-9s\,^1S$	1—0
1177,48	4	7,42	17,95	$3p\,^1P^\circ-8d\,^1D$	1—2
1158,14	1	7,42	18,12	$3p\,^1P^\circ-10s\,^1S$	1—0
1157,13	3	7,42	18,13	$3p\,^1P^\circ-9d\,^1D$	1—2
1152,14	4	—	—	—	—
1142,97	2	7,42	18,26	$3p\,^1P^\circ-10d\,^1D$	1—2
1049,93	2	4,66	16,47	$3p\,^3P^\circ-5d\,^3D$	2—1, 2, 3
1048,53	1	4,64	16,47	$3p\,^3P^\circ-5d\,^3D$	1—1, 2
990,88	1	4,66	17,17	$3p\,^3P^\circ-7s\,^3S$	2—1
955,99	1	4,66	17,63	$3p\,^3P^\circ-8s\,^3S$	2—1
935,20	1	0,00	13,26	$3s^2\,^1S-4p\,^1P^\circ$	0—1

Al III, ground state $1s^2\,2s^2\,2p^6\,3s\,^2S_{1/2}$
Ionization potential 229 453,99 cm^{-1}; 28,447 eV

λ, Å	I	E_H, eV	E_B, eV	Transition	J
5722,65	6	15,64	17,80	$4s\,^2S-4p\,^2P^\circ$	$1/2-1/2$
5696,47	8	15,64	17,81	$4s\,^2S-4p\,^2P^\circ$	$1/2-3/2$
5260,91	0	23,54	25,89	$5f\,^2F^\circ-7d\,^2D$	$5/2,\ 7/2-3/2,\ 5/2$
5172,6	1	23,54	25,94	$5g\,^2G-7f\,^2F^\circ$	$7/2,\ 9/2-5/2,\ 7/2$
5163,90	7	23,54	25,94	$5g\,^2G-7h\,^2H^\circ$	$7/2,\ 9/2-9/2,\ 11/2$
5150,86	6	23,54	25,94	$5f\,^2F^\circ-7g\,^2G$	$5/2,\ 7/2-7/2,\ 9/2$
4903,71	4	23,41	25,94	$5d\,^2D-7f\,^2F^\circ$	$3/2,\ 5/2-5/2,\ 7/2$
4701,65	6	20,78	23,41	$4f\,^2F^\circ-5d\,^2D$	$5/2,\ 7/2-3/2,\ 5/2$

λ, Å	I	E_H, eV	E_B, eV	Transition	J
4529,176	10	17,81	20,55	$4p\ ^2P^\circ$—$4d\ ^2D$	$^3/_2$—$^5/_2$
4528,911	1	17,81	20,55	$4p\ ^2P^\circ$—$4d\ ^2D$	$^3/_2$—$^3/_2$
4512,535	8	17,80	20,55	$4p\ ^2P^\circ$—$4d\ ^2D$	$^1/_2$—$^3/_2$
4490,90	2	20,78	23,54	$4f\ ^2F^\circ$—$5f\ ^2F^\circ$	$^5/_2$, $^7/_2$—$^5/_2$, $^7/_2$
4479,968	4	20,78	23,54	$4f\ ^2F^\circ$—$5g\ ^2G$	$^7/_2$—$^7/_2$, $^9/_2$
4479,891	3	20,78	23,54	$4f\ ^2F^\circ$—$5g\ ^2G$	$^5/_2$—$^7/_2$
4364,59	2	22,12	24,96	$5p\ ^2P^\circ$—$6d\ ^2D$	$^3/_2$—$^3/_2$, $^5/_2$
4357,2	—	22,12	24,96	$5p\ ^2P^\circ$—$6d\ ^2D$	$^1/_2$—$^3/_2$
4199,00	0,5	—	—	—	—
4188,88	0,5	23,54	26,49	$5f\ ^2F^\circ$—$8d\ ^2D$	$^5/_2$, $^7/_2$—$^3/_2$, $^5/_2$
4150,138	8	20,55	23,54	$4d\ ^2D$—$5f\ ^2F^\circ$	$^3/_2$—$^5/_2$
4149,917	1	20,55	23,54	$4d\ ^2D$—$5f\ ^2F^\circ$	$^5/_2$—$^5/_2$
4149,897	10	20,55	23,54	$4d\ ^2D$—$5f\ ^2F^\circ$	$^5/_2$—$^7/_2$
4142,15	2	23,54	26,53	$5f\ ^2F^\circ$—$8g\ ^2G$	$^5/_2$, $^7/_2$—$^7/_2$, $^9/_2$
4141,25	0	23,54	26,42	$5f\ ^2F^\circ$—$8h\ ^2H^\circ$	$^7/_2$—$^9/_2$
4140,21	2	—	—	—	—
3980,56	2	23,41	26,50	$5d\ ^2D$—$8f\ ^2F^\circ$	$^3/_2$, $^5/_2$—$^5/_2$, $^7/_2$
3713,103	15	17,81	21,15	$4p\ ^2P^\circ$—$5s\ ^2S$	$^3/_2$—$^1/_2$
3702,086	10	17,80	21,15	$4p\ ^2P^\circ$—$5s\ ^2S$	$^1/_2$—$^1/_2$
3658,3	1	23,54	26,92	$5g\ ^2G$—$9h\ ^2H^\circ$	$^7/_2$, $^9/_2$—$^9/_2$, $^{11}/_2$
3612,352	15	14,37	17,80	$3d\ ^2D$—$4p\ ^2P^\circ$	$^3/_2$—$^1/_2$
3601,916	1	14,37	17,81	$3d\ ^2D$—$4p\ ^2P^\circ$	$^3/_2$—$^3/_2$
3601,623	20	14,37	17,81	$3d\ ^2D$—$4p\ ^2P^\circ$	$^5/_2$—$^3/_2$
3287,37	1	22,12	25,89	$5p\ ^2P^\circ$—$7d\ ^2D$	$^3/_2$—$^3/_2$, $^5/_2$
3283,114	0,5	22,12	25,89	$5p\ ^2P^\circ$—$7d\ ^2D$	$^1/_2$—$^3/_2$
2961,06	1,5	20,76	24,96	$4f\ ^2F^\circ$—$6d\ ^2D$	$^5/_2$—$^3/_2$, $^5/_2$
2909,77	2	20,76	25,03	$4f\ ^2F^\circ$—$6f\ ^2F^\circ$	$^7/_2$—$^5/_2$, $^7/_2$
2907,05	10	20,76	25,04	$4f\ ^2F^\circ$—$6g\ ^2G$	$^5/_2$—$^7/_2$
2906,34	3	—	—	—	—
2762,815	9	20,55	25,03	$4d\ ^2D$—$6f\ ^2F^\circ$	$^5/_2$—$^5/_2$, $^7/_2$
2760,48	1	20,55	25,04	$4d\ ^2D$—$6f\ ^2F^\circ$	$^3/_2$—$^5/_2$
2422,44	5	20,76	25,89	$4f\ ^2F^\circ$—$7d\ ^2D$	$^5/_2$—$^3/_2$, $^5/_2$
2400,33	3	20,76	25,94	$4f\ ^2F^\circ$—$7f\ ^2F^\circ$	$^7/_2$—$^5/_2$, $^7/_2$
2398,98	5	20,76	25,94	$4f\ ^2F^\circ$—$7g\ ^2G$	$^7/_2$—$^7/_2$, $^9/_2$
2398,76	5	20,76	25,94	$4f\ ^2F^\circ$—$7g\ ^2G$	$^5/_2$—$^7/_2$
2299,47	3	20,80	25,94	$4d\ ^2D$—$7f\ ^2F^\circ$	$^3/_2$—$^5/_2$
2298,36	0	20,55	25,94	$4d\ ^2D$—$7f\ ^2F^\circ$	$^5/_2$—$^7/_2$
2213,56	2	17,81	23,41	$4p\ ^2P^\circ$—$5d\ ^2D$	$^3/_2$—$^3/_2$, $^5/_2$
2209,66	1	17,81	23,41	$4p\ ^2P^\circ$—$5d\ ^2D$	$^1/_2$—$^3/_2$
2154,635	1	20,78	26,52	$4f\ ^2F^\circ$—$8g\ ^2G$	$^5/_2$, $^7/_2$—$^7/_2$, $^9/_2$
1935,83	10	14,37	20,78	$3d\ ^2D$—$4f\ ^2F^\circ$	$^3/_2$, $^5/_2$—$^5/_2$, $^7/_2$
1862,749	10	0,00	6,65	$3s\ ^2S$—$3p\ ^2P^\circ$	$^1/_2$—$^1/_2$
1854,715	10	0,00	6,65	$3s\ ^2S$—$3p\ ^2P^\circ$	$^1/_2$—$^3/_2$
1611,849	8	6,65	14,37	$3p\ ^2P^\circ$—$3d\ ^2D$	$^3/_2$—$^3/_2$, $^5/_2$
1605,776	8	6,65	14,37	$3p\ ^2P^\circ$—$3d\ ^2D$	$^1/_2$—$^3/_2$
1384,144	5	6,65	15,64	$3p\ ^2P^\circ$—$4s\ ^2S$	$^3/_2$—$^1/_2$
1379,670	3	6,65	15,64	$3p\ ^2P^\circ$—$4s\ ^2S$	$^1/_2$—$^1/_2$
1352,92	1	14,37	23,54	$3d\ ^2D$—$5f\ ^2F^\circ$	$^3/_2$, $^5/_2$—$^5/_2$, $^7/_2$
1162,66	0	14,37	25,04	$3d\ ^2D$—$6f\ ^2F^\circ$	$^3/_2$, $^5/_2$—$^5/_2$, $^7/_2$
893,905	5	6,65	20,55	$3p\ ^2P^\circ$—$4d\ ^2D$	$^3/_2$—$^3/_2$, $^5/_2$
892,056	4	6,65	20,55	$3p\ ^2P^\circ$—$4d\ ^2D$	$^1/_2$—$^3/_2$
856,768	5	6,65	21,15	$3p\ ^2P^\circ$—$5s\ ^2S$	$^3/_2$—$^1/_2$
855,040	4	6,65	21,15	$3p\ ^2P^\circ$—$5s\ ^2S$	$^1/_2$—$^1/_2$
726,948	3	6,65	23,73	$3p\ ^2P^\circ$—$6s\ ^2S$	$^3/_2$—$^1/_2$
725,716	2	6,65	23,73	$3p\ ^2P^\circ$—$6s\ ^2S$	$^1/_2$—$^1/_2$
696,212	4	0,00	17,81	$3s\ ^2S$—$4p\ ^2P^\circ$	$^1/_2$—$^1/_2$
695,817	5	0,00	17,81	$3s\ ^2S$—$4p\ ^2P^\circ$	$^1/_2$—$^3/_2$

λ, Å	I	E_H, eV	E_B, eV	Transition	J
671,198	2	6,65	25,16	$3p\ ^2P^\circ$—$7s\ ^2S$	$^3/_2$—$^1/_2$
670,144	1	6,65	25,16	$3p\ ^2P^\circ$—$7s\ ^2S$	$^1/_2$—$^1/_2$
560,390	7	0,00	22,15	$3s\ ^2S$—$5p\ ^2P^\circ$	$^1/_2$—$^1/_2$, $^3/_2$
511,215	4	0,00	24,25	$3s\ ^2S$—$6p\ ^2P^\circ$	$^1/_2$—$^1/_2$, $^3/_2$
486,95	1	0,00	25,45	$3s\ ^2S$—$7p\ ^2P^\circ$	$^1/_2$—$^3/_2$, $^3/_2$

Al IV, ground state $1s^2 2s^2 2p^6\ ^1S_0$

Ionization potential 967783 cm^{-1}; 119,983 eV

λ, Å	I	E_H, eV	E_B, eV	Transition	J
1881,19	1	76,68	83,27	3s [1$^1/_2$]°—3p [$^1/_2$]	1—1
1818,55	1	76,45	83,27	3s [1$^1/_2$]°—3p [$^1/_2$]	2—1
1639,00	2	77,45	85,02	3s′ [$^1/_2$]°—3p [1$^1/_2$]	1—2
1589,27	1	76,86	84,66	3s′ [$^1/_2$]°—3p [1$^1/_2$]	0—1
1584,45	2	77,45	85,28	3s′ [$^1/_2$]°—3p′ [1$^1/_2$]	1—2
1582,04	3	76,68	84,51	3s [1$^1/_2$]°—3p [2$^1/_2$]	1—2
1564,14	1	77,45	85,38	3s′ [$^1/_2$]°—3p′ [$^1/_2$]	1—1
1557,24	5	76,45	84,41	3s [1$^1/_2$]°—3p [2$^1/_2$]	2—3
1553,00	1	76,68	84,66	3s [1$^1/_2$]°—3p [1$^1/_2$]	1—1
1537,52	2	76,45	84,51	3s [1$^1/_2$]°—3p [2$^1/_2$]	2—2
1526,15	1	77,45	85,57	3s′ [$^1/_2$]°—3p′ [$^1/_2$]	1—0
1486,87	1	76,68	85,02	3s [1$^1/_2$]°—3p [1$^1/_2$]	1—2
1481,46	0	76,86	85,23	3s′ [$^1/_2$]°—3p′ [1$^1/_2$]	0—1
1449,70	0	76,68	85,23	3s [1$^1/_2$]°—3p′ [1$^1/_2$]	1—1
1447,47	2	76,45	85,02	3s [1$^1/_2$]°—3p [1$^1/_2$]	2—2
1441,81	1	76,68	85,28	3s [1$^1/_2$]°—3p′ [1$^1/_2$]	1—2
1431,93	2	76,68	85,34	3s [1$^1/_2$]°—3p [$^1/_2$]	1—0
1425,00	0	76,68	85,38	3s [1$^1/_2$]°—3p′ [$^1/_2$]	1—1
1417,58	0	85,38	94,12	3p′ [$^1/_2$]—3d [$^1/_2$]°	1—0
1412,24	0	76,45	85,23	3s [1$^1/_2$]°—3p′ [1$^1/_2$]	2—1
1409,52	0	85,38	94,17	3p′ [$^1/_2$]—3d [$^1/_2$]°	1—1
1404,72	2	76,45	85,28	3s [1$^1/_2$]°—3p′ [1$^1/_2$]	2—2
1388,77	2	76,45	85,38	3s [1$^1/_2$]°—3p′ [$^1/_2$]	2—1
1359,49	0	85,23	94,35	3p′ [1$^1/_2$]—3d [1$^1/_2$]°	1—2
1353,73	0	85,02	94,17	3p [1$^1/_2$]—3d [$^1/_2$]°	2—1
1302,13	2	85,57	95,10	3p′ [$^1/_2$]—3d [1$^1/_2$]°	0—1
1283,48	0	84,51	94,17	3p [2$^1/_2$]—3d [$^1/_2$]°	2—1
1272,70	3	85,02	94,76	3p [1$^1/_2$]—3d [2$^1/_2$]°	2—3
1264,14	3	85,38	95,18	3p′ [$^1/_2$]—3d′ [1$^1/_2$]°	1—2
1262,51	1	85,28	95,10	3p′ [1$^1/_2$]—3d [1$^1/_2$]°	2—1
1257,58	3	85,28	95,13	3p′ [1$^1/_2$]—3d′ [2$^1/_2$]°	2—3
1251,35	1	85,28	95,18	3p′ [1$^1/_2$]—3d′ [1$^1/_2$]°	2—2
1248,76	2	85,23	95,16	3p′ [1$^1/_2$]—3d′ [2$^1/_2$]°	1—2
1240,83	3	84,51	94,50	3p [2$^1/_2$]—3d [3$^1/_2$]°	2—3
1240,18	2	84,66	94,65	3p [1$^1/_2$]—3d [2$^1/_2$]°	1—2
1237,14	4	84,41	94,43	3p [2$^1/_2$]—3d [3$^1/_2$]°	3—4
1229,94	0	85,02	95,10	3p [1$^1/_2$]—3d [1$^1/_2$]°	2—1
1228,30	1	84,41	94,50	3p [2$^1/_2$]—3d [3$^1/_2$]°	3—3
1219,19	0	85,02	95,18	3p [1$^1/_2$]—3d′ [1$^1/_2$]°	2—2
1216,78	1	85,38	95,57	3p′ [$^1/_2$]—3d′ [1$^1/_2$]°	1—1
1211,80	0	85,34	95,57	3p [$^1/_2$]—3d′ [1$^1/_2$]°	0—1
1198,47	1	84,41	94,76	3p [2$^1/_2$]—3d [2$^1/_2$]°	3—3
1167,35	0	84,51	95,13	3p [2$^1/_2$]—3d′ [2$^1/_2$]°	2—3

λ, Å	I	E_H, eV	E_B, eV	Transition	J
1161,85	0	84,51	95,18	$3p\ [2^1/_2]-3d'\ [1^1/_2]^\circ$	2—2
1156,21	1	84,41	95,13	$3p\ [2^1/_2]-3d'\ [2^1/_2]^\circ$	3—3
1150,85	0	84,41	95,18	$3p\ [2^1/_2]-3d'\ [1^1/_2]^\circ$	3—2
1142,03	1	83,27	94,12	$3p\ [^1/_2]-3d\ [^1/_2]^\circ$	1—0
1136,80	3	83,27	94,17	$3p\ [^1/_2]-3d\ [^1/_2]^\circ$	1—1
1118,80	4	83,27	94,35	$3p\ [^1/_2]-3d\ [1^1/_2]^\circ$	1—2
161,686	14	0,00	76,68	$2p^6\ ^1S-3s\ [1^1/_2]^\circ$	0—1
160,073	16	0,00	77,45	$2p^6\ ^1S-3s'\ [^1/_2]^\circ$	0—1
131,652	3	0,00	94,17	$2p^6\ ^1S-3d\ [^1/_2]^\circ$	0—1
130,403	11	0,00	95,10	$2p^6\ ^1S-3d\ [1^1/_2]^\circ$	0—1
129,729	12	0,00	95,57	$2p^6\ ^1S-3d'\ [1^1/_2]^\circ$	0—1
124,543	6	0,00	99,55	$2p^6\ ^1S-4s\ [1^1/_2]^\circ$	0—1
124,034	8	0,00	99,55	$2p^6\ ^1S-4s'\ [^1/_2]^\circ$	0—1
117,377	0	0,00	105,62	$2p^6\ ^1S-4d\ [^1/_2]^\circ$	0—1
116,920	5	0,00	106,04	$2p^6\ ^1S-4d\ [1^1/_2]^\circ$	0—1
116,459	7	0,00	106,46	$2p^6\ ^1S-4d'\ [1^1/_2]^\circ$	0—1
114,759	0	0,00	108,03	$2p^6\ ^1S-5s\ [1^1/_2]^\circ$	0—1
114,329	0	0,00	108,44	$2p^6\ ^1S-5s'\ [^1/_2]^\circ$	0—1
111,780	0	0,00	110,91	$2p^6\ ^1S-5d\ [^1/_2]^\circ$	0—1
111,590	1	0,00	111,10	$2p^6\ ^1S-5d\ [1^1/_2]^\circ$	0—1
111,200	1	0,00	111,50	$2p^6\ ^1S-5d'\ [1^1/_2]^\circ$	0—1
108,907	0	0,00	113,84	$2p^6\ ^1S-6d\ [1^1/_2]^\circ$	0—1
108,535	0	0,00	114,23	$2p^6\ ^1S-6d'\ [1^1/_2]^\circ$	0—1

Al V, ground state $1s^2\ 2s^2\ 2p^5\ ^2P^0_{3/2}$

Ionization potential 1 240 600 cm^{-1}; 153,806 eV

λ, Å	I	E_H, eV	E_B, eV	Transition	J
281,397	14	0,42	44,48	$2p^5\ ^2P^\circ-2p^6\ ^2S$	$^1/_2-^1/_2$
278,699	16	0,00	44,48	$2p^5\ ^2P^\circ-2p^6\ ^2S$	$^3/_2-^1/_2$
136,668	8	44,48	136,79	$2p^6\ ^2S-3s'''\ ^2P^\circ$	$^1/_2-^3/_2$
133,242	2	0,42	93,47	$2p^5\ ^2P^\circ-3s\ ^4P$	$^1/_2-^3/_2$
133,013	4	0,00	93,21	$2p^5\ ^2P^\circ-3s\ ^4P$	$^3/_2-^5/_2$
		0,42	93,63	$2p^5\ ^2P^\circ-3s\ ^4P$	$^1/_2-^1/_2$
132,630	10	0,00	93,47	$2p^5\ ^2P^\circ-3s\ ^4P$	$^3/_2-^3/_2$
131,441	20	0,42	94,75	$2p^5\ ^2P^\circ-3s\ ^2P$	$^1/_2-^3/_2$
131,003	20	0,42	95,06	$2p^5\ ^2P^\circ-3s\ ^2P$	$^1/_2-^1/_2$
130,848	20	0,00	94,75	$2p^5\ ^2P^\circ-3s\ ^2P$	$^3/_2-^3/_2$
130,413	20	0,00	95,06	$2p^5\ ^2P^\circ-3s\ ^2P$	$^3/_2-^1/_2$
126,065	15	0,42	98,77	$2p^5\ ^2P^\circ-3s'\ ^2D$	$^1/_2-^3/_2$
125,525	15	0,00	98,77	$2p^5\ ^2P^\circ-3s'\ ^2D$	$^3/_2-^5/_2,\ ^3/_2$
118,984	6	0,42	104,62	$2p^5\ ^2P^\circ-3s''\ ^2S$	$^1/_2-^1/_2$
118,500	10	0,00	104,62	$2p^5\ ^2P^\circ-3s''\ ^2S$	$^3/_2-^1/_2$
109,021	3	0,42	114,14	$2p^5\ ^2P^\circ-3d\ ^4D$	$^1/_2-^3/_2$
108,851	1	0,42	114,32	$2p^5\ ^2P^\circ-3d\ ^4P$	$^1/_2-^3/_2$
108,707	6	0,00	114,04	$2p^5\ ^2P^\circ-3d\ ^4D$	$^3/_2-^5/_2$
108,616	1	0,00	114,14	$2p^5\ ^2P^\circ-3d\ ^4D$	$^3/_2-^3/_2$
108,462	10	0,42	114,73	$2p^5\ ^2P^\circ-3d\ ^2D$	$^1/_2-^3/_2$
108,446	3	0,00	114,32	$2p^5\ ^2P^\circ-3d\ ^4P$	$^3/_2-^3/_2$
108,404	5	0,42	114,79	$2p^5\ ^2P^\circ-3d\ ^2P$	$^1/_2-^1/_2$
108,385	10	0,00	114,39	$2p^5\ ^2P^\circ-3d\ ^4P$	$^3/_2-^5/_2$
108,315	4	0,00	114,46	$2p^5\ ^2P^\circ-3d\ ^2F$	$^3/_2-^5/_2$
108,112	12	0,42	115,10	$2p^5\ ^2P^\circ-3d\ ^2P$	$^1/_2-^3/_2$
108,057	12	0,00	114,73	$2p^5\ ^2P^\circ-3d\ ^2D$	$^3/_2-^3/_2$

λ, Å	I	E_H, eV	E_B, eV	Transition	J
108,004	5	0,00	114,79	$2p^5\ ^2P^\circ - 3d\ ^2P$	$^3/_2 - ^1/_2$
107,945	20	0,00	114,85	$2p^5\ ^2P^\circ - 3d\ ^2D$	$^3/_2 - ^5/_2$
107,711	6	0,00	115,10	$2p^5\ ^2P^\circ - 3d\ ^2P$	$^3/_2 - ^3/_2$
104,495	8	0,42	119,07	$2p^5\ ^2P^\circ - 3d'\ ^2P$	$^1/_2 - ^1/_2$
104,447	10	0,42	119,12	$2p^5\ ^2P^\circ - 3d'\ ^2S$	$^1/_2 - ^1/_2$
104,363	10	0,42	119,22	$2p^5\ ^2P^\circ - 3d'\ ^2P$	$^1/_2 - ^3/_2$
104,180	14	0,42	119,43	$2p^5\ ^2P^\circ - 3d'\ ^2D$	$^1/_2 - ^3/_2$
104,121	3	0,00	119,07	$2p^5\ ^2P^\circ - 3d'\ ^2P$	$^3/_2 - ^1/_2$
104,073	10	0,00	119,12	$2p^5\ ^2P^\circ - 3d'\ ^2S$	$^3/_2 - ^1/_2$
103,990	4	0,00	119,22	$2p^5\ ^2P^\circ - 3d'\ ^2P$	$^3/_2 - ^3/_2$
103,881	14	0,00	119,35	$2p^5\ ^2P^\circ - 3d'\ ^2D$	$^3/_2 - ^5/_2$
103,805	10	0,00	119,43	$2p^5\ ^2P^\circ - 3d'\ ^2D$	$^3/_2 - ^3/_2$
99,769	0,5	0,42	124,69	$2p^5\ ^2P^\circ - 4s\ ^2P$	$^1/_2 - ^3/_2$
99,616	7	0,42	124,88	$2p^5\ ^2P^\circ - 3d''\ ^2D$	$^1/_2 - ^3/_2$
99,544	2	0,42	124,97	$2p^5\ ^2P^\circ - 4s\ ^2P$	$^1/_2 - ^1/_2$
99,427	4	0,00	124,69	$2p^5\ ^2P^\circ - 4s\ ^2P$	$^3/_2 - ^3/_2$
99,290	10	0,00	124,86	$2p^5\ ^2P^\circ - 3d''\ ^2D$	$^3/_2 - ^3/_2,\ ^5/_2$
99,200	1	0,00	124,97	$2p^5\ ^2P^\circ - 4s\ ^2P$	$^3/_2 - ^1/_2$
99,150	1	0,42	129,37	$2p^5\ ^2P^\circ - 4s'\ ^2D$	$^1/_2 - ^3/_2$
95,835	2	0,00	129,36	$2p^5\ ^2P^\circ - 4s'\ ^2D$	$^3/_2 - ^3/_2,\ ^5/_2$
94,187	2	0,42	132,06	$2p^5\ ^2P^\circ - 4d\ ^2P$	$^1/_2 - ^1/_2$
94,160	2	0,42	132,09	$2p^5\ ^2P^\circ - 4d\ ^2D$	$^1/_2 - ^3/_2$
94,117	2,5	0,00	131,73	$2p^5\ ^2P^\circ - 4d\ ^4D$	$^3/_2 - ^5/_2$
93,981	2	0,00	131,92	$2p^5\ ^2P^\circ - 4d\ ^4P$	$^3/_2 - ^5/_2$
93,955	6	0,42	132,38	$2p^5\ ^2P^\circ - 4d\ ^2P$	$^1/_2 - ^3/_2$
93,855	4	0,00	132,09	$2p^5\ ^2P^\circ - 4d\ ^2D$	$^3/_2 - ^3/_2$
93,755	7	0,00	132,24	$2p^5\ ^2P^\circ - 4d\ ^2D$	$^3/_2 - ^5/_2$
91,750	1	0,42	135,13	$2p^5\ ^2P^\circ - 4s''\ ^2S$	$^1/_2 - ^1/_2$
90,982	1	0,42	136,69	$2p^5\ ^2P^\circ - 4d'\ ^2S$	$^1/_2 - ^1/_2$
90,914	4	0,42	136,79	$2p\ ^2P^\circ - 4d'\ ^2P$	$^1/_2 - ^3/_2$
90,701	4	0,00	136,69	$2p^5\ ^2P^\circ - 4d'\ ^2S$	$^3/_2 - ^1/_2$
90,646	2	0,00	136,77	$2p^5\ ^2P^\circ - 4d'\ ^2D$	$^3/_2 - ^5/_2$
90,630	5	0,00	136,79	$2p^5\ ^2P^\circ - 4d'\ ^2P$	$^3/_2 - ^3/_2$
88,817	1	0,42	140,01	$2p^5\ ^2P^\circ - 5d\ ^2P$ $2p^5\ ^2P^\circ - 5d\ ^2D$	$^1/_2 - ^1/_2$ $^1/_2 - ^3/_2$
88,688	4	0,42	140,01	$2p^5\ ^2P^\circ - 5d\ ^2D$	$^3/_2 - ^5/_2$
88,636	2	0,42	140,30	$2p^5\ ^2P^\circ - 5d\ ^2P$	$^1/_2 - ^3/_2$
88,539	8	0,00	140,01	$2p^5\ ^2P^\circ - 5d\ ^2D$	$^3/_2 - ^3/_2$
88,425	2	0,00	140,21	$2p^5\ ^2P^\circ - 5d\ ^2D$	$^3/_2 - ^5/_2$
87,279	1	0,42	142,48	$2p^5\ ^2P^\circ - 4d''\ ^2D$	$^1/_2 - ^3/_2$
87,020	2	0,00	142,47	$2p^5\ ^2P^\circ - 4d''\ ^2D$	$^3/_2 - ^3/_2,\ ^5/_2$
85,865	2	0,42	144,81	$2p^5\ ^2P^\circ - 5d'\ ^2P$	$^1/_2 - ^3/_2$
85,804	7	—	—	—	—
85,662	1	0,00	144,73	$2p^5\ ^2P^\circ - 5d'\ ^2S$	$^3/_2 - ^1/_2$

Al VI, ground state $1s^2\ 2s^2\ 2p^4\ ^3P_2$
Ionization potential 1 536 300 cm^{-1}; 190,466 eV

λ, Å	I	E_H, eV	E_B, eV	Transition	J
312,241	6	0,34	40,04	$2p^4\ ^3P - 2p^5\ ^3P^\circ$	1—2
310,908	6	0,47	40,35	$2p^4\ ^3P - 2p^5\ ^3P^\circ$	0—1
309,852	6	0,34	40,35	$2p^4\ ^3P - 2p^5\ ^3P^\circ$	1—1
309,596	8	0,00	40,04	$2p^4\ ^3P - 2p^5\ ^3P^\circ$	2—2
308,560	6	0,34	40,52	$2p^4\ ^3P - 2p^5\ ^3P^\circ$	1—0

λ, Å	I	E_H, eV	E_B, eV	Transition	J
307,248	7	0,00	40,35	$2p^4\,{}^3P$—$2p^5\,{}^3P^\circ$	2—1
275,350	6	10,99	56,02	$2p^4\,{}^1S$—$2p^5\,{}^1P^\circ$	0—1
243,760	12	5,16	56,02	$2p^4\,{}^1D$—$2p^5\,{}^1P^\circ$	2—1
113,756	1	40,35	149,34	$2p^5\,{}^3P^\circ$—$3s'''\,{}^3P$	1—2
113,623	1	40,35	149,45	$2p^5\,{}^3P^\circ$—$3s'''\,{}^3P$	1—1
113,437	3	40,04	149,34	$2p^5\,{}^3P^\circ$—$3s'''\,{}^3P$	2—2
113,314	1	40,04	149,45	$2p^5\,{}^3P^\circ$—$3s'''\,{}^3P$	2—1
109,974	4	0,47	113,21	$2p^4\,{}^3P$—$3s\,{}^3S^\circ$	0—1
109,843	12	0,34	113,21	$2p^4\,{}^3P$—$3s\,{}^3S^\circ$	1—1
109,514	20	0,00	113,21	$2p^4\,{}^3P$—$3s\,{}^3S^\circ$	2—1
109,284	7	10,99	124,43	$2p^4\,{}^1S$—$3s''\,{}^1P^\circ$	0—1
107,620	14	5,16	120,35	$2p^4\,{}^1D$—$3s'\,{}^1D^\circ$	2—2
104,466	8	0,47	119,15	$2p^4\,{}^3P$—$3s'\,{}^3D$	0—1
104,344	16	0,34	119,15	$2p^4\,{}^3P$—$3s'\,{}^3D^\circ$	1—1, 2
104,047	20	0,00	119,15	$2p^4\,{}^3P$—$3s'\,{}^3D^\circ$	2—1, 2, 3
103,940	6	5,16	124,43	$2p^4\,{}^1D$—$3s''\,{}^1P^\circ$	2—1
101,027	3	0,47	123,19	$2p^4\,{}^3P$—$3s''\,{}^3P^\circ$	0—1
100,919	4	0,34	123,19	$2p^4\,{}^3P$—$3s''\,{}^3P^\circ$	1—0, 1, 2
100,894	4	0,34	123,21	$2p^4\,{}^3P$—$3s''\,{}^3P^\circ$	1—2
100,639	2	0,00	123,19	$2p^4\,{}^3P$—$3s''\,{}^3P^\circ$	2—1
100,616	12	0,00	123,21	$2p^4\,{}^3P$—$3s''\,{}^3P^\circ$	2—2
96,673	1	40,35	168,60	$2p^5\,{}^3P^\circ$—$3s^V\,{}^3S$	1—1
95,436	2	10,99	140,90	$2p^4\,{}^1S$—$3d'\,{}^1P^\circ$	0—1
92,970	5	0,47	133,83	$2p^4\,{}^3P$—$3d\,{}^3D^\circ$	0—1
92,875	10	0,34	133,83	$2p^4\,{}^3P$—$3d\,{}^3D^\circ$	1—1, 2
92,636	4	0,00	133,83	$2p^4\,{}^3P$—$3d\,{}^3D^\circ$	2—2
92,626	15	0,00	133,85	$2p^4\,{}^3P$—$3d\,{}^3D^\circ$	2—3
91,332	10	5,16	140,90	$2p^4\,{}^1D$—$3d'\,{}^1P^\circ$	2—1
90,858	12	5,16	141,61	$2p^4\,{}^1D$—$3d'\,{}^1D^\circ$	2—2
90,200	20	5,16	142,60	$2p^4\,{}^1D$—$3d'\,{}^1F^\circ$	2—3
88,688	4	5,16	144,95	$2p^4\,{}^1D$—$3d''\,{}^1D^\circ$	2—2
88,539	8	5,16	145,18	$2p^4\,{}^1D$—$3d''\,{}^1P^\circ$	2—1
88,469	5	0,47	140,61	$2p^4\,{}^3P$—$3d'\,{}^3D^\circ$	0—1
88,376	15	0,34	140,61	$2p^4\,{}^3P$—$3d'\,{}^3D^\circ$	1—1, 2
88,325	2	0,00	140,36	$2p^4\,{}^3P$—$3d'\,{}^3F^\circ$	2—3
88,273	15	5,16	145,61	$2p^4\,{}^1D$—$3d''\,{}^1F^\circ$	2—3
88,170	20	0,00	140,61	$2p^4\,{}^3P$—$3d'\,{}^3D^\circ$	2—1, 2, 3
87,887	5	0,47	140,61	$2p^4\,{}^3P$—$3d'\,{}^3D^\circ$	0—1
87,866	7	0,34	141,44	$2p^4\,{}^3P$—$3d'\,{}^3P^\circ$	1—2
87,802	5	0,34	141,54	$2p^4\,{}^3P$—$3d'\,{}^3P^\circ$	1—1
87,783	5	0,34	141,57	$2p^4\,{}^3P$—$3d'\,{}^3P^\circ$	1—0
87,655	13	0,00	141,44	$2p^4\,{}^3P$—$3d'\,{}^3P^\circ$	2—2
87,629	2	0,47	141,96	$2p^4\,{}^3P$—$3d'\,{}^3S^\circ$	0—1
87,592	10	0,00	141,54	$2p^4\,{}^3P$—$3d'\,{}^3P^\circ$	2—1
87,544	7	0,34	141,96	$2p^4\,{}^3P$—$3d'\,{}^3S^\circ$	1—1
87,334	8	0,00	141,96	$2p^4\,{}^3P$—$3d'\,{}^3S^\circ$	2—1
86,147	4	0,47	144,37	$2p^4\,{}^3P$—$3d''\,{}^3P^\circ$	0—1
86,097	3	0,34	144,34	$2p^4\,{}^3P$—$3d''\,{}^3P^\circ$	1—0
86,020	3	0,34	144,47	$2p^4\,{}^3P$—$3d''\,{}^3P^\circ$	1—2
85,817	7	0,00	144,47	$2p^4\,{}^3P$—$3d''\,{}^3P^\circ$	2—2
85,764	8	0,34	144,89	$2p^4\,{}^3P$—$3d''\,{}^3F^\circ$	1—2
85,724	6	0,00	144,62	$2p^4\,{}^3P$—$3d''\,{}^3F^\circ$	2—3
85,622	6	0,34	145,13	$2p^4\,{}^3P$—$3d''\,{}^3D^\circ$	1—2
85,569	4	0,00	144,89	$2p^4\,{}^3P$—$3d''\,{}^3F^\circ$	2—2
85,515	20	0,00	144,98	$2p^4\,{}^3P$—$3d''\,{}^3D^\circ$	2—3
85,423	2	0,00	145,13	$2p^4\,{}^3P$—$3d''\,{}^3D^\circ$	2—2
82,267	1	0,34	151,04	$2p^4\,{}^3P$—$4s\,{}^3S^\circ$	1—1
82,082	1,5	0,00	151,04	$2p^4\,{}^3P$—$4s\,{}^3S^\circ$	2—1

λ, Å	I	E_H, eV	E_B, eV	Transition	J
81,738	1	10,99	162,67	$2p^4\,^1S—4s''\,^1P^\circ$	0—1
80,770	1,5	5,16	158,65	$2p^4\,^1D—4s'\,^1D^\circ$	2—2
78,459	1,5	0,00	158,02	$2p^4\,^3P—4s'\,^3D^\circ$	2—1, 2, 3
78,178	1	0,47	159,06	$2p^4\,^3P—4d\,^3D^\circ$	0—1
78,112	2	0,34	159,06	$2p^4\,^3P—4d\,^3D^\circ$	1—1, 2
77,945	10	0,00	159,06	$2p^4\,^3P—4d\,^3D^\circ$	2—1, 2, 3
76,953	1	5,16	166,26	$2p^4\,^1D—4d'\,^1P^\circ$	2—1
76,697	4	5,16	166,80	$2p^4\,^1D—4d'\,^1D^\circ$	2—2
76,618	4	5,16	166,97	$2p^4\,^1D—4d'\,^1F^\circ$	2—3
74,892	2	5,16	170,70	$2p^4\,^1D—4d''\,^1F^\circ$	2—3
74,813	1	0,34	166,06	$2p^4\,^3P—4d'\,^3D^\circ$	1—1, 2
74,656	5	0,00	166,06	$2p^4\,^3P—4d'\,^3D^\circ$	2—1, 2, 3
74,592	3	0,34	166,54	$2p^4\,^3P—4d'\,^3P^\circ$	1—2
74,504	1	0,34	166,75	$2p^4\,^3P—4d'\,^3S^\circ$	1—1
74,444	6	0,00	166,54	$2p^4\,^3P—4d'\,^3P^\circ$	2—2
74,346	1	0,00	166,75	$2p^4\,^3P—4d'\,^3S^\circ$	2—1
73,076	2	0,34	170,00	$2p^4\,^3P—4d''\,^3P^\circ$	1—2
72,926	2	0,00	170,00	$2p^4\,^3P—4d''\,^3P^\circ$	2—2
72,865	1	0,34	170,49	$2p^4\,^3P—4d''\,^3D^\circ$	1—2
72,810	5	0,00	170,28	$2p^4\,^3P—4d''\,^3D^\circ$	2—3
68,293	0	0,00	181,72	$2p^4\,^3P—5d''\,^3P^\circ$	2—2
68,167	1	0,00	181,87	$2p^4\,^3P—5d''\,^3D^\circ$	2—3

Unclassified Lines of Aluminum

λ, Å		Presumed classification	λ, Å	I	Presumed classification
977,4	4	—	78,508	3	—
904,5	5	—	78,938	6	—
718,9	3	—	76,853	4	—
678,0	5	—	76,794	4	—
95,720	4	—	72,674	6	—
94,970	3	—	69,631	4	—
91,023	10	—	69,165	4	—

SILICON, Z = 14

Si I, ground state $1s^2\,2s^2\,2p^6\,3s^2\,3p^2\,{}^3P_0$

Ionization potential 65 747,5 cm^{-1}; 8,151 eV

λ, Å	I	E_H, eV	E_B, eV	Transition	J
25854,38	6	5,62	6,10	$3p^3\,{}^3D^\circ$—$4p\,{}^3P$	3—2
22062,71	12	6,73	7,29	$3d\,{}^3D^\circ$—$4f\,{}^3F$	3—4
21879,35	8	6,72	7,29	$3d\,{}^3D^\circ$—$4f\,{}^1F$	2—3
21819,69	5	6,72	7,29	$3d\,{}^3D^\circ$—$4f\,{}^3F$	2—3
21779,77	9	6,72	7,29	$3d\,{}^3D^\circ$—$4f\,{}^3F$	1—2
21354,24	21	6,22	6,80	$4p\,{}^1D$—$5s\,{}^1P^\circ$	2—1
20917,13	12	6,73	7,32	$3d\,{}^3D^\circ$—$4f\,{}^3G$	3—4
19928,88	31	6,10	6,72	$4p\,{}^3P$—$3d\,{}^3D^\circ$	2—2
19722,50	110	6,10	6,73	$4p\,{}^3P$—$3d\,{}^3D^\circ$	2—3
19508,13	14	6,08	6,72	$4p\,{}^3P$—$3d\,{}^3D^\circ$	1—1
19506,12	5	6,10	6,73	$4p\,{}^3P$—$5s\,{}^3P^\circ$	2—1
19493,38	13	6,08	6,72	$4p\,{}^3P$—$3d\,{}^3D^\circ$	1—1
19432,97	48	6,08	6,72	$4p\,{}^3P$—$3d\,{}^3D^\circ$	1—2
19385,94	15	6,08	6,72	$4p\,{}^3P$—$3d\,{}^3D^\circ$	0—1
19283,29	8	6,08	6,73	$4p\,{}^3P$—$5s\,{}^3P^\circ$	1—0
19030,79	5	6,08	6,73	$4p\,{}^3P$—$5s\,{}^3P^\circ$	1—1
18914,48	8	6,08	6,73	$4p\,{}^3P$—$5s\,{}^3P^\circ$	0—1
18722,90	26	6,10	6,76	$4p\,{}^3P$—$5s\,{}^3P^\circ$	2—2
18422,72	7	6,62	7,29	$3d\,{}^1F^\circ$—$4f\,{}^3F$	3—4
17617,00	9	6,62	7,32	$3d\,{}^1F^\circ$—$4f\,{}^3G$	3—4
17327,29	28	6,62	7,33	$3d\,{}^1F^\circ$—$4f\,{}^3/_2\,[{}^9/_2]$	3—4
16680,77	29	5,98	6,73	$4p\,{}^3D$—$3d\,{}^3D^\circ$	3—3
16381,55	16	5,96	6,72	$4p\,{}^3D$—$3d\,{}^3D^\circ$	2—2
16380,12	8	5,86	6,62	$4p\,{}^1P$—$3d\,{}^1P^\circ$	1—1
16241,84	7	5,96	6,73	$4p\,{}^3D$—$3d\,{}^3D^\circ$	2—3
16215,68	11	5,95	6,72	$4p\,{}^3D$—$3d\,{}^3D^\circ$	1—1
16163,71	60	5,95	6,72	$4p\,{}^3D$—$3d\,{}^3D^\circ$	1—2
16094,80	20	5,96	6,73	$4p\,{}^3D$—$5s\,{}^3P^\circ$	2—1
16060,03	95	5,95	6,73	$4p\,{}^3D$—$5s\,{}^3P^\circ$	1—0
15960,04	40	5,98	6,76	$4p\,{}^3D$—$5s\,{}^3P^\circ$	3—2
15888,39	190	5,08	5,86	$4s\,{}^1P^\circ$—$4p\,{}^1P$	1—1
15884,41	5	5,95	6,73	$4p\,{}^3D$—$5s\,{}^3P^\circ$	1—1
15833,58	7	6,22	7,01	$4p\,{}^1D$—$4d\,{}^1D^\circ$	2—2
15557,81	7	5,96	6,76	$4p\,{}^3D$—$5s\,{}^3P^\circ$	2—2
14224,54	6	5,08	5,95	$4s\,{}^1P^\circ$—$4p\,{}^3D$	1—1
13711,36	5	6,12	7,03	$4p\,{}^3S$—$3p^3\,{}^3P^\circ$	1—2
13693,85	8	6,12	7,03	$4p\,{}^3S$—$3p^3\,{}^3P^\circ$	1—1
13309,04	5	6,10	7,03	$4p\,{}^3P$—$3p^3\,{}^3P^\circ$	2—1
13287,58	9	4,93	5,86	$4s\,{}^3P$—$4p\,{}^1P^\circ$	1—1
13176,90	11	5,86	6,80	$4p\,{}^1P$—$5s\,{}^1P^\circ$	1—1
12395,82	6	4,95	5,95	$4s\,{}^3P^\circ$—$4p\,{}^3D$	2—1
12270,68	120	4,95	5,96	$4s\,{}^3P^\circ$—$4p\,{}^3D$	2—2
12103,50	5	4,93	5,95	$4s\,{}^3P^\circ$—$4p\,{}^3D$	1—1
12031,48	10	4,95	5,98	$4s\,{}^3P^\circ$—$4p\,{}^3D$	2—3
11991,52	5	4,92	5,95	$4s\,{}^3P^\circ$—$4p\,{}^3D$	0—1
11984,18	10	4,93	5,96	$4s\,{}^3P^\circ$—$4p\,{}^3D$	1—2
11289,83	15	6,19	7,29	$3d\,{}^3F^\circ$—$4f\,{}^3F$	3—4
11187,588	16	6,18	7,29	$3d\,{}^3F^\circ$—$4f\,{}^3F$	2—3
11130,03	12	6,21	7,32	$3d\,{}^3F^\circ$—$4f\,{}^3G$	4—4
11017,9648	80	6,21	7,33	$3d\,{}^3F^\circ$—$4f\,{}^3/_2\,[{}^9/_2]$	4—5
10984,527	20	6,19	7,32	$3d\,{}^3F^\circ$—$4f\,{}^3G$	3—3
10982,061	30	6,19	7,32	$3d\,{}^3F^\circ$—$4f\,{}^3G$	3—4
10979,308	80	4,95	6,08	$4s\,{}^3P^\circ$—$4p\,{}^3P$	2—1

λ, Å	I	E_H, eV	E_B, eV	Transition	J
10885,336	30	6,18	7,32	3d $^3F^\circ$—4f ^3G	2—3
10882,802	30	5,98	7,12	4p ^3D—4d $^3F^\circ$	3—3
10869,5408	130	5,08	6,22	4s $^1P^\circ$—4p ^1D	1—2
10868,79	30	6,19	7,33	3d $^3F^\circ$—4f $^3/_2$ [$^9/_2$]	3—4
10843,854	60	5,86	7,00	4p ^1P—4d $^1D^\circ$	1—2
10827,091	140	4,95	6,10	4s $^3P^\circ$—4p ^3P	2—2
10796,06	7	6,18	7,33	3d $^3F^\circ$—4f ^1D	2—2
10786,8560	80	4,93	6,08	4s $^3P^\circ$—4p ^3P	1—0
10784,5597	30	5,96	7,11	4p ^3D—4d $^3F^\circ$	2—2
10749,3837	60	4,93	6,08	4s $^3P^\circ$—4p ^3P	1—1
10727,4076	30	5,98	7,14	4p ^3D—4d $^3F^\circ$	3—4
10694,2510	30	5,96	7,12	4p ^3D—4d $^3F^\circ$	2—3
10689,719	25	5,95	7,11	4p ^3D—4d $^2F^\circ$	1—2
10660,9748	120	4,92	6,08	4s $^3P^\circ$—4p ^3P	0—1
10627,6467	20	5,86	7,03	4p ^1P—4d $^3P^\circ$	1—2
10603,431	120	4,93	6,10	4s $^3P^\circ$—4p ^3P	1—2
10585,1412	120	4,95	6,12	4s $^3P^\circ$—4p ^3S	2—1
10371,269	30	4,93	6,12	4s $^3P^\circ$—4p ^3S	1—1
10288,942	10	4,92	6,12	4s $^3P^\circ$—4p ^3S	0—1
9891,72	10	6,12	7,38	4p ^3S—6s $^3P^\circ$	1—2
9887,06	10	6,22	7,48	4p ^1D—5d $^1D^\circ$	2—2
9689,39	10	6,10	7,38	4p ^3P—6s $^3P^\circ$	2—2
9585,92	10	4,93	6,22	4s $^3P^\circ$—4p ^1D	1—2
9570,65	8	6,08	7,38	4p ^3P—6s $^3P^\circ$	1—2
9505,19	20	6,12	7,43	4p ^3S—5d $^3P^\circ$	1—2
9421,78	15	6,12	7,44	4p ^3S—5d $^3P^\circ$	1—1
9413,506	100	5,08	6,40	4s $^1P^\circ$—4p ^1S	1—0
9387,33	10	6,21	7,53	3d $^3F^\circ$—6p [$^3/_2$, $^3/_2$]	4—3
9318,22	10	6,10	7,43	4p ^3P—5d $^3P^\circ$	2—2
9253,67	15	6,26	7,60	3d $^3P^\circ$—5f ^1D	2—2
9208,35	15	6,08	7,43	4p ^3P—5d $^3P^\circ$	1—2
9021,58	10	6,27	7,64	3d $^3P^\circ$—5f $^3/_2$ [$^3/_2$]	1—2
9008,51	15	6,26	7,64	3d $^3P^\circ$—5f ^3D	2—3
8949,10	10	5,96	7,35	4p ^3D—6s $^3P^\circ$	2—1
8925,30	10	5,95	7,34	4p ^3D—6s $^3P^\circ$	1—0
8892,7277	20	5,98	7,38	4p ^3D—6s $^3P^\circ$	3—2
8790,3889	35	6,19	7,60	3d $^3F^\circ$—5f ^3F	3—4
8780,747	11	6,22	7,63	4p ^1D—5d $^3D^\circ$	2—3
8766,422	14	5,96	7,38	4p ^3D—6s $^3P^\circ$	2—2
8752,009	100	5,87	7,29	3d $^1D^\circ$—4f ^1F	2—3
8751,174	10	5,87	7,29	3d $^1D^\circ$—4f ^3F	2—2
8742,4509	75	5,87	7,29	3d $^1D^\circ$—4f ^3F	2—3
8728,0110	40	6,18	7,60	3d $^3F^\circ$—5f ^3F	2—3
8680,079	11	5,86	7,29	4p ^1P—4d $^1P^\circ$	1—1
8648,4622	50	6,21	7,64	3d $^3F^\circ$—5f $^1/_2$ [$^7/_2$]	4—3
8647,114	15	—	—	—	—
8606,014	8	6,19	7,50	3d $^3F^\circ$—6p [$^1/_2$, $^3/_2$]	3—2
8597,0470	20	6,19	7,63	3d $^3F^\circ$—5f ^3G	3—3
8595,962	25	6,19	7,63	3d $^3F^\circ$—5f ^3G	3—4
8556,7803	120	5,87	7,32	3d $^1D^\circ$—4f ^3G	2—3
8536,1645	40	6,18	7,63	3d $^3F^\circ$—5f ^3G	2—3
8502,2207	60	5,87	7,33	3d $^1D^\circ$—4f ^3D	2—3
8501,547	40	5,87	7,33	3d $^1D^\circ$—4f ^1D	2—2
8492,078	15	5,86	7,32	4p ^1P—4d $^3D^\circ$	1—1
8443,982	40	5,87	7,34	3d $^1D^\circ$—4f ^3D	2—2
8435,24	8	4,93	6,40	4s $^3P^\circ$—4p ^1S	1—0
8338,328	20	5,86	7,35	4p ^1P—6s $^3P^\circ$	1—1

λ, Å	I	E_H, eV	E_B, eV	Transition	J
8317,39	15	5,61	7,10	$3p^3\ {}^3D^\circ - 5p\ {}^3P$	1—0
8306,710	25	5,62	7,10	$3p^3\ {}^3D^\circ - 5p\ {}^3P$	2—1
8294,675	13	5,61	7,10	$3p^3\ {}^3D^\circ - 5p\ {}^3P$	1—1
8230,642	35	5,62	7,12	$3p^3\ {}^3D^\circ - 5p\ {}^3P$	3—2
8215,15	10	6,26	7,77	$3d\ {}^3P^\circ - 6f\ {}^1/_2\ [{}^5/_2]$	2—3
8171,288	25	6,10	7,62	$4p\ {}^3P - 5d\ {}^1F^\circ$	2—3
8162,170	15	5,62	7,13	$3p^3\ {}^3D^\circ - 5p\ {}^3S$	2—1
8154,872	15	6,10	7,62	$4p\ {}^3P - 5d\ {}^3D^\circ$	2—2
8140,55	15	5,95	7,48	$4p\ {}^3D - 5d\ {}^1D^\circ$	1—2
8093,241	70	5,86	7,39	$4p\ {}^1P - 6s\ {}^1P^\circ$	1—1
8071,285	25	6,10	7,63	$4p\ {}^3P - 5d\ {}^3D^\circ$	2—3
8070,598	25	6,08	7,62	$4p\ {}^3P - 5d\ {}^3D^\circ$	1—2
8035,619	35	5,98	7,53	$4p\ {}^3D - 5d\ {}^3F^\circ$	3—3
8026,950	25	6,12	7,67	$4p\ {}^3S - 7s\ {}^1P^\circ$	1—1
7975,579	13	6,12	7,67	$3d\ {}^3D^\circ - 5p\ {}^1D$	2—2
7970,306	35	5,96	7,52	$4p\ {}^3D - 5d\ {}^3F^\circ$	2—2
7944,0011	140	5,98	7,54	$4p\ {}^3D - 5d\ {}^3F^\circ$	3—4
7932,3490	120	5,96	7,53	$4p\ {}^3D - 5d\ {}^3F^\circ$	2—3
7925,850	15	6,22	7,79	$4p\ {}^1D - 6d\ {}^1F^\circ$	2—3
7918,3857	90	5,95	7,52	$4p\ {}^3D - 5d\ {}^3F^\circ$	1—2
7913,432	25	5,86	7,43	$4p\ {}^1P - 5d\ {}^3P^\circ$	1—2
7912,383	20	6,10	7,66	$4p\ {}^3P - 7s\ {}^3P^\circ$	2—2
7849,967	30	6,19	7,77	$3d\ {}^3F^\circ - 6f\ {}^1/_2\ [{}^7/_2]$	3—4
7800,008	30	6,18	7,77	$3d\ {}^3F^\circ - 6f\ {}^1/_2\ [{}^7/_2]$	2—3
7742,71	40	6,21	7,81	$3d\ {}^3F^\circ - 6f\ {}^3/_2\ [{}^9/_2]$	4—5
7680,2668	100	5,86	7,48	$4p\ {}^1P - 5d\ {}^1D^\circ$	1—2
7640,31	20	6,18	7,80	$3d\ {}^3F^\circ - 6f\ {}^3/_2\ [{}^7/_2]$	2—3
7482,19	25	5,86	7,52	$4p\ {}^1P - 5d\ {}^3F^\circ$	1—2
7455,36	25	5,96	7,63	$4p\ {}^3D - 5d\ {}^3D^\circ$	2—1
7424,60	85	5,62	7,29	$3p^3\ {}^3D^\circ - 4f\ {}^3F$	3—3
7423,4969	425	5,62	7,29	$3p^3\ {}^3D^\circ - 4f\ {}^3F$	3—4
7415,9462	275	5,62	7,29	$3p^3\ {}^3D^\circ - 4f\ {}^1F$	2—3
7415,35	40	5,62	7,29	$3p^3\ {}^3D^\circ - 4f\ {}^3F$	2—2
7409,0818	200	5,62	7,29	$3p^3\ {}^3D^\circ - 4f\ {}^3F$	2—3
7405,774	375	5,61	7,29	$3p^3\ {}^3D^\circ - 4f\ {}^3F$	1—2
7395,52	15	5,95	7,63	$4p\ {}^3D - 7s\ {}^3P^\circ$	1—0
7373,00	35	5,98	7,66	$4p\ {}^3D - 7s\ {}^3P^\circ$	3—2
7290,26	55	5,62	7,32	$3p^3\ {}^3D^\circ - 4f\ {}^3G$	3—3
7289,1730	400	5,62	7,32	$3p^3\ {}^3D^\circ - 4f\ {}^3G$	3—4
7282,81	40	6,21	7,91	$3d\ {}^3F^\circ - 7f\ {}^3/_2\ [{}^9/_2]$	4—5
7275,294	160	5,62	7,32	$3p^3\ {}^3D^\circ - 4f\ {}^3G$	2—3
7250,625	180	5,62	7,33	$3p^3\ {}^3D^\circ - 4f\ {}^3D$	3—3
7250,14	25	5,62	7,33	$3p^3\ {}^3D^\circ - 4f\ {}^1D$	3—2
7235,82	60	5,62	7,33	$3p^3\ {}^3D^\circ - 4f\ {}^3D$	2—3
7235,326	100	5,62	7,33	$3p^3\ {}^3D^\circ - 4f\ {}^1D$	2—2
7226,206	100	5,61	7,33	$3p^3\ {}^3D^\circ - 4f\ {}^1D$	1—2
7208,21	25	5,62	7,34	$3p^3\ {}^3D^\circ - 4f\ {}^3D$	3—2
7193,90	30	5,62	7,34	$3p^3\ {}^3D^\circ - 4f\ {}^3D$	2—1
7193,58	65	5,62	7,34	$3p^3\ {}^3D^\circ - 4f\ {}^3D$	2—2
7184,89	70	5,61	7,34	$3p^3\ {}^3D^\circ - 4f\ {}^3D$	1—1
7184,57	20	5,61	7,34	$3p^3\ {}^3D^\circ - 4f\ {}^3D$	1—2
7165,545	200	5,87	7,60	$3d\ {}^1D^\circ - 5f\ {}^1D$	2—2
7164,69	70	5,87	7,60	$3d\ {}^1D^\circ - 5f\ {}^3F$	2—3
7034,903	250	5,87	7,63	$3d\ {}^1D^\circ - 5f\ {}^3G$	2—3
7026,62	25	5,86	7,63	$4p\ {}^1P - 5d\ {}^3D^\circ$	1—1
7017,646	90	5,87	7,64	$3d\ {}^1D^\circ - 5f\ {}^3D$	2—3
7017,28	30	5,87	7,64	$3d\ {}^1D^\circ - 5f\ {}^3D$	2—2
7016,74	10	5,96	7,73	$4p\ {}^3D - 6d\ {}^3F^\circ$	2—2

λ, Å	I	E_H, eV	E_B, eV	Transition	J
7005,883	180	5,98	7,75	$4p\ ^3D—6d\ ^3F^\circ$	3—4
7003,5665	180	5,96	7,73	$4p\ ^3D—6d\ ^3F^\circ$	2—3
6992,88	15	5,87	7,64	$3d\ ^1D^\circ—5f\ ^3/_2\ [^3/_2]$	2—2
6976,523	80	5,95	7,73	$4p\ ^3D—6d\ ^3F^\circ$	1—2
6867,22	20	5,86	7,67	$4p\ ^1P—6d\ ^3P^\circ$	1—2
6848,568	30	5,86	7,67	$4p\ ^1P—7s\ ^1P^\circ$	1—1
6741,64	30	5,98	7,82	$4p\ ^3D—8s\ [^3/_2,\ ^1/_2]^\circ$	3—2
6721,853	100	5,86	7,71	$4p\ ^1P—6d\ ^1D^\circ$	1—2
6635,65	25	5,86	7,72	$4p\ ^1P—6d\ ^3F^\circ$	1—2
6583,71	15	5,95	7,84	$4p\ ^3D—7d\ ^1D^\circ$	1—2
6560,556	25	5,96	7,85	$4p\ ^3D—7d\ ^3F^\circ$	2—3
6555,4624	45	5,98	7,87	$4p\ ^3D—7d\ ^3F^\circ$	3—4
6527,1989	45	5,87	7,77	$3d\ ^1D^\circ—6f\ ^1/_2\ [^7/_2]$	2—3
6526,609	45	5,87	7,77	$3d\ ^1D^\circ—6f\ ^1/_2\ [^5/_2]$	2—3
6518,73	20	5,95	7,85	$4p\ ^3D—4d\ ^3F^\circ$	1—3
6452,29	20	5,62	7,54	$3p^3\ ^3D^\circ—6p\ [^3/_2,\ ^1/_2]$	3—2
6437,79	8	5,86	7,79	$4p\ ^1P—8s\ ^3P^\circ$	1—1
6414,97	25	5,87	7,80	$3d\ ^1D^\circ—6f\ ^3/_2\ [^7/_2]$	2—3
6407,27	15	5,87	7,80	$3d\ ^1D^\circ—6f\ ^3/_2\ [^5/_2]$	2—3
6394,67	15	5,87	7,81	$3d\ ^1D^\circ—6f\ ^3/_2\ [^3/_2]$	2—2
6331,954	45	5,08	7,04	$4s\ ^1P^\circ—5p\ ^1P$	1—1
6279,35	15	5,86	7,84	$4p\ ^1P—7d\ ^1D^\circ$	1—2
6254,85	20	5,62	7,60	$3p^3\ ^3D^\circ—5f\ ^3F$	3—3
6254,1876	180	5,62	7,60	$3p^3\ ^3D^\circ—5f\ ^3F$	3—4
6253,60	15	5,08	7,06	$4s\ ^1P^\circ—5p\ ^3D$	1—1
6244,468	125	5,62	7,60	$3p^3\ ^3D^\circ—5f\ ^1D$	2—2
6243,8129	125	5,62	7,60	$3p^3\ ^3D^\circ—5f\ ^3F$	2—3
6238,2871	40	5,08	7,07	$4s\ ^1P^\circ—5p\ ^3D$	1—2
6237,3199	160	5,61	7,60	$3p^3\ ^3D^\circ—5f\ ^3F$	1—2
6155,70	20	5,62	7,63	$3p^3\ ^3D^\circ—5f\ ^3G$	3—3
6155,1338	160	5,62	7,63	$3p^3\ ^3D^\circ—5f\ ^3G$	3—4
6145,0151	100	5,62	7,63	$3p^3\ ^3D^\circ—5f\ ^3G$	2—3
6142,487	100	5,62	7,64	$3p^3\ ^3D^\circ—5f\ ^3D$	3—3
6131,850	90	5,62	7,64	$3p^3\ ^3D^\circ—5f\ ^3D$	2—3
6131,574	85	5,62	7,64	$3p^3\ ^3D^\circ—5f\ ^3D$	2—2
6125,0207	90	5,61	7,64	$3p^3\ ^3D^\circ—5f\ ^3D$	1—1
6112,926	10	5,62	7,64	$3p^3\ ^3D^\circ—5f\ ^3/_2\ [^3/_2]$	2—2
6106,605	15	5,61	7,64	$3p^3\ ^3D^\circ—5f\ ^3/_2\ [^3/_2]$	1—1
6091,92	15	5,87	7,91	$3d\ ^1D^\circ—7f\ ^3/_2\ [^7/_2]$	2—3
6087,80	10	5,87	7,91	$3d\ ^1D^\circ—7f\ ^3/_2\ [^5/_2]$	2—3
6067,624	20	5,08	7,12	$4s\ ^1P^\circ—5p\ ^3P$	1—2
5948,545	200	5,08	7,17	$4s\ ^1P^\circ—5p\ ^1D$	1—2
5873,764	40	4,93	7,04	$4s\ ^3P^\circ—5p\ ^1P$	1—1
5797,8591	100	4,95	7,09	$4s\ ^3P^\circ—5p\ ^3D$	2—3
5793,0714	90	4,93	7,07	$4s\ ^3P^\circ—5p\ ^3D$	1—2
5780,3839	70	4,92	7,06	$4s\ ^3P^\circ—5p\ ^3D$	0—1
5772,1453	70	5,08	7,23	$4s\ ^1P^\circ—5p\ ^1S$	1—0
5762,9769	45	5,62	7,77	$3p^3\ ^3D^\circ—6f\ ^1/_2\ [^7/_2]$	3—4
5754,2195	45	4,95	7,10	$4s\ ^3P^\circ—5p\ ^3P$	2—1
5753,625	45	5,62	7,77	$3p^3\ ^3D^\circ—6f\ ^1/_2\ [^5/_2]$	2—3
5747,6670	45	5,61	7,77	$3p^3\ ^3D^\circ—6f\ ^1/_2\ [^5/_2]$	1—2
5708,397	160	4,95	7,12	$4s\ ^3P^\circ—5p\ ^3P$	2—2
5701,1048	90	4,93	7,10	$4s\ ^3P^\circ—5p\ ^3P$	1—0
5690,4251	100	4,93	7,10	$4s\ ^3P^\circ—5p\ ^3P$	1—1
5684,4843	120	4,95	7,13	$4s\ ^3P^\circ—5p\ ^3S$	2—1
5675,418	20	5,62	7,80	$3p^3\ ^3D^\circ—6f\ ^3/_2\ [^7/_2]$	3—4
5669,743	10	5,62	7,80	$3p^3\ ^3D^\circ—6f\ ^3/_2\ [^5/_2]$	3—3

λ, Å	I	E_H, eV	E_B, eV	Transition	J
5666,677	10	5,62	7,80	$3p^3\ ^3D^\circ$—$6f\ ^3/_2\ [^7/_2]$	2—3
5665,5536	80	4,92	7,10	$4s\ ^3P^\circ$—$5p\ ^3P$	0—1
5660,683	13	5,62	7,80	$3p^3\ ^3D^\circ$—$6f\ ^3/_2\ [^5/_2]$	2—3
5660,502	10	5,62	7,80	$3p^3\ ^3D^\circ$—$6f\ ^3/_2\ [^5/_2]$	2—2
5654,924	15	5,61	7,80	$3p^3\ ^3D^\circ$—$6f\ ^3/_2\ [^5/_2]$	1—2
5645,960	90	4,93	7,12	$4s\ ^3P^\circ$—$5p\ ^3P$	1—2
5622,2214	30	4,93	7,13	$4s\ ^3P^\circ$—$5p\ ^3S$	1—1
5621,607	15	5,08	7,29	$4s\ ^1P^\circ$—$4f\ ^3F$	1—2
5602,875	20	4,95	7,17	$4s\ ^3P^\circ$—$5p\ ^1D$	2—2
5517,535	35	5,08	7,33	$4s\ ^1P^\circ$—$4f\ ^1D$	1—2
5493,23	40	5,08	7,34	$4s\ ^1P^\circ$—$4f\ ^3D$	1—2
5421,61	10	5,62	7,91	$3p^3\ ^3D^\circ$—$7f\ ^3/_2\ [^7/_2]$	3—3
5421,168	10	5,62	7,91	$3p^3\ ^3D^\circ$—$7f\ ^3/_2\ [^7/_2]$	3—4
5156,023	8	5,08	7,49	$4s\ ^1P^\circ$—$6p\ [^1/_2,\ ^1/_2]$	1—1
5128,031	10	5,08	7,50	$4s\ ^1P^\circ$—$6p\ [^1/_2,\ ^3/_2]$	1—1
5125,598	10	5,08	7,50	$4s\ ^1P^\circ$—$6p\ [^1/_2,\ ^3/_2]$	1—2
5006,0607	40	5,08	7,56	$4s\ ^1P^\circ$—$6p\ [^3/_2,\ ^3/_2]$	1—2
4947,6067	30	5,08	7,59	$4s\ ^1P^\circ$—$6p\ [^3/_2,\ ^3/_2]$	1—0
4851,540	13	5,08	7,64	$4s\ ^1P^\circ$—$5f\ ^3D$	1—2
4839,861	11	5,08	7,64	$4s\ ^1P^\circ$—$5f\ ^3/_2\ [^3/_2]$	1—2
4823,31	10	4,93	7,50	$4s\ ^3P^\circ$—$6p\ [^1/_2,\ ^3/_2]$	1—1
4821,1666	15	4,93	7,50	$4s\ ^3P^\circ$—$6p\ [^1/_2,\ ^3/_2]$	1—2
4805,4402	20	4,92	7,50	$4s\ ^3P^\circ$—$6p\ [^1/_2,\ ^3/_2]$	0—1
4792,324	80	4,95	7,54	$4s\ ^3P^\circ$—$6p\ [^3/_2,\ ^1/_2]$	2—2
4792,212	35	4,93	7,52	$4s\ ^3P^\circ$—$6p\ [^1/_2,\ ^1/_2]$	1—0
4782,9905	50	4,95	7,54	$4s\ ^3P^\circ$—$6p\ [^3/_2,\ ^1/_2]$	2—1
4772,7847	25	4,93	7,53	$4s\ ^3P^\circ$—$6p\ [^3/_2,\ ^3/_2]$	1—1
4758,972	13	4,95	7,56	$4s\ ^3P^\circ$—$6p\ [^3/_2,\ ^3/_2]$	2—2
4755,2756	25	4,92	7,53	$4s\ ^3P^\circ$—$6p\ [^3/_2,\ ^3/_2]$	0—1
4747,9936	25	4,93	7,54	$4s\ ^3P^\circ$—$6p\ [^3/_2,\ ^1/_2]$	1—2
4706,76	8	5,08	7,72	$4s\ ^1P^\circ$—$7p\ [^1/_2,\ ^3/_2]$	1—2
4638,17	15	—	—	—	—
4627,383	18	5,08	7,76	$4s\ ^1P^\circ$—$7p\ [^3/_2,\ ^3/_2]$	1—2
4434,69	10	4,92	7,71	$4s\ ^3P^\circ$—$7p\ [^1/_2,\ ^3/_2]$	0—1
4430,470	10	4,95	7,75	$4s\ ^3P^\circ$—$7p\ [^3/_2,\ ^1/_2]$	2—2
4425,49	10	4,95	7,75	$4s\ ^3P^\circ$—$7p\ [^3/_2,\ ^1/_2]$	2—1
4392,59	10	4,93	7,75	$4s\ ^3P^\circ$—$7p\ [^3/_2,\ ^1/_2]$	1—2
4102,9359	70	1,91	4,93	$3p^2\ ^1S$—$4s\ ^3P^\circ$	0—1
3905,5227	300	1,91	5,08	$3p^2\ ^1S$—$4s\ ^1P^\circ$	0—1
3020,0044	75	0,03	4,13	$3p^2\ ^3P$—$3p^3\ ^5S^\circ$	2—2
3006,7387	50	0,01	4,13	$3p^2\ ^3P$—$3p^3\ ^5S^\circ$	1—2
2987,6453	150	0,78	4,93	$3p^2\ ^1D$—$4s\ ^3P^\circ$	2—1
2970,3547	55	0,78	4,95	$3p^2\ ^1D$—$4s\ ^3P^\circ$	2—2
2881,5792	1000	0,78	5,08	$3p^2\ ^1D$—$4s\ ^1P^\circ$	2—1
2842,3345	15	1,91	6,27	$3p^2\ ^1S$—$3d\ ^3P^\circ$	0—1
2631,2819	190	1,91	6,62	$3p^2\ ^1S$—$3d\ ^1P^\circ$	0—1
2577,1514	45	1,91	6,72	$3p^2\ ^1S$—$3d\ ^3D^\circ$	0—1
2568,6407	85	1,91	6,73	$3p^2\ ^1S$—$5s\ ^3P^\circ$	0—1
2564,8242	20	0,78	5,61	$3p^2\ ^1D$—$3p^3\ ^3D^\circ$	2—1
2563,6787	30	0,78	5,62	$3p^2\ ^1D$—$3p^3\ ^3D^\circ$	2—2
2532,3814	110	1,91	6,80	$3p^2\ ^1S$—$5s\ ^1P^\circ$	0—1
2528,5086	450	0,03	4,93	$3p^2\ ^3P$—$4s\ ^3P^\circ$	2—1
2524,1079	425	0,01	4,92	$3p^2\ ^3P$—$4s\ ^3P^\circ$	1—0
2519,2023	350	0,01	4,93	$3p^2\ ^1D$—$4s\ ^3P^\circ$	1—1
2516,1125	500	0,03	4,95	$3p^2\ ^3P$—$4s\ ^3P^\circ$	2—2
2514,3161	375	0,00	4,93	$3p^2\ ^3P$—$4s\ ^3P^\circ$	0—1
2506,8973	425	0,01	4,95	$3p^2\ ^1D$—$4s\ ^3P^\circ$	1—2
2452,1180	70	0,03	5,08	$3p^2\ ^3P$—$4s\ ^1P^\circ$	2—1

λ, Å	I	E_H, eV	E_B, eV	Transition	J
2443,3643	65	0,01	5,08	$3p^2\,{}^3P-4s\,{}^1P^\circ$	1—1
2438,7674	65	0,00	5,08	$3p^2\,{}^3P-4s\,{}^1P^\circ$	0—1
2435,1545	300	0,78	5,87	$3p^2\,{}^1D-3d\,{}^1D^\circ$	2—2
2303,0585	55	1,91	7,29	$3p^2\,{}^1S-4d\,{}^1P^\circ$	0—1
2295,401	10	0,78	6,18	$3p^2\,{}^1D-3d\,{}^3F^\circ$	2—2
2291,034	35	0,78	6,19	$3p^2\,{}^1D-3d\,{}^3F^\circ$	2—3
2289,6074	20	1,91	7,32	$3p^2\,{}^1S-4d\,{}^3D^\circ$	0—1
2278,281	10	1,91	7,35	$3p^2\,{}^1S-6s\,{}^3P^\circ$	0—1
2259,587	10	1,91	7,39	$3p^2\,{}^1S-6s\,{}^1P^\circ$	0—1
2218,9148	50	0,03	5,61	$3p^2\,{}^3P-3p^3\,{}^3D^\circ$	2—1
2218,0569	120	0,03	5,62	$3p^2\,{}^3P-3p^3\,{}^3D^\circ$	2—2
2216,6688	120	0,03	5,62	$3p^2\,{}^3P-3p^3\,{}^3D^\circ$	2—3
2211,7441	110	0,01	5,61	$3p^2\,{}^3P-3p^3\,{}^3D^\circ$	1—1
2210,8940	115	0,01	5,62	$3p^2\,{}^3P-3p^3\,{}^3D^\circ$	1—2
2207,9783	110	0,00	5,61	$3p^2\,{}^3P-3p^3\,{}^3D^\circ$	0—1
2177,432	10	1,91	7,60	$3p^2\,{}^1S-5d\,{}^1P^\circ$	0—1
2147,911	50	1,91	7,68	$3p^2\,{}^1S-6d\,{}^3P^\circ$	0—1
2124,1225	100	0,78	6,62	$3p^2\,{}^1D-3d\,{}^1F^\circ$	2—3
2122,994	15	0,78	6,62	$3p^2\,{}^1D-3d\,{}^1P^\circ$	2—1
2121,1945	10	0,03	5,87	$3p^2\,{}^3P-3d\,{}^1D^\circ$	2—2
2114,631	30	0,01	5,87	$3p^2\,{}^3P-3d\,{}^1D^\circ$	1—2
2103,213	30	1,91	7,80	$3p^2\,{}^1S-6d\,{}^3D^\circ$	0—1
2094,211	10	1,91	7,83	$3p^2\,{}^1S-8s\,{}^1P^\circ$	0—1
2084,4669	10	0,78	6,73	$3p^2\,{}^1D-3d\,{}^3D^\circ$	2—3
2065,516	30	4,13	10,16	$3p^3\,{}^5S^\circ-4s\,{}^5P$	2—3
2061,192	40	4,13	10,14	$3p^3\,{}^5S^\circ-4s\,{}^5P$	2—2
2058,136	15	0,78	6,80	$3p^2\,{}^1D-5s\,{}^1P^\circ$	2—1
2054,828	50	4,13	10,13	$3p^3\,{}^5S^\circ-4s\,{}^5P$	2—1
2010,974	30	0,03	6,19	$3p^2\,{}^3P-3d\,{}^3F^\circ$	2—3
2008,439	15	0,01	6,18	$3p^2\,{}^3P-3d\,{}^3F^\circ$	1—2
1991,848	50	0,78	7,00	$3p^2\,{}^1D-4d\,{}^1D^\circ$	2—2
1988,9950	15	0,03	6,26	$3p^2\,{}^3P-3d\,{}^3P^\circ$	2—2
1986,3637	10	0,03	6,27	$3p^2\,{}^3P-3d\,{}^3P^\circ$	2—1
1984,434	30	0,78	7,03	$3p^2\,{}^1D-3p^3\,{}^3P^\circ$	2—2
1983,2341	20	0,01	6,26	$3p^2\,{}^3P-3d\,{}^3P^\circ$	1—2
1980,6203	15	0,01	6,27	$3p^2\,{}^3P-3d\,{}^3P^\circ$	1—1
1979,2062	15	0,01	6,27	$3p^2\,{}^3P-3d\,{}^3P^\circ$	1—0
1977,5982	15	0,00	6,27	$3p^2\,{}^3P-3d\,{}^3P^\circ$	0—1
1954,966	100	0,78	7,12	$3p^2\,{}^1D-4d\,{}^3F^\circ$	2—3
1904,660	50	0,78	7,29	$3p^2\,{}^1D-4d\,{}^1P^\circ$	2—1
1901,331	1000	0,78	7,30	$3p^2\,{}^1D-4d\,{}^1F^\circ$	2—3
1893,245	200	0,78	7,33	$3p^2\,{}^1D-4d\,{}^3D^\circ$	2—3
1887,700	200	0,78	7,35	$3p^2\,{}^1D-6s\,{}^3P^\circ$	2—1
1881,851	200	0,03	6,62	$3p^2\,{}^3P-3d\,{}^1F^\circ$	2—3
1880,953	20	0,03	6,62	$3p^2\,{}^3P-3d\,{}^1P^\circ$	2—1
1875,809	100	0,01	6,62	$3p^2\,{}^3P-3d\,{}^1P^\circ$	1—1
1874,838	500	0,78	7,39	$3p^2\,{}^1D-6s\,{}^1P^\circ$	2—1
1873,100	100	0,00	6,62	$3p^2\,{}^3P-3d\,{}^1P^\circ$	0—1
1853,148	50	0,03	6,72	$3p^2\,{}^3P-3d\,{}^3D^\circ$	2—1
1852,464	200	0,03	6,72	$3p^2\,{}^3P-3d\,{}^3D^\circ$	2—2
1851,791	30	0,78	7,48	$3p^2\,{}^1D-5d\,{}^1D^\circ$	2—2
1850,668	500	0,03	6,73	$3p^2\,{}^3P-3d\,{}^3D^\circ$	2—3
1848,737	100	0,03	6,73	$3p^2\,{}^3P-5s\,{}^3P^\circ$	2—1
1848,144	200	0,01	6,72	$3p^2\,{}^3P-3d\,{}^3D^\circ$	1—1
1847,468	400	0,01	6,72	$3p^2\,{}^3P-3d\,{}^3D^\circ$	1—2
1846,103	100	0,01	6,73	$3p^2\,{}^3P-5s\,{}^3P^\circ$	1—0
1845,510	300	0,00	6,72	$3p^2\,{}^3P-3d\,{}^3D^\circ$	0—1

λ, Å	I	E_H, eV	E_B, eV	Transition	J
1843,765	200	0,01	6,73	$3p^2\ ^3P—5s\ ^3P^\circ$	1—1
1841,440	200	0,03	6,76	$3p^2\ ^3P—5s\ ^3P^\circ$	2—2
1841,146	100	0,00	6,73	$3p^2\ ^3P—5s\ ^3P^\circ$	0—1
1838,006	30	0,78	7,53	$3p^2\ ^1D—5d\ ^3F^\circ$	2—3
1836,506	200	0,01	6,76	$3p^2\ ^3P—5s\ ^3P^\circ$	1—2
1829,893	20	0,03	6,80	$3p^2\ ^3P—5s\ ^1P^\circ$	2—1
1822,452	50	0,00	6,80	$3p^2\ ^3P—5s\ ^1P^\circ$	0—1
1814,068	500	0,78	7,62	$3p^2\ ^1D—5d\ ^1F^\circ$	2—3
1809,092	50	0,78	7,63	$3p^2\ ^1D—5d\ ^3D^\circ$	2—3
1799,122	100	0,78	7,67	$3p^2\ ^1D—7s\ ^1P^\circ$	2—1
1797,343	15	0,78	7,68	$3p^2\ ^1D—6d\ ^3P^\circ$	2—1
1790,292	20	0,78	7,71	$3p^2\ ^1D—6d\ ^1D^\circ$	2—2
1783,232	50	0,78	7,73	$3p^2\ ^1D—6d\ ^3F^\circ$	2—3
1776,826	100	0,03	7,00	$3p^2\ ^3P—4d\ ^1D^\circ$	2—2
1770,922	100	0,03	7,03	$3p^2\ ^3P—3p^3\ ^3P^\circ$	2—2
1770,629	30	0,03	7,03	$3p^2\ ^3P—3p^3\ ^3P^\circ$	2—1
1769,762	15	0,78	7,79	$3p^2\ ^1D—6d\ ^1F^\circ$	2—3
1766,346	20	0,01	7,03	$3p^2\ ^3P—3p^3\ ^3P^\circ$	1—2
1766,060	30	0,01	7,03	$3p^2\ ^3P—3p^3\ ^3P^\circ$	1—1
1765,60	30	0,78	7,80	$3p^2\ ^1D—6d\ ^3D^\circ$	2—3
1765,030	40	0,01	7,03	$3p^2\ ^3P—3p^3\ ^3P^\circ$	1—0
1763,664	50	0,00	7,03	$3p^2\ ^3P—3p^3\ ^3P^\circ$	0—1
1759,601	20	0,78	7,83	$3p^2\ ^1D—8s\ ^1P^\circ$	2—1
1753,112	30	0,78	7,85	$3p^2\ ^1D—7d\ ^3F^\circ$	2—3
1747,404	50	0,03	7,12	$3p^2\ ^3P—4d\ ^3F^\circ$	2—3
1745,332	15	0,01	7,11	$3p^2\ ^3P—4d\ ^3F^\circ$	1—2
1743,884	30	0,78	7,89	$3p^2\ ^1D—7d\ ^1F^\circ$	2—3
1740,378	20	0,78	7,90	$3p^2\ ^1D—7d\ ^3D^\circ$	2—3
1734,769	10	0,78	7,93	$3p^2\ ^1D—8d\ ^3F^\circ$	2—3
1704,434	50	0,03	7,30	$3p^2\ ^3P—4d\ ^1F^\circ$	2—3
1702,862	30	0,01	7,29	$3p^2\ ^3P—4d\ ^1P^\circ$	1—1
1700,626	30	0,00	7,29	$3p^2\ ^3P—4d\ ^1P^\circ$	0—1
1700,423	50	0,03	7,32	$3p^2\ ^3P—4d\ ^3D^\circ$	2—2
1699,716	10	0,03	7,32	$3p^2\ ^3P—4d\ ^3D^\circ$	2—1
1697,938	200	0,03	7,33	$3p^2\ ^3P—4d\ ^3D^\circ$	2—3
1696,203	200	0,01	7,32	$3p^2\ ^3P—4d\ ^3D^\circ$	1—2
1695,507	50	0,01	7,32	$3p^2\ ^3P—4d\ ^3D^\circ$	1—1
1693,461	20	0,03	7,35	$3p^2\ ^3P—6s\ ^3P^\circ$	2—1
1693,292	50	0,00	7,32	$3p^2\ ^3P—4d\ ^3D^\circ$	0—1
1690,786	30	0,01	7,34	$3p^2\ ^3P—6s\ ^3P^\circ$	1—0
1689,290	50	0,01	7,35	$3p^2\ ^3P—6s\ ^3P^\circ$	1—1
1687,095	15	0,00	7,35	$3p^2\ ^3P—6s\ ^3P^\circ$	0—1
1686,815	30	0,03	7,38	$3p^2\ ^3P—6s\ ^3P^\circ$	2—2
1682,675	30	0,01	7,38	$3p^2\ ^3P—6s\ ^3P^\circ$	1—2
1676,818	10	0,00	7,39	$3p^2\ ^3P—6s\ ^1P^\circ$	0—1
1675,198	200	0,03	7,43	$3p^2\ ^3P—5d\ ^3P^\circ$	2—2
1672,593	100	0,03	7,44	$3p^2\ ^3P—5d\ ^3P^\circ$	2—1
1671,111	20	0,01	7,43	$3p^2\ ^3P—5d\ ^3P^\circ$	1—2
1668,517	100	0,01	7,44	$3p^2\ ^3P—5d\ ^3P^\circ$	1—1
1667,618	100	0,01	7,44	$3p^2\ ^3P—5d\ ^3P^\circ$	1—0
1666,369	50	0,00	7,44	$3p^2\ ^3P—5d\ ^3P^\circ$	0—1
1664,521	30	0,03	7,48	$3p^2\ ^3P—5d\ ^1D^\circ$	2—2
1660,484	10	0,01	7,48	$3p^2\ ^3P—5d\ ^1D^\circ$	1—2
1653,351	30	0,03	7,53	$3p^2\ ^3P—5d\ ^3F^\circ$	2—3
1651,013	20	0,01	7,52	$3p^2\ ^3P—5d\ ^3F^\circ$	1—2
1640,257	20	4,13	11,71	$3p^3\ ^5S^\circ—3d\ ^5P$	2—1
1638,274	10	4,13	11,70	$3p^3\ ^5S^\circ—3d\ ^5P$	2—2
1633,978	50	0,03	7,62	$3p^2\ ^3P—5d\ ^1F^\circ$	2—2

λ, Å	I	E_H, eV	E_B, eV	Transition	J
1633,318	15	0,03	7,62	$3p^2\ ^3P—5d\ ^3D°$	2—2
1633,203	15	0,01	7,60	$3p^2\ ^3P—5d\ ^1P°$	1—1
1631,134	75	0,00	7,60	$3p^2\ ^3P—5d\ ^1P°$	0—1
1629,921	200	0,03	7,63	$3p^2\ ^3P—5d\ ^3D°$	2—3
1629,426	300	0,01	7,62	$3p^2\ ^3P—5d\ ^3D°$	1—2
1625,707	30	0,00	7,63	$3p^2\ ^3P—5d\ ^3D°$	0—1
1625,534	15	0,01	7,64	$3p^2\ ^3P—7s\ ^3P°$	1—1
1622,867	100	0,03	7,67	$3p^2\ ^3P—6d\ ^3P°$	2—2
1620,389	20	0,03	7,68	$3p^2\ ^3P—6d\ ^3P°$	2—1
1619,531	10	0,01	7,66	$3p^2\ ^3P—7s\ ^3P°$	1—2
1616,571	20	0,01	7,68	$3p^2\ ^3P—6d\ ^3P°$	1—1
1615,937	20	0,01	7,68	$3p^2\ ^3P—5d\ ^3P°$	1—0
1614,650	10	0,03	7,71	$3p^2\ ^3P—6d\ ^1D°$	2—2
1614,557	10	0,00	7,68	$3p^2\ ^3P—6d\ ^3P°$	0—1
1608,900	10	0,03	7,73	$3p^2\ ^3P—6d\ ^3F°$	2—3
1597,950	30	0,03	7,79	$3p^2\ ^3P—6d\ ^1F°$	2—3
1595,760	20	0,01	7,78	$3p^2\ ^3P—8s\ ^3P°$	1—0
1594,927	50	0,03	7,80	$3p^2\ ^3P—6d\ ^3P°$	2—2
1594,548	150	0,03	7,80	$3p^2\ ^3P—6d\ ^3D°$	2—3
1592,409	50	0,01	7,80	$3p^2\ ^3P—6d\ ^3D°$	1—2
1586,783	20	0,00	7,81	$3p^2\ ^3P—6d\ ^3P°$	0—1
1586,133	15	0,01	7,83	$3p^2\ ^3P—6d\ ^3P°$	1—0
1580,303	10	0,01	7,85	$3p^2\ ^3P—7d\ ^3F°$	1—3
1576,817	15	0,03	7,89	$3p^2\ ^3P—7d\ ^1F°$	2—3
1575,115	20	0,01	7,88	$3p^2\ ^3P—9s\ [^1/_2,\ ^1/_2]°$	1—0
1574,817	50	0,03	7,90	$3p^2\ ^3P—7d\ ^3P°$	2—2
1573,874	50	0,03	7,90	$3p^2\ ^3P—7d\ ^3D°$	2—3
1573,650	20	0,03	7,91	$3p^2\ ^3P—7d\ ^3P°$	2—1
1571,377	10	0,01	7,90	$3p^2\ ^3P—7d\ ^3D°$	1—2
1569,322	10	0,03	7,93	$3p^2\ ^3P—8d\ ^3F°$	2—3
1568,172	15	—	—	—	—
1567,703	10	0,01	7,92	$3p^2\ ^3P—9s\ [^3/_2,\ ^1/_2]°$	1—2
1564,589	10	0,01	7,93	$3p^2\ ^3P—8d\ ^3F°$	1—2
1561,982	10	0,03	7,96	$3p^2\ ^3P—8d\ ^3P°$	2—2
1561,792	10	0,01	7,95	$3p^2\ ^3P—10s\ [^1/_2,\ ^1/_2]°$	1—0
1560,739	15	0,03	7,97	$3p^2\ ^3P—8d\ ^3D°$	2—3
1560,067	15	—	—	—	—
1258,795	50	0,03	9,88	$3p^2\ ^3P—3p^3\ ^3S°$	2—1
1256,490	40	0,01	9,88	$3p^2\ ^3P—3p^3\ ^3S°$	1—1
1255,276	10	0,00	9,88	$3p^2\ ^3P—3p^3\ ^3S°$	0—1

Si II, ground state $1s^2\,2s^2\,2p^6\,3s^2\,3p\ ^2P^0_{1/2}$
Ionization potential 131 838,4 cm⁻¹; 16,345 eV

λ, Å	I	E_H, eV	E_B, eV	Transition	J
9412,72	100	12,84	14,15	$4f\ ^2F°—5g\ ^2G$	—
8044,50	15	13,93	15,47	$5d\ ^2D—8f\ ^2F°$	—
7911,47	10	14,10	15,67	$5f\ ^2F°—9g\ ^2G$	—
7849,72	500	12,52	14,10	$4d\ ^2D—5f\ ^2F°$	$^5/_2—^7/_2$
7848,80	400	12,52	14,10	$4d\ ^2D—5f\ ^2F°$	$^3/_2—^5/_2$
7312,29	3	14,10	15,80	$5f\ ^2F°—10g\ ^2G$	—
7125,84	4	12,88	14,62	$5p^2\ P°—7s\ ^2S$	$^3/_2—^1/_2$

λ, Å	I	E_H, eV	E_B, eV	Transition	J
6829,82	40	12,88	14,69	$5p\ ^2P^\circ—6d\ ^2D$	$^3/_2—^5/_2$
6818,45	20	12,88	14,69	$5p\ ^2P^\circ—6d\ ^2D$	$^1/_2—^3/_2$
6751,88	5	14,50	16,34	$4s'\ ^4P^\circ—4p'\ ^4D$	$^3/_2—^1/_2$
6750,28	20	14,53	16,36	$4s'\ ^4P^\circ—4p'\ ^4D$	$^5/_2—^5/_2$
6717,04	50	14,50	16,35	$4s'\ ^4P^\circ—4p'\ ^4D$	$^3/_2—^3/_2$
6699,38	20	14,49	16,35	$4s'\ ^4P^\circ—4p'\ ^4D$	$^1/_2—^1/_2$
6679,65	3	12,84	14,69	$4f\ ^2F^\circ—6d\ ^2D$	—
6671,88	100	14,53	16,39	$4s'\ ^4P^\circ—4p'\ ^4D$	$^5/_2—^7/_2$
6665,00	15	14,49	16,35	$4s'\ ^4P^\circ—4p'\ ^4D$	$^1/_2—^3/_2$
6660,52	50	14,50	16,36	$4s'\ ^4P^\circ—4p'\ ^4D$	$^3/_2—^5/_2$
6371,359	1000	8,12	10,07	$4s\ ^2S—4p\ ^2P^\circ$	$^1/_2—^1/_2$
6347,103	1000	8,12	10,07	$4s\ ^2S—4p\ ^2P^\circ$	$^1/_2—^3/_2$
6239,630	100	12,84	14,82	$4f\ ^2F^\circ—6g\ ^2G$	—
6160,16	5	16,64	18,65	$4p'\ ^4P—5s'\ ^4P^\circ$	$^5/_2—^3/_2$
6155,17	5	16,62	18,63	$4p'\ ^4P—5s'\ ^4P^\circ$	$^3/_2—^1/_2$
6131,76	4	16,61	18,63	$4p'\ ^4P—5s'\ ^4P^\circ$	$^1/_2—^1/_2$
6086,67	10	16,61	18,65	$4p'\ ^4P—5s'\ ^4P^\circ$	$^1/_2—^3/_2$
6080,06	20	16,64	18,68	$4p'\ ^4P—5s'\ ^4P^\circ$	$^5/_2—^5/_2$
6067,45	10	—	—	—	—
6030,27	5	16,62	18,68	$4p'\ ^4P—5s'\ ^4P^\circ$	$^3/_2—^5/_2$
6019,76	4	16,62	18,65	$4p'\ ^4P—5s'\ ^4P^\circ$	$^3/_2—^3/_2$
5978,929	500	10,07	12,15	$4p\ ^2P^\circ—5s\ ^2S$	$^3/_2—^1/_2$
5957,561	500	10,07	12,15	$4p\ ^2P^\circ—5s\ ^2S$	$^1/_2—^1/_2$
5915,220	150	14,53	16,62	$4s'\ ^4P^\circ—4p'\ ^4P$	$^5/_2—^3/_2$
5868,404	300	14,53	16,64	$4s'\ ^4P^\circ—4p'\ ^4P$	$^5/_2—^5/_2$
5867,483	10	14,50	16,61	$4s'\ ^4P^\circ—4p'\ ^4P$	$^3/_2—^1/_2$
5846,134	50	14,50	16,62	$4s'\ ^4P^\circ—4p'\ ^4P$	$^3/_2—^3/_2$
5827,801	30	14,49	16,61	$4s'\ ^4P^\circ—4p'\ ^4P$	$^1/_2—^1/_2$
5806,738	200	14,49	16,62	$4s'\ ^4P^\circ—4p'\ ^4P$	$^1/_2—^3/_2$
5800,468	150	14,50	16,64	$4s'\ ^4P^\circ—4p'\ ^4P$	$^3/_2—^5/_2$
5794,90	30	16,45	18,59	$4p'\ ^2P—4d'\ ^2D^\circ$	$^3/_2—^5/_2$
5785,73	30	16,44	18,58	$4p'\ ^2P—4d'\ ^2D^\circ$	$^1/_2—^3/_2$
5706,370	100	14,17	16,34	$3d'\ ^4F^\circ—4p'\ ^4D$	$^3/_2—^1/_2$
5701,374	200	14,17	16,35	$3d'\ ^4F^\circ—4p'\ ^4D$	$^5/_2—^3/_2$
5688,811	300	14,18	16,36	$3d'\ ^4F^\circ—4p'\ ^4D$	$^7/_2—^5/_2$
5681,44	30	14,17	16,35	$3d'\ ^4F^\circ—4p'\ ^4D$	$^3/_2—^3/_2$
5669,562	1000	14,20	16,38	$3d'\ ^4F^\circ—4p'\ ^4D$	$^9/_2—^7/_2$
5660,656	150	14,17	16,36	$3d'\ ^4F^\circ—4p'\ ^4D$	$^5/_2—^5/_2$
5639,478	200	14,53	16,72	$4s'\ ^4P^\circ—4p'\ ^4S$	$^5/_2—^3/_2$
5632,973	100	14,18	16,38	$3d'\ ^4F^\circ—4p'\ ^4D$	$^7/_2—^7/_2$
5605,351	3	14,17	16,38	$3d'\ ^4F^\circ—4p'\ ^4D$	$^5/_2—^7/_2$
5576,661	150	14,50	16,72	$4s'\ ^4P^\circ—4p'\ ^4S$	$^3/_2—^3/_2$
5575,973	5	12,88	15,10	$5p\ ^2P^\circ—8s\ ^2S$	$^3/_2—^1/_2$
5540,74	100	14,49	16,72	$4s'\ ^4P^\circ—4p'\ ^4S$	$^1/_2—^3/_2$
5496,45	200	16,72	18,98	$4p'\ ^4S—4d'\ ^4P^\circ$	$^3/_2—^5/_2$
5478,73	5	16,64	18,90	$4p'\ ^4P—4d'\ ^4D^\circ$	$^5/_2—^5/_2$
5469,450	30	12,88	15,14	$5p\ ^2P^\circ—7d\ ^2D$	$^3/_2—^5/_2$
5469,21	100	16,72	18,99	$4p'\ ^4S—4d'\ ^4P^\circ$	$^3/_2—^3/_2$
5466,868	500	12,52	14,79	$4d\ ^2D—6f\ ^2F^\circ$	$^5/_2—^7/_2$
5466,432	500	12,52	14,79	$4d\ ^2D—6f\ ^2F^\circ$	$^3/_2—^5/_2$
5462,146	10	12,88	15,14	$5p\ ^2P^\circ—7d\ ^2D$	$^1/_2—^3/_2$
5456,45	100	16,64	18,91	$4p'\ ^4P—4d'\ ^4D^\circ$	$^5/_2—^7/_2$
5454,49	15	16,72	19,00	$4p'\ ^4S—4d'\ ^4P^\circ$	$^3/_2—^1/_2$
5447,26	20	16,62	18,90	$4p'\ ^4P—4d'\ ^4D^\circ$	$^3/_2—^3/_2$
5438,62	100	16,62	18,90	$4p'\ ^4P—4d'\ ^4D^\circ$	$^3/_2—^5/_2$
5432,89	15	16,61	18,89	$4p'\ ^4P—4d'\ ^4D^\circ$	$^1/_2—^1/_2$
5428,92	15	16,61	18,90	$4p'\ ^4P—4d'\ ^4D^\circ$	$^1/_2—^3/_2$
5417,24	15	16,36	18,65	$4p'\ ^4D—5s'\ ^4P^\circ$	$^5/_2—^3/_2$

λ, Å	I	E_H, eV	E_B, eV	Transition	J
5415,64	5	16,35	18,63	$4p'\,^4D-5s'\,^4P^\circ$	3/2—1/2
5405,34	100	16,38	18,68	$4p'\,^4D-5s'\,^4P^\circ$	7/2—5/2
5393,18	3	16,34	18,63	$4p'\,^4D-5s'\,^4P^\circ$	1/2—1/2
5380,48	5	16,35	18,65	$4p'\,^4D-5s'\,^4P^\circ$	3/2—3/2
5354,89	5	16,36	18,68	$4p'\,^4D-5s'\,^4P^\circ$	5/2—5/2
5295,19	30	16,64	18,98	$4p'\,^4P-4d'\,^4P^\circ$	5/2—5/2
5269,74	3	16,64	18,99	$4p'\,^4P-4d'\,^4P^\circ$	5/2—3/2
5257,64	3	16,62	18,98	$4p'\,^4P-4d'\,^4P^\circ$	3/2—5/2
5240,31	5	16,38	18,75	$4p'\,^4D-4d'\,^4F^\circ$	7/2—7/2
5219,37	10	16,36	18,74	$4p'\,^4D-4d'\,^4F^\circ$	5/2—5/2
		16,62	19,00	$4p'\,^4P-4d'\,^4P^\circ$	3/2—1/2
5202,413	500	16,38	18,77	$4p'\,^4D-4d'\,^4F^\circ$	7/2—9/2
		16,35	18,73	$4p'\,^4D-4d'\,^4F^\circ$	3/2—3/2
5192,86	200	16,36	18,75	$4p'\,^4D-4d'\,^4F^\circ$	5/2—7/2
5185,535	100	12,84	15,23	$4f\,^2F^\circ-7g\,^2G$	—
5185,25	100	16,35	18,74	$4p'\,^4D-4d'\,^4F^\circ$	3/2—5/2
5181,90	100	16,34	18,73	$4p'\,^4D-4d'\,^4F^\circ$	1/2—3/2
5056,314	30	10,07	12,52	$4p\,^2P^\circ-4d\,^2D$	3/2—3/2
5055,981	1000	10,07	12,52	$4p\,^2P^\circ-4d\,^2D$	3/2—5/2
5041,026	1000	10,07	12,52	$4p\,^2P^\circ-4d\,^2D$	1/2—3/2
4932,80	20	16,77	19,29	$3p^3\,^2D^\circ-4f'\,^2F$	3/2—5/2
4908,18	5	16,77	19,29	$3p^3\,^2D^\circ-4f'\,^4F$	5/2—7/2
4906,99	20	16,38	18,91	$4p'\,^4D-4d'\,^4D^\circ$	7/2—7/2
4902,65	3	12,88	15,41	$5p\,^2P^\circ-9s\,^2S$	3/2—1/2
4883,20	15	16,36	18,90	$4p'\,^4D-4d'\,^4D^\circ$	5/2—5/2
4861,095	10	12,52	15,07	$4d\,^2D-4s'\,^2P^\circ$	5/2—3/2
4850,550	5	12,88	15,43	$5p\,^2P^\circ-8d\,^2D$	3/2—5/2
4845,26	3	16,34	18,89	$4p'\,^4D-4d'\,^4D^\circ$	1/2—1/2
4792,29	5	16,45	19,03	$4p'\,^2P-4d'\,^2P^\circ$	3/2—3/2
4782,89	3	16,44	19,03	$4p'\,^2P-4d'\,^2P^\circ$	1/2—1/2
4776,20	3	16,38	18,98	$4p'\,^4D-4d'\,^4P^\circ$	7/2—5/2
4673,273	20	12,84	15,49	$4f\,^2F^\circ-8g\,^2G$	—
4621,721	150	12,52	15,21	$4d\,^2D-7f\,^2F^\circ$	5/2—7/2
4621,418	100	12,52	15,21	$4d\,^2D-7f\,^2F^\circ$	3/2—5/2
4376,957	5	12,84	15,67	$4f\,^2F^\circ-9g\,^2G$	—
4259,202	5	12,15	15,05	$5s\,^2S-4s'\,^2P^\circ$	1/2—1/2
4232,864	10	12,15	15,07	$5s\,^2S-4s'\,^2P^\circ$	1/2—3/2
4200,898	40	12,52	15,47	$4d\,^2D-8f\,^2F^\circ$	5/2—7/2
4200,657	30	12,52	15,47	$4a\,^2D-8f\,^2F^\circ$	3/2—5/2
4198,133	50	13,49	16,44	$3d'\,^2D^\circ-4p'\,^2P$	3/2—1/2
4190,724	100	13,49	16,45	$3d'\,^2D^\circ-4p'\,^2P$	5/2—3/2
4187,137	5	12,84	15,80	$4f\,^2F^\circ-10g\,^2G$	—
4183,345	10	13,49	16,45	$3d'\,^2D^\circ-4p'\,^2P$	3/2—3/2
4130,893	500	9,84	12,84	$3d\,^2D-4f\,^2F^\circ$	5/2—7/2
4128,067	300	9,84	12,84	$3d\,^2D-4f\,^2F^\circ$	3/2—5/2
4076,781	15	9,84	12,88	$3d\,^2D-5p\,^2P^\circ$	3/2—1/2
4075,451	20	9,84	12,88	$3d\,^2D-5p\,^2P^\circ$	5/2—3/2
4072,711	3	9,84	12,88	$3d\,^2D-5p\,^2P^\circ$	3/2—3/2
4016,22	5	16,35	19,44	$3d'\,^2F^\circ-4f'\,^4G$	7/2—9/2
3998,01	10	16,32	19,43	$3d'\,^2F^\circ-4f'\,^2G$	5/2—7/2
3991,77	15	16,35	19,46	$3d'\,^2F^\circ-4f'\,^2G$	7/2—9/2
3977,46	10	16,32	19,44	$3d'\,^2F^\circ-4f'\,^4G$	5/2—7/2
3955,74	10	16,35	19,49	$3d'\,^2F^\circ-4f'\,^2D$	7/2—5/2
3954,507	10	12,52	15,66	$4d\,^2D-9f\,^2F^\circ$	5/2—7/2
3954,296	5	12,52	15,66	$4d\,^2D-9f\,^2F^\circ$	3/2—5/2
3919,00	5	16,32	19,49	$3d'\,^2F^\circ-4f'\,^2D$	5/2—3/2
3862,595	200	6,86	10,07	$3p^2\,^2D-4p\,^2P^\circ$	3/2—1/2
3856,017	500	6,86	10,07	$3p^2\,^2D-4p\,^2P^\circ$	5/2—3/2
3853,664	100	6,86	10,07	$3p^2\,^2D-4p^2\,P^\circ$	3/2—3/2

λ, Å	I	E_H, eV	E_B, eV	Transition	J
3339,819	500	10,07	13,78	$4p\ ^2P°—6s\ ^2S$	$^3/_2—^1/_2$
3333,139	300	10,07	13,78	$4p\ ^2P°—6s\ ^2S$	$^1/_2—^1/_2$
3223,01	20	15,45	19,29	$3d'\ ^4P°—4f'\ ^4F$	$^3/_2—^5/_2$
3220,44	10	15,44	19,29	$3d'\ ^4P°—4f'\ ^2F$	$^5/_2—^7/_2$
3217,99	15	15,44	19,29	$3d'\ ^4P°—4f'\ ^4F$	$^5/_2—^5/_2$
3214,66	75	15,44	19,30	$3d'\ ^4P°—4f'\ ^4F$	$^5/_2—^7/_2$
3210,025	200	10,07	13,93	$4p\ ^2P°—5d\ ^2D$	$^3/_2—^5/_2$
3203,872	100	10,07	13,93	$4p\ ^2P°—5d\ ^2D$	$^1/_2—^3/_2$
3202,49	20	15,43	19,30	$3d'\ ^4D°—4f'\ ^4F$	$^7/_2—^7/_2$
3199,514	200	15,43	19,30	$3d'\ ^4D°—4f'\ ^4F$	$^7/_2—^9/_2$
3195,41	100	15,41	19,29	$3d'\ ^4D°—4f'\ ^4F$	$^1/_2—^3/_2$
3194,69	50	15,41	19,29	$3d'\ ^4D°—4f'\ ^2F$	$^5/_2—^7/_2$
3194,21	50	15,41	19,29	$3d'\ ^4D°—4f'\ ^4F$	$^3/_2—^3/_2$
3193,09	150	15,41	19,29	$3d'\ ^4D°—4f'\ ^4F$	$^3/_2—^5/_2$
3192,25	50	15,41	19,29	$3d'\ ^4D°—4f'\ ^4F$	$^5/_2—^5/_2$
3188,97	150	15,41	19,30	$3d'\ ^4D°—4f'\ ^4F$	$^5/_2—^7/_2$
3185,99	10	15,07	18,96	$4s'\ ^2P°—4p'\ ^2S$	$^3/_2—^1/_2$
3149,92	20	16,38	20,32	$4p'\ ^4D—5d'\ ^4F°$	$^7/_2—^9/_2$
3053,184	150	15,44	19,50	$3d'\ ^4P°—4f'\ ^4D$	$^5/_2—^7/_2$
3048,30	50	15,45	19,51	$3d'\ ^4P°—4f'\ ^4D$	$^3/_2—^5/_2$
3045,77	10	15,45	19,52	$3d'\ ^4P°—4f'\ ^4D$	$^1/_2—^3/_2$
3043,85	10	15,44	19,51	$3d'\ ^4P°—4f'\ ^4D$	$^5/_2—^5/_2$
3043,692	100	{15,45	19,52	$3d'\ ^4P°—4f'\ ^4D$	$^3/_2—^3/_2$
		{15,45	19,53	$3d'\ ^4P°—4f'\ ^4D$	$^1/_2—^1/_2$
3042,191	30	15,43	19,50	$3d'\ ^4D°—4f'\ ^4D$	$^7/_2—^7/_2$
3041,573	20	15,45	19,53	$3d'\ ^4P°—4f'\ ^4D$	$^3/_2—^1/_2$
3039,21	3	15,44	19,52	$3d'\ ^4P°—4f'\ ^4D$	$^5/_2—^3/_2$
3030,000	100	15,41	19,50	$3d'\ ^4D°—4f'\ ^4D$	$^5/_2—^7/_2$
3021,55	20	15,41	19,51	$3d'\ ^4D°—4f'\ ^4D$	$^3/_2—^5/_2$
3015,980	3	15,41	19,52	$3d'\ ^4D°—4f'\ ^4D$	$^1/_2—^1/_2$
3014,920	3	15,41	19,52	$3d'\ ^4D°—4f'\ ^4D$	$^3/_2—^1/_2$
2905,692	500	9,84	14,10	$3d^2\ D—5f\ ^2F°$	$^5/_2—^7/_2$
2904,283	300	9,84	14,10	$3d\ ^2D—5f\ ^2F°$	$^3/_2—^5/_2$
2887,511	10	9,84	14,13	$3d\ ^2D—6p\ ^2P°$	$^5/_2—^3/_2$
2887,358	5	9,84	14,13	$3d\ ^2D—6p\ ^2P°$	$^3/_2—^1/_2$
2834,472	3	10,41	14,79	$3p^2\ ^2P—7p\ ^2P°$	$^3/_2—^3/_2$
2820,580	2	10,39	14,78	$3p^2\ ^2P—7p\ ^2P°$	$^1/_2—^1/_2$
2726,702	5	10,07	14,62	$4p\ ^2P°—7s\ ^2S$	$^3/_2—^1/_2$
2682,210	10	10,07	14,69	$4p\ ^2P°—6d\ ^2D$	$^3/_2—^5/_2$
2677,906	3	10,07	14,69	$4p\ ^2P°—6d\ ^2D$	$^1/_2—^3/_2$
2659,781	5	10,41	15,07	$3p^2\ ^2P—4s'\ ^2P°$	$^3/_2—^3/_2$
2655,803	3	10,39	15,05	$3p^2\ ^2P—4s'\ ^2P°$	$^1/_2—^1/_2$
2554,530	10	10,41	15,27	$3p^2\ ^2P—8p\ ^2P°$	$^3/_2—^3/_2$
2544,046	3	10,39	15,26	$3p^2\ ^2P—8p\ ^2P°$	$^1/_2—^1/_2$
2501,970	5	9,84	14,79	$3d\ ^2D—6f\ ^2F°$	$^5/_2—^7/_2$
2500,928	3	9,84	14,79	$3d\ ^2D—6f\ ^2F°$	$^3/_2—^5/_2$
2428,45	10	14,20	19,30	$3d'\ ^4F°—4f'\ ^4F$	$^9/_2—^9/_2$
2423,42	3	14,18	19,30	$3d'\ ^4F°—4f'\ ^4F$	$^7/_2—^7/_2$
2421,72	3	14,18	19,30	$3d'\ ^4F°—4f'\ ^4F$	$^7/_2—^9/_2$
2420,19	3	14,17	19,29	$3d'\ ^4F°—4f'\ ^4F$	$^5/_2—^5/_2$
2374,255	5	9,84	15,05	$3d\ ^2D—4s'\ ^2P°$	$^3/_2—^1/_2$
2366,972	5	9,84	15,07	$3d\ ^2D—4s'\ ^2P°$	$^5/_2—^3/_2$
2366,053	5	9,84	15,07	$3d\ ^2D—4s'\ ^2P°$	$^3/_2—^3/_2$
2364,33	3	14,20	19,44	$3d'\ ^4F°—4f'\ ^4G$	$^9/_2—^9/_2$
2360,59	5	14,17	19,42	$3d'\ ^4F°—4f'\ ^4G$	$^5/_2—^5/_2$
2360,20	10	14,17	19,44	$3d'\ ^4F°—4f'\ ^2G$	$^5/_2—^7/_2$
2357,97	50	14,18	19,44	$3d'\ ^4F°—4f'\ ^4G$	$^7/_2—^7/_2$
2357,18	30	14,17	19,42	$3d'\ ^4F°—4f'\ ^4G$	$^3/_2—^5/_2$
2356,295	100	14,20	19,46	$3d'\ ^4F°—4f'\ ^4G$	$^9/_2—^{11}/_2$

λ, Å	I	E_H, eV	E_B, eV	Transition	J
2353,09	20	14,17	19,44	$3d'\ ^4F^\circ$—$4f'\ ^4G$	$^5/_2$—$^7/_2$
2350,174	20	0,04	5,31	$3p\ ^2P^\circ$—$3p^2\ ^4P$	$^3/_2$—$^1/_2$
2349,54	10	14,18	19,46	$3d'\ ^4F^\circ$—$4f'\ ^2G$	$^7/_2$—$^9/_2$
2344,203	10	0,04	5,32	$3p\ ^2P^\circ$—$3p^2\ ^4P$	$^3/_2$—$^3/_2$
2334,606	30	0,04	5,34	$3p\ ^2P^\circ$—$3p^2\ ^4P$	$^3/_2$—$^5/_2$
2334,404	30	0,00	5,31	$3p\ ^2P^\circ$—$3p^2\ ^4P$	$^1/_2$—$^1/_2$
2136,560	50	13,49	19,29	$3d'\ ^2D^\circ$—$4f'\ ^2F$	$^5/_2$—$^7/_2$
2136,402	30	13,49	19,29	$3d'\ ^2D^\circ$—$4f'\ ^2F$	$^3/_2$—$^5/_2$
2133,99	10	13,49	19,29	$3d'\ ^2D^\circ$—$4f'\ ^2F$	$^5/_2$—$^7/_2$
2072,701	200	6,86	12,84	$3p^2\ ^2D^\circ$—$4f\ ^2F^\circ$	$^5/_2$—$^7/_2$
2072,016	200	6,86	12,84	$3p^2\ ^2D$—$4f\ ^2F^\circ$	$^3/_2$—$^5/_2$
2059,014	50	6,86	12,88	$3p^2\ ^2D$—$5p\ ^2P^\circ$	$^3/_2$—$^1/_2$
2058,646	50	6,86	12,88	$3p^2\ ^2D$—$5p\ ^2P^\circ$	$^5/_2$—$^3/_2$
2016,654	3	9,50	15,65	$3p^2\ ^2S$—$3d'\ ^2P^\circ$	$^1/_2$—$^3/_2$
1949,564	100	10,41	16,77	$3p^2\ ^2P$—$3p^3\ ^2D^\circ$	$^3/_2$—$^5/_2$
1949,331	10	10,41	16,77	$3p^2\ ^2P$—$3p^3\ ^2D^\circ$	$^3/_2$—$^3/_2$
1945,504	3	10,07	16,44	$4p\ ^2P^\circ$—$4p'\ ^2P$	$^1/_2$—$^1/_2$
1944,586	15	10,07	16,45	$4p\ ^2P^\circ$—$4p'\ ^2P$	$^3/_2$—$^3/_2$
1941,667	50	10,39	16,77	$3p^2\ ^2P$—$3p^3\ ^2D^\circ$	$^1/_2$—$^3/_2$
1910,621	50	9,84	16,32	$3d\ ^2D$—$3d'\ ^2F^\circ$	$^3/_2$—$^5/_2$
1905,878	3	12,52	19,03	$4d\ ^2D$—$4d'\ ^2P^\circ$	$^3/_2$—$^1/_2$
1904,326	5	12,52	19,03	$4d\ ^2D$—$4d'\ ^2P^\circ$	$^5/_2$—$^3/_2$
1902,459	100	9,84	16,35	$3d\ ^2D$—$3d'\ ^2F^\circ$	$^5/_2$—$^7/_2$
1870,782	3	6,86	13,49	$3p^2\ ^2D$—$3d\ ^2D^\circ$	$^5/_2$—$^3/_2$
1870,227	15	6,86	13,49	$3p^2\ ^2D$—$3d\ ^2D^\circ$	$^3/_2$—$^3/_2$
1869,317	20	6,86	13,49	$3p^2\ ^2D$—$3d\ ^2D^\circ$	$^5/_2$—$^5/_2$
1817,445	10	0,04	6,86	$3p\ ^2P^\circ$—$3p^2\ ^2D$	$^3/_2$—$^3/_2$
1816,921	200	0,04	6,86	$3p\ ^2P^\circ$—$3p^2\ ^2D$	$^3/_2$—$^5/_2$
1808,003	150	0,00	6,86	$3p\ ^2P^\circ$—$3p^2\ ^2D$	$^1/_2$—$^3/_2$
1787,538	8	9,84	16,77	$3d\ ^2D$—$3p^3\ ^2D^\circ$	$^5/_2$—$^5/_2$
1786,817	4	9,84	16,77	$3d\ ^2D$—$3p^3\ ^2D^\circ$	$^3/_2$—$^3/_2$
1711,296	20	6,86	14,10	$3p^2\ ^2D$—$5f\ ^2F^\circ$	$^5/_2$—$^7/_2$
1710,826	10	6,86	14,10	$3p^2\ ^2D$—$5f\ ^2F^\circ$	$^3/_2$—$^5/_2$
1704,967	2	6,86	14,13	$3p^2\ ^2D$—$6p\ ^2P^\circ$	$^3/_2$—$^1/_2$
1661,059	3	10,39	17,85	$3p^2\ ^2P$—$3p^3\ ^2P^\circ$	$^1/_2$—$^1/_2$
1564,066	5	6,86	14,78	$3p^2\ ^2D$—$7p\ ^2P^\circ$	$^3/_2$—$^1/_2$
1563,765	10	6,86	14,79	$3p^2\ ^2D$—$7p\ ^2P^\circ$	$^5/_2$—$^3/_2$
1562,845	15	6,86	14,79	$3p^2\ ^2D$—$6f\ ^2F^\circ$	$^5/_2$—$^7/_2$
1562,451	10	6,86	14,79	$3p^2\ ^2D$—$6f\ ^2F^\circ$	$^3/_2$—$^5/_2$
1533,445	1000	0,04	8,12	$3p\ ^2P^\circ$—$4s\ ^2S$	$^3/_2$—$^1/_2$
1526,719	500	0,00	8,12	$3p\ ^2P^\circ$—$4s\ ^2S$	$^1/_2$—$^1/_2$
1518,221	5	10,41	18,58	$3p^2\ ^2P$—$4d'\ ^2D^\circ$	$^3/_2$—$^3/_2$
1516,910	60	10,41	18,59	$3p^2\ ^2P$—$4d'\ ^2D^\circ$	$^3/_2$—$^5/_2$
1513,570	30	10,39	18,58	$3p^2\ ^2P$—$4d'\ ^2D^\circ$	$^1/_2$—$^3/_2$
1512,072	50	6,86	15,05	$3p^2\ ^2D$—$4s\ ^2P^\circ$	$^3/_2$—$^1/_2$
1509,101	100	6,86	15,07	$3p^2\ ^2D$—$4s\ ^2P^\circ$	$^5/_2$—$^3/_2$
1508,741	3	6,86	15,07	$3p^2\ ^2D$—$4s\ ^2P^\circ$	$^3/_2$—$^3/_2$
1485,513	100	9,50	17,85	$3p^2\ ^2S$—$3p^3\ ^2P^\circ$	$^1/_2$—$^3/_2$
1485,224	30	6,86	15,21	$3p^2\ ^2D$—$7f\ ^2F^\circ$	$^5/_2$—$^7/_2$
1485,024	90	9,50	17,85	$3p^2\ ^2S$—$3p^3\ ^2P^\circ$	$^1/_2$—$^1/_2$
1484,873	15	6,86	15,21	$3p^2\ ^2D$—$7f\ ^2F^\circ$	$^3/_2$—$^5/_2$
1475,188	5	6,86	15,26	$3p^2\ ^2D$—$8p\ ^2P^\circ$	$^3/_2$—$^1/_2$
1474,649	15	6,86	15,27	$3p^2\ ^2D$—$8p\ ^2P^\circ$	$^5/_2$—$^3/_2$
1438,931	4	6,86	15,47	$3p^2\ ^2D$—$8f\ ^2F^\circ$	$^5/_2$—$^7/_2$
1417,781	5	9,84	18,58	$3d\ ^2D$—$4d'\ ^2D^\circ$	$^3/_2$—$^3/_2$
1416,972	10	9,84	18,59	$3d\ ^2D$—$4d'\ ^2D^\circ$	$^5/_2$—$^5/_2$
1410,219	20	6,86	15,65	$3p^2\ ^2D$—$3d'\ ^2P^\circ$	$^5/_2$—$^3/_2$
1409,073	10	6,86	15,65	$3p^2\ ^2D$—$3d'\ ^2P^\circ$	$^3/_2$—$^1/_2$

λ, Å	I	E_H, eV	E_B, eV	Transition	J
1404,478	6	6,86	15,68	$3p^2\ {}^2D—10p\ {}^2P^\circ$	$^5/_2—^3/_2$
1403,783	5	6,86	15,69	$3p^2\ {}^2D—10p\ {}^2P^\circ$	$^3/_2—^1/_2$
1353,718	100	5,34	14,50	$3p^2\ {}^4P—4s'\ {}^4P^\circ$	$^5/_2—^3/_2$
1352,635	100	5,32	14,49	$3p^2\ {}^4P—4s'\ {}^4P^\circ$	$^3/_2—^1/_2$
1350,658	20	5,31	14,49	$3p^2\ {}^4P—4s'\ {}^4P^\circ$	$^1/_2—^1/_2$
1350,520	20	5,32	14,50	$3p^2\ {}^4P—4s'\ {}^4P^\circ$	$^3/_2—^3/_2$
1350,057	150	5,34	14,53	$3p^2\ {}^4P—4s'\ {}^4P^\circ$	$^5/_2—^5/_2$
1348,543	100	5,31	14,50	$3p^2\ {}^4P—4s'\ {}^4P^\circ$	$^1/_2—^3/_2$
1346,873	100	5,32	14,53	$3p^2\ {}^4P—4s'\ {}^4P^\circ$	$^3/_2—^5/_2$
1309,458	20	6,86	16,32	$3p^2\ {}^2D—3d'\ {}^2F^\circ$	$^3/_2—^5/_2$
1309,274	200	0,04	9,50	$3p\ {}^2P^\circ—3p^2\ {}^2S$	$^3/_2—^1/_2$
1305,590	50	6,86	16,35	$3p^2\ {}^2D—3d'\ {}^2F^\circ$	$^5/_2—^7/_2$
1304,369	100	0,00	9,50	$3p\ {}^2P^\circ—3p^2\ {}^2S$	$^1/_2—^1/_2$
1265,023	200	0,04	9,84	$3p\ {}^2P^\circ—3d\ {}^2D$	$^3/_2—^3/_2$
1264,730	2000	0,04	9,84	$3p\ {}^2P^\circ—3d\ {}^2D$	$^3/_2—^5/_2$
1260,418	1000	0,00	9,84	$3p\ {}^2P^\circ—3d\ {}^2D$	$^1/_2—^3/_2$
1251,164	200	5,34	15,25	$3p^2\ {}^4P—3p^3\ {}^4S^\circ$	$^5/_2—^3/_2$
1250,433	150	6,86	16,77	$3p^2\ {}^2D—3p^3\ {}^2D^\circ$	$^5/_2—^5/_2$
1250,089	100	6,86	16,77	$3p^2\ {}^2D—3p^3\ {}^2D^\circ$	$^3/_2—^3/_2$
1248,426	150	5,32	15,25	$3p^2\ {}^4P—3p^3\ {}^4S^\circ$	$^3/_2—^3/_2$
1246,738	100	5,31	15,25	$3p^2\ {}^4P—3p^3\ {}^4S^\circ$	$^1/_2—^3/_2$
1235,920	10	—	—	—	—
1231,406	5	5,34	15,41	$3p^2\ {}^4P—3d'\ {}^4D^\circ$	$^5/_2—^5/_2$
1229,388	200	5,34	15,43	$3p^2\ {}^4P—3d'\ {}^4D^\circ$	$^5/_2—^7/_2$
1228,746	150	5,32	15,41	$3p^2\ {}^4P—3d'\ {}^4D^\circ$	$^3/_2—^5/_2$
1228,617	25	5,32	15,41	$3p^2\ {}^4P—3d'\ {}^4D^\circ$	$^3/_2—^3/_2$
1228,437	10	5,32	15,41	$3p^2\ {}^4P—3d'\ {}^4D^\circ$	$^3/_2—^1/_2$
1227,604	100	5,34	15,44	$3p^2\ {}^4P—3d'\ {}^4P^\circ$	$^5/_2—^5/_2$
1226,986	40	5,31	15,41	$3p^2\ {}^4P—3d'\ {}^4D^\circ$	$^1/_2—^3/_2$
1226,887	20	5,34	15,45	$3p^2\ {}^4P—3d'\ {}^4D^\circ$	$^5/_2—^3/_2$
1226,814	50	5,31	15,41	$3p^2\ {}^4P—3d'\ {}^4D^\circ$	$^1/_2—^1/_2$
1224,972	10	5,32	15,44	$3p^2\ {}^4P—3d'\ {}^4P^\circ$	$^3/_2—^5/_2$
1224,252	20	5,32	15,45	$3p^2\ {}^4P—3d'\ {}^4P^\circ$	$^3/_2—^3/_2$
1223,907	20	5,32	15,45	$3p^2\ {}^4P—3d'\ {}^4P^\circ$	$^3/_2—^1/_2$
1222,635	5	5,31	15,45	$3p^2\ {}^4P—3d'\ {}^4P^\circ$	$^1/_2—^3/_2$
1216,117	10	—	—	—	—
1197,389	100	0,04	10,39	$3p\ {}^2P^\circ—3p^2\ {}^2P$	$^3/_2—^1/_2$
1194,496	250	0,04	10,41	$3p\ {}^2P^\circ—3p^2\ {}^2P$	$^3/_2—^3/_2$
1193,284	200	0,00	10,39	$3p\ {}^2P^\circ—3p^2\ {}^2P$	$^1/_2—^1/_2$
1190,418	100	0,00	10,41	$3p\ {}^2P^\circ—3p^2\ {}^2P$	$^1/_2—^3/_2$
1127,907	40	6,86	17,85	$3p^2\ {}^2D—3p^3\ {}^2P^\circ$	$^5/_2—^3/_2$
1127,442	20	6,86	17,85	$3p^2\ {}^2D—3p^3\ {}^2P^\circ$	$^3/_2—^1/_2$
1057,503	15	6,86	18,58	$3p^2\ {}^2D—4d'\ {}^2D^\circ$	$^3/_2—^3/_2$
1057,050	30	6,86	18,59	$3p^2\ {}^2D—4d'\ {}^2D^\circ$	$^5/_2—^5/_2$
1023,693	50	0,04	12,15	$3p\ {}^2P^\circ—5s\ {}^2S$	$^3/_2—^1/_2$
1020,699	25	0,00	12,15	$3p\ {}^2P^\circ—5s\ {}^2S$	$^1/_2—^1/_2$
992,675	200	0,04	12,52	$3p\ {}^2P^\circ—4d\ {}^2D$	$^3/_2—^5/_2$
989,867	100	0,00	12,52	$3p\ {}^2P^\circ—4d\ {}^2D$	$^1/_2—^3/_2$
931,667	5	5,34	18,65	$3p^2\ {}^4P—5s'\ {}^4P^\circ$	$^5/_2—^3/_2$
931,200	5	5,32	18,63	$3p^2\ {}^4P—5s'\ {}^4P^\circ$	$^3/_2—^1/_2$
929,810	20	5,34	18,68	$3p^2\ {}^4P—5s'\ {}^4P^\circ$	$^5/_2—^5/_2$
928,297	5	5,32	18,68	$3p^2\ {}^4P—5s'\ {}^4P^\circ$	$^3/_2—^5/_2$
913,853	20	5,34	18,91	$3p^2\ {}^4P—4d'\ {}^4D^\circ$	$^5/_2—^7/_2$
913,264	3	5,32	18,90	$3p^2\ {}^4P—4d'\ {}^4D^\circ$	$^3/_2—^3/_2$
913,012	10	5,32	18,90	$3p^2\ {}^4P—4d'\ {}^4D^\circ$	$^3/_2—^5/_2$
912,459	5	5,31	18,89	$3p^2\ {}^4P—4d'\ {}^4D^\circ$	$^1/_2—^1/_2$
912,375	5	5,31	18,90	$3p^2\ {}^4P—4d'\ {}^4D^\circ$	$^1/_2—^3/_2$

λ, Å	I	E_H, eV	E_B, eV	Transition	J
909,209	3	5,34	18,98	$3p^2\,^4P$—$4d'\,^4P^\circ$	$^5/_2$—$^5/_2$
901,735	20	0,04	13,78	$3p\,^2P^\circ$—$6s\,^2S$	$^3/_2$—$^1/_2$
899,405	10	0,00	13,78	$3p\,^2P^\circ$—$6s\,^2S$	$^1/_2$—$^1/_2$
891,999	200	0,04	13,93	$3p\,^2P^\circ$—$5d\,^2D$	$^3/_2$—$^5/_2$
889,722	100	0,00	13,93	$3p\,^2P^\circ$—$5d\,^2D$	$^1/_2$—$^3/_2$
850,142	10	0,04	14,62	$3p\,^2P^\circ$—$7s\,^2S$	$^3/_2$—$^1/_2$
848,074	5	0,00	14,62	$3p\,^2P^\circ$—$7s\,^2S$	$^1/_2$—$^1/_2$
845,774	40	0,04	14,69	$3p\,^2P^\circ$—$6d\,^2D$	$^3/_2$—$^5/_2$
843,718	20	0,00	14,69	$3p\,^2P^\circ$—$6d\,^2D$	$^1/_2$—$^3/_2$
822,844	5	0,04	15,10	$3p\,^2P^\circ$—$8s\,^2S$	$^3/_2$—$^1/_2$
821,450	2	5,34	20,43	$3p^2\,^4P$—$5d'\,^4D^\circ$	$^5/_2$—$^7/_2$
820,918	3	0,00	15,10	$3p\,^2P^\circ$—$8s\,^2S$	$^1/_2$—$^1/_2$
820,516	20	0,04	15,14	$3p\,^2P^\circ$—$7d\,^2D$	$^3/_2$—$^5/_2$
818,590	2	0,00	15,14	$3p\,^2P^\circ$—$7d\,^2D$	$^1/_2$—$^3/_2$
805,101	10	0,04	15,43	$3p\,^2P^\circ$—$8d\,^2D$	$^3/_2$—$^3/_2$, $^5/_2$
755,362	2	0,04	16,45	$3p\,^2P^\circ$—$4p\,^2P$	$^3/_2$—$^3/_2$

Si III, ground state $1s^2\,2s^2\,2p^6\,3s^2\,^1S_0$
Ionization potential 270139,3 cm^{-1}; 33,491 eV

λ, Å	I	E_H, eV	E_B, eV	Transition	J
9799,906	2	25,33	26,59	$4d\,^1D$—$5p\,^1P^\circ$	2—1
9323,899	3	20,55	21,88	$3d\,^1D$—$4p\,^1P^\circ$	2—1
9173,267	2	30,23	31,58	$6f\,^1F^\circ$—$8d\,^1D$	3—2
8728,019	3	33,16	34,58	$4d'\,^1D^\circ$—$4f'$ [$3^1/_2$]	2—3
8341,931	2	29,25	30,73	$4s'\,^1P^\circ$—$4p'\,^1D$	1—2
8292,615	3	30,08	31,57	$6h\,^{3,1}H^\circ$—$8i\,^{3,1}I$	—
8271,944	6	26,65	28,15	$5p\,^3P^\circ$—$5d\,^3D$	0—1
8271,377	5	26,65	28,15	$5p\,^3P^\circ$—$5d\,^3D$	1—1
8269,324	8	26,65	28,15	$5p\,^3P^\circ$—$5d\,^3D$	1—2
8265,640	5	26,65	28,15	$5p\,^3P^\circ$—$5d\,^3D$	2—2
8262,568	9	26,65	28,15	$5p\,^3P^\circ$—$5d\,^3D$	2—3
8212,05	2	30,05	31,56	$6f\,^3F^\circ$—$8g\,^3G$	—
8194,71	3	30,06	31,57	$6g\,^{3,1}G$—$8h\,^{3,1}H^\circ$	5——
8194,18	3	30,06	31,57	$6g\,^{3,1}G$—$8h\,^{3,1}H^\circ$	3, 4——
8191,679	8	28,54	30,06	$5f\,^3F^\circ$—$6g\,^3G$	4—5
8191,16	6	28,54	30,06	$5f\,^3F^\circ$—$6g\,^3G$	3—4
8190,431	7	28,54	30,06	$5f\,^3F^\circ$—$6g\,^1G$	3—4
		28,54	30,06	$5f\,^3F^\circ$—$6g\,^3G$	2—3
8103,448	11	28,55	30,08	$5g\,^{3,1}G$—$6h\,^{3,1}H^\circ$	5——
8102,862	9	28,55	30,08	$5g\,^{3,1}G$—$6h\,^{3,1}H^\circ$	3, 4——
7612,356	12	26,59	28,22	$5p\,^1P^\circ$—$5d\,^1D$	1—2
7497,286	3	26,81	28,46	$3d'\,^3P^\circ$—$6s\,^3S$	1—1
7466,322	9	24,99	26,65	$4d\,^3D$—$5p\,^3P^\circ$	3—2
7465,669	4	24,99	26,65	$4d\,^3D$—$5p\,^3P^\circ$	2—2
7462,624	8	24,99	26,65	$4d\,^3D$—$5p\,^3P^\circ$	1, 2—1
7461,890	5	24,99	26,65	$4d\,^3D$—$5p\,^3P^\circ$	1—0
7442,327	4	26,80	28,46	$3d'\,^3P^\circ$—$6s\,^3S$	2—1
7408,467	3	28,12	29,79	$4s'\,^3P^\circ$—$6d\,^3D$	2—3
7267,090	2	28,08	29,79	$4s'\,^3P^\circ$—$6d\,^3D$	1—2
7151,08	2	30,23	31,96	$6f\,^1F^\circ$—$9d\,^1D$	3—2
7047,58	5	28,35	30,11	$3d'\,^1P^\circ$—$4p'\,^1P$	1—1
6851,65	7	24,99	26,80	$4d\,^3D$—$3d'\,^3P^\circ$	3—2
6851,18	3	24,99	26,80	$4d\,^3D$—$3d'\,^3P^\circ$	2—2

λ, Å	I	E_{H}, eV	E_{B}, eV	Transition	J
6834,38	2	26,65	28,46	$5p\ ^{3}P^{\circ}$—$6s\ ^{3}S$	0—1
6834,08	4	26,65	28,46	$5p\ ^{3}P^{\circ}$—$6s\ ^{3}S$	1—1
6831,560	6	26,65	28,46	$5p\ ^{3}P^{\circ}$—$6s\ ^{3}S$	2—1
6805,244	4	24,99	26,81	$4d\ ^{3}D$—$3d'\ ^{3}P^{\circ}$	1, 2—1
6776,623	2	24,99	26,82	$4d\ ^{3}D$—$3d'\ ^{3}P^{\circ}$	1—0
6535,163	2	30,08	31,97	$6h\ ^{3,1}H^{\circ}$—$9i\ ^{3,1}I$	—
6524,357	6	28,15	30,05	$5d\ ^{3}D$—$6f\ ^{3}F^{\circ}$	3—3, 4
6522,626	4	28,15	30,05	$5d\ ^{3}D$—$6f\ ^{3}F^{\circ}$	2—2, 3
6521,485	3	28,15	30,05	$5d\ ^{3}D$—$6f\ ^{3}F^{\circ}$	1—2
6471,86	2	30,06	31,97	$6g\ ^{3,1}G$—$9h\ ^{3,1}H^{\circ}$	5——
6314,459	7	26,59	28,56	$5p\ ^{1}P^{\circ}$—$6s\ ^{1}S$	1—0
6173,712	3	19,72	21,73	$4s\ ^{1}S$—$4p\ ^{3}P^{\circ}$	0—1
6169,835	3	28,22	30,23	$5d\ ^{1}D$—$6f\ ^{1}F^{\circ}$	2—3
6152,556	2	29,05	31,07	$6p\ ^{1}P^{\circ}$—$7d\ ^{1}D$	1—2
5898,788	10	27,96	30,06	$5f\ ^{1}F^{\circ}$—$6g\ ^{1}G$	3—4
5810,187	3	28,22	30,35	$5d\ ^{1}D$—$7p\ ^{1}P^{\circ}$	2—1
5739,733	20	19,72	21,88	$4s\ ^{1}S$—$4p\ ^{1}P^{\circ}$	0—1
5716,289	8	25,98	28,15	$4f\ ^{3}F^{\circ}$—$5d\ ^{3}D$	4—3
5704,598	7	25,98	28,15	$4f\ ^{3}F^{\circ}$—$5d\ ^{3}D$	3—2
5703,121	4	25,98	28,15	$4f\ ^{3}F^{\circ}$—$5d\ ^{3}D$	3—3
5696,50	7	25,97	28,15	$4f\ ^{3}F^{\circ}$—$5d\ ^{3}D$	2—1
5695,522	3	25,97	28,15	$4f\ ^{3}F^{\circ}$—$5d\ ^{3}D$	2—2
5601,461	2	28,15	30,36	$5d\ ^{3}D$—$7p\ ^{3}P^{\circ}$	1—0
5600,952	3	28,15	30,36	$5d\ ^{3}D$—$7p\ ^{3}P^{\circ}$	2—1
5599,246	4	28,15	30,36	$5d\ ^{3}D$—$7p\ ^{3}P^{\circ}$	3—2
5539,926	3	28,12	30,35	$4s'\ ^{3}P^{\circ}$—$4p'\ ^{3}D$	2—2
5490,114	3	28,08	30,34	$4s'\ ^{3}P^{\circ}$—$4p'\ ^{3}D$	1—1
5473,045	7	28,12	30,38	$4s'\ ^{3}P^{\circ}$—$4p'\ ^{3}D$	2—3
5451,961	4	28,06	30,34	$4s'\ ^{3}P^{\circ}$—$4p'\ ^{3}D$	0—1
5451,462	6	28,08	30,35	$4s'\ ^{3}P^{\circ}$—$4p'\ ^{3}D$	1—2
5303,415	2	28,54	30,88	$5f\ ^{3}F^{\circ}$—$7d\ ^{3}D$	4—3
5197,264	5	28,35	30,73	$3d'\ ^{1}P^{\circ}$—$4p'\ ^{1}D$	1—2
5135,110	3	28,35	30,76	$3d'\ ^{1}P^{\circ}$—$4p'\ ^{3}P$	1—2
5114,116	8	28,54	30,97	$5f\ ^{3}F^{\circ}$—$7g\ ^{3}G$	4—5
5113,76	7	28,54	30,97	$5f\ ^{3}F^{\circ}$—$7g\ ^{3}G$	3—4
		28,54	30,97	$5f\ ^{3}F^{\circ}$—$7g\ ^{3}G$	2—3
5091,419	10	28,55	30,98	$5g\ ^{3,1}G$—$7h\ ^{3,1}H^{\circ}$	—
4912,332	4	25,56	28,08	$5s\ ^{3}S$—$4s'\ ^{3}P^{\circ}$	1—1
4842,57	5	25,56	28,12	$5s\ ^{3}S$—$4s'\ ^{3}P^{\circ}$	1—2
4828,968	18	25,98	28,55	$4f\ ^{3}F^{\circ}$—$5g\ ^{3}G$	4—4, 5
4819,718	16	25,98	28,55	$4f\ ^{3}F^{\circ}$—$5g\ ^{3}G$	3—3, 4
4813,330	15	25,97	28,55	$4f\ ^{3}F^{\circ}$—$5g\ ^{3}G$	2—3
4800,428	8	25,77	28,35	$5s\ ^{1}S$—$3d'\ ^{1}P^{\circ}$	0—1
4734,627	2	28,12	30,73	$4s'\ ^{3}P^{\circ}$—$4p'\ ^{1}D$	2—2
4730,519	7	28,12	30,74	$4s'\ ^{3}P^{\circ}$—$4p'\ ^{3}P$	2—1
4716,651	16	25,33	27,96	$4d\ ^{1}D$—$5f\ ^{1}F^{\circ}$	2—3
4683,797	7	28,08	30,72	$4s'\ ^{3}P^{\circ}$—$4p'\ ^{3}P$	1—0
4683,022	9	28,12	30,76	$4s'\ ^{3}P^{\circ}$—$4p'\ ^{3}P$	2—2
4665,869	8	28,08	30,74	$4s'\ ^{3}P^{\circ}$—$4p'\ ^{3}P$	1—1
4638,277	7	28,06	30,74	$4s'\ ^{3}P^{\circ}$—$4p'\ ^{3}P$	0—1
4619,657	7	28,08	30,76	$4s'\ ^{3}P^{\circ}$—$4p'\ ^{3}P$	1—2
4574,759	20	19,01	21,72	$4s\ ^{3}S$—$4p\ ^{3}P^{\circ}$	1—0
4567,823	25	19,01	21,73	$4s\ ^{3}S$—$4p\ ^{3}P^{\circ}$	1—1
4553,996	8	28,12	30,84	$4s'\ ^{3}P^{\circ}$—$4p'\ ^{3}S$	2—1
4552,616	30	19,01	21,74	$4s\ ^{3}S$—$4p\ ^{3}P^{\circ}$	1—2
4494,048	6	28,08	30,84	$4s'\ ^{3}P^{\circ}$—$4p'\ ^{3}S$	1—1
4482,884	3	28,11	30,88	$4s'\ ^{3}P^{\circ}$—$7d\ ^{3}D$	2—3
4468,452	2	28,06	30,84	$4s'\ ^{3}P^{\circ}$—$4p'\ ^{3}S$	0—1
4423,556	4	30,38	33,18	$4p'\ ^{3}D$—$4d'\ ^{3}F^{\circ}$	3—4

λ, Å	I	E_H, eV	E_B, eV	Transition	J
4406,721	8	28,15	30,96	$5d\ ^3D-7f\ ^3F^\circ$	3—3, 4
4405,901	6	28,15	30,96	$5d\ ^3D-7f\ ^3F^\circ$	2—2, 3
4405,351	4	28,15	30,96	$5d\ ^3D-7f\ ^3F^\circ$	1—2
4377,626	8	25,39	28,22	$4f\ ^1F^\circ-5d\ ^1D$	3—2
4356,821	4	34,58	37,42	$4f'\ [3^1/_2]-5g'\ [4^1/_2]^\circ$	4—5
4356,100	2	34,58	37,42	$4f'\ [3^1/_2]-5g'\ [3^1/_2]^\circ$	3—3
4355,525	3	34,58	37,42	$4f'\ [3^1/_2]-5g'\ [4^1/_2]^\circ$	4—4
4355,281	3	34,58	37,42	$4f'\ [3^1/_2]-5g'\ [3^1/_2]^\circ$	3—4
4352,810	2	34,57	37,42	$4f'\ [2^1/_2]-5g'\ [3^1/_2]^\circ$	3—3
4351,974	2	34,57	37,42	$4f'\ [2^1/_2]-5g'\ [3^1/_2]^\circ$	3—4
4341,400	8	29,25	32,10	$4s'\ ^1P^\circ-4p'\ ^1S$	1—0
4338,501	9	19,02	21,88	$3p^2\ ^1S-4p\ ^1P^\circ$	0—1
4211,679	2	28,54	31,49	$5f\ ^3F^\circ-8d\ ^3D$	4—3
4115,504	5	27,96	30,97	$5f\ ^1F^\circ-7g\ ^1G$	3—4
4111,512	3	28,54	31,56	$5f\ ^3F^\circ-8g\ ^3G$	4—5
4111,255	2	28,54	31,56	$5f\ ^3F^\circ-8g\ ^3G$	3—4, 3
4102,422	8	25,33	28,35	$4d\ ^1D-3d'\ ^1P^\circ$	2—1
4101,86	5	28,55	31,57	$5g\ ^{3,1}G-8h\ ^{3,1}H^\circ$	—
4064,113	2	30,11	33,16	$4p'\ ^1P-4d'\ ^1D^\circ$	1—2
4030,752	2	24,99	28,06	$4d\ ^3D-4s'\ ^3P^\circ$	1—0
4010,192	4	24,99	28,08	$4d\ ^3D-4s'\ ^3P^\circ$	1, 2—1
3981,238	5	{ 27,96	31,07	$5f\ ^1F^\circ-7d\ ^1D$	3—2
		21,88	24,99	$4p\ ^1P^\circ-4d\ ^3D$	1—1, 2
3963,838	6	24,99	28,12	$4d\ ^3D-4s'\ ^3P^\circ$	2, 3—2
3956,66	2	26,65	29,78	$5p\ ^3P^\circ-6d\ ^3D$	0—1
3953,080	4	26,65	29,79	$5p\ ^3P^\circ-6d\ ^3D$	1—2
3947,488	6	26,65	29,79	$5p\ ^3P^\circ-6d\ ^3D$	2—3
3924,468	20	25,39	28,55	$4f\ ^1F^\circ-5g\ ^1G$	3—4
3842,458	7	26,59	29,82	$5p\ ^1P^\circ-6d\ ^1D$	1—2
3806,544	30	21,74	24,99	$4p\ ^3P^\circ-4d\ ^3D$	2—2, 3
3796,114	25	21,73	24,99	$4p\ ^3P^\circ-4d\ ^3D$	1—1, 2
3791,41	20	21,73	24,99	$4p\ ^3P^\circ-4d\ ^3D$	0—1
3682,25	2	26,65	30,02	$5p\ ^3P^\circ-7s\ ^3S$	0—1
3682,15	5	26,65	30,02	$5p\ ^3P^\circ-7s\ ^3S$	1—1
3681,402	7	26,65	30,02	$5p\ ^3P^\circ-7s\ ^3S$	2—1
3676,731	3	28,22	31,59	$5d\ ^1D-8f\ ^1F^\circ$	2—3
3662,366	2	26,95	30,34	$3d'\ ^3D^\circ-4p'\ ^3D$	2—1
3655,112	4	26,95	30,34	$3d'\ ^3D^\circ-4p'\ ^3D$	1—1
3651,721	2	26,96	30,35	$3d'\ ^3D^\circ-4p'\ ^3D$	3—2
3645,123	6	26,95	30,35	$3d'\ ^3D^\circ-4p'\ ^3D$	2—2
3639,445	5	28,15	31,55	$5d\ ^3D-8f\ ^3F^\circ$	3—3, 4
3638,898	3	28,15	31,55	$5d\ ^3D-8f\ ^3F^\circ$	2—2, 3
3638,524	2	28,15	31,55	$5d\ ^3D-8f\ ^3F^\circ$	1—2
3637,943	2	26,95	30,35	$3d'\ ^3D^\circ-4p'\ ^3D$	1—2
3622,538	8	26,96	30,38	$3d'\ ^3D^\circ-4p'\ ^3D$	3—3
3619,581	3	28,55	31,97	$5g\ ^3G-9h\ ^{3,1}H^\circ$	5——
3590,465	20	21,88	25,33	$4p\ ^1P^\circ-4d\ ^1D$	1—2
3580,050	3	24,69	28,15	$3d'\ ^3F^\circ-5d\ ^3D$	4—3
3569,673	8	26,59	30,07	$5p\ ^1P^\circ-7s\ ^1S$	1—0
3563,11	2	24,67	28,15	$3d'\ ^3F^\circ-5d\ ^3D$	3—2
3525,939	9	26,59	30,11	$5p\ ^1P^\circ-4p'\ ^1P$	1—1
3486,911	15	24,99	28,54	$4d\ ^3D-5f\ ^3F^\circ$	—
3440,37	5	27,96	31,56	$5f\ ^1F^\circ-8g\ ^1G$	3—4
3439,242	3	21,73	25,33	$4p\ ^3P^\circ-4d\ ^1D$	1—2
3346,717	2	26,65	30,35	$5p\ ^3P^\circ-4p'\ ^3D$	1—2
3321,578	4	26,65	30,38	$5p\ ^3P^\circ-4p'\ ^3D$	2—3
3279,258	7	26,95	30,73	$3d'\ ^3D^\circ-4p'\ ^3P$	1—0
3276,264	10	26,95	30,74	$3d'\ ^3D^\circ-4p'\ ^3P$	2—1

λ, Å	I	E_H, eV	E_B, eV	Transition	J
3270,456	6	26,95	30,74	$3d'\ ^3D^\circ - 4p'\ ^3P$	1—1
3258,664	12	26,96	30,76	$3d'\ ^3D^\circ - 4p'\ ^3P$	3—2
3254,800	7	25,98	29,79	$4f\ ^3F^\circ - 6d\ ^3D$	4—3
3253,741	5	25,98	29,79	$4f\ ^3F^\circ - 6d\ ^3D$	3—2
3253,401	7	26,95	30,76	$3d'\ ^3D^\circ - 4p'\ ^3P$	2—2
3253,117	4	25,97	29,78	$4f\ ^3F^\circ - 6d\ ^3D$	2—1
3241,622	15	21,74	25,56	$4p\ ^3P^\circ - 5s\ ^3S$	2—1
3233,934	14	21,73	25,56	$4p\ ^3P^\circ - 5s\ ^3S$	1—1
3230,499	12	21,72	25,56	$4p\ ^3P^\circ - 5s\ ^3S$	0—1
3216,249	7	25,33	29,18	$4d\ ^1D - 3d'\ ^1F^\circ$	2—3
3210,554	15	24,69	28,55	$3d'\ ^3F^\circ - 5g\ ^3G$	4—4, 5
3196,504	14	24,67	28,55	$3d'\ ^3F^\circ - 5g\ ^3G$	3—3, 4
		24,67	28,55	$3d'\ ^3F^\circ - 5g\ ^1G$	3—4
3186,022	13	24,66	28,55	$3d'\ ^3F^\circ - 5g\ ^3G$	2—3
3185,125	16	21,88	25,77	$4p\ ^1P^\circ - 5s\ ^1S$	1—0
3165,38	2	26,81	30,73	$3d'\ ^3P^\circ - 4p'\ ^3P$	1—0
3163,281	2	26,82	30,74	$3d'\ ^3P^\circ - 4p'\ ^3P$	0—1
3161,610	8	25,33	29,25	$4d\ ^1D - 4s'\ ^1P^\circ$	2—1
3147,371	7	26,80	30,74	$3d'\ ^3P^\circ - 4p'\ ^3P$	2—1
3135,906	3	26,81	30,76	$3d'\ ^3P^\circ - 4p'\ ^3P$	1—2
3126,267	6	26,80	30,76	$3d'\ ^3P^\circ - 4p'\ ^3P$	2—2
3096,826	16	17,72	21,72	$3d\ ^3D - 4p\ ^3P^\circ$	1—0
3093,65	5	17,72	21,73	$3d\ ^3D - 4p\ ^3P^\circ$	1—1
3093,424	20	17,72	21,73	$3d\ ^3D - 4p\ ^3P^\circ$	2—1
3086,46	6	17,72	21,74	$3d\ ^3D - 4p\ ^3P^\circ$	2—2
3086,236	25	17,72	21,74	$3d\ ^3D - 4p\ ^3P^\circ$	3—2
3083,363	2	26,82	30,84	$3d'\ ^3P^\circ - 4p'\ ^3S$	0—1
3077,523	4	26,81	30,84	$3d'\ ^3P^\circ - 4p'\ ^3S$	1—1
3068,238	7	26,80	30,84	$3d'\ ^3P^\circ - 4p'\ ^3S$	2—1
3046,284	3	24,99	29,06	$4d\ ^3D - 6p\ ^3P^\circ$	1—0
3045,076	5	24,99	29,06	$4d\ ^3D - 6p\ ^3P^\circ$	1, 2—1
3043,932	7	24,99	29,06	$4d\ ^3D - 6p\ ^3P^\circ$	3—2
3040,933	9	25,98	30,06	$4f\ ^3F^\circ - 6g\ ^3G$	4—4, 5
3037,287	8	25,98	30,06	$4f\ ^3F^\circ - 6g\ ^3G$	3—3, 4
3034,732	6	25,97	30,06	$4f\ ^3F^\circ - 6g\ ^3G$	2—3
3032,66	4	26,65	30,74	$5p\ ^3P^\circ - 4p'\ ^3P$	2—1
3013,091	5	26,65	30,76	$5p\ ^3P^\circ - 4p'\ ^3P$	2—2
2980,519	5	17,72	21,88	$3d\ ^3D - 4p\ ^1P^\circ$	2—1
2959,67	3	26,65	30,84	$5p\ ^3P^\circ - 4p'\ ^3S$	0, 1—1
2959,150	5	26,65	30,84	$5p\ ^3P^\circ - 4p'\ ^3S$	2—1
2875,09	2	26,65	30,96	$5p\ ^3P^\circ - 8s\ ^3S$	0, 1—1
2874,626	4	26,65	30,96	$5p\ ^3P^\circ - 8s\ ^3S$	2—1
2839,622	5	25,97	30,34	$4f\ ^3F^\circ - 4p'\ ^3D$	2—1
2831,490	7	25,98	30,35	$4f\ ^3F^\circ - 4p'\ ^3D$	3—2
2817,110	9	25,98	30,38	$4f\ ^3F^\circ - 4p'\ ^3D$	4—3
2655,512	14	25,39	30,06	$4f\ ^1F^\circ - 6g\ ^1G$	3—4
2640,788	11	25,42	30,11	$3d'\ ^1D^\circ - 4p'\ ^1P$	2—1
2559,210	14	20,55	25,39	$3d\ ^1D - 4f\ ^1F^\circ$	2—3
2546,093	10	20,55	25,42	$3d\ ^1D - 3d'\ ^1D^\circ$	2—2
2541,818	25	10,28	15,15	$3p\ ^1P^\circ - 3p^2\ ^1D$	1—2
2483,196	6	25,98	30,97	$4f\ ^3F^\circ - 7g\ ^3G$	3—4
2481,508	3	25,97	30,97	$4f\ ^3F^\circ - 7g\ ^3G$	2—3
2449,484	11	24,99	30,05	$4d\ ^3D - 6f\ ^3F^\circ$	—
2429,35	7	24,69	29,79	$3d'\ ^3F^\circ - 6d\ ^3D$	4—3
2329,931	2	25,42	30,73	$3d'\ ^1D^\circ - 4p'\ ^1D$	2—2
2308,191	10	24,69	30,06	$3d'\ ^3F^\circ - 6g\ ^3G$	4—4, 5
2306,42	2	24,99	30,36	$4d\ ^3D - 7p\ ^3P^\circ$	3—2
2300,930	8	24,67	30,06	$3d'\ ^3F^\circ - 6g\ ^3G$	3—3, 4
2296,873	10	29,18	34,58	$3d'\ ^1F^\circ - 4f'\ [3^1/_2]$	3—3

λ, A	I	E_H, eV	E_B, eV	Transition	J
2295,476	6	24,66	30,06	$3d'\ ^3F^\circ$—$6g\ ^3G$	2—3
2222,01	2	25,39	30,97	$4f\ ^1F^\circ$—$7g\ ^1G$	3—4
2182,049	3	24,66	30,34	$3d'\ ^3F^\circ$—$4p'\ ^3D$	2—1
2180,836	4	24,67	30,35	$3d'\ ^3F^\circ$—$4p'\ ^3D$	3—2
2176,894	5	24,69	30,38	$3d'\ ^3F^\circ$—$4p'\ ^3D$	4—3
2157,280	2	29,25	34,99	$4s'\ ^1P^\circ$—$5p'\ ^1P$	1—1
2075,04	2	24,99	30,96	$4d\ ^3D$—$7f\ ^3F^\circ$	—
2049,913	2	20,55	26,59	$3d\ ^1D$—$5p\ ^1P^\circ$	2—1
1892,030	3	0,00	6,55	$3s^2\ ^1S$—$3p\ ^3P^\circ$	0—1
1842,547	9	15,15	21,88	$3p^2\ ^1D$—$4p\ ^1P^\circ$	2—1
1803,023	3	19,72	26,59	$4s\ ^1S$—$5p\ ^1P^\circ$	0—1
1709,018	3	—	—	—	—
1699,808	5	—	—	—	—
1673,315	7	20,55	27,96	$3d\ ^1D$—$5f\ ^1F^\circ$	2—3
1588,950	2	20,55	28,35	$3d\ ^1D$—$3d'\ ^1P^\circ$	2—1
1513,533	2	29,18	37,37	$3d'\ ^1F^\circ$—$5f'\ [3^1/_2]$	3—3
1506,060	6	21,88	30,11	$4p\ ^1P^\circ$—$4p'\ ^1P$	1—1
1501,870	9	17,72	25,97	$3d\ ^3D$—$4f\ ^3F^\circ$	1, 2, 3—2
1501,191	10	17,72	25,98	$3d\ ^3D$—$4f\ ^3F^\circ$	2, 3—3
1500,241	12	17,72	25,98	$3d\ ^3D$—$4f\ ^3F^\circ$	3—4
1496,172	7	—	—	—	—
1457,253	5	20,55	29,05	$3d\ ^1D$—$6p\ ^1P^\circ$	2—1
1447,196	6	6,59	15,15	$3p\ ^3P^\circ$—$3p^2\ ^1D$	2—2
1441,732	5	6,55	15,15	$3p\ ^3P^\circ$—$3p^2\ ^1D$	1—2
1439,391	2	21,73	30,34	$4p\ ^3P^\circ$—$4p'\ ^3D$	1—1
1438,702	2	21,72	30,34	$4p\ ^3P^\circ$—$4p'\ ^3D$	0—1
1438,228	2	21,73	30,35	$4p\ ^3P^\circ$—$4p'\ ^3D$	2—2
1436,724	4	21,73	30,35	$4p\ ^3P^\circ$—$4p'\ ^3D$	1—2
1436,166	7	19,72	28,35	$4s\ ^1S$—$3d'\ ^1P^\circ$	0—1
1435,776	8	20,55	29,18	$3d\ ^1D$—$3d'\ ^1F^\circ$	2—3
1433,690	6	21,74	30,38	$4p\ ^3P^\circ$—$4p'\ ^3D$	2—3
1424,775	2	20,55	29,25	$3d\ ^1D$—$4s'\ ^1P^\circ$	2—1
1417,237	13	10,28	19,02	$3p\ ^1P^\circ$—$3p^2\ ^1S$	1—0
1387,994	5	17,72	26,65	$3d\ ^3D$—$5p\ ^3P^\circ$	—
1377,238	2	21,73	30,73	$4p\ ^3P^\circ$—$4p'\ ^3P$	1—0
1377,082	3	21,74	30,74	$4p\ ^3P^\circ$—$4p'\ ^3P$	2—1
1375,688	2	21,73	30,74	$4p\ ^3P^\circ$—$4p'\ ^3P$	1—1
1375,083	2	21,72	30,74	$4p\ ^3P^\circ$—$4p'\ ^3P$	0—1
1373,030	5	21,74	30,76	$4p\ ^3P^\circ$—$4p'\ ^3P$	2—2
1371,652	3	21,73	30,76	$4p\ ^3P^\circ$—$4p'\ ^3P$	1—2
1369,437	5	19,01	28,06	$4s\ ^3S$—$4s'\ ^3P^\circ$	1—0
1367,049	7	19,01	28,08	$4s\ ^3S$—$4s'\ ^3P^\circ$	1—1
1365,253	8	17,72	26,80	$3d\ ^3D$—$3d'\ ^3P^\circ$	1, 2, 3—2
1363,459	7	17,72	26,81	$3d\ ^3D$—$3d'\ ^3P^\circ$	1, 2—1
1362,366	5	17,72	26,82	$3d\ ^3D$—$3d'\ ^3P^\circ$	1—0
1361,597	8	19,01	28,12	$4s\ ^3S$—$4s'\ ^3P^\circ$	1—2
1343,388	6	17,72	26,95	$3d\ ^3D$—$3d'\ ^3D^\circ$	1, 2—1
1342,392	7	17,72	26,95	$3d\ ^3D$—$3d'\ ^3D^\circ$	1, 2, 3—2
1341,465	8	17,72	26,96	$3d\ ^3D$—$3d'\ ^3D^\circ$	2, 3—3
1313,867	3	—	—	—	—
1312,590	13	10,28	19,72	$3p\ ^1P^\circ$—$4s\ ^1S$	1—0
1303,320	16	6,59	16,10	$3p\ ^3P^\circ$—$3p^2\ ^3P$	2—1
1301,146	14	6,55	16,08	$3p\ ^3P^\circ$—$3p^2\ ^3P$	1—0
1298,960	18	6,59	16,13	$3p\ ^3P^\circ$—$3p^2\ ^3P$	2—2
1298,891	15	6,55	16,10	$3p\ ^3P^\circ$—$3p^2\ ^3P$	1—1
1296,726	14	6,54	16,10	$3p\ ^3P^\circ$—$3p^2\ ^3P$	0—1
1294,543	17	6,55	16,13	$3p\ ^3P^\circ$—$3p^2\ ^3P$	1—2
1280,354	6	20,55	30,23	$3d\ ^1D$—$6f\ ^1F^\circ$	2—3

λ, Å	I	E_H, eV	E_B, eV	Transition	J
1235,431	7	19,02	29,05	$3p^2\ {}^1S-6p\ {}^1P^\circ$	0—1
1212,011	2	19,02	29,25	$3p^2\ {}^1S-4s'\ {}^1P^\circ$	0—1
1210,456	10	15,15	25,39	$3p^2\ {}^1D-4f\ {}^1F^\circ$	2—3
1207,517	9	15,15	25,42	$3p^2\ {}^1D-3d'\ {}^1D^\circ$	2—2
1206,533	30	10,28	20,55	$3p\ {}^1P^\circ-3d\ {}^1D$	1—2
1206,510	30	0,00	10,27	$3s^2\ {}^1S-3p\ {}^1P^\circ$	0—1
1182,018	3	20,55	31,04	$3d\ {}^1D-7f\ {}^1F^\circ$	2—3
1178,004	8	16,13	26,65	$3p^2\ {}^3P-5p\ {}^3P^\circ$	2—1, 2
1174,432	6	16,10	26,65	$3p^2\ {}^3P-5p\ {}^3P^\circ$	1—2
1174,369	5	16,10	26,65	$3p^2\ {}^3P-5p\ {}^3P^\circ$	1—0, 1
1172,529	4	16,08	26,65	$3p^2\ {}^3P-5p\ {}^3P^\circ$	0—1
1161,579	8	16,13	26,80	$3p^2\ {}^3P-3d'\ {}^3P^\circ$	2—2
1160,255	6	16,13	26,81	$3p^2\ {}^3P-3d'\ {}^3P^\circ$	2—1
1158,102	7	16,10	26,80	$3p^2\ {}^3P-3d'\ {}^3P^\circ$	1—2
1156,782	4	16,10	26,81	$3p^2\ {}^3P-3d'\ {}^3P^\circ$	1—1
1155,957	6	16,10	26,82	$3p^2\ {}^3P-3d'\ {}^3P^\circ$	1—0
1154,998	6	16,08	26,81	$3p^2\ {}^3P-3d'\ {}^3P^\circ$	0—1
1145,22	8	17,72	28,54	$3d\ {}^3D-5f\ {}^3F^\circ$	3—3, 4
1145,177	7	17,72	28,54	$3d\ {}^3D-5f\ {}^3F^\circ$	1, 2, 3—2
		17,72	28,54	$3d\ {}^3D-5f\ {}^3F^\circ$	2—3
1144,959	6	16,13	26,95	$3p^2\ {}^3P-3d'\ {}^3D^\circ$	2—2
1144,306	8	16,13	26,96	$3p^2\ {}^3P-3d'\ {}^3D^\circ$	2—3
1142,282	6	16,10	26,95	$3p^2\ {}^3P-3d'\ {}^3D^\circ$	1—1
1141,580	7	16,10	26,95	$3p^2\ {}^3P-3d'\ {}^3D^\circ$	1—2
1140,545	6	16,08	26,95	$3p^2\ {}^3P-3d'\ {}^3D^\circ$	0—1
1117,686	4	—	—	—	—
1113,228	18	6,59	17,72	$3p\ {}^3P^\circ-3d\ {}^3D$	2—1, 2, 3
1109,970	16	6,55	17,72	$3p\ {}^3P^\circ-3d\ {}^3D$	1—1, 2
1108,368	14	6,54	17,72	$3p\ {}^3P^\circ-3d\ {}^3D$	0—1
1083,210	6	15,15	26,59	$3p^2\ {}^1D-5p\ {}^1P^\circ$	2—1
1079,112	4	—	—	—	—
1076,589	10	—	—	—	—
1076,253	3	—	—	—	—
1053,289	10	—	—	—	—
1037,053	7	16,13	28,08	$3p^2\ {}^3P-4s'\ {}^3P^\circ$	2—1
1035,657	3	16,10	28,06	$3p^2\ {}^3P-4s'\ {}^3P^\circ$	1—0
1034,287	4	16,10	28,08	$3p^2\ {}^3P-4s'\ {}^3P^\circ$	1—1
1033,920	8	16,13	28,12	$3p^2\ {}^3P-4s'\ {}^3P^\circ$	2—2
1032,851	3	16,08	28,08	$3p^2\ {}^3P-4s'\ {}^3P^\circ$	0—1
1031,169	7	16,10	28,12	$3p^2\ {}^3P-4s'\ {}^3P^\circ$	1—2
1005,365	7	17,72	30,05	$3d\ {}^3D-6f\ {}^3F^\circ$	—
997,389	16	6,59	19,01	$3p\ {}^3P^\circ-4s\ {}^3S$	2—1
994,787	13	6,55	19,01	$3p\ {}^3P^\circ-4s\ {}^3S$	1—1
993,519	10	6,54	19,01	$3p\ {}^3P^\circ-4s\ {}^3S$	0—1
967,946	9	15,15	27,96	$3p^2\ {}^1D-5f\ {}^1F^\circ$	2—3
939,093	7	15,15	28,35	$3p^2\ {}^1D-3d'\ {}^1P^\circ$	2—1
936,060	3	17,72	30,96	$3d\ {}^3D-7f\ {}^3F^\circ$	—
883,398	5	15,15	29,18	$3p^2\ {}^1D-3d'\ {}^1F^\circ$	2—3
823,408	9	10,28	25,33	$3p\ {}^1P^\circ-4d\ {}^1D$	1—2
800,066	5	10,28	25,77	$3p\ {}^1P^\circ-5s\ {}^1S$	1—0
690,689	2	10,28	28,22	$3p\ {}^1P^\circ-5d\ {}^1D$	1—2
678,055	2	10,28	28,56	$3p\ {}^1P^\circ-6s\ {}^1S$	1—0
673,477	5	6,59	24,99	$3p\ {}^3P^\circ-4d\ {}^3D$	2—1, 2, 3
672,293	4	6,55	24,99	$3p\ {}^3P^\circ-4d\ {}^3D$	1—1, 2
671,718	2	6,54	24,99	$3p\ {}^3P^\circ-4d\ {}^3D$	0—1
653,332	8	6,59	25,56	$3p\ {}^3P^\circ-5s\ {}^3S$	2—1
652,223	6	6,55	25,56	$3p\ {}^3P^\circ-5s\ {}^3S$	1—1
651,668	4	6,54	25,56	$3p\ {}^3P^\circ-5s\ {}^3S$	0—1

λ, Å	I	E_H, eV	E_B, eV	Transition	J
566,613	8	0,00	21,88	$3s^2\ ^1S$—$4p\ ^1P^\circ$	0—1
565,698	2	6,55	28,46	$3p\ ^3P^\circ$—$6s\ ^3S$	1—1
565,289	1	6,54	28,46	$3p\ ^3P^\circ$—$6s\ ^3S$	0—1
466,129	4	0,00	26,59	$3s^2\ ^1S$—$5p\ ^1P^\circ$	0—1

Si IV, ground state $1s^2 2s^2 2p^6 3s\ ^2S_{1/2}$
Ionization potential 364 093,1 cm^{-1}; 45,139 eV

λ, Å	I	E_H, eV	E_B, eV	Transition	J
9018,162	1	32,90	34,28	$5s\ ^2S$—$5p\ ^2P^\circ$	$^1/_2$—$^1/_2$
8957,245	2	32,90	34,29	$5s\ ^2S$—$5p\ ^2P^\circ$	$^1/_2$—$^3/_2$
8240,606	1	39,08	40,59	$6f\ ^2F^\circ$—$7d\ ^2D$	—
7752,905	1	39,09	40,69	$6g\ ^2G$—$7f\ ^2F^\circ$	—
7730,469	1	39,09	40,69	$6h\ ^2H^\circ$—$7g\ ^2G$	—
7725,64	2	39,09	40,70	$6h\ ^2H^\circ$—$7h\ ^2H^\circ$	—
7723,818	6	39,09	40,70	$6h\ ^2H^\circ$—$7i\ ^2I$	—
7718,785	5	39,09	40,70	$6g\ ^2G$—$7h\ ^2H^\circ$	—
7678,748	4	39,08	40,69	$6f\ ^2F^\circ$—$7g\ ^2G$	—
7654,555	4	37,90	39,52	$6p\ ^2P^\circ$—$7s\ ^2S$	$^3/_2$—$^1/_2$
7630,497	2	37,89	39,52	$6p\ ^2P^\circ$—$7s\ ^2S$	$^1/_2$—$^1/_2$
7068,410	4	36,14	37,89	$5d\ ^2D$—$6p\ ^2P^\circ$	$^3/_2$—$^1/_2$
7047,939	6	[illegible]	37,90	$5d\ ^2D$—$6p\ ^2P^\circ$	$^3/_2$, $^5/_2$—$^3/_2$
6998,358	3	38,92	40,69	$6d\ ^2D$—$7f\ ^2F^\circ$	$^5/_2$—$^7/_2$
6701,207	7	34,29	36,14	$5p\ ^2P^\circ$—$5d\ ^2D$	$^3/_2$—$^3/_2$, $^5/_2$
6667,556	5	34,28	36,14	$5p\ ^2P^\circ$—$5d\ ^2D$	$^1/_2$—$^3/_2$
5309,493	1	38,92	41,25	$6d\ ^2D$—$8p\ ^2P^\circ$	$^3/_2$—$^1/_2$
5304,971	2	38,92	41,25	$6d\ ^2D$—$8p\ ^2P^\circ$	$^3/_2$, $^5/_2$—$^3/_2$
4950,105	3	36,41	38,92	$5f\ ^2F^\circ$—$6d\ ^2D$	—
4673,297	2	36,43	39,08	$5g\ ^2G$—$6f\ ^2F^\circ$	—
4667,14	1	39,08	41,74	$6f\ ^2F^\circ$—$8g\ ^2G$	—
4656,92	3	36,43	39,09	$5g\ ^2G$—$6g\ ^2G$	—
4654,323	10	36,43	39,09	$5g\ ^2G$—$6h\ ^2H^\circ$	—
4647,45	1	36,41	39,08	$5f\ ^2F^\circ$—$6f\ ^2F^\circ$	—
4631,241	9	36,41	39,09	$5f\ ^2F^\circ$—$6g\ ^2G$	—
4628,62	3	36,41	39,09	$5f\ ^2F^\circ$—$6h\ ^2H^\circ$	—
4611,27	0	37,90	40,59	$6p\ ^2P^\circ$—$7d\ ^2D$	$^3/_2$—$^5/_2$, $^3/_2$
4411,652	0	37,15	39,96	$6s\ ^2S$—$7p\ ^2P^\circ$	$^1/_2$—$^3/_2$
4403,734	2	38,92	41,73	$6d\ ^2D$—$8f\ ^2F^\circ$	—
4328,175	5	34,29	37,15	$5p\ ^2P^\circ$—$6s\ ^2S$	$^3/_2$—$^1/_2$
4314,104	3	34,28	37,15	$5p\ ^2P^\circ$—$6s\ ^2S$	$^1/_2$—$^1/_2$
4212,407	7	36,14	39,08	$5d\ ^2D$—$6f\ ^2F^\circ$	—
4116,097	9	24,05	27,06	$4s\ ^2S$—$4p\ ^2P^\circ$	$^1/_2$—$^1/_2$
4088,854	10	24,05	27,08	$4s\ ^2S$—$4p\ ^2P^\circ$	$^1/_2$—$^3/_2$
4038,057	2	37,90	40,97	$6p\ ^2P^\circ$—$8s\ ^2S$	$^3/_2$—$^1/_2$
4031,39	1	37,89	40,97	$6p\ ^2P^\circ$—$8s\ ^2S$	$^1/_2$—$^1/_2$
3773,151	6	31,00	34,28	$4d\ ^2D$—$5p\ ^2P^\circ$	$^3/_2$—$^1/_2$
3762,435	8	31,00	34,29	$4d\ ^2D$—$5p\ ^2P^\circ$	$^3/_2$, $^5/_2$—$^3/_2$
3244,192	1	36,14	39,96	$5d\ ^2D$—$7p\ ^2P^\circ$	$^3/_2$—$^1/_2$
3165,710	9	27,08	31,00	$4p\ ^2P^\circ$—$4d\ ^2D$	$^3/_2$—$^5/_2$, $^3/_2$
3149,561	7	27,06	31,00	$4p\ ^2P^\circ$—$4d\ ^2D$	$^1/_2$—$^3/_2$
2971,522	1	36,41	40,59	$5f\ ^2F^\circ$—$7d\ ^2D$	—
2904,470	2	36,43	40,70	$5g\ ^2G$—$7h\ ^2H^\circ$	—
2895,131	3	36,41	40,69	$5f\ ^2F^\circ$—$7g\ ^2G$	—
2723,812	3	36,14	40,69	$5d\ ^2D$—$7f\ ^2F^\circ$	—

λ, Å	I	E_H, eV	E_B, eV	Transition	J
2677,57	1	34,29	38,92	$5p\ ^2P^\circ$—$6d\ ^2D$	$^3/_2$—$^5/_2$, $^3/_2$
2675,249	4	31,51	36,14	$4f\ ^2F^\circ$—$5d\ ^2D$	$^7/_2$—$^5/_2$
2675,120	4	31,51	36,14	$4f\ ^2F^\circ$—$5d\ ^2D$	$^5/_2$—$^3/_2$, $^5/_2$
2672,193	1	34,28	38,92	$5p\ ^2P^\circ$—$6d\ ^2D$	$^1/_2$—$^3/_2$
2517,506	7	31,51	36,43	$4f\ ^2F^\circ$—$5g\ ^2G$	—
2485,378	1	32,90	37,89	$5s\ ^2S$—$6p\ ^2P$	$^1/_2$—$^1/_2$
2482,816	2	32,90	37,90	$5s\ ^2S$—$6p\ ^2P$	$^1/_2$—$^3/_2$
2370,985	3	34,29	39,52	$5p\ ^2P^\circ$—$7s\ ^2S$	$^3/_2$—$^1/_2$
2366,755	2	34,28	39,52	$5p\ ^2P^\circ$—$7s\ ^2S$	$^1/_2$—$^1/_2$
2328,56	2	36,41	41,74	$5f\ ^2F^\circ$—$8g\ ^2G$	—
2287,041	5	31,00	36,41	$4d\ ^2D$—$5f\ ^2F^\circ$	—
2127,467	4	27,08	32,90	$4p\ ^2P^\circ$—$5s\ ^2S$	$^3/_2$—$^1/_2$
2120,179	3	27,06	32,90	$4p\ ^2P^\circ$—$5s\ ^2S$	$^1/_2$—$^1/_2$
1729,997	5	—	—	—	—
1727,377	5	19,88	27,06	$3d\ ^2D$—$4p\ ^2P^\circ$	$^3/_2$—$^1/_2$
1726,006	4	—	—	—	—
1722,534	6	19,88	27,08	$3d\ ^2D$—$4p\ ^2P^\circ$	$^3/_2$, $^5/_2$—$^3/_2$
1673,374	150	—	—	—	—
1642,168	4	—	—	—	—
1634,607	1	31,51	39.09	$4f\ ^2F^\circ$—$6g\ ^2G$	—
1430,791	1	—	—	—	—
1410,950	1	—	—	—	—
1407,701	3	—	—	—	—
1402,769	12	0,00	8.84	$3s\ ^2S$—$3p\ ^2P^\circ$	$^1/_2$—$^1/_2$
1393,755	15	0,00	8,90	$3s\ ^2S$—$3p\ ^2P^\circ$	$^1/_2$—$^3/_2$
1291,969	30	—	—	—	—
1280,336	20	—	—	—	—
1128,340	10	8,90	19,88	$3p\ ^2P^\circ$—$3d\ ^2D$	$^3/_2$—$^5/_2$
1128,325	10	8,90	19,88	$3p\ ^2P^\circ$—$3d\ ^2D$	$^3/_2$—$^3/_2$
1122,485	8	8,84	19,88	$3p\ ^2P^\circ$—$3d\ ^2D$	$^1/_2$—$^3/_2$
1117,990	4	—	—	—	—
1100,050	1	—	—	—	—
1098,408	1	—	—	—	—
1077,325	1	—	—	—	—
1066,650	8	19,88	31,51	$3d\ ^2D$—$4f\ ^2F^\circ$	$^3/_2$—$^5/_2$
1066,636	8	19,88	31,51	$3d\ ^2D$—$4f\ ^2F^\circ$	$^5/_2$—$^5/_2$
1066,614	8	19,88	31,51	$3d\ ^2D$—$4f\ ^2F^\circ$	$^5/_2$—$^7/_2$
1056,582	12	—	—	—	—
1051,596	70	—	—	—	—
984,889	1	—	—	—	—
933,420	5	—	—	—	—
894,347	3	—	—	—	—
854,789	7	—	—	—	—
849,248	1	—	—	—	—
840,901	5	—	—	—	—
836,126	1	—	—	—	—
818,129	8	8,90	24,05	$3p\ ^2P^\circ$—$4s\ ^2S$	$^3/_2$—$^1/_2$
815,049	7	8,84	24,05	$3p\ ^2P^\circ$—$4s\ ^2S$	$^1/_2$—$^1/_2$
749,941	5	19,88	36,41	$3d\ ^2D$—$5f\ ^2F^\circ$	—
730,241	2	—	—	—	—
712,688	1	—	—	—	—
688,935	2	—	—	—	—
683,568	1	—	—	—	—
683,135	1	—	—	—	—
681,911	1	—	—	—	—
680,679	1	—	—	—	—
676,981	1	—	—	—	—
671,354	5	—	—	—	—

λ, Å	I	E_H, eV	E_B, eV	Transition	J
659,854	1	—	—	—	—
645,764	2	19,88	39,08	$3d\ ^2D—6f\ ^2F°$	—
639,208	1	—	—	—	—
635,312	10	—	—	—	—
633,183	3	—	—	—	—
629,750	2	—	—	—	—
629,363	2	—	—	—	—
612,082	2	—	—	—	—
608,904	1	—	—	—	—
561,689	1	—	—	—	—
542,296	5	—	—	—	—
516,348	3	8,90	32,90	$3p\ ^2P°—5s\ ^2S$	$^3/_2—^1/_2$
515,118	2	8,84	32,90	$3p\ ^2P°—5s\ ^2S$	$^1/_2—^1/_2$
458,155	3	0,00	27,06	$3s\ ^2S—4p\ ^2P°$	$^1/_2—^1/_2$
457,818	4	0,00	27,08	$3s\ ^2S—4p\ ^2P°$	$^1/_2—^3/_2$

Si V, ground state $1s^2\,2s^2\,2p^6\,^1S_0$
Ionization potential 1 345 100 cm^{-1}; 166,762 eV

λ, Å	I	E_H, eV	E_B, eV	Transition	J
118,968	20	0,00	104,21	$2p^6\ ^1S—3s\ [1^1/_2]°$	0—1
117,860	20	0,00	105,19	$2p^6\ ^1S—3s'\ [^1/_2]°$	0—1
98,209	2	0,00	126,24	$2p^6\ ^1S—3d\ [^1/_2]°$	0—1
97,143	10	0,00	127,62	$2p^6\ ^1S—3d\ [1^1/_2]°$	0—1
96,439	15	0,00	128,56	$2p^6\ ^1S—3d'\ [1^1/_2]°$	0—1
90,852	4	0,00	136,46	$2p^6\ ^1S—4s\ [1^1/_2]°$	0—1
90,453	4	0,00	137,06	$2p^6\ ^1S—4s'\ [^1/_2]°$	0—1
85,576	6	0,00	144,87	$2p^6\ ^1S—4d\ [1^1/_2]°$	0—1
85,175	10	0,00	145,56	$2p^6\ ^1S—4d'\ [1^1/_2]°$	0—1
81,113	2	0,00	152,85	$2p^6\ ^1S—5d\ [1^1/_2]°$	0—1
80,807	2	0,00	153,42	$2p^6\ ^1S—5d'\ [1^1/_2]°$	0—1
78,903	1	0,00	157,13	$2p^6\ ^1S—6d\ [1^1/_2]°$	0—1
78,611	1	0,00	157,71	$2p^6\ ^1S—6d'\ [1^1/_2]°$	0—1

Si VI, ground state $1s^2\,2s^2\,2p^5\,^2P^0{}_{3/2}$
Ionization potential 1 654 800 cm^{-1}; 205,157 eV

λ, Å	I	E_H, eV	E_B, eV	Transition	J
249,125	8	0,63	50,40	$2p^5\ ^2P°—2p^6\ ^2S$	$^1/_2—^1/_2$
246,001	8	0,00	50,40	$2p^5\ ^2P°—2p^6\ ^2S$	$^3/_2—^1/_2$
103,163	2	50,40	170,57	$2p^6\ ^2S—3s'''\,^2P°$	$^1/_2—^3/_2$
102,846	1	50,40	170,94	$2p^6\ ^2S—3s'''\,^2P°$	$^1/_2—^1/_2$
101,160	0	0,63	123,19	$2p^5\ ^2P°—3s\ ^4P$	$^1/_2—^3/_2$
100,963	1	0,00	122,79	$2p^5\ ^2P°—3s\ ^4P$	$^3/_2—^5/_2$
100,640	10	0,00	123,19	$2p^5\ ^2P°—3s\ ^4P$	$^3/_2—^3/_2$
99,966	10	0,63	124,65	$2p^5\ ^2P°—3s\ ^2P$	$^1/_2—^3/_2$
99,598	10	0,63	125,11	$2p^5\ ^2P°—3s\ ^2P$	$^1/_2—^1/_2$
99,460	15	0,00	124,65	$2p^5\ ^2P°—3s\ ^2P$	$^3/_2—^3/_2$

λ, Å	I	E_H eV	E_B, eV	Transition	J
99,095	10	0,00	125,11	$2p^5\,{}^2P^\circ - 3s\,{}^2P$	$^3/_2-{}^1/_2$
96,488	10	0,63	129,12	$2p^5\,{}^2P^\circ - 3s\,{}^2D$	$^1/_2-{}^3/_2$
96,020	10	0,00	129,12	$2p^5\,{}^2P^\circ - 3s\,{}^2D$	$^3/_2-{}^3/_2,\ {}^5/_2$
91,798	4	0,63	135,69	$2p^5\,{}^2P^\circ - 3s''\,{}^2S$	$^1/_2-{}^1/_2$
91,369	4	0,00	135,69	$2p^5\,{}^2P^\circ - 3s''\,{}^2S$	$^3/_2-{}^1/_2$
84,082	12	0,63	148,07	$2p^5\,{}^2P^\circ - 3d\,{}^4F$	$^1/_2-{}^3/_2$
83,965	0	0,63	148,28	$2p^5\,{}^2P^\circ - 3d\,{}^4P$	$^1/_2-{}^3/_2$
83,802	6	0,00	147,94	$2p^5\,{}^2P^\circ - 3d\,{}^4F$	$^3/_2-{}^5/_2$
83,729	1	0,00	148,07	$2p^5\,{}^2P^\circ - 3d\,{}^4F$	$^3/_2-{}^3/_2$
83,684	1	0,00	148,15	$2p^5\,{}^2P^\circ - 3d\,{}^4P$	$^3/_2-{}^1/_2$
83,639	3	0,63	148,86	$2p^5\,{}^2P^\circ - 3d\,{}^2P$	$^1/_2-{}^1/_2$
83,611	8	0,00	148,28	$2p^5\,{}^2P^\circ - 3d\,{}^4P$	$^3/_2-{}^3/_2$
		0,63	148,91	$2p^5\,{}^2P^\circ - 3d\,{}^2D$	$^1/_2-{}^3/_2$
83,526	8	0,00	148,43	$2p^5\,{}^2P^\circ - 3d\,{}^4P$	$^3/_2-{}^5/_2$
83,358	8	0,63	149,36	$2p^5\,{}^2P^\circ - 3d\,{}^2P$	$^1/_2-{}^3/_2$
83,283	1	0,00	148,86	$2p^5\,{}^2P^\circ - 3d\,{}^2P$	$^3/_2-{}^1/_2$
83,258	5	0,00	148,91	$2p^5\,{}^2P^\circ - 3d\,{}^2D$	$^3/_2-{}^3/_2$
83,128	15	0,00	149,14	$2p^5\,{}^2P^\circ - 3d\,{}^2D$	$^3/_2-{}^5/_2$
83,006	4	0,00	149,36	$2p^5\,{}^2P^\circ - 3d\,{}^2P$	$^3/_2-{}^3/_2$
81,030	7	0,63	153,63	$2p^5\,{}^2P^\circ - 3d'\,{}^2P$	$^1/_2-{}^1/_2$
80,908	8	0,63	153,86	$2p^5\,{}^2P^\circ - 3d'\,{}^2S$	$^1/_2-{}^1/_2$
80,821	8	0,63	154,03	$2p^5\,{}^2P^\circ - 3d'\,{}^2P$	$^1/_2-{}^3/_2$
80,725	10	0,63	154,21	$2p^5\,{}^2P^\circ - 3d'\,{}^2D$	$^1/_2-{}^3/_2$
80,698	10	0,00	153,63	$2p^5\,{}^2P^\circ - 3d'\,{}^2P$	$^3/_2-{}^1/_2$
80,577	12	0,00	153,86	$2p^5\,{}^2P^\circ - 3d'\,{}^2S$	$^3/_2-{}^1/_2$
80,501	10	0,00	154,01	$2p^5\,{}^2P^\circ - 3d'\,{}^2D$	$^3/_2-{}^5/_2$
80,491	5	0,00	154,03	$2p^5\,{}^2P^\circ - 3d'\,{}^2P$	$^3/_2-{}^3/_2$
80,449	10	0,00	154,11	$2p^5\,{}^2P^\circ - 3d'\,{}^2F$	$^3/_2-{}^5/_2$
80,395	5	0,00	154,21	$2p^5\,{}^2P^\circ - 3d'\,{}^2D$	$^3/_2-{}^3/_2$
77,718	6	0,63	160,15	$2p^5\,{}^2P^\circ - 3d''\,{}^2D$	$^1/_2-{}^3/_2$
77,429	10	0,00	160,12	$2p^5\,{}^2P^\circ - 3d''\,{}^2D$	$^3/_2-{}^5/_2$
75,587	1	0,00	164,02	$2p^5\,{}^2P^\circ - 4s\,{}^4P$	$^3/_2-{}^3/_2$
75,486	1	0,63	164,88	$2p^5\,{}^2P^\circ - 4s\,{}^2P$	$^1/_2-{}^3/_2$
75,193	4	0,00	164,88	$2p^5\,{}^2P^\circ - 4s\,{}^2P$	$^3/_2-{}^3/_2$
72,896	1	0,00	170,07	$2p^5\,{}^2P^\circ - 4s'\,{}^2D$	$^3/_2-{}^5/_2,\ {}^3/_2$
71,718	0	0,63	173,50	$2p^5\,{}^2P^\circ - 4d\,{}^4F$	$^1/_2-{}^3/_2$
71,644	0	0,63	173,68	$2p^5\,{}^2P^\circ - 4d\,{}^4P$	$^1/_2-{}^3/_2$
71,561	1	0,63	173,88	$2p^5\,{}^2P^\circ - 4d\,{}^2P$	$^1/_2-{}^1/_2$
71,534	1	0,63	173,95	$2p^5\,{}^2P^\circ - 4d\,{}^2D$	$^1/_2-{}^3/_2$
71,474	1	0,00	173,46	$2p^5\,{}^2P^\circ - 4d\,{}^4F$	$^3/_2-{}^5/_2$
71,384	4	0,00	173,68	$2p^5\,{}^2P^\circ - 4d\,{}^4P$	$^3/_2-{}^3/_2$
71,366	3	0,63	174,35	$2p^5\,{}^2P^\circ - 4d\,{}^2P$	$^1/_2-{}^3/_2$
71,340	1	0,00	173,78	$2p^5\,{}^2P^\circ - 4d\,{}^4P$	$^3/_2-{}^5/_2$
71,304	0	0,00	173,88	$2p^5\,{}^2P^\circ - 4d\,{}^2P$	$^3/_2-{}^1/_2$
71,273	2	0,00	173,95	$2p^5\,{}^2P^\circ - 4d\,{}^2D$	$^3/_2-{}^3/_2$
71,181	5	0,00	174,17	$2p^5\,{}^2P^\circ - 4d\,{}^2D$	$^3/_2-{}^5/_2$
69,448	2	0,63	179,15	$2p^5\,{}^2P^\circ - 4d'\,{}^2P$	$^1/_2-{}^3/_2$
69,421	1	0,63	179,22	$2p^5\,{}^2P^\circ - 4d'\,{}^2D$	$^1/_2-{}^3/_2$
69,236	5	0,00	179,07	$2p^5\,{}^2P^\circ - 4d'\,{}^2S$	$^3/_2-{}^1/_2$
69,204	1	0,00	179,15	$2p^5\,{}^2P^\circ - 4d'\,{}^2D$	$^3/_2-{}^5/_2$
66,796	0	0,00	185,61	$2p^5\,{}^2P^\circ - 4d''\,{}^2D$	$^3/_2-{}^5/_2$
66,772	0	0,00	185,67	$2p^5\,{}^2P^\circ - 5d\,{}^2D$	$^3/_2-{}^5/_2$
65,211	0	0,63	190,75	$2p^5\,{}^2P^\circ - 5d'\,{}^2P$	$^1/_2-{}^3/_2$
65,004	0	0,00	190,72	$2p^5\,{}^2P^\circ - 5d'\,{}^2S$	$^3/_2-{}^1/_2$

CHLORINE, Z = 17

Cl I, ground state $1s^2\,2s^2\,2p^6\,3s^2\,3p^5\ ^2P^\circ_{3/2}$

Ionization potential 107 995,46 cm^{-1}; 13,017 eV

λ, Å	I	E_H, eV	E_B, eV	Transition	J
25323,66	6	10,43	10,92	$4p\ ^4D^\circ$—$3d\ ^4D$	$^5/_2$—$^5/_2$
25047,31	6	—	—	—	—
24470,02	100	10,40	10,90	$4p\ ^4D^\circ$—$3d\ ^4D$	$^7/_2$—$^7/_2$
24146,23	4	—	—	—	—
23956,10	11	—	—	—	—
23882,69	18	10,40	10,92	$4p\ ^4D^\circ$—$3d\ ^4D$	$^7/_2$—$^5/_2$
23188,84	12	11,29	11,83	$3d\ ^2F$—$5p\ ^2D^\circ$	$^7/_2$—$^5/_2$
23038,78	17	11,22	11,76	$3d\ ^4F$—$5p\ ^4D^\circ$	$^7/_2$—$^5/_2$
22891,90	4	—	—	—	—
22721,74	5	—	—	—	—
22688,70	12	11,39	11,94	$3d\ ^2P$—$5p\ ^2P^\circ$	$^3/_2$—$^3/_2$
22522,80	6	—	—	—	—
22288,52	9	11,30	11,85	$5s\ ^4P$—$5p\ ^4S^\circ$	$^5/_2$—$^3/_2$
22026,68	40	11,18	11,74	$3d\ ^4F$—$5p\ ^4D^\circ$	$^9/_2$—$^7/_2$
21902,34	14	11,25	11,82	$3d\ ^4F$—$5p\ ^4D^\circ$	$^5/_2$—$^3/_2$
21830,38	10	11,30	11,86	$5s\ ^4P$—$5p\ ^2D^\circ$	$^5/_2$—$^3/_2$
21582,40	12	11,37	11,94	$3d\ ^4P$—$5p\ ^2P^\circ$	$^3/_2$—$^3/_2$
20725,44	56	10,33	10,93	$4p\ ^4P^\circ$—$3d\ ^4D$	$^1/_2$—$^3/_2$
20370,12	85	10,33	10,94	$4p\ ^4P^\circ$—$3d\ ^4D$	$^1/_2$—$^1/_2$
20199,36	227	10,31	10,92	$4p\ ^4P^\circ$—$3d\ ^4D$	$^3/_2$—$^5/_2$
19766,78	185	10,31	10,93	$4p\ ^4P^\circ$—$3d\ ^4D$	$^3/_2$—$^3/_2$
19755,28	717	10,28	10,90	$4p\ ^4P^\circ$—$3d\ ^4D$	$^5/_2$—$^7/_2$
19443,27	6	10,31	10,94	$4p\ ^4P^\circ$—$3d\ ^4D$	$^3/_2$—$^1/_2$
19370,30	227	10,28	10,92	$4p\ ^4P^\circ$—$3d\ ^4D$	$^5/_2$—$^5/_2$
18971,55	21	10,28	10,93	$4p\ ^4P^\circ$—$3d\ ^4D$	$^5/_2$—$^3/_2$
18742,79	22	10,63	11,29	$4p\ ^4S^\circ$—$5s\ ^4P$	$^3/_2$—$^1/_2$
18541,37	74	10,63	11,30	$4p\ ^4S^\circ$—$5s\ ^4P$	$^3/_2$—$^5/_2$
18382,27	40	10,63	11,30	$4p\ ^4S^\circ$—$5s\ ^4P$	$^3/_2$—$^3/_2$
17825,76	5	10,63	11,32	$4p\ ^4S^\circ$—$3d\ ^2D$	$^3/_2$—$^5/_2$
17767,65	7	10,59	11,29	$4p\ ^2P^\circ$—$5s\ ^4P$	$^3/_2$—$^1/_2$
17586,44	60	10,59	11,30	$4p\ ^2P^\circ$—$5s\ ^4P$	$^3/_2$—$^5/_2$
17519,72	4	—	—	—	—
17443,93	46	10,59	11,30	$4p\ ^2P^\circ$—$5s\ ^4P$	$^3/_2$—$^3/_2$
17226,30	27	10,63	11,35	$4p\ ^4S^\circ$—$3d\ ^2D$	$^3/_2$—$^3/_2$
17119,13	28	10,49	11,22	$4p\ ^2D^\circ$—$3d\ ^4F$	$^5/_2$—$^7/_2$
16941,45	10	10,59	11,32	$4p\ ^2P^\circ$—$3d\ ^2D$	$^3/_2$—$^5/_2$
16871,76	18	10,63	11,36	$4p\ ^4S^\circ$—$3d\ ^4P$	$^3/_2$—$^5/_2$
16813,82	14	10,63	11,37	$4p\ ^4S^\circ$—$3d\ ^4P$	$^3/_2$—$^3/_2$
16800,73	10	—	—	—	—
16759,73	6	—	—	—	—
16671,38	55	—	—	—	—
16624,76	4	11,36	12,11	$3d\ ^4P$—$(^3P_2)\ 4f\ [3]^\circ$	$^5/_2$—$^5/_2$, $^7/_2$
16385,70	7	10,54 10,49	11,29 11,25	$4p\ ^2D^\circ$—$3d\ ^2F$ $4p\ ^2D^\circ$—$3d\ ^4F$	$^3/_2$—$^7/_2$ $^5/_2$—$^5/_2$
16293,39	15	11,35	12,11	$3d\ ^2D$—$(^3P_2)\ 4f\ [3]^\circ$	$^3/_2$—$^5/_2$
16286,18	39	10,54	11,30	$4p\ ^2D^\circ$—$5s\ ^4P$	$^3/_2$—$^5/_2$
16284,18	7	11,35	12,11	$3d\ ^2P$—$(^3P_2)\ 4f\ [2]^\circ$	$^3/_2$—$^3/_2$, $^5/_2$
16214,99	10	11,35	12,11	$3d\ ^2D$—$(^3P_2)\ 4f\ [2]^\circ$	$^3/_2$—$^5/_2$
16198,47	259	10,63	11,39	$4p\ ^4S^\circ$—$3d\ ^2P$	$^3/_2$—$^3/_2$
16189,88	14	11,35	12,12	$3d\ ^2P$—$(^3P_2)\ 4f\ [1]^\circ$	$^3/_2$—$^3/_2$
16179,12	10	11,43	12,20	$3d\ ^2P$—$(^3P_1)\ 4f\ [2]^\circ$	$^1/_2$—$^3/_2$
16077,62	129	10,59	11,36	$4p\ ^2P^\circ$—$3d\ ^4P$	$^3/_2$—$^5/_2$
16067,35	10	10,65	11,42	$4p\ ^2P^\circ$—$5s\ ^2P$	$^1/_2$—$^3/_2$

λ, Å	I	E_H, eV	E_B, eV	Transition	J
16060,41	10	—	—	—	—
16024,95	25	10,59	11,37	$4p\ ^2P^\circ - 3d\ ^4P$	$^3/_2 - ^3/_2$
15970,49	283	10,50	11,28	$4p\ ^4D^\circ - 3d\ ^4F$	$^1/_2 - ^3/_2$
15959,97	735	10,47	11,25	$4p\ ^4D^\circ - 3d\ ^4F$	$^3/_2 - ^5/_2$
15928,92	342	10,65	11,43	$4p\ ^2P^\circ - 3d\ ^2P$	$^1/_2 - ^1/_2$
15883,34	277	10,57	11,35	$4p\ ^2S^\circ - 3d\ ^2D$	$^1/_2 - ^3/_2$
15869,66	2780	10,40	11,18	$4p\ ^4D^\circ - 3d\ ^4F$	$^7/_2 - ^9/_2$
15818,41	193	10,54	11,32	$4p\ ^2D^\circ - 3d\ ^2F$	$^3/_2 - ^5/_2$
15808,54	25	11,32	12,11	$3d\ ^2F - (^3P_2)\ 4f\ [4]^\circ$	$^5/_2 - ^7/_2$
15792,00	21	11,32	12,11	$3d\ ^2F - (^3P_2)\ 4f\ [3]^\circ$	$^5/_2 - ^5/_2,\ ^7/_2$
15730,06	1487	10,43	11,22	$4p\ ^4D^\circ - 3d\ ^4F$	$^5/_2 - ^7/_2$
15717,70	4	11,32	12,11	$3d\ ^2F - (^3P_2)\ 4f\ [2]^\circ$	$^5/_2 - ^5/_2$
15668,64	7	10.50	11,29	$4p\ ^4D^\circ - 5s\ ^4P$	$^1/_2 - ^1/_2$
15615,16	7	—	—	—	—
15608,08	18	10,63	11,42	$4p\ ^4S^\circ - 5s\ ^2P$	$^3/_2 - ^3/_2$
15580,66	5	10,63	11,42	$4p\ ^4S^\circ - 3d\ ^4P$	$^3/_2 - ^1/_2$
15520,29	1094	10,49	11,29	$4p\ ^2D^\circ - 3d\ ^2F$	$^5/_2 - ^7/_2$
15477,78	15	10,63	11,43	$4p\ ^4S^\circ - 3d\ ^2P$	$^3/_2 - ^1/_2$
15467,59	169	10,47	11,28	$4p\ ^4D^\circ - 3d\ ^4F$	$^3/_2 - ^3/_2$
15465,07	381	10,59	11,39	$4p\ ^2P^\circ - 3d\ ^2P$	$^3/_2 - ^3/_2$
15435,14	27	11,39	12,20	$3d\ ^2D - (^3P_1)\ 4f\ [4]^\circ$	$^5/_2 - ^7/_2$
15416,07	32	10,50	11,30	$4p\ ^4D^\circ - 5s\ ^4P$	$^1/_2 - ^3/_2$
15382,31	17	11,43	12,24	$3d\ ^2P - (^3P_0)\ 4f\ [3]^\circ$	$^1/_2 - ^5/_2$?
15373,88	23	—	—	—	—
15351,42	2	11,39	12,20	$3d\ ^2D - (^3P_1)\ 4f\ [3]^\circ$	$^5/_2 - ^7/_2,\ ^5/_2$
15320,46	7	10,49	11,30	$4p\ ^2D^\circ - 5s\ ^4P$	$^5/_2 - ^3/_2$
15309,08	28	11,30	12,11	$5s\ ^4P - (^3P_2)\ 4f\ [2]^\circ$	$^3/_2 - ^3/_2,\ ^5/_2$
15269,83	8	11,30	12,11	$5s\ ^4P - (^3P_2)\ 4f\ [3]^\circ$	$^5/_2 - ^5/_2,\ ^7/_2$
15262,98	150	10,54	11,35	$4p\ ^2D^\circ - 3d\ ^2D$	$^3/_2 - ^3/_2$
15225,72	13	11,30	12,12	$5s\ ^4P - (^3P_2)\ 4f\ [1]^\circ$	$^3/_2 - ^3/_2$
15203,46	15	—	—	—	—
15199,65	22	11,29	12,11	$3d\ ^2F - (^3P_2)\ 4f\ [4]^\circ$	$^7/_2 - ^9/_2$
15183,97	8	—	—	—	—
15181,94	5	11,29	12,11	$3d\ ^2F - (^3P_2)\ 4f\ [3]^\circ$	$^7/_2 - ^7/_2,\ ^5/_2$
15161,15	145	10,40	11,22	$4p\ ^4D^\circ - 3d\ ^4F$	$^7/_2 - ^7/_2$
15108,04	269	10,43	11,25	$4p\ ^4D^\circ - 3d\ ^4F$	$^5/_2 - ^5/_2$
15094,12	48	11,29	12,11	$3d\ ^2F - (^3P_2)\ 4f\ [5]^\circ$	$^7/_2 - ^9/_2$
15051,60	4	10,47	11,30	$4p\ ^4D^\circ - 5s\ ^4P$	$^3/_2 - ^5/_2$
14987,69	29	11,29	12,12	$5s\ ^4P - (^3P_2)\ 4f\ [1]^\circ$	$^1/_2 - ^3/_2$
14983,51	95	10,54	11,36	$4p\ ^2D^\circ - 3d\ ^4P$	$^3/_2 - ^5/_2$
14955,33	78	10,65	11,48	$4p\ ^2P^\circ - 5s\ ^2P$	$^1/_2 - ^1/_2$
14947,73	43	10,47	11,30	$4p\ ^4D^\circ - 5s\ ^4P$	$^3/_2 - ^3/_2$
14938,14	108	10,54	11,37	$4p\ ^2D^\circ - 3d\ ^4P$	$^3/_2 - ^3/_2$
		11,37	12,19	$3d\ ^4P - (^3P_1)\ 4f\ [2]^\circ$	$^3/_2 - ^3/_2,\ ^5/_2$
14931,70	294	10,49	11,32	$4p\ ^2D^\circ - 3d\ ^2D$	$^5/_2 - ^5/_2$
14924,95	7	—	—	—	—
14918,68	6	—	—	—	—
14901,33	10	10,59	11,42	$4p\ ^2P^\circ - 3d\ ^4P$	$^3/_2 - ^1/_2$
		10,49	11,32	$4p\ ^2D^\circ - 3d\ ^2F$	$^5/_2 - ^5/_2$
14892,33	3	11,36	12,20	$3d\ ^4P - (^3P_1)\ 4f\ [2]^\circ$	$^5/_2 - ^3/_2,\ ^5/_2$
14863,53	5	11,28	12,11	$3d\ ^4F - (^3P_2)\ 4f\ [3]^\circ$	$^3/_2 - ^5/_2$
14806,75	82	10,59	11,43	$4p\ ^2P^\circ - 3d\ ^2P$	$^3/_2 - ^1/_2$
14798,50	5	11,28	12,11	$3d\ ^4F - (^3P_2)\ 4f\ [2]^\circ$	$^3/_2 - ^3/_2,\ ^5/_2$
14792,29	50	11,36	12,20	$3d\ ^4P - (^3P_1)\ 4f\ [3]^\circ$	$^5/_2 - ^7/_2$
14731,37	45	11,39	12,23	$3d\ ^2D - (^3P_0)\ 4f\ [3]^\circ$	$^5/_2 - ^5/_2,\ ^7/_2$
14682,82	16	11,35	12,20	$3d\ ^2P - (^3P_1)\ 4f\ [2]^\circ$	$^3/_2 - ^3/_2,\ ^5/_2$
14576,78	3	10,47	11,32	$4p\ ^4D^\circ - 3d\ ^2D$	$^3/_2 - ^5/_2$
14626,35	9	11,35	12,20	$3d\ ^2D - (^3P_1)\ 4f\ [2]^\circ$	$^5/_2 - ^3/_2,\ ^5/_2$

λ, Å	I	E_H, eV	E_B, eV	Transition	J
14556,68	25	10,63	11,48	$4p\ ^4S^\circ - 5s\ ^2P$	$3/2 - 1/2$
14529,13	4	11,35	12,20	$3d\ ^2D - (^3P_1)\ 4f\ [3]^\circ$	$5/2 - 5/2$
14508,63	16	10,49	11,35	$4p\ ^2D^\circ - 3d\ ^2D$	$5/2 - 3/2$
14497,41	60	10,57	11,42	$4p\ ^2S^\circ - 5s\ ^2P$	$1/2 - 3/2$
14450,44	95	11,25	12,11	$3d\ ^4F - (^3P_2)\ 4f\ [4]^\circ$	$5/2 - 7/2$
14436,26	13	11,25	12,11	$3d\ ^4F - (^3P_2)\ 4f\ [3]^\circ$	$5/2 - 7/2,\ 5/2$
14384,93	4	10,57	11,43	$4p\ ^2S^\circ - 3d\ ^2P$	$1/2 - 1/2$
14369,71	148	10,13	11,29	$4p\ ^4D^\circ - 3d\ ^2F$	$5/2 - 7/2$
14297,53	2	10,50	11,37	$4p\ ^4D^\circ - 3d\ ^4P$	$1/2 - 3/2$
14292,07	73	10,43	11,30	$4p\ ^4D^\circ - 5s\ ^4P$	$5/2 - 5/2$
14255,80	3	10,49	11,36	$4p\ ^2D^\circ - 3d\ ^4P$	$5/2 - 5/2$
14221,48	2	11,32	12,20	$3d\ ^2F - (^3P_1)\ 4f\ [2]^\circ$	$5/2 - 3/2,\ 5/2$
14214,95	5	11,36	12,23	$3d\ ^4P - (^3P_0)\ 4f\ [3]^\circ$	$5/2 - 7/2$
14198,27	48	10,43	11,30	$4p\ ^4D^\circ - 5s\ ^4P$	$5/2 - 3/2$
14173,84	11	10,47	11,35	$4p\ ^4D^\circ - 3d\ ^2D$	$3/2 - 3/2$
14129,80	14	11,32	12,20	$3d\ ^2F - (^3P_1)\ 4f\ [3]^\circ$	$5/2 - 7/2$
14122,44	4	—	—	—	—
13983,32	4	—	—	—	—
13978,14	120	10,54	11,42	$4p\ ^2D^\circ - 5s\ ^2P$	$3/2 - 3/2$
13961,45	19	10,59	11,48	$4p\ ^2P^\circ - 5s\ ^2P$	$3/2 - 1/2$
13956,82	2	10,54	11,42	$4p\ ^2D^\circ - 3d\ ^4P$	$3/2 - 1/2$
13932,97	15	10,47	11,36	$4p\ ^4D^\circ - 3d\ ^4P$	$3/2 - 5/2$
13923,92	20	11,22	12,11	$3d\ ^4F - (^3P_2)\ 4f\ [4]^\circ$	$7/2 - 7/2$
13911,08	2	11,22	12,11	$3d\ ^4F - (^3P_2)\ 4f\ [3]^\circ$	$7/2 - 5/2$
13893,10	110	10,47	11,37	$4p\ ^4D^\circ - 3d\ ^4P$	$3/2 - 3/2$
13885,14	7	11,30	12,20	$5s\ ^4P - (^3P_1)\ 4f\ [2]^\circ$	$5/2 - 3/2,\ 5/2$
13853,31	13	10,43	11,32	$4p\ ^4D^\circ - 3d\ ^2D$	$5/2 - 5/2$
13837,58	125	11,22	12,11	$3d\ ^4F - (^3P_2)\ 4f\ [5]^\circ$	$7/2 - 9/2$
13827,67	9	—	—	—	—
13821,72	525	10,40	11,30	$4p\ ^4D^\circ - 5s\ ^4P$	$7/2 - 5/2$
13802,82	11	—	—	—	—
13772,48	50	10,49	11,39	$4p\ ^2D^\circ - 3d\ ^2P$	$5/2 - 3/2$
13710,06	2	11,30	12,20	$5s\ ^4P - (^3P_1)\ 4f\ [3]^\circ$	$5/2 - 5/2,\ 7/2$
13706,12	5	11,29	12,20	$3d\ ^2F - (^3P_1)\ 4f\ [4]^\circ$	$7/2 - 9/2$
13602,16	11	11,32	12,23	$3d\ ^2F - (^3P_0)\ 4f\ [3]^\circ$	$5/2 - 5/2,\ 7/2$
13586,00	6	10,57	11,48	$4p\ ^2S^\circ - 5s\ ^2P$	$1/2 - 1/2$
13578,45	28	—	—	—	—
13498,30	160	10,43	11,35	$4p\ ^4D^\circ - 3d\ ^2D$	$5/2 - 3/2$
13469,98	9	10,47	11,39	$4p\ ^4D^\circ - 3d\ ^2P$	$3/2 - 3/2$
13465,13	2	11,28	12,20	$3d\ ^4F - (^3P_1)\ 4f\ [2]^\circ$	$3/2 - 3/2,\ 5/2$
13419,89	90	10,40	11,32	$4p\ ^4D^\circ - 3d\ ^2D$	$7/2 - 5/2$
13396,04	95	10,50	11,42	$4p\ ^4D^\circ - 3d\ ^4P$	$1/2 - 1/2$
13382,46	30	11,28	12,20	$3d\ ^4F - (^3P_1)\ 4f\ [3]^\circ$	$3/2 - 5/2$
13378,04	33	11,18	12,11	$3d\ ^4F - (^3P_2)\ 4f\ [4]^\circ$	$9/2 - 9/2$
13346,76	550	10,49	11,42	$4p\ ^2D^\circ - 5s\ ^2P$	$5/2 - 3/2$
13296,01	310	11,18	12,11	$3d\ ^4F - (^3P_2)\ 4f\ [5]^\circ$	$9/2 - 11/2$
13243,83	350	10,43	11,37	$4p\ ^4D^\circ - 3d\ ^4P$	$5/2 - 3/2$
13213,42	7	11,30	12,23	$5s\ ^4P - (^3P_0)\ 4f\ [3]^\circ$	$5/2 - 5/2,\ 7/2$
13208,29	20	10,28	11,22	$4p\ ^4P^\circ - 3d\ ^4F$	$5/2 - 7/2$
13182,58	8	10,33	11,28	$4p\ ^4P^\circ - 3d\ ^4F$	$1/2 - 3/2$
13168,90	13	—	—	—	—
13129,66	100	10,54	11,48	$4p\ ^2D^\circ - 5s\ ^2P$	$3/2 - 1/2$
13122,59	16	10,31	11,25	$4p\ ^4P^\circ - 3d\ ^4F$	$3/2 - 5/2$
13107,98	4	—	—	—	—
13095,13	49	11,25	12,20	$3d\ ^4F - (^3P_1)\ 4f\ [4]^\circ$	$5/2 - 7/2$
13062,73	5	—	—	—	—
13059,70	4	10,47	11,42	$4p\ ^4D^\circ - 5s\ ^2P$	$3/2 - 3/2$

λ, Å	I	E_H, eV	E_B, eV	Transition	J
13040,99	125	10,47	11,42	$4p\ ^4D^\circ - 3d\ ^4P$	$^3/_2 - ^1/_2$
13034,59	9	11,25	12,20	$3d\ ^4F - (^3P_1)\ 4f\ [3]^\circ$	$^5/_2 - ^5/_2,\ ^7/_2$
12976,77	20	10,33	11,29	$4p\ ^4P^\circ - 5s\ ^4P$	$^1/_2 - ^1/_2$
12908,57	24	11,28	12,23	$3d\ ^4F - (^3P_0)\ 4f\ [3]^\circ$	$^3/_2 - ^5/_2$
12872,10	39	10,40	11,36	$4p\ ^4D^\circ - 3d\ ^4P$	$^7/_2 - ^5/_2$
12869,80	4	—	—	—	—
12859,16	13	10,43	11,39	$4p\ ^4D^\circ - 3d\ ^2P$	$^5/_2 - ^3/_2$
12803,05	37	10,33	11,30	$4p\ ^4P^\circ - 5s\ ^4P$	$^1/_2 - ^3/_2$
12795,90	12	—	—	—	—
12661,75	47	11,22	12,20	$3d\ ^4F - (^3P_1)\ 4f\ [4]^\circ$	$^7/_2 - ^9/_2$
12621,35	47	—	—	—	—
12594,15	142	10,31	11,29	$4p\ ^4P^\circ - 5s\ ^4P$	$^3/_2 - ^1/_2$
12585,09	10	11,25	12,24	$3d\ ^4F - (^3P_0)\ 4f\ [3]^\circ$	$^5/_2 - ^5/_2,\ ^7/_2$
12563,58	38	—	—	—	—
12502,69	63	10,31	11,30	$4p\ ^4P^\circ - 5s\ ^4P$	$^3/_2 - ^5/_2$
12430,13	12	10,31	11,30	$4p\ ^4P^\circ - 5s\ ^4P$	$^3/_2 - ^3/_2$
12389,03	4	—	—	—	—
12280,55	16	—	—	—	—
12236,26	5	—	—	—	—
12231,13	16	10,33	11,35	$4p\ ^4P^\circ - 3d\ ^2D$	$^1/_2 - ^3/_2$
12229,52	4	—	—	—	—
12179,72	77	10,28	11,30	$4p\ ^4P^\circ - 5s\ ^4P$	$^5/_2 - ^5/_2$
12172,95	60	10,31	11,32	$4p\ ^4P^\circ - 3d\ ^2D$	$^3/_2 - ^5/_2$
12110,96	60	10,28	11,30	$4p\ ^4P^\circ - 5s\ ^4P$	$^5/_2 - ^3/_2$
12021,67	172	10,33	11,37	$4p\ ^4P^\circ - 3d\ ^4P$	$^1/_2 - ^3/_2$
11866,50	195	10,28	11,32	$4p\ ^4P^\circ - 3d\ ^2D$	$^5/_2 - ^5/_2$
11720,55	180	10,31	11,36	$4p\ ^4P^\circ - 3d\ ^4P$	$^3/_2 - ^5/_2$
11692,73	85	10,31	11,37	$4p\ ^4P^\circ - 3d\ ^4P$	$^3/_2 - ^3/_2$
11636,22	4	—	—	—	—
11598,74	5	10,28	11,35	$4p\ ^4P^\circ - 3d\ ^2D$	$^5/_2 - ^3/_2$
11579,91	9	—	—	—	—
11577,24	11	—	—	—	—
11573,48	2	11,35	12,42	$3d\ ^2P - (^3P_2)\ 5f\ [1]^\circ$	$^3/_2 - ^1/_2,\ ^3/_2$
11436,34	1000	10,28	11,36	$4p\ ^4P^\circ - 3d\ ^4P$	$^5/_2 - ^5/_2$
11409,68	269	10,28	11,37	$4p\ ^4P^\circ - 3d\ ^4P$	$^5/_2 - ^3/_2$
11392,66	231	10,33	11,42	$4p\ ^4P^\circ - 5s\ ^2P$	$^1/_2 - ^1/_2$
11378,01	45	10,33	11,42	$4p\ ^4P^\circ - 3d\ ^4P$	$^1/_2 - ^1/_2$
11373,93	5	—	—	—	—
11331,08	5	11,32	12,42	$3d\ ^2F - (^3P_2)\ 5f\ [4]^\circ$	$^5/_2 - ^7/_2$
11326,53	6	11,32	12,42	$3d\ ^2D - 100166^\circ$	$^5/_2 - ^5/_2$
11306,70	3	11,32	12,42	$3d\ ^2D - 100184^\circ$	$^5/_2 - ^7/_2$
11302,26	4	—	—	—	—
11286,39	9	—	—	—	—
11151,25	6	—	—	—	—
11122,97	300	10,28	11,39	$4p\ ^4P^\circ - 3d\ ^2P$	$^5/_2 - ^3/_2$
11096,70	56	10,31	11,42	$4p\ ^4P^\circ - 5s\ ^2P$	$^3/_2 - ^3/_2$
11093,76	6	11,30	12,42	$5s\ ^4P - 100181^\circ$	$^3/_2 - ^3/_2$
11093,04	6	—	—	—	—
11082,93	206	10,31	11,42	$4p\ ^4P^\circ - 3d\ ^4P$	$^3/_2 - ^3/_2$
11072,10	3	11,30	12,42	$5s\ ^4P - (^3P_2)\ 5f\ [1]^\circ$	$^3/_2 - ^3/_2,\ ^1/_2$
11063,58	2	11,42	12,54	$5s\ ^2P - 101172^\circ$	$^3/_2 - ^3/_2,\ ^5/_2$
11055,22	2	11,30	12,42	$5s\ ^4P - (^3P_2)\ 5f\ [3]^\circ$	$^5/_2 - ^7/_2,\ ^5/_2$
11014,52	3	11,29	12,42	$3d\ ^2F - (^3P_2)\ 5f\ [4]^\circ$	$^7/_2 - ^9/_2,\ ^7/_2$
10986,71	13	11,29	12,42	$3d\ ^2F - 100184^\circ$	$^7/_2 - ^7/_2$
10945,43	5	11,29	12,42	$5s\ ^4P - (^3P_2)\ 5f\ [1]^\circ$	$^1/_2 - ^3/_2,\ ^1/_2$
10841,55	100	{ 10,28 11,28	11,42 12,42	$4p\ ^4P^\circ - 5s\ ^2P$ $3d\ ^4F - 100166^\circ$	$^5/_2 - ^3/_2$ $^3/_2 - ^5/_2$
10831,68	9	11,36	12,51	$3d\ ^4P - (^3P_1)\ 5f\ [3]^\circ$	$^5/_2 -$ —

λ, Å	I	E_H, eV	E_B, eV	Transition	J
10822,20	3	10,33	11,48	$4p\ ^4P^\circ$—$5s\ ^2P$	$^1/_2$—$^1/_2$
10785,12	7	11,39	12,54	$3d\ ^2P$—101172°	$^3/_2$—$^3/_2$, $^5/_2$
10717,84	2	11,35	12,50	$3d\ ^2D$—$(^3P_1)\ 5f\ [2]^\circ$	$^3/_2$— —
10690,94	14	—	—	—	—
10681,99	7	—	—	—	—
10620,37	7	—	—	—	—
10600,53	18	10,94	12,11	$3d\ ^4D$—$(^3P_2)\ 4f\ [2]^\circ$	$^1/_2$—$^3/_2$
10560,89	4	10,94	12,12	$3d\ ^4D$—$(^3P_2)\ 4f\ [1]^\circ$	$^1/_2$—$^1/_2$
10554,96	8	10,31	11,48	$4p\ ^4P^\circ$—$5s\ ^2P$	$^3/_2$—$^1/_2$
10539,18	44	10,93	12,11	$3d\ ^4D$—$(^3P_2)\ 4f\ [3]^\circ$	$^3/_2$—$^5/_2$
10506,72	33	10,93	12,11	$3d\ ^4D$—$(^3P_2)\ 4f\ [2]^\circ$	$^3/_2$—$^3/_2$
10472,38	3	11,32	12,51	$3d\ ^2F$—$(^3P_1)\ 5f\ [3]^\circ$	$^5/_2$— —
10467,86	7	10,93	12,12	$3d\ ^4D$—$(^3P_2)\ 4f\ [1]^\circ$	$^3/_2$—$^1/_2$
10427,54	44	10,92	12,11	$3d\ ^4D$—$(^3P_2)\ 4f\ [4]^\circ$	$^5/_2$—$^7/_2$
10420,26	105	10,92	12,11	$3d\ ^4D$—$(^3P_2)\ 4f\ [3]^\circ$	$^5/_2$—$^7/_2$
10420,05	105	10,92	12,11	$3d\ ^4D$—$(^3P_2)\ 4f\ [3]^\circ$	$^5/_2$—$^5/_2$
10392,51	331	9,28	10,47	$4s\ ^2P$—$4p\ ^4D^\circ$	$^1/_2$—$^3/_2$
10387,97	34	10,92	12,11	$3d\ ^4D$—$(^3P_2)\ 4f\ [2]^\circ$	$^5/_2$—$^5/_2$
10350,02	2	10,92	12,12	$3d\ ^4D$—$(^3P_2)\ 4f\ [1]^\circ$	$^5/_2$—$^3/_2$
10329,77	5	11,22	12,42	$3d\ ^4F$—$(^3P_2)\ 5f\ [4]^\circ$	$^7/_2$—$^7/_2$, $^9/_2$
10320,08	205	10,91	12,11	$3d\ ^4D$—$(^3P_2)\ 4f\ [4]^\circ$	$^7/_2$—$^9/_2$
10312,16	44	10,91	12,11	$3d\ ^4D$—$(^3P_2)\ 4f\ [3]^\circ$	$^7/_2$—$^7/_2$, $^5/_2$
10305,50	22	—	—	—	—
10280,07	4	10,91	12,11	$3d\ ^4D$—$(^3P_2)\ 4f\ [2]^\circ$	$^7/_2$—$^5/_2$
10221,12	10	—	—	—	—
10091,64	40	9,20	10,43	$4s\ ^2P$—$4p\ ^4D^\circ$	$^3/_2$—$^5/_2$
10002,25	4	11,18	12,42	$3d\ ^4F$—100184°	$^9/_2$—$^7/_2$
9875,95	50	9,28	10,54	$4s\ ^2P$—$4p\ ^2D^\circ$	$^1/_2$—$^3/_2$
9815,74	3	10,93	12,20	$3d\ ^4D$—$(^3P_1)\ 4f\ [2]^\circ$	$^3/_2$—$^5/_2$
9808,46	5	10,43	11,69	$4s'\ ^2D$—$4p'\ ^2P^\circ$	$^3/_2$—$^3/_2$
9806,90	25	10,43	11,69	$4s'\ ^2D$—$4p'\ ^2P^\circ$	$^5/_2$—$^3/_2$
9744,33	30	9,20	10,47	$4s\ ^2P$—$4p\ ^4D^\circ$	$^3/_2$—$^3/_2$
9702,35	40	9,03	10,31	$4s\ ^4P$—$4p\ ^4P^\circ$	$^1/_2$—$^3/_2$
9669,54	5	10,92	12,20	$3d\ ^4D$—$(^3P_1)\ 4f\ [3]^\circ$	$^5/_2$—$^5/_2$, $^7/_2$
9661,90	20	10,43	11,71	$4s'\ ^2D$—$4p'\ ^2P^\circ$	$^3/_2$—$^1/_2$
9632,37	20	9,28	10,57	$4s\ ^2P$—$4p\ ^2S^\circ$	$^1/_2$—$^1/_2$
9609,06	35	10,91	12,20	$3d\ ^4D$—$(^3P_1)\ 4f\ [4]^\circ$	$^7/_2$—$^9/_2$
9592,20	75	9,20	10,49	$4s\ ^2P$—$4p\ ^2D^\circ$	$^3/_2$—$^5/_2$
9588,01	5	11,25	12,54	$3d\ ^4F$—101172°	$^5/_2$—$^3/_2$, $^5/_2$
9584,77	50	8,98	10,28	$4s\ ^4P$—$4p\ ^4P^\circ$	$^3/_2$—$^5/_2$
9576,43	8	10,91	12,20	$3d\ ^4D$—$(^3P_1)\ 4f\ [3]^\circ$	$^7/_2$—$^7/_2$
9571,30	5	11,29	12,59	$3d\ ^2F$—101530°	$^7/_2$—$^7/_2$
9554,96	4	9,20	10,50	$4s\ ^2P$—$4p\ ^4D^\circ$	$^3/_2$—$^1/_2$
9516,94	15	10,93	12,23	$3d\ ^4D$—$(^3P_0)\ 4f\ [3]^\circ$	$^3/_2$—$^5/_2$
9486,89	25	9,03	10,33	$4s\ ^4P$—$4p\ P^\circ$	$^1/_2$—$^1/_2$
9481,93	3	10,43	11,74	$4s'\ ^2D$—$5p\ ^4P^\circ$	$^5/_2$—$^3/_2$
9452,06	75	9,28	10,59	$4s\ ^2P$—$4p\ ^2P^\circ$	$^1/_2$—$^3/_2$
9419,82	40	10,92	12,23	$3d\ ^4D$—$(^3P_0)\ 4f\ [3]^\circ$	$^5/_2$—$^5/_2$, $^7/_2$
9414,07	2	11,39	12,71	$3d\ ^2P$—102522°	$^3/_2$—$^3/_2$, $^5/_2$
9393,81	50	8,98	10,31	$4s\ ^4P$—$4p\ ^4P^\circ$	$^3/_2$—$^3/_2$
9288,82	60	9,20	10,54	$4s\ ^2P$—$4p\ ^2D^\circ$	$^3/_2$—$^3/_2$
9212,39	3	10,43	11,77	$4s'\ ^2D$—$5p\ ^4P^\circ$	$^3/_2$—$^1/_2$
9197,49	25	9,28	10,63	$4s\ ^2P$—$4p\ ^4S^\circ$	$^1/_2$—$^3/_2$
9191,67	60	8,98	10,33	$4s\ ^4P$—$4p\ ^4P^\circ$	$^3/_2$—$^1/_2$
9121,10	75	8,92	10,28	$4s\ ^4P$—$4p\ ^4P^\circ$	$^5/_2$—$^5/_2$
9073,15	50	9,20	10,57	$4s\ ^2P$—$4p\ ^2S^\circ$	$^3/_2$—$^1/_2$
9069,66	25	10,43	11,79	$4s'\ ^2D$—$4p'\ ^2F^\circ$	$^3/_2$—$^5/_2$
9068,39	3	10,43	11,79	$4s'\ ^2D$—$4p'\ ^2F^\circ$	$^5/_2$—$^5/_2$

λ, Å	I	E_H, eV	E_B, eV	Transition	J
9050,10	4	11,22	12,59	$3d\ ^4F$—101530°	7/2—7/2
9045,40	40	9,28	10,65	$4s\ ^2P$—$4p\ ^2P^\circ$	1/2—1/2
9038,96	30	10,43	11,80	$4s'\ ^2D$—$4p'\ ^2F^\circ$	5/2—7/2
8980,10	2	11,29	12,67	$3d\ ^2F$—102218°	7/2—7/2
8948,01	50	8,92	10,31	$4s\ ^4P$—$4p\ ^4P^\circ$	5/2—3/2
8931,20	2	10,43	11,82	$4s'\ ^2D$—$5p\ ^4D^\circ$	5/2—3/2
8912,88	40	9,20	10,59	$4s\ ^2P$—$4p\ ^2P^\circ$	3/2—3/2
8841,70	15	11,48	12,88	$5s\ ^2P$—103905°	1/2—3/2
8815,28	30	11,18	12,59	$3d\ ^4F$—101530°	9/2—7/2
8713,69	3	11,25	12,67	$3d\ ^4F$—102218°	5/2—7/2
8711,58	1	10,43	11,85	$4s'\ ^2D$—$5p\ ^2S^\circ$	3/2—1/2
8700,44	5	11,42	12,85	$3d\ ^4P$—103638°	1/2—3/2
8686,28	30	9,20	10,63	$4s\ ^2P$—$4p\ ^4S^\circ$	3/2—3/2
8641,75	3	10,43	11,86	$4p\ ^4D^\circ$—$4d\ ^4D$	5/2—7/2
8634,84	2	11,28	12,71	$3d\ ^4F$—102522°	3/2—3/2, 5/2
8628,61	4	10,43	11,86	$4s'\ ^2D$—$5p\ ^2D^\circ$	5/2—3/2
8585,96	100	8,98	10,43	$4s\ ^4P$—$4p\ ^4D^\circ$	3/2—5/2
8577,98	7	10,43	11,87	$4p\ ^4D^\circ$—$4d\ ^4D$	5/2—5/2
8575,25	75	9,03	10,47	$4s\ ^4P$—$4p\ ^4D^\circ$	1/2—3/2
8550,46	20	9,20	10,65	$4s\ ^2P$—$4p\ ^2P^\circ$	3/2—1/2
8519,72	8	11,22	12,67	$3d\ ^4F$—102218°	7/2—7/2
8497,32	5	10,43	11,88	$4p\ ^4D^\circ$—$4d\ ^4D$	5/2—3/2
8488,85	2	11,25	12,71	$3d\ ^4F$—102522°	5/2—3/2, 5/2
8472,96	3	10,63	12,09	$4p\ ^4S^\circ$—$4d\ ^2D$	3/2—5/2
8467,32	25	10,40	11,86	$4p\ ^4D^\circ$—$4d\ ^4D$	7/2—7/2
8428,25	100	9,03	10,50	$4s\ ^4P$—$4p\ ^4D^\circ$	1/2—1/2
8406,20	10	10,40	11,87	$4p\ ^4D^\circ$—$4d\ ^4D$	7/2—5/2
8403,70	1	10,59	12,07	$4p\ ^2P^\circ$—$4d\ ^4P$	3/2—5/2
8397,61	3	—	—	—	—
8392,20	15	10,94	12,42	$3d\ ^4D$—100181°	1/2—3/2
8383,58	4	—	—	—	—
8375,95	150	8,92	10,40	$4s\ ^4P$—$4p\ ^4D^\circ$	5/2—7/2
8358,28	6	11,37	12,85	$3d\ ^4P$—103638°	3/2—3/2
8343,90	50	10,93	12,42	$3d\ ^4D$—100166°	3/2—5/2
		11,36	12,85	$3d\ ^4P$—103638°	5/2—3/2
8333,27	5000	10,93	12,42	$3d\ ^4D$—100181°	3/2—3/2
		8,98	10,47	$4s\ ^4P$—$4p\ ^4D^\circ$	3/2—3/2
8304,69	5	10,59	12,08	$4p\ ^2P^\circ$—$6s\ ^4P$	3/2—3/2
8286,67	5	—	—	—	—
8280,95	7	10,63	12,13	$4p\ ^4S^\circ$—$4d\ ^2D$	3/2—3/2
8273,79	7	10,49	11,99	$4p\ ^2D^\circ$—$4d\ ^4F$	5/2—7/2
8271,70	7	—	—	—	—
8269,15	60	10,92	12,42	$3d\ ^4D$—100166°	5/2—5/2
8267,97	3	10,59	12,09	$4p\ ^2P^\circ$—$4d\ ^2D$	3/2—5/2
8258,64	4	10,92	12,42	$3d\ ^4D$—100181°	5/2—3/2
8221,73	75	8,98	10,49	$4s\ ^4P$—$4p\ ^2D^\circ$	3/2—5/2
8220,40	60	9,03	10,54	$4s\ ^4P$—$4p\ ^2D^\circ$	1/2—3/2
8212,00	100	8,92	10,43	$4s\ ^4P$—$4p\ ^4D^\circ$	5/2—5/2
8206,40	2	10,65	12,16	$4p\ ^2P^\circ$—$6s\ ^4P$	1/2—1/2
8203,76	12	10,49	12,00	$4p\ ^2D^\circ$—$4d\ ^2F$	5/2—7/2
8200,95	25	10,90	12,42	$3d\ ^4D$—100166°	7/2—5/2
8200,20	35	10,43	11,94	$4s'\ ^2D$—$5p\ ^2P^\circ$	3/2—3/2
8199,02	35	10,43	11,94	$4s'\ ^2D$—$5p\ ^2P^\circ$	5/2—3/2
8194,35	50	8,98	10,50	$4s\ ^4P$—$4p\ ^4D^\circ$	3/2—1/2
8170,09	10	10,57	12,08	$4p\ ^2S^\circ$—$6s\ ^4P$	1/2—3/2
8161,52	12	11,36	12,88	$3d\ ^4P$—103905°	5/2—3/2
8129,55	2	10,49	12,02	$4p\ ^2D^\circ$—$4d\ ^4F$	5/2—5/2
8121,40	5	10,63	12,15	$4p\ ^4S^\circ$—$4d\ ^4P$	3/2—3/2
8117,75	4	—	—	—	—

λ, Å	I	E_H, eV	E_B, eV	Transition	J
8094,76	12	10,54	12,07	$4p\ ^2D^\circ-4d\ ^4P$	$^3/_2-^5/_2$
8087,69	20	10,43	11,96	$4s'\ ^2D-4p'\ ^2D^\circ$	$^3/_2-^5/_2$
8086,67	75	10,43	11,96	$4s'\ ^2D-4p'\ ^2D^\circ$	$^5/_2-^5/_2$
8085,54	60	10,43	11,96	$4s'\ ^2D-4p'\ ^2D^\circ$	$^3/_2-^3/_2$
8084,48	35	10,43	11,96	$4s'\ ^2D-4p'\ ^2D^\circ$	$^5/_2-^3/_2$
8051,08	20	9,03	10,57	$4s\ ^4P-4p\ ^2S^\circ$	$^1/_2-^1/_2$
8023,30	18	10,47	12,02	$4p\ ^4D^\circ-4d\ ^4F$	$^3/_2-^5/_2$
8015,57	45	10,43	11,97	$4s'\ ^2D-5p\ ^2P^\circ$	$^3/_2-^1/_2$
7997,80	50	8,98	10,54	$4s\ ^4P-4p\ ^2D^\circ$	$^3/_2-^3/_2$
7985,80	4	—	—	—	—
7980,58	15	8,92	10,47	$4s\ ^4P-4p\ ^4D^\circ$	$^5/_2-^3/_2$
7976,95	25	10,33	11,88	$4p\ ^4P^\circ-4d\ ^4D$	$^1/_2-^3/_2$
7974,72	20	10,49	12,05	$4p\ ^2D^\circ-4d\ ^2F$	$^5/_2-^5/_2$
7968,66	3	10,54	12,09	$4p\ ^2D^\circ-4d\ ^2D$	$^3/_2-^5/_2$
7952,49	15	10,50	12,06	$4p\ ^4D^\circ-4d\ ^4F$	$^1/_2-^3/_2$
7940,65	2	10,49	12,05	$4p\ ^2D^\circ-6s\ ^4P$	$^5/_2-^5/_2$
7938,90	8	—	—	—	—
7933,85	50	10,40	11,96	$4p\ ^4D^\circ-4d\ ^4F$	$^7/_2-^9/_2$
7924,62	100	9,03	10,59	$4s\ ^4P-4p\ ^2P^\circ$	$^1/_2-^3/_2$
7915,09	25	10,33	11,90	$4p\ ^4P^\circ-4d\ ^4D$	$^1/_2-^1/_2$
7899,28	45	10,31	11,87	$4p\ ^4P^\circ-4d\ ^4D$	$^3/_2-^5/_2$
7898,10	5	10,59	12,16	$4p\ ^2P^\circ-6s\ ^4P$	$^3/_2-^1/_2$
7893,33	10	—	—	—	—
7886,00	6	—	—	—	—
7878,22	75	8,92	10,49	$4s\ ^4P-4p\ ^2D^\circ$	$^5/_2-^5/_2$
7872,50	1	10,47	12,05	$4p\ ^4D^\circ-4d\ ^2F$	$^3/_2-^5/_2$
7870,68	1	10,43	12,00	$4p\ ^4D^\circ-4d\ ^2F$	$^5/_2-^7/_2$
7839,42	8	10,47	12,05	$4p\ ^4D^\circ-6s\ ^4P$	$^3/_2-^5/_2$
7837,40	6	8,98	10,57	$4s\ ^4P-4p\ ^2S^\circ$	$^3/_2-^1/_2$
7830,76	30	10,31	11,88	$4p\ ^4P^\circ-4d\ ^4D$	$^3/_2-^3/_2$
7825,80	3	10,47	12,06	$4p\ ^4D^\circ-4d\ ^4F$	$^3/_2-^3/_2$
7821,35	45	10,28	11,86	$4p\ ^4P^\circ-4d\ ^4D$	$^5/_2-^7/_2$
7815,34	1	10,50	12,08	$4p\ ^4D^\circ-6s\ ^4P$	$^1/_2-^3/_2$
7810,06	3	10,57	12,15	$4p\ ^2S^\circ-4d\ ^4P$	$^1/_2-^3/_2$
7802,27	6	10,43	12,02	$4p\ ^4D^\circ-4d\ ^4F$	$^5/_2-^5/_2$
7798,59	5	10,54	12,13	$4p\ ^2D^\circ-4d\ ^2D$	$^3/_2-^3/_2$
7790,56	5	10,49	12,08	$4p\ ^2D^\circ-6s\ ^4P$	$^5/_2-^3/_2$
7787,75	4	10,40	11,99	$4p\ ^4D^\circ-4d\ ^4F$	$^7/_2-^7/_2$
7777,82	10	10,47	12,07	$4p\ ^4D^\circ-4d\ ^4P$	$^3/_2-^5/_2$
7771,10	12	10,31	11,90	$4p\ ^4P^\circ-4d\ ^4D$	$^3/_2-^1/_2$
7769,18	30	10,28	11.87	$4p\ ^4P^\circ-4d\ ^4D$	$^5/_2-^5/_2$
7754,78	6	—	—	—	—
7744,94	125	9,03	10,63	$4s\ ^4P-4p\ ^4S^\circ$	$^1/_2-^3/_2$
7717,57	100	8,98	10,59	$4s\ ^4P-4p\ ^2P^\circ$	$^3/_2-^3/_2$
7702,89	10	10,28	11,88	$4p\ ^4P^\circ-4d\ ^4D$	$^5/_2-^3/_2$
7697,40	8	10,93	12,54	$3d\ ^4D-101172^\circ$	$^3/_2-^3/_2,\ ^5/_2$
7692,97	7	10,47	12,08	$4p\ ^4D^\circ-6s\ ^4P$	$^3/_2-^3/_2$
7672,44	25	8,92	10,54	$4s\ ^4P-4p\ ^2D^\circ$	$^5/_2-^3/_2$
7659,50	5	{ 10,43	12,05	$4p\ ^4D^\circ-4d\ ^2F$	$^5/_2-^5/_2$
		10,40	12,02	$4p\ ^4D^\circ-4d\ ^4F$	$^7/_2-^5/_2$
7656,86	6	10,54	12,15	$4p\ ^2D^\circ-4d\ ^4P$	$^3/_2-^3/_2$
7633,72	20	10,92	12,54	$3d\ ^4D-101172^\circ$	$^5/_2-^3/_2,\ ^5/_2$
7628,35	1	10,43	12,05	$4p\ ^4D^\circ-6s\ ^4P$	$^5/_2-^5/_2$
7561,19	4	—	—	—	—
7547,06	100	8,98	10,63	$4s\ ^4P-4p\ ^4S^\circ$	$^3/_2-^3/_2$
7496,56	1	10,93	12,59	$3d\ ^4D-101520^\circ$	$^3/_2-^5/_2$
7492,12	10	10,40	12,05	$4p\ ^4D^\circ-6s\ ^4P$	$^7/_2-^5/_2$
7489,46	8	10,43	12,08	$4p\ ^4D^\circ-6s\ ^4P$	$^5/_2-^3/_2$
7462,40	8	10,49	12,15	$4p\ ^2D^\circ-4d\ ^4P$	$^5/_2-^3/_2$

λ, Å	I	E_H, eV	E_B, eV	Transition	J
7459,42	3	10,43	12,09	$4p\ ^4D^\circ$—$4d\ ^2D$	5/2—5/2
7454,08	2	10,50	12,16	$4p\ ^4D^\circ$—$6s\ ^4P$	1/2—1/2
7444,32	3	8,98	10,65	$4s\ ^4P$—$4p\ ^2P^\circ$	3/2—1/2
7436,13	10	10,92	12,59	$3d\ ^4D$—101520°	5/2—5/2
7435,71	7	10,40	12,07	$4p\ ^4D^\circ$—$4d\ ^4P$	7/2—5/2
7414,10	90	8,92	10,59	$4s\ ^4P$—$4p\ ^2P$	5/2—3/2
7382,47	3	—	—	—	—
7380,92	4	10,90	12,59	$3d\ ^4D$—101520°	7/2—5/2
7372,65	1	10,47	12,15	$4p\ ^4D^\circ$—$4d\ ^4P$	3/2—3/2
7342,74	3	10,47	12,16	$4p\ ^4D^\circ$—$6s\ ^4P$	3/2—1/2
7342,00	1	10,65	12,34	$4p\ ^2P^\circ$—$4d\ ^2P$	1/2—3/2
7329,33	3	10,40	12,09	$4p\ ^4D^\circ$—$4d\ ^2D$	7/2—5/2
7256,63	125	8,92	10,63	$4s\ ^4P$—$4p\ ^4S^\circ$	5/2—3/2
7252,48	3	—	—	—	—
7244,76	3	10,63	12,34	$4p\ ^4S^\circ$—$4d\ ^2P$	3/2—3/2
7194,94	5	10,33	12,06	$4p\ ^4P^\circ$—$4d\ ^4F$	1/2—3/2
7185,68	2	10,43	12,15	$4p\ ^4D^\circ$—$4d\ ^4P$	5/2—3/2
7146,38	5	—	—	—	—
7127,35	3	10,28	12,02	$4p\ ^4P^\circ$—$4d\ ^4F$	5/2—5/2
7094,20	8	10,59	12,34	$4p\ ^2P^\circ$—$4d\ ^2P$	3/2—3/2
7086,80	25	10,31	12,05	$4p\ ^4P^\circ$—$6s\ ^4P$	3/2—5/2
7082,35	3	10,33	12,08	$4p\ ^4P^\circ$—$6s\ ^4P$	1/2—3/2
7075,64	3	10,31	12,06	$4p\ ^4P^\circ$—$4d\ ^4F$	3/2—3/2
7060,18	2	10,47	12,23	$4p\ ^4D^\circ$—$4d\ ^4P$	3/2—1/2
7058,25	1	10,57	12,32	$4p\ ^2S^\circ$—$5d\ ^4D$	1/2—1/2
7036,30	5	10,31	12,07	$4p\ ^4P^\circ$—$4d\ ^4P$	3/2—5/2
7019,30	4	10,90	12,67	$3d\ ^4D$—102218°	7/2—7/2
7011,24	3	—	—	—	—
7008,00	10	10,28	12,05	$4p\ ^4P^\circ$—$4d\ ^2F$	5/2—5/2
7006,30	4	10,59	12,36	$4p\ ^2P^\circ$—$4d\ ^2P$	3/2—1/2
6995,88	12	10,57	12,34	$4p\ ^2S^\circ$—$4d\ ^2P$	1/2—3/2
6981,85	25	10,28	12,05	$4p\ ^4P^\circ$—$6s\ ^4P$	5/2—5/2
6979,60	3	10,59	12,37	$4p\ ^2P^\circ$—$5d\ ^4F$	3/2—5/2
6977,00	5	—	—	—	—
6966,80	8	10,31	12,08	$4p\ ^4P^\circ$—$6s\ ^4P$	3/2—3/2
6962,50	6	—	—	—	—
6932,90	25	10,28	12,07	$4p\ ^4P^\circ$—$4d\ ^4P$	5/2—5/2
6925,35	6	—	—	—	—
6924,40	5	—	—	—	—
6920,31	3	10,92	12,71	$3d\ ^4D$—102522°	5/2—3/2, 5/2
6910,32	6	10,57	12,36	$4p\ ^2S^\circ$—$4d\ ^2P$	1/2—1/2
6872,85	6	10,54	12,34	$4p\ ^2D^\circ$—$4d\ ^2P$	3/2—3/2
6865,36	8	10,28	12,08	$4p\ ^4P^\circ$—$6s\ ^4P$	5/2—3/2
6854,45	10	—	—	—	—
6841,74	6	10,63	12,44	$4p\ ^4S^\circ$—$5d\ ^2D$	3/2—3/2
6840,23	15	10,28	12,09	$4p\ ^4P^\circ$—$4d\ ^2D$	5/2—5/2
6837,60	5	10,49	12,31	$4p\ ^2D^\circ$—$5d\ ^4D$	5/2—5/2
6816,50	2	10,50	12,32	$4p\ ^4D^\circ$—$5d\ ^4D$	1/2—3/2
6811,50	3	10,31	12,13	$4p\ ^4P^\circ$—$4d\ ^2D$	3/2—3/2
6810,04	15	10,33	12,15	$4p\ ^4P^\circ$—$4d\ ^4P$	1/2—3/2
6791,92	3	10,50	12,32	$4p\ ^4D^\circ$—$5d\ ^4D$	1/2—1/2
6790,20	2	10,54	12,36	$4p\ ^2D^\circ$—$4d\ ^2P$	3/2—1/2
6784,45	1	10,33	12,16	$4p\ ^4P^\circ$—$6s\ ^4P$	1/2—1/2
6765,20	3	10,54	12,37	$4p\ ^2D^\circ$—$5d\ ^4F$	3/2—5/2
6762,30	2	10,47	12,31	$4p\ ^4D^\circ$—$5d\ ^4D$	3/2—5/2
6757,75	5	—	—	—	—
6751,54	4	10,59	12,43	$4p\ ^2P^\circ$—$5d\ ^2D$	3/2—5/2
6730,24	5	10,63	12,47	$4p\ ^4S^\circ$—$5d\ ^2F$	3/2—5/2

λ, Å	I	E_H, eV	E_B, eV	Transition	J
6723,40	4	10,47	12,32	4p $^4D^\circ$—5d ^4D	3/2—3/2
6714,65	4	10,28	12,13	4p $^4P^\circ$—4d ^2D	5/2—3/2
6709,90	15	—	—	—	—
6703,20	6	10,31	12,15	4p $^4P^\circ$—4d ^4P	3/2—3/2
6699,40	3	10,47	12,32	4p $^4D^\circ$—5d ^4D	3/2—1/2
6697,45	2	10,57	12,41	4p $^2S^\circ$—5d ^4P	1/2—1/2
6678,39	10	10,31	12,16	4p $^4P^\circ$—6s ^4P	3/2—1/2
6643,10	1	10,47	12,34	4p $^4D^\circ$—4d ^2P	3/2—3/2
6609,26	7	10,28	12,15	4p $^4P^\circ$—4d ^4P	5/2—3/2
6604,57	7	10,43	12,31	4p $^4D^\circ$—5d ^4D	5/2—5/2
6600,10	3	10,59	12,47	4p $^2P^\circ$—5d ^2F	3/2—5/2
6567,35	3	10,43	12,32	4p $^4D^\circ$—5d ^4D	5/2—3/2
6565,90	1	10,47	12,36	4p $^4D^\circ$—4d ^2P	3/2—1/2
6551,28	3	—	—	—	—
6550,80	3	{ 10,54	12,43	4p $^2D^\circ$—5d ^2D	3/2—5/2
		10,50	12,39	4p $^4D^\circ$—5d ^4F	1/2—3/2
6542,40	8	10,47	12,37	4p $^4D^\circ$—5d ^4F	3/2—5/2
6536,10	1	10,59	12,49	4p $^2P^\circ$—5d ^4F	3/2—1/2
6531,39	20	10,40	12,30	4p $^4D^\circ$—5d ^4D	7/2—7/2
6509,00	2	10,54	12,44	4p $^2D^\circ$—5d ^2D	3/2—3/2
6502,21	5	10,40	12,31	4p $^4D^\circ$—5d ^4D	7/2—5/2
6490,55	1	10,43	12,34	4p $^4D^\circ$—4d ^2P	5/2—3/2
6471,45	3	—	—	—	—
6464,60	1	10,47	12,39	4p $^4D^\circ$—5d ^4F	3/2—3/2
6457,14	1	10,50	12,41	4p $^4D^\circ$—5d ^4P	1/2—1/2
6450,30	12	10,49	12,42	4p $^2D^\circ$—5d ^2F	5/2—7/2
6443,76	2	10,31	12,23	4p $^4P^\circ$—4d ^4P	3/2—1/2
6434,79	15	10,43	12,36	4p $^4D^\circ$—5d ^4F	5/2—7/2
6425,61	8	—	—	—	—
6408,05	7	{ 10,54	12,47	4p $^2D^\circ$—5d ^2F	3/2—5/2
		10,49	12,43	4p $^2D^\circ$—5d ^2D	5/2—5/2
6398,63	20	10,40	12,34	4p $^4D^\circ$—5d ^4F	7/2—9/2
6394,75	4	10,43	12,37	4p $^4D^\circ$—5d ^4F	5/2—5/2
6376,28	3	—	—	—	—
6373,37	1	10,47	12,41	4p $^4D^\circ$—5d ^4P	3/2—1/2
6367,98	3	10,49	12,44	4p $^2D^\circ$—5d ^2D	5/2—3/2
6341,66	10	10,47	12,43	4p $^4D^\circ$—5d ^2D	3/2—5/2
6334,96	4	—	—	—	—
6326,74	3	10,57	12,53	4p $^2S^\circ$—6d ^4D	1/2—3/2
6321,59	6	—	—	—	—
6280,20	5	10,43	12,40	4p $^4D^\circ$—3	5/2—3/2
6252,26	8	10,33	12,32	4p $^4P^\circ$—5d ^4D	1/2—3/2
6242,54	4	10,43	12,42	4p $^4D^\circ$—5d ^2F	5/2—7/2
6231,48	8	10,33	12,32	4p $^4P^\circ$—5d ^4D	1/2—1/2
6226,39	3	10,49	12,48	4p $^2D^\circ$—5d ^2P	5/2—3/2
6211,55	6	10,40	12,40	4p $^4D^\circ$—5d ^4P	7/2—5/2
6194,72	15	10,31	12,31	4p $^4P^\circ$—5d ^4D	3/2—5/2
6165,38	1	10,43	12,44	4p $^4D^\circ$—5d ^2D	5/2—3/2
6162,05	12	10,31	12,32	4p $^4P^\circ$—5d ^4D	3/2—3/2
6151,34	2	10,40	12,42	4p $^4D^\circ$—5d ^2F	7/2—7/2
6141,79	4	10,31	12,32	4p $^4P^\circ$—5d ^4D	3/2—1/2
6140,21	25	10,28	12,30	4p $^4P^\circ$—5d ^4D	5/2—7/2
6114,37	15	10,28	12,31	4p $^4P^\circ$—5d ^4D	5/2—5/2
6082,53	4	10,28	12,32	4p $^4P^\circ$—5d ^4D	5/2—3/2
6019,71	6	10,33	12,39	4p $^4P^\circ$—2	1/2—1/2
5991,42	4	10,33	12,40	4p $^4P^\circ$—3	1/2—3/2
5948,40	4	10,33	12,41	4p $^4P^\circ$—5d ^4P	1/2—1/2
5934,00	1	10,28	12,37	4p $^4P^\circ$—5d ^4F	5/2—5/2

λ, Å	I	E_H, eV	E_B, eV	Transition	J
5930,35	6	10,43 10,31	12,52 12,40	$4p\ ^4D^\circ$—$6d\ ^4D$ $4p\ ^4P^\circ$—$5d\ ^4P$	$^5/_2$—$^7/_2$ $^3/_2$—$^5/_2$
5908,22	2	10,43	12,53	$4p\ ^4D^\circ$—$6d\ ^4D$	$^5/_2$—$^5/_2$
5866,75	4	10,31	12,41	$4p\ ^4P^\circ$—$5d\ ^4P$	$^3/_2$—$^1/_2$
5856,70	8	10,28	12,40	$4p\ ^4P^\circ$—$5d\ ^4P$	$^5/_2$—$^5/_2$
5847,68	7	10,40	12,52	$4p\ ^4D^\circ$—$6d\ ^4D$	$^7/_2$—$^7/_2$
5846,70	8	—	—	—	—
5844,15	6	10,31	12,42	$4p\ ^4P^\circ$—$5d\ ^4P$	$^3/_2$—$^3/_2$
5839,85	4	10,31	12,43	$4p\ ^4P^\circ$—$5d\ ^2D$	$^3/_2$—$^5/_2$
5826,24	2	10,40	12,53	$4p\ ^4D^\circ$—$6d\ ^4D$	$^7/_2$—$^5/_2$
5806,76	2	10,31	12,44	$4p\ ^4D^\circ$—$5d\ ^2D$	$^3/_2$—$^3/_2$
5802,84	5	10,28	12,42	$4p\ ^4P^\circ$—$5d\ ^2F$	$^5/_2$—$^7/_2$
5799,88	12	—	—	—	—
5796,26	15	—	—	—	—
5774,72	4	—	—	—	—
5772,58	5	10,28	12,42	$4p\ ^4P^\circ$—$5d\ ^4P$	$^5/_2$—$^3/_2$
5768,30	2	10,28	12,43	$4p\ ^4P^\circ$—$5d\ ^2D$	$^5/_2$—$^5/_2$
5765,55	3	10,33	12,48	$4p\ ^4P^\circ$—$5d\ ^2P$	$^1/_2$—$^3/_2$
5726,16	5	10,31	12,47	$4p\ ^4P^\circ$—$5d\ ^2F$	$^3/_2$—$^5/_2$
5686,28	1	10,33	12,51	$4p\ ^4P^\circ$—$6d\ ^4D$	$^1/_2$—$^1/_2$
5654,20	1	10,33	12,53	$4p\ ^4P^\circ$—$6d\ ^4D$	$^1/_2$—$^3/_2$
5620,72	1	10,28	12,48	$4p\ ^4P^\circ$—$5d\ ^2P$	$^5/_2$—$^3/_2$
5580,45	3	10,31	12,53	$4p\ ^4P^\circ$—$6d\ ^4D$	$^3/_2$—$^3/_2$
5578,13	3	10,31	12,53	$4p\ ^4P^\circ$—$6d\ ^4D$	$^3/_2$—$^5/_2$
5532,13	8	10,28	12,52	$4p\ ^4P^\circ$—$6d\ ^4D$	$^5/_2$—$^7/_2$
5512,95	4	10,28	12,53	$4p\ ^4P^\circ$—$6d\ ^4D$	$^5/_2$—$^5/_2$
5493,14	3	—	—	—	—
5140,35	5	9,28	11,69	$4s\ ^2P$—$4p'\ ^2P^\circ$	$^1/_2$—$^3/_2$
5099,80	8	9,28	11,71	$4s\ ^2P$—$4p'\ ^2P^\circ$	$^1/_2$—$^1/_2$
4976,62	10	9,20	11,69	$4s\ ^2P$—$4p'\ ^2P^\circ$	$^3/_2$—$^3/_2$
4971,77	—	9,28	11,77	$4s\ ^2P$—$5p\ ^4P^\circ$	$^1/_2$—$^1/_2$
4938,59	—	9,20	11,71	$4s\ ^2P$—$4p'\ ^2P^\circ$	$^3/_2$—$^1/_2$
4891,52	—	9,20	11,74	$4s\ ^2P$—$5p\ ^4P^\circ$	$^3/_2$—$^3/_2$
4852,70	8	9,20	11,76	$4s\ ^2P$—$5p\ ^4D^\circ$	$^3/_2$—$^5/_2$
4818,64	2	9,28	11,85	$4s\ ^2P$—$5p\ ^4S^\circ$	$^1/_2$—$^3/_2$
4818,42	3	9,20	11,77	$4s\ ^2P$—$5p\ ^4P^\circ$	$^3/_2$—$^1/_2$
4796,76	2	9,28	11,86	$4s\ ^2P$—$5p\ ^2D^\circ$	$^1/_2$—$^3/_2$
4740,71	10	9,20	11,82	$4s\ ^2P$—$5p\ ^4D^\circ$	$^3/_2$—$^3/_2$
4721,24	8	9,20	11,83	$4s\ ^2P$—$5p\ ^2D^\circ$	$^3/_2$—$^5/_2$
4691,53	12	9,20	11,84	$4s\ ^2P$—$5p\ ^4D^\circ$	$^3/_2$—$^1/_2$
4677,76	7	9,20	11,85	$4s\ ^2P$—$5p\ ^2S^\circ$	$^3/_2$—$^1/_2$
4674,40	2	9,20	11,85	$4s\ ^2P$—$5p\ ^4S^\circ$	$^3/_2$—$^3/_2$
4661,22	18	9,28	11,94	$4s\ ^2P$—$5p\ ^2P^\circ$	$^1/_2$—$^3/_2$
4654,05	10	9,20	11,86	$4s\ ^2P$—$5p\ ^2D^\circ$	$^3/_2$—$^3/_2$
4623,96	10	9,28	11,96	$4s\ ^2P$—$4p'\ ^2D^\circ$	$^1/_2$—$^3/_2$
4601,00	20	9,28	11,97	$4s\ ^2P$—$5p\ ^2P^\circ$	$^1/_2$—$^1/_2$
4580,47	3	8,98	11,69	$4s\ ^4P$—$4p'\ ^2P^\circ$	$^3/_2$—$^3/_2$
4578,17	4	9,03	11,74	$4s\ ^4P$—$5p\ ^4P^\circ$	$^1/_2$—$^3/_2$
4548,26	—	8,98	11,71	$4s\ ^4P$—$4p'\ ^2P^\circ$	$^3/_2$—$^1/_2$
4526,20	30	9,20	11,94	$4s\ ^2P$—$5p\ ^2P^\circ$	$^3/_2$—$^3/_2$
4491,05	10	9,20	11,96	$4s\ ^2P$—$4p'\ ^2D^\circ$	$^3/_2$—$^3/_2$
4475,31	15	8,98	11,76	$4s\ ^4P$—$5p\ ^4D^\circ$	$^3/_2$—$^5/_2$
4469,37	18	9,20	11,97	$4s\ ^2P$—$5p\ ^2P^\circ$	$^3/_2$—$^1/_2$
4446,11	4	8,98	11,77	$4s\ ^4P$—$5p\ ^4P^\circ$	$^3/_2$—$^1/_2$
4445,83	4	9,03	11,82	$4s\ ^4P$—$5p\ ^4D^\circ$	$^1/_2$—$^3/_2$
4438,48	20	8,92	11,71	$4s\ ^4P$—$5p\ ^4P^\circ$	$^5/_2$—$^5/_2$
4403,03	15	8,92	11,74	$4s\ ^4P$—$5p\ ^4P^\circ$	$^5/_2$—$^3/_2$

λ, Å	I	E_H, eV	E_B, eV	Transition	J
4402,58	4	9,03	11,84	$4s\,^4P-5p\,^4D^\circ$	$1/2-1/2$
4390,38	7	9,03	11,85	$4s\,^4P-5p\,^2S^\circ$	$1/2-1/2$
4389,76	25	8,92	11,74	$4s\,^4P-5p\,^4D^\circ$	$5/2-7/2$
4387,55	6	9,03	11,85	$4s\,^4P-5p\,^4S^\circ$	$1/2-3/2$
4379,90	20	8,98	11,82	$4s\,^4P-5p\,^4D^\circ$	$3/2-3/2$
4371,55	5	8,92	11,76	$4s\,^4P-5p\,^4D^\circ$	$5/2-5/2$
4369,52	15	9,03	11,86	$4s\,^4P-5p\,^2D^\circ$	$1/2-3/2$
4363,30	20	8,98	11,83	$4s\,^4P-5p\,^2D^\circ$	$3/2-5/2$
4337,80	1	8,98	11,84	$4s\,^4P-5p\,^4D^\circ$	$3/2-1/2$
4326,12	—	8,98	11,85	$4s\,^4P-5p\,^2S^\circ$	$3/2-1/2$
4323,35	20	8,98	11,85	$4s\,^4P-5p\,^4S^\circ$	$3/2-3/2$
4280,43	2	8,92	11,82	$4s\,^4P-5p\,^4D^\circ$	$5/2-3/2$
4264,59	5	8,92	11,83	$4s\,^4P-5p\,^2D^\circ$	$5/2-5/2$
4226,44	15	8,92	11,85	$4s\,^4P-5p\,^4S^\circ$	$5/2-3/2$
4209,68	12	8,92	11,86	$4s\,^4P-5p\,^2D^\circ$	$5/2-3/2$
4147,20	2	8,98	11,97	$4s\,^4P-5p\,^2P^\circ$	$3/2-1/2$
4139,00	1	9,28	12,28	$4s\,^2P-6p\,^4D^\circ$	$1/2-3/2$
4104,78	3	8,92	11,94	$4s\,^4P-5p\,^2P^\circ$	$5/2-3/2$
4032,14	5	9,20	12,28	$4s\,^2P-6p\,^4D^\circ$	$3/2-3/2$
4005,51	—	9,28	12,38	$4s\,^2P-6p\,^2P^\circ$	$1/2-3/2$
3992,81	2	—	—	—	—
3944,79	3	9,20	12,34	$4s\,^2P-1^\circ$	$3/2-3/2$
3768,05	—	8,98	12,28	$4s\,^4P-6p\,^4D^\circ$	$3/2-3/2$
3703,03	—	9,03	12,38	$4s\,^4P-6p\,^2P^\circ$	$1/2-3/2$
3694,15	—	8,92	12,28	$4s\,^4P-6p\,^4D^\circ$	$5/2-3/2$
3691,58	—	8,98	12,34	$4s\,^4P-1^\circ$	$3/2-3/2$
1396,527	8	0,11	8,98	$3p\,^2P^\circ-4s\,^4P$	$1/2-3/2$
1389,961	6	0,11	9,03	$3p\,^2P^\circ-4s\,^4P$	$1/2-1/2$
1389,688	6	0,00	8,92	$3p\,^2P^\circ-4s\,^4P$	$3/2-5/2$
1379,529	11	0,00	8,98	$3p\,^2P^\circ-4s\,^4P$	$3/2-3/2$
1373,118	4	0,00	9,03	$3p\,^2P^\circ-4s\,^4P$	$3/2-1/2$
1363,449	10	0,11	9,20	$3p^5\,^2P^\circ-4s\,^2P$	$1/2-3/2$
1351,657	10	0,11	9,28	$3p^5\,^2P^\circ-4s\,^2P$	$1/2-1/2$
1347,238	12	0,00	9,20	$3p^5\,^2P^\circ-4s\,^2P$	$3/2-3/2$
1335,723	9	0,00	9,28	$3p^5\,^2P^\circ-4s\,^2P$	$3/2-1/2$
1201,358	11	0,11	10,43	$3p^5\,^2P^\circ-4s'\,^2D$	$1/2-3/2$
1188,768	12	0,00	10,43	$3p^5\,^2P^\circ-4s'\,^2D$	$3/2-3/2,\ 5/2$

Cl II, ground state $1s^2\,2s^2\,2p^6\,3s^2\,3p^4\,^3P_2$
Ionization potential 192000 cm^{-1}; 23,80 eV

λ, Å	I	E_H, eV	E_B, eV	Transition	J
9483,00	2	18,73	20,04	$3d'\,^3D^\circ-4p''\,^3P$	3—2
8820,70	5				
8391,96	3	14,86	16,33	$3d\,^3D^\circ-4p\,^3P$	1—1
8382,76	5	14,86	16,34	$3d\,^3D^\circ-4p\,^3P$	1—0
8361,81	8	14,85	16,33	$3d\,^3D^\circ-4p\,^3P$	2—1
8360,63	15	14,85	16,34	$3d\,^3D^\circ-4p\,^3P$	3—2
8353,00	2	14,85	16,34	$3d\,^3D^\circ-4p\,^3P$	2—2
7644,80	4	18,74	20,36	$3d'\,^3D^\circ-1$	1—2
		19,78	21,40	$4p''\,^1D-4d'\,^3F^\circ$	2—3
7620,51	4	18,73	20,36	$3d'\,^3D^\circ-1$	3—2
7578,07	10	18,72	20,36	$3d'\,^3D^\circ-1$	2—2

λ, Å	I	E_H, eV	E_B, eV	Transition	J
7565,53	18				
7389,28	7	18,68	20,36	$3d'\ {}^3P^\circ—1$	2—2
7147,80	3	18,10	19,84	$3d''\ {}^3P^\circ—{}^3P$	2—1
7074,98	4	18,10	19,85	$3d''\ {}^3P^\circ—{}^3P$	2—2
6993,27	2	16,39	18,16	$3d'\ {}^3G^\circ—4p'\ {}^3D$	4—3
6952,13	25	17,90	19,68	$3d''\ {}^3F^\circ—4p''\ {}^3D$	2—1
6930,45	4	17,90	19,68	$3d''\ {}^3F^\circ—4p''\ {}^3D$	2—2
6850,21	40	17,87	19,68	$3d''\ {}^3F^\circ—4p''\ {}^3D$	3—2
6841,86	10	17,87	19,69	$3d''\ {}^3F^\circ—4p''\ {}^3D$	3—3
6831,62	30	17,19	19,00	$4s''\ {}^1P^\circ—4p'\ {}^1D$	1—2
6759,42	35	17,85	19,69	$3d''\ {}^3F^\circ—4p''\ {}^3D$	4—3
6713,43	40	16,39	18,23	$3d'\ {}^3G^\circ—4p'\ {}^3F$	3—2
6686,04	45	16,39	18,24	$3d'\ {}^3G^\circ—4p'\ {}^3F$	4—3
6681,03	15	16,39	18,24	$3d'\ {}^3G^\circ—4p'\ {}^3F$	3—3
6661,68	75	16,39	18,25	$3d'\ {}^3G^\circ—4p'\ {}^3F$	5—4
6653,75	25	16,39	18,25	$3d'\ {}^3G^\circ—4p'\ {}^3F$	4—4
6522,38	10	18,16	20,06	$4p'\ {}^3D—4d\ {}^3D^\circ$	3—3
6478,07	2	16,39	18,30	$3d'\ {}^3G^\circ—4p'\ {}^1F$	4—3
6475,38	2	18,16	20,07	$4p'\ {}^3D—4d\ {}^3D^\circ$	3—2
6465,32	3	18,14	20,06	$4p'\ {}^3D—4d\ {}^3D^\circ$	2—3
6419,25	8	18,14	20,07	$4p'\ {}^3D—4d\ {}^3D^\circ$	2—2
6417,59	2	18,14	20,07	$4p'\ {}^3D—4d\ {}^3D^\circ$	1—2
6399,41	10	18,10	20,04	$3d''\ {}^3P^\circ—4p''\ {}^3P$	2—2
6391,30	3	18,10	20,04	$3d''\ {}^3P^\circ—4p''\ {}^3P$	2—1
6385,51	2	18,14	20,08	$4p'\ {}^3D—4d\ {}^3D^\circ$	2—1
6384,13	5	18,14	20,08	$4p'\ {}^3D—4d\ {}^3D^\circ$	1—1
6243,00	2	13,96	15,95	$4s\ {}^3S^\circ—4p\ {}^5P$	1—1
6227,18	6	13,96	15,95	$4s\ {}^3S^\circ—4p\ {}^5P$	1—2
6094,65	100	16,00	18,03	$4s'\ {}^1D^\circ—4p'\ {}^1P$	2—1
5922,33	7	—	—	—	—
5790,50	25	16,00	18,14	$4s'\ {}^1D^\circ—4p'\ {}^3D$	2—1
5634,84	18	15,84	18,03	$3d'\ {}^1P^\circ—4p'\ {}^1P$	1—1
5568,81	15	19,00	21,23	$4p'\ {}^1D—5s'\ {}^1D^\circ$	2—2
5535,39	5	16,00	18,24	$4s'\ {}^1D^\circ—4p'\ {}^3F$	2—3
5457,47	30	13,68	15,95	$3d\ {}^5D^\circ—4p\ {}^5P$	0—1
5457,02	75	13,67	15,95	$3d\ {}^5D^\circ—4p\ {}^5P$	1—1
5456,27	50	13,67	15,95	$3d\ {}^5D^\circ—4p\ {}^5P$	2—1
5444,99	10	13,67	15,95	$3d\ {}^5D^\circ—4p\ {}^5P$	1—2
5444,25	60	13,67	15,95	$3d\ {}^5D^\circ—4p\ {}^5P$	2—2
5443,42	100	13,67	15,95	$3d\ {}^5D^\circ—4p\ {}^5P$	3—2
5424,36	25	13,67	15,96	$3d\ {}^5D^\circ—4p\ {}^5P$	2—3
5423,52	100	13,67	15,96	$3d\ {}^5D^\circ—4p\ {}^5P$	3—3
5423,25	150	13,67	15,96	$3d\ {}^5D^\circ—4p\ {}^5P$	4—3
5414,20	2	17,39	19,68	$3d''\ {}^1D^\circ—4p''\ {}^3D$	2—1
5392,12	100	16,00	18,30	$4s'\ {}^1D^\circ—4p'\ {}^1F$	2—3
5356,14	10	—	—	—	—
5338,92	5	15,71	18,03	$4s'\ {}^3D^\circ—4p'\ {}^1P$	2—1
5333,70	15	15,71	18,03	$4s'\ {}^3D^\circ—4p'\ {}^1P$	1—1
5285,48	30	16,34	18,68	$4p\ {}^3P—3d'\ {}^3P^\circ$	2—2
5249,22	3	16,34	18,70	$4p\ {}^3P—3d'\ {}^3P^\circ$	2—1
5245,69	4	16,33	18,70	$4p\ {}^3P—3d'\ {}^3P^\circ$	1—1
5221,34	75	13,96	16,33	$4s\ {}^3S^\circ—4p\ {}^3P$	1—1
5217,93	150	13,96	16,34	$4s\ {}^3S^\circ—4p\ {}^3P$	1—2
		13,96	16,34	$4s\ {}^3S^\circ—4p\ {}^3P$	1—0
5193,03	10	16,34	18,72	$4p\ {}^3P—3d'\ {}^3D^\circ$	2—2
5189,70	25	16,33	18,72	$4p\ {}^3P—3d'\ {}^3D^\circ$	1—2
5175,85	20	17,39	19,78	$3d''\ {}^1D^\circ—4p''\ {}^1D$	2—2
		19,00	21,40	$4p'\ {}^1D—4d'\ {}^3F^\circ$	2—2

λ, Å	I	E_H, eV	E_B, eV	Transition	J
5173,15	25	16,34	18,73	$4p\ ^3P$—$3d'\ ^3D^\circ$	2—3
5162,34	10	16,34	18,74	$4p\ ^3P$—$3d'\ ^3D^\circ$	0—1
5158,79	8	16,33	18,74	$4p\ ^3P$—$3d'\ ^3D^\circ$	1—1
5113,36	40	15,72	18,14	$4s'\ ^3D^\circ$—$4p'\ ^3D$	3—2
5104,08	25	15,71	18,14	$4s'\ ^3D^\circ$—$4p'\ ^3D$	2—1
5103,04	125	15,71	18,14	$4s'\ ^3D^\circ$—$4p'\ ^3D$	2—2
5099,30	100	15,71	18,14	$4s'\ ^3D^\circ$—$4p'\ ^3D$	1—1
5098,34	20	15,71	18,14	$4s'\ ^3D^\circ$—$4p'\ ^3D$	1—2
5078,25	150	15,72	18,16	$4s'\ ^3D^\circ$—$4p'\ ^3D$	3—3
5068,10	10	15,71	18,16	$4s'\ ^3D^\circ$—$4p'\ ^3D$	2—3
4995,52	60	15,68	18,16	$3d'\ ^3F^\circ$—$4p'\ ^3D$	4—3
4970,12	50	15,65	18,14	$3d'\ ^3F^\circ$—$4p'\ ^3D$	3—2
4943,24	15	17,28	19,78	$3d''\ ^1P^\circ$—$4p''\ ^1D$	1—2
4936,99	25	15,65	18,16	$3d'\ ^3F^\circ$—$4p'\ ^3D$	3—3
4931,76	2	15,72	18,23	$4s'\ ^3D^\circ$—$4p'\ ^3F$	3—2
4925,17	15	15,63	18,14	$3d'\ ^3F^\circ$—$4p'\ ^3D$	2—1
4924,83	10	17,09	19,61	$4s''\ ^3P^\circ$—$4p''\ ^3S$	2—1
4924,28	18	15,63	18,14	$3d'\ ^3F^\circ$—$4p'\ ^3D$	2—2
4922,14	20	15,71	18,23	$4s'\ ^3D^\circ$—$4p'\ ^3F$	2—2
4917,72	125	15,71	18,23	$4s'\ ^3D^\circ$—$4p'\ ^3F$	1—2
4914,32	12	15,72	18,24	$4s'\ ^3D^\circ$—$4p'\ ^3F$	3—3
4907,17	15	17,08	19,61	$4s''\ ^3P^\circ$—$4p''\ ^3S$	1—1
4904,76	135	15,71	18,24	$4s'\ ^3D^\circ$—$4p'\ ^3F$	2—3
4898,94	7	17,08	19,61	$4s''\ ^3P^\circ$—$4p''\ ^3S$	0—1
4896,77	200	15,72	18,25	$4s'\ ^3D^\circ$—$4p'\ ^3F$	3—4
4891,62	4	15,63	18,16	$3d'\ ^3F^\circ$—$4p'\ ^3D$	2—3
4877,70	5	18,60	21,14	$4p'\ ^3P$—$5s'\ ^3D^\circ$	0—1
4857,04	10	18,59	21,14	$4p'\ ^3P$—$5s'\ ^3D^\circ$	1—2
4847,07	4	17,48	20,04	$3d''\ ^3D^\circ$—$4p''\ ^3P$	2—2
4842,44	8	17,48	20,04	$3d''\ ^3D^\circ$—$4p''\ ^3P$	2—1
4836,79	20	15,68	18,24	$3d'\ ^3F^\circ$—$4p'\ ^3F$	4—3
4833,50	2	20,04	22,61	$4p''\ ^3P$—$5s''\ ^3P^\circ$	1—0
4829,23	3	20,04	22,61	$4p''\ ^3P$—$5s''\ ^3P^\circ$	0—1
4821,87	2	16,00	18,57	$4s'\ ^1D^\circ$—$4p'\ ^3P$	2—2
4820,95	4	{ 18,57	21,14	$4p'\ ^3P$—$5s'\ ^3D^\circ$	2—2
		20,04	22,61	$4p''\ ^3P$—$5s''\ ^3P^\circ$	2—1
4819,79	25	15,68	18,25	$3d'\ ^3F^\circ$—$4p'\ ^3F$	4—4
4819,46	200	13,37	15,95	$4s\ ^5S^\circ$—$4p\ ^5P$	2—1
4811,57	12	18,57	21,15	$4p'\ ^3P$—$5s'\ ^3D^\circ$	2—3
4810,06	225	13,37	15,95	$4s\ ^5S^\circ$—$4p\ ^5P$	2—2
4809,05	9	19,00	21,58	$4p'\ ^1D$—$4d'\ ^1F^\circ$	2—3
4807,68	5	20,04	22,62	$4p''\ ^3P$—$5s''\ ^3P^\circ$	1—2
4803,16	2	20,04	22,62	$4p''\ ^3P$—$5s''\ ^3P^\circ$	2—2
4798,40	15	15,65	18,23	$3d'\ ^3F^\circ$—$4p'\ ^3F$	3—2
4794,54	250	13,37	15,96	$4s\ ^5S^\circ$—$4p\ ^5P$	2—3
4792,04	12	15,71	18,30	$4s'\ ^3D^\circ$—$4p'\ ^1F$	2—3
4785,44	50	17,09	19,68	$4s''\ ^3P^\circ$—$4p''\ ^3D$	2—2
4781,82	50	15,65	18,24	$3d'\ ^3F^\circ$—$4p'\ ^3F$	3—3
4781,32	75	17,09	19,69	$4s''\ ^3P^\circ$—$4p''\ ^3D$	2—3
4778,93	45	17,08	19,68	$4s''\ ^3P^\circ$—$4p''\ ^3D$	1—1
4776,38	5	17,45	20,04	$3d''\ ^3D^\circ$—$4p''\ ^3P$	1—0
4771,66	20	17,19	19,78	$4s''\ ^1P^\circ$—$4p''\ ^1D$	1—2
4771,09	40	17,08	19,68	$4s''\ ^3P^\circ$—$4p''\ ^3D$	0—1
4768,68	150	17,08	19,68	$4s''\ ^3P^\circ$—$4p''\ ^3D$	1—2
4765,30	10	15,65	18,25	$3d'\ ^3F^\circ$—$4p'\ ^3F$	3—4
4755,64	50	15,63	18,23	$3d'\ ^3F^\circ$—$4p'\ ^3F$	2—2
4753,49	8	18,60	21,21	$4p'\ ^3P$—$5d\ ^3D^\circ$	0—1
4748,67	20	18,59	21,20	$4p'\ ^3P$—$5d\ ^3D^\circ$	1—2

λ, Å	I	E_H, eV	E_B, eV	Transition	J
4740,40	150	17,39	20,00	$3d''\,^1D^\circ$—$4p''\,^1P$	2—1
4739,42	10	15,63	18,24	$3d'\,^3F^\circ$—$4p'\,^3F$	2—3
4738,41	10	18,59	21,21	$4p'\,^3P$—$5d\,^3D^\circ$	1—1
4721,43	25	18,57	21,20	$4p'\,^3P$—$5d\,^3D^\circ$	2—3
4714,28	8	18,57	21,20	$4p'\,^3P$—$5d\,^3D^\circ$	2—2
4676,73	8	17,19	19,84	$4s''\,^1P^\circ$—^3P	1—1
		17,39	20,04	$3d''\,^1D^\circ$—$4p''\,^3P$	2—2
4624,36	6				
4592,29	2	17,08	19,78	$4s''\,^3P^\circ$—$4p''\,^1D$	1—2
4585,03	15	16,34	19,04	$4p\,^3P$—$3d'\,^3S^\circ$	2—1
4584,28	20				
4582,40	8	16,33	19,04	$4p\,^3P$—$3d'\,^3S^\circ$	1—1
4572,13	100	16,34	19,05	$4p\,^3P$—$5s\,^3S^\circ$	2—1
		16,34	19,05	$4p\,^3P$—$5s\,^3S^\circ$	0—1
4569,42	50	16,33	19,05	$4p\,^3P$—$5s\,^3S^\circ$	1—1
4544,48	10	17,28	20,00	$3d''\,^1P^\circ$—$4p''\,^1P$	1—1
4540,29	6	—	—	—	—
4539,25	6	—	—	—	—
4536,78	20	17,08	19,82	$4s''\,^3P^\circ$—^3P	1—0
4534,34	5	—	—	—	—
4519,19	18	17,09	19,84	$4s''\,^3P^\circ$—^3P	2—1
4504,27	20	17,08	19,84	$4s''\,^3P^\circ$—^3P	1—1
4497,30	18	19,85	22,61	^3P—$5s''\,^3P^\circ$	2—1
		17,08	19,84	$4s''\,^3P^\circ$—^3P	0—1
4490,00	50	17,09	19,85	$4s''\,^3P^\circ$—^3P	2—2
4482,02	10	19,85	22,62	^3P—$5s''\,^3P^\circ$	2—2
4475,28	20	17,08	19,85	$4s''\,^3P^\circ$—^3P	1—2
		19,84	22,61	^3P—$5s''\,^3P^\circ$	1—0
4468,48	2	19,84	22,61	^3P—$5s''\,^3P^\circ$	1—1
4453,32	3	19,84	22,62	^3P—$5s''\,^3P^\circ$	1—2
4436,96	3	19,82	22,61	^3P—$5s''\,^3P^\circ$	0—1
4399,14	15	17,19	20,00	$4s''\,^1P^\circ$—$4p''\,^1P$	1—1
4372,91	80	17,52	20,36	$3d''\,^3D^\circ$—1	3—2
4343,62	100	15,72	18,57	$4s'\,^3D^\circ$—$4p'\,^3P$	3—2
4336,26	45	15,71	18,57	$4s'\,^3D^\circ$—$4p'\,^3P$	2—2
4332,80	9	15,71	18,57	$4s'\,^3D^\circ$—$4p'\,^3P$	1—2
4309,06	50	17,48	20,36	$3d''\,^3D^\circ$—1	2—2
4307,42	75	15,71	18,59	$4s'\,^3D^\circ$—$4p'\,^3P$	2—1
4304,07	40	15,71	18,59	$4s'\,^3D^\circ$—$4p'\,^3P$	1—1
4291,76	50	15,71	18,60	$4s'\,^3D^\circ$—$4p'\,^3P$	1—0
4276,51	30	18,25	21,15	$4p'\,^3F$—$5s'\,^3D^\circ$	4—3
4270,61	25	18,24	21,14	$4p'\,^3F$—$5s'\,^3D^\circ$	3—2
4261,22	20	18,23	21,14	$4p'\,^3F$—$5s'\,^3D^\circ$	2—1
4259,52	35	17,45	20,36	$3d''\,^3D^\circ$—1	1—2
		17,09	20,00	$4s''\,^3P^\circ$—$4p''\,^1P$	2—1
4257,54	4	18,23	21,14	$4p'\,^3F$—$5s'\,^3D^\circ$	2—2
4253,51	75	15,96	18,87	$4p\,^5P$—$5s\,^5S^\circ$	3—2
4241,38	60	15,95	18,87	$4p\,^5P$—$5s\,^5S^\circ$	2—2
4235,49	25	19,68	22,61	$4p''\,^3D$—$5s''\,^3P^\circ$	2—1
		18,30	21,23	$4p'\,^1F$—$5s'\,^1D^\circ$	3—2
4234,09	50	15,95	18,87	$4p\,^5P$—$5s\,^5S^\circ$	1—2
4233,60	4	19,68	22,61	$4p''\,^3D$—$5s''\,^3P^\circ$	1—0
4227,37	4	19,68	22,61	$4p''\,^3D$—$5s''\,^3P^\circ$	1—1
4224,92	15	19,69	22,62	$4p''\,^3D$—$5s''\,^3P^\circ$	3—2
4221,80	3	19,68	22,62	$4p''\,^3D$—$5s''\,^3P^\circ$	2—2
4218,76	4	—	—	—	—
4208,03	30	17,09	20,04	$4s''\,^3P^\circ$—$4p''\,^3P$	2—2
4205,07	10	18,25	21,20	$4p'\,^3F$—$5d\,^3D^\circ$	4—3

λ, Å	I	$E_{н}$, eV	$E_{в}$, eV	Transition	J
4204,54	18	17,09	20,04	$4s'' \, {}^3P^\circ - 4p'' \, {}^3P$	2—1
4195,11	18	17,08	20,04	$4s'' \, {}^3P^\circ - 4p'' \, {}^3P$	1—2
4192,24	6	18,24	21,20	$4p' \, {}^3F - 5d \, {}^3D^\circ$	3—3
4191,59	15	17,08	20,04	$4s'' \, {}^3P^\circ - 4p'' \, {}^3P$	1—1
4188,82	15	17,08	20,04	$4s'' \, {}^3P^\circ - 4p'' \, {}^3P$	1—0
4187,06	2	13,37	16,33	$4s \, {}^5S^\circ - 4p \, {}^3P$	2—1
4186,63	5	18,24	21,20	$4p' \, {}^3F - 5d \, {}^3D^\circ$	3—2
4185,61	20	17,08	20,04	$4s'' \, {}^3P^\circ - 4p'' \, {}^3P$	0—1
4184,89	7	13,37	16,34	$4s \, {}^5S^\circ - 4p \, {}^3P$	2—2
4181,05	4		—	—	—
4179,61	2	18,23	21,20	$4p' \, {}^3F - 5d \, {}^3D^\circ$	2—3
4170,66	8	15,06	18,03	$3d' \, {}^1D^\circ - 4p' \, {}^1P$	2—1
4166,10	4	18,23	21,21	$4p' \, {}^3F - 5d \, {}^3D^\circ$	2—1
4157,98	5	—	—	—	—
4157,82	25	—	—	—	—
4156,15	7	—	—		—
4153,98	2	18,16	21,14	$4p' \, {}^3D - 5s' \, {}^3D^\circ$	3—2
4147,09	30	18,16	21,15	$4p' \, {}^3D - 5s' \, {}^3D^\circ$	3—3
4143,04	5	20,04	23,03	$4p'' \, {}^3P - 4d'' \, {}^3P^\circ$	2—2
4134,31	4	18,14	21,14	$4p' \, {}^3D - 5s' \, {}^3D^\circ$	2—1
4133,66	20	18,14	21,14	$4p' \, {}^3D - 5s' \, {}^3D^\circ$	1—1
4132,48	200	16,00	19,00	$4s' \, {}^1D^\circ - 4p' \, {}^1D$	2—2
4130,86	25	18,14	21,14	$4p' \, {}^3D - 5s' \, {}^3D^\circ$	2—2
4130,22	8	18,14	21,14	$4p' \, {}^3D - 5s' \, {}^3D^\circ$	1—2
4125,96	3	20,04	23,04	$4p'' \, {}^3P - 4d'' \, {}^3D^\circ$	2—3
4124,00	12	18,14	21,15	$4p' \, {}^3D - 5s' \, {}^3D^\circ$	2—3
4118,84	4	19,61	22,62	$4p'' \, {}^3S - 5s'' \, {}^3P^\circ$	1—2
4079,88	15	18,16	21,20	$4p' \, {}^3D - 5d \, {}^3D^\circ$	3—3
4077,93	4	19,85	22,89	${}^3P - 4d'' \, {}^3F^\circ$	2—2
4074,51	6	18,16	21,20	$4p' \, {}^3D - 5d \, {}^3D^\circ$	3—2
4057,52	6	18,14	21,20	$4p' \, {}^3D - 5d \, {}^3D^\circ$	2—3
4055,46	4	—	—		—
4054,18	9	19,84	22,89	${}^3P - 4d'' \, {}^3F^\circ$	1—2
4052,22	12	18,14	21,20	$4p' \, {}^3D - 5d \, {}^3D^\circ$	2—2
4051,58	4	18,14	21,20	$4p' \, {}^3D - 5d \, {}^3D^\circ$	1—2
4044,65	4	18,14	21,21	$4p' \, {}^3D - 5d \, {}^3D^\circ$	2—1
4044,09	9	18,14	21,21	$4p' \, {}^3D - 5d \, {}^3D^\circ$	1—1
4040,64	9	18,16	21,23	$4p' \, {}^3D - 5s' \, {}^1D^\circ$	3—2
4036,53	10	18,60	21,67	$4p' \, {}^3P - 4d' \, {}^3D^\circ$	0—1
4025,68	7	18,59	21,67	$4p' \, {}^3P - 4d' \, {}^3D^\circ$	1—1
4020,06	15	18,59	21,67	$4p' \, {}^3P - 4d' \, {}^3D^\circ$	1—2
4018,24	3	18,14	21,23	$4p' \, {}^3D - 5s' \, {}^1D^\circ$	1—2
3995,24	6	18,57	21,67	$4p' \, {}^3P - 4d' \, {}^3D^\circ$	2—2
3990,19	20	18,57	21,68	$4p' \, {}^3P - 4d' \, {}^3D^\circ$	2—3
3988,17	4	18,03	21,14	$4p' \, {}^1P - 5s' \, {}^3D^\circ$	1—2
3981,94	15	—	—	—	—
3971,18	7	—	—	—	—
3954,21	20	19,00	22,13	$4p' \, {}^1D - 4d' \, {}^1D^\circ$	2—2
3949,96	10	16,34	19,47	$4p \, {}^3P - 3p^5 \, 4s \, {}^3P^\circ$	2—2
3928,63	5	18,24	21,40	$4p' \, {}^3F - 4d' \, {}^3F^\circ$	3—2
3927,88	6	18,25	21,40	$4p' \, {}^3F - 4d' \, {}^3F^\circ$	4—3
3921,75	3	15,08	18,24	$3d' \, {}^1F^\circ - 4p' \, {}^3F$	3—3
3917,57	18	18,23	21,40	$4p' \, {}^3F - 4d' \, {}^3F^\circ$	2—2
3916,70	20	18,24	21,40	$4p' \, {}^3F - 4d' \, {}^3F^\circ$	3—3
3915,82	3	15,84	19,00	$3d' \, {}^1P^\circ - 4p' \, {}^1D$	1—2
3913,92	30	18,25	21,42	$4p' \, {}^3F - 4d' \, {}^3F^\circ$	4—4
3910,60	2	15,08	18,25	$3d' \, {}^1F^\circ - 4p' \, {}^3F$	3—4
3905,80	4	18,23	21,40	$4p' \, {}^3F - 4d' \, {}^3F^\circ$	2—3

λ, Å	I	E_H, eV	E_B, eV	Transition	J
3902,84	9	18,24	21,42	$4p'\,^3F-4d'\,^3F^\circ$	3—4
3901,89	5	19,85	23,03	$^3P-4d''\,^3P^\circ$	2—2
3901,12	4	14,86	18,03	$3d\,^3D^\circ-4p'\,^1P$	1—1
3899,27	4	18,30	21,48	$4p'\,^1F-4d'\,^3G^\circ$	3—4
3894,55	2	14,85	18,03	$3d\,^3D^\circ-4p'\,^1P$	2—1
3886,63	4	19,85	23,04	$^3P-4d''\,^3D^\circ$	2—3
3883,80	12	18,03	21,23	$4p'\,^1P-5s'\,^1D^\circ$	1—2
3868,62	40	19,69	22,89	$4p''\,^3D-4d''\,^3F^\circ$	3—4
3864,60	15	19,69	22,89	$4p''\,^3D-4d''\,^3F^\circ$	3—3
3861,95	20	19,68	22,89	$4p''\,^3D-4d''\,^3F^\circ$	2—3
3861,40	50	15,96	19,17	$4p\,^5P-4d\,^5D^\circ$	3—2
3860,98	100	15,96	19,17	$4p\,^5P-4d\,^5D^\circ$	3—3
3860,80	150	15,96	19,17	$4p\,^5P-4d\,^5D^\circ$	3—4
3860,05	2	16,34	19,55	$4p\,^3P-3p^54s\,^3P^\circ$	2—1
		16,34	19,55	$4p\,^3P-3p^54s\,^3P^\circ$	0—1
3859,17	7	19,84	23,05	$^3P-4d''\,^3P^\circ$	1—1
3854,75	15	19,68	22,89	$4p''\,^3D-4d''\,^3F^\circ$	1—2
3851,69	30	15,95	19,17	$4p\,^5P-4d\,^5D^\circ$	2—1
3851,38	75	15,95	19,17	$4p\,^5P-4d\,^5D^\circ$	2—2
3850,97	100	15,95	19,17	$4p\,^5P-4d\,^5D^\circ$	2—3
3849,33	3	15,08	18,30	$3d'\,^1F^\circ-4p'\,^1F$	3—3
3845,84	30	15,95	19,17	$4p\,^5P-4d\,^5D^\circ$	1—0
3845,69	75	15,95	19,17	$4p\,^5P-4d\,^5D^\circ$	1—1
3845,42	50	15,95	19,17	$4p\,^5P-4d\,^5D^\circ$	1—2
3843,26	100	17,28	20,50	$3d''\,^1P^\circ-4p''\,^1S$	1—0
3838,37	20	18,25	21,48	$4p'\,^3F-4d'\,^3G^\circ$	4—4
3833,40	200	18,25	21,48	$4p'\,^3F-4d'\,^3G^\circ$	4—5
3830,80	15	18,24	21,48	$4p'\,^3F-4d'\,^3G^\circ$	3—3
3829,27	15	15,06	18,30	$3d'\,^1D^\circ-4p'\,^1F$	2—3
3827,62	150	18,24	21,48	$4p'\,^3F-4d'\,^3G^\circ$	3—4
3820,25	100	18,23	21,48	$4p'\,^3F-4d'\,^3G^\circ$	2—3
3818,40	30	18,16	21,40	$4p'\,^3D-4d'\,^3F^\circ$	3—3
3810,10	30	18,14	21,40	$4p'\,^3D-4d'\,^3F^\circ$	2—2
3809,51	40	18,14	21,40	$4p'\,^3D-4d'\,^3F^\circ$	1—2
3805,24	75	18,16	21,42	$4p'\,^3D-4d'\,^3F^\circ$	3—4
3798,80	50	18,14	21,40	$4p'\,^3D-4d'\,^3F^\circ$	2—3
3793,75	25	—	—	—	—
3781,23	30	18,30	21,58	$4p'\,^1F-4d'\,^1F^\circ$	3—3
3776,20	4	15,72	19,00	$4s'\,^3D^\circ-4p'\,^1D$	3—2
3774,25	25	14,86	18,14	$3d\,^3D^\circ-4p'\,^3D$	1—1
3773,68	20	14,86	18,14	$3d\,^3D^\circ-4p'\,^3D$	1—2
3769,13	20	14,85	18,14	$3d\,^3D^\circ-4p'\,^3D$	3—2
3768,13	18	14,85	18,14	$3d\,^3D^\circ-4p'\,^3D$	2—1
3767,57	30	14,85	18,14	$3d\,^3D^\circ-4p'\,^3D$	2—2
3756,92	2	19,00	22,30	$4p'\,^1D-4d'\,^1P^\circ$	2—1
3750,00	30	14,85	18,16	$3d\,^3D^\circ-4p'\,^3D$	3—3
3748,46	15	14,85	18,16	$3d\,^3D^\circ-4p'\,^3D$	2—3
3738,76	4	17,19	20,50	$4s''\,^1P^\circ-4p''\,^1S$	1—0
3733,73	10	18,16	21,48	$4p'\,^3D-4d'\,^3G^\circ$	3—4
3717,94	15	18,14	21,48	$4p'\,^3D-4d'\,^3G^\circ$	2—3
3705,54	2	19,69	23,03	$4p''\,^3D-4d''\,^3P^\circ$	3—2
3691,88	5	19,69	23,04	$4p''\,^3D-4d''\,^3D^\circ$	3—3
3688,44	15	18,03	21,40	$4p'\,^1P-4d'\,^3F^\circ$	1—2
3673,83	18	14,86	18,23	$3d\,^3D^\circ-4p'\,^3F$	1—2
3669,46	2	14,85	18,23	$3d\,^3D^\circ-4p'\,^3F$	3—2
3668,03	20	14,85	18,23	$3d\,^3D^\circ-4p'\,^3F$	2—2
3659,84	18	14,85	18,24	$3d\,^3D^\circ-4p'\,^3F$	3—3
3658,38	20	14,85	18,24	$3d\,^3D^\circ-4p'\,^3F$	2—3

λ, Å	I	E_H, eV	E_B, eV	Transition	J
3650,13	30	14,85	18,25	$3d\ ^3D^\circ—4p'\ ^3F$	3—4
3648,07	10	18,60	22,00	$4p'\ ^3P—4d'\ ^3S^\circ$	0—1
3639,19	18	18,59	22,00	$4p'\ ^3P—4d'\ ^3S^\circ$	1—1
3623,79	9	19,61	23,03	$4p''\ ^3S—4d''\ ^3P^\circ$	1—2
3618,88	15	18,57	22,00	$4p'\ ^3P—4d'\ ^3S^\circ$	2—1
3615,09	10	18,25	21,68	$4p'\ ^3F—4d'\ ^3D^\circ$	4—3
3610,07	12	—	—	—	—
3609,75	4	18,24	21,67	$4p'\ ^3F—4d'\ ^3D^\circ$	3—2
3605,61	7	18,24	21,68	$4p'\ ^3F—4d'\ ^3D^\circ$	3—3
3605,39	5	19,61	23,05	$4p''\ ^3S—4d''\ ^3P^\circ$	1—1
3604,92	3	18,23	21,67	$4p'\ ^3F—4d'\ ^3D^\circ$	2—1
3604,51	15	18,60	22,04	$4p'\ ^3P—4d'\ ^3P^\circ$	0—1
3603,72	10	18,59	22,03	$4p'\ ^3P—4d'\ ^3P^\circ$	1—0
3600,42	5	18,23	21,67	$4p'\ ^3F—4d'\ ^3D^\circ$	2—2
3595,82	8	18,59	22,04	$4p'\ ^3P—4d'\ ^3P^\circ$	1—1
3587,78	12	18,59	22,05	$4p'\ ^3P—4d'\ ^3P^\circ$	1—2
3576,00	15	18,57	22,04	$4p'\ ^3P—4d'\ ^3P^\circ$	2—1
3568,04	20	18,57	22,05	$4p'\ ^3P—4d'\ ^3P^\circ$	2—2
3526,13	30	18,16	21,67	$4p'\ ^3D—4d'\ ^3D^\circ$	3—2
3522,14	40	18,16	21,68	$4p'\ ^3D—4d'\ ^3D^\circ$	3—3
3513,69	12	18,14	21,67	$4p'\ ^3D—4d'\ ^3D^\circ$	2—1
3513,22	35	18,14	21,67	$4p'\ ^3D—4d'\ ^3D^\circ$	1—1
3509,39	40	18,14	21,67	$4p'\ ^3D—4d'\ ^3D^\circ$	2—2
3508,94	12	18,14	21,67	$4p'\ ^3D—4d'\ ^3D^\circ$	1—2
3505,44	12	18,14	21,68	$4p'\ ^3D—4d'\ ^3D^\circ$	2—3
3479,82	30	—	—	—	—
3448,14	4	—	—	—	—
3409,92	5	18,03	21,67	$4p'\ ^1P—4d'\ ^3D^\circ$	1—1
3405,89	3	18,03	21,67	$4p'\ ^1P—4d'\ ^3D^\circ$	1—2
3353,39	125	14,34	18,03	$3p^5\ ^1P^\circ—4p'\ ^1P$	1—1
3350,07	4	—	—	—	—
3337,20	3	14,86	18,57	$3d\ ^3D^\circ—4p'\ ^3P$	1—2
3333,64	40	14,85	18,57	$3d\ ^3D^\circ—4p'\ ^3P$	3—2
3332,42	15	14,85	18,57	$3d\ ^3D^\circ—4p'\ ^3P$	2—2
3329,12	150	16,34	20,06	$4p\ ^3P—4d\ ^3D^\circ$	2—3
3320,14	30	14,86	18,59	$3d\ ^3D^\circ—4p'\ ^3P$	1—1
3316,86	50	16,34	20,07	$4p\ ^3P—4d\ ^3D^\circ$	2—2
3315,44	100	16,33	20,07	$4p\ ^3P—4d\ ^3D^\circ$	1—2
		14,85	18,59	$3d\ ^3D^\circ—4p'\ ^3P$	2—1
3312,78	15	14,86	18,60	$3d\ ^3D^\circ—4p'\ ^3P$	1—0
3307,90	50	16,34	20,08	$4p\ ^3P—4d\ ^3D^\circ$	0—1
3306,45	40	16,33	20,08	$4p\ ^3P—4d\ ^3D^\circ$	1—1
3276,81	40	16,00	19,78	$4s'\ ^1D^\circ—4p''\ ^1D$	2—2
3231,75	12	18,30	22,13	$4p'\ ^1F—4d'\ ^1D^\circ$	3—2
3222,55	7	—	—	—	—
3203,05	20	—	—	—	—
3202,12	6	—	—	—	—
3189,04	20	18,16	22,05	$4p'\ ^3D—4d'\ ^3P^\circ$	3—2
3187,42	5	18,14	22,03	$4p'\ ^3D—4d'\ ^3P^\circ$	1—0
3181,70	7	18,14	22,04	$4p'\ ^3D—4d'\ ^3P^\circ$	2—1
3181,26	5	18,14	22,04	$4p'\ ^3D—4d'\ ^3P^\circ$	1—1
3180,43	7	—	—	—	—
3176,95	5	—	—	—	—
3175,30	6	18,14	22,05	$4p'\ ^3D—4d'\ ^3P^\circ$	2—2
3173,66	20	—	—	—	—
3172,56	6	—	—	—	—
3170,23	15	—	—	—	—
3169,45	7	—	—	—	—
3161,44	20	15,08	19,00	$3d'\ ^1F^\circ—4p'\ ^1D$	3—2

λ, Å	I	E_H, eV	E_B, eV	Transition	J
3160,52	10	—	—	—	—
3147,86	20	15,06	19,00	$3d'\,^1D^\circ-4p'\,^1D$	2—2
3125,96	5	15,71	19,68	$4s'\,^3D^\circ-4p''\,^3D$	2—1
3125,44	6	15,72	19,68	$4s'\,^3D^\circ-4p''\,^3D$	3—2
3124,28	6	15,71	19,68	$4s'\,^3D^\circ-4p''\,^3D$	1—1
3123,72	15	15,72	19,69	$4s'\,^3D^\circ-4p''\,^3D$	3—3
3121,62	10	15,71	19,68	$4s'\,^3D^\circ-4p''\,^3D$	2—2
3119,82	12	15,71	19,69	$4s'\,^3D^\circ-4p''\,^3D$	2—3
		15,71	19,68	$4s'\,^3D^\circ-4p''\,^3D$	1—2
3096,72	25	16,00	20,00	$4s'\,^1D^\circ-4p''\,^1P$	2—1
3092,90	8	—	—	—	—
3092,22	50	15,68	19,69	$3d'\,^3F^\circ-4p''\,^3D$	4—3
3071,35	40	15,65	19,68	$3d'\,^3F^\circ-4p''\,^3D$	3—2
3069,66	5	15,65	19,69	$3d'\,^3F^\circ-4p''\,^3D$	3—3
3058,00	40	15,63	19,68	$3d'\,^3F^\circ-4p''\,^3D$	2—1
3053,74	10	15,63	19,68	$3d'\,^3F^\circ-4p''\,^3D$	2—2
3045,00	10	15,71	19,78	$4s'\,^3D^\circ-4p''\,^1D$	2—2
3037,98	35	17,52	21,60	$3d''\,^3D^\circ-2$	3—2
3022,93	30	18,03	22,13	$4p'\,^1P-4d'\,^1D^\circ$	1—2
3018,82	12	15,71	19,82	$4s'\,^3D^\circ-{}^3P$	1—0
3006,98	20	17,48	21,60	$3d''\,^3D^\circ-2$	2—2
3006,05	20	15,71	19,84	$4s'\,^3D^\circ-{}^3P$	2—1
3004,39	10	15,71	19,84	$4s'\,^3D^\circ-{}^3P$	1—1
2996,63	40	15,72	19,85	$4s'\,^3D^\circ-{}^3P$	3—2
2993,09	8	15,71	19,85	$4s'\,^3D^\circ-{}^3P$	2—2
2982,78	18	17,45	21,60	$3d''\,^3D^\circ-2$	1—2
2980,90	4	18,14	22,30	$4p'\,^3D-4d'\,^1P^\circ$	1—1
2980,47	2	15,63	19,78	$3d'\,^3F^\circ-4p''\,^1D$	2—2
2978,48	7	—	—	—	—
2973,46	2	15,84	20,00	$3d'\,^1P^\circ-4p''\,^1P$	1—1
2972,63	5	—	—	—	—
2964,21	2	13,96	18,14	$4s\,^3S^\circ-4p'\,^3D$	1—2
2950,35	5	19,00	23,20	$4p'\,^1D-6s'\,^1D^\circ$	2—2
2934,60	5	—	—	—	—
2912,06	15	—	—	—	—
2906,25	20	18,03	22,30	$4p'\,^1P-4d'\,^1P^\circ$	1—1
2902,45	4	—	—	—	—
2887,41	4	11,65	15,95	$3p^5\,^3P^\circ-4p\,^5P$	1—1
		15,71	20,00	$4s'\,^3D^\circ-4p''\,^1P$	1—1
2884,01	2	11,65	15,95	$3p^5\,^3P^\circ-4p\,^5P$	1—2
2876,42	5	—	—	—	—
2868,41	10	15,72	20,04	$4s'\,^3D^\circ-4p''\,^3P$	3—2
2865,21	4	15,71	20,04	$4s'\,^3D^\circ-4p''\,^3P$	2—2
2863,55	7	15,71	20,04	$4s'\,^3D^\circ-4p''\,^3P$	2—1
2862,06	5	15,71	20,04	$4s'\,^3D^\circ-4p''\,^3P$	1—1
2860,71	5	15,71	20,04	$4s'\,^3D^\circ-4p''\,^3P$	1—0
2844,28	4	—	—	—	—
2835,59	3	11,58	15,95	$3p^5\,^3P^\circ-4p\,^5P$	2—1
2832,33	4	11,58	15,95	$3p^5\,^3P^\circ-4p\,^5P$	2—2
2800,27	4	—	—	—	—
2799,60	4	—	—	—	—
2763,88	10	—	—	—	—
2758,69	5	—	—	—	—
2754,10	25	—	—	—	—
2751,52	5	—	—	—	—
2719,61	4	13,67	18,23	$3d\,^5D^\circ-4p'\,^3F$	3—2
2714,38	8	13,67	18,24	$3d\,^5D^\circ-4p'\,^3F$	3—3

λ, Å	I	E_H, eV	E_B, eV	Transition	J
2712,77	4	—	—	—	—
2709,82	2	18,59	23,16	$4p'\ ^3P$—$6s'\ ^3D^\circ$	1—1
2709,60	4	—	—	—	—
2709,03	10	13,67	18,25	$3d\ ^5D^\circ$—$4p'\ ^3F$	4—4
2706,76	4	—	—	—	—
2698,56	2	18,57	23,16	$4p'\ ^3P$—$6s'\ ^3D^\circ$	2—1
2694,63	3	18,57	23,17	$4p'\ ^3P$—$6s'\ ^3D^\circ$	2—3
2689,39	6	—	—	—	—
2688,04	150	13,96	18,57	$4s\ ^3S^\circ$—$4p'\ ^3P$	1—2
2679,37	5	—	—	—	—
2676,95	100	13,96	18,59	$4s\ ^3S^\circ$—$4p'\ ^3P$	1—1
2672,19	50	13,96	18,60	$4s\ ^3S^\circ$—$4p'\ ^3P$	1—0
2671,43	6	11,70	16,33	$3p^5\ ^3P^\circ$—$4p\ ^3P$	0—1
2667,36	40	16,34	20,98	$4p\ ^3P$—$6s\ ^3S^\circ$	2—1
		16,34	20,98	$4p\ ^3P$—$6s\ ^3S^\circ$	0—1
2666,46	20	16,33	20,98	$4p\ ^3P$—$6s\ ^3S^\circ$	1—1
2658,74	100	14,34	19,00	$3p^5\ ^1P^\circ$—$4p'\ ^1D$	1—2
2648,19	10	—	—	—	—
2647,79	5	11,65	16,33	$3p^5\ ^3P^\circ$—$4p\ ^3P$	1—1
2646,88	25	11,65	16,34	$3p^5\ ^3P^\circ$—$4p\ ^3P$	1—2
		11,65	16,34	$3p^5\ ^3P^\circ$—$4p\ ^3P$	1—0
2642,28	4	—	—	—	—
2634,95	12	15,08	19,78	$3d'\ ^1F^\circ$—$4p''\ ^1D$	3—2
2631,33	2	15,65	20,36	$3d'\ ^3F^\circ$—1	3—2
2630,20	4	—	—	—	—
2621,87	4	—	—	—	—
2619,80	4	18,16	22,89	$4p'\ ^3D$—$4d''\ ^3F^\circ$	3—4
2615,13	10	—	—	—	—
2614,65	5	—	—	—	—
2608,72	2	18,14	22,89	$4p'\ ^3D$—$4d''\ ^3F^\circ$	2—3
2608,24	2	18,14	22,89	$4p'\ ^3D$—$4d''\ ^3F^\circ$	1—2
2605,67	5	—	—	—	—
2604,18	8	11,58	16,33	$3p^5\ ^3P^\circ$—$4p\ ^3P$	2—1
2603,36	10	11,58	16,34	$3p^5\ ^3P^\circ$—$4p\ ^3P$	2—2
2580,40	4	—	—	—	—
2571,10	8	14,86	19,68	$3d\ ^3D^\circ$—$4p''\ ^3D$	1—1
2568,25	3	14,85	19,68	$3d\ ^3D^\circ$—$4p''\ ^3D$	2—1
2568,13	4	14,86	19,68	$3d\ ^3D^\circ$—$4p''\ ^3D$	1—2
2566,01	5	14,85	19,68	$3d\ ^3D^\circ$—$4p''\ ^3D$	3—2
2565,29	15	14,85	19,68	$3d\ ^3D^\circ$—$4p''\ ^3D$	2—2
2564,84	20	14,85	19,69	$3d\ ^3D^\circ$—$4p''\ ^3D$	3—3
2564,13	6	14,85	19,69	$3d\ ^3D^\circ$—$4p''\ ^3D$	2—3
2549,85	50	16,34	21,20	$4p\ ^3P$—$5d\ ^3D^\circ$	2—3
2547,76	12	16,34	21,20	$4p\ ^3P$—$5d\ ^3D^\circ$	2—2
2546,94	20	16,33	21,20	$4p\ ^3P$—$5d\ ^3D^\circ$	1—2
2544,84	15	16,34	21,21	$4p\ ^3P$—$5d\ ^3D^\circ$	0—1
2543,98	10	16,33	21,21	$4p\ ^3P$—$5d\ ^3D^\circ$	1—1
2518,15	4	18,25	23,17	$4p'\ ^3F$—$6s'\ ^3D^\circ$	4—3
2515,92	3	18,24	23,17	$4p'\ ^3F$—$6s'\ ^3D^\circ$	3—2
2514,01	3	14,85	19,78	$3d\ ^3D^\circ$—$4p''\ 'D$	3—2
2512,41	2	18,23	23,16	$4p'\ ^3F$—$6s'\ ^3D^\circ$	2—1
2511,33	3	18,23	23,17	$4p'\ ^3F$—$6s'\ ^3D^\circ$	2—2
2502,75	40	15,96	20,91	$4p\ ^5P$—$6s\ ^5S^\circ$	3—2
2498,53	30	15,95	20,91	$4p\ ^5P$—$6s\ ^5S^\circ$	2—2
2496,04	20	15,95	20,91	$4p\ ^5P$—$6s\ ^5S^\circ$	1—2
2472,69	3	18,16	23,17	$4p'\ ^3D$—$6s'\ ^3D^\circ$	3—3
2466,72	2	18.14	23,17	$4p'\ ^3D$—$6s'\ ^3D^\circ$	2—2
2459,86	10	—	—	—	—
2452,30	10	—	—	—	—

λ, Å	I	E_H, eV	E_B eV	Transition	J
2445,34	20	—	—	—	—
2444,12	7	—	—	—	—
2440,49	4	—	—	—	—
2440,33	5	—	—	—	—
2434,10	50	15,96 15,96	21,05 21,05	$4p\,^5P-5d\,^5D^\circ$ $4p\,^5P-5d\,^5D^\circ$	3—4 3—3
2430,16	30	15,95 15,95	21,05 21,05	$4p\,^5P-5d\,^5D^\circ$ $4p\,^5P-5d\,^5D^\circ$	2—3 2—2
2428,02	10	—	—	—	—
2427,79	20	15,95	21,05	$4p\,^5P-5d^5D^\circ$	1—2
2424,01	10	—	—	—	—
2419,85	4	—	—	—	—
2412,48	10	—	—	—	—
2407,10	5	—	—	—	—
2404,59	5	—	—	—	—
2403,87	3	—	—	—	—
2399,85	3	—	—	—	—
2398,91	2	18,03	23,20	$4p'\,^1P-6s'\,^1D^\circ$	1—2
2340,60	2	15,06	20,36	$3d'\,^1D^\circ-1$	2—2
2323,02	4	—	—	—	—
2322,00	1	16,34	21,67	$4p\,^3P-4d'\,^3D^\circ$	2—2
2321,28	1	16,33	21,67	$4p\,^3P-4d'\,^3D^\circ$	1—2
2320,25	2	16,34	21,68	$4p\,^3P-4d'\,^3D^\circ$	2—3
2288,17	7	—	—	—	—
2276,25	4	—	—	—	—
2253,16	30	14,86	20,36	$3d\,^3D^\circ-1$	1—2
2251,50	40	14,85	20,36	$3d\,^3D^\circ-1$	3—2
2250,96	20	14,85	20,36	$3d\,^3D^\circ-1$	2—2
2109,37	2	13,96	19,84	$4s\,^3S^\circ-^3P$	1—1
2102,99	3	13,96	19,85	$4s\,^3S^\circ-^3P$	1—2
1923,35	4	11,70	18,14	$3p^5\,^3P^\circ-4p'\,^3D$	0—1
1883,14	3	11,58	18,16	$3p^5\,^3P^\circ-4p'\,^3D$	2—3
1791,91	4	11,65	18,57	$3p^5\,^3P^\circ-4p'\,^3P$	1—2
1787,10	3	11,65	18,59	$3p^5\,^3P^\circ-4p'\,^3P$	1—1
1785,06	1	11,65	18,60	$3p^5\,^3P^\circ-4p'\,^3P$	1—0
1772,01	3	11,58	18,57	$3p^5\,^3P^\circ-4p'\,^3P$	2—2
1767,24	1	11,58	18,59	$3p^5\,^3P^\circ-4p'\,^3P$	2—1
1558,05	1	11,65	19,61	$3p^5\,^3P^\circ-4p''\,^3S$	1—1
1528,91	1	11,58	19,68	$3p^5\,^3P^\circ-4p''\,^3D$	2—2
1471,06	2	11,58	20,00	$3p^5\,^3P^\circ-4p''\,^1P$	2—1
1223,71	2	1,44	11,58	$3p^4\,^1D-3p^5\,^3P^\circ$	2—2
1079,08	15	0,09	11,58	$3p^4\,^3P-3p^5\,^3P^\circ$	1—2
1075,24	7	0,12	11,65	$3p^4\,^3P-3p^5\,^3P^\circ$	0—1
1071,76	10	0,09	11,65	$3p^4\,^3P-3p^5\,^3P^\circ$	1—1
1071,05	20	0,00	11,58	$3p^4\,^3P-3p^5\,^3P^\circ$	2—2
1067,94	4	0,09	11,70	$3p^4\,^3P-3p^5\,^3P^\circ$	1—0
1063,83	10	0,00	11,65	$3p^4\,^3P-3p^5\,^3P^\circ$	2—1
961,49	10	1,44	14,34	$3p^4\,^1D-3p^5\,^1P^\circ$	2—1
914,90	2	0,12	13,67	$3p^4\,^3P-3d\,^5D^\circ$	0—1
895,95	3	0,12	13,96	$3p^4\,^3P-4s\,^3S^\circ$	0—1
893,56	3	0,09	13,96	$3p^4\,^3P-4s\,^3S^\circ$	1—1
888,07	4	0,00	13,96	$3p^4\,^3P-4s\,^3S^\circ$	2—1
864,67	5	0,00	14,34	$3p^4\,^3P-3p^5\,^1P^\circ$	2—1
851,70	7	1,44	16,00	$3p^4\,^1D-4s'\,^1D^\circ$	2—2
841,41	4	0,12	14,86	$3p^4\,^3P-3d\,^3D^\circ$	0—1
839,63	2	0,09	14,85	$3p^4\,^3P-3d\,^3D^\circ$	1—2
839,30	2	0,09	14,86	$3p^4\,^3P-3d\,^3D^\circ$	1—1
834,67	10	0,00	14,85	$3p^4\,^3P-3d\,^3D^\circ$	2—3
827,85	1	0,09	15,06	$3p^4\,^3P-3d'\,^1D^\circ$	1—2

λ, Å	I	E_H, eV	E_B, eV	Transition	J
795,36	2	0,12	15,71	$3p^4\ ^3P—4s'\ ^3D°$	0—1
793,47	3	0,09	15,71	$3p^4\ ^3P—4s'\ ^3D°$	1—1
793,34	3	0,09	15,71	$3p^4\ ^3P—4s'\ ^3D°$	1—2
792,19	2	1,44	17,09	$3p^4\ ^1D—4s''\ ^3P°$	2—2
789,01	7	0,00	15,71	$3p^4\ ^3P—4s'\ ^3D°$	2—2
788,75	4	0,00	15,72	$3p^4\ ^3P—4s'\ ^3D°$	2—3
787,62	3	1,44	17,19	$3p^4\ ^1D—4s''\ ^1P°$	2—1
787,15	1	0,09	15,84	$3p^4\ ^3P—3d'\ ^1P°$	1—1
777,55	3	1,44	17,39	$3p^4\ ^1D—3d''\ ^1D°$	2—2
730,92	3	0,12	17,08	$3p^4\ ^3P—4s''\ ^3P°$	0—1
729,52	2	0,09	17,08	$3p^4\ ^3P—4s''\ ^3P°$	1—0
729,39	3	0,09	17,08	$3p^4\ ^3P—4s''\ ^3P°$	1—1
728,94	3	0,09	17,09	$3p^4\ ^3P—4s''\ ^3P°$	1—2
725,64	2	0,00	17,08	$3p^4\ ^3P—4s''\ ^3P°$	2—1
725,27	3	0,00	17,09	$3p^4\ ^3P—4s''\ ^3P°$	2—2
719,26	1	1,44	18,68	$3p^4\ ^1D—3d'\ ^3P°$	2—2
717,15	2	1,44	18,73	$3p^4\ ^1D—3d'\ ^3D°$	2—3
715,58	3	0,12	17,45	$3p^4\ ^3P—3d''\ ^3D°$	0—1
714,03	2	0,09	17,45	$3p^4\ ^3P—3d''\ ^3D°$	1—1
712,66	3	0,09	17,48	$3p^4\ ^3P—3d''\ ^3D°$	1—2
709,16	2	0,00	17,48	$3p^4\ ^3P—3d''\ ^3D°$	2—2
707,43	4	0,00	17,52	$3p^4\ ^3P—3d''\ ^3D°$	2—3
687,55	1	1,44	19,47	$3p^4\ ^1D—3p^54s\ ^3P°$	2—2
667,49	1	0,12	18,70	$3p^4\ ^3P—3d'\ ^3P°$	0—1
666,17	2	0,09	18,70	$3p^4\ ^3P—3d'\ ^3P°$	1—1
666,08	3	0,12	18,74	$3p^4\ ^3P—3d'\ ^3D°$	0—1
665,21	1	0,09	18,72	$3p^4\ ^3P—3d'\ ^3D°$	1—2
664,67	2	0,09	18,74	$3p^4\ ^3P—3d'\ ^3D°$	1—1
663,67	2	0,00	18,68	$3p^4\ ^3P—3d'\ ^3P°$	2—2
663,08	2	0,00	18,70	$3p^4\ ^3P—3d'\ ^3P°$	2—1
662,15	1	0,00	18,72	$3p^4\ ^3P—3d'\ ^3D°$	2—2
661,82	2	0,00	18,73	$3p^4\ ^3P—3d'\ ^3D°$	2—3
655,09	1	0,12	19,05	$3p^4\ ^3P—5s\ ^3S°$	0—1
653,80	1	0,09	19,05	$3p^4\ ^3P—5s\ ^3S°$	1—1
651,13	1	0,00	19,04	$3p^4\ ^3P—3d'\ ^3S°$	2—1
650,88	1	0,00	19,05	$3p^4\ ^3P—5s\ ^3S°$	2—1
639,42	2	0,09	19,47	$3p^4\ ^3P—3p^54s\ ^3P°$	1—2
638,23	2	0,12	19,55	$3p^4\ ^3P—3p^54s\ ^3P°$	0—1
637,06	1	0,09	19,55	$3p^4\ ^3P—3p^54s\ ^3P°$	1—1
636,62	2	0,00	19,47	$3p^4\ ^3P—3p^54s\ ^3P°$	2—2
635,87	2	0,09	19,58	$3p^4\ ^3P—3p^54s\ ^3P°$	1—0
634,24	1	0,00	19,55	$3p^4\ ^3P—3p^54s\ ^3P°$	2—1
626,70	1	1,44	21,23	$3p^4\ ^1D—5s'\ ^1D°$	2—2
621,12	4	0,12	20,08	$3p^4\ ^3P—4d\ ^3D°$	0—1
620,28	1	0,09	20,07	$3p^4\ ^3P—4d\ ^3D°$	1—2
618,02	2	0,00	20,06	$3p^4\ ^3P—4d\ ^3D°$	2—3
617,61	1	0,00	20,07	$3p^4\ ^3P—4d\ ^3D°$	2—2
617,27	0	0,00	20,08	$3p^4\ ^3P—4d\ ^3D°$	2—1
612,73	0	1,44	21,68	$3p^4\ ^1D—4d'\ ^3D°$	2—3
599,19	0	1,44	22,13	$3p^4\ ^1D—4d'\ ^1D°$	2—2
594,49	0	1,44	22,30	$3p^4\ ^1D—4d'\ ^1P°$	2—1
589,82	0	0,12	21,14	$3p^4\ ^3P—5s'\ ^3D°$	0—1
588,77	0	0,09	21,14	$3p^4\ ^3P—5s'\ ^3D°$	1—2
586,25	0	0,00	21,15	$3p^4\ ^3P—5s'\ ^3D°$	2—3
584,10	1	0,00	21,23	$3p^4\ ^3P—5s'\ ^1D°$	2—2
575,30	0	0,12	21,67	$3p^4\ ^3P—4d'\ ^3D°$	0—1
574,37	3	0,09	21,67	$3p^4\ ^3P—4d'\ ^3D°$	1—2
571,95	1	0,00	21,68	$3p^4\ ^3P—4d'\ ^3D°$	2—3

λ, Å	I	E_H, eV	E_B, eV	Transition	J
566,77	0	0,12	22,00	$3p^4\ ^3P—4d'\ ^3S^\circ$	0—1
565,75	0	0,12	22,04	$3p^4\ ^3P—4d'\ ^3P^\circ$	0—1
563,58	0	0,00	22,00	$3p^4\ ^3P—4d'\ ^3S^\circ$	2—1
562,54	0	0,00	22,04	$3p^4\ ^3P—4d'\ ^3P^\circ$	2—1
562,28	3	0,00	22,05	$3p^4\ ^3P—4d'\ ^3P^\circ$	2—2
558,14	1	0,09	22,30	$3p^4\ ^3P—4d'\ ^1P^\circ$	1—1

Cl III, ground state $1s^2\,2s^2\,2p^6\,3s^2\,3p^3\ ^4S^\circ_{3/2}$
Ionization potential 321936 cm⁻¹; 39,912 eV

λ, Å	I	E_H, eV	E_B, eV	Transition	J
4971,64	0	23,04	25,53	$3d\ ^2P—4p\ ^2D^\circ$	$^3/_2—^5/_2$
4863,75	1	24,32	26,87	$3d'\ ^2F—4p'\ ^2F^\circ$	$^7/_2—^7/_2$
4854,37	0	23,36	25,92	$4s'\ ^2D—4p\ ^2P^\circ$	$^3/_2—^1/_2$
4808,00	1	23,36	25,93	$4s'\ ^2D—4p\ ^2P^\circ$	$^5/_2—^3/_2$
4703,14	3	24,21	26,84	$3d'\ ^2D—4p'\ ^2F^\circ$	$^3/_2—^5/_2$
4695,07	1	22,29	24,93	$3d\ ^4P—4p\ ^4D^\circ$	$^1/_2—^1/_2$
4669,50	0	22,27	24,93	$3d\ ^4P—4p\ ^4D^\circ$	$^3/_2—^1/_2$
4638,96	2	22,29	24,96	$3d\ ^4P—4p\ ^4D^\circ$	$^1/_2—^3/_2$
4635,83	0	24,17	26,84	$3d'\ ^2D—4p'\ ^2F^\circ$	$^5/_2—^5/_2$
4613,78	2	22,27	24,96	$3d\ ^4P—4p\ ^4D^\circ$	$^3/_2—^3/_2$
4608,21	5	24,32	27,01	$3d'\ ^2F—4p'\ ^2D^\circ$	$^7/_2—^5/_2$
4604,43	0	24,32	27,01	$3d'\ ^2F—4p'\ ^2D^\circ$	$^5/_2—^5/_2$
4596,22	4	24,17	26,87	$3d'\ ^2D—4p'\ ^2F^\circ$	$^5/_2—^7/_2$
4591,10	4	24,32	27,02	$3d'\ ^2F—4p'\ ^2D^\circ$	$^5/_2—^3/_2$
4523,33	4	22,27	25,01	$3d\ ^4P—4p\ ^4D^\circ$	$^3/_2—^5/_2$
4489,17	1	22,25	25,01	$3d\ ^4P—4p\ ^4D^\circ$	$^5/_2—^5/_2$
4414,90	2	24,21	27,02	$3d'\ ^2D—4p'\ ^2D^\circ$	$^3/_2—^3/_2$
4380,57	2	23,09	25,92	$3d\ ^2P—4p\ ^2P^\circ$	$^1/_2—^1/_2$
4370,91	4	22,25	25,09	$3d\ ^4P—4p\ ^4D^\circ$	$^5/_2—^7/_2$
4369,60	2	24,67	27,51	$3d'\ ^2P—4p'\ ^2P^\circ$	$^3/_2—^1/_2$
4364,79	3	22,69	25,53	$3d\ ^2D—4p\ ^2D^\circ$	$^5/_2—^5/_2$
4354,03	2	22,57	25,42	$3d\ ^2D—4p\ ^2D^\circ$	$^3/_2—^3/_2$
4353,73	2	23,09	25,93	$3d\ ^2P—4p\ ^2P^\circ$	$^1/_2—^3/_2$
4341,47	2	24,65	27,51	$3d'\ ^2P—4p'\ ^2P^\circ$	$^1/_2—^1/_2$
4324,66	2	24,67	27,54	$3d'\ ^2P—4p'\ ^2P^\circ$	$^3/_2—^3/_2$
4308,42	1	23,04	25,92	$3d\ ^2P—4p\ ^2P^\circ$	$^3/_2—^1/_2$
4297,04	0	24,65	27,54	$3d'\ ^2P—4p'\ ^2P^\circ$	$^1/_2—^3/_2$
4282,46	4	23,04	25,93	$3d\ ^2P—4p\ ^2P^\circ$	$^3/_2—^3/_2$
4124,00	1	22,29	25,29	$3d\ ^4P—4p\ ^4P^\circ$	$^1/_2—^1/_2$
4106,83	5	22,29	25,31	$3d\ ^4P—4p\ ^4P^\circ$	$^1/_2—^3/_2$
4104,23	5	22,27	25,29	$3d\ ^4P—4p\ ^4P^\circ$	$^3/_2—^1/_2$
4087,00	4	22,27	25,31	$3d\ ^4P—4p\ ^4P^\circ$	$^3/_2—^3/_2$
4059,07	6	22,25	25,31	$3d\ ^4P—4p\ ^4P^\circ$	$^5/_2—^3/_2$
4018,50	6	22,27	25,36	$3d\ ^4P—4p\ ^4P^\circ$	$^3/_2—^5/_2$
3991,50	7	22,25	25,36	$3d\ ^4P—4p\ ^4P^\circ$	$^5/_2—^5/_2$
3958,39	0	22,29	25,42	$3d\ ^4P—4p\ ^2D^\circ$	$^1/_2—^3/_2$
3925,87	5	22,20	25,36	$4s\ ^2P—4p\ ^4P^\circ$	$^3/_2—^5/_2$
3881,73	3	22,11	25,31	$4s\ ^2P—4p\ ^4P^\circ$	$^1/_2—^3/_2$
3850,81	4	22,20	25,42	$4s\ ^2P—4p\ ^2D^\circ$	$^3/_2—^3/_2$
3824,47	4	22,69	25,93	$3d\ ^2D—4p\ ^2P^\circ$	$^5/_2—^3/_2$
3822,02	4	22,29	25,53	$3d\ ^4P—4p\ ^4S^\circ$	$^1/_2—^3/_2$
3804,83	3	22,27	25,53	$3d\ ^4P—4p\ ^4S^\circ$	$^3/_2—^3/_2$
3803,57	3	22,27	25,53	$3d\ ^4P—4p\ ^2D^\circ$	$^3/_2—^5/_2$

λ, Å	I	E_H, eV	E_B, eV	Transition	J
3779,35	5	22,25	25,53	$3d\ ^4P—4p\ ^2D°$	$^5/_2—^5/_2$
3764,42	2	27,54	30,83	$4p'\ ^2P°—4d\ ^2D$	$^3/_2—^5/_2$
3759,10	3	24,21	27,51	$3d'\ ^2D—4p'\ ^2P°$	$^3/_2—^1/_2$
3748,81	8	22,11	25,42	$4s\ ^2P—4p\ ^2D°$	$^1/_2—^3/_2$
3741,70	3	21,65	24,96	$4s\ ^4P—4p\ ^4D°$	$^5/_2—^3/_2$
3725,74	3	24,21	27,54	$3d'\ ^2D—4p'\ ^2P°$	$^3/_2—^3/_2$
3725,46	1	27,01	30,34	$4p'\ ^2D°—4d\ ^2F$	$^5/_2—^7/_2$
3720,45	8	22,20	25,53	$4s\ ^2P—4p\ ^2D°$	$^3/_2—^5/_2$
3707,34	6	22,57	25,92	$3d\ ^2D—4p\ ^2P°$	$^3/_2—^1/_2$
3705,45	6	21,58	24,93	$4s\ ^4P—4p\ ^4D°$	$^3/_2—^1/_2$
3688,10	2	22,57	25,93	$3d\ ^2D—4p\ ^2P°$	$^3/_2—^3/_2$
3683,39	5	24,17	27,54	$3d'\ ^2D—4p'\ ^2P°$	$^5/_2—^3/_2$
3682,05	7	21,65	25,01	$4s\ ^4P—4p\ ^4D°$	$^5/_2—^5/_2$
3670,28	7	21,58	24,96	$4s\ ^4P—4p\ ^4D°$	$^3/_2—^3/_2$
3661,48	1	26,84	30,23	$4p'\ ^2F°—4d\ ^2F$	$^5/_2—^5/_2$
3656,95	7	21,54	24,93	$4s\ ^4P—4p\ ^4D°$	$^1/_2—^1/_2$
3622,69	7	21,54	24,96	$4s\ ^4P—4p\ ^4D°$	$^1/_2—^3/_2$
3612,85	8	21,58	25,01	$4s\ ^4P—4p\ ^4D°$	$^3/_2—^5/_2$
3602,10	9	21,65	25,09	$4s\ ^4P—4p\ ^4D°$	$^5/_2—^7/_2$
3573,69	2	26,87	30,34	$4p'\ ^2F°—4d'\ ^2F$	$^7/_2—^7/_2$
3560,68	8	23,36	26,84	$4s'\ ^2D—4p'\ ^2F°$	$^3/_2—^5/_2$
3553,35	1	23,36	26,84	$4s'\ ^2D—4p'\ ^2F°$	$^5/_2—^5/_2$
3530,03	9	23,36	26,87	$4s'\ ^2D—4p'\ ^2F°$	$^5/_2—^7/_2$
3400,15	2	23,36	27,01	$4s'\ ^2D—4p'\ ^2D°$	$^3/_2—^5/_2$
3393,45	8	23,36	27,01	$4s'\ ^2D—4p'\ ^2D°$	$^5/_2—^5/_2$
3392,89	8	23,36	27,02	$4s'\ ^2D—4p'\ ^2D°$	$^3/_2—^3/_2$
3387,60	6	21,65	25,31	$4s\ ^4P—4p\ ^4P°$	$^5/_2—^3/_2$
3386,22	5	23,36	27,02	$4s'\ ^2D—4p'\ ^2D°$	$^5/_2—^3/_2$
3340,42	9	{ 21,65	25,36	$4s\ ^4P—4p\ ^4P°$	$^5/_2—^5/_2$
		{ 21,58	25,29	$4s\ ^4P—4p\ ^4P°$	$^3/_2—^1/_2$
3336,16	5	22,30	25,92	$4s\ ^2P—4p\ ^2P°$	$^3/_2—^1/_2$
3329,06	8	21,58	25,31	$4s\ ^4P—4p\ ^4P°$	$^3/_2—^3/_2$
3320,57	7	22,20	25,93	$4s\ ^2P—4p\ ^2P°$	$^3/_2—^3/_2$
3300,95	3	21,54	25,29	$4s\ ^4P—4p\ ^4P°$	$^1/_2—^1/_2$
3289,80	7	21,54	25,31	$4s\ ^4P—4p\ ^4P°$	$^1/_2—^3/_2$
3283,41	6	21,58	25,36	$4s\ ^4P—4p\ ^4P°$	$^3/_2—^5/_2$
3265,45	0	27,02	30,81	$4p'\ ^2D°—4d\ ^2D$	$^3/_2—^3/_2$
3259,32	6	22,11	25,92	$4s\ ^2P—4p\ ^2P°$	$^1/_2—^1/_2$
3245,05	2	27,01	30,83	$4p'\ ^2D°—4d\ ^2D$	$^5/_2—^5/_2$
3244,44	5	22,11	25,93	$4s\ ^2P—4p\ ^2P°$	$^1/_2—^3/_2$
3230,78	1	21,58	25,42	$4s\ ^4P—4p\ ^2D°$	$^3/_2—^3/_2$
3193,84	0	21,54	25,42	$4s\ ^4P—4p\ ^2D°$	$^1/_2—^3/_2$
3191,45	9	21,65	25,53	$4s\ ^4P—4p\ ^4S°$	$^5/_2—^3/_2$
3190,58	4	21,65	25,53	$4s\ ^4P—4p\ ^2D°$	$^5/_2—^5/_2$
3139,34	8	21,58	25,53	$4s\ ^4P—4p\ ^4S°$	$^3/_2—^3/_2$
3123,74	1	26,84	30,81	$4p\ ^2F°—4d\ ^2D$	$^5/_2—^3/_2$
3104,46	6	21,54	25,53	$4s\ ^4P—4p\ ^4S°$	$^1/_2—^3/_2$
2991,82	3	23,36	27,51	$4s'\ ^2D—4p'\ ^2P°$	$^3/_2—^1/_2$
2970,67	4	23,36	27,54	$4s'\ ^2D—4p'\ ^2P°$	$^3/_2—^3/_2$
2965,56	6	23,36	27,54	$4s'\ ^2D—4p'\ ^2P°$	$^5/_2—^3/_2$
2949,1	1	25,93	30,14	$4p\ ^2P°—4d\ ^4P$	$^3/_2—^3/_2$
2805,17	2	25,53	29,95	$4p\ ^4S°—4d\ ^4D$	$^3/_2—^3/_2$
2796,37	1	25,53	29,96	$4p\ ^4S°—4d\ ^4D$	$^3/_2—^5/_2$
2769,3	3	25,53	30,01	$4p\ ^2D°—4d\ ^4D$	$^5/_2—^7/_2$
2727,7	2	25,42	29,96	$4p\ ^2D°—4d\ ^4D$	$^3/_2—^5/_2$
2724,03	5	27,02	31,57	$4p'\ ^2D°—4d'\ ^2D$	$^3/_2—^3/_2$
2717,62	2	27,54	32,10	$4p'\ ^2P°—5s'\ ^2D$	$^3/_2—^5/_2$
2714,37	2	27,01	31,57	$4p'\ ^2D°—4d'\ ^2D$	$^5/_2—^5/_2$

λ, Å	I	$E_{\text{н}}$, eV	$E_{\text{в}}$, eV	Transition	J
2710,37	7	25,53	30,10	$4p\ ^4S^\circ - 4d\ ^4P$	$3/2 - 5/2$
2699,79	1	27,51	32,10	$4p'\ ^2P^\circ - 5s'\ ^2P$	$5/2 - 3/2$
		25,36	29,95	$4p\ ^4P^\circ - 4d\ ^4D$	$5/2 - 3/2$
2691,52	5	25,53	30,14	$4p\ ^4S^\circ - 4d\ ^4P$	$3/2 - 3/2$
		25,36	29,96	$4p\ ^4P^\circ - 4d\ ^4D$	$5/2 - 5/2$
2685,40	4	27,02	31,63	$4p'\ ^2D^\circ - 4d'\ ^2F$	$3/2 - 5/2$
2684,76	5	27,01	31,62	$4p'\ ^2D^\circ - 4d'\ ^2F$	$5/2 - 7/2$
2682,40	3	25,53	30,15	$4p\ ^4S^\circ - 4d\ ^4P$	$3/2 - 1/2$
2680,88	2	27,01	31,63	$4p'\ ^2D^\circ - 4d'\ ^2F$	$5/2 - 5/2$
2675,4	2	25,09	29,72	$4p\ ^4D^\circ - 4d\ ^4F$	$7/2 - 5/2$
2669,52	3	25,31	29,95	$4p\ ^4P^\circ - 4d\ ^4D$	$3/2 - 3/2$
2665,54	6	25,36	30,01	$4p\ ^4P^\circ - 4d\ ^4D$	$5/2 - 7/2$
2663,20	3	25,29	29,95	$4p\ ^4P^\circ - 4d\ ^4D$	$1/2 - 1/2$
2662,29	3	25,29	29,95	$4p\ ^4P^\circ - 4d\ ^4D$	$1/2 - 3/2$
2661,65	5	25,31	29,96	$4p\ ^4P^\circ - 4d\ ^4D$	$3/2 - 5/2$
2651,19	3	25,09	29,76	$4p\ ^4D^\circ - 4d\ ^4F$	$7/2 - 7/2$
2633,18	5	25,01	29,72	$4p\ ^4D^\circ - 4d\ ^4F$	$5/2 - 5/2$
2632,67	5	26,87	31,57	$4p'\ ^2F^\circ - 4d'\ ^2D$	$7/2 - 5/2$
2624,71	3	26,84	31,57	$4p'\ ^2F^\circ - 4d'\ ^2D$	$5/2 - 3/2$
2620,05	4	26,84	31,57	$4p'\ ^2F^\circ - 4d'\ ^2D$	$5/2 - 5/2$
2618,78	4	24,96	29,69	$4p\ ^4D^\circ - 4d\ ^4F$	$3/2 - 3/2$
2616,97	4	25,09	29,82	$4p\ ^4D^\circ - 4d\ ^4F$	$7/2 - 9/2$
2611,45	2	25,36	30,10	$4p\ ^4P^\circ - 4d\ ^4P$	$5/2 - 5/2$
2609,50	4	25,01	29,76	$4p\ ^4D^\circ - 4d\ ^4F$	$5/2 - 7/2$
2605,04	2	26,87	31,62	$4p'\ ^2F^\circ - 4d'\ ^2F$	$7/2 - 7/2$
2603,59	5	24,96	29,72	$4p\ ^4D^\circ - 4d\ ^4F$	$3/2 - 5/2$
2601,16	4	24,93	29,69	$4p\ ^4D^\circ - 4d\ ^4F$	$1/2 - 3/2$
2593,97	2	25,36	30,14	$4p\ ^4P^\circ - 4d\ ^4P$	$5/2 - 3/2$
2592,45	2	26,84	31,62	$4p'\ ^2F^\circ - 4d'\ ^2F$	$5/2 - 7/2$
2588,80	3	26,84	31,63	$4p'\ ^2F^\circ - 4d'\ ^2F$	$5/2 - 5/2$
2580,67	6	25,53	30,34	$4p\ ^2D^\circ - 4d\ ^2F$	$5/2 - 7/2$
2578,26	5	22,20	27,01	$4s\ ^2P - 4p'\ ^2D^\circ$	$3/2 - 5/2$
2577,13	5	25,42	30,23	$4p\ ^2D^\circ - 4d\ ^2F$	$3/2 - 5/2$
2574,13	0	22,20	27,02	$4s\ ^2P - 4p'\ ^2D^\circ$	$3/2 - 3/2$
2566,23	1	25,31	30,14	$4p\ ^4P^\circ - 4d\ ^4P$	$3/2 - 3/2$
2562,52	1	25,53	30,37	$4p\ ^4S^\circ - 5s\ ^4P$	$3/2 - 1/2$
2559,50	3	25,29	30,14	$4p\ ^4P^\circ - 4d\ ^4P$	$1/2 - 3/2$
2557,9	3	25,31	30,15	$4p\ ^4P^\circ - 4d\ ^4P$	$3/2 - 1/2$
2542,65	2	25,09	29,96	$4p\ ^4D^\circ - 4d\ ^4D$	$7/2 - 5/2$
2540,84	3	25,92	30,81	$4p\ ^2P^\circ - 4d\ ^2D$	$3/2 - 3/2$
2533,95	1	25,53	30,42	$4p\ ^4S^\circ - 5s\ ^4P$	$3/2 - 3/2$
2532,48	5	25,93	30,83	$4p\ ^2P^\circ - 4d\ ^2D$	$3/2 - 5/2$
2531,76	5	25,92	30,81	$4p\ ^2P^\circ - 4d\ ^2D$	$1/2 - 3/2$
2528,08	5	22,11	27,02	$4s\ ^2P - 4p'\ ^2D^\circ$	$1/2 - 3/2$
2519,45	5	25,09	30,01	$4p\ ^4D^\circ - 4d\ ^4D$	$7/2 - 7/2$
2510,92	4	25,01	29,95	$4p\ ^4D^\circ - 4d\ ^4D$	$5/2 - 3/2$
2504,23	5	25,01	29,96	$4p\ ^4D^\circ - 4d\ ^4D$	$5/2 - 5/2$
2490,3	5	25,36	30,34	$4p\ ^4P^\circ - 4d\ ^2F$	$5/2 - 7/2$
2486,91	5	25,53	30,52	$4p\ ^4S^\circ - 5s\ ^4P$	$3/2 - 5/2$
2485,1	3	24,96	29,95	$4p\ ^4D^\circ - 4d\ ^4D$	$3/2 - 1/2$
2484,27	4	24,96	29,95	$4p\ ^4D^\circ - 4d\ ^4D$	$3/2 - 3/2$
2481,77	2	25,01	30,01	$4p\ ^4D^\circ - 4d\ ^4D$	$5/2 - 7/2$
2477,29	2	24,96	29,96	$4p\ ^4D^\circ - 4d\ ^4D$	$3/2 - 5/2$
2471,07	5	25,09	30,10	$4p\ ^4D^\circ - 4d\ ^4P$	$7/2 - 5/2$
2469,20	5	24,93	29,95	$4p\ ^4D^\circ - 4d\ ^4D$	$1/2 - 1/2$
2468,37	3	24,93	29,95	$4p\ ^4D^\circ - 4d\ ^4D$	$1/2 - 3/2$
2448,58	6	25,31	30,37	$4p\ ^4P^\circ - 5s\ ^4P$	$3/2 - 1/2$
2447,14	6	25,36	30,42	$4p\ ^4P^\circ - 5s\ ^4P$	$5/2 - 3/2$
2442,47	5	25,29	30,37	$4p\ ^4P^\circ - 5s\ ^4P$	$1/2 - 1/2$

λ, Å	I	E_H, eV	E_B, eV	Transition	J
2439,69	5	27,02	32,10	$4p'\,{}^2D^\circ-5s'\,{}^2D$	3/2—3/2
2436,1	5	27,01	32,10	$4p'\,{}^2D^\circ-5s'\,{}^2D$	5/2—5/2
2435,1	2	25,01	30,10	$4p\,{}^4D^\circ-4d\,{}^4P$	5/2—5/2
2422,47	4	25,31	30,42	$4p\,{}^4P^\circ-5s\,{}^4P$	3/2—3/2
2419,5	5	25,01	30,14	$4p\,{}^4D^\circ-4d\,{}^4P$	5/2—3/2
2416,42	7	25,29	30,42	$4p\,{}^4P^\circ-5s\,{}^4P$	1/2—3/2
2403,32	5	25,36	30,52	$4p\,{}^4P^\circ-5s\,{}^4P$	5/2—5/2
2394,73	5	24,96	30,14	$4p\,{}^4D^\circ-4d\,{}^4P$	3/2—3/2
2387,3	3	24,96	30,15	$4p\,{}^4D^\circ-4d\,{}^4P$	3/2—1/2
2379,47	5	25,31	30,52	$4p\,{}^4P^\circ-5s\,{}^4P$	3/2—5/2
2372,7	0	24,93	30,15	$4p\,{}^4D^\circ-4d\,{}^4P$	1/2—1/2
2370,37	6	26,87	32,10	$4p'\,{}^2F^\circ-5s'\,{}^2D$	7/2—5/2
2359,67	6	26,84	32,10	$4p'\,{}^2F^\circ-5s'\,{}^2D$	5/2—3/2
2347,7	2	25,53	30,81	$4p\,{}^2D^\circ-4d\,{}^2D$	5/2—3/2
2340,64	6	25,53	30,83	$4p\,{}^2D^\circ-4d\,{}^2D$	5/2—5/2
2336,45	5	22,20	27,51	$4s\,{}^2P-4p'\,{}^2P^\circ$	3/2—1/2
2323,50	6	22,20	27,54	$4s\,{}^2P-4p'\,{}^2P^\circ$	3/2—3/2
2298,51	5	22,11	27,51	$4s\,{}^2P-4p'\,{}^2P^\circ$	1/2—1/2
		25,42	30,81	$4p\,{}^2D^\circ-4d\,{}^2D$	3/2—3/2
2291,81	4	25,42	30,83	$4p\,{}^2D^\circ-4d\,{}^2D$	3/2—5/2
		24,96	30,37	$4p\,{}^4D^\circ-5s\,{}^4P$	3/2—1/2
2291,38	4	25,01	30,42	$4p\,{}^4D^\circ-5s\,{}^4P$	5/2—3/2
2286,0	3	22,11	27,54	$4s\,{}^2P-4p'\,{}^2P^\circ$	1/2—3/2
2283,93	7	25,09	30,52	$4p\,{}^4D^\circ-5s\,{}^4P$	7/2—5/2
2278,34	5	24,93	30,37	$4p\,{}^4D^\circ-5s\,{}^4P$	1/2—1/2
2272,8	1	25,36	30,81	$4p\,{}^4P^\circ-4d\,{}^2D$	5/2—3/2
2268,95	5	24,96	30,42	$4p\,{}^4D^\circ-5s\,{}^4P$	3/2—3/2
2266,08	2	25,36	30,83	$4p\,{}^4P^\circ-4d\,{}^2D$	5/2—5/2
2255,64	2	24,93	30,42	$4p\,{}^4D^\circ-5s\,{}^4P$	1/2—3/2
2253,07	7	25,01	30,52	$4p\,{}^4D^\circ-5s\,{}^4P$	5/2—5/2
2231,16	3	24,96	30,52	$4p\,{}^4D^\circ-5s\,{}^4P$	3/2—5/2
2034,88	3	18,84	24,93	$3d\,{}^4D-4p\,{}^4D^\circ$	1/2—1/2
2032,14	3	18,83	24,93	$3d\,{}^4D-4p\,{}^4D^\circ$	3/2—1/2
2024,21	3	18,84	24,96	$3d\,{}^4D-4p\,{}^4D^\circ$	1/2—3/2
2021,46	3	18,83	24,96	$3d\,{}^4D-4p\,{}^4D^\circ$	3/2—3/2
2020,19	3	18,83	24,96	$3d\,{}^4D-4p\,{}^4D^\circ$	5/2—3/2
2011,34	1	25,93	32,10	$4p\,{}^2P^\circ-5s'\,{}^2D$	3/2—5/2
2006,84	4	18,84	25,01	$3d\,{}^4D-4p\,{}^4D^\circ$	7/2—5/2
2003,97	0	18,83	25,01	$3d\,{}^4D-4p\,{}^4D^\circ$	3/2—5/2
2002,72	3	18,83	25,01	$3d\,{}^4D-4p\,{}^4D^\circ$	5/2—5/2
1983,61	5	18,84	25,09	$3d\,{}^4D-4p\,{}^4D^\circ$	7/2—7/2
1979,46	3	18,83	25,09	$3d\,{}^4D-4p\,{}^4D^\circ$	5/2—7/2
1920,32	4	18,84	25,29	$3d\,{}^4D-4p\,{}^4P^\circ$	1/2—1/2
1917,87	4	18,83	25,29	$3d\,{}^4D-4p\,{}^4P^\circ$	3/2—1/2
1916,53	4	18,84	25,31	$3d\,{}^4D-4p\,{}^4P^\circ$	1/2—3/2
1914,09	3	18,83	25,31	$3d\,{}^4D-4p\,{}^4P^\circ$	3/2—3/2
1912,90	4	18,83	25,31	$3d\,{}^4D-4p\,{}^4P^\circ$	5/2—3/2
1901,61	5	18,84	25,36	$3d\,{}^4D-4p\,{}^4P^\circ$	7/2—5/2
1897,85	3	18,83	25,36	$3d\,{}^4D-4p\,{}^4P^\circ$	5/2—5/2
1889,06	0	25,53	32,10	$4p\,{}^2D^\circ-5s'\,{}^2D$	5/2—5/2
1880,10	3	18,83	25,42	$3d\,{}^4D-4p\,{}^2D^\circ$	5/2—3/2
1852,11	2	18,84	25,53	$3d\,{}^4D-4p\,{}^4S^\circ$	1/2—3/2
1849,64	0	18,83	25,53	$3d\,{}^4D-4p\,{}^4S^\circ$	3/2—3/2
1848,74	0	18,83	25,53	$3d\,{}^4D-4p\,{}^4S^\circ$	5/2—3/2
1833,31	4	18,17	24,93	$3d\,{}^4F-4p\,{}^4D^\circ$	3/2—1/2
1832,08	4	18,19	24,96	$3d\,{}^4F-4p\,{}^4D^\circ$	5/2—3/2
1828,40	5	18,23	25,01	$3d\,{}^4F-4p\,{}^4D^\circ$	7/2—5/2
1824,59	3	18,17	24,96	$3d\,{}^4F-4p\,{}^4D^\circ$	3/2—3/2

λ, Å	I	E_H, eV	E_B, eV	Transition	J
1822,50	6	18,29	25,09	$3d\,{}^4F-4p\,{}^4D^\circ$	$9/2-7/2$
1817,73	4	18,19	25,01	$3d\,{}^4F-4p\,{}^4D^\circ$	$5/2-5/2$
1810,26	1	18,17	25,01	$3d\,{}^4F-4p\,{}^4D^\circ$	$3/2-5/2$
1808,51	4	18,23	25,09	$3d\,{}^4F-4p\,{}^4D^\circ$	$7/2-7/2$
1797,98	2	18,19	25,09	$3d\,{}^4F-4p\,{}^4D^\circ$	$5/2-7/2$
1689,50	1	18,19	25,53	$3d\,{}^4F-4p\,{}^4S^\circ$	$5/2-3/2$
1683,18	0	18,17	25,53	$3d\,{}^4F-4p\,{}^4S^\circ$	$3/2-3/2$
1015,023	7	0,00	12,21	$3p^3\,{}^4S^\circ-3p^4\,{}^4P$	$3/2-5/2$
1008,777	6	0,00	12,29	$3p^3\,{}^4S^\circ-3p^4\,{}^4P$	$3/2-3/2$
1005,280	5	0,00	12,33	$3p^3\,{}^4S^\circ-3p^4\,{}^4P$	$3/2-1/2$
953,40	2	12,29	25,29	$3p^4\,{}^4P-4p\,{}^4P^\circ$	$3/2-1/2$
948,72	1	12,29	25,36	$3p^4\,{}^4P-4p\,{}^4P^\circ$	$3/2-5/2$
946,97	1	12,21	25,31	$3p^4\,{}^4P-4p\,{}^4P^\circ$	$5/2-3/2$
943,22	1	12,21	25,36	$3p^4\,{}^4P-4p\,{}^4P^\circ$	$5/2-5/2$
939,31	0	12,33	25,53	$3p^4\,{}^4P-4p\,{}^4S^\circ$	$1/2-3/2$
936,28	1	12,29	25,53	$3p^4\,{}^4P-4p\,{}^4S^\circ$	$3/2-3/2$
930,94	1	12,21	25,53	$3p^4\,{}^4P-4p\,{}^4S^\circ$	$5/2-3/2$
747,553	1	2,25	18,83	$3p^3\,{}^2D^\circ-3d\,{}^4D$	$5/2-3/2$
747,415	1	2,24	18,83	$3p^3\,{}^2D^\circ-3d\,{}^4D$	$3/2-5/2$
746,864	1	2,24	18,84	$3p^3\,{}^2D^\circ-3d\,{}^4D$	$3/2-1/2$
673,598	1	3,71	22,11	$3p^3\,{}^2P^\circ-4s\,{}^2P$	$3/2-1/2$
673,127	3	3,70	22,11	$3p^3\,{}^2P^\circ-4s\,{}^2P$	$1/2-1/2$
670,383	3	3,71	22,20	$3p^3\,{}^2P^\circ-4s\,{}^2P$	$3/2-3/2$
669,949	2	3,70	22,20	$3p^3\,{}^2P^\circ-4s\,{}^2P$	$1/2-3/2$
657,320	2	3,71	22,57	$3p^3\,{}^2P^\circ-3d\,{}^2D$	$3/2-3/2$
656,772	2	3,70	22,57	$3p^3\,{}^2P^\circ-3d\,{}^2D$	$1/2-3/2$
653,013	2	3,71	22,69	$3p^3\,{}^2P^\circ-3d\,{}^2D$	$3/2-5/2$
641,304	1	3,71	23,04	$3p^3\,{}^2P^\circ-3d\,{}^2P$	$3/2-3/2$
640,928	1	3,70	23,04	$3p^3\,{}^2P^\circ-3d\,{}^2P$	$1/2-3/2$
639,757	1	3,71	23,09	$3p^3\,{}^2P^\circ-3d\,{}^2P$	$3/2-1/2$
631,006	1	3,71	23,36	$3p^3\,{}^2P^\circ-4s'\,{}^2D$	$3/2-5/2$
630,746	1	3,71	23,36	$3p^3\,{}^2P^\circ-4s'\,{}^2D$	$3/2-3/2$
630,380	1	3,70	23,36	$3p^3\,{}^2P^\circ-4s'\,{}^2P$	$1/2-3/2$
623,768	3	2,24	22,11	$3p^3\,{}^2D^\circ-4s\,{}^2P$	$3/2-1/2$
621,280	4	2,25	22,20	$3p^3\,{}^2D^\circ-4s\,{}^2P$	$5/2-3/2$
621,027	3	2,24	22,20	$3p^3\,{}^2D^\circ-4s\,{}^2P$	$3/2-3/2$
619,025	1	2,25	22,27	$3p^3\,{}^2D^\circ-3d\,{}^4P$	$5/2-3/2$
609,901	0	2,25	22,57	$3p^3\,{}^2D^\circ-3d\,{}^2D$	$5/2-3/2$
609,673	4	2,24	22,57	$3p^3\,{}^2D^\circ-3d\,{}^2D$	$3/2-3/2$
606,345	5	2,25	22,69	$3p^3\,{}^2D^\circ-3d\,{}^2D$	$5/2-5/2$
606,100	2	2,24	22,69	$3p^3\,{}^2D^\circ-3d\,{}^2D$	$3/2-5/2$
605,855	1	3,71	24,17	$3p^3\,{}^2P^\circ-3d'\,{}^2D$	$3/2-5/2$
596,240	4	2,25	23,04	$3p^3\,{}^2D^\circ-3d\,{}^2P$	$5/2-3/2$
595,990	3	2,24	23,04	$3p^3\,{}^2D^\circ-3d\,{}^2P$	$3/2-3/2$
594,636	4	2,24	23,09	$3p^3\,{}^2D^\circ-3d\,{}^2P$	$3/2-1/2$
591,962	2	3,71	24,65	$3p^3\,{}^2P^\circ-3d'\,{}^2P$	$3/2-1/2$
591,646	4	3,70	24,65	$3p^3\,{}^2P^\circ-3d'\,{}^2P$	$1/2-1/2$
591,428	4	3,71	24,67	$3p^3\,{}^2P^\circ-3d'\,{}^2P$	$3/2-3/2$
591,118	3	3,70	24,67	$3p^3\,{}^2P^\circ-3d'\,{}^2P$	$1/2-3/2$
587,295	4	2,25	23,36	$3p^3\,{}^2D^\circ-4s'\,{}^2D$	$5/2-5/2$
587,078	3	2,24	23,36	$3p^3\,{}^2D^\circ-4s'\,{}^2D$	$3/2,\ 5/2-3/2,\ 5/2$
586,874	4	2,24	23,36	$3p^3\,{}^2D^\circ-4s'\,{}^2D$	$3/2-3/2$
575,582	3	0,00	21,54	$3p^3\,{}^4S^\circ-4s\,{}^4P$	$3/2-1/2$
574,408	3	0,00	21,58	$3p^3\,{}^4S^\circ-4s\,{}^4P$	$3/2-3/2$
572,693	4	0,00	21,65	$3p^3\,{}^4S^\circ-4s\,{}^4P$	$3/2-5/2$
565,480	4	2,25	24,17	$3p^3\,{}^2D^\circ-3d'\,{}^2D$	$5/2-5/2$
565,272	3	2,24	24,17	$3p^3\,{}^2D^\circ-3d'\,{}^2D$	$3/2-5/2$

λ, Å	I	E_H, eV	E_B, eV	Transition	J
564,514	2	2,25	24,21	$3p^3\ ^2D^\circ$—$3d'\ ^2D$	5/2—3/2
564,287	4	2,24	24,21	$3p^3\ ^2D^\circ$—$3d'\ ^2D$	3/2—3/2
561,738	7	2,25	24,32	$3p^3\ ^2D^\circ$—$3d'\ ^2F$	5/2—5/2
561,680	7	2,25	24,32	$3p^3\ ^2D^\circ$—$3d'\ ^2F$	5/2—7/2
561,530	7	2,24	24,32	$3p^3\ ^2D^\circ$—$3d'\ ^2F$	3/2—5/2
560,636	1	0,00	22,11	$3p^3\ ^4S^\circ$—$4s\ ^2P$	3/2—1/2
558,385	1	0,00	22,20	$3p^3\ ^4S^\circ$—$4s\ ^2P$	3/2—3/2
557,118	7	0,00	22,25	$3p^3\ ^4S^\circ$—$3d\ ^4P$	3/2—5/2
556,605	7	0,00	22,27	$3p^3\ ^4S^\circ$—$3d\ ^4P$	3/2—3/2
556,232	6	0,00	22,29	$3p^3\ ^4S^\circ$—$3d\ ^4P$	3/2—1/2
552,908	2	2,25	24,67	$3p^3\ ^2D^\circ$—$3d'\ ^2P$	5/2—3/2
457,444	0	3,71	30,81	$3p^3\ ^2P^\circ$—$4d\ ^2D$	3/2—3/2
457,245	2	3,70	30,81	$3p^3\ ^2P^\circ$—$4d\ ^2D$	1/2—3/2
457,169	3	3,71	30,83	$3p^3\ ^2P^\circ$—$4d\ ^2D$	3/2—5/2
442,947	2	2,24	30,23	$3p^3\ ^2D^\circ$—$4d\ ^2F$	3/2—5/2
441,398	3	2,25	30,34	$3p^3\ ^2D^\circ$—$4d\ ^2F$	5/2—7/2
433,774	0	2,25	30,83	$3p^3\ ^2D^\circ$—$4d\ ^2D$	5/2—5/2
433,664	0	2,24	30,83	$3p^3\ ^2D^\circ$—$4d\ ^2D$	3/2—5/2
422,713	1	2,24	31,57	$3p^3\ ^2D^\circ$—$4d'\ ^2D$	3/2, 5/2—3/2, 5/2
421,990	3	2,25	31,62	$3p^3\ ^2D^\circ$—$4d'\ ^2F$	5/2—7/2
421,771	3	2,24	31,63	$3p^3\ ^2D^\circ$—$4d'\ ^2F$	3/2—5/2
415,333	1	2,25	32,10	$3p^3\ ^2D^\circ$—$5s'\ ^2D$	5/2—5/2
415,196	1	2,24	32,10	$3p^3\ ^2D^\circ$—$5s'\ ^2D$	3/2—3/2
411,812	4	0,00	30,10	$3p^3\ ^4S^\circ$—$4d\ ^4P$	3/2—5/2
411,373	4	0,00	30,14	$3p^3\ ^4S^\circ$—$4d\ ^4P$	3/2—3/2
411,163	3	0,00	30,15	$3p^3\ ^4S^\circ$—$4d\ ^4P$	3/2—1/2
407,513	0	0,00	30,42	$3p^3\ ^4S^\circ$—$5s\ ^4P$	3/2—3/2
406,274	1	0,00	30,52	$3p^3\ ^4S^\circ$—$5s\ ^4P$	3/2—5/2

Cl IV, ground state $1s^2\,2s^2\,2p^6\,3s^2\,3p^2\ ^3P_0$

Ionization potential 431226 cm^{-1}; 53,462 eV

λ, Å	I	E_H, eV	E_B, eV	Transition	J
3167,87	2	26,84	30,75	$4s\ ^3P^\circ$—$4p\ ^3D$	2—2
3106,09	1	26,70	30,69	$4s\ ^3P^\circ$—$4p\ ^3D$	1—1
3076,68	6	26,84	30,87	$4s\ ^3P^\circ$—$4p\ ^3D$	2—3
3071,36	3	26,66	30,69	$4s\ ^3P^\circ$—$4p\ ^3D$	0—1
3063,13	5	26,70	30,75	$4s\ ^3P^\circ$—$4p\ ^3D$	1—2
2835,4	4	26,84	31,21	$4s\ ^3P^\circ$—$4p\ ^3P$	2—1
2782,47	7	26,84	31,29	$4s\ ^3P^\circ$—$4p\ ^3P$	2—2
2770,64	4	26,70	31,18	$4s\ ^3P^\circ$—$4p\ ^3P$	1—0
2751,23	5	26,70	31,21	$4s\ ^3P^\circ$—$4p\ ^3P$	1—1
2724,03	5	26,66	31,21	$4s\ ^3P^\circ$—$4p\ ^3P$	0—1
2701,36	4	26,70	31,29	$4s\ ^3P^\circ$—$4p\ ^3P$	1—2
1651,21	1	23,18	30,69	$3d\ ^3D^\circ$—$4p\ ^3D$	1—1
1648,04	0	23,23	30,75	$3d\ ^3D^\circ$—$4p\ ^3D$	3—2
1643,40	1	23,21	30,75	$3d\ ^3D^\circ$—$4p\ ^3D$	2—2
1638,95	0	23,18	30,75	$3d\ ^3D^\circ$—$4p\ ^3D$	1—2
1622,86	2	23,23	30,87	$3d\ ^3D^\circ$—$4p\ ^3D$	3—3
1617,43	1	31,29	38,96	$4p\ ^3P$—$5s\ ^3P^\circ$	2—2
1551,27	1	23,18	31,18	$3d\ ^3D^\circ$—$4p\ ^3P$	1—0
1549,15	2	23,21	31,21	$3d\ ^3D^\circ$—$4p\ ^3P$	2—1
1545,19	2	23,18	31,21	$3d\ ^3D^\circ$—$4p\ ^3P$	1—1

λ, Å	I	E_H, eV	E_B, eV	Transition	J
1539,30	2	30,75	38,80	$4p\ ^3D—5s\ ^3P°$	2—1
1537,21	3	23,23	31,29	$3d\ ^3D°—4p\ ^3P$	3—2
1533,25	1	23,21	31,29	$3d\ ^3D°—4p\ ^3P$	2—2
1532,19	1	30,87	38,96	$4p\ ^3D—5s\ ^3P°$	3—2
1529,28	0	23,18	31,29	$3d\ ^3D°—4p\ ^3P$	1—2
1440,95	0	22,60	31,18	$3d\ ^3P°—4p\ ^3P$	1—0
1426,89	1	22,52	31,21	$3d\ ^3P°—4p\ ^3P$	2—1
1421,97	1	22,57	31,29	$3d\ ^3P°—4p\ ^3P$	1—2
1413,39	1	22,52	31,29	$3d\ ^3P—4p\ ^3P$	2—2
985,749	4	0,16	12,74	$3p^2\ ^3P—3p^3\ ^3D°$	2—2
984,952	7	0,16	12,75	$3p^2\ ^3P—3p^3\ ^3D°$	2—3
977,901	4	0,06	12,74	$3p^2\ ^3P—3p^3\ ^3D°$	1—1
977,560	6	0,06	12,74	$3p^2\ ^3P—3p^3\ ^3D°$	1—2
973,212	5	0,00	12,74	$3p^2\ ^3P—3p^3\ ^3D°$	0—1
840,933	4	0,16	14,91	$3p^2\ ^3P—3p^3\ ^3P°$	2—2
840,808	6	0,16	14,91	$3p^2\ ^3P—3p^3\ ^3P°$	2—1
834,967	5	0,06	14,91	$3p^2\ ^3P—3p^3\ ^3P°$	1—2
834,840	5	0,06	14,91	$3p^2\ ^3P—3p^3\ ^3P°$	1—1
834,659	3	0,06	14,91	$3p^2\ ^3P—3p^3\ ^3P°$	1—0
831,431	4	0,00	14,91	$3p^2\ ^3P—3p^3\ ^3P°$	0—1
776,91	0	14,91	30,87	$3p^3\ ^3P°—4p\ ^3D$	2—3
756,563	1	4,04	20,42	$3p^2\ ^1S—3p^3\ ^3S°$	0—1
745,205	4	4,04	20,67	$3p^2\ ^1S—3p^3\ ^1P°$	0—1
684,490	0	12,74	30,87	$3p^3\ ^3D°—4p\ ^3D$	3—3
672,428	0	12,74	31,18	$3p^3\ ^3D°—4p\ ^3P$	1—0
668,770	2	12,75	31,29	$3p^3\ ^3D°—4p\ ^3P$	3—2
662,454	3	1,71	20,42	$3p^2\ ^1D—3p^3\ ^3S°$	2—1
653,696	4	1,71	20,67	$3p^2\ ^1D—3p^3\ ^1P°$	2—1
612,070	4	0,16	20,42	$3p^2\ ^3P—3p^3\ ^3S°$	2—1
608,903	4	0,06	20,42	$3p^2\ ^3P—3p^3\ ^3S°$	1—1
607,088	3	0,00	20,42	$3p^2\ ^3P—3p^3\ ^3S°$	0—1
604,590	5	0,16	20,67	$3p^2\ ^3P—3p^3\ ^1P°$	2—1
601,499	5	0,06	20,67	$3p^2\ ^3P—3p^3\ ^1P°$	1—1
599,733	2	0,00	20,67	$3p^2\ ^3P—3p^3\ ^1P°$	0—1
554,619	7	0,16	22,52	$3p^2\ ^3P—3d\ ^3P°$	2—2
553,297	6	0,16	22,57	$3p^2\ ^3P—3d\ ^3P°$	2—1
552,017	7	0,06	22,52	$3p^2\ ^3P—3d\ ^3P°$	1—2
550,706	3	0,06	22,57	$3p^2\ ^3P—3d\ ^3P°$	1—1
550,020	4	0,06	22,60	$3p^2\ ^3P—3d\ ^3P°$	1—0
549,219	5	0,00	22,57	$3p^2\ ^3P—3d\ ^3P°$	0—1
538,595	4	0,16	23,18	$3p^2\ ^3P—3d\ ^3D°$	2—1
538,119	6	0,16	23,21	$3p^2\ ^3P—3d\ ^3D°$	2—2
537,606	9	0,16	23,23	$3p^2\ ^3P—3d\ ^3D°$	2—3
536,150	6	0,06	23,18	$3p^2\ ^3P—3d\ ^3D°$	1—1
535,666	7	0,06	23,21	$3p^2\ ^3P—3d\ ^3D°$	1—2
535,039	4	4,04	27,21	$3p^2\ ^1S—4s\ ^1P°$	0—1
534,727	8	0,00	23,18	$3p^2\ ^3P—3d\ ^3D°$	0—1
486,172	8	1,71	27,21	$3p^2\ ^1D—4s\ ^1P°$	2—1
467,194	3	0,16	26,70	$3p^2\ ^3P—4s\ ^3P°$	2—1
466,132	3	0,06	26,66	$3p^2\ ^3P—4s\ ^3P°$	1—0
465,350	3	0,06	26,70	$3p^2\ ^3P—4s\ ^3P°$	1—1
464,861	4	0,16	26,84	$3p^2\ ^3P—4s\ ^3P°$	2—2
464,292	3	0,00	26,70	$3p^2\ ^3P—4s\ ^3P°$	0—1
463,011	3	0,06	26,84	$3p^2\ ^3P—4s\ ^3P°$	1—2
440,245	2	—	—	—	—
439,255	3	—	—	—	—
437,825	4	—	—	—	—
331,835	2	1,71	39,07	$3p^2\ ^3P—5s\ ^1P°$	2—1

λ, Å	I	E_H, eV	E_B, eV	Transition	J
320,881	1	0,16	38,80	$3p^2\,^3P—5s\,^3P^\circ$	2—1
320,250	1	0,06	38,77	$3p^2\,^3P—5s\,^3P^\circ$	1—0
319,993	0	0,06	38,80	$3p^2\,^3P—5s\,^3P^\circ$	1—1
319,616	3	0,16	38,96	$3p^3\,^3P—5s\,^3P^\circ$	2—2
319,513	1	0,00	38,80	$3p^2\,^3P—5s\,^3P^\circ$	0—1
318,750	1	0,06	38,96	$3p^2\,^3P—5s\,^3P^\circ$	1—2

Cl V, ground state $1s^2\,2s^2\,2p^6\,3s^2\,3p\,^2P^0_{1/2}$
Ionization potential 547000 cm⁻¹; 67,81 eV

λ, Å	I	E_H, eV	E_B, eV	Transition	J
894,910	1	0,18	14,04	$3p\,^2P^\circ—3p^2\,^2D$	$^3/_2—^3/_2$
894,340	4	0,18	14,05	$3p\,^2P^\circ—3p^2\,^2D$	$^3/_2—^5/_2$
883,127	4	0,00	14,04	$3p\,^2P^\circ—3p^2\,^2D$	$^1/_2—^3/_2$
688,933	4	0,18	18,18	$3p\,^2P^\circ—3p^2\,^2S$	$^3/_2—^1/_2$
683,171	4	10,83	28,98	$3p^2\,^4P—3p^3\,^4S^\circ$	$^5/_2—^3/_2$
681,924	4	0,00	18,18	$3p\,^2P^\circ—3p^2\,^2S$	$^1/_2—^1/_2$
679,257	3	10,73	28,98	$3p^2\,^4P—3p^3\,^4S^\circ$	$^3/_2—^3/_2$
676,785	3	10,66	28,98	$3p^2\,^4P—3p^3\,^4S^\circ$	$^1/_2—^3/_2$
639,226	3	0,18	19,58	$3p\,^2P^\circ—3p^3\,^2P$	$^3/_2—^1/_2$
635,323	4	0,18	19,70	$3p\,^2P^\circ—3p^2\,^2P$	$^3/_2—^3/_2$
633,186	4	0,00	19,58	$3p\,^2P^\circ—3p^2\,^2P$	$^1/_2—^1/_2$
629,354	3	0,00	19,70	$3p\,^2P^\circ—3p^2\,^2P$	$^1/_2—^3/_2$
554,210	1	—	—	—	—
551,643	1	—	—	—	—
551,117	2	—	—	—	—
547,630	10	10,83	33,47	$3p^2\,^4P—3d\,^4P^\circ$	$^5/_2—^5/_2$
546,329	6	10,83	33,52	$3p^2\,^4P—3d\,^4P^\circ$	$^5/_2—^3/_2$
545,114	10	10,73	33,47	$3p^2\,^4P—3d\,^4P^\circ$	$^3/_2—^5/_2$
543,818	1	10,73	33,52	$3p^2\,^4P—3d\,^4P^\circ$	$^3/_2—^3/_2$
542,868	4	10,73	33,57	$3p^2\,^4P—3d\,^4P^\circ$	$^3/_2—^1/_2$
542,395	3	0,18	23,04	$3p\,^2P—3d\,^2D$	$^3/_2—^3/_2$
542,297	6	0,18	23,05	$3p\,^2P—3d\,^2D$	$^3/_2—^5/_2$
542,229	8	10,66	33,52	$3p^2\,^4P—3d\,^4P^\circ$	$^1/_2—^3/_2$
541,284	3	10,66	33,57	$3p^2\,^4P—3d\,^4P^\circ$	$^1/_2—^1/_2$
539,441	0	10,83	33,82	$3p^2\,^4P—3d\,^4D^\circ$	$^5/_2—^3/_2$
538,977	3	10,83	33,84	$3p^2\,^4P—3d\,^4D^\circ$	$^5/_2—^5/_2$
538,681	4	10,83	33,85	$3p^2\,^4P—3d\,^4D^\circ$	$^5/_2—^7/_2$
538,032	5	0,00	23,04	$3p\,^2P^\circ—3d\,^2D$	$^1/_2—^3/_2$
537,461	3	10,73	33,80	$3p^2\,^4P—3d\,^4D^\circ$	$^3/_2—^1/_2$
537,006	4	10,73	33,82	$3p^2\,^4P—3d\,^4D^\circ$	$^3/_2—^3/_2$
536,532	3	10,73	33,84	$3p^2\,^4P—3d\,^4D^\circ$	$^3/_2—^5/_2$
535,916	2	10,66	33,80	$3p^2\,^4P—3d\,^4D^\circ$	$^1/_2—^1/_2$
535,455	2	10,66	33,82	$3p^2\,^4P—3d\,^4D^\circ$	$^1/_2—^3/_2$
392,433	5	0,18	31,78	$3p\,^2P^\circ—4s\,^2S$	$^3/_2—^1/_2$
390,148	4	0,00	31,78	$3p\,^2P^\circ—4s\,^2S$	$^1/_2—^1/_2$
375,103	2	10,83	43,89	$3p^2\,^4P—4s\,^4P^\circ$	$^5/_2—^3/_2$
374,662	1	10,73	43,82	$3p^2\,^4P—4s\,^4P^\circ$	$^3/_2—^1/_2$
373,911	0	10,66 10,73	43,82 43,89	$3p^2\,^4P—4s\,^4P^\circ$ $3p^2\,^4P—4s\,^4P^\circ$	$^1/_2—^1/_2$ $^3/_2—^3/_2$
373,776	3	10,83	44,00	$3p^2\,^4P—4s\,^4P^\circ$	$^5/_2—^5/_2$
373,165	2	10,66	43,89	$4s^2\,^4P—4s\,^4P^\circ$	$^1/_2—^3/_2$
372,589	2	10,73	44,00	$3p^2\,^4P—4s\,^4P^\circ$	$^3/_2—^5/_2$
287,327	3	0,18	43,33	$3p\,^2P^\circ—4d\,^2D$	$^3/_2—^3/_2$, $^5/_2$
286,127	2	0,00	43,33	$3p\,^2P^\circ—4d\,^2D$	$^1/_2—^3/_2$
237,231	2	0,18	52,45	$3p\,^2P^\circ—5d\,^2D$	$^3/_2—^5/_2$
236,435	1	0,00	52,44	$3p\,^2P^\circ—5d\,^2D$	$^1/_2—^3/_2$

Cl VI, ground state $1s^2 2s^2 2p^6 3s^2\ {}^1S_0$

Ionization potential 780000 cm^{-1}; 96,70 eV

λ, Å	I	E_H, eV	E_B, eV	Transition	J
736,762	3	12,38	29,21	$3p\ {}^3P^\circ—3p^2\ {}^3P$	2—1
733,891	3	12,24	29,13	$3p\ {}^3P^\circ—3p^2\ {}^3P$	1—0
730,311	4	12,24	29,21	$3p\ {}^3P^\circ—3p^2\ {}^3P$	1—1
		12,38	29,36	$3p\ {}^3P^\circ—3p^2\ {}^3P$	2—2
727,537	3	12,17	29,21	$3p\ {}^3P^\circ—3p^2\ {}^3P$	0—1
724,129	3	12,24	29,36	$3p\ {}^3P^\circ—3p^2\ {}^3P$	1—2
671,37	4	0,00	18,47	$3s^2\ {}^1S—3p\ {}^1P^\circ$	0—1
580,444	2	29,36	50,72	$3p^2\ {}^3P—3d\ {}^3P^\circ$	2—2
577,444	1	29,36	50,83	$3p^2\ {}^3P—3d\ {}^3P^\circ$	2—1
576,419	2	29,21	50,72	$3p^2\ {}^3P—3d\ {}^3P^\circ$	1—2
571,435	0	29,36	51,05	$3p^2\ {}^3P—3d\ {}^3D^\circ$	2—1
571,376	1	29,13	50,83	$3p^2\ {}^3P—3d\ {}^3P^\circ$	0—1
570,881	0	29,21	50,93	$3p^2\ {}^3P—3d\ {}^3P^\circ$	1—0
570,529	2	29,36	51,09	$3p^2\ {}^3P—3d\ {}^3D^\circ$	2—2
570,025	4	29,36	51,11	$3p^2\ {}^3P—3d\ {}^3D^\circ$	2—3
567,479	1	29,21	51,05	$3p^2\ {}^3P—3d\ {}^3D^\circ$	1—1
566,630	2	29,21	51,09	$3p^2\ {}^3P—3d\ {}^3D^\circ$	1—2
565,480	6	29,13	51,05	$3p^2\ {}^3P—3d\ {}^3D^\circ$	0—1
555,580	3	12,38	34,70	$3p\ {}^3P^\circ—3d\ {}^3D$	2—2
555,485	20	12,38	34,70	$3p\ {}^3P^\circ—3d\ {}^3D$	2—3
552,053	2	12,24	34,69	$3p\ {}^3P^\circ—3d\ {}^3D$	1—1
551,992	10	12,24	34,70	$3p\ {}^3P^\circ—3d\ {}^3D$	1—2
550,355	5	12,17	34,69	$3p\ {}^3P—3d\ {}^3D$	0—1
399,995	8	34,70	65,69	$3d\ {}^3D—4f\ {}^3F^\circ$	3—2, 3, 4
399,957	7	34,70	65,69	$3d\ {}^3D—4f\ {}^3F^\circ$	2—2, 3
399,938	5	34,69	65,69	$3d\ {}^3D—4f\ {}^3F^\circ$	1—2
325,161	25	12,38	50,51	$3p\ {}^3P^\circ—4s\ {}^3S$	2—1
323,936	20	12,24	50,51	$3p\ {}^3P^\circ—4s\ {}^3S$	1—1
323,356	15	12,17	50,51	$3p\ {}^3P^\circ—4s\ {}^3S$	0—1
243,883	3	12,38	63,21	$3p\ {}^3P^\circ—4d\ {}^3D$	2—2
243,854	12	12,38	63,22	$3p\ {}^3P^\circ—4d\ {}^3D$	2—3
243,208	2	12,24	63,21	$3p\ {}^3P^\circ—4d\ {}^3D$	1—1
243,194	8	12,24	63,21	$3p\ {}^3P^\circ—4d\ {}^3D$	1—2
242,885	5	12,17	63,21	$3p\ {}^3P^\circ—4d\ {}^3D$	0—1
195,227	3	12,38	75,88	$3p\ {}^3P^\circ—5d\ {}^3D$	2—3
194,796	2	12,24	75,88	$3p\ {}^3P^\circ—5d\ {}^3D$	1—2

ARGON, Z = 18

Ar I, ground state $1s^2 2s^2 2p^6 3s^2 3p^6\ {}^1S_0$

Ionization potential 127109,9 cm^{-1}; 15,759 eV

λ, Å	I	E_H, eV	E_B, eV	Transition	J
25660,9	65	14,25	14,74	$5s' [^1/_2]^\circ$—$4p' [^1/_2]$	1—0
25504,4	35	14,09	14,58	$5s [1^1/_2]^\circ$—$5p [^1/_2]$	1—0
25125,08	23	14,01	14,51	$3d [3^1/_2]^\circ$—$5p [2^1/_2]$	3—2
23966,68	30	13,33	13,84	$4p' [^1/_2]$—$3d [^1/_2]^\circ$	1—0
23845,13	56	13,98	14,50	$3d [3^1/_2]^\circ$—$5p [2^1/_2]$	4—3
23133,22	35	13,33	13,86	$4p' [^1/_2]$—$3d [^1/_2]^\circ$	1—1
22113,2	10	13,90	14,46	$3d [1^1/_2]^\circ$—$5p [^1/_2]$	2—1
22077,20	53	13,30	13,86	$4p' [1^1/_2]$—$3d [^1/_2]^\circ$	2—1
22039,57	9	13,28	13,84	$4p' [1^1/_2]$—$3d [^1/_2]^\circ$	1—0
21534,16	58	13,33	13,90	$4p' [^1/_2]$—$3d [1^1/_2]^\circ$	1—2
21333,27	15	13,28	13,86	$4p' [1^1/_2]$—$3d [^1/_2]^\circ$	1—1
20986,10	155	13,27	13,86	$4p [^1/_2]$—$3d [^1/_2]^\circ$	0—1
20811,14	22	13,90	14,50	$3d [1^1/_2]^\circ$—$5p [2^1/_2]$	2—3
20733,35	11	14,30	14,90	$3d' [1^1/_2]^\circ$—$4f [1^1/_2]$	1—1
20647,17	16	13,86	14,46	$3d [^1/_2]^\circ$—$5p [^1/_2]$	1—1
20616,21	356	13,30	13,90	$4p' [1^1/_2]$—$3d [1^1/_2]^\circ$	1—2
20568,5	8	13,90	14,51	$3d [1^1/_2]^\circ$—$5p [2^1/_2]$	2—2
20316,82	23	13,48	14,09	$4p' [^1/_2]$—$5p [1^1/_2]^\circ$	0—1
20069,6	7	14,06	14,66	$3d [2^1/_2]^\circ$—$5p' [1^1/_2]$	2—1
20025,90	7	13,84	14,46	$3d [^1/_2]^\circ$—$5p [^1/_2]$	0—1
19965,75	37	13,28	13,90	$4p' [1^1/_2]$—$3d [1^1/_2]^\circ$	1—2
19944,8	7	13,90	14,52	$3d [1^1/_2]^\circ$—$5p [1^1/_2]$	2—1
19817,54	75	13,90	14,53	$3d [1^1/_2]^\circ$—$5p [1^1/_2]$	2—2
18632,17	13	13,86	14,53	$3d [^1/_2]^\circ$—$5p [1^1/_2]$	1—2
18570,53	8	14,23	14,90	$3d' [1^1/_2]^\circ$—$4f [1^1/_2]$	2—2, 1
18564,74	8	14,24	14,90	$3d' [2^1/_2]^\circ$—$4f [4^1/_2]$	3—4
18429,27	40	14,23	14,91	$3d' [1^1/_2]^\circ$—$4f [2^1/_2]$	2—3, 2
18427,68	26	13,48	14,15	$4p' [^1/_2]$—$3d [1^1/_2]^\circ$	0—1
18417,91	27	14,24	14,91	$3d' [2^1/_2]^\circ$—$4f [3^1/_2]$	3—4, 3
17914,43	10	—	—	—	—
17887,35	15	14,21	14,91	$3d' [2^1/_2]^\circ$—$4f [2^1/_2]$	2—3, 2
17823,90	61	14,21	14,91	$3d' [2^1/_2]^\circ$—$4f [3^1/_2]$	2—3
17444,93	128	{ 13,30	14,01	$4p' [1^1/_2]$—$3d [3^1/_2]^\circ$	2—3
		13,15	13,86	$4p' [1^1/_2]$—$3d [^1/_2]^\circ$	1—1
16940,39	100	13,17	13,90	$4p [1^1/_2]$—$3d [1^1/_2]^\circ$	2—2
16739,94	5	13,33	14,07	$4p' [^1/_2]$—$5s [1^1/_2]^\circ$	1—2
16549,81	6	14,15	14,90	$3d [1^1/_2]^\circ$—$4f [1^1/_2]$	1—1, 2
16520,14	9	13,15	13,90	$4p [1^1/_2]$—$3d [1^1/_2]^\circ$	1—2
16436,92	18	14,15	14,91	$3d [1^1/_2]^\circ$—$4f [2^1/_2]$	1—2
16122,97	27	13,09	13,86	$4p [2^1/_2]$—$3d [^1/_2]^\circ$	2—1
15989,34	20	13,48	14,25	$4p' [^1/_2]$—$5s' [^1/_2]^\circ$	0—1
15899,93	20	14,30	15,08	$3d' [1^1/_2]^\circ$—$4f' [2^1/_2]$	1—2
15883,21	50	13,28	14,06	$4p' [1^1/_2]$—$3d [2^1/_2]^\circ$	1—2
15816,64	16	13,90	14,69	$3d [1^1/_2]^\circ$—$5p' [^1/_2]$	2—1
15402,58	10	14,10	14,90	$3d [2^1/_2]^\circ$—$4f [4^1/_2]$	3—4
15353,51	2	13,28	14,09	$4p' [1^1/_2]$—$5s [1^1/_2]^\circ$	1—1
15349,52	10	14,10	14,91	$3d [2^1/_2]^\circ$—$4f [2^1/_2]$	3—3
15329,56	5	13,09	13,90	$4p [2^1/_2]$—$3d [2^1/_2]^\circ$	2—2
15302,26	75	14,10	14,91	$3d [2^1/_2]^\circ$—$4f [3^1/_2]$	3—3, 4
15172,33	22	13,27	14,09	$4p [^1/_2]$—$5s' [1^1/_2]^\circ$	0—1
15046,42	70	13,48	14,30	$4p' [^1/_2]$—$3d' [1^1/_2]^\circ$	0—1
15030,71	42	13,33	14,15	$4p' [^1/_2]$—$3d [1^1/_2]^\circ$	1—2
14786,29	2	14,07	14,91	$5s [1^1/_2]^\circ$—$4f [2^1/_2]$	2—3

λ, Å	I	E_H, eV	E_B, eV	Transition	J
14739,11	3	13,17	14,01	$4p\,[1^1/_2]-3d\,[3^1/_2]^\circ$	2—3
14692,39	5	14,06	14,91	$3d\,[2^1/_2]^\circ-4f\,[2^1/_2]$	2—2, 3
14649,97	60	14,06	14,91	$3d\,[2^1/_2]^\circ-4f\,[3^1/_2]$	2—3, 4
14634,11	80	14,24	15,08	$3d'\,[2^1/_2]^\circ-4f'\,[3^1/_2]$	3—3, 4
14596,27	40	14,23	15,08	$3d'\,[1^1/_2]^\circ-4f'\,[2^1/_2]$	2—3
14577,51	10	13,30	14,15	$4p'\,[1^1/_2]-3d\,[1^1/_2]^\circ$	2—2
14257,46	50	14,21	15,08	$3d'\,[2^1/_2]^\circ-4f'\,[3^1/_2]$	2—3, 4
14249,93	7	13,28	14,15	$4p'\,[1^1/_2]-3d\,[1^1/_2]^\circ$	1—1
14093,61	120	13,27	14,15	$4p\,[^1/_2]-3d\,[1^1/_2]^\circ$	0—1
13992,59	10	13,33	14,21	$4p'\,[^1/_2]-3d'\,[2^1/_2]^\circ$	1—2
13910,83	150	14,01	14,90	$3d\,[3^1/_2]^\circ-4f\,[4^1/_2]$	3—4
13907,41	12	13,17	14,06	$4p\,[1^1/_2]-3d\,[2^1/_2]^\circ$	2—2
13866,97	20	14,01	14,91	$3d\,[3^1/_2]^\circ-4f\,[2^1/_2]$	3—3, 2
13828,79	20	14,01	14,91	$3d\,[3^1/_2]^\circ-4f\,[3^1/_2]$	3—3, 4
13825,99	30	13,17	14,07	$4p\,[1^1/_2]-5s\,[1^1/_2]^\circ$	2—2
13718,77	1000	13,08	13,98	$4p\,[2^1/_2]-3d\,[3^1/_2]^\circ$	3—4
13678,53	300	13,33	14,23	$4p'\,[^1/_2]-3d'\,[1^1/_2]^\circ$	1—2
13622,38	500	13,15	14,06	$4p\,[1^1/_2]-3d\,[2^1/_2]^\circ$	1—2
13599,18	55	13,30	14,21	$4p'\,[1^1/_2]-3d'\,[2^1/_2]^\circ$	2—2
13573,60	25	13,33	14,24	$4p'\,[^1/_2]-5s'\,[^1/_2]^\circ$	1—0
13543,75	15	13,15	14,07	$4p\,[1^1/_2]-5s\,[1^1/_2]^\circ$	1—2
13503,99	850	13,09	14,01	$4p\,[2^1/_2]-3d\,[3^1/_2]^\circ$	2—3
13499,24	50	13,17	14,09	$4p\,[1^1/_2]-5s\,[1^1/_2]^\circ$	2—1
13406,57	250	13,98	14,90	$3d\,[3^1/_2]^\circ-4f\,[4^1/_2]$	4—5
13367,38	800	13,17	14,10	$4p\,[1^1/_2]-3d\,[2^1/_2]^\circ$	2—3
13330,32	7	13,98	14,91	$3d\,[3^1/_2]^\circ-4f\,[3^1/_2]$	4—3, 4
13313,39	600	13,28	14,21	$4p'\,[1^1/_2]-3d^1\,[2^1/_2]^\circ$	1—2
13302,37	3	13,30	14,23	$4p'\,[1^1/_2]-3d'\,[1^1/_2]^\circ$	2—2
13273,05	750	13,30	14,24	$4p'\,[1^1/_2]-3d'\,[2^1/_2]^\circ$	2—3
13231,37	120	13,15	14,09	$4p\,[1^1/_2]-5s\,[1^1/_2]^\circ$	1—1
13228,49	200	13,08	14,01	$4p\,[2^1/_2]-3d\,[3^1/_2]^\circ$	3—3
13214,70	150	12,91	13,84	$4p\,[^1/_2]-3d\,[^1/_2]^\circ$	1—0
13028,27	5	13,28	14,23	$4p'\,[1^1/_2]-3d'\,[1^1/_2]^\circ$	1—2
13008,47	200	13,30	14,25	$4p'\,[1^1/_2]-5s'\,[^1/_2]^\circ$	2—1
12956,59	250	12,91	13,86	$4p\,[^1/_2]-3d\,[^1/_2]^\circ$	1—1
12933,33	60	13,28	14,24	$4p'\,[1^1/_2]-5s'\,[^1/_2]^\circ$	1—0
12802,68	300	13,09	14,06	$4p\,[2^1/_2]-3d\,[2^1/_2]^\circ$	2—2
12746,31	40	13,28	14,25	$4p'\,[^1/_2]-5s'\,[^1/_2]^\circ$	1—1
12733,59	75	13,09	14,07	$4p\,[2^1/_2]-5s\,[1^1/_2]^\circ$	2—2
12702,39	150	13,33	14,30	$4p'\,[^1/_2]-3d'\,[1^1/_2]^\circ$	1—1
12639,01	2	13,17	14,15	$4p\,[1^1/_2]-3d\,[1^1/_2]^\circ$	2—1
12621,82	6	13,27	14,25	$4p\,[^1/_2]-5s'\,[^1/_2]^\circ$	0—1
12596,27	5	14,10	15,08	$3d\,[2^1/_2]^\circ-4f'\,[3^1/_2]$	3—3, 4
12554,44	5	13,08	14,06	$4p\,[2^1/_2]-3d\,[2^1/_2]^\circ$	3—2
12487,63	700	13,08	14,07	$4p\,[2^1/_2]-5s\,[1^1/_2]^\circ$	3—2
12456,05	400	13,09	14,09	$4p\,[2^1/_2]-5s\,[1^1/_2]^\circ$	2—1
12439,19	500	12,91	13,90	$4p\,[^1/_2]-3d\,[1^1/_2]^\circ$	1—2
12419,39	20	13,90	14,90	$3d\,[1^1/_2]^\circ-4f\,[1^1/_2]$	2—2
12402,88	400	13,15	14,15	$4p\,[1^1/_2]-3d\,[1^1/_2]^\circ$	1—1
12356,82	100	13,90	14,91	$3d\,[1^1/_2]^\circ-4f\,[2^1/_2]$	2—3
12343,72	150	13,09	14,10	$4p\,[2^1/_2]-3d\,[2^1/_2]^\circ$	2—3
12151,57	15	14,06	15,08	$3d\,[2^1/_2]^\circ-4f'\,[3^1/_2]$	2—3
12139,79	100	13,28	14,30	$4p'\,[1^1/_2]-3d'\,[1^1/_2]^\circ$	1—1
12112,20	300	13,08	14,10	$4p\,[2^1/_2]-3d\,[2^1/_2]^\circ$	3—3
12026,63	5	13,27	14,30	$4p\,[^1/_2]-3d'\,[1^1/_2]^\circ$	0—1
11943,50	25	13,86	14,90	$3d\,[^1/_2]^\circ-4f\,[1^1/_2]$	1—1, 2
11896,60	3	13,17	14,21	$4p\,[1^1/_2]-3d'\,[2^1/_2]^\circ$	2—2
11884,47	5	13,86	14,91	$3d\,[^1/_2]^\circ-4f\,[2^1/_2]$	1—2

λ, Å	I	E_H, eV	E_B, eV	Transition	J
11733,26	20	13,84	14,90	$3d\,[^1/_2]^\circ - 4f\,[1^1/_2]$	0—1
11719,51	30	13,09	14,15	$4p\,[2^1/_2] - 3d\,[1^1/_2]^\circ$	2—1
11708,22	3	14,15	15,21	$3d\,[1^1/_2]^\circ - 5f\,[1^1/_2]$	1—1, 2
11687,61	5	13,15	14,21	$4p\,[1^1/_2] - 3d'\,[2^1/_2]^\circ$	1—2
11678,47	4	14,15	15,21	$3d\,[1^1/_2]^\circ - 5f\,[2^1/_2]$	1—2
11668,72	100	13,17	14,23	$4p\,[1^1/_2] - 3d'\,[1^1/_2]^\circ$	2—2
11580,39	8	14,01	15,08	$3d\,[3^1/_2]^\circ - 4f'\,[3^1/_2]$	3—3, 4
11488,12	150	11,83	12,91	$4s'\,[^1/_2]^\circ - 4p\,[^1/_2]$	1—1
11467,57	30	13,15	14,23	$4p\,[1^1/_2] - 3d'\,[1^1/_2]^\circ$	1—2
11441,83	80	13,17	14,25	$4p\,[1^1/_2] - 5s'\,[^1/_2]^\circ$	2—1
11398,63	7	14,30	15,39	$3d'\,[1^1/_2]^\circ - 5f'\,[2^1/_2]$	1—2
11393,66	50	13,15	14,24	$4p\,[1^1/_2] - 5s'\,[^1/_2]^\circ$	1—0
11248,33	8	13,15	14,25	$4p\,[1^1/_2] - 5s'\,[^1/_2]^\circ$	1—1
11209,67	1	14,10	15,20	$3d\,[2^1/_2]^\circ - 6p'\,[1^1/_2]$	3—2
11195,37	2	13,90	15,01	$3d\,[1^1/_2]^\circ - 6p\,[^1/_2]$	2—1
11133,86	20	14,10	15,21	$3d\,[2^1/_2]^\circ - 5f\,[4^1/_2]$	3—4
11118,75	20	14,10	15,21	$3d\,[2^1/_2]^\circ - 5f\,[2^1/_2]$	3—3, 2
11106,44	60	14,10	15,21	$3d\,[2^1/_2]^\circ - 5f\,[3^1/_2]$	3—3, 4
11078,87	200	13,09	14,21	$4p\,[2^1/_2] - 3d'\,[2^1/_2]^\circ$	2—2
11043,13	2	13,90	15,03	$3d\,[1^1/_2]^\circ - 6p\,[2^1/_2]$	2—2
11028,60	1	14,09	15,21	$5s\,[1^1/_2]^\circ - 5f\,[2^1/_2]$	1—2
10977,30	1	14,15	15,28	$3d\,[1^1/_2]^\circ - 7p\,[1^1/_2]$	1—1
10964,00	2	13,90	15,03	$3d\,[1^1/_2]^\circ - 6p\,[1^1/_2]$	2—1
10950,74	120	13,17	14,30	$4p\,[1^1/_2] - 3d'\,[1^1/_2]^\circ$	2—1
10947,90	20	13,90	15,03	$3d\,[1^1/_2]^\circ - 6p\,[1^1/_2]$	2—2
10895,9	1	14,06	15,20	$3d\,[2^1/_2]^\circ - 6p'\,[^1/_2]$	2—1
10892,37	30	13,08	14,21	$4p\,[2^1/_2] - 3d'\,[2^1/_2]^\circ$	3—2
10885,9	2	14,06	15,20	$3d\,[2^1/_2]^\circ - 6p'\,[^1/_2]$	2—1
10880,96	150	13,09	14,23	$4p\,[2^1/_2] - 3d'\,[1^1/_2]^\circ$	2—2
10861,04	25	13,09	14,24	$4p\,[2^1/_2] - 3d'\,[2^1/_2]^\circ$	2—3
10845,43	2	14,07	14,21	$5s\,[1^1/_2]^\circ - 5f\,[1^1/_2]$	2—1, 2
10837,39	1	14,24	15,38	$3d'\,[2^1/_2]^\circ - 6f\,[4^1/_2]$	3—4
10831,88	1	14,24	15,38	$3d'\,[2^1/_2]^\circ - 6f\,[2^1/_2]$	2—3, 2
10824,00	1	{ 14,23	15,38	$3d'\,[1^1/_2]^\circ - 6f\,[1^1/_2]$	2—2, 1
		14,24	15,38	$3d'\,[2^1/_2]^\circ - 6f\,[3^1/_2]$	2—3
10822,74	1	14,15	15,30	$3d\,[1^1/_2]^\circ - 7p\,[^1/_2]$	1—0
10820,18	6	14,07	15,21	$5s\,[1^1/_2]^\circ - 5f\,[2^1/_2]$	2—3, 2
10812,16	1	14,23	15,38	$3d'\,[1^1/_2]^\circ - 6f\,[2^1/_2]$	2—3, 2
10807,04	5	13,86	15,01	$3d\,[^1/_2]^\circ - 6p\,[^1/_2]$	1—1
10795,91	2	14,06	15,21	$3d\,[2^1/_2]^\circ - 5f\,[1^1/_2]$	2—1, 2
10773,35	30	13,15	14,30	$4p\,[1^1/_2] - 3d'\,[1^1/_2]^\circ$	1—1
10770,35	15	14,06	15,21	$3d\,[2^1/_2]^\circ - 5f\,[2^1/_2]$	2—3, 2
10759,13	60	14,06	15,21	$3d\,[2^1/_2]^\circ - 5f\,[3^1/_2]$	2—3
10733,87	50	14,24	15,39	$3d'\,[2^1/_2]^\circ - 5f'\,[3^1/_2]$	3—3, 4
10732,10	2	14,24	15,39	$3d'\,[2^1/_2]^\circ - 5f'\,[2^1/_2]$	3—3, 2
10722,22	6	12,91	14,06	$4p\,[^1/_2] - 3d\,[2^1/_2]^\circ$	1—2
10712,77	40	14,23	15,39	$3d'\,[1^1/_2]^\circ - 5f'\,[2^1/_2]$	2—3, 2
10700,98	80	13,08	14,23	$4p\,[2^1/_2] - 3d'\,[1^1/_2]^\circ$	3—2
10683,40	50	13,09	14,25	$4p\,[2^1/_2] - 5s'\,[^1/_2]^\circ$	2—1
10681,78	200	13,08	14,24	$4p\,[2^1/_2] - 3d'\,[2^1/_2]^\circ$	3—3
10673,55	500	12,91	14,07	$4p\,[^1/_2] - 5s\,[1^1/_2]^\circ$	1—2
10634,25	5	13,84	15,01	$3d\,[^1/_2]^\circ - 6p\,[^1/_2]$	0—1
10623,38	2	14,21	15,38	$3d'\,[2^1/_2]^\circ - 6f\,[2^1/_2]$	2—3, 2
10615,7	1	14,21	15,38	$3d'\,[2^1/_2]^\circ - 6f\,[3^1/_2]$	2—3
10591,23	2	13,86	15,03	$3d\,[^1/_2]^\circ - 6p\,[1^1/_2]$	1—1
10576,18	4	13,86	15,03	$3d\,[^1/_2]^\circ - 6p\,[1^1/_2]$	1—2
10529,32	50	14,21	15,39	$3d'\,[2^1/_2]^\circ - 5f'\,[3^1/_2]$	2—3
10527,34	2	14,21	15,39	$3d'\,[2^1/_2]^\circ - 5f'\,[2^1/_2]$	2—3, 2
10506,47	100	13,90	15,08	$3d\,[1^1/_2]^\circ - 4f'\,[2^1/_2]$	2—3, 2

λ, A	I	E_H, eV	E_B, eV	Transition	J
10478,10	200	12,91	14,09	4p [1/2]—5s [1 1/2]°	1—1
10470,051	500	11,72	12,91	4s′ [1/2]°—4p [1/2]	0—1
10357,6	1	13,86	15,06	3d [1/2]°—6p [1/2]	1—0
10332,76	60	14,01	15,21	3d [3 1/2]°—5f [4 1/2]	3—4
10319,62	2	14,01	15,21	3d [3 1/2]°—5f [2 1/2]	3—3, 2
10309,15	20	14,01	15,21	3d [3 1/2]°—5f [3 1/2]	3—3, 4
10266,79	1	14,07	15,27	5s [1 1/2]°—7p [2 1/2]	2—3
10254,04	10	13,09	14,30	4p [2 1/2]—3d′ [1 1/2]°	2—1
10208,7	1	14,07	15,28	5s [1 1/2]°—7p [1 1/2]	2—1
10206,9	1	14,06	15,28	3d [2 1/2]°—7p [2 1/2]	2—2
10171,2	1	14,06	15,28	3d [2 1/2]°—7p [1 1/2]	2—1
10163,45	30	13,86	15,08	3d [1/2]°—4f′ [2 1/2]	1—2
10104,82	4	14,15	15,38	3d [1 1/2]°—6f [1 1/2]	1—1, 2
10094,32	8	14,15	15,38	3d [1 1/2]°—6f [2 1/2]	1—2
10069,04	50	13,48	14,71	4p′ [1/2]—4d [1/2]°	0—1
10052,10	150	13,98	15,21	3d [3 1/2]°—5f [4 1/2]	4—4, 5
10039,75	2	13,98	15,21	3d [3 1/2]°—5f [2 1/2]	4—3
10029,70	40	13,98	15,21	3d [3 1/2]°—5f [3 1/2]	4—3, 4
10007,61	3	14,15	15,39	3d [1 1/2]°—5f′ [2 1/2]	1—2
9994,94	1	14,21	15,45	3d′ [2 1/2]°—7p′ [1 1/2]	2—1
9951,88	20	12,91	14,15	4p [1/2]—3d [1 1/2]°	1—1
9937,80	1	14,23	15,48	3d′ [1 1/2]°—7f [2 1/2]	2—3, 2
9882,18	6	14,30	15,56	3d′ [1 1/2]°—6f′ [2 1/2]	1—2
9815,22	1	14,01	15,27	3d [3 1/2]°—7p [2 1/2]	3—3
9800,92	4	14,01	15,28	3d [3 1/2]°—7p [2 1/2]	3—2
9784,5010	1000	11,83	13,09	4s′ [1/2]°—4p [2 1/2]	1—2
9774,79	1	14,21	15,48	3d′ [2 1/2]°—4f [3 1/2]	2—3
9677,80	8	14,10	15,38	3d [2 1/2]°—6f [4 1/2]	3—4
9673,39	6	14,10	15,38	3d [2 1/2]°—6f [2 1/2]	3—3, 2
9666,86	50	14,10	15,38	3d [2 1/2]°—6f [3 1/2]	3—3, 4
9657,7841	1500	11,62	12,91	4s [1 1/2]°—4p [1/2]	1—1
9595,09	4	14,10	15,39	3d [2 1/2]°—5f′ [3 1/2]	3—3, 4
9593,67	1	14,10	15,39	3d [2 1/2]°—5f′ [2 1/2]	3—3, 2
9561,60	5	13,98	15,27	3d [3 1/2]°—7p [2 1/2]	4—3
9555,2	4	13,90	15,20	3d [1 1/2]°—6p′ [1 1/2]	2—1
9547,73	2	13,90	15,20	3d [1 1/2]°—6p′ [1/2]	2—1
9486,02	3	12,91	14,21	4p [1/2]—3d′ [2 1/2]°	1—2
9478,39	50	13,90	15,21	3d [1 1/2]°—5f [1 1/2]	2—1, 2
9459,09	100	13,90	15,21	3d [1 1/2]°—5f [2 1/2]	2—3, 2
9446,57	2	14,07	15,38	5s [1 1/2]°—6f [2 1/2]	2—3, 2
9408,66	3	14,06	15,38	3d [2 1/2]°—6f [2 1/2]	2—3, 2
9402,69	20	14,06	15,38	3d [2 1/2]°—6f [3 1/2]	2—3
9377,63	5	14,24	15,56	{ 3d′ [2 1/2]°—6f′ [2 1/2]	3—3, 2
				3d′ [2 1/2]°—6f′ [3 1/2]	3—3, 4
9362,50	4	14,23	15,56	{ 3d′ [1 1/2]°—6f′ [2 1/2]	2—3, 2
				3d′ [1 1/2]°—6f [3 1/2]	2—3
9354,218	200	11,83	13,15	4s′ [1/2]°—4p [1 1/2]	1—1
9340,59	3	12,91	14,23	4p [1/2]—3d′ [1 1/2]°	1—2
9334,80	8	{ 14,15	15,48	3d [1 1/2]°—7f [1 1/2]	1—1, 2
		14,06	15,39	3d [2 1/2]°—5f′ [3 1/2]	2—3
9333,32	1	14,06	15,39	3d [2 1/2]°—5f′ [2 1/2]	2—3, 2
9328,08	2	14,15	15,48	3d [1 1/2]°—7f [2 1/2]	2—2
9291,58	100	12,91	14,24	4p [1/2]—5s′ [1/2]°	1—0
9242,17	1	13,86	15,20	3d [1/2]°—6p′ [1 1/2]	1—2
9224,4955	1000	11,83	13,17	4s′ [1/2]°—4p [1 1/2]	1—2
9221,08	5	14,21	15,56	3d′ [2 1/2]°—6f′ [3 1/2]	2—3
9198,61	50	13,86	15,21	3d [1/2]°—5f [1 1/2]	1—1, 2
9194,637	150	12,91	14,25	4p [1/2]—5s′ [1/2]°	1—1

λ, Å	I	E_H, eV	E_B, eV	Transition	J
9180,17	6	13,86	15,21	$3d\ [^1/_2]°—5f\ [2^1/_2]$	1—2
9122,9660	500	11,55	12,91	$4s\ [1^1/_2]°—4p\ [^1/_2]$	2—1
9111,3	1	13,86	15,22	$3d\ [^1/_2]°—6p'\ [^1/_2]$	1—0
9075,42	60	13,33	14,69	$4p'\ [^1/_2]—4d\ [^1/_2]°$	1—0
9073,34	50	13,84	15,21	$3d\ [^1/_2]°—5f\ [1^1/_2]$	0—1
9066,77	40	14,01	15,38	$3d\ [3^1/_2]°—6f\ [4^1/_2]$	3—4
9057,51	2	13,48	14,85	$4p'\ [0^1/_2]—6s\ [1^1/_2]°$	0—1
9057,23	4	14,01	15,38	$3d\ [3^1/_2]—6f\ [3^1/_2]$	3—3, 4
8994,09	10	14,01	15,39	$3d\ [3^1/_2]°—5f'\ [3^1/_2]$	3—3, 4
8992,84	1	14,01	15,39	$3d\ [3^1/_2]°—5f'\ [2^1/_2]$	3—3, 2
8988,20	3	13,90	15,28	$3d\ [1^1/_2]°—7p\ [1^1/_2]$	2—2
8970,98	2	14,10	15,48	$3d\ [2^1/_2]°—7f\ [4^1/_2]$	3—4
8967,39	2	14,10	15,48	$3d\ [2^1/_2]°—7f\ [2^1/_2]$	3—3, 2
8964,48	10	14,10	15,48	$3d\ [2^1/_2]°—7f\ [3^1/_2]$	3—3, 4
8962,19	40	13,33	14,71	$4p'\ [^1/_2]—4d\ [^1/_2]°$	1—1
8895,42	1	14,15	15,54	$3d\ [1^1/_2]°—8f\ [1^1/_2]$	1—1, 2
8891,70	1	14,15	15,55	$3d\ [1^1/_2]°—8f\ [2^1/_2]$	1—2
8874,84	4	12,91	14,30	$4p\ [^1/_2]—3d'\ [1^1/_2]°$	1—1
8849,97	150	13,98	15,38	$3d\ [3^1/_2]°—6f\ [4^1/_2]$	4—4, 5
8846,17	1	13,98	15,38	$3d\ [3^1/_2]°—6f\ [2^1/_2]$	4—3
8840,82	20	13,98	15,38	$3d\ [3^1/_2]°—6f\ [3^1/_2]$	4—3, 4
8840,39	3	—	—	—	—
8819,37	1	14,15	15,56	$3d\ [1^1/_2]°—6f'\ [2^1/_2]$	1—2
8805,16	3	13,86	15,27	$3d\ [^1/_2]°—7p\ [^1/_2]$	1—1
8799,082	100	13,30	14,71	$4p'\ [1^1/_2]—4d\ [^1/_2]°$	2—1
8784,59	30	13,28	14,69	$4p'\ [1^1/_2]—4d\ [^1/_2]°$	1—0
8761,6907	200	13,33	14,74	$4p'\ [^1/_2]—4d\ [1^1/_2]°$	1—2
8741,26	1	13,86	15,28	$3d\ [^1/_2]°—7p\ [1^1/_2]$	1—1
8739,51	3	14,06	15,48	$3d\ [2^1/_2]°—7f\ [2^1/_2]$	2—3, 2
8736,63	20	14,06	15,48	$3d\ [2^1/_2]°—7f\ [3^1/_2]$	2—3
8736,19	2	13,86	15,28	$3d\ [^1/_2]°—7p\ [1^1/_2]$	1—2
8713,79	5	14,27	15,66	$3d'\ [2^1/_2]°—7f'\ [2^1/_2]$	3—3, 2
8700,95	3	14,23	15,66	$3d'\ [1^1/_2]°—7f'\ [2^1/_2]$	2—3, 2
8690,12	2	13,84	15,27	$3d\ [^1/_2]°—7p\ [^1/_2]$	0—1
8678,43	60	13,28	14,71	$4p'\ [1^1/_2]—4d\ [^1/_2]°$	1—1
8667,9438	400	11,72	13,15	$4s'\ [^1/_2]°—2p\ [1^1/_2]$	0—1
8642,89	1	13,86	15,30	$3d\ [^1/_2]°—7p\ [^1/_2]$	1—0
8620,4602	100	13,27	14,71	$4p\ [^1/_2]—4d\ [^1/_2]°$	0—1
8605,7790	150	13,30	14,74	$4p'\ [1^1/_2]—4d\ [1^1/_2]°$	2—2
8579,49	4	14,21	15,66	$3d'\ [2^1/_2]°—7f'\ [3^1/_2]$	2—3
8578,06	5	11,83	13,27	$4s'\ [^1/_2]°—4p\ [^1/_2]$	1—0
8565,13	1	14,10	15,55	$3d\ [2^1/_2]°—8f\ [4^1/_2]$	3—4
8563,38	1	14,10	15,55	$3d\ [2^1/_2]°—8f\ [2^1/_2]$	3—3, 2
8561,38	3	14,10	15,55	$3d\ [2^1/_2]°—8f\ [3^1/_2]$	3—3, 4
8521,4428	2000	11,83	13,28	$4s'\ [^1/_2]°—4p'\ [1^1/_2]$	1—1
8496,64	2	14,10	15,56	{ $3d\ [2^1/_2]°—6f'\ [2^1/_2]$ $3d\ [2^1/_2]°—6f'\ [3^1/_2]$	3—3, 4 3—3, 4
8490,30	40	13,28	14,74	$4p'\ [^1/_2]—4d\ [1^1/_2]°$	1—2
8443,44	20	14,01	15,48	$3d\ [3^1/_2]°—7f\ [4^1/_2]$	3—4
8440,26	1	14,01	15,48	$3d\ [3^1/_2]°—7f\ [2^1/_2]$	3—3, 2
8437,71	6	14,01	15,48	$3d\ [3^1/_2]°—7f\ [3^1/_2]$	3—3, 4
8424,6473	2500	11,62	13,09	$4s\ [1^1/_2]°—4p\ [2^1/_2]$	1—2
8408,2094	3000	11,83	13,30	$4s'\ [^1/_2]°—4p'\ [1^1/_2]$	1—2
8399,35	20	13,90	15,38	$3d\ [1^1/_2]°—6f\ [1^1/_2]$	2—1, 2
8392,28	80	13,90	15,38	$3d\ [1^1/_2]°—6f\ [2^1/_2]$	2—3, 2
8384,73	60	13,30	14,78	$4p'\ [1^1/_2]—4d\ [3^1/_2]°$	2—3
8367,03	3	13,33	14,81	$4p'\ [^1/_2]—4d\ [2^1/_2]°$	1—2
8355,30	1	14,06	15,55	$3d\ [2^1/_2]°—8f\ [2^1/_2]$	2—3, 2

λ, Å	I	E_H, eV	E_B, eV	Transition	J
8353,50	4	14,06	15,55	$3d\ [2^1/_2]^\circ - 8f\ [3^1/_2]$	2—3
8332,21	20	13,90	15,39	$3d\ [1^1/_2]^\circ - 5f'\ [2^1/_2]$	2—3, 2
8305,02	1	14,10	15,59	$3d\ [2^1/_2]^\circ - 9f\ [2^1/_2]$	3—3, 2
8303,79	1	14,10	15,55	$3d\ [2^1/_2]^\circ - 9f\ [3^1/_2]$	3—3, 4
8291,88	8	14,06	15,56	$3d\ [2^1/_2]^\circ - 6f'\ [3^1/_2]$	2—3
8264,5221	1500	11,83	13,33	$4s'\ [^1/_2]^\circ - 4p'\ [^1/_2]$	1—1
8255,07	50	13,98	15,48	$3d\ [3^1/_2]^\circ - 7f\ [4^1/_2]$	4—4, 5
8249,58	4	13,98	15,48	$3d\ [3^1/_2]^\circ - 7f\ [3^1/_2]$	4—3, 4
8224,72	6	13,30	14,81	$4p'\ [1^1/_2] - 4d\ [2^1/_2]^\circ$	2—2
8203,42	20	13,33	14,84	$4p'\ [^1/_2] - 6s\ [1^1/_2]^\circ$	1—2
8178,96	20	13,86	15,38	$3d\ [^1/_2]^\circ - 6f\ [1^1/_2]$	1—1, 2
8178,84	40	—	—	—	—
8171,95	10	13,86	15,38	$3d\ [^1/_2]^\circ - 6f\ [2^1/_2]$	1—2
8151,86	3	13,33	14,85	$4p'\ [^1/_2] - 6s\ [1^1/_2]^\circ$	1—1
8143,54	10	13,30	14,82	$4p'\ [1^1/_2] - 4d\ [2^1/_2]^\circ$	2—2, 3
8119,18	50	13,28	14,81	$4p'\ [1^1/_2] - 4d\ [2^1/_2]^\circ$	1—2
8115,3108	5000	11,55	13,08	$4s\ [1^1/_2]^\circ - 4p\ [2^1/_2]$	2—3
8103,6920	2000	11,62	13,15	$4s\ [1^1/_2]^\circ - 4p\ [1^1/_2]$	1—1
8094,06	20	13,33	14,86	$4p'\ [^1/_2] - 4d\ [1^1/_2]^\circ$	1—1
8089,93	5	14,01	15,55	$3d\ [3^1/_2]^\circ - 8f\ [4^1/_2]$	3—4
8079,68	20	13,84	15,38	$3d\ [^1/_2]^\circ - 6f\ [1^1/_2]$	0—1
		14,01	15,55	$3d\ [3^1/_2]^\circ - 8f\ [3^1/_2]$	3—3, 4
8066,60	20	13,30	14,84	$4p'\ [1^1/_2] - 6s\ [1^1/_2]^\circ$	2—2
8053,305	100	13,17	14,71	$4p\ [1^1/_2] - 4d\ [^1/_2]^\circ$	2—1
8046,13	50	13,15	14,69	$4p\ [1^1/_2] - 4d\ [^1/_2]^\circ$	1—0
8037,23	20	13,48	15,02	$4p'\ [^1/_2] - 6s'\ [^1/_2]^\circ$	0—1
8021,9	2	14,01	15,56	$3d\ [3^1/_2]^\circ - 6f'\ [3^1/_2]$	3—3, 4
8014,7853	800	11,55	13,09	$4s\ [1^1/_2]^\circ - 4p\ [2^1/_2]$	2—2
8006,1566	600	11,62	13,17	$4s\ [1^1/_2]^\circ - 4p\ [1^1/_2]$	1—2
7965,08	3	13,28	14,84	$4p'\ [1^1/_2] - 6s\ [1^1/_2]^\circ$	1—2
7960,84	2	13,30	14,86	$4p'\ [1^1/_2] - 4d\ [1^1/_2]^\circ$	2—1
7956,99	10	13,15	14,71	$4p\ [1^1/_2] - 4d\ [^1/_2]^\circ$	1—1
7948,1755	400	11,72	13,28	$4s'\ [^1/_2]^\circ - 4p'\ [1^1/_2]$	0—1
7916,45	20	13,28	14,85	$4p'\ [1^1/_2] - 6s\ [1^1/_2]^\circ$	1—1
7910,23	4	13,98	15,55	$3d\ [3^1/_2]^\circ - 8f\ [4^1/_2]$	4—4, 5
7906,91	1	13,98	15,55	$3d\ [3^1/_2]^\circ - 8f\ [3^1/_2]$	4—3, 4
7891,0777	100	13,17	14,74	$4p\ [1^1/_2] - 4d\ [1^1/_2]^\circ$	2—2
7868,20	40	13,27	14,85	$4p\ [^1/_2] - 6s\ [1^1/_2]^\circ$	0—1
7861,91	15	13,28	14,86	$4p'\ [1^1/_2] - 4d\ [1^1/_2]^\circ$	1—1
7860,44	2	13,90	15,48	$3d\ [1^1/_2]^\circ - 7f\ [1^1/_2]$	2—1, 2
7855,73	8	13,90	15,48	$3d\ [1^1/_2]^\circ - 7f\ [2^1/_2]$	3—3, 2
7853,29	1	14,01	15,59	$3d\ [3^1/_2]^\circ - 9f\ [4^1/_2]$	3—4
7814,33	10	13,27	14,86	$4p\ [^1/_2] - 4d\ [1^1/_2]^\circ$	0—1
7798,55	30	13,15	14,74	$4p\ [1^1/_2] - 4d\ [1^1/_2]^\circ$	1—2
7724,2064	200	11,72	13,33	$4s'\ [^1/_2]^\circ - 4p'\ [^1/_2]$	0—1
7723,7599	200	11,55	13,15	$4s\ [1^1/_2]^\circ - 4p\ [1^1/_2]$	2—1
7704,81	20	13,17	14,78	$4p\ [1^1/_2] - 4d\ [3^1/_2]^\circ$	2—3
7690,10	2	13,98	15,59	$3d\ [3^1/_2]^\circ - 9f\ [4^1/_2]$	4—4, 5
7670,04	50	13,09	14,71	$4p\ [2^1/_2] - 4d\ [^1/_2]^\circ$	2—1
7667,03	4	13,86	15,48	$3d\ [^1/_2]^\circ - 7f\ [1^1/_2]$	1—1, 2
7662,3	2	13,86	15,48	$3d\ [^1/_2]^\circ - 7f\ [2^1/_2]$	1—2
7635,1056	500	11,55	13,17	$4s\ [1^1/_2]^\circ - 4p\ [1^1/_2]$	2—2
7628,86	50	13,33	14,95	$4p'\ [^1/_2] - 4d'\ [2^1/_2]^\circ$	1—2
7618,33	30	13,33	14,95	$4p'\ [^1/_2] - 4d'\ [1^1/_2]^\circ$	1—2
7514,6514	200	11,62	13,27	$4s\ [1^1/_2]^\circ - 4p\ [^1/_2]$	1—0
7510,42	10	13,30	14,95	$4p'\ [1^1/_2] - 4d'\ [2^1/_2]^\circ$	2—2
7503,8685	700	11,83	13,48	$4s'\ [^1/_2]^\circ - 4p'\ [^1/_2]$	1—0
7484,24	15	13,15	14,81	$4p\ [1^1/_2] - 4d\ [2^1/_2]^\circ$	1—2
7471,1676	4	11,62	13,28	$4s\ [1^1/_2]^\circ - 4p'\ [1^1/_2]$	1—1

λ, Å	I	E_H, eV	E_B, eV	Transition	J
7436,25	10	13,08	14,74	$4p\ [2^1/_2]—4d\ [1^1/_2]^\circ$	3—2
7435,33	30	13,17	14,84	$4p\ [1^1/_2]—6s\ [1^1/_2]^\circ$	2—2
7425,290	12	13,30	14,97	$4p'\ [1^1/_2]—4d'\ [2^1/_2]^\circ$	2—3
7422,26	6	13,28	14,95	$4p'\ [1^1/_2]—4d'\ [2^1/_2]^\circ$	1—2
7412,334	15	13,28	14,95	$4p'\ [1^1/_2]—4d'\ [1^1/_2]^\circ$	1—2
7392,97	15	13,17	14,85	$4p\ [1^1/_2]—6s\ [1^1/_2]^\circ$	2—1
7383,9796	400	11,62	13,30	$4s\ [1^1/_2]^\circ—4p'\ [1^1/_2]$	1—2
7372,1189	100	13,08	14,76	$4p\ [2^1/_2]—4d\ [3^1/_2]^\circ$	3—4
7353,316	100	13,09	14,78	$4p\ [2^1/_2]—4d\ [3^1/_2]^\circ$	3—4
		13,15	14,84	$4p\ [1^1/_2]—6s\ [1^1/_2]^\circ$	1—2
7350,78	6	13,33	15,01	$4p'\ [^1/_2]—6s'\ [^1/_2]^\circ$	1—0
7345,34	1	13,17	14,86	$4p\ [1^1/_2]—4d\ [1^1/_2]^\circ$	2—1
7316,0068	30	13,33	15,02	$4p'\ [^1/_2]—6s'\ [^1/_2]^\circ$	1—1
7311,724	100	13,15	14,85	$4p\ [1^1/_2]—6s\ [1^1/_2]^\circ$	1—1
7284,44	6	13,30	15,00	$4p'\ [1^1/_2]—4d'\ [1^1/_2]^\circ$	2—1
7272,9349	100	11,62	13,33	$4s\ [1^1/_2]^\circ—4p'\ [^1/_2]$	1—1
7270,66	10	13,08	14,78	$4p\ [2^1/_2]—4d\ [3^1/_2]^\circ$	3—3
7267,20	2	13,48	15,18	$4p'\ [^1/_2]—7s\ [1^1/_2]^\circ$	0—1
7265,173	3	13,15	14,86	$4p\ [1^1/_2]—4d\ [1^1/_2]^\circ$	1—1
7229,93	4	13,09	14,81	$4p\ [2^1/_2]—4d\ [2^1/_2]^\circ$	2—2
7206,9812	100	13,30	15,02	$4p'\ [1^1/_2]—6s'\ [^1/_2]^\circ$	2—1
7202,55	2	13,28	15,00	$4p'\ [1^1/_2]—4d'\ [1^1/_2]^\circ$	1—1
7176,34	4	—	—	—	—
7162,57	8	13,27	15,00	$4p\ [^1/_2]—4d'\ [1^1/_2]^\circ$	0—1
7158,83	30	13,28	15,01	$4p'\ [1^1/_2]—6s'\ [^1/_2]^\circ$	1—0
7147,0408	30	11,55	13,28	$4s\ [1^1/_2]^\circ—4p'\ [1^1/_2]$	2—1
7125,825	30	13,28	15,02	$4p'\ [1^1/_2]—6s'\ [^1/_2]^\circ$	1—1
7107,4777	200	13,09	14,84	$4p\ [2^1/_2]—6s\ [1^1/_2]^\circ$	2—2
7086,70	15	13,27	15,02	$4p\ [^1/_2]—6s'\ [^1/_2]^\circ$	0—1
7068,73	30	13,09	14,85	$4p\ [2^1/_2]—6s\ [1^1/_2]^\circ$	2—1
7067,2175	400	11,55	13,30	$4s\ [1^1/_2]^\circ—4p'\ [1^1/_2]$	2—2
7030,2519	100	13,09	14,84	$4p\ [2^1/_2]—6s\ [1^1/_2]^\circ$	3—2
6992,17	4	13,33	15,10	$4p'\ [^1/_2]—5d\ [^1/_2]^\circ$	1—0
6965,4304	400	11,55	13,33	$4s\ [1^1/_2]^\circ—4p'\ [^1/_2]$	2—1
6960,23	20	13,17	14,95	$4p\ [1^1/_2]—4d'\ [2^1/_2]^\circ$	2—2
6951,46	20	13,17	14,95	$4p\ [1^1/_2]—4d'\ [1^1/_2]^\circ$	2—2
6937,6658	100	12,91	14,69	$4p\ [^1/_2]—4d\ [^1/_2]^\circ$	1—0
6925,010	2	13,33	15,12	$4p'\ [^1/_2]—5d\ [^1/_2]^\circ$	1—1
6888,1704	100	13,15	14,95	$4p\ [1^1/_2]—4d'\ [2^1/_2]^\circ$	1—2
6887,10	20	13,17	14,97	$4p\ [1^1/_2]—4d'\ [2^1/_2]^\circ$	2—3
6879,59	40	13,15	14,95	$4p\ [1^1/_2]—4d'\ [1^1/_2]^\circ$	1—2
6871,2898	150	12,91	14,71	$4p\ [^1/_2]—4d\ [^1/_2]^\circ$	1—1
6851,884	4	13,33	15,14	$4p'\ [^1/_2]—5d\ [1^1/_2]^\circ$	1—2
6827,2529	30	13,30	15,12	$4p'\ [1^1/_2]—5d\ [^1/_2]^\circ$	2—1
6818,291	4	13,28	15,10	$4p'\ [1^1/_2]—5d\ [^1/_2]^\circ$	1—0
6779,933	4	13,48	15,31	$4p'\ [^1/_2]—6d\ [^1/_2]^\circ$	0—1
6766,6134	100	13,17	15,00	$4p\ [1^1/_2]—4d'\ [1^1/_2]^\circ$	2—1
6756,10	100	13,30	15,14	$4p'\ [1^1/_2]—5d\ [1^1/_2]^\circ$	2—2
6754,30	8	13,28	15,12	$4p'\ [1^1/_2]—5d\ [^1/_2]^\circ$	1—1
6752,8347	100	12,91	14,74	$4p\ [^1/_2]—4d\ [1^1/_2]^\circ$	1—2
6722,893	4	13,30	15,15	$4p'\ [1^1/_2]—5d\ [3^1/_2]^\circ$	2—3
6719,2193	100	13,27	15,12	$4p\ [^1/_2]—5d\ [^1/_2]^\circ$	0—1
6698,8752	100	13,17	15,02	$4p\ [1^1/_2]—6s'\ [^1/_2]^\circ$	2—1
6698,474	6	13,15	15,00	$4p\ [1^1/_2]—4d'\ [1^1/_2]^\circ$	1—1
6689,91	2	13,33	15,18	$4p'\ [^1/_2]—7s\ [1^1/_2]^\circ$	1—2
6684,73	6	13,28	15,14	$4p'\ [1^1/_2]—5d\ [1^1/_2]^\circ$	2—2
6677,2812	30	11,62	13,48	$4s\ [1^1/_2]^\circ—4p'\ [^1/_2]$	1—0
6672,10	2	13,09	14,95	$4p\ [2^1/_2]—4d'\ [2^1/_2]^\circ$	2—2

λ, Å	I	E_H, eV	E_B, eV	Transition	J
6664,0533	100	13,09	14,95	$4p\ [2^1/_2]-4d'\ [1^1/_2]^\circ$	2—2
6660,6784	100	13,15	15,01	$4p\ [1^1/_2]-6s'\ [^1/_2]^\circ$	1—0
6656,88	6	13,33	15,19	$4p'\ [^1/_2]-5d\ [1^1/_2]^\circ$	1—1
6632,087	8	13,15	15,02	$4p\ [1^1/_2]-6s'\ [^1/_2]^\circ$	1—1
6604,8542	30	13,09	14,97	$4p\ [2^1/_2]-4d'\ [2^1/_2]^\circ$	2—3
6604,02	2	13,08	14,95	$4p\ [2^1/_2]-4d'\ [2^1/_2]^\circ$	3—2
6598,684	6	13,30	15,18	$4p'\ [1^1/_2]-7s\ [1^1/_2]^\circ$	2—2
		13,28	15,16	$4p'\ [1^1/_2]-5d\ [2^1/_2]^\circ$	1—2
6596,1155	8	13,08	14,95	$4p\ [2^1/_2]-4d'\ [1^1/_2]^\circ$	3—2
6594,66	2	13,48	15,36	$4p'\ [^1/_2]-7s'\ [^1/_2]^\circ$	0—1
6581,60	2	13,30	15,18	$4p'\ [1^1/_2]-7s\ [1^1/_2]^\circ$	2—1
6571,37	2	13,48	15,37	$4p'\ [^1/_2]-6d\ [1^1/_2]^\circ$	0—1
6538,1137	30	13,08	14,95	$4p\ [2^1/_2]-4d'\ [1^1/_2]^\circ$	3—2
6530,52	1	13,28	15,18	$4p\ [2^1/_2]-7s\ [1^1/_2]^\circ$	3—2
6513,848	8	13,28	15,18	$4p'\ [1^1/_2]-7s\ [1^1/_2]^\circ$	1—1
6499,109	6	13,28	15,19	$4p'\ [1^1/_2]-5d\ [1^1/_2]^\circ$	1—1
6493,971	15	13,09	15,00	$4p\ [2^1/_2]-4d'\ [1^1/_2]^\circ$	2—1
6481,141	8	13,27	15,18	$4p\ [^1/_2]-7s\ [1^1/_2]^\circ$	0—1
6466,5505	20	13,27	15,19	$4p\ [^1/_2]-5d\ [1^1/_2]^\circ$	0—1
6431,559	15	13,09	15,02	$4p\ [2^1/_2]-6s'\ [^1/_2]^\circ$	2—1
6416,3075	100	12,91	14,84	$4p\ [^1/_2]-6s\ [1^1/_2]^\circ$	1—2
6384,7189	100	12,91	14,85	$4p\ [^1/_2]-6s\ [1^1/_2]^\circ$	1—1
6369,5783	30	13,17	15,12	$4p\ [1^1/_2]-5d\ [^1/_2]^\circ$	2—1
6364,8945	20	13,15	15,10	$4p\ [1^1/_2]-5d\ [^1/_2]^\circ$	1—0
6349,20	2	12,91	14,86	$4p\ [^1/_2]-4d\ [1^1/_2]^\circ$	1—1
6309,14	8	13,15	15,12	$4p\ [1^1/_2]-5d\ [^1/_2]^\circ$	1—1
6307,6598	30	13,17	15,14	$4p\ [1^1/_2]-5d\ [1^1/_2]^\circ$	2—2
6296,8762	20	13,33	15,30	$4p'\ [^1/_2]-5d'\ [1^1/_2]^\circ$	1—2
6278,652	6	13,17	15,15	$4p\ [1^1/_2]-5d\ [3^1/_2]^\circ$	2—3
6259,41	1	13,33	15,31	$4p'\ [^1/_2]-6d\ [^1/_2]^\circ$	1—1
6248,4064	15	13,15	15,14	$4p\ [1^1/_2]-5d\ [1^1/_2]^\circ$	1—2
6243,3958	6	13,33	15,31	$4p'\ [^1/_2]-6d\ [^1/_2]^\circ$	1—0
6230,928	4	13,17	15,16	$4p\ [1^1/_2]-5d\ [2^1/_2]^\circ$	2—2
6215,9423	60	13,30	15,30	$4p'\ [1^1/_2]-5d'\ [1^1/_2]^\circ$	2—2
6212,5044	100	13,17	15,17	$4p\ [1^1/_2]-5d\ [2^1/_2]^\circ$	2—3
6179,41	4	13,30	15,31	$4p'\ [1^1/_2]-6d\ [^1/_2]^\circ$	2—1
6173,0980	100	13,15	15,16	$4p\ [1^1/_2]-5d\ [2^1/_2]^\circ$	1—2
6170,1761	100	13,17	15,18	$4p\ [1^1/_2]-7s\ [1^1/_2]^\circ$	2—2
6165,123	8	13,30	15,31	$4p'\ [1^1/_2]-5d'\ [2^1/_2]$	2—2
6155,2393	60	13,17	15,18	$4p\ [1^1/_2]-7s\ [1^1/_2]^\circ$	2—1
		13,28	15,30	$4p'\ [1^1/_2]-5d'\ [1^1/_2]^\circ$	1—2
6145,4432	100	13,30	15,32	$4p'\ [1^1/_2]-5d'\ [2^1/_2]^\circ$	2—3
6142,05	1	13,17	15,19	$4p\ [1^1/_2]-5d\ [1^1/_2]^\circ$	2—1
6128,726	8	13,33	15,35	$4p'\ [^1/_2]-5d'\ [1^1/_2]^\circ$	1—1
6127,416	15	13,09	15,12	$4p\ [2^1/_2]-5d\ [^1/_2]^\circ$	2—1
6121,86	1	13,33	15,35	$4p'\ [^1/_2]-6d\ [2^1/_2]^\circ$	1—2
6119,662	2	13,28	15,31	$4p'\ [^1/_2]-6d\ [^1/_2]^\circ$	1—1
6113,463	8	13,15	15,18	$4p\ [1^1/_2]-7s\ [1^1/_2]^\circ$	1—2
6105,6354	60	13,28	15,31	$4p\ [1^1/_2]-5d'\ [2^1/_2]^\circ$	1—2
6104,60	6	13,28	15,31	$4p'\ [1^1/_2]-6d\ [^1/_2]^\circ$	1—0
		13,33	15,36	$4p'\ [^1/_2]-7s'\ [^1/_2]^\circ$	1—0
6101,16	6	13,33	15,36	$4p'\ [^1/_2]-7s'\ [^1/_2]^\circ$	1—1
6098,8046	60	13,15	14,18	$4p\ [1^1/_2]-7s\ [1^1/_2]^\circ$	1—1
6093,33	1	13,33	15,36	$4p'\ [^1/_2]-8s\ [1^1/_2]^\circ$	1—2
6090,7865	10	13,27	15,31	$4p\ [^1/_2]-6d\ [^1/_2]^\circ$	0—1
		13,48	15,51	$4p'\ [^1/_2]-8d\ [^1/_2]^\circ$	0—1
6085,86	2	13,15	15,19	$4p\ [1^1/_2]-5d\ [1^1/_2]^\circ$	1—1
6081,245	4	13,33	15,35	$4p'\ [^1/_2]-6d\ [1^1/_2]^\circ$	1—1
6064,758	6	13,30	15,35	$4p'\ [1^1/_2]-6d\ [3^1/_2]^\circ$	2—3

λ, Å	I	E_H, eV	E_B, eV	Transition	J
6059,3735	100	12,91	14,95	$4p\ [^1/_2]-4d'\ [2^1/_2]^\circ$	1—2
6052,7234	30	12,91	14,95	$4p\ [^1/_2]-4d'\ [1^1/_2]^\circ$	1—2
6045,34	1	13,30	15,35	$4p'\ [1^1/_2]-6d\ [2^1/_2]^\circ$	2—2
6043,2254	100	13,09	15,15	$4p\ [2^1/_2]-5d\ [3^1/_2]^\circ$	2—3
		13,30	15,35	$4p'\ [1^1/_2]-6d\ [2^1/_2]^\circ$	2—3
6032,1291	60	13,08	15,13	$4p\ [2^1/_2]-5d\ [3^1/_2]^\circ$	3—4
6025,1515	10	13,30	15,36	$4p'\ [1^1/_2]-7s'\ [^1/_2]^\circ$	2—1
6017,53	1	13,30	15,36	$4p'\ [1^1/_2]-8s\ [1^1/_2]^\circ$	2—2
6013,6790	6	13,08	15,14	$4p\ [2^1/_2]-5d\ [1^1/_2]^\circ$	3—2
6005,7246	4	13,30	15,37	$4p\ [1^1/_2]-8s\ [1^1/_2]^\circ$	2—1
5999,0004	20	13,09	15,16	$4p\ [2^1/_2]-5d\ [2^1/_2]^\circ$	2—2
5994,66	2	13,28	15,35	$4p'\ [1^1/_2]-5d\ [1^1/_2]^\circ$	1—1
5988,11	2	13,28	15,35	$4p'\ [1^1/_2]-6d\ [2^1/_2]^\circ$	1—2
5987,3027	40	13,08	15,15	$4p\ [2^1/_2]-5d\ [3^1/_2]^\circ$	3—3
5981,90	5	13,09	15,17	$4p\ [2^1/_2]-5d\ [2^1/_2]^\circ$	2—3
5971,6036	5	13,28	15,36	$4p'\ [1^1/_2]-7s'\ [^1/_2]^\circ$	1—0
5968,31	1	13,28	15,36	$4p'\ [1^1/_2]-7s'\ [^1/_2]^\circ$	1—1
5949,2595	10	13,28	15,35	$4p'\ [1^1/_2]-6d\ [1^1/_2]^\circ$	1—1
5943,89	2	13,08	15,16	$4p\ [2^1/_2]-5d\ [2^1/_2]^\circ$	3—2
5942,6722	40	13,09	15,18	$4p\ [2^1/_2]-7s\ [1^1/_2]^\circ$	2—2
5940,86	2	13,27	15,36	$4p\ [^1/_2]-7s'\ [^1/_2]^\circ$	0—1
5928,8124	200	13,09	15,18	$4p\ [2^1/_2]-7s\ [1^1/_2]^\circ$	2—1
5927,13	10	13,08	15,17	$4p\ [2^1/_2]-5d\ [2^1/_2]^\circ$	3—3
5916,58	5	13,09	15,19	$4p\ [2^1/_2]-5d\ [1^1/_2]^\circ$	2—1
5912,0861	500	12,91	15,00	$4p\ [^1/_2]-4d'\ [1^1/_2]^\circ$	1—1
5888,5851	300	13,08	15,18	$4p\ [2^1/_2]-7s\ [1^1/_2]^\circ$	3—2
5882,6250	100	12,91	15,01	$4p\ [^1/_2]-6s'\ [^1/_2]^\circ$	1—0
5870,26	2	13,33	15,44	$4p'\ [^1/_2]-7d\ [^1/_2]^\circ$	1—0
5860,3118	60	12,91	15,02	$4p\ [^1/_2]-6s'\ [^1/_2]^\circ$	1—1
5843,74	2	13,33	15,45	$4p'\ [^1/_2]-7d\ [1^1/_2]^\circ$	1—2
5834,2660	60	13,17	15,30	$4p\ [1^1/_2]-5d'\ [1^1/_2]^\circ$	2—2
5802,0809	40	13,17	15,31	$4p\ [1^1/_2]-6d\ [^1/_2]^\circ$	2—1
5790,39	5	13,30	15,44	$4p'\ [1^1/_2]-7d\ [^1/_2]^\circ$	2—1
5789,477	20	13,17	15,31	$4p\ [1^1/_2]-5d'\ [2^1/_2]^\circ$	2—2
5783,541	40	13,15	15,30	$4p\ [1^1/_2]-5d'\ [1^1/_2]^\circ$	1—2
5774,00	40	13,30	15,45	$4p'\ [1^1/_2]-7d\ [1^1/_2]^\circ$	2—2
5772,1160	100	13,17	15,32	$4p\ [1^1/_2]-5d'\ [2^1/_2]^\circ$	2—3
5758,84	5	13,30	15,45	$4p'\ [1^1/_2]-7d\ [3^1/_2]^\circ$	2—3
5747,18	2	13,28	15,44	$4p'\ [1^1/_2]-7d\ [^1/_2]^\circ$	1—0
5739,5207	500	13,15	15,31	$4p\ [1^1/_2]-5d'\ [2^1/_2]^\circ$	1—2
5738,416	20	13,15	15,31	$4p\ [1^1/_2]-6d\ [^1/_2]^\circ$	1—0
5737,96	5	13,28	15,44	$4p'\ [1^1/_2]-7d\ [^1/_2]^\circ$	1—1
5712,48	1	13,27	15,44	$4p\ [^1/_2]-7d\ [^1/_2]^\circ$	0—1
5700,874	60	13,17	15,35	$4p\ [1^1/_2]-6d\ [3^1/_2]^\circ$	2—3
5693,10	1	13,28	15,46	$4p'\ [1^1/_2]-7d\ [2^1/_2]^\circ$	1—2
5689,91	200	13,33	15,51	$4p'\ [1^1/_2]-6d'\ [1^1/_2]^\circ$	1—2
5689,64	200	13,17	15,35	$4p\ [1^1/_2]-5d'\ [1^1/_2]^\circ$	2—1
5687,40	20	—	—	—	—
5683,73	40	13,17	15,35	$4p\ [1^1/_2]-6d\ [2^1/_2]^\circ$	2—2
5681,9014	500	13,17	15,35	$4p\ [1^1/_2]-6d\ [2^1/_2]^\circ$	2—3
5674,73	1	13,33	15,51	$4p'\ [^1/_2]-6d'\ [2^1/_2]^\circ$	1—2
5667,40	1	13,28	15,47	$4p'\ [1^1/_2]-9s\ [1^1/_2]^\circ$	1—2
		13,33	15,51	$4p'\ [^1/_2]-8d\ [^1/_2]^\circ$	1—1
5665,82	5	13,17	15,36	$4p\ [1^1/_2]-7s'\ [^1/_2]^\circ$	2—1
5663,80	1	13,28	15,47	$4p'\ [1^1/_2]-9s\ [1^1/_2]^\circ$	1—1
5662,00	5	13,28	15,47	$4p'\ [1^1/_2]-7d\ [1^1/_2]^\circ$	1—1
5659,1278	500	13,17	15,36	$4p\ [1^1/_2]-8s\ [1^1/_2]^\circ$	2—2
5650,7054	1500	12,91	15,10	$4p\ [^1/_2]-5d\ [^1/_2]^\circ$	1—0
5648,66	200	13,17	15,37	$4p\ [1^1/_2]-8s\ [1^1/_2]^\circ$	2—1

λ, Å	I	E_H, eV	E_B, eV	Transition	J
5641,34	60	13,15	15,35	$4p\,[1^1/_2]—5d'\,[1^1/_2]^\circ$	1—1
5639,11	100	13,27	15,47	$4p\,[^1/_2]—9s\,[1^1/_2]^\circ$	0—1
5637,29	20	13,27	15,47	$4p\,[^1/_2]—7d\,[1^1/_2]^\circ$	0—1
5635,575	60	13,15	15,35	$4p\,[1^1/_2]—6d\,[2^1/_2]^\circ$	1—2
5630,44	10	13,09	15,30	$4p\,[2^1/_2]—5d'\,[1^1/_2]^\circ$	2—2
5623,778	60	13,30	15,51	$4p'\,[1/^1_2]—6d'\,[1^1/_2]^\circ$	2—2
5620,89	60	13,15	15,36	$4p\,[1^1/_2]—7s'\,[^1/_2]^\circ$	1—0
5620,636	2	13,33	15,53	$4p'\,[^1/_2]—6d'\,[1^1/_2]^\circ$	1—1
5619,00	5	—	—	—	—
5618,010	60	13,15	15,36	$4p\,[1^1/_2]—7s'\,[^1/_2]^\circ$	1—1
5611,35	20	13,15	15,36	$4p\,[1^1/_2]—8s\,[1^1/_2]^\circ$	1—2
5608,90	20	13,30	15,51	$4p'\,[1^1/_2]—6d'\,[2^1/_2]^\circ$	2—2
5606,7341	500	12,91	15,12	$4p\,[^1/_2]—5d\,[^1/_2]^\circ$	1—1
5605,25	5	13,33	15,53	$4p'\,[^1/_2]—8d\,[1^1/_2]^\circ$	1—1
5604,36	20	13,33	15,54	$4p'\,[^1/_2]—8s'\,[^1/_2]^\circ$	1—0
5601,85	2	13,30	15,51	$4p'\,[1^1/_2]—8d\,[^1/_2]^\circ$	2—1
5601,08	60	13,15	15,35	$4p\,[1^1/_2]—6d\,[1^1/_2]^\circ$	1—1
5600,43	40	13,09	15,31	$4p\,[2^1/_2]—6d\,[^1/_2]^\circ$	2—1
5598,50	20	13,33	15,54	$4p'\,[^1/_2]—8s'\,[^1/_2]^\circ$	1—1
5597,4783	500	13,30	15,51	$4p'\,[1^1/_2]—6d'\,[2^1/_2]^\circ$	2—2
5591,75	5	—	—	—	—
5588,7213	500	13,09	15,31	$4p\,[2^1/_2]—5d'\,[2^1/_2]^\circ$	2—2
5581,83	60	13,08	15,30	$4p\,[2^1/_2]—5d'\,[1^1/_2]^\circ$	3—2
5574,20	5	13,28	15,51	$4p'\,[1^1/_2]—6d'\,[1^1/_2]^\circ$	1—2
5572,5428	500	13,09	15,32	$4p\,[2^1/_2]—5d'\,[2^1/_2]^\circ$	2—3
5565,96	5	—	—	—	—
5560,22	10	13,30	15,53	$4p'\,[1^1/_2]—8d\,[3^1/_2]^\circ$	2—3
5559,62	200	13,28	15,51	$4p'\,[1^1/_2]—6d'\,[2^1/_2]^\circ$	1—2
5558,7031	500	12,91	15,14	$4p\,[^1/_2]—5d\,[1^1/_2]^\circ$	1—2
5553,40	2	13,30	15,53	$4p'\,[1^1/_2]—8d\,[2^1/_2]^\circ$	2—2
5552,76	10	13,30	15,53	$4p'\,[1^1/_2]—8d\,[2^1/_2]^\circ$	2—2
5542,73	2	13,28	15,51	$4p'\,[1^1/_2]—8d\,[^1/_2]^\circ$	1—1
5541,46	2	13,30	15,54	$4p'\,[1^1/_2]—10s\,[1^1/_2]^\circ$	2—2
5540,90	40	13,08	15,31	$4p\,[2^1/_2]—5d'\,[2^1/_2]^\circ$	3—2
		13,30	15,53	$4p'\,[1^1/_2]—8d\,[1^1/_2]^\circ$	2—1
5534,45	60	13,30	15,54	$4p'\,[1^1/_2]—8s'\,[^1/_2]^\circ$	2—1
5528,93	40	13,27	15,51	$4p\,[^1/_2]—8d\,[^1/_2]^\circ$	0—1
5524,9598	300	13,08	15,32	$4p\,[2^1/_2]—5d'\,[2^1/_2]^\circ$	3—3
5523,70	5	13,33	15,57	$4p'\,[^1/_2]—9d\,[^1/_2]^\circ$	1—0
5518,20	5	13,33	15,57	$4p'\,[^1/_2]—9d\,[^1/_2]^\circ$	1—1
5507,63	10	13,28	15,53	$4p'\,[1^1/_2]—6d'\,[1^1/_2]^\circ$	1—1
5506,1149	500	13,09	15,35	$4p\,[2^1/_2]—6d\,[3^1/_2]^\circ$	2—3
5505,18	10	13,28	15,53	$4p'\,[1^1/_2]—8d\,[2^1/_2]^\circ$	1—2
5499,00	10	12,91	15,16	$4p\,[^1/_2]—5d\,[2^1/_2]^\circ$	1—2
5495,876	1000	13,08	15,33	$4p\,[2^1/_2]—6d\,[3^1/_2]^\circ$	3—4
5493,49	20	13,28	15,54	$4p'\,[1^1/_2]—10s\,[1^1/_2]^\circ$	1—2
5492,06	40	13,28	15,54	$4p'\,[1^1/_2]—8s'\,[^1/_2]^\circ$	1—0
5490,122	60	13,09	15,35	$4p\,[2^1/_2]—6d\,[2^1/_2]^\circ$	2—2
5488,46	2	13,09	15,35	$4p\,[2^1/_2]—6d\,[2^1/_2]^\circ$	2—3
5486,47	20	13,28	15,54	$4p'\,[1^1/_2]—8s'\,[^1/_2]^\circ$	1—1
5483,32	10	13,27	15,54	$4p\,[^1/_2]—8s'\,[^1/_2]^\circ$	0—1
5473,455	500	13,09	15,36	$4p\,[2^1/_2]—7s'\,[^1/_2]^\circ$	2—1
5469,65	20	—	—	—	—
5467,1626	60	13,09	15,36	$4p\,[2^1/_2]—8s\,[1^1/_2]^\circ$	2—2
5459,61	20	13,08	15,35	$4p\,[2^1/_2]—6d\,[3^1/_2]^\circ$	3—3
5457,75	10	13,17	15,44	$4p\,[1^1/_2]—7d\,[^1/_2]^\circ$	2—1
5457,4158	200	13,09	15,35	$4p\,[2^1/_2]—6d\,[1^1/_2]^\circ$	2—1
5456,01	5	13,30	15,57	$4p'\,[1^1/_2]—9d\,[^1/_2]^\circ$	2—1
5451,6539	500	12,91	15,18	$4p\,[^1/_2]—7s\,[1^1/_2]^\circ$	1—2

λ, Å	I	E_H, eV	E_B, eV	Transition	J
5448,61	10	13,30	15,58	$4p'[1^1/_2]-9d[1^1/_2]^\circ$	2—2
5443,88	20	13,08	15,35	$4p[2^1/_2]-6d[2^1/_2]^\circ$	3—2
5443,21	100	13,17	15,45	$4p[1^1/_2]-7d[1^1/_2]^\circ$	2—2
5442,22	500	13,08	15,35	$4p[2^1/_2]-6d[2^1/_2]^\circ$	3—2
5439,9903	500	12,91	15,18	$4p[^1/_2]-7s[1^1/_2]^\circ$	1—1
5435,83	1	13,30	15,58	$4p'[1^1/_2]-9d[2^1/_2]^\circ$	2—3
5433,48	1	13,33	15,61	$4p'[^1/_2]-10d[^1/_2]^\circ$	1—0
5432,60	1	13,33	15,61	$4p'[^1/_2]-10d[^1/_2]^\circ$	1—1
5430,27	10	13,33	15,61	$4p'[^1/_2]-10d[1^1/_2]^\circ$	1—2
5429,69	20	12,91 13,17	15,19 15,45	$4p[^1/_2]-5d[1^1/_2]^\circ$ $4p[1^1/_2]-7d[3^1/_2]^\circ$	1—1 2—3
5427,39	1	13,30	15,59	$4p'[1^1/_2]-11s[1^1/_2]^\circ$	2—2
5422,55	2	13,33	15,61	$4p'[^1/_2]-10d[1^1/_2]^\circ$	1—2
5421,3536	500	13,08	15,36	$4p[2^1/_2]-8s[1^1/_2]^\circ$	3—2
5417,22	10	13,17	15,46	$4p[1^1/_2]-7d[2^1/_2]^\circ$	2—2
5413,32	10	13,15	15,44	$4p[1^1/_2]-7d[^1/_2]^\circ$	1—1
5410,4750	500	13,17	15,46	$4p[1^1/_2]-7d[2^1/_2]^\circ$	2—3
5409,34	1	13,28	15,57	$4p'[1^1/_2]-9d[^1/_2]^\circ$	1—1
5402,08	1	13,28	15,58	$4p'[1^1/_2]-9d[1^1/_2]^\circ$	1—2
5399,01	20	13,15	15,45	$4p[1^1/_2]-7d[1^1/_2]^\circ$	1—2
5393,971	200	13,17	15,47	$4p[1^1/_2]-9s[1^1/_2]^\circ$	2—2
5390,72	40	13,17	15,47	$4p[1^1/_2]-9s[1^1/_2]^\circ$	2—1
5389,10	40	13,17	15,47	$4p[1^1/_2]-7d[1^1/_2]^\circ$	2—1
5387,37	40	13,33	15,63	$4p'[^1/_2]-7d'[2^1/_2]^\circ$	1—2
5386,79	1	13,27	15,57	$4p[^1/_2]-9d[^1/_2]^\circ$	0—1
5373,4951	500	13,15	15,46	$4p[1^1/_2]-7d[2^1/_2]^\circ$	1—2
5372,29	1	13,30	15,61	$4p'[1^1/_2]-10d[^1/_2]^\circ$	2—1
5369,97	5	13,30	15,61	$4p'[1^1/_2]-10d[1^1/_2]^\circ$	2—2
5362,48	1	13,30	15,61	$4p'[^1/_2]-10d[3^1/_2]^\circ$	2—3
5356,49	10	—	—	—	—
5353,46	20	11,55	13,86	$4s[1^1/_2]^\circ-3d[^1/_2]^\circ$	2—1
5350,58	20	13,15	15,47	$4p[1^1/_2]-9s[1^1/_2]^\circ$	1—2
5347,412	200	13,15	15,47	$4p[1^1/_2]-9s[1^1/_2]^\circ$	1—1
5345,81	20	13,15	15,47	$4p[1^1/_2]-7d[1^1/_2]^\circ$	1—1
5344,28	5	13,33	15,65	$4p'[^1/_2]-9s'[^1/_2]^\circ$	1—0
5341,78	10	13,33	15,65	$4p'[^1/_2]-9s'[^1/_2]^\circ$	1—1
5328,02	20	13,30	15,63	$4p'[1^1/_2]-7d'[1^1/_2]^\circ$	2—2
5327,07	1	13,28	15,61	$4p'[1^1/_2]-10d[^1/_2]^\circ$	1—1
5324,80	5	13,28	15,61	$4p'[1^1/_2]-10d[1^1/_2]^\circ$	1—2
5317,726	60	13,30	15,63	$4p'[1^1/_2]-7d'[2^1/_2]^\circ$	2—2
5309,517	200	13,17	15,51	$4p[1^1/_2]-6d'[1^1/_2]^\circ$	2—2
5305,17	1	13,27	15,61	$4p[^1/_2]-10d[^1/_2]^\circ$	0—1
5296,91	1	13,30	15,64	$4p'[1^1/_2]-11d[2^1/_2]^\circ$	2—3
5296,32	5	13,17	15,51	$4p[1^1/_2]-6d'[2^1/_2]^\circ$	2—2
5290,00	20	13,17	15,51	$4p[1^1/_2]-8d[^1/_2]^\circ$	2—1
5286,071	60	13,17	15,52	$4p[1^1/_2]-6d'[2^1/_2]^\circ$	2—3
5283,43	20	13,28 13,30	15,63 15,65	$4p'[1^1/_2]-7d'[2^1/_2]^\circ$ $4p'[1^1/_2]-9s'[^1/_2]^\circ$	1—2 2—1
5280,40	60	—	—	—	—
5279,05	20	13,09	15,44	$4p[2^1/_2]-7d[^1/_2]^\circ$	2—1
5267,48	2	13,15	15,51	$4p[1^1/_2]-6d'[1^1/_2]^\circ$	1—2
5263,02	2	11,55	13,90	$4s[1^1/_2]^\circ-3d[1^1/_2]^\circ$	2—1
5254,4710	60	13,15	15,51	$4p[1^1/_2]-6d'[2^1/_2]^\circ$	1—2
5252,7890	300	13,09	15,45	$4p[2^1/_2]-7d[3^1/_2]^\circ$	2—3
5249,20	40	13,17	15,53	$4p[1^1/_2]-6d'[1^1/_2]^\circ$	2—1
5248,18	1	13,15	15,51	$4p[1^1/_2]-8d[^1/_2]^\circ$	1—1
5246,76	5	13,17	15,53	$4p[1^1/_2]-8d[2^1/_2]^\circ$	2—2
5246,24	40	13,17	15,53	$4p[1^1/_2]-8d[2^1/_2]^\circ$	2—3
5242,13	2	13,28	15,65	$4p'[1^1/_2]-9s'[^1/_2]^\circ$	1—0

λ, Å	I	E_H, eV	E_B, eV	Transition	J
5241,091	60	13,09	15,46	$4p\,[2^1/_2]-7d\,[2^1/_2]^\circ$	2—2
5239,71	2	13,28	15,65	$4p'\,[1^1/_2]-9s'\,[^1/_2]^\circ$	1—0
5236,21	20	13,17	15,54	$4p\,[1^1/_2]-10s\,[1^1/_2]^\circ$	2—2
5234,74	5	13,09	15,46	$4p\,[2^1/_2]-7d\,[2^1/_2]^\circ$	2—3
5229,86	40	13,17	15,54	$4p\,[1^1/_2]-8s'\,[^1/_2]^\circ$	2—1
5222,90	20	13,08	15,45	$4p\,[2^1/_2]-7d\,[1^1/_2]^\circ$	3—2
5221,2729	500	13,08	15,45	$4p\,[2^1/_2]-7d\,[3^1/_2]^\circ$	3—4
5219,30	40	13,09	15,47	$4p\,[2^1/_2]-9s\,[1^1/_2]^\circ$	2—2
5216,28	60	13,09	15,47	$4p\,[2^1/_2]-9s\,[1^1/_2]^\circ$	2—1
5214,774	200	13,09	15,47	$4p\,[2^1/_2]-7d\,[1^1/_2]^\circ$	2—1
5210,492	200	13,08	15,45	$4p\,[2^1/_2]-7d\,[3^1/_2]^\circ$	3—3
5208,04	10	13,15	15,53	$4p\,[1^1/_2]-6d'\,[1^1/_2]^\circ$	1—1
5207,17	10	13,15	15,54	$4p\,[1^1/_2]-8s'\,[^1/_2]^\circ$	1—1
5205,79	10	13,15	15,53	$4p\,[1^1/_2]-8d\,[2^1/_2]^\circ$	1—2
5198,96	2	13,08	15,46	$4p\,[2^1/_2]-7d\,[2^1/_2]^\circ$	3—2
5195,29	1	13,15	15,54	$4p\,[1^1/_2]-10s\,[1^1/_2]^\circ$	1—2
5194,77	20	13,15	15,53	$4p\,[1^1/_2]-8d\,[1^1/_2]^\circ$	1—1
5194,02	5	13,15	15,54	$4p\,[1^1/_2]-8s'\,[^1/_2]^\circ$	1—0
5192,72	60	13,08	15,46	$4p\,[2^1/_2]-7d\,[2^1/_2]^\circ$	3—3
5187,7507	800	12,91	15,30	$4p\,[^1/_2]-5d'\,[1^1/_2]^\circ$	1—2
5177,540	40	13,08	15,47	$4p\,[2^1/_2]-9s\,[1^1/_2]^\circ$	3—2
5162,2858	500	12,91	15,31	$4p\,[^1/_2]-6d\,[^1/_2]^\circ$	1—1
5159,69	10	13,17	15,57	$4p\,[1^1/_2]-9d\,[1^1/_2]^\circ$	2—1
5153,11	20	13,17	15,58	$4p\,[1^1/_2]-9d\,[1^1/_2]^\circ$	2—2
5151,3943	200	12,91	15,31	$4p\,[^1/_2]-6d\,[^1/_2]^\circ$	1—0
5141,81	20	13,17	15,58	$4p\,[1^1/_2]-9d\,[2^1/_2]^\circ$	2—3
5134,17	2	13,17	15,59	$4p\,[1^1/_2]-11s\,[1^1/_2]^\circ$	2—2
5132,61	1	13,17	15,59	$4p\,[1^1/_2]-11s\,[1^1/_2]^\circ$	2—1
5127,802	60	13,09	15,51	$4p\,[2^1/_2]-6d'\,[2^1/_2]^\circ$	2—2
5124,72	1	13,15	15,57	$4p\,[1^1/_2]-9d\,[^1/_2]^\circ$	1—0
5121,88	5	13,09	15,51	$4p\,[2^1/_2]-8d\,[^1/_2]^\circ$	2—1
5120,01	1	13,15	15,57	$4p\,[1^1/_2]-9d\,[^1/_2]^\circ$	1—1
5118,2057	60	13,09	15,52	$4p\,[2^1/_2]-6d'\,[2^1/_2]^\circ$	2—3
5113,50	1	13,15	15,58	$4p\,[1^1/_2]-9d\,[1^1/_2]^\circ$	1—2
5104,74	20	13,15	15,58	$4p\,[1^1/_2]-9d\,[2^1/_2]^\circ$	1—2
5099,64	5	13,08	15,51	$4p\,[2^1/_2]-6d'\,[1^1/_2]^\circ$	3—2
5098,97	20	11,55	13,98	$4s\,[1^1/_2]^\circ-3d\,[3^1/_2]^\circ$	2—4
5094,84	1	13,15	15,59	$4p\,[1^1/_2]-11s\,[1^1/_2]^\circ$	1—2
5093,32	10	13,15	15,59	$4p\,[1^1/_2]-11s\,[1^1/_2]^\circ$	1—1
5087,085	60	13,09	15,53	$4p\,[2^1/_2]-8d\,[3^1/_2]^\circ$	2—3
5084,79	1	13,17	15,61	$4p\,[1^1/_2]-10d\,[^1/_2]^\circ$	2—1
5082,74	20	13,09	15,54	$4p\,[2^1/_2]-8s'\,[^1/_2]^\circ$	2—1
5081,44	10	13,09	15,53	$4p\,[2^1/_2]-8d\,[2^1/_2]^\circ$	2—2
5078,03	40	13,08	15,52	$4p\,[2^1/_2]-6d'\,[2^1/_2]^\circ$	3—3
5076,03	1	13,17	15,61	$4p\,[1^1/_2]-10d\,[3^1/_2]^\circ$	2—3
5073,0758	200	12,91	15,35	$4p\,[^1/_2]-5d'\,[1^1/_2]^\circ$	1—1
5071,30	5	13,09	15,54	$4p\,[2^1/_2]-10s\,[1^1/_2]^\circ$	2—2
5070,99	40	13,09	15,53	$4p\,[2^1/_2]-8d\,[1^1/_2]^\circ$	2—1
		13,09	15,54	$4p\,[2^1/_2]-10s\,[1^1/_2]^\circ$	2—1
5069,66	5	13,17	15,62	$4p\,[1^1/_2]-10d\,[2^1/_2]^\circ$	2—3
5068,39	5	12,91	15,35	$4p\,[^1/_2]-6d\,[2^1/_2]^\circ$	1—2
5065,48	5	13,09	15,54	$4p\,[2^1/_2]-8s'\,[^1/_2]^\circ$	2—1
5063,99	5	13,17	15,62	$4p\,[1^1/_2]-12s\,[1^1/_2]^\circ$	2—2
5062,72	1	13,17	15,62	$4p\,[1^1/_2]-12s\,[1^1/_2]^\circ$	2—1
5060,0793	500	13,08	15,52	$4p\,[2^1/_2]-8d\,[3^1/_2]^\circ$	3—4
5056,53	200	12,91	15,36	$4p\,[^1/_2]-7s'\,[^1/_2]^\circ$	1—0
5054,1783	300	12,91	15,36	$4p\,[^1/_2]-7s'\,[^1/_2]^\circ$	1—1
5048,8130	500	12,91	15,36	$4p\,[^1/_2]-8s\,[1^1/_2]^\circ$	1—2
5047,30	2	13,08	15,53	$4p\,[2^1/_2]-8d\,[3^1/_2]^\circ$	3—3

λ, Å	I	E_H, eV	E_B, eV	Transition	J
5047,00	1	13,15	15,61	4p [1¹/₂]—10d [¹/₂]°	1—0
5044,15	2	13,15	15,61	4p [1¹/₂]—10d [1¹/₂]°	1—2
5041,23	10	13,08	15,53	4p [2¹/₂]—8d [2¹/₂]°	3—3
5040,51	10	12,91	15,35	4p [¹/₂]—6d [1¹/₂]°	1—1
5035,88	5	13,17	15,63	4p [1¹/₂]—7d′ [2¹/₂]°	2—3
5034,25	10	13,15	15,61	4p [1¹/₂]—10d [2¹/₂]°	1—2
5032,026	60	13,08	15,54	4p [2¹/₂]—10s [1¹/₂]°	3—2
5029,64	5	11,55	14,01	4s [1¹/₂]°—3d [3¹/₂]°	2—3
5025,74	1	13,15	15,62	4p [1¹/₂]—12s [1¹/₂]°	1—2
5024,50	1	13,08	15,59	4p [2¹/₂]—11s [1¹/₂]°	1—1
5017,25	5	13,17	15,64	4p [1¹/₂]—11d [2¹/₂]°	2—3
5013,47	1	13,17	15,64	4p [1¹/₂]—13s [1¹/₂]°	2—2
5007,09	2	13,15	15,63	4p [1¹/₂]—7d′ [2¹/₂]°	1—2
5006,84	2	11,83	14,30	4s′ [¹/₂]°—3d′ [1¹/₂]°	1—1
5005,13	1	13,17	15,65	4p [1¹/₂]—9s′ [¹/₂]°	2—1
5004,318	20	—	—	—	—
4999,65	1	13,09	15,57	4p [2¹/₂]—9d [¹/₂]°	2—1
4991,66	1	13,15	15,63	4p [1¹/₂]—11d [¹/₂]°	1—0
4989,948	80	13,09	15,58	4p [2¹/₂]—9d [3¹/₂]°	2—3
4985,09	10	13,09	15,58	4p [2¹/₂]—9d [2¹/₂]°	2—2
4982,81	1	13,09	15,58	4p [2¹/₂]—9d [2¹/₂]°	2—3
4979,05	1	13,17	15,66	4p [1¹/₂]—12d [2¹/₂]°	2—3
4975,66	2	13,09	15,59	4p [2¹/₂]—11s [1¹/₂]°	2—2
4974,18	10	13,09	15,59	4p [2¹/₂]—11s [1¹/₂]°	2—1
4973,53	5	—	—	—	—
4969,88	1	13,15	15,65	4p [1¹/₂]—9s′ [¹/₂]°	1—0
4956,750	100	13,08	15,58	4p [2¹/₂]—9d [3¹/₂]°	3—4
4955,21	2	13,08	15,58	4p [2¹/₂]—9d [1¹/₂]°	3—2
4951,75	10	13,08	15,58	4p [2¹/₂]—9d [3¹/₂]°	3—3
4949,64	1	13,17	15,68	4p [1¹/₂]—13d [2¹/₂]°	2—3
4944,80	5	13,08	15,58	4p [2¹/₂]—9d [2¹/₂]°	3—3
4937,718	30	13,08	15,59	4p [2¹/₂]—11s [1¹/₂]°	3—2
4929,16	2	13,09	15,61	4p [2¹/₂]—10d [¹/₂]°	2—1
4921,042	80	13,09	15,61	4p [2¹/₂]—10d [3¹/₂]°	3—3
4917,85	5	13,09	15,61	4p [2¹/₂]—10d [2¹/₂]°	2—2
4915,03	1	13,09	15,62	4p [2¹/₂]—10d [2¹/₂]°	2—3
4909,71	2	13,09	15,62	4p [2¹/₂]—12s [1¹/₂]°	2—2
4908,52	10	13,09	15,62	4p [2¹/₂]—12s [1¹/₂]°	2—1
4901,26	2	11,62	14,15	4s [1¹/₂]°—3d [1¹/₂]°	1—1
4894,6909	150	12,91	15,44	4p [¹/₂]—7d [¹/₂]°	1—0
4890,19	1	13,08	15,61	4p [2¹/₂]—10d [1¹/₂]°	3—2
4887,9478	200	12,91	15,44	4p [¹/₂]—7d [¹/₂]°	1—1
4886,29	30	13,08	15,61	4p [2¹/₂]—10d [3¹/₂]°	3—4
4883,86	5	13,08	15,61	4p [2¹/₂]—10d [3¹/₂]°	3—3
4883,27	30	13,09	15,63	4p [2¹/₂]—7d′ [2¹/₂]°	2—3
4877,96	1	13,08	15,62	4p [2¹/₂]—10d [2¹/₂]°	3—3
4876,2619	200	12,91	15,45	4p [¹/₂]—7d [1¹/₂]°	1—2
4872,73	10	13,08	15,62	4p [2¹/₂]—12s [1¹/₂]°	3—2
4867,84	10	13,08	15,64	4p [2¹/₂]—11d [3¹/₂]°	3—3
4865,91	1	13,09	15,64	4p [2¹/₂]—11d [2¹/₂]°	2—3
4862,16	1	13,09	15,64	4p [2¹/₂]—13s [1¹/₂]°	2—2
4859,44	5	11,55	14,10	4s [1¹/₂]°—3d [2¹/₂]°	2—3
4855,37	1	12,91	15,46	4p [¹/₂]—7d [2¹/₂]°	1—2
4854,37	1	13,09	15,65	4p [2¹/₂]—9s′ [¹/₂]°	2—1
4846,73	5	13,08	15,63	4p [2¹/₂]—7d′ [2¹/₂]°	3—3
4836,697	150	12,91	15,47	4p [¹/₂]—9s [1¹/₂]°	1—2
4835,97	30	13,08	15,64	4p [2¹/₂]—11d [3¹/₂]°	3—3
4834,10	30	12,91	15,47	4p [¹/₂]—9s [1¹/₂]°	1—1

λ, Å	I	E_H, eV	E_B, eV	Transition	J
4832,79	5	12,91	15,47	4p [1/2]—7d [11/2]°	1—1
4832,38	5	13,09	15,66	4p [21/2]—12d [31/2]°	2—3
4830,54	1	13,08	15,64	4p [21/2]—11d [11/2]°	3—2
4829,47	2	13,08	15,64	4p [21/2]—11d [21/2]°	3—3
4825,97	2	13,08	15,64	4p [21/2]—13s [11/2]°	3—2
4804,33	5	13,09	15,67	4p [21/2]—13d [31/2]°	2—3
4798,742	30	13,08	15,66	4p [21/2]—12d [31/2]°	3—4
4796,57	1	13,08	15,66	4p [21/2]—12d [31/2]°	3—3
4794,10	1	13,08	15,66	4p [21/2]—12d [21/2]°	3—3
4791,15	2	13,08	15,66	4p [21/2]—14s [11/2]°	3—2
4770,34	2	13,08	15,67	4p [21/2]—13d [31/2]°	3—4
4768,6750	150	12,91	15,51	4p [1/2]—6d′ [11/2]°	1—2
4752,9404	150	12,91	15,51	4p [1/2]—8d [1/2]°	1—1
4748,23	5	13,08	15,69	4p [21/2]—14d [31/2]°	3—4
4746,823	80	12,91	15,52	4p [1/2]—8d [1/2]°	1—0
4730,66	5	—	—	—	—
4727,48	5	—	—	—	—
4724,10	5	—	—	—	—
4719,94	20	12,91	15,53	4p [1/2]—6d′ [11/2]°	1—1
4719,22	2	12,91	15,54	4p [1/2]—8s′ [1/2]°	1—1
4718,10	2	12,91	15,53	4p [1/2]—8d [21/2]°	1—2
4709,50	30	12,91	15,54	4p [1/2]—10s [11/2]°	1—2
4709,08	10	12,91	15,53	4p [1/2]—8d [11/2]°	1—1
4708,46	2	12,91	15,54	4p [1/2]—8s′ [1/2]°	1—0
4704,35	2	12,91	15,54	4p [1/2]—8s′ [1/2]°	1—1
4702,3155	1200	11,83	14,46	4s′ [1/2]°—5p [1/2]	1—1
4651,388	20	12,91	15,57	4p [1/2]—9d [1/2]°	1—0
4647,493	40	12,91	15,57	4p [1/2]—9d [1/2]°	1—1
4642,148	80	12,91	15,58	4p [1/2]—9d [11/2]°	1—2
4628,4409	1000	11,83	14,51	4s′ [1/2]°—5p [21/2]	1—2
4626,78	30	12,91	15,59	4p [1/2]—11s [11/2]°	1—2
4625,46	10	12,91	15,59	4p [1/2]—11s [11/2]°	1—1
4596,0964	1000	11,83	14,52	4s′ [1/2]°—5p [11/2]	1—1
4589,288	80	11,83	14,53	4s′ [1/2]°—5p [11/2]	1—2
4587,21	5	12,91	15,61	4p [1/2]—10d [1/2]°	1—0
4586,610	10	12,91	15,61	4p [1/2]—10d [1/2]°	1—1
4584,958	10	12,91	15,63	4p [1/2]—7d′ [11/2]°	1—2
4569,69	2	12,91	15,62	4p [1/2]—12s [11/2]°	1—2
4568,64	2	12,91	15,62	4p [1/2]—12s [11/2]°	1—1
4564,82	4	—	—	—	—
4554,319	15	12,91	15,63	4p [1/2]—7d′ [21/2]°	1—2
4544,746	30	—	—	—	—
4541,60	20	12,91	15,63	4p [1/2]—11d [1/2]°	1—0
4534,78	20	—	—	—	—
4523,35	1	12,91	15,65	4p [1/2]—9s′ [1/2]°	1—0
4522,3238	800	11,72	14,46	4s′ [1/2]°—5p [1/2]	0—1
4510,7335	1000	11,83	14,58	4s′ [1/2]°—5p [1/2]	1—0
4509,87	4	—	—	—	—
4507,45	1	12,91	15,66	4p [1/2]—12d [1/2]°	1—0
4505,16	3	—	—	—	—
4480,87	5	—	—	—	—
4479,31	5	—	—	—	—
4474,72	5	—	—	—	—
4461,46	5	—	—	—	—
4460,53	10	—	—	—	—
4456,61	3	—	—	—	—
4448,88	3	—	—	—	—

λ, Å	I	E_H, eV	E_B, eV	Transition	J
4445,84	5	12,91	15,70	$4p\ [^1/_2]—8d'\ [1^1/_2]°$	1—2
4423,996	80	11,72	14,52	$4s'\ [^1/_2]°—5p\ [1^1/_2]$	0—1
4368,36	5	—	—	—	—
4367,87	10	—	—	—	—
4363,7957	80	11,62	14,46	$4s\ [1^1/_2]°—5p\ [^1/_2]$	1—1
4345,167	1000	11,83	14,66	$4s'\ [^1/_2]°—5p'\ [1^1/_2]$	1—1
4335,3381	800	11,83	14,69	$4s'\ [^1/_2]°—5p'\ [^1/_2]$	1—1
4333,5612	1000	11,83	14,69	$4s'\ [^1/_2]°—5p'\ [1^1/_2]$	1—2
4310,47	20	—	—	—	—
4300,1011	1200	11,62	14,51	$4s\ [1^1/_2]°—5p\ [2^1/_2]$	1—2
4299,24	5	—	—	—	—
4294,97	20	—	—	—	—
4289,09	5	—	—	—	—
4272,1690	1200	11,62	14,52	$4s\ [1^1/_2]°—5p\ [1^1/_2]$	1—1
4271,24	5	—	—	—	—
4266,2868	1200	11,62	14,53	$4s\ [1^1/_2]°—5p\ [1^1/_2]$	1—2
4265,52	2	—	—	—	—
4259,3617	1200	11,83	14,74	$4s'\ [^1/_2]°—5p'\ [^1/_2]$	1—0
4258,59	5	—	—	—	—
4254,95	10	—	—	—	—
4251,1850	800	11,55	14,46	$4s\ [1^1/_2]°—5p\ [^1/_2]$	2—1
4250,41	3	—	—	—	—
4249,37	20	—	—	—	—
4243,57	20	—	—	—	—
4200,6746	1200	11,55	14,50	$4s\ [1^1/_2]°—5p\ [2^1/_2]$	2—3
4198,3176	1200	11,62	14,58	$4s\ [1^1/_2]°—5p\ [^1/_2]$	1—0
4191,0288	1200	11,72	14,66	$4s'\ [^1/_2]°—5p'\ [1^1/_2]$	0—1
4190,7138	600	11,55	14,51	$4s\ [1^1/_2]°—5p\ [2^1/_2]$	2—2
4181,8837	1000	11,72	14,69	$4s'\ [^1/_2]°—5p'\ [^1/_2]$	0—1
4176,33	20	—	—	—	—
4175,40	10	—	—	—	—
4168,70	3	—	—	—	—
4168,41	3	—	—	—	—
4164,1795	1000	11,55	14,52	$4s\ [1^1/_2]°—5p\ [1^1/_2]$	2—1
4158,5906	1200	11,55	14,53	$4s\ [1^1/_2]°—5p\ [1^1/_2]$	2—2
4152,54	20	—	—	—	—
4054,5253	80	11,62	14,66	$4s\ [1^1/_2]°—5p'\ [1^1/_2]$	1—1
4045,9658	150	11,62	14,69	$4s\ [1^1/_2]°—5p'\ [^1/_2]$	1—1
4044,4185	1200	11,62	14,69	$4s\ [1^1/_2]°—5p'\ [1^1/_2]$	1—2
4032,97	20	11,83	14,90	$4s'\ [^1/_2]°—4f\ [1^1/_2]$	1—1, 2
3979,7149	10	11,62	14,74	$4s\ [1^1/_2]°—5p'\ [^1/_2]$	1—0
3948,9785	2000	11,85	14,69	$4s\ [1^1/_2]°—5p'\ [^1/_2]$	2—1
3947,5048	1000	11,55	14,69	$4s\ [1^1/_2]°—5p'\ [1^1/_2]$	2—2
3899,878	100	11,72	14,90	$4s'\ [^1/_2]°—4f\ [1^1/_2]$	0—1
3894,6603	300	11,83	15,01	$4s'\ [^1/_2]°—6p\ [^1/_2]$	1—1
3876,080	10	11,83	15,03	$4s'\ [^1/_2]°—6p\ [2^1/_2]$	1—2
3866,2752	5	11,83	15,03	$4s'\ [^1/_2]°—6p\ [1^1/_2]$	1—1
3864,2669	10	11,83	15,03	$4s'\ [^1/_2]°—6p\ [1^1/_2]$	1—2
3834,6788	800	11,83	15,06	$4s'\ [^1/_2]°—6p\ [^1/_2]$	1—0
3781,3570	300	11,62	14,90	$4s\ [1^1/_2]°—4f\ [1^1/_2]$	1—1, 2
3775,4408	10	11,62	14,91	$4s\ [1^1/_2]°—4f\ [2^1/_2]$	1—2
3770,3698	400	11,72	15,01	$4s'\ [^1/_2]°—6p\ [^1/_2]$	0—1
3743,7653	100	11,72	15,03	$4s'\ [^1/_2]°—6p\ [1^1/_2]$	0—1
3696,5082	20	11,55	14,90	$4s\ [1^1/_2]°—4f\ [1^1/_2]$	2—1, 2
3690,8960	300	11,55	14,91	$4s\ [1^1/_2]°—4f\ [2^1/_2]$	2—3, 2
3675,2367	300	11,83	15,20	$4s'\ [^1/_2]°—6p'\ [1^1/_2]$	1—1
3674,05	2	11,83	15,20	$4s'\ [^1/_2]°—6p'\ [^1/_2]$	1—1
3670,6693	300	11,83	15,20	$4s'\ [^1/_2]°—6p'\ [1^1/_2]$	1—2

λ, Å	I	E_H, eV	E_B, eV	Transition	J
3663,76	5	11,83	15,21	$4s'\,[^1/_2]^\circ - 5f\,[1^1/_2]$	1—1, 2
3659,5305	100	11,62	15,01	$4s\,[1^1/_2]^\circ - 6p\,[^1/_2]$	1—1
3649,8330	800	11,83	15,22	$4s'\,[^1/_2]^\circ - 6p'\,[^1/_2]$	1—0
3643,1169	100	11,62	15,03	$4s\,[1^1/_2]^\circ - 6p\,[2^1/_2]$	1—2
3634,4605	300	11,62	15,03	$4s\,[1^1/_2]^\circ - 6p\,[1^1/_2]$	1—1
3632,6837	300	11,62	15,03	$4s\,[1^1/_2]^\circ - 6p\,[1^1/_2]$	1—2
3606,5224	1000	11,62	15,06	$4s\,[1^1/_2]^\circ - 6p\,[^1/_2]$	1—0
3599,7116	20	11,83	15,27	$4s'\,[^1/_2]^\circ - 7p\,[^1/_2]$	1—1
3593,418	—	11,83	15,28	$4s'\,[^1/_2]^\circ - 7p\,[2^1/_2]$	1—2
3588,97	2	11,83	15,28	$4s'\,[^1/_2]^\circ - 7p\,[1^1/_2]$	1—1
3588,11	3	11,83	15,28	$4s'\,[^1/_2]^\circ - 7p\,[1^1/_2]$	1—2
3582,6971	30	11,62	15,08	$4s\,[1^1/_2]^\circ - 4f'\,[2^1/_2]$	1—2
3572,2960	300	11,83	15,30	$4s'\,[^1/_2]^\circ - 7p\,[^1/_2]$	1—0
3567,6562	300	11,55	15,02	$4s\,[1^1/_2]^\circ - 6p\,[2^1/_2]$	2—3
3564,2955	100	11,55	15,03	$4s\,[1^1/_2]^\circ - 6p\,[2^1/_2]$	2—2
		11,72	15,20	$4s'\,[^1/_2]^\circ - 6p'\,[1^1/_2]$	0—1
3563,2864	100	11,72	15,20	$4s'\,[^1/_2]^\circ - 6p'\,[^1/_2]$	0—1
3556,0076	100	11,55	15,03	$4s\,[1^1/_2]^\circ - 6p\,[1^1/_2]$	2—1
3554,3056	300	11,55	15,03	$4s\,[1^1/_2]^\circ - 6p\,[1^1/_2]$	2—2
3553,58	15	11,72	15,21	$4s'\,[^1/_2]^\circ - 5f\,[1^1/_2]$	0—1
3506,4807	30	11,55	15,08	$4s\,[1^1/_2]^\circ - 4f'\,[2^1/_2]$	2—3, 2
3493,2747	20	11,72	15,27	$4s'\,[^1/_2]^\circ - 7p\,[^1/_2]$	0—1
3490,50	3	11,83	15,38	$4s'\,[^1/_2]^\circ - 6f\,[1^1/_2]$	1—1, 2
3483,17	5	11,72	15,28	$4s'\,[^1/_2]^\circ - 7p\,[1^1/_2]$	0—1
3465,15	2	11,62	15,20	$4s\,[1^1/_2]^\circ - 6p'\,[1^1/_2]$	2—1
3464,08	1	11,62	15,20	$4s\,[1^1/_2]^\circ - 6p'\,[^1/_2]$	2—1
3461,0785	300	11,62	15,20	$4s\,[1^1/_2]^\circ - 6p'\,[1^1/_2]$	1—2
3457,81	3	11,83	15,41	$4s'\,[^1/_2]^\circ - 8p\,[^1/_2]$	1—1
3454,944	20	11,62	15,21	$4s\,[1^1/_2]^\circ - 5f\,[1^1/_2]$	1—1, 2
3452,32	3	11,62	15,21	$4s\,[1^1/_2]^\circ - 5f\,[2^1/_2]$	1—2
3449,52	2	11,83	15,42	$4s'\,[^1/_2]^\circ - 8p\,[1^1/_2]$	1—2
3442,58	10	11,83	15,43	$4s'\,[^1/_2]^\circ - 8p\,[^1/_2]$	1—0
		11,62	15,22	$4s\,[1^1/_2]^\circ - 6p'\,[^1/_2]$	1—0
3418,51	3	11,83	15,45	$4s'\,[^1/_2]^\circ - 7p'\,[1^1/_2]$	1—1
3417,68	3	11,83	15,45	$4s'\,[^1/_2]^\circ - 7p'\,[^1/_2]$	1—1
3416,80	5	11,83	15,45	$4s'\,[^1/_2]^\circ - 7p'\,[1^1/_2]$	1—2
3406,1804	30	11,83	15,47	$4s'\,[^1/_2]^\circ - 7p'\,[^1/_2]$	1—0
3397,920	20	11,62	15,27	$4s\,[1^1/_2]^\circ - 7p\,[^1/_2]$	1—1
3393,7522	250	11,83	15,48	$4s'\,[^1/_2]^\circ - 7f\,[1^1/_2]$	1—1, 2
		11,55	15,20	$4s\,[1^1/_2]^\circ - 6p'\,[^1/_2]$	2—1
3392,7812	100	11,55	15,20	$4s\,[1^1/_2]^\circ - 6p'\,[1^1/_2]$	2—1
3392,31	3	11,62	15,28	$4s\,[1^1/_2]^\circ - 7p\,[2^1/_2]$	1—2
3390,29	3	11,72	15,38	$4s'\,[^1/_2]^\circ - 6f\,[1^1/_2]$	0—1
3389,854	20	11,55	15,20	$4s\,[1^1/_2]^\circ - 6p'\,[1^1/_2]$	2—2
3388,365	20	11,62	15,28	$4s\,[1^1/_2]^\circ - 7p\,[1^1/_2]$	1—1
3387,600	20	11,62	15,28	$4s\,[1^1/_2]^\circ - 7p\,[1^1/_2]$	1—2
3383,98	2	11,55	15,21	$4s\,[1^1/_2]^\circ - 5f\,[1^1/_2]$	1—2, 1
3381,49	20	11,55	15,21	$4s\,[1^1/_2]^\circ - 5f\,[2^1/_2]$	2—3, 2
3373,4823	300	11,62	15,30	$4s\,[1^1/_2]^\circ - 7p\,[^1/_2]$	1—0
3372,88	3	11,83	15,50	$4s'\,[^1/_2]^\circ - 9p\,[^1/_2]$	1—1
3368,84	1	11,83	15,51	$4s'\,[^1/_2]^\circ - 9p\,[1^1/_2]$	1—2
3363,47	20	11,83	15,51	$4s'\,[^1/_2]^\circ - 9p\,[^1/_2]$	1—0
3359,48	10	11,72	15,41	$4s'\,[^1/_2]^\circ - 8p\,[^1/_2]$	0—1
3352,20	1	11,72	15,42	$4s'\,[^1/_2]^\circ - 8p\,[1^1/_2]$	0—1
3333,84	2	11,83	15,54	$4s'\,[^1/_2]^\circ - 8f\,[1^1/_2]$	1—1, 2
3325,5006	100	11,55	15,27	$4s\,[1^1/_2]^\circ - 7p\,[2^1/_2]$	2—3
3323,825	30	11,55	15,28	$4s\,[1^1/_2]^\circ - 7p\,[2^1/_2]$	2—2
3322,44	5	11,72	15,45	$4s'\,[^1/_2]^\circ - 7p'\,[1^1/_2]$	0—1

λ, Å	I	E_H, eV	E_B, eV	Transition	J
3321,58	5	11,72	15,45	$4s'\,[^1/_2]^\circ-7p'\,[^1/_2]$	0—1
3320,67	2	11,83	15,56	$4s'\,[^1/_2]^\circ-10p\,[^1/_2]$	1—1
3320,06	3	11,55	15,28	$4s\,[1^1/_2]^\circ-7p\,[1^1/_2]$	2—1
3319,3446	300	11,55	15,28	$4s\,[1^1/_2]^\circ-7p\,[1^1/_2]$	2—2
3317,54	1	11,83	15,56	$4s'\,[^1/_2]^\circ-10p\,[1^1/_2]$	1—2
3314,49	2	11,83	15,57	$4s'\,[^1/_2]^\circ-10p\,[^1/_2]$	1—0
3310,47	3	—	—	—	—
3300,39	20	11,62	15,38	$4s\,[1^1/_2]^\circ-6f\,[1^1/_2]$	1—1, 2
3299,26	2	11,62	15,38	$4s\,[1^1/_2]^\circ-6f\,[2^1/_2]$	1—2
3299,02	1	11,72	15,48	$4s'\,[^1/_2]^\circ-7f\,[1^1/_2]$	0—1
3289,95	3	11,62	15,39	$4s\,[1^1/_2]^\circ-5f'\,[2^1/_2]$	1—2
3289,39	3	11,83	15,60	$4s'\,[^1/_2]^\circ-8p'\,[1^1/_2]$	1—2
3285,10	2	11,83	15,60	$4s'\,[^1/_2]^\circ-8p'\,[^1/_2]$	1—0
3283,74	1	11,83	15,60	$4s'\,[^1/_2]^\circ-11p\,[^1/_2]$	1—1
3282,70	1	11,83	15,60	$4s'\,[^1/_2]^\circ-11p\,[1^1/_2]$	1—2
3279,25	3	11,72	15,50	$4s'\,[^1/_2]^\circ-9p\,[^1/_2]$	0—1
3278,93	3	11,83	15,61	$4s'\,[^1/_2]^\circ-11p\,[^1/_2]$	1—0
3275,72	2	11,72	15,51	$4s'\,[^1/_2]^\circ-9p\,[1^1/_2]$	0—1
3271,16	10	11,62	15,41	$4s\,[1^1/_2]^\circ-8p\,[^1/_2]$	1—1
3266,34	1	11,62	15,42	$4s\,[1^1/_2]^\circ-8p\,[2^1/_2]$	1—2
3264,29	3	11,62	15,42	$4s\,[1^1/_2]^\circ-8p\,[1^1/_2]$	1—1
3263,78	3	11,62	15,42	$4s\,[1^1/_2]^\circ-8p\,[1^1/_2]$	1—2
3257,585	100	11,62	15,43	$4s\,[1^1/_2]^\circ-8p\,[^1/_2]$	1—0
3256,20	2	11,83	15,63	$4s'\,[^1/_2]^\circ-12p\,[^1/_2]$	1—0
3242,40	2	11,72	15,54	$4s'\,[^1/_2]^\circ-8f\,[1^1/_2]$	0—1, 2
3238,49	1	11,83	15,65	$4s'\,[^1/_2]^\circ-13p\,[^1/_2]$	1—0
3235,57	2	11,55	15,38	$4s\,[1^1/_2]^\circ-6p\,[1^1/_2]$	2—1, 2
3234,491	100	11,62	15,45	$4s\,[1^1/_2]^\circ-7p'\,[1^1/_2]$	1—2
		11,55	15,38	$4s\,[1^1/_2]^\circ-6f\,[2^1/_2]$	2—2
3229,91	3	11,72	15,56	$4s'\,[^1/_2]^\circ-10p\,[^1/_2]$	0—1
3225,58	20	11,55	15,39	$4s\,[1^1/_2]^\circ-5f'\,[2^1/_2]$	2—3, 2
3213,84	2	11,62	15,48	$4s\,[1^1/_2]^\circ-7f\,[1^1/_2]$	1—1, 2
3212,99	2	11,62	15,48	$4s\,[1^1/_2]^\circ-7f\,[2^1/_2]$	1—2
3211,99	2	11,83	15,61	$4s'\,[^1/_2]^\circ-9p'\,[^1/_2]$	1—0
3207,50	10	11,55	15,41	$4s\,[1^1/_2]^\circ-8p\,[^1/_2]$	2—1
3203,66	10	11,55	15,42	$4s\,[1^1/_2]^\circ-8p\,[2^1/_2]$	2—3
3202,85	5	11,55	15,42	$4s\,[1^1/_2]^\circ-8p\,[2^1/_2]$	2—2
3201,12	3	11,72	15,59	$4s'\,[^1/_2]^\circ-8p'\,[1^1/_2]$	0—1
3200,84	2	11,55	15,42	$4s\,[1^1/_2]^\circ-8p\,[1^1/_2]$	2—1
3200,39	100	11,55	15,42	$4s\,[1^1/_2]^\circ-8p\,[1^1/_2]$	2—2
3195,12	5	11,62	15,50	$4s\,]1^1/_2]^\circ-9p\,[^1/_2]$	1—1
3194,93	1	11,72	15,60	$4s'\,[^1/_2]^\circ-11p\,[^1/_2]$	0—1
3191,72	2	11,62	15,51	$4s\,[1^1/_2]^\circ-9p\,[1^1/_2]$	1—1
3191,50	2	11,62	15,51	$4s\,[1^1/_2]^\circ-9p\,[1^1/_2]$	1—2
3186,63	5	11,62	15,51	$4s\,[1^1/_2]^\circ-9p\,[^1/_2]$	1—0
3173,71	2	11,55	15,45	$4s\,[1^1/_2]^\circ-7p'\,[1^1/_2]$	2—1
3172,961	150	11,55	15,45	$4s\,[1^1/_2]^\circ-7p'\,[^1/_2]$	2—1
3172,18	5	11,55	15,45	$4s\,[1^1/_2]^\circ-7p'\,[1^1/_2]$	2—2
3160,06	5	11,62	15,54	$4s\,[1^1/_2]^\circ-8f\,[1^1/_2]$	1—1, 2
3159,55	1	11,62	15,55	$4s\,[1^1/_2]^\circ-8f\,[2^1/_2]$	1—2
3152,29	3	11,55	15,48	$4s\,[1^1/_2]^\circ-7f\,[1^1/_2]$	2—1, 2
3151,52	3	11,55	15,48	$4s\,[1^1/_2]^\circ-7f\,[2^1/_2]$	2—3, 2
3150,42	1	11,62	15,56	$4s\,[1^1/_2]^\circ-6f'\,[2^1/_2]$	1—2
3148,20	1	11,62	15,56	$4s\,[1^1/_2]^\circ-10p\,[^1/_2]$	1—1
3145,63	1	11,62	15,56	$4s\,[1^1/_2]^\circ-10p\,[1^1/_2]$	1—1
3145,42	1	11,62	15,56	$4s\,[1^1/_2]^\circ-10p\,[1^1/_2]$	1—2
3142,60	3	11,62	15,57	$4s\,[1^1/_2]^\circ-10p\,[^1/_2]$	1—0
3134,27	2	11,55	15,50	$4s\,[1^1/_2]^\circ-9p\,[^1/_2]$	2—1

λ, Å	I	E_H, eV	E_B, eV	Transition	J
3132,87	3	11,55	15,50	$4s\,[1^1/_2]^\circ$—$9p\,[2^1/_2]$	2—3
3132,31	2	11,55	15,50	$4s\,[1^1/_2]^\circ$—$9p\,[2^1/_2]$	2—2
3131,04	2	11,55	15,51	$4s\,[1^1/_2]^\circ$—$9p\,[1^1/_2]$	2—1
3130,80	20	11,55	15,51	$4s\,[1^1/_2]^\circ$—$9p\,[1^1/_2]$	2—2
3120,06	3	11,62	15,60	$4s\,[1^1/_2]^\circ$—$8p'\,[1^1/_2]$	1—2
3117,85	3	—	—	—	—
3116,63	3	—	—	—	—
3116,22	1	11,62	15,60	$4s\,[1^1/_2]^\circ$—$8p'\,[^1/_2]$	1—0
3114,96	1	11,62	15,60	$4s\,[1^1/_2]^\circ$—$11p\,[^1/_2]$	1—1
3114,10	1	11,62	15,60	$4s\,[1^1/_2]^\circ$—$11p\,[1^1/_2]$	1—2, 1
3110,66	3	11,62	15,61	$4s\,[1^1/_2]^\circ$—$11p\,[^1/_2]$	1—0
3100,09	5	11,55	15,55	$4s\,[1^1/_2]^\circ$—$8f\,[2^1/_2]$	2—3, 2
3092,97	1	11,62	15,63	$4s\,[1^1/_2]^\circ$—$12p\,[^1/_2]$	1—1
3091,32	2	11,55	15,56	$4s\,[1^1/_2]^\circ$—$6f'\,[2^1/_2]$	2—3
3090,18	1	11,62	15,63	$4s\,[1^1/_2]^\circ$—$12p\,[^1/_2]$	1—0
3089,17	2	11,55	15,56	$4s\,[1^1/_2]^\circ$—$10p\,[^1/_2]$	2—1
3087,81	1	11,55	15,56	$4s\,[1^1/_2]^\circ$—$10p\,[2^1/_2]$	2—3
3087,31	1	11,55	15,56	$4s\,[1^1/_2]^\circ$—$10p\,[2^1/_2]$	2—2
3086,47	2	11,55	15,56	$4s\,[1^1/_2]^\circ$—$10p\,[1^1/_2]$	2—1
3074,15	1	11,62	15,65	$4s\,[1^1/_2]^\circ$—$13p\,[^1/_2]$	1—0
3065,73	1	11,55	15,59	$4s\,[1^1/_2]^\circ$—$9f\,[2^1/_2]$	2—3
3063,44	5	11,55	15,59	$4s\,[1^1/_2]^\circ$—$8p'\,[^1/_2]$	2—1
3062,82	1	11,55	15,59	$4s\,[1^1/_2]^\circ$—$8p'\,[1^1/_2]$	2—1
3062,06	3	11,55	15,60	$4s\,[1^1/_2]^\circ$—$8p'\,[1^1/_2]$	2—2
3056,28	3	11,55	15,60	$4s\,[1^1/_2]^\circ$—$11p\,[1^1/_2]$	2—2
1066,660	15	0,00	11,62	$3p^6\,{}^1S$—$4s\,[1^1/_2]^\circ$	0—1
1048,218	25	0,00	11,83	$3p^6\,{}^1S$—$4s'\,[^1/_2]^\circ$	0—1
894,310	4	0,00	13,86	$3p^6\,{}^1S$—$3d\,[^1/_2]^\circ$	0—1
879,949	3	0,00	14,09	$3p^6\,{}^1S$—$5s\,[1^1/_2]^\circ$	0—1
876,063	4	0,00	14,15	$3p^6\,{}^1S$—$3d\,[1^1/_2]^\circ$	0—1
869,754	2	0,00	14,25	$3p^6\,{}^1S$—$5s'\,[^1/_2]^\circ$	0—1
866,805	4	0,00	14,30	$3p^6\,{}^1S$—$3d'\,[1^1/_2]^\circ$	0—1
842,808	2	0,00	14,71	$3p^6\,{}^1S$—$4d\,[^1/_2]^\circ$	0—1
835,003	6	0,00	14,85	$3p^6\,{}^1S$—$6s\,[1^1/_2]^\circ$	0—1
834,397	6	0,00	14,86	$3p^6\,{}^1S$—$4d\,[1^1/_2]^\circ$	0—1
826,371	2	0,00	15,00	$3p^6\,{}^1S$—$4d'\,[1^1/_2]^\circ$	0—1
825,348	1	0,00	15,02	$3p^6\,{}^1S$—$6s'\,[^1/_2]^\circ$	0—1
820,129	—	0,00	15,12	$3p^6\,{}^1S$—$5d\,[^1/_2]^\circ$	0—1
816,466	4	0,00	15,18	$3p^6\,{}^1S$—$7s\,[1^1/_2]^\circ$	0—1
816,233	4	0,00	15,19	$3p^6\,{}^1S$—$5d\,[1^1/_2]^\circ$	0—1
809,933	2	0,00	15,31	$3p^6\,{}^1S$—$6d\,[^1/_2]^\circ$	0—1
807,702	2	0,00	15,35	$3p^6\,{}^1S$—$6d\,[1^1/_2]^\circ$	0—1
807,220	2	0,00	15,36	$3p^6\,{}^1S$—$7s'\,[^1/_2]^\circ$	0—1
806,875	2	0,00	15,37	$3p^6\,{}^1S$—$8s\,[1^1/_2]^\circ$	0—1
801,359	1	0,00	15,47	$3p^6\,{}^1S$—$7d\,[1^1/_2]^\circ$	0—1
799,137	0	0,00	15,51	$3p^6\,{}^1S$—$8d\,[^1/_2]^\circ$	0—1
797,744	1	0,00	15,54	$3p^6\,{}^1S$—$8d'\,[^1/_2]^\circ$	0—1

Ar II, **ground state** $1s^2\,2s^2\,2p^6\,3s^2\,3p^5\,{}^2P^0_{3/2}$

Ionization potential 222 848,2 cm^{-1}; 27,628 eV

λ, Å	I	E_H, eV	E_B, eV	Transition	J
11253,496	1	22,70	23,80	$5s\,{}^2P$—$4p''\,{}^2P^\circ$	$^3/_2$—$^3/_2$
11173,266	2	22,59	23,70	$5s\,{}^4P$—$5p\,{}^4S^\circ$	$^3/_2$—$^3/_2$

λ, Å	I	E_H, eV	E_B, eV	Transition	J
11068,44	1	23,57	24,69	$5p\,{}^4D^\circ - 6s\,{}^4P$	$3/2 - 3/2$
11067,929	2	23,26	24,38	$4d\,{}^2F - ({}^3P_0)4f\,[3]^\circ$	$5/2 - 7/2$
10982,382	2	24,82	25,95	$4d'\,{}^2F - ({}^1D)4f\,[4]^\circ$	$5/2 - 7/2$
10974,33	1	24,81	25,94	$4d'\,{}^2F - ({}^1D)4f\,[3]^\circ$	$7/2 - 7/2$
10973,80	2	23,08	24,21	$4d\,{}^4P - ({}^3P_2)4f\,[1]^\circ$	$1/2 - 1/2$
10954,260	2	23,08	24,21	$4d\,{}^4P - ({}^3P_2)4f\,[1]^\circ$	$1/2 - 3/2$
10923,438	7	18,73	19,87	$3d\,{}^2D - 4p\,{}^2P^\circ$	$5/2 - 3/2$
10916,67	1	23,51	24,65	$5p\,{}^4D^\circ - 6s\,{}^4P$	$5/2 - 5/2$
10869,698	2	23,01	24,15	$4d\,{}^4F - ({}^3P_2)4f\,[4]^\circ$	$7/2 - 7/2$
10867,87	1	23,17	24,31	$4d\,{}^4P - ({}^3P_1)4f\,[2]^\circ$	$5/2 - 5/2$
10867,343	3	24,81	25,95	$4d'\,{}^2F - ({}^1D)4f\,[4]^\circ$	$7/2 - 9/2$
10829,452	3	18,66	19,80	$3d\,{}^2D - 4p\,{}^2P^\circ$	$3/2 - 1/2$
10817,858	1	24,76	25,90	$4d'\,{}^2D - ({}^1D)4f\,[2]^\circ$	$5/2 - 5/2$
10812,901	12	18,62	19,76	$3d\,{}^2F - 4p\,{}^2D^\circ$	$5/2 - 3/2$
10785,13	1	24,79	25,94	$4d'\,{}^2D - ({}^1D)4f\,[3]^\circ$	$3/2 - 5/2$
10764,378	8	23,16	24,31	$4d\,{}^2F - ({}^3P_1)4f\,[4]^\circ$	$7/2 - 9/2$
10720,530	1	18,45	19,61	$4s'\,{}^2D - 4p\,{}^4D^\circ$	$5/2 - 3/2$
10683,050	12	18,33	19,49	$3d\,{}^4P - 4p\,{}^4D^\circ$	$5/2 - 7/2$
10660,99	2	23,65	24,81	$5p\,{}^4D^\circ - 6s\,{}^4P$	$1/2 - 1/2$
10639,86	1	24,74	25,90	$4d'\,{}^2P - ({}^1D)4f\,[2]^\circ$	$3/2 - 5/2$
10638,121	8	23,48	24,65	$5p\,{}^4D^\circ - 6s\,{}^4P$	$7/2 - 5/2$
10619,458	7	23,17	24,34	$4d\,{}^4P - ({}^3P)4f\,[3]^\circ$	$5/2 - 7/2$
10614,01	1	23,57	24,74	$5p\,{}^4D^\circ - 4d'\,{}^2P$	$3/2 - 3/2$
10580,83	2	23,62	24,79	$5p\,{}^2D^\circ - 5d\,{}^4D$	$5/2 - 5/2$
10555,90	1	23,62	24,79	$5p\,{}^2D^\circ - 4d'\,{}^2D$	$5/2 - 3/2$
10541,552	5	23,51	24,69	$5p\,{}^4D^\circ - 6s\,{}^4P$	$5/2 - 3/2$
10535,52	2	23,16	24,34	$4d\,{}^2F - ({}^3P_1)4f\,[3]^\circ$	$7/2 - 7/2$
10519,510	9	23,01	24,19	$4d\,{}^4F - ({}^3P_2)4f\,[5]^\circ$	$7/2 - 9/2$
10500,212	6	23,62	24,80	$5p\,{}^2D^\circ - 6s\,{}^2P$	$5/2 - 3/2$
10495,941	2	23,62	24,80	$5p\,{}^2P^\circ - 6s\,{}^2P$	$3/2 - 3/2$
10467,173	20	18,49	19,68	$3d\,{}^2F - 4p\,{}^2D^\circ$	$7/2 - 5/2$
10447,771	2	24,76	25,94	$4d'\,{}^2D - ({}^1D)4f\,[3]^\circ$	$5/2 - 7/2$
10442,57	1	23,57	24,76	$5p\,{}^4D^\circ - 4d'\,{}^2D$	$3/2 - 5/2$
10440,511	6	22,51	23,70	$5s\,{}^4P - 5p\,{}^4S^\circ$	$5/2 - 3/2$
10410,53	2	23,62	24,81	$5p\,{}^2P^\circ - 6s\,{}^4P$	$3/2 - 1/2$
10401,510	1	23,68	24,87	$5p\,{}^2D^\circ - 6s\,{}^2P$	$3/2 - 1/2$
10392,604	5	23,12	24,31	$4d\,{}^4P - ({}^3P_1)4f\,[2]^\circ$	$3/2 - 5/2$
10383,900	1	23,62	24,81	$5p\,{}^2D^\circ - 4d'\,{}^2F$	$5/2 - 7/2$
10325,34	1	18,06	19,26	$3d\,{}^2P - 4p\,{}^4P^\circ$	$3/2 - 3/2$
10305,616	1	23,48	24,69	$5p\,{}^4P^\circ - 6s\,{}^4P$	$1/2 - 3/2$
10299,077	5	22,95	24,15	$4d\,{}^4F - ({}^3P_2)4f\,[4]^\circ$	$9/2 - 9/2$
10273,689	5	23,17	24,38	$4d\,{}^4P - ({}^3P_0)4f\,[3]^\circ$	$5/2 - 7/2$
10268,320	2	25,19	26,40	$5p'\,{}^2F^\circ - 6s'\,{}^2D$	$7/2 - 5/2$
10230,845	4	23,44	24,65	$5p\,{}^4P^\circ - 6s\,{}^4P$	$3/2 - 5/2$
10220,980	1	25,19	26,40	$5p'\,{}^2F^\circ - 6s'\,{}^2D$	$5/2 - 3/2$
10203,917	5	18,33	19,55	$3d\,{}^4P - 4p\,{}^4D^\circ$	$5/2 - 5/2$
10138,408	1	23,70	24,92	$5p\,{}^4S^\circ - 5d\,{}^4P$	$3/2 - 1/2$
10111,595	8	20,27	21,50	$3d'\,{}^2F - 4p'\,{}^2D^\circ$	$7/2 - 5/2$
10110,660	3	18,45	19,68	$4s'\,{}^2D - 4p\,{}^2D^\circ$	$5/2 - 5/2$
10093,016	1	23,08	24,31	$4d\,{}^4P - ({}^3P_1)4f\,[2]^\circ$	$1/2 - 3/2$
10022,278	4	23,10	24,34	$4d\,{}^4F - ({}^3P_1)4f\,[3]^\circ$	$3/2 - 5/2$
9993,874	4	24,21	25,45	$({}^3P_2)4f\,[1]^\circ - ({}^3P_2)5g\,[2]$	$3/2 - 5/2$
9989,02	1	21,35	22,59	$4p'\,{}^2P^\circ - 5s\,{}^4P$	$3/2 - 3/2$
9988,39	1	23,48	24,73	$5p\,{}^4P^\circ - 4d'\,{}^2P$	$1/2 - 1/2$
9977,825	3	24,21	25,45	$({}^3P_2)4f\,[1]^\circ - ({}^3P_2)5g\,[2]$	$1/2 - 3/2$
9972,313	1	23,51	24,76	$5p\,{}^4D^\circ - 4d'\,{}^2D$	$5/2 - 5/2$
9967,045	12	22,95	24,19	$4d\,{}^4F - ({}^3P_2)4f\,[5]^\circ$	$9/2 - 11/2$
9965,41	1	24,34	25,58	$({}^3P_1)4f\,[3]^\circ - ({}^3P_1)5g\,[3]$	$7/2 - 7/2$

λ, Å	I	E_H, eV	E_B, eV	Transition	J
9962,314	1	22,95	24,19	$4d\ {}^4F-({}^3P_2)4f\ [5]^\circ$	9/2—9/2
9952,809	5	23,07	24,31	$4d\ {}^4F-({}^3P_1)4f\ [4]^\circ$	5/2—7/2
9951,087	4	23,40	24,65	$5p\ {}^4P^\circ-6s\ {}^4P$	5/2—5/2
9949,151	7	20,24	21,49	$3d'\ {}^2F-4p'\ {}^2D^\circ$	5/2—3/2
9935,046	1	22,26	23,51	$3d''\ {}^2D-5p\ {}^4D^\circ$	5/2—5/2
9931,680	1	25,94	27,20	$({}^1D)4f\ [3]^\circ-({}^1D)5g\ [4]$	5/2—7/2
9928,830	1	25,94	27,20	$({}^1D)4f\ [3]^\circ-({}^1D)5g\ [4]$	7/2—9/2
9916,144	4	24,34	25,59	$({}^3P_1)4f\ [3]^\circ-({}^3P_1)5g\ [4]$	5/2—7/2
9914,246	2	24,19	25,44	$({}^3P_2)4f\ [5]^\circ-({}^3P_2)5g\ [5]$	9/2—9/2
9909,712	2	24,19	25,44	$({}^3P_2)4f\ [5]^\circ-({}^3P_2)5g\ [5]$	11/2—11/2
9906,394	5	18,62	19,87	$3d\ {}^2F-4p\ {}^2P^\circ$	5/2—3/2
9905,880	5	24,34	25,59	$({}^3P_1)4f\ [3]^\circ-({}^3P_1)5g\ [4]$	7/2—9/2
9904,29	1	20,24	21,50	$3d'\ {}^2F-4p'\ {}^2D^\circ$	5/2—5/2
9854,065	8	24,19	25,45	$({}^3P_2)4f\ [5]^\circ-({}^3P_2)5g\ [6]$	9/2—11/2
9849,460	10	24,19	25,45	$({}^3P_2)4f\ [5]^\circ-({}^3P_2)5g\ [6]$	11/2—13/2
9837,170	3	23,12	24,38	$4d\ {}^4P-({}^3P_0)4f\ [3]^\circ$	3/2—5/2
9829,856	3	18,29	19,55	$3d\ {}^4P-4p\ {}^4D^\circ$	3/2—5/2
9825,843	1	22,31	23,57	$3d''\ {}^2D-5p\ {}^4D^\circ$	3/2—3/2
9824,642	2	23,70	24,96	$5p\ {}^4S^\circ-5d\ {}^4P$	3/2—3/2
9819,18	1	23,80	25,06	$4p''\ {}^2P^\circ-5d\ {}^2F$	3/2—5/2
9814,424	3	24,38	25,64	$({}^3P_0)4f\ [3]^\circ-({}^3P_0)5g\ [4]$	5/2—7/2
9803,697	4	24,38	25,64	$({}^3P_0)4f\ [3]^\circ-({}^3P_0)5g\ [4]$	7/2—9/2
9802,019	4	24,62	25,89	$4d'\ {}^2G-({}^1D)4f\ [5]^\circ$	9/2—11/2
9793,239	3	24,62	25,89	$4d'\ {}^2G-({}^1D)4f\ [5]^\circ$	7/2—9/2
9783,100	3	24,18	25,45	$({}^3P_2)4f\ [2]^\circ-({}^3P_2)5g\ [3]$	5/2—7/2
9773,575	4	24,31	25,58	$({}^3P_1)4f\ [4]^\circ-({}^3P_1)5g\ [5]$	7/2—9/2
9771,833	2	23,07	24,34	$4d\ {}^4F-({}^3P_1)4f\ [3]^\circ$	5/2—7/2
9761,847	1	23,07	24,34	$4d\ {}^4F-({}^3P_1)4f\ [3]^\circ$	5/2—5/2
9758,644	4	24,31	25,58	$({}^3P_1)4f\ [4]^\circ-({}^3P_1)5g\ [5]$	9/2—11/2
9756,157	3	24,31	25,58	$({}^3P_1)4f\ [2]^\circ-({}^3P_1)5g\ [3]$	5/2—7/2
9750,145	3	24,18	25,45	$({}^3P_2)4f\ [2]^\circ-({}^3P_2)5g\ [3]$	3/2—5/2
9743,460	2	24,31	25,58	$({}^3P_1)4f\ [2]^\circ-({}^3P_1)5g\ [3]$	3/2—5/2
9739,770	2	22,31	23,58	$3d''\ {}^2D-5p\ {}^2P^\circ$	3/2—1/2
9734,554	1	24,18	25,45	$({}^3P_2)4f\ [2]^\circ-({}^3P_2)5g\ [2]$	5/2—5/2
9713,117	2	23,10	24,38	$4d\ {}^4F-({}^3P_0)4f\ [3]^\circ$	3/2—5/2
9711,779	1	24,31	25,59	$({}^3P_1)4f\ [4]^\circ-({}^3P_1)5g\ [4]$	9/2—9/2
9701,961	1	24,18	25,45	$({}^3P_2)4f\ [2]^\circ-({}^3P_2)5g\ [2]$	3/2—3/2
9701,515	1	23,51	24,79	$5p\ {}^4D^\circ-5d\ {}^4D$	5/2—5/2
9678,812	2	24,16	25,44	$({}^3P_2)4f\ [3]^\circ-({}^3P_2)5g\ [5]$	7/2—9/2
9676,287	3	24,16	25,44	$({}^3P_2)4f\ [3]^\circ-({}^3P_2)5g\ [4]$	7/2—9/2
9655,974	3	24,16	25,44	$({}^3P_2)4f\ [3]^\circ-({}^3P_2)5g\ [4]$	5/2—7/2
9643,312	1	24,16	25,45	$({}^3P_2)4f\ [3]^\circ-({}^3P_2)5g\ [3]$	7/2—7/2
9641,190	2	23,62	24,90	$5p\ {}^2D^\circ-5d\ {}^4F$	5/2—7/2
9623,235	2	24,16	25,45	$({}^3P_2)4f\ [3]^\circ-({}^3P_2)5g\ [3]$	5/2—5/2
9622,068	4	24,15	25,44	$({}^3P_2)4f\ [4]^\circ-({}^3P_2)5g\ [5]$	7/2—9/2
9619,575	3	{ 24,15	25,44	$({}^3P_2)4f\ [4]^\circ-({}^3P_2)5g\ [4]$	7/2—9/2
		{ 24,15	25,44	$({}^3P_2)4f\ [4]^\circ-({}^3P_2)5g\ [4]$	7/2—7/2
9612,508	2	25,89	27,18	$({}^1D)4f\ [5]^\circ-({}^1D)5g\ [6]$	9/2, 11/2—11/2, 13/2
9601,933	6	24,15	25,44	$({}^3P_2)4f\ [4]^\circ-({}^3P_2)5g\ [5]$	9/2—11/2
9599,325	2	24,15	25,44	$({}^3P_2)4f\ [4]^\circ-({}^3P_2)5g\ [4]$	9/2—9/2
9586,996	2	24,15	25,45	$({}^3P_2)4f\ [4]^\circ-({}^3P_2)5g\ [3]$	7/2—7/2
9553,631	5	23,48	24,78	$5p\ {}^4D^\circ-5d\ {}^4D$	7/2—7/2
9540,664	5	23,01	24,31	$4d\ {}^4F-({}^3P_1)4f\ [4]^\circ$	7/2—9/2
9535,640	3	23,51	24,81	$5p\ {}^4D^\circ-4d'\ {}^2F$	5/2—7/2
9526,39	1	23,01	24,31	$4d\ {}^4F-({}^3P_1)4f\ [4]^\circ$	7/2—7/2
9508,440	3	22,26	23,57	$3d''\ {}^2D-5p\ {}^4D^\circ$	5/2—3/2
9480,871	1	23,48	24,79	$5p\ {}^4D^\circ-5d\ {}^4D$	7/2—5/2
9475,239	4	18,45	19,76	$4s'\ {}^2D-4p\ {}^2D^\circ$	5/2—3/2
9436,22	1	23,48	24,80	$5p\ {}^4P^\circ-6s\ {}^2P$	1/2—3/2

λ, Å	I	E_H, eV	E_B, eV	Transition	J
9420,484	4	23,65	24,96	$5p\ ^4D^\circ$—$5d\ ^4P$	1/2—3/2
		23,70	25,02	$5p\ ^4S^\circ$—$5d\ ^4P$	3/2—5/2
9418,582	1	18,66	19,97	$3d\ ^2D$—$4p\ ^2S^\circ$	3/2—1/2
9374,163	3	18,29	19,61	$3d\ ^4P$—$4p\ ^4D^\circ$	3/2—3/2
9360,466	1	23,01	24,34	$4d\ ^4F$—$(^3P_1)4f\ [3]^\circ$	7/2—7/2
9344,793	2	25,31	26,63	$5p'\ ^2D^\circ$—$5d'\ ^2F$	5/2—7/2
9331,05	1	23,48	24,81	$5p\ ^4P^\circ$—$5d\ ^4D$	1/2—3/2
9313,51	1	24,62	25,95	$4d'\ ^2G$—$(^1D)4f\ [4]^\circ$	9/2—9/2
9305,87	1	24,62	25,95	$4d'\ ^2G$—$(^1D)4f\ [4]^\circ$	7/2—7/2
9279,712	4	18,43	19,76	$4s'\ ^2D$—$4p\ ^2D^\circ$	3/2—3/2
9258,78	1	25,31	26,64	$5p'\ ^2D^\circ$—$5d'\ ^2F$	3/2—5/2
9252,628	2	22,84	24,18	$4d\ ^4D$—$(^3P_2)4f\ [2]^\circ$	1/2—3/2
9219,001	2	23,62	24,96	$5p\ ^2P^\circ$—$5d\ ^4F$	3/2—5/2
9210,39	1	18,33	19,68	$3d\ ^4P$—$4p\ ^2D^\circ$	5/2—5/2
9192,605	3	22,81	24,16	$4d\ ^4D$—$(^3P_2)4f\ [3]^\circ$	3/2—5/2
9191,17	1	21,35	22,70	$4p'\ ^2P^\circ$—$5s\ ^2P$	3/2—3/2
9168,917	1	22,82	24,18	$3d'\ ^2S$—$(^3P_2)4f\ [2]^\circ$	1/2—3/2
9159,030	2	22,26	23,62	$3d''\ ^2D$—$5p\ ^2P^\circ$	5/2—3/2
9156,049	3	23,44	24,79	$5p\ ^4P^\circ$—$5d\ ^4D$	3/2—5/2
9150,82	1	18,29	19,64	$3d\ ^4P$—$4p\ ^4D^\circ$	3/2—1/2
9106,573	4	23,48	24,84	$5p\ ^4D^\circ$—$5d\ ^4F$	7/2—9/2
9098,58	2	23,65	25,01	$5p\ ^4D^\circ$—$5d\ ^4F$	1/2—3/2
9095,099	3	25,19	26,55	$5p'\ ^2F^\circ$—$5d'\ ^2G$	7/2—9/2
9079,707	2	22,81	24,18	$4d\ ^4D$—$(^3P_2)4f\ [2]^\circ$	3/2—3/2
9068,023	5	22,79	24,15	$4d\ ^4D$—$(^3P_2)4f\ [4]^\circ$	5/2—7/2
9060,749	3	25,19	26,55	$5p'\ ^2F^\circ$—$5d'\ ^2G$	5/2—7/2
9051,236	1	22,81	24,18	$4d\ ^4D$—$(^3P_2)4f\ [2]^\circ$	3/2—5/2
9035,915	3	22,79	24,16	$4d\ ^4D$—$(^3P_2)4f\ [3]^\circ$	5/2—5/2
9031,35	1	23,44	24,81	$5p\ ^4P^\circ$—$6s\ ^4P$	3/2—1/2
9017,596	7	18,43	19,80	$4s'\ ^2D$—$4p\ ^2P^\circ$	3/2—1/2
9014,938	1	22,84	24,21	$4d\ ^4D$—$(^3P_2)4f\ [1]^\circ$	1/2—1/2
9008,455	6	23,62	24,99	$5p\ ^2D^\circ$—$5d\ ^2F$	5/2—7/2
8997,803	4	23,44	24,81	$5p\ ^4P^\circ$—$5d\ ^4D$	3/2—3/2
8995,865	7	23,40	24,78	$5p\ ^4P^\circ$—$5d\ ^4D$	5/2—7/2
8986,615	6	22,77	24,15	$4d\ ^4D$—$(^3P_2)4f\ [4]^\circ$	7/2—9/2
8971,365	4	23,68	25,06	$5p\ ^2D^\circ$—$5d\ ^2F$	3/2—5/2
8968,947	4	22,77	24,15	$4d\ ^4D$—$(^3P_2)4f\ [4]^\circ$	7/2—7/2
8937,530	1	22,77	24,16	$4d\ ^4D$—$(^3P_2)4f\ [3]^\circ$	7/2—5/2
8935,448	1	22,82	24,21	$3d'\ ^2S$—$(^3P_2)4f\ [1]^\circ$	1/2—1/2
8931,326	5	23,40	24,79	$5p\ ^4P^\circ$—$5d\ ^4D$	5/2—5/2
8926,819	1	22,79	24,18	$4d\ ^4D$—$(^3P_2)4f\ [2]^\circ$	5/2—3/2
8926,074	3	18,25	19,64	$3d\ ^4P$—$4p\ ^4D^\circ$	1/2—1/2
8920,198	2	22,77	24,16	$4d\ ^4D$—$(^3P_2)4f\ [3]^\circ$	7/2—7/2
8915,522	1	23,62	25,01	$5p\ ^2P^\circ$—$5d\ ^4F$	3/2—3/2
8905,650	6	23,51	24,90	$5p\ ^4D^\circ$—$5d\ ^4F$	5/2—7/2
8904,512	1	18,29	19,68	$3d\ ^4P$—$4p\ ^2D^\circ$	3/2—5/2
8899,297	3	22,79	24,18	$4d\ ^4D$—$(^3P_2)4f\ [2]^\circ$	5/2—5/2
8895,144	1	23,57	24,96	$5p\ ^4D^\circ$—$5d\ ^4P$	3/2—3/2
8890,147	4	23,57	24,96	$5p\ ^4D^\circ$—$5d\ ^4F$	3/2—5/2
8870,216	1	23,62	25,02	$5p\ ^2D^\circ$—$5d\ ^4P$	5/2—5/2
8867,170	2	23,62	25,02	$5p\ ^2P^\circ$—$5d\ ^4P$	3/2—5/2
8850,695	1	22,81	24,21	$4d\ ^4D$—$(^3P_2)4f\ [1]^\circ$	3/2—1/2
8842,527	1	24,18	25,58	$(^3P_2)4f\ [2]^\circ$—$(^3P_1)5g\ [3]$	5/2—7/2
8838,009	1	22,81	24,21	$4d\ ^4D$—$(^3P_2)4f\ [1]^\circ$	3/2—3/2
8803,860	1	22,77	24,18	$4d\ ^4D$—$(^3P_2)4f\ [2]^\circ$	7/2—5/2
8796,142	5	23,44	24,84	$5p\ ^4P^\circ$—$5d\ ^4D$	3/2—1/2
8790,555	1	23,40	24,81	$5p\ ^4P^\circ$—$4d'\ ^2F$	5/2—7/2
8771,855	15	18,45	19,87	$4s'\ ^2D$—$4p\ ^2P^\circ$	5/2—3/2

λ, Å	I	E_H, eV	E_B, eV	Transition	J
8768,215	1	23,89	25,31	$4d\ ^2D—5p'\ ^2D^\circ$	3/2—3/2
8754,009	2	22,26	23,68	$3d''\ ^2D—5p\ ^2D^\circ$	5/2—3/2
8719,374	3	23,48	24,90	$5p\ ^4D^\circ—5d\ ^4F$	7/2—7/2
8716,947	1	23,40	24,82	$5p\ ^4P^\circ—4d'\ ^2F$	5/2—5/2
8693,086	2	22,79	24,21	$4d\ ^4D—(^3P_2)4f\ [1]^\circ$	5/2—3/2
8674,767	1	23,58	25,01	$5p\ ^2P^\circ—5d\ ^4F$	1/2—3/2
8657,390	1	24,15	25,58	$(^3P_2)4f\ [4]^\circ—(^3P_1)5g\ [5]$	9/2—11/2
8631,102	1	23,44	24,87	$5p\ ^4P^\circ—6s\ ^2P$	3/2—1/2
8623,804	5	23,48	24,92	$5p\ ^4P^\circ—5d\ ^4P$	1/2—1/2
8607,611	2	23,57	25,01	$5p\ ^4D^\circ—5d\ ^4F$	3/2—3/2
8604,016	6	18,43	19,87	$4s'\ ^2D—4p\ ^2P^\circ$	3/2—3/2
8592,624	3	25,19	26,63	$5p'\ ^2F^\circ—5d'\ ^2F$	7/2—7/2
8585,262	3	23,62	25,06	$5p\ ^2D^\circ—5d\ ^2F$	5/2—5/2
8562,550	2	23,57	25,02	$5p\ ^4D^\circ—5d\ ^4P$	3/2—5/2
8547,023	4	23,51	24,96	$5p\ ^4D^\circ—5d\ ^4F$	5/2—5/2
8500,997	2	25,19	26,64	$5p'\ ^2F^\circ—5d'\ ^2F$	5/2—5/2
8411,88	1	22,84	24,31	$4d\ ^4D—(^3P_1)4f\ [2]^\circ$	1/2—3/2
8395,734	3	23,48	24,96	$5p\ ^4P^\circ—5d\ ^4P$	1/2—3/2
8376,079	2	17,74	19,22	$3d\ ^4F—4p\ ^4P^\circ$	5/2—5/2
8363,074	2	23,51	24,99	$5p\ ^4D^\circ—5d\ ^2F$	5/2—7/2
8346,420	1	21,35	22,84	$4p'\ ^2P^\circ—4d\ ^4D$	3/2—1/2
8345,183	2	17,77	19,26	$3d\ ^4F—4p\ ^4P^\circ$	3/2—3/2
8342,630	1	22,82	24,31	$3d'\ ^2S—(^3P_1)4f\ [2]^\circ$	1/2—3/2
8338,384	1	23,44	24,92	$5p\ ^4P^\circ—5d\ ^4P$	3/2—1/2
8327,907	2	18,06	19,55	$3d\ ^2P—4p\ ^4D^\circ$	3/2—5/2
8296,723	1	23,57	25,06	$5p\ ^4D^\circ—5d\ ^2F$	3/2—5/2
8259,521	2	22,81	24,31	$4d\ ^4D—(^3P_1)4f\ [2]^\circ$	3/2—5/2
8217,817	1	18,25	19,76	$3d\ ^4P—4p\ ^2D^\circ$	1/2—3/2
8190,258	1	18,45	19,97	$4s'\ ^2D—4p\ ^4S^\circ$	5/2—3/2
8165,405	3	17,74	19,26	$3d\ ^4F—4p\ ^4P^\circ$	5/2—3/2
8150,647	1	23,89	25,41	$4d\ ^2D—(^3P_2)5f\ [3]^\circ$	3/2—5/2
8110,65	1	17,69	19,22	$3d\ ^4F—4p\ ^4P^\circ$	7/2—5/2
8083,75	1	23,89	25,42	$4d\ ^2D—(^3P_2)5f\ [2]^\circ$	3/2—5/2
8044,308	2	22,77	24,31	$4d\ ^4D—(^3P_1)4f\ [4]^\circ$	7/2—9/2
8036,853	2	23,87	25,41	$4d\ ^2D—(^3P_2)5f\ [3]^\circ$	5/2—7/2
8034,625	1	23,80	25,34	$4p''\ ^2P^\circ—5d\ ^2D$	3/2—3/2
8017,542	2	18,43	19,97	$4s'\ ^2D—4p\ ^2S^\circ$	3/2—1/2
7992,90	1	22,79	24,34	$4d\ ^4D—(^3P_1)4f\ [3]^\circ$	5/2—7/2
7983,61	1	23,87	25,42	$4d\ ^2D—(^3P_2)5f\ [2]^\circ$	5/2—5/2
7927,35	2	23,80	25,36	$4p''\ ^2P^\circ—5d\ ^2D$	3/2—5/2
7915,813	1	22,77	24,34	$4d\ ^4D—(^3P_1)4f\ [3]^\circ$	7/2—7/2
7904,770	2	22,81	24,38	$4d\ ^4D—(^3P_0)4f\ [3]^\circ$	3/2—5/2
7849,397	3	18,29	19,87	$3d\ ^4P—4p\ ^2P^\circ$	3/2—3/2
7846,555	2	23,48	25,06	$5p\ ^4D^\circ—5d\ ^2F$	7/2—5/2
		23,44	25,02	$5p\ ^4P^\circ—5d\ ^4P$	3/2—5/2
7802,252	1	24,19	25,78	$(^3P_2)4f\ [5]^\circ—6d\ ^4F$	11/2—9/2
7795,410	2	22,79	24,38	$4d\ ^4D—(^3P_0)4f\ [3]^\circ$	5/2—7/2
7757,003	1	23,85	25,44	$4p''\ ^2P^\circ—5d\ ^2P$	1/2—3/2
7753,28	1	24,15	25,75	$(^3P_2)4f\ [4]^\circ—6d\ ^4D$	9/2—7/2
7683,458	1	18,25	19,87	$3d\ ^4P—4p\ ^2P^\circ$	1/2—3/2
7681,49	1	24,19	25,80	$(^3P_2)4f\ [5]^\circ—6d\ ^4F$	9/2—7/2
7680,948	2	23,40	25,02	$5p\ ^4P^\circ—5d\ ^4P$	5/2—5/2
7654,031	2	18,06	19,68	$3d\ ^2P—4p\ ^2D^\circ$	3/2—5/2
7618,03	1	19,80	21,43	$4p\ ^2P^\circ—3d'\ ^2D$	1/2—3/2
7589,320	15	18,33	19,97	$3d\ ^4P—4p\ ^4S^\circ$	5/2—3/2
7505,153	1	19,97	21,62	$4p\ ^2S^\circ—3d'\ ^2P$	1/2—3/2
7455,996	2	23,89	25,55	$4d\ ^2D—(^3P_1)5f\ [2]^\circ$	3/2—5/2
7440,491	4	19,76	21,43	$4p\ ^2D^\circ—3d'\ ^2D$	3/2—3/2

λ, Å	I	E_H, eV	E_B, eV	Transition	J
7428,574	2	17,94	19,61	$3d\ ^2P—4p\ ^4D^\circ$	$^1/_2—^3/_2$
7419,341	1	23,67	25,34	$5p\ ^2S^\circ—5d\ ^2D$	$^1/_2—^3/_2$
7380,433	15	18,29	19,97	$3d\ ^4P—4p\ ^4S^\circ$	$^3/_2—^3/_2$
7358,338	2	18,29	19,97	$3d\ ^4P—4p\ ^2S^\circ$	$^3/_2—^1/_2$
7355,180	2	23,87	25,56	$4d\ ^2D—(^3P_1)5f\ [4]^\circ$	$^5/_2—^7/_2$
7348,049	7	19,68	21,37	$4p\ ^2D^\circ—3d'\ ^2D$	$^5/_2—^5/_2$
7284,236	4	18,06	19,76	$3d\ ^2P—4p\ ^2D^\circ$	$^3/_2—^3/_2$
7280,454	2	19,97	21,67	$4p\ ^2S^\circ—3d'\ ^2P$	$^1/_2—^1/_2$
7233,546	15	18,25	19,97	$3d\ ^4P—4p\ ^4S^\circ$	$^1/_2—^3/_2$
7182,098	2	23,89	25,62	$4d\ ^2D—(^3P_0)5f\ [3]^\circ$	$^3/_2—^5/_2$
7121,740	4	18,06	19,80	$3d\ ^2P—4p\ ^2P^\circ$	$^3/_2—^1/_2$
7101,190	1	23,62	25,36	$5p\ ^2P^\circ—5d\ ^2D$	$^3/_2—^5/_2$
7090,560	1	19,68	21,43	$4p\ ^2D^\circ—3d'\ ^2D$	$^5/_2—^3/_2$
7077,024	5	17,74	19,49	$3d\ ^4F—4p\ ^4D^\circ$	$^5/_2—^7/_2$
7054,993	3	19,87	21,62	$4p\ ^2P^\circ—3d'\ ^2P$	$^3/_2—^3/_2$
6990,122	5	17,77	19,55	$3d\ ^4F—4p\ ^4D^\circ$	$^3/_2—^5/_2$
6985,708	1	23,57	25,34	$5p\ ^4D^\circ—5d\ ^2D$	$^3/_2—^3/_2$
6900,880	2	23,63	25,42	$4d\ ^2P—(^3P_2)5f\ [2]^\circ$	$^3/_2—^5/_2$
6886,618	20	17,69	19,49	$3d\ ^4F—4p\ ^4D^\circ$	$^7/_2—^7/_2$
6863,535	20	17,74	19,55	$3d\ ^4F—4p\ ^4D^\circ$	$^5/_2—^5/_2$
6861,270	15	18,06	19,87	$3d\ ^2P—4p\ ^2P^\circ$	$^3/_2—^3/_2$
6846,540	1	23,63	25,44	$4d\ ^2P—(^3P_2)5f\ [1]^\circ$	$^3/_2—^3/_2$
6839,584	4	21,67	23,48	$3d'\ ^2P—5p\ ^4P^\circ$	$^1/_2—^1/_2$
6818,371	8	19,55	21,37	$4p\ ^4D^\circ—3d'\ ^2D$	$^5/_2—^5/_2$
6808,532	9	17,94	19,76	$3d\ ^2P—4p\ ^2D^\circ$	$^1/_2—^3/_2$
6799,288	3	19,80	21,62	$4p\ ^2P^\circ—3d'\ ^2P$	$^1/_2—^3/_2$
6756,548	20	17,77	19,61	$3d\ ^4F—4p\ ^4D^\circ$	$^3/_2—^3/_2$
6696,296	4	22,31	24,16	$3d''\ ^2D—(^3P_2)4f\ [3]^\circ$	$^3/_2—^5/_2$
6684,307	50	17,69	19,55	$3d\ ^4F—4p\ ^4D^\circ$	$^7/_2—^5/_2$
6666,356	15	17,94	19,80	$3d\ ^2P—4p\ ^2P^\circ$	$^1/_2—^1/_2$
6657,499	2	19,76	21,62	$4p\ ^2D^\circ—3d'\ ^2P$	$^3/_2—^3/_2$
6653,583	1	23,58	25,44	$5p\ ^2P^\circ—5d\ ^2P$	$^1/_2—^3/_2$
6643,716	100	17,63	19,49	$3d\ ^4F—4p\ ^4D^\circ$	$^9/_2—^7/_2$
6639,743	30	17,77	19,64	$3d\ ^4F—4p\ ^4D^\circ$	$^3/_2—^1/_2$
6638,226	50	17,74	19,61	$3d\ ^4F—4p\ ^4D^\circ$	$^5/_2—^3/_2$
6620,977	6	22,31	24,18	$3d''\ ^2D—(^3P_2)4f\ [2]^\circ$	$^3/_2—^5/_2$
6614,354	6	19,80	21,67	$4p\ ^2P^\circ—3d'\ ^2P$	$^1/_2—^1/_2$
6611,196	2	23,55	25,42	$4d\ ^2P—(^3P_2)5f\ [2]^\circ$	$^1/_2—^3/_2$
6564,170	3	22,26	24,15	$3d''\ ^2D—(^3P_2)4f\ [4]^\circ$	$^5/_2—^7/_2$
6557,724	2	21,62	23,51	$3d'\ ^2P—5p\ ^4D^\circ$	$^3/_2—^5/_2$
6551,498	3	23,55	25,44	$4d\ ^2P—(^3P_2)5f\ [1]^\circ$	$^1/_2—^3/_2$
6547,350	3	21,26	24,16	$3d''\ ^2D—(^3P_2)4f\ [3]^\circ$	$^5/_2—^5/_2$
6540,409	2	21,67	23,57	$3d'\ ^2P—5p\ ^4D^\circ$	$^1/_2—^3/_2$
6532,927	2	23,80	25,70	$4p''\ ^2P^\circ—7s\ ^4P$	$^3/_2—^3/_2$
6509,089	6	17,77	19,68	$3d\ ^4F—4p\ ^2D^\circ$	$^3/_2—^5/_2$
6508,184	6	24,21	26,12	$(^3P_2)4f\ [1]^\circ—(^3P_2)6g\ [2]$	$^3/_2—^5/_2$
6506,138	3	22,31	24,21	$3d''\ ^2D—(^3P_2)4f\ [1]^\circ$	$^3/_2—^3/_2$
6502,157	3	21,67	23,58	$3d'\ ^2P—5p\ ^2P^\circ$	$^1/_2—^1/_2$
6501,348	4	24,21	26,12	$(^3P_2)4f\ [1]^\circ—(^3P_2)6g\ [2]$	$^1/_2—^3/_2$
6500,216	12	18,06	19,97	$3d\ ^2P—4p\ ^4S^\circ$	$^3/_2—^3/_2$
6483,076	20	18,06	19,97	$3d\ ^2P—4p\ ^2S^\circ$	$^3/_2—^1/_2$
6480,085	2	19,76	21,67	$4p\ ^2D^\circ—3d'\ ^2P$	$^3/_2—^1/_2$
6475,312	4	22,26	24,18	$3d''\ ^2D—(^3P_2)4f\ [2]^\circ$	$^5/_2—^5/_2$
6472,431	6	24,34	26,25	$(^3P_1)4f\ [3]^\circ—(^3P_1)6g\ [4]$	$^5/_2—^7/_2$
6468,050	7	24,34	26,25	$(^3P_1)4f\ [3]^\circ—(^3P_1)6g\ [4]$	$^7/_2—^9/_2$
6458,403	2	24,19	26,11	$(^3P_2)4f\ [5]^\circ—(^3P_2)6g\ [5]$	$^9/_2—^9/_2$
6456,489	3	24,19	26,11	$(^3P_2)4f\ [5]^\circ—(^3P_2)6g\ [5]$	$^{11}/_2—^{11}/_2$
6445,117	1	23,63	25,55	$4d\ ^2P—(^3P_1)5f\ [2]^\circ$	$^3/_2—^3/_2$

λ, Å	I	E_H, eV	E_B, eV	Transition	J
6443,858	8	24,19	26,11	$({}^3P_2)4f\ [5]^\circ-({}^3P_2)6g\ [6]$	$9/2-11/2$
6441,908	9	24,19	26,11	$({}^3P_2)4f\ [5]^\circ-({}^3P_2)6g\ [6]$	$11/2-13/2$
6437,604	8	17,94	19,87	$3d\ {}^2P-4p\ {}^2P^\circ$	$1/2-3/2$
6433,683	1	23,17	25,10	$4d\ {}^4P-6p\ {}^4D^\circ$	$5/2-7/2$
6424,144	1	23,26	25,19	$4d\ {}^2F-5p'\ {}^2F^\circ$	$5/2-5/2$
6422,903	6	24,38	26,31	$({}^3P_0)4f\ [3]^\circ-({}^3P_0)6g\ [4]$	$5/2-7/2$
6418,354	8	24,38	26,31	$({}^3P_0)4f\ [3]^\circ-({}^3P_0)6g\ [4]$	$7/2-9/2$
6417,417	1	24,18	26,11	$({}^3P_2)4f\ [2]^\circ-({}^3P_2)6g\ [4]$	$5/2-7/2$
6408,904	6	24,18	26,11	$({}^3P_2)4f\ [2]^\circ-({}^3P_2)6g\ [3]$	$5/2-7/2$
6403,004	6	24,31	26,25	$({}^3P_1)4f\ [4]^\circ-({}^3P_1)6g\ [5]$	$7/2-9/2$
6399,215	15	17,74	19,68	$3d\ {}^4F-4p\ {}^2D^\circ$	$5/2-5/2$
6397,184	2	24,18	26,12	$({}^3P_2)4f\ [2]^\circ-({}^3P_2)6g\ [2]$	$5/2-5/2$
6396,614	7	24,31	26,25	$({}^3P_1)4f\ [4]^\circ-({}^3P_1)6g\ [5]$	$9/2-11/2$
6394,723	5	24,18	26,11	$({}^3P_2)4f\ [2]^\circ-({}^3P_2)6g\ [3]$	$3/2-5/2$
6393,803	6	24,31	26,25	$({}^3P_1)4f\ [2]^\circ-({}^3P_1)6g\ [3]$	$5/2-7/2$
6391,117	2	24,31	26,25	$({}^3P_1)4f\ [4]^\circ-({}^3P_1)6g\ [4]$	$7/2-7/2$
		24,31	26,25	$({}^3P_1)4f\ [4]^\circ-({}^3P_1)6g\ [4]$?	$7/2-9/2$
6388,335	5	24,31	26,25	$({}^3P_1)4f\ [2]^\circ-({}^3P_1)6g\ [3]$	$3/2-5/2$
6383,095	2	24,18	26,12	$({}^3P_2)4f\ [2]^\circ-({}^3P_2)6g\ [2]$	$3/2-3/2$
6382,696	3	23,63	25,57	$4d\ {}^2P-({}^3P_1)5f\ [3]^\circ$	$3/2-5/2$
6375,945	3	19,68	21,62	$4p\ {}^2D^\circ-3d'\ {}^2P$	$5/2-3/2$
6369,128	2	21,62	23,57	$3d'\ {}^2P-5p\ {}^4D^\circ$	$3/2-3/2$
6365,440	1	22,26	24,21	$3d''\ {}^2D-({}^3P_2)4f\ [1]^\circ$	$5/2-3/2$
6357,668	5	24,16	26,11	$({}^3P_2)4f\ [3]^\circ-({}^3P_2)6g\ [5]$	$7/2-9/2$
6357,025	6	24,16	26,11	$({}^3P_2)4f\ [3]^\circ-({}^3P_2)6g\ [4]$	$7/2-9/2$
6348,601	2	24,16	26,11	$({}^3P_2)4f\ [3]^\circ-({}^3P_2)6g\ [3]$	$7/2-7/2$
6348,227	6	24,16	26,11	$({}^3P_2)4f\ [3]^\circ-({}^3P_2)6g\ [4]$	$5/2-7/2$
6339,897	3	24,16	26,11	$({}^3P_2)4f\ [3]^\circ-({}^3P_2)6g\ [3]$	$5/2-5/2$
6333,142	5	24,15	26,11	$({}^3P_2)4f\ [4]^\circ-({}^3P_2)6g\ [5]$	$7/2-9/2$
6332,832	1	21,62	23,58	$3d'\ {}^2P-5p\ {}^2P^\circ$	$3/2-1/2$
6332,499	5	24,15	26,11	$({}^3P_2)4f\ [4]^\circ-({}^3P_2)6g\ [4]$	$7/2-9/2$
6328,474	1	24,16	26,12	$({}^3P_2)4f\ [3]^\circ-({}^3P_2)6g\ [2]$	$5/2-3/2$
6326,117	2	25,89	27,85	$({}^1D)4f\ [5]^\circ-({}^1D)6g\ [6]$	$11/2,\ 9/2-13/2,\ 11/2$
6324,414	8	24,15	26,11	$({}^3P_2)4f\ [4]^\circ-({}^3P_2)6g\ [5]$	$9/2-11/2$
6323,735	3	24,15	26,11	$({}^3P_2)4f\ [4]^\circ-({}^3P_2)6g\ [4]$	$9/2-9/2$
6315,40	1	24,15	26,11	$({}^3P_2)4f\ [4]^\circ-({}^3P_2)6g\ [3]$	$9/2-7/2$
6295,446	2	23,89	25,86	$4d\ {}^2D-({}^1D)4f\ [1]^\circ$	$3/2-3/2$
6282,823	1	21,67	23,65	$3d'\ {}^2P-5p\ {}^4D^\circ$	$1/2-1/2$
6277,425	2	23,70	25,67	$5p\ {}^4S^\circ-7s\ {}^4P$	$3/2-5/2$
6249,975	1	23,85	25,83	$4p''\ {}^2P^\circ-7s\ {}^4P$	$1/2-1/2$
6243,125	25	17,69	19,68	$3d\ {}^4F-4p\ {}^2D^\circ$	$7/2-5/2$
6239,713	7	17,77	19,76	$3d\ {}^4F-4p\ {}^2D^\circ$	$3/2-3/2$
6232,892	2	23,63	25,62	$4d\ {}^2P-({}^3P_0)5f\ [3]^\circ$	$3/2-5/2$
6210,420	1	21,62	23,62	$3d'\ {}^2P-5p\ {}^2P^\circ$	$3/2-3/2$
6208,935	2	21,62	23,62	$3d'\ {}^2P-5p\ {}^2D^\circ$	$3/2-5/2$
6206,463	1	23,70	25,70	$5p\ {}^4S^\circ-7s\ {}^4P$	$3/2-3/2$
6201,099	6	21,67	23,67	$3d'\ {}^2P-5p\ {}^2S^\circ$	$1/2-1/2$
6192,301	2	22,31	24,31	$3d''\ {}^2D-({}^3P_1)4f\ [2]^\circ$	$3/2-3/2$
6187,136	6	21,31	24,31	$3d''\ {}^2D-({}^3P_1)4f\ [2]^\circ$	$3/2-5/2$
6183,024	2	23,55	25,55	$4d\ {}^2P-({}^2P_1)5f\ [2]^\circ$	$1/2-3/2$
6174,378	3	21,67	23,68	$3d'\ {}^3P-5p\ {}^2D^\circ$	$1/2-3/2$
6172,290	40	19,11	21,13	$3d'\ {}^2G-4p'\ {}^2F^\circ$	$7/2-5/2$
6167,628	3	23,89	25,90	$4d\ {}^2D-({}^1D)4f\ [2]^\circ$	$3/2-5/2$
6166,790	3	21,43	23,44	$3d'\ {}^2D-5p\ {}^4P^\circ$	$3/2-3/2$
		23,89	25,90	$4d\ {}^2D-({}^1D)4f\ [2]^\circ$	$3/2-3/2$
6142,615	2	23,80	25,82	$4p''\ {}^2P^\circ-7s\ {}^2P$	$3/2-3/2$
6138,660	12	21,14	23,16	$4p'\ {}^2F^\circ-4d\ {}^2F$?	$7/2-7/2$
		17,74	19,76	$3d\ {}^4F-4p\ {}^2D^\circ$	$5/2-3/2$

λ, Å	I	E_H, eV	E_B, eV	Transition	J
6124,571	3	21,62	23,65	$3d'\ ^2P-5p\ ^4D^\circ$	$^3/_2-^1/_2$
6123,368	15	19,11	21,14	$3d'\ ^2G-4p'\ ^2F^\circ$	$^7/_2-^7/_2$
6120,102	5	17,77	19,80	$3d\ ^4F-4p\ ^2P^\circ$	$^3/_2-^1/_2$
6118,724	4	17,94	19,97	$3d\ ^2P-4p\ ^4S^\circ$	$^1/_2-^3/_2$
6114,929	50	19,11	21,14	$3d'\ ^2G-4p'\ ^2F^\circ$	$^9/_2-^7/_2$
6111,742	2	23,87	25,90	$4d\ ^2D-(^3P_2)7p\ [2]^\circ$	$^5/_2-^5/_2$
6109,15	2	23,87	25,90	$4d\ ^2D-(^1D)4f\ [2]^\circ$	$^5/_2-^5/_2$
6103,546	12	17,94	19,97	$3d\ ^2P-4p\ ^2S^\circ$	$^1/_2-^1/_2$
6102,765	1	23,16	25,19	$4d\ ^2F-5p'\ ^2F^\circ$	$^7/_2-^7/_2$
6101,925	3	22,31	24,34	$3d''\ ^2D-(^3P_1)4f\ [3]^\circ$	$^3/_2-^5/_2$
6084,507	2	21,37	23,40	$3d'\ ^2D-5p\ ^4P^\circ$	$^5/_2-^5/_2$
6083,875	2	23,85	25,88	$4p''\ ^2P^\circ-7s\ ^2P$	$^1/_2-^1/_2$
6077,431	6	17,26	19,30	$4s\ ^2P-4p\ ^4P^\circ$	$^1/_2-^1/_2$
6049,072	6	22,26	24,31	$3d''\ ^2D-(^3P_1)4f\ [4]^\circ$	$^5/_2-^7/_2$
6046,894	8	21,62	23,67	$3d'\ ^2P-5p\ ^2S^\circ$	$^3/_2-^1/_2$
6044,468	7	23,89	25,94	$4d\ ^2D-(^1D)4f\ [3]^\circ$	$^3/_2-^5/_2$
6030,844	1	23,70	25,76	$5p\ ^4S^\circ-6d\ ^4D$	$^3/_2-^5/_2$
6028,220	1	23,62	25,67	$5p\ ^2D^\circ-7s\ ^4P$	$^5/_2-^5/_2$
6027,248	5	21,49	23,55	$4p'\ ^2D^\circ-4d\ ^2P$	$^3/_2-^1/_2$
6019,493	4	21,43	23,48	$3d'\ ^2D-5p\ ^4P^\circ$	$^3/_2-^1/_2$
6003,170	1	19,61	21,67	$4p\ ^4D^\circ-3d'\ ^2P$	$^3/_2-^1/_2$
5989,339	8	23,87	25,94	$4d\ ^2D-(^1D)4f\ [3]^\circ$	$^5/_2-^7/_2$
5988,288	3	23,87	25,94	$4d\ ^2D-(^1D)4f\ [3]^\circ$	$^5/_2-^5/_2$
5985,920	7	22,31	24,38	$3d''\ ^2D-(^3P_0)4f\ [3]^\circ$	$^3/_2-^5/_2$
5984,454	3	21,37	23,44	$3d'\ ^2D-5p\ ^4P^\circ$	$^5/_2-^3/_2$
5977,995	4	22,26	24,34	$3d''\ ^2D-(^3P_1)4f\ [3]^\circ$	$^5/_2-^5/_2$
5975,945	2	24,18	26,25	$(^3P_2)4f\ [2]^\circ-(^3P_1)6g\ [3]$	$^3/_2-^5/_2$
5973,314	2	19,55	21,62	$4p\ ^4D^\circ-3d'\ ^2P$	$^5/_2-^3/_2$
5965,031	3	21,62	23,70	$3d'\ ^2P-5p\ ^4S^\circ$	$^3/_2-^3/_2$
5953,820	2	23,80	25,88	$4p''\ ^2P^\circ-7s\ ^2P$	$^3/_2-^1/_2$
5950,905	6	17,14	19,22	$4s\ ^2P-4p\ ^4P^\circ$	$^3/_2-^5/_2$
5941,825	4	21,43	23,51	$3d'\ ^2D-5p\ ^4D^\circ$	$^3/_2-^5/_2$
5936,64	1	23,17	25,26	$4d\ ^4P-6p\ ^2P^\circ$	$^5/_2-^3/_2$
5935,792	2	23,70	25,79	$5p\ ^4S^\circ-6d\ ^4D$	$^3/_2-^3/_2$
5904,291	2	24,15	26,25	$(^3P_2)4f\ [4]^\circ-(^3P_1)6g\ [5]$	$^9/_2-^{11}/_2$
		24,15	26,25	$(^3P_2)4f\ [4]^\circ-(^3P_1)6g\ [5]$	$^9/_2-^9/_2$
5886,088	3	19,26	21,37	$4p\ ^4P^\circ-3d'\ ^2D$	$^3/_2-^5/_2$
5870,443	4	22,26	24,38	$3d''\ ^2D-(^3P_0)4f\ [3]^\circ$	$^5/_2-^7/_2$
5866,598	2	22,26	24,38	$3d''\ ^2D-(^3P_0)4f\ [3]^\circ$	$^5/_2-^5/_2$
5853,10	2	23,70	25,82	$5p\ ^4S^\circ-7s\ ^2P$	$^3/_2-^3/_2$
5843,781	7	17,14	19,26	$4s\ ^2P-4p\ ^4P^\circ$	$^3/_2-^3/_2$
5840,048	1	21,43	23,55	$4p'\ ^2P^\circ-4d\ ^2P$	$^1/_2-^1/_2$
5838,96	1	19,30	21,43	$4p\ ^4P^\circ-3d'\ ^2D$	$^1/_2-^3/_2$
5828,059	3	21,67	23,80	$3d'\ ^2P-4p''\ ^2P^\circ$	$^1/_2-^3/_2$
5826,036	2	23,70	25,83	$5p\ ^4S^\circ-7s\ ^4P$	$^3/_2-^1/_2$
5822,114	3	23,57	25,70	$5p\ ^4D^\circ-7s\ ^4P$	$^3/_2-^3/_2$
5816,272	2	21,13	23,26	$4p'\ ^2F^\circ-4d\ ^2F$?	$^5/_2-^5/_2$
5812,746	6	21,50	23,63	$4p'\ ^2D^\circ-4d\ ^2P$	$^5/_2-^3/_2$
5807,596	1	24,19	26,32	$(^3P_2)4f\ [5]^\circ-7d\ ^4F$	$^9/_2-^7/_2$
5800,46	1	23,62	25,76	$5p\ ^2D^\circ-6d\ ^4D$	$^5/_2-^5/_2$
5799,734	3	23,68	25,82	$5p\ ^2D^\circ-7s\ ^2P$	$^3/_2-^3/_2$
5786,560	5	21,43	23,57	$3d'\ ^2D-5p\ ^4D^\circ$	$^3/_2-^3/_2$
5781,268	2	19,22	21,37	$4p\ ^4P^\circ-3d'\ ^2D$	$^5/_2-^5/_2$
5776,374	2	23,67	25,82	$5p\ ^2S^\circ-7s\ ^2P$	$^1/_2-^3/_2$
5774,697	1	23,70	25,85	$5p\ ^4S^\circ-6d\ ^4P$	$^3/_2-^1/_2$
5772,326	5	21,37	23,51	$3d'\ ^2D-5p\ ^4D^\circ$	$^5/_2-^5/_2$
5766,542	2	22,95	25,10	$4d\ ^4F-6p\ ^4D^\circ$	$^9/_2-^7/_2$
5756,600	3	21,43	23,58	$3d'\ ^2D-5p\ ^2P^\circ$	$^3/_2-^1/_2$

λ, Å	I	E_H, eV	E_B, eV	Transition	J
5753,54	1	24,15	26,31	$(^3P_2)\,4f\,[4]^\circ$—$(^3P_0)\,6g\,[4]$	$^7/_2$—$^9/_2$
5743,278	2	23,26	25,41	$4d\,^2F$—$(^3P_2)5f\,[3]^\circ$	$^5/_2$—$^7/_2$
5732,694	1	23,80	25,96	$4p''\,^2P^\circ$—$6d\,^2D$	$^3/_2$—$^3/_2$
5732,210	1	23,51	25,67	$5p\,^4D^\circ$—$7s\,^4P$	$^5/_2$—$^5/_2$
5724,325	5	17,14	19,30	$4s\,^2P$—$4p\,^4P^\circ$	$^3/_2$—$^1/_2$
5716,029	1	23,26	25,42	$4d\,^2F$—$(^3P_2)\,5f\,[2]^\circ$	$^5/_2$—$^5/_2$
5711,453	1	23,70	25,87	$5p\,^4S^\circ$—$6d\,^4F$	$^3/_2$—$^5/_2$
5708,616	1	21,67	23,85	$3d'\,^2P$—$4p''\,^2P^\circ$	$^1/_2$—$^1/_2$
5707,215	1	23,65	25,82	$5p\,^4D^\circ$—$7s\,^2P$	$^1/_2$—$^3/_2$
5704,371	3	23,70	25,87	$5p\,^4S^\circ$—$6d\,^4P$	$^3/_2$—$^3/_2$
5691,650	8	21,62	23,80	$3d'\,^2P$—$4p''\,^2P^\circ$	$^3/_2$—$^3/_2$
5681,480	2	23,65	25,83	$5p\,^4D^\circ$—$7s\,^4P$	$^1/_2$—$^1/_2$
		23,70	25,88	$5p\,^4S^\circ$—$7s\,^2P$	$^3/_2$—$^1/_2$
5672,952	7	23,51	25,70	$5p\,^4D^\circ$—$7s\,^4P$	$^5/_2$—$^3/_2$
		23,80	25,98	$4p''\,^2P^\circ$—$6d\,^2P$	$^3/_2$—$^3/_2$
5655,236	2	21,43	23,62	$3d'\,^2D$—$5p\,^2P^\circ$	$^3/_2$—$^3/_2$
5654,450	8	23,48	25,67	$5p\,^4D^\circ$—$7s\,^4P$	$^7/_2$—$^5/_2$
5654,020	2	21,43	23,62	$3d'\,^2D$—$5p\,^2D^\circ$	$^3/_2$—$^5/_2$
5642,413	2	17,77	19,97	$3d\,^4F$—$4p\,^2S^\circ$	$^3/_2$—$^1/_2$
5635,882	5	23,62	25,82	$5p\,^2D^\circ$—$7s\,^2P$	$^5/_2$—$^3/_2$
5634,661	2	23,62	25,82	$5p\,^2P^\circ$—$7s\,^2P$	$^3/_2$—$^3/_2$
5631,381	1	23,89	26,09	$4d\,^2D$—$(^3P_2)\,6f\,[3]^\circ$	$^3/_2$—$^5/_2$
5631,160	3	23,68	25,88	$5p\,^2D^\circ$—$7s\,^2P$	$^3/_2$—$^1/_2$
5625,684	6	21,37	23,57	$3d'\,^2D$—$5p\,^4D^\circ$	$^5/_2$—$^3/_2$
5624,005	1	21,43	23,63	$4p'\,^2P^\circ$—$4d\,^2P$	$^1/_2$—$^3/_2$
5611,667	2	23,89	26,10	$4d\,^2D$—$(^3P_2)\,6f\,[2]^\circ$	$^3/_2$—$^5/_2$
5609,578	2	23,62	25,83	$5p\,^2P^\circ$—$7s\,^4P$	$^3/_2$—$^1/_2$
5603,932	2	23,48	25,70	$5p\,^4P^\circ$—$7s\,^4P$	$^1/_2$—$^3/_2$
5593,52	1	23,89	26,11	$4d\,^2D$—$(^3P_2)\,6f\,[1]^\circ$	$^3/_2$—$^3/_2$
5592,200	1	23,01	25,23	$4d\,^4F$—$6p\,^2D^\circ$	$^7/_2$—$^5/_2$
5582,61	1	23,87	26,09	$4d\,^2D$—$(^3P_2)\,6f\,[3]^\circ$	$^5/_2$—$^5/_2$
5578,518	6	23,87	26,09	$4d\,^2D$—$(^3P_2)\,6f\,[3]^\circ$	$^5/_2$—$^7/_2$
5577,689	8	21,62	23,85	$3d'\,^2P$—$4p''\,^2P^\circ$	$^3/_2$—$^1/_2$
5563,196	2	23,87	26,10	$4d\,^2D$—$(^3P_2)\,6f\,[2]^\circ$	$^5/_2$—$^5/_2$
5554,050	8	23,63	25,86	$4d\,^2P$—$(^1D)\,4f\,[1]^\circ$	$^3/_2$—$^1/_2$
		23,63	25,86	$4d\,^2P$—$(^1D)\,4f\,[1]^\circ$	$^3/_2$—$^3/_2$
5545,045	6	23,70	25,94	$5p\,^4S^\circ$—$6d\,^4P$	$^3/_2$—$^5/_2$
5543,880	1	23,65	25,88	$5p\,^4D^\circ$—$7s\,^2P$?	$^1/_2$—$^1/_2$
5537,290	5	23,44	25,67	$5p\,^4P^\circ$—$7s\,^4P$	$^3/_2$—$^5/_2$
5535,51	1	23,17	25,41	$4d\,^4P$—$(^3P_2)\,5f\,[4]^\circ$	$^5/_2$—$^7/_2$
5525,856	2	23,51	25,76	$5p\,^4D^\circ$—$6d\,^4D$?	$^5/_2$—$^5/_2$
5523,690	2	23,70	25,94	$5p\,^4S^\circ$—$6d\,^4F$	$^3/_2$—$^3/_2$
5521,74	1	23,17	25,41	$4d\,^4P$—$(^3P_2)\,5f\,[3]^\circ$	$^5/_2$—$^7/_2$
5519,337	4	21,43	23,67	$3d'\,^2D$—$5p\,^2S^\circ$	$^3/_2$—$^1/_2$
5516,668	2	23,16	25,41	$4d\,^2F$—$(^3P_2)\,5f\,[4]^\circ$	$^7/_2$—$^9/_2$
5514,367	4	—	—	—	—
5513,303	1	23,58	25,83	$5p\,^2P^\circ$—$7s\,^4P$	$^1/_2$—$^1/_2$
5507,753	4	23,80	26,05	$4p''\,^2P^\circ$—$6d\,^2D$	$^3/_2$—$^5/_2$
5504,917	3	22,82	25,07	$3d'\,^2S$—$6p\,^4P^\circ$	$^1/_2$—$^3/_2$
5503,256	3	23,62	25,87	$5p\,^2P^\circ$—$6d\,^4F$	$^3/_2$—$^5/_2$
5501,480	2	21,37	23,62	$3d'\,^2D$—$5p\,^2P^\circ$	$^5/_2$—$^3/_2$
5500,334	7	21,37	23,62	$3d'\,^2D$—$5p\,^2D^\circ$	$^5/_2$—$^5/_2$
5498,972	2	23,16	25,41	$4d\,^2F$—$(^3P_2)\,5f\,[3]^\circ$	$^7/_2$—$^7/_2$
5498,185	8	21,43	23,68	$3d'\,^2D$—$5p\,^2D^\circ$	$^3/_2$—$^3/_2$
5497,123	3	23,68	25,94	$5p\,^2D^\circ$—$6d\,^4P$	$^3/_2$—$^5/_2$
5490,667	1	23,70	25,96	$5p\,^4S^\circ$—$6d\,^2F$	$^3/_2$—$^5/_2$
5486,102	1	23,57	25,83	$5p\,^4D^\circ$—$7s\,^4P$	$^3/_2$—$^1/_2$
5484,311	1	22,84	25,10	$4d\,^4D$—$6p\,^4P^\circ$	$^1/_2$—$^1/_2$

λ, Å	I	E_H, eV	E_B, eV	Transition	J
5481,997	1	23,44	25,70	$5p\,^4P^\circ—7s\,^4P$	$^3/_2—^3/_2$
5472,642	1	22,81	25,07	$4d\,^4D—6p\,^4P^\circ$	$^3/_2—^3/_2$
5470,307	2	22,84	25,10	$4d\,^4D—6p\,^4D^\circ$	$^1/_2—^3/_2$
5469,105	6	23,48	25,75	$5p\,^4D^\circ—6d\,^4D$	$^7/_2—^7/_2$
5466,440	5	23,16	25,43	$4d\,^2F—(^3P_2)\,5f\,[5]^\circ$	$^7/_2—^9/_2$
5456,382	5	23,63	25,90	$4d\,^2P—(^3P_2)\,7p\,[2]^\circ$	$^3/_2—^5/_2$
5454,307	9	23,63	25,90	$4d\,^2P—(^1D)\,4f\,[2]^\circ$	$^3/_2—^5/_2$
		23,40	25,67	$5p\,^4P^\circ—7s\,^4P$	$^5/_2—^5/_2$
5453,634	5	23,63	25,90	$4d\,^2P—(^1D)\,4f\,[2]^\circ$	$^3/_2—^3/_2$
		23,48	25,76	$5p\,^4D^\circ—6d\,^4D$	$^7/_2—^5/_2$
5447,556	2	22,79	25,06	$4d\,^4D—6p\,^4P^\circ$	$^5/_2—^5/_2$
5443,681	5	23,68	25,96	$5p\,^2D^\circ—6d\,^2F$	$^3/_2—^5/_2$
5440,932	2	22,82	25,10	$3d'\,^2S—6p\,^4D^\circ$	$^1/_2—^3/_2$
5439,676	3	21,35	23,63	$4p'\,^2P^\circ—4d\,^2P$	$^3/_2—^3/_2$
5432,94	1	23,68	25,96	$5p\,^2D^\circ—6d\,^2D$	$^3/_2—^3/_2$
5416,710	1	22,79	25,07	$4d\,^4D—6p\,^4P^\circ$	$^5/_2—^3/_2$
5412,434	2	23,67	25,96	$5p\,^2S^\circ—6d\,^2D$	$^1/_2—^3/_2$
5411,646	3	22,77	25,06	$4d\,^4D—6p\,^4P^\circ$	$^7/_2—^5/_2$
5409,401	1	22,81	25,10	$4d\,^4D—6p\,^4D^\circ$	$^3/_2—^3/_2$
5407,348	7	23,51	25,80	$5p\,^4D^\circ—6d\,^4F$	$^5/_2—^7/_2$
5402,604	8	23,62	25,91	$5p\,^2D^\circ—6d\,^2F$	$^5/_2—^7/_2$
5402,170	1	22,80	25,10	$5s\,^2P—6p\,^4P^\circ$	$^1/_2—^1/_2$
5400,62	1	23,40	25,70	$5p\,^4P^\circ—7s\,^4P$	$^5/_2—^3/_2$
5397,522	9	23,48	25,78	$5p\,^4D^\circ—6d\,^4F$	$^7/_2—^9/_2$
5394,81	1	23,26	25,55	$4d\,^2F—(^3P_1)\,5f\,[2]^\circ$	$^5/_2—^5/_2$
5393,603	5	23,65	25,94	$5p\,^4D^\circ—6d\,^4F$	$^1/_2—^3/_2$
5386,519	2	23,26	25,56	$4d\,^2F—(^3P_1)\,5f\,[4]^\circ$	$^5/_2—^7/_2$
5384,378	5	23,57	25,87	$5p\,^4D^\circ—6d\,^4F$	$^3/_2—^5/_2$
5382,330	2	23,48	25,79	$5p\,^4P^\circ—6d\,^4D$	$^1/_2—^3/_2$
5379,163	1	23,68	25,98	$5p\,^2D^\circ—6d\,^2P$	$^3/_2—^3/_2$
5378,078	3	23,57	25,87	$5p\,^4D^\circ—6d\,^4P$	$^3/_2—^3/_2$
5376,636	3	24,21	26,52	$(^3P_2)\,4f\,[1]^\circ—(^3P_2)\,7g\,[2]$	$^3/_2—^5/_2$
5372,007	2	24,21	26,52	$(^3P_2)\,4f\,[1]^\circ—(^3P_2)\,7g\,[2]$	$^1/_2—^3/_2$
5365,485	1	22,79	25,10	$4d\,^4D—6p\,^4D^\circ$	$^5/_2—^7/_2$
5364,142	1	23,10	25,41	$4d\,^4F—(^3P_2)\,5f\,[3]^\circ$	$^3/_2—^5/_2$
5359,069	2	23,67	25,98	$5p\,^2S^\circ—6d\,^2P$	$^1/_2—^3/_2$
5358,616	2	23,26	25,57	$4d\,^2F—(^3P_1)\,5f\,[3]^\circ$	$^5/_2—^7/_2$
5358,363	6	23,55	25,86	$4d\,^2P—(^1D)\,4f\,[1]^\circ$	$^1/_2—^1/_2$
5354,82	1	24,34	26,65	$(^3P_1)\,4f\,[3]^\circ—(^3P_1)\,7g\,[3]$	$^7/_2—^7/_2$
5351,449	4	24,34	26,65	$(^3P_1)\,4f\,[3]^\circ—(^3P_1)\,7g\,[4]$	$^5/_2—^7/_2$
5349,717	2	23,62	25,94	$5p\,^2D^\circ—6d\,^4P$	$^5/_2—^5/_2$
5348,604	2	23,62	25,94	$5p\,^2P^\circ—6d\,^4P$	$^3/_2—^5/_2$
5348,283	3	24,34	26,65	$(^3P_1)\,4f\,[3]^\circ—(^3P_1)\,7g\,[4]$	$^7/_2—^9/_2$
5345,609	1	19,30	21,62	$4p\,^4P^\circ—3d'\,^2P$	$^1/_2—^3/_2$
5344,534	5	23,44	25,76	$5p\,^4P^\circ—6d\,^4D$	$^3/_2—^5/_2$
5339,33	1	23,12	25,44	$4d\,^4P—(^3P_2)\,5f\,[1]^\circ$	$^3/_2—^3/_2$
5338,106	1	23,48	25,80	$5p\,^4D^\circ—6d\,^4F$	$^7/_2—^7/_2$
5335,916	1	24,19	26,51	$(^3P_2)\,4f\,[5]^\circ—(^3P_2)\,7g\,[5]$	$^{11}/_2—^{11}/_2$
5331,034	4	24,19	26,52	$(^3P_2)\,4f\,[5]^\circ—(^3P_2)\,7g\,[6]$	$^9/_2—^{11}/_2$
5330,664	3	22,77	25,10	$4d\,^4D—6p\,^4D^\circ$	$^7/_2—^7/_2$
5329,712	5	24,19	26,52	$(^3P_2)\,4f\,[5]^\circ—(^3P_2)\,7g\,[6]$	$^{11}/_2—^{13}/_2$
5315,214	3	24,38	26,71	$(^3P_0)\,4f\,[3]^\circ—(^3P_0)\,7g\,[4]$	$^5/_2—^7/_2$
5314,258	2	23,48	25,82	$5p\,^4P^\circ—7s\,^2P$	$^1/_2—^3/_2$
5312,002	6	24,38	26,71	$(^3P_0)\,4f\,[3]^\circ—(^3P_0)\,7g\,[4]$	$^7/_2—^9/_2$
5308,074	5	21,37	23,70	$3d'\,^2D—5p\,^4S^\circ$	$^5/_2—^3/_2$
5305,690	6	24,18	26,51	$(^3P_2)\,4f\,[2]^\circ—(^3P_2)\,7g\,[3]$	$^5/_2—^7/_2$
		19,97	22,31	$4p\,^2S^\circ—3d''\,^2D$	$^1/_2—^3/_2$
5300,761	4	24,31	26,65	$(^3P_1)\,4f\,[4]^\circ—(^3P_1)\,7g\,[5]$	$^7/_2—^9/_2$

λ, Å	I	E_H, eV	E_B, eV	Transition	J
5299,075	3	23,62	25,96	$5p\ ^2D^\circ—6d\ ^2F$	5/2—5/2
5297,993	3	23,62	25,96	$5p\ ^2P^\circ—6d\ ^2F$	3/2—5/2
5296,791	3	23,89	26,23	$4d\ ^2D—(^3P_1)\ 6f\ [2]^\circ$	3/2—5/2
5296,386	4	24,31	26,65	$(^3P_1)\ 4f\ [4]^\circ—(^3P_1)\ 7g\ [5]$	9/2—11/2
5295,892	3	24,18	26,51	$(^3P_2)\ 4f\ [2]^\circ—(^3P_2)\ 7g\ [3]$	3/2—5/2
5293,821	4	24,31	26,65	$(^3P_1)\ 4f\ [2]^\circ—(^3P_1)\ 7g\ [3]$	5/2—7/2
5290,035	3	24,31	26,65	$(^3P_1)\ 4f\ [2]^\circ—(^3P_1)\ 7g\ [3]$	3/2—5/2
5288,634	1	23,07	25,41	$4d\ ^4F—(^3P_2)\ 5f\ [3]^\circ$	5/2—5/2
5286,895	15	17,26	19,61	$4s\ ^2P—4p\ ^4D^\circ$	1/2—3/2
5285,438	2	22,82	25,17	$3d'\ ^2S—5p'\ ^2P^\circ$	1/2—3/2
5283,437	1	23,07	25,41	$4d\ ^4F—(^3P_2)\ 5f\ [3]^\circ$	5/2—7/2
5281,628	7	23,40	25,75	$5p\ ^4P^\circ—6d\ ^4D$	5/2—7/2
5273,580	1	23,89	26,24	$4d\ ^2D—(^3P_1)\ 6f\ [3]^\circ$	3/2—5/2
5269,755	4	23,44	25,79	$5p\ ^4P^\circ—6d\ ^4D$	3/2—3/2
5268,246	3	24,16	26,51	$(^3P_2)\ 4f\ [3]^\circ—(^3P_2)\ 7g\ [5]$	7/2—9/2
5267,958	3	24,16	26,51	$(^3P_2)\ 4f\ [3]^\circ—(^3P_2)\ 7g\ [4]$	7/2—9/2
5267,168	6	23,40	25,76	$5p\ ^4P^\circ—6d\ ^4D$	5/2—5/2
5264,783	6	23,55	25,90	$4d\ ^2P—(^1D)\ 4f\ [2]^\circ$	1/2—3/2
5264,305	1	24,16	26,51	$(^3P_2)\ 4f\ [3]^\circ—(^3P_2)\ 7g\ [3]$	7/2—7/2
5261,903	4	24,16	26,51	$(^3P_2)\ 4f\ [3]^\circ—(^3P_2)\ 7g\ [4]$	5/2—7/2
5259,174	1	23,08	25,44	$4d\ ^4P—(^3P_2)\ 5f\ [1]^\circ$	1/2—1/2
5258,223	1	24,16	26,51	$(^3P_2)\ 4f\ [3]^\circ—(^3P_2)\ 7g\ [3]$	5/2—5/2
5256,569	3	23,51	25,87	$5p\ ^4D^\circ—6d\ ^4F$	5/2—5/2
5255,677	1	22,81	25,17	$4d\ ^4D—5p'\ ^2P^\circ$	3/2—3/2
5252,138	1	23,26	25,62	$4d\ ^2F—(^3P_0)\ 5f\ [3]^\circ$	5/2—7/2
5251,400	3	24,15	26,51	$(^3P_2)\ 4f\ [4]^\circ—(^3P_2)\ 7g\ [5]$	7/2—9/2
5251,108	3	24,15	26,51	$(^3P_2)\ 4f\ [4]^\circ—(^3P_2)\ 7g\ [4]$	7/2—9/2
5249,547	4	23,48	25,85	$5p\ ^4P^\circ—6d\ ^4P$	1/2—1/2
5247,986	3	23,87	26,23	$4d\ ^2D—(^3P_1)\ 6f\ [4]^\circ$	5/2—7/2
5247,469	1	24,15	26,51	$(^3P_2)\ 4f\ [4]^\circ—(^3P_2)\ 7g\ [3]$	7/2—7/2
5245,389	5	24,15	26,51	$(^3P_2)\ 4f\ [4]^\circ—(^3P_2)\ 7g\ [5]$	9/2—11/2
5241,786	2	23,58	25,94	$5p\ ^2P^\circ—6d\ ^4F$	1/2—3/2
5236,853	1	23,62	25,98	$5p\ ^2P^\circ—6d\ ^2P$	3/2—3/2
5236,231	2	23,57	25,94	$5p\ ^4D^\circ—6d\ ^4P$?	3/2—5/2
5235,564	4	—	—	—	—
5230,523	3	19,30	21,67	$4p\ ^4P^\circ—3d'\ ^2P$	1/2—1/2
		23,68	26,05	$5p\ ^2D^\circ—6d\ ^2D$	3/2—5/2
5221,854	3	21,43	23,80	$3d'\ ^2D—4p''\ ^2P^\circ$	3/2—3/2
5219,589	1	22,84	25,21	$4d\ ^4D—6p\ ^2D^\circ$	1/2—3/2
5216,816	8	21,50	23,87	$4p'\ ^2D^\circ—4d\ ^2D$	5/2—5/2
5204,440	3	23,44	25,82	$5p\ ^4P^\circ—7s\ ^2P$	3/2—3/2
		21,49	23,87	$4p'\ ^2D^\circ—4d\ ^2D$	3/2—5/2
5202,201	2	23,58	25,96	$5p\ ^2P^\circ—6d\ ^2D$	1/2—3/2
5192,810	3	22,82	25,21	$3d'\ ^2S—6p\ ^2D^\circ$	1/2—3/2
5191,364	4	23,48	25,87	$5p\ ^4P^\circ—6d\ ^4P$	1/2—3/2
5176,233	10	18,73	21,13	$3d\ ^2D—4p'\ ^2F^\circ$	5/2—5/2
5175,426	2	23,16	25,56	$4d\ ^2F—(^3P_1)\ 5f\ [4]^\circ$	7/2—9/2
5165,774	8	19,87	22,26	$4p\ ^2P^\circ—3d''\ ^2D$	3/2—5/2
5162,742	5	21,49	23,89	$4p'\ ^2D^\circ—4d\ ^2D$	3/2—3/2
5159,505	1	22,70	25,10	$5s\ ^2P—6p\ ^4D^\circ$	3/2—3/2
5145,319	25	17,14	19,55	$4s\ ^2P—4p\ ^4D^\circ$	3/2—5/2
5141,790	20	18,73	21,14	$3d\ ^2D—4p'\ ^2F^\circ$	5/2—7/2
5132,145	2	23,01	25,43	$4d\ ^4F—(^3P_2)\ 5f\ [5]^\circ$	7/2—9/2
5131,106	1	23,40	25,82	$5p\ ^4P^\circ—7s\ ^2P$	5/2—3/2
5129,083	2	23,57	25,98	$5p\ ^4D^\circ—6d\ ^2P$	3/2—3/2
5125,765	8	21,43	23,85	$3d'\ ^2D—4p''\ ^2P^\circ$	3/2—1/2
5122,972	2	—	—	—	—
5121,486	2	23,87	26,29	$4d\ ^2D—(^3P_0)\ 6f\ [3]^\circ$	5/2—5/2

λ, Å	I	E_H, eV	E_B, eV	Transition	J
5096,851	1	23,62	26,05	$5p\,^2D^\circ$—$6d\,^2D$	$^5/_2$—$^5/_2$
5095,845	2	23,62	26,05	$5p\,^2P^\circ$—$6d\,^2D$	$^3/_2$—$^5/_2$
5092,174	1	23,44	25,87	$5p\,^4P^\circ$—$6d\,^4F$	$^3/_2$—$^5/_2$
5090,496	10	21,37	23,80	$3d'\,^2D$—$4p''\,^2P^\circ$	$^5/_2$—$^3/_2$
5062,036	30	16,81	19,26	$4s\,^4P$—$4p\,^4P^\circ$	$^1/_2$—$^3/_2$
5059,394	2	—	—	—	—
5042,416	1	22,77	25,23	$4d\,^4D$—$6p\,^2D^\circ$	$^7/_2$—$^5/_2$
5024,778	1	21,43	23,89	$4p'\,^2P^\circ$—$4d\,^2D$	$^1/_2$—$^3/_2$
5017,629	10	17,14	19,61	$4s\,^2P$—$4p\,^4D^\circ$	$^3/_2$—$^3/_2$
5017,160	20	18,66	21,13	$3d\,^2D$—$4p'\,^2F^\circ$	$^3/_2$—$^5/_2$
5014,92	1	23,63	26,10	$4d\,^2P$—$(^3P_2)\,6f\,[2]^\circ$	$^3/_2$—$^5/_2$
5009,334	30	16,75	19,22	$4s\,^4P$—$4p\,^4P^\circ$	$^3/_2$—$^5/_2$
4996,782	2	22,95	25,43	$4d\,^4F$—$(^3P_2)\,5f\,[5]^\circ$	$^9/_2$—$^{11}/_2$
4993,746	2	23,57	26,05	$5p\,^4D^\circ$—$6d\,^2D$	$^3/_2$—$^5/_2$
4972,157	15	16,81	19,30	$4s\,^4P$—$4p\,^4P^\circ$	$^1/_2$—$^1/_2$
4965,073	25	17,26	19,76	$4s\,^2P$—$4p\,^2D^\circ$	$^1/_2$—$^3/_2$
4959,478	1	23,44	25,94	$5p\,^4P^\circ$—$6d\,^4P$	$^3/_2$—$^5/_2$
4955,111	7	21,67	24,18	$3d'\,^2P$—$(^3P_2)\,4f\,[2]^\circ$	$^1/_2$—$^3/_2$
4952,924	2	17,14	19,64	$4s\,^2P$—$4p\,^4D^\circ$	$^3/_2$—$^1/_2$
4949,398	5	19,76	22,26	$4p\,^2D^\circ$—$3d''\,^2D$	$^3/_2$—$^5/_2$
4942,915	6	19,80	22,31	$4p\,^2P^\circ$—$3d''\,^2D$	$^1/_2$—$^3/_2$
4936,083	3	18,62	21,13	$3d\,^2F$—$4p'\,^2F^\circ$	$^5/_2$—$^5/_2$
4933,206	25	16,75	19,26	$4s\,^4P$—$4p\,^4P^\circ$	$^3/_2$—$^3/_2$
4914,309	2	21,35	23,87	$4p'\,^2P^\circ$—$4d\,^2D$	$^3/_2$—$^5/_2$
4904,753	12	18,62	21,14	$3d\,^2F$—$4p'\,^2F^\circ$	$^5/_2$—$^7/_2$
4889,033	15	17,26	19,80	$4s\,^2P$—$4p\,^2P^\circ$	$^1/_2$—$^1/_2$
4888,263	5	21,62	24,16	$3d'\,^2P$—$(^3P_2)\,4f\,[3]^\circ$	$^3/_2$—$^5/_2$
4882,233	10	21,67	24,21	$3d'\,^2P$—$(^3P_2)\,4f\,[1]^\circ$	$^1/_2$—$^3/_2$
4879,860	30	17,14	19,68	$4s\,^2P$—$4p\,^2D^\circ$	$^3/_2$—$^5/_2$
4867,557	5	19,76	22,31	$4p\,^2D^\circ$—$3d''\,^2D$	$^3/_2$—$^3/_2$
4865,919	12	19,97	22,51	$4p\,^4S^\circ$—$5s\,^4P$	$^3/_2$—$^5/_2$
4856,156	1	21,62	24,18	$3d'\,^2P$—$(^3P_2)\,4f\,[2]^\circ$	$^3/_2$—$^3/_2$
4847,815	25	16,75	19,30	$4s\,^4P$—$4p\,^4P^\circ$	$^3/_2$—$^1/_2$
4806,017	35	16,64	19,22	$4s\,^4P$—$4p\,^4P^\circ$	$^5/_2$—$^5/_2$
4792,090	6	19,68	22,26	$4p\,^2D^\circ$—$3d''\,^2D$	$^5/_2$—$^5/_2$
4786,155	5	21,62	24,21	$3d'\,^2P$—$(^3P_2)\,4f\,[1]^\circ$	$^3/_2$—$^3/_2$
4764,862	25	17,26	19,87	$4s\,^2P$—$4p\,^2P^\circ$	$^1/_2$—$^3/_2$
4757,215	1	22,80	25,41	$5s\,^2P$—$6p\,^2S^\circ$	$^1/_2$—$^1/_2$
4735,905	25	16,64	19,26	$4s\,^4P$—$4p\,^4P^\circ$	$^5/_2$—$^3/_2$
4732,056	12	18,73	21,35	$3d\,^2D$—$4p'\,^2P^\circ$	$^5/_2$—$^3/_2$
4730,664	3	19,97	22,59	$4p\,^2S^\circ$—$5s\,^4P$	$^1/_2$—$^3/_2$
4726,859	25	17,14	19,76	$4s\,^2P$—$4p\,^2D^\circ$	$^3/_2$—$^3/_2$
4721,594	12	19,97	22,59	$4p\,^4S^\circ$—$5s\,^4P$	$^3/_2$—$^3/_2$
4710,823	7	18,49	21,13	$3d\,^2F$—$4p'\,^2F^\circ$	$^7/_2$—$^5/_2$
4703,359	9	21,67	24,31	$3d'\,^2P$—$(^3P_1)\,4f\,[2]^\circ$	$^1/_2$—$^3/_2$
4682,277	10	18,49	21,14	$3d\,^2F$—$4p'\,^2F^\circ$	$^7/_2$—$^7/_2$
4681,494	2	19,87	22,51	$4p\,^2P^\circ$—$5s\,^4P$	$^3/_2$—$^5/_2$
4666,260	1	19,61	22,26	$4p\,^4D^\circ$—$3d''\,^2D$	$^3/_2$—$^5/_2$
4657,893	25	17,14	19,80	$4s\,^2P$—$4p\,^2P^\circ$	$^3/_2$—$^1/_2$
4637,233	12	18,45	21,13	$4s'\,^2D$—$4p'\,^2F^\circ$	$^5/_2$—$^5/_2$
4614,10	1	21,62	24,31	$3d'\,^2P$—$(^3P_1)\,4f\,[2]^\circ$	$^3/_2$—$^3/_2$
4611,245	3	21,62	24,31	$3d'\,^2P$—$(^3P_1)\,4f\,[2]^\circ$	$^3/_2$—$^5/_2$
4609,560	25	18,45	21,14	$4s'\,^2D$—$4p'\,^2F^\circ$	$^5/_2$—$^7/_2$
4600,02	1	20,74	23,44	$4s''\,^2S$—$5p\,^4P^\circ$	$^1/_2$—$^3/_2$
4598,760	10	18,66	21,35	$3d\,^2D$—$4p'\,^2P^\circ$	$^3/_2$—$^3/_2$
4589,896	25	18,43	21,13	$4s'\,^2D$—$4p'\,^2F^\circ$	$^3/_2$—$^5/_2$
4587,895	3	17,26	19,97	$4s\,^2P$—$4p\,^4S^\circ$	$^1/_2$—$^3/_2$
4579,346	25	17,26	19,97	$4s\,^2P$—$4p\,^2S^\circ$	$^1/_2$—$^1/_2$
4572,894	2	19,97	22,68	$4p\,^2S^\circ$—$5s\,^4P$	$^1/_2$—$^1/_2$

λ, Å	I	E_H, eV	E_B, eV	Transition	J
4564,415	7	19,97	22,68	$4p\ ^4S^\circ—5s\ ^4P$	$^3/_2—^1/_2$
4563,751	7	21,62	24,34	$3d'\ ^2P—(^3P_1)\ 4f\ [3]^\circ$	$^3/_2—^5/_2$
4561,018	3	19,55	22,26	$4p\ ^4D^\circ—3d''\ ^2D$	$^5/_2—^5/_2$
4547,760	4	19,87	22,59	$4p\ ^2P^\circ—5s\ ^4P$	$^3/_2—^3/_2$
4545,045	25	17,14	19,87	$4s\ ^2P—4p\ ^2P^\circ$	$^3/_2—^3/_2$
4543,871	3	19,97	22,70	$4p\ ^2S^\circ—5s\ ^2P$	$^1/_2—^3/_2$
4538,713	1	21,14	23,87	$4p'\ ^2F^\circ—4d\ ^2D$	$^7/_2—^5/_2$
4537,648	7	21,43	24,16	$3d'\ ^2D—(^3P_2)\ 4f\ [3]^\circ$	$^3/_2—^5/_2$
4535,492	6	19,97	22,70	$4p\ ^4S^\circ—5s\ ^2P$	$^3/_2—^3/_2$
4530,553	7	18,62	21,35	$3d\ ^2F—4p'\ ^2P^\circ$	$^5/_2—^3/_2$
4517,526	3	20,74	23,48	$4s''\ ^2S—5p\ ^4P^\circ$	$^1/_2—^1/_2$
4516,095	1	22,51	25,26	$5s\ ^4P—6p\ ^2P^\circ$	$^5/_2—^3/_2$
4509,957	2	21,43	24,18	$3d'\ ^2D—(^3P_2)\ 4f\ [2]^\circ$	$^3/_2—^3/_2$
4503,08	1	19,76	22,51	$4p\ ^2D^\circ—5s\ ^4P$	$^3/_2—^5/_2$
4502,931	7	21,43	24,18	$3d'\ ^2D—(^3P_2)\ 4f\ [2]^\circ$	$^3/_2—^5/_2$
4498,543	7	21,62	24,38	$3d'\ ^2P—(^3P_0)\ 4f\ [3]^\circ$	$^3/_2—^5/_2$
4490,988	8	18,73	21,49	$3d\ ^2D—4p'\ ^2D^\circ$	$^5/_2—^3/_2$
4481,810	15	18,73	21,50	$3d\ ^2D—4p'\ ^2D^\circ$	$^5/_2—^5/_2$
4474,759	10	18,66	21,43	$3d\ ^2D—4p'\ ^2P^\circ$	$^3/_2—^1/_2$
4460,560	12	16,44	19,22	$3d\ ^4D—4p\ ^4P^\circ$	$^3/_2—^5/_2$
4458,885	1	23,08	25,86	$4d\ ^4P—(^1D)\ 4f\ [1]^\circ$	$^1/_2—^3/_2$
4456,55	1	23,16	25,94	$4d\ ^2F—(^1D)\ 4f\ [3]^\circ$	$^7/_2—^7/_2$
4449,517	4	21,43	24,21	$3d'\ ^2D—(^3P_2)\ 4f\ [1]^\circ$	$^3/_2—^3/_2$
4448,881	8	21,50	24,28	$4p'\ ^2D^\circ—5s'\ ^2D$	$^5/_2—^5/_2$
4448,459	3	21,50	24,28	$4p'\ ^2D^\circ—5s'\ ^2D$	$^5/_2—^3/_2$
4445,848	4	21,37	24,15	$3d'\ ^2D—(^3P_2)\ 4f\ [4]^\circ$	$^5/_2—^7/_2$
4440,122	4	19,80	22,59	$4p\ ^2P^\circ—5s\ ^4P$	$^1/_2—^3/_2$
4439,878	4	21,49	24,28	$4p'\ ^2D^\circ—5s'\ ^2D$	$^3/_2—^5/_2$
4439,463	7	21,49	24,28	$4p'\ ^2D^\circ—5s'\ ^2D$	$^3/_2—^3/_2$
4438,808	1	23,16	25,95	$4d\ ^2F—(^1D)\ 4f\ [4]^\circ$	$^7/_2—^9/_2$
4438,117	3	21,37	24,16	$3d'\ ^2D—(^3P_2)\ 4f\ [3]^\circ$	$^5/_2—^5/_2$
4433,841	10	21,37	24,16	$3d'\ ^2D—(^3P_2)\ 4f\ [3]^\circ$	$^5/_2—^7/_2$
4431,004	15	16,42	19,22	$3d\ ^4D—4p\ ^4P^\circ$	$^5/_2—^5/_2$
4430,192	20	16,81	19,61	$4s\ ^4P—4p\ ^4D^\circ$	$^1/_2—^3/_2$
4426,005	25	16,75	19,55	$4s\ ^4P—4p\ ^4D^\circ$	$^3/_2—^5/_2$
4420,912	12	16,46	19,26	$3d\ ^4D—4p\ ^4P^\circ$	$^1/_2—^3/_2$
4412,905	3	18,33	21,14	$3d\ ^4P—4p'\ ^2F^\circ$	$^5/_2—^7/_2$
4406,469	2	23,51	26,32	$5p\ ^4D^\circ—7d\ ^4F$	$^5/_2—^7/_2$
4404,903	5	21,37	24,18	$3d'\ ^2D—(^3P_2)\ 4f\ [2]^\circ$	$^5/_2—^5/_2$
4401,744	2	19,87	22,68	$4p\ ^2P^\circ—5s\ ^4P$	$^3/_2—^1/_2$
4400,988	20	16,41	19,22	$3d\ ^4D—4p\ ^4P^\circ$	$^7/_2—^5/_2$
4400,099	18	16,44	19,26	$3d\ ^4D—4p\ ^4P^\circ$	$^3/_2—^3/_2$
4394,622	3	19,97	22,79	$4p\ ^4S^\circ—4d\ ^4D$	$^3/_2—^5/_2$
4386,962	1	23,62	26,44	$5p\ ^2D^\circ—7d\ ^2F?$	$^5/_2—^7/_2$
4385,058	10	20,74	23,57	$4s''\ ^2S—5p\ ^4D^\circ$	$^1/_2—^3/_2$
4383,754	8	17,14	19,97	$4s\ ^2P—4p\ ^4S^\circ$	$^3/_2—^3/_2$
4382,934	3	23,48	26,31	$5p\ ^4D^\circ—7d\ ^4F$	$^7/_2—^9/_2$
4379,879	5	19,97	22,80	$4p\ ^2S^\circ—5s\ ^2P$	$^1/_2—^1/_2$
4379,667	20	16,81	19,64	$4s\ ^4P—4p\ ^4D^\circ$	$^1/_2—^1/_2$
4379,226	3	19,76	22,59	$4p\ ^2D^\circ—5s\ ^4P$	$^3/_2—^3/_2$
4375,948	12	17,14	19,97	$4s\ ^2P—4p\ ^2S^\circ$	$^3/_2—^1/_2$
4374,857	6	19,87	22,70	$4p\ ^2P^\circ—5s\ ^2P$	$^3/_2—^3/_2$
4372,491	5	19,68	22,51	$4p\ ^2D^\circ—5s\ ^4P$	$^5/_2—^5/_2$
4371,329	20	16,42	19,26	$3d\ ^4D—4p\ ^4P^\circ$	$^5/_2—^3/_2$
4370,751	15	18,66	21,49	$3d\ ^2D—4p'\ ^2D^\circ$	$^3/_2—^3/_2$
4367,829	10	20,74	23,58	$4s''\ ^2S—5p\ ^2P^\circ$	$^1/_2—^1/_2$
4362,065	10	18,66	21,50	$3d\ ^2D—4p'\ ^2D^\circ$	$^3/_2—^5/_2$
4358,490	3	19,97	22,81	$4p\ ^4S^\circ—4d\ ^4D$	$^3/_2—^3/_2$

λ, Å	I	E_H, eV	E_B, eV	Transition	J
4352,204	15	16,46	19,30	$3d\,^4D-4p\,^4P^\circ$	$1/2-1/2$
4348,063	50	16,64	19,49	$4s\,^4P-4p\,^4D^\circ$	$5/2-7/2$
4338,228	2	19,97	22,82	$4p\,^4S^\circ-3d'\,^2S$	$3/2-1/2$
4337,070	8	21,43	24,28	$4p'\,^2P^\circ-5s'\,^2D$	$1/2-3/2$
4332,031	15	16,44	19,30	$3d\,^4D-4p\,^4P^\circ$	$3/2-1/2$
4331,199	25	16,75	19,61	$4s\,^4P-4p\,^4D^\circ$	$3/2-3/2$
4319,636	2	19,97	22,84	$4p\,^4S^\circ-4d\,^4D$?	$3/2-1/2$
4309,236	9	20,74	23,62	$4s''\,^2S-5p\,^2P^\circ$	$1/2-3/2$
4309,090	8	18,62	21,49	$3d\,^2F-4p'\,^2D^\circ$	$5/2-3/2$
4300,650	12	18,62	21,50	$3d\,^2F-4p'\,^2D^\circ$	$5/2-5/2$
4300,449	2	21,43	24,31	$3d'\,^2D-(^3P_1)\,4f\,[2]^\circ$	$3/2-3/2$
4297,964	7	21,43	24,31	$3d'\,^2D-(^3P_1)\,4f\,[2]^\circ$	$3/2-5/2$
4286,343	2	23,40	26,29	$5p\,^4P^\circ-7d\,^4D$	$5/2-7/2$
4282,896	12	16,75	19,64	$4s\,^4P-4p\,^4D^\circ$	$3/2-1/2$
4279,909	2	—	—	—	—
4277,524	20	18,45	21,35	$4s'\,^2D-4p'\,^2P^\circ$	$5/2-3/2$
4275,158	8	19,80	22,70	$4p\,^2P^\circ-5s\,^2P$	$1/2-3/2$
4267,730	3	20,74	23,65	$4s''\,^2S-5p\,^4D^\circ$	$1/2-1/2$
4267,490	3	19,61	22,51	$4p\,^4D^\circ-5s\,^4P$	$3/2-5/2$
4266,528	25	16,64	19,55	$4s\,^4P-4p\,^4D^\circ$	$5/2-5/2$
4256,663	1	21,43	24,34	$3d'\,^2D-(^3P_1)\,4f\,[3]^\circ$	$3/2-5/2$
4255,600	4	19,68	22,59	$4p\,^2D^\circ-5s\,^4P$	$5/2-3/2$
4243,640	2	19,87	22,79	$4p\,^2P^\circ-4d\,^4D$	$3/2-5/2$
4237,223	12	18,43	21,35	$4s'\,^2D-4p'\,^2P^\circ$	$3/2-3/2$
4229,872	8	20,74	23,67	$4s''\,^2S-5p\,^2S^\circ$	$1/2-1/2$
4228,162	20	16,75	19,68	$4s\,^4P-4p\,^2D^\circ$	$3/2-5/2$
4226,988	10	21,35	24,28	$4p'\,^2P^\circ-5s'\,^2D$	$3/2-5/2$
4226,607	5	21,35	24,28	$4p'\,^2P^\circ-5s'\,^2D$	$3/2-3/2$
4222,640	10	19,87	22,80	$4p\,^2P^\circ-5s\,^2P$	$3/2-1/2$
4218,667	11	19,76	22,70	$4p\,^2D^\circ-5s\,^2P$	$3/2-3/2$
4217,433	9	20,74	23,68	$4s''\,^2S-5p\,^2D^\circ$	$1/2-3/2$
4214,854	2	23,16	26,10	$4d\,^2F-(^3P_2)\,6f\,[5]^\circ$	$7/2-9/2$
4210,950	1	21,37	24,31	$3d'\,^2D-(^3P_1)\,4f\,[2]^\circ$	$5/2-3/2$
4209,944	1	19,87	22,81	$4p\,^2P^\circ-4d\,^4D$	$3/2-3/2$
4203,410	11	21,37	24,31	$3d'\,^2D-(^3P_1)\,4f\,[4]^\circ$	$5/2-7/2$
4201,971	12	16,81	19,76	$4s\,^4P-4p\,^2D^\circ$	$1/2-3/2$
4201,551	4	19,64	22,59	$4p\,^4D^\circ-5s\,^4P$	$1/2-3/2$
4199,892	6	21,43	24,38	$3d'\,^2D-(^3P_0)\,4f\,[3]^\circ$	$3/2-5/2$
4189,653	10	20,74	23,70	$4s''\,^2S-5p\,^4S^\circ$	$1/2-3/2$
4179,302	12	19,55	22,51	$4p\,^4D^\circ-5s\,^4P$	$5/2-5/2$
4178,371	12	16,64	19,61	$4s\,^4P-4p\,^4D^\circ$	$5/2-3/2$
4168,967	4	21,37	24,34	$3d'\,^2D-(^3P_1)\,4f\,[3]^\circ$	$5/2-5/2$
4156,090	12	19,61	22,59	$4p\,^4D^\circ-5s\,^4P$	$3/2-3/2$
4147,377	2	16,81	19,80	$4s\,^4P-4p\,^2P^\circ$	$1/2-1/2$
4144,240	1	23,67	26,66	$5p\,^2S^\circ-5s''\,^2S$	$1/2-1/2$
4131,730	15	18,43	21,43	$4s'\,^2D-4p'\,^2P^\circ$	$3/2-1/2$
4129,693	4	19,80	22,80	$4p\,^2P^\circ-5s\,^2P$	$1/2-1/2$
4128,643	9	18,49	21,50	$3d\,^2F-4p'\,^2D^\circ$	$7/2-5/2$
4124,058	1	19,26	22,26	$4p\,^4P^\circ-3d''\,^2D$	$3/2-5/2$
4116,377	6	21,37	24,38	$3d'\,^2D-(^3P_0)\,4f\,[3]^\circ$	$5/2-7/2$
4114,487	2	21,37	24,38	$3d'\,^2D-(^3P_0)\,4f\,[3]^\circ$	$5/2-5/2$
4112,819	8	16,75	19,76	$4s\,^4P-4p\,^2D^\circ$	$3/2-3/2$
4103,913	20	19,49	22,51	$4p\,^4D^\circ-5s\,^4P$	$7/2-5/2$
		19,68	22,70	$4p\,^2D^\circ-5s\,^2P$	$5/2-3/2$
4099,458	2	19,80	22,82	$4p\,^2P^\circ-3d'\,^2S$	$1/2-1/2$
4097,190	1	22,84	25,86	$4d\,^4D-(^1D)\,4f\,[1]^\circ$	$1/2-1/2$
4097,138	3	22,84	25,86	$4d\,^4D-(^1D)\,4f\,[1]^\circ$	$1/2-3/2$
4082,393	15	16,64	19,68	$4s\,^4P-4p\,^2D^\circ$	$5/2-5/2$

λ, Å	I	E_H, eV	E_B, eV	Transition	J
4080,686	4	22,82	25,86	$3d'\,^2S-(^1D)\,4f\,[1]^\circ$	$1/2-1/2$
4080,645	6	22,82	25,86	$3d'\,^2S-(^1D)\,4f\,[1]^\circ$	$1/2-3/2$
4079,582	12	18,45	21,49	$4s'\,^2D-4p'\,^2D^\circ$	$5/2-3/2$
4076,939	9	19,76	22,80	$4p\,^2D^\circ-5s\,^2P$	$3/2-1/2$
4076,638	12	19,64	22,68	$4p\,^4D^\circ-5s\,^4P$	$1/2-1/2$
4072,385	12	19,55	22,59	$4p\,^4D-5s\,^4P$	$5/2-3/2$
4072,006	25	18,45	21,50	$4s'\,^2D-4p'\,^2D^\circ$	$5/2-5/2$
4070,789	2	23,62	26,66	$5p\,^2P^\circ-5s''\,^2S$	$3/2-1/2$
4065,113	4	19,76	22,81	$4p\,^2D^\circ-4d\,^4D$	$3/2-3/2$
4057,672	1	16,81	19,87	$4s\,^4P-4p\,^2P^\circ$	$1/2-3/2$
4053,540	1	19,64	22,70	$4p\,^4D^\circ-5s\,^2P$	$1/2-3/2$
4052,923	12	20,74	23,80	$4s''\,^2S-4p''\,^2P^\circ$	$1/2-3/2$
4047,480	1	19,76	22,82	$4p\,^2D^\circ-3d'\,^2S$	$3/2-1/2$
4045,677	2	18,29	21,35	$3d\,^4P-4p'\,^2P^\circ$	$3/2-3/2$
4042,896	15	18,43	21,49	$4s'\,^2D-4p'\,^2D^\circ$	$3/2-3/2$
4042,190	3	18,06	21,13	$3d\,^2P-4p'\,^2F^\circ$	$3/2-5/2$
4038,807	15	16,42	19,49	$3d\,^4D-4p\,^4D^\circ$	$5/2-7/2$
4035,459	12	18,43	21,50	$4s'\,^2D-4p'\,^2D^\circ$	$3/2-5/2$
4033,818	12	19,61	22,68	$4p\,^4D^\circ-5s\,^4P$	$3/2-1/2$
4031,378	2	19,76	22,84	$4p\,^2D^\circ-4d\,^4D$	$3/2-1/2$
4019,843	2	23,58	26,66	$5p\,^2P^\circ-5s''\,^2S$	$1/2-1/2$
4013,858	25	16,41	19,49	$3d\,^4D-4p\,^4D^\circ$	$7/2-7/2$
4011,202	6	19,61	22,70	$4p\,^4D^\circ-5s\,^2P$	$3/2-3/2$
4007,632	2	19,68	22,77	$4p\,^2D^\circ-4d\,^4D$	$5/2-7/2$
4005,362	2	23,57	26,66	$5p\,^4D^\circ-5s''\,^2S$	$3/2-1/2$
4001,135	1	18,25	21,35	$3d\,^4P-4p'\,^2P^\circ$	$1/2-3/2$
3999,248	1	22,31	25,41	$3d''\,^2D-6p\,^2S^\circ$	$3/2-1/2$
3994,789	10	20,74	23,85	$4s''\,^2S-4p''\,^2P^\circ$	$1/2-1/2$
		19,97	23,07	$4p\,^4S^\circ-4d\,^4F$?	$3/2-5/2$
3992,053	12	16,44	19,55	$3d\,^4D-4p\,^4D^\circ$	$3/2-5/2$
3988,158	9	19,68	22,79	$4p\,^2D^\circ-4d\,^4D$	$5/2-5/2$
3979,356	12	19,97	23,08	$4p\,^4S^\circ-4d\,^4P$	$3/2-1/2$
3974,753	9	16,64	19,76	$4s\,^4P-4p\,^2D^\circ$	$5/2-3/2$
3974,478	10	16,75	19,87	$4s\,^4P-4p\,^2P^\circ$	$3/2-3/2$
3968,360	20	16,42	19,55	$3d\,^4D-4p\,^4D^\circ$	$5/2-5/2$
3958,382	6	19,68	22,81	$4p\,^2D^\circ-4d\,^4D$	$5/2-3/2$
3952,729	9	19,97	23,10	$4p\,^4S^\circ-4d\,^4F$	$3/2-3/2$
3946,096	12	21,14	24,28	$4p'\,^2F^\circ-5s'\,^2D$	$7/2-5/2$
3944,272	15	16,41	19,55	$3d\,^4D-4p\,^4D^\circ$	$7/2-5/2$
3938,843	1	19,97	23,12	$4p\,^2S^\circ-4d\,^4P$	$1/2-3/2$
3935,275	1	22,26	25,41	$3d''\,^2D-(^3P_2)\,5f\,[3]^\circ$	$5/2-7/2$
3933,17	2	19,55	22,70	$4p\,^4D^\circ-5s\,^2P$	$5/2-3/2$
3932,548	15	19,97	23,12	$4p\,^4S^\circ-4d\,^4P$	$3/2-3/2$
3931,235	12	16,46	19,61	$3d\,^4D-4p\,^4D^\circ$	$1/2-3/2$
3928,629	25	16,81	19,97	$4s\,^4P-4p\,^4S^\circ$	$1/2-3/2$
3926,03	7	21,13	24,28	$4p'\,^2F^\circ-5s'\,^2D$	$5/2-5/2$
3925,722	10	21,13	24,28	$4p'\,^2F^\circ-5s'\,^2D$	$5/2-3/2$
3923,556	1	23,48	26,64	$5p\,^4D^\circ-8d\,^4F$	$7/2-9/2$
3922,528	1	19,64	22,80	$4p\,^4D^\circ-5s\,^2P$	$1/2-1/2$
3922,359	2	16,81	19,97	$4s\,^4P-4p\,^2S^\circ$	$1/2-1/2$
3917,766	4	18,33	21,50	$3d\,^4P-4p'\,^2D^\circ$	$5/2-5/2$
3914,768	12	16,44	19,61	$3d\,^4D-4p\,^4D^\circ$	$3/2-3/2$
3911,572	10	19,64	28,81	$4p\,^4D^\circ-4d\,^4D$	$1/2-3/2$
3900,624	11	19,61	22,79	$4p\,^4D^\circ-4d\,^4D$	$3/2-5/2$
3895,250	1	19,64	22,82	$4p\,^4D^\circ-3d'\,^2S$	$1/2-1/2$
3891,984	15	16,42	19,61	$3d\,^4D-4p\,^4D^\circ$	$5/2-3/2$
3891,400	12	16,46	19,64	$3d\,^4D-4p\,^4D^\circ$	$1/2-1/2$
3880,335	6	19,64	22,84	$4p\,^4D^\circ-4d\,^4D$	$1/2-1/2$
3875,264	12	16,44	19,64	$3d\,^4D-4p\,^4D^\circ$	$3/2-1/2$

λ, Å	I	$E_н$, eV	$E_в$, eV	Transition	J
3872,143	11	19,61	22,81	$4p\ ^4D^\circ$—$4d\ ^4D$	$^3/_2$—$^3/_2$
3869,614	2	19,87	23,07	$4p\ ^2P^\circ$—$4d\ ^4F$	$^3/_2$—$^5/_2$
3868,524	20	19,97	23,17	$4p\ ^4S^\circ$—$4d\ ^4P$	$^3/_2$—$^5/_2$
3861,348	3	18,29	21,50	$3d\ ^4P$—$4p'\ ^2D^\circ$	$^3/_2$—$^5/_2$
3856,127	1	19,61	22,82	$4p\ ^4D^\circ$—$3d'\ ^2S$	$^3/_2$—$^1/_2$
3855,160	4	19,87	23,08	$4p\ ^2P^\circ$—$4d\ ^4P$	$^3/_2$—$^1/_2$
3850,578	30	16,75	19,97	$4s\ ^4P$—$4p\ ^4S^\circ$	$^3/_2$—$^3/_2$
3846,860	2	—	—	—	—
3845,406	10	16,64	19,87	$4s\ ^4P$—$4p\ ^2P^\circ$	$^5/_2$—$^3/_2$
3844,735	9	19,55	22,77	$4p\ ^4D^\circ$—$4d\ ^4D$	$^5/_2$—$^7/_2$
3844,565	4	16,75	19,97	$4s\ ^4P$—$4p\ ^2S^\circ$	$^3/_2$—$^1/_2$
3841,518	6	19,61	22,84	$4p\ ^4D^\circ$—$4d\ ^4D$	$^3/_2$—$^1/_2$
3838,239	1	23,40	26,63	$5p\ ^4P^\circ$—$8d\ ^4D$	$^5/_2$—$^7/_2$
3830,515	4	21,49	24,73	$4p'\ ^2D^\circ$—$4d'\ ^2P$	$^3/_2$—$^1/_2$
3830,390	10	16,44	19,68	$3d\ ^4D$—$4p\ ^2D^\circ$	$^3/_2$—$^5/_2$
3830,165	1	19,87	23,10	$4p\ ^2P^\circ$—$4d\ ^4F$	$^3/_2$—$^3/_2$
3826,807	12	19,55	27,79	$4p\ ^4D^\circ$—$4d\ ^4D$	$^5/_2$—$^5/_2$
3825,676	8	21,50	24,74	$4p'\ ^2D^\circ$—$4d'\ ^2P$	$^5/_2$—$^3/_2$
3823,254	3	20,27	23,51	$3d'\ ^2F$—$5p\ ^4D^\circ$	$^7/_2$—$^5/_2$
3819,017	5	21,49	24,74	$4p'\ ^2D^\circ$—$4d'\ ^2P$	$^3/_2$—$^3/_2$
3811,212	2	19,87	23,12	$4p\ ^2P^\circ$—$4d\ ^4P$	$^3/_2$—$^3/_2$
3809,456	15	19,26	22,51	$4p\ ^4P^\circ$—$5s\ ^4P$	$^3/_2$—$^5/_2$
3808,577	11	16,42	19,68	$3d\ ^4D$—$4p\ ^2D^\circ$	$^5/_2$—$^5/_2$
3803,172	10	21,50	24,76	$4p'\ ^2D^\circ$—$4d'\ ^2D$	$^5/_2$—$^5/_2$
3799,381	10	19,55	22,81	$4p\ ^4D^\circ$—$4d\ ^4D$	$^5/_2$—$^3/_2$
3796,599	8	21,49	24,76	$4p'\ ^2D^\circ$—$4d'\ ^2D$	$^3/_2$—$^5/_2$
3793,226	1	20,24	23,51	$3d'\ ^2F$—$5p\ ^4D^\circ$	$^5/_2$—$^5/_2$
3786,383	12	16,41	19,68	$3d\ ^4D$—$4p\ ^2D^\circ$	$^7/_2$—$^5/_2$
3780,841	25	19,49	22,77	$4p\ ^4D^\circ$—$4d\ ^4D$	$^7/_2$—$^7/_2$
3777,529	2	19,80	23,08	$4p\ ^2P^\circ$—$4d\ ^4P$	$^1/_2$—$^1/_2$
3774,522	4	21,50	24,78	$4p'\ ^2D^\circ$—$5d\ ^4D$	$^5/_2$—$^7/_2$
3770,516	10	19,30	22,59	$4p\ ^4P^\circ$—$5s\ ^4P$	$^1/_2$—$^3/_2$
3766,118	11	18,06	21,35	$3d\ ^2P$—$4p'\ ^2P^\circ$	$^3/_2$—$^3/_2$
3765,269	20	19,22	22,51	$4p\ ^4P^\circ$—$5s\ ^4P$	$^5/_2$—$^5/_2$
3763,504	12	19,49	22,79	$4p\ ^4D^\circ$—$4d\ ^4D$	$^7/_2$—$^5/_2$
3763,111	2	21,50	24,79	$4p'\ ^2D^\circ$—$5d\ ^4D$	$^5/_2$—$^5/_2$
3756,671	4	21,49	24,79	$4p'\ ^2D^\circ$—$5d\ ^4D$	$^3/_2$—$^5/_2$
3754,052	6	21,43	24,73	$4p'\ ^2P^\circ$—$4d'\ ^2P$	$^1/_2$—$^1/_2$
3753,521	9	{ 19,80	23,10	$4p\ ^2P^\circ$—$4d\ ^4F$	$^1/_2$—$^3/_2$
		{ 21,49	24,79	$4p'\ ^2D^\circ$—$4d'\ ^2D$	$^3/_2$—$^3/_2$
3751,330	1	22,79	26,09	$4d\ ^4D$—$(^3P_2)\ 6f\ [4]^\circ$	$^5/_2$—$^7/_2$
3751,047	2	19,87	23,17	$4p\ ^2P^\circ$—$4d\ ^4P$	$^3/_2$—$^5/_2$
3750,485	5	16,46	19,76	$3d\ ^4D$—$4p\ ^2D^\circ$	$^1/_2$—$^3/_2$
3746,915	5	19,76	23,07	$4p\ ^2D^\circ$—$4d\ ^4F$	$^3/_2$—$^5/_2$
3746,452	6	21,49	24,80	$4p'\ ^2D^\circ$—$6s\ ^2P$	$^3/_2$—$^3/_2$
3744,274	1	22,31	25,62	$3d''\ ^2D$—$(^3P_0)\ 5f\ [3]^\circ$	$^3/_2$—$^5/_2$
3737,893	15	21,50	24,81	$4p'\ ^2D^\circ$—$4d'\ ^2F$	$^5/_2$—$^7/_2$
3735,495	6	{ 21,49	24,81	$4p'\ ^2D^\circ$—$6s\ ^4P$?	$^3/_2$—$^1/_2$
		{ 16,44	19,76	$3d\ ^4D$—$4p\ ^2D^\circ$	$^3/_2$—$^3/_2$
3729,310	30	16,64	19,97	$4s\ ^4P$—$4p\ ^4S^\circ$	$^5/_2$—$^3/_2$
3724,521	8	21,50	24,82	$4p'\ ^2D^\circ$—$4d'\ ^2F$	$^5/_2$—$^5/_2$
3720,428	9	19,26	22,59	$4p\ ^4P^\circ$—$5s\ ^4P$	$^3/_2$—$^3/_2$
3718,208	12	21,49	24,82	$4p'\ ^2D^\circ$—$4d'\ ^2F$	$^3/_2$—$^5/_2$
3717,174	10	19,68	23,01	$4p\ ^2D^\circ$—$4d\ ^4F$	$^5/_2$—$^7/_2$
3714,737	6	16,42	19,76	$3d\ ^4D$—$4p\ ^2D^\circ$	$^5/_2$—$^3/_2$
3713,019	2	21,35	24,69	$4p'\ ^2P^\circ$—$6s\ ^4P$	$^3/_2$—$^3/_2$
3709,918	6	19,76	23,10	$4p\ ^2D^\circ$—$4d\ ^4F$	$^3/_2$—$^3/_2$
3706,937	5	16,46	19,80	$3d\ ^4D$—$4p\ ^2P^\circ$	$^1/_2$—$^1/_2$
3703,550	1	23,16	25,51	$4d\ ^2F$—$(^3P_2)7f\ [5]^\circ$	$^7/_2$—$^9/_2$

λ, Å	I	E_H, eV	E_B, eV	Transition	J
3702,005	2	20,27	23,62	$3d'\,^2F-5p\,^2D^\circ$	$7/2-5/2$
3694,643	2	21,49	24,84	$4p'\,^2D^\circ-5d\,^4D$	$3/2-1/2$
3692,126	1	19,76	23,12	$4p\,^2D^\circ-4d\,^4P$	$3/2-3/2$
3682,547	7	18,06	21,43	$3d\,^2P-4p'\,^2P^\circ$	$3/2-1/2$
3680,064	9	21,43	24,79	$4p'\,^2P^\circ-4d'\,^2D$	$1/2-3/2$
3678,274	10	19,22	22,59	$4p\,^4P^\circ-5s\,^4P$	$5/2-3/2$
3673,266	5	21,43	24,80	$4p'\,^2P^\circ-6s\,^2P$	$1/2-3/2$
3671,005	4	21,35	24,73	$4p'\,^2P^\circ-4d'\,^2P$	$3/2-1/2$
3669,605	9	19,30	22,68	$4p\,^4P^\circ-5s\,^4P$	$1/2-1/2$
3660,439	10	21,35	24,74	$4p'\,^2P^\circ-4d'\,^2P$	$3/2-3/2$
3657,218	2	21,43	24,81	$4p'\,^2P^\circ-5d\,^4D$	$1/2-3/2$
3656,051	10	19,68	23,07	$4p\,^2D^\circ-4d\,^4F$	$5/2-5/2$
3655,281	12	19,87	23,26	$4p\,^2P^\circ-4d\,^2F$	$3/2-5/2$
3650,891	7	19,30	22,70	$4p\,^4P^\circ-5s\,^2P$	$1/2-3/2$
3639,830	12	21,35	24,76	$4p'\,^2P^\circ-4d'\,^2D$	$3/2-5/2$
3637,031	10	21,50	24,90	$4p'\,^2D^\circ-5d\,^4F$	$5/2-7/2$
3635,636	3	19,76	23,17	$4p\,^2D^\circ-4d\,^4P$	$3/2-5/2$
3634,814	7	19,94	21,35	$3d\,^2P-4p'\,^2P^\circ$	$1/2-3/2$
3623,444	2	21,43	24,84	$4p'\,^2P^\circ-5d\,^4D$	$1/2-1/2$
3622,140	12	19,26	22,68	$4p\,^4P^\circ-5s\,^4P$	$3/2-1/2$
3621,012	3	16,44	19,87	$3d\,^4D-4p\,^2P^\circ$	$3/2-3/2$
3620,807	3	19,68	23,10	$4p\,^2D^\circ-4d\,^4F$	$5/2-3/2$
3611,812	5	18,06	21,49	$3d\,^2P-4p'\,^2D^\circ$	$3/2-3/2$
3611,365	1	21,49	24,92	$4p'\,^2D^\circ-5d\,^4P$	$3/2-1/2$
3607,401	2	20,24	23,68	$3d'\,^2F-5p\,^2D^\circ$	$5/2-3/2$
3605,883	12	18,06	21,50	$3d\,^2P-4p'\,^2D^\circ$	$3/2-5/2$
3603,905	4	{ 19,26	22,70	$4p\,^4P^\circ-5s\,^2P$	$3/2-3/2$
		19,68	23,12	$4p\,^2D^\circ-4d\,^4P$	$5/2-3/2$
3603,462	4	19,64	23,08	$4p\,^4D^\circ-4d\,^4P$	$1/2-1/2$
3601,512	4	16,42	19,87	$3d\,^4D-4p\,^2P^\circ$	$5/2-3/2$
3600,219	3	21,35	24,79	$4p'\,^2P^\circ-4d'\,^2D$	$3/2-3/2$
3588,448	30	19,49	22,95	$4p\,^4D^\circ-4d\,^4F$	$7/2-9/2$
3582,362	20	19,61	23,07	$4p\,^4D^\circ-4d\,^4F$	$3/2-5/2$
3581,608	18	19,64	23,10	$4p\,^4D^\circ-4d\,^4F$	$1/2-3/2$
3578,357	5	21,35	24,81	$4p'\,^2P^\circ-5d\,^4D$	$3/2-3/2$
3576,611	25	19,55	23,01	$4p\,^4D^\circ-4d\,^4F$	$5/2-7/2$
3575,761	1	21,50	24,96	$4p'\,^2D^\circ-5d\,^4F$	$5/2-5/2$
3570,746	1	21,49	24,96	$4p'\,^2D^\circ-5d\,^4P$	$3/2-3/2$
3569,940	3	{ 19,61	23,08	$4p\,^4D^\circ-4d\,^4P$	$3/2-1/2$
		21,49	24,96	$4p'\,^2D^\circ-5d\,^4F$	$3/2-5/2$
3565,033	12	19,64	23,12	$4p\,^4D^\circ-4d\,^4P$	$1/2-3/2$
3564,33	7	19,22	22,70	$4p\,^4P^\circ-5s\,^2P$	$5/2-3/2$
3562,194	7	21,14	24,62	$4p'\,^2F^\circ-4d'\,^2G$	$7/2-7/2$
3561,031	20	21,14	24,62	$4p'\,^2F^\circ-4d'\,^2G$	$7/2-9/2$
3559,508	25	19,68	23,16	$4p\,^2D^\circ-4d\,^2F$	$5/2-7/2$
3556,906	7	17,94	21,43	$3d\,^2P-4p'\,^2P^\circ$	$1/2-1/2$
3550,030	5	19,68	23,17	$4p\,^2D^\circ-4d\,^4P$	$5/2-5/2$
3548,519	15	19,61	23,10	$4p\,^4D^\circ-4d\,^4F$	$3/2-3/2$
3545,842	18	21,13	24,62	$4p'\,^2F^\circ-4d'\,^2G$	$5/2-7/2$
3545,597	18	19,76	23,26	$4p\,^2D^\circ-4d\,^2F$	$3/2-5/2$
3543,149	7	21,50	24,99	$4p'\,^2D^\circ-5d\,^2F$	$5/2-7/2$
3535,319	18	19,30	22,81	$4p\,^4P^\circ-4d\,^4D$	$1/2-3/2$
3532,233	1	19,61	23,12	$4p\,^4D^\circ-4d\,^4P$	$3/2-3/2$
3531,178	2	16,46	19,97	$3d\,^4D-4p\,^4S^\circ$	$1/2-3/2$
3521,977	4	19,30	22,82	$4p\,^4P^\circ-3d'\,^2S$	$1/2-1/2$
3521,555	1	21,50	25,02	$4p'\,^2D^\circ-5d\,^4P$	$5/2-5/2$
3521,263	12	19,49	23,01	$4p\,^4D^\circ-4d\,^4F$	$7/2-7/2$
3519,996	15	19,55	23,07	$4p\,^4D^\circ-4d\,^4F$	$5/2-5/2$

λ, Å	I	E_H, eV	E_B, eV	Transition	J
3517,894	6	16,44	19,97	$3d\,^4D-4p\,^4S^\circ$	$^3/_2-^3/_2$
3514,388	20	19,26	22,79	$4p\,^4P^\circ-4d\,^4D$	$^3/_2-^5/_2$
3509,783	10	19,30	22,84	$4p\,^4P^\circ-4d\,^4D$	$^1/_2-^1/_2$
3499,481	7	16,42	19,97	$3d\,^4D-4p\,^4S^\circ$	$^5/_2-^3/_2$
3495,775	1	21,62	25,17	$3d'\,^2P-5p'\,^2P^\circ$	$^3/_2-^3/_2$
3491,538	25	19,22	22,77	$4p\,^4P^\circ-4d\,^4D$	$^5/_2-^7/_2$
3491,243	20	19,26	22,81	$4p\,^4P^\circ-4d\,^4D$	$^3/_2-^3/_2$
3490,884	8	17,94	21,49	$3d\,^2P-4p'\,^2D^\circ$	$^1/_2-^3/_2$
3488,188	1	22,31	25,86	$3d''\,^2D-(^1D)\,4f\,[1]^\circ$	$^3/_2-^1/_2$
3487,318	3	19,55	23,10	$4p\,^4D^\circ-4d\,^4F$	$^5/_2-^3/_2$
3480,511	9	19,61	23,17	$4p\,^4D^\circ-4d\,^4P$	$^3/_2-^5/_2$
3478,236	5	19,26	22,82	$4p\,^4P^\circ-3d'\,^2S$	$^3/_2-^1/_2$
3476,749	20	19,22	22,79	$4p\,^4P^\circ-4d\,^4D$	$^5/_2-^5/_2$
3471,600	3	19,55	23,12	$4p\,^4D^\circ-4d\,^4P$	$^5/_2-^3/_2$
3470,264	4	21,49	25,06	$4p'\,^2D^\circ-5d\,^2F$	$^3/_2-^5/_2$
3466,343	8	19,49	23,07	$4p\,^4D^\circ-4d\,^4F$	$^7/_2-^5/_2$
		19,26	22,84	$4p\,^4P^\circ-4d\,^4D$	$^3/_2-^1/_2$
3465,787	4	19,97	23,55	$4p\,^2S^\circ-4d\,^2P$	$^1/_2-^1/_2$
3464,132	10	19,68	23,26	$4p\,^2D^\circ-4d\,^2F$	$^5/_2-^5/_2$
3454,098	12	19,22	22,81	$4p\,^4P^\circ-4d\,^4D$	$^5/_2-^3/_2$
3448,281	1	22,31	25,90	$3d''\,^2D-(^1D)\,4f\,[2]^\circ$	$^3/_2-^3/_2$
3447,290	1	22,26	25,86	$3d''\,^2D-(^1D)\,4f\,[1]^\circ$	$^5/_2-^3/_2$
3439,094	1	23,55	27,15	$4d\,^2P-(^1D)\,5f\,[2]^\circ$	$^1/_2-^3/_2$
3435,773	1	21,67	25,28	$3d'\,^2P-6p\,^4S^\circ$	$^1/_2-^3/_2$
3433,369	1	23,16	26,77	$4d\,^2F-(^3P_2)\,8f\,[5]^\circ$	$^7/_2-^9/_2$
3432,585	3	21,13	24,74	$4p'\,^2F^\circ-4d'\,^2P$	$^5/_2-^3/_2$
3431,737	2	21,35	24,96	$4p'\,^2P^\circ-5d\,^4P$	$^3/_2-^3/_2$
3430,990	3	21,35	24,96	$4p'\,^2P^\circ-5d\,^4F$	$^3/_2-^5/_2$
3430,417	9	19,55	23,16	$4p\,^4D^\circ-4d\,^2F$	$^5/_2-^7/_2$
3429,617	7	21,14	24,76	$4p'\,^2F^\circ-4d'\,^2D$	$^7/_2-^5/_2$
3421,615	8	19,55	23,17	$4p\,^4D^\circ-4d\,^4P$	$^5/_2-^5/_2$
3416,560	1	23,01	26,64	$4d\,^4F-(^3P_1)\,7f\,[4]^\circ$	$^7/_2-^9/_2$
3414,462	4	21,13	24,76	$4p'\,^2F^\circ-4d'\,^2D$	$^5/_2-^5/_2$
3409,699	2	22,31	25,94	$3d''\,^2D-(^1D)\,4f\,[3]^\circ$	$^3/_2-^5/_2$
3409,413	1	22,26	25,90	$3d''\,^2D-(^3P_2)\,7p\,[2]^\circ$	$^5/_2-^5/_2$
3408,612	2	22,26	25,90	$3d''\,^2D-(^1D)\,4f\,[2]^\circ$	$^5/_2-^5/_2$
3406,298	3	21,14	24,78	$4p'\,^2F^\circ-5d\,^4D$	$^7/_2-^7/_2$
3397,900	5	19,61	23,26	$4p\,^4D^\circ-4d\,^2F$	$^3/_2-^5/_2$
3397,608	1	21,43	25,07	$3d'\,^2D-6p\,^4P^\circ$	$^3/_2-^3/_2$
3397,002	1	21,14	24,79	$4p'\,^2F^\circ-5d\,^4D$	$^7/_2-^5/_2$
3388,533	10	19,97	23,63	$4p\,^2S^\circ-4d\,^2P$	$^1/_2-^3/_2$
3383,865	1	19,97	23,63	$4p\,^4S^\circ-4d\,^2P$	$^3/_2-^3/_2$
3382,133	3	21,13	24,79	$4p'\,^2F^\circ-5d\,^4D$	$^5/_2-^5/_2$
3381,063	1	21,35	25,02	$4p'\,^2P^\circ-5d\,^4P$	$^3/_2-^5/_2$
3379,577	4	21,13	24,79	$4p'\,^2F^\circ-4d'\,^2D$	$^5/_2-^3/_2$
3379,458	4	19,49	23,16	$4p\,^4D^\circ-4d\,^2F$	$^7/_2-^7/_2$
3378,442	1	21,43	25,10	$3d'\,^2D-6p\,^4P^\circ$	$^3/_2-^1/_2$
3376,443	12	21,14	24,81	$4p'\,^2F^\circ-4d'\,^2F$	$^7/_2-^7/_2$
3373,842	3	21,13	24,80	$4p'\,^2F^\circ-6s\,^2P$	$^5/_2-^3/_2$
3370,925	8	22,26	25,94	$3d''\,^2D-(^1D)\,4f\,[3]^\circ$	$^5/_2-^7/_2$
		19,49	23,17	$4p\,^4D^\circ-4d\,^4P$	$^7/_2-^5/_2$
3366,586	6	19,87	23,55	$4p\,^2P^\circ-4d\,^2P$	$^3/_2-^1/_2$
3365,536	8	21,14	24,82	$4p'\,^2F^\circ-4d'\,^2F$	$^7/_2-^5/_2$
3364,362	2	21,43	25,11	$3d'\,^2D-6p\,^4D^\circ$	$^3/_2-^5/_2$
3363,300	2	21,62	25,31	$3d'\,^2P-5p'\,^2D^\circ$	$^3/_2-^5/_2$
3361,752	6	21,13	24,81	$4p'\,^2F^\circ-4d'\,^2F$	$^5/_2-^7/_2$
3350,933	12	21,13	24,82	$4p'\,^2F^\circ-4d'\,^2F$	$^5/_2-^5/_2$
3347,694	1	21,14	24,84	$4p'\,^2F^\circ-5d\,^4F$	$^7/_2-^9/_2$
3341,746	6	19,55	23,26	$4p\,^4D^\circ-4d\,^2F$	$^5/_2-^5/_2$

λ, Å	I	E_H, eV	E_B, eV	Transition	J
3341,507	3	21,37	25,07	$3d'\ ^2D—6p\ ^4P°$	$^5/_2—^3/_2$
3338,828	5	21,35	25,06	$4p'\ ^2P°—5d\ ^2F$	$^3/_2—^5/_2$
3337,116	2	22,79	26,50	$4d\ ^4D—(^3P_2)\ 7f\ [4]°$	$^5/_2—^7/_2$
3324,228	2	22,77	26,50	$4d\ ^4D—(^3P_2)\ 7f\ [4]°$	$^7/_2—^9/_2$
3320,375	2	21,67	25,41	$3d'\ ^2P—6p\ ^2S°$	$^1/_2—^1/_2$
3317,825	3	21,37	25,10	$3d'\ ^2D—6p\ ^4D°$	$^5/_2—^3/_2$
3312,936	3	—	—	—	—
3309,343	5	21,37	25,11	$3d'\ ^2D—6p\ ^4D°$	$^5/_2—^5/_2$
3307,229	9	19,80	23,55	$4p\ ^2P°—4d\ ^2P$	$^1/_2—^1/_2$
3306,445	5	21,67	25,42	$3d'\ ^2P—(^3P_2)\ 5f\ [2]°$	$^1/_2—^3/_2$
3298,418	2	23,01	26,77	$4d\ ^4F—(^3P_2)\ 8f\ [5]°$	$^7/_2—^9/_2$
3297,020	2	21,43	25,19	$3d'\ ^2D—5p'\ ^2F°$	$^3/_2—^5/_2$
3293,921	9	21,14	24,90	$4p'\ ^2F°—5d\ ^4F$	$^7/_2—^7/_2$
3293,641	10	19,87	23,63	$4p\ ^2P°—4d\ ^2P$	$^3/_2—^3/_2$
3291,441	6	21,67	25,44	$3d'\ ^2P—(^3P_2)\ 5f\ [1]°$	$^1/_2—^3/_2$
3281,703	12	19,30	23,08	$4p\ ^4P°—4d\ ^4P$	$^1/_2—^1/_2$
3279,937	4	21,13	24,90	$4p'\ ^2F°—5d\ ^4F$	$^5/_2—^7/_2$
3276,085	3	21,43	25,21	$3d'\ ^2D—6p\ ^2D°$	$^3/_2—^3/_2$
3275,639	4	21,62	25,41	$3d'\ ^2P—6p\ ^2S°$	$^3/_2—^1/_2$
3273,316	6	19,76	23,55	$4p\ ^2D°—4d\ ^2P$	$^3/_2—^1/_2$
3270,474	5	21,62	25,41	$4d'\ ^2P—(^3P_2)\ 5f\ [3]°$	$^3/_2—^5/_2$
3268,987	5	19,22	23,01	$4p\ ^4P°—4d\ ^4F$	$^5/_2—^7/_2$
3263,572	12	19,30	23,10	$4p\ ^4P°—4d\ ^4F$	$^1/_2—^3/_2$
3262,083	2	21,62	25,42	$3d'\ ^2P—(^3P_2)\ 5f\ [2]°$	$^3/_2—^3/_2$
3259,656	6	21,62	25,42	$3d'\ ^2P—(^3P_2)\ 5f\ [2]°$	$^3/_2—^5/_2$
3258,894	2	17,69	21,50	$3d\ ^4F—4p'\ ^2D°$	$^7/_2—^5/_2$
3253,918	3	19,26	23,07	$4p\ ^4P°—4d\ ^4F$	$^3/_2—^5/_2$
3249,801	15	19,30	23,12	$4p\ ^4P°—4d\ ^4P$	$^1/_2—^3/_2$
3247,481	3	21,62	25,44	$3d'\ ^2P—(^3P_2)\ 5f\ [1]°$	$^3/_2—^3/_2$
3243,689	14	19,26	23,08	$4p\ ^4P°—4d\ ^4P$	$^3/_2—^1/_2$
3241,708	2	22,95	26,77	$4d\ ^4F—(^3P_2)\ 8f\ [5]°$	$^9/_2—^{11}/_2$
3236,809	6	19,80	23,63	$4p\ ^2P°—4d\ ^2P$	$^1/_2—^3/_2$
3235,175	3	21,43	25,26	$3d'\ ^2D—6p\ ^2P°$	$^3/_2—^3/_2$
3230,680	2	21,13	24,96	$4p'\ ^2F°—5d\ ^4P$	$^5/_2—^3/_2$
3230,021	4	21,13	24,96	$4p'\ ^2F°—5d\ ^4F$	$^5/_2—^5/_2$
3225,973	6	19,26	23,10	$4p\ ^4P°—4d\ ^4F$	$^3/_2—^3/_2$
3222,393	6	21,50	25,34	$4p'\ ^2D°—5d\ ^2D$	$^5/_2—^3/_2$
3221,625	7	19,22	23,07	$4p\ ^4P°—4d\ ^4F$	$^5/_2—^5/_2$
3217,669	5	21,49	25,34	$4p'\ ^2D°—5d\ ^2D$	$^3/_2—^3/_2$
3216,729	8	21,14	24,99	$4p'\ ^2F°—5d\ ^2F$	$^7/_2—^7/_2$
3215,688	3	21,43	25,28	$3d'\ ^2D—6p\ ^4S°$	$^3/_2—^3/_2$
3212,516	9	19,26	23,12	$4p\ ^4P°—4d\ ^4P$	$^3/_2—^3/_2$
3207,655	2	21,37	25,23	$3d'\ ^2D—6p\ ^2D°$	$^5/_2—^5/_2$
3207,577	4	21,49	25,36	$4p'\ ^2D°—5d\ ^2P$	$^3/_2—^1/_2$
3204,996	8	21,50	25,36	$4p'\ ^2D°—5d\ ^2D$	$^5/_2—^5/_2$
3204,318	9	19,76	23,63	$4p\ ^2D°—4d\ ^2P$	$^3/_2—^3/_2$
3203,392	3	21,13	24,99	$4p'\ ^2F°—5d\ ^2F$	$^5/_2—^7/_2$
3198,920	2	21,14	25,02	$4p'\ ^2F°—5d\ ^4P$	$^7/_2—^5/_2$
3195,752	5	21,67	25,55	$3d'\ ^2P—(^3P_1)\ 5f\ [2]°$	$^1/_2—^3/_2$
3195,574	2	21,43	25,31	$3d'\ ^2D—5p'\ ^2D°$	$^3/_2—^3/_2$
3194,598	4	20,27	24,15	$3d'\ ^2F—(^3P_2)\ 4f\ [4]°$	$^7/_2—^9/_2$
3194,229	9	19,22	23,10	$4p\ ^4P°—4d\ ^4F$	$^5/_2—^3/_2$
3193,512	1	21,43	25,31	$3d'\ ^2D—5p'\ ^2D°$	$^3/_2—^5/_2$
3192,363	3	20,27	24,15	$3d'\ ^2F—(^3P_2)\ 4f\ [4]°$	$^7/_2—^7/_2$
3188,369	1	20,27	24,16	$3d'\ ^2F—(^3P_2)\ 4f\ [3]°$	$^7/_2—^5/_2$
3186,169	5	20,27	24,16	$3d'\ ^2F—(^3P_2)\ 4f\ [3]°$	$^7/_2—^7/_2$
3185,734	3	21,13	25,02	$4p'\ ^2F°—5d\ ^4P$	$^5/_2—^5/_2$
3181,038	12	19,22	23,12	$4p\ ^4P°—5d\ ^4P$	$^5/_2—^3/_2$

λ, Å	I	E_H, eV	E_B, eV	Transition	J
3172,855	3	19,97	23,87	$4p\,^4S^\circ-4d\,^2D$	$3/2-5/2$
3171,403	3	20,24	24,15	$3d'\,^2F-(^3P_2)\,4f\,[4]^\circ$	$5/2-7/2$
3169,667	15	19,26	23,17	$4p\,^4P^\circ-4d\,^4P$	$3/2-5/2$
3167,464	3	20,24	24,16	$3d'\,^2F-(^3P_2)\,4f\,[3]^\circ$	$5/2-5/2$
3165,288	6	20,24	24,16	$3d'\,^2F-(^3P_2)\,4f\,[3]^\circ$	$5/2-7/2$
3163,535	2	21,43	25,34	$4p'\,^2P^\circ-5d\,^2D$	$1/2-3/2$
3161,456	8	20,27	24,19	$3d'\,^2F-(^3P_2)\,4f\,[5]^\circ$	$7/2-9/2$
3161,369	7	19,97	23,89	$4p\,^2S^\circ-4d\,^2D$	$1/2-3/2$
3154,289	2	21,62	25,55	$3d'\,^2P-(^3P_1)\,5f\,[2]^\circ$	$3/2-3/2$
3153,782	4	21,43	25,36	$4p'\,^2P^\circ-5d\,^2P$	$1/2-1/2$
3152,613	3	21,62	25,55	$3d'\,^2P-(^3P_1)\,5f\,[2]^\circ$	$3/2-5/2$
3150,510	4	20,24	24,18	$3d'\,^2F-(^3P_2)\,4f\,[2]^\circ$	$5/2-5/2$
3148,202	5	21,13	25,06	$4p'\,^2F^\circ-5d\,^2F$	$5/2-5/2$
3146,422	4	19,22	23,16	$4p\,^4P^\circ-4d\,^2F$	$5/2-7/2$
3145,900	2	21,37	25,31	$3d'\,^2D-5p'\,^2D^\circ$	$5/2-3/2$
3143,891	3	21,37	25,31	$3d'\,^2D-5p'\,^2D^\circ$	$5/2-5/2$
3140,963	2	21,50	25,44	$4p'\,^2D^\circ-5d\,^2P$	$5/2-3/2$
3139,257	4	21,62	25,57	$3d'\,^2P-(^3P_1)\,5f\,[3]^\circ$	$3/2-5/2$
3139,015	12	19,22	23,17	$4p\,^4P^\circ-4d\,^4P$	$5/2-5/2$
3137,629	3	19,68	23,63	$4p\,^2D^\circ-4d\,^2P$	$5/2-3/2$
3136,481	3	21,49	25,44	$4p'\,^2D^\circ-5d\,^2P$	$3/2-3/2$
3124,268	1	20,24	24,21	$3d'\,^2F-(^3P_2)\,4f\,[1]^\circ$	$5/2-3/2$
3114,378	3	21,43	25,41	$3d'\,^2D-6p\,^2S^\circ$	$3/2-1/2$
3109,711	4	21,43	25,41	$3d'\,^2D-(^3P_2)\,5f\,[3]^\circ$	$3/2-5/2$
3108,801	2	17,14	21,13	$4s\,^2P-4p'\,^2F^\circ$	$3/2-5/2$
3104,359	5	21,35	25,34	$4p'\,^2P^\circ-5d\,^2D$	$3/2-3/2$
3102,953	1	22,77	26,77	$4d\,^4D-(^3P_2)\,8f\,[4]^\circ$	$7/2-9/2$
3102,585	4	21,62	25,62	$3d'\,^2P-(^3P_0)\,5f\,[3]^\circ$	$3/2-5/2$
3101,004	2	19,26	23,26	$4p\,^4P^\circ-4d\,^2F$	$3/2-5/2$
3099,923	5	21,43	25,42	$3d'\,^2D-(^3P_2)\,5f\,[2]^\circ$	$3/2-5/2$
3094,960	4	21,35	25,36	$4p'\,^2P^\circ-5d\,^2P$	$3/2-1/2$
3093,403	10	19,87	23,87	$4p\,^2P^\circ-4d\,^2D$	$3/2-5/2$
3088,910	3	21,43	25,44	$3d'\,^2D-(^3P_2)\,5f\,[1]^\circ$	$3/2-3/2$
3088,209	7	21,35	25,36	$4p'\,^2P^\circ-5d\,^2D$	$3/2-5/2$
3085,026	5	21,43	25,44	$4p'\,^2P^\circ-5d\,^2P$	$1/2-3/2$
3083,193	1	19,61	23,63	$4p\,^4D^\circ-4d\,^2P$	$3/2-3/2$
3082,979	5	21,43	25,45	$4p'\,^2P^\circ-4d'\,^2S$	$1/2-1/2$
3066,889	6	20,27	24,31	$3d'\,^2F-(^3P_1)\,4f\,[4]^\circ$	$7/2-9/2$
3065,120	3	21,37	25,41	$3d'\,^2D-(^3P_2)\,5f\,[4]^\circ$	$5/2-7/2$
3062,643	3	21,37	25,41	$3d'\,^2D-(^3P_2)\,5f\,[3]^\circ$	$5/2-5/2$
3060,909	8	21,37	25,41	$3d'\,^2D-(^3P_2)\,5f\,[3]^\circ$	$5/2-7/2$
3053,151	5	21,37	25,42	$3d'\,^2D-(^3P_2)\,5f\,[2]^\circ$	$5/2-5/2$
3050,043	1	20,24	24,31	$3d'\,^2F-(^3P_1)\,4f\,[2]^\circ$	$5/2-3/2$
3048,784	2	20,24	24,31	$3d'\,^2F-(^3P_1)\,4f\,[2]^\circ$	$5/2-5/2$
3048,021	2	20,27	24,34	$3d'\,^2F-(^3P_1)\,4f\,[3]^\circ$	$7/2-7/2$
3046,079	5	20,24	24,31	$3d'\,^2F-(^3P_1)\,4f\,[4]^\circ$	$5/2-7/2$
3042,463	1	21,37	25,44	$3d'\,^2D-(^3P_2)\,5f\,[1]^\circ$	$5/2-3/2$
3036,887	2	19,55	23,63	$4p\,^4D^\circ-4d\,^2P$	$5/2-3/2$
3033,510	10	17,26	21,35	$4s\,^2P-4p'\,^2P^\circ$	$1/2-3/2$
3028,914	8	19,80	23,89	$4p\,^2P^\circ-4d\,^2D$	$1/2-3/2$
		20,24	24,34	$3d'\,^2F-(^3P_1)\,4f\,[3]^\circ$	$5/2-7/2$
3028,721	3	21,35	25,44	$4p'\,^2P^\circ-5d\,^2P$	$3/2-3/2$
3026,745	5	21,35	25,45	$4p'\,^2P^\circ-4d'\,^2S$	$3/2-1/2$
3014,481	6	19,76	23,87	$4p\,^2D^\circ-4d\,^2D$	$3/2-5/2$
3004,486	2	21,43	25,55	$3d'\,^2D-(^3P_1)\,5f\,[2]^\circ$	$3/2-3/2$
3002,961	6	21,43	25,55	$3d'\,^2D-(^3P_1)\,5f\,[2]^\circ$	$3/2-5/2$
3000,442	9	19,76	23,89	$4p\,^2D^\circ-4d\,^2D$	$3/2-3/2$
3000,110	5	20,24	24,38	$3d'\,^2F-(^3P_0)\,4f\,[3]^\circ$	$5/2-7/2$

λ, Å	I	E_H, eV	E_B, eV	Transition	J
2999,110	2	20,24	24,38	$3d'\ ^2F-(^3P_0)\ 4f\ [3]^\circ$	$^5/_2-^5/_2$
2990,843	2	21,43	25,57	$3d'\ ^2D-(^3P_1)\ 5f\ [3]^\circ$	$^3/_2-^5/_2$
2979,051	15	17,26	21,43	$4s\ ^2P-4p'\ ^2P^\circ$	$^1/_2-^1/_2$
2960,260	5	21,67	25,86	$3d'\ ^2P-(^1D)\ 4f\ [1]^\circ$	$^1/_2-^1/_2$
		21,67	25,86	$3d'\ ^2P-(^1D)\ 4f\ [1]^\circ$	$^1/_2-^3/_2$
2957,532	3	21,43	25,62	$3d'\ ^2D-(^3P_0)\ 5f\ [3]^\circ$	$^3/_2-^5/_2$
2956,541	4	21,37	25,56	$3d'\ ^2D-(^3P_1)\ 5f\ [4]^\circ$	$^5/_2-^7/_2$
2955,388	10	19,68	23,87	$4p\ ^2D^\circ-4d\ ^2D$	$^5/_2-^5/_2$
2947,275	2	21,37	25,57	$3d'\ ^2D-(^3P_1)\ 5f\ [3]^\circ$	$^5/_2-^5/_2$
2942,892	20	17,14	21,35	$4s\ ^2P-4p'\ ^2P^\circ$	$^3/_2-^3/_2$
2941,893	1	19,68	23,89	$4p\ ^2D^\circ-4d\ ^2D$	$^5/_2-^3/_2$
2935,538	3	21,14	25,36	$4p'\ ^2F^\circ-5d\ ^2D$	$^7/_2-^5/_2$
2932,589	8	17,26	21,49	$4s\ ^2P-4p'\ ^2D^\circ$	$^1/_2-^3/_2$
2931,483	9	21,67	25,90	$3d'\ ^2P-(^1D)\ 4f\ [2]^\circ$	$^1/_2-^3/_2$
2924,642	10	21,62	25,86	$3d'\ ^2P-(^1D)\ 4f\ [1]^\circ$	$^3/_2-^3/_2$
2915,967	1	19,64	23,89	$4p\ ^4D^\circ-4d\ ^2D$	$^1/_2-^3/_2$
2915,593	4	21,37	25,62	$3d'\ ^2D-(^3P_0)\ 5f\ [3]^\circ$	$^5/_2-^7/_2$
2914,932	1	21,37	25,62	$3d'\ ^2D-(^3P_0)\ 5f\ [3]^\circ$	$^5/_2-^5/_2$
2897,332	6	21,62	25,90	$3d'\ ^2P-(^3P_2)\ 7p\ [2]^\circ$	$^3/_2-^5/_2$
2896,753	10	21,62	25,90	$3d'\ ^2P-(^1D)\ 4f\ [2]^\circ$	$^3/_2-^5/_2$
2896,564	2	21,62	25,90	$3d'\ ^2P-(^1D)\ 4f\ [2]^\circ$	$^3/_2-^3/_2$
2893,985	1	19,61	23,89	$4p\ ^4D^\circ-4d\ ^2D$	$^3/_2-^3/_2$
2891,612	18	17,14	21,43	$4s\ ^2P-4p'\ ^2P^\circ$	$^3/_2-^1/_2$
2879,327	4	—	—	—	—
2874,583	3	19,97	24,28	$4p\ ^2S^\circ-5s'\ ^2D$	$^1/_2-^3/_2$
2871,399	1	19,97	24,28	$4p\ ^4S^\circ-5s'\ ^2D$	$^3/_2-^5/_2$
2871,022	1	21,13	25,44	$4p'\ ^2F^\circ-5d\ ^2P$	$^5/_2-^3/_2$
2869,283	1	21,62	25,94	$3d'\ ^2P-(^1D)\ 4f\ [3]^\circ$	$^3/_2-^5/_2$
2865,841	4	19,55	23,87	$4p\ ^4D^\circ-4d\ ^2D$	$^5/_2-^5/_2$
2860,742	3	20,74	25,07	$4s''\ ^2S-6p\ ^4P^\circ$	$^1/_2-^3/_2$
2847,816	3	17,14	21,49	$4s\ ^2P-4p'\ ^2D^\circ$	$^3/_2-^3/_2$
2847,146	2	20,74	25,10	$4s''\ ^2S-6p\ ^4P^\circ$	$^1/_2-^1/_2$
2844,129	4	17,14	21,50	$4s\ ^2P-4p'\ ^2D^\circ$	$^3/_2-^5/_2$
2843,369	3	20,74	25,10	$4s''\ ^2S-6p\ ^4D^\circ$	$^1/_2-^3/_2$
2806,168	5	19,87	24,28	$4p\ ^2P^\circ-5s'\ ^2D$	$^3/_2-^5/_2$
2805,990	1	19,87	24,28	$4p\ ^2P^\circ-5s'\ ^2D$	$^3/_2-^3/_2$
2800,919	1	21,67	26,10	$3d'\ ^2P-(^3P_2)\ 6f\ [2]^\circ$	$^1/_2-^3/_2$
2795,425	2	21,43	25,86	$3d'\ ^2D-(^1D)\ 4f\ [1]^\circ$	$^3/_2-^1/_2$
		21,43	25,86	$3d'\ ^2D-(^1D)\ 4f\ [1]^\circ$?	$^3/_2-^3/_2$
2795,289	2	21,67	26,11	$3d'\ ^2P-(^3P_2)\ 6f\ [1]^\circ$	$^1/_2-^3/_2$
2774,099	2	20,74	25,21	$4s''\ ^2S-6p\ ^2D^\circ$	$^1/_2-^3/_2$
2772,740	2	21,62	26,09	$3d'\ ^2P-(^3P_2)\ 6f\ [3]^\circ$	$^3/_2-^5/_2$
2769,748	4	21,43	25,90	$3d'\ ^2D-(^1D)\ 4f\ [2]^\circ$	$^3/_2-^3/_2$
2767,945	2	21,62	26,10	$3d'\ ^2P-(^3P_2)\ 6f\ [2]^\circ$	$^3/_2-^5/_2$
2764,648	4	19,80	24,28	$4p\ ^2P^\circ-5s'\ ^2D$	$^1/_2-^3/_2$
2763,520	1	21,62	26,11	$3d'\ ^2P-(^3P_2)\ 6f\ [1]^\circ$	$^3/_2-^3/_2$
2757,304	3	21,37	25,86	$3d'\ ^2D-(^1D)\ 4f\ [1]^\circ$	$^5/_2-^3/_2$
2754,864	2	16,64	21,14	$4s\ ^4P-4p'\ ^2F^\circ$	$^5/_2-^7/_2$
2744,797	6	21,43	25,94	$3d'\ ^2D-(^1D)\ 4f\ [3]^\circ$	$^3/_2-^5/_2$
2741,962	1	21,35	25,87	$4p'\ ^2P^\circ-6d\ ^4F$	$^3/_2-^5/_2$
2741,067	2	19,76	24,28	$4p\ ^2D^\circ-5s'\ ^2D$	$^3/_2-^5/_2$
2740,912	1	19,76	24,28	$4p\ ^2D^\circ-5s'\ ^2D$	$^3/_2-^3/_2$
2740,333	1	21,35	25,87	$4p'\ ^2P^\circ-6d\ ^4P$	$^3/_2-^3/_2$
2733,022	4	21,37	25,90	$3d'\ ^2D-(^3P_2)\ 7p\ [2]^\circ$	$^5/_2-^5/_2$
2732,504	6	21,37	25,90	$3d'\ ^2D-(^1D)\ 4f\ [2]^\circ$	$^5/_2-^5/_2$
2732,335	1	21,37	25,90	$3d'\ ^2D-(^1D)\ 4f\ [2]^\circ$	$^5/_2-^3/_2$
2731,639	1	21,43	25,96	$4p'\ ^2P^\circ-6d\ ^2D$	$^1/_2-^3/_2$
2720,184	2	21,67	26,23	$3d'\ ^2P-(^3P_1)\ 6f\ [2]^\circ$	$^1/_2-^3/_2$
2716,860	2	—	—	—	—

λ, Å	I	E_H, eV	E_B, eV	Transition	J
2708,272	6	21,37	25,94	$3d'\ ^2D-(^1D)\ 4f\ [3]^\circ$	$^5/_2-^7/_2$
2708,052	2	21,37	25,94	$3d'\ ^2D-(^1D)\ 4f\ [3]^\circ$	$^5/_2-^5/_2$
2701,719	2	—	—	—	—
2692,596	5	16,75	21,35	$4s\ ^4P-4p'\ ^2P^\circ$	$^3/_2-^3/_2$
2690,033	2	21,35	25,96	$4p'\ ^2P^\circ-6d\ ^2F$	$^3/_2-^5/_2$
2689,093	2	21,62	26,23	$3d'\ ^2P-(^3P_1)\ 6f\ [2]^\circ$	$^3/_2-^5/_2$
2687,395	1	21,35	25,96	$4p'\ ^2P^\circ-6d\ ^2D$	$^3/_2-^3/_2$
2686,322	2	16,81	21,43	$4s\ ^4P-4p'\ ^2P^\circ$	$^1/_2-^1/_2$
2683,094	3	21,62	26,24	$3d'\ ^2P-(^3P_1)\ 6f\ [3]^\circ$	$^3/_2-^5/_2$
2674,170	2	21,35	25,98	$4p'\ ^2P^\circ-6d\ ^2P$	$^3/_2-^3/_2$
2656,303	2	21,43	26,09	$3d'\ ^2D-(^3P_2)\ 6f\ [3]^\circ$	$^3/_2-^5/_2$
2654,056	2	21,62	26,29	$3d'\ ^2P-(^3P_0)\ 6f\ [3]^\circ$	$^3/_2-^5/_2$
2652,899	1	21,43	26,10	$3d'\ ^2D-(^3P_2)\ 6f\ [2]^\circ$	$^3/_2-^3/_2$
2651,906	2	21,43	26,10	$3d'\ ^2D-(^3P_2)\ 6f\ [2]^\circ$	$^3/_2-^5/_2$
2649,599	2	16,75	21,43	$4s\ ^4P-4p'\ ^2P^\circ$	$^3/_2-^1/_2$
2647,844	1	21,43	26,11	$3d'\ ^2D-(^3P_2)\ 6f\ [1]^\circ$	$^3/_2-^3/_2$
2647,247	6	19,97	24,65	$4p\ ^4S^\circ-6s\ ^4P$	$^3/_2-^5/_2$
2636,906	2	21,35	26,05	$4p'\ ^2P^\circ-6d\ ^2D$	$^3/_2-^5/_2$
2636,354	2	16,42	21,13	$3d\ ^4D-4p'\ ^2F^\circ$	$^5/_2-^5/_2$
2634,001	2	18,73	23,44	$3d\ ^2D-5p\ ^4P^\circ$	$^5/_2-^3/_2$
2627,397	3	19,97	24,69	$4p\ ^2S^\circ-6s\ ^4P$	$^1/_2-^3/_2$
		16,42	21,14	$3d\ ^4D-4p'\ ^2F^\circ$	$^5/_2-^7/_2$
2625,711	1	16,41	21,13	$3d\ ^4D-4p'\ ^2F^\circ$	$^7/_2-^5/_2$
2624,593	3	19,97	24,69	$4p\ ^4S^\circ-6s\ ^4P$	$^3/_2-^3/_2$
2623,090	1	21,37	26,09	$3d'\ ^2D-(^3P_2)\ 6f\ [4]^\circ$	$^5/_2-^7/_2$
2621,879	1	21,37	26,09	$3d'\ ^2D-(^3P_2)\ 6f\ [3]^\circ$	$^5/_2-^5/_2$
2620,985	4	21,37	26,09	$3d'\ ^2D-(^3P_2)\ 6f\ [3]^\circ$	$^5/_2-^7/_2$
2617,596	2	19,55	24,28	$4p\ ^4D^\circ-5s'\ ^2D$	$^5/_2-^5/_2$
		21,37	26,10	$3d'\ ^2D-(^3P_2)\ 6f\ [2]^\circ$	$^5/_2-^5/_2$
2616,811	3	16,41	21,14	$3d\ ^4D-4p'\ ^2F^\circ$	$^7/_2-^7/_2$
2600,956	3	19,97	24,74	$4p\ ^2S^\circ-4d'\ ^2P$	$^1/_2-^3/_2$
2592,178	1	18,66	23,44	$3d\ ^2D-5p\ ^4P^\circ$	$^3/_2-^3/_2$
2592,074	1	18,73	23,51	$3d\ ^2D-5p\ ^4D^\circ$	$^5/_2-^5/_2$
2591,696	1	19,87	24,65	$4p\ ^2P^\circ-6s\ ^4P$	$^3/_2-^5/_2$
2580,360	1	21,43	26,23	$3d'\ ^2D-(^3P_1)\ 6f\ [2]^\circ$	$^3/_2-^3/_2$
2579,428	2	21,43	26,23	$3d'\ ^2D-(^3P_1)\ 6f\ [2]^\circ$	$^3/_2-^5/_2$
2570,411	4	19,97	24,79	$4p\ ^2S^\circ-4d'\ ^2D$	$^1/_2-^3/_2$
		18,62	23,44	$3d\ ^2F-5p\ ^4P^\circ$	$^5/_2-^3/_2$
2569,984	3	19,87	24,69	$4p\ ^2P^\circ-6s\ ^4P$	$^3/_2-^3/_2$
2569,202	4	19,97	24,79	$4p\ ^4S^\circ-5d\ ^4D$	$^3/_2-^5/_2$
2567,727	1	19,97	24,79	$4p\ ^4S^\circ-4d'\ ^2D$	$^3/_2-^3/_2$
2567,095	1	19,97	24,80	$4p\ ^2S^\circ-6s\ ^2P$	$^1/_2-^3/_2$
2565,782	3	18,66	23,48	$3d\ ^2D-5p\ ^4P^\circ$	$^3/_2-^1/_2$
2564,416	4	19,97	24,80	$4p\ ^4S^\circ-6s\ ^2P$	$^3/_2-^3/_2$
2562,090	6	18,73	23,57	$3d\ ^2D-5p\ ^4D^\circ$	$^5/_2-^3/_2$
2561,954	1	19,97	24,81	$4p\ ^2S^\circ-6s\ ^4P$	$^1/_2-^1/_2$
2560,853	1	20,27	25,11	$3d'\ ^2F-6p\ ^4D^\circ$	$^7/_2-^5/_2$
2559,281	3	19,97	24,81	$4p\ ^4S^\circ-6s\ ^4P$	$^3/_2-^1/_2$
2556,586	4	19,97	24,81	$4p\ ^4S^\circ-5d\ ^4D$	$^3/_2-^3/_2$
2553,400	2	16,64	21,50	$4s\ ^4P-4p'\ ^2D^\circ$	$^5/_2-^5/_2$
2551,571	1	18,66	23,51	$3d\ ^2D-5p\ ^4D^\circ$	$^3/_2-^5/_2$
2549,788	3	19,87	24,73	$4p\ ^2P^\circ-4d'\ ^2P$	$^3/_2-^1/_2$
2547,184	2	21,43	26,29	$3d'\ ^2D-(^3P_0)\ 6f\ [3]^\circ$	$^3/_2-^5/_2$
2546,866	2	—	—	—	—
2545,642	3	21,37	26,23	$3d'\ ^2D-(^3P_1)\ 6f\ [4]^\circ$	$^5/_2-^7/_2$
2544,685	6	19,87	24,74	$4p\ ^2P^\circ-4d'\ ^2P$	$^3/_2-^3/_2$
2540,037	3	19,97	24,84	$4p\ ^4S^\circ-5d\ ^4D$	$^3/_2-^1/_2$
2536,018	7	19,76	24,65	$4p\ ^2D^\circ-6s\ ^4P$	$^3/_2-^5/_2$
		18,73	23,62	$3d\ ^2D-5p\ ^2P^\circ$	$^5/_2-^3/_2$

λ, Å	I	E_H, eV	E_B, eV	Transition	J
2535,758	1	18,73	23,62	3d ^2D—5p $^2D^\circ$	5/2—5/2
2535.250	3	19,80	24,69	4p $^2P^\circ$—6s ^4P	1/2—3/2
2534,712	7	19,87	24,76	4p $^2P^\circ$—4d′ ^2D	3/2—5/2
2530,423	1	18,62	23,51	3d ^2F—5p $^4D^\circ$	5/2—5/2
2528,679	3	19,97	24,87	4p $^2S^\circ$—6s ^2P	1/2—1/
2528,318	4	21,50	26,40	4p′ $^2D^\circ$—6s′ ^2D	5/2—5/2
2526,076	2	19,97	24,87	4p $^4S^\circ$—6s ^2P	3/2—1/2
2525,479	4	21,49	26,40	4p′ $^2D^\circ$—6s′ ^2D	3/2—3/2
2522,497	4	18,66	23,57	3d ^2D—5p $^4D^\circ$	3/2—3/2
2516,791	6	18,66	23,58	3d ^2D—5p $^2P^\circ$	3/2—1/2
2515,598	4	19,80	24,73	4p $^2P^\circ$—4d′ ^2P	1/2—1/2
2515,272	3	19,76	24,69	4p $^2D^\circ$—6s ^4P	3/2—3/2
2512,260	4	19,87	24,80	4p $^2P^\circ$—6s ^2P	3/2—3/2
2510,624	3	19,80	24,74	4p $^2P^\circ$—4d′ ^2P	1/2—3/2
2508,548	1	20,24	25,19	3d′ ^2F—5p′ $^2F^\circ$	5/2—5/2
2507,333	3	19,87	24,81	4p $^2P^\circ$—6s ^4P	3/2—1/2
2504,738	3	19,87	24,81	4p $^2P^\circ$—5d ^4D	3/2—3/2
2503,935	4	18,73	23,68	3d ^2D—5p $^2D^\circ$	5/2—3/2
2501,836	4	18,62	23,57	3d ^2F—5p $^4D^\circ$	5/2—3/2
2500,397	5	19,97	24,92	4p $^4S^\circ$—5d ^4P	3/2—1/2
2499,527	4	19,87	24,82	4p $^2P^\circ$—4d′ ^2F	3/2—5/2
2497,221	3	18,66	23,62	3d ^2D—5p $^2P^\circ$	3/2—3/2
2495,920	2	19,76	24,73	4p $^2D^\circ$—4d′ ^2P	3/2—1/2
2494,114	4	18,73	23,70	3d ^2D—5p $^4S^\circ$	5/2—3/2
		19,68	24,65	4p $^2D^\circ$—6s ^4P	5/2—5/2
2492,013	3	21,43	26,40	4p′ $^2P^\circ$—6s′ ^2D	1/2—3/2
2491,036	4	19,76	24,74	4p $^2D^\circ$—4d′ ^2P	3/2—3/2
2486,906	3	18,45	23,44	4s′ ^2D—5p $^4P^\circ$	5/2—3/2
2483,225	2	18,66	23,65	3d ^2D—5p $^4D^\circ$	3/2—1/2
2482,151	4	19,80	24,79	4p $^2P^\circ$—4d′ ^2D	1/2—3/2
2481,478	5	19,76	24,76	4p $^2D^\circ$—4d′ ^2D	3/2—5/2
2480,858	6	19,97	24,96	4p $^4S^\circ$—5d ^4P	3/2—3/2
2480,467	3	19,97	24,96	4p $^4S^\circ$—5d ^4F	3/2—5/2
2479,055	5	19,80	24,80	4p $^2P^\circ$—6s ^2P	1/2—3/2
2476,970	2	18,62	23,62	3d ^2F—5p $^2P^\circ$	5/2—3/2
2475,462	4	19,87	24,87	4p $^2P^\circ$—6s ^2P	3/2—1/2
2474,252	1	19,80	24,81	4p $^2P^\circ$—6s ^4P	1/2—1/2
2473,998	4	19,68	24,69	4p $^2D^\circ$—6s ^4P	5/2—3/2
2470,355	3	18,66	23,67	3d ^2D—5p $^2S^\circ$	3/2—1/2
2469,876	2	18,49	23,51	3d ^2F—5p $^4D^\circ$	7/2—5/2
2462,998	2	19,76	24,79	4p $^2D^\circ$—4d′ ^2D	3/2—3/2
2461,203	1	19,11	24,15	3d′ ^2G—($^3P_2$) 4f [4]°	7/2—7/2
2460,635	2	20,27	25,31	3d′ ^2F—5p′ $^2D^\circ$	7/2—5/2
2459,953	4	19,76	24,80	4p $^2D^\circ$—6s ^2P	3/2—3/2
2457,954	2	19,97	25,01	4p $^4S^\circ$—5d ^4F	3/2—3/2
2457,525	1	19,11	24,16	3d′ ^2G—($^3P_2$) 4f [3]°	7/2—7/2
2456,266	2	19,80	24,84	4p $^2P^\circ$—5d ^4D	1/2—1/2
2455,628	1	19,64	24,69	4p $^4D^\circ$—6s ^4P	1/2—3/2
2455,235	1	19,76	24,81	4p $^2D^\circ$—6s ^4P	3/2—1/2
2455,080	5	21,35	26,40	4p′ $^2P^\circ$—6s′ ^2D	3/2—5/2
2454,270	8	19,97	25,02	4p $^4S^\circ$—5d ^4P	3/2—5/2
2452,743	2	19,76	24,81	4p $^2D^\circ$—5d ^4D	3/2—3/2
2450,541	1	19,68	24,74	4p $^2D^\circ$—4d′ ^2P	5/2—3/2
2449,407	2	20,24	25,31	3d′ ^2F—5p′ $^2D^\circ$?	5/2—3/2
2449,179	2	18,43	23,48	4s′ ^2D—5p $^4P^\circ$	3/2—1/2
2447,743	2	19,76	24,82	4p $^2D^\circ$—4d′ ^2F	3/2—5/2
2446,355	1	18,62	23,68	3d ^2F—5p $^2D^\circ$	5/2—3/2
2444,828	3	18,33	23,40	3d ^4P—5p $^4P^\circ$	5/2—5/2
2443,219	2	19,80	24,87	4p $^2P^\circ$—6s ^2P	1/2—1/2

λ, Å	I	E_H, eV	E_B, eV	Transition	J
2442,794	2	19,11	24,19	$3d'\ ^2G-(^3P_2)\ 4f\ [5]^\circ$	7/2—9/2
2441,288	2	19,68	24,76	$4p\ ^2D^\circ-4d'\ ^2D$	5/2—5/2
2440,028	4	19,61	24,69	$4p\ ^4D^\circ-6s\ ^4P$	3/2—3/2
2437,517	2	19,76	24,84	$4p\ ^2D^\circ-5d\ ^4D$	3/2—1/2
2437,200	1	19,64	24,73	$4p\ ^4D^\circ-4d'\ ^2P$	1/2—1/2
2434,364	2	16,41	21,50	$3d\ ^4D-4p'\ ^2D^\circ$	7/2—5/2
2431,923	1	19,97	25,06	$4p\ ^4S^\circ-5d\ ^2F$	3/2—5/2
2431,62	2	19,87	24,96	$4p\ ^2P^\circ-5d\ ^4F$?	3/2—5/2
2430,032	5	19,55	24,65	$4p\ ^4D^\circ-6s\ ^4P$	5/2—5/2
2429,446	2	19,68	24,78	$4p\ ^2D^\circ-5d\ ^4D$	5/2—7/2
2428,523	1	18,33	23,44	$3d\ ^4P-5p\ ^4P^\circ$	5/2—3/2
2424,659	4	19,76	24,87	$4p\ ^2D^\circ-6s\ ^2P$	3/2—1/2
2423,528	5	19,68	24,79	$4p\ ^2D^\circ-4d'\ ^2D$?	5/2—3/2
2422,695	4	18,29	23,40	$3d\ ^4P-5p\ ^4P^\circ$	3/2—5/2
		18,45	23,57	$4s'\ ^2D-5p\ ^4D^\circ$	5/2—3/2
2422,089	2	21,50	26,61	$4p'\ ^2D^\circ-5d'\ ^2D$	5/2—5/2
2421,502	3	20,74	25,86	$4s''\ ^2S-(^1D)\ 4f\ [1]^\circ$	1/2—3/2
2420,457	6	19,68	24,80	$4p\ ^2D^\circ-6s\ ^2P$	5/2—3/2
2419,413	1	21,49	26,61	$4p'\ ^2D^\circ-5d'\ ^2D$	3/2—5/2
2419,164	1	19,80	24,92	$4p\ ^2P^\circ-5d\ ^4P$	1/2—1/2
2418,704	1	18,49	23,62	$3d\ ^2F-5p\ ^2D^\circ$	7/2—5/2
2417,214	2	19,61	24,74	$4p\ ^4D^\circ-4d'\ ^2P$	3/2—3/2
2414,224	5	19,68	24,81	$4p\ ^2D^\circ-4d'\ ^2F$	5/2—7/2
2413,486	1	19,68	24,81	$4p\ ^2D^\circ-5d\ ^4D$	5/2—3/2
2412,910	1	20,27	25,41	$3d'\ ^2F-(^3P_2)\ 5f\ [4]^\circ$	7/2—9/2
2412,461	4	21,50	26,63	$4p'\ ^2D^\circ-5d'\ ^2F$	5/2—7/2
2410,94	6	19,55	24,69	$4p\ ^4D^\circ-6s\ ^4P$	5/2—3/2
2409,702	1	18,43	23,57	$4s'\ ^2D-5p\ ^4D^\circ$	3/2—3/2
2409,503	2	20,27	25,41	$3d'\ ^2F-(^3P_2)\ 5f\ [3]^\circ$	7/2—7/2
2408,943	1	18,66	23,80	$3d\ ^2D-4p''\ ^2P^\circ$	3/2—3/2
2408,207	2	19,61	24,76	$4p\ ^4D^\circ-4d'\ ^2D$	3/2—5/2
2407,862	2	21,50	26,64	$4p'\ ^2D^\circ-5d'\ ^2F$	5/2—5/2
2406,647	5	18,33	23,48	$3d\ ^4P-5p\ ^4D^\circ$	5/2—7/2
2405,776	2	19,64	24,79	$4p\ ^4D^\circ-4d'\ ^2D$	1/2—3/2
2405,228	5	21,49	26,64	$4p'\ ^2D^\circ-5d'\ ^2F$	3/2—5/2
2404,352	9	19,49	24,65	$4p\ ^4D^\circ-6s\ ^4P$	7/2—5/2
2403,237	5	20,27	25,43	$3d'\ ^2F-(^3P_2)\ 5f\ [5]^\circ$	7/2—9/2
2399,851	3	—	—	—	—
2399,372	2	18,45	23,62	$4s'\ ^2D-5p\ ^2P^\circ$	5/2—3/2
2398,372	5	19,64	24,81	$4p\ ^4D^\circ-6s\ ^4P$	1/2—1/2
2397,548	2	20,24	25,41	$3d'\ ^2F-(^3P_2)\ 5f\ [3]^\circ$	5/2—7/2
2390,878	2	18,25	23,44	$3d\ ^4P-5p\ ^4P^\circ$	1/2—3/2
2388,268	1	18,66	23,85	$3d\ ^2D-4p''\ ^2P^\circ$	3/2—1/2
2387,933	3	19,61	24,80	$4p\ ^4D^\circ-6s\ ^2P$	3/2—3/2
2385,936	1	19,11	24,31	$3d'\ ^2G-(^3P_1)\ 4f\ [4]^\circ$	7/2—9/2
2384,969	4	19,87	25,06	$4p\ ^2P^\circ-5d\ ^2F$	3/2—5/2
2383,934	2	18,29	23,48	$3d\ ^4P-5p\ ^4P^\circ$	3/2—1/2
2383,486	6	19,61	24,81	$4p\ ^4D^\circ-6s\ ^4P$	3/2—1/2
2382,955	1	19,76	24,96	$4p\ ^2D^\circ-5d\ ^4P$	3/2—3/2
2381,138	3	19,61	24,81	$4p\ ^4D^\circ-5d\ ^4D$	3/2—3/2
2379,863	3	19,55	24,76	$4p\ ^4D^\circ-4d'\ ^2D$	5/2—5/2
2376,430	1	19,61	24,82	$4p\ ^4D^\circ-4d'\ ^2F$	3/2—5/2
2371,718	4	19,68	24,90	$4p\ ^2D^\circ-5d\ ^4F$	5/2—7/2
2371,662	1	18,29	23,51	$3d\ ^4P-5p\ ^4D^\circ$	3/2—5/2
2369,916	2	—	—	—	—
2369,187	2	19,64	24,87	$4p\ ^4D^\circ-6s\ ^2P$	1/2—1/2
2368,612	1	19,55	24,78	$4p\ ^4D^\circ-5d\ ^4D$	5/2—7/2
2367,248	1	18,33	23,57	$3d\ ^4P-5p\ ^4D^\circ$	5/2—3/2

λ, A	I	E_H, eV	E_B, eV	Transition	J
2366,778	2	19,61	24,84	$4p\,^4D^\circ$—$5d\,^4D$	3/2—1/2
2364,112	5	19,55	24,79	$4p\,^4D^\circ$—$5d\,^4D$	5/2—5/2
2362,866	1	19,55	24,79	$4p\,^4D^\circ$—$4d'\,^2D$	5/2—3/2
2362,083	1	18,43	23,67	$4s'\,^2D$—$5p\,^2S^\circ$	3/2—1/2
2360,058	4	19,55	24,80	$4p\,^4D^\circ$—$6s\,^2P$	5/2—3/2
2358,408	2	19,76	25,02	$4p\,^2D^\circ$—$5d\,^4P$	3/2—5/2
2357,589	5	21,14	26,40	$4p'\,^2F^\circ$—$6s'\,^2D$	7/2—5/2
2354,793	1	21,35	26,61	$4p'\,^2P^\circ$—$5d'\,^2D$	3/2—5/2
2354,135	6	19,55	24,81	$4p\,^4D^\circ$—$4d'\,^2F$	5/2—7/2
2353,426	3	19,55	24,81	$4p\,^4D^\circ$—$5d\,^4D$	5/2—3/2
2352,731	2	—	—	—	—
2350,486	5	21,13	26,40	$4p'\,^2F^\circ$—$6s'\,^2D$	5/2—3/2
2348,910	1	21,37	26,64	$3d'\,^2D$—$(^3P_1)\,7f\,[4]^\circ$	5/2—7/2
2346,570	2	19,64	24,92	$4p\,^4D^\circ$—$5d\,^4P$	1/2—1/2
		18,29	23,57	$3d\,^4P$—$5p\,^4D^\circ$	3/2—3/2
2344,204	6	19,49	24,78	$4p\,^4D^\circ$—$5d\,^4D$	7/2—7/2
2339,795	4	19,49	24,79	$4p\,^4D^\circ$—$5d\,^4D$	7/2—5/2
2337,780	6	19,76	25,06	$4p\,^2D^\circ$—$5d\,^2F$	3/2—5/2
2333,036	2	20,24	25,56	$3d'\,^2F$—$(^3P_1)\,5f\,[4]^\circ$	5/2—7/2
2332,895	1	21,35	26,66	$4p'\,^2P^\circ$—$5s''\,^2S$	3/2—1/2
2331,452	8	18,25	23,57	$3d\,^4P$—$5p\,^4D^\circ$	1/2—3/2
		19,68	24,99	$4p\,^2D^\circ$—$5d\,^2F$	5/2—7/2
2329,357	1	19,64	24,96	$4p\,^4D^\circ$—$5d\,^4P$	1/2—3/2
2327,784	2	20,24	25,57	$3d'\,^2F$—$(^3P_1)\,5f\,[3]^\circ$	5/2—7/2
2324,427	3	18,29	23,62	$3d\,^4P$—$5p\,^2D^\circ$	3/2—5/2
2322,081	2	19,68	25,02	$4p\,^2D^\circ$—$5d\,^4P$	5/2—5/2
2317,745	5	18,45	23,80	$4s'\,^2D$—$4p''\,^2P^\circ$	5/2—3/2
2316,299	8	19,49	24,84	$4p\,^4D^\circ$—$5d\,^4F$	7/2—9/2
2315,306	3	19,61	24,96	$4p\,^4D^\circ$—$5d\,^4P$	3/2—3/2
2314,970	6	19,61	24,96	$4p\,^4D^\circ$—$5d\,^4F$	3/2—5/2
2313,720	7	19,55	24,90	$4p\,^4D^\circ$—$5d\,^4F$	5/2—7/2
2309,860	2	18,25	23,62	$3d\,^4P$—$5p\,^2P^\circ$	1/2—3/2
2309,148	6	19,64	25,01	$4p\,^4D^\circ$—$5d\,^4F$	1/2—3/2
2307,456	2	20,04	25,62	$3d'\,^2F$—$(^3P_0)\,5f\,[3]^\circ$	5/2—7/2
2307,266	2	19,97	25,34	$4p\,^2S^\circ$—$5d\,^2D$	1/2—3/2
2305,859	2	18,43	23,80	$4s'\,^2D$—$4p''\,^2P^\circ$	3/2—3/2
2302,077	4	19,68	25,06	$4p\,^2D^\circ$—$5d\,^2F$	5/2—5/2
		19,97	25,36	$4p\,^2S^\circ$—$5d\,^2P$	1/2—1/2
2301,825	3	19,30	24,69	$4p\,^4P^\circ$—$6s\,^4P$	1/2—3/2
2300,179	5	19,26	24,65	$4p\,^4P^\circ$—$6s\,^4P$	3/2—5/2
2297,879	2	18,25	23,65	$3d\,^4P$—$5p\,^4D^\circ$	1/2—1/2
2295,349	3	19,61	25,01	$4p\,^4D^\circ$—$5d\,^4F$	3/2—3/2
2292,130	4	19,61	25,02	$4p\,^4D^\circ$—$5d\,^4P$	3/2—5/2
2290,425	3	19,49	24,90	$4p\,^4D^\circ$—$5d\,^4F$	7/2—7/2
2289,771	5	21,14	26,55	$4p'\,^2F^\circ$—$5d'\,^2G$	7/2—9/2
2289,376	2	18,29	23,70	$3d\,^4P$—$5p\,^4S^\circ$	3/2—3/2
2288,765	4	19,55	24,96	$4p\,^4D^\circ$—$5d\,^4F$	5/2—5/2
2286,925	4	18,43	23,85	$4s'\,^2D$—$4p''\,^2P^\circ$	3/2—1/2
2285,801	4	18,73	24,15	$3d\,^2D$—$(^3P_2)\,4f\,[4]^\circ$	5/2—7/2
2285,612	1	19,30	24,73	$4p\,^4P^\circ$—$4d'\,^2P$	1/2—1/2
2283,994	7	19,22	24,65	$4p\,^4P^\circ$—$6s\,^4P$	5/2—5/2
2283,753	1	18,73	24,16	$3d\,^2D$—$(^3P_2)\,4f\,[3]^\circ$	5/2—5/2
2283,243	7	21,13	26,55	$4p'\,^2F^\circ$—$5d'\,^2G$	5/2—7/2
		18,25	23,68	$3d\,^4P$—$5p\,^2D^\circ$	1/2—3/2
2282,621	8	18,73	24,16	$3d\,^2D$—$(^3P_2)\,4f\,[3]^\circ$	5/2—7/2
2281,512	1	19,30	24,74	$4p\,^4P^\circ$—$4d'\,^2P$	1/2—3/2
2275,358	3	19,55	24,99	$4p\,^4D^\circ$—$5d\,^2F$	5/2—7/2
2275,054	1	18,25	23,70	$3d\,^4P$—$5p\,^4S^\circ$	1/2—3/2
2274,923	3	18,73	24,18	$3d\,^2D$—$(^3P_2)\,4f\,[2]^\circ$	5/2—5/2

λ, Å	I	E_H, eV	E_B, eV	Transition	J
2272,640	2	19,61	25,06	$4p\ ^4D^\circ - 5d\ ^2F$	$^3/_2 - ^5/_2$
2267,111	2	19,22	24,69	$4p\ ^4P^\circ - 6s\ ^4P$	$^5/_2 - ^3/_2$
		19,26	24,73	$4p\ ^4P^\circ - 4d'\ ^2P$	$^3/_2 - ^1/_2$
2266,441	2	19,55	25,02	$4p\ ^4D^\circ - 5d\ ^4P$	$^5/_2 - ^5/_2$
2265,215	4	19,97	25,44	$4p\ ^2S^\circ - 5d\ ^2P$	$^1/_2 - ^3/_2$
2263,068	2	19,26	24,74	$4p\ ^4P^\circ - 4d'\ ^2P$	$^3/_2 - ^3/_2$
2262,632	2	21,67	27,15	$3d'\ ^2P - (^1D)\ 5f\ [2]^\circ$	$^1/_2 - ^3/_2$
2258,342	1	21,13	26,61	$4p'\ ^2F^\circ - 5d'\ ^2D$	$^5/_2 - ^5/_2$
2257,965	1	19,30	24,79	$4p\ ^4P^\circ - 4d'\ ^2D$	$^1/_2 - ^3/_2$
2256,545	3	21,14	26,63	$4p'\ ^2F^\circ - 5d'\ ^2F$	$^7/_2 - ^7/_2$
2255,408	3	19,30	24,80	$4p\ ^4P^\circ - 6s\ ^2P$	$^1/_2 - ^3/_2$
2255,178	1	19,26	24,76	$4p\ ^4P^\circ - 4d'\ ^2D$	$^3/_2 - ^5/_2$
2254,283	5	19,87	25,36	$4p\ ^2P^\circ - 5d\ ^2D$	$^3/_2 - ^5/_2$
2252,248	6	18,66	24,16	$3d\ ^2D - (^3P_2)\ 4f\ [3]^\circ$	$^3/_2 - ^5/_2$
2251,403	2	19,30	24,81	$4p\ ^4P^\circ - 6s\ ^4P$?	$^1/_2 - ^1/_2$
2249,658	1	18,06	23,57	$3d\ ^2P - 5p\ ^4D^\circ$	$^3/_2 - ^3/_2$
2249,347	3	19,30	24,81	$4p\ ^4P^\circ - 5d\ ^4D$	$^1/_2 - ^3/_2$
2245,975	3	21,13	26,64	$4p'\ ^2F^\circ - 5d'\ ^2F$	$^5/_2 - ^5/_2$
2245,410	2	18,66	24,18	$3d\ ^2D - (^3P_2)\ 4f\ [2]^\circ$	$^3/_2 - ^3/_2$
2245,116	2	18,06	23,58	$3d\ ^2P - 5p\ ^2P^\circ$	$^3/_2 - ^1/_2$
2244,080	1	19,49	25,02	$4p\ ^4D^\circ - 5d\ ^4P$	$^7/_2 - ^5/_2$
2243,662	5	18,66	24,18	$3d\ ^2D - (^3P_2)\ 4f\ [2]^\circ$	$^3/_2 - ^5/_2$
2241,858	2	21,62	27,15	$3d'\ ^2P - (^1D)\ 5f\ [2]^\circ$	$^3/_2 - ^5/_2$
2241,028	6	19,26	24,79	$4p\ ^4P^\circ - 5d\ ^4D$	$^3/_2 - ^5/_2$
2239,906	1	19,26	24,79	$4p\ ^4P^\circ - 4d'\ ^2D$	$^3/_2 - ^3/_2$
2237,721	2	18,62	24,15	$3d\ ^2F - (^3P_2)\ 4f\ [4]^\circ$	$^5/_2 - ^7/_2$
2237,385	1	19,26	24,80	$4p\ ^4P^\circ - 6s\ ^2P$	$^3/_2 - ^3/_2$
2236,527	3	19,30	24,84	$4p\ ^4P^\circ - 5d\ ^4D$	$^1/_2 - ^1/_2$
2235,904	2	19,80	25,34	$4p\ ^2P^\circ - 5d\ ^2D$	$^1/_2 - ^3/_2$
2235,760	3	18,62	24,16	$3d\ ^2F - (^3P_2)\ 4f\ [3]^\circ$	$^5/_2 - ^5/_2$
2234,673	6	18,62	24,16	$3d\ ^2F - (^3P_2)\ 4f\ [3]^\circ$	$^5/_2 - ^7/_2$
2233,478	4	19,26	24,81	$4p\ ^4P^\circ - 6s\ ^4P$	$^3/_2 - ^1/_2$
2231,423	5	19,26	24,81	$4p\ ^4P^\circ - 5d\ ^4D$	$^3/_2 - ^3/_2$
2231,024	1	19,80	25,36	$4p\ ^2P^\circ - 5d\ ^2P$	$^1/_2 - ^1/_2$
2230,317	3	18,66	24,21	$3d\ ^2D - (^3P_2)\ 4f\ [1]^\circ$	$^3/_2 - ^3/_2$
2229,648	8	19,22	24,78	$4p\ ^4P^\circ - 5d\ ^4D$	$^5/_2 - ^7/_2$
2227,298	5	18,62	24,18	$3d\ ^2F - (^3P_2)\ 4f\ [2]^\circ$	$^5/_2 - ^5/_2$
		19,26	24,82	$4p\ ^4P^\circ - 4d'\ ^2F$?	$^3/_2 - ^5/_2$
2225,662	6	19,22	24,79	$4p\ ^4P^\circ - 5d\ ^4D$	$^5/_2 - ^5/_2$
2224,550	1	19,22	24,79	$4p\ ^4P^\circ - 4d'\ ^2D$	$^5/_2 - ^3/_2$
2222,066	3	18,73	24,31	$3d\ ^2D - (^3P_1)\ 4f\ [2]^\circ$	$^5/_2 - ^3/_2$
		19,22	24,80	$4p\ ^4P^\circ - 6s\ ^2P$	$^5/_2 - ^3/_2$
2221,352	1	19,87	25,45	$4p\ ^2P^\circ - 4d'\ ^2S$	$^3/_2 - ^1/_2$
2220,347	2	19,76	25,34	$4p\ ^2D^\circ - 5d\ ^2D$	$^3/_2 - ^3/_2$
2219,962	6	18,73	24,31	$3d\ ^2D - (^3P_1)\ 4f\ [4]^\circ$	$^5/_2 - ^7/_2$
2218,805	4	19,26	24,84	$4p\ ^4P^\circ - 5d\ ^4D$	$^3/_2 - ^1/_2$
2218,375	1	18,06	23,65	$3d\ ^2P - 5p\ ^4D^\circ$	$^3/_2 - ^1/_2$
2216,190	4	19,22	24,81	$4p\ ^4P^\circ - 5d\ ^4D$	$^5/_2 - ^3/_2$
2214,147	1	18,62	24,21	$3d\ ^2F - (^3P_2)\ 4f\ [1]^\circ$	$^5/_2 - ^3/_2$
2210,883	3	18,73	24,34	$3d\ ^2D - (^3P_1)\ 4f\ [3]^\circ$	$^5/_2 - ^7/_2$
2210,321	2	18,73	24,34	$3d\ ^2D - (^3P_1)\ 4f\ [3]^\circ$	$^5/_2 - ^5/_2$
2205,738	4	19,30	24,92	$4p\ ^4P^\circ - 5d\ ^4P$	$^1/_2 - ^1/_2$
2204,698	1	18,06	23,68	$3d\ ^2P - 5p\ ^2D^\circ$	$^3/_2 - ^3/_2$
2202,135	1	17,94	23,57	$3d\ ^2P - 5p\ ^4D^\circ$	$^1/_2 - ^3/_2$
2201,573	1	20,27	25,90	$3d'\ ^2F - (^3P_2)\ 7p\ [2]^\circ$	$^7/_2 - ^5/_2$
2201,242	2	20,27	25,90	$3d'\ ^2F - (^1D)\ 4f\ [2]^\circ$	$^7/_2 - ^5/_2$
2197,786	1	17,94	23,58	$3d\ ^2P - 5p\ ^2P^\circ$	$^1/_2 - ^1/_2$
2196,389	1	19,80	25,44	$4p\ ^2P^\circ - 5d\ ^2P$	$^1/_2 - ^3/_2$

λ, Å	I	E_H, eV	E_B, eV	Transition	J
2195,445	5	18,73	24,38	$3d\ ^2D-(^3P_0)\ 4f\ [3]^\circ$	5/2—7/2
2194,907	2	18,73	24,38	$3d\ ^2D-(^3P_0)\ 4f\ [3]^\circ$	5/2—5/2
2192,224	2	18,66	24,31	$3d\ ^2D-(^3P_1)\ 4f\ [2]^\circ$	3/2—3/2
2191,579	4	20,24	25,90	$3d'\ ^2F-(^3P_2)\ 7p\ [2]^\circ$	5/2—5/2
		18,66	24,31	$3d\ ^2D-(^3P_1)\ 4f\ [2]^\circ$	3/2—5/2
2191,287	5	18,49	24,15	$3d\ ^2F-(^3P_2)\ 4f\ [4]^\circ$	7/2—9/2
		20,24	25,90	$3d'\ ^2F-(^1D)\ 4f\ [2]^\circ$	5/2—5/2
2190,511	4	19,30	24,96	$4p\ ^4P^\circ-5d\ ^4P$	1/2—3/2
2190,235	2	18,49	24,15	$3d\ ^2F-(^3P_2)\ 4f\ [4]^\circ$	7/2—7/2
2188,492	3	19,26	24,92	$4p\ ^4P^\circ-5d\ ^4P$	3/2—1/2
2187,320	6	18,49	24,16	$3d\ ^2F-(^3P_2)\ 4f\ [3]^\circ$	7/2—7/2
2185,489	5	20,27	25,94	$3d'\ ^2F-(^1D)\ 4f\ [3]^\circ$	7/2—7/2
2181,378	1	19,76	25,44	$4p\ ^2D^\circ-5d\ ^2P$	3/2—3/2
2181,211	7	20,27	25,95	$3d'\ ^2F-(^1D)\ 4f\ [4]^\circ$	7/2—9/2
2180,789	1	18,66	24,34	$3d\ ^2D-(^3P_1)\ 4f\ [3]^\circ$	3/2—5/2
2180,247	1	18,49	24,18	$3d\ ^2F-(^3P_2)\ 4f\ [2]^\circ$	7/2—5/2
2180,089	2	19,68	25,36	$4p\ ^2D^\circ-5d\ ^2D$	5/2—5/2
2176,387	1	17,74	23,44	$3d\ ^4F-5p\ ^4P^\circ$	5/2—3/2
2175,959	2	18,62	24,31	$3d\ ^2F-(^3P_1)\ 4f\ [2]^\circ$	5/2—5/2
2175,636	10	20,24	25,94	$3d'\ ^2F-(^1D)\ 4f\ [3]^\circ$	5/2—7/2
		18,49	24,19	$3d\ ^2F-(^3P_2)\ 4f\ [5]^\circ$	7/2—9/2
		20,24	25,94	$3d'\ ^2F-(^1D)\ 4f\ [3]^\circ$	5/2—5/2
2174,585	5	18,62	24,31	$3d\ ^2F-(^3P_1)\ 4f\ [4]^\circ$	5/2—7/2
2174,190	2	18,45	24,15	$4s'\ ^2D-(^3P_2)\ 4f\ [4]^\circ$	5/2—7/2
2173,209	1	19,26	24,96	$4p\ ^4P^\circ-5d\ ^4F$	3/2—5/2
2172,637	2	19,30	25,01	$4p\ ^4P^\circ-5d\ ^4F$	1/2—3/2
2172,341	2	18,45	24,16	$4s'\ ^2D-(^3P_2)\ 4f\ [3]^\circ$	5/2—5/2
2171,418	5	20,24	25,95	$3d'\ ^2F-(^1D)\ 4f\ [4]^\circ$	5/2—7/2
2171,312	3	18,45	24,16	$4s'\ ^2D-(^3P_2)\ 4f\ [3]^\circ$	5/2—7/2
2171,038	1	19,97	25,67	$4p\ ^4S^\circ-7s\ ^4P$	3/2—5/2
2170,914	1	17,69	23,40	$3d\ ^4F-5p\ ^4P^\circ$	7/2—5/2
2165,821	6	18,62	24,34	$3d\ ^2F-(^3P_1)\ 4f\ [3]^\circ$	5/2—7/2
2164,351	2	18,45	24,18	$4s'\ ^2D-(^3P_2)\ 4f\ [2]^\circ$	5/2—5/2
2162,292	2	17,94	23,67	$3d\ ^2P-5p\ ^2S^\circ$	1/2—1/2
2161,895	3	18,43	24,16	$4s'\ ^2D-(^3P_2)\ 4f\ [3]^\circ$	3/2—5/2
2159,046	2	19,22	24,96	$4p\ ^4P^\circ-5d\ ^4P$	5/2—3/2
		17,94	23,68	$3d\ ^2P-5p\ ^2D^\circ$	1/2—3/2
2158,883	2	18,06	23,80	$3d\ ^2P-4p''\ ^2P^\circ$	3/2—3/2
2158,755	2	19,22	24,96	$4p\ ^4P^\circ-5d\ ^4F$	5/2—5/2
2155,588	1	18,43	24,18	$4s'\ ^2D-(^3P_2)\ 4f\ [2]^\circ$	3/2—3/2
2153,980	3	18,43	24,18	$4s'\ ^2D-(^3P_2)\ 4f\ [2]^\circ$	3/2—5/2
2153,068	3	19,26	25,02	$4p\ ^4P^\circ-5d\ ^4P$	3/2—5/2
2151,052	6	18,62	24,38	$3d\ ^2F-(^3P_0)\ 4f\ [3]^\circ$	5/2—7/2
2150,537	2	18,62	24,38	$3d\ ^2F-(^3P_0)\ 4f\ [3]^\circ$	5/2—5/2
2147,681	2	17,74	23,51	$3d\ ^4F-5p\ ^4D^\circ$	5/2—5/2
2146,823	2	19,22	24,99	$4p\ ^4P^\circ-5d\ ^2F$	5/2—7/2
2143,884	3	13,48	19,26	$3p^6\ ^2S-4p\ ^4P^\circ$	1/2—3/2
2142,263	2	18,06	23,85	$3d\ ^2P-4p''\ ^2P^\circ$	3/2—1/2
2141,682	2	18,43	24,21	$4s'\ ^2D-(^3P_2)\ 4f\ [1]^\circ$	3/2—3/2
		19,22	25,01	$4p\ ^4P^\circ-5d\ ^4F$	5/2—3/2
2140,747	2	17,69	23,48	$3d\ ^4F-5p\ ^4D^\circ$	7/2—7/2
2138,882	3	19,22	25,02	$4p\ ^4P^\circ-5d\ ^4P$	5/2—5/2
2130,429	6	18,49	24,31	$3d\ ^2F-(^3P_1)\ 4f\ [4]^\circ$	7/2—9/2
2129,810	3	17,69	23,51	$3d\ ^4F-5p\ ^4D^\circ$	7/2—5/2
2129,427	4	18,33	24,15	$3d\ ^4P-(^3P_2)\ 4f\ [4]^\circ$	5/2—7/2
2127,646	3	18,33	24,16	$3d\ ^4P-(^3P_2)\ 4f\ [3]^\circ$	5/2—5/2
2127,050	2	17,74	23,57	$3d\ ^4F-5p\ ^4D^\circ$	5/2—3/2
2126,668	4	18,33	24,16	$3d\ ^4P-(^3P_2)\ 4f\ [3]^\circ$	5/2—7/2
2125,706	1	20,27	26,10	$3d'\ ^2F-(^3P_2)\ 6f\ [5]^\circ$	7/2—9/2

λ, Å	I	E_H, eV	E_B, eV	Transition	J
2125,272	1	19,87	25,70	$4p\ ^2P°—7s\ ^4P$	$^3/_2—^3/_2$
2121,542	1	18,33	24,18	$3d\ ^4P—(^3P_2)\ 4f\ [2]°$	$^5/_2—^3/_2$
2121,306	2	18,49	24,34	$3d\ ^2F—(^3P_1)\ 4f\ [3]°$	$^7/_2—^7/_2$
2119,985	3	18,33	24,18	$3d\ ^4P—(^3P_2)\ 4f\ [2]°$	$^5/_2—^5/_2$
2118,948	1	20,24	26,09	$3d'\ ^2F—(^3P_2)\ 6f\ [3]°$	$^5/_2—^7/_2$
2117,934	1	19,97	25,82	$4p\ ^4S°—7s\ ^2P$	$^3/_2—^3/_2$
2116,687	5	17,63	23,48	$3d\ ^4F—5p\ ^4D°$	$^9/_2—^7/_2$
2115,090	1	17,94	23,80	$3d\ ^2P—4p''\ ^2P°$	$^1/_2—^3/_2$
2114,532	1	18,45	24,31	$4s'\ ^2D—(^3P_1)\ 4f\ [4]°$	$^5/_2—^7/_2$
2110,896	2	18,29	24,16	$3d\ ^4P—(^3P_2)\ 4f\ [3]°$	$^3/_2—^5/_2$
2110,747	2	17,77	23,65	$3d\ ^4F—5p\ ^4D°$	$^3/_2—^1/_2$
2109,046	2	17,74	23,62	$3d\ ^4F—5p\ ^2P°$	$^5/_2—^3/_2$
2108,886	1	17,74	23,62	$3d\ ^4F—5p\ ^2D°$	$^5/_2—^5/_2$
2108,068	2	18,33	24,21	$3d\ ^4P—(^3P_2)\ 4f\ [1]°$	$^5/_2—^3/_2$
2106,537	1	18,43	24,31	$4s'\ ^2D—(^3P_1)\ 4f\ [2]°$	$^3/_2—^3/_2$
2106,247	1	18,45	24,34	$4s'\ ^2D—(^3P_1)\ 4f\ [3]°$	$^5/_2—^7/_2$
2105,935	3	18,43	24,31	$4s'\ ^2D—(^3P_1)\ 4f\ [2]°$	$^3/_2—^5/_2$
2104,885	3	18,29	24,18	$3d\ ^4P—(^3P_2)\ 4f\ [2]°$	$^3/_2—^3/_2$
2103,353	5	18,29	24,18	$3d\ ^4P—(^3P_2)\ 4f\ [2]°$	$^3/_2—^5/_2$
2101,467	1	19,80	25,70	$4p\ ^2P°—7s\ ^4P$	$^1/_2—^3/_2$
2098,123	1	19,97	25,87	$4p\ ^4S°—6d\ ^4P$	$^3/_2—^3/_2$
2096,808	1	19,97	25,88	$4p\ ^2S°—7s\ ^2P$	$^1/_2—^1/_2$
2095,976	1	18,43	24,34	$4s'\ ^2D—(^3P_1)\ 4f\ [3]°$	$^3/_2—^5/_2$
2092,764	3	18,25	24,18	$3d\ ^4P—(^3P_2)\ 4f\ [2]°$	$^1/_2—^3/_2$
2092,337	3	18,29	24,21	$3d\ ^4P—(^3P_2)\ 4f\ [1]°$	$^3/_2—^1/_2$
2091,627	5	18,29	24,21	$3d\ ^4P—(^3P_2)\ 4f\ [1]°$	$^3/_2—^3/_2$
2087,718	1	19,76	25,70	$4p\ ^2D°—7s\ ^4P$	$^3/_2—^3/_2$
2086,816	2	17,74	23,68	$3d\ ^4F—5p\ ^2D°$	$^5/_2—^3/_2$
2082,109	3	18,43	24,38	$4s'\ ^2D—(^3P_0)\ 4f\ [3]°$	$^3/_2—^5/_2$
2080,357	5	18,25	24,21	$3d\ ^4P—(^3P_0)\ 4f\ [1]°$	$^1/_2—^1/_2$
2079,654	4	18,25	24,21	$3d\ ^4P—(^3P_2)\ 4f\ [1]°$	$^1/_2—^3/_2$
2076,178	2	19,97	25,94	$4p\ ^4S°—6d\ ^4P$	$^3/_2—^5/_2$
2074,003	1	18,33	24,31	$3d\ ^4P—(^3P_1)\ 4f\ [2]°$	$^5/_2—^3/_2$
2073,426	4	18,33	24,31	$3d\ ^4P—(^3P_1)\ 4f\ [2]°$	$^5/_2—^5/_2$
2064,212	5	18,33	24,34	$3d\ ^4P—(^3P_1)\ 4f\ [3]°$	$^5/_2—^7/_2$
2063,761	2	18,33	24,34	$3d\ ^4P—(^3P_1)\ 4f\ [3]°$	$^5/_2—^5/_2$
2060,079	1	19,87	25,88	$4p\ ^2P°—7s\ ^2P$	$^3/_2—^1/_2$
2059,190	1	19,68	25,70	$4p\ ^2D°—7s\ ^4P$	$^5/_2—^3/_2$
2058,087	3	18,29	24,31	$3d\ ^4P—(^3P_1)\ 4f\ [2]°$	$^3/_2—^3/_2$
2057,514	5	18,29	24,31	$3d\ ^4P—(^3P_1)\ 4f\ [2]°$	$^3/_2—^5/_2$
2050,794	5	18,33	24,38	$3d\ ^4P—(^3P_0)\ 4f\ [3]°$	$^5/_2—^7/_2$
2050,324	2	18,33	24,38	$3d\ ^4P—(^3P_0)\ 4f\ [3]°$	$^5/_2—^5/_2$
2047,995	2	18,29	24,34	$3d\ ^4P—(^3P_1)\ 4f\ [3]°$	$^3/_2—^5/_2$
2046,492	4	18,25	24,31	$3d\ ^4P—(^3P_1)\ 4f\ [2]°$	$^1/_2—^3/_2$
2042,355	3	19,11	25,19	$3d'\ ^2G—5p'\ ^2F°$	$^7/_2—^5/_2$
2039,490	3	19,11	25,19	$3d'\ ^2G—5p'\ ^2F°$	$^9/_2—^7/_2$
2034,760	2	18,29	24,38	$3d\ ^4P—(^3P_0)\ 4f\ [3]°$	$^3/_2—^5/_2$
2032,173	3	18,06	24,16	$3d\ ^2P—(^3P_2)\ 4f\ [3]°$	$^3/_2—^5/_2$
2028,558	1	19,76	25,87	$4p\ ^2D°—6d\ ^4F$	$^3/_2—^5/_2$
2026,602	2	18,06	24,18	$3d\ ^2P—(^3P_2)\ 4f\ [2]°$	$^3/_2—^3/_2$
2025,183	3	18,06	24,18	$3d\ ^2P—(^3P_2)\ 4f\ [2]°$	$^3/_2—^5/_2$
2024,733	2	{ 19,76	25,88	$4p\ ^2D°—7s\ ^2P$	$^3/_2—^1/_2$
		{ 19,22	25,34	$4p\ ^4P°—5d\ ^2D$	$^5/_2—^3/_2$
2023,118	2	19,68	25,80	$4p\ ^2D°—6d\ ^4F$	$^5/_2—^7/_2$
2022,73	1	19,55	25,67	$4p\ ^4D°—7s\ ^4P$	$^5/_2—^5/_2$
2018,754	2	19,68	25,82	$4p\ ^2D°—7s\ ^2P$	$^5/_2—^3/_2$
2015,319	2	19,55	25,70	$4p\ ^4D°—7s\ ^4P$	$^5/_2—^3/_2$
2014,311	1	18,06	24,21	$3d\ ^2P—(^3P_2)\ 4f\ [1]°$	$^3/_2—^3/_2$
2007,178	1	19,76	25,94	$4p\ ^2D°—6d\ ^4P$	$^3/_2—^5/_2$

λ, Å	I	E_H, eV	E_B, eV	Transition	J
2004,914	3	19,49	25,67	$4p\ ^4D^\circ$—$7s\ ^4P$	$^7/_2$—$^5/_2$
2003,903	2	19,87	26,05	$4p\ ^2P^\circ$—$6d\ ^2D$	$^3/_2$—$^5/_2$
2003,325	1	19,64	25,83	$4p\ ^4D^\circ$—$7s\ ^4P$	$^1/_2$—$^1/_2$
2000,000	2	19,76	25,96	$4p\ ^2D^\circ$—$6d\ ^2F$	$^3/_2$—$^5/_2$
1988,620	3	19,68	25,91	$4p\ ^2D^\circ$—$6d\ ^2F$	$^5/_2$—$^7/_2$
		17,94	24,18	$3d\ ^2P$—$(^3P_2)\ 4f\ [2]^\circ$	$^1/_2$—$^3/_2$
1983,831	1	18,06	24,31	$3d\ ^2P$—$(^3P_1)\ 4f\ [2]^\circ$	$^3/_2$—$^3/_2$
1983,296	1	18,06	24,31	$3d\ ^2P$—$(^3P_1)\ 4f\ [2]^\circ$	$^3/_2$—$^5/_2$
1981,74	1	19,49	25,75	$4p\ ^4D^\circ$—$6d\ ^4D$?	$^7/_2$—$^7/_2$
1981,394	2	19,55	25,80	$4p\ ^4D^\circ$—$6d\ ^4F$	$^5/_2$—$^7/_2$
1979,988	1	19,61	25,87	$4p\ ^4D^\circ$—$6d\ ^4F$	$^3/_2$—$^5/_2$
1976,765	3	17,94	24,21	$3d\ ^2P$—$(^3P_2)\ 4f\ [1]^\circ$	$^1/_2$—$^3/_2$
1974,467	3	18,06	24,34	$3d\ ^2P$—$(^3P_2)\ 4f\ [3]^\circ$	$^3/_2$—$^5/_2$
1973,4837	2	13,48	19,76	$3p^6\ ^2S$—$4p\ ^2D^\circ$	$^1/_2$—$^3/_2$
1972,270	2	19,49	25,78	$4p\ ^4D^\circ$—$6d\ ^4F$	$^7/_2$—$^9/_2$
1966,952	1	19,64	25,94	$4p\ ^4D^\circ$—$6d\ ^4F$	$^1/_2$—$^3/_2$
1962,164	3	18,06	24,38	$3d\ ^2P$—$(^3P_0)\ 4f\ [3]^\circ$	$^3/_2$—$^5/_2$
1961,3610	4	13,48	19,80	$3p^6\ ^2S$—$4p\ ^2P^\circ$	$^1/_2$—$^1/_2$
1946,800	2	17,94	24,31	$3d\ ^2P$—$(^3P_1)\ 4f\ [2]^\circ$	$^1/_2$—$^3/_2$
1945,111	1	17,14	23,51	$4s\ ^2P$—$5p\ ^4D^\circ$	$^3/_2$—$^5/_2$
1941,0724	3	13,48	19,87	$3p^6\ ^2S$—$4p\ ^2P^\circ$	$^1/_2$—$^3/_2$
1937,042	1	17,77	24,18	$3d\ ^4F$—$(^3P_2)\ 4f\ [2]^\circ$	$^3/_2$—$^3/_2$
1933,694	2	17,74	24,15	$3d\ ^4F$—$(^3P_2)\ 4f\ [4]^\circ$	$^5/_2$—$^7/_2$
1932,231	2	17,74	24,16	$3d\ ^4F$—$(^3P_2)\ 4f\ [3]^\circ$	$^5/_2$—$^5/_2$
1931,421	1	17,74	24,16	$3d\ ^4F$—$(^3P_2)\ 4f\ [3]^\circ$	$^5/_2$—$^7/_2$
1920,016	2	17,69	24,15	$3d\ ^4F$—$(^3P_2)\ 4f\ [4]^\circ$	$^7/_2$—$^9/_2$
1919,197	3	17,69	24,15	$3d\ ^4F$—$(^3P_2)\ 4f\ [4]^\circ$	$^7/_2$—$^7/_2$
1907,989	4	17,69	24,19	$3d\ ^4F$—$(^3P_2)\ 4f\ [5]^\circ$	$^7/_2$—$^9/_2$
1900,638	4	17,63	24,15	$3d\ ^4F$—$(^3P_2)\ 4f\ [4]^\circ$	$^9/_2$—$^9/_2$
1899,834	1	17,63	24,15	$3d\ ^4F$—$(^3P_2)\ 4f\ [4]^\circ$	$^9/_2$—$^7/_2$
1899,271	1	19,22	25,75	$4p\ ^4P^\circ$—$6d\ ^4D$	$^5/_2$—$^7/_2$
1889,029	6	17,63	24,19	$3d\ ^4F$—$(^3P_2)\ 4f\ [5]^\circ$	$^9/_2$—$^{11}/_2$
1888,788	4	17,77	24,34	$3d\ ^4F$—$(^3P_1)\ 4f\ [3]^\circ$	$^3/_2$—$^5/_2$
1886,387	4	17,74	24,31	$3d\ ^4F$—$(^3P_1)\ 4f\ [4]^\circ$	$^5/_2$—$^7/_2$
1879,788	2	17,74	24,34	$3d\ ^4F$—$(^3P_1)\ 4f\ [4]^\circ$	$^5/_2$—$^7/_2$
1879,419	1	17,74	24,34	$3d\ ^4F$—$(^3P_1)\ 4f\ [3]^\circ$	$^5/_2$—$^5/_2$
1877,523	4	17,77	24,38	$3d\ ^4F$—$(^3P_0)\ 4f\ [3]^\circ$	$^3/_2$—$^5/_2$
1873,140	6	17,69	24,31	$3d\ ^4F$—$(^3P_1)\ 4f\ [4]^\circ$	$^7/_2$—$^9/_2$
1872,582	1	17,69	24,34	$3d\ ^4F$—$(^3P_1)\ 4f\ [4]^\circ$	$^7/_2$—$^7/_2$
1868,660	3	17,74	24,38	$3d\ ^4F$—$(^3P_0)\ 4f\ [3]^\circ$	$^5/_2$—$^7/_2$
1866,093	1	17,69	24,34	$3d\ ^4F$—$(^3P_1)\ 4f\ [4]^\circ$	$^7/_2$—$^7/_2$
1862,856	1	16,75	23,40	$4s\ ^4P$—$5p\ ^4P^\circ$?	$^3/_2$—$^5/_2$
1834,039	2	16,64	23,40	$4s\ ^4P$—$5p\ ^4P^\circ$	$^5/_2$—$^5/_2$
1831,525	5	19,12	25,89	$3d'\ ^2G$—$(^1D)\ 4f\ [5]^\circ$	$^7/_2$—$^9/_2$
1830,771	5	19,12	25,89	$3d'\ ^2G$—$(^1D)\ 4f\ [5]^\circ$	$^9/_2$—$^{11}/_2$
1823,207	1	18,62	25,42	$3d\ ^2F$—$(^3P_2)\ 5f\ [3]^\circ$	$^5/_2$—$^7/_2$
1813,772	1	19,12	25,95	$3d'\ ^2G$—$(^1D)\ 4f\ [4]^\circ$	$^7/_2$—$^7/_2$
1813,009	1	19,12	25,95	$3d'\ ^2G$—$(^1D)\ 4f\ [4]^\circ$	$^9/_2$—$^9/_2$
1791,561	1	18,49	25,42	$3d\ ^2F$—$(^3P_2)\ 5f\ [3]^\circ$	$^7/_2$—$^7/_2$
1788,101	3	18,49	25,43	$3d\ ^2F$—$(^3P_2)\ 5f\ [5]^\circ$	$^7/_2$—$^9/_2$
1785,669	1	18,62	25,56	$3d\ ^2F$—$(^3P_1)\ 5f\ [4]^\circ$	$^5/_2$—$^7/_2$
1782,587	1	18,62	25,57	$3d\ ^2F$—$(^3P_1)\ 5f\ [3]^\circ$	$^5/_2$—$^5/_2$
1776,670	1	16,42	23,40	$3d\ ^4D$—$5p\ ^4P^\circ$	$^5/_2$—$^5/_2$
1771,829	2	16,41	23,40	$3d\ ^4D$—$5p\ ^4P^\circ$	$^7/_2$—$^5/_2$
1770,652	1	18,62	25,62	$3d\ ^2F$—$(^3P_0)\ 5f\ [3]^\circ$	$^5/_2$—$^7/_2$
1768,042	1	16,42	23,43	$3d\ ^4D$—$5p\ ^4P^\circ$	$^5/_2$—$^3/_2$
1755,810	1	18,49	25,56	$3d\ ^2F$—$(^3P_1)\ 5f\ [4]^\circ$	$^7/_2$—$^9/_2$
1751,679	2	16,41	23,48	$3d\ ^4D$—$5p\ ^4D^\circ$	$^7/_2$—$^7/_2$

λ, Å	I	E_H, eV	E_B, eV	Transition	J
1736,830	1	18,29	25,43	$3d\,^4P-(^3P_2)\,5f\,[2]^\circ$	$^3/_2-^5/_2$
1733,362	1	18,29	25,44	$3d\,^4P-(^3P_2)\,5f\,[1]^\circ$	$^3/_2-^3/_2$
1729,262	1	18,25	25,42	$3d\,^4P-(^3P_2)\,5f\,[2]^\circ$	$^1/_2-^3/_2$
1729,075	1	18,73	25,90	$3d\,^2D-(^1D)\,4f\,[2]^\circ$ $3d\,^2D-(^1D)\,4f\,[2]^\circ$	$^5/_2-^5/_2$ $^5/_2-^3/_2$
1725,138	1	18,25	25,44	$3d\,^4P-(^3P_2)\,5f\,[1]^\circ$	$^1/_2-^3/_2$
1719,346	2	18,73	25,94	$3d\,^2D-(^1D)\,4f\,[3]^\circ$ $3d\,^2D-(^1D)\,4f\,[3]^\circ$	$^5/_2-^7/_2$ $^5/_2-^5/_2$
1718,680	1	16,41	23,62	$3d\,^4D-5p\,^2D^\circ$	$^7/_2-^5/_2$
1713,215	2	18,33	25,57	$3d\,^4P-(^3P_1)\,5f\,[3]^\circ$	$^5/_2-^7/_2$
1705,977	1	18,29	25,55	$3d\,^4P-(^3P_1)\,5f\,[2]^\circ$	$^3/_2-^5/_2$
1702,186	1	18,33	25,62	$3d\,^4P-(^3P_0)\,5f\,[3]^\circ$	$^5/_2-^7/_2$
1701,358	1	18,66	25,94	$3d\,^2D-(^1D)\,4f\,[3]^\circ$	$^3/_2-^5/_2$
1662,253	1	18,49	25,95	$3d\,^2F-(^1D)\,4f\,[4]^\circ$ $3d\,^2F-(^1D)\,4f\,[4]^\circ$	$^7/_2-^7/_2$ $^7/_2-^9/_2$
1653,322	1	17,94	25,44	$3d\,^2P-(^3P_2)\,5f\,[1]^\circ$	$^1/_2-^3/_2$
1650,531	1	18,06	25,57	$3d\,^2P-(^3P_1)\,5f\,[3]^\circ$	$^3/_2-^5/_2$
1649,299	1	18,43	25,94	$4s'\,^2D-(^1D)\,4f\,[3]^\circ$	$^3/_2-^5/_2$
1640,335	1	18,06	25,62	$3d\,^2P-(^3P_0)\,5f\,[3]^\circ$	$^3/_2-^5/_2$
1629,834	1	18,49	26,10	$3d\,^2F-(^3P_2)\,6f\,[5]^\circ$	$^7/_2-^9/_2$
1628,825	1	17,94	25,55	$3d\,^2P-(^3P_1)\,5f\,[2]^\circ$	$^1/_2-^3/_2$
1616,972	1	17,74	25,41	$3d\,^4F-(^3P_2)\,5f\,[4]^\circ$	$^5/_2-^7/_2$
1607,168	1	17,69	25,41	$3d\,^4F-(^3P_2)\,5f\,[4]^\circ$	$^7/_2-^9/_2$
1606,927	4	16,44	24,16	$3d\,^4D-(^3P_2)\,4f\,[3]^\circ$	$^3/_2-^5/_2$
1606,197	3	16,46	24,18	$3d\,^4D-(^3P_2)\,4f\,[2]^\circ$	$^1/_2-^3/_2$
1604,083	5	16,42	24,15	$3d\,^4D-(^3P_2)\,4f\,[4]^\circ$	$^5/_2-^7/_2$
1603,443	4	16,44	24,18	$3d\,^4D-(^3P_2)\,4f\,[2]^\circ$	$^3/_2-^3/_2$
1603,074	4	16,42	24,16	$3d\,^4D-(^3P_2)\,4f\,[3]^\circ$	$^5/_2-^5/_2$
1602,893	2	17,69	25,43	$3d\,^4F-(^3P_2)\,5f\,[5]^\circ$	$^7/_2-^9/_2$
1602,554	2	16,44 16,42	24,18 24,16	$3d\,^4D-(^3P_2)\,4f\,[2]^\circ$ $3d\,^4D-(^3P_2)\,4f\,[3]^\circ$	$^3/_2-^5/_2$ $^5/_2-^7/_2$
1600,694	6	16,41	24,15	$3d\,^4D-(^3P_2)\,4f\,[4]^\circ$	$^7/_2-^9/_2$
1600,133	4	16,41	24,15	$3d\,^4D-(^3P_2)\,4f\,[4]^\circ$	$^7/_2-^7/_2$
1599,597	1	16,42	24,18	$3d\,^4D-(^3P_2)\,4f\,[2]^\circ$	$^5/_2-^3/_2$
1599,125	1	16,41	24,16	$3d\,^4D-(^3P_2)\,4f\,[3]^\circ$	$^7/_2-^5/_2$
1598,872	1	16,46	24,21	$3d\,^4D-(^3P_2)\,4f\,[1]^\circ$	$^1/_2-^1/_2$
1598,724	2	16,42	24,18	$3d\,^4D-(^3P_2)\,4f\,[2]^\circ$	$^5/_2-^5/_2$
1598,561	1	16,41	24,16	$3d\,^4D-(^3P_2)\,4f\,[3]^\circ$	$^7/_2-^7/_2$
1596,141	1	16,44	24,21	$3d\,^4D-(^3P_2)\,4f\,[1]^\circ$	$^3/_2-^1/_2$
1595,734	1	16,44	24,21	$3d\,^4D-(^3P_2)\,4f\,[1]^\circ$	$^3/_2-^3/_2$
1594,787	1	16,41	24,18	$3d\,^4D-(^3P_2)\,4f\,[2]^\circ$	$^7/_2-^5/_2$
1593,581	2	17,63	25,41	$3d\,^4F-(^3P_2)\,5f\,[4]^\circ$	$^9/_2-^9/_2$
1591,933	1	16,42	24,21	$3d\,^4D-(^3P_2)\,4f\,[1]^\circ$	$^5/_2-^3/_2$
1590,229	2	17,77	25,57	$3d\,^4F-(^3P_1)\,5f\,[3]^\circ$	$^3/_2-^5/_2$
1589,463	5	17,63	25,43	$3d\,^4F-(^3P_2)\,5f\,[5]^\circ$	$^9/_2-^{11}/_2$
1586,256	2	17,74	25,56	$3d\,^4F-(^3P_1)\,5f\,[4]^\circ$	$^5/_2-^7/_2$
1583,83	1	17,74	25,57	$3d\,^4F-(^3P_1)\,5f\,[3]^\circ$	$^5/_2-^7/_2$
1580,960	1	18,06	25,90	$3d\,^2P-(^1D)\,4f\,[2]^\circ$ $3d\,^2P-(^1D)\,4f\,[2]^\circ$	$^3/_2-^5/_2$ $^3/_2-^3/_2$
1580,768	2	17,77	25,62	$3d\,^4F-(^3P_0)\,5f\,[3]^\circ$	$^3/_2-^5/_2$
1578,812	3	16,46	24,31	$3d\,^4D-(^3P_1)\,4f\,[2]^\circ$	$^1/_2-^3/_2$
1576,897	3	17,69	25,56	$3d\,^4F-(^3P_1)\,5f\,[4]^\circ$	$^7/_2-^9/_2$
1575,815	3	16,44	24,31	$3d\,^4D-(^3P_1)\,4f\,[2]^\circ$	$^3/_2-^5/_2$
1574,992	6	13,48	21,35	$3p^6\,^2S-4p'\,^2P^\circ$	$^1/_2-^3/_2$
1574,402	1	17,74	25,62	$3d\,^4F-(^3P_0)\,5f\,[3]^\circ$	$^5/_2-^7/_2$
1571,390	1	16,42	24,31	$3d\,^4D-(^3P_1)\,4f\,[4]^\circ$	$^5/_2-^7/_2$
1567,987	4	16,41	24,31	$3d\,^4D-(^3P_1)\,4f\,[4]^\circ$	$^7/_2-^9/_2$
1566,812	1	16,42	24,34	$3d\,^4D-(^3P_1)\,4f\,[4]^\circ$	$^5/_2-^7/_2$

λ, Å	I	E_H, eV	E_B, eV	Transition	J
1565,377	1	17,94	25,86	{ $3d\ ^2P$—(^1D) $4f$ [1]° $3d\ ^2P$—(^1D) $4f$ [1]°	1/2—1/2 1/2—3/2
1563,036	1	16,41	24,34	$3d\ ^4D$—($^3P_1$) $4f$ [4]°	7/2—7/2
1562,441	2	16,44	24,38	$3d\ ^4D$—($^3P_0$) $4f$ [3]°	3/2—5/2
1560,184	4	13,48	21,43	$3p^6\ ^2S$—$4p'\ ^2P$°	1/2—1/2
1559,072	3	16,42	24,37	$3d\ ^4D$—($^3P_0$) $4f$ [3]°	5/2—7/2
1557,302	1	17,94	25,90	$3d\ ^2P$—(^1D) $4f$ [2]°	1/2—3/2
1547,354	1	13,48	21,49	$3p^6\ ^2S$—$4p'\ ^2D$°	1/2—3/2
1544,711	2	19,12	27,14	$3d'\ ^2G$—(^1D) $5f$ [5]°	7/2—9/2
1544,177	2	19,12	27,14	$3d'\ ^2G$—(^1D) $5f$ [5]°	9/2—11/2
1474,537	1	17,69	26,10	$3d\ ^4F$—($^3P_2$) $6f$ [5]°	7/2—9/2
1472,594	1	16,64	25,06	$4s\ ^4P$—$6p\ ^4P$°	5/2—5/2
1466,524	1	16,64	25,10	$4s\ ^4P$—$6p\ ^4D$°	5/2—7/2
1465,153	1	17,63	26,09	$3d\ ^4F$—($^3P_2$) $6f$ [4]°	9/2—9/2
1464,176	1	17,77	26,24	$3d\ ^4F$—($^3P_1$) $6f$ [3]°	3/2—5/2
1463,155	2	17,63	26,10	$3d\ ^4F$—($^3P_2$) $6f$ [5]°	9/2—11/2
1459,875	1	17,74	26,23	$3d\ ^4F$—($^3P_1$) $6f$ [4]°	5/2—7/2
1455,484	1	17,77	26,29	$3d\ ^4F$—($^3P_0$) $6f$ [3]°	3/2—5/2
1451,879	1	17,69	26,23	$3d\ ^4F$—($^3P_1$) $6f$ [4]°	7/2—9/2
1396,231	1	17,63	26,51	$3d\ ^4F$—($^3P_2$) $7f$ [5]°	9/2—11/2
1382,765	1	16,46	25,42	$3d\ ^4D$—($^3P_2$) $5f$ [2]°	1/2—3/2
1382,228	2	16,44	25,41	$3d\ ^4D$—($^3P_2$) $5f$ [3]°	3/2—5/2
1380,723	1	16,44	25,42	$3d\ ^4D$—($^3P_2$) $5f$ [2]°	3/2—3/2
1379,884	3	16,42	25,41	$3d\ ^4D$—($^3P_2$) $5f$ [4]°	5/2—7/2
1379,377	1	16,42	25,41	$3d\ ^4D$—($^3P_2$) $5f$ [3]°	5/2—5/2
1377,211	4	16,41	25,41	$3d\ ^4D$—($^3P_2$) $5f$ [4]°	7/2—9/2
1376,956	1	16,41	25,41	$3d\ ^4D$—($^3P_2$) $5f$ [4]°	7/2—7/2
1363,031	2	16,46	25,55	$3d\ ^4D$—($^3P_1$) $5f$ [2]°	1/2—3/2
1360,735	1	16,44	25,55	$3d\ ^4D$—($^3P_1$) $5f$ [2]°	3/2—5/2
1354,912	2	16,41	25,56	$3d\ ^4D$—($^3P_1$) $5f$ [4]°	7/2—9/2
1351,330	1	16,44	25,62	$3d\ ^4D$—($^3P_0$) $5f$ [3]°	3/2—5/2
1348,745	1	16,42	25,62	$3d\ ^4D$—($^3P_0$) $5f$ [3]°	5/2—7/2
1284,793	1	16,44	26,09	$3d\ ^4D$—($^3P_2$) $6f$ [3]°	3/2—5/2
1282,620	1	16,42	26,09	$3d\ ^4D$—($^3P_2$) $6f$ [4]°	5/2—7/2
1280,225	1	16,41	26,09	$3d\ ^4D$—($^3P_2$) $6f$ [4]°	5/2—9/2
1268,483	1	16,46	26,23	$3d\ ^4D$—($^3P_1$) $6f$ [2]°	1/2—3/2
932,0528	10	0,18	13,48	$3p^5\ ^2P$°—$3p^6\ ^2S$	1/2—1/2
919,7815	10	0,00	13,48	$3p^5\ ^2P$°—$3p^6\ ^2S$	3/2—1/2
762,200	3	0,18	16,44	$3p^5\ ^2P$°—$3d\ ^4D$	1/2—3/2
754,824	3	0,00	16,42	$3p^5\ ^2P$°—$3d\ ^4D$	3/2—5/2
748,198	4	0,18	16,75	$3p^5\ ^2P$°—$4s\ ^4P$	1/2—3/2
745,323	7	0,18	16,81	$3p^5\ ^2P$°—$4s\ ^4P$	1/2—1/2
744,925	8	0,00	16,64	$3p^5\ ^2P$°—$4s\ ^4P$	3/2—5/2
740,270	10	0,00	16,75	$3p^5\ ^2P$°—$4s\ ^4P$	3/2—3/2
737,457	1	0,00	16,81	$3p^5\ ^2P$°—$4s\ ^4P$	3/2—1/2
730,929	5	0,18	17,14	$3p^5\ ^2P$°—$4s\ ^2P$	1/2—3/2
725,550	4	0,18	17,26	$3p^5\ ^2P$°—$4s\ ^2P$	1/2—1/2
723,361	5	0,00	17,14	$3p^5\ ^2P$°—$4s\ ^2P$	3/2—3/2
718,091	4	0,00	17,26	$3p^5\ ^2P$°—$4s\ ^2P$	3/2—1/2
704,523	4	0,18	17,77	$3p^5\ ^2P$°—$3d\ ^4F$	1/2—3/2
698,771	4	0,00	17,74	$3p^5\ ^2P$°—$3d\ ^4F$	3/2—5/2
697,940	2	0,18	17,94	$3p^5\ ^2P$°—$3d\ ^2P$	1/2—1/2
697,489	2	0,00	17,77	$3p^5\ ^2P$°—$3d\ ^4F$	3/2—3/2
693,301	2	0,18	18,06	$3p^5\ ^2P$°—$3d\ ^2P$	1/2—3/2
691,038	1	0,00	17,94	$3p^5\ ^2P$°—$3d\ ^2P$	3/2—1/2
686,489	2	0,00	18,06	$3p^5\ ^2P$°—$3d\ ^2P$	3/2—3/2
679,400	6	0,18	18,43	$3p^5\ ^2P$°—$4s'\ ^2D$	1/2—3/2
679,221	3	0,00	18,25	$3p^5\ ^2P$°—$3d\ ^4P$	3/2—1/2

λ, Å	I	E_H, eV	E_B, eV	Transition	J
677,951	5	0,00	18,29	$3p^5\ {}^2P^\circ - 3d\ {}^4P$	$^3/_2 - {}^3/_2$
676,241	6	0,00	18,33	$3p^5\ {}^2P^\circ - 3d\ {}^4P$	$^3/_2 - {}^5/_2$
672,856	2	0,00	18,43	$3p^5\ {}^2P^\circ - 4s'\ {}^2D$	$^3/_2 - {}^3/_2$
671,852	6	0,00	18,45	$3p^5\ {}^2P^\circ - 4s'\ {}^2D$	$^3/_2 - {}^5/_2$
670,948	5	0,18	18,66	$3p^5\ {}^2P^\circ - 3d\ {}^2D$	$^1/_2 - {}^3/_2$
666,010	6	0,00	18,62	$3p^5\ {}^2P^\circ - 3d\ {}^2F$	$^3/_2 - {}^5/_2$
664,563	4	0,00	18,66	$3p^5\ {}^2P^\circ - 3d\ {}^2D$	$^3/_2 - {}^3/_2$
661,869	5	0,00	18,73	$3p^5\ {}^2P^\circ - 3d\ {}^2D$	$^3/_2 - {}^5/_2$
612,371	5	0,00	20,24	$3p^5\ {}^2P^\circ - 3d'\ {}^2F$	$^3/_2 - {}^5/_2$
602,858	2	0,18	20,74	$3p^5\ {}^2P^\circ - 4s''\ {}^2S$	$^1/_2 - {}^1/_2$
597,701	2	0,00	20,74	$3p^5\ {}^2P^\circ - 4s''\ {}^2S$	$^3/_2 - {}^1/_2$
583,437	2	0,18	21,43	$3p^5\ {}^2P^\circ - 3d'\ {}^2D$	$^1/_2 - {}^3/_2$
580,264	3	0,00	21,37	$3p^5\ {}^2P^\circ - 3d'\ {}^2D$	$^3/_2 - {}^5/_2$
578,604	2	0,00	21,43	$3p^5\ {}^2P^\circ - 3d'\ {}^2D$	$^3/_2 - {}^3/_2$
578,107	2	0,18	21,62	$3p^5\ {}^2P^\circ - 3d'\ {}^2P$	$^1/_2 - {}^3/_2$
576,738	2	0,18	21,67	$3p^5\ {}^2P^\circ - 3d'\ {}^2P$	$^1/_2 - {}^1/_2$
573,364	2	0,00	21,62	$3p^5\ {}^2P^\circ - 3d'\ {}^2P$	$^3/_2 - {}^3/_2$
572,014	2	0,00	21,67	$3p^5\ {}^2P^\circ - 3d'\ {}^2P$	$^3/_2 - {}^1/_2$
560,224	2	0,18	22,31	$3p^5\ {}^2P^\circ - 3d''\ {}^2D$	$^1/_2 - {}^3/_2$
556,817	2	0,00	22,27	$3p^5\ {}^2P^\circ - 3d''\ {}^2D$	$^3/_2 - {}^5/_2$
555,764	1	0,00	22,31	$3p^5\ {}^2P^\circ - 3d''\ {}^2D$	$^3/_2 - {}^3/_2$
553,123	1	0,18	22,59	$3p^5\ {}^2P^\circ - 5s\ {}^4P$	$^1/_2 - {}^3/_2$
550,896	1	0,18	22,68	$3p^5\ {}^2P^\circ - 5s\ {}^4P$	$^1/_2 - {}^1/_2$
550,481	1	0,18	22,70	$3p^5\ {}^2P^\circ - 5s\ {}^2P$	$^1/_2 - {}^3/_2$
548,781	2	0,00	22,59	$3p^5\ {}^2P^\circ - 5s\ {}^4P$	$^3/_2 - {}^3/_2$
547,456	2	0,18	22,82	$3p^5\ {}^2P^\circ - 3d'\ {}^2S$	$^1/_2 - {}^1/_2$
547,166	2	0,18	22,84	$3p^5\ {}^2P^\circ - 4d\ {}^4D$	$^1/_2 - {}^1/_2$
546,175	2	0,00	22,70	$3p^5\ {}^2P^\circ - 5s\ {}^2P$	$^3/_2 - {}^3/_2$
543,730	2	0,00	22,80	$3p^5\ {}^2P^\circ - 5s\ {}^2P$	$^3/_2 - {}^1/_2$
543,205	2	0,00	22,82	$3p^5\ {}^2P^\circ - 3d'\ {}^2S$	$^3/_2 - {}^1/_2$
542,911	2	0,00	22,84	$3p^5\ {}^2P^\circ - 4d\ {}^4D$	$^3/_2 - {}^1/_2$
537,140	1	0,00	23,08	$3p^5\ {}^2P^\circ - 4d\ {}^4P$	$^3/_2 - {}^1/_2$
533,082	1	0,00	23,26	$3p^5\ {}^2P^\circ - 4d\ {}^2F$	$^3/_2 - {}^5/_2$
530,494	1	0,18	23,55	$3p^5\ {}^2P^\circ - 4d\ {}^2P$	$^1/_2 - {}^1/_2$
526,497	1	0,00	23,55	$3p^5\ {}^2P^\circ - 4d\ {}^2P$	$^3/_2 - {}^1/_2$
524,683	1	0,00	23,63	$3p^5\ {}^2P^\circ - 4d\ {}^2P$	$^3/_2 - {}^3/_2$
522,791	1	0,18	23,89	$3p^5\ {}^2P^\circ - 4d\ {}^2D$	$^1/_2 - {}^3/_2$
519,329	1	0,00	23,87	$3p^5\ {}^2P^\circ - 4d\ {}^2D$	$^3/_2 - {}^5/_2$
518,910	1	0,00	23,89	$3p^5\ {}^2P^\circ - 4d\ {}^2D$	$^3/_2 - {}^3/_2$
514,310	1	0,18	24,28	$3p^5\ {}^2P^\circ - 5s'\ {}^2D$	$^1/_2 - {}^3/_2$
510,554	1	0,00	24,28	$3p^5\ {}^2P^\circ - 5s'\ {}^2D$ $3p^5\ {}^2P^\circ - 5s'\ {}^2D$	$^3/_2 - {}^5/_2$ $^3/_2 - {}^3/_2$
505,013	0,5	0,18	24,73	$3p^5\ {}^2P^\circ - 4d'\ {}^2P$	$^1/_2 - {}^1/_2$
503,649	0,5	0,18	24,79	$3p^5\ {}^2P^\circ - 4d'\ {}^2P$	$^1/_2 - {}^3/_2$
502,157	1	0,00	24,69	$3p^5\ {}^2P^\circ - 6s\ {}^4P$	$^3/_2 - {}^3/_2$
501,184	0,5	0,00	24,74	$3p^5\ {}^2P^\circ - 4d'\ {}^2D$	$^3/_2 - {}^3/_2$
500,798	0,5	0,00	24,76	$3p^5\ {}^2P^\circ - 4d'\ {}^2D$	$^3/_2 - {}^5/_2$
496,650	0,5	0,00	24,96	$3p^5\ {}^2P^\circ - 5d\ {}^4F$? $3p^5\ {}^2P^\circ - 5d\ {}^4P$?	$^3/_2 - {}^5/_2$ $^3/_2 - {}^3/_2$
494,686	0,5	0,00	25,06	$3p^5\ {}^2P^\circ - 5d\ {}^2F$?	$^3/_2 - {}^5/_2$
490,698	0,5	0,18	25,36	$3p^5\ {}^2P^\circ - 5d\ {}^2D$	$^1/_2 - {}^3/_2$
489,196	0,5	0,00	25,34	$3p^5\ {}^2P^\circ - 5d\ {}^2P$	$^3/_2 - {}^3/_2$
488,782	0,5	0,00	25,36	$3p^5\ {}^2P^\circ - 5d\ {}^2D$	$^3/_2 - {}^5/_2$

Ar III, ground state $1s^2\ 2s^2\ 2p^6\ 3s^2\ 3p^4\ {}^3P_2$
Ionization potential 329 965,80 cm^{-1}; 40,908 eV

λ, Å	I	E_H, eV	E_B, eV	Transition	J
4198,83	3	25,73	28,68	$4s''\ {}^3P^\circ$—$4p'\ {}^3P$	1—2
4173,69	2	25,75	28,72	$4s''\ {}^3P^\circ$—$4p'\ {}^3P$	0—1
4149,03	3	25,73	28,72	$4s''\ {}^3P^\circ$—$4p'\ {}^3P$	1—1
4146,70	5	25,69	28,68	$4s''\ {}^3P^\circ$—$4p'\ {}^3P$	2—2
4127,19	4	25,73	28,73	$4s''\ {}^3P^\circ$—$4p'\ {}^3P$	1—0
4098,19	4	25,69	28,72	$4s''\ {}^3P^\circ$—$4p'\ {}^3P$	2—1
4059,89	3	26,60	29,65	$3d''\ {}^3P^\circ$—$4p''\ {}^3S$	0—1
4023,60	6	26,57	29,65	$3d''\ {}^3P^\circ$—$4p''\ {}^3S$	1—1
3960,53	8	26,53	29,65	$3d''\ {}^3P^\circ$—$4p''\ {}^3S$	2—1
3907,84	7	26,60	29,77	$3d''\ {}^3P^\circ$—$4p''\ {}^3D$	0—1
3874,22	4	26,57	29,77	$3d''\ {}^3P^\circ$—$4p''\ {}^3D$	1—1
3858,32	10	26,57	29,79	$3d''\ {}^3P^\circ$—$4p''\ {}^3D$	1—2
3815,70	1	26,53	29,77	$3d''\ {}^3P^\circ$—$4p''\ {}^3D$	2—1
3800,25	6	26,53	29,79	$3d''\ {}^3P^\circ$—$4p''\ {}^3D$	2—2
3795,37	20	26,53	29,79	$3d''\ {}^3P^\circ$—$4p''\ {}^3D$	2—3
3514,18	6	22,40	25,93	$4s\ {}^3S^\circ$—$4p\ {}^3P$	1—1
3511,69	5	24,38	27,91	$4s'\ {}^3D^\circ$—$4p'\ {}^3D$	3—2
3511,12	8	22,40	25,93	$4s\ {}^3S^\circ$—$4p\ {}^3P$	1—2
3509,33	5	22,40	25,93	$4s\ {}^3S^\circ$—$4p\ {}^3P$	1—0
3503,58	15	24,38	27,91	$4s'\ {}^3D^\circ$—$4p'\ {}^3D$	2—2
3502,70	6	24,38	27,91	$4s'\ {}^3D^\circ$—$4p'\ {}^3D$	2—1
3500,58	5	24,37	27,91	$4s'\ {}^3D^\circ$—$4p'\ {}^3D$	1—2
3499,67	12	24,37	27,91	$4s'\ {}^3D^\circ$—$4p'\ {}^3D$	1—1
3498,31	6	{ 26,60	30,14	$3d''\ {}^3P^\circ$—$4p''\ {}^3P$	0—1
		{ 26,57	30,12	$3d''\ {}^3P^\circ$—$4p''\ {}^3P$	1—0
3497,10	4	26,23	29,77	$3d''\ {}^3D^\circ$—$4p''\ {}^3D$	1—1
3484,12	3	26,23	29,79	$3d''\ {}^3D^\circ$—$4p''\ {}^3D$	1—2
3480,55	20	24,38	27,94	$4s'\ {}^3D^\circ$—$4p'\ {}^3D$	3—3
3472,61	6	24,38	27,94	$4s'\ {}^3D^\circ$—$4p'\ {}^3D$	2—3
3471,32	9	26,57	30,14	$3d''\ {}^3P^\circ$—$4p''\ {}^3P$	1—1
3438,04	8	26,57	30,18	$3d''\ {}^3P^\circ$—$4p''\ {}^3P$	1—2
3430,03	2	26,16	29,77	$3d''\ {}^3D^\circ$—$4p''\ {}^3D$	2—1
3424,25	9	26,53	30,14	$3d''\ {}^3P^\circ$—$4p''\ {}^3P$	2—1
3417,49	7	26,16	29,79	$3d''\ {}^3D^\circ$—$4p''\ {}^3D$	2—2
3413,53	6	26,16	29,79	$3d''\ {}^3D^\circ$—$4p''\ {}^3D$	2—3
3391,85	15	26,53	30,18	$3d''\ {}^3P^\circ$—$4p''\ {}^3P$	2—2
3361,28	7	24,38	28,06	$4s'\ {}^3D^\circ$—$4p'\ {}^3F$	2—2
3358,49	15	24,37	28,06	$4s'\ {}^3D^\circ$—$4p'\ {}^3F$	1—2
3352,11	4	24,38	28,08	$4s'\ {}^3D^\circ$—$4p'\ {}^3F$	3—3
3344,72	20	24,38	28,08	$4s'\ {}^3D^\circ$—$4p'\ {}^3F$	2—3
3336,13	25	24,38	28,10	$4s'\ {}^3D^\circ$—$4p'\ {}^3F$	3—4
3327,34	4	26,06	29,79	$3d''\ {}^3D^\circ$—$4p''\ {}^3D$	3—2
3323,59	9	26,06	29,79	$3d''\ {}^3D^\circ$—$4p''\ {}^3D$	3—3
3311,25	15	21,62	25,36	$4s\ {}^5S^\circ$—$4p\ {}^5P$	2—1
3301,88	20	21,62	25,37	$4s\ {}^5S^\circ$—$4p\ {}^5P$	2—2
3285,85	25	21,62	25,39	$4s\ {}^5S^\circ$—$4p\ {}^5P$	2—3
3187,90	6	26,23	30,12	$3d''\ {}^3D^\circ$—$4p''\ {}^3P$	1—0
3171,64	2	25,75	29,65	$4s''\ {}^3P^\circ$—$4p''\ {}^3S$	0—1
3157,42	5	25,73	29,65	$4s''\ {}^3P^\circ$—$4p''\ {}^3S$	1—1
3127,90	7	25,69	29,65	$4s''\ {}^3P^\circ$—$4p''\ {}^3S$	2—1
3110,41	7	26,16	30,14	$3d''\ {}^3D^\circ$—$4p''\ {}^3P$	2—1
3083,64	3	26,16	30,18	$3d''\ {}^3D^\circ$—$4p''\ {}^3P$	2—2
3078,15	10	25,75	29,77	$4s''\ {}^3P^\circ$—$4p''\ {}^3D$	0—1
3064,77	10	25,73	29,77	$4s''\ {}^3P^\circ$—$4p''\ {}^3D$	1—1
3054,82	12	25,73	29,79	$4s''\ {}^3P^\circ$—$4p''\ {}^3D$	1—2
3036,96	3	25,69	29,77	$4s''\ {}^3P^\circ$—$4p''\ {}^3D$	2—1

λ, Å	I	E_H, eV	E_B, eV	Transition	J
3027,16	5	25,69	29,79	$4s''\,{}^3P^\circ$—$4p''\,{}^3D$	2—2
3024,05	12	25,69	29,79	$4s''\,{}^3P^\circ$—$4p''\,{}^3D$	2—3
3010,02	10	26,06	30,18	$3d''\,{}^3D^\circ$—$4p''\,{}^3P$	3—2
2884,12	9	24,38	28,68	$4s'\,{}^3D^\circ$—$4p'\,{}^3P$	3—2
2878,72	5	24,38	28,68	$4s'\,{}^3D^\circ$—$4p'\,{}^3P$	2—2
2876,65	1	24,37	28,68	$4s'\,{}^3D^\circ$—$4p'\,{}^3P$	1—2
2855,29	8	24,38	28,72	$4s'\,{}^3D^\circ$—$4p'\,{}^3P$	2—1
2853,23	6	24,37	28,72	$4s'\,{}^3D^\circ$—$4p'\,{}^3P$	1—1
2842,88	7	24,37	28,73	$4s'\,{}^3D^\circ$—$4p'\,{}^3P$	1—0
2824,66	6	25,73	30,12	$4s''\,{}^3P^\circ$—$4p''\,{}^3P$	1—0
2818,26	6	25,75	30,14	$4s''\,{}^3P^\circ$—$4p''\,{}^3P$	0—1
2807,02	4	25,73	30,14	$4s''\,{}^3P^\circ$—$4p''\,{}^3P$	1—1
2785,23	5	25,73	30,18	$4s''\,{}^3P^\circ$—$4p''\,{}^3P$	1—2
2783,65	5	25,69	30,14	$4s''\,{}^3P^\circ$—$4p''\,{}^3P$	2—1
2762,23	7	25,69	30,18	$4s''\,{}^3P^\circ$—$4p''\,{}^3P$	2—2
2743,89	3	23,40	27,91	$3d'\,{}^3D^\circ$—$4p'\,{}^3D$	3—2
2724,84	10	23,40	27,94	$3d'\,{}^3D^\circ$—$4p'\,{}^3D$	3—3
2685,63	6	28,73	33,35	$4p'\,{}^3P$—$4d'\,{}^3D^\circ$	0—1
2678,38	9	23,29	27,91	$3d'\,{}^3D^\circ$—$4p'\,{}^3D$	2—2
2677,87	3	23,29	27,91	$3d'\,{}^3D^\circ$—$4p'\,{}^3D$	2—1
2676,46	4	28,72	33,35	$4p'\,{}^3P$—$4d'\,{}^3D^\circ$	1—1
2674,02	8	28,72	33,35	$4p'\,{}^3P$—$4d'\,{}^3D^\circ$	1—2
2660,22	3	23,29	27,94	$3d'\,{}^3D^\circ$—$4p'\,{}^3D$	2—3
2656,17	1	28,68	33,35	$4p'\,{}^3P$—$4d'\,{}^3D^\circ$	2—1
2654,63	10	28,68	33,35	$4p'\,{}^3P$—$4d'\,{}^3D^\circ$	2—3
2653,77	4	28,68	33,35	$4p'\,{}^3P$—$4d'\,{}^3D^\circ$	2—2
2645,47	2	23,40	28,08	$3d'\,{}^3D^\circ$—$4p'\,{}^3F$	3—3
2632,40	4	23,21	27,91	$3d'\,{}^3D^\circ$—$4p'\,{}^3D$	1—2
2631,90	7	23,21	27,91	$3d'\,{}^3D^\circ$—$4p'\,{}^3D$	1—1
2617,26	1	25,38	30,12	$3d'\,{}^3S^\circ$—$4p''\,{}^3P$	1—0
2613,95	3	23,17	27,91	$3d'\,{}^3F^\circ$—$4p'\,{}^3D$	2—2
2613,44	3	23,17	27,91	$3d'\,{}^3F^\circ$—$4p'\,{}^3D$	2—1
2602,12	1	25,38	30,14	$3d'\,{}^3S^\circ$—$4p''\,{}^3P$	1—1
2597,25	3	23,14	27,91	$3d'\,{}^3F^\circ$—$4p'\,{}^3D$	3—2
2594,41	1	23,29	28,08	$3d'\,{}^3D^\circ$—$4p'\,{}^3F$	2—3
2583,39	3	25,38	30,18	$3d'\,{}^3S^\circ$—$4p''\,{}^3P$	1—2
2580,17	2	23,14	27,94	$3d'\,{}^3F^\circ$—$4p'\,{}^3D$	3—3
2563,29	5	23,11	27,94	$3d'\,{}^3F^\circ$—$4p'\,{}^3D$	4—3
2533,92	3	23,17	28,06	$3d'\,{}^3F^\circ$—$4p'\,{}^3F$	2—2
2524,48	1	23,17	28,08	$3d'\,{}^3F^\circ$—$4p'\,{}^3F$	2—3
2518,26	2	23,14	28,06	$3d'\,{}^3F^\circ$—$4p'\,{}^3F$	3—2
2508,91	3	23,14	28,08	$3d'\,{}^3F^\circ$—$4p'\,{}^3F$	3—3
2506,69	5	28,72	33,66	$4p'\,{}^3P$—$4d'\,{}^3P^\circ$	1—2
2504,42	4	28,73	33,68	$4p'\,{}^3P$—$4d'\,{}^3P^\circ$	0—1
2499,96	1	23,14	28,10	$3d'\,{}^3F^\circ$—$4p'\,{}^3F$	3—4
2496,40	5	28,72	33,68	$4p'\,{}^3P$—$4d'\,{}^3P^\circ$	1—1
2494,90	6	28,72	33,68	$4p'\,{}^3P$—$4d'\,{}^3P^\circ$	1—0
2492,95	3	23,11	28,08	$3d'\,{}^3F^\circ$—$4p'\,{}^3F$	4—3
2488,86	12	28,68	33,66	$4p'\,{}^3P^\circ$—$4d'\,{}^3P^\circ$	2—2
2485,63	2	28,08	33,07	$4p'\,{}^3F$—$4d'\,{}^3F^\circ$	3—2
2484,87	2	28,10	33,09	$4p'\,{}^3F$—$4d'\,{}^3F^\circ$	4—3
2484,11	6	23,11	28,10	$3d'\,{}^3F^\circ$—$4p'\,{}^3F$	4—4
2479,76	3	28,73	33,73	$4p'\,{}^3P$—$4d'\,{}^3S^\circ$	0—1
2478,79	6	28,68	33,68	$4p'\,{}^3P$—$4d'\,{}^3P^\circ$	2—1
2476,55	6	28,06	33,07	$4p'\,{}^3F$—$4d'\,{}^3F^\circ$	2—2
2476,10	7	{ 28,73	33,74	$4p'\,{}^3P$—$5s'\,{}^3D^\circ$	0—1
		{ 28,08	33,09	$4p'\,{}^3F$—$4d'\,{}^3F^\circ$	3—3
2472,95	8	28,10	33,11	$4p'\,{}^3F$—$4d'\,{}^3F^\circ$	4—4

λ, Å	I	E_H, eV	E_B, eV	Transition	J
2471,92	6	28,72	33,73	$4p'\ ^3P-4d'\ ^3S^\circ$	1—1
2468,30	2	28,72	33,74	$4p'\ ^3P-5s'\ ^3D^\circ$	1—1
2467,10	3	28,06	33,09	$4p'\ ^3F-4d'\ ^3F^\circ$	2—3
2464,62	5	28,72	33,75	$4p'\ ^3P-5s'\ ^3D^\circ$	1—2
2464,26	4	28,08	33,11	$4p'\ ^3F-4d'\ ^3F^\circ$	3—4
2454,63	6	28,68	33,73	$4p'\ ^3P-4d'\ ^3S^\circ$	2—1
2447,43	1	28,68	33,75	$4p'\ ^3P-5s'\ ^3D^\circ$	2—2
2443,69	7	28,68	33,75	$4p'\ ^3P-5s'\ ^3D^\circ$	2—3
2427,20	4	28,10	33,21	$4p'\ ^3F-4d'\ ^3G^\circ$	4—4
2424,49	2	25,39	30,50	$4p\ ^5P-4d\ ^5D^\circ$	3—2
2424,27	6	25,39	30,50	$4p\ ^5P-4d\ ^5D^\circ$	3—3
2423,93	12	25,39	30,50	$4p\ ^5P-4d\ ^5D$	3—4
2423,52	12	28,10	30,21	$4p'\ ^3F-4d'\ ^3G^\circ$	4—5
2421,81	4	28,08	33,20	$4p'\ ^3F-4d'\ ^3G^\circ$	3—3
2418,82	10	28,08	33,21	$4p'\ ^3F-4d'\ ^3G^\circ$	3—4
2416,00	3	25,37	30,50	$4p\ ^5P-4d\ ^5D^\circ$	2—1
2415,84	4	25,37	30,50	$4p\ ^5P-4d\ ^5D^\circ$	2—2
2415,61	7	25,37	30,50	$4p\ ^5P-4d\ ^5D^\circ$	2—3
2413,20	10	28,06	33,20	$4p'\ ^3F-4d'\ ^3G^\circ$	2—3
2411,01	5	25,36	30,50	$4p\ ^5P-4d\ ^5D^\circ$	1—1
2410,80	4	25,36	30,50	$4p\ ^5P-4d\ ^5D^\circ$	1—2
2410,34	4	27,94	33,09	$4p'\ ^3D-4d'\ ^3F^\circ$	3—3
2404,98	6	27,91	33,07	$4p'\ ^3D-4d'\ ^3F^\circ$	1—2
2404,50	3	27,91	33,07	$4p'\ ^3D-4d'\ ^3F^\circ$	2—2
2399,15	12	27,94	33,11	$4p'\ ^3D-4d'\ ^3F^\circ$	3—4
2395,63	10	27,91	33,09	$4p'\ ^3D-4d'\ ^3F^\circ$	2—3
2360,26	9	28,10	33,35	$4p'\ ^3F-4d'\ ^3D^\circ$	4—3
2352,33	5	28,08	33,35	$4p'\ ^3F-4d'\ ^3D^\circ$	3—3
2351,67	7	28,08	33,35	$4p'\ ^3F-4d'\ ^3D^\circ$	3—2
2345,42	5	28,06	33,35	$4p'\ ^3F-4d'\ ^3D^\circ$	2—1
2345,17	9	23,40	28,68	$3d'\ ^3D^\circ-4p'\ ^3P$	3—2
2343,56	3	28,06	33,35	$4p'\ ^3F-4d'\ ^3D^\circ$	2—2
2319,37	10	25,93	31,27	$4p\ ^3P-4d\ ^3D^\circ$	2—2
2319,13	10	25,93	31,28	$4p\ ^3P-4d\ ^3D^\circ$	0—1
2318,35	2	25,93	31,28	$4p\ ^3P-4d\ ^3D^\circ$	2—1
2318,04	12	25,93	31,27	$4p\ ^3P-4d\ ^3D^\circ$	1—2
2317,47	15	25,93	31,28	$4p\ ^3P-4d\ ^3D^\circ$	2—3
2317,00	9	25,93	31,28	$4p\ ^3P-4d\ ^3D^\circ$	1—1
2302,92	6	25,93	31,31	$4p\ ^3P-5s\ ^3S^\circ$	0—1
2302,17	15	25,93	31,31	$4p\ ^3P-5s\ ^3S^\circ$	2—1
2300,85	10	25,93	31,31	$4p\ ^3P-5s\ ^3S^\circ$	1—1
2297,15	5	23,29	28,68	$3d'\ ^3D^\circ-4p'\ ^3P$	2—2
2296,24	4	24,38	29,77	$4s'\ ^3D^\circ-4p''\ ^3D$	2—1
2294,91	5	24,37	29,77	$4s'\ ^3D^\circ-4p''\ ^3D$	1—1
2294,05	3	24,38	29,79	$4s'\ ^3D^\circ-4p''\ ^3D$	3—2
2293,03	12	27,94	33,35	$4p'\ ^3D-4d'\ ^3D^\circ$	3—3
2292,39	5	27,94	33,35	$4p'\ ^3D-4d'\ ^3D^\circ$	3—2
2292,25	4	24,38	29,79	$4s'\ ^3D^\circ-4p''\ ^3D$	3—3
2290,61	6	24,38	29,79	$4s'\ ^3D^\circ-4p''\ ^3D$	2—2
2289,31	4	24,37	29,79	$4s'\ ^3D^\circ-4p''\ ^3D$	1—2
2288,82	2	24,38	29,79	$4s'\ ^3D^\circ-4p''\ ^3D$	2—3
2282,21	7	23,29	28,72	$3d'\ ^3D^\circ-4p'\ ^3P$	2—1
2281,22	7	27,91	33,35	$4p'\ ^3D-4d'\ ^3D$	1—1
2280,85	5	27,91	33,35	$4p'\ ^3D-4d'\ ^3D$	2—1
2279,68	4	27,91	33,35	$4p'\ ^3D-4d'\ ^3D^\circ$	2—3
2279,47	3	27,91	33,35	$4p'\ ^3D-4d'\ ^3D^\circ$	1—2
2279,10	10	27,91	33,35	$4p'\ ^3D-4d'\ ^3D^\circ$	2—2
2248,73	7	23,21	28,72	$3d'\ ^3D^\circ-4p'\ ^3P$	1—1

λ, Å	I	E_H, eV	E_B, eV	Transition	J
2242,29	6	23,21	28,73	$3d'\ ^3D°—4p'\ ^3P$	1—0
2192,06	15	28,10	33,75	$4p'\ ^3F—5s'\ ^3D°$	4—3
2188,22	10	28,08	33,75	$4p'\ ^3F—5s'\ ^3D°$	3—2
2184,06	8	28,06	33,74	$4p'\ ^3F—5s'\ ^3D°$	2—1
2177,22	25	25,39	31,08	$4p\ ^5P—5s\ ^5S°$	3—2
2170,23	20	25,37	31,08	$4p\ ^5P—5s\ ^5S°$	2—2
2168,26	10	27,94	33,66	$4p'\ ^3D—4d'\ ^3P°$	3—2
2166,19	15	25,36	31,08	$4p\ ^5P—5s\ ^5S°$	1—2
2157,53	3	24,37	30,12	$4s'\ ^3D°—4p''\ ^3P$	1—0
2156,38	3	27,91	33,66	$4p'\ ^3D—4d'\ ^3P°$	2—2
2149,07	3	27,91	33,68	$4p'\ ^3D—4d'\ ^3P°$	1—1
2148,73	8	27,91	33,68	$4p'\ ^3D—4d'\ ^3P°$	2—1
2148,38	5	24,38	30,14	$4s'\ ^3D°—4p''\ ^3P$	2—1
2147,95	6	27,91	33,68	$4p'\ ^3D—4d'\ ^3P°$	1—0
2138,59	10	24,38	30,18	$4s'\ ^3D°—4p''\ ^3P$	3—2
2136,73	3	27,94	33,75	$4p'\ ^3D—5s'\ ^3D°$	3—2
2133,87	15	27,94	33,75	$4p'\ ^3D—5s'\ ^3D°$	3—3
2128,22	6	27,91	33,74	$4p'\ ^3D—5s'\ ^3D°$	1—1
2127,89	3	27,91	33,74	$4p'\ ^3D—5s'\ ^3D°$	2—1
2125,50	3	27,91	33,75	$4p'\ ^3D—5s'\ ^3D°$	1—2
2125,16	10	27,91	33,75	$4p'\ ^3D—5s'\ ^3D°$	2—2
2122,34	8	27,91	33,75	$4p'\ ^3D—5s'\ ^3D°$	2—3
1973,780	4	22,40	28,63	$4s\ ^3S°—4p'\ ^3P$	1—2
1962,74	2	22,40	28,72	$4s\ ^3S°—4p'\ ^3P$	1—1
1957,83	1	22,40	28,73	$4s\ ^3S°—4p'\ ^3P$	1—0
1919,515	4	19,47	25,93	$3d\ ^3D°—4p\ ^3P$	1—1
1918,667	4	19,47	25,93	$3d\ ^3D°—4p\ ^3P$	1—2
1918,06	1	19,47	25,93	$3d\ ^3D°—4p\ ^3P$	1—0
1915,564	7	19,46	25,93	$3d\ ^3D°—4p\ ^3P$	2—1
1914,653	3	19,46	25,93	$3d\ ^3D°—4p\ ^3P$	2—2
1914,398	9	19,45	25,93	$3d\ ^3D°—4p\ ^3P$	3—2
1843,19	2	—	—	—	—
1839,43	3	—	—	—	—
1836,42	5	—	—	—	—
1675,637	4	17,96	25,36	$3d\ ^5D°—4p\ ^5P$	2—1
1675,484	7	17,96	25,36	$3d\ ^5D°—4p\ ^5P$	1, 0—1
1673,425	7	17,96	25,37	$3d\ ^5D°—4p\ ^5P$	3—2
1673,241	3	17,96	25,37	$3d\ ^5D°—4p\ ^5P$	2—2
1673,14	1	17,96	25,37	$3d\ ^5D°—4p\ ^5P$	1—2
1669,671	7	17,97	25,39	$3d\ ^5D°—4p\ ^5P$	4—3
1669,304	5	17,96	25,39	$3d\ ^5D°—4p\ ^5P$	3—3
1669,10	1	17,96	25,39	$3d\ ^5D°—4p\ ^5P$	2—3
1468,006	2	19,47	27,91	$3d\ ^3D°—4p'\ ^3D$	1—2
1467,841	3	19,47	27,91	$3d\ ^3D°—4p'\ ^3D$	1—1
1465,712	3	19,46	27,91	$3d\ ^3D°—4p'\ ^3D$	2—2
1465,532	2	19,46	27,91	$3d\ ^3D°—4p'\ ^3D$	{ 2—1 3—2
1460,234	2	19,46	27,94	$3d\ ^3D°—4p'\ ^3D$	2—3
1460,077	4	19,45	27,94	$3d\ ^3D°—4p'\ ^3D$	3—3
1205,95	1	4,12	14,23	$3p^4\ ^1S—3p^5\ ^3P°$	0—1
1002,095	3	1,74	14,11	$3p^4\ ^1D—3p^5\ ^3P°$	2—2
887,404	10	0,14	14,11	$3p^4\ ^3P—3p^5\ ^3P°$	1—2
883,179	9	0,19	14,23	$3p^4\ ^3P—3p^5\ ^3P°$	0—1
879,622	8	0,14	14,23	$3p^4\ ^3P—3p^5\ ^3P°$	1—1
878,728	12	0,00	14,11	$3p^4\ ^3P—3p^5\ ^3P°$	2—2
875,534	9	0,14	14,30	$3p^4\ ^3P—3p^5\ ^3P°$	1—0
871,099	10	0,00	14,23	$3p^4\ ^3P—3p^5\ ^3P°$	2—1
769,152	12	1,74	17,86	$3p^4\ ^1D—3p^5\ ^1P°$	2—1
699,72	1	1,74	19,45	$3p^4\ ^1D—3d\ ^3D°$	2—3

λ, Å	I	E_H, eV	E_B, eV	Transition	J
697,74	2	0,19	17,96	$3p^4\ ^3P—3d\ ^5D^\circ$	0—1
695,537	6	0,14	17,96	$3p^4\ ^3P—3d\ ^5D^\circ$	1—0, 1, 2
690,170	8	0,00	17,96	$3p^4\ ^3P—3d\ ^5D^\circ$	2—1, 2, 3
676,241	6	—	—	—	—
643,256	9	0,19	19,47	$3p^4\ ^3P—3d\ ^3D^\circ$	0—1
641,808	12	0,14	19,46	$3p^4\ ^3P—3d\ ^3D^\circ$	1—2
641,364	5	0,14	19,47	$3p^4\ ^3P—3d\ ^3D^\circ$	1—1
637,282	20	0,00	19,46	$3p^4\ ^3P—3d\ ^3D^\circ$	2—3, 2
636,818	3	0,00	19,47	$3p^4\ ^3P—3d\ ^3D^\circ$	2—1
623,767	5	—	—	—	—
604,152	10	—	—	—	—
579,212	3	1,74	23,14	$3p^4\ ^1D—3d'\ ^3F^\circ$	2—3
578,386	4	1,74	23,17	$3p^4\ ^1D—3d'\ ^3F^\circ$	2—2
577,153	3	0,14	21,62	$3p^4\ ^3P—4s\ ^5S^\circ$	1—2
573,468	4	0,00	21,62	$3p^4\ ^3P—4s\ ^5S^\circ$	2—2
558,321	5	0,19	22,40	$3p^4\ ^3P—4s\ ^3S^\circ$	0—1
556,893	6	0,14	22,40	$3p^4\ ^3P—4s\ ^3S^\circ$	1—1
553,470	9	0,00	22,40	$3p^4\ ^3P—4s\ ^3S^\circ$	2—1
538,788	6	0,19	23,21	$3p^4\ ^3P—3d'\ ^3D^\circ$	0—1
537,459	6	0,14	23,21	$3p^4\ ^3P—3d'\ ^3D^\circ$	1—1
536,745	8	—	—	—	—
535,580	7	0,14	23,29	$3p^4\ ^3P—3d'\ ^3D^\circ$	1—2
534,26	1	0,00	23,21	$3p^4\ ^3P—3d'\ ^3D^\circ$	2—1
532,413	7	0,00	23,29	$3p^4\ ^3P—3d'\ ^3D^\circ$	2—2
529,900	9	0,00	23,40	$3p^4\ ^3P—3d'\ ^3D^\circ$	2—3
512,769	7	0,19	24,37	$3p^4\ ^3P—4s'\ ^3D^\circ$	0—1
511,565	7	0,14	24,37	$3p^4\ ^3P—4s'\ ^3D^\circ$	1—1
511,497	8	0,14	24,38	$3p^4\ ^3P—4s'\ ^3D^\circ$	1—2
508,655	2	0,00	24,37	$3p^4\ ^3P—4s'\ ^3D^\circ$	2—2
508,595	4	—	—	—	—
508,434	9	0,00	24,38	$3p^4\ ^3P—4s'\ ^3D^\circ$	2—3
492,228	3	0,19	25,38	$3p^4\ ^3P—3d'\ ^3S^\circ$	0—1
491,121	4	0,14	25,38	$3p^4\ ^3P—3d'\ ^3S^\circ$	1—1
490,68	3	—	—	—	—
488,452	7	0,00	25,38	$3p^4\ ^3P—3d'\ ^3S^\circ$	2—1
487,988	7	—	—	—	
487,025	7	—	—	—	—
485,515	4	0,19	25,73	$3p^4\ ^3P—4s''\ ^3P^\circ$	0—1
485,150	6	0,14	25,69	$3p^4\ ^3P—4s''\ ^3P^\circ$	1—2
484,445	5	0,14	25,73	$3p^4\ ^3P—4s''\ ^3P^\circ$	1—1
484,116	5	0,14	25,75	$3p^4\ ^3P—4s''\ ^3P^\circ$	1—0
482,548	8	0,00	25,69	$3p^4\ ^3P—4s''\ ^3P^\circ$	2—2
481,848	6	0,00	25,73	$3p^4\ ^3P—4s''\ ^3P^\circ$	2—1
476,432	7	0,14	26,16	$3p^4\ ^3P—3d''\ ^3D^\circ$	1—2
473,918	6	0,00	26,16	$3p^4\ ^3P—3d''\ ^3D^\circ$	2—2
473,025	6	0,00	26,23	$3p^4\ ^3P—3d''\ ^3D^\circ$	2—1
469,968	4	0,19	26,57	$3p^4\ ^3P—3d''\ ^3D^\circ$	0—1
469,831	4	0,14	26,53	$3p^4\ ^3P—3d''\ ^3D^\circ$	1—2
468,956	3	0,14	26,57	$3p^4\ ^3P—3d''\ ^3D^\circ$	1—1
468,467	4	0,14	26,60	$3p^4\ ^3P—3d''\ ^3D^\circ$	1—0
467,390	6	0,00	26,53	$3p^4\ ^3P—3d''\ ^3D^\circ$	2—2
466,530	5	0,00	26,57	$3p^4\ ^3P—3d''\ ^3D^\circ$	2—1
398,86	1	0,00	31,08	$3p^4\ ^3P—5s\ ^5S^\circ$	2—2
397,67	1	0,14	31,31	$3p^4\ ^3P—5s\ ^3S^\circ$	1—1
396,38	4	0,00	31,28	$3p^4\ ^3P—4d\ ^3D^\circ$	2—3
395,92	1	0,00	31,31	$3p^4\ ^3P—5s\ ^3S^\circ$	2—1

Ar IV, ground state $1s^2\,2s^2\,2p^6\,3s^2\,3p^3\;{}^4S^0_{3/2}$

Ionization potential 482400 cm^{-1}; 59,806 eV

λ, Å	I	E_H, eV	E_B, eV	Transition	J
5838,01	—	—	—	—	—
5847,03	—	—	—	—	—
5830,04	—	—	—	—	—
5739,88	—	—	—	—	—
4922,50	—	—	—	—	—
4894,53	—	—	—	—	—
4871,78	—	—	—	—	—
4808,66	—	—	—	—	—
4697,87	—	—	—	—	—
4690,9	—	—	—	—	—
4639,36	—	—	—	—	—
4417,30	—	—	—	—	—
4364,80	—	—	—	—	—
4229,81	—	—	—	—	—
4182,97	—	—	—	—	—
4089,04	—	—	—	—	—
3908,43	—	—	—	—	—
3858,46	—	—	—	—	—
3800,42	—	—	—	—	—
3777,52	—	—	—	—	—
3750,79	—	—	—	—	—
3713,19	—	—	—	—	—
3692,5	—	—	—	—	—
3424,43	—	—	—	—	—
3393,35	—	—	—	—	—
3391,86	—	—	—	—	—
3324,78	—	—	—	—	—
3157,60	—	—	—	—	—
3134,90	3	31,91	35,86	$4s\,^2P$—$4p\,^4P^\circ$	$^3/_2$—$^3/_2$
3125,98	—	—	—	—	—
3077,40	8	31,91	35,93	$4s\,^2P$—$4p\,^4P^\circ$	$^3/_2$—$^5/_2$
3065,11	—	—	—	—	—
3039,75	—	—	—	—	—
3037,98	6	31,91	35,99	$4s\,^2P$—$4p\,^2D^\circ$	$^3/_2$—$^3/_2$
3016,15	5	31,75	35,86	$4s\,^2P$—$4p\,^4P^\circ$	$^1/_2$—$^3/_2$
2985,04	—	—	—	—	—
2926,33	11	31,75	35,99	$4s\,^2P$—$4p\,^2D^\circ$	$^1/_2$—$^3/_2$
2918,28	3	31,24	35,49	$4s\,^4P$—$4p\,^4D^\circ$	$^5/_2$—$^3/_2$
2913,00	12	31,91	36,16	$4s\,^2P$—$4p\,^2D^\circ$	$^3/_2$—$^5/_2$
2874,40	6	31,24	35,55	$4s\,^4P$—$4p\,^4D^\circ$	$^5/_2$—$^5/_2$
2851,94	4	31,11	35,45	$4s\,^4P$—$4p\,^4D^\circ$	$^3/_2$—$^1/_2$
2830,25	10	31,11	35,49	$4s\,^4P$—$4p\,^4D^\circ$	$^3/_2$—$^3/_2$
2809,44	16	31,24	35,65	$4s\,^4P$—$4p\,^4D^\circ$	$^5/_2$—$^7/_2$
2797,11	7	31,02	35,45	$4s\,^4P$—$4d\,^4D^\circ$	$^1/_2$—$^1/_2$
2788,96	14	31,11	35,55	$4s\,^4P$—$4p\,^4D^\circ$	$^3/_2$—$^5/_2$
2785,39	—	—	—	—	—
2784,47	12	33,24	37,70	$4s'\,^2D$—$4p'\,^2F^\circ$	$^3/_2$—$/^5_2$
2782,92	3	33,24	37,70	$4s'\,^2D$—$4p'\,^2F^\circ$	$^5/_2$—$^5/_2$
2776,26	10	31,02	34,49	$4s\,^4P$—$4p\,^4D^\circ$	$^1/_2$—$^3/_2$
2757,92	14	33,24	37,74	$4s'\,^2D$—$4p'\,^2F^\circ$	$^5/_2$—$^7/_2$
2682,63	9	31,24	35,86	$4s\,^4P$—$4p\,^4P^\circ$	$^5/_2$—$^3/_2$
2640,34	15	31,24	35,93	$4s\,^4P$—$4p\,^4P^\circ$	$^5/_2$—$^5/_2$
2626,32	2	33,24	37,97	$4s'\,^2D$—$4p'\,^2D^\circ$	$^3/_2$—$^5/_2$
2624,92	12	33,24	37,97	$4s'\,^2D$—$4p'\,^2D^\circ$	$^5/_2$—$^5/_2$
2621,36	12	33,24	37,98	$4s'\,^2D$—$4p'\,^2D^\circ$	$^3/_2$—$^3/_2$

λ, A	I	E_H, eV	E_B, eV	Transition	J
2619,98	6	33,24	37,98	$4s'\ ^2D—4p'\ ^2D°$	$^5/_2—^3/_2$
2615,68	12	31,11	35,85	$4s\ ^4P—4p\ ^4P°$	$^3/_2—^1/_2$
2611,24	3	31,24	35,99	$4s\ ^4P—4p\ ^2D°$	$^5/_2—^3/_2$
2608,44	7	31,91	36,66	$4s\ ^2P—4p\ ^2P$	$^3/_2—^1/_2$
2608,06	10	31,11	35,86	$4s\ ^4P—4p\ ^4P°$	$^3/_2—^3/_2$
2599,47	12	31,91	36,67	$4s\ ^2P—4p\ ^2P°$	$^3/_2—^3/_2$
2569,53	7	31,02	35,85	$4s\ ^4P—4p\ ^4P°$	$^1/_2—^1/_2$
2568,07	10	31,11	35,93	$4s\ ^4P—4p\ ^4P°$	$^3/_2—^5/_2$
2562,17	12	31,02	35,86	$4s\ ^4P—4p\ ^4P°$	$^1/_2—^3/_2$
2540,55	4	31,11	35,99	$4s\ ^4P—4p\ ^2D°$	$^3/_2—^3/_2$
2534,1	—	—	—	—	—
2525,69	9	31,75	36,66	$4s\ ^2P—4p\ ^2P°$	$^1/_2—^1/_2$
2518,40	6	31,24	36,16	$4s\ ^4P—4p\ ^2D°$	$^5/_2—^5/_2$
2517,28	5	31,75	36,67	$4s\ ^2P—4p\ ^2P°$	$^1/_2—^3/_2$
2513,28	12	31,24	36,17	$4s\ ^4P—4p\ ^4S°$	$^5/_2—^3/_2$
2496,93	3	31,02	35,99	$4s\ ^4P—4p\ ^2D°$	$^1/_2—^3/_2$
2467,3	—	—	—	—	—
2452,58	4	31,11	36,16	$4s\ ^4P—4p\ ^2D°$	$^3/_2—^5/_2$
2447,71	8	31,11	36,17	$4s\ ^4P—4p\ ^4S°$	$^3/_2—^3/_2$
2407,20	6	31,02	36,17	$4s\ ^4P—4p\ ^4S°$	$^1/_2—^3/_2$
2368,15	3	31,91	37,14	$4s\ ^2P—4p\ ^2S°$	$^3/_2—^1/_2$
2351,8	—	—	—	—	—
2345,4	—	—	—	—	—
2299,72	4	31,75	37,14	$4s\ ^2P—4p\ ^2S°$	$^1/_2—^1/_2$
1197,84	1	4,34	14,69	$3p^3\ ^2P°—3p^4\ ^4P$	$^3/_2—^3/_2$
1190,354	2	4,34	14,76	$3p^3\ ^2P°—3p^4\ ^4P$	$^3/_2—^1/_2$
1187,80	1	4,32	14,76	$3p^3\ ^2P°—3p^4\ ^4P$	$^1/_2—^1/_2$
1037,931	1	2,63	14,58	$3p^3\ ^2D°—3p^4\ ^4P$	$^5/_2—^5/_2$
901,804	2	4,34	18,09	$3p^3\ ^2P°—3p^4\ ^2D$	$^3/_2—^3/_2$
901,168	9	4,34	18,10	$3p^3\ ^2P°—3p^4\ ^2D$	$^3/_2—^5/_2$
900,362	5	4,32	18,09	$3p^3\ ^2P°—3p^4\ ^2D$	$^1/_2—^3/_2$
850,602	25	0,00	14,58	$3p^3\ ^4S°—3p^4\ ^4P$	$^3/_2—^5/_2$
843,772	20	0,00	14,69	$3p^3\ ^4S°—3p^4\ ^4P$	$^3/_2—^3/_2$
840,029	15	0,00	14,76	$3p^3\ ^4S°—3p^4\ ^4P$	$^3/_2—^1/_2$
801,913	5	2,63	18,09	$3p^3\ ^2D°—3p^4\ ^2D$	$^5/_2—^3/_2$
801,409	10	2,63	18,10	$3p^3\ ^2D°—3p^4\ ^2D$	$^5/_2—^5/_2$
801,086	10	2,61	18,09	$3p^3\ ^2D°—3p^4\ ^2D$	$^3/_2—^3/_2$
800,573	5	2,61	18,10	$3p^3\ ^2D°—3p^4\ ^2D$	$^3/_2—^5/_2$
761,470	5	4,34	20,62	$3p^3\ ^2P°—3p^4\ ^2P$	$^3/_2—^3/_2$
760,439	3	4,32	20,62	$3p^3\ ^2P°—3p^4\ ^2P$	$^1/_2—^3/_2$
755,212	3	4,34	20,76	$3p^3\ ^2P°—3p^4\ ^2P$	$^3/_2—^1/_2$
754,205	4	4,32	20,76	$3p^3\ ^2P°—3p^4\ ^2P$	$^1/_2—^1/_2$
700,277	8	4,34	22,05	$3p^3\ ^2P°—3p^4\ ^2S$	$^3/_2—^1/_2$
699,408	6	4,32	22,05	$3p^3\ ^2P°—3p^4\ ^2S$	$^1/_2—^1/_2$
689,007	12	2,63	20,62	$3p^3\ ^2D°—3p^4\ ^2P$	$^5/_2—^3/_2$
688,392	7	2,61	20,62	$3p^3\ ^2D°—3p^4\ ^2P$	$^3/_2—^3/_2$
683,278	10	2,61	20,76	$3p^3\ ^2D°—3p^4\ ^2P$	$^3/_2—^1/_2$
399,634	3	0,00	31,02	$3p^3\ ^4S°—4s\ ^4P$	$^3/_2—^1/_2$
398,546	4	0,00	31,11	$3p^3\ ^4S°—4s\ ^4P$	$^3/_2—^3/_2$
396,869	4	0,00	31,24	$3p^3\ ^4S°—4s\ ^4P$	$^3/_2—^5/_2$

Ar V, ground state $1s^2\,2s^2\,2p^6\,3s^2\,3p^2\,{}^3P_0$

Ionization potential 605100 cm⁻¹; 75,02 eV

λ, Å	I	E_H, eV	E_B, eV	Transition	J
836,126	2	0,25	15,08	$3p^2\,{}^3P$—$3p^3\,{}^3D^\circ$	2—1
835,792	1	0,25	15,09	$3p^2\,{}^3P$—$3p^3\,{}^3D^\circ$	2—2
834,878	4	0,25	15,10	$3p^2\,{}^3P$—$3p^3\,{}^3D^\circ$	2—3
827,349	3	0,09	15,08	$3p^2\,{}^3P$—$3p^3\,{}^3D^\circ$	1—1
827,055	5	0,09	15,09	$3p^2\,{}^3P$—$3p^3\,{}^3D^\circ$	1—2
822,159	4	0,00	15,08	$3p^2\,{}^3P$—$3p^3\,{}^3D^\circ$	0—1
715,645	3	0,25	17,58	$3p^2\,{}^3P$—$3p^3\,{}^3P^\circ$	2—2
715,599	4	0,25	17,58	$3p^2\,{}^3P$—$3p^3\,{}^3P^\circ$	2—1
709,195	5	0,09	17,58	$3p^2\,{}^3P$—$3p^3\,{}^3P^\circ$	1—1, 0
705,353	3	0,00	17,58	$3p^2\,{}^3P$—$3p^3\,{}^3P^\circ$	0—1
558,481	5	2,02	24,22	$3p^2\,{}^1D$—$3p^3\,{}^1P^\circ$	2—1
527,693	6	0,25	23,75	$3p^2\,{}^3P$—$3p^3\,{}^3S^\circ$	2—1
524,189	5	0,09	23,75	$3p^2\,{}^3P$—$3p^3\,{}^3S^\circ$	1—1
522,090	3	0,00	23,75	$3p^2\,{}^3P$—$3p^3\,{}^3S^\circ$	0—1
517,250	0	0,25	24,22	$3p^2\,{}^3P$—$3p^3\,{}^1P^\circ$	2—1
513,914	1	0,09	24,22	$3p^2\,{}^3P$—$3p^3\,{}^1P^\circ$	1—1
511,886	0	0,00	24,22	$3p^2\,{}^3P$—$3p^3\,{}^1P^\circ$	0—1
463,938	7	0,25	26,97	$3p^2\,{}^3P$—$3d\,{}^3P^\circ$	2—2
462,415	3	0,25	27,06	$3p^2\,{}^3P$—$3d\,{}^3P^\circ$	2—1
461,227	6	0,09	26,97	$3p^2\,{}^3P$—$3d\,{}^3P^\circ$	1—2
459,728	1	0,09	27,06	$3p^2\,{}^3P$—$3d\,{}^3P^\circ$	1—1
458,975	2	0,09	27,11	$3p^2\,{}^3P$—$3d\,{}^3P^\circ$	1—0
458,121	3	0,00	27,06	$3p^2\,{}^3P$—$3d\,{}^3P^\circ$	0—1
450,079	1	0,25	27,80	$3p^2\,{}^3P$—$3d\,{}^3D^\circ$	2—1
449,493	4	0,25	27,83	$3p^2\,{}^3P$—$3d\,{}^3D^\circ$	2—2
449,065	18	0,25	27,86	$3p^2\,{}^3P$—$3d\,{}^3D^\circ$	2—3
447,527	4	0,09	27,80	$3p^2\,{}^3P$—$3d\,{}^3D^\circ$	1—1
446,949	8	0,09	27,83	$3p^2\,{}^3P$—$3d\,{}^3D^\circ$	1—2
445,997	5	0,00	27,80	$3p^2\,{}^3P$—$3d\,{}^3D^\circ$	0—1
350,878	3	2,02	37,35	$3p^2\,{}^1D$—$4s\,{}^1P^\circ$	2—1
339,886	3	0,25	36,73	$3p^2\,{}^3P$—$4s\,{}^3P^\circ$	2—1
339,009	3	0,09	36,67	$3p^2\,{}^3P$—$4s\,{}^3P^\circ$	1—0
338,426	2	0,09	36,73	$3p^2\,{}^3P$—$4s\,{}^3P^\circ$	1—1
337,998	6	0,25	36,93	$3p^2\,{}^3P$—$4s\,{}^3P^\circ$	2—2
337,555	3	0,00	36,73	$3p^2\,{}^3P$—$4s\,{}^3P^\circ$	0—1
336,555	3	0,09	36,93	$3p^2\,{}^3P$—$4s\,{}^3P^\circ$	1—2

Ar VI, ground state $1s^2\,2s^2\,2p^6\,3s^2\,3p\,{}^2P^0_{1/2}$

Ionization potential 736 600 cm⁻¹; 91,32 eV

λ, Å	I	E_H, eV	E_B, eV	Transition	J
767,71	1	0,27	16,42	$3p\,{}^2P^\circ$—$3p^2\,{}^2D$	3/2—3/2
767,06	2	0,27	16,44	$3p\,{}^2P^\circ$—$3p^2\,{}^2D$	3/2—5/2
754,93	1	0,00	16,42	$3p\,{}^3P^\circ$—$3p^2\,{}^2D$	1/2—3/2
596,694	4	0,27	21,05	$3p\,{}^2P^\circ$—$3p^2\,{}^2S$	3/2—1/2
594,096	3	12,65	33,52	$3p^2\,{}^4P$—$3p^3\,{}^4S^\circ$	5/2—3/2
589,783	2	12,50	33,52	$3p^2\,{}^4P$—$3p^3\,{}^4S^\circ$	3/2—3/2
588,921	5	0,00	21,05	$3p\,{}^2P^\circ$—$3p^2\,{}^2S$	1/2—1/2
587,006	1	12,40	33,52	$3p^2\,{}^4P$—$3p^3\,{}^4S^\circ$	1/2—3/2
555,639	4	0,27	22,59	$3p\,{}^2P^\circ$—$3p^2\,{}^2P$	3/2—1/2
551,371	8	0,27	22,76	$3p\,{}^2P^\circ$—$3p^2\,{}^2P$	3/2—3/2

λ, Å	I	E_H, eV	E_B, eV	Transition	J
548,905	5	0,00	22,59	$3p\ ^2P^\circ$—$3p^2\ ^2P$	$^1/_2$—$^1/_2$
544,731	4	0,00	22,76	$3p\ ^2P^\circ$—$3p^2\ ^2P$	$^1/_2$—$^3/_2$
466,932	4	12,65	39,20	$3p^2\ ^4P$—$3d\ ^4P^\circ$	$^5/_2$—$^5/_2$
465,586	2	12,65	39,28	$3p^2\ ^4P$—$3d\ ^4P^\circ$	$^5/_2$—$^3/_2$
464,257	4	12,50	39,20	$3p^2\ ^4P$—$3d\ ^4P^\circ$	$^3/_2$—$^5/_2$
462,146	4	0,27	27,10	$3p\ ^2P^\circ$—$3d\ ^2D$	$^3/_2$—$^3/_2$
462,007	25	0,27	27,11	$3p\ ^2P^\circ$—$3d\ ^2D$	$^3/_2$—$^5/_2$
461,898	1	12,50	39,34	$3p^2\ ^4P$—$3d\ ^4P^\circ$	$^3/_2$—$^1/_2$
461,227	6	12,40	39,28	$3p^2\ ^4P$—$3d\ ^4P^\circ$	$^1/_2$—$^3/_2$
460,202	1	12,40	39,34	$3p^2\ ^4P$—$3d\ ^4P^\circ$	$^1/_2$—$^1/_2$
460,058	1	12,65	39,60	$3p^2\ ^4P$—$3d\ ^4D^\circ$	$^5/_2$—$^3/_2$
459,603	3	12,65	39,62	$3p^2\ ^4P$—$3d\ ^4D^\circ$	$^5/_2$—$^5/_2$
459,320	10	12,65	39,64	$3p^2\ ^4P$—$3d\ ^4D^\circ$	$^5/_2$—$^7/_2$
458,039	1	12,50	39,56	$3p^2\ ^4P$—$3d\ ^4D^\circ$	$^3/_2$—$^1/_2$
457,475	20	{ 0,00	27,10	$3p\ ^2P^\circ$—$3d\ ^2D$	$^1/_2$—$^3/_2$
		12,50	39,60	$3p^2\ ^4P$—$3d\ ^4D^\circ$	$^3/_2$—$^3/_2$
457,007	5	12,50	39,62	$3p^2\ ^4P$—$3d\ ^4D^\circ$	$^3/_2$—$^5/_2$
456,375	3	12,40	39,56	$3p^2\ ^4P$—$3d\ ^4D^\circ$	$^1/_2$—$^1/_2$
455,813	2	12,40	39,60	$3p^2\ ^4P$—$3d\ ^4D^\circ$	$^1/_2$—$^3/_2$
294,052	6	0,27	42,44	$3p\ ^2P^\circ$—$4s\ ^2S$	$^3/_2$—$^1/_2$
292,154	5	0,00	42,44	$3p\ ^2P^\circ$—$4s\ ^2S$	$^1/_2$—$^1/_2$
283,164	3	12,50	56,28	$3p^2\ ^4P$—$4s\ ^4P^\circ$	$^3/_2$—$^1/_2$
282,556	1	12,50	56,37	$3p^2\ ^4P$—$4s\ ^4P^\circ$	$^3/_2$—$^3/_2$
282,423	6	12,65	56,55	$3p^2\ ^4P$—$4s\ ^4P^\circ$	$^5/_2$—$^5/_2$
281,915	4	12,40	56,37	$3p^2\ ^4P$—$4s\ ^4P^\circ$	$^1/_2$—$^3/_2$
281,433	3	12,50	56,55	$3p^2\ ^4P$—$4s\ ^4P^\circ$	$^3/_2$—$^5/_2$
220,946	5	0,27	56,39	$3p\ ^2P^\circ$—$4d\ ^2D$	$^3/_2$—$^5/_2$
219,896	3	0,00	56,38	$3p\ ^2P^\circ$—$4d\ ^2D$	$^1/_2$—$^3/_2$
180,719	3	0,27	68,88	$3p\ ^3P^\circ$—$5d\ ^2D$	$^3/_2$—$^5/_2$
180,074	2	0,00	68,85	$3p\ ^2P^\circ$—$5d\ ^2D$	$^1/_2$—$^3/_2$

Ar VII, ground state $1s^2\ 2s^2\ 2p^6\ 3s^2\ ^1S_0$

Ionization potential 1 000 400 cm^{-1}; 124,03 eV

λ, Å	I	E_H, eV	E_B, eV	Transition	J
644,388	2	14,33	33,57	$3p\ ^3P^\circ$—$3p^2\ ^3P$	2—1
641,318	2	14,12	33,45	$3p\ ^3P^\circ$—$3p^2\ ^3P$	1—0
637,466	1	14,12	33,57	$3p\ ^3P^\circ$—$3p^2\ ^3P$	1—1
637,052	4	14,33	33,79	$3p\ ^3P^\circ$—$3p^2\ ^3P$	2—2
634,208	2	14,02	33,57	$3p\ ^3P^\circ$—$3p^2\ ^3P$	0—1
630,306	2	14,12	33,79	$3p\ ^3P^\circ$—$3p^2\ ^3P$	1—2
585,754	15	0,00	21,17	$3s^2\ ^1S$—$3p\ ^1P^\circ$	0—1
479,485	2	14,33	40,19	$3p\ ^3P^\circ$—$3d\ ^3D$	2—2
479,379	12	14,33	40,19	$3p\ ^3P^\circ$—$3d\ ^3D$	2—3
475,733	2	14,12	40,18	$3p\ ^3P^\circ$—$3d\ ^3D$	1—1
475,656	8	14,12	40,19	$3p\ ^3P^\circ$—$3d\ ^3D$	1—2
463,938	4	14,02	40,18	$3p\ ^3P^\circ$—$3d\ ^3D$	0—1
297,701	6	40,19	81,84	$3p\ ^3D$—$4f\ ^3F^\circ$	3—4, 3, 2
297,658	4	40,19	81,84	$3p\ ^3D$—$4f\ ^3F^\circ$	2—3, 2
297,621	3	40,18	81,84	$3p\ ^3D$—$4f\ ^3F^\circ$	1—2
250,940	7	14,33	63,73	$3p\ ^3P^\circ$—$4s\ ^3S$	2—1
249,886	5	14,12	63,73	$3p\ ^3P^\circ$—$4s\ ^3S$	1—1
249,384	2	14,02	63,73	$3p\ ^3P^\circ$—$4s\ ^3S$	0—1

λ, Å	I	E_H, eV	E_B, eV	Transition	J
192,635	7	14,33	78,69	$3p\ ^3P^\circ$—$4d\ ^3D$	2—3
192,041	5	14,12	78,68	$3p\ ^3P^\circ$—$4d\ ^3D$	1—2
191,759	3	14,02	78,67	$3p\ ^3P^\circ$—$4d\ ^3D$	0—1
176,566	10	0,00	70,22	$3s^2\ ^1S$—$4p\ ^1P^\circ$	0—1
152,259	3	14,33	95,75	$3p\ ^3P^\circ$—$5d\ ^3D$	2—3
151,876	2	14,12	95,75	$3p\ ^3P^\circ$—$5d\ ^3D$	1—2
151,698	1	14,02	95,75	$3p\ ^3P^\circ$—$5d\ ^3D$	0—1

Ar VIII, ground state $1s^2\ 2s^2\ 2p^6\ 3s\ ^2S_{1/2}$
Ionization potential 1 157 400 cm^{-1}; 143,49 eV

λ, Å	I	E_H, eV	E_B, eV	Transition	J
713,81	5	0,00	17,36	$3s\ ^2S$—$3p\ ^2P^\circ$	$^1/_2$—$^1/_2$
700,24	10	0,00	17,70	$3s\ ^2S$—$3p\ ^2P^\circ$	$^1/_2$—$^3/_2$
526,870	1	17,70	41,23	$3p\ ^2P^\circ$—$3d\ ^2D$	$^3/_2$—$^3/_2$
526,457	5	17,70	41,25	$3p\ ^2P^\circ$—$3d\ ^2D$	$^3/_2$—$^5/_2$
519,429	3	17,36	41,23	$3p\ ^2P^\circ$—$3d\ ^2D$	$^1/_2$—$^3/_2$
338,222	0	41,23	77,89	$3d\ ^2D$—$4p\ ^2P^\circ$	$^3/_2$—$^1/_2$
337,257	2	41,25	78,01	$3d\ ^2D$—$4p\ ^2P^\circ$	$^5/_2$—$^3/_2$
260,332	6	41,25	88,87	$3d\ ^2D$—$4f\ ^2F^\circ$	$^5/_2$—$^7/_2$
260,253	4	41,23	88,87	$3d\ ^2D$—$4f\ ^2F^\circ$	$^3/_2$—$^5/_2$
230,875	7	17,70	71,40	$3p\ ^3P^\circ$—$4s\ ^2S$	$^3/_2$—$^1/_2$
229,437	5	17,36	71,40	$3p\ ^3P^\circ$—$4s\ ^2S$	$^1/_2$—$^1/_2$
184,315	5	41,25	108,51	$3d\ ^2D$—$5f\ ^2F^\circ$	$^5/_2$—$^7/_2$
184,273	3	41,23	108,51	$3d\ ^2D$—$5f\ ^2F^\circ$	$^3/_2$—$^5/_2$
180,254	15	17,70	86,48	$3p\ ^2P^\circ$—$4d\ ^2D$	$^3/_2$—$^5/_2$
179,400	10	17,36	86,47	$3p\ ^2P^\circ$—$4d\ ^2D$	$^1/_2$—$^3/_2$
159,175	5	0,00	77,89	$3s\ ^2S$—$4p\ ^2P^\circ$	$^1/_2$—$^1/_2$
158,923	8	0,00	78,01	$3s\ ^2S$—$4p\ ^2P^\circ$	$^1/_2$—$^3/_2$
149,333	3	17,70	100,72	$3p\ ^2P^\circ$—$5s\ ^2S$	$^3/_2$—$^1/_2$
148,725	2	17,36	100,72	$3p\ ^2P^\circ$—$5s\ ^2S$	$^1/_2$—$^1/_2$
138,440	5	17,70	107,25	$3p\ ^2P^\circ$—$5d\ ^2D$	$^3/_2$—$^5/_2$
137,926	3	17,36	107,25	$3p\ ^2P^\circ$—$5d\ ^2D$	$^1/_2$—$^3/_2$
123,033	1	17,70	118,47	$3p\ ^2P^\circ$—$6d\ ^2D$	$^3/_2$—$^5/_2$
122,624	0	17,36	118,47	$3p\ ^2P^\circ$—$6d\ ^2D$	$^1/_2$—$^3/_2$
120,157	1	0,00	103,18	$3s\ ^2S$—$5p\ ^2P^\circ$	$^1/_2$—$^1/_2$
120,093	2	0,00	103,23	$3s\ ^2S$—$5p\ ^2P^\circ$	$^1/_2$—$^3/_2$

POTASSIUM, Z = 19

K I, ground state $1s^2\,2s^2\,2p^6\,3s^2\,3p^6\,4s\,^2S_{1/2}$

Ionization potential 35 009,78 **cm⁻¹**; 4,340 **eV**

λ, Å	I	E_H, eV	E_B, eV	Transition	J
85100	10	3,60	3,74	$6p\,^2P^\circ$—$5d\,^2D$	3/2—3/2, 5/2
84520	10	3,60	3,74	$6p\,^2P^\circ$—$5d\,^2D$	1/2—3/2
74260	10	3,80	3,96	$5g\,^2G$—$6h\,^2H^\circ$	—
		3,79	3,96	$5f\,^2F^\circ$—$6g\,^2G$	—
64610	10	3,40	3,60	$6s\,^2S$—$6p\,^2P^\circ$	1/2—1/2
64310	10	3,40	3,60	$6s\,^2S$—$6p\,^2P^\circ$	1/2—3/2
62360	20	3,40	3,60	$3d\,^2D$—$6p\,^2P^\circ$	3/2—1/2
62030	20	3,40	3,60	$3d\,^3D$—$6p\,^2P^\circ$	1/2, 3/2—3/2
40115,5	60	—	—	—	—
37370,7	10	—	—	—	—
37354,3	40	3,06	3,40	$5p\,^2P^\circ$—$4d\,^2D$	3/2—3/2, 5/2
37075,6	30	3,06	3,40	$5p\,^2P^\circ$—$4d\,^2D$	1/2—3/2
36626,4	30	3,06	3,40	$5p\,^2P^\circ$—$6s\,^2S$	1/2—1/2
36372,7	10	3,06	3,40	$5p\,^2P^\circ$—$6s\,^2S$	3/2—1/2
31596,8	40	2,67	3,06	$3d\,^2D$—$5p\,^2P^\circ$	3/2—1/2
31395	80	2,67	3,06	$3d\,^2D$—$5p\,^2P^\circ$	3/2, 5/2—3/2
27215,0	10	2,61	3,06	$5s\,^2S$—$5p\,^2P^\circ$	1/2—1/2
27065,6	20	2,61	3,06	$5s\,^3S$—$5p\,^2P^\circ$	1/2—3/2
15168,40	16	2,67	3,49	$3d\,^2D$—$4f\,^2F^\circ$	3/2—5/2
15163,08	—	2,67	3,49	$3d\,^2D$—$4f\,^2F^\circ$	5/2—7/2
13397,09	—	2,67	3,60	$3d\,^2D$—$6p\,^2P^\circ$	3/2—1/2
13377,86	—	2,67	3,60	$3d\,^2D$—$6p\,^2P^\circ$	5/2—3/2
12522,11	15	1,62	2,61	$4p\,^2P^\circ$—$5s\,^2S$	3/2—1/2
12432,24	16	1,61	2,61	$4p\,^2P^\circ$—$5s\,^2S$	1/2—1/2
11772,83	17	1,62	2,67	$4p\,^2P^\circ$—$3d\,^2D$	3/2—5/2
11769,62	16	1,62	2,67	$4p\,^2P^\circ$—$3d\,^2D$	3/2—3/2
11690,21	17	1,61	2,67	$4p\,^2P^\circ$—$3d\,^2D$	1/2—3/2
11022,67	16	2,67	3,80	$3d\,^2D$—$5f\,^2F^\circ$	3/2—5/2
11019,87	17	2,67	3,80	$3d\,^2D$—$5f\,^2F^\circ$	5/2—7/2
10487,11	8	2,67	3,85	$3d\,^2D$—$7p\,^2P^\circ$	3/2—1/2
10482,15	5	2,67	3,85	$3d\,^2D$—$7p\,^2P^\circ$	3/2—3/2
10479,63	9	2,67	3,85	$3d\,^2D$—$7p\,^2P^\circ$	5/2—3/2
9954,141	5	2,61	3,85	$5s\,^2S$—$7p\,^2P^\circ$	1/2—1/2
9949,668	6	2,61	3,85	$5s\,^2S$—$7p\,^2P^\circ$	1/2—3/2
9597,829	14	2,67	3,96	$3d\,^2D$—$6f\,^2F^\circ$	3/2—5/2
9595,704	15	2,67	3,96	$3d\,^2D$—$6f\,^2F^\circ$	5/2—7/2
9351,590	6	2,67	4,00	$3d\,^2D$—$8p\,^2P^\circ$	3/2—1/2
9349,248	3	2,67	4,00	$3d\,^2D$—$8p\,^2P^\circ$	3/2—3/2
9347,235	7	2,67	4,00	$3d\,^2D$—$8p\,^2P^\circ$	5/2—3/2
8925,436	4	2,61	4,00	$5s\,^2S$—$8p\,^2P^\circ$	1/2—1/2
8923,312	5	2,61	4,00	$5s\,^2S$—$8p\,^2P^\circ$	1/2—3/2
8904,017	12	2,67	4,06	$3d\,^2D$—$7f\,^2F^\circ$	3/2—5/2
8902,188	13	2,67	4,06	$3d\,^2D$—$7f\,^2F^\circ$	5/2—7/2
8767,053	3	2,67	4,08	$3d\,^2D$—$9p\,^2P^\circ$	3/2—1/2
8763,955	4	2,67	4,08	$3d\,^2D$—$9p\,^2P^\circ$	5/2—3/2
8505,112	10	2,67	4,13	$3d\,^2D$—$8f\,^2F^\circ$	3/2—5/2
8503,449	11	2,67	4,13	$3d\,^2D$—$8f\,^2F^\circ$	5/2—7/2
8419,996	1	2,67	4,14	$3d\,^2D$—$10p\,^2P^\circ$	3/2—1/2
8417,535	2	2,67	4,14	$3d\,^2D$—$10p\,^2P^\circ$	5/2—3/2
8391,44	—	2,61	4,08	$5s\,^2S$—$9p\,^2P^\circ$	1/2—1/2
8390,223	3	2,61	4,08	$5s\,^2S$—$9p\,^2P^\circ$	1/2—3/2
8251,743	8	2,67	4,17	$3d\,^2D$—$9f\,^2F^\circ$	3/2—5/2
8250,180	9	2,67	4,17	$3d\,^2D$—$9f\,^2F^\circ$	5/2—7/2
8079,618	6	2,67	4,20	$3d\,^2D$—$10f\,^2F^\circ$	3/2—5/2

λ, Å	I	E_H, eV	E_B, eV	Transition	J
8078,114	7	2,67	4,20	$3d\ ^2D—10f\ ^2F^\circ$	$^5/_2—^7/_2$
7956,832	4	2,67	4,23	$3d\ ^2D—11f\ ^2F^\circ$	$^3/_2—^5/_2$
7955,371	5	2,67	4,23	$3d\ ^2D—11f\ ^2F^\circ$	$^5/_2—^7/_2$
7698,959	24	0,00	1,61	$4s\ ^2S—4p\ ^2P^\circ$	$^1/_2—^1/_2$
7664,899	25	0,00	1,62	$4s\ ^2S—4p\ ^2P^\circ$	$^1/_2—^3/_2$
6964,672	12	1,62	3,40	$4p\ ^2P^\circ—4d\ ^2D$	$^3/_2—^5/_2$
6964,18	7	1,62	3,40	$4p\ ^2P^\circ—4d\ ^2D$	$^3/_2—^3/_2$
6938,767	20	1,62	3,40	$4p\ ^2P^\circ—6s\ ^2S$	$^3/_2—^1/_2$
6936,284	12	1,61	3,40	$4p\ ^2P^\circ—4d\ ^2D$	$^1/_2—^3/_2$
6911,084	19	1,61	3,40	$4p\ ^2P^\circ—6s\ ^2S$	$^1/_2—^1/_2$
5831,887	17	1,62	3,74	$4p\ ^2P^\circ—5d\ ^2D$	$^3/_2—^5/_2$
5812,148	15	1,61	3,74	$4p\ ^2P^\circ—5d\ ^2D$	$^1/_2—^3/_2$
5801,752	17	1,62	3,75	$4p\ ^2P^\circ—7s\ ^2S$	$^3/_2—^1/_2$
5782,384	16	1,61	3,75	$4p\ ^2P^\circ—7s\ ^2S$	$^1/_2—^1/_2$
5359,574	14	1,62	3,93	$4p\ ^2P^\circ—6d\ ^2D$	$^3/_2—^5/_2$
5342,970	12	1,61	3,93	$4p\ ^2P^\circ—6d\ ^2D$	$^1/_2—^3/_2$
5339,688	13	1,62	3,94	$4p\ ^2P^\circ—8s\ ^2S$	$^3/_2—^1/_2$
5323,276	12	1,61	3,94	$4p\ ^2P^\circ—8s\ ^2S$	$^1/_2—^1/_2$
5112,249	12	1,62	4,04	$4p\ ^2P^\circ—7d\ ^2D$	$^3/_2—^5/_2$
5099,200	11	1,62	4,05	$4p\ ^2P^\circ—9s\ ^2S$	$^3/_2—^1/_2$
5097,171	11	1,61	4,04	$4p\ ^2P^\circ—7d\ ^2D$	$^1/_2—^3/_2$
5084,226	10	1,61	4,05	$4p\ ^2P^\circ—9s\ ^2S$	$^1/_2—^1/_2$
4965,031	10	1,62	4,11	$4p\ ^2P^\circ—8d\ ^2D$	$^3/_2—^5/_2$
4956,146	9	1,62	4,12	$4p\ ^2P^\circ—10s\ ^2S$	$^3/_2—^1/_2$
4950,815	9	1,61	4,11	$4p\ ^2P^\circ—8d\ ^2D$	$^1/_2—^3/_2$
4942,015	8	1,61	4,12	$4p\ ^2P^\circ—10s\ ^2S$	$^1/_2—^1/_2$
4869,757	9	1,62	4,16	$4p\ ^2P^\circ—9d\ ^2D$	$^3/_2—^5/_2$
4863,483	8	1,62	4,17	$4p\ ^2P^\circ—11s\ ^2S$	$^3/_2—^1/_2$
4856,090	8	1,61	4,16	$4p\ ^2P^\circ—9d\ ^2D$	$^1/_2—^3/_2$
4849,865	7	1,61	4,17	$4p\ ^2P^\circ—11s\ ^2S$	$^1/_2—^1/_2$
4804,348	8	1,62	4,20	$4p\ ^2P^\circ—10d\ ^2D$	$^3/_2—^5/_2$
4799,754	6	1,62	4,20	$4p\ ^2P^\circ—12s\ ^2S$	$^3/_2—^1/_2$
4791,049	7	1,61	4,20	$4p\ ^2P^\circ—10d\ ^2D$	$^1/_2—^3/_2$
4786,491	5	1,61	4,20	$4p\ ^2P^\circ—12s\ ^2S$	$^1/_2—^1/_2$
4757,389	7	1,62	4,22	$4p\ ^2P^\circ—11d\ ^2D$	$^3/_2—^5/_2$
4753,934	5	1,62	4,22	$4p\ ^2P^\circ—13s\ ^2S$	$^3/_2—^1/_2$
4744,345	6	1,61	4,22	$4p\ ^2P^\circ—11d\ ^2D$	$^1/_2—^3/_2$
4740,914	4	1,61	4,22	$4p\ ^2P^\circ—13s\ ^2S$	$^1/_2—^1/_2$
4642,373	11	0,00	2,67	$4s\ ^2S—3d\ ^2D$	$^1/_2—^5/_2$
4641,876	10	0,00	2,67	$4s\ ^2S—3d\ ^2D$	$^1/_2—^3/_2$
4047,206	17	0,00	3,06	$4s\ ^2S—5p\ ^2P^\circ$	$^1/_2—^1/_2$
4044,136	18	0,00	3,07	$4s\ ^2S—5p\ ^2P^\circ$	$^1/_2—^3/_2$
3648,981	4	0,00	3,40	$4s\ ^2S—4d\ ^2D$	$^1/_2—^5/_2$
3648,841	3	0,00	3,40	$4s\ ^2S—4d\ ^2D$	$^1/_2—^3/_2$
3447,375	10	0,00	3,60	$4s\ ^2S—6p\ ^2P^\circ$	$^1/_2—^1/_2$
3446,372	11	0,00	3,60	$4s\ ^2S—6p\ ^2P^\circ$	$^1/_2—^3/_2$
3217,621	6	0,00	3,85	$4s\ ^2S—7p\ ^2P^\circ$	$^1/_2—^1/_2$
3217,155	7	0,00	3,85	$4s\ ^2S—7p\ ^2P^\circ$	$^1/_2—^3/_2$
3102,043	3	0,00	4,00	$4s\ ^2S—8p\ ^2P^\circ$	$^1/_2—^1/_2$
3101,790	4	0,00	4,00	$4s\ ^2S—8p\ ^2P^\circ$	$^1/_2—^3/_2$
3034,920	2	0,00	4,08	$4s\ ^2S—9p\ ^2P^\circ$	$^1/_2—^1/_2$
3034,761	1	0,00	4,08	$4s\ ^2S—9p\ ^2P^\circ$	$^1/_2—^3/_2$
2992,223	1	0,00	4,14	$4s\ ^2S—10p\ ^2P^\circ$	$^1/_2—^1/_2$
2992,118	1	0,00	4,14	$4s\ ^2S—10p\ ^2P^\circ$	$^1/_2—^3/_2$

K II, ground state $1s^2\,2s^2\,2p^6\,3s^2\,3p^6\ {}^1S_0$

Ionization potential 256 637 cm^{-1}; 31,817 eV

λ, Å	I	E_H, eV	E_B, eV	Transition	J
6595,00	2	21,27	23,14	3d [2½]°—4p [2½]	2—2
6427,96	5	21,18	23,11	3d [3½]°—4p [2½]	3—3
6307,29	7	21,18	23,14	3d [3½]°—4p [2½]	3—2
6246,59	6	21,27	23,25	3d [3½]°—4p [1½]	2—1
6120,27	8	—	—	—	—
6012,41	1	21,27	23,33	3d [2½]°—4p [1½]	2—2
5969,64	2	20,64	22,71	4s′ [½]°—4p [½]	1—1
5772,32	4	21,18	23,33	3d [3½]°—4p [1½]	3—2
5642,73	5	21,27	23,46	3d [2½]°—4p′ [1½]	2—1
5536,01	3	20,47	22,71	4s′ [½]°—4p [½]	0—1
5512,69	2	21,27	23,51	3d [2½]°—4p′ [1½]	2—2
5488,06	2	24,15	26,41	4p′ [½]—5s [1½]°	0—1
5470,13	6	20,45	22,71	3d [1½]—4p [½]	2—1
5310,24	5	21,18	23,51	3d [3½]°—4p′ [1½]	3—2
5056,27	7	20,26	22,71	3d [½]°—4p [½]	0—1
5017,34	1	20,64	23,11	4s′ [½]°—4p [2½]	1—3
5005,60	8	20,24	22,71	4s [1½]°—4p [½]	1—1
4943,24	6	20,64	23,14	4s′ [½]°—4p [2½]	1—2
4938,75	3	24,15	26,66	4p′ [½]—5s′ [½]°	0—1
4829,23	9	20,15	22,71	4s [1½]°—4p [½]	2—1
4774,92	4	20,64	23,25	4s′ [½]°—4p [1½]	1—1
4659,38	5	20,45	23,11	3d [1½]°—4p [2½]	2—3
4608,45	8	20,64	23,33	4s′ [½]°—4p [1½]	1—2
4595,65	5	20,45	23,14	3d [1½]°—4p [2½]	2—2
4505,33	6	20,39	23,14	3d [½]°—4p [2½]	1—2
4466,65	5	20,47	23,25	4s′ [½]°—4p [1½]	0—1
4455,00	2	23,57	26,35	4p′ [½]—5s [1½]°	1—2
4423,73	3	20,45	23,25	3d [1½]°—4p [1½]	2—1
4388,16	7	20,64	23,46	4s′ [½]°—4p′ [1½]	1—1
4383,80	1	23,53	26,35	4p [½]—5s [1½]°	0—2
4374,87	1	—	—	—	—
4362,96	3	23,51	26,35	4p′ [1½]—5s [1½]°	2—2
4340,03	5	20,39	23,25	3d [½]°—4p [1½]	1—1
4317,85	2	—	—	—	—
4309,10	7	20,64	23,51	4s′ [½]°—4p′ [1½]	1—2
4305,265	1	23,53	26,41	4p [½]—5s [1½]°	0—1
4305,00	7	20,45	23,33	3d [1½]°—4p [1½]	2—2
4288,70	4	20,64	23,53	4s′ [½]°—4p [½]	1—0
4284,89	3	23,51	26,41	4p′ [1½]—5s [1½]°	2—1
4263,40	7	20,24	23,14	4s [1½]°—4p [2½]	1—2
4225,67	7	20,39	23,33	3d [½]°—4p [1½]	1—2
4222,97	7	20,64	23,57	4s′ [½]°—4p′ [½]	1—1
4209,49	4	23,46	26,41	4p′ [1½]—5s [1½]	1—1
4186,24	8	20,15	23,11	4s [1½]°—4p [2½]	2—3
4149,19	7	20,47	23,46	4s′ [½]°—4p′ [1½]	0—1
4134,72	7	20,15	23,14	4s [1½]°—4p [2½]	2—2
4114,99	6	20,24	23,25	4s [1½]°—4p [1½]	1—1
4112,14	4	20,45	23,45	3d [1½]°—4p′ [1½]	2—1
4093,69	5	23,33	26,35	4p [1½]—5s [1½]°	2—2
4065,23	4	23,57	26,62	4p′ [½]—5s′ [½]°	1—0
4042,59	6	20,45	23,51	3d [1½]°—4p′ [1½]	2—2
4039,69	4	20,39	23,46	3d [½]°—4p′ [1½]	1—1
4024,88	4	23,33	26,41	4p [1½]—5s [1½]°	2—1
4017,52	4	23,57	26,66	4p′ [½]—5s′ [½]°	1—1
4012,10	5	20,24	23,33	4s [1½]°—4p [1½]	1—2

λ, Å	I	E_H, eV	E_B, eV	Transition	J
4001,24	7	20,47	23,57	4s′ [1/2]°—4p′ [1/2]	0—1
3995,10	6	20,15	23,25	4s [1 1/2]°—4p [1 1/2]	2—1
3981,80	4	23,25	26,35	4p [1 1/2]—5s [1 1/2]°	1—2
3972,58	6	20,39	23,51	3d [1/2]°—4p′ [1 1/2]	1—2
3966,72	6	20,45	23,57	3d [1 1/2]°—4p′ [1/2]	2—1
3959,84	3	23,53	26,66	4p [1/2]—5s′ [1/2]°	0—1
3956,10	3	23,57	26,71	4p′ [1/2]—4d [1/2]°	1—1
3955,21	6	20,39	23,53	3d [1/2]°—4p [1/2]	1—0
3942,53	6	23,51	26,66	4p′ [1 1/2]—5s′ [1/2]°	2—1
3934,46	5	—	—	—	—
3926,36	5	23,25	26,41	4p [1 1/2]—5s [1 1/2]°	1—1
3923,00	5	23,46	26,62	4p′ [1 1/2]—5s′ [1/2]°	1—0
3900,11	3	23,53	26,71	4p [1/2]—4d [1/2]°	0—1
3899,28	3	20,39	23,57	3d [1/2]°—4p′ [1/2]	1—1
3897,92	8	20,15	23,33	4s [1 1/2]°—4p [1 1/2]	2—2
3886,84	2	23,57	26,76	4p′ [1/2]—4d [1 1/2]°	1—2
3883,42	3	23,51	26,71	4p′ [1 1/2]—4d [1/2]°	2—1
3878,62	4	23,46	26,66	4p′ [1 1/2]—5s′ [1/2]°	1—1
3873,74	5	20,26	23,46	3d [1/2]°—4p′ [1 1/2]	0—1
3861,41	3	23,14	26,35	4p [2 1/2]—5s [1 1/2]°	3—2
3844,02	1	20,24	23,46	4s [1 1/2]°—4p′ [1 1/2]	1—1
3821,30	3	23,46	26,71	4p′ [1 1/2]—4d [1/2]°	1—1
3817,50	7	23,11	26,35	4p [2 1/2]—5s [1 1/2]°	3—2
3816,56	6	23,51	26,76	4p′ [1 1/2]—4d [1 1/2]°	2—2
3800,14	6	23,14	26,41	4p [2 1/2]—5s [1 1/2]°	2—1
3783,19	6	20,24	23,51	4s [1 1/2]°—4p′ [1 1/2]	1—2
3767,36	6	20,24	23,53	4s [1 1/2]°—4p [1/2]	1—0
3756,62	3	23,46	26,76	4p′ [1 1/2]—4d [1 1/2]°	1—2
3744,42	5	20,26	23,57	3d [1/2]°—4p′ [1/2]	0—1
3739,13	5	20,15	23,46	4s [1 1/2]°—4p′ [1 1/2]	2—1
3721,34	5	23,33	26,66	4p [1 1/2]—5s′ [1/2]°	2—1
3716,60	5	20,24	23,57	4s [1 1/2]°—4p′ [1/2]	1—1
3681,54	6	20,15	23,51	4s [1 1/2]°—4p′ [1 1/2]	2—2
3676,05	3	23,25	26,62	4p [1 1/2]—5s′ [1/2]°	1—0
3668,60	3	23,35	26,71	4p [1 1/2]—4d [1/2]°	2—1
3647,95	2	23,51	26,91	4p′ [1 1/2]—4d′ [?]°	2—2
3637,00	3	23,25	26,66	4p [1 1/2]—5s′ [1/2]°	1—1
3626,42	4	—	—	—	—
3618,49	6	20,15	23,57	4s [1 1/2]°—4p′ [1/2]	2—1
3608,88	5	23,33	26,76	4p [1 1/2]—4d [1 1/2]°	2—2
3593,22	2	23,46	26,91	4p′ [1 1/2]—4d [?]°	1—2
3586,60	2	23,25	26,71	4p [1 1/2]—4d [1/2]°	1—1
3562,15	4	23,51	26,99	4p′ [1 1/2]—4d [3 1/2]°	2—3
3530,75	7	20,64	24,15	4s′ [1/2]°—4p′ [1/2]	1—0
3529,53	3	23,25	26,76	4p [1 1/2]—4d [1 1/2]°	1—2
3528,51	1	23,14	26,66	4p [2 1/2]—5s′ [1/2]°	2—1
3481,11	6	23,14	26,71	4p [2 1/2]—4d [1/2]°	2—1
3457,85	2	23,33	26,91	4p [1 1/2]—4d [?]°	2—2
3440,05	7	23,57	27,18	4p′ [1/2]—4d [2 1/2]°	1—2
3427,13	2	23,14	26,76	4p [2 1/2]—4d [1 1/2]°	2—2
3421,83	4	—	—	—	—
3416,76	2	—	—	—	—
3404,24	6	22,71	26,35	4p [1/2]—5s [1 1/2]°	1—2
3392,63	3	23,11	26,76	4p [2 1/2]—4d [1 1/2]°	3—2
3384,86	6	23,51	27,18	4p′ [1 1/2]—4d [2 1/2]°	2—2
3380,62	6	23,33	26,99	4p [1 1/2]—4d [3 1/2]°	2—3
3373,60	6	—	—	—	—
3356,51	2	22,71	26,41	4p [1/2]—5s [1 1/2]°	1—1

λ, Å	I	E_H, eV	E_B, eV	Transition	J
3345,32	6	—	—	—	—
3337,67	1	23,46	27,18	$4p'\ [1^1/_2]-4d\ [2^1/_2]^\circ$	1—2
3301,60	3	20,39	24,15	$3d\ [^1/_2]^\circ-4p'\ [^1/_2]$	1—0
3290,65	5	23,14	26,91	$4p\ [2^1/_2]-4d\ [?]^\circ$	2—2
3258,81	3	23,11	26,91	$4p\ [2^1/_2]-4d\ [?]^\circ$	3—2
3253,98	3	—	—	—	—
3220,60	4	23,33	27,18	$4p\ [1^1/_2]-4d\ [2^1/_2]^\circ$	2—2
3190,07	5	23,11	26,99	$4p\ [2^1/_2]-4d\ [3^1/_2]^\circ$	3—3
3189,28	2	—	—	—	—
3171,81	2	22,71	26,62	$4p\ [^1/_2]-5s'\ [^1/_2]^\circ$	1—0
3169,80	3	20,24	24,15	$4s\ [1^1/_2]^\circ-4p'\ [^1/_2]$	1—0
3157,15	2	23,25	27,18	$4p\ [1^1/_2]-4d\ [2^1/_2]^\circ$	1—2
3142,75	2	22,71	26,66	$4p\ [^1/_2]-5s'\ [^1/_2]^\circ$	1—1
3105,00	6	22,71	26,71	$4p\ [^1/_2]-4d\ [^1/_2]^\circ$	1—1
3075,00	3	23,14	27,18	$4p\ [2^1/_2]-4d\ [2^1/_2]^\circ$	2—2
3062,18	5	22,71	26,76	$4p\ [^1/_2]-4d\ [1^1/_2]^\circ$	1—2
3047,16	2	23,11	27,18	$4p\ [2^1/_2]-4d\ [2^1/_2]^\circ$	3—2
3030,43	2	23,57	27,66	$4p'\ [^1/_2]-4d'\ [1^1/_2]^\circ$	1—1
2950,88	2	23,46	27,66	$4p'\ [1^1/_2]-4d'\ [1^1/_2]^\circ$	1—1
2808,99	3	23,25	27,66	$4p\ [1^1/_2]-4d'\ [1^1/_2]^\circ$	1—1
2777,89	2	22,71	27,18	$4p\ [^1/_2]-4d\ [2^1/_2]^\circ$	1—2
2743,55	4	23,14	27,66	$4p\ [2^1/_2]-4d'\ [1^1/_2]^\circ$	2—1
2550,02	6	—	—	—	—
2504,60	3	22,71	27,66	$4p\ [^1/_2]-4d'\ [1^1/_2]^\circ$	1—1
2358,70	1	—	—	—	—
2342,30	3	—	—	—	—
2296,79	1	—	—	—	—
2265,04	5	—	—	—	—
2255,29	3	21,27	26,76	$3d\ [2^1/_2]^\circ-4d\ [1^1/_2]^\circ$	2—2
2243,0	2	21,18	26,71	$3d\ [3^1/_2]^\circ-4d\ [^1/_2]^\circ$	3—1
2210,53	4	—	—	—	—
2190,00	6	—	—	—	—
2105,45	1	—	—	—	—
1764,0	0	20,15	27,18	$4s\ [1^1/_2]^\circ-4d\ [2^1/_2]^\circ$	2—2
1725,0	3	20,47	27,66	$4s'\ [^1/_2]^\circ-4d'\ [1^1/_2]^\circ$	0—1
615,40	2	—	—	—	—
612,621	4	0,00	20,24	$3p^6\ {}^1S-4s\ [1^1/_2]^\circ$	0—1
607,931	5	0,00	20,39	$3p^6\ {}^1S-3d\ [^1/_2]^\circ$	0—1
600,765	6	0,00	20,64	$3p^6\ {}^1S-4s'\ [^1/_2]^\circ$	0—1
574,634	1	—	—	—	—
550,323	1	—	—	—	—
546,123	3		—	—	—
539,731	3	—	—	—	—
518,249	3	—	—	—	—
515,653	3	—	—	—	—
495,144	6	—	—	—	—
485,626	2	—	—	—	—
485,513	3	—	—	—	—
485,084	5	—	—	—	—
476,029	2	—	—	—	—
469,499	—	0,00	26,41	$3p^6\ {}^1S-5s\ [1^1/_2]^\circ$	0—1
465,078	1	0,00	26,66	$3p^6\ {}^1S-5s'\ [^1/_2]^\circ$	0—1
446,830	5	—	—	—	—
441,812	5	—	—	—	—
429,923	3	—	—	—	—
429,656	3	—	—	—	—
261,200	1	—	—	—	—
261,028	1	—	—	—	—

K III, ground state $1s^2\,2s^2\,2p^6\,3s^2\,3p^5\ {}^2P^o_{3/2}$

Ionization potential 369000 cm⁻¹; 45,747 eV

λ, Å	*I*	E_H, eV	E_B, eV	Transition	*J*
3885,501	1	26,37	29,56	$4s\ {}^2P-4p\ {}^4P^o$	$^3/_2-^1/_2$
3513,88	5	25,97	29,50	$4s\ {}^4P-4p\ {}^4P^o$	$^1/_2-^3/_2$
3481,11	6	26,37	29,93	$4s\ {}^2P-4p\ {}^4D^o$	$^3/_2-^5/_2$
3468,32	6	25,87	29,45	$4s\ {}^4P-4p\ {}^4P^o$	$^3/_2-^5/_2$
3421,83	4	26,56	30,18	$4s\ {}^2P-4p\ {}^2D^o$	$^1/_2-^3/_2$
3420,82	6	25,87	29,50	$4s\ {}^4P-4p\ {}^4P^o$	$^3/_2-^3/_2$
3364,22	6	26,56	30,24	$4s\ {}^2P-4p\ {}^2P^o$	$^1/_2-^3/_2$
3322,40	6	25,72	29,45	$4s\ {}^4P-4p\ {}^4P^o$	$^5/_2-^5/_2$
3289,06	6	26,37	30,14	$4s\ {}^2P-4p\ {}^2D^o$	$^3/_2-^5/_2$
3278,79	6	25,72	29,50	$4s\ {}^4P-4p\ {}^4P^o$	$^5/_2-^3/_2$
3209,34	6	26,56	30,42	$4s\ {}^2P-4p\ {}^2P^o$	$^1/_2-^1/_2$
3201,95	6	26,37	30,24	$4s\ {}^2P-4p\ {}^2P^o$	$^3/_2-^3/_2$
3056,84	5	25,97	30,02	$4s\ {}^4P-4p\ {}^4D^o$	$^1/_2-^3/_2$
3052,07	6	25,87	29,93	$4s\ {}^4P-4p\ {}^4D^o$	$^3/_2-^5/_2$
2992,24	6	25,72	29,86	$4s\ {}^4P-4p\ {}^4D^o$	$^5/_2-^7/_2$
2986,20	5	25,87	30,02	$4s\ {}^4P-4p\ {}^4D^o$	$^3/_2-^3/_2$
2948,94	0	26,37	30,58	$4s\ {}^2P-4p\ {}^4S^o$	$^3/_2-^3/_2$
2938,45	5	25,72	29,93	$4s\ {}^4P-4p\ {}^4D^o$	$^5/_2-^5/_2$
2898,90	1	25,97	30,24	$4s\ {}^4P-4p\ {}^2P^o$	$^1/_2-^3/_2$
2736,96	0	25,72	30,24	$4s\ {}^4P-4p\ {}^2P^o$	$^5/_2-^3/_2$
2689,90	5	25,97	30,58	$4s\ {}^4P-4p\ {}^4S^o$	$^1/_2-^3/_2$
2635,11	5	25,87	30,58	$4s\ {}^4P-4p\ {}^4S^o$	$^3/_2-^3/_2$
2550,02	6	25,72	30,58	$4s\ {}^4P-4p\ {}^4S^o$	$^5/_2-^3/_2$
874,045	3	—	—	—	—
873,865	2	—	—	—	—
872,313	4	—	—	—	—
778,528	7	0,27	16,19	$3p^5\ {}^2P^o-3p^6\ {}^2S$	$^1/_2-^1/_2$
765,644	6	0,00	16,19	$3p^5\ {}^2P^o-3p^6\ {}^2S$	$^3/_2-^1/_2$
765,314	4	—	—	—	—
708,838	4	—	—	—	—
696,202	1	—	—	—	—
695,820	3	—	—	—	—
582,140	1	—	—	—	—
529,796	8	0,27	23,67	$3p^5\ {}^2P^o-3d\ {}^2D$	$^1/_2-^3/_2$
527,565	1	—	—	—	—
523,792	5	0,00	23,67	$3p^5\ {}^2P^o-3d\ {}^2D$	$^3/_2-^3/_2$
520,611	10	0,00	23,81	$3p^5\ {}^2P^o-3d\ {}^2D$	$^3/_2-^5/_2$
515,514	4	—	—	—	—
514,943	2	—	—	—	—
497,104	15	0,00	24,94	$3p^5\ {}^2P^o-3d\ {}^2F$	$^3/_2-^5/_2$
484,200	1	0,27	25,87	$3p^5\ {}^2P^o-4s\ {}^4P$	$^1/_2-^3/_2$
483,972	3	—	—	—	—
482,408	2	0,27	25,97	$3p^5\ {}^2P^o-4s\ {}^4P$	$^1/_2-^1/_2$
482,107	2	0,00	25,72	$3p^5\ {}^2P^o-4s\ {}^4P$	$^3/_2-^5/_2$
480,965	1	—	—	—	—
479,185	8	0,00	25,87	$3p^5\ {}^2P^o-4s\ {}^4P$	$^3/_2-^3/_2$
474,920	9	0,27	26,37	$3p^5\ {}^2P^o-4s\ {}^2P$	$^1/_2-^3/_2$
471,569	15	0,27	26,56	$3p^5\ {}^2P^o-4s\ {}^2P$	$^1/_2-^1/_2$
470,089	20	0,00	26,37	$3p^5\ {}^2P^o-4s\ {}^2P$	$^3/_2-^3/_2$
466,793	15	0,00	26,56	$3p^5\ {}^2P^o-4s\ {}^2P$	$^3/_2-^1/_2$
448,595	15	0,27	27,91	$3p^5\ {}^2P^o-4s'\ {}^2D$	$^1/_2-^3/_2$
444,344	15	0,00	27,90	$3p^5\ {}^2P^o-4s'\ {}^2D$	$^3/_2-^5/_2$
442,913	3	—	—	—	—
442,043	2	—	—	—	—
440,429	15	—	—	—	—

λ, Å	I	E_H, eV	E_B, eV	Transition	J
438,869	4	—	—	—	—
437,216	3	—	—	—	—
435,676	10	—	—	—	—
434,722	15	—	—	—	—
418,623	6	0,27	29,88	$3p^5\ ^2P^\circ—3d'\ ^2P$	$^1/_2—^3/_2$
417,535	6	0,27	29,96	$3p^5\ ^2P^\circ—4s''\ ^2S$	$^1/_2—^1/_2$
416,001	6	0,27	30,07	$3p^5\ ^2P^\circ—3d'\ ^2P$	$^1/_2—^1/_2$
414,870	6	0,00	29,88	$3p^5\ ^2P^\circ—3d'\ ^2P$	$^3/_2—^3/_2$
413,792	10	0,00	29,96	$3p^5\ ^2P^\circ—4s''\ ^2S$	$^3/_2—^1/_2$
412,289	5	0,00	30,07	$3p^5\ ^2P^\circ—3d'\ ^2P$	$^3/_2—^1/_2$
410,102	8	0,27	30,50	$3p^5\ ^2P^\circ—3d'\ ^2D$	$^1/_2—^3/_2$
409,737	8	—	—	—	—
408,959	8	0,00	30,32	$3p^5\ ^2P^\circ—3d'\ ^2D$	$^3/_2—^5/_2$
406,484	6	0,00	30,50	$3p^5\ ^2P^\circ—3d'\ ^2D$	$^3/_2—^3/_2$
402,907	6	—	—	—	—
402,104	4	0,27	31,10	$3p^5\ ^2P^\circ—3d'\ ^2S$	$^1/_2—^1/_2$
400,210	8	0,27	31,25	$3p^5\ ^2P^\circ—1$	$^1/_2$——
398,633	3	0,00	31,10	$3p^5\ ^2P^\circ—3d'\ ^2S$	$^3/_2—^1/_2$
396,763	0	0,00	31,25	$3p^5\ ^2P^\circ—1$	$^3/_2$——
391,918	4	—	—	—	—
390,114	5	—	—	—	—
387,372	2	—	—	—	—
382,229	6	0,27	32,70	$3p^5\ ^2P^\circ—5s\ ^2P$	$^1/_2—^1/_2$
380,477	5	0,00	32,58	$3p^5\ ^2P^\circ—5s\ ^2P$	$^3/_2—^3/_2$
379,877	6	—	—	—	—
379,118	8	0,00	32,70	$3p^5\ ^2P^\circ—5s\ ^2P$	$^3/_2—^1/_2$
347,999	3	0,27	35,89	$3p^5\ ^2P^\circ—5s'\ ^2D$	$^1/_2—^3/_2$
345,545	2	0,00	35,88	$3p^5\ ^2P^\circ—5s'\ ^2D$	$^3/_2—^5/_2$
345,405	2	0,00	35,89	$3p^5\ ^2P^\circ—5s'\ ^2D$	$^3/_2—^3/_2$
345,197	2	—	—	—	—
344,635	4	—	—	—	—
344,270	4	—	—	—	—
341,924	6	—	—	—	—
331,416	1	0,27	37,68	$3p^5\ ^2P^\circ—3d''\ ^2D$	$^1/_2—^3/_2$
330,684	5	0,00	37,49	$3p^5\ ^2P^\circ—3d''\ ^2D$	$^3/_2—^5/_2$
329,053	2	0,00	37,68	$3p^5\ ^2P^\circ—3d''\ ^2D$	$^3/_2—^3/_2$
328,933	3	—	—	—	—
328,845	2	—	—	—	—
327,605	1	0,27	38,11	$3p^5\ ^2P^\circ—2$	$^1/_2$——
325,278	0	0,00	38,11	$3p^5\ ^2P^\circ—2$	$^3/_2$——
240,758	3	—	—	—	—
239,010	2	—	—	—	—
203,890	3	—	—	—	—
203,823	2	—	—	—	—

K IV, ground state $1s^2\ 2s^2\ 2p^6\ 3s^2\ 3p^4\ ^3P_2$

Ionization potential 491 300 cm^{-1}; 60,909 eV

λ, Å	I	E_H, eV	E_B, eV	Transition	J
1025,742	3	—	—	—	—
892,621	2	—	—	—	—
855,815	2	—	—	—	—
754,673	8	0,21	16,64	$4p^4\ ^3P—3p^5\ ^3P^\circ$	1—2
754,194	3	4,78	21,22	$3p^4\ ^1S—3p^5\ ^1P^\circ$	0—1

λ, Å	I	E_H, eV	E_B, eV	Transition	J
749,993	6	0,29	16,82	$3p^4\ ^3P—3p^5\ ^3P°$	0—1
746,350	8	0,21	16,82	$3p^4\ ^3P—3p^5\ ^3P°$	1—1
745,264	10	0,00	16,64	$3p^4\ ^3P—3p^5\ ^3P°$	2—2
741,950	10	0,21	16,92	$3p^4\ ^3P—3p^5\ ^3P°$	1—0
737,144	10	0,00	16,82	$3p^4\ ^3P—3p^5\ ^3P°$	2—1
725,848	1	—	—	—	—
705,641	3	—	—	—	—
687,495	6	—	—	—	—
672,941	5	—	—	—	—
646,188	15	2,03	21,22	$3p^4\ ^1D—3p^5\ ^1P°$	2—1
605,908	1	—	—	—	—
605,316	1	—	—	—	—
591,434	1	—	—	—	—
591,311	1	—	—	—	—
591,237	1	—	—	—	—
543,973	1	—	—	—	—
543,640	2	—	—	—	—
528,879	1	0,29	16,92	$3p^4\ ^3P—3d\ ^3D°$	0—1
527,617	3	0,21	23,70	$3p^4\ ^3P—3d\ ^3D°$	1—2
527,064	2	0,21	23,73	$3p^4\ ^3P—3d\ ^3D°$	1—1
526,448	4	0,00	23,55	$3p^4\ ^3P—3d\ ^3D°$	2—3
523,001	5	0,00	23,70	$3p^4\ ^3P—3d\ ^2D°$	2—2
506,029	2	—	—	—	—
505,761	1	—	—	—	—
500,125	3	—	—	—	—
500,047	2	—	—	—	—
499,993	2	—	—	—	—
485,359	2	2,03	27,58	$3p^4\ ^1D—3d'\ ^1F°$	2—3
456,328	8	2,03	29,20	$3p^4\ ^1D—3d'\ ^1P°$	2—1
445,607	4	0,21	28,03	$3p^4\ ^3P—3d'\ ^3P°$	1—1
443,567	6	0,00	27,95	$3p^4\ ^3P—3d'\ ^3P°$	2—2
442,518	2	0,21	28,22	$3p^4\ ^3P—3d\ ^3P°$	1—0
442,300	4	0,00	28,03	$3p^4\ ^3P—3d'\ ^3P°$	2—1
440,905	4	—	—	—	—
417,280	3	2,03	31,74	$3p^4\ ^1D—3d''\ ^3P°$	2—2
408,076	5	2,03	32,41	$3p^4\ ^1D—3d''\ ^1P°$	2—1
405,773	2	2,03	32,59	$3p^4\ ^1D—3d''\ ^3D°$	2—2
404,412	3	2,03	32,69	$3p^4\ ^1D—3d''\ ^3D°$	2—1
393,142	10	0,21	31,74	$3p^4\ ^3P—3d''\ ^3P°$	1—2
392,467	4	0,29	31,88	$3p^4\ ^3P—3d''\ ^3P°$	0—1
392,274	2	4,78	36,38	$3p^4\ ^1S—4s''\ ^3P°$	0—1
391,462	4	0,21	31,88	$3p^4\ ^3P—3d''\ ^3P°$	1—1
390,574	6	0,00	31,74	$3p^4\ ^3P—3d''\ ^3P°$	2—2
390,415	5	0,29	31,96	$3p^4\ ^3P—3d''\ ^3P°$	1—0
389,069	5	2,03	33,90	$3p^4\ ^1D—3d''\ ^1D°$	2—2
388,920	3	0,00	31,88	$3p^4\ ^3P—3d''\ ^3P°$	1—2
384,956	3	0,21	32,41	$3p^4\ ^3P—3d''\ ^1P°$	1—1
384,095	5	0,00	32,28	$3p^4\ ^3P—4s\ ^3S°$	2—1
382,906	6	0,21	32,59	$3p^4\ ^3P—3d''\ ^3D°$	1—2
382,646	4	0,29	32,69	$3p^4\ ^3P—3d''\ ^3D°$	0—1
382,487	3	2,03	34,45	$3p^4\ ^1D—4s'\ ^3D°$	2—2
		0,00	32,41	$3p^4\ ^3P—3d''\ ^1P°$	2—1
381,702	4	0,21	32,69	$3p^4\ ^3P—3d''\ ^3D°$	1—1
380,477	5	0,00	32,59	$3p^4\ ^3P—3d''\ ^3D°$	2—2
379,279	2	0,00	32,69	$3p^4\ ^3P—3d''\ ^3D°$	2—1
375,955	6	2,03	35,01	$3p^4\ ^1D—4s'\ ^1D°$	2—2
363,021	3	0,29	34,44	$3p^4\ ^3P—4s'\ ^3D°$	0—1
362,154	3	0,21	34,44	$3p^4\ ^3P—4s'\ ^3D°$	1—1
362,085	5	0,21	34,45	$3p^4\ ^3P—4s'\ ^3D°$	1—2

λ, Å	I	E_H, eV	E_B, eV	Transition	J
360,568	2	2,03	36,41	$3p^4\ ^1D—4s''\ ^3P^\circ$	2—2
359,907	4	0,00	34,45	$3p^4\ ^3P—4s'\ ^3D^\circ$	2—1, 2
359,730	6	0,00	34,46	$3p^4\ ^3P—4s'\ ^3D^\circ$	2—3
356,260	3	0,21	35,01	$3p^4\ ^3P—4s'\ ^1D^\circ$	1—2
354,927	6	2,03	36,96	$3p^4\ ^1D—4s''\ ^1P^\circ$	2—1
354,139	2	0,00	35,01	$3p^4\ ^3P—4s'\ ^1D^\circ$	2—2
343,468	3	0,29	36,38	$3p^4\ ^3P—4s''\ ^3P^\circ$	0—1
342,805	2	0,21	36,37	$3p^4\ ^3P—4s''\ ^3P^\circ$	1—0
342,703	2	0,21	36,38	$3p^4\ ^3P—4s''\ ^3P^\circ$	1—1
342,410	3	0,21	36,41	$3p^4\ ^3P—4s''\ ^3P^\circ$	1—2
340,745	3	0,00	36,38	$3p^4\ ^3P—4s'\ ^3P^\circ$	2—1
340,462	6	0,00	36,41	$3p^4\ ^3P—4s''\ ^3P^\circ$	2—2
283,765	2	—	—	—	—
274,552	3	—	—	—	—
273,546	1	0,29	45,61	$3p^4\ ^3P—5s\ ^3S^\circ$	0—1
273,065	2	0,21	45,61	$3p^4\ ^3P—5s\ ^3S^\circ$	1—1
271,820	3	0,00	45,61	$3p^4\ ^3P—5s\ ^3S^\circ$	2—1
263,819	2	—	—	—	—
263,716	2	—	—	—	—
170,227	1	—	—	—	—
166,246	1	—	—	—	—
166,163	0	—	—	—	—

K V, ground state $1s^2\ 2s^2\ 2p^6\ 3s^2\ 3p^3\ ^4S^0_{1^1/_2}$

Ionization potential 666 000 cm^{-1}; 82,6 eV

λ, Å	I	E_H, eV	E_B, eV	Transition	J
1027,174	2	—	—	—	—
1021,332	2	—	—	—	—
917,498	1	—	—	—	—
898,953	3	—	—	—	—
881,405	3	—	—	—	—
874,985	1	—	—	—	—
874,883	1	—	—	—	—
869,965	3	—	—	—	—
868,552	1	—	—	—	—
868,140	1	—	—	—	—
867,921	1	—	—	—	—
854,771	1	—	—	—	—
854,416	2	—	—	—	—
839,439	1	—	—	—	—
830,785	1	—	—	—	—
826,395	1	—	—	—	—
825,559	1	4,97	19,98	$3p^3\ ^2P^\circ—3p^4\ ^2D$	$^3/_2—^3/_2$
823,358	3	4,93	19,98	$3p^3\ ^2P^\circ—3p^4\ ^2D$	$^1/_2—^3/_2$
823,047	3	4,97	20,03	$3p^3\ ^2P^\circ—3p^4\ ^2D$	$^3/_2—^5/_2$
821,568	1	—	—	—	—
810,893	1	—	—	—	—
810,215	1	—	—	—	—
809,673	2	—	—	—	—
806,577	2	—	—	—	—
806,373	2	—	—	—	—
803,826	2	—	—	—	—
802,122	2	—	—	—	

λ, Å	I	E_H, eV	E_B, eV	Transition	J
786,464	2	—	—	—	—
784,713	2	—	—	—	—
771,456	3	—	—	—	—
771,376	3	4,97	21,04	$3p^3\ ^2P^\circ$—1	$^3/_2$——
770,287	3	4,97	21,06	$3p^3\ ^2P^\circ$—2	$^3/_2$——
769,402	2	4,93	21,04	$3p^3\ ^2P^\circ$—1	$^1/_2$——
758,559	1	—	—	—	—
757,112	4	—	—	—	—
750,381	1	—	—	—	—
750,230	3	—	—	—	—
743,292	2	—	—	—	—
731,858	2	0,00	16,94	$3p^3\ ^4S^\circ—3p^4\ ^4P$	$^3/_2—^5/_2$
724,420	8	0,00	17,11	$3p^3\ ^4S^\circ—3p^4\ ^4P$	$^3/_2—^3/_2$
720,432	6	0,00	17,21	$3p^3\ ^4S^\circ—3p^4\ ^4P$	$^3/_2—^1/_2$
713,041	1	—	—	—	—
695,042	3	—	—	—	—
694,477	2	—	—	—	—
644,963	0	4,93	24,15	$3p^3\ ^2P^\circ—3p^4\ ^2P$	$^1/_2—^3/_2$
639,982	2	4,97	24,34	$3p^3\ ^2P^\circ—3p^4\ ^2P$	$^3/_2—^1/_2$
638,668	5	4,93	24,34	$3p^3\ ^2P^\circ—3p^4\ ^2P$	$^1/_2—^1/_2$
625,404	4	—	—	—	—
603,429	8	4,97	25,51	$3p^3\ ^2P^\circ—3p^4\ ^2S$	$^3/_2—^1/_2$
602,269	5	4,93	25,51	$3p^3\ ^2P^\circ—3p^4\ ^2S$	$^1/_2—^1/_2$
586,322	8	3,00	24,15	$3p^3\ ^2D^\circ—3p^4\ ^2P$	$^5/_2—^3/_2$
585,510	5	2,98	24,15	$3p^3\ ^2D^\circ—3p^4\ ^2P$	$^3/_2—^3/_2$
580,319	7	2,98	24,34	$3p^3\ ^2D^\circ—3p^4\ ^2P$	$^3/_2—^1/_2$
544,627	1	—	—	—	—
544,537	1	—	—	—	—
536,216	2	—	—	—	—
535,287	1	—	—	—	—
534,873	2	—	—	—	—
515,320	1	—	—	—	—
509,601	1	—	—	—	—
483,745	4	0,00	25,63	$3p^3\ ^4S^\circ—3d\ ^4F$	$^3/_2—^3/_2$
482,706	4	0,00	25,68	$3p^3\ ^4S^\circ—3d\ ^4F$	$^3/_2—^5/_2$
468,447	2	—	—	—	—
462,596	1	—	—	—	—
459,005	3	—	—	—	—
456,328	4	4,97	32,14	$3p^3\ ^2P^\circ—3d\ ^2P$	$^3/_2—^3/_2$
455,670	1	4,93	32,14	$3p^3\ ^2P^\circ—3d\ ^2P$	$^1/_2—^3/_2$
452,900	3	4,97	32,34	$3p^3\ ^2P^\circ—3d\ ^2P$	$^3/_2—^1/_2$
452,227	2	4,93	32,34	$3p^3\ ^2P^\circ—3d\ ^2P$	$^1/_2—^1/_2$
449,708	4	0,00	27,57	$3p^3\ ^4S^\circ—3d\ ^4D$	$^3/_2—^5/_2$
449,013	3	0,00	27,61	$3p^3\ ^4S^\circ—3d\ ^4D$	$^3/_2—^3/_2$
447,085	3	—	—	—	—
445,878	1	—	—	—	—
438,647	2	4,93	33,23	$3p^3\ ^2P^\circ$—3	$^3/_2$——
438,023	5	4,97	33,23	$3p^3\ ^2P^\circ$—3	$^1/_2$——
425,588	7	3,00	32,14	$3p^3\ ^2D^\circ—3d\ ^2P$	$^5/_2—^3/_2$
425,159	5	2,98	32,14	$3p^3\ ^2D^\circ—3d\ ^2P$	$^3/_2—^3/_2$
422,178	5	2,98	32,34	$3p^3\ ^2D^\circ—3d\ ^2P$	$^3/_2—^1/_2$
421,446	3	—	—	—	—
419,731	1	3,00	32,54	$3p^3\ ^2D^\circ—3d\ ^2F$	$^5/_2—^5/_2$
419,310	2	2,98	32,54	$3p^3\ ^2D^\circ—3d\ ^2F$	$^3/_2—^5/_2$
419,045	2	3,00	32,59	$3p^3\ ^2D^\circ—3d\ ^2F$	$^5/_2—^7/_2$
415,793	4	3,00	32,82	$3p^3\ ^2D^\circ—3d\ ^2D$	$^5/_2—^5/_2$
415,465	3	3,00	32,85	$3p^3\ ^2D^\circ—3d\ ^2D$	$^5/_2—^3/_2$
415,052	5	2,98	32,85	$3p^3\ ^2D^\circ—3d\ ^2D$	$^3/_2—^3/_2$

λ, Å	I	E_H, eV	E_B, eV	Transition	J
414,465	3	4,93	34,84	$3p\ ^2P^\circ—3d'\ ^2D$	$^1/_2—^3/_2$
412,080	6	—	—	—	—
399,754	4	3,00	34,02	$2p^3\ ^2D^\circ—4$	$^5/_2—$
399,400	3	2,98	34,02	$2p^3\ ^2D^\circ—4$	$^3/_2—$
398,878	4	4,97	36,05	$3p^3\ ^2P^\circ—3d'\ ^2P$	$^3/_2—^3/_2$
398,363	4	4,93	36,05	$3p^3\ ^2P^\circ—3d'\ ^2P$	$^1/_2—^3/_2$
394,909	3	4,93	36,32	$3p^3\ ^2P^\circ—3d'\ ^2S$	$^1/_2—^3/_2$
389,428	2	3,00	34,84	$3p^3\ ^2D^\circ—3d'\ ^2D$	$^5/_2—^3/_2$
389,069	5	2,98	34,84	$3p^3\ ^2D^\circ—3d'\ ^2D$	$^3/_2—^3/_2$
387,800	6	0,00	31,97	$3p^3\ ^4S^\circ—3d\ ^4P$	$^3/_2—^5/_2$
386,710	4	—	—	—	—
385,689	5	0,00	32,14	$3p^3\ ^4S^\circ—4d\ ^4P$	$^3/_2—^3/_2$
385,020	1	0,00	32,20	$3p^3\ ^4S^\circ—4d\ ^4P$	$^3/_2—^1/_2$
384,516	2	—	—	—	—
384,400	2	—	—	—	—
383,318	2	—	—	—	—
378,219	3	4,97	37,75	$3p^3\ ^2P^\circ—3d''\ ^2D$	$^3/_2—^3/_2$
377,763	5	4,93	37,75	$3p^3\ ^2P^\circ—3d''\ ^2D$	$^1/_2—^3/_2$
376,061	3	4,97	37,93	$3p^3\ ^2P^\circ—3d''\ ^2D$	$^3/_2—^5/_2$
374,939	5	—	—	—	—
373,318	3	—	—	—	—
373,074	2	0,00	33,23	$3p^3\ ^4S^\circ—3$	$^3/_2——$
372,462	4	3,00	36,29	$3p^3\ ^2D^\circ—3d'\ ^2F$	$^5/_2—^5/_2$
372,148	10	2,98	36,29	$3p^3\ ^2D^\circ—3d'\ ^2F$	$^3/_2—^5/_2$
370,580	3	—	—	—	—
370,523	3	—	—	—	—
354,927	6	3,00	37,93	$3p^3\ ^2D^\circ—3d''\ ^2D$	$^5/_2—^5/_2$
354,627	0	2,98	37,93	$3p^3\ ^2D^\circ—3d''\ ^2D$	$^3/_2—^5/_2$
352,750	2	3,00	38,15	$3p^3\ ^2D^\circ—5$	$^5/_2——$
352,463	2	2,98	38,15	$3p^3\ ^2D^\circ—5$	$^3/_2——$
349,793	3	3,00	38,45	$3p^3\ ^2D^\circ—6$	$^5/_2——$
349,504	4	2,98	38,45	$3p^3\ ^2D^\circ—6$	$^3/_2——$
331,168	1	—	—	—	—
329,307	0	4,97	42,61	$3p^3\ ^2P^\circ—4s\ ^2P$	$^3/_2—^1/_2$
328,973	2	4,93	42,61	$3p^3\ ^2P^\circ—4s\ ^2P$	$^1/_2—^1/_2$
327,376	4	4,97	42,84	$3p^3\ ^2P^\circ—4s\ ^2P$	$^3/_2—^3/_2$
327,031	2	4,93	42,84	$3p^3\ ^2P^\circ—4s\ ^2P$	$^1/_2—^3/_2$
318,969	1	—	—	—	—
315,537	3	4,97	44,26	$3p^3\ ^2P^\circ—4s'\ ^2D$	$^3/_2—^5/_2$
315,181	4	4,93	44,26	$3p^3\ ^2P^\circ—4s'\ ^2D$	$^1/_2—^3/_2$
312,770	5	2,98	42,61	$3p^3\ ^2D^\circ—4s\ ^2P$	$^3/_2—^1/_2$
311,243	6	3,00	42,84	$3p^3\ ^2D^\circ—4s\ ^2P$	$^5/_2—^3/_2$
300,503	6	3,00	44,26	$3p^3\ ^2D^\circ—4s'\ ^2D$	$^5/_2—^5/_2$
300,252	7	2,98	44,26	$3p^3\ ^2D^\circ—4s'\ ^2D$	$^3/_2—^3/_2$
297,064	5	0,00	41,73	$3p^3\ ^4S^\circ—4s\ ^4P$	$^3/_2—^1/_2$
296,169	6	0,00	41,86	$3p^3\ ^4S^\circ—4s\ ^4P$	$^3/_2—^3/_2$
294,836	6	0,00	42,05	$3p^3\ ^4S^\circ—4s\ ^4P$	$^3/_2—^5/_2$
285,850	2	—	—	—	—
285,734	1	—	—	—	—
282,355	3	—	—	—	—
277,394	3	—	—	—	—
264,478	2	—	—	—	—
264,339	2	—	—	—	—
246,235	1	—	—	—	—
232,673	1	—	—	—	—
214,351	2	—	—	—	—
213,121	1	—	—	—	—

K VI, ground state $1s^2\,2s^2\,2p^6\,3s^2\,3p^2\;{}^3P_0$
Ionization potential 804 513 cm^{-1}; 99,741 eV

λ, Å	I	E_H, eV	E_B, eV	Transition	J
1000,056	2	—	—	—	—
982,115	2	—	—	—	—
968,518	6	—	—	—	—
938,287	2	—	—	—	—
930,318	1	—	—	—	—
929,374	1	—	—	—	—
918,581	2	—	—	—	—
882,184	2	—	—	—	—
872,006	1	—	—	—	—
776,957	4	—	—	—	—
771,103	5	—	—	—	—
770,022	1	—	—	—	—
757,199	4	—	—	—	—
753,877	3	—	—	—	—
747,848	2	—	—	—	—
739,177	1	—	—	—	—
725,309	2	0,36	17,46	$3p^2\,{}^3P—3p^3\,{}^3D^\circ$	2—2
724,420	8	0,36	17,48	$3p^2\,{}^3P—3p^3\,{}^3D^\circ$	2—3
716,272	2	0,14	17,45	$3p^2\,{}^3P—3p^3\,{}^3D^\circ$	1—1
715,999	5	0,14	17,46	$3p^2\,{}^3P—3p^3\,{}^3D^\circ$	1—2
712,728	1	—	—	—	—
710,932	1	—	—	—	—
710,519	4	0,00	17,45	$3p^2\,{}^3P—3p^3\,{}^3D^\circ$	0—1
668,864	3	—	—	—	—
663,134	3	—	—	—	—
661,402	3	—	—	—	—
659,852	3	—	—	—	—
657,931	3	—	—	—	—
657,327	3	—	—	—	—
651,324	1	—	—	—	—
637,195	1	—	—	—	—
630,940	1	—	—	—	—
630,302	2	—	—	—	—
627,560	2	—	—	—	—
623,016	8	0,36	20,26	$3p^2\,{}^3P—3p^3\,{}^3P^\circ$	2—2, 1
616,136	6	0,14	20,26	$3p^2\,{}^3P—3p^3\,{}^3P^\circ$	1—0, 1,
612,272	1	—	—	—	—
611,862	3	0,00	20,26	$3p^2\,{}^3P—3p^3\,{}^3P^\circ$	0—1
519,372	5	—	—	—	—
501,649	2	2,35	27,07	$3p^2\,{}^1D—3p^3\,{}^3S^\circ$	2—1
491,062	2	—	—	—	—
490,423	2	—	—	—	—
488,120	10	2,35	27,75	$3p^2\,{}^1D—3p^3\,{}^1P^\circ$	2—1
481,313	2	—	—	—	—
480,397	1	—	—	—	—
473,207	2	—	—	—	—
464,270	10	0,36	27,07	$3p^2\,{}^3P—3p^3\,{}^3S^\circ$	2—1
461,737	3	—	—	—	—
460,438	8	0,14	27,07	$3p^2\,{}^3P—3p^3\,{}^3S^\circ$	1—1
458,048	7	0,00	27,07	$3p^2\,{}^3P—3p^3\,{}^3S^\circ$	0—1
457,323	1	—	—	—	—
452,667	3	0,36	27,75	$3p^2\,{}^3P—3p^3\,{}^1P^\circ$	2—1
451,320	2	—	—	—	—
449,013	3	0,14	27,75	$3p^2\,{}^3P—3p^3\,{}^1P^\circ$	1—1
446,009	4	—	—	—	—

λ, Å	I	E_H, eV	E_B, eV	Transition	J
441,370	3	—	—	—	—
429,438	2	—	—	—	—
428,538	5	—	—	—	—
428,315	2	—	—	—	—
426,338	2	—	—	—	—
420,807	4	—	—	—	—
418,160	2	—	—	—	—
416,509	1	—	—	—	—
412,790	1	—	—	—	—
406,102	2	—	—	—	—
405,675	2	—	—	—	—
405,475	1	—	—	—	—
405,333	2	—	—	—	—
405,178	2	—	—	—	—
404,684	4	—	—	—	—
400,951	5	0,36	31,28	$3p^2\ ^3P—3d\ ^3P^\circ$	2—2
399,073	2	0,36	31,43	$3p^2\ ^3P—3d\ ^3P^\circ$	2—1
398,087	4	0,14	31,28	$3p^2\ ^3P—3d\ ^3P^\circ$	1—2
396,235	4	0,14	31,43	$3p^2\ ^3P—3d\ ^3P^\circ$	1—1
395,395	5	0,14	31,50	$3p^2\ ^3P—3d\ ^3P^\circ$	1—0
394,477	3	0,00	31,42	$3p^2\ ^3P—3d\ ^3P^\circ$	0—1
389,750	2	—	—	—	—
389,531	2	—	—	—	—
388,485	2	—	—	—	—
388,233	4	—	—	—	—
386,505	2	—	—	—	—
377,263	2	—	—	—	—
372,774	4	—	—	—	—
370,115	2	—	—	—	—
368,030	2	—	—	—	—
367,378	2	—	—	—	—
365,614	3	—	—	—	—
357,685	3	—	—	—	—
357,645	3	—	—	—	—
356,615	3	—	—	—	—
356,372	1	—	—	—	—
355,800	1	—	—	—	—
355,603	1	—	—	—	—
355,469	1	—	—	—	—
350,164	2	—	—	—	—
338,161	3	—	—	—	—
335,175	3	—	—	—	—
303,023	1	—	—	—	—
302,657	2	—	—	—	—
293,438	2	—	—	—	—
293,332	3	—	—	—	—
293,050	2	—	—	—	—
289,241	1	—	—	—	—
259,609	2	0,36	48,12	$3p^2\ ^3P—4s\ ^3P^\circ$	2—1
258,873	3	0,14	48,03	$3p^2\ ^3P—4s\ ^3P^\circ$	1—0
258,411	1	0,14	48,12	$3p^2\ ^3P—4s\ ^3P^\circ$	1—1
258,018	4	0,36	48,41	$3p^2\ ^3P—4s\ ^3P^\circ$	2—2
257,657	2	0,00	48,12	$3p^2\ ^3P—4s\ ^3P^\circ$	0—1
256,831	3	0,14	48,41	$3p^2\ ^3P—4s\ ^3P^\circ$	1—2
253,106	1	—	—	—	—
252,965	1	—	—	—	—
247,777	1	—	—	—	—
205,862	2	—	—	—	—

λ, Å	I	E_H, eV	E_B, eV	Transition	J
205,772	2	—	—	—	—
200,341	1	—	—	—	—
196,978	1	—	—	—	—
185,883	2	—	—	—	—
181,265	1	—	—	—	—
157,433	1	—	—	—	—

Unclassified Lines of Potassium

λ, Å	I	Presumed classification	λ, Å	I	Presumed classification
2240,89	4	—	1668,7	4	—
1895,7	2	—	1659,7	4	—
1892,7	2	—	1033,875	2	—
1890,9	2	—	1032,768	2	—
1887,9	2	—	774,738	3	—
1843,9	2	—	748,393	2	—
1840,2	2	—	747,677	3	—
1837,1	2	—	688,085	3	—
1787,4	4	—	534,059	2	—
1773,5	6	—	528,519	2	—
1770,8	6	—	520,493	3	—
1765,4	3	—	446,926	2	—
1752,4	4	—	368,580	2	—
1749,3	8	—	355,133	2	—
1741,2	4	—	353,455	3	—
1739,4	2	—	353,325	3	—
1721,6	3	—	308,129	2	—
1719,8	3	—	306,620	2	—
1709,4	3	—	303,690	2	—
1704,5	9	—	267,036	2	—
1703,5	9	—	266,938	2	—
1698,9	10	—	266,344	4	—
1675,1	2	—	247,708	3	—
1672,9	3	—	247,561	2	—
1669,3	4	—	247,202	2	—

CALCIUM, Z = 20

Ca I, ground state $1s^2\,2s^2\,2p^6\,3s^2\,3p^6\,4s^2\,{}^1S_0$

Ionization potential 49 304,80 cm^{-1}; 6,113 eV

λ, Å	I	E_H, eV	E_B, eV	Transition	J
22651,30	30	4,68	5,23	$4d\,^3D-4f\,^3F^\circ$	3—2, 3,4
22625,51	20	4,68	5,23	$4d\,^3D-4f\,^3F^\circ$	2—2, 3
22608,39	10	4,68	5,23	$4d\,^3D-4f\,^3F^\circ$	1—2
19961,37	40	3,91	4,53	$5s\,^3S-5p\,^3P^\circ$	1—0
19932,94	100	3,91	4,53	$5s\,^3S-5p\,^3P^\circ$	1—1
19916,34	50	1,90	2,52	$4p\,^3P^\circ-3d\,^3D$	2—1
19861,70	500	1,90	2,52	$4p\,^3P^\circ-3d\,^3D$	2—2
19852,96	250	3,91	4,53	$5s\,^3S-5p\,^3P^\circ$	1—2
19815,14	30	4,62	5,25	$4d\,^1D-4f\,^1F^\circ$	2—3
19776,67	2000	1,90	2,53	$4p\,^3P^\circ-3d\,^3D$	2—3
19505,62	500	1,89	2,52	$4p\,^3P^\circ-3d\,^3D$	1—1
19452,82	1500	1,89	2,52	$4p\,^3P^\circ-3d\,^3D$	1—2
19309,43	500	1,88	2,52	$4p\,^3P^\circ-3d\,^3D$	0—1
19114,83	5	4,74	5,39	$4p'\,^3D^\circ-3d^2\,^3F$	3—2, 3
19045,86	90	4,74	5,39	$4p'\,^3D^\circ-3d^2\,^3F$	3—4
19021,39	4	4,74	5,39	$4p'\,^3D^\circ-3d^2\,^3F$	2—2
18969,71	60	4,74	5,39	$4p'\,^3D^\circ-3d^2\,^3F$	2—3
18924,96	30	4,73	5,39	$4p'\,^3D^\circ-3d^2\,^3F$	1—2
16202,94	10	4,53	5,30	$5p\,^3P^\circ-5d\,^3D$	2—1, 2
16195,33	150	4,53	5,30	$5p\,^3P^\circ-5d\,^3D$	2—3
16156,04	100	4,55	5,32	$4p'\,^1P^\circ-5d\,^1D$	1—2
16149,79	70	4,53	5,30	$5p\,^3P^\circ-5d\,^3D$	1—1, 2
16135,80	20	4,53	5,30	$5p\,^3P^\circ-5d\,^3D$	0—1
13167,75	24	4,45	5,39	$4p'\,^3F^\circ-3d^2\,^3F$	4—3
13134,96	400	4,45	5,39	$4p'\,^3F^\circ-3d^2\,^3F$	4—4
13086,26	50	4,44	5,39	$4p'\,^1D^\circ-3d^2\,^3F$	2—2
13061,84	8	4,44	5,39	$4p'\,^1D^\circ-3d^2\,^3F$	2—3
13057,82	20	4,44	5,39	$4p'\,^3F^\circ-3d^2\,^3F$	3—2
13033,41	300	4,44	5,39	$4p'\,^3F^\circ-3d^2\,^3F$	3—3
13001,37	20	4,44	5,39	$4p'\,^3F^\circ-3d^2\,^3F$	3—4
12909,07	200	4,43	5,39	$4p'\,^3F^\circ-3d^2\,^3F$	2—2
12885,21	15	4,43	5,39	$4p'\,^3F^\circ-3d^2\,^3F$	2—3
12826,60	25	3,91	4,88	$5s\,^3S-4p'\,^3P^\circ$	1—0
12823,46	100	3,91	4,88	$5s\,^3S-4p'\,^3P^\circ$	1—1
12815,69	400	3,91	4,88	$5s\,^3S-4p'\,^3P^\circ$	1—2
10879,78	4	4,88	6,02	$4p'\,^3P^\circ-3d^2\,^3P$	1—0
10869,37	3	4,88	6,02	$4p'\,^3P^\circ-3d^2\,^3P$	2—1
10863,72	2	4,88	6,02	$4p'\,^3P^\circ-3d^2\,^3P$	1—1
10861,51	3	4,88	6,02	$4p'\,^3P^\circ-3d^2\,^3P$	0—1
10838,77	10	4,88	6,02	$4p'\,^3P^\circ-3d^2\,^3P$	2—2
10833,12	4	4,88	6,02	$4p'\,^3P^\circ-3d^2\,^3P$	1—2
10343,85	500	2,93	4,13	$4p\,^1P^\circ-5s\,^1S$	1—0
9701,81	20	4,74	6,02	$4p'\,^3D^\circ-3d^2\,^3P$	3—2
9688,60	15	4,74	6,02	$4p'\,^3D^\circ-3d^2\,^3P$	2—1
9676,25	5	4,73	6,02	$4p'\,^3D^\circ-3d^2\,^3P$	1—0
9664,29	3	4,74	6,02	$4p'\,^3D^\circ-3d^2\,^3P$	2—2
9663,58	2	4,73	6,02	$4p'\,^3D^\circ-3d^2\,^3P$	1—1
7468,41	3	—	—	—	—
7326,146	400	2,93	4,62	$4p\,^1P^\circ-4d\,^1D$	1—2
7202,194	200	2,71	4,43	$3d\,^1D-4p'\,^3F^\circ$	2—2
7148,147	500	2,71	4,44	$3d\,^1D-4p'\,^1D^\circ$	2—2
6798,51	6	2,71	4,53	$3d\,^1D-5p\,^3P^\circ$	2—1

λ, Å	I	E_H, eV	E_B, eV	Transition	J
6717,685	500	2,71	4,55	$3d\ ^1D-4p'\ ^1P^\circ$	2—1
6709,86	1	2,93	4,78	$4p\ ^1P^\circ-4p^2\ ^3P$	1—2
6572,781	50	0,00	1,89	$4s^2\ ^1S-4p\ ^3P^\circ$	0—1
6508,742	1	2,53	4,43	$3d\ ^3D-4p'\ ^3F^\circ$	3—2
6499,649	30	2,52	4,43	$3d\ ^3D-4p'\ ^3F^\circ$	2—2
6493,780	80	2,52	4,43	$3d\ ^3D-4p'\ ^3F^\circ$	1—2
6471,660	40	2,53	4,44	$3d\ ^3D-4p'\ ^3F^\circ$	3—3
6464,70	1	2,53	4,44	$3d\ ^3D-4p'\ ^1D^\circ$	3—2
6462,566	125	2,52	4,44	$3d\ ^3D-4p'\ ^3F^\circ$	2—3
6455,600	10	2,52	4,44	$3d\ ^3D-4p'\ ^1D^\circ$	2—2
6449,810	50	2,52	4,44	$3d\ ^3D-4p'\ ^1D^\circ$	1—2
6439,073	150	2,53	4,45	$3d\ ^3D-4p'\ ^3F^\circ$	3—4
6405,89	3	—	—	—	—
6395,16	3	—	—	—	—
6361,79	5	4,45	6,40	$4p'\ ^3F^\circ-4d'\ ^3G$	4—5
6343,29	4	4,44	6,39	$4p'\ ^3F^\circ-4d'\ ^3G$	3—4
6318,11	3	4,43	6,39	$4p'\ ^3F^\circ-4d'\ ^3G$	2—3
6169,559	40	2,53	4,53	$3d\ ^3D-5p\ ^3P^\circ$	3—2
6169,055	25	2,52	4,53	$3d\ ^3D-5p\ ^3P^\circ$	2—1
6166,443	15	2,52	4,53	$3d\ ^3D-5p\ ^3P^\circ$	1—0
6163,758	10	2,52	4,53	$3d\ ^3D-5p\ ^3P^\circ$	1—1
6162,172	150	1,90	3,91	$4p\ ^3P^\circ-5s\ ^3S$	2—1
6161,289	10	2,52	4,53	$3d\ ^3D-5p\ ^3P^\circ$	2—2
6156,10	1	2,52	4,53	$3d\ ^3D-5p\ ^3P^\circ$	1—2
6122,219	100	1,89	3,91	$4p\ ^3P^\circ-5s\ ^3S$	1—1
6102,722	80	1,88	3,91	$4p\ ^3P^\circ-5s\ ^3S$	0—1
5867,572	1	2,93	5,04	$4p\ ^1P^\circ-4p^2\ ^1S$	1—0
5857,454	100	2,93	5,05	$4p\ ^1P^\circ-4p^2\ ^1D$	1—2
5764,32	3	—	—	—	—
5761,88	1	4,45	6,60	$4p'\ ^3F^\circ-4d'\ ^3F$	4—3
5757,69	4	4,45	6,60	$4p'\ ^3F^\circ-4d'\ ^3F$	4—4
5746,81	2	4,44	6,60	$4p'\ ^3F^\circ-4d'\ ^3F$	3—2
5743,28	3				
5735,74	3	4,44	6,60	$4p'\ ^3F^\circ-4d'\ ^3F$	3—3
5731,70	1	4,44	6,60	$4p'\ ^3F^\circ-4d'\ ^3F$	3—4
5717,99	4	4,43	6,60	$4p'\ ^3F^\circ-4d'\ ^3F$	2—2
5707,03	1	4,43	6,60	$4p'\ ^3F^\circ-4d'\ ^3F$	2—3
5688,47	4	—	—	—	—
5682,88	3	—	—	—	—
5602,846	25	2,52	4,73	$3d\ ^3D-4p'\ ^3D^\circ$	2—1
5601,285	30	2,53	4,74	$3d\ ^3D-4p'\ ^3D^\circ$	3—2
5598,487	50	2,52	4,73	$3d\ ^3D-4p'\ ^3D^\circ$	1—1
5594,468	60	2,52	4,74	$3d\ ^3D-4p'\ ^3D^\circ$	2—2
5590,120	20	2,52	4,74	$3d\ ^3D-4p'\ ^3D^\circ$	1—2
5588,757	80	2,53	4,74	$3d\ ^3D-4p'\ ^3D^\circ$	3—3
5581,971	25	2,52	4,74	$3d\ ^3D-4p'\ ^3D^\circ$	2—3
5512,979	20	2,93	5,18	$4p\ ^1P^\circ-6s\ ^1S$	1—0
5349,472	25	2,71	5,03	$3d\ ^1D-4p'\ ^1F^\circ$	2—3
5270,270	60	2,53	4,88	$3d\ ^3D-4p'\ ^3P^\circ$	3—2
5265,557	40	2,52	4,88	$3d\ ^3D-4p'\ ^3P^\circ$	2—1
5264,239	20	2,52	4,88	$3d\ ^3D-4p'\ ^3P^\circ$	2—2
5262,244	25	2,52	4,88	$3d\ ^3D-4p'\ ^3P^\circ$	1—0
5261,706	20	2,52	4,88	$3d\ ^3D-4p'\ ^3P^\circ$	1—1
5260,375	2	2,52	4,88	$3d\ ^3D-4p'\ ^3P^\circ$	1—2
5188,848	50	2,93	5,32	$4p\ ^1P^\circ-5d\ ^1D$	1—2
5041,620	40	2,71	5,17	$3d\ ^1D-5p\ ^1P^\circ$	2—1
4878,132	50	2,71	5,25	$3d\ ^1D-4f\ ^1F^\circ$	2—3
4847,296	2	2,93	5,49	$4p\ ^1P^\circ-7s\ ^1S$	1—0

λ, Å	I	E_H, eV	E_B, eV	Transition	J
4685,265	12	2,93	5,58	$4p\ ^1P^\circ-6d\ ^1D$	1—2
4585,923	2	2,53	5,23	$3d\ ^3D-4f\ ^3F^\circ$	3—3
4585,871	50	2,53	5,23	$3d\ ^3D-4f\ ^3F^\circ$	3—4
4581,402	40	2,52	5,23	$3d\ ^3D-4f\ ^3F^\circ$	2—3
4578,558	30	2,52	5,23	$3d\ ^3D-4f\ ^3F^\circ$	1—2
4526,935	30	2,71	5,45	$3d\ ^1D-6p\ ^1P^\circ$	2—1
4512,282	5	2,53	5,27	$3d\ ^3D-6p\ ^3P^\circ$	3—2
4509,446	3	2,52	5,27	$3d\ ^3D-6p\ ^3P^\circ$	2—1
4507,854	1	2,52	5,27	$3d\ ^3D-6p\ ^3P^\circ$	2—2
4507,417	1	2,52	5,27	$3d\ ^3D-6p\ ^3P^\circ$	1—0
4506,624	1	2,52	5,27	$3d\ ^3D-6p\ ^3P^\circ$	1—1
4505,00	0	2,52	5,27	$3d\ ^3D-6p\ ^3P^\circ$	1—2
4456,612	10	1,90	4,68	$4p\ ^3P^\circ-4d\ ^3D$	2—1
4455,887	40	1,90	4,68	$4p\ ^3P^\circ-4d\ ^3D$	2—2
4454,781	80	1,90	4,68	$4p\ ^3P^\circ-4d\ ^3D$	2—3
4435,688	40	1,89	4,68	$4p\ ^3P^\circ-4d\ ^3D$	1—1
4434,960	60	1,89	4,68	$4p\ ^3P^\circ-4d\ ^3D$	1—2
4425,441	50	1,88	4,68	$4p\ ^3P^\circ-4d\ ^3D$	0—1
4355,096	25	2,71	5,55	$3d\ ^1D-5f\ ^1F^\circ$	2—3
4318,652	45	1,90	4,77	$4p\ ^3P^\circ-4p^2\ ^3P$	2—1
4307,741	45	1,89	4,76	$4p\ ^3P^\circ-4p^2\ ^3P$	1—0
4302,527	60	1,90	4,78	$4p\ ^3P^\circ-4p^2\ ^3P$	2—2
4298,986	30	1,89	4,77	$4p\ ^3P^\circ-4p^2\ ^3P$	1—1
4289,364	40	1,88	4,77	$4p\ ^3P^\circ-4p^2\ ^3P$	0—1
4283,010	40	1,89	4,78	$4p\ ^3P^\circ-4p^2\ ^3P$	1—2
4240,456	6	2,71	5,63	$3d\ ^1D-7p\ ^1P^\circ$	2—1
4226,728	500	0,00	2,93	$4s^2\ ^1S-4p\ ^1P^\circ$	0—1
4108,554	10	2,71	5,73	$3d\ ^1D-6f\ ^1F^\circ$	2—3
4098,533	15	2,53	5,55	$3d\ ^3D-5f\ ^3F^\circ$	3—4
4094,930	12	2,52	5,55	$3d\ ^3D-5f\ ^3F^\circ$	2—3
4092,633	8	2,52	5,55	$3d\ ^3D-5f\ ^3F^\circ$	1—2
4058,912	1	2,71	5,76	$3d\ ^1D-8p\ ^1P^\circ$	2—1
3973,707	12	1,90	5,02	$4p\ ^3P^\circ-6s\ ^3S$	2—1
3972,570	1	2,71	5,83	$3d\ ^1D-7f\ ^1F^\circ$	2—3
3957,053	10	1,89	5,02	$4p\ ^3P^\circ-6s\ ^3S$	1—1
3948,901	6	1,88	5,02	$4p\ ^3P^\circ-6s\ ^3S$	0—1
3923,50	0	1,89	5,04	$4p\ ^3P^\circ-4p^2\ ^1S$	1—0
3889,141	1	2,71	5,90	$3d\ ^1D-8f\ ^1F^\circ$	2—3
3875,807	4	2,53	5,72	$3d\ ^3D-6f\ ^3F^\circ$	3—2, 3, 4
3872,552	3	2,52	5,72	$3d\ ^3D-6f\ ^3F^\circ$	2—2, 3
3870,508	2	2,52	5,72	$3d\ ^3D-6f\ ^3F^\circ$	1—2
3761,72	0	1,89	5,18	$4p\ ^3P^\circ-6s\ ^1S$	1—0
3753,367	1	2,53	5,83	$3d\ ^3D-7f\ ^3F^\circ$	3—2, 3, 4
3750,349	1	2,52	5,83	$3d\ ^3D-7f\ ^3F^\circ$	2—2, 3
3748,374	1	2,52	5,83	$3d\ ^3D-7f\ ^3F^\circ$	1—2
3678,240	3	2,53	5,90	$3d\ ^3D-8f\ ^3F^\circ$	3—2, 3, 4
3675,307	2	2,53	5,90	$3d\ ^3D-8f\ ^3F^\circ$	2—2, 3
3673,448	1	2,52	5,90	$3d\ ^3D-8f\ ^3F^\circ$	1—2
3644,990	2	1,90	5,30	$4p\ ^3P^\circ-5d\ ^3D$	2—1
3644,765	15	1,90	5,30	$4p\ ^3P^\circ-5d\ ^3D$	2—2
3644,410	40	1,90	5,30	$4p\ ^3P^\circ-5d\ ^3D$	2—3
3630,974	15	1,89	5,30	$4p\ ^3P^\circ-5d\ ^3D$	1—1
3630,748	30	1,89	5,30	$4p\ ^3P^\circ-5d\ ^3D$	1—2
3624,111	20	1,88	5,30	$4p\ ^3P^\circ-5d\ ^3D$	0—1
3487,598	12	1,90	5,45	$4p\ ^3P^\circ-7s\ ^3S$	2—1
3474,763	8	1,89	5,45	$4p\ ^3P^\circ-7s\ ^3S$	1—1
3468,476	4	1,88	5,45	$4p\ ^3P^\circ-7s\ ^3S$	0—1
3362,28	0	1,90	5,59	$4p\ ^3P^\circ-6d\ ^3D$	2—1

λ, Å	I	E_H, eV	E_B, eV	Transition	J
3362,131	35	1,90	5,59	$4p\ ^3P^\circ—6d\ ^3D$	2—2
3361,918	35	1,90	5,59	$4p\ ^3P^\circ—6d\ ^3D$	2—3
3350,361	25	1,89	5,59	$4p\ ^3P^\circ—6d\ ^3D$	1—1
3350,209	25	1,89	5,59	$4p\ ^3P^\circ—6d\ ^3D$	1—2
3344,513	8	1,88	5,59	$4p\ ^3P^\circ—6d\ ^3D$	0—1
3286,067	4	1,90	5,67	$4p\ ^3P^\circ—8s\ ^3S$	2—1
3274,661	2	1,89	5,67	$4p\ ^3P^\circ—8s\ ^3S$	1—1
3269,090	1	1,88	5,67	$4p\ ^3P^\circ—8s\ ^3S$	0—1
3226,129	8	1,90	5 74	$4p\ ^3P^\circ—7d\ ^3D$	2—2
3225,896	8	1,90	5,74	$4p\ ^3P^\circ—7d\ ^3D$	2—3
3215,334	5	1,89	5,74	$4p\ ^3P^\circ—7d\ ^3D$	1—1
3215,145	5	1,89	5,74	$4p\ ^3P^\circ—7d\ ^3D$	1—2
3209,930	2	1,88	5,74	$4p\ ^3P^\circ—7d\ ^3D$	0—1
3180,521	1	1,90	5,80	$4p\ ^3P^\circ—9s\ ^3S$	2—1
3169,854	1	1,89	5,80	$4p\ ^3P^\circ—9s\ ^3S$	1—1
3164,618	1	1,88	5,80	$4p\ ^3P^\circ—9s\ ^3S$	0—1
3151,280	4	1,90	5,83	$4p\ ^3P^\circ—8d\ ^3D$	2—2
3150,738	4	1,90	5,83	$4p\ ^3P^\circ—8d\ ^3D$	2—3
3141,164	3	1,89	5,83	$4p\ ^3P^\circ—8d\ ^3D$	1—1
3140,782	3	1,89	5,83	$4p\ ^3P^\circ—8d\ ^3D$	1—2
3136,003	1	1,88	5,83	$4p\ ^3P^\circ—8d\ ^3D$	0—1
3117,656	1	1,90	5,87	$4p\ ^3P^\circ—10s\ ^3S$	2—1
3108,58	3	—	—	—	—
3107,388	1	1,89	5,87	$4p\ ^3P^\circ—10s\ ^3S$	1—1
3102,36	0	1,88	5,87	$4p\ ^3P^\circ—10s\ ^3S$	0—1
3080,826	2	—	—	—	—
3009,205	5	1,90	6,02	$4p\ ^3P^\circ—3d^2\ ^3P$	2—1
3006,858	6	1,90	6,02	$4p\ ^3P^\circ—3d^2\ ^3P$	2—2
3000,863	5	1,89	6,02	$4p\ ^3P^\circ—3d^2\ ^3P$	1—0
2999,641	4	1,89	6,02	$4p\ ^3P^\circ—3d^2\ ^3P$	1—1
2997,309	5	1,89	6,02	$4p\ ^3P^\circ—3d^2\ ^3P$	1—2
2994,958	5	1,88	6,02	$4p\ ^3P^\circ—3d^2\ ^3P$	0—1
2772,80	1	1,90	6,37	$4p\ ^3P^\circ—4d'\ ^3D$	2—2
2770,79	3	1,90	6,37	$4p\ ^3P^\circ—4d'\ ^3D$	2—3
2766,13	1	1,89	6,37	$4p\ ^3P^\circ—4d'\ ^3D$	1—1
2764,60	2	1,89	6,37	$4p\ ^3P^\circ—4d'\ ^3D$	1—2
2762,05	2	1,88	6,37	$4p\ ^3P^\circ—4d'\ ^3D$	0—1
2757,40	2	1,90	6,39	$4p\ ^3P^\circ—4d'\ ^3S$	2—1
2749,34	1	1,89	6,39	$4p\ ^3P^\circ—4d'\ ^3S$	1—1
2745,49	1	1,88	6,39	$4p\ ^3P^\circ—4d'\ ^3S$	0—1
2734,82	2	0,00	4,53	$4s^2\ ^1S—5p\ ^3P^\circ$	0—1
2721,645	10	0,00	4,55	$4s^2\ ^1S—4p'\ ^1P^\circ$	0—1
2617,66	3	0,00	4,73	$4s^2\ ^1S—4p'\ ^3D^\circ$	0—1
2565,20	2	1,90	6,73	$4p\ ^3P^\circ—4d'\ ^3P$	2—1
2564,09	3	1,90	6,73	$4p\ ^3P^\circ—4d'\ ^3P$	2—2
2558,60	2	1,89	6,73	$4p\ ^3P^\circ—4d'\ ^3P$	1—0
2558,20	2	1,89	6,73	$4p\ ^3P^\circ—4d'\ ^3P$	1—1
2557,18	2	1,89	6,73	$4p\ ^3P^\circ—4d'\ ^3P$	1—2
2554,82	2	1,88	6,73	$4p\ ^3P^\circ—4d'\ ^3P$	0—1
2541,40	0	0,00	4,88	$4s^2\ ^1S—4p'\ ^3P^\circ$	0—1
2398,559	2	0,00	5,17	$4s^2\ ^1S—5p\ ^1P^\circ$	0—1
2275,471	1	0,00	5,45	$4s^2\ ^1S—6p\ ^1P^\circ$	0—1
2200,728	1	0,00	5,63	$4s^2\ ^1S—7p\ ^1P^\circ$	0—1
2150,78	1	0,00	5,76	$4s^2\ ^1S—8p\ ^1P^\circ$	0—1

Ca II, ground state $1s^2 2s^2 2p^6 3s^2 3p^6 4s\ ^2S_{1/2}$
Ionization potential 95 748,0 cm^{-1}; 11,870 eV

λ, Å	I	E_H, eV	E_B, eV	Transition	J
11949,72	1	6,47	7,50	$5s\ ^2S-5p\ ^2P^\circ$	$1/2-1/2$
11838,99	2	6,47	7,51	$5s\ ^2S-5p\ ^2P^\circ$	$1/2-3/2$
9931,39	9	7,51	8,76	$5p\ ^2P^\circ-6s\ ^2S$	$3/2-1/2$
9890,63	11	8,44	9,69	$4f\ ^2F^\circ-5g\ ^2G$	$5/2,\ 7/2-7/2,\ 9/2$
9854,74	8	7,50	8,76	$5p\ ^2P^\circ-6s\ ^2S$	$1/2-1/2$
8927,36	11	7,05	8,44	$4d\ ^2D-4f\ ^2F^\circ$	$5/2-7/2$
8912,07	10	7,05	8,44	$4d\ ^2D-4f\ ^2F^\circ$	$3/2-5/2$
8662,140	16	1,69	3,12	$3d\ ^2D-4p\ ^2P^\circ$	$3/2-1/2$
8542,089	17	1,70	3,15	$3d\ ^2D-4p\ ^2P^\circ$	$5/2-3/2$
8498,018	13	1,69	3,15	$3d\ ^2D-4p\ ^2P^\circ$	$3/2-3/2$
8254,725	7	7,51	9,02	$5p\ ^2P^\circ-5d\ ^2D$	$3/2-3/2$
8248,797	11	7,51	9,02	$5p\ ^2P^\circ-5d\ ^2D$	$3/2-5/2$
8201,720	10	7,50	9,02	$5p\ ^2P^\circ-5d\ ^2D$	$1/2-3/2$
8020,504	2	8,44	9,98	$4f\ ^2F^\circ-6d\ ^2D$	$5/2-3/2$
8017,502	2	8,44	9,98	$4f\ ^2F^\circ-6d\ ^2D$	$7/2-5/2$
6456,874	8	8,44	10,36	$4f\ ^2F^\circ-6g\ ^2G$	$5/2,\ 7/2-7/2,\ 9/2$
5923,69	1	8,44	10,53	$4f\ ^2F^\circ-7d\ ^2D$	$5/2-3/2$
5922,72	1	8,44	10,53	$4f\ ^2F^\circ-7d\ ^2D$	$7/2-5/2$
5339,189	5	8,44	10,76	$4f\ ^2F^\circ-7g\ ^2G$	$5/2,\ 7/2-7/2,\ 9/2$
5307,223	7	7,51	9,85	$5p\ ^2P^\circ-7s\ ^2S$	$3/2-1/2$
5285,268	6	7,50	9,85	$5p\ ^2P^\circ-7s\ ^2S$	$1/2-1/2$
5021,138	4	7,51	9,98	$5p\ ^2P^\circ-6d\ ^2D$	$3/2-3/2$
5019,971	8	7,51	9,98	$5p\ ^2P^\circ-6d\ ^2D$	$3/2-5/2$
5001,479	7	7,50	9,98	$5p\ ^2P^\circ-6d\ ^2D$	$1/2-3/2$
4799,973	4	8,44	11,02	$4f\ ^2F^\circ-8g\ ^2G$	$5/2,\ 7/2-7/2,\ 9/2$
4721,028	4	7,05	9,67	$4d\ ^2D-5f\ ^2F^\circ$	$5/2-7/2$
4716,736	3	7,05	9,67	$4d\ ^2D-5f\ ^2F^\circ$	$3/2-5/2$
4489,178	2	8,44	11,20	$4f\ ^2F^\circ-9g\ ^2G$	$5/2,\ 7/2-7/2,\ 9/2$
4479,226	1	6,47	9,23	$5s\ ^2S-6p\ ^2P^\circ$	$1/2-1/2$
4472,043	2	6,47	9,24	$5s\ ^2S-6p\ ^2P^\circ$	$1/2-3/2$
4220,074	5	7,51	10,45	$5p\ ^2P^\circ-8s\ ^2S$	$3/2-1/2$
4206,175	4	7,50	10,45	$5p\ ^2P^\circ-8s\ ^2S$	$1/2-1/2$
4110,279	3	7,51	10,53	$5p\ ^2P^\circ-7d\ ^2D$	$3/2-3/2$
4109,816	6	7,51	10,53	$5p\ ^2P^\circ-7d\ ^2D$	$3/2-5/2$
4097,102	5	7,50	10,53	$5p\ ^2P^\circ-7d\ ^2D$	$1/2-3/2$
3968,468	22	0,00	3,12	$4s\ ^2S-4p\ ^2P^\circ$	$1/2-1/2$
3933,663	23	0,00	3,15	$4s\ ^2S-4p\ ^2P^\circ$	$1/2-3/2$
3758,386	3	7,05	10,35	$4d\ ^2D-6f\ ^2F^\circ$	$5/2-7/2$
3755,668	2	7,05	10,35	$4d\ ^2D-6f\ ^2F^\circ$	$3/2-5/2$
3739,375	1	7,50	10,82	$5p\ ^2P^\circ-9s\ ^2S$	$1/2-1/2$
3736,901	18	3,15	6,47	$4p\ ^2P^\circ-5s\ ^2S$	$3/2-1/2$
3706,026	17	3,12	6,47	$4p\ ^2P^\circ-5s\ ^2S$	$1/2-1/2$
3694,355	1	7,51	10,87	$5p\ ^2P^\circ-8d\ ^2D$	$3/2-3/2$
3694,108	4	7,51	10,87	$5p\ ^2P^\circ-8d\ ^2D$	$3/2-5/2$
3683,696	3	7,50	10,87	$5p\ ^2P^\circ-8d\ ^2D$	$1/2-3/2$
3495,156	1	7,51	11,06	$5p\ ^2P^\circ-10s\ ^2S$	$3/2-1/2$
3461,871	2	7,51	11,09	$5p\ ^2P^\circ-9d\ ^2D$	$3/2-5/2$
3452,657	1	7,50	11,09	$5p\ ^2P^\circ-9d\ ^2D$	$1/2-3/2$
3347,035	1	7,05	10,75	$4d\ ^2D-7f\ ^2F^\circ$	$5/2-7/2$
3181,275	15	3,15	7,05	$4p\ ^2P^\circ-4d\ ^2D$	$3/2-3/2$
3179,332	18	3,15	7,05	$4p\ ^2P^\circ-4d\ ^2D$	$3/2-5/2$
3158,869	17	3,12	7,05	$4p\ ^2P^\circ-4d\ ^2D$	$1/2-3/2$
2573,09	3	—	—	—	—
2208,611	3	3,15	8,76	$4p\ ^2P^\circ-6s\ ^2S$	$3/2-1/2$
2197,787	2	3,12	8,76	$4p\ ^2P^\circ-6s\ ^2S$	$1/2-1/2$

λ, Å	I	E_H, eV	E_B, eV	Transition	J
2132,304	1	1,69	7,50	3d ^2D—5p $^2P^\circ$	3/2—1/2
2131,505	2	1,70	7,51	3d ^2D—5p $^2P^\circ$	5/2—3/2
2128,750	0	1,69	7,51	3d ^2D—5p $^2P^\circ$	3/2—3/2
2113,146	1	3,15	9,02	4p $^2P^\circ$—5d ^2D	3/2—3/2
2112,757	2	3,15	9,02	4p $^2P^\circ$—5d ^2D	3/2—5/2
2103,235	2	3,12	9,02	4p $^2P^\circ$—5d ^2D	1/2—3/2
1850,691	2	3,15	9,85	4p $^2P^\circ$—7s ^2S	3/2—1/2
1843,088	1	3,12	9,85	4p $^2P^\circ$—7s ^2S	1/2—1/2
1840,061	8	1,70	8,44	3d ^2D—4f $^2F^\circ$	5/2—5/2, 7/2
1838,008	7	1,69	8,44	3d ^2D—4f $^2F^\circ$	3/2—5/2
1814,647	1	3,15	9,98	4p $^2P^\circ$—6d ^2D	3/2—3/2
1814,495	1	3,15	9,98	4p $^2P^\circ$—6d ^2D	3/2—5/2
1807,337	1	3,12	9,98	4p $^2P^\circ$—6d ^2D	1/2—3/2
1698,183	—	3,15	10,45	4p $^2P^\circ$—8s ^2S	3/2—1/2
1691,779	1	3,12	10,45	4p $^2P^\circ$—8s ^2S	1/2—1/2
1680,129	1	3,15	10,53	4p $^2P^\circ$—7d ^2D	3/2—3/2
1680,051	—	3,15	10,53	4p $^2P^\circ$—7d ^2D	3/2—5/2
1673,860	1	3,12	10,53	4p $^2P^\circ$—7d ^2D	1/2—3/2
1651,991	1	0,00	7,50	4s ^2S—5p $^2P^\circ$	1/2—1/2
1649,858	2	0,00	7,51	4s ^2S—5p $^2P^\circ$	1/2—3/2
1644,441	0	1,70	9,24	3d ^2D—6p $^2P^\circ$	5/2—3/2
1643,770	—	1,69	9,23	3d ^2D—6p $^2P^\circ$	3/2—1/2
1642,802	—	1,69	9,24	3d ^2D—6p $^2P^\circ$	3/2—3/2
1554,642	4	1,70	9,67	3d ^2D—5f $^2F^\circ$	5/2—5/2, 7/2
1553,176	4	1,69	9,67	3d ^2D—5f $^2F^\circ$	3/2—5/2
1433,749	—	1,70	10,35	3d ^2D—6f $^2F^\circ$	5/2—5/2, 7/2
1432,503	1	1,69	10,35	3d ^2D—6f $^2F^\circ$	3/2—5/2
1370,6	3	—	—	—	—
1369,1	3	—	—	—	—
1342,535	—	0,00	9,23	4s ^2S—6p $^2P^\circ$	1/2—1/2
1341,889	1	0,00	9,24	4s ^2S—6p $^2P^\circ$	1/2—3/2

Ca III, ground state $1s^2\,2s^2\,2p^6\,3s^2\,3p^6\,{}^1S_0$

Ionization potential 413 127 cm^{-1}; 51,218 eV

λ, Å	I	E_H, eV	E_B, eV	Transition	J
4081,74	5	30,71	33,75	4s′ [1/2]°—4p [1/2]	1—1
3761,62	6	30,45	33,75	4s′ [1/2]°—4p [1/2]	0—1
3537,75	7	30,24	33,75	4s [1 1/2]°—4p [1/2]	1—1
3372,68	8	30,07	33,75	4s [1 1/2]°—4p [1/2]	2—1
3367,81	5	30,71	34,39	4s′ [1/2]°—4p [2 1/2]	1—2
3233,02	4	30,71	34,54	4s′ [1/2]°—4p [1 1/2]	1—1
3119,66	8	30,71	34,68	4s′ [1/2]°—4p [1 1/2]	1—2
3028,66	6	30,45	34,54	4s′ [1/2]°—4p [1 1/2]	0—1
2989,30	6	30,71	34,86	4s′ [1/2]°—4p′ [1 1/2]	1—1
2988,61	7	30,24	34,39	4s [1 1/2]°—4p [2 1/2]	1—2
2924,33	8	30,71	34,95	4s′ [1/2]°—4p′ [1 1/2]	1—2
2907,90	2	30,71	34,97	4s′ [1/2]°—4p [1/2]	1—0
2899,78	9	30,07	34,34	4s [1 1/2]°—4p [2 1/2]	2—3
2881,80	7	30,24	34,54	4s [1 1/2]°—4p [1 1/2]	1—1
2869,95	7	30,07	34,39	4s [1 1/2]°—4p [2 1/2]	2—2
2866,57	7	30,71	35,03	4s′ [1/2]°—4p′ [1/2]	1—1
2813,88	7	30,45	34,85	4s′ [1/2]°—4p′ [1 1/2]	0—1

λ, Å	I	E_H, eV	E_B, eV	Transition	
2791,63	6	30,24	34,68	$4s\ [1^1/_2]^\circ - 4p\ [1^1/_2]$	1—2
2771,27	4	30,07	34,54	$4s\ [1^1/_2]^\circ - 4p\ [1^1/_2]$	2—1
2704,87	6	30,45	35,03	$4s'\ [^1/_2]^\circ - 4p'\ [^1/_2]$	0—1
2687,78	8	30,07	34,68	$4s\ [1^1/_2]^\circ - 4p\ [1^1/_2]$	2—2
2686,73	3	30,24	34,85	$4s\ [1^1/_2]^\circ - 4p'\ [1^1/_2]$	1—1
2634,17	6	30,24	34,95	$4s\ [1^1/_2]^\circ - 4p'\ [1^1/_2]$	1—2
2620,82	6	30,24	34,97	$4s\ [1^1/_2]^\circ - 4p\ [^1/_2]$	1—0
2590,34	2	30,07	34,85	$4s\ [1^1/_2]^\circ - 4p'\ [1^1/_2]$	2—1
2587,09	3	30,24	35,03	$4s\ [1^1/_2]^\circ - 4p'\ [^1/_2]$	1—1
2541,49	6	30,07	34,95	$4s\ [1^1/_2]^\circ - 4p'\ [1^1/_2]$	2—2
2497,67	5	30,07	35,03	$4s\ [1^1/_2]^\circ - 4p'\ [^1/_2]$	2—1
2472,52	1	35,03	40,04	$4p'\ [^1/_2] - 4d\ [^1/_2]^\circ$	1—1
2442,54	1	34,97	40,04	$4p\ [^1/_2] - 4d\ [^1/_2]^\circ$	0—1
2431,08	1	34,95	40,04	$4p'\ [1^1/_2] - 4d\ [^1/_2]^\circ$	2—1
2393,20	3	34,95	40,13	$4p'\ [1^1/_2] - 4d\ [1^1/_2]^\circ$	2—2
2351,40	1	34,85	40,13	$4p'\ [1^1/_2] - 4d\ [1^1/_2]^\circ$	1—2
2310,87	0	34,68	40,04	$4p\ [1^1/_2] - 4d\ [^1/_2]^\circ$	2—1
2276,54	2	34,68	40,13	$4p\ [1^1/_2] - 4d\ [1^1/_2]^\circ$	2—2
2256,33	0	34,95	40,44	$4p'\ [1^1/_2] - 4d\ [3^1/_2]^\circ$	2—3
2252,65	2	34,54	40,04	$4p\ [1^1/_2] - 4d\ [^1/_2]^\circ$	1—1
2244,31	2	28,87	34,39	$3d'\ [1^1/_2]^\circ - 4p\ [2^1/_2]$	1—2
2219,87	1	34,54	40,13	$4p\ [1^1/_2] - 4d\ [1^1/_2]^\circ$	1—2
2219,22	2	34,85	40,44	$4p'\ [1^1/_2] - 4d\ [3^1/_2]^\circ$	1—3
2204,40	3	35,03	40,65	$4p'\ [^1/_2] - 5s\ [1^1/_2]^\circ$	1—2
2191,27	2	34,39	40,04	$4p\ [2^1/_2] - 4d\ [^1/_2]^\circ$	2—1
2183,30	3	28,87	34,54	$3d'\ [1^1/_2]^\circ - 4p\ [1^1/_2]$	1—1
2171,60	5	34,95	40,65	$4p'\ [1^1/_2] - 5s\ [1^1/_2]^\circ$	2—2
2163,51	4	34,95	40,68	$4p'\ [1^1/_2] - 4d\ [2^1/_2]^\circ$	2—2
2160,40	2	34,39	40,13	$4p\ [2^1/_2] - 4d\ [1^1/_2]^\circ$	2—2
2152,47	6	34,68	40,44	$4p\ [1^1/_2] - 4d\ [3^1/_2]^\circ$	2—3
2143,81	1	34,34	40,13	$4p\ [2^1/_2] - 4d\ [1^1/_2]^\circ$	3—2
2140,39	6	34,95	40,74	$4p'\ [1^1/_2] - 5s\ [1^1/_2]^\circ$	2—1
2131,49	3	28,87	34,68	$3d'\ [1^1/_2]^\circ - 4p\ [1^1/_2]$	1—2
2129,20	6	34,85	40,68	$4p'\ [1^1/_2] - 4d\ [2^1/_2]^\circ$	1—2
2098,56	5	27,84	33,74	$3d\ [1^1/_2]^\circ - 4p\ [^1/_2]$	1—1
2074,89	2	34,68	40,65	$4p\ [1^1/_2] - 5s\ [1^1/_2]^\circ$	2—2
2069,55	2	28,87	34,85	$3d'\ [1^1/_2]^\circ - 4p'\ [1^1/_2]$	1—1
2067,64	2	34,68	40,68	$4p\ [1^1/_2] - 4d\ [2^1/_2]^\circ$	2—2
2062,17	3	35,03	41,04	$4p'\ [^1/_2] - 5s'\ [^1/_2]^\circ$	1—0
2056,64	3	28,32	34,34	$3d'\ [2^1/_2]^\circ - 4p\ [2^1/_2]$	3—3
2048,36	3	34,39	40,44	$4p\ [2^1/_2] - 4d\ [3^1/_2]^\circ$	2—3
2047,14	3	35,03	41,09	$4p'\ [^1/_2] - 5s'\ [^1/_2]^\circ$	1—1
2046,65	4	34,68	40,74	$4p\ [1^1/_2] - 5s\ [1^1/_2]^\circ$	2—1
2041,61	4	28,32	34,39	$3d'\ [2^1/_2]^\circ - 4p\ [2^1/_2]$	3—2
2038,35	3	28,87	34,95	$3d'\ [1^1/_2]^\circ - 4p'\ [1^1/_2]$	1—2
2033,46	4	34,34	40,44	$4p\ [2^1/_2] - 4d\ [3^1/_2]^\circ$	3—3
2030,01	1	28,87	34,97	$3d'\ [1^1/_2]^\circ - 4p\ [^1/_2]$	1—0
2027,72	2	34,54	40,65	$4p\ [1^1/_2] - 5s\ [1^1/_2]^\circ$	1—2
2026,68	2	34,97	41,09	$4p\ [^1/_2] - 5s'\ [^1/_2]^\circ$	0—1
2020,83	3	34,54	40,68	$4p\ [1^1/_2] - 4d\ [2^1/_2]^\circ$	1—2
2018,77	4	34,95	41,09	$4p'\ [1^1/_2] - 5s'\ [^1/_2]^\circ$	2—1
2014,10	3	28,19	34,34	$3d'\ [1^1/_2]^\circ - 4p\ [2^1/_2]$	2—3
2009,90	1	28,87	35,03	$3d'\ [1^1/_2]^\circ - 4p'\ [^1/_2]$	1—1
2003,09	3	34,85	41,04	$4p'\ [1^1/_2] - 5s'\ [^1/_2]^\circ$	1—0
2000,96	4	34,54	40,74	$4p\ [1^1/_2] - 5s\ [1^1/_2]^\circ$	1—1
1999,79	4	28,19	34,39	$3d'\ [1^1/_2]^\circ - 4p\ [2^1/_2]$	2—2
1989,61	2	34,85	41,09	$4p'\ [1^1/_2] - 5s'\ [^1/_2]^\circ$	1—1
1978,63	3	34,39	40,65	$4p\ [2^1/_2] - 5s\ [1^1/_2]^\circ$	2—2

λ, Å	I	E_H, eV	E_B, eV	Transition	J
1972,01	1	34,39	40,68	$4p\ [2^1/_2]-4d\ [2^1/_2]^\circ$	2—2
1968,03	5	33,75	40,04	$4p\ [^1/_2]-4d\ [^1/_2]^\circ$	1—1
1964,70	5	34,34	40,65	$4p\ [2^1/_2]-5s\ [1^1/_2]^\circ$	3—2
1958,18	2	34,34	40,68	$4p\ [2^1/_2]-4d\ [2^1/_2]^\circ$	3—2
1953,06	4	28,00	34,34	$3d'\ [2^1/_2]^\circ-4p\ [2^1/_2]$	2—3
		34,39	40,74	$4p\ [2^1/_2]-5s\ [1^1/_2]^\circ$	2—1
1952,16	3	28,19	34,54	$3d'\ [1^1/_2]^\circ-4p\ [1^1/_2]$	2—1
1948,31	5	28,32	34,68	$3d'\ [2^1/_2]^\circ-4p\ [1^1/_2]$	3—2
1943,12	6	33,75	40,13	$4p\ [^1/_2]-4d\ [1^1/_2]^\circ$	1—2
1939,72	4	28,00	34,39	$3d'\ [2^1/_2]^\circ-4p\ [2^1/_2]$	2—2
1935,79	3	34,68	41,09	$4p\ [1^1/_2]-5s'\ [^1/_2]^\circ$	2—1
1910,17	4	28,19	34,68	$3d'\ [1^1/_2]^\circ-4p\ [1^1/_2]$	2—2
1907,46	2	34,54	41,04	$4p\ [1^1/_2]-5s'\ [^1/_2]^\circ$	1—0
1894,17	3	28,00	34,54	$3d'\ [2^1/_2]^\circ-4p\ [1^1/_2]$	2—1
1892,92	1	27,84	34,39	$3d\ [1^1/_2]^\circ-4p\ [2^1/_2]$	1—2
1872,39	5	34,95	41,57	$4p'\ [1^1/_2]-4d'\ [2^1/_2]^\circ$	2—3
1870,28	6	28,32	34,95	$3d'\ [2^1/_2]^\circ-4p'\ [1^1/_2]$	3—2
1860,50	3	28,19	34,85	$3d'\ [1^1/_2]^\circ-4p'\ [1^1/_2]$	2—1
1854,72	6	28,00	34,68	$3d'\ [2^1/_2]^\circ-4p\ [1^1/_2]$	2—2
1849,51	2	27,84	34,54	$3d\ [1^1/_2]^\circ-4p\ [1^1/_2]$	1—1
1828,43	1	28,19	34,95	$3d'\ [1^1/_2]^\circ-4p\ [^1/_2]$	2—0
1812,17	5	27,84	34,68	$3d\ [1^1/_2]^\circ-4p\ [1^1/_2]$	1—2
		28,19	35,03	$3d'\ [1^1/_2]^\circ-4p'\ [^1/_2]$	2—1
1807,91	5	28,00	34,85	$3d'\ [2^1/_2]^\circ-4p'\ [1^1/_2]$	2—1
1800,24	4	34,68	41,57	$4p\ [1^1/_2]-4d'\ [2^1/_2]^\circ$	2—2
1794,31	4	33,74	40,65	$4p\ [^1/_2]-5s\ [1^1/_2]^\circ$	1—2
1783,92	4	28,00	34,95	$3d'\ [2^1/_2]^\circ-4p'\ [1^1/_2]$	2—2
1773,29	3	33,74	40,74	$4p\ [^1/_2]-5s\ [1^1/_2]^\circ$	1—1
1762,14	3	28,00	35,03	$3d'\ [2^1/_2]^\circ-4p'\ [^1/_2]$	2—1
1744,61	3	27,84	34,95	$3d\ [1^1/_2]^\circ-4p'\ [1^1/_2]$	1—2
1738,56	3	27,84	34,97	$3d\ [1^1/_2]^\circ-4p\ [^1/_2]$	1—0
1726,88	1	34,39	41,57	$4p\ [2^1/_2]-4d'\ [2^1/_2]^\circ$	2—3
1716,23	1	34,34	41,57	$4p\ [2^1/_2]-4d\ [2^1/_2]^\circ$	3—3
1698,95	1	33,74	41,04	$4p\ [^1/_2]-5s'\ [^1/_2]^\circ$	1—0
1688,81	1	33,74	41,09	$4p\ [^1/_2]-5s'\ [^1/_2]^\circ$	1—1
1595,24	1	26,57	34,34	$3d\ [2^1/_2]^\circ-4p\ [2^1/_2]$	2—3
1586,19	4	26,57	34,39	$3d\ [2^1/_2]^\circ-4p\ [2^1/_2]$	2—2
1571,31	5	26,45	34,34	$3d\ [3^1/_2]^\circ-4p\ [2^1/_2]$	3—3
1562,50	6	26,45	34,39	$3d\ [3^1/_2]^\circ-4p\ [2^1/_2]$	3—2
1555,48	4	26,57	34,54	$3d\ [2^1/_2]^\circ-4p\ [1^1/_2]$	2—1
1528,89	0	26,57	34,68	$3d\ [2^1/_2]^\circ-4p\ [1^1/_2]$	2—2
1506,94	3	26,45	34,68	$3d\ [3^1/_2]^\circ-4p\ [1^1/_2]$	3—2
1496,92	2	26,57	34,85	$3d\ [2^1/_2]^\circ-4p'\ [1^1/_2]$	2—1
1484,92	4	25,39	33,74	$3d\ [1^1/_2]^\circ-4p\ [^1/_2]$	2—1
1480,55	2	26,57	34,95	$3d\ [2^1/_2]^\circ-4p'\ [1^1/_2]$	2—2
1463,41	4	25,27	33,74	$3d\ [^1/_2]^\circ-4p\ [^1/_2]$	1—1
1459,87	3	26,45	34,95	$3d\ [3^1/_2]^\circ-4p'\ [1^1/_2]$	3—2
1385,39	2	25,39	34,34	$3d\ [1^1/_2]^\circ-4p\ [2^1/_2]$	2—3
1359,81	1	25,27	34,39	$3d\ [^1/_2]^\circ-4p\ [2^1/_2]$	1—2
1355,37	1	25,39	34,54	$3d\ [1^1/_2]^\circ-4p\ [1^1/_2]$	2—1
1334,94	3	25,39	34,68	$3d\ [1^1/_2]^\circ-4p\ [1^1/_2]$	2—2
1317,60	2	25,27	34,68	$3d\ [^1/_2]^\circ-4p\ [1^1/_2]$	1—2
1310,58	1	25,39	34,85	$3d\ [1^1/_2]^\circ-4p'\ [1^1/_2]$	2—1
1297,96	3	25,39	34,95	$3d\ [1^1/_2]^\circ-4p'\ [1^1/_2]$	2—2
1281,50	2	25,27	34,95	$3d\ [^1/_2]^\circ-4p'\ [1^1/_2]$	1—2
1278,38	2	25,27	34,97	$3d\ [^1/_2]^\circ-4p\ [^1/_2]$	1—0
1270,54	2	25,27	35,03	$3d\ [^1/_2]^\circ-4p'\ [^1/_2]$	1—1
1035,366	4	—	—	—	—
1034,848	3	—	—	—	—

λ, Å	I	$E_н$, eV	$E_в$, eV	Transition	J
1031,760	4	—	—	—	—
1028,560	4	—	—	—	—
1019,371	2	—	—	—	—
1014,998	3	—	—	—	—
998,397	3	—	—	—	—
987,336	5	—	—	—	—
984,935	4	—	—	—	—
977,544	2	—	—	—	—
917,278	2	—	—	—	—
916,917	2	—	—	—	—
906,615	2	—	—	—	—
894,351	4	—	—	—	—
890,892	2	—	—	—	—
883,159	3	—	—	—	—
864,695	4	—	—	—	—
856,791	4	—	—	—	—
856,635	4	—	—	—	—
850,966	3	—	—	—	—
846,611	3	—	—	—	—
840,921	3	—	—	—	—
772,641	2	—	—	—	—
740,555	3	—	—	—	—
696,206	3	—	—	—	—
695,824	4	—	—	—	—
688,907	3	—	—	—	—
686,190	2	—	—	—	—
681,908	4	—	—	—	—
653,720	4	—	—	—	—
639,230	6	—	—	—	—
633,187	5	—	—	—	—
629,355	4	—	—	—	—
621,279	3	—	—	—	—
612,062	5	—	—	—	—
610,740	3	—	—	—	—
608,895	3	—	—	—	—
607,069	3	—	—	—	—
606,333	2	—	—	—	—
603,622	3	—	—	—	—
596,223	2	—	—	—	—
594,640	3	—	—	—	—
591,449	3	—	—	—	—
587,311	3	—	—	—	—
586,880	2	—	—	—	—
575,580	6	—	—	—	—
574,398	3	—	—	—	—
572,691	4	—	—	—	—
570,291	2	—	—	—	—
561,670	3	—	—	—	—
561,518	3	—	—	—	—
557,112	3	—	—	—	—
556,583	3	—	—	—	—
556,215	2	—	—	—	—
554,615	3	—	—	—	—
550,004	3	—	—	—	—
549,201	4	—	—	—	—
545,091	2	—	—	—	—
536,132	4	—	—	—	—
534,715	4	—	—	—	—

λ, Å	I	E_H, eV	E_B, eV	Transition	J
528,286	8	—	—	—	—
490,546	2	0,00	25,27	$3p^6\,{}^1S—3d\,[{}^1/_2]^\circ$	0—1
478,305	4	—	—	—	—
464,830	3	—	—	—	—
409,971	18	0,00	30,24	$3p^6\,{}^1S—4s\,[1{}^1/_2]^\circ$	0—1
403,732	20	0,00	30,71	$3p^6\,{}^1S—4s'\,[{}^1/_2]^\circ$	0—1
396,382	3	—	—	—	—
392,420	2	—	—	—	—
369,647	5	—	—	—	—
368,303	3	—	—	—	—
357,973	8	—	—	—	—
344,219	2	—	—	—	—
340,389	3	—	—	—	—
304,330	3	0,00	40,74	$3p^6\,{}^1S—5s\,[1{}^1/_2]^\circ$	0—1
301,741	4	0,00	41,08	$3p^6\,{}^1S—5s'\,[{}^1/_2]^\circ$	1—1
248,636	4	—	—	—	—
242,384	3	—	—	—	—

Ca IV, ground state $1s^2\,2s^2\,2p^6\,3s^2\,3p^5\,{}^2P^0_{3/2}$

Ionization potential 542000 cm^{-1}; 67,196 eV

λ, Å	I	E_H, eV	E_B, eV	Transition	J
1030,273	4	—	—	—	—
1029,566	3	—	—	—	—
1027,309	5	—	—	—	—
1024,339	5	—	—	—	—
1023,820	4	—	—	—	—
997,579	7	—	—	—	—
994,311	6	—	—	—	—
892,671	3	—	—	—	—
858,855	3	—	—	—	—
775,526	3	—	—	—	—
669,725	10	0,39	18,90	$3p^5\,{}^2P^\circ—3p^6\,{}^2S$	$^1/_2—^1/_2$
656,038	15	0,00	18,90	$3p^5\,{}^2P^\circ—3p^6\,{}^2S$	$^3/_2—^1/_2$
635,318	8	—	—	—	—
565,463	3	18,90	40,82	$3p^6\,{}^2S—4p\,{}^2P^\circ$	$^1/_2—^3/_2$
538,967	2	—	—	—	—
538,648	3	—	—	—	—
537,004	2	—	—	—	—
536,790	2	—	—	—	—
536,531	2	—	—	—	—
535,647	4	—	—	—	—
456,981	5	0,39	27,52	$3p^5\,{}^2P^\circ—3d\,{}^4F$	$^1/_2—^3/_2$
450,565	10	0,00	27,52	$3p^5\,{}^2P^\circ—3d\,{}^4F$	$^3/_2—^3/_2$
445,018	1	0,39	28,25	$3p^5\,{}^2P^\circ—3d\,{}^4D$	$^1/_2—^3/_2$
444,766	3	—	—	—	—
443,821	15	0,39	28,32	$3p^5\,{}^2P^\circ—3d\,{}^2D$	$^1/_2—^3/_2$
439,700	5	0,00	28,20	$3p^5\,{}^2P^\circ—3d\,{}^4D$	$^3/_2—^5/_2$
438,930	4	0,00	28,25	$3p^5\,{}^2P^\circ—3d\,{}^4D$	$^3/_2—^3/_2$
437,773	5	0,00	28,32	$3p^5\,{}^2P^\circ—3d\,{}^2D$	$^3/_2—^3/_2$
437,271	2	0,00	28,35	$3p^5\,{}^2P^\circ—3d\,{}^4D$	$^3/_2—^1/_2$
434,570	12	0,00	28,53	$3p^5\,{}^2P^\circ—3d\,{}^2D$	$^3/_2—^5/_2$
374,744	5	0,00	33,08	$3p^5\,{}^2P^\circ—3d\,{}^2F$	$^3/_2—^5/_2$
345,130	4	—	—	—	—

λ, Å	I	E_H, eV	E_B, eV	Transition	J
344,958	4	0,39	36,33	$3p^5\,{}^2P^\circ$—$4s\,{}^4P$	$^1/_2$—$^3/_2$
343,933	5	—	—	—	—
343,438	4	0,39	36,49	$3p^5\,{}^2P^\circ$—$4s\,{}^4P$	$^1/_2$—$^1/_2$
343,203	6	0,00	36,12	$3p^5\,{}^2P^\circ$—$4s\,{}^4P$	$^3/_2$—$^5/_2$
342,447	5	—	—	—	—
341,455	4	—	—	—	—
341,284	4	0,00	36,33	$3p^5\,{}^2P^\circ$—$4s\,{}^4P$	$^3/_2$—$^3/_2$
340,286	4	—	—	—	—
339,800	5	0,00	36,49	$3p^5\,{}^2P^\circ$—$4s\,{}^4P$	$^3/_2$—$^1/_2$
338,929	5	0,39	36,97	$3p^5\,{}^2P^\circ$—$4s\,{}^2P$	$^1/_2$—$^3/_2$
338,828	4	—	—	—	—
336,555	15	0,39	37,22	$3p^5\,{}^2P^\circ$—$4s\,{}^2P$	$^1/_2$—$^1/_2$
335,374	25	0,00	36,97	$3p^5\,{}^2P^\circ$—$4s\,{}^2P$	$^3/_2$—$^3/_2$
333,057	2	0,00	37,22	$3p^5\,{}^2P^\circ$—$4s\,{}^2P$	$^3/_2$—$^1/_2$
332,808	3	0,39	37,64	$3p^5\,{}^2P^\circ$—$3d'\,{}^2D$	$^1/_2$—$^3/_2$
332,531	5	0,39	37,64	$3p^5\,{}^2P^\circ$—$3d'\,{}^2S$	$^1/_2$—$^1/_2$
331,991	5	—	—	—	—
331,442	4	—	—	—	—
329,391	3	0,00	37,64	$3p^5\,{}^2P^\circ$—$3d'\,{}^2D$	$^3/_2$—$^3/_2$
329,116	5	0,00	37,67	$3p^5\,{}^2P^\circ$—$3d'\,{}^2S$	$^3/_2$—$^1/_2$
328,577	1	—	—	—	—
321,593	10	0,39	38,94	$3p^5\,{}^2P^\circ$—$4s'\,{}^2D$	$^1/_2$—$^3/_2$
318,392	4	0,00	38,94	$3p^5\,{}^2P^\circ$—$4s'\,{}^2D$	$^3/_2$—$^3/_2$
318,093	15	0,00	38,98	$3p^5\,{}^2P^\circ$—$4s'\,{}^2D$	$^3/_2$—$^5/_2$
304,910	3	—	—	—	—
299,315	4	0,39	41,81	$3p^5\,{}^2P^\circ$—$4s''\,{}^2S$	$^1/_2$—$^1/_2$
296,958	6	—	—	—	—
296,554	5	0,00	41,81	$3p^5\,{}^2P^\circ$—$4s''\,{}^2S$	$^3/_2$—$^1/_2$
251,354	3	0,39	49,71	$3p^5\,{}^2P^\circ$—$5s'\,{}^2D$	$^1/_2$—$^3/_2$
250,153	3	0,00	49,56	$3p^5\,{}^2P^\circ$—$5s'\,{}^2D$	$^3/_2$—$^5/_2$
249,408	3	0,00	49,71	$3p^5\,{}^2P^\circ$—$5s'\,{}^2D$	$^3/_2$—$^3/_2$

Ca V, ground state $1s^2\,2s^2\,2p^6\,3s^2\,3p^4\,{}^3P_2$

Ionization potential 680 800 cm^{-1}; 84,39 eV

λ, Å	I	E_H, eV	E_B, eV	Transition	J
1021,139	3	—	—	—	—
1014,162	4	—	—	—	—
1012,613	3	—	—	—	—
1009,638	3	—	—	—	—
1001,544	3	—	—	—	—
1000,310	6	—	—	—	—
994,946	3	—	—	—	—
987,680	5	—	—	—	—
975,825	4	—	—	—	—
973,437	6	—	—	—	—
968,236	3	—	—	—	—
966,466	6	—	—	—	—
962,896	2	—	—	—	—
842,950	3	—	—	—	—
821,583	1	—	—	—	—
810,937	4	—	—	—	—
803,325	1	—	—	—	—
779,919	4	—	—	—	—

λ, Å	I	E_H, eV	E_B, eV	Transition	J
779,824	3	—	—	—	—
774,354	3	—	—	—	—
774,088	5	—	—	—	—
748,409	3	—	—	—	—
730,257	5	—	—	—	—
727,646	1	—	—	—	—
676,775	1	—	—	—	—
671,365	6	—	—	—	—
656,763	6	0,30	19,17	$3p^4\ ^3P—3p^5\ ^3P^\circ$	1—2
651,550	5	0,41	19,43	$3p^4\ ^3P—3p^5\ ^3P^\circ$	0—1
647,876	5	0,30	19,43	$3p^4\ ^3P—3p^5\ ^3P^\circ$	1—1
646,570	8	0,00	19,17	$3p^4\ ^3P—3p^5\ ^3P^\circ$	2—2
643,118	6	0,30	19,57	$3p^4\ ^3P—3p^5\ ^3P^\circ$	1—0
637,928	8	0,00	19,43	$3p^4\ ^3P—3p^5\ ^3P^\circ$	2—1
594,239	1	—	—	—	—
593,472	1	—	—	—	—
593,404	1	—	—	—	—
558,602	10	—	—	—	—
555,482	5	—	—	—	—
551,103	2	—	—	—	—
549,070	3	—	—	—	—
542,290	10	—	—	—	—
530,303	6	—	—	—	—
528,746	3	—	—	—	—
509,293	2	—	—	—	—
476,606	2	—	—	—	—
446,036	1	—	—	—	—
445,933	1	—	—	—	—
425,000	15	—	—	—	—
387,077	5	5,44	37,46	$3p^4\ ^1S—3d''\ ^1P^\circ$	0—1
381,606	3	—	—	—	—
380,396	5	0,41	33,00	$3p^4\ ^3P—3d\ ^3D^\circ$	0—1
379,765	3	—	—	—	—
379,138	2	0,30	33,00	$3p^4\ ^3P—3d\ ^3D^\circ$	1—1
377,181	5	—	—	—	—
376,279	3	0,30	33,25	$3p^4\ ^3P—3d\ ^3D^\circ$	1—2
375,333	3	—	—	—	—
374,000	4	—	—	—	—
372,904	6	0,00	33,25	$3p^4\ ^3P—3d\ ^3D^\circ$	2—2
371,225	6	0,00	33,40	$3p^4\ ^3P—3d\ ^3D^\circ$	2—3
356,246	5	2,33	37,14	$3p^4\ ^1D—3d''\ ^3P^\circ$	2—1
352,915	9	2,33	37,46	$3p^4\ ^1D—3d''\ ^1P^\circ$	2—1
343,640	4	2,33	38,41	$3p^4\ ^1D—3d''\ ^3D^\circ$	2—2
338,056	5	0,30	36,97	$3p^4\ ^3P—3d''\ ^3P^\circ$	1—2
337,541	4	0,41	37,14	$3p^4\ ^3P—3d''\ ^3P^\circ$	0—1
336,554	4	0,30	37,14	$3p^4\ ^3P—3d''\ ^3P^\circ$	1—1
335,344	5	0,00	36,97	$3p^4\ ^3P—3d''\ ^3P^\circ$	2—2
334,545	6	0,41	37,46	$3p^4\ ^3P—3d''\ ^1P^\circ$	0—1
334,135	3	—	—	—	—
333,857	4	0,00	37,14	$3p^4\ ^3P—3d''\ ^3P^\circ$	2—1
333,570	3	0,30	37,46	$3p^4\ ^3P—3d''\ ^1P^\circ$	1—1
333,438	4	—	—	—	—
330,937	6	0,00	37,46	$3p^4\ ^3P—3d''\ ^1P^\circ$	2—1
325,282	5	0,30	38,41	$3p^4\ ^3P—3d''\ ^3D^\circ$	1—2
325,020	3	0,41	38,55	$3p^4\ ^3P—3d''\ ^3D^\circ$	0—1
324,477	5	—	—	—	—
324,110	3	0,30	38,55	$3p^4\ ^3P—3d''\ ^3D^\circ$	1—1
323,223	6	0,00	38,36	$3p^4\ ^3P—3d''\ ^3D^\circ$	2—3
322,757	5	0,00	38,41	$3p^4\ ^3P—3d''\ ^3D^\circ$	2—2

λ, Å	I	E_H, eV	E_B, eV	Transition	J
322,166	10	—	—	—	—
321,609	6	0,00	38,55	$3p^4\,{}^3P$—$3d''\,{}^3D^\circ$	2—1
301,139	0	2,33	43,51	$3p^4\,{}^1D$—$4s\,{}^3S^\circ$	2—1
287,657	3	0,41	43,51	$3p^4\,{}^3P$—$4s\,{}^3S^\circ$	0—1
286,965	9	5,44	48,63	$3p^4\,{}^1S$—$4s''\,{}^1P^\circ$	0—1
286,947	5	0,30	43,51	$3p^4\,{}^3P$—$4s\,{}^3S^\circ$	1—1
284,948	5	0,00	43,51	$3p^4\,{}^3P$—$4s\,{}^3S^\circ$	2—1
284,794	2	2,33	45,86	$3p^4\,{}^1D$—$4s'\,{}^3D^\circ$	2—3
280,992	8	2,33	46,46	$3p^4\,{}^1D$—$4s'\,{}^1D^\circ$	2—2
272,982	4	0,41	45,82	$3p^4\,{}^3P$—$4s'\,{}^3D^\circ$	0—1
272,336	3	0,30	45,82	$3p^4\,{}^3P$—$4s'\,{}^3D^\circ$	1—1
272,265	5	0,30	45,83	$3p^4\,{}^3P$—$4s'\,{}^3D^\circ$	1—2
271,440	1	2,33	48,00	$3p^4\,{}^1D$—$4s''\,{}^3P^\circ$	2—1
271,141	4	2,33	48,06	$3p^4\,{}^1D$—$4s''\,{}^3P^\circ$	2—2
270,570	2	0,00	45,82	$3p^4\,{}^3P$—$4s'\,{}^3D^\circ$	2—1
270,494	3	0,00	45,83	$3p^4\,{}^3P$—$4s'\,{}^3D^\circ$	2—2
270,305	6	0,00	45,86	$3p^4\,{}^3P$—$4s'\,{}^3D^\circ$	2—3
268,583	2	0,30	46,46	$3p^4\,{}^3P$—$4s'\,{}^1D^\circ$	1—2
267,772	8	2,33	48,63	$3p^4\,{}^1D$—$4s''\,{}^1P^\circ$	2—1
266,863	3	0,00	46,46	$3p^4\,{}^3P$—$4s'\,{}^1D^\circ$	2—2
260,446	3	0,41	48,00	$3p^4\,{}^3P$—$4s''\,{}^3P^\circ$	0—1
259,978	3	0,30	47,98	$3p^4\,{}^3P$—$4s''\,{}^3P^\circ$	1—0
259,856	3	0,30	48,00	$3p^4\,{}^3P$—$4s''\,{}^3P^\circ$	1—1
259,576	3	0,30	48,06	$3p^4\,{}^3P$—$4s''\,{}^3P^\circ$	1—2
258,251	3	0,00	48,00	$3p^4\,{}^3P$—$4s''\,{}^3P^\circ$	2—1
257,976	5	0,00	48,06	$3p^4\,{}^3P$—$4s''\,{}^3P^\circ$	2—2
200,860	3	0,41	62,13	$3p^4\,{}^3P$—$5s\,{}^3S^\circ$	0—1
200,512	5	0,30	62,13	$3p^4\,{}^3P$—$5s\,{}^3S^\circ$	1—1
199,890	3	5,44	67,46	$3p^4\,{}^1S$—$5s''\,{}^1P^\circ$	0—1
199,553	6	0,00	62,13	$3p^4\,{}^3P$—$5s\,{}^3S^\circ$	2—1
197,648	2	2,33	65,06	$3p^4\,{}^1D$—$5s'\,{}^3D^\circ$	2—2
197,531	2	2,33	65,09	$3p^4\,{}^1D$—$5s'\,{}^3D^\circ$	2—3
196,970	5	2,33	65,28	$3p^4\,{}^1D$—$5s'\,{}^1D^\circ$	2—2
191,801	2	0,41	65,04	$3p^4\,{}^3P$—$5s'\,{}^3D^\circ$	0—1
191,480	2	0,30	65,04	$3p^4\,{}^3P$—$5s'\,{}^3D^\circ$	1—1
191,439	3	0,30	65,06	$3p^4\,{}^3P$—$5s'\,{}^3D^\circ$	1—2
190,558	3	0,00	65,06	$3p^4\,{}^3P$—$5s'\,{}^3D^\circ$	2—2
190,457	5	0,00	65,09	$3p^4\,{}^3P$—$5s'\,{}^3D^\circ$	2—3
190,363	4	2,33	67,46	$3p^4\,{}^1D$—$5s''\,{}^1P^\circ$	2—1
185,540	2	0,41	67,23	$3p^4\,{}^3P$—$5s''\,{}^3P^\circ$	0—1
185,288	1	0,30	67,23	$3p^4\,{}^3P$—$5s''\,{}^3P^\circ$	1—0, 1
185,102	2	0,30	67,28	$3p^4\,{}^3P$—$5s''\,{}^3P^\circ$	1—2
184,415	1	0,00	67,23	$3p^4\,{}^3P$—$5s''\,{}^3P^\circ$	2—1
184,280	3	0,00	67,28	$3p^4\,{}^3P$—$5s''\,{}^3P^\circ$	2—2

Ca VI, ground state $1s^2\,2s^2\,2p^6\,3s^2\,3p^3\,{}^4S^0_{3/2}$

Ionization potential 879 000 cm^{-1}; 109 eV

λ, Å	I	E_H, eV	E_B, eV	Transition	J
1032,612	2	—	—	—	—
1021,508	4	—	—	—	—
1018,346	3	—	—	—	—
975,055	3	—	—	—	—
969,652	6	—	—	—	—

λ, Å	I	E_H, eV	E_B, eV	Transition	J
916,682	2	—	—	—	—
860,827	2	—	—	—	—
854,923	3	—	—	—	—
817,058	2	—	—	—	—
816,805	2	—	—	—	—
811,480	1	—	—	—	—
778,718	1	—	—	—	—
777,508	1	—	—	—	—
775,966	3	—	—	—	—
774,532	3	—	—	—	—
772,389	1	—	—	—	—
770,928	1	—	—	—	—
776,522	1	5,62	21,79	$3p^3\ ^2P^\circ$—1	$^3/_2$—$^5/_2$
765,154	6	—	—	—	—
764,358	3	5,62	21,84	$3p^3\ ^2P^\circ$—2	$^3/_2$—$^3/_2$
763,344	2	—	—	—	—
689,538	3	—	—	—	—
685,807	4	—	—	—	—
674,278	2	5,62	24,00	$3p^3\ ^2P^\circ$—$3p^4\ ^2D$	$^3/_2$—$^5/_2$
674,046	1	—	—	—	—
641,883	2	0,00	19,31	$3p^3\ ^4S^\circ$—$3p^4\ ^4P$	$^3/_2$—$^5/_2$
633,815	2	0,00	19,56	$3p^3\ ^4S^\circ$—$3p^4\ ^4P$	$^3/_2$—$^3/_2$
629,594	2	0,00	19,69	$3p^3\ ^4S^\circ$—$3p^4\ ^4P$	$^3/_2$—$^1/_2$
617,517	4	—	—	—	—
614,015	3	—	—	—	—
602,389	0	3,40	23,98	$3p^3\ ^2D^\circ$—$3p^4\ ^2D$	$^5/_2$—$^3/_2$
601,700	5	3,40	24,00	$3p^3\ ^2D^\circ$—$3p^4\ ^2D$	$^5/_2$—$^5/_2$
600,917	6	3,35	23,98	$3p^3\ ^2D^\circ$—$3p^4\ ^2D$	$^3/_2$—$^3/_2$
590,396	3	—	—	—	—
587,872	1	—	—	—	—
587,604	2	—	—	—	—
581,466	3	—	—	—	—
579,775	2	—	—	—	—
578,732	4	—	—	—	—
564,275	2	—	—	—	—
562,250	3	5,62	27,67	$3p^3\ ^2P^\circ$—$3p^4\ ^2P$	$^3/_2$—$^3/_2$
547,898	3	—	—	—	—
537,613	6	5,62	28,68	$3p^3\ ^2P^\circ$—$3p^4\ ^2S$	$^3/_2$—$^1/_2$
536,008	0	5,55	28,68	$3p^3\ ^2P^\circ$—$3p^4\ ^2S$	$^1/_2$—$^1/_2$
505,199	8	—	—	—	—
400,824	3	5,62	36,55	$3p^3\ ^2P^\circ$—$3d\ ^2P$	$^3/_2$—$^3/_2$
399,925	0	5,55	36,55	$3p^3\ ^2P^\circ$—$3d\ ^2P$	$^1/_2$—$^3/_2$
396,917	2	5,62	36,85	$3p^3\ ^2P^\circ$—$3d\ ^2P$	$^3/_2$—$^1/_2$
396,055	2	5,55	36,85	$3p^3\ ^2P^\circ$—$3d\ ^2P$	$^1/_2$—$^1/_2$
390,137	3	—	—	—	—
387,080	4	5,62	37,65	$3p^3\ ^2P^\circ$—3	$^3/_2$—
386,254	1	5,55	37,65	$3p^3\ ^2P^\circ$—3	$^1/_2$—
386,106	1	—	—	—	—
385,941	1	—	—	—	—
385,091	2	—	—	—	—
384,172	3	—	—	—	—
384,028	2	—	—	—	—
383,505	2	—	—	—	—
381,849	2	—	—	—	—
381,464	2	—	—	—	—
380,003	1	—	—	—	—
378,745	1	—	—	—	—
378,653	1	—	—	—	—
378,551	1	3,40	36,10	$3p^3\ ^2D^\circ$—$3d\ ^2F$	$^3/_2$—$^5/_2$
373,997	7	3,40	36,55	$3p^3\ ^2D^\circ$—$3d\ ^2P$	$^5/_2$—$^3/_2$

λ, Å	I	E_H, eV	E_B, eV	Transition	J
373,700	3	—	—	—	—
373,418	5	3,35	36,55	$3p^3\ ^2D°—3d\ ^2P$	$^3/_2—^3/_2$
370,022	7	3,35	36,85	$3p^3\ ^2D°—3d\ ^2P$	$^3/_2—^1/_2$
367,371	2	—	—	—	—
363,525	2	5,62	39,72	$3p^3\ ^2P°—3d'\ ^2S$	$^3/_2—^1/_2$
362,788	1	5,55	39,72	$3p^3\ ^2P°—3d'\ ^2S$	$^1/_2—^1/_2$
362,612	4	5,62	39,81	$3p^3\ ^2P°—3d'\ ^2D$	$^3/_2—^5/_2$
361,645	2	—	—	—	—
361,234	2	5,55	39,87	$3p^3\ ^2P°—3d'\ ^2D$	$^1/_2—^3/_2$
361,114	4	—	—	—	—
358,153	3	—	—	—	—
350,394	1	—	—	—	—
349,494	1	—	—	—	—
348,927	1	—	—	—	—
348,650	1	5,62	41,18	$3p^3\ ^2P°—3d'\ ^2P$	$^3/_2—^1/_2$
347,967	3	5,55	41,18	$3p^3\ ^2P°—3d'\ ^2P$	$^1/_2—^1/_2$
347,431	1	—	—	—	—
347,334	1	—	—	—	—
347,005	4	5,62	41,35	$3p^3\ ^2P°—3d'\ ^2P$	$^3/_2—^3/_2$
346,335	2	5,55	41,35	$3p^3\ ^2P°—3d'\ ^2P$	$^1/_2—^3/_2$
340,528	8	3,40	39,81	$3p^3\ ^2D°—3d'\ ^2D$	$^5/_2—^5/_2$
340,037	4	3,35	39,81	$3p^3\ ^2D°—3d'\ ^2D$	$^3/_2—^5/_2$
339,940	4	3,40	39,87	$3p^3\ ^2D°—3d'\ ^2D$	$^5/_2—^3/_2$
339,463	6	3,35	39,87	$3p^3\ ^2D°—3d'\ ^2D$	$^3/_2—^3/_2$
329,298	3	—	—	—	—
327,806	4	—	—	—	—
327,175	2	—	—	—	—
321,110	1	—	—	—	—
320,445	2	—	—	—	—
316,947	3	5,62	44,73	$3p^3\ ^2P°—3d''\ ^2D$	$^3/_2—^3/_2$
316,389	0	5,55	44,73	$3p^3\ ^2P°—3d''\ ^2D$	$^1/_2—^3/_2$
316,115	3	—	—	—	—
291,976	1	—	—	—	—
251,816	1	5,62	54,85	$3p^3\ ^2P°—4s\ ^2P$	$^3/_2—^1/_2$
251,465	4	5,55	54,85	$3p^3\ ^2P°—4s\ ^2P$	$^1/_2—^1/_2$
250,265	4	5,62	55,16	$3p^3\ ^2P°—4s\ ^2P$	$^3/_2—^3/_2$
249,914	3	5,55	55,16	$3p^3\ ^2P°—4s\ ^2P$	$^1/_2—^3/_2$
242,631	5	5,62	56,71	$3p^3\ ^2P°—4s'\ ^2D$	$^3/_2—^5/_2$
242,592	3	5,62	56,72	$3p^3\ ^2P°—4s'\ ^2D$	$^3/_2—^3/_2$
242,265	3	5,55	56,72	$3p^3\ ^2P°—4s'\ ^2D$	$^1/_2—^3/_2$
240,721	6	3,35	54,85	$3p^3\ ^2D°—4s\ ^2P$	$^3/_2—^1/_2$
239,535	7	3,40	55,16	$3p^3\ ^2D°—4s\ ^2P$	$^5/_2—^3/_2$
239,296	0	3,35	55,16	$3p^3\ ^2D°—4s\ ^2P$	$^3/_2—^3/_2$
232,531	5	3,40	56,71	$3p^3\ ^2D°—4s\ ^2D$	$^5/_2—^5/_2$
232,282	6	3,35	56,72	$3p^3\ ^2D°—4s\ ^2D$	$^3/_2—^3/_2$
230,495	5	0,00	53,79	$3p^3\ ^4S°—4s\ ^4P$	$^3/_2—^1/_2$
229,734	7	0,00	53,97	$3p^3\ ^4S°—4s\ ^4P$	$^3/_2—^3/_2$
228,628	7	0,00	54,23	$3p^3\ ^4S°—4s\ ^4P$	$^3/_2—^5/_2$

Unclassified Lines of Calcium

λ, Å	I	Expected assignment	λ, Å	I	Expected assignment
1667,7	30	—	486,160	3	—
1545,7	3	—	485,636	3	—
1402,9	10	—	484,368	3	—
1400,7	4	—	480,471	3	—
1393,4	8	—	461,085	5	—
552,005	4	—	397,178	4	—
542,842	3	—	209,723	3	—

TITANIUM, Z = 22

Ti I, ground state $1s^2\,2s^2\,2p^6\,3s^2\,3p^6\,3d^2\,4s^2\ ^3F_2$

Ionization potential 55 138, cm^{-1}; 6,836 eV

λ, Å	I	E_H, eV	E_B, eV	Transition	J
11973,88	6	1,46	2,50	$b\,^3F-z\,^3D^\circ$	4—3
11949,58	5	1,44	2,48	$b\,^3F-z\,^3D^\circ$	3—2
11892,85	5	1,43	2,47	$b\,^3F-z\,^3D^\circ$	2—1
11797,24	3	1,43	2,48	$b\,^3F-z\,^3D^\circ$	2—2
11780,54	4	1,44	2,50	$b\,^3F-z\,^3D^\circ$	3—3
11539,50	5	—	—	—	—
11403,89	8	—	—	—	—
11381,53	7	—	—	—	—
11292,43	6	3,14	4,24	$y\,^3D^\circ-d\,^3P$	1—0
11246,88	8	3,15	4,26	$y\,^3D^\circ-d\,^3P$	2—1
11243,90	10	3,18	4,28	$y\,^3D^\circ-d\,^3P$	3—2
11230,91	5	—	—	—	—
11095,79	5	3,14	4,26	$y\,^3D^\circ-d\,^3P$	1—1
11057,58	3	3,16	4,28	$z\,^3P^\circ-d\,^3P$	2—2
10990,70	3	3,15	4,28	$y\,^3D^\circ-d\,^3P$	2—2
10896,10	8	3,35	4,49	$x\,^3F^\circ-b\,^3G$	4—5
10833,66	3	3,72	4,86	$x\,^5D^\circ-f\,^5F$	3—3
10820,31	5	3,33	4,48	$x\,^3F^\circ-b\,^3G$	3—4
10817,35	5	{ 3,71	4,85	$x\,^5D^\circ-f\,^5F$	2—2
		3,73	4,87	$x\,^5D^\circ-f\,^5F$	4—4
10793,65	3	3,32	4,47	$x\,^3F^\circ-b\,^3G$	2—3
10781,34	3	2,25	3,40	$b\,^3P-x\,^3D^\circ$	2—2
10774,92	12	{ 3,70	4,85	$x\,^5D^\circ-f\,^5F$	0—1
		0,82	1,97	$a\,^5F-z\,^5G^\circ$	2—2
10756,90	5	{ 2,24	3,39	$b\,^3P-x\,^3D^\circ$	1—1
		3,70	4,85	$x\,^5D^\circ-f\,^5F$	1—2
10741,77	7	3,71	4,86	$x\,^5D^\circ-f\,^5F$	2—3
10732,89	8	0,83	1,98	$a\,^5F-z\,^5G^\circ$	3—3
10731,11	6	3,72	4,87	$x\,^5D^\circ-f\,^5F$	3—4
10726,33	18	0,81	1,97	$a\,^5F-z\,^5G^\circ$	1—2
10689,52	15	3,73	4,89	$x\,^5D^\circ-f\,^5F$	4—5
10677,04	10	0,84	2,00	$a\,^5F-z\,^5G^\circ$	4—4
10661,61	20	0,82	1,98	$a\,^5F-z\,^5G^\circ$	2—3
10607,78	10	0,85	2,02	$a\,^5F-z\,^5G^\circ$	5—5
10584,66	25	0,83	2,00	$a\,^5F-z\,^5G^\circ$	3—4
10565,97	5	2,24	3,41	$a\,^3H-y\,^3G^\circ$	4—3
10553,02	8	2,25	3,42	$a\,^3H-y\,^3G^\circ$	5—4
10551,81	3	1,89	3,06	$a\,^3G-z\,^1G^\circ$	5—4
10496,14	30	0,84	2,00	$a\,^5F-z\,^5G^\circ$	4—5
10460,07	10	2,26	3,44	$a\,^3H-y\,^3G^\circ$	6—5
10396,85	25	0,85	2,04	$a\,^5F-z\,^5G^\circ$	5—6
10257,30	3	3,72	4,93	$x\,^3G^\circ-f\,^3F$	5—4
10189,26	3	1,46	2,68	$b\,^3F-z\,^3G^\circ$	4—4
10179,92	3	3,89	5,11	$w\,^3G^\circ-e\,^3G$	3—3
10170,60	3	1,44	2,66	$b\,^3F-z\,^3G^\circ$	3—3
10147,09	4	3,92	5,14	$w\,^3G^\circ-e\,^3G$	5—5
10145,48	8	—	—	—	—
10120,90	10	2,17	3,40	$a\,^3D-x\,^3D^\circ$	3—2
10119,20	3	3,90	5,13	$w\,^3G^\circ-e\,^3G$	4—4
10066,47	8	2,16	3,39	$a\,^3D-x\,^3D^\circ$	2—1
10059,87	12	1,43	2,66	$b\,^3F-z\,^3G^\circ$	2—3
10057,69	25	2,17	3,41	$a\,^3D-x\,^3D^\circ$	3—3
10050,11	5	—	—	—	—
10048,78	12	1,44	2,68	$b\,^3F-z\,^3G^\circ$	3—4
10034,45	15	1,46	2,70	$b\,^3F-z\,^3G^\circ$	4—5

λ, Å	I	E_H, eV	E_B, eV	Transition	J
10011,72	15	2,15	3,39	$a\,^3D—x\,^3D°$	1—1
10003,02	25	2,16	3,40	$a\,^3D—x\,^3D°$	2—2
9997,94	15	1,87	3,11	$a\,^3G—y\,^3F°$	3—2
9981,16	5	—	—	—	—
9948,98	8	2,15	3,40	$a\,^3D—x\,^3D°$	1—2
9941,33	8	2,16	3,41	$a\,^3D—x\,^3D°$	2—3
9927,35	20	1,88	3,13	$a\,^3G—y\,^3F°$	4—3
9879,41	3	1,87	3,13	$a\,^3G—y\,^3F°$	3—3
9832,15	25	1,89	3,15	$a\,^3G—y\,^3F°$	5—4
9813,45	5	2,32	3,58	$z\,^5D°—a\,^5D$	4—3
9787,67	50	0,83	2,09	$a\,^5F—z\,^5F°$	3—2
9783,59	20	0,82	2,08	$a\,^5F—z\,^5F°$	2—1
9783,30	40	0,84	2,10	$a\,^5F—z\,^5F°$	4—3
9770,28	40	0,85	2,12	$a\,^5F—z\,^5F°$	5—4
9768,22	5	2,31	3,57	$z\,^5D°—a\,^5D$	3—2
9746,86	15	2,32	3,59	$z\,^5D°—a\,^5D$	4—4
9743,60	50	0,81	2,08	$a\,^5F—z\,^5F°$	1—1
9737,77	5	2,30	3,57	$z\,^5D°—a\,^5D$	2—1
9728,36	60	0,82	2,09	$a\,^5F—z\,^5F°$	2—2
9718,96	25	1,50	2,78	$a\,^1G—z\,^1F°$	4—3
9717,00	10	2,31	3,58	$z\,^5D°—a\,^5D$	3—3
9715,51	3	2,29	3,57	$z\,^5D°—a\,^5D$	1—0
9705,64	80	0,83	2,10	$a\,^5F—z\,^5F°$	3—3
9702,86	3	2,30	3,57	$z\,^5D°—a\,^5D$	2—2
9688,86	30	0,81	2,09	$a\,^5F—z\,^5F°$	1—2
9678,98	3	2,29	3,57	$z\,^5D°—a\,^5D$	0—1
9675,55	90	0,84	2,12	$a\,^5F—z\,^5F°$	4—4
9663,19	3	2,29	3,57	$z\,^5D°—a\,^5D$	1—2
9661,42	10	2,17	3,46	$a\,^3D—z\,^5P°$	3—3
9647,40	50	0,82	2,10	$a\,^5F—z\,^5F°$	2—3
9638,28	100	0,85	2,13	$a\,^5F—z\,^5F°$	5—5
9606,77	3	3,58	4,87	$y\,^5F°—f\,^5F$	5—4
9599,53	50	0,83	2,12	$a\,^5F—z\,^5F°$	3—4
9590,15	3	3,21	4,51	$y\,^5D°—e\,^5F$	4—5
9588,77	4	3,57	4,86	$y\,^5F°—f\,^5F$	4—3
9570,08	4	3,56	4,85	$y\,^5F°—f\,^5F$	3—2
9546,07	50	0,84	2,13	$a\,^5F—z\,^5F°$	4—5
9511,80	8	3,55	4,85	$y\,^5F°—f\,^5F$	2—2
9511,55	10	3,55	4,85	$y\,^5F°—f\,^5F$	1—1
9510,81	12	3,56	4,86	$y\,^5F°—f\,^5F$	3—3
9508,49	20	3,57	4,87	$y\,^5F°—f\,^5F$	4—4
9506,04	25	3,58	4,89	$y\,^5F°—f\,^5F$	5—5
9453,22	3	3,55	4,86	$y\,^5F°—f\,^5F$	2—3
9431,77	3	3,56	4,87	$y\,^5F°—f\,^5F$	3—4
9312,48	4	3,32	4,65	$x\,^3F°—e\,^3F$	2—2
9285,04	5	3,33	4,67	$x\,^3F°—e\,^3F$	3—3
9257,62	7	3,35	4,69	$x\,^3F°—e\,^3F$	4—4
9246,14	10	3,15	4,49	$y\,^3F°—b\,^3G$	4—5
9167,53	8	3,13	4,48	$y\,^3F°—b\,^3G$	3—4
9123,14	5	3,11	4,47	$y\,^3F°—b\,^3G$	2—3
9090,70	25	1,75	3,11	$a\,^5P—z\,^5S°$	3—2
9027,32	15	1,74	3,11	$a\,^5P—z\,^5S°$	2—2
8989,44	12	1,73	3,11	$a\,^5P—z\,^5S°$	1—2
8863,09	3	3,71	5,11	$x\,^3G°—e\,^3G$	3—3
8821,14	12	1,75	3,15	$a\,^5P—y\,^3D°$	3—2
8819,39	8	1,07	2,47	$a\,^3P—z\,^3D°$	2—1
8794,40	8	4,30	5,71	$z\,^1H°—e\,^1G$	5—4
8778,66	30	1,75	3,16	$a\,^5P—z\,^3P°$	3—2

λ, Å	I	E_H, eV	E_B, eV	Transition	J
8766,64	75	1,07	2,48	$a\,^3P - z\,^3D^\circ$	2—2
8761,44	15	1,74	3,15	$a\,^5P - y\,^3D^\circ$	2—2
8737,31	7	3,72	5,14	$x\,^3G^\circ - e\,^3G$	5—5
8734,70	75	1,05	2,47	$a\,^3P - z\,^3D^\circ$	1—1
8725,76	6	1,73	3,15	$a\,^5P - y\,^3D^\circ$	1—2
8719,56	30	1,74	3,16	$a\,^5P - z\,^3P^\circ$	2—2
8692,34	100	1,05	2,47	$a\,^3P - z\,^3D^\circ$	0—1
8682,99	125	1,05	2,48	$a\,^3P - z\,^3D^\circ$	1—2
8675,38	150	1,07	2,50	$a\,^3P - z\,^3D^\circ$	2—3
8641,47	40	—	—	—	—
8636,38	18	—	—	—	—
8629,33	18	—	—	—	—
8618,44	20	2,24	3,68	$b\,^3P - w\,^3D^\circ$	1—1
8618,14	15	—	—	—	—
8612,91	7	1,74	3,18	$a\,^5P - y\,^5D^\circ$	2—1
8600,98	25	{ 2,25 1,73	3,69 3,17	$b\,^3P - w\,^3D^\circ$ $a\,^5P - y\,^5D^\circ$	2—2 1—0
8598,18	60	2,27	3,71	$b\,^1G - x\,^3G^\circ$	4—3
8578,40	15	1,73	3,18	$a\,^5P - y\,^5D^\circ$	1—1
8569,72	50	2,23	3,68	$b\,^3P - w\,^3D^\circ$	0—1
8565,45	25	1,74	3,19	$a\,^5P - y\,^5D^\circ$	2—2
8550,54	25	1,75	3,20	$a\,^5P - y\,^5D^\circ$	3—3
8548,07	100	1,87	3,32	$a\,^3G - x\,^3F^\circ$	3—2
8539,36	60	2,24	3,69	$b\,^3P - w\,^3D^\circ$	1—2
8531,36	15	1,73	3,19	$a\,^5P - y\,^5D^\circ$	1—2
8526,36	8	—	—	—	—
8525,99	8	—	—	—	—
8518,37	100	1,88	3,33	$a\,^3G - x\,^3F^\circ$	4—3
8518,05	60	2,13	3,59	$z\,^5F^\circ - a\,^5D$	5—4
8496,03	60	{ 2,25 3,70	3,71 5,16	$b\,^3P - w\,^3D^\circ$ $a\,^1F - v\,^1F^\circ$	2—3 3—3
8495,51	15	2,24	3,70	$b\,^3P - x\,^5D^\circ$	1—0
8494,42	30	1,74	3,20	$a\,^5P - y\,^5D^\circ$	2—3
8483,16	25	1,87	3,33	$a\,^3G - x\,^3F^\circ$	3—3
8468,46	100	1,89	3,35	$a\,^3G - x\,^3F^\circ$	5—4
8467,15	75	2,12	3,58	$z\,^5F^\circ - a\,^5D$	4—3
8460,96	7	—	—	—	—
8457,10	40	1,75	3,21	$a\,^5P - y\,^5D^\circ$	3—4
8450,89	75	2,25	3,72	$a\,^3H - x\,^3G^\circ$	5—4
8442,98	20	2,25	3,72	$b\,^3P - x\,^5D^\circ$	2—3
8438,93	75	2,26	3,72	$a\,^3H - x\,^3G^\circ$	6—5
8435,68	300	0,84	2,31	$a\,^5F - z\,^5D^\circ$	4—3
8434,98	300	0,85	2,32	$a\,^5F - z\,^5D^\circ$	5—4
8426,50	200	0,83	2,30	$a\,^5F - z\,^5D^\circ$	3—2
8424,41	50	2,10	3,57	$z\,^5F^\circ - a\,^5D$	3—2
8423,10	20	1,88	3,35	$a\,^3G - x\,^3F^\circ$	4—4
8418,70	10	—	—	—	—
8417,54	25	2,12	3,59	$z\,^5F^\circ - a\,^5D$	4—4
8416,97	60	2,24	3,71	$a\,^3H - x\,^3G^\circ$	4—3
8412,36	150	0,82	2,29	$a\,^5P - z\,^5D^\circ$	2—1
8402,54	5	2,25	3,72	$a\,^3H - x\,^3G^\circ$	5—5
8396,93	90	0,81	2,29	$a\,^5F - z\,^5D^\circ$	1—0
8389,48	25	2,09	3,57	$z\,^5F^\circ - a\,^5D$	2—1
8382,82	90	0,81	2,29	$a\,^5F - z\,^5D^\circ$	1—1
8382,54	100	0,82	2,30	$a\,^5F - z\,^5D^\circ$	2—2
8377,90	100	0,83	2,31	$a\,^5F - z\,^5D^\circ$	3—3
7996,53	3	3,34	4,89	$y\,^5G^\circ - f\,^5F$	6—5

λ, Å	I	E_H, eV	E_B, eV	Transition	J
7978,88	4	1,89	3,44	$a\,^3G—y\,^3G^\circ$	5—5
		3,32	4,87	$y\,^5G^\circ—f\,^5F$	5—4
7949,17	3	1,50	3,06	$a\,^1G—z\,^1G^\circ$	4—4
7440,60	3	2,26	3,92	$a\,^3H—w\,^3G^\circ$	6—5
7357,74	3	1,44	3,13	$b\,^3F—y\,^3F^\circ$	3—3
7344,72	4	1,46	3,15	$b\,^3F—y\,^3F^\circ$	4—4
7251,74	8	1,43	3,14	$b\,^3F—y\,^3D^\circ$	2—1
7244,86	10	1,44	3,15	$b\,^3F—y\,^3D^\circ$	3—2
7216,20	5	1,44	3,16	$b\,^3F—z\,^3P^\circ$	3—2
7209,44	20	1,46	3,18	$b\,^3F—y\,^3D^\circ$	4—3
7038,80	6	2,34	4,11	$c\,^3P—x\,^3P^\circ$	2—2
6861,47	6	2,27	4,07	$b\,^1G—y\,^1F^\circ$	4—3
6743,124	10	0,90	2,74	$a\,^1D—z\,^1D^\circ$	2—2
6599,112	12	0,90	2,78	$a\,^1D—z\,^1F^\circ$	2—3
6575,180	3	2,58	4,46	$a\,^1H—y\,^1G^\circ$	5—4
6556,066	25	1,46	3,35	$b\,^3F—x\,^3F^\circ$	4—4
6554,236	20	1,44	3,33	$b\,^3F—x\,^3F^\circ$	3—3
6546,276	20	1,43	3,32	$b\,^3F—x\,^3F^\circ$	2—2
6508,135	3	1,43	3,33	$b\,^3F—x\,^3F^\circ$	2—3
6497,689	3	1,44	3,35	$b\,^3F—x\,^3F^\circ$	3—4
6366,354	8	1,46	3,41	$b\,^3F—x\,^3D^\circ$	4—3
6359,896	8	0,05	2,00	$a\,^3F—z\,^5G^\circ$	4—4
6336,104	8	1,44	3,40	$b\,^3F—x\,^3D^\circ$	3—2
6325,22	10	0,02	1,98	$a\,^3F—z\,^5G^\circ$	3—3
6318,027	5	1,43	3,39	$b\,^3F—x\,^3D^\circ$	2—1
6312,240	10	1,46	3,42	$b\,^3F—y\,^3G^\circ$	4—4
6303,754	10	1,44	3,41	$b\,^3F—y\,^3G^\circ$	3—3
6296,646	12	0,00	1,97	$a\,^3F—z\,^5G^\circ$	2—2
6273,389	6	0,02	2,00	$a\,^3F—z\,^5G^\circ$	3—4
6261,101	35	1,43	3,41	$b\,^3F—y\,^3G^\circ$	2—3
6258,706	50	1,46	3,44	$b\,^3F—y\,^3G^\circ$	4—5
6258,103	40	1,44	3,42	$b\,^3F—y\,^3G^\circ$	3—4
6221,41	8	2,66	4,65	$z\,^3G^\circ—e\,^3F$	3—2
6220,460	12	2,68	4,67	$z\,^3G^\circ—e\,^3F$	4—3
6215,212	20	2,70	4,69	$z\,^3G^\circ—e\,^3F$	5—4
6186,14	3	2,17	4,18	$a\,^3D—w\,^3F^\circ$	3—4
6146,225	3	1,87	3,89	$a\,^3G—w\,^3G^\circ$	3—3
6126,217	20	1,07	3,09	$a\,^3P—z\,^3S^\circ$	2—1
6121,008	3	1,88	3,90	$a\,^3G—w\,^3G^\circ$	4—4
6098,665	7	3,06	5,09	$z\,^1G^\circ—e\,^1F$	4—3
6092,814	4	1,89	3,92	$a\,^3G—w\,^3G^\circ$	5—5
6091,175	20	2,27	4,30	$b\,^1G—z\,^1H^\circ$	4—5
6085,228	20	1,05	3,09	$a\,^3P—z\,^3S^\circ$	1—1
6064,631	9	1,05	3,09	$a\,^3P—z\,^3S^\circ$	0—1
5999,668	8	2,24	4,30	$a\,^3H—z\,^1H^\circ$	4—5
5999,003	4	2,17	4,24	$a\,^3D—v\,^3F^\circ$	3—4
5978,543	25	1,87	3,95	$a\,^3G—z\,^3H^\circ$	3—4
5965,828	30	1,88	3,96	$a\,^3G—z\,^3H^\circ$	4—5
5953,162	30	1,89	3,97	$a\,^3G—z\,^3H^\circ$	5—6
5941,755	12	1,05	3,14	$a\,^3P—y\,^3D^\circ$	1—1
5937,806	6	1,07	3,15	$a\,^3P—y\,^3D^\circ$	2—2
5929,27	3	3,58	5,67	$y\,^5F^\circ—f\,^5G$	5—5
5924,42	3	3,87	5,96	$v\,^3D^\circ—f\,^5F$	1—1
5922,112	18	1,05	3,14	$a\,^3P—y\,^3D^\circ$	0—1
5919,06	10	2,25	4,34	$b\,^3P—x\,^1D^\circ$	2—2
5918,548	10	1,07	3,16	$a\,^3P—z\,^3P^\circ$	2—2
5916,18	5	2,78	4,87	$z\,^1F^\circ—f\,^5F$	3—4
5915,123	9	2,26	4,42	$a\,^3H—y\,^3H^\circ$	6—6

λ, Å	I	E_H, eV	E_B, eV	Transition	J
5914,93	5	3,87 3,57	5,96 5,66	$v\,^3D^\circ-j\,^5F$ $y\,^5F^\circ-f\,^5G$	3—2 4—3
5911,55	3	3,19	5,28	$y\,^5D^\circ-e\,^5P$	2—1
5907,83	4	3,57	5,67	$y\,^5F^\circ-f\,^5G$	4—4
5906,35	5	1,46 3,20	3,56 5,30	$b\,^3F-y\,^5F^\circ$ $y\,^5D^\circ-e\,^5P$	4—3 3—2
5904,35	4	3,21	5,31	$y\,^5D^\circ-e\,^5P$	4—3
5903,317	5	1,07	3,17	$a\,^3P-z\,^3P^\circ$	2—1
5900,80	4	3,95	6,05	$z\,^3H^\circ-k\,^5F$	4—5
5899,295	25	1,05	3,15	$a\,^3P-y\,^3D^\circ$	1—2
5889,96	8	3,90 2,24	5,99 4,34	$w\,^3G^\circ-j\,^5F$ $b\,^3P-x\,^1D^\circ$	4—5 1—2
5889,05	3	3,18	5,28	$y\,^5D^\circ-e\,^5P$	1—1
5885,05	3	3,56	5,66	$y\,^5F^\circ-f\,^5G$	3—3
5880,306	5	1,05	3,16	$a\,^3P-z\,^3P^\circ$	1—2
5874,42	3	3,87	5,98	$v\,^3D^\circ-j\,^5F$	2—3
5872,36	10	3,34 3,19	5,47 5,30	$y\,^5G^\circ-f\,^5H$ $y\,^5D^\circ-e\,^5P$	6—7 2—2
5871,18	6	2,49	4,60	$a\,^1P-w\,^3P^\circ$	1—0
5869,23	3	1,07 3,94	3,18 6,06	$a\,^3P-y\,^5D^\circ$ $y\,^3P^\circ-e\,^3D$	2—1 2—3
5866,453	35	1,07	3,18	$a\,^3P-y\,^3D^\circ$	2—3
5865,32	6	3,57 1,05	5,68 3,17	$y\,^5F^\circ-f\,^3H$ $a\,^3P-z\,^3P^\circ$	4—5 1—1
5859,71	4	3,34 3,20	5,45 5,31	$y\,^5G^\circ-f\,^5H$ $y\,^5D^\circ-e\,^5P$	6—5 3—3
5847,12	10	1,07	3,19	$a\,^3P-y\,^5D^\circ$	2—2
5839,73	4	1,46	3,58	$b\,^3F-y\,^5F^\circ$	4—5
5838,03	12	3,06	5,18	$z\,^1G^\circ-e\,^5G$	4—4
5837,34	6	3,55	5,67	$y\,^5F^\circ-h\,^5F$	2—1
5829,86	5	1,44	3,57	$b\,^3F-y\,^5F^\circ$	3—4
5827,28	3	1,74	3,87	$a\,^5P-v\,^3D^\circ$	2—1
5823,679	3	2,27	4,40	$b\,^1G-y\,^3H^\circ$	4—4
5819,96	8	1,74	3,87	$a\,^5P-v\,^3D^\circ$	2—3
5817,88	6	3,90	6,03	$w\,^3G^\circ-k\,^5F$	4—4
5816,86	6	3,30	5,44	$y\,^5G^\circ-f\,^5H$	4—3
5810,08	3	1,05	3,19	$a\,^3P-y\,^5D^\circ$	1—2
5807,23	8	3,20	5,31	$y\,^5D^\circ-e\,^5P$	3—3
5804,265	5	3,34	5,47	$y\,^5G^\circ-f\,^5H$	6—7
5798,44	4	3,72	5,86	$x\,^3G^\circ-g\,^5G$	5—5
5791,26	7	2,49	4,63	$a\,^1P-w\,^3P^\circ$	1—2
5788,08	5	2,27	4,41	$b\,^1G-y\,^3H^\circ$	4—5
5785,979	5	3,32	5,46	$y\,^5G^\circ-f\,^5H$	5—6
5785,66	25	3,29	5,44	$y\,^5G^\circ-f\,^5H$	3—3
5784,38	3	3,20	5,34	$y\,^5D^\circ-g\,^5F$	3—2
5783,68	3	3,15	5,30	$y\,^3D^\circ-e\,^5P$	2—2
5780,70	12	3,14 2,25	5,28 4,39	$y\,^3D^\circ-e\,^5P$ $b\,^3P-y\,^3S^\circ$	1—1 2—1
5776,96	3	3,21 3,59	5,36 5,73	$y\,^5D^\circ-g\,^5F$ $a\,^5D-u\,^1G^\circ$	4—4 4—4
5774,54	13	2,25	4,40	$a\,^3H-y\,^3H^\circ$	5—4
5774,037	5	3,30	5,45	$y\,^5G^\circ-f\,^5H$	4—5
5771,28	6	4,51	6,65	$y\,^5P^\circ-e\,^1P$	2—1
5766,330	4	3,29	5,44	$y\,^5G^\circ-f\,^5H$	3—4
5763,52	3	2,51 1,74	4,66 3,89	$b\,^1D-v\,^3G^\circ$ $a\,^5P-w\,^3G^\circ$	2—3 2—3
5762,295	4	3,28	5,44	$y\,^5G^\circ-f\,^5H$	2—3
5756,45	6	2,26	4,41	$a\,^3H-y\,^3H^\circ$	6—5

λ, Å	I	E_H, eV	E_B, eV	Transition	J
5751,74	4	3,19	5,34	$y\,^5D^\circ-g\,^5F$	2—2
		2,78	4,93	$z\,^1F^\circ-f\,^3F$	3—4
5751,05	3	1,75	3,90	$a\,^5P-w\,^3G^\circ$	3—4
5745,07	8	2,32	4,48	$z\,^5D^\circ-e\,^5F$	4—3
5744,47	5	3,21	5,37	$y\,^5D^\circ-g\,^5F$	4—5
5739,975	4	2,24	4,40	$a\,^3H-y\,^3H^\circ$	4—4
5739,464	9	2,25	4,41	$a\,^3H-y\,^3H^\circ$	5—5
5735,77	3	3,17	5,34	$y\,^5D^\circ-g\,^5F$	0—1
5734,95	3	2,25	4,41	$b\,^3P-w\,^5D^\circ$	2—2
5734,24	10	2,30	4,46	$z\,^5D^\circ-e\,^5F$	2—1
		3,20	5,36	$y\,^5D^\circ-g\,^5F$	3—4
5731,08	4	2,23	4,39	$b\,^3P-y\,^3S^\circ$	0—1
5730,51	3	2,51	4,67	$b\,^1D-u\,^3F^\circ$	2—2
		3,18	5,34	$y\,^5D^\circ-g\,^5F$	1—2
		2,74	4,90	$z\,^1D^\circ-f\,^3F$	2—2
5728,25	4	2,09	4,26	$z\,^5F^\circ-d\,^3P$	2—1
5721,80	4	2,33	4,50	$c\,^3P-y\,^5P^\circ$	1—1
5720,445	3	2,29	4,46	$z\,^5D^\circ-e\,^5F$	1—1
5716,450	4	2,30	4,46	$z\,^5D^\circ-e\,^5F$	2—2
5715,123	9	2,26	4,42	$a\,^3H-y\,^3H^\circ$	6—6
5713,895	3	2,29	4,46	$z\,^5D^\circ-e\,^5F$	0—1
5711,852	4	2,31	4,48	$z\,^5D^\circ-e\,^5F$	3—3
5710,68	3	2,32	4,49	$z\,^5D^\circ-b\,^3G$	4—5
5709,95	3	3,11	5,28	$z\,^5S^\circ-e\,^5P$	2—1
5709,33	4	2,08	4,26	$z\,^5F^\circ-d\,^3P$	1—1
5708,199	3	2,32	4,49	$z\,^5D^\circ-e\,^5F$	4—4
5706,85	5	3,93	6,04	$y\,^3P^\circ-e\,^3D$	1—2
		3,87	6,04	$v\,^3D^\circ-e\,^3D$	3—2
5705,43	4	2,24	4,41	$a\,^3H-y\,^3H^\circ$	4—5
5702,666	6	2,29	4,46	$z\,^5D^\circ-e\,^5F$	1—2
5702,11	6	2,48	4,65	$z\,^3D^\circ-e\,^3F$	2—2
5701,66	7	3,87	6,04	$v\,^3D^\circ-e\,^3D$	2—2,
		2,50	4,67	$z\,^3D^\circ-e\,^3F$	3—3
5692,53	3	2,70	4,87	$z\,^3G^\circ-f\,^5F$	5—4
5689,465	10	2,30	4,48	$z\,^5D^\circ-e\,^5F$	2—3
5687,52	4	2,15	4,33	$a\,^3D-y\,^1P^\circ$	1—0
5683,80	3	3,18	5,36	$y\,^3D^\circ-g\,^5F$	3—4
5681,08	6	3,30	5,49	$y\,^5G^\circ-f\,^5D$	4—3
5675,413	9	2,31	4,49	$z\,^5D^\circ-e\,^5F$	3—4
5673,45	10	3,11	5,30	$z\,^5S^\circ-e\,^5P$	2—2
5669,76	5	3,13	5,31	$y\,^3F^\circ-e\,^5P$	3—3
		3,87	6,04	$v\,^3D^\circ-e\,^3D$	2—2
		3,87	6,06	$v\,^3D^\circ-e\,^3D$	3—3
5662,891	4	2,48	4,67	$z\,^3D^\circ-e\,^3F$	2—3
5662,154	12	2,32	4,51	$z\,^5D^\circ-e\,^5F$	4—5
5659,104	3	0,90	3,09	$a\,^1D-z\,^3S^\circ$	2—1
5656,51	4	2,70	4,89	$z\,^3G^\circ-f\,^5F$	5—5
5656,09	4	3,87	6,06	$v\,^3D^\circ-h\,^5D$	1—1
5654,78	8	2,66	4,85	$z\,^3G^\circ-f\,^5F$	3—2
5648,570	5	2,50	4,69	$z\,^3D^\circ-e\,^3F$	3—4
5644,137	18	2,27	4,46	$b\,^1G-y\,^1G^\circ$	4—4
5638,52	12	3,87	6,06	$v\,^3D^\circ-h\,^5D$	1—2
5634,73	3	3,57	5,77	$y\,^5F^\circ-h\,^3F$	4—4
5630,29	6	3,11	5,31	$z\,^5S^\circ-e\,^5P$	2—3
5627,45	3	3,94	6,15	$y\,^3P^\circ-f\,^3D$	2—2
		3,32	5,53	$x\,^3F^\circ-e\,^1D$	2—2
5623,58	5	1,74	3,94	$a\,^5P-y\,^3P^\circ$	2—2
		2,58	4,78	$a\,^1H-z\,^3I^\circ$	5—5
5618,32	5	1,50	3,71	$a\,^1G-w\,^3D^\circ$	4—3
5597,69	3	—	—	—	—

λ, Å	I	$E_н$, eV	$E_в$, eV	Transition	J
5565,478	9	2,24	4,46	$a\,^3H - y\,^1G^\circ$	4—4
5514,536	25	1,44	3,69	$b\,^3F - w\,^3D^\circ$	3—2
5514,350	20	1,43	3,68	$b\,^3F - w\,^3D^\circ$	2—1
5512,529	25	1,46	3,71	$b\,^3F - w\,^3D^\circ$	4—3
5503,897	8	2,58	4,83	$a\,^1H - x\,^1G^\circ$	5—4
5490,151	12	1,46	3,72	$b\,^3F - x\,^5D^\circ$	4—3
5488,210	5	2,40	4,65	$z\,^3F^\circ - e\,^3F$	2—2
5481,862	5	1,43	3,69	$b\,^3F - w\,^3D^\circ$	2—2
5481,426	6	2,41	4,67	$z\,^3F^\circ - e\,^3F$	3—3
5477,695	8	2,43	4,69	$z\,^3F^\circ - e\,^3F$	4—4
5474,228	6	1,46	3,72	$b\,^3F - x\,^3G^\circ$	4—5
5471,198	5	1,44	3,71	$b\,^3F - w\,^3D^\circ$	3—3
5460,502	4	0,05	2,32	$a\,^3F - z\,^5D^\circ$	4—4
5453,646	3	1,44	3,72	$b\,^3F - x\,^3G^\circ$	3—4
5429,139	6	2,34	4,63	$c\,^3P - w\,^3P^\circ$	2—2
5426,256	3	0,02	2,31	$a\,^3F - z\,^5D^\circ$	3—3
5409,609	6	1,89	4,18	$a\,^3G - w\,^3F^\circ$	5—4
5397,093	4	1,88	4,18	$a\,^3G - w\,^3F^\circ$	4—3
5389,996	3	1,87	4,17	$a\,^3G - w\,^3F^\circ$	3—2
5369,65	4	—	—	—	—
5351,072	4	2,78	5,09	$z\,^1F^\circ - e\,^1F$	3—3
5298,429	4	2,51	4,84	$b\,^1D - x\,^1P^\circ$	2—1
5297,236	6	1,87	4,21	$a\,^3G - v\,^3F^\circ$	3—2
5295,781	4	1,07	3,41	$a\,^3P - x\,^3D^\circ$	2—3
5283,441	8	1,88	4,22	$a\,^3G - v\,^3F^\circ$	4—3
5282,378	3	1,05	3,40	$a\,^3P - x\,^3D^\circ$	1—2
5265,967	10	1,89	4,24	$a\,^3G - v\,^3F^\circ$	5—4
5263,483	3	2,13	4,49	$z\,^5F^\circ - e\,^5F$	5—4
5259,976	3	2,74	5,09	$z\,^1D^\circ - e\,^1F$	2—3
5255,811	5	2,12	4,48	$z\,^5F^\circ - e\,^5F$	4—3
5252,105	8	0,05	2,41	$a\,^3F - z\,^3F^\circ$	4—3
5247,293	5	2,10	4,46	$z\,^5F^\circ - e\,^5F$	3—2
5246,574	3	0,84	3,20	$a\,^5F - y\,^5D^\circ$	4—3
5238,560	6	{ 0,85 { 2,09	3,21 4,46	$a\,^5F - y\,^5D^\circ$ $z\,^5F^\circ - e\,^5F$	5—4 2—1
5224,928	8	2,12	4,49	$z\,^5F^\circ - e\,^5F$	4—4
5224,558	6	2,10	4,48	$z\,^5F^\circ - e\,^5F$	3—3
5224,301	15	2,13	4,51	$z\,^5F^\circ - e\,^5F$	5—5
5223,623	6	2,09	4,46	$z\,^5F^\circ - e\,^5F$	2—2
5222,685	6	2,08	4,46	$z\,^5F^\circ - e\,^5F$	1—1
5219,697	8	0,02	2,40	$a\,^3F - z\,^3F^\circ$	3—2
5212,371	3	2,25	4,63	$b\,^3P - w\,^3P^\circ$	2—2
5210,386	40	0,05	2,43	$a\,^3F - z\,^3F^\circ$	4—4
5207,852	3	2,08	4,46	$z\,^5F^\circ - e\,^5F$	1—2
5206,059	5	2,49	4,87	$a\,^1P - w\,^1D^\circ$	1—2
5201,096	4	2,09	4,48	$z\,^5F^\circ - e\,^5F$	2—3
5194,043	4	2,10	4,49	$z\,^5F^\circ - e\,^5F$	3—4
5192,971	35	0,02	2,41	$a\,^3F - z\,^3F^\circ$	3—3
5186,329	3	2,12	4,51	$z\,^5F^\circ - e\,^5F$	4—5
5173,742	30	0,00	2,40	$a\,^3F - z\,^3F^\circ$	2—2
5152,185	10	0,02	2,43	$a\,^3F - z\,^3F^\circ$	3—4
5147,483	10	0,00	2,41	$a\,^3F - z\,^3F^\circ$	2—3
5145,465	12	1,46	3,87	$b\,^3F - v\,^3D^\circ$	4—3
5120,430	12	2,58	5,00	$a\,^1H - z\,^1I^\circ$	5—6
5113,448	10	1,44	3,87	$b\,^3F - v\,^3D^\circ$	3—2
5109,427	4	1,44	3,87	$b\,^3F - v\,^3D^\circ$	3—3
5087,055	8	1,43	3,87	$b\,^3F - v\,^3D^\circ$	2—1
5085,333	4	1,43	3,87	$b\,^3F - v\,^3D^\circ$	2—2

λ, Å	I	E_H, eV	E_B, eV	Transition	J
5071,475	7	1,46	3,90	$b\ ^3F—w\ ^3G^\circ$	4—4
5069,351	5	2,15	4,60	$a\ ^3D—w\ ^3P^\circ$	1—0
5068,332	3	2,66	5,11	$z\ ^3G^\circ—e\ ^3G$	3—3
5065,985	7	1,44	3,89	$b\ ^3F—w\ ^3G^\circ$	3—3
5064,654	25	0,05	2,50	$a\ ^3F—z\ ^3D^\circ$	4—3
5064,068	4	2,70	5,14	$z\ ^3G^\circ—e\ ^3G$	5—5
5062,112	7	2,16	4,61	$a\ ^3D—w\ ^3P^\circ$	2—1
5054,070	3	{ 2,68	5,13	$z\ ^3G^\circ—e\ ^3G$	4—4
		1,88	4,33	$a\ ^1S—y\ ^1P^\circ$	0—1
5052,879	8	2,17	4,63	$a\ ^3D—w\ ^3P^\circ$	3—2
5045,400	5	0,85	3,30	$a\ ^5F—y\ ^5G^\circ$	5—4
5043,578	7	0,84	3,29	$a\ ^5F—y\ ^5G^\circ$	4—3
5040,642	6	0,83	3,28	$a\ ^5F—y\ ^5G^\circ$	3—2
5039,959	22	0,02	2,48	$a\ ^3F—z\ ^3D^\circ$	3—2
5038,400	25	1,43	3,89	$b\ ^3F—w\ ^3G^\circ$	2—3
5036,468	25	1,44	3,90	$b\ ^3F—w\ ^3G^\circ$	3—4
5035,908	25	1,46	3,92	$b\ ^3F—w\ ^3G^\circ$	4—5
5025,570	18	2,04	4,51	$z\ ^5G^\circ—e\ ^5F$	6—5
5024,842	20	0,82	3,28	$a\ ^5F—y\ ^5G^\circ$	2—2
5022,871	25	0,83	3,29	$a\ ^5F—y\ ^5G^\circ$	3—3
5020,028	25	0,84	3,30	$a\ ^5F—y\ ^5G^\circ$	4—4
5016,162	20	0,85	3,32	$a\ ^5F—y\ ^5G^\circ$	5—5
5014,277	25	0,81	3,28	$a\ ^5F—y\ ^5G^\circ$	1—2
5014,185	25	0,00	2,47	$a\ ^3F—z\ ^3D^\circ$	2—1
5013,284	18	2,02	4,49	$z\ ^5G^\circ—e\ ^5F$	5—4
5009,652	7	0,02	2,50	$a\ ^3F—z\ ^3D^\circ$	3—3
5007,209	40	0,82	3,29	$a\ ^5F—y\ ^5G^\circ$	2—3
5000,991	10	2,00	4,48	$z\ ^5G^\circ—e\ ^5F$	4—3
4999,504	45	0,83	3,30	$a\ ^5F—y\ ^5G^\circ$	3—4
4997,099	8	0,00	2,48	$a\ ^3F—z\ ^3D^\circ$	2—2
4991,067	50	0,84	3,32	$a\ ^5F—y\ ^5G^\circ$	4—5
4989,140	10	1,98	4,46	$z\ ^5G^\circ—e\ ^5F$	3—2
4981,732	60	0,85	3,34	$a\ ^5F—y\ ^5G^\circ$	5—6
4978,191	10	1,97	4,46	$z\ ^5G^\circ—e\ ^5F$	2—1
4977,731	5	2,02	4,51	$z\ ^5G^\circ—e\ ^5F$	5—5
4975,344	10	2,51	5,00	$b\ ^1D—w\ ^1F^\circ$	2—3
4973,051	6	2,00	4,49	$z\ ^5G^\circ—e\ ^5F$	4—4
4968,566	6	1,98	4,48	$z\ ^5G^\circ—e\ ^5F$	3—3
4964,713	5	1,97	4,46	$z\ ^5G^\circ—e\ ^5F$	2—2
4948,183	3	2,17	4,68	$a\ ^3D—u\ ^3F^\circ$	3—3
4941,562	3	2,16	4,67	$a\ ^3D—u\ ^3F^\circ$	2—2
4938,283	8	2,58	5,09	$a\ ^1H—y\ ^1H^\circ$	5—5
4937,719	4	0,81	3,32	$a\ ^5F—x\ ^3F^\circ$	1—2
4928,342	12	2,15	4,67	$a\ ^3D—u\ ^3F^\circ$	1—2
4926,148	4	0,82	3,33	$a\ ^5F—x\ ^3F^\circ$	2—3
4925,396	5	1,88	4,40	$a\ ^3G—y\ ^3H^\circ$	4—4
4921,768	12	2,17	4,69	$a\ ^3D—u\ ^3F^\circ$	3—4
4919,867	12	2,16	4,68	$a\ ^3D—u\ ^3F^\circ$	2—3
4915,236	5	1,89	4,41	$a\ ^3G—y\ ^3H^\circ$	5—5
4913,616	20	1,87	4,40	$a\ ^3G—y\ ^3H^\circ$	3—4
4899,910	20	1,88	4,41	$a\ ^3G—y\ ^3H^\circ$	4—5
4885,082	20	1,89	4,42	$a\ ^3G—y\ ^3H^\circ$	5—6
4880,922	3	2,15	4,69	$a\ ^3D—u\ ^3D^\circ$	1—1
4870,129	20	2,25	4,79	$a\ ^3H—z\ ^3I^\circ$	5—6
4868,264	18	2,24	4,78	$a\ ^3H—z\ ^3I^\circ$	4—5
4864,187	4	2,16	4,71	$a\ ^3D—u\ ^3D^\circ$	2—2
4856,012	20	2,26	4,81	$a\ ^3H—z\ ^3I^\circ$	6—7
4848,487	8	2,17	4,73	$a\ ^3D—u\ ^3D^\circ$	3—3
4840,874	25	0,90	3,46	$a\ ^1D—y\ ^1D^\circ$	2—2

λ, Å	I	E_H, eV	E_B, eV	Transition	J
4836,125	6	2,27	4,83	$b\,^{1}G - x\,^{1}G^{\circ}$	4—4
4825,445	3	2,32	4,89	$z\,^{5}D^{\circ} - f\,^{5}F$	4—5
4820,410	20	1,50	4,07	$a\,^{1}G - y\,^{1}F^{\circ}$	4—3
4812,240	5	2,34	4,92	$c\,^{3}P - s\,^{3}D^{\circ}$	2—2
4811,074	4	1,89	4,46	$a\,^{3}G - y\,^{1}G^{\circ}$	5—4
4808,531	5	3,06	5,64	$z\,^{1}G^{\circ} - e\,^{1}H$	4—5
4805,416	12	2,34	4,92	$c\,^{3}P - s\,^{3}D^{\circ}$	2—3
4799,797	12	2,27	4,85	$b\,^{1}G - x\,^{3}H^{\circ}$	4—4
4797,983	5	2,33	4,92	$c\,^{3}P - s\,^{3}D^{\circ}$	1—1
4796,210	6	2,33	4,92	$c\,^{3}P - s\,^{3}D^{\circ}$	0—1
4792,482	10	2,33	4,92	$c\,^{3}P - s\,^{3}D^{\circ}$	1—2
4781,718	6	0,85	3,44	$a\,^{5}F - y\,^{3}G^{\circ}$	5—5
4778,259	10	2,24	4,83	$a\,^{3}H - x\,^{1}G^{\circ}$	4—4
4771,103	3	0,83	3,42	$a\,^{5}F - y\,^{3}G^{\circ}$	3—4
4769,775	4	2,26	4,85	$a\,^{3}H - x\,^{3}H^{\circ}$	6—5
4766,330	4	2,25	4,85	$a\,^{3}H - x\,^{3}H^{\circ}$	5—4
4759,272	25	2,26	4,86	$a\,^{3}H - x\,^{3}H^{\circ}$	6—6
4758,913	4	0,84	3,44	$a\,^{5}F - y\,^{3}G^{\circ}$	4—5
4758,120	25	2,25	4,85	$a\,^{3}H - x\,^{3}H^{\circ}$	5—5
4747,680	3	2,25	4,86	$a\,^{3}H - x\,^{3}H^{\circ}$	5—6
4742,791	20	2,24	4,85	$a\,^{3}H - x\,^{3}H^{\circ}$	4—4
4742,129	3	2,15	4,77	$a\,^{3}D - t\,^{3}F^{\circ}$	1—2
4734,682	3	2,24	4,85	$a\,^{3}H - x\,^{3}H^{\circ}$	4—5
4733,426	6	2,16	4,78	$a\,^{3}D - t\,^{3}F^{\circ}$	2—3
4731,172	9	2,17	4,79	$a\,^{3}D - t\,^{3}F^{\circ}$	3—4
4723,171	10	1,07	3,69	$a\,^{3}P - w\,^{3}D^{\circ}$	2—2
4722,603	10	1,05	3,68	$a\,^{3}P - w\,^{3}D^{\circ}$	1—1
4715,295	4	0,05	2,68	$a\,^{3}F - z\,^{3}G^{\circ}$	4—4
4710,186	18	{ 1,05	3,68	$a\,^{3}P - w\,^{3}D^{\circ}$	0—1
		{ 2,17	4,81	$a\,^{3}D - t\,^{3}D^{\circ}$	3—3
4698,86	6	2,16	4,80	$a\,^{3}D - t\,^{3}D^{\circ}$	2—2
4698,766	20	1,05	3,69	$a\,^{3}P - w\,^{3}D^{\circ}$	1—2
4696,923	4	2,15	4,79	$a\,^{3}D - t\,^{3}D^{\circ}$	1—1
4693,670	5	0,02	2,66	$a\,^{3}F - z\,^{3}G^{\circ}$	3—3
4691,336	20	1,07	3,71	$a\,^{3}P - w\,^{3}D^{\circ}$	2—3
4690,827	3	1,07	3,71	$a\,^{3}P - x\,^{3}G^{\circ}$	2—3
4688,392	3	3,09	5,73	$z\,^{3}S^{\circ} - e\,^{3}P$	1—2
4686,921	4	2,15	4,80	$a\,^{3}D - t\,^{3}D^{\circ}$	1—2
4681,908	30	0,05	2,70	$a\,^{3}F - z\,^{3}G^{\circ}$	4—5
4675,118	10	1,07	3,72	$a\,^{3}F - x\,^{5}D^{\circ}$	2—3
4667,585	25	0,02	2,68	$a\,^{3}F - z\,^{3}G^{\circ}$	3—4
4656,468	25	0,00	2,66	$a\,^{3}F - z\,^{3}G^{\circ}$	2—3
4656,048	6	1,75	4,41	$a\,^{5}P - w\,^{5}D^{\circ}$	3—2
4655,712	3	2,34	5,01	$c\,^{3}P - v\,^{3}P^{\circ}$	2—1
4650,016	10	1,74	4,40	$a\,^{5}P - w\,^{5}D^{\circ}$	2—1
4645,193	12	1,73	4,40	$a\,^{5}P - w\,^{5}D^{\circ}$	1—0
4639,944	15	1,73	4,40	$a\,^{5}P - w\,^{5}D^{\circ}$	1—1
4639,669	15	1,75	4,42	$a\,^{5}P - w\,^{5}D^{\circ}$	3—3
4639,369	18	1,74	4,41	$a\,^{5}P - w\,^{5}D^{\circ}$	2—2
4637,887	8	2,34	5,02	$c\,^{3}P - v\,^{3}P^{\circ}$	2—2
4635,539	3	2,33	5,01	$c\,^{3}P - v\,^{3}P^{\circ}$	0—1
4629,336	15	1,73	4,41	$a\,^{5}P - w\,^{5}D^{\circ}$	1—2
4623,098	25	1,74	4,42	$a\,^{5}P - w\,^{5}D^{\circ}$	2—3
4619,525	3	2,33	5,02	$c\,^{3}P - v\,^{3}P^{\circ}$	1—2
4617,269	30	1,75	4,43	$a\,^{5}P - w\,^{5}D^{\circ}$	3—4
4599,23	5	—	—	—	—
4570,906	3	2,40	5,11	$z\,^{3}F^{\circ} - e\,^{3}G$	2—3
4563,427	5	2,43	5,14	$z\,^{3}F^{\circ} - e\,^{3}G$	4—5

λ, Å	I	E_H, eV	E_B, eV	Transition	J
4562,637	6	0,02	2,74	$a\,^3F - z\,^1D^\circ$	3—2
4559,920	6	1,46	4,18	$b\,^3F - w\,^3F^\circ$	4—4
4555,486	30	0,85	3,57	$a\,^5F - y\,^5F^\circ$	5—4
4555,069	3	2,41	5,13	$z\,^3F^\circ - e\,^3G$	3—4
4552,453	35	0,84	3,56	$a\,^5F - y\,^5F^\circ$	4—3
4548,764	35	0,83	3,55	$a\,^5F - y\,^5F^\circ$	3—2
4544,688	30	0,82	3,55	$a\,^5F - y\,^5F^\circ$	2—1
4536,051	40	0,81	3,55	$a\,^5F - y\,^5F^\circ$	1—1
4535,920	40	0,82	3,55	$a\,^5F - y\,^5F^\circ$	2—2
4535,574	50	0,83	3,56	$a\,^5F - y\,^5F^\circ$	3—3
4534,782	60	0,84	3,57	$a\,^5F - y\,^5F^\circ$	4—4
4533,238	80	0,85	3,58	$a\,^5F - y\,^5F^\circ$	5—5
4527,455	4	0,02	2,74	$a\,^3F - z\,^1D^\circ$	2—2
4527,305	35	0,81	3,55	$a\,^5F - y\,^5F^\circ$	1—2
4522,798	40	0,82	3,56	$a\,^5F - y\,^5F^\circ$	2—3
4518,700	8	1,43	4,17	$b\,^3F - w\,^3F^\circ$	2—2
4518,022	50	0,83	3,57	$a\,^5F - y\,^5F^\circ$	3—4
4512,734	40	0,84	3,58	$a\,^5F - y\,^5F^\circ$	4—5
4503,762	4	2,13	4,89	$z\,^5F^\circ - f\,^5F$	5—5
4497,709	3	2,12	4,87	$z\,^5F^\circ - f\,^5F$	4—4
4496,146	20	1,75	4,51	$a\,^5P - y\,^5P^\circ$	3—2
4492,540	3	2,10	4,86	$z\,^5F^\circ - f\,^5F$	3—3
4489,089	20	1,74	4,50	$a\,^5P - y\,^5P^\circ$	2—1
4482,688	10	1,46	4,22	$b\,^3F - v\,^3F^\circ$	4—3
4481,261	30	1,75	4,51	$a\,^5P - y\,^5P^\circ$	3—3
4480,600	5	1,74	4,51	$a\,^5P - y\,^5P^\circ$	2—2
4479,724	9	1,73	4,50	$a\,^5P - y\,^5P^\circ$	1—1
4474,852	8	{ 1,44	4,21	$b\,^3F - v\,^3F^\circ$	3—2
		{ 2,10	4,87	$z\,^5F^\circ - f\,^5F$	3—4
4471,238	20	1,73	4,51	$a\,^5P - y\,^5P^\circ$	1—2
4465,807	20	1,74	4,51	$a\,^5P - y\,^5P^\circ$	2—3
4463,539	8	1,89	4,66	$a\,^3G - v\,^3G^\circ$	5—4
4463,391	8	1,88	4,66	$a\,^3G - v\,^3G^\circ$	4—3
4462,089	3	0,00	2,78	$a\,^3F - z\,^1F^\circ$	2—3
4457,428	40	1,46	4,24	$b\,^3F - v\,^3F^\circ$	4—4
4455,321	30	1,44	4,22	$b\,^3F - v\,^3F^\circ$	3—3
4453,708	20	1,87	4,66	$a\,^3G - v\,^3G^\circ$	3—3
4453,312	30	1,43	4,21	$b\,^3F - v\,^3F^\circ$	2—2
4450,896	25	1,88	4,66	$a\,^3G - v\,^3G^\circ$	4—4
4449,143	30	1,89	4,67	$a\,^3G - v\,^3G^\circ$	5—5
4443,20	3	2,74	5,53	$z\,^1D^\circ - e\,^1D$	2—2
4441,272	4	1,87	4,66	$a\,^3G - v\,^3G^\circ$	3—4
4440,345	10	1,87	4,66	$a\,^3G - x\,^1F^\circ$	3—3
4436,586	4	1,88	4,67	$a\,^3G - v\,^3G^\circ$	4—5
4434,003	15	{ 1,43	4,22	$b\,^3F - v\,^3F^\circ$	2—3
		{ 1,87	4,67	$a\,^3G - u\,^3F^\circ$	3—2
4433,578	3	2,40	5,19	$z\,^3F^\circ - g\,^3F$	2—2
4431,284	4	2,23	5,03	$b\,^3P - r\,^3D^\circ$	0—1
4430,366	7	1,44	4,24	$b\,^3F - v\,^3F^\circ$	3—4
4430,023	3	2,41	5,21	$z\,^3F^\circ - g\,^3F$	3—3
4427,098	40	1,50	4,30	$a\,^1G - z\,^1H^\circ$	4—5
4426,054	10	1,88	4,68	$a\,^3G - u\,^3F^\circ$	4—3
4425,840	3	1,07	3,87	$a\,^3P - v\,^3D^\circ$	2—2
4422,823	10	1,07	3,87	$a\,^3P - v\,^3D^\circ$	2—3
4421,754	6	2,24	5,04	$b\,^3P - r\,^3D^\circ$	1—2
4417,274	15	1,89	4,69	$a\,^3G - u\,^3F^\circ$	5—4
4416,535	4	1,87	4,68	$a\,^3G - u\,^3F^\circ$	3—3
4404,911	5	1,88	4,69	$a\,^3G - u\,^3F^\circ$	4—4

λ, Å	I	E_H, eV	E_B, eV	Transition	J
4404,397	5	1,05	3,87	$a\,^3P-v\,^3D^\circ$	1—2
4404,276	10	2,25	5,06	$b\,^3P-x\,^3S^\circ$	2—1
		2,25	5,06	$b\,^3P-r\,^3D^\circ$	2—3
4393,925	8	2,27	5,09	$b\,^1G-y\,^1H^\circ$	4—5
4388,077	3	2,24	5,06	$b\,^3P-x\,^3S^\circ$	1—1
4372,383	3	2,49	5,32	$a\,^1P-w\,^1P^\circ$	1—1
4369,682	5	2,58	5,41	$a\,^1H-v\,^1G^\circ$	5—4
4360,487	4	2,17	5,02	$a\,^3D-v\,^3P^\circ$	3—2
4354,064	3	2,16	5,01	$a\,^3D-v\,^3P^\circ$	2—1
4346,104	5	2,24	5,09	$a\,^3H-y\,^1H^\circ$	4—5
4326,359	9	0,83	3,69	$a\,^5F-w\,^3D^\circ$	3—2
4325,134	9	2,25	5,11	$a\,^3H-u\,^3G^\circ$	5—4
4321,655	8	2,24	5,10	$a\,^3H-u\,^3G^\circ$	4—3
4318,631	10	2,26	5,13	$a\,^3H-u\,^3G^\circ$	6—5
4314,801	25	0,84	3,71	$a\,^5F-w\,^3D^\circ$	4—3
4314,74	25	0,82	3,69	$a\,^5F-w\,^3D^\circ$	2—2
4314,356	5	0,84	3,71	$a\,^5F-x\,^3G^\circ$	4—3
4305,910	60	0,85	3,73	$a\,^5F-x\,^5D^\circ$	5—4
4301,089	50	0,84	3,72	$a\,^5F-x\,^5D^\circ$	4—3
4300,566	50	0,83	3,71	$a\,^5F-x\,^5D^\circ$	3—2
4299,636	15	0,83	3,71	$a\,^5F-w\,^3D^\circ$	3—3
4299,229	15	1,75	4,63	$a\,^5P-y\,^5S^\circ$	3—2
4298,664	40	0,82	3,70	$a\,^5F-x\,^5D^\circ$	2—1
4295,751	22	0,81	3,70	$a\,^5F-x\,^5D^\circ$	1—0
4291,214	5	1,74	4,63	$a\,^5P-w\,^3P^\circ$	2—2
		0,84	3,72	$a\,^5F-x\,^3G^\circ$	4—5
4290,933	22	0,81	3,70	$a\,^5F-x\,^5D^\circ$	1—1
4289,919	3	2,17	5,06	$a\,^3D-r\,^3D^\circ$	3—3
4289,068	25	0,82	3,71	$a\,^5F-x\,^5D^\circ$	2—2
4288,161	3	1,05	3,94	$a\,^3P-y\,^3P^\circ$	1—2
		0,82	3,71	$a\,^5F-w\,^3D^\circ$	2—3
4287,405	22	0,84	3,73	$a\,^5F-x\,^5D^\circ$	4—4
4286,006	25	0,83	3,72	$a\,^5F-x\,^5D^\circ$	3—3
4284,988	8	1,74	4,63	$a\,^5P-y\,^5S^\circ$	2—2
4282,702	12	1,87	4,77	$a\,^3G-t\,^3F^\circ$	3—2
4281,371	10	0,81	3,71	$a\,^5F-x\,^5D^\circ$	1—2
4278,829	3	2,31	5,20	$z\,^5D^\circ-e\,^5D$	3—2
4278,231	7	2,58	5,48	$a\,^1H-x\,^1H^\circ$	5—5
4276,441	8	1,73	4,63	$a\,^5P-y\,^5S^\circ$	1—2
4274,584	15	1,88	4,78	$a\,^3G-t\,^3F^\circ$	4—3
		0,82	3,72	$a\,^5F-x\,^5D^\circ$	2—3
4272,440	8	0,83	3,73	$a\,^5F-x\,^5D^\circ$	3—4
4270,139	7	2,32	5,22	$z\,^5D^\circ-g\,^3F$	4—4
4266,227	3	2,30	5,20	$z\,^5D^\circ-e\,^5D$	2—2
4265,723	4	1,87	4,78	$a\,^3G-t\,^3F^\circ$	3—3
4265,273	3	2,29	5,19	$z\,^5D^\circ-e\,^5D$	0—1
4263,134	15	1,89	4,79	$a\,^3G-t\,^3F^\circ$	5—4
4261,609	5	2,31	5,21	$z\,^5D^\circ-e\,^5D$	3—3
4258,523	4	2,29	5,20	$z\,^5D^\circ-e\,^5D$	1—2
4256,025	8	2,32	5,23	$z\,^5D^\circ-e\,^5D$	4—4
4251,618	3	1,88	4,79	$a\,^3G-t\,^3F^\circ$	4—4
4249,114	5	2,30	5,21	$z\,^5D^\circ-e\,^5D$	2—3
4237,889	7	2,51	5,43	$b\,^1D-u\,^1D^\circ$	2—2
4227,654	5	2,49	5,42	$a\,^1P-v\,^1D^\circ$	1—2
4224,795	5	2,78	5,71	$z\,^1F^\circ-e\,^1G$	3—4
4211,729	4	2,49	5,43	$a\,^1P-u\,^1D^\circ$	1—2
4203,465	8	2,25	5,20	$b\,^3P-u\,^3P^\circ$	2—2
4200,752	6	2,25	5,20	$b\,^3P-u\,^3P^\circ$	2—1
4188,694	5	2,24	5,20	$b\,^3P-u\,^3P^\circ$	1—2

λ, Å	I	E_H, eV	E_B, eV	Transition	J
4186,119	25	1,50	4,46	$a\ ^1G—y\ ^1G°$	4—4
4183,294	4	2,24	5,20	$b\ ^3P—u\ ^3P°$	1—0
4180,86	3	1,88	4,84	$a\ ^1S—x\ ^1P°$	0—1
4174,472	3	2,23	5,20	$b\ ^3P—u\ ^3P°$	0—1
4171,018	8	2,15	5,12	$a\ ^3D—s\ ^3F°$	1—2
4169,330	7	1,89	4,86	$a\ ^3G—x\ ^3H°$	5—6
4166,311	6	1,88	4,85	$a\ ^3G—x\ ^3H°$	4—5
4164,134	4	1,87	4,85	$a\ ^3G—x\ ^3H°$	3—4
4159,634	9	2,16	5,14	$a\ ^3D—s\ ^3F°$	2—3
4150,963	10	2,17	5,16	$a\ ^3D—s\ ^3F°$	3—4
4150,557	3	2,30	5,28	$z\ ^5D°—e\ ^5P$	2—1
4143,280	3	2,29	5,28	$z\ ^5D°—e\ ^5P$	1—1
4143,048	7	2,31	5,30	$z\ ^5D°—e\ ^5P$	3—2
4137,284	10	2,32	5,31	$z\ ^5D°—e\ ^5P$	4—3
4131,244	4	2,30	5,30	$z\ ^5D°—e\ ^5P$	2—2
4127,531	15	2,70	5,70	$z\ ^3G°—f\ ^3H$	5—6
4123,559	10	2,68	5,68	$z\ ^3G°—f\ ^3H$	4—5
4123,287	5	2,78	5,78	$z\ ^1F°—f\ ^1F$	3—3
4122,143	10	2,66	5,67	$z\ ^3G°—f\ ^3H$	3—4
4112,708	20	0,05	3,06	$a\ ^3F—z\ ^1G°$	4—4
4099,166	8	2,17	5,20	$a\ ^3D—u\ ^3P°$	3—2
4082,456	20	1,07	4,10	$a\ ^3P—x\ ^3P°$	2—1
4079,708	6	2,16	5,20	$a\ ^3D—u\ ^3P°$	2—2
4078,471	30	1,07	4,11	$a\ ^3P—x\ ^3P°$	2—2
4077,143	4	2,16	5,20	$a\ ^3D—u\ ^3P°$	2—1
4076,370	4	0,02	3,06	$a\ ^3F—z\ ^1G°$	3—4
4074,356	3	2,32	5,36	$z\ ^5D°—g\ ^5F$	4—4
4068,981	4	2,74	5,78	$z\ ^1D°—f\ ^1F$	2—3
4068,144	3	2,15	5,20	$a\ ^3D—u\ ^3P°$	1—1
4065,094	15	1,05	4,10	$a\ ^3P—x\ ^3P°$	1—0
4064,203	15	1,05	4,10	$a\ ^3P—x\ ^3P°$	1—1
4060,263	20	1,05	4,11	$a\ ^3P—x\ ^3P°$	1—2
4058,139	7	2,32	5,37	$z\ ^5D°—g\ ^5F$	4—5
4057,612	5	2,31	5,36	$z\ ^5D°—g\ ^5F$	3—4
4055,011	20	1,05	4,10	$a\ ^3P—x\ ^3P°$	0—1
4040,310	4	2,12	5,18	$z\ ^5F°—e\ ^5G$	4—4
4035,828	10	2,17	5,25	$a\ ^3D—q\ ^3D°$	3—3
4034,884	5	2,15	5,23	$a\ ^3D—q\ ^3D°$	1—1
4033,883	6	2,16	5,23	$a\ ^3D—q\ ^3D°$	2—2
4032,628	3	2,70	5,77	$z\ ^3G°—h\ ^3F$	5—4
4031,753	3	2,10	5,18	$z\ ^5F°—e\ ^5G$	3—3
4030,512	25	2,13	5,21	$z\ ^5F°—e\ ^5G$	5—6
4026,539	25	2,12	5,20	$z\ ^5F°—e\ ^5G$	4—5
4024,573	35	0,05	3,13	$a\ ^3F—y\ ^3F°$	4—3
4021,812	25	2,10	5,18	$z\ ^5F°—e\ ^5G$	3—4
4017,771	15	2,09	5,18	$z\ ^5F°—e\ ^5G$	2—3
4016,943	3	2,16	5,25	$a\ ^3D—q\ ^3D°$	2—3
4016,264	6	2,13	5,22	$z\ ^5F°—g\ ^3F$	5—4
4015,377	12	2,08	5,17	$z\ ^5F°—e\ ^5G$	1—2
4013,587	12	2,13	5,22	$z\ ^5F°—e\ ^5H$	5—6
4012,786	3	2,12	5,21	$z\ ^5F°—g\ ^3F$	4—3
4011,534	3	0,00	3,09	$a\ ^3F—z\ ^3S°$	2—1
4009,653	15	0,02	3,11	$a\ ^3F—z\ ^5S°$	3—2
4008,926	35	0,02	3,11	$a\ ^3F—y\ ^3F°$	3—2
4008,046	9	2,12	5,21	$z\ ^5F°—e\ ^5H$	4—5
4007,195	3	2,09	5,19	$z\ ^5F°—e\ ^5H$	2—3
4005,952	6	2,10	5,20	$z\ ^5F°—e\ ^5H$	3—4
4003,789	10	2,13	5,23	$z\ ^5F°—e\ ^5D$	5—4
4002,466	9	2,12	5,21	$z\ ^5F°—e\ ^5D$	4—3

λ, Å	I	$E_{\text{н}}$, eV	$E_{\text{в}}$, eV	Transition	J
3999,336	7	2,10	5,20	$z\,^5F^\circ—e\,^5D$	3—2
3998,635	100	0,05	3,15	$a\,^3F—y\,^3F^\circ$	4—4
3994,683	4	2,10	5,21	$z\,^5F^\circ—g\,^3F$	3—3
		2,09	5,19	$z\,^5F^\circ—e\,^5D$	2—1
3989,758	80	0,02	3,13	$a\,^3F—y\,^3F^\circ$	3—3
3984,313	3	2,10	5,21	$z\,^5F^\circ—e\,^5D$	3—3
3982,478	30	0,00	3,11	$a\,^3F—z\,^5S^\circ$	2—2
3981,761	70	0,00	3,11	$a\,^3F—y\,^3F^\circ$	2—2
3964,269	35	0,02	3,15	$a\,^3F—y\,^3F^\circ$	3—4
3962,851	35	0,00	3,13	$a\,^3F—y\,^3F^\circ$	2—3
3958,206	80	0,05	3,18	$a\,^3F—y\,^3D^\circ$	4—3
3956,336	60	0,02	3,15	$a\,^3F—y\,^3D^\circ$	3—2
3948,670	60	0,00	3,14	$a\,^3F—y\,^3D^\circ$	2—1
3947,770	40	0,02	3,16	$a\,^3F—z\,^3P^\circ$	3—2
3934,527	50	0,02	3,18	$a\,^3F—y\,^3D^\circ$	3—3
3934,228	9	0,05	3,20	$a\,^3F—y\,^5D^\circ$	4—3
3929,875	40	0,00	3,15	$a\,^3F—y\,^3D^\circ$	2—2
3926,319	10	2,58	5,73	$a\,^1H—u\,^1G^\circ$	5—4
3924,51	50	0,02	3,18	$a\,^3F—y\,^3D^\circ$	3—3
3921,423	30	0,00	3,16	$a\,^3F—z\,^3P^\circ$	2—2
3919,822	5	1,50	4,66	$a\,^1G—x\,^1F^\circ$	4—3
3915,879	3	0,02	3,19	$a\,^3F—y\,^5D^\circ$	3—2
3914,751	5	0,00	3,17	$a\,^3F—z\,^3P^\circ$	2—1
3914,334	35	0,05	3,21	$a\,^3F—y\,^5D^\circ$	4—4
3911,185	8	2,04	5,21	$z\,^5G^\circ—e\,^5G$	6—6
3904,785	40	0,90	4,07	$a\,^1D—y\,^1F^\circ$	2—3
3900,958	12	0,02	3,20	$a\,^3F—y\,^5D^\circ$	3—3
3898,487	8	0,00	3,18	$a\,^3F—y\,^3D^\circ$	2—3
3895,243	30	2,04	5,22	$z\,^5G^\circ—e\,^5H$	6—6
3889,948	6	0,00	3,19	$a\,^3F—y\,^5D^\circ$	2—2
3888,020	4	2,00	5,18	$z\,^5G^\circ—e\,^5G$	4—4
3882,892	20	2,04	5,23	$z\,^5G^\circ—e\,^5H$	6—7
3882,313	10	2,02	5,21	$z\,^5G^\circ—e\,^5H$	5—5
3882,147	15	2,02	5,21	$z\,^5G^\circ—e\,^5G$	5—6
3881,399	4	0,02	3,21	$a\,^3F—y\,^5D^\circ$	3—4
3875,262	20	0,00	3,20	$a\,^3F—y\,^5D^\circ$	2—3
		2,00	5,20	$z\,^5G^\circ—e\,^5G$	4—5
3873,203	10	2,00	5,20	$z\,^5G^\circ—e\,^5H$	4—4
3869,275	5	1,97	5,17	$z\,^5G^\circ—e\,^5G$	2—2
3868,397	10	1,98	5,18	$z\,^5G^\circ—e\,^5G$	3—4
3867,739	3	1,98	5,19	$z\,^5G^\circ—e\,^5H$	3—3
3866,446	15	2,02	5,22	$z\,^5G^\circ—e\,^5H$	5—6
3862,823	10	1,97	5,18	$z\,^5G^\circ—e\,^5G$	2—3
3858,133	15	2,00	5,21	$z\,^5G^\circ—e\,^5H$	4—5
3853,719	10	1,98	5,20	$z\,^5G^\circ—e\,^5H$	3—4
3853,038	10	1,97	5,19	$z\,^5G^\circ—e\,^5H$	2—3
3846,436	6	—	—	—	—
3842,61	3	2,13	5,36	$z\,^5F^\circ—g\,^5F$	5—4
3836,763	5	—	—	—	—
3833,91	3	1,46	4,69	$b\,^3F—u\,^3F^\circ$	4—4
3828,180	3	2,13	5,37	$z\,^5F^\circ—g\,^5F$	5—5
3817,639	5	2,10	5,35	$z\,^5F^\circ—g\,^5F$	3—3
3814,855	4	2,09	5,34	$z\,^5F^\circ—g\,^5F$	2—2
3811,385	4	1,87	5,12	$a\,^3G—s\,^3F^\circ$	3—2
3801,093	3	1,88	5,14	$a\,^3G—s\,^3F^\circ$	4—3
3798,276	6	1,43	4,69	$b\,^3F—u\,^3D^\circ$	2—1
3795,903	7	1,44	4,71	$b\,^3F—u\,^3D^\circ$	3—2
3789,293	8	1,46	4,73	$b\,^3F—u\,^3D^\circ$	4—3

λ, Å	I	E_H, eV	E_B, eV	Transition	J
3786,253	3	1,89	5,16	$a\,^3G - s\,^3F^\circ$	5—4
3786,043	20	0,90	4,17	$a\,^1D - z\,^1P^\circ$	2—1
3771,652	25	0,05	3,33	$a\,^3F - x\,^3F^\circ$	4—3
3766,445	3	1,05	4,34	$a\,^3P - x\,^1D^\circ$	1—2
3766,043	20	0,90	4,17	$a\,^1D - z\,^1P^\circ$	2—1
3753,623	25	0,02	3,32	$a\,^3F - x\,^3F^\circ$	3—2
3752,860	80	0,05	3,35	$a\,^3F - x\,^3F^\circ$	4—4
3748,101	6	1,87	5,18	$a\,^3G - w\,^3H^\circ$	3—4
3741,059	60	0,02	3,33	$a\,^3F - x\,^3F^\circ$	3—3
3738,901	5	1,88	5,19	$a\,^3G - w\,^3H^\circ$	4—5
3733,767	4	1,89	5,21	$a\,^3G - w\,^3H^\circ$	5—6
3729,806	50	0,00	3,32	$a\,^3F - x\,^3F^\circ$	2—2
3725,155	20	1,07	4,39	$a\,^3P - y\,^3S^\circ$	2—1
3724,570	20	1,50	4,83	$a\,^1G - x\,^1G^\circ$	4—4
3722,568	15	0,02	3,35	$a\,^3F - x\,^3F^\circ$	3—4
3717,393	20	0,00	3,33	$a\,^3F - x\,^3F^\circ$	2—3
3709,963	20	1,05	4,39	$a\,^3P - y\,^3S^\circ$	1—1
3708,625	4	2,43	5,77	$z\,^3F^\circ - h\,^3F$	4—4
3707,549	10	2,02	5,36	$z\,^5G^\circ - g\,^5F$	5—4
3704,295	15	1,46	4,81	$b\,^3F - t\,^3D^\circ$	4—3
3702,291	10	1,05	4,39	$a\,^3P - y\,^3S^\circ$	0—1
3698,183	3	2,25	5,60	$b\,^3P - t\,^3P^\circ$	2—2
3694,445	10	1,44	4,80	$b\,^3F - t\,^3D^\circ$	3—2
3689,916	15	0,05	3,41	$a\,^3F - x\,^3D^\circ$	4—3
3687,354	5	0,05	3,41	$a\,^3F - y\,^3G^\circ$	4—3
3671,672	20	0,05	3,42	$a\,^3F - y\,^3G^\circ$	4—4
3668,965	15	0,02	3,40	$a\,^3F - x\,^3D^\circ$	3—2
3660,631	12	0,02	3,41	$a\,^3F - x\,^3D^\circ$	3—3
3658,097	20	0,02	3,41	$a\,^3F - y\,^3G^\circ$	3—3
3654,592	15	0,00	3,39	$a\,^3F - x\,^3D^\circ$	2—1
3653,497	100	0,05	3,44	$a\,^3F - y\,^3G^\circ$	4—5
3646,198	12	0,00	3,40	$a\,^3F - x\,^3D^\circ$	2—2
3642,675	80	0,02	3,42	$a\,^3F - y\,^3G^\circ$	3—4
3637,966	10	0,00	3,41	$a\,^3F - x\,^3D^\circ$	2—3
3635,462	80	0,00	3,41	$a\,^3F - y\,^3G^\circ$	2—3
3635,202	8	0,05	3,46	$a\,^3F - z\,^5P^\circ$	4—3
3633,458	5	—	—	—	
3626,085	4	0,02	3,44	$a\,^3F - z\,^5P^\circ$	3—2
3610,154	12	0,90	4,33	$a\,^1D - y\,^1P^\circ$	2—1
3606,786	4	0,02	3,46	$a\,^3F - z\,^5P^\circ$	3—3
3604,284	8	0,02	3,46	$a\,^3F - y\,^1D^\circ$	3—2
3598,714	15	0,90	4,34	$a\,^1D - x\,^1D^\circ$	2—2
3578,25	3	1,46	4,92	$b\,^3F - s\,^3D^\circ$	4—3
3574,245	8	2,27	5,73	$b\,^1G - u\,^1G^\circ$	4—4
3558,51	6	—	—	—	—
3547,029	15	1,50	5,00	$a\,^1G - w\,^1F^\circ$	4—3
3542,51	3	2,24	5,73	$a\,^3H - u\,^1G^\circ$	4—4
3525,161	3	1,87	5,39	$a\,^3G - r\,^3F^\circ$	3—2
3516,838	3	1,88	5,40	$a\,^3G - r\,^3F^\circ$	4—3
3511,626	3	0,02	3,55	$a\,^3F - y\,^5F^\circ$	3—2
3506,643	8	0,05	3,58	$a\,^3F - y\,^5F^\circ$	4—5
3499,099	8	1,07	4,61	$a\,^3P - w\,^3P^\circ$	2—1
3495,754	6	1,05	4,60	$a\,^3P - w\,^3P^\circ$	1—0
3493,280	4	0,02	3,57	$a\,^3F - y\,^5F^\circ$	3—4
3485,689	6	1,05	4,61	$a\,^3P - w\,^3P^\circ$	1—1
3481,675	3	2,50	6,06	$z\,^3D^\circ - e\,^3D$	3—3
3481,136	3	2,48	6,04	$z\,^3D^\circ - e\,^3D$	2—2

λ, Å	I	E_H, eV	E_B, eV	Transition	J
3480,525	12	1,07	4,63	$a\,^{3}P - w\,^{3}P^{\circ}$	2—2
3478,918	6	1,05	4,61	$a\,^{3}P - w\,^{3}P^{\circ}$	0—1
3476,452	3	1,07	4,63	$a\,^{3}P - y\,^{5}S^{\circ}$	2—2
3467,260	6	1,05	4,63	$a\,^{3}P - w\,^{3}P^{\circ}$	1—2
3458,020	3	0,84	4,42	$a\,^{5}F - w\,^{5}D^{\circ}$	4—3
3457,494	4	0,85	4,43	$a\,^{5}F - w\,^{5}D^{\circ}$	5—4
3456,661	6	1,50	5,09	$a\,^{1}G - y\,^{1}H^{\circ}$	4—5
3444,403	3	1,43	5,03	$b\,^{3}F - r\,^{3}D^{\circ}$	2—1
3443,644	5	1,44	5,04	$b\,^{3}F - r\,^{3}D^{\circ}$	3—2
3439,305	8	1,46	5,06	$b\,^{3}F - r\,^{3}D^{\circ}$	4—3
3428,956	4	1,89	5,50	$a\,^{3}G - t\,^{3}G^{\circ}$	5—5
3415,993	5	2,43	6,06	$z\,^{3}F^{\circ} - e\,^{3}D$	4—3
3411,67	5	2,41	6,04	$z\,^{3}F^{\circ} - e\,^{3}D$	3—2
3405,094	5	1,05	4,69	$a\,^{3}P - u\,^{3}D^{\circ}$	1—1
3403,369	4	1,07	4,71	$a\,^{3}P - u\,^{3}D^{\circ}$	2—2
3400,162	3	2,40	6,04	$z\,^{3}F^{\circ} - e\,^{3}D$	2—1
3398,634	8	1,05	4,69	$a\,^{3}P - u\,^{3}D^{\circ}$	0—1
3392,713	10	1,50	5,16	$a\,^{1}G - v\,^{1}F^{\circ}$	4—3
3390,682	10	1,05	4,71	$a\,^{3}P - u\,^{3}D^{\circ}$	1—2
3385,944	40	0,05	3,71	$a\,^{3}F - w\,^{3}D^{\circ}$	4—3
3385,664	12	0,05	3,71	$a\,^{3}F - x\,^{3}G^{\circ}$	4—3
3382,312	15	1,07	4,73	$a\,^{3}P - u\,^{3}D^{\circ}$	2—3
3379,216	15	0,05	3,72	$a\,^{3}F - x\,^{3}G^{\circ}$	4—4
3377,577	30	0,02	3,69	$a\,^{3}F - w\,^{3}D^{\circ}$	3—2
3377,485	20	0,05	3,72	$a\,^{3}F - x\,^{5}D^{\circ}$	4—3
3375,706	3	2,31	5,98	$z\,^{5}D^{\circ} - j\,^{5}F$	3—3
3371,447	80	0,05	3,72	$a\,^{3}F - x\,^{3}G^{\circ}$	4—5
3370,436	40	0,00	3,68	$a\,^{3}F - w\,^{3}D^{\circ}$	2—1
3367,881	3	2,30	5,98	$z\,^{5}D^{\circ} - j\,^{5}F$	2—3
3366,176	5	2,04	5,72	$z\,^{5}G^{\circ} - h\,^{5}F$	6—5
3362,100	3	2,31	5,99	$z\,^{5}D^{\circ} - j\,^{5}F$	3—4
3361,835	10	0,02	3,71	$a\,^{3}F - x\,^{5}D^{\circ}$	3—2
3361,263	40	0,02	3,71	$a\,^{3}F - w\,^{3}D^{\circ}$	3—3
3360,990	10	0,02	3,71	$a\,^{3}F - x\,^{3}G^{\circ}$	3—3
3358,479	8	2,32	6,01	$z\,^{5}D^{\circ} - j\,^{5}F$	4—5
3358,271	10	0,00	3,69	$a\,^{3}F - w\,^{3}D^{\circ}$	2—2
3354,634	60	0,02	3,72	$a\,^{3}F - x\,^{3}G^{\circ}$	3—4
3352,937	6	0,02	3,72	$a\,^{3}F - x\,^{5}D^{\circ}$	3—3
3348,535	5	0,00	3,70	$a\,^{3}F - x\,^{5}D^{\circ}$	2—1
3342,151	6	0,00	3,71	$a\,^{3}F - w\,^{3}D^{\circ}$	2—3
3341,875	50	0,00	3,71	$a\,^{3}F - x\,^{3}G^{\circ}$	2—3
3325,229	3	2,12	5,84	$z\,^{5}F^{\circ} - g\,^{5}G$	4—4
3325,155	3	2,13	5,86	$z\,^{5}F^{\circ} - g\,^{5}G$	5—5
3324,754	4	{ 2,09	5,82	$z\,^{5}F^{\circ} - g\,^{5}G$	2—2
		2,10	5,83	$z\,^{5}F^{\circ} - g\,^{5}G$	3—3
3323,803	4	2,27	6,00	$b\,^{1}G - u\,^{1}F^{\circ}$	4—3
3321,588	8	1,07	4,80	$a\,^{3}P - t\,^{3}D^{\circ}$	2—2
3318,362	4	2,08	5,82	$z\,^{5}F^{\circ} - g\,^{5}G$	1—2
3314,523	8	1,05	4,79	$a\,^{3}P - t\,^{3}D^{\circ}$	1—1
3314,422	10	1,07	4,81	$a\,^{3}P - t\,^{3}D^{\circ}$	2—3
3312,690	5	2,10	5,84	$z\,^{5}F^{\circ} - g\,^{5}G$	3—4
3309,730	6	2,12	5,86	$z\,^{5}F^{\circ} - g\,^{5}G$	4—5
3309,501	15	1,05	4,80	$a\,^{3}P - t\,^{3}D^{\circ}$	1—2
3308,391	10	1,05	4,79	$a\,^{3}P - t\,^{3}D^{\circ}$	0—1
3306,879	10	2,13	5,88	$z\,^{5}F^{\circ} - g\,^{5}G$	5—6
3299,413	10	0,90	4,66	$a\,^{1}D - v\,^{3}G^{\circ}$	2—3
3294,903	6	2,32	6,08	$z\,^{5}D^{\circ} - h\,^{5}D$	4—4
3292,078	20	0,90	4,66	$a\,^{1}D - x\,^{1}F^{\circ}$	2—3
3278,922	12	0,90	4,68	$a\,^{1}D - u\,^{3}F^{\circ}$	2—3

λ, Å	I	E_H, eV	E_B, eV	Transition	J
3274,047	5	1,46	5,25	$b\,^3F-q\,^3D^\circ$	4—3
3270,562	3	1,44	5,23	$b\,^3F-q\,^3D^\circ$	3—2
3260,259	3	1,07	4,87	$a\,^3P-w\,^1D^\circ$	2—2
3248,602	15	1,05	4,87	$a\,^3P-w\,^1D^\circ$	1—2
3243,803	4	0,05	3,87	$a\,^3F-v\,^3D^\circ$	4—3
3243,513	3	2,04	5,86	$z\,^5G^\circ-g\,^5G$	6—5
3238,224	4	2,02	5,84	$z\,^5G^\circ-g\,^5G$	5—4
3232,791	3	2,00	5,83	$z\,^5G^\circ-g\,^5G$	4—3
3226,128	12	2,04	5,88	$z\,^5G^\circ-g\,^5G$	6—6
3223,519	10	2,02	5,86	$z\,^5G^\circ-g\,^5G$	5—5
3222,741	3	0,02	3,87	$a\,^3F-v\,^3D^\circ$	3—2
3221,381	10	2,00	5,84	$z\,^5G^\circ-g\,^5G$	4—4
3219,212	8	1,98	5,83	$z\,^5G^\circ-g\,^5G$	3—3
3217,942	8	1,97	5,82	$z\,^5G^\circ-g\,^5G$	2—2
3216,203	3	1,07	4,92	$a\,^3P-s\,^3D^\circ$	2—2
3214,240	12	0,05	3,90	$a\,^3F-w\,^3G^\circ$	4—4
3213,145	8	{ 1,07 2,13	4,92 5,99	$a\,^3P-s\,^3D^\circ$ $z\,^5F^\circ-j\,^5F$	2—3 5—4
3209,030	4	1,97	5,83	$z\,^5G^\circ-g\,^5G$	2—3
3207,897	5	1,98	5,84	$z\,^5G^\circ-g\,^5G$	3—4
3207,337	5	1,05	4,92	$a\,^3P-s\,^3D^\circ$	1—1
3206,825	5	2,00	5,86	$z\,^5G^\circ-g\,^5G$	4—5
3206,344	5	2,02	5,88	$z\,^5G^\circ-g\,^5G$	5—6
3205,848	5	0,00	3,87	$a\,^3F-v\,^3D^\circ$	2—1
3204,870	6	1,05	4,92	$a\,^3P-g\,^3D^\circ$	1—2
3203,828	15	0,02	3,89	$a\,^3F-w\,^3G^\circ$	3—3
3201,594	5	1,05	4,92	$a\,^3P-s\,^3D^\circ$	0—1
3199,915	100	0,05	3,92	$a\,^3F-w\,^3G^\circ$	4—5
3191,994	80	0,02	3,90	$a\,^3F-w\,^3G^\circ$	3—4
3186,451	60	0,00	3,89	$a\,^3F-w\,^3G^\circ$	2—3
3179,291	3	0,90	4,80	$a\,^1D-t\,^3D^\circ$	2—2
3172,731	4	0,90	4,81	$a\,^1D-t\,^3D^\circ$	2—3
3170,925	3	0,05	3,96	$a\,^3F-z\,^3H^\circ$	4—5
3147,268	3	2,12	6,06	$z\,^5F^\circ-e\,^3D$	4—3
3146,260	3	2,04	5,98	$z\,^5G^\circ-h\,^5G$	6—6
3143,350	12	2,04	5,98	$z\,^5G^\circ-g\,^5H$	6—7
3141,670	10	2,13	6,08	$z\,^5F^\circ-h\,^5D$	5—4
3141,537	15	0,00	4,84	$a\,^1D-x\,^1P^\circ$	2—1
3139,87	10	2,02	5,96	$z\,^5G^\circ-g\,^5H$	5—6
3135,069	8	2,00	5,95	$z\,^5G^\circ-g\,^5H$	4—5
3132,707	6	2,00	5,95	$z\,^5G^\circ-h\,^5G$	4—4
3130,804	15	—	—	—	—
3130,175	8	1,98	5,94	$z\,^5G^\circ-g\,^5H$	3—4
3129,075	7	2,12	6,08	$z\,^5F^\circ-h\,^5D$	4—3
3128,640	8	{ 1,07 2,10	5,03 6,06	$a\,^3P-r\,^3D^\circ$ $z\,^5F^\circ-h\,^5D$	2—1 3—2
3127,90	5	{ 2,12 1,98	6,08 5,94	$z\,^5F^\circ-h\,^5D$ $z\,^5G^\circ-h\,^5G$	4—4 3—3
3127,684	8	1,97	5,93	$z\,^5G^\circ-g\,^5H$	2—3
3123,769	20	2,04	6,01	$z\,^5G^\circ-j\,^5F$	6—5
3123,074	15	0,90	4,87	$a\,^1D-w\,^1D^\circ$	2—2
3120,212	3	2,09	6,06	$z\,^5F^\circ-h\,^5D$	2—2
3119,725	15	1,50	5,48	$a\,^1G-x\,^1H^\circ$	4—5
3118,130	15	2,02	5,99	$z\,^5G^\circ-j\,^5F$	5—4
3117,899	5	1,05	5,03	$a\,^3P-r\,^3D^\circ$	1—1
3117,455	6	1,07	5,04	$a\,^3P-r\,^3D^\circ$	2—2
3114,092	20	2,00	5,98	$z\,^5G^\circ-j\,^5F$	4—3
3112,482	8	1,05	5,03	$a\,^3P-r\,^3D^\circ$	0—1

λ, Å	I	E_H, eV	E_B, eV	Transition	J
3111,283	10	1,98	5,96	$z\,^5G^\circ - j\,^5F$	3—2
3109,581	8	—	—	—	—
3107,468	12	1,97	5,96	$z\,^5G^\circ - j\,^5F$	2—1
3106,806	8	1,05	5,04	$a\,^3P - r\,^3D^\circ$	1—2
3102,517	3	2,00	5,99	$z\,^5G^\circ - j\,^5F$	4—4
3101,526	4	1,98	5,98	$z\,^5G^\circ - j\,^5F$	3—3
3100,666	12	{ 1,07	5,06	$a\,^3P - x\,^3S^\circ$	2—1
		1,07	5,06	$a\,^3P - r\,^3D^\circ$	2—3
3093,813	3	2,04	6,05	$z\,^5G^\circ - k\,^5F$	6—5
3090,137	8	1,05	5,06	$a\,^3P - x\,^3S^\circ$	1—1
3084,819	4	1,05	5,06	$a\,^3P - x\,^3S^\circ$	0—1
3042,535	3	0,00	4,07	$a\,^3F - y\,^1F^\circ$	2—3
3002,730	3	{ 0,05	4,18	$a\,^3F - w\,^3F^\circ$	4—3
		1,43	5,56	$b\,^3F - g\,^3F^\circ$	2—2
3000,892	20	0,05	4,18	$a\,^3F - w\,^3F^\circ$	4—4
2990,981	3	1,44	5,59	$b\,^3F - n\,^3D^\circ$	3—2
2990,488	3	1,46	5,60	$b\,^3F - n\,^3D^\circ$	4—3
2990,036	3	1,43	5,57	$b\,^3F - n\,^3D^\circ$	2—1
2985,464	3	0,02	4,17	$a\,^3F - w\,^3F^\circ$	3—2
2983,290	20	0,02	4,18	$a\,^3F - w\,^3F^\circ$	3—3
2974,926	4	1,07	5,23	$a\,^3P - q\,^3D^\circ$	2—2
2970,554	4	1,05	5,23	$a\,^3P - q\,^3D^\circ$	1—1
2970,372	10	0,00	4,17	$a\,^3F - w\,^3F^\circ$	2—2
2968,226	4	0,00	4,18	$a\,^3F - w\,^3F^\circ$	2—3
2967,218	25	0,05	4,22	$a\,^3F - v\,^3F^\circ$	4—3
2965,686	15	1,07	5,25	$a\,^3P - q\,^3D^\circ$	2—3
2965,681	8	1,05	5,23	$a\,^3P - q\,^3D^\circ$	0—1
2965,231	6	1,05	5,23	$a\,^3P - q\,^3D^\circ$	1—2
2959,98	5	1,07	5,25	$a\,^3P - p\,^3D^\circ$	2—3
2959,71	3	1,05	5,24	$a\,^3P - p\,^3D^\circ$	1—2
2956,795	25	0,02	4,21	$a\,^3F - v\,^3F^\circ$	3—2
2956,18	70	0,05	4,24	$a\,^3F - v\,^3F^\circ$	4—4
2948,38	60	0,02	4,22	$a\,^3F - v\,^3F^\circ$	3—3
2947,700	3	0,90	5,10	$a\,^1D - u\,^3G^\circ$	2—3
2941,963	60	0,00	4,21	$a\,^3F - v\,^3F^\circ$	2—2
2937,293	25	0,02	4,24	$a\,^3F - v\,^3F^\circ$	3—4
2933,526	25	0,00	4,22	$a\,^3F - v\,^3F^\circ$	2—3
2928,313	30	1,50	5,73	$a\,^1G - u\,^1G^\circ$	4—4
2912,082	40	0,90	5,16	$a\,^1D - v\,^1F^\circ$	2—3
2905,655	5	—	—	—	—
2809,154	5	1,05	5,46	$a\,^3P - o\,^3D^\circ$	1—2
2805,694	6	1,07	5,48	$a\,^3P - o\,^3D^\circ$	2—3
2802,498	15	0,90	5,32	$a\,^1D - w\,^1P^\circ$	2—1
2758,066	20	1,50	6,00	$a\,^1G - u\,^1F^\circ$	4—3
2757,374	6	1,07	5,56	$a\,^3P - w\,^3S^\circ$	2—1
2749,031	5	1,05	5,56	$a\,^3P - w\,^3S^\circ$	1—1
2744,838	5	1,05	5,56	$a\,^3P - w\,^3S^\circ$	0—1
2742,297	15	0,90	5,42	$a\,^1D - v\,^1D^\circ$	2—2
2739,808	15	1,07	5,59	$a\,^3P - t\,^3P^\circ$	2—1
2735,613	6	0,90	5,43	$a\,^1D - u\,^1D^\circ$	2—2
2735,283	10	1,05	5,58	$a\,^3P - t\,^3P^\circ$	1—0
2733,264	30	1,07	5,60	$a\,^3P - t\,^3P^\circ$	2—2
2731,583	7	1,05	5,59	$a\,^3P - t\,^3P^\circ$	1—1
2731,141	4	1,07	5,60	$a\,^3P - n\,^3D^\circ$	2—3
2727,420	8	1,05	5,59	$a\,^3P - t\,^3P^\circ$	0—1
2725,081	10	1,05	5,60	$a\,^3P - t\,^3P^\circ$	1—2
2688,826	10	—	—	—	—
2685,137	3	0,05	4,66	$a\,^3F - v\,^3G^\circ$	4—4
2684,795	5	—	—	—	—

λ, Å	I	E_H, eV	E_B, eV	Transition	J
2679,923	20	0,05	4,67	$a\ ^3F-v\ ^3G^\circ$	4—5
2669,592	15	0,02	4,66	$a\ ^3F-v\ ^3G^\circ$	3—4
2661,962	10	0,00	4,66	$a\ ^3F-v\ ^3G^\circ$	2—3
2657,178	10	0,00	4,66	$a\ ^3F-x\ ^1F^\circ$	2—3
2654,921	5	0,00	4,67	$a\ ^3F-u\ ^3F^\circ$	2—2
2646,631	40	0,05	4,73	$a\ ^3F-u\ ^3D^\circ$	4—3
2644,253	40	0,02	4,71	$a\ ^3F-u\ ^3D^\circ$	3—2
2641,089	40	0,00	4,69	$a\ ^3F-u\ ^3D^\circ$	2—1
2632,414	15	0,00	4,71	$a\ ^3F-u\ ^3D^\circ$	2—2
2619,940	10	0,05	4,78	$a\ ^3F-t\ ^3F^\circ$	4—3
2611,476	8	0,02	4,77	$a\ ^3F-t\ ^3F^\circ$	3—2
2611,288	25	0,05	4,79	$a\ ^3F-t\ ^3F^\circ$	4—4
2605,121	25	0,02	4,78	$a\ ^3F-t\ ^3F^\circ$	3—3
2604,883	3	0,05	4,81	$a\ ^3F-t\ ^3D^\circ$	4—3
2599,885	25	0,00	4,77	$a\ ^3F-t\ ^3F^\circ$	2—2
2596,564	10	0,02	4,79	$a\ ^3F-t\ ^3F^\circ$	3—4
2590,247	5	0,02	4,81	$a\ ^3F-t\ ^3D^\circ$	3—3
2586,274	3	0,00	4,79	$a\ ^3F-t\ ^3D^\circ$	2—1
2580,803	5	—	—	—	—
2541,910	20	0,05	4,92	$a\ ^3F-s\ ^3D^\circ$	4—3
2527,980	5	0,02	4,92	$a\ ^3F-s\ ^3D^\circ$	3—3
2520,534	10	0,00	4,92	$a\ ^3F-s\ ^3D^\circ$	2—1
2519,017	8	0,00	4,92	$a\ ^3F-s\ ^3D^\circ$	2—2
2470,987	3	0,05	5,06	$a\ ^3F-r\ ^3D^\circ$	4—3
2440,976	10	0,05	5,13	$a\ ^3F-u\ ^3G^\circ$	4—5
2434,067	3	0,05	5,14	$a\ ^3F-s\ ^3F^\circ$	4—3
2433,211	6	0,02	5,11	$a\ ^3F-u\ ^3G^\circ$	3—4
2424,247	10	0,05	5,16	$a\ ^3F-s\ ^3F^\circ$	4—4
2421,296	10	0,02	5,14	$a\ ^3F-s\ ^3F^\circ$	3—3
2418,362	10	0,00	5,12	$a\ ^3F-s\ ^3F^\circ$	2—2
2384,516	4	0,05	5,25	$a\ ^3F-q\ ^3D^\circ$	4—3
2378,145	3	0,02	5,23	$a\ ^3F-q\ ^3D^\circ$	3—2
2305,665	12	0,05	5,42	$a\ ^3F-r\ ^3F^\circ$	4—4
2302,730	10	0,02	5,40	$a\ ^3F-r\ ^3F^\circ$	3—3
2299,852	10	0,00	5,39	$a\ ^3F-r\ ^3F^\circ$	2—2
2294,200	3	0,02	5,42	$a\ ^3F-r\ ^3F^\circ$	3—4
2293,745	3	0,00	5,40	$a\ ^3F-r\ ^3F^\circ$	2—3
2279,964	12	0,05	5,48	$a\ ^3F-o\ ^3D^\circ$	4—3
2276,703	10	0,02	5,46	$a\ ^3F-o\ ^3D^\circ$	3—2
2273,280	8	0,00	5,45	$a\ ^3F-o\ ^3D^\circ$	2—1
2272,613	8	0,05	5,50	$a\ ^3F-t\ ^3G^\circ$	4—5
2264,020	5	—	—	—	—
2244,690	7	0,05	5,57	$a\ ^3F-q\ ^3F^\circ$	4—3
2238,750	8	0,05 0,02	5,58 5,56	$a\ ^3F-q\ ^3F^\circ$ $a\ ^3F-q\ ^3F^\circ$	4—4 3—2
2238,20	6	—	—	—	—
2233,809	8	0,02	5,57	$a\ ^3F-q\ ^3F^\circ$	3—3
2230,492	7	0,05	5,60	$a\ ^3F-n\ ^3D^\circ$	4—3
2230,244	4	0,00	5,56	$a\ ^3F-q\ ^3F^\circ$	2—2
2229,67	7	—	—	—	—
2226,798	6	0,02	5,59	$a\ ^3F-n\ ^3D^\circ$	3—2
2225,11	8	—	—	—	—
2223,199	7	0,00	5,57	$a\ ^3F-n\ ^3D^\circ$	2—1
2219,75	5	0,02	5,60	$a\ ^3F-n\ ^3D^\circ$	3—3
2218,38	5	0,00	5,59	$a\ ^3F-n\ ^3D^\circ$	2—2
2143,52	6	—	—	—	—
2142,05	5	—	—	—	—
2139,41	5	—	—	—	—

λ, Å	I	E_H, eV	E_B, eV	Transition	J
2126,89	5	—	—	—	—
2123,50	7	—	—	—	—
2121,90	6	—	—	—	—
2117,01	6	—	—	—	—

Ti II, ground state $1s^2\,2s^2\,2p^6\,3s^2\,3p^6\,3d^2\,4s\ ^4F_{3/2}$
Ionization potential 110 000 cm^{-1}; 13,637 eV

λ, Å	I	E_H, eV	E_B, eV	Transition	J
6717,911	1	3,12	4,97	$c\ ^2D—b\ ^2F^\circ$	$^7/_2—^9/_2$
6680,26	1	3,09	4,95	$c\ ^2D—b\ ^2F^\circ$	$^3/_2—^5/_2$
6559,580	1	2,05	3,94	$b\ ^2P—z\ ^2D^\circ$	$^1/_2—^3/_2$
6491,61	2	2,06	3,97	$b\ ^2P—z\ ^2D^\circ$	$^3/_2—^5/_2$
6212,30	1	2,65	4,64	$a\ ^2S—z\ ^2S^\circ$	$^1/_2—^1/_2$
5473,517	1	2,65	4,91	$a\ ^2S—z\ ^2P^\circ$	$^1/_2—^3/_2$
5452,03	1	2,65	4,92	$a\ ^2S—z\ ^2P^\circ$	$^1/_2—^1/_2$
5422,47	1	1,57	3,86	$a\ ^2H—z\ ^4F^\circ$	$^9/_2—^7/_2$
5396,30	1	1,58	3,88	$a\ ^2H—z\ ^4F^\circ$	$^{11}/_2—^9/_2$
5381,020	1	1,57	3,87	$b\ ^2D—z\ ^2F^\circ$	$^3/_2—^5/_2$
5336,809	4	1,58	3,90	$b\ ^2D—z\ ^2F^\circ$	$^5/_2—^7/_2$
5268,62	1	2,60	4,95	$b\ ^2F—y\ ^2F^\circ$	$^5/_2—^5/_2$
5226,534	5	1,57	3,94	$b\ ^2D—z\ ^2D^\circ$	$^3/_2—^3/_2$
5188,700	6	1,58	3,97	$b\ ^2D—z\ ^2D^\circ$	$^5/_2—^5/_2$
5185,90	2	1,89	4,28	$b\ ^2G—z\ ^2G^\circ$	$^7/_2—^7/_2$
5129,143	1	1,89	4,31	$b\ ^2G—z\ ^2G^\circ$	$^9/_2—^9/_2$
5072,30	2	3,12	5,57	$c\ ^2D—x\ ^2D^\circ$	$^5/_2—^5/_2$
4839,251	1	2,65	5,21	$a\ ^2S—z\ ^4P^\circ$	$^1/_2—^1/_2$
4805,105	2	2,06	4,64	$b\ ^2P—z\ ^2S^\circ$	$^3/_2—^1/_2$
4798,535	2	1,08	3,66	$a\ ^2D—z\ ^4G^\circ$	$^3/_2—^5/_2$
4779,986	1	2,05	4,64	$b\ ^2P—z\ ^2S^\circ$	$^1/_2—^1/_2$
4764,535	1	1,24	3,84	$a\ ^2P—z\ ^4F^\circ$	$^3/_2—^5/_2$
4762,77	1	1,08	3,69	$a\ ^2D—z\ ^4G^\circ$	$^5/_2—^7/_2$
4719,515	1	1,24	3,87	$b\ ^4P—z\ ^2F^\circ$	$^5/_2—^5/_2$
4655,75	1	1,16	3,82	$a\ ^4P—z\ ^4F^\circ$	$^1/_2—^3/_2$
4636,345	1	1,16	3,84	$a\ ^4P—z\ ^4F^\circ$	$^3/_2—^5/_2$
4629,29	1	1,18	3,86	$a\ ^4P—z\ ^4F^\circ$	$^5/_2—^7/_2$
4589,961	2	1,24	3,94	$a\ ^2P—z\ ^2D^\circ$	$^3/_2—^3/_2$
4583,443	1	1,16	3,87	$a\ ^4P—z\ ^2F^\circ$	$^3/_2—^5/_2$
4580,458	1	1,23	3,94	$b\ ^4P—z\ ^2D^\circ$	$^3/_2—^3/_2$
4571,971	50	1,57	4,28	$a\ ^2H—z\ ^2G^\circ$	$^9/_2—^7/_2$
4563,761	30	1,22	3,94	$a\ ^2P—z\ ^2D^\circ$	$^1/_2—^3/_2$
4549,622	60	1,58	4,31	$a\ ^2H—z\ ^2G^\circ$	$^{11}/_2—^9/_2$
4533,966	30	1,25	3,97	$a\ ^2P—z\ ^2D^\circ$	$^3/_2—^5/_2$
4501,270	40	1,12	3,87	$a\ ^2G—z\ ^2F^\circ$	$^7/_2—^5/_2$
4488,319	15	3,12	5,88	$c\ ^2D—x\ ^2F^\circ$	$^5/_2—^7/_2$
4468,493	50	1,13	3,90	$a\ ^2G—z\ ^2F^\circ$	$^9/_2—^7/_2$
4450,487	10	1,08	3,87	$a\ ^2D—z\ ^2F^\circ$	$^5/_2—^5/_2$
4443,802	50	1,08	3,87	$a\ ^2D—z\ ^2F^\circ$	$^3/_2—^5/_2$
4417,718	40	1,16	3,97	$a\ ^4P—z\ ^2D^\circ$	$^3/_2—^5/_2$
4411,080	15	3,09	5,90	$c\ ^2D—x\ ^2F^\circ$	$^3/_2—^5/_2$
4399,767	35	1,24	4,05	$a\ ^2P—z\ ^4D^\circ$	$^3/_2—^5/_2$
4395,848	2	1,24	4,06	$b\ ^4P—z\ ^4D^\circ$	$^5/_2—^7/_2$
4395,031	60	1,08	3,90	$a\ ^2D—z\ ^2F^\circ$	$^5/_2—^7/_2$
4394,057	2	1,22	4,04	$a\ ^2P—z\ ^4D^\circ$	$^1/_2—^3/_2$

λ, Å	I	E_H, eV	E_B, eV	Transition	J
4386,858	10	2,60	5,42	$b\,^2F-y\,^2G^\circ$	5/2—7/2
4367,657	15	2,59	5,43	$b\,^2F-y\,^2G^\circ$	7/2—9/2
4344,291	2	1,08	3,94	$a\,^2D-z\,^2D^\circ$	5/2—3/2
4337,916	50	1,08	3,94	$a\,^2D-z\,^2D^\circ$	3/2—3/2
4314,979	40	1,16	4,03	$a\,^4P-z\,^4D^\circ$	1/2—1/2
4312,861	35	1,18	4,05	$a\,^4P-z\,^4D^\circ$	5/2—5/2
4307,900	40	1,16	4,04	$a\,^4P-z\,^4D^\circ$	3/2—3/2
4301,928	15	1,16	4,04	$a\,^4P-z\,^4D^\circ$	1/2—3/2
4300,052	60	1,18	4,06	$a\,^4P-z\,^4D^\circ$	5/2—7/2
4294,101	10	1,08	3,97	$a\,^2D-z\,^2D^\circ$	5/2—5/2
4290,222	50	1,16	4,05	$a\,^4P-z\,^4D^\circ$	3/2—5/2
4287,893	2	1,08	3,97	$a\,^2D-z\,^2D^\circ$	3/2—5/2
4174,088	2	2,60	5,57	$b\,^2F-x\,^2D^\circ$	5/2—5/2
4171,897	30	2,60	5,57	$b\,^2F-x\,^2D^\circ$	5/2—3/2
4163,644	40	2,59	5,57	$b\,^2F-x\,^2D^\circ$	7/2—5/2
4053,814	3	1,89	4,95	$b\,^2G-y\,^2F^\circ$	7/2—5/2
4028,332	7	1,89	4,97	$b\,^2G-y\,^2F^\circ$	9/2—7/2
4025,136	2	0,57	3,69	$a\,^2F-z\,^4G^\circ$	7/2—7/2
4012,372	4	0,57	3,66	$a\,^2F-z\,^4G^\circ$	5/2—5/2
3932,007	2	1,13	4,28	$a\,^2G-z\,^2G^\circ$	9/2—7/2
3913,464	60	1,12	4,28	$a\,^2G-z\,^2G^\circ$	7/2—7/2
3900,546	70	1,13	4,31	$a\,^2G-z\,^2G^\circ$	9/2—9/2
3814,580	4	0,57	3,82	$a\,^2F-z\,^4F^\circ$	5/2—3/2
3813,390	2	0,57	3,86	$a\,^2F-z\,^4F^\circ$	7/2—7/2
3796,899	2	0,57	3,84	$a\,^2F-z\,^4F^\circ$	5/2—5/2
3776,062	6	1,58	4,86	$b\,^2D-y\,^2D^\circ$	5/2—3/2
3774,650	3	0,57	3,86	$a\,^2F-z\,^4F^\circ$	5/2—7/2
3761,866	15	2,59	5,88	$b\,^2F-x\,^2F^\circ$	7/2—7/2
3761,320	200	0,57	3,87	$a\,^2F-z\,^2F^\circ$	5/2—5/2
3759,291	200	0,57	3,90	$a\,^2F-z\,^2F^\circ$	7/2—7/2
3757,684	30	1,57	4,86	$b\,^2D-y\,^2D^\circ$	3/2—3/2
3748,010	10	2,60	5,90	$b\,^2F-x\,^2F^\circ$	5/2—5/2
3741,633	50	1,58	4,89	$b\,^2D-y\,^2D^\circ$	5/2—5/2
3721,632	15	0,57	3,90	$a\,^2F-z\,^2F^\circ$	5/2—7/2
3706,219	20	1,57	4,91	$b\,^2D-z\,^2P^\circ$	3/2—3/2
3685,192	250	0,57	3,97	$a\,^2F-z\,^2D^\circ$	7/2—5/2
		0,57	3,94	$a\,^2F-z\,^2D^\circ$	5/2—3/2
3679,673	3	1,58	4,95	$b\,^2D-y\,^2F^\circ$	5/2—5/2
3662,237	40	1,57	4,95	$b\,^2D-y\,^2F^\circ$	3/2—5/2
3659,765	60	1,58	4,97	$b\,^2D-y\,^2F^\circ$	5/2—7/2
3641,330	100	1,24	4,64	$a\,^2P-z\,^2S^\circ$	3/2—1/2
3624,826	70	1,22	4,64	$a\,^2P-z\,^2S^\circ$	1/2—1/2
3596,048	60	0,57	4,05	$a\,^2F-z\,^4D^\circ$	7/2—5/2
3593,093	2	1,58	5,03	$b\,^2D-y\,^4D^\circ$	5/2—5/2
3587,130	12	0,57	4,06	$a\,^2F-z\,^4D^\circ$	7/2—7/2
3573,737	20	0,57	4,04	$a\,^2F-z\,^4D^\circ$	5/2—3/2
3566,00	6	1,16	4,64	$a\,^4P-z\,^2S^\circ$	3/2—1/2
3565,326	3	1,58	5,06	$b\,^2D-y\,^4D^\circ$	5/2—7/2
3561,575	3	0,57	4,05	$a\,^2F-z\,^4D^\circ$	5/2—5/2
3535,408	40	2,06	5,57	$b\,^2P-x\,^2D^\circ$	3/2—5/2
3533,868	2	2,06	5,57	$b\,^2P-x\,^2D^\circ$	3/2—3/2
3520,253	20	2,05	5,57	$b\,^2P-x\,^2D^\circ$	1/2—3/2
3510,840	60	1,89	5,42	$b\,^2G-y\,^2G^\circ$	7/2—7/2
3509,844	3	1,89	5,42	$b\,^2G-y\,^2G^\circ$	9/2—7/2
3504,890	80	1,89	5,43	$b\,^2G-y\,^2G^\circ$	9/2—9/2
3500,340	2	0,12	3,66	$b\,^4F-z\,^4G^\circ$	5/2—5/2
3492,39	3	4,28	7,83	$z\,^2G^\circ-e\,^2F$	7/2—5/2
3491,053	10	0,11	3,66	$b\,^4F-z\,^4G^\circ$	3/2—5/2
3489,739	2	0,13	3,69	$b\,^4F-z\,^4G^\circ$	7/2—7/2

λ, Å	I	E_H, eV	E_B, eV	Transition	J
3483,80	4	4,31	7,87	$z\,^2G^\circ—e\,^2F$	9/2—7/2
3477,181	15	0,12	3,69	$b\,^4F—z\,^4G^\circ$	5/2—7/2
3465,562	3	2,06	5,64	$b\,^2P—y\,^2P^\circ$	3/2—1/2
3461,496	20	0,13	3,72	$b\,^4F—z\,^4G^\circ$	7/2—9/2
3456,390	20	2,06	5,65	$b\,^2P—y\,^2P^\circ$	3/2—3/2
3452,470	4	2,05	5,64	$b\,^2P—y\,^2P^\circ$	1/2—1/2
3444,306	30	0,15	3,75	$b\,^4F—z\,^4G^\circ$	9/2—11/2
3416,957	2	1,24	4,86	$a\,^2P—y\,^2D^\circ$	3/2—3/2
3409,809	4	0,03	3,66	$a\,^4F—z\,^4G^\circ$	7/2—5/2
3407,205	3	0,05	3,69	$a\,^4F—z\,^4G^\circ$	9/2—7/2
3402,422	8	1,22	4,86	$a\,^2P—y\,^2D^\circ$	1/2—3/2
3394,574	40	0,01	3,66	$a\,^4F—z\,^4G^\circ$	5/2—5/2
3388,755	8	1,24	4,89	$a\,^2P—y\,^2D^\circ$	3/2—5/2
3387,834	50	0,03	3,69	$a\,^4F—z\,^4G^\circ$	7/2—7/2
3383,761	125	0,00	3,66	$a\,^4F—z\,^4G^\circ$	3/2—5/2
3380,278	30	0,05	3,72	$a\,^4F—z\,^4G^\circ$	9/2—9/2
3374,352	8	1,24	4,91	$a\,^2P—z\,^2P^\circ$	3/2—3/2
3372,800	100	0,01	3,69	$a\,^4F—z\,^4G^\circ$	5/2—7/2
3372,208	10	0,57	4,28	$a\,^2F—z\,^2G^\circ$	7/2—7/2
3369,212	2	1,23	4,91	$b\,^4P—z\,^2P^\circ$	3/2—3/2
3366,176	8	1,24	4,92	$a\,^2P—z\,^2P^\circ$	3/2—1/2
3361,213	125	0,03	3,72	$a\,^4F—z\,^4G^\circ$	7/2—9/2
3352,071	5	1,22	4,92	$a\,^2P—z\,^2P^\circ$	1/2—1/2
3349,399	125	0,05	3,75	$a\,^4F—z\,^4G^\circ$	9/2—11/2
3349,035	75	0,57	4,31	$a\,^2F—z\,^2G^\circ$	7/2—9/2
3348,844	10	0,12	3,82	$b\,^4F—z\,^4F^\circ$	5/2—3/2
3346,724	15	0,13	3,84	$b\,^4F—z\,^4F^\circ$	7/2—5/2
3343,770	10	0,15	3,86	$b\,^4F—z\,^4F^\circ$	9/2—7/2
3341,875	100	0,57	4,28	$a\,^2F—z\,^2G^\circ$	5/2—7/2
3340,344	35	0,11	3,82	$b\,^4F—z\,^4F^\circ$	3/2—3/2
3337,85	2	1,24	4,95	$a\,^2P—y\,^2F^\circ$	3/2—5/2
3335,192	40	0,12	3,84	$b\,^4F—z\,^4F^\circ$	5/2—5/2
3332,111	30	1,24	4,96	$b\,^4P—z\,^4S^\circ$	5/2—3/2
3329,455	70	0,13	3,86	$b\,^4F—z\,^4F^\circ$	7/2—7/2
3326,762	20	0,12	3,84	$b\,^4F—z\,^4F^\circ$	3/2—5/2
3322,936	75	0,15	3,88	$b\,^4F—z\,^4F^\circ$	9/2—9/2
3321,700	25	1,23	4,96	$b\,^4P—z\,^4S^\circ$	3/2—3/2
3318,024	10	0,12	3,86	$b\,^4F—z\,^4F^\circ$	5/2—7/2
3315,324	10	1,22	4,96	$b\,^4P—z\,^4S^\circ$	1/2—3/2
3308,806	8	0,13	3,88	$b\,^4F—z\,^4F^\circ$	7/2—9/2
3301,71	2	1,16	4,92	$a\,^4P—z\,^2P^\circ$	3/2—1/2
3288,575	5	1,23	5,00	$b\,^4P—y\,^4D^\circ$	3/2—1/2
3288,428	5	1,24	5,01	$b\,^4P—y\,^4D^\circ$	5/2—3/2
3287,64	40	1,89	5,66	$b\,^2G—z\,^2H^\circ$	7/2—9/2
3282,329	20	1,23	5,01	$b\,^4P—y\,^4D^\circ$	1/2—1/2
3279,995	4	1,12	4,89	$a\,^2G—y\,^2D^\circ$	7/2—5/2
3278,922	35	1,08	4,86	$a\,^2D—y\,^2D^\circ$	5/2—3/2
3278,290	30	1,23	5,01	$b\,^4P—y\,^4D^\circ$	3/2—3/2
3276,774	5	1,18	4,96	$a\,^4P—z\,^4S^\circ$	5/2—3/2
3275,293	3	1,08	4,86	$a\,^2D—y\,^2D^\circ$	3/2—3/2
3272,080	25	1,22	5,01	$b\,^4P—y\,^4D^\circ$	1/2—3/2
3271,625	25	1,24	5,03	$b\,^4P—y\,^4D^\circ$	5/2—5/2
3263,686	4	1,16	4,96	$a\,^4P—z\,^4S^\circ$	3/2—3/2
3261,596	60	1,89	5,69	$b\,^2G—z\,^2H^\circ$	9/2—11/2
		1,23	5,03	$b\,^4P—y\,^4D^\circ$	3/2—5/2
3260,259	3	1,16	4,96	$a\,^4P—z\,^4S^\circ$	1/2—3/2
3254,250	30	0,05	3,86	$a\,^4F—z\,^4F^\circ$	9/2—7/2
3252,914	40	0,03	3,84	$a\,^3F—z\,^4F^\circ$	7/2—5/2

λ, Å	I	E_H, eV	E_B, eV	Transition	J
3251,911	30	0,01	3,82	$a\ ^4F-z\ ^4F^\circ$	$^5/_2-^3/_2$
3249,370	2	1,08	4,89	$a\ ^2D-y\ ^2D^\circ$	$^3/_2-^5/_2$
3248,602	50	1,24	5,06	$b\ ^4P-y\ ^4D^\circ$	$^5/_2-^7/_2$
3241,984	60	0,00	3,82	$a\ ^4F-z\ ^4F^\circ$	$^3/_2-^3/_2$
3239,664	30	1,08	4,91	$a\ ^2D-z\ ^2P^\circ$	$^5/_2-^3/_2$
3239,037	60	0,01	3,83	$a\ ^4F-z\ ^4F^\circ$	$^5/_2-^5/_2$
3236,573	70	0,03	3,86	$a\ ^4F-z\ ^4F^\circ$	$^7/_2-^7/_2$
3236,122	20	1,08	4,91	$a\ ^2D-z\ ^2P^\circ$	$^3/_2-^3/_2$
3234,517	75	0,05	3,88	$a\ ^4F-z\ ^4F^\circ$	$^9/_2-^9/_2$
3232,280	30	1,12	4,95	$a\ ^2G-y\ ^2F^\circ$	$^7/_2-^5/_2$
3231,315	4	0,13	3,97	$b\ ^4F-z\ ^2D^\circ$	$^7/_2-^5/_2$
3229,397	35	1,13	4,97	$a\ ^2G-y\ ^2F^\circ$	$^9/_2-^7/_2$
3229,193	40	0,00	3,84	$a\ ^4F-z\ ^4F^\circ$	$^3/_2-^5/_2$
3228,605	30	1,08	4,92	$a\ ^2D-z\ ^2P^\circ$	$^3/_2-^1/_2$
3226,771	2	0,03	3,87	$a\ ^4F-z\ ^2F^\circ$	$^7/_2-^5/_2$
3224,241	35	1,58	5,43	$a\ ^2H-y\ ^2G^\circ$	$^{11}/_2-^9/_2$
3222,843	35	0,01	3,86	$a\ ^4F-z\ ^4F^\circ$	$^5/_2-^7/_2$
3218,270	25	1,57	5,42	$a\ ^2H-y\ ^2G^\circ$	$^9/_2-^7/_2$
3217,056	30	0,03	3,88	$a\ ^4F-z\ ^4F^\circ$	$^7/_2-^9/_2$
3214,750	4	0,05	3,90	$a\ ^4F-z\ ^2F^\circ$	$^9/_2-^7/_2$
3203,435	3	0,00	3,87	$a\ ^4F-z\ ^2F^\circ$	$^3/_2-^5/_2$
3202,535	40	1,08	4,95	$a\ ^2D-y\ ^2F^\circ$	$^3/_2-^5/_2$
3197,518	2	0,03	3,90	$a\ ^4F-z\ ^2F^\circ$	$^7/_2-^7/_2$
3195,717	3	1,08	4,96	$a\ ^2D-z\ ^4S^\circ$	$^5/_2-^3/_2$
3194,76	6	—	—	—	—
3194,56	8	3,88	7,76	$z\ ^4F^\circ-e\ ^4F$	$^9/_2-^9/_2$
3194,26	5	3,86	7,74	$z\ ^4F^\circ-e\ ^4F$	$^7/_2-^7/_2$
3192,68	4	3,84	7,72	$z\ ^4F^\circ-e\ ^4F$	$^5/_2-^5/_2$
3192,26	2	1,08	4,96	$a\ ^2D-z\ ^4S^\circ$	$^3/_2-^3/_2$
3190,874	30	1,08	4,97	$a\ ^2D-y\ ^4F^\circ$	$^5/_2-^7/_2$
3189,52	5	3,82	7,71	$z\ ^4F^\circ-e\ ^4F$	$^3/_2-^3/_2$
3184,09	2	0,01	3,90	$a\ ^4F-z\ ^2F^\circ$	$^5/_2-^7/_2$
3182,57	6	3,94	7,83	$z\ ^2D^\circ-e\ ^2F$	$^3/_2-^5/_2$
3181,84	8	3,97	7,87	$z\ ^2D^\circ-e\ ^2F$	$^5/_2-^7/_2$
3180,225	2	3,82	7,72	$z\ ^4F^\circ-e\ ^4F$	$^3/_2-^5/_2$
3178,630	3	3,84	7,74	$z\ ^4F^\circ-e\ ^4F$	$^5/_2-^7/_2$
3175,66	2	3,86	7,76	$z\ ^4F^\circ-e\ ^4F$	$^7/_2-^9/_2$
3174,80	5	—	—	—	—
3168,519	40	0,15	4,06	$b\ ^4F-z\ ^4D^\circ$	$^9/_2-^7/_2$
3164,91	8		—	—	—
3162,570	35	0,13	4,05	$b\ ^4F-z\ ^4D^\circ$	$^7/_2-^5/_2$
3161,755	30	0,12	4,04	$b\ ^4F-z\ ^4D^\circ$	$^5/_2-^3/_2$
3161,205	25	0,11	4,03	$b\ ^4F-z\ ^4D^\circ$	$^3/_2-^1/_2$
3157,397	2	0,01	3,97	$a\ ^4F-z\ ^2D^\circ$	$^5/_2-^3/_2$
3155,670	12	0,13	4,06	$b\ ^4F-z\ ^4D^\circ$	$^7/_2-^7/_2$
3154,195	12	0,11	4,04	$b\ ^4F-z\ ^4D^\circ$	$^3/_2-^3/_2$
3152,251	15	0,12	4,05	$b\ ^4F-z\ ^4D^\circ$	$^5/_2-^5/_2$
3148,033	12	0,00	3,94	$a\ ^4F-z\ ^2D^\circ$	$^3/_2-^3/_2$
3143,756	10	0,03	3,97	$a\ ^4F-z\ ^2D^\circ$	$^7/_2-^5/_2$
3130,804	15	0,01	3,97	$a\ ^4F-z\ ^2D^\circ$	$^5/_2-^5/_2$
3128,640	10	3,90	7,87	$z\ ^2P^\circ-e\ ^2F$	$^7/_2-^7/_2$
3127,883	10	3,87	7,83	$z\ ^2F^\circ-e\ ^2F$	$^5/_2-^5/_2$
3122,065	2	1,24	5,21	$a\ ^2P-z\ ^4P^\circ$	$^3/_2-^1/_2$
3119,800	15	1,24	5,22	$b\ ^4P-z\ ^4P^\circ$	$^5/_2-^3/_2$
3118,85	2	1,08	5,06	$a\ ^2D-y\ ^4D^\circ$	$^5/_2-^7/_2$
3117,669	20	1,23	5,21	$b\ ^4P-z\ ^4P^\circ$	$^3/_2-^1/_2$
3112,050	10	1,22	5,21	$b\ ^4P-z\ ^4P^\circ$	$^1/_2-^1/_2$
3110,620	20	1,23	5,22	$b\ ^4P-z\ ^4P^\circ$	$^3/_2-^3/_2$

λ, Å	I	E_H, eV	E_B, eV	Transition	J
3110,095	8	1,58	5,57	$b\,^2D—x\,^2D^\circ$	$^5/_2—^5/_2$
3106,234	35	1,24	5,23	$b\,^4P—z\,^4P^\circ$	$^5/_2—^5/_2$
3105,084	20	1,22	5,22	$b\,^4P—z\,^4P^\circ$	$^1/_2—^3/_2$
3104,593	3	1,89	5,88	$b\,^2G—x\,^2F^\circ$	$^7/_2—^7/_2$
3103,804	50	1,89	5,88	$b\,^2G—x\,^2F^\circ$	$^9/_2—^7/_2$
3102,975	2	1,22	5,22	$a\,^2P—z\,^4P^\circ$	$^1/_2—^3/_2$
3102,975	2	1,22	5,22	$a\,^2P—z\,^4P^\circ$	$^1/_2—^3/_2$
3097,186	25	1,23	5,23	$b\,^4P—z\,^4P^\circ$	$^3/_2—^5/_2$
3096,424	2	1,57	5,57	$b\,^2D—x\,^2D^\circ$	$^3/_2—^3/_2$
3090,051	8	3,75	7,76	$z\,^4G^\circ—e\,^4F$	$^{11}/_2—^9/_2$
3089,401	15	1,89	5,90	$b\,^2G—x\,^2F^\circ$	$^7/_2—^5/_2$
3088,027	75	0,05	4,06	$a\,^4F—z\,^4D^\circ$	$^9/_2—^7/_2$
3081,575	5	3,72	7,76	$z\,^4G^\circ—e\,^4F$	$^9/_2—^7/_2$
3078,645	50	0,03	4,05	$a\,^4F—z\,^4D^\circ$	$^7/_2—^5/_2$
3075,225	40	0,01	4,04	$a\,^4F—z\,^4D^\circ$	$^5/_2—^3/_2$
3072,971	40	0,00	4,03	$a\,^4F—z\,^4D^\circ$	$^3/_2—^1/_2$
3072,107	30	0,03	4,06	$a\,^4F—z\,^4D^\circ$	$^7/_2—^7/_2$
3071,242	15	1,18	5,22	$a\,^4P—z\,^4P^\circ$	$^5/_2—^3/_2$
3066,514	3	1,16	5,21	$a\,^4P—z\,^4P^\circ$	$^3/_2—^1/_2$
3066,354	20	0,00	4,04	$a\,^4F—z\,^4D^\circ$	$^3/_2—^3/_2$
3066,220	30	0,01	4,05	$a\,^4F—z\,^4D^\circ$	$^5/_2—^5/_2$
3063,502	4	1,16	5,21	$a\,^4P—z\,^4P^\circ$	$^1/_2—^1/_2$
3063,280	2	3,66	7,71	$z\,^4G^\circ—e\,^4F$	$^5/_2—^3/_2$
3059,741	{ 6	1,16	5,22	$a\,^4P—z\,^4P^\circ$	$^3/_2—^3/_2$
	4	0,01	4,06	$a\,^4F—z\,^4D^\circ$	$^5/_2—^7/_2$
3058,090	50	1,18	5,23	$a\,^4P—z\,^4P^\circ$	$^5/_2—^5/_2$
3057,395	10	0,00	3,97	$a\,^4F—z\,^2D^\circ$	$^3/_2—^5/_2$
3056,740	15	1,16	5,22	$a\,^4P—z\,^4P^\circ$	$^1/_2—^3/_2$
3048,766	6	1,58	5,65	$b\,^2D—y\,^2P^\circ$	$^5/_2—^3/_2$
3046,685	30	1,16	5,23	$a\,^4P—z\,^4P^\circ$	$^3/_2—^5/_2$
3045,085	5	—	—	—	—
3043,851	5	1,57	5,64	$b\,^2D—y\,^2P^\circ$	$^3/_2—^1/_2$
3038,706	6	1,58	5,66	$a\,^2H—z\,^2H^\circ$	$^{11}/_2—^9/_2$
3029,730	35	1,57	5,66	$a\,^2H—z\,^2H^\circ$	$^9/_2—^9/_2$
3023,86	12	4,28	8,33	$z\,^2G^\circ—e\,^2G$	$^7/_2—^7/_2$
3022,820	15	4,31	8,41	$z\,^2G^\circ—e\,^2G$	$^9/_2—^9/_2$
3017,187	50	1,58	5,69	$a\,^2H—z\,^2H^\circ$	$^{11}/_2—^{11}/_2$
3008,322	2	1,57	5,69	$a\,^2H—z\,^2H^\circ$	$^9/_2—^{11}/_2$
2995,75	5	—	—	—	—
2990,17	10	3,97	8,12	$z\,^2D^\circ—f\,^2F$	$^5/_2—^7/_2$
2979,20	10	3,94	8,10	$z\,^2D^\circ—f\,^2F$	$^3/_2—^5/_2$
2977,80	7	—	—	—	—
2958,98	50	4,28	8,47	$z\,^2G^\circ—e\,^2H$	$^7/_2—^9/_2$
2958,30	2	3,88	8,07	$z\,^4F^\circ—e\,^4G$	$^9/_2—^9/_2$
2954,76	60	4,31	8,50	$z\,^2G^\circ—e\,^2H$	$^9/_2—^{11}/_2$
2952,10	4	3,86	8,06	$z\,^4F^\circ—e\,^4G$	$^7/_2—^7/_2$
2945,47	50	3,88	8,09	$z\,^4F^\circ—e\,^4G$	$^9/_2—^{11}/_2$
2943,12	12	3,90	8,12	$z\,^2F^\circ—f\,^2F$	$^7/_2—^7/_2$
2941,993	50	3,86	8,07	$z\,^4F^\circ—e\,^4G$	$^7/_2—^9/_2$
2941,39	8	—	—	—	—
2938,69	30	3,84	8,06	$z\,^4F^\circ—e\,^4G$	$^5/_2—^7/_2$
2936,17	30	3,82	9,04	$z\,^4F^\circ—e\,^4G$	$^3/_2—^5/_2$
2931,27	40	3,87	8,10	$z\,^2F^\circ—f\,^2F$	$^5/_2—^5/_2$
2928,69	15	—	—	—	—
2927,87	2	3,88	8,11	$z\,^4F^\circ—e\,^4H$	$^9/_2—^{11}/_2$
2926,75	10	3,88	8,12	$z\,^4F^\circ—f\,^2F$	$^9/_2—^7/_2$
2918,77	2	3,87	8,12	$z\,^2F^\circ—f\,^2F$	$^5/_2—^7/_2$
2916,09	10	—	—	—	—

λ, Å	I	E_H, eV	E_B, eV	Transition	J
2914,89	10	—	—	—	—
2913,34	10	—	—	—	—
2913,08	1	0,03	4,28	$a\ ^4F—z\ ^2G^\circ$	$^7/_2—^7/_2$
2909,912	7	0,05	4,31	$a\ ^4F—z\ ^2G^\circ$	$^9/_2—^9/_2$
2908,14	4	—	—	—	—
2906,69	20	—	—	—	—
2891,050	15	0,57	4,89	$a\ ^2F—y\ ^2D^\circ$	$^7/_2—^5/_2$
2890,59	8	—	—	—	—
2888,923	15	0,57	4,86	$a\ ^2F—y\ ^2D^\circ$	$^5/_2—^3/_2$
2888,62	10	—	—	—	—
2887,456	2	1,13	5,42	$a\ ^2G—y\ ^2G^\circ$	$^9/_2—^7/_2$
2884,099	70	1,13	5,43	$a\ ^2G—y\ ^2G^\circ$	$^9/_2—^9/_2$
2880,28	3	1,58	5,88	$b\ ^2D—x\ ^2F^\circ$	$^5/_2—^7/_2$
2877,418	60	1,12	5,42	$a\ ^2G—y\ ^2G^\circ$	$^7/_2—^7/_2$
2875,79	10	—	—	—	—
2875,39	15	—	—	—	—
2874,08	2	1,12	5,43	$a^2G—y\ ^2G^\circ$	$^7/_2—^9/_2$
2870,04	25	—	—	—	—
2868,732	15	0,57	4,89	$a\ ^2F—y\ ^2D^\circ$	$^5/_2—^5/_2$
2862,34	30	1,24	5,57	$a\ ^2P—x\ ^2D^\circ$	$^3/_2—^5/_2$
2861,99	20	—	—	—	—
2861,291	3	1,24	5,57	$a\ ^2P—x\ ^2D^\circ$	$^3/_2—^3/_2$
2860,79	4	—	—	—	—
2858,399	8	0,57	4,91	$a\ ^2F—z\ ^2P^\circ$	$^5/_2—^3/_2$
2857,79	15	—	—	—	—
2856,616	2	1,57	5,90	$b\ ^2D—x\ ^2F^\circ$	$^3/_2—^5/_2$
2856,24	25	3,75	8,09	$z\ ^4G^\circ—e\ ^4G$	$^{11}/_2—^{11}/_2$
2853,922	10	0,57	4,95	$a\ ^2F—y\ ^2F^\circ$	$^7/_2—^5/_2$
2851,087	20	1,22	5,57	$a\ ^2P—x\ ^2D^\circ$	$^1/_2—^3/_2$
2846,09	15	3,72	8,07	$z\ ^4G^\circ—e\ ^4G$	$^9/_2—^9/_2$
2844,09	2	3,69	8,04	$z\ ^4G^\circ—e\ ^4G$	$^7/_2—^5/_2$
2841,914	30	0,57	4,97	$a\ ^2F—y\ ^2F^\circ$	$^7/_2—^7/_2$
2839,70	15	3,75	8,11	$z\ ^4G^\circ—e\ ^4H$	$^{11}/_2—^{11}/_2$
2836,60	15	3,69	8,06	$z\ ^4G^\circ—e\ ^4G$	$^7/_2—^7/_2$
2834,14	10	3,72	8,09	$z\ ^4G^\circ—e\ ^4G$	$^9/_2—^{11}/_2$
2832,158	20	0,57	4,95	$a\ ^2F—y\ ^2F^\circ$	$^5/_2—^5/_2$
2828,87	30	3,72	8,10	$z\ ^4G^\circ—e\ ^4H$	$^9/_2—^9/_2$
2828,80	30	3,66	8,04	$z\ ^4G^\circ—e\ ^4G$	$^5/_2—^5/_2$
2828,150	60	3,75	8,13	$z\ ^4G^\circ—e\ ^4H$	$^{11}/_2—^{13}/_2$
2827,22	10	3,69	8,07	$z\ ^4G^\circ—e\ ^4G$	$^7/_2—^9/_2$
2821,41	8	3,66	8,06	$z\ ^4G^\circ—e\ ^4G$	$^5/_2—^7/_2$
2820,36	4	0,57	4,96	$a\ ^2F—y\ ^2F^\circ$	$^5/_2—^7/_2$
2819,99	8	3,69	8,08	$z\ ^4G^\circ—e\ ^4H$	$^7/_2—^7/_2$
2817,838	60	3,72	8,11	$z\ ^4G^\circ—e\ ^4H$	$^9/_2—^{11}/_2$
2815,57	2	1,16	5,57	$a\ ^4P—x\ ^2D^\circ$	$^3/_2—^5/_2$
2810,276	50	3,69	8,10	$z\ ^4G^\circ—e\ ^4H$	$^7/_2—^9/_2$
2806,407	5	1,22	5,64	$a\ ^2P—y\ ^2P^\circ$	$^1/_2—^1/_2$
2805,00	40	3,66	8,08	$z\ ^4G^\circ—e\ ^4H$	$^5/_2—^7/_2$
2800,65	30	3,88	8,31	$z\ ^4F^\circ—e\ ^4D$	$^9/_2—^7/_2$
2790,62	3	3,86	8,30	$z\ ^4F^\circ—e\ ^4D$	$^7/_2—^5/_2$
2788,00	8	3,84	8,28	$z\ ^4F^\circ—e\ ^4D$	$^5/_2—^3/_2$
2785,99	6	3,86	8,31	$z\ ^4F^\circ—e\ ^4D$	$^7/_2—^7/_2$
2784,648	3	0,57	5,06	$a\ ^2F—y\ ^4D^\circ$	$^7/_2—^7/_2$
2782,30	2	3,82	8,28	$z\ ^4F^\circ—e\ ^4D$	$^3/_2—^1/_2$
2780,55	5	0,57	5,03	$a\ ^2F—y\ ^4D^\circ$	$^5/_2—^5/_2$
2778,48	2	{ 3,82	8,28	$z\ ^4F^\circ—e\ ^4D$	$^3/_2—^3/_2$
		3,84	8,30	$z\ ^4F^\circ—e\ ^4D$	$^5/_2—^5/_2$
2764,821	10	1,08	5,57	$a\ ^2D—x\ ^2D^\circ$	$^5/_2—^5/_2$

λ, Å	I	E_H, eV	E_B, eV	Transition	J
2762,22	2	1,08	5,57	$a\,^2D-x\,^2D^\circ$	$^3/_2-^5/_2$
2761,291	7	1,08	5,57	$a\,^2D-x\,^2D^\circ$	$^3/_2-^3/_2$
2758,35	2	4,04	8,54	$z\,^4D^\circ-f\,^4F$	$^3/_2-^5/_2$
2757,62	3	4,05	8,55	$z\,^4D^\circ-f\,^4F$	$^5/_2-^7/_2$
2752,85	4	4,06	8,56	$z\,^4D^\circ-f\,^4F$	$^7/_2-^9/_2$
2751,70	50	3,86	8,41	$z\,^4F^\circ-e\,^2G$	$^7/_2-^9/_2$
2746,70	30	3,87	8,38	$z\,^2F^\circ-e\,^2G$	$^5/_2-^7/_2$
2742,30	8	—	—	—	—
2738,70	3	2,06	6,59	$b\,^2P-x\,^2P^\circ$	$^3/_2-^3/_2$
2730,95	6	2,05	6,59	$b\,^2P-x\,^2P^\circ$	$^1/_2-^1/_2$
2725,79	3	1,12	5,66	$a\,^2G-z\,^2H^\circ$	$^7/_2-^9/_2$
2719,39	2	1,08	5,64	$a\,^2D-y\,^2P^\circ$	$^3/_2-^1/_2$
2717,304	3	1,13	5,69	$a\,^2G-z\,^2H^\circ$	$^9/_2-^{11}/_2$
2716,20	4	1,08	5,65	$a\,^2D-y\,^2P^\circ$	$^5/_2-^3/_2$
2698,52	30	—	—	—	—
2646,08	50	3,88	8,56	$z\,^4F^\circ-f\,^4F$	$^9/_2-^9/_2$
2642,15	20	3,86	8,55	$z\,^4F^\circ-f\,^4F$	$^7/_2-^7/_2$
2638,70	10	3,84	8,54	$z\,^4F^\circ-f\,^4F$	$^5/_2-^5/_2$
2635,60	5	3,82	8,53	$z\,^4F^\circ-f\,^4F$	$^3/_2-^3/_2$
2604,11	2	0,13	4,89	$b\,^4F-y\,^2D^\circ$	$^7/_2-^5/_2$
2572,648	5	—	—	—	—
2571,036	20	0,57	5,43	$a\,^2F-y\,^2G^\circ$	$^7/_2-^9/_2$
2555,988	10	0,57	5,42	$a\,^2F-y\,^2G^\circ$	$^5/_2-^7/_2$
2535,881	10	0,11	5,00	$b\,^4F-y\,^4D^\circ$	$^3/_2-^1/_2$
2534,640	20	0,12	5,01	$b\,^4F-y\,^4D^\circ$	$^5/_2-^3/_2$
2531,266	20	0,13	5,03	$b\,^4F-y\,^4D^\circ$	$^7/_2-^5/_2$
2525,619	30	0,15	5,06	$b\,^4F-y\,^4D^\circ$	$^9/_2-^7/_2$
2524,655	8	0,12	5,03	$b\,^4F-y\,^4D^\circ$	$^5/_2-^5/_2$
2517,448	2	0,13	5,06	$b\,^4F-y\,^4D^\circ$	$^7/_2-^7/_2$
2510,90	2	0,12	5,06	$b\,^4F-y\,^4D^\circ$	$^5/_2-^7/_2$
2498,94	2	0,57	5,57	$a\,^2F-x\,^2D^\circ$	$^7/_2-^5/_2$
2478,64	5	0,01	5,01	$a\,^4F-y\,^4D^\circ$	$^5/_2-^3/_2$
2477,21	2	0,03	5,03	$a\,^4F-y\,^4D^\circ$	$^7/_2-^5/_2$
2474,22	2	0,05	5,06	$a\,^4F-y\,^4D^\circ$	$^9/_2-^7/_2$
2450,44	6	1,58	6,64	$b\,^2D-w\,^2D^\circ$	$^5/_2-^5/_2$
2447,92	2	1,58	6,64	$b\,^2D-w\,^2D^\circ$	$^5/_2-^3/_2$
2442,67	2	1,57	6,64	$b\,^2D-w\,^2D^\circ$	$^3/_2-^5/_2$
2440,21	5	1,57	6,64	$b\,^2D-w\,^2D^\circ$	$^3/_2-^3/_2$
2357,82	2	1,23	6,49	$b\,^4P-x\,^4D^\circ$	$^3/_2-^1/_2$
2355,17	2	1,24	6,51	$b\,^4P-x\,^4D^\circ$	$^3/_2-^3/_2$
2354,12	3	—	—	—	—
2350,67	2	1,23	6,50	$b\,^4P-x\,^4D^\circ$	$^3/_2-^3/_2$
2349,97	3	1,23	6,51	$b\,^4P-x\,^4D^\circ$	$^3/_2-^5/_2$
2347,46	2	1,22	6,50	$b\,^4P-x\,^4D^\circ$	$^1/_2-^3/_2$
2342,31	3	—	—	—	—
2341,23	3	—	—	—	—
2334,54	3	—	—	—	—
2269,14	3	1,89	7,32	$b\,^2G-w\,^2F^\circ$	$^7/_2-^5/_2$
2261,23	3	1,89	7,37	$b\,^2G-w\,^2F^\circ$	$^9/_2-^7/_2$
2253,26	2	3,09	8,60	$c\,^2D-v\,^2D^\circ$	$^3/_2-^3/_2$
2250,09	2	3,12	8,63	$c\,^2D-v\,^2D^\circ$	$^5/_2-^5/_2$
2227,14	2	1,08	6,64	$a\,^2D-w\,^2D^\circ$	$^3/_2-^3/_2$
2162,68	4	1,24	6,97	$b\,^4P-y\,^4P^\circ$	$^5/_2-^3/_2$
2159,50	3	1,23	6,97	$b\,^4P-y\,^4P^\circ$	$^3/_2-^1/_2$
2159,09	5	1,24	6,98	$b\,^4P-y\,^4P^\circ$	$^5/_2-^5/_2$
2158,29	2	1,23	6,97	$b\,^4P-y\,^4P^\circ$	$^3/_2-^3/_2$
2155,58	4	1,22	6,97	$b\,^4P-y\,^4P^\circ$	$^1/_2-^3/_2$
2154,70	4	1,23	6,98	$b\,^4P-y\,^4P^\circ$	$^3/_2-^5/_2$

λ, Å	I	E_H, eV	E_B, eV	Transition	J
2054,54	3	0,57	6,64	$a\,^2F—w\,^2D°$	$^7/_2—^5/_2$
2041,49	3	0,57	6,64	$a\,^2F—w\,^2D°$	$^5/_2—^3/_2$
1909,74	2	0,01	6,50	$a\,^4F—x\,^4D°$	$^5/_2—^3/_2$
1909,33	2	0,01	6,51	$a\,^4F—x\,^4D°$	$^5/_2—^5/_2$
1908,29	3	0,03	6,53	$a\,^4F—x\,^4D°$	$^7/_2—^7/_2$
1906,30	3	0,00	6,50	$a\,^4F—x\,^4D°$	$^3/_2—^3/_2$

Ti III, ground state $1s^2\,2s^2\,2p^6\,3s^2\,3p^6\,3d^2\,^3F_2$
Ionization potential 227 000 cm^{-1}; 28,143 eV

λ, Å	I	E_H, eV	E_B, eV	Transition	J
4215,55	5	—	—	—	—
4207,54	3	—	—	—	—
4204,95	2	—	—	—	—
4200,11	2	—	—	—	—
2984,76	10	5,17	9,32	$b\,^1D—z\,^1D°$	2—2
2819,02	1	5,17	9,57	$b\,^1D—z\,^3D°$	2—2
2798,95	1	5,17	9,60	$b\,^1D—z\,^3F°$	2—2
2798,73	0	5,17	9,60	$b\,^1D—z\,^3D°$	2—3
2773,73	1	5,17	9,64	$b\,^1D—z\,^3F°$	2—3
2718,64	0	4,76	9,32	$a\,^3D—z\,^1D°$	3—2
2701,95	1	4,74	9,32	$a\,^3D—z\,^1D°$	2—2
2692,15	1	4,72	9,32	$a\,^3D—z\,^1D°$	1—2
2580,43	5	4,76	9,57	$a\,^3D—z\,^3D°$	3—2
2576,43	5	4,74	9,55	$a\,^3D—z\,^3D°$	2—1
2567,53	8	4,72	9,55	$a\,^3D—z\,^3D°$	1—1
2565,42	8	4,74	9,57	$a\,^3D—z\,^3D°$	2—2
2563,42	15	4,76	9,60	$a\,^3D—z\,^3D°$	3—3
2556,58	1	4,72	9,57	$a\,^3D—z\,^3D°$	1—2
2548,69	1	4,74	9,60	$a\,^3D—z\,^3F°$	2—2
2548,55	1	4,74	9,60	$a\,^3D—z\,^3D°$	2—3
2547,98	0	5,17	10,03	$b\,^1D—z\,^3P°$	2—1
2542,41	1	4,76	9,64	$a\,^3D—z\,^3F°$	3—3
2540,02	15	4,72	9,60	$a\,^3D—z\,^3F°$	1—2
2527,80	15	4,74	9,64	$a\,^3D—z\,^3F°$	2—3
2516,01	20	4,76	9,69	$a\,^3D—z\,^3F°$	3—4
2413,97	15	5,17	10,30	$b\,^1D—z\,^1F°$	2—3
2375,02	6	5,17	10,39	$b\,^1D—z\,^1P°$	2—1
2346,78	6	4,76	10,05	$a\,^3D—z\,^3P°$	3—2
2339,01	5	4,74	10,03	$a\,^3D—z\,^3P°$	2—1
2334,33	3	4,74	10,05	$a\,^3D—z\,^3P°$	2—2
2331,67	3	4,72	10,03	$a\,^3D—z\,^3P°$	1—1
2331,35	3	4,72	10,04	$a\,^3D—z\,^3P°$	1—0
2327,04	1	4,72	10,05	$a\,^3D—z\,^3P°$	1—2
2237,82	1		—	—	—
2199,30	1	—	—	—	—
1957,02	0	9,69	16,02	$z\,^3F°—e\,^3G$	4—4
1948,79	5	9,69	16,05	$z\,^3F°—e\,^3G$	4—5
1941,40	4	9,64	16,02	$z\,^3F°—e\,^3G$	3—4
1935,18	3	9,60	16,00	$z\,^3F°—e\,^3G$	2—3
1929,34	1	10,05	16,47	$z\,^3P°—e\,^3S$	2—1
1926,18	0	10,03	16,47	$z\,^3P°—e\,^3S$	1—1
1901,31	3	9,60	16,12	$z\,^3D°—e\,^3D$	3—3
1897,27	0	9,57	16,10	$z\,^3D°—e\,^3D$	2—2

λ, Å	I	E_H, eV	E_B, eV	Transition	J
1832,21	0	10,05	16,83	$z\,^3P^\circ - e\,^3P$	2—1
1831,31	0	10,03	16,80	$z\,^3P^\circ - e\,^3P$	1—0
1829,42	0	10,03	16,83	$z\,^3P^\circ - e\,^3P$	0, 1—1
1828,14	1	10,05	16,83	$z\,^3P^\circ - e\,^3P$	2—2
1825,30	0	10,03	16,83	$z\,^3P^\circ - e\,^3P$	1—2
1811,09	2	9,69	16,54	$z\,^3F^\circ - e\,^3F$	4—4
1797,69	1	9,64	16,54	$z\,^3F^\circ - e\,^3F$	3—4
1797,10	0	9,60	16,50	$z\,^3F^\circ - e\,^3F$	2—2
1792,56	2	9,60	16,51	$z\,^3D^\circ - e\,^3F$	3—3
1788,86	1	9,57	16,50	$z\,^3D^\circ - e\,^3F$	2—2
1787,32	2	9,60	16,54	$z\,^3D^\circ - e\,^3F$	3—4
1784,36	1	9,57	16,51	$z\,^3D^\circ - e\,^3F$	2—3
1783,58	1	9,55	16,50	$z\,^3D^\circ - e\,^3F$	1—2
1715,24	0	9,60	16,83	$z\,^3D^\circ - e\,^3P$	3—2
1506,07	10	1,31	9,55	$a\,^3P - z\,^3D^\circ$	1—1
1504,91	5	1,33	9,57	$a\,^3P - z\,^3D^\circ$	2—2
1504,59	10	1,31	9,55	$a\,^3P - z\,^3D^\circ$	0—1
1502,36	10	1,31	9,57	$a\,^3P - z\,^3D^\circ$	1—2
1499,17	20	1,33	9,60	$a\,^3P - z\,^3D^\circ$	2—3
1498,65	30	1,05	9,32	$a\,^1D - z\,^1D^\circ$	2—2
1496,59	1	1,31	9,60	$a\,^3P - z\,^3F^\circ$	1—2
1495,08	1	1,74	10,03	$a\,^1S - z\,^3P^\circ$	0—1
1491,98	5	1,33	9,64	$a\,^3P - z\,^3F^\circ$	2—3
1455,22	40	1,79	10,30	$a\,^1G - z\,^1F^\circ$	4—3
1450,29	2	1,05	9,60	$a\,^1D - z\,^3D^\circ$	2—3
1433,85	2	1,74	10,39	$a\,^1S - z\,^1P^\circ$	0—1
1424,14	20	1,33	10,03	$a\,^3P - z\,^3P^\circ$	2—1
1422,41	25	1,33	10,05	$a\,^3P - z\,^3P^\circ$	2—2
1421,69	20	1,31	10,03	$a\,^3P - z\,^3P^\circ$	1—0, 1
1420,42	15	1,31	10,03	$a\,^3P - z\,^3P^\circ$	0—1
1420,04	15	1,31	10,05	$a\,^3P - z\,^3P^\circ$	1—2
1339,72	2	1,05	10,30	$a\,^1D - z\,^1F^\circ$	2—3
1327,60	15	1,05	10,39	$a\,^1D - z\,^1P^\circ$	2—1
1298,95	40	0,02	9,57	$a\,^3F - z\,^3D^\circ$	3—2
1298,67	50	0,05	9,60	$a\,^3F - z\,^3D^\circ$	4—3
		0,00	9,55	$a\,^3F - z\,^3D^\circ$	2—1
1295,91	30	0,00	9,57	$a\,^3F - z\,^3D^\circ$	2—2
1294,67	50	0,02	9,60	$a\,^3F - z\,^3D^\circ$	3—2, 3
1293,26	30	0,05	9,64	$a\,^3F - z\,^3F^\circ$	4—3
1291,64	20	0,00	9,60	$a\,^3F - z\,^3F^\circ$	2—2
1289,32	30	0,02	9,64	$a\,^3F - z\,^3F^\circ$	3—3
1286,38	40	0,05	9,69	$a\,^3F - z\,^3F^\circ$	4—4
1282,49	3	0,02	9,69	$a\,^3F - z\,^3F^\circ$	3—4
1008,08	0	4,72	17,02	$a\,^3D - y\,^3P^\circ$	1—0
1007,15	1	4,74	17,05	$a\,^3D - y\,^3P^\circ$	2—1
1005,75	0	4,72	17,05	$a\,^3D - y\,^3P^\circ$	1—1
1004,68	2	4,76	17,11	$a\,^3D - y\,^3P^\circ$	3—2
1002,23	0	4,74	17,11	$a\,^3D - y\,^3P^\circ$	2—2

Ti IV, ground state $1s^2\,2s^2\,2p^6\,3s^2\,3p^6\,3d\,^2D_{3/2}$
Ionization potential 348 817,8 cm^{-1}; 43,245 eV

λ, Å	I	E_H, eV	E_B, eV	Transition	J
5492,43	6	26,33	28,59	$5s\,^2S - 5p\,^2P^\circ$	$^1/_2 - ^1/_2$
5398,82	8	26,33	28,63	$5s\,^2S - 5p\,^2P^\circ$	$^1/_2 - ^3/_2$

λ, Å	I	E_H, eV	E_B, eV	Transition	J
4647,40	3	{ 34,53	37,19	$5g\ ^2G—6h\ ^2H^\circ$	$^9/_2—^{11}/_2$
		{ 34,53	37,19	$5g\ ^2G—6h\ ^2H^\circ$	$^7/_2—^9/_2$
4403,54	2	29,27	32,09	$4f\ ^2F^\circ—5d\ ^2D$	$^5/_2—^3/_2$
4397,37	2	29,28	32,10	$4f\ ^2F^\circ—5d\ ^2D$	$^7/_2—^5/_2$
3576,44	4	28,63	32,10	$5p\ ^2P^\circ—5d\ ^2D$	$^3/_2—^5/_2$
3541,44	3	28,59	32,09	$5p\ ^2P^\circ—5d\ ^2D$	$^1/_2—^3/_2$
2957,50	4	24,40	28,59	$4d\ ^2D—5p\ ^2P^\circ$	$^3/_2—^1/_2$
2937,52	5	24,41	28,63	$4d\ ^2D—5p\ ^2P^\circ$	$^5/_2—^3/_2$
2930,14	1	24,40	28,63	$4d\ ^2D—5p\ ^2P^\circ$	$^3/_2—^3/_2$
2900,02	0	34,53	38,80	$5g\ ^2G—7h\ ^2H$	—
2862,67	1	28,63	32,96	$5p\ ^2P^\circ—6s\ ^2S$	$^3/_2—^1/_2$
2836,98	0	28,59	32,96	$5p\ ^2P^\circ—6s\ ^2S$	$^1/_2—^1/_2$
2547,30	3	24,41	29,27	$4d\ ^2D—4f\ ^2F^\circ$	$^5/_2—^5/_2$
2546,85	12	24,41	29,28	$4d\ ^2D—4f\ ^2F^\circ$	$^5/_2—^7/_2$
2541,75	8	24,40	29,27	$4d\ ^2D—4f\ ^2F^\circ$	$^3/_2—^5/_2$
2359,51	5	29,28	34,53	$4f\ ^2F^\circ—5g\ ^2G$	$^7/_2—^7/_2$, $^9/_2$
2359,11	5	29,27	34,53	$4f\ ^2F^\circ—5g\ ^2G$	$^5/_2—^7/_2$
2103,08	10	9,97	15,86	$4s\ ^2S—4p\ ^2P^\circ$	$^1/_2—^1/_2$
2067,50	15	9,97	15,96	$4s\ ^2S—4p\ ^2P^\circ$	$^1/_2—^3/_2$
1469,21	15	15,96	24,40	$4p\ ^2P^\circ—4d\ ^2D$	$^3/_2—^3/_2$
1467,25	30	15,96	24,41	$4p\ ^2P^\circ—4d\ ^2D$	$^3/_2—^5/_2$
1451,75	30	15,86	24,40	$4p\ ^2P^\circ—4d\ ^2D$	$^1/_2—^3/_2$
1195,25	5	15,96	26,33	$4p\ ^2P^\circ—5s\ ^2S$	$^3/_2—^1/_2$
1183,63	5	15,86	26,33	$4p\ ^2P^\circ—5s\ ^2S$	$^1/_2—^1/_2$
781,78	20	0,00	15,86	$3d\ ^2D—4p\ ^2P^\circ$	$^3/_2—^1/_2$
779,14	20	0,05	15,96	$3d\ ^2D—4p\ ^2P^\circ$	$^5/_2—^3/_2$
776,82	10	0,00	15,96	$3d\ ^2D—4p\ ^2P^\circ$	$^3/_2—^3/_2$
729,39	0	15,96	32,96	$4p\ ^2P^\circ—6s\ ^2S$	$^3/_2—^1/_2$
424,28	3	0,05	29,28	$3d\ ^2D—4f\ ^2F^\circ$	$^5/_2—^7/_2$
423,58	4	0,05	29,27	$3d\ ^2D—4f\ ^2F^\circ$	$^5/_2—^5/_2$

Ti V, ground state $1s^2\,2s^2\,2p^6\,3s^2\,3p^6\ ^1S_0$
Ionization potential 805 500 cm^{-1}; 99,864 eV

λ, Å	I	E_H, eV	E_B, eV	Transition	J
228,898	75	0,00	54,16	$3p^6\ ^1S—4s\ ^3P^\circ$	0—1
225,337	100	0,00	55,02	$3p^6\ ^1S—4s\ ^1P^\circ$	0—1
164,450	6	0,00	75,39	$3p^6\ ^1S—5s\ ^3P^\circ$	0—1
163,140	5	0,00	75,99	$3p^6\ ^1S—5s\ ^1P^\circ$	0—1

Ti VI, ground state $1s^2\,2s^2\,2p^6\,3s^2\,3p^5\ ^2P^0_{3/2}$
Ionization potential 966 000 cm^{-1}; 119,762 eV

λ, Å	I	E_H, eV	E_H, eV	Transition	J
524,11	10	0,72	24,38	$3p^5\ ^2P^\circ—3p^6\ ^2S$	$^1/_2—^1/_2$
508,58	12	0,00	24,38	$3p^5\ ^2P^\circ—3p^6\ ^2S$	$^3/_2—^1/_2$
201,862	5	0,00	61,42	$3p^5\ ^2P^\circ—4s\ ^4P$	$^3/_2—^3/_2$
201,313	5	0,72	62,31	$3p^5\ ^2P^\circ—4s\ ^2P$	$^1/_2—^3/_2$
199,759	6	0,72	62,79	$3p^5\ ^2P^\circ—4s\ ^2P$	$^1/_2—^1/_2$

λ, Å	I	E_H, eV	E_B, eV	Transition	J
198,974	8	0,00	62,31	$3p^5\ ^2P^\circ$—$4s\ ^2P$	$^3/_2$—$^3/_2$
197,455	5	0,00	62,79	$3p^5\ ^2P^\circ$—$4s\ ^2P$	$^3/_2$—$^1/_2$
194,900	7	0,72	64,32	$3p^5\ ^2P^\circ$—$4s'\ ^2D$	$^1/_2$—$^3/_2$
192,747	8	0,00	64,32	$3p^5\ ^2P^\circ$—$4s'\ ^2D$	$^3/_2$—$^5/_2$
192,705	1	0,00	64,34	$3p^5\ ^2P^\circ$—$4s'\ ^2D$	$^3/_2$—$^3/_2$
184,104	4	0,72	68,06	$3p^5\ ^2P^\circ$—$4s''\ ^2S$	$^1/_2$—$^1/_2$
182,148	5	0,00	68,06	$3p^5\ ^2P^\circ$—$4s''\ ^2S$	$^3/_2$—$^1/_2$

IRON, Z = 26

Fe I, ground state $1s^2 2s^2 2p^6 3s^2 3p^6 3d^6 4s^2\ {}^5D_4$

Ionization potential 63 700 cm⁻¹; 7,897 eV

λ, Å	I	E_H, eV	E_B, eV	Transition	J
11973,067	8	2,18	3,21	$a\ {}^5P—z\ {}^5D^\circ$	3—4
11882,861	7	2,20	3,24	$a\ {}^5P—z\ {}^5D^\circ$	2—3
11783,275	6	2,83	3,88	$b\ {}^3P—z\ {}^3D^\circ$	2—3
11689,988	8	2,22	3,28	$a\ {}^5P—z\ {}^5D^\circ$	1—1
11638,279	7	2,18	3,24	$a\ {}^5P—z\ {}^5D^\circ$	3—3
11593,600	5	2,22	3,29	$a\ {}^5P—z\ {}^5D^\circ$	1—0
11439,129	15	2,84	3,92	$b\ {}^3P—z\ {}^3D^\circ$	1—2
11422,335	6	2,20	3,28	$a\ {}^5P—z\ {}^5D^\circ$	2—1
11119,809	10	2,84	3,95	$b\ {}^3P—z\ {}^3D^\circ$	1—1
10863,60	5	4,73	5,87	$y\ {}^3D^\circ—e\ {}^5F$	3—4
10532,21	10	3,93	5,10	$z\ {}^3D^\circ—X$	2—2
10469,59	20	3,89	5,07	$z\ {}^3D^\circ—X$	3—3
10395,811	8	2,18	3,36	$a\ {}^5P—z\ {}^5F^\circ$	3—4
10216,351	100	4,73	5,94	$y\ {}^3D^\circ—e\ {}^3F$	3—4
10145,601	80	4,79	6,01	$y\ {}^3D^\circ—e\ {}^3F$	2—3
10065,080	60	4,83	6,06	$y\ {}^3D^\circ—e\ {}^3F$	1—2
9889,082	40	5,03	6,28	$x\ {}^5F^\circ—e\ {}^5G$	4—5
9861,793	30	5,06	6,31	$x\ {}^5F^\circ—e\ {}^5G$	3—4
9800,335	20	5,08	6,34	$x\ {}^5F^\circ—e\ {}^5G$	2—3
9763,913	15	5,03	6,30	$x\ {}^5F^\circ—e\ {}^7F$	4—5
9763,450	15	5,10	6,38	$x\ {}^5F^\circ—e\ {}^5G$	1—2
9738,624	200	4,98	6,26	$x\ {}^5F^\circ—e\ {}^5G$	5—6
9653,143	20	4,73	6,01	$y\ {}^3D^\circ—e\ {}^3F$	3—3
9626,562	30	5,03	6,31	$x\ {}^5F^\circ—e\ {}^5G$	4—4
9569,960	40	4,99	6,28	$x\ {}^5F^\circ—e\ {}^5G$	5—5
9414,14	20	5,06	6,37	$x\ {}^5F^\circ—f\ {}^5F$	3—4
9372,904	6	2,56	3,87	$b\ {}^3F—z\ {}^3F^\circ$	4—4
9350,44	10	4,56	5,87	$y\ {}^3F^\circ—e\ {}^5F$	4—4
9259,05	15	4,91	6,24	$x\ {}^5D^\circ—f\ {}^5D$	4—4
9258,31	20	4,56	5,94	$y\ {}^3F^\circ—e\ {}^3F$	3—4
9210,033	6	2,85	4,19	$b\ {}^3P—y\ {}^5D^\circ$	1—2
9118,892	25	2,84	4,19	$b\ {}^3P—y\ {}^5D^\circ$	2—2
9089,415	30	2,96	4,31	$b\ {}^3G—z\ {}^5G^\circ$	5—5
9088,324	50	2,85	4,21	$b\ {}^3P—z\ {}^3P^\circ$	1—2
9079,599	8	4,65	6,01	$y\ {}^3F^\circ—e\ {}^3F$	2—3
9024,47	15	4,91	6,28	$x\ {}^5D^\circ—e\ {}^5G$	4—5
9012,098	30	4,99	6,36	$x\ {}^5F^\circ—g\ {}^5D$	5—4
8999,564	200	2,83	4,21	$b\ {}^3P—z\ {}^3P^\circ$	2—2
8975,408	10	2,99	4,37	$b\ {}^3G—z\ {}^5G^\circ$	4—4
8945,204	20	5,03	6,41	$x\ {}^5F^\circ—g\ {}^5D$	4—3
8866,961	150	4,55	5,94	$y\ {}^3F^\circ—e\ {}^3F$	4—4
8838,433	30	2,87	4,26	$b\ {}^3P—z\ {}^3P^\circ$	0—1
8824,227	250	2,20	3,60	$a\ {}^5P—z\ {}^5P^\circ$	2—3
8804,624	10	2,29	3,69	$a\ {}^3P—z\ {}^5P^\circ$	2—1
8793,376	120	4,61	6,01	$y\ {}^3F^\circ—e\ {}^3F$	3—3
8764,00	100	4,65	6,06	$y\ {}^3F^\circ—e\ {}^3F$	2—2
8757,192	25	2,85	4,26	$b\ {}^3P—z\ {}^3P^\circ$	1—1
8710,29	20	4,91	6,33	$x\ {}^5D^\circ—f\ {}^5F$	4—5
8688,632	1500	2,18	3,60	$a\ {}^5P—z\ {}^5P^\circ$	3—3
8674,751	60	2,84	4,26	$b\ {}^3P—z\ {}^3P^\circ$	2—1
8661,907	600	2,22	3,65	$a\ {}^5P—z\ {}^5P^\circ$	1—2
8621,612	10	2,96	4,39	$b\ {}^3G—z\ {}^3G^\circ$	5—5
8611,807	40	2,85	4,29	$b\ {}^3P—z\ {}^3P^\circ$	1—0

λ, Å	I	E_H, eV	E_B, eV	Transition	J
8582,267	15	3,00	4,44	$b\,^3G$—$z\,^3G^\circ$	4—4
8515,08	20	3,02	4,47	$b\,^3G$—$z\,^3G^\circ$	3—3
8514,075	150	2,20	3,65	$a\,^5P$—$z\,^5P^\circ$	2—2
8468,413	300	2,23	3,68	$a\,^5P$—$z\,^5P^\circ$	1—1
8439,603	20	4,56	6,01	$y\,^3F^\circ$—$e\,^3F$	4—3
8387,780	1200	2,18	3,65	$a\,^5P$—$z\,^5P^\circ$	3—2
8365,642	25	3,26	4,73	$a\,^3D$—$y\,^3D^\circ$	3—3
8339,431	80	4,45	5,92	$z\,^3G^\circ$—$e\,^5F$	4—3
8331,941	200	4,40	5,87	$z\,^3G^\circ$—$e\,^5F$	5—4
8327,063	1200	2,20	3,68	$a\,^5P$—$z\,^5P^\circ$	2—1
8293,527	20	3,31	4,79	$a\,^3D$—$y\,^3D^\circ$	2—2
8248,151	30	4,38	5,88	$z\,^5G^\circ$—$e\,^5F$	4—4
8239,130	8	2,43	3,93	$a\,^3P$—$z\,^3D^\circ$	1—2
8232,347	50	4,42	5,92	$z\,^5G^\circ$—$e\,^5F$	3—3
8220,406	150	4,32	5,82	$z\,^5G^\circ$—$e\,^5F$	6—5
8207,667	40	4,45	5,97	$z\,^5G^\circ$—$e\,^5F$	2—2
8198,951	80	4,43	5,94	$z\,^3G^\circ$—$e\,^3F$	4—4
8085,20	200	4,45	5,97	$z\,^5G^\circ$—$e\,^5F$	2—1
8047,60	15	0,87	2,40	$a\,^5F$—$z\,^7D^\circ$	5—5
8046,073	600	4,42	5,95	$z\,^5G^\circ$—$e\,^5F$	3—2
8028,341	50	4,47	6,01	$z\,^3G^\circ$—$e\,^3F$	3—3
7998,972	700	4,37	5,92	$z\,^5G^\circ$—$e\,^5F$	4—3
7994,473	20	—	—	—	—
7945,878	600	4,39	5,94	$z\,^3G^\circ$—$e\,^3F$	5—4
7941,09	10	3,28	4,83	$a\,^3D$—$y\,^3D^\circ$	1—1
7937,166	700	4,31	5,87	$z\,^5G^\circ$—$e\,^5F$	5—4
7912,867	6	0,86	2,42	$a\,^5F$—$z\,^7D^\circ$	5—4
7832,224	400	4,44	6,01	$z\,^3G^\circ$—$e\,^3F$	4—3
7808,04	6	—	—	—	—
7780,586	300	4,48	6,06	$z\,^3G^\circ$—$e\,^3F$	3—2
7748,278	125	2,96	4,55	$b\,^3G$—$y\,^3F^\circ$	5—4
7710,390	25	4,22	5,82	$y\,^5F^\circ$—$e\,^5F$	4—5
7664,301	80	3,01	4,61	$b\,^3G$—$y\,^3F^\circ$	4—3
7661,223	30	4,26	5,87	$y\,^5F^\circ$—$e\,^5F$	3—4
7653,76	6	4,80	6,42	$y\,^3D^\circ$—$e\,^3D$	2—2
7620,538	25	4,73	6,36	$y\,^3D^\circ$—$e\,^3D$	3—3
7586,044	150	4,31	5,94	$z\,^5G^\circ$—$e\,^3F$	5—4
7583,797	50	3,02	4,65	$b\,^3G$—$y\,^3F^\circ$	3—2
7568,925	30	4,28	5,92	$y\,^5F^\circ$—$e\,^5F$	2—3
7531,171	60	4,37	6,01	$z\,^5G^\circ$—$e\,^3F$	4—3
7511,045	800	4,18	5,82	$y\,^5F^\circ$—$e\,^5F$	5—5
7507,28	8	4,42	6,06	$z\,^5G^\circ$—$e\,^3F$	3—2
7495,088	400	4,22	5,87	$y\,^5F^\circ$—$e\,^5F$	4—4
7491,678	12	4,30	5,95	$y\,^5F^\circ$—$e\,^5F$	1—2
7445,776	200	4,26	5,92	$y\,^5F^\circ$—$e\,^5F$	3—3
7411,178	100	4,28	5,95	$y\,^5F^\circ$—$e\,^5F$	2—2
7389,425	80	4,30	5,97	$y\,^5F^\circ$—$e\,^5F$	1—1
7386,402	8	4,91	6,59	$x\,^5D^\circ$—$f\,^5P$	4—3
7293,068	15	4,25	5,95	$y\,^5F^\circ$—$e\,^5F$	3—2
7288,760	10	4,22	5,92	$y\,^5F^\circ$—$e\,^5F$	4—3
7239,885	8	4,22	5,92	$z\,^3P^\circ$—$e\,^5F$	2—3
7223,668	12	3,02	4,73	$c\,^3P$—$y\,^3D^\circ$	2—3
7207,406	500	4,15	5,87	$y\,^5D^\circ$—$e\,^5F$	3—4
7187,341	800	4,10	5,82	$y\,^5D^\circ$—$e\,^5F$	4—5
7164,469	250	4,19	5,92	$y\,^5D^\circ$—$y\,^5F$	2—3
7130,942	150	4,22	5,95	$y\,^5D^\circ$—$e\,^5F$	1—2
7090,404	40	4,23	5,97	$y\,^5D^\circ$—$e\,^5F$	0—1

λ, Å	I	E_H, eV	E_B, eV	Transition	J
7068,413	40	4,08	5,82	$c\,^3F—w\,^3D^\circ$	4—3
7038,251	40	4,22	5,97	$y\,^5D^\circ—e\,^5F$	1—1
7024,649	10	4,57	6,33	$y\,^5P^\circ—f\,^7D$	3—2
7022,976	50	4,19	5,96	$y\,^5D^\circ—e\,^5F$	2—2
7016,436	60	4,15	5,92	$y\,^5D^\circ—e\,^5F$	3—3
7016,075	20	2,42	4,19	$a\,^3P—y\,^5D^\circ$	1—2
6999,902	30	4,10	5,87	$y\,^5D^\circ—e\,^5F$	4—4
6978,856	100	2,49	4,26	$a\,^3P—z\,^3P^\circ$	0—1
6951,261	25	4,56	6,34	$y\,^5P^\circ—e\,^7F$	3—3
6945,208	150	2,43	4,21	$a\,^3P—z\,^3P^\circ$	1—2
6916,702	60	4,16	5,94	$y\,^5D^\circ—e\,^3F$	3—4
6885,77	20	4,65	6,44	$y\,^3F^\circ—e\,^3D$	2—1
6858,164	40	4,62	6,42	$y\,^3F^\circ—e\,^3D$	3—2
6855,179	150	4,56	6,37	$y\,^5P^\circ—g\,^5D$	3—4
6843,671	60	4,55	6,36	$y\,^3F^\circ—e\,^3D$	4—3
6841,349	80	4,61	6,42	$y\,^5P^\circ—g\,^5D$	2—3
6828,610	50	4,64	6,46	$y\,^5P^\circ—g\,^5D$	1—2
6810,25	20	4,61	6,43	$y\,^5P^\circ—e\,^5P$	2—3
6806,851	10	2,73	4,55	$a\,^3G—y\,^3F^\circ$	4—4
6750,155	100	2,42	4,26	$a\,^3P—z\,^3P^\circ$	1—1
6726,478	20	4,62	6,45	$y\,^5P^\circ—e\,^5P$	2—1
6705,117	15	4,62	6,46	$y\,^5P^\circ—e\,^5P$	2—2
6703,574	10	2,77	4,61	$a\,^3G—y\,^3F^\circ$	3—3
6677,994	600	2,70	4,55	$a\,^3G—y\,^3F^\circ$	5—4
6663,444	80	2,43	4,29	$a\,^3P—z\,^3P^\circ$	1—0
6633,772	50	4,56	6,43	$y\,^5P^\circ—e\,^5P$	3—3
6609,117	30	2,57	4,44	$b\,^3F—z\,^3G^\circ$	4—4
6597,607	15	4,80	6,68	$y\,^3D^\circ—g\,^5F$	2—3
6593,875	60	2,43	4,31	$a\,^3H—z\,^5G^\circ$	5—5
6592,920	300	2,73	4,61	$a\,^3G—y\,^3F^\circ$	4—3
6575,024	30	2,59	4,47	$b\,^3F—z\,^3G^\circ$	3—3
6569,224	50	4,73	6,62	$y\,^3D^\circ—g\,^5F$	3—4
6546,245	200	2,76	4,65	$a\,^3G—y\,^3F^\circ$	3—2
6518,374	20	2,84	4,73	$b\,^3P—y\,^3D^\circ$	2—3
6498,950	5	0,96	2,86	$a\,^5F—z\,^7F^\circ$	3—3
6496,456	20	4,79	6,70	$y\,^3D^\circ—f\,^3D$	2—2
6494,985	1000	2,40	4,31	$a\,^3H—z\,^5G^\circ$	6—5
6481,877	20	2,28	4,19	$a\,^3P—y\,^5D^\circ$	2—2
6475,632	12	2,56	4,47	$b\,^3F—z\,^3G^\circ$	4—3
6469,214	15	4,84	6,75	$y\,^3D^\circ—f\,^3D$	1—1
6462,730	30	2,46	4,37	$a\,^3H—z\,^5G^\circ$	4—4
6430,852	300	2,19	4,11	$a\,^5P—y\,^5D^\circ$	3—4
6421,355	200	2,28	4,21	$a\,^3P—z\,^3P^\circ$	2—2
6419,977	30	4,74	6,67	$y\,^3D^\circ—f\,^3D$	3—3
6411,659	400	3,65	5,58	$z\,^5P^\circ—e\,^5D$	2—3
6408,028	60	3,68	5,62	$z\,^5P^\circ—e\,^5D$	1—2
6400,013	800	3,60	5,54	$z\,^5P^\circ—e\,^5D$	3—4
6393,605	400	2,43	4,37	$a\,^3H—z\,^5G^\circ$	5—4
6335,335	10	2,21	4,16	$a\,^5P—y\,^5D^\circ$	2—3
6318,022	10	2,46	4,42	$a\,^3H—z\,^5G^\circ$	4—3
6301,510	15	3,66	5,62	$z\,^5P^\circ—e\,^5D$	2—2
6254,263	6	2,29	4,26	$a\,^3P—z\,^3P^\circ$	2—1
6252,561	20	2,40	4,39	$a\,^3H—z\,^3G^\circ$	6—5
6246,329	15	3,61	5,58	$z\,^5P^\circ—e\,^5D$	3—3
6230,728	25	2,57	4,55	$b\,^3F—y\,^3F^\circ$	4—4
6191,561	20	2,44	4,44	$a\,^3H—z\,^3G^\circ$	5—4
6137,697	18	2,60	4,61	$b\,^3F—y\,^3F^\circ$	3—3
6136,621	20	2,45	4,47	$a\,^3H—z\,^3G^\circ$	4—3

λ, Å	I	E_H, eV	E_B, eV	Transition	J
6065,487	15	2,62	4,65	$b\,^3F—y\,^3F^\circ$	2—2
6024,063	15	4,56	6,61	$y\,^3F^\circ—f\,^5G$	4—5
6020,179	10	4,61	6,67	$y\,^3F^\circ—f\,^5G$	3—4
6016,655	5	3,54	5,60	$a\,^1D—x\,^3D^\circ$	2—3
6008,576	9	3,88	5,94	$z\,^3D^\circ—e\,^3F$	3—4
6003,034	8	3,88	5,94	$z\,^3F^\circ—e\,^3F$	4—4
5987,055	6	4,80	6,87	$y\,^3D^\circ—e\,^3P$	2—1
5984,804	8	4,73	6,80	$y\,^3D^\circ—e\,^3P$	3—2
5983,704	6	4,55	6,62	$y\,^3F^\circ—g\,^5F$	4—4
5930,186	8	4,66	6,74	$y\,^3F^\circ—e\,^3G$	2—3
5914,162	8	4,62	6,70	$y\,^3F^\circ—f\,^3D$	3—2
5862,363	8	4,55	6,65	$y\,^3F^\circ—e\,^3G$	4—5
5763,013	10	4,21	6,35	$z\,^3P^\circ—e\,^3D$	2—3
5753,136	5	4,26	6,42	$z\,^3P^\circ—e\,^3D$	1—2
5709,3864	10	3,37	5,54	$z\,^5F^\circ—e\,^5D$	4—4
5701,551	7	2,56	4,73	$b\,^3F—y\,^3D^\circ$	4—3
5662,525	6	4,19	6,37	$y\,^5F^\circ—g\,^5D$	5—4
5658,8247	10	3,40	5,58	$z\,^5F^\circ—e\,^5D$	3—3
5624,5501	10	3,42	5,62	$z\,^5F^\circ—e\,^5D$	2—2
5615,6521	50	3,33	5,54	$z\,^5F^\circ—e\,^5D$	5—4
5602,9529	10	3,43	5,64	$z\,^5F^\circ—e\,^5D$	1—1
5586,7634	40	3,37	5,58	$z\,^5F^\circ—e\,^5D$	4—3
5576,106	10	3,43	5,65	$z\,^5F^\circ—e\,^5D$	1—0
5572,8501	30	3,40	5,62	$z\,^5F^\circ—e\,^5D$	3—2
5569,6256	20	3,42	5,64	$z\,^5F^\circ—e\,^5D$	2—1
5563,604	5	4,19	6,42	$y\,^5D^\circ—g\,^5D$	2—3
5554,887	5	4,55	6,78	$y\,^3F^\circ—f\,^3F$	4—4
5506,7824	18	0,99	3,24	$a\,^5F—z\,^5D^\circ$	2—3
5501,4686	12	0,95	3,21	$a\,^5F—z\,^5D^\circ$	3—4
5497,5196	15	1,01	3,26	$a\,^5F—z\,^5D^\circ$	1—2
5487,138	8	4,14	6,40	$c\,^3F—t\,^5D^\circ$	3—2
5476,571	10	4,10	6,36	$y\,^5D^\circ—g\,^5D$	4—4
5473,920	5	4,16	6,42	$y\,^5D^\circ—g\,^5D$	3—3
5463,283	10	4,47	6,70	$z\,^3G^\circ—e\,^3G$	4—4
5455,6131	40	1,01	3,28	$a\,^5F—z\,^5D^\circ$	1—1
5446,920	40	0,99	3,26	$a\,^5F—z\,^5D^\circ$	2—2
5445,037	15	4,39	6,66	$z\,^3G^\circ—e\,^3G$	5—5
5434,5268	30	1,01	3,29	$a\,^5F—z\,^5D^\circ$	1—0
5429,6999	40	0,96	3,24	$a\,^5F—z\,^5D^\circ$	3—3
5424,076	45	4,32	6,61	$z\,^5G^\circ—e\,^5H$	6—7
5415,207	35	4,39	6,68	$z\,^3G^\circ—e\,^3H$	5—6
5410,909	15	4,47	6,76	$z\,^3G^\circ—e\,^3H$	3—4
5405,7781	40	0,99	3,28	$a\,^5F—z\,^5D^\circ$	2—1
5404,148	30	4,44	6,73	$z\,^3G^\circ—e\,^3H$	4—5
5400,503	5	4,37	6,67	$z\,^5G^\circ—f\,^5G$	4—4
5397,1311	40	0,91	3,21	$a\,^5F—z\,^5D^\circ$	4—4
5393,1752	10	3,24	5,54	$z\,^5D^\circ—e\,^5D$	3—4
5383,371	35	4,31	6,62	$z\,^5G^\circ—e\,^5H$	5—6
5371,4926	50	0,96	3,26	$a\,^5F—z\,^5D^\circ$	3—2
5369,957	25	4,37	6,68	$z\,^5G^\circ—e\,^5H$	4—5
5367,460	20	4,42	6,73	$z\,^5G^\circ—e\,^5H$	3—4
5364,883	15	4,45	6,76	$z\,^5G^\circ—e\,^5H$	2—3
5341,0255	20	1,61	3,92	$a\,^3F—z\,^3D^\circ$	2—2
5339,9371	12	3,26	5,58	$z\,^5D^\circ—e\,^5D$	2—3
5328,5336	15	1,56	3,88	$a\,^3F—z\,^3D^\circ$	3—3
5328,0418	50	0,91	3,24	$a\,^5F—z\,^5D^\circ$	4—3
5324,182	30	3,21	5,54	$z\,^5D^\circ—e\,^5D$	4—4

λ, Å	I	E_H, eV	E_B, eV	Transition	J
5307,3633	5	1,61	3,95	$a\,^3F—z\,^3F°$	2—3
5302,3073	10	3,28	5,62	$z\,^5D°—e\,^5D$	1—2
5283,6283	18	3,25	5,59	$z\,^5D°—e\,^5D$	3—3
5281,7970	10	3,04	5,38	$z\,^7P°—e\,^7D$	2—3
5270,3602	30	1,61	3,96	$a\,^3F—z\,^3D°$	2—1
5269,5402	60	0,86	3,21	$a\,^5F—z\,^5D°$	5—4
5266,564	30	3,00	5,35	$z\,^7P°—e\,^7D$	3—4
5263,3134	8	3,26	5,62	$z\,^5D°—e\,^5D$	2—2
5250,6490	6	2,20	4,56	$a\,^5P—y\,^5P°$	2—3
5242,4955	5	3,63	5,99	$a\,^1I—z\,^1H°$	6—5
5232,9474	40	2,94	5,31	$z\,^7P°—e\,^7D$	4—5
5227,1911	40	1,55	3,92	$a\,^3F—z\,^3D°$	3—2
5226,8686	15	3,04	5,41	$z\,^7P°—e\,^7D$	2—2
5217,3964	5	3,21	5,58	$z\,^5D°—e\,^5D$	4—3
5216,2770	10	1,61	3,99	$a\,^3F—z\,^3F°$	2—2
5215,1871	6	3,26	5,64	$z\,^5D°—e\,^5D$	2—1
5208,6007	7	3,24	5,62	$z\,^5D°—e\,^5D$	3—2
5204,5840	5	0,09	2,47	$a\,^5D—z\,^7D°$	2—2
5202,3395	8	2,18	4,56	$a\,^5P—y\,^5P°$	3—3
5195,478	8	4,22	6,61	$y\,^5D°—f\,^5G$	4—5
5194,9441	10	1,56	3,95	$a\,^3F—z\,^3F°$	3—3
5192,3509	30	3,00	5,38	$z\,^7P°—e\,^7D$	3—3
5191,4615	20	3,03	5,42	$z\,^7P°—e\,^7D$	2—1
5171,5987	20	1,48	3,88	$a\,^3F—z\,^3F°$	4—4
5167,4905	40	1,48	3,89	$a\,^3F—z\,^3D°$	4—3
5166,2841	4	0,00	2,40	$a\,^5D—z\,^7D°$	4—5
5162,288	10	4,18	6,58	$y\,^5F°—g\,^5F$	5—5
5150,8425	6	0,99	3,39	$a\,^5F—z\,^5F°$	2—3
5142,9320	6	0,95	3,36	$a\,^5F—z\,^5F°$	3—4
5139,4702	20	2,94	5,35	$z\,^7P°—e\,^7D$	4—4
5139,2578	10	3,00	5,41	$z\,^7P°—e\,^7D$	3—2
5137,388	6	4,17	6,59	$y\,^5F°—h\,^5D$	5—4
5133,680	20	4,17	6,59	$y\,^5F°—f\,^5G$	5—6
5127,3624	5	0,91	3,33	$a\,^5F—z\,^5F°$	4—5
5125,130	6	4,22	6,64	$y\,^5F°—h\,^5D$	4—3
5123,7231	6	1,01	3,43	$a\,^5F—z\,^5F°$	1—1
5110,4139	10	0,00	2,42	$a\,^5D—z\,^7D°$	4—4
5107,4505	6	0,99	3,41	$a\,^5F—z\,^5F°$	2—2
5098,7030	8	2,18	4,61	$a\,^5P—y\,^5P°$	3—2
5096,995	6	4,28	6,72	$y\,^5F°—f\,^5G$	2—3
5090,789	6	4,25	6,69	$y\,^5F°—h\,^5D$	3—2
5083,3413	7	0,95	3,39	$a\,^5F—z\,^5F°$	3—3
5074,760	10	4,22	6,66	$y\,^5F°—e\,^3G$	4—5
5068,7730	10	2,94	5,38	$z\,^7P°—e\,^7D$	4—3
5065,016	6	4,26	6,70	$y\,^5F°—e\,^3G$	3—4
5051,6379	10	0,91	3,36	$a\,^5F—z\,^5F°$	4—4
5049,8253	15	2,28	4,73	$a\,^3P—y\,^3D°$	2—3
5041,7585	10	1,48	3,94	$a\,^3F—z\,^3F°$	4—3
5041,0747	7	0,95	3,41	$a\,^5F—z\,^5F°$	3—2
5036,294	6	—	—	—	—
5031,901	8	4,37	6,83	$z\,^5G°—f\,^3F$	4—3
5028,131	4	3,57	6,03	$a\,^1H—y\,^1G°$	5—4
5022,250	6	3,98	6,44	$z\,^3F°—e\,^3D$	2—1
5014,959	10	3,94	6,41	$z\,^3F°—e\,^3D$	3—2
5012,0712	12	0,86	3,33	$a\,^5F—z\,^5F°$	5—5
5006,1254	20	2,83	5,30	$z\,^7F°—e\,^7D$	5—5
5005,725	10	3,88	6,35	$z\,^3D°—e\,^3D$	3—3
5002,7998	6	3,39	5,87	$z^5\,F°—e\,^5F$	3—4

λ, Å	I	E_H, eV	E_B, eV	Transition	J
5001,871	12	3,88	6,36	$z\,^3F^\circ - e\,^3D$	4—3
4994,1323	8	0,91	3,39	$a\,^5F - z\,^5F^\circ$	4—3
4988,963	6	4,16	6,64	$y\,^5D^\circ - h\,^5D$	3—3
4985,5539	7	2,86	5,35	$z\,^7F^\circ - e\,^7D$	3—4
4985,260	7	3,92	6,41	$z\,^3D^\circ - e\,^3D$	2—2
4983,855	6	4,16	6,59	$y\,^5D^\circ - h\,^5D$	4—4
4983,258	5	4,15	6,64	$y\,^5D^\circ - f\,^5P$	3—2
4982,507	8	4,10	6,59	$y\,^5D^\circ - f\,^5P$	4—3
4973,108	3	3,95	6,44	$z\,^3D^\circ - e\,^3D$	1—1
4966,0968	8	3,33	5,82	$z\,^5F^\circ - e\,^5F$	5—5
4957,6059	60	2,83	5,31	$z\,^7F^\circ - e\,^7D$	6—5
4957,3054	20	2,85	5,35	$z\,^7F^\circ - e\,^7D$	4—4
4939,6896	4	0,86	3,36	$a\,^5F - z\,^5F^\circ$	5—4
4938,8206	10	2,87	5,38	$z\,^7F^\circ - e\,^7D$	2—3
4924,7753	3	2,28	4,79	$a\,^3P - y\,^3D^\circ$	2—2
4920,5096	60	2,83	5,35	$z\,^7F^\circ - e\,^7D$	5—4
4919,0003	30	2,86	5,38	$z\,^7F^\circ - e\,^7D$	3—3
4910,025	2	3,39	5,92	$z\,^5F^\circ - e\,^5F$	3—3
4903,3169	12	2,88	5,41	$z\,^7F^\circ - e\,^7D$	1—2
4891,4989	50	2,84	5,37	$z\,^7F^\circ - e\,^7D$	4—3
4890,7616	25	2,87	5,41	$z\,^7F^\circ - e\,^7D$	2—2
4878,2182	12	2,87	5,42	$z\,^7F^\circ - e\,^7D$	0—1
4872,1444	20	2,88	5,42	$z\,^7F^\circ - e\,^7D$	1—1
4871,3244	25	2,86	5,41	$z\,^7F^\circ - e\,^7D$	3—2
4859,7480	15	2,87	5,42	$z\,^7F^\circ - e\,^7D$	2—1
4791,248	5	3,27	5,86	$a\,^3D - w\,^3D^\circ$	1—1
4789,6537	7	3,54	6,13	$a\,^1D - z\,^1D^\circ$	2—2
4786,8106	5	3,01	5,60	$c\,^3P - x\,^3D^\circ$	2—3
4736,7807	12	3,21	5,82	$z\,^5D^\circ - e\,^5F$	4—5
4707,2807	8	3,24	5,87	$z\,^5D^\circ - e\,^5F$	3—4
4691,4144	6	3,00	5,63	$b\,^3G - y\,^3G^\circ$	4—4
4678,852	7	3,60	6,24	$z\,^5P^\circ - f\,^5D$	3—4
4668,1422	6	3,26	5,92	$z\,^5D^\circ - e\,^5F$	2—3
4667,459	6	3,60	6,25	$z\,^5P^\circ - e\,^7P$	3—4
4647,4370	6	2,95	5,61	$b\,^3G - y\,^3G^\circ$	5—5
4637,518	3	3,28	5,95	$z\,^5D^\circ - e\,^5F$	1—2
4625,0527	3	3,24	5,92	$z\,^5D^\circ - e\,^5F$	3—3
4619,298	3	3,60	6,28	$z\,^5P^\circ - f\,^5D$	3—2
4611,289	5	3,65	6,33	$z\,^5P^\circ - e\,^5S$	2—2
4602,9446	9	1,48	4,18	$a\,^3F - y\,^5F^\circ$	4—5
4592,6547	5	1,56	4,26	$a\,^3F - y\,^5F^\circ$	3—3
4547,8505	4	3,54	6,26	$a\,^1D - z\,^1F^\circ$	2—3
4531,152	8	1,48	4,22	$a\,^3F - y\,^5F^\circ$	4—4
4528,6175	18	2,18	4,91	$a\,^5P - x\,^5D^\circ$	3—4
4525,146	5	3,60	6,34	$z\,^5P^\circ - e\,^5S$	3—2
4494,5669	12	2,20	4,95	$a\,^5P - x\,^5D^\circ$	2—3
4489,7416	3	0,12	2,89	$a\,^5D - z\,^7F^\circ$	0—1
4484,225	4	3,60	6,36	$z\,^5P^\circ - g\,^5D$	3—4
4482,2563	6	2,23	4,99	$a\,^5P - x\,^5D^\circ$	1—2
4482,1720	4	0,11	2,87	$a\,^5D - z\,^7F^\circ$	1—2
4476,0206	10	2,84	5,61	$b\,^3P - x\,^3D^\circ$	1—2
4469,380	5	3,65	6,43	$z\,^5P^\circ - e\,^5P$	2—3
4466,5542	12	2,83	5,60	$b\,^3P - x\,^3D^\circ$	2—3
4461,6544	8	0,09	2,86	$a\,^5D - z\,^7F^\circ$	2—3
4459,1213	10	2,18	4,95	$a\,^5P - x\,^5D^\circ$	3—3
4454,3835	5	2,83	5,61	$b\,^3P - x\,^3D^\circ$	2—2
4447,7212	9	2,23	5,01	$a\,^5P - x\,^5D^\circ$	1—1

λ, Å	I	E_H, eV	E_B, eV	Transition	J
4443,1963	7	2,86	5,64	$b\,^3P-x\,^3D^\circ$	0—1
4442,3428	12	2,20	4,99	$a\,^5P-x\,^5D^\circ$	2—2
4433,221	3	3,65	6,44	$z\,^5P^\circ-e\,^5P$	2—1
4430,6175	6	2,23	5,02	$a\,^5P-x\,^5D^\circ$	1—0
4427,3118	10	0,05	2,85	$a\,^5D-z\,^7F^\circ$	3—4
4422,5703	6	2,84	5,64	$b\,^3P-x\,^3D^\circ$	1—1
4415,1250	20	1,61	4,42	$a\,^3F-z\,^5G^\circ$	2—3
4408,4176	6	2,20	5,01	$a\,^5P-x\,^5D^\circ$	2—1
4407,7130	5	2,18	4,99	$a\,^5P-x\,^5D^\circ$	3—2
4404,7525	30	1,56	4,37	$a\,^3F-z\,^5G^\circ$	3—4
4390,9542	4	3,01	5,84	$b\,^3G-z\,^3H^\circ$	3—4
4388,411	4	3,60	6,43	$z\,^5P^\circ-e\,^5P$	3—3
4387,8959	4	3,07	5,89	$c\,^3P-y\,^3S^\circ$	1—1
4383,5473	45	1,48	4,31	$a\,^3F-z\,^5G^\circ$	4—5
4375,9318	9	0,00	2,83	$a\,^5D-z\,^7F^\circ$	4—5
4369,7745	7	3,05	5,88	$a\,^1G-z\,^1G^\circ$	4—4
4367,5811	5	2,99	5,82	$b\,^3G-z\,^3H^\circ$	4—5
4352,7371	9	2,22	5,07	$a\,^5P-z\,^5S^\circ$	1—2
4337,0484	10	1,56	4,42	$a\,^3F-z\,^5G^\circ$	3—3
4327,098	3	3,54	6,41	$a\,^1D-y\,^1D^\circ$	2—2
4325,7647	35	1,61	4,47	$a\,^3F-z\,^3G^\circ$	2—3
4315,0872	10	2,20	5,07	$a\,^5P-z\,^5S^\circ$	2—2
4309,3771	4	2,95	5,82	$b\,^3G-z\,^3H^\circ$	5—6
4307,9048	35	1,56	4,44	$a\,^3F-z\,^3G^\circ$	3—4
4299,2409	18	2,42	5,31	$z\,^7D^\circ-e\,^7D$	4—5
4298,040	3	3,05	5,93	$a\,^1G-x\,^3G^\circ$	4—5
4294,1271	15	1,48	4,37	$a\,^3F-z\,^5G^\circ$	4—4
4285,4453	3	3,23	6,12	$b\,^3H-y\,^3H^\circ$	6—6
4282,4057	12	2,18	5,07	$a\,^5P-z\,^5S^\circ$	3—2
4271,7634	35	1,48	4,39	$a\,^3F-z\,^3G^\circ$	4—5
4271,1589	20	2,45	5,35	$z\,^7D^\circ-e\,^7D$	3—4
4267,83	5	3,11	6,01	$c\,^3P-x\,^3P^\circ$	0—1
4260,4794	35	2,40	5,31	$z\,^7D^\circ-e\,^7D$	5—5
4250,7896	25	1,56	4,47	$a\,^3F-z\,^3G^\circ$	3—3
4250,1248	25	2,47	5,38	$z\,^7D^\circ-e\,^7D$	2—3
4248,2275	4	3,07	5,98	$c\,^3P-x\,^3P^\circ$	1—2
4247,433	12	3,37	6,29	$z\,^5F^\circ-e\,^5G$	4—5
4245,2594	6	2,85	5,77	$b\,^3P-z\,^3S^\circ$	0—1
4238,821	10	3,40	6,31	$z\,^5F^\circ-e\,^5G$	3—4
4235,9433	25	2,42	5,35	$z\,^7D^\circ-e\,^7D$	4—4
4233,6089	18	2,48	5,41	$z\,^7D^\circ-e\,^7D$	1—2
4227,432	30	3,33	6,26	$z\,^5F^\circ-e\,^5G$	5—6
4225,465	6	3,42	6,35	$z\,^5F^\circ-e\,^5G$	2—3
4224,176	6	3,37	6,31	$z\,^5F^\circ-e\,^7F$	4—5
4222,2181	12	2,45	5,38	$z\,^7D^\circ-e\,^7D$	3—3
4219,3641	12	3,57	6,51	$a\,^1H-y\,^3I^\circ$	5—6
4217,555	7	3,43	6,37	$z\,^5F^\circ-e\,^5G$	1—2
4216,1854	8	0,00	2,94	$a\,^5D-z\,^7P^\circ$	4—4
4213,650	5	2,84	5,78	$b\,^3P-y\,^3P^\circ$	1—0
4210,3497	15	2,48	5,42	$z\,^7D^\circ-e\,^7D$	1—1
4208,615	3	3,39	6,34	$z\,^5F^\circ-e\,^7F$	3—3
4206,6985	3	0,05	3,00	$a\,^5D-z\,^7P^\circ$	3—3
4203,9867	10	2,84	5,79	$b\,^3P-y\,^3P^\circ$	1—2
4202,0320	30	1,48	4,44	$a\,^3F-z\,^3G^\circ$	4—4
4199,0981	20	3,04	5,99	$a\,^1G-z\,^1H^\circ$	4—5
4198,3098	20	2,40	5,35	$z\,^7D^\circ-e\,^7D$	5—4
4196,214	4	3,39	6,35	$z\,^5F^\circ-e\,^5G$	3—3
4195,337	5	3,33	6,29	$z\,^5F^\circ-e\,^5G$	5—5

λ, Å	I	E_H, eV	E_B, eV	Transition	J
4191,4358	15	2,47	5,42	$z\,^7D°—e\,^7D$	2—1
4187,8015	20	2,42	5,38	$z\,^7D°—e\,^7D$	4—3
4187,0436	20	2,45	5,41	$z\,^7D°—e\,^7D$	3—2
4184,8941	10	2,83	5,79	$b\,^3P—y\,^3P°$	2—2
4181,7571	15	2,83	5,79	$b\,^3P—u\,^5D°$	2—3
4177,595	4	0,92	3,87	$a\,^5F—z\,^3F°$	4—4
4176,572	7	3,37	6,34	$z\,^5F°—f\,^5F$	4—5
4175,6386	10	2,84	5,81	$b\,^3P—u\,^5D°$	1—2
4174,9137	5	0,92	3,88	$a\,^5F—z\,^3D°$	4—3
4158,798	5	3,43	6,40	$z\,^5F°—f\,^5F$	1—2
4157,791	8	3,42	6,41	$z\,^5F°—f\,^5F$	2—3
4156,8021	12	2,83	5,81	$b\,^3P—u\,^5D°$	2—2
4154,812	9	3,37	6,34	$z\,^5F°—e\,^7G$	4—5
4154,5021	12	2,83	5,81	$b\,^3P—y\,^3P°$	2—1
4153,910	10	3,39	6,38	$z\,^5F°—f\,^5F$	3—4
4149,370	5	3,33	6,31	$z\,^5F°—e\,^7G$	5—6
4147,6719	10	1,48	4,47	$a\,^3F—z\,^3G°$	4—3
4143,8703	30	1,56	4,55	$a\,^3F—y\,^3F°$	3—4
4143,4174	15	3,05	6,03	$a\,^1G—y\,^1G°$	4—4
4137,004	7	3,41	6,40	$a\,^1P—y\,^1D°$	1—2
4134,6798	12	2,83	5,82	$b\,^3P—w\,^3D°$	2—3
4132,903	8	2,84	5,84	$b\,^3P—w\,^3D°$	1—2
4132,0603	25	1,61	4,61	$a\,^3F—y\,^3F°$	2—3
4127,6113	7	2,86	5,86	$b\,^3P—w\,^3D°$	0—1
4121,8050	5	2,83	5,82	$b\,^3P—x\,^3F°$	2—3
4118,5484	15	3,57	6,58	$a\,^1H—z\,^1I°$	5—6
4109,8053	9	2,84	5,86	$b\,^3P—w\,^3D°$	1—1
4107,4917	12	2,83	5,84	$b\,^3P—u\,^5D°$	2—1
4098,187	4	3,24	6,26	$z\,^5D°—f\,^5D$	3—3
4085,324	4	3,24	6,27	$z\,^5D°—e\,^7P$	3—3
4084,499	6	3,33	6,36	$z\,^5F°—g\,^5D$	5—4
4076,637	8	3,21	6,24	$z\,^5D°—f\,^5D$	4—4
4071,7399	40	1,61	4,65	$a\,^3F—y\,^3F°$	2—2
4067,982	8	3,21	6,25	$z\,^5D°—e\,^7P$	4—4
4066,979	6	2,83	5,87	$b\,^3P—1°$	2—2
4063,5963	45	1,56	4,61	$a\,^3F—y\,^3F°$	3—3
4062,4440	10	2,84	5,89	$b\,^3P—y\,^3S°$	1—1
4045,8147	60	1,48	4,55	$a\,^3F—y\,^3F°$	4—4
4044,6125	6	2,83	5,89	$b\,^3P—y\,^3S°$	2—1
4030,492	6	3,21	6,29	$z\,^5D°—e\,^5G$	4—5
4024,739	6	3,24	6,31	$z\,^5D°—e\,^5G$	3—4
4021,8696	12	2,76	5,84	$a\,^3G—z\,^3H°$	3—4
4017,152	6	3,05	6,13	$a\,^1G—v\,^3G°$	4—5
4014,534	10	3,57	6,66	$a\,^1H—y\,^1H°$	5—5
4009,7154	10	2,22	5,31	$a\,^5P—x\,^5P°$	1—2
4007,2735	6	2,76	5,84	a $^3G—x\,^3F°$	3—2
4005,2440	25	1,56	4,65	$a\,^3F—y\,^3F°$	3—2
3998,0554	10	2,69	5,79	$a\,^3G—u\,^5D°$	5—4
3997,3952	15	2,73	5,82	$a\,^3G—z\,^3H°$	4—5
3986,172	5	3,25	6,36	$a\,^3D—v\,^3F°$	3—4
3985,388	3	3,30	6,41	$a\,^3D—y\,^1D°$	2—2
3983,9593	10	2,73	5,83	$a\,^3G—x\,^3F°$	4—3
3981,7743	7	2,73	5,84	$a\,^3G—z\,^3H°$	4—4
3977,7437	12	2,20	5,31	$a\,^5P—x\,^5P°$	2—2
3971,3250	9	2,69	5,81	$a\,^3G—x\,^3F°$	5—4
3969,2595	30	1,48	4,61	$a\,^3F—y\,^3F°$	4—3
3967,4234	8	3,29	6,43	$b\,^3H—u\,^3G°$	4—3

λ, Å	I	E_H, eV	E_B, eV	Transition	J
3966,6304	10	2,73	5,87	$a\,^3G - z\,^1G^\circ$	3—4
3966,0645	10	1,61	4,73	$a\,^3F - y\,^3D^\circ$	2—3
3963,109	6	3,28	6,41	$z\,^5D^\circ - f\,^5F$	1—2
3956,681	12	2,68	5,81	$a\,^3G - z\,^3H^\circ$	5—6
3956,4574	9	3,23	6,37	$b\,^3H - u\,^3G^\circ$	6—5
3952,606	8	2,68	5,82	$a\,^3G - z\,^3H^\circ$	5—5
3951,168	9	3,27	6,41	$a\,^3D - y\,^1D^\circ$	1—2
3949,9558	10	2,18	5,31	$a\,^5P - x\,^5P^\circ$	3—2
3948,7778	10	3,26	6,41	$b\,^3H - u\,^3G^\circ$	5—4
3948,107	6	3,25	6,38	$z\,^5D^\circ - f\,^5F$	3—4
3942,4418	6	2,84	5,98	$b\,^3P - x\,^3P^\circ$	1—2
3940,8797	5	0,96	4,11	$a\,^5F - y\,^5D^\circ$	3—4
3935,8143	8	2,83	5,98	$b\,^3P - v\,^5F^\circ$	2—2
3933,605	10	{ 3,07	6,22	$c\,^3P - w\,^3P^\circ$	1—2
		3,26	6,41	$z\,^5D^\circ - f\,^5F$	2—1
3930,2981	25	0,09	3,24	$a\,^5D - z\,^5D^\circ$	2—3
3927,9216	30	0,11	3,26	$a\,^5D - z\,^5D^\circ$	1—2
3925,947	6	2,86	6,01	$b\,^3P - x\,^3P^\circ$	0—1
3922,9134	25	0,05	3,21	$a\,^5D - z\,^5D^\circ$	3—4
3920,2601	20	0,12	3,28	$a\,^5D - z\,^5D^\circ$	0—1
3918,646	6	3,01	6,17	$b\,^3G - v\,^3G^\circ$	3—3
3917,1834	8	0,99	4,16	$a\,^5F - y\,^5D^\circ$	2—3
3916,733	6	3,23	6,40	$b\,^3H - 6^\circ$	6—5
3913,6339	4	2,28	5,44	$a\,^3P - w\,^5D^\circ$	2—3
3907,9371	4	2,76	5,92	$a\,^3G - w\,^5G^\circ$	3—2
3906,4814	8	0,11	3,28	$a\,^5D - z\,^5D^\circ$	1—1
3903,9011	5	2,99	6,16	$b\,^3G - y\,^3H^\circ$	4—4
3902,9484	20	1,56	4,73	$a\,^3F - y\,^3D^\circ$	3—3
3899,7086	30	0,09	3,26	$a\,^5D - z\,^5D^\circ$	2—2
3898,012	10	1,01	4,19	$a\,^5F - y\,^5D^\circ$	1—2
3897,895	8	2,69	5,87	$a\,^3G - w\,^5G^\circ$	5—6
3895,6579	25	0,11	3,29	$a\,^5D - z\,^5D^\circ$	1—0
3893,3935	7	2,95	6,13	$b\,^3G - v\,^3G^\circ$	5—5
3891,929	5	3,41	6,60	$a\,^1P - z\,^1P^\circ$	1—1
3888,517	20	1,60	4,79	$a\,^3F - y\,^3D^\circ$	2—2
3886,2839	40	0,05	3,24	$a\,^5D - z\,^5D^\circ$	3—3
3885,5165	5	2,42	5,61	$a\,^3P - x\,^3D^\circ$	1—2
3878,663	8	2,44	5,63	$a\,^3H - y\,^3G^\circ$	4—3
3878,5745	100	0,09	3,28	$a\,^5D - z\,^5D^\circ$	2—1
3878,0206	60	0,96	4,16	$a\,^5F - y\,^5D^\circ$	3—3
3873,7624	8	2,43	5,63	$a\,^3H - y\,^3G^\circ$	5—4
3872,5032	60	0,99	4,19	$a\,^5F - y\,^5D^\circ$	2—2
3871,7513	4	2,95	6,14	$b\,^3G - y\,^3H^\circ$	5—5
3869,5615	4	2,73	5,93	$a\,^3G - x\,^3G^\circ$	4—5
3867,2184	7	3,02	6,22	$c\,^3P - w\,^3P^\circ$	2—2
3865,5256	30	1,01	4,22	$a\,^5F - y\,^5D^\circ$	1—1
3859,9132	300	0,00	3,21	$a\,^5D - z\,^5D^\circ$	4—4
3859,2143	10	2,40	5,61	$a\,^3H - y\,^3G^\circ$	6—5
3856,373	50	0,05	3,26	$a\,^5D - z\,^5D^\circ$	3—2
3852,5752	6	2,18	5,39	$a\,^5P - w\,^5D^\circ$	3—4
3850,8193	12	0,99	4,21	$a\,^5F - z\,^3P^\circ$	2—2
3849,9694	40	1,01	4,23	$a\,^5F - y\,^5D^\circ$	1—0
3846,8023	8	3,25	6,48	$a\,^3D - t\,^3D^\circ$	3—3
3845,1706	5	2,42	5,64	$a\,^3P - x\,^3D^\circ$	1—1
3843,2596	8	3,05	6,27	$a\,^1G - z\,^1F^\circ$	4—3
3841,0499	80	1,61	4,83	$a\,^3F - y\,^3D^\circ$	2—1
3840,4397	80	0,99	4,22	$a\,^5F - y\,^5D^\circ$	2—1
3839,2584	7	3,05	6,27	$a\,^1G - x\,^1G^\circ$	4—4
3836,333	4	3,30	6,53	$a\,^3D - t\,^3D^\circ$	2—2

λ, Å	I	E_H, eV	E_B, eV	Transition	J
3834,2244	100	0,96	4,19	$a\,^5F—y\,^5D°$	3—2
3833,3103	5	2,56	5,79	$b\,^3F—u\,^5D°$	4—4
3827,8256	75	1,56	4,79	$a\,^3F—y\,^3D°$	3—2
3825,8834	200	0,91	4,16	$a\,^5F—y\,^5D°$	4—3
3824,4455	50	0,00	3,24	$a\,^5D—z\,^5D°$	4—3
3821,1807	10	3,27	6,51	$b\,^3H—y\,^3I°$	5—6
3820,4274	250	0,86	4,11	$a\,^5F—y\,^5D°$	5—4
3815,8430	100	1,48	4,73	$a\,^3F—y\,^3D°$	4—3
3812,9658	40	0,96	4,21	$a\,^5F—z\,^3P°$	3—2
3808,7306	4	2,56	5,81	$b\,^3F—x\,^3F°$	4—4
3807,5392	7	2,22	5,47	$a\,^5P—w\,^5D°$	1—2
3806,6992	10	3,25	6,53	$b\,^3H—w\,^3H°$	5—5
3805,3450	12	3,30	6,56	$b\,^3H—y\,^3I°$	4—5
3799,5498	50	0,96	4,22	$a\,^5F—y\,^5F°$	3—4
3798,5134	40	0,91	4,18	$a\,^5F—y\,^5F°$	4—5
3797,517	12	3,22	6,50	$b\,^3H—w\,^3H°$	6—6
3795,0045	60	0,99	4,26	$a\,^5F—y\,^5F°$	2—3
3794,340	8	2,45	5,72	$a\,^3H—z\,^3I°$	4—5
3790,0943	12	0,99	4,26	$a\,^5F—z\,^3P°$	2—1
3787,8825	50	1,01	4,28	$a\,^5F—y\,^5F°$	1—2
3786,6781	8	1,01	4,29	$a\,^5F—z\,^3P°$	1—0
3786,176	4	2,83	6,10	$b\,^3P—v\,^3D°$	2—2
3785,951	6	2,43	5,70	$a\,^3H—z\,^3I°$	5—6
3779,446	4	2,56	5,84	$b\,^3F—x\,^3F°$	4—3
3776,4553	6	2,18	5,46	$a\,^5P—w\,^5F°$	3—4
3774,8266	5	2,22	5,50	$a\,^5P—w\,^5D°$	1—1
3767,1939	80	1,01	4,30	$a\,^5F—y\,^5F°$	1—1
3765,5414	20	3,23	6,52	$b\,^3H—y\,^3I°$	6—7
3763,7910	100	0,99	4,28	$a\,^5F—y\,^5F°$	2—2
3760,5335	6	2,22	5,51	$a\,^5P—y\,^5S°$	1—2
3760,052	8	2,40	5,70	$a\,^3H—z\,^3I°$	6—7
3758,2350	150	0,96	4,26	$a\,^5F—y\,^5F°$	3—3
3753,6134	8	2,18	5,47	$a\,^5P—w\,^5D°$	3—2
3749,4875	200	0,91	4,22	$a\,^5F—y\,^5F°$	4—4
3748,492	7	3,57	6,87	$a\,^1H—v\,^3H°$	5—6
3748,2639	60	0,11	3,41	$a\,^5D—z\,^5F°$	1—2
3746,929	6	3,01	6,31	$z\,^7P°—f\,^7D$	3—3
3745,9013	40	0,12	3,43	$a\,^5D—z\,^5F°$	0—1
3745,5623	100	0,09	3,39	$a\,^5D—z\,^5F°$	2—3
3743,468	6	3,57	6,88	$a\,^1H—x\,^1H°$	5—5
3743,3640	20	0,99	4,30	$a\,^5F—y\,^5F°$	2—1
3738,3078	10	3,27	6,58	$b\,^3H—z\,^1I°$	5—6
3737,1333	150	0,05	3,36	$a\,^5D—z\,^5F°$	3—4
3735,330	6	2,94	6,25	$z\,^7P°—e\,^7P$	4—4
3734,8659	300	0,86	4,18	$a\,^5F—y\,^5F°$	5—5
3733,3191	40	0,11	3,43	$a\,^5D—z\,^5F°$	1—1
3732,399	10	2,20	5,51	$a\,^5P—y\,^5S°$	2—2
3727,6211	50	0,96	4,28	$a\,^5F—y\,^5F°$	3—2
3726,925	6	3,04	6,37	$z\,^7P°—e\,^7F$	2—2
3724,3796	8	2,28	5,60	$a\,^3P—x\,^3D°$	2—3
3722,5642	50	0,09	3,41	$a\,^5D—z\,^5F°$	2—2
3719,9367	250	0,00	3,33	$a\,^5D—z\,^5F°$	4—5
3716,448	12	2,94	6,27	$z\,^7P°—e\,^7P$	4—3
3709,2484	75	0,91	4,26	$a\,^5F—y\,^5F°$	4—3
3707,9216	8	2,17	5,51	$a\,^5P—y\,^5S°$	3—2
3707,823	20	0,09	3,43	$a\,^5D—z\,^5F°$	2—1
3707,049	8	3,00	6,34	$z\,^7P°—e\,^7F$	3—3

λ, Å	I	E_H, eV	E_B, eV	Transition	J
3705,5674	100	0,05	3,39	$a\ ^5D-z\ ^5F^\circ$	3—3
3704,4635	10	2,69	6,03	$a\ ^3G-y\ ^1G^\circ$	5—4
3701,090	20	3,00	6,35	$z\ ^7P^\circ-e\ ^7F$	3—4
3695,054	8	2,59	5,94	$b\ ^3F-v\ ^5F^\circ$	3—4
3694,010	20	3,04	6,40	$z\ ^7P^\circ-e\ ^7S$	2—3
3689,463	12	2,94	6,30	$z\ ^7P^\circ-f\ ^7D$	4—4
3687,4589	40	0,86	4,22	$a\ ^5F-y\ ^5F^\circ$	5—4
3686,003	15	2,94	6,31	$z\ ^7P^\circ-e\ ^7F$	4—5
3684,1102	15	2,73	6,09	$a\ ^3G-v\ ^3D^\circ$	4—3
3683,0562	10	0,05	3,41	$a\ ^5D-z\ ^5F^\circ$	3—2
3682,209	20	3,54	6,91	$a\ ^1D-w\ ^1D^\circ$	2—2
3679,9152	40	0,00	3,36	$a\ ^5D-z\ ^5F^\circ$	4—4
3678,8620	3	2,43	5,79	$a\ ^3P-y\ ^3P^\circ$	1—0
3677,6309	12	2,75	6,11	$a\ ^3G-w\ ^3F^\circ$	3—2
3676,3135	6	2,56	5,93	$b\ ^3F-x\ ^3G^\circ$	4—5
3670,071	3	2,95	6,33	$b\ ^3G-x\ ^3H^\circ$	5—6
3670,028	3	2,84	6,22	$b\ ^3P-w\ ^3P^\circ$	1—2
3669,5229	10	2,73	6,10	$a\ ^3G-w\ ^3F^\circ$	4—3
3659,521	8	2,45	5,84	$a\ ^3H-z\ ^3H^\circ$	4—4
3651,4699	20	2,76	6,15	$a\ ^3G-v\ ^3G^\circ$	3—4
3649,5090	12	2,69	6,08	$a\ ^3G-w\ ^3F^\circ$	5—4
3647,8439	100	0,91	4,31	$a\ ^5F-z\ ^5G^\circ$	4—5
3645,825	6	3,11	6,51	$c\ ^3P-u\ ^3D^\circ$	0—1
3640,3918	15	2,73	6,13	$a\ ^3G-v\ ^3G^\circ$	4—5
3638,2998	12	2,76	6,16	$a\ ^3G-y\ ^3H^\circ$	3—4
3634,334	6	2,95	6,35	$z\ ^7P^\circ-e\ ^5G$	4—3
3631,4646	125	0,96	4,37	$a\ ^5F-z\ ^5G^\circ$	3—4
3631,096	7	2,83	6,24	$z\ ^5F^\circ-f\ ^7D$	5—5
3625,148	6	2,83	6,25	$z\ ^7F^\circ-f\ ^5D$	5—4
3623,1878	8	2,40	5,82	$a\ ^3H-z\ ^3H^\circ$	6—6
3622,005	12	2,76	6,17	$a\ ^3G-v\ ^3G^\circ$	3—3
3621,4640	15	2,73	6,14	$a\ ^3G-y\ ^3H^\circ$	4—5
3618,7694	125	0,99	4,42	$a\ ^5F-z\ ^5G^\circ$	2—3
3617,788	12	3,02	6,45	$c\ ^3P-u\ ^3D^\circ$	2—3
3612,074	8	2,83	6,26	$z\ ^7F^\circ-e\ ^5G$	5—6
3610,162	20	2,81	6,24	$z\ ^7F^\circ-e\ ^7F$	6—6
3608,8609	100	1,01	4,45	$a\ ^5F-z\ ^5G^\circ$	1—2
3606,6821	20	2,69	6,12	$a\ ^3G\quad y\ ^3H^\circ$	5 6
3605,458	15	2,73	6,16	$a\ ^3G-y\ ^3H^\circ$	4—4
3603,2068	10	2,69	6,13	$a\ ^3G-v\ ^3G^\circ$	5—5
3594,636	3	2,85	6,30	$z\ ^7F^\circ-f\ ^7D$	4—4
3589,1063	8	0,86	4,31	$a\ ^5F-z\ ^5G^\circ$	5—5
3586,9861	30	0,99	4,45	$a\ ^5F-z\ ^5G^\circ$	2—2
3586,114	10	3,23	6,69	$b\ ^3H-t\ ^3G^\circ$	6—5
3585,7068	20	0,91	4,37	$a\ ^5F-z\ ^5G^\circ$	4—4
3585,3206	30	0,96	4,42	$a\ ^5F-z\ ^5G^\circ$	3—3
3584,6627	8	2,69	6,14	$a\ ^3G-y\ ^3H^\circ$	5—5
3581,195	250	0,86	4,32	$a\ ^5F-z\ ^5G^\circ$	5—6
3571,996	6	2,83	6,31	$z\ ^7F^\circ-e\ ^7F$	5—5
3570,258	20	2,80	6,27	$z\ ^7F^\circ-e\ ^7G$	6—7
3570,0996	100	0,91	4,39	$a\ ^5F-z\ ^3G^\circ$	4—5
3565,3807	60	0,96	4,44	$a\ ^5F-z\ ^3G^\circ$	3—4
3558,5170	30	0,99	4,47	$a\ ^5F-z\ ^3G^\circ$	2—3
3556,883	7	2,85	6,34	$z\ ^7F^\circ-f\ ^5F$	4—5
3554,929	40	2,83	6,32	$z\ ^7F^\circ-e\ ^7G$	5—6
3553,741	6	3,57	7,06	$a\ ^1H-v\ ^1G^\circ$	5—4
3542,078	15	2,85	6,37	$z\ ^7F^\circ-e\ ^7G$	3—4
3541,086	15	2,85	6,35	$z\ ^7F^\circ-e\ ^7G$	4—5

λ, Å	I	E_H, eV	E_B, eV	Transition	J
3540,118	3	2,86	6,37	$z\,^{7}F^{\circ}$—$g\,^{5}D$	3—4
3536,557	15	2,88	6,38	$z\,^{7}F^{\circ}$—$e\,^{7}G$	2—3
3533,202	10	2,89	6,39	$z\,^{7}F^{\circ}$—$e\,^{7}G$	1—2
3529,820	6	2,89	6,39	$z\,^{7}F^{\circ}$—$e\,^{7}G$	1—1
3527,797	4	2,85	6,37	$z\,^{7}F^{\circ}$—$e\,^{7}G$	4—4
3526,1676	15	0,95	4,47	$a\,^{5}F$—$z\,^{3}G^{\circ}$	3—3
3521,2630	25	0,91	4,44	$a\,^{5}F$—$z\,^{3}G^{\circ}$	4—4
3513,8196	30	0,86	4,39	$a\,^{5}F$—$z\,^{3}G^{\circ}$	5—5
3497,8420	40	0,11	3,65	$a\,^{5}D$—$z\,^{5}P^{\circ}$	1—2
3497,108	10	2,18	5,72	$a\,^{5}P$—$w\,^{5}P^{\circ}$	3—3
3495,2879	8	2,56	6,10	$b\,^{3}F$—$w\,^{3}F^{\circ}$	4—3
3490,5749	100	0,05	3,60	$a\,^{5}D$—$z\,^{5}P^{\circ}$	3—3
3485,3418	7	2,20	5,75	$a\,^{5}P$—$w\,^{5}P^{\circ}$	2—1
3476,7036	40	0,12	3,68	$a\,^{5}D$—$z\,^{5}P^{\circ}$	0—1
3475,651	6	2,18	5,72	$a\,^{5}P$—$w\,^{5}P^{\circ}$	3—2
3475,4511	70	0,09	3,65	$a\,^{5}D$—$z\,^{5}P^{\circ}$	2—2
3471,3460	6	2,27	5,83	$a\,^{3}P$—$u\,^{5}D^{\circ}$	2—1
3465,8621	60	0,11	3,68	$a\,^{5}D$—$z\,^{5}P^{\circ}$	1—1
3452,2760	10	0,96	4,55	$a\,^{5}F$—$y\,^{3}F^{\circ}$	3—4
3451,9166	10	2,22	5,81	$a\,^{5}P$—$u\,^{5}D^{\circ}$	1—2
3450,3304	10	2,22	5,81	$a\,^{5}P$—$y\,^{3}P^{\circ}$	1—1
3447,2797	8	2,20	5,79	$a\,^{5}P$—$y\,^{3}P^{\circ}$	2—2
3445,1508	20	2,20	5,79	$a\,^{5}P$—$u\,^{5}D^{\circ}$	2—3
3443,8775	50	0,09	3,68	$a\,^{5}D$—$z\,^{5}P^{\circ}$	2—1
3440,9899	75	0,05	3,65	$a\,^{5}D$—$z\,^{5}P^{\circ}$	3—2
3440,6069	150	0,00	3,60	$a\,^{5}D$—$z\,^{5}P^{\circ}$	4—3
3428,1948	8	2,20	5,81	$a\,^{5}P$—$u\,^{5}D^{\circ}$	2—2
3427,1213	20	2,18	5,79	$a\,^{5}P$—$u\,^{5}D^{\circ}$	3—4
3424,2861	10	2,18	5,79	$a\,^{5}P$—$u\,^{5}D^{\circ}$	3—3
3422,6583	7	2,22	5,84	$a\,^{5}P$—$w\,^{3}D^{\circ}$	1—2
3418,512	10	2,22	5,84	$a\,^{5}P$—$u\,^{5}D^{\circ}$	1—0
3417,8428	12	2,22	5,84	$a\,^{5}P$—$u\,^{5}D^{\circ}$	1—1
3413,1339	15	2,20	5,82	$a\,^{5}P$—$w\,^{3}D^{\circ}$	2—3
3407,4611	20	2,18	5,81	$a\,^{5}P$—$x\,^{3}F^{\circ}$	3—4
3406,8021	6	2,22	5,86	$a\,^{5}P$—$w\,^{3}D^{\circ}$	1—1
3404,3557	6	2,20	5,83	$a\,^{5}P$—$x\,^{3}F^{\circ}$	2—3
3402,262	5	3,23	6,88	$b\,^{3}H$—$v\,^{3}H^{\circ}$	6—6
3401,5200	6	0,91	4,56	$a\,^{5}F$—$y\,^{5}P^{\circ}$	4—3
3399,3356	15	2,20	5,84	$a\,^{5}P$—$w\,^{3}D^{\circ}$	2—2
3396,9774	4	0,96	4,60	$a\,^{5}F$—$y\,^{5}P^{\circ}$	3—2
3394,5854	5	2,20	5,84	$a\,^{5}P$—$u\,^{3}D^{\circ}$	2—1
3392,6540	15	2,18	5,82	$a\,^{5}P$—$w\,^{3}D^{\circ}$	3—3
3392,3058	8	2,20	5,85	$a\,^{5}P$—$x\,^{3}F^{\circ}$	2—2
3383,9808	8	2,18	5,83	$a\,^{5}P$—$x\,^{3}F^{\circ}$	3—3
3380,1117	8	2,76	6,42	$a\,^{3}G$—$u\,^{3}G^{\circ}$	3—3
3379,0206	6	2,17	5,83	$a\,^{5}P$—$w\,^{3}D^{\circ}$	3—2
3378,685	6	2,69	6,36	$a\,^{3}G$—$v\,^{3}F^{\circ}$	5—4
3370,7852	10	2,69	6,37	$a\,^{3}G$—$u\,^{3}G^{\circ}$	5—5
3369,549	8	2,73	6,41	$a\,^{3}G$—$u\,^{3}G^{\circ}$	4—4
3355,2287	6	3,30	7,00	$b\,^{3}H$—$u\,^{3}H^{\circ}$	4—4
3347,9271	6	2,28	5,98	$a\,^{3}P$—$v\,^{5}F^{\circ}$	2—2
3341,905	5	2,68	6,38	$a\,^{3}G$—6°	5—5
3340,5666	6	2,28	5,98	$a\,^{3}P$—$x\,^{3}P^{\circ}$	2—2
3337,6664	6	2,69	6,41	$a\,^{3}G$—$u\,^{3}G^{\circ}$	5—4
3335,7699	4	2,84	6,56	$b\,^{3}P$—$v\,^{3}P^{\circ}$	1—2
3334,2201	4	2,43	6,14	$a\,^{3}H$—$y\,^{3}H^{\circ}$	5—5
3331,613	4	2,43	6,15	$a\,^{3}H$—$v\,^{3}G^{\circ}$	5—4

λ, Å	I	E_H, eV	E_B, eV	Transition	J
3328,8667	5	3,26	6,99	$b\,^3H—u\,^3H^\circ$	5—5
3325,462	4	2,45	6,17	$a\,^3H—v\,^3G^\circ$	4—3
3324,5385	4	2,40	6,13	$a\,^3H—v\,^3G^\circ$	6—5
3323,7375	7	2,82	6,56	$b\,^3P—v\,^3P^\circ$	2—2
3322,477	5	2,94	6,67	$z\,^7P^\circ—g\,^7D$	4—5
3317,121	4	2,28	6,01	$a\,^3P—x\,^3P^\circ$	2—1
3314,7420	7	3,30	7,04	$a\,^3D—u\,^3F^\circ$	2—3
3310,342	4	2,95	6,68	$b\,^3G—t\,^3G^\circ$	5—5
3306,498	6	2,27	6,00	$a\,^3P—u\,^5D^\circ$	2—2
3306,354	20	2,22	5,97	$a\,^5P—v\,^5P^\circ$	1—2
3305,9719	20	2,20	5,94	$a\,^5P—v\,^5P^\circ$	2—3
3298,1331	6	2,22	5,98	$a\,^5P—v\,^5F^\circ$	1—2
3292,5910	8	2,22	5,98	$a\,^5P—v\,^5P^\circ$	1—1
3292,023	8	3,25	7,02	$a\,^3D—u\,^3F^\circ$	3—4
3290,9899	5	2,22	5,98	$a\,^5P—x\,^3P^\circ$	1—2
3286,7541	20	2,18	5,94	$a\,^5P—v\,^5P^\circ$	3—3
3284,5888	5	2,20	5,97	$a\,^5P—v\,^5P^\circ$	2—2
3280,2613	8	3,30	7,08	$b\,^3H—x\,^3I^\circ$	4—5
3278,734	4	2,42	6,20	$a\,^3P—w\,^3P^\circ$	1—1
3276,4713	4	2,20	5,98	$a\,^5P—v\,^5F^\circ$	2—2
3271,0014	15	2,20	5,98	$a\,^5P—v\,^5P^\circ$	2—1
3268,236	5	2,22	6,01	$a\,^5P—x\,^3P^\circ$	1—1
3265,6182	15	2,18	5,97	$a\,^5P—v\,^5P^\circ$	3—2
3265,0473	8	0,09	3,89	$a\,^5D—z\,^3D^\circ$	2—3
3259,994	6	2,45	6,24	$z\,^7D^\circ—f\,^5D$	3—4
3257,5940	8	2,18	5,98	$a\,^5P—v\,^5F^\circ$	3—2
3254,3628	10	3,27	7,08	$b\,^3H—x\,^3I^\circ$	5—6
3253,602	4	3,25	7,06	$a\,^3D—v\,^1G^\circ$	3—4
3251,235	8	2,20	6,00	$a\,^5P—w\,^3G^\circ$	2—3
3248,206	10	2,45	6,26	$z\,^7D^\circ—f\,^5D$	3—3
3246,962	6	2,20	6,01	$a\,^5P—x\,^3P^\circ$	2—1
3246,005	8	0,11	3,93	$a\,^5D—z\,^3D^\circ$	1—2
3244,190	5	2,42	6,24	$z\,^7D^\circ—f\,^7D$	4—5
3239,436	15	2,42	6,24	$z\,^7D^\circ—f\,^5D$	4—4
3236,2231	8	0,05	3,88	$a\,^5D—z\,^3F^\circ$	3—4
3234,6138	7	0,05	3,89	$a\,^5D—z\,^3D^\circ$	3—3
3233,971	12	2,42	6,25	$z\,^7D^\circ—e\,^7P$	4—4
3233,054	8	3,24	7,07	$b\,^3H—x\,^3I^\circ$	6—7
3230,967	10	2,45	6,29	$z\,^7D^\circ—f\,^5D$	3—2
3230,211	6	2,47	6,31	$z\,^7D^\circ—e\,^7P$	2—2
3228,254	5	2,47	6,31	$z\,^7D^\circ—f\,^5D$	2—1
3227,798	15	2,42	6,26	$z\,^7D^\circ—f\,^5D$	4—3
3225,789	20	2,40	6,23	$z\,^7D^\circ—e\,^7F$	5—6
3222,069	20	2,40	6,24	$z\,^7D^\circ—f\,^7D$	5—5
3219,581	12	2,45	6,30	$z\,^7D^\circ—f\,^7D$	3—4
3217,380	10	2,40	6,24	$z\,^7D^\circ—f\,^5D$	5—4
3215,940	12	2,47	6,33	$z\,^7D^\circ—f\,^7D$	2—2
3214,3964	8	0,09	3,95	$a\,^5D—z\,^3F^\circ$	2—3
3214,040	20	2,45	6,31	$z\,^7D^\circ—f\,^7D$	3—3
		3,36	7,22	$z\,^5F^\circ—g\,^5G$	4—5
3211,992	10	2,40	6,25	$z\,^7D^\circ—e\,^7P$	5—4
3210,834	10	2,47	6,33	$z\,^7D^\circ—f\,^7D$	2—1
3210,236	8	2,42	6,29	$z\,^7D^\circ—e\,^5G$	4—5
3209,297	12	2,81	6,67	$z\,^7F^\circ—g\,^7D$	6—5
		3,42	7,28	$z\,^5F^\circ—g\,^5G$	2—3
3208,475		3,43	7,29	$z\,^5F^\circ—g\,^5G$	1—2
3205,400	15	2,48	6,35	$z\,^7D^\circ—e\,^7F$	1—1

λ, Å	I	E_H, eV	E_B, eV	Transition	J
3200,475	15	2,47	6,34	$z\,^7D^\circ - e\,^7F$	2—3
3199,525	15	2,42	6,30	$z\,^7D^\circ - f\,^7D$	4—4
3196,997	20	0,05	3,92	$a\,^5D - z\,^3D^\circ$	3—2
3196,930	20	2,42	6,30	$z\,^7D^\circ - e\,^7F$	4—5
3193,314	8	2,47	6,35	$z\,^7D^\circ - e\,^5G$	2—3
3193,2268	10	0,00	3,88	$a\,^5D - z\,^3F^\circ$	4—4
3192,802	8	2,48	6,36	$z\,^7D^\circ - e\,^7F$	1—2
3191,6599	7	0,00	3,89	$a\,^5D - z\,^3D^\circ$	4—3
3188,821	7	2,48	6,37	$z\,^7D^\circ - e\,^5G$	1—2
3188,571	4	2,40	6,29	$z\,^7D^\circ - e\,^5G$	5—5
3184,8955	7	0,05	3,95	$a\,^5D - z\,^3F^\circ$	3—3
3182,9798	4	2,20	6,09	$a\,^5P - v\,^3D^\circ$	2—3
3180,7562	5	0,09	3,99	$a\,^5D - z\,^3F^\circ$	2—2
3180,226	20	2,45	6,35	$z\,^7D^\circ - e\,^7F$	3—4
3178,015	10	2,40	6,30	$z\,^7D^\circ - f\,^7D$	5—4
3175,447	12	2,40	6,31	$z\,^7D^\circ - e\,^7F$	5—5
3166,438	6	2,56	6,48	$b\,^3F - t\,^3D^\circ$	4—3
3165,861	4	2,45	6,37	$z\,^7D^\circ - e\,^7G$	3—4
3161,949	8	2,40	6,31	$z\,^7D^\circ - e\,^7G$	5—6
3160,658	10	2,42	6,35	$z\,^7D^\circ - e\,^7F$	4—4
3157,887	6	2,47	6,40	$z\,^7D^\circ - e\,^7S$	2—3
3157,040	8	2,42	6,35	$z\,^7D^\circ - e\,^7G$	4—5
3156,274	5	3,24	7,17	$z\,^5D^\circ - i\,^5D$	3—3
3153,206	5	2,45	6,38	$z\,^7D^\circ - f\,^5F$	3—4
3151,351	10	2,73	6,66	$a\,^3G - y\,^1H^\circ$	4—5
3144,495	6	2,47	6,41	$z\,^7D^\circ - f\,^5F$	2—2
3143,9896	8	3,21	7,15	$z\,^5D^\circ - i\,^5D$	4—4
3142,453	6	2,45	6,40	$z\,^7D^\circ - e\,^7S$	3—3
3140,391	5	3,24	7,19	$z\,^5D^\circ - i\,^5D$	3—2
3134,1115	10	0,95	4,91	$a\,^5F - x\,^5D^\circ$	3—4
3129,3349	5	1,49	5,44	$a\,^3F - w\,^5D^\circ$	4—3
3126,175	8	—	—	—	—
3125,654	15	{ 0,99	4,95	$a\,^5F - x\,^5D^\circ$	2—3
		2,40	6,36	$z\,^7D^\circ - e\,^7G$	5—4
3120,4364	6	2,45	6,43	$a\,^3H - u\,^3G^\circ$	4—3
3119,4956	6	2,43	6,41	$a\,^3H - u\,^3G^\circ$	5—4
3116,6337	12	1,01	4,99	$a\,^5F - x\,^5D^\circ$	1—2
3100,6667	20	0,96	4,95	$a\,^5F - x\,^5D^\circ$	3—3
3100,3054	20	0,99	4,99	$a\,^5F - x\,^5D^\circ$	2—2
3099,9695	15	0,91	4,91	$a\,^5F - x\,^5D^\circ$	4—4
3099,8968	20	1,01	5,01	$a\,^5F - x\,^5D^\circ$	1—1
3098,192	6	2,69	6,69	$a\,^3G - t\,^3G^\circ$	5—5
3091,5786	20	1,01	5,02	$a\,^5F - x\,^5D^\circ$	1—0
3083,7430	20	0,99	5,01	$a\,^5F - x\,^5D^\circ$	2—1
3078,018	4	0,96	4,98	$a\,^5F - y\,^7D^\circ$	3—3
3075,7214	25	0,96	4,99	$a\,^5F - x\,^5D^\circ$	3—2
3068,1749	8	1,61	5,65	$a\,^3F - x\,^3D^\circ$	2—1
3067,2457	30	0,91	4,95	$a\,^5F - x\,^5D^\circ$	4—3
3067,1196	8	1,61	5,65	$a\,^3F - y\,^3G^\circ$	2—3
3059,0871	100	0,05	4,11	$a\,^5D - y\,^5D^\circ$	3—4
3057,4471	40	0,86	4,91	$a\,^5F - x\,^5D^\circ$	5—4
3055,2638	12	1,56	5,62	$a\,^3F - x\,^3D^\circ$	3—2
3053,070	5	2,42	6,49	$a\,^3P - u\,^3D^\circ$	1—2
3047,6060	100	0,09	4,16	$a\,^5D - y\,^5D^\circ$	2—3
3045,078	5	0,91	4,98	$a\,^5F - y\,^7P^\circ$	4—3
3042,6667	15	0,99	5,06	$a\,^5F - x\,^5F^\circ$	2—3
3042,0215	15	1,01	5,08	$a\,^5F - x\,^5F^\circ$	1—2
3041,7401	15	0,96	5,03	$a\,^5F - x\,^5F^\circ$	3—4

λ, Å	I	E_H, eV	E_B, eV	Transition	J
3041,6396	10	1,56	5,62	$a\ ^3F—y\ ^3G^\circ$	3—4
3040,4281	15	0,91	4,99	$a\ ^5F—x\ ^5F^\circ$	4—5
3037,3901	80	0,11	4,19	$a\ ^5D—y\ ^5D^\circ$	1—2
3031,639	15	1,01	5,10	$a\ ^5F—x\ ^5F^\circ$	1—1
3031,215	12	2,45	6,54	$a\ ^3H—w\ ^3H^\circ$	4—4
3030,1494	15	2,43	6,53	$a\ ^3H—w\ ^3H^\circ$	5—5
3026,4637	15	0,99	5,08	$a\ ^5F—x\ ^5F^\circ$	2—2
3025,8442	50	0,12	4,22	$a\ ^5D—y\ ^5D^\circ$	0—1
3025,638	15	2,40	6,50	$a\ ^3H—w\ ^3H^\circ$	6—6
3024,0337	15	0,11	4,21	$a\ ^5D—z\ ^3P^\circ$	1—2
3021,0743	150	0,05	4,16	$a\ ^5D—y\ ^5D^\circ$	3—3
3020,6405	200	0,00	4,11	$a\ ^5D—y\ ^5D^\circ$	4—4
3020,4918	100	0,09	4,19	$a\ ^5D—y\ ^5D^\circ$	2—2
3018,9848	15	0,96	5,06	$a\ ^5F—x\ ^5F^\circ$	3—3
3017,6288	15	0,11	4,22	$a\ ^5D—y\ ^5D^\circ$	1—1
3016,185	12	0,99	5,10	$a\ ^5F—x\ ^5F^\circ$	2—1
3011,482	7	2,76	6,87	$a\ ^3G—v\ ^3H^\circ$	3—4
3009,5707	25	0,91	5,03	$a\ ^5F—x\ ^5F^\circ$	4—4
3008,1399	60	0,11	4,23	$a\ ^5D—y\ ^5D^\circ$	1—0
3007,1469	8	1,48	5,60	$a\ ^3F—x\ ^3D^\circ$	4—3
3003,0323	10	0,96	5,08	$a\ ^5F—x\ ^5F^\circ$	3—2
3000,9489	100	0,09	4,22	$a\ ^5D—y\ ^5D^\circ$	2—1
3000,4527	8	1,48	5,61	$a\ ^3F—y\ ^3G^\circ$	4—5
2999,5125	30	0,86	4,99	$a\ ^5F—x\ ^5F^\circ$	5—5
2994,4281	100	0,05	4,19	$a\ ^5D—y\ ^5D^\circ$	3—2
2994,385	5	2,42	6,56	$a\ ^3P—v\ ^3P^\circ$	1—2
2991,637	5	—	—	—	—
2990,3933	6	2,73	6,87	$a\ ^3G—v\ ^3H^\circ$	4—5
2987,2923	10	0,91	5,06	$a\ ^5F—x\ ^5F^\circ$	4—3
2986,4569	3	0,11	4,26	$a\ ^5D—z\ ^3P^\circ$	1—1
2984,785	10	0,86	5,01	$a\ ^5F—y\ ^7P^\circ$	5—4
2983,5714	125	0,00	4,16	$a\ ^5D—y\ ^5D^\circ$	4—3
2981,854	6	2,18	6,34	$a\ ^5P—t\ ^5D^\circ$	3—4
2981,4459	20	0,05	4,21	$a\ ^5D—z\ ^3P^\circ$	3—2
2980,539	5	2,76	6,92	$a\ ^3G—w\ ^1F^\circ$	3—3
2976,131	5	2,28	6,45	$a\ ^3P—u\ ^3D^\circ$	2—3
2973,2368	60	0,05	4,22	$a\ ^5D—y\ ^5F^\circ$	3—4
2973,1336	60	0,09	4,26	$a\ ^5D—y\ ^5F^\circ$	2—3
2972,279	3	2,20	6,36	$a\ ^5P—t\ ^5D^\circ$	2—3
2970,106	40	0,11	4,28	$a\ ^5D—y\ ^5F^\circ$	1—2
2969,4759	10	0,86	5,03	$a\ ^5F—x\ ^5F^\circ$	5—4
2969,3606	5	0,11	4,28	$a\ ^5D—z\ ^3P^\circ$	1—0
2966,8997	125	0,00	4,18	$a\ ^5D—y\ ^5F^\circ$	4—5
2965,2561	20	0,12	4,30	$a\ ^5D—y\ ^5F^\circ$	0—1
2959,9929	10	2,69	6,88	$a\ ^3G—v\ ^3H^\circ$	5—6
2959,682	5	2,81	7,00	$z\ ^7F^\circ—1$	6—5
2957,3660	30	0,11	4,30	$a\ ^5D—y\ ^5F^\circ$	1—1
2954,6543	5	2,28	6,48	$a\ ^3P—t\ ^3D^\circ$	2—3
2953,9411	50	0,09	4,28	$a\ ^5D—y\ ^5F^\circ$	2—2
2953,486	5	2,76	6,95	$a\ ^3G—s\ ^3G^\circ$	3—3
2950,243	20	2,17	6,37	$a\ ^5P—5^\circ$	3—3
2947,8773	60	0,05	4,26	$a\ ^5D—y\ ^5F^\circ$	3—3
2941,3438	15	0,09	4,30	$a\ ^5D—y\ ^5F^\circ$	2—1
2937,811	10	2,20	6,41	$a\ ^5P—7^\circ$	2—2
2936,9049	60	0,00	4,22	$a\ ^5D—y\ ^5F^\circ$	4—4
2929,121	6	3,31	7,54	$b\ ^3H—t\ ^3H^\circ$	4—4
2929,0085	25	0,05	4,28	$a\ ^5D—y\ ^5F^\circ$	3—2
2923,852	7	2,69	6,93	$a\ ^3G—s\ ^3G^\circ$	5—5

λ, Å	I	E_H, eV	E_B, eV	Transition	J
2920,6915	5	2,48	6,72	$a\ ^3P—t\ ^5P^\circ$	0—1
2918,023	10	3,24	7,48	$b\ ^3H—t\ ^3H^\circ$	6—6
2912,1589	20	0,00	4,26	$a\ ^5D—y\ ^5F^\circ$	4—3
2907,520	5	2,73	6,99	$a\ ^3G—u\ ^3H^\circ$	4—5
2901,915	5	2,40	6,67	$z\ ^7D^\circ—g\ ^7D$	5—5
2901,3820	5	1,56	5,82	$a\ ^3F—w\ ^3D^\circ$	3—3
2898,355	5	—	—	—	—
2895,036	8	1,56	5,84	$a\ ^3F—x\ ^3F^\circ$	3—3
2894,5055	10	2,28	6,56	$a\ ^3P—v\ ^3P^\circ$	2—2
2887,807	5	2,68	6,98	$a\ ^3G—u\ ^3H^\circ$	5—6
2877,3021	8	1,48	5,79	$a\ ^3F—u\ ^5D^\circ$	4—4
2875,3034	5	1,48	5,79	$a\ ^3F—u\ ^5D^\circ$	4—3
2874,173	10	0,00	4,31	$a\ ^5D—z\ ^5G^\circ$	4—5
2872,3346	7	0,96	5,27	$a\ ^5F—x\ ^5P^\circ$	3—3
2869,3083	10	0,05	4,37	$a\ ^5D—z\ ^5G^\circ$	3—4
2866,6264	7	0,99	5,31	$a\ ^5F—x\ ^5P^\circ$	2—2
2863,8644	8	0,09	4,42	$a\ ^5D—z\ ^5G^\circ$	2—3
2863,4311	8	1,48	5,81	$a\ ^3F—x\ ^3F^\circ$	4—4
2858,8970	5	0,11	4,45	$a\ ^5D—z\ ^5G^\circ$	1—2
2851,7979	15	1,01	5,35	$a\ ^5F—y\ ^5G^\circ$	1—2
2845,5959	8	0,95	5,31	$a\ ^5F—x\ ^5P^\circ$	3—2
2843,9775	20	0,99	5,35	$a\ ^5F—y\ ^5G^\circ$	2—3
2843,6314	10	0,91	5,27	$a\ ^5F—x\ ^5P^\circ$	4—3
2840,4229	6	0,05	4,42	$a\ ^5D—z\ ^5G^\circ$	3—3
2838,1205	10	0,99	5,35	$a\ ^5F—y\ ^5G^\circ$	2—2
2835,4574	6	0,00	4,37	$a\ ^5D—z\ ^5G^\circ$	4—4
2832,4364	25	0,96	5,33	$a\ ^5F—y\ ^5G^\circ$	3—4
2828,8094	7	0,99	5,37	$a\ ^5F—z\ ^5H^\circ$	2—3
2827,8931	5	0,05	4,43	$a\ ^5D—z\ ^3G^\circ$	3—4
2825,687	6	0,00	4,39	$a\ ^5D—z\ ^3G^\circ$	4—5
2825,557	20	0,96	5,34	$a\ ^5F—z\ ^5H^\circ$	3—4
2823,2767	20	0,96	5,35	$a\ ^5F—y\ ^5G^\circ$	3—3
2817,5047	6	0,96	5,35	$a\ ^5F—y\ ^5G^\circ$	3—2
2813,2877	30	0,91	5,32	$a\ ^5F—y\ ^5G^\circ$	4—5
2806,985	20	0,91	5,33	$a\ ^5F—z\ ^5H^\circ$	4—5
2804,5212	20	0,91	5,33	$a\ ^5F—y\ ^5G^\circ$	4—4
2797,7765	15	0,91	5,34	$a\ ^5F—z\ ^5H^\circ$	4—4
2788,105	30	0,86	5,30	$a\ ^5F—y\ ^5G^\circ$	5—6
2787,9331	5	1,49	5,93	$a\ ^3F—x\ ^3G^\circ$	4—5
2778,2214	20	0,86	5,32	$a\ ^5F—y\ ^5G^\circ$	5—5
2772,1107	20	0,09	4,56	$a\ ^5D—y\ ^5P^\circ$	2—3
2769,2985	6	2,41	6,88	$a\ ^3H—v\ ^3H^\circ$	6—6
2767,5232	20	0,91	5,39	$a\ ^5F—w\ ^5D^\circ$	4—4
2763,108	4	0,99	5,47	$a\ ^5F—w\ ^5F^\circ$	2—3
2762,0275	15	0,96	5,44	$a\ ^5F—w\ ^5D^\circ$	3—3
2761,785	18	2,42	6,90	$a\ ^3P—w\ ^1D^\circ$	1—2
2759,817	5	1,01	5,50	$a\ ^5F—w\ ^5F^\circ$	1—1
2757,3170	10	1,01	5,50	$a\ ^5F—w\ ^5D^\circ$	1—1
2756,3295	20	0,11	4,61	$a\ ^5D—y\ ^5P^\circ$	1—2
2756,2677	20	0,05	4,55	$a\ ^5D—y\ ^3F^\circ$	3—4
2750,878	5	2,17	6,68	$a\ ^5P—10^\circ$	3—3
2750,1415	25	0,05	4,56	$a\ ^5D—y\ ^5P^\circ$	3—3
2746,982	20	0,87	5,38	$a\ ^5F—z\ ^5H^\circ$	5—6
2744,5287	8	0,99	5,50	$a\ ^5F—w\ ^5D^\circ$	2—1
2744,0691	10	0,12	4,64	$a\ ^5D—y\ ^5P^\circ$	0—1
2742,4064	30	0,09	4,61	$a\ ^5D—y\ ^5P^\circ$	2—2
2742,2554	20	0,95	5,47	$a\ ^5F—w\ ^5D^\circ$	3—2
2737,3108	20	0,11	4,64	$a\ ^5D—y\ ^5P^\circ$	1—1

λ, Å	I	E_H, eV	E_B, eV	Transition	J
2735,614	8	2,20	6,73	$a\,^5P—t\,^5P°$	2—1
2735,4762	8	0,91	5,44	$a\,^5F—w\,^5D°$	4—3
2733,5816	15	0,86	5,39	$a\,^5F—w\,^5D°$	5—4
2728,0212	5	0,91	5,45	$a\,^5F—w\,^5F°$	4—4
2726,054	6	1,01	5,58	$a\,^5F—v\,^5D°$	1—0
2724,9542	10	0,95	5,50	$a\,^5F—v\,^5D°$	3—4
2723,5786	15	0,09	4,64	$a\,^5D—y\,^5P°$	2—1
2720,9035	40	0,05	4,61	$a\,^5D—y\,^5P°$	3—2
2719,025	60	{ 0,00	4,56	$a\,^5D—y\,^5P°$	4—3
		2,59	7,15	$b\,^3F—t\,^3F°$	3—3
2718,4365	6	0,99	5,55	$a\,^5F—v\,^5D°$	2—1
2711,6560	4	0,91	5,48	$a\,^5F—w\,^5F°$	4—5
2707,507	20	—	—	—	—
2706,5829	8	0,96	5,53	$a\,^5F—v\,^5D°$	3—2
2699,1075	6	0,91	5,50	$a\,^5F—v\,^5D°$	4—4
2696,284	5	2,41	7,00	$z\,^7D°—1$	5—5
2694,538	5	2,39	6,99	$z\,^7D°—2$	5—4
2689,2130	8	0,91	5,52	$a\,^5F—v\,^5D°$	4—3
2679,0626	10	0,86	5,48	$a\,^5F—w\,^5F°$	5—5
2666,8133	8	0,86	5,50	$a\,^5F—v\,^5D°$	5—4
2647,5588	5	0,05	4,73	$a\,^5D—y\,^3D°$	3—3
2644,000	8	1,01	5,70	$a\,^5F—x\,^5G°$	1—2
2641,6468	4	0,91	5,60	$a\,^5F—x\,^3D°$	4—3
2635,8100	8	0,99	5,69	$a\,^5F—x\,^5G°$	2—3
2632,2382	4	0,99	5,70	$a\,^5F—x\,^5G°$	2—2
2629,587	5	0,12	4,83	$a\,^5D—y\,^3D°$	0—1
2623,532	5	0,96	5,68	$a\,^5F—x\,^5G°$	3—4
2618,0191	5	0,96	5,69	$a\,^5F—x\,^5G°$	3—3
2606,8286	6	0,91	5,66	$a\,^5F—x\,^5G°$	4—5
2605,6578	6	0,86	5,61	$a\,^5F—y\,^3G°$	5—5
2599,570	6	0,91	5,68	$a\,^5F—x\,^5G°$	4—4
2587,999	8	—	—	—	—
2584,5370	8	0,86	5,65	$a\,^5F—x\,^5G°$	5—6
2582,299	6	—	—	—	—
2569,601	6	0,86	5,68	$a\,^5F—x\,^5G°$	5—4
2553,193	7	—	—	—	—
2551,094	8	—	—	—	—
2549,6142	10	0,05	4,91	$a\,^5D—x\,^5D°$	3—4
2545,9795	10	0,09	4,95	$a\,^5D—x\,^5D°$	2—3
2543,920	6	2,60	7,46	$b\,^3F—r\,^3G°$	3—4
2542,101	6	2,61	7,48	$b\,^3F—r\,^3G°$	2—3
2540,9734	10	0,11	4,99	$a\,^5D—x\,^5D°$	1—2
2539,3576	7	0,91	5,79	$a\,^5F—u\,^5D°$	4—3
2535,6086	8	0,12	5,01	$a\,^5D—x\,^5D°$	0—1
2529,1361	10	0,09	4,99	$a\,^5D—x\,^5D°$	2—2
2527,4358	15	0,05	4,95	$a\,^5D—x\,^5D°$	3—3
2527,16	5	—	—	—	—
2525,022	7	—	—	—	—
2524,2939	8	0,11	5,02	$a\,^5D—x\,^5D°$	1—0
2523,658	6	—	—	—	—
2523,11	5	—	—	—	—
2522,8505	40	0,00	4,91	$a\,^5D—x\,^5D°$	4—4
2522,488	6	0,91	5,82	$a\,^5F—z\,^3H°$	4—5
2521,9197	7	0,91	5,83	$a\,^5F—w\,^3D°$	4—3
2519,6305	10	1,01	5,92	$a\,^5F—w\,^5G°$	1—2
2518,1029	12	0,09	5,01	$a\,^5D—x\,^5D°$	2—1
2517,659	8	0,99	5,91	$a\,^5F—w\,^5G°$	2—3
2512,363	5	0,05	4,98	$a\,^5D—y\,^7P°$	3—3

λ, Å	I	E_H, eV	E_B, eV	Transition	J
2510,8362	15	0,05	4,99	$a\,^5D—x\,^5D°$	3—2
2507,899	6	0,95	5,89	$a\,^5F—w\,^5G°$	3—4
2505,485	5	—	—	—	—
2501,1332	20	0,00	4,95	$a\,^5D—x\,^5D°$	4—3
2498,895	10	0,05	5,01	$a\,^5D—y\,^7P°$	3—4
2496,5343	6	0,91	5,87	$a\,^5F—w\,^5G°$	4—5
2493,999	6	0,95	5,92	$a\,^5F—x\,^3G°$	3—4
2491,984	8	2,56	7,53	$b\,^3F—t\,^3H°$	4—4
2491,1562	20	0,11	5,08	$a\,^5D—x\,^5F°$	1—2
2490,6454	30	0,09	5,06	$a\,^5D—x\,^5F°$	2—3
2489,751	15	0,12	5,10	$a\,^5D—x\,^5F°$	0—1
2488,950	6	2,56	7,54	$b\,^3F—q\,^3G°$	4—3
2488,1437	40	0,05	5,03	$a\,^5D—x\,^5F°$	3—4
2487,064	12	1,01	5,99	$a\,^5F—v\,^5F°$	1—1
2486,693	10	0,95	5,94	$a\,^5F—v\,^5F°$	3—4
2486,371	10	0,00	4,96	$a\,^5D—y\,^7P°$	4—3
2485,985	10	0,91	5,89	$a\,^5F—w\,^5G°$	4—4
2484,187	15	0,11	5,10	$a\,^5D—x\,^5F°$	1—1
2483,531	10	0,99	5,98	$a\,^5F—v\,^5F°$	2—2
2483,2718	60	0,00	4,99	$a\,^5D—x\,^5F°$	4—5
2479,7774	20	0,09	5,08	$a\,^5D—x\,^5F°$	2—2
2479,4813	6	0,99	5,99	$a\,^5F—x\,^3P°$	2—2
2474,8151	8	0,95	5,96	$a\,^5F—v\,^5F°$	3—3
2472,909	12	0,05	5,06	$a\,^5D—x\,^5F°$	3—3
2472,8962	5	0,09	5,10	$a\,^5D—x\,^5F°$	2—1
2465,150	6	0,92	5,94	$a\,^5F—v\,^5F°$	4—4
2463,728	6	0,96	5,99	$a\,^5F—x\,^3P°$	3—2
2462,6483	10	0,00	5,03	$a\,^5D—x\,^5F°$	4—4
2457,5980	6	0,86	5,90	$a\,^5F—v\,^5F°$	5—5
2454,706	6	2,58	7,43	$b\,^3F—r\,^3G°$	4—5
2445,210	6	0,87	5,92	$a\,^5F—x\,^3G°$	5—4
2443,8728	20	0,87	5,93	$a\,^5F—x\,^3G°$	5—5
2442,567	20	2,43	7,50	$a\,^3H—t\,^3H°$	5—5
2440,107	15	2,46	7,53	$a\,^3H—t\,^3H°$	4—4
2439,744	25	2,41	7,48	$a\,^3H—t\,^3H°$	6—6
2436,344	10	—	—	—	—
2431,025	20	—	—	—	—
2389,9732	25	0,09	5,27	$a\,^5D—x\,^5P°$	2—3
2374,5192	10	0,12	5,34	$a\,^5D—x\,^5P°$	0—1
2373,618	20	0,05	5,27	$a\,^5D—x\,^5P°$	3—3
2371,4313	15	0,09	5,31	$a\,^5D—x\,^5P°$	2—2
2369,4567	8	0,11	5,34	$a\,^5D—x\,^5P°$	1—1
2350,408	5	0,00	5,27	$a\,^5D—x\,^5P°$	4—3
2320,3585	40	0,05	5,39	$a\,^5D—w\,^5D°$	3—4
2313,1048	40	0,09	5,44	$a\,^5D—w\,^5D°$	2—3
2308,9999	30	0,11	5,47	$a\,^5D—w\,^5D°$	1—2
2304,627	5	0,99	6,37	$a\,^5F—t\,^5D°$	2—3
2303,5815	20	0,11	5,49	$a\,^5D—w\,^5F°$	1—2
2303,422	15	0,12	5,50	$a\,^5D—w\,^5F°$	0—1
2301,6849	20	0,12	5,50	$a\,^5D—w\,^5D°$	0—1
2301,173	6	—	—	—	—
2300,139	30	0,09	5,47	$a\,^5D—w\,^5F°$	2—3
2299,2209	25	0,09	5,47	$a\,^5D—w\,^5D°$	2—2
2298,662	6	0,11	5,50	$a\,^5D—w\,^5F°$	1—1
2298,1699	10	0,00	5,39	$a\,^5D—w\,^5D°$	4—4
2297,7877	35	0,05	5,44	$a\,^5D—w\,^5D°$	3—3
2296,9279	15	0,11	5,50	$a\,^5D—w\,^5D°$	1—1
2294,406	25	0,11	5,51	$a\,^5D—w\,^5D°$	1—0

λ, Å	I	E_H, eV	E_B, eV	Transition	J
2293,8482	25	0,09	5,49	$a\,^5D—w\,^5F°$	2—2
2292,5249	30	0,05	5,45	$a\,^5D—w\,^5F°$	3—4
2291,120	15	0,99	6,40	$a\,^5F—u\,^5F°$	2—3
2290,545	9	0,99	6,40	$a\,^5F—t\,^5D°$	2—2
2289,031	10	1,01	6,43	$a\,^5F—u\,^5F°$	1—2
2287,630	15	0,91	6,33	$a\,^5F—t\,^5D°$	4—4
2287,2505	30	0,09	5,50	$a\,^5D—w\,^5D°$	2—1
2284,0864	40	0,05	5,47	$a\,^5D—w\,^5D°$	3—2
2283,6557	12	0,11	5,53	$a\,^5D—v\,^5D°$	1—2
2283,3045	9	0,12	5,55	$a\,^5D—v\,^5D°$	0—1
2283,079	9	1,01	6,44	$a\,^5F—t\,^5D°$	1—0
2280,222	8	0,99	6,43	$a\,^5F—u\,^5F°$	2—2
2279,924	10	0,09	5,52	$a\,^5D—v\,^5D°$	2—3
2277,672	12	0,95	6,40	$a\,^5F—u\,^5F°$	3—3
2277,096	9	0,95	6,40	$a\,^5F—t\,^5D°$	3—2
2276,0263	12	0,00	5,44	$a\,^5D—w\,^5D°$	4—3
2275,187	6	0,11	5,55	$a\,^5D—v\,^5D°$	1—0
2274,088	9	0,99	6,44	$a\,^5F—u\,^5F°$	2—1
2272,816	8	0,91	6,37	$a\,^5F—t\,^5D°$	4—3
2272,0703	15	0,05	5,50	$a\,^5D—v\,^5D°$	3—4
2271,778	40	0,91	6,37	$a\,^5F—u\,^5F°$	4—4
2270,8628	18	0,00	5,48	$a\,^5D—w\,^5F°$	4—4
2269,0990	18	0,09	5,55	$a\,^5D—v\,^5D°$	2—1
2267,466	15	0,86	6,32	$a\,^5F—u\,^5F°$	5—5
2267,080	9	0,05	5,51	$a\,^5D—y\,^5S°$	3—2
2266,903	10	0,95	6,43	$a\,^5F—u\,^5F°$	3—2
2265,0546	20	0,05	5,52	$a\,^5D—v\,^5D°$	3—3
2264,390	45	0,86	6,34	$a\,^5F—t\,^5D°$	5—4
2263,474	6	0,00	5,47	$a\,^5D—w\,^5F°$	4—3
2259,5109	15	0,00	5,48	$a\,^5D—w\,^5F°$	4—5
2255,859	45	0,91	6,41	$a\,^5F—u\,^5P°$	4—3
2251,8749	12	0,11	5,61	$a\,^5D—x\,^3D°$	1—2
2250,7911	10	0,00	5,50	$a\,^5D—v\,^5D°$	4—4
2248,857	25	0,86	6,37	$a\,^5F—u\,^5F°$	5—4
2245,6536	15	0,09	5,60	$a\,^5D—x\,^3D°$	2—3
2242,579	15	0,09	5,61	$a\,^5D—x\,^3D°$	2—2
2231,2138	15	0,05	5,60	$a\,^5D—x\,^3D°$	3—3
2229,0735	5	0,09	5,65	$a\,^5D—x\,^3D°$	2—1
2228,1722	10	0,05	5,61	$a\,^5D—x\,^3D°$	3—2
2222,763	7	1,49	7,06	$a\,^3F—v\,^1G°$	4—4
2211,2364	7	0,09	5,69	$a\,^5D—x\,^5G°$	2—3
2210,6894	9	0,00	5,60	$a\,^5D—x\,^3D°$	4—3
2207,0692	6	0,00	5,61	$a\,^5D—y\,^3G°$	4—5
2200,722	15	0,11	5,74	$a\,^5D—w\,^5P°$	1—2
2200,370	10	0,12	5,75	$a\,^5D—w\,^5P°$	0—1
2196,0428	50	0,11	5,75	$a\,^5D—w\,^5P°$	1—1
2191,838	60	0,09	5,74	$a\,^5D—w\,^5P°$	2—2
2191,2052	10	0,12	5,78	$a\,^5D—z\,^3S°$	0—1
2187,1950	40	0,09	5,75	$a\,^5D—w\,^5P°$	2—1
2186,485	40	0,05	5,72	$a\,^5D—w\,^5P°$	3—3
2178,090	35	0,05	5,74	$a\,^5D—w\,^5P°$	3—2
2176,8414	6	0,12	5,81	$a\,^5D—y\,^3P°$	0—1
2173,2146	8	0,11	5,81	$a\,^5D—u\,^5D°$	1—2
2172,581	6	0,11	5,81	$a\,^5D—y\,^3P°$	1—1
2171,2976	40	0,09	5,79	$a\,^5D—u\,^5D°$	2—3
2166,773	100	0,00	5,72	$a\,^5D—w\,^5P°$	4—3
2165,860	20	—	—	—	—

λ, Å	I	E_H, eV	E_B, eV	Transition	J
2164,547	7	0,09	5,81	$a\,^5D—u\,^5D°$	2—2
2163,8633	6	0,12	5,85	$a\,^5D—u\,^5D°$	0—1
2163,368	10	—	—	—	—
2161,5802	5	0,11	5,84	$a\,^5D—w\,^3D°$	1—2
2158,49	6	—	—	—	—
2157,795	5	0,05	5,79	$a\,^5D—u\,^5D°$	3—3
2153,0075	5	0,09	5,84	$a\,^5D—w\,^3D°$	2—2
2119,125	5	0,05	5,89	$a\,^5D—w\,^5G°$	3—4
2115,1697	20	0,09	5,94	$a\,^5D—v\,^5P°$	2—3
2114,488	25	0,11	5,97	$a\,^5D—v\,^5P°$	1—2
2113,08	20	0,86	6,73	$a\,^5F—t\,^3G°$	5—4
2112,966	25	0,12	5,98	$a\,^5D—v\,^5P°$	0—1
2111,274	20	—	—	—	—
2110,233	30	0,12	5,99	$a\,^5D—v\,^5F°$	0—1
2109,861	25	—	—	—	—
2108,959	30	0,11	5,98	$a\,^5D—v\,^5P°$	1—1
2108,302	12	0,11	5,98	$a\,^5D—x\,^3P°$	1—2
2108,139	12	0,00	5,87	$a\,^5D—w\,^5G°$	4—5
2106,380	25	0,09	5,97	$a\,^5D—v\,^5P°$	2—2
2106,260	20	0,11	5,99	$a\,^5D—v\,^5F°$	1—1
2103,048	25	0,09	5,98	$a\,^5D—v\,^5F°$	2—2
2102,910	20	0,12	6,01	$a\,^5D—x\,^3P°$	0—1
2102,3542	30	0,05	5,94	$a\,^5D—v\,^5P°$	3—3
2100,7984	30	0,09	5,98	$a\,^5D—v\,^5P°$	2—1
2100,144	10	0,09	5,98	$a\,^5D—x\,^3P°$	2—2
2098,953	25	0,11	6,01	$a\,^5D—x\,^3P°$	1—1
2098,081	15	0,09	5,99	$a\,^5D—v\,^5F°$	2—1
2093,660	40	0,05	5,97	$a\,^5D—v\,^5P°$	3—2
2090,862	20	0,09	6,01	$a\,^5D—x\,^3P°$	2—1
2090,380	30	0,05	5,98	$a\,^5D—v\,^5F°$	3—2
2087,525	25	0,05	5,98	$a\,^5D—x\,^3P°$	3—2
2084,117	50	0,00	5,94	$a\,^5D—v\,^5P°$	4—3
2041,204	25	—	—	—	—
2017,090	15	—	—	—	—
2016,512	5	0,91	7,06	$a\,^5F—v\,^1G°$	4—4
2007,215	15	—	—	—	—
2006,260	15	—	—	—	—
1980,129	25	0,00	6,32	$a\,^5D—u\,^5F°$	4—5
1964,043	20	0,09	6,40	$a\,^5D—u\,^5F°$	2—3
1963,629	15	0,09	6,40	$a\,^5D—t\,^5D°$	2—2
1963,110	25	0,11	6,43	$a\,^5D—u\,^5F°$	1—2
1962,871	20	0,05	6,37	$a\,^5D—t\,^5D°$	3—3
1962,746	15	0,11	6,43	$a\,^5D—t\,^5D°$	1—1
1962,100	30	0,05	6,37	$a\,^5D—u\,^5F°$	3—4
1962,031	25	0,12	6,41	$a\,^5D—u\,^5F°$	0—1
1961,236	20	0,09	6,41	$a\,^5D—u\,^5P°$	2—3
1960,129	30	0,00	6,32	$a\,^5D—u\,^5F°$	4—5
1958,739	15	0,11	6,44	$a\,^5D—t\,^5D°$	1—0
1958,598	30	0,11	6,44	$a\,^5D—u\,^5F°$	1—1
1957,831	25	0,00	6,34	$a\,^5D—t\,^5D°$	4—4
1956,026	30	0,09	6,43	$a\,^5D—u\,^5F°$	2—2
1955,690	20	0,09	6,43	$a\,^5D—t\,^5D°$	2—1
1952,997	20	0,05	6,40	$a\,^5D—u\,^5F°$	3—3
1952,596	30	0,05	6,40	$a\,^5D—t\,^5D°$	3—2
1952,262	20	0,11	6,46	$a\,^5D—u\,^5P°$	1—1
1951,556	25	0,09	6,44	$a\,^5D—u\,^5F°$	2—1
1950,223	20	0,05	6,41	$a\,^5D—u\,^5P°$	3—3
1946,978	25	0,00	6,37	$a\,^5D—t\,^5D°$	4—3

λ, Å	I	E_H, eV	E_B, eV	Transition	J
1946,219	10	0,00	6,37	$a\,^5D-u\,^5F^\circ$	4—4
1945,294	25	0,09	6,46	$a\,^5D-u\,^5P^\circ$	2—1
1945,070	20	0,05	6,43	$a\,^5D-u\,^5F^\circ$	3—2
1940,649	25	0,05	6,44	$a\,^5D-u\,^5P^\circ$	3—2
1937,274	35	0,00	6,40	$a\,^5D-u\,^5F^\circ$	4—3
1934,528	25	0,00	6,41	$a\,^5D-u\,^5P^\circ$	4—3
1888,32	12	0,09	6,65	$a\,^5D-y\,^1F^\circ$	2—3
1887,761	14	0,05	6,62	$a\,^5D-t\,^5P^\circ$	3—3
1880,14	5	0,09	6,68	$a\,^5D-10^\circ$	2—3
1876,421	10	0,12	6,73	$a\,^5D-t\,^5P^\circ$	0—1
1873,259	15	0,11	6,73	$a\,^5D-t\,^5P^\circ$	1—1
1873,052	12	0,00	6,62	$a\,^5D-t\,^5P^\circ$	4—3
1872,359	15	0,09	6,71	$a\,^5D-t\,^5P^\circ$	2—2
1866,815	10	0,09	6,73	$a\,^5D-t\,^5P^\circ$	2—1
1866,07	12	0,05	6,70	$a\,^5D-11^\circ$	3—3
1862,318	15	0,05	6,71	$a\,^5D-t\,^5P^\circ$	3—2
1855,58	15	0,00	6,68	$a\,^5D-10^\circ$	4—3

Fe II, ground state $1s^2\,2s^2\,2p^6\,3s^2\,3p^6\,3d^6\,4s\,^6D_{9/2}$
Ionization potential 130 524 cm^{-1}; 6,187 eV

λ, Å	I	E_H, eV	E_B, eV	Transition	J
7711,73	15	3,90	5,51	$b\,^4D-z\,^4D^\circ$	$^7/_2$—$^7/_2$
7515,88	6	3,90	5,55	$b\,^4D-z\,^4D^\circ$	$^7/_2$—$^5/_2$
7462,38	20	3,88	5,55	$b\,^4D-z\,^4D^\circ$	$^5/_2$—$^5/_2$
7445,34	6	3,88	5,55	$b\,^4D-z\,^4D^\circ$	$^3/_2$—$^5/_2$
7376,46	20	—	—	—	—
7334,66	8	7,26	8,95	$d\,^2G-x\,^2H^\circ$	$^9/_2$—$^{11}/_2$
7320,70	40	3,88	5,58	$b\,^4D-z\,^4D^\circ$	$^5/_2$—$^3/_2$
7310,24	6	3,88	5,58	$b\,^4D-z\,^4D^\circ$	$^1/_2$—$^3/_2$
7307,957	50	3,88	5,58	$b\,^4D-z\,^4D^\circ$	$^3/_2$—$^3/_2$
7287,36	6	6,22	7,92	$c\,^4F-y\,^4G^\circ$	$^9/_2$—$^{11}/_2$
7264,99	10	6,22	7,92	$c\,^4F-y\,^4G^\circ$	$^7/_2$ $^9/_2$
7224,51	12	3,88	5,60	$b\,^4D-z\,^4D^\circ$	$^1/_2$—$^1/_2$
7222,39	8	3,88	5,60	$b\,^4D-z\,^4D^\circ$	$^3/_2$—$^1/_2$
7193,23	8	6,21	7,94	$c\,^4F-y\,^4G^\circ$	$^5/_2$—$^7/_2$
7134,99	5	6,20	7,94	$c\,^4F-y\,^4G^\circ$	$^3/_2$—$^5/_2$
7067,44	20	—	—	—	—
6627,28	5	7,26	9,14	$d\,^2G-w\,^2H^\circ$	$^7/_2$—$^9/_2$
6588,69	5	—	—	—	—
6517,01	5	—	—	—	—
6516,053	20	2,89	4,79	$a\,^6S-z\,^6D^\circ$	$^5/_2$—$^7/_2$
6506,33	5	—	—	—	—
6493,05	8	—	—	—	—
6456,376	200	3,90	5,82	$b\,^4D-z\,^4P^\circ$	$^7/_2$—$^5/_2$
6446,43	20	6,21	8,14	$c\,^4F-x\,^4G^\circ$	$^7/_2$—$^9/_2$
6442,93	6	—	—	—	—
6432,654	8	2,89	4,82	$a\,^6S-z\,^6D^\circ$	$^5/_2$—$^5/_2$
6416,905	20	3,88	5,82	$b\,^4D-z\,^4P^\circ$	$^5/_2$—$^5/_2$
6385,473	5	—	—	—	—
6383,753	15	—	—	—	—
6331,969	12	6,21	8,16	$c\,^4F-x\,^4G^\circ$	$^5/_2$—$^7/_2$
6305,318	15	6,22	8,17	$c\,^4F-x\,^4F^\circ$	$^9/_2$—$^9/_2$
6247,562	80	3,88	5,87	$b\,^4D-z\,^4P^\circ$	$^5/_2$—$^3/_2$

Å, λ	I	E_H, eV	E_B, eV	Transition	J
6236,375	20	3,88	5,87	$b\,^4D—z\,^4P^\circ$	$^3/_2—^3/_2$
6179,378	5	5,56	7,56	$c\,^2F—z\,^2D^\circ$	$^7/_2—^5/_2$
6175,158	15	6,22	8,21	$c\,^4F—x\,^4F^\circ$	$^7/_2—^7/_2$
6149,238	20	3,88	5,90	$b\,^4D—z\,^4P^\circ$	$^1/_2—^1/_2$
6147,735	30	3,88	5,90	$b\,^4D—z\,^4P^\circ$	$^3/_2—^1/_2$
6103,54	8	6,21	8,23	$c\,^4F—x\,^4F^\circ$	$^5/_2—^5/_2$
6084,11	5	3,20	5,22	$a\,^4G—z\,^6F^\circ$	$^9/_2—^7/_2$
6045,497	6	6,21	8,24	$c\,^4F—x\,^4F^\circ$	$^3/_2—^3/_2$
5991,383	10	3,15	5,22	$a\,^4G—z\,^6F^\circ$	$^{11}/_2—^9/_2$
5962,4	30	—	—	—	—
5903,6	8	—	—	—	—
5891,36	8	7,27	9,37	$d\,^2G—w\,^2F^\circ$	$^9/_2—^7/_2$
5567,815	10	—	—	—	—
5466,94	20	—	—	—	—
5427,832	30	—	—	—	—
5362,864	5	3,20	5,51	$a\,^4G—z\,^4D^\circ$	$^9/_2—^7/_2$
5316,609	8	3,15	5,48	$a\,^4G—z\,^4F^\circ$	$^{11}/_2—^9/_2$
5284,092	5	2,89	5,23	$a\,^6S—z\,^6F^\circ$	$^5/_2—^7/_2$
5275,994	7	3,20	5,54	$a\,^4G—z\,^4F^\circ$	$^9/_2—^7/_2$
5234,620	7	3,22	5,59	$a\,^4G—z\,^4F^\circ$	$^7/_2—^5/_2$
5197,569	6	3,23	5,61	$a\,^4G—z\,^4F^\circ$	$^5/_2—^3/_2$
5169,030	12	2,89	5,29	$a\,^6S—z\,^6P^\circ$	$^5/_2—^7/_2$
5156,10	6	—	—	—	—
5136,795	6	2,84	5,25	$b\,^4F—z\,^6F^\circ$	$^5/_2—^3/_2$
5100,95	15	—	—	—	—
5018,434	12	2,89	5,36	$a\,^6S—z\,^6P^\circ$	$^5/_2—^5/_2$
4923,916	12	2,89	5,41	$a\,^6S—z\,^6P^\circ$	$^5/_2—^3/_2$
4648,933	10	2,58	5,25	$b\,^4P—z\,^6F^\circ$	$^5/_2—^5/_2$
4629,336	7	2,80	5,48	$b\,^4F—z\,^4F^\circ$	$^9/_2—^9/_2$
4583,848	11	2,80	5,51	$b\,^4F—z\,^4D^\circ$	$^9/_2—^7/_2$
4555,890	8	2,83	5,54	$b\,^4F—z\,^4F^\circ$	$^7/_2—^7/_2$
4549,470	10	2,83	5,57	$b\,^4F—z\,^4D^\circ$	$^7/_2—^5/_2$
4522,634	9	2,84	5,60	$b\,^4F—z\,^4D^\circ$	$^5/_2—^3/_2$
4520,225	7	2,80	5,54	$b\,^4F—z\,^4F^\circ$	$^9/_2—^7/_2$
4515,337	7	2,84	5,59	$b\,^4F—z\,^4F^\circ$	$^5/_2—^5/_2$
4508,283	8	2,85	5,60	$b\,^4F—z\,^4D^\circ$	$^3/_2—^1/_2$
4491,401	5	2,85	5,61	$b\,^4F—z\,^4F^\circ$	$^3/_2—^3/_2$
4416,817	7	2,78	5,58	$b\,^4P—z\,^4D^\circ$	$^1/_2—^3/_2$
4385,381	7	2,78	5,60	$b\,^4P—z\,^4D^\circ$	$^1/_2—^1/_2$
4351,764	9	2,70	5,55	$b\,^4P—z\,^4D^\circ$	$^3/_2—^5/_2$
4303,166	8	2,70	5,58	$b\,^4P—z\,^4D^\circ$	$^3/_2—^3/_2$
4296,567	6	2,70	5,59	$b\,^4P—z\,^4F^\circ$	$^3/_2—^5/_2$
4233,168	11	2,58	5,51	$b\,^4P—z\,^4D^\circ$	$^5/_2—^7/_2$
4178,855	8	2,58	5,54	$b\,^4P—z\,^4F^\circ$	$^5/_2—^7/_2$
4173,450	8	2,58	5,55	$b\,^4P—z\,^4D^\circ$	$^5/_2—^5/_2$
4066,328	12	7,70	10,74	$z\,^2D^\circ—e\,^4F$	$^3/_2—^3/_2$
4024,552	5	4,50	7,57	$b\,^2D—z\,^2D^\circ$	$^5/_2—^5/_2$
3759,460	6	4,74	8,03	$c\,^2D—z\,^2P^\circ$	$^3/_2—^1/_2$
3748,489	8	4,73	8,03	$c\,^2D—z\,^2P^\circ$	$^5/_2—^3/_2$
3624,890	5	4,62	8,03	$a\,^2S—z\,^2P^\circ$	$^1/_2—^1/_2$
3621,273	6	4,62	8,03	$a\,^2S—z\,^2P^\circ$	$^1/_2—^3/_2$
3614,873	5	4,16	7,58	$c\,^2G—z\,^4H^\circ$	$^7/_2—^7/_2$
3494,672	5	2,28	5,82	$a\,^2P—z\,^4P^\circ$	$^3/_2—^5/_2$
3493,474	10	4,15	7,69	$c\,^2G—z\,^2G^\circ$	$^9/_2—^9/_2$
3468,680	8	4,16	7,72	$c\,^2G—z\,^2G^\circ$	$^7/_2—^7/_2$
3456,928	5	3,90	7,48	$b\,^4D—y\,^4P^\circ$	$^7/_2—^5/_2$
3436,112	5	3,96	7,56	$b\,^2F—z\,^2D^\circ$	$^7/_2—^5/_2$
3416,021	5	2,28	5,90	$a\,^2P—z\,^4P^\circ$	$^3/_2—^1/_2$

λ, Å	I	E_H, eV	E_B, eV	Transition	J
3388,134	12	3,90	7,56	$b\,^4D—z\,^4H^\circ$	$^7/_2—^9/_2$
3323,068	8	3,97	7,70	$b\,^2F—z\,^4G^\circ$	$^7/_2—^9/_2$
3297,888	5	3,94	7,70	$b\,^2F—z\,^2D^\circ$	$^5/_2—^3/_2$
3295,814	6	1,07	4,84	$a\,^4D—z\,^6D^\circ$	$^3/_2—^3/_2$
3289,347	7	3,81	7,58	$b\,^2G—z\,^4H^\circ$	$^7/_2—^7/_2$
3281,300	7	1,04	4,82	$a\,^4D—z\,^6D^\circ$	$^5/_2—^5/_2$
3277,346	9	0,98	4,77	$a\,^4D—z\,^6D^\circ$	$^7/_2—^9/_2$
3276,606	5	3,94	7,72	$b\,^2F—z\,^2G^\circ$	$^5/_2—^7/_2$
3259,048	10	3,90	7,69	$b\,^4D—y\,^4F^\circ$	$^7/_2—^9/_2$
3258,773	10	3,89	7,69	$b\,^4D—y\,^4F^\circ$	$^5/_2—^7/_2$
3255,890	8	0,98	4,79	$a\,^4D—z\,^6D^\circ$	$^7/_2—^7/_2$
3252,437	5	3,90	7,71	$b\,^4D—z\,^2D^\circ$	$^5/_2—^3/_2$
3247,213	9	3,88	7,70	$b\,^4D—y\,^4F^\circ$	$^3/_2—^5/_2$
3243,724	8	4,15	7,97	$c\,^2G—z\,^2F^\circ$	$^9/_2—^7/_2$
3237,819	8	3,89	7,71	$b\,^4D—y\,^4F^\circ$	$^1/_2—^3/_2$
3237,402	5	3,88	7,71	$b\,^4D—y\,^4F^\circ$	$^3/_2—^3/_2$
3232,791	7	4,16	7,98	$c\,^2G—z\,^2F^\circ$	$^7/_2—^5/_2$
3231,702	5	3,88	7,72	$b\,^4D—z\,^2G^\circ$	$^5/_2—^7/_2$
3227,747	13	1,67	5,51	$a\,^4P—z\,^4D^\circ$	$^5/_2—^7/_2$
3213,314	13	1,70	5,55	$a\,^4P—z\,^4D^\circ$	$^3/_2—^5/_2$
3210,451	10	1,73	5,58	$a\,^4P—z\,^4D^\circ$	$^1/_2—^3/_2$
3196,076	10	1,67	5,54	$a\,^4P—z\,^4F^\circ$	$^5/_2—^7/_2$
3193,809	11	1,73	5,60	$a\,^4P—z\,^4D^\circ$	$^1/_2—^1/_2$
3192,926	9	1,67	5,55	$a\,^4P—z\,^4D^\circ$	$^5/_2—^5/_2$
3187,293	8	4,15	8,03	$c\,^2G—y\,^2G^\circ$	$^9/_2—^9/_2$
3186,741	11	1,70	5,58	$a\,^4P—z\,^4D^\circ$	$^3/_2—^3/_2$
3185,315	5	1,72	5,61	$a\,^4P—z\,^4F^\circ$	$^1/_2—^3/_2$
3183,108	8	1,70	5,59	$a\,^4P—z\,^4F^\circ$	$^3/_2—^5/_2$
3180,164	7	4,74	8,63	$c\,^2D—y\,^2F^\circ$	$^3/_2—^5/_2$
3179,504	8	4,73	8,62	$c\,^2D—y\,^2F^\circ$	$^5/_2—^7/_2$
3177,535	10	3,91	7,80	$b\,^4D—x\,^4D^\circ$	$^7/_2—^7/_2$
3170,346	6	1,70	5,60	$a\,^4P—z\,^4D^\circ$	$^3/_2—^1/_2$
3167,859	11	3,82	7,72	$b\,^2G—z\,^2G^\circ$	$^7/_2—^7/_2$
3163,091	5	1,67	5,59	$a\,^4P—z\,^4F^\circ$	$^5/_2—^5/_2$
3162,800	8	4,16	8,07	$c\,^2G—y\,^2G^\circ$	$^7/_2—^7/_2$
3161,949	5	1,70	5,61	$a\,^4P—z\,^4F^\circ$	$^3/_2—^3/_2$
3154,206	12	3,77	7,69	$b\,^2G—z\,^2G^\circ$	$^9/_2—^9/_2$
3144,758	5	3,91	7,84	$b\,^4D—x\,^4D^\circ$	$^7/_2—^5/_2$
3135,364	9	3,89	7,86	$b\,^4D—x\,^4D^\circ$	$^5/_2—^5/_2$
3116,590	6	3,89	7,86	$b\,^4D—x\,^4D^\circ$	$^5/_2—^3/_2$
3114,293	7	3,89	7,86	$b\,^4D—x\,^4D^\circ$	$^3/_2—^3/_2$
3105,548	5	3,89	7,88	$b\,^4D—x\,^4D^\circ$	$^1/_2—^1/_2$
3105,168	5	3,89	7,88	$b\,^4D—x\,^4D^\circ$	$^3/_2—^1/_2$
3096,296	5	3,96	7,96	$b\,^2F—z\,^2F^\circ$	$^7/_2—^7/_2$
3078,698	8	5,82	9,85	$z\,^4P^\circ—e\,^4D$	$^5/_2—^7/_2$
3077,168	10	4,08	8,10	$a\,^2I—z\,^2H^\circ$	$^{13}/_2—^{11}/_2$
3076,455	6	5,87	9,90	$z\,^4P^\circ—e\,^4D$	$^3/_2—^5/_2$
3065,315	6	3,95	7,98	$b\,^2F—z\,^2F^\circ$	$^5/_2—^5/_2$
3062,233	9	4,08	8,13	$a\,^2I—z\,^2H^\circ$	$^{11}/_2—^9/_2$
3056,802	5	4,08	8,13	$a\,^2I—x\,^4G^\circ$	$^{13}/_2—^{11}/_2$
3049,011	5	5,87	9,94	$z\,^4P^\circ—e\,^4D$	$^3/_2—^3/_2$
3044,843	5	3,96	8,03	$b\,^2F—y\,^2G^\circ$	$^7/_2—^9/_2$
3036,986	5	5,82	9,89	$z\,^4P^\circ—e\,^4D$	$^5/_2—^5/_2$
3020,001	10	4,08	8,18	$a\,^2I—x\,^4F^\circ$	$^{11}/_2—^9/_2$
3002,649	13	1,70	5,82	$a\,^4P—z\,^4P^\circ$	$^3/_2—^5/_2$
3002,330	5	3,95	8,07	$b\,^2F—y\,^2G^\circ$	$^5/_2—^7/_2$
3000,059	5	3,81	7,94	$b\,^2G—y\,^4G^\circ$	$^7/_2—^5/_2$
2997,301	7	4,49	8,61	$b\,^2D—y\,^2F^\circ$	$^5/_2—^7/_2$

λ, Å	I	E_H, eV	E_B, eV	Transition	J
2985,550	13	1,73	5,87	$a\,^4P—z\,^4P°$	$^1/_2—^3/_2$
2984,830	15	1,67	5,82	$a\,^4P—z\,^4P°$	$^5/_2—^5/_2$
2982,062	8	4,48	8,63	$b\,^2D—y\,^2F°$	$^3/_2—^5/_2$
2979,352	8	1,09	5,25	$a\,^4D—z\,^6F°$	$^1/_2—^3/_2$
2975,938	5	1,09	5,26	$a\,^4D—z\,^6F°$	$^1/_2—^1/_2$
2970,682	5	3,76	7,94	$b\,^2G—y\,^4G°$	$^9/_2—^7/_2$
2970,513	5	1,07	5,25	$a\,^4D—z\,^6F°$	$^3/_2—^5/_2$
2969,934	8	3,81	7,98	$b\,^2G—z\,^2F°$	$^7/_2—^5/_2$
2965,037	10	1,70	5,87	$a\,^4P—z\,^4P°$	$^3/_2—^3/_2$
2964,629	9	1,72	5,90	$a\,^4P—z\,^4P°$	$^1/_2—^1/_2$
2964,131	7	3,38	7,57	$a\,^2F—z\,^4G°$	$^7/_2—^5/_2$
2961,272	5	1,07	5,26	$a\,^4D—z\,^6F°$	$^3/_2—^1/_2$
2959,599	7	3,38	7,57	$a\,^2F—z\,^2D°$	$^7/_2—^5/_2$
2953,778	11	1,04	5,22	$a\,^4D—z\,^6F°$	$^5/_2—^7/_2$
2949,178	10	3,76	7,97	$b\,^2G—z\,^2F°$	$^9/_2—^7/_2$
2947,658	13	1,67	5,87	$a\,^4P—z\,^4P°$	$^5/_2—^3/_2$
2944,398	13	1,70	5,90	$a\,^4P—z\,^4P°$	$^3/_2—^1/_2$
2939,506	5	1,04	5,25	$a\,^4D—z\,^6F°$	$^5/_2—^3/_2$
2926,587	12	0,98	5,22	$a\,^4D—z\,^6F°$	$^7/_2—^9/_2$
2922,024	5	3,91	8,14	$b\,^4D—x\,^4G°$	$^7/_2—^9/_2$
2902,456	5	3,76	8,03	$b\,^2G—y\,^2G°$	$^9/_2—^9/_2$
2897,262	8	3,42	7,70	$a\,^2F—z\,^2D°$	$^5/_2—^3/_2$
2895,215	7	3,91	8,18	$b\,^4D—x\,^4F°$	$^7/_2—^9/_2$
2894,778	7	3,26	7,55	$b\,^2H—z\,^4H°$	$^9/_2—^{11}/_2$
2888,093	5	3,19	7,49	$b\,^2P—y\,^4P°$	$^3/_2—^5/_2$
2885,928	5	4,07	8,37	$a\,^2I—y\,^2H°$	$^{13}/_2—^{11}/_2$
2883,702	10	3,24	7,54	$b\,^2H—z\,^4H°$	$^{11}/_2—^{13}/_2$
2880,756	9	0,98	5,29	$a\,^4D—z\,^6P°$	$^7/_2—^7/_2$
2876,802	7	3,38	7,69	$a\,^2F—y\,^4F°$	$^7/_2—^7/_2$
2875,346	8	3,38	7,69	$a\,^2F—z\,^2G°$	$^7/_2—^9/_2$
2873,401	10	3,81	8,13	$b\,^2G—z\,^2H°$	$^7/_2—^9/_2$
2872,382	9	3,26	7,58	$b\,^2H—z\,^4H°$	$^9/_2—^7/_2$
2871,125	6	3,24	7,56	$b\,^2H—z\,^4H°$	$^{11}/_2—^9/_2$
2871,059	6	3,20	7,51	$a\,^4G—z\,^4G°$	$^9/_2—^{11}/_2$
2868,874	5	1,04	5,36	$a\,^4D—z\,^6P°$	$^5/_2—^5/_2$
2864,973	5	3,90	8,23	$b\,^4D—x\,^4F°$	$^7/_2—^7/_2$
2858,343	11	3,22	7,56	$a\,^4G—z\,^4G°$	$^7/_2—^7/_2$
2857,171	7	3,88	8,23	$b\,^4D—x\,^4F°$	$^5/_2—^7/_2$
2856,928	8	5,51	9,85	$z\,^4D°—e\,^4D$	$^7/_2—^7/_2$
2856,392	5	5,36	9,70	$z\,^6P°—e\,^6D$	$^5/_2—^7/_2$
2856,141	7	3,20	7,54	$a\,^4G—z\,^4G°$	$^9/_2—^9/_2$
2855,670	9	3,22	7,56	$a\,^4G—z\,^4H°$	$^7/_2—^9/_2$
2849,606	7	3,20	7,54	$a\,^4G—z\,^4H°$	$^9/_2—^{11}/_2$
2848,899	5	4,08	8,42	$a\,^2I—y\,^2H°$	$^{11}/_2—^9/_2$
2848,332	7	5,59	9,94	$z\,^4F°—e\,^4D$	$^5/_2—^3/_2$
2848,122	7	5,55	9,90	$z\,^4D°—e\,^4D$	$^5/_2—^5/_2$
2848,046	8	3,23	7,58	$a\,^4G—z\,^4H°$	$^5/_2—^7/_2$
2843,485	5	3,88	8,24	$b\,^4D—x\,^4F°$	$^3/_2—^5/_2$
2840,756	8	3,77	8,13	$b\,^2G—x\,^4G°$	$^9/_2—^{11}/_2$
2840,647	9	3,33	7,69	$b\,^2P—z\,^2D°$	$^1/_2—^3/_2$
2840,342	7	3,15	7,51	$a\,^4G—z\,^4G°$	$^{11}/_2—^{11}/_2$
2839,819	6	5,29	9,65	$z\,^6P°—e\,^6D$	$^7/_2—^9/_2$
2839,535	7	5,48	9,85	$z\,^4F°—e\,^4D$	$^9/_2—^7/_2$
2837,300	5	3,26	7,63	$b\,^2H—z\,^4I°$	$^9/_2—^{11}/_2$
2835,716	9	3,19	7,57	$b\,^2P—z\,^4G°$	$^3/_2—^5/_2$
2833,100	5	5,36	9,73	$z\,^6P°—e\,^6D$	$^5/_2—^5/_2$
2831,562	11	3,19	7,57	$b\,^2P—z\,^2D°$	$^3/_2—^5/_2$

λ, Å	I	E_H, eV	E_B, eV	Transition	J
2828,634	6	3,24	7,62	$b\ ^2H—z\ ^4I^\circ$	11/2—9/2
2827,431	5	3,24	7,63	$b\ ^2H—z\ ^4I^\circ$	11/2—13/2
2813,613	5	3,22	7,62	$a\ ^4G—z\ ^4I^\circ$	7/2—9/2
2809,806	7	5,29	9,70	$z\ ^6P^\circ—e\ ^6D$	7/2—7/2
2805,791	5	3,38	7,80	$a\ ^2F—x\ ^4D^\circ$	7/2—7/2
2799,286	7	3,26	7,69	$b\ ^2H—y\ ^4F^\circ$	9/2—7/2
2797,914	5	3,26	7,70	$b\ ^2H—z\ ^2G^\circ$	9/2—9/2
2793,888	7	3,20	7,63	$a\ ^4G—z\ ^4I^\circ$	9/2—11/2
2785,213	8	5,20	9,65	$z\ ^6F^\circ—e\ ^6D$	11/2—9/2
2783,696	12	3,24	7,70	$b\ ^2H—z\ ^2G^\circ$	11/2—9/2
2779,299	11	3,26	7,72	$b\ ^2H—z\ ^2G^\circ$	9/2—7/2
2777,892	5	3,24	7,70	$b\ ^2H—y\ ^4F^\circ$	11/2—9/2
2776,923	5	5,22	9,69	$z\ ^6F^\circ—e\ ^6D$	7/2—7/2
2774,691	7	3,33	7,80	$b\ ^2P—y\ ^4D^\circ$	1/2—3/2
2771,184	5	3,76	8,23	$b\ ^2G—y\ ^4H^\circ$	9/2—11/2
2770,508	5	3,15	7,62	$a\ ^4G—z\ ^4I^\circ$	11/2—9/2
2769,354	9	3,15	7,63	$a\ ^4G—z\ ^4I^\circ$	11/2—13/2
2769,153	6	3,22	7,69	$a\ ^4G—z\ ^2G^\circ$	7/2—9/2
2768,934	8	1,07	5,55	$a\ ^4D—z\ ^4D^\circ$	3/2—5/2
2767,503	13	3,24	7,72	$b\ ^2H—z\ ^2I^\circ$	11/2—13/2
2761,813	9	1,09	5,58	$a\ ^4D—z\ ^4D^\circ$	1/2—3/2
2757,025	5	3,20	7,69	$a\ ^4G—y\ ^4F^\circ$	9/2—7/2
2756,512	5	3,23	7,72	$a\ ^4G—z\ ^2G^\circ$	5/2—7/2
2755,737	15	0,98	5,48	$a\ ^4D—z\ ^4F^\circ$	7/2—9/2
2754,907	6	5,23	9,73	$z\ ^6F^\circ—e\ ^6D$	7/2—5/2
2753,287	12	3,26	7,77	$b\ ^2H—z\ ^2I^\circ$	9/2—11/2
2751,123	6	3,19	7,70	$b\ ^2P—z\ ^2D^\circ$	3/2—3/2
2749,484	12	1,09	5,60	$a\ ^4D—z\ ^4D^\circ$	1/2—1/2
2749,324	14	1,04	5,55	$a\ ^4D—z\ ^6F^\circ$	5/2—7/2
2749,184	13	1,07	5,58	$a\ ^4D—z\ ^4D^\circ$	3/2—3/2
2746,483	14	1,07	5,59	$a\ ^4D—z\ ^4F^\circ$	3/2—5/2
2743,196	14	1,09	5,61	$a\ ^4D—z\ ^4F^\circ$	1/2—3/2
2741,397	6	3,42	7,94	$a\ ^2F—y\ ^4G^\circ$	5/2—5/2
2739,546	15	0,98	5,51	$a\ ^4D—z\ ^4D^\circ$	7/2—7/2
2736,968	12	1,07	5,60	$a\ ^4D—z\ ^4D^\circ$	3/2—1/2
2730,738	11	1,07	5,61	$a\ ^4D—z\ ^4F^\circ$	3/2—3/2
2727,539	13	1,04	5,58	$a\ ^4D—z\ ^4D^\circ$	5/2—3/2
2727,382	8	3,15	7,60	$a\ ^4G—z\ ^2G^\circ$	11/2—9/2
2724,885	9	1,04	5,59	$a\ ^4D—z\ ^4F^\circ$	5/2—5/2
2722,737	5	6,22	10,77	$c\ ^4P—v\ ^4D^\circ$	5/2—7/2
2722,040	5	3,39	7,94	$a\ ^2F—y\ ^4G^\circ$	7/2—7/2
2719,296	5	4,50	9,05	$b\ ^2D—x\ ^2F^\circ$	5/2—7/2
2718,639	5	6,22	10,77	$c\ ^4F—v\ ^4D^\circ$	9/2—7/2
2716,218	9	3,42	7,99	$a\ ^2F—z\ ^2F^\circ$	5/2—5/2
2714,412	13	0,99	5,55	$a\ ^4D—z\ ^4D^\circ$	7/2—5/2
2712,388	6	3,20	7,77	$a\ ^4G—z\ ^4I^\circ$	9/2—11/2
2711,845	9	3,15	7,72	$a\ ^4G—z\ ^2I^\circ$	11/2—13/2
2709,056	7	3,19	7,77	$b\ ^2P—y\ ^4D^\circ$	3/2—5/2
2707,132	6	4,48	9,06	$b\ ^2D—x\ ^2F^\circ$	3/2—5/2
2706,566	7	4,50	9,07	$b\ ^2D—y\ ^2P^\circ$	5/2—3/2
2703,989	10	3,39	7,97	$a\ ^2F—z\ ^2F^\circ$	7/2—7/2
2697,462	5	4,48	9,07	$b\ ^2D—y\ ^2P^\circ$	3/2—1/2
2692,836	5	0,99	5,58	$a\ ^4D—z\ ^4F^\circ$	7/2—5/2
2692,597	10	3,77	8,36	$b\ ^2G—y\ ^2H^\circ$	9/2—11/2
2684,751	10	3,82	8,42	$b\ ^2G—y\ ^2H^\circ$	7/2—9/2
2666,635	10	3,42	8,07	$a\ ^2F—y\ ^2G^\circ$	5/2—7/2
2664,664	10	3,38	8,03	$a\ ^2F—y\ ^2G^\circ$	7/2—9/2
2658,251	5	3,97	8,63	$b\ ^2F—y\ ^2F^\circ$	7/2—7/2

λ, Å	I	E_H, eV	E_B, eV	Transition	J
2650,492	4	5,82	10,49	$z\,^4P^\circ - f\,^4D$	$^5/_2 - ^7/_2$
2649,464	4	6,79	11,47	$d\,^2F - u\,^2G^\circ$	$^5/_2 - ^7/_2$
2639,553	5	3,33	8,03	$b\,^2P - z\,^2P^\circ$	$^1/_2 - ^1/_2$
2637,644	6	3,33	8,04	$b\,^2P - z\,^2P^\circ$	$^1/_2 - ^3/_2$
2633,194	5	4,73	9,44	$c\,^2D - x\,^2P^\circ$	$^5/_2 - ^3/_2$
2631,609	8	2,80	7,51	$b\,^4F - z\,^4G^\circ$	$^9/_2 - ^{11}/_2$
2631,323	13	0,08	4,79	$a\,^6D - z\,^6D^\circ$	$^5/_2 - ^7/_2$
2631,051	13	0,11	4,82	$a\,^6D - z\,^6D^\circ$	$^3/_2 - ^5/_2$
2630,066	8	2,85	7,57	$b\,^4F - z\,^4G^\circ$	$^3/_2 - ^5/_2$
2629,587	8	2,84	7,56	$b\,^4F - z\,^4G^\circ$	$^5/_2 - ^7/_2$
2628,292	13	0,12	4,84	$a\,^6D - z\,^6D^\circ$	$^1/_2 - ^3/_2$
2626,500	6	2,85	7,57	$b\,^4F - z\,^2D^\circ$	$^3/_2 - ^5/_2$
2625,666	13	0,05	4,77	$a\,^6D - z\,^6D^\circ$	$^7/_2 - ^9/_2$
2625,490	9	4,08	8,80	$a\,^2I - z\,^2K^\circ$	$^{11}/_2 - ^{13}/_2$
2623,721	5	2,84	7,57	$b\,^4F - z\,^4G^\circ$	$^5/_2 - ^5/_2$
2621,668	10	0,12	4,85	$a\,^6D - z\,^6D^\circ$	$^1/_2 - ^1/_2$
2620,695	7	2,83	7,56	$b\,^4F - z\,^4G^\circ$	$^7/_2 - ^7/_2$
2620,407	6	0,11	4,84	$a\,^6D - z\,^6D^\circ$	$^3/_2 - ^3/_2$
2619,076	7	2,80	7,54	$b\,^4F - z\,^4G^\circ$	$^9/_2 - ^9/_2$
2617,616	12	0,08	4,82	$a\,^6D - z\,^6D^\circ$	$^5/_2 - ^5/_2$
2613,823	13	0,11	4,85	$a\,^6D - z\,^6D^\circ$	$^3/_2 - ^1/_2$
2611,872	13	0,05	4,79	$a\,^6D - z\,^6D^\circ$	$^7/_2 - ^7/_2$
2611,072	6	1,08	5,82	$a\,^4D - z\,^4P^\circ$	$^3/_2 - ^5/_2$
2609,122	5	3,97	8,71	$b\,^2F - x\,^2G^\circ$	$^7/_2 - ^9/_2$
2607,087	13	0,08	4,81	$a\,^6D - z\,^6D^\circ$	$^5/_2 - ^5/_2$
2606,504	7	4,50	9,25	$b\,^2D - x\,^2D^\circ$	$^5/_2 - ^5/_2$
2605,416	6	3,23	7,98	$a\,^4G - z\,^2F^\circ$	$^5/_2 - ^5/_2$
2605,303	6	4,48	9,24	$b\,^2D - x\,^2D^\circ$	$^3/_2 - ^3/_2$
2605,040	6	5,57	10,32	$c\,^2F - v\,^2G^\circ$	$^5/_2 - ^7/_2$
2599,396	14	0,00	4,77	$a\,^6D - z\,^6D^\circ$	$^9/_2 - ^9/_2$
2598,369	14	0,05	4,82	$a\,^6D - z\,^6D^\circ$	$^7/_2 - ^5/_2$
2593,726	7	1,10	5,87	$a\,^4D - z\,^4P^\circ$	$^1/_2 - ^3/_2$
2592,779	9	4,08	8,85	$a\,^2I - z\,^2K^\circ$	$^{13}/_2 - ^{15}/_2$
2591,543	10	1,04	5,82	$a\,^4D - z\,^4P^\circ$	$^5/_2 - ^5/_2$
2587,948	7	4,16	8,94	$c\,^2G - x\,^2H^\circ$	$^7/_2 - ^9/_2$
2585,876	13	0,00	4,79	$a\,^6D - z\,^6D^\circ$	$^9/_2 - ^7/_2$
2582,585	10	1,08	5,87	$a\,^4D - z\,^4P^\circ$	$^3/_2 - ^3/_2$
2577,923	9	1,09	5,90	$a\,^4D - z\,^4P^\circ$	$^1/_2 - ^1/_2$
2576,865	7	4,15	8,96	$c\,^2G - x\,^2H^\circ$	$^9/_2 - ^{11}/_2$
2574,368	9	2,58	7,40	$b\,^4P - z\,^4S^\circ$	$^5/_2 - ^3/_2$
2570,841	7	3,81	8,63	$b\,^2G - y\,^2F^\circ$	$^7/_2 - ^5/_2$
2570,525	5	5,95	10,77	$d\,^2D - v\,^4D^\circ$	$^5/_2 - ^7/_2$
2568,405	6	2,78	7,60	$b\,^4P - y\,^4P^\circ$	$^1/_2 - ^3/_2$
2566,912	9	1,07	5,90	$a\,^4D - z\,^4P^\circ$	$^3/_2 - ^1/_2$
2566,218	5	5,56	10,39	$c\,^2F - v\,^2G^\circ$	$^7/_2 - ^9/_2$
2563,474	12	1,05	5,87	$a\,^4D - z\,^4P^\circ$	$^5/_2 - ^3/_2$
2562,534	13	0,99	5,82	$a\,^4D - z\,^4P^\circ$	$^7/_2 - ^5/_2$
2560,272	7	3,20	8,04	$b\,^2P - z\,^2P^\circ$	$^3/_2 - ^3/_2$
2559,912	5	3,42	8,26	$a\,^2F - y\,^4H^\circ$	$^5/_2 - ^7/_2$
2559,774	5	3,23	8,07	$a\,^4G - y\,^2G^\circ$	$^5/_2 - ^7/_2$
2555,447	7	2,85	7,70	$b\,^4F - y\,^4F^\circ$	$^3/_2 - ^5/_2$
2555,066	5	2,84	7,69	$b\,^4F - y\,^4F^\circ$	$^5/_2 - ^7/_2$
2550,680	8	3,25	8,11	$b\,^2H - z\,^2H^\circ$	$^{11}/_2 - ^{11}/_2$
2550,020	8	3,27	8,13	$b\,^2H - z\,^2H^\circ$	$^9/_2 - ^9/_2$
2549,453	8	2,84	7,70	$b\,^4F - y\,^4F^\circ$	$^5/_2 - ^5/_2$
2549,399	8	2,85	7,71	$b\,^4F - y\,^4F^\circ$	$^3/_2 - ^3/_2$
2549,082	7	3,76	8,62	$b\,^2G - y\,^2F^\circ$	$^9/_2 - ^7/_2$

λ, Å	I	E_H, eV	E_B, eV	Transition	J
2548,925	5	4,08	8,94	$a\,^2I—x\,^2H°$	$^{11}/_2—^9/_2$
2548,741	7	2,70	7,56	$b\,^4P—y\,^4P°$	$^3/_2—^1/_2$
2548,590	6	2,67	7,54	$a\,^4H—z\,^4G°$	$^9/_2—^9/_2$
2547,330	5	2,69	7,56	$a\,^4H—z\,^4G°$	$^7/_2—^7/_2$
2546,667	8	2,83	7,69	$b\,^4F—y\,^4F°$	$^7/_2—^7/_2$
2545,215	7	2,69	7,56	$a\,^4H—z\,^4H°$	$^7/_2—^9/_2$
2544,972	6	2,70	7,57	$b\,^4P—z\,^2D°$	$^3/_2—^5/_2$
2543,431	5	2,84	7,71	$b\,^4F—y\,^4F°$	$^5/_2—^3/_2$
2543,384	9	2,67	7,54	$a\,^4H—z\,^4H°$	$^9/_2—^{11}/_2$
2542,733	5	3,33	8,21	$b\,^2P—z\,^2S°$	$^1/_2—^1/_2$
2541,831	7	2,69	7,57	$a\,^4H—z\,^4G°$	$^7/_2—^5/_2$
2541,096	7	2,83	7,70	$b\,^4F—y\,^4F°$	$^7/_2—^5/_2$
2540,666	6	2,83	7,70	$b\,^4F—y\,^4F°$	$^7/_2—^9/_2$
2539,003	10	2,63	7,51	$a\,^4H—z\,^4G°$	$^{13}/_2—^{11}/_2$
2538,997	8	2,67	7,56	$a\,^4H—z\,^4G°$	$^9/_2—^7/_2$
2538,809	9	2,66	7,54	$a\,^4H—z\,^4G°$	$^{11}/_2—^9/_2$
2538,500	5	2,69	7,57	$a\,^4H—z\,^2D°$	$^7/_2—^5/_2$
2538,205	6	4,07	8,95	$a\,^2I—x\,^2H°$	$^{13}/_2—^{11}/_2$
2537,142	5	4,77	9,64	$z\,^6D°—e\,^6D$	$^9/_2—^9/_2$
2536,817	9	2,67	7,56	$a\,^4H—z\,^4H°$	$^9/_2—^9/_2$
2536,673	7	3,24	8,13	$b\,^2H—x\,^4G°$	$^{11}/_2—^{11}/_2$
2535,480	7	2,80	7,69	$b\,^4F—y\,^4F°$	$^9/_2—^7/_2$
2534,416	9	2,69	7,58	$a\,^4H—z\,^4H°$	$^7/_2—^7/_2$
2533,627	10	2,66	7,55	$a\,^4H—z\,^4H°$	$^{11}/_2—^{11}/_2$
2530,103	6	2,83	7,72	$b\,^4F—z\,^2G°$	$^7/_2—^7/_2$
2529,547	10	2,70	7,60	$b\,^4P—y\,^4P°$	$^3/_2—^3/_2$
2529,545	10	2,80	7,70	$b\,^4F—y\,^4F°$	$^9/_2—^9/_2$
2529,221	5	3,24	8,14	$b\,^2H—x\,^4G°$	$^{11}/_2—^9/_2$
2529,080	5	4,74	9,64	$c\,^2D—v\,^2F°$	$^3/_2—^5/_2$
2527,102	6	2,66	7,56	$a\,^4H—z\,^4H°$	$^{11}/_2—^9/_2$
2526,295	9	2,58	7,49	$b\,^4P—y\,^4P°$	$^5/_2—^5/_2$
2526,071	5	2,67	7,58	$a\,^4H—z\,^4H°$	$^9/_2—^7/_2$
2525,387	10	2,63	7,54	$a\,^4H—z\,^4H°$	$^{13}/_2—^{13}/_2$
2521,814	7	4,16	9,07	$c\,^2G—w\,^2G°$	$^7/_2—^7/_2$
2521,089	7	3,42	8,33	$a\,^2F—y\,^2D°$	$^5/_2—^3/_2$
2519,044	7	3,38	8,30	$a\,^2F—y\,^2D°$	$^7/_2—^5/_2$
2517,120	6	2,78	7,70	$b\,^4P—z\,^2D°$	$^1/_2—^3/_2$
2514,383	7	3,81	8,74	$b\,^2G—x\,^2G°$	$^7/_2—^7/_2$
2512,513	5	4,48	9,41	$b\,^2D—w\,^2F°$	$^3/_2—^5/_2$
2511,759	10	2,69	7,62	$a\,^4H—z\,^4I°$	$^7/_2—^9/_2$
2509,122	5	3,24	8,18	$b\,^2H—x\,^4F°$	$^{11}/_2—^9/_2$
2506,093	7	3,20	8,14	$a\,^4G—x\,^4G°$	$^9/_2—^9/_2$
2503,870	7	3,77	8,72	$b\,^2G—x\,^2G°$	$^9/_2—^9/_2$
2503,560	5	2,85	7,80	$b\,^4F—y\,^4D°$	$^3/_2—^3/_2$
2503,325	7	3,15	8,10	$a\,^4G—z\,^2H°$	$^{11}/_2—^{11}/_2$
2502,390	7	3,22	8,17	$a\,^4G—x\,^4G°$	$^7/_2—^7/_2$
2500,919	5	4,73	9,68	$c\,^2D—v\,^2F°$	$^5/_2—^7/_2$
2498,894	10	2,67	7,63	$a\,^4H—z\,^4I°$	$^9/_2—^{11}/_2$
2497,820	7	2,84	7,80	$b\,^4F—y\,^4D°$	$^5/_2—^3/_2$
2493,261	12	2,63	7,60	$a\,^4H—z\,^4I°$	$^{13}/_2—^{15}/_2$
2493,180	12	2,66	7,63	$a\,^4H—z\,^4I°$	$^{11}/_2—^{13}/_2$
2491,392	6	3,20	8,17	$a\,^4G—x\,^4G°$	$^9/_2—^7/_2$
2490,856	6	2,83	7,80	$b\,^4F—x\,^4D°$	$^7/_2—^7/_2$
2489,822	8	3,15	8,13	$a\,^4G—x\,^4G°$	$^{11}/_2—^{11}/_2$
2487,356	5	5,41	10,39	$z\,^6P°—32$	$^3/_2—^5/_2$
2486,345	7	3,20	8,18	$a\,^4G—x\,^4F°$	$^9/_2—^9/_2$
2484,243	5	3,24	8,23	$b\,^2H—y\,^4H°$	$^{11}/_2—^{13}/_2$
2482,654	8	3,15	8,14	$a\,^4G—x\,^4G°$	$^{11}/_2—^9/_2$

λ, Å	I	E_H, eV	E_B, eV	Transition	J
2482,115	8	2,63	7,63	$a\,^4H - z\,^4I^\circ$	$13/2 - 13/2$
2480,158	8	2,80	7,80	$b\,^4F - x\,^4D^\circ$	$9/2 - 7/2$
2478,568	6	2,84	7,84	$b\,^4F - x\,^4D^\circ$	$5/2 - 5/2$
2474,765	6	3,22	8,23	$a\,^4G - x\,^4F^\circ$	$7/2 - 7/2$
2473,314	6	2,78	7,79	$b\,^4P - y\,^4D^\circ$	$1/2 - 1/2$
2472,426	5	2,85	7,86	$b\,^4F - x\,^4D^\circ$	$3/2 - 3/2$
2470,658	7	2,83	7,84	$b\,^4F - x\,^4D^\circ$	$7/2 - 5/2$
2469,712	8	5,29	10,30	$z\,^6P^\circ - e\,^6F$	$7/2 - 7/2$
2467,732	6	5,41	10,43	$z\,^6P^\circ - 34$	$3/2 - 3/2$, $1/2$
2466,811	7	2,84	7,86	$b^4\,F - x\,^4D^\circ$	$5/2 - 3/2$
2466,670	7	2,85	7,88	$b\,^4F - x\,^4D^\circ$	$3/2 - 1/2$
2465,200	7	2,77	7,80	$b\,^4P - y\,^4D^\circ$	$1/2 - 3/2$
2464,907	7	3,23	8,25	$a\,^4G - x\,^4F^\circ$	$5/2 - 3/2$
2464,007	7	3,21	8,23	$a\,^4G - x\,^4F^\circ$	$9/2 - 7/2$
2463,900	5	5,36	10,39	$z\,^6P^\circ - 32$	$5/2 - 5/2$
2463,280	6	3,15	8,18	$a\,^4G - x\,^4F^\circ$	$11/2 - 9/2$
2461,857	8	3,23	8,25	$a\,^4G - y\,^4H^\circ$	$7/2 - 9/2$
2461,282	8	3,23	8,26	$a\,^4G - y\,^4H^\circ$	$5/2 - 7/2$
2460,453	5	5,48	10,52	$z\,^4F^\circ - e\,^4G$	$9/2 - 11/2$
2458,964	5	3,88	8,92	$b\,^4D - w\,^4P^\circ$	$5/2 - 3/2$
2458,782	8	3,21	8,24	$a\,^4G - y\,^4H^\circ$	$9/2 - 11/2$
2455,892	10	5,41	10,45	$z\,^6P^\circ - e\,^6P$	$3/2 - 3/2$
2454,567	6	4,08	9,13	$a\,^2I - w\,^2H^\circ$	$13/2 - 11/2$
2453,935	25	5,22	10,27	$z\,^6F^\circ - e\,^6F$	$9/2 - 11/2$
2453,747	15	5,23	10,28	$z\,^6F^\circ - e\,^6F$	$7/2 - 9/2$
2450,134	5	5,25	10,30	$z\,^6F^\circ - e\,^6F$	$5/2 - 7/2$
2447,747	6	4,08	9,14	$a\,^2I - w\,^2H^\circ$	$11/2 - 9/2$
2446,462	5	2,66	7,72	$a\,^4H - z\,^2I^\circ$	$11/2 - 13/2$
2446,405	25	5,22	10,28	$z\,^6F^\circ - e\,^6F$	$9/2 - 9/2$
2445,558	7	2,70	7,77	$b\,^4P - y\,^4D^\circ$	$3/2 - 5/2$
2445,114	40	5,20	10,27	$z\,^6F^\circ - e\,^6F$	$11/2 - 11/2$
2444,512	8	2,58	7,65	$b\,^4P - y\,^4D^\circ$	$5/2 - 7/2$
2444,274	10	5,23	10,30	$z\,^6F^\circ - e\,^6F$	$7/2 - 7/2$
2443,842	15	5,25	10,32	$z\,^6F^\circ - e\,^6F$	$3/2 - 5/2$
2439,860	8	5,25	10,32	$z\,^6F^\circ - e\,^6F$	$5/2 - 5/2$
2439,300	8	3,15	8,23	$a\,^4G - y\,^4H^\circ$	$11/2 - 13/2$
2437,632	20	5,20	10,28	$z\,^6F^\circ - e\,^6F$	$11/2 - 9/2$
2437,100	5	5,26	10,34	$z\,^6F^\circ - e\,^6F$	$1/2 - 3/2$
2436,987	10	5,22	10,30	$z\,^6F^\circ - e\,^6F$	$9/2 - 7/2$
2436,615	20	5,36	10,44	$z\,^6P^\circ - e\,^6P$	$5/2 - 5/2$
2434,988	25	5,29	10,37	$z\,^6P^\circ - 30$	$7/2 - 7/2$
2434,944	7	2,85	7,94	$b\,^4F - y\,^4G^\circ$	$3/2 - 5/2$
2434,822	5	5,25	10,34	$z\,^6F^\circ - e\,^6F$	$3/2 - 3/2$
2434,733	7	4,08	9,17	$a\,^2I - y\,^2I^\circ$	$11/2 - 11/2$
2434,229	20	5,29	10,38	$z\,^6P^\circ - e\,^6P$	$7/2 - 7/2$
2434,052	15	5,23	10,32	$z\,^6F^\circ - e\,^6F$	$7/2 - 5/2$
2432,267	7	2,84	7,94	$b\,^4F - y\,^4G^\circ$	$5/2 - 7/2$
2430,876	10	5,25	10,34	$z\,^6F^\circ - e\,^6F$	$5/2 - 3/2$
2430,072	7	2,83	7,93	$b\,^4F - y\,^4G^\circ$	$7/2 - 9/2$
2429,148	10	5,29	10,39	$z\,^6P^\circ - 32$	$7/2 - 5/2$
2428,970	6	5,25	10,36	$z\,^6F^\circ - e\,^6F$	$3/2 - 1/2$
2428,367	6	3,90	9,01	$b\,^4D - w\,^4F^\circ$	$7/2 - 9/2$
2424,143	8	2,80	7,92	$b\,^4F - y\,^4G^\circ$	$9/2 - 11/2$
2417,866	6	3,24	8,37	$b\,^2H - y\,^2H^\circ$	$11/2 - 11/2$
2413,309	9	0,12	5,26	$a\,^6D - z\,^6F^\circ$	$1/2 - 3/2$
2411,066	9	0,12	5,26	$a\,^6D - z\,^6F^\circ$	$1/2 - 1/2$
2410,517	9	0,11	5,25	$a\,^6D - z\,^6F^\circ$	$3/2 - 5/2$

λ, Å	I	E_H, eV	E_B, eV	Transition	J
2406,658	9	0,11	5,26	$a\,^6D - z\,^6F^\circ$	$^3/_2 - ^3/_2$
2404,882	9	0,08	5,24	$a\,^6D - z\,^6F^\circ$	$^5/_2 - ^7/_2$
2404,430	7	0,11	5,26	$a\,^6D - z\,^6F^\circ$	$^3/_2 - ^1/_2$
2402,450	8	5,22	10,38	$z\,^6F^\circ - e\,^6P$	$^9/_2 - ^7/_2$
2399,239	9	0,08	5,25	$a\,^6D - z\,^6F^\circ$	$^5/_2 - ^5/_2$
2399,237	9	0,38	5,55	$a\,^4F - z\,^4D^\circ$	$^3/_2 - ^5/_2$
2395,625	9	0,05	5,22	$a\,^6D - z\,^6F^\circ$	$^7/_2 - ^9/_2$
2395,408	7	0,05	5,26	$a\,^6D - z\,^6F^\circ$	$^7/_2 - ^3/_2$
2388,627	9	0,05	5,24	$a\,^6D - z\,^6F^\circ$	$^7/_2 - ^7/_2$
2384,386	7	0,38	5,58	$a\,^4F - z\,^4D^\circ$	$^3/_2 - ^3/_2$
2383,242	7	0,35	5,55	$a\,^4F - z\,^4D^\circ$	$^5/_2 - ^5/_2$
2382,039	9	0,00	5,20	$a\,^6D - z\,^6F^\circ$	$^9/_2 - ^{11}/_2$
2380,759	7	0,08	5,29	$a\,^6D - z\,^6P^\circ$	$^5/_2 - ^7/_2$
2379,275	7	0,30	5,51	$a\,^6F - z\,^4D^\circ$	$^7/_2 - ^7/_2$
2376,435	5	5,20	10,42	$z\,^6F^\circ - e\,^6G$	$^{11}/_2 - ^{13}/_2$
2375,192	7	0,38	5,60	$a\,^4F - z\,^4D^\circ$	$^3/_2 - ^1/_2$
2373,6250	8	0,00	5,22	$a\,^6D - z\,^6F^\circ$	$^9/_2 - ^9/_2$
2370,496	5	0,38	5,61	$a\,^4F - z^4F^\circ$	$^3/_2 - ^3/_2$
2369,960	5	5,22	10,45	$z\,^6F^\circ - e\,^6G$	$^9/_2 - ^{11}/_2$
2368,595	7	0,35	5,58	$a\,^4F - z\,^4D^\circ$	$^5/_2 - ^3/_2$
2366,595	5	0,35	5,59	$a\,^4F - z\,^4F^\circ$	$^5/_2 - ^5/_2$
2364,826	8	0,05	5,29	$a\,^6D - z\,^6P^\circ$	$^7/_2 - ^7/_2$
2362,019	6	0,30	5,55	$a\,^4F - z\,^4F^\circ$	$^7/_2 - ^7/_2$
2360,293	8	0,30	5,55	$a\,^4F - z\,^4D^\circ$	$^7/_2 - ^5/_2$
2359,997	8	0,23	5,48	$a\,^4F - z\,^4F^\circ$	$^9/_2 - ^9/_2$
2359,111	8	5,25	10,50	$z\,^6F^\circ - e\,^6G$	$^5/_2 - ^7/_2$
2359,104	8	0,11	5,36	$a\,^6D - z\,^6P^\circ$	$^3/_2 - ^5/_2$
2354,889	5	0,35	5,61	$a\,^4F - z\,^4F^\circ$	$^5/_2 - ^3/_2$
2354,466	5	2,67	7,94	$a\,^4H - y\,^4G^\circ$	$^9/_2 - ^7/_2$
2351,198	5	2,66	7,93	$a\,^4H - y\,^4G^\circ$	$^{11}/_2 - ^9/_2$
2348,303	8	0,08	5,36	$a\,^6D - z\,^6P^\circ$	$^5/_2 - ^5/_2$
2348,099	8	0,23	5,51	$a\,^4F - z\,^4D^\circ$	$^9/_2 - ^7/_2$
2345,327	5	2,63	7,92	$a\,^4H - y\,^4G^\circ$	$^{13}/_2 - ^{11}/_2$
2344,278	8	0,12	5,41	$a\,^6D - z\,^6P^\circ$	$^1/_2 - ^3/_2$
2343,959	6	0,30	5,59	$a\,^4F - z\,^4F^\circ$	$^7/_2 - ^5/_2$
2343,492	8	0,00	5,29	$a\,^6D - z\,^6P^\circ$	$^9/_2 - ^7/_2$
2338,005	8	0,11	5,41	$a\,^6D - z\,^6P^\circ$	$^3/_2 - ^3/_2$
2332,796	8	0,05	5,36	$a\,^6D - z\,^6P^\circ$	$^7/_2 - ^5/_2$
2331,306	7	0,23	5,55	$a\,^4F - z\,^4F^\circ$	$^9/_2 - ^7/_2$
2327,394	7	0,08	5,41	$a\,^6D - z\,^6P^\circ$	$^5/_2 - ^3/_2$
2257,788	25	4,82	10,30	$z\,^6D^\circ - e\,^6F$	$^5/_2 - ^7/_2$
2256,897	10	4,84	10,32	$z\,^6D^\circ - e\,^6F$	$^3/_2 - ^5/_2$
2255,691	50	4,79	10,28	$z\,^6D^\circ - e\,^6F$	$^7/_2 - ^9/_2$
2254,066	8	4,85	10,34	$z\,^6D^\circ - e\,^6F$	$^1/_2 - ^3/_2$
2251,831	80	4,77	10,27	$z\,^6D^\circ - e^6F$	$^9/_2 - ^{11}/_2$
2249,181	25	4,84	10,34	$z\,^6D^\circ - e\,^6F$	$^3/_2 - ^3/_2$
2249,175	10	0,00	5,51	$a\,^6D - z\,^4D^\circ$	$^9/_2 - ^7/_2$
2249,063	30	4,82	10,32	$z\,^6D^\circ - e\,^6F$	$^5/_2 - ^5/_2$
2247,692	35	4,79	10,30	$z\,^6D^\circ - e\,^6F$	$^7/_2 - ^7/_2$
2245,505	45	4,77	10,28	$z\,^6D^\circ - e\,^6F$	$^9/_2 - ^9/_2$
2244,216	8	4,84	10,36	$z\,^6D^\circ - e\,^6F$	$^3/_2 - ^1/_2$
2241,426	20	4,82	10,34	$z\,^6D^\circ - e\,^6F$	$^5/_2 - ^3/_2$
2239,047	25	4,79	10,32	$z\,^6D^\circ - e\,^6F$	$^7/_2 - ^5/_2$
2237,577	20	4,77	10,30	$z\,^6D^\circ - e\,^6F$	$^9/_2 - ^7/_2$
2231,512	10	4,84	10,39	$z\,^6D^\circ - 32$	$^3/_2 - ^5/_2$
2228,761	30	4,82	10,37	$z\,^6D^\circ - 30$	$^5/_2 - ^7/_2$
2220,453	6	4,85	10,43	$z\,^6D^\circ - 34$	$^1/_2 - ^3/_2, ^1/_2$
2220,388	25	2,52	8,10	$a\,^2H - z\,^2H^\circ$	$^{11}/_2 - ^{11}/_2$

λ, Å	I	E_H, eV	E_B, eV	Transition	J
2219,889	20	2,66	8,23	$a\,^4H-y\,^4H^\circ$	$^{11}/_2-^{11}/_2$
2218,289	30	4,79	10,38	$z\,^6D^\circ-e\,^6P$	$^7/_2-^7/_2$
2215,094	10	4,82	10,41	$z\,^6D^\circ-33$	$^5/_2-^3/_2$
2214,059	20	4,79	10,39	$z\,^6D^\circ-32$	$^7/_2-^5/_2$
2213,679	20	2,63	8,23	$a\,^4H-y\,^4H^\circ$	$^{13}/_2-^{13}/_2$
2211,243	12	3,88	9,48	$b\,^4D-x\,^2P^\circ$	$^1/_2-^1/_2$
2211,112	5	2,63	8,23	$a\,^4H-y\,^4H^\circ$	$^{13}/_2-^{11}/_2$
2210,952	5	2,52	8,12	$a\,^2H-z\,^2H^\circ$	$^{11}/_2-^9/_2$
2209,049	20	4,77	10,37	$z\,^6D^\circ-30$	$^9/_2-^7/_2$
2208,419	30	4,77	10,38	$z\,^6D^\circ-e\,^6P$	$^9/_2-^7/_2$
2206,153	8	4,84	10,45	$z\,^6D^\circ-e\,^6P$	$^3/_2-^3/_2$
2201,595	5	4,82	10,44	$z\,^6D^\circ-e\,^6P$	$^5/_2-^5/_2$
2197,273	5	3,33	8,98	$b\,^2P-w\,^4D^\circ$	$^1/_2-^1/_2$
2192,674	5	3,33	8,99	$b\,^2P-w\,^4D^\circ$	$^1/_2-^3/_2$
2191,935	10	4,78	10,44	$z\,^6D^\circ-e\,^6P$	$^7/_2-^5/_2$
2187,868	15	2,64	8,30	$a\,^2D-y\,^2D^\circ$	$^3/_2-^5/_2$
2187,678	10	2,03	7,69	$a\,^2G-y\,^4F^\circ$	$^7/_2-^7/_2$
2187,444	12	3,38	9,04	$a\,^2F-x\,^2F^\circ$	$^7/_2-^7/_2$
2185,622	8	3,38	9,05	$a\,^2F-x\,^2F^\circ$	$^7/_2-^5/_2$
2183,803	10	3,26	8,93	$b\,^2H-x\,^2H^\circ$	$^9/_2-^9/_2$
2183,468	8	2,58	8,25	$a\,^2H-y\,^4H^\circ$	$^9/_2-^9/_2$
2183,301	12	2,03	7,70	$a\,^2G-y\,^4F^\circ$	$^7/_2-^9/_2$
2181,407	5	4,85	10,53	$z\,^6D^\circ-e\,^6G$	$^1/_2-^3/_2$
2181,137	8	4,84	10,51	$z\,^6D^\circ-e\,^6G$	$^3/_2-^5/_2$
2180,8692	12	4,77	10,45	$z\,^6D^\circ-e\,^6G$	$^9/_2-^{11}/_2$
2180,255	12	4,82	10,50	$z\,^6D^\circ-e\,^6G$	$^5/_2-^7/_2$
2177,025	10	2,34	8,03	$a\,^2P-z\,^2P^\circ$	$^1/_2-^1/_2$
2176,826	20	4,84	10,53	$z\,^6D^\circ-e\,^6G$	$^3/_2-^3/_2$
2175,445	25	2,03	7,72	$a\,^2G-z\,^2G^\circ$	$^7/_2-^7/_2$
2174,849	8	2,64	8,33	$a\,^2D-y\,^2D^\circ$	$^3/_2-^3/_2$
2173,720	15	1,70	7,40	$a\,^4P-z\,^4S^\circ$	$^3/_2-^3/_2$
2173,220	20	3,26	8,96	$b\,^2H-w\,^4F^\circ$	$^9/_2-^7/_2$
2172,989	15	2,54	8,24	$a\,^2D-x\,^4F^\circ$	$^5/_2-^5/_2$
2172,679	8	4,82	10,52	$z\,^6D^\circ-f\,^4D$	$^5/_2-^5/_2$
2170,193	5	4,84	10,55	$z\,^6D^\circ-f\,^4D$	$^1/_2-^1/_2$
2169,994	12	3,34	9,07	$b\,^2P-y\,^2P^\circ$	$^1/_2-^1/_2$
2169,950	12	4,77	10,48	$z\,^6D^\circ-e\,^6G$	$^9/_2-^9/_2$
2169,431	10	4,82	10,53	$z\,^6D^\circ-e\,^6G$	$^5/_2-^3/_2$
2168,925	8	3,24	8,95	$b\,^2H-x\,^2H^\circ$	$^{11}/_2-^{11}/_2$
2167,880	12	3,23	8,94	$a\,^4G-w\,^4F^\circ$	$^5/_2-^3/_2$
2167,401	12	2,52	8,23	$a\,^2H-y\,^4H^\circ$	$^{11}/_2-^{11}/_2$
2166,198	20	3,22	8,93	$a\,^4G-x\,^2H^\circ$	$^7/_2-^9/_2$
2165,555	10	2,84	8,56	$b\,^4F-x\,^4P^\circ$	$^5/_2-^5/_2$
2164,558	25	3,23	8,95	$a\,^4G-w\,^4F^\circ$	$^5/_2-^5/_2$
2164,339	20	4,82	10,54	$z\,^6D^\circ-f\,^4D$	$^5/_2-^3/_2$
2163,370	20	4,79	10,52	$z\,^6D^\circ-f\,^4D$	$^7/_2-^5/_2$
2162,023	20	1,96	7,69	$a\,^2G-z\,^2G^\circ$	$^9/_2-^9/_2$
2161,582	20	2,52	8,26	$a\,^2H-y\,^4H^\circ$	$^{11}/_2-^9/_2$
2161,313	20	3,34	9,07	$b\,^2P-y\,^2P^\circ$	$^1/_2-^1/_2$
2161,161	15	3,22	8,95	$a\,^4G-w\,^4F^\circ$	$^7/_2-^5/_2$
2159,199	25	1,96	7,70	$a\,^2G-y\,^4F^\circ$	$^9/_2-^9/_2$
2159,152	10	0,08	5,81	$a\,^6D-z\,^4P^\circ$	$^5/_2-^5/_2$
2158,518	25	1,96	7,70	$a\,^2G-y\,^4F^\circ$	$^9/_2-^9/_2$
2155,839	12	3,22	8,97	$a\,^4G-w\,^4F^\circ$	$^7/_2-^7/_2$
2153,373	12	2,28	8,03	$a\,^2P-z\,^2P^\circ$	$^3/_2-^1/_2$
2153,281	5	3,19	8,95	$b\,^2P-w\,^4P^\circ$	$^3/_2-^1/_2$
2152,488	25	2,58	8,33	$b\,^4P-y\,^2D^\circ$	$^5/_2-^3/_2$

λ, Å	I	E_H, eV	E_B, eV	Transition	J
2152,373	12	2,28	8,03	$a\,^2P—z\,^2P°$	$^3/_2—^1/_2$
2151,774	25	2,28	8,04	$a\,^2P—z\,^2P°$	$^3/_2—^3/_2$
2150,762	10	3,24	9,00	$b\,^2H—w\,^4F°$	$^{11}/_2—^9/_2$
2150,618	20	2,54	8,30	$a\,^2D—y\,^2D°$	$^5/_2—^5/_2$
2147,719	15	3,20	8,96	$a\,^4G—w\,^4F°$	$^9/_2—^7/_2$
2146,058	10	0,05	5,82	$a\,^6D—z\,^4P°$	$^7/_2—^5/_2$
2139,6987	25	0,08	5,87	$a\,^6D—z\,^4P°$	$^5/_2—^3/_2$
2138,103	20	2,54	8,33	$a\,^2D—y\,^2D°$	$^5/_2—^3/_2$
2137,735	15	0,11	5,90	$a\,^6D—z\,^4P°$	$^3/_2—^1/_2$
2136,519	20	3,26	9,07	$b\,^2H—w\,^2G°$	$^9/_2—^7/_2$
2133,990	8	3,20	9,00	$a\,^4G—w\,^4F°$	$^9/_2—^9/_2$
2130,548	12	3,24	9,06	$b\,^2H—w\,^2G°$	$^{11}/_2—^9/_2$
2130,259	15	1,67	7,49	$a\,^4P—y\,^4P°$	$^5/_2—^5/_2$
2127,967	10	3,81	9,64	$b\,^2G—v\,^2F°$	$^7/_2—^5/_2$
2119,050	12	2,52	8,36	$a\,^2H—y\,^2H°$	$^{11}/_2—^{11}/_2$
2118,195	8	2,58	8,42	$a\,^2H—y\,^2H°$	$^9/_2—^9/_2$
2117,633	25	3,15	9,01	$a\,^4G—w\,^4D°$	$^{11}/_2—^9/_2$
2116,960	25	3,15	9,00	$a\,^4G—w\,^4F°$	$^{11}/_2—^9/_2$
2110,724	15	2,34	8,22	$a\,^2P—z\,^2S°$	$^1/_2—^1/_2$
2110,240	25	3,82	9,69	$b\,^2G—v\,^2F°$	$^7/_2—^7/_2$
2109,613	25	3,20	9,07	$b\,^2P—y\,^2P°$	$^3/_2—^3/_2$
2109,097	10	3,26	9,14	$b\,^2H—w\,^2H°$	$^9/_2—^9/_2$
2108,139	15	1,69	7,57	$a\,^4P—z\,^2D°$	$^3/_2—^5/_2$
2107,555	10	3,24	9,13	$b\,^2H—w\,^2H°$	$^{11}/_2—^{11}/_2$
2098,181	25	2,52	8,42	$a\,^2H—y\,^2H°$	$^{11}/_2—^9/_2$
2097,512	25	1,69	7,60	$a\,^4P—y\,^4P°$	$^3/_2—^3/_2$
2093,683	35	3,76	9,69	$b\,^2G—v\,^2F°$	$^9/_2—^7/_2$
2087,527	25	2,34	8,21	$a\,^2P—z\,^2S°$	$^1/_2—^1/_2$
2080,912	20	2,03	7,99	$a\,^2G—z\,^2F°$	$^7/_2—^5/_2$
2078,164	8	1,97	7,92	$a\,^2G—y\,^4G°$	$^9/_2—^9/_2$
2077,507	12	2,64	8,60	$a\,^2D—x\,^4P°$	$^3/_2—^1/_2$
2075,683	5	2,28	8,24	$a\,^2P—x\,^4F°$	$^3/_2—^5/_2$
2074,195	8	1,96	7,94	$a\,^2G—y\,^4G°$	$^9/_2—^7/_2$
2073,147	8	1,73	7,70	$a\,^4P—z\,^2D°$	$^1/_2—^3/_2$
2071,821	10	2,28	8,25	$a\,^2P—x\,^4F°$	$^3/_2—^3/_2$
2070,330	8	3,38	9,37	$a\,^2F—w\,^2F°$	$^7/_2—^7/_2$
2069,952	10	3,42	9,41	$a\,^2F—w\,^2F°$	$^5/_2—^5/_2$
2067,917	20	2,64	8,64	$a\,^2D—y\,^2F°$	$^3/_2—^5/_2$
2066,005	15	2,34	8,34	$a\,^2P—y\,^2D°$	$^1/_2—^3/_2$
2064,335	25	1,96	7,97	$a\,^2G—z\,^2F°$	$^9/_2—^7/_2$
2063,672	25	1,96	7,97	$a\,^2G—z\,^2F°$	$^9/_2—^7/_2$
2057,332	12	1,67	7,69	$a\,^4P—y\,^4F°$	$^5/_2—^7/_2$
2055,270	20	2,28	8,31	$a\,^2P—y\,^2D°$	$^3/_2—^5/_2$
2051,688	25	2,03	8,08	$a\,^2G—y\,^2G°$	$^7/_2—^7/_2$
2048,492	5	2,58	8,63	$a\,^2H—y\,^2F°$	$^9/_2—^7/_2$
2041,345	25	1,96	8,04	$a\,^2G—y\,^2G°$	$^9/_2—^9/_2$
2040,687	25	1,96	8,04	$a\,^2G—y\,^2G°$	$^9/_2—^9/_2$
2036,435	20	2,54	8,63	$a\,^2D—y\,^2F°$	$^5/_2—^7/_2$
2033,064	25	2,03	8,12	$a\,^2G—z\,^2H°$	$^7/_2—^9/_2$
2029,182	8	1,96	8,08	$a\,^2G—y\,^2G°$	$^9/_2—^7/_2$
2027,778	5	2,84	8,96	$b\,^4F—w\,^4F°$	$^5/_2—^5/_2$
2020,739	25	1,67	7,80	$a\,^4P—x\,^4D°$	$^5/_2—^7/_2$
2019,427	25	1,96	8,10	$a\,^2G—z\,^2H°$	$^9/_2—^{11}/_2$
2017,090	15	1,73	7,86	$a\,^4P—x\,^4D°$	$^1/_2—^3/_2$
2016,154	10	2,83	8,99	$b\,^4F—w\,^4D°$	$^3/_2—^5/_2$
2016,092	10	2,84	8,98	$b\,^4F—w\,^4D°$	$^5/_2—^3/_2$
2015,500	20	1,70	7,84	$a\,^4P—x\,^4D°$	$^3/_2—^5/_2$
2013,268	15	1,73	7,88	$a\,^4P—x\,^4D°$	$^1/_2—^1/_2$

λ, Å	I	E_H, eV	E_B, eV	Transition	J
2010,688	25	2,58	8,74	$a\,^2H—x\,^2G^\circ$	$^9/_2—^7/_2$
2007,711	12	1,70	7,86	$a\,^4P—x\,^4D^\circ$	$^3/_2—^3/_2$
2007,452	15	1,67	7,84	$a\,^4P—x\,^4D^\circ$	$^5/_2—^5/_2$
2007,013	12	2,83	9,01	$b\,^4F—w\,^4D^\circ$	$^7/_2—^5/_2$
2000,368	30	2,52	8,71	$a\,^2H—x\,^2G^\circ$	$^{11}/_2—^9/_2$
1999,430	10	2,80	9,01	$b\,^4F—w\,^4F^\circ$	$^9/_2—^9/_2$
1994,857	20	3,19	9,41	$b\,^2P—w\,^2F^\circ$	$^3/_2—^5/_2$
1993,289	8	1,96	8,18	$a\,^2G—x\,^4F^\circ$	$^9/_2—^9/_2$
1964,330	12	2,69	9,00	$a\,^4H—w\,^4D^\circ$	$^7/_2—^5/_2$
1963,110	25	2,69	9,01	$a\,^4H—w\,^4D^\circ$	$^7/_2—^7/_2$
1958,121	5	2,67	9,01	$a\,^4H—w\,^4F^\circ$	$^9/_2—^9/_2$
1948,372	10	2,58	8,94	$a\,^2H—x\,^2H^\circ$	$^9/_2—^9/_2$
1938,899	8	2,85	9,25	$b\,^4F—x\,^2D^\circ$	$^3/_2—^5/_2$
1936,781	20	2,03	8,42	$a\,^2G—y\,^2H^\circ$	$^7/_2—^9/_2$
1935,296	15	1,96	8,36	$a\,^2G—y\,^2H^\circ$	$^9/_2—^{11}/_2$
1932,477	15	2,64	9,06	$a\,^2D—x\,^2F^\circ$	$^3/_2—^5/_2$
1925,987	20	2,52	8,96	$a\,^2H—z\,^2H^\circ$	$^{11}/_2—^{11}/_2$
1922,797	20	2,54	8,99	$a\,^2D—w\,^4D^\circ$	$^5/_2—^3/_2$
1917,337	15	1,96	8,42	$a\,^2G—y\,^2H^\circ$	$^9/_2—^9/_2$
1910,669	8	2,58	9,07	$a\,^2H—w\,^2G^\circ$	$^9/_2—^7/_2$
1904,787	15	2,54	9,05	$a\,^2D—x\,^2F^\circ$	$^5/_2—^7/_2$
1898,538	10	2,54	9,08	$a\,^2D—y\,^2P^\circ$	$^5/_2—^3/_2$
1895,675	10	2,52	9,06	$a\,^2H—w\,^2G^\circ$	$^{11}/_2—^9/_2$
1894,006	10	2,58	9,13	$a\,^2H—w\,^2H^\circ$	$^9/_2—^{11}/_2$
1888,729	20	2,58	9,14	$a\,^2H—w\,^2H^\circ$	$^9/_2—^9/_2$
1880,976	20	2,58	9,17	$a\,^2H—y\,^2I^\circ$	$^9/_2—^{11}/_2$
1877,462	20	2,52	9,13	$a\,^2H—w\,^2H^\circ$	$^{11}/_2—^{11}/_2$
1876,835	15	2,03	8,63	$a\,^2G—y\,^2F^\circ$	$^7/_2—^5/_2$
1876,173	8	2,64	9,25	$a\,^2D—x\,^2D^\circ$	$^3/_2—^3/_2$
1875,536	15	4,48	11,08	$b\,^2D—3^\circ$	$^3/_2—^5/_2$
1864,743	20	2,52	9,17	$a\,^2H—y\,^2I^\circ$	$^{11}/_2—^{13}/_2$
1860,040	20	1,96	8,62	$a\,^2G—y\,^2F^\circ$	$^9/_2—^7/_2$
1859,744	15	0,98	7,65	$a\,^4D—y\,^4D^\circ$	$^7/_2—^7/_2$
1857,935	12	0,08	6,75	$a\,^6D—z\,^8P^\circ$	$^5/_2—^5/_2, ^7/_2$
1848,768	12	2,54	9,25	$a\,^2D—x\,^2D^\circ$	$^5/_2—^5/_2$
1848,231	5	0,05	6,75	$a\,^6D—z\,^8P^\circ$	$^7/_2—^5/_2, ^7/_2$
1846,581	12	2,03	8,74	$a\,^2G—x\,^2G^\circ$	$^7/_2—^7/_2$
1844,590	5	5,48	12,20	$z\,^4F—e\,^4H$	$^9/_2—^9/_2$
1841,701	10	1,04	7,77	$a\,^4D—y\,^4D^\circ$	$^5/_2—^5/_2$
1835,869	15	1,96	8,71	$a\,^2G—x\,^2G^\circ$	$^9/_2—^9/_2$
1809,316	10	2,64	9,49	$a\,^2D—x\,^2P^\circ$	$^3/_2—^1/_2$
1798,163	10	2,54	9,43	$a\,^2D—x\,^2P^\circ$	$^5/_2—^3/_2$
1793,371	10	2,03	8,94	$a\,^2G—x\,^2H^\circ$	$^7/_2—^9/_2$
1787,997	35	2,89	9,82	$a\,^6S—x\,^6P^\circ$	$^5/_2—^3/_2$
1786,738	40	2,89	9,83	$a\,^6S—x\,^6P^\circ$	$^5/_2—^5/_2$
1785,262	40	2,89	9,83	$a\,^6S—x\,^6P^\circ$	$^5/_2—^7/_2$
1772,518	15	1,96	8,96	$a\,^2G—x\,^2H^\circ$	$^9/_2—^{11}/_2$
1761,379	25	2,03	9,06	$a\,^2G—w\,^2G^\circ$	$^7/_2—^7/_2$
1760,415	20	1,96	9,01	$a\,^2G—w\,^4F^\circ$	$^9/_2—^9/_2$
1746,816	20	1,96	9,06	$a\,^2G—w\,^2G^\circ$	$^9/_2—^9/_2$
1732,253	15	6,21	13,37	$c\,^4F—20^\circ$	$^9/_2—^7/_2$
1731,038	10	2,28	9,44	$a\,^2P—x\,^2P^\circ$	$^3/_2—^3/_2$
1726,394	12	0,38	7,57	$a\,^4F—z\,^4G^\circ$	$^3/_2—^5/_2$
1725,402	5	4,49	11,68	$b\,^2D—14^\circ$	$^5/_2—^5/_2$
1724,963	8	0,30	7,49	$a\,^4F—y\,^4P^\circ$	$^7/_2—^5/_2$
1724,847	8	0,38	7,57	$a\,^4F—z\,^2D^\circ$	$^3/_2—^5/_2$
1720,621	20	0,35	7,54	$a\,^4F—z\,^4G^\circ$	$^5/_2—^7/_2$

λ, Å	I	E_H, eV	E_B, eV	Transition	J
1720,042	10	1,72	8,92	$a\ ^4P-w\ ^4P^\circ$	$1/2-3/2$
1715,507	12	1,69	8,91	$a\ ^4P-w\ ^4P^\circ$	$3/2-5/2$
1713,002	20	0,30	7,54	$a\ ^4F-z\ ^4G^\circ$	$7/2-9/2$
1709,678	15	1,67	8,91	$a\ ^4P-w\ ^4P^\circ$	$5/2-5/2$
1708,627	8	0,30	7,54	$a\ ^4F-z\ ^4G^\circ$	$7/2-7/2$
1702,045	25	0,23	7,51	$a\ ^4F-z\ ^4G^\circ$	$9/2-11/2$
1696,800	8	0,23	7,54	$a\ ^4F-z\ ^4G^\circ$	$9/2-9/2$
1691,289	8	0,38	7,71	$a\ ^4F-y\ ^4F^\circ$	$3/2-3/2$
1690,781	8	1,66	8,99	$a\ ^4P-w\ ^4D^\circ$	$5/2-5/2$
1689,821	10	1,66	9,00	$a\ ^4P-w\ ^4D^\circ$	$5/2-7/2$
1686,475	8	0,30	7,65	$a\ ^4F-y\ ^4D^\circ$	$7/2-7/2$
1685,953	5	0,35	7,70	$a\ ^4F-y\ ^4F^\circ$	$5/2-5/2$
1685,457	8	0,30	7,65	$a\ ^4F-y\ ^4D^\circ$	$7/2-7/2$
1679,388	15	2,02	9,40	$a\ ^2G-w\ ^2F^\circ$	$7/2-5/2$
1674,716	10	0,38	7,79	$a\ ^4F-y\ ^4D^\circ$	$3/2-1/2$
1673,470	15	1,96	9,37	$a\ ^2G-w\ ^2F^\circ$	$9/2-7/2$
1670,759	25	0,23	7,65	$a\ ^4F-y\ ^4D^\circ$	$9/2-7/2$
1663,226	15	0,35	7,80	$a\ ^4F-y\ ^4D^\circ$	$5/2-3/2$
1659,487	20	0,30	7,77	$a\ ^4F-y\ ^4D^\circ$	$7/2-5/2$
1658,785	15	0,23	7,71	$a\ ^4F-y\ ^4F^\circ$	$9/2-9/2$
1654,484	5	0,38	7,89	$a\ ^4F-x\ ^4D^\circ$	$3/2-1/2$
1654,105	5	1,09	8,59	$a\ ^4D-x\ ^4P^\circ$	$1/2-3/2$
1650,709	20	1,09	8,60	$a\ ^4D-x\ ^4P^\circ$	$1/2-1/2$
1649,583	20	1,07	8,59	$a\ ^4D-x\ ^4P^\circ$	$3/2-3/2$
1649,444	15	0,35	7,87	$a\ ^4F-x\ ^4D^\circ$	$5/2-3/2$
1647,161	25	1,04	8,57	$a\ ^4D-x\ ^4P^\circ$	$5/2-5/2$
1646,187	20	1,07	8,60	$a\ ^4D-x\ ^4P^\circ$	$3/2-1/2$
1643,588	15	0,30	7,85	$a\ ^4F-x\ ^4D^\circ$	$7/2-5/2$
1642,187	5	3,38	10,93	$a\ ^2F-1^\circ$	$7/2-7/2$
1641,761	25	1,04	8,59	$a\ ^4D-x\ ^4P^\circ$	$5/2-3/2$
1640,167	12	0,38	7,95	$a\ ^4F-y\ ^4G^\circ$	$3/2-5/2$
1639,403	30	0,12	7,68	$a\ ^6D-y\ ^6P^\circ$	$1/2-3/2$
1637,400	15	0,23	7,80	$a\ ^4F-x\ ^4D^\circ$	$9/2-7/2$
1636,334	30	0,11	7,68	$a\ ^6D-y\ ^6P^\circ$	$3/2-3/2$
1635,389	35	0,98	8,57	$a\ ^4D-x\ ^4P^\circ$	$7/2-5/2$
1634,353	20	0,11	7,69	$a\ ^6D-y\ ^6P^\circ$	$3/2-5/2$
1633,907	15	0,35	7,93	$a\ ^4F-y\ ^4G^\circ$	$5/2-7/2$
1631,124	30	0,08	7,68	$a\ ^6D-y\ ^6P^\circ$	$5/2-3/2$
1629,155	30	0,08	7,69	$a\ ^6D-y\ ^6P^\circ$	$5/2-5/2$
1625,919	15	0,08	7,71	$a\ ^6D-y\ ^6P^\circ$	$5/2-7/2$
1625,525	20	0,30	7,92	$a\ ^4F-y\ ^4G^\circ$	$7/2-9/2$
1623,102	8	0,30	7,93	$a\ ^4F-y\ ^4G^\circ$	$7/2-7/2$
1621,685	30	0,05	7,69	$a\ ^6D-y\ ^6P^\circ$	$7/2-5/2$
1618,464	25	0,05	7,71	$a\ ^6D-y\ ^6P^\circ$	$7/2-7/2$
1612,814	20	0,23	7,92	$a\ ^4F-y\ ^4G^\circ$	$9/2-11/2$
1610,933	15	0,23	7,92	$a\ ^4F-y\ ^4G^\circ$	$9/2-9/2$
1608,446	35	0,00	7,71	$a\ ^6D-y\ ^6P^\circ$	$9/2-7/2$
1602,588	12	3,94	11,68	$b\ ^2F-14^\circ$	$5/2-5/2$
1588,295	10	0,38	8,19	$a\ ^4F-x\ ^4G^\circ$	$3/2-5/2$
1584,954	15	0,35	8,17	$a\ ^4F-x\ ^4G^\circ$	$5/2-7/2$
1581,243	8	0,35	8,19	$a\ ^4F-z\ ^4G^\circ$	$5/2-5/2$
1580,635	25	0,30	8,14	$a\ ^4F-x\ ^4G^\circ$	$7/2-9/2$
1574,931	20	0,38	8,25	$a\ ^4F-x\ ^4F^\circ$	$3/2-3/2$
1573,831	5	0,35	8,22	$a\ ^4F-x\ ^4F^\circ$	$5/2-7/2$
1570,248	20	0,35	8,24	$a\ ^4F-x\ ^4F^\circ$	$5/2-5/2$
1569,670	12	0,23	8,14	$a\ ^4F-x\ ^4G^\circ$	$9/2-11/2$
1568,031	8	0,35	8,25	$a\ ^4F-x\ ^4F^\circ$	$5/2-3/2$
1566,825	20	0,23	8,15	$a\ ^4F-x\ ^4G^\circ$	$9/2-9/2$

λ, Å	I	E_H, eV	E_B, eV	Transition	J
1563,790	25	0,30	8,23	$a\,^4F - x\,^4F^\circ$	$^7/_2 - ^7/_2$
1559,106	20	0,23	8,18	$a\,^4F - x\,^4F^\circ$	$^9/_2 - ^9/_2$
1558,706	10	0,38	8,33	$a\,^4F - y\,^2D^\circ$	$^3/_2 - ^3/_2$
1558,543	10	0,35	8,30	$a\,^4F - y\,^2D^\circ$	$^5/_2 - ^5/_2$
1495,311	15	2,89	11,38	$a\,^6S - w\,^6P^\circ$	$^5/_2 - ^3/_2$
1476,054	10	2,80	11,20	$b\,^4F - 6^\circ$	$^9/_2 - ^9/_2,\ ^7/_2$
1473,834	20	2,89	11,29	$a\,^6S - w\,^6P^\circ$	$^5/_2 - ^7/_2$
1465,043	20	2,89	11,34	$a\,^6S - w\,^6P^\circ$	$^5/_2 - ^5/_2$
1424,747	12	0,30	8,98	$a\,^4F - w\,^4D^\circ$	$^7/_2 - ^5/_2$
1424,047	8	0,30	9,01	$a\,^4F - w\,^4D^\circ$	$^7/_2 - ^7/_2$
1417,744	20	2,64	11,38	$a\,^2D - w\,^6P^\circ$	$^3/_2 - ^3/_2$
1413,707	25	0,98	9,75	$a\,^4D - w\,^2D^\circ$	$^7/_2 - ^5/_2$
1412,834	12	0,23	9,01	$a\,^4F - w\,^4D^\circ$	$^9/_2 - ^7/_2$
1397,581	12	4,62	13,41	$a\,^2S - 23^\circ$	$^1/_2 - ^3/_2$
1381,250	10	2,70	11,68	$b\,^4P - 14^\circ$	$^3/_2 - ^5/_2$
1364,590	12	1,97	11,05	$a\,^2G - 2^\circ$	$^9/_2 - ^7/_2$
1362,771	20	2,58	11,68	$b\,^4P - 14^\circ$	$^5/_2 - ^5/_2$
1360,870	5	2,28	11,38	$a\,^2P - w\,^6P^\circ$	$^3/_2 - ^3/_2$
1296,088	20	1,70	11,27	$a\,^4P - 8^\circ$	$^3/_2 - ^3/_2,\ ^1/_2$
1294,914	12	1,70	11,27	$a\,^4P - 9^\circ$	$^3/_2 - ^3/_2$
1291,594	15	1,67	11,26	$a\,^4P - 9^\circ$	$^5/_2 - ^3/_2$
1290,204	15	1,67	11,28	$a\,^4P - 10^\circ$	$^5/_2 - ^3/_2$
1275,801	20	0,11	9,82	$a\,^6D - x\,^6P^\circ$	$^3/_2 - ^3/_2$
1275,154	15	0,11	9,83	$a\,^6D - x\,^6P^\circ$	$^3/_2 - ^5/_2$
1272,638	15	0,08	9,82	$a\,^6D - x\,^6P^\circ$	$^5/_2 - ^3/_2$
1272,001	25	0,08	9,83	$a\,^6D - x\,^6P^\circ$	$^5/_2 - ^5/_2$
1267,437	25	0,05	9,82	$a\,^6D - x\,^6P^\circ$	$^7/_2 - ^5/_2$
1266,694	20	0,05	9,83	$a\,^6D - x\,^6P^\circ$	$^7/_2 - ^7/_2$
1260,542	20	0,00	9,83	$a\,^6D - x\,^6P^\circ$	$^9/_2 - ^7/_2$
1233,660	8	3,38	13,44	$a\,^2F - 26^\circ$	$^7/_2 - ^7/_2$
1220,882	5	1,04	11,19	$a\,^4D - 5^\circ$	$^5/_2 - ^7/_2$
1214,409	10	0,98	11,19	$a\,^4D - 5^\circ$	$^7/_2 - ^7/_2$
1213,764	20	1,07	11,29	$a\,^4D - 11^\circ$	$^3/_2 - ^5/_2$
1213,149	20	0,98	11,20	$a\,^4D - 6^\circ$	$^7/_2 - ^9/_2,\ ^7/_2$
1171,606	8	2,58	13,28	$b\,^4P - 17^\circ$	$^5/_2 - ^7/_2,\ ^5/_2$
1165,269	12	1,04	11,68	$a\,^4D - 14^\circ$	$^5/_2 - ^5/_2$
1159,347	20	0,98	11,68	$a\,^4D - 14^\circ$	$^7/_2 - ^5/_2$
1154,401	20	0,12	10,85	$a\,^6D - y\,^6F^\circ$	$^1/_2 - ^3/_2$
1153,955	15	0,12	10,86	$a\,^6D - y\,^6F^\circ$	$^1/_2 - ^1/_2$
1153,281	20	0,11	10,86	$a\,^6D - y\,^6F^\circ$	$^3/_2 - ^5/_2$
1152,882	20	0,11	10,85	$a\,^6D - y\,^6F^\circ$	$^3/_2 - ^3/_2$
1152,440	15	0,12	10,86	$a\,^6D - y\,^6F^\circ$	$^3/_2 - ^1/_2$
1151,163	25	0,08	10,85	$a\,^6D - y\,^6F^\circ$	$^5/_2 - ^7/_2$
1150,689	20	0,08	10,85	$a\,^6D - y\,^6F^\circ$	$^5/_2 - ^5/_2$
1150,292	20	0,08	10,85	$a\,^6D - y\,^6F^\circ$	$^5/_2 - ^3/_2$
1148,693	8	2,58	13,37	$b\,^4P - 20^\circ$	$^5/_2 - ^7/_2$
1148,295	30	0,05	10,84	$a\,^6D - y\,^6F^\circ$	$^7/_2 - ^9/_2$
1147,413	25	0,05	10,85	$a\,^6D - y\,^6F^\circ$	$^7/_2 - ^7/_2$
1146,963	15	0,05	10,85	$a\,^6D - y\,^6F^\circ$	$^7/_2 - ^5/_2$
1144,946	35	0,00	10,82	$a\,^6D - y\,^6F^\circ$	$^9/_2 - ^{11}/_2$
1144,052	5	2,58	13,42	$b\,^4P - 24^\circ$	$^5/_2 - ^7/_2$
1143,235	25	0,00	10,84	$a\,^6D - y\,^6F^\circ$	$^9/_2 - ^9/_2$
1142,334	25	0,00	10,84	$a\,^6D - y\,^6F^\circ$	$^9/_2 - ^7/_2$
1138,642	25	0,05	10,93	$a\,^6D - 1^\circ$	$^7/_2 - ^7/_2$
1138,039	5	0,30	11,19	$a\,^4F - 5^\circ$	$^7/_2 - ^7/_2$
1133,678	25	0,00	10,93	$a\,^6D - 1^\circ$	$^9/_2 - ^7/_2$
1133,413	25	0,35	11,29	$a\,^4F - 11^\circ$	$^5/_2 - ^5/_2$

λ, Å	I	E_H, eV	E_B, eV	Transition	J
1130,428	25	0,08	11,05	$a\ ^6D—2°$	$^5/_2—^7/_2$
1129,777	12	0,23	11,20	$a\ ^4F—6°$	$^9/_2—^9/_2,\ ^7/_2$
1128,909	20	0,11	11,09	$a\ ^6D—3°$	$^3/_2—^5/_2$
1128,530	10	2,88	13,87	$a\ ^6S—29°$	$^5/_2—^5/_2$
1128,180	5	0,30	11,29	$a\ ^4F—11°$	$^7/_2—^5/_2$
1128,074	25	0,12	11,11	$a\ ^6D—4°$	$^1/_2—^3/_2$
1126,850	20	0,05	11,05	$a\ ^6D—2°$	$^7/_2—^7/_2$
1126,603	20	0,11	11,11	$a\ ^6D—4°$	$^3/_2—^3/_2$
1126,425	20	0,08	11,09	$a\ ^6D—3°$	$^5/_2—^5/_2$
1124,134	20	0,08	11,11	$a\ ^6D—4°$	$^5/_2—^3/_2$
1122,858	25	0,05	11,09	$a\ ^6D—3°$	$^7/_2—^5/_2$
1121,987	25	0,00	11,04	$a\ ^6D—2°$	$^9/_2—^7/_2$
1112,086	35	0,12	11,26	$a\ ^6D—9°$	$^1/_2—^3/_2$
1111,114	15	0,05	11,20	$a\ ^6D—6°$	$^7/_2—^9/_2,\ ^7/_2$
1106,362	5	0,00	11,20	$a^6\ D—6°$	$^9/_2—^9/_2,\ ^7/_2$
1106,215	15	0,08	11,29	$a\ ^6D—11°$	$^5/_2—^5/_2$
1102,385	8	0,11	11,34	$a\ ^6D—w\ ^6P°$	$^3/_2—^5/_2$
1101,538	20	0,05	11,29	$a\ ^6D—w\ ^6P°$	$^7/_2—^7/_2$
1100,525	20	0,12	11,38	$a\ ^6D—w\ ^6P°$	$^1/_2—^3/_2$
1100,026	20	0,08	11,35	$a\ ^6D—w\ ^6P°$	$^5/_2—^5/_2$
1099,117	25	0,11	11,38	$a\ ^6D—w\ ^6P°$	$^3/_2—^3/_2$
1096,886	30	0,00	11,29	$a\ ^6D—w\ ^6P°$	$^9/_2—^7/_2$
1096,793	20	0,08	11,38	$a\ ^6D—w\ ^6P°$	$^5/_2—^3/_2$
1096,616	20	0,05	11,35	$a\ ^6D—w\ ^6P°$	$^7/_2—^5/_2$
1071,596	30	0,08	11,64	$a\ ^6D—13°$	$^5/_2—^7/_2$
1069,038	15	0,08	11,67	$a\ ^6D—14°$	$^5/_2—^5/_2$
1068,356	30	0,05	11,64	$a\ ^6D—13°$	$^7/_2—^7/_2$
1063,982	15	0,00	11,64	$a\ ^6D—13°$	$^9/_2—^7/_2$
1062,758	20	0,08	11,75	$a\ ^6D—15°$	$^5/_2—^7/_2$
1059,571	20	0,05	11,75	$a\ ^6D—15°$	$^7/_2—^7/_2$
1055,269	25	0,00	11,75	$a\ ^6D—15°$	$^9/_2—^7/_2$
1015,520	20	1,04	13,25	$a\ ^4D—16°$	$^5/_2—^7/_2$
1015,083	10	1,07	13,29	$a\ ^4D—18°$	$^3/_2—^5/_2$
1012,417	25	1,04	13,28	$a\ ^4D—17°$	$^5/_2—^7/_2,\ ^5/_2$
1012,088	20	1,04	13,29	$a\ ^4D—18°$	$^5/_2—^5/_2$
1011,037	25	0,98	13,25	$a\ ^4D—16°$	$^7/_2—^7/_2$
1007,975	25	0,98	13,28	$a\ ^4D—17°$	$^7/_2—^7/_2,\ ^5/_2$
1007,057	20	0,98	13,29	$a\ ^4D—18°$	$^7/_2—^5/_2$
995,829	8	0,98	13,44	$a\ ^4D—26°$	$^7/_2—^7/_2$
952,470	10	0,23	13,25	$a\ ^4F—16°$	$^9/_2—^7/_2$
945,095	25	0,30	13,42	$a\ ^4F—24°$	$^7/_2—^7/_2$
943,910	15	0,30	13,44	$a\ ^4F—26°$	$^7/_2—^7/_2$
943,267	12	0,23	13,37	$a\ ^4F—20°$	$^9/_2—^7/_2$
941,660	12	0,08	13,25	$a\ ^6D—16°$	$^5/_2—^7/_2$
939,159	20	0,05	13,25	$a\ ^6D—16°$	$^7/_2—^7/_2$
938,967	10	0,08	13,28	$a\ ^6D—17°$	$^5/_2—^7/_2,\ ^5/_2$
936,484	8	0,05	13,28	$a\ ^6D—17°$	$^7/_2—^7/_2,\ ^5/_2$
932,687	30	0,08	13,37	$a\ ^6D—20°$	$^5/_2—^7/_2$
932,244	30	0,11	13,40	$a\ ^6D—22°$	$^3/_2—^5/_2$
931,709	10	0,11	13,41	$a\ ^6D—23°$	$^3/_2—^3/_2$
931,142	25	0,12	13,44	$a\ ^6D—25°$	$^1/_2—^3/_2$
930,558	30	0,08	13,40	$a\ ^6D—22°$	$^5/_2—^5/_2$
930,219	30	0,05	13,37	$a\ ^6D—20°$	$^7/_2—^7/_2$
930,165	30	0,11	13,44	$a\ ^6D—25°$	$^3/_2—^3/_2$
930,030	30	0,08	13,41	$a\ ^6D—23°$	$^5/_2—^3/_2$
929,612	30	0,08	13,42	$a\ ^6D—24°$	$^5/_2—^7/_2$
929,538	30	0,05	13,38	$a\ ^6D—21°$	$^7/_2—^9/_2,\ ^7/_2$
928,470	20	0,08	13,44	$a\ ^6D—25°$	$^5/_2—^3/_2$

λ, Å	I	E_H, eV	E_B, eV	Transition	J
928,107	30	0,05	13,40	$a\ ^6D—22°$	$^7/_2—^5/_2$
927,632	8	0,12	13,49	$a\ ^6D—27°$	$^1/_2—^3/_2$
927,176	30	0,05	13,42	$a\ ^6D—24°$	$^7/_2—^7/_2$
926,900	25	0,00	13,37	$a\ ^6D—20°$	$^9/_2—^7/_2$
926,618	10	0,11	13,49	$a\ ^6D—27°$	$^3/_2—^3/_2$
926,220	60	0,00	13,38	$a\ ^6D—21°$	$^9/_2—^7/_2$, $^9/_2$
924,970	15	0,08	13,49	$a\ ^6D—27°$	$^5/_2—^3/_2$
923,884	30	0,00	13,42	$a\ ^6D—24°$	$^9/_2—^7/_2$
900,360	5	0,11	13,87	$a\ ^6D—29°$	$^3/_2—^5/_2$

Fe III, ground state $1s^2\,2s^2\,2p^6\,3s^2\,3p^6\,3d^6\ ^5D_4$

Ionization potential 247 200 cm^{-1}; 30,647 eV

λ, Å	I	E_H, eV	E_B, eV	Transition	J
6185,26	9	22,94	24,94	$e\ ^5P—w\ ^5D°$	2—3
6169,74	9	22,74	24,75	$e\ ^3G—u\ ^3H°$	5—6
6056,36	9	22,54	24,59	$f\ ^5G—x\ ^5G°$	3—3
6054,18	11	22,54	24,59	$f\ ^5G—x\ ^5G°$	4—4
6048,72	11	22,54	24,59	$f\ ^5G—x\ ^5G°$	5—5
6036,56	13	22,54	24,59	$f\ ^5G—x\ ^5G°$	6—6
6032,59	7	18,81	20,87	$e\ ^5S—w\ ^5P°$	2—3
5999,54	5	18,81	20,88	$e\ ^5S—w\ ^5P°$	2—2
5953,62	6	18,79	20,87	$e\ ^5D—w\ ^5P°$	4—3
5929,69	18	18,51	20,60	$e\ ^7S—y\ ^7P°$	3—2
5920,13	7	18,79	20,88	$e\ ^5D—w\ ^5P°$	3—2
5891,91	6	18,51	20,61	$e\ ^7S—y\ ^7P°$	3—3
5833,93	10	18,51	20,63	$e\ ^7S—y\ ^7P°$	3—4
5363,80	8	—	—	—	—
5302,99	6	18,26	20,60	$e\ ^7D—y\ ^7P°$	2—2
5282,29	7	18,26	20,61	$e\ ^7D—y\ ^7P°$	4—3
5276,47	7	18,26	20,61	$e\ ^7D—y\ ^7P°$	3—3
5243,31	10	18,27	20,63	$e\ ^7D—y\ ^7P°$	5—4
5149,33	7	—	—	—	—
5100,706	10	—	—	—	—
5030,75	6	—	—	—	—
5002,02	8	—	—	—	—
4559,09	6	—	—	—	—
4431,02	7	8,25	11,04	$a\ ^5P—z\ ^5P°$	2—3
4419,59	10	8,24	11,04	$a\ ^5P—z\ ^5P°$	3—3
4395,76	6	8,26	11,08	$a\ ^5P—z\ ^5P°$	1—2
4372,81	20	22,91	25,74	$w\ ^5F°—e\ ^5G$	1—2
4310,36	12	22,87	25,74	$z\ ^7F°—e\ ^5G$	6—7
4304,77	10	22,86	25,74	$z\ ^7F°—e\ ^7G$	5—6
4296,85	10	22,86	25,74	$z\ ^7F°—e\ ^7G$	4—5
4286,16	10	22,85	25,74	$z\ ^7F°—e\ ^7G$	3—4
4271,47	6	—	—	—	—
4249,95	7	—	—	—	—
4243,85	8	—	—	—	—
4235,54	10	—	—	—	—
4222,39	8	—	—	—	—
4210,87	10	—	—	—	—
4200,38	6	—	—	—	—
4200,06	6	—	—	—	—
4189,10	7	—	—	—	—

λ, Å	I	E_H, eV	E_B, eV	Transition	J
4174,27	10	—	—	—	—
4166,84	9	20,63	23,61	$y\ ^7P^\circ - f\ ^7D$	4—4
4164,73	20	20,63	23,61	$y\ ^7P^\circ - f\ ^7D$	4—5
4154,98	8	—	—	—	—
4140,48	6	20,61	23,60	$y\ ^7P^\circ - f\ ^7D$	3—2
4139,35	8	20,61	23,60	$y\ ^7P^\circ - f\ ^7D$	3—3
4137,76	10	20,61	23,61	$y\ ^7P^\circ - f\ ^7D$	3—4
4122,78	8	20,60	23,60	$y\ ^7P^\circ - f\ ^7D$	2—1
4122,02	8	20,60	23,60	$y\ ^7P^\circ - f\ ^7D$	2—2
4120,90	8	20,60	23,60	$y\ ^7P^\circ - f\ ^7D$	2—3
4081,00	7	20,63	23,67	$y\ ^7P^\circ - f\ ^7S$	4—3
3968,71	8	20,88	24,00	$w\ ^5P^\circ - f\ ^5D$	2—3
3954,33	12	20,87	24,00	$w\ ^5P^\circ - f\ ^5D$	3—4
3603,88	9	11,21	14,65	$d\ ^3F - z\ ^3F^\circ$	2—2
3600,94	10	11,22	14,66	$d\ ^3F - z\ ^3F^\circ$	3—3
3586,04	9	11,22	14,67	$d\ ^3F - z\ ^3F^\circ$	4—4
3501,76	8	11,12	14,66	$c\ ^3G - z\ ^3F^\circ$	3—3
3500,28	7	11,58	15,12	$e\ ^3F - z\ ^3G^\circ$	4—5
3499,59	7	11,13	14,67	$c\ ^3G - z\ ^3F^\circ$	4—4
3396,70	8	11,02	14,67	$b\ ^3H - z\ ^3H^\circ$	6—6
3382,18	6	14,17	17,83	$d\ ^3G - v\ ^3G^\circ$	5—5
3360,87	6	14,17	17,86	$d\ ^3G - v\ ^3G^\circ$	4—4
3347,70	8	10,99	14,70	$b\ ^3H - z\ ^3H^\circ$	5—5
3339,38	10	10,37	14,08	$a\ ^5F - z\ ^5G^\circ$	1—2
3329,89	7	10,99	14,71	$b\ ^3H - z\ ^3H^\circ$	4—4
3305,22	10	10,33	14,08	$a\ ^5F - z\ ^5G^\circ$	2—3
3292,04	8	10,32	14,08	$a\ ^5F - z\ ^5G^\circ$	3—3
3288,81	15	10,32	14,09	$a\ ^5F - z\ ^5G^\circ$	3—4
3280,56	6	10,31	14,09	$a\ ^5F - z\ ^5G^\circ$	4—4
3276,08	15	10,31	14,09	$a\ ^5F - z\ ^5G^\circ$	4—5
3273,53	6	10,31	14,09	$a\ ^5F - z\ ^5G^\circ$	5—5
3266,88	20	10,31	14,10	$a\ ^5F - z\ ^5G^\circ$	5—6
3262,46	6	14,17	17,97	$d\ ^3G - x\ ^1G^\circ$	5—4
3218,34	6	14,62	18,48	$d\ ^1G - w\ ^1G^\circ$	4—4
3215,63	8	10,23	14,08	$c\ ^3D - z\ ^5G^\circ$	1—2
3204,76	6	10,22	14,08	$c\ ^3D - z\ ^5G^\circ$	2—3
3178,01	10	11,22	15,12	$d\ ^3F - z\ ^3G^\circ$	3—4
3175,99	10	11,22	15,12	$d\ ^3F - z\ ^3G^\circ$	4—5
3174,09	10	11,21	15,12	$d\ ^3F - z\ ^3G^\circ$	2—3
3136,43	10	11,22	15,17	$d\ ^3F - z\ ^3D^\circ$	4—3
3120,847	20	11,15	15,12	$c\ ^3G - z\ ^3G^\circ$	5—4
3111,609	8	10,31	14,29	$a\ ^5F - z\ ^5H^\circ$	4—5
3110,841	8	11,22	15,20	$d\ ^3F - z\ ^3D^\circ$	3—2
3107,978	6	11,13	15,11	$c\ ^3G - z\ ^3G^\circ$	4—5
3084,07	6	11,22	15,24	$d\ ^3F - y\ ^5D^\circ$	3—2
3054,138	6	10,37	14,43	$a\ ^5F - z\ ^5D^\circ$	1—1
3027,006	6	11,00	15,12	$b\ ^3H - z\ ^3G^\circ$	6—5
3023,83	8	10,33	14,43	$a\ ^5F - z\ ^5D^\circ$	2—2
3018,789	6	10,33	14,44	$a\ ^5F - z\ ^5D^\circ$	2—3
3015,260	7	10,31	14,42	$a\ ^5F - z\ ^5F^\circ$	4—5
3013,167	20	10,31	14,42	$a\ ^5F - z\ ^5F^\circ$	5—5
3007,793	6	10,32	14,44	$a\ ^5F - z\ ^5D^\circ$	3—3
3007,275	20	18,79	22,91	$e\ ^5D - w\ ^5F^\circ$	4—4
3001,617	12	10,31	14,44	$a\ ^5F - z\ ^5F^\circ$	4—4
2977,222	6	10,33	14,50	$a\ ^5F - z\ ^5F^\circ$	2—1
2963,230	8	10,32	14,50	$a\ ^5F - z\ ^5F^\circ$	3—2
2958,286	6	10,46	14,65	$c\ ^3F - z\ ^3F^\circ$	2—2
2948,388	8	10,31	14,51	$a\ ^5F - z\ ^5F^\circ$	4—3

λ, Å	I	E_H, eV	E_B, eV	Transition	J
2939,55	7	10,22	14,43	$c\,^3D - z\,^5D^\circ$	2—2
2923,902	8	10,43	14,67	$c\,^3F - z\,^3F^\circ$	4—4
2907,701	12	11,22	15,48	$d\,^3F - y\,^3D^\circ$	4—3
2907,497	10	10,31	14,58	$a\,^5F - z\,^5D^\circ$	5—4
2905,80	8	10,23	14,49	$c\,^3D - z\,^5S^\circ$	1—2
2904,431	12	11,22	15,49	$d\,^3F - y\,^3D^\circ$	3—2
2902,47	9	10,23	14,50	$c\,^3D - z\,^5F^\circ$	1—1
2895,076	8	11,21	15,49	$d\,^3F - y\,^3D^\circ$	2—1
2858,664	7	11,22	15,55	$d\,^3F - y\,^3F^\circ$	4—4
2850,288	7	13,13	17,48	$d\,^3D - v\,^3F^\circ$	3—4
2818,624	6	13,58	17,98	$c\,^1D - x\,^1F^\circ$	2—3
2813,241	10	11,14	15,55	$c\,^3G - y\,^3F^\circ$	5—4
2803,441	6	11,13	15,55	$c\,^3G - y\,^3F^\circ$	4—4
2788,258	6	11,13	15,58	$c\,^3G - y\,^3F^\circ$	4—3
2773,306	8	13,58	18,05	$c\,^1D - x\,^1D^\circ$	2—2
2705,117	7	18,26	22,84	$e\,^7D - z\,^7F^\circ$	2—2
2701,13	8	18,26	22,85	$e\,^7D - z\,^7F^\circ$	3—3
2700,045	8	18,26	22,85	$e\,^7D - z\,^7F^\circ$	2—3
2698,414	7	18,26	22,86	$e\,^7D - z\,^7F^\circ$	4—4
2696,905	7	18,26	22,86	$e\,^7D - z\,^7F^\circ$	3—4
2695,314	9	18,26	22,86	$e\,^7D - z\,^7F^\circ$	4—5
2695,150	10	18,27	22,87	$e\,^7D - z\,^7F^\circ$	5—6
2678,810	6	12,03	16,66	$c\,^1F - z\,^1G^\circ$	3—4
2646,751	6	10,32	15,00	$a\,^5F - y\,^5F^\circ$	3—3
2645,39	9	10,43	15,12	$c\,^3F - z\,^3G^\circ$	4—5
2617,149	8	11,59	16,33	$c\,^1G - z\,^1H^\circ$	4—5
2608,112	7	10,31	15,06	$a\,^5F - y\,^5F^\circ$	5—5
2595,622	8	9,90	14,67	$a\,^3I - z\,^3H^\circ$	7—6
2584,038	6	11,58	16,37	$e\,^3F - x\,^3F^\circ$	3—3
2582,37	8	9,90	14,70	$a\,^3I - z\,^3H^\circ$	6—5
2574,838	7	9,90	14,71	$a\,^3I - z\,^3H^\circ$	5—4
2551,098	6	11,47	16,33	$a\,^1H - z\,^1H^\circ$	5—5
2511,418	6	10,31	15,24	$a\,^5F - y\,^5D^\circ$	5—4
2447,374	7	11,59	16,66	$c\,^1G - z\,^1G^\circ$	4—4
2438,174	8	5,08	10,17	$a\,^5S - z\,^7P^\circ$	2—2
2418,568	7	5,08	10,21	$a\,^5S - z\,^7P^\circ$	2—3
2403,551	6	10,99	16,15	$b\,^3H - z\,^3I^\circ$	4—5
2389,533	8	11,47	16,66	$a\,^1H - z\,^1G^\circ$	5—4
2363,51	7	11,13	16,37	$c\,^3G - x\,^3F^\circ$	4—3
2338,961	10	9,54	14,84	$b\,^3D - z\,^3P^\circ$	3—2
2336,768	10	11,13	16,43	$c\,^3G - y\,^3H^\circ$	4—5
2329,905	9	9,55	14,87	$b\,^3D - z\,^3P^\circ$	1—1
2326,948	10	11,12	16,45	$c\,^3G - y\,^3H^\circ$	3—4
2324,358	8	13,13	18,46	$d\,^3D - w\,^3P^\circ$	2—1
2321,71	10	11,46	16,80	$a\,^1H - z\,^1I^\circ$	5—6
2319,466	8	11,59	16,94	$c\,^1G - w\,^3F^\circ$	4—4
2319,220	10	9,55	14,90	$b\,^3D - z\,^3P^\circ$	1—0
2315,70	10	9,14	14,49	$c\,^3P - z\,^5S^\circ$	2—2
2303,012	7	11,58	16,96	$e\,^3F - w\,^3F^\circ$	4—3
2302,808	8	12,03	17,41	$c\,^1F - y\,^1F^\circ$	3—3
2295,859	15	10,90	16,30	$b\,^1F - z\,^1D^\circ$	3—2
2293,056	10	13,13	18,54	$d\,^3D - w\,^3P^\circ$	3—2
2291,850	6	13,13	18,54	$d\,^3D - w\,^3P^\circ$	2—2
2278,432	6	11,22	16,66	$d\,^3F - z\,^1G^\circ$	3—4
2277,820	8	11,22	16,66	$d\,^3F - z\,^1G^\circ$	4—4
2276,870	8	9,56	15,00	$b\,^3D - y\,^5F^\circ$	2—3
2274,00	8	12,03	17,48	$c\,^1F - v\,^3F^\circ$	3—4

λ, Å	I	E_H, eV	E_B, eV	Transition	J
2267,42	10	11,46	16,94	$a\,^{1}H - w\,^{3}F^{\circ}$	5—4
2261,592	12	10,90	16,38	$b\,^{1}F - x\,^{3}F^{\circ}$	3—2
2260,547	7	8,77	14,25	$b\,^{3}G - z\,^{5}H^{\circ}$	3—3
2257,406	8	9,54	15,03	$b\,^{3}D - y\,^{5}F^{\circ}$	3—4
2243,405	8	8,77	14,29	$b\,^{3}G - z\,^{5}H^{\circ}$	4—5
2241,54	12	10,77	16,30	$b\,^{1}D - z\,^{1}D^{\circ}$	2—2
2238,155	10	11,58	17,12	$e\,^{3}F - x\,^{3}G^{\circ}$	4—5
2235,908	10	11,58	17,12	$e\,^{3}F - x\,^{3}G^{\circ}$	3—4
2235,699	6	11,58	17,12	$e\,^{3}F - x\,^{3}G^{\circ}$	4—4
2233,654	6	11,22	16,76	$d\,^{3}F - y\,^{5}G^{\circ}$	3—4
2232,690	10	11,58	17,13	$e\,^{3}F - x\,^{3}G^{\circ}$	2—3
2232,548	8	10,90	16,45	$b\,^{1}F - y\,^{3}H^{\circ}$	3—4
2232,430	10	8,76	14,31	$b\,^{3}G - z\,^{5}H^{\circ}$	5—6
2229,267	10	11,22	16,77	$d\,^{3}F - y\,^{5}G^{\circ}$	4—5
2227,848	7	9,15	14,72	$c\,^{3}P - y\,^{5}P^{\circ}$	1—2
2221,830	10	9,14	14,72	$c\,^{3}P - y\,^{5}P^{\circ}$	2—2
2210,073	6	10,77	16,37	$b\,^{1}D - x\,^{3}F^{\circ}$	2—3
2208,85	10	10,77	16,38	$b\,^{1}D - x\,^{3}F^{\circ}$	2—2
2202,458	8	9,54	15,17	$b\,^{3}D - z\,^{3}D^{\circ}$	3—3
2195,532	6	11,13	16,77	$c\,^{3}G - y\,^{5}G^{\circ}$	4—5
2191,215	8	8,76	14,42	$b\,^{3}G - z\,^{5}F^{\circ}$	5—5
2186,876	6	11,22	16,88	$d\,^{3}F - x\,^{5}F^{\circ}$	4—5
2183,980	6	8,76	14,44	$b\,^{3}G - z\,^{5}F^{\circ}$	5—4
2180,410	12	9,15	14,84	$c\,^{3}P - z\,^{3}P^{\circ}$	1—2
2179,258	6	9,54	15,23	$b\,^{3}D - y\,^{5}D^{\circ}$	3—3
2174,658	15	9,14	14,84	$c\,^{3}P - z\,^{3}P^{\circ}$	2—2
2173,829	7	9,54	15,24	$b\,^{3}D - y\,^{5}D^{\circ}$	3—4
2171,045	12	9,17	14,87	$c\,^{3}P - z\,^{3}P^{\circ}$	0—1
2166,952	12	9,15	14,87	$c\,^{3}P - z\,^{3}P^{\circ}$	1—1
2161,270	10	9,14	14,87	$c\,^{3}P - z\,^{3}P^{\circ}$	2—1
2160,655	6	11,58	17,32	$e\,^{3}F - w\,^{3}G^{\circ}$	2—3
2158,472	12	11,59	17,34	$c\,^{1}G - y\,^{1}G^{\circ}$	4—4
2157,710	12	9,15	14,90	$c\,^{3}P - z\,^{3}P^{\circ}$	1—0
2152,706	6	11,58	17,34	$e\,^{3}F - y\,^{1}G^{\circ}$	4—4
2151,776	15	10,90	16,66	$b\,^{1}F - z\,^{1}G^{\circ}$	3—4
2147,904	7	8,66	14,43	$b\,^{5}D - z\,^{5}D^{\circ}$	2—1
2146,339	6	10,99	16,76	$b\,^{3}H - y\,^{5}G^{\circ}$	4—4
2146,062	8	8,66	14,43	$b\,^{5}D - z\,^{5}D^{\circ}$	2—2
2145,616	6	8,65	14,43	$b\,^{5}D - z\,^{5}D^{\circ}$	1—1
2144,743	7	10,34	16,12	$b\,^{1}I - z\,^{3}I^{\circ}$	6—7
2144,282	8	8,64	14,42	$b\,^{5}D - z\,^{5}F^{\circ}$	4—5
2143,827	7	8,66	14,44	$b\,^{5}D - z\,^{5}F^{\circ}$	3—4
2143,470	8	8,66	14,44	$b\,^{5}D - z\,^{5}D^{\circ}$	3—3
2143,045	7	9,54	15,34	$b\,^{3}D - y\,^{5}P^{\circ}$	2—3
2137,365	8	8,64	14,44	$b\,^{5}D - z\,^{5}F^{\circ}$	4—4
2134,861	9	10,34	16,15	$b\,^{1}I - z\,^{3}I^{\circ}$	6—5
2123,590	8	10,46	16,30	$c\,^{3}F - z\,^{1}D^{\circ}$	2—2
2118,567	6	8,65	14,50	$b\,^{5}D - z\,^{5}F^{\circ}$	1—2
2116,588	7	8,66	14,51	$b\,^{5}D - z\,^{5}F^{\circ}$	2—3
2113,891	6	12,03	17,89	$c\,^{1}F - t\,^{3}F^{\circ}$	3—4
2107,324	10	8,77	14,65	$b\,^{3}G - z\,^{3}F^{\circ}$	3—2
2103,799	12	8,77	14,66	$b\,^{3}G - z\,^{3}F^{\circ}$	4—3
2100,961	8	11,22	17,11	$d\,^{3}F - x\,^{3}G^{\circ}$	4—5
2099,332	6	11,22	17,12	$d\,^{3}F - x\,^{3}G^{\circ}$	3—4
2097,692	12	8,76	14,67	$b\,^{3}G - z\,^{3}F^{\circ}$	5—4
2097,480	15	8,76	14,67	$b\,^{3}G - z\,^{3}H^{\circ}$	5—6
2096,430	6	8,66	14,58	$b\,^{5}D - z\,^{5}D^{\circ}$	3—4
2092,945	6	11,21	17,13	$d\,^{3}F - x\,^{3}G^{\circ}$	2—3

λ, Å	I	E_H, eV	E_B, eV	Transition	J
2091,312	7	9,56	15,48	$b\,^3D - y\,^3D^\circ$	2—2
2090,240	6	8,64	14,57	$b\,^5D - z\,^5D^\circ$	4—4
2090,139	12	8,77	14,70	$b\,^3G - z\,^3H^\circ$	4—5
2090,053	7	11,12	17,05	$c\,^3G - x\,^3H^\circ$	3—4
2089,089	6	9,56	15,49	$b\,^3D - y\,^3D^\circ$	2—1
2087,907	7	9,56	15,49	$b\,^3D - y\,^3D^\circ$	1—1
2087,132	8	9,54	15,48	$b\,^3D - y\,^3D^\circ$	3—3
2084,349	10	8,77	14,71	$b\,^3G - z\,^3H^\circ$	3—4
2083,530	6	11,13	17,08	$c\,^3G - x\,^3H^\circ$	4—5
2078,989	14	5,08	11,04	$a\,^5S - z\,^5P^\circ$	2—3
2070,539	8	10,34	16,33	$b\,^1I - z\,^1H^\circ$	6—5
2068,243	12	5,08	11,08	$a\,^5S - z\,^5P^\circ$	2—2
2067,302	6	11,15	17,14	$c\,^3G - x\,^3H^\circ$	5—6
2061,751	9	9,54	15,55	$b\,^3D - y\,^3F^\circ$	3—4
2061,552	10	5,08	11,09	$a\,^5S - z\,^5P^\circ$	2—1
2059,677	7	9,56	15,58	$b\,^3D - y\,^3F^\circ$	2—3
2058,560	8	10,34	16,36	$b\,^1I - z\,^1K^\circ$	6—7
2057,058	6	9,55	15,58	$b\,^3D - y\,^3F^\circ$	1—2
2056,145	7	9,14	15,17	$c\,^3P - z\,^3D^\circ$	2—3
2055,855	6	10,43	16,46	$c\,^3F - x\,^3F^\circ$	4—4
2050,739	7	8,64	14,68	$b\,^5D - y\,^5P^\circ$	4—3
2049,384	7	9,15	15,20	$c\,^3P - z\,^3D^\circ$	1—2
2039,507	6	11,47	17,55	$a\,^1H - y\,^1I^\circ$	5—6
2008,494	6	8,26	14,43	$a\,^5P - z\,^5D^\circ$	1—1
2004,143	8	8,25	14,43	$a\,^5P - z\,^5D^\circ$	2—2
2000,228	9	9,90	16,10	$a\,^3I - z\,^3I^\circ$	5—6
1999,588	9	9,90	16,10	$a\,^3I - z\,^3I^\circ$	6—6
1996,420	12	7,87	14,08	$a\,^5G - z\,^5G^\circ$	2—2
1995,563	12	7,87	14,08	$a\,^5G - z\,^5G^\circ$	3—3
1995,266	7	7,87	14,08	$a\,^5G - z\,^5G^\circ$	4—3
1994,073	13	7,87	14,09	$a\,^5G - z\,^5G^\circ$	4—4
1993,262	7	7,87	14,09	$a\,^5G - z\,^5G^\circ$	5—4
1992,858	6	10,46	16,68	$c\,^3F - y\,^3G^\circ$	2—3
1992,196	9	9,90	16,12	$a\,^3I - z\,^3I^\circ$	6—7
1992,017	9	9,90	16,12	$a\,^3I - z\,^3I^\circ$	7—7
1991,613	14	7,87	14,09	$a\,^5G - z\,^5G^\circ$	5—5
1989,975	7	7,86	14,09	$a\,^5G - z\,^5G^\circ$	6—5
1987,503	15	7,86	14,10	$a\,^5G - z\,^5G^\circ$	6—6
1984,288	9	9,90	16,15	$a^3\,I - z\,^3I^\circ$	5—5
1984,027	7	10,21	16,46	$c\,^3D - x\,^3F^\circ$	3—4
1982,805	8	8,24	14,49	$a\,^5P - z\,^5S^\circ$	3—2
1982,076	6	8,25	14,50	$a\,^5P - z\,^5F^\circ$	2—2
1976,126	8	8,24	14,51	$a\,^5P - z\,^5F^\circ$	3—3
1966,740	8	10,99	17,30	$b\,^3H - y\,^3I^\circ$	4—5
1965,309	8	10,50	16,80	$c\,^3F - y\,^3G^\circ$	3—4
1964,776	8	9,90	16,21	$a\,^3I - z\,^3K^\circ$	5—6
1964,260	7	8,65	14,96	$b\,^5D - y\,^5F^\circ$	1—1
1964,169	8	9,90	16,21	$a\,^3I - z\,^3K^\circ$	6—6
1961,230	6	8,66	14,98	$b\,^5D - y\,^5F^\circ$	2—2
1960,318	13	9,90	16,22	$a\,^3I - z\,^3K^\circ$	7—8
1959,324	8	8,65	14,98	$b\,^5D - y\,^5F^\circ$	1—2
1958,583	11	8,24	14,57	$a\,^5P - z\,^5D^\circ$	3—4
1957,938	6	11,59	17,92	$c\,^1G - y\,^1H^\circ$	4—5
1954,975	8	11,00	17,34	$b\,^3H - y\,^3I^\circ$	5—6
1954,223	10	8,66	15,00	$b\,^5D - y\,^5F^\circ$	2—3
1953,488	10	9,90	16,24	$a\,^3I - z\,^3K^\circ$	6—7
1953,322	13	8,77	15,12	$b\,^3G - z\,^3G^\circ$	3—3

λ, Å	I	E_H, eV	E_B, eV	Transition	J
1952,648	11	8,77	15,12	$b\,^3G-z\,^3G^\circ$	4—4
1951,007	12	8,76	15,12	$b\,^3G-z\,^3G^\circ$	5—5
1950,334	10	11,02	17,38	$b\,^3H-y\,^3I^\circ$	6—7
1945,342	12	8,66	15,03	$b\,^5D-y\,^5F^\circ$	3—4
1943,481	14	7,87	14,25	$a\,^5G-z\,^5H^\circ$	2—3
1940,018	8	8,64	15,03	$b\,^5D-y\,^5F^\circ$	4—4
1938,901	10	11,22	17,61	$d\,^3F-u\,^3F^\circ$	4—4
1937,345	14	7,87	14,27	$a\,^5G-z\,^5H^\circ$	3—4
1931,507	14	8,64	15,06	$b\,^5D-y\,^5F^\circ$	4—5
1930,387	15	7,87	14,29	$a\,^5G-z\,^5H^\circ$	4—5
1926,304	18	3,73	10,16	$a\,^7S-z\,^7P^\circ$	3—2
1926,013	10	8,25	14,68	$a\,^5P-y\,^5P^\circ$	2—3
1924,532	6	9,54	15,98	$b\,^3D-y\,^3P^\circ$	3—2
1923,877	7	8,24	14,68	$a\,^5P-y\,^5P^\circ$	3—3
1923,003	7	10,32	16,77	$a\,^5F-y\,^5G^\circ$	3—4
1922,789	15	7,87	14,32	$a\,^5G-z\,^5H^\circ$	5—6
1918,480	7	10,50	16,96	$c\,^3F-w\,^3D^\circ$	3—2
1918,284	7	8,26	14,72	$a\,^5P-y\,^5P^\circ$	1—2
1917,960	6	10,22	16,68	$c\,^3D-x\,^3P^\circ$	2—1
1917,453	9	10,34	16,81	$b\,^1I-z\,^1I^\circ$	6—6
1917,351	8	10,31	16,77	$a\,^5F-y\,^5G^\circ$	4—5
1915,083	15	7,86	14,34	$a\,^5G-z\,^5H^\circ$	6—7
1914,056	19	3,73	10,21	$a\,^7S-z\,^7P^\circ$	3—3
1911,338	7	11,47	17,95	$a\,^1H-x\,^1H^\circ$	5—5
1910,401	6	8,25	14,74	$a\,^5P-y\,^5P^\circ$	2—1
1907,577	10	9,90	16,40	$a\,^3I-y\,^3H^\circ$	7—6
1906,814	6	10,23	16,73	$c\,^3D-y\,^5G^\circ$	1—2
1906,457	6	10,43	16,94	$c\,^3F-w\,^3F^\circ$	4—4
1902,402	6	10,22	16,73	$c\,^3D-x\,^3D^\circ$	2—3
1901,096	9	10,31	16,83	$a\,^5F-y\,^5G^\circ$	5—6
1898,870	6	9,14	15,67	$c\,^3P-z\,^3S^\circ$	2—1
1896,803	9	9,90	16,44	$a\,^3I-y\,^3H^\circ$	6—5
1895,456	20	3,73	10,27	$a\,^7S-z\,^7P^\circ$	3—4
1893,981	11	9,90	16,45	$a\,^3I-y\,^3H^\circ$	5—4
1890,669	13	7,86	14,42	$a\,^5G-z\,^5F^\circ$	6—5
1887,471	8	7,87	14,44	$a\,^5G-z\,^5F^\circ$	4—4
1887,197	8	7,87	14,44	$a\,^5G-z\,^5D^\circ$	4—3
1886,757	12	7,87	14,44	$a\,^5G-z\,^5F^\circ$	5—4
1885,125	9	10,31	16,88	$a\,^5F-x\,^5F^\circ$	5—5
1884,596	8	8,66	15,23	$b\,^5D-y\,^5D^\circ$	3—2
1882,047	10	8,65	15,24	$b\,^5D-y\,^5D^\circ$	1—1
1877,989	12	8,64	15,24	$b\,^5D-y\,^5D^\circ$	4—4
1872,214	6	11,15	17,77	$c\,^3G-w\,^3H^\circ$	5—6
1871,152	9	7,87	14,50	$a\,^5G-z\,^5F^\circ$	2—1
1869,828	10	7,87	14,50	$a\,^5G-z\,^5F^\circ$	3—2
1866,305	9	7,87	14,51	$a\,^5G-z\,^5F^\circ$	4—3
1865,202	7	12,03	18,68	$c\,^1F-w\,^1F^\circ$	3—3
1856,690	7	8,66	15,33	$b\,^5D-y\,^5P^\circ$	2—2
1854,826	9	8,66	15,34	$b\,^5D-y\,^5P^\circ$	3—3
1852,677	6	11,15	17,84	$c\,^3G-v\,^3G^\circ$	5—5
1851,261	6	11,57	18,27	$e\,^3F-u\,^3G^\circ$	4—5
1849,407	7	10,31	17,01	$a\,^5F-x\,^5D^\circ$	5—4
1845,521	7	10,32	17,04	$a\,^5F-x\,^5D^\circ$	3—3
1844,547	6	10,99	17,72	$b\,^3H-w\,^3H^\circ$	5—5
1838,309	7	11,02	17,77	$b\,^3H-w\,^3H^\circ$	6—6
1775,983	6	8,25	15,23	$a\,^5P-y\,^5D^\circ$	2—3
1770,554	6	8,24	15,24	$a\,^5P-y\,^5D^\circ$	3—4
1611,763	7	11,09	18,79	$z\,^5P^\circ-e\,^5D$	1—2

λ, Å	I	E_H, eV	E_B, eV	Transition	J
1611,763	7	11,09	18,79	$z\,^5P^\circ - e\,^5D$	1—1, 0
1607,723	9	11,07	18,79	$z\,^5P^\circ - e\,^5D$	2—3, 2
1601,289	6	11,04	18,79	$z\,^5P^\circ - e\,^5D$	3—3, 2
1601,211	10	11,04	18,79	$z\,^5P^\circ - e\,^5D$	3—4
1595,597	6	11,04	18,81	$z\,^5P^\circ - e\,^5S$	3—2
1550,862	8	10,27	18,26	$z\,^7P^\circ - e\,^7D$	4—4
1550,196	12	10,27	18,27	$z\,^7P^\circ - e\,^7D$	4—5
1539,128	8	10,21	18,26	$z\,^7P^\circ - e\,^7D$	3—3
1538,632	10	10,21	18,26	$z\,^7P^\circ - e\,^7D$	3—4
1531,864	7	10,17	18,26	$z\,^7P^\circ - e\,^7D$	2—1
1531,644	8	10,17	18,26	$z\,^7P^\circ - e\,^7D$	2—2
1531,293	6	10,17	18,26	$z\,^7P^\circ - e\,^7D$	2—3
1505,166	10	10,27	18,51	$z\,^7P^\circ - e\,^7S$	4—3
1493,640	9	10,21	18,51	$z\,^7P^\circ - e\,^7S$	3—3
1486,265	7	10,17	18,51	$z\,^7P^\circ - e\,^7S$	2—3
1131,194	7	0,12	11,07	$a\,^5D - z\,^5P^\circ$	1—2
1129,19	7	0,12	11,09	$a\,^5D - z\,^5P^\circ$	1—1
1128,72	7	0,09	11,07	$a\,^5D - z\,^5P^\circ$	2—2
1128,02	8	0,05	11,04	$a\,^5D - z\,^5P^\circ$	3—3
1126,72	6	0,09	11,09	$a\,^5D - z\,^5P^\circ$	2—1
1124,883	9	0,05	11,07	$a\,^5D - z\,^5P^\circ$	3—2
1122,526	9	0,00	11,04	$a\,^5D - z\,^5P^\circ$	4—3
1066,181	10	3,05	14,67	$a\,^3G - z\,^3F^\circ$	5—4
1066,143	10	3,05	14,67	$a\,^3G - z\,^3H^\circ$	5—6
1063,872	8	3,82	15,47	$a\,^3D - y\,^3D^\circ$	3—3
1061,708	6	3,80	15,48	$a\,^3D - y\,^3D^\circ$	2—2
1038,355	6	2,71	14,65	$a\,^3F - z\,^3F^\circ$	2—2
1035,768	6	2,69	14,66	$a\,^3F - z\,^3F^\circ$	3—3
1032,123	8	2,66	14,67	$a\,^3F - z\,^3F^\circ$	4—4
1030,924	6	3,09	15,12	$a\,^3G - z\,^3G^\circ$	4—4
1026,790	6	3,05	15,12	$a\,^3G - z\,^3G^\circ$	5—5
1019,789	6	3,82	15,98	$a\,^3D - y\,^3P^\circ$	3—2
1018,286	8	3,81	15,98	$a\,^3D - y\,^3P^\circ$	2—2
1017,745	8	2,52	14,70	$a\,^3H - z\,^3H^\circ$	5—5
1017,254	9	2,49	14,68	$a\,^3H - z\,^3H^\circ$	6—6
997,599	6	2,70	15,12	$a\,^3F - z\,^3G^\circ$	3—4
997,081	7	2,41	14,84	$a\,^3P - z\,^3P^\circ$	2—2
995,150	6	2,66	15,12	$a\,^3F - z\,^3G^\circ$	4—5
994,724	6	3,12	15,58	$a\,^3G - y\,^3F^\circ$	3—2
993,080	7	3,09	15,57	$a\,^3G - y\,^3F^\circ$	4—3
991,829	6	3,83	16,33	$a\,^1G - z\,^1H^\circ$	4—5
991,232	9	2,66	15,17	$a\,^3F - z\,^3D^\circ$	4—3
990,800	6	2,69	15,20	$a\,^3F - z\,^3D^\circ$	3—2
985,824	8	2,54	15,11	$a\,^3H - z\,^3G^\circ$	4—3
983,877	10	2,52	15,12	$a\,^3H - z\,^3G^\circ$	5—4
981,373	10	2,48	15,12	$a\,^3H - z\,^3G^\circ$	6—5
967,197	6	2,66	15,48	$a\,^3F - y\,^3D^\circ$	4—3
961,901	7	4,44	17,32	$a\,^1D - y\,^1D^\circ$	2—2
950,344	10	3,76	16,81	$a\,^1I - z\,^1I^\circ$	6—6
934,703	7	2,41	15,67	$a\,^3P - z\,^3S^\circ$	2—1
910,961	6	2,54	16,15	$a\,^3H - z\,^3I^\circ$	4—5
905,338	7	2,52	16,21	$a\,^3H - z\,^3K^\circ$	5—6
899,417	8	3,76	17,55	$a\,^1I - y\,^1I^\circ$	6—6
892,417	6	3,05	16,94	$a\,^3G - w\,^3F^\circ$	5—4
891,442	8	2,54	16,45	$a\,^3H - y\,^3H^\circ$	4—4
891,172	10	2,49	16,40	$a\,^3H - y\,^3H^\circ$	6—6
890,755	9	2,52	16,44	$a\,^3H - y\,^3H^\circ$	5—5

λ, Å	I	E_H, eV	E_B, eV	Transition	J
883,688	6	3,09	17,12	$a\,^3G - x\,^3G^\circ$	4—4
881,088	7	3,05	17,11	$a\,^3G - x\,^3G^\circ$	5—5
880,949	6	2,66	16,73	$a\,^3F - x\,^3D^\circ$	4—3
880,447	6	2,69	16,78	$a\,^3F - x\,^3D^\circ$	3—2
873,462	8	3,76	17,96	$a\,^1I - x\,^1H^\circ$	6—5
861,832	10	0,05	14,44	$a\,^5D - z\,^5F^\circ$	3—4
861,761	8	0,05	14,44	$a\,^5D - z\,^5D^\circ$	3—3
859,838	6	2,41	16,82	$a\,^3P - w\,^3D^\circ$	2—3
859,721	8	0,00	14,42	$a\,^5D - z\,^5F^\circ$	4—5
859,626	6	0,09	14,51	$a\,^5D - z\,^5F^\circ$	2—3
858,602	6	0,00	14,44	$a\,^5D - z\,^5F^\circ$	4—4
854,367	6	2,54	17,05	$a\,^3H - x\,^3H^\circ$	4—4
851,992	6	3,09	17,64	$a\,^3G - u\,^3F^\circ$	4—3
851,842	6	3,12	17,67	$a\,^3G - u\,^3F^\circ$	3—2
851,332	7	2,52	17,08	$a\,^3H - x\,^3H^\circ$	5—5
851,150	7	3,04	17,61	$a\,^3G - u\,^3F^\circ$	5—4
847,924	6	0,12	14,74	$a\,^5D - y\,^5P^\circ$	1—1
847,700	6	3,09	17,72	$a\,^3G - w\,^3H^\circ$	4—5
847,578	7	0,09	14,72	$a\,^5D - y\,^5P^\circ$	2—2
847,425	8	0,05	14,68	$a\,^5D - y\,^5P^\circ$	3—3
846,534	6	0,09	14,74	$a\,^5D - y\,^5P^\circ$	2—1
845,925	7	2,49	17,14	$a\,^3H - x\,^3H^\circ$	6—6
845,408	9	0,05	14,72	$a\,^5D - y\,^5P^\circ$	3—2
844,284	10	0,00	14,68	$a\,^5D - y\,^5P^\circ$	4—3
842,020	6	3,04	17,77	$a\,^3G - w\,^3H^\circ$	5—6
838,048	8	2,52	17,31	$a\,^3H - w\,^3G^\circ$	5—4
837,439	7	2,49	17,29	$a\,^3H - w\,^3G^\circ$	6—5
836,521	7	2,66	17,48	$a\,^3F - v\,^3F^\circ$	4—4
834,944	6	3,83	18,67	$a\,^1G - w\,^1F^\circ$	4—3
827,777	6	0,05	15,03	$a\,^5D - y\,^5F^\circ$	3—4
823,257	6	0,00	15,06	$a\,^5D - y\,^5F^\circ$	4—5
817,038	7	0,05	15,23	$a\,^5D - y\,^5D^\circ$	3—3
816,273	6	0,05	15,24	$a\,^5D - y\,^5D^\circ$	3—4
816,163	6	0,12	15,30	$a\,^5D - y\,^5D^\circ$	1—0
814,242	6	0,12	15,32	$a\,^5D - y\,^5P^\circ$	1—1
813,382	10	0,00	15,24	$a\,^5D - y\,^5D^\circ$	4—4
811,284	8	0,05	15,33	$a\,^5D - y\,^5P^\circ$	3—2
810,940	7	0,05	15,34	$a\,^5D - y\,^5P^\circ$	3—3
808,840	8	2,54	16,68	$a\,^3H - y\,^3G^\circ$	4—3
807,855	8	2,52	17,86	$a\,^3H - v\,^3G^\circ$	5—4
807,547	9	2,49	17,83	$a\,^3H - v\,^3G^\circ$	6—5
728,810	6	0,00	17,01	$a\,^5D - x\,^5D^\circ$	4—4

Fe IV, ground state $1s^2\,2s^2\,2p^6\,3s^2\,3p^6\,3d^5\,^6S_{5/2}$
Ionization potential 460 278 cm⁻¹; 57,063 eV

λ, Å	I	E_H, eV	E_B, eV	Transition	J
1825,55	8	14,68	21,47	$4s\,^4D - 4p\,^4P^\circ$	$3/2 - 1/2$
1822,72	2	14,72	21,52	$4s\,^4D - 4p\,^4P^\circ$	$5/2 - 3/2$
1819,29	1	14,66	21,47	$4s\,^4D - 4p\,^4P^\circ$	$1/2 - 1/2$
1815,61	25	14,78	21,60	$4s\,^4D - 4p\,^4P^\circ$	$7/2 - 5/2$
1812,53	1	14,68	21,52	$4s\,^4D - 4p\,^4P^\circ$	$3/2 - 3/2$
1801,53	5	14,72	21,60	$4s\,^4D - 4p\,^4P^\circ$	$5/2 - 5/2$
1688,44	2	14,78	22,12	$4s\,^4D - 4p\,^4F^\circ$	$7/2 - 7/2$

λ, Å	I	E_H, eV	E_B, eV	Transition	J
1681,45	25	14,78	22,15	$4s\,^4D—4p\,^4F^\circ$	$^7/_2—^9/_2$
1680,86	1	14,72	22,10	$4s\,^4D—4p\,^4F^\circ$	$^5/_2—^5/_2$
1675,78	25	14,72	22,12	$4s\,^4D—4p\,^4F^\circ$	$^5/_2—^7/_2$
1674,89	5	14,68	22,08	$4s\,^4D—4p\,^4F^\circ$	$^3/_2—^3/_2$
1672,18	5	14,68	22,10	$4s\,^4D—4p\,^4F^\circ$	$^3/_2—^5/_2$
1669,73	2	14,66	22,08	$4s\,^4D—4p\,^4F^\circ$	$^1/_2—^3/_2$
1663,52	10	13,44	20,89	$4s\,^6D—4p\,^6F^\circ$	$^3/_2—^3/_2$
1663,21	10	13,40	20,85	$4s\,^6D—4p\,^6F^\circ$	$^1/_2—^1/_2$
1662,26	20	13,48	20,94	$4s\,^6D—4p\,^6F^\circ$	$^5/_2—^5/_2$
1660,07	20	13,54	21,00	$4s\,^6D—4p\,^6F^\circ$	$^7/_2—^7/_2$
1656,61	15	13,40	20,89	$4s\,^6D—4p\,^6F^\circ$	$^1/_2—^3/_2$
1656,25	10	13,61	21,10	$4s\,^6D—4p\,^6F^\circ$	$^9/_2—^9/_2$
1652,85	20	13,44	20,94	$4s\,^6D—4p\,^6F^\circ$	$^3/_2—^5/_2$
1647,05	45	13,48	21,00	$4s\,^6D—4p\,^6F^\circ$	$^5/_2—^7/_2$
1640,03	65	13,54	21,10	$4s\,^6D—4p\,^6F^\circ$	$^7/_2—^9/_2$
1630,99	75	13,61	21,21	$4s\,^6D—4p\,^6F^\circ$	$^9/_2—^{11}/_2$
1579,73	3	13,48	21,33	$4s\,^6D—4p\,^6D^\circ$	$^5/_2—^3/_2$
1574,68	8	13,40	21,28	$4s\,^6D—4p\,^6D^\circ$	$^1/_2—^1/_2$
1571,21	10	13,44	21,33	$4s\,^6D—4p\,^6D^\circ$	$^3/_2—^3/_2$
1566,54	3	13,61	21,52	$4s\,^6D—4p\,^6D^\circ$	$^9/_2—^7/_2$
1565,05	2	13,40	21,33	$4s\,^6D—4p\,^6D^\circ$	$^1/_2—^3/_2$
1563,30	10	—	—	—	—
1560,26	15	—	—	—	—
1559,08	15	13,54	21,49	$4s\,^6D—4p\,^6P^\circ$	$^7/_2—^5/_2$
1556,48	15	13,61	21,58	$4s\,^6D—4p\,^6P^\circ$	$^9/_2—^7/_2$
1555,01	1	—	—	—	—
1552,11	15	13,54	21,52	$4s\,^6D—4p\,^6D^\circ$	$^7/_2—^7/_2$
1547,58	15	—	—	—	—
1546,03	8	—	—	—	—
1542,15	15	13,54	21,58	$4s\,^6D—4p\,^6P^\circ$	$^7/_2—^7/_2$
1540,77	1	13,48	21,52	$4s\,^6D—4p\,^6D^\circ$	$^5/_2—^7/_2$
1538,67	25	13,61	21,92	$4s\,^6D—4p\,^6D^\circ$	$^9/_2—^9/_2$
1524,67	15	13,54	21,92	$4s\,^6D—4p\,^6D^\circ$	$^7/_2—^9/_2$
1487,35	5	—	—	—	—
1485,48	12	—	—	—	—
1479,65	38	14,78	23,16	$4s\,^4D—4p\,^4D^\circ$	$^7/_2—^7/_2$
1477,69	5	19,83	28,22	$4s^2\,^4F—4sp\,^4G^\circ$	$^9/_2—^9/_2$
1475,67	28	—	—	—	—
1474,52	2	14,68	23,09	$4s\,^4D—4p\,^4D^\circ$	$^3/_2—^1/_2$
1472,13	35	14,68	23,10	$4s\,^4D—4p\,^4D^\circ$	$^3/_2—^3/_2$
1470,54	2	14,66	23,09	$4s\,^4D—4p\,^4D^\circ$	$^1/_2—^1/_2$
1469,92	20	14,72	23,16	$4s\,^4D—4p\,^4D^\circ$	$^5/_2—^7/_2$
1469,04	37	—	—	—	—
1468,11	2	14,66	23,10	$4s\,^4D—4p\,^4D^\circ$	$^1/_2—^3/_2$
1464,81	40	19,83	28,29	$4s^2\,^4F—4sp\,^4G^\circ$	$^9/_2—^{11}/_2$
1463,25	4	19,65	28,12	$4s^2\,^4F—4sp\,^4G^\circ$	$^5/_2—^5/_2$
1459,92	40	19,72	28,22	$4s^2\,^4F—4sp\,^4G^\circ$	$^7/_2—^9/_2$
1455,66	25	19,65	28,16	$4s^2\,^4F—4sp\,^4G^\circ$	$^5/_2—^7/_2$
1453,67	15	19,59	28,12	$4s^2\,^4F—4sp\,^4G^\circ$	$^3/_2—^5/_2$
1280,43	10	21,21	30,89	$4p\,^6F—4d\,^6G^\circ$	$^{11}/_2—^{11}/_2$
1273,49	2	21,10	30,83	$4p\,^6F—4d\,^6G^\circ$	$^9/_2—^9/_2$
1271,08	15	21,21	30,97	$4p\,^6F—4d\,^6G^\circ$	$^{11}/_2—^{13}/_2$
1268,40	2	—	—	—	—
1265,28	15	21,10	30,89	$4p\,^6F^\circ—4d\,^6G$	$^9/_2—^{11}/_2$
1263,47	15	20,94	30,75	$4p\,^6F^\circ—4d\,^6G$	$^5/_2—^5/_2$
1261,72	10	—	—	—	—
1259,54	30	20,94	30,90	$4p\,^6F^\circ—4d\,^6G$	$^5/_2—^7/_2$
1258,68	2	—	—	—	—

λ, Å	I	E_H, eV	E_B, eV	Transition	J
1257,29	6	20,89	30,75	$4p\ ^6F^\circ$—$4d\ ^6G$	$^3/_2$—$^5/_2$
1254,80	10	20,85	30,64	$4p\ ^6P^\circ$—$4d\ ^6G$	$^1/_2$—$^3/_2$
587,6	2	—	—	—	—
576,8	40	—	—	—	—
574,5	50	—	—	—	—
526,60	60	0,00	23,5	$3d^5\ ^6S$—$4p\ ^6P^\circ$	$^5/_2$—$^3/_2$
526,28	75	0,00	23,6	$3d^5\ ^6S$—$4p\ ^6P^\circ$	$^5/_2$—$^5/_2$
525,68	100	0,00	23,6	$3d^5\ ^6S$—$4p\ ^6P^\circ$	$^5/_2$—$^7/_2$

Fe V, ground state $1s^2\,2s^2\,2p^6\,3s^2\,3p^6\,3d^4\ ^5D_0$

Ionization potential 78 eV

λ, Å	I	E_H, eV	E_B, eV	Transition	J
1554,17	1	24,40	32,38	$b\ ^3F$—$z\ ^3D^\circ$	4—3
1550,80	2	24,29	32,29	$b\ ^3F$—$z\ ^3D^\circ$	3—2
1544,50	3	24,20	32,13	$b\ ^3F$—$z\ ^3F^\circ$	2—3
1543,66	2	24,20	32,23	$b\ ^3F$—$z\ ^3D^\circ$	2—1
1533,27	2	24,20	32,29	$b\ ^3F$—$z\ ^3D^\circ$	2—2
1532,70	4	24,29	32,38	$b\ ^3F$—$z\ ^3D^\circ$	3—3
1479,49	4	24,40	32,78	$b\ ^3F$—$z\ ^3G^\circ$	4—4
1465,37	3	23,20	31,66	$a\ ^5F$—$z\ ^5G^\circ$	3—3
1464,73	6	24,40	32,87	$b\ ^3F$—$z\ ^3G^\circ$	4—5
1462,67	3	23,11	31,66	$a\ ^5F$—$z\ ^5G^\circ$	1—3
1460,86	2	23,27	31,76	$a\ ^5F$—$z\ ^5G^\circ$	4—4
1459,85	5	24,29	31,76	$b\ ^3F$—$y\ ^5G^\circ$	3—4
1456,23	5	23,15	31,66	$a\ ^5F$—$z\ ^5G^\circ$	2—3
1455,59	5	24,20	31,66	$b\ ^3F$—$z\ ^5G^\circ$	2—3
1454,71	3	23,36	31,88	$a\ ^5F$—$z\ ^5G^\circ$	5—5
1448,91	6	23,20	31,76	$a\ ^5F$—$z\ ^5G^\circ$	3—4
1440,59	7	23,27	31,88	$a\ ^5F$—$z\ ^5G^\circ$	4—5
1430,61	8	23,36	32,02	$a\ ^5F$—$z\ ^5G^\circ$	5—6
1420,24	3	24,40	33,13	$b\ ^3F$—$z\ ^3F^\circ$	4—3
1409,51	7	23,36	32,15	$a\ ^5F$—$z\ ^5D^\circ$	5—4
1409,19	6	23,20	32,00	$a\ ^5F$—$z\ ^5D^\circ$	3—2
1408,19	1	23,15	31,95	$a\ ^5F$—$z\ ^5D^\circ$	2—1
1406,78	7	24,40	33,22	$b\ ^3F$—$z\ ^3F^\circ$	4—4
1402,45	6	24,29	33,13	$b\ ^3F$—$z\ ^3F^\circ$	3—3
1400,30	4	24,20	33,05	$b\ ^3F$—$z\ ^3F^\circ$	2—2
1397,99	3	23,20	32,07	$a\ ^5F$—$z\ ^5D^\circ$	3—3
1394,77	3	23,11	32,00	$a\ ^5F$—$z\ ^5D^\circ$	1—2
1389,97	0	23,15	32,07	$a\ ^5F$—$z\ ^5D^\circ$	2—3
1389,05	1	24,29	33,22	$b\ ^3F$—$z\ ^3F^\circ$	3—4
1388,07	5	24,20	33,13	$b\ ^3F$—$z\ ^3F^\circ$	2—3
1386,33	0	23,36	32,30	$a\ ^5F$—$z\ ^5F^\circ$	5—4
1385,32	2	23,20	32,15	$a\ ^5F$—$z\ ^5D^\circ$	3—4
1384,75	1	23,20	32,16	$a\ ^5F$—$z\ ^5F^\circ$	3—2
1384,17	1	23,27	32,23	$a\ ^5F$—$z\ ^5F^\circ$	4—3
1380,18	2	23,11	32,10	$a\ ^5F$—$z\ ^5F^\circ$	1—1
1376,45	6	23,36	32,16	$a\ ^5F$—$z\ ^5F^\circ$	5—2
1373,68	6	23,27	32,30	$a\ ^5F$—$z\ ^5F^\circ$	4—4
1371,00	4	23,11	32,16	$a\ ^5F$—$z\ ^5F^\circ$	1—2
1365,73	3	23,15	32,23	$a\ ^5F$—$z\ ^5F^\circ$	2—3
1365,14	3	23,20	32,29	$a\ ^5F$—$z\ ^3D^\circ$	3—2

λ, Å	I	E_H, eV	E_B, eV	Transition	J
1363,72	3	23,27	32,36	$a\ ^5F—z\ ^5F°$	4—5
1363,00	4	23,20	32,30	$a\ ^5F—z\ ^5F°$	3—4
1361,42	5	23,27	32,38	$a\ ^5F—z\ ^3D°$	4—3
1359,41	1	23,11	32,23	$a\ ^5F—z\ ^3D°$	1—1
1357,28	1	23,15	32,29	$a\ ^5F—z\ ^3D°$	2—2
1303,59	1	23,27	32,78	$a\ ^5F—z\ ^3G°$	4—4
1302,99	1	23,20	32,72	$a\ ^5F—z\ ^3G°$	3—3
432,919	1	3,32	31,95	$a\ ^3F—z\ ^5D°$	2—1
432,340	3	3,33	32,00	$a\ ^3F—z\ ^5D°$	3—2
431,541	3	3,34	32,07	$a\ ^3F—z\ ^5D°$	4—3
430,624	2	3,28	32,07	$a\ ^3P—z\ ^5D°$	2—3
430,053	1	3,33	32,16	$a\ ^3F—z\ ^5F°$	3—2
429,206	1	3,34	32,23	$a\ ^3F—z\ ^5F°$	4—3
428,909	5	3,10	32,00	$a\ ^3P—z\ ^5D°$	1—2
428,763	5	3,32	32,23	$a\ ^3F—z\ ^3D°$	2—1
428,292	0	3,28	32,23	$a\ ^3P—z\ ^5F°$	2—3
428,131	3	3,33	32,29	$a\ ^3F—z\ ^3D°$	3—2
428,000	0	3,32	32,29	$a\ ^3F—z\ ^3D°$	2—2
427,918	2	2,98	31,95	$a\ ^3P—z\ ^5D°$	0—1
427,782	1	3,74	32,72	$a\ ^3G—z\ ^3G°$	4—3
427,442	2	3,28	32,29	$a\ ^3P—z\ ^3D°$	2—2
427,320	1	3,77	32,78	$a\ ^3G—z\ ^3G°$	5—4
427,190	3	3,70	32,72	$a\ ^3G—z\ ^3G°$	3—3
426,969	3	3,34	32,38	$a\ ^3F—z\ ^3D°$	4—3
426,814	4	3,74	32,78	$a\ ^3G—z\ ^3G°$	4—4
426,745	3	3,33	32,38	$a\ ^3F—z\ ^3D°$	3—3
426,609	1	3,10	32,16	$a\ ^3P—z\ ^5F°$	1—2
426,097	5	3,77	32,87	$a\ ^3G—z\ ^3G°$	5—5
426,045	5	3,28	32,38	$a\ ^3P—z\ ^3D°$	2—3
425,840	0	2,98	32,10	$a\ ^3P—z\ ^5F°$	0—1
425,589	1	3,74	32,87	$a\ ^3G—z\ ^3G°$	4—5
425,476	1	3,10	32,23	$a\ ^3P—z\ ^3D°$	1—1
424,733	3	3,10	32,29	$a\ ^3P—z\ ^3D°$	1—2
423,833	2	2,98	32,23	$a\ ^3P—z\ ^3D°$	0—1
422,287	6	3,70	33,05	$a\ ^3G—z\ ^3F°$	3—2
421,765	4	3,74	33,13	$a\ ^3G—z\ ^3F°$	4—3
421,682	2	3,32	32,72	$a\ ^3F—z\ ^3G°$	2—3
421,045	5	3,77	33,22	$a\ ^3G—z\ ^3F°$	5—4
420,874	2	3,33	33,22	$a\ ^3F—z\ ^3G°$	3—4
420,546	5	3,74	33,22	$a\ ^3G—z\ ^3F°$	4—4
419,915	3	3,09	32,87	$a\ ^3H—z\ ^3G°$	4—5
418,457	5	3,09	32,72	$a\ ^3H—z\ ^3G°$	4—3
418,033	6	3,13	32,78	$a\ ^3H—z\ ^3G°$	5—4
417,516	0	3,09	32,78	$a\ ^3H—z\ ^3G°$	4—4
417,382	6	3,16	32,87	$a\ ^3H—z\ ^3G°$	6—5
417,048	1	3,33	33,05	$a\ ^3F—z\ ^3F°$	3—2
416,910	2	3,32	33,05	$a\ ^3F—z\ ^3F°$	2—2
416,208	5	3,34	33,13	$a\ ^3F—z\ ^3F°$	4—3
415,972	3	3,33	33,13	$a\ ^3F—z\ ^3F°$	3—3
415,825	1	3,32	33,13	$a\ ^3F—z\ ^3F°$	2—3
415,006	4	3,34	33,22	$a\ ^3F—z\ ^3F°$	4—4
414,790	1	3,33	33,22	$a\ ^3F—z\ ^3F°$	3—4
402,197	1	3,74	34,56	$a\ ^3G—y\ ^3G°$	4—3
401,639	2	3,70	34,56	$a\ ^3G—y\ ^3G°$	3—3
401,030	2	3,74	34,65	$a\ ^3G—y\ ^3G°$	4—4
400,625	4	3,77	34,72	$a\ ^3G—y\ ^3G°$	5—5
396,902	0	3,33	34,56	$a\ ^3F—y\ ^3G°$	3—3
396,773	3	3,32	31,66	$a\ ^3F—z\ ^3G°$	2—3

λ, Å	I	E_H, eV	E_B, eV	Transition	J
395,789	2	3,33	31,76	$a\ ^3F-z\ ^3G^\circ$	3—4
395,155	3	3,34	31,88	$a\ ^3F-z\ ^3G^\circ$	4—5
393,911	4	3,09	31,66	$a\ ^3H-z\ ^3G^\circ$	4—3
393,270	5	3,13	31,76	$a\ ^3H-z\ ^3G^\circ$	5—4
392,907	6	3,16	32,02	$a\ ^3H-z\ ^3G^\circ$	6—6
388,607	2	0,10	32,00	$a\ ^5D-z\ ^5D^\circ$	3—2
388,500	2	0,16	32,07	$a\ ^5D-z\ ^5D^\circ$	4—3
387,983	3	0,00	31,95	$a\ ^5D-z\ ^5D^\circ$	0—1
387,775	4	0,10	32,07	$a\ ^5D-z\ ^5D^\circ$	3—3
387,616	4	0,02	32,00	$a\ ^5D-z\ ^5D^\circ$	1—2
387,500	6	0,16	32,15	$a\ ^5D-z\ ^5D^\circ$	4—4
387,199	5	0,05	32,07	$a\ ^5D-z\ ^5D^\circ$	2—3
386,897	4	0,05	32,10	$a\ ^5D-z\ ^5F^\circ$	2—1
386,783	4	0,10	32,15	$a\ ^5D-z\ ^5D^\circ$	3—4
386,737	3	0,10	32,16	$a\ ^5D-z\ ^5F^\circ$	3—2
386,585	1	0,16	32,23	$a\ ^5D-z\ ^5F^\circ$	4—3
386,483	1	0,02	32,10	$a\ ^5D-z\ ^5F^\circ$	1—1
386,256	0	0,00	32,10	$a\ ^5D-z\ ^5F^\circ$	0—1
386,156	4	0,05	32,16	$a\ ^5D-z\ ^5F^\circ$	2—2
385,869	5	0,10	32,23	$a\ ^5D-z\ ^5F^\circ$	3—3
385,740	5	0,16	32,29	$a\ ^5D-z\ ^5F^\circ$	4—4
385,023	4	0,10	32,29	$a\ ^5D-z\ ^5F^\circ$	3—4
384,957	6	0,16	32,36	$a\ ^5D-z\ ^5F^\circ$	4—5
384,826	1	0,02	32,23	$a\ ^5D-z\ ^3D^\circ$	1—1
384,610	2	0,00	32,23	$a\ ^5D-z\ ^3D^\circ$	0—1
384,212	3	0,02	32,29	$a\ ^5D-z\ ^3D^\circ$	1—2
383,484	3	0,05	32,38	$a\ ^5D-z\ ^3D^\circ$	2—3
381,881	4	3,77	36,24	$a\ ^3G-z\ ^3G^\circ$	5—5
381,671	0	3,77	36,25	$a\ ^3G-z\ ^3G^\circ$	5—4
381,467	0	3,74	36,24	$a\ ^3G-z\ ^3G^\circ$	4—5
381,260	3	3,74	36,25	$a\ ^3G-z\ ^3G^\circ$	4—4
381,152	2	3,74	36,26	$a\ ^3G-z\ ^3G^\circ$	4—3
380,664	3	3,70	36,26	$a\ ^3G-z\ ^3G^\circ$	3—3
379,294	3	0,10	36,25	$a\ ^5D-z\ ^3G^\circ$	3—4
379,032	1	0,16	36,24	$a\ ^5D-z\ ^3G^\circ$	4—5
374,864	5	3,16	36,24	$a\ ^3H-z\ ^3G^\circ$	6—5
374,464	2	3,13	36,24	$a\ ^3H-z\ ^3G^\circ$	5—5
374,240	4	3,13	36,25	$a\ ^3H-z\ ^3G^\circ$	5—4
373,720	5	3,09	36,26	$a\ ^3H-z\ ^3G^\circ$	4—3
366,001	3	0,05	33,92	$a\ ^5D-z\ ^5P^\circ$	2—1
365,858	6	0,10	33,98	$a\ ^5D-z\ ^5P^\circ$	3—2
365,634	3	0,02	33,92	$a\ ^5D-z\ ^5P^\circ$	1—1
365,440	6	0,16	34,08	$a\ ^5D-z\ ^5P^\circ$	4—3
365,339	3	0,05	33,99	$a\ ^5D-z\ ^5P^\circ$	2—2
364,973	3	0,02	33,99	$a\ ^5D-z\ ^5P^\circ$	1—2
364,795	4	0,10	34,08	$a\ ^5D-z\ ^5P^\circ$	3—3
364,292	3	0,05	34,08	$a\ ^5D-z\ ^5P^\circ$	2—3

Fe VI, ground state $1s^2 2s^2 2p^6 3s^2 3p^6 3d^3\ ^4F_{3/2}$
Ionization potential 102 eV

λ, Å	I	E_H, eV	E_B, eV	Transition	J
318,364	3	—	—	—	—
317,319	3	—	—	—	—

λ, Å	I	E_H, eV	E_B, eV	Transition	J
315,506	3	—	—	—	—
315,027	4	—	—	—	—
314,814	1	2,55	41,94	$a\ ^2G—z\ ^4G^\circ$	$^7/_2—^5/_2$
314,299	3	2,64	42,32	$a\ ^2G—z\ ^4G^\circ$	$^9/_2—^7/_2$
312,263	7	3,56	43,26	$a\ ^2H—z\ ^2G^\circ$	$^9/_2—^7/_2$
311,702	7	3,62	43,39	$a\ ^2H—z\ ^2G^\circ$	$^{11}/_2—^9/_2$
311,236	2	3,56	43,39	$a\ ^2H—z\ ^2G^\circ$	$^9/_2—^9/_2$
311,138	1	2,35	42,47	$a\ ^4P—z\ ^5F^\circ$	$^3/_2—^5/_2$
310,807	0	2,43	42,32	$a\ ^4P—z\ ^4F^\circ$	$^5/_2—^7/_2$
310,601	4	2,55	42,47	$a\ ^2G—z\ ^5F^\circ$	$^7/_2—^5/_2$
310,274	5	2,64	42,60	$a\ ^2G—z\ ^5F^\circ$	$^9/_2—^7/_2$
309,627	1	2,43	42,47	$a\ ^4P—z\ ^5F^\circ$	$^5/_2—^5/_2$
308,993	3	2,35	42,47	$a\ ^4P—z\ ^5F^\circ$	$^3/_2—^5/_2$
308,960	3	2,55	42,68	$a\ ^2G—z\ ^4D^\circ$	$^7/_2—^5/_2$
308,664	5	2,43	42,60	$a\ ^4P—z\ ^2F^\circ$	$^5/_2—^7/_2$
308,534	4	2,64	42,82	$a\ ^2G—z\ ^4D^\circ$	$^9/_2—^7/_2$
308,383	2	2,35	42,55	$a\ ^4P—z\ ^4D^\circ$	$^3/_2—^3/_2$
308,187	2	2,32	42,55	$a\ ^4P—z\ ^4D^\circ$	$^1/_2—^3/_2$
308,007	3	2,35	42,60	$a\ ^4P—z\ ^4D^\circ$	$^3/_2—^1/_2$
307,884	0	2,55	42,82	$a\ ^2G—z\ ^4D^\circ$	$^7/_2—^7/_2$
307,800	3	2,32	42,60	$a\ ^4P—z\ ^4D^\circ$	$^1/_2—^1/_2$
307,404	3	2,55	42,88	$a\ ^2G—z\ ^2D^\circ$	$^7/_2—^5/_2$
307,375	4	2,35	42,68	$a\ ^4P—z\ ^4D^\circ$	$^3/_2—^5/_2$
307,013	2	2,35	42,73	$a\ ^4P—z\ ^2D^\circ$	$^3/_2—^3/_2$
306,922	5	2,43	42,82	$a\ ^4P—z\ ^4D^\circ$	$^5/_2—^7/_2$
306,823	2	2,32	42,73	$a\ ^4P—z\ ^2D^\circ$	$^1/_2—^3/_2$
306,460	1	2,43	42,88	$a\ ^4P—z\ ^2D^\circ$	$^5/_2—^5/_2$
305,837	1	2,35	42,88	$a\ ^4P—z\ ^2D^\circ$	$^3/_2—^5/_2$
305,200	4	2,64	43,26	$a\ ^2G—z\ ^2G^\circ$	$^9/_2—^7/_2$
304,551	7	2,55	43,26	$a\ ^2G—z\ ^2G^\circ$	$^7/_2—^7/_2$
304,221	7	2,64	43,39	$a\ ^2G—z\ ^2G^\circ$	$^9/_2—^9/_2$
303,558	4	2,55	43,39	$a\ ^2G—z\ ^2G^\circ$	$^7/_2—^9/_2$
300,997	2	3,53	44,72	$A—y\ ^4D^\circ$	$——^5/_2$
299,803	1	3,56	44,91	$a\ ^2H—y\ ^4D^\circ$	$^9/_2—^7/_2$
299,579	1	3,53	44,91	$A—y\ ^4D^\circ$	$——^7/_2$
297,568	8	3,62	45,28	$a\ ^2H—y\ ^2G^\circ$	$^{11}/_2—^9/_2$
297,308	7	3,56	45,26	$a\ ^3H—y\ ^2G^\circ$	$^9/_2—^7/_2$
297,131	2	3,56	45,28	$a\ ^2H—y\ ^2G^\circ$	$^9/_2—^9/_2$
296,988	6	2,35	44,10	$a\ ^4P—z\ ^4S^\circ$	$^3/_2—^3/_2$
296,808	5	2,32	44,10	$a\ ^4P—z\ ^4S^\circ$	$^1/_2—^3/_2$
296,723	3	3,53	45,31	$A—z\ ^4P$	$——^5/_2$
296,317	1	0,25	42,09	$a\ ^4F—z\ ^4G^\circ$	$^9/_2—^3/_2$
295,634	4	0,00	41,94	$a\ ^4F—z\ ^4G^\circ$	$^3/_2—^5/_2$
295,042	2	0,25	42,27	$a\ ^4F—z\ ^4G^\circ$	$^9/_2—^9/_2$
295,014	4	0,06	41,94	$a\ ^4F—z\ ^4G^\circ$	$^5/_2—^5/_2$
294,960	4	0,06	42,09	$a\ ^4F—z\ ^4F^\circ$	$^5/_2—^3/_2$
294,850	4	0,15	42,19	$a\ ^4F—z\ ^4F^\circ$	$^7/_2—^5/_2$
294,665	4	0,25	42,32	$a\ ^4F—z\ ^4F^\circ$	$^9/_2—^7/_2$
294,520	7	0,00	42,09	$a\ ^4F—z\ ^4F^\circ$	$^3/_2—^3/_2$
294,339	5	0,15	42,27	$a\ ^4F—z\ ^4G^\circ$	$^7/_2—^9/_2$
294,265	7	0,06	42,19	$a\ ^4F—z\ ^4F^\circ$	$^5/_2—^5/_2$
294,040	0	2,55	44,72	$a\ ^2G—y\ ^4D^\circ$	$^7/_2—^5/_2$
293,966	8	0,15	42,32	$a\ ^4F—z\ ^4F^\circ$	$^7/_2—^7/_2$
293,820	1	0,00	42,19	$a\ ^4F—z\ ^4F^\circ$	$^3/_2—^5/_2$
293,745	8	0,25	42,45	$a\ ^4F—z\ ^4F^\circ$	$^9/_2—^9/_2$
293,488	4	0,25	42,49	$a\ ^4F—z\ ^4G^\circ$	$^9/_2—^{11}/_2$
293,384	4	0,06	42,32	$a\ ^4F—z\ ^4F^\circ$	$^5/_2—^7/_2$
293,292	4	2,64	44,91	$a\ ^2G—y\ ^4D^\circ$	$^9/_2—^7/_2$

λ, Å	I	E_H, eV	E_B, eV	Transition	J
293,214	1	2,32	44,60	$a\,^4P—y\,^4D^\circ$	$^1/_2—^3/_2$
293,046	2	0,15	42,45	$a\,^4F—z\,^4F^\circ$	$^7/_2—^9/_2$
292,925	5	0,15	42,47	$a\,^4F—z\,^5F^\circ$	$^7/_2—^5/_2$
292,736	7	0,25	42,60	$a\,^4F—z\,^5F^\circ$	$^9/_2—^7/_2$
292,597	4	2,35	44,60	$a\,^4P—y\,^4D^\circ$	$^3/_2—^5/_2$
292,343	1	0,06	42,47	$a\,^4F—z\,^5F^\circ$	$^5/_2—^5/_2$
292,038	2	0,15	42,60	$a\,^4F—z\,^5F^\circ$	$^7/_2—^7/_2$
291,931	0	0,00	42,47	$a\,^4F—z\,^5F^\circ$	$^3/_2—^5/_2$
291,829	5	2,43	44,91	$a\,^4P—y\,^4D^\circ$	$^5/_2—^7/_2$
291,800	5	0,06	42,45	$a\,^4F—z\,^4F^\circ$	$^5/_2—^9/_2$
291,632	2	3,62	46,13	$a\,^2H—z\,^2H^\circ$	$^{11}/_2—^9/_2$
291,473	5	0,15	42,68	$a\,^4F—z\,^4D^\circ$	$^7/_2—^5/_2$
291,229	6	3,56	46,13	$a\,^2H—z\,^2H^\circ$	$^9/_2—^9/_2$
291,184	6	0,25	42,60	$a\,^4F—z\,^5F^\circ$	$^9/_2—^7/_2$
291,020	5	0,00	42,60	$a\,^4F—z\,^4D^\circ$	$^3/_2—^1/_2$
290,890	2	0,06	42,68	$a\,^4F—z\,^4D^\circ$	$^5/_2—^5/_2$
290,737	4	2,64	45,28	$a\,^2G—y\,^2G^\circ$	$^9/_2—^9/_2$
290,577	4	0,06	42,73	$a\,^4F—z\,^2D^\circ$	$^5/_2—^3/_2$
290,499	2	0,15	42,82	$a\,^4F—z\,^4D^\circ$	$^7/_2—^7/_2$
290,302	5	2,55	45,26	$a\,^2G—y\,^2G^\circ$	$^7/_2—^7/_2$
290,271	6	3,62	46,33	$a\,^2H—z\,^2H^\circ$	$^{11}/_2—^{11}/_2$
290,146	4	0,00	42,73	$a\,^4F—z\,^2D^\circ$	$^3/_2—^3/_2$
290,089	4	0,15	42,88	$a\,^4F—z\,^2D^\circ$	$^7/_2—^5/_2$
290,038	4	2,43	45,18	$a\,^4P—z\,^4P^\circ$	$^5/_2—^3/_2$
289,851	4	2,35	45,18	$a\,^4P—z\,^4P^\circ$	$^3/_2—^3/_2$
289,672	2	2,32	45,13	$a\,^4P—z\,^4P^\circ$	$^1/_2—^1/_2$
289,520	4	0,06	42,88	$a\,^4F—z\,^2D^\circ$	$^5/_2—^5/_2$
289,468	3	2,35	45,18	$a\,^4P—z\,^4P^\circ$	$^3/_2—^3/_2$
289,302	4	2,32	45,18	$a\,^4P—z\,^4P^\circ$	$^1/_2—^3/_2$
289,112	5	2,43	45,31	$a\,^4P—z\,^4P^\circ$	$^5/_2—^5/_2$
288,551	4	2,35	45,31	$a\,^4P—z\,^4P^\circ$	$^3/_2—^5/_2$
287,333	1	0,25	43,39	$a\,^4F—z\,^2G^\circ$	$^9/_2—^9/_2$
284,504	4	2,55	45,26	$a\,^2G—y\,^2G^\circ$	$^7/_2—^9/_2$
283,770	5	2,64	46,33	$a\,^2G—z\,^2H^\circ$	$^9/_2—^{11}/_2$
278,471	3	0,00	44,52	$a\,^4F—y\,^4D^\circ$	$^3/_2—^1/_2$
278,339	5	0,06	44,60	$a\,^4F—y\,^4D^\circ$	$^5/_2—^3/_2$
278,149	5	0,15	44,72	$a\,^4F—y\,^4D^\circ$	$^7/_2—^5/_2$
277,951	3	0,00	44,60	$a\,^4F—y\,^4D^\circ$	$^3/_2—^3/_2$
277,610	2	0,06	44,72	$a\,^4F—y\,^4D^\circ$	$^5/_2—^5/_2$
277,569	6	0,25	44,91	$a\,^4F—y\,^4D^\circ$	$^9/_2—^7/_2$
276,947	3	0,15	44,91	$a\,^4F—y\,^4D^\circ$	$^7/_2—^7/_2$

COPPER, Z = 29

Cu I, ground state $1s^2\,2s^2\,2p^6\,3s^2\,3p^6\,3d^{10}\,4s\ ^2S_{1/2}$

Ionization potential 62317,2 cm^{-1}; 7,726 eV

λ, Å	I	E_H, eV	E_B, eV	Transition	J
18229	5	6,19	6,87	$4d\,^2D$—$4f\,^2F^\circ$	5/2—7/2
18194	7	6,19	6,87	$4d\,^2D$—$4f\,^2F^\circ$	3/2—5/2
16653	4	6,12	6,87	$5p\,^2P^\circ$—$5d\,^2D$	1/2, 3/2—3/2
16008	5	5,35	6,12	$5s\,^2S$—$5p\,^2P^\circ$	1/2—1/2, 3/2
11118,2	1	5,08	6,19	$4p'\,^4P^\circ$—$4d\,^2D$	1/2—3/2
10883,3	1	6,12	7,26	$5p\,^2P^\circ$—$8s\,^2S$	1/2, 3/2—1/2
10771,7	2	7,28	8,43	$4p''\,^2D^\circ$—$5s''\,^2D$	3/2—3/2
10179,2	1	4,97	6,19	$4p'\,^4P^\circ$—$4d\,^2D$	3/2—3/2
10172,00	2	4,97	6,19	$4p'\,^4P^\circ$—$4d\,^2D$	3/2—5/2
10146,78	10	7,21	8,43	$4p''\,^2F^\circ$—$5s''\,^2D$	5/2—3/2
10124,5	5	6,79	8,02	$6p\,^2P^\circ$—$5s'\,^2D$	3/2—5/2
9739,6	4	6,12	7,39	$5p\,^2P^\circ$—$9s\,^2S$	1/2, 3/2—1/2
9530,3	5	7,02	8,32	$4p''\,^2D^\circ$—$5s''\,^2D$	5/2—5/2
9472,4	2	7,12	8,43	$7p\,^2P^\circ$—$5s''\,^2D$	1/2—3/2
9263,54	3	5,69	7,03	$4p'\,^2P^\circ$—$7s\,^2S$	3/2—1/2
8996,2	20	6,95	8,32	$4p''\,^2F^\circ$—$5s''\,^2D$	7/2—5/2
8584,0	10	5,35	6,79	$5s\,^2S$—$6p\,^2P^\circ$	1/2—3/2
8408,15	20	5,35	6,82	$5s\,^2S$—$6p\,^2P^\circ$	1/2—1/2
8092,634	2000	3,82	5,35	$4p\,^2P^\circ$—$5s\,^2S$	3/2—1/2
7933,130	1500	3,79	5,35	$4p\,^2P^\circ$—$5s\,^2S$	1/2—1/2
7570,09	200	5,35	6,98	$5s\,^2S$—$4p''\,^2P^\circ$	1/2—3/2
7452,5	2	7,28	8,94	$4p''\,^2D^\circ$—$4d'\,^2D$	3/2—3/2
7427,2	5	7,28	8,94	$4p''\,^2D^\circ$—$4d'\,^2F$	3/2—5/2
7193,56	50	7,21	8,93	$4p''\,^2F^\circ$—$4d'\,^2G$	5/2—7/2
7154,29	5	7,21	8,94	$4p''\,^2F^\circ$—$4d'\,^4F$	5/2—7/2
7124,66	5	7,21	8,94	$4p''\,^2F^\circ$—$4d'\,^2F$	5/2—5/2
7039,37	25	6,12	7,88	$5p\,^2P^\circ$—$5s'\,^4D$	3/2, 1/2—3/2
		7,02	8,78	$4p''\,^2D^\circ$—$4d'\,^2S$	5/2—1/2
7000,05	2	5,35	7,12	$5s\,^2S$—$7p\,^2P^\circ$	1/2—1/2
6968,34	5	7,02	8,80	$4p''\,^2D^\circ$—$4d'\,^4S$	5/2—3/2
6935,82	5	7,28	9,06	$4p''\,^2D^\circ$—$4d'\,^4D$	3/2—3/2
6920,06	50	7,02	8,81	$4p''\,^2D^\circ$—$4d'\,^2D$	5/2—5/2
6905,94	100	7,02	8,82	$4p''\,^2D^\circ$—$4d'\,^2F$	5/2—7/2
6890,90	10	7,12	8,92	$7p\,^2P^\circ$—$4d'\,^4P$	1/2—3/2
		7,28	9,07	$4p''\,^2D^\circ$—$4d'\,^4G$	3/2—5/2
6889,92	10	6,98	8,78	$4p''\,^2P^\circ$—$4d'\,^2S$	3/2—1/2
6881,94	10	7,02	8,82	$4p''\,^2D^\circ$—$4d'\,^4P$	5/2—5/2
6840,99	3	7,28	9,09	$4p''\,^2D^\circ$—$4d'\,^4F$	3/2—5/2
6821,86	2	6,98	8,80	$4p''\,^2P^\circ$—$4d'\,^4S$	3/2—3/2
6775,64	2	6,98	8,81	$4p''\,^2P^\circ$—$4d'\,^2D$	3/2—5/2
6741,12	100	6,95	8,78	$4p''\,^2F^\circ$—$4d'\,^2G$	7/2—9/2
6672,23	10	7,21	9,06	$4p''\,^2F^\circ$—$4d'\,^4G$	5/2—7/2
6634,7	2	6,95	8,81	$4p''\,^2F^\circ$—$4d'\,^2D$	7/2—5/2
6629,67	5	7,21	9,07	$4p''\,^2F^\circ$—$4d'\,^4G$	5/2—5/2
6621,61	30	6,95	8,82	$4p''\,^2F^\circ$—$4d'\,^2F$	7/2—7/2
6485,18	5	7,02	8,93	$4p''\,^2D^\circ$—$4d'\,^4D$	5/2—5/2
6474,20	10	7,02	8,94	$4p''\,^2D^\circ$—$4d'\,^4F$	5/2—7/2
6325,45	5	5,72	7,74	$4p'\,^2D^\circ$—$5s'\,^4D$	3/2—7/2
6268,30	20	6,95	8,92	$4p''\,^2F^\circ$—$4d'\,^4G$	7/2—9/2
6223,66	4	6,95	8,94	$4p''\,^2F^\circ$—$4d'\,^4F$	7/2—7/2
6221,11	2	6,79	8,78	$6p\,^2P^\circ$—$4d'\,^2S$	3/2—1/2
6147,31	20	—	—	—	—
6032,33	2	7,28	9,33	$4p''\,^2D^\circ$—$4d''\,^2P$	3/2—3/2
5966,59	3	7,28	9,35	$4p''\,^2D^\circ$—$4d''\,^2D$	3/2—5/2

λ, Å	I	E_H, eV	E_B, eV	Transition	J
5856,94	5	5,69	7,80	$4p'\,{}^2P^\circ - 5s'\,{}^4D$	$3/2-5/2$
5851,1	2	7,21	9,32	$4p''\,{}^2F^\circ - 4d''\,{}^2G$	$5/2-7/2$
5782,132	1500	1,64	3,79	$4s^2\,{}^2D - 4p\,{}^2P^\circ$	$3/2-1/2$
5732,325	75	5,57	7,74	$4p'\,{}^2F^\circ - 5s'\,{}^4D$	$7/2-7/2$
5727,96	5	7,21	9,37	$4p''\,{}^2F^\circ - 4d''\,{}^2F$	$5/2-7/2$
5700,240	1500	1,64	3,82	$4s^2\,{}^2D - 4p\,{}^2P^\circ$	$3/2-3/2$
5646,5	2	5,69	7,88	$4p'\,{}^2P^\circ - 5s'\,{}^4D$	$3/2-3/2$
5554,935	100	5,51	7,74	$4p'\,{}^4P^\circ - 5s'\,{}^4D$	$5/2-7/2$
5535,78	50	5,78	8,02	$4p'\,{}^2D^\circ - 5s'\,{}^2D$	$5/2-5/2$
5463,138	150	5,72	7,99	$4p'\,{}^2D^\circ - 5s'\,{}^4D$	$3/2-1/2$
5432,05	250	5,52	7,80	$4p'\,{}^4D^\circ - 5s'\,{}^4D$	$3/2-5/2$
5408,34	100	5,72	8,02	$4p'\,{}^2D^\circ - 5s'\,{}^2D$	$3/2-5/2$
5391,62	450	5,51	7,80	$4p'\,{}^4D^\circ - 5s'\,{}^4D$	$5/2-5/2$
5376,867	5	5,69	7,99	$4p'\,{}^2P^\circ - 5s'\,{}^4D$	$3/2-1/2$
5360,030	200	5,69	7,99	$4p'\,{}^2P^\circ - 5s'\,{}^4D$	$3/2-1/2$
5357,33	3	5,78	8,09	$4p'\,{}^2D^\circ - 5s'\,{}^2D$	$5/2-3/2$
5354,95	250	5,57	7,88	$4p'\,{}^4D^\circ - 5s'\,{}^4D$	$1/2-3/2$
5352,666	300	5,42	7,74	$4p'\,{}^2F^\circ - 5s'\,{}^4D$	$5/2-7/2$
5323,78	3	5,69	8,02	$4p'\,{}^2P^\circ - 5s'\,{}^2D$	$3/2-5/2$
5292,517	1650	5,39	7,74	$4p'\,{}^4D^\circ - 5s'\,{}^4D$	$7/2-7/2$
5283,530	5	7,02	9,37	$4p''\,{}^2D^\circ - 4d''\,{}^2F$	$5/2-7/2$
5250,52	500	5,52	7,88	$4p'\,{}^4D^\circ - 5s'\,{}^4D$	$3/2-3/2$
5237,65	10	5,72	8,09	$4p'\,{}^2D^\circ - 5s'\,{}^2D$	$3/2-3/2$
5220,070	500	3,82	6,19	$4p\,{}^2P^\circ - 4d\,{}^2D$	$3/2-3/2$
5218,202	2500	3,82	6,19	$4p\,{}^2P^\circ - 4d\,{}^2D$	$3/2-5/2$
5212,780	140	5,51	7,88	$4p'\,{}^4D^\circ - 5s'\,{}^4D$	$5/2-3/2$
5200,87	500	5,42	7,80	$4p'\,{}^2F^\circ - 5s'\,{}^4D$	$5/2-5/2$
5158,36	50	5,69	8,09	$4p'\,{}^2P^\circ - 5s'\,{}^2D$	$3/2-3/2$
5153,235	2000	3,79	6,19	$4p\,{}^2P^\circ - 4d\,{}^2D$	$1/2-3/2$
5144,120	550	5,39	7,80	$4p'\,{}^4D^\circ - 5s'\,{}^4D$	$7/2-5/2$
5142,7	10	5,68	8,09	$4p'\,{}^2P^\circ - 5s'\,{}^2D$	$1/2-3/2$
5115,49	10	6,95	9,37	$4p''\,{}^2F^\circ - 4d''\,{}^2F$	$7/2-7/2$
5111,913	300	5,57	7,99	$4p'\,{}^4D^\circ - 5s'\,{}^4D$	$1/2-1/2$
5105,541	1500	1,39	3,82	$4s^2\,{}^2D - 4p\,{}^2P^\circ$	$5/2-3/2$
5076,173	100	5,57	8,02	$4p'\,{}^2F^\circ - 5s'\,{}^2D$	$7/2-5/2$
5034,36	100	5,42	7,88	$4p'\,{}^2F^\circ - 5s'\,{}^4D$	$5/2-3/2$
5016,611	400	5,52	7,99	$4p'\,{}^4D^\circ - 5s'\,{}^4D$	$3/2-1/2$
4866,10	75	5,78	8,32	$4p'\,{}^2D^\circ - 5s''\,{}^2D$	$5/2-5/2$
4842,290	25	5,24	7,80	$4p'\,{}^4F^\circ - 5s'\,{}^4D$	$3/2-5/2$
4797,042	20	5,15	7,74	$4p'\,{}^4F^\circ - 5s'\,{}^4D$	$5/2-7/2$
4794,00	150	5,51	8,09	$4p'\,{}^4D^\circ - 5s'\,{}^2D$	$5/2-3/2$
4776,22	20	5,42	8,02	$4p'\,{}^2F^\circ - 5s'\,{}^2D$	$5/2-5/2$
4767,49	75	5,72	8,32	$4p'\,{}^2D^\circ - 5s''\,{}^2D$	$3/2-5/2$
4704,594	450	5,10	7,74	$4p'\,{}^4F^\circ - 5s'\,{}^4D$	$7/2-7/2$
4701,71	10	5,69	8,32	$4p'\,{}^2P^\circ - 5s''\,{}^2D$	$3/2-5/2$
4697,490	350	5,24	7,88	$4p'\,{}^4F^\circ - 5s'\,{}^4D$	$3/2-3/2$
4677,340	3	5,78	8,43	$4p'\,{}^2D^\circ - 5s''\,{}^2D$	$5/2-3/2$
4674,72	500	5,15	7,80	$4p'\,{}^4F^\circ - 5s'\,{}^4D$	$5/2-5/2$
4651,124	2000	5,07	7,74	$4p'\,{}^4F^\circ - 5s'\,{}^4D$	$9/2-7/2$
4642,58	150	5,42	8,09	$4p'\,{}^2F^\circ - 5s'\,{}^2D$	$5/2-3/2$
4586,97	1300	5,10	7,80	$4p'\,{}^4F^\circ - 5s'\,{}^4D$	$7/2-5/2$
4539,695	800	5,15	7,88	$4p'\,{}^4F^\circ - 5s'\,{}^4D$	$5/2-3/2$
4530,785	800	3,82	6,55	$4p\,{}^2P^\circ - 6s\,{}^2S$	$3/2-1/2$
4525,112	40	5,69	8,43	$4p'\,{}^2P^\circ - 5s''\,{}^2D$	$3/2-3/2$
4513,192	50	5,68	8,43	$4p'\,{}^2P^\circ - 5s''\,{}^2D$	$1/2-3/2$
4509,374	400	5,24	7,99	$4p'\,{}^4F^\circ - 5s'\,{}^4D$	$3/2-1/2$
4507,35	200	5,57	8,32	$4p'\,{}^2F^\circ - 5s''\,{}^2D$	$7/2-5/2$
4480,350	500	3,79	6,55	$4p\,{}^2P^\circ - 6s\,{}^2S$	$1/2-1/2$

λ, Å	I	E_H, eV	E_B, eV	Transition	J
4415,54	200	5,08	7,88	$4p'\,^4P^\circ-5s'\,^4D$	$1/2-3/2$
4397,0	10	5,51	8,32	$4p'\,^4D^\circ-5s''\,^2D$	$5/2-5/2$
4378,20	550	4,97	7,80	$4p'\,^4P^\circ-5s'\,^4D$	$3/2-5/2$
4354,74	10	5,24	8,09	$4p'\,^4F^\circ-5s'\,^2D$	$3/2-3/2$
4336,00	10	5,57	8,43	$4p'\,^4D^\circ-5s''\,^2D$	$1/2-3/2$
4328,68	20	5,15	8,02	$4p'\,^4F^\circ-5s'\,^2D$	$5/2-5/2$
4275,107	950	4,84	7,74	$4p'\,^4P^\circ-5s'\,^4D$	$5/2-7/2$
4267,204	2	5,52	8,43	$4p'\,^4D^\circ-5s''\,^2D$	$3/2-3/2$
4259,401	150	4,97	7,88	$4p'\,^4P^\circ-5s'\,^4D$	$3/2-3/2$
4253,390	20	—	—	—	—
4248,956	150	5,08	7,99	$4p'\,^4P^\circ-5s'\,^4D$	$1/2-1/2$
4242,26	30	5,51	8,43	$4p'\,^4D^\circ-5s''\,^2D$	$5/2-3/2$
4230,9	5	5,39	8,32	$4p'\,^4D^\circ-5s''\,^2D$	$7/2-5/2$
4218,8	2	5,15	8,09	$4p'\,^4F^\circ-5s'\,^2D$	$5/2-3/2$
4177,758	100	4,84	7,80	$4p'\,^4P^\circ-5s'\,^4D$	$5/2-5/2$
4123,287	30	5,42	8,43	$4p'\,^2F^\circ-5s''\,^2D$	$5/2-3/2$
4121,74	10	5,78	8,78	$4p'\,^2D^\circ-4d'\,^2P$	$5/2-3/2$
4111,4	3	5,08	8,09	$4p'\,^4P^\circ-5s'\,^2D$	$1/2-3/2$
4104,218	25	4,97	7,99	$4p'\,^4P^\circ-5s'\,^4D$	$3/2-1/2$
4080,534	15	5,78	8,81	$4p'\,^2D^\circ-4d'\,^2D$	$5/2-5/2$
4075,572	50	5,78	8,82	$4p'\,^2D^\circ-4d'\,^2F$	$5/2-7/2$
4073,224	20	4,97	8,02	$4p'\,^4P^\circ-5s'\,^2D$	$3/2-5/2$
4069,53	6	4,84	7,88	$4p'\,^4P^\circ-5s'\,^4D$	$5/2-3/2$
4063,238	650	3,82	6,87	$4p\,^2P^\circ-5d\,^2D$	$3/2-3/2$
4062,641	2000	3,82	6,87	$4p\,^2P^\circ-5d\,^2D$	$3/2-5/2$
4056,78	35	3,82	6,87	$4p\,^2P^\circ-4f\,^2F^\circ$	$3/2-7/2$
4056,38	35	3,82	6,87	$4p\,^2P^\circ-4f\,^2F^\circ$	$3/2-5/2$
4052,380	2	5,78	8,84	$4p'\,^2D^\circ-4d'\,^4D$	$5/2-7/2$
4050,617	20	5,72	8,78	$4p'\,^2D^\circ-4d'\,^2P$	$3/2-3/2$
4027,026	10	5,72	8,80	$4p'\,^2D^\circ-4d'\,^4S$	$3/2-3/2$
4022,629	1250	3,79	6,87	$4p\,^2P^\circ-5d\,^2D$	$1/2-3/2$
4015,8	10	3,79	6,87	$4p\,^2P^\circ-4f\,^2F^\circ$	$1/2-5/2$
4010,836	8	5,72	8,81	$4p'\,^2D^\circ-4d'\,^2D$	$3/2-5/2$
4003,028	15	5,69	8,78	$4p'\,^2P^\circ-4d'\,^2P$	$3/2-3/2$
3998,018	3	5,72	8,82	$4p'\,^2D^\circ-4d'\,^4P$	$3/2-5/2$
3979,954	5	5,69	8,80	$4p'\,^2P^\circ-4d'\,^4S$	$3/2-3/2$
3975,7	5	4,97	8,09	$4p'\,^4P^\circ-5s'\,^2D$	$3/2-3/2$
3964,16	5	5,69	8,81	$4p'\,^2P^\circ-4d'\,^2D$	$3/2-5/2$
3951,616	2	5,69	8,82	$4p'\,^2P^\circ-4d'\,^4P$	$3/2-5/2$
3946,938	3	5,78	8,92	$4p'\,^2D^\circ-4d'\,^4P$	$5/2-3/2$
3933,027	5	5,78	8,93	$4p'\,^2D^\circ-4d'\,^2G$	$5/2-7/2$
3925,274	8	5,78	8,93	$4p'\,^2D^\circ-4d'\,^4D$	$5/2-5/2$
3921,267	5	5,78	8,94	$4p'\,^2D^\circ-4d'\,^4F$	$5/2-7/2$
3899,22	8	4,84	8,02	$4p'\,^4P^\circ-5s'\,^2D$	$5/2-5/2$
3888,40	4	5,72	8,91	$4p'\,^2D^\circ-4d'\,^4P$	$3/2-1/2$
3885,92	3	6,12	9,31	$5p\,^2P^\circ-4d''\,^2P$	$3/2,\ 1/2-1/2$
3881,714	5	5,72	8,92	$4p'\,^2D^\circ-4d'\,^4P$	$3/2-3/2$
3862,781	5	6,12	9,33	$5p\,^2P^\circ-4d''\,^2P$	$3/2,\ 1/2-3/2$
3861,747	250	3,82	7,03	$4p\,^2P^\circ-7s\,^2S$	$3/2-1/2$
3860,898	5	5,72	8,93	$4p'\,^2D^\circ-4d'\,^4D$	$3/2-5/2$
3860,472	600	5,57	8,78	$4p'\,^2F^\circ-4d'\,^2G$	$7/2-9/2$
3844,51	4	5,69	8,91	$4p'\,^2P^\circ-4d'\,^4P$	$3/2-1/2$
3837,976	5	5,69	8,92	$4p'\,^2P^\circ-4d'\,^4P$	$3/2-3/2$
3825,047	100	3,79	7,03	$4p\,^2P^\circ-7s\,^2S$	$1/2-1/2$
3820,884	60	5,57	8,82	$4p'\,^2F^\circ-4d'\,^2F$	$7/2-7/2$
3817,490	5	5,69	8,93	$4p'\,^2P^\circ-4d'\,^4D$	$3/2-5/2$
3813,542	10	5,57	8,82	$4p'\,^2F^\circ-4d'\,^4P$	$7/2-5/2$
3811,95	8	5,69	8,94	$4p'\,^2P^\circ-4d'\,^2D$	$3/2-3/2$

λ, Å	I	E_H, eV	E_B, eV	Transition	J
3805,232	100	5,69	8,94	$4p'\ ^2P°—4d'\ ^2P$	$^3/_2—^1/_2$
3803,49	5	5,68	8,94	$4p'\ ^2P°—4d'\ ^2D$	$^1/_2—^3/_2$
3800,502	30	5,57	8,84	$4p'\ ^2F°—4d'\ ^4D$	$^7/_2—^7/_2$
3799,88	10	5,52	8,78	$4p'\ ^4D°—4d'\ ^2P$	$^3/_2—^3/_2$
3797,245	8	5,57	8,84	$4p'\ ^2F°—4d'\ ^4F$	$^7/_2—^9/_2$
3785,49	5	5,15	8,43	$4p'\ ^4F°—5s''\ ^2D$	$^5/_2—^3/_2$
3780,045	5	5,51	8,78	$4p'\ ^4D°—4d'\ ^2P$	$^5/_2—^3/_2$
3779,067	2	5,52	8,80	$4p'\ ^4D°—4d'\ ^4S$	$^3/_2—^3/_2$
3771,904	100	5,78	9,06	$4p'\ ^2D°—4d'\ ^4G$	$^5/_2—^7/_2$
3764,837	5	5,52	8,81	$4p'\ ^4D°—4d'\ ^2D$	$^3/_2—^5/_2$
3759,492	60	5,51	8,80	$4p'\ ^4D°—4d'\ ^4S$	$^5/_2—^3/_2$
3758,296	5	5,78	9,07	$4p'\ ^2D°—4d'\ ^4G$	$^5/_2—^5/_2$
3753,519	8	5,52	8,82	$4p'\ ^4D°—4d'\ ^4P$	$^3/_2—^5/_2$
3745,356	20	5,51	8,81	$4p'\ ^4D°—4d'\ ^2D$	$^5/_2—^5/_2$
3743,363	3	5,78	9,09	$4p'\ ^2D°—4d'\ ^4F$	$^5/_2—^5/_2$
3741,242	450	5,51	8,82	$4p'\ ^4D°—4d'\ ^2F$	$^5/_2—^7/_2$
3734,180	200	5,51	8,82	$4p'\ ^4D°—4d'\ ^4P$	$^5/_2—^5/_2$
3721,666	8	5,51	8,84	$4p'\ ^4D°—4d'\ ^4D$	$^5/_2—^7/_2$
3720,771	150	1,64	4,97	$4s^2\ ^2D—4p'\ ^4P°$	$^3/_2—^3/_2$
3712,009	30	5,72	9,06	$4p'\ ^2D°—4d'\ ^4D$	$^3/_2—^3/_2$
3707,12	4	5,57	8,91	$4p'\ ^4D°—4d'\ ^4P$	$^1/_2—^1/_2$
3701,070	5	5,57	8,92	$4p'\ ^4D°—4d'\ ^4P$	$^1/_2—^3/_2$
3700,536	250	5,57	8,92	$4p'\ ^2F°—4d'\ ^4G$	$^7/_2—^9/_2$
3699,097	10	5,72	9,07	$4p'\ ^2D°—4d'\ ^4G$	$^3/_2—^5/_2$
3695,358	8	5,57	8,93	$4p'\ ^2F°—4d'\ ^2G$	$^7/_2—^7/_2$
3687,708	40	3,82	7,18	$4p\ ^2P°—6d\ ^2D$	$^3/_2—^3/_2$
3687,438	400	3,82	7,18	$4p\ ^2P°—6d\ ^2D$	$^3/_2—^5/_2$
3684,930	200	5,57	8,94	$4p'\ ^2F°—4d'\ ^4F$	$^7/_2—^7/_2$
3684,672	450	5,72	9,09	$4p'\ ^2D°—4d'\ ^4F$	$^3/_2—^5/_2$
3676,878	50	5,57	8,94	$4p'\ ^4D°—4d'\ ^2D$	$^1/_2—^3/_2$
3671,953	100	5,69	9,06	$4p'\ ^2P°—4d'\ ^4D$	$^3/_2—^3/_2$
3665,735	125	5,42	8,80	$4p'\ ^2F°—4d'\ ^4S$	$^5/_2—^3/_2$
3664,08	5	5,68	9,06	$4p'\ ^2P°—4d'\ ^4D$	$^1/_2—^3/_2$
3659,353	125	5,69	9,07	$4p'\ ^2P°—4d'\ ^4G$	$^3/_2—^5/_2$
3656,785	125	5,52	8,91	$4p'\ ^4D°—4d'\ ^4P$	$^3/_2—^1/_2$
3655,859	600	5,39	8,78	$4p'\ ^4D°—4d'\ ^2G$	$^7/_2—^9/_2$
3654,243	200	3,79	7,18	$4p\ ^2P°—6d\ ^2D$	$^1/_2—^3/_2$
3652,34	100	5,42	8,81	$4p'\ ^2F°—4d'\ ^2D$	$^5/_2—^5/_2$
3650,855	5	5,52	8,92	$4p'\ ^4D°—4d'\ ^4P$	$^3/_2—^3/_2$
3648,383	125	5,42	8,82	$4p'\ ^2F°—4d'\ ^2F$	$^5/_2—^7/_2$
3645,232	250	5,69	9,09	$4p'\ ^2P°—4d'\ ^4F$	$^3/_2—^5/_2$
3643,632	5	5,69	9,09	$4p'\ ^2P°—4d'\ ^4F$	$^3/_2—^3/_2$
3641,693	50	5,42	8,82	$4p'\ ^2F°—4d'\ ^4P$	$^5/_2—^5/_2$
3635,916	250	5,68	9,09	$4p'\ ^2P°—4d'\ ^4F$	$^1/_2—^3/_2$
3632,558	50	5,51	8,92	$4p'\ ^4D°—4d'\ ^4P$	$^5/_2—^3/_2$
3632,308	5	5,52	8,93	$4p'\ ^4D°—4d'\ ^4D$	$^3/_2—^5/_2$
3629,771	10	5,42	8,84	$4p'\ ^2F°—4d'\ ^4D$	$^5/_2—^7/_2$
3627,32	125	5,52	8,94	$4p'\ ^4D°—4d'\ ^2D$	$^3/_2—^3/_2$
3624,236	100	5,39	8,81	$4p'\ ^4D°—4d'\ ^2D$	$^7/_2—^5/_2$
3621,245	600	5,52	8,94	$4p'\ ^4D°—4d'\ ^2F$	$^3/_2—^5/_2$
3620,352	225	5,39	8,82	$4p'\ ^4D°—4d'\ ^2F$	$^7/_2—^7/_2$
3614,218	200	5,51	8,93	$4p'\ ^4D°—4d'\ ^4D$	$^5/_2—^5/_2$
3613,761	600	5,39	8,82	$4p'\ ^4D°—4d'\ ^4P$	$^7/_2—^5/_2$
3610,809	200	5,51	8,94	$4p'\ ^4D°—4d'\ ^4F$	$^5/_2—^7/_2$
3609,295	200	1,64	5,08	$4s^2\ ^2D—4p'\ ^4P°$	$^3/_2—^1/_2$
3602,032	1400	5,39	8,84	$4p'\ ^4D°—4d'\ ^4D$	$^7/_2—^7/_2$
3599,132	1400	5,39	8,84	$4p'\ ^4D°—4d'\ ^4F$	$^7/_2—^9/_2$

λ, Å	I	E_H, eV	E_B, eV	Transition	J
3598,011	10	3,82	7,26	$4p\ ^2P^\circ-8s\ ^2S$	$3/2-1/2$
3594,023	30	1,39	4,84	$4s^2\ ^2D-4p'\ ^4P^\circ$	$5/2-5/2$
3566,131	5	3,79	7,26	$4p\ ^2P^\circ-8s\ ^2S$	$1/2-1/2$
3546,433	15	5,57	9,06	$4p'\ ^4D^\circ-4d'\ ^4D$	$1/2-3/2$
3544,963	125	5,42	8,92	$4p'\ ^2F^\circ-4d'\ ^4P$	$5/2-3/2$
3533,746	500	5,42	8,93	$4p'\ ^2F^\circ-4d'\ ^2G$	$5/2-7/2$
3530,383	2000	1,64	5,15	$4s^2\ ^2D-4p'\ ^4F^\circ$	$3/2-5/2$
3527,482	500	5,42	8,93	$4p'\ ^2F^\circ-4d'\ ^4D$	$5/2-5/2$
3524,231	1250	5,42	8,94	$4p'\ ^2F^\circ-4d'\ ^4F$	$5/2-7/2$
3520,031	500	5,57	9,09	$4p'\ ^4D^\circ-4d'\ ^4F$	$1/2-3/2$
3517,039	100	5,42	8,94	$4p'\ ^2F^\circ-4d'\ ^2F$	$5/2-5/2$
3512,121	650	5,39	8,92	$4p'\ ^4D^\circ-4d'\ ^4G$	$7/2-9/2$
3511,985	10	3,82	7,35	$4p\ ^2P^\circ-7d\ ^2D$	$3/2-3/2$
3511,835	50	3,82	7,35	$4p\ ^2P^\circ-7d\ ^2D$	$3/2-5/2$
3507,407	5	5,39	8,93	$4p'\ ^4D^\circ-4d'\ ^2G$	$7/2-7/2$
3501,529	3	5,24	8,78	$4p'\ ^4F^\circ-4d'\ ^2P$	$3/2-3/2$
3501,251	5	5,39	8,93	$4p'\ ^4D^\circ-4d'\ ^4D$	$7/2-5/2$
3500,324	50	5,52	9,06	$4p'\ ^4D^\circ-4d'\ ^2D$	$3/2-3/2$
3498,938	3	5,78	9,32	$4p'\ ^2D^\circ-6s'\ ^2D$	$5/2-5/2$
3498,063	125	5,39	8,94	$4p'\ ^4D^\circ-4d'\ ^4F$	$7/2-7/2$
3488,858	100	5,52	9,07	$4p'\ ^4D^\circ-4d'\ ^4G$	$3/2-5/2$
3487,566	60	5,78	9,33	$4p'\ ^2D^\circ-4d''\ ^2P$	$5/2-3/2$
3483,761	1250	5,51	9,06	$4p'\ ^4D^\circ-4d'\ ^4G$	$5/2-7/2$
3481,614	5	3,79	7,35	$4p\ ^2P^\circ-7d\ ^2D$	$1/2-3/2$
3475,999	750	5,52	9,09	$4p'\ ^4D^\circ-4d'\ ^4F$	$3/2-5/2$
3474,578	5	5,52	9,09	$4p'\ ^4D^\circ-4d'\ ^4F$	$3/2-3/2$
3472,141	200	5,51	9,07	$4p'\ ^4D^\circ-4d'\ ^4G$	$5/2-5/2$
3471,748	2	5,24	8,81	$4p'\ ^4F^\circ-4d'\ ^2D$	$3/2-5/2$
3466,24	25	5,78	9,35	$4p'\ ^2D^\circ-4d''\ ^2D$	$5/2-3/2$
3465,401	50	5,78	9,35	$4p'\ ^2D^\circ-4d''\ ^2D$	$5/2-5/2$
3463,499	5	3,82	7,39	$4p\ ^2P^\circ-9s\ ^2S$	$3/2-1/2$
3459,428	25	5,51	9,09	$4p'\ ^4D^\circ-4d'\ ^4F$	$5/2-5/2$
3457,850	750	1,39	4,97	$4s^2\ ^2D-4p'\ ^4P^\circ$	$5/2-3/2$
3454,686	200	5,78	9,36	$4p'\ ^2D^\circ-4d''\ ^2F$	$5/2-5/2$
3450,332	750	5,78	9,37	$4p'\ ^2D^\circ-4d''\ ^2F$	$5/2-7/2$
3447,590	3	5,72	9,32	$4p'\ ^2D^\circ-6s'\ ^2D$	$3/2-5/2$
3440,507	250	1,64	5,24	$4s^2\ ^2D-4p'\ ^4F^\circ$	$3/2-3/2$
3436,543	5	5,72	9,33	$4p'\ ^2D^\circ-4d''\ ^2P$	$3/2-3/2$
3433,972	3	3,79	7,39	$4p\ ^2P^\circ-9s\ ^2S$	$1/2-1/2$
3422,10	15	5,72	9,35	$4p'\ ^2D^\circ-4d''\ ^2S$	$3/2-1/2$
3420,166	8	5,69	9,31	$4p'\ ^2P^\circ-4d''\ ^2P$	$3/2-1/2$
3415,80	200	5,72	9,35	$4p'\ ^2D^\circ-4d''\ ^2D$	$3/2-3/2$
3414,017	5	3,82	7,45	$4p\ ^2P^\circ-8d\ ^2D$	$3/2-5/2$
3413,343	200	5,68	9,31	$4p'\ ^2P^\circ-4d''\ ^2P$	$1/2-1/2$
3413,107	10	5,15	8,78	$4p'\ ^4F^\circ-4d'\ ^2P$	$5/2-3/2$
3404,66	125	5,72	9,36	$4p'\ ^2D^\circ-4d''\ ^2F$	$3/2-5/2$
3403,107	5	5,42	9,06	$4p'\ ^2F^\circ-4d'\ ^4G$	$5/2-7/2$
3402,244	225	5,69	9,33	$4p'\ ^2P^\circ-4d''\ ^2P$	$3/2-3/2$
3396,324	10	5,15	8,80	$4p'\ ^4F^\circ-4d'\ ^4S$	$5/2-3/2$
3395,476	60	5,68	9,33	$4p'\ ^2P^\circ-4d''\ ^2P$	$1/2-3/2$
3392,016	8	5,42	9,07	$4p'\ ^2F^\circ-4d'\ ^4G$	$5/2-5/2$
3388,07	8	5,69	9,35	$4p'\ ^2P^\circ-4d''\ ^2S$	$3/2-1/2$
3385,394	2	3,79	7,45	$4p\ ^2P^\circ-8d\ ^2D$	$1/2-3/2$
3384,80	15	5,15	8,81	$4p'\ ^4F^\circ-4d'\ ^2D$	$5/2-5/2$
3381,421	200	5,15	8,82	$4p'\ ^4F^\circ-4d'\ ^2F$	$5/2-7/2$
3381,124	60	5,69	9,35	$4p'\ ^2P^\circ-4d''\ ^2D$	$3/2-5/2$
3379,864	3	5,42	9,09	$4p'\ ^2F^\circ-4d'\ ^4F$	$5/2-5/2$
3379,653	5	5,24	8,91	$4p'\ ^4F^\circ-4d'\ ^4P$	$3/2-1/2$

λ, Å	I	E_H, eV	E_B, eV	Transition	J
3378,707	2	5,39	9,06	$4p'\,{}^4D^\circ-4d'\,{}^4G$	$7/2-7/2$
		5,51	9,17	$4p'\,{}^4D^\circ-6s'\,{}^4D$	$5/2-7/2$
3375,672	30	5,15	8,82	$4p'\,{}^4F^\circ-4d'\,{}^4P$	$5/2-5/2$
3375,18	8	5,68	9,35	$4p'\,{}^2P^\circ-4d''\,{}^2D$	$1/2-3/2$
3365,342	750	5,10	8,78	$4p'\,{}^4F^\circ-4d'\,{}^2G$	$7/2-9/2$
3362,12	2	5,78	9,46	$4p'\,{}^2D^\circ-6s'\,{}^2D$	$5/2-3/2$
3358,74	2	5,24	8,93	$4p'\,{}^4F^\circ-4d'\,{}^4D$	$3/2-5/2$
3358,27	2	5,52	9,21	$4p'\,{}^4D^\circ-6s'\,{}^4D$	$3/2-5/2$
3354,474	60	5,24	8,94	$4p'\,{}^4F^\circ-4d'\,{}^2D$	$3/2-3/2$
3353,466	10	3,82	7,51	$4p\,{}^2P^\circ-9d\,{}^2D$	$3/2-5/2$
3349,279	450	5,24	8,94	$4p'\,{}^4F^\circ-4d'\,{}^2F$	$3/2-5/2$
3342,77	5	5,51	9,21	$4p'\,{}^4D^\circ-6s'\,{}^4D$	$5/2-5/2$
3342,454	5	5,08	8,78	$4p'\,{}^4P^\circ-4d'\,{}^2P$	$1/2-3/2$
3337,845	1500	1,39	5,10	$4s^2\,{}^2D-4p'\,{}^4F^\circ$	$5/2-7/2$
3335,215	400	5,10	8,82	$4p'\,{}^4F^\circ-4d'\,{}^2F$	$7/2-7/2$
3329,636	225	5,10	8,82	$4p'\,{}^4F^\circ-4d'\,{}^4P$	$7/2-5/2$
3325,812	3	3,79	7,51	$4p\,{}^2P^\circ-9d\,{}^2D$	$1/2-3/2$
3325,328	3	5,08	8,80	$4p'\,{}^4P^\circ-4d'\,{}^4S$	$1/2-3/2$
3319,682	150	5,10	8,84	$4p'\,{}^4F^\circ-4d'\,{}^4D$	$7/2-7/2$
3317,218	750	5,10	8,84	$4p'\,{}^4F^\circ-4d'\,{}^4F$	$7/2-9/2$
3310,987	8	5,57	9,31	$4p'\,{}^4D^\circ-4d''\,{}^2P$	$1/2-1/2$
3309,558	4	5,57	9,32	$4p'\,{}^2F^\circ-6s'\,{}^2D$	$7/2-5/2$
3307,948	2500	5,07	8,82	$4p'\,{}^4F^\circ-4d'\,{}^4G$	$9/2-11/2$
3305,530	4	5,68	9,43	$4p'\,{}^2P^\circ-6s'\,{}^4D$	$1/2-1/2$
3302,787	4	5,42	9,17	$4p'\,{}^2F^\circ-6s'\,{}^4D$	$5/2-7/2$
3294,168	5	5,57	9,33	$4p'\,{}^4D^\circ-4d''\,{}^2P$	$1/2-3/2$
3293,815	2	5,78	9,54	$4p'\,{}^2D^\circ-5d'\,{}^2D$	$5/2-5/2$
3292,965	450	5,07	8,84	$4p'\,{}^4F^\circ-4d'\,{}^4D$	$9/2-7/2$
3292,827	650	1,39	5,15	$4s^2\,{}^2D-4p'\,{}^4F^\circ$	$5/2-5/2$
3292,393	125	5,15	8,92	$4p'\,{}^4F^\circ-4d'\,{}^4P$	$5/2-3/2$
3290,541	1500	5,07	8,84	$4p'\,{}^4F^\circ-4d'\,{}^4F$	$9/2-9/2$
3286,193	2	3,79	7,56	$4p\,{}^2P^\circ-10d\,{}^2D$	$1/2-3/2$
3282,716	1400	5,15	8,93	$4p'\,{}^4F^\circ-4d'\,{}^2G$	$5/2-7/2$
3279,815	2000	1,64	5,42	$4s^2\,{}^2D-4p'\,{}^2F^\circ$	$3/2-5/2$
3277,310	650	5,15	8,93	$4p'\,{}^4F^\circ-4d'\,{}^4D$	$5/2-5/2$
3273,957	10000	0,00	3,79	$4s\,{}^2S-4p\,{}^2P^\circ$	$1/2-1/2$
3268,278	650	5,15	8,94	$4p'\,{}^4F^\circ-4d'\,{}^2F$	$5/2-5/2$
3266,023	650	5,57	9,37	$4p'\,{}^2F^\circ-4d''\,{}^2F$	$7/2-7/2$
3252,220	650	4,97	8,78	$4p'\,{}^4P^\circ-4d'\,{}^2P$	$3/2-3/2$
3247,540	10000	0,00	3,82	$4s\,{}^2S-4p\,{}^2P^\circ$	$1/2-3/2$
3243,164	1500	5,10	8,92	$4p'\,{}^4F^\circ-4d'\,{}^4G$	$7/2-9/2$
3239,16	150	5,10	8,93	$4p'\,{}^4F^\circ-4d'\,{}^2G$	$7/2-7/2$
3235,713	650	5,24	9,07	$4p'\,{}^4F^\circ-4d'\,{}^4G$	$3/2-5/2$
3233,899	450	5,10	8,93	$4p'\,{}^4F^\circ-4d'\,{}^4D$	$7/2-5/2$
3231,178	650	5,10	8,94	$4p'\,{}^4F^\circ-4d'\,{}^4F$	$7/2-7/2$
3226,602	150	5,08	8,92	$4p'\,{}^4P^\circ-4d'\,{}^4P$	$1/2-3/2$
3226,541	50	4,97	8,81	$4p'\,{}^4P^\circ-4d'\,{}^2D$	$3/2-5/2$
3225,698	5	5,52	9,36	$4p'\,{}^4D^\circ-4d''\,{}^2F$	$3/2-5/2$
3225,088	2	5,10	8,94	$4p'\,{}^4F^\circ-4d'\,{}^2F$	$7/2-5/2$
3224,664	450	5,24	9,09	$4p'\,{}^4F^\circ-4d'\,{}^4F$	$3/2-5/2$
3223,435	400	5,24	9,09	$4p'\,{}^4F^\circ-4d'\,{}^4F$	$3/2-3/2$
3221,35	8	5,51	9,35	$4p'\,{}^4D^\circ-4d''\,{}^2D$	$5/2-3/2$
3220,65	8	5,51	9,35	$4p'\,{}^4D^\circ-4d''\,{}^2D$	$5/2-5/2$
3218,204	5	4,97	8,82	$4p'\,{}^4P^\circ-4d'\,{}^4P$	$3/2-5/2$
3217,64	10	5,07	8,92	$4p'\,{}^4F^\circ-4d'\,{}^4G$	$9/2-9/2$
3211,43	30	5,51	9,36	$4p'\,{}^4D^\circ-4d''\,{}^2F$	$5/2-5/2$
3209,498	4	5,57	9,43	$4p'\,{}^4D^\circ-6s'\,{}^4D$	$1/2-1/2$
3208,231	1400	1,64	5,51	$4s^2\,{}^2D-4p'\,{}^4D^\circ$	$3/2-5/2$

λ, Å	I	E_H, eV	E_B, eV	Transition	J
3194,099	1500	1,64	5,52	$4s^2\,{}^2D-4p'\,{}^4D^\circ$	3/2—3/2
3192,22	2	5,42	9,30	$4p'\,{}^2F^\circ-6s'\,{}^4D$	5/2—3/2
3179,343	2	5,42	9,32	$4p'\,{}^2F^\circ-6s'\,{}^2D$	5/2—5/2
3175,67	60	5,42	9,32	$4p'\,{}^2F^\circ-4d''\,{}^2G$	5/2—7/2
3171,663	5	5,52	9,43	$4p'\,{}^4D^\circ-6s'\,{}^4D$	3/2—1/2
3169,681	500	5,15	9,06	$4p'\,{}^4F^\circ-4d'\,{}^4G$	5/2—7/2
3160,047	25	5,15	9,07	$4p'\,{}^4F^\circ-4d'\,{}^4G$	5/2—5/2
3156,629	450	1,64	5,57	$4s^2\,{}^2D-4p'\,{}^4D^\circ$	3/2—1/2
3151,62	8	5,42	9,35	$4p'\,{}^2F^\circ-4d''\,{}^2D$	5/2—5/2
3149,508	30	5,15	9,09	$4p'\,{}^4F^\circ-4d'\,{}^4F$	5/2—5/2
3148,57	2	5,78	9,71	$4p'\,{}^2D^\circ-6s''\,{}^2D$	5/2—5/2
3148,333	3	5,15	9,09	$4p'\,{}^4F^\circ-4d'\,{}^4F$	5/2—3/2
3146,821	450	4,97	8,91	$4p'\,{}^4P^\circ-4d'\,{}^4P$	3/2—1/2
3142,797	8	5,42	9,36	$4p'\,{}^2F^\circ-4d''\,{}^2F$	5/2—5/2
3142,444	750	4,97	8,92	$4p'\,{}^4P^\circ-4d'\,{}^4P$	3/2—3/2
3140,312	400	4,84	8,78	$4p'\,{}^4P^\circ-4d'\,{}^2P$	5/2—3/2
3137,72	5	5,57	9,52	$4p'\,{}^2F^\circ-5d'\,{}^2G$	7/2—9/2
3131,33	5	5,51	9,46	$4p'\,{}^4D^\circ-6s'\,{}^2D$	5/2—3/2
3128,701	650	4,97	8,93	$4p'\,{}^4P^\circ-4d'\,{}^4D$	3/2—5/2
3126,109	1400	4,84	8,80	$4p'\,{}^4P^\circ-4d'\,{}^4S$	5/2—3/2
3120,435	50	4,97	8,94	$4p'\,{}^4P^\circ-4d'\,{}^2P$	3/2—1/2
3118,355	5	5,39	9,37	$4p'\,{}^4D^\circ-4d''\,{}^2F$	7/2—7/2
3116,348	400	4,84	8,81	$4p'\,{}^4P^\circ-4d'\,{}^2D$	5/2—5/2
3113,482	50	4,84	8,82	$4p'\,{}^4P^\circ-4d'\,{}^2F$	5/2—7/2
3108,605	2000	4,84	8,82	$4p'\,{}^4P^\circ-4d'\,{}^4P$	5/2—5/2
3108,452	600	5,08	9,06	$4p'\,{}^4P^\circ-4d'\,{}^4D$	1/2—3/2
3099,928	1250	4,84	8,84	$4p'\,{}^4P^\circ-4d'\,{}^4D$	5/2—7/2
3093,989	1500	1,39	5,39	$4s^2\,{}^2D-4p'\,{}^4D^\circ$	5/2—7/2
3088,132	125	5,08	9,09	$4p'\,{}^4P^\circ-4d'\,{}^4F$	1/2—3/2
3086,47	2	5,78	9,79	$4p'\,{}^2D^\circ-5d'\,{}^4G$	5/2—7/2
3084,96	2	5,72 5,52 5,78	9,74 9,54 9,79	$4p'\,{}^2D^\circ-6s''\,{}^2D$ $4p'\,{}^4D^\circ-5d'\,{}^2D$ $4p'\,{}^2D^\circ-5d'\,{}^4D$	3/2—3/2 3/2—5/2 5/2—3/2
3073,798	1400	1,39	5,42	$4s^2\,{}^2D-4p'\,{}^2F^\circ$	5/2—5/2
3071,96	2	5,51	9,54	$4p'\,{}^4D^\circ-5d'\,{}^2D$	5/2—5/2
3070,97	5	5,51	9,54	$4p'\,{}^4D^\circ-5d'\,{}^2F$	5/2—7/2
3068,906	15	1,64	5,68	$4s^2\,{}^2D-4p'\,{}^2P^\circ$	3/2—1/2
3066,011	3	5,42	9,46	$4p'\,{}^2F^\circ-6s'\,{}^2D$	5/2—3/2
3063,411	2500	1,64	5,69	$4s^2\,{}^2D-4p'\,{}^2P^\circ$	3/2—3/2
3060,84	2	5,51	9,55	$4p'\,{}^4D^\circ-5d'\,{}^4D$	5/2—7/2
3057,36	8	5,69	9,74	$4p'\,{}^2P^\circ-6s''\,{}^2D$	3/2—3/2
3053,38	10	5,24	9,30	$4p'\,{}^4F^\circ-6s'\,{}^4D$	3/2—3/2
3052,554	15	5,15	9,21	$4p'\,{}^4F^\circ-6s'\,{}^4D$	5/2—5/2
3051,901	2	5,68	9,74	$4p'\,{}^2P^\circ-6s''\,{}^2D$	1/2—3/2
3044,028	20	5,10	9,17	$4p'\,{}^4F^\circ-6s'\,{}^4D$	7/2—7/2
3039,488	10	5,57	9,65	$4p'\,{}^2F^\circ-5d'\,{}^4G$	7/2—9/2
3036,101	2500	1,64	5,72	$4s^2\,{}^2D-4p'\,{}^2D^\circ$	3/2—3/2
3034,555	3	5,72	9,81	$4p'\,{}^2D^\circ-5d'\,{}^4F$	3/2—5/2
3033,480	2	5,57	9,66	$4p'\,{}^2F^\circ-5d'\,{}^4F$	7/2—7/2
3030,258	10	4,97	9,06	$4p'\,{}^4P^\circ-4d'\,{}^4D$	3/2—3/2
3029,60	2	4,84	8,93	$4p'\,{}^4P^\circ-4d'\,{}^2G$	5/2—7/2
3027,82	5	5,57	9,66	$4p'\,{}^4D^\circ-5d'\,{}^2D$	1/2—3/2
3024,994	100	3,79 4,84	7,88 8,93	$4p\,{}^2P^\circ-5s'\,{}^4D$ $4p'\,{}^4P^\circ-4d'\,{}^4D$	1/2—3/2 5/2—5/2
3022,608	300	4,84	8,94	$4p'\,{}^4P^\circ-4d'\,{}^4F$	5/2—7/2
3021,544	300	5,07	9,17	$4p'\,{}^4F^\circ-6s'\,{}^4D$	9/2—7/2
3018,09	2	5,69	9,79	$4p'\,{}^2P^\circ-5d'\,{}^4D$	3/2—3/2
3014,848	30	5,10	9,21	$4p'\,{}^4F^\circ-6s'\,{}^4D$	7/2—5/2

λ, Å	I	E_H, eV	E_B, eV	Transition	J
3013,510	2	5,69	9,80	$4p'\ {}^2P^\circ - 5d'\ {}^4G$	$3/2-5/2$
3012,005	250	4,97	9,09	$4p'\ {}^4P^\circ - 4d'\ {}^4F$	$3/2-5/2$
3010,838	2000	1,39	5,51	$4s^2\ {}^2D - 4p'\ {}^4D^\circ$	$5/2-5/2$
3008,12	5	5,24	9,36	$4p'\ {}^4F^\circ - 4d''\ {}^2F$	$3/2-5/2$
		5,42	9,54	$4p'\ {}^2F^\circ - 5d'\ {}^2F$	$5/2-7/2$
3002,281	10	—	—	—	—
3001,774	2	5,42	9,55	$4p'\ {}^2F^\circ - 5d'\ {}^4P$	$5/2-5/2$
3001,24	5	5,39	9,52	$4p'\ {}^4D^\circ - 5d'\ {}^2G$	$7/2-9/2$
2998,384	150	1,39	5,52	$4s^2\ {}^2D - 4p'\ {}^4D^\circ$	$5/2-3/2$
2997,364	2000	1,64	5,78	$4s^2\ {}^2D - 4p'\ {}^2D^\circ$	$3/2-5/2$
2994,13	5	5,69	9,83	$4p'\ {}^2P^\circ - 7s'\ {}^2D$	$3/2-5/2$
		5,52	9,66	$4p'\ {}^4D^\circ - 5d'\ {}^2D$	$3/2-3/2$
2991,780	15	5,52	9,67	$4p'\ {}^4D^\circ - 5d'\ {}^2F$	$3/2-5/2$
2989,010	2	5,39	9,54	$4p'\ {}^4D^\circ - 5d'\ {}^2F$	$7/2-7/2$
2985,926	10	5,15	9,30	$4p'\ {}^4F^\circ - 6s'\ {}^4D$	$5/2-3/2$
2984,267	5	5,51	9,66	$4p'\ {}^4D^\circ - 5d'\ {}^4D$	$5/2-5/2$
2983,038	3	5,51	9,66	$4p'\ {}^4D^\circ - 5d'\ {}^4F$	$5/2-7/2$
2982,765	8	5,39	9,55	$4p'\ {}^4D^\circ - 5d'\ {}^4P$	$7/2-5/2$
2982,123	3	5,57	9,73	$4p'\ {}^2F^\circ - 7s'\ {}^4D$	$7/2-5/2$
2979,380	25	5,39	9,55	$4p'\ {}^4D^\circ - 5d'\ {}^4D$	$7/2-7/2$
2978,295	30	5,39	9,56	$4p'\ {}^4D^\circ - 5d'\ {}^4F$	$7/2-9/2$
2974,675	10	5,15	9,32	$4p'\ {}^4F^\circ - 6s'\ {}^2D$	$5/2-5/2$
2961,165	2500	1,39	5,57	$4s^2\ {}^2D - 4p'\ {}^2F^\circ$	$5/2-7/2$
2951,21	5	3,82	8,02	$4p\ {}^2P^\circ - 5s'\ {}^2D$	$3/2-5/2$
2945,23	3	5,51	9,71	$4p'\ {}^4D^\circ - 6s''\ {}^2D$	$5/2-5/2$
		5,78	9,99	$4p'\ {}^2D^\circ - 6d'\ {}^4D$	$5/2-5/2$
2939,453	2	5,15	9,37	$4p'\ {}^4F^\circ - 4d''\ {}^2F$	$5/2-7/2$
2938,868	15	5,10	9,32	$4p'\ {}^4F^\circ - 6s'\ {}^2D$	$7/2-5/2$
2937,766	2	5,24	9,46	$4p'\ {}^4F^\circ - 6s'\ {}^2D$	$3/2-3/2$
2933,060	20	4,84	9,06	$4p'\ {}^4P^\circ - 4d'\ {}^4G$	$5/2-7/2$
2931,699	10	5,08	9,30	$4p'\ {}^4P^\circ - 6s'\ {}^4D$	$1/2-3/2$
2930,416	5	5,42	9,65	$4p'\ {}^2F^\circ - 5d'\ {}^4P$	$5/2-3/2$
2926,057	10	5,08	9,31	$4p'\ {}^4P^\circ - 4d''\ {}^2P$	$1/2-1/2$
2925,439	30	5,42	9,66	$4p'\ {}^2F^\circ - 5d'\ {}^2G$	$5/2-7/2$
2924,882	10	5,42	9,66	$4p'\ {}^2F^\circ - 5d'\ {}^4D$	$5/2-5/2$
2923,704	80	5,42	9,66	$4p'\ {}^2F^\circ - 5d'\ {}^4F$	$5/2-7/2$
2923,212	20	4,97	9,21	$4p'\ {}^4P^\circ - 6s'\ {}^4D$	$3/2-5/2$
2022,830	10	5,57	9,81	$4p'\ {}^4D^\circ - 5d'\ {}^4F$	$1/2-3/2$
2920,296	10	5,42	9,67	$4p'\ {}^2F^\circ - 5d'\ {}^2F$	$5/2-5/2$
2912,916	2	5,08	9,33	$4p'\ {}^4P^\circ - 4d''\ {}^2P$	$1/2-3/2$
2911,215	30	5,39	9,65	$4p'\ {}^4D^\circ - 5d'\ {}^4G$	$7/2-9/2$
2905,662	5	5,39	9,66	$4p'\ {}^4D^\circ - 5d'\ {}^4F$	$7/2-7/2$
2891,64	30	5,52	9,81	$4p'\ {}^4D^\circ - 5d'\ {}^4F$	$3/2-5/2$
2890,84	50	5,51	9,79	$4p'\ {}^4D^\circ - 5d'\ {}^4G$	$5/2-7/2$
2885,408	5	5,51	9,80	$4p'\ {}^4D^\circ - 5d'\ {}^4G$	$5/2-5/2$
		5,24	9,54	$4p'\ {}^4F^\circ - 5d'\ {}^2D$	$3/2-5/2$
2882,934	1500	1,39	5,69	$4s^2\ {}^2D - 4p'\ {}^2P^\circ$	$5/2-3/2$
2879,743	2	5,78	10,08	$4p'\ {}^2D^\circ - 5d''\ {}^2D$	$5/2-3/2$
2878,86	5	3,79	8,09	$4p\ {}^2P^\circ - 5s'\ {}^2D$	$1/2-3/2$
		5,57	9,87	$4p'\ {}^4D^\circ - 6d'\ {}^4S$	$1/2-3/2$
2877,101	5	5,39	9,70	$4p'\ {}^4D^\circ - 7s'\ {}^4D$	$7/2-7/2$
2876,025	2	5,42	9,73	$4p'\ {}^2F^\circ - 7s'\ {}^4D$	$5/2-5/2$
2875,67	10	5,78	10,09	$4p'\ {}^2D^\circ - 5d''\ {}^2F$	$5/2-5/2$
2875,240	2	5,15	9,46	$4p'\ {}^4F^\circ - 6s'\ {}^2D$	$5/2-3/2$
2874,560	20	5,78	10,09	$4p'\ {}^2D^\circ - 5d''\ {}^2F$	$5/2-7/2$
2869,80	2	5,39	9,71	$4p'\ {}^4D^\circ - 6s''\ {}^2D$	$7/2-5/2$
2868,470	10	5,42	9,74	$4p'\ {}^2F^\circ - 6s''\ {}^2D$	$5/2-3/2$
		5,51	9,83	$4p'\ {}^4D^\circ - 7s'\ {}^4D$	$5/2-3/2$

λ, Å	I	E_H, eV	E_B, eV	Transition	J
2862,07	5	4,97	9,30	$4p'\,^4P^\circ—6s'\,^4D$	3/2—3/2
2858,734	200	1,39	5,72	$4s^2\,^2D—4p'\,^2D^\circ$	5/2—3/2
2858,225	50	4,84	9,17	$4p'\,^4P^\circ—6s'\,^4D$	5/2—7/2
2856,660	2	4,97	9,33	$4p'\,^4P^\circ—4d''\,^2P$	3/2—1/2
2851,743	15	4,97	9,32	$4p'\,^4P^\circ—6s'\,^2D$	3/2—5/2
2846,478	15	5,08	9,43	$4p'\,^4P^\circ—6s'\,^4D$	1/2—1/2
2844,842	10	5,72	10,08	$4p'\,^2D^\circ—5d''\,^2D$	3/2—3/2
2844,160	15	4,97	9,33	$4p'\,^4P^\circ—4d''\,^2P$	3/2—3/2
2840,92	10	5,72	10,09	$4p'\,^2D^\circ—5d''\,^2F$	3/2—5/2
2834,30	2	4,97	9,35	$4p'\,^4P^\circ—4d''\,^2S$	3/2—1/2
2832,49	5	4,84	9,21	$4p'\,^4P^\circ—6s'\,^4D$	5/2—5/2
2830,93	3	—	—	—	—
2829,42	5	4,97	9,35	$4p'\,^4P^\circ—4d''\,^2D$	3/2—5/2
2824,370	1250	1,39	5,78	$4s^2\,^2D—4p'\,^2D^\circ$	5/2—5/2
2818,68	4	5,15	9,55	$4p'\,^4F^\circ—5d'\,^4P$	5/2—5/2
2813,558	2	5,42	9,83	$4p'\,^2F^\circ—7s'\,^4D$	5/2—3/2
2812,74	2	5,42	9,83	$4p'\,^2F^\circ—7s'\,^2D$	5/2—5/2
		5,57	9,83	$4p'\,^2F^\circ—7s'\,^2D$	7/2—5/2
		5,57	9,98	$4p'\,^2F^\circ—6d'\,^4G$	7/2—9/2
2805,71	5	5,24	9,66	$4p'\,^4F^\circ—5d'\,^2D$	3/2—3/2
2803,686	10	5,24	9,67	$4p'\,^4F^\circ—5d'\,^2F$	3/2—5/2
2802,556	10	5,10	9,52	$4p'\,^4F^\circ—5d'\,^2G$	7/2—9/2
2793,485	2	5,52	9,96	$4p'\,^4D^\circ—7s'\,^4D$	3/2—1/2
2791,951	5	5,10	9,54	$4p'\,^4F^\circ—5d'\,^2F$	7/2—7/2
2786,496	10	5,10	9,55	$4p'\,^4F^\circ—5d'\,^4P$	7/2—5/2
2783,551	20	5,10	9,55	$4p'\,^4F^\circ—5d'\,^4D$	7/2—7/2
2782,592	20	5,10	9,56	$4p'\,^4F^\circ—5d'\,^4F$	7/2—9/2
2768,878	125	5,07	9,55	$4p'\,^4F^\circ—5d'\,^4G$	9/2—11/2
2766,371	2500	1,64	6,12	$4s^2\,^2D—5d\,^2P^\circ$	3/2—3/2, 1/2
2764,762	5	5,07	9,55	$4p'\,^4F^\circ—5d'\,^4D$	9/2—7/2
2763,809	15	5,07	9,56	$4p'\,^4F^\circ—5d'\,^4F$	9/2—9/2
2760,25	2	4,97	9,46	$4p'\,^4P^\circ—6s'\,^2D$	3/2—3/2
2755,69	5	5,15	9,65	$4p'\,^4F^\circ—5d'\,^4P$	5/2—3/2
2751,810	10	5,57	10,08	$4p'\,^2F^\circ—5d''\,^2D$	7/2—5/2
2751,29	10	5,15	9,66	$4p'\,^4F^\circ—5d'\,^2G$	5/2—7/2
2750,786	5	5,15	9,66	$4p'\,^4F^\circ—5d'\,^4D$	5/2—5/2
2749,734	2	5,15	9,66	$4p'\,^4F^\circ—5d'\,^4F$	5/2—7/2
2746,713	20	5,15	9,67	$4p'\,^4F^\circ—5d'\,^2F$	5/2—5/2
2745,452	20	5,57	10,09	$4p'\,^2F^\circ—5d''\,^2F$	7/2—7/2
2737,608	2	4,84	9,36	$4p'\,^4P^\circ—4d''\,^2F$	5/2—5/2
2734,858	10	4,84	9,37	$4p'\,^4P^\circ—4d''\,^2F$	5/2—7/2
2723,953	30	5,10	9,65	$4p'\,^4F^\circ—5d'\,^4G$	7/2—9/2
2722,702	5	4,97	9,53	$4p'\,^4P^\circ—5d'\,^2P$	3/2—3/2
2720,62	2	5,10	9,66	$4p'\,^4F^\circ—5d'\,^2G$	7/2—7/2
2720,199	15	5,24	9,80	$4p'\,^4F^\circ—5d'\,^4G$	3/2—5/2
2719,097	15	5,10	9,66	$4p'\,^4F^\circ—5d'\,^4F$	7/2—7/2
2718,847	2	5,52	10,08	$4p'\,^4D^\circ—5d''\,^2D$	3/2—3/2
2715,543	20	5,24	9,81	$4p'\,^4F^\circ—5d'\,^4F$	3/2—5/2
2715,35	5	5,24	9,81	$4p'\,^4F^\circ—5d'\,^4F$	3/2—3/2
2714,54	2	5,42	9,99	$4p'\,^2F^\circ—6d'\,^2G$	5/2—7/2
2714,00	2	4,97	9,54	$4p'\,^4P^\circ—5d'\,^2D$	3/2—5/2
2705,18	2	5,24	9,83	$4p'\,^4F^\circ—7s'\,^4D$	3/2—3/2
2702,65	10	5,07	9,66	$4p'\,^4F^\circ—5d'\,^2G$	9/2—7/2
		5,08	9,66	$4p'\,^4P^\circ—5d'\,^2D$	1/2—3/2
2694,080	5	5,10	9,70	$4p'\,^4F^\circ—7s'\,^4D$	7/2—7/2
2676,428	20	5,07	9,70	$4p'\,^4F^\circ—7s'\,^4D$	9/2—7/2
2672,05	5	—	—	—	—
2671,204	20	5,15	9,79	$4p'\,^4F^\circ—5d'\,^4G$	5/2—7/2

λ, Å	I	E_H, eV	E_B, eV	Transition	J
2666,59	2	5,15	9,80	$4p'\,{}^4F^\circ-5d'\,{}^4G$	$^5/_2-{}^5/_2$
2652,065	2	5,15	9,83	$4p'\,{}^4F^\circ-7s'\,{}^4D$	$^5/_2-{}^3/_2$
2651,693	10	4,97	9,65	$4p'\,{}^4P^\circ-5d'\,{}^4P$	$^3/_2-{}^1/_2$
2649,840	30	4,97	9,65	$4p'\,{}^4P^\circ-5d'\,{}^4P$	$^3/_2-{}^3/_2$
2645,303	20	4,97	9,66	$4p'\,{}^4P^\circ-5d'\,{}^4D$	$^3/_2-{}^5/_2$
2641,550	5	4,97	9,67	$4p'\,{}^4P^\circ-5d'\,{}^2F$	$^3/_2-{}^5/_2$
2634,933	30	4,84	9,54	$4p'\,{}^4P^\circ-5d'\,{}^4S$	$^5/_2-{}^3/_2$
2630,004	20	4,84	9,55	$4p'\,{}^4P^\circ-5d'\,{}^4P$	$^5/_2-{}^5/_2$
2627,365	20	4,84	9,55	$4p'\,{}^4P^\circ-5d'\,{}^4D$	$^5/_2-{}^7/_2$
2626,678	10	5,08	9,79	$4p'\,{}^4P^\circ-5d'\,{}^4D$	$^1/_2-{}^3/_2$
2622,875	5	5,10	9,83	$4p'\,{}^4F^\circ-7s'\,{}^2D$	$^7/_2-{}^5/_2$
2618,366	2500	1,39	6,12	$4s^2\,{}^2D-5p\,{}^2P^\circ$	$^5/_2-{}^3/_2$
2605,26	3	4,97	9,73	$4p'\,{}^4P^\circ-7s'\,{}^4D$	$^3/_2-{}^5/_2$
2593,65	2	5,10	9,88	$4p'\,{}^4F^\circ-6d'\,{}^4F$	$^7/_2-{}^9/_2$
2592,627	200	—	—	—	—
2580,57	5	1,39	6,19	$4s^2\,{}^2D-4d\,{}^2D$	$^5/_2-{}^5/_2$
2579,29	20	5,07	9,88	$4p'\,{}^4F^\circ-6d'\,{}^4G$	$^9/_2-{}^{11}/_2$
2577,12	2	5,07	9,88	$4p'\,{}^4F^\circ-6d'\,{}^4F$	$^9/_2-{}^9/_2$
2570,800	10	4,84	9,66	$4p'\,{}^4P^\circ-5d'\,{}^4D$	$^5/_2-{}^5/_2$
2569,888	10	4,84	9,66	$4p'\,{}^4P^\circ-5d'\,{}^4F$	$^5/_2-{}^7/_2$
2567,330	2	4,97	9,80	$4p'\,{}^4P^\circ-5d'\,{}^4G$	$^3/_2-{}^5/_2$
2563,955	3	5,15	9,99	$4p'\,{}^4F^\circ-6d'\,{}^2G$	$^5/_2-{}^7/_2$
2563,553	3	5,15	9,99	$4p'\,{}^4F^\circ-6d'\,{}^4F$	$^5/_2-{}^7/_2$
2563,167	10	4,97	9,81	$4p'\,{}^4P^\circ-5d'\,{}^4F$	$^3/_2-{}^5/_2$
2553,29	2	4,97	9,83	$4p'\,{}^4P^\circ-7s'\,{}^2D$	$^3/_2-{}^5/_2$
2547,48	10	4,84	9,70	$4p'\,{}^4P^\circ-7s'\,{}^4D$	$^5/_2-{}^7/_2$
2540,38	5	5,10	9,98	$4p'\,{}^4F^\circ-6d'\,{}^4G$	$^7/_2-{}^9/_2$
2536,86	2	5,10	9,99	$4p'\,{}^4F^\circ-6d'\,{}^4F$	$^7/_2-{}^7/_2$
2536,67	2	5,07	9,96	$4p'\,{}^4F^\circ-8s'\,{}^4D$	$^9/_2-{}^7/_2$
2536,03	2	5,10	9,99	$4p'\,{}^4F^\circ-6d'\,{}^2F$	$^7/_2-{}^5/_2$
2494,89	10	3,82	8,78	$4p\,{}^2P^\circ-4d'\,{}^2P$	$^3/_2-{}^3/_2$
2492,146	2000	0,00	4,97	$4s\,{}^2S-4p'\,{}^4P^\circ$	$^1/_2-{}^3/_2$
2479,754	10	3,82	8,81	$4p\,{}^2P^\circ-4d'\,{}^2D$	$^3/_2-{}^5/_2$
2474,818	5	3,82	8,82	$4p\,{}^2P^\circ-4d'\,{}^4P$	$^3/_2-{}^5/_2$
2460,93	5	4,84	9,87	$4p'\,{}^4P^\circ-6d'\,{}^4S$	$^5/_2-{}^3/_2$
2458,88	5	4,84	9,88	$4p'\,{}^4P^\circ-6d'\,{}^4P$	$^5/_2-{}^5/_2$
2457,74	5	4,84	9,88	$4p'\,{}^4P^\circ-6d'\,{}^4D$	$^5/_2-{}^7/_2$
2441,637	1000	0,00	5,08	$4s\,{}^2S-4p'\,{}^4P^\circ$	$^1/_2-{}^1/_2$
2416,605	5	3,82	8,94	$4p\,{}^2P^\circ-4d'\,{}^2P$	$^3/_2-{}^1/_2$
2415,197	5	3,79	8,92	$4p\,{}^2P^\circ-4d'\,{}^4P$	$^1/_2-{}^3/_2$
2406,665	1500	1,64	6,79	$4s^2\,{}^2D-6p\,{}^2P^\circ$	$^3/_2-{}^3/_2$
2404,864	2	3,79	8,94	$4p\,{}^2P^\circ-4d'\,{}^2D$	$^1/_2-{}^3/_2$
2392,627	2500	1,64	6,82	$4s^2\,{}^2D-6p\,{}^2P^\circ$	$^3/_2-{}^1/_2$
2363,220	5	0,00	5,24	$4s\,{}^2S-4p'\,{}^4F^\circ$	$^1/_2-{}^3/_2$
2348,352	2	3,79	9,06	$4p\,{}^2P^\circ-4d'\,{}^4D$	$^1/_2-{}^3/_2$
2319,561	500	1,64	6,98	$4s^2\,{}^2D-4p''\,{}^2P^\circ$	$^3/_2-{}^3/_2$
2303,116	1000	1,64	7,02	$4s^2\,{}^2D-4p''\,{}^2D^\circ$	$^3/_2-{}^5/_2$
2293,842	2500	1,39	6,79	$4s^2\,{}^2D-6p\,{}^2P^\circ$	$^5/_2-{}^3/_2$
2263,079	2200	1,64	7,12	$4s^2\,{}^2D-7p\,{}^2P^\circ$	$^3/_2-{}^1/_2$
2260,528	1300	1,39	6,87	$4s^2\,{}^2D-4f\,{}^2F^\circ$	$^5/_2-{}^7/_2$
2247,503	2	3,82	9,33	$4p\,{}^2P^\circ-4d''\,{}^2P$	$^3/_2-{}^3/_2$
2244,265	2300	0,00	5,52	$4s\,{}^2S-4p'\,{}^4D^\circ$	$^1/_2-{}^3/_2$
2238,454	1100	1,64	7,18	$4s^2\,{}^2D-5f\,{}^2F^\circ$	$^3/_2-{}^5/_2$
2237,34	5	—	—	—	—
2236,278	900	1,64	7,18	$4s^2\,{}^2D-7p\,{}^2P^\circ$	$^3/_2-{}^3/_2$
2230,084	2500	1,39	6,95	$4s^2\,{}^2D-4p''\,{}^2F^\circ$	$^5/_2-{}^7/_2$
2227,775	1600	1,64	7,21	$4s^2\,{}^2D-4p''\,{}^2F^\circ$	$^3/_2-{}^5/_2$
2225,697	2100	0,00	5,57	$4s\,{}^2S-4p'\,{}^4D^\circ$	$^1/_2-{}^1/_2$

λ, Å	I	E_H, eV	E_B, eV	Transition	J
2215,654	1000	1,64	7,24	$4s^2\,{}^2D-4p''\,{}^2P^\circ$	3/2—1/2
2214,581	1600	1,39	6,98	$4s^2\,{}^2D-4p''\,{}^2P^\circ$	5/2—3/2
2205,65	5	—	—	—	—
2199,752	1300	1,64	7,28	$4s^2\,{}^2D-4p''\,{}^2D^\circ$	3/2—3/2
2199,583	1700	1,39	7,02	$4s^2\,{}^2D-4p''\,{}^2D^\circ$	5/2—5/2
2181,720	1700	0,00	5,68	$4s\,{}^2S-4p'\,{}^2P^\circ$	1/2—1/2
2178,944	1600	0,00	5,69	$4s\,{}^2S-4p'\,{}^2P^\circ$	1/2—3/2
2171,817	200	1,64	7,35	$4s^2\,{}^2D-8p\,{}^2P^\circ$	3/2—3/2
2169,562	300	1,64	7,35	$4s^2\,{}^2D-8p\,{}^2P^\circ$	3/2—1/2
2165,093	1300	0,00	5,72	$4s\,{}^2S-4p'\,{}^2D^\circ$	1/2—3/2
2149,40	10	—	—	—	—
2142,72	5	—	—	—	—
2140,56	2	1,39	7,18	$4s^2\,{}^2D-5f\,{}^2F^\circ$	5/2—5/2
2138,533	500	1,39	7,18	$4s^2\,{}^2D-7p\,{}^2P^\circ$	5/2—3/2
2130,762	50	1,39	7,21	$4s^2\,{}^2D-4p''\,{}^2F^\circ$	5/2—5/2
2124,35	5	3,82	9,65	$4p\,{}^2P^\circ-5d'\,{}^4P$	3/2—3/2
2113,26	2	3,79	9,65	$4p\,{}^2P^\circ-5d'\,{}^4P$	1/2—3/2
2105,112	800	1,39	7,28	$4s^2\,{}^2D-4p''\,{}^2D^\circ$	5/2—3/2
2079,529	20	1,39	7,35	$4s^2\,{}^2D-8p\,{}^2P^\circ$	5/2—3/2
2068,321	5	3,82	9,81	$4p\,{}^2P^\circ-5d'\,{}^4F$	3/2—5/2
2045,62	5	1,39	7,45	$4s^2\,{}^2D-9p\,{}^2P^\circ$	5/2—3/2
2024,335	200	0,00	6,12	$4s\,{}^2S-5p\,{}^2P^\circ$	1/2—3/2, 1/2
1825,348	100	0,00	6,79	$4s\,{}^2S-6p\,{}^2P^\circ$	1/2—3/2
1817,334 1817,265	} 20	0,00	6,82	$4s\,{}^2S-6p\,{}^2P^\circ$	1/2—1/2
1774,820	200	0,00	6,98	$4s\,{}^2S-4p''\,{}^2P^\circ$	1/2—3/2
1764,540	10	0,00	7,03	$4s\,{}^2S-7s\,{}^2S$	1/2—1/2
1749,202	2	1,64	8,73	$4s^2\,{}^2D-5p'\,{}^4F^\circ$	3/2—5/2
1741,574	50	0,00	7,12	$4s\,{}^2S-7p\,{}^2P^\circ$	1/2—1/2
1732,674	20	1,64	8,80	$4s^2\,{}^2D-5p'\,{}^2F^\circ$	3/2—5/2
1731,32	2	1,64	8,80	$4s^2\,{}^2D-5p'\,{}^4P^\circ$	3/2—1/2
1730,576	10	1,64	8,81	$4s^2\,{}^2D-5p'\,{}^4D^\circ$	3/2—3/2
1725,664	50	0,00	7,18	$4s\,{}^2S-7p\,{}^2P^\circ$	1/2—3/2
1713,364	50	0,00	7,24	$4s\,{}^2S-4p''\,{}^2P^\circ$	1/2—1/2
1709,396	2	1,64	8,89	$4s^2\,{}^2D-5p'\,{}^2D^\circ$	3/2—5/2
1707,391	5	0,00	7,26	$4s\,{}^2S-8s\,{}^2S$	1/2—1/2
1703,843	30	0,00	7,28	$4s\,{}^2S-4p''\,{}^2D^\circ$	1/2—3/2
1701,292	10	1,64	8,93	$4s^2\,{}^2D-5p'\,{}^2D^\circ$	3/2—3/2
1692,654	5	1,39	8,71	$4s^2\,{}^2D-5p'\,{}^4P^\circ$	5/2—3/2
1691,076	30	1,39	8,72	$4s^2\,{}^2D-5p'\,{}^4F^\circ$	5/2—7/2
1688,865	15	1,39	8,73	$4s^2\,{}^2D-5p'\,{}^4F^\circ$	5/2—5/2
1688,093	30	1,39	8,73	$4s^2\,{}^2D-5p'\,{}^4D^\circ$	5/2—7/2
1687,043	20	0,00	7,35	$4s\,{}^2S-8p\,{}^2P^\circ$	1/2—3/2
1685,682	25	0,00	7,35	$4s\,{}^2S-8p\,{}^2P^\circ$	1/2—1/2
1684,674	20	1,39	8,75	$4s^2\,{}^2D-5p'\,{}^4D^\circ$	5/2—5/2
1673,440	5	1,39	8,80	$4s^2\,{}^2D-5p'\,{}^2F^\circ$	5/2—5/2
1671,484	3	1,39	8,81	$4s^2\,{}^2D-5p'\,{}^4D^\circ$	5/2—3/2
1664,708	10	0,00	7,45	$4s\,{}^2S-9p\,{}^2P^\circ$	1/2—3/2
1664,303	10	0,00	7,45	$4s\,{}^2S-9p\,{}^2P^\circ$	1/2—1/2
1655,318	30	1,39	8,88	$4s^2\,{}^2D-5p'\,{}^2F^\circ$	5/2—7/2
1651,721	20	1,39	8,89	$4s^2\,{}^2D-5p'\,{}^2D^\circ$	5/2—5/2
1650,301	5	0,00	7,51	$4s\,{}^2S-10p\,{}^2P^\circ$	1/2—3/2
1650,119	5	0,00	7,51	$4s\,{}^2S-10p\,{}^2P^\circ$	1/2—1/2
1640,474	5	0,00	7,56	$4s\,{}^2S-11p\,{}^2P^\circ$	1/2—3/2, 1/2
1632,326	5	1,64	9,24	$4s^2\,{}^2D-5p''\,{}^2D^\circ$	3/2—5/2, 3/2
1621,316	20	1,64	9,29	$4s^2\,{}^2D-5p''\,{}^2F^\circ$	3/2—5/2
1616,940	20	1,64	9,31	$4s^2\,{}^2D-5p''\,{}^2P^\circ$	3/2—1/2
1585,871	5	1,39	9,21	$4s^2\,{}^2D-5p''\,{}^2P^\circ$	5/2—3/2
1583,799	15	1,39	9,22	$4s^2\,{}^2D-5p''\,{}^2F^\circ$	5/2—7/2
1579,658	5	1,39	9,24	$4s^2\,{}^2D-5p''\,{}^2D^\circ$	5/2—5/2, 3/2

Cu II, **ground state** $1s^2\,2s^2\,2p^6\,3s^2\,3p^6\,3d^{10}\,{}^1S_0$

Ionization potential 163 665,6 cm⁻¹; 20,291 eV

λ, Å	I	E_H, eV	E_B, eV	Transition	J
10166,91	15	16,89	18,11	$4f\,{}^3G^\circ$—$5g\,{}^3H$	4—5
10162,88	1	16,89	18,11	$4f\,{}^3G^\circ$—$5g\,{}^3G$	4—4
10080,47	10	17,14	18,37	$4f\,{}^3G^\circ$—$5g\,{}^3H$	3—4
10055,02	30	16,88	18,11	$4f\,{}^3G^\circ$—$5g\,{}^3H$	5—6
10051,12	3	16,88	18,11	$4f\,{}^3G^\circ$—$5g\,{}^3G$	5—5
10049,88	1	17,14	18,37	$4f\,{}^1G^\circ$—$5g\,{}^1G$	4—4
10038,19	15	17,14	18,37	$4f\,{}^1G^\circ$—$5g\,{}^1H$	4—5
10036,32	5	16,88	18,11	$4f\,{}^3F^\circ$—$5g\,{}^3F$	4—4
10026,93	1	16,88	18,11	$4f\,{}^3F^\circ$—$5g\,{}^3H$	4—5
10023,05	30	16,88	18,11	$4f\,{}^3F^\circ$—$5g\,{}^3G$	4—5
10006,68	10	17,13	18,37	$4f\,{}^3F^\circ$—$5g\,{}^3G$	2—3
9994,32	1	17,13	18,37	$4f\,{}^1F^\circ$—$5g\,{}^1F$	3—3
9960,46	15	17,13	18,37	$4f\,{}^1F^\circ$—$5g\,{}^1G$	3—4
9960,07	10	15,32	16,56	$5p\,{}^1D^\circ$—$6s\,{}^3D$	2—3
		16,87	18,11	$4f\,{}^3F^\circ$—$5g\,{}^1D$	3—2
9939,05	20	16,86	18,11	$4f\,{}^3D^\circ$—$5g\,{}^1D$	2—2
		16,87	18,11	$4f\,{}^3F^\circ$—$5g\,{}^3F$	3—3
9926,10	10	16,87	18,11	$4f\,{}^3F^\circ$—$5g\,{}^3G$	3—4
9925,67	20	17,12	18,37	$4f\,{}^3H^\circ$—$5g\,{}^3I$	4—5
9918,05	15	16,86	18,11	$4f\,{}^3D^\circ$—$5g\,{}^3F$	2—3
9916,52	30	17,12	18,37	$4f\,{}^1H^\circ$—$5g\,{}^1I$	5—6
9915,20	1	16,86	18,11	$4f\,{}^3D^\circ$—$5g\,{}^3D$	3—3
9905,44	2	16,86	18,11	$4f\,{}^1P^\circ$—$5g\,{}^3D$	1—2
9894,44	5	16,86	18,11	$4f\,{}^3D^\circ$—$5g\,{}^3F$	3—4
9893,04	5	17,11	18,37	$4f\,{}^3D^\circ$—$5g\,{}^3F$	1—2
		17,12	18,37	$4f\,{}^3H^\circ$—$5g\,{}^3H$	4—4
9884,09	10	16,86	18,11	$4f\,{}^1P^\circ$—$5g\,{}^1D$	1—2
		17,12	18,37	$4f\,{}^1H^\circ$—$5g\,{}^1H$	5—5
9881,57	15	16,86	18,11	$4f\,{}^3D^\circ$—$5g\,{}^3G$	3—4
9868,20	15	17,11	18,37	$4f\,{}^1D^\circ$—$5g\,{}^1F$	2—3
9864,26	40	16,85	18,11	$4f\,{}^3H^\circ$—$5g\,{}^3I$	5—6
9861,41	50	16,85	18,11	$4f\,{}^3H^\circ$—$5g\,{}^3I$	6—7
9858,87	3	16,85	18,11	$4f\,{}^3P^\circ$—$5g\,{}^3D$	2—2
9850,58	3	16,85	18,11	$4f\,{}^3P^\circ$—$5g\,{}^3D$	0—1
9837,94	25	16,85	18,11	$4f\,{}^3P^\circ$—$5g\,{}^3D$	2—3
9830,90	5	16,85	18,11	$4f\,{}^3H^\circ$—$5g\,{}^3H$	5—5
9829,06	3	15,32	16,58	$5p\,{}^1D^\circ$—$6s\,{}^3D$	2—2
9828,06	5	16,85	18,11	$4f\,{}^3H^\circ$—$5g\,{}^3H$	6—6
9813,35	20	16,84	18,11	$4f\,{}^3P^\circ$—$5g\,{}^3D$	1—2
9735,94	15	16,84	18,11	$4p'\,{}^3G^\circ$—$5g\,{}^3H$	4—5
9732,28	3	16,84	18,11	$4p'\,{}^3G^\circ$—$5g\,{}^3G$	4—4
9688,71	10	13,68	14,96	$5s\,{}^1D$—$5p\,{}^3F^\circ$	2—3
9512,43	2	13,68	14,99	$5s\,{}^1D$—$4p''\,{}^1D^\circ$	2—2
9473,36	1	13,68	14,99	$5s\,{}^1D$—$5p\,{}^3P^\circ$	2—1
9463,71	3	16,84	18,15	$4f\,{}^3P^\circ$—$6d\,{}^3P$	1—1
9451,59	2	15,25	16,56	$5p\,{}^1F^\circ$—$6s\,{}^3D$	3—3
9332,04	5	13,68	15,01	$5s\,{}^1D$—$4p''\,{}^1F^\circ$	2—3
9226,86	1	13,65	14,99	$5s\,{}^3D$—$5p\,{}^3P^\circ$	1—1
9205,40	20	15,23	16,58	$5p\,{}^1P^\circ$—$6s\,{}^3D$	1—2
9103,33	10	15,22	16,58	$5p\,{}^3F^\circ$—$6s\,{}^3D$	2—2
8609,26	3	15,12	16,56	$5p\,{}^3D^\circ$—$6s\,{}^3D$	2—3
8606,64	1	13,68	15,12	$5s\,{}^1D$—$5p\,{}^3D^\circ$	2—2
8511,04	40	15,12	16,58	$5p\,{}^3D^\circ$—$6s\,{}^3D$	2—2
8503,46	15	13,43	14,89	$5s\,{}^3D$—$5p\,{}^3P^\circ$	2—2
8477,26	10	9,12	10,59	$4p\,{}^1P^\circ$—$4s^2\,{}^1D$	1—2
8283,21	60	15,07	16,56	$5p\,{}^3D^\circ$—$6s\,{}^3D$	3—3

λ, Å	I	E_H, eV	E_B, eV	Transition	J
8277,60	50	13,39	14,89	$5s\ ^3D—5p\ ^3P^\circ$	3—2
8256,90	5	15,32	16,82	$5p\ ^1D^\circ—6s\ ^3D$	2—1
8235,30	10	13,65	15,15	$5s\ ^3D—5p\ ^3P^\circ$	1—0
8192,28	30	{ 15,07	16,58	$5p\ ^3D^\circ—6s\ ^3D$	3—2
		{ 15,32	16,83	$5p\ ^1D^\circ—6s\ ^1D$	2—2
8095,55	40	13,43	14,96	$5s\ ^3D—5p\ ^3F^\circ$	2—3
8088,58	20	15,29	16,82	$5p\ ^3D^\circ—6s\ ^3D$	1—1
8075,46	2	13,68	15,22	$5s\ ^1D—5p\ ^3F^\circ$	2—2
8026,45	10	15,29	16,83	$5p\ ^3D^\circ—6s\ ^1D$	1—2
7996,72	10	13,68	15,23	$5s\ ^1D—5p\ ^1P^\circ$	2—1
7988,17	60	15,01	16,56	$4p''\ ^1F^\circ—6s\ ^3D$	3—3
7972,01	8	13,43	14,99	$5s\ ^3D—4p''\ ^1D^\circ$	2—2
7944,42	25	13,43	14,99	$5s\ ^3D—5p\ ^3P^\circ$	2—1
7902,57	25	13,68	15,25	$5s\ ^1D—5p\ ^1F^\circ$	2—3
7895,83	20	13,65	15,22	$5s\ ^3D—5p\ ^3F^\circ$	1—2
7890,56	3	13,39	14,96	$5s\ ^3D—5p\ ^3F^\circ$	3—3
7860,58	5	14,99	16,56	$4p''\ ^1D^\circ—6s\ ^3D$	2—3
7845,03	25	{ 13,43	15,01	$5s\ ^3D—4p''\ ^1F^\circ$	2—3
		{ 15,25	16,83	$5p\ ^1F^\circ—6s\ ^1D$	3—2
7825,66	50	13,39	14,98	$5s\ ^3D—5p\ ^3F^\circ$	3—4
7820,57	5	13,65	15,23	$5s\ ^3D—5p\ ^1P^\circ$	1—1
7812,33	10	15,23	16,82	$5p\ ^1P^\circ—6s\ ^3D$	1—1
7807,66	75	14,98	16,56	$5p\ ^3F^\circ—6s\ ^3D$	4—3
7805,19	25	14,99	16,58	$5p\ ^3P^\circ—6s\ ^3D$	1—2
7778,74	30	14,99	16,58	$4p''\ ^1D^\circ—6s\ ^3D$	2—2
7754,37	10	15,23	16,83	$5p\ ^1P^\circ—6s\ ^1D$	1—2
7744,09	5	14,96	16,56	$5p\ ^3F^\circ—6s\ ^3D$	3—3
7738,68	30	15,22	16,82	$5p\ ^3F^\circ—6s\ ^3D$	2—1
7726,64	5	13,68	15,29	$5s\ ^1D—5p\ ^3D^\circ$	2—1
7664,70	75	14,96	16,58	$5p\ ^3F^\circ—6s\ ^3D$	3—2
7652,36	30	13,39	15,01	$5s\ ^3D—4p''\ ^1F^\circ$	3—3
7579,87	10	13,43	15,07	$5s\ ^3D—5p\ ^3D^\circ$	2—3
7579,02	30	13,68	15,32	$5s\ ^1D—5p\ ^1D^\circ$	2—2
7562,01	25	13,65	15,29	$5s\ ^3D—5p\ ^3D^\circ$	1—1
7438,15	15	15,15	16,82	$5p\ ^3P^\circ—6s\ ^3D$	0—1
7433,85	5	15,18	16,84	$4d\ ^1S—4f\ ^3P^\circ$	0—1
7420,70	8	13,65	15,32	$5s\ ^3D—5p\ ^1D^\circ$	1—2
7404,34	100	14,89	16,56	$5p\ ^3P^\circ—6s\ ^3D$	2—3
7399,89	20	13,39	15,07	$5s\ ^3D—5p\ ^3D^\circ$	3—3
7382,18	10	15,18	16,86	$4d\ ^1S—4f\ ^1P^\circ$	0—1
7331,74	15	14,89	16,58	$5p\ ^3P^\circ—6s\ ^3D$	2—2
7326,02	15	13,43	15,12	$5s\ ^3D—5p\ ^3D^\circ$	2—2
7306,60	12	15,12	16,82	$5p\ ^3D^\circ—6s\ ^3D$	2—1
7255,83	20	15,12	16,83	$5p\ ^3D^\circ—6s\ ^1D$	2—2
7194,92	15	8,86	10,59	$4p\ ^3D^\circ—4s^2\ ^1D$	2—2
7022,75	2	15,07	16,83	$5p\ ^3D^\circ—6s\ ^1D$	3—2
6872,43	3	8,78	10,59	$4p\ ^3D^\circ—4s^2\ ^1D$	3—2
6823,40	3	16,95	18,77	$5d\ ^3G—6f\ ^3H^\circ$	5—6
6809,90	4	15,01	16,83	$4p''\ ^1F^\circ—6s\ ^1D$	3—2
6806,60	4	14,76	16,58	$4p''\ ^3F^\circ—6s\ ^3D$	2—2
6780,40	3	14,99	16,82	$5p\ ^3P^\circ—6s\ ^3D$	1—1
6770,70	8	9,12	10,95	$4p\ ^1P^\circ—4s^2\ ^3P$	1—2
6758,55	8	15,12	16,96	$5p\ ^3D^\circ—5d\ ^3P$	2—1
6737,64	5	14,99	16,83	$5p\ ^3P^\circ—6s\ ^1D$	1—2
6660,99	8	9,09	10,95	$4p\ ^1D^\circ—4s^2\ ^3P$	2—2
6649,22	2	15,32	17,18	$5p\ ^1D^\circ—5d\ ^1P$	2—1
6641,41	10	15,12	16,99	$5p\ ^3D^\circ—5d\ ^3F$	2—3
6631,85	2	14,96	16,83	$5p\ ^3F^\circ—6s\ ^1D$	3—2

λ, Å	I	E_H, eV	E_B, eV	Transition	J
6624,29	8	15,12	16,99	$5p\ ^3D^\circ$—$5d\ ^3D$	2—2
6564,50	10	15,07	16,95	$5p\ ^3D^\circ$—$5d\ ^3P$	3—2
6555,05	5	9,09	10,99	$4p\ ^1D^\circ$—$4s^2\ ^3P$	2—1
6551,58	2	9,06	10,95	$4p\ ^3D^\circ$—$4s^2\ ^3P$	1—2
6541,93	2	14,34	16,23	$4d\ ^3P$—$4p^{\mathrm{IV}\,5}D^\circ$	1—2, 1
6530,30	8	14,34	16,23	$4d\ ^3P$—$4p^{\mathrm{IV}\,5}D^\circ$	2—2, 1
6521,14	14	9,12	11,02	$4p\ ^1P^\circ$—$4s^2\ ^3P$	1—0
6494,04	30	15,07	16,97	$5p\ ^3D^\circ$—$5d\ ^3D$	3—3
6484,46	20	14,99	16,90	$5p\ ^3P^\circ$—$5d\ ^3S$	1—1
6481,46	15	15,32	17,23	$5p\ ^1D^\circ$—$5d\ ^1D$	2—2
6470,152	50	14,65	16,56	$4p''\ ^3G^\circ$—$6s\ ^3D$	3—3
6466,60	3	14,99	16,90	$4p''\ ^1D^\circ$—$5d\ ^3S$	2—1
6457,54	3	17,14	19,06	$4f\ ^3G^\circ$—$7d\ ^3G$	3—3
		16,23	18,15	$4p^{\mathrm{IV}\,5}D^\circ$—$6d\ ^3P$	2, 1—1
6448,49	10	9,06	10,99	$4p\ ^3D^\circ$—$4s^2\ ^3P$	1—1
6443,47	5	16,25	18,17	$4p''\ ^1F^\circ$—$6d\ ^3F$	3—4
6441,698	40	15,07	16,99	$5p\ ^3D^\circ$—$5d\ ^3F$	3—4
6432,78	3	14,34	16,27	$4d\ ^3P$—$4p^{\mathrm{IV}\,5}D^\circ$	1—0
		13,39	15,32	$5s\ ^3D$—$5p\ ^1D^\circ$	3—2
6423,90	30	15,32	17,25	$5p\ ^1D^\circ$—$5d\ ^1F$	2—3
6414,62	20	14,65	16,58	$4p''\ ^3G^\circ$—$6s\ ^3D$	3—2
6411,18	10	15,29	17,22	$5p\ ^3D^\circ$—$5d\ ^3D$	1—1
6403,70	5	16,88	18,81	$4f\ ^3F^\circ$—$7d\ ^3F$	4—4
		15,18	17,11	$4d\ ^1S$—$4f\ ^3D^\circ$	0—1
6377,84	20	15,01	16,95	$4p''\ ^1F^\circ$—$5d\ ^3P$	3—2
6373,27	5	15,01	16,96	$4p''\ ^1F^\circ$—$5d\ ^3G$	3—4
6357,45	15	15,23	17,18	$5p\ ^1P^\circ$—$5d\ ^1P$	1—1
6318,00	3	9,06	11,02	$4p\ ^3D^\circ$—$4s^2\ ^3P$	1—0
6312,83	20	15,29	17,25	$5p\ ^3D^\circ$—$5d\ ^3F$	1—2
6311,292	30	15,01	16,97	$4p''\ ^1F^\circ$—$5d\ ^3D$	3—3
6305,956	15	14,99	16,96	$5p\ ^3P^\circ$—$5d\ ^3P$	1—1
6300,988	40	15,25	17,22	$5p\ ^1F^\circ$—$5d\ ^1G$	3—4
6288,72	5	14,99	16,96	$4p''\ ^1D^\circ$—$5d\ ^3P$	2—1
6276,708	10	—	—	—	—
6276,624	10	14,59	16,56	$4p''\ ^3F^\circ$—$6s\ ^3D$	4—3
6273,330	60	14,98	16,95	$5p\ ^3F^\circ$—$5d\ ^3G$	4—5
6261,826	40	15,01	16,99	$4p''\ ^1F^\circ$—$5d\ ^3F$	3—4
6257,86	5	14,98	16,96	$5p\ ^3F^\circ$—$5d\ ^3G$	4—4
6219,818	30	15,22	17,21	$5p\ ^3F^\circ$—$5d\ ^3G$	2—3
6216,910	60	14,96	16,96	$5p\ ^3F^\circ$—$5d\ ^3G$	3—4
6208,46	15	15,25	17,25	$5p\ ^1F^\circ$—$5d\ ^1F$	3—3
6204,27	15	15,23	17,23	$5p\ ^1P^\circ$—$5d\ ^1D$	1—2
6198,11	5	14,98	16,97	$5p\ ^3F^\circ$—$5d\ ^3D$	4—3
6188,69	20	14,99	16,99	$5p\ ^3P^\circ$—$5d\ ^3D$	1—2
6186,860	20	14,99	16,99	$4p''\ ^1D^\circ$—$5d\ ^3F$	2—3
6172,020	20	14,99	16,99	$4p''\ ^1D^\circ$—$5d\ ^3D$	2—2
6158,00	5	14,96	16,97	$5p\ ^3F^\circ$—$5d\ ^3D$	3—3
6154,24	30	14,89	16,90	$5p\ ^3P^\circ$—$5d\ ^3S$	2—1
6150,42	20	14,98	16,99	$5p\ ^3F^\circ$—$5d\ ^3F$	4—4
6114,468	20	14,96	16,99	$5p\ ^3F^\circ$—$5d\ ^3F$	3—3
6110,90	5	14,96	16,99	$5p\ ^3F^\circ$—$5d\ ^3F$	3—4
6107,45	10	15,15	17,18	$5p\ ^3P^\circ$—$5d\ ^1P$	0—1
6105,97	5	15,22	17,25	$5p\ ^3F^\circ$—$5d\ ^1F$	2—3
6100,01	5	14,96	16,99	$5p\ ^3F^\circ$—$5d\ ^3D$	3—2
6097,33	10	15,22	17,25	$5p\ ^3F^\circ$—$5d\ ^3F$	2—2
6080,320	30	8,92	10,95	$4p\ ^1F^\circ$—$4s^2\ ^3P$	3—2
6072,25	5	14,52	16,56	$4p''\ ^3D^\circ$—$6s\ ^3D$	2—3
6023,25	10	14,52	16,58	$4p''\ ^3D^\circ$—$6s\ ^3D$	2—2

λ, Å	I	E_H, eV	E_B, eV	Transition	J
6013,40	8	14,76	16,82	$4p''\,^3F^\circ-6s\,^3D$	2—1
6000,104	40	14,89	16,95	$5p\,^3P^\circ-5d\,^3P$	2—2
5995,59	10	15,15	17,22	$5p\,^3P^\circ-5d\,^3D$	0—1
5993,27	8	14,89	16,96	$5p\,^3P^\circ-5d\,^3P$	2—1
5988,30	25	14,99	17,06	$5p\,^3P^\circ-5d\,^3P$	1—0
		14,20	16,27	$4d\,^3S-4p^{IV}\,^5D^\circ$	1—0
5979,20	3	14,76	16,83	$4p''\,^3F^\circ-6s\,^1D$	2—2
5941,168	50	14,89	16,97	$5p\,^3P^\circ-5d\,^3D$	2—3
5937,59	5	15,12	17,21	$5p\,^3D^\circ-5d\,^3G$	2—3
5926,90	3	8,86	10,95	$4p\,^3D^\circ-4s^2\,^3P$	2—2
5901,21	5	8,49	10,59	$4p\,^3F^\circ-4s^2\,^1D$	3—2
5897,986	25	14,46	16,56	$4p''\,^3F^\circ-6s\,^3D$	3—3
5858,63	5	13,43	15,55	$5s\,^3D-4p''\,^1P^\circ$	2—1
5851,93	2	14,46	16,58	$4p''\,^3F^\circ-6s\,^3D$	3—2
5842,67	4	8,86	10,99	$4p\,^3D^\circ-4s^2\,^3P$	2—1
5833,68	5	15,12	17,25	$5p\,^3D^\circ-5d\,^1F$	2—3
5826,02	10	15,12	17,25	$5p\,^3D^\circ-5d\,^3F$	2—2
5806,00	25	14,43	16,56	$4p''\,^3D^\circ-6s\,^3D$	3—3
5761,37	2	14,43	16,58	$4p''\,^3D^\circ-6s\,^3D$	3—2
5759,43	5	15,07	17,22	$5p\,^3D^\circ-5d\,^1G$	3—4
5721,78	20	8,42	10,59	$4p\,^3P^\circ-4s^2\,^1D$	1—2
5692,41	2	15,97	18,15	$4p^{IV}\,^5P^\circ-6d\,^3P$	2—2
5689,86	5	14,65	16,83	$4d\,^1D-4p'\,^3D^\circ$	2—3
5682,42	20	14,64	16,82	$4p''\,^3D^\circ-6s\,^3D$	1—1
5664,47	3	14,69	16,88	$4d\,^1F-4f\,^3F^\circ$	3—4
5641,30	20	15,23	17,43	$5p\,^1P^\circ-5d\,^1S$	1—0
5635,57	2	14,43	16,63	$4d\,^3F-4p'\,^3G^\circ$	4—5
5633,14	3	14,65	16,85	$4d\,^1D-4f\,^3P^\circ$	2—2
5615,20	5	15,01	17,22	$4p''\,^1F^\circ-5d\,^1G$	3—4
5593,73	5	14,65	16,87	$4d\,^1D-4f\,^3F^\circ$	2—3
5534,98	3	14,99	17,23	$5p\,^3P^\circ-5d\,^1D$	1—2
5482,65	3	14,30	16,56	$4p''\,^3G^\circ-6s\,^3D$	4—3
5469,63	3	14,61	16,88	$4d\,^1G-4f\,^3G^\circ$	4—5
5437,36	2	14,61	16,89	$4d\,^1G-4f\,^3G^\circ$	4—4
5393,96	3	14,52	16,82	$4p''\,^3D^\circ-6s\,^3D$	2—1
5390,45	5	14,70	16,99	$4d\,^3F-4p'\,^3G^\circ$	2—3
5376,85	3	14,69	16,99	$4d\,^1F-4p'\,^3G^\circ$	3—3
5368,42	10	14,65	16,96	$4p''\,^3G^\circ-5d\,^3G$	3—4
5365,62	5	14,53	16,84	$4d\,^1P-4f\,^3P^\circ$	1—1
5276,522	15	14,60	16,95	$4d\,^3G-4p^{IV}\,^3H^\circ$	3—4
5269,988	30	8,23	10,59	$4p\,^3P^\circ-4s^2\,^1D$	2—2
5245,36	10	14,59	16,95	$4p''\,^3F^\circ-5d\,^3G$	4—5
5229,58	3	14,70	17,07	$4d\,^3F-4p'\,^3F^\circ$	2—2
5207,128	20	14,61	17,00	$4d\,^1G-4p^{VI}\,^1H^\circ$	4—5
5183,364	20	14,45	16,84	$4d\,^3P-4f\,^3P^\circ$	0—1
5175,89	2	14,20	16,59	$4d\,^3S-4p^{V}\,^3P^\circ$	1—2
5158,090	10	14,45	16,86	$4d\,^3P-4d\,^1P^\circ$	0—1
		14,70	17,10	$4d\,^3F-4p'\,^3D^\circ$	2—1
5124,461	20	14,42	16,84	$4d\,^3F-4p'\,^3G^\circ$	3—4
5120,745	20	14,43	16,85	$4d\,^3D-4f\,^3P^\circ$	2—2
5108,331	3	14,43	16,86	$4d\,^3D-4f\,^1P^\circ$	2—1
5100,08	10	14,43	16,86	$4d\,^3D-4f\,^3D^\circ$	2—3
5093,792	20	14,43	16,86	$4d\,^3D-4f\,^3D^\circ$	2—2
		14,42	16,86	$4d\,^3F-4p^{V}\,^1D^\circ$	3—2
5088,932	10	14,43	16,87	$4d\,^3D-4f\,^3F^\circ$	2—2
5088,487	10	14,69	17,13	$4d\,^1F-4f\,^1F^\circ$	3—3
5088,260	30	14,43	16,87	$4d\,^3D-4f\,^3F^\circ$	2—3
5083,991	15	14,42	16,86	$4d\,^3F-4f\,^3D^\circ$	3—3

λ, Å	I	E_H, eV	E_B, eV	Transition	J
5083,991	15	14,43	16,87	$4d\ ^3F-4f\ ^3F^\circ$	4—3
5077,805	5	14,42	16,86	$4d\ ^3F-4f\ ^3D^\circ$	3—2
5072,293	20	14,42	16,87	$4d\ ^3F-4f\ ^3F^\circ$	3—3
5067,082	30	14,70	17,14	$4d\ ^3F-4f\ ^3G^\circ$	2—3
5065,448	40	14,69	17,14	$4d\ ^1F-4f\ ^1G^\circ$	3—4
5060,635	30	8,54	10,99	$4p\ ^3P^\circ-4s^2\ ^3P$	0—1
5058,897	30	14,43	16,88	$4d\ ^3F-4f\ ^3F^\circ$	4—4
5051,778	60	14,43	16,88	$4d\ ^3F-4f\ ^3G^\circ$	4—5
5047,343	10	14,42	16,88	$4d\ ^3F-4f\ ^3F^\circ$	3—4
5041,322	10	14,39	16,85	$4d\ ^3D-4f\ ^3P^\circ$	3—2
5039,002	10	14,65	17,11	$4d\ ^1D-4f\ ^1D^\circ$	2—2
5030,778	2	14,39	16,86	$4d\ ^3D-4p\text{V}\,^1D^\circ$	3—2
5024,027	5	14,43	16,89	$4d\ ^3F-4f\ ^3G^\circ$	4—4
5021,285	20	14,39	16,86	$4d\ ^3D-4f\ ^3D^\circ$	3—3
5020,139	5	8,49	10,95	$4p\ ^3F^\circ-4s^2\ ^3P$	3—2
5015,207	10	14,39	16,86	$4d\ ^3D-4f\ ^3D^\circ$	3—2
5012,611	20	14,42	16,89	$4d\ ^3F-4f\ ^3G^\circ$	3—4
5009,833	20	14,39	16,87	$4d\ ^3D-4f\ ^3F^\circ$	3—3
5006,787	30	14,65	17,13	$4d\ ^1D-4f\ ^1F^\circ$	2—3
4985,503	40	14,39	16,88	$4d\ ^3D-4f\ ^3F^\circ$	3—4
4985,136	2	14,43	16,92	$4d\ ^3D-4p'\ ^3F^\circ$	2—3
4980,006	10	14,62	17,11	$4d\ ^3D-4f\ ^1D^\circ$	1—2
4974,151	10	14,65	17,14	$4d\ ^1D-4f\ ^3G^\circ$	2—3
4973,689	10	14,62	17,11	$4d\ ^3D-4f\ ^3D^\circ$	1—1
4969,812	3	14,42	16,92	$4d\ ^3F-4p'\ ^3F^\circ$	3—3
4955,964	5	14,34	16,84	$4d\ ^3G-4p'\ ^3G^\circ$	4—4
4953,733	50	14,61	17,12	$4d\ ^1G-4f\ ^1H^\circ$	4—5
4951,627	12	14,39	16,89	$4d\ ^3D-4f\ ^3G^\circ$	3—4
4951,454	3	14,61	17,12	$4d\ ^1G-4f\ ^3H^\circ$	4—4
4949,479	3	14,34	16,84	$4d\ ^3P-4f\ ^3P^\circ$	1—1
4943,020	20	14,34	16,84	$4d\ ^3P-4f\ ^3P^\circ$	2—1
4940,060	5	14,34	16,85	$4d\ ^3P-4f\ ^3P^\circ$	1—0
4937,967	15	14,34	16,85	$4d\ ^3P-4f\ ^3P^\circ$	1—2
4937,196	20	14,62	17,13	$4d\ ^3D-4f\ ^3F^\circ$	1—2
4931,653	100	14,34	16,85	$4d\ ^3D-4f\ ^3H^\circ$	4—5
4931,483	20	14,34	16,85	$4d\ ^3P-4f\ ^3P^\circ$	2—2
4926,390	20	14,34	16,86	$4d\ ^3P-4f\ ^1P^\circ$	1—1
4921,461	3	14,34	16,86	$4d\ ^3P-4p\text{V}\,^1D^\circ$	2—2
4920,031	5	14,34	16,86	$4d\ ^3P-4f\ ^1P^\circ$	2—1
4918,373	30	14,60	17,12	$4d\ ^3G-4f\ ^3H^\circ$	3—4
4915,821	15	14,61	17,14	$4d\ ^1G-4f\ ^1G^\circ$	4—4
4912,909	20	14,34	16,86	$4d\ ^3P-4f\ ^3D^\circ$	1—2
4912,362	15	14,34	16,86	$4d\ ^3P-4f\ ^3D^\circ$	2—3
4909,726	100	14,33	16,85	$4d\ ^3G-4f\ ^3H^\circ$	5—6
4909,032	5	14,33	16,85	$4d\ ^3G-4f\ ^3H^\circ$	5—5
4906,548	20	14,34	16,86	$4d\ ^3P-4f\ ^3D^\circ$	2—2
4901,412	25	14,34	16,87	$4d\ ^3P-4f\ ^3F^\circ$	2—3
4896,396	3	14,53	17,07	$4d\ ^1P-4p'\ ^3F^\circ$	1—2
4889,690	30	8,42	10,95	$4p\ ^3P^\circ-4s^2\ ^3P$	1—2
4883,761	3	14,34	16,88	$4d\ ^3G-4f\ ^3F^\circ$	4—4
4883,217	3	14,60	17,14	$4d\ ^3G-4f\ ^1G^\circ$	3—4
4873,291	15	14,60	17,14	$4d\ ^3G-4f\ ^3G^\circ$	3—3
4861,548	2	14,33	16,88	$4d\ ^3G-4f\ ^3F^\circ$	5—4
4854,966	30	14,33	16,88	$4d\ ^3G-4f\ ^3G^\circ$	5—5
4851,248	15	14,34	16,89	$4d\ ^3G-4f\ ^3G^\circ$	4—4
4847,368	2	14,20	16,75	$4d\ ^3S-4p\text{V}\,^3D^\circ$	1—1
4832,236	30	8,42	10,99	$4p\ ^3P^\circ-4s^2\ ^3P$	1—1
4812,940	40	14,53	17,11	$4d\ ^1P-4f\ ^1D^\circ$	1—2

λ, Å	I	E_H, eV	E_B, eV	Transition	J
4807,039	10	14,53	17,11	$4d\ ^1P—4f\ ^3D°$	1—1
4805,651	3	14,34	16,92	$4d\ ^3P—4p'\ ^3F°$	2—3
4766,729	5	14,20	16,80	$4d\ ^3S—4p^V\ ^3P°$	1—0
4758,421	30	8,42	11,02	$4p\ ^3P°—4s^2\ ^3P$	1—0
4753,458	3	14,34	16,95	$4d\ ^3G—4p^{VI}\ ^3H°$	4—4
4687,770	5	15,32	17,96	$5p\ ^1D°—7s\ ^3D$	2—2
4681,990	50	14,20	16,84	$4d\ ^3S—4f\ ^3P°$	1—1
4673,555	30	14,20	16,85	$4d\ ^3S—4f\ ^3P°$	1—0
4671,686	40	14,20	16,85	$4d\ ^3S—4f\ ^3P°$	1—2
4667,297	15	14,34	17,00	$4d\ ^3G—4p^{VI}\ ^1H°$	4—5
4662,638	15	14,20	16,86	$4d\ ^3S—4p^V\ ^1D°$	1—2
4661,350	15	14,20	16,86	$4d\ ^3S—4f\ ^1P°$	1—1
4660,294	8	14,45	17,11	$4d\ ^3P—4f\ ^3D°$	0—1
4649,266	10	14,20	16,86	$4d\ ^3S—4f\ ^3D°$	1—2
4608,457	5	14,43	17,12	$4d\ ^3F—4f\ ^1H°$	4—5
4597,942	5	14,43	17,13	$4d\ ^3D—4f\ ^1F°$	2—3
4596,903	10	14,42	17,12	$4d\ ^3F—4f\ ^3H°$	3—4
4555,922	100	8,23	10,95	$4p\ ^3P°—4s^2\ ^3P$	2—2
4541,032	25	15,23	17,96	$5p\ ^1P°—7s\ ^3D$	1—2
4540,335	10	—	—	—	—
4540,207	10	—	—	—	—
4516,050	5	15,22	17,96	$5p\ ^3F°—7s\ ^3D$	2—2
4505,997	75	8,23	10,99	$4p\ ^3P°—4s^2\ ^3P$	2—1
4462,684	3	14,34	17,12	$4d\ ^3G—4f\ ^1H°$	4—5
4444,823	3	14,34	17,13	$4d\ ^3P—4f\ ^1F°$	2—3
4380,76	2	14,65	17,48	$4d\ ^1D—6p\ ^1P°$	2—1
4378,430	8	15,12	17,95	$5p\ ^3D°—7s\ ^3D$	2—3
4365,362	30	15,12	17,96	$5p\ ^3D°—7s\ ^3D$	2—2
4292,469	30	15,07	17,95	$5p\ ^3D°—7s\ ^3D$	3—3
4291,10	2	13,39	16,28	$5s\ ^3D—4p^{IV}\ ^5D°$	3—3
4285,239	10	15,32	18,21	$5p\ ^1D°—7s\ ^3D$	2—1
4279,959	20	15,07	17,96	$5p\ ^3D°—7s\ ^3D$	3—2
4276,044	30	15,32	18,22	$5p\ ^1D°—7s\ ^1D$	2—2
4255,59	3	14,30	17,11	$4d\ ^3S—4f\ ^1D°$	1—2
4243,35	2	15,23	18,15	$5p\ ^1P°—6d\ ^3P$	1—1
4239,448	25	15,29	18,21	$5p\ ^3D°—7s\ ^3D$	1—1
4230,444	10	15,29	18,22	$5p\ ^3D°—7s\ ^1D$	1—2
4227,936	30	8,92	11,85	$4p\ ^1F°—4s^2\ ^1G$	3—4
4216,89	3	15,23	18,17	$5p\ ^1P°—6d\ ^3D$	1—2
4211,861	30	15,01	17,95	$4p''\ ^1F°—7s\ ^3D$	3—3
4179,512	30	15,25	18,22	$5p\ ^1F°—7s\ ^1D$	3—2
4176,11	5	14,99	17,95	$4p''\ ^1D°—7s\ ^3D$	2—3
4171,858	25	14,99	17,96	$5p\ ^3P°—7s\ ^3D$	1—2
4164,288	20	14,99	17,96	$4p''\ ^1D°—7s\ ^3D$	2—2
4162,296	8	15,23	18,21	$5p\ ^1P°—7s\ ^3D$	1—1
4161,155	30	14,98	17,95	$5p\ ^3F°—7s\ ^3D$	4—3
4153,623	10	15,23	18,22	$5p\ ^1P°—7s\ ^1D$	1—2
4143,020	5	14,96	17,95	$5p\ ^3F°—7s\ ^3D$	3—3
4141,296	20	15,22	18,21	$5p\ ^3F°—7s\ ^3D$	2—1
4132,62	3	15,22	18,22	$5p\ ^3F°—7s\ ^1D$	2—2
4131,359	35	14,96	17,96	$5p\ ^3F°—7s\ ^3D$	3—2
4068,090	5	15,12	18,17	$5p\ ^3D°—6d\ ^3F$	2—3
4065,009	3	15,12	18,17	$5p\ ^3D°—6d\ ^3D$	2—2
4053,658	10	15,15	18,21	$5p\ ^3P°—7s\ ^3D$	0—1
4043,751	35	14,89	17,95	$5p\ ^3P°—7s\ ^3D$	2—3
4043,502	75	8,78	11,85	$4p\ ^3D°—4s^2\ ^1G$	3—4
4032,642	3	14,89	17,96	$5p\ ^3P°—7s\ ^3D$	2—2
4014,18	2	15,12	18,21	$5p\ ^3D°—7s\ ^3D$	2—1

λ, Å	I	E_H, eV	E_B, eV	Transition	J
4006,159	3	15,12	18,22	$5p\ ^3D^\circ$—$7s\ ^1D$	2—2
4003,470	2	15,07	18,16	$5p\ ^3D^\circ$—$6d\ ^3D$	3—3
3993,295	5	15,07	18,17	$5p\ ^3D^\circ$—$6d\ ^3F$	3—4
3987,021	3	15,32	18,43	$5p\ ^1D^\circ$—$6d\ ^1F$	2—3
3945,749	2	15,01	18,15	$4p''\ ^1F^\circ$—$6d\ ^3P$	3—2
3945,570	5	15,29	18,43	$5p\ ^3D^\circ$—$6d\ ^3F$	1—2
3933,260	3	15,01	18,16	$4p''\ ^1F^\circ$—$6d\ ^3D$	3—3
3923,438	3	15,01	18,17	$4p''\ ^1F^\circ$—$6d\ ^3F$	3—4
3920,641	5	15,25	18,41	$5p\ ^1F^\circ$—$6d\ ^1G$	3—4
3903,163	15	14,98	18,15	$5p\ ^3F^\circ$—$6d\ ^3G$	4—5
3896,682	3	14,99	18,17	$5p\ ^3P^\circ$—$6d\ ^3D$	1—2
3892,913	5	14,99	18,17	$4p''\ ^1D^\circ$—$6d\ ^3F$	2—3
3891,114	2	15,23	18,42	$5p\ ^1P^\circ$—$6d\ ^1D$	1—2
3890,073	3	14,99	18,17	$4p''\ ^1D^\circ$—$6d\ ^3D$	2—2
3884,523	5	15,22	18,41	$5p\ ^3F^\circ$—$6d\ ^3G$	2—3
3884,120	10	14,96	18,15	$5p\ ^3F^\circ$—$6d\ ^3G$	3—4
3879,387	5	14,98	18,17	$5p\ ^3F^\circ$—$6d\ ^3F$	4—4
3868,358	3	14,76	17,96	$4p''\ ^3F^\circ$—$7s\ ^3D$	2—2
3866,291	2	15,01	18,22	$4p''\ ^1F^\circ$—$7s\ ^1D$	3—2
3864,121	5	14,96	18,17	$5p\ ^3F^\circ$—$6d\ ^3F$	3—3
3850,03	2	14,99	18,21	$5p\ ^3P^\circ$—$7s\ ^3D$	1—1
3849,570	2	13,39	16,61	$5s\ ^3D$—$4p^V\ ^3D^\circ$	3—3
3842,577	4	14,99	18,22	$5p\ ^3P^\circ$—$7s\ ^1D$	1—2
3836,150	4	14,99	18,22	$4p''\ ^1D^\circ$—$7s\ ^1D$	2—2
3826,908	5	14,89	18,13	$5p\ ^3P^\circ$—$6d\ ^3S$	2—1
3818,869	3	14,99	18,24	$5p\ ^3P^\circ$—$6d\ ^3P$	1—0
3797,832	5	14,89	18,15	$5p\ ^3P^\circ$—$6d\ ^3P$	2—2
3786,261	5	14,89	18,16	$5p\ ^3P^\circ$—$6d\ ^3D$	2—3
3748,207	3	14,65	17,95	$4p''\ ^3G^\circ$—$7s\ ^3D$	3—3
3738,637	3	14,65	17,96	$4p''\ ^3G^\circ$—$7s\ ^3D$	3—2
3686,555	100	8,49	11,85	$4p\ ^3F^\circ$—$4s^2\ ^1G$	3—4
3682,428	10	14,59	17,95	$4p''\ ^3F^\circ$—$7s\ ^3D$	4—3
3602,227	2	14,52	17,96	$4p''\ ^3D^\circ$—$7s\ ^3D$	2—2
3548,742	3	14,46	17,95	$4p''\ ^3F^\circ$—$7s\ ^3D$	3—3
3384,948	3	14,70	18,36	$4d\ ^3F$—$5f\ ^3F^\circ$	2—2
3380,717	10	14,70	18,36	$4d\ ^3F$—$5f\ ^3G^\circ$	2—3
3379,961	3	14,69	18,36	$4d\ ^1F$—$5f\ ^1F^\circ$	3—3
3378,512	3	14,43	18,10	$4d\ ^3D$—$5f\ ^3D^\circ$	2—3
3377,706	5	14,43	18,10	$4d\ ^3D$—$5f\ ^3D^\circ$	2—2
3374,953	20	14,69	18,36	$4d\ ^1F$—$5f\ ^1G^\circ$	3—4
3373,594	15	14,43	18,10	$4d\ ^3D$—$5f\ ^3F^\circ$	2—3
3371,412	8	14,43	18,10	$4d\ ^3F$—$5f\ ^3F^\circ$	4—4
3370,457	30	14,43	18,10	$4d\ ^3F$—$5f\ ^3G^\circ$	4—5
3366,560	5	14,42	18,10	$4d\ ^3F$—$5f\ ^3F^\circ$	3—3
3366,269	10	14,42	18,10	$4d\ ^3F$—$5f\ ^3F^\circ$	3—4
3365,65	15	14,42	18,10	$4d\ ^3F$—$5f\ ^3G^\circ$	3—4
3352,044	8	14,65	18,35	$4d\ ^1D$—$5f\ ^1D^\circ$	2—2
3349,463	5	14,39	18,09	$4d\ ^3D$—$5f\ ^3P^\circ$	3—2
3343,743	20	14,65 14,39	18,36 18,10	$4d\ ^1D$—$5f\ ^1F^\circ$ $4d\ ^3D$—$5f\ ^3D^\circ$	2—3 3—3
3339,084	3	14,65	18,36	$4d\ ^1D$—$5f\ ^3G^\circ$	2—3
3338,937	3	14,39	18,10	$4d\ ^3D$—$5f\ ^3F^\circ$	3—3
3338,647	10	14,39	18,10	$4d\ ^3D$—$5f\ ^3F^\circ$	3—4
3325,812	8	14,62	18,35	$4d\ ^3D$—$5f\ ^1D^\circ$	1—2
3323,735	5	—	—	—	—
3317,140	5	14,62	18,36	$4d\ ^3D$—$5f\ ^3F^\circ$	1—2
3316,279	20	14,34	18,09	$4d\ ^1G$—$5f\ ^1H^\circ$	4—5
3303,516	5	14,34 14,34	18,09 18,09	$4d\ ^3P$—$5f\ ^3P^\circ$ $4d\ ^3P$—$5f\ ^3P^\circ$	2—1 1—2

λ, Å	I	E_H, eV	E_B, eV	Transition	J
3301,228	40	14,34	18,09	$4d\,^3G-5f\,^3H^\circ$	4—5
3300,885	20	14,60	18,35	$4d\,^3G-5f\,^3H^\circ$	3—4
3300,644	10	14,34	18,09	$4d\,^3P-5f\,^3P^\circ$	2—2
3300,444	5	14,34	18,10	$4d\,^3P-5f\,^1P^\circ$	1—1
3297,199	10	14,34	18,10	$4d\,^3P-5f\,^3D^\circ$	1—2
3295,103	15	14,34	18,10	$4d\,^3P-5f\,^3D^\circ$	2—3
3294,336	3	14,34	18,10	$4d\,^3P-5f\,^3D^\circ$	2—2
3293,334	2	14,60	18,36	$4d\,^3G-5f\,^3G^\circ$	3—3
3292,124	10	14,34	18,10	$4d\,^3G-5f\,^3G^\circ$	4—4
3290,422	50	14,33	18,09	$4d\,^3G-5f\,^3H^\circ$	5—6
		14,34	18,10	$4d\,^3P-5f\,^3F^\circ$	2—3
3281,696	10	14,33	18,10	$4d\,^3G-5f\,^3G^\circ$	5—5
3250,469	10	14,53	18,35	$4d\,^1P-5f\,^1D^\circ$	1—2
3238,83	5	14,98	18,80	$5p\,^3F^\circ-7d\,^3G$	4—5
3218,77	3	14,96	18,81	$5p\,^3F^\circ-7d\,^3F$	3—3
3186,017	5	14,20	18,09	$4d\,^3S-5f\,^3P^\circ$	1—0
3185,729	2	13,39	17,28	$5s\,^3D-6p\,^3F^\circ$	3—4
3184,843	15	14,20	18,09	$4d\,^3S-5f\,^3P^\circ$	1—1
3182,175	10	14,20	18,09	$4d\,^3S-5f\,^3P^\circ$	1—2
3179,793	5	14,45	18,35	$4d\,^3P-5f\,^3D^\circ$	0—1
3177,965	3	14,89	18,79	$5p\,^3P^\circ-7d\,^3S$	2—1
3166,56	5	14,89	18,80	$5p\,^3P^\circ-7d\,^3P$	2—2
3162,03	3	14,89	18,81	$5p\,^3P^\circ-7d\,^3D$	2—3
3158,64	5	13,65	17,57	$5s\,^3D-6p\,^3F^\circ$	1—2
3152,883	3	14,42	18,35	$4d\,^3F-5f\,^3H^\circ$	3—4
3151,049	10	13,39	17,32	$5s\,^3D-6p\,^3D^\circ$	3—3
3150,538	3	14,43	18,36	$4d\,^3F-5f\,^3G^\circ$	4—3
		10,95	14,89	$4s^2\,^3P-5p\,^3P^\circ$	2—2
2986,33	2	14,20	18,35	$4d\,^3S-5f\,^1D^\circ$	1—2
2945,368	2	11,02	15,23	$4s^2\,^3P-5p\,^1P^\circ$	0—1
2884,1955	60	9,09	13,39	$4p\,^1D^\circ-5s\,^3D$	2—3
2877,6996	40	9,12	13,43	$4p\,^1P^\circ-5s\,^3D$	1—2
2857,746	5	9,09	13,43	$4p\,^1D^\circ-5s\,^3D$	2—2
2848,72	2	14,42	18,77	$4d\,^3F-6f\,^3F^\circ$	3—3
2840,489	2	10,95	15,32	$4s^2\,^3P-5p\,^1D^\circ$	2—2
2837,3685	50	9,06	13,43	$4p\,^3D^\circ-5s\,^3D$	1—2
2830,31	2	14,39	18,77	$4d\,^3D-6f\,^3D^\circ$	3—3
2810,80	3	14,61	19,02	$4d\,^1G-6f\,^1H^\circ$	4—5
2799,69	2	14,60	19,02	$4d\,^3G-6f\,^3H^\circ$	3—4
2799,536	5	14,34	18,77	$4d\,^3G-6f\,^3H^\circ$	4—5
2797,44	2	14,34	18,77	$4d\,^3P-6f\,^3P^\circ$	2—2
2797,26	2	14,34	18,77	$4d\,^3P-6f\,^3D^\circ$	1—2
2795,31	2	14,34	18,77	$4d\,^3P-6f\,^3D^\circ$	2—3
2791,798	10	14,33	18,77	$4d\,^3G-6f\,^3H^\circ$	5—6
2769,6693	50	8,92	13,39	$4p\,^1F^\circ-5s\,^3D$	3—3
2745,2710	20	8,92	13,43	$4p\,^1F^\circ-5s\,^3D$	3—2
2739,7658	8	9,12	13,65	$4p\,^1P^\circ-5s\,^3D$	1—1
2737,3422	10	8,86	13,39	$4p\,^3D^\circ-5s\,^3D$	2—3
2731,93	2	10,59	15,12	$4s^2\,^1D-5p\,^3D^\circ$	2—2
2721,6771	25	9,09	13,65	$4p\,^1D^\circ-5s\,^3D$	2—1
2718,775	35	9,12	13,68	$4p\,^1P^\circ-5s\,^1D$	1—2
2713,5079	50	8,86	13,43	$4p\,^3D^\circ-5s\,^3D$	2—2
2711,88	5	14,20	18,77	$4d\,^3S-6f\,^3P^\circ$	1—2
2703,184	30	9,06	13,65	$4p\,^3D^\circ-5s\,^3D$	1—1
2700,963	30	9,09	13,68	$4p\,^1D^\circ-5s\,^1D$	2—2
2689,2998	50	8,78	13,39	$4p\,^3D^\circ-5s\,^3D$	3—3
2676,08	2	10,59	15,22	$4s^2\,^1D-5p\,^3F^\circ$	2—2
2666,2910	20	8,78	13,43	$4p\,^3D^\circ-5s\,^3D$	3—2
2648,60	3	8,87	13,55	$4s^2\,^3F-4p'\,^5D^\circ$	3—3

λ, Å	I	E_H, eV	E_B, eV	Transition	J
2620,6663	5	8,66	13,39	4p $^3F^\circ$—5s ^3D	2—3
2614,41	8	8,64	13,38	4s^2 ^3F—4p′ $^5D^\circ$	4—4
2600,2711	20	8,92	13,68	4p $^1F^\circ$—5s ^1D	3—2
2598,8125	20	8,66	13,43	4p $^3F^\circ$—5s ^3D	2—2
2590,5290	15	8,86	13,65	4p $^3D^\circ$—5s ^3D	2—1
2571,7563	10	8,86	13,68	4p $^3D^\circ$—5s ^1D	2—2
2553,32	3	9,02	13,87	4s^2 ^3F—4p′ $^5G^\circ$	2—3
2544,8055	100	8,52	13,39	4p $^3F^\circ$—5s ^3D	4—3
2529,3048	50	8,78	13,68	4p $^3D^\circ$—5s ^1D	3—2
2526,5929	25	8,49	13,39	4p $^3F^\circ$—5s ^3D	3—3
2518,95	8	8,87	13,79	4s^2 ^3F—4p′ $^5G^\circ$	3—4
2506,2732	30	8,49	13,43	4p $^3F^\circ$—5s ^3D	3—2
2489,664	5	3,26	8,23	4s ^1D—4p $^3P^\circ$	2—2
2485,794	20	8,66	13,65	4p $^3F^\circ$—5s ^3D	2—1
2473,3339	20	8,42	13,43	4p $^3P^\circ$—5s ^3D	1—2
2468,5006	5	8,66	13,68	4p $^3F^\circ$—5s ^1D	2—2
2448,21	5	11,02	16,09	4s^2 $^3P^\circ$—4p^{IV} $^5P^\circ$	0—1
2443,32	6	9,12	14,20	4p $^1P^\circ$—4d ^3S	1—1
2442,67	15	8,64	13,72	4s^2 ^3F—4p′ $^5G^\circ$	4—5
2428,92	8	9,09	14,20	4p $^1D^\circ$—4d ^3S	2—1
2424,428	50	8,54	13,65	4p $^3P^\circ$—5s ^3D	0—1
2414,84	5	10,95	16,09	4s^2 ^3P—4p^{IV} $^5P^\circ$	2—1
2414,18	5	9,06	14,20	4p $^3D^\circ$—4d ^3S	1—1
2403,3378	100	8,23	13,39	4p $^3P^\circ$—5s ^3D	2—3
2400,1138	20	3,26	8,42	4s ^1D—4p $^3P^\circ$	2—1
2384,9441	10	8,23	13,43	4p $^3P^\circ$—5s ^3D	2—2
2384,80	5	8,49	13,68	4p $^3F^\circ$—5s ^1D	3—2
2379,39	3	11,02	16,23	4s^2 ^3P—4p^{IV} $^5D^\circ$	0—2, 1
2378,83	10	10,95	16,16	4s^2 ^3P—4p‴ $^3P^\circ$	2—2
2378,39	2	9,02	14,23	4s^2 ^3F—4p′ $^5F^\circ$	2—1
2376,29	50	9,12	14,34	4p $^1P^\circ$—4d ^3P	1—1
2370,7474	20	8,42	13,65	4p $^3P^\circ$—5s ^3D	1—1
2369,8897	100	3,26	8,49	4s ^1D—4p $^3F^\circ$	2—3
2364,15	3	9,09	14,34	4p $^1D^\circ$—4d ^3P	2—2
2362,68	2	9,09	14,34	4p $^1D^\circ$—4d ^3P	2—1
2361,19	3	10,99	16,23	4s^2 ^3P—4p^{IV} $^5D^\circ$	1—2, 1
2356,6410	10	2,98	8,23	4s ^0D—4p $^3P^\circ$	1—2
2355,0141	15	8,42	13,68	4p $^3P^\circ$—5s ^1D	1—2
2353,96	2	8,87	14,13	4s^2 ^3F—4p′ $^5F^\circ$	3—3
2348,74	15	9,06	14,34	4p $^3D^\circ$—4d ^3P	1—1
2342,17	3	10,95	16,25	4s^2 ^3P—4p‴ $^1F^\circ$	2—3
2339,73	3	9,09	14,39	4p $^1D^\circ$—4d ^3D	2—3
2336,17	20	9,12	14,43	4p $^1P^\circ$—4d ^3D	1—2
2323,01	8	9,09	14,43	4p $^1D^\circ$—4d ^3D	2—2
2309,51	6	9,06	14,43	4p $^3D^\circ$—4d ^3D	1—2
2299,47	7	9,06	14,45	4p $^3D^\circ$—4d ^3P	1—0
2294,3683	40	2,83	8,23	4s ^3D—4p $^3P^\circ$	2—2
2290,998	15	9,12	14,53	4p $^1P^\circ$—4d ^1P	1—1
2286,642	15	8,92	14,34	4p $^1F^\circ$—4d ^3P	3—2
2278,33	7	9,09	14,53	4p $^1D^\circ$—4d ^1P	2—1
2276,2582	35	2,98	8,42	4s ^3D—4p $^3P^\circ$	1—1
2274,74	3	8,23	13,68	4p $^3P^\circ$—5s ^1D	2—2
2265,36	7	9,06	14,53	4p $^3D^\circ$—4d ^1P	1—1
2263,780	35	8,92	14,39	4p $^1F^\circ$—4d ^3D	3—3
2263,212	8	8,86	14,34	4p $^3D^\circ$—4d ^3P	2—1
2254,975	6	9,12	14,62	4p $^1P^\circ$—4d ^3D	1—1
2253,00	2	10,59	16,09	4s^2 ^1D—4p^{IV} $^5P^\circ$	2—1
2251,84	2	9,09	14,60	4p $^1D^\circ$—4d ^3G	2—3

λ, Å	I	E_H, eV	E_B, eV	Transition	J
2248,960	25	8,92	14,43	$4p\,^1F^\circ—4d\,^3F$	3—4
2247,002	75	2,72	8,23	$4s\,^3D—4p\,^3P^\circ$	3—2
2243,10	6	9,12	14,65	$4p\,^1P^\circ—4d\,^1D$	1—2
2242,6183	50	3,26	8,78	$4s\,^1D—4p\,^3D^\circ$	2—3
2242,14	6	8,86	14,39	$4p\,^3D^\circ—4d\,^3D$	2—3
2231,571	30	8,78	14,34	$4p\,^3D^\circ—4d\,^3P$	3—2
2230,948	30	9,09	14,65	$4p\,^1D^\circ—4d\,^1D$	2—2
2230,40	10	8,78	14,34	$4p\,^3D^\circ—4d\,^3G$	3—4
2230,087	30	9,06	14,62	$4p\,^3D^\circ—4d\,^3D$	1—1
2229,850	30	8,83	14,42	$4p\,^3D^\circ—4d\,^3F$	2—3
2228,8700	40	2,98	8,54	$4s\,^3D—4p\,^3P^\circ$	1—0
2226,773	40	8,86	14,43	$4p\,^3D^\circ—4d\,^3D$	2—2
2224,701	15	9,12	14,70	$4p\,^1P^\circ—4d\,^3F$	1—2
2221,65	2	10,59	16,16	$4s^2\,^1D—4p'''\,^3P^\circ$	2—2
2218,504	25	9,06	14,65	$4p\,^3D^\circ—4d\,^1D$	1—2
2218,1082	50	2,83	8,42	$4s\,^3D—4p\,^3P^\circ$	2—1
2215,100	35	9,09	14,69	$4p\,^1D^\circ—4d\,^1F$	2—3
2212,741	10	9,09	14,70	$4p\,^1D^\circ—4d\,^3F$	2—2
2210,2684	60	3,26	8,86	$4s\,^1D—4p\,^3D^\circ$	2—2
2209,795	30	8,78	14,39	$4p\,^3D^\circ—4d\,^3D$	3—3
2200,498	25	9,06	14,70	$4p\,^3D^\circ—4d\,^3F$	1—2
2195,674	25	8,78	14,43	$4p\,^3D^\circ—4d\,^3F$	3—4
2192,2678	75	2,83	8,49	$4s\,^3D—4p\,^3F^\circ$	2—3
2189,6323	50	3,26	8,92	$4s\,^1D—4p\,^1F^\circ$	2—3
		8,64	14,30	$4s^2\,^3F—4p''\,^3G^\circ$	4—4
		10,59	16,25	$4s^2\,^1D—4p'''\,^1F^\circ$	2—3
2189,36	3	8,54	14,20	$4p\,^3P^\circ—4d\,^3S$	0—1
2182,85	6	8,66	14,34	$4p\,^3F^\circ—4d\,^3P$	2—1
2181,41	4	8,92	14,60	$4p\,^1F^\circ—4d\,^3G$	3—3
2180,74	10	8,64	14,33	$4s^2\,^3F—4p''\,^3G^\circ$	4—5
2179,399	60	2,98	8,66	$4s\,^3D—4p\,^3F^\circ$	1—2
2174,968	35	8,92	14,61	$4p\,^1F^\circ—4d\,^1G$	3—4
2161,314	20	8,86	14,60	$4p\,^3D^\circ—4d\,^3G$	2—3
2151,801	25	8,66	14,42	$4d\,^3F^\circ—4d\,^3F$	2—3
2148,9838	60	2,72	8,49	$4s\,^3D—4p\,^3F^\circ$	3—3
2146,91	8	8,92	14,69	$4p\,^1F^\circ—4d\,^1F$	3—3
2145,48	15	8,42	14,20	$4p\,^3P^\circ—4d\,^3S$	1—1
2144,70	2	8,92	14,70	$4p\,^1F^\circ—4d\,^3F$	3—2
		8,87	14,65	$4s^2\,^3F—4p''\,^3G^\circ$	3—3
2135,9815	75	2,72	8,52	$4s\,^3D—4p\,^3F^\circ$	3—4
2134,3413	35	8,52	14,33	$4p\,^3F^\circ—4d\,^3G$	4—5
		3,26	9,06	$4s\,^1D—4p\,^3D^\circ$	2—1
2131,23	2	8,78	14,60	$4p\,^3D^\circ—4d\,^3G$	3—3
2130,08	8	8,52	14,34	$4p\,^3F^\circ—4d\,^3G$	4—4
2126,0449	50	2,83	8,66	$4s\,^3D—4p\,^3F^\circ$	2—2
2125,24	4	8,86	14,70	$4p\,^3D^\circ—4d\,^3F$	2—2
2125,098	8	8,78	14,61	$4p\,^3D^\circ—4d\,^1G$	3—4
2122,9793	50	3,26	9,09	$4s\,^1D—4p\,^1D^\circ$	2—2
2118,38	2	8,49	14,34	$4p\,^3F^\circ—4d\,^3P$	3—2
2117,300	35	8,49	14,34	$4p\,^3F^\circ—4d\,^3G$	3—4
2112,1001	30	3,26	9,12	$4s\,^1D—4p\,^1P^\circ$	2—1
2111,30	6	8,52	14,39	$4p\,^3F^\circ—4d\,^3D$	4—3
2106,39	2	8,87	14,75	$4s^2\,^3F—4p''\,^1G^\circ$	3—4
2104,7971	40	2,98	8,86	$4s\,^3D—4p\,^3D^\circ$	1—2
2098,72	2	8,49	14,39	$4p\,^3F^\circ—4d\,^3D$	3—3
2098,386	15	8,52	14,43	$4p\,^3F^\circ—4d\,^3F$	4—4
2094,77	2	8,42	14,34	$4p\,^3P^\circ—4d\,^3P$	1—2
2093,606	10	8,42	14,34	$4p\,^3P^\circ—4d\,^3P$	1—1
2087,930	35	8,49	14,42	$4p\,^3F^\circ—4d\,^3F$	3—3
		8,66	14,60	$4p\,^3F^\circ—4d\,^3G$	2—3

λ, Å	I	E_H, eV	E_B, eV	Transition	J
2085,295	8	2,72	8,66	$4s\ ^3D—4p\ ^3F^\circ$	3—2
2084,33	2	8,64	14,59	$4s^2\ ^3F—4p''\ ^3F^\circ$	4—4
2082,92	2	2,83	8,78	$4s\ ^3D—4p\ ^3D^\circ$	2—3
2080,03	2	8,66	14,62	$4p\ ^3F^\circ—4d\ ^3D$	2—1
2078,646	40	8,23	14,20	$4p\ ^3P^\circ—4d\ ^3S$	2—1
2069,92	2	8,66	14,65	$4p\ ^3F^\circ—4d\ ^1D$	2—2
2066,25	8	8,54	14,53	$4p\ ^3P^\circ—4d\ ^1P$	0—1
2062,41	10	8,42	14,43	$4p\ ^3P^\circ—4d\ ^3D$	1—2
2054,9795	50	2,83	8,86	$4s\ ^3D—4p\ ^3D^\circ$	2—2
2054,43	6	8,42	14,45	$4p\ ^3P^\circ—4d\ ^3P$	1—0
2054,27	4	8,66	14,70	$4p\ ^3F^\circ—4d\ ^3F$	2—2
2047,65	8	9,12	15,18	$4p\ ^1P^\circ—4d\ ^1S$	1—0
2043,8031	60	2,72	8,78	$4s\ ^3D—4p\ ^3D^\circ$	3—3
2037,1269	30	2,83	8,92	$4s\ ^3D—4p\ ^1F^\circ$	2—3
2035,8539	30	2,98	9,06	$4s\ ^3D—4p\ ^3D^\circ$	1—1
2031,023	15	8,23	14,34	$4p\ ^3P^\circ—4d\ ^3P$	2—2
2025,4902	8	2,98	9,09	$4s\ ^3D—4p\ ^1D^\circ$	1—2
2016,885	8	2,72	8,86	$4s\ ^3D—4p\ ^3D^\circ$	3—2
2015,576	5	2,98	9,12	$4s\ ^3D—4p\ ^1P^\circ$	1—1
2012,96	15	8,23	14,39	$4p\ ^3P^\circ—4d\ ^3D$	2—3
1999,7000	60	2,72	8,92	$4s\ ^3D—4p\ ^1F^\circ$	3—3
1989,2116	30	2,83	9,06	$4s\ ^3D—4p\ ^3D^\circ$	2—1
1979,3124	50	2,83	9,09	$4s\ ^3D—4p\ ^1D^\circ$	2—2
1977,02	15	9,02	15,29	$4s^2\ ^3F—5p\ ^3D^\circ$	2—1
1970,489	15	2,83	9,12	$4s\ ^3D—4p\ ^1P^\circ$	2—1
1967,99	2	8,23	14,53	$4p\ ^3P^\circ—4d\ ^1P$	2—1
1957,51	20	8,64	14,98	$4s^2\ ^3F—5p\ ^3F^\circ$	4—4
1952,56	5	8,87	15,22	$4s^2\ ^3F—5p\ ^3F^\circ$	3—2
1946,49	10	8,64	15,01	$4s^2\ ^3F—4p''\ ^1F^\circ$	4—3
1944,586	25	2,72	9,09	$4s\ ^3D—4p\ ^1D^\circ$	3—2
1929,74	25	8,64	15,07	$4s^2\ ^3F—5p\ ^3D^\circ$	4—3
		10,59	17,01	$4s^2\ ^1D—4p^{V}\ ^1P^\circ$	2—1
1922,13	5	8,87	15,32	$4s^2\ ^3F—5p\ ^1D^\circ$	3—2
1920,665	5	8,23	14,69	$4p\ ^3P^\circ—4d\ ^1F$	2—3
1807,84	15	11,02	17,88	$4s^2\ ^3P—4p^{IV}\ 3^\circ$	0—1
1800,95	2	8,64	15,53	$4s^2\ ^3F—4p'''\ ^3D^\circ$	4—3
1790,65	5	9,02	15,94	$4s^2\ ^3F—4p'''\ ^3F^\circ$	2—2
1753,27	15	8,87	15,94	$4s^2\ ^3F—4p'''\ ^3F^\circ$	3—2
		9,02	16,09	$4s^2\ ^3F—4p^{IV}\ ^5P^\circ$	2—1
1744,50	20	8,87	15,97	$4s^2\ ^3F—4p^{IV}\ ^5P^\circ$	3—2
		10,99	18,09	$4s^2\ ^3P—5f\ ^3P^\circ$	1—2
1736,54	10	8,87	16,01	$4s^2\ ^3F—4p^{IV}\ ^5P^\circ$	3—3
1734,21	3	9,02	16,16	$4s^2\ ^3F—4p'''\ ^3P^\circ$	2—2
1717,72	15	9,02	16,23	$4s^2\ ^3F—4p^{IV}\ ^5D^\circ$	2—2, 1
1699,09	30	8,64	15,94	$4s^2\ ^3F—4p'''\ ^3F^\circ$	4—3
		8,87	16,16	$4s^2\ ^3F—4p'''\ ^3P^\circ$	3—2
1683,15	40	8,64	16,01	$4s^2\ ^3F—4p^{IV}\ ^5P^\circ$	4—3
		8,87	16,23	$4s^2\ ^3F—4p^{IV}\ ^5D^\circ$	3—2, 1
1672,77	10	8,87	16,28	$4s^2\ ^3F—4p^{IV}\ ^5D^\circ$	3—3
1663,0017	30	9,12	16,58	$4p\ ^1P^\circ—6s\ ^3D$	1—2
1660,0009	20	9,09	16,56	$4p\ ^1D^\circ—6s\ ^3D$	2—3
1656,3216	20	9,09	16,58	$4p\ ^1D^\circ—6s\ ^3D$	2—2
1649,4573	25	9,06	16,58	$4p\ ^3D^\circ—6s\ ^3D$	1—2
1636,61	10	9,02	16,59	$4s^2\ ^3F—4p^{V}\ ^3P^\circ$	2—2
1630,27	25	8,64	16,25	$4s^2\ ^3F—4p'''\ ^1F^\circ$	4—3
1623,17	30	8,64	16,28	$4s^2\ ^3F—4p^{IV}\ ^5D^\circ$	4—3
1622,44	40	9,02	16,66	$4s^2\ ^3F—4p^{V}\ ^3P^\circ$	2—1

λ, Å	I	E_H, eV	E_B, eV	Transition	J
1621,4256	60	8,92	16,56	$4p\,^1F^\circ$—$6s\,^3D$	3—3
1617,9151	20	8,92	16,58	$4p\,^1F^\circ$—$6s\,^3D$	3—2
1611,1180	10	9,12	16,82	$4p\,^1P^\circ$—$6s\,^3D$	1—1
1610,2964	15	8,86	16,56	$4p\,^3D^\circ$—$6s\,^3D$	2—3
1608,6396	25	9,12	16,83	$4p\,^1P^\circ$—$6s\,^1D$	1—2
1606,8338	40	8,86	16,58	$4p\,^3D^\circ$—$6s\,^3D$	2—2
1605,274	30	8,87	16,59	$4s^2\,^3F$—$4p^V\,^3P^\circ$	3—2
1604,8474	20	9,09	16,82	$4p\,^1D^\circ$—$6s\,^3D$	2—1
1602,3882	40	9,09	16,83	$4p\,^1D^\circ$—$6s\,^1D$	2—2
1602,250	15	9,02	16,75	$4s^2\,^3F$—$4p^V\,^3D^\circ$	2—1
1598,4024	40	9,06	16,82	$4p\,^3D^\circ$—$6s\,^3D$	1—1
1593,5557	60	8,78	16,56	$4p\,^3D^\circ$—$6s\,^3D$	3—3
1590,1646	40	8,78	16,58	$4p\,^3D^\circ$—$6s\,^3D$	3—2
1583,683	50	8,87	16,70	$4s^2\,^3F$—$4p^V\,^3D^\circ$	3—2
1582,849	10	9,12	16,96	$4p\,^1P^\circ$—$5d\,^3P$	1—1
1581,991	40	8,87	16,71	$4s^2\,^3F$—$4p'\,^3F^\circ$	3—4
1580,628	30	9,02	16,86	$4s^2\,^3F$—$4f\,^3D^\circ$	2—3
1580,025	15	9,02	16,86	$4s^2\,^3F$—$4f\,^3D^\circ$	2—2
1579,492	30	9,02	16,87	$4s^2\,^3F$—$4f\,^3F^\circ$	2—3
1575,349	5	9,12	16,99	$4p\,^1P^\circ$—$5d\,^3D$	1—2
1570,568	3	9,06	16,96	$4p\,^3D^\circ$—$5d\,^3P$	1—1
1570,035	2	9,09	16,99	$4p\,^1D^\circ$—$5d\,^3F$	2—3
1569,426	10	9,02	16,92	$4s^2\,^3F$—$4p'\,^3F^\circ$	2—3
1569,2123	10	8,66	16,56	$4p\,^3F^\circ$—$6s\,^3D$	2—3
1566,4151	40	8,92	16,83	$4p\,^1F^\circ$—$6s\,^1D$	3—2
1565,9240	40	8,66	16,58	$4p\,^3F^\circ$—$6s\,^3D$	2—2
1563,189	5	9,06	16,99	$4p\,^3D^\circ$—$5d\,^3D$	1—2
1558,3446	30	8,86	16,82	$4p\,^3D^\circ$—$6s\,^3D$	2—1
1557,583	20	8,87	16,83	$4s^2\,^3F$—$4p'\,^3D^\circ$	3—3
1555,698	50	8,64	16,61	$4s^2\,^3F$—$4p^V\,^3D^\circ$	4—3
1555,134	40	8,87	16,84	$4s^2\,^3F$—$4p'\,^3G^\circ$	3—4
1553,893	25	9,02	16,99	$4s^2\,^3F$—$4p'\,^3G^\circ$	2—3
1552,641	50	8,64	16,63	$4s^2\,^3F$—$4p'\,^3G^\circ$	4—5
1551,379	30	8,87	16,86	$4s^2\,^3F$—$4f\,^3D^\circ$	3—3
1550,644	30	9,02	17,01	$4s^2\,^3F$—$4p^V\,^1P^\circ$	2—1
1550,292	3	8,87	16,87	$4s^2\,^3F$—$4f\,^3F^\circ$	3—3
1547,950	10	8,87	16,88	$4s^2\,^3F$—$4f\,^3F^\circ$	3—4
1544,674	40	8,87	16,89	$4s^2\,^3F$—$4f\,^3G^\circ$	3—4
1541,7031	75	8,52	16,56	$4p\,^3F^\circ$—$6s\,^3D$	4—3
1540,589	30	8,87	16,92	$4s^2\,^3F$—$4p'\,^3F^\circ$	3—3
1540,3889	30	8,78	16,83	$4p\,^3D^\circ$—$6s\,^1D$	3—2
1540,231	20	9,02	17,07	$4s^2\,^3F$—$4p'\,^3F^\circ$	2—2
1538,488	10	8,92	16,97	$4p\,^1F^\circ$—$5d\,^3D$	3—3
		9,12	17,18	$4p\,^1P^\circ$—$5d\,^1P$	1—1
1537,560	50	8,64	16,71	$4s^2\,^3F$—$4p'\,^3F^\circ$	4—4
1535,515	15	8,92	16,99	$4p\,^1F^\circ$—$5d\,^3F$	3—4
1535,0024	25	8,49	16,56	$4p\,^3F^\circ$—$6s\,^3D$	3—3
1533,976	25	9,02	17,10	$4s^2\,^3F$—$4p'\,^3D^\circ$	2—1
1532,124	30	8,87	16,96	$4s^2\,^3F$—$4p'\,^3D^\circ$	3—2
1531,8557	50	8,49	16,58	$4p\,^3F^\circ$—$6s\,^3D$	3—2
		9,02	17,11	$4s^2\,^3F$—$4f\,^1D^\circ$	2—2
		8,86	16,96	$4p\,^3D^\circ$—$5d\,^3P$	2—1
1528,782	2	9,02	17,13	$4s^2\,^3F$—$4f\,^1F^\circ$	2—3
1527,801	5	9,02	17,13	$4s^2\,^3F$—$4f\,^3F^\circ$	2—2
1526,969	5	9,06	17,18	$4p\,^3D^\circ$—$5d\,^1P$	1—1
1525,794	30	9,02	17,14	$4s^2\,^3F$—$4f\,^3G^\circ$	2—3

λ, Å	I	E_H, eV	E_B, eV	Transition	J
1525,653	10	8,87	16,99	$4s^2\,^3F—4p'\,^3G^\circ$	3—3
		9,09	17,22	$4p\,^1D^\circ—5d\,^3D$	2—1
		9,12	17,25	$4p\,^1P^\circ—5d\,^3F$	1—2
1524,857	20	8,86	16,99	$4p\,^3D^\circ—5d\,^3D$	2—2
1523,740	10	9,09	17,23	$4p\,^1D^\circ—5d\,^1D$	2—2
1522,575	15	9,02	17,16	$4s^2\,^3F—4p^{VI}\,^3F^\circ$	2—3
1520,543	20	9,09	17,25	$4p\,^1D^\circ—5d\,^1F$	2—3
1519,8370	60	8,66	16,82	$4p\,^3F^\circ—6s\,^3D$	2—1
		9,06	17,22	$4p\,^3D^\circ—5d\,^3D$	1—1
1519,4917	50	8,42	16,58	$4p\,^3P^\circ—6s\,^3D$	1—2
1517,930	10	9,06	17,23	$4p\,^3D^\circ—5d\,^1D$	1—2
1517,6312	20	8,66	16,83	$4p\,^3F^\circ—6s\,^1D$	2—2
1517,162	10	8,78	16,95	$4p\,^3D^\circ—5d\,^3P$	3—2
1516,902	5	8,78	16,96	$4p\,^3D^\circ—5d\,^3G$	3—4
1514,492	50	8,64	16,83	$4s^2\,^3F—4p'\,^3D^\circ$	4—3
1514,238	10	9,06	17,25	$4p\,^3D^\circ—5d\,^3F$	1—2
1513,360	20	8,78	16,97	$4p\,^3D^\circ—5d\,^3D$	3—3
1512,457	20	8,87	17,07	$4s^2\,^3F—4p'\,^3F^\circ$	3—2
1512,174	20	8,64	16,84	$4s^2\,^3F—4p'\,^3G^\circ$	4—4
1510,502	35	8,78	16,99	$4p\,^3D^\circ—5d\,^3F$	3—4
1508,627	30	8,64	16,86	$4s^2\,^3F—4f\,^3D^\circ$	4—3
1508,175	25	9,02	17,24	$4s^2\,^3F—4p^{VI}\,^3F^\circ$	2—2
1505,848	5	8,87	17,10	$4s^2\,^3F—4p^{VI}\,^3F^\circ$	3—4
1505,384	20	8,64	16,88	$4s^2\,^3F—4f\,^3F^\circ$	4—4
1504,755	25	8,64	16,88	$4s^2\,^3F—4f\,^3G^\circ$	4—5
1503,368	15	9,02	17,26	$4s^2\,^3F—6p\,^3P^\circ$	2—1
1501,333	10	9,02	17,27	$4s^2\,^3F—6p\,^3F^\circ$	2—3
1499,510	10	8,87	17,14	$4s^2\,^3F—4f\,^1G^\circ$	3—4
1498,566	3	8,87	17,14	$4s^2\,^3F—4f\,^3G^\circ$	3—3
1496,6860	35	8,54	16,82	$4p\,^3P^\circ—6s\,^3D$	0—1
1495,426	25	8,87	17,16	$4s^2\,^3F—4p^{VI}\,^3F^\circ$	3—3
1494,658	5	8,66	16,96	$4p\,^3F^\circ—5d\,^3P$	2—1
1493,359	25	8,92	17,22	$4p\,^1F^\circ—5d\,^1G$	3—4
1492,837	30	9,02	17,32	$4s^2\,^3F—6p\,^3D^\circ$	2—2
1492,684	10	9,12	17,43	$4p\,^1P^\circ—5d\,^1S$	1—0
1492,149	10	9,02	17,32	$4s^2\,^3F—6p\,^3D^\circ$	2—3
1488,6373	75	8,23	16,56	$4p\,^3P^\circ—6s\,^3D$	2—3
		8,66	16,99	$4p\,^3F—5d\,^3F$	2—3
1485,6777	40	8,23	16,58	$4p\,^3P^\circ—6s\,^3D$	2—2
		8,49	16,83	$4p\,^3F^\circ—6s\,^1D$	3—2
1485,318	20	8,86	17,21	$4p\,^3D^\circ—5d\,^3G$	2—3
1481,541	20	8,87	17,24	$4s^2\,^3F—4p^{VI}\,^3F^\circ$	3—2
1478,230	2	8,86	17,25	$4p\,^3D^\circ—5d\,^3F$	2—2
1475,846	30	9,02	17,42	$4s^2\,^3F—4p^{IV}\,1^\circ$	2—1
1474,934	20	8,87	17,27	$4s^2\,^3F—6p\,^3F^\circ$	3—3
1473,9788	25	8,42	16,83	$4p\,^3P^\circ—6s\,^1D$	1—2
1473,531	15	8,87	17,28	$4s^2\,^3F—6p\,^3F^\circ$	3—4
1472,399	20	0,00	8,42	$3d^{10}\,^1S—4p\,^3P^\circ$	0—1
1470,697	40	8,52	16,95	$4p\,^3F^\circ—5d\,^3G$	4—5
1469,691	15	8,78	17,22	$4p\,^3D^\circ—5d\,^1G$	3—4
1466,751	5	8,87	17,32	$4s^2\,^3F—6p\,^3D^\circ$	3—2
1466,519	10	8,52	16,97	$4p\,^3F^\circ—5d\,^3D$	4—3
1466,067	20	8,87	17,32	$4s^2\,^3F—6p\,^3D^\circ$	3—3
1465,542	15	8,64	17,10	$4s^2\,^3F—4p^{VI}\,^3F^\circ$	4—4
1463,771	50	8,49	16,96	$4p\,^3F^\circ—5d\,^3G$	3—4
		8,52	16,99	$4p\,^3F^\circ—5d\,^3F$	4—4
1461,556	15	8,42	16,90	$4p\,^3P^\circ—5d\,^3S$	1—1
1459,412	25	9,02	17,51	$4s^2\,^3F—6p\,^3D^\circ$	2—1

λ, Å	I	$E_{\text{н}}$, eV	$E_{\text{в}}$, eV	Transition	J
1458,004	30	8,49	16,99	$4p\ ^3F^\circ$—$5d\ ^3F$	3—3
1457,175	10	8,49	16,99	$4p\ ^3F^\circ$—$5d\ ^3D$	3—2
1455,655	3	8,64	17,16	$4s^2\ ^3F$—$4p^{VI}\ ^3F^\circ$	4—3
1452,291	20	8,42	16,96	$4p\ ^3P^\circ$—$5d\ ^3P$	1—1
1450,307	25	8,66	17,21	$4p\ ^3F^\circ$—$5d\ ^3G$	2—3
1449,056	20	9,02	17,57	$4s^2\ ^3F$—$6p\ ^3F^\circ$	2—2
1445,982	20	8,42	17,00	$4p\ ^3P^\circ$—$5d\ ^3D$	1—2
1444,1305	2	8,23	16,82	$4p\ ^3P^\circ$—$6s\ ^3D$	2—1
1443,541	10	8,66	17,25	$4p\ ^3F^\circ$—$5d\ ^3F$	2—2
1442,1389	15	8,23	16,83	$4p\ ^3P^\circ$—$6s\ ^1D$	2—2
4436,233	15	8,64	17,27	$4s^2\ ^3F$—$6p\ ^3F^\circ$	4—3
1435,312	10	8,87	17,51	$4s^2\ ^3F$—$6p\ ^1F^\circ$	3—3
1134,916	25	8,64	17,28	$4s^2\ ^3F$—$6p\ ^3F^\circ$	4—4
1434,758	15	8,42	17,06	$4p\ ^3P^\circ$—$5d\ ^3P$	1—0
1433,837	10	8,54	17,18	$4p\ ^3P^\circ$—$5d\ ^1P$	0—1
1430,243	40	8,23	16,90	$4p\ ^3P^\circ$—$5d\ ^3S$	2—1
1428,366	15	8,87	17,55	$4s^2\ ^3F$—$6p\ ^1D^\circ$	3—2
1427,835	20	8,64	17,32	$4s^2\ ^3F$—$6p\ ^3D^\circ$	4—3
1427,589	10	8,54	17,22	$4p\ ^3P^\circ$—$5d\ ^3D$	0—1
1421,760	25	8,23	16,95	$4p\ ^3P^\circ$—$5d\ ^3P$	2—2
1421,382	5	8,23	16,96	$4p\ ^3P^\circ$—$5d\ ^3P$	2—1
1419,742	2	8,49	17,22	$4p\ ^3F^\circ$—$5d\ ^1G$	3—4
1418,423	25	8,23	16,97	$4p\ ^3P^\circ$—$5d\ ^3D$	2—3
1414,897	10	8,42	17,18	$4p\ ^3P^\circ$—$5d\ ^1P$	1—1
1408,811	2	8,42	17,22	$4p\ ^3P^\circ$—$5d\ ^3D$	1—1
1407,160	15	8,42	17,23	$4p\ ^3P^\circ$—$5d\ ^1D$	1—2
1402,776	15	9,12	17,96	$4p\ ^1P^\circ$—$7s\ ^3D$	1—2
1399,355	3	9,09	17,95	$4p\ ^1D^\circ$—$7s\ ^3D$	2—3
1398,636	10	8,64	17,51	$4s^2\ ^3F$—$6p\ ^1F^\circ$	4—3
1393,126	10	9,06	17,96	$4p\ ^3D^\circ$—$7s\ ^3D$	1—2
1375,520	3	8,23	17,27	$4p\ ^3P^\circ$—$5d\ ^1F$	2—3
1371,840	20	8,92	17,95	$4p\ ^1F^\circ$—$7s\ ^3D$	3—3
1370,558	2	8,92	17,96	$4p\ ^1F^\circ$—$7s\ ^3D$	3—2
1367,952	25	0,00	9,06	$3d^{10}\ ^1S$—$4p\ ^3D^\circ$	0—1
1363,501	5	9,12	18,22	$4p\ ^1P^\circ$—$7s\ ^1D$	1—2
1362,598	20	8,86	17,96	$4p\ ^3D^\circ$—$7s\ ^3D$	2—2
1359,935	5	9,09	18,21	$4p\ ^1D^\circ$—$7s\ ^3D$	2—1
1359,010	20	9,09	18,22	$4p\ ^1D^\circ$—$7s\ ^1D$	2—2
1358,764	30	0,00	9,12	$3d^{10}\ ^1S$—$4p\ ^1P^\circ$	0—1
1355,304	15	9,06	18,21	$4p\ ^3D^\circ$—$7s\ ^3D$	1—1
1351,837	25	8,78	17,95	$4p\ ^3D^\circ$—$7s\ ^3D$	3—3
1350,592	15	8,78	17,96	$4p\ ^3D^\circ$—$7s\ ^3D$	3—2
1340,909	3	8,92	18,16	$4p\ ^1F^\circ$—$6d\ ^3D$	3—3
1339,769	5	8,92	18,17	$4p\ ^1F^\circ$—$6d\ ^3F$	3—4
1334,556	2	8,86	18,15	$4p\ ^3D^\circ$—$6d\ ^3P$	2—1
1333,054	20	8,92	18,22	$4p\ ^1F^\circ$—$7s\ ^1D$	3—2
		8,66	17,96	$4p\ ^3F^\circ$—$7s\ ^3D$	2—2
1332,222	5	8,86	18,17	$4p\ ^3D^\circ$—$6d\ ^3F$	2—3
1328,416	5	9,09	18,43	$4p\ ^1D^\circ$—$6d\ ^1F$	2—3
1326,394	10	8,86	18,21	$4p\ ^3D^\circ$—$7s\ ^3D$	2—1
1325,511	3	8,86	18,22	$4p\ ^3D^\circ$—$7s\ ^1D$	2—2
1323,811	6	9,06	18,43	$4p\ ^3D^\circ$—$6d\ ^3F$	1—2
1323,187	3	8,78	18,15	$4p\ ^3D^\circ$—$6d\ ^3P$	3—2
1322,627	6	9,12	18,50	$4p\ ^1P^\circ$—$6d\ ^1S$	1—0
1321,788	5	8,78	18,16	$4p\ ^3D^\circ$—$6d\ ^3D$	3—3
1320,687	10	8,78	18,17	$4p\ ^3D^\circ$—$6d\ ^3F$	3—4
1314,335	30	8,52	17,95	$4p\ ^3F^\circ$—$7s\ ^3D$	4—3
1314,147	15	8,78	18,22	$4p\ ^3D^\circ$—$7s\ ^1D$	3—2
1309,463	15	8,49	17,95	$4p\ ^3F^\circ$—$7s\ ^3D$	3—3

λ, Å	I	E_H, eV	E_B, eV	Transition	J
1308,296	30	8,49	17,96	$4p\ ^3F^\circ - 7s\ ^3D$	3—2
1305,554	5	8,92	18,41	$4p\ ^1F^\circ - 6d\ ^1G$	3—4
1303,979	2	8,66	18,17	$4p\ ^3F^\circ - 6d\ ^3F$	2—3
1303,656	2	8,66	18,17	$4p\ ^3F^\circ - 6d\ ^3D$	2—2
1299,267	10	8,42	17,96	$4p\ ^3P^\circ - 7s\ ^3D$	1—2
1298,394	15	8,66	18,21	$4p\ ^3F^\circ - 7s\ ^3D$	2—1
1297,549	2	8,66	18,22	$4p\ ^3F^\circ - 7s\ ^1D$	2—2
1287,464	15	8,52	18,15	$4p\ ^3F^\circ - 6d\ ^3G$	4—5
1284,868	8	8,52	18,17	$4p\ ^3F^\circ - 6d\ ^3F$	4—4
1282,450	15	8,49	18,15	$4p\ ^3F^\circ - 6d\ ^3G$	3—4
1281,458	8	8,54	18,22	$4p\ ^3P^\circ - 7s\ ^3D$	0—1
1281,098	3	8,49	18,16	$4p\ ^3F^\circ - 6d\ ^3D$	3—3
1280,265	5	8,49	18,17	$4p\ ^3F^\circ - 6d\ ^3F$	3—3
1275,570	30	8,23	17,95	$4p\ ^3P^\circ - 7s\ ^3D$	2—3
1274,463	3	8,23	17,96	$4p\ ^3P^\circ - 7s\ ^3D$	2—2
1274,069	3	8,49	18,22	$4p\ ^3F^\circ - 7s\ ^1D$	3—2
1273,704	2	8,42	18,15	$4p\ ^3P^\circ - 6d\ ^3P$	1—1
1272,036	8	8,66	18,41	$4p\ ^3F^\circ - 6d\ ^3G$	2—3
1271,326	2	8,42	18,17	$4p\ ^3P^\circ - 6d\ ^3D$	1—2
1266,308	10	8,42	18,21	$4p\ ^3P^\circ - 7s\ ^3D$	1—1
1265,504	15	8,42	18,22	$4p\ ^3P^\circ - 7s\ ^1D$	1—2
1262,928	3	8,42	18,24	$4p\ ^3P^\circ - 6d\ ^3P$	1—0
1253,179	5	8,23	18,13	$4p\ ^3P^\circ - 6d\ ^3S$	2—1
1250,045	10	8,23	18,15	$4p\ ^3P^\circ - 6d\ ^3P$	2—2
1248,790	5	8,23	18,16	$4p\ ^3P^\circ - 6d\ ^3D$	2—3
1241,961	2	8,23	18,22	$4p\ ^3P^\circ - 7s\ ^1D$	2—2
1201,626	2	8,49	18,80	$4p\ ^3F^\circ - 7d\ ^3G$	3—4
1192,261	2	8,66	19,06	$4p\ ^3F^\circ - 7d\ ^3G$	2—3
1185,899	2	8,23	18,69	$4p\ ^3P^\circ - 8s\ ^3D$	2—3
1162,610	3	2,72	13,38	$4s\ ^3D - 4p'\ ^5D^\circ$	3—4
1157,871	8	2,98	13,68	$4s\ ^3D - 4p'\ ^5D^\circ$	1—2
1157,021	5	2,83	13,55	$4s\ ^3D - 4p'\ ^5D^\circ$	2—3
1147,762	8	2,98	13,78	$4s\ ^3D - 4p'\ ^5D^\circ$	1—1
1144,853	30	2,72	13,55	$4s\ ^3D - 4p'\ ^5D^\circ$	3—3
1142,642	20	2,83	13,68	$4s\ ^3D - 4p'\ ^5D^\circ$	2—2
1123,226	5	2,83	13,87	$4s\ ^3D - 4p'\ ^5G^\circ$	2—3
1119,945	15	2,72	13,79	$4s\ ^3D - 4p'\ ^5G^\circ$	3—4
1112,407	5	—	—	—	—
1106,446	3	3,26	14,46	$4s\ ^1D - 4p''\ ^3F^\circ$	2—3
1105,182	5	2,98	14,19	$4s\ ^3D - 4p'\ ^5F^\circ$	1—2
1097,049	25	2,83	14,13	$4s\ ^3D - 4p'\ ^5F^\circ$	2—3
1094,401	30	2,72	14,05	$4s\ ^3D - 4p'\ ^5F^\circ$	3—4
1091,288	5	2,83	14,19	$4s\ ^3D - 4p'\ ^5F^\circ$	2—2
1089,236	3	3,26	14,64	$4s\ ^1D - 4p''\ ^3D^\circ$	2—1
1088,393	20	3,26	14,65	$4s\ ^1D - 4p''\ ^3G^\circ$	2—3
1086,110	5	2,72	14,13	$4s\ ^3D - 4p'\ ^5F^\circ$	3—3
1073,738	30	1,98	14,52	$4s\ ^3D - 4p''\ ^3D^\circ$	1—2
1070,308	15	2,72	14,30	$4s\ ^3D - 4p''\ ^3G^\circ$	3—4
1069,193	50	2,83	14,43	$4s\ ^3D - 4p''\ ^3D^\circ$	2—3
1066,133	20	2,83	14,46	$4s\ ^3D - 4p''\ ^3F^\circ$	2—3
1065,7822	20	3,26	14,89	$4s\ ^1D - 5p\ ^3P^\circ$	2—2
1063,003	60	2,98	14,64	$4s\ ^3D - 4p''\ ^3D^\circ$	1—1
1060.630	60	2,83	14,52	$4s\ ^3D - 4p''\ ^3D^\circ$	2—2
1059,0960	60	3,26	14,96	$4s\ ^1D - 5p\ ^3F^\circ$	2—3
1058,796	40	2,72	14,43	$4s\ ^3D - 4p''\ ^3D^\circ$	3—3
1056,9545	60	3,26	14,99	$4s\ ^1D - 4p''\ ^1D^\circ$	2—2
1055,795	40	2,72	14,46	$4s\ ^3D - 4p''\ ^3F^\circ$	3—3
1054,6903	60	3,26	15,01	$4s\ ^1D - 4p''\ ^1F^\circ$	2—3

λ, Å	I	E_H, eV	E_B, eV	Transition	J
1052,170	20	2,98	14,76	$4s\,^3D—4p''\,^3F^\circ$	1—2
1050,399	10	2,72	14,52	$4s\,^3D—4p''\,^3D^\circ$	3—2
1050,153	10	2,83	14,64	$4s\,^3D—4p''\,^3D^\circ$	2—1
1049,7556	50	3,26	15,07	$4s\,^1D—5p\,^3D^\circ$	2—3
1049,363	20	2,83	14,65	$4s\,^3D—4p''\,^3G^\circ$	2—3
1044,7434	80	3,26	15,12	$4s\,^1D—5p\,^3D^\circ$	2—2
1044,516	80	2,72	14,59	$4s\,^3D—4p''\,^3F^\circ$	3—4
1039,569	60	2,83	14,76	$4s\,^3D—4p''\,^3F^\circ$	2—2
1039,345	60	2,72	14,65	$4s\,^3D—4p''\,^3G^\circ$	3—3
1036,4695	60	3,26	15,22	$4s\,^1D—5p\,^3F^\circ$	2—2
1035,1631	8	3,26	15,23	$4s\,^1D—5p\,^1P^\circ$	2—1
1033,5679	10	3,26	15,25	$4s\,^1D—5p\,^1F^\circ$	2—3
1031,7661	8	2,98	14,99	$4s\,^3D—5p\,^3P^\circ$	1—1
1030,261	20	2,72	14,75	$4s\,^3D—4p''\,^1G^\circ$	3—4
1029,747	10	2,72	14,76	$4s\,^3D—4p''\,^3F^\circ$	3—2
1028,3281	25	2,83	14,89	$4s\,^3D—5p\,^3P^\circ$	2—2
1027,8312	50	3,26	15,32	$4s\,^1D—5p\,^1D^\circ$	2—2
1022,1021	5	2,83	14,96	$4s\,^3D—5p\,^3F^\circ$	2—3
1020,1075	15	2,83	14,99	$4s\,^3D—4p''\,^1D^\circ$	2—2
1019,6545	15	2,83	14,99	$4s\,^3D—5p\,^3P^\circ$	2—1
1018,7075	50	2,72	14,89	$4s\,^3D—5p\,^3P^\circ$	3—2
1018,0643	15	{ 2,83	15,01	$4s\,^3D—4p''\,^1F^\circ$	2—3
		2,98	15,15	$4s\,^3D—5p\,^3P^\circ$	1—0
1012,6834	3	2,98	15,22	$4s\,^3D—5p\,^3F^\circ$	1—2
1012,5972	25	2,72	14,96	$4s\,^3D—5p\,^3F^\circ$	3—3
1011,4362	2	2,98	15,23	$4s\,^3D—5p\,^1P^\circ$	1—1
1010,6395	3	2,72	14,99	$4s\,^3D—4p''\,^1D^\circ$	3—2
1010,453	10	3,26	15,53	$4s\,^1D—4p'''\,^3D^\circ$	2—3
1010,267	30	3,26	15,53	$4s\,^1D—4p'''\,^3D^\circ$	2—2
1008,7284	30	{ 2,83	15,12	$4s\,^3D—5p\,^3D^\circ$	2—2
		3,26	15,55	$4s\,^1D—4p'''\,^1P^\circ$	2—1
1008,5692	30	2,72	15,01	$4s\,^3D—4p''\,^1F^\circ$	3—3
1004,0557	30	2,72	15,07	$4s\,^3D—5p\,^3D^\circ$	3—3
1001,0130	8	2,83	15,22	$4s\,^3D—5p\,^3F^\circ$	2—2
999,7944	5	2,83	15,23	$4s\,^3D—5p\,^1P^\circ$	2—1
998,3063	8	2,83	15,25	$4s\,^3D—5p\,^1F^\circ$	2—3
992,9533	25	2,83	15,32	$4s\,^3D—5p\,^1D^\circ$	2—2
989,2368	8	2,72	15,25	$4s\,^3D—5p\,^1F^\circ$	3—3
987,656	10	2,98	15,53	$4s\,^3D—4p'''\,^3D^\circ$	1—2
984,530	10	2,98	15,57	$4s\,^3D—4p'''\,^3D^\circ$	1—1
979,418	5	3,26	15,91	$4s\,^1D—4p^{IV}\,^5S^\circ$	2—2
977,567	25	3,26	15,94	$4s\,^1D—4p'''\,^3F^\circ$	2—3
976,708	10	2,83	15,53	$4s\,^3D—4p'''\,^3D^\circ$	2—3
976,540	10	2,83	15,53	$4s\,^3D—4p'''\,^3D^\circ$	2—2
974,759	20	3,26	15,97	$4s\,^1D—4p^{IV}\,^5P^\circ$	2—2
973,508	2	2,83	15,57	$4s\,^3D—4p'''\,^3D^\circ$	2—1
972,263	2	3,26	16,01	$4s\,^1D—4p^{IV}\,^5P^\circ$	2—3
968,037	25	2,72	15,53	$4s\,^3D—4p'''\,^3D^\circ$	3—3
966,231	3	3,26	16,09	$4s\,^1D—4p^{IV}\,^5P^\circ$	2—1
960,409	20	3,26	16,16	$4s\,^1D—4p'''\,^3P^\circ$	2—2
958,149	40	{ 3,26	16,20	$4s\,^1D—4p'''\,^1D^\circ$	2—2
		2,98	15,91	$4s\,^3D—4p^{IV}\,^5S^\circ$	1—2
956,286	25	2,98	15,94	$4s\,^3D—4p'''\,^3F^\circ$	1—2
955,321	5	3,26	16,23	$4s\,^1D—4p^{IV}\,^5D^\circ$	2—2, 1
954,378	20	3,26	16,25	$4s\,^1D—4p'''\,^1P^\circ$	2—3
951,413	5	—	—	—	—
947,700	2	2,83	15,91	$4s\,^3D—4p^{IV}\,^5S^\circ$	2—2
945,976	50	2,83	15,94	$4s\,^3D—4p'''\,^3F^\circ$	2—3

λ, Å	I	E_H, eV	E_B, eV	Transition	J
945,860	40	2,83	15,94	$4s\,{}^3D-4p'''\,{}^3F^\circ$	2—2
945,524	60	2,98	16,09	$4s\,{}^3D-4p^{IV}\,{}^5P^\circ$	1—1
943,328	60	2,83	15,97	$4s\,{}^3D-4p^{IV}\,{}^5P^\circ$	2—2
939,522	10	2,72	15,91	$4s\,{}^3D-4p^{IV}\,{}^5S^\circ$	3—2
937,814	5	2,72	15,94	$4s\,{}^3D-4p'''\,{}^3F^\circ$	3—3
935,892	60	2,72	15,97	$4s\,{}^3D-4p'''\,{}^3F^\circ$	3—4
935,35	20	2,83	16,09	$4s\,{}^3D-4p^{IV}\,{}^5P^\circ$	2—1
935,25	40	2,72	15,97	$4s\,{}^3D-4p^{IV}\,{}^5P^\circ$	3—2
935,074	60	2,98	16,23	$4s\,{}^3D-4p^{IV}\,{}^5D^\circ$	1—2, 1
932,940	60	2,72	16,01	$4s\,{}^3D-4p^{IV}\,{}^5P^\circ$	3—3
929,897	5	2,83	16,16	$4s\,{}^3D-4p'''\,{}^3P^\circ$	2—2
929,732	2	3,26	16,59	$4s\,{}^1D-4p^{V}\,{}^3P^\circ$	2—2
925,125	30	2,83	16,23	$4s\,{}^3D-4p^{IV}\,{}^5D^\circ$	2—2, 1
		3,26	16,66	$4s\,{}^1D-4p^{V}\,{}^3P^\circ$	2—1
924,239	50	2,83	16,25	$4s\,{}^3D-4p'''\,{}^1F^\circ$	2—3
922,411	20	3,26	16,70	$4s\,{}^1D-4p^{V}\,{}^3D^\circ$	2—2
922,017	60	2,72	16,16	$4s\,{}^3D-4p'''\,{}^3P^\circ$	3—2
917,303	20	2,72	16,23	$4s\,{}^3D-4p^{IV}\,{}^5D^\circ$	3—2
914,209	80	2,72	16,28	$4s\,{}^3D-4p^{IV}\,{}^5D^\circ$	3—3
912,414	3	3,26	16,84	$4s\,{}^1D-4f\,{}^3P^\circ$	2—1
910,518	15	2,98	16,59	$4s\,{}^3D-4p^{V}\,{}^3P^\circ$	1—2
906,109	40	2,98	16,66	$4s\,{}^3D-4p^{V}\,{}^3P^\circ$	1—1
901,071	60	2,83	16,59	$4s\,{}^3D-4p^{V}\,{}^3P^\circ$	2—2
899,791	50	2,98	16,75	$4s\,{}^3D-4p^{V}\,{}^3D^\circ$	1—1
		2,83	16,61	$4s\,{}^3D-4p^{V}\,{}^3D^\circ$	2—3
897,790	15	3,26	17,07	$4s\,{}^1D-4p'\,{}^3F^\circ$	2—2
896,970	40	2,98	16,80	$4s\,{}^3D-4p^{V}\,{}^3P^\circ$	1—0
896,753	60	2,83	16,66	$4s\,{}^3D-4p^{V}\,{}^3P^\circ$	2—1
894,226	40	2,83	16,70	$4s\,{}^3D-4p^{V}\,{}^3D^\circ$	2—2
893,674	80	2,72	16,59	$4s\,{}^3D-4p^{V}\,{}^3P^\circ$	3—2
892,411	50	2,72	16,61	$4s\,{}^3D-4p^{V}\,{}^3D^\circ$	3—3
890,567	60	2,83	16,75	$4s\,{}^3D-4p^{V}\,{}^3D^\circ$	2—1
886,946	60	2,72	16,70	$4s\,{}^3D-4p^{V}\,{}^3D^\circ$	3—2
886,515	10	2,98	16,96	$4s\,{}^3D-4p'\,{}^3D^\circ$	1—2
885,842	25	2,83	16,83	$4s\,{}^3D-4p'\,{}^3D^\circ$	2—3
884,824	5	2,83	16,84	$4s\,{}^3D-4f\,{}^3P^\circ$	2—1
884,430	8	3,26	17,27	$4s\,{}^1D-6p\,{}^3F^\circ$	2—3
884,127	10	2,83	16,86	$4s\,{}^3D-4p^{V}\,{}^1D^\circ$	2—2
883,837	5	2,83	16,86	$4s\,{}^3D-4f\,{}^3D^\circ$	2—3
883,282	5	2,98	17,01	$4s\,{}^3D-4p^{V}\,{}^1P^\circ$	1—1
880,325	5	2,83	16,92	$4s\,{}^3D-4p'\,{}^3F^\circ$	2—3
879,906	2	2,98	17,07	$4s\,{}^3D-4p'\,{}^3F^\circ$	1—2
878,696	50	2,72	16,83	$4s\,{}^3D-4p'\,{}^3D^\circ$	3—3
877,839	15	2,98	17,10	$4s\,{}^3D-4p'\,{}^3D^\circ$	1—1
877,559	20	2,83	16,96	$4s\,{}^3D-4p'\,{}^3D^\circ$	2—2
877,007	25	2,72	16,86	$4s\,{}^3D-4p^{V}\,{}^1D^\circ$	3—2
876,719	20	2,72	16,86	$4s\,{}^3D-4f\,{}^3D^\circ$	3—3
873,264	15	2,72	16,92	$4s\,{}^3D-4p'\,{}^3F^\circ$	3—3
871,064	8	2,83	17,07	$4s\,{}^3D-4p'\,{}^3F^\circ$	2—2
870,544	8	2,72	16,96	$4s\,{}^3D-4p'\,{}^3D^\circ$	3—2
869,336	25	2,98	17,24	$4s\,{}^3D-4p^{VI}\,{}^3F^\circ$	1—2
869,062	10	2,83	17,10	$4s\,{}^3D-4p'\,{}^3D^\circ$	2—1

λ, Å	I	E_H, eV	E_B, eV	Transition	J
867,726	8	2,98	17,26	$4s\,^3D-6p\,^3P^\circ$	1—1
866,440	5	2,83	17,14	$4s\,^3D-4f\,^3G^\circ$	2—3
865,383	40	2,83	17,16	$4s\,^3D-4p^{VI}\,^3F^\circ$	2—3
864,199	10	2,72	17,07	$4s\,^3D-4p'\,^3F^\circ$	3—2
		2,98	17,32	$4s\,^3D-6p\,^3D^\circ$	1—2
862,011	40	2,72	17,10	$4s\,^3D-4p^{VI}\,^3F^\circ$	3—4
858,482	25	2,83	17,27	$4s\,^3D-6p\,^3F^\circ$	2—3
		2,98	17,42	$4s\,^3D-4p^{IV}\,1^\circ$	1—1
		2,72	17,16	$4s\,^3D-4p^{VI}\,^3F^\circ$	3—3
855,701	10	2,83	17,32	$4s\,^3D-6p\,^3D^\circ$	2—2
855,474	5	2,83	17,32	$4s\,^3D-6p\,^3D^\circ$	2—3
852,898	3	2,98	17,51	$4s\,^3D-6p\,^3D^\circ$	1—1
851,76	2	2,72	17,27	$4s\,^3D-6p\,^3F^\circ$	3—3
851,300	25	2,72	17,28	$4s\,^3D-6p\,^3F^\circ$	3—4
850,76	2	2,98	17,55	$4s\,^3D-6p\,^1D^\circ$	1—2
849,354	3	2,98	17,57	$4s\,^3D-6p\,^3F^\circ$	1—2
848,806	15	2,72	17,32	$4s\,^3D-6p\,^3D^\circ$	3—3
844,910	5	2,83	17,51	$4s\,^3D-6p\,^1F^\circ$	2—3
844,616	3	2,83	17,51	$4s\,^3D-6p\,^3D^\circ$	2—1
842,483	3	2,83	17,55	$4s\,^3D-6p\,^1D^\circ$	2—2
841,105	2	2,83	17,57	$4s\,^3D-6p\,^3F^\circ$	2—2
826,995	30	0,00	14,99	$3d^{10}\,^1S-5p\,^3P^\circ$	0—1
823,800	2	2,83	17,88	$4s\,^3D-4p^{IV}\,3^\circ$	2—1
813,882	20	0,00	15,23	$3d^{10}\,^1S-5p\,^1P^\circ$	0—1
810,997	15	0,00	15,29	$3d^{10}\,^1S-5p\,^3D^\circ$	0—1
806,550	3	3,26	18,63	$4s\,^1D-4p^{VII}\,^3P^\circ$	2—2
797,452	10	0,00	15,55	$3d^{10}\,^1S-4p'''\,^1P^\circ$	0—1
779,300	8	2,72	18,63	$4s\,^3D-4p^{VII}\,^3P^\circ$	3—2
736,031	25	0,00	16,84	$3d^{10}\,^1S-4f\,^3P^\circ$	0—1
735,519	20	0,00	16,86	$3d^{10}\,^1S-4f\,^1P^\circ$	0—1
724,487	15	0,00	17,11	$3d^{10}\,^1S-4f\,^3D^\circ$	0—1
718,171	10	0,00	17,26	$3d^{10}\,^1S-6p\,^3P^\circ$	0—1
709,303	10	0,00	17,48	$3d^{10}\,^1S-6p\,^1P^\circ$	0—1
685,396	2	0,00	18,09	$3d^{10}\,^1S-5f\,^3P^\circ$	0—1
685,139	8	0,00	18,10	$3d^{10}\,^1S-5f\,^1P^\circ$	0—1
675,601	2	0,00	18,35	$3d^{10}\,^1S-5f\,^3D^\circ$	0—1

Cu III, **ground state $1s^2\,2s^2\,2p^6\,3s^2\,3p^6\,3d^9\,^2D_{5/2}$**

Ionization potential 297 100 cm^{-1}; 36,834 eV

λ, Å	I	E_H, eV	E_B, eV	Transition	J
2822,05	1	11,04	15,43	$a\,^2G-z\,^2G^\circ$	$^9/_2-^9/_2$
2812,96	5	11,04	15,72	$a\,^2G-z\,^4F^\circ$	$^9/_2-^7/_2$
2698,46	3	11,04	15,63	$a\,^2G-z\,^2G^\circ$	$^7/_2-^7/_2$
2696,39	6	11,04	15,63	$a\,^2G-z\,^2G^\circ$	$^9/_2-^7/_2$
2643,92	40	11,04	15,72	$a\,^2G-z\,^2F^\circ$	$^9/_2-^7/_2$
2641,54	8	11,04	15,73	$a\,^2G-z\,^2D^\circ$	$^7/_2-^5/_2$
2609,31	50	9,99	14,74	$a\,^4P-z\,^4D^\circ$	$^5/_2-^7/_2$
2522,36	25	11,04	15,95	$a\,^2G-z\,^2F^\circ$	$^7/_2-^5/_2$
2497,58	20	9,99	14,95	$a\,^4P-z\,^4D^\circ$	$^5/_2-^5/_2$
2482,34	30	9,99	14,95	$a\,^4P-z\,^4D^\circ$	$^3/_2-^5/_2$
2438,47	25	9,67	14,74	$b\,^2D-z\,^4D^\circ$	$^5/_2-^7/_2$
2412,32	15	9,97	15,11	$a\,^4P-z\,^4D^\circ$	$^1/_2-^3/_2$

λ, Å	I	E_H, eV	E_B, eV	Transition	J
2412,08	4	10,59	15,73	$a\,^2P-z\,^2D^\circ$	$^3/_2-^5/_2$
2405,49	20	9,96	15,11	$a\,^4P-z\,^4D^\circ$	$^3/_2-^3/_2$
2391,73	10	9,77	14,95	$b\,^2D-z\,^4D^\circ$	$^3/_2-^5/_2$
2368,15	20	9,97	15,20	$a\,^4P-z\,^4D^\circ$	$^1/_2-^1/_2$
2363,21	8	10,68	15,92	$a\,^2P-z\,^2D^\circ$	$^1/_2-^3/_2$
2361,56	10	9,96	15,20	$a\,^4P-z\,^4D^\circ$	$^3/_2-^1/_2$
2346,17	40	9,67	14,95	$b\,^2D-z\,^4D^\circ$	$^5/_2-^5/_2$
2320,28	8	9,77	15,11	$b\,^2D-z\,^4D^\circ$	$^3/_2-^3/_2$
2315,10	4	—	—	—	—
2312,31	5	10,59	15,95	$a\,^2P-z\,^2F^\circ$	$^3/_2-^5/_2$
2279,45	2	9,77	15,20	$b\,^2D-z\,^4D^\circ$	$^3/_2-^1/_2$
2277,43	4	9,67	15,11	$b\,^2D-z\,^4D^\circ$	$^5/_2-^3/_2$
2271,69	5	9,99	15,44	$a\,^4P-z\,^4F^\circ$	$^5/_2-^7/_2$
2077,81	2	9,77	15,73	$b\,^2D-z\,^2D^\circ$	$^3/_2-^5/_2$
2043,37	5	9,67	15,73	$b\,^2D-z\,^2D^\circ$	$^5/_2-^5/_2$
2000,78	3	11,04	17,23	$a\,^2G-y\,^2F^\circ$	$^9/_2-^7/_2$
1928,715	2	8,31	14,74	$a\,^2F-z\,^4D^\circ$	$^7/_2-^7/_2$
1882,250	2	10,68	17,27	$a\,^2P-z\,^2P^\circ$	$^1/_2-^1/_2$
1867,747	50	8,55	15,19	$a\,^2F-z\,^4G^\circ$	$^5/_2-^7/_2$
1858,685	1	11,04	17,71	$a\,^2G-y\,^4D^\circ$	$^9/_2-^7/_2$
1840,917	200	8,31	15,04	$a\,^2F-z\,^4G^\circ$	$^7/_2-^9/_2$
1826,339	10	10,59	17,38	$a\,^2P-z\,^2P^\circ$	$^3/_2-^3/_2$
1820,339	5	—	—	—	—
1798,761	5	8,55	15,44	$a\,^2F-z\,^4F^\circ$	$^5/_2-^7/_2$
1787,902	1	9,99	16,92	$a\,^4P-z\,^4P^\circ$	$^5/_2-^3/_2$
1783,935	5	9,99	16,94	$a\,^4P-z\,^4P^\circ$	$^5/_2-^5/_2$
1783,799	20	9,97	16,92	$a\,^4P-z\,^4P^\circ$	$^1/_2-^3/_2$
1780,062	5	9,96	16,92	$a\,^4P-z\,^4P^\circ$	$^3/_2-^3/_2$
1776,136	20	9,96	16,94	$a\,^4P-z\,^4P^\circ$	$^3/_2-^5/_2$
1773,697	1	10,68	17,67	$a\,^2P-y\,^4D^\circ$	$^1/_2-^3/_2$
1772,478	2	8,55	15,54	$a\,^2F-z\,^4F^\circ$	$^5/_2-^5/_2$
		10,68	17,67	$a\,^2P-y\,^4D^\circ$	$^1/_2-^1/_2$
1768,869	200	8,31	15,32	$a\,^2F-z\,^4F^\circ$	$^7/_2-^9/_2$
1766,219	2	9,97	16,99	$a\,^4P-z\,^4P^\circ$	$^1/_2-^1/_2$
1762,557	30	9,96	16,99	$a\,^4P-z\,^4P^\circ$	$^3/_2-^1/_2$
1761,155	20	8,55	15,59	$a\,^2F-z\,^4F^\circ$	$^5/_2-^3/_2$
1760,586	10	7,69	14,74	$a\,^4F-z\,^4D^\circ$	$^7/_2-^7/_2$
1755,012	20	10,59	17,66	$a\,^2P-y\,^4D^\circ$	$^3/_2-^5/_2$
1750,391	500	8,55	15,63	$a\,^2F-z\,^2G^\circ$	$^5/_2-^7/_2$
1741,378	500	8,31	15,43	$a\,^2F-z\,^2G^\circ$	$^7/_2-^9/_2$
1741,135	30	7,83	14,95	$a\,^4F-z\,^4D^\circ$	$^5/_2-^5/_2$
1739,508	300	11,04	18,17	$a\,^2G-z\,^2H^\circ$	$^7/_2-^9/_2$
1738,648	10	11,04	18,17	$a\,^2G-z\,^2H^\circ$	$^9/_2-^9/_2$
1738,145	30	9,99	17,12	$a\,^4P-y\,^2F^\circ$	$^5/_2-^5/_2$
1737,893	30	8,31	15,44	$a\,^2F-z\,^4F^\circ$	$^7/_2-^7/_2$
1732,998	5	9,77	16,92	$b\,^2D-z\,^4P^\circ$	$^3/_2-^3/_2$
1728,139	200	8,55	15,72	$a\,^2F-z\,^2F^\circ$	$^5/_2-^7/_2$
1726,275	5	8,55	15,73	$a\,^2F-z\,^2D^\circ$	$^5/_2-^5/_2$
1724,810	10	7,92	15,11	$a\,^4F-z\,^4D^\circ$	$^3/_2-^3/_2$
1722,379	1000	7,54	14,74	$a\,^4F-z\,^4D^\circ$	$^9/_2-^7/_2$
1717,134	5	—	—	—	—
1716,400	10	9,77	16,99	$b\,^2D-z\,^4P^\circ$	$^3/_2-^1/_2$
1713,346	5	8,31	15,54	$a\,^2F-z\,^4F^\circ$	$^7/_2-^5/_2$
1711,437	30	9,99	17,23	$a\,^4P-y\,^2F^\circ$	$^5/_2-^7/_2$
1711,257	30	9,99	17,23	$a\,^4P-y\,^2D^\circ$	$^5/_2-^3/_2$
1709,036	700	7,69	14,95	$a\,^4F-z\,^4D^\circ$	$^7/_2-^5/_2$
1708,958	200	9,67	16,92	$b\,^2D-z\,^4P^\circ$	$^5/_2-^3/_2$
1707,500	5	9,97	17,23	$a\,^4P-y\,^2D^\circ$	$^1/_2-^3/_2$
1705,633	400	11,04	18,30	$a\,^2G-z\,^2H^\circ$	$^9/_2-^{11}/_2$

λ, Å	I	E_H, eV	E_B, eV	Transition	J
1705,333	300	9,67	16,94	$b\,^2D-z\,^4P^\circ$	$5/2-5/2$
1704,072	10	9,96	17,23	$a\,^4P-y\,^2D^\circ$	$3/2-3/2$
1702,994	500	7,83	15,11	$a\,^4F-z\,^4D^\circ$	$5/2-3/2$
1702,349	30	10,68	17,96	$a\,^2P-x\,^2D^\circ$	$1/2-3/2$
1702,190	300	10,59	17,88	$a\,^2P-x\,^2D^\circ$	$3/2-5/2$
1702,102	400	7,92	15,20	$a\,^4F-z\,^4D^\circ$	$3/2-1/2$
1701,023	400	11,04	18,32	$a\,^2G-x\,^2F^\circ$	$9/2-7/2$
1696,202	15	9,96	17,27	$a\,^4P-z\,^2P^\circ$	$3/2-1/2$
1692,706	300	8,31	15,63	$a\,^2F-z\,^2G^\circ$	$7/2-7/2$
1689,051	200	9,99	17,33	$a\,^4P-y\,^2D^\circ$	$5/2-5/2$
1688,618	100	10,68	18,02	$a\,^2P-y\,^2P^\circ$	$1/2-3/2$
1687,134	600	7,69	15,04	$a\,^4F-z\,^4G^\circ$	$7/2-9/2$
1686,214	300	9,77	17,12	$b\,^2D-y\,^2F^\circ$	$3/2-5/2$
1684,642	500	7,83	15,19	$a\,^4F-z\,^4G^\circ$	$5/2-7/2$
1682,695	30	10,59	17,96	$a\,^2P-x\,^2D^\circ$	$3/2-3/2$
1682,044	10	9,96	17,33	$a\,^4P-y\,^2D^\circ$	$3/2-5/2$
1681,481	300	8,55	15,92	$a\,^2F-z\,^2D^\circ$	$5/2-3/2$
1679,151	400	7,92	15,30	$a\,^4F-z\,^4G^\circ$	$3/2-5/2$
1677,373	200	11,04	18,43	$a\,^2G-x\,^2F^\circ$	$7/2-5/2$
1676,469	15	9,99	17,38	$a\,^4P-z\,^2P^\circ$	$5/2-3/2$
1674,602	500	8,55	15,95	$a\,^2F-z\,^2F^\circ$	$5/2-5/2$
1671,886	500	8,31	15,72	$a\,^2F-z\,^2F^\circ$	$7/2-7/2$
1670,140	500	8,31	15,73	$a\,^2F-z\,^2D^\circ$	$7/2-5/2$
1669,273	10	10,59	18,02	$a\,^2P-y\,^2P^\circ$	$3/2-3/2$
1660,887	30	9,77	17,23	$b\,^2D-y\,^2D^\circ$	$3/2-3/2$
1658,472	200	7,83	15,30	$a\,^4F-z\,^4G^\circ$	$5/2-5/2$
1654,574	300	7,69	15,19	$a\,^4F-z\,^4G^\circ$	$7/2-7/2$
1653,399	10	9,77	17,27	$b\,^2D-z\,^2P^\circ$	$3/2-1/2$
1652,010	300	7,54	15,04	$a\,^4F-z\,^4G^\circ$	$9/2-9/2$
1651,758	15	10,68	18,18	$a\,^2P-y\,^2P^\circ$	$1/2-1/2$
1642,208	2000	7,54	15,09	$a\,^4F-z\,^4G^\circ$	$9/2-11/2$
1639,960	10	9,77	17,33	$b\,^2D-y\,^2D^\circ$	$3/2-5/2$
1638,956	300	9,67	17,23	$b\,^2D-y\,^2F^\circ$	$5/2-7/2$
1633,192	1	10,59	18,18	$a\,^2P-y\,^2P^\circ$	$3/2-1/2$
1629,301	1	7,69	15,30	$a\,^4F-z\,^4G^\circ$	$7/2-5/2$
1628,295	300	7,83	15,44	$a\,^4F-z\,^4F^\circ$	$5/2-7/2$
1628,088	50	9,77	17,38	$b\,^2D-z\,^2P^\circ$	$3/2-3/2$
1626,411	200	7,69	15,32	$a\,^4F-z\,^4F^\circ$	$7/2-9/2$
1626,139	200	7,92	15,54	$a\,^4F-z\,^4F^\circ$	$3/2-5/2$
1625,500	1	10,68	18,31	$a\,^2P-z\,^2S^\circ$	$1/2-1/2$
1621,723	3	8,31	15,95	$a\,^2F-z\,^2F^\circ$	$7/2-5/2$
1620,776	1	7,54	15,19	$a\,^4F-z\,^4G^\circ$	$9/2-7/2$
1618,408	5	9,67	17,33	$b\,^2D-y\,^2D^\circ$	$5/2-5/2$
1616,607	300	7,92	15,59	$a\,^4F-z\,^4F^\circ$	$3/2-3/2$
1616,160	15	9,99	17,66	$a\,^4P-y\,^4D^\circ$	$5/2-5/2$
1610,571	75	9,97	17,67	$a\,^4P-y\,^4D^\circ$	$1/2-3/2$
1609,757	100	9,96	17,66	$a\,^4P-y\,^4D^\circ$	$3/2-5/2$
1609,599	50	9,97	17,67	$a\,^4P-y\,^4D^\circ$	$1/2-1/2$
1607,542	100	{ 9,96	17,67	$a\,^4P-y\,^4D^\circ$	$3/2-3/2$
		10,59	18,31	$a\,^2P-z\,^2S^\circ$	$3/2-1/2$
1606,837	10	9,67	17,38	$b\,^2D-z\,^2P^\circ$	$5/2-3/2$
1606,730	300	7,83	15,54	$a\,^4F-z\,^4F^\circ$	$5/2-5/2$
1605,969	300	9,99	17,71	$a\,^4P-y\,^4D^\circ$	$5/2-7/2$
1603,146	400	7,69	15,43	$a\,^4F-z\,^2G^\circ$	$7/2-9/2$
1600,194	500	7,69	15,44	$a\,^4F-z\,^4F^\circ$	$7/2-7/2$
1597,418	10	7,83	15,59	$a\,^4F-z\,^4F^\circ$	$5/2-3/2$

λ, Å	I	E_H, eV	E_B, eV	Transition	J
1593,758	1000	7,54	15,32	$a\,{}^4F—z\,{}^4F^\circ$	$9/2—9/2$
1588,551	3	7,83	15,63	$a\,{}^4F—z\,{}^2G^\circ$	$5/2—7/2$
1579,353	15	7,69	15,54	$a\,{}^4F—z\,{}^4F^\circ$	$7/2—5/2$
1571,390	1	7,54	15,43	$a\,{}^4F—z\,{}^2G^\circ$	$9/2—9/2$
1571,154	3	9,77	17,66	$b\,{}^2D—y\,{}^4D^\circ$	$3/2—5/2$
1570,202	30	7,83	15,72	$a\,{}^4F—z\,{}^2F^\circ$	$5/2—7/2$
1568,655	2	7,83	15,73	$a\,{}^4F—z\,{}^2D^\circ$	$5/2—5/2$
1568,564	2	7,54	15,44	$a\,{}^4F—z\,{}^4F^\circ$	$9/2—7/2$
1565,194	5	9,96	17,88	$a\,{}^4P—x\,{}^2D^\circ$	$3/2—5/2$
1561,790	3	7,69	15,63	$a\,{}^4F—z\,{}^2G^\circ$	$7/2—7/2$
1549,203	10	7,92	15,92	$a\,{}^4F—z\,{}^2D^\circ$	$3/2—3/2$
1548,867	300	11,04	19,04	$a\,{}^2G—y\,{}^2G^\circ$	$7/2—7/2$
1544,110	2	11,04	19,07	$a\,{}^2G—y\,{}^2G^\circ$	$7/2—9/2$
1544,062	2	7,69	15,72	$a\,{}^4F—z\,{}^2F^\circ$	$7/2—7/2$
1543,438	500	11,04	19,07	$a\,{}^2G—y\,{}^2G^\circ$	$9/2—9/2$
1543,180	2	9,99	18,02	$a\,{}^4P—y\,{}^2P^\circ$	$5/2—3/2$
1542,562	2	7,69	15,73	$a\,{}^4F—z\,{}^2D^\circ$	$7/2—5/2$
1541,970	40	9,67	17,71	$b\,{}^2D—y\,{}^4D^\circ$	$5/2—7/2$
1531,588	1	7,83	15,92	$a\,{}^4F—z\,{}^2D^\circ$	$5/2—3/2$
1522,580	15	—	—	—	—
1520,546	20	—	—	—	—
1502,107	1	9,77	18,02	$b\,{}^2D—y\,{}^2P^\circ$	$3/2—3/2$
1486,904	10	9,99	18,32	$a\,{}^4P—x\,{}^2F^\circ$	$5/2—7/2$
1486,659	25	9,99	18,33	$a\,{}^4P—z\,{}^4S^\circ$	$5/2—3/2$
1484,010	5	9,67	18,02	$b\,{}^2D—y\,{}^2P^\circ$	$5/2—3/2$
1483,831	15	9,97	18,33	$a\,{}^4P—z\,{}^4S^\circ$	$1/2—3/2$
1481,243	20	9,96	18,33	$a\,{}^4P—z\,{}^4S^\circ$	$3/2—3/2$
1460,915	10	15,73	24,22	$z\,{}^2D^\circ—5$	$5/2—7/2$
1458,021	6	—	—	—	—
1455,200	3	15,95	24,47	$z\,{}^2F^\circ—e\,{}^2F$	$5/2—5/2$
1451,478	1	15,73	24,27	$z\,{}^2D^\circ—e\,{}^2F$	$5/2—7/2$
1450,165	3	15,72	24,27	$z\,{}^2F^\circ—e\,{}^2F$	$7/2—7/2$
1444,692	1	15,95	24,53	$z\,{}^2F^\circ—9$	$5/2—3/2,\ 5/2$
1440,446	3	15,63	24,24	$z\,{}^2G^\circ—6$	$7/2—7/2,\ 9/2$
1437,645	3	15,44	24,07	$z\,{}^4F^\circ—e\,{}^4F$	$7/2—7/2$
1436,994	15	15,43	24,06	$z\,{}^2G^\circ—3$	$9/2—7/2,\ 9/2$
1436,376	1	15,95	24,58	$z\,{}^2F^\circ—10$	$5/2—7/2$
1432,275	3	15,32	23,97	$z\,{}^4F^\circ—e\,{}^4F$	$9/2—9/2$
1431,901	3	9,67	18,32	$b\,{}^2D—x\,{}^2F^\circ$	$5/2—7/2$
1431,671	10	9,67	18,33	$b\,{}^2D—z\,{}^4S^\circ$	$5/2—3/2$
1430,969	3	9,77	18,43	$b\,{}^2D—x\,{}^2F^\circ$	$3/2—5/2$
1430,373	3	15,72	24,39	$z\,{}^2F^\circ—7$	$7/2—5/2$
1429,201	5	15,32	23,99	$z\,{}^4F^\circ—1$	$9/2—7/2,\ 9/2$
1428,081	5	8,55	17,23	$a\,{}^2F—y\,{}^2D^\circ$	$5/2—3/2$
1425,282	1	15,73	24,43	$z\,{}^2D^\circ—8$	$5/2—5/2,\ 7/2$
1425,079	2	15,54	24,24	$z\,{}^4F^\circ—e\,{}^4F$	$5/2—5/2$
1424,020	5	15,72	24,43	$z\,{}^2F^\circ—8$	$7/2—5/2,\ 7/2$
1423,504	10	—	—	—	—
1418,811	5	15,73	24,47	$z\,{}^2D^\circ—f\,{}^2F$	$5/2—7/2$
		15,32	24,06	$z\,{}^4F^\circ—3$	$9/2—7/2,\ 9/2$
1417,538	10	15,72	24,47	$z\,{}^2F^\circ—f\,{}^2F$	$7/2—7/2$
1417,124	2	15,32	24,07	$z\,{}^4F^\circ—e\,{}^4F$	$9/2—7/2$
1417,060	2	15,72	24,47	$z\,{}^2F^\circ—e\,{}^2F$	$7/2—5/2$
1415,478	1	15,63	24,39	$z\,{}^2G^\circ—7$	$7/2—5/2$
1414,431	3	15,59	24,35	$z\,{}^4F^\circ—e\,{}^4F$	$3/2—3/2$
1412,794	5	15,32	24,09	$z\,{}^4F^\circ—e\,{}^4G$	$9/2—11/2$

λ, Å	I	E_H, eV	E_B, eV	Transition	J
1412,724	5	15,44	24,22	$z\,^4F^\circ$—5	7/2—7/2
1409,248	1	15,44	24,24	$z\,^4F^\circ$—6	7/2—7/2, 9/2
		15,63	24,43	$z\,^2G^\circ$—8	7/2—5/2, 7/2
1408,536	1	15,44	24,24	$z\,^4F^\circ$—$e\,^4F$	7/2—5/2
1408,310	1	15,73	24,53	$z\,^2D^\circ$—9	5/2—3/2, 5/2
1407,196	3	15,54	24,35	$z\,^4F^\circ$—$e\,^4F$	5/2—3/2
1407,139	5	8,31	17,12	$a\,^2F$—$y\,^2F^\circ$	7/2—5/2
1403,763	1	8,55	17,38	$a\,^2F$—$z\,^2P^\circ$	5/2—3/2
1403,181	10	—	—	—	—
1402,917	1	15,63	24,47	$z\,^2G^\circ$—$f\,^2F$	7/2—7/2
1402,435	3	15,63	24,47	$z\,^2G^\circ$—$e\,^2F$	7/2—5/2
1402,250	5	—	—	—	—
1401,655	5	15,43	24,27	$z\,^2G^\circ$—$e\,^2F$	9/2—7/2
1401,602	5	—	—	—	—
1401,376	2	15,54	24,39	$z\,^4F^\circ$—7	5/2—5/2
1399,190	5	15,72	24,58	$z\,^2F^\circ$—10	7/2—7/2
1398,379	5	15,32	24,18	$z\,^4F^\circ$—4	9/2—7/2, 9/2
1396,417	1	15,19	24,07	$z\,^4G^\circ$—$e\,^4F$	7/2—7/2
1395,274	10	15,09	23,97	$z\,^4G^\circ$—$e\,^4F$	11/2—9/2
1390,306	10				
1389,528	5	8,31	17,23	$a\,^2F$—$y\,^2F^\circ$	7/2—7/2
		15,32	24,24	$z\,^4F^\circ$—6	9/2—7/2, 9/2
1388,276	1	15,04	23,97	$z\,^4G^\circ$—$e\,^4F$	9/2—9/2
1386,714	1	15,30	24,24	$z\,^4G^\circ$—$e\,^4F$	5/2—5/2
1385,921	3	15,59	24,53	$z\,^4F^\circ$—9	3/2—3/2, 5/2
1385,380	1	15,04	23,99	$z\,^4G^\circ$—1	9/2—7/2, 9/2
		15,44	24,39	$z\,^4F^\circ$—7	7/2—5/2
1384,929	3	15,63	24,58	$z\,^2G^\circ$—10	7/2—7/2
1384,840	5	—	—	—	—
1384,324	5	15,32	24,27	$z\,^4F^\circ$—$e\,^2F$	9/2—7/2
1382,561	5	—	—	—	—
1379,379	1	15,44	24,43	$z\,^4F^\circ$—8	7/2—5/2, 7/2
1378,238	1	15,19	24,18	$z\,^4G^\circ$—4	7/2—7/2, 9/2
1377,504	30	—	—	—	—
1376,807	30	15,09	24,09	$z\,^4G^\circ$—$e\,^4G$	11/2—11/2
1375,621	5	15,04	24,06	$z\,^4G^\circ$—3	9/2—7/2, 9/2
1374,758	3	8,31	17,33	$a\,^2F$—$y\,^2D^\circ$	7/2—5/2
1374,033	3	15,04	24,07	$z\,^4G^\circ$—$e\,^4F$	9/2—7/2
1372,899	5	15,19	24,22	$z\,^4G^\circ$—5	7/2—7/2
1371,144	10	15,43	24,47	$z\,^2G^\circ$—$f\,^2F$	9/2—7/2
1369,988	1	15,04	24,09	$z\,^4G^\circ$—$e\,^4G$	9/2—11/2
1369,612	5	15,19	24,24	$z\,^4G^\circ$—6	7/2—7/2, 9/2
1368,923	2	15,19	24,24	$z\,^4G^\circ$—$e\,^4F$	7/2—5/2
1367,646	5	—	—	—	—
1358,440	2	15,30	24,43	$z\,^4G^\circ$—8	5/2—5/2, 7/2
1356,424	5	15,04	24,18	$z\,^4G^\circ$—4	9/2—7/2, 9/2
1353,964	2	15,43	24,58	$z\,^2G^\circ$—10	9/2—7/2
1351,271	3	15,04	24,22	$z\,^4G^\circ$—5	9/2—7/2
1349,441	5	—	—	—	—
1348,077	1	15,04	24,24	$z\,^4G^\circ$—6	9/2—7/2, 9/2
1347,048	3	15,19	24,39	$z\,^4G^\circ$—7	7/2—5/2
1346,062	5	—	—	—	—
1343,730	5	—	—	—	—
1343,032	2	15,30	24,53	$z\,^4G^\circ$—9	5/2—3/2, 5/2
1342,193	3	14,74	23,97	$z\,^4D^\circ$—$e\,^4F$	7/2—9/2
1339,497	5	14,74	23,99	$z\,^4D^\circ$—1	7/2—7/2, 9/2
1337,572	5	—	—	—	—
1332,985	15	—	—	—	—

λ, Å	I	E_H, eV	E_B, eV	Transition	J
1330,365	2	14,74	24,06	$z\,^4D^\circ$—3	$^7/_2$—$^7/_2$, $^9/_2$
1327,178	5	—	—	—	—
1324,033	5	—	—	—	—
1318,582	2	7,83	17,23	$a\,^4F$—$y\,^2F^\circ$	$^5/_2$—$^7/_2$
1316,143	5	—	—	—	—
1312,400	10	14,74	24,18	$z\,^4D^\circ$—4	$^7/_2$—$^7/_2$, $^9/_2$
1307,595	3	14,95	24,43	$z\,^4D^\circ$—8	$^5/_2$—$^5/_2$, $^7/_2$
1271,839	2	7,92	17,67	$a\,^4F$—$y\,^4D^\circ$	$^3/_2$—$^3/_2$
1271,234	5	7,92	17,67	$a\,^4F$—$y\,^4D^\circ$	$^3/_2$—$^1/_2$
1259,937	10	7,83	17,67	$a\,^4F$—$y\,^4D^\circ$	$^5/_2$—$^3/_2$
1254,717	3	8,55	18,43	$a\,^2F$—$x\,^2F^\circ$	$^5/_2$—$^5/_2$
1244,377	10	7,69	17,66	$a\,^4F$—$y\,^4D^\circ$	$^7/_2$—$^5/_2$
1238,325	1	7,69	17,71	$a\,^4F$—$y\,^4D^\circ$	$^7/_2$—$^7/_2$
1237,776	3	8,31	18,32	$a\,^2F$—$x\,^2F^\circ$	$^7/_2$—$^7/_2$
1219,290	5	7,54	17,71	$a\,^4F$—$y\,^4D^\circ$	$^9/_2$—$^7/_2$
829,343	5	0,00	14,95	$a\,^2D$—$z\,^4D^\circ$	$^5/_2$—$^5/_2$
808,583	20	0,26	15,59	$a\,^2D$—$z\,^4F^\circ$	$^3/_2$—$^3/_2$
802,841	150	0,00	15,44	$a\,^2D$—$z\,^4F^\circ$	$^5/_2$—$^7/_2$
801,154	200	0,26	15,73	$a\,^2D$—$z\,^2D^\circ$	$^3/_2$—$^5/_2$
797,566	100	0,00	15,54	$a\,^2D$—$z\,^4F^\circ$	$^5/_2$—$^5/_2$
795,258	2	0,00	15,59	$a\,^2D$—$z\,^4F^\circ$	$^5/_2$—$^3/_2$
793,065	100	0,00	15,63	$a\,^2D$—$z\,^2G^\circ$	$^5/_2$—$^7/_2$
791,371	300	0,26	15,92	$a\,^2D$—$z\,^2D^\circ$	$^3/_2$—$^3/_2$
789,840	200	0,26	15,95	$a\,^2D$—$z\,^2F^\circ$	$^3/_2$—$^5/_2$
788,462	300	0,00	15,72	$a\,^2D$—$z\,^2F^\circ$	$^5/_2$—$^7/_2$
788,073	400	0,00	15,73	$a\,^2D$—$z\,^2D^\circ$	$^5/_2$—$^5/_2$
778,603	50	0,00	15,92	$a\,^2D$—$z\,^2D^\circ$	$^5/_2$—$^3/_2$
777,125	200	0,00	15,95	$a\,^2D$—$z\,^2F^\circ$	$^5/_2$—$^5/_2$
743,970	30	0,26	16,92	$a\,^2D$—$z\,^4P^\circ$	$^3/_2$—$^3/_2$
743,303	20	0,26	16,94	$a\,^2D$—$z\,^4P^\circ$	$^3/_2$—$^5/_2$
735,224	100	0,26	17,12	$a\,^2D$—$y\,^2F^\circ$	$^3/_2$—$^5/_2$
732,688	5	0,00	16,92	$a\,^2D$—$z\,^4P^\circ$	$^5/_2$—$^3/_2$
732,026	100	0,00	16,94	$a\,^2D$—$z\,^4P^\circ$	$^5/_2$—$^5/_2$
730,365	150	0,26	17,23	$a\,^2D$—$y\,^2D^\circ$	$^3/_2$—$^3/_2$
728,906	2	0,26	17,27	$a\,^2D$—$z\,^2P^\circ$	$^3/_2$—$^1/_2$
726,295	10	0,26	17,33	$a\,^2D$—$y\,^2D^\circ$	$^3/_2$—$^5/_2$
723,958	20	0,26	17,38	$a\,^2D$—$z\,^2P^\circ$	$^3/_2$—$^3/_2$
719,506	150	0,00	17,23	$a\,^2D$—$y\,^2F^\circ$	$^5/_2$—$^7/_2$
715,530	200	0,00	17,33	$a\,^2D$—$y\,^2D^\circ$	$^5/_2$—$^5/_2$
713,262	10	0,00	17,38	$a\,^2D$—$z\,^2P^\circ$	$^5/_2$—$^3/_2$
712,473	15	0,26	17,66	$a\,^2D$—$y\,^4D^\circ$	$^3/_2$—$^5/_2$
712,040	5	0,26	17,67	$a\,^2D$—$y\,^4D^\circ$	$^3/_2$—$^3/_2$
711,834	3	0,26	17,67	$a\,^2D$—$y\,^4D^\circ$	$^3/_2$—$^1/_2$
703,622	15	0,26	17,88	$a\,^2D$—$x\,^2D^\circ$	$^3/_2$—$^5/_2$
702,112	20	0,00	17,66	$a\,^2D$—$y\,^4D^\circ$	$^5/_2$—$^5/_2$
701,692	15	0,00	17,67	$a\,^2D$—$y\,^4D^\circ$	$^5/_2$—$^3/_2$
700,271	150	0,26	17,96	$a\,^2D$—$x\,^2D^\circ$	$^3/_2$—$^3/_2$
700,182	20	0,00	17,71	$a\,^2D$—$y\,^4D^\circ$	$^5/_2$—$^7/_2$
697,930	20	0,26	18,02	$a\,^2D$—$y\,^2P^\circ$	$^3/_2$—$^3/_2$
693,510	50	0,00	17,88	$a\,^2D$—$x\,^2D^\circ$	$^5/_2$—$^5/_2$
691,557	100	0,26	18,18	$a\,^2D$—$y\,^2P^\circ$	$^3/_2$—$^1/_2$
690,250	75	0,00	17,96	$a\,^2D$—$x\,^2D^\circ$	$^5/_2$—$^3/_2$
687,987	100	0,00	18,02	$a\,^2D$—$y\,^2P^\circ$	$^5/_2$—$^3/_2$
686,903	15	0,26	18,31	$a\,^2D$—$z\,^2S^\circ$	$^3/_2$—$^1/_2$
682,171	200	0,26	18,43	$a\,^2D$—$x\,^2F^\circ$	$^3/_2$—$^5/_2$
676,564	300	0,00	18,32	$a\,^2D$—$x\,^2F^\circ$	$^5/_2$—$^7/_2$
672,659	50	0,00	18,43	$a\,^2D$—$x\,^2F^\circ$	$^5/_2$—$^5/_2$

Unclassified Lines of Copper

λ, Å	I	Expected assignment
1411,69	30	Cu IV
476,201	20	
472,347	20	
467,106	15	
464,824	20	
464,640	20	
463,712	20	
459,881	15	
453,425	40	
453,130	20	
452,654	30	
451,152	25	
450,015	25	
448,420	25	
446,995	25	
444,999	15	
379,326	25	
377,756	15	
361,838	15	
361,220	25	
360,618	30	
359,873	50	
358,865	90	
357,897	100	
355,425	20	
353,031	15	
349,964	30	
348,413	15	
347,854	30	
346,004	60	
345,368	90	
342,713	80	
342,432	20	
341,483	25	
341,183	20	
339,887	30	
339,420	15	
338,314	25	
336,279	25	
335,919	25	
335,470	15	
335,016	20	
334,204	30	
333,562	30	
332,893	100	
329,851	30	
329,805	30	
329,047	100	
328,831	20	
328,737	40	
328,536	25	
328,412	50	
327,620	20	
327,383	15	
326,575	20	
325,687	20	
325,038	20	
324,607	50	
324,485	70	
323,816	60	
322,617	15	
312,505	20	
310,727	15	
310,380	20	
299,217	15	
298,901	20	
282,440	20	
281,744	40	
281,492	50	
278,128	20	
274,601	15	
273,417	20	
272,424	15	
271,443	20	
270,740	20	
270,298	30	
269,653	20	
269,044	40	
268,773	50	
268,309	60	
267,562	25	
266,061	20	
265,641	50	
264,414	20	
264,029	15	
263,760	30	
262,938	30	
262,442	20	
261,806	20	
261,606	30	
260,967	25	
260,245	25	
259,871	25	
259,558	20	
259,199	20	
258,927	80	
258,265	45	
258,004	15	
257,626	20	
257,315	20	
256,898	40	
256,365	30	
255,417	35	
255,214	45	
254,772	70	
254,510	50	
253,786	15	
253,465	15	
253,083	15	
252,780	75	
252,223	15	
251,947	20	
251,670	20	

λ, Å	I	Expected assignment	λ, Å	I	Expected assignment
251,278	20		209,648	30	
250,400	30		209,241	30	
249,415	15		208,902	60	
249,213	20		208,502	15	
248,426	50		207,925	35	
247,742	25		207,733	15	
241,583	15		207,282	20	
235,299	30		206,842	35	
225,497	25		206,355	60	
222,378	15		205,610	30	
217,743	20		205,278	60	
216,454	25		204,725	40	
216,063	50		204,056	15	
215,611	50		203,432	60	
214,206	20		203,010	60	
211,707	20		202,065	20	
211,109	50		201,615	20	
210,612	15		201,329	15	
210,217	15				

KRYPTON, Z = 36

Kr I, ground state $1s^2\,2s^2\,2p^6\,3s^2\,3p^6\,3d^{10}\,4s^2\,4p^6\ {}^1S_0$

Ionization potential 112915,2 cm⁻¹; 13,999 eV

λ, Å	I	E_H, eV	E_B, eV	Transition	J
25233,78	70	11,55	12,04	$5p\,[1^1/_2]-4d\,[^1/_2]^\circ$	2—1
24292,17	38	12,35	12,86	$4d\,[1^1/_2]^\circ-6p\,[^1/_2]$	1—0
24260,45	28	11,53	12,04	$5p\,[1^1/_2]-4d\,[^1/_2]^\circ$	1—1
23502,37	17	12,26	12,78	$4d\,[2^1/_2]^\circ-6p\,[2^1/_2]$	2—2
23340,44	65	12,28	12,82	$4d\,[2^1/_2]^\circ-6p\,[1^1/_2]$	3—2
22485,79	38	12,26	12,81	$4d\,[2^1/_2]^\circ-6p\,[1^1/_2]$	2—1
21900,51	2250	11,55	12,11	$5p\,[1^1/_2]-4d\,[1^1/_2]^\circ$	2—2
21165,46	319	11,53	12,11	$5p\,[1^1/_2]-4d\,[1^1/_2]^\circ$	1—2
20423,97	142	12,18	12,78	$4d\,[3^1/_2]^\circ-6p\,[2^1/_2]$	3—2
20419,00	1	12,82	13,42	$6p\,[1^1/_2]-6d\,[1^1/_2]^\circ$	2—2
20209,87	84	12,26	12,87	$5p'\,[^1/_2]-5d\,[^1/_2]^\circ$	0—1
18797,59	40	12,12	12,78	$4d\,[3^1/_2]^\circ-6p\,[2^1/_2]$	4—3
18787,73	10	12,14	12,80	$5p'\,[1^1/_2]-4d'\,[1^1/_2]^\circ$	2—2
18785,45	37	10,64	11,30	$5s'\,[^1/_2]^\circ-5p\,[^1/_2]$	1—1
18695,91	62	12,14	12,80	$5p'\,[^1/_2]-4d'\,[1^1/_2]^\circ$	1—2
18581,19	30	11,44	12,11	$5p\,[2^1/_2]-4d\,[1^1/_2]^\circ$	2—2
18418,82	4	12,11	12,78	$4d\,[1^1/_2]^\circ-6p\,[2^1/_2]$	2—3
18184,43	15	12,14	12,82	$5p'\,[1^1/_2]-4d'\,[2^1/_2]^\circ$	2—2
18167,12	1500	11,44	12,12	$5p\,[2^1/_2]-4d\,[3^1/_2]^\circ$	3—4
18098,46	10	12,14	12,82	$5p'\,[^1/_2]-4d'\,[2^1/_2]^\circ$	1—2
18001,71	400	11,67	12,35	$5p\,[^1/_2]-4d\,[1^1/_2]^\circ$	0—1
17842,70	270	11,30	12,00	$5p\,[^1/_2]-4d\,[^1/_2]^\circ$	1—0
17770,21	4	12,11	12,81	$4d\,[1^1/_2]^\circ-6p\,[1^1/_2]$	2—1
17630,44	4	12,10	12,80	$5p'\,[1^1/_2]-4d'\,[1^1/_2]^\circ$	1—2
17616,57	37	12,11	12,82	$4d\,[1^1/_2]^\circ-6p\,[1^1/_2]$	2—2
17404,67	32	11,55	12,26	$5p\,[1^1/_2]-4d'\,[1^1/_2]^\circ$	2—2
17367,98	360	12,14	12,86	$5p'\,[1^1/_2]-4d'\,[2^1/_2]^\circ$	2—3
17230,21	10	12,04	12,76	$4d\,[^1/_2]^\circ-6p\,[^1/_2]$	1—1
17098,76	300	12,10	12,82	$5p'\,[1^1/_2]-4d'\,[2^1/_2]^\circ$	1—2
17070,04	10	12,14	12,87	$5p'\,[1^1/_2]-5d\,[^1/_2]^\circ$	2—1
16994,36	10	12,14	12,87	$5p'\,[^1/_2]-5d\,[^1/_2]^\circ$	1—1
16935,71	800	11,53	12,26	$5p\,[1^1/_2]-4d\,[2^1/_2]^\circ$	1—2
16896,58	700	11,30	12,04	$5p\,[^1/_2]-4d\,[^1/_2]^\circ$	1—1
16890,40	1000	11,44	12,18	$5p\,[2^1/_2]-4d\,[3^1/_2]^\circ$	2—3
16853,45	480	11,44	12,18	$5p\,[2^1/_2]-4d\,[3^1/_2]^\circ$	3—3
16784,65	950	11,55	12,28	$5p\,[1^1/_2]-4d\,[2^1/_2]^\circ$	2—3
16726,48	70	10,56	11,30	$5s'\,[^1/_2]^\circ-5p\,[^1/_2]$	0—1
16573,10	16	12,26	13,00	$5p'\,[^1/_2]-4d'\,[1^1/_2]^\circ$	0—1
16465,29	15	12,38	13,14	$6s\,[1^1/_2]^\circ-4f\,[1^1/_2]$	1—2
16347,31	5	12,00	12,76	$4d\,[^1/_2]^\circ-6p\,[^1/_2]$	0—1
16315,58	12	12,38	13,14	$6s\,[1^1/_2]^\circ-4f\,[2^1/_2]$	1—2
16109,46	3	12,10	12,87	$5p'\,[1^1/_2]-5d\,[^1/_2]^\circ$	1—1
16052,31	2	12,04	12,81	$4d\,[^1/_2]^\circ-6p\,[1^1/_2]$	1—1
15925,64	6	12,04	12,82	$4d\,[^1/_2]^\circ-6p\,[1^1/_2]$	1—2
15890,52	25	12,26	13,03	$5p'\,[^1/_2]-6s'\,[^1/_2]^\circ$	0—1
15823,40	2	12,35	13,14	$4d\,[1^1/_2]^\circ-4f\,[1^1/_2]$	1—1
15820,10	35	12,35	13,14	$4d\,[1^1/_2]^\circ-4f\,[1^1/_2]$	1—2
15771,44	1	12,35	13,14	$6s\,[1^1/_2]^\circ-4f\,[1^1/_2]$	2—2
15680,94	75	12,35	13,14	$4d\,[1^1/_2]^\circ-4f\,[2^1/_2]$	1—2
15634,98	7	12,35	13,14	$6s\,[1^1/_2]^\circ-4f\,[2^1/_2]$	2—3
15474,02	65	10,64	11,44	$5s\,[^1/_2]^\circ-5p\,[2^1/_2]$	1—2
15433,63	4	12,10	12,90	$5p'\,[1^1/_2]-5d\,[^1/_2]^\circ$	1—0
15371,89	350	11,55	12,35	$5p\,[1^1/_2]-6s\,[1^1/_2]^\circ$	2—2

λ, Å	I	E_H, eV	E_B, eV	Transition	J
15335,29	850	11,30	12,11	$5p\,[^1/_2]-4d\,[1^1/_2]^\circ$	1—2
15326,87	35	11,55	12,35	$5p\,[1^1/_2]-4d\,[1^1/_2]^\circ$	2—1
15239,85	900	11,44	12,26	$5p\,[2^1/_2]-4d\,[2^1/_2]^\circ$	2—2
15209,52	42	11,44	12,26	$5p\,[2^1/_2]-4d\,[2^1/_2]^\circ$	3—2
15005,57	25	11,53	12,35	$5p\,[1^1/_2]-6s\,[1^1/_2]^\circ$	1—2
14973,74	8	12,04	12,86	$4d\,[^1/_2]^\circ-6p\,[^1/_2]$	1—0
14961,76	110	11,53	12,35	$5p\,[1^1/_2]-4d\,[1^1/_2]^\circ$	1—1
14765,64	230	11,55	12,38	$5p\,[1^1/_2]-6s\,[1^1/_2]^\circ$	2—1
14762,83	250	11,44	12,28	$5p\,[2^1/_2]-4d\,[2^1/_2]^\circ$	2—3
14734,46	900	11,44	12,28	$5p\,[2^1/_2]-4d\,[2^1/_2]^\circ$	3—3
14715,55	2	12,26	13,10	$5p'\,[^1/_2]-5d\,[1^1/_2]^\circ$	0—1
14469,33	30	12,28	13,14	$4d\,[2^1/_2]^\circ-4f\,[4^1/_2]$	3—4
14426,93	1100	11,53	12,38	$5p\,[1^1/_2]-6s\,[1^1/_2]^\circ$	1—1
14402,58	80	12,28	13,14	$4d\,[2^1/_2]^\circ-4f\,[2^1/_2]$	3—3
14401,35	30	12,28	13,14	$4d\,[2^1/_2]^\circ-4f\,[2^1/_2]$	3—2
		12,14	13,00	$5p'\,[1^1/_2]-4d'\,[1^1/_2]^\circ$	2—1
14347,82	400	12,28	13,15	$4d\,[2^1/_2]^\circ-4f\,[3^1/_2]$	3—3, 4
14341,25	9	12,14	13,01	$5p'\,[1^1/_2]-5d\,[2^1/_2]^\circ$	2—3
14156,62	15	12,14	13,02	$5p'\,[1^1/_2]-5d\,[1^1/_2]^\circ$	2—2
14104,27	40	12,14	13,02	$5p'\,[^1/_2]-5d\,[1^1/_2]^\circ$	1—2
13974,15	70	12,26	13,14	$4d\,[2^1/_2]^\circ-4f\,[2^1/_2]$	2—2
13939,13	85	12,14	13,03	$5p'\,[^1/_2]-6s'\,[^1/_2]^\circ$	1—0
13924,00	270	12,26	13,15	$4d\,[2^1/_2]^\circ-4f\,[3^1/_2]$	2—3
13882,64	240	12,14	13,03	$5p'\,[1^1/_2]-6s'\,[^1/_2]^\circ$	2—1
13832,57	50	12,14	13,03	$5p'\,[^1/_2]-6s'\,[^1/_2]^\circ$	1—1
13800,03	3	12,14	13,04	$5p'\,[^1/_2]-5d\,[2^1/_2]^\circ$	1—2
13763,72	6	12,14	13,04	$5p'\,[1^1/_2]-5d\,[2^1/_2]^\circ$	2—3
13738,86	400	10,64	11,55	$5s'\,[^1/_2]^\circ-5p\,[1^1/_2]$	1—2
13711,23	100	12,10	13,00	$5p'\,[1^1/_2]^\circ-4d'\,[1^1/_2]^\circ$	1—1
13658,38	360	11,44	12,35	$5p\,[2^1/_2]-6s\,[1^1/_2]^\circ$	2—2
13634,22	1700	11,44	12,35	$5p\,[2^1/_2]-6s\,[1^1/_2]^\circ$	2—3
13622,28	800	11,44	12,35	$5p\,[2^1/_2]-4d\,[1^1/_2]^\circ$	2—1
13337,52	55	12,10	13,03	$5p'\,[1^1/_2]-6s'\,[^1/_2]^\circ$	1—0
13304,30	5	12,35	13,28	$6s\,[1^1/_2]^\circ-7p\,[2^1/_2]$	2—3
13240,52	75	12,10	13,03	$5p'\,[1^1/_2]-6s'\,[^1/_2]^\circ$	1—1
13210,56	10	12,10	13,04	$5p'\,[1^1/_2]-5d\,[2^1/_2]^\circ$	1—2
13177,38	850	11,44	12,38	$5p\,[2^1/_2]-6s\,[1^1/_2]^\circ$	2—1
13022,05	15	12,86	13,81	$4d'\,[2^1/_2]^\circ-4f'\,[3^1/_2]$	3—3, 4
12985,08	12	11,30	12,26	$5p\,[^1/_2]-4d\,[2^1/_2]^\circ$	1—2
12977,98	2	12,14	13,10	$5p'\,[1^1/_2]-5d\,[1^1/_2]^\circ$	2—1
12934,48	1	12,14	13,10	$5p'\,[^1/_2]-5d\,[1^1/_2]^\circ$	1—1
12879,00	500	12,18	13,14	$4d\,[3^1/_2]^\circ-4f\,[4^1/_2]$	3—4
12861,89	55	10,56	11,53	$5s'\,[^1/_2]^\circ-5p\,[1^1/_2]$	0—1
12825,08	5	12,18	13,14	$4d\,[3^1/_2]^\circ-4f\,[2^1/_2]$	3—2
		12,18	13,14	$4d\,[3^1/_2]^\circ-4f\,[2^1/_2]$	3—3
12782,39	100	12,18	13,15	$4d\,[3^1/_2]^\circ-4f\,[3^1/_2]$	3—3, 4
12598,19	15	12,82	13,81	$4d'\,[2^1/_2]^\circ-4f\,[3^1/_2]$	2—3
12321,48	9	12,80	13,81	$4d'\,[1^1/_2]^\circ-4f'\,[2^1/_2]$	2—3
12240,81	2	12,28	13,29	$4d\,[2^1/_2]^\circ-7p\,[1^1/_2]$	3—2
12229,23	4	12,10	13,11	$5p'\,[1^1/_2]-7s\,[1^1/_2]^\circ$	1—1
12204,39	700	12,12	13,14	$4d\,[3^1/_2]^\circ-4f\,[4^1/_2]$	4—5
12156,97	2	12,12	13,14	$4d\,[3^1/_2]^\circ-4f\,[2^1/_2]$	4—3
12123,47	40	10,64	11,67	$5s'\,[^1/_2]^\circ-5p\,[^1/_2]$	1—0
12117,81	100	12,12	13,15	$4d\,[3^1/_2]^\circ-4f\,[3^1/_2]$	4—3, 4
12077,42	115	12,11	13,14	$4d\,[1^1/_2]^\circ-4f\,[1^1/_2]$	2—2
11997,08	480	12,11	13,14	$4d\,[1^1/_2]^\circ-4f\,[2^1/_2]$	2—3
11996,00	25	12,11	13,14	$4d\,[1^1/_2]^\circ-4f\,[2^1/_2]$	2—2

λ, Å	I	E_H, eV	E_B, eV	Transition	J
11819,43	2000	11,30	12,35	5p [1/2]—6s [1 1/2]°	1—2
11792,25	120	11,30	12,35	5p [1/2]—4d [1 1/2]°	1—1
11655,8	1	12,38	13,45	6s [1 1/2]°—5f [1 1/2]	1—2
11611,6	1	12,38	13,45	6s [1 1/2]°—5f [2 1/2]	1—2
11457,52	80	11,30	12,38	5p [1/2]—6s [1 1/2]°	1—1
11339,44	1	12,26	13,35	5p′ [1/2]—6d [1/2]°	0—1
11328,51	4	12,35	13,45	4d [1 1/2]°—5f [1 1/2]	1—2
11316,1	1	12,35	13,45	4d [1 1/2]°—5f [1 1/2]	1—1
11303,8	1	12,35	13,45	6s [1 1/2]°—5f [1 1/2]	2—2
11262,71	2	12,35	13,45	6s [1 1/2]°—5f [2 1/2]	2—2
11259,16	50	12,04	13,14	4d [1/2]°—4f [1 1/2]	1—1
11257,74	80	12,04	13,14	4d [1/2]°—4f [1 1/2]	1—2
11214,58	5	12,18	13,28	4d [3 1/2]°—7p [2 1/2]	3—2, 3
11187,13	40	12,04	13,14	4d [1/2]°—4f [2 1/2]	1—2
10874,92	100	12,00	13,14	4d [1/2]°—4f [1 1/2]	0—1
10729,43	2	12,11	13,27	4d [1 1/2]°—7p [1/2]	2—1
10699,33	20	12,12	13,28	4d [3 1/2]°—7p [2 1/2]	4—3
10647,63	1	12,28	13,45	4d [2 1/2]°—5f [1 1/2]	3—2
10626,70	8	12,28	13,45	4d [2 1/2]°—5f [4 1/2]	3—4
10608,43	20	12,28	13,45	4d [2 1/2]°—5f [2 1/2]	3—2
10593,01	100	12,28	13,45	4d [2 1/2]°—5f [3 1/2]	3—3, 4
10575,50	2	12,11	13,28	4d [1 1/2]°—7p [2 1/2]	2—2, 3
10486,29	2	12,11	13,29	4d [1 1/2]°—7p [1 1/2]	2—1
10458,56	6	12,11	13,29	4d [1 1/2]°—7p [1 1/2]	2—2
10374,44	10	12,26	13,45	4d [2 1/2]°—5f [2 1/2]	2—2
10360,37	100	12,26	13,45	4d [2 1/2]°—5f [3 1/2]	2—3, 4
10322,88	2	12,14	13,34	5p′ [1/2]—6d [1/2]°	1—0
10296,93	80	11,67	12,87	5p [1/2]—5d [1/2]°	0—1
10273,6	2	12,14	13,35	5p′ [1 1/2]—6d [1/2]°	2—1
10147,68	10	12,14	13,36	5p′ [1 1/2]—6d [1 1/2]°	2—2
10120,96	30	12,14	13,36	5p′ [1/2]—6d [1 1/2]°	1—2
10077,66	10	12,04	13,27	4d [1/2]°—7p [1/2]	1—1
10065,96	10	12,14	13,37	5p′ [1 1/2]—6d [3 1/2]°	2—3
10054,86	2	12,38	13,62	6s [1 1/2]°—6f [1 1/2]	1—2
10038,65	3	12,38	13,62	6s [1 1/2]°—6f [2 1/2]	1—2
9989,3	1	12,10	13,34	5p′ [1 1/2]—6d [1/2]°	1—0
9917,60	3	12,10	13,34	5p′ [1 1/2]—6d [1/2]°	1—1
9916,37	4	12,28	13,53	4d [2 1/2]°—8p [1 1/2]	3—2
9897,08	2	12,14	13,40	5p′ [1 1/2]—6d [2 1/2]°	2—3
9862,95	4	12,04	13,29	4d [1/2]°—7p [1 1/2]	1—1
9856,24	500	11,55	12,80	5p [1 1/2]—4d′ [1 1/2]°	2—2
9838,33	5	12,04	13,29	4d [1/2]°—7p [1 1/2]	1—2
9810,27	2	12,35	13,62	4d [1 1/2]°—6f [1 1/2]	1—2
9794,89	3	12,35	13,62	4d [1 1/2]°—6f [2 1/2]	1—2
9768,69	2	12,00	13,27	4d [1/2]°—7p [1/2]	0—1
9751,759	2000	10,03	11,30	5s [1 1/2]°—5p [1/2]	1—1
9743,11	50	12,18	13,45	4d [3 1/2]°—5f [4 1/2]	3—4
9727,51	2	12,18	13,45	4d [3 1/2]°—5f [2 1/2]	3—2
9722,78	1	12,26	13,53	4d [2 1/2]°—8p [1 1/2]	2—1
9714,85	15	12,18	13,45	4d [3 1/2]°—5f [3 1/2]	3—3, 4
9704,22	50	11,53	12,80	5p [1 1/2]—4d′ [1 1/2]°	1—2
9687,83	10	11,55	12,82	5p [1 1/2]—4d′ [2 1/2]°	2—2
9682,26	2	12,04	13,32	4d [1/2]°—7p [1/2]	1—0
9669,03	1	12,14	13,42	5p′ [1/2]—6d [1 1/2]°	1—1
9615,63	3	12,10	13,39	5p′ [1 1/2]—6d [2 1/2]°	1—2
9540,89	30	11,53	12,82	5p [1 1/2]—4d′ [2 1/2]°	1—2
9532,3	1	12,26	13,56	5p′ [1/2]—7d [1/2]°	0—1

λ, Å	I	E_H, eV	E_B, eV	Transition	J
9450,88	20	11,55	12,86	5p [1¹/₂]—4d' [2¹/₂]°	2—3
9362,03	100	11,55	12,87	5p [1¹/₂]—5d [¹/₂]°	2—1
9352,23	100	12,12	13,45	4d [3¹/₂]°—5f [4¹/₂]	4—5, 4
9337,9	1	12,12	13,45	4d [3¹/₂]°—5f [2¹/₂]	4—3
9326,03	10	12,12	13,45	4d [3¹/₂]°—5f [3¹/₂]	4—3, 4
9299,40	1	12,11	13,44	4d [1¹/₂]°—6p' [1¹/₂]	2—1
9287,87	1	12,28	13,62	4d [2¹/₂]°—6f [4¹/₂]	3—4
9279,9	2	12,28	13,62	4d [2¹/₂]°—6f [2¹/₂]	3—2, 3
9273,02	8	12,28	13,62	4d [2¹/₂]°—6f [3¹/₂]	3—3, 4
9270,96	10	12,11	13,45	4d [1¹/₂]°—5f [1¹/₂]	1—2
9262,69	1	12,11	13,45	4d [1¹/₂]°—5f [1¹/₂]	1—1
9243,54	30	—	—	—	—
9243,00	1	12,11	13,45	4d [1¹/₂]°—5f [2¹/₂]	1—2
9234,16	1	12,11	13,45	4d [1¹/₂]°—6p' [¹/₂]	1—1
9224,83	1	11,53	12,87	5p [1¹/₂]—5d [¹/₂]°	1—1
9188,69	2	12,11	13,46	4d [1¹/₂]°—6p' [1¹/₂]	1—2
9122,49	20	11,44	12,80	5p [2¹/₂]—4d' [1¹/₂]°	2—2
9111,69	20	11,44	12,80	5p [2¹/₂]—4d' [1¹/₂]°	3—2
9100,58	1	12,26	13,62	4d [2¹/₂]°—6f [2¹/₂]	2—2, 3
9094,33	4	12,26	13,62	4d [2¹/₂]°—6f [3¹/₂]	2—3, 4
9044,47	3	11,67	13,03	5p [¹/₂]—6s' [¹/₂]°	0—1
8999,19	30	11,53	12,90	5p [1¹/₂]—5d [¹/₂]°	1—0
8977,99	50	11,44	12,82	5p [2¹/₂]—4d' [2¹/₂]°	2—2
8967,53	10	11,44	12,82	5p [2¹/₂]—4d' [2¹/₂]°	3—2
8928,6920	2000	9,91	11,30	5s [1¹/₂]°—5p [¹/₂]	2—1
8870,32	4	—	—	—	—
8842,46	3	12,12	13,53	4d [3¹/₂]°—8p [2¹/₂]	4—3
8805,78	20	12,04	13,44	4d [¹/₂]°—6p' [1¹/₂]	1—1
8780,25	30	12,04	13,45	4d [¹/₂]°—5f [1¹/₂]	1—2
8776,7490	6000	10,03	11,44	5s [1¹/₂]°—5p [2¹/₂]	1—2
8774,05	50	11,44	12,86	5p [2¹/₂]—4d' [2¹/₂]°	2—3
8773,00	4	12,04	13,45	4d [¹/₂]°—5f [1¹/₂]	1—1
8764,112	150	11,44	12,86	5p [2¹/₂]—4d' [2¹/₂]°	3—3
8755,20	30	12,04	13,45	4d [¹/₂]°—5f [2¹/₂]	1—2
8747,29	2	12,04	13,45	4d [¹/₂]°—6p' [¹/₂]	1—1
8746,43	3	12,14	13,56	5p' [1¹/₂]—7d [1¹/₂]°	2—2
8742,49	1	12,26	13,67	5p' [¹/₂]—8d [¹/₂]°	0—1
8726,54	8	12,14	13,56	5p' [¹/₂]—7d [1¹/₂]°	1—2
8722,17	1	12,11	13,53	4d [1¹/₂]°—8p [1¹/₂]	2—1
8713,62	2	12,11	13,53	4d [1¹/₂]°—8p [1¹/₂]	2—2
8697,50	40	11,44	12,87	5p [2¹/₂]—5d [¹/₂]°	2—1
8673,48	2	12,14	13,57	5p' [1¹/₂]—7d [3¹/₂]°	2—3
8651,49	8	11,67	13,10	5p [¹/₂]—5d [1¹/₂]°	0—1
8632,81	1	12,28	13,72	4d [2¹/₂]°—7f [4¹/₂]	3—4, 5
8631,5	1	12,14	13,58	5p' [1¹/₂]—7d [2¹/₂]°	2—2
8628,70	1	12,28	13,72	4d [2¹/₂]°—7f [2¹/₂]	3—3, 2
8624,82	4	12,28	13,72	4d [2¹/₂]°—7f [3¹/₂]	3—3, 4
8610,67	5	12,18	13,62	4d [3¹/₂]°—6f [1¹/₂]	3—2
8605,85	40	12,18	13,62	4d [3¹/₂]°—6f [4¹/₂]	3—4
8599,4	1	12,18	13,62	4d [3¹/₂]°—6f [2¹/₂]	3—2
8593,1	10	12,18	13,62	4d [3¹/₂]°—6f [3¹/₂]	3—3, 4
8569,02	20	12,00	13,44	4d [¹/₂]°—6p' [1¹/₂]	0—1
8560,89	50	11,67	13,11	5p [¹/₂]—7s [1¹/₂]°	0—1
8537,93	40	12,00	13,45	4d [¹/₂]°—5f [1¹/₂]	0—1
8508,8700	3000	10,64	12,10	5s' [¹/₂]°—5p' [1¹/₂]	1—1
8498,21	30	11,55	13,00	5p [1¹/₂]—4d' [1¹/₂]°	2—1
8477,20	2	11,55	13,01	5p [1¹/₂]—5d [3¹/₂]°	2—3
8469,96	2	12,26	13,72	4d [2¹/₂]°—7f [3¹/₂]	2—3, 4

λ, Å	I	E_H, eV	E_B, eV	Transition	J
8412,428	100	11,55	13,02	$5p\ [1^1/_2]-5d\ [1^1/_2]^\circ$	2—2
8384,90	15	11,53	13,00	$5p\ [1^1/_2]-4d'\ [1^1/_2]^\circ$	1—1
8375,93	5	—	—	—	—
8332,73	1	12,04	13,52	$4d\ [^1/_2]^\circ-8p\ [^1/_2]$	1—1
8321,09	2	—	—	—	—
8303,20	10	11,55	13,04	$5p\ [1^1/_2]-5d\ [2^1/_2]^\circ$	2—2
8301,39	20	11,53	13,02	$5p\ [1^1/_2]-5d\ [1^1/_2]^\circ$	1—2
8298,1077	5000	10,03	11,53	$5s\ [1^1/_2]^\circ-5p\ [1^1/_2]$	1—1
8287,56	4	12,12	13,62	$4d\ [3^1/_2]^\circ-6f\ [3^1/_2]$	4—3, 4
8281,0495	1500	10,64	12,14	$5s'\ [^1/_2]^\circ-5p'\ [^1/_2]$	1—1
8272,355	100	11,55	13,04	$5p\ [1^1/_2]-5d\ [2^1/_2]^\circ$	2—3
8263,2398	3000	10,64	12,14	$5s'\ [^1/_2]^\circ-5p'\ [1^1/_2]$	1—2
8228,89	10	12,11	13,62	$4d\ [1^1/_2]^\circ-6f\ [1^1/_2]$	2—2
8222,69	6	12,14	13,65	$5p'\ [1^1/_2]-5d'\ [1^1/_2]^\circ$	2—2
8218,40	80	12,11	13,62	$4d\ [1^1/_2]^\circ-6f\ [2^1/_2]$	2—3
8212,24	5	12,11	13,62	$4d\ [1^1/_2]^\circ-6f\ [3^1/_2]$	2—3
8210,1	1	12,14	13,65	$5p'\ [1^1/_2]-5d'\ [2^1/_2]^\circ$	2—2
8206,62	40	11,53	13,03	$5p\ [1^1/_2]-6s'\ [^1/_2]^\circ$	1—1
8205,22	20	12,14	13,65	$5p'\ [^1/_2]-5d'\ [1^1/_2]^\circ$	1—2
8195,070	50	11,53	13,04	$5p\ [1^1/_2]-5d\ [2^1/_2]^\circ$	1—2
8192,4	2	12,14	13,65	$5p'\ [^1/_2]-5d'\ [2^1/_2]^\circ$	1—2
8190,0543	3000	10,03	11,55	$5s\ [1^1/_2]^\circ-5p\ [1^1/_2]$	1—2
8144,96	15	11,30	12,82	$5p\ [^1/_2]-4d'\ [2^1/_2]^\circ$	1—2
8132,98	60	12,14	13,67	$5p'\ [1^1/_2]-5d'\ [2^1/_2]^\circ$	2—3
8112,900	6000	9,91	11,44	$5s\ [1^1/_2]^\circ-5p\ [2^1/_2]$	2—3
8104,3642	4000	9,91	11,44	$5s\ [1^1/_2]^\circ-5p\ [2^1/_2]$	2—2
8104,02	500	11,44	12,97	$5p\ [2^1/_2]-5d\ [3^1/_2]^\circ$	3—4
8059,5038	1500	10,56	12,10	$5s'\ [^1/_2]^\circ-5p'\ [1^1/_2]$	0—1
8040,50	8	12,18	13,72	$4d\ [3^1/_2]^\circ-7f\ [4^1/_2]$	3—4
8033,52	2	12,18	13,72	$4d\ [3^1/_2]^\circ-7f\ [3^1/_2]$	3—3, 4
7993,12	5	12,10	13,73	$5p'\ [1^1/_2]-5d'\ [1^1/_2]^\circ$	1—1
7990,78	2	—	—	—	—
7982,406	100	11,55	13,10	$5p\ [1^1/_2]-7s\ [1^1/_2]^\circ$	2—2
7981,82	30	11,55	13,10	$5p\ [1^1/_2]-5d\ [1^1/_2]^\circ$	2—1
7981,19	20	12,10	13,65	$5p'\ [1^1/_2]-5d'\ [2^1/_2]^\circ$	1—2
7962,62	1	12,11	13,67	$4d\ [1^1/_2]^\circ-9p\ [1^1/_2]$	2—2
7957,67	2	12,14	13,70	$5p'\ [1^1/_2]-8d\ [2^1/_2]^\circ$	2—3
7946,99	20	11,44	13,00	$5p\ [2^1/_2]-4d\ [1^1/_2]^\circ$	2—1
7938,34	2	12,14	13,70	$5p'\ [^1/_2]-8d\ [1^1/_2]^\circ$	1—2
7928,5996	180	11,44	13,01	$5p\ [2^1/_2]-5d\ [3^1/_2]^\circ$	2—3
7920,47	40	11,44	13,01	$5p\ [2^1/_2]-5d\ [3^1/_2]^\circ$	3—3
7913,4242	200	11,30	12,87	$5p\ [^1/_2]-5d\ [^1/_2]^\circ$	1—1
7904,62	30	11,55	13,11	$5p\ [1^1/_2]-7s\ [1^1/_2]^\circ$	2—1
7882,36	10	11,53	13,10	$5p\ [1^1/_2]-7s\ [1^1/_2]^\circ$	1—2
7881,76	30	11,53	13,10	$5p\ [1^1/_2]-5d\ [1^1/_2]^\circ$	1—1
7871,93	2	11,44	13,02	$5p\ [2^1/_2]-5d\ [1^1/_2]^\circ$	2—2
7863,91	20	11,44	13,02	$5p\ [2^1/_2]-5d\ [1^1/_2]^\circ$	3—2
7854,8215	800	10,56	12,14	$5s'\ [^1/_2]^\circ-5p'\ [^1/_2]$	0—1
7840,40	4	12,04	13,62	$4d\ [^1/_2]^\circ-6f\ [1^1/_2]$	1—1
7840,01	8	12,04	13,62	$4d\ [^1/_2]^\circ-6f\ [1^1/_2]$	1—2
7830,21	2	12,04	13,62	$4d\ [^1/_2]^\circ-6f\ [2^1/_2]$	1—2
7806,52	15	11,53	13,11	$5p\ [1^1/_2]-7s\ [1^1/_2]^\circ$	1—1
7786,66	2	11,44	13,03	$5p\ [2^1/_2]-6s'\ [^1/_2]^\circ$	2—1
7776,28	15	11,44	13,04	$5p\ [2^1/_2]-5d\ [2^1/_2]^\circ$	2—2
7772,40	5	12,12	13,72	$4d\ [3^1/_2]^\circ-7f\ [4^1/_2]$	4—4, 5
7768,43	5	11,44	13,04	$5p\ [2^1/_2]-5d\ [2^1/_2]^\circ$	3—2
7765,89	1	12,12	13,72	$4d\ [3^1/_2]^\circ-7f\ [3^1/_2]$	4—3, 4

λ, Å	I	E_H, eV	E_B, eV	Transition	J
7749,16	3	11,44	13,04	$5p\ [2^1/_2]-5d\ [2^1/_2]^\circ$	2—3
7746,828	50	11,30	12,90	$5p\ [^1/_2]-5d\ [^1/_2]^\circ$	1—0
7741,39	10	11,44	13,04	$5p\ [2^1/_2]-5d\ [2^1/_2]^\circ$	3—3
7712,94	1	12,18	13,79	$4d\ [3^1/_2]^\circ-8f\ [4^1/_2]$	3—4
7708,96	1	12,11	13,72	$4d\ [1^1/_2]^\circ-7f\ [1^1/_2]$	2—2
7703,41	2	12,11	13,72	$4d\ [1^1/_2]^\circ-7f\ [2^1/_2]$	2—3
7694,5393	500	9,91	11,53	$5s\ [1^1/_2]^\circ-5p\ [1^1/_2]$	2—1
7685,2460	400	10,64	12,26	$5s'\ [^1/_2]^\circ-5p'\ [^1/_2]$	1—0
7652,16	4	12,00	13,62	$4d\ [^1/_2]^\circ-6f\ [1^1/_2]$	0—1
7601,5443	2000	9,91	11,55	$5s\ [1^1/_2]^\circ-5p\ [1^1/_2]$	2—2
7587,4130	1000	10,03	11,67	$5s\ [1^1/_2]^\circ-5p\ [^1/_2]$	1—0
7550,63	3	—	—	—	—
7543,10	3	—	—	—	—
7494,15	30	11,44	13,10	$5p\ [2^1/_2]-7s\ [1^1/_2]^\circ$	2—2
7493,58	20	11,44	13,10	$5p\ [2^1/_2]-5d\ [1^1/_2]^\circ$	2—1
7486,862	100	11,44	13,10	$5p\ [2^1/_2]-7s\ [1^1/_2]^\circ$	3—2
7465,01	3	12,12	13,79	$4d\ [3^1/_2]^\circ-7f\ [4^1/_2]$	4—4, 5
7459,70	1	12,12	13,79	$4d\ [3^1/_2]^\circ-7f\ [3^1/_2]$	4—3, 4
7425,54	60	11,44	13,11	$5p\ [2^1/_2]-7s\ [1^1/_2]^\circ$	2—1
7402,70	1	12,11	13,79	$4d\ [1^1/_2]^\circ-7f\ [2^1/_2]$	2—2
7367,02	2	12,04	13,72	$4d\ [^1/_2]^\circ-7f\ [1^1/_2]$	1—1
7366,80	2	12,04	13,72	$4d\ [^1/_2]^\circ-7f\ [1^1/_2]$	1—2
7362,83	4	—	—	—	—
7361,34	1	12,04	13,72	$4d\ [^1/_2]^\circ-7f\ [2^1/_2]$	1—2
7359,96	5	11,67	13,35	$5p\ [^1/_2]-6d\ [^1/_2]^\circ$	0—1
7355,48	4	—	—	—	—
7341,16	2	—	—	—	—
7334,33	4	—	—	—	—
7327,00	5	—	—	—	—
7301,25	5	12,11	13,81	$4d\ [1^1/_2]^\circ-4f'\ [2^1/_2]$	2—3
7287,262	80	11,30	13,00	$5p\ [^1/_2]-4d'\ [1^1/_2]^\circ$	1—1
7268,28	1	12,12	13,83	$4d\ [3^1/_2]^\circ-9f\ [4^1/_2]$	4—5, 4
7234,58	2	—	—	—	—
7227,34	2	—	—	—	—
7224,103	100	11,30	13,02	$5p\ [^1/_2]-5d\ [1^1/_2]^\circ$	1—2
7200,59	2	12,00	13,72	$4d\ [^1/_2]^\circ-7f\ [1^1/_2]$	0—1
7180,47	3	—	—	—	—
7152,21	5	11,30	13,03	$5p\ [^1/_2]-6s'\ [^1/_2]^\circ$	1—1
7143,45	8	11,30	13,04	$5p\ [^1/_2]-5d\ [2^1/_2]^\circ$	1—2
7133,67	1	12,12	13,86	$4d\ [3^1/_2]^\circ-10f\ [4^1/_2]$	4—5, 4
7089,51	1	12,04	13,78	$4d\ [^1/_2]^\circ-8f\ [1^1/_2]$	1—1
7086,43	1	12,04	13,79	$4d\ [^1/_2]^\circ-8f\ [2^1/_2]$	1—2
7057,27	10	11,67	13,42	$5p\ [^1/_2]-6d\ [1^1/_2]^\circ$	0—1
7008,62	2	—	—	—	—
7001,62	2	—	—	—	—
7000,79	7	11,67	13,44	$5p\ [^1/_2]-8s\ [1^1/_2]^\circ$	0—1
6993,05	2	12,04	13,81	$4d\ [^1/_2]^\circ-4f'\ [2^1/_2]$	1—2
6935,38	2	12,00	13,78	$4d\ [^1/_2]^\circ-8f\ [1^1/_2]$	0—1
6911,29	2	12,04	13,83	$4d\ [^1/_2]^\circ-9f\ [1^1/_2]$	1—1
6904,68	100	11,30	13,10	$5p\ [^1/_2]-7s\ [1^1/_2]^\circ$	1—2
6904,22	15	11,30	13,10	$5p\ [^1/_2]-5d\ [1^1/_2]^\circ$	1—1
6869,63	20	11,55	13,35	$5p\ [1^1/_2]-6d\ [^1/_2]^\circ$	2—1
6862,82	3	—	—	—	—
6853,32	2	—	—	—	—
6846,40	20	11,30	13,11	$5p\ [^1/_2]-7s\ [1^1/_2]^\circ$	1—1
6829,09	8	11,53	13,34	$5p\ [1^1/_2]-6d\ [^1/_2]^\circ$	1—0
6813,10	50	11,55	13,36	$5p\ [1^1/_2]-6d\ [1^1/_2]^\circ$	2—2
6795,40	4	11,53	13,35	$5p\ [1^1/_2]-6b\ [^1/_2]^\circ$	1—1

λ, Å	I	E_H, eV	E_B, eV	Transition	J
6789,21	1	12,04	13,86	$4d\ [{}^1/_2]^\circ - 10f\ [1{}^1/_2]$	1—1, 2
6776,15	3	11,55	13,37	$5p\ [1{}^1/_2] - 6d\ [3{}^1/_2]^\circ$	2—3
6764,51	2	12,00	13,83	$4d\ [{}^1/_2]^\circ - 9f\ [1{}^1/_2]$	0—1
6740,10	20	11,53	13,36	$5p\ [1{}^1/_2] - 6d\ [1{}^1/_2]^\circ$	1—2
6723,36	4	11,55	13,39	$5p\ [1{}^1/_2] - 6d\ [2{}^1/_2]^\circ$	2—2
6699,228	60	11,55	13,40	$5p\ [1{}^1/_2] - 6d\ [2{}^1/_2]^\circ$	2—3
6652,239	40	11,53	13,39	$5p\ [1{}^1/_2] - 6d\ [2{}^1/_2]^\circ$	1—2
6647,94	2	12,00	13,86	$4d\ [{}^1/_2]^\circ - 10f\ [1{}^1/_2]$	0—1
6612,38	2	—	—	—	—
6605,12	2	11,55	13,42	$5p\ [1{}^1/_2] - 6d\ [1{}^1/_2]^\circ$	2—1
6576,42	20	11,55	13,43	$5p\ [1{}^1/_2] - 8s\ [1{}^1/_2]^\circ$	2—2
6555,69	6	11,55	13,44	$5p\ [1{}^1/_2] - 8s\ [1{}^1/_2]^\circ$	2—1
6555,56	2	11,67	13,56	$5p\ [{}^1/_2] - 7d\ [{}^1/_2]^\circ$	0—1
6536,55	8	11,53	13,42	$5p\ [1{}^1/_2] - 6d\ [1{}^1/_2]^\circ$	1—1
6508,37	3	11,53	13,43	$5p\ [1{}^1/_2] - 8s\ [1{}^1/_2]^\circ$	1—2
6504,89	10	11,44	13,35	$5p\ [2{}^1/_2] - 6d\ [{}^1/_2]^\circ$	2—1
6488,07	15	11,53	13,44	$5p\ [1{}^1/_2] - 8s\ [1{}^1/_2]^\circ$	1—1
6456,2910	200	11,44	13,36	$5p\ [2{}^1/_2] - 6d\ [3{}^1/_2]^\circ$	3—4
6454,19	1	11,44	13,36	$5p\ [2{}^1/_2] - 6d\ [1{}^1/_2]^\circ$	2—2
6448,78	10	11,44	13,36	$5p\ [2{}^1/_2] - 6d\ [1{}^1/_2]^\circ$	3—2
6421,0283	100	11,44	13,37	$5p\ [2{}^1/_2] - 6d\ [3{}^1/_2]^\circ$	2—3
6415,65	20	11,44	13,37	$5p\ [2{}^1/_2] - 6d\ [3{}^1/_2]^\circ$	3—3
6410,17	5	11,67	13,60	$5p\ [{}^1/_2] - 7d\ [1{}^1/_2]^\circ$	0—1
6373,58	30	11,44	13,39	$5p\ [2{}^1/_2] - 6d\ [2{}^1/_2]^\circ$	2—2
6373,19	1	11,67	13,61	$5p\ [{}^1/_2] - 9s\ [1{}^1/_2]^\circ$	0—1
6368,26	4	11,44	13,39	$5p\ [2{}^1/_2] - 6d\ [2{}^1/_2]^\circ$	3—2
6351,90	8	11,44	13,40	$5p\ [2{}^1/_2] - 6d\ [2{}^1/_2]^\circ$	2—3
6346,66	20	11,44	13,40	$5p\ [2{}^1/_2] - 6d\ [2{}^1/_2]^\circ$	3—3
6267,33	2	11,44	13,42	$5p\ [2{}^1/_2] - 6d\ [1{}^1/_2]^\circ$	2—1
6241,39	10	11,44	13,43	$5p\ [2{}^1/_2] - 8s\ [1{}^1/_2]^\circ$	2—2
6236,3520	30	11,44	13,43	$5p\ [2{}^1/_2] - 8s\ [1{}^1/_2]^\circ$	3—2
6222,71	20	11,44	13,44	$5p\ [2{}^1/_2] - 8s\ [1{}^1/_2]^\circ$	2—1
6172,08	2	11,67	13,67	$5p\ [{}^1/_2] - 8d\ [{}^1/_2]^\circ$	0—1
6163,65	7	11,55	13,56	$5p\ [1{}^1/_2] - 7d\ [{}^1/_2]^\circ$	2—1
6151,38	20	11,55	13,56	$5p\ [1{}^1/_2] - 7d\ [1{}^1/_2]^\circ$	2—2
6115,23	3	11,55	13,57	$5p\ [1{}^1/_2] - 7d\ [3{}^1/_2]^\circ$	2—3
6108,34	3	11,53	13,55	$5p\ [1{}^1/_2] - 7d\ [{}^1/_2]^\circ$	1—0
6103,86	1	11,53	13,56	$5p\ [1{}^1/_2] - 7d\ [{}^1/_2]^\circ$	1—1
6094,31	2	11,55	13,58	$5p\ [1{}^1/_2] - 7d\ [2{}^1/_2]^\circ$	2—2
6091,81	6	11,53	13,56	$5p\ [1{}^1/_2] - 7d\ [1{}^1/_2]^\circ$	1—2
6088,00	2	11,67	13,70	$5p\ [{}^1/_2] - 8d\ [1{}^1/_2]^\circ$	0—1
6082,8630	40	11,30	13,34	$5p\ [{}^1/_2] - 6d\ [{}^1/_2]^\circ$	1—0
6075,24	20	11,55	13,59	$5p\ [1{}^1/_2] - 7d\ [2{}^1/_2]^\circ$	2—3
6056,1280	60	11,30	13,35	$5p\ [{}^1/_2] - 6d\ [{}^1/_2]^\circ$	1—1
6049,35	3	11,67	13,71	$5p\ [{}^1/_2] - 10s\ [1{}^1/_2]^\circ$	0—1
6035,82	15	11,53	13,58	$5p\ [1{}^1/_2] - 7d\ [2{}^1/_2]^\circ$	2—2
6012,1570	50	11,55	13,61	$5p\ [1{}^1/_2] - 9s\ [1{}^1/_2]^\circ$	2—2
		11,30	13,36	$5p\ [{}^1/_2] - 6d\ [1{}^1/_2]^\circ$	1—2
6002,19	3	11,55	13,61	$5p\ [1{}^1/_2] - 9s\ [1{}^1/_2]^\circ$	2—1
5993,8506	60	10,03	12,10	$5s\ [1{}^1/_2]^\circ - 5p'\ [1{}^1/_2]$	1—1
5977,65	4	11,53	13,60	$5p\ [1{}^1/_2] - 7d\ [1{}^1/_2]^\circ$	1—1
5955,14	2	11,53	13,61	$5p\ [1{}^1/_2] - 9s\ [1{}^1/_2]^\circ$	1—2
5945,44	5	11,53	13,61	$5p\ [1{}^1/_2] - 9s\ [1{}^1/_2]^\circ$	1—1
5942,13	2	11,30	13,39	$5p\ [{}^1/_2] - 6d\ [2{}^1/_2]^\circ$	1—2
5887,68	3	11,55	13,65	$5p\ [1{}^1/_2] - 5d'\ [1{}^1/_2]^\circ$	2—2
5881,18	2	11,55	13,65	$5p\ [1{}^1/_2] - 5d'\ [2{}^1/_2]^\circ$	2—2
5879,9000	50	10,03	12,14	$5s\ [1{}^1/_2]^\circ - 5p'\ [{}^1/_2]$	1—1
5870,9153	3000	10,03	12,14	$5s\ [1{}^1/_2]^\circ - 5p'\ [1{}^1/_2]$	1—2

λ, Å	I	E_H, eV	E_B, eV	Transition	J
5866,7514	50	10,64	12,76	5s′ [1/2]°—6p [1/2]	1—1
5857,32	1	11,44	13,56	5p [2 1/2]—7d [1 1/2]°	2—2
5852,86	5	11,44	13,56	5p [2 1/2]—7d [1 1/2]°	3—2
5849,66	2	11,30	13,42	5p [1/2]—6d [1 1/2]°	1—1
5841,44	4	11,55	13,67	5p [1 1/2]—5d′ [2 1/2]°	2—3
5832,8600	100	11,44	13,57	5p [2 1/2]—7d [3 1/2]°	3—4
5827,07	20	11,30	13,43	5p [1/2]—8s [1 1/2]°	1—2
5824,50	40	11,44	13,57	5p [2 1/2]—7d [3 1/2]°	2—3
5823,51	3	11,55	13,67	5p [1 1/2]—8d [1/2]°	2—1
5820,10	15	11,44	13,57	5p [2 1/2]—7d [3 1/2]°	3—3
5810,80	8	11,30	13,44	5p [1/2]—8s [1 1/2]°	1—1
5805,53	20	11,44	13,58	5p [2 1/2]—7d [2 1/2]°	2—2
5801,17	2	11,44	13,58	5p [2 1/2]—7d [2 1/2]°	3—2
5788,24	7	11,44	13,59	5p [2 1/2]—7d [2 1/2]°	2—3
5787,29	6	10,64	12,78	5s′ [1/2]°—6p [2 1/2]	1—2
5783,89	10	11,44	13,59	5p [2 1/2]—7d [2 1/2]°	3—3
5775,56	2	10,03	12,18	5s [2 1/2]°—4d [3 1/2]°	1—3
5762,90	4	11,55	13,70	5p [1 1/2]—8d [3 1/2]°	2—3
5755,04	2	11,53	13,67	5p [1 1/2]—8d [1/2]°	1—1
5754,33	1	11,55	13,70	5p [1 1/2]—8d [2 1/2]°	2—2
5750,57	10	11,55	13,70	5p [1 1/2]—8d [2 1/2]°	2—3
5749,02	5	11,55	13,70	5p [1 1/2]—8d [1 1/2]°	2—2
5730,86	4	11,44	13,61	5p [2 1/2]—9s [1 1/2]°	2—2
5726,59	20	11,44	13,61	5p [2 1/2]—9s [1 1/2]°	3—2
5723,56	15	10,64	12,81	5s′ [1/2]°—6p [1 1/2]	1—1
5721,88	10	11,44	13,61	5p [2 1/2]—9s [1 1/2]°	2—1
5717,61	3	11,55	13,71	5p [1 1/2]—10s [1 1/2]°	2—2
5714,11	2	11,55	13,71	5p [1 1/2]—10s [1 1/2]°	2—1
5707,5188	40	10,64	12,82	5s′ [1/2]°—6p [1 1/2]	1—2
5702,19	10	11,53	13,70	5p [1 1/2]—8d [2 1/2]°	1—2
5696,95	1	11,53	13,70	5p [1 1/2]—8d [1 1/2]°	1—2
5696,54	3	11,53	13,70	5p [1 1/2]—8d [1 1/2]°	1—1
5672,4519	50	9,91	12,10	5s [1 1/2]°—5p′ [1 1/2]	2—1
5666,09	1	11,53	13,71	5p [1 1/2]—10s [1 1/2]°	1—2
5662,67	3	11,53	13,71	5p [1 1/2]—10s [1 1/2]°	1—1
5649,5625	100	10,56	12,76	5s′ [1/2]°—6p [1/2]	0—1
5611,82	4	11,44	13,65	5p [2 1/2]—5d′ [2 1/2]°	2—2
5608,37	3	9,91	12,12	5s [1 1/2]°—4d [3 1/2]°	2—4
5607,72	1	11,44	13,65	5p [2 1/2]—5d′ [2 1/2]°	3—2
5591,41	2	11,55	13,76	5p [1 1/2]—7s′ [1/2]°	2—1
5580,3890	80	10,64	12,86	5s′ [1/2]°—6p [1/2]	1—0
5577,64	3	11,55	13,77	5p [1 1/2]—9d [1 1/2]°	2—2
5575,6	10	—	—	—	—
5573,13	2	10,03	12,26	5s [1 1/2]°—5p′ [1/2]	1—0
5570,2890	2000	9,91	12,14	5s [1 1/2]°—5p′ [1/2]	2—1
5562,2254	500	9,91	12,14	5s [1 1/2]°—5p′ [1 1/2]	2—2
5559,26	2	11,44	13,67	5p [2 1/2]—8d [1/2]°	2—1
5544,4	1	11,55	13,78	5p [1 1/2]—11s [1 1/2]°	2—2
5542,10	1	11,53	13,75	5p [1 1/2]—9d [1/2]°	1—1
5539,4	1	11,55	13,78	5p [1 1/2]—11s [1 1/2]°	2—1
5528,63	2	11,53	13,77	5p [1 1/2]—9d [1 1/2]°	1—2
5521,17	3	11,53	13,77	5p [1 1/2]—9d [2 1/2]°	1—2
5520,52	40	11,44	13,69	5p [2 1/2]—8d [3 1/2]°	3—4
5516,66	20	10,56	12,81	5s′ [1/2]°—6p [1 1/2]	0—1
5511,16	1	11,53	13,78	5p [1 1/2]—9d [1 1/2]°	1—1
5504,34	20	11,30	13,55	5p [1/2]—7d [1/2]°	1—0
5504,02	15	11,44	13,70	5p [2 1/2]—8d [3 1/2]°	2—3
5500,71	50	11,30	13,56	5p [1/2]—7d [1/2]°	1—1

λ, Å	I	E_H, eV	E_B, eV	Transition	J
5496,21	3	11,44	13,70	$5p\ [2^1/_2]-8d\ [2^1/_2]^\circ$	2—2
5492,77	1	11,44	13,70	$5p\ [2^1/_2]-8d\ [2^1/_2]^\circ$	2—3
5491,33	2	11,44	13,70	$5p\ [2^1/_2]-8d\ [1^1/_2]^\circ$	2—2
5490,94	50	11,53	13,78	$5p\ [1^1/_2]-11s\ [1^1/_2]^\circ$	1—1
		11,30	13,56	$5p\ [^1/_2]-7d\ [1^1/_2]^\circ$	1—2
5488,86	5	11,44	13,70	$5p\ [2^1/_2]-8d\ [2^1/_2]^\circ$	3—3
5487,46	1	11,44	13,70	$5p\ [2^1/_2]-8d\ [1^1/_2]^\circ$	3—2
5476,58	2	9,91	12,18	$5s\ [1^1/_2]^\circ-4d\ [3^1/_2]^\circ$	2—3
5462,65	2	11,44	13,71	$5p\ [2^1/_2]-10s\ [1^1/_2]^\circ$	2—2
5461,37	1	11,55	13,81	$5p\ [1^1/_2]-10d\ [^1/_2]^\circ$	2—1
5459,47	4	11,44	13,71	$5p\ [2^1/_2]-10s\ [1^1/_2]^\circ$	2—1
5458,80	7	11,44	13,71	$5p\ [2^1/_2]-10s\ [1^1/_2]^\circ$	3—2
5456,39	2	11,55	13,82	$5p\ [1^1/_2]-10d\ [1^1/_2]^\circ$	2—2
5447,86	3	11,55	13,82	$5p\ [1^1/_2]-10d\ [2^1/_2]^\circ$	2—3
5445,43	1	11,30	13,58	$5p\ [^1/_2]-7d\ [2^1/_2]^\circ$	1—2
5431,77	1	11,55	13,83	$5p\ [1^1/_2]-12s\ [1^1/_2]^\circ$	2—2
5414,42	1	11,53	13,81	$5p\ [1^1/_2]-10d\ [^1/_2]^\circ$	1—1
5409,44	1	11,53	13,82	$5p\ [1^1/_2]-10d\ [1^1/_2]^\circ$	1—2
5403,03	2	11,53	13,82	$5p\ [1^1/_2]-10d\ [2^1/_2]^\circ$	1—2
5379,64	15	11,30	13,61	$5p\ [^1/_2]-9s\ [1^1/_2]^\circ$	1—2
5372,57	1	11,55	13,85	$5p\ [1^1/_2]-11d\ [1^1/_2]^\circ$	2—2
5371,74	2	11,30	13,61	$5p\ [^1/_2]-9s\ [1^1/_2]^\circ$	1—1
5365,91	1	11,55	13,86	$5p\ [1^1/_2]-11d\ [2^1/_2]^\circ$	2—3
5347,37	2	11,44	13,75	$5p\ [2^1/_2]-9d\ [^1/_2]^\circ$	2—1
5339,13	20	11,44	13,76	$5p\ [2^1/_2]-9d\ [3^1/_2]^\circ$	3—4
5334,78	10	11,44	13,77	$5p\ [2^1/_2]-9d\ [1^1/_2]^\circ$	2—2
5331,08	2	11,44	13,77	$5p\ [2^1/_2]-9d\ [1^1/_2]^\circ$	3—2
5327,87	2	11,44	13,77	$5p\ [2^1/_2]-9d\ [2^1/_2]^\circ$	2—2
5325,70	1	11,44	13,77	$5p\ [2^1/_2]-9d\ [2^1/_2]^\circ$	2—3
5322,02	2	11,44	13,77	$5p\ [2^1/_2]-9d\ [2^1/_2]^\circ$	3—3
5304,43	1	11,44	13,78	$5p\ [2^1/_2]-11s\ [1^1/_2]^\circ$	2—2
5300,74	3	11,44	13,78	$5p\ [2^1/_2]-11s\ [1^1/_2]^\circ$	3—2
5299,79	2	11,44	13,78	$5p\ [2^1/_2]-11s\ [1^1/_2]^\circ$	2—1
5279,84	9	11,30	13,65	$5p\ [^1/_2]-5d'\ [1^1/_2]^\circ$	1—2
5274,61	4	11,30	13,65	$5p\ [^1/_2]-5d'\ [2^1/_2]^\circ$	1—2
5232,06	2	9,91	12,28	$5s\ [1^1/_2]^\circ-4d\ [2^1/_2]^\circ$	2—3
5228,18	20	11,30	13,67	$5p\ [^1/_2]-8d\ [^1/_2]^\circ$	1—1
5223,57	5	11,44	13,82	$5p\ [2^1/_2]-10d\ [3^1/_2]^\circ$	3—4
5222,38	3	11,44	13,82	$5p\ [2^1/_2]-10d\ [3^1/_2]^\circ$	2—3
5218,84	1	11,44	13,82	$5p\ [2^1/_2]-10d\ [3^1/_2]^\circ$	3—3
5217,78	1	11,44	13,82	$5p\ [2^1/_2]-10d\ [2^1/_2]^\circ$	2—2
5215,81	8	11,30	13,68	$5p\ [^1/_2]-8d\ [^1/_2]^\circ$	1—0
5212,41	1	11,44	13,82	$5p\ [2^1/_2]-10d\ [2^1/_2]^\circ$	3—3
5198,97	1	11,44	13,83	$5p\ [2^1/_2]-12s\ [1^1/_2]^\circ$	2—1
5197,82	1	11,44	13,83	$5p\ [2^1/_2]-12s\ [1^1/_2]^\circ$	3—2
5172,36	2	11,30	13,70	$5p\ [^1/_2]-8d\ [2^1/_2]^\circ$	1—2
5168,06	4	11,30	13,70	$5p\ [^1/_2]-8d\ [1^1/_2]^\circ$	1—2
5167,73	1	11,30	13,70	$5p\ [^1/_2]-8d\ [1^1/_2]^\circ$	1—1
5145,39	1	11,44	13,85	$5p\ [2^1/_2]-11d\ [3^1/_2]^\circ$	2—3
5145,04	2	11,44	13,85	$5p\ [2^1/_2]-11d\ [3^1/_2]^\circ$	3—4
5142,7	4	11,30	13,71	$5p\ [^1/_2]-10s\ [1^1/_2]^\circ$	1—1
5139,9	1	11,30	13,71	$5p\ [^1/_2]-10s\ [1^1/_2]^\circ$	1—2
5109,81	2	—	—	—	—
5090,36	1	11,44	13,88	$5p\ [2^1/_2]-12d\ [3^1/_2]^\circ$	2—3
5089,12	2	11,44	13,88	$5p\ [2^1/_2]-12d\ [3^1/_2]^\circ$	3—4
5058,08	4	11,30	13,75	$5p\ [^1/_2]-9d\ [^1/_2]^\circ$	1—0
5047,74	1	11,44	13,90	$5p\ [2^1/_2]-13d\ [3^1/_2]^\circ$	3—4
5040,34	7	11,30	13,75	$5p\ [^1/_2]-9d\ [^1/_2]^\circ$	1—1

λ, Å	I	E_H, eV	E_B, eV	Transition	J
5029,15	5	11,30	13,77	5p [1/2]—9d [1 1/2]°	1—2
5002,14	2	11,30	13,78	5p [1/2]—11s [1 1/2]°	1—2
4969,36	15	10,64	13,14	5s' [1/2]°—4f [1 1/2]	1—2
4969,08	20	10,64	13,14	5s' [1/2]°—4f [1 1/2]	1—1
4955,27	15	10,64	13,14	5s' [1/2]°—4f [2 1/2]	1—2
4938,38	2	11,30	13,81	5p [1/2]—10d [1/2]°	1—0
4934,48	4	11,30	13,81	5p [1/2]—10d [1/2]°	1—1
4930,38	4	11,30	13,82	5p [1/2]—10d [1 1/2]°	1—2
4910,39	2	11,30	13,83	5p [1/2]—12s [1 1/2]°	1—2
4867,24	1	11,30	13,85	5p [1/2]—11d [1/2]°	1—0
4864,91	2	11,30	13,85	5p [1/2]—11d [1/2]°	1—1
4861,84	2	11,30	13,85	5p [1/2]—11d [1 1/2]°	1—2
4861,31	4	—	—	—	—
4812,6367	40	10,56	13,14	5s' [1/2]°—4f [1 1/2]	0—1
4810,51	3	—	—	—	—
4724,89	20	10,64	13,27	5s' [1/2]°—7p [1/2]	1—1
4722,16	3	—	—	—	—
4694,84	4	10,64	13,28	5s' [1/2]°—7p [2 1/2]	1—2
4677,16	1	10,64	13,29	5s' [1/2]°—7p [1 1/2]	1—1
4671,61	10	10,64	13,29	5s' [1/2]°—7p [1 1/2]	1—2
4636,14	20	10,64	13,32	5s' [1/2]°—7p [1/2]	1—0
4550,2985	40	10,03	12,76	5s [1 1/2]°—6p [1/2]	1—1
4538,06	3	10,56	13,29	5s' [1/2]°—7p [1 1/2]	0—1
4502,3546	600	10,03	12,78	5s [1 1/2]°—6p [2 1/2]	1—2
4463,6901	800	10,03	12,81	5s [1 1/2]°—6p [1 1/2]	1—1
4453,9177	600	10,03	12,82	5s [1 1/2]°—6p [1 1/2]	1—2
4425,1908	100	10,64	13,44	5s' [1/2]°—6p' [1 1/2]	1—1
4418,7626	50	10,64	13,45	5s' [1/2]°—5f [1 1/2]	1—2
4416,8338	20	10,64	13,45	5s' [1/2]°—5f [1 1/2]	1—1
4412,39	6	10,64	13,45	5s' [1/2]°—5f [2 1/2]	1—2
4410,3685	50	10,64	13,45	5s' [1/2]°—6p' [1/2]	1—1
4399,9670	200	10,64	13,46	5s' [1/2]°—6p' [1 1/2]	1—2
4380,11	2	—	—	—	—
4376,1219	800	10,03	12,86	5s [1 1/2]°—6p [1/2]	1—0
4362,6424	500	9,91	12,76	5s [1 1/2]°—6p [1/2]	2—1
4354,23	2	—	—	—	—
4353,90	2	—	—	—	—
4351,3602	100	10,64	13,40	5s' [1/2]°—6p' [1/2]	1—0
4349,55	2	—	—	—	—
4319,5798	1000	9,91	12,78	5s [1 1/2]°—6p [2 1/2]	2—3
4318,5523	400	9,91	12,78	5s [1 1/2]°—6p [2 1/2]	2—2
4302,4455	10	10,64	13,52	5s' [1/2]°—8p [1/2]	1—1
4300,4877	50	10,56	13,44	5s' [1/2]°—6p' [1 1/2]	0—1
4292,64	6	10,56	13,45	5s' [1/2]°—5f [1 1/2]	0—1
4290,78	4	—	—	—	—
4288,02	5	10,64	13,53	5s' [1/2]°—8p [1 1/2]	1—2
4286,4875	40	10,56	13,45	5s' [1/2]°—6p' [1/2]	0—1
4282,9686	100	9,91	12,81	5s [1 1/2]°—6p [1 1/2]	2—1
4273,9700	1000	9,91	12,82	5s [1 1/2]°—6p [1 1/2]	2—2
4263,2881	20	10,64	13,55	5s' [1/2]°—8p [1/2]	1—0
4184,4726	20	10,56	13,52	5s' [1/2]°—8p [1/2]	0—1
4172,83	3	10,56	13,53	5s' [1/2]°—8p [1 1/2]	0—1
4167,28	5	10,64	13,62	5s' [1/2]°—6f [1 1/2]	1—1
4164,48	2	10,64	13,62	5s' [1/2]°—6f [2 1/2]	1—2
4108,43	3	10,64	13,66	5s' [1/2]°—9p [1/2]	1—1
4097,84	1	10,64	13,67	5s' [1/2]°—9p [1 1/2]	1—2
4086,90	2	10,64	13,68	5s' [1/2]°—9p [1/2]	1—0
4056,57	3	10,56	13,62	5s' [1/2]°—6f [1 1/2]	0—1

λ, Å	I	E_H, eV	E_B, eV	Transition	J
4029,66	2	10,64	13,72	$5s'\ [^1/_2]^\circ$—$7f\ [1^1/_2]$	1—1
4028,03	1	10,64	13,72	$5s'\ [^1/_2]^\circ$—$7f\ [2^1/_2]$	1—2
4000,72	2	10,56	13,66	$5s'\ [^1/_2]^\circ$—$9p\ [^1/_2]$	0—1
3994,82	3	10,64	13,75	$5s'\ [^1/_2]^\circ$—$10p\ [^1/_2]$	1—1
3991,2581	10	10,03	13,14	$5s\ [1^1/_2]^\circ$—$4f\ [1^1/_2]$	1—1
3991,0797	20	10,03	13,14	$5s\ [1^1/_2]^\circ$—$4f\ [1^1/_2]$	1—2
3982,1699	6	10,03	13,14	$5s\ [1^1/_2]^\circ$—$4f\ [2^1/_2]$	1—2
3926,05	1	10,56	13,72	$5s'\ [^1/_2]^\circ$—$7f\ [1^1/_2]$	0—1
3892,69	1	10,56	13,75	$5s'\ [^1/_2]^\circ$—$10p\ [^1/_2]$	0—1
3846,12	2	9,91	13,14	$5s\ [1^1/_2]^\circ$—$4f\ [1^1/_2]$	2—1
3845,9778	15	9,91	13,14	$5s\ [1^1/_2]^\circ$—$4f\ [1^1/_2]$	2—2
3837,81	30	9,91	13,14	$5s\ [1^1/_2]^\circ$—$4f\ [2^1/_2]$	2—3
3812,2155	20	10,03	13,28	$5s\ [1^1/_2]^\circ$—$7p\ [2^1/_2]$	1—2
3800,5437	30	10,03	13,29	$5s\ [1^1/_2]^\circ$—$7p\ [1^1/_2]$	1—1
3796,8839	20	10,03	13,29	$5s\ [1^1/_2]^\circ$—$7p\ [1^1/_2]$	1—2
3773,4241	50	10,03	13,32	$5s\ [1^1/_2]^\circ$—$7p\ [^1/_2]$	1—0
3698,0452	6	9,91	13,27	$5s\ [1^1/_2]^\circ$—$7p\ [^1/_2]$	2—1
3679,58	100	9,91	13,28	$5s\ [1^1/_2]^\circ$—$7p\ [2^1/_2]$	2—3, 2
3668,7363	10	9,91	13,29	$5s\ [1^1/_2]^\circ$—$7p\ [1^1/_2]$	2—1
3665,3259	80	9,91	13,29	$5s\ [1^1/_2]^\circ$—$7p\ [1^1/_2]$	2—2
3632,4896	4	10,03	13,44	$5s\ [1^1/_2]^\circ$—$6p'\ [1^1/_2]$	1—1
3628,1570	10	10,03	13,45	$5s\ [1^1/_2]^\circ$—$5f\ [1^1/_2]$	1—2
3626,91	2	10,03	13,45	$5s\ [1^1/_2]^\circ$—$5f\ [1^1/_2]$	1—1
3623,84	1	10,03	13,45	$5s\ [1^1/_2]^\circ$—$5f\ [2^1/_2]$	1—2
3622,53	1	10,03	13,45	$5s\ [1^1/_2]^\circ$—$6p'\ [^1/_2]$	1—1
3615,4755	20	10,03	13,46	$5s\ [1^1/_2]^\circ$—$6p'\ [1^1/_2]$	1—2
3549,44	1	10,03	13,52	$5s\ [1^1/_2]^\circ$—$8p\ [^1/_2]$	1—1
3546,46	3	10,03	13,53	$5s\ [1^1/_2]^\circ$—$8p\ [2^1/_2]$	1—2
3540,9538	5	10,03	13,53	$5s\ [1^1/_2]^\circ$—$8p\ [1^1/_2]$	1—1
3539,5416	5	10,03	13,53	$5s\ [1^1/_2]^\circ$—$8p\ [1^1/_2]$	1—2
3522,6747	15	10,03	13,55	$5s\ [1^1/_2]^\circ$—$8p\ [^1/_2]$	1—0
3511,8963	4	9,91	13,44	$5s\ [1^1/_2]^\circ$—$6p'\ [1^1/_2]$	2—1
3507,84	3	9,91	13,45	$5s\ [1^1/_2]^\circ$—$5f\ [1^1/_2]$	2—2
3506,66	3	9,91	13,45	$5s\ [1^1/_2]^\circ$—$5f\ [1^1/_2]$	2—1
3503,8981	15	9,91	13,45	$5s\ [1^1/_2]^\circ$—$5f\ [2^1/_2]$	2—3
3502,5537	20	9,91	13,45	$5s\ [1^1/_2]^\circ$—$6p'\ [^1/_2]$	2—1
3495,9900	10	9,91	13,46	$5s\ [1^1/_2]^\circ$—$6p'\ [1^1/_2]$	2—2
3460,13	2	—	—	—	—
3456,87	3	10,03	13,62	$5s\ [1^1/_2]^\circ$—$6f\ [1^1/_2]$	1—1
3454,90	1	10,03	13,62	$5s\ [1^1/_2]^\circ$—$6f\ [2^1/_2]$	1—2
3434,1423	8	9,91	13,52	$5s\ [1^1/_2]^\circ$—$8p\ [^1/_2]$	2—1
3431,7217	20	9,91	13,53	$5s\ [1^1/_2]^\circ$—$8p\ [2^1/_2]$	2—3
3431,45	2	9,91	13,53	$5s\ [1^1/_2]^\circ$—$8p\ [2^1/_2]$	2—2
3426,27	2	9,91	13,53	$5s\ [1^1/_2]^\circ$—$8p\ [1^1/_2]$	2—1
3424,9433	15	9,91	13,53	$5s\ [1^1/_2]^\circ$—$8p\ [1^1/_2]$	2—2
3412,80	1	10,03	13,66	$5s\ [1^1/_2]^\circ$—$9p\ [2^1/_2]$	1—2
3409,89	2	10,03	13,67	$5s\ [1^1/_2]^\circ$—$9p\ [1^1/_2]$	1—1
3408,97	2	10,03	13,67	$5s\ [1^1/_2]^\circ$—$9p\ [1^1/_2]$	1—2
3401,40	5	10,03	13,68	$5s\ [1^1/_2]^\circ$—$9p\ [^1/_2]$	1—0
3361,74	2	10,03	13,72	$5s\ [1^1/_2]^\circ$—$7f\ [1^1/_2]$	1—1
3347,50	2	9,91	13,62	$5s\ [1^1/_2]^\circ$—$6f\ [1^1/_2]$	2—1
3345,73	4	9,91	13,62	$5s\ [1^1/_2]^\circ$—$6f\ [2^1/_2]$	2—3
3337,17	1	10,03	13,75	$5s\ [1^1/_2]^\circ$—$10p\ [^1/_2]$	1—1
3334,47	1	10,03	13,75	$5s\ [1^1/_2]^\circ$—$10p\ [2^1/_2]$	1—2
3332,47	1	10,03	13,75	$5s\ [1^1/_2]^\circ$—$10p\ [1^1/_2]$	1—2
3328,00	2	10,03	13,76	$5s\ [1^1/_2]^\circ$—$10p\ [^1/_2]$	1—0
3306,17	7	9,91	13,66	$5s\ [1^1/_2]^\circ$—$9p\ [2^1/_2]$	2—3, 2

λ, Å	I	E_H, eV	E_B, eV	Transition	J
3302,54	10	10,03	13,78	$5s\ [1^1/_2]^\circ - 8f\ [1^1/_2]$	1—2
		9,91	13,67	$5s\ [1^1/_2]^\circ - 9p\ [1^1/_2]$	2—2
3280,59	1	10,03	13,81	$5s\ [1^1/_2]^\circ - 11p\ [^1/_2]$	1—0
3258,00	1	9,91	13,72	$5s\ [1^1/_2]^\circ - 7f\ [1^1/_2]$	2—2
3257,10	1	9,91	13,72	$5s\ [1^1/_2]^\circ - 7f\ [2^1/_2]$	2—3
3232,80	2	9,91	13,75	$5s\ [1^1/_2]^\circ - 10p\ [2^1/_2]$	2—3, 2
3230,68	2	9,91	13,75	$5s\ [1^1/_2]^\circ - 10p\ [1^1/_2]$	2—2
3186,01	1	9,91	13,80	$5s\ [1^1/_2]^\circ - 11p\ [2^1/_2]$	2—3, 2
3184,53	1	9,91	13,81	$5s\ [1^1/_2]^\circ - 11p\ [1^1/_2]$	2—2
1235,839	13	0,00	10,03	$4p^6\ {}^1S - 5s\ [1^1/_2]^\circ$	0—1
1164,868	4	0,00	10,64	$4p^6\ {}^1S - 5s'\ [^1/_2]^\circ$	0—1
1134,89	3	—	—	—	—
1134,15	3	—	—	—	—
1030,020	2	0,00	12,04	$4p^6\ {}^1S - 4d\ [^1/_2]^\circ$	0—1
1003,542	2	0,00	12,35	$4p^6\ {}^1S - 4d\ [1^1/_2]^\circ$	0—1
1001,048	2	0,00	12,38	$4p^6\ {}^1S - 6s\ [1^1/_2]^\circ$	0—1
963,34	1	0,00	12,87	$4p^6\ {}^1S - 5d\ [^1/_2]^\circ$	0—1
953,42	1	0,00	13,00	$4p^6\ {}^1S - 4d'\ [1^1/_2]^\circ$	0—1
951,06	0	0,00	13,03	$4p^6\ {}^1S - 6s'\ [^1/_2]^\circ$	0—1
946,52	1	0,00	13,10	$4p^6\ {}^1S - 5d\ [1^1/_2]^\circ$	0—1
945,45	1	0,00	13,11	$4p^6\ {}^1S - 7s\ [1^1/_2]^\circ$	0—1

Kr II, ground state $1s^2\,2s^2\,2p^6\,3s^2\,3p^6\,3d^{10}\,4s^2\,4p^5\ {}^2P^0_{3/2}$
Ionization potential 198 182,00 cm^{-1}; 24,570 eV

λ, Å	I	E_H, eV	E_B, eV	Transition	J
10659,5	1	17,37	18,54	$5p\ {}^2P^\circ - 4d'\ {}^2D$	$^3/_2 - {}^5/_2$
10639,34	6	17,37	18,54	$5p\ {}^2D^\circ - 4d'\ {}^2D$	$^5/_2 - {}^5/_2$
10562,84	4	17,65	18,62	$5p\ {}^2S^\circ - 4d'\ {}^2P$	$^1/_2 - {}^3/_2$
10431,84	2	—	—	—	—
10428,40	10	—	—	—	—
10389,28	8	16,18	17,37	$4d\ {}^4P - 5p\ {}^2P^\circ$	$^1/_2 - {}^3/_2$
10361,15	100	20,86	22,06	$5f\ {}^2F^\circ - 5d'\ {}^2F$	$^7/_2 - {}^7/_2$
		16,18	17,38	$4d\ {}^4P - 5p\ {}^4D^\circ$	$^1/_2 - {}^1/_2$
10221,46	1000	15,62	16,83	$4d\ {}^4F - 5p\ {}^4D^\circ$	$^9/_2 - {}^7/_2$
10177,41	3	17,65	18,87	$5p\ {}^2S^\circ - 4d'\ {}^2P$	$^1/_2 - {}^1/_2$
10167,61	10	17,60	18,82	$5p\ {}^2D^\circ - 4d'\ {}^2P$	$^3/_2 - {}^3/_2$
10157,07	2	20,02	21,24	$5d\ {}^4D - 4^\circ$	$^5/_2 - {}^5/_2$
10127,74	4	19,96	21,19	$5d\ {}^4P - 3^\circ$	$^5/_2 - {}^3/_2$
10042,27	20	20,60	21,83	$1^\circ - 5d'\ {}^2D$	$^3/_2 - {}^3/_2$
10017,97	20	—	—	—	—
9966,67	5	16,83	18,08	$5p\ {}^4P^\circ - 5s''\ {}^2S$	$^1/_2 - {}^1/_2$
9954,75	20	17,37	18,62	$5p\ {}^2P^\circ - 4d'\ {}^2D$	$^3/_2 - {}^3/_2$
9892,97	10	17,57	18,82	$5p\ {}^4S^\circ - 4d'\ {}^2D$	$^3/_2 - {}^3/_2$
9851,40	3	—	—	—	—
9833,8	5	—	—	—	—
9826,58	100	—	—	—	—
9823,39	100	—	—	—	—
9803,14	500	—	—	—	—
9800,6	5	—	—	—	—
9795,1	2	—	—	—	—
9777,6	2	—	—	—	—
9770,1	2	—	—	—	—
9739,4	2	—	—	—	—

λ, Å	I	E_H, eV	E_B, eV	Transition	J
9720,6	3	—	—	—	—
9717,16	10	—	—	—	—
9711,60	200	—	—	—	—
9693,27	2	—	—	—	—
9672,90	6	—	—	—	—
9663,34	200	16,29	17,57	$4d\ ^4P$—$5p\ ^4S^\circ$	$^5/_2$—$^3/_2$
9622,5	3	—	—	—	—
9619,61	400	—	—	—	—
9613,80	100	20,89	22,17	$5f\ ^2F^\circ$—$5g\ ^4G$	$^5/_2$—$^5/_2$
9605,80	500	—	—	—	—
9594,24	100	{ 16,09	17,38	$4d\ ^4F$—$5p\ ^4D^\circ$	$^3/_2$—$^1/_2$
		16,08	17,37	$4d\ ^4F$—$5p\ ^2D^\circ$	$^5/_2$—$^5/_2$
9577,52	500	16,08	17,37	$4d\ ^4F$—$5p\ ^2P^\circ$	$^5/_2$—$^3/_2$
9564,32	5	—	—	—	—
9561,26	2	—	—	—	—
9552,85	10	—	—	—	—
9549,4	2	—	—	—	—
9543,64	10	20,88	22,17	$5f\ ^4F^\circ$—$5g\ ^4G$	$^3/_2$—$^5/_2$
9520,23	4	—	—	—	—
9504,70	100	15,85	17,16	$5s'\ ^2D$—$5p\ ^4D^\circ$	$^5/_2$—$^3/_2$
9500,60	100	—	—	—	—
9476,4	5	—	—	—	—
9475,06	100	20,86	22,17	$5f\ ^2F^\circ$—$5g\ ^4G$	$^7/_2$—$^9/_2$
9470,93	200	20,86	22,17	$5f\ ^2F^\circ$—$5g\ ^4G$	$^7/_2$—$^7/_2$
9461,67	3	—	—	—	—
9440,02	100	20,15	21,47	$6s\ ^2P$—$5f'\ ^2F^\circ$	$^1/_2$—$^5/_2$
9437,21	20	18,62	19,94	$5p'\ ^2P^\circ$—$5d\ ^4P$	$^3/_2$—$^1/_2$
9430,25	5	20,86	22,17	$5f\ ^2F^\circ$—$5g\ ^4G$	$^7/_2$—$^5/_2$
9414,94	100	16,29	17,60	$4d\ ^4P$—$5p\ ^2D^\circ$	$^5/_2$—$^3/_2$
9413,32	3	—	—	—	—
9402,82	200	20,85	22,17	$5f\ ^4F^\circ$—$5g\ ^4D$	$^5/_2$—$^7/_2$
9388,08	50	—	—	—	—
9363,6	1	20,09	21,42	$6s\ ^2P$—7°	$^3/_2$—$^5/_2$
9361,95	300	17,17	18,49	2—$5p'\ ^2F^\circ$	$^3/_2$—$^5/_2$
9349,08	100	20,84	22,17	$5f\ ^4F^\circ$—$5g\ ^4G$	$^7/_2$—$^9/_2$
9345,11	100	20,84	22,17	$5f\ ^4F^\circ$—$5g\ ^4G$	$^7/_2$—$^7/_2$
9337,73	2	—	—	—	—
9330,66	5	20,09	21,42	$6s\ ^2P$—$5f'\ ^2D^\circ$	$^3/_2$—$^5/_2$
9326,19	4	—	—	—	—
9320,99	200	20,84	22,17	$5f\ ^4F^\circ$—$5g\ ^4G$	$^9/_2$—$^9/_2$
9317,84	30	20,84	22,17	$5f\ ^4F^\circ$—$5g\ ^4G$	$^9/_2$—$^7/_2$
9305,76	1	20,84	22,17	$5f\ ^4F^\circ$—$5g\ ^4G$	$^7/_2$—$^5/_2$
9296,1	60	—	—	—	—
9293,82	500	20,84	22,17	$5f\ ^4F^\circ$—$5g\ ^4G$	$^9/_2$—$^{11}/_2$
9289,95	20	—	—	—	—
9271,99	50	15,82	17,16	$5s'\ ^2D$—$5p\ ^4D^\circ$	$^3/_2$—$^3/_2$
9269,38	2	—	—	—	—
9266,17	2	21,33	22,67	$4d''\ ^2D$—11°	$^3/_2$—$^5/_2$
9262,93	2	18,87	20,21	$5p'\ ^2P^\circ$—$5d\ ^4D$	$^1/_2$—$^1/_2$
9245,45	20	—	—	—	—
9238,48	500	16,87	18,21	$5p\ ^4D^\circ$—$4d'\ ^2G$	$^5/_2$—$^7/_2$
9233,18	50	16,23	17,57	$4d\ ^4P$—$5p\ ^4S^\circ$	$^3/_2$—$^3/_2$
9207,27	8	—	—	—	—
9196,7	1	21,32	22,67	$4d''\ ^2D$—11°	$^5/_2$—$^5/_2$
9181,23	10	—	—	—	—
9175,42	40	—	—	—	—
9164,04	3	—	—	—	—
9157,82	2	—	—	—	—
9133,4	8	—	—	—	—

λ, Å	I	E_H, eV	E_B, eV	Transition	J
9131,21	6	—	—	—	—
9115,00	20	—	—	—	—
9099,72	15	—	—	—	—
9096,49	4	—	—	—	—
9094,5	2	—	—	—	—
9090,98	3	—	—	—	—
9087,18	2	—	—	—	—
9044,55	10	19,57	20,94	$6s\,^4P-5p''\,^2P^\circ$	$3/2-3/2$
9039,95	20	—	—	—	—
9025,67	10	17,25	18,62	$5p\,^2P^\circ-4d'\,^2D$	$1/2-3/2$
9006,15	10	16,23	17,60	$4d\,^4P-5p\,^2D^\circ$	$3/2-3/2$
8999,11	6	16,83	18,21	$5p\,^4D^\circ-4d'\,^2G$	$7/2-7/2$
8978,70	15	17,16	18,54	$5p\,^4D^\circ-4d'\,^2D$	$3/2-5/2$
8908,26	3	—	—	—	—
8840,09	4	18,56	19,96	$5p'\,^2F^\circ-5d\,^4P$	$7/2-5/2$
8833,42	3	—	—	—	—
8829,38	5	20,02 20,01	21,42 21,42	$5d\,^4D-5f'\,^2D^\circ$ $5d\,^4P-7^\circ$	$5/2-5/2$ $3/2-5/2$
8810,10	2	—	—	—	—
8804,65	3	20,06	21,47	$5d\,^4D-5f'\,^2F^\circ$	$3/2-5/2$
8798,65	3	—	—	—	—
8717,31	2	20,21	21,63	$5d\,^4D-5f'\,^2P^\circ$	$1/2-1/2$
8707,61	8	16,18	17,60	$4d\,^4P-5p\,^2D^\circ$	$1/2-3/2$
8690,19	100	15,82	17,25	$5s'\,^2D-5p\,^2P^\circ$	$3/2-1/2$
8674,26	2	16,65	18,08	$5p\,^4P^\circ-5s''\,^2S$	$3/2-1/2$
8651,50	5	—	—	—	—
8619,34	1	18,62	20,06	$5p'\,^2P^\circ-5d\,^4D$	$3/2-3/2$
8613,58	2	—	—	—	—
8595,91	4	—	—	—	—
8563,59	2	18,62	20,07	$5p'\,^2P^\circ-6s\,^4P$	$3/2-1/2$
8551,33	2	17,37	18,82	$5p\,^2P^\circ-4d'\,^2P$	$3/2-3/2$
8537,98	3	17,37	18,82	$5p\,^2P^\circ-4d'\,^2P$	$5/2-3/2$
8523,88	3	—	—	—	—
8473,31	100	17,16	18,62	$5p\,^4D^\circ-4d'\,^2D$	$3/2-3/2$
8464,92	4	20,07	21,53	$6s\,^4P-5f'\,^2P^\circ$	$1/2-3/2$
8432,37	1	18,62	20,09	$5p'\,^2P^\circ-6s'\,^2P$	$3/2-3/2$
8411,14	1	20,06	21,53	$5d\,^4D-5f'\,^2P^\circ$	$3/2-3/2$
8378,87	2	—	—	—	—
8333,14	2	—	—	—	—
8202,72	200	15,86	17,37	$4d\,^4F-5p\,^2D^\circ$	$7/2-5/2$
8178,68	2	—	—	—	—
8157,25	10	15,85	17,37	$5s'\,^2D-5p\,^2D^\circ$	$5/2-5/2$
8147,70	1	21,15	22,67	$6s'\,^2D-11^\circ$	$5/2-5/2$
8145,15	100	15,85	17,37	$5s'\,^2D-5p\,^2P^\circ$	$5/2-3/2$
8142,17	1	20,01 18,49	21,53 20,02	$5d\,^4P-5f'\,^2P^\circ$ $5p'\,^2F^\circ-5d\,^4D$	$3/2-3/2$ $5/2-5/2$
8132,96	6	—	—	—	—
8130,03	10	16,08	17,60	$4d\,^4F-5p\,^2D^\circ$	$5/2-3/2$
8123,44	4	—	—	—	—
8095,96	3	—	—	—	—
7993,22	200	15,10	16,65	$4d\,^4D-5p\,^4P^\circ$	$1/2-3/2$
7973,62	120	15,82	17,37	$5s'\,^2D-5p\,^2P^\circ$	$3/2-3/2$
7957,07	3	15,82	17,38	$5s'\,^2D-5p\,^4D^\circ$	$3/2-1/2$
7931,41	40	17,00	18,56	$4d\,^2F-5p'\,^2F^\circ$	$5/2-7/2$
7956,52	30	17,25	18,82	$5p\,^2P^\circ-4d'\,^2P$	$1/2-3/2$
7791,90	6	18,62	20,21	$5p'\,^2P^\circ-5d\,^4D$	$3/2-2/2$
7781,97	100	—	—	—	—
7749,16	1	18,87	20,47	$5p'\,^2P^\circ-5d\,^2P$	$1/2-3/2$
7735,69	250	15,00	16,60	$4d\,^4D-5p\,^4P^\circ$	$3/2-5/2$

λ, Å	I	E_H, eV	E_B, eV	Transition	J
7641,16	150	17,25	18,87	$5p\ ^2P^\circ$—$4d'\ ^2P$	$^1/_2$—$^1/_2$
7635,13	5	20,01	21,63	$5d\ ^4P$—$5f'\ ^2P^\circ$	$^3/_2$—$^1/_2$
7629,46	5	17,00	18,62	$4d\ ^2F$—$5p'\ ^2P^\circ$	$^5/_2$—$^3/_2$
7615,69	3	—	—	—	—
7515,48	20	—	—	—	—
7524,46	300	15,00	16,65	$4d\ ^4D$—$5p\ ^4P^\circ$	$^3/_2$—$^3/_2$
7467,99	6	—	—	—	—
7435,78	200	16,87	18,54	$5p\ ^4D^\circ$—$4d'\ ^2D$	$^5/_2$—$^5/_2$
7434,74	15	17,16	18,82	$5p\ ^4D^\circ$—$4d'\ ^2P$	$^3/_2$—$^3/_2$
7407,02	400	14,93	16,60	$4d\ ^4D$—$5p\ ^4P^\circ$	$^5/_2$—$^5/_2$
7359,97	3	—	—	—	—
7301,29	4	—	—	—	—
7289,78	400	14,90	16,60	$4d\ ^4D$—$5p\ ^4P^\circ$	$^7/_2$—$^5/_2$
7272,97	4	17,17	18,87	2—$5p'\ ^2P^\circ$	$^3/_2$—$^1/_2$
7241,56	2	17,16	18,87	$5p\ ^4D^\circ$—$4d'\ ^2P$	$^3/_2$—$^1/_2$
7233,52	1	17,17	18,88	2—$5p'\ ^2D^\circ$	$^3/_2$—$^5/_2$
7213,13	250	{ 14,93	16,65	$4d\ ^4D$—$5p\ ^4P^\circ$	$^5/_2$—$^3/_2$
		15,85	17,57	$5s'\ ^2D$—$5p\ ^4S^\circ$	$^5/_2$—$^3/_2$
7156,81	1	18,87	20,60	$4d'\ ^2P$—1°	$^1/_2$—$^3/_2$
7139,99	60	15,10	16,83	$4d\ ^4D$—$5p\ ^4P^\circ$	$^1/_2$—$^1/_2$
7078,44	3	15,82	17,57	$5s'\ ^2D$—$5p\ ^4S^\circ$	$^3/_2$—$^3/_2$
7073,97	60	15,85	17,60	$5s'\ ^2D$—$5p\ ^2D^\circ$	$^5/_2$—$^3/_2$
7022,56	2	18,62	20,39	$5p'\ ^2P^\circ$—$6s''\ ^2S$	$^3/_2$—$^1/_2$
6977,95	3	18,82	20,60	$4d'\ ^2P$—1°	$^3/_2$—$^3/_2$
6944,06	10	{ 16,83	18,62	$5p\ ^4P^\circ$—$4d'\ ^2D$	$^1/_2$—$^3/_2$
		15,82	17,60	$5s'\ ^2D$—$5p\ ^2D^\circ$	$^3/_2$—$^3/_2$
6870,85	40	16,69	18,49	$4d\ ^2P$—$5p'\ ^2F^\circ$	$^3/_2$—$^5/_2$
6771,22	50	15,82	17,65	$5s'\ ^2D$—$5p\ ^2S^\circ$	$^3/_2$—$^1/_2$
6764,43	80	15,00	16,83	$5s\ ^2P$—$5p\ ^4P^\circ$	$^1/_2$—$^1/_2$
6763,61	100	15,00	16,83	$4d\ ^4D$—$5p\ ^4P^\circ$	$^3/_2$—$^1/_2$
6634,36	15	15,00	16,87	$4d\ ^4D$—$5p\ ^4D^\circ$	$^3/_2$—$^5/_2$
6627,96	2	17,60	19,47	$5p\ ^2D^\circ$—$6s\ ^4P$	$^3/_2$—$^5/_2$
6624,22	2	17,00	18,87	$4d\ ^2F$—$5p'\ ^2D^\circ$	$^5/_2$—$^3/_2$
6605,00	15	16,68	18,56	$4d\ ^2D$—$5p'\ ^2F^\circ$	$^5/_2$—$^7/_2$
6602,90	10	17,00	18,87	$4d\ ^2F$—$5p'\ ^2P^\circ$	$^5/_2$—$^1/_2$
6570,07	150	17,00	18,88	$4d\ ^2F$—$5p'\ ^2D^\circ$	$^5/_2$—$^5/_2$
6565,32	6	16,65	18,54	$5p\ ^4P^\circ$—$4d'\ ^2D$	$^3/_2$—$^5/_2$
6510,95	100	14,93	16,83	$4d\ ^4D$—$5p\ ^4D^\circ$	$^5/_2$—$^7/_2$
6510,14	8	17,57	19,47	$5p\ ^4S^\circ$—$6s\ ^4P$	$^3/_2$—$^5/_2$
6493,7	2	20,82	22,73	$5d\ ^2D$—13°	$^3/_2$—$^5/_2$
6470,89	50	14,69	16,60	$5s\ ^2P$—$5p\ ^4P^\circ$	$^3/_2$—$^5/_2$
6440,74	5	17,65	19,57	$5p\ ^2S^\circ$—$6s\ ^4P$	$^1/_2$—$^3/_2$
6420,18	300	14,90	16,83	$4d\ ^4D$—$5p\ ^4D^\circ$	$^7/_2$—$^7/_2$
6416,61	60	16,69	18,62	$4d\ ^2P$—$5p'\ ^2P^\circ$	$^3/_2$—$^3/_2$
6412,53	4	16,60	18,54	$5p\ ^4P^\circ$—$4d'\ ^2D$	$^5/_2$—$^5/_2$
6409,84	10	18,88	20,82	$5p'\ ^2D^\circ$—$5d\ ^2D$	$^5/_2$—$^3/_2$
6404,69	3	—	—	—	—
6394,28	4	16,68	18,62	$4d\ ^2D$—$5p'\ ^2P^\circ$	$^5/_2$—$^3/_2$
6391,14	30	14,93	16,87	$4d\ ^4D$—$5p\ ^4D^\circ$	$^5/_2$—$^5/_2$
6344,61	4	16,87	18,82	$5p\ ^4D^\circ$—$4d'\ ^2P$	$^5/_2$—$^3/_2$
6322,42	4	14,69	16,65	$5s\ ^2P$—$5p\ ^4P^\circ$	$^3/_2$—$^3/_2$
6303,66	100	14,90	16,87	$4d\ ^4D$—$5p\ ^4D^\circ$	$^7/_2$—$^5/_2$
6290,96	3	16,65	18,62	$5p\ ^4P^\circ$—$4d'\ ^2D$	$^3/_2$—$^3/_2$
6257,84	4	18,62	20,60	$4d'\ ^2D$—1°	$^3/_2$—$^3/_2$
6230,74	10	18,63	18,82	$5p\ ^4P^\circ$—$4d'\ ^2P$	$^1/_2$—$^3/_2$
6228,14	1	19,47	21,46	$6s\ ^4P$—$5f'\ ^2F^\circ$	$^5/_2$—$^7/_2$
6196,14	3	—	—	—	—
6185,35	7	17,57	19,57	$5p\ ^4S^\circ$—$6s\ ^4P$	$^3/_2$—$^3/_2$

λ, Å	I	$E_н$, eV	$E_в$, eV	Transition	J
6171,77	6	18,87	20,88	$4d'\,^2P-5f\,^4F^\circ$	$^1/_2-^3/_2$
6168,80	50	16,48	18,49	$4d\,^2D-5p'\,^2F^\circ$	$^3/_2-^5/_2$
6150,54	1	16,60	18,62	$5p\,^4P^\circ-4d'\,^2D$	$^5/_2-^3/_2$
6119,56	10	—	—	—	—
6112,61	4	20,70 18,82	22,73 20,85	$5d\,^2P-13^\circ$ $4d'\,^2P-5f\,^4F^\circ$	$^1/_2-^5/_2$ $^3/_2-^5/_2$
6107,61	5	—	—	—	—
6094,50	30	16,83	18,87	$5p\,^4P^\circ-4d'\,^2P$	$^1/_2-^1/_2$
6079,71	20	18,87	20,91	$4d'\,^2P-5f\,^2D^\circ$	$^1/_2-^3/_2$
6047,13	1	18,87	20,92	$4d'\,^2P-5f\,^4D^\circ$	$^1/_2-^3/_2$
6046,06	10	—	—	—	—
6040,7	10	—	—	—	—
6038,1	1	18,82	20,88	$4d'\,^2P-5f\,^4F^\circ$	$^3/_2-^3/_2$
6022,39	40	15,10	17,16	$4d\,^2D-5p\,^4D^\circ$	$^1/_2-^3/_2$
6009,99	10	18,82	20,89	$4d'\,^2P-5f\,^2F^\circ$	$^3/_2-^5/_2$
6008,10	3	18,54	20,60	$4d'\,^2D-1^\circ$	$^5/_2-^3/_2$
5992,22	200	14,58	16,65	$5s\,^4P-5p\,^4P^\circ$	$^1/_2-^3/_2$
5974,82	2	16,48	18,56	$1-5p'\,^2F^\circ$	$^5/_2-^7/_2$
5967,54	15	18,87	20,94	$4d'\,^2P-5p''\,^2P^\circ$	$^1/_2-^3/_2$
5949,93	3	18,82	20,91	$4d'\,^2P-5f\,^2D^\circ$	$^3/_2-^3/_2$
5941,82	4	—	—	—	—
5935,03	8	18,62	20,71	$5p'\,^2P^\circ-5d\,^4F$	$^3/_2-^3/_2$
5918,81	2	18,82	20,92	$4d'\,^2P-5f\,^4D^\circ$	$^3/_2-^3/_2$
5911,72	10	—	—	—	—
5900,89	8	17,37	19,47	$5p\,^2P^\circ-6s\,^4P$	$^3/_2-^5/_2$
5894,56	8	17,37	19,47	$5p\,^2D^\circ-6s\,^4P$	$^5/_2-^5/_2$
5860,75	10	—	—	—	—
5854,04	4	—	—	—	—
5842,49	1	18,82	20,94	$4d'\,^2P-5p''\,^2P^\circ$	$^3/_2-^3/_2$
5801,81	1	16,48	18,62	$1-5p'\,^2P^\circ$	$^5/_2-^3/_2$
5800,16	6	16,48	18,62	$4d\,^2D-5p'\,^2P^\circ$	$^3/_2-^3/_2$
5777,72	2	18,70	20,84	$4d'\,^2F-5f\,^4F^\circ$	$^5/_2-^7/_2$
5773,5	1	18,62	20,77	$5p'\,^2P^\circ-5d\,^2F$	$^3/_2-^5/_2$
5771,41	100	15,10	17,25	$4d\,^4D-5p\,^2P^\circ$	$^1/_2-^1/_2$
5755,60	2	18,70	20,85	$4d'\,^2F-5f\,^4F^\circ$	$^5/_2-^5/_2$
5752,98	60	15,00	17,16	$5s\,^2P-5p\,^4D^\circ$	$^1/_2-^3/_2$
5749,27	5	18,88	21,04	$5p'\,^2D^\circ-5d\,^2D$	$^5/_2-^5/_2$
5699,84	10	16,65 16,32	18,82 18,49	$5p\,^4P^\circ-4d'\,^2P$ $4d\,^2F-5p'\,^2F^\circ$	$^3/_2-^3/_2$ $^7/_2-^5/_2$
5692,11	5	—	—	—	—
5690,35	200	16,69	18,87	$4d\,^2P-5p'\,^2D^\circ$	$^3/_2-^3/_2$
5681,89	400	14,69	16,87	$5s\,^2P-5p\,^4D^\circ$	$^3/_2-^5/_2$
5674,52	30	16,69	18,87	$4d\,^2P-5p'\,^2P^\circ$	$^3/_2-^1/_2$
5672,78	40	16,68	18,87	$4d\,^2D-5p'\,^2D^\circ$	$^5/_2-^3/_2$
5664,85	1	18,70	20,89	$4d'\,^2F-5f\,^2F^\circ$	$^5/_2-^5/_2$
5650,37	10	16,69	18,88	$4d\,^2P-5p'\,^2D^\circ$	$^3/_2-^5/_2$
5648,38	1	20,47	22,67	$5d\,^2P-11^\circ$	$^3/_2-^5/_2$
5645,00	1	18,62	20,82	$5p'\,^2P-5d\,^2D$	$^3/_2-^3/_2$
5641,07	3	17,38	19,57	$5p\,^4D^\circ-6s\,^4P$	$^1/_2-^3/_2$
5633,02	100	16,68 17,37	18,88 19,57	$4d\,^2D-5p'\,^2D^\circ$ $5p\,^2P^\circ-6s\,^4P$	$^5/_2-^5/_2$ $^3/_2-^3/_2$
5627,02	1	17,37	19,57	$5p\,^2D^\circ-6s\,^4P$	$^5/_2-^3/_2$
5617,63	2	16,29	18,49	$4d\,^4D-5p'\,^2F^\circ$	$^5/_2-^5/_2$
5585,4	1	16,65	18,87	$5p\,^4P^\circ-4d'\,^2P$	$^3/_2-^1/_2$
5584,4	1	16,60	18,82	$5p\,^4P^\circ-4d'\,^2P$	$^5/_2-^3/_2$
5568,65	100	14,93	17,16	$4d\,^4D-5p\,^4D^\circ$	$^5/_2-^3/_2$
5552,99	100	18,62	20,85	$4d'\,^2D-5f\,^4F^\circ$	$^3/_2-^5/_2$
5546,11	5	20,39	22,62	$6s''\,^2S-8^\circ$	$^1/_2-^3/_2$

λ, Å	I	E_H, eV	E_B, eV	Transition	J
5541,65	4	20,39	22,62	$6s''\ ^2S-9^\circ$	1/2—1/2
5532,29	5	16,32	18,56	$4d\ ^2F-5p'\ ^2F^\circ$	7/2—7/2
5523,47	30	15,00	17,25	$5s\ ^2P-5p\ ^2P^\circ$	1/2—1/2
5522,94	60	15,00	17,25	$4d\ ^4D-5p\ ^2P^\circ$	3/2—1/2
5499,54	50	14,58	16,83	$5s\ ^4P-5p\ ^4P^\circ$	1/2—1/2
5491,43	4	18,62	20,88	$4d'\ ^2D-5f\ ^4F^\circ$	1/2—1/2
5476,46	4	18,88	21,15	$5p'\ ^2D^\circ-6s'\ ^2D$	5/2—5/2
547,549	1	18,58	20,84	$4d'\ ^2F-5f\ ^4F^\circ$	7/2—7/2
546,817	200	18,62	20,89	$4d'\ ^2D-5f\ ^2F^\circ$	3/2—5/2
5449,61	2	18,49	20,77	$5p'\ ^2F^\circ-5d\ ^2F$	5/2—5/2
5446,34	80	15,10	17,37	$4d\ ^4D-5p\ ^2P^\circ$	1/2—3/2
5439,38	1	18,87	21,15	$5p'\ ^2D^\circ-6s'\ ^2D$	3/2—5/2
5438,63	40	15,10	17,38	$4d\ ^4D-5p\ ^4D^\circ$	1/2—1/2
5433,24	2	18,58	20,86	$4d'\ ^2F-5f\ ^2F^\circ$	7/2—7/2
5423,56	1	17,65	19,94	$5p\ ^2S^\circ-5d\ ^4P$	1/2—1/2
5418,43	30	18,62	20,91	$4d'\ ^2D-5f\ ^2D^\circ$	3/2—3/2
5373,19	3	18,58	20,89	$4d'\ ^2F-5f\ ^2F^\circ$	7/2—5/2
5355,45	10	18,54	20,85	$4d'\ ^2D-5f\ ^4F^\circ$	5/2—5/2
5346,76	60	17,16	19,47	$5p\ ^4D^\circ-6s\ ^4P$	3/2—5/2
5333,41	500	18,54	20,86	$4d'\ ^2D-5f\ ^2F^\circ$	5/2—7/2
5329,15	4	18,62	20,94	$4d'\ ^2D-5p''\ ^2P^\circ$	1/2—3/2
5322,77	60	17,25	19,57	$5p\ ^2P^\circ-6s\ ^4P$	1/2—3/2
5317,41	30	17,60	19,94	$5p\ ^2D^\circ-5d\ ^4P$	3/2—1/2
5310,26	4	16,29	18,62	$4d\ ^4P-5p'\ ^2P^\circ$	5/2—3/2
5308,66	200	14,27	16,60	$5s\ ^4P-5p\ ^4P^\circ$	3/2—5/2
5297,8	1	18,54	20,88	$4d'\ ^2D-5f\ ^4F^\circ$	5/2—3/2
5276,50	100	18,54	20,89	$4d'\ ^2D-5f\ ^2F^\circ$	5/2—5/2
5256,75	30	17,60	19,96	$5p\ ^2D^\circ-5d\ ^4P$	3/2—5/2
5249,06	4	17,65	20,01	$5p\ ^2S^\circ-5d\ ^4P$	1/2—3/2
5245,25	4	—	—	—	—
5241,29	2	17,57	19,94	$5p\ ^4S^\circ-5d\ ^4P$	3/2—1/2
5230,15	3	18,54	20,91	$4d'\ ^2D-5f\ ^2D^\circ$	5/2—3/2
5229,52	60	15,00	17,37	$4d\ ^4D-5p\ ^2D^\circ$	3/2—5/2
5225,05	3	15,00	17,37	$5s\ ^2P-5p\ ^2P^\circ$	1/2—3/2
5224,56	7	15,00	17,37	$4d\ ^4D-5p\ ^2P^\circ$	3/2—3/2
5217,93	12	15,00	17,38	$5s\ ^2P-5p\ ^4D^\circ$	1/2—1/2
5217,45	30	15,00	17,38	$4d\ ^4D-5p\ ^4D^\circ$	3/2—1/2
5208,32	500	14,27	16,65	$5s\ ^4P-5p\ ^4P^\circ$	3/2—3/2
5201,56	2	16,48	18,87	$1-5p'\ ^2D^\circ$	5/2—3/2
5200,22	60	16,48	18,87	$4d\ ^2D-5p'\ ^2D^\circ$	3/2—3/2
5186,99	60	16,48	18,87	$4d\ ^2D-5p'\ ^2P^\circ$	3/2—1/2
5182,30	1	17,57	19,96	$5p\ ^4S^\circ-5d\ ^4P$	3/2—5/2
5177,71	6	16,23	18,62	$4d\ ^4P-5p'\ ^2P^\circ$	3/2—3/2
5166,80	80	16,48	18,88	$4d\ ^2D-5p'\ ^2D^\circ$	3/2—5/2
5152,01	3	—	—	—	—
5149,61	3	17,60	20,01	$5p\ ^2D^\circ-5d\ ^4P$	3/2—3/2
5145,28	4	16,09	18,49	$4d\ ^4F-5p'\ ^2F^\circ$	3/2—5/2
5143,05	60	17,65	20,06	$5p\ ^2S^\circ-5d\ ^4D$	1/2—3/2
5141,10	1	20,21	22,62	$5d\ ^4D-9^\circ$	1/2—1/2
5133,52	1	16,08	18,49	$4d\ ^4F-5p'\ ^2F^\circ$	5/2—5/2
5125,73	400	17,16	19,57	$5p\ ^4D^\circ-6s\ ^4P$	3/2—3/2
5123,16	15	17,65	20,07	$5p\ ^2S^\circ-6s\ ^4P$	1/2—1/2
5086,52	250	18,88	21,32	$5p'\ ^2D^\circ-4d''\ ^2D$	5/2—5/2
5078,19	2	17,57	20,01	$5p\ ^4S^\circ-5d\ ^4P$	3/2—3/2
5077,23	40	14,93	17,37	$4d\ ^4D-5p\ ^2D^\circ$	5/2—5/2
5075,92	4	17,65	20,09	$5p\ ^2S^\circ-6s\ ^2P$	1/2—3/2
5072,55	40	14,93	17,37	$4d\ ^4D-5p\ ^2P^\circ$	5/2—3/2
5067,41	3	—	—	—	—
5067,22	1	17,57	20,02	$5p\ ^4S^\circ-5d\ ^4D$	3/2—5/2

λ, Å	I	E_H, eV	E_B, eV	Transition	J
5065,58	20	18,88	21,33	$5p'\,{}^2D^\circ-4d''\,{}^2D$	${}^5/_2-{}^3/_2$
5054,53	30	20,21	22,66	$5d\,{}^4D-10^\circ$	${}^1/_2-{}^3/_2$
		18,87	21,32	$5p'\,{}^2D^\circ-4d''\,{}^2D$	${}^3/_2-{}^5/_2$
5047,52	4	17,60	20,06	$5p\,{}^2D^\circ-5d\,{}^4D$	${}^3/_2-{}^3/_2$
5046,31	80	18,87	21,33	$5p'\,{}^2P^\circ-4d''\,{}^2D$	${}^1/_2-{}^3/_2$
5033,85	100	18,87	21,33	$5p'\,{}^2D^\circ-4d''\,{}^2D$	${}^3/_2-{}^3/_2$
5028,36	30	17,60	20,07	$5p\,{}^2D^\circ-6s\,{}^4P$	${}^3/_2-{}^1/_2$
5022,40	200	14,69	17,16	$5s\,{}^2P-5p\,{}^4D^\circ$	${}^3/_2-{}^3/_2$
5021,88	100	14,90	17,37	$4d\,{}^4D-5p\,{}^2D^\circ$	${}^7/_2-{}^5/_2$
5020,43	4	20,15	22,62	$6s\,{}^2P-9^\circ$	${}^1/_2-{}^1/_2$
5015,71	1	20,21	22,68	$5d\,{}^4D-12^\circ$	${}^1/_2-{}^3/_2$
5013,29	100	15,10	17,57	$4d\,{}^4D-5p\,{}^4S^\circ$	${}^1/_2-{}^3/_2$
4998,54	5	18,56	21,04	$5p'\,{}^2F^\circ-5d\,{}^2D$	${}^7/_2-{}^5/_2$
4997,22	1	16,08	18,56	$4d\,{}^4F-5p'\,{}^2F^\circ$	${}^5/_2-{}^7/_2$
4982,83	50	17,60	20,09	$5p\,{}^2D^\circ-6s\,{}^2P$	${}^3/_2-{}^3/_2$
4978,89	100	17,57	20,06	$5p\,{}^4S^\circ-5d\,{}^4D$	${}^3/_2-{}^3/_2$
4960,25	100	17,57	20,07	$5p\,{}^4S^\circ-6s\,{}^4P$	${}^3/_2-{}^1/_2$
4948,50	50	17,65	20,15	$5p'\,{}^2S^\circ-6s\,{}^2P$	${}^1/_2-{}^1/_2$
4945,59	300	15,10	17,60	$4d\,{}^4D-5p\,{}^2D^\circ$	${}^1/_2-{}^3/_2$
4937,97	1	20,15	22,66	$6s\,{}^2P-10^\circ$	${}^1/_2-{}^3/_2$
4915,94	100	17,57	20,09	$5p\,{}^4S^\circ-6s\,{}^2P$	${}^3/_2-{}^3/_2$
4914,62	2	18,08	20,60	$5s''\,{}^2S-1^\circ$	${}^1/_2-{}^3/_2$
4908,34	2	18,62	21,15	$5p'\,{}^2P^\circ-6s'\,{}^2D$	${}^3/_2-{}^5/_2$
4897,2	3	—	—	—	—
4875,63	1	16,08	18,62	$4d\,{}^4F-5p'\,{}^2P^\circ$	${}^5/_2-{}^3/_2$
4870,14	20	18,87	21,41	$4d'\,{}^2P-5f'\,{}^2D^\circ$	${}^1/_2-{}^1/_2$
4857,20	150	15,10	17,65	$4d\,{}^4D-5p\,{}^2S^\circ$	${}^1/_2-{}^1/_2$
4852,61	2	20,07	22,62	$6s\,{}^4P-9^\circ$	${}^1/_2-{}^1/_2$
4846,60	700	14,69	17,25	$5s\,{}^2P-5p\,{}^2P^\circ$	${}^3/_2-{}^1/_2$
4845,14	2	17,38	19,94	$5p\,{}^4D^\circ-5d\,{}^4P$	${}^1/_2-{}^1/_2$
4839,04	4	17,37	19,94	$5p\,{}^2P^\circ-5d\,{}^4P$	${}^3/_2-{}^1/_2$
4836,56	20	17,65	20,21	$5p\,{}^2S^\circ-5d\,{}^4D$	${}^1/_2-{}^1/_2$
4833,68	4	16,32	18,88	$4d\,{}^2F-5p'\,{}^2D^\circ$	${}^7/_2-{}^5/_2$
4832,07	800	14,27	16,83	$5s\,{}^4P-5p\,{}^4P^\circ$	${}^3/_2-{}^1/_2$
4825,18	300	15,00	17,57	$5s\,{}^2P-5p\,{}^4S^\circ$	${}^1/_2-{}^3/_2$
4811,76	300	14,58	17,16	$5s\,{}^4P-5p\,{}^4D^\circ$	${}^1/_2-{}^3/_2$
4802,97	4	16,29	18,87	$4d\,{}^4P-5p'\,{}^2D^\circ$	${}^5/_2-{}^3/_2$
4796,33	60	17,57	20,15	$5p\,{}^4S^\circ-6s\,{}^2P$	${}^3/_2-{}^1/_2$
4791,15	3	18,56	21,15	$5p\,{}^2F^\circ-6s'\,{}^2D$	${}^7/_2-{}^5/_2$
4788,76	5	17,37	19,96	$5p\,{}^2P^\circ-5d\,{}^4P$	${}^3/_2-{}^5/_2$
4784,8	1	17,37	19,96	$5p\,{}^2D^\circ-5d\,{}^4P$	${}^5/_2-{}^5/_2$
4774,46	2	16,29	18,88	$4d\,{}^4P-5p'\,{}^2D^\circ$	${}^5/_2-{}^5/_2$
4773,01	40	18,82	21,42	$4d'\,{}^2P-5f'\,{}^2D^\circ$	${}^3/_2-{}^5/_2$
4765,74	1000	14,27	16,87	$5s\,{}^4P-5p\,{}^4D^\circ$	${}^3/_2-{}^5/_2$
4762,43	300	15,00	17,60	$5s\,{}^2P-5p\,{}^2D^\circ$	${}^1/_2-{}^3/_2$
4758,77	1	16,87	19,47	$5p\,{}^4D^\circ-6s\,{}^4P$	${}^5/_2-{}^5/_2$
4752,02	100	17,60	20,21	$5p\,{}^2D^\circ-5d\,{}^4D$	${}^3/_2-{}^1/_2$
4739,00	3000	13,99	16,60	$5s\,{}^4P-5p\,{}^4D^\circ$	${}^5/_2-{}^5/_2$
4706,31	3	15,86	18,49	$4d\,{}^4F-5p'\,{}^2F^\circ$	${}^7/_2-{}^5/_2$
4705,44	2	17,38	20,01	$5p\,{}^4D^\circ-5d\,{}^4P$	${}^1/_2-{}^3/_2$
4699,69	30	17,37	20,01	$5p\,{}^2P^\circ-5d\,{}^4P$	${}^3/_2-{}^3/_2$
4695,66	50	17,37	20,01	$5p\,{}^2D^\circ-5d\,{}^4P$	${}^5/_2-{}^3/_2$
4694,44	200	16,23	18,87	$4d\,{}^4P-5p'\,{}^2D^\circ$	${}^3/_2-{}^3/_2$
		16,83	19,47	$5p\,{}^4D^\circ-6s\,{}^4P$	${}^7/_2-{}^5/_2$
4691,28	100	17,57	20,21	$5p\,{}^4S^\circ-5d\,{}^4D$	${}^3/_2-{}^1/_2$
		15,85	18,49	$5s'\,{}^2D-5p'\,{}^2F^\circ$	${}^5/_2-{}^5/_2$
4688,3	3	18,88	21,53	$5p'\,{}^2D^\circ-7s\,{}^4P$	${}^5/_2-{}^3/_2$
4687,28	10	18,82	21,47	$4d'\,{}^2P-5f'\,{}^2F^\circ$	${}^3/_2-{}^5/_2$
4686,30	8	17,37	20,02	$5p\,{}^2D^\circ-5d\,{}^4D$	${}^5/_2-{}^5/_2$

λ, Å	I	E_H, eV	E_B, eV	Transition	J
4683,68	5	16,23	18,87	$4d\,^4P—5p'\,^2P^\circ$	3/2—1/2
4680,41	500	20,02	22,66	$5d\,^4D—10^\circ$	5/2—3/2
		15,00	17,65	$5s\,^2P—5p\,^2S^\circ$	1/2—1/2
4673,80	3	20,02	22,67	$5d\,^4D—11^\circ$	5/2—5/2
4672,09	2	18,49	21,15	$5p'\,^2F^\circ—6s\,^2D$	5/2—5/2
4658,87	2000	13,99	16,65	$5s\,^4P—5p\,^4P^\circ$	5/2—3/2
4650,17	30	14,58	17,25	$5s\,^4P—5p\,^2P^\circ$	1/2—1/2
4637,66	1	20,00	22,67	$5d\,^4D—11^\circ$	7/2—5/2
4635,42	8	14,93	17,60	$4d\,^4D—5p\,^2D^\circ$	5/2—3/2
4633,88	800	15,82	18,49	$5s'\,^2D—5p'\,^2F^\circ$	3/2—5/2
4619,99	5	17,38	20,06	$5p\,^4D^\circ—5d\,^4D$	1/2—3/2
4619,15	1000	14,69	17,37	$5s\,^2P—5p\,^2D^\circ$	3/2—5/2
4615,28	500	14,69	17,37	$5s\,^2P—5p\,^2P^\circ$	3/2—3/2
4614,50	15	17,37	20,06	$5p\,^2P^\circ—5d\,^4D$	3/2—3/2
4613,79	2	19,94	22,62	$5d\,^4P—8^\circ$	1/2—3/2
4610,65	60	19,94	22,62	$5d\,^4P—9^\circ$	1/2—1/2
		17,37	20,06	$5p\,^2D^\circ—5d\,^4D$	5/2—3/2
4609,72	20	14,69	17,38	$5s\,^2P—5p\,^4D^\circ$	3/2—1/2
4608,48	1	17,25	19,94	$5p\,^2P^\circ—5d\,^4P$	1/2—1/2
4604,02	60	17,38	20,07	$5p\,^4D^\circ—6s\,^4P$	1/2—1/2
4601,42	1	16,18	18,87	$4d\,^4P—5p'\,^2P^\circ$	1/2—1/2
4598,49	50	17,37	20,07	$5p\,^2P^\circ—6s\,^4P$	3/2—1/2
4592,80	150	18,62	21,32	$5p'\,^2P^\circ—4d''\,^2D$	3/2—5/2
4591,50	1	15,86	18,56	$4d\,^4F—5p'\,^2F^\circ$	7/2—7/2
4582,85	300	16,87	19,57	$5p\,^4D^\circ—6s\,^4P$	5/2—3/2
4580,11	2	19,96	22,67	$5d\,^4P—11^\circ$	5/2—5/2
4577,20	800	15,85	18,56	$5s'\,^2D—5p'\,^2F^\circ$	5/2—7/2
4575,8	1	18,62	21,33	$5p'\,^2P^\circ—4d''\,^2D$	3/2—3/2
4573,33	30	18,82	21,53	$4d'\,^2P—5f'\,^2P^\circ$	3/2—3/2
4565,82	1	17,38	20,09	$5p\,^4D^\circ—6s\,^2P$	1/2—3/2
4564,9	1	20,01	22,73	$5d\,^4P—13^\circ$	3/2—5/2
4560,38	3	17,37	20,09	$5p\,^2P^\circ—6s\,^2P$	3/2—3/2
4556,61	200	17,37	20,09	$5p\,^2D^\circ—6s\,^2P$	5/2—3/2
4552,77	3	18,70	21,42	$4d'\,^2F—5f'\,^2D^\circ$	5/2—5/2
4528,62	3	17,65	20,39	$5p\,^2S^\circ—6s''\,^2S$	1/2—1/2
4523,14	400	16,83	19,57	$5p\,^4P^\circ—6s\,^4P$	1/2—3/2
4489,88	400	18,56	21,32	$5p'\,^2F^\circ—4d''\,^2D$	7/2—5/2
4488,22	3	—	—	—	—
4481,85	50	17,25	20,01	$5p\,^2P^\circ—5d\,^4P$	1/2—3/2
4479,86	5	15,85	18,62	$4d'\,^2P—5f'\,^2P^\circ$	5/2—3/2
4475,00	800	15,85	18,62	$6s'\,^2D—5p'\,^2P^\circ$	5/2—3/2
4460,45	1	17,37	20,15	$5p\,^2D^\circ—5d\,^4F$	5/2—7/2
4459,99	8	17,16	19,94	$5p\,^4D^\circ—5d\,^4P$	3/2—1/2
4457,25	40	17,37	20,15	$5p\,^2P^\circ—6s\,^2P$	3/2—1/2
4454,37	10	17,60	20,39	$5p\,^2D^\circ—6s''\,^2S$	3/2—1/2
4453,21	50	18,08	20,86	$5s''\,^2S—5p''\,^2P^\circ$	1/2—1/2
4450,34	4	—	—	—	—
4443,72	3	16,09	18,87	$4d\,^4F—5p'\,^2P^\circ$	3/2—1/2
4436,81	600	14,58	17,37	$5s\,^4P—5p\,^2P^\circ$	1/2—3/2
4431,67	500	14,58	17,38	$5s\,^4P—5p\,^4D^\circ$	1/2—1/2
4428,9	1	16,09	18,88	$4d\,^4F—5p'\,^2D^\circ$	3/2—5/2
4422,70	100	15,82	18,62	$5s'\,^2D—5p'\,^2P^\circ$	3/2—3/2
4420,16	1	16,08	18,88	$4d\,^4F—5p'\,^2D^\circ$	5/2—5/2
4417,24	40	17,16	19,96	$5p\,^4D^\circ—5d\,^4P$	3/2—5/2
4408,89	40	18,82	21,63	$4d'\,^2P—5f'\,^2P^\circ$	3/2—1/2
4404,33	30	17,25	20,06	$5p\,^2P^\circ—5d\,^4D$	1/2—3/2
4400,87	100	17,57	20,39	$5p\,^4S^\circ—6s''\,^2S$	3/2—1/2
4399,39	15	—	—	—	—
4389,72	20	17,25	20,07	$5p\,^2P^\circ—6s\,^4P$	1/2—1/2

λ, Å	I	E_H, eV	E_B, eV	Transition	J
4388,90	3	17,65	20,47	$5p\,{}^2S^\circ-5d\,{}^2P$	$^1/_2-{}^3/_2$
4386,54	300	16,65	19,47	$5p\,{}^4P^\circ-6s\,{}^4P$	$^3/_2-{}^5/_2$
4385,27	50	18,49	21,32	$5p'\,{}^2F^\circ-4d''\,{}^2D$	$^5/_2-{}^5/_2$
4381,52	100	18,08	20,91	$5s''\,{}^2S-5f\,{}^2D^\circ$	$^1/_2-{}^3/_2$
4377,71	40	—	—	—	—
4371,25	20	17,38	20,21	$5p\,{}^4D^\circ-5d\,{}^4D$	$^1/_2-{}^1/_2$
4369,69	200	18,49	21,33	$5p'\,{}^2F^\circ-4d''\,{}^2D$	$^5/_2-{}^3/_2$
4366,26	6	17,37	20,21	$5p\,{}^2P^\circ-5d\,{}^4D$	$^3/_2-{}^1/_2$
4364,61	4	18,08	20,92	$5s''\,{}^2S-5f\,{}^4D^\circ$	$^1/_2-{}^3/_2$
4355,47	3000	13,99	16,83	$5s\,{}^4P-5p\,{}^4D^\circ$	$^5/_2-{}^7/_2$
4351,02	40	18,62	21,47	$4d'\,{}^2D-5f'\,{}^2F^\circ$	$^3/_2-{}^5/_2$
4341,33	8	17,16	20,01	$5p\,{}^4D^\circ-5d\,{}^4P$	$^3/_2-{}^3/_2$
4333,34	50	17,16	20,02	$5p\,{}^4D^\circ-5d\,{}^4D$	$^3/_2-{}^5/_2$
4331,24	80	17,60	20,47	$5p\,{}^2D^\circ-5d\,{}^4F$	$^3/_2-{}^5/_2$
4322,98	150	18,08	20,94	$5s''\,{}^2S-5p''\,{}^2P^\circ$	$^1/_2-{}^3/_2$
4319,12	4	17,60	20,47	$5p\,{}^2D^\circ-5d\,{}^2P$	$^3/_2-{}^3/_2$
4317,81	500	16,60	19,47	$5p\,{}^4P^\circ-6s\,{}^4P$	$^5/_2-{}^5/_2$
4305,81	3	—	—	—	—
4301,53	40	13,99	16,87	$5s\,{}^4P-5p\,{}^4D^\circ$	$^5/_2-{}^5/_2$
4300,49	200	14,69	17,57	$5s\,{}^2P-5p\,{}^4S^\circ$	$^3/_2-{}^3/_2$
4295,21	8	—	—	—	—
4292,92	600	14,27	17,16	$5s\,{}^4P-5p\,{}^4D^\circ$	$^3/_2-{}^3/_2$
4287,45	4	—	—	—	—
4285,40	4	—	—	—	—
4280,61	5	17,57	20,47	$5p\,{}^4S^\circ-5d\,{}^4F$	$^3/_2-{}^5/_2$
4273,48	4	—	—	—	—
4268,81	100	17,57	20,47	$5p\,{}^4S^\circ-5d\,{}^2P$	$^3/_2-{}^3/_2$
4268,57	60	17,16	20,06	$5p\,{}^4D^\circ-5d\,{}^4D$	$^3/_2-{}^3/_2$
4260,85	5	17,25	20,15	$5p\,{}^2P^\circ-6s\,{}^2P$	$^1/_2-{}^1/_2$
4259,44	80	—	—	—	—
4254,85	100	17,16	20,07	$5p\,{}^4D^\circ-6s\,{}^4P$	$^3/_2-{}^1/_2$
4252,67	50	18,62	21,53	$4d'\,{}^2D-5f'\,{}^2P^\circ$	$^3/_2-{}^3/_2$
4250,58	150	14,69	17,60	$5s\,{}^2P-5p\,{}^2D^\circ$	$^3/_2-{}^3/_2$
4236,64	100	16,65	19,57	$5p\,{}^4P^\circ-6s\,{}^4P$	$^3/_2-{}^3/_2$
4229,21	8	—	—	—	—
4228,79	20	18,54	21,47	$4d'\,{}^2D-5f'\,{}^2F^\circ$	$^5/_2-{}^5/_2$
4222,20	20	17,16	20,09	$5p\,{}^4D^\circ-6s\,{}^2P$	$^3/_2-{}^3/_2$
4217,88	2	15,62	18,56	$4d\,{}^4F-5p'\,{}^2F^\circ$	$^9/_2-{}^7/_2$
4210,67	25	—	—	—	—
4240,31	3	—	—	—	—
4201,42	30	18,88	21,83	$5p'\,{}^2D^\circ-5d'\,{}^2D$	$^5/_2-{}^3/_2$
4185,12	50	14,69	17,65	$5s\,{}^2P-5p\,{}^2S^\circ$	$^3/_2-{}^1/_2$
4179,58	20	18,87	21,83	$5p'\,{}^2D^\circ-5d'\,{}^2D$	$^3/_2-{}^3/_2$
4177,02	3	—	—	—	—
4172,51	20	16,60	19,57	$5p\,{}^4P^\circ-6s\,{}^4P$	$^5/_2-{}^3/_2$
4163,82	2	14,27	17,25	$5s\,{}^4P-5p\,{}^2P^\circ$	$^3/_2-{}^1/_2$
4159,00	4	—	—	—	—
4145,12	250	14,58	17,57	$5s\,{}^4P-5p\,{}^4S^\circ$	$^1/_2-{}^3/_2$
4139,11	100	—	—	—	—
4137,96	50	—	—	—	—
4135,86	3	18,54	21,53	$4d'\,{}^2D-5f'\,{}^2P^\circ$	$^5/_2-{}^3/_2$
4133,68	5	17,16	20,15	$5p\,{}^4D^\circ-6s\,{}^2P$	$^3/_2-{}^1/_2$
4118,14	30	17,38	20,39	$5p\,{}^4D^\circ-6s''\,{}^2S$	$^1/_2-{}^1/_2$
4113,73	8	17,37	20,39	$5p\,{}^2P^\circ-6s''\,{}^2S$	$^3/_2-{}^1/_2$
4110,16	5	18,62	21,63	$4d'\,{}^2D-5f'\,{}^2P^\circ$	$^3/_2-{}^1/_2$
4109,23	100	15,85	18,87	$5s'\,{}^2D-5p'\,{}^2D^\circ$	$^5/_2-{}^3/_2$
4099,71	3	15,86	18,88	$4d\,{}^4F-5p'\,{}^2D^\circ$	$^7/_2-{}^5/_2$
4098,72	250	14,58	17,60	$5s\,{}^4P-5p\,{}^2D^\circ$	$^1/_2-{}^3/_2$

λ, Å	I	E_H, eV	E_B, eV	Transition	J
4088,33	500	15,85	18,88	$5s'\,^{2}D-5p'\,^{2}D^{\circ}$	$^{5}/_{2}-{}^{5}/_{2}$
4081,40	1	18,21	21,25	$4d'\,^{2}G-5^{\circ}$	$^{7}/_{2}-{}^{5}/_{2}$
4066,09	6	—	—	—	—
4065,11	300	15,82	18,87	$5s'\,^{2}D-5p'\,^{2}D^{\circ}$	$^{3}/_{2}-{}^{3}/_{2}$
4057,01	300	15,82	18,87	$5s'\,^{2}D-5p'\,^{2}P^{\circ}$	$^{3}/_{2}-{}^{1}/_{2}$
4050,42	50	17,65	20,71	$5p\,^{2}S^{\circ}-5d\,^{4}F$	$^{1}/_{2}-{}^{3}/_{2}$
4044,67	80	15,82	18,88	$5s'\,^{2}D-5p'\,^{2}D^{\circ}$	$^{3}/_{2}-{}^{5}/_{2}$
4037,83	30	14,58	17,65	$5s\,^{4}P-5p\,^{2}D^{\circ}$	$^{1}/_{2}-{}^{1}/_{2}$
4008,48	10	17,37	20,47	$5p\,^{2}P^{\circ}-5d\,^{4}F$	$^{3}/_{2}-{}^{5}/_{2}$
4008,08	25	16,87	19,96	$5p\,^{4}D^{\circ}-5d\,^{4}P$	$^{5}/_{2}-{}^{5}/_{2}$
4005,57	30	17,37	20,47	$5p\,^{2}D^{\circ}-5d\,^{4}F$	$^{5}/_{2}-{}^{5}/_{2}$
3997,95	100	17,60	20,70	$5p\,^{2}D^{\circ}-3$	$^{3}/_{2}-{}^{5}/_{2}$
		17,37	20,47	$5p\,^{2}P^{\circ}-5d\,^{2}P$	$^{3}/_{2}-{}^{3}/_{2}$
3996,69	3	16,83	19,94	$5p\,^{4}P^{\circ}-5d\,^{4}P$	$^{1}/_{2}-{}^{1}/_{2}$
3994,83	100	14,27	17,37	$5s\,^{4}P-5p\,^{2}D^{\circ}$	$^{3}/_{2}-{}^{5}/_{2}$
3991,94	15	14,27	17,37	$5s\,^{4}P-5p\,^{2}P^{\circ}$	$^{3}/_{2}-{}^{3}/_{2}$
3990,66	15	17,60	20,71	$5p\,^{2}D^{\circ}-5d\,^{4}F$	$^{3}/_{2}-{}^{3}/_{2}$
3987,78	25	14,27	17,38	$5s\,^{4}P-5p\,^{4}D^{\circ}$	$^{3}/_{2}-{}^{1}/_{2}$
3987,09	5	18,08	21,19	$5s''\,^{2}S-3^{\circ}$	$^{1}/_{2}-{}^{3}/_{2}$
3964,89	30	16,87	20,00	$5p\,^{4}D^{\circ}-5d\,^{4}D$	$^{5}/_{2}-{}^{7}/_{2}$
3962,34	10	17,57	20,70	$5p\,^{4}S^{\circ}-5d\,^{2}P^{\circ}$	$^{3}/_{2}-{}^{1}/_{2}$
		16,83	19,96	$5p\,^{4}D^{\circ}-5d\,^{4}P$	$^{7}/_{2}-{}^{5}/_{2}$
3954,78	90	17,57	20,70	$5p\,^{4}S^{\circ}-3$	$^{3}/_{2}-{}^{5}/_{2}$
		17,37	20,50	$5p\,^{2}D^{\circ}-5d\,^{2}F$	$^{5}/_{2}-{}^{7}/_{2}$
3953,59	20	13,51	16,65	$4p^{6}\,^{2}S-5p\,^{4}P^{\circ}$	$^{1}/_{2}-{}^{3}/_{2}$
3947,66	5	17,57	20,71	$5p\,^{4}S^{\circ}-5d\,^{4}F$	$^{3}/_{2}-{}^{3}/_{2}$
3945,83	1	17,25	30,39	$5p\,^{2}P^{\circ}-6s''\,^{2}S$	$^{1}/_{2}-{}^{1}/_{2}$
3945,48	5	16,87	20,01	$5p\,^{4}D^{\circ}-5d\,^{4}P$	$^{5}/_{2}-{}^{3}/_{2}$
3942,93	20	—	—	—	—
3940,92	5	18,88	22,03	$5p'\,^{2}D^{\circ}-5d'\,^{2}P$	$^{5}/_{2}-{}^{3}/_{2}$
3938,88	20	16,87	20,02	$5p\,^{4}D^{\circ}-5d\,^{4}D$	$^{5}/_{2}-{}^{5}/_{2}$
3929,26	20	18,87	22,03	$5p'\,^{2}P^{\circ}-5d'\,^{2}P$	$^{1}/_{2}-{}^{3}/_{2}$
3921,68	6	18,87	22,03	$5p'\,^{2}D^{\circ}-5d'\,^{2}P$	$^{3}/_{2}-{}^{3}/_{2}$
3920,14	200	16,83	20,00	$5p\,^{4}D^{\circ}-5d\,^{4}D$	$^{7}/_{2}-{}^{7}/_{2}$
3917,64	50	18,87	22,03	$5p'\,^{2}D^{\circ}-5d'\,^{2}F$	$^{3}/_{2}-{}^{5}/_{2}$
3916,90	3	17,60	20,77	$5p\,^{2}D^{\circ}-5d\,^{2}F$	$^{3}/_{2}-{}^{5}/_{2}$
3912,88	5	17,65	20,82	$5p\,^{2}S^{\circ}-5d\,^{2}D$	$^{1}/_{2}-{}^{3}/_{2}$
3912,59	70	13,99	17,17	$5s\,^{4}P-5p\,^{4}D^{\circ}$	$^{5}/_{2}-{}^{3}/_{2}$
3906,25	150	18,88	22,06	$5p'\,^{2}D^{\circ}-5d\,^{2}F$	$^{5}/_{2}-{}^{7}/_{2}$
3901,15	10	16,83	20,01	$5p\,^{4}P^{\circ}-5d\,^{4}P$	$^{1}/_{2}-{}^{3}/_{2}$
3894,71	60	16,83	20,02	$5p\,^{4}D^{\circ}-5d\,^{4}D$	$^{7}/_{2}-{}^{5}/_{2}$
3887,54	5	18,87	22,06	$5p'\,^{2}P^{\circ}-7s\,^{2}P$	$^{1}/_{2}-{}^{3}/_{2}$
3885,28	1	16,87	20,06	$5p\,^{4}D^{\circ}-5d\,^{4}D$	$^{5}/_{2}-{}^{3}/_{2}$
3880,07	2	18,87	22,06	$5p'\,^{2}D^{\circ}-7s\,^{2}P$	$^{3}/_{2}-{}^{3}/_{2}$
3875,44	150	17,57	20,77	$5p\,^{4}S^{\circ}-5d\,^{2}F$	$^{3}/_{2}-{}^{5}/_{2}$
3858,78	5	18,62	21,83	$5p'\,^{2}P^{\circ}-5d'\,^{2}D$	$^{3}/_{2}-{}^{3}/_{2}$
3857,32	20	17,60	20,82	$5p\,^{2}D^{\circ}-5d\,^{2}D$	$^{3}/_{2}-{}^{3}/_{2}$
3846,83	5	16,87	20,09	$5p\,^{4}D^{\circ}-6s\,^{2}P$	$^{5}/_{2}-{}^{3}/_{2}$
3844,45	50	18,88	22,11	$5p'\,^{2}D^{\circ}-5d'\,^{2}D$	$^{5}/_{2}-{}^{3}/_{2}$
3842,28	20	16,83	20,06	$5p\,^{4}P^{\circ}-5d\,^{4}D$	$^{1}/_{2}-{}^{3}/_{2}$
3839,37	4	17,25	20,47	$5p\,^{2}P^{\circ}-5d\,^{2}P$	$^{1}/_{2}-{}^{3}/_{2}$
3836,54	30	17,16	20,39	$5p\,^{4}D^{\circ}-6s''\,^{2}S$	$^{3}/_{2}-{}^{1}/_{2}$
3831,17	2	16,83	20,07	$5p\,^{4}P-6s\,^{4}P$	$^{1}/_{2}-{}^{1}/_{2}$
3826,15	2	18,87	22,11	$5p'\,^{2}D^{\circ}-5d'\,^{2}D$	$^{3}/_{2}-{}^{5}/_{2}$
3817,11	15	17,57	20,82	$5p\,^{4}S^{\circ}-5d\,^{2}D$	$^{3}/_{2}-{}^{3}/_{2}$
3806,52	1	18,21	21,47	$4d'\,^{2}G-5f'\,^{2}F^{\circ}$	$^{7}/_{2}-{}^{5}/_{2}$
3806,17	8	18,87	22,13	$5p'\,^{2}P^{\circ}-7s\,^{2}P$	$^{1}/_{2}-{}^{1}/_{2}$
3804,67	30	16,83	20,09	$5p\,^{4}P^{\circ}-6s\,^{2}P$	$^{1}/_{2}-{}^{3}/_{2}$

λ, Å	I	E_H, eV	E_B, eV	Transition	J
3783,13	500	16,83	20,11	$5p\,^4D°—5d\,^4F$	$^7/_2—^9/_2$
3778,09	500	16,87	20,15	$5p\,^4D°—5d\,^4F$	$^5/_2—^7/_2$
3771,34	30	16,65	19,94	$5p\,^4P°—5d\,^4P$	$^3/_2—^1/_2$
3765,88	2	18,87	22,17	$5p'\,^2P—5d'\,^2P$	$^1/_2—^1/_2$
3758,93	6	18,87	22,17	$5p'\,^2D—5d'\,^2P$	$^3/_2—^1/_2$
3754,24	80	14,27	17,57	$5s\,^4P—5p\,^4S°$	$^3/_2—^3/_2$
3744,80	150	17,16	20,47	$5p\,^4D°—5d\,^4F$	$^3/_2—^5/_2$
3741,69	200	18,49	21,81	$5p'\,^2F°—5d'\,^2G$	$^5/_2—^7/_2$
3740,73	6	16,65	19,96	$5p\,^4P°—5d\,^4P$	$^3/_2—^5/_2$
3735,78	40	17,16	20,47	$5p\,^4D°—5d\,^2P$	$^3/_2—^3/_2$
3732,92	6	13,51	16,83	$4p^6\,^2S—5p\,^4P°$	$^1/_2—^1/_2$
3732,61	15	16,83	20,15	$5p\,^4P°—6s\,^2P°$	$^1/_2—^1/_2$
3731,67	2	17,38	20,70	$5p\,^4D°—5d\,^2P$	$^1/_2—^1/_2$
3728,04	7	17,37	20,70	$5p\,^2P°—5d\,^2P$	$^3/_2—^1/_2$
3721,35	150	17,38	20,71	$5p\,^2P°—3$	$^1/_2—^3/_2$
3718,63	200	17,38	20,71	$5p\,^4D°—5d\,^4F$	$^1/_2—^3/_2$
3718,02	300	18,56	21,89	$5p'\,^2F°—5d'\,^2G$	$^7/_2—^9/_2$
3716,15	4	14,27	17,60	$5s\,^4P—5p\,^2D°$	$^3/_2—^3/_2$
3715,04	12	17,37	20,71	$5p\,^2P°—5d\,^4F$	$^3/_2—^3/_2$
3712,48	1	17,37	20,71	$5p\,^2D°—5d\,^4F$	$^5/_2—^3/_2$
3711,27	1	18,49	21,83	$5p'\,^2F°—5d'\,^2D$	$^5/_2—^3/_2$
3690,65	30	16,60	19,96	$5p\,^4P°—5d\,^4P$	$^5/_2—^5/_2$
3686,15	80	16,65	20,01	$5p\,^4P°—5d\,^4P$	$^3/_2—^3/_2$
3680,37	100	16,65	20,02	$5p\,^4P°—5d\,^4D$	$^3/_2—^5/_2$
3678,66	7	—	—	—	—
3669,01	150	—	—	—	—
3668,59	6	16,83	20,21	$5p\,^4P°—5d\,^4D$	$^1/_2—^1/_2$
3666,01	5	14,27	17,65	$5s\,^4P—5p\,^2S°$	$^3/_2—^1/_2$
3663,44	20	13,99	17,37	$5s\,^4P—5p\,^2D°$	$^5/_2—^5/_2$
3661,00	15	13,99	17,37	$5s\,^4P—5p\,^2P°$	$^5/_2—^3/_2$
3653,97	250	16,60	20,00	$5p\,^4P°—5d\,^4D$	$^5/_2—^7/_2$
3651,02	25	17,37	29,77	$5p\,^2P°—5d\,^2F$	$^7/_2—^5/_2$
3648,61	40	17,37	20,77	$5p\,^2D°—5d\,^2F$	$^3/_2—^5/_2$
3637,93	4	18,62	22,03	$5p'\,^2P°—5d'\,^2P$	$^3/_2—^3/_2$
3637,48	20	16,60	20,01	$5p\,^4P°—5d\,^4P$	$^5/_2—^3/_2$
3634,42	3	18,62	22,03	$5p'\,^2P°—5d'\,^2F$	$^3/_2—^5/_2$
3633,54	3	16,65	20,06	$5p\,^4P°—5d\,^4D$	$^3/_2—^3/_2$
3631,87	200	16,60	20,02	$5p\,^4P°—5d\,^4D$	$^5/_2—^5/_2$
3623,61	30	16,65	20,07	$5p\,^4P°—6s\,^4P$	$^3/_2—^1/_2$
3607,88	100	17,60	21,04	$5p\,^2D°—5d\,^2D$	$^3/_2—^5/_2$
3602,12	2	18,62	22,06	$5p'\,^2P°—7s\,^2P$	$^3/_2—^3/_2$
3599,90	40	16,65	20,09	$5p\,^4P°—6s\,^2P$	$^3/_2—^3/_2$
3599,21	25	17,37	20,82	$5p\,^2P°—5d\,^2D$	$^3/_2—^3/_2$
3596,86	2	17,37	20,82	$5p\,^2D°—5d\,^2D$	$^5/_2—^3/_2$
3589,65	70	17,25	20,70	$5p\,^2P°—5d\,^2P$	$^1/_2—^1/_2$
3586,25	12	16,60	20,06	$5p\,^4P°—5d\,^4D$	$^5/_2—^3/_2$
3577,60	4	17,25	20,71	$5p\,^2P°—5d\,^4F$	$^1/_2—^3/_2$
3572,68	15	17,57	21,04	$5p\,^4S°—5d\,^2D$	$^3/_2—^5/_2$
3569,68	2	18,56	22,03	$5p'\,^2F°—5d'\,^2F$	$^7/_2—^5/_2$
3555,54	8	18,62	22,11	$5p'\,^2P°—5d'\,^2D$	$^3/_2—^5/_2$
3553,49	20	16,60	20,09	$5p\,^4P°—6s\,^2P$	$^5/_2—^3/_2$
3548,71	6	15,00	18,49	$4d\,^4D—5p'\,^2F°$	$^3/_2—^5/_2$
3544,54	30	18,56	22,06	$5p'\,^2F°—5d'\,^2F$	$^7/_2—^7/_2$
3544,14	30	17,65	21,15	$5p\,^2S°—6s'\,^2D$	$^1/_2—^3/_2$
3535,35	50	16,65	20,15	$5p\,^4P°—6s\,^2P$	$^3/_2—^1/_2$
3527,42	3	—	—	—	—
3517,37	5	15,10	18,62	$4d\,^4D—5p'\,^2P°$	$^1/_2—^3/_2$
3503,25	50	18,49	22,03	$5p'\,^2F°—5d'\,^2F$	$^5/_2—^5/_2$
3498,92	2	17,16	20,70	$5p\,^4D°—5d\,^2P$	$^3/_2—^1/_2$

λ, Å	I	E_H, eV	E_B, eV	Transition	J
3498,50	4	17,60	21,15	$5p\ ^2D^\circ - 6s'\ ^2D$	$3/2 - 5/2$
3497,45	3	18,62	22,17	$5p'\ ^2P^\circ - 5d'\ ^2P$	$3/2 - 1/2$
3493,57	2	18,56	22,11	$5p'\ ^2F^\circ - 5d'\ ^2D$	$7/2 - 5/2$
3493,04	8	17,16	20,70	$5p\ ^4D^\circ - 3$	$3/2 - 5/2$
3488,65	30	16,83	20,39	$5p\ ^4P^\circ - 6s''\ ^2S$	$1/2 - 1/2$
3487,49	7	17,16	20,71	$5p\ ^4D^\circ - 5d\ ^4F$	$3/2 - 3/2$
3479,00	3	18,49	22,06	$5p'\ ^2F^\circ - 5d'\ ^2F$	$5/2 - 3/2$
3477,89	5	16,65	20,21	$5p\ ^4P^\circ - 5d\ ^4D$	$3/2 - 1/2$
		14,93	18,49	$4d\ ^4D - 5p'\ ^2F^\circ$	$5/2 - 5/2$
3475,31	3	—	—	—	—
3470,05	30	17,25	20,82	$5p\ ^2P^\circ - 5d\ ^2D$	$1/2 - 3/2$
3465,41	6	17,57	21,15	$5p\ ^4S^\circ - 6s'\ ^2D$	$3/2 - 5/2$
3460,09	50	13,99	17,57	$5s\ ^4P - 5p\ ^4S^\circ$	$5/2 - 3/2$
3453,46	3	—	—	—	—
3446,51	50	16,87	20,47	$5p\ ^4D^\circ - 5d\ ^4F$	$5/2 - 5/2$
3443,29	5	—	—	—	—
3438,88	3	16,87	20,47	$5p\ ^4D^\circ - 5d\ ^2P$	$5/2 - 3/2$
3431,03	8	—	—	—	—
3429,91	3	18,49	22,11	$5p'\ ^2F^\circ - 5d'\ ^2D$	$5/2 - 5/2$
3427,71	30	13,99	17,60	$5s\ ^4P - 5p\ ^2D^\circ$	$5/2 - 3/2$
3423,73	20	15,00	18,62	$5s\ ^2P - 5p'\ ^2P^\circ$	$1/2 - 3/2$
3414,80	10	14,93	18,56	$4d\ ^4D - 5p'\ ^2F^\circ$	$5/2 - 7/2$
3412,67	1	16,83	20,47	$5p\ ^4D^\circ - 5d\ ^4F$	$7/2 - 5/2$
3405,16	80	16,83	20,47	$5p\ ^4P^\circ - 5d\ ^2P$	$1/2 - 3/2$
3402,79	2	13,51	17,16	$4p^6\ ^2S - 5p\ ^4D^\circ$	$1/2 - 3/2$
3389,67	5	14,90	18,56	$4d\ ^4D - 5p'\ ^2F^\circ$	$7/2 - 7/2$
3387,11	7	—	—	—	—
3385,23	15	17,16	20,82	$5p\ ^4D^\circ - 5d\ ^2D$	$3/2 - 3/2$
3381,11	20	17,37	21,04	$5p\ ^2P^\circ - 5d\ ^2D$	$3/2 - 5/2$
3379,03	15	17,37	21,04	$5p\ ^2D^\circ - 5d\ ^2D$	$5/2 - 5/2$
3375,78	3	16,83	20,50	$5p\ ^4D^\circ - 5d\ ^2F$	$7/2 - 7/2$
3357,58	2	14,93	18,62	$4d\ ^4D - 5p'\ ^2P^\circ$	$5/2 - 3/2$
3335,16	4	17,60	21,32	$5p\ ^2D^\circ - 4d''\ ^2D$	$3/2 - 5/2$
3326,13	1	17,60	21,33	$5p\ ^2D^\circ - 4d''\ ^2D$	$3/2 - 3/2$
3321,16	8	13,51	17,25	$4p^6\ ^2S - 5p\ ^2P^\circ$	$1/2 - 1/2$
3315,72	15	16,65	20,39	$5p\ ^4P^\circ - 6s''\ ^2S$	$3/2 - 1/2$
3302,28	4	—	—	—	—
3301,75	5	18,87	22,62	$4d'\ ^2P - 8^\circ$	$1/2 - 3/2$
3300,18	4	18,87	22,62	$4d'\ ^2P - 9^\circ$	$1/2 - 1/2$
3295,29	3	—	—	—	—
3287,69	2	17,38	21,15	$5p\ ^4D^\circ - 6s'\ ^2D$	$1/2 - 3/2$
3287,38	2	15,10	18,87	$4d\ ^4D - 5p'\ ^2D^\circ$	$1/2 - 3/2$
3282,08	15	15,10	18,87	$4d\ ^4D - 5p'\ ^2P^\circ$	$1/2 - 1/2$
3264,33	5	18,87	22,66	$4d'\ ^2P - 10^\circ$	$1/2 - 3/2$
3263,12	1	18,82	22,62	$4d'\ ^2P - 8^\circ$	$3/2 - 3/2$
3261,58	8	18,82	22,62	$4d'\ ^2P - 9^\circ$	$3/2 - 1/2$
3256,67	4	14,69	18,49	$5s\ ^2P - 5p'\ ^2F^\circ$	$3/2 - 5/2$
3248,03	6	18,87	22,68	$4d'\ ^2P - 12^\circ$	$1/2 - 3/2$
3247,00	12	16,65	20,47	$5d\ ^4P^\circ - 5d\ ^4F$	$3/2 - 5/2$
3240,20	2	16,65	20,47	$5d\ ^4P^\circ - 5d\ ^2P$	$3/2 - 3/2$
3232,15	2	16,87	20,70	$5p\ ^4D^\circ - 3$	$5/2 - 5/2$
3226,57	5	18,82	22,66	$4d'\ ^2P - 10^\circ$	$3/2 - 3/2$
3223,52	12	18,82	22,67	$4d'\ ^2P - 11^\circ$	$3/2 - 5/2$
3223,00	6	17,00	20,84	$4d\ ^2F - 5f\ ^4F^\circ$	$5/2 - 7/2$
3220,25	6	—	—	—	—
3216,25	7	17,00	20,85	$4d\ ^2F - 5f\ ^4F^\circ$	$5/2 - 5/2$
3210,89	7	13,51	17,37	$4p^6\ ^2S - 5p\ ^2P^\circ$	$1/2 - 3/2$
3210,64	2	18,82	22,68	$4d'\ ^2P - 12^\circ$	$3/2 - 3/2$
3209,7	7	16,60	20,47	$5p\ ^4P^\circ - 5d\ ^4F$	$5/2 - 5/2$

λ, Å	I	E_H, eV	E_B, eV	Transition	J
3208,28	40	13,51	17,38	$4p^6\,{}^2S-5p\,{}^4D^\circ$	$1/2-1/2$
		17,00	20,86	$4d\,{}^2F-5f\,{}^2F^\circ$	$5/2-5/2$
3207,29	1	16,83	20,70	$5p\,{}^4P^\circ-5d\,{}^2P$	$1/2-1/2$
3205,44	2	15,00	18,87	$5s\,{}^2P-5p'\,{}^2D^\circ$	$1/2-3/2$
3205,26	4	15,00	18,87	$4d\,{}^4D-5p'\,{}^2D^\circ$	$3/2-3/2$
3202,54	15	16,60	20,47	$5p\,{}^4P^\circ-5d\,{}^2P$	$5/2-3/2$
3200,40	50	15,00	18,87	$5s\,{}^2P^\circ-5p'\,{}^2P^\circ$	$1/2-1/2$
3197,65	4	16,83	20,71	$5p\,{}^4P^\circ-5d\,{}^4F$	$1/2-3/2$
3195,50	2	17,00	20,88	$4d\,{}^2F-5f\,{}^4F^\circ$	$5/2-3/2$
3192,54	2	17,60	21,49	$5p\,{}^2D^\circ-7s\,{}^4P$	$3/2-5/2$
		15,00	18,88	$4d\,{}^4D-5p'\,{}^2D^\circ$	$3/2-5/2$
3187,61	4	17,00	20,89	$4d\,{}^2F-5f\,{}^2F^\circ$	$5/2-5/2$
3181,25	5	—	—	—	—
3178,92	1	16,87	20,77	$5p\,{}^4D^\circ-5d\,{}^2F$	$5/2-5/2$
3176,94	15	17,25	21,15	$5p\,{}^2P^\circ-6s'\,{}^2D$	$1/2-3/2$
3175,67	40	18,82	22,73	$4d'\,{}^2F-13^\circ$	$3/2-5/2$
3170,63	2	17,00	20,91	$4d\,{}^2F-5f\,{}^2D^\circ$	$5/2-3/2$
3164,94	3	16,68	20,60	$4d\,{}^2D-1^\circ$	$5/2-3/2$
		17,57	21,49	$5p\,{}^4S^\circ-7s\,{}^4P$	$3/2-5/2$
3150,93	80	14,69	18,62	$5s\,{}^2P-5p'\,{}^2P^\circ$	$3/2-3/2$
3147,39	1	14,93	18,87	$4d\,{}^4D-5p'\,{}^2D^\circ$	$5/2-3/2$
3140,44	3	17,37	21,32	$5p\,{}^2P^\circ-4d''\,{}^2D$	$3/2-5/2$
3139,86	4	17,00	20,94	$4d\,{}^2F-5p''\,{}^2P^\circ$	$5/2-3/2$
3139,58	20	16,87	20,82	$5p\,{}^4D^\circ-5d\,{}^2D$	$5/2-3/2$
2135,10	8	17,38	21,33	$5p\,{}^4D^\circ-4d''\,{}^2D$	$1/2-3/2$
		14,93	18,88	$4d\,{}^4D-5p'\,{}^2D^\circ$	$5/2-5/2$
3132,84	4	—	—	—	—
3126,02	6	—	—	—	—
3113,92	2	14,90	18,88	$4d\,{}^4D-5p'\,{}^2D^\circ$	$7/2-5/2$
3111,45	2	16,83	20,82	$5p\,{}^4P^\circ-5d\,{}^2D$	$1/2-3/2$
3105,68	1	17,16	21,15	$5p\,{}^4D^\circ-6s'\,{}^2D$	$3/2-3/2$
3096,52	20	18,62	22,62	$4d'\,{}^2D-8^\circ$	$3/2-3/2$
3095,14	30	18,62	22,62	$4d'\,{}^2D-9^\circ$	$3/2-1/2$
3066,72	2	14,58	18,62	$5s\,{}^4P-5p'\,{}^2P^\circ$	$1/2-3/2$
3063,57	3	18,62	22,66	$4d'\,{}^2D-10^\circ$	$3/2-3/2$
3061,51	6	—	—	—	—
3060,84	30	16,65	20,70	$4d'\,{}^2D-11^\circ$	$3/2-5/2$
3056,01	30	16,65	20,70	$5p\,{}^4P^\circ-3$	$3/2-5/2$
3055,31	3	13,51	17,57	$4p^6\,{}^2S-5p\,{}^4S^\circ$	$1/2-3/2$
3051,75	1	16,65	20,71	$5p\,{}^4P^\circ-5d\,{}^4F$	$3/2-3/2$
3049,23	8	18,62	22,68	$4d'\,{}^2D-12^\circ$	$3/2-3/2$
3038,38	3	—	—	—	—
3034,16	2	18,54	22,62	$4d'\,{}^2D-8^\circ$	$5/2-3/2$
		17,25	21,33	$5p\,{}^2P^\circ-4d''\,{}^2D$	$1/2-3/2$
3032,77	5	—	—	—	—
3031,59	5	—	—	—	—
3030,01	4	13,51	17,60	$4p^6\,{}^2S-5p\,{}^2D^\circ$	$1/2-3/2$
3022,49	5	16,60	20,70	$5p\,{}^4P^\circ-3$	$5/2-5/2$
3018,30	1	16,60	20,71	$5p\,{}^4P^\circ-5d\,{}^4F$	$5/2-3/2$
3017,65	20	18,62	22,73	$4d'\,{}^2D-13^\circ$	$3/2-5/2$
3012,00	1	17,37	21,49	$5p\,{}^2D^\circ-7s\,{}^4P$	$5/2-5/2$
3008,42	8	16,65	20,77	$5p\,{}^4P^\circ-5d\,{}^2F$	$3/2-5/2$
3002,48	2	18,54	22,66	$4d'\,{}^2D-10^\circ$	$5/2-3/2$
2999,84	40	18,54	22,67	$4d'\,{}^2D-11^\circ$	$5/2-5/2$
2996,60	20	13,51	17,65	$4p^6\,{}^2S-5p\,{}^2S^\circ$	$1/2-1/2$
2988,69	3	18,54	22,68	$4d'\,{}^2D-12^\circ$	$5/2-3/2$
2985,33	4	—	—	—	—
2983,94	2	17,37	21,53	$5p\,{}^2P^\circ-7s\,{}^4P$	$3/2-3/2$

λ, Å	I	E_H, eV	E_B, eV	Transition	J
2982,34	1	17,37	21,53	$5p\,^2D^\circ—7s\,^4P$	5/2—3/2
2979,81	20	16,68	20,84	$4d\,^2D—5f\,^4F^\circ$	5/2—7/2
2978,87	25	16,69	20,85	$4d\,^2P—5f\,^4F^\circ$	3/2—5/2
2976,28	3	17,16	21,32	$5p\,^4D^\circ—4d''\,^2D$	3/2—5/2
2975,92	3	16,60	20,77	$5p\,^4P^\circ—5d\,^2F$	5/2—5/2
2974,04	25	16,68	20,85	$4d\,^2D—5f\,^4F^\circ$	5/2—5/2
2972,34	2	16,87	21,04	$5p\,^4D^\circ—5d\,^2D$	5/2—5/2
2971,80	4	16,69	20,86	$4d\,^2P—5p''\,^2P^\circ$	3/2—1/2
2967,25	80	16,68	20,86	$4d\,^2D—5f\,^2F^\circ$	5/2—7/2
2966,13	3	—	—	—	—
2965,11	2	14,69	18,87	$5s\,^2P—5p'\,^2D^\circ$	3/2—3/2
2961,05	4	16,69	20,88	$4d\,^2P—5f\,^4F^\circ$	3/2—3/2
2960,78	5	14,69	18,87	$5s\,^2P—5p'\,^2P^\circ$	3/2—1/2
2960,14	40	—	—	—	—
2958,35	20	18,54	22,73	$4d'\,^2D—13^\circ$	5/2—5/2
2956,30	3	16,68	20,88	$4d\,^2D—5f\,^4F^\circ$	5/2—3/2
2954,28	12	14,69	18,88	$5s\,^2P—5p'\,^2D^\circ$	3/2—5/2
		16,69	20,89	$4d\,^2P—5f\,^2F^\circ$	3/2—5/2
2950,21	30	—	—	—	—
2949,54	15	16,68	20,89	$4d\,^2D—5f\,^2F^\circ$	5/2—5/2
2939,70	2	16,69	20,91	$4d\,^2P—5f\,^2D^\circ$	3/2—3/2
2932,06	1	16,69	20,92	$4d\,^2P—5f\,^4D^\circ$	3/2—3/2
2930,40	2	17,60	21,83	$5p\,^2D^\circ—5d'\,^2D$	3/2—3/2
2921,92	4	—	—	—	—
2913,23	4	16,69	20,94	$4d\,^2P—5p''\,^2P^\circ$	3/2—3/2
2908,63	5	16,68	20,94	$4d\,^2D—5p''\,^2P^\circ$	5/2—3/2
2907,15	1	17,57	21,83	$5p\,^4S^\circ—5d'\,^2D$	3/2—3/2
2895,92	1	16,32	20,60	$4d\,^2F—1^\circ$	7/2—3/2
2894,63	2	17,25	21,53	$5p\,^2P^\circ—7s\,^4P$	1/2—3/2
2884,21	2	17,17	21,47	$2—5f'\,^2F^\circ$	3/2—5/2
2873,72	4	16,83	21,15	$5p\,^4P^\circ—6s'\,^2D$	1/2—3/2
2862,17	2	17,16	21,49	$5p\,^4D^\circ—7s\,^4P$	3/2—5/2
2847,36	25	14,27	18,62	$5s\,^4P—5p'\,^2P^\circ$	3/2—3/2
2844,46	20	16,48	20,84	$1—5f\,^4F^\circ$	5/2—7/2
2839,20	2	16,48	20,85	$1—5f\,^4F^\circ$	5/2—5/2
2838,79	20	16,48	20,85	$4d\,^2D—5f\,^4F^\circ$	3/2—5/2
2835,35	8	17,16	21,53	$5p\,^4D^\circ—7s\,^4P$	3/2—3/2
2833,00	100	16,48	20,86	$1—5f\,^2F^\circ$	5/2—7/2
2832,39	2	16,48	20,86	$4d\,^2D—5p''\,^2P^\circ$	3/2—1/2
2830,43	3	17,65	22,03	$5p\,^2S^\circ—5d'\,^2P$	1/2—3/2
2823,03	2	16,48	20,88	$1—5f\,^4F^\circ$	5/2—3/2
2822,63	5	16,65	21,04	$5p\,^4P^\circ—5d\,^2D$	3/2—5/2
		16,48	20,88	$4d\,^2D—5f\,^4F^\circ$	3/2—3/2
2816,87	30	16,48	20,89	$1—5f\,^2F^\circ$	5/2—5/2
2816,46	60	16,48	20,89	$4d\,^2D—5f\,^2F^\circ$	3/2—5/2
2808,72	1	17,65	22,06	$5p\,^2S^\circ—7s\,^2P$	1/2—3/2
2803,60	4	16,48	20,91	$1—5f\,^2D^\circ$	5/2—3/2
2803,20	20	16,48	20,91	$4d\,^2D—5f\,^2D^\circ$	3/2—3/2
2801,23	2	17,60	22,03	$5p\,^2D^\circ—5d'\,^2P$	3/2—3/2
2800,98	2	17,65	22,07	$5p\,^2S^\circ—7s\,^4P$	1/2—1/2
2796,26	2	16,48	20,92	$4d\,^2D—5f\,^4D^\circ$	3/2—3/2
2795,81	30	—	—	—	—
2789,83	3	—	—	—	—
2779,97	1	17,60	22,06	$5p\,^2D^\circ—7s\,^2P$	3/2—3/2
		17,57	22,03	$5p\,^4S^\circ—5d'\,^2P$	3/2—3/2
2779,51	4	16,48	20,94	$1—5p''\,^2P^\circ$	5/2—3/2
2779,11	20	16,48	20,94	$4d\,^2D—5p''\,^2P^\circ$	3/2—3/2
2778,99	2	17,37	21,83	$5p\,^2P^\circ—5d'\,^2D$	3/2—3/2

λ, Å	λ	E_H, eV	E_B, eV	Transition	J
2777,96	1	17,57	22,03	$5p\ ^4S^\circ-5d'\ ^2F$	$3/2-5/2$
2774,59	3	17,00	21,46	$4d\ ^2F-5f'\ ^2F^\circ$	$5/2-7/2$
2772,60	10	17,00	21,47	$4d\ ^2F-5f'\ ^2F^\circ$	$5/2-5/2$
2759,02	4	17,57	22,06	$5p\ ^4S^\circ-7s\ ^2P$	$3/2-3/2$
2751,59	5	17,57	22,07	$5p\ ^4S^\circ-7s\ ^4P$	$3/2-1/2$
2747,41	2	—	—	—	—
2746,31	15	—	—	—	—
2744,64	1	17,65	22,17	$5p\ ^2S^\circ-5d'\ ^2P$	$1/2-1/2$
2742,56	40	16,32	20,84	$4d\ ^2F-5f\ ^4F^\circ$	$7/2-9/2$
2740,11	1	16,32	20,84	$4d\ ^2F-5f\ ^4F^\circ$	$7/2-7/2$
2738,13	1	17,60	22,13	$5p\ ^2D^\circ-7s\ ^2P$	$3/2-1/2$
2733,26	50	—	—	—	—
2732,33	4	17,00	21,53	$4d\ ^2F-5f'\ ^2P^\circ$	$5/2-3/2$
2729,46	30	16,32	20,86	$4d\ ^2F-5f\ ^2F^\circ$	$7/2-7/2$
2719,90	5	—	—	—	—
2717,70	1	17,57	22,13	$5p\ ^4S^\circ-7s\ ^2P$	$3/2-1/2$
2717,18	1	17,60	22,17	$5p\ ^2D^\circ-5d'\ ^2P$	$3/2-1/2$
2716,16	10	16,29	20,85	$4d\ ^4P-5f\ ^2F^\circ$	$5/2-5/2$
2714,49	3	16,32	20,89	$4d\ ^2F-5f\ ^2F^\circ$	$7/2-5/2$
2712,40	80	16,32	20,89	$4d\ ^2F-2^\circ$	$7/2-7/2$
2711,11	2	13,99	18,56	$5s\ ^4P-5p'\ ^2F^\circ$	$5/2-7/2$
2710,27	3	16,29	20,86	$4d\ ^4P-5p''\ ^2P^\circ$	$5/2-1/2$
2701,34	15	16,29	20,89	$4d\ ^4P-5f\ ^4F^\circ$	$5/2-3/2$
		17,25	21,83	$5p\ ^2P^\circ-5d'\ ^2D$	$1/2-3/2$
2700,60	3	—	—	—	—
2695,70	30	16,29	20,89	$4d\ ^4P-5f\ ^2F^\circ$	$5/2-5/2$
2691,20	2	14,27	18,87	$5s\ ^4P-5p'\ ^2P^\circ$	$3/2-1/2$
2688,37	4	—	—	—	—
2683,55	15	16,29	20,91	$4d\ ^4P-5f\ ^2D^\circ$	$5/2-3/2$
2677,20	6	16,29	20,92	$4d\ ^4P-5f\ ^4D^\circ$	$5/2-3/2$
2675,31	4	16,23	20,86	$4d\ ^4P-5p''\ ^2P^\circ$	$3/2-1/2$
2672,79	3	17,00	21,63	$4d\ ^2F-5f'\ ^2P^\circ$	$5/2-1/2$
2666,61	6	16,23	20,88	$4d\ ^4P-5f\ ^4F^\circ$	$3/2-3/2$
2664,37	4	17,38	22,03	$5p\ ^4D^\circ-5d'\ ^2P$	$1/2-3/2$
2664,00	8	16,83	21,49	$5p\ ^4P^\circ-7s\ ^4P$	$1/2-5/2$
2662,57	2	17,37	22,03	$5p\ ^2P^\circ-5d'\ ^2P$	$3/2-3/2$
2661,47	5	16,29	20,94	$4d\ ^4P-5p''\ ^2P^\circ$	$5/2-3/2$
2661,22	1	17,37	22,03	$5p\ ^2D^\circ-5d'\ ^2P$	$5/2-3/2$
		16,23	20,89	$4d\ ^4P-5f\ ^2F^\circ$	$3/2-5/2$
2660,97	8	16,87	21,53	$5p\ ^4D^\circ-7s\ ^4P$	$5/2-3/2$
2659,60	2	17,38	22,03	$5p\ ^2D^\circ-5d'\ ^2F$	$5/2-5/2$
2656,38	15	—	—	—	—
2653,95	6	16,18	20,85	$4d\ ^4P-5f\ ^4F^\circ$	$1/2-5/2$
2649,67	4	17,16	21,83	$5p\ ^4D^\circ-5d'\ ^2D$	$3/2-3/2$
2649,27	20	16,23	20,91	$4d\ ^4P-5f\ ^2D^\circ$	$3/2-3/2$
2648,15	20	—	—	—	—
2643,06	20	16,23	20,92	$4d\ ^4P-5f\ ^4D^\circ$	$3/2-3/2$
2642,08	4	17,37	22,06	$5p\ ^2D^\circ-7s\ ^2P$	$5/2-3/2$
2640,74	2	16,83	21,53	$5p\ ^4P^\circ-7s\ ^4P$	$1/2-3/2$
2638,32	2	17,38	22,07	$5p\ ^4D^\circ-7s\ ^4P$	$1/2-1/2$
2636,51	3	17,37	22,07	$5p\ ^2P^\circ-7s\ ^4P$	$3/2-1/2$
2634,41	6	16,18	20,89	$4d\ ^4P-5f\ ^2F^\circ$	$1/2-5/2$
2627,75	7	16,23	20,94	$4d\ ^4P-5p''\ ^2P^\circ$	$3/2-3/2$
		16,60	21,32	$5p\ ^4P^\circ-4d''\ ^2D$	$5/2-5/2$
2627,22	3	—	—	—	—
2624,78	6	—	—	—	—
2622,82	2	16,18	20,91	$4d\ ^4P-5f\ ^2D^\circ$	$1/2-3/2$
2620,65	6	—	—	—	—

λ, Å	I	E_H, eV	E_B, eV	Transition	J
2620,44	40	16,69	21,42	$4d\ ^2P—5f'\ ^2D^\circ$	$^3/_2—^5/_2$
2616,71	10	{ 16,18	20,92	$4d\ ^4P—5f\ ^4D^\circ$	$^1/_2—^3/_2$
		{ 16,68	21,42	$4d\ ^2D—5f'\ ^2D^\circ$	$^5/_2—^5/_2$
2610,98	10	—	—	—	—
2610,76	1	15,85	20,60	$5s'\ ^2D—1^\circ$	$^5/_2—^3/_2$
2605,41	1	17,37	22,13	$5p\ ^2P^\circ—7s\ ^2P^\circ$	$^3/_2—^1/_2$
2602,11	7	16,08	20,84	$4d\ ^4F—5f\ ^4F^\circ$	$^5/_2—^7/_2$
2597,73	7	16,08	20,85	$4d\ ^4F—5f\ ^2F^\circ$	$^5/_2—^5/_2$
2596,73	5	—	—	—	—
2595,36	4	16,09	20,86	$4d\ ^4F—5p''\ ^2P^\circ$	$^3/_2—^1/_2$
2594,40	4	16,69	21,47	$4d\ ^2P—5f'\ ^2F^\circ$	$^3/_2—^5/_2$
2592,48	60	16,08	20,86	$4d\ ^4F—5f\ ^2F^\circ$	$^5/_2—^7/_2$
2591,25	1	17,25	22,03	$5p\ ^2P^\circ—5d'\ ^2P$	$^1/_2—^3/_2$
2590,74	2	16,68	21,47	$4d\ ^2D—5f'\ ^2F^\circ$	$^5/_2—^5/_2$
2589,08	30	—	—	—	—
2584,15	3	16,08	20,88	$4d\ ^4F—5f\ ^4F^\circ$	$^5/_2—^3/_2$
2581,74	5	—	—	—	—
2578,98	2	16,08	20,89	$4d\ ^4F—5f\ ^2F^\circ$	$^5/_2—^5/_2$
2572,03	10	—	—	—	—
2566,61	1	17,25	22,07	$5p\ ^2P^\circ—7s\ ^4P^\circ$	$^1/_2—^1/_2$
2562,04	1	16,08	20,92	$4d\ ^4F—5f\ ^4D^\circ$	$^5/_2—^3/_2$
2561,94	3	16,65	21,49	$5p\ ^4P^\circ—7s\ ^4P$	$^3/_2—^5/_2$
2559,10	8	16,69	21,53	$4d\ ^2P—5f'\ ^2P^\circ$	$^3/_2—^3/_2$
2556,36	6	—	—	—	—
2555,91	6	—	—	—	—
2538,34	5	16,60	21,49	$5p\ ^4P^\circ—7s\ ^4P$	$^5/_2—^5/_2$
2531,73	1	13,99	18,88	$5s\ ^4P—5p'\ ^2D^\circ$	$^5/_2—^5/_2$
2527,16	3	—	—	—	—
2517,95	8	—	—	—	—
2511,74	3	16,48	21,42	$1—5f'\ ^2D^\circ$	$^5/_2—^5/_2$
2510,56	5	16,87	21,81	$5p\ ^4D^\circ—5d'\ ^2G$	$^5/_2—^7/_2$
2506,56	5	—	—	—	—
2503,87	7	16,32	21,27	$4d\ ^2F—6^\circ$	$^7/_2—^7/_2$
2494,66	1	17,37	21,83	$5p\ ^4D^\circ—5d'\ ^2D$	$^5/_2—^3/_2$
2489,39	8	16,48	21,46	$1—5f'\ ^2F^\circ$	$^5/_2—^7/_2$
2487,62	4	15,86	20,84	$4d\ ^4F—5f\ ^4F^\circ$	$^7/_2—^7/_2$
2487,50	3	16,48	21,47	$4d\ ^2D—5f'\ ^2F^\circ$	$^3/_2—^5/_2$
2483,62	1	15,86	20,85	$4d\ ^4F—5f\ ^4F^\circ$	$^7/_2—^5/_2$
2478,85	3	15,86	20,86	$4d\ ^4F—5f\ ^2F^\circ$	$^7/_2—^7/_2$
2474,69	2	15,85	20,86	$5s'\ ^2D—5f\ ^2F^\circ$	$^5/_2—^7/_2$
2470,45	10	—	—	—	—
2464,77	100	15,86	20,89	$4d\ ^4F—2^\circ$	$^7/_2—^7/_2$
2463,27	2	15,82	20,85	$5s'\ ^2D—5f\ ^4F^\circ$	$^3/_2—^5/_2$
2462,33	2	15,85	20,89	$5s'\ ^2D—5f\ ^2F^\circ$	$^5/_2—^5/_2$
2456,07	6	—	—	—	—
2455,31	2	16,48	21,53	$1—5f'\ ^2P^\circ$	$^5/_2—^3/_2$
2455,04	2	16,48	21,53	$4d\ ^2D—5f'\ ^2P^\circ$	$^3/_2—^3/_2$
2446,44	8	15,82	20,89	$5s'\ ^2D—5f\ ^2F^\circ$	$^3/_2—^5/_2$
2432,74	8	—	—	—	—
2428,35	20	—	—	—	—
2426,36	10	13,51	18,62	$4p^6\ ^2S—5p'\ ^2P^\circ$	$^1/_2—^3/_2$
2418,41	4	—	—	—	—
2414,94	2	16,29	21,42	$4d\ ^4P—5f'\ ^2D^\circ$	$^7/_2—^7/_2$
2414,89	10	—	—	—	—
2413,81	10	—	—	—	—
2409,06	5	16,32	21,46	$4d\ ^2F—5f'\ ^2F^\circ$	$^7/_2—^7/_2$
2408,52	5	—	—	—	—
2392,78	10	16,29	21,47	$4d\ ^4P—5f'\ ^2F^\circ$	$^5/_2—^5/_2$

λ, Å	I	$E_н$, eV	$E_в$, eV	Transition	J
2390,50	4	16,23	21,41	$4d\ ^4P—5f'\ ^2D°$	$^3/_2—^3/_2$
2375,52	20	15,62	20,84	$4d\ ^4F—5f\ ^4F°$	$^9/_2—^9/_2$
2373,68	4	15,62	20,84	$4d\ ^4F—5f\ ^4F°$	$^9/_2—^7/_2$
2368,94	3	16,18	21,41	$4d\ ^4P—5f'\ ^2D°$	$^1/_2—^3/_2$
2365,52	3	16,23	21,47	$4d\ ^4P—5f'\ ^2F°$	$^3/_2—^5/_2$
2362,74	6	16,29	21,53	$4d\ ^4P—5f'\ ^2P°$	$^5/_2—^3/_2$
2353,68	50	—	—	—	—
2352,86	2	15,62	20,89	$4d\ ^4F—2°$	$^9/_2—^7/_2$
2344,38	10	16,18	21,47	$4d\ ^4P—5f'\ ^2F°$	$^1/_2—^5/_2$
2316,32	10	—	—	—	—
2315,52	8	16,18	21,53	$4d\ ^4P—5f'\ ^2P°$	$^1/_2—^3/_2$
2314,24	6	—	—	—	—
2312,00	6	13,51	18,87	$4p^6\ ^2S—5p'\ ^2P°$	$^1/_2—^1/_2$
2302,67	3	16,09	21,47	$4d\ ^4F—5f'\ ^2F°$	$^3/_2—^5/_2$
2301,73	6	16,08	21,46	$4d\ ^4F—5f'\ ^2F°$	$^5/_2—^7/_2$
2300,38	6	16,08	21,47	$4d\ ^4F—5f'\ ^2F°$	$^5/_2—^5/_2$
2287,79	30	15,85	21,27	$5s'\ ^2D—6°$	$^5/_2—^7/_2$
2283,07	30	—	—	—	—
2273,24	8	—	—	—	—
2272,55	1	16,08	21,53	$4d\ ^4F—5f'\ ^2P°$	$^5/_2—^3/_2$
2250,32	8	—	—	—	—
2245,39	10	—	—	—	—
2237,15	4	—	—	—	—
2227,92	30	15,85	21,42	$5s'\ ^2D—7°$	$^5/_2—^5/_2$
2212,96	5	15,82	21,42	$5s'\ ^2D—5f'\ ^2D°$	$^3/_2—^5/_2$
2212,29	6	—	—	—	—
2211,71	5	15,86	21,46	$4d\ ^4F—5f'\ ^2F°$	$^7/_2—^7/_2$
2208,41	1	15,85	21,46	$5s'\ ^2D—5f'\ ^2F°$	$^5/_2—^7/_2$
2185,52	5	—	—	—	—
2177,79	3	—	—	—	—
2164,38	4	—	—	—	—
2162,50	3	—	—	—	—
2145,08	10	15,10	20,88	$4d\ ^4D—5f\ ^4F°$	$^1/_2—^3/_2$
2133,85	2	15,10	20,91	$4d\ ^4D—5f\ ^2D°$	$^1/_2—^3/_2$
2129,80	1	15,10	20,92	$4d\ ^4D—5f\ ^4D°$	$^1/_2—^3/_2$
2123,48	3	—	—	—	—
2118,83	12	15,00	20,85	$4d\ ^4D—5f\ ^4F°$	$^3/_2—^5/_2$
2109,81	5	15,00	20,88	$4d\ ^4D—5f\ ^4F°$	$^3/_2—^3/_2$
2098,97	1	15,00	20,91	$4d\ ^4D—5f\ ^2D°$	$^3/_2—^3/_2$
2096,24	15	14,93	20,84	$4d\ ^4D—5f\ ^4F°$	$^5/_2—^3/_2$
2095,02	1	15,00	20,92	$4d\ ^4D—5f\ ^4D°$	$^3/_2—^1/_2$
2093,37	3	14,93	20,85	$4d\ ^4D—5f\ ^4F°$	$^5/_2—^5/_2$
2088,16	20	14,90	20,84	$4d\ ^4D—5f\ ^4F°$	$^7/_2—^9/_2$
2086,73	5	14,90	20,84	$4d\ ^4D—5f\ ^4F°$	$^7/_2—^7/_2$
2084,54	1	14,93	20,88	$4d\ ^4D—5f\ ^4F°$	$^5/_2—^3/_2$
2083,87	1	14,90	20,85	$4d\ ^4D—5f\ ^4F°$	$^7/_2—^5/_2$
2080,53	1	14,90	20,84	$4d\ ^4D—5f\ ^2F°$	$^7/_2—^7/_2$
964,962	30	0,67	13,51	$4p^5\ ^2P°—4p^6\ ^2S$	$^1/_2—^1/_2$
917,434	20	0,00	13,51	$4p^5\ ^2P°—4p^6\ ^2S$	$^3/_2—^1/_2$
911,384	25	0,67	14,27	$4p^5\ ^2P°—5s\ ^4P$	$^1/_2—^3/_2$
890,982	20	0,67	14,58	$4p^5\ ^2P°—5s\ ^4P$	$^1/_2—^1/_2$
886,302	30	0,00	13,99	$4p^5\ ^2P°—5s\ ^4P$	$^3/_2—^5/_2$
884,144	30	0,67	14,69	$4p^5\ ^2P°—5s\ ^2P$	$^1/_2—^3/_2$
868,869	25	0,00	14,69	$4p^5\ ^2P°—5s\ ^4P$	$^3/_2—^3/_2$
864,812	20	0,67	15,00	{ $4p^5\ ^2P°—5s\ ^2P$ $4p^5\ ^2P°—4d\ ^4D$	$^1/_2—^1/_2$ $^1/_2—^3/_2$
859,040	20	0,67	15,10	$4p^5\ ^2P°—4d\ ^4D$	$^1/_2—^1/_2$
850,318	6	0,00	14,58	$4p^5\ ^2P°—5s\ ^4P$	$^3/_2—^1/_2$

λ, Å	I	E_H, eV	E_B, eV	Transition	J
844,058	25	0,00	14,69	$4p^5\,{}^2P^\circ$—$5s\,{}^2P$	$^3/_2$—$^3/_2$
830,377	18	0,00	14,93	$4p^5\,{}^2P^\circ$—$4d\,{}^4D$	$^3/_2$—$^5/_2$
826,432	22	0,00	15,00	$4p^5\,{}^2P^\circ$—$5s\,{}^2P$ $4p^5\,{}^2P^\circ$—$4d\,{}^4D$	$^3/_2$—$^1/_2$ $^3/_2$—$^3/_2$
821,161	20	0,00	15,10	$4p^5\,{}^2P^\circ$—$4d\,{}^4D$	$^3/_2$—$^1/_2$
818,147	25	0,67	15,82	$4p^5\,{}^2P^\circ$—$5s'\,{}^2S$	$^1/_2$—$^3/_2$
799,083	9	0,67	16,18	$4p^5\,{}^2P^\circ$—$4d\,{}^4P$	$^1/_2$—$^1/_2$
796,678	6	0,67	16,23	$4p^5\,{}^2P^\circ$—$4d\,{}^4P$	$^1/_2$—$^3/_2$
783,715	20	0,00 0,67	15,82 16,48	$4p^5\,{}^2P^\circ$—$5s'\,{}^2D$ $4p^5\,{}^2P^\circ$—$4d\,{}^2D$	$^3/_2$—$^3/_2$ $^1/_2$—$^3/_2$
782,084	25	0,00	15,85	$4p^5\,{}^2P^\circ$—$5s'\,{}^2D$	$^3/_2$—$^5/_2$
773,684	18	0,67	16,69	$4p^5\,{}^2P^\circ$—$4d\,{}^2P$	$^1/_2$—$^3/_2$
771,024	18	0,00	16,08	$4p^5\,{}^2P^\circ$—$4d\,{}^4F$	$^3/_2$—$^5/_2$
766,202	9	0,00	16,18	$4p^5\,{}^2P^\circ$—$4d\,{}^4P$	$^3/_2$—$^1/_2$
763,976	11	0,00	16,23	$4p^5\,{}^2P^\circ$—$4d\,{}^4P$	$^3/_2$—$^3/_2$
761,050	18	0,00	16,29	$4p^5\,{}^2P^\circ$—$4d\,{}^4P$	$^3/_2$—$^5/_2$
752,051	30	0,00	16,48	$4p^5\,{}^2P^\circ$—$4d\,{}^2D$	$^3/_2$—$^3/_2$
743,122	9	0,00	16,68	$4p^5\,{}^2P^\circ$—$4d\,{}^2D$	$^3/_2$—$^5/_2$
742,821	9	0,00	16,69	$4p^5\,{}^2P^\circ$—$4d\,{}^2P$	$^3/_2$—$^3/_2$
729,402	20	0,00	17,00	$4p^5\,{}^2P^\circ$—$4d\,{}^2F$	$^3/_2$—$^5/_2$
722,036	50	0,00	17,17	$4p^5\,{}^2P^\circ$—2	$^3/_2$—$^3/_2$
712,036	8	0,67	18,08	$4p^5\,{}^2P^\circ$—$5s''\,{}^2S$	$^1/_2$—$^1/_2$
690,557	11	0,67	18,62	$4p^5\,{}^2P^\circ$—$4d'\,{}^2D$	$^1/_2$—$^3/_2$
685,812	11	0,00	18,08	$4p^5\,{}^2P^\circ$—$5s''\,{}^2D$	$^3/_2$—$^1/_2$
682,791	16	0,67	18,82	$4p^5\,{}^2P^\circ$—$4d'\,{}^2P$	$^1/_2$—$^3/_2$
681,119	16	0,67	18,87	$4p^5\,{}^2P^\circ$—$4d'\,{}^2P$	$^1/_2$—$^1/_2$
668,827	20	0,00	18,54	$4p^5\,{}^2P^\circ$—$4d'\,{}^2D$	$^3/_2$—$^5/_2$
665,870	9	0,00	18,62	$4p^5\,{}^2P^\circ$—$4d'\,{}^2D$	$^3/_2$—$^3/_2$
663,039	20	0,00	18,70	$4p^5\,{}^2P^\circ$—$4d'\,{}^2F$	$^3/_2$—$^5/_2$
658,637	5	0,00	18,82	$4p^5\,{}^2P^\circ$—$4d'\,{}^2P$	$^3/_2$—$^3/_2$
657,088	13	0,00	18,87	$4p^5\,{}^2P^\circ$—$4d'\,{}^2P$	$^3/_2$—$^1/_2$
655,677	5	0,67	19,57	$4p^5\,{}^2P^\circ$—$6s\,{}^4P$	$^1/_2$—$^3/_2$
643,404	9	0,67	19,94	$4p^5\,{}^2P^\circ$—$5d\,{}^4P$	$^1/_2$—$^1/_2$
640,870	5	0,67	20,01	$4p^5\,{}^2P^\circ$—$5d\,{}^4P$	$^1/_2$—$^3/_2$
639,263	5	0,67	20,06	$4p^5\,{}^2P^\circ$—$5d\,{}^4D$	$^1/_2$—$^3/_2$
638,952	5	0,67	20,07	$4p^5\,{}^2P^\circ$—$6s\,{}^4P$	$^1/_2$—$^1/_2$
638,214	4	0,67	20,09	$4p^5\,{}^2P^\circ$—$6s\,{}^2P$	$^1/_2$—$^3/_2$
636,154	3	0,67	20,15	$4p^5\,{}^2P^\circ$—$6s\,{}^2P$	$^1/_2$—$^1/_2$
634,265	4	0,67	20,21	$4p^5\,{}^2P^\circ$—$5d\,{}^4D$	$^1/_2$—$^1/_2$
633,375	5	0,00	19,57	$4p^5\,{}^2P^\circ$—$6s\,{}^4P$	$^3/_2$—$^3/_2$
621,910	5	0,00	19,94	$4p^5\,{}^2P^\circ$—$5d\,{}^4P$	$^3/_2$—$^1/_2$
621,071	5	0,00	19,96	$4p^5\,{}^2P^\circ$—$5d\,{}^4P$	$^3/_2$—$^5/_2$
619,548	2	0,00	20,01	$4p^5\,{}^2P^\circ$—$5d\,{}^4P$	$^3/_2$—$^3/_2$
619,379	2	0,00	20,02	$4p^5\,{}^2P^\circ$—$5d\,{}^4D$	$^3/_2$—$^5/_2$
618,879	3	0,67	20,70	$4p^5\,{}^2P^\circ$—$5d\,{}^2P$	$^1/_2$—$^1/_2$
618,515	2	0,67	20,71	$4p^5\,{}^2P^\circ$—$5d\,{}^4F$	$^1/_2$—$^3/_2$
618,042	4	0,00	20,06	$4p^5\,{}^2P^\circ$—$5d\,{}^4D$	$^3/_2$—$^3/_2$
617,750	4	0,00	20,07	$4p^5\,{}^2P^\circ$—$6s\,{}^4P$	$^3/_2$—$^1/_2$
617,068	6	0,00	20,09	$4p^5\,{}^2P^\circ$—$6s\,{}^2P$	$^3/_2$—$^3/_2$
615,225	4	0,67	20,82	$4p^5\,{}^2P^\circ$—$5d\,{}^2D$	$^1/_2$—$^3/_2$
615,134	4	0,00	20,15	$4p^5\,{}^2P^\circ$—$6s\,{}^2P$	$^3/_2$—$^1/_2$
613,336	4	0,00	20,21	$4p^5\,{}^2P^\circ$—$5d\,{}^4D$	$^3/_2$—$^1/_2$
608,124	5	0,00	20,39	$4p^5\,{}^2P^\circ$—$6s''\,{}^2S$	$^3/_2$—$^1/_2$
605,776	5	0,00	20,47	$4p^5\,{}^2P^\circ$—$5d\,{}^4F$	$^3/_2$—$^5/_2$
605,536	5	0,00	20,47	$4p^5\,{}^2P^\circ$—$5d\,{}^2P$	$^3/_2$—$^3/_2$
605,316	5	0,67	21,15	$4p^5\,{}^2P^\circ$—$6s'\,{}^2D$	$^1/_2$—$^3/_2$
599,944	4	0,67	21,33	$4p^5\,{}^2P^\circ$—$4d''\,{}^2D$	$^1/_2$—$^3/_2$
598,968	3	0,00	20,70	$4p^5\,{}^2P^\circ$—$5d\,{}^2P$	$^3/_2$—$^1/_2$

λ, Å	I	E_H, eV	E_B, eV	Transition	J
598,791	3	0,00	20,70	$4p^5\ {}^2P^\circ$—3	$3/2$—$5/2$
598,66	0	0,00	20,71	$4p^5\ {}^2P^\circ$—$5d\ {}^4F$	$3/2$—$3/2$
596,944	4	0,00	20,77	$4p^5\ {}^2P^\circ$—$5d\ {}^2F$	$3/2$—$5/2$
595,530	7	0,00	20,82	$4p^5\ {}^2P^\circ$—$5d\ {}^2D$	$3/2$—$3/2$
589,262	5	0,00	21,04	$4p^5\ {}^2P^\circ$—$5d\ {}^2D$	$3/2$—$5/2$
586,269	1	0,00	21,15	$4p^5\ {}^2P^\circ$—$6s'\ {}^2D$	$3/2$—$5/2$, $3/2$
585,684	2	0,67	21,83	$4p^5\ {}^2P^\circ$—$5d'\ {}^2D$	$1/2$—$3/2$
581,496	3	0,00	21,32	$4p^5\ {}^2P^\circ$—$4d''\ {}^2D$	$3/2$—$5/2$
581,22	1	0,00	21,33	$4p^5\ {}^2P^\circ$—$4d''\ {}^2D$	$3/2$—$3/2$
580,342	3	0,67	22,03	$4p^5\ {}^2P^\circ$—$5d'\ {}^2P$	$1/2$—$3/2$
579,11	0	0,67	22,07	$4p^5\ {}^2P^\circ$—$7s\ {}^4P$	$1/2$—$1/2$
577,01	0	0,00	21,49	$4p^5\ {}^2P^\circ$—$7s\ {}^4P$	$3/2$—$5/2$
576,647	4	0,67	22,17	$4p^5\ {}^2P^\circ$—$5d'\ {}^2P$	$1/2$—$1/2$
575,902	2	0,00	21,53	$4p^5\ {}^2P^\circ$—$7s\ {}^4P$	$3/2$—$3/2$
560,788	3	0,00	22,11	$4p^5\ {}^2P^\circ$—$5d'\ {}^2D$	$3/2$—$5/2$
559,320	4	0,00	22,17	$4p^5\ {}^2P^\circ$—$5d'\ {}^2D$	$3/2$—$1/2$

Kr III, ground state $1s^2 2s^2 2p^6 3s^2 3p^6 3d^{10} 4s^2 4p^4\ {}^3P_2$

Ionization potential 298 020 cm^{-1}; 36,947 eV

λ, Å	I	E_H, eV	E_B, eV	Transition	J
7353,42	1	24,01	25,69	$4d''\ {}^3P^\circ$—$5p''\ {}^3D$	2—1
7057,45	2	20,51	22,27	$4d'\ {}^1D^\circ$—$5p\ {}^3P$	2—1
6977,95	3	23,38	25,15	$4d''\ {}^1D^\circ$—$5p'\ {}^1D$	2—2
6818,13	1	23,34	25,15	$4d''\ {}^1P^\circ$—$5p'\ {}^1D$	1—2
6793,53	3	22,35	24,17	$5s''\ {}^3P^\circ$—$5p'\ {}^3F$	2—3
6728,41	1	24,01	25,85	$4d''\ {}^3P^\circ$—$5p''\ {}^3D$	2—2
6683,55	1	24,01	25,86	$4d''\ {}^3P^\circ$—$5p''\ {}^3S$	2—1
6651,75	10	20,46	22,33	$5s'\ {}^3D^\circ$—$5p\ {}^3P$	3—2
6602,90	10	22,68	24,56	$4d''\ {}^1F^\circ$—$5p'\ {}^3P$	3—2
6444,70	1	21,72	23,65	$4d''\ {}^3F^\circ$—$5p'\ {}^3D$	2—1
6395,09	2	24,01	25,95	$4d''\ {}^3P^\circ$—$5p''\ {}^1P$	2—1
6310,22	10	22,60	24,56	$4d''\ {}^3D^\circ$—$5p'\ {}^3P$	1—2
6250,98	5	20,29	22,27	$5s'\ {}^3D^\circ$—$5p\ {}^3P$	2—1
6164,76	1	24,01	26,02	$4d''\ {}^3P^\circ$—$5p''\ {}^3D$	2—3
6110,81	5	20,24	22,27	$5s'\ {}^3D^\circ$—$5p\ {}^3P$	1—1
6078,38	10	20,29	22,33	$5s'\ {}^3D^\circ$—$5p\ {}^3P$	2—2
6050,11	3	22,60	24,65	$4d''\ {}^3D^\circ$—$5p'\ {}^3P$	1—0
6037,17	10	22,60	24,65	$4d''\ {}^3D^\circ$—$5p'\ {}^3P$	1—1
5935,03	8	22,47	24,56	$5s''\ {}^1P^\circ$—$5p'\ {}^3P$	1—2
5891,72	5	20,24	22,35	$5s'\ {}^3D^\circ$—$5p\ {}^3P$	1—0
5873,50	1	23,58	25,69	$4d''\ {}^3P^\circ$—$5p''\ {}^3D$	1—1
5715,80	1	21,72	23,89	$4d''\ {}^3F^\circ$—$5p'\ {}^3D$	2—2
5597,32	5	22,35	24,56	$5s''\ {}^3P^\circ$—$5p'\ {}^3P$	2—2
5501,43	10	21,92	24,17	$4d''\ {}^3D^\circ$—$5p'\ {}^3F$	2—3
5477,66	2	21,63	23,89	$4d'\ {}^3D^\circ$—$5p'\ {}^3D$	3—2
5475,49	1	21,38	23,65	$4d'\ {}^3D^\circ$—$5p'\ {}^3D$	2—1
5438,20	2	23,58	25,86	$4d''\ {}^3P^\circ$—$5p''\ {}^3S$	1—1
5412,19	5	24,01	26,30	$4d''\ {}^3P^\circ$—$5p''\ {}^1D$	2—2
5389,12	1	24,65	26,95	$5p'\ {}^3P$—$5d\ {}^3D^\circ$	1—2
5381,39	2	22,35	24,65	$5s''\ {}^3P^\circ$—$5p'\ {}^3P$	2—1
5371,40	4	24,01	26,32	$4d''\ {}^3P^\circ$—$5p''\ {}^3P$	2—1
5362,11	1	21,72	24,03	$4d''\ {}^3F^\circ$—$5p'\ {}^3F$	2—2
5349,77	2	21,92	24,23	$4d''\ {}^3D^\circ$—$5p'\ {}^1F$	2—3

λ, Å	I	E_H, eV	E_B, eV	Transition	J
5338,20	2	21,32	23,65	$4d'\ ^1P^\circ$—$5p'\ ^3D$	1—1
5263,18	1	21,68	24,03	$4d''\ ^3F^\circ$—$5p'\ ^3D$	3—3
5257,83	2	23,34	25,69	$4d''\ ^1P^\circ$—$5p''\ ^3D$	1—1
5188,68	1	24,56	26,95	$5p'\ ^3P$—$5d\ ^3D^\circ$	2—2
5160,09	1	21,63	24,03	$4d'\ ^3D^\circ$—$5p'\ ^3D$	3—3
5152,01	3	21,63	24,03	$4d'\ ^3D^\circ$—$5p'\ ^3F$	3—2
5151,68	2	24,01	26,41	$4d''\ ^3P^\circ$—$5p''\ ^3P$	2—2
5110,98	1	23,58	26,01	$4d''\ ^3P^\circ$—$5p''\ ^3P$	1—0
5069,96	4	21,45	23,89	$4d'\ ^3S^\circ$—$5p'\ ^3D$	1—2
5061,46	2	21,72	24,17	$4d''\ ^3P^\circ$—$5p'\ ^3P$	2—3
5042,86	2	21,19	23,65	$5s'\ ^1D^\circ$—$5p'\ ^3D$	2—1
5018,72	2	21,70	24,17	$4d''\ ^3F^\circ$—$5p''\ ^3F$	4—3
5016,45	20	22,68	25,15	$4d''\ ^1F^\circ$—$5p'\ ^1D$	3—2
4988,52	10	23,38	25,86	$4d''\ ^1D^\circ$—$5p''\ ^3S$	2—1
4977,08	2	19,84	22,33	$4d'\ ^3G^\circ$—$5p\ ^3P$	3—2
4965,78	2	21,68	24,17	$4d''\ ^3F^\circ$—$5p'\ ^3F$	3—3
4940,21	2	21,38	23,89	$4d'\ ^3D^\circ$—$5p'\ ^3D$	2—2
4906,28	6	23,34	25,86	$4d''\ ^1P^\circ$—$5p''\ ^3S$	1—1
4892,21	5	21,70	24,23	$4d''\ ^3F^\circ$—$5p'\ ^1F$	4—3
4873,87	1	21,63	24,17	$4d'\ ^3D^\circ$—$5p'\ ^3F$	3—3
4845,62	2	21,70	24,26	$4d''\ ^3F^\circ$—$5p'\ ^3F$	4—4
4841,9	1	21,68	24,23	$4d''\ ^3F^\circ$—$5p'\ ^1F$	3—3
4826,08	2	23,38	25,95	$4d''\ ^1D^\circ$—$5p''\ ^1P$	2—1
4789,74	7	21,45	24,03	$4d'\ ^3S^\circ$—$5p'\ ^3F$	1—2
4754,48	6	21,63	24,23	$4d'\ ^3D^\circ$—$5p'\ ^1F$	3—3
4749,00	2	23,34	25,95	$4d''\ ^1P^\circ$—$5p''\ ^1P$	1—1
4729,72	4	21,45	24,07	$4d'\ ^3S^\circ$—$5p'\ ^1P$	1—1
4710,48	10	21,63	24,26	$4d'\ ^3D^\circ$—$5p'\ ^3F$	3—4
4693,65	3	23,38	26,02	$4d''\ ^1D^\circ$—$5p''\ ^3D$	2—3
4673,80	3	21,38	24,03	$4d'\ ^3D^\circ$—$5p'\ ^3F$	2—2
4621,40	1	22,47	26,15	$5s''\ ^1P^\circ$—$5p'\ ^1D$	1—2
4565,51	1	23,58	26,30	$4d''\ ^3P^\circ$—$5p''\ ^1D$	1—2
4537,25	6	21,92	24,65	$4d''\ ^3D^\circ$—$5p'\ ^3P$	2—1
4536,46	10	23,58	26,32	$4d''\ ^3P^\circ$—$5p''\ ^3P$	1—1
4518,64	2	21,32	24,07	$4d'\ ^1P^\circ$—$5p'\ ^1P$	1—1
4443,72	3	21,38	24,17	$4d'\ ^3D^\circ$—$5p'\ ^3F$	2—3
4443,28	15	21,10	23,89	$4d'\ ^3D^\circ$—$5p'\ ^3D$	1—2
4378,68	8	23,58	26,41	$4d''\ ^3P^\circ$—$5p''\ ^3P$	1—2
4360,63	1	21,19	24,03	$5s'\ ^1D^\circ$—$5p'\ ^3D$	2—3
4344,24	8	21,38	24,23	$4d'\ ^3D^\circ$—$5p'\ ^1F$	2—3
4305,20	9	21,19	24,07	$5s'\ ^1D^\circ$—$5p'\ ^1P$	2—1
4294,83	10	21,68	24,56	$4d''\ ^3F^\circ$—$5p'\ ^3P$	3—2
4244,33	5	23,38	26,30	$4d''\ ^1D^\circ$—$5p''\ ^1D$	2—2
4233,72	1	21,72	24,65	$4d''\ ^3F^\circ$—$5p'\ ^3P$	2—1
4232,82	2	22,92	25,85	$4d''\ ^3D^\circ$—$5p''\ ^3D$	3—2
4226,58	25	21,20	24,03	$4d'\ ^3D^\circ$—$5p'\ ^3F$	1—2
4225,92	20	21,63	24,56	$4d'\ ^3D^\circ$—$5p'\ ^3P$	3—2
4195,91	1	25,95	28,90	$5p''\ ^1P$—$6s'\ ^3D^\circ$	1—2
4184,59	2	23,34	26,30	$4d''\ ^1P^\circ$—$5p''\ ^1D$	1—2
4171,79	15	18,79	21,76	$5s\ ^3S^\circ$—$5p\ ^5P$	1—1
4160,21	4	23,34	26,32	$4d''\ ^1P^\circ$—$5p''\ ^3P$	1—1
4154,46	40	20,29	24,17	$5s'\ ^1D^\circ$—$5p'\ ^3F$	2—3
4131,33	40	18,79	21,79	$5s\ ^3S^\circ$—$5p\ ^5P$	1—2
4067,37	50	21,19	24,23	$5s'\ ^1D^\circ$—$5p'\ ^1F$	2—3
4027,17	1	23,34	26,41	$4d''\ ^1P^\circ$—$5p''\ ^3F$	1—2
4002,61	15	22,92	26,02	$4d''\ ^3D^\circ$—$5p''\ ^3D$	3—3
3979,05	3	21,45	24,56	$4d'\ ^3S^\circ$—$5p'\ ^3P$	1—2

λ, Å	I	E_H, eV	E_B, eV	Transition	J
3957,67	25	20,51	23,65	$4d'\,^1D°—5p'\,^3D$	2—1
3938,53	4	19,18	22,33	$4d'\,^3F°—5p\,^3P$	3—2
3913,90	3	22,68	25,85	$4d''\,^1F°—5p''\,^3D$	3—2
3898,70	10	21,38	24,56	$4d'\,^3D°—5p'\,^3P$	2—2
3874,04	3	21,45	24,65	$4d'\,^3S°—5p'\,^3P$	1—0
3868,70	40	21,45	24,65	$4d'\,^3S°—5p'\,^3P$	1—1
3847,49	3	22,47	25,69	$5s''\,^1P°—5p''\,^3D$	1—1
3835,37	2	19,04	22,27	$4d'\,^3F°—5p\,^3P$	2—1
3829,57	1	21,92	25,15	$4d''\,^3D°—5p'\,^1D$	2—2
3809,16	7	22,60	25,85	$4d''\,^3D°—5p''\,^3D$	1—2
3792,70	15	21,38	24,65	$4d'\,^3D°—5p'\,^3P$	2—1
3769,69	2	19,04	22,33	$4d'\,^3F°—5p\,^3P$	2—2
3726,32	5	21,32	24,65	$4d'\,^1P°—5p'\,^3P$	1—1
3699,98	2	22,60	25,95	$4d''\,^3D°—5p''\,^1P$	1—1
3696,69	5	18,44	21,79	$4d\,^3D°—5p\,^5P$	3—2
3690,65	30	20,29	23,65	$5s'\,^3D°—5p'\,^3D$	2—1
3674,23	4	21,19	24,56	$5s'\,^1D°—5p'\,^3P$	2—2
3671,14	1	22,92	26,30	$4d''\,^3D°—5p''\,^1D$	3—2
3670,23	4	20,51	23,89	$4d'\,^1D°—5p'\,^3D$	2—2
3655,77	1	22,47	25,86	$5s''\,^1P°—4d''\,^3S$	1—1
3641,34	30	20,24	23,65	$5s'\,^3D°—5p'\,^3D$	1—1
3632,5	1	22,60	26,01	$4d''\,^3D°—5p''\,^3P$	1—0
3615,82	20	20,46	23,89	$5s'\,^3D°—5p'\,^3D$	3—2
3611,06	5	21,72	25,15	$4d''\,^3F°—5p'\,^1D$	2—2
3603,96	2	18,32	21,76	$4d\,^3D°—5p\,^5P$	2—1
3598,04	1	18,44	21,88	$4d\,^3D°—5p\,^5P$	3—3
3582,48	5	21,10	24,56	$4d'\,^3D°—5p'\,^3P$	1—2
3579,95	2	21,19	24,65	$5s'\,^1D°—5p'\,^3P$	2—1
3567,72	15	22,47	25,95	$5s''\,^1P°—5p''\,^1P$	1—1
3564,23	100	18,79	22,27	$5s\,^3S°—5p\,^3P$	1—1
3562,09	2	21,68	25,15	$4d''\,^3F°—5p'\,^1D$	3—2
3549,42	20	22,92	26,41	$4d''\,^3D°—5p''\,^3P$	3—2
3537,20	2	22,35	25,85	$5s''\,^3P°—5p''\,^3D$	2—2
3524,78	5	22,35	25,86	$5s''\,^3P°—5p''\,^3S$	2—1
3521,11	4	20,51	24,03	$4d'\,^1D°—5p'\,^3F$	2—2
3514,55	15	21,63	25,15	$4d'\,^3D°—5p'\,^1D$	3—2
3507,42	200	18,79	22,33	$5s\,^3S°—5p\,^3P$	1—2
3497,13	10	21,10	24,65	$4d'\,^3D°—5p'\,^3P$	1—0
3492,80	8	21,10	24,65	$4d'\,^3D°—5p'\,^3P$	1—1
3488,59	100	20,51	24,07	$4d'\,^1D°—5p'\,^1P$	2—1
3485,08	1	25,95	29,50	$5p''\,^1P—5d'\,1°$	1—1
3474,65	70	20,46	24,03	$5s'\,^3D°—5p'\,^3D$	3—3
3471,02	3	20,46	24,03	$5s'\,^3D°—5p'\,^3F$	3—2
3448,71	10	22,10	25,69	$5s''\,^3P°—5p''\,^3D$	1—1
3446,85	8	22,10	25,69	$5s''\,^3P°—5p''\,^3D$	0—1
3442,86	6	22,35	25,95	$5s''\,^3P°—5p''\,^1P$	2—1
3439,46	100	20,29	23,89	$5s'\,^3D°—5p'\,^3D$	2—2
3428,83	10	22,68	26,30	$4d''\,^1F°—5p''\,^1D$	3—2
3396,58	15	20,24	23,89	$5s'\,^3D°—5p'\,^3D$	1—2
3388,93	20	20,51	24,17	$4d'\,^1D°—5p'\,^3F$	2—3
3374,96	40	22,35	26,02	$5s''\,^3P°—5p''\,^3D$	2—3
3351,93	100	18,07	21,76	$5s\,^5S°—5p\,^5P$	2—1
3348,17	10	22,60	26,30	$4d''\,^3D°—5p''\,^1D$	1—2
3342,48	50	20,46	24,17	$5s'\,^3D°—5p'\,^3F$	3—3
3332,50	10	22,60	26,32	$4d''\,^3D°—5p''\,^3P$	1—1
3330,76	60	20,51	24,23	$4d'\,^1D°—5p'\,^1F$	2—3
3325,75	200	18,07	21,79	$5s\,^5S°—5p\,^5P$	2—2
3311,47	50	20,29	24,03	$5s'\,^3D°—5p'\,^3D$	2—3

λ, Å	I	E_H, eV	E_B, eV	Transition	J
3308,73	1	25,15	28,90	$5p'\,{}^1D$—$6s'\,{}^3D^\circ$	2—2
3308,16	20	20,29	24,03	$5s'\,{}^3D^\circ$—$5p'\,{}^3F$	2—2
3304,75	30	22,10	25,85	$5s''\,{}^3P^\circ$—$5p''\,{}^3D$	1—2
3293,88	4	22,10	25,86	$5s''\,{}^3P^\circ$—$5p''\,{}^3S$	1—1
3292,21	1	22,10	25,86	$5s''\,{}^3P^\circ$—$5p''\,{}^3S$	0—1
3285,89	30	20,46	24,23	$5s'\,{}^3D^\circ$—$5p'\,{}^1F$	3—3
3285,25	3	21,38	25,15	$4d'\,{}^3D^\circ$—$5p'\,{}^1D$	2—2
3279,42	2	20,29	24,07	$5s'\,{}^3D^\circ$—$5p'\,{}^1P$	2—1
3271,65	30	18,48	22,27	$4d\,{}^3D^\circ$—$5p\,{}^3P$	1—1
3268,48	100	20,24	24,03	$5s'\,{}^3D^\circ$—$5p'\,{}^3F$	1—2
3264,81	150	20,46	24,26	$5s'\,{}^3D^\circ$—$5p'\,{}^3F$	3—4
3246,62	5	22,60	26,41	$4d''\,{}^3D^\circ$—$5p''\,{}^3P$	1—2
3245,69	300	18,07	21,88	$5s\,{}^5S^\circ$—$5p\,{}^5P$	2—3
3240,44	40	20,24	24,07	$5s'\,{}^3D^\circ$—$5p'\,{}^1P$	1—1
3239,52	40	22,47	26,30	$5s''\,{}^1P^\circ$—$5p''\,{}^1D$	1—2
3235,21	2	21,32	25,15	$4d'\,{}^1P^\circ$—$5p'\,{}^1D$	1—2
3224,85	20	22,47	26,32	$5s''\,{}^1P^\circ$—$5p''\,{}^3P$	1—1
3223,74	3	18,48	22,33	$4d\,{}^3D^\circ$—$5p\,{}^3P$	1—2
3222,24	10	22,10	25,95	$5s''\,{}^3P^\circ$—$5p''\,{}^1P$	1—1
3220,62	20	22,10	25,95	$5s''\,{}^3P^\circ$—$5p''\,{}^1P$	0—1
3191,21	80	20,29	24,17	$5s'\,{}^3D^\circ$—$5p'\,{}^3F$	2—3
3189,11	100	18,24	22,33	$4d\,{}^3D^\circ$—$5p\,{}^3P$	3—2
3170,93	20	22,10	26,01	$5s''\,{}^3P$—$5p''\,{}^3P$	1—0
3156,63	1	25,15	29,08	$5p'\,{}^1D$—$6s'\,{}^3D^\circ$	2—3
3151,75	10	21,92	25,85	$4d''\,{}^3D^\circ$—$5p''\,{}^3D$	2—2
3144,32	9	22,47	26,41	$5s''\,{}^1P^\circ$—$5p''\,{}^3P$	1—2
3141,88	20	21,92	25,86	$4d''\,{}^3D^\circ$—$5p''\,{}^3S$	2—1
3141,35	60	18,32	22,27	$4d\,{}^3D^\circ$—$5p\,{}^3P$	2—1
3139,58	15	20,29	24,23	$5s'\,{}^3D^\circ$—$5p'\,{}^1F$	2—3
3136,20	10	22,35	26,30	$5s''\,{}^3P^\circ$—$5p''\,{}^1D$	2—2
3124,39	100	21,19	25,15	$5s'\,{}^1D^\circ$—$5p'\,{}^1D$	2—2
3122,46	20	22,35	26,32	$5s''\,{}^3P^\circ$—$5p''\,{}^3P$	2—1
3120,61	30	21,72	25,69	$4d''\,{}^3F^\circ$—$5p''\,{}^3D$	2—1
3112,25	60	20,19	24,17	$4d'\,{}^1G^\circ$—$5p'\,{}^3F$	4—3
3097,16	40	18,32	22,33	$4d\,{}^3D^\circ$—$5p\,{}^3P$	2—2
3063,13	60	20,19	24,23	$4d'\,{}^1G^\circ$—$5p'\,{}^1P$	4—3
3062,43	3	20,51	24,56	$4d'\,{}^1D^\circ$—$5p'\,{}^3P$	2—2
3056,72	30	19,84	23,89	$4d'\,{}^3G^\circ$—$5p'\,{}^3D$	3—2
3046,93	50	22,35	26,41	$5s''\,{}^3P^\circ$—$5p''\,{}^3P$	2—2
3044,80	6	20,19	24,26	$4d'\,{}^1G^\circ$—$5p'\,{}^3F$	4—4
3024,45	80	20,46	24,56	$5s'\,{}^3D^\circ$—$5p'\,{}^3P$	3—2
3022,30	50	21,92	26,02	$4d''\,{}^3D^\circ$—$5p''\,{}^3D$	2—3
3002,24	6	21,72	25,85	$4d''\,{}^3F^\circ$—$5p''\,{}^3D$	2—2
2996,60	20	20,51	24,65	$4d'\,{}^1D^\circ$—$5p'\,{}^3P$	2—1
2993,27	2	21,72	25,86	$4d''\,{}^3F^\circ$—$5p''\,{}^3S$	2—1
2992,22	60	19,89	24,03	$4d'\,{}^3G^\circ$—$5p'\,{}^3D$	4—3
2968,31	20	21,68	25,85	$4d''\,{}^3F^\circ$—$5p''\,{}^3D$	3—2
2955,20	3	19,84	24,03	$4d'\,{}^3G^\circ$—$5p'\,{}^3D$	3—3
2952,56	50	19,84	24,03	$4d'\,{}^3G^\circ$—$5p'\,{}^3F$	3—2
2952,09	4	22,10	26,30	$5s''\,{}^3P^\circ$—$5p''\,{}^1D$	1—2
2948,13	10	18,07	22,27	$5s\,{}^5S^\circ$—$5p\,{}^3P$	2—1
2939,91	15	22,10	26,32	$5s''\,{}^3P^\circ$—$5p''\,{}^3P$	1—1
2938,56	4	22,10	26,32	$5s''\,{}^3P^\circ$—$5p''\,{}^3P$	0—1
2935,23	20	21,63	25,85	$4d'\,{}^3D^\circ$—$5p''\,{}^3D$	3—2
2934,00	10	21,72	25,95	$4d''\,{}^3F^\circ$—$5p''\,{}^1P$	2—1
2917,67	10	21,45	25,69	$4d'\,{}^3S^\circ$—$5p''\,{}^3D$	1—1
2915,78	6	24,65	28,90	$5p'\,{}^3P$—$6s'\,{}^3D^\circ$	1—2

λ, Å	I	E_H, eV	E_B, eV	Transition	J
2909,17	30	18,07	22,33	$5s\ ^5S^\circ—5p\ ^3P$	2—2
2900,04	20	20,29	24,56	$5s'\ ^3D^\circ—5p'\ ^3P$	2—2
2895,92	1	24,65	28,93	$5p'\ ^3P—5d'\ ^1D^\circ$	1—2
2893,68	40	19,89	24,17	$4d'\ ^3G^\circ—5p'\ ^3F$	4—3
2892,18	100	19,97	24,26	$4d'\ ^3G^\circ—5p'\ ^3F$	5—4
2884,55	2	21,72	26,02	$4d''\ ^3F^\circ—5p''\ ^3D$	2—3
2874,24	2	21,38	25,69	$4d'\ ^3D^\circ—5p''\ ^3D$	2—1
2872,85	5	22,10	26,41	$5s''\ ^3P^\circ—5p''\ ^3P$	1—2
2870,61	50	21,70	26,02	$4d''\ ^3F^\circ—5p''\ ^3D$	4—3
2859,05	4	19,84	24,17	$4d'\ ^3G^\circ—5p'\ ^3F$	3—3
2856,09	5	24,56	28,90	$5p'\ ^3P—6s'\ ^3D^\circ$	2—2
2853,22	2	21,68	26,02	$4d''\ ^3F^\circ—5p''\ ^3D$	3—3
2851,16	30	19,89	24,23	$4d'\ ^3G^\circ—5p'\ ^1F$	4—3
2841,00	30	20,29	24,65	$5s'\ ^3D^\circ—5p'\ ^3P$	2—1
2835,94	6	21,32	25,69	$4d'\ ^1P^\circ—5p''\ ^3D$	1—1
2829,41	6	21,92	26,30	$4d''\ ^3D^\circ—5p''\ ^1D$	2—2
2822,63	6	21,63	26,02	$4d'\ ^3D^\circ—5p''\ ^3D$	3—3
2820,95	4	22,33	26,72	$5p\ ^3P—6s\ ^5S^\circ$	2—2
2817,53	2	19,84	24,23	$4d'\ ^3G^\circ—5p'\ ^1F$	3—3
2814,48	15	20,24	24,65	$5s'\ ^3D^\circ—5p'\ ^3P$	1—0
2813,97	15	21,45	25,85	$4d'\ ^3S^\circ—5p''\ ^3D$	1—2
2811,67	25	21,19	24,65	$5s'\ ^3D^\circ—5p'\ ^3P$	1—1
2806,07	20	21,45	25,86	$4d'\ ^3S^\circ—5p''\ ^3S$	1—1
2785,26	2	22,27	26,72	$5p\ ^3P—6s\ ^5S^\circ$	1—2
2768,54	4	26,41	30,89	$5p''\ ^3P—6s''\ ^3P^\circ$	2—1
2765,90	2	21,38	25,86	$4d'\ ^3D^\circ—5p''\ ^3S$	2—1
2756,53	8	21,92	26,41	$4d''\ ^3D^\circ—5p''\ ^3P$	2—2
2750,36	10	21,19	25,69	$5s'\ ^1D^\circ—5p''\ ^3D$	2—1
2744,05	2	22,33	26,84	$5p\ ^3P—5d\ ^5D^\circ$	2—1
2743,03	3	22,33	26,84	$5p\ ^3P—5d\ ^5D^\circ$	2—2
2742,05	5	24,56	29,08	$5p'\ ^3P—6s'\ ^3D^\circ$	2—3
2741,84	2	22,33	26,85	$5p\ ^3P—5d\ ^5D^\circ$	2—3
2730,41	5	21,32	25,86	$4d'\ ^1P^\circ—5p''\ ^3S$	1—1
2715,19	7	21,38	25,95	$4d'\ ^3D^\circ—5p''\ ^1P$	2—1
2710,27	2	22,27	26,84	$5p\ ^3P—5d\ ^5D^\circ$	1—1
2709,02	1	26,32	30,89	$5p''\ ^3P—6s''\ ^3P^\circ$	1—1
2708,34	1	21,72	26,30	$4d''\ ^3F^\circ—5p''\ ^1D$	2—2
2698,71	2	26,30	30,89	$5p''\ ^1D—6s''\ ^3P^\circ$	2—1
2698,07	3	21,72	26,32	$4d''\ ^3F^\circ—5p''\ ^3P$	2—1
2697,30	25	17,17	21,76	$4d\ ^5D^\circ—5p\ ^5P$	2—1
2696,59	25	17,17	21,76	$4d\ ^5D^\circ—5p\ ^5P$	1—1
2694,81	20	17,16	21,76	$4d\ ^5D^\circ—5p\ ^5P$	0—1
2691,86	4	—	—	—	—
2690,23	15	19,04	23,65	$4d'\ ^3F^\circ—5p'\ ^3D$	2—1
2681,19	40	17,17	21,79	$4d\ ^5D^\circ—5p\ ^5P$	3—2
2680,72	7	22,33	26,05	$5p\ ^3P—5d\ ^3D^\circ$	2—2
2680,32	30	17,17	21,79	$4d\ ^5D^\circ—5p\ ^5P$	2—2
2679,62	15	17,17	21,79	$4d\ ^5D^\circ—5p\ ^5P$	1—2
2676,00	8	24,65	29,28	$5p'\ ^3P—6s'\ ^1D^\circ$	1—2
2672,79	3	21,38	26,02	$4d'\ ^3D^\circ—5p''\ ^3D$	2—3
2670,67	20	20,51	25,15	$4d'\ ^1D^\circ—5p'\ ^1D$	2—2
2658,00	2	21,19	25,85	$5s'\ ^1D^\circ—5p''\ ^3D$	2—2
2653,66	4	21,63	26,30	$4d'\ ^3D^\circ—5p''\ ^1D$	3—2
2650,96	1	21,19	25,86	$5s\ ^1D^\circ—5p''\ ^3S$	2—1
2648,69	10	19,35	24,03	$4d'\ ^3F^\circ—5p'\ ^3D$	4—3
2648,43	4	22,27	26,95	$5p\ ^3P—5d\ ^3D^\circ$	1—2
2641,74	2	20,46	25,15	$5s'\ ^3D^\circ—5p'\ ^1D$	3—2
2641,00	4	26,41	31,11	$5p''\ ^3P—6s''\ ^3P^\circ$	2—2

λ, Å	I	E_H, eV	E_B, eV	Transition	J
2639,76	60	17,19	21,88	$4d\ ^5D^\circ-5p\ ^5P$	4—3
2630,66	15	19,18	23,89	$4d'\ ^3F^\circ-5p'\ ^3D$	3—2
2628,90	25	17,17	21,88	$4d\ ^5D^\circ-5p\ ^5P$	3—3
2628,08	6	17,17	21,88	$4d\ ^5D^\circ-5p\ ^5P$	2—3
2625,64	2	24,56	29,28	$5p'\ ^3P-6s'\ ^1D^\circ$	2—2
2623,11	1	19,84	24,56	$4d'\ ^3G^\circ-5p'\ ^3P$	3—2
2615,19	3	21,68	26,41	$4d''\ ^3F^\circ-5p''\ ^3P$	3—2
2609,66	1	21,10	25,85	$4d'\ ^3D^\circ-5p''\ ^3D$	1—2
2604,35	8	21,19	25,95	$5s'\ ^1D^\circ-5p''\ ^1P$	2—1
2589,47	3	21,63	26,41	$4d'\ ^3D^\circ-5p''\ ^3P$	3—2
2586,78	3	26,32	31,11	$5p''\ ^3P-6s''\ ^3P^\circ$	1—2
2571,19	6	19,35	24,17	$4d'\ ^3F^\circ-5p'\ ^3F$	4—3
2570,48	10	24,26	29,08	$5p'\ ^3F-6s'\ ^3D^\circ$	4—3
2563,25	30	21,88	26,72	$5p\ ^5P-6s\ ^5S^\circ$	3—2
2558,00	5	21,10	25,95	$4d'\ ^3D^\circ-5p''\ ^1P$	1—1
2557,55	1	24,23	29,08	$5p'\ ^1F-6s'\ ^3D^\circ$	3—3
2555,13	10	19,18	24,03	$4d'\ ^3F^\circ-5p'\ ^3D$	3—3
2554,25	8	19,04	23,89	$4d'\ ^3F^\circ-5p'\ ^3D$	2—2
2553,81	1	24,65	29,50	$5p'\ ^3P-5d'\ 1^\circ$	1—1
2553,16	8	19,18	24,03	$4d'\ ^3F^\circ-5p'\ ^3F$	3—2
2551,49	2	24,65	29,50	$5p'\ ^3P-5d'\ 1^\circ$	0—1
2549,51	2	22,33	27,19	$5p\ ^3P-5d\ ^3D^\circ$	2—1
2548,60	4	24,07	28,93	$5p'\ ^1P-5d\ ^1D^\circ$	1—2
2546,67	1	24,03	28,90	$5p'\ ^3F-6s'\ ^3D^\circ$	2—2
2546,36	1	20,29	25,15	$5s'\ ^3D^\circ-5p'\ ^1D$	2—2
2544,72	3	24,03	28,90	$5p'\ ^3D-6s'\ ^3D^\circ$	3—2
2537,57	6	19,35	24,23	$4d'\ ^3F^\circ-5p'\ ^1F$	4—3
2537,16	1	26,41	31,30	$5p''\ ^3P-6s''\ ^1P^\circ$	2—1
2531,46	1	24,03	28,93	$5p'\ ^3F-5d'\ ^1D^\circ$	2—2
2529,52	1	24,03	28,93	$5p'\ ^3D-5d'\ ^1D^\circ$	3—2
2525,51	2	21,10	26,01	$4d'\ ^3D^\circ-5p''\ ^3P$	1—0
2524,97	10	19,35	24,26	$4d'\ ^3F^\circ-5p'\ ^3F$	4—4
2524,27	2	24,17	29,08	$5p'\ ^3F-6s'\ ^3D^\circ$	3—3
2520,32	2	22,27	27,19	$5p\ ^3P-5d\ ^3D^\circ$	1—1
2519,29	6	24,26	29,18	$5p'\ ^3F-5d'\ ^1G^\circ$	4—4
2515,42	10	21,79	26,72	$5p\ ^5P-6s\ ^5S^\circ$	2—2
2512,92	6	24,65	29,58	$5p'\ ^3P-5d'\ 2^\circ$	1—2
2511,92	1	21,38	26,32	$4d'\ ^3D^\circ-5p''\ ^3P$	2—1
2507,84	1	24,56	29,50	$5p'\ ^3P-5d'\ 1^\circ$	2—1
2506,86	5	24,23	29,18	$5p'\ ^1F-5d'\ ^1G^\circ$	3—4
2500,64	8	21,76	26,72	$5p\ ^5P-6s\ ^5S^\circ$	1—2
2498,77	3	21,88	26,84	$5p\ ^5P-5d\ ^5D^\circ$	3—2
2497,71	15	21,88	26,85	$5p\ ^5P-5d\ ^5D^\circ$	3—3
2494,01	40	21,88	26,85	$5p\ ^5P-5d\ ^5D^\circ$	3—4
2491,35	2	21,32	26,30	$4d'\ ^1P^\circ-5p''\ ^1D$	1—2
2487,03	1	26,32	31,30	$5p''\ ^3P-6s''\ ^1P^\circ$	1—1
2482,99	2	19,04	24,03	$4d'\ ^3F^\circ-5p'\ ^3D$	2—3
2481,04	1	19,04	24,03	$4d'\ ^3F^\circ-5p'\ ^3F$	2—2
2478,37	2	26,30	31,30	$5p''\ ^1D-6s''\ ^1P^\circ$	2—1
2474,90	3	24,17	29,18	$5p'\ ^3F-5d'\ ^1G^\circ$	3—4
2473,96	4	23,89	28,90	$5p'\ ^3D-6s'\ ^3D^\circ$	2—2
2468,43	6	24,56	29,58	$5p'\ ^3P-5d'\ 2^\circ$	2—2
2462,76	3	21,38	26,41	$4d'\ ^3D^\circ-5p''\ ^3P$	2—2
2459,63	5	23,89	23,93	$5p'\ ^3D-5d'\ ^1D^\circ$	2—2
2457,72	10	22,33	27,37	$5p\ ^3P-5d\ ^3D^\circ$	2—3
2454,12	3	21,79	26,84	$5p\ ^5P-5d\ ^5D^\circ$	2—1
2453,74	1	24,03	29,08	$5p'\ ^3D-6s'\ ^3D^\circ$	3—3

λ, Å	I	E_H, eV	E_B, eV	Transition	J
2453,28	8	21,79	26,84	$5p\ ^5P-5d\ ^5D^\circ$	2—2
2452,29	10	21,79	26,85	$5p\ ^5P-5d\ ^5D^\circ$	2—3
2451,52	4	19,18	24,23	$4d'\ ^3F^\circ-5p'\ ^1F$	3—3
2440,89	5	21,76	26,84	$5p\ ^5P-5d\ ^5D^\circ$	1—0
2440,05	6	21,76	26,84	$5p\ ^5P-5d\ ^5D^\circ$	1—1
2439,78	1	19,18	24,26	$4d'\ ^3F^\circ-5p'\ ^3F$	3—4
2439,21	6	21,76	26,84	$5p\ ^5P-5d\ ^5D^\circ$	1—2
2434,64	2	21,32	26,41	$4d'\ ^1P^\circ-5p''\ ^3P$	1—2
2431,04	1	18,79	23,89	$5s\ ^3S^\circ-5p'\ ^3D$	1—2
2428,92	1	17,17	22,27	$4d\ ^5D^\circ-5p\ ^3P$	1—1
2427,48	1	17,16	22,27	$4d\ ^5D^\circ-5p\ ^3P$	0—1
2414,78	1	19,04	24,17	$4d'\ ^3F^\circ-5p'\ ^3F$	2—3
2407,10	10	24,03	29,18	$5p'\ ^3D-5d'\ ^1G^\circ$	3—4
2403,65	3	17,17	22,33	$4d\ ^5D^\circ-5p\ ^3P$	3—4
2403,29	1	21,79	26,95	$5p\ ^5P-5d\ ^3D^\circ$	2—2
2402,96	3	17,17	22,33	$4d\ ^5D^\circ-5p\ ^3P$	2—2
2402,10	2	17,17	22,33	$4d\ ^5D^\circ-5p\ ^3P$	1—2
2401,58	1	25,95	31,11	$5p''\ ^1P-6s\ ^3P^\circ$	1—2
2400,10	4	18,48	24,65	$4d\ ^3D^\circ-5p'\ ^3D$	1—1
2393,94	40	22,33	27,50	$5p\ ^3P-6s\ ^3S^\circ$	2—1
2387,90	1	23,89	29,08	$5p'\ ^3D-6s'\ ^3D^\circ$	2—3
2376,69	1	24,07	29,28	$5p'\ ^1P-6s'\ ^1D^\circ$	1—2
2368,19	4	22,27	27,50	$5p\ ^3P-6s\ ^3S^\circ$	1—1
2364,70	1	18,79	24,03	$5s\ ^3S^\circ-5p'\ ^3F$	1—2
2363,26	3	25,86	31,11	$5p''\ ^3S-6s''\ ^3P^\circ$	1—2
2361,82	4	24,03	29,28	$5p'\ ^3F-6s'\ ^1D^\circ$	2—2
2360,14	3	24,03	29,28	$5p'\ ^3D-6s'\ ^1D^\circ$	3—2
2358,48	3	23,65	28,90	$5p'\ ^3D-6s'\ ^3D^\circ$	1—2
2345,45	6	23,65	28,93	$5p'\ ^3D-5d'\ ^1D^\circ$	1—2
2329,22	3	18,32	23,65	$4d\ ^3D^\circ-5p'\ ^3D$	2—1
2322,32	1	20,51	25,85	$4d'\ ^1D^\circ-5p''\ ^3D$	2—2
2317,87	1	24,23	29,58	$5p'\ ^1F-5d'\ 2^\circ$	3—2
2303,00	2	19,18	24,56	$4d'\ ^3F^\circ-5p'\ ^3P$	3—2
2299,15	3	23,89	29,28	$5p'\ ^3D-6s'\ ^1D^\circ$	2—2
2291,28	3	18,48	23,89	$4d\ ^3D^\circ-5p'\ ^3D^\circ$	1—2
2290,52	1	24,17	29,58	$5p'\ ^3F-5d'\ 2^\circ$	3—2
2279,79	4	24,07	29,50	$5p'\ ^1P-5d'\ 1^\circ$	1—1
		25,86	31,30	$5p''\ ^3S-6s''\ ^1P^\circ$	1—1
2273,76	3	18,44	23,89	$4d\ ^3D^\circ-5p'\ ^3D$	3—2
2259,76	6	21,88	27,37	$5p\ ^5P-5d\ ^3D^\circ$	3—3
2232,35	1	24,03	29,58	$5p'\ ^3D-5d'\ 2^\circ$	3—2
2230,69	1	20,46	26,02	$5s'\ ^3D^\circ-5p''\ ^3D$	3—3
2219,14	1	18,48	24,07	$4d\ ^3D^\circ-5p'\ ^1P$	1—1
2215,60	2	18,44	24,03	$4d\ ^3D^\circ-5p'\ ^3F$	3—2
2172,25	1	18,32	24,03	$4d\ ^3D^\circ-5p'\ ^3D$	2—3
2170,83	2	18,32	24,03	$4d\ ^3D^\circ-5p'\ ^3F$	2—2
2162,50	3	18,44	24,17	$4d\ ^3D^\circ-5p'\ ^3F$	3—3
2158,43	1	18,32	24,07	$4d\ ^3D^\circ-5p'\ ^1P$	2—1
2148,58	2	18,79	24,56	$5s\ ^3S^\circ-5p'\ ^3P$	1—2
2142,49	1	20,51	26,30	$4d'\ ^1D^\circ-5p''\ ^1D$	2—2
2138,70	1	18,44	24,23	$4d\ ^3D^\circ-5p'\ ^1F$	3—3
2129,75	1	18,44	24,26	$4d\ ^3D^\circ-5p'\ ^3F$	3—4
2116,00	1	18,79	24,65	$5s\ ^3S^\circ-5p'\ ^3P$	1—1
1923,88	0	17,59	24,03	$4p^5\ ^1P^\circ-5p'\ ^3F$	1—2
1914,086	3	17,59	24,07	$4p^5\ ^1P^\circ-5p'\ ^1P$	1—1
1721,637	1	15,07	22,27	$4p^5\ ^3P^\circ-5p\ ^3P$	0—1
1659,809	2	14,80	22,27	$4p^5\ ^3P^\circ-5p\ ^3P$	1—1
1647,359	2	14,80	22,33	$4p^5\ ^3P^\circ-5p\ ^3P$	1—2

λ, Å	I	E_H, eV	E_B, eV	Transition	J
1638,816	3	17,59	25,15	$4p^5\,{}^1P°—5p'\,{}^1D$	1—2
1569,886	2	14,37	22,27	$4p^5\,{}^3P°—5p\,{}^3P$	2—1
1558,802	3	14,37	22,33	$4p^5\,{}^3P°—5p\,{}^3P$	2—2
1483,429	2	17,59	25,95	$4p^5\,{}^1P°—5p''\,{}^1P$	1—1
1423,553	1	17,59	26,30	$4p^5\,{}^1P°—5p''\,{}^1D$	1—2
1400,90	1	14,80	23,65	$4p^5\,{}^3P°—5p'\,{}^3D$	1—1
1377,833	2	15,07	24,07	$4p^5\,{}^3P°—5p'\,{}^1P$	0—1
1363,853	2	14,80	23,89	$4p^5\,{}^3P°—5p'\,{}^3D$	1—2
1342,678	1	14,80	24,03	$4p^5\,{}^3P°—5p'\,{}^3F$	1—2
1302,586	2	14,37	23,89	$4p^5\,{}^3P°—5p'\,{}^3D$	2—2
1293,988	3	15,07	24,65	$4p^5\,{}^3P°—5p'\,{}^3P$	0—1
1283,798	3	14,37	24,03	$4p^5\,{}^3P°—5p'\,{}^3D$	2—3
1283,313	3	14,37	24,03	$4p^5\,{}^3P°—5p'\,{}^3F$	2—2
1278,943	1	14,37	24,07	$4p^5\,{}^3P°—5p'\,{}^1P$	2—1
1270,204	5	14,80	24,56	$4p^5\,{}^3P°—5p'\,{}^3P$	1—2
1265,315	4	14,37	24,17	$4p^5\,{}^3P°—5p'\,{}^3F$	2—3
1259,309	3	14,80	24,65	$4p^5\,{}^3P°—5p'\,{}^3P$	1—0
1258,745	3	14,80	24,65	$4p^5\,{}^3P°—5p'\,{}^3P$	1—1
1216,896	5	14,37	24,56	$4p^5\,{}^3P°—5p'\,{}^3P$	2—2
1206,346	5	14,37	24,65	$4p^5\,{}^3P°—5p'\,{}^3P$	2—1
1158,724	6	4,10	14,80	$4p^4\,{}^1S—4p^5\,{}^3P°$	0—1
987,281	18	1,82	14,37	$4p^4\,{}^1D—4p^5\,{}^3P°$	2—2
954,774	4	1,82	14,80	$4p^4\,{}^1D—4p^5\,{}^3P°$	2—1
919,143	2	4,10	17,59	$4p^4\,{}^1S—4p^5\,{}^1P°$	0—1
897,801	40	0,56	14,37	$4p^4\,{}^3P—4p^5\,{}^3P°$	1—2
876,674	22	0,66	14,80	$4p^4\,{}^3P—4p^5\,{}^3P°$	0—1
870,825	20	0,56	14,80	$4p^4\,{}^3P—4p^5\,{}^3P°$	1—1
862,578	35	0,00	14,37	$4p^4\,{}^3P—4p^5\,{}^3P°$	2—2
854,733	25	0,56	15,07	$4p^4\,{}^3P—4p^5\,{}^3P°$	1—0
837,666	22	0,66	14,80	$4p^4\,{}^3P—4p^5\,{}^3P°$	2—1
785,968	25	1,82	17,59	$4p^4\,{}^1D—4p^5\,{}^1P°$	2—1
768,104	1	4,10	20,24	$4p^4\,{}^1S—5s'\,{}^3D°$	0—1
750,986	4	0,66	17,17	$4p^4\,{}^3P—4d\,{}^5D°$	0—1
746,834	5	0,56	17,16	$4p^4\,{}^3P—4d\,{}^5D°$	1—0
746,695	7	0,56	17,17	$4p^4\,{}^3P—4d\,{}^5D°$	1—2, 1
745,763	3	1,82	18,44	$4p^4\,{}^1D—4d\,{}^3D°$	2—3
743,870	3	1,82	18,48	$4p^4\,{}^1D—4d\,{}^3D°$	2—1
732,259	4	0,66	17,59	$4p^4\,{}^3P—4p^5\,{}^1P°$	0—1
730,264	3	1,82	18,79	$4p^4\,{}^1D—5s\,{}^3S°$	2—1
722,036	50	0,00	17,17	$4p^4\,{}^3P—4d\,{}^5D°$	2—3, 2, 1
719,85	1	1,82	19,04	$4p^4\,{}^1D—4d'\,{}^3F°$	2—2
714,772	2	4,10	21,45	$4p^4\,{}^1S—4d'\,{}^3S°$	0—1
713,999	7	1,82	19,18	$4p^4\,{}^1D—4d'\,{}^3F°$	2—3
708,356	8	0,56	18,07	$4p^4\,{}^3P—5s\,{}^5S°$	1—2
704,838	4	0,00	17,59	$4p^4\,{}^3P—4p^5\,{}^1P°$	2—1
698,037	20	0,56	18,32	$4p^4\,{}^3P—4d\,{}^3D°$	1—2
695,604	15	0,66	18,48	$4p^4\,{}^3P—4d\,{}^3D°$	0—1
691,919	18	0,56	18,48	$4p^4\,{}^3P—4d\,{}^3D°$	1—1
687,979	11	1,82	19,84	$4p^4\,{}^1D—4d'\,{}^3G°$	2—3
686,254	20	0,00	18,07	$4p^4\,{}^3P—5s\,{}^5S°$	2—2
683,666	18	0,66	18,79	$4p^4\,{}^3P—5s\,{}^3S°$	0—1
680,119	22	0,56	18,79	$4p^4\,{}^3P—5s\,{}^3S°$	1—1
676,564	25	0,00	18,32	$4p^4\,{}^3P—4d\,{}^3D°$	2—2
674,828	8	4,10	22,47	$4p^4\,{}^1S—5s''\,{}^1P°$	0—1
672,826	7	1,82	20,24	$4p^4\,{}^1D—5s'\,{}^3D°$	2—1
672,330	25	0,00	18,44	$4p^4\,{}^3P—4d\,{}^3D°$	2—3
671,175	7	1,82	20,29	$4p^4\,{}^1D—5s'\,{}^3D°$	2—2
671,058	7	0,56	19,04	$4p^4\,{}^3P—4d'\,{}^3F°$	1—2

λ, Å	I	E_H, eV	E_B, eV	Transition	J
670,813	3	0,00	18,48	$4p^4\,{}^3P$—$4d\,{}^3D°$	2—1
670,300	4	4,10	22,60	$4p^4\,{}^1S$—$4d''\,{}^3D°$	0—1
664,844	11	1,82	20,46	$4p^4\,{}^1D$—$5s'\,{}^3D°$	2—3
663,039	20	1,82	20,51	$4p^4\,{}^1D$—$4d\,{}^1D°$	2—2
659,716	22	0,00	18,79	$4p^4\,{}^3P$—$5s\,{}^3S°$	2—1
651,198	8	0,00	19,04	$4p^4\,{}^3P$—$4d'\,{}^3F°$	2—2
646,417	20	0,00	19,18	$4p^4\,{}^3P$—$4d'\,{}^3F°$	2—3
644,521	1	4,10	23,34	$4p^4\,{}^1S$—$4d''\,{}^1P°$	0—1
642,84	1	1,82	21,10	$4p^4\,{}^1D$—$4d'\,{}^3D°$	2—1
639,981	15	1,82	21,19	$4p^4\,{}^1D$—$5s'\,{}^1D°$	2—2
636,348	1	4,10	23,58	$4p^4\,{}^1S$—$4d''\,{}^3P°$	0—1
633,631	5	1,82	21,38	$4p^4\,{}^1D$—$4d'\,{}^3D°$	2—2
633,082	7	0,66	20,24	$4p^4\,{}^3P$—$5s'\,{}^3D°$	0—1
631,550	7	1,82	21,45	$4p^4\,{}^1D$—$4d'\,{}^3S°$	2—1
630,037	15	0,56	20,24	$4p^4\,{}^3P$—$5s'\,{}^3D°$	1—1
628,581	15	0,56	20,29	$4p^4\,{}^3P$—$5s'\,{}^3D°$	1—2
625,758	13	1,82	21,63	$4p^4\,{}^1D$—$4d'\,{}^3D°$	2—3
625,011	9	0,00	19,84	$4p^4\,{}^3P$—$4d'\,{}^3G°$	2—3
624,268	3	1,82	21,68	$4p^4\,{}^1D$—$4d''\,{}^3F°$	2—3
622,795	11	1,82	21,72	$4p^4\,{}^1D$—$4d''\,{}^3F°$	2—2
621,448	8	0,56	20,51	$4p^4\,{}^3P$—$4d'\,{}^1D°$	1—2
616,728	5	1,82	21,92	$4p^4\,{}^1D$—$4d''\,{}^3D°$	2—2
612,485	6	0,00	20,24	$4p^4\,{}^3P$—$5s'\,{}^3D°$	2—1
611,187	8	1,82	22,10	$4p^4\,{}^1D$—$5s''\,{}^3P°$	2—1
611,100	9	0,00	20,29	$4p^4\,{}^3P$—$5s'\,{}^3D°$	2—2
606,460	9	0,66	21,10	$4p^4\,{}^3P$—$4d'\,{}^3D°$	0—1
605,862	9	0,00	20,46	$4p^4\,{}^3P$—$5s''\,{}^3D°$	2—3
604,355	4	0,00	20,51	$4p^4\,{}^3P$—$4d'\,{}^1D°$	2—2
603,849	6	1,82	22,35	$4p^4\,{}^1D$—$5s''\,{}^3P°$	2—2
603,666	7	0,56	21,10	$4p^4\,{}^3P$—$4d'\,{}^3D°$	1—1
601,134	7	0,56	21,19	$4p^4\,{}^3P$—$5s'\,{}^1D°$	1—2
600,167	5	1,82	22,47	$4p^4\,{}^1D$—$5s''\,{}^1P°$	2—1
599,944	4	0,66	21,32	$4p^4\,{}^3P$—$4d'\,{}^1P°$	0—1
597,194	6	0,56	21,32	$4p^4\,{}^3P$—$4d'\,{}^1P°$	1—1
596,576	6	1,82	22,60	$4p^4\,{}^1D$—$4d''\,{}^3D°$	2—1
596,401	6	0,66	21,45	$4p^4\,{}^3P$—$4d'\,{}^3S°$	0—1
595,530	7	0,56	21,38	$4p^4\,{}^3P$—$4d'\,{}^3D°$	1—2
594,090	9	1,82	22,68	$4p^4\,{}^1D$—$4d''\,{}^1F°$	2—3
593,699	7	0,56	21,45	$4p^4\,{}^3P$—$4d'\,{}^3S°$	1—1
587,543	4	0,00	21,10	$4p^4\,{}^3P$—$4d'\,{}^3D°$	2—1
587,374	4	1,82	22,92	$4p^4\,{}^1D$—$4d''\,{}^3D°$	2—3
585,950	8	0,56	21,72	$4p^4\,{}^3P$—$4d''\,{}^3F°$	1—2
585,140	8	0,00	21,19	$4p^4\,{}^3P$—$5s'\,{}^1D°$	2—2
580,577	6	0,56	21,92	$4p^4\,{}^3P$—$4d''\,{}^3D°$	1—2
579,823	6	0,00	21,38	$4p^4\,{}^3P$—$4d'\,{}^3D°$	2—2
578,220	5	0,66	22,10	$4p^4\,{}^3P$—$5s''\,{}^3P°$	0—1
578,09	0	0,00	21,45	$4p^4\,{}^3P$—$4d'\,{}^3S°$	2—1
576,076	4	1,82	23,34	$4p^4\,{}^1D$—$4d''\,{}^1P°$	2—1
575,716	5	0,56	22,10	$4p^4\,{}^3P$—$5s''\,{}^3P°$	1—1, 0
574,956	5	1,82	23,38	$4p^4\,{}^1D$—$4d''\,{}^1D°$	2—2
573,228	13	0,00	21,63	$4p^4\,{}^3P$—$4d'\,{}^3D°$	2—3
571,983	15	0,00	21,68	$4p^4\,{}^3P$—$4d''\,{}^3F°$	2—3
570,738	4	0,00	21,72	$4p^4\,{}^3P$—$4d''\,{}^3F°$	2—2
569,156	7	0,56	22,35	$4p^4\,{}^3P$—$5s''\,{}^3P°$	1—2
565,879	4	0,56	22,47	$4p^4\,{}^3P$—$5s''\,{}^1P°$	1—1
565,640	5	0,00	21,92	$4p^4\,{}^3P$—$4d''\,{}^3D°$	2—2
565,124	4	0,66	22,60	$4p^4\,{}^3P$—$4d''\,{}^3D°$	0—1
562,690	5	0,56	22,60	$4p^4\,{}^3P$—$4d''\,{}^3D°$	1—1

λ, Å	I	E_H, eV	E_B, eV	Transition	J
560,986	5	0,00	22,10	$4p^4\,^3P—5s''\,^3P°$	2—1
558,634	15	1,82	24,01	$4p^4\,^1D—4d''\,^3P°$	2—2
554,794	7	0,00	22,35	$4p^4\,^3P—5s''\,^3P°$	2—2
551,689	4	0,00	22,47	$4p^4\,^3P—5s''\,^1P°$	2—1
548,652	5	0,00	22,60	$4p^4\,^3P—4d''\,^3D°$	2—1
546,686	5	0,66	23,34	$4p^4\,^3P—4d''\,^1P°$	0—1
546,547	6	0,00	22,68	$4p^4\,^3P—4d''\,^1F°$	2—3
544,413	5	0,56	23,34	$4p^4\,^3P—4d''\,^1P°$	1—1
543,420	8	0,56	23,38	$4p^4\,^3P—4d''\,^1D°$	1—2
540,860	4	0,00	22,92	$4p^4\,^3P—4d''\,^3D°$	2—3
540,788	5	0,66	23,58	$4p^4\,^3P—4d''\,^3P°$	0—1
538,544	8	0,56	23,58	$4p^4\,^3P—4d''\,^3P°$	1—1
531,255	4	0,00	23,34	$4p^4\,^3P—4d''\,^1P°$	2—1
530,306	6	0,00	23,38	$4p^4\,^3P—4d''\,^1D°$	2—2
528,811	4	0,56	24,01	$4p^4\,^3P—4d''\,^3P°$	1—2
525,687	4	0,00	23,58	$4p^4\,^3P—4d''\,^3P°$	2—1
516,384	4	0,00	24,01	$4p^4\,^3P—4d''\,^3P°$	2—2

Kr IV, ground state $1s^2\,2s^2\,2p^6\,3s^2\,3p^6\,3d^{10}\,4s^2\,4p^3\,^4S^0_{3/2}$

λ, Å	I	E_H, eV	E_B, eV	Transition	J
3934,29	5	22,63	25,78	$4d\ 12—5p\ ^4P°$	$^3/_2—^1/_2$
3860,58	5	22,63	25,84	$4d\ 12—5p\ ^4P°$	$^3/_2—^3/_2$
3809,30	3	22,63	25,89	$4d\ 12—5p\ ^4P°$	$^3/_2—^5/_2$
3261,70	3	22,04	25,84	$4d\ 11—5p\ ^4P°$	$^3/_2,\ ^5/_2—^3/_2$
3224,99	6	22,04	25,89	$4d\ 11—5p\ ^4P°$	$^3/_2,\ ^5/_2—^5/_2$
3199,91	2	21,45	25,32	$4d\ 9—5p\ ^4D°$	$^1/_2,\ ^3/_2—^3/_2$
3142,01	3	21,38	25,32	$5s\,^4P—5p\ ^4D°$	$^5/_2—^3/_2$
2983,22	2	20,98	25,13	$5s\,^4P—5p\ ^4D°$	$^3/_2—^1/_2$
2859,3	3	21,45	25,78	$4d\ 9—5p\ ^4D°$	$^1/_2,\ ^3/_2—^1/_2$
2856,2	2	21,38	25,72	$5s\,^4P—5p\ ^4D°$	$^5/_2—^5/_2$
2853,0	5	20,98	25,32	$5s\,^4P—5p\ ^4D°$	$^3/_2—^3/_2$
2836,08	3	21,52	25,89	$4d\ 10—5p\ ^4P°$	$^3/_2,\ ^5/_2—^5/_2$
2829,60	3	21,52	25,90	$4d\ 10—5p\ ^4S°$	$^3/_2,\ ^5/_2—^3/_2$
2774,70	6	21,38	25,84	$5s\,^4P—5p\ ^4P°$	$^5/_2—^3/_2$
2748,18	8	21,38	25,89	$5s\,^4P—5p\ ^4P°$	$^5/_2—^5/_2$
2742,13	2	21,38	25,90	$5s\,^4P—5p\ ^4S°$	$^5/_2—^3/_2$
2736,65	2	20,79	25,32	$4d\ 4—5p\ ^4D°$	$^5/_2—^3/_2$
2733,36	2	21,52	26,05	$4d\ 10—5p\ 1°$	$^3/_2,\ ^5/_2—^3/_2,\ ^5/_2$
2730,55	3	20,59	25,13	$5s\,^4P—5p\ ^4D°$	$^1/_2—^1/_2$
2673,0	2	21,26	25,90	$4d\ 8—5p\ ^4S°$	$^5/_2—^3/_2$
2651,6	2	21,38	26,05	$5s\,^4P—5p\ 1°$	$^5/_2—^3/_2,\ ^5/_2$
2621,11	7	20,59	25,32	$5s\,^4P—5p\ ^4D°$	$^1/_2—^3/_2$
2615,3	8	20,98	25,72	$5s\,^4P—5p\ ^4D°$	$^3/_2—^5/_2$
2609,5	10	21,38	26,13	$5s\,^4P—5p\ ^4D°$	$^5/_2—^7/_2$
2606,17	5	20,57	25,32	$4d\ 2—5p\ ^4D°$	$^3/_2,\ ^5/_2—^3/_2$
2586,9	3	21,26	26,05	$4d\ 8—5p\ 1°$	$^5/_2—^3/_2,\ ^5/_2$
2579,0	2	20,98	25,78	$5s\,^4P—5p\ ^4P°$	$^3/_2—^1/_2$
2558,08	4	21,26	26,11	$4d\ 8—5p\ 2°$	$^5/_2—^3/_2,\ ^5/_2$
2547,0	6	{ 20,98	25,84	$5s\,^4P—5p\ ^4P°$	$^3/_2—^3/_2$
		21,26	26,13	$4d\ 8—5p\ ^4D°$	$^5/_2—^7/_2$
2546,0	5	21,18	26,05	$4d\ 7—5p\ 1°$	$^3/_2,\ ^5/_2—^3/_2,\ ^5/_2$

λ, Å	I	E_H, eV	E_B, eV	Transition	J
2546,0	5	20,85	25,72	$4d$ 6—$5p$ $^4D^\circ$	5/2, 7/2—5/2
2524,5	5	20,98	25,89	$5s$ ^4P—$5p$ $^4P^\circ$	3/2—5/2
2519,38	6	20,98	25,90	$5s$ ^4P—$5p$ $^4S^\circ$	3/2—3/2
2518,02	5	21,18	26,11	$4d$ 7—$5p$ 2°	3/2, 5/2—3/2, 5/2
2517,0	4	20,79	25,72	$4d$ 4—$5p$ $^4D^\circ$	5/2—5/2
2510,2	2	20,38	25,32	$4d$ 1—$5p$ $^4D^\circ$	3/2, 5/2—3/2
2474,06	5	21,38	26,39	$5s$ ^4P—$5p$ 3°	5/2—3/2, 5/2
2459,74	6	20,85	25,89	$4d$ 6—$5p$ $^4P^\circ$	5/2, 7/2—5/2
2451,7	4	20,83	25,89	$4d$ 5—$5p$ $^4P^\circ$	5/2, 7/2—5/2
2442,68	5	20,98	26,05	$5s$ ^4P—$5p$ 1°	3/2—3/2, 5/2
2432,8	1	20,79	25,89	$4d$ 4—$5p$ $^4P^\circ$	5/2—5/2
2428,04	3	20,79	25,90	$4d$ 4—$5p$ $^4S^\circ$	5/2—3/2
2416,9	4	20,59	25,72	$4d$ 3—$5p$ $^4D^\circ$	3/2, 5/2—5/2
		20,98	26,11	$5s$ ^4P—$5p$ 2°	3/2—3/2, 5/2
2406,42	2	20,57	25,72	$4d$ 2—$2p$ $^4D^\circ$	3/2, 5/2—5/2
2388,05	3	20,59	25,78	$5s$ ^4P—$5p$ $^4P^\circ$	1/2—1/2
2360,4	2	20,59	25,84	$4d$ 3—$5p$ $^4P^\circ$	3/2, 5/2—3/2
2358,5	3	20,59	25,84	$4d$ 3—$5p$ $^4P^\circ$	3/2, 5/2—3/2
2348,27	4	20,85	26,13	$4d$ 6—$5p$ $^4D^\circ$	5/2, 7/2—7/2
2340,93	2	20,83	26,13	$4d$ 5—$5p$ $^4D^\circ$	5/2, 7/2—7/2
2339,15	2	20,59	25,89	$4d$ 3—$5p$ $^4P^\circ$	3/2, 5/2—5/2
2336,75	4	20,59	25,90	$5s$ ^4P—$5p$ $^4S^\circ$	1/2—3/2
2329,3	3	20,57	25,89	$4d$ 2—$5p$ $^4P^\circ$	3/2, 5/2—5/2
2324,85	1	20,57	25,90	$4d$ 2—$5p$ $^4S^\circ$	3/2, 5/2—3/2
2323,57	1	20,79	26,13	$4d$ 4—$5p$ $^4D^\circ$	5/2—7/2
2291,26	6	20,98	26,39	$5s$ ^4P—$5p$ 3°	3/2—3/2, 5/2
2259,42	1	20,57	26,05	$4d$ 2—$5p$ 1°	3/2, 5/2—3/2, 5/2
2252,54	2	20,38	25,89	$4d$ 1—$5p$ $^4P^\circ$	3/2, 5/2—5/2
2237,34	3	20,57	26,11	$4d$ 2—$5p$ 2°	3/2, 5/2—3/2, 5/2
842,035	22	0,00	14,72	$4p^3$ $^4S^\circ$—$4p^4$ $^4P^\circ$	3/2—5/2
816,822	18	0,00	15,18	$4p^3$ $^4S^\circ$—$4p^4$ $^4P^\circ$	3/2—3/2
805,763	7	0,00	15,39	$4p^3$ $^4S^\circ$—$4p^4$ $^4P^\circ$	3/2—1/2

Kr V, ground state $1s^2\,2s^2\,2p^6\,3s^2\,3p^6\,3d^{10}\,4s^2\,4p^2\,{}^3P_0$

λ, Å	I	E_H, eV	E_B, eV	Transition	J
708,85	8	—	—	—	—
472,16	3	—	—	—	—

Kr VI, ground state $1s^2\,2s^2\,2p^6\,3s^2\,3p^6\,3d^{10}\,4s^2\,4p\,{}^2P^0_{1/2}$

λ, Å	I	E_H, eV	E_B, eV	Transition	J
742,83	8	—	—	$4p$ $^2P^\circ$—$4p^2$ ^2D	3/2—5/2
705,84	8	—	—	$4p$ $^2P^\circ$—$4p^2$ ^2D	1/2—3/2
580,63	2	—	—	$4p$ $^2P^\circ$—$4p^2$ ^2P	3/2—1/2
569,13	5	—	—	$4p$ $^2P^\circ$—$4p^2$ ^2P	3/2—3/2
554,52	5	—	—	$4p$ $^2P^\circ$—$4p^2$ ^2P	1/2—1/2

λ, Å	I	E_H, eV	E_B, eV	Transition	J
544,03	5	—	—	$4p\ ^2P^\circ—4p^2\ ^2P$	$^1/_2—^3/_2$
465,27	6	—	—	$4p\ ^2P^\circ—4d\ ^2D$	$^3/_2—^5/_2$
450,20	2	—	—	$4p\ ^2P^\circ—4d\ ^2D$	$^1/_2—^3/_2$

Kr VII, ground state $1s^2\,2s^2\,2p^6\,3s^2\,3p^6\,3d^{10}\,4s^2\,^1S_0$

λ, Å	I	E_H, eV	E_B, eV	Transition	J
618,67	1	—	—	$4p\ ^3P^\circ—4p^2\ ^3P$	2—2
585,37	8	—	—	$4s^2\ ^1S—4p\ ^1P^\circ$	0—1

Kr VIII, ground state $1s^2\,2s^2\,2p^6\,3s^2\,3p^6\,3d^{10}\,4s\ ^2S_{1/2}$

λ, Å	I	E_H, eV	E_B, eV	Transition	J
695,91	8	—	—	$4s\ ^2S—4p\ ^2P^\circ$	$^1/_2—^1/_2$
651,57	10	—	—	$4s\ ^2S—4p\ ^2P^\circ$	$^1/_2—^3/_2$

XENON, Z = 54

Xe I, ground state $1s^2\,2s^2\,2p^6\,3s^2\,3p^6\,3d^{10}\,4s^2\,4p^6\,4d^{10}\,5s^2\,5p^6\,{}^1S_0$

Ionization potential 97 834,4 cm^{-1}; 12,129 eV

λ, Å	I	E_H, eV	E_B, eV	Transition	J
26511,1	30	9,93	10,40	6p [1/2]—5d [1 1/2]°	0—1
26272,0	60	9,69	10,16	6p [2 1/2]—5d [2 1/2]°	2—2
24825,3	20	9,72	10,22	6p [2 1/2]—5d [2 1/2]°	3—3
23195,5	10	9,69	10,22	6p [2 1/2]—5d [2 1/2]°	2—3
20262,2	6	9,79	10,40	6p [1 1/2]—5d [1 1/2]°	1—1
18788,0	3	9,93	10,60	6p [1/2]—7s [1 1/2]°	0—1
17325,5	5	9,69	10,40	6p [2 1/2]—5d [1 1/2]°	2—1
16727,52	50	9,82	10,56	6p [1 1/2]—7s [1 1/2]°	2—2
16052,02	50	9,82	10,60	6p [1 1/2]—7s [1 1/2]°	2—1
15418,01	110	9,79	10,60	6p [1 1/2]—7s [1 1/2]°	1—1
14732,38	200	9,72	10,56	6p [2 1/2]—7s [1 1/2]°	3—2
14659,84	5	10,16	11,00	5d [2 1/2]°—7p [1 1/2]	2—1
14364,90	20	10,40	11,26	5d [1 1/2]°—4f [1 1/2]	1—2
14241,39	40	10,40	11,27	5d [1 1/2]°—4f [2 1/2]	1—2
14142,09	80	9,69	10,56	6p [2 1/2]—7s [1 1/2]°	2—2
13656,48	150	9,96	10,60	6p [2 1/2]—7s [1 1/2]°	2—1
13543,16	5	10,04	10,95	5d [3 1/2]°—7p [2 1/2]	3—2
12623,32	300	9,58	10,56	6p [1/2]—7s [1 1/2]°	1—2
12590,00	26	9,92	10,90	5d [1/2]°—7p [1/2]	1—1
12451,21	2	9,96	10,95	5d [1 1/2]°—7p [2 1/2]	2—2
12258,10	6	9,89	10,90	5d [1/2]°—7p [1/2]	0—1
12235,14	80	9,58	10,60	6p [1/2]—7s [1 1/2]°	1—1
12084,82	20	9,94	10,97	5d [3 1/2]°—7p [2 1/2]	4—3
11952,57	10	9,96	10,99	5d [1 1/2]°—7p [1 1/2]	2—2
11951,1	1	9,92	10,95	5d [1/2]°—7p [2 1/2]	1—2
11911,44	3	9,92	10,96	5d [1/2]°—6p' [1 1/2]	1—1
11874,36	1	10,22	11,26	5d [2 1/2]°—4f [1 1/2]	3—2
11857,86	2	9,93	10,98	6p [1/2]—6d [1/2]°	0—1
11857,00	30	10,22	11,26	5d [2 1/2]°—4f [4 1/2]	3—4
11793,04	40	10,22	11,27	5d [2 1/2]°—4f [2 1/2]	3—3
11742,01	90	10,22	11,27	5d [2 1/2]°—4f [3 1/2]	3—4
11614,08	25	0,80	10,96	5d [1/2]°—6p' [1 1/2]	0—1
11537,4	1	10,40	11,47	5d [1 1/2]°—8p [1/2]	1—0
11491,22	15	9,92	10,99	5d [1/2]°—7p [1 1/2]	1—2
11415,04	15	9,92	11,00	5d [1/2]°—7p [1 1/2]	1—1
11309,56	5	9,96	11,05	5d [1 1/2]°—6p' [1 1/2]	2—2
11289,10	10	9,92	11,01	5d [1/2]°—7p [1/2]	1—0
11214,89	5	10,16	11,26	5d [2 1/2]°—4f [1 1/2]	2—1
11175,5	1	10,56	11,67	7s [1 1/2]°—9p [2 1/2]	2—3
11162,67	10	9,96	11,07	5d [1 1/2]°—6p' [1/2]	2—1
11141,09	50	9,89	11,00	5d [1/2]°—7p [1 1/2]	0—1
11130,81	8	10,16	11,27	5d [2 1/2]°—4f [2 1/2]	2—3
11127,20	100	10,16	11,27	5d [2 1/2]°—4f [2 1/2]	2—2
11085,25	250	10,16	11,27	5d [2 1/2]°—4f [3 1/2]	2—3
10895,32	200	9,92	11,05	5d [1/2]°—6p' [1 1/2]	1—2
10838,34	1000	8,44	9,58	6s [1 1/2]°—6p [1/2]	1—1
10758,86	100	9,92	11,07	5d [1/2]°—6p' [1/2]	1—1
10706,78	150	9,82	10,98	6p [1 1/2]—6d [1/2]°	2—1
10549,76	20	10,40	11,58	5d [1 1/2]°—5f [1 1/2]	1—2
10527,84	40	9,82	11,00	6p [1 1/2]—6d [1 1/2]°	2—2
10515,15	10	9,89	11,07	5d [1/2]°—6p' [1/2]	0—1
10507,91	6	10,40	11,58	5d [1 1/2]°—5f [2 1/2]	1—2
10484,83	8	9,79	10,97	6p [1 1/2]—6d [1/2]°	1—0

λ, Å	I	E_H, eV	E_B, eV	Transition	J
10420,52	1	9,79	10,98	$6p\,[1^1/_2]-6d\,[^1/_2]^\circ$	1—1
10251,07	20	9,79	11,00	$6p\,[1^1/_2]-6d\,[1^1/_2]^\circ$	1—2
10188,36	10	9,82	11,04	$6p\,[1^1/_2]-6d\,[3^1/_2]^\circ$	2—3
10125,47	20	9,92	11,14	$5d\,[^1/_2]^\circ-6p'\,[^1/_2]$	1—0
10119,8	1	10,04	11,26	$5d\,[3^1/_2]^\circ-4f\,[1^1/_2]$	3—2
10107,34	80	10,04	11,26	$5d\,[3^1/_2]^\circ-4f\,[4^1/_2]$	3—4
10084,79	20	9,93	11,16	$6p\,[^1/_2]-6d\,[1^1/_2]^\circ$	0—1
10060,96	10	10,04	11,27	$5d\,[3^1/_2]^\circ-4f\,[2^1/_2]$	3—3
10057,96	5	10,04	11,27	$5d\,[3^1/_2]^\circ-4f\,[2^1/_2]$	3—2
10023,72	50	10,04	11,27	$5d\,[3^1/_2]^\circ-4f\,[3^1/_2]$	3—3, 4
9966,58	10	9,82	11,06	$6p\,[1^1/_2]-6d\,[2^1/_2]^\circ$	2—2
9923,192	3000	8,44	9,69	$6s\,[1^1/_2]^\circ-6p\,[2^1/_2]$	1—2
9799,699	2000	8,31	9,58	$6s\,[1^1/_2]^\circ-6p\,[^1/_2]$	2—1
9718,16	100	9,79	11,06	$6p\,[1^1/_2]-6d\,[2^1/_2]^\circ$	1—2
9710,03	2	10,16	11,43	$5d\,[2^1/_2]^\circ-8p\,[2^1/_2]$	2—2
9700,99	20	9,72	11,00	$6p\,[2^1/_2]-6d\,[1^1/_2]^\circ$	3—2
9685,32	150	9,82	11,10	$6p\,[1^1/_2]-6d\,[2^1/_2]^\circ$	2—3
9605,80	3	10,16	11,45	$5d\,[2^1/_2]^\circ-8p\,[1^1/_2]$	2—1
9585,14	20	9,69	10,98	$6p\,[2^1/_2]-6d\,[^1/_2]^\circ$	2—1
9513,379	200	9,72	11,02	$6p\,[2^1/_2]-6d\,[3^1/_2]^\circ$	3—4
9505,78	10	9,96	11,26	$5d\,[1^1/_2]^\circ-4f\,[1^1/_2]$	2—1
9497,07	40	9,96	11,26	$5d\,[1^1/_2]^\circ-4f\,[1^1/_2]$	2—2
9487,76	4	—	—	—	—
9445,34	80	9,96	11,27	$5d\,[1^1/_2]^\circ-4f\,[2^1/_2]$	2—3
9442,68	20	9,96	11,27	$5d\,[1^1/_2]^\circ-4f\,[2^1/_2]$	2—2
9441,46	20	9,69	11,00	$6p\,[2^1/_2]-6d\,[1^1/_2]^\circ$	2—2
9412,01	60	9,72	11,04	$6p\,[2^1/_2]-6d\,[3^1/_2]^\circ$	3—3
9374,76	100	9,94	11,26	$5d\,[3^1/_2]^\circ-4f\,[4^1/_2]$	4—5
9374,02	10	9,94	11,26	$5d\,[3^1/_2]^\circ-4f\,[4^1/_2]$	4—4
9334,08	3	9,94	11,27	$5d\,[3^1/_2]^\circ-4f\,[2^1/_2]$	4—3
9306,64	40	9,57	10,90	$6s'\,[^1/_2]^\circ-7p\,[^1/_2]$	1—1
9301,95	30	9,94	11,27	$5d\,[3^1/_2]^\circ-4f\,[3^1/_2]$	4—3, 4
9245,18	3	9,93	11,27	$6p\,[^1/_2]-8s\,[1^1/_2]^\circ$	0—1
9222,39	5	9,72	11,06	$6p\,[2^1/_2]-6d\,[2^1/_2]^\circ$	3—2
9216,51	1	10,40	11,75	$5d\,[1^1/_2]^\circ-6f\,[1^1/_2]$	1—2
9211,38	25	9,92	11,26	$5d\,[^1/_2]^\circ-4f\,[1^1/_2]$	1—1
9203,20	30	9,92	11,26	$5d\,[^1/_2]^\circ-4f\,[1^1/_2]$	1—2
9197,18	2	10,40	11,75	$5d\,[1^1/_2]^\circ-6f\,[2^1/_2]$	1—2
9167,52	100	9,69	11,04	$6p\,[2^1/_2]-6d\,[3^1/_2]^\circ$	2—3
9162,654	500	8,44	9,79	$6s\,[1^1/_2]^\circ-6p\,[1^1/_2]$	1—1
9158,38	2	—	—	—	—
9152,12	20	9,92	11,27	$5d\,[^1/_2]^\circ-4f\,[2^1/_2]$	1—2
9141,8	2	10,22	11,58	$5d\,[2^1/_2]^\circ-5f\,[1^1/_2]$	3—2
9131,59	3	10,22	11,58	$5d\,[2^1/_2]^\circ-5f\,[4^1/_2]$	3—4
9112,24	4	10,22	11,58	$5d\,[2^1/_2]^\circ-5f\,[2^1/_2]$	3—3
9096,13	50	10,22	11,58	$5d\,[2^1/_2]^\circ-5f\,[3^1/_2]$	3—3, 4
9045,446	400	8,31	9,69	$6s\,[1^1/_2]^\circ-6p\,[2^1/_2]$	2—2
9032,18	50	9,89	11,26	$5d\,[^1/_2]^\circ-4f\,[1^1/_2]$	0—1
9025,98	30	9,79	11,16	$6p\,[1^1/_2]-6d\,[1^1/_2]^\circ$	1—1
8987,57	200	9,69	11,06	$6p\,[2^1/_2]-6d\,[2^1/_2]^\circ$	2—2
8981,05	100	9,72	11,10	$6p\,[2^1/_2]-6d\,[2^1/_2]^\circ$	3—3
8952,78	50	9,57	10,95	$6s'\,[^1/_2]^\circ-7p\,[2^1/_2]$	1—2
8952,254	1000	8,44	9,82	$6s\,[1^1/_2]^\circ-6p\,[1^1/_2]$	1—2
8930,83	200	9,57	10,96	$6s'\,[^1/_2]^\circ-6p'\,[1^1/_2]$	1—1
8908,73	200	9,58	10,97	$6p\,[^1/_2]-6d\,[^1/_2]^\circ$	1—0
8885,71	10	10,04	11,43	$5d\,[3^1/_2]^\circ-8p\,[2^1/_2]$	3—2
8862,32	300	9,58	10,98	$6p\,[^1/_2]-6d\,[^1/_2]^\circ$	1—1
8851,44	1	10,04	11,44	$5d\,[3^1/_2]^\circ-8p\,[2^1/_2]$	3—3
8819,412	5000	8,31	9,72	$6s\,[1^1/_2]^\circ-6p\,[2^1/_2]$	2—3

λ, Å	I	E_H, eV	E_B, eV	Transition	J
8758,20	100	9,69	11,10	$6p\ [2^1/_2]$—$6d\ [2^1/_2]^\circ$	2—3
8739,39	300	9,58	11,00	$6p\ [^1/_2]$—$6d\ [1^1/_2]^\circ$	1—2
8711,54	2	10,16	11,58	$5d\ [2^1/_2]^\circ$—$5f\ [2^1/_2]$	2—3
8709,64	40	10,16	11,58	$5d\ [2^1/_2]^\circ$—$5f\ [2^1/_2]$	2—2
8696,86	200	10,16	11,58	$5d\ [2^1/_2]^\circ$—$5f\ [3^1/_2]$	2—3
8692,20	100	9,57	10,99	$6s'\ [^1/_2]^\circ$—$7p\ [1^1/_2]$	1—2
8648,54	250	9,57	11,00	$6s'\ [^1/_2]^\circ$—$7p\ [1^1/_2]$	1—1
8624,24	80	9,82	11,26	$6p\ [1^1/_2]$—$8s\ [1^1/_2]^\circ$	2—2
8576,01	200	9,57	11,01	$6s'\ [^1/_2]^\circ$—$7p\ [^1/_2]$	1—0
8564,7	1	10,40	11,85	$5d\ [1^1/_2]^\circ$—$7f\ [1^1/_2]$	1—2
8553,97	2	10,40	11,85	$5d\ [1^1/_2]^\circ$—$7f\ [2^1/_2]$	1—2
8530,10	30	9,82	11,27	$6p\ [1^1/_2]$—$8s\ [1^1/_2]^\circ$	2—1
8522,55	30	9,45	10,90	$6s'\ [^1/_2]^\circ$—$7p\ [^1/_2]$	0—1
8450,37	1	9,96	11,42	$5d\ [1^1/_2]^\circ$—$8p\ [^1/_2]$	2—1
8437,55	10	9,79	11,26	$6p\ [1^1/_2]$—$8s\ [1^1/_2]^\circ$	1—2
8409,190	2000	8,31	9,79	$6s\ [1^1/_2]^\circ$—$6p\ [1^1/_2]$	2—1
8402,03	5	9,96	11,43	$5d\ [1^1/_2]^\circ$—$8p\ [2^1/_2]$	2—1
8392,37	20	9,69	11,16	$6p\ [2^1/_2]$—$6d\ [1^1/_2]^\circ$	2—1
8372,79	5	9,82	11,30	$6p\ [1^1/_2]$—$5d'\ [2^1/_2]^\circ$	2—2
8371,38	3	9,96	11,44	$5d\ [1^1/_2]^\circ$—$8p\ [2^1/_2]$	2—3
8349,05	40	9,58	11,06	$6p\ [^1/_2]$—$6d\ [2^1/_2]^\circ$	1—2
8347,45	60	9,79	11,27	$6p\ [1^1/_2]$—$8s\ [1^1/_2]^\circ$	1—1
8346,823	2000	9,57	11,05	$6s'\ [^1/_2]^\circ$—$6p'\ [1^1/_2]$	1—2
8324,58	20	9,93	11,42	$6p\ [^1/_2]$—$7d\ [^1/_2]^\circ$	0—1
8323,90	2	9,96	11,45	$5d\ [1^1/_2]^\circ$—$8p\ [1^1/_2]$	2—1
8297,71	15	9,96	11,45	$5d\ [1^1/_2]^\circ$—$8p\ [1^1/_2]$	2—2
8280,1163	7000	8,44	9,93	$6s\ [1^1/_2]^\circ$—$6p\ [^1/_2]$	1—0
8266,519	500	9,97	11,07	$6s'\ [^1/_2]^\circ$—$6p'\ [^1/_2]$	1—1
8231,6348	10000	8,31	9,82	$6s\ [1^1/_2]^\circ$—$6p\ [1^1/_2]$	2—2
8206,341	700	9,45	10,96	$6s'\ [^1/_2]^\circ$—$6p'\ [1^1/_2]$	0—1
8196,73	2	9,79	11,30	$6p\ [1^1/_2]$—$5d'\ [2^1/_2]^\circ$	1—2
8182,93	1	10,40	11,92	$5d\ [1^1/_2]^\circ$—$8f\ [2^1/_2]$	1—2
8171,02	100	9,82	11,34	$6p\ [1^1/_2]$—$5d'\ [1^1/_2]^\circ$	2—2
		9,92	11,43	$5d\ [^1/_2]^\circ$—$8p\ [2^1/_2]$	1—2
8165,37	2	—	—	—	—
8123,29	2	10,22	11,75	$5d\ [2^1/_2]^\circ$—$6f\ [1^1/_2]$	3—2
8118,29	15	10,22	11,75	$5d\ [2^1/_2]^\circ$—$6f\ [4^1/_2]$	3—4
8109,40	15	10,22	11,75	$5d\ [2^1/_2]^\circ$—$6f\ [2^1/_2]$	3—3
8107,91	6	10,22	11,75	$5d\ [2^1/_2]^\circ$—$6f\ [2^1/_2]$	3—2
8101,98	100	10,22	11,75	$5d\ [2^1/_2]^\circ$—$6f\ [3^1/_2]$	3—3, 4
8097,24	3	9,92	11,45	$5d\ [^1/_2]^\circ$—$8p\ [1^1/_2]$	1—1
8073,99	1	9,89	11,42	$5d\ [^1/_2]^\circ$—$8p\ [^1/_2]$	0—1
8064,94	2	10,04	11,58	$5d\ [3^1/_2]^\circ$—$5f\ [1^1/_2]$	3—2
8061,340	150	9,72	11,26	$6p\ [2^1/_2]$—$8s\ [1^1/_2]^\circ$	3—2
8057,258	200	10,04	11,58	$5d\ [3^1/_2]^\circ$—$5f\ [4^1/_2]$	3—4
8042,18	15	10,04	11,58	$5d\ [3^1/_2]^\circ$—$5f\ [2^1/_2]$	3—3
8040,56	10	10,04	11,58	$5d\ [3^1/_2]^\circ$—$5f\ [2^1/_2]$	3—2
8029,67	100	10,04	11,58	$5d\ [3^1/_2]^\circ$—$5f\ [3^1/_2]$	3—3, 4
8003,26	10	9,79	11,34	$6p\ [1^1/_2]$—$5d'\ [1^1/_2]^\circ$	1—2
7976,03	8	9,82	11,37	$6p\ [1^1/_2]$—$5d'\ [2^1/_2]^\circ$	2—3
7967,341	500	9,45	11,00	$6s'\ [^1/_2]^\circ$—$7p\ [1^1/_2]$	0—1
7954,22	4	9,92	11,47	$5d\ [^1/_2]^\circ$—$8p\ [^1/_2]$	1—0
7937,41	40	9,93	11,49	$6p\ [^1/_2]$—$7d\ [1^1/_2]^\circ$	0—1
7887,395	300	9,57	11,14	$6s'\ [^1/_2]^\circ$—$6p'\ [^1/_2]$	1—0
7881,320	100	9,69	11,26	$6p\ [2^1/_2]$—$8s\ [1^1/_2]^\circ$	2—2
7841,23	15	9,72	11,30	$6p\ [2^1/_2]$—$5d'\ [2^1/_2]^\circ$	3—2
7832,98	10	9,58	11,16	$6p\ [^1/_2]$—$6d\ [1^1/_2]^\circ$	1—1
7802,651	100	9,69	11,27	$6p\ [2^1/_2]$—$8s\ [1^1/_2]^\circ$	2—1
7790,53	1	10,16	11,75	$5d\ [2^1/_2]^\circ$—$6f\ [2^1/_2]$	2—3

λ, Å	I	E_H, eV	E_B, eV	Transition	J
7789,42	15	10,16	11,75	$5d\,[2^1/_2]^\circ - 6f\,[2^1/_2]$	2—2
7783,66	50	10,16	11,75	$5d\,[2^1/_2]^\circ - 6f\,[3^1/_2]$	2—3
7740,31	40	9,82	11,42	$6p\,[1^1/_2] - 7d\,[^1/_2]^\circ$	2—1
7670,81	1	9,69	11,30	$6p\,[2^1/_2] - 5d'\,[2^1/_2]^\circ$	2—2
7666,61	10	9,96	11,57	$5d\,[1^1/_2]^\circ - 5f\,[1^1/_2]$	2—1
7664,56	30	9,96	11,58	$5d\,[1^1/_2]^\circ - 5f\,[1^1/_2]$	2—2
7664,02	10	9,72	11,34	$6p\,[2^1/_2] - 5d'\,[1^1/_2]^\circ$	3—2
7643,91	100	9,96	11,58	$5d\,[1^1/_2]^\circ - 5f\,[2^1/_2]$	2—3
7642,30	—	9,96	11,58	$5d\,[1^1/_2]^\circ - 5f\,[2^1/_2]$	2—2
7642,025	500	9,45	11,07	$6s'\,[^1/_2]^\circ - 6p'\,[^1/_2]$	0—1
7609,82	3	10,22	11,85	$5d\,[2^1/_2]^\circ - 7f\,[4^1/_2]$	3—4
7608,46	5	10,04	11,67	$5d\,[3^1/_2]^\circ - 9p\,[2^1/_2]$	3—2
7604,97	2	10,22	11,85	$5d\,[3^1/_2]^\circ - 7f\,[2^1/_2]$	3—3
7600,77	10	10,22	11,85	$5d\,[2^1/_2]^\circ - 7f\,[3^1/_2]$	3—3, 4
7594,36	1	10,04	11,67	$5d\,[3^1/_2]^\circ - 9p\,[2^1/_2]$	3—3
7589,61	6	9,79	11,42	$6p\,[1^1/_2] - 7d\,[^1/_2]^\circ$	1—1
7584,680	200	9,94	11,58	$5d\,[3^1/_2]^\circ - 5f\,[4^1/_2]$	4—5
7584,29	10	9,94	11,58	$5d\,[3^1/_2]^\circ - 5f\,[4^1/_2]$	4—4
7570,93	6	9,94	11,98	$5d\,[3^1/_2]^\circ - 5f\,[2^1/_2]$	4—3
7559,79	40	9,94	11,58	$5d\,[3^1/_2]^\circ - 5f\,[3^1/_2]$	4—3, 4
7514,96	3	9,93	11,58	$6p\,[^1/_2] - 9s\,[1^1/_2]^\circ$	0—1
7514,54	8	9,79	11,44	$6p\,[1^1/_2] - 7d\,[^1/_2]^\circ$	1—0
7501,13	20	9,69	11,34	$6p\,[2^1/_2] - 5d'\,[1^1/_2]^\circ$	2—2
7492,23	20	9,72	11,37	$6p\,[2^1/_2] - 5d'\,[2^1/_2]^\circ$	3—3
7474,01	25	9,92	11,57	$5d\,[^1/_2]^\circ - 5f\,[1^1/_2]$	1—1
7472,01	40	9,92	11,58	$5d\,[^1/_2]^\circ - 5f\,[1^1/_2]$	1—2
7451,00	25	9,92	11,58	$5d\,[^1/_2]^\circ - 5f\,[2^1/_2]$	1—2
7441,94	20	9,82	11,49	$6p\,[1^1/_2] - 7d\,[3^1/_2]^\circ$	2—3
7424,05	20	9,82	11,49	$6p\,[1^1/_2] - 7d\,[2^1/_2]^\circ$	2—2
7405,77	3	9,93	11,61	$6p\,[^1/_2] - 5d'\,[1^1/_2]^\circ$	0—1
7404,51	12	9,82	11,49	$6p\,[1^1/_2] - 7d\,[1^1/_2]^\circ$	2—1
7400,41	30	9,82	11,50	$6p\,[1^1/_2] - 7d\,[1^1/_2]^\circ$	2—2
7393,793	150	9,82	11,50	$6p\,[1^1/_2] - 7d\,[2^1/_2]^\circ$	2—3
7386,002	100	9,58	11,26	$6p\,[^1/_2] - 8s\,[1^1/_2]^\circ$	1—2
7355,58	40	9,89	11,57	$5d\,[^1/_2]^\circ - 5f\,[1^1/_2]$	0—1
7336,480	50	9,69	11,37	$6p\,[2^1/_2] - 5d'\,[2^1/_2]^\circ$	2—3
7323,05	2	10,16	11,85	$5d\,[2^1/_2]^\circ - 7f\,[2^1/_2]$	2—2
7312,452	80	9,57	11,26	$6s'\,[^1/_2]^\circ - 4f\,[1^1/_2]$	1—1
7319,94	15	10,16	11,85	$5d\,[2^1/_2]^\circ - 7f\,[3^1/_2]$	2—3
7316,87	20	9,58	11,27	$6p\,[^1/_2] - 8s\,[1^1/_2]^\circ$	1—1
7316,272	70	9,57	11,26	$6s'\,[^1/_2]^\circ - 4f\,[1^1/_2]$	1—2
7313,01	1	10,22	11,91	$5d\,[2^1/_2]^\circ - 8f\,[4^1/_2]$	3—4
7307,37	5	10,22	11,92	$5d\,[2^1/_2]^\circ - 8f\,[3^1/_2]$	3—3, 4
7285,301	60	9,79	11,49	$6p\,[1^1/_2] - 7d\,[2^1/_2]^\circ$	1—2
7283,961	40	9,57	11,27	$6s'\,[^1/_2]^\circ - 4f\,[2^1/_2]$	1—2
7266,49	25	9,79	11,49	$6p\,[1^1/_2] - 7d\,[1^1/_2]^\circ$	1—1
7262,54	20	9,79	11,50	$6p\,[1^1/_2] - 7d\,[1^1/_2]^\circ$	1—2
7257,94	60	10,04	11,75	$5d\,[3^1/_2]^\circ - 6f\,[4^1/_2]$	3—4
7250,87	5	10,04	11,75	$5d\,[2^1/_2]^\circ - 6f\,[3^1/_2]$	3—3
		9,96	11,67	$5d\,[1^1/_2]^\circ - 9p\,[2^1/_2]$	2—2
7249,92	2	10,04	11,75	$5d\,[3^1/_2]^\circ - 6f\,[2^1/_2]$	3—2
7244,94	20	10,04	11,75	$5d\,[3^1/_2]^\circ - 6f\,[3^1/_2]$	3—3, 4
7238,20	3	9,96	11,67	$5d\,[1^1/_2]^\circ - 9p\,[2^1/_2]$	2—3
7220,24	1	9,96	11,67	$5d\,[1^1/_2]^\circ - 9p\,[1^1/_2]$	2—1
7209,14	5	9,96	11,68	$5d\,[1^1/_2]^\circ - 9p\,[1^1/_2]$	2—2
7200,79	15	9,58	11,30	$6p\,[^1/_2] - 5d'\,[2^1/_2]^\circ$	1—2
7172,70	10	9,94	11,67	$5d\,[3^1/_2]^\circ - 9p\,[2^1/_2]$	4—3
7136,57	15	9,69	11,42	$6p\,[2^1/_2] - 7d\,[^1/_2]^\circ$	2—1

λ, Å	I	E_H, eV	E_B, eV	Transition	J
7119,598	500	9,72	11,46	6*p* [2½]—7*d* [3½]°	3—4
7078,46	1	9,92	11,67	5*d* [½]°—9*p* [2½]	1—2
7051,06	3	9,58	11,34	6*p* [½]—5*d*′ [1½]°	1—2
7049,36	1	10,16	11,92	5*d* [2½]°—8*f* [2½]	2—2
7049,07	1	9,92	11,67	5*d* [½]°—9*p* [1½]	1—1
7047,37	30	9,82	11,58	6*p* [1½]—9*s* [1½]°	2—2
7035,53	20	9,82	11,58	6*p* [1½]—9*s* [1½]°	2—1
7034,80	3	—	—	—	—
7019,02	30	9,72	11,49	6*p* [2½]—7*d* [3½]°	3—3
7003,10	4	9,72	11,49	6*p* [2½]—7*d* [2½]°	3—2
6991,65	1	9,92	11,69	5*d* [½]°—9*p* [½]	1—0
6982,05	30	9,72	11,50	6*p* [2½]—7*d* [1½]°	3—2
6976,182	100	9,72	11,50	6*p* [2½]—7*d* [2½]°	3—3
6936,69	8	9,96	11,74	5*d* [1½]°—6*f* [1½]	2—1
6935,62	50	9,96	11,74	5*d* [1½]°—6*f* [1½]	2—1
6925,53	100	9,96	11,75	5*d* [1½]°—6*f* [2½]	2—3
6924,67	15	9,96	11,75	5*d* [1½]°—6*f* [2½]	2—2
6922,22	8	9,79	11,58	6*p* [1½]—9*s* [1½]°	1—2
6910,82	30	9,79	11,58	6*p* [1½]—9*s* [1½]°	1—1
6882,155	300	9,69	11,49	6*p* [2½]—7*d* [3½]°	2—3
6872,107	100	9,94	11,75	5*d* [3½]°—6*f* [4½]	4—5
6866,838	50	9,69	11,49	6*p* [2½]—7*d* [2½]°	2—2
6865,58	5	9,94	11,75	5*d* [3½]°—6*f* [2½]	4—3
6863,20	20	9,93	11,74	6*p* [½]—8*d* [1½]°	0—1
6860,19	40	9,94	11,75	5*d* [3½]°—6*f* [3½]	4—3, 4
6850,13	30	9,69	11,49	6*p* [2½]—7*d* [1½]°	2—1
6848,82	50	10,04	11,85	5*d* [3½]°—7*f* [4½]	3—4
6846,613	60	9,69	11,50	6*p* [2½]—7*d* [1½]°	2—2
6844,84	2	10,04	11,85	5*d* [3½]°—7*f* [2½]	3—3
6844,27	1	10,04	11,85	5*d* [3½]°—7*f* [2½]	3—2
6841,50	20	10,04	11,85	5*d* [3½]°—7*f* [3½]	3—3, 4
6840,96	8	9,69	11,50	6*p* [2½]—7*d* [2½]°	2—3
6827,315	200	9,45	11,26	6*s*′ [½]°—4*f* [1½]	0—1
6818,38	15	9,79	11,61	6*p* [1½]—5*d*′ [1½]°	1—1
6815,64	12	9,93	11,75	6*p* [½]—10*s* [1½]°	0—1
6778,60	40	9,92	11,74	5*d* [½]°—6*f* [1½]	1—1
6777,57	50	9,92	11,75	5*d* [½]°—6*f* [1½]	1—2
6767,12	10	9,92	11,75	5*d* [½]°—6*f* [2½]	1—2
6728,008	200	9,96	11,80	5*d* [1½]°—10*p* [2½]	2—2
		9,58	11,42	6*p* [½]—7*d* [½]°	1—1
6706,46	1	9,96	11,81	5*d* [1½]°—10*p* [1½]	2—2
6681,036	20	9,89	11,74	5*d* [½]°—6*f* [1½]	0—1
6678,972	25	9,57	11,42	6*s*′ [½]°—8*p* [½]	1—1
6668,920	150	9,58	11,44	6*p* [½]—7*d* [½]°	1—0
6666,965	60	9,72	11,58	6*p* [2½]—9*s* [1½]°	3—2
6664,85	4	—	—	—	—
6657,92	20	9,82	11,68	6*p* [1½]—8*d* [½]°	2—1
6648,75	3	9,57	11,43	6*s*′ [½]°—8*p* [2½]	1—2
6632,464	50	9,82	11,69	6*p* [1½]—8*d* [1½]°	2—2
6630,44	2	9,82	11,69	6*p* [1½]—8*d* [3½]°	2—3
6608,87	10	9,82	11,70	6*p* [1½]—8*d* [2½]°	2—2
6607,41	30	10,04	11,91	5*d* [3½]°—8*f* [4½]	3—4
6602,87	4	10,04	11,92	5*d* [3½]°—8*f* [3½]	3—3, 4
6595,561	100	9,82	11,70	6*p* [1½]—8*d* [2½]°	2—3
6590,86	8	9,79	11,67	6*p* [1½]—8*d* [½]°	1—0
6583,27	20	9,57	11,45	6*s*′ [½]°—8*p* [1½]	1—2
6560,65	4	9,96	11,85	5*d* [1½]°—7*f* [1½]	2—2
6559,97	25	9,96	11,85	5*d* [1½]°—7*f* [1½]	2—2
6554,196	50	9,96	11,85	5*d* [1½]°—7*f* [2½]	2—3

λ, Å	I	E_H, eV	E_B, eV	Transition	J
6553,66	4	9,96	11,85	$5d\,[1^1/_2]^\circ{-}7f\,[2^1/_2]$	2—1
6546,12	20	9,79	11,68	$6p\,[1^1/_2]{-}8d\,[^1/_2]^\circ$	1—1
6543,360	40	9,69	11,58	$6p\,[2^1/_2]{-}9s\,[1^1/_2]^\circ$	2—2
6533,159	100	9,69	11,58	$6p\,[2^1/_2]{-}9s\,[1^1/_2]^\circ$	2—1
6521,508	40	9,79	11,69	$6p\,[1^1/_2]{-}8d\,[1^1/_2]^\circ$	1—2
6507,50	3	8,31	10,22	$6s\,[1^1/_2]^\circ{-}5d\,[2^1/_2]^\circ$	2—3
6504,18	200	9,57	11,47	$6s'\,[^1/_2]^\circ{-}8p\,[^1/_2]$	1—0
		9,94	11,85	$5d\,[3^1/_2]^\circ{-}7f\,[4^1/_2]$	4—5, 4
6500,37	15	9,93	11,84	$6p\,[^1/_2]{-}9d\,[1^1/_2]^\circ$	0—1
6498,718	100	9,79	11,70	$6p\,[1^1/_2]{-}8d\,[2^1/_2]^\circ$	1—2
6497,43	30	9,94	11,85	$5d\,[3^1/_2]^\circ{-}7f\,[3^1/_2]$	4—3, 4
6487,765	120	9,58	11,49	$6p\,[^1/_2]{-}7d\,[2^1/_2]^\circ$	1—2
6472,841	150	9,58	11,49	$6p\,[^1/_2]{-}7d\,[1^1/_2]^\circ$	1—1
6469,705	300	9,58	11,50	$6p\,[^1/_2]{-}7d\,[1^1/_2]^\circ$	1—2
6461,50	3	9,93	11,85	$6p\,[^1/_2]{-}11s\,[1^1/_2]^\circ$	0—1
6451,79	10	10,04	11,96	$5d\,[3^1/_2]^\circ{-}9f\,[4^1/_2]$	3—4
6450,48	7	9,69	11,61	$6p\,[2^1/_2]{-}5d'\,[1^1/_2]^\circ$	2—3
6448,70	2	10,04	11,96	$5d\,[3^1/_2]^\circ{-}9f\,[3^1/_2]$	3—3, 4
6430,155	20	9,82	11,75	$6p\,[1^1/_2]{-}10s\,[1^1/_2]^\circ$	2—2
6418,98	30	9,92	11,85	$5d\,[^1/_2]^\circ{-}7f\,[1^1/_2]$	1—1
		9,82	11,75	$6p\,[1^1/_2]{-}10s\,[1^1/_2]^\circ$	1—1
6418,41	30	9,92	11,85	$5d\,[^1/_2]^\circ{-}7f\,[1^1/_2]$	1—2
6412,38	10	9,92	11,85	$5d\,[^1/_2]^\circ{-}7f\,[2^1/_2]$	1—2
6355,77	20	9,79	11,74	$6p\,[1^1/_2]{-}8d\,[1^1/_2]^\circ$	1—1
6344,98	2	10,04	11,99	$5d\,[3^1/_2]^\circ{-}10f\,[4^1/_2]$	3—4
6337,58	8	9,96	11,91	$5d\,[1^1/_2]^\circ{-}8f\,[1^1/_2]$	2—2
6333,97	40	9,96	11,92	$5d\,[1^1/_2]^\circ{-}8f\,[2^1/_2]$	2—3
6331,50	20	9,89	11,85	$5d\,[^1/_2]^\circ{-}7f\,[1^1/_2]$	0—1
6325,81	2	9,79	11,75	$6p\,[1^1/_2]{-}10s\,[1^1/_2]^\circ$	1—2
6318,062	500	9,72	11,68	$6p\,[2^1/_2]{-}8d\,[3^1/_2]^\circ$	3—4
6314,97	15	9,79	11,75	$6p\,[1^1/_2]{-}10s\,[1^1/_2]$	1—1
6294,45	15	9,72	11,69	$6p\,[2^1/_2]{-}8d\,[1^1/_2]^\circ$	3—2
6292,649	50	9,72	11,69	$6p\,[2^1/_2]{-}8d\,[3^1/_2]^\circ$	3—3
6286,011	100	9,94	11,91	$5d\,[3^1/_2]^\circ{-}8f\,[4^1/_2]$	4—5
6284,38	2	9,94	11,92	$5d\,[3^1/_2]^\circ{-}8f\,[2^1/_2]$	4—3
6281,81	5	9,94	11,92	$5d\,[3^1/_2]^\circ{-}8f\,[3^1/_2]$	4—3
6276,99	4	9,93	11,91	$6p\,[^1/_2]{-}10d\,[1^1/_2]^\circ$	0—1
6273,23	10	9,72	11,70	$6p\,[2^1/_2]{-}8d\,[2^1/_2]^\circ$	3—2
6268,34	1	10,04	12,02	$5d\,[3^1/_2]^\circ{-}11f\,[4^1/_2]$	3—4
6265,301	40	9,45	11,42	$6s'\,[^1/_2]^\circ{-}8p\,[^1/_2]$	0—1
6261,212	50	9,72	11,70	$6p\,[2^1/_2]{-}8d\,[2^1/_2]^\circ$	3—3
6242,09	8	9,82	11,81	$6p\,[1^1/_2]{-}9d\,[^1/_2]^\circ$	2—1
6224,169	40	9,82	11,81	$6p\,[1^1/_2]{-}9d\,[1^1/_2]^\circ$	2—2
6220,84	1	9,82	11,81	$6p\,[1^1/_2]{-}9d\,[3^1/_2]^\circ$	2—3
6209,11	3	9,82	11,82	$6p\,[1^1/_2]{-}9d\,[2^1/_2]^\circ$	2—2
6206,297	20	9,69	11,68	$6p\,[2^1/_2]{-}8d\,[^1/_2]^\circ$	2—1
6205,75	4	9,92	11,91	$5d\,[^1/_2]^\circ{-}8f\,[1^1/_2]$	1—1
6205,35	6	9,92	11,91	$5d\,[^1/_2]^\circ{-}8f\,[1^1/_2]$	1—2
6201,49	3	9,92	11,92	$5d\,[^1/_2]^\circ{-}8f\,[2^1/_2]$	1—2
6200,890	60	9,82	11,82	$6p\,[1^1/_2]{-}9d\,[2^1/_2]^\circ$	2—3
6198,260	100	9,58	11,58	$6p\,[^1/_2]{-}9s\,[1^1/_2]^\circ$	1—2
6195,49	1	9,45	11,45	$6s'\,[^1/_2]^\circ{-}8p\,[1^1/_2]$	0—1
6193,89	1	9,96	11,96	$5d\,[1^1/_2]^\circ{-}9f\,[1^1/_2]$	2—2
6191,40	4	9,96	11,96	$5d\,[1^1/_2]^\circ{-}9f\,[2^1/_2]$	2—3
6189,10	20	9,58	11,58	$6p\,[^1/_2]{-}9s\,[1^1/_2]^\circ$	1—1
6184,16	3	9,69	11,69	$6p\,[2^1/_2]{-}8d\,[1^1/_2]^\circ$	2—2
6182,420	300	9,69	11,69	$6p\,[2^1/_2]{-}8d\,[3^1/_2]^\circ$	2—3
6179,665	120	9,57	11,57	$6s'\,[^1/_2]^\circ{-}5f\,[1^1/_2]$	1—1
6178,302	150	9,57	11,58	$6s'\,[^1/_2]^\circ{-}5f\,[1^1/_2]$	1—2

λ, Å	I	E_H, eV	E_B, eV	Transition	J
6163,935	80	9,57	11,58	$6s'$ [$^1/_2$]°—$5f$ [$2^1/_2$]	1—2
6163,660	90	9,69	11,70	$6p$ [$2^1/_2$]—$8d$ [$2^1/_2$]°	2—2
6162,16	3	9,79	11,80	$6p$ [$1^1/_2$]—$9d$ [$^1/_2$]°	1—0
6152,069	20	9,69	11,70	$6p$ [$2^1/_2$]—$8d$ [$2^1/_2$]°	2—3
6144,97	20	9,94	11,96	$5d$ [$3^1/_2$]°—$9f$ [$4^1/_2$]	4—5, 4
6143,70	4	9,79	11,81	$6p$ [$1^1/_2$]—$9d$ [$^1/_2$]°	1—1
6142,13	1	9,94	11,96	$5d$ [$3^1/_2$]°—$9f$ [$3^1/_2$]	4—3, 4
6131,47	1	9,93	11,96	$6p$ [$^1/_2$]—$11d$ [$1^1/_2$]°	0—1
6126,36	15	9,79	11,81	$6p$ [$1^1/_2$]—$9d$ [$1^1/_2$]°	1—2
6123,91	5	9,89	11,91	$5d$ [$^1/_2$]°—$8f$ [$1^1/_2$]	0—1
6114,86	10	9,58	11,61	$6p$ [$^1/_2$]—$5d'$ [$1^1/_2$]°	1—1
6111,951	40	9,72	11,75	$6p$ [$2^1/_2$]—$10s$ [$1^1/_2$]°	3—2
6111,759	30	9,79	11,82	$6p$ [$1^1/_2$]—$9d$ [$2^1/_2$]°	1—2
6108,37	8	9,82	11,85	$6p$ [$1^1/_2$]—$11s$ [$1^1/_2$]°	2—2
6103,88	3	9,82	11,85	$6p$ [$1^1/_2$]—$11s$ [$1^1/_2$]°	2—1
6095,15	1	9,96	11,99	$5d$ [$1^1/_2$]°—$10f$ [$1^1/_2$]	2—1, 2
6093,38	3	9,96	11,99	$5d$ [$1^1/_2$]°—$10f$ [$2^1/_2$]	2—3
6067,77	1	9,92	11,96	$5d$ [$^1/_2$]°—$9f$ [$1^1/_2$]	1—1
6067,52	2	9,92	11,96	$5d$ [$^1/_2$]°—$9f$ [$1^1/_2$]	1—2
6064,91	1	9,92	11,96	$5d$ [$^1/_2$]°—$9f$ [$2^1/_2$]	1—2
6048,00	6	9,94	11,99	$5d$ [$3^1/_2$]°—$10f$ [$4^1/_2$]	4—5, 4
6043,38	10	9,79	11,84	$6p$ [$1^1/_2$]—$9d$ [$1^1/_2$]°	1—1
6034,92	2	9,69	11,74	$6p$ [$2^1/_2$]—$8d$ [$1^1/_2$]°	2—1
6031,36	1	9,93	11,99	$6p$ [$^1/_2$]—$12d$ [$1^1/_2$]°	0—1
6026,76	4	9,82	11,89	$6p$ [$1^1/_2$]—$7s'$ [$^1/_2$]°	2—1
6022,89	1	9,96	12,02	$5d$ [$1^1/_2$]°—$11f$ [$2^1/_2$]	2—3
6014,10	1	9,79	11,85	$6p$ [$1^1/_2$]—$11s$ [$1^1/_2$]°	1—2
6009,78	8	9,79	11,85	$6p$ [$1^1/_2$]—$11s$ [$1^1/_2$]°	1—1
6007,909	15	9,69	11,75	$6p$ [$2^1/_2$]—$10s$ [$1^1/_2$]°	2—2
5998,115	30	9,69	11,75	$6p$ [$2^1/_2$]—$10s$ [$1^1/_2$]°	2—1
5989,18	20	9,82	11,89	$6p$ [$1^1/_2$]—$10d$ [$1^1/_2$]°	2—2
5986,23	4	9,82	11,89	$6p$ [$1^1/_2$]—$10d$ [$^1/_2$]°	2—1
5979,42	1	9,82	11,89	$6p$ [$1^1/_2$]—$10d$ [$2^1/_2$]°	2—2
5978,29	2	9,94	12,02	$5d$ [$3^1/_2$]°—$11f$ [$4^1/_2$]	4—5, 4
5974,152	40	9,82	11,90	$6p$ [$1^1/_2$]—$10d$ [$2^1/_2$]°	2—3
5972,82	1	9,92	11,99	$5d$ [$^1/_2$]°—$10f$ [$1^1/_2$]	1—1, 2
5970,41	1	9,92	11,99	$5d$ [$^1/_2$]°—$10f$ [$2^1/_2$]	1—2
5934,172	100	9,72	11,81	$6p$ [$2^1/_2$]—$9d$ [$3^1/_2$]°	3—4
5931,241	80	9,58	11,67	$6p$ [$^1/_2$]—$8d$ [$^1/_2$]°	1—0
5925,56	6	9,72	11,81	$6p$ [$2^1/_2$]—$9d$ [$1^1/_2$]°	3—2
5922,550	20	9,72	11,81	$6p$ [$2^1/_2$]—$9d$ [$3^1/_2$]°	3—3
5921,85	10	9,57	11,66	$6s'$ [$^1/_2$]°—$9p$ [$^1/_2$]	1—1
5916,65	4	9,82	11,92	$6p$ [$1^1/_2$]—$12s$ [$1^1/_2$]°	2—2
5911,90	5	9,72	11,82	$6p$ [$2^1/_2$]—$9d$ [$2^1/_2$]°	3—2
5906,76	3	9,57	11,67	$6s'$ [$^1/_2$]°—$9p$ [$2^1/_2$]	1—2
5904,462	20	9,72	11,82	$6p$ [$2^1/_2$]—$9d$ [$2^1/_2$]°	3—3
5898,56	8	9,79	11,89	$6p$ [$1^1/_2$]—$10d$ [$1^1/_2$]°	1—2
5895,62	2	9,79	11,89	$6p$ [$1^1/_2$]—$10d$ [$^1/_2$]°	1—1
5894,988	100	9,58	11,68	$6p$ [$^1/_2$]—$8d$ [$^1/_2$]°	1—1
5889,12	20	9,79	11,89	$6p$ [$1^1/_2$]—$10d$ [$2^1/_2$]°	1—2
5878,92	6	9,57	11,68	$6s'$ [$^1/_2$]°—$9p$ [$1^1/_2$]	1—2
5875,018	100	9,58	11,69	$6p$ [$^1/_2$]—$8d$ [$1^1/_2$]°	1—2
5856,509	15	9,58	11,70	$6p$ [$^1/_2$]—$8d$ [$2^1/_2$]°	1—2
5849,85	3	9,79	11,91	$6p$ [$1^1/_2$]—$10d$ [$1^1/_2$]°	1—1
5846,21	2	9,57	11,69	$6s'$ [$^1/_2$]°—$9p$ [$^1/_2$]	1—0
5845,46	1	9,82	11,94	$6p$ [$1^1/_2$]—$11d$ [$^1/_2$]°	2—1
5843,43	5	9,69	11,81	$6p$ [$2^1/_2$]—$9d$ [$^1/_2$]°	2—1
5840,83	4	9,82	11,94	$6p$ [$1^1/_2$]—$11d$ [$1^1/_2$]°	2—2

λ, Å	I	E_H, eV	E_B, eV	Transition	J
5830,63	20	9,82	11,95	$6p\ [1^1/_2]-11d\ [2^1/_2]^\circ$	2—3
5827,72	1	9,69	11,81	$6p\ [2^1/_2]-9d\ [1^1/_2]^\circ$	2—2
5824,800	150	9,69	11,81	$6p\ [2^1/_2]-9d\ [3^1/_2]^\circ$	2—3
5823,890	300	9,45	11,57	$6s'\ [^1/_2]^\circ-5f\ [1^1/_2]$	0—1
5820,52	25	9,72	11,85	$6p\ [2^1/_2]-11s\ [1^1/_2]^\circ$	3—2
5814,505	60	9,69	11,82	$6p\ [2^1/_2]-9d\ [2^1/_2]^\circ$	2—2
5807,311	15	9,69	11,82	$6p\ [2^1/_2]-9d\ [2^1/_2]^\circ$	2—3
5792,26	1	9,82	11,96	$6p\ [1^1/_2]-13s\ [1^1/_2]^\circ$	2—2
5754,60	1	9,79	11,94	$6p\ [1^1/_2]-11d\ [1^1/_2]^\circ$	1—2
5748,20	8	9,79	11,95	$6p\ [1^1/_2]-11d\ [2^1/_2]^\circ$	1—2
5740,73	1	9,82	11,98	$6p\ [1^1/_2]-12d\ [1^1/_2]^\circ$	2—2
5740,17	6	9,58	11,74	$6p\ [^1/_2]-8d\ [1^1/_2]^\circ$	1—1
5733,48	4	9,82	11,98	$6p\ [1^1/_2]-12d\ [2^1/_2]^\circ$	2—3
5726,10	4	9,69	11,85	$6p\ [2^1/_2]-11s\ [1^1/_2]^\circ$	2—2
5723,26	1	9,79	11,96	$6p\ [1^1/_2]-11d\ [1^1/_2]^\circ$	1—1
5722,14	15	9,69	11,85	$6p\ [2^1/_2]-11s\ [1^1/_2]^\circ$	2—1
5716,252	80	9,72	11,89	$6p\ [2^1/_2]-10d\ [3^1/_2]^\circ$	3—4
5715,716	70	9,58	11,75	$6p\ [^1/_2]-10s\ [1^1/_2]^\circ$	1—2
5712,21	2	9,72	11,89	$6p\ [2^1/_2]-10d\ [1^1/_2]^\circ$	3—2
5709,80	10	9,72	11,89	$6p\ [2^1/_2]-10d\ [3^1/_2]^\circ$	3—3
5706,87	3	9,58	11,75	$6p\ [^1/_2]-10s\ [1^1/_2]^\circ$	1—1
5703,34	1	9,72	11,89	$6p\ [2^1/_2]-10d\ [2^1/_2]^\circ$	3—2
5698,54	8	9,72	11,90	$6p\ [2^1/_2]-10d\ [2^1/_2]^\circ$	3—3
5696,479	80	9,57	11,74	$6s'\ [^1/_2]^\circ-6f\ [1^1/_2]$	1—1
5695,750	100	9,57	11,75	$6s'\ [^1/_2]^\circ-6f\ [1^1/_2]$	1—2
5688,373	40	9,57	11,75	$6s'\ [^1/_2]^\circ-6f\ [2^1/_2]$	1—2
5664,46	1	9,82	12,01	$6p\ [1^1/_2]-13d\ [3^1/_2]^\circ$	2—3
5654,31	1	9,69	11,88	$6p\ [2^1/_2]-7s'\ [^1/_2]^\circ$	2—1
5652,84	2	9,79	11,98	$6p\ [1^1/_2]-12d\ [2^1/_2]^\circ$	1—2
5646,19	5	9,72	11,92	$6p\ [2^1/_2]-12s\ [1^1/_2]^\circ$	3—2
5621,24	1	9,69	11,89	$6p\ [2^1/_2]-10d\ [1^1/_2]^\circ$	2—2
5618,878	80	9,69	11,89	$6p\ [2^1/_2]-10d\ [3^1/_2]^\circ$	2—3
5612,65	15	9,69	11,89	$6p\ [2^1/_2]-10d\ [2^1/_2]^\circ$	2—2
5607,99	3	9,69	11,90	$6p\ [2^1/_2]-10d\ [2^1/_2]^\circ$	2—3
5594,37	6	9,45	11,66	$6s'\ [^1/_2]^\circ-9p\ [^1/_2]$	0—1
5585,18	1	9,79	12,01	$6p\ [1^1/_2]-13d\ [2^1/_2]^\circ$	1—2
5581,784	50	9,58	11,80	$6p\ [^1/_2]-9d\ [^1/_2]^\circ$	1—0
5579,28	40	9,72	11,94	$6p\ [2^1/_2]-11d\ [3^1/_2]^\circ$	3—4
5575,27	2	9,72	11,94	$6p\ [2^1/_2]-11d\ [3^1/_2]^\circ$	3—3
5567,77	2	9,72	11,95	$6p\ [2^1/_2]-11d\ [2^1/_2]^\circ$	3—3
5566,615	100	9,58	11,81	$6p\ [^1/_2]-9d\ [^1/_2]^\circ$	1—1
5566,22	5	—	—	—	—
5563,50	2	9,57	11,80	$6s'\ [^1/_2]^\circ-10p\ [^1/_2]$	1—1
5557,28	2	9,69	11,92	$6p\ [2^1/_2]-12s\ [1^1/_2]^\circ$	2—2
5555,06	1	9,57	11,80	$6s'\ [^1/_2]^\circ-10p\ [2^1/_2]$	1—2
5553,10	3	9,69	11,92	$6p\ [2^1/_2]-12s\ [1^1/_2]^\circ$	2—1
5552,385	80	9,58	11,81	$6p\ [^1/_2]-9d\ [1^1/_2]^\circ$	1—2
5540,38	3	9,58	11,82	$6p\ [^1/_2]-9d\ [2^1/_2]^\circ$	1—2
		9,57	11,81	$6s'\ [^1/_2]^\circ-10p\ [1^1/_2]$	1—2
5532,78	2	9,72	11,96	$6p\ [2^1/_2]-13s\ [1^1/_2]^\circ$	3—2
5523,05	3	9,57	11,81	$6s'\ [^1/_2]^\circ-10p\ [^1/_2]$	1—0
5488,555	20	9,69	11,94	$6p\ [2^1/_2]-11d\ [3^1/_2]^\circ$	2—3
5487,03	6	9,72	11,98	$6p\ [2^1/_2]-12d\ [3^1/_2]^\circ$	3—4
5484,46	4	9,69	11,95	$6p\ [2^1/_2]-11d\ [2^1/_2]^\circ$	2—2
5484,16	1	9,72	11,98	$6p\ [2^1/_2]-12d\ [3^1/_2]^\circ$	3—3
5481,33	1	9,69	11,95	$6p\ [2^1/_2]-11d\ [2^1/_2]^\circ$	2—3
5479,12	1	9,72	11,98	$6p\ [2^1/_2]-12d\ [2^1/_2]^\circ$	3—3
5460,037	15	9,58	11,85	$6p\ [^1/_2]-11s\ [1^1/_2]^\circ$	1—2

λ, Å	I	E_H, eV	E_B, eV	Transition	J
5456,45	2	9,58	11,85	$6p\,[{}^1/_2]-11s\,[1^1/_2]^\circ$	1—1
5454,54	1	9,72	11,99	$6p\,[2^1/_2]-14s\,[1^1/_2]^\circ$	3—2
5444,87	1	9,69	11,96	$6p\,[2^1/_2]-13s\,[1^1/_2]^\circ$	2—1
5440,39	15	9,57	11,85	$6s'\,[{}^1/_2]^\circ-7f\,[1^1/_2]$	1—1
5439,923	30	9,57	11,85	$6s'\,[{}^1/_2]^\circ-7f\,[1^1/_2]$	1—2
5435,60	5	9,57	11,85	$6s'\,[{}^1/_2]^\circ-7f\,[2^1/_2]$	1—2
5421,76	2	9,72	12,01	$6p\,[2^1/_2]-13d\,[3^1/_2]^\circ$	3—4
5418,02	5	—	—	—	—
5400,45	4	9,69	11,98	$6p\,[2^1/_2]-12d\,[3^1/_2]^\circ$	2—3
5397,63	1	9,72	12,02	$6p\,[2^1/_2]-15s\,[1^1/_2]^\circ$	3—2
		9,69	11,98	$6p\,[2^1/_2]-12d\,[2^1/_2]^\circ$	2—2
5394,738	20	9,58	11,89	$6p\,[{}^1/_2]-10d\,[{}^1/_2]^\circ$	1—0
5392,795	100	9,45	11,74	$6s'\,[{}^1/_2]^\circ-6f\,[1^1/_2]$	0—1
5373,74	1	9,72	12,03	$6p\,[2^1/_2]-14d\,[3^1/_2]^\circ$	3—4
5367,03	6	—	—	—	—
5364,626	30	9,58	11,89	$6p\,[{}^1/_2]-10d\,[1^1/_2]^\circ$	1—2
5362,244	15	9,58	11,89	$6p\,[{}^1/_2]-10d\,[{}^1/_2]^\circ$	1—1
5356,80	1	9,58	11,89	$6p\,[{}^1/_2]-10d\,[2^1/_2]^\circ$	1—2
5337,89	2	9,69	12,01	$6p\,[2^1/_2]-13d\,[3^1/_2]^\circ$	2—3
5335,91	1	9,69	12,01	$6p\,[2^1/_2]-13d\,[2^1/_2]^\circ$	2—2
5306,37	3	9,58	11,92	$6p\,[{}^1/_2]-12s\,[1^1/_2]^\circ$	1—2
5286,38	3	9,57	11,91	$6s'\,[{}^1/_2]^\circ-8f\,[1^1/_2]$	1—1
5286,11	4	9,57	11,91	$6s'\,[{}^1/_2]^\circ-8f\,[1^1/_2]$	1—2
5283,30	2	9,57	11,92	$6s'\,[{}^1/_2]^\circ-8f\,[2^1/_2]$	1—2
5273,48	1	9,45	11,80	$6s'\,[{}^1/_2]^\circ-10p\,[{}^1/_2]$	0—1
5251,89	2	—	—	—	—
5248,98	4	9,58	11,94	$6p\,[{}^1/_2]-11d\,[{}^1/_2]^\circ$	1—1
5245,27	4	9,58	11,94	$6p\,[1^1/_2]-11d\,[1^1/_2]^\circ$	1—2
5206,07	1	9,58	11,96	$6p\,[{}^1/_2]-13s\,[1^1/_2]^\circ$	1—2
5185,85	2	9,57	11,96	$6s'\,[{}^1/_2]^\circ-9f\,[1^1/_2]$	1—1, 2
5167,30	1	9,58	11,98	$6p\,[{}^1/_2]-12d\,[{}^1/_2]^\circ$	1—1
5164,39	1	9,58	11,98	$6p\,[{}^1/_2]-12d\,[1^1/_2]^\circ$	1—2
5162,711	10	9,45	11,85	$6s'\,[{}^1/_2]^\circ-7f\,[1^1/_2]$	0—1
5116,46	2	9,57	11,99	$6s'\,[{}^1/_2]^\circ-10f\,[1^1/_2]$	1—2, 1
5028,2796	200	8,44	10,90	$6s\,[1^1/_2]^\circ-7p\,[{}^1/_2]$	1—1
5023,88	3	9,45	11,91	$6s'\,[{}^1/_2]^\circ-8f\,[1^1/_2]$	0—1
4923,1522	500	8,44	10,95	$6s\,[1^1/_2]^\circ-7p\,[2^1/_2]$	1—2
4916,508	500	8,44	10,96	$6s\,[1^1/_2]^\circ-6p'\,[1^1/_2]$	1—1
4843,294	300	8,44	10,99	$6s\,[1^1/_2]^\circ-7p\,[1^1/_2]$	1—2
4829,709	400	8,44	11,00	$6s\,[1^1/_2]^\circ-7p\,[1^1/_2]$	1—1
4807,019	500	8,44	11,01	$6s\,[1^1/_2]^\circ-7p\,[{}^1/_2]$	1—0
4792,6192	150	8,31	10,90	$6s\,[1^1/_2]^\circ-7p\,[1^1/_2]$	2—1
4734,1524	600	8,44	11,05	$6s\,[1^1/_2]^\circ-6p'\,[1^1/_2]$	1—2
4708,21	5	8,44	11,07	$6s\,[1^1/_2]^\circ-6p'\,[{}^1/_2]$	1—1
4697,020	300	8,31	10,95	$6s\,[1^1/_2]^\circ-7p\,[2^1/_2]$	2—2
4690,9711	100	8,31	10,96	$6s\,[1^1/_2]^\circ-6p'\,[1^1/_2]$	2—1
4671,226	2000	8,31	10,97	$6s\,[1^1/_2]^\circ-7p\,[2^1/_2]$	2—3
4624,2757	1000	8,31	10,99	$6s\,[1^1/_2]^\circ-7p\,[1^1/_2]$	2—2
4611,8896	100	8,31	11,00	$6s\,[1^1/_2]^\circ-7p\,[1^1/_2]$	2—1
4582,7474	300	8,44	11,14	$6s\,[1^1/_2]^\circ-6p'\,[{}^1/_2]$	1—0
4576,60	2	8,31	11,02	$6s\,[1^1/_2]^\circ-6d\,[3^1/_2]^\circ$	2—4
4524,6805	400	8,31	11,05	$6s\,[1^1/_2]^\circ-6p'\,[1^1/_2]$	2—2
4500,9772	500	8,31	11,07	$6s\,[1^1/_2]^\circ-6p'\,[{}^1/_2]$	2—1
4385,7693	70	8,44	11,26	$6s\,[1^1/_2]^\circ-4f\,[1^1/_2]$	1—1
4383,9092	100	8,44	11,26	$6s\,[1^1/_2]^\circ-4f\,[1^1/_2]$	1—2
4372,287	20	8,44	11,27	$6s\,[1^1/_2]^\circ-4f\,[2^1/_2]$	1—2
4205,404	10	8,31	11,26	$6s\,[1^1/_2]^\circ-4f\,[1^1/_2]$	2—1
4203,6945	50	8,31	11,26	$6s\,[1^1/_2]^\circ-4f\,[1^1/_2]$	2—2
4193,5296	150	8,31	11,27	$6s\,[1^1/_2]^\circ-4f\,[2^1/_2]$	2—3

λ, Å	I	E_H, eV	E_B, eV	Transition	J
4193,01	20	8,31	11,27	$6s\,[1^1/_2]^\circ - 4f\,[2^1/_2]$	2—2
4146,78	2	8,44	11,42	$6s\,[1^1/_2]^\circ - 8p\,[^1/_2]$	1—1
4135,1337	20	8,44	11,43	$6s\,[1^1/_2]^\circ - 8p\,[2^1/_2]$	1—2
4116,1151	80	8,44	11,45	$6s\,[1^1/_2]^\circ - 8p\,[1^1/_2]$	1—1
4109,7093	60	8,44	11,45	$6s\,[1^1/_2]^\circ - 8p\,[1^1/_2]$	1—2
4078,8207	100	8,44	11,47	$6s\,[1^1/_2]^\circ - 8p\,[^1/_2]$	1—0
3985,202	30	8,31	11,42	$6s\,[1^1/_2]^\circ - 8p\,[1^1/_2]$	2—1
3974,417	40	8,31	11,43	$6s\,[1^1/_2]^\circ - 8p\,[2^1/_2]$	2—2
3967,541	200	8,31	11,44	$6s\,[1^1/_2]^\circ - 8p\,[2^1/_2]$	2—3
3956,85	6	8,31	11,45	$6s\,[1^1/_2]^\circ - 8p\,[1^1/_2]$	2—1
3950,925	120	8,31	11,45	$6s\,[1^1/_2]^\circ - 8p\,[1^1/_2]$	2—2
3948,72	10	8,44	11,57	$6s\,[1^1/_2]^\circ - 5f\,[1^1/_2]$	1—1
3948,163	60	8,44	11,58	$6s\,[1^1/_2]^\circ - 5f\,[1^1/_2]$	1—2
3942,29	2	8,44	11,58	$6s\,[1^1/_2]^\circ - 5f\,[2^1/_2]$	1—2
3835,6	2	8,44	11,67	$6s\,[1^1/_2]^\circ - 9p\,[2^1/_2]$	1—2
3826,86	15	8,44	11,67	$6s\,[1^1/_2]^\circ - 9p\,[1^1/_2]$	1—1
3823,74	10	8,44	11,68	$6s\,[1^1/_2]^\circ - 9p\,[1^1/_2]$	1—2
3809,84	30	8,44	11,69	$6s\,[1^1/_2]^\circ - 9p\,[^1/_2]$	1—0
3801,90	3	8,31	11,57	$6s\,[1^1/_2]^\circ - 5f\,[1^1/_2]$	2—1
3801,39	30	8,31	11,58	$6s\,[1^1/_2]^\circ - 5f\,[1^1/_2]$	2—2
3796,30	40	8,31	11,58	$6s\,[1^1/_2]^\circ - 5f\,[2^1/_2]$	2—3
3795,95	3	8,31	11,58	$6s\,[1^1/_2]^\circ - 5f\,[2^1/_2]$	2—2
3745,69	4	8,44	11,74	$6s\,[1^1/_2]^\circ - 6f\,[1^1/_2]$	1—1
3745,38	10	8,44	11,75	$6s\,[1^1/_2]^\circ - 6f\,[1^1/_2]$	1—2
3742,22	1	8,44	11,75	$6s\,[1^1/_2]^\circ - 6f\,[2^1/_2]$	1—2
3702,74	2	8,31	11,66	$6s\,[1^1/_2]^\circ - 9p\,[^1/_2]$	2—1
3696,82	4	8,31	11,67	$6s\,[1^1/_2]^\circ - 9p\,[2^1/_2]$	2—2
3693,49	40	8,31	11,67	$6s\,[1^1/_2]^\circ - 9p\,[2^1/_2]$	2—3
3688,80	1	8,31	11,67	$6s\,[1^1/_2]^\circ - 9p\,[1^1/_2]$	2—1
3685,90	40	8,31	11,68	$6s\,[1^1/_2]^\circ - 9p\,[1^1/_2]$	2—2
3679,31	4	8,44	11,80	$6s\,[1^1/_2]^\circ - 10p\,[1^1/_2]$	1—1
3677,54	2	8,44	11,81	$6s\,[1^1/_2]^\circ - 10p\,[1^1/_2]$	1—2
3669,91	10	8,44	11,81	$6s\,[1^1/_2]^\circ - 10p\,[^1/_2]$	1—0
3633,06	6	8,44	11,85	$6s\,[1^1/_2]^\circ - 7f\,[1^1/_2]$	1—2
3613,06	8	8,31	11,75	$6s\,[1^1/_2]^\circ - 6f\,[1^4/_2]$	2—2
3610,32	15	8,31	11,75	$6s\,[1^1/_2]^\circ - 6f\,[2^1/_2]$	2—3
3592,80	2	8,44	11,89	$6s\,[1^1/_2]^\circ - 11p\,[1^1/_2]$	1—1
3591,67	1	8,44	11,89	$6s\,[1^1/_2]^\circ - 11p\,[1^1/_2]$	1—2
3587,02	4	8,44	11,89	$6s\,[1^1/_2]^\circ - 11p\,[^1/_2]$	1—0
3563,80	3	8,44	11,91	$6s\,[1^1/_2]^\circ - 8f\,[1^1/_2]$	1—2
3555,92	1	8,31	11,80	$6s\,[1^1/_2]^\circ - 10p\,[2^1/_2]$	2—2
3554,04	10	8,31	11,80	$6s\,[1^1/_2]^\circ - 10p\,[2^1/_2]$	2—3
3549,86	10	8,31	11,81	$6s\,[1^1/_2]^\circ - 10p\,[1^1/_2]$	2—2
3537,35	1	8,44	11,94	$6s\,[1^1/_2]^\circ - 12p\,[1^1/_2]$	1—1
3536,61	1	8,44	11,94	$6s\,[1^1/_2]^\circ - 12p\,[1^1/_2]$	1—2
3533,48	2	8,44	11,94	$6s\,[1^1/_2]^\circ - 12p\,[^1/_2]$	1—0
3517,90	2	8,44	11,96	$6s\,[1^1/_2]^\circ - 9f\,[1^1/_2]$	1—2
3508,42	2	8,31	11,85	$6s\,[1^1/_2]^\circ - 7f\,[1^1/_2]$	2—2
3506,74	5	8,31	11,85	$6s\,[1^1/_2]^\circ - 7f\,[2^1/_2]$	2—3
3496,86	1	8,44	11,98	$6s\,[1^1/_2]^\circ - 13p\,[^1/_2]$	1—0
3472,36	4	8,31	11,88	$6s\,[1^1/_2]^\circ - 11p\,[2^1/_2]$	2—3
3469,81	4	8,31	11,89	$6s\,[1^1/_2]^\circ - 11p\,[1^1/_2]$	2—2
3443,83	1	8,31	11,91	$6s\,[1^1/_2]^\circ - 8f\,[1^1/_2]$	2—2
3442,66	3	8,31	11,92	$6s\,[1^1/_2]^\circ - 8f\,[2^1/_2]$	2—3
3420,00	2	8,31	11,94	$6s\,[1^1/_2]^\circ - 12p\,[2^1/_2]$	2—3
3418,37	2	8,31	11,94	$6s\,[1^1/_2]^\circ - 12p\,[1^1/_2]$	2—2
3400,79	1	8,31	11,96	$6s\,[1^1/_2]^\circ - 9f\,[1^1/_2]$	2—2
3400,07	2	8,31	11,96	$6s\,[1^1/_2]^\circ - 9f\,[2^1/_2]$	2—3

λ, Å	I	E_H, eV	E_B, eV	Transition	J
3384,36	1	8,31	11,98	$6s\ [1^1/_2]^\circ$—$13p\ [2^1/_2]$	2—3
3383,20	1	8,31	11,98	$6s\ [1^1/_2]^\circ$—$13p\ [1^4/_2]$	2—2
3370,34	1	8,31	11,99	$6s\ [1^1/_2]^\circ$—$10f\ [2^1/_2]$	2—3
3358,96	1	8,31	12,00	$6s\ [1^1/_2]^\circ$—$14p\ [2^1/_2]$	2—3
3358,17	1	8,31	12,01	$6s\ [1^1/_2]^\circ$—$14p\ [1^1/_2]$	2—2
3348,63	1	8,31	12,02	$6s\ [1^1/_2]^\circ$—$11f\ [2^1/_2]$	2—3
3340,04	1	8,31	12,03	$6s\ [1^1/_2]^\circ$—$15p\ [2^1/_2]$	2—3
1469,610	5	0,00	8,44	$5p^6\ ^1S$—$6s\ [1^1/_2]^\circ$	0—1
1295,587	8	0,00	9,57	$5p^6\ ^1S$—$6s'\ [^1/_2]^\circ$	0—1
1277,50	6	—	—	—	—
1250,203	2	0,00	9,92	$5p^6\ ^1S$—$5d\ [^1/_2]^\circ$	0—1
1192,040	2	0,00	10,40	$5p^6\ ^1S$—$5d\ [1^1/_2]^\circ$	0—1
1100,46	15	—	—	—	—
1088,94	10	—	—	—	—
1067,10	5	—	—	—	—
1027,04	10	—	—	—	—

Xe II, ground state $1s^2\,2s^2\,2p^6\,3s^2\,3p^6\,3d^{10}\,4s^2\,4p^6\,4d^{10}\,5s^2\,5p^5\ ^2P^\circ_{3/2}$

Ionization potential 171 068,4 cm⁻¹; 21,208 eV

λ, Å	I	E_H, eV	E_B, eV	Transition	J
10220,8	3	—	—	—	—
10206,9	1	—	—	—	—
10095,7	1	—	—	—	—
10054,2	1	—	—	—	—
9990,9	2	—	—	—	—
9983,4	1	—	—	—	—
9908,9	2	—	—	—	—
9895,8	1	—	—	—	—
9865,56	6	—	—	—	—
9837,8	2	—	—	—	—
9820,90	2	—	—	—	—
9810,28	2	—	—	—	—
9774,8	1	14,48	15,75	$6p\ ^4D^\circ$—$5d'\ ^2P$	$^3/_2$—$^1/_2$
9744,8	1	15,75	17,02	$5d'\ ^2P$—1°	$^1/_2$—$^3/_2$
9734,0	3	—	—	—	—
9706,2	2	—	—	—	—
9698,68	50	13,20	14,48	$5d\ ^4P$—$6p\ ^4D^\circ$	$^5/_2$—$^3/_2$
9641,6	4	—	—	—	—
9630,95	3	—	—	—	—
9615,71	4	—	—	—	—
9604,50	7	13,97	15,26	$5d'\ ^2D$—$6p\ ^2D^\circ$	$^5/_2$—$^5/_2$
9591,35	50	12,59	13,89	$5d\ ^4F$—$6p\ ^4P^\circ$	$^5/_2$—$^5/_2$
9577,70	2	—	—	—	—
9475,23	3	13,97	15,28	$5d'\ ^2D$—$6p\ ^2P^\circ$	$^5/_2$—$^3/_2$
9464,3	10	—	—	—	—
9447,6	1	14,76	16,08	$5d''\ ^2D$—$6p'\ ^2P^\circ$	$^5/_2$—$^3/_2$
9407,57	1	—	—	—	—
9400,59	15	12,54	13,86	$6s\ ^4P$—$6p\ ^4P^\circ$	$^1/_2$—$^3/_2$
9331,67	4	12,74	14,07	$6s\ ^2P$—$6p\ ^4D^\circ$	$^3/_2$—$^5/_2$
9304,77	1	14,48	15,81	$6p\ ^4D^\circ$—$5d'\ ^2P$	$^3/_2$—$^3/_2$
9298,7	2	—	—	—	—
9288,4	5	16,02	17,36	$5d'\ ^2S$—13°	$^1/_2$—$^1/_2$

λ, Å	I	E_H, eV	E_B, eV	Transition	J
9265,67	10	14,07	15,41	$6p\ ^4D°—5d''\ ^2D$	$^5/_2—^3/_2$
9259,60	1	—	—	—	—
9244,15	2	—	—	—	—
9238,59	2	—	—	—	—
9226,39	7	13,14	14,48	$5d\ ^2P—6p\ ^4D°$	$^1/_2—^3/_2$
9193,8	2	12,74	14,09	$6s\ ^2P—6p\ ^4P°$	$^3/_2—^1/_2$
9136,6	5	—	—	—	—
9106,24	1	14,76	16,12	$5d''\ ^2D—6p'\ ^2F°$	$^5/_2—^7/_2$
9068,0	2	—	—	—	—
8902,66	5	—	—	—	—
8881,48	2	—	—	—	—
8869,40	2	—	—	—	—
8855,74	5	—	—	—	—
8839,9	3	—	—	—	—
8804,61	30	—	—	—	—
8796,92	2	14,00	15,41	$6s'\ ^2D—6p\ ^2D°$	$^3/_2—^3/_2$
8785,88	4	—	—	—	—
8760,14	6	—	—	—	—
8752,14	7	—	—	—	—
8716,19	50	13,06	14,48	$5d\ ^2P—6p\ ^4D°$	$^3/_2—^3/_2$
8655,72	3	—	—	—	—
8636,4	2	—	—	—	—
8628,94	25	13,97	15,41	$5d'\ ^2D—6p\ ^2D°$	$^5/_2—^3/_2$
8604,23	50	—	—	—	—
8584,0	1	14,00	15,44	$6s'\ ^2D—6p\ ^2P°$	$^3/_2—^1/_2$
8566,7	2	—	—	—	—
8515,19	50	—	—	—	—
8500,96	2	—	—	—	—
8482,64	5	13,80	15,26	$5d'\ ^2D—6p\ ^2D°$	$^3/_2—^5/_2$
8467,8	1	15,28	16,74	$6p\ ^2P°—7s\ ^4P$	$^3/_2—^1/_2$
8446,6	2	—	—	—	—
8378,3	5	16,39	17,87	$6p'\ ^2D°—6d\ ^4P$	$^5/_2—^3/_2$
8366,4	30	—	—	—	—
8351,3	3	15,81	17,29	$5d'\ ^2P—9°$	$^3/_2—^3/_2$
8347,24	100	12,59	14,07	$5d\ ^4F—6p\ ^4D°$	$^5/_2—^5/_2$
8329,44	30	—	—	—	—
8317,10	40	—	—	—	—
8316,2	10	—	—	—	—
8297,55	100	—	—	—	—
8285,70	15	13,58	15,08	$5d\ ^2D—6p\ ^4S°$	$^5/_2—^3/_2$
8282,85	15	—	—	—	—
8262,73	30	—	—	—	—
8260,81	5	—	—	—	—
8256,40	20	—	—	—	—
8251,30	2	—	—	—	—
8245,37	4	—	—	—	—
8214,85	20	12,59	14,10	$5d\ ^4F—6p\ ^4D°$	$^5/_2—^7/$
8213,50	2	—	—	—	—
8186,9	10	—	—	—	—
8167,55	10	—	—	—	—
8151,80	100	—	—	—	—
8144,8	3	15,02	16,55	$6p\ ^2S°—4$	$^1/_2—^3/_2,\ ^1/_2$
8142,13	5	—	—	—	—
8136,83	30	15,08	16,60	$6p\ ^4S°—6$	$^3/_2—^5/_2$
8131,40	20	—	—	—	—
8120,16	30	—	—	—	—
8115,94	50	—	—	—	—
8098,55	12	—	—	—	—

λ, A	I	E_H, eV	E_B, eV	Transition	J
8095,13	10	—	—	—	—
8080,31	50	—	—	—	—
8070,97	50	—	—	—	—
8047,28	20	—	—	—	—
8038,26	100	—	—	—	—
8035,40	20	—	—	—	—
8031,64	100	—	—	—	—
8028,0	1	15,28	16,82	$6p\ ^2P^\circ$—$6d\ ^4D$	$^3/_2$—$^3/_2$
8023,85	50	—	—	—	—
8020,07	5	14,48	16,02	$6p\ ^4D^\circ$—$5d'\ ^2S$	$^3/_2$—$^1/_2$
8014,26	50	—	—	—	—
8008,45	300	—	—	—	—
8005,8	2	15,75	17,29	$5d'\ ^2P$—9°	$^1/_2$—$^3/_2$
8001,95	10	—	—	—	—
7996,5	3	14,93	16,48	$6p\ ^4D^\circ$—2	$^1/_2$—$^1/_2$
7992,34	100	—	—	—	—
7991,5	5	13,86	15,41	$6p\ ^4P^\circ$—$5d''\ ^2D$	$^3/_2$—$^3/_2$
7987,99	40	12,54	14,09	$6s\ ^4P$—$6p\ ^4P^\circ$	$^1/_2$—$^1/_2$
7981,1	100	—	—	—	—
7976,4	3	12,92	14,48	$6s\ ^2P$—$6p\ ^4D^\circ$	$^1/_2$—$^3/_2$
7974,76	20	—	—	—	—
7942,54	100	—	—	—	—
7920,48	10	—	—	—	—
7897,7	5	15,81	17,38	$5d'\ ^2P$—17°	$^3/_2$—$^1/_2$
7889,4	50	—	—	—	—
7882,71	20	15,81	17,38	$5d'\ ^2P$—$6p''\ ^2P^\circ$	$^3/_2$—$^3/_2$
7862,7	3	16,08	17,65	$6p'\ ^2P^\circ$—$7s\ ^2P$	$^3/_2$—$^3/_2$
7828,28	20	—	—	—	—
7818,31	10	—	—	—	—
7805,8	1	15,81	17,40	$5d'\ ^2P$—19°	$^3/_2$—$^3/_2$
7787,04	100	14,76	16,36	$5d''\ ^2D$—$6p'\ ^2D^\circ$	$^5/_2$—$^3/_2$
7777,1	10	—	—	—	—
7774,18	4	—	—	—	—
7772,12	20	—	—	—	—
7712,42	30	13,80	15,41	$5d'\ ^2D$—$6p\ ^2D^\circ$	$^3/_2$—$^3/_2$
7670,66	200	13,31	14,93	$5d\ ^4P$—$6p\ ^4D^\circ$	$^3/_2$—$^1/_2$
7618,57	100	14,76	16,39	$5d''\ ^2D$—$6p'\ ^2D^\circ$	$^5/_2$—$^5/_2$
7548,45	300	13,80	15,44	$5d'\ ^2D$ $6p\ ^2P^\circ$	$^3/_2$—$^1/_2$
7530,70	50	13,38	15,02	$5d\ ^2D$—$6p\ ^2S^\circ$	$^3/_2$—$^1/_2$
7508,6	1	16,74	18,40	$7s\ ^4P$—31°	$^1/_2$—$^3/_2$
7503,00	3	15,75	17,40	$5d'\ ^2P$—19°	$^1/_2$—$^3/_2$
7495,36	50	14,09	15,75	$6p\ ^4P^\circ$—$5d'\ ^2P$	$^1/_2$—$^1/_2$
7410,14	4	15,44	17,12	$6p\ ^2P^\circ$—12	$^1/_2$—$^3/_2$
7400,5	4	13,25	14,93	$5d\ ^4P$—$6p\ ^4D^\circ$	$^1/_2$—$^1/_2$
7378,38	30	13,58	15,26	$5d\ ^2D$—$6p\ ^2D^\circ$	$^5/_2$—$^5/_2$
7343,37	30	15,26	16,95	$6p\ ^2D^\circ$—10	$^5/_2$—$^5/_2$
7339,30	300	13,39	15,08	$6s'\ ^2D$—$6p\ ^4S^\circ$	$^5/_2$—$^3/_2$
7301,80	200	13,58	15,28	$5d\ ^2D$—$6p\ ^2P^\circ$	$^5/_2$—$^3/_2$
7284,34	100	13,38	15,08	$5d\ ^2D$—$6p\ ^4S^\circ$	$^3/_2$—$^3/_2$
7279,75	4	16,39	18,09	$6p'\ ^2D^\circ$—$6d\ ^2D$	$^5/_2$—$^3/_2$
7276,47	4	—	—	—	—
7258,6	2	15,41	17,12	$6p\ ^2D^\circ$—12	$^3/_2$—$^3/_2$
7245,38	2	13,31	15,02	$5d\ ^4P$—$6p\ ^2S^\circ$	$^3/_2$—$^1/_2$
7215,97	20	14,09	15,81	$6p\ ^4P^\circ$—$5d'\ ^2P$	$^1/_2$—$^3/_2$
7164,83	800	14,25	15,98	$5d'\ ^2F$—$6p'\ ^2F^\circ$	$^7/_2$—$^5/_2$
7149,03	300	12,74	14,48	$6s\ ^2P$—$6p\ ^4D^\circ$	$^3/_2$—$^3/_2$
7147,50	100	—	—	—	—
7143,81	8	—	—	—	—

λ, Å	I	E_H, eV	E_B, eV	Transition	J
7133,27	10	14,07	15,81	$6p\ ^4D^\circ—5d'\ ^2P$	$^5/_2—^3/_2$
7100,8	2	15,08	16,82	$6p\ ^4S^\circ—6d\ ^4D$	$^3/_2—^3/_2$
7082,15	200	14,23	15,98	$5d'\ ^2F—6p'\ ^2F^\circ$	$^5/_2—^5/_2$
7075,0	2	—	—	—	—
7072,43	4	—	—	—	—
7052,57	3	—	—	—	—
7017,06	80	13,31	15,08	$5d\ ^4P—6p\ ^4S^\circ$	$^3/_2—^3/_2$
7003,96	50	13,25	15,02	$5d\ ^4P—6p\ ^2S^\circ$	$^1/_2—^1/_2$
6990,88	2000	12,32	14,10	$5d\ ^4F—6p\ ^4D^\circ$	$^3/_2—^7/_2$
6942,11	1000	—	—	—	—
6910,22	100	13,14	14,03	$5d\ ^2P—6p\ ^4D^\circ$	$^1/_2—^1/_2$
6890,41	3	—	—	—	—
6876,69	3	—	—	—	—
6873,2	10	—	—	—	—
6805,74	1000	12,25	14,07	$5d\ ^4F—6p\ ^4D^\circ$	$^7/_2—^5/_2$
6790,37	80	13,25	15,08	$5d\ ^2P—6p\ ^4S^\circ$	$^1/_2—^3/_2$
6788,71	100	13,39	15,41	$6s'\ ^2D—6p\ ^2D^\circ$	$^5/_2—^3/_2$
6702,25	80	14,23	16,08	$5d'\ ^2F—6p'\ ^2P^\circ$	$^5/_2—^3/_2$
6694,32	400	12,01	13,86	$5d\ ^4D—6p\ ^4P^\circ$	$^1/_2—^3/_2$
6691,22	1	15,08	16,93	$6p\ ^4S^\circ—8$	$^3/_2—^5/_2$
6663,1	2	16,36	18,22	$6p'\ ^2D^\circ—14$	$^3/_2—^5/_2$
6642,9	1	16,43	18,30	$7s\ ^4P—29^\circ$	$^5/_2—^3/_2$
6638,85	2	15,38	17,25	$6s''\ ^2S—5^\circ$	$^1/_2—^3/_2$
6634,13	6	15,44	17,31	$6p\ ^2P^\circ—6d\ ^4F$	$^1/_2—^3/_2$
6632,44	2	—	—	—	—
6620,02	200	13,06	14,93	$5d\ ^2P—6p\ ^4D^\circ$	$^3/_2—^1/_2$
6618,40	50	13,39	15,26	$6s'\ ^2D—6p\ ^2D^\circ$	$^5/_2—^5/_2$
6614,96	10	15,38	17,26	$6s''\ ^2S—7^\circ$	$^1/_2—^3/_2$
6613,31	4	—	—	—	—
6598,84	80	14,25	16,12	$5d'\ ^2F—6p'\ ^2F^\circ$	$^7/_2—^7/_2$
6597,25	300	13,20	15,08	$5d\ ^4P—6p\ ^4S^\circ$	$^5/_2—^3/_2$
6595,01	800	—	—	—	—
6573,68	30	13,38	15,26	$5d\ ^2D—6p\ ^2D^\circ$	$^3/_2—^5/_2$
6569,13	5	13,86	15,75	$6p\ ^4P^\circ—5d'\ ^2P$	$^3/_2—^1/_2$
6563,19	15	13,14	15,02	$5d\ ^2P—6p\ ^2S^\circ$	$^1/_2—^1/_2$
6556,70	4	{ 13,39	15,28	$6s'\ ^2D—6p\ ^2P^\circ$	$^5/_2—^3/_2$
		12,59	14,48	$5d\ ^4F—6p\ ^4D^\circ$	$^5/_2—^3/_2$
6528,65	200	14,23	16,12	$5d'\ ^2F—6p'\ ^2F^\circ$	$^5/_2—^7/_2$
6515,48	1	{ 16,46	18,36	$6p'\ ^2P^\circ—6d'\ ^2D$	$^1/_2—^3/_2$
		15,41	17,31	$5d''\ ^2D—11^\circ$	$^3/_2—^5/_2$
6512,83	300	13,38	15,28	$5d\ ^2D—6p\ ^2P^\circ$	$^3/_2—^3/_2$
6479,69	2	15,38	17,29	$6s''\ ^2S—9^\circ$	$^1/_2—^3/_2$
6461,48	3	—	—	—	—
6442,3	1	15,98	17,90	$6p'\ ^2F^\circ—6d\ ^2D$	$^5/_2—^5/_2$
6426,73	2	16,08	18,00	$6p'\ ^2P^\circ—6d\ ^2P^\circ$	$^3/_2—^1/_2$
6421,47	1	13,89	15,81	$6p\ ^4P^\circ—5d'\ ^2P$	$^5/_2—^3/_2$
6418,58	20	14,09	16,02	$6p\ ^4P^\circ—5d'\ ^2S$	$^1/_2—^1/_2$
6397,99	60	12,54	14,48	$6s\ ^4P—6p\ ^4D^\circ$	$^1/_2—^3/_2$
6375,28	100	13,14	15,08	$5d\ ^2P—6p\ ^4S^\circ$	$^1/_2—^3/_2$
6362,8	2	—	—	—	—
6356,35	500	15,41	17,36	$5d''\ ^2D—15^\circ$	$^3/_2—^5/_2$
6353,25	50	{ 13,86	15,81	$6p\ ^4P^\circ—5d'\ ^2P$	$^3/_2—^3/_2$
		14,48	16,43	$6p\ ^4D^\circ—7s\ ^4P$	$^3/_2—^5/_2$
6343,96	300	11,91	13,86	$5d\ ^4D—6p\ ^4P^\circ$	$^3/_2—^3/_2$
6325,17	2	—	—	—	—
6311,46	5	15,28	17,24	$6p\ ^2P^\circ—6d\ ^4F$	$^3/_2—^5/_2$
6305,01	1	16,43	18,40	$7s\ ^4P—31^\circ$	$^5/_2—^3/_2$
6300,86	100	13,06	15,02	$5d\ ^2P—6p\ ^2S^\circ$	$^3/_2—^1/_2$

λ, Å	I	E_H, eV	E_B, eV	Transition	J
6298,31	20	13,31	15,28	$5d\ ^4P-6p\ ^2P^\circ$	$^3/_2-^3/_2$
6296,39	10	16,39	18,36	$6p'\ ^2D^\circ-6d'\ ^2D$	$^5/_2-^3/_2$
6284,41	50	15,41	17,38	$5d''\ ^2D-6p''\ ^2P^\circ$	$^3/_2-^3/_2$
6277,54	300	11,91	13,89	$5d\ ^4D-6p\ ^4P^\circ$	$^3/_2-^5/_2$
6270,82	400	14,00	15,98	$6s'\ ^2D-6p'\ ^2F^\circ$	$^3/_2-^5/_2$
6255,32	2	15,26	17,24	$6p\ ^2D^\circ-6d\ ^4F$	$^5/_2-^5/_2$
6235,40	1	15,41	17,40	$5d''\ ^2D-19^\circ$	$^3/_2-^3/_2$
6234,04	10	16,48	18,47	$2-33^\circ$	$^1/_2-^3/_2$
6206,16	200	—	—	—	—
6203,45	1	15,38	17,38	$6s''\ ^2S-17^\circ$	$^1/_2-^1/_2$
6196,63	4	14,48	16,48	$6p\ ^4D^\circ-2$	$^3/_2-^1/_2$
6194,07	300	15,38	17,38	$6s''\ ^2S-6p''\ ^2P^\circ$	$^1/_2-^3/_2$
6185,93	1	16,36	18,09	$6p'\ ^2D^\circ-6d'\ ^2D$	$^3/_2-^3/_2$
6185,03	15	13,97	15,98	$5d'\ ^2D-6p'\ ^2F^\circ$	$^5/_2-^5/_2$
6184,57	20	12,92	14,93	$6s\ ^2P-6p\ ^4D^\circ$	$^1/_2-^1/_2$
6155,28	1	16,12	18,14	$6p'\ ^2F^\circ-7s'\ ^2D$	$^5/_2-^5/_2$
6146,45	50	15,38	17,40	$6s''\ ^2S-19^\circ$	$^1/_2-^3/_2$
6143,40	1	16,08	18,09	$6p'\ ^2P^\circ-6d\ ^2D$	$^3/_2-^3/_2$
6127,44	2	13,06	15,08	$5d\ ^2P-6p\ ^4S^\circ$	$^3/_2-^3/_2$
6115,08	50	13,25	15,28	$5d\ ^4P-6p\ ^2P^\circ$	$^1/_2-^3/_2$
6101,43	200	13,38	15,41	$5d\ ^2D-6p\ ^2D^\circ$	$^3/_2-^3/_2$
6097,59	1000	11,83	13,86	$5d\ ^4D-6p\ ^4P^\circ$	$^5/_2-^3/_2$
6093,56	300	14,48	16,51	$6p\ ^4D^\circ-7s\ ^4P$	$^3/_2-^3/_2$
6083,21	1	15,08	17,12	$6p\ ^4S^\circ-12$	$^3/_2-^3/_2$
6051,15	1000	11,83	13,89	$5d\ ^4D-6p\ ^4P^\circ$	$^7/_2-^5/_2$
6048,53	5	15,26	17,31	$6p\ ^2D^\circ-6d\ ^4F$	$^5/_2-^3/_2$
6036,20	500	11,83	13,89	$5d\ ^4D-6p\ ^4P^\circ$	$^5/_2-^5/_2$
6024,77	3	15,41	17,47	$6p\ ^2D^\circ-6d\ ^4P$	$^3/_2-^1/_2$
6008,92	100	13,20	15,26	$5d\ ^4P-6p\ ^2D^\circ$	$^5/_2-^5/_2$
5998,3	1	13,38	15,44	$5d\ ^2D-6p\ ^2P^\circ$	$^3/_2-^1/_2$
5991,86	1	15,41	17,48	$5d''\ ^2D-21^\circ$	$^3/_2-^3/_2$
5988,44	1	—	—	—	—
5976,46	1000	11,79	13,86	$6s\ ^4P-6p\ ^4P^\circ$	$^3/_2-^3/_2$
5971,13	200	14,00	16,08	$6s'\ ^2D-6p'\ ^2P^\circ$	$^3/_2-^3/_2$
5958,03	50	13,20	15,28	$5d\ ^4P-6p\ ^2P^\circ$	$^5/_2-^3/_2$
5945,53	300	12,01	14,09	$5d\ ^4D-6p\ ^4P^\circ$	$^1/_2-^1/_2$
5934,55	2	—	—	—	—
5921,50	2	15,02	17,12	$6p\ ^2S^\circ-12$	$^1/_3-^3/_2$
5917,44	50	11,79	13,89	$6s\ ^4P-6p\ ^4P^\circ$	$^3/_2-^5/_2$
5912,80	5	13,31	15,41	$5d\ ^4P-6p\ ^2D^\circ$	$^3/_2-^3/_2$
5909,67	30	15,38	17,48	$6s''\ ^2S-21^\circ$	$^1/_2-^3/_2$
5905,13	200	12,92	15,02	$6s\ ^2P-6p\ ^2S^\circ$	$^1/_2-^1/_2$
5893,29	150	16,12	18,22	$6p'\ ^2F^\circ-14$	$^7/_2-^5/_2$
		13,97	16,08	$5d'\ ^2D-6p'\ ^2P^\circ$	$^5/_2-^3/_2$
5859,47	2	16,46	18,57	$6p'\ ^2P^\circ-7s'\ ^2D$	$^1/_2-^3/_2$
5855,47	1	15,98	18,09	$6p'\ ^2F^\circ-6d\ ^2D$	$^5/_2-^3/_2$
5846,69	2	—	—	—	—
5835,5	5	14,48	16,60	$6p\ ^4D^\circ-6$	$^3/_2-^5/_2$
5821,57	1	14,23	16,36	$5d'\ ^2F-6p'\ ^2D^\circ$	$^5/_2-^3/_2$
5815,96	50	13,31	15,44	$5d\ ^4P-6p\ ^2P^\circ$	$^3/_2-^1/_2$
5809,5	1	17,24	19,38	$6d\ ^4F-39^\circ$	$^5/_2-^3/_2$
5791,98	1	16,08	18,22	$6p'\ ^2P^\circ-14$	$^3/_2-^5/_2$
5776,39	100	13,14	15,28	$5d\ ^2P-6p\ ^2P^\circ$	$^1/_2-^3/_2$
5758,65	100	13,97	16,12	$5d'\ ^2D-6p'\ ^4F^\circ$	$^5/_2-^7/_2$
5754,18	2	—	—	—	—
5752,56	10	12,92	15,08	$6s\ ^2P-6p\ ^4S^\circ$	$^1/_2-^3/_2$
5751,03	200	13,25	15,41	$5d\ ^2P-6p\ ^2D^\circ$	$^1/_2-^3/_2$
5746,88	5	—	—	—	—

λ, Å	I	E_H, eV	E_B, eV	Transition	J
5726,91	200	13,86	16,02	$6p\ ^4P^\circ - 5d'\ ^2S$	$3/2-1/2$
		14,23	16,39	$5d'\ ^2F - 6p'\ ^2D^\circ$	$5/2-5/2$
5716,19	100	16,39	18,56	$6p'\ ^2D^\circ - 6d'\ ^2D$	$5/2-5/2$
5699,61	100	13,80	15,98	$5d'\ ^2D - 6p'\ ^2F^\circ$	$3/2-5/2$
5686,49	2	—	—	—	—
5681,87	1	16,39	18,57	$6p'\ ^2D^\circ - 7s'\ ^2D$	$5/2-3/2$
5675,15	1	12,74	14,93	$6s\ ^2P - 6p\ ^4D^\circ$	$3/2-1/2$
5670,96	50	15,28	17,47	$6p\ ^2P^\circ - 6d\ ^4P$	$3/2-1/2$
5667,56	300	11,91	14,09	$5d\ ^4D - 6p\ ^4P^\circ$	$3/2-1/2$
5664,02	3	—	—	—	—
5659,38	150	13,14	15,44	$5d\ ^2P - 6p\ ^2P^\circ$	$1/2-1/2$
5633,24	3	—	—	—	—
5624,78	1	16,36	18,56	$6p'\ ^2D^\circ - 6d'\ ^2D$	$3/2-5/2$
5616,67	150	13,06	15,26	$5d\ ^2P - 6p\ ^2D^\circ$	$3/2-5/2$
5612,89	1	15,44	17,65	$6p\ ^2P^\circ - 7s\ ^2P$	$1/2-3/2$
5594,87	4	—	—	—	—
5591,61	2	16,36	18,57	$6p'\ ^2D^\circ - 7s'\ ^2D$	$3/2-3/2$
5583,5	2	—	—	—	—
5581,93	2	—	—	—	—
5572,19	50	—	—	—	—
5554,99	3	—	—	—	—
5551,50	2	—	—	—	—
5531,07	400	11,83	14,07	$5d\ ^4D - 6p\ ^4D^\circ$	$7/2-5/2$
5525,59	50	15,41	17,65	$6p\ ^2D^\circ - 7s\ ^2P$	$3/2-3/2$
5518,56	1	11,83	14,07	$5d\ ^4D - 6p\ ^4D^\circ$	$5/2-5/2$
5509,20	2	—	—	—	—
5507,46	2	—	—	—	—
5495,07	20	—	—	—	—
5472,61	500	11,83	14,10	$5d\ ^4D - 6p\ ^4D^\circ$	$7/2-7/2$
5469,58	20	14,48	16,74	$6p\ ^4D^\circ - 7s\ ^4P$	$3/2-1/2$
5460,39	300	11,83	14,10	$5d\ ^4D - 6p\ ^4D^\circ$	$5/2-7/2$
5450,90	20	13,80	16,08	$5d'\ ^2D - 6p'\ ^2P^\circ$	$3/2-3/2$
5450,45	100	13,14	15,41	$5d\ ^2P - 6p\ ^2D^\circ$	$1/2-3/2$
5445,52	150	15,44	17,72	$6p\ ^2P^\circ - 7s\ ^2P$	$1/2-1/2$
5438,96	400	12,74	15,02	$6s\ ^2P - 6p\ ^2S^\circ$	$3/2-1/2$
5428,07	2	16,08	18,36	$6p'\ ^2P^\circ - 6d'\ ^2D$	$3/2-3/2$
5419,15	2000	11,79	14,07	$6s\ ^4P - 6p\ ^4D^\circ$	$3/2-5/2$
5418,2	2	—	—	—	—
5415,36	50	15,02	17,31	$6p\ ^2S^\circ - 6d\ ^4F$	$1/2-3/2$
5372,39	300	11,79	14,09	$6s\ ^4P - 6p\ ^4P^\circ$	$3/2-1/2$
5368,07	100	13,14	15,44	$5d\ ^2P - 6p\ ^2P^\circ$	$1/2-1/2$
5363,27	150	15,41	17,72	$6p\ ^2D^\circ - 7s\ ^2P$	$3/2-5/2$
5339,38	1100	11,54	13,86	$6s\ ^4P - 6p\ ^4P^\circ$	$5/2-3/2$
		14,48	16,80	$6p\ ^4D^\circ - 6d\ ^4D$	$3/2-7/2$
5327,90	3	—	—	—	—
5313,87	800	14,10	16,43	$6p\ ^4D^\circ - 7s\ ^4P$	$7/2-5/2$
5309,27	200	12,74	15,08	$6s\ ^2P - 6p\ ^4S^\circ$	$3/2-3/2$
5292,22	1000	11,54	13,89	$6s\ ^4P - 6p\ ^4P^\circ$	$5/2-5/2$
5291,3	2	—	—	—	—
5282,46	2	14,48	16,82	$6p\ ^4D^\circ - 6d\ ^4D$	$3/2-3/2$
5268,31	50	13,06	15,41	$5d\ ^2D - 6p\ ^2D^\circ$	$3/2-3/2$
5261,95	200	14,00	16,36	$6s'\ ^2D - 6p'\ ^2D^\circ$	$3/2-3/2$
5260,44	200	12,92	15,28	$6s\ ^2P - 6p\ ^2P^\circ$	$1/2-3/2$
5259,89	30	14,07	16,43	$6p\ ^4D^\circ - 7s\ ^4P$	$5/2-5/2$
5247,75	20	—	—	—	—
5226,90	2	16,02	18,40	$5d'\ ^2S - 31^\circ$	$1/2-3/2$
5226,62	20	15,28	17,65	$6p\ ^2P^\circ - 7s\ ^2P$	$3/2-3/2$
5201,88	2	15,98	18,36	$6p'\ ^2F^\circ - 6d'\ ^2D$	$5/2-3/2$
5201,42	20	13,97	16,36	$6d'\ ^2D - 6p'\ ^2D^\circ$	$5/2-3/2$

λ, Å	I	E_H, eV	E_B, eV	Transition	J
5199,9	1	14,93	17,31	$6p\ ^4D^\circ$—$6d\ ^4F$	$^1/_2$—$^3/_2$
5194,92	5	14,09	16,48	$6p\ ^4P^\circ$—2	$^1/_2$—$^1/_2$
5192,10	80	15,08	17,47	$6p\ ^4S^\circ$—$6d\ ^4P$	$^3/_2$—$^1/_2$
5191,37	300	12,54	14,93	$6s\ ^4P$—$6p\ ^4D^\circ$	$^1/_2$—$^1/_2$
5188,11	200	15,26	17,65	$6p\ ^2D^\circ$—$7s\ ^2P$	$^5/_2$—$^3/_2$
5184,48	50	14,00	16,39	$6s'\ ^2D$—$6p'\ ^2D^\circ$	$^3/_2$—$^5/_2$
5178,82	50	13,58	15,98	$5d\ ^2D$—$6p'\ ^2F^\circ$	$^5/_2$—$^5/_2$
5125,70	30	13,97	16,39	$6d'\ ^2D$—$6p'\ ^2D^\circ$	$^5/_2$—$^5/_2$
5122,42	200	14,09	16,51	$6p\ ^4P^\circ$—$7s\ ^4P$	$^1/_2$—$^3/_2$
5117,76	2	—	—	—	—
5108,58	2	15,44	17,87	$6p\ ^2P^\circ$—$6d\ ^4P$	$^1/_2$—$^3/_2$
5099,59	5	—	—	—	—
5092,02	60	16,12	18,56	$6p'\ ^2F^\circ$—$6d'\ ^2D$	$^7/_2$—$^5/_2$
5081,07	30	15,28	17,77	$6p\ ^2P^\circ$—$7s\ ^2P$	$^3/_2$—$^1/_2$
5080,62	600	14,07	16,51	$6p\ ^4D^\circ$—$7s\ ^4P$	$^5/_2$—$^3/_2$
5069,82	10	—	—	—	—
5066,33	3	—	—	—	—
5052,54	30	14,48	16,93	$6p\ ^4D^\circ$—8	$^3/_2$—$^5/_2$
5044,92	150	14,00	16,46	$6s'\ ^2D$—$6p'\ ^2P^\circ$	$^3/_2$—$^1/_2$
5036,15	3	15,41	17,87	$6p\ ^2D^\circ$—$6d\ ^4P$	$^3/_2$—$^3/_2$
5018,75	1	14,76	17,23	$5d''\ ^2D$—3°	$^5/_2$—$^7/_2$
5012,83	50	14,48	16,95	$6p\ ^4D^\circ$—10	$^3/_2$—$^5/_2$
5001,01	3	—	—	—	—
4993,93	5	17,81	18,29	$5d'\ ^2P$—27°	$^3/_2$—$^5/_2$
4993,03	10	12,54	15,02	$6s\ ^4P$—$6p\ ^2S^\circ$	$^1/_2$—$^1/_2$
4991,17	100	16,08	18,56	$6p'\ ^2P^\circ$—$6d'\ ^2D$	$^3/_2$—$^5/_2$
4988,77	300	12,92	15,41	$6s\ ^2P$—$6p\ ^2D^\circ$	$^1/_2$—$^3/_2$
4974,87	2	15,41	17,90	$6p\ ^2D^\circ$—$6d\ ^2D$	$^3/_2$—$^5/_2$
4974,41	1	14,76	17,26	$5d''\ ^2D$—7°	$^5/_2$—$^3/_2$
4972,71	400	13,58	16,08	$5d\ ^2D$—$6p'\ ^2P^\circ$	$^5/_2$—$^3/_2$
4971,71	200	—	—	—	—
4965,00	4	16,08	18,57	$6p'\ ^2P^\circ$—$7s'\ ^2D$	$^3/_2$—$^3/_2$
4962,8	1	15,28	17,78	$6p\ ^2P^\circ$—$6d\ ^4P$	$^3/_2$—$^5/_2$
4946,72	1	14,10	16,60	$6p\ ^4D^\circ$—6	$^7/_2$—$^5/_2$
4921,48	800	12,74	15,26	$6s\ ^2P$—$6p\ ^2D^\circ$	$^3/_2$—$^5/_2$
4919,66	200	12,92	15,44	$6s\ ^2P$—$6p\ ^2P^\circ$	$^1/_2$—$^1/_2$
4905,20	2	—	—	—	—
4899,9	1	14,07	16,60	$6p\ ^4D^\circ$—6	$^5/_2$—$^5/_2$
4890,09	300	11,54	14,07	$6s\ ^4P$—$6p\ ^4D^\circ$	$^5/_2$—$^5/_2$
4887,30	300	12,74	15,28	$6s\ ^2P$—$6p\ ^2P^\circ$	$^3/_2$—$^3/_2$
4885,19	4	—	—	—	—
4884,15	100	14,93	17,47	$6p\ ^4D^\circ$—$6d\ ^4P$	$^1/_2$—$^1/_2$
4883,53	600	12,54	15,08	$6s\ ^4P$—$6p\ ^4S^\circ$	$^1/_2$—$^3/_2$
4876,50	500	13,58	16,12	$5d\ ^2D$—$6p'\ ^2F^\circ$	$^5/_2$—$^7/_2$
4862,54	800	13,89	16,43	$6p\ ^4P^\circ$—$7s\ ^4P$	$^5/_2$—$^5/_2$
4853,77	40	13,80	16,36	$5d'\ ^2D$—$6p'\ ^2D^\circ$	$^3/_2$—$^3/_2$
4844,33	2000	11,54	14,10	$6s\ ^4P$—$6p\ ^4D^\circ$	$^5/_2$—$^7/_2$
4840,87	1	15,44	18,00	$6p\ ^2P^\circ$—$6d\ ^2P$	$^1/_2$—$^1/_2$
4823,41	300	13,86	16,43	$6p\ ^4P^\circ$—$7s\ ^4P$	$^3/_2$—$^5/_2$
4818,02	200	11,91	14,48	$5d\ ^4D$—$6p\ ^4D^\circ$	$^3/_2$—$^3/_2$
4817,22	40	15,08	17,65	$6p\ ^4S^\circ$—$7s\ ^2P$	$^3/_2$—$^3/_2$
4806,92	3	—	—	—	—
4799,45	15	15,98	18,56	$6p'\ ^2F^\circ$—$6d'\ ^2D$	$^5/_2$—$^5/_2$
4796,53	6	—	—	—	—
4795,40	3	15,81	18,40	$5d'\ ^2P$—31°	$^3/_2$—$^3/_2$
4790,20	3	—	—	—	—
4787,77	100	13,80	16,39	$5d'\ ^2D$—$6p'\ ^2D^\circ$	$^3/_2$—$^5/_2$
4786,65	10	15,28	17,87	$6p\ ^2P^\circ$—$6d\ ^4P$	$^3/_2$—$^3/_2$

λ, Å	I	E_H, eV	E_B, eV	Transition	J
4779,18	80	11,27	13,86	$5p^6\ ^2S$—$6p\ ^4P^\circ$	$^1/_2$—$^3/_2$
4775,76	8	15,41	18,00	$6p\ ^2D^\circ$—$6d\ ^2P$	$^3/_2$—$^1/_2$
4775,18	5	15,98	18,57	$6p'\ ^2F^\circ$—$7s'\ ^2D$	$^5/_2$—$^3/_2$
4773,19	80	14,76	17,36	$5d''\ ^2D$—15°	$^5/_2$—$^5/_2$
4769,05	150	13,38	15,98	$5d\ ^2D$—$6p'\ ^2F^\circ$	$^3/_2$—$^5/_2$
4732,51	15	13,86 14,76	16,48 17,38	$6p\ ^4P^\circ$—2 $5d''\ ^2D$—$6p''\ ^2P^\circ$	$^3/_2$—$^1/_2$ $^5/_2$—$^3/_2$
4731,19	100	15,28	17,90	$6p\ ^2P^\circ$—$6d\ ^2D$	$^3/_2$—$^5/_2$
4721,00	2	—	—	—	—
4715,18	100	15,02	17,65	$6p\ ^2S^\circ$—$7s\ ^2P$	$^1/_2$—$^3/_2$
4712,63	40	—	—	—	—
4708,92	8	13,89	16,51	$6p\ ^4P^\circ$—$7s\ ^4P$	$^5/_2$—$^3/_2$
4706,96	2	—	—	—	—
4704,67	10	14,76	17,40	$5d''\ ^2D$—19°	$^5/_2$—$^3/_2$
4699,62	3	15,26	17,90	$6p\ ^2D^\circ$—$6d\ ^2D$	$^5/_2$—$^5/_2$
4698,01	300	14,48	17,12	$6p\ ^4D^\circ$—12	$^3/_2$—$^3/_2$
4693,34	15	15,08	17,72	$6p\ ^4S^\circ$—$7s\ ^2P$	$^3/_2$—$^1/_2$
4679,45	3	15,75	18,40	$5d'\ ^2P$—31°	$^1/_2$—$^3/_2$
4678,31	2	15,44	18,09	$6p\ ^2P^\circ$—$6d\ ^2D$	$^1/_2$—$^3/_2$
4676,75	2	—	—	—	—
4676,46	200	—	—	—	—
4674,56	40	11,83	14,48	$5d\ ^4D$—$6p\ ^4D^\circ$	$^5/_2$—$^3/_2$
4672,20	100	13,86	16,51	$6p\ ^4P^\circ$—$7s\ ^4P$	$^3/_2$—$^3/_2$
4668,49	100	13,80	16,46	$5d'\ ^2D$—$6p'\ ^2P^\circ$	$^3/_2$—$^1/_2$
4666,28	40	15,81	18,47	$5d'\ ^2P$—33°	$^3/_2$—$^3/_2$
4653,00	40	13,31	15,98	$5d\ ^4P$—$6p'\ ^2F^\circ$	$^3/_2$—$^5/_2$
4651,94	200	12,74	15,41	$6s\ ^2P$—$6p\ ^2D^\circ$	$^3/_2$—$^3/_2$
4649,17	2	—	—	—	—
4633,30	50	12,59	15,26	$5d\ ^4F$—$6p\ ^2D^\circ$	$^5/_2$—$^5/_2$
4620,11	2	—	—	—	—
4617,50	90	15,41	18,09	$6p\ ^2D^\circ$—$6d\ ^2D$	$^3/_2$—$^3/_2$
4615,50	200	13,39	16,08	$6s'\ ^2D$—$6p'\ ^2P^\circ$	$^5/_2$—$^3/_2$
4615,06	100	13,86	16,55	$6p\ ^4P^\circ$—4	$^3/_2$—$^1/_2$, $^3/_2$
4603,03	600	11,79	14,48	$6s\ ^4P$—$6p\ ^4D^\circ$	$^3/_2$—$^3/_2$
4596,30	1	15,02	17,72	$6p\ ^2S^\circ$—$7s\ ^2P$	$^1/_2$—$^1/_2$
4593,70	6	13,38	16,08	$5d\ ^2D$—$6p'\ ^2P^\circ$	$^3/_2$—$^3/_2$
4592,05	300	15,08	17,78	$6p\ ^4S^\circ$—$6d\ ^4P$	$^3/_2$—$^5/_2$
4585,48	500	14,10	16,80	$6p\ ^4D^\circ$—$6d\ ^4D$	$^7/_2$—$^7/_2$
4580,70	80	16,39	19,10	$6p\ ^2D^\circ$—$6d'\ ^2F$	$^5/_2$—$^5/_2$
4577,06	200	14,10	16,80	$6p\ ^4D^\circ$—$6d\ ^4D$	$^7/_2$—$^5/_2$
4571,85	30	—	—	—	—
4569,12	4	—	—	—	—
4563,00	2	—	—	—	—
4555,94	200	14,48	17,20	$6p\ ^4D^\circ$—20	$^3/_2$—$^1/_2$
4550,79	10	15,28	18,00	$6p\ ^2P^\circ$—$6d\ ^2P$	$^3/_2$—$^1/_2$
4545,23	400	14,07	16,80	$6p\ ^4D^\circ$—$6d\ ^4D$	$^5/_2$—$^7/_2$
4540,89	400	15,41	18,14	$6p\ ^2D^\circ$—$7s'\ ^2D$	$^3/_2$—$^5/_2$
4536,92	80	14,09	16,82	$6p\ ^4P^\circ$—$6d\ ^4D$	$^1/_2$—$^3/_2$
4532,49	200	13,39	16,12	$6s'\ ^2D$—$6p'\ ^2F^\circ$	$^5/_2$—$^7/_2$
4524,21	200	12,54	15,28	$6s\ ^4P$—$6p\ ^2P^\circ$	$^1/_2$—$^3/_2$
4521,86	100	16,36	19,10	$6p'\ ^2D^\circ$—$6d'\ ^2F$	$^3/_2$—$^5/_2$
4519,69	3	—	—	—	—
4511,80	2	—	—	—	—
4507,11	5	—	—	—	—
4488,60	4	—	—	—	—
4485,95	20	13,31	16,08	$5d\ ^4P$—$6p'\ ^2P^\circ$	$^3/_2$—$^3/_2$
4480,86	500	14,48	17,24	$6p\ ^4D^\circ$—$6d\ ^4F$	$^3/_2$—$^5/_2$
4473,85	4	—	—	—	—

λ, Å	I	E_H, eV	E_B, eV	Transition	J
4470,90	30	13,58	16,36	$5d\ ^2D—6p'\ ^2D°$	$^5/_2—^3/_2$
4464,60	1	13,20	15,98	$5d\ ^4P—6p'\ ^2F°$	$^5/_2—^5/_2$
4462,19	1000	—	—	—	—
4448,13	500	—	—	—	—
4440,95	50	15,08	17,87	$6p\ ^4S°—6d\ ^4P$	$^3/_2—^3/_2$
4427,52	2	—	—	—	—
4416,07	150	15,41	18,22	$6p\ ^2D°—14$	$^3/_2—^5/_2$
4414,84	300	13,58	16,39	$5d\ ^2D—6p'\ ^2D°$	$^5/_2—^5/_2$
4406,88	200	15,28	18,09	$6p\ ^2P°—6d\ ^2D$	$^3/_2—^3/_2$
4395,77	500	14,23	17,05	$5d'\ ^2F—4f'\ ^2F$	$^5/_2—^7/_2$
4393,20	500	15,08	17,90	$6p\ ^4S°—6d\ ^2D$	$^3/_2—^5/_2$
4384,93	60	11,27	14,09	$5p^6\ ^2S—6p\ ^4P°$	$^1/_2—^1/_2$
4379,44	10	15,26	18,09	$6p\ ^2D°—6d\ ^2D$	$^5/_2—^3/_2$
4373,78	100	14,48	17,31	$6p\ ^4D°—6d\ ^4F$	$^3/_2—^3/_2$
4372,46	2	14,10	16,93	$6p\ ^4D°—8$	$^7/_2—^5/_2$
4369,20	200	14,09	16,93	$6p\ ^4P°—18$	$^1/_2—^1/_2$
4367,05	30	14,10	16,93	$6p\ ^4D°—6d\ ^4F$	$^7/_2—^7/_2$
4360,32	2	—	—	—	—
4342,56	6	14,10	16,95	$6p\ ^4D°—10$	$^7/_2—^5/_2$
4337,07	30	15,28	18,14	$6p\ ^2P°—7s'\ ^2D$	$^3/_2—^5/_2$
4335,81	10	14,07	16,93	$6p\ ^4D°—8$	$^5/_2—^5/_2$
4330,52	1000	14,07	16,93	$6p\ ^4D°—6d\ ^4F$	$^5/_2—^7/_2$
4321,82	40	12,54	15,41	$6s\ ^4P—6p\ ^2D°$	$^1/_2—^3/_2$
4310,51	500	15,26	18,14	$6p\ ^2D°—7s'\ ^2D$	$^5/_2—^5/_2$
4306,21	1	15,41	18,29	$5d''\ ^2D—25°$	$^3/_2—^3/_2$
4296,75	2	15,41	18,30	$5d''\ ^2D—29°$	$^3/_2—^3/_2$
4296,40	500	13,86	16,74	$6p\ ^4P°—7s\ ^4P$	$^3/_2—^1/_2$
4269,84	40	12,54	15,44	$6s\ ^4P—6p\ ^2P°$	$^1/_2—^1/_2$
4263,57	5	15,38	18,29	$6s''\ ^2S—25°$	$^1/_2—^3/_2$
4263,44	30	—	—	—	—
4251,57	100	15,44	18,36	$6p\ ^2P°—6d'\ ^2D$	$^1/_2—^3/_2$
4245,38	500	13,89	16,80	$6p\ ^4P°—6d\ ^4D$	$^5/_2—^7/_2$
4244,41	30	13,06	15,98	$5d\ ^2P—6p'\ ^2F°$	$^3/_2—^5/_2$
4243,88	10	12,01	14,93	$5d\ ^4D—6p\ ^4D°$	$^1/_2—^1/_2$
4238,25	500	13,89	16,80	$6p\ ^4P°—6d\ ^4D$	$^5/_2—^5/_2$
4223,00	400	15,28	18,22	$6p\ ^2P°—14$	$^3/_2—^5/_2$
4215,60	200	11,54	14,48	$6s\ ^4P—6p\ ^4D°$	$^5/_2—^3/_2$
4214,69	6	13,14	16,08	$5d\ ^2P—6p'\ ^2P°$	$^1/_2—^3/_2$
4213,72	400	14,93	17,87	$6p\ ^4D°—6d\ ^4P$	$^1/_2—^3/_2$
4209,47	200	13,89	16,82	$6p\ ^4P°—6d\ ^4D$	$^5/_2—^3/_2$
4208,48	400	13,86	16,80	$6p\ ^4P°—6d\ ^4D$	$^3/_2—^5/_2$
4203,22	5	16,43	19,38	$7s\ ^4P—39°$	$^5/_2—^3/_2$
4201,25	15	15,41	18,36	$6p\ ^2D°—6d'\ ^2D$	$^3/_2—^3/_2$
4197,81	10	15,26	18,22	$6p\ ^2D°—14$	$^5/_2—^5/_2$
4193,15	500	—	—	—	—
4180,10	1000	13,86	16,82	$6p\ ^4P°—6d\ ^4D$	$^3/_2—^3/_2$
4170,99	8	16,12	19,10	$6p'\ ^2F°—6d'\ ^2F$	$^7/_2—^5/_2$
4162,16	60	13,38	16,36	$5d\ ^2D—6p'\ ^2D°$	$^3/_2—^3/_2$
4158,04	200	15,02	18,00	$6p\ ^2S°—6d\ ^2P$	$^1/_2—^1/_2$
4156,17	2	—	—	—	—
4148,19	2	14,48	17,47	$6p\ ^4D°—6d\ ^4P$	$^3/_2—^1/_2$
4138,81	3	—	—	—	—
4131,01	20	13,39	16,39	$6s'\ ^2D—6p'\ ^2D°$	$^5/_2—^5/_2$
4121,86	5	14,23	17,23	$5d'\ ^2F—3°$	$^5/_2—^7/_2$
4113,52	2	13,38	16,39	$5d\ ^2D—6p'\ ^2D°$	$^3/_2—^5/_2$
4113,26	2	15,38	18,40	$6s''\ ^2S—31°$	$^1/_2—^3/_2$
4112,14	30	15,08	18,09	$6p\ ^4S°—6d\ ^2D$	$^3/_2—^3/_2$
4110,41	30	12,01	15,02	$5d\ ^4D—6p\ ^2S°$	$^1/_2—^1/_2$

λ, Å	I	E_H, eV	E_B, eV	Transition	J
4104,95	40	13,06	16,08	5d ²P—6p′ ²P°	3/2—3/2
4103,10	8	16,08	19,10	6p′ ²P°—6d′ ²F	3/2—5/2
4100,97	1	14,23	17,25	5d′ ²F—5°	5/2—3/2
4100,34	20	11,91	14,93	5d ⁴D—6p ⁴D°	3/2—1/2
4098,89	100	14,09	17,12	6p ⁴P°—12	1/2—3/2
4091,88	3	14,23	17,26	5d′ ²F—7°	5/2—3/2
4073,50	15	13,31	16,36	5d ⁴P—6p′ ²D°	3/2—3/2
4072,10	6	14,07	17,12	6p ⁴D°—12	5/2—3/2
4062,12	6	13,89	16,93	6p ⁴P°—8	5/2—5/2
4061,06	3	12,92	15,98	6s ²P—6p′ ²F°	1/2—5/2
4057,46	200	13,89	16,93	6p ⁴P°—6d ⁴F	5/2—7/2
4051,27	10	15,08	18,14	6p ⁴S°—7s′ ²D	3/2—5/2
4044,90	8	—	—	—	—
4044,64	6	13,86	16,92	6p ⁴P°—16	3/2—3/2
4039,69	1	14,23	17,29	5d′ ²F—9°	5/2—3/2
4037,59	200	15,02	18,09	6p ²S°—6d ²D	1/2—3/2
4037,29	100	13,86	16,93	6p ⁴P°—18	3/2—1/2
4035,87	1	12,01	15,08	5d ⁴D—6p ⁴S°	1/2—3/2
4029,82	1	14,93	18,00	6p ⁴D°—6d ²P	1/2—1/2
4027,97	3	—	—	—	—
4026,20	5	15,28	18,36	6p ²P°—6d′ ²D	3/2—3/2
4025,19	30	13,38	16,46	5d ²D—6p′ ²P°	3/2—1/2
4017,86	2	15,38	18,47	6s″ ²S—33°	1/2—3/2
4016,56	2	—	—	—	—
4002,35	80	15,41	18,51	5d″ ²D—35 °	3/2—5/2
4000,55	5	—	—	—	—
3996,05	3	13,25	16,36	5d ⁴P—6p′ ²D°	1/2—3/2
3990,33	60	14,09	17,20	6p ⁴P°—20	1/2—1/2
3980,41	2	14,25	17,36	5d′ ²F—15°	7/2—5/2
3978,98	2	15,02	18,14	6p ²S°—7s′ ²D	1/2—5/2
3975,59	4	11,91	15,02	5d ⁴D—6p ²S°	3/2—1/2
3972,58	50	15,98	19,10	6p′ ²F°—6d′ ²F	5/2—5/2
3954,73	20	14,23	17,36	5d′ ²F—15°	5/2—5/2
3951,61	5	15,08	18,22	6p ⁴S°—14	3/2—5/2
3943,57	20	11,79	14,93	6s ⁴P—6p ⁴D°	3/2—1/2
3942,21	3	13,31	16,46	5d ⁴P—6p′ ²P°	3/2—1/2
3938,92	15	15,41	18,56	5d″ ²D—37°	3/2—5/2
3937,66	2	14,10	17,24	6p ⁴D°—6d ⁴F	7/2—5/2
3933,22	1	12,92	16,08	6s ²P—6p′ ²P°	1/2—3/2
3926,80	1	14,23	17,38	5d′ ²F—6p′ ²P°	5/2—3/2
3918,57	2	15,41	18,57	6p ²D°—7s′ ²D	3/2—3/2
3916,60	1	14,93	18,09	6p ⁴D°—6d ²D	1/2—3/2
3907,91	100	14,07	17,24	6p ⁴D°—6d ⁴F	5/2—5/2
3905,85	10	11,91	15,08	5d ⁴D—6p ⁴S°	3/2—3/2
3905,34	1	14,48	17,65	6p ⁴D°—7s ³P	3/2—3/2
3885,45	4	—	—	—	—
3885,00	20	13,20	16,39	5d ⁴P—6p′ ²D°	5/2—5/2
3869,63	20	13,25	16,46	5d ⁴P—6p′ ²P°	1/2—1/2
3858,53	20	11,27	14,48	5p⁶ ²S—6p ⁴D°	1/2—3/2
3849,87	50	14,09	17,31	6p ⁴P°—6d ²F	1/2—3/2
3848,58	6	13,14	16,36	6p ²P—6p′ ²D°	1/2—3/2
3829,77	10	13,89	17,12	6p ⁴P°—12	5/2—3/2
3826,27	2	14,07	17,31	6p ⁴D°—6d ⁴F	5/2—3/2
3823,35	2	—	—	—	—
3811,05	40	11,83	15,08	5d ⁴D—6p ⁴S°	5/2—3/2
3807,29	10	14,00	17,26	6s′ ²D—7°	3/2—3/2
3800,99	15	13,97	17,23	5d′ ²D—3°	5/2—7/2
3787,32	3	12,01	15,28	5d ⁴D—6p ²P°	1/2—3/2

λ, Å	I	E_H, eV	E_B, eV	Transition	J
3783,23	10	13,97	17,25	$5d'\ ^2D$—5°	$^5/_2$—$^3/_2$
3780,70	1	15,28	18,56	$6p\ ^2P°$—$6d'\ ^2D$	$^3/_2$—$^5/_2$
3778,78	1	15,08	18,36	$6p\ ^4S°$—$6d'\ ^2D$	$^3/_2$—$^3/_2$
3775,49	1	13,97	17,26	$5d'\ ^2D$—7°	$^5/_2$—$^3/_2$
3770,12	3	—	—	—	—
3763,37	15	11,79	15,08	$6s\ ^4P$—$6p\ ^4S°$	$^3/_2$—$^3/_2$
3762,26	10	—	—	—	—
3762,05	3	14,00	17,29	$6s'\ ^2D$—9°	$^3/_2$—$^3/_2$
3756,87	10	13,06	16,36	$5d\ ^2P$—$6p'\ ^2D°$	$^3/_2$—$^3/_2$
3737,20	5	—	—	—	—
3731,18	20	13,14	16,46	$5d\ ^2P$—$6p'\ ^2P°$	$^1/_2$—$^1/_2$
3720,80	40	12,74	16,08	$6s\ ^2P$—$6p'\ ^2P°$	$^3/_2$—$^3/_2$
3717,20	20	13,06	16,39	$5d\ ^2P$—$6p'\ ^2D°$	$^3/_2$—$^5/_2$
3715,69	2	15,02	18,36	$6p\ ^2S°$—$6d'\ ^2D$	$^1/_2$—$^3/_2$
3711,64	20	13,86	17,20	$6p\ ^4P°$—20	$^3/_2$—$^1/_2$
3691,84	1	11,91	15,26	$5d\ ^4D$—$6p\ ^2D°$	$^3/_2$—$^5/_2$
3690,74	1	14,00	17,36	$6s'\ ^2D$—13°	$^3/_2$—$^1/_2$
3674,04	1	14,09	17,47	$6p\ ^4P°$—$6d\ ^4P$	$^1/_2$—$^1/_2$
3672,57	20	11,91	15,28	$5d\ ^4D$—$6p\ ^2P°$	$^3/_2$—$^3/_2$
3663,93	5	14,00	17,38	$6s'\ ^2D$—$6p''\ ^2P°$	$^3/_2$—$^3/_2$
3661,70	20	13,86	17,24	$6p\ ^4P°$—$6d\ ^4F$	$^3/_2$—$^5/_2$
3658,44	6	13,97	17,36	$5d'\ ^2D$—15°	$^5/_2$—$^5/_2$
3657,74	5	12,59	15,98	$5d\ ^4F$—$6p'\ ^2F°$	$^5/_2$—$^5/_2$
3644,91	5	13,06	16,46	$5d\ ^2P$—$6p'\ ^2P°$	$^3/_2$—$^1/_2$
3644,43	5	12,01	15,41	$5d\ ^4D$—$6p\ ^2D°$	$^1/_2$—$^3/_2$
3634,48	1	13,97	17,38	$5d'\ ^2D$—$6p''\ ^2P°$	$^5/_2$—$^3/_2$
3621,98	3	14,48	17,90	$6p\ ^4D°$—$6d\ ^2D$	$^3/_2$—$^5/_2$
3612,37	20	11,83	15,26	$5d\ ^4D$—$6p\ ^2D°$	$^7/_2$—$^5/_2$
3611,52	1	13,89	17,31	$6p\ ^4P°$—$6d\ ^4F$	$^5/_2$—$^3/_2$
3607,41	8	12,01	15,44	$5d\ ^4D$—$6p\ ^2P°$	$^1/_2$—$^1/_2$
3604,83	3	—	—	—	—
3589,88	1	13,86	17,31	$6p\ ^4P°$—$6d\ ^4F$	$^3/_2$—$^3/_2$
3588,62	6	11,83	15,28	$5d\ ^4D$—$6p\ ^2P°$	$^5/_2$—$^3/_2$
3564,30	20	11,79	15,26	$6s\ ^4P$—$6p\ ^2D°$	$^3/_2$—$^5/_2$
3562,50	1	14,00	17,48	$6s'\ ^2D$—21°	$^3/_2$—$^3/_2$
3561,75	1	15,08	18,56	$6p\ ^4S°$—$6d'\ ^2D$	$^3/_2$—$^5/_2$
3548,60	2	13,80	17,29	$5d'\ ^2D$—9°	$^3/_2$—$^3/_2$
3546,29	1	11,79	15,28	$6s\ ^4P$—$6p\ ^2P°$	$^3/_2$—$^3/_2$
3538,08	2	11,91	15,41	$5d\ ^4D$—$6p\ ^2D°$	$^3/_2$—$^3/_2$
3534,61	1	13,97	17,48	$5d'\ ^2D$—21°	$^5/_2$—$^3/_2$
3530,21	3	13,80	17,31	$5d'\ ^2D$—11°	$^3/_2$—$^5/_2$
3514,58	8	—	—	—	—
3508,88	20	12,92	16,46	$6s\ ^2P$—$6p'\ ^2P°$	$^1/_2$—$^1/_2$
3506,56	15	12,54	16,08	$6s\ ^4P$—$6p'\ ^2P°$	$^1/_2$—$^3/_2$
3504,25	1	12,59	16,12	$5d\ ^4F$—$6p'\ ^2F°$	$^5/_2$—$^7/_2$
3503,15	15	11,91	15,44	$5d\ ^4D$—$6p\ ^2P°$	$^3/_2$—$^1/_2$
3501,77	20	—	—	—	—
3500,36	30	11,54	15,08	$6s\ ^4P$—$6p\ ^4S°$	$^5/_2$—$^3/_2$
3485,23	1	13,80	17,36	$5d'\ ^2D$—13°	$^3/_2$—$^1/_2$
3482,21	2	14,09	17,65	$6p\ ^4P°$—$7s\ ^2P$	$^1/_2$—$^3/_2$
3474,23	20	15,81	19,38	$5d'\ ^2P$—39°	$^3/_2$—$^3/_2$
3464,17	1	13,80	17,36	$5d'\ ^2D$—17°	$^3/_2$—$^1/_2$
3462,81	1	14,07	17,65	$6p\ ^4D°$—$7s\ ^2P$	$^5/_2$—$^3/_2$
3461,26	100	13,80	17,38	$5d'\ ^2D$—$6p''\ ^2P°$	$^3/_2$—$^3/_2$
3460,08	8	11,83	15,41	$5d\ ^4D$—$6p\ ^2D°$	$^5/_2$—$^3/_2$
3446,34	25	13,80	17,40	$5d'\ ^2D$—19°	$^3/_2$—$^3/_2$
3440,75	4	—	—	—	—

λ, Å	I	E_H, eV	E_B, eV	Transition	J
3437,73	3	—	—	—	—
3436,48	1	13,86	17,47	6p $^4P^\circ$—6d ^4P	3/2—1/2
3432,49	1	12,74	16,36	6s ^2P—6p′ $^2D^\circ$	3/2—3/2
3420,73	40	11,79	15,41	6s ^4P—6p $^2D^\circ$	3/2—3/2
3417,04	1	14,09	17,72	6p $^4P^\circ$—7s ^2P	1/2—1/2
3413,20	6	14,76	18,40	5d″ ^2D—31°	5/2—3/2
3409,49	8	—	—	—	—
3399,37	1	12,74	16,39	6s ^2P—6p′ $^2D^\circ$	3/2—5/2
3395,50	3	13,58	17,23	5d ^2D—3°	5/2—7/2
3388,05	2	11,79	15,44	6s ^4P—6p $^2P^\circ$	3/2—1/2
3386,30	2	14,48	18,14	6p $^4D^\circ$—7s′ ^2D	3/2—5/2
3384,13	40	11,27	14,93	5p^6 ^2S—6p $^4D^\circ$	1/2—1/2
3381,34	1	13,58	17,25	5d ^2D—5°	5/2—3/2
3375,16	3	13,58	17,26	5d ^2D—7°	5/2—3/2
3373,92	2	—	—	—	—
3366,72	300	14,10	17,78	6p $^4D^\circ$—6d ^4P	7/2—5/2
3350,44	6	—	—	—	—
3347,27	3	14,76	18,47	5d″ ^2D—33°	5/2—3/2
3344,97	4	14,07	17,78	6p $^4D^\circ$—6d ^4P	5/2—5/2
3338,80	4	12,74	16,46	6s ^2P—6p′ $^2P^\circ$	3/2—1/2
3327,46	15	11,54	15,26	6s ^4P—6p $^2D^\circ$	5/2—5/2
3316,39	6	14,48	18,22	6p $^4D^\circ$—14°	3/2—5/2
3313,48	2	—	—	—	—
3311,80	2	11,54	15,28	6s ^4P—6p $^2P^\circ$	5/2—3/2
3310,85	1	14,76	18,51	5d″ ^2D—35°	5/2—5/2
3310,38	3	—	—	—	—
3309,39	2	—	—	—	—
3298,72	6	11,27	15,02	5p^6 ^2S—6p $^2S^\circ$	1/2—1/2
3281,26	12	14,09	17,87	6p $^4P^\circ$—6d ^4P	1/2—3/2
3280,48	8	—	—	—	—
3274,94	4	—	—	—	—
3272,91	60	14,00	17,79	6s′ ^2D—23°	3/2—3/2
3268,08	1	13,86	17,65	6p $^4P^\circ$—7s ^2P	3/2—3/2
3267,34	3	14,76	18,56	5d″ ^2D—37°	5/2—5/2
3267,05	6	—	—	—	—
3266,08	4	—	—	—	—
3262,02	4	13,58	17,38	5d ^2D—6p″ $^2P^\circ$	5/2—3/2
3260,73	2	12,32	16,12	5d ^4F—6p′ $^2F^\circ$	9/2—7/2
3259,36	12	12,59	16,39	5d ^4F—6p′ $^2D^\circ$	5/2—5/2
3250,56	25	11,27	15,08	5p^6 ^2S—6p $^4S^\circ$	1/2—3/2
3250,04	2	—	—	—	—
3249,35	1	{ 12,54	16,36	6s ^4P—6p′ $^2D^\circ$	1/2—3/2
		13,97	17,79	5d′ ^2D—23°	5/2—3/2
3247,74	6	—	—	—	—
3233,23	1	15,26	19,10	6p $^2D^\circ$—6d′ ^2F	5/2—5/2
3229,03	4	—	—	—	—
3225,08	15	13,39	17,23	6s′ ^2D—3°	5/2—7/2
3212,29	5	13,39	17,25	6s′ ^2D—5°	5/2—3/2
3206,72	1	13,39	17,26	6s′ ^2D—7°	5/2—3/2
3202,04	10	11,54	15,41	6s ^4P—6p $^2D^\circ$	5/2—3/2
3201,68	3	13,38	17,25	5d ^2D—5°	3/2—1/2
3196,22	25	—	—	—	—
3193,75	1	14,48	18,36	6p $^4D^\circ$—6d′ ^2D	3/2—3/2
3181,39	3	13,58	17,48	5d ^2D—21°	5/2—3/2
3175,64	80	—	—	—	—
3175,25	6	—	—	—	—
3174,59	1	13,39	17,29	6s′ ^2D—9°	5/2—3/2
3168,67	3	14,09	18,00	6p $^4P^\circ$—6d ^2P	1/2—1/2

λ, Å	I	E_H, eV	E_B, eV	Transition	J
3165,27	6	12,54	16,46	$6s\,^4P-6p'\,^2P^\circ$	1/2—1/2
3164,44	4	—	—	—	—
3164,23	6	13,38	17,29	$5d\,^2D-9^\circ$	3/2—3/2
3162,93	25	13,86	17,78	$6p\,^4P^\circ-6d\,^4P$	3/2—5/2
3159,75	4	13,39	17,31	$6s'\,^2D-11^\circ$	5/2—5/2
3148,99	5	13,31	17,25	$5d\,^4P-5^\circ$	3/2—3/2
3145,02	4	—	—	—	—
3143,62	6	13,31	17,26	$5d\,^4P-7^\circ$	3/2—3/2
3130,40	3	—	—	—	—
3128,40	1	13,06	17,02	$5d\,^2P-1^\circ$	3/2—3/2
3124,02	12	15,41	19,38	$5d''\,^2D-39^\circ$	3/2—3/2
3121,87	250	13,39	17,36	$6s'\,^2D-15^\circ$	5/2—5/2
3116,78	2	—	—	—	—
3112,74	20	13,31	17,29	$5d\,^4P-9^\circ$	3/2—3/2
3107,82	20	—	—	—	—
3104,40	70	13,39	17,38	$6s'\,^2D-6p''\,^2P^\circ$	5/2—3/2
3102,73	3	—	—	—	—
3101,51	50	15,38	19,38	$6s''\,^2S-39^\circ$	1/2—3/2
3098,50	1	13,31	17,31	$5d\,^4P-11^\circ$	3/2—5/2
3098,21	2	14,09	18,09	$6p\,^4P^\circ-6d\,^2D$	1/2—3/2
3096,90	8	13,38	17,38	$5d\,^2D-17^\circ$	3/2—1/2
3094,53	30	13,38	17,38	$5d\,^2D-6p''\,^2P^\circ$	3/2—3/2
3092,41	15	13,39	17,40	$6s'\,^2D-19^\circ$	5/2—3/2
3090,47	1	13,86	17,87	$6p\,^4P^\circ-6d\,^4P$	3/2—3/2
3088,92	3	—	—	—	—
3087,34	1	11,27	15,28	$5p^6\,^2S-6p\,^2P^\circ$	1/2—3/2
3082,87	2	14,07	18,09	$6p\,^4D^\circ-6d\,^2D$	5/2—3/2
3082,62	20	13,38	17,40	$5d\,^2D-19^\circ$	3/2—3/2
3073,17	2	13,20	17,23	$5d\,^4P-3^\circ$	5/2—7/2
3071,39	6	—	—	—	—
3067,30	30	13,25	17,29	$5d\,^4P-9^\circ$	1/2—3/2
3066,60	1	14,10	18,14	$6p\,^4D-7s'\,^2D$	7/2—5/2
3061,54	12	13,20	17,25	$5d\,^4P-5^\circ$	5/2—3/2
3056,49	20	13,20	17,26	$5d\,^4P-7^\circ$	5/2—3/2
3050,98	3	14,23	18,29	$5d'\,^2F-25^\circ$	5/2—3/2
3048,92	3	—	—	—	—
3048,50	2	14,23	18,29	$5d'\,^2F-27^\circ$	5/2—5/2
3048,17	5	—	—	—	—
3047,76	8	—	—	—	—
3046,27	25	14,23	18,30	$5d'\,^2F-29^\circ$	5/2—3/2
3045,25	30	13,31	17,38	$5d\,^4P-6p''\,^2P^\circ$	3/2—3/2
3044,75	10	11,91	15,98	$5d\,^4D-6p'\,^2F^\circ$	3/2—5/2
3042,12	12	—	—	—	—
3037,35	6	14,48	18,56	$6p\,^4D^\circ-6d'\,^2D$	3/2—5/2
3036,80	30	15,41	19,49	$5d''\,^2D-41^\circ$	3/2—3/2
3033,71	10	13,31	17,40	$5d\,^4P-19^\circ$	3/2—3/2
3033,11	6	—	—	—	—
3027,63	2	14,48 12,92	18,57 17,02	$6p\,^4D^\circ-7s'\,^2D$ $6s\,^2P-1^\circ$	3/2—3/2 1/2—3/2
3027,27	3	13,20	17,29	$5d\,^4P-9^\circ$	5/2—3/2
3022,10	2	—	—	—	—
3020,29	2	—	—	—	—
3019,78	2	13,25	17,36	$5d\,^4P-13^\circ$	1/2—1/2
3017,43	100	—	—	—	—
3015,52	20	15,38	19,49	$6s''\,^2S-41^\circ$	1/2—3/2
3013,82	2	13,20	17,31	$5d\,^4P-11^\circ$	5/2—5/2
3012,88	1	13,25	17,25	$5d\,^4P-5^\circ$	1/2—3/2
3006,97	2	—	—	—	—
3003,98	40	13,25	17,38	$5d\,^4P-17^\circ$	1/2—1/2

λ, Å	I	E_H, eV	E_B, eV	Transition	J
2999,21	15	—	—	—	—
2991,73	3	14,07 11,27	18,22 15,41	$6p\ ^4D^\circ$—14 $5p^6\ ^2S$—$6p\ ^2D^\circ$	$^5/_2$—$^5/_2$ $^1/_2$—$^3/_2$
2990,54	12	13,25	17,40	$5d\ ^4P$—19°	$^1/_2$—$^3/_2$
2986,82	8	11,83	15,98	$5d\ ^4D$—$6p'\ ^2F^\circ$	$^5/_2$—$^5/_2$
2986,18	10	—	—	—	—
2982,23	2	—	—	—	—
2979,32	300	13,20	17,36	$5d\ ^4P$—15°	$^5/_2$—$^5/_2$
2977,90	5	—	—	—	—
2976,39	8	—	—	—	—
2974,86	20	13,31	17,48	$5d\ ^4P$—21°	$^3/_2$—$^3/_2$
2972,31	8	11,91	16,08	$5d\ ^4D$—$6p'\ ^2P^\circ$	$^3/_2$—$^3/_2$
2969,80	12	—	—	—	—
2969,23	3	—	—	—	—
2966,74	1	11,27	15,44	$5p^6\ ^2S$—$6p\ ^2P^\circ$	$^1/_2$—$^1/_2$
2964,19	12	—	—	—	—
2963,41	50	13,20	17,38	$5d\ ^4P$—$6p''\ ^2P^\circ$	$^5/_2$—$^3/_2$
2955,84	2	—	—	—	—
2954,78	2	—	—	—	—
2952,48	2	13,20	17,40	$5d\ ^4P$—19°	$^5/_2$—$^3/_2$
2951,58	2	13,06	17,26	$5d\ ^2P$—7°	$^3/_2$—$^3/_2$
2949,77	4	—	—	—	—
2944,61	4	—	—	—	—
2943,41	4	—	—	—	—
2942,10	20	—	—	—	—
2941,38	8	—	—	—	—
2939,72	5	—	—	—	—
2935,86	60	—	—	—	—
2934,80	2	13,14	17,36	$5d\ ^2P$—13°	$^1/_2$—$^1/_2$
2933,34	1	13,25	17,48	$5d\ ^4P$—21°	$^1/_2$—$^3/_2$
2927,58	2	13,86	18,09	$6p\ ^4P^\circ$—$6d\ ^2D$	$^3/_2$—$^3/_2$
2924,38	2	13,06	17,29	$5d\ ^2P$—9°	$^3/_2$—$^3/_2$
2923,95	6	—	—	—	—
2923,03	1	14,23	18,47	$5d'\ ^2F$—33°	$^5/_2$—$^3/_2$
2919,87	40	13,14	17,38	$5d\ ^2P$—17°	$^1/_2$—$^1/_2$
2910,64	1	13,89	18,14	$6p\ ^4P^\circ$—$7s'\ ^2D$	$^5/_2$—$^5/_2$
2910,27	3	—	—	—	—
2907,18	80	13,14	17,40	$5d\ ^2P$—19°	$^1/_2$—$^3/_2$
2905,10	2	14,09	18,36	$6p\ ^4P^\circ$—$6d'\ ^2D$	$^1/_2$—$^3/_2$
2904,18	3	—	—	—	—
2902,68	3	—	—	—	—
2895,22	150	14,23	18,51	$5d'\ ^2F$—35°	$^5/_2$—$^5/_2$
2889,07	10	11,79	16,08	$6s\ ^4P$—$6p'\ ^2P^\circ$	$^3/_2$—$^3/_2$
2887,12	10	11,83	16,12	$5d\ ^4D$—$6p'\ ^2F^\circ$	$^7/_2$—$^7/_2$
2883,71	12	11,83	16,12	$5d\ ^4D$—$6p'\ ^2F^\circ$	$^5/_2$—$^7/_2$
2881,14	1	13,06	17,36	$5d\ ^2P$—13°	$^3/_2$—$^1/_2$
2871,24	50	—	—	—	—
2867,36	2	13,97	18,30	$5d'\ ^2D$—29°	$^5/_2$—$^3/_2$
2866,76	5	13,06	17,38	$5d\ ^2P$—17°	$^3/_2$—$^1/_2$
2864,73	150	13,06	17,38	$5d\ ^2P$—$6p''\ ^2P^\circ$	$^3/_2$—$^3/_2$
2861,90	20	14,23	18,56	$5d'\ ^2F$—37°	$^5/_2$—$^5/_2$
2856,65	2	—	—	—	—
2854,53	60	13,06	17,40	$5d\ ^2P$—19°	$^3/_2$—$^3/_2$
2853,11	1	13,14	17,48	$5d\ ^2P$—21°	$^1/_2$—$^3/_2$
2852,39	3	—	—	—	—
2850,95	3	12,01	16,36	$5d\ ^4D$—$6p'\ ^2D^\circ$	$^1/_2$—$^3/_2$
2849,66	8	—	—	—	—
2846,48	3	—	—	—	—

λ, Å	I	E_H, eV	E_B, eV	Transition	J
2845,92	8	—	—	—	—
2844,45	5	—	—	—	—
2839,57	2	—	—	—	—
2838,85	3	—	—	—	—
2836,16	1	12,92	17,29	$6s\ ^2P$—9°	1/2—3/2
2832,46	2	—	—	—	—
2832,00	2	—	—	—	—
2827,90	2	—	—	—	—
2826,94	5	—	—	—	—
2820,06	4	14,00	18,40	$6s'\ ^2D$—31°	3/2—3/2
2819,02	1	13,39	17,79	$6s'\ ^2D$—23°	5/2—3/2
2808,56	4	—	—	—	—
2807,55	2	—	—	—	—
2803,02	5	—	—	—	—
2802,50	1	13,97	18,40	$5d'\ ^2D$—31°	5/2—3/2
2797,65	30	12,59	17,02	$5d\ ^4F$—1°	5/2—3/2
2796,49	2	—	—	—	—
2792,52	1	11,54	15,98	$6s\ ^4P$—$6p'\ ^2F^\circ$	5/2—5/2
2789,52	2	—	—	—	—
2785,42	3	11,91	16,36	$5d\ ^4D$—$6p'\ ^2D^\circ$	3/2—3/2
2782,73	2	—	—	—	—
2774,86	15	14,00	18,47	$6s'\ ^2D$—33°	3/2—3/2
2773,55	5	—	—	—	—
2770,41	2	12,92	17,40	$6s\ ^2P$—19°	1/2—3/2
2767,00	1	14,09	18,57	$6p\ ^4P^\circ$—$7s'\ ^2D$	1/2—3/2
2763,56	1	11,91	16,39	$5d\ ^4D$—$6p'\ ^2D^\circ$	3/2—5/2
2762,77	2	14,07	18,56	$6p\ ^4D^\circ$—$6d'\ ^2D$	5/2—5/2
2758,36	1	13,80	18,30	$5d'\ ^2D$—29°	3/2—3/2
2757,86	40	13,97	18,47	$5d'\ ^2D$—33°	5/2—3/2
2744,04	2	—	—	—	—
2743,16	2	—	—	—	—
2734,14	50	13,25	17,79	$5d\ ^4P$—23°	1/2—3/2
2733,15	25	13,97	18,51	$5d'\ ^2D$—35°	5/2—5/2
2731,46	1	11,54	16,08	$6s\ ^4P$—$6p'\ ^2P^\circ$	5/2—3/2
2723,40	1	11,91	16,46	$5d\ ^4D$—$6p'\ ^2P^\circ$	3/2—1/2
2721,28	1	12,92	17,48	$6s\ ^2P$—21°	1/2—3/2
2718,79	1	11,83	16,39	$5d\ ^4D$—$6p'\ ^2D^\circ$	7/2—5/2
2717,35	30	—	—	—	—
2715,76	3	11,83	16,39	$5d\ ^4D$—$6p'\ ^2D^\circ$	5/2—5/2
2703,44	10	13,97	18,56	$5d'\ ^2D$—37°	5/2—5/2
2702,34	2	13,20	17,79	$5d\ ^4P$—23°	5/2—3/2
2702,22	2	11,54	16,12	$6s\ ^4P$—$6p'\ ^2F^\circ$	5/2—7/2
2691,40	1	11,79	16,39	$6s\ ^4P$—$6p'\ ^2D^\circ$	3/2—5/2
2687,03	5	—	—	—	—
2686,14	3	14,76	19,38	$5d''\ ^2D$—39°	5/2—3/2
2677,18	50	—	—	—	—
2672,22	4	12,74	17,38	$6s\ ^2P$—$6p''\ ^2P^\circ$	3/2—3/2
2668,02	5	12,59	17,23	$5d\ ^4F$—3°	5/2—7/2
2663,29	3	12,74	17,40	$6s\ ^2P$—19°	3/2—3/2
2659,28	1	12,59	17,25	$5d\ ^4F$—5°	5/2—3/2
2657,00	5	13,80	18,47	$5d'\ ^2D$—33°	3/2—3/2
2655,39	2	—	—	—	—
2634,20	2	—	—	—	—
2633,88	2	13,58	18,29	$5d\ ^2D$—25°	5/2—3/2
2631,25	2	11,27	15,98	$5p^6\ ^2S$—$6p'\ ^2F^\circ$	1/2—5/2
2630,40	6	13,58	18,30	$5d\ ^2D$—29°	5/2—3/2
2629,54	5	—	—	—	—
2621,39	2	—	—	—	—

λ, Å	I	E_H, eV	E_B, eV	Transition	J
2607,52	1	12,54	17,29	$6s\ ^4P$—9°	$^1/_2$—$^3/_2$
2606,93	5	—	—	—	—
2605,54	50	—	—	—	—
2598,42	2	—	—	—	—
2597,01	4	12,59	17,36	$5d\ ^4F$—15°	$^5/_2$—$^5/_2$
2596,86	5	—	—	—	—
2584,88	1	12,59	17,38	$5d\ ^4F$—$6p''\ ^2P°$	$^5/_2$—$^3/_2$
2576,97	15	11,27	16,08	$5p^6\ ^2S$—$6p'\ ^2P°$	$^1/_2$—$^3/_2$
2561,48	2	12,54	17,38	$6s\ ^4P$—17°	$^1/_2$—$^1/_2$
2560,89	3	—	—	—	—
2554,20	1	11,54	16,39	$6s\ ^4P$—$6p'\ ^2D°$	$^5/_2$—$^5/_2$
2551,70	3	12,54	17,40	$6s\ ^4P$—19°	$^1/_2$—$^3/_2$
2546,37	3	—	—	—	—
2538,02	3	13,58	18,47	$5d\ ^2D$—33°	$^5/_2$—$^3/_2$
2531,36	3	—	—	—	—
2530,18	2	13,39	18,29	$6s'\ ^2D$—25°	$^5/_2$—$^3/_2$
2528,49	6	13,39	18,29	$6s'\ ^2D$—27°	$^5/_2$—$^5/_2$
2526,98	12	13,39	18,30	$6s'\ ^2D$—29°	$^5/_2$—$^3/_2$
2526,79	12	—	—	—	—
2524,46	3	12,32	17,23	$5d\ ^4F$—3°	$^9/_2$—$^7/_2$
2519,17	6	—	—	—	—
2516,12	12	—	—	—	—
2514,29	5	—	—	—	—
2506,86	8	—	—	—	—
2491,78	5	13,58	18,56	$5d\ ^2D$—37°	$^5/_2$—$^5/_2$
2490,76	20	13,31	18,29	$5d\ ^4P$—25°	$^3/_2$—$^3/_2$
2489,11	50	13,31	18,29	$5d\ ^4P$—27°	$^3/_2$—$^5/_2$
2478,82	4	—	—	—	—
2475,89	100	—	—	—	—
2470,18	5	13,38	18,40	$5d\ ^2D$—31°	$^3/_2$—$^3/_2$
2469,46	5	—	—	—	—
2468,43	5	—	—	—	—
2466,60	2	—	—	—	—
2444,40	2	—	—	—	—
2441,60	2	13,39	18,47	$6s'\ ^2D$—33°	$^5/_2$—$^3/_2$
2438,76	1	13,31	18,40	$5d\ ^4P$—31°	$^3/_2$—$^3/_2$
2435,47	6	13,38	18,47	$5d\ ^2D$—33°	$^3/_2$—$^3/_2$
2432,72	12	13,20	18,30	$5d\ ^4P$—29°	$^5/_2$—$^3/_2$
2425,05	40	—	—	—	—
2422,94	10	—	—	—	—
2422,12	2	13,39	18,51	$6s'\ ^2D$—35°	$^5/_2$—$^5/_2$
2421,27	20	—	—	—	—
2410,72	7	13,25	18,40	$5d\ ^4P$—31°	$^1/_2$—$^3/_2$
2409,74	40	—	—	—	—
2405,92	3	14,23	19,38	$5d'\ ^2F$—39°	$^5/_2$—$^3/_2$
2401,79	2	13,14	18,30	$5d\ ^2P$—29°	$^1/_2$—$^3/_2$
2398,76	4	13,39	18,56	$6s'\ ^2D$—37°	$^5/_2$—$^5/_2$
2392,33	2	—	—	—	—
2392,15	2	—	—	—	—
2387,75	4	11,27	16,46	$5p^6\ ^2S$—$6p'\ ^2P°$	$^1/_2$—$^1/_2$
2386,14	2	—	—	—	—
2385,85	1	13,20	18,40	$5d\ ^4P$—31°	$^5/_2$—$^3/_2$
2369,62	4	—	—	—	—
2368,68	5	11,79	17,02	$6s\ ^4P$—1°	$^3/_2$—$^3/_2$
2362,50	1	12,54	17,79	$6s\ ^4P$—23°	$^1/_2$—$^3/_2$
2356,72	4	—	—	—	—
2356,25	1	13,14	18,40	$5d\ ^2P$—31°	$^1/_2$—$^3/_2$
2353,52	1	13,20	18,47	$5d\ ^4P$—33°	$^5/_2$—$^3/_2$

λ, Å	I	E_H, eV	E_B, eV	Transition	J
2351,56	4	—	—	—	—
2351,18	4	—	—	—	—
2344,47	12	12,01	17,29	$5d\,^4D$—9°	$^1/_2$—$^3/_2$
2342,18	3	—	—	—	—
2335,42	2	13,20	18,51	$5d\,^4P$—35°	$^5/_2$—$^5/_2$
2319,70	7	11,91	17,25	$5d\,^4D$—5°	$^3/_2$—$^3/_2$
2316,80	10	11,91	17,26	$5d\,^4D$—7°	$^3/_2$—$^3/_2$
2313,70	5	13,20	18,56	$5d\,^4P$—37°	$^5/_2$—$^5/_2$
2307,28	3	12,01	17,38	$5d\,^4D$—17°	$^1/_2$—$^1/_2$
2304,60	1	14,00	19,38	$6s'\,^2D$—39°	$^3/_2$—$^3/_2$
2299,98	6	11,91	17,29	$5d\,^4D$—9°	$^3/_2$—$^3/_2$
2299,36	2	12,01	17,40	$5d\,^4D$—19°	$^1/_2$—$^3/_2$
2296,52	30	—	—	—	—
2294,57	15	11,83	17,23	$5d\,^4D$—3°	$^7/_2$—$^7/_2$
2292,40	20	11,83	17,23	$5d\,^4D$—3°	$^5/_2$—$^7/_2$
2290,84	2	13,06	18,47	$5d\,^2P$—33°	$^3/_2$—$^3/_2$
2285,94	8	11,83	17,25	$5d\,^4D$—5°	$^5/_2$—$^3/_2$
2285,24	2	—	—	—	—
2268,72	1	11,79	17,25	$6s\,^4P$—5°	$^3/_2$—$^3/_2$
2266,80	3	11,83	17,29	$5d\,^4D$—9°	$^5/_2$—$^3/_2$
2265,94	2	11,79	17,26	$6s\,^4P$—7°	$^3/_2$—$^3/_2$
2265,62	3	12,92	18,40	$6s\,^2P$—31°	$^1/_2$—$^3/_2$
2264,20	2	11,91	17,38	$5d\,^4D$—17°	$^3/_2$—$^1/_2$
2262,95	2	11,91	17,38	$5d\,^4D$—$6p''\,^2P^\circ$	$^3/_2$—$^3/_2$
2259,22	1	11,83	17,31	$5d\,^4D$—11°	$^5/_2$—$^5/_2$
2256,56	1	11,91	17,40	$5d\,^4D$—19°	$^3/_2$—$^3/_2$
2249,86	4	11,79	17,29	$6s\,^4P$—9°	$^3/_2$—$^3/_2$
2241,86	2	11,83	17,36	$5d\,^4D$—15°	$^7/_2$—$^5/_2$
2230,79	1	11,83	17,38	$5d\,^4D$—$6p''\,^2P^\circ$	$^5/_2$—$^3/_2$
1972,6	5	—	—	—	—
1244,756	5	1,31	11,27	$5p^5\,^2P^\circ$—$5p^6\,^2S$	$^1/_2$—$^1/_2$
1183,053	7	1,31	11,79	$5p^5\,^2P^\circ$—$6s\,^4P$	$^1/_2$—$^3/_2$
1169,63	2	1,31	11,91	$5p^5\,^2P^\circ$—$5d\,^4P$	$^1/_2$—$^3/_2$
1158,474	5	1,31	12,01	$5p^5\,^2P^\circ$—$5d\,^4P$	$^1/_2$—$^1/_2$
1100,432	10	0,00	11,27	$5p^5\,^2P^\circ$—$5p^6\,^2S$	$^3/_2$—$^1/_2$
1083,860	5	1,31	12,74	$5p^5\,^2P^\circ$—$6s\,^2P$	$^1/_2$—$^3/_2$
1074,476	15	0,00	11,54	$5p^5\,^2P^\circ$—$6s\,^4P$	$^3/_2$—$^5/_2$
1051,920	10	0,00	11,79	$5p^5\,^2P^\circ$—$6s\,^4P$	$^3/_2$—$^3/_2$
1048,272	8	0,00	11,83	$5p^5\,^2P^\circ$—$5d\,^4D$	$^3/_2$—$^5/_2$
1041,306	9	0,00	11,91	$5p^5\,^2P^\circ$—$5d\,^4D$	$^3/_2$—$^3/_2$
1037,680	6	1,31	13,25	$5p^5\,^2P^\circ$—$5d\,^4P$	$^1/_2$—$^1/_2$
1032,438	4	0,00	12,01	$5p^5\,^2P^\circ$—$5d\,^4D$	$^3/_2$—$^1/_2$
976,678	6	1,31	14,00	$5p^5\,^2P^\circ$—$6s'\,^2D$	$^1/_2$—$^3/_2$
972,769	7	0,00	12,74	$5p^5\,^2P^\circ$—$6s\,^2P$	$^3/_2$—$^3/_2$
971,84	3	—	—	—	—
939,16	12	—	—	—	—
935,405	2	0,00	13,25	$5p^5\,^2P^\circ$—$5d\,^4P$	$^3/_2$—$^1/_2$
931,25	10	—	—	—	—
925,866	5	0,00	13,39	$5p^5\,^2P^\circ$—$6s'\,^2D$	$^3/_2$—$^5/_2$
912,71	8	—	—	—	—
887,24	6	—	—	—	—
885,54	3	0,00	14,00	$5p^5\,^2P^\circ$—$6s'\,^2D$	$^3/_2$—$^3/_2$
880,802	5	1,31	15,38	$5p^5\,^2P^\circ$—$6s''\,^2S$	$^1/_2$—$^1/_2$
871,42	6	—	—	—	—
854,71	6	—	—	—	—
805,95	5	—	—	—	—
804,45	5	—	—	—	—
803,066	3	1,31	16,74	$5p^5\,^2P^\circ$—$7s\,^4P$	$^1/_2$—$^1/_2$
787,31	6	—	—	—	—
740,406	3	0,00	16,74	$5p^5\,^2P^\circ$—$7s\,^4P$	$^3/_2$—$^1/_2$

Xe III, **ground state** $1s^2 2s^2 2p^6 3s^2 3p^6 3d^{10} 4s^2 4p^6 4d^{10} 5s^2 5p^4\ ^3P_2$

Ionization potential **259 089 cm⁻¹; 32,121 eV**

λ, Å	I	E_H, eV	E_B, eV	Transition	J
7653,8	1	19,70	21,32	$5d''\ 23°—6p'\ ^1D$	2—2
7460,82	5	17,19	18,85	$6s'\ ^3D°—6p\ ^3P$	3—2
7448,9	1	20,06	21,72	$5d''\ 27°—6p''\ ^3D$	2—1
7298,93	1	18,01	19,71	$5d'\ ^1D°—6p'\ ^3D$	2—1
7185,92	2	17,13	18,85	$5d'\ ^3D°—6p\ ^3P$	1—2
7174,90	2	16,91	18,61	$5d'\ 41°—6p\ ^3P$	2—1
7049,34	2	18,40	20,16	$5d''\ 17°—6p'\ ^3F$	3—3
7043,94	4	19,08	20,84	$6s''\ ^3P°—6p'\ ^3P$	2—1
6818,12	1	17,13	18,94	$5d'\ ^3D°—6p\ ^3P$	1—0
6371,65	1	16,91	18,85	$5d'\ 41°—6p\ ^3P$	2—2
6268,30	2	17,73	19,71	$6s'\ ^1D°—6p'\ ^3D$	2—1
6260,16	2	18,65	20,63	$5d''\ 19°—6p'\ ^3D$	2—3
6238,24	60	18,40	20,39	$5d''\ 17°—6p'\ ^1F$	3—3
6221,66	25	18,39	20,39	$5d''\ 15°—6p'\ ^1F$	2—3
5857,61	10	16,52	18,63	$6s\ ^3D°—6p\ ^3P$	1—1
5780,55	2	18,01	20,16	$5d'\ ^1D°—6p'\ ^3F$	2—3
5761,96	2	19,17	21,32	$6s''\ ^1P°—6p'\ ^1D$	1—2
5748,71	12	16,70	18,85	$6s'\ ^3D°—6p\ ^3P$	2—2
5701,31	6	17,75	19,92	$5d'\ ^3D°—6p'\ ^3F$	3—2
5666,46	1	17,73	19,92	$6s'\ ^1D°—6p'\ ^3F$	2—2
5566,02	2	18,40	20,63	$5d''\ 17°—6p'\ ^3D$	3—3
5552,83	12	18,39	20,63	$5d''\ 15°—6p'\ ^3D$	2—3
5524,39	40	19,08	21,32	$6s''\ P°—6p'\ ^1D$	2—2
5510,55	1	18,40	20,65	$5d''\ 17°—6p'\ ^3F$	3—4
5454,30	1	20,40	22,67	$6p'\ 4—6d\ ^5D°$	1—0
5413,56	12	18,40	20,69	$5d''\ 17°—6p'\ ^3P$	3—2
5401,04	50	18,39	20,69	$5d''\ 15°—6p'\ ^3P$	2—2
5384,17	2	19,76	22,06	$5d''\ 25°—6p''\ ^3D$	1—2
5371,09	1	15,91	18,22	$5d'\ ^3F°—6p\ ^5P$	2—2
5367,06	30	17,61	19,92	$5d'\ ^3D°—6p'\ ^3F$	2—2
5310,99	6	16,52	18,85	$6s'\ ^3D°—6p\ ^3P$	1—2
5238,95	60	18,32	20,69	$5d''\ 13°—6p'\ ^3P$	1—2
5233,16	3	17,75	20,12	$5d'\ ^3D°—6p'\ ^3F$	3—2
5223,66	20	18,01	20,39	$5d'\ ^1D°—6p'\ ^1F$	2—3
5143,03	4	17,75	20,16	$5d'\ ^3D°—6p'\ ^3F$	3—3
5114,57	1	17,73	20,16	$6s'\ ^1D°—6p'\ ^3F$	2—3
5107,38	20	16,52	18,94	$6s'\ ^3D°—6p\ ^3P$	1—0
5070,53	1	18,39	20,84	$5d''\ 15°—6p'\ ^3P$	2—1
5008,55	10	17,45	19,92	$5d'\ ^3S°—6p'\ ^3F$	1—2
4927,53	3	18,32	20,84	$5d''\ 13°—6p'\ ^3P$	1—1
4869,47	40	17,61	20,16	$5d'\ ^3D°—6p'\ ^3F$	2—3
4794,48	12	17,13	19,71	$5d'\ ^3D°—6p'\ ^3D$	1—1
4743,89	4	18,01	20,63	$5d'\ ^1D°—6p'\ ^3D$	2—3
4723,57	30	15,57	18,20	$6s\ ^3S°—6p\ ^5P$	1—1
4697,49	2	17,75	20,39	$5d'\ ^3D°—6p'\ ^1F$	3—3
4685,17	1	19,08	21,72	$6s''\ ^3P°—6p''\ ^3D$	2—1
4683,53	60	15,57	18,22	$6s''\ ^3S°—6p\ ^5P$	1—2
4673,66	30	17,73	20,39	$6s'\ ^1D°—6p'\ ^1F$	2—3
4657,78	9	17,73	20,40	$6s'\ ^1D°—6p'\ ^1P$	2—1
4643,63	1	17,45	20,12	$5d'\ ^3S°—6p'\ ^3D$	1—2
4632,68	2	18,01	20,69	$5d'\ ^1D°—6p'\ ^3P$	2—2
4631,50	2	18,65	21,32	$5d''\ 19°—6p'\ ^1D$	2—2
4537,33	30	17,19	19,92	$6s'\ ^3D°—6p'\ ^3F$	3—2
4525,67	1	15,46	18,20	$5d\ ^3D°—6p\ ^5P$	2—1
4503,46	10	20,06	22,81	$5d''\ 27°—6p''\ 32$	2—1

λ, Å	I	E_H, eV	E_B, eV	Transition	J
4488,81	2	15,46	18,22	5d $^3D°$—6p ^5P	2—2
4468,15	1	17,61	20,39	5d′ $^3D°$—6p′ ^1F	2—3
4453,61	8	17,61	20,40	5d′ $^3D°$—6p′ 4	2—1
4434,16	50	17,13	19,92	5d′ $^3D°$—6p′ ^3F	1—2
4425,25	1	21,57	24,37	6p ^3D—35°	3—2, 3
4417,81	1	16,91	19,71	5d′ 41°—6p′ ^3D	2—1
4395,12	4	19,76	22,58	5d″ 25°—6p″ 28	1—1
4387,52	4	18,01	20,84	5d′ $^1D°$—6p′ ^3P	2—1
4357,66	1	15,64	18,48	5d′ ^3F—6p ^5P	3—3
4309,33	6	17,75	20,62	5d′ $^3D°$—4f′ 6	3—4
4308,00	10	19,70	22,58	5d″ 23°—6p″ 28	2—1
4305,86	2	17,75	20,63	5d $^3D°$—6p′ ^3D	3—3
4285,89	30	17,73	20,63	6s′ $^1D°$—6p′ ^3D	2—3
4274,13	1	19,17	22,07	6s′ $^1P°$—6p″ 26	1—1
4272,60	20	17,75	20,65	5d′ $^3D°$—6p′ ^3F	3—4
4240,24	10	18,40	21,32	5d″ 17°—6p′ ^1D	3—2
4235,82	1	17,19	20,12	6s′ $^3D°$—6p′ ^3D	3—2
4232,66	1	18,39	21,32	5d″ 15°—6p′ ^1D	2—2
4216,75	10	15,91	18,85	5d′ $^3F°$—6p ^3P	2—2
4214,04	20	17,75	20,69	5d′ $^3D°$—6p′ ^3P	3—2
4209,62	10	18,78	21,72	6s″ $^3P°$—6p″ ^3D	1—1
4203,92	10	17,45	20,40	5d′ $^3S°$—6p′ 4	1—1
4194,88	5	{ 19,76	22,72	5d″ 25°—6p″ $^3P°$	1—0
		17,73	20,69	6s′ $^1D°$—6p′ ^3P	2—2
4176,53	20	17,19	20,16	6s′ $^3D°$—6p′ ^3F	3—3
4167,66	1	21,32	24,30	6p′ ^1D—31°	2—2
4154,65	2	19,08	22,06	6s″ $^3P°$—6p″ ^3D	2—2
4152,03	5	20,06	23,05	5d″ 27°—6p″ 36	2—1, 2
4145,73	100	17,13	20,12	5d′ $^3D°$—6p′ ^3D	1—2
4142,01	10	19,08	22,07	6s″ $^3P°$—6p″ 26	2—1
4132,42	3	18,32	21,32	5d″ 13°—6p′ ^1D	1—2
4112,34	1	17,61	20,63	5d′ $^3D°$—6p″ 8	2—3
4110,06	10	16,91	19,92	5d′ 41°—6p′ ^3F	2—2
4109,07	100	16,70	19,71	6s′ $^3D°$—6p′ ^3D	2—1
4060,88	2	21,32	24,38	6p′ ^1D—37°	2—2, 3
4060,43	60	19,76	22,81	5d″ 25°—6p″ 32	1—1
4050,05	200	15,57	18,63	6s $^3S°$—6p ^3P	1—1
4043,21	20	18,66	21,72	6s″ $^3P°$—6p″ ^3D	0—1
4028,58	10	17,61	20,69	5d′ $^3D°$—6p′ ^3P	2—2
4021,62	4	15,12	18,20	5d $^3D°$—6p ^5P	1—1
3992,85	20	17,73	20,84	6s′ $^1D°$—6p′ ^3P	2—1
3985,96	8	19,70	22,81	5d″ 23°—6p″ 32	2—1
3969,91	4	19,76	22,88	5d″ 25°—6p″ 34	1—1
3950,56	300	15,06	18,20	6s $^5S°$—6p ^5P	2—1
3922,53	500	15,06	18,22	6s $^5S°$—6p ^5P	2—2
3915,30	4	18,40	21,57	5d″ 17°—6p″ ^3D	3—3
3903,70	4	15,46	18,63	5d $^3D°$—6p ^3P	2—1
3880,46	60	16,52	19,71	6s′ $^3D°$—6p′ ^3D	1—1
3877,80	200	17,19	20,39	6s′ $^3D°$—6p′ ^1F	3—3
3861,05	10	16,91	20,12	5d′ 41°—6p′ ^3D	2—2
3854,30	10	15,64	18,85	5d′ $^3F°$—6p ^3P	3—2
3841,88	20	17,61	20,84	5d′ $^3D°$—6p′ ^3P	2—1
3841,52	100	16,70	19,92	6s′ $^3D°$—6p′ ^3F	2—2
3829,77	20	21,57	24,80	6p″ ^3D—39°	3—3
3791,67	12	17,13	20,40	5d′ $^3D°$—6p′ 4	1—1
3780,98	300	15,57	18,85	6s $^3S°$—6p ^3P	1—2
3776,30	40	18,78	22,06	6s″ $^3P°$—6p″ ^3D	1—2
3772,53	20	19,76	23,05	5d″ 25°—6p″ 36	1—1, 2

λ, Å	I	E_H, eV	E_B, eV	Transition	J
3765,85	10	18,78	22,07	$6s''\,^3P^\circ$—$6p''$ 26	1—1
3745,72	25	18,01	21,32	$5d'\,^1D^\circ$—$6p'\,^1D$	2—2
3708,15	4	19,70	23,05	$5d''$ 23°—$6p''$ 36	2—1, 2
3676,63	50	15,57	18,94	$6s\,^2S^\circ$—$6p\,^3P$	1—0
3654,63	20	17,45	20,84	$5d'\,^3S^\circ$—$6p'\,^3P$	1—1
3653,12	3	15,46	18,85	$5d\,^3D^\circ$—$6p\,^3P$	2—2
3644,14	5	18,32	21,72	$5d''$ 13°—$6p''\,^3D$	1—1
3641,00	15	16,52	19,92	$6s'\,^3D^\circ$—$6p'\,^3F$	1—2
3636,03	3	19,17	22,58	$6s''\,^1P^\circ$—$6p''$ 28	1—1
3632,14	20	18,66	22,07	$6s''\,^3P^\circ$—$6p''$ 26	0—1
3628,57	3	18,65	22,06	$5d''$ 19°—$6p''\,^3D$	2—2
3624,05	600	15,06	18,48	$6s\,^5S^\circ$—$6p\,^5P$	2—3
3623,13	40	16,70	20,12	$6s'\,^3D^\circ$—$6p'\,^3D$	2—2
3618,90	4	18,65	22,07	$5d''$ 19°—$6p''$ 26	2—1
3609,44	20	17,19	20,62	$6s'\,^3D^\circ$—$4f'$ 6	3—4
3607,01	40	17,19	20,63	$6s'\,^3D^\circ$—$6p'\,^3D$	3—3
3601,89	6	14,76	18,20	$5p^5\,^1P^\circ$—$6p\,^5P$	1—1
3592,00	5	15,03	18,48	$5d\,^3D^\circ$—$6p\,^5P$	3—3
3583,64	80	17,19	20,65	$6s'\,^3D^\circ$—$6p'\,^3F$	3—4
3579,69	100	16,70	20,16	$6s'\,^3D^\circ$—$6p'\,^3F$	2—3
3561,38	40	16,91	20,39	$5d'$ 41°—$6p'\,^1F$	2—3
3552,13	50	16,91	20,40	$5d'$ 41°—$6p'$ 4	2—1
3542,33	50	17,19	20,69	$6s'\,^3D^\circ$—$6p'\,^3P$	3—2
3539,96	20	19,08	22,58	$6s''\,^3P^\circ$—$6p''$ 28	2—1
3522,83	80	15,12	18,63	$5d\,^3D^\circ$—$6p\,^3P$	1—1
3497,89	4	19,17	22,72	$6s''\,^1P^\circ$—$6p''\,^3P$	1—0
3479,11	1	17,13	20,69	$5d'\,^3D^\circ$—$6p'\,^3P$	1—2
3468,19	40	15,06	18,63	$6s\,^5S^\circ$—$6p\,^3P$	2—1
3467,20	25	17,75	21,32	$5d'\,^3D^\circ$—$6p'\,^1D$	3—2
3454,25	70	17,73	21,32	$6s'\,^1D^\circ$—$6p'\,^1D$	2—2
3444,23	60	16,52	20,12	$6s'\,^3D^\circ$—$6p'\,^3D$	1—2
3435,78	4	20,69	24,30	$6p'\,^3P$—31°	2—2
3403,89	8	19,17	22,81	$6s''\,^1P^\circ$—$6p''$ 32	1—1
3379,02	5	18,39	22,06	$5d''$ 15°—$6p''\,^3D$	2—2
3377,09	2	20,63	24,30	$6p'\,^3D$—31°	3—2
3370,65	4	18,39	22,07	$5d''$ 15°—$6p''$ 26	2—1
3362,81	3	20,69	24,38	$6p'\,^3P$—37°	2—2, 3
3357,98	30	16,70	20,39	$6s'\,^3D^\circ$—$6p'\,^1F$	2—3
3349,76	12	16,70	20,40	$6s'\,^3D^\circ$—$6p'$ 4	2—1
3344,97	4	20,63	24,33	$6p'\,^3D$—33°	3—2, 3
3340,06	10	{ 19,17	22,88	$6s''\,^1P^\circ$—$6p''$ 34	1—1
		18,01	21,72	$5d'\,^1D^\circ$—$6p''\,^3D$	2—1
3338,98	25	17,13	20,84	$5d'\,^3D^\circ$—$6p'\,^3P$	1—1
3334,26	1	20,65	24,37	$6p'\,^3F$—35°	4—3
3331,65	40	16,91	20,63	$5d'$ 41°—$6p'\,^3D$	2—3
3319,53	2	19,08	22,81	$6s''\,^3P^\circ$—$6p''$ 32	2—1
3317,44	2	15,12	18,85	$5d\,^3D^\circ$—$6p\,^3P$	1—2
3314,87	10	18,32	22,06	$5d''$ 13°—$6p''\,^3D$	1—2
3314,30	1	20,63	24,37	$6p'\,^3D$—35°	3—2, 3
3306,80	10	18,32	22,07	$5d''$ 13°—$6p''$ 26	1—1
3301,60	20	18,85	22,61	$6p\,^3P$—$7s\,^5S^\circ$	2—2
3287,92	30	18,85	22,62	$6p\,^3P$—$6d\,^5D^\circ$	2—3
3285,89	10	18,85	22,62	$6p\,^3P$—$6d\,^5D^\circ$	2—2
3284,70	3	16,38	20,16	$5d'\,^3G^\circ$—$6p'\,^3F$	4—3
3278,48	8	18,85	22,63	$6p\,^3P$—$6d\,^5D^\circ$	2—1
3276,39	8	16,91	20,69	$5d'$ 41°—$6p'\,^3P$	2—2
3268,96	80	15,06	18,85	$6s\,^5S^\circ$—$6p\,^3P$	2—2
3246,84	10	17,75	21,57	$5d'\,^3D^\circ$—$6p''\,^3D$	3—3
3242,86	100	15,03	18,85	$5d\,^3D^\circ$—$6p\,^3P$	3—2

λ, Å	I	E_H, eV	E_B, eV	Transition	J
3236,84	25	15,12	18,94	$5d\ ^3D°—6p\ ^3P$	1—0
3199,22	4	19,17	23,05	$6s''\ ^1P°—6p''\ 36$	1—1, 2
3196,51	25	14,76	18,63	$5p^5\ ^1P°—6p\ ^3P$	1—1
3185,24	40	20,40	24,29	$6p'\ 4—29°$	1—2
3177,19	5	20,40	24,30	$6p'\ 4—31°$	1—2
3169,82	5	20,39	24,30	$6p'\ ^1F—31°$	3—2
3160,70	2	18,66	22,58	$6s''\ ^3P°—6p''\ 28$	0—1
3152,98	8	16,70	20,63	$6s'\ ^3D°—6p'\ ^3D$	2—3
3151,82	10	16,91	20,84	$5d'\ 41°—6p'\ ^3P$	2—1
3150,69	20	18,65	22,58	$5d''\ 19°—6p''\ 28$	2—1
3124,61	1	19,08	23,05	$6s''\ ^3P°—6p''\ 36$	2—1, 2
3114,46	12	20,39	24,37	$6p'\ ^1F—35°$	3—2, 3
3106,33	30	17,73	21,72	$6s'\ ^1D°—6p''\ ^3D$	2—1
3103,47	3	16,70	20,69	$6s'\ ^3D°—6p'\ ^3P$	2—2
3099,91	8	18,63	22,63	$6p\ ^3P—6d\ ^5D°$	1—1
3091,06	50	15,91	19,92	$5d'\ ^3F°—6p'\ ^3F$	2—2
3083,54	40	16,14	20,16	$5d'\ ^3G°—6p'\ ^3F$	3—3
3073,49	10	18,78	22,81	$6s''\ ^3P°—6p''\ 32$	1—1
3054,49	15	18,01	22,07	$5d'\ ^1D°—6p''\ 26$	2—1
3026,52	8	14,76	18,85	$5p^5\ ^1P°—6p\ ^3P$	1—2
3023,80	100	{ 16,52	20,62	$6s'\ ^3D°—6p'\ ^3P$	1—0
		{ 18,48	22,58	$6p\ ^5P—6d\ ^5D°$	3—4
3014,18	6	17,61	21,72	$5d'\ ^3D°—6p''\ ^3D$	2—1
3004,32	30	18,48	22,61	$6p\ ^5P—7s\ ^5S°$	3—2
3001,85	10	20,16	24,29	$6p'\ ^3F—29°$	3—2
2994,69	8	20,16	24,30	$6p'\ ^3F—31°$	3—2
2992,91	40	18,48	22,62	$6p\ ^5P—6d\ ^5D°$	3—3
2991,45	8	16,70	20,84	$6s'\ ^3D°—6p'\ ^3P$	2—1
2991,25	10	18,48	22,62	$6p\ ^5P—6d\ ^5D°$	3—2
2984,63	15	20,65	24,80	$6p'\ ^3F—39°$	4—3
2971,24	8	16,52	20,69	$6s'\ ^3D°—6p'\ ^3P$	1—2
2969,45	4	20,16	24,33	$6p'\ ^3F—33°$	3—2, 3
2968,56	10	20,63	24,80	$6p'\ ^3D—39°$	3—3
2966,97	10	20,62	24,80	$4f'\ 6—39°$	4—3
2964,98	15	20,12	24,30	$6p'\ ^3D—31°$	2—2
2948,06	40	15,91	20,12	$5d'\ ^3F°—6p'\ ^3D$	2—2
2947,53	40	16,42	20,62	$5d'\ ^1G°—4f'\ 6$	4—4
2945,25	60	20,16	24,37	$6p'\ ^3F—35°$	3—2, 3
2940,22	40	20,12	24,33	$6p'\ ^3D—33°$	2—2, 3
2939,13	10	20,16	24,38	$6p'\ ^3F—37°$	3—2, 3
2932,74	25	13,97	18,24	$5d\ ^5D°—6p\ ^5P$	0—1
2930,29	20	16,42	20,65	$5d'\ ^1G°—6p'\ ^3F$	4—4
2923,51	25	16,38	20,62	$5d'\ ^3G°—4f'\ 6$	4—4
2917,59	20	16,14	20,39	$5d'\ ^3G°—6p'\ ^1F$	3—3
2914,12	20	15,46	19,71	$5d\ ^3D°—6p'\ ^3D$	2—1
2911,90	40	13,94	18,20	$5d\ ^5D°—6p\ ^5P$	1—1
2911,47	2	18,32	22,58	$3d''\ 13°—6p''\ 28$	1—1
2906,56	50	16,38	20,65	$5d'\ ^3G°—6p'\ ^3F$	4—4
2899,57	1	15,84	20,12	$5d'\ ^1F°—6p'\ ^3D$	3—2
2897,69	2	17,45	21,72	$5d'\ ^3S°—6p\ ^3D$	1—1
2896,63	30	13,94	18,22	$5d\ ^5D°—6p\ ^5P$	1—2
2891,71	25	15,64	19,92	$5d'\ ^3F°—6p'\ ^3F$	3—2
2872,73	2	17,75	22,06	$5d'\ ^3D°—6p''\ ^3D$	3—2
2871,68	30	15,84	20,16	$5d'\ ^1F°—6p'\ ^3F$	3—3
2868,42	1	16,52	20,84	$6s'\ ^3D°—6p'\ ^3P$	1—1
2863,86	1	17,73	22,06	$6s'\ ^1D°—6p''\ ^3D$	2—2
2862,41	30	13,87	18,20	$5d\ ^5D°—6p\ ^5P$	2—1
2857,81	1	17,73	22,07	$6s'\ ^1D°—6p''\ 26$	2—1

λ, Å	I	E_H, eV	E_B, eV	Transition	J
2850,25	2	15,57	19,92	$6s\ ^3S^\circ - 6p'\ ^3F$	1—2
2847,66	40	13,87	18,22	$5d\ ^5D^\circ - 6p\ ^5P$	2—2
2839,57	2	19,92	24,29	$6p'\ ^3F - 29^\circ$	2—2
2838,85	3	18,94	23,31	$6p\ ^3P - 7s\ ^3S^\circ$	0—1
2833,18	6	19,92	24,30	$6p'\ ^3F - 31^\circ$	2—2
2832,95	6	17,19	21,57	$6s'\ ^3D^\circ - 6p''\ ^3D$	3—3
2827,45	30	13,84	18,22	$5d\ ^5D^\circ - 6p\ ^5P$	3—2
2826,05	20	18,22	22,61	$6p\ ^5P - 7s\ ^5S^\circ$	2—2
2815,94	40	18,22	22,62	$6p\ ^5P - 6d\ ^5D^\circ$	2—3
2814,47	30	18,22	22,62	$6p\ ^5P - 6d\ ^5D^\circ$	2—2
2811,67	8	18,20	22,61	$6p\ ^5P - 7s\ ^5S^\circ$	1—2
2810,52	1	19,92	24,33	$6p'\ ^3F - 33^\circ$	2—2, 3
2809,07	8	18,22	22,63	$6p\ ^5P - 6d\ ^5D^\circ$	2—1
2807,25	10	20,39	24,80	$6p'\ ^1F - 39^\circ$	3—3
2806,39	3	16,91	21,32	$5d'\ 41^\circ - 6p'\ ^1D$	2—2
2805,08	2	18,39	22,81	$5d''\ 15^\circ - 6p''\ 32$	2—1
2800,22	20	18,20	22,62	$6p\ ^5P - 6d\ ^5D^\circ$	1—2
2794,86	20	18,20	22,63	$6p\ ^5P - 6d\ ^5D^\circ$	1—1
2783,37	12	19,92	24,38	$6p'\ ^3F - 37^\circ$	2—2, 3
2779,64	5	18,85	23,31	$6p\ ^3P - 7s\ ^3S^\circ$	2—1
		17,61	22,07	$5d'\ ^3D^\circ - 6p''\ 26$	2—1
2776,96	10	15,46	19,92	$5d\ ^3D^\circ - 6p'\ ^3F$	2—2
2772,41	10	18,20	22,67	$6p\ ^5P - 6d\ ^5D^\circ$	1—0
2766,20	5	15,64	20,12	$5d'\ ^3F^\circ - 6p'\ ^3D$	3—2
2763,00	1	16,14	20,62	$5d'\ ^3G^\circ - 4f'\ 6$	3—4
2761,60	12	18,39	22,88	$5d''\ 15^\circ - 6p''\ 34$	2—1
		16,14	20,63	$5d'\ ^3G^\circ - 6p'\ ^3D$	3—3
2760,76	6	18,32	22,81	$5d''\ 13^\circ - 6p''\ 32$	1—1
2747,88	8	16,14	20,63	$5d'\ ^3G^\circ - 6p'\ ^3F$	3—4
2740,80	12	15,64	20,16	$5d'\ ^3F^\circ - 6p'\ ^3F$	3—3
2728,22	4	15,57	20,12	$6s\ ^2S^\circ - 6p'\ ^3D$	1—2
2727,22	4	15,84	20,39	$5d'\ ^1F^\circ - 6p'\ ^1F$	3—3
2696,50	8	15,12	19,71	$5d\ ^3D^\circ - 6p'\ ^3D$	1—1
2687,03	5	13,87	18,46	$5d\ ^5D^\circ - 6p\ ^5P$	2—3
2685,58	2	17,45	22,06	$5d'\ ^3S^\circ - 6p''\ ^3D$	1—2
2678,54	1	16,70	21,32	$6s'\ ^3D^\circ - 6p'\ ^1D$	2—2
2669,00	10	13,84	18,48	$5d\ ^5D^\circ - 6p\ ^5P$	3—3
2661,00	1	15,46	20,12	$5d\ ^3D^\circ - 6p'\ ^3D$	2—2
2658,26	3	13,97	18,63	$5d\ ^5D^\circ - 6p\ ^3P$	0—1
2650,20	1	18,63	23,31	$6p\ ^3P - 7s\ ^3S^\circ$	1—1
2641,12	5	13,94	18,63	$5d\ ^5D^\circ - 6p\ ^3P$	1—1
2637,54	3	15,46	20,16	$5d\ ^3D^\circ - 6p'\ ^3F$	2—3
2624,52	1	18,32	23,05	$5d''\ 13^\circ - 6p''\ 36$	1—1, 2
2608,90	6	15,64	20,39	$5d'\ ^3F^\circ - 6p'\ ^1F$	3—3
2591,69	4	15,84	20,62	$5d'\ ^1F^\circ - 4f'\ 6$	3—4
2590,45	2	15,84	20,63	$5d'\ ^1F^\circ - 6p'\ ^3D$	3—3
2578,62	2	15,12	19,92	$5d\ ^3D^\circ - 6p'\ ^3F$	1—2
2578,36	5	15,84	20,63	$5d'\ ^1F^\circ - 6p'\ ^3F$	3—4
2572,30	1	16,91	21,72	$5d'\ 41^\circ - 6p''\ ^3D$	2—1
2570,26	1	15,57	20,40	$6s\ ^3S^\circ - 6p'\ 4$	1—1
2533,31	2	15,03	19,92	$5d\ ^3D^\circ - 6p'\ ^3F$	3—2
2524,09	1	13,94	18,85	$5d\ ^5D^\circ - 6p\ ^3P$	1—2
2515,14	1	15,46	20,39	$5d\ ^3D^\circ - 6p'\ ^1F$	2—3
2510,52	3	15,46	20,40	$5d\ ^3D^\circ - 6p'\ 4$	2—1
2501,04	4	14,76	19,71	$5p^5\ ^1P^\circ - 6p'\ ^3D$	1—1
2486,69	3	13,87	18,85	$5d\ ^5D^\circ - 6p\ ^3P$	2—2
2483,43	1	15,64	20,63	$5d'\ ^3F^\circ - 6p'\ ^3D$	3—3
2472,34	1	15,64	20,65	$5d'\ ^3F^\circ - 6p'\ ^3F$	3—4

λ, Å	I	E_H, eV	E_B, eV	Transition	J
2471,28	3	13,84	18,85	$5d\ ^5D^\circ$—$6p\ ^3P$	3—2
2463,02	1	18,01	23,05	$5d'\ ^1D^\circ$—$6p''$ 36	2—1, 2
2452,62	3	15,64	20,69	$5d'\ ^3F^\circ$—$6p'\ ^3P$	3—2
2447,64	1	13,57	18,63	$5p^5\ ^3P^\circ$—$6p\ ^3P$	0—1
2436,48	5	15,03	20,12	$5d\ ^3D^\circ$—$6p'\ ^3D$	3—2
2416,73	4	15,03	20,16	$5d\ ^3D^\circ$—$6p'\ ^3F$	3—3
2414,52	1	17,45	22,58	$5d'\ ^3S^\circ$—$6p''$ 28	1—1
2403,76	1	16,91	20,06	$5d'\ 41^\circ$—$6p''\ ^3D$	2—2
2312,29	1	14,76	20,12	$5p^5\ ^1P^\circ$—$6p'\ ^3D$	1—2
2303,73	1	15,46	20,84	$5d\ ^3D^\circ$—$6p'\ ^3P$	2—1
2235,35	1	16,52	22,06	$6s'\ ^3D^\circ$—$6p''\ ^3D$	1—2
1978,702	3	15,46	21,72	$5d\ ^3D^\circ$—$6p''\ ^3D$	2—1
1921,630	2	12,18	18,63	$5p^5\ ^3P^\circ$—$6p\ ^3P$	2—1
1896,904	5	15,03	21,57	$5d\ ^3D^\circ$—$6p''\ ^3D$	3—3
1874,907	5	15,46	22,07	$5d\ ^3D^\circ$—$6p''$ 26	2—1
1854,365	6	13,94	20,63	$5d\ ^5D^\circ$—$6p'\ ^3D$	1—3
1834,254	4	13,87	20,63	$5d\ ^5D^\circ$—$6p'\ ^3D$	2—3
1826,477	7	13,84	20,62	$5d\ ^5D^\circ$—$4f'$ 6	3—4
1825,858	3	13,84	20,63	$5d\ ^5D^\circ$—$6p'\ ^3D$	3—3
1819,845	6	13,84	20,65	$5d\ ^5D^\circ$—$6p'\ ^3F$	3—4
1817,381	2	13,87	20,69	$5d\ ^5D^\circ$—$6p'\ ^3P$	2—2
1804,109	2	12,84	19,71	$5p^5\ ^3P^\circ$—$6p'\ ^3D$	1—1
1750,749	2	12,84	19,92	$5p^5\ ^3P^\circ$—$6p'\ ^3F$	1—2
1579,498	5	12,84	20,69	$5p^5\ ^3P^\circ$—$6p'\ ^3P$	1—2
1562,563	4	12,18	20,12	$5p^5\ ^3P^\circ$—$6p'\ ^3D$	2—2
1554,438	1	12,18	20,16	$5p^5\ ^3P^\circ$—$6p'\ ^3F$	2—3
1549,975	2	12,84	20,84	$5p^5\ ^3P^\circ$—$6p'\ ^3P$	1—1
1511,121	3	12,18	20,39	$5p^5\ ^3P^\circ$—$6p'\ ^1F$	2—3
1509,454	2	12,18	20,40	$5p^5\ ^3P^\circ$—$6p'$ 4	2—1
1468,180	4	12,18	20,63	$5p^5\ ^3P^\circ$—$6p'\ ^3D$	2—3
1457,356	3	12,18	20,69	$5p^5\ ^3P^\circ$—$6p'\ ^3P$	2—2
1432,204	4	12,18	20,84	$5p^5\ ^3P^\circ$—$6p\ ^3P$	2—1
1393,496	1	12,84	21,72	$5p^5\ ^3P^\circ$—$6p''\ ^3D$?	1—1
1356,364	3	12,18	21,32	$5p^5\ ^3P^\circ$—$6p'\ ^1D$	2—2
1253,645	2	12,18	22,07	$5p^5\ ^3P^\circ$—$6p''$ 26	2—1
1232,074	25	2,12	12,18	$5p^4\ ^1D$—$5p^5\ ^3P^\circ$	2—2
1225,089	3	4,64	14,76	$5p^4\ ^1S$—$5p^5\ ^1P^\circ$	0—1
1183,053	8	4,64	15,12	$5p^4\ ^1S$—$5d\ ^3D^\circ$	0—1
1156,480	9	2,12	12,84	$5p^4\ ^1D$—$5p^5\ ^3P^\circ$	2—1
1130,344	30	1,21	12,18	$5p^4\ ^3P$—$5p^5\ ^3P^\circ$	1—2
1066,391	12	1,21	12,84	$5p^4\ ^3P$—$5p^5\ ^3P^\circ$	1—1
1058,128	2	2,12	13,84	$5p^4\ ^1D$—$5d\ ^5D^\circ$	2—3
1055,328	5	2,12	13,87	$5p^4\ ^1D$—$5d\ ^5D^\circ$	2—2
1048,754	3	2,12	13,94	$5p^4\ ^1D$—$5d\ ^5D^\circ$	2—1
1047,801	10	1,01	12,84	$5p^4\ ^3P$—$5p^5\ ^3P^\circ$	0—1
1017,680	35	0,00	12,18	$5p^4\ ^3P$—$5p^5\ ^3P^\circ$	2—2
1010,376	3	4,64	16,91	$5p^4\ ^1S$—$5d'\ 41^\circ$	0—2?
1003,370	35	1,21	13,57	$5p^4\ ^3P$—$5p^5\ ^3P^\circ$	1—0
981,088	7	2,12	14,76	$5p^4\ ^1D$—$5p^5\ ^1P^\circ$	2—1
974,124	8	1,21	13,94	$5p^4\ ^3P$—$5d\ ^5D^\circ$	1—1
971,818	8	1,21	13,97	$5p^4\ ^3P$—$5d\ ^5D^\circ$	1—0
965,540	10	0,00	12,84	$5p^4\ ^3P$—$5p^5\ ^3P^\circ$	2—1
960,325	2	2,12	15,03	$5p^4\ ^1D$—$5d\ ^3D^\circ$	2—3
958,585	4	1,01	13,94	$5p^4\ ^3P$—$5d\ ^5D^\circ$	0—1
953,975	3	2,12	15,12	$5p^4\ ^1D$—$5d\ ^3D^\circ$	2—1
917,257	4	2,12	15,64	$5p^4\ ^1D$—$5d'\ ^3F^\circ$	2—3
915,488	3	1,21	14,76	$5p^4\ ^3P$—$5p^5\ ^1P^\circ$	1—1
901,746	7	1,01	14,76	$5p^4\ ^3P$—$5p^5\ ^1P^\circ$	0—1
898,873	8	2,12	15,91	$5p^4\ ^1D$—$5d'\ ^3F^\circ$	2—2

λ, Å	I	E_H, eV	E_B, eV	Transition	J
896,003	20	0,00	13,84	$5p^4\ {}^3P - 5d\ {}^5D^\circ$	2—3
895,406	4	1,21	15,06	$5p^4\ {}^3P - 6s\ {}^5S^\circ$	1—2
893,989	20	0,00	13,87	$5p^4\ {}^3P - 5d\ {}^5D^\circ$	2—2
891,833	9	1,21	15,12	$5p^4\ {}^3P - 5d\ {}^3D^\circ$	1—1
889,276	15	0,00	13,94	$5p^4\ {}^3P - 5d\ {}^5D^\circ$	2—1
878,790	8	1,01	15,12	$5p^4\ {}^3P - 5d\ {}^3D^\circ$	0—1
870,346	6	1,21	15,46	$5p^4\ {}^3P - 5d\ {}^3D^\circ$	1—2
863,386	8	1,21	15,57	$5p^4\ {}^3P - 6s\ {}^3S^\circ$	1—1
861,071	5	2,12	16,52	$5p^4\ {}^1D - 6s'\ {}^3D^\circ$	2—1
852,950	25	4,64	19,17	$5p^4\ {}^1S - 6s''\ {}^1P^\circ$	0—1
851,147	8	1,01	15,57	$5p^4\ {}^3P - 6s\ {}^3S^\circ$	0—1
850,572	5	2,12	16,70	$5p^4\ {}^1D - 6s'\ {}^3D^\circ$	2—2
840,162	7	0,00	14,76	$5p^4\ {}^3P - 5p^5\ {}^1P^\circ$	2—1
838,449	3	2,12	16,91	$5p^4\ {}^1D - 5d'\ 41^\circ$	2—2
826,134	4	2,12	17,13	$5p^4\ {}^1D - 5d'\ {}^3D^\circ$	2—1
824,881	30	0,00	15,03	$5p^4\ {}^3P - 5d\ {}^3D^\circ$	2—3
823,210	25	0,00	15,06	$5p^4\ {}^3P - 6s\ {}^5S^\circ$	2—2
822,647	4	2,12	17,19	$5p^4\ {}^1D - 6s'\ {}^3D^\circ$	2—3
820,166	4	0,00	15,12	$5p^4\ {}^3P - 5d\ {}^3D^\circ$	2—1
810,119	7	1,21	16,52	$5p^4\ {}^3P - 6s'\ {}^3D^\circ$	1—1
808,860	3	2,12	17,45	$5p^4\ {}^1D - 5d'\ {}^3S^\circ$	2—1
801,980	15	0,00	15,46	$5p^4\ {}^3P - 5d\ {}^3D^\circ$	2—2
800,819	3	1,21	16,70	$5p^4\ {}^3P - 6s'\ {}^3D^\circ$	1—2
800,228	2	2,12	17,61	$5p^4\ {}^1D - 5d'\ {}^3D^\circ$	2—1
799,338	8	1,01	16,52	$5p^4\ {}^3P - 6s'\ {}^3D^\circ$	0—1
796,070	12	0,00	15,57	$5p^4\ {}^3P - 6s\ {}^3S^\circ$	2—1
793,977	8	2,12	17,73	$5p^4\ {}^1D - 6s'\ {}^1D^\circ$	2—2
793,292	8	2,12	17,75	$5p^4\ {}^1D - 5d'\ {}^3D^\circ$	2—3
792,896	15	0,00	15,64	$5p^4\ {}^3P - 5d'\ {}^3F^\circ$	2—3
790,064	5	1,21	16,91	$5p^4\ {}^3P - 5d'\ 41^\circ$	1—2
780,030	7	2,12	18,01	$5p^4\ {}^1D - 5d'\ {}^1D^\circ$	2—2
779,781	5	1,01	16,91	$5p^4\ {}^3P - 5d'\ 41^\circ$	0—2?
779,126	25	1,21	17,13	$5p^4\ {}^3P - 5d'\ {}^3D^\circ$	1—1
		0,00	15,91	$5p^4\ {}^3P - 5d'\ {}^3F^\circ$	2—2
769,143	10	1,01	17,13	$5p^4\ {}^3P - 5d'\ {}^3D^\circ$	0—1
765,120	7	2,12	18,32	$5p^4\ {}^1D - 5d''\ 13^\circ$	2—1
763,736	7	1,21	17,45	$5p^4\ {}^3P - 5d'\ {}^3S^\circ$	1—1
761,790	5	2,12	18,39	$5p^4\ {}^1D - 5d''\ 15^\circ$	2—2
756,031	10	1,21	17,61	$5p^4\ {}^3P - 5d'\ {}^3D^\circ$	1—2
754,144	5	1,01	17,45	$5p^4\ {}^3P - 5d'\ {}^3S^\circ$	0—1
750,447	8	1,21	17,73	$5p^4\ {}^3P - 6s'\ {}^1D^\circ$	1—2
750,155	8	2,12	18,65	$5p^4\ {}^1D - 5d''\ 19^\circ$	2—2
744,141	6	2,12	18,78	$5p^4\ {}^1D - 6s''\ {}^3P^\circ$	2—1
742,566	15	0,00	16,70	$5p^4\ {}^3P - 6s'\ {}^3D^\circ$	2—2
737,979	7	1,21	18,01	$5p^4\ {}^3P - 5d'\ {}^1D^\circ$	1—2
733,314	10	0,00	16,91	$5p^4\ {}^3P - 5d'\ 41^\circ$	2—2
731,030	15	2,12	19,08	$5p^4\ {}^1D - 6s''\ {}^3P^\circ$	2—2
727,058	9	2,12	19,17	$5p^4\ {}^1D - 6s''\ {}^1P^\circ$	2—1
724,623	3	1,21	18,32	$5p^4\ {}^3P - 5d''\ 13^\circ$	1—1
723,873	5	0,00	17,13	$5p^4\ {}^3P - 5d'\ {}^3D$	2—1
721,630	4	1,21	18,39	$5p^4\ {}^3P - 5d''\ 15^\circ$	1—2
721,199	10	0,00	17,19	$5p^4\ {}^3P - 6s'\ {}^3D^\circ$	2—3
715,986	4	1,01	18,32	$5p^4\ {}^3P - 5d''\ 13^\circ$	0—1
711,190	4	1,21	18,65	$5p^4\ {}^3P - 5d''\ 19^\circ$	1—2
710,677	5	1,21	18,66	$5p^4\ {}^3P - 6s''\ {}^3P^\circ$	1—0
710,576	5	0,00	17,45	$5p^4\ {}^3P - 5d'\ {}^3S^\circ$	2—1
705,783	5	1,21	18,78	$5p^4\ {}^3P - 6s''\ {}^3P^\circ$	1—1
705,096	12	2,12	19,70	$5p^4\ {}^1D - 5d''\ 23^\circ$	2—2
703,906	9	0,00	17,61	$5p^4\ {}^2P - 5d'\ {}^3D^\circ$	2—2

λ, Å	I	E_H, eV	E_B, eV	Transition	J
702,799	8	2,12	19,76	$5p^4\,^1D$—$5d''$ 25°	2—1
699,070	5	0,00	17,73	$5p^4\,^3P$—$6s'\,^1D°$	2—2
698,541	20	0,00	17,75	$5p^4\,^3P$—$5d'\,^3D°$	2—3
697,526	8	1,01	18,78	$5p^4\,^3P$—$6s''\,^3P°$	0—1
693,972	10	1,21	19,08	$5p^4\,^3P$—$6s''\,^3P°$	1—2
691,036	7	2,12	20,06	$5p^4\,^1D$—$5d''$ 27°	2—2
690,397	7	1,21	19,17	$5p^4\,^3P$—$6s''\,^1P°$	1—1
688,231	4	0,00	18,01	$5p^4\,^3P$—$5d'\,^1D°$	2—2
682,564	7	1,01	19,17	$5p^4\,^3P$—$6s''\,^1P°$	0—1
676,606	9	0,00	18,32	$5p^4\,^3P$—$5d''$ 13°	2—1
673,996	9	0,00	18,39	$5p^4\,^3P$—$5d''$ 15°	2—2
673,813	9	0,00	18,40	$5p^4\,^3P$—$5d''$ 17°	2—3
670,55	2	1,21	19,70	$5p^4\,^3P$—$5d''$ 23°	1—2
668,473	4	1,21	19,76	$5p^4\,^3P$—$5d''$ 25°	1—1
664,877	6	0,00	18,65	$5p^4\,^3P$—$5d''$ 19°	2—2
661,124	4	1,01	19,76	$5p^4\,^3P$—$5d''$ 25°	0—1
660,124	8	0,00	18,78	$5p^4\,^3P$—$6s''\,^3P°$	2—1
657,828	8	1,21	20,06	$5p^4\,^3P$—$5d''$ 27°	1—2
646,667	5	0,00	19,17	$5p^4\,^3P$—$6s''\,^1P°$	2—1
629,217	7	0,00	19,70	$5p^4\,^3P$—$5d''$ 23°	2—2
627,393	4	0,00	19,76	$5p^4\,^3P$—$5d''$ 25°	2—1

Xe IV, ground state $1s^2\,2s^2\,2p^6\,3s^2\,3p^6\,3d^{10}\,4s^2\,4p^6\,4d^{10}\,5s^2\,5p^3\,^4S^0_{3/2}$

λ, Å	I	E_H, eV	E_B, eV	Transition	J
1249,28	—	5,19	15,12	$5p^3\,^2P°$—121929	3/2—3/2
1217,26	—	2,17	12,36	$5p^3\,^2D°$—99644	5/2—3/2, 5/2
1196,29	—	5,19	15,56	$5p^3\,^2P°$—125475	3/2—3/2, 5/2
1157,46	—	1,64	12,36	$5p^3\,^2D°$—99664	3/2—3/2, 5/2
1118,42	—	2,17	13,26	$5p^3\,^2D°$—106924	5/2—3/2, 5/2
1097,16	—	5,19	16,49	$5p^3\,^2P°$—133027	3/2—3/2
1067,74	—	1,64	13,26	$5p^3\,^2D°$—106924	3/2—3/2, 5/2
1065,04	—	3,48	15,12	$5p^3\,^2P°$—121929	1/2—3/2
1041,81	—	1,64	13,55	$5p^3\,^2D°$—109255	3/2—1/2
1026,28	—	3,48	15,56	$5p^3\,^2P°$—125475	1/2—3/2
1003,37	—	0,00	12,36	$5p^3\,^4S°$—99664	3/2—3/2, 5/2
957,70	—	2,17	15,12	$5p^3\,^2D°$—121929	5/2—3/2
952,46	—	3,48	16,49	$5p^3\,^2P°$—133027	1/2—3/2
935,24	—	0,00	13,26	$5p^3\,^4S°$—106924	3/2—3/2, 5/2
926,24	—	2,17	15,56	$5p^3\,^2D°$—125475	5/2—3/2, 5/2
920,29	—	1,64	15,12	$5p^3\,^2D°$—121929	3/2—3/2
915,29	—	0,00	13,55	$5p^3\,^4S°$—109255	3/2—1/2
891,21	—	1,64	15,56	$5p^3\,^2D°$—125475	3/2—3/2, 5/2
865,69	—	2,17	16,49	$5p^3\,^2D°$—133027	5/2—3/2
851,29	—	2,17	16,73	$5p^3\,^2D°$—134981	5/2—3/2, 5/2
840,46	—	2,17	16,92	$5p^3\,^2D°$—136495	5/2—3/2, 5/2
835,01	—	1,64	16,49	$5p^3\,^2D°$—133027	3/2—3/2
821,60	—	1,64	16,73	$5p^3\,^2D°$—134981	3/2—3/2, 5/2
820,15	—	0,00	15,12	$5p^3\,^4S°$—121929	3/2—3/2
811,51	—	1,64	16,92	$5p^3\,^2D°$—136495	3/2—3/2, 5/2
796,97	—	0,00	15,56	$5p^3\,^4S°$—125475	3/2—3/2, 5/2
783,73	—	2,17	17,99	$5p^3\,^2D°$—145107	5/2—3/2

λ, Å	I	E_H, eV	E_B, eV	Transition	J
758,50	—	1,64	17,99	$5p^3\ ^2D^\circ$—145107	$^3/_2$—$^3/_2$
751,73	—	0,00	16,49	$5p^3\ ^4S^\circ$—133027	$^3/_2$—$^3/_2$
740,85	—	0,00	16,73	$5p^3\ ^4S^\circ$—134981	$^3/_2$—$^3/_2$, $^5/_2$
732,63	—	0,00	16,92	$5p^3\ ^4S^\circ$—136495	$^3/_2$—$^3/_2$, $^5/_2$
689,15	—	0,00	17,99	$5p^3\ ^4S^\circ$—145107	$^3/_2$—$^3/_2$
683,97	—	0,00	18,13	$5p^3\ ^4S^\circ$—146205	$^3/_2$—$^3/_2$, $^5/_2$
672,57	—	0,00	18,43	$5p^3\ ^4S^\circ$—148684	$^3/_2$—$^3/_2$, $^5/_2$
647,12	—	0,00	19,16	$5p^3\ ^4S^\circ$—154532	$^3/_2$—$^1/_2$
630,48	—	0,00	19,66	$5p^3\ ^4S^\circ$—158610	$^3/_2$—$^1/_2$
626,40	—	0,00	19,79	$5p^3\ ^4S^\circ$—159643	$^3/_2$—$^3/_2$, $^5/_2$
619,44	—	0,00	20,01	$5p^3\ ^4S^\circ$—161435	$^3/_2$—$^1/_2$
611,26	—	0,00	20,28	$5p^3\ ^4S^\circ$—163596	$^3/_2$—$^3/_2$
602,43	—	0,00	20,58	$5p^3\ ^4S^\circ$—165995	$^3/_2$—$^3/_2$
598,06	—	0,00	20,73	$5p^3\ ^4S^\circ$—167208	$^3/_2$—$^1/_2$
587,78	—	0,00	21,09	$5p^3\ ^4S^\circ$—170132	$^3/_2$—$^3/_2$, $^5/_2$
586,54	—	0,00	21,14	$5p^3\ ^4S^\circ$—170491	$^3/_2$—$^3/_2$, $^5/_2$
577,30	—	0,00	21,48	$5p^3\ ^4S^\circ$—173221	$^3/_2$—$^3/_2$
568,04	—	0,00	21,83	$5p^3\ ^4S^\circ$—176043	$^3/_2$—$^3/_2$, $^5/_2$
558,66	—	0,00	22,19	$5p^3\ ^4S^\circ$—179001	$^3/_2$—$^3/_2$, $^5/_2$

Note. No experimental data on the spectrum of Xe IV have yet been published. In the table we therefor give wavelengths calculated from energy levels [10].

Xe V, ground state $1s^2\,2s^2\,2p^6\,3s^2\,3p^6\,3d^{10}\,4s^2\,4p^6\,4d^{10}\,5s^2\,5p^2\ ^3P$

λ, Å	I	E_H, eV	E_B, eV	Transition	J
682,56	3	—	—	—	—

Xe VI, ground state $1s^2\,2s^2\,2p^6\,3s^2\,3p^6\,3d^{10}\,4s^2\,4p^6\,4d^{10}\,5s^2\,5p\ ^2P^\circ_{1/2}$

λ, Å	I	E_H, eV	E_B, eV	Transition	J
880,04	2	—	—	—	—
800,84	2	—	—	—	—
599,84	3	—	—	—	—

Xe VII, ground state $1s^2\,2s^2\,2p^6\,3s^2\,3p^6\,3d^{10}\,4s^2\,4p^6\,4d^{10}\,5s^2\ ^1S_0$

λ, Å	I	E_H, eV	E_B, eV	Transition	J
995,50	3	—	—	$5s^2\ ^1S$—$5p\ ^3P^\circ$	0—1
723,71	3	—	—	$5p\ ^3P^\circ$—$5p^2\ ^3P$	2—2
698,02	10	—	—	$5s^2\ ^1S$—$5p\ ^1P^\circ$	0—1
566,04	2	—	—	$5p\ ^3P^\circ$—$5d\ ^3D$	2—3
531,18	1	—	—	$5p\ ^3P^\circ$—$5d\ ^3D$	1—2

Xe VIII, ground state $1s^2 2s^2 2p^6 3s^2 3p^6 3d^{10} 4s^2 4p^6 4d^{10} 5s\ ^2S_{1/2}$

λ, Å	I	E_H, eV	E_B, eV	Transition	J
858,59	3	—	—	$5s\ ^2S—5p\ ^2P^\circ$	$^1/_2—^1/_2$
740,44	7	—	—	$5s\ ^2S—5p\ ^2P^\circ$	$^1/_2—^3/_2$
562,55	2	—	—	$5p\ ^2P^\circ—5d\ ^2D$	$^3/_2—^5/_2$
517,00	2	—	—	$5p\ ^2P^\circ—5d\ ^2D$	$^1/_2—^3/_2$

Unclassified Lines of Xenon Belonging to Xe I or Xe II [11]

λ, Å	I	λ, Å	I	λ, Å	I
1170,43	3	784,09	2	682,82	7
1152,19	7	774,53	4	680,58	2
1110,62	3	750,76	3	673,87	3
1085,47	2	726,99	2	665,09	4
959,22	2	718,89	3	643,19	2
842,35	5	716,09	2	640,36	2
839,73	2	690,40	1	639,99	2
793,53	4	686,73	1		

CESIUM, Z = 55

Cs I, ground state: $1s^2 2s^2 2p^6 3s^2 3p^6 3d^{10} 4s^2 4p^6 4d^{10} 5s^2 5p^6 6s\ ^2S_{1/2}$

Ionization potential 31 406,45 cm⁻¹; 3,894 eV

λ, Å	I	E_H, eV	E_B, eV	Transition	J
74250	10	3,35	3,52	$5g\ ^2G—6h\ ^2H°$	$^7/_2$, $^9/_2—^9/_2$, $^{11}/_2$
71930	13	3,01	3,19	$8s\ ^2S—8p\ ^2P°$	$^1/_2—^1/_2$
68070	15	3,01	3,20	$8s\ ^2S—8p\ ^2P°$	$^1/_2—^3/_2$
42202,3	4	2,72	3,01	$7p\ ^2P°—8s\ ^2S$	$^3/_2—^1/_2$
39398,5	30	3,03	3,35	$4f\ ^2F°—5g\ ^2G$	$^5/_2$, $^7/_2—^7/_2$, $^9/_2$
39180,1	10	2,70	3,01	$7p\ ^2P°—8s\ ^2S$	$^1/_2—^1/_2$
36127,7	20	1,45	1,80	$6p\ ^2P°—5d\ ^2D$	$^3/_2—^3/_2$
34900	25	1,45	1,81	$6p\ ^2P°—5d\ ^2D$	$^3/_2—^5/_2$
30952	22	2,30	2,70	$7s\ ^2S—7p\ ^2P°$	$^1/_2—^1/_2$
30102	30	1,38	1,80	$6p\ ^2P°—5d\ ^2D$	$^1/_2—^3/_2$
29308	3	2,30	2,72	$7s\ ^2S—7p\ ^2P°$	$^1/_2—^3/_2$
24373	9,5	2,72	3,23	$7p\ ^2P°—7d\ ^2D$	$^3/_2—^3/_2$
24248	80	2,72	3,23	$7p\ ^2P°—7d\ ^2D$	$^3/_2—^5/_2$
23340	50	2,70	3,23	$7p\ ^2P°—7d\ ^2D$	$^1/_2—^3/_2$
23032	15	2,80	3,34	$6d\ ^2D—5f\ ^2F°$	$^5/_2—^5/_2$, $^7/_2$
22949	0,2	3,03	3,57	$4f\ ^2F°—9d\ ^2D$	$^5/_2—^3/_2$
22909	0,6	3,03	3,57	$4f\ ^2F°—9d\ ^2D$	$^5/_2$, $^7/_2—^5/_2$
22810	10	2,80	3,34	$6d\ ^2D—5f\ ^2F°$	$^3/_2—^5/_2$
20140	6	2,72	3,34	$7p\ ^2P°—9s\ ^2S$	$^3/_2—^1/_2$
19809	1,3	2,80	3,43	$6d\ ^2D—9p\ ^2P°$	$^3/_2—^1/_2$
19800	1,7	2,80	3,43	$6d\ ^2D—9p\ ^2P°$	$^5/_2—^3/_2$
19624	0,2	2,80	3,43	$6d\ ^2D—9p\ ^2P°$	$^3/_2—^3/_2$
19430	3	2,70	3,34	$7p\ ^2P°—9s\ ^2S$	$^1/_2—^1/_2$
17549	4	2,80	3,51	$6d\ ^2D—6f\ ^2F°$	$^5/_2—^5/_2$, $^7/_2$
17417	4	2,80	3,51	$6d\ ^2D—6f\ ^2F°$	$^3/_2—^5/_2$
17046,8	—	2,72	3,45	$7p\ ^2P°—8d\ ^2D$	$^3/_2—^3/_2$
17015	9	2,72	3,45	$7p\ ^2P°—8d\ ^2D$	$^3/_2—^5/_2$
16540	4	2,70	3,45	$7p\ ^2P°—8d\ ^2D$	$^1/_2—^3/_2$
14694,93	1000	1,45	2,30	$6p\ ^2P°—7s\ ^2S$	$^3/_2—^1/_2$
13939	5	2,30	3,19	$7s\ ^2S—8p\ ^2P°$	$^1/_2—^1/_2$
13779	12	2,30	3,20	$7s\ ^2S—8p\ ^2P°$	$^1/_2—^3/_2$
13758,83	36	1,80	2,70	$5d\ ^2D—7p\ ^2P°$	$^3/_2—^1/_2$
13602,57	36	1,81	2,72	$5d\ ^2D—7p\ ^2P°$	$^5/_2—^3/_2$
13588,31	290	1,38	2,30	$6p\ ^2P°—7s\ ^2S$	$^1/_2—^1/_2$
13424,32	30	1,80	2,72	$5d\ ^2D—7p\ ^2P°$	$^3/_2—^3/_2$
10985	1,5	2,30	3,43	$7s\ ^2S—9p\ ^2P°$	$^1/_2—^1/_2$
10930	7	2,30	3,43	$7s\ ^2S—9p\ ^2P°$	$^1/_2—^3/_2$
10123,6025	1200	1,81	3,03	$5d\ ^2D—4f\ ^2F°$	$^5/_2—^7/_2$
10123,415	200	1,81	3,03	$5d\ ^2D—4f\ ^2F°$	$^5/_2—^5/_2$
10024,3595	1000	1,81	3,03	$5d\ ^2D—4f\ ^2F°$	$^5/_2—^3/_2$
9208,5382	200	1,45	2,80	$6p\ ^2P°—6d\ ^2D$	$^3/_2—^3/_2$
9172,3217	1000	1,45	2,80	$6p\ ^2P°—6d\ ^2D$	$^3/_2—^5/_2$
8943,483	2000	0,00	1,38	$6s\ ^2S—6p\ ^2P°$	$^1/_2—^1/_2$
8761,415	500	1,38	2,80	$6p\ ^2P°—6d\ ^2D$	$^1/_2—^3/_2$
8521,149	4000	0,00	1,45	$6s\ ^2S—6p\ ^2P°$	$^1/_2—^3/_2$
8079,0332	1000	1,81	3,34	$5d\ ^2D—5f\ ^2F°$	$^5/_2—^7/_2$
8078,923	100	1,81	3,34	$5d\ ^2D—5f\ ^2F°$	$^5/_2—^5/_2$
8053,35	100	1,81	3,35	$5d\ ^2D—5g\ ^2G$	$^5/_2—^7/_2$, $^9/_2$
8015,7235	200	1,80	3,34	$5d\ ^2D—5f\ ^2F°$	$^3/_2—^5/_2$
7990,68	100	1,80	3,35	$5d\ ^2D—5g\ ^2G$	$^3/_2—^7/_2$, $^9/_2$
7943,8820	800	1,45	3,01	$6p\ ^2P°—8s\ ^2S$	$^3/_2—^1/_2$
7608,9032	500	1,38	3,01	$6p\ ^2P°—8s\ ^2S$	$^1/_2—^1/_2$

λ, Å	I	E_H, eV	E_B, eV	Transition	J
7279,9570	500	1,81	3,51	$5d\,^2D-6f\,^2F^\circ$	$^5/_2-^7/_2$
7279,895	100	1,81	3,51	$5d\,^2D-6f\,^2F^\circ$	$^5/_2-^5/_2$
7270,70	15	1,81	3,51	$5d\,^2D-6g\,^2G$	$^5/_2-^7/_2,\ ^9/_2$
7228,5356	500	1,80	3,51	$5d\,^2D-6f\,^2F^\circ$	$^3/_2-^5/_2$
7219,70	15	1,80	3,51	$5d\,^2D-6g\,^2G$	$^3/_2-^7/_2,\ ^9/_2$
6983,4912	25	1,45	3,23	$6p\,^2P^\circ-7d\,^2D$	$^3/_2-^3/_2$
6973,2966	500	1,45	3,23	$6p\,^2P^\circ-7d\,^2D$	$^3/_2-^5/_2$
6895,005	—	0,00	1,80	$6s\,^2S-5d\,^2D$	$^1/_2-^3/_2$
6870,4552	200	1,81	3,61	$5d\,^2D-7f\,^2F^\circ$	$^5/_2-^7/_2$
6870,419	2	1,81	3,61	$5d\,^2D-7f\,^2F^\circ$	$^5/_2-^5/_2$
6848,906	—	0,00	1,81	$6s\,^2S-5d\,^2D$	$^1/_2-^5/_2$
6824,6520	200	1,80	3,61	$5d\,^2D-7f\,^2F^\circ$	$^3/_2-^5/_2$
6723,2943	500	1,38	3,23	$6p\,^2P^\circ-7d\,^2D$	$^1/_2-^3/_2$
6628,6605	35	1,81	3,68	$5d\,^2D-8f\,^2F^\circ$	$^5/_2-^7/_2$
6586,5096	500	1,45	3,34	$6p\,^2P^\circ-9s\,^2S$	$^3/_2-^1/_2$
6586,022	35	1,80	3,68	$5d\,^2D-8f\,^2F^\circ$	$^3/_2-^5/_2$
6472,6226	15	1,81	3,72	$5d\,^2D-9f\,^2F^\circ$	$^5/_2-^7/_2$
6431,9693	15	1,80	3,72	$5d\,^2D-9f\,^2F^\circ$	$^3/_2-^5/_2$
6365,5235	2	1,81	3,76	$5d\,^2D-10f\,^2F^\circ$	$^5/_2-^7/_2$
6354,5548	200	1,38	3,34	$6p\,^2P^\circ-9s\,^2S$	$^1/_2-^1/_2$
6326,2055	2	1,80	3,76	$5d\,^2D-10f\,^2F^\circ$	$^3/_2-^5/_2$
6288,5975	2	1,81	3,78	$5d\,^2D-11f\,^2F^\circ$	$^5/_2-^7/_2$
6250,2206	2	1,80	3,78	$5d\,^2D-11f\,^2F^\circ$	$^3/_2-^5/_2$
6231,31	1	1,81	3,80	$5d\,^2D-12f\,^2F^\circ$	$^5/_2-^7/_2$
6217,5986	15	1,45	3,45	$6p\,^2P^\circ-8d\,^2D$	$^3/_2-^3/_2$
6213,0998	100	1,45	3,45	$6p\,^2P^\circ-8d\,^2D$	$^3/_2-^5/_2$
6193,66	—	1,80	3,80	$5d\,^2D-12f\,^2F^\circ$	$^3/_2-^5/_2$
6187,54	—	1,81	3,81	$5d\,^2D-13f\,^2F^\circ$	$^5/_2-^7/_2$
6153,24	—	1,81	3,82	$5d\,^2D-14f\,^2F^\circ$	$^5/_2-^7/_2$
6150,38	—	1,80	3,81	$5d\,^2D-13f\,^2F^\circ$	$^3/_2-^5/_2$
6116,52	—	1,80	3,82	$5d\,^2D-14f\,^2F^\circ$	$^3/_2-^5/_2$
6034,0895	35	1,45	3,51	$6p\,^2P^\circ-10s\,^2S$	$^3/_2-^1/_2$
6010,4905	50	1,38	3,45	$6p\,^2P^\circ-8d\,^2D$	$^1/_2-^3/_2$
5847,64	—	1,45	3,57	$6p\,^2P^\circ-9d\,^2D$	$^3/_2-^3/_2$
5845,1410	30	1,45	3,57	$6p\,^2P^\circ-9d\,^2D$	$^3/_2-^5/_2$
5838,8347	—	1,38	3,51	$6p\,^2P^\circ-10s\,^2S$	$^1/_2-^1/_2$
5745,7244	—	1,45	3,61	$6p\,^2P^\circ-11s\,^2S$	$^3/_2-^1/_2$
5664,0183	15	1,38	3,57	$6p\,^2P^\circ-9d\,^2D$	$^1/_2-^3/_2$
5636,67	—	1,45	3,65	$6p\,^2P^\circ-10d\,^2D$	$^3/_2-^3/_2$
5635,2123	10	1,45	3,65	$6p\,^2P^\circ-10d\,^2D$	$^3/_2-^5/_2$
5573,6740	—	1,45	3,68	$6p\,^2P^\circ-12s\,^2S$	$^3/_2-^1/_2$
5568,4078	—	1,38	3,61	$6p\,^2P^\circ-11s\,^2S$	$^1/_2-^1/_2$
5503,8524	—	1,45	3,71	$6p\,^2P^\circ-11d\,^2D$	$^3/_2-^3/_2$
5502,8843	—	1,45	3,71	$6p\,^2P^\circ-11d\,^2D$	$^3/_2-^5/_2$
5465,9443	5	1,38	3,65	$6p\,^2P^\circ-10d\,^2D$	$^1/_2-^3/_2$
5461,9231	—	1,45	3,72	$6p\,^2P^\circ-13s\,^2S$	$^3/_2-^1/_2$
5414,28	—	1,45	3,74	$6p\,^2P^\circ-12d\,^2D$	$^3/_2-^3/_2$
5413,6145	—	1,45	3,74	$6p\,^2P^\circ-12d\,^2D$	$^3/_2-^5/_2$
5406,6672	—	1,38	3,68	$6p\,^2P^\circ-12s\,^2S$	$^1/_2-^1/_2$
5350,8	—	1,45	3,77	$6p\,^2P^\circ-13d\,^2D$	$^3/_2-^3/_2$
5350,3512	—	1,45	3,77	$6p\,^2P^\circ-13d\,^2D$	$^3/_2-^5/_2$
5340,9418	—	1,38	3,71	$6p\,^2P^\circ-11d\,^2D$	$^1/_2-^3/_2$
5303,7766	—	1,45	3,79	$6p\,^2P^\circ-14d\,^2D$	$^3/_2-^5/_2$
5301,40	—	1,38	3,72	$6p\,^2P^\circ-13s\,^2S$	$^1/_2-^1/_2$
5256,5633	—	1,38	3,74	$6p\,^2P^\circ-12d\,^2D$	$^1/_2-^3/_2$
5196,7343	—	1,38	3,77	$6p\,^2P^\circ-13d\,^2D$	$^1/_2-^3/_2$
5152,6813	—	1,38	3,79	$6p\,^2P^\circ-14d\,^2D$	$^1/_2-^3/_2$
4593,172	1000	0,00	2,70	$6s\,^2S-7p\,^2P^\circ$	$^1/_2-^1/_2$

λ, Å	I	E_H, eV	E_B, eV	Transition	J
4555,280	2000	0,00	2,72	$6s\ ^2S-7p\ ^2P^\circ$	$^1/_2-^3/_2$
4425,726	—	0,00	2,80	$6s\ ^2S-6d\ ^2D$	$^1/_2-^3/_2$
4417,344	—	0,00	2,80	$6s\ ^2S-6d\ ^2D$	$^1/_2-^5/_2$
3888,610	150	0,00	3,19	$6s\ ^2S-8p\ ^2P^\circ$	$^1/_2-^1/_2$
3876,146	300	0,00	3,20	$6s\ ^2S-8p\ ^2P^\circ$	$^1/_2-^3/_2$
3617,295	60	0,00	3,43	$6s\ ^2S-9p\ ^2P^\circ$	$^1/_2-^1/_2$
3611,459	200	0,00	3,43	$6s\ ^2S-9p\ ^2P^\circ$	$^1/_2-^3/_2$
3480,063	50	0,00	3,56	$6s\ ^2S-10p\ ^2P^\circ$	$^1/_2-^1/_2$
3476,814	100	0,00	3,56	$6s\ ^2S-10p\ ^2P^\circ$	$^1/_2-^3/_2$
3399,983	30	0,00	3,64	$6s\ ^2S-11p\ ^2P^\circ$	$^1/_2-^1/_2$
3397,969	60	0,00	3,65	$6s\ ^2S-11p\ ^2P^\circ$	$^1/_2-^3/_2$
3348,825	15	0,00	3,70	$6s\ ^2S-12p\ ^2P^\circ$	$^1/_2-^1/_2$
3347,494	30	0,00	3,70	$6s\ ^2S-12p\ ^2P^\circ$	$^1/_2-^3/_2$
3314,059	5	0,00	3,74	$6s\ ^2S-13p\ ^2P^\circ$	$^1/_2-^1/_2$
3313,124	10	0,00	3,74	$6s\ ^2S-13p\ ^2P^\circ$	$^1/_2-^3/_2$
3289,290	2	0,00	3,77	$6s\ ^2S-14p\ ^2P^\circ$	$^1/_2-^1/_2$
3288,605	4	0,00	3,77	$6s\ ^2S-14p\ ^2P^\circ$	$^1/_2-^3/_2$
3270,980	1	0,00	3,79	$6s\ ^2S-15p\ ^2P^\circ$	$^1/_2-^1/_2$
3270,477	2	0,00	3,79	$6s\ ^2S-15p\ ^2P^\circ$	$^1/_2-^3/_2$
3183,132	—	0,00	3,89	Limit of series	

Cs II, ground state: $1s^2\,2s^2\,2p^6\,3s^2\,3p^6\,3d^{10}\,4s^2\,4p^6\,4d^{10}\,5s^2\,5p^6\ ^1S_0$

Ionization potential 202 263 cm⁻¹; 25,076 eV

λ, Å	I	E_H, eV	E_B, eV	Transition	J
6955,519	20	14,10	15,88	$5d\ [2^1/_2]^\circ-6p\ [2^1/_2]$	3—2
6536,440	15	13,98	15,88	$5d\ [2^1/_2]^\circ-6p\ [2^1/_2]$	2—2
6506,254	5	17,77	19,68	$6p\ [^1/_2]-158717^\circ$	1—2
6495,528	15	14,10	16,01	$5d\ [2^1/_2]^\circ-6p\ [2^1/_2]$	3—3
6419,541	10	13,75	15,68	$5d\ [1^1/_2]^\circ-6p\ [^1/_2]$	1—1
6128,619	20	13,98	16,01	$5d\ [2^1/_2]^\circ-6p\ [2^1/_2]$	2—3
6076,738	2	16,51	18,55	$6p\ [^1/_2]-6d\ [^1/_2]^\circ$	0—1
5925,651	60	13,91	16,01	$5d\ [3^1/_2]^\circ-6p\ [2^1/_2]$	4—3
5863,701	5	14,10	16,21	$5d\ [2^1/_2]^\circ-6p\ [1^1/_2]$	3—2
5831,159	60	13,75	15,88	$5d\ [1^1/_2]^\circ-6p\ [2^1/_2]$	1—2
5814,181	25	13,98	16,12	$5d\ [2^1/_2]^\circ-6p\ [1^1/_2]$	2—1
5579,033	2	15,33	17,55	$5d'\ [1^1/_2]^\circ-6p'\ [1^1/_2]$	1—1
5563,019	125	13,98	16,21	$5d\ [2^1/_2]^\circ-6p\ [1^1/_2]$	2—2
5419,687	60	16,21	18,50	$6p\ [1^1/_2]-7s\ [1^1/_2]^\circ$	2—2
5407,35	2	—	—	—	—
5370,979	80	13,38	15,68	$6s\ [1^1/_2]^\circ-6p\ [^1/_2]$	1—1
5358,53	500	17,92	20,23	$6p'\ [^1/_2]-7s'\ [^1/_2]^\circ$	0—1
5349,16	25	15,23	17,55	$6s'\ [^1/_2]^\circ-6p'\ [1^1/_2]$	1—1
5306,609	25	16,21	18,55	$6p\ [1^1/_2]-6d\ [^1/_2]^\circ$	2—1
5274,044	40	13,34	15,68	$5d\ [^1/_2]^\circ-6p\ [^1/_2]$	0—1
5263,21	2	17,78	20,13	$6p'\ [1^1/_2]-162388^\circ$	2—3
5249,373	80	13,75	16,12	$5d\ [1^1/_2]^\circ-6p\ [1^1/_2]$	1—1
5227,002	200	13,31	15,68	$6s\ [1^1/_2]^\circ-6p\ [^1/_2]$	2—1
5209,62	15	—	—	—	—
5096,604	40	16,12	18,55	$6p\ [1^1/_2]-6d\ [^1/_2]^\circ$	1—1
5081,77	10	17,77	20,21	$6p'\ [^1/_2]-7s'\ [^1/_2]^\circ$	1—0
5080,10	5	—	—	—	—
5070,684	2	15,33	17,77	$5d'\ [1^1/_2]^\circ-6p'\ [^1/_2]$	1—1
5059,866	25	15,33	17,78	$5d'\ [1^1/_2]^\circ-6p'\ [1^1/_2]$	1—2

λ, Å	I	E_H, eV	E_B, eV	Transition	J
5052,696	25	13,43	15,88	$5d\,[1^1/_2]^\circ - 6p\,[2^1/_2]$	2—2
5043,800	80	13,75	16,21	$5d\,[1^1/_2]^\circ - 6p\,[1^1/_2]$	1—2
5041,828	5	17,77	20,23	$6p'\,[^1/_2] - 7s'\,[^1/_2]^\circ$	1—1
5012,979	5	17,92	20,39	$6p'\,[^1/_2] - 164465^\circ$	0—1
4972,593	25	16,01	18,50	$6p\,[2^1/_2] - 7s\,[1^1/_2]^\circ$	3—2
4952,835	30	13,38	15,88	$6s\,[1^1/_2]^\circ - 6p\,[2^1/_2]$	1—2
4943,01	10	—	—	—	—
4925,744	5	13,17	15,68	$5d\,[^1/_2]^\circ - 6p\,[^1/_2]$	1—1
4879,95	2	15,23	17,77	$6s'\,[^1/_2]^\circ - 6p'\,[^1/_2]$	1—1
4870,024	30	15,23	17,78	$6s'\,[^1/_2]^\circ - 6p'\,[1^1/_2]$	1—2
4830,161	30	13,31	15,88	$6s\,[1^1/_2]^\circ - 6p\,[2^1/_2]$	2—2
4806,924	5	17,55	20,13	$6p'\,[1^1/_2] - 162352^\circ$	1—0
4786,363	15	15,33	17,92	$5d'\,[1^1/_2]^\circ - 6p'\,[^1/_2]$	1—0
4763,616	25	—	—	—	—
4749,132	10	17,78	20,39	$6p'\,[1^1/_2] - 164444^\circ$	2—2
4744,60	5	17,78	20,39	$6p'\,[1^1/_2] - 164465^\circ$	2—1
4739,665	20	17,77	20,39	$6p'\,[^1/_2] - 164444^\circ$	1—2
4732,975	20	15,88	18,50	$6p\,[2^1/_2] - 7s\,[1^1/_2]^\circ$	2—2
4726,684	5	14,93	17,55	$5d'\,[2^1/_2]^\circ - 6p'\,[1^1/_2]$	2—1
4701,793	25	17,78	20,41	$6p'\,[1^1/_2] - 164656^\circ$	2—1
4695,610	10	17,92	20,56	$6p'\,[^1/_2] - 165813^\circ$	0—1
4692,482	5	17,77	20,41	$6p'\,[^1/_2] - 164656^\circ$	1—1
4670,280	20	16,21	18,86	$6p\,[1^1/_2] - 6d\,[1^1/_2]^\circ$	2—1
4656,538	12	17,55	20,21	$6p'\,[1^1/_2] - 7s'\,[^1/_2]^\circ$	1—0
4646,508	25	15,88	18,55	$6p\,[2^1/_2] - 6d\,[^1/_2]^\circ$	2—1
4640,333	5	—	—	—	—
4623,091	20	17,55	20,23	$6p'\,[1^1/_2] - 7s'\,[^1/_2]^\circ$	1—1
4620,59	10	—	—	—	—
4616,13	15	15,23	17,92	$6s'\,[^1/_2]^\circ - 6p'\,[^1/_2]$	1—0
4609,99	10	13,43	16,12	$5d\,[1^1/_2]^\circ - 6p\,[1^1/_2]$	2—1
4603,755	60	13,31	16,01	$6s\,[1^1/_2]^\circ - 6p\,[2^1/_2]$	2—3
4597,673	10	—	—	—	—
4571,786	15	13,17	15,88	$5d\,[^1/_2]^\circ - 6p\,[2^1/_2]$	1—2
4566,983	15	14,84	17,55	$5d'\,[1^1/_2]^\circ - 6p'\,[1^1/_2]$	2—1
4538,942	30	16,21	18,94	$6p\,[1^1/_2] - 6d\,[1^1/_2]^\circ$	2—2
4526,725	35	13,38	16,12	$6s\,[1^1/_2]^\circ - 6p\,[1^1/_2]$	1—1
4525,59	2	—	—	—	—
4522,846	15	—	—	—	—
4515,495	10	13,14	15,88	$5d\,[3^1/_2]^\circ - 6p\,[2^1/_2]$	3—2
4506,834	10	16,12	18,86	$6p\,[1^1/_2] - 6d\,[1^1/_2]^\circ$	1—1
4506,705	15	—	—	—	—
4501,525	35	13,75	16,51	$5d\,[1^1/_2]^\circ - 6p\,[^1/_2]$	1—0
4493,660	10	—	—	—	—
4459,185	15	17,78	20,56	$6p'\,[1^1/_2] - 165813^\circ$	2—1
4457,680	15	13,34	16,12	$5d\,[^1/_2]^\circ - 6p\,[1^1/_2]$	0—1
4453,44	15	17,78	20,56	$6p'\,[1^1/_2] - 165843^\circ$	2—1
4450,785	2	13,43	16,21	$5d\,[1^1/_2]^\circ - 6p\,[1^1/_2]$	2—2
4444,004	10	—	—	—°	—
4436,06	2	16,21	19,00	$6p\,[1^1/_2] - 6d\,[2^1/_2]^\circ$?	2—2
4435,708	20	—	—	—	—
4424,046	10	13,31	16,12	$6s\,[1^1/_2]^\circ - 6p\,[1^1/_2]$	2—1
4410,208	20	—	—	—	—
4405,253	35	15,68	18,50	$6p\,[^1/_2] - 7s\,[1^1/_2]^\circ$	1—2
4403,854	20	—	—	—	—
4399,495	20	17,78	20,59	$6p'\,[1^1/_2] - 166117^\circ$	2—3
4396,909	15	17,78	20,60	$6p'\,[1^1/_2] - 166131^\circ$	2—2
4388,764	10	17,77	20,60	$6p'\,[^1/_2] - 166131^\circ$	1—2
4386,566	2	16,21	19,04	$6p\,[1^1/_2] - 6d\,[2^1/_2]^\circ$?	2—3

λ, Å	I	E_H, eV	E_B, eV	Transition	J
4384,428	25	16,12	18,94	6p [1¹/₂]—6d [1¹/₂]°	1—2
4373,018	30	13,38	16,21	6s [1¹/₂]°—6p [1¹/₂]	1—2
4363,69	2	17,55	20,39	6p′ [1¹/₂]—164465°	1—1
4363,275	50	16,21	19,05	6p [1¹/₂]—6d [3¹/₂]°?	2—3
4356,575	2	14,93	17,77	5d′ [2¹/₂]°—6p′ [¹/₂]	2—1
4348,620	2	14,93	17,78	5d′ [2¹/₂]°—6p′ [1¹/₂]	2—2
4330,239	20	15,68	18,55	6p [¹/₂]—6d [¹/₂]°	1—1
4327,580	10	17,55	20,41	6p′ [1¹/₂]—164656°	1—1
4316,992	2	13,14	16,01	5d [3¹/₂]°—6p [2¹/₂]	3—3
4307,942	8	17,78	20,65	6p′ [1¹/₂]—166600°	2—3
4306,48	10	—	—	—	—
4300,636	30	16,51	19,39	6p [¹/₂]—156399°	0—1
4292,008	12	17,78	20,66	6p′ [1¹/₂]—166687°	2—2
4288,350	35	16,12	19,00	6p [1¹/₂]—6d [2¹/₂]°?	1—2
4284,229	2	17,77	20,66	6p′ [¹/₂]—166687°	1—2
4277,100	50	13,31	16,21	6s [1¹/₂]°—6p [1¹/₂]	2—2
4271,84	10	—	—	—	—
4241,973	10	17,78	20,70	6p′ [1¹/₂]—166961°	2—2
4234,408	20	17,77	20,70	6p′ [¹/₂]—166961°	1—2
4232,188	25	17,78	20,70	6p′ [1¹/₂]—167015°	2—3
4227,28	5	—	—	—	—
4221,119	15	16,01	18,94	6p [2¹/₂]—6d [1¹/₂]°	3—2
4220,571	2	14,84	17,77	5d′ [1¹/₂]°—6p′ [¹/₂]	2—1
4213,129	30	14,84	17,78	5d′ [1¹/₂]°—6p′ [1¹/₂]	2—2
4193,198	8	—	—	—	—
4186,249	5	—	—	—	—
4158,610	18	17,78	20,76	6p′ [1¹/₂]—167434°	2—1
4151,267	20	17,77	20,76	6p′ [¹/₂]—167434°	1—1
4132,003	10	16,01	19,00	6p [2¹/₂]—6d [2¹/₂]°?	3—2
4121,210	15	17,55	20,56	6p′ [1¹/₂]—165813°	1—1
4119,288	8	—	—	—	—
4108,232	5	—	—	—	—
4102,01	1	—	—	—	—
4073,364	8	13,17	16,21	5d [¹/₂]°—6p [1¹/₂]	1—2
4068,773	30	16,01	19,05	6p [2¹/₂]—6d [3¹/₂]°?	3—3
4067,958	30	17,55	20,60	6p′ [1¹/₂]—166131°	1—2
4053,956	15	17,92	20,97	6p′ [¹/₂]—169183°	0—1
4047,184	20	15,88	18,94	6p [2¹/₂]—6d [1¹/₂]°	2—2
4028,43	2	13,14	16,21	5d [3¹/₂]°—6p [1¹/₂]	3—2
3993,863	4	—	—	—	—
3978,000	10	17,55	20,66	6p′ [1¹/₂]—166687°	1—2
3967,212	4	—	—	—	—
3965,187	25	15,88	19,00	6p [2¹/₂]—6d [2¹/₂]°?	2—2
3959,495	20	13,38	16,51	6s [1¹/₂]°—6p [¹/₂]	1—0
3925,583	25	15,88	19,04	6p [2¹/₂]—6d [2¹/₂]°?	2—3
3906,933	20	15,88	19,05	6p [2¹/₂]—6d [3¹/₂]°?	2—3
3900,09	4	16,21	19,39	6p [1¹/₂]—156399°	2—1
3896,978	7	15,68	18,86	6p [¹/₂]—6d [1¹/₂]°	1—1
3870,164	4	17,77	20,97	6p′ [¹/₂]—169183°	1—1
3848,27	2	—	—	—	—
3805,412	2	—	—	—	—
3805,096	25	15,68	18,94	6p [¹/₂]—6d [1¹/₂]°	1—2
3785,424	20	16,12	19,39	6p [1¹/₂]—156399°	1—1
3751,402	4	—	—	—	—
3734,337	10	—	—	—	—
3732,539	4	15,68	19,00	6p [¹/₂]—6d [2¹/₂]°?	1—2
3699,20	10	—	—	—	—
3687,64	4	17,78	21,14	6p′ [1¹/₂]—170504°	2—2
3680,101	4	17,55	20,92	6p′ [1¹/₂]—168721°	1—0

λ, A	I	E_H, eV	E_B, eV	Transition	J
3651,073	4	—	—	—	—
3630,620	2	—	—	—	—
3618,549	2	17,55	20,97	6p′ [1½]—169183°	1—1
3576,570	2	16,21	19,68	6p [1½]—158717°	2—2
3566,11	2	17,55	21,02	6p′ [1½]—169588°	1—2
3565,111	10	—	—	—	—
3559,68	10	—	—	—	—
3531,376	4	15,88	19,39	6p [2½]—156399°	2—1
3475,973	2	13,98	17,55	5d [2½]°—6p′ [1½]	2—1
3459,185	15	—	—	—	—
3429,49	3	17,78	21,39	6p′ [1½]—172544°	2—3
3396,60	2	17,78	21,43	6p′ [1½]—172826°	2—3
3376,261	2	16,01	19,68	6p [2½]—158717°	3—2
3368,555	30	14,10	17,78	5d [2½]°—6p′ [1½]	3—2
3345,00	2	17,77	21,48	6p′ [½]—173244°	1—0
3329,428	10	16,51	20,23	6p [½]—7s′ [½]°	0—1
3271,626	20	13,98	17,77	5d [2½]°—6p′ [½]	2—1
3267,135	30	13,98	17,78	5d [2½]°—6p′ [1½]	2—2
3265,924	30	13,75	17,55	5d [1½]°—6p′ [1½]	1—1
3263,982	5	15,88	19,68	6p [2½]—158717°	2—2
3180,94	10	17,92	21,81	6p′ [½]—175951°	0—1
3173,355	5	16,51	20,41	6p [½]—164656°	0—1
3161,333	2	16,21	20,13	6p [1½]—162388°	2—3
3154,75	4	17,55	21,48	6p′ [1½]—173244°	1—0
3095,86	6	17,55	21,55	6p′ [1½]—173837°	1—1
3092,31	10	15,33	19,34	5d′ [1½]°—7p [½]	1—1
3089,053	5	16,12	20,13	6p [1½]—162352°	1—0
3084,875	5	13,75	17,77	5d [1½]°—6p′ [½]	1—1
3080,874	6	13,75	17,78	5d [1½]°—6p′ [1½]	1—2
3078,07	6	15,33	19,35	5d′ [1½]°—7p [2½]	1—2
3066,60	10	17,77	21,81	6p′ [½]—175951°	1—1
3060,976	5	16,51	20,56	6p [½]—165813°	0—1
3020,37	4	15,23	19,34	6s′ [½]°—7p [½]	1—1
3012,041	8	16,12	20,23	6p [1½]—7s′ [½]°	1—1
3006,75	—	15,23	19,35	6s′ [½]°—7p [2½]	1—2
3001,271	10	—	—	—	—
2990,85	2	15,33	19,47	5d′ [1½]°—7p [1½]	1—1
2977,258	3	13,75	17,92	5d [1½]°—6p′ [½]	1—0
2970,851	5	13,38	17,55	6s [1½]°—6p′ [1½]	1—1
2968,383	5	16,21	20,39	6p [1½]—164444°	2—2
2949,800	5	17,77	21,97	6p′ [½]—10s [1½]°	1—1
		15,23	19,43	6s′ [½]°—7p [1½]	1—2
2942,25	8	—	—	—	—
2940,953	20	13,34	17,55	5d [½]°—6p′ [1½]	0—1
2931,09	20	15,33	19,56	5d′ [1½]°—7p [½]	1—0
2926,274	1	13,31	17,55	6s [1½]°—6p′ [1½]	2—1
2914,652	8	15,88	20,13	6p [2½]—162388°	2—3
2899,75	8	16,12	20,39	6p [1½]—164465°	1—1
2883,745	5	16,12	20,41	6p [1½]—164656°	1—1
2881,19	15	15,17	19,47	6s′ [½]°—7p [1½]	0—1
2866,37	8	15,23	19,56	6s′ [½]°—7p [½]	1—0
2852,415	8	13,43	17,77	5d [1½]°—6p′ [½]	2—1
2848,955	3	13,43	17,78	5d [1½]°—6p′ [1½]	2—2
2847,655	1	—	—	—	—
2847,24	3	19,05	23,40	6d [3½]°?—188791	3—2
2846,193	10	—	—	—	—
2837,28	00	19,04	23,40	6d [2½]°?—188791	3—2
2829,423	5	13,17	17,55	5d [½]°—6p′ [1½]	1—1
2829,045	5	16,01	20,39	6p [2½]—164444°	3—2

λ, Å	I	E_H, eV	E_B, eV	Transition	J
2827,91	00	16,21	20,59	$6p\ [1^1/_2]—166117^\circ$	2—3
2826,802	1	16,21	20,60	$6p\ [1^1/_2]—166131^\circ$	2—2
2820,268	5	13,38	17,77	$6s\ [1^1/_2]^\circ—6p'\ [^1/_2]$	1—1
2816,943	20	19,00	23,40	$6d\ [2^1/_2]^\circ?—188791$	2—2
2799,41	10	14,93	19,35	$5d'\ [2^1/_2]^\circ—7p\ [2^1/_2]$	2—2
2794,50	10	—	—	—	—
2793,316	5	13,34	17,77	$5d\ [^1/_2]^\circ—6p'\ [^1/_2]$	0—1
2789,797	10	16,21	20,65	$6p\ [1^1/_2]—166600^\circ$	2—3
2788,24	10	16,12	20,56	$6p\ [1^1/_2]—165843^\circ$	1—1
2784,666	3	—	—	—	—
2780,065	3	13,31	17,77	$6s\ [1^1/_2]^\circ—6p'\ [^1/_2]$	2—1
2776,99	15	18,94	23,40	$6d\ [1^1/_2]^\circ—188791$	2—2
2766,095	5	16,12	20,60	$6p\ [1^1/_2]—166131^\circ$	1—2
2761,97	8	16,21	20,70	$6p\ [1^1/_2]—166961^\circ$	2—2
2757,81	7	16,21	20,70	$6p\ [1^1/_2]—167015^\circ$	2—3
2749,839	8	14,93	19,43	$5d'\ [2^1/_2]^\circ—7p\ [1^1/_2]$	2—2
2748,23	15	15,88	20,39	$6p\ [2^1/_2]—164465^\circ$	2—1
2740,73	15	—	—	—	—
2733,879	5	15,88	20,41	$6p\ [2^1/_2]—164656^\circ$	2—1
2730,065	5	18,86	23,40	$6d\ [1^1/_2]^\circ—188791$	1—2
2726,802	1	15,68	20,23	$6p\ [^1/_2]—7s'\ [^1/_2]^\circ$	1—1
2726,30	0	16,21	20,76	$6p\ [1^1/_2]—167434^\circ$	2—1
2724,21	10	16,12	20,66	$6p\ [1^1/_2]—166687^\circ$	1—2
2717,86	1	16,01	20,57	$6p\ [2^1/_2]—165890^\circ$	3—2
2703,95	3	16,12	20,70	$6p\ [1^1/_2]—166961^\circ$	1—2
2701,19	4	16,01	20,59	$6p\ [2^1/_2]—166117^\circ$	3—3
2689,412	5	13,17	17,78	$5d\ [^1/_2]^\circ—6p'\ [1^1/_2]$	1—2
2686,60	10	16,51	21,12	$6p\ [^1/_2]—170363^\circ$	0—1
2673,24	6	14,84	19,47	$5d'\ [1^1/_2]^\circ—7p\ [1^1/_2]$	2—1
2671,17	4	19,05	23,69	$6d\ [3^1/_2]^\circ?—191103$	3—3
2669,792	10	19,04	23,68	$6d\ [2^1/_2]^\circ?—191002$	3—2
		16,12	20,76	$6p\ [1^1/_2]—167434^\circ$	1—1
2666,358	1	16,01	20,65	$6p\ [2^1/_2]—166600^\circ$	3—3
2662,62	1	19,04	23,69	$6d\ [2^1/_2]^\circ?—191103$	3—3
2660,24	5	16,01	20,66	$6p\ [2^1/_2]—166687^\circ$	3—2
2651,71	12	19,00	23,68	$6d\ [2^1/_2]^\circ?—191002$	2—2
2648,07	10	15,88	20,56	$6p\ [2^1/_2]—165843^\circ$	2—1
2644,69	5	19,00	23,69	$6d\ [2^1/_2]^\circ?—191103$	2—3
		15,88	20,57	$6p\ [2^1/_2]—165890^\circ$	2—2
2640,92	3	16,01	20,70	$6p\ [2^1/_2]—166961^\circ$	3—2
2637,14	8	16,01	20,70	$6p\ [2^1/_2]—167015^\circ$	3—3
2635,882	1	15,68	20,39	$6p\ [^1/_2]—164444^\circ$	1—2
2628,86	2	15,88	20,59	$6p\ [2^1/_2]—166117^\circ$	2—3
2627,952	5	15,88	20,60	$6p\ [2^1/_2]—166131^\circ$	2—2
2616,27	10	18,94	23,68	$6d\ [1^1/_2]^\circ—191002$	2—2
2610,140	1	13,17	17,92	$5d\ [^1/_2]^\circ—6p'\ [^1/_2]$	1—0
2609,44	15	18,94	23,69	$6d\ [1^1/_2]^\circ—191103$	2—3
2595,886	3	15,88	20,65	$6p\ [2^1/_2]—166600^\circ$	2—3
2590,09	10	15,88	20,66	$6p\ [2^1/_2]—166687^\circ$	2—2
2581,05	0	16,12	20,92	$6p\ [1^1/_2]—168721^\circ$	1—0
2576,74	10	19,05	23,86	$6d\ [3^1/_2]^\circ?—192475$	3—2
2575,07	3	16,21	21,02	$6p\ [1^1/_2]—169588^\circ$	2—2
2574,54	10	18,86	23,68	$6d\ [1^1/_2]^\circ—191002$	1—2
2573,03	30	19,05	23,87	$6d\ [3^1/_2]^\circ?—192530$	3—3
2571,79	2	15,88	20,70	$6p\ [2^1/_2]—166961^\circ$	2—2
2568,69	15	19,04	23,86	$6d\ [2^1/_2]^\circ?—192475$	3—2
2568,17	10	15,88	20,70	$6p\ [2^1/_2]—167015^\circ$	2—3
2565,02	1	19,04	23,87	$6d\ [2^1/_2]^\circ?—192530$	3—3
2552,00	2	19,00	23,86	$6d\ [2^1/_2]^\circ?—192475$	2—2

λ, Å	I	$E_н$, eV	$E_в$, eV	Transition	J
2551,17	10	18,55	23,40	$6d\,[^{1}/_{2}]^{\circ}$—188791	1—2
2550,65	7	16,12	20,97	$6p\,[1^{1}/_{2}]$—169183°	1—1
2548,43	2	19,00	23,87	$6d\,[2^{1}/_{2}]^{\circ}$?—192530	2—3
2542,18	1	15,68	20,56	$6p\,[^{1}/_{2}]$—165843°	1—1
2540,83	3	15,88	20,76	$6p\,[2^{1}/_{2}]$—167434°	2—1
2539,174	5	—	—	—	—
2539,08	10	19,00	23,89	$6d\,[2^{1}/_{2}]^{\circ}$?—192675	2—1
2524,64	0	16,21	21,12	$6p\,[1^{1}/_{2}]$—170363°	2—1
2523,66	4	15,68	20,60	$6p\,[^{1}/_{2}]$—166131°	1—2
2519,17	00	18,94	23,86	$6d\,[1^{1}/_{2}]^{\circ}$—192475	2—2
2515,72	10	18,94	23,87	$6d\,[1^{1}/_{2}]^{\circ}$—192530	2—3
		16,21	21,14	$6p\,[1^{1}/_{2}]$—170504°	2—2
2506,53	1	18,94	23,89	$6d\,[1^{1}/_{2}]^{\circ}$—192675	2—1
2488,74	1	15,68	20,66	$6p\,[^{1}/_{2}]$—166687°	1—2
2480,41	6	18,86	23,86	$6d\,[1^{1}/_{2}]^{\circ}$—192475	1—2
2476,07	10	16,12	21,12	$6p\,[1^{1}/_{2}]$—170363°	1—1
2471,88	1	15,68	20,70	$6p\,[^{1}/_{2}]$—166961°	1—2
2469,58	0	16,01	21,02	$6p\,[2^{1}/_{2}]$—169588°	3—2
2468,12	3	18,86	23,89	$6d\,[1^{1}/_{2}]^{\circ}$—192675	1—1
2457,32	1	16,51	21,55	$6p\,[^{1}/_{2}]$—173837°	0—1
2443,24	5	15,68	20,76	$6p\,[^{1}/_{2}]$—167434°	1—1
2432,71	5	15,88	20,97	$6p\,[2^{1}/_{2}]$—169183°	2—1
2414,89	8	18,55	23,68	$6d\,[^{1}/_{2}]^{\circ}$—191002	1—2
		16,01	21,14	$6p\,[2^{1}/_{2}]$—170504°	3—2
2408,96	0	15,88	21,02	$6p\,[2^{1}/_{2}]$—169588°	2—2
2392,86	15	16,21	21,39	$6p\,[1^{1}/_{2}]$—172544°	2—3
2376,80	0	16,21	21,43	$6p\,[1^{1}/_{2}]$—172826°	2—3
2368,75	0	15,68	20,92	$6p\,[^{1}/_{2}]$—168721°	1—0
2364,81	10	15,88	21,12	$6p\,[2^{1}/_{2}]$—170363°	2—1
2357,85	5	14,10	19,35	$5d\,[2^{1}/_{2}]^{\circ}$—$7p\,[2^{1}/_{2}]$	3—2
2356,95	0	15,88	21,14	$6p\,[2^{1}/_{2}]$—170504°	2—2
2354,44	10	15,33	20,59	$5d'\,[1^{1}/_{2}]^{\circ}$—$7p'\,[1^{1}/_{2}]$	1—1
2343,13	8	15,68	20,97	$6p\,[^{1}/_{2}]$—169183°	1—1
2335,90	00	16,51	21,81	$6p\,[^{1}/_{2}]$—175951°	0—1
		14,10	19,40	$5d\,[2^{1}/_{2}]^{\circ}$—$7p\,[2^{1}/_{2}]$	3—3
2321,07	10	18,55	23,89	$6d\,[^{1}/_{2}]^{\circ}$—192675	1—1
		15,68	21,02	$6p\,[^{1}/_{2}]$—169588°	1—2
		16,21	21,55	$6p\,[1^{1}/_{2}]$—173837°	2—1
2315,68	6	13,98	19,34	$5d\,[2^{1}/_{2}]^{\circ}$—$7p\,[^{1}/_{2}]$	2—1
2312,47	0	15,23	20,59	$6s'\,[^{1}/_{2}]^{\circ}$—$7p'\,[1^{1}/_{2}]$	1—1
2311,16	0	16,12	21,48	$6p\,[1^{1}/_{2}]$—173244°	1—0
2307,71	5	13,98	19,35	$5d\,[2^{1}/_{2}]^{\circ}$—$7p\,[2^{1}/_{2}]$	2—2
2286,68	5	16,01	21,43	$6p\,[2^{1}/_{2}]$—172826°	3—3
		13,98	19,40	$5d\,[2^{1}/_{2}]^{\circ}$—$7p\,[2^{1}/_{2}]$	2—3
2281,50	00	15,33	20,76	$5d'\,[1^{1}/_{2}]^{\circ}$—$7p'\,[^{1}/_{2}]$	1—1
2280,02	2	15,68	21,12	$6p\,[^{1}/_{2}]$—170363°	1—1
2279,96	2	16,12	21,55	$6p\,[1^{1}/_{2}]$—173837°	1—1
2273,98	0	13,98	19,43	$5d\,[2^{1}/_{2}]^{\circ}$—$7p\,[1^{1}/_{2}]$	2—2
2273,83	20	19,05	24,50	$6d\,[3^{1}/_{2}]^{\circ}$?—197642	3—3
2272,76	0	15,68	21,14	$6p\,[^{1}/_{2}]$—170504°	1—2
2267,61	20	19,04	24,50	$6d\,[2^{1}/_{2}]^{\circ}$?—197642	3—3
2267,29	3	16,51	21,97	$6p\,[^{1}/_{2}]$—$10s\,[1^{1}/_{2}]^{\circ}$	0—1
2258,35	5	13,98	19,47	$5d\,[2^{1}/_{2}]^{\circ}$—$7p\,[1^{1}/_{2}]$	2—1
2257,82	12	13,91	19,40	$5d\,[3^{1}/_{2}]^{\circ}$—$7p\,[2^{1}/_{2}]$	4—3
2254,58	15	19,00	24,50	$6d\,[2^{1}/_{2}]^{\circ}$?—197642	2—3
2248,80	0	15,88	21,39	$6p\,[2^{1}/_{2}]$—172544°	2—3
2242,05	2	15,23	20,76	$6s'\,[^{1}/_{2}]^{\circ}$—$7p'\,[^{1}/_{2}]$	1—1
2234,57	0	15,88	21,43	$6p\,[2^{1}/_{2}]$—172826°	2—3
2228,88	10	18,94	24,50	$6d\,[1^{1}/_{2}]^{\circ}$—197642	2—3

λ, Å	I	E_H, eV	E_B, eV	Transition	J
2227,01	00	15,33	20,89	$5d'\ [1^1/_2]^\circ - 7p'\ [1^1/_2]$	1—2
2220,51	9	13,75	19,34	$5d\ [1^1/_2]^\circ - 7p\ [^1/_2]$	1—1
2217,91	3	15,17	20,76	$6s'\ [^1/_2]^\circ - 7p'\ [^1/_2]$	0—1
2213,15	5	13,75	19,35	$5d\ [1^1/_2]^\circ - 7p\ [2^1/_2]$	1—2
2212,40	0	16,21	21,81	$6p\ [1^1/_2] - 175951^\circ$	2—1
2189,47	10	15,23	20,89	$6s'\ [^1/_2]^\circ - 7p'\ [1^1/_2]$	1—2
2187,87	3	14,93	20,59	$5d'\ [2^1/_2]^\circ - 7p'\ [1^1/_2]$	2—1
2182,14	5	13,75	19,43	$5d\ [1^1/_2]^\circ - 7p\ [1^1/_2]$	1—2
2179,60	10	—	—	—	—
2177,61	3	—	—	—	—
2167,70	3	13,75	19,47	$5d\ [1^1/_2]^\circ - 7p\ [1^1/_2]$	1—1
2153,06	0	14,84	20,59	$5d'\ [1^1/_2]^\circ - 7p'\ [1^1/_2]$	2—1
2146,75	10	—	—	—	—
2139,48	2	15,68	21,48	$6p\ [^1/_2] - 173244^\circ$	1—0
2136,17	3	13,75	19,56	$5d\ [1^1/_2]^\circ - 7p\ [^1/_2]$	1—0
2133,77	0	19,04	24,84	$6d\ [2^1/_2]^\circ ? - 200406$	3—3
2124,80	0	14,93	20,76	$5d'\ [2^1/_2]^\circ - 7p'\ [^1/_2]$	2—1
2122,27	1	19,00	24,84	$6d\ [2^1/_2]^\circ ? - 200406$	2—3
2115,55	0	16,12	21,97	$6p\ [1^1/_2] - 10s\ [1^1/_2]^\circ$	1—1
2112,65	5	15,68	21,55	$6p\ [^1/_2] - 173837^\circ$	1—1
2099,50	4	18,94	24,84	$6d\ [1^1/_2]^\circ - 200406$	2—3
2097,52	0	13,43	19,34	$5d\ [1^1/_2]^\circ - 7p\ [^1/_2]$	2—1
2091,97	8	14,84	20,76	$5d'\ [1^1/_2]^\circ - 7p'\ [^1/_2]$	2—1
2088,71	8	15,88	21,81	$6p\ [2^1/_2] - 175951^\circ$	2—1
2080,05	8	13,38	19,34	$6s\ [1^1/_2]^\circ - 7p\ [^1/_2]$	1—1
2077,43	8	14,93	20,89	$5d'\ [2^1/_2]^\circ - 7p'\ [1^1/_2]$	2—2
2073,60	1	13,43	19,40	$5d\ [1^1/_2]^\circ - 7p\ [2^1/_2]$	2—3
2063,13	00	13,43	19,43	$5d\ [1^1/_2]^\circ - 7p\ [1^1/_2]$	2—2
2058,10	3	13,31	19,34	$6s\ [1^1/_2]^\circ - 7p\ [^1/_2]$	2—1
2051,75	0	13,31	19,35	$6s\ [1^1/_2]^\circ - 7p\ [2^1/_2]$	2—2
2046,25	00	13,38	19,43	$6s\ [1^1/_2]^\circ - 7p\ [1^1/_2]$	1—2
2046,00	00	14,84	20,89	$5d'\ [1^1/_2]^\circ - 7p'\ [1^1/_2]$	2—2
2035,15	7	13,31	19,40	$6s\ [1^1/_2]^\circ - 7p\ [2^1/_2]$	2—3
2033,78	0	15,88	21,97	$6p\ [2^1/_2] - 10s\ [1^1/_2]^\circ$	2—1
2025,05	5	13,31	19,43	$6s\ [1^1/_2]^\circ - 7p\ [1^1/_2]$	2—2
2022,29	0	15,68	21,81	$6p\ [^1/_2] - 175951^\circ$	1—1
2005,83	00	13,38	19,56	$6s\ [1^1/_2]^\circ - 7p\ [^1/_2]$	1—0
1971,57	3	15,68	21,97	$6p\ [^1/_2] - 10s\ [1^1/_2]^\circ$	1—1
1968,21	2	13,14	19,43	$5d\ [3^1/_2]^\circ - 7p\ [1^1/_2]$	3—2
1966,91	2	13,17	19,47	$5d\ [^1/_2]^\circ - 7p\ [1^1/_2]$	1—1
1501,3	5	—	—	—	—
1191,55	8	14,10	24,50	$5d\ [2^1/_2]^\circ - 197642$	3—3
1178,65	10	13,98	24,50	$5d\ [2^1/_2]^\circ - 197642$	2—3
926,75	20	0,00	13,38	$5p^6\ {}^1S - 6s\ [1^1/_2]^\circ$	0—1
901,34	20	0,00	13,75	$5p^6\ {}^1S - 5d\ [1^1/_2]^\circ$	0—1
813,85	20	0,00	15,23	$5p^6\ {}^1S - 6s'\ [^1/_2]^\circ$	0—1
808,77	20	0,00	15,33	$5p^6\ {}^1S - 5d'\ [1^1/_2]^\circ$	0—1
668,43	12	0,00	18,55	$5p^6\ {}^1S - 6d\ [^1/_2]^\circ$	0—1
657,15	5	0,00	18,86	$5p^6\ {}^1S - 6d\ [1^1/_2]^\circ$	0—1
639,42	12	0,00	19,39	$5p^6\ {}^1S - 156399^\circ$	0—1
612,82	7	0,00	20,23	$5p^6\ {}^1S - 7s'\ [^1/_2]^\circ$	0—1
607,98	1	0,00	20,39	$5p^6\ {}^1S - 164465^\circ$	0—1
607,31	3	0,00	20,41	$5p^6\ {}^1S - 164656^\circ$	0—1
602,95	1	0,00	20,56	$5p^6\ {}^1S - 165843^\circ$	0—1
591,08	3	0,00	20,97	$5p^6\ {}^1S - 169183^\circ$	0—1
575,34	1	0,00	21,55	$5p^6\ {}^1S - 173837^\circ$	0—1
564,25	3	0,00	21,97	$5p^6\ {}^1S - 10s\ [1^1/_2]^\circ$	0—1

Cs III, ground state $1s^2\,2s^2\,2p^6\,3s^2\,3p^6\,3d^{10}\,4s^2\,4p^6\,4d^{10}\,5s^2\,5p^5\;{}^2P^{\circ}_{3/2}$

Ionization potential 279 000 cm^{-1}*; 34,6 eV*

λ, Å	I	E_H, eV	E_B, eV	Transition	J
877,9	7	1,72	15,84	$5p^5\,{}^2P^{\circ}$—$5p^6\,{}^2S$	1/2—1/2
817,9	1	1,72	16,88	$5p^5\,{}^2P^{\circ}$—136146	1/2— —
802,8	00	1,72	17,16	$5p^5\,{}^2P^{\circ}$—138453	1/2— —
782,6	3	0,00	15,84	$5p^5\,{}^2P^{\circ}$—$5p^6\,{}^2S$	3/2—1/2
758,9	1	1,72	18,06	$5p^5\,{}^2P^{\circ}$—145655	1/2— —
734,5	1	0,00	16,88	$5p^5\,{}^2P^{\circ}$—136146	3/2— —
722,2	2	0,00	17,16	$5p^5\,{}^2P^{\circ}$—138453	3/2— —
708,4	1	1,72	19,22	$5p^5\,{}^2P^{\circ}$—155040	1/2— —
686,5	0	0,00	18,06	$5p^5\,{}^2P^{\circ}$—145655	3/2— —
649,4	0	1,72	20,81	$5p^5\,{}^2P^{\circ}$—167872	1/2— —
645,0	4	0,00	19,22	$5p^5\,{}^2P^{\circ}$—155040	3/2— —
595,7	2	{ 0,00	20,81	$5p^5\,{}^2P^{\circ}$—167872	3/2— —
		{ 1,72	22,53	$5p^5\,{}^2P^{\circ}$—181758	1/2— —
592,9	0	1,72	22,63	$5p^5\,{}^2P^{\circ}$—182546	1/2— —
571,7	1	1,72	23,40	$5p^5\,{}^2P^{\circ}$—188794	1/2— —
550,2	2	0,00	22,53	$5p^5\,{}^2P^{\circ}$—181758	3/2— —
547,8	2	0,00	22,63	$5p^5\,{}^2P^{\circ}$—182546	3/2— —
529,7	0	0,00	23,40	$5p^5\,{}^2P^{\circ}$—188794	3/2— —

*Calculated data, see [17].

Unclassified Lines of Cesium Belonging to Cs II or Cs III [11, 12, 14, 16]

λ, Å	I	λ, Å	I	λ, Å	I
69310 Cs I?	15	4864,24	10	4447,649	10
7248,99	2	4851,583	8	4440,26	15
7229,01	35	4835,03	15	4397,994	10
7205,99	2	4825,42	10	4368,77	10
7188,32	2	4804,61	10	4367,66	10
7160,88	2	4768,41	10	4359,02	10
7149,554	10	4758,92	10	4335,411	8
7130,532	5	4757,87	10	4326,315	10
7121,18	2	4733,06	20	4312,778	10
7085,72	2	4728,18	10	4297,514	10
6979,681	15	4716,19	10	4282,59	10
6892,42	2	4674,89	10	4281,31	10
6825,22	15	4651,1	10	4272,87	10
6724,476	15	4616,28	15	4268,89	10
6646,564	15	4616,01	10	4264,675	50
6386,94	25	4610,505	10	4219,516	5
5984,393	15	4599,22	15	4173,533	15
5566,7	40	4572,611	10	4163,243	15
5507,174	15	4543,71	10	4081,471	10
5402,793	40	4534,64	10	4043,422	20
5349,31	15	4532,500	10	4039,841	50
5348,95	25	4531,45	15	4035,83	15
5209,44	15	4522,36	15	4025,67	10
5081,773	15	4496,758	15	4023,582	10
5001,641	2	4469,09	2	4014,99	10

λ, Å	I	λ, Å	I	λ, Å	I
4010,54	10	3450,36	6	3130,7	4
4006,772	10	3443,88	4	3129,1	4
4006,537	30	3430,4	4	3125,3	4
4001,682	20	3418,11	6	3118,35	4
3974,239	6	3411,313	10	3112,18	4
3955,923	10	3406,626	10	3109,3	4
3929,46	2	3397,187	6	3097,38	10
3921,69	4	3393,25	3	3094,82	4
3913,37	2	3389,15	6	3091,6	4
3904,806	4	3379,0	6	3088,9	4
3900,82	4	3364,52	4	3072,7	4
3893,09	4	3358,8	4	3069,73	4
3892,206	4	3357,687	6	3067,8	4
3864,367	4	3353,88	4	3067,2	4
3864,249	6	3349,445	10	3063,7	4
3861,489	4	3344,004	10	3062,7	4
3837,449	4	3340,574	10	3061,24	6
3819,61	4	3330,33	4	3060,12	6
3797,908	4	3324,5	4	3058,6	6
3729,980	4	3322,8	4	3056,04	6
3724,9	4	3315,498	10	3054,56	4
3710,774	4	3311,52	4	3054,20	4
3699,475	10	3303,72	4	3053,5	4
3680,454	4	3299,86	6	3050,8	6
3661,391	6	3282,1	3	3045,9	4
3655,73	4	3278,26	4	3042,3	4
3641,40	4	3275,68	4	3039,31	4
3641,332	5	3268,314	10	3032,41	4
3634,75	6	3263,06	4	3031,5	4
3624,56	4	3262,29	6	3030,35	4
3622,691	6	3255,35	10	3029,15	4
3618,161	6	3250,58	6	3028,25	4
3614,989	4	3247,5	4	3020,9	4
3608,285	10	3242,28	10	3015,8	4
3605,535	4	3234,16	6	3002,88	6
3602,852	8	3227,2	4	2999,513	8
3600,73	10	3219,1	4	2998,20	2
3598,97	4	3213,7	6	2997,2	2
3597,73	6	3209,65	10	2996,15	2
3597,430	10	3207,07	4	2995,34	20
3592,48	4	3204,27	4	2986,89	2
3581,3	4	3201,09	4	2985,3	2
3573,24	4	3198,7	4	2983,91	2
3569,28	4	3195,5	4	2982,5	2
3541,45	4	3193,6	4	2982,03	2
3533,364	6	3192,1	4	2976,81	2
3518,15	6	3189,20	4	2975,65	2
3516,03	4	3178,61	10	2975,13	2
3514,022	6	3172,56	10	2972,8	2
3504,85	4	3169,73	4	2969,0	8
3503,67	4	3153,88	6	2965,4	2
3479,25	4	3152,30	6	2965,0	8
3470,92	4	3151,14	6	2962,8	2
3469,81	4	3149,36	10	2962,4	2
3465,20	4	3145,2	4	2951,59	2
3463,425	6	3141,46	4	2947,85	2
3457,18	4	3134,8	4	2944,1	2
3455,48	4			2938,5	20

λ, Å	I	λ,	I	λ, Å	I
2924,48	2	2721,6	2	2483,0	2
2922,21	2	2719,0	2	2480,7	2
2921,83	2	2715,8	2	2477,58	20
2921,03	20	2714,0	2	2466,8	2
2915,24	2	2711,6	2	2466,3	2
2910,82	2	2710,5	2	2462,0	2
2906,17	2	2709,0	2	2459,23	2
2901,1	2	2706,79	20	2455,80	8
2895,32	2	2705,3	2	2443,2	2
2894,85	2	2704,1	2	2439,8	2
2893,81	2	2700,30	8	2437,1	2
2891,75	2	2699,16	8	2432,6	8
2886,67	20	2691,83	2	2427,65	20
2884,42	8	2681,99	8	2426,41	8
2879,25	8	2681,34	8	2425,15	20
2877,29	8	2678,92	20	2422,9	2
2875,30	8	2677,01	2	2421,4	2
2872,35	8	2674,62	2	2420,06	2
2871,32	2	2674,0	2	2415,0	2
2868,33	8	2668,76	8	2414,77	2
2866,90	2	2658,71	2	2411,98	2
2865,45	2	2656,83	2	2406,9	2
2862,40	8	2650,7	20	2401,7	2
2860,85	8	2646,20	8	2396,86	2
2859,32	20	2642,63	20	2394,92	2
2857,83	2	2641,0	2	2393,6	2
2854,45	8	2637,6	2	2390,02	4
2851,23	20	2634,17	2	2387,26	3
2850,4	2	2627,84	8	2380,15	3
2845,67	20	2621,1	2	2379,54	2
2844,48	2	2619,22	2	2375,82	2
2841,5	2	2614,62	8	2373,4	2
2838,09	20	2613,6	2	2364,827	2
2835,01	8	2612,1	2	2359,23	2
2824,12	8	2605,40	20	2357,9	2
2823,03	8	2603,72	2	2356,12	2
2819,28	2	2600,36	20	2354,42	2
2817,98	20	2598,7	2	2351,911	2
2815,33	2	2596,95	20	2351,74	8
2810,82	20	2591,17	20	2344,38	4
2809,91	8	2589,3	2	2340,47	2
2792,16	8	2582,5	2	2337,88	4
2788,81	2	2560,37	8	2332,42	8
2787,02	8	2554,8	2	2321,1	6
2784,10	8	2546,1	2	2284,60	2
2780,81	2	2543,92	20	2272,79	2
2779,9	8	2538,67	2	2256,10	3
2779,1	8	2533,44	20	2246,56	2
2774,46	8	2528,8	8	2233,28	3
2769,5	2	2525,68	20	2229,12	5
2764,42	20	2520,8	2	2211,30	2
2755,20	20	2512,1	2	2209,61	3
2751,11	2	2511,51	8	2200,68	8
2734,85	2	2502,2	2	2197,15	6
2731,8	2	2496,9	2	2187,88	3
2727,80	2	2495,04	20	2186,31	3
2723,95	8	2489,5	2	2182,13	5
		2485,42	20		

λ, Å	I	λ, Å	I	λ, Å	I
2163,54	8	1981,5	1	1910,2	0
2141,30	10	1976,8	1	1908,1	1
2131,85	10	1974,5	1	1900,7	1
2127,69	6	1962,3	0	1897,7	0
2109,11	6	1961,4	2	1897,1	0
2101,49	8	1955,3	1	1896,8	0
2091,83	8	1945,1	1	1889,2	6
2083,75	8	1942,3	4	1884,0	6
2083,01	7	1941,2	0	1877,6	1
2077,14	8	1938,8	2	1873,2	2
2076,29	8	1937,4	1	1861,3	0
2058,47	3	1936,0	0	1859,3	2
2058,00	1	1935,2	8	1840,6	2
2057,61	3	1930,9	1	1675,5	1
2056,30	4	1925,0	0	1673,2	1
1996,5	5	1923,4	2	1669,5	1
1991,9	1	1920,0	1		
1990,1	1	1919,3	1		
1985,1	1	1915,6	3		
1983,7	1	1914,6	4		

Section IV

Summary Table of Spectral Lines Arranged According to Wavelength

λ	Symbol	I
190569	H	—
123684	H	3
113057	H	—
90850	Na I	40
90480	Na I	30
85100	K I	10
84520	K I	10
75004,5	H	—
74260	K I	10
74577,6	H	20
74430	Na I	10
74250	Cs I	10
71930	Cs I	13
69310	Cs I?	15
68070	Cs I	15
64610	K I	10
64310	K I	10
62360	K I	20
62030	K I	20
46524,7	H	3
42202,3	Cs I	4
40511,4	H	120
40449	Na I	80
40115,5	K I	60
39398,5	Cs I	30
39180,1	Cs I	10
37370,7	K I	10
37354,3	K I	40
37075,6	K I	30
36626,4	K I	30
36372,7	K I	10
36127,7	Cs I	20
34900	Cs I	25
31596,8	K I	40
31395	K I	80
30952	Cs I	22
30102	Cs I	30
29308	Cs I	3
27215,0	K I	10
27065,6	K I	20
26877,82	Li I	8
26511,1	Xe I	30
26392,9	Mg I	5
26272,0	Xe I	60
26251,3	H	40
25854,38	Si I	6
25842,20	C I	1
25833,66	C I	1
25706,03	C I	1
25697,56	C I	1
25660,9	Ar I	65
25504,4	Ar I	35
25323,66	Cl I	6
25233,78	Kr I	70
25125,08	Ar I	23
25047,31	Cl I	6
24935,6	Ne I	7
24825,3	Xe I	20
24777,7	Ne I	15

λ	Symbol	I
24470,02	Cl I	100
24464,66	Li I	6
24458,7	Ne I	36
24448,5	Ne I	20
24373	Cs I	9,5
24366,4	Ne I	95
24292,17	Kr I	38
24260,45	Kr I	28
24248,9	Ne I	32
24248	Cs I	80
24161,5	Ne I	25
24146,23	Cl I	4
24097,8	Ne I	11
23978,4	Ne I	68
23966,68	Ar I	30
23956,2	Ne I	47
23956,10	Cl I	11
23951,3	Ne I	119
23882,69	Cl I	18
23845,13	Ar I	56
23709,4	Ne I	62
23636,3	Ne I	205
23565,6	Ne I	40
23502,37	Kr I	17
23379,13	Na I	240
23372,1	Ne I	62
23348,41	Na I	237
23340,44	Kr I	65
23340	Cs I	50
23260,7	Ne I	45
23195,5	Xe I	10
23188,84	Cl I	35
23133,22	Ar I	35
23101,0	Ne I	62
23038,78	Cl I	17
23032	Cs I	15
22949	Cs I	0,2
22909	Cs I	0,6
22906,56	C I	1
22891,90	Cl I	4
22810	Cs I	10
22721,74	Cl I	5
22688,70	Cl I	12
22662,5	Ne I	15
22651,30	Ca I	30
22625,51	Ca I	20
22608,39	Ca I	10
22529,7	Ne I	105
22522,80	Cl I	6
22485,79	Kr I	38
22468,4	Ne I	8
22428,2	Ne I	15
22288,52	Cl I	9
22245,3	Ne I	12
22113,2	Ar I	10
22083,67	Na I	276
22077,20	Ar I	53

λ	Symbol	I	λ	Symbol	I
22062,71	Si I	1	19727,33	Al I	18
22056,44	Na I	300	19722,50	Si I	110
22039,57	Ar I	9	19721,99	C I	23
22026,68	Cl I	40	19624	Cs I	0,2
21902,34	Cl I	14	19574,0	Ne I	10
21900,51	Kr I	2250	19543,13	He I	65
21879,35	Si I	8	19508,13	Si I	14
21830,38	Cl I	10	19506,12	Si I	5
21819,69	Si I	5	19505,62	Ca I	500
21779,77	Si I	9	19493,38	Si I	13
21707,4	Ne I	25	19452,82	Ca I	1500
21655,2	H	13	19443,27	Cl I	6
21582,40	Cl I	12	19432,97	Si I	48
21534,16	Ar I	58	19430	Cs I	3
21432,11	Mg II	5	19385,94	Si I	15
21368,91	Mg II	7	19370,30	Cl I	227
21354,24	Si I	21	19309,43	Ca I	500
21333,27	Ar I	15	19283,29	Si I	8
21259,89	C I	8	19274,78	Li I	4
21211,55	C I	2	19114,83	Ca I	5
21191,41	C I	4	19089,37	He I	550
21165,46	Kr I	319	19045,86	Ca I	90
21163,75	Al I	13	19021,39	Ca I	4
21132,04	He I	40	19030,79	Si I	5
21121,31	He I	150	18971,55	Cl I	21
21120,04	He I	150	18969,71	Ca I	60
21093,04	Al I	12	18926,54	C I	3
21040,9	Ne I	27	18924,96	Ca I	30
20986,10	Ar I	155	18914,48	Si I	8
20917,13	Si I	12	18844,42	C I	2
20811,14	Ar I	22	18797,59	Kr I	40
20733,35	Ar I	11	18788,0	Xe I	3
20725,44	Cl I	56	18787,73	Kr I	10
20647,17	Ar I	16	18785,45	Kr I	37
20616,21	Ar I	356	18751,1	H	700
20581,30	He I	10000	18751,01	N I	2
20568,5	Ar I	8	18746,0	D	700
20423,97	Kr I	142	18744,3	T	700
			18742,79	Cl I	22
20419,00	Kr I	1	18722,90	Si I	26
20370,12	Cl I	85	18703,09	Li I	7
20350,6	Ne I	10	18703,09	He I	1500
20316,82	Ar I	23	18695,91	Kr I	62
20262,2	Xe I	6	18685,96	He I	3600
20209,87	Kr I	84	18670,00	N I	4
20199,36	Cl I	227	18658,16	N	32
20140	Cs I	6	18632,17	Ar I	13
20069,6	Ar I	7	18630,19	N I	13
20025,90	Ar I	7	18624,94	Ne I	20
19965,75	Ar I	37	18622,68	Mg II	25
19961,37	Ca I	40	18618,69	Ne I	15
19944,8	Ar I	7	18597,30	Ne I	120
19932,94	Ca I	100	18591,12	Ne I	25
19928,88	Si I	31	18587,24	N I	13
19916,34	Ca I	50	18581,19	Kr I	30
19861,70	Ca I	500	18574,80	Mg II	20
19852,96	Ca I	250	18570,53	Ar I	8
19817,54	Ar I	75	18566,75	N I	4
19815,14	Ca I	30	18564,74	Ar I	8
19809	Cs I	1,3	18555,55	He I	6
19800	Cs I	1,7	18541,37	Cl I	74
19776,67	Ca I	2000	18475,79	Ne I	3
19766,78	Cl I	185	18465,25	Na I	2
19755,28	Cl I	717	18458,58	Ne I	10

λ	Symbol	I
18429,27	Ar I	40
18427,68	Ar I	26
18422,72	Si I	7
18422,43	Ne I	110
18418,82	Kr I	4
18417,91	Ar I	27
18403,16	Ne I	60
18390,10	Ne I	180
18385,17	Ne I	160
18382,27	Cl I	40
18359,21	Ne I	6
18320,67	C I	8
18304,00	Ne I	140
18282,58	Ne I	200
18276,59	Ne I	260
18251,58	N I	11
18243,63	O I	22
18240,54	N I	13
18229,66	N I	60
18229	Cu I	5
18226,57	Ne I	10
18221,12	C I	8
18220,76	Ne I	15
18210,56	N I	32
18199,13	N I	8
18194	Cu I	7
18184,43	Kr I	15
18171,60	N I	13
18169,74	N I	13
18167,12	Kr I	1500
18139,80	C I	13
18116,27	N I	6
18108,61	N I	12
18098,46	Kr I	10
18097,71	N I	10
18082,71	Ne I	130
18049,56	N I	33
18034,86	C I	5
18035,49	Ne I	20
18030,47	C I	2
18029,95	N I	30
18021,21	O I	23
18001,71	Kr I	400
17979,89	N I	51
17966,12	C I	2
17959,24	C I	3
17936,55	N I	17
17925,70	N I	8
17918,38	C I	4
17918,06	N I	7
17914,43	Ar I	10
17887,35	Ar I	15
17878,26	N I	100
17852,09	N I	10
17842,70	Kr I	270
17826,33	C I	4
17825,76	Cl I	5
17823,90	Ar I	61
17814,03	C I	3
17787,27	N I	8
17770,21	Kr I	4
17768,94	C I	3
17767,65	Cl I	7
17717,72	Mg II	15
17699,09	Al I	13
17643,98	N I	42
17637,38	C I	3
17636,83	N I	8
17630,44	Kr I	4
17617,00	Si I	9
17616,57	Kr I	37
17586,44	Cl I	60
17584,86	N I	100
17551,6	Li I	20
17549	Cs I	4
17546,05	Li I	7
17531,99	N I	18
17519,72	Cl I	4
17516,58	N I	125
17505,64	C I	3
17480,41	N I	27
17474,16	N I	32
17455,97	C I	2
17448,60	C I	11
17444,93	Ar I	128
17443,93	Cl I	46
17436,22	N I	24
17429,23	N I	16
17417	Cs I	4
17404,67	Kr I	32
17385,13	N I	12
17367,98	Kr I	360
17367,55	N I	23
17338,56	C I	10
17327,29	Si I	28
17326,86	N I	16
17325,5	Xe I	5
17323,51	C I	2
17291,81	N I	6
17282,04	N I	4
17274,99	C I	3
17269,17	N I	11
17234,48	C I	2
17230,21	Kr I	10
17226,30	Cl I	27
17219,55	N I	10
17119,13	Cl I	28
17108,66	Mg I	30
17098,76	Kr I	300
17070,04	Kr I	10
17046,8	Cs I	—
17015	Cs I	9
17003,15	He I	200
17002,38	He I	1800
16994,36	Kr I	10
16941,45	Cl I	10
16940,39	Ar I	100
16935,71	Kr I	800
16896,58	Kr I	700
16890,40	Kr I	1000
16871,76	Cl I	18
16853,45	Kr I	480
16813,82	Cl I	14
16800,73	Cl I	10
16784,65	Kr I	950
16763,36	Al I	9
16759,73	Cl I	6
16750,56	Al I	12
16739,84	Ar I	5
16727,52	Xe I	50

λ	Symbol	I
16726,48	Kr I	70
16718,96	Al I	11
16680,77	Si I	29
16671,38	Cl I	55
16653	Cu I	4
16624,76	Cl I	4
16573,10	Kr I	16
16549,81	Ar I	6
16540	Cs I	4
16520,14	Ar I	9
16465,29	Kr I	15
16436,92	Ar I	18
16388,85	Na I	27
16385,70	Cl I	7
16381,55	Si I	16
16380,12	Si I	8
16373,85	Na I	30
16347,31	Kr I	5
16315,58	Kr I	12
16293,39	Cl I	15
16286,18	Cl I	39
16284,18	Cl I	7
16241,84	Si I	7
16215,68	Si I	11
16214,99	Cl I	10
16202,94	Ca I	10
16198,47	Cl I	259
16195,33	Ca I	150
16189,88	Cl I	14
16179,12	Cl I	10
16163,71	Si I	60
16156,04	Ca I	100
16149,79	Ca I	70
16135,80	Ca I	20
16122,97	Ar I	27
16109,46	Kr I	3
16094,80	Si I	20
16077,62	Cl I	129
16067,35	Cl I	10
16060,41	Cl I	10
16060,03	Si I	95
16052,31	Kr I	2
16052,02	Xe I	50
16024,95	Cl I	25
16021,64	C I	3
16008	Cu I	5
16004,81	C I	2
15989,34	Ar I	20
15970,49	Cl I	283
15960,04	Si I	40
15959,97	Cl I	735
15928,92	Cl I	342
15925,64	Kr I	6
15899,93	Ar I	20
15890,52	Kr I	25
15888,39	Si I	190
15884,41	Si I	5
15883,34	Cl I	277
15883,21	Ar I	50
15869,63	Cl I	2780
15833,58	Si I	7
15823,40	Kr I	2
15820,10	Kr I	35
15818,41	Cl I	193
15816,64	Ar I	16
15808,54	Cl I	25
15792,00	Cl I	21
15771,44	Kr I	1
15771,10	N I	22
15765,84	Mg I	10
15748,99	Mg I	8
15740,71	Mg I	6
15730,06	Cl I	1487
15717,70	Cl I	4
15688,86	N I	54
15680,94	Kr I	75
15668,64	Cl I	7
15634,98	Kr I	7
15615,16	Cl I	7
15608,08	Cl I	18
15582,27	N I	200
15580,66	Cl I	5
15557,81	Si I	7
15520,29	Cl I	1094
15496,13	N I	34
15477,78	Cl I	15
15474,02	Kr I	65
15467,59	Cl I	169
15465,07	Cl I	381
15435,14	Cl I	27
15433,63	Kr I	4
15418,01	Xe I	110
15416,07	Cl I	32
15402,58	Ar I	10
15382,31	Cl I	17
15373,88	Cl I	23
15371,89	Kr I	350
15353,51	Ar I	2
15351,42	Cl I	2
15349,52	Ar I	10
15335,29	Kr I	850
15329,56	Ar I	5
15326,87	Kr I	35
15320,46	Cl I	7
15309,08	Cl I	28
15302,26	Ar I	75
15296,83	Cl I	8
15262,98	Cl I	150
15239,85	Kr I	900
15234,4	Ne I	2
15225,72	Cl I	13
15209,52	Kr I	42
15203,46	Cl I	15
15199,65	Cl I	22
15183,97	Cl I	8
15181,94	Cl I	5
15172,33	Ar I	22

λ	Symbol	I
15168,40	K I	16
15163,08	K I	—
15161,15	Cl I	145
15146,66	N I	75
15108,04	Cl I	269
15102,29	NI	26
15094,96	NI	75
15094,12	Cl I	48
15083,66	He I	60
15051,60	Cl I	4
15050,88	N I	80
15047,70	Mg I	25
15046,42	Ar I	70
15040,24	Mg I	30
15030,71	Ar I	42
15024,99	Mg I	35
15005,57	Kr I	25
14987,69	Cl I	29
14983,51	Cl I	95
14973,74	Kr I	8
14966,60	NI	180
14965,4	N I	25
14961,76	Kr I	110
14955,33	Cl I	78
14952,07	NI	15
14947,73	Cl I	43
14938,14	Cl I	108
14931,70	Cl I	294
14924,95	Cl I	7
14918,68	Cl I	6
14901,33	Cl I	10
14892,33	Cl I	3
14877,62	Mg I	28
14868,87	N I	100
14863,53	Cl I	5
14806,75	Cl I	82
14798,50	Cl I	5
14792,29	Cl I	50
14786,29	Ar I	2
14782,98	CI	4
14779,73	Na I	36
14767,48	Na I	1155
14765,64	Kr I	230
14762,83	Kr I	250
14757,07	N I	300
14739,11	Ar I	3
14734,46	Kr I	900
14732,38	Xe I	200
14731,37	Cl I	45
14715,55	Kr I	2
14694,93	Cs I	1000
14692,39	Ar I	5
14682,82	Cl I	6
14681,04	N I	55
14659,84	Xe I	5
14649,97	Ar I	60
14637,03	CI	2
14634,11	Ar I	80
14626,35	Cl I	9
14604.64	NI	27
14598,42	N I	17
14596,27	Ar I	40
14577,51	Ar I	12
14576,78	Cl I	3
14556,68	Cl I	25
14548,55	N I	10
14542,50	CI	179
14529,13	Cl I	4
14522,81	N I	36
14508,63	Cl I	16
14497,41	Cl I	60
14469,33	Kr I	30
14454,62	NI	29
14450,44	Cl I	95
14442,24	CI	13
14436,26	Cl I	13
14426,93	Kr I	1100
14429,03	CI	12
14420,12	CI	61
14403,25	CI	16
14402,58	Kr I	80
14401,35	Kr I	30
14399,65	CI	38
14384,93	Cl I	4
14369,71	Cl I	148
14364,90	Xe I	20
14347,82	Kr I	400
14341,25	Kr I	9
14313,21	N I	80
14297,53	Cl I	2
14292,07	Cl I	73
14257,46	Ar I	50
14255,80	Cl I	3
14249,93	Ar I	7
14241,39	Xe I	40
14224,54	Si I	6
14221,48	Cl I	2
14214,95	Cl I	5
14198,27	Cl I	48
14173,84	Cl I	11
14156,62	Kr I	15
14142,09	Xe I	80
14129,80	Cl I	14
14122,44	Cl I	4
14104,27	Kr I	40
14093,61	Ar I	120
13992,59	Ar I	10
13983,32	Cl I	4
13978,14	Cl I	120
13974,15	Kr I	70
13961,45	Cl I	19
13956,82	Cl I	2
13939,13	Kr I	85
13939	Cs I	5
13932,97	Cl I	15
13924,00	Kr I	270
13923,92	Cl I	20
13911,08	Cl I	2
13910,83	Ar I	150
13907,41	Ar I	12
13893,10	Cl I	110
13885,14	Cl I	7
13882,64	Kr I	240
13866,97	Ar I	20
13863,31	Cl I	13
13837,58	Cl I	125

λ	Symbol	I
13832,57	Kr I	50
13828,79	Ar I	20
13827,67	Cl I	9
13825,99	Ar I	30
13821,72	Cl I	525
13802,82	Cl I	11
13800,03	Kr I	3
13779	Cs I	12
13772,48	Cl I	50
13765,29	C I	1
13763,72	Kr I	6
13758,83	Cs I	36
13743,93	C I	3
13741,86	C I	1
13738,86	Kr I	400
13718,77	Ar I	1000
13711,36	Si I	5
13711,23	Kr I	100
13710,06	Cl I	2
13706,12	Cl I	5
13705,41	C I	1
13697,81	C I	6
13693,85	Si I	8
13686,03	NI	14
13678,53	Ar I	300
13668,60	N I	65
13658,38	Kr I	360
13656,48	Xe I	150
13651,63	NI	60
13649,3	N I	10
13634,22	Kr I	1700
13624,18	N I	350
13622,38	Ar I	500
13622,28	Kr I	800
13615,56	NI	35
13602,57	Cs I	36
13602,16	Cl I	11
13602,27	N I	190
13599,18	Ar I	55
13588,55	NI	155
13588,31	Cs I	290
13587,73	N I	200
13586,00	Cl I	6
13581,35	C I	5
13581,33	N I	1200
13578,45	Cl I	28
13573,60	Ar I	25
13559,66	C I	12
13557,75	Li I	4
13544,61	NI	65
13543,75	Ar I	15
13534,64	N I	60
13543,16	Xe I	5
13503,99	Ar I	850
13502,27	C I	20
13499,24	Ar I	50
13498,30	Cl I	160
13498,30	Cl I	160
13469,98	Cl I	9
13465,13	Cl I	2
13464,53	N I	185
13448,12	NI	21
13429,61	N I	670
13424,32	Cs I	30
13419,89	Cl I	90
13406,57	Ar I	250
13397,09	K I	—
13396,04	Cl I	95
13382,46	Cl I	30
13378,04	Cl I	33
13377,86	K I	—
13367,38	Ar I	800
13346,76	Cl I	550
13337,52	Kr I	55
13330,32	Ar I	7
13313,39	Ar I	600
13309,04	Si I	5
13304,30	Kr I	5
13302,37	Ar I	3
13296,01	Cl I	310
13287,58	Si I	9
13273,05	Ar I	750
13243,83	Cl I	350
13240,52	Kr I	75
13231,37	Ar I	120
13228,49	Ar I	200
13214,70	Ar I	150
13213,42	Cl I	7
13210,56	Kr I	10
13208,29	Cl I	20
13182,58	Cl I	8
13177,38	Kr I	850
13176,90	Si I	11
13168,90	Cl I	13
13167,75	Ca I	24
13165,11	O I	24
13164,85	O I	26
13163,89	O I	25
13150,76	Al I	14
13134,96	Ca I	400
13129,66	Cl I	100
13123,41	Al I	15
13122,59	Cl I	16
13107,98	Cl I	4
13095,13	Cl I	49
13086,26	Ca I	50
13062,73	Cl I	5
13061,84	Ca I	8
13059,70	Cl I	4
13057,82	Ca I	20
13040,99	Cl I	125
13034,59	Cl I	9
13033,41	Ca I	300
13028,27	Ar I	5
13022,05	Kr I	15
13008,47	Ar I	200
13001,37	Ca I	20
12985,08	Kr I	12
12977,98	Kr I	2
12976,77	Cl I	20
12968,44	He I	50
12956,59	Ar I	250
12934,48	Kr I	1
12933,33	Ar I	60
12912,4	Ne I	2
12909,07	Ca I	200
12908,57	Cl I	24
12897,32	N I	51

λ	Symbol	I	λ	Symbol	I
12885,21	Ca I	15	12464,2	N I	5
12879,00	Kr I	500	12464,02	O I	21
12872,10	Cl I	39	12461,25	N I	680
12869,80	Cl I	4	12459,49	Ne I	2
12861,89	Kr I	55	12456,05	Ar I	400
12859,16	Cl I	13	12451,21	Xe I	2
12845,95	He I	30	12439,19	Ar I	500
12826,60	Ca I	25	12438,40	N I	195
12825,08	Kr I	5	12432,24	K I	16
12823,46	Ca I	100	12430,13	Cl I	12
12818,05	H	140	12428,81	N I	6
12815,69	Ca I	400	12419,39	Ar I	20
12814,56	D	140	12404,27	N I	98
12813,40	T	140	12402,88	Ar I	400
12803,05	Cl I	37	12395,82	Si I	6
12802,68	Ar I	300	12391,9	N I	5
12795,90	Cl I	12	12389,03	Cl I	4
12793,31	Li I	5	12384,83	N I	12
12790,27	He I	125	12381,65	N I	375
12784,79	He I	400	12356,82	Ar I	100
12782,39	Kr I	100	12343,72	Ar I	150
12778,5	N I	5	12322,76	N I	350
12771,51	N I	15	12321,48	Kr I	9
12757,26	Al I	4	12298,55	N I	120
12749,83	Al I	21	12288,97	N I	260
12747,65	Al I	2	12280,55	Cl I	16
12746,31	Ar I	40	12270,80	N I	20
12733,59	Ar I	75	12270,68	Si I	120
12730,68	N I	35	12261,28	N I	27
12708,89	N I	30	12258,10	Xe I	6
12702,39	Ar I	150	12250,11	N I	11
12690,1	Ne I	2	12240,81	Kr I	2
12679,17	Na I	83	12236,26	Cl I	5
12662,16	N I	27	12235,14	Xe I	80
12661,75	Cl I	47	12231,32	N I	75
12639,01	Ar I	2	12237,67	Li I	4
12623,32	Xe I	300	12231,13	Cl I	16
12621,82	Ar I	6	12229,52	Cl I	4
12621,35	Cl I	47	12229,23	Kr I	4
12614,10	C I	26	12210,17	N I	12
12601,48	C I	8	12204,39	Kr I	700
12598,19	Kr I	15	12203,93	N I	150
12596,27	Ar I	5	12186,82	N I	480
12594,8	Ne I	1	12179,72	Cl I	77
12594,15	Cl I	142	12172,95	Cl I	60
12590,00	Xe I	26	12156,97	Kr I	2
12585,09	Cl I	10	12151,57	Ar I	15
12581,59	C I	6	12142,16	N I	12
12581,00	N I	27	12139,79	Ar I	100
12578,8	N I	3	12129,97	N I	170
12575,99	N I	8	12124,60	N I	35
12570,04	O I	20	12123,47	Kr I	40
12569,04	C I	5	12117,81	Kr I	100
12564,4	N I	4	12110,96	Cl I	60
12563,58	Cl I	38	12112,20	Ar I	300
12562,12	C I	6	12109,30	N I	25
12557,66	N I	14	12106,59	N I	45
12554,44	Ar I	5	12103,50	Si I	5
12549,48	C I	5	12084,82	Xe I	20
12527,51	He I	100	12083,66	Mg I	30
12522,11	K I	15	12077,42	Kr I	115
12502,69	Cl I	63	12074,51	N I	230
12487,63	Ar I	700	12066,38	Ne I	15
12469,62	N I	1350	12031,48	Si I	10

λ	Symbol	I	λ	Symbol	I
12026,63	Ar I	5	11658,85	C I	13
12021,67	Cl I	172	11655,8	Kr I	1
11998,36	N I	110	11669,63	C I	24
11997,08	Kr I	480	11651,45	N I	2
11996,00	Kr I	25	11647,99	C I	5
11991,52	Si I	5	11638,279	Fe I	7
11984,99	Ne I	10	11636,22	Cl I	4
11984,18	Si I	10	11628,83	C I	23
11973,88	Ti I	6	11626,40	He II	—
11973,067	Fe I	8	11625,173	N I	3
11969,48	He I	—	11620,14	Mg II	3
11969,07	He I	220	11619,29	C I	12
11952,57	Xe I	10	11614,18	Ne I	80
11951,1	Xe I	1	11614,08	Xe I	25
11949,72	Ca II	1	11611,6	Kr I	1
11949,58	Ti I	5	11601,62	Ne I	25
11943,50	Ar I	25	11600,56	Mg II	3
11911,44	Xe I	3	11598,74	Cl I	5
11896,60	Ar I	3	11593,600	Fe I	5
11895,75	C I	30	11580,39	Ar I	8
11892,91	C I	17	11579,91	Cl I	9
11892,85	Ti I	5	11577,24	Cl I	11
11884,47	Ar I	5	11573,48	Cl I	2
11882,861	Fe I	7	11566,114	N I	4
11879,59	C I	8	11557,17	F I	5
11874,36	Xe I	1	11544,65	F I	2
11866,50	Cl I	195	11539,50	Ti I	5
11862,99	C I	5	11537,4	Xe I	1
11857,86	Xe I	2	11536,41	Ne I	50
11857,00	Xe I	30	11525,11	Ne I	90
11848,73	C I	6	11522,82	Ne I	150
11838,99	Ca II	2	11491,22	Xe I	15
11828,18	Mg I	45	11488,12	Ar I	150
11819,43	Kr I	2000	11480,22	F I	1
11801,08	C I	7	11473,70	F I	3
11797,24	Ti I	3	11467,57	Ar I	30
11793,04	Xe I	40	11457,52	Kr I	80
11792,25	Kr I	120	11441,83	Ar I	80
11789,93	Ne I	10	11439,129	Fe I	15
11789,11	Ne I	50	11436,34	Cl I	1000
11783,275	Fe I	6	11422,335	Fe I	6
11780,54	Ti I	4	11415,04	Xe I	15
11777,54	C I	11	11414,20	F I	1,5
11772,83	K I	17	11409,68	Cl I	269
11769,62	K I	16	11409,24	Ne I	100
11766,87	Ne I	60	11403,89	Ti I	8
11754,76	C I	114	11403,78	Na I	12
11753,32	C I	142	11398,63	Ar I	7
11748,22	C I	82	11393,66	Ar I	50
11742,01	Xe I	90	11392,66	Cl I	231
11733,26	Ar I	20	11390,53	Ne I	1
11720,55	Cl I	180	11381,53	Ti I	7
11719,51	Ar I	30	11381,45	Na I	11
11708,22	Ar I	3	11378,01	Cl I	45
11692,73	Cl I	85	11373,93	Cl I	5
11690,21	K I	17	11366,80	Ne I	3
11689,988	Fe I	8	11339,44	Kr I	1
11688,08	Ne I	10	11333,60	Ne I	3
11687,61	Ar I	5			
11678,47	Ar I	4			
11674,14	C I	7			
11652,91	C I	5			
11668,72	Ar I	100			
11659,68	C I	47			

λ	Symbol	I	λ	Symbol	I
11331,08	Cl I	5	11287,022	O I	21
11330,285	C I	6	11286,914	O I	24
11329,56	Ne I	1	11286,39	Cl I	9
11328,51	Kr I	4	11286,344	O I	23
11326,53	Cl I	6	11266,198	N I	3
11323,169	N I	3	11262,71	Kr I	2
11316,1	Kr I	1	11259,16	Kr I	50
11313,891	N I	4	11257,74	Kr I	80
11309,56	Xe I	5	11256,35	Mg II	4
11306,70	Cl I	3	11255,93	Mg II	5
11304,47	Ne I	2	11254,881	Al I	15
11303,96	Ne I	5	11253,496	Ar II	1
11303,8	Kr I	1	11253,190	Al I	14
11302,376	O I	23	11248,33	Ar I	8
11302,26	Cl I	4	11246,88	Ti I	8
11298,45	Ne I	1	11243,90	Ti I	10
11297,682	O I	22	11237,582	N I	2
11295,104	O I	21	11230,91	Ti I	5
11294,238	N I	2	11227,076	N I	3
11293,00	Ne I	2	11225,90	He I	—
11292,43	Ti I	6	11214,89	Xe I	5
11291,657	N I	5	11214,58	Kr I	5
11289,83	Si I	15	11209,67	Ar I	1
11289,10	Xe I	10	11197,21	Na I	2
11287,318	O I	21	11195,37	Ar I	2

λ	Symbol	I
11190,19	Na I	1
11187,588	Si I	16
11187,13	Kr I	40
11180,114	N I	1
11177,59	Ne I	300
11175,5	Xe I	1
11173,266	Ar II	2
11162,67	Xe I	10
11160,29	Ne I	10
11151,25	Cl I	6
11143,09	Ne I	300
11141,09	Xe I	50
11138,55	Ne I	4
11134,62	Ne I	4
11133,86	Ar I	20
11130,81	Xe I	8
11130,03	Si I	12
11127,20	Xe I	100
11122,97	Cl I	300
11120,37	Ne I	5
11119,809	Fe I	10
11118,75	Ar I	20
11118,2	Cu I	1
11106,44	Ar I	60
11096,70	Cl I	56
11095,79	Ti I	5
11093,76	Cl I	6
11093,04	Cl I	6
11085,25	Xe I	250
11082,93	Cl I	206
11078,87	Ar I	200
11072,10	Cl I	3
11068,44	Ar II	1
11067,929	Ar II	2
11063,58	Cl I	2
11060,88	Ne I	2
11057,58	Ti I	3
11055,22	Cl I	2
11049,80	Ne I	20
11045,00	He I	8
11044,06	Ne I	15
11043,13	Ar I	2
11033,661	Mg I	14
11032,103	Mg I	15
11032,09	Li I	1
11028,60	Ar I	1
11022,67	K I	16
11020,93	Ne I	10
11019,87	K I	17
11017,9648	Si I	80
11014,52	Cl I	3
11013,07	He I	8
10996,56	He I	3
10990,70	Ti I	3
10986,71	Cl I	13
10985	Cs I	1,5
10984,527	Si I	20
10982,382	Ar II	2
10982,061	Si I	30
10979,308	Si I	80
10977,30	Ar I	1
10976,06	LiI	0
10974,33	Ar II	1
10973,80	Ar II	2
10965,450	Mg I	28
10964,00	Ar I	2
10957,304	Mg I	27
10954,260	Ar II	2
10953,320	Mg I	25
10951	Mg II	10
10950,74	Ar I	120
10947,90	Ar I	20
10945,43	Cl I	5
10940,37	F I	4
10938,09	H	28
10935,11	D	28
10934,12	T	28
10930	Cs I	7
10924,81	F I	2,5
10923,438	Ar II	7
10919,07	LiI	3
10916,98	He I	25
10916,67	Ar II	1
10915,27	Mg II	7
10914,23	Mg II	10
10912,92	He I	60
10902,16	He I	1
10896,10	Ti I	8
10895,9	Ar I	1
10895,32	Xe I	200
10892,37	Ar I	30
10891,733	Al I	11
10888,53	Ne I	8
10886,35	Ne I	5
10885,9	Ar I	2
10885,336	Si I	30
10884,60	N I	2
10883,3	Cu I	1
10883,28	F I	2,5
10882,802	Si I	30
10880,96	Ar I	150
10879,78	Ca I	4
10879,19	N I	1
10874,92	Kr I	100
10872,975	Al I	10
10869,698	Ar II	2
10869,5408	Si I	130
10869,37	Ca I	3
10868,79	Si I	30
10867,87	Ar II	1
10867,343	Ar II	3
10863,72	Ca I	2
10863,60	Fe I	5
10862,31	F I	20
10861,51	Ca I	3
10861,04	Ar I	25
10845,43	Ar I	2
10844,54	Ne I	200

λ	Symbol	I
10843,854	Si I	60
10841,55	Cl I	100
10838,77	Ca I	10
10838,34	Xe I	1000
10838,30	Ne I	3
10837,39	Ar I	1
10834,87	Na I	8
10833,66	Ti I	3
10833,12	Ca I	4
10831,88	Ar I	1
10831,68	Cl I	9
10830,337	He I	25000
10830,33	Ne I	4
10830,248	He I	15000
10829,452	Ar II	3
10829,088	He I	5000
10827,091	Si I	140
10824,00	Ar I	1
10822,74	Ar I	1
10822,20	Cl I	3
10820,31	Ti I	5
10820,18	Ar I	6
10819,95	Ne I	5
10817,858	Ar II	1
10817,35	Ti I	5
10814,83	Ne I	4
10812,901	Ar II	12
10811,085	Mg I	35
10812,16	Ar I	1
10808,22	Ne I	7
10807,04	Ar I	5
10806,43	Ne I	5
10798,12	Ne I	150
10796,06	Si I	7
10795,91	Ar I	2
10793,65	Ti I	3
10789,37	Ne I	2
10786,8560	Si I	80
10786,770	Al I	4
10785,13	Ar II	1
10785,12	Cl I	7
10784,5597	Si I	30
10782,045	Al I	9
10781,34	Ti I	3
10780,57	Ne I	6
10774,993	N I	3
10774,92	Ti I	12
10773,35	Ar I	30
10771,7	Cu I	2
10770,35	Ar I	15
10769,43	F I	4
10768,364	Al I	8
10766,15	Ne I	10
10764,378	Ar II	8
10764,09	Ne I	12
10760,34	Ne I	1
10759,13	Ar I	60
10758,86	Xe I	100
10758,28	Ne I	2
10757,888	N I	7
10756,90	Ti I	5
10753,985	C I	2
10753,530	O I	17
10749,3837	Si I	60
10749,29	Na I	9
10746,44	Na I	10
10741,77	Ti I	7
10733,87	Ar I	50
10732,89	Ti I	8
10732,10	Ar I	2
10731,11	Ti I	6
10730,510	N I	4
10729,533	C I	6
10729,43	Kr I	2
10727,4076	Si I	30
10726,33	Ti I	18
10722,22	Ar I	6
10720,530	Ar II	1
10717,954	N I	6
10717,84	Cl I	2
10713,550	N I	8
10712,77	Ar I	40
10707,333	C I	6
10706,78	Xe I	150
10700,98	Ar I	80
10699,33	Kr I	20
10694,2510	Si I	30
10693,167	N I	3
10691,250	C I	10
10690,94	Cl I	14
10690,48	Ne I	6
10689,719	Si I	25
10689,52	Ti I	15
10685,345	C I	6
10683,40	Ar I	50
10683,082	C I	8
10683,050	Ar II	12
10681,99	Cl I	7
10681,78	Ar I	200
10677,04	Ti I	10
10675,940	O I	16
10675,725	O I	17
10673,80	Ne I	2
10673,55	Ar I	500
10667,65	He I	15
10661,61	Ti I	20
10660,99	Ar II	2
10660,9748	Si I	120
10659,5	Kr II	1
10653,034	N I	8
10647,63	Kr I	1
10643,981	N I	6
10639,86	Ar II	1
10639,34	Kr II	6

λ	Symbol	I
10638,121	Ar II	8
10634,25	Ar I	5
10627,6467	Si I	20
10626,70	Kr I	8
10623,38	Ar I	2
10623,177	N I	5
10620,63	Ne I	40
10620,37	Cl I	7
10619,458	Ar II	7
10615,7	Ar I	1
10614,01	Ar II	1
10608,43	Kr I	20
10607,78	Ti I	10
10603,431	Si I	120
10600,53	Cl I	18
10596,958	N I	6
10593,01	Kr I	100
10592,28	F I	2
10591,905	N I	5
10591,23	Ar I	2
10588,71	F I	5
10585,1412	Si I	120
10584,66	Ti I	25
10580,83	Ar II	2
10576,18	Ar I	4
10575,50	Kr I	2
10572,28	Na I	3
10566,00	Na I	1
10565,97	Ti I	5
10563,339	N I	5
10562,84	Kr II	4
10562,43	Ne I	200
10560,89	Cl I	4
10555,90	Ar II	1
10554,96	Cl I	8
10553,02	Ti I	8
10551,81	Ti I	3
10549,76	Xe I	20
10549,635	N I	8
10546,76	N II	4
10541,552	Ar II	5
10541,226	C I	4
10539,554	N I	10
10539,18	Cl I	44
10535,52	Ar II	2
10533,775	N I	5
10532,21	Fe I	10
10529,32	Ar I	50
10527,84	Xe I	40
10527,34	Ar I	2
10520,574	N I	8
10519,510	Ar II	9
10515,15	Xe I	10
10513,403	N I	7
10510,60	Li I	3
10507,91	Xe I	6
10506,998	N I	8
10506,72	Cl I	33
10506,47	Ar I	100
10500,266	N I	6
10500,212	Ar II	6
10496,14	Ti I	30
10495,941	Ar II	2
10490,21	F I	1
10487,11	K I	8
10486,29	Kr I	2
10484,83	Xe I	8
10482,15	K I	5
10479,63	K I	9
10478,10	Ar I	200
10472,38	Cl I	3
10470,051	Ar I	500
10469,59	Fe I	20
10467,86	Cl I	7
10467,173	Ar II	20
10460,07	Ti I	10
10458,56	Kr I	6
10447,771	Ar II	2
10442,57	Ar II	1
10440,511	Ar II	6
10432,53	Ne I	3
10431,92	F I	1
10431,84	Kr II	2
10428,40	Kr II	10
10427,54	Cl I	44
10426,29	F I	6
10420,52	Xe I	1
10420,26	Cl I	105
10420,05	Cl I	105
10417,29	F I	7
10410,53	Ar II	2
10401,510	Ar II	1
10396,85	Ti I	25
10395,811	Fe I	8
10392,604	Ar II	5
10392,51	Cl I	331
10392,23	Mg II	6
10391,76	Mg II	5
10389,28	Kr II	8
10387,97	Cl I	34
10383,900	Ar II	1
10380,84	F I	7
10374,44	Kr I	10
10371,269	Si I	30
10361,15	Kr II	100
10360,37	Kr I	100
10357,6	Ar I	1
10350,02	Cl I	2
10343,85	Ca I	500
10332,95	F I	2,5
10332,76	Ar I	60
10329,77	Cl I	5
10325,34	Ar II	1
10322,88	Kr I	2
10320,08	Cl I	205
10319,62	Ar I	2

λ	Symbol	I
10312,16	Cl I	44
10311,54	He I	7
10311,23	He I	50
10309,15	Ar I	20
10305,616	Ar II	1
10305,50	Cl I	22
10299,077	Ar II	5
10296,93	Kr I	80
10295,40	Ne I	80
10293,01	F I	3,5
10288,942	Si I	10
10287,96	F I	1,5
10285,45	F I	15
10280,07	Cl I	4
10273,689	Ar II	5
10273,6	Kr I	2
10270,75	F I	4
10268,320	Ar II	2
10266,79	Ar I	1
10257,30	Ti I	3
10254,04	Ar I	10
10251,07	Xe I	20
10245,70	Ne I	7
10241,98	F I	4
10233,06	He I	2
10230,845	Ar II	4
10226,82	F I	3
10224,6	Ne I	2
10222,50	F I	2
10221,46	Kr II	1000
10221,12	Cl I	10
10220,980	Ar II	1
10220,8	Xe II	3
10216,351	Fe I	100
10210,73	Ne I	2
10209,57	F I	4
10208,7	Ar I	1
10206,9	Ar I	1
10206,9	Xe II	1
10203,917	Ar II	5
10199,98	N I	2
10189,26	Ti I	3
10188,36	Xe I	10
10186,15	F I	5
10179,92	Ti I	3
10179,2	Cu I	1
10177,41	Kr II	3
10172,00	Cu I	2
10171,2	Ar I	1
10170,60	Ti I	3
10167,61	Kr II	10
10167,252	O I	10
10166,91	Cu II	15
10166,79	N I	3
10164,849	N I	7
10163,50	F I	3
10163,45	Ar I	30
10162,88	Cu II	1
10157,07	Kr II	2
10147,68	Kr I	10
10147,274	N I	8
10147,09	Ti I	4
10146,78	Cu I	10
10145,601	Fe I	80
10145,48	Ti I	8
10138,50	He I	5
10138,408	Ar II	1
10128,285	N I	7
10127,74	Kr II	4
10126,27	N II	5
10125,47	Xe I	20
10124,5	Cu I	5
10123,871	CI	6
10123,61	He II	—
10123,6025	Cs I	1200
10123,415	Cs I	200
10120,96	Kr I	30
10120,90	Ti I	10
10119,8	Xe I	1
10119,20	Ti I	3
10118,49	N II	4
10114,644	N I	13
10112,484	N I	12
10111,595	Ar II	8
10110,660	Ar II	3
10108,895	N I	11
10107,34	Xe I	80
10107,19	Al II	4
10105,147	N I	10
10104,82	Ar I	4
10095,7	Xe II	1
10094,32	Ar I	8
10093,016	Ar II	1
10092,16	Mg II	14
10091,64	Cl I	40
10091,53	Ne I	3
10087,13	F I	6
10084,79	Xe I	20
10080,47	Cu II	10
10077,66	Kr I	10
10077,32	Al II	1
10076,29	Al II	6
10074,17	F I	1
10072,04	He I	3
10070,12	N II	6
10069,04	Ar I	50
10066,47	Ti I	8
10065,96	Kr I	10
10065,15	N II	7
10065,08	Fe I	60
10064,25	F I	4
10060,96	Xe I	10
10059,87	Ti I	12
10057,96	Xe I	5
10057,69	Ti I	25
10055,02	Cu II	30

λ	Symbol	I	λ	Symbol	I
10054,86	Kr I	2	9972,313	Ar II	1
10054,259	N I	4	9970,92	F I	4
10054,2	Xe II	1	9969,34	N II	7
10052,10	Ar I	150	9967,045	Ar II	12
10051,12	Cu II	3	9966,67	Kr II	5
10050,11	Ti I	5	9966,58	Xe I	10
10049,88	Cu II	1	9965,736	N I	3
10049,38	H	6	9965,41	Ar II	1
10048,78	Ti I	12	9963,55	Ne I	6
10046,64	D	6	9962,314	Ar II	1
10045,73	T	6	9961,86	N II	6
10042,27	Kr II	20	9961,281	Na I	7
10039,75	Ar I	2	9960,46	Cu II	15
10038,9	Ne I	2	9960,07	Cu II	10
10038,65	Kr I	3	9955,09	Li I	2
10038,19	Cu II	15	9954,75	Kr II	20
10038,03	F I	4	9954,74	Ca II	8
10037,1	Ne I	2	9954,141	K I	5
10036,32	Cu II	5	9952,809	Ar II	5
10035,45	N II	7	9951,88	Ar I	20
10034,45	Ti I	15	9951,087	Ar II	4
10032,81	Li I	2	9949,668	K I	6
10031,16	He I	10	9949,151	Ar II	7
10029,70	Ar I	40	9948,98	Ti I	8
10027,73	He I	30	9947,94	Ne I	15
10026,93	Cu II	1	9947,066	N I	4
10024,3595	Cs I	1000	9944,9	Ne I	2
10023,72	Xe I	50	9944,1	Ne I	7
10023,27	N II	8	9941,33	Ti I	8
10023,05	Cu II	30	9939,05	Cu II	20
10022,278	Ar II	4	9938,85	Ne I	15
10017,97	Kr II	20	9937,80	Ar I	1
10017,822	N I	5	9936,83	Ne I	10
10011,72	Ti I	15	9935,046	Ar II	1
10008,55	Ne I	4	9931,680	Ar II	1
10007,61	Ar I	3	9931,474	N I	5
10007,31	Ne I	30	9931,39	Ca II	9
10006,68	Cu II	10	9928,830	Ar II	1
10005,54	Ne I	20	9927,35	Ti I	20
10003,055	N I	5	9926,10	Cu II	10
10003,02	Ti I	25	9925,67	Cu II	20
10002,25	Cl I	4	9923,192	Xe I	3000
9997,94	Ti I	15	9918,52	Ne I	4
9997,750	N I	4	9918,05	Cu II	15
9994,94	Ar I	1	9917,60	Kr I	3
9994,32	Cu II	1	9916,52	Cu II	30
9993,874	Ar II	4	9916,37	Kr I	4
9993,209	Mg I	18	9916,144	Ar II	4
9990,9	Xe II	2	9915,20	Cu II	1
9989,3	Kr I	1	9915,13	Ne I	20
9989,02	Ar II	1	9914,246	Ar II	2
9988,39	Ar II	1	9909,712	Ar II	2
9986,475	Mg I	17	9909,220	N I	2
9983,4	Xe II	1	9908,9	Xe II	2
9983,20	Mg I	15	9906,394	Ar II	5
9981,16	Ti I	5	9905,880	Ar II	5
9980,424	N I	3	9905,65	F I	1,5
9977,825	Ar II	3	9905,54	N I	0
9974,2	Ne I	2			

λ	Symbol	I	λ	Symbol	I
9905,44	Cu II	2	9825,847	O I	13
9904,29	Ar II	1	9825,843	Ar II	1
9902,65	F I	12	9824,642	Ar II	2
9902,31	Ne I	30	9823,42	Ne I	5
9900,58	Ne I	40	9823,39	Kr II	100
9899,06	Ne I	2	9822,754	N I	7
9897,30	Ne I	3	9822,11	F I	15
9897,08	Kr I	2	9820,90	Xe II	2
9895,8	Xe II	1	9819,18	Ar II	1
9894,44	Cu II	5	9815,74	Cl I	3
9893,04	Cu II	5	9815,22	Ar I	1
9892,97	Kr II	10	9814,424	Ar II	3
9892,18	Ar I	6	9814,026	N I	4
9891,743	O I	13	9813,45	Ti I	5
9891,72	Si I	10	9813,35	Cu II	20
9891,09	N II	7	9810,28	Xe II	2
9890,63	Ca II	11	9810,27	Kr I	2
9889,082	Fe I	40	9810,018	N I	5
9887,39	N II	6	9808,46	Cl I	5
9887,06	Si I	10	9806,90	Cl I	25
9884,09	Cu II	10	9803,697	Ar II	4
9883,58	F I	8	9803,14	Kr II	500
9883,369	N I	3	9802,019	Ar II	4
9881,57	Cu II	15	9800,92	Ar I	4
9879,41	Ti I	3	9800,6	Kr II	5
9875,95	Cl I	50	9800,335	Fe I	20
9875,90	Ne I	2	9799,906	Si III	2
9872,159	N I	6	9799,699	Xe I	2000
9868,21	N II	5	9798,565	N I	5
9868,20	Cu II	15	9795,1	Kr II	2
9865,56	Xe II	6	9794,89	Kr I	3
9865,41	N II	6	9794,80	F I	6
9864,26	Cu II	40	9794,01	N II	3
9863,332	N I	9	9793,239	Ar II	3
9862,95	Kr I	4	9788,298	N I	4
9861,793	Fe I	30	9788,1	Ne I	2
9861,41	Cu II	50	9787,67	Ti I	50
9858,87	Cu II	3	9786,788	N I	4
9856,24	Kr I	500	9784,5010	Ar I	1000
9854,74	Ca II	8	9783,59	Ti I	20
9854,065	Ar II	8	9783,30	Ti I	40
9851,40	Kr II	3	9783,100	Ar II	3
9850,58	Cu II	3	9777,6	Kr II	2
9849,460	Ar II	10	9776,904	N I	4
9838,33	Kr I	5	9774,8	Xe II	1
9837,94	Cu II	25	9774,79	Ar I	1
9837,8	Xe II	2	9773,575	Ar II	4
9837,47	Ne I	20	9771,833	Ar II	2
9837,170	Ar II	3	9770,28	Ti I	40
9834,623	N I	6	9770,1	Kr II	2
9833,8	Kr II	5	9768,69	Kr I	2
9832,15	Ti I	25	9768,22	Ti I	5
9830,90	Cu II	5	9763,913	Fe I	15
9829,856	Ar II	3	9763,450	Fe I	15
9829,06	Cu II	3	9761,847	Ar II	1
9828,06	Cu II	5	9760,65	O I	5
9826,58	Kr II	100	9760,57	Ne	2
9826,002	O I	12			

λ	Symbol	I
9758,644	Ar II	4
9756,157	Ar II	3
9753,57	F I	1,5
9751,759	Kr I	2000
9750,145	Ar II	3
9746,86	Ti I	15
9744,8	Xe II	1
9744,33	Cl I	30
9743,60	Ti I	50
9743,460	Ar II	2
9743,11	Kr I	50
9741,49	O I	4
9741,43	N II	4
9741,3	Ne I	1
9739,770	Ar II	2
9739,6	Cu I	4
9739,4	Kr II	2
9738,624	Fe I	200
9737,77	Ti I	5
9737,75	N II	4
9736,70	F I	9
9735,94	Cu II	15
9734,554	Ar II	1
9734,34	F I	25
9734,0	Xe II	3
9732,28	Cu II	3
9728,36	Ti I	60
9728,2	Ne I	1
9727,51	Kr I	2
9724,8	Ne I	1
9722,78	Kr I	1
9722,36	N II	1
9720,6	Kr II	3
9720,57	F I	1
9718,96	Ti I	25
9718,66	N II	1
9718,16	Xe I	100
9717,73	C III	2
9717,16	Kr II	10
9717,00	Ti I	10
9715,51	Ti I	3
9715,11	C III	5
9714,85	Kr I	15
9713,117	Ar II	2
9711,779	Ar II	1
9711,60	Kr II	200
9710,03	Xe I	2
9706,44	C III	2
9706,2	Xe II	2
9705,64	Ti I	80
9705,39	C III	3
9704,22	Kr I	50
9702,86	Ti I	3
9702,60	He I	15
9702,40	Ne I	3
9702,35	Cl I	40
9701,961	Ar II	1
9701,81	Ca I	20
9701,515	Ar II	1
9701,12	C III	2
9700,99	Xe I	20
9699,40	F I	7
9698,68	Xe II	50
9696,77	N II	1
9694,01	N I	1
9694,0	N II	1
9693,27	Kr II	2
9689,39	Si I	10
9688,86	Ti I	30
9688,71	Cu II	10
9688,60	Ca I	15
9687,83	Kr I	10
9686,37	Li I	2
9685,32	Xe I	150
9682,26	Kr I	2
9682,19	He I	1
9678,98	Ti I	3
9678,812	Ar II	2
9677,80	Ar I	8
9677,41	O I	1
9676,287	Ar II	3
9676,25	Ca I	5
9675,55	Ti I	90
9673,39	Ar I	6
9672,90	Kr II	6
9669,54	Cl I	5
9669,03	Kr I	1
9666,86	Ar I	50
9665,424	Ne I	1000
9664,29	Ca I	3
9663,58	Ca I	2
9663,34	Kr II	200
9663,19	Ti I	3
9662,04	F I	12
9661,90	Cl I	20
9661,42	Ti I	10
9658,44	C I	10
9657,7841	Ar I	1500
9655,974	Ar II	3
9653,143	Fe I	20
9647,40	Ti I	50
9643,312	Ar II	1
9642,2	Ne I	1
9641,6	Xe II	4
9641,190	Ar II	2
9638,28	Ti I	100
9632,435	Mg II	11
9632,37	Cl I	20
9631,888	Mg II	12
9630,95	Xe II	3
9626,562	Fe I	30
9625,64	He I	3
9623,235	Ar II	2
9622,5	Kr II	3
9622,068	Ar II	4
9620,80	C I	9

λ	Symbol	I
9619,61	Kr II	400
9619,575	Ar II	3
9615,71	Xe II	4
9615,63	Kr I	3
9613,80	Kr II	100
9612,508	Ar II	2
9609,06	Cl I	35
9606,77	Ti I	3
9605,80	Kr II	500
9605,80	Xe I	3
9604,50	Xe II	7
9603,42	He I	5
9603,03	C I	7
9601,933	Ar II	6
9599,53	Ti I	50
9599,325	Ar II	2
9597,829	K I	14
9595,704	K I	15
9595,09	Ar I	4
9594,24	Kr II	100
9593,67	Ar I	1
9592,20	Cl I	75
9592,19	Ne I	5
9591,35	Xe II	50
9590,15	Ti I	3
9588,77	Ti I	4
9588,01	Cl I	5
9586,996	Ar II	2
9585,92	Si I	10
9585,14	Xe I	20
9584,79	Ne I	3
9584,77	Cl I	50
9581,42	Li II	—
9577,70	Xe II	2
9577,52	Kr II	500
9576,43	Cl I	8
9574,80	F I	3
9573,99	Ne I	2
9571,30	Cl I	5
9570,65	Si I	8
9570,08	Ti I	4
9569,960	Fe I	40
9564,32	Kr II	5
9561,60	Ar I	5
9561,26	Kr II	2
9555,2	Ar I	4
9554,96	Cl I	4
9553,631	Ar II	5
9552,99	FI	0,7
9552,89	He I	2
9552,85	Kr II	10
9552,30	F I	0,7
9549,4	Kr II	2
9547,73	Ar I	2
9547,40	Ne I	300
9546,07	Ti I	50
9545,974	H	5
9543,64	Kr II	10
9543,376	D	5
9542,509	T	5
9540,89	Kr I	30
9540,664	Ar II	5
9535,640	Ar II	3
9534,167	Ne I	500
9532,3	Kr I	1
9530,73	Li I	1
9530,3	Cu I	5
9529,27	He I	5
9526,39	Ar II	1
9526,17	He I	15
9522,01	O I	4
9520,23	Kr II	4
9516,94	Cl I	15
9516,87	He I	3
9516,60	He I	20
9513,379	Xe I	200
9512,43	Cu II	2
9511,80	Ti I	8
9511,55	Ti I	10
9510,81	Ti I	12
9508,440	Ar II	3
9508,49	Ti I	20
9508,4	Ne I	5
9506,59	Ne I	3
9506,04	Ti I	25
9505,78	Xe I	10
9505,67	O I	5
9505,433	Mg I	5
9505,30	F I	25
9505,19	Si I	20
9504,70	Kr II	100
9503,108	Mg I	7
9502,454	Mg I	8
9500,60	Kr II	100
9499,39	O I	0
9498,04	O I	8
9497,9	Ne	2
9497,07	Xe I	40
9492,76	O I	1
9487,76	Xe I	4
9487,49	O I	6
9486,89	Cl I	25
9486,680	Ne I	500
9486,02	Ar I	3
9483,00	Cl II	2
9481,93	Cl I	3
9480,871	Ar II	1
9480,73	N II	1
9478,39	Ar I	50
9476,4	Kr II	5
9475,239	Ar II	4
9475,23	Xe II	3
9475,06	Kr II	100
9473,36	Cu II	1
9472,4	Cu I	2
9470,93	Kr II	200

λ	Symbol	I
9467,81	Ne I	2
9465,938	Na I	6
9464,3	Xe II	10
9464,23	N I	1
9463,71	Cu II	3
9463,61	He I	50
9461,67	Kr II	3
9460,676	N I	10
9459,21	Ne I	300
9459,09	Ar I	100
9454,0	Ne I	1
9453,50	N II	1
9453,22	Ti I	3
9452,08	Ne I	10
9452,06	Cl I	75
9451,59	Cu II	2
9450,88	Kr I	20
9447,6	Xe II	1
9446,57	Ar I	2
9445,34	Xe I	80
9445,26	Ne I	3
9443,8	Ne I	2
9442,82	N II	3
9442,68	Xe I	20
9441,46	Xe I	20
9440,02	Kr II	100
9439,40	N II	1
9438,783	Mg I	20
9437,21	Kr II	20
9436,22	Ar II	1
9433,67	F I	200
9432,94	Ne I	40
9432,764	Mg I	19
9431,77	Ti I	3
9431,20	N II	1
9430,25	Kr II	5
9429,814	Mg I	17
9425,38	Ne I	500
9421,78	Si I	15
9420,484	Ar II	4
9419,82	Cl I	40
9418,582	Ar II	1
9414,964	Mg I	25
9414,94	Kr II	100
9414,14	Fe I	20
9414,07	Cl I	2
9413,506	Si I	100
9413,32	Kr II	3
9412,72	Si II	100
9412,32	Ne I	4
9412,01	Xe I	60
9410,75	Ne I	6
9408,66	Ar I	3
9407,57	Xe II	1
9405,73	C I	16
9405,75	Ne I	8
9402,82	Kr II	200
9402,69	Ar I	20
9400,59	Xe II	15
9399,24	O I	1
9393,81	Cl I	50
9393,8	Ne	2
9392,789	N I	15
9389,47	F I	0,8
9388,08	Kr II	50
9387,33	Si I	10
9386,805	N I	14
9386,75	F I	2,5
9384,96	F I	40
9377,63	Ar I	5
9377,2	Ne I	5
9376,71	Li I	1
9374,76	Xe I	100
9374,163	Ar II	3
9374,02	Xe I	10
9373,28	Ne I	200
9372,904	Fe I	6
9368,02	Ne	2
9363,6	Kr II	1
9362,50	Ar I	4
9362,03	Kr I	100
9361,95	Kr II	300
9360,466	Ar II	1
9358,37	C III	1
9354,218	Ar I	200
9353,3	Ne I	3
9352,23	Kr I	100
9351,590	K I	6
9350,44	Fe I	10
9349,248	K I	3
9349,08	Kr II	100
9347,235	K I	7
9345,11	Kr II	100
9344,93	He II	—
9344,793	Ar II	2
9340,59	Ar I	3
9340,544	Mg II	10
9340,5	Ne I	2
9337,9	Kr I	1
9337,73	Kr II	2
9334,80	Ar I	8
9334,08	Xe I	3
9333,32	Ar I	1
9332,04	Cu II	5
9331,979	Al II	2
9331,67	Xe II	4
9331,546	Al II	3
9331,05	Ar II	1
9330,66	Kr II	5
9328,08	Ar I	2
9327,545	Mg II	10
9326,66	Ne	2
9326,52	Ne I	600
9326,19	Kr II	4
9326,03	Kr I	10
9325,84	N II	0

λ	Symbol	I
9323,899	Si III	3
9320,99	Kr II	200
9318,22	Si I	10
9317,84	Kr II	30
9314,34	F I	60
9313,98	Ne I	300
9313,51	Ar II	1
9312,48	Ti I	4
9310,58	Ne I	150
9306,64	Xe I	40
9305,87	Ar II	1
9305,76	Kr II	1
9304,77	Xe II	1
9301,95	Xe I	30
9300,85	Ne I	600
9299,40	Kr I	1
9298,7	Xe II	2
9296,1	Kr II	60
9293,82	Kr II	500
9291,58	Ar I	100
9290,747	Al II	5
9290,649	Al II	6
9289,95	Kr II	20
9288,82	Cl I	60
9288,550	Al II	2
9288,4	Xe II	5
9288,145	Al II	3
9287,87	Kr I	1
9286,794	Al II	2
9286,578	Al II	1
9285,04	Ti I	5
9281,06	N II	3
9279,9	Kr I	2
9279,712	Ar II	4
9275,53	Ne I	100
9273,02	Kr I	8
9271,99	Kr II	50
9270,96	Kr I	10
9269,38	Kr II	2
9266,61	N II	1
9266,17	Kr II	2
9266,006	O I	24
9265,938	O I	21
9265,67	Xe II	10
9263,54	Cu I	3
9262,93	Kr II	2
9262,774	O I	23
9262,69	F I	8
9262,69	Kr I	1
9262,671	O I	22
9262,584	O I	19
9260,935	O I	20
9260,845	O I	21
9260,806	O I	20
9259,60	Xe II	1
9259,05	Fe I	15
9258,78	Ar II	1
9258,31	Fe I	20
9257,62	Ti I	7
9255,778	Mg I	30
9253,98	N II	1
9253,67	Si I	15
9252,628	Ar II	2
9249,41	Al II	1
9246,499	Mg I	12
9246,14	Ti I	10
9245,45	Kr II	20
9245,18	Xe I	3
9244,57	F I	15
9244,266	Mg II	13
9244,15	Xe II	2
9243,54	Kr I	30
9243,00	Kr I	1
9242,17	Ar I	1
9242,02	N II	2
9238,59	Xe II	2
9238,48	Kr II	500
9235,38	F I	50
9234,16	Kr I	1
9233,18	Kr II	50
9232,85	F I	6
9229,40	F I	2,5
9229,017	H	4
9226,86	Cu II	1
9226,67	Ne I	200
9226,50	D	4
9226,39	Xe II	7
9225,667	T	4
9224,83	Kr I	1
9224,4955	Ar I	1000
9223,05	F I	6
9222,39	Xe I	5
9221,88	Ne I	150
9221,59	Ne I	200
9221,08	Ar I	5
9220,05	Ne I	400
9219,001	Ar II	2
9218,248	Mg II	14
9217,32	Li I	2
9217,10	N II	2
9216,51	Xe I	1
9214,61	Li I	1
9212,9	Ne I	2
9212,39	Cl I	3
9211,38	Xe I	25
9210,39	Ar II	1
9210,337	He I	10
9210,033	Fe I	6
9208,5382	Cs I	200
9208,35	Si I	15
9208,001	N I	8
9207,59	N I	3
9207,27	Kr II	8
9205,40	Cu II	20
9203,20	Xe I	30

λ	Symbol	I	λ	Symbol	I
9201,76	Ne I	600	9121,00	N II	1
9198,61	Ar I	50	9118,892	Fe I	25
9197,49	Cl I	25	9115,00	Kr II	20
9197,18	Xe I	2	9112,24	Xe I	4
9196,7	Kr II	1	9111,80	C I	10
9194,637	Ar I	150	9111,69	Kr I	20
9193,8	Xe II	2	9111,3	Ar I	1
9192,605	Ar II	3	9107,87	F I	10
9191,8	Ne I	3	9106,573	Ar II	4
9191,67	Cl I	60	9106,24	Xe II	1
9191,65	F I	10	9103,53	Ne I	3
9191,17	Ar II	1	9103,33	Cu II	10
9188,69	Kr I	2	9102,33	F I	50
9187,84	N I	3	9102,1	Ne I	1
9187,449	N I	9	9100,58	Kr I	1
9182,83	C I	4	9099,72	Kr II	15
9181,23	Kr II	10	9098,58	Ar II	2
9180,17	Ar I	6	9097,49	F I	3,5
9178,68	F I	350	9096,49	Kr II	4
9175,42	Kr II	40	9096,17	N II	1
9174,52	He I	2	9096,13	Xe I	50
9173,267	Si III	2	9095,099	Ar II	3
9172,3217	Cs I	1000	9094,83	C I	12
9172,14	Al I	4	9094,5	Kr II	2
9170,86	Al I	5	9094,33	Kr I	4
9169,76	F I	3	9092,93	N II	0
9168,917	Ar II	1	9090,98	Kr II	3
9167,53	Ti I	8	9090,70	Ti I	25
9167,52	Xe I	100	9089,906	Al I	5
9165,938	Na I	6	9089,45	NII	1
9164,04	Kr II	3	9089,415	Fe I	30
9163,261	Al I	2	9088,51	C I	9
9162,654	Xe I	500	9088,324	Fe I	50
9162,33	F I	2	9087,18	Kr II	2
9159,030	Ar II	2	9079,707	Ar II	2
9158,38	Xe I	2	9079,599	Fe I	8
9157,82	Kr II	2	9078,28	C I	8
9156,049	Ar II	3	9075,42	Ar I	60
9156,02	O I	4	9073,34	Ar I	50
9153,878	Na I	4	9073,15	Cl I	50
9152,12	Xe I	20	9073,04	Ne I	8
9151,78	F I	180	9069,7	Ne	2
9150,82	Ar II	1	9069,66	Cl I	25
9148,68	Ne I	600	9069,61	N II	1
9146,40	N II	2	9068,39	Cl I	3
9141,8	Xe I	2	9068,023	Ar II	5
9139,950	Al I	6	9068,0	Xe II	2
9136,6	Xe II	5	9066,77	Ar I	40
9133,4	Kr II	8	9063,78	N II	0
9132,53	F I	1,4	9063,27	He I	6
9131,59	Xe I	3	9062,47	C I	8
9131,21	Kr II	6	9061,43	C I	9
9123,14	Ti I	5	9060,749	Ar II	3
9122,9660	Ar I	500	9060,472	N I	10
9122,63	F I	40	9057,51	Ar I	2
9122,49	Kr I	20	9057,23	Ar I	4
9121,14	Ne I	20	9052,54	Ne I	6
9121,10	Cl I	75			

λ	Symbol	I
9051,236	Ar II	1
9050,10	Cl I	4
9049,890	N I	12
9049,47	N I	5
9049,06	Ne I	3
9046,8	Ne I	1
9045,878	N I	13
9045,446	Xe I	400
9045,40	Cl I	40
9045,4	Ne	2
9044,55	Kr II	10
9044,47	Kr I	3
9042,10	F I	400
9039,95	Kr II	20
9039,0	Ne I	3
9038,96	Cl I	30
9036,98	Ne I	6
9035,915	Ar II	3
9032,18	Xe I	50
9032,04	N II	1
9031,35	Ar II	1
9028,918	N I	9
9027,32	Ti I	15
9025,98	Xe I	30
9025,67	Kr II	10
9025,49	F I	350
9024,47	Fe I	15
9021,58	Si I	10
9018,162	Si IV	1
9017,596	Ar II	7
9015,19	F I	0,7
9014,938	Ar II	1
9014,911	H	3
9012,457	D	3
9012,098	Fe I	30
9011,639	T	3
9010,39	N II	1
9008,51	Si I	15
9008,455	Ar II	6
9006,19	F I	50
9006,15	Kr II	10
8999,564	Fe I	200
8999,19	Kr I	30
8999,11	Kr II	6
8997,803	Ar II	4
8997,156	Mg I	10
8996,978	He I	2
8996,2	Cu I	20
8995,865	Ar II	7
8994,99	Ar I	10
8992,84	Ar I	1
8991,692	Mg I	9
8989,44	Ti I	12
8989,026	Mg I	7
8988,58	Ne I	200
8988,20	Ar I	3
8987,57	Xe I	200
8986,615	Ar II	6
8986,15	N II	4
8983,84	Si I	4
8983,65	F I	3,5
8983,28	N II	3
8981,18	F I	12
8981,05	Xe I	100
8980,10	Cl I	2
8978,70	Kr II	15
8977,99	Kr I	50
8975,408	Fe I	10
8971,365	Ar II	4
8971,36	N II	1
8970,98	Ar I	2
8968,947	Ar II	4
8968,6	Ne I	2
8967,53	Kr I	10
8967,39	Ar I	2
8964,48	Ar I	10
8963,66	F I	4
8962,34	Ne I	3
8962,19	Ar I	40
8960,75	C I	2
8957,245	Si IV	2
8952,78	Xe I	50
8952,254	Xe I	1000
8949,10	Si I	10
8948,12	Ne I	7
8948,01	Cl I	50
8945,204	Fe I	20
8943,483	Cs I	2000
8942,962	Na I	2
8941,47	Ne I	6
8937,530	Ar II	1
8936,61	F I	0,7
8935,448	Ar II	1
8931,326	Ar II	5
8931,20	Cl I	2
8930,83	Xe I	200
8930,04	N II	1
8929,24	Ne I	10
8928,6920	Kr I	2000
8927,4	Ne I	2
8927,36	Ca II	11
8926,819	Ar II	1
8926,074	Ar II	3
8925,504	Al I	4
8925,436	K I	4
8925,30	Si I	10
8923,569	Mg I	20
8923,555	Al I	9
8923,312	K I	5
8921,14	Li I	0
8920,198	Ar II	2
8919,4987	Ne I	300
8916,89	F I	2
8915,522	Ar II	1
8915,44	Ne I	3
8914,74	He I	2

λ	Symbol	I
8914,43	F I	7
8913,0	Ne I	3
8912,900	Al I	7
8912,88	Cl I	40
8912,78	F I	300
8912,07	Ca II	10
8910,27	F I	140
8908,73	Xe I	200
8908,26	Kr II	3
8905,650	Ar II	6
8904,512	Ar II	1
8904,34	C I	2
8904,017	K I	12
8903,20	C I	1
8902,66	Xe II	5
8902,188	K I	13
8900,92	F I	1000
8899,92	F I	60
8899,297	Ar II	3
8895,6	Ne I	2
8895,42	Ar I	1
8895,144	Ar II	1
8893,32	N II	1
8892,7277	Si I	20
8892,22	Ne I	10
8891,70	Ar I	1
8890,67	C I	2
8890,147	Ar II	4
8885,71	Xe I	10
8881,48	Xe II	2
8874,84	Ar I	4
8873,39	C I	3
8870,32	Kr I	4
8870,216	Ar II	1
8869,40	Xe II	2
8867,170	Ar II	2
8866,961	Fe I	150
8865,7562	Ne I	500
8865,3057	Ne I	100
8863,09	Ti I	3
8862,787	H	2
8862,32	Xe I	300
8860,374	D	2
8859,570	T	2
8858,39	Al II	1
8855,74	Xe II	5
8855,40	N II	0
8853,8669	Ne I	700
8851,44	Xe I	1
8850,659	Ar II	1
8849,97	Ar I	150
8849,06	F I	70
8846,46	N II	1
8846,17	Ar I	1
8844,502	F I	120
8842,527	Ar II	1
8842,46	Kr I	3
8842,1	Ne	2
8841,70	Cl I	15
8841,277	Al I	10
8840,82	Ar I	20
8840,39	Ar I	3
8840,09	Kr II	4
8839,9	Xe II	3
8838,433	Fe I	30
8838,009	Ar II	1
8835,082	Mg II	11
8833,42	Kr II	3
8831,232	F I	100
8830,9078	Ne I	50
8829,38	Kr II	5
8828,909	Al I	8
8824,323	Mg II	10
8824,227	Fe I	250
8821,14	Ti I	12
8820,70	Cl II	5
8820,45	O I	15
8820,36	Ne I	6
8819,60	O I	5
8819,56	N II	2
8819,412	Xe I	5000
8819,39	Ti I	8
8819,37	Ar I	1
8815,28	Cl I	30
8810,10	Kr II	2
8807,582	F I	900
8806,757	Mg I	50
8805,78	Kr I	20
8805,16	Ar I	3
8804,65	Kr II	3
8804,624	Fe I	10
8804,61	Xe II	30
8803,860	Ar II	1
8799,9	C II	0
8799,36	F I	70
8799,082	Ar I	100
8798,65	Kr II	3
8796,92	Xe II	2
8796,142	Ar II	5
8794,40	Ti I	8
8793,8	C II	1
8793,376	Fe I	120
8792,51	Ne I	30
8792,50	F I	35
8790,3889	Si I	35
8790,555	Ar II	1
8785,88	Xe II	4
8785,63	F I	14
8784,59	Ar I	30
8783,7539	Ne I	1000
8782,014	Ne I	50
8780,747	Si I	11
8780,6223	Ne I	1200
8780,25	Kr I	30

λ	Symbol	I	λ	Symbol	I
8778,75	Ne I	150	8725,76	Ti I	6
8778,66	Ti I	30	8722,17	Kr I	1
8777,73	F I	120	8719,56	Ti I	30
8776,7490	Kr I	6000	8719,374	Ar II	3
8776,74	He I	2	8718,841	N I	14
8774,05	Kr I	50	8717,825	Mg I	13
8773,896	Al I	14	8717,31	Kr II	2
8773,00	Kr I	4	8716,947	Ar II	1
8772,95	N II	3	8716,19	Xe II	50
8772,866	Al I	13	8714,52	Ne I	5
8771,855	Ar II	15	8713,79	Ar I	5
8771,6592	Ne I	400	8713,69	Cl I	3
8768,215	Ar II	1	8713,62	Kr I	2
8767,55	Ne I	15	8712,689	Mg I	12
8767,053	K I	3	8711,708	N I	15
8766,422	Si I	14	8711,58	Cl I	1
8766,64	Ti I	75	8711,54	Xe I	2
8766,61	F I	10	8710,54	N II	6
8764,112	Kr I	150	8710,29	Fe I	20
8764,00	Fe I	100	8710,175	Mg I	10
8763,955	K I	4	8709,64	Xe I	40
8763,39	N II	1	8707,61	Kr II	8
8761,6907	Ar I	200	8704,1132	Ne I	200
8761,44	Ti I	15	8703,255	N I	14
8761,415	Cs I	500	8700,95	Ar I	3
8760,14	Xe II	6	8700,44	Cl I	5
8758,20	Xe I	100	8699,002	N II	5
8757,192	Fe I	25	8697,79	N II	3
8755,20	Kr I	30	8697,50	Kr I	40
8754,009	Ar II	2	8696,86	Xe I	200
8753,08	C I	3	8696,71	C II	5
8752,009	Si I	100	8694,900	N II	4
8752,14	Xe II	7	8693,086	Ar II	2
8751,174	Si I	10	8692,34	Ti I	100
8750,475	H	—	8692,20	Xe I	100
8748,093	D	—	8690,19	Kr II	100
8747,357	N I	9	8690,12	Ar I	2
8747,298	T	—	8688,632	Fe I	1500
8747,29	Kr I	2	8687,430	N II	5
8746,43	Kr I	3	8686,28	Cl I	30
8745,657	Mg II	11	8686,161	N I	14
8742,4509	Si I	75	8683,400	N I	16
8742,49	Kr I	1	8682,99	Ti I	125
8741,26	Ar I	1	8682,56	C II	8
8739,51	Ar I	3	8681,9216	Ne I	500
8739,39	Xe I	300	8680,31	Al II	3
8737,31	Ti I	7	8680,270	N I	17
8737,270	F I	140	8680,079	Si I	11
8736,63	Ar I	20	8679,4898	Ne I	500
8736,19	Ar I	2	8678,43	Ar I	60
8736,021	Mg I	17	8676,076	N II	7
8734,990	Mg II	10	8675,38	Ti I	150
8734,70	Ti I	75	8675,28	Al II	1
8732,80	F I	6	8674,92	Al II	2
8728,909	N I	10	8674,767	Ar II	1
8728,019	Si III	3	8674,751	Fe I	60
8728,0110	Si I	40	8674,26	Kr II	2
8726,54	Kr I	8			

λ	Symbol	I	λ	Symbol	I
8673,48	Kr I	2	8610,67	Kr I	5
8672,62	F I	35	8609,26	Cu II	3
8671,28	Al II	1	8707,611	Ar II	2
8667,9438	Ar I	400	8606,64	Cu II	1
8667,71	F I	1	8606,06	F I	6
8665,22	C III	3	8606,014	Si I	8
8665,021	H	—	8605,85	Kr I	40
8664,63	F I	6	8605,7790	Ar I	150
8663,65	C III	2	8604,47	F I	1,4
8662,140	Ca II	16	8604,32	N II	3
8661,907	Fe I	600	8604,23	Xe II	50
8660,52	N II	3	8604,016	Ar II	6
8657,390	Ar II	1	8600,98	Ti I	25
8656,93	F I	0,8	8599,4	Kr I	1
8655,869	N I	14	8598,394	H	—
8655,72	Xe II	3	8598,18	Ti I	60
8655,5206	Ne I	400	8597,0470	Si I	20
8654,3837	Ne I	1500	8595,962	Si I	25
8653,38	N II	3	8595,91	Kr II	4
8652,6	C III	1	8594,005	N I	15
8651,50	Kr II	5	8593,1	Kr I	10
8651,49	Kr I	8	8592,624	Ar II	3
8650,889	Na I	6	8591,2583	Ne I	400
8649,922	Na I	7	8586,00	O I	2
8648,4622	Si I	50	8585,96	Cl I	100
8648,54	Xe I	250	8585,262	Ar II	3
8647,114	Si I	15	8584,0	Cu I	10
8647,0400	Ne I	300	8584,0	Xe II	1
8642,89	Ar I	1	8582,91	Ne I	60
8641,75	Cl I	3	8582,267	Fe I	15
8641,47	Ti I	40	8579,49	Ar I	4
8640,7	Al II	8	8578,40	Ti I	15
8638,31	N II	3	8578,06	Ar I	5
8636,4	Xe II	2	8577,98	Cl I	7
8636,38	Ti I	18	8576,01	Xe I	200
8635,31	Ne I	50	8575,25	Cl I	75
8634,84	Cl I	2	8571,3535	Ne I	100
8634,6472	Ne I	600	8569,72	Ti I	50
8632,81	Kr I	1	8569,02	Kr I	20
8631,5	Kr I	1	8567,735	N I	14
8631,102	Ar II	1	8566,7	Xe II	2
8629,33	Ti I	18	8565,45	Ti I	25
8629,238	N I	16	8565,13	Ar I	1
8628,94	Xe II	25	8564,7	Xe I	1
8628,70	Kr I	1	8563,59	Kr II	2
8628,61	Cl I	4	8563,38	Ar I	1
8624,82	Kr I	4	8562,550	Ar II	2
8624,24	Xe I	80	8561,38	Ar I	3
8623,804	Ar II	5	8560,89	Kr I	50
8621,612	Fe I	10	8556,7803	Si I	120
8620,4602	Ar I	100	8553,97	Xe I	2
8619,34	Kr II	1	8551,33	Kr II	2
8618,44	Ti I	20	8550,54	Ti I	25
8618,14	Ti I	15	8550,46	Cl I	20
8613,58	Kr II	2	8548,07	Ti I	100
8612,91	Ti I	7	8547,023	Ar II	4
8612,58	F I	6	8545,384	H	—
8611,807	Fe I	40			

λ	Symbol	I	λ	Symbol	I
8544,6952	Ne I	60	8468,46	Ti I	100
8542,089	Ca II	17	8468,413	Fe I	300
8539,36	Ti I	60	8467,8	Xe II	1
8537,98	Kr II	3	8467,32	Cl I	25
8537,93	Kr I	40	8467,256	H	—
8537,04	F I	7	8467,15	Ti I	75
8536,26	C I	1	8466,483	Mg I	2
8536,1645	Si I	40	8465,352	Li I	4
8531,36	Ti I	15	8464,92	Kr II	4
8530,10	Xe I	30	8463,3569	Ne I	150
8526,36	Ti I	8	8460,96	Ti I	7
8525,99	Ti I	8	8457,10	Ti I	40
8523,88	Kr II	3	8450,89	Ti I	75
8522,55	Xe I	30	8450,37	Xe I	1
8521,4428	Ar I	2000	8446,758	O I	29
8521,149	Cs I	4000	8446,6	Xe II	2
8519,72	Cl I	8	8446,359	O I	30
8518,37	Ti I	100	8446,250	O I	27
8518,05	Ti I	60	8443,982	Si I	40
8515,19	Xe II	50	8443,44	Ar I	20
8515,08	Fe I	20	8442,98	Ti I	20
8514,075	Fe I	150	8440,26	Ar I	1
8511,04	Cu II	40	8439,603	Fe I	20
8510,45	C I	1	8438,93	Ti I	75
8508,8700	Kr I	3000	8438,742	N II	11
8508,66	O I	2	8437,958	H	—
8505,112	K I	10	8437,71	Ar I	6
8503,46	Cu II	15	8437,55	Xe I	10
8503,449	K I	11	8435,68	Ti I	300
8502,487	H	—	8435,24	Si I	8
8502,2207	Si I	60	8434,98	Ti I	300
8501,547	Si I	40	8432,37	Kr II	1
8500,997	Ar II	2	8430,88	C I	1
8500,96	Xe II	2	8429,128	O I	1
8500,32	C III	10	8428,342	O I	2
8498,21	Kr I	30	8428,25	Cl I	100
8498,018	Ca II	13	8426,50	Ti I	200
8497,32	Cl I	5	8426,326	O I	4
8496,64	Ar I	2	8424,780	O I	1
8496,03	Ti I	60	8424,6473	Ar I	2500
8495,51	Ti I	15	8424,41	Ti I	50
8495,3591	Ne I	500	8423,10	Ti I	20
8494,42	Ti I	30	8420,968	O I	1
8492,078	Si I	15	8419,996	K I	1
8490,30	Ar I	40	8418,70	Ti I	10
8488,85	Cl I	2	8418,4265	Ne I	400
8484,4424	Ne I	80	8417,54	Ti I	25
8483,16	Ti I	25	8417,535	K I	2
8482,64	Xe II	5	8417,161	Ne I	100
8477,26	Cu II	10	8416,97	Ti I	60
8477,20	Kr I	2	8414,49	C II	1
8473,694	Mg I	7	8413,42	C II	2
8473,31	Kr II	100	8413,321	H	—
8472,96	Cl I	3	8412,428	Kr I	100
8470,72	Ne I	5	8412,36	Ti I	150
8469,96	Kr I	2	8411,88	Ar II	1
8468,845	Mg I	5	8411,14	Kr II	1
			8409,190	Xe I	2000

λ	Symbol	I
8408,2094	Ar I	3000
8408,15	Cu I	20
8406,20	Cl I	10
8403,70	Cl I	1
8402,54	Ti I	5
8402,03	Xe I	5
8399,35	Ar I	20
8397,61	Cl I	3
8396,93	Ti I	90
8395,734	Ar II	3
8392,400	H	—
8392,37	Xe I	20
8392,28	Ar I	80
8392,20	Cl I	15
8391,96	Cl II	3
8391,44	K I	—
8390,223	K I	3
8389,48	Ti I	25
8387,780	Fe I	1200
8384,90	Kr I	15
8384,73	Ar I	60
8383,58	Cl I	4
8382,82	Ti I	90
8382,76	Cl II	5
8382,54	Ti I	100
8378,87	Kr II	2
8378,3	Xe II	3
8377,90	Ti I	100
8377,6062	Ne I	800
8376,41	Ne I	200
8376,079	Ar II	2
8375,95	Cl I	150
8375,93	Kr I	5
8374,478	H	—
8372,79	Xe I	5
8371,38	Xe I	3
8367,03	Ar I	3
8366,4	Xe II	30
8365,7464	Ne I	150
8365,642	Fe I	25
8363,52	Al II	8
8363,30	Al II	1
8363,074	Ar II	2
8361,81	Cl II	8
8361,69	He I	10
8360,63	Cl II	15
8359,57	Al II	9
8359,23	Al II	1
8359,006	H	—
8358,72	C III	2
8358,28	Cl I	6
8357,86	C III	2
8355,30	Ar I	1
8354,35	Al II	10
8353,50	Ar I	4
8353,00	Cl II	2
8351,3	Xe II	3
8349,05	Xe I	40
8347,94	C III	5
8347,45	Xe I	60
8347,24	Xe II	100
8346,823	Xe I	2000
8346,420	Ar II	1
8346,120	Mg I	15
8345,556	F I	120
8345,553	H	—
8345,183	Ar II	2
8343,90	Cl I	50
8342,630	Ar II	1
8341,931	Si III	2
8341,59	C III	6
8339,431	Fe I	80
8338,328	Si I	20
8338,384	Ar II	1
8335,15	C I	13
8333,785	H	—
8333,27	Cl I	5000
8333,14	Kr II	2
8332,99	C III	7
8332,73	Kr I	1
8332,21	Ar I	20
8331,941	Fe I	200
8329,44	Xe II	30
8327,907	Ar II	2
8327,063	Fe I	1200
8324,58	Xe I	20
8323,90	Xe I	2
8323,428	H	—
8321,09	Kr I	2
8317,39	Si I	15
8317,10	Xe II	40
8316,2	Xe II	10
8314,262	H	—
8310,264	Mg I	10
8306,710	Si I	25
8305,596	Mg I	9
8306,115	H	—
8305,02	Ar I	1
8304,69	Cl I	5
8303,79	Ar I	1
8303,313	Mg I	7
8303,20	Kr I	10
8302,40	F I	600
8301,54	Ne I	150
8301,39	Kr I	20
8300,3248	Ne I	600
8298,837	H	—
8298,581	F I	2000
8298,1077	Kr I	5000
8297,71	Xe I	15
8297,55	Xe II	100
8296,723	Ar II	1
8296,51	C III	1
8296,205	N II	4
8294,675	Si I	13

λ	Symbol	I	λ	Symbol	I
8293,527	Fe I	20	8235,30	Cu II	10
8292,615	Si III	3	8234,639	Mg II	11
8292,309	H	—	8233,194	Mg II	7
8291,88	Ar I	8	8233,085	O I	13
8287,56	Kr I	4	8232,347	Fe I	50
8286,67	Cl I	5	8232,19	F I	500
8286,434	H	—	8231,6348	Xe I	10000
8285,70	Xe II	15	8230,773	F I	3000
8283,21	Cu II	60	8230,642	Si I	35
8282,85	Xe II	15	8230,016	O I	10
8281,125	H	—	8228,89	Kr I	10
8281,0495	Kr I	1500	8227,680	O I	10
8280,95	Cl I	7	8224,72	Ar I	6
8280,1163	Xe I	7000	8223,121	N I	13
8278,44	F I	2	8222,924	Mg II	7
8277,60	Cu II	50	8222,69	Kr I	6
8276,310	H	—	8221,84	O I	15
8274,615	F I	1500	8221,829	O I	20
8273,79	Cl I	7	8221,73	Cl I	75
8272,355	Kr I	100	8220,406	Fe I	150
8272,26	C III	1	8220,40	Cl I	60
8271,944	Si III	6	8218,40	Kr I	80
8271,70	Cl I	7	8217,817	Ar II	1
8271,377	Si III	5	8216,317	N I	15
8269,324	Si III	8	8215,15	Si I	10
8269,15	Cl I	60	8214,85	Xe II	20
8267,97	Cl I	3	8214,726	F I	2500
8267,117	Ne I	80	8213,989	Mg II	10
8266,519	Xe I	500	8213,50	Xe II	2
8266,0788	Ne I	200	8213,034	Mg I	20
8265,640	Si III	5	8212,24	Kr I	5
8264,5221	Ar I	1500	8212,05	Si III	2
8263,2398	Kr I	3000	8212,00	Cl I	100
8262,73	Xe II	30	8210,708	N I	11
8262,568	Si III	9	8210,1	Kr I	1
8262,49	F I	12	8209,839	Mg I	10
8260,81	Xe II	5	8208,634	F I	350
8259,521	Ar II	2	8207,667	Fe I	40
8259,3795	Ne I	150	8206,62	Kr I	40
8258,64	Cl I	4	8206,40	Cl I	2
8256,90	Cu II	5	8206,341	Xe I	700
8256,40	Xe II	20	8205,22	Kr I	20
8255,62	C III	1	8203,76	Cl I	12
8255,07	Ar I	50	8203,42	Ar I	20
8254,725	Ca II	7	8202,72	Kr II	200
8251,743	K I	8	8201,766	N I	7
8251,30	Xe II	2	8201,720	Ca II	10
8250,180	K I	9	8201,43	N I	2
8249,58	Ar I	4	8200,95	Cl I	25
8248,797	Ca II	11	8200,357	N I	10
8248,6812	Ne I	30	8200,20	Cl I	35
8248,151	Fe I	30	8199,02	Cl I	35
8245,37	Xe II	4	8198,951	Fe I	80
8242,374	N I	13	8197,734	F I	60
8240,606	Si IV	1	8196,73	Xe I	2
8239,130	Fe I	8	8196,48	C III	10
8236,77	He II	—	8195,070	Kr I	50
8235,408	O I	5			

λ	Symbol	I
8194,8237	Na I	9
8194,7905	Na I	1
8194,71	Si III	3
8194,35	Cl I	50
8194,18	Si III	3
8192,4	Kr I	2
8192,28	Cu II	30
8191,679	Si III	8
8191,241	F I	300
8191,16	Si III	6
8190,431	Si III	7
8190,258	Ar II	1
8190,0543	Kr I	3000
8188,005	N I	13
8186,9	Xe II	10
8184,852	N I	13
8183,2556	Na I	5
8182,93	Xe I	1
8179,339	F I	600
8178,96	Ar I	20
8178,84	Ar I	40
8178,68	Kr II	2
8174,50	N I	1
8171,95	Ar I	10
8171,288	Si I	25
8171,02	Xe I	100
8170,09	Cl I	10
8167,55	Xe II	10
8166,51	N I	2
8166,235	N I	8
8165,405	Ar II	3
8165,37	Xe I	2
8162,170	Si I	15
8161,52	Cl I	12
8160,15	Al II	3
8159,51	F I	300
8159,132	Mg I	2
8157,25	Kr II	10
8154,872	Si I	15
8154,644	Mg I	1
8151,86	Ar I	3
8151,80	Xe II	100
8150,66	N I	1
8150,647	Ar II	1
8147,70	Kr II	1
8145,15	Kr II	100
8144,96	Kr I	15
8144,8	Xe II	3
8143,54	Ar I	10
8142,17	Kr II	1
8142,13	Xe II	5
8140,55	Si I	15
8136,83	Xe II	30
8136,4061	Ne I	300
8132,98	Kr I	60
8132,96	Kr II	6
8131,40	Xe II	20
8130,03	Kr II	10
8129,55	Cl I	2
8129,26	F I	600
8129,170	N I	3
8128,908	Ne I	60
8126,56	F I	350
8126,378	Li I	300
8123,44	Kr II	4
8123,29	Xe I	2
8121,40	Cl I	5
8120,434	Mg II	8
8120,16	Xe II	30
8119,72	Al II	1,5
8119,18	Ar I	50
8118,5495	Ne I	100
8118,29	Xe I	15
8117,75	Cl I	4
8115,94	Xe II	50
8115,3108	Ar I	5000
8115,220	Mg II	9
8112,900	Kr I	6000
8110,65	Ar II	1
8109,46	Xe I	15
8107,91	Xe I	6
8105,631	N I	2
8104,3642	Kr I	4000
8104,02	Kr I	500
8103,6920	Ar I	2000
8103,448	Si III	11
8102,862	Si III	9
8101,98	Xe I	100
8098,724	Mg I	10
8098,55	Xe II	12
8097,24	Xe I	3
8095,96	Kr II	3
8095,55	Cu II	40
8095,13	Xe II	10
8094,76	Cl I	12
8094,06	Ar I	20
8093,241	Si I	70
8093,08	Ne I	2
8092,634	Cu I	2000
8089,93	Ar I	5
8088,58	Cu II	20
8087,69	Cl I	20
8086,67	Cl I	75
8085,54	Cl I	60
8085,20	Fe I	200
8084,48	Cl I	35
8083,75	Ar II	1
8083,80	C I	5
8082,4576	Ne I	200
8080,31	Xe II	50
8079,68	Ar I	20
8079,618	K I	6
8079,0332	Cs I	1000
8078,923	Cs I	100
8078,48	C I	4

λ	Symbol	I	λ	Symbol	I
8078,114	K I	7	8024,11	Ne I	2
8077,521	F I	350	8023,85	Xe II	50
8076,64	C II	8	8023,30	Cl I	18
8076,298	Al I	2	8021,9	Ar I	2
8076,06	Ne I	1	8021,26	C I	3
8075,519	F I	900	8021,14	C III	1
8075,46	Cu II	2	8020,504	Ca II	2
8075,353	Al I	8	8020,07	Xe II	5
8073,99	Xe I	1	8018,12	F I	8
8071,285	Si I	25	8018,56	C I	1
8070,97	Xe II	50	8017,542	Ar II	2
8070,598	Si I	25	8017,502	Ca II	2
8070,42	C I	3	8015,7235	Cs I	200
8066,60	Ar I	20	8015,57	Cl I	45
8065,968	Al I	6	8014,7853	Ar I	800
8064,94	Xe I	2	8014,26	Xe II	50
8062,78	C II	6	8011,05	F I	5
8062,36	C I	3	8009,05	F I	1
8062,12	C II	5	8008,45	Xe II	300
8061,340	Xe I	150	8007,79	F I	15
8059,5038	Kr I	1500	8006,1566	Ar I	600
8058,62	C I	8	8005,8	Xe II	2
8057,258	Xe I	200	8003,26	Xe I	10
8054,232	Mg I	7	8003,186	Al I	7
8053,35	Cs I	100	8001,95	Xe II	10
8053,305	Ar I	100	7998,972	Fe I	700
8051,08	Cl I	20	7997,80	Cl I	50
8049,854	Mg I	5	7996,72	Cu II	10
8048,32	C II	3	7996,53	Ti I	3
8047,73	Mg I	3	7996,5	Xe II	3
8047,60	Fe I	15	7995,074	O I	15
8047,28	Xe II	20	7994,473	Fe I	20
8046,13	Ar I	50	7993,42	C I	3
8046,073	Fe I	600	7993,22	Kr II	200
8045,33	C I	4	7993,12	Kr I	5
8044,50	Si II	15	7993,048	Al I	5
8044,308	Ar II	2	7992,90	Ar II	1
8042,18	Xe I	15	7992,53	C I	0
8041,79	Ne I	2	7992,34	Xe II	100
8040,931	F I	1000	7991,5	Xe II	5
8040,56	Xe I	10	7990,78	Kr I	2
8040,50	Kr I	8	7990,68	Cs I	100
8039,39	C II	6	7988,17	Cu II	60
8038,26	Xe II	100	7987,99	Xe II	40
8037,76	C II	5	7987,89	C I	2
8037,23	Ar I	20	7987,333	O I	11
8036,853	Ar II	2	7986,977	O I	13
8035,619	Si I	35	7985,80	Cl I	4
8035,40	Xe II	20	7983,61	Ar II	1
8034,625	Ar II	1	7982,406	Kr I	100
8033,52	Kr I	2	7982,398	O I	11
8031,64	Xe II	100	7981,941	O I	10
8029,67	Xe I	100	7981,82	Kr I	30
8028,86	C II	2	7981,19	Kr I	20
8028,341	Fe I	50	7981,1	Xe II	100
8028,0	Xe II	1	7980,58	Cl I	15
8026,950	Si I	25	7978,96	F I	5
8026,45	Cu II	10			

λ	Symbol	I	λ	Symbol	I
7978,88	Ti I	4	7930,806	Mg I	7
7976,95	Cl I	25	7929,65	F I	4
7976,4	Xe II	3	7928,5996	Kr I	180
7976,03	Xe I	8	7927,35	Ar II	2
7975,579	Si I	13	7927,1172	Ne I	40
7974,76	Xe II	20	7925,850	Si I	15
7974,72	Cl I	20	7924,62	Cl I	100
7973,62	Kr II	120	7920,48	Xe II	10
7972,01	Cu II	8	7920,47	Kr I	40
7970,306	Si I	35	7918,3857	Si I	90
7968,66	Cl I	3	7916,45	Ar I	20
7967,341	Xe I	500	7915,813	Ar II	1
7965,08	Ar I	3	7915,419	N I	7
7962,62	Kr I	1	7915,09	Cl I	25
7960,84	Ar I	2	7913,432	Si I	25
7957,67	Kr I	2	7913,4242	Kr I	200
7957,07	Kr II	3	7912,867	Fe I	6
7956,99	Ar I	10	7912,383	Si I	20
7956,832	K I	4	7911,47	Si II	10
7956,32	F I	300	7910,23	Ar I	4
7955,371	K I	5	7906,91	Ar I	1
7954,22	Xe I	4	7904,770	Ar II	2
7954,09	F I	60	7904,62	Kr I	30
7952,66	F I	2	7902,57	Cu II	25
7952,49	Cl I	15	7899,28	Cl I	45
7952,19	C I	3	7899,27	N I	3
7952,182	O I	9	7898,985	N I	8
7951,35	C I	1	7898,558	F I	500
7950,824	O I	10	7898,10	Cl I	5
7949,17	Ti I	3	7897,7	Xe II	5
7948,52	F I	40	7897,62	N II	4
7948,1755	Ar I	400	7896,368	Mg II	13
7947,566	O I	10	7895,83	Cu II	20
7947,204	O I	3	7893,33	Cl I	10
7946,99	Kr I	20	7891,0777	Ar I	100
7945,878	Fe I	600	7890,56	Cu II	3
7944,60	C I	3	7889,62	F I	8
7944,42	Cu II	25	7889,4	Xe II	50
7944,16	Ne I	20	7887,395	Xe I	300
7944,0011	Si I	140	7886,31	O I	4
7943,8820	Cs I	800	7886,00	Cl I	6
7943,1805	Ne I	200	7882,71	Xe II	20
7943,178	O I	6	7882,36	Kr I	10
7942,54	Xe II	100	7881,76	Kr I	30
7941,09	Fe I	10	7881,667	Mg I	2
7940,65	Cl I	2	7881,320	Xe I	100
7939,49	O I	1	7879,18	F I	300
7938,90	Cl I	8	7878,22	Cl I	75
7938,34	Kr I	2	7877,051	Mg II	12
7937,41	Xe I	40	7875,56	F I	18
7937,166	Fe I	700	7872,50	Cl I	1
7936,9946	Ne I	70	7871,93	Kr I	2
7936,314	F I	350	7870,68	Cl I	1
7933,85	Cl I	50	7868,20	Ar I	40
7933,130	Cu I	1500	7863,91	Kr I	20
7932,3490	Si I	120	7862,7	Xe II	3
7931,41	Kr II	40	7861,91	Ar I	15
7930,93	F I	220			

λ	Symbol	I
7860,89	C I	8
7860,58	Cu II	5
7860,44	Ar I	2
7855,73	Ar I	8
7854,8215	Kr I	800
7853,29	Ar I	1
7852,86	C I	4
7849,967	Si I	30
7849,72	Si II	500
7849,397	Ar II	3
7848,80	Si II	400
7848,25	C I	4
7846,555	Ar II	2
7845,03	Cu II	25
7841,23	Xe I	15
7840,40	Kr I	4
7840,28	C I	2
7840,04	Ne I	1
7840,01	Kr I	8
7839,42	Cl I	8
7839,0550	Ne I	30
7837,40	Cl I	6
7837,11	C I	3
7836,134	Al I	12
7835,309	Al I	11
7833,06	Ne I	7
7832,98	Xe I	10
7832,63	C I	3
7832,224	Fe I	400
7830,76	Cl I	30
7830,21	Kr I	2
7828,28	Xe II	20
7825,80	Cl I	3
7825,66	Cu II	50
7823,72	Al II	2
7822,59	F I	80
7821,35	Cl I	45
7820,79	F I	6
7820,57	Cu II	5
7818,31	Xe II	10
7816,15	He I	5
7815,83	Al II	1
7815,34	Cl I	1
7814,33	Ar I	10
7813,76	F I	10
7812,33	Cu II	10
7811,135	Mg I	3
7810,237	Na I	3
7810,06	Cl I	3
7809,781	Na I	4
7808,04	Fe I	6
7807,66	Cu II	75
7806,52	Kr I	15
7805,8	Xe II	1
7805,19	Cu II	25
7802,651	Xe I	100
7802,27	Cl I	6
7802,252	Ar II	1
7800,212	F I	15000
7800,008	Si I	30
7798,59	Cl I	5
7798,55	Ar I	30
7796,00	C III	4
7795,410	Ar II	2
7791,90	Kr II	6
7790,978	Mg II	4
7790,56	Cl I	5
7790,53	Xe I	1
7789,42	Xe I	15
7787,75	Cl I	4
7787,04	Xe II	100
7786,66	Kr I	2
7786,500	Mg II	5
7783,66	Xe I	50
7781,97	Kr II	100
7780,586	Fe I	300
7780,42	C III	3
7778,74	Cu II	30
7777,82	Cl I	10
7777,1	Xe II	10
7776,28	Kr I	15
7775,388	O I	26
7774,18	Xe II	4
7774,166	O I	27
7772,40	Kr I	5
7772,12	Xe II	20
7771,943	O I	28
7771,10	Cl I	12
7769,18	Cl I	30
7768,43	Kr I	5
7765,89	Kr I	1
7762,237	N II	10
7759,297	Mg I	1
7757,003	Ar II	1
7756,52	Kr II	30
7754,78	Cl I	6
7754,696	F I	18000
7754,37	Cu II	10
7753,28	Ar II	1
7752,905	Si IV	1
7749,16	Kr I	3
7749,16	Kr II	1
7748,278	Fe I	125
7746,828	Kr I	50
7746,343	Mg I	1
7744,94	Cl I	125
7744,09	Cu II	5
7742,71	Si I	40
7741,39	Kr I	10
7740,31	Xe I	40
7738,68	Cu II	30
7735,69	Kr II	250
7730,469	Si IV	1
7726,64	Cu II	5
7726,2	C IV	6

λ	Symbol	I
7725,64	Si IV	2
7724,6281	Ne I	10
7724,2064	Ar I	200
7723,818	Si IV	6
7723,7599	Ar I	200
7722,614	Mg I	1
7718,785	Si IV	5
7717,57	Cl I	100
7712,94	Kr I	1
7712,42	Xe II	30
7711,73	Fe II	15
7710,390	Fe I	25
7708,96	Kr I	1
7707,43	C III	6
7706,77	O I	5
7704,81	Ar I	20
7703,41	Kr I	2
7702,96	N IV	4
7702,89	Cl I	10
7698,959	K I	24
7697,40	Cl I	8
7694,5393	Kr I	500
7692,97	Cl I	7
7692,50	C I	2
7691,550	Mg I	15
7690,165	Mg I	8
7690,10	Ar I	2
7685,2460	Kr I	400
7685,20	C I	4
7683,458	Ar II	1
7681,49	Ar II	1
7680,948	Ar II	2
7680,2668	Si I	100
7878,748	Si IV	4
7672,44	Cl I	25
7670,85	Ne I	5
7670,81	Xe I	1
7670,66	Xe II	200
7670,04	Ar I	50
7667,03	Ar I	4
7666,61	Xe I	10
7665,48	O I	1
7664,899	K I	25
7664,70	Cu II	75
7664,56	Xe I	30
7664,301	Fe I	80
7664,02	Xe I	10
7663,45	O I	3
7662,43	C I	5
7662,3	Ar I	2
7661,223	Fe I	30
7659,902	Mg I	17
7659,50	Cl I	5
7659,152	Mg I	19
7657,603	Mg I	20
7656,86	Cl I	6
7654,555	Si IV	4
7654,031	Ar II	2
7653,8	Xe III	1

λ	Symbol	I
7653,76	Fe I	6
7652,36	Cu II	30
7652,16	Kr I	4
7644,80	Cl II	4
7643,91	Xe I	100
7642,30	Xe I	—
7642,025	Xe I	500
7641,16	Kr II	150
7640,31	Si I	20
7639,99	O I	1
7635,33	Al II	2
7635,13	Kr II	5
7635,1056	Ar I	500
7633,72	Cl I	20
7630,497	Si IV	2
7629,46	Kr II	5
7628,86	Ar I	50
7628,35	Cl I	1
7628,0	N V	—
7627,85	Al II	1
7625,94	C III	2
7621,33	Ne I	5
7620,538	Fe I	25
7620,51	Cl II	4
7619	N V	—
7618,57	Xe II	100
7618,46	N V	5
7618,33	Ar I	30
7618,03	Ar II	1
7615,69	Kr II	3
7615,339	Al I	1
7615,2	N V	—
7614,820	Al I	7
7612,65	C III	7
7612,356	Si III	12
7609,82	Xe I	3
7608,9032	Cs I	500
7608,46	Xe I	5
7607,170	F I	7000
7606,159	Al I	5
7604,97	Xe I	2
7601,5443	Kr I	2000
7600,77	Xe I	10
7600,2	N V	—
7595,29	C III	2
7594,36	Xe I	1
7592,74	He II	—
7592,28	C III	5
7589,61	Xe I	6
7589,558	Mg II	3
7589,320	Ar II	15
7587,4130	Kr I	1000
7586,40	C III	4
7586,044	Fe I	150
7584,680	Xe I	200
7584,29	Xe I	10
7583,797	Fe I	50
7582,40	N IV	2
7582,169	Li I	3
7580,764	Mg II	4

λ	Symbol	I
7579,87	Cu II	10
7579,02	Cu II	30
7578,16	C III	4
7578,07	Cl II	10
7576,68	C III	2
7573,384	F I	5000
7572,06	Ne I	5
7570,93	Xe I	6
7570,09	Cu I	200
7568,925	Fe I	30
7565,53	Cl II	18
7563,214	Al I	3
7562,01	Cu II	25
7561,19	Cl I	4
7559,79	Xe I	40
7554,162	Al I	1
7552,235	F I	5000
7550,63	Kr I	3
7548,45	Xe II	300
7547,06	Cl I	100
7544,0439	Ne I	100
7543,10	Kr I	3
7535,7739	Ne I	300
7531,171	Fe I	60
7530,70	Xe II	50
7530,60	C II	2
7524,46	Kr II	300
7519,86	C II	4
7519,50	C II	7
7515,88	Fe II	6
7515,48	Kr II	20
7514,96	Xe I	3
7514,919	F I	900
7514,6514	Ar I	200
7514,54	Xe I	8
7511,045	Fe I	800
7510,42	Ar I	10
7508,90	C II	3
7508,6	Xe II	1
7507,28	Fe I	8
7505,67	C I	1
7505,31	C II	2
7505,153	Ar II	1
7503,8685	Ar I	700
7503,00	Xe II	3
7501,13	Xe I	20
7497,286	Si III	3
7496,56	Cl I	1
7495,36	Xe II	50
7495,088	Fe I	400
7494,15	Kr I	30
7493,58	Kr I	20
7492,23	Xe I	20
7492,12	Cl I	10
7491,678	Fe I	12
7489,46	Cl I	8
7489,155	F I	2500
7488,8712	Ne I	500

λ	Symbol	I
7486,862	Kr I	100
7486,52	C III	3
7486,225	Mg I	5
7484,24	Ar I	15
7483,44	C I	3
7482,723	F I	2200
7482,19	Si I	25
7480,652	O I	8
7479,148	O I	8
7477,264	O I	7
7476,54	F I	70
7476,473	O I	12
7476,18	C I	2
7474,01	Xe I	25
7473,30	C I	1
7473,226	O I	5
7472,4383	Ne I	50
7472,01	Xe I	40
7471,374	O I	2
7471,37	Al II	1
7471,1676	Ar I	4
7470,09	C I	1
7468,41	Ca I	3
7468,309	N I	16
7467,99	Kr II	6
7466,322	Si III	9
7465,669	Si III	4
7465,645	F I	4000
7465,45	C I	1
7465,01	Kr I	3
7462,624	Si III	8
7462,40	Cl I	8
7462,38	Fe II	20
7461,890	Si III	5
7460,82	Xe III	5
7459,70	Kr I	1
7459,42	Cl I	3
7455,996	Ar II	2
7455,36	Si I	25
7454,08	Cl I	2
7452,5	Cu I	2
7451,00	Xe I	25
7449,12	Al II	5
7448,9	Xe III	1
7445,776	Fe I	200
7445,34	Fe II	6
7444,32	Cl I	3
7442,327	Si III	4
7442,299	N I	15
7441,94	Xe I	20
7440,60	Ti I	3
7440,491	Ar II	4
7438,8981	Ne I	300
7438,15	Cu II	15
7438	O V	—
7436,25	Ar I	10
7436,13	Cl I	10

λ	Symbol	I	λ	Symbol	I
7435,78	Kr II	200	7362,83	Kr I	4
7435,71	Cl I	7	7362,297	Al I	9
7435,33	Ar I	30	7361,568	Al I	8
7434,74	Kr II	15	7361,34	Kr I	1
7433,85	Cu II	5	7359,97	Kr II	3
7428,574	Ar II	2	7359,96	Kr I	5
7427,2	Cu I	5	7358,338	Ar II	2
7425,54	Kr I	60	7357,74	Ti I	3
7425,290	Ar I	12	7355,58	Xe I	40
7424,60	Si I	85	7355,48	Kr I	4
7424.05	Xe I	20	7355,180	Ar II	2
7423,639	N I	14	7353,96	C III	0
7423,4969	Si I	425	7353,42	Kr III	1
7422,26	Ar I	6	7353,316	Ar I	100
7420,70	Cu II	8	7350,78	Ar I	6
7419,341	Ar II	1	7348,049	Ar II	7
7415,9462	Si I	275	7345,34	Ar I	1
7415,35	Si I	40	7344,72	Ti I	4
7414,10	Cl I	90	7343,37	Xe II	30
7412,334	Ar I	15	7342,74	Cl I	3
7411,178	Fe I	100	7342,00	Cl I	1
7410,14	Xe II	4	7341,16	Kr I	2
7409,0818	Si I	200	7339,30	Xe II	300
7408,467	Si III	3	7336,480	Xe I	50
7407,02	Kr II	400	7334,66	Fe II	8
7405,774	Si I	375	7334,33	Kr I	4
7405,77	Xe I	3	7331,957	F I	5000
7404,51	Xe I	12	7331,74	Cu II	15
7404,34	Cu II	100	7329,33	Cl I	3
7402,70	Kr I	1	7327,8	N V	—
7400,5	Xe II	4	7327,00	Kr I	5
7400,41	Xe I	30	7326,146	Ca I	400
7399,89	Cu II	20	7326,02	Cu II	15
7398,688	F I	10000	7325,57	Ne I	15
7395,52	Si I	15	7323,05	Xe I	2
7393,793	Xe I	150	7320,70	Fe II	40
7392,97	Ar I	15	7319,94	Xe I	15
7389,425	Fe I	80	7319,33	F I	6
7389,28	Cl II	7	7316,87	Xe I	20
7387,685	Mg I	12	7316,272	Xe I	70
7387,004	Mg I	5	7316,0068	Ar I	30
7386,402	Fe I	8	7314,303	F I	700
7386,002	Xe I	100	7313,77	F I	40
7383,9796	Ar I	400	7313,01	Xe I	1
7382,47	Cl I	3	7312,452	Xe I	80
7382,18	Cu II	10	7312,29	Si II	3
7380,92	Cl I	4	7311,724	Ar I	100
7380,433	Ar II	15	7311,019	F I	15000
7378,38	Xe II	30	7310,24	Fe II	6
7376,46	Fe II	20	7309,033	F I	1000
7373,491	Na I	1	7307,957	Fe II	50
7373,229	Na I	2	7307,93	Ne I	15
7373,00	Si I	35	7307,37	Xe I	5
7372,65	Cl I	1	7306,60	Cu II	12
7372,1189	Ar I	100	7304,82	Ne I	30
7367,02	Kr I	2	7301,80	Xe II	200
7366,80	Kr I	2	7301,29	Kr II	4
7364,73	C I	3			

λ	Symbol	I
7301,25	Kr I	5
7298,98	F I	150
7298,93	Xe III	1
7293,068	Fe I	15
7291,060	Mg I	10
7290,26	Si I	55
7289,78	Kr II	400
7289,1730	Si I	400
7288,760	Fe I	10
7287,36	Fe II	6
7287,262	Kr I	80
7286,11	C I	0
7285,301	Xe I	60
7284,44	Ar I	6
7284,34	Xe II	100
7284,236	Ar II	4
7283,961	Xe I	40
7282,81	Si I	40
7281,349	He I	500
7280,454	Ar II	2
7279,9570	Cs I	500
7279,895	Cs I	100
7279,75	Xe II	4
7276,47	Xe II	4
7275,294	Si I	160
7272,97	Kr II	4
7272,9349	Ar I	100
7270,70	Cs I	15
7270,66	Ar I	10
7268,28	Kr I	1
7267,20	Ar I	2
7267,090	Si III	2
7266,49	Xe I	25
7265,173	Ar I	3
7264,99	Fe II	10
7262,54	Xe I	20
7258,6	Xe II	2
7257,94	Xe I	60
7256,63	Cl I	125
7256,53	N II	2
7255,83	Cu II	20
7254,529	O I	17
7254,447	O I	20
7254,154	O I	19
7252,48	Cl I	3
7251,74	Ti I	8
7250,87	Xe I	5
7250,625	Si I	180
7250,14	Si I	25
7249,92	Xe I	2
7248,99	Cs	2
7245,38	Xe II	2
7245,1665	Ne I	1000
7244,94	Xe I	20
7244,86	Ti I	10
7244,76	Cl I	3
7241,56	Kr II	2

λ	Symbol	I
7241,32	C I	2
7240,12	F I	2
7239,885	Fe I	8
7238,20	Xe I	3
7237,17	C II	7
7236,42	C II	20
7235,82	Si I	60
7235,326	Si I	100
7234,58	Kr I	2
7233,546	Ar II	15
7233,52	Kr II	1
7231,32	C II	18
7229,93	Ar I	4
7229,01	Cs	35
7228,5356	Cs I	500
7227,34	Kr I	2
7226,206	Si I	100
7224,51	Fe II	12
7224,24	C I	1
7224,103	Kr I	100
7223,668	Fe I	12
7222,39	Fe II	8
7220,24	Xe I	1
7219,70	Cs I	15
7216,20	Ti I	5
7216,03	C I	0
7215,97	Xe II	20
7215,06	N II	3
7213,13	Kr II	250
7212,29	C III	1
7211,836	Si III	1
7210,52	C III	2
7209,44	Ti I	20
7209,14	Xe I	5
7208,21	Si I	25
7207,406	Fe I	500
7206,9812	Ar I	100
7205,99	Cs	2
7202,55	Ar I	2
7202,360	F I	15000
7202,26	C I	2
7202,194	Ca I	200
7200,79	Xe I	15
7200,59	Kr I	2
7194,94	Cl I	5
7194,92	Cu II	15
7193,90	Si I	30
7193,58	Si I	65
7193,56	Cu I	50
7193,23	Fe II	8
7193,172	Mg I	10
7188,32	Cs	2
7188,20	N II	2
7187,341	Fe I	800
7185,92	Xe III	2
7185,68	Cl I	2
7184,89	Si I	70
7184,57	Si I	20

λ	Symbol	I
7182,098	Ar II	2
7180,47	Kr I	3
7177,50	He II	—
7176,34	Ar I	4
7174,90	Xe III	2
7173,9380	Ne I	1000
7172,70	Xe I	10
7166,676	Mg II	2
7165,545	Si I	200
7164,83	Xe II	800
7164,69	Si I	70
7164,469	Fe I	250
7162,57	Ar I	8
7160,88	Cs	2
7158,83	Ar I	30
7157,360	O I	7
7156,81	Kr II	1
7156,80	O I	12
7154,29	Cu I	5
7152,21	Kr I	5
7151,08	Si III	2
7149,554	Cs	10
7149,03	Xe II	300
7148,147	Ca I	500
7147,80	Cl II	3
7147,50	Xe II	100
7147,0408	Ar I	30
7146,38	Cl I	5
7144,19	C II	1
7143,81	Xe II	8
7143,45	Kr I	8
7139,99	Kr II	60
7138,87	N II	4
7138,70	Ne I	30
7136,57	Xe I	15
7135,040	Li I	1
7134,99	Fe II	5
7134,11	C II	6
7133,67	Kr I	1
7133,27	Xe II	10
7132,45	C II	1
7132,11	C I	1
7130,942	Fe I	150
7130,532	Cs	5
7129,18	N IV	—
7127,890	F I	30000
7127,35	Cl I	3
7127,27	N IV	1
7125,84	Si II	4
7125,825	Ar I	30
7125,73	C II	7
7125,37	Ne	3
7124,66	Cu I	5
7122,98	N IV	5
7122,20	C I	1
7121,740	Ar II	4
7121,18	Cs	2
7119,90	C II	12
7119,67	C I	7
7119,598	Xe I	500
7116,99	C I	8
7115,63	C II	10
7115,19	C I	9
7113,04	C II	7
7113,18	C I	9
7112,48	C II	6
7112,2	Ne I	10
7111,48	C I	7
7111,30	N IV	1
7109,61	F I	3
7109,40	N IV	3
7108,94	C I	3
7107,4777	Ar I	200
7103,28	N IV	1
7101,190	Ar II	1
7100,8	Xe II	2
7100,12	C I	5
7094,20	Cl I	8
7093,25	C I	3
7090,560	Ar II	1
7090,404	Fe I	40
7089,51	Kr I	1
7087,83	C I	4
7086,80	Cl I	25
7086,70	Ar I	15
7086,43	Kr I	1
7085,72	Cs	2
7085,51	C I	0
7084,644	Al I	6
7083,968	Al I	5
7082,35	Cl I	3
7082,15	Xe II	200
7078,46	Xe I	1
7078,44	Kr II	3
7077,024	Ar II	5
7076,48	C I	2
7075,64	Cl I	3
7075,0	Xe II	2
7074,98	Cl II	4
7074,86	C I	1
7073,97	Kr II	60
7072,43	Xe II	4
7068,73	Ar I	30
7068,413	Fe I	40
7068,410	Si IV	4
7067,44	Fe II	20
7067,2175	Ar I	400
7065,707	He I	300
7065,190	He I	2500
7064,42	Ne I	2
7063,70	C II	8
7063,624	Al II	3
7060,409	Mg I	8
7060,18	Cl I	2
7059,1079	Ne I	200

λ	Symbol	I
7058,25	Cl I	1
7057,45	Kr III	2
7057,27	Kr I	10
7056,87	C I	0
7056,56	Al II	4
7054,993	Ar II	3
7053,09	C II	6
7052,57	Xe II	3
7051,2937	Ne I	70
7051,06	Xe I	3
7049,36	Xe I	1
7049,34	Xe III	2
7049,07	Xe I	1
7047,939	Si IV	6
7047,58	Si III	5
7047,37	Xe I	30
7046,26	C II	4
7043,94	Xe III	4
7042,056	Al II	5
7039,37	Cu I	25
7038,80	Ti I	6
7038,251	Fe I	40
7037,469	F I	45000
7037,25	C III	7
7036,30	Cl I	5
7035,53	Xe I	20
7034,903	Si I	250
7034,80	Xe I	3
7032,4128	Ne I	1000
7030,2519	Ar I	100
7026,62	Si I	25
7025,52	O I	3
7024,649	Fe I	10
7024,0500	Ne I	500
7022,976	Fe I	50
7022,75	Cu II	2
7022,56	Kr II	2
7019,30	Cl I	4
7019,02	Xe I	30
7017,646	Si I	90
7017,28	Si I	30
7017,06	Xe II	80
7016,74	Si I	10
7016,436	Fe I	60
7016,075	Fe I	20
7014,73	N II	2
7013,98	N II	2
7011,24	Cl I	3
7008,62	Kr I	2
7008,00	Cl I	10
7006,30	Cl I	4
7005,883	Si I	180
7004,06	O IV	—
7003,96	Xe II	50
7003,5665	Si I	180
7003,10	Xe I	4
7002,228	O I	17
7001,915	O I	15
7001,62	Kr I	2
7000,79	Kr I	7
7000,05	Cu I	2
6999,902	Fe I	30
6998,358	Si IV	3
6995,88	Cl I	12
6993,27	Cl II	2
6993,05	Kr I	2
6992,88	Si I	15
6992,17	Ar I	4
6991,65	Xe I	1
6990,88	Xe II	2000
6990,122	Ar II	5
6985,708	Ar II	1
6983,4912	Cs I	25
6982,69	F I	1
6982,05	Xe I	30
6982,02	N I	00
6981,85	Cl I	25
6979,681	Cs	15
6979,60	Cl I	3
6979,10	N I	1
6978,856	Fe I	100
6977,95	Kr II	3
6977,95	Kr III	3
6977,00	Cl I	5
6976,523	Si I	80
6976,182	Xe I	100
6975,64	N II	4
6973,2966	Cs I	500
6972,674	Mg I	5
6971,71	F I	1
6968,34	Cu I	5
6966,81	N II	3
6966,80	Cl I	8
6966,349	F I	4000
6965,430	Ar I	400
6965,404	Mg I	6
6964,672	K I	12
6964,18	K I	7
6962,50	Cl I	6
6962,31	C I	0
6960,23	Ar I	20
6955,519	Cs II	20
6952,13	Cl II	25
6951,50	N I	1
6951,46	Ar I	20
6951,261	Fe I	25
6945,22	N I	4
6945,208	Fe I	150
6944,06	Kr II	10
6942,11	Xe II	1000
6941,752	N II	5
6938,767	K I	20
6937,6658	Ar I	100
6936,69	Xe I	8
6936,284	K I	12
6935,82	Cu I	5

λ	Symbol	I	λ	Symbol	I
6935,62	Xe I	50	6872,107	Xe I	100
6935,38	Kr I	2	6872,05	C III	4
6932,90	Cl I	25	6871,2898	Ar I	150
6931,39	O IV	—	6870,85	Kr II	40
6930,45	Cl II	4	6870,4552	Cs I	200
6929,4672	Ne I	1000	6870,419	Cs I	2
6926,90	N I	1	6870,215	F I	8000
6925,53	Xe I	100	6869,74	O II	1
6925,35	Cl I	6	6869,63	Kr I	20
6925,010	Ar I	2	6869,580	N II	4
6924,67	Xe I	15	6868,80	C III	1
6924,40	Cl I	5	6867,22	Si I	20
6922,22	Xe I	8	6866,838	Xe I	50
6920,31	Cl I	3	6865,58	Xe I	5
6920,06	Cu I	50	6865,36	Cl I	8
6919,96	Al II	1	6863,535	Ar II	20
6917,93	Al II	1	6863,20	Xe I	20
6916,702	Fe I	60	6862,82	Kr I	3
6911,29	Kr I	2	6862,71	C III	3
6911,084	K I	19	6861,47	Ti I	6
6910,82	Xe I	30	6861,270	Ar II	15
6910,75	O II	3	6860,19	Xe I	40
6910,32	Cl I	6	6858,164	Fe I	40
6910,22	Xe II	100	6857,27	C III	2
6909,816	F I	6000	6857,030	N II	3
6909,0	O V	—	6856,030	F I	50 000
6908,11	O II	2	6855,179	Fe I	150
6906,54	O II	4	6854,45	Cl I	10
6905,94	Cu I	100	6853,70	C III	1
6904,68	Kr I	100	6853,32	Kr I	2
6904,22	Kr I	15	6851,884	Ar I	4
6902,475	F I	15000	6851,65	Si III	7
6900,880	Ar II	2	6851,20	C III	1
6899,64	C III	1	6851,18	Si III	3
6895,29	O II	5	6850,21	Cl II	40
6895,005	Cs I	—	6850,13	Xe I	30
6894,898	Mg I	4	6848,906	Cs I	—
6892,42	Cs	2	6848,82	Xe I	50
6890,90	Cu I	10	6848,568	Si I	30
6890,88	He II	—	6847,237	N II	4
6890,41	Xe II	3	6846,97	O II	1
6889,92	Cu I	10	6846,613	Xe I	60
6888,1704	Ar I	100	6846,540	Ar II	1
6887,834	N II	5	6846,40	Kr I	20
6887,10	Ar I	20	6844,84	Xe I	2
6886,618	Ar II	20	6844,27	Xe I	1
6885,77	Fe I	20	6843,671	Fe I	60
6885,07	O II	1	6841,86	Cl II	10
6882,155	Xe I	300	6841,74	Cl I	6
6881,94	Cu I	10	6841,50	Xe I	20
6881,09	C III	1	6841,349	Fe I	80
6879,59	Ar I	40	6840,99	Cu I	3
6878,5	O V	—	6840,96	Xe I	8
6876,69	Xe II	3	6840,23	Cl I	15
6874,30	N I	1	6839,584	Ar II	4
6873,2	Xe II	10	6837,74	F I	5
6872,85	Cl I	6	6837,60	Cl I	5
6872,43	Cu II	3			

λ	Symbol	I	λ	Symbol	I
6837,094	Al II	3	6790,20	Cl I	2
6834,38	Si III	2	6789,8	O V	—
6834,264	F I	9000	6789,21	Kr I	1
6834,094	N II	6	6788,71	Xe II	100
6834,08	Si III	4	6787,851	Mg II	8
6831,62	Cl II	30	6787,22	C II	6
6831,560	Si III	6	6784,45	Cl I	1
6830,1	O V	8	6783,90	C II	10
6829,82	Si II	40	6781,451	Mg II	7
6829,09	Kr I	8	6780,61	C II	5
6828,610	Fe I	50	6780,40	Cu II	3
6828,12	C I	6	6779,933	Ar I	4
6827,315	Xe I	200	6779,93	C II	8
6827,2529	Ar I	30	6778,60	Xe I	40
6825,56	F I	15	6777,57	Xe I	50
6825,22	Cs	15	6776,623	Si III	2
6824,6520	Cs I	200	6776,15	Kr I	3
6823,40	Cu II	3	6775,64	Cu I	2
6823,382	Al II	2	6774,93	C III	0
6821,86	Cu I	2	6773,984	F I	7000
6819,4	O V	—	6773,37	C III	1
6819,270	Mg II	8	6771,22	Kr II	50
6818,45	Si II	20	6770,70	Cu II	8
6818,38	Xe I	15	6767,12	Xe I	10
6818,371	Ar II	8	6767,007	F I	50
6818,291	Ar I	4	6766,8	O V	—
6818,13	Kr III	1	6766,6134	Ar I	100
6818,12	Xe III	1	6766,54	F I	5
6816,827	Al II	1	6765,20	Cl I	3
6816,50	Cl I	2	6764,51	Kr I	2
6815,64	Xe I	12	6764,43	Kr II	80
6813,10	Kr I	50	6763,61	Kr II	100
6812,860	Mg II	7	6763,325	F I	5
6812,29	C II	3	6762,934	F I	70
6811,67	F I	1	6762,30	Cl I	2
6811,50	Cl I	3	6762,17	C III	4
6810,25	Fe I	20	6759,586	Ne I	15
6810,04	Cl I	15	6759,42	Cl II	35
6809,989	N II	7	6758,60	N I	4
6809,90	Cu II	4	6758,55	Cu II	8
6808,532	Ar II	9	6757,75	Cl I	5
6806,851	Fe I	10	6756,548	Ar II	20
6806,85	F I	10	6756,10	Ar I	100
6806,60	Cu II	4	6755,16	C II	3
6805,74	Xe II	1000	6754,30	Ar I	8
6805,244	Si III	4	6752,8347	Ar I	100
6801,31	N II	1	6752,40	N I	4
6800,68	C II	7	6751,88	Si II	5
6799,288	Ar II	3	6751,54	Cl I	4
6798,51	Ca I	6	6750,55	C II	8
6798,11	C II	3	6750,28	Si II	20
6795,528	F I	1500	6750,155	Fe I	100
6795,40	Kr I	4	6744,38	C III	7
6793,82	N I	00	6743,124	Ti I	10
6793,53	Kr III	3	6742,43	C II	3
6791,92	Cl I	3	6742,24	C III	5
6791,47	C II	7	6741,90	F I	1
6790,37	Xe II	80			

λ	Symbol	I	λ	Symbol	I
6741,64	Si I	30	6698,474	Ar I	6
6741,29	N I	3	6697,45	Cl I	2
6741,12	Cu I	100	6696,296	Ar II	4
6740,10	Kr I	20	6696,023	Al I	13
6738,62	C II	6	6694,32	Xe II	400
6738,058	Ne I	70	6691,22	Xe II	1
6737,76	F I	6	6690,481	F I	1800
6737,64	Cu II	5	6689,91	Ar I	2
6734,00	C II	2	6688,79	C I	4
6733,58	C II	2	6686,04	Cl II	45
6733,48	N I	6	6684,73	Ar I	6
6731,07	C II	5	6684,307	Ar II	50
6731,04	C III	6	6683,95	C I	4
6730,24	Cl I	5	6683,55	Kr III	1
6728,41	Kr III	1	6683,26	He II	—
6728,008	Xe I	200	6681,036	Xe I	20
6727,39	C III	6	6681,03	Cl II	15
6727,19	C II	4	6680,26	Ti II	1
6726,538	O I	6	6679,65	Si II	3
6726,478	Fe I	20	6678,972	Xe I	25
6726,283	O I	9	6678,39	Cl I	10
6724,56	C II	2	6678,2764	Ne I	500
6724,476	Cs	15	6678,19	O II	0
6723,65	C II	1	6678,151	He I	1000
6723,40	Cl I	4	6677,994	Fe I	600
6723,36	Kr I	4	6677,2812	Ar I	30
6723,2943	Cs I	500	6674,11	C I	4
6723,12	N I	9	6672,23	Cu I	10
6722,893	Ar I	4	6672,10	Ar I	2
6721,853	Si I	100	6671,88	Si II	100
6721,35	O II	5	6671,84	C I	5
6719,2193	Ar I	100	6668,920	Xe I	150
6717,911	Ti II	1	6667,556	Si IV	5
6717,685	Ca I	500	6667,00	F I	7
6717,2	N V	—	6666,965	Xe I	60
6717,0428	Ne I	70	6666,94	O II	1
6717,04	Si II	50	6666,8967	Ne I	100
6714,65	Cl I	4	6666,75	N I	0
6713,43	Cl II	40	6666,356	Ar II	15
6713,12	N I	1	6665,00	Si II	15
6711,29	C I	1	6664,85	Xe I	4
6709,90	Cl I	15	6664,0533	Ar I	100
6709,86	Ca I	1	6663,444	Fe I	80
6708,81	N I	4	6663,1	Xe II	2
6708,282	F I	400	6663,04	C I	4
6707,807	Li I	1000	6662,73	C I	3
6706,46	Xe I	1	6661,68	Cl II	75
6706,20	N I	4	6660,99	Cu II	8
6705,117	Fe I	15	6660,6784	Ar I	100
6703,574	Fe I	10	6660,62	F I	12
6703,20	Cl I	6	6660,52	Si II	50
6702,25	Xe II	80	6657,92	Xe I	20
6701,207	Si IV	7	6657,499	Ar II	2
6699,40	Cl I	3	6655,51	C I	6
6699,38	Si II	20	6656,88	Ar I	6
6699,228	Kr I	60	6656,510	N I	1
6698,8752	Ar I	100	6654,61	C I	3
6698,673	Al I	11			

λ	Symbol	I	λ	Symbol	I
6653,95	C I	1	6614,96	Xe II	10
6653,78	O I	5	6614,354	Ar II	6
6653,75	Cl II	25	6613,622	N II	5
6653,583	Ar II	1	6613,31	Xe II	4
6653,458	N I	5	6612,38	Kr I	2
6652,239	Kr I	40	6611,35	C I	4
6652,0925	Ne I	150	6611,196	Ar II	2
6651,97	F I	0,5	6611,04	F I	5
6651,75	Kr III	10	6610,565	N II	13
6650,405	F I	400	6609,55	F I	2,5
6649,51	F I	6	6609,26	Cl I	7
6649,22	Cu II	2	6609,117	Fe I	30
6648,75	Xe I	3	6608,87	Xe I	10
6647,94	Kr I	2	6607,73	F I	2,5
6646,564	Cs	15	6607,41	Xe I	30
6646,510	N I	2	6606,77	N I	00
6644,963	N I	9	6605,79	C I	1
6643,716	Ar II	100	6605,12	Kr I	2
6643,10	Cl I	1	6605,00	Kr II	15
6642,9	Xe II	1	6604,86	F I	0,6
6641,41	Cu II	10	6604,8542	Ar I	30
6640,90	O II	4	6604,57	Cl I	7
6640,80	Ne I	5	6604,02	Ar I	2
6640,012	Ne I	10	6602,907	Ne I	100
6639,743	Ar II	30	6602,90	Kr II	10
6638,85	Xe II	2	6602,90	Kr III	10
6638,226	Ar II	50	6602,87	Xe I	4
6636,938	N I	4	6602,42	C I	2
6635,65	Si I	25	6600,10	Cl I	3
6634,789	N II	3	6599,725	F I	6
6634,7	Cu I	2	6599,112	Ti I	12
6634,36	Kr II	15	6598,9529	Ne I	1000
6634,13	Xe II	6	6598,84	Xe II	80
6633,772	Fe I	50	6598,684	Ar I	6
6632,464	Xe I	50	6597,607	Fe I	15
6632,44	Xe II	2	6597,25	Xe II	300
6632,087	Ar I	8	6596,90	F I	0,8
6631,85	Cu II	2	6596,85	C I	1
6630,834	Mg I	2	6596,85	F I	8
6630,44	Xe I	2	6596,1155	Ar I	8
6629,795	N II	7	6595,666	N II	3
6629,67	Cu I	5	6595,561	Xe I	100
6628,6605	Cs I	35	6595,24	C I	1
6627,96	Kr II	2	6595,01	Xe II	800
6627,62	O II	3	6595,00	K II	2
6627,28	Fe II	5	6594,66	Ar I	2
6627,02	N I	0	6593,875	Fé I	60
6624,29	Cu II	8	6592,920	Fe I	300
6624,22	Kr II	2	6591,45	C I	1
6622,543	N I	3	6590,86	Xe I	8
6622,05	C II	1	6589,21	F I	4
6621,61	Cu I	30	6588,69	Fe II	5
6620,977	Ar II	6	6587,61	C I	8
6620,569	Mg II	6	6586,5096	Cs I	500
6620,440	Mg II	5	6586,27	C I	2
6620,02	Xe II	200	6586,022	Cs I	35
6618,40	Xe II	50	6583,71	Si I	15
6617,23	C I	0			

λ	Symbol	I	λ	Symbol	I
6583,36	F I	2	6546,245	Fe I	200
6583,27	Xe I	20	6546,12	Xe I	20
6582,88	C II	15	6545,973	Mg II	11
6581,60	Ar I	2	6545,530	N II	3
6580,389	F I	300	6544,162	N II	4
6578,871	F I	12	6543,360	Xe I	40
6578,77	C I	2	6542,40	Cl I	8
6578,05	C II	18	6541,93	Cu II	2
6576,42	Kr I	20	6540,409	Ar II	2
6575,180	Ti I	3	6538,1137	Ar I	30
6575,024	Fe I	30	6536,55	Kr I	8
6573,68	Xe II	30	6536,440	Cs II	15
6573,61	F I	5	6536,10	Cl I	1
6572,781	Ca I	50	6535,163	Si III	2
6571,37	Ar I	2	6533,159	Xe I	100
6570,07	Kr II	150	6532,927	Ar II	2
6569,694	F I	450	6532,8824	Ne I	100
6569,224	Fe I	50	6532,550	N II	5
6569,14	F I	7	6531,39	Cl I	20
6569,13	Xe II	5	6530,52	Ar I	1
6568,71	C I	2	6530,30	Cu II	8
6567,35	Cl I	3	6528,65	Xe II	200
6565,90	Cl I	1	6527,1989	Si I	45
6565,32	Kr II	6	6527,16	He II	—
6564,50	Cu II	10	6526,609	Si I	45
6564,20	N II	3	6524,357	Si III	6
6564,170	Ar II	3	6522,626	Si III	4
6563,59	F I	7	6522,39	N II	2
6563,19	Xe II	15	6522,38	Cl II	10
6562,849	H	2000	6521,508	Xe I	40
6562,725	H	1000	6521,485	Si III	3
6561,78	N II	3	6521,14	Cu II	14
6561,032	D	3000	6518,73	Si I	20
6560,65	Xe I	4	6518,374	Fe I	20
6560,556	Si I	25	6517,01	Fe II	5
6560,435	T	3000	6516,053	Fe II	20
6560,203	N II	3	6515,48	Xe II	1
6560,099	He II	100	6513,848	Ar I	8
6559,97	Xe I	25	6512,83	Xe II	300
6559,580	Ti II	1	6512,71	F I	12
6557,724	Ar II	2	6510,95	Kr II	100
6556,70	Xe II	4	6510,14	Kr II	8
6556,066	Ti I	25	6509,089	Ar II	6
6555,69	Kr I	6	6509,00	Cl I	2
6555,56	Kr I	2	6508,742	Ca I	1
6555,4624	Si I	45	6508,37	Kr I	3
6555,05	Cu II	5	6508,184	Ar II	6
6554,47	N II	3	6508,135	Ti I	3
6554,236	Ti I	20	6507,50	Xe I	3
6554,196	Xe I	50	6506,5279	Ne I	100
6553,93	F I	6	6506,45	N I	0
6553,66	Xe I	4	6506,33	Fe II	6
6551,58	Cu II	2	6506,254	Cs II	5
6551,498	Ar II	3	6506,138	Ar II	3
6551,28	Cl I	3	6504,89	Kr I	10
6550,80	Cl I	3	6504,608	N II	6
6547,350	Ar II	3	6504,18	Xe I	200
6546,276	Ti I	20			

λ	Symbol	I
6502,21	Cl I	5
6502,157	Ar II	3
6502,08	F I	18
6501,348	Ar II	4
6500,37	Xe I	15
6500,216	Ar II	12
6499,649	Ca I	30
6499,52	N I	3
6499,109	Ar I	6
6498,950	Fe I	5
6498,718	Xe I	100
6497,689	Ti I	3
6497,43	Xe I	30
6496,456	Fe I	20
6495,528	Cs II	15
6494,985	Fe I	1000
6494,04	Cu II	30
6493,971	Ar I	15
6493,780	Ca I	80
6493,7	Kr II	2
6493,05	Fe II	8
6491,79	N II	2
6491,61	Ti II	2
6491,28	N I	3
6490,55	Cl I	1
6488,07	Kr I	15
6487,765	Xe I	120
6487,55	N III	0
6485,18	Cu I	5
6484,88	N I	9
6484,46	Cu II	20
6483,75	N I	3
6483,076	Ar II	20
6482,74	N I	9
6482,053	N II	13
6481,877	Fe I	20
6481,73	N I	2
6481,46	Cu II	15
6481,141	Ar I	8
6480,50	N I	0
6480,085	Ar II	2
6479,69	Xe II	2
6478,69	N III	2
6478,45	F I	8
6478,07	Cl II	2
6476,39	F I	7
6475,632	Fe I	12
6475,38	Cl II	2
6475,312	Ar II	4
6474,20	Cu I	10
6472,841	Xe I	150
6472,6226	Cs I	15
6472,431	Ar II	6
6471,86	Si III	2
6471,660	Ca I	40
6471,45	Cl I	3
6471,03	N I	1
6470,89	Kr II	50
6470,152	Cu II	50
6469,705	Xe I	300
6469,214	Fe I	15
6468,77	N III	00
6468,32	N I	4
6468,050	Ar II	7
6466,86	N III	4
6466,60	Cu II	3
6466,5505	Ar I	20
6465,32	Cl II	3
6464,70	Ca I	1
6464,60	Cl I	1
6463,50	F I	70
6463,03	N III	2
6462,730	Fe I	30
6462,566	Ca I	125
6461,95	C II	5
6461,50	Xe I	3
6461,48	Xe II	3
6460,33	C III	0
6458,403	Ar II	2
6457,93	N I	3
6457,69	N II	0
6457,54	Cu II	3
6457,14	Cl I	1
6457,06	F I	0,8
6456,874	Ca II	8
6456,489	Ar II	3
6456,376	Fe II	200
6456,2910	Kr I	200
6455,975	O I	19
6455,600	Ca I	10
6454,77	C II	1
6454,445	O I	18
6454,19	Kr I	1
6453,95	N III	3
6453,602	O I	17
6453,32	F I	10
6452,75	N I	1
6452,29	Si I	20
6451,79	Xe I	10
6450,78	N III	2
6450,48	Xe I	7
6450,30	Cl I	12
6449,810	Ca I	50
6448,78	Kr I	10
6448,70	Xe I	2
6448,49	Cu II	10
6448,49	N I	0
6448,14	F I	0,6
6447,69	F I	6
6446,43	Fe II	20
6445,117	Ar II	1
6445,05	N III	2
6444,7118	Ne I	150
6444,70	Kr III	1
6443,858	Ar II	8

λ	Symbol	I	λ	Symbol	I
6443,76	Cl I	2	6413,55	C I	3
6443,47	Cu II	5	6412,53	Kr II	4
6442,93	Fe II	6	6412,38	Xe I	10
6442,3	Xe II	1	6411,659	Fe I	400
6441,908	Ar II	9	6411,18	Cu II	10
6441,70	N I	5	6410,17	Kr I	5
6441,698	Cu II	40	6409,84	Kr II	10
6440,95	N I	3	6409,753	Ne I	150
6440,74	Kr II	5	6408,904	Ar II	6
6439,073	Ca I	150	6408,05	Cl I	7
6437,79	Si I	8	6408,028	Fe I	60
6437,604	Ar II	8	6407,27	Si I	15
6437,01	N I	4	6406,619	Mg I	6
6434,79	Cl I	15	6406,44	He II	—
6434,11	F I	15	6405,89	Ca I	3
6433,683	Ar II	1	6405,171	F I	60
6433,45	N II	1	6404,69	Kr II	3
6432,78	Cu II	3	6403,70	Cu II	5
6432,654	Fe II	8	6403,004	Ar II	6
6431,9693	Cs I	15	6402,2460	Ne I	2000
6431,559	Ar I	15	6401,076	Ne I	100
6430,852	Fe I	300	6400,013	Fe I	800
6430,155	Xe I	20	6399,41	Cl II	10
6428,05	N I	00	6399,215	Ar II	15
6427,96	K II	5	6399,16	N II	2
6426,73	Xe II	2	6398,63	Cl I	20
6425,61	Cl I	8	6397,99	Xe II	60
6424,144	Ar II	1	6397,98	C I	5
6423,90	Cu II	30	6397,184	Ar II	2
6422,93	N I	3	6396,614	Ar II	7
6422,903	Ar II	6	6395,16	Ca I	3
6422,87	F I	2,5	6395,09	Kr III	2
6422,43	F I	7	6394,75	Cl I	4
6421,7108	Ne I	100	6394,723	Ar II	5
6421,47	Xe II	1	6394,67	Si I	15
6421,355	Fe I	200	6394,28	Kr II	4
6421,0283	Kr I	100	6393,803	Ar II	6
6420,47	N I	3	6393,605	Fe I	400
6420,18	Kr II	300	6391,30	Cl II	3
6419,977	Fe I	30	6391,14	Kr II	30
6419,541	Cs II	10	6391,117	Ar II	2
6419,25	Cl II	8	6389,87	C I	2
6418,98	Xe I	30	6388,335	Ar II	5
6418,58	Xe II	20	6386,94	Cs	25
6418,41	Xe I	30	6385,72	C II	1
6418,354	Ar II	8	6385,51	Cl II	2
6417,59	Cl II	2	6385,473	Fe II	5
6417,54	C I	2	6385,17	F I	10
6417,417	Ar II	1	6384,7189	Ar I	100
6417,05	N I	2	6384,31	N II	2
6416,905	Fe II	20	6384,13	Cl II	5
6416,61	Kr II	60	6383,753	Fe II	15
6416,31	F I	18	6383,095	Ar II	2
6416,3075	Ar I	100	6382,9914	Ne I	1000
6415,65	Kr I	20	6382,696	Ar II	3
6414,97	Si I	25	6380,77	N IV	8
6414,62	Cu II	20	6379,615	N II	9
6413,651	F I	8000			

λ	Symbol	I	λ	Symbol	I
6378,79	C I	0	6339,897	Ar II	3
6377,84	Cu II	20	6337,58	Xe I	8
6376,28	Cl I	3	6337,20	C I	1
6375,945	Ar II	3	6336,104	Ti I	8
6375,28	Xe II	100	6335,70	C I	0
6374,292	O I	4	6335,70	Al II	5
6373,58	Kr I	30	6335,335	Fe I	10
6373,37	Cl I	1	6334,96	Cl I	4
6373,33	F I	2	6334,4279	Ne I	1000
6373,27	Cu II	5	6333,97	Xe I	40
6373,19	Kr I	1	6333,142	Ar II	5
6371,77	F I	2	6332,832	Ar II	1
6371,65	Xe III	1	6332,499	Ar II	5
6371,359	Si II	1000	6331,969	Fe II	12
6369,5783	Ar I	30	6331,954	Si I	45
6369,128	Ar II	2	6331,50	Xe I	20
6368,26	Kr I	4	6330,901	Ne I	150
6367,98	Cl I	3	6328,6	O V	—
6367,43	F I	10	6328,474	Ar II	1
6366,354	Ti I	8	6328,39	N II	5
6366,282	O I	3	6328,1646	Ne I	300
6365,84	F I	0,8	6326,74	Cl I	3
6365,5235	Cs I	2	6326,2055	Cs I	2
6365,440	Ar II	1	6326,117	Ar II	2
6365,013	Ne I	100	6325,81	Xe I	2
6364,8945	Ar I	20	6325,45	Cu I	5
6363,34	F I	8	6325,22	Ti I	10
6362,8	Xe II	2	6325,17	Xe II	2
6361,79	Ca I	5	6324,682	O I	3
6359,896	Ti I	8	6324,414	Ar II	8
6357,668	Ar II	5	6323,735	Ar II	3
6357,569	N II	5	6323,283	O I	1
6357,45	Cu II	15	6322,42	Kr II	4
6357,025	Ar II	6	6321,70	N I	00
6356,545	N II	6	6321,59	Cl I	6
6356,35	Xe II	500	6319,493	Mg I	7
6355,77	Xe I	20	6319,236	Mg I	9
6354,5548	Cs I	200	6318,80	N II	1
6353,25	Xe II	50	6318,716	Mg I	10
6351,90	Kr I	8	6318,11	Ca I	3
6351,8618	Ne I	100	6318,062	Xe I	500
6350,76	C III	2	6318,027	Ti I	5
6349,20	Ar I	20	6318,022	Fe I	10
6348,601	Ar II	2	6318,00	Cu II	3
6348,508	F I	10000	6315,40	Ar II	1
6348,227	Ar II	6	6314,97	Xe I	15
6347,103	Si II	1000	6314,74	F I	0,8
6346,962	Mg II	9	6314,459	Si III	7
6346,86	N II	5	6313,6921	Ne I	150
6346,737	Mg II	10	6313,61	F I	3
6346,66	Kr I	20	6312,83	Cu II	20
6344,98	Xe I	20	6312,240	Ti I	10
6344,61	Kr II	4	6311,46	Xe II	5
6343,96	Xe II	300	6311,292	Cu II	30
6343,29	Ca I	4	6310,8	He II	—
6342,32	C I	2	6310,22	Kr III	10
6341,66	Cl I	10	6309,14	Ar I	8
6340,569	N II	7			

λ	Symbol	I
6307,6598	Ar I	30
6307,29	K II	7
6305,956	Cu II	15
6305,318	Fe II	15
6305,01	Xe II	1
6304,7892	Ne I	100
6303,754	Ti I	10
6303,68	N I	0
6303,66	Kr II	100
6301,510	Fe I	15
6300,988	Cu II	40
6300,86	Xe II	100
6298,31	Xe II	20
6296,8762	Ar I	20
6296,646	Ti I	12
6296,39	Xe II	10
6295,446	Ar II	2
6295,20	C II	0
6294,45	Xe I	15
6293,7447	Ne I	100
6292,649	Xe I	50
6292,37	C I	2
6290,96	Kr II	3
6290,01	C II	1
6288,72	Cu II	5
6288,5975	Cs I	2
6286,011	Xe I	100
6285,78	N I	1
6285,70	N II	2
6284,56	C II	0
6284,41	Xe II	50
6284,38	Xe I	2
6284,322	N II	6
6282,823	Ar II	1
6281,81	Xe I	5
6280,20	Cl I	5
6279,35	Si I	15
6279,028	F I	9
6278,652	Ar I	6
6277,54	Xe II	300
6277,425	Ar II	2
6276,99	Xe I	4
6276,708	Cu II	10
6276,624	Cu II	10
6276,039	Ne I	50
6275,79	C II	1
6275,43	N I	1
6273,389	Ti I	6
6273,330	Cu II	60
6273,23	Xe I	10
6273,018	Ne I	70
6272,83	N I	1
6270,82	Xe II	400
6268,34	Xe I	1
6268,30	Cu I	20
6268,30	Xe III	2
6267,33	Kr I	2
6266,89	O I	3

λ	Symbol	I
6266,4950	Ne I	1000
6265,301	Xe I	40
6264,346	O I	3
6263,696	F I	18
6261,826	Cu II	40
6261,55	O I	6
6261,212	Xe I	50
6261,101	Ti I	35
6260,16	Xe III	2
6259,59	C II	4
6259,41	Ar I	1
6259,22	O I	0
6258,796	Ne I	100
6258,706	Ti I	50
6258,103	Ti I	40
6257,86	Cu II	5
6257,84	Kr II	4
6257,18	C II	2
6256,84	O I	4
6256,750	Mg I	7
6256,54	C II	2
6255,32	Xe II	2
6254,85	Si I	20
6254,690	F I	80
6254,263	Fe I	6
6254,1876	Si I	180
6253,84	C II	2
6253,60	Si I	15
6252,732	Ne I	2
6252,561	Fe I	20
6252,26	Cl I	8
6250,98	Kr III	5
6250,74	C II	4
6250,2206	Cs I	2
6249,975	Ar II	1
6249,593	Ne I	5
6248,4064	Ar I	15
6247,562	Fe II	80
6246,7294	Ne I	100
6246,59	K II	6
6246,57	C II	1
6246,329	Fe I	15
6244,468	Si I	125
6243,8129	Si I	125
6243,3958	Ar I	6
6243,347	Al II	10
6243,125	Ar II	25
6243,00	Cl II	2
6242,70	C I	1
6242,54	Cl I	4
6242,412	N II	7
6242,09	Xe I	8
6241,39	Kr I	10
6240,4	Li I	2
6239,713	Ar II	7
6239,651	F I	13000
6239,630	Si II	100

λ	Symbol	I
6238,2871	Si I	40
6238,24	Xe III	60
6237,3199	Si I	160
6237,27	C I	1
6236,375	Fe II	20
6236,3520	Kr I	30
6235,40	Xe II	1
6234,04	Xe II	10
6233,8	He II	—
6232,892	Ar II	2
6231,759	Al II	7
6231,48	Cl I	8
6231,31	Cs I	1
6230,928	Ar I	40
6230,74	Kr II	10
6230,728	Fe I	25
6230,11	F I	3
6228,14	Kr II	1
6227,18	Cl II	6
6226,39	Cl I	3
6226,193	Al II	5
6225,742	Ne I	50
6225,356	F I	18
6224,169	Xe I	40
6223,66	Cu I	4
6222,71	Kr I	20
6221,66	Xe III	25
6221,41	Ti I	8
6221,11	Cu I	2
6220,84	Xe I	1
6220,460	Ti I	12
6219,89	N IV	4
6219,818	Cu II	30
6218,67	N II	0
6217,5986	Cs I	15
6217,2813	Ne I	1000
6216,910	Cu II	60
6215,9423	Ar I	60
6215,43	N IV	3
6215,212	Ti I	20
6213,8758	Ne I	150
6213,0998	Cs I	100
6212,5044	Ar I	100
6212,41	N IV	1
6212,30	Ti II	1
6212,249	F I	18
6211,55	Cl I	6
6210,87	F I	400
6210,420	Ar II	1
6209,11	Xe I	3
6208,935	Ar II	2
6208,46	Cu II	15
6208,440	Mg I	3
6206,463	Ar II	1
6206,297	Xe I	20
6206,16	Xe II	200
6205,7775	Ne I	100
6205,75	Xe I	400
6205,56	C III	5
6205,35	Xe I	6
6204,27	Cu II	15
6203,45	Xe II	1
6202,981	Ne I	15
6201,70	Al II	9
6201,52	Al II	10
6201,49	Xe I	3
6201,099	Ar II	6
6200,890	Xe I	60
6198,260	Xe I	100
6198,11	Cu II	5
6196,63	Xe II	4
6196,14	Kr II	3
6195,49	Xe I	1
6194,72	Cl I	15
6194,07	Xe II	300
6193,89	Xe I	1
6193,66	Cs I	—
6193,0663	Ne I	50
6192,301	Ar II	2
6191,561	Fe I	20
6191,40	Xe I	4
6189,10	Xe I	20
6189,0649	Ne I	70
6188,69	Cu II	20
6187,54	Cs I	—
6187,136	Ar II	6
6186,860	Cu II	20
6186,14	Ti I	3
6185,93	Xe II	1
6185,35	Kr II	7
6185,26	Fe III	9
6185,03	Xe II	15
6184,57	Xe II	20
6184,16	Xe I	3
6183,68	N II	0
6183,42	Al II	10
6183,169	Ne I	5
6183,024	Ar II	2
6182,45	Al II	7
6182,420	Xe I	300
6182,28	Al II	8
6182,1460	Ne I	150
6181,68	Al II	6
6181,57	Al II	5
6179,665	Xe I	120
6179,41	Ar I	4
6179,378	Fe II	5
6178,302	Xe I	150
6175,291	Ne I	50
6175,158	Fe II	15
6174,8829	Ne I	70
6174,378	Ar II	3
6173,712	Si III	3
6173,313	N II	7
6173,0980	Ar I	100
6172,821	Ne I	15
6172,290	Ar II	40
6172,08	Kr I	2

λ	Symbol	I
6172,020	Cu II	20
6171,77	Kr II	6
6170,6	He II	—
6170,1761	Ar I	100
6170,166	N II	6
6169,835	Si III	3
6169,74	Fe III	9
6169,559	Ca I	40
6169,055	Ca I	25
6168,80	Kr II	50
6167,755	N II	8
6167,628	Ar II	3
6166,790	Ar II	3
6166,628	F I	25
6166,443	Ca I	15
6165,38	Cl I	1
6165,123	Ar I	8
6164,76	Kr III	1
6164,136	F I	1,5
6163,96	C III	0
6163,935	Xe I	80
6163,758	Ca I	10
6163,660	Xe I	90
6163,65	Kr I	7
6163,5939	Ne I	1000
6162,172	Ca I	150
6162,16	Xe I	3
6162,05	Cl I	12
6161,289	Ca I	10
6160,7470	Na I	2
6160,16	Si II	5
6159,97	C III	0
6158,183	O I	21
6158,00	Cu II	5
6156,765	O I	20
6156,68	C III	3
6156,145	Ne I	50
6156,10	Ca I	1
6155,975	O I	19
6155,70	Si I	20
6155,359	F I	1
6155,28	Xe II	1
6155,2393	Ar I	60
6155,17	Si II	5
6155,1338	Si I	160
6155,09	C III	2
6154,24	Cu II	30
6154,2253	Na I	1
6154,13	C III	1
6153,24	Cs I	—
6152,556	Si III	2
6152,55	F I	2
6152,069	Xe I	20
6151,43	C II	4
6151,38	Kr I	20
6151,34	Cl I	2
6150,755	N II	4
6150,54	Kr II	1
6150,42	Cu II	20
6150,38	Cs I	—
6150,303	Ne I	100
6149,76	F I	800
6149,238	Fe II	20
6149,23	C III	0
6147,81	C III	0
6147,735	Fe II	30
6147,31	Cu I	20
6146,45	Xe II	50
6146,225	Ti I	3
6145,4432	Ar I	100
6145,029	F I	8
6145,0151	Si I	100
6144,97	Xe I	20
6143,70	Xe I	4
6143,40	Xe II	1
6143,0623	Ne I	1000
6142,615	Ar II	2
6142,508	Ne I	100
6142,487	Si I	100
6142,13	Xe I	1
6142,05	Ar I	1
6141,79	Cl I	4
6140,21	Cl I	25
6138,660	Ar II	12
6137,697	Fe I	18
6136,894	N II	4
6136,621	Fe I	20
6133,220	F I	70
6131,850	Si I	90
6131,76	Si II	4
6131,574	Si I	85
6131,47	Xe I	1
6131,43	F I	1
6129,93	F I	1
6128,726	Ar I	8
6128,619	Cs II	20
6128,4598	Ne I	100
6127,49	F I	0,6
6127,44	Xe II	2
6127,416	Ar I	15
6126,36	Xe I	15
6126,217	Ti I	20
6125,0207	Si I	90
6124,571	Ar II	3
6123,91	Xe I	5
6123,368	Ar II	15
6122,219	Ca I	100
6121,86	Ar I	1
6121,008	Ti I	3
6120,82	C I	2
6120,27	K II	8
6120,102	Ar II	5
6119,662	Ar I	2
6119,56	Kr II	10
6119,23	N IV	—

λ	Symbol	I
6118,724	Ar II	4
6118,027	Ne I	15
6117,222	F I	10
6116,52	Cs I	—
6115,85	C I	2
6115,23	Kr I	3
6115,08	Xe II	50
6114,929	Ar II	50
6114,86	Xe I	10
6114,468	Cu II	20
6114,37	Cl I	15
6113,463	Ar I	8
6113,15	C I	1
6112,926	Si I	10
6112,61	Kr II	4
6111,951	Xe I	40
6111,759	Xe I	30
6111,742	Ar II	2
6110,90	Cu II	5
6110,81	Kr III	5
6109,15	Ar II	2
6108,53	C I	2
6108,37	Xe I	8
6108,34	Kr I	3
6107,65	C I	1
6107,61	Kr II	5
6107,45	Cu II	10
6106,605	Si I	15
6106,398	O I	4
6105,97	Cu II	5
6105,6354	Ar I	60
6104,60	Ar I	6
6103,88	Xe I	3
6103,86	Kr I	1
6103,611	Li I	500
6103,546	Ar II	12
6103,34	Fe II	8
6102,765	Ar II	1
6102,722	Ca I	80
6102,56	C II	4
6101,925	Ar II	3
6101,43	Xe II	200
6101,16	Ar I	6
6100,46	C I	4
6100,03	C I	2
6100,01	Cu II	5
6098,92	C I	1
6098,8046	Ar I	60
6098,665	Ti I	7
6098,51	C II	9
6098,34	F I	25
6097,59	Xe II	1000
6097,33	Cu II	10
6096,1630	Ne I	300
6095,29	C II	7
6095,15	Xe I	1
6094,65	Cl II	100
6094,50	Kr II	30
6094,31	Kr I	2
6094,30	C I	0
6093,56	Xe II	300
6093,38	Xe I	3
6093,33	Ar I	1
6092,84	C I	1
6092,814	Ti I	4
6091,92	Si I	15
6091,81	Kr I	6
6091,175	Ti I	20
6090,902	F I	6
6090,7865	Ar I	10
6088,61	F I	2,5
6088,00	Kr I	2
6087,80	Si I	10
6086,69	C I	0
6086,67	Si II	10
6085,86	Ar I	2
6085,228	Ti I	20
6084,507	Ar II	2
6084,11	Fe II	5
6083,875	Ar II	2
6083,21	Xe II	1
6082,8630	Kr I	40
6082,53	Cl I	4
6081,245	Ar I	4
6080,320	Cu II	30
6080,113	F I	100
6080,06	Si II	20
6079,77	C I	1
6079,71	Kr II	20
6078,40	C I	2
6078,38	Kr III	10
6077,431	Ar II	6
6076,738	Cs II	2
6075,83	N I	3
6075,24	Kr I	20
6074,3377	Ne I	1000
6073,17	Al II	3
6072,25	Cu II	5
6070,83	C I	1
6068,37	Al II	1
6068,25	C I	0
6067,77	Xe I	1
6067,624	Si I	20
6067,52	Xe I	2
6067,45	Si II	10
6066,40	Al II	2
6065,487	Fe I	15
6065,00	N II	3
6064,91	Xe I	1
6064,758	Ar I	6
6064,631	Ti I	9
6064,5359	Ne I	50
6062,09	C I	0
6061,06	Al II	6
6059,3735	Ar I	100

λ	Symbol	I	λ	Symbol	I
6056,36	Fe III	9	6022,39	Kr II	40
6056,1280	Kr I	60	6021,91	F I	10
6054,18	Fe III	11	6020,179	Fe I	10
6052,7234	Ar I	30	6019,87	C I	0
6052,19	F I	1	6019,76	Si II	4
6051,15	Xe II	1000	6019,71	Cl I	6
6050,11	Kr III	3	6019,493	Ar II	4
6049,35	Kr I	3	6018,47	F I	1
6049,072	Ar II	6	6017,70	N I	2
6048,80	F I	1	6017,53	Ar I	1
6048,72	Fe III	11	6016,655	Fe I	5
6048,53	Xe II	5	6016,45	C I	6
6048,00	Xe I	6	6015,828	F I	150
6047,54	F I	900	6015,40	N I	1
6047,13	Kr II	1	6014,85	C I	9
6046,894	Ar II	8	6014,10	Xe I	1
6046,494	O I	10	6014,03	F I	40
6046,438	O I	13	6013,6790	Ar I	6
6046,232	O I	12	6013,40	Cu II	8
6046,1348	Ne I	50	6013,22	C I	10
6046,06	Kr II	10	6012,41	K II	1
6045,497	Fe II	6	6012,24	C I	5
6045,34	Ar I	1	6012,1570	Kr I	50
6044,79	C I	0	6010,68	C I	7
6044,468	Ar II	7	6010,4905	Cs I	50
6043,38	Xe I	10	6009,99	Kr II	10
6043,2254	Ar I	100	6009,78	Xe I	8
6042,46	C I	1	6008,92	Xe II	100
6042,013	Ne I	15	6008,576	Fe I	9
6040,7	Kr II	10	6008,48	N I	10
6039,17	C I	0	6008,10	Kr II	3
6038,944	F I	18	6007,909	Xe I	15
6038,1	Kr II	1	6007,18	C I	6
6038,04	F I	80	6006,38	Al II	10
6037,96	C II	0	6006,03	C I	9
6037,17	Kr III	10	6005,7246	Ar I	4
6036,56	Fe III	13	6003,67	C I	1
6036,20	Xe II	500	6003,170	Ar II	1
6035,82	Kr I	15	6003,034	Fe I	8
6034,92	Xe I	2	6002,98	C I	4
6034,0895	Cs I	35	6002,19	Kr I	3
6033,34	F I	8	6001,81	Al II	4
6032,59	Fe III	7	6001,78	F I	8
6032,33	Cu I	2	6001,18	Al II	1
6032,1291	Ar I	60	6001,13	C I	8
6031,36	Xe I	1	6000,9275	Ne I	100
6030,844	Ar II	1	6000,104	Cu II	40
6030,27	Si II	5	5999,83	Al II	2
6029,9971	Ne I	1000	5999,753	F I	15
6029,95	F I	20	5999,668	Ti I	8
6028,220	Ar II	1	5999,54	Fe III	5
6027,248	Ar II	5	5999,47	N I	6
6026,76	Xe I	4	5999,0004	Ar I	20
6025,1515	Ar I	10	5999,003	Ti I	4
6024,77	Xe II	3	5998,3	Xe II	1
6024,063	Fe I	15	5998,115	Xe I	30
6023,25	Cu II	10	5996,06	C I	2
6022,89	Xe I	1			

λ	Symbol	I
5995,59	Cu II	10
5995,28	O I	3
5994,66	Ar I	2
5994,425	F I	50
5993,8506	Kr I	60
5993,27	Cu II	8
5993,18	O I	1
5992,22	Kr II	200
5991,93	O I	2
5991,86	Xe II	1
5991,6532	Ne I	75
5991,42	Cl I	4
5991,383	Fe II	10
5991,34	O I	1
5989,40	CI	1
5989,339	Ar II	8
5989,18	Xe I	20
5989,03	C I	2
5988,44	Xe II	1
5988,30	Cu II	25
5988,288	Ar II	3
5988,11	Ar I	2
5987,9074	Ne I	150
5987,3027	Ar I	40
5987,055	Fe I	6
5986,635	F I	30
5986,23	Xe I	4
5985,920	Ar II	7
5984,94	F I	1,5
5984,804	Fe I	8
5984,454	Ar II	3
5984,393	Cs	15
5984,26	C I	3
5983,704	Fe I	6
5982,67	C I	2
5982,401	Ne I	8
5982,27	C I	0
5981,90	Ar I	5
5981,22	C I	1
5979,42	Xe I	1
5979,20	Cu II	3
5978,929	Si II	500
5978,543	Ti I	25
5978,29	Xe I	2
5977,995	Ar II	4
5977,65	Kr I	4
5976,46	Xe II	1000
5975,945	Ar II	2
5975,5340	Ne I	600
5974,82	Kr II	2
5974,6273	Ne I	500
5974,152	Xe I	40
5973,314	Ar II	2
5972,82	Xe I	1
5972,59	C I	0
5972,05	Al II	5
5971,6036	Ar I	5
5971,13	Xe II	200
5970,73	C I	0
5970,41	Xe I	1
5969,64	K II	2
5969,33	C I	4
5969,056	F I	1
5968,31	Ar I	1
5967,54	Kr II	15
5966,59	Cu I	3
5966,171	Ne I	35
5965,828	Ti I	30
5965,4710	Ne I	500
5965,28	F I	70
5965,031	Ar II	3
5963,99	C I	4
5962,4	Fe II	30
5962,166	F I	3
5961,6228	Ne I	70
5960,901	N II	4
5959,187	F I	25
5958,583	O I	13
5958,386	O I	12
5958,03	Xe II	50
5957,561	Si II	500
5956,87	F I	2
5955,14	Kr I	2
5954,276	N II	5
5953,820	Ar II	2
5953,62	Fe III	6
5953,162	Ti I	30
5952,388	N II	8
5952,13	C I	2
5950,905	Ar II	6
5950,147	F I	12
5950,04	C I	1
5949,93	Kr II	3
5949,2595	Ar I	10
5948,545	Si I	200
5948,40	Cl I	4
5947,61	C I	1
5945,53	Xe II	300
5945,44	Kr I	5
5944,8342	Ne I	500
5943,89	Ar I	2
5943,499	Mg II	4
5943,39	C I	0
5942,6722	Ar I	40
5942,13	Kr I	2
5941,825	Ar II	4
5941,82	Kr II	4
5941,755	Ti I	12
5941,653	N II	12
5941,179	F I	5
5941,168	Cu II	50
5940,86	Ar I	2
5940,697	F I	5
5940,240	N II	8
5940,10	C I	0
5939,319	Ne I	50
5938,629	Mg II	3
5937,806	Ti I	6
5937,59	Cu II	5

λ	Symbol	I	λ	Symbol	I
5937,56	F I	2,5	5911,72	Kr II	10
5936,64	Ar II	1	5911,55	Ti I	3
5935,792	Ar II	2	5909,67	Xe II	30
5935,03	Kr II	8	5908,22	Cl I	2
5935,03	Kr III	8	5907,83	Ti I	4
5934,55	Xe II	2	5907,21	C II	6
5934,458	Ne I	75	5906,76	Xe I	3
5934,172	Xe I	100	5906,4294	Ne I	50
5934,00	Cl I	1	5906,35	Ti I	5
5933,958	Ne I	8	5905,13	Xe II	200
5931,779	N II	11	5904,462	Xe I	20
5931,39	F I	6	5904,35	Ti I	4
5931,241	Xe I	80	5904,291	Ar II	2
5930,35	Cl I	6	5903,6	Fe II	8
5930,186	Fe I	8	5903,317	Ti I	5
5929,69	Fe III	18	5903,06	F I	0,8
5929,27	Ti I	3	5902,783	Ne I	5
5928,8124	Ar I	200	5902,4623	Ne I	50
5928,233	Mg II	4	5902,097	Ne I	3
5927,811	N II	9	5901,21	Cu II	5
5927,60	F I	1	5900,89	Kr II	8
5927,13	Ar I	10	5900,80	Ti I	4
5926,90	Cu II	3	5899,83	N II	1
5925,651	Cs II	60	5899,295	Ti I	25
5925,56	Xe I	6	5898,788	Si III	10
5924,42	Ti I	3	5898,56	Xe I	8
5923,69	Ca II	1	5898,406	Ne I	20
5923,366	Mg II	3	5897,986	Cu II	25
5922,72	Ca II	1	5897,25	N II	2
5922,709	Ne I	25	5895,9236	Na I	16
5922,550	Xe I	20	5895,62	Xe I	2
5922,33	Cl II	7	5894,988	Xe I	100
5922,112	Ti I	18	5894,56	Kr II	8
5921,85	Xe I	10	5894,07	C III	3
5921,50	Xe II	2	5893,29	Xe II	150
5920,13	Fe III	7	5893,15	N II	3
5919,45	C II	3	5892,00	C I	1
5919,06	Ti I	10	5891,91	Fe III	6
5919,037	Ne I	8	5891,74	F I	8
5918,9068	Ne I	250	5891,72	Kr III	5
5918,81	Kr II	2	5891,59	C II	12
5918,548	Ti I	10	5891,36	Fe II	8
5918,158	Mg II	6	5889,96	Ti I	8
5917,44	Xe II	50	5889,9504	Na I	32
5916,65	Xe I	4	5889,77	C II	15
5916,58	Ar I	5	5889,52	C I	2
5916,429	Mg II	7	5889,27	C II	6
5916,18	Ti I	5	5889,12	Xe I	20
5915,220	Si II	150	5889,05	Ti I	3
5915,123	Ti I	9	5888,5851	Ar I	300
5914,93	Ti I	5	5887,68	Kr I	3
5914,64	C II	4	5886,088	Ar II	3
5914,162	Fe I	8	5885,05	Ti I	3
5913,6327	Ne I	250	5882,6250	Ar I	100
5912,80	Xe II	5	5881,8950	Ne I	1000
5912,58	C I	0	5881,18	Kr I	2
5912,0861	Ar I	500	5880,54	C III	1
5911,90	Xe I	5			

λ	Symbol	I
5880,306	Ti I	5
5879,9000	Kr I	50
5878,92	Xe I	60
5877,31	C I	2
5875,966	He I	1000
5875,621	He I	7500
5875,018	Xe I	100
5874,42	Ti I	3
5873,764	Si I	40
5873,50	Kr III	1
5872,8275	Ne I	500
5872,36	Ti I	10
5872,149	Ne I	75
5871,69	C III	2
5871,18	Ti I	6
5870,971	Ne I	3
5870,9153	Kr I	3000
5870,66	C I	3
5870,443	Ar II	4
5870,26	Ar I	2
5869,23	Ti I	3
5868,4183	Ne I	75
5868,404	Si II	300
5867,81	Al II	3
5867,572	Ca I	1
5867,483	Si II	10
5866,7514	Kr I	50
5866,75	Cl I	4
5866,598	Ar II	2
5866,453	Ti I	35
5865,32	Ti I	6
5864,95	C I	0
5863,701	Cs II	5
5863,24	C III	1
5862,363	Fe I	8
5861,53	Al II	7
5860,75	Kr II	10
5860,3118	Ar I	60
5859,71	Ti I	4
5859,47	Xe II	2
5858,63	Cu II	5
5858,35	C III	1
5857,61	Xe III	10
5857,454	Ca I	100
5857,32	Kr I	1
5856,94	Cu I	5
5856,70	Cl I	8
5856,509	Xe I	15
5856,23	N I	1
5856,04	C II	5
5855,47	Xe II	1
5854,16	N I	2
5854,04	Kr II	4
5853,62	Al II	5
5853,10	Ar II	2
5853,06	F I	2,5
5852,86	Kr I	5
5852,4878	Ne I	2000
5851,93	Cu II	2
5851,1	Cu I	2
5850,25	C I	0
5849,85	Xe I	3
5849,66	Kr I	2
5847,68	Cl I	7
5847,64	Cs I	—
5847,12	Ti I	10
5847,03	Ar IV	—
5846,70	Cl I	8
5846,69	Xe II	2
5846,35	C I	0
5846,21	Xe I	2
5846,134	Si II	50
5845,46	Xe I	1
5845,1410	Cs I	30
5844,15	Cl I	6
5843,84	N IV	—
5843,781	Ar II	7
5843,74	Ar I	2
5843,61	C II	2
5843,43	Xe I	5
5842,67	Cu II	4
5842,49	Kr II	1
5841,44	Kr I	4
5841,01	N I	2
5840,83	Xe I	4
5840,048	Ar II	1
5839,85	Cl I	4
5839,820	Mg I	3
5839,73	Ti I	4
5838,96	Ar II	1
5838,8347	Cs I	—
5838,03	Ti I	12
5838,01	Ar IV	—
5837,34	Ti I	6
5836,35	C II	4
5835,5	Xe II	5
5834,71	N I	1
5834,2660	Ar I	60
5833,93	Fe III	10
5833,68	Cu II	5
5832,8600	Kr I	100
5831,887	K I	17
5831,159	Cs II	60
5830,63	Xe I	20
5830,04	Ar IV	—
5829,86	Ti I	5
5829,53	N I	6
5828,91	Ne I	75
5828,059	Ar II	3
5827,85	C II	2
5827,801	Si II	30
5827,72	Xe I	1
5827,48	F I	0,6
5827,28	Ti I	3
5827,07	Kr I	20
5826,44	N IV	—

λ	Symbol	I	λ	Symbol	I
5826,42	C III	1	5800,468	Si II	150
5826,24	Cl I	2	5800,46	Ar II	1
5826,036	Ar II	2	5800,23	C I	3
5826,02	Cu II	10	5800,16	Kr II	6
5824,800	Xe I	150	5799,88	Cl I	12
5824,64	C I	1	5799,734	Ar II	3
5824,50	Kr I	40	5798,90	C I	0
5823,890	Xe I	300	5798,44	Ti I	4
5823,679	Ti I	3	5797,8591	Si I	100
5823,51	Kr I	3	5796,26	Cl I	15
5823,14	C II	2	5795,08	N IV	—
5822,114	Ar II	3	5794,90	Si II	30
5821,57	Xe II	1	5794,46	C I	3
5820,52	Xe I	25	5793,51	N I	1
5820,1558	Ne I	500	5793,12	C I	7
5820,10	Kr I	15	5793,0714	Si I	90
5819,96	Ti I	8	5792,26	Xe I	1
5819,50	C I	1	5791,98	Xe II	1
5818,30	C II	2	5791,77	C II	1
5817,88	Ti I	6	5791,26	Ti I	7
5817,70	C I	0	5790,50	Cl II	25
5816,86	Ti I	6	5790,39	Ar I	5
5816,645	Ne I	50	5789,477	Ar I	20
5816,48	N I	2	5788,24	Kr I	7
5816,272	Ar II	2	5788,08	Ti I	5
5815,96	Xe II	50	5787,29	Kr I	6
5814,505	Xe I	60	5786,560	Ar II	5
5814,181	Cs II	25	5785,979	Ti I	5
5813,51	C I	1	5785,73	Si II	30
5812,746	Ar II	6	5785,66	Ti I	25
5812,400	Ne I	15	5785,560	Mg I	4
5812,30	N IV	—	5785,45	F I	0,5
5812,148	K I	15	5785,312	Mg I	5
5811,98	C IV	9	5784,38	Ti I	3
5811,4066	Ne I	300	5783,89	Kr I	10
5810,80	Kr I	8	5783,68	Ti I	3
5810,187	Si III	3	5783,541	Ar I	40
5810,08	Ti I	3	5782,384	K I	16
5809,5	Xe II	1	5782,132	Cu I	1500
5807,596	Ar II	1	5781,268	Ar II	2
5807,311	Xe I	15	5780,70	Ti I	12
5807,23	Ti I	8	5780,55	Xe III	2
5806,76	Cl I	2	5780,3839	Si I	70
5806,738	Si II	200	5777,72	Kr II	2
5806,00	Cu II	25	5776,96	Ti I	3
5805,80	C I	3	5776,39	Xe II	100
5805,53	Kr I	20	5776,374	Ar II	2
5805,19	C I	4	5775,56	Kr I	2
5804,4496	Ne I	500	5774,72	Cl I	4
5804,265	Ti I	5	5774,697	Ar II	1
5804,098	Ne I	75	5774,54	Ti I	13
5802,84	Cl I	5	5774,037	Ti I	5
5802,0809	Ar I	40	5774,00	Ar I	40
5801,81	Kr II	1	5773,5	Kr II	1
5801,752	K I	17	5772,58	Cl I	5
5801,33	C IV	10	5772,326	Ar II	5
5801,17	Kr I	2	5772,32	K II	4
5800,59	C I	6			

λ	Symbol	I
5772,1453	Si I	70
5772,1160	Ar I	100
5771,66	C III	2
5771,41	Kr II	100
5771,28	Ti I	6
5770,307	Ne I	50
5768,30	Cl I	2
5767,440	N II	7
5766,542	Ar II	2
5766,330	Ti I	4
5765,55	Cl I	3
5764,4188	Ne I	700
5764,32	Ca I	3
5764,063	Ne I	3
5763,52	Ti I	3
5763,013	Fe I	10
5762,9769	Si I	45
5762,90	Kr I	4
5762,295	Ti I	4
5761,96	Xe III	2
5761,88	Ca I	1
5761,37	Cu II	2
5760,5885	Ne I	70
5759,43	Cu II	5
5758,84	Ar I	5
5758,65	Xe II	100
5757,69	Ca I	4
5756,600	Ar II	3
5756,45	Ti I	6
5755,60	Kr II	2
5755,04	Kr I	2
5754,60	Xe I	1
5754,33	Kr I	1
5754,2195	Si I	45
5754,18	Xe II	2
5753,625	Si I	45
5753,54	Ar II	1
5753,136	Fe I	5
5752,98	Kr II	60
5752,64	N I	4
5752,56	Xe II	10
5751,74	Ti I	4
5751,05	Ti I	3
5751,03	Xe II	200
5750,57	Kr I	10
5750,424	O I	5
5749,27	Kr II	5
5749,02	Kr I	5
5748,71	Xe III	12
5748,650	Ne I	70
5748,2985	Ne I	500
5748,20	Xe I	8
5747,6670	Si I	45
5747,36	N I	2
5747,296	N II	8
5747,18	Ar I	2
5746,88	Xe II	5
5746,81	Ca I	2
5745,7244	Cs I	—
5745,07	Ti I	8
5744,47	Ti I	5
5743,28	Ca I	3
5743,278	Ar II	2
5740,73	Xe I	1
5740,65	N I	2
5740,17	Xe I	6
5739,975	Ti I	4
5739,88	Ar IV	—
5739,733	Si III	20
5739,5207	Ar I	500
5739,464	Ti I	9
5738,416	Ar I	20
5737,96	Ar I	5
5736,94	N IV	—
5735,77	Ti I	3
5735,74	Ca I	3
5735,63	N I	1
5734,95	Ti I	3
5734,39	F I	3
5734,24	Ti I	10
5733,48	Xe I	4
5732,694	Ar II	1
5732,325	Cu I	75
5732,210	Ar II	1
5731,70	Ca I	1
5731,103	O I	3
5731,08	Ti I	4
5730,86	Kr I	4
5730,65	N II	5
5730,51	Ti I	3
5728,25	Ti I	4
5727,96	Cu I	5
5726,91	Xe II	200
5726,59	Kr I	20
5726,16	Cl I	5
5726,10	Xe I	4
5724,325	Ar II	5
5723,56	Kr I	15
5723,26	Xe I	1
5722,65	Al III	6
5722,59	F I	6
5722,14	Xe I	15
5721,88	Kr I	10
5721,80	Ti I	4
5721,78	Cu II	20
5720,78	C I	2
5720,613	O I	1
5720,445	Ti I	3
5719,532	Ne I	75
5719,2248	Ne I	500
5719,16	F I	1,5
5718,899	Ne I	150
5717,99	Ca I	4
5717,61	Kr I	3
5716,450	Ti I	4

λ	Symbol	I	λ	Symbol	I
5716,289	Si III	8	5696,479	Xe I	80
5716,252	Xe I	80	5696,47	Al III	8
5716,19	Xe II	100	5695,92	C III	12
5716,029	Ar II	1	5695,750	Xe I	100
5715,80	Kr III	1	5695,522	Si III	3
5715,716	Xe I	70	5694,30	C II	2
5715,339	Ne I	35	5693,11	C I	3
5715,123	Ti I	9	5693,10	Ar I	1
5714,11	Kr I	2	5692,53	Ti I	3
5713,895	Ti I	3	5692,41	Cu II	2
5713,56	C II	0	5692,11	Kr II	5
5712,51	C II	1	5691,650	Ar II	8
5712,48	Ar I	1	5690,4251	Si I	100
5712,21	Xe I	2	5690,35	Kr II	200
5711,852	Ti I	4	5689,91	Ar I	200
5711,453	Ar II	1	5689,86	Cu II	5
5711,0880	Mg I	30	5689,8163	Ne I	150
5710,766	N II	10	5689,64	Ar I	200
5710,68	Ti I	3	5689,465	Ti I	10
5709,95	Ti I	3	5689,14	F I	18
5709,80	Xe I	10	5688,811	Si II	300
5709,3864	Fe I	10	5688,47	Ca I	4
5709,33	Ti I	4	5688,373	Xe I	40
5708,616	Ar II	1	5688,2046	Na I	9
5708,397	Si I	160	5688,1934	Na I	1
5708,199	Ti I	3	5687,52	Ti I	4
5708,03	C II	0	5687,40	Ar I	20
5707,62	F I	2	5686,49	Xe II	2
5707,5188	Kr I	40	5686,28	Cl I	1
5707,31	F I	25	5686,213	N II	10
5707,215	Ar II	1	5685,74	F I	8
5707,03	Ca I	1	5684,647	Ne I	25
5706,87	Xe I	3	5684,4843	Si I	120
5706,85	Ti I	5	5683,80	Ti I	3
5706,370	Si II	100	5683,73	Ar I	40
5705,43	Ti I	4	5682,88	Ca I	3
5704,598	Si III	7	5682,6333	Na I	5
5704,371	Ar II	3	5682,42	Cu II	20
5703,34	Xe I	1	5681,9014	Ar I	500
5703,121	Si III	4	5681,89	Kr II	400
5702,666	Ti I	6	5681,87	Xe II	1
5702,19	Kr I	10	5681,480	Ar II	2
5702,11	Ti I	6	5681,44	Si II	30
5701,66	Ti I	7	5681,08	Ti I	6
5701,551	Fe I	7	5679,562	N II	14
5701,374	Si II	200	5676,019	N II	11
5701,31	Xe III	6	5675,418	Si I	20
5701,16	C II	2	5675,413	Ti I	9
5701,1048	Si I	90	5675,15	Xe II	1
5700,874	Ar I	60	5674,73	Ar I	1
5700,82	F I	25	5674,52	Kr II	30
5700,240	Cu I	1500	5674,39	F I	8
5699,84	Kr II	10	5673,45	Ti I	10
5699,61	Xe II	100	5672,952	Ar II	7
5698,54	Xe I	8	5672,78	Kr II	40
5696,95	Kr I	1	5672,4519	Kr I	50
5696,54	Kr I	3	5671,668	F I	90
5696,50	Si III	7			

λ	Symbol	I	λ	Symbol	I
5670,96	Xe II	50	5648,07	C II	10
5669,8	Na I	3	5646,5	Cu I	2
5669,76	Ti I	5	5646,254	F I	8
5669,743	Si I	10	5646,19	Xe I	5
5669,562	Si II	1000	5645,960	Si I	90
5668,96	C I	7	5645,00	Kr II	1
5667,56	Xe II	300	5644,137	Ti I	18
5667,532	F I	40	5642,73	K II	5
5667,40	Ar I	1	5642,413	Ar II	2
5667,04	N I	1	5641,34	Ar I	60
5666,677	Si I	10	5641,30	Cu II	20
5666,627	N II	12	5641,07	Kr II	3
5666,46	Xe III	1	5640,55	C II	8
5666,09	Kr I	1	5639,478	Si II	200
5665,82	Ar I	5	5639,11	Ar I	100
5665,5536	Si I	80	5638,52	Ti I	12
5664,85	Kr II	1	5637,29	Ar I	20
5664,47	Cu II	3	5636,67	Cs I	—
5664,46	Xe I	1	5635,882	Ar II	5
5664,02	Xe II	3	5635,575	Ar I	60
5664,0183	Cs I	15	5635,57	Cu II	2
5663,80	Ar I	1	5635,2123	Cs I	10
5662,891	Ti I	4	5634,84	Cl II	18
5662,67	Kr I	3	5634,73	Ti I	3
5662,5489	Ne I	75	5634,661	Ar II	2
5662,525	Fe I	6	5633,24	Xe II	3
5662,47	C II	12	5633,14	Cu II	3
5662,154	Ti I	12	5633,02	Kr II	100
5662,00	Ar I	5	5632,973	Si II	100
5661,106	F I	7	5631,72	N II	1
5660,683	Si I	13	5631,381	Ar II	1
5660,656	Si II	150	5631,160	Ar II	3
5660,502	Si I	10	5630,44	Ar I	10
5659,38	Xe II	150	5630,29	Ti I	6
5659,15	F I	15	5629,93	C I	1
5659,1278	Ar I	500	5627,45	Ti I	3
5659,104	Ti I	3	5627,02	Kr II	1
5658,8247	Fe I	10	5626,93	F I	12
5656,6588	Ne I	500	5625,684	Ar II	6
5656,51	Ti I	4	5625,43	N I	2
5656,09	Ti I	4	5624,78	Xe II	1
5656,030	Ne I	75	5624,5501	Fe I	10
5655,236	Ar II	2	5624,06	F I	20
5654,924	Si I	15	5624,005	Ar II	1
5654,78	Ti I	8	5623,778	Ar I	60
5654,450	Ar II	8	5623,58	Ti I	5
5654,31	Xe I	1	5623,20	N I	4
5654,20	Cl I	1	5622,2214	Si I	30
5654,020	Ar II	2	5621,607	Si I	15
5652,84	Xe I	2	5621,24	Xe I	1
5652,601	F I	0,5	5620,89	Ar I	60
5652,5664	Ne I	75	5620,72	Cl I	1
5650,7054	Ar I	1500	5620,636	Ar I	2
5650,37	Kr II	10	5619,00	Ar I	5
5649,5625	Kr I	100	5618,878	Xe I	80
5648,66	Ar I	200	5618,32	Ti I	5
5648,570	Ti I	5	5618,18	N I	1
5648,38	Kr II	1			

λ	Symbol	I
5618,010	Ar I	60
5617,63	Kr II	2
5616,67	Xe II	150
5616,54	N I	5
5616,49	C I	0
5615,65	Fe I	50
5615,20	Cu II	5
5614,81	C I	0
5613,19	Al II	3
5612,89	Xe II	1
5612,65	Xe I	15
5611,82	Kr I	4
5611,667	Ar II	2
5611,36	N I	1
5611,35	Ar I	20
5609,578	Ar II	2
5608,90	Ar I	20
5608,37	Kr I	3
5607,99	Xe I	3
5607,72	Kr I	1
5607,66	F I	0,5
5607,51	O V	—
5606,7341	Ar I	500
5605,351	Si II	3
5605,25	Ar I	5
5604,36	Ar I	20
5604,28	N I	0
5604,11	O V	—
5603,932	Ar II	2
5603,73	C I	0
5602,9529	Fe I	10
5602,875	Si I	20
5602,846	Ca I	25
5602	O VI	—
5601,85	Ar I	2
5601,461	Si III	2
5601,285	Ca I	30
5601,08	Ar I	60
5600,952	Si III	3
5600,54	N I	0
5600,43	Ar I	40
5599,246	Si III	4
5598,50	Ar I	20
5598,487	Ca I	50
5597,90	O V	—
5597,69	Ti I	3
5597,4783	Ar I	500
5597,32	Kr III	5
5597,30	C I	1
5594,87	Xe II	4
5594,468	Ca I	60
5594,37	Xe I	6
5593,73	Cu II	5
5593,52	Ar II	1
5593,23	Al II	10
5592,37	O III	6
5592,200	Ar II	1
5592,01	O III	—
5591,75	Ar I	5
5591,734	F I	1
5591,61	Xe II	2
5591,41	Kr I	2
5591,15	Ne I	8
5590,120	Ca I	20
5589,378	Ne I	50
5589,31	F II	0
5588,757	Ca I	80
5588,7213	Ar I	500
5587,888	F I	1
5586,7634	Fe I	40
5585,905	Ne I	5
5585,4	Kr II	1
5585,18	Xe I	1
5584,4	Kr II	1
5583,5	Xe II	2
5583,29	O V	—
5582,61	Ar II	1
5581,971	Ca I	25
5581,93	Xe II	2
5581,83	Ar I	60
5581,784	Xe I	50
5580,45	Cl I	3
5580,3890	Kr I	80
5579,93	O V	—
5579,28	Xe I	40
5579,033	Cs II	2
5578,518	Ar II	6
5578,13	Cl I	3
5577,689	Ar II	8
5577,64	Kr I	3
5577,33	F I	10
5576,661	Si II	150
5576,106	Fe I	10
5576,049	Ne I	35
5575,973	Si II	5
5575,6	Kr I	10
5575,27	Xe I	2
5574,20	Ar I	5
5573,6740	Cs I	—
5573,13	Kr I	2
5572,8501	Fe I	30
5572,5428	Ar I	500
5572,19	Xe II	50
5572,00	O V	—
5570,2890	Kr I	2000
5570,216	F I	2,5
5569,6256	Fe I	20
5568,81	Cl II	15
5568,65	Kr II	100
5568,4078	Cs I	—
5567,815	Fe II	10
5567,77	Xe I	2
5567,63	N I	1
5566,7	Cs	40
5566,615	Xe I	100
5566,22	Xe I	5

λ	Symbol	I
5566,02	Xe III	2
5565,96	Ar I	5
5565,478	Ti I	9
5565,25	N II	3
5564,37	N I	9
5563,84	N I	3
5563,604	Fe I	5
5563,50	Xe I	2
5563,196	Ar II	2
5563,047	Ne I	75
5563,019	Cs II	125
5562,7662	Ne I	500
5562,441	Ne I	150
5562,2254	Kr I	500
5560,37	N I	9
5560,22	Ar I	10
5559,62	Ar I	200
5559,26	Kr I	2
5559,087	Ne I	35
5558,7031	Ar I	500
5557,948	Al I	8
5557,44	N I	2
5557,28	Xe I	2
5557,063	Al I	10
5555,06	Xe I	1
5555,003	O I	9
5554,99	Xe II	3
5554,935	Cu I	100
5554,887	Fe I	5
5554,832	O I	8
5554,050	Ar II	8
5553,53	F I	0,7
5553,40	Ar I	2
5553,17	C I	1
5553,10	Xe I	3
5552,99	Kr II	100
5552,83	Xe III	12
5552,76	Ar I	10
5552,67	N II	4
5552,43	F I	12
5552,385	Xe I	80
5551,922	N II	5
5551,59	C I	5
5551,50	Xe II	2
5551,03	C I	2
5548,90	C I	1
5548,24	C I	0
5547,27	C I	3
5546,74	F I	0,7
5546,11	Kr II	5
5545,11	N I	3
5545,07	C I	6
5545,045	Ar II	6
5544,4	Kr I	1
5543,880	Ar II	1
5543,82	C I	0
5543,471	N II	5
5542,73	Ar I	2
5542,10	Kr I	1
5541,65	Kr II	4
5541,46	Ar I	2
5540,90	Ar I	40
5540,76	C I	2
5540,74	Si II	100
5540,52	F I	18
5540,38	Xe I	3
5540,36	N I	1
5540,059	N II	4
5539,926	Si III	3
5539,4	Kr I	1
5539,33	F I	6
5538,651	Ne I	50
5538,61	F I	0,5
5537,61	C II	3
5537,290	Ar II	5
5536,01	K II	3
5535,78	Cu I	50
5535,51	Ar II	1
5535,39	Cl II	5
5535,37	N I	1
5535,363	N II	8
5535,35	C II	5
5534,98	Cu II	3
5534,81	C I	1
5534,862	F I	2
5534,45	Ar I	60
5533,6788	Ne I	75
5532,78	Xe I	2
5532,29	Kr II	5
5532,13	Cl I	8
5532,0	Na I	2
5531,07	Xe II	400
5530,244	N II	7
5529,78	C I	1
5528,93	Ar I	40
5528,63	Kr I	2
5528,4047	Mg I	40
5526,84	C I	2
5526,239	N II	5
5525,856	Ar II	2
5525,59	Xe II	50
5524,9598	Ar I	300
5524,39	Xe III	40
5523,70	Ar I	5
5523,690	Ar II	2
5523,47	Kr II	30
5523,05	Xe I	3
5522,94	Kr II	60
5521,74	Ar II	1
5521,17	Kr I	3
5520,63	Ne I	3
5520,52	Kr I	40

λ	Symbol	I	λ	Symbol	I
5519,337	Ar II	4	5493,23	Si I	40
5518,56	Xe II	1	5493,22	N II	1
5518,20	Ar I	5	5493,14	Cl I	3
5517,535	Si I	35	5492,8	O I	3
5516,668	Ar II	2	5492,77	Kr I	1
5516,66	Kr I	20	5492,43	Ti IV	6
5516,64	C I	0	5492,06	Ar I	40
5514,536	Ti I	25	5491,43	Kr II	4
5514,367	Ar II	4	5491,33	Kr I	2
5514,350	Ti I	20	5490,94	Kr I	50
5513,303	Ar II	1	5490,667	Ar II	1
5512,979	Ca I	20	5490,16	C II	1
5512,95	Cl I	4	5490,151	Ti I	12
5512,770	O I	8	5490,122	Ar I	60
5512,69	K II	2	5490,114	Si III	3
5512,603	O I	7	5488,95	C II	1
5512,529	Ti I	25	5488,86	Kr I	5
5511,485	Ne I	15	5488,555	Xe I	20
5511,176	Ne I	3	5488,46	Ar I	2
5511,16	Kr I	1	5488,210	Ti I	5
5510,55	Xe III	1	5488,06	K II	2
5509,597	Mg I	2	5487,46	Kr I	1
5509,20	Xe II	2	5487,138	Fe I	8
5508,11	O III	1	5487,03	Xe I	6
5507,753	Ar II	4	5486,6	O I	3
5507,63	Ar I	10	5486,47	Ar I	20
5507,46	Xe II	2	5486,102	Ar II	1
5507,339	Ne I	25	5485,90	C II	2
5507,174	Cs	15	5484,7	Li II	10
5506,7824	Fe I	18	5484,46	Xe I	4
5506,1149	Ar I	500	5484,311	Ar II	1
5505,18	Ar I	10	5484,16	Xe I	1
5504,917	Ar II	3	5483,35	C II	1
5504,34	Kr I	20	5483,32	Ar I	10
5504,02	Kr I	15	5482,65	Cu II	3
5503,897	Ti I	8	5481,997	Ar II	1
5503,8524	Cs I	—	5481,862	Ti I	5
5503,256	Ar II	3	5481,426	Ti I	6
5502,8843	Cs I	—	5481,33	Xe I	1
5502,88	Al II	3	5480,062	N II	7
5501,480	Ar II	2	5479,12	Xe I	1
5501,4686	Fe I	12	5478,73	Si II	5
5501,43	Kr III	10	5478,59	C II	4
5500,71	Kr I	50	5478,096	N II	7
5500,334	Ar II	7	5477,695	Ti I	8
5499,54	Kr II	50	5477,66	Kr III	2
5499,00	Ar I	10	5476,58	Kr I	2
5498,972	Ar II	2	5476,571	Fe I	10
5498,185	Ar II	8	5476,46	Kr II	4
5497,5196	Fe I	15	5475,49	Kr II	1
5497,123	Ar II	3	5475,49	Kr III	1
5496,45	Si II	200	5475,29	N II	4
5496,21	Kr I	3	5474,228	Ti I	6
5495,876	Ar I	1000	5473,920	Fe I	5
5495,666	N II	10	5473,7	O V	—
5495,07	Xe II	20	5473,517	Ti II	1
5494,4158	Ne I	50	5473,455	Ar I	500
5493,49	Ar I	20	5473,045	Si III	7

λ	Symbol	I
5472,642	Ar II	1
5472,61	Xe II	500
5471,198	Ti I	5
5470,307	Ar II	2
5470,13	K II	6
5469,65	Ar I	20
5469,63	Cu II	3
5469,58	Xe II	20
5469,450	Si II	30
5469,21	Si II	100
5469,105	Ar II	6
5468,17	Kr II	200
5467,1626	Ar I	60
5466,94	Fe II	20
5466,868	Si II	500
5466,440	Ar II	5
5466,432	Si II	500
5465,9443	Cs I	5
5464,136	Mg II	2
5463,283	Fe I	10
5463,138	Cu I	150
5462,65	Kr I	2
5462,592	N II	7
5462,146	Si II	10
5461,9231	Cs I	—
5461,37	Kr I	1
5460,502	Ti I	4
5460,39	Xe II	300
5460,037	Xe I	15
5460,019	Mg II	1
5459,61	Ar I	20
5459,47	Kr I	4
5458,80	Kr I	7
5457,75	Ar I	10
5457,47	Cl II	30
5457,4158	Ar I	200
5457,02	Cl II	75
5456,45	Si II	100
5456,45	Xe I	2
5456,39	Kr I	2
5456,382	Ar II	5
5456,27	Cl II	50
5456,01	Ar I	5
5455,6131	Fe I	40
5454,54	Xe I	1
5454,49	Si II	15
5454,307	Ar II	9
5454,30	Xe III	1
5454,221	N II	7
5453,646	Ti I	3
5453,634	Ar II	5
5452,5	O IV	—
5452,083	N II	7
5452,03	Ti II	1
5451,961	Si III	4
5451,6539	Ar I	500
5451,462	Si III	6
5451,259	Mg II	1
5450,90	Xe II	20
5450,45	Xe II	100
5450,05	F I	1
5449,61	Kr II	2
5448,61	Ar I	10
5448,5091	Ne I	150
5447,86	Kr I	3
5447,556	Ar II	2
5447,26	Si II	20
5447,120	Ne I	8
5446,920	Fe I	40
5446,34	Kr II	80
5445,52	Xe II	150
5445,43	Kr I	1
5445,037	Fe I	15
5444,99	Cl II	10
5444,87	Xe I	1
5444,25	Cl II	60
5443,88	Ar I	20
5443,681	Ar II	5
5443,42	Cl II	100
5443,21	Ar I	100
5442,22	Ar I	500
5440,932	Ar II	2
5440,39	Xe I	15
5439,9903	Ar I	500
5439,923	Xe I	30
5439,676	Ar II	3
5439,38	Kr II	1
5438,96	Xe II	400
5438,63	Kr II	40
5438,62	Si II	100
5438,20	Kr III	2
5437,36	Cu II	2
5436,861	O I	11
5435,83	Ar I	1
5435,775	O I	10
5435,60	Xe I	5
5435,176	O I	9
5434,5268	Fe I	30
5434,039	Mg II	4
5433,6513	Ne I	250
5433,48	Ar I	1
5433,24	Kr II	2
5432,94	Ar II	1
5432,89	Si II	15
5432,60	Ar I	1
5432,05	Cu I	250
5431,77	Kr I	1
5431,5	O V	—
5430,27	Ar I	10
5429,6999	Fe I	40
5429,69	Ar I	20
5429,139	Ti I	6
5428,92	Si II	15
5428,07	Xe II	2

λ	Symbol	I	λ	Symbol	I
5427,832	Fe II	30	5405,34	Si II	100
5427,39	Ar I	1	5404,87	O I	3
5426,5	O IV	—	5404,148	Fe I	30
5426,256	Ti I	3	5403,03	Kr I	2
5424,36	Cl II	25	5402,793	Cs	40
5424,076	Fe I	45	5402,604	Ar II	8
5423,56	Kr II	1	5402,170	Ar II	1
5423,52	Cl II	100	5401,543	Mg II	9
5423,25	Cl II	150	5401,450	N I	4
5422,55	Ar I	2	5401,04	Xe III	50
5422,47	Ti II	1	5400,62	Ar II	1
5421,76	Xe I	2	5400,5616	Ne I	2000
5421,61	Si I	10	5400,503	Fe I	5
5421,3536	Ar I	500	5400,45	Xe I	4
5421,168	Si I	10	5399,01	Ar I	20
5420,155	Ne I	50	5398,82	Ti IV	8
5419,687	Cs II	60	5397,718	F I	1
5419,15	Xe II	2000	5397,63	Xe I	1
5418,5584	Ne I	150	5397,522	Ar II	9
5418,43	Kr II	30	5397,131	Fe I	40
5418,2	Xe II	2	5397,093	Ti I	4
5418,02	Xe I	5	5396,30	Ti II	1
5417,4	O V	—	5394,81	Ar II	1
5417,24	Si II	15	5394,738	Xe I	20
5417,22	Ar I	10	5393,971	Ar I	200
5416,710	Ar II	1	5393,96	Cu II	3
5415,64	Si II	5	5393,603	Ar II	5
5415,36	Xe II	50	5393,18	Si II	3
5415,207	Fe I	35	5393,175	Fe I	10
5414,42	Kr I	1	5392,795	Xe I	100
5414,28	Cs I	—	5392,12	Cl II	100
5414,20	Cl II	2	5391,62	Cu I	450
5413,6145	Cs I	—	5390,72	Ar I	40
5413,56	Xe III	12	5390,68	N II	1
5413,32	Ar I	10	5390,45	Cu II	5
5412,900	F I	0,6	5389,996	Ti I	3
5412,649	Ne I	250	5389,12	Kr III	1
5412,434	Ar II	2	5389,10	Ar I	40
5412,19	Kr III	5	5388,48	Al II	1
5411,881	N I	5	5387,37	Ar I	40
5411,646	Ar II	3	5386,79	Ar I	1
5411,524	He II	50	5386,519	Ar II	2
5410,909	Fe I	15	5385,88	F I	0,6
5410,76	O I	4	5384,378	Ar II	5
5410,4750	Ar I	500	5384,17	Xe III	2
5410,15	F I	4	5383,71	N II	2
5410,12	Ne I	5	5383,371	Fe I	35
5410	O VI	—	5383,250	Ne I	25
5409,609	Ti I	6	5382,330	Ar II	2
5409,44	Kr I	1	5381,39	Kr III	2
5409,401	Ar II	1	5381,03	F I	0,8
5409,34	Ar I	1	5381,020	Ti II	1
5408,87	O I	3	5380,48	Si II	5
5408,59	O I	4	5380,34	C I	10
5408,34	Cu I	100	5379,64	Kr I	15
5407,35	Cs II	2	5379,163	Ar II	1
5407,348	Ar II	7	5378,45	N I	0
5406,6672	Cs I	—	5378,078	Ar II	3
5405,7781	Fe I	40			

λ	Symbol	I	λ	Symbol	I
5376,867	Cu I	5	5357,33	Cu I	3
5376,85	Cu II	3	5356,80	Xe I	1
5376,636	Ar II	3	5356,77	N I	5
5376,0	O V	—	5356,49	Ar I	10
5374,9774	Ne I	50	5356,14	Cl II	10
5374,19	Kr II	3	5355,45	Kr II	10
5373,74	Xe I	1	5355,422	Ne I	150
5373,4951	Ar I	500	5355,176	Ne I	150
5372,66	N I	3	5354,95	Cu I	250
5372,57	Kr I	1	5354,89	Si II	5
5372,39	Xe II	300	5354,82	Ar II	1
5372,3110	Ne I	75	5353,513	Ne I	5
5372,29	Ar I	1	5353,46	Ar I	20
5372,007	Ar II	2	5353,12	C III	0
5371,84	Al II	1	5352,666	Cu I	300
5371,74	Kr I	2	5352,1	O V	—
5371,4926	Fe I	50	5351,449	Ar II	4
5371,40	Kr III	4	5351,220	N II	4
5371,10	N I	1	5351,072	Ti I	4
5371,09	Xe III	1	5350,8	Cs I	—
5370,979	Cs II	80	5350,58	Ar I	20
5370,10	F I	5	5350,3512	Cs I	—
5369,97	Ar I	5	5349,77	Kr III	2
5369,957	Fe I	25	5349,717	Ar II	2
5369,65	Ti I	4	5349,472	Ca I	25
5368,58	C II	1	5349,31	Cs	15
5368,42	Cu II	10	5349,204	Ne I	150
5368,07	Xe II	100	5349,16	Cs II	25
5367,67	C II	1	5348,95	Cs	25
5367,460	Fe I	20	5348,604	Ar II	2
5367,27	N I	1	5348,283	Ar II	3
5367,06	Xe III	30	5347,412	Ar I	200
5367,03	Xe I	6	5347,37	Kr I	2
5366,222	Ne I	25	5346,76	Kr II	60
5365,91	Kr I	1	5345,977	Mg I	1
5365,62	Cu II	5	5345,84	C III	1
5365,485	Ar II	1	5345,81	Ar I	20
5364,883	Fe I	15	5345,609	Ar II	1
5364,626	Xe I	30	5344,534	Ar II	5
5364,142	Ar II	1	5344,28	Ar I	5
5363,80	Fe III	8	5344,23	N I	00
5363,27	Xe II	150	5343,3	O V	—
5362,864	Fe II	5	5343,2834	Ne I	600
5362,48	Ar I	1	5342,970	K I	12
5362,42	O IV	—	5342,80	F I	1
5362,248	Ne I	25	5342,700	Ne I	1
5362,244	Xe I	15	5342,40	C II	2
5362,11	Kr III	1	5341,78	Ar I	10
5360,442	Ne I	35	5341,46	C III	0
5360,030	Cu I	200	5341,0938	Ne I	1000
5360,0121	Ne I	150	5341,0255	Fe I	20
5359,95	C III	2	5340,9418	Cs I	—
5359,574	K I	14	5340,213	N II	3
5359,069	Ar II	2	5339,9371	Fe I	12
5358,616	Ar II	2	5339,85	C II	1
5358,53	Cs II	500	5339,688	K I	13
5358,363	Ar II	6	5339,38	Xe II	1000
5358,020	Ne I	10			

λ	Symbol	I	λ	Symbol	I
5339,33	Ar II	1	5314,45	N III	2
5339,189	Ca II	5	5314,258	Ar II	2
5339,13	Kr I	20	5313,87	Xe II	800
5338,92	Cl II	5	5313,419	N II	2
5338,732	N II	4	5312,32	Al II	5
5338,20	Kr III	2	5312,002	Ar II	6
5338,106	Ar II	1	5310,99	Xe III	6
5337,89	Xe I	2	5310,76	Al II	2
5337,42	C III	0	5310,52	N I	1
5336,809	Ti II	4	5310,26	Kr II	4
5335,916	Ar II	1	5310,24	K II	5
5335,91	Xe I	1	5309,517	Ar I	200
5335,710	Ne I	10	5309,493	Si IV	1
5334,79	C II	6	5309,48	N I	1
5334,78	Kr I	10	5309,27	Xe II	200
5334,42	N I	1	5308,66	Kr II	200
5333,70	Cl II	15	5308,074	Ar II	5
5333,41	Kr II	500	5307,3633	Fe I	5
5333,323	Ne I	50	5307,223	Ca II	7
5332,89	C II	4	5306,84	C I	2
5331,08	Kr I	2	5306,609	Cs II	25
5331,034	Ar II	4	5306,37	Xe I	3
5330,7775	Ne I	600	5306,32	C I	0
5330,739	O I	13	5305,690	Ar II	6
5330,664	Ar II	3	5305,32	O IV	15
5329,712	Ar II	5	5305,17	Ar I	1
5329,685	O I	12	5305,10	C III	2
5329,15	Kr II	4	5304,971	Si IV	2
5329,101	O I	11	5304,9	N I	1
5328,70	N I	5	5304,7580	Ne I	70
5328,5336	Fe I	15	5304,43	Kr I	1
5328,0418	Fe I	50	5303,7766	Cs I	—
5328,02	Ar I	20	5303,415	Si III	2
5327,90	Xe II	3	5302,99	Fe III	6
5327,87	Kr I	2	5302,35	C I	1
5327,76	N II	1	5302,3073	Fe I	10
5327,07	Ar I	1	5301,40	Cs I	—
5326,3968	Ne I	75	5300,84	C I	1
5325,70	Kr I	1	5300,761	Ar II	4
5324,80	Ar I	5	5300,74	Kr I	3
5324,61	Al II	4	5300,55	C I	3
5324,182	Fe I	30	5300,12	C I	1
5323,78	Cu I	3	5299,79	Kr I	2
5323,276	K I	12	5299,075	Ar II	3
5322,77	Kr II	60	5299,044	O I	5
5322,02	Kr I	2	5298,93	N III	1
5320,953	N II	4	5298,887	O I	4
5320,550	Ne I	2	5298,429	Ti I	4
5320,203	N II	3	5298,1891	Ne I	150
5317,726	Ar I	60	5298	O VI	—
5317,46	C I	1	5297,993	Ar II	3
5317,41	Kr II	30	5297,86	N III	1
5316,98	F I	0,5	5297,8	Kr II	1
5316,806	Ne I	25	5297,236	Ti I	6
5316,609	Fe II	8	5296,93	C I	0
5316,07	Al II	7	5296,91	Ar I	1
5315,97	F I	3			
5315,69	C I	0			
5315,214	Ar II	3			
5314,781	Ne I	30			

λ	Symbol	I
5296,791	Ar II	3
5296,386	Ar II	4
5296,32	Ar I	5
5295,892	Ar II	3
5295,781	Ti I	4
5295,19	Si II	30
5293,821	Ar II	4
5292,75	N I	0
5292,517	Cu I	1650
5292,22	Xe II	1000
5292	O VI	—
5291,3	Xe II	2
5291,22	C I	1
5291	O VI	—
5290,09	C II	0
5290,035	Ar II	3
5290,00	Ar I	20
5289	O VI	—
5288,634	Ar II	1
5288,32	C I	0
5286,895	Ar II	15
5286,47	C II	1
5286,38	Xe I	3
5286,11	Xe I	4
5286,071	Ar I	60
5285,85	Al II	6
5285,57	F I	0,8
5285,48	Cl II	30
5285,438	Ar II	2
5285,268	Ca II	6
5284,092	Fe II	5
5283,77	Al II	8
5283,6283	Fe I	18
5283,530	Cu I	5
5283,441	Ti I	8
5283,437	Ar II	1
5283,43	Ar I	20
5283,30	Xe I	2
5282,52	N III	00
5282,46	Xe II	2
5282,378	Ti I	3
5282,29	Fe III	7
5281,7970	Fe I	10
5281,628	Ar II	7
5281,18	N I	3
5280,40	Ar I	60
5280,24	C I	2
5280,21	Al II	6
5280,0853	Ne I	50
5279,84	Kr I	9
5279,05	Ar I	20
5279,01	F I	12
5279	O VI	—
5278,62	Al II	3
5277,68	Al II	2
5276,81	Al II	2
5276,522	Cu II	15
5276,50	Kr II	100
5276,47	Fe III	7
5276,42	Al II	2
5275,994	Fe II	7
5275,121	O I	4
5274,968	O I	2
5274,61	Kr I	4
5274,044	Cs II	40
5274,0393	Ne I	40
5273,580	Ar II	1
5273,49	N V	—
5273,48	Xe I	1
5272,60	N III	1
5272,53	C III	6
5270,59	N III	1
5270,3602	Fe I	30
5270,270	Ca I	60
5269,988	Cu II	30
5269,755	Ar II	4
5269,74	Si II	3
5269,5402	Fe I	60
5268,96	C I	4
5268,62	Ti II	1
5268,31	Xe II	50
5268,246	Ar II	3
5268,06	O III	5
5267,958	Ar II	3
5267,48	Ar I	1
5267,168	Ar II	6
5266,564	Fe I	30
5265,967	Ti I	10
5265,557	Ca I	40
5264,783	Ar II	6
5264,368	Mg II	7
5264,305	Ar II	1
5264,239	Ca I	20
5264,215	Mg II	8
5263,483	Ti I	3
5263,3134	Fe I	8
5263,21	Cs II	2
5263,18	Kr III	1
5263,02	Ar I	2
5262,244	Ca I	25
5261,95	Xe II	200
5261,903	Ar II	4
5261,706	Ca I	20
5260,91	Al III	0
5260,91	N III	1
5260,57	N II	2
5260,44	Xe II	200
5260,375	Ca I	2
5259,976	Ti I	3
5259,89	Xe II	30
5259,71	C II	5
5259,174	Ar II	1
5259,06	C II	5
5258,223	Ar II	1

λ	Symbol	I	λ	Symbol	I
5257,84	Kr III	2	5236,21	Ar I	20
5257,64	Si II	3	5235,564	Ar II	4
5257,24	C II	7	5234,74	Ar I	5
5256,75	Kr II	30	5234,620	Fe II	7
5256,569	Ar II	3	5234,0271	Ne I	50
5256,5633	Cs I	—	5233,16	Xe III	3
5256,09	C II	2	5232,9474	Fe I	40
5255,811	Ti I	5	5232,06	Kr I	2
5255,677	Ar II	1	5230,523	Ar II	3
5254,4710	Ar I	60	5230,41	F I	15
5253,58	C III	5	5230,15	Kr II	3
5253,57	C II	4	5229,86	Ar I	40
5252,7890	Ar I	300	5229,58	Cu II	3
5252,138	Ar II	1	5229,52	Kr II	60
5252,105	Ti I	8	5228,18	Kr I	20
5251,89	Xe I	2	5227,1911	Fe I	40
5251,400	Ar II	3	5227,002	Cs II	200
5251,108	Ar II	3	5226,96	F I	3
5250,6490	Fe I	6	5226,90	Xe II	2
5250,52	Cu I	500	5226,8686	Fe I	15
5249,547	Ar II	400	5226,69	N IV	3
5249,51	C II	2	5226,62	Xe II	20
5249,373	Cs II	80	5226,534	Ti II	5
5249,22	Cl II	3	5225,05	Kr II	3
5249,20	Ar I	40	5224,928	Ti I	8
5249,11	C III	4	5224,558	Ti I	6
5249,06	Kr II	4	5224,56	Kr II	7
5248,98	Xe I	4	5224,301	Ti I	15
5248,18	Ar I	1	5223,57	Kr I	5
5247,986	Ar II	3	5223,66	Xe III	20
5247,75	Xe II	20	5223,623	Ti I	6
5247,469	Ar II	1	5222,90	Ar I	20
5247,293	Ti I	5	5222,685	Ti I	6
5246,76	Ar I	5	5222,38	Kr I	3
5246,574	Ti I	3	5222,3517	Ne I	50
5246,24	Ar I	40	5221,854	Ar II	3
5245,69	Cl II	4	5221,42	F I	0,5
5245,61	N IV	—	5221,34	Cl II	75
5245,389	Ar II	5	5221,2729	Ar I	500
5245,36	Cu II	10	5220,070	Cu I	500
5245,27	Xe I	4	5219,697	Ti I	8
5245,25	Kr II	4	5219,589	Ar II	1
5244,67	C III	3	5219,37	Si II	10
5244,28	F I	0,8	5219,30	Ar I	40
5243,31	Fe III	10	5218,84	Kr I	1
5242,4955	Fe I	5	5218,202	Cu I	2500
5242,13	Ar I	2	5217,93	Cl II	150
5241,786	Ar II	2	5217,93	Kr II	12
5241,29	Kr II	2	5217,78	Kr I	1
5241,091	Ar I	60	5217,45	Kr II	30
5240,31	Si II	5	5217,3964	Fe I	5
5239,71	Ar I	2	5216,816	Ar II	8
5238,95	Xe III	60	5216,28	Ar I	60
5238,69	F I	2	5216,2770	Fe I	10
5238,560	Ti I	6	5215,81	Kr I	8
5237,65	Cu I	10	5215,1871	Fe I	6
5236,853	Ar II	1	5214,774	Ar I	200
5236,231	Ar I	2			

λ	Symbol	I
5214,339	Ne I	35
5212,780	Cu I	140
5212,41	Kr I	1
5212,371	Ti I	3
5210,5672	Ne I	50
5210,492	Ar I	200
5210,386	Ti I	40
5209,62	Cs II	15
5209,44	Cs	15
5208,8648	Ne I	70
5208,6007	Fe I	7
5208,32	Kr II	500
5208,04	Ar I	10
5207,96	F I	1
5207,852	Ti I	3
5207,17	Ar I	10
5207,128	Cu II	20
5206,73	O II	5
5206,565	Ne I	3
5206,07	Xe I	1
5206,059	Ti I	5
5205,79	Ar I	10
5205,15	N IV	3
5205,11	N II	0
5204,5840	Fe I	5
5204,440	Ar II	3
5204,29	N IV	5
5203,8962	Ne I	150
5202,413	Si II	500
5202,3395	Fe I	8
5202,201	Ar II	2
5201,88	Xe II	2
5201,71	N I	2
5201,56	Kr II	2
5201,42	Xe II	20
5201,096	Ti I	4
5200,87	Cu I	500
5200,40	N IV	4
5200,22	Kr II	60
5199,9	Xe II	1
5199,48	N II	1
5198,97	Kr I	1
5198,96	Ar I	2
5197,82	Kr I	1
5197,569	Fe II	6
5197,264	Si III	5
5196,7343	Cs I	—
5195,478	Fe I	8
5195,29	Ar I	1
5194,9441	Fe I	10
5194,92	Xe II	5
5194,77	Ar I	20
5194,043	Ti I	4
5194,02	Ar I	5
5193,2227	Ne I	150
5193,1302	Ne I	150
5193,03	Cl II	10
5192,971	Ti I	35
5192,86	Si II	200
5192,810	Ar II	3
5192,72	Ar I	60
5192,3509	Fe I	30
5192,10	Xe II	80
5191,97	N II	2
5191,4615	Fe I	20
5191,37	Xe II	300
5191,364	Ar II	4
5191,322	Ne I	35
5190,56	O II	3
5190,380	N II	4
5189,70	Cl II	25
5189,51	N I	1
5189,27	F I	4
5188,848	Ca I	50
5188,700	Ti II	6
5188,68	Kr III	1
5188,6122	Ne I	150
5188,11	Xe II	200
5187,7507	Ar I	800
5187,1	N I	1
5186,99	Kr II	60
5186,41	F I	0,8
5186,329	Ti I	3
5186,200	N II	2
5185,90	Ti II	2
5185,85	Xe I	2
5185,535	Si II	100
5185,25	Si II	100
5184,964	N II	4
5184,48	Xe II	50
5183,6042	Mg I	45
5183,364	Cu II	20
5183,200	N II	4
5182,320	Ne I	2
5182,30	Kr II	1
5181,90	Si II	100
5181,47	N I	0
5180,352	N II	4
5179,52	N II	7
5179,35	N II	7
5178,82	Xe II	50
5177,71	Kr II	6
5177,540	Ar I	40
5177,060	N II	4
5176,563	N II	2
5176,233	Ar II	10
5176,00	O II	2
5175,891	N II	6
5175,89	Cu II	2
5175,85	Cl II	20
5175,426	Ar II	2
5174,463	N II	4
5173,742	Ti I	30
5173,386	N II	5
5173,16	F II	2
5173,15	Cl II	25
5172,970	N II	3
5172,6843	Mg I	44

λ	Symbol	I
5172,6	Al III	1
5172,36	Kr I	2
5172,346	N II	4
5171,5987	Fe I	20
5171,45	N II	4
5171,30	N II	2
5170,168	N II	4
5169,45	N I	1
5169,030	Fe II	12
5168,99	N II	1
5168,06	Kr I	4
5168,056	N II	4
5167,73	Kr I	1
5167,4905	Fe I	40
5167,3216	Mg I	42
5167,30	Xe I	1
5166,80	Kr II	80
5166,2841	Fe I	4
5165,774	Ar II	8
5164,39	Xe I	1
5163,90	Al III	7
5163,474	Ne I	10
5162,78	N I	1
5162,742	Ar II	5
5162,711	Xe I	10
5162,34	Cl II	10
5162,288	Fe I	10
5162,2858	Ar I	500
5160,09	Kr III	1
5160,02	O II	4
5159,92	C I	1
5159,69	Ar I	10
5159,505	Ar II	1
5158,902	Ne I	50
5158,79	Cl II	8
5158,36	Cu I	50
5158,187	Al II	1
5158,090	Cu II	10
5156,667	Ne I	50
5156,10	Fe II	6
5156,023	Si I	8
5155,29	C I	1
5154,4271	Ne I	50
5153,57	C I	2
5153,4024	Na I	2
5153,235	Cu I	2000
5153,11	Ar I	20
5152,6813	Cs I	—
5152,185	Ti I	10
5152,01	Kr II	3
5152,01	Kr III	3
5151,9610	Ne I	75
5151,68	Kr III	2
5151,3943	Ar I	200
5151,09	C II	13
5150,86	Al III	6
5150,8425	Fe I	6
5150,077	Ne I	35
5149,61	Kr II	3
5149,33	Fe III	7
5148,8381	Na I	1
5147,483	Ti I	10
5146,06	O I	5
5145,654	Al II	1,5
5145,465	Ti I	12
5145,39	Kr I	1
5145,319	Ar II	25
5145,28	Kr II	4
5145,16	C II	15
5145,122	Ne I	35
5145,04	Kr I	2
5145,011	Ne I	500
5144,998	Al II	1
5144,9384	Ne I	500
5144,120	Cu I	550
5143,49	C II	12
5143,2665	Ne I	5
5143,05	Kr II	60
5143,03	Xe III	4
5142,9320	Fe I	6
5142,7	Cu I	10
5142,7	Kr I	4
5141,81	Ar I	20
5141,790	Ar II	20
5141,10	Kr II	1
5140,35	Cl I	5
5139,9	Kr I	1
5139,4702	Fe I	20
5139,2578	Fe I	10
5139,17	C II	9
5137,388	Fe I	6
5137,26	C II	7
5136,795	Fe II	6
5135,110	Si III	3
5134,17	Ar I	2
5133,680	Fe I	20
5133,52	Kr II	1
5133,28	C II	12
5132,94	C II	12
5132,61	Ar I	1
5132,145	Ar II	2
5131,106	Ar II	1
5130,53	O I	3
5129,143	Ti II	1
5129,083	Ar II	2
5128,280	Ne I	2
5128,031	Si I	10
5127,802	Ar I	60
5127,3624	Fe I	5
5126,93	C II	2
5125,765	Ar II	8
5125,73	Kr II	400
5125,70	Xe II	30
5125,598	Si I	10
5125,20	C II	4
5125,130	Fe I	6
5124,72	Ar I	1

λ	Symbol	I
5124,461	Cu II	20
5123,723	Fe I	6
5123,16	Kr II	15
5122,972	Ar II	2
5122,42	Xe II	200
5122,337	Ne I	150
5122,257	Ne I	150
5122,15	C II	2
5121,88	Ar I	5
5121,866	Ne I	2
5121,82	C II	5
5121,486	Ar II	2
5120,745	Cu II	20
5120,506	Ne I	25
5120,430	Ti I	12
5120,10	C II	3
5120,01	Ar I	1
5119,45	C II	4
5118,2057	Ar I	60
5117,76	Xe II	2
5117,011	Ne I	35
5116,75	C II	2
5116,5032	Ne I	150
5116,46	Xe I	2
5115,49	Cu I	10
5114,57	Xe III	1
5114,26	C II	4
5114,2	O V	—
5114,116	Si III	8
5113,76	Si III	7
5113,69	C II	4
5113,6724	Ne I	75
5113,50	Ar I	1
5113,448	Ti I	10
5113,36	Cl II	40
5112,249	K I	12
5112	O VI	—
5111,913	Cu I	300
5110,98	Kr III	1
5110,4139	Fe I	10
5110,269	F I	1
5109,81	Kr I	2
5109,427	Ti I	4
5108,58	Xe II	2
5108,331	Cu II	3
5107,943	Al I	4
5107,91	C II	1
5107,520	Al I	6
5107,4505	Fe I	6
5107,38	Xe III	20
5105,541	Cu I	1500
5104,74	Ar I	20
5104,701	Ne I	35
5104,437	N II	5
5104,08	Cl II	25
5103,04	Cl II	125
5100,95	Fe II	15
5100,706	Fe III	10
5100,34	Al II	1
5100,08	Cu II	10
5099,80	Cl I	8
5099,64	Ar I	5
5099,59	Xe II	5
5099,30	Cl II	100
5099,200	K I	11
5099,042	Ne I	25
5098,97	Ar I	20
5098,7030	Fe I	8
5098,34	Cl II	20
5097,171	K I	11
5096,995	Fe I	6
5096,851	Ar II	1
5096,604	Cs II	40
5095,845	Ar II	2
5095,58	N II	1
5094,84	Ar I	1
5093,792	Cu II	20
5093,65	Al II	2
5093,32	Ar I	10
5092,174	Ar II	1
5092,02	Xe II	60
5091,419	Si III	10
5090,789	Fe I	6
5090,496	Ar II	10
5090,36	Kr I	1
5090,321	Ne I	8
5089,63	C I	0
5089,12	Kr I	2
5088,932	Cu II	10
5088,487	Cu II	10
5088,260	Cu II	30
5087,085	Ar I	60
5087,055	Ti I	8
5086,52	Kr II	250
5085,333	Ti I	4
5085,02	Al II	4
5084,79	Ar I	1
5084,226	K I	10
5083,991	Cu II	15
5083,968	Ne I	25
5083,3413	Fe I	7
5082,74	Ar I	20
5081,773	Cs	15
5081,77	Cs II	10
5081,44	Ar I	10
5081,360	Ne I	15
5081,07	Xe II	30
5080,62	Xe II	600
5080,3852	Ne I	150
5080,10	Cs II	5
5078,762	Ne I	15
5078,352	F I	3
5078,25	Cl II	150
5078,19	Kr II	2
5078,03	Ar I	40

λ	Symbol	I	λ	Symbol	I
5077,805	Cu II	5	5056,314	Si II	30
5077,23	Kr II	40	5056,27	K II	7
5076,59	C I	1	5055,981	Si II	1000
5076,581	Ne I	35	5054,53	Kr II	30
5076,173	Cu I	100	5054,1783	Ar I	300
5076,03	Ar I	1	5054,070	Ti I	3
5075,92	Kr II	4	5053,52	C I	2
5074,760	Fe I	10	5052,930	Ne I	25
5074,201	Ne I	35	5052,879	Ti I	8
5074,062	Ne I	3	5052,696	Cs II	25
5073,590	N II	5	5052,54	Xe II	30
5073,0758	Ar I	200	5052,17	C I	8
5072,55	Kr II	40	5051,778	Cu II	60
5072,30	Ti II	2	5051,6379	Fe I	10
5072,293	Cu II	20	5049,8253	Fe I	15
5071,62	N IV	—	5049,24	C II	2
5071,475	Ti I	7	5048,8130	Ar I	500
5071,30	Ar I	5	5047,74	Kr I	1
5070,99	Ar I	40	5047,738	He I	50
5070,684	Cs II	2	5047,70	O I	5
5070,53	Xe III	1	5047,52	Kr II	4
5069,96	Kr III	4	5047,343	Cu II	10
5069,82	Xe II	10	5047,30	Ar I	2
5069,802	Mg II	3	5047,11	C II	3
5069,66	Ar I	5	5047,00	Ar I	1
5069,351	Ti I	5	5046,608	Ne I	3
5068,937	Mg II	4	5046,51	N II	2
5068,7730	Fe I	10	5046,31	Kr II	80
5068,39	Ar I	5	5045,816	Ne I	15
5068,332	Ti I	3	5045,400	Ti I	5
5068,10	Cl II	10	5045,100	N II	11
5067,41	Kr II	3	5044,98	C II	1
5067,22	Kr II	1	5044,92	Xe II	150
5067,082	Cu II	30	5044,35	C II	5
5067,0	N V	—	5044,15	Ar I	2
5066,33	Xe II	3	5043,800	Cs II	80
5065,985	Ti I	7	5043,578	Ti I	7
5065,58	Kr II	20	5042,86	Kr III	2
5065,48	Ar I	5	5042,853	Ne I	15
5065,448	Cu II	40	5042,416	Ar II	1
5065,016	Fe I	6	5041,828	Cs II	5
5064,654	Ti I	25	5041,80	C I	6
5064,15	C I	0	5041,76	C II	2
5064,068	Ti I	4	5041,7585	Fe I	10
5063,99	Ar I	5	5041,620	Ca I	40
5062,72	Ar I	1	5041,598	Ne I	1
5062,112	Ti I	7	5041,48	C I	6
5062,036	Ar II	30	5041,322	Cu II	10
5061,46	Kr III	2	5041,23	Ar I	10
5060,635	Cu II	30	5041,0747	Fe I	7
5060,0793	Ar I	500	5041,026	Si II	1000
5059,866	Cs II	25	5040,74	C II	2
5059,66	C I	0	5040,72	N II	3
5059,394	Ar II	2	5040,69	F I	0,6
5059,150	Ne I	2	5040,642	Ti I	6
5058,897	Cu II	30	5040,51	Ar I	10
5058,08	Kr I	4			
5057,68	C I	0			
5056,53	Ar I	200			

λ	Symbol	I	λ	Symbol	I
5040,34	Kr I	7	5018,783	O I	5
5040,13	C I	4	5018,75	Xe II	1
5039,959	Ti I	22	5018,72	Kr III	2
5039,07	C I	7	5018,434	Fe II	12
5039,002	Cu II	10	5018,39	C IV	2
5038,400	Ti I	25	5018,06	C I	2
5037,8	Li II	6	5017,76	C I	1
5037,7512	Ne I	500	5017,629	Ar II	10
5037,577	Ne I	3	5717,34	K II	1
5037,16	O I	15	5017,25	Ar I	5
5036,468	Ti I	25	5017,160	Ar II	20
5036,294	Fe I	6	5017,09	C I	3
5036,15	Xe II	3	5016,611	Cu I	400
5035,989	Ne I	35	5016,58	C IV	1
5035,91	C II	5	5016,45	Kr III	20
5035,908	Ti I	25	5016,387	N II	9
5035,88	Ar I	5	5016,162	Ti I	20
5034,36	Cu I	100	5015,71	Kr II	1
5034,25	Ar I	10	5015,6779	He I	500
5033,85	Kr II	100	5015,207	Cu II	10
5032,07	C II	7	5015,187	Ne I	5
5032,026	Ar I	60	5014,959	Fe I	10
5031,901	Fe I	6	5014,92	Ar II	1
5031,483	Ne I	2	5014,277	Ti I	25
5031,3504	Ne I	250	5014,185	Ti I	25
5030,778	Cu II	2	5013,47	Ar I	1
5030,75	Fe III	6	5013,29	Kr II	100
5029,64	Ar I	5	5013,284	Ti I	18
5029,15	Kr I	5	5012,979	Cs II	5
5028,81	N II	1	5012,83	Xe II	50
5028,36	Kr II	30	5012,611	Cu II	20
5028,2796	Xe I	200	5012,28	C I	2
5028,131	Fe I	4	5012,0712	Fe I	12
5025,74	Ar I	1	5012,029	N II	6
5025,662	N II	9	5012,00	C I	2
5025,570	Ti I	18	5011,30	N II	5
5024,92	C I	3	5011,003	Ne I	25
5024,842	Ti I	20	5010,620	N II	10
5024,778	Ar II	1	5009,833	Cu II	20
5024,50	Ar I	1	5009,652	Ti I	7
5024,027	Cu II	5	5009,334	Ar II	30
5023,88	Xe I	3	5008,55	Xe III	10
5023,85	C I	7	5007,325	N II	11
5023,048	N II	5	5007,209	Ti I	40
5022,871	Ti I	25	5007,09	Ar I	2
5022,870	Ne I	25	5006,84	Ar I	2
5022,40	Kr II	200	5006,787	Cu II	30
5022,250	Fe I	6	5006,1254	Fe I	20
5022,06	N II	0	5006,0607	Si I	40
5021,88	Kr II	100	5005,725	Fe I	10
5021,285	Cu II	20	5005,60	K II	8
5021,138	Ca II	4	5005,333	Ne I	50
5020,43	Kr II	4	5005,1587	Ne I	500
5020,217	O I	7	5005,149	N II	14
5020,139	Cu II	5	5005,13	Ar I	1
5020,028	Ti I	25	5004,318	Ar I	20
5019,971	Ca II	8	5003,88	N II	0
5019,291	O I	6			

λ	Symbol	I
5003,561	Ne I	2
5002,7998	Fe I	6
5002,703	N II	9
5002,14	Kr I	2
5002,02	Fe III	8
5001,98	F II	3
5001,871	Fe I	12
5001,641	Cs	2
5001,479	Ca II	7
5001,477	N II	12
5001,136	N II	11
5001,01	Xe II	3
5000,991	Ti I	10
5000,97	Al II	3
5000,395	Ne I	3
4999,65	Ar I	1
4999,504	Ti I	45
4998,54	Kr II	5
4998,502	Ne I	10
4997,482	Ne I	15
4997,227	N II	4
4997,22	Kr II	1
4997,099	Ti I	8
4996,782	Ar II	2
4996,209	Ne I	2
4995,52	Cl II	60
4994,930	Ne I	150
4994,363	N II	10
4994,1323	Fe I	8
4993,93	Xe II	5
4993,746	Ar II	2
4993,03	Xe II	10
4991,66	Ar I	1
4991,41	C I	0
4991,240	N II	5
4991,17	Xe II	100
4991,067	Ti I	50
4989,948	Ar I	80
4989,31	F I	2
4989,140	Ti I	10
4988,963	Fe I	6
4988,77	Xe II	300
4988,52	Kr III	10
4987,367	N II	8
4986,43	F I	2,5
4985,5539	Fe I	7
4985,503	Cu II	40
4985,260	Fe I	7
4985,136	Cu II	2
4985,09	Ar I	10
4983,855	Fe I	6
4983,258	Fe I	5
4982,83	Kr II	50
4982,8134	Na I	2
4982,81	Ar I	1
4982,507	Fe I	8
4981,732	Ti I	60
4981,54	F I	1,5
4980,45	F I	0,8
4980,006	Cu II	10
4979,625	Ne I	5
4979,05	Ar I	1
4978,89	Kr II	100
4978,5414	Na I	1
4978,191	Ti I	10
4977,731	Ti I	5
4977,08	Kr III	2
4976,62	Cl I	10
4975,961	Ne I	10
4975,66	Ar I	2
4975,344	Ti I	10
4974,87	Xe II	2
4974,760	Ne I	50
4974,41	Xe II	1
4974,18	Ar I	10
4974,151	Cu II	10
4973,689	Cu II	10
4973,538	Ne I	100
4973,53	Ar I	5
4973,108	Fe I	3
4973,051	Ti I	6
4972,71	Xe II	400
4972,593	Cs II	25
4972,157	Ar II	15
4971,77	Cl I	—
4971,720	Li I	50
4971,71	Xe II	200
4971,64	Cl III	0
4970,12	Cl II	50
4969,88	Ar I	1
4969,812	Cu II	3
4969,8	Ne II	0
4969,36	Kr I	15
4969,08	Kr I	20
4968,793	O I	8
4968,566	Ti I	6
4967,882	O I	7
4967,378	O I	6
4966,0968	Fe I	8
4965,78	Kr III	2
4965,073	Ar II	25
4965,031	K I	10
4965,00	Xe II	4
4964,73	C II	4
4964,713	Ti I	5
4962,8	Xe II	1
4962,10	Al II	3
4960,65	F I	6
4960,25	Kr II	100
4959,92	C II	1
4959,478	Ar II	1
4958,85	F I	3
4958,67	C II	1
4957,6059	Fe I	60
4957,3054	Fe I	20
4957,122	Ne I	150

λ	Symbol	I
4957,0335	Ne I	1000
4956,750	Ar I	100
4956,146	K I	9
4956,01	F I	1
4955,964	Cu II	5
4955,78	O II	3
4955,41	F I	1,5
4955,382	Ne I	150
4955,27	Kr I	15
4955,21	Ar I	2
4955,111	Ar II	7
4953,85	C II	3
4953,733	Cu II	50
4952,924	Ar II	2
4952,835	Cs II	30
4952,20	F I	2,5
4951,75	Ar I	10
4951,627	Cu II	12
4951,454	Cu II	3
4951,27	N V	—
4950,815	K I	9
4950,16	F I	1
4950,105	Si IV	3
4949,64	Ar I	1
4949,479	Cu II	3
4949,398	Ar II	5
4948,50	Kr II	50
4948,183	Ti I	3
4947,6067	Si I	30
4946,72	Xe II	1
4945,59	Kr II	300
4945	N V	—
4944,9899	Ne I	100
4944,80	Ar I	5
4944,56	N V	9
4943,58	C I	0
4943,24	Cl II	15
4943,24	K II	6
4943,06	O II	7
4943,020	Cu II	20
4943,01	Cs II	10
4943	N V	—
4942,915	Ar II	6
4942,02	C I	0
4942,015	K I	8
4941,562	Ti I	3
4941,12	O II	5
4940,21	Kr III	2
4940,060	Cu II	5
4939,6896	Fe I	4
4939,0457	Ne I	100
4938,8206	Fe I	10
4938,75	K II	3
4938,59	Cl I	—
4938,38	Kr I	2
4938,283	Ti I	8
4937,97	Kr II	1
4937,967	Cu II	15
4937,719	Ti I	4
4937,718	Ar I	30
4937,196	Cu II	20
4936,99	Cl II	25
4936,083	Ar II	3
4935,03	N I	10
4934,48	Kr I	4
4933,8	N V	—
4933,25	F II	5
4933,206	Ar II	25
4932,80	Si II	20
4932,05	C I	8
4931,76	Cl II	2
4931,653	Cu II	100
4931,483	Cu II	20
4930,944	Ne I	50
4930,45	F I	1
4930,38	Kr I	4
4929,16	Ar I	2
4928,83	F I	2
4928,342	Ti I	12
4928,235	Ne I	70
4927,53	Xe III	3
4926,67	F I	0,8
4926,40	C I	0
4926,390	Cu II	20
4926,148	Ti I	4
4925,744	Cs II	5
4925,396	Ti I	5
4925,17	Cl II	15
4924,83	Cl II	10
4924,7753	Fe I	3
4924,60	O II	6
4924,28	Cl II	18
4923,916	Fe II	12
4923,1522	Xe I	500
4922,68	C I	1
4922,50	Ar IV	—
4922,3	Ne II	0
4922,14	Cl II	20
4921,9310	He I	100
4921,768	Ti I	12
4921,48	Xe II	800
4921,461	Cu II	3
4921,042	Ar I	80
4920,5096	Fe I	60
4920,031	Cu II	5
4919,867	Ti I	12
4919,66	Xe II	200
4919,0003	Fe I	30
4918,98	Al II	3
4918,373	Cu II	30
4917,85	Ar I	5
4917,72	Cl II	125
4916,508	Xe I	500
4915,94	Kr II	100
4915,821	Cu II	15
4915,236	Ti I	5

λ	Symbol	I	λ	Symbol	I
4915,03	Ar I	1	4890,09	Xe II	300
4914,90	N I	5	4889,690	Cu II	30
4914,62	Kr II	2	4889,033	Ar II	15
4914,32	Cl II	12	4888,91	C I	1
4914,309	Ar II	2	4888,37	F I	0,8
4913,616	Ti I	20	4888,365	Ne I	5
4912,909	Cu II	20	4888,263	Ar II	5
4912,362	Cu II	15	4887,9478	Ar I	200
4912,332	Si III	4	4887,30	Xe II	300
4910,39	Kr I	2	4886,30	N I	2
4910,025	Fe I	2	4886,29	Ar I	30
4909,726	Cu II	100	4885,19	Xe II	4
4909,71	Ar I	2	4885,084	Ne I	100
4909,032	Cu II	5	4885,082	Ti I	20
4908,52	Ar I	10	4884,9170	Ne I	1000
4908,34	Kr II	2	4884,25	F I	2,5
4908,18	Si II	5	4884,15	Xe II	100
4907,17	Cl II	15	4884,14	N III	1
4906,99	Si II	20	4883,86	Ar I	5
4906,88	O II	5	4883,761	Cu II	3
4906,548	Cu II	20	4883,53	Xe II	600
4906,28	Kr III	6	4883,403	Ne I	15
4905,20	Xe II	2	4883,27	Ar I	30
4904,76	Cl II	135	4883,217	Cu II	3
4904,753	Ar II	12	4883,20	Si II	15
4903,71	Al III	4	4882,233	Ar II	10
4903,3169	Fe I	12	4881,81	N III	0
4902,77	Al II	5	4881,79	N I	1
4902,65	Si II	3	4881,3	Li II	3
4901,412	Cu II	25	4880,922	Ti I	3
4901,26	Ar I	2	4880,46	F I	0,6
4900,625	Ti I	7	4879,95	Cs II	2
4899,910	Ti I	20	4879,860	Ar II	30
4899,9	Xe II	1	4878,2182	Fe I	12
4899,64	Al II	3	4878,132	Ca I	50
4899,013	Ne I	50	4877,96	Ar I	1
4898,94	Cl II	7	4877,70	Cl II	5
4898,76	Al II	5	4876,50	Xe II	500
4898,63	C I	1	4876,2619	Ar I	200
4898,52	Al II	2	4875,63	Kr II	1
4898,06	F I	1	4874,02	F I	1,5
4897,924	Ne I	70	4873,87	Kr III	1
4897,2	Kr II	3	4873,58	N III	2
4896,77	Cl II	200	4873,291	Cu II	15
4896,71	N III	0	4872,73	Ar I	10
4896,396	Cu II	3	4872,1444	Fe I	20
4895,111	N II	8	4871,78	Ar IV	—
4894,6909	Ar I	150	4871,58	O II	5
4894,53	Ar IV	—	4871,3244	Fe I	25
4893,43	C I	0	4870,14	Kr II	20
4892,228	Ne I	10	4870,129	Ti I	20
4892,21	Kr III	5	4870,024	Cs II	30
4892,1007	Ne I	500	4869,757	K I	9
4891,62	Cl II	4	4869,47	Xe III	40
4891,52	Cl I	—	4868,845	Mg II	2
4891,4989	Fe I	50	4868,69	F I	0,5
4890,93	O II	4	4868,268	Ne I	70
4890,7616	Fe I	25	4868,264	Ti I	18
4890,65	C I	2			
4890,19	Ar I	1			

λ	Symbol	I	λ	Symbol	I
4867,84	Ar I	10	4851,501	Ne I	60
4867,557	Ar II	5	4851,248	Cu II	15
4867,24	Kr I	1	4851,082	Mg II	7
4867,18	N III	5	4850,550	Si II	5
4867,07	C II	2	4849,865	K I	7
4867,010	Ne I	70	4849,530	Ne I	30
4866,476	Ne I	80	4849,4	Ne II	0
4866,33	F I	2	4848,487	Ti I	8
4866,10	Cu I	75	4847,815	Ar II	25
4865,919	Ar II	12	4847,38	N I	2
4865,91	Ar I	1	4847,368	Cu II	2
4865,501	Ne I	100	4847,296	Ca I	2
4864,95	O II	3	4847,13	F I	0,5
4864,91	Kr I	2	4847,07	Cl II	4
4864,351	Ne I	30	4846,73	Ar I	5
4864,24	Cs	10	4846,60	Kr II	700
4864,187	Ti I	4	4845,767	Ne I	5
4863,75	Cl III	1	4845,62	Kr III	2
4863,483	K I	8	4845,26	Si II	3
4863,0800	Ne I	100	4845,145	Ne I	15
4862,57	C II	4	4845,14	Kr II	2
4862,54	Xe II	800	4845,01	O II	1
4862,16	Ar I	1	4844,33	Xe II	2000
4861,84	Kr I	2	4843,45	O II	0
4861,548	Cu II	2	4843,294	Xe I	300
4861,332	H	500	4843,26	O II	1
4861,33	N III	4	4842,941	Ne I	50
4861,31	Kr I	4	4842,57	Si III	5
4861,095	Si II	10	4842,566	Ne I	10
4861,03	O II	3	4842,44	Cl II	8
4860,170	N II	4	4842,290	Cu I	25
4860,029	D	500	4841,9	Kr III	1
4859,7480	Fe I	15	4840,874	Ti I	25
4859,604	Ne I	15	4840,87	Xe II	1
4859,6	C III	0	4839,861	Si I	11
4859,595	T	500	4839,251	Ti II	1
4859,44	Ar I	5	4839,04	Kr II	4
4859,37	F II	7	4837,93	N I	1
4859,323	He II	7	4837,3139	Ne I	500
4858,88	N III	3	4836,79	Cl II	20
4858,74	N III	2	4836,76	C I	0
4857,20	Kr II	150	4836,697	Ar I	150
4857,04	Cl II	10	4836,56	Kr II	20
4856,76	O II	2	4836,125	Ti I	6
4856,49	O II	2	4835,97	Ar I	30
4856,156	Ar II	1	4835,03	Cs	15
4856,090	K I	8	4834,10	Ar I	30
4856,012	Ti I	20	4833,68	Kr II	4
4855,37	Ar I	1	4833,50	Cl II	2
4854,966	Cu II	30	4832,79	Ar I	5
4854,37	Ar I	1	4832,38	Ar I	5
4854,37	Cl III	0	4832,236	Cu II	30
4853,77	Xe II	40	4832,07	Kr II	800
4852,70	Cl I	8	4831,16	N I	1
4852,6571	Ne I	100	4830,54	Ar I	1
4852,61	Kr II	2	4830,161	Cs II	30
4851,583	Cs	8	4829,709	Xe I	400
4851,540	Si I	13	4829,47	Ar I	2

λ	Symbol	I
4829,288	Ne I	5
4829,23	Cl II	3
4829,23	K II	9
4828,968	Si III	18
4827,587	Ne I	300
4827,3444	Ne I	1000
4826,80	C I	3
4826,08	Kr III	2
4825,97	Ar I	2
4825,529	Ne I	50
4825,445	Ti I	3
4825,42	Cs	10
4825,18	Kr II	300
4823,93	O IV	—
4823,41	Xe II	300
4823,370	Ne I	50
4823,31	Si I	10
4823,174	Ne I	100
4821,9236	Ne I	300
4821,87	Cl II	2
4821,1666	Si I	15
4820,95	Cl II	4
4820,410	Ti I	20
4819,937	Ne I	70
4819,79	Cl II	25
4819,718	Si III	16
4819,46	Cl II	200
4818,789	Ne I	150
4818,64	Cl I	2
4818,42	Cl I	3
4818,02	Xe II	200
4817,6386	Ne I	300
4817,37	C I	4
4817,22	Xe II	40
4816,900	Ne I	1
4814,338	Ne I	50
4813,330	Si III	15
4813,07	O IV	1
4812,940	Cu II	40
4812,92	C I	2
4812,6367	Kr I	40
4812,240	Ti I	5
4811,76	Kr II	300
4811,57	Cl II	12
4811,074	Ti I	4
4810,634	Ne I	100
4810,51	Kr I	3
4810,306	N II	4
4810,0640	Ne I	150
4810,06	Cl II	225
4809,500	Ne I	10
4809,05	Cl II	9
4808,66	Ar IV	—
4808,531	Ti I	5
4808,00	Cl III	1
4807,68	Cl II	5
4807,039	Cu II	10
4807,019	Xe I	500
4806,924	Cs II	5
4806,92	Xe II	3
4806,017	Ar II	35
4805,651	Cu II	3
4805,4402	Si I	20
4805,416	Ti I	12
4805,105	Ti II	2
4804,61	Cs	10
4804,348	K I	8
4804,33	Ar I	5
4803,289	N II	10
4803,225	Ne I	1
4803,16	Cl II	2
4802,981	O I	4
4802,97	Kr II	4
4802,70	C II	1
4802,363	Ne I	10
4802,132	O I	3
4801,80	O I	2
4801,076	Ne I	2
4800,77	O IV	—
4800,428	Si III	8
4800,111	Ne I	15
4799,973	Ca II	4
4799,797	Ti I	12
4799,754	K I	6
4799,45	Xe II	15
4799,2	O IV	10
4798,742	Ar I	30
4798,535	Ti II	2
4798,40	Cl II	15
4798,25	O IV	5
4797,983	Ti I	5
4797,042	Cu I	20
4796,76	Cl I	2
4796,57	Ar I	1
4796,53	Xe II	6
4796,33	Kr II	60
4796,210	Ti I	6
4796,08	C I	0
4795,88	C I	0
4795,62	Ne II	2
4795,40	Xe II	3
4794,54	Cl II	250
4794,48	Xe III	12
4794,22	O IV	2
4794,10	Ar I	1
4794,00	Cu I	150
4793,66	C III	2
4793,650	N II	4
4792,65	C I	0
4792,6192	Xe I	150
4792,482	Ti I	10
4792,324	Si I	80
4792,29	Si II	5
4792,212	Si I	35
4792,090	Ar II	6

λ	Symbol	I	λ	Symbol	I
4792,04	Cl II	12	4773,01	Kr II	40
4791,71	C I	0	4772,913	O I	4
4791,248	Fe I	5	4772,7847	Si I	25
4791,15	Ar I	2	4772,57	O IV	2
4791,15	Kr II	3	4772,448	O I	3
4791,049	K I	7	4771,75	C I	8
4790,728	Ne I	30	4771,66	Cl II	20
4790,218	Ne I	500	4771,103	Ti I	3
4790,20	Xe II	3	4771,09	Cl II	40
4789,74	Kr III	7	4770,34	Ar I	2
4789,6537	Fe I	7	4770,03	C I	5
4789,600	Ne I	100	4769,775	Ti I	4
4788,9270	Ne I	1000	4769,05	Xe II	150
4788,8	Li II	8	4768,68	Cl II	150
4788,76	Kr II	5	4768,6750	Ar I	150
4788,131	N II	8	4768,41	Cs	10
4787,77	Xe II	100	4767,49	Cu I	75
4787,7	O IV	3	4766,729	Cu II	5
4786,8106	Fe I	5	4766,68	C I	4
4786,7	C IV	0	4766,330	Ti I	4
4786,65	Xe II	10	4765,74	Kr II	1000
4786,491	K I	5	4765,30	Cl II	10
4786,4	O IV	20	4764,862	Ar II	25
4786,363	Cs II	15	4764,535	Ti II	1
4786,155	Ar II	5	4763,616	Cs II	25
4785,88	C IV	1	4762,77	Ti II	1
4785,44	Cl II	50	4762,54	C I	5
4784,8	Kr II	1	4762,43	Kr II	300
4784,022	Ne I	2	4762,31	C I	5
4783,80	C I	1	4762,10	N IV	—
4783,43	O IV	4	4759,272	Ti I	25
4782,9905	Si I	50	4758,972	Si I	13
4782,89	Si II	3	4758,92	Cs	10
4781,95	Ne II	1	4758,913	Ti I	4
4781,82	Cl II	50	4758,77	Kr II	1
4781,718	Ti I	6	4758,728	Ne I	150
4781,32	Cl II	75	4758,421	Cu II	30
4781,239	Ne I	2	4758,120	Ti I	25
4781,190	N II	4	4757,87	Cs	10
4780,884	Ne I	30	4757,389	K I	7
4780,338	Ne I	300	4757,215	Ar II	1
4779,986	Ti II	1	4755,64	Cl II	50
4779,722	N II	7	4755,2756	Si I	25
4779,18	Xe II	80	4754,48	Kr III	6
4779,09	O IV	2	4754,440	Ne I	100
4778,93	Cl II	45	4753,934	K I	5
4778,259	Ti I	10	4753,49	Cl II	8
4776,38	Cl II	5	4753,458	Cu II	3
4776,22	Cu I	20	4753,13	N I	2
4776,20	Si II	3	4753,123	Ne I	1
4775,91	C I	6	4752,9404	Ar I	150
4775,76	Xe II	8	4752,7320	Ne I	500
4775,18	Xe II	5	4752,70	O II	2
4774,92	K II	4	4752,50	N IV	—
4774,46	Kr II	2	4752,02	Kr II	100
4774,241	N II	4	4751,8218	Na I	2
4773,752	O I	5	4751,802	Ne I	30
4773,19	Xe II	80	4751,34	O II	4

λ	Symbol	I	λ	Symbol	I
4751	O VI	—	4731,172	Ti I	9
4750,686	Ne I	30	4730,664	Ar II	3
4750,26	N I	2	4730,66	Ar I	5
4749,73	N V	—	4730,519	Si III	7
4749,5754	Ne I	300	4730,24	Ne II	0,5
4749,132	Cs II	10	4730,16	C III	1
4749,00	Kr III	2	4730,0285	Mg I	10
4748,67	Cl II	20	4729,72	Kr III	4
4748,23	Ar I	5	4728,18	Cs	10
4747,9936	Si I	25	4727,48	Ar I	5
4747,9410	Na I	1	4727,41	C II	2
4747,680	Ti I	3	4726,859	Ar II	25
4747,28	C II	2	4726,684	Cs II	5
4746,823	Ar I	80	4725,145	Ne I	70
4744,92	K II	4	4724,89	Kr I	20
4744,77	C II	5	4724,33	C III	1
4744,60	Cs II	5	4724,162	Ne I	5
4744,345	K I	6	4724,10	Ar I	5
4744,04	N I	3	4723,810	Ne I	70
4743,89	Xe III	4	4723,57	Xe III	30
4742,90	N I	2	4723,171	Ti I	10
4742,791	Ti I	20	4722,714	Ne I	15
4742,57	C I	2	4722,603	Ti I	10
4742,129	Ti I	3	4722,16	Kr I	3
4741,71	O II	3	4722,150	Ne I	5
4740,914	K I	4	4721,594	Ar II	12
4740,71	Cl I	10	4721,57	N II	2
4740,40	Cl II	150	4721,536	Ne I	70
4740,26	N IV	—	4721,43	Cl II	25
4739,712	Mg II	5	4721,24	Cl I	8
4739,665	Cs II	20	4721,028	Ca II	4
4739,588	Mg II	6	4721,00	Xe II	2
4739,42	Cl II	10	4719,94	Ar I	20
4739,00	Kr II	3000	4719,515	Ti II	1
4738,47	C I	3	4719,37	Ne II	1,5
4738,41	Cl II	10	4719,22	Ar I	2
4738,21	C I	1	4718,38	N II	4
4738,0	F II	0	4718,10	Ar I	2
4737,97	C II	3	4717,608	Ne I	70
4736,7807	Fe I	12	4716,736	Ca II	3
4735,905	Ar II	25	4716,651	Si III	16
4735,46	C II	2	4716,19	Cs	10
4735,17	C I	2	4715,3466	Ne I	1500
4734,682	Ti I	3	4715,295	Ti I	4
4734,627	Si III	2	4715,246	Ne I	30
4734,60	C II	2	4715,18	Xe II	100
4734,37	F II	2	4715,132	Ne I	30
4734,26	C I	5	4714,336	Ne I	70
4734,1524	Xe I	600	4714,28	Cl II	8
4733,426	Ti I	6	4713,376	He I	20
4733,06	Cs	20	4713,1455	He I	150
4732,975	Cs II	20	4712,800	Ne I	10
4732,53	Ne II	1	4712,63	Xe II	40
4732,51	Xe II	15	4712,135	Ne I	15
4732,056	Ar II	12	4712,07	N II	2
4731,22	N I	1	4712,066	Ne I	1000
4731,19	Xe II	100	4710,823	Ar II	7

λ	Symbol	I
4710,48	Kr III	10
4710,478	Ne I	30
4710,186	Ti I	18
4710,0669	Ne I	1000
4710,04	Ne II	2
4710,04	O II	5
4709,59	N II	2
4709,50	Ar I	30
4709,08	Ar I	10
4708,92	Xe II	8
4708,8619	Ne I	1200
4708,46	Ar I	2
4708,21	Xe I	5
4707,80	O II	0
4707.31	N IV	4
4707,2807	Fe I	8
4706,96	Xe II	2
4706,76	Si I	8
4706,40	N II	2
4706,31	Kr II	3
4705,44	Kr II	2
4705,355	O II	8
4704,67	Xe II	10
4704,594	Cu I	450
4704,3949	Ne I	1500
4704,35	Ar I	2
4704,24	N II	2
4703,359	Ar II	9
4703,18	O II	3
4703,14	Cl III	3
4703,06	F I	1
4702,9909	Mg I	30
4702,526	Ne I	150
4702,51	N II	2
4702,3155	Ar I	1200
4701,793	Cs II	25
4701,76	O II	0
4701,71	Cu I	10
4701,65	Al III	6
4701,23	O II	2
4701,2	Ne II	0
4700,469	Ne I	5
4700,1	Ne II	0
4700,04	N II	2
4699,69	Kr II	30
4699,62	Xe II	3
4699,21	O II	7
4698,99	O II	3
4698,86	Ti I	6
4698,766	Ti I	20
4698,55	N II	1
4698,48	O II	1
4698,01	Xe II	300
4697,87	Ar IV	—
4697,490	Cu I	350
4697,49	Xe III	2
4697,020	Xe I	300
4697,00	F I	0,6
4696,943	Ne I	5
4696,923	Ti I	4
4696,36	O II	2
4695,89	N II	2
4695,66	Kr II	50
4695,610	Cs II	10
4695,363	Ne I	20
4695,07	Cl III	1
4694,84	Kr I	4
4694,637	N II	6
4694,44	Kr II	200
4693,670	Ti I	5
4693,65	Kr III	3
4693,34	Xe II	15
4692,482	Cs II	5
4691,580	Ne I	15
4691,53	Cl I	12
4691,47	O II	1
4691,4144	Fe I	6
4691,336	Ti I	20
4691,28	Kr II	100
4690,9711	Xe I	100
4690,97	O II	0
4690,9	Ar IV	—
4690,827	Ti I	3
4688,392	Ti I	3
4688,3	Kr II	3
4688,191	Ne I	2
4687,770	Cu II	5
4687,6724	Ne I	100
4687,28	Kr II	10
4686,921	Ti I	4
4686,30	Kr II	8
4685,74	N I	3
4685,682	He II	300
4685,4	C IV	1
4685,265	Ca I	12
4685,17	Xe III	1
4683,797	Si III	7
4683,764	Ne I	30
4683,68	Kr II	5
4683,53	Xe III	60
4683,238	Ne I	5
4683,022	Si III	9
4682,910	Ne I	10
4682,277	Ar II	10
4682,146	Ne I	20
4681,990	Cu II	50
4681,930	Ne I	20
4681,908	Ti I	30
4681,494	Ar II	2
4681,200	Ne I	50
4680,41	Kr II	500
4680,363	Ne I	100
4679,45	Xe II	3
4679,135	Ne I	150
4678,852	Fe I	7
4678,604	Ne I	50

λ	Symbol	I
4678,31	Xe II	2
4678,218	Ne I	300
4678,14	N II	6
4677,76	Cl I	7
4677,7	Li II	8
4677,340	Cu I	30
4677,16	Kr I	1
4677,00	O II	0
4676,75	Xe II	2
4676,73	Cl II	8
4676,46	Xe II	200
4676,234	O II	8
4675,118	Ti I	10
4674,909	N II	5
4674,89	Cs	10
4674,72	Cu I	500
4674,56	Xe II	40
4674,40	Cl I	2
4673,95	C III	6
4673,80	Kr II	3
4673,80	Kr III	3
4673,75	O II	4
4673,70	O I	3
4673,66	Xe III	30
4673,555	Cu II	30
4673,297	Si IV	2
4673,273	Si II	20
4672,75	O I	3
4672,20	Xe II	100
4672,09	Kr II	2
4671,8	Li II	4
4671,686	Cu II	40
4671,61	Kr I	10
4671,226	Xe I	2000
4670,884	Ne I	70
4670,280	Cs II	20
4669,77	N I	3
4669,53	O II	0
4669,50	Cl III	0
4669,33	O II	0
4668,5595	Na I	2
4668,49	Xe II	100
4668,1422	Fe I	6
4667,585	Ti I	25
4667,459	Fe I	6
4667,356	Ne I	100
4667,297	Cu II	15
4667,206	N II	5
4667,14	Si IV	1
4666,8	Al II	11
4666,654	Ne I	50
4666,28	Xe II	40
4666,260	Ar II	1
4665,869	Si III	8
4665,86	C III	8
4664,8107	Na I	1
4663,64	C III	6
4663,518	Ne I	20
4663,092	Ne I	40
4663,054	Al II	10
4662,638	Cu II	15
4661,635	O II	9
4661,350	Cu II	15
4661,22	Cl I	18
4661,1054	Ne I	150
4660,294	Cu II	8
4660,05	N I	2
4659,38	K II	5
4659,06	C III	5
4658,87	Kr II	2000
4658,30	C IV	9
4657,893	Ar II	25
4657,78	Xe III	9
4657,72	N I	1
4656,92	Si IV	3
4656,65	N I	1
4656,538	Cs II	12
4656,468	Ti I	25
4656,3936	Ne I	300
4656,048	Ti I	6
4655,75	Ti II	1
4655,712	Ti I	3
4655,359	O I	3
4654,558	O I	2
4654,532	N II	5
4654,323	Si IV	10
4654,118	O I	1
4654,05	Cl I	10
4653,699	Ne I	50
4653,00	Xe II	40
4652,101	Ne I	30
4652,06	C III	5
4651,94	Xe II	200
4651,47	C III	11
4651,388	Ar I	20
4651,124	Cu I	2000
4651,1	Cs	10
4651,08	N I	1
4651,01	C III	5
4650,841	O II	6
4650,646	Al II	1,5
4650,544	Al II	2
4650,25	C III	13
4650,17	Kr II	30
4650,016	Ti I	10
4649,904	Ne I	70
4649,266	Cu II	10
4649,17	Xe II	2
4649,139	O II	10
4648,933	Fe II	10
4648,62	Al II	1
4647,493	Ar I	40
4647,45	Si IV	1
4647,4370	Fe I	6
4647,42	C III	14
4647,40	Ti IV	3
4647,34	Ne II	0,5

λ	Symbol	I
4646,508	Cs II	25
4646	C IV	—
4645,885	Ne I	1
4645,42	F I	2
4645,4180	Ne I	300
4645,193	Ti I	12
4644,833	Ne I	40
4644,43	F I	1
4643,931	Ne I	2
4643,63	Xe III	1
4643,182	Ne I	5
4643,085	N II	11
4642,58	Cu I	150
4642,373	K I	11
4642,148	Ar I	80
4641,90	N III	7
4641,876	K I	10
4641,811	O II	9
4640,64	N III	10
4640,443	Ne I	70
4640,384	Al II	3,5
4640,362	Al II	4
4640,333	Cs II	5
4639,944	Ti I	15
4639,833	Al II	1,5
4639,725	Al II	2
4639,669	Ti I	15
4639,591	Ne I	30
4639,384	Al II	1
4639,369	Ti I	18
4639,36	Ar IV	—
4638,96	Cl III	2
4638,91	C II	2
4638,854	O II	6
4638,277	Si III	7
4638,17	Si I	15
4637,887	Ti I	8
4637,66	Kr II	1
4637,63	C II	1
4637,518	Fe I	3
4637,233	Ar II	12
4636,974	Ne I	50
4636,634	Ne I	70
4636,345	Ti II	1
4636,14	Kr I	20
4636,125	Ne I	70
4636,0	Li I	1
4635,83	Cl III	0
4635,7	Al II	1
4635,539	Ti I	3
4635,42	Kr II	8
4634,73	Ne II	2
4634,16	N III	8
4633,88	Kr II	800
4633,30	Xe II	50
4632,68	Xe III	2
4631,50	Xe III	2
4631,405	Mg II	1

λ	Symbol	I
4631,241	Si IV	9
4630,878	Mg II	2
4630,543	N II	14
4629,98	C II	2
4629,7	Al II	1
4629,336	Fe II	7
4629,336	Ti I	15
4629,29	Ti II	1
4628,62	Si IV	3
4628,460	Ne I	30
4628,4409	Ar I	1000
4628,3113	Ne I	150
4627,85	Ne II	3
4627,799	Ne I	2
4627,44	C II	1
4627,383	Si I	18
4626,78	Ar I	30
4625,61	N I	1
4625,56	C II	3
4625,46	Ar I	10
4625,0527	Fe I	3
4624,36	Cl II	6
4624,2757	Xe I	1000
4623,96	Cl I	10
4623,098	Ti I	25
4623,091	Cs II	20
4621,721	Si II	150
4621,418	Si II	100
4621,40	Kr III	1
4621,394	N II	10
4621,299	Mg I	3
4621,28	O II	0
4620,59	Cs II	10
4620,11	Xe II	2
4619,99	Kr II	5
4619,98	N V	10
4619,657	Si III	7
4619,525	Ti I	3
4619,4	N V	—
4619,298	Fe I	3
4619,23	C II	8
4619,15	Kr II	1000
4618,40	C II	6
4617,837	Ne I	70
4617,50	Xe II	90
4617,269	Ti I	30
4616,911	Ne I	5
4616,28	Cs	15
4616,13	Cs II	15
4616,01	Cs	10
4615,98	Ne II	4
4615,50	Xe II	200
4615,28	Kr II	500
4615,06	Xe II	100
4614,50	Kr II	15
4614,391	Ne I	100
4614,10	Ar II	1
4613,866	N II	9
4613,79	Kr II	2
4613,78	Cl III	2

λ	Symbol	I	λ	Symbol	I
4613,67	O II	1	4592,6547	Fe I	5
4613,11	O II	0	4592,29	Cl II	2
4612,89	Ne II	1	4592,05	Xe II	300
4611,8896	Xe I	100	4591,50	Kr II	1
4611,289	Fe I	5	4591,10	Cl III	4
4611,27	Si IV	0	4590,971	O II	9
4611,245	Ar II	3	4589,961	Ti II	2
4610,65	Kr II	60	4589,896	Ar II	25
4610,505	Cs	10	4589,89	O I	3
4610,14	O II	3	4589,75	Al II	4
4609,99	Cs II	10	4589,689	Al II	1
4609,910	Ne I	150	4589,288	Ar I	80
4609,72	Kr II	20	4588,98	O I	2
4609,7	Al II	1	4588,194	Al II	5
4609,560	Ar II	25	4588,13	Ne II	3
4609,42	O II	4	4587,895	Ar II	3
4609,365	Ne I	30	4587,6	C III	0
4608,48	Kr II	1	4587,21	Ar I	5
4608,457	Cu II	5	4586,97	Cu I	1300
4608,45	K II	8	4586,610	Ar I	10
4608,21	Cl III	5	4586,145	Ne I	2
4608,085	N II	3	4585,923	Ca I	2
4607,157	N II	10	4585,876	Ne I	10
4606,33	N IV	6	4585,871	Ca I	50
4604,938	Ne I	5	4585,82	Al II	6
4604,680	Ne I	1	4585,48	Xe II	500
4604,43	Cl III	0	4585,03	Cl II	15
4604,095	Ne I	15	4584,958	Ar I	10
4604,02	Kr II	60	4584,28	Cl II	20
4603,755	Cs II	60	4583,848	Fe II	11
4603,73	N V	12	4583,443	Ti II	1
4603,2	N V	—	4582,980	Ne I	5
4603,03	Xe II	600	4582,85	Kr II	300
4602,9446	Fe I	9	4582,7474	Xe I	300
4602,871	Li I	100	4582,556	Ne I	15
4602,53	N II	3	4582,4521	Ne I	150
4602,11	O II	2	4582,40	Cl II	8
4602,02	Li I	1	4582,105	Ne I	15
4601,480	N II	11	4582,035	Ne I	150
4601,42	Kr II	1	4581,402	Ca I	40
4601,00	Cl I	20	4580,70	Xe II	80
4600,11	Ne II	1	4580,47	Cl I	3
4600,02	Ar II	1	4580,458	Ti II	1
4599,23	Ti I	5	4580,35	Ne II	3
4599,22	Cs	15	4580,11	Kr II	2
4598,760	Ar II	10	4579,346	Ar II	25
4598,49	Kr II	50	4578,558	Ca I	30
4597,942	Cu II	5	4578,17	Cl I	4
4597,673	Cs II	10	4577,66	O I	3
4596,903	Cu II	10	4577,20	Kr II	800
4596,30	Xe II	1	4577,06	Xe II	200
4596,22	Cl III	4	4576,79	O I	2
4596,174	O II	8	4576,60	Xe I	2
4596,0964	Ar I	1000	4576,1	F II	0
4595,65	K II	5	4575,858	Ne I	20
4595,249	Ne I	50	4575,8	Kr II	1
4593,70	Xe II	6	4575,0620	Ne I	300
4593,3	C III	1			
4593,243	Ne I	50			
4593,172	Cs I	1000			
4592,85	F VI	1			
4592,80	Kr II	150			

λ	Symbol	I	λ	Symbol	I
4574,759	Si III	20	4554,28	O V	0
4574,49	Ne II	1	4554,21	N I	1
4573,759	Ne I	30	4553,996	Si III	8
4573,557	Ne I	50	4553,38	N I	1
4573,33	Kr II	30	4553,16	Ne II	4
4573,066	Ne I	5	4552,77	Kr II	3
4572,894	Ar II	2	4552,616	Si III	30
4572,611	Cs	10	4552,598	Ne I	30
4572,13	Cl II	100	4552,527	N II	7
4571,971	Ti II	50	4552,453	Ti I	35
4571,85	Xe II	30	4550,79	Xe II	10
4571,786	Cs II	15	4550,640	Ne I	1
4571,0956	Mg I	28	4550,2985	Kr I	40
4570,906	Ti I	3	4549,622	Ti II	60
4569,69	Ar I	2	4549,470	Fe II	10
4569,50	O III	1	4548,764	Ti I	35
4569,42	Cl II	50	4548,26	Cl I	—
4569,12	Xe II	4	4547,8505	Fe I	4
4569,01	Ne II	5	4547,760	Ar II	4
4568,64	Ar I	2	4547,728	Ne I	15
4568,5	O IV	—	4547,34	N III	10
4567,845	Ne I	10	4547,218	Ne I	10
4567,823	Si III	25	4546,36	N III	3
4567,139	Ne I	15	4545,729	Ne I	1
4566,983	Cs II	15	4545,23	Xe II	400
4566,830	Ne I	40	4545,186	Na I	8
4565,888	Ne I	60	4545,045	Ar II	25
4565,82	Kr II	1	4544,80	N III	0
4565,51	Kr III	1	4544,746	Ar I	30
4565,49	Ne II	1	4544,688	Ti I	30
4564,9	Kr II	1	4544,502	Ne I	50
4564,83	F I	0,6	4544,48	Cl II	10
4564,82	Ar I	4	4544,11	Ne II	1
4564,764	N II	3	4543,871	Ar II	3
4564,415	Ar II	7	4543,71	Cs	10
4563,761	Ti II	30	4541,633	Na I	7
4563,751	Ar II	7	4541,60	Ar I	20
4563,427	Ti I	5	4541,59	He II	5
4563,00	Xe II	2	4541,032	Cu II	25
4562,637	Ti I	6	4540,89	Xe II	400
4562,05	Ne II	1	4540,380	Ne I	50
4561,018	Ar II	3	4540,335	Cu II	10
4560,38	Kr II	3	4540,29	Cl II	6
4559,920	Ti I	6	4540,207	Cu II	10
4559,09	Fe III	6	4539,695	Cu I	800
4556,698	Ne I	2	4539,25	Cl II	6
4556,61	Kr II	200	4539,168	Ne I	50
4555,94	Xe II	200	4538,942	Cs II	30
4555,922	Cu II	100	4538,713	Ar II	1
4555,890	Fe II	8	4538,293	Ne I	300
4555,486	Ti I	30	4538,06	Kr I	3
4555,392	Ne I	30	4537,7545	Ne I	1000
4555,30	O III	0	4537,683	Ne I	300
4555,280	Cs I	2000	4537,648	Ar II	7
4555,069	Ti I	3	4537,33	Xe III	30
4554,824	Ne I	40	4537,25	Kr III	6
4554,561	Ne I	5	4536,92	Xe II	80
4554,415	Ne I	10	4536,78	Cl II	20
4554,319	Ar I	15			

λ	Symbol	I
4536,46	Kr III	10
4536,312	Ne I	150
4536,051	Ti I	40
4535,920	Ti I	40
4535,574	Ti I	50
4535,492	Ar II	6
4535,47	Ne II	3
4535,11	N III	2
4534,782	Ti I	60
4534,78	Ar I	20
4534,66	Ne II	2
4534.64	Cs	10
4534,57	N III	3
4534,34	Cl II	5
4534,291	Mg II	6
4533,966	Ti II	30
4533,238	Ti I	80
4532.500	Cs	10
4532,49	Xe II	200
4532,395	Ne I	1
4531.45	Cs	15
4531,152	Fe I	8
4530,84	N III	1
4530,785	Cu I	800
4530,553	Ar II	7
4530,410	N II	9
4529.7	O III	00
4529.61	N JV	—
4529,476	Ne I	30
4529,176	Al III	10
4528,911	Al III	1
4528,62	Kr II	3
4528,6175	Fe I	18
4527,973	Ne I	1
4527,86	N III	0
4527,725	Ne I	15
4527,455	Ti I	4
4527,305	Ti I	35
4526,935	Ca I	30
4526,725	Cs II	35
4526,685	Ne I	15
4526,20	Cl I	30
4526,177	Ne I	50
4525,764	Ne I	70
4525,67	Xe III	1
4525,59	Cs II	2
4525,146	Fe I	5
4525,112	Cu I	40
4524,6805	Xe I	400
4524,21	Xe II	200
4523,60	N III	4
4523,35	Ar I	1
4523,33	Cl III	4
4523,14	Kr II	400
4522,846	Cs II	15
4522,798	Ti I	40
4522,66	Ne II	4
4522,634	Fe II	9
4522.36	Cs	15
4522,3238	Ar I	800
4522.2	O V	—
4521,86	Xe II	100

λ	Symbol	I
4520,225	Fe II	7
4519,69	Xe II	3
4519,19	Cl II	18
4518,700	Ti I	8
4518,64	Kr III	2
4518,18	N III	3
4518,022	Ti I	50
4517,79	Ne II	2
4517,736	Ne I	100
4517,526	Ar II	3
4517,29	Ne	2
4516,936	Ne I	50
4516,77	C III	6
4516,095	Ar II	1
4516,050	Cu II	5
4515,78	C III	5
4515,495	Cs II	10
4515,411	Ne I	30
4515,337	Fe II	7
4515,33	C III	3
4515,022	Ne I	2
4514,891	Ne I	70
4514,89	N III	7
4514,80	Ne II	2
4513,192	Cu I	50
4512,734	Ti I	40
4512,535	Al III	8
4512,282	Ca I	5
4511,80	Xe II	2
4511,509	Ne I	20
4511,37	Ne II	4
4511,29	Ne II	2
4510,92	N III	6
4510,7335	Ar I	1000
4510,170	Ne I	15
4509,957	Ar II	2
4509,87	Ar I	4
4509,446	Ca I	3
4509,374	Cu I	400
4508,77	N II	2
4508,283	Fe II	8
4508,21	Ne II	3
4507,854	Ca I	1
4507,557	N II	6
4507,45	Ar I	1
4507,417	Ca I	1
4507,35	Cu I	200
4507,11	Xe II	5
4506,834	Cs II	10
4506,705	Cs II	15
4506,624	Ca I	1
4506,50	O II	2
4505,997	Cu II	75
4505,33	K II	6
4505,16	Ar I	3
4505,00	Ca I	0
4504,27	Cl II	20

λ	Symbol	I
4503,762	Ti I	4
4503,53	N I	1
4503,46	Xe III	10
4503,08	Ar II	1
4502,931	Ar II	7
4502,52	Ne II	0,5
4502,3546	Kr I	600
4502,27	N I	2
4501,525	Cs II	25
4501,270	Ti II	40
4500,9772	Xe I	500
4500,182	Ne I	50
4499,843	Ne I	5
4499,000	Ne I	2
4498,94	Ne II	5
4498,543	Ar II	7
4497,709	Ti I	3
4497,658	Na I	11
4497,45	N I	1
4497,30	Cl II	18
4496,758	Cs	15
4496,146	Ti I	20
4495,86	N IV	—
4494,67	N I	5
4494,5669	Fe I	12
4494,177	Na I	10
4494,048	Si III	6
4493,699	Ne I	50
4493,660	Cs II	10
4493,108	Ne I	5
4492,689	Ne I	15
4492,540	Ti I	3
4492,412	Ne I	30
4492,40	N I	7
4492,132	Ne I	5
4491,838	Ne I	50
4491,771	Ne I	80
4491,401	Fe II	5
4491,25	O II	3
4491,05	Cl I	10
4490,988	Ar II	8
4490,90	Al III	2
4490,00	Cl II	50
4489,88	Kr II	400
4489,7416	Fe I	3
4489,48	O II	1
4489,178	Ca II	2
4489,17	Cl III	1
4489,089	Ti I	20
4488,81	Xe III	2
4488,60	Xe II	4
4488,319	Ti II	15
4488,22	Kr II	3
4488,17	O II	2
4488,12	N II	2
4488,0926	Ne I	300
4488,09	O II	2
4487,72	O II	0
4485,95	Xe II	20
4484,225	Fe I	4
4483,190	Ne I	150
4482,884	Si III	3
4482,688	Ti I	10
4482,2563	Fe I	6
4482,1720	Fe I	4
4482,02	Cl II	10
4481,85	Kr II	50
4481,810	Ar II	15
4481,327	Mg II	13
4481,261	Ti I	30
4481,130	Mg II	14
4480,87	Ar I	5
4480,86	Xe II	500
4480,823	Ne I	15
4480,600	Ti I	5
4480,350	Cu I	500
4479,968	Al III	4
4479,92	N IV	—
4479,891	Al III	3
4479,86	Kr II	5
4479,724	Ti I	9
4479,31	Ar I	5
4479,226	Ca II	1
4478,83	C I	4
4478,59	C I	4
4478,32	C I	2
4477,88	O II	2
4477,691	N II	2
4477,47	C I	4
4476,08	O II	0
4476,0206	Fe I	10
4475,656	Ne I	100
4475,31	Cl I	15
4475,28	Cl II	20
4475,22	Ne II	1
4475,131	Ne I	5
4475,00	Kr II	800
4474,95	O III	1
4474,852	Ti I	8
4474,759	Ar II	10
4474,72	Ar I	5
4473,85	Xe II	4
4472,4	O IV	—
4472,246	Ne I	1
4472,043	Ca II	2
4471,682	He I	120
4471,52	Ne II	3
4471,479	He I	1000
4471,238	Ti I	20
4470,971	Ne I	5
4470,90	Xe II	30
4469,41	O II	4
4469,380	Fe I	5
4469,37	Cl I	18
4469,32	O II	3
4469,09	Cs	2
4468,91	Ne II	5

λ	Symbol	I
4468,493	Ti II	50
4468,48	Cl II	2
4468,452	Si III	2
4468,15	Xe III	1
4467,88	O II	4
4467,491	Ne I	1
4467,31	C I	2
4466,8120	Ne I	70
4466,65	K II	5
4466,5542	Fe I	12
4466,503	Ne I	2
4466,48	C I	5
4466,32	O II	2
4466,28	O II	4
4466,045	Ne I	5
4465,807	Ti I	20
4465,651	Ne I	50
4465,527	N II	2
4465,40	O II	4
4464,68	C I	2
4464,60	Xe II	1
4464,45	C I	2
4463,89	C I	2
4463,6901	Kr I	800
4463,539	Ti I	8
4463,391	Ti I	8
4462,856	Ne I	2
4462,684	Cu II	3
4462,19	Xe II	1000
4462,089	Ti I	3
4461,6544	Fe I	8
4461,56	O III	1
4461,50	C I	1
4461,46	Ar I	5
4461,30	C I	1
4460,560	Ar II	12
4460,53	Ar I	10
4460,45	Kr II	1
4460,175	Ne I	100
4459,99	Kr II	8
4459,933	N II	3
4459,185	Cs II	15
4459,1213	Fe I	10
4458,885	Ar II	1
4457,680	Cs II	15
4457,428	Ti I	40
4457,25	Kr II	40
4456,95	Ne II	5
4456,612	Ca I	10
4456,61	Ar I	3
4456,55	Ar II	1
4456,380	Ne I	1
4455,887	Ca I	40
4455,564	Ne I	15
4455,321	Ti I	30
4455,00	K II	2
4454,781	Ca I	80
4454,3835	Fe I	5
4454,37	Kr II	10
4454,285	Ne I	5
4453,9177	Kr I	600
4453,708	Ti I	20
4453,61	Xe III	8
4453,528	Ne I	1
4453,44	Cs II	15
4453,324	Ne I	2
4453,32	Cl II	3
4453,312	Ti I	30
4453,253	Ne I	5
4453,21	Kr II	50
4452,983	Ne I	15
4452,55	Ne II	1
4452,377	O II	6
4450,896	Ti I	25
4450,785	Cs II	2
4450,487	Ti II	10
4450,34	Kr II	4
4449,517	Ar II	4
4449,143	Ti I	30
4448,881	Ar II	8
4448,88	Ar I	3
4448,459	Ar II	3
4448,21	O II	6
4448,13	Xe II	500
4447,82	O III	0
4447,8	Al II	3
4447,7212	Fe I	9
4447,649	Cs	10
4447,18	F II	12
4447,033	N II	12
4446,71	F II	10
4446,538	Ne I	1
4446,51	F II	6
4446,46	Ne II	3
4446,11	Cl I	4
4445,848	Ar II	8
4445,84	Ar I	5
4445,83	Cl I	4
4445,671	Ne I	1
4444,978	Ne I	30
4444,823	Cu II	3
4444,004	Cs II	10
4443,802	Ti II	50
4443,72	Kr II	3
4443,72	Kr III	3
4443,28	Kr III	15
4443,20	Ti I	3
4443,1963	Fe I	7
4443,08	C III	2
4443,05	O II	5
4442,67	Ne II	3
4442,3428	Fe I	12
4442,018	N II	6
4441,49	C IV	3
4441,272	Ti I	4
4440,95	Xe II	50

λ	Symbol	I	λ	Symbol	I
4440,890	Ne I	1	4427,755	Ne I	30
4440,812	Ne I	2	4427,52	Xe II	2
4440,363	Ne I	15	4427,3118	Fe I	10
4440,345	Ti I	10	4427,236	N II	5
4440.34	C IV	2	4427,098	Ti I	40
4440,26	Cs	15	4426,054	Ti I	10
4440,122	Ar II	4	4426,005	Ar II	25
4440,1	O III	0	4425,840	Ti I	3
4439,95	Ne II	2	4425,726	Cs I	—
4439,878	Ar II	4	4425,49	Si I	10
4439,463	Ar II	7	4425,441	Ca I	50
4439,30	Ne II	3	4425,400	Ne I	150
4438,808	Ar II	1	4425,25	Xe III	1
4438,48	Cl I	20	4425,1908	Kr I	100
4438,117	Ar II	3	4424,8096	Ne I	300
4437,551	He I	15	4424,046	Cs II	10
4436,96	Cl II	3	4423,996	Ar I	80
4436,81	Kr II	600	4423,73	K II	3
4436,598	Mg II	4	4423,556	Si III	4
4436,586	Ti I	4	4423,246	Na I	7
4436,486	Mg II	5	4422,823	Ti I	10
4436,06	Cs II	2	4422,70	Kr II	100
4435,708	Cs II	20	4422,5703	Fe I	6
4435,688	Ca I	40	4422,5205	Ne I	300
4435,094	Ne I	5	4421,754	Ti I	6
4434,960	Ca I	60	4421,559	Ne I	50
4434,69	Si I	10	4421,38	Ne II	3
4434,43	O III	2	4420,912	Ar II	12
4434,16	Xe III	50	4420,558	Ne I	1
4434,003	Ti I	15	4420,16	Kr II	1
4433,990	Mg II	9	4419,885	Na I	6
4433,841	Ar II	10	4419,59	Fe III	10
4433,7239	Ne I	70	4418,7626	Kr I	50
4433,578	Ti I	3	4417,82	N II	1
4433,475	N II	5	4417,81	Xe III	1
4433,398	Ne I	10	4417,718	Ti II	40
4433,221	Fe I	3	4417,344	Cs I	—
4432,735	N II	8	4417,30	Ar IV	—
4432,526	Ne I	20	4417,3	F II	3
4432,26	Ne II	1	4417,274	Ti I	15
4431,816	N II	3	4417,24	Kr II	40
4431,67	Kr II	500	4417,07	N II	4
4431,67	Ne II	1	4416,972	O II	8
4431,284	Ti I	4	4416,8338	Kr I	20
4431,02	Fe III	7	4416,817	Fe II	7
4431,004	Ar II	15	4416,817	Ne I	50
4430,90	Ne II	4	4416,77	Ne II	2
4430,6175	Fe I	6	4416,535	Ti I	4
4430,470	Si I	10	4416,07	Xe II	150
4430,366	Ti I	7	4415,54	Cu I	200
4430,192	Ar II	20	4415,141	Ne I	5
4430,023	Ti I	3	4415,1250	Fe I	20
4429,60	Ne II	2	4414,909	O II	10
4429,410	Ne I	1	4414,90	Cl III	2
4428,9	Kr II	1	4414,84	Xe II	300
4428,54	Ne II	6	4414,37	O II	1
4427,994	Mg II	8	4413,561	Ne I	15
4427,981	Ne I	15			
4427,964	N II	4			

λ	Symbol	I	λ	Symbol	I
4413,255	C II	1	4395,95	O II	7
4413,20	Ne II	4	4395,848	Ti II	2
4412,905	Ar II	3	4395,77	Xe II	500
4412,54	Ne II	2	4395,76	Fe III	6
4412,39	Kr I	6	4395,556	Ne I	50
4412,285	Ne I	20	4395,306	Ne I	1
4411,652	Si IV	0	4395,12	Xe III	4
4411,506	C II	7	4395,031	Ti II	60
4411,163	C II	6	4395,008	Ne I	1
4411,080	Ti II	15	4394,773	Ne I	15
4410,3685	Kr I	50	4394,622	Ar II	3
4410,208	Cs II	20	4394,370	Ne I	15
4409,979	C II	5	4394,057	Ti II	2
4409,84	Mg I	1	4393,925	Ti I	8
4409,620	Ne I	20	4393,340	Na I	9
4409,30	Ne II	7	4393,20	Xe II	500
4409,161	C II	2	4392,59	Si I	10
4408,89	Kr II	40	4391,94	Ne II	7
4408,4176	Fe I	6	4390,9542	Fe I	4
4407,7130	Fe I	5	4390,564	Mg II	10
4406,88	Xe II	200	4390,38	Cl I	7
4406,721	Si III	8	4390,029	Na I	8
4406,469	Ar II	2	4389,76	Cl I	25
4406,02	O II	1	4389,72	Kr II	20
4405,901	Si III	6	4388,90	Kr II	3
4405,582	Ne I	2	4388,764	Cs II	10
4405,351	Si III	4	4388,411	Fe I	4
4405,253	Cs II	35	4388,16	K II	7
4404,911	Ti I	5	4388,077	Ti I	3
4404,903	Ar II	5	4388,016	C III	4
4404,7525	Fe I	30	4387,9294	He I	50
4404,397	Ti I	5	4387,8959	Fe I	4
4404,33	Kr II	30	4387,55	Cl I	6
4404,276	Ti I	10	4387,52	Xe III	4
4403,854	Cs II	20	4386,962	Ar II	1
4403,734	Si IV	2	4386,858	Ti II	10
4403,54	Ti IV	2	4386,566	Cs II	2
4403,03	Cl I	15	4386,54	Kr II	300
4402,985	Ne I	1	4385,7693	Xe I	70
4402,58	Cl I	4	4385,381	Fe II	7
4402,580	Ne I	1	4385,27	Kr II	50
4402,374	Ne I	2	4385,058	Ar II	10
4401,744	Ar II	2	4385,00	Ne II	2
4400,988	Ar II	20	4384,93	Xe II	60
4400,87	Kr II	100	4384,637	Mg II	9
4400,099	Ar II	18	4384,428	Cs II	25
4399,9670	Kr I	200	4384,08	Ne II	1
4399,767	Ti II	35	4383,9092	Xe I	100
4399,495	Cs II	20	4383,80	K II	1
4399,39	Kr II	15	4383,754	Ar II	8
4399,14	Cl II	15	4383,5473	Fe I	45
4398,136	Ne I	5	4383,544	C III	2
4397,994	Cs	10	4382,934	Ar II	3
4397,94	Ne II	6	4382,898	C III	3
4397,37	Ti IV	2	4381,52	Kr II	100
4397,175	Ne I	1	4381,220	Ne I	30
4397,0	Cu I	10	4380,76	Cu II	2
4396,909	Cs II	15			
4395,969	Ne I	1			

λ	Symbol	I
4380,57	Cl III	2
4380,375	Mg I	6
4380,11	Kr I	2
4379,952	C III	2
4379,90	Cl I	20
4379,879	Ar II	5
4379,667	Ar II	20
4379,55	O II	3
4379,50	Ne II	6
4379,481	C III	2
4379,44	Xe II	10
4379,226	Ar II	3
4379,09	N III	10
4378,68	Kr III	8
4378,430	Cu II	8
4378,41	O II	0
4378,20	Cu I	550
4378,01	O II	0
4377,95	Ne II	2
4377,754	Ne I	2
4377,71	Kr II	40
4377,626	Si III	8
4376,957	Si II	5
4376,562	C II	5
4376,1219	Kr I	800
4375,948	Ar II	12
4375,9318	Fe I	9
4375,009	C II	4
4374,997	Ne I	2
4374,98	N II	2
4374,87	K II	1
4374,857	Ar II	6
4374,272	C II	9
4373,78	Xe II	100
4373,018	Cs II	30
4372,91	Cl II	80
4372,81	Fe III	20
4372,491	Ar II	5
4372,487	C II	7
4372,46	Xe II	2
4372,383	Ti I	3
4372,350	C II	6
4372,287	Xe I	20
4372,157	Ne I	30
4371,796	Ne I	2
4371,65	O II	2
4371,55	Cl I	5
4371,37	C I	6
4371,329	Ar II	20
4371,25	Kr II	20
4370,91	Cl III	4
4370,751	Ar II	15
4370,661	C II	1
4369,857	C II	2
4369,7745	Fe I	7
4369,77	Ne II	5
4369,69	Kr II	200
4369,682	Ti I	5
4369,60	Cl III	2
4369,52	Cl I	15
4369,28	O II	4
4369,20	Xe II	200
4368,77	Cs	10
4368,36	Ar I	5
4368,30	O I	10
4368,263	C II	4
4368,047	C II	1
4367,87	Ar I	10
4367,829	Ar II	10
4367,66	Cs	10
4367,657	Ti II	15
4367,5811	Fe I	5
4367,50	C III	3
4367,05	Xe II	30
4366,896	O II	7
4366,26	Kr II	6
4365,72	Ne II	2
4365,705	Ne I	1
4365,362	Cu II	30
4364,80	Ar IV	—
4364,79	Cl III	3
4364,61	Kr II	4
4364,59	Al III	2
4363,7957	Ar I	80
4363,69	Cs II	2
4363,524	Ne I	70
4363,30	Cl I	20
4363,275	Cs II	50
4363,228	Ne I	2
4362,96	K II	3
4362,690	Ne I	30
4362,6424	Kr I	500
4362,065	Ar II	10
4361,87	C III	4
4360,63	Kr III	1
4360,487	Ti I	4
4360,32	Xe II	2
4359,38	O II	1
4359,02	Cs	10
4358,90	C III	2
4358,816	Ne I	2
4358,490	Ar II	3
4358,40	O II	0
4358,27	N I	10
4357,918	Ne I	5
4357,66	Xe III	1
4357,298	Ne I	2
4357,25	O II	0
4357,2	Al III	—
4356,821	Si III	4
4356,807	Al II	1,5
4356,711	Al II	2
4356,575	Cs II	2
4356,100	Si III	2
4355,525	Si III	3
4355,47	Kr II	3000
4355,41	C I	1
4355,281	Si III	3

λ	Symbol	I	λ	Symbol	I
4355,096	Ca I	25	4340,468	H	200
4354,74	Cu I	10	4340,420	Ne I	2
4354,529	Mg I	6	4340,36	O II	2
4354,23	Kr I	2	4340,256	Ne I	2
4354,064	Ti I	3	4340,03	K II	5
4354,03	Cl III	2	4339,78	Ne II	1
4353,90	Kr I	2	4339,52	N III	3
4353,73	Cl III	2	4339,287	D	200
4353,66	N III	2	4338,893	T	200
4353,60	O II	1	4338,67	He II	3
4352,810	Si III	2	4338,501	Si III	9
4352,7371	Fe I	9	4338,228	Ar II	2
4352,204	Ar II	15	4338,200	Ne I	2
4351,974	Si III	2	4337,916	Ti II	50
4351,9056	Mg I	20	4337,80	Cl I	1
4351,764	Fe II	9	4337,070	Ar II	8
4351,3602	Kr I	100	4337,07	Xe II	30
4351,269	O II	6	4337,0484	Fe I	10
4351,02	Kr II	40	4336,865	O II	6
4349,55	Kr I	2	4336,48	N I	5
4349,426	O II	8	4336,26	Cl II	45
4348,97	C I	1	4336,221	Ne I	50
4348,620	Cs II	2	4336,00	Cu I	10
4348,36	N III	5	4335,81	Xe II	10
4348,063	Ar II	50	4335,53	N III	4
4347,802	Al II	3,5	4335,411	Cs	8
4347,785	Al II	4	4335,3381	Ar I	800
4347,425	O II	6	4334,29	O II	0
4347,316	Al II	1,5	4334,1267	Ne I	70
4347,223	Al II	2	4334,0	N V	—
4346,918	Al II	1	4333,5612	Ar I	1000
4346,12	Ne II	1	4333,34	Kr II	50
4346,104	Ti I	5	4332,80	Cl II	9
4346,036	Ne I	15	4332,76	O II	1
4345,762	Ne I	1	4332,031	Ar II	15
4345,562	O II	7	4331,945	Mg II	3
4345,479	Ne I	2	4331,89	O II	2
4345,167	Ar I	1000	4331,47	O II	0
4344,736	Na I	5	4331,24	Kr II	80
4344,42	O II	0	4331,199	Ar II	25
4344,31	C I	1	4331,13	O II	0
4344,291	Ti II	2	4330,52	Xe II	1000
4344,24	Kr III	8	4330,44	N III	2
4344,11	C I	0	4330,239	Cs II	20
4343,62	Cl II	100	4330,14	N III	2
4343,41	N I	1	4328,68	Cu I	20
4343,36	O II	0	4328,62	O II	2
4342,83	O II	1	4328,175	Si IV	5
4342,56	Xe II	6	4328,15	N III	3
4342,40	C I	0	4327,89	O II	0
4342,00	O II	4	4327,580	Cs II	10
4341,64	C I	2	4327,48	O II	3
4341,489	Na I	4	4327,265	Ne I	10
4341,47	Cl III	2	4327,098	Fe I	3
4341,42	Ne II	2	4326,359	Ti I	9
4341,400	Si III	8	4326,315	Cs	10
4341,33	Kr II	8	4326,156	C II	5
			4326,12	Cl I	—
			4325,827	C II	4

λ	Symbol	I
4325,77	O II	3
4325,7647	Fe I	35
4325,7	Li II	3
4325,560	C III	8
4325,15	Ne II	1
4325,134	Ti I	9
4324,66	Cl III	2
4324,615	Na I	7
4323,93	N III	2
4323,35	Cl I	20
4323,102	C II	3
4322,98	Kr II	150
4322,66	Ne II	1
4322,26	Ne II	2
4321,99	N I	1
4321,82	Xe II	40
4321,655	Ti I	8
4321,647	C II	3
4321,492	Ne I	2
4321,400	Na I	6
4321,37	N III	1
4319,93	O II	1
4319,636	Ar II	2
4319,631	O II	8
4319,5798	Kr I	1000
4319,511	Ne I	1
4319,12	Kr II	4
4318,834	Ne I	5
4318,652	Ca I	45
4318,631	Ti I	10
4318,600	C II	5
4318,5523	Kr I	400
4317,85	K II	2
4317,81	Kr II	500
4317,70	N I	5
4317,65	O II	0
4317,260	C II	8
4317,139	O II	8
4316,992	Cs II	2
4316,008	Ne I	15
4315,80	O II	00
4315,44	C III	3
4315,35	O II	0
4315,0872	Fe I	10
4314,979	Ti II	40
4314,801	Ti I	25
4314,74	Ti I	25
4314,695	Ne I	30
4314,356	Ti I	5
4314,110	Ne I	1
4314,104	Si IV	3
4313,43	O II	1
4313,11	N I	4
4313,100	C II	6
4312,861	Ti II	35
4312,10	O II	0
4310,51	XeII	500
4310,47	Ar I	20

λ	Symbol	I
4310,36	Fe III	12
4310,130	Ne I	2
4309,3771	Fe I	4
4309,33	Xe III	6
4309,236	Ar II	9
4309,10	K II	7
4309,090	Ar II	8
4309,06	Cl II	50
4308,96	O II	1
4308,42	Cl III	1
4308,00	Xe III	10
4307,942	Cs II	8
4307,9048	Fe I	35
4307,900	Ti II	40
4307,741	Ca I	45
4307,59	C II	2
4307,42	Cl II	75
4307,31	O II	1
4307,20	Al II	3
4306,48	Cs II	10
4306,33	C II	1
4306,2625	Ne I	70
4306,21	Xe II	1
4305,910	Ti I	60
4305,86	Xe III	2
4305,81	Kr II	3
4305,53	O II	0
4305,46	N I	6
4305,265	K II	1
4305,20	Kr III	9
4305,00	K II	7
4304,77	Fe III	10
4304,07	Cl II	40
4303,955	Ne I	5
4303,82	O II	5
4303,695	Ne I	1
4303,248	Ne I	30
4303,166	Fe II	8
4303,06	O II	0
4302,81	O II	0
4302,527	Ca I	60
4302,4455	Kr I	10
4301,928	Ti II	15
4301,53	Kr II	40
4301,089	Ti I	50
4300,650	Ar II	12
4300,636	Cs II	30
4300,566	Ti I	50
4300,49	Kr II	200
4300,4877	Kr I	50
4300,449	Ar II	2
4300,1011	Ar I	1200
4300,052	Ti II	60
4299,636	Ti I	15
4299,2409	Fe I	18
4299,24	Ar I	5
4299,229	Ti I	15

λ	Symbol	I
4299,177	F II	10
4298,986	Ca I	30
4298,664	Ti I	40
4298,040	Fe I	3
4297,964	Ar II	7
4297,514	Cs	10
4297,04	Cl III	0
4296,85	Fe III	10
4296,75	Xe II	2
4296,567	Fe II	6
4296,40	Xe II	500
4295,920	C II	4
4295,751	Ti I	22
4295,21	Kr II	8
4294,97	Ar I	20
4294,83	Kr III	10
4294,82	O II	3
4294,76	N III	0
4294,1271	Fe I	15
4294,101	Ti II	10
4292,92	Kr II	600
4292,64	Kr I	6
4292,469	Cu II	30
4292,23	O II	0
4292,008	Cs II	12
4291,976	Ne I	2
4291,819	C II	3
4291,76	Cl II	50
4291,25	O II	1
4291,214	Ti I	5
4291,10	Cu II	2
4291,006	Na I	3
4290,933	Ti I	22
4290,80	N III	3
4290,78	Kr I	4
4290,55	N III	1
4290,40	Ne II	6
4290,222	Ti II	50
4280,010	Ti I	3
4289,876	C II	2
4289,799	Ne I	2
4289,364	Ca I	40
4289,09	Ar I	5
4289,068	Ti I	25
4288,83	O II	1
4288,72	N III	1
4288,70	K II	4
4288,541	Ne I	5
4288,350	Cs II	35
4288,21	N III	0
4288,161	Ti I	3
4288,02	Kr I	5
4287,893	Ti II	2
4287,838	Na I	2
4287,45	Kr II	4
4287,405	Ti I	22
4286,4875	Kr I	40
4286,343	Ar II	2
4286,16	Fe III	10

λ	Symbol	I
4286,006	Ti I	25
4285,89	Xe III	30
4285,704	C II	3
4285,70	O II	3
4285,4453	Fe I	3
4285,40	Kr II	4
4285,239	Cu II	10
4284,988	Ti I	8
4284,92	N I	2
4284,89	K II	3
4284,51	N III	1
4284,229	Cs II	2
4283,75	O II	0
4283,242	Ne I	10
4283,13	O II	0
4283,010	Ca I	40
4282,9686	Kr I	100
4282,96	O II	1
4282,896	Ar II	12
4282,82	O II	0
4282,702	Ti I	12
4282,59	Cs	10
4282,46	Cl III	4
4282,4057	Fe I	12
4282,20	N I	1
4281,40	O II	0
4281,39	N I	2
4281,371	Ti I	10
4281,31	Cs	10
4281,25	O II	0
4280,61	Kr II	5
4280,43	Cl I	2
4279,959	Cu II	20
4279,909	Ar II	2
4279,279	Ne I	15
4278,89	F II	4
4278,850	Ne I	5
4278,829	Ti I	3
4278,231	Ti I	7
4277,90	O II	1
4277,524	Ar II	20
4277,51	F II	6
4277,40	O II	1
4277,100	Cs II	50
4276,787	Na I	5
4276,71	O II	1
4276,64	O II	3
4276,51	Cl II	30
4276,441	Ti I	8
4276,21	O II	0
4276,044	Cu II	30
4275,90	O II	0
4275,5598	Ne I	70
4275,52	O II	4
4275,21	F II	8
4275,167	Ne I	1
4275,158	Ar II	8
4275,107	Cu I	950
4274,656	Ne I	50

λ	Symbol	I	λ	Symbol	I
4274,584	Ti I	15	4258,523	Ti I	4
4274,13	O II	00	4257,894	C III	2
4274,13	Xe III	1	4257,82	Ne II	3
4273,9700	Kr I	1000	4257,54	Cl II	4
4273,642	Na I	4	4257,25	Ne II	1
4273,48	Kr II	4	4256,663	Ar II	1
4273,17	O II	0	4256,498	Ne I	2
4273,107	Li I	10	4256,455	C III	1
4272,87	Cs	10	4256,025	Ti I	8
4272,60	Xe III	20	4255,600	Ar II	4
4272,440	Ti I	8	4255,59	Cu II	3
4272,1690	Ar I	1200	4255,42	C III	1
4271,84	Cs II	10	4254,95	Ar I	10
4271,7634	Fe I	35	4254,85	Kr II	100
4271,47	Fe III	6	4254,7	N I	4
4271,24	Ar I	5	4253,98	O II	4
4271,1589	Fe I	20	4253,74	O II	4
4270,61	Cl II	25	4253,51	Cl II	75
4270,267	Ne I	50	4253,390	Cu I	20
4270,139	Ti I	7	4253,28	N I	4
4269,84	Xe II	40	4252,775	Ne I	2
4269,724	Ne I	70	4252,67	Kr II	50
4269,02	C I	6	4252,520	Na I	2
4268,89	Cs	10	4252,418	Ne I	2
4268,81	Kr II	1000	4251,618	Ti I	3
4268,57	Kr II	60	4251,57	Xe II	100
4268,009	Ne I	70	4251,1850	Ar I	800
4267,83	Fe I	5	4250,7896	Fe I	25
4267,730	Ar II	3	4250,68	Ne II	4
4267,724	Ne I	5	4250,58	Kr II	150
4267,490	Ar II	3	4250,41	Ar I	3
4267,286	Ne I	1	4250,1248	Fe I	25
4267,258	C II	20	4249,95	Fe III	7
4267,204	Cu I	2	4249,538	Ne I	2
4267,003	C II	18	4249,410	Na I	1
4266,528	Ar II	25	4249,37	Ar I	20
4266,2868	Ar I	1200	4249,114	Ti I	5
4266,227	Ti I	3	4248,956	Cu I	150
4265,723	Ti I	4	4248,2275	Fe I	4
4265,52	Ar I	2	4247,433	Fe I	12
4265,273	Ti I	3	4247,31	N II	1
4264,675	Cs	50	4247,308	C III	4
4264,59	Cl I	5	4247,20	N II	1
4263,57	Xe II	5	4246,16	F II	15
4263,44	Xe II	30	4245,38	Xe II	500
4263,40	K II	7	4245,2594	Fe I	6
4263,2881	Kr I	20	4244,41	Xe II	30
4263,134	Ti I	15	4244,33	Kr III	5
4262,479	Ne I	2	4244,17	Ne II	0
4262,3	O IV	—	4243,88	Xe II	10
4261,609	Ti I	5	4243,85	Fe III	8
4261,22	Cl II	20	4243,640	Ar II	2
4260,85	Kr II	5	4243,57	Ar I	20
4260,4794	Fe I	35	4243,35	Cu II	2
4259,739	Ne I	1	4242,543	Mg II	2
4259,52	Cl II	35	4242,489	N II	3
4259,44	Kr II	80	4242,445	Mg II	3
4259,401	Cu I	150			
4259,3617	Ar I	1200			
4259,202	Si II	5			
4258,59	Ar I	5			

λ	Symbol	I	λ	Symbol	I
4242,26	Cu I	30	4227,28	Cs II	5
4242,20	Ne II	1	4226,988	Ar II	10
4242,082	Na I	3	4226,812	Al II	6
4241,973	Cs II	10	4226,728	Ca I	500
4241,784	N II	10	4226,607	Ar II	5
4241,38	Cl II	60	4226,58	Kr III	25
4241,240	N II	3	4226,44	Cl I	15
4240,75	Al II	3	4225,92	Kr III	20
4240,456	Ca I	6	4225,67	K II	7
4240,24	Xe III	10	4225,465	Fe I	6
4239,95	Ne II	2	4225,12	F II	4
4239,5	O III	00	4224,92	Cl II	15
4239,448	Cu II	25	4224,795	Ti I	5
4238,987	Na I	2	4224,74	N I	4
4238,821	Fe I	10	4224,57	Ne II	1
4238,25	Xe II	500	4224,176	Fe I	6
4237,889	Ti I	7	4223,36	C I	3
4237,223	Ar II	12	4223,16	C I	4
4237,05	N II	7	4223,04	N I	5
4236,91	N II	8	4223,00	Xe II	400
4236,64	Kr II	100	4222,97	K II	7
4235,9433	Fe I	25	4222,78	O I	5
4235,82	Xe III	1	4222,640	Ar II	10
4235,54	Fe III	10	4222,47	C I	3
4235,49	Cl II	25	4222,39	Fe III	8
4234,408	Cs II	20	4222,2181	Fe I	12
4234,09	Cl II	50	4222,20	Kr II	20
4233,86	Ne II	3	4221,80	Cl II	3
4233,72	Kr III	1	4221,554	Ne I	1
4233,6089	Fe I	18	4221,119	Cs II	15
4233,60	Cl II	4	4220,92	Ne II	2
4233,32	O I	7	4220,79	N I	2
4233,168	Fe II	11	4220,571	Cs II	2
4232,864	Si II	10	4220,074	Ca II	5
4232,82	Kr III	2	4219,76	Ne II	6
4232,66	Xe III	1	4219,516	Cs	5
4232,323	Ne I	1	4219,3641	Fe I	12
4232,188	Cs II	25	4218,8	Cu I	2
4231,60	Ne II	4	4218,76	Cl II	4
4230,9	Cu I	5	4218,667	Ar II	11
4230,444	Cu II	10	4217,88	Kr II	2
4230,35	N I	4	4217,555	Fe I	7
4229,872	Ar II	8	4217,433	Ar II	9
4229,81	Ar IV	—	4217,15	Ne II	3
4229,59	N I	2	4217,09	O I	4
4229,21	Kr II	8	4216,89	Cu II	3
4228,79	Kr II	20	4216,75	Xe III	10
4228,33	C I	5	4216,1854	Fe I	8
4228,162	Ar II	20	4215,92	N I	2
4227,982	Al II	4	4215,69	N III	3
4227,936	Cu II	30	4215,60	Xe II	200
4227,923	Al II	1,5	4215,55	Ti III	5
4227,743	N II	8	4214,854	Ar II	2
4227,654	Ti I	5	4214,73	N I	5
4227,493	Al II	5	4214,69	Xe II	6
4227,432	Fe I	30	4214,04	Xe III	20
4227,406	Al II	2	4213,72	Xe II	400
4227,37	Cl II	4	4213,650	Fe I	5
			4213,2	O V	—

λ	Symbol	I
4213,129	Cs II	30
4213,07	C I	2
4212,694	Ne I	20
4212,497	Mg I	2
4212,407	Si IV	7
4212,33	C I	4
4211,861	Cu II	30
4211,82	C I	2
4211,729	Ti I	4
4211,679	Si III	2
4211,61	C I	1
4211,12	C I	2
4210,950	Ar II	1
4210,87	Fe III	10
4210,67	Kr II	25
4210,3497	Fe I	15
4209,944	Ar II	1
4209,91	C I	0
4209,71	C I	0
4209,68	Cl I	12
4209,62	Xe III	10
4209,49	K II	4
4209,47	Xe II	200
4209,09	N II	0
4209,05	N I	1
4208,615	Fe I	3
4208,48	Xe II	400
4208,03	Cl II	30
4207,87	F II	2
4207,54	Ti III	3
4207,50	N II	3
4207,442	F II	5
4207,162	F II	7
4206,6985	Fe I	3
4206,51	N II	2
4206,43	Ne II	2
4206,29	N I	1
4206,175	Ca II	4
4206,11	N II	1
4205,65	N I	2
4205,404	Xe I	10
4205,096	Mg I	2
4205,07	Cl II	10
4204,95	Ti III	2
4204,54	Cl II	18
4204,31	Kr II	3
4203,9867	Fe I	10
4203,92	Xe III	10
4203,6945	Xe I	50
4203,465	Ti I	8
4203,410	Ar II	11
4203,270	Ne I	2
4203,22	Xe II	5
4202,40	Al II	2
4202,0320	Fe I	30
4201,971	Ar II	12
4201,551	Ar II	4
4201,42	Kr II	30
4201,35	N II	1
4201,25	Xe II	15
4200,898	Si II	40
4200,752	Ti I	6
4200,6746	Ar I	1200
4200,657	Si II	30
4200,38	Fe III	6
4200,11	Ti III	2
4200,06	Fe III	6
4200,02	N III	6
4199,980	N II	5
4199,892	Ar II	6
4199,87	He II	2
4199,0981	Fe I	20
4199,00	Al III	0,5
4198,83	Ar III	3
4198,3176	Ar I	1200
4198,3098	Fe I	20
4198,3	Na I	10
4198,133	Si II	50
4198,099	Ne I	70
4197,81	Xe II	10
4196,72	O II	1
4196,415	Ne I	15
4196,26	O II	00
4196,214	Fe I	4
4195,974	N II	3
4195,91	Kr III	1
4195,70	N III	5
4195,337	Fe I	5
4195,11	Cl II	18
4194,88	Xe III	5
4193,5296	Xe I	150
4193,49	N I	3
4193,482	Mg II	2
4193,198	Cs II	8
4193,15	Xe II	500
4193,01	Xe I	20
4192,62	F II	2
4192,50	O II	2
4192,24	Cl II	6
4191,59	Cl II	15
4191,4358	Fe I	15
4191,0288	Ar I	1200
4190,724	Si II	100
4190,7138	Ar I	600
4190,29	Ti II	1
4189,788	O II	10
4189,653	Ar II	10
4189,10	Fe III	7
4188,88	Al III	0,5
4188,82	Cl II	15
4188,694	Ti I	5
4187,8015	Fe I	20
4187,137	Si II	5
4187,06	Cl II	2

λ	Symbol	I
4187,06	N I	1
4187,0436	Fe I	20
4186,900	C III	9
4186,63	Cl II	5
4186,249	Cs II	5
4186,24	K II	8
4186,119	Ti I	25
4185,61	Cl II	20
4185,456	O II	8
4185,12	Kr II	50
4184,8941	Fe I	10
4184,89	Cl II	7
4184,59	Kr III	2
4184,4726	Kr I	20
4183,345	Si II	10
4183,294	Ti I	4
4182,97	Ar IV	—
4182,60	N IV	—
4181,8837	Ar I	1000
4181,7571	Fe I	15
4181,10	N II	2
4181,05	Cl II	4
4180,86	Ti I	3
4180,10	Xe II	1000
4179,674	N II	5
4179,61	Cl II	2
4179,58	Kr II	20
4179,512	Cu II	30
4179,302	Ar II	12
4178,86	N II	0
4178,855	Fe II	8
4178,371	Ar II	12
4178,2	O V	—
4177,758	Cu I	100
4177,595	Fe I	4
4177,109	Mg I	2
4177,02	Kr II	3
4176,572	Fe I	7
4176,53	Xe III	20
4176,33	Ar I	20
4176,161	N II	8
4176,11	Cu II	5
4175,6386	Fe I	10
4175,488	Ne I	40
4175,40	Ar I	10
4175,223	Ne I	60
4174,9137	Fe I	5
4174,472	Ti I	3
4174,369	Ne I	70
4174,27	Fe III	10
4174,088	Ti II	2
4173,966	Ne I	2
4173,69	Ar III	2
4173,67	N IV	—
4173,572	N II	3
4173,533	Cs	15
4173,450	Fe II	8
4173,089	C III	2
4172,83	Kr I	3
4172,51	Kr II	20
4171,897	Ti II	30
4171,858	Cu II	25
4171,79	Kr III	15
4171,607	N II	6
4171,018	Ti I	8
4170,99	Xe II	8
4170,66	Cl II	8
4169,38	N II	1
4169,330	Ti I	7
4169,230	O II	4
4168,967	Ar II	4
4168,967	He I	3
4168,70	Ar I	3
4168,424	Al II	1
4168,41	Ar I	3
4167,66	Xe III	1
4167,28	Kr I	5
4167,2712	Mg I	15
4166,95	C III	1
4166,84	Fe III	9
4166,64	N I	1
4166,311	Ti I	6
4166,10	Cl II	4
4166,091	Ne I	30
4165,101	Mg I	4
4164,802	Ne I	50
4164,73	Fe III	20
4164,48	Kr I	2
4164,288	Cu II	20
4164,1795	Ar I	1000
4164,134	Ti I	4
4163,82	Kr II	2
4163,644	Ti II	40
4163,26	C III	2
4163,243	Cs	15
4162,86	C III	7
4162,296	Cu II	8
4162,16	Xe II	60
4161,155	Cu II	30
4161,14	N II	1
4160,50	N II	2
4160,263	Al II	3
4160,239	Al II	2,5
4160,21	Kr III	4
4159,809	Al II	1
4159,725	Al II	1,5
4159,634	Ti I	9
4159,450	Al II	1
4159,00	Kr II	4
4158,76	O V	0
4158,798	Fe I	5
4158,610	Cs II	18
4158,5906	Ar I	1200
4158,04	Xe II	200
4157,98	Cl II	5
4157,82	Cl II	25
4157,791	Fe I	8

λ	Symbol	I
4157,01	N II	3
4156,8021	Fe I	12
4156,76	C III	2
4156,54	O II	3
4156,49	C III	6
4156,39	N II	1
4156,3	Li II	1
4156,17	Xe II	2
4156,15	Cl II	7
4156,090	Ar II	12
4154,98	Fe III	8
4154,812	Fe I	9
4154,77	N II	2
4154,65	Xe III	2
4154,5021	Fe I	12
4154,46	Kr III	40
4153,98	Cl II	2
4153,910	Fe I	10
4153,623	Cu II	10
4153,37	C I	0
4153,302	O II	7
4153.2	O V	—
4152,54	Ar I	20
4152,512	C III	5
4152,03	Xe III	5
4151,46	N I	12
4151,267	Cs II	20
4150,963	Ti I	10
4150,67	Ne II	3
4150,557	Ti I	3
4150,138	Al III	8
4149,917	Al III	1
4149,897	Al III	10
4149,370	Fe I	5
4149,19	K II	7
4149,03	Ar III	3
4148,4	Li I	—
4148,19	Xe II	2
4147,98	C I	1
4147,6719	Fe I	10
4147,377	Ar II	2
4147,20	Cl I	2
4147,09	Cl II	30
4146,97	C I	0
4146,78	Xe I	2
4146,70	Ar III	5
4146,26	C I	2
4146,09	O II	3
4145,90	O II	0
4145,78	N I	2
4145,776	N II	6
4145,73	Xe III	100
4145,12	Kr II	250
4144,240	Ar II	1
4143,8703	Fe I	30
4143,77	O II	2
4143,761	He I	10
4143,52	O II	1
4143,4174	Fe I	15
4143,4	N I	—
4143,280	Ti I	3
4143,048	Ti I	7
4143,04	Cl II	5
4143,020	Cu II	5
4142,24	O II	0
4142,15	Al III	2
4142,08	O II	1
4142,01	Xe III	10
4141,96	O II	1
4141,296	Cu II	20
4141,25	Al III	0
4140,74	O II	0
4140,48	Fe III	6
4140,21	Al III	2
4139,35	Fe III	8
4139,11	Kr II	100
4139,00	Cl I	1
4138,81	Xe II	3
4137,96	Kr II	50
4137,76	Fe III	10
4137,63	N I	7
4137,284	Ti I	10
4137,004	Fe I	7
4135,9	O V	—
4135,86	Kr II	3
4135,1337	Xe I	20
4134,72	K II	7
4134,6798	Fe I	12
4134,31	Cl II	4
4133,68	Kr II	5
4133,672	N II	5
4133,66	Cl II	20
4133,65	Ne II	3
4132,903	Fe I	8
4132,806	O II	6
4132,62	Cu II	3
4132.598	Li I	50
4132,48	Cl II	200
4132,42	Xe III	3
4132,0603	Fe I	25
4132,003	Cs II	10
4131,782	N II	4
4131,730	Ar II	15
4131,359	Cu II	35
4131,33	Kr III	40
4131,244	Ti I	4
4131,054	Ne I	70
4131,01	Xe II	20
4130,893	Si II	500
4130,86	Cl II	25
4130,512	Ne I	20
4130,22	Cl II	8
4129,693	Ar II	4
4129,34	O II	2
4129,16	N I	1

λ	Symbol	I
4128,643	Ar II	9
4128,072	Ne I	30
4128,067	Si II	300
4127,6113	Fe I	7
4127,531	Ti I	15
4127,19	Ar III	4
4126,96	F II	2
4126,941	Ne I	2
4125,96	Cl II	3
4125,4	O V	—
4124,078	N II	4
4124,058	Ar II	1
4124,00	Cl II	12
4124,00	Cl III	1
4123,90	O V	2
4123,559	Ti I	10
4123,287	Ti I	5
4123,287	Cu I	30
4123,069	Na II	3
4122,78	Fe III	8
4122,143	Ti I	10
4122,02	Fe III	8
4121,86	Xe II	5
4121,843	C III	5
4121,8050	Fe I	5
4121,74	Cu I	10
4121,7	O V	—
4121,48	O II	4
4121,210	Cs II	15
4120,90	Fe III	8
4120,815	He I	60
4120,554	O II	2
4120,279	O II	3
4119,288	Cs II	8
4119,221	O II	8
4119,219	F II	7
4119,2	O V	—
4118,84	Cl II	4
4118,756	F II	3
4118,5484	Fe I	15
4118,14	Kr II	30
4118,10	Ne II	0
4117,008	F II	5
4116,547	F II	7
4116,377	Ar II	6
4116,1151	Xe I	80
4116,097	Si IV	9
4115,504	Si III	5
4114,99	K II	6
4114,95	Na II	3
4114,487	Ar II	2
4114,36	N II	0
4113,972	N I	6
4113,82	O II	1
4113,73	Kr II	8
4113,52	Xe II	2
4113,26	Xe II	2
4112,975	F II	5
4112,885	Ne I	10
4112,819	Ar II	8
4112,734	F II	4
4112,708	Ti I	20
4112,34	Xe III	1
4112,14	K II	4
4112,14	Xe II	30
4112,100	Ne I	15
4112,029	O II	4
4111,882	Ne I	1
4111,512	Si III	3
4111,4	Cu I	3
4111,255	Si III	2
4110,83	N II	2
4110,795	O II	3
4110,41	Xe II	30
4110,279	Ca II	3
4110,20	O II	1
4110,16	Kr II	5
4110,06	Xe III	10
4110,04	N II	3
4109,959	N I	12
4109,816	Ca II	6
4109,8053	Fe I	9
4109,7093	Xe I	60
4109,54	Mg II	3
4109,23	Kr II	100
4109,173	F II	8
4109,07	Xe III	100
4108,75	O II	0
4108,554	Ca I	10
4108,43	Kr I	3
4108,232	Cs II	5
4107,4917	Fe I	12
4107,07	O II	1
4106,83	Cl III	5
4106,03	O II	0
4105,000	O II	7
4104,95	Xe II	40
4104,78	Cl I	3
4104,743	O II	5
4104,23	Cl III	5
4104,218	Cu I	25
4103,913	Ar II	20
4103,871	F II	7
4103,724	F II	7
4103,525	F II	15
4103,37	N III	9
4103,217	F II	5
4103,10	Xe II	8
4103,085	F II	10
4103,017	O II	5
4102,9359	Si I	70
4102,422	Si III	8
4102,18	N I	2
4102,01	Cs II	1
4101,86	Si III	5

λ	Symbol	I	λ	Symbol	I
4101,737	H	100	4082,40	C I	1
4100,97	Xe II	1	4082,393	Ar II	15
4100,621	D	100	4082,270	N II	5
4100,34	Xe II	20	4081,833	Mg I	2
4100,30	Ne II	1	4081,74	Ca III	5
4100,249	T	100	4081,471	Cs	10
4100,04	He II	2	4081,40	Kr II	1
4099,951	N I	9	4081,10	O III	1
4099,77	Mg I	2	4081,00	Fe III	7
4099,71	Kr II	3	4080,686	Ar II	4
4099,458	Ar II	2	4080,645	Ar II	6
4099,166	Ti I	8	4080,534	Cu I	15
4098,89	Xe II	100	4080,48	Ne II	2
4098,77	Ne II	4	4080,148	Ne I	50
4098,72	Kr II	250	4079,88	Cl II	15
4098,533	Ca I	15	4079,708	Ti I	6
4098,27	O II	0	4079,582	Ar II	12
4098,19	Ar III	4	4079,359	Ne I	2
4098,187	Fe I	4	4078,862	O II	4
4097,84	Kr I	1	4078,8207	Xe I	100
4097,31	N III	10	4078,471	Ti I	30
4097,3	N II	—	4077,93	Cl II	4
4097,260	O II	4	4077,778	C II	4
4097,190	Ar II	1	4077,625	C II	2
4097,138	Ar II	3	4077,143	Ti I	4
4097,102	Ca II	5	4076,939	Ar II	9
4096,58	N II	0	4076,908	N II	3
4096,543	O II	3	4076,83	C II	0
4096,18	O II	0	4076,781	Si II	15
4095,904	N II	4	4076,638	Ar II	12
4095,63	O II	0	4076,637	Fe I	8
4094,930	Ca I	12	4076,526	C II	4
4094,18	O II	0	4076,370	Ti I	4
4093,90	Mg II	1	4076,251	C II	3
4093,69	K II	5	4076,142	C II	5
4092,940	O II	8	4075,868	O II	10
4092,633	Ca I	8	4075,851	C II	12
4091,88	Xe II	3	4075,572	Cu I	50
4089,295	O II	4	4075,451	Si II	20
4089,04	Ar IV	—	4075,395	C II	4
4088,854	Si IV	10	4075,059	Mg I	3
4088,33	Kr II	500	4074,845	C II	8
4087,60	Na II	0	4074,518	C II	10
4087,303	N II	3	4074,51	Cl II	6
4087,16	O II	2	4074,356	Ti I	3
4087,00	Cl III	4	4073,90	O III	0
4086,90	Kr I	2	4073,50	Xe II	15
4086,69	Ne II	1	4073,364	Cs II	8
4085,324	Fe I	4	4073,33	C I	1
4085,124	O II	3	4073,224	Cu I	20
4084,66	O II	1	4073,042	N II	6
4084,499	Fe I	6	4072,711	Si II	3
4083,919	F II	6	4072,64	C I	3
4083,907	O II	2	4072,385	Ar II	12
4083,16	C I	1	4072,164	O II	8
4082,98	C I	1	4072,10	Xe II	6
4082,89	N II	1	4072,006	Ar II	25
4082,456	Ti I	20	4071,7399	Fe I	40

λ	Symbol	I	λ	Symbol	I
4071,20	O II	0	4057,46	Xe II	200
4070,97	C I	2	4057,01	Kr II	300
4070,789	Ar II	2	4056,90	N II	4
4070,261	C III	9	4056,78	Cu I	35
4069,897	O II	6	4056,57	Kr I	3
4069,634	O II	4	4056,38	Cu I	35
4069,53	Cu I	6	4056,062	C III	7
4069,389	Ne I	5	4055,46	Cl II	4
4069,243	Ne I	30	4055,011	Ti I	20
4068,981	Ti I	4	4054,689	Mg I	2
4068,912	C III	9	4054,55	O II	00
4068,835	Ne I	30	4054,5253	Ar I	80
4068,773	Cs II	30	4054,18	Cl II	9
4068,144	Ti I	3	4054,10	O II	0
4068,090	Cu II	5	4053,956	Cs II	15
4067,982	Fe I	8	4053,814	Ti II	3
4067,958	Cs II	30	4053,658	Cu II	10
4067,940	C III	8	4053,540	Ar II	1
4067,37	Kr III	50	4052,923	Ar II	12
4066,979	Fe I	6	4052,380	Cu I	2
4066,75	C I	2	4052,22	Cl II	12
4066,328	Fe II	12	4051,58	Cl II	4
4066,09	Kr II	6	4051,27	Xe II	10
4065,25	C I	4	4050,617	Cu I	20
4065,23	K II	4	4050,42	Kr II	50
4065,113	Ar II	4	4050,05	Xe III	200
4065,11	Kr II	300	4048,22	O II	1
4065,094	Ti I	15	4047,480	Ar II	1
4065,009	Cu II	3	4047,206	K I	17
4064,829	Ne I	15	4047,184	Cs II	20
4064,27	C I	3	4046,15	O II	00
4064,203	Ti I	15	4045,9658	Ar I	1150
4064,113	Si III	2	4045,8147	Fe I	60
4064,036	Ne I	50	4045,677	Ar II	2
4063,5963	Fe I	45	4045,662	Ne I	2
4063,58	C I	2	4044,96	O II	0
4063,238	Cu I	650	4044,90	Xe II	8
4062,90	Ne II	3	4044,777	N II	4
4062,90	O II	1	4044,67	Kr II	80
4062,641	Cu I	2000	4044,65	Cl II	4
4062,4440	Fe I	10	4044,64	Xe II	6
4062,12	Xe II	6	4044,6125	Fe I	6
4061,06	Xe II	3	4044,4185	Ar I	1200
4061,00	O II	2	4044,136	K I	18
4060,98	O II	2	4044,09	Cl II	9
4060,88	Xe III	2	4043,751	Cu II	35
4060,58	O II	3	4043,529	N II	9
4060,43	Xe III	60	4043,502	Cu II	75
4060,263	Ti I	20	4043,422	Cs	20
4059,89	Ar III	3	4043,21	Xe III	20
4059,07	Cl III	6	4042,896	Ar II	15
4058,912	Ca I	1	4042,642	Ne I	50
4058,139	Ti I	7	4042,59	K II	6
4057,759	N IV	8	4042,327	Ne I	10
4057,672	Ar II	1	4042,190	Ar II	3
4057,612	Ti I	5	4041,311	N II	11
4057,52	Cl II	6	4041,31	O II	0
4057,5052	Mg I	10	4040,64	Cl II	9

λ	Symbol	I
4040,310	Ti I	4
4039,841	Cs	50
4039,69	K II	4
4039,69	Xe II	1
4039,345	N II	2
4038,807	Ar II	15
4038,057	Si IV	2
4037,96	N II	1
4037,83	Kr II	30
4037,696	Ne I	5
4037,615	Ne I	15
4037,59	Xe II	200
4037,35	N I	1
4037,29	Xe II	100
4037,262	Ne I	5
4036,53	Cl II	10
4035,87	Xe II	1
4035,83	Cs	15
4035,828	Ti I	10
4035,459	Ar II	12
4035,09	O II	0
4035,080	N II	9
4034,884	Ti I	5
4033,883	Ti I	6
4033,818	Ar II	12
4033,64	N I	1
4033,23	C I	0
4033,18	O II	0
4032,97	Ar I	20
4032,642	Cu II	3
4032,628	Ti I	3
4032,14	Cl I	5
4031,80	C I	3
4031,753	Ti I	3
4031,39	Si IV	1
4031,378	Ar II	2
4030,752	Si III	2
4030,512	Ti I	25
4030,492	Fe I	6
4029,82	Xe II	1
4029,66	Kr I	2
4029,41	C I	4
4028,58	Xe III	10
4028,43	Cs II	2
4028,36	C I	2
4028,332	Ti II	7
4028,03	Kr I	1
4027,97	Xe II	3
4027,17	Kr III	1
4027,026	Cu I	10
4026,539	Ti I	25
4026,5	Al II	5
4026,40	O II	0
4026,359	He I	25
4026,20	Xe II	5
4026,1912	He I	250
4026,075	N II	7
4025,68	Cl II	7
4025,67	Cs	10
4025,60	He II	—
4025,495	F II	15
4025,22	C I	1
4025,19	Xe II	30
4025,136	Ti II	2
4025,010	F II	10
4024,88	K II	4
4024,739	Fe I	6
4024,727	F II	20
4024,573	Ti I	35
4024,552	Fe II	5
4024,04	O II	1
4023,973	He I	2
4023,60	Ar III	6
4023,582	Cs	10
4022,84	C I	3
4022,629	Cu I	1250
4022,12	C I	2
4021,8696	Fe I	12
4021,812	Ti I	25
4021,62	Xe III	4
4021,167	C II	3
4020,06	Cl II	15
4020,015	Ne I	2
4019,843	Ar II	2
4018,50	Cl III	6
4018,24	Cl II	3
4017,86	Xe II	2
4017,771	Ti I	15
4017,52	K II	4
4017,278	C II	5
4017,152	Fe I	6
4016,943	Ti I	3
4016,56	Xe II	2
4016,264	Ti I	6
4016,22	Si II	5
4015,8	Cu I	10
4015,377	Ti I	12
4014,99	Cs	10
4014,534	Fe I	10
4014,18	Cu II	2
4013,995	Ne I	2
4013,858	Ar II	25
4013,80	Mg II	2
4013,752	Ne I	1
4013,587	Ti I	12
4012,786	Ti I	3
4012,372	Ti II	4
4012,10	K II	5
4011,534	Ti I	3
4011,202	Ar II	6
4010,99	N I	2
4010,836	Cu I	8
4010,54	Cs	10
4010,192	Si III	4
4009,93	C I	4
4009,884	C II	7
4009,7154	Fe I	10
4009,653	Ti I	15
4009,58	Al II	1
4009,268	He I	5
4008,926	Ti I	35

λ	Symbol	I	λ	Symbol	I
4008,48	Kr II	10	3994,683	Ti I	4
4008,08	Kr II	25	3993,863	Cs II	4
4008,046	Ti I	9	3993,295	Cu II	5
4007,632	Ar II	2	3992,85	Xe III	20
4007,2735	Fe I	6	3992,81	Cl I	2
4007,195	Ti I	3	3992,210	F I	2
4006,772	Cs	10	3992,053	Ar II	12
4006,537	Cs	30	3991,94	Kr II	15
4006,159	Cu II	3	3991,77	Si II	15
4005,952	Ti I	6	3991,50	Cl III	7
4005,57	Kr II	30	3991,2581	Kr I	10
4005,51	Cl I	—	3991,0797	Kr I	20
4005,362	Ar II	2	3990,66	Kr II	15
4005,2440	Fe I	25	3990,33	Xe II	60
4003,789	Ti I	10	3990,19	Cl II	20
4003,64	N III	4	3989,758	Ti I	80
4003,470	Cu II	2	3988,17	Cl II	4
4003,028	Cu I	15	3988,158	Ar II	9
4002,98	C I	2	3987,78	Kr II	25
4002,61	Kr III	15	3987,09	Kr II	5
4002,466	Ti I	9	3987,021	Cu II	3
4002,35	Xe II	80	3986,88	C I	1
4001,682	Cs	20	3986,7533	Mg I	8
4001,65	N I	1	3986,172	Fe I	5
4001,56	C III	0	3985,96	Xe III	8
4001,24	K II	7	3985,520	Li I	5
4001,135	Ar II	1	3985,46	O II	0
4000,72	Kr I	2	3985,388	Fe I	3
4000,55	Xe II	5	3985,202	Xe I	30
3999,98	N I	4	3984,313	Ti I	3
3999,92	C III	0	3984,253	Ne I	7
3999,86	Ne II	1	3984,212	Mg I	1
3999,336	Ti I	7	3984,065	Ne I	2
3999,263	Ne I	1	3983,9593	Fe I	10
3999,248	Ar II	1	3982,719	O II	5
3998,69	N III	3	3982,478	Ti I	30
3998,635	Ti I	100	3982,1699	Kr I	6
3998,594	Ne I	1	3981,94	Cl II	15
3998,0554	Fe I	10	3981,80	K II	4
3998,018	Cu I	3	3981,7743	Fe I	7
3998,01	Si II	10	3981,761	Ti I	70
3997,95	Kr II	100	3981,238	Si III	5
3997,3952	Fe I	15	3980,56	Al III	2
3997,14	C I	1	3980,41	Xe II	2
3996,97	C I	0	3980,323	C II	8
3996,69	Kr II	3	3979,954	Cu I	5
3996,49	C I	0	3979,7149	Ar I	10
3996,381	Al II	3	3979,356	Ar II	12
3996,159	Al II	4	3979,05	Kr III	3
3996,075	Al II	1	3978,98	Xe II	2
3996,05	Xe II	3	3978,759	C II	4
3995,86	Al II	5	3978,000	Cs II	10
3995,721	Ne I	1	3977,7437	Fe I	12
3995,24	Cl II	6	3977,46	Si II	10
3995,17	O IV	2	3977,269	C II	5
3995,10	K II	6	3977,10	O IV	1
3994,998	N II	15	3975,953	C II	1
3994,86	N I	3			
3994,83	Kr II	100			
3994,82	Kr I	3			
3994,789	Ar II	10			

λ	Symbol	I	λ	Symbol	I
3975,7	Cu I	5	3959,495	Cs II	20
3975,59	Xe II	4	3958,39	Cl III	0
3975,341	C II	2	3958,382	Ar II	6
3974,791	F II	6	3958,206	Ti I	80
3974,753	Ar II	9	3957,67	Kr III	25
3974,66	O IV	—	3957,20	N I	3
3974,478	Ar II	10	3957,053	Ca I	10
3974,417	Xe I	40	3956,85	Xe I	6
3974,239	Cs	6	3956,82	O IV	—
3973,760	C II	7	3956,681	Fe I	12
3973,707	Ca I	12	3956,66	Si III	2
3973,263	O II	10	3956,4574	Fe I	9
3972,670	F II	4	3956,336	Ti I	60
3972,58	K II	6	3956,10	K II	3
3972,58	Xe II	50	3955,923	Cs	10
3972,570	Ca I	1	3955,851	N II	10
3972,439	C II	6	3955,74	Si II	10
3972,411	F II	2	3955,21	K II	6
3972,047	F II	6	3954,78	Kr II	90
3971,626	F II	3	3954,73	Xe II	20
3971,574	C II	2	3954,596	O I	5
3971,3250	Fe I	9	3954,507	Si II	10
3971,18	Cl II	7	3954,687	O I	10
3970,386	C II	4	3954,372	O II	7
3970,074	H	80	3954,33	Fe III	12
3969,95	N I	1	3954,296	Si II	5
3969,91	Xe III	4	3954,21	Cl II	20
3969,520	C II	3	3953,95	C II	0
3969,2595	Fe I	30	3953,799	F I	2
3968,995	D	80	3953,59	Kr II	20
3968,92	C II	0	3953,080	Si III	4
3968,71	Fe III	8	3953,056	O I	2
3968,637	T	80	3952,982	O I	1
3968,468	Ca II	22	3952,729	Ar II	9
3968,43	He II	—	3952,679	C II	1
3968,360	Ar II	20	3952,606	Fe I	8
3967,541	Xe I	200	3952,26	F II	2
3967,441	O II	1	3952,21	N I	3
3967,4234	Fe I	8	3952,058	C II	9
3967,212	Cs II	4	3951,987	O I	3
3966,72	K II	6	3951,616	Cu I	2
3966,6304	Fe I	10	3951,61	Xe II	5
3966,14	F I	1,5	3951,168	Fe I	9
3966,0645	Fe I	10	3950,925	Xe I	120
3965,187	Cs II	25	3950,56	Xe III	300
3964,89	Kr II	30	3949,96	Cl II	10
3964,7289	He I	100	3949,9558	Fe I	10
3964,269	Ti I	35	3949,530	C II	4
3964,16	Cu I	5	3949,373	C II	1
3963,838	Si III	6	3948,9785	Ar I	2000
3963,13	O II	0	3948,901	Ca I	6
3963,109	Fe I	6	3948,7778	Fe I	10
3962,851	Ti I	35	3948,72	Xe I	10
3962,34	Kr II	10	3948,670	Ti I	60
3961,59	O III	8	3948,563	F I	5
3961,5200	Al I	26	3948,333	C II	6
3961,40	C I	3	3948,163	Xe I	60
3960,53	Ar III	8	3948,107	Fe I	6
3959,84	K II	3	3947,770	Ti I	40

λ	Symbol	I
3947,715	C II	6
3947,66	Kr II	5
3947,594	O I	4
3947,5048	Ar I	1000
3947,489	O I	7
3947,488	Si III	6
3947,301	O I	10
3947,079	C II	2
3946,938	Cu I	3
3946,429	C II	1
3946,278	C II	5
3946,096	Ar II	12
3945,83	Kr II	1
3945,749	Cu II	2
3945,65	F II	4
3945,570	Cu II	5
3945,48	Kr II	5
3945,29	O IV	—
3945,197	C II	4
3945,048	O II	5
3945,003	C II	5
3944,79	Cl I	3
3944,33	F II	6
3944,272	Ar II	15
3944,193	C II	3
3944,0058	Al I	24
3943,57	Xe II	20
3943,540	Ne I	2
3942,93	Kr II	20
3942,78	N III	1
3942,53	K II	6
3942,4418	Fe I	6
3942,29	Xe I	2
3942,22	C I	3
3942,21	Xe II	3
3942,19	Ne II	3
3942,14	O IV	—
3941,52	F II	3
3941,23	N II	1
3940,92	Kr II	5
3940,8797	Fe I	5
3940,66	N II	2
3939,57	N II	4
3939,03	F II	7
3938,92	Xe II	15
3938,88	Kr II	20
3938,843	Ar II	1
3938,53	Kr III	4
3938,52	N III	4
3938,400	Mg I	6
3937,66	Xe II	2
3935,912	He I	2
3935,8143	Fe I	8
3935,677	Al I	4
3935,275	Ar II	1
3935,00	F II	3
3935,0	O III	2
3934,89	C IV	1
3934,527	Ti I	50
3934,46	K II	5
3934,41	N III	3
3934,29	C IV	2
3934,29	Kr IV	5
3934,262	F I	5
3934,228	Ti I	9
3933,663	Ca II	23
3933,605	Fe I	10
3933,260	Cu II	3
3933,22	Xe II	1
3933,17	Ar II	2
3933,11	F I	3
3933,027	Cu I	5
3932,548	Ar II	15
3932,007	Ti II	2
3931,996	Al I	5
3931,235	Ar II	12
3930,689	F I	8
3930,63	O IV	—
3930,2981	Fe I	25
3929,875	Ti I	40
3929,46	Cs	2
3929,26	Kr II	20
3928,63	Cl II	5
3928,629	Ar II	25
3927,9216	Fe I	30
3927,88	Cl II	6
3926,80	Xe II	1
3926,58	O II	—
3926,534	He I	7
3926,36	K II	5
3926,319	Ti I	10
3926,05	Kr I	1
3926,03	Ar II	7
3925,947	Fe I	6
3925,87	Cl III	5
3925,722	Ar II	10
3925,583	Cs II	25
3925,274	Cu I	8
3924,51	Ti I	50
3924,468	Si III	20
3923,556	Ar II	1
3923,50	Ca I	0
3923,438	Cu II	3
3923,00	K II	5
3922,9134	Fe I	25
3922,53	Xe III	500
3922,528	Ar II	1
3922,359	Ar II	2
3921,75	Cl II	3
3921,69	Cs	4
3921,68	Kr II	6
3921,6	Li I	—
3921,423	Ti I	30
3921,267	Cu I	5
3920,693	C II	18
3920,641	Cu II	5
3920,2601	Fe I	20

λ	Symbol	I	λ	Symbol	I
3920,14	Kr II	200	3901,12	Cl II	4
3919,822	Ti I	5	3900,958	Ti I	12
3919,287	O II	6	3900,82	Cs	4
3919,00	Si II	5	3900,680	Al II	10
3918,999	N II	9	3900,624	Ar II	11
3918,978	C II	15	3900,546	Ti II	70
3918,646	Fe I	6	3900,11	K II	3
3918,57	Xe II	2	3900,09	Cs II	4
3917,766	Ar II	4	3899,878	Ar I	100
3917,64	Kr II	50	3899,723	Ne I	2
3917,57	Cl II	18	3899,7086	Fe I	30
3917,1834	Fe I	8	3899,542	Mg I	1
3916,90	Kr II	3	3899,28	K II	3
3916,733	Fe I	6	3899,27	Cl II	4
3916,70	Cl II	20	3899,22	Cu I	8
3916,60	Xe II	1	3898,833	F II	6
3915,879	Ti I	3	3898,725	F II	2
3915,82	Cl II	3	3898,70	Kr III	10
3915,30	Xe III	4	3898,487	Ti I	8
3915,329	Li I	25	3898,478	F I	5
3914,768	Ar II	12	3898,120	Mg I	4
3914,751	Ti I	5	3898,012	Fe I	10
3914,334	Ti I	35	3897,92	K II	8
3913,92	Cl II	30	3897,895	Fe I	8
3913,90	Kr III	3	3896,978	Cs II	7
3913,6339	Fe I	4	3896,682	Cu II	3
3913,464	Ti II	60	3896,66	F II	3
3913,37	Cs	2	3896,30	O II	1
3912,88	Kr II	5	3896,12	F II	1
3912,59	Kr II	70	3895,572	Mg I	3
3912,088	O II	2	3895,6579	Fe I	25
3911,960	O II	10	3895,250	Ar II	1
3911,572	Ar II	10	3895,243	Ti I	30
3911,185	Ti I	8	3894,71	Kr II	60
3910,60	Cl II	2	3894,6603	Ar I	300
3908,43	Ar IV	—	3894,55	Cl II	2
3907,9371	Fe I	4	3893,53	O II	2
3907,91	Xe II	100	3893,3935	Fe I	7
3907,84	Ar III	7	3893,304	Mg I	2
3907,45	O II	4	3893,09	Cs	4
3906,933	Cs II	20	3893,04	F II	2
3906,4814	Fe I	8	3892,913	Cu II	5
3906,25	Kr II	150	3892,69	Kr I	1
3905,85	Xe II	10	3892,206	Cs	4
3905,80	Cl II	4	3891,984	Ar II	15
3905,5227	Si I	300	3891,906	Mg I	2
3905,34	Xe II	1	3891,929	Fe I	5
3904,806	Cs	4	3891,400	Ar II	12
3904,785	Ti I	40	3891,114	Cu II	2
3903,9011	Fe I	5	3890,241	Mg I	3
3903,859	Mg I	4	3890,073	Cu II	3
3903,819	F II	4	3889,948	Ti I	6
3903,70	Xe III	4	3889,475	C III	1
3903,163	Cu II	15	3889,427	Ne I	5
3902,9484	Fe I	20	3889,144	C III	6
3902,84	Cl II	9	3889,141	Ca I	1
3901,955	F II	5	3889,051	H	60
3901,89	Cl II	5	3888,648	He I	5000
3901,852	F II	2	3888,610	Cs I	150
3901,15	Kr II	10	3888,517	Fe I	20

λ	Symbol	I
3888,40	Cu I	4
3888,020	Ti I	4
3887,993	D	60
3887,640	T	60
3887,54	Kr II	5
3887,134	Ne I	1
3886,84	K II	2
3886,63	Cl II	4
3886,2839	Fe I	40
3885,941	C III	5
3885,92	Cu I	3
3885,5165	Fe I	5
3885,501	K III	1
3885,45	Xe II	4
3885,28	Kr II	1
3885,00	Xe II	20
3884,523	Cu II	5
3884,120	Cu II	10
3883,824	C II	1
3883,816	C III	4
3883,80	Cl II	12
3883,42	K II	3
3883,15	O II	3
3882,892	Ti I	20
3882,698	Ne I	2
3882,45	O II	1
3882,313	Ti I	10
3882,197	O II	7
3882,147	Ti I	15
3881,73	Cl III	3
3881,714	Cu I	5
3881,399	Ti I	4
3880,588	C II	7
3880,46	Xe III	60
3880,335	Ar II	6
3880,07	Kr II	2
3879,640	C II	7
3879,387	Cu II	5
3878,663	Fe I	8
3878,62	K II	4
3878,5745	Fe I	100
3878,306	Mg I	3
3878,181	He I	3
3878,028	C II	7
3878,0206	Fe I	60
3877,80	Xe III	200
3876,664	C II	12
3876,408	C II	12
3876,187	C II	12
3876,146	Cs I	300
3876,080	Ar I	10
3876,055	C II	9
3875,82	O II	4
3875,807	Ca I	4
3875,44	Kr II	150
3875,264	Ar II	12
3875,262	Ti I	20
3874,666	C II	2

λ	Symbol	I
3874,22	Ar III	4
3874,10	O II	2
3874,04	Kr III	3
3873,7624	Fe I	8
3873,74	K II	5
3873,203	Ti I	10
3873,067	C II	0
3872,552	Ca I	3
3872,5032	Fe I	60
3872,45	O II	1
3872,143	Ar II	11
3871,791	He I	5
3871,7513	Fe I	4
3871,669	C II	7
3870,508	Ca I	2
3870,164	Cs II	4
3869,63	Xe II	20
3869,614	Ar II	2
3869,5615	Fe I	4
3869,275	Ti I	5
3869,10	N I	4
3868,874	C II	6
3868,70	Kr III	40
3868,62	Cl II	40
3868,524	Ar II	20
3868,397	Ti I	10
3868,358	Cu II	3
3867,739	Ti I	3
3867,630	He I	5
3867,475	He I	30
3867,2184	Fe I	7
3866,446	Ti I	15
3866,291	Cu II	2
3866,2752	Ar I	5
3866,160	Al II	2
3865,5256	Fe I	30
3864,68	O II	1
3864,60	Cl II	15
3864,45	O II	1
3864,367	Cs	4
3864,2669	Ar I	10
3864,249	Cs	6
3864,13	O II	7
3864,121	Cu II	5
3863,50	O II	2
3862,823	Ti I	10
3862,781	Cu I	5
3862,595	Si II	200
3862,181	C II	2
3861,95	Cl II	20
3861,747	Cu I	250
3861,489	Cs	4
3861,41	K II	3
3861,40	Cl II	50
3861,348	Ar II	3
3861,05	Xe III	10
3860,98	Cl II	100
3860,898	Cu I	5
3860,80	Cl II	150
3860,58	Kr IV	5

λ	Symbol	I	λ	Symbol	I
3860,472	Cu I	600	3846,436	Ti I	6
3860,05	Cl II	2	3846,12	Kr I	2
3859,9132	Fe I	300	3845,9778	Kr I	15
3859,33	Al II	3	3845,84	Cl II	30
3859,2143	Fe I	10	3845,69	Cl II	75
3859,17	Cl II	7	3845,42	Cl II	50
3858,860	Mg I	2	3845,406	Ar II	10
3858,78	Kr II	5	3845,1706	Fe I	5
3858,53	Xe II	20	3844,735	Ar II	9
3858,46	Ar IV	—	3844,565	Ar II	4
3858,32	Ar III	10	3844,51	Cu I	4
3858,133	Ti I	15	3844,45	Kr II	50
3857,32	Kr II	20	3844,02	K II	1
3857,18	O II	4	3843,58	O II	3
3856,62	C II	0	3843,26	Cl II	100
3856,373	Fe I	50	3843,2596	Fe I	8
3856,16	O II	5	3842,82	O II	3
3856,127	Ar II	1	3842,61	Ti I	3
3856,057	N II	6	3842,577	Cu II	4
3856,017	Si II	500	3842,458	Si III	7
3855,160	Ar II	4	3842,317	Al II	1
3855,100	N II	5	3842,28	Kr II	20
3854,965	Mg I	1	3842,213	Al II	2
3854,75	Cl II	15	3842,183	N II	5
3854,30	Xe III	10	3842,037	Al II	3
3853,960	Mg I	2	3841,88	Xe III	20
3853,719	Ti I	10	3841,52	Xe III	100
3853,664	Si II	100	3841,518	Ar II	6
3853,038	Ti I	10	3841,0499	Fe I	80
3852,5752	Fe I	6	3840,48	Ne II	1
3851,69	Cl II	30	3840,4397	Fe I	80
3851,667	F II	10	3839,37	Kr II	4
3851,47	O II	0	3839,2584	Fe I	7
3851,38	Cl II	75	3838,374	N II	8
3851,04	O II	3	3838,37	Cl II	20
3850,97	Cl II	100	3838,2943	Mg I	20
3850,8193	Fe I	12	3838,2918	Mg I	20
3850,81	Cl III	4	3838,239	Ar II	1
3850,81	O II	2	3838,15	Li I	3
3850,578	Ar II	30	3838,100	He I	2
3850,385	Mg II	7	3837,976	Cu I	5
3850,03	Cu II	2	3837,81	Kr I	30
3849,987	F II	15	3837,68	F II	0,5
3849,9694	Fe I	40	3837,449	Cs	4
3849,87	Xe II	50	3836,763	Ti I	5
3849,570	Cu II	2	3836,683	C II	2
3849,33	Cl II	3	3836,54	Kr II	30
3848,914	Mg I	1	3836,333	Fe I	4
3848,58	Xe II	6	3836,150	Cu II	4
3848,27	Cs II	2	3835,730	C II	6
3848,209	Mg II	8	3835,6	Xe I	2
3847,89	O II	3	3835,386	H	40
3847,49	Kr III	3	3835,37	Kr III	2
3847,409	N II	5	3834,84	N I	2
3847,086	F II	20	3834,6788	Ar I	800
3846,860	Ar II	2	3834,342	D	40
3846,83	Kr II	5	3834,24	N I	4
3846,8023	Fe I	8	3834,24	O VI	1

λ	Symbol	I	λ	Symbol	I
3834,2244	Fe I	100	3820,4274	Fe I	250
3833,994	T	40	3820,25	Cl II	100
3833,91	Ti I	3	3819,758	He I	10
3833,554	He I	4	3819,61	Cs	4
3833,40	Cl II	200	3819,6072	He I	100
3833,35	C I	3	3819,017	Ar II	5
3833,3103	Fe I	5	3818,869	Cu II	3
3833,10	O II	3	3818,52	F II	2
3832,3037	Mg I	20	3818,44	Ne II	6
3832,2996	Mg I	18	3818,40	Cl II	30
3831,743	C II	8	3818,27	N I	2
3831,17	Kr II	2	3817,639	Ti I	5
3830,80	Cl II	15	3817,50	K II	7
3830,515	Ar II	4	3817,490	Cu I	5
3830,45	O II	4	3817,11	Kr II	15
3830,390	Ar II	10	3816,72	O III	1
3830,39	N I	9	3816,56	K II	6
3830,26	O I	—	3815,8430	Fe I	100
3830,165	Ar II	1	3815,70	Ar III	1
3829,793	N II	6	3814,855	Ti I	4
3829,77	Ne II	7	3814,65	F II	4
3829,77	Xe II	10	3814,580	Ti II	4
3829,77	Xe III	20	3813,542	Cu I	10
3829,57	Kr III	1	3813,390	Ti II	2
3829,3549	Mg I	36	3812,9658	Fe I	40
3829,27	Cl II	15	3812,2155	Kr I	20
3828,85	C I	2	3811,95	Cu I	8
3828,180	Ti I	3	3811,385	Ti I	4
3827,8256	Fe I	75	3811,35	O VI	2
3827,68	F II	0,5	3811,212	Ar II	2
3827,62	Cl II	150	3811,05	Xe II	40
3826,908	Cu II	5	3810,99	O III	2
3826,86	Xe I	15	3810,10	Cl II	30
3826,807	Ar II	12	3809,84	Xe I	30
3826,27	Xe II	2	3809,51	Cl II	40
3826,15	Kr II	2	3809,456	Ar II	15
3825,8834	Fe I	200	3809,30	Kr IV	3
3825,676	Ar II	8	3809,16	Kr III	7
3825,530	O I	—	3808,7306	Fe I	4
3825,249	O I	4	3808,577	Ar II	11
3825,090	O I	3	3807,5392	Fe I	7
3825,047	Cu I	100	3807,29	Xe II	10
3824,47	Cl III	4	3806,6992	Fe I	10
3824,4455	Fe I	50	3806,544	Si III	30
3824,425	O I	3	3806,52	Kr II	1
3823,95	N IV	0	3806,30	Ne II	2
3823,74	Xe I	10	3806,17	Kr II	8
3823,469	O I	10	3805,90	F II	5
3823,35	Xe II	2	3805,740	He I	3
3823,254	Ar II	3	3805,412	Cs II	2
3823,19	Ne II	1	3805,3450	Fe I	12
3822,63	O I	—	3805,24	Cl II	75
3822,07	N I	6	3805,232	Cu I	100
3822,02	Cl III	4	3805,096	Cs II	25
3821,68	O II	4	3804,83	Cl III	3
3821,30	K II	3	3804,67	Kr II	30
3821,1807	Fe I	10	3804,31	C I	2
3820,884	Cu I	60	3803,57	Cl III	3

λ	Symbol	I	λ	Symbol	I
3803,49	Cu I	5	3786,253	Ti I	3
3803,172	Ar II	10	3786,261	Cu II	5
3803,14	O II	6	3786,176	Fe I	4
3801,90	Xe I	3	3786,043	Ti I	20
3801,39	Xe I	30	3785,97	F II	3
3801,093	Ti I	3	3785,951	Fe I	6
3801,09	F II	3	3785,49	Cu I	5
3800,99	Xe II	15	3785,424	Cs II	20
3800,5437	Kr I	30	3785,01	O II	0
3800,502	Cu I	30	3784,862	He I	2
3800,42	Ar IV	—	3783,92	Ne	1
3800,25	Ar III	6	3783,23	Xe II	10
3800,14	K II	6	3783,19	K II	6
3800,02	Ne II	5	3783,13	Kr II	500
3799,88	Cu I	10	3782,31	Ne	1
3799,5498	Fe I	50	3781,63	F II	2
3799,381	Ar II	10	3781,3570	Ar I	300
3798,80	Cl II	50	3781,23	Cl II	30
3798,5134	Fe I	40	3780,98	Xe III	300
3798,46	F II	3	3780,841	Ar II	25
3798,276	Ti I	6	3780,70	Xe II	1
3797,908	Cs	4	3780,045	Cu I	5
3797,900	H	20	3779,446	Fe I	4
3797,832	Cu II	5	3779,35	Cl III	5
3797,517	Fe I	12	3779,23	N III	—
3797,245	Cu I	8	3779,067	Cu I	2
3796,899	Ti II	2	3778,78	Xe II	1
3796,8839	Kr I	20	3778,09	Kr II	500
3796,866	D	20	3777,60	O II	4
3796,599	Ar II	8	3777,529	Ar II	2
3796,522	T	20	3777,52	Ar IV	—
3796,30	Xe I	40	3777,16	Ne II	8
3796,114	Si III	25	3776,4553	Fe I	6
3795,95	Xe I	3	3776,30	Xe III	40
3795,903	Ti I	7	3776,20	Cl II	4
3795,37	Ar III	20	3776,062	Ti II	6
3795,0045	Fe I	60	3775,49	Xe II	1
3794,72	Li I	10	3775,4408	Ar I	10
3794,60	F II	2	3774,8266	Fe I	5
3794,48	O II	3	3774,650	Ti II	3
3794,340	Fe I	8	3774,522	Ar II	4
3793,75	Cl II	25	3774,38	O IV	—
3793,68	C I	2	3774,25	Cl II	25
3793,226	Ar II	1	3774,00	O III	6
3792,87	N III	1	3773,68	Cl II	20
3792,70	Kr III	15	3773,4241	Kr I	50
3792,40	F II	4	3773,151	Si IV	6
3792,12	F II	1	3772,53	Xe III	20
3791,67	Xe III	12	3771,904	Cu I	100
3791,41	Si III	20	3771,652	Ti I	25
3791,26	O III	6	3771,64	Ne	1
3790,96	Ne II	3	3771,45	N III	2
3790,0943	Fe I	12	3771,34	Kr II	30
3789,293	Ti I	8	3771,08	N III	7
3787,8825	Fe I	50	3770,632	H	15
3787,32	Xe II	3	3770,516	Ar II	10
3786,6781	Fe I	8	3770,37	N III	—
3786,383	Ar II	12			
3786,29	Ne	2			

λ	Symbol	I
3770,3698	Ar I	400
3770,12	Xe II	3
3769,69	Kr III	2
3769,654	Ne I	5
3769,606	D	15
3769,449	Ne I	7
3769,264	T	15
3769,13	Cl II	20
3768,784	He I	2
3768,13	Cl II	18
3768,05	Cl I	—
3768,047	Ne I	5
3767,57	Cl II	30
3767,36	K II	6
3767,1939	Fe I	80
3766,445	Ti I	3
3766,29	Ne II	8
3766,118	Ar II	11
3766,043	Ti I	20
3765,88	Kr II	2
3765,85	Xe III	10
3765,819	Ne I	5
3765,5414	Fe I	20
3765,269	Ar II	20
3764,837	Cu I	5
3764,42	Cl III	2
3763,96	C I	0
3763,7910	Fe I	100
3763,504	Ar II	12
3763,37	Xe II	15
3763,111	Ar II	2
3762,63	O II	5
3762,62	N III	—
3762,435	Si IV	8
3762,26	Xe II	10
3762,25	C I	2
3762,05	Xe II	3
3761,866	Ti II	15
3761,72	Ca I	0
3761,62	Ca III	6
3761,6	O V	—
3761,320	Ti II	200
3760,5335	Fe I	6
3760,052	Fe I	8
3759,87	O III	9
3759,492	Cu I	60
3759,460	Fe II	6
3759,291	Ti II	200
3759,10	Cl III	3
3758,93	Kr II	6
3758,45	O IV	0
3758,386	Ca II	3
3758,296	Cu I	5
3758,2350	Fe I	150
3757,84	C I	1
3757,684	Ti II	30
3757,66	N III	—
3757,60	N III	—
3757,21	O III	5
3757,05	C I	3
3756,92	Cl II	2
3756,87	Xe II	10
3756,671	Ar II	4
3756,62	K II	3
3756,52	C I	2
3756,107	He I	1
3755,82	O IV	—
3755,668	Ca II	2
3755,12	C I	1
3754,67	O III	7
3754,62	N III	6
3754,24	Kr II	80
3754,215	Ne I	50
3754,052	Ar II	6
3753,83	Ne II	5
3753,623	Ti I	25
3753,6134	Fe I	8
3753,521	Ar II	9
3753,519	Cu I	8
3753,367	Ca I	1
3753,3	F II	0
3753,10	Al II	1
3752,860	Ti I	80
3752,65	N III	3
3752,36	F II	2
3751,402	Cs II	4
3751,330	Ar II	1
3751,26	Ne II	5
3751,047	Ar II	2
3750,79	Ar IV	—
3750,485	Ar II	5
3750,349	Ca I	1
3750,154	H	10
3750,00	Cl II	30
3749,49	O II	9
3749,4875	Fe I	200
3748,81	Cl III	8
3748,492	Fe I	7
3748,489	Fe II	8
3748,46	Cl II	15
3748,374	Ca I	1
3748,2639	Fe I	60
3748,207	Cu II	3
3748,101	Ti I	6
3748,010	Ti II	10
3747,54	N IV	6
3747,1	O V	—
3746,929	Fe I	6
3746,915	Ar II	5
3746,452	Ar II	6
3745,9013	Fe I	40
3745,83	N III	4
3745,72	Xe III	25
3745,69	Xe I	4
3745,5623	Fe I	100
3745,38	Xe I	10

λ	Symbol	I	λ	Symbol	I
3745,356	Cu I	20	3732,61	Kr II	15
3744,80	Kr II	150	3732,539	Cs II	4
3744,73	O IV	0	3732,399	Fe I	10
3744,66	Ne II	4	3732,35	C I	2
3744,42	K II	5	3732,13	O III	1
3744,274	Ar II	1	3731,950	Al II	1
3743,7653	Ar I	100	3731,67	Kr II	2
3743,468	Fe I	6	3731,18	Xe II	20
3743,3640	Fe I	20	3729,980	Cs	4
3743,363	Cu I	3	3729,806	Ti I	50
3742,85	C I	1	3729,70	O III	1
3742,22	Xe I	1	3729,34	O II	2
3741,70	Cl III	3	3729,310	Ar II	30
3741,69	Kr II	200	3729,03	C I	1
3741,69	O II	0	3729,03	O IV	3
3741,633	Ti II	50	3728,82	O III	1
3741,44	C I	2	3728,49	O III	0
3741,242	Cu I	450	3728,04	Kr II	7
3741,059	Ti I	60	3727,6211	Fe I	50
3740,79	C I	0	3727,33	O II	8
3740,73	Kr II	6	3727,08	Ne II	9
3740,60	Ne	1	3726,925	Fe I	6
3739,92	O II	6	3726,4	O V	—
3739,60	F II	4	3726,32	Kr III	5
3739,375	Ca II	1	3725,81	O IV	2
3739,13	K II	5	3725,74	Cl III	3
3738,901	Ti I	5	3725,46	Cl III	1
3738,76	Cl II	4	3725,30	O III	3
3738,637	Cu II	3	3725,155	Ti I	20
3738,3078	Fe I	10	3724,9	Cs	4
3738,003	Al II	3	3724,570	Ti I	20
3737,893	Ar II	15	3724,521	Ar II	8
3737,20	Xe II	5	3724,3796	Fe I	8
3737,19	C I	0	3722,568	Ti I	15
3737,1333	Fe I	150	3722,5642	Fe I	50
3736,901	Ca II	18	3722	O IV	2
3736,78	O IV	4	3721,95	O III	1
3735,94	O II	3	3721,940	H	6
3735,78	C I	1	3721,86	Ne II	2
3735,78	Kr II	40	3721,666	Cu I	8
3735,495	Ar II	6	3721,632	Ti II	15
3735,330	Fe I	6	3721,35	Kr II	150
3734,94	Ne II	7	3721,34	K II	5
3734,8659	Fe I	300	3720,86	O III	3
3734,80	O III	1	3720,80	Xe II	40
3734,567	Al II	1	3720,771	Cu I	150
3734,51	C I	0	3720,45	Cl III	8
3734,370	H	8	3720,428	Ar II	9
3734,337	Cs II	10	3719,9367	Fe I	250
3734,180	Cu I	200	3718,7	Li I	5
3734	O IV	3	3718,63	Kr II	200
3733,910	Al II	2	3718,208	Ar II	12
3733,767	Ti I	4	3718,02	Kr II	300
3733,73	Cl II	10	3717,94	Cl II	15
3733,3191	Fe I	40	3717,5	O V	–
3733,010	He I	3	3717,393	Ti I	20
3732,92	Kr II	6	3717,20	Xe II	20
3732,865	He I	10	3717,174	Ar II	10
			3716,60	K II	5
			3716,448	Fe I	12

λ	Symbol	I	λ	Symbol	I
3716,15	Kr II	4	3702,75	O III	5
3715,69	Xe II	2	3702,74	Xe I	2
3715,08	O III	6	3702,291	Ti I	10
3715,04	Kr II	12	3702,2	O V	—
3714,737	Ar II	6	3702,086	Al III	10
3714,03	O III	2	3702,005	Ar II	2
3714,43	N IV	—	3701,81	Ne II	4
3713,19	Ar IV	—	3701,2247	Ne I	40
3713,103	Al III	15	3701,090	Fe I	20
3713,084	Ne II	10	3701,070	Cu I	5
3713,019	Ar II	2	3700,536	Cu I	250
3712,75	O II	7	3699,98	Kr III	2
3712,733	F I	0,8	3699,475	Cs	10
3712,48	O III	2	3699,20	Cs II	10
3712,48	Kr II	1	3699,097	Cu I	10
3712,04	C I	1	3699,0	O V	—
3712,009	Cu I	30	3698,70	O III	5
3711,974	H	5	3698,183	Ti I	3
3711,64	Xe II	20	3698,0452	Kr I	6
3711,27	Kr II	1	3697,154	H	3
3711,074	Na II	6	3697,09	Ne II	2
3710,774	Cs	4	3696,82	Xe I	4
3710,365	F II	4	3696,69	Kr III	5
3709,963	Ti I	20	3696,5082	Ar I	20
3709,918	Ar II	6	3695,37	O III	4
3709,64	Ne II	7	3695,358	Cu I	8
3709,52	O III	2	3695,054	Fe I	8
3709,2484	Fe I	75	3694,643	Ar II	2
3708,625	Ti I	4	3694,445	Ti I	10
3708,15	Xe III	4	3694,355	Ca II	1
3707,9216	Fe I	8	3694,197	Ne II	10
3707,823	Fe I	20	3694,15	Cl I	—
3707,549	Ti I	10	3694,15	N IV	—
3707,34	Cl III	6	3694,108	Ca II	4
3707,24	O III	6	3694,010	Fe I	20
3707,12	Cu I	4	3693,49	Xe I	40
3707,049	Fe I	8	3692,5	Ar IV	—
3706,937	Ar II	5	3692,44	O I	7
3706,63	F II	0,5	3692,126	Ar II	1
3706,219	Ti II	20	3691,88	Cl II	5
3706,026	Ca II	17	3691,859	F I	2
3705,93	F I	1,5	3691,84	Xe II	1
3705,5674	Fe I	100	3691,58	Cl I	—
3705,56	C I	1	3691,557	H	2
3705,54	Cl II	2	3690,8960	Ar I	300
3705,45	Cl III	6	3690,74	Xe II	1
3705,148	He I	3	3690,65	Kr II	30
3705,005	He I	30	3690,65	Kr III	30
3704,73	O III	3	3690,2	O V	—
3704,51	F II	8	3690,018	F I	4
3704,4635	Fe I	10	3689,95	N IV	—
3704,295	Ti I	15	3689,916	Ti I	15
3703,855	H	4	3689,6	C IV	2
3703,71	C III	4	3689,463	Fe I	12
3703,550	Ar II	1	3689,40	F I	1
3703,37	O III	5	3688,80	Xe I	1
3703,3	O V	—	3688,44	Cl II	15
3703,217	Al II	4	3688,10	Cl III	2
3703,03	Cl I	—			

λ	Symbol	I
3687,88	N I	2
3687,708	Cu I	40
3687,64	Cs II	4
3687,4589	Fe I	40
3687,438	Cu I	400
3687,354	Ti I	5
3686,833	H	—
3686,555	Cu II	100
3686,15	Kr II	80
3686,003	Fe I	15
3685,90	Xe I	40
3685,7351	Ne I	100
3685,192	Ti II	250
3684,930	Cu I	200
3684,672	Cu I	450
3684,1102	Fe I	15
3684,1	Li II	3
3683,696	Ca II	3
3683,39	Cl III	5
3683,0562	Fe I	10
3682,810	H	—
3682,547	Ar II	7
3682,428	Cu II	10
3682,25	Si III	2
3682,2421	Ne I	100
3682,209	Fe I	20
3682,15	Si III	5
3682,05	Cl III	7
3681,54	K II	6
3681,402	Si III	7
3681,10	N I	3
3680,454	Cs	4
3680,37	Kr II	100
3680,101	Cs II	4
3680,064	Ar II	9
3679,9152	Fe I	40
3679,80	Ne II	2
3679,673	Ti II	3
3679,67	F II	5
3679,58	Kr I	100
3679,355	H	—
3679,31	Xe I	4
3678,8620	Fe I	3
3678,66	Kr II	7
3678,274	Ar II	10
3678,240	Ca I	3
3677,6309	Fe I	12
3677,54	Xe I	2
3676,878	Cu I	50
3676,731	Si III	3
3676,63	Xe III	50
3676,365	H	—
3676,3135	Fe I	6
3676,05	K II	3
3675,307	Ca I	2
3675,2367	Ar I	300
3674,23	Kr III	4
3674,05	Ar I	2
3674,04	Xe II	1
3673,83	Cl II	18
3673,761	H	—
3673,448	Ca I	1
3673,266	Ar II	5
3672,85	F I	2,5
3672,57	Xe II	20
3671,953	Cu I	100
3671,672	Ti I	20
3671,478	H	—
3671,14	Kr III	1
3671,005	Ar II	4
3670,6693	Ar I	300
3670,4	Li I	3
3670,28	Cl III	7
3670,23	Kr III	4
3670,071	Fe I	3
3670,028	Fe I	3
3669,91	Xe I	10
3669,605	Ar II	9
3669,5229	Fe I	10
3669,466	H	—
3669,46	Cl II	2
3669,01	Kr II	150
3668,965	Ti I	15
3668,9	F II	0
3668,7363	Kr I	10
3668,60	C I	1
3668,60	K II	3
3668,59	Kr II	6
3668,174	F I	12
3668,03	Cl II	20
3667,757	F I	4
3667,684	H	—
3666,097	H	—
3666,01	Kr II	5
3665,735	Cu I	125
3665,3259	Kr I	80
3664,679	H	—
3664,112	Ne II	9
3664,08	Cu I	5
3663,93	Xe II	5
3663,76	Ar I	5
3663,44	Kr II	20
3663,406	H	—
3662,366	Si III	2
3662,258	H	—
3662,237	Ti II	40
3661,793	F I	3
3661,70	Xe II	20
3661,48	Cl III	1
3661,391	Cs	6
3661,221	H	—
3661,00	Kr II	15
3660,631	Ti I	12
3660,439	Ar II	10
3660,270	H	—
3659,93	Ne II	3
3659,84	Cl II	18

λ	Symbol	I	λ	Symbol	I
3659,765	Ti II	60	3646,84	O III	2
3659,5305	Ar I	100	3646,198	Ti I	12
3659,521	Fe I	8	3645,825	Fe I	6
3659,353	Cu I	125	3645,232	Cu I	250
3658,44	Xe II	6	3645,20	O III	1
3658,38	Cl II	20	3645,123	Si III	6
3658,3	Al III	1	3644,990	Ca I	2
3658,097	Ti I	20	3644,91	Xe II	5
3657,74	Xe II	5	3644,86	Ne II	4
3657,218	Ar II	2	3644,765	Ca I	15
3657,187	F I	3	3644,43	Xe II	5
3656,95	Cl III	7	3644,410	Ca I	40
3656,785	Cu I	125	3644,14	Xe III	5
3656,50	F II	0,5	3643,89	Ne II	5
3656,051	Ar II	10	3643,632	Cu I	5
3655,859	Cu I	600	3643,1169	Ar I	100
3655,77	Kr III	1	3642,798	F II	7
3655,73	Cs	4	3642,675	Ti I	80
3655,281	Ar II	12	3641,985	F II	8
3655,112	Si III	4	3641,693	Cu I	50
3655,00	Al II	8	3641,40	Cs	4
3654,979	Al II	4	3641,34	Kr III	30
3654,63	Xe III	20	3641,332	Cs	5
3654,592	Ti I	15	3641,330	Ti II	100
3654,243	Cu I	200	3641,011	F II	3
3653,97	Kr II	250	3641,00	Xe III	15
3653,497	Ti I	100	3640,891	F II	9
3653,12	Xe III	3	3640,3918	Fe I	15
3653,00	O III	1	3639,830	Ar II	12
3652,982	F I	1,5	3639,445	Si III	5
3652,82	C I	0	3639,19	Cl II	18
3652,34	Cu I	100	3638,898	Si III	3
3652,130	He I	2	3638,70	O III	3
3651,990	He I	7	3638,524	Si III	2
3651,721	Si III	2	3638,2998	Fe I	12
3651,4699	Fe I	20	3637,966	Ti I	10
3651,174	F I	2	3637,943	Si III	2
3651,090	Al II	4	3637,93	Kr II	4
3651,073	Cs II	4	3637,48	Kr II	20
3651,064	Al II	6	3637,031	Ar II	10
3651,02	Kr II	25	3637,00	K II	3
3650,891	Ar II	7	3636,03	Xe III	3
3650,855	Cu I	5	3635,916	Cu I	250
3650,70	O III	0	3635,636	Ar II	3
3650,19	N I	5	3635,462	Ti I	80
3650,13	Cl II	30	3635,202	Ti I	8
3649,8330	Ar I	800	3634,814	Ar II	7
3649,5090	Fe I	12	3634,75	Cs	6
3649,221	Al II	1	3634,48	Xe II	1
3649,20	O III	00	3634,4605	Ar I	300
3649,182	Al II	1,5	3634,42	Kr II	3
3648,981	K I	4	3634,369	He I	2
3648,841	K I	3	3634,334	Fe I	6
3648,62	C I	0	3634,232	He I	15
3648,61	Kr II	40	3633,6643	Ne I	100
3648,383	Cu I	125	3633,54	Kr II	3
3648,07	Cl II	10	3633,458	Ti I	5
3647,95	K II	2	3633,06	Xe I	6
3647,8439	Fe I	100	3632,75	Ne II	2
			3632 6837	Ar I	300

λ	Symbol	I
3632,558	Cu I	50
3632,5	Kr III	1
3632,4896	Kr I	4
3632,308	Cu I	5
3632,14	Xe III	20
3631,87	Kr II	200
3631,4646	Fe I	125
3631,266	Na II	8
3631,096	Fe I	7
3630,974	Ca I	15
3630,776	F I	1,5
3630,748	Ca I	30
3630,620	Cs II	2
3629,963	F I	4
3629,771	Cu I	10
3628,57	Xe III	3
3628,1570	Kr I	10
3628,06	Ne II	4
3627,63	Mg I	4
3627,32	Cu I	125
3626,91	Kr I	2
3626,42	K II	4
3626,085	Ti I	4
3625,61	C I	0
3625,148	Fe I	6
3624,890	Fe II	5
3624,826	Ti II	70
3624,56	Cs	4
3624,236	Cu I	100
3624,111	Ca I	20
3624,05	Xe III	600
3623,84	Kr I	1
3623,79	Cl II	9
3623,61	Kr II	30
3623,444	Ar II	2
3623,1878	Fe I	8
3623,13	Xe III	40
3622,691	Cs	6
3622,69	Cl III	7
3622,538	Si III	8
3622,53	Kr I	1
3622,140	Ar II	12
3622,005	Fe I	12
3622	O VI	—
3621,98	Xe II	3
3621,4640	Fe I	15
3621,273	Fe II	6
3621,245	Cu I	600
3621,012	Ar II	3
3620,807	Ar II	3
3620,789	F I	1
3620,352	Cu I	225
3619,581	Si III	3
3618,90	Xe III	4
3618,88	Cl II	15
3618,7694	Fe I	125
3618,549	Cs II	2
3618,49	K II	6
3618,161	Cs	6
3617,788	Fe I	12
3617,295	Cs I	60
3615,858	N II	2
3615,82	Kr III	20
3615,583	Mg II	3
3615,4755	Kr I	20
3615,09	Cl II	10
3614,989	Cs	4
3614,873	Fe II	5
3614,218	Cu I	200
3613,781	Mg II	4
3613,761	Cu I	600
3613,643	He I	30
3613,06	Xe I	8
3612,85	Cl III	8
3612,37	Xe II	20
3612,352	Al III	15
3612,35	Ne II	3
3612,074	Fe I	8
3611,812	Ar II	5
3611,52	Xe II	1
3611,459	Cs I	200
3611,365	Ar II	1
3611,06	Kr III	5
3610,809	Cu I	200
3610,32	Xe I	15
3610,162	Fe I	20
3610,154	Ti I	12
3610,07	Cl II	12
3609,808	F I	1,5
3609,75	Cl II	4
3609,625	C III	6
3609,56	C I	0
3609,44	Xe III	20
3609,295	Cu I	200
3609,1787	Ne I	50
3609,097	N II	4
3609,063	C III	5
3608,89	F II	3
3608,88	K II	5
3608,8609	Fe I	100
3608,81	C III	3
3608,70	C I	1
3608,285	Cs	10
3607,94	C I	0
3607,88	Kr II	100
3607,41	Xe II	8
3607,401	Ar II	2
3607,32	F II	3
3607,01	Xe III	40
3606,80	F II	4
3606,786	Ti I	4
3606,6821	Fe I	20
3606,5224	Ar I	1000
3605,883	Ar II	12
3605,61	Cl II	7
3605,535	Cs	4
3605,458	Fe I	15
3605,39	Cl II	5

λ	Symbol	I	λ	Symbol	I
3604,92	Cl II	3	3593,418	Ar I	—
3604,83	Xe II	3	3593,22	K II	2
3604,51	Cl II	15	3593,093	Ti II	2
3604,401	F I	2	3592,80	Xe I	2
3604,284	Ti I	8	3592,48	Cs	4
3603,96	Kr III	2	3592,00	Xe III	5
3603,95	C I	0	3591,67	Xe I	1
3603,905	Ar II	4	3590,862	C II	8
3603,88	Fe III	9	3590,63	F II	7
3603,72	Cl II	10	3590,47	Ne II	2
3603,72	F II	6	3590,465	Si III	20
3603,53	C I	2	3589,88	Xe II	1
3603,462	Ar II	4	3589,657	C II	9
3603,44	C I	1	3589,65	Kr II	70
3603,2068	Fe I	10	3589,345	F II	6
3602,852	Cs	8	3589,1063	Fe I	8
3602,85	F II	8	3588,97	Ar I	2
3602,227	Cu II	2	3588,915	C II	5
3602,12	Kr II	2	3588,62	Xe II	6
3602,10	Cl III	9	3588,448	Ar II	30
3602,032	Cu I	1400	3588,11	Ar I	3
3601,916	Al III	1	3587,980	F II	5
3601,89	Xe III	6	3587,78	Cl II	12
3601,623	Al III	20	3587,657	C II	6
3601,512	Ar II	4	3587,441	Al II	7
3601,47	C I	0	3587,42	F II	3
3601,403	F II	7	3587,405	He I	2
3600,94	Fe III	10	3587,327	Al II	2
3600,73	Cs	10	3587,270	He I	10
3600,42	Cl II	5	3587,176	Al II	1
3600,219	Ar II	3	3587,13	F II	3
3600,1694	Ne I	100	3587,130	Ti II	12
3599,90	Kr II	40	3587,057	Al II	8
3599,7116	Ar I	20	3587,02	Xe I	4
3599,448	He I	2	3586,9861	Fe I	30
3599,314	He I	5	3586,908	Al II	3,5
3599,21	Kr II	25	3586,692	Al II	2
3599,132	Cu I	1400	3586,60	K II	2
3598,97	Cs	4	3586,546	Al II	9
3598,714	Ti I	15	3586,25	Kr II	12
3598,704	F II	7	3586,114	Fe I	10
3598,04	Kr III	1	3586,04	Fe III	9
3598,011	Cu I	10	3585,809	C II	3
3597,73	Cs	6	3585,7068	Fe I	20
3597,50	Al II	2	3585,3206	Fe I	30
3597,430	Cs	10	3584,977	C II	7
3596,86	Kr II	2	3584,6627	Fe I	8
3596,048	Ti II	60	3583,64	Xe III	80
3595,917	F II	5	3582,6971	Ar I	30
3595,82	Cl II	8	3582,48	Kr III	5
3595,46	C I	0	3582,362	Ar II	20
3595,14	C I	0	3581,763	C II	3
3594,636	Fe I	3	3581,608	Ar II	18
3594,18	Ne II	4	3581,3	Cs	4
3594,103	F I	6	3581,195	Fe I	250
3594,023	Cu I	30	3580,050	Si III	3
3593,640	Ne I	300	3579,95	Kr III	2
3593,597	N II	5			
3593,5263	Ne I	500			

λ	Symbol	I
3579,69	Xe III	100
3578,357	Ar II	5
3578,25	Ti I	3
3577,60	Kr II	4
3577,23	F II	2
3576,611	Ar II	25
3576,570	Cs II	2
3576,44	Ti IV	4
3576,00	Cl II	15
3575,761	Ar II	1
3574,92	F II	3
3574,64	Ne II	5
3574,346	F I	1,5
3574,245	Ti I	8
3574,23	Ne II	0
3573,737	Ti II	20
3573,69	Cl III	2
3573,24	Cs	4
3572,68	Kr II	15
3572,2960	Ar I	300
3571,996	Fe I	6
3571,68	F II	3
3571,26	Ne II	4
3570,746	Ar II	1
3570,258	Fe I	20
3570,0996	Fe I	100
3569,940	Ar II	3
3569,68	Kr II	2
3569,673	Si III	8
3569,47	F II	2
3569,28	Cs	4
3568,53	Ne II	6
3568,04	Cl II	20
3567,72	Kr III	15
3567,6562	Ar I	300
3567	O IV	2
3566,131	Cu I	5
3566,11	Cs II	2
3566,00	Ti II	6
3565,84	Ne II	4
3565,3807	Fe I	60
3565,326	Ti II	3
3565,111	Cs II	10
3565,033	Ar II	12
3564,33	Ar II	7
3564,30	Xe II	20
3564,2955	Ar II	100
3564,23	Kr III	100
3563,87	F II	0
3563,80	Xe I	3
3563,36	O IV	2
3563,2864	Ar I	100
3563,11	Si III	2
3562,979	He I	4
3562,9551	Ne I	15
3562,50	Xe II	1
3562,194	Ar II	7
3562,15	K II	4

λ	Symbol	I
3562,09	Kr III	2
3561,75	Xe II	1
3561,575	Ti II	3
3561,38	Xe III	40
3561,23	Ne II	4
3561,031	Ar II	20
3560,68	Cl III	8
3560,42	O IV	1
3559,68	Cs II	10
3559,508	Ar II	25
3558,5170	Fe I	30
3558,51	Ti I	6
3557,84	Ne II	4
3556,90	O III	1
3556,906	Ar II	7
3556,883	Fe I	6
3556,0076	Ar I	100
3555,92	Xe I	1
3555,54	Kr II	8
3554,929	Fe I	40
3554,547	He I	1
3554,415	He I	7
3554,39	Ne II	1
3554,3056	Ar I	300
3554,04	Xe I	10
3553,741	Fe I	6
3553,58	Ar I	15
3553,49	Kr II	20
3553,366	Mg II	8
3553,35	Cl III	1
3552,13	Xe III	50
3552,00	Al II	1
3551,52	Ne II	1
3550,030	Ar II	5
3549,86	Xe I	10
3549,516	Mg II	7
3549,44	Kr I	1
3549,42	Kr III	20
3548,742	Cu II	3
3548,71	Kr II	6
3548,69	Xe II	2
3548,519	Ar II	15
3548,5	F II	0,5
3547,029	Ti I	15
3546,6	F II	0
3546,46	Kr I	3
3546,433	Cu I	15
3546,29	Xe II	1
3546,22	Ne II	1
3546,06	F II	1
3545,842	Ar II	18
3545,62	N I	2
3545,597	Ar II	18
3545,5	F II	0,5
3544,963	Cu I	125
3544,54	Kr II	30
3544,392	F II	3

λ	Symbol	I
3544,14	Kr II	30
3543,149	Ar II	7
3542,90	Ne II	7
3542,51	Ti I	3
3542,33	Xe III	50
3542,28	Ne II	2
3542,078	Fe I	15
3541,937	F II	8
3541,765	F II	9
3541,45	Cs	4
3541,44	Ti IV	3
3541,086	Fe I	15
3540,9538	Kr I	5
3540,118	Fe I	3
3539,96	Xe III	20
3539,94	Ne II	0,5
3539,5416	Kr I	5
3539,45	F II	1
3538,813	Mg II	8
3538,474	F II	3
3538,08	Xe II	2
3537,99	Ne II	3
3537,75	Ca III	7
3537,35	Xe I	1
3537,20	Kr III	2
3536,838	F II	7
3536,809	He I	3
3536,61	Xe I	1
3536,557	Fe I	15
3535,408	Ti II	40
3535,35	Kr II	50
3535,319	Ar II	18
3535,162	F II	4
3534,972	Mg II	7
3534,61	Xe II	1
3533,97	O II	00
3533,868	Ti II	2
3533,746	Cu I	500
3533,48	Xe I	2
3533,364	Cs	6
3533,202	Fe I	10
3533,043	Na II	10
3532,65	N I	4
3532,233	Ar II	1
3531,376	Cs II	4
3531,178	Ar II	2
3530,75	K II	7
3530,491	He I	5
3530,383	Cu I	2000
3530,21	Xe II	3
3530,03	Cl III	9
3529,820	Fe I	6
3529,53	K II	3
3528,51	K II	1
3527,797	Fe I	4
3527,482	Cu I	500
3527,42	Kr II	3
3526,1676	Fe I	15
3526,13	Cl II	30
3525,939	Si III	9
3525,161	Ti I	3
3524,78	Kr III	5
3524,231	Cu I	1250
3522,883	F II	6
3522,83	Xe III	80
3522,72	Ne II	1
3522,6747	Kr I	15
3522,14	Cl II	40
3521,977	Ar II	4
3521,555	Ar II	1
3521,263	Ar II	12
3521,2630	Fe I	25
3521,11	Kr III	4
3520,9	O IV	—
3520,4714	Ne I	1000
3520,253	Ti II	20
3520,031	Cu I	500
3519,996	Ar II	15
3518,31	C I	0
3518,15	Cs	6
3517,90	Xe I	2
3517,894	Ar II	6
3517,37	Kr II	5
3517,317	He I	2
3517,039	Cu I	100
3516,92	O II	0
3516,838	Ti I	3
3516,03	Cs	4
3515,602	Mg I	4
3515,1900	Ne I	200
3514,80	C I	2
3514,58	Xe II	8
3514,55	Kr III	15
3514,388	Ar II	20
3514,18	Ar III	6
3514,022	Cs	6
3513,88	K III	5
3513,8196	Fe I	30
3513,69	Cl II	12
3513,22	Cl II	35
3512,512	He I	4
3512,121	Cu I	650
3511,985	Cu I	10
3511,8963	Kr I	4
3511,835	Cu I	50
3511,69	Ar III	5
3511,626	Ti I	3
3511,12	Ar III	8
3510,840	Ti II	60
3510,7207	Ne I	50
3509,844	Ti II	3
3509,783	Ar II	10
3509,39	Cl II	40
3509,33	Ar III	5
3509	O VI	—

λ	Symbol	I	λ	Symbol	I
3508,94	Cl II	12	3497,89	Xe III	4
3508,88	Xe II	20	3497,8420	Fe I	40
3508,42	Xe I	2	3497,45	Kr II	3
3507,84	Kr I	3	3497,13	Kr III	10
3507,42	Kr III	200	3497,108	Fe I	10
3507,407	Cu I	5	3497,10	Ar III	4
3506,74	Xe I	5	3496,86	Xe I	1
3506,66	Kr I	3	3496,27	O II	1
3506,643	Ti I	8	3495,9900	Kr I	10
3506,56	Xe II	15	3495,775	Ar II	1
3506,4807	Ar I	30	3495,754	Ti I	6
3506,02	O II	0	3495,44	O II	0
3505,763	F II	4	3495,2879	Fe I	8
3505,614	F II	15	3495,156	Ca II	1
3505,508	F II	6	3494,672	Fe II	5
3505,44	Cl II	12	3494,66	O II	00
3504,890	Ti II	80	3493,57	Kr II	2
3504,85	Cs	4	3493,474	Fe II	10
3504,25	Xe II	1	3493,280	Ti I	4
3503,8981	Kr I	15	3493,2747	Ar I	20
3503,67	Cs	4	3493,215	F II	5
3503,61	Ne II	5	3493,04	Kr II	8
3503,58	Ar III	15	3492,80	Kr III	8
3503,25	Kr II	50	3492,39	Ti II	3
3503,15	Xe II	15	3492,24	O IV	0
3503,095	F II	12	3491,538	Ar II	25
3502,954	F II	8	3491,243	Ar II	20
3502,859	F II	4	3491,053	Ti II	10
3502,70	Ar III	6	3490,884	Ar II	8
3502,5537	Kr I	20	3490,685	He I	2
3502,379	He I	2	3490,5749	Fe I	100
3502,2	O IV	—	3490,50	Ar I	3
3501,77	Xe II	20	3489,84	O IV	1
3501,76	Fe III	8	3489,739	Ti II	2
3501,67	O II	00	3488,858	Cu I	100
3501,562	F II	5	3488,65	Kr II	30
3501,529	Cu I	3	3488,59	Kr III	100
3501,487	F II	3	3488,188	Ar II	1
3501,416	F II	10	3488,18	O II	0
3501,251	Cu I	5	3487,723	He I	2
3501,2154	Ne I	200	3487,598	Ca I	12
3500,58	Ar III	5	3487,566	Cu I	60
3500,5	O II	00	3487,49	Kr II	7
3500,36	Xe II	30	3487,318	Ar II	3
3500,340	Ti II	2	3486,911	Si III	15
3500,324	Cu I	50	3485,689	Ti I	6
3500,28	Fe III	7	3485,3418	Fe I	7
3499,67	Ar III	12	3485,23	Xe II	1
3499,59	Fe III	7	3485,08	Kr III	1
3499,481	Ar II	7	3484,96	N IV	13
3499,099	Ti I	8	3484,12	Ar III	3
3498,938	Cu I	3	3483,80	Ti II	4
3498,92	Kr II	2	3483,761	Cu I	1250
3498,645	He I	3	3483,17	Ar I	14
3498,50	Kr II	4	3482,99	N IV	5
3498,31	Ar III	6	3482,628	Al I	5
3498,0632	Ne I	100	3482,21	Xe II	2
3498,063	Cu I	125			

λ	Symbol	I	λ	Symbol	I
3481,96	Ne II	6	3470,81	O II	8
3481,675	Ti I	3	3470,42	O II	5
3481,614	Cu I	5	3470,264	Ar II	4
3481,136	Ti I	3	3470,05	Kr II	30
3481,11	K II	6	3469,81	Cs	4
3481,11	K III	6	3469,81	Xe I	4
3480,75	Ne II	2	3468,680	Fe II	8
3480,55	Ar III	20	3468,476	Ca I	4
3480,525	Ti I	12	3468,32	K III	6
3480,511	Ar II	9	3468,19	Xe III	40
3480,063	Cs I	50	3467,260	Ti I	6
3479,82	Cl II	30	3467,20	Xe III	25
3479,806	Al I	5	3466,90	O III	0
3479,53	Ne II	1	3466,5781	Ne I	200
3479,25	Cs	4	3466,343	Ar II	8
3479,11	Xe III	1	3466,24	Cu I	25
3479,00	Kr II	3	3466,15	O III	2
3478,957	He I	2	3465,8621	Fe I	60
3478,918	Ti I	6	3465,787	Ar II	4
3478,71	N IV	15	3465,562	Ti II	3
3478,236	Ar II	6	3465,41	Kr II	6
3477,89	Kr II	5	3465,401	Cu I	50
3477,69	Ne II	3	3465,20	Cs	4
3477,181	Ti II	15	3465,15	Ar I	2
3476,814	Cs I	100	3464,3385	Ne I	100
3476,749	Ar II	20	3464,17	Xe II	1
3476,7036	Fe I	40	3464,132	Ar II	10
3476,452	Ti I	3	3464,08	Ar I	1
3476,2	F II	0	3463,499	Cu I	5
3475,999	Cu I	750	3463,425	Cs	6
3475,973	Cs II	2	3463,37	N IV	6
3475,68	F II	2	3462,81	Xe II	1
3475,651	Fe I	6	3462,494	Na II	3
3475,4511	Fe I	70	3461,871	Ca II	2
3475,31	Kr II	3	3461,496	Ti II	20
3475,26	O III	—	3461,36	N IV	2
3475,25	Ne II	1	3461,26	Xe II	100
3474,94	O II	1	3461,0785	Ar I	300
3474,800	F II	7	3460,5235	Ne I	100
3474,763	Ca I	8	3460,13	Kr I	2
3474,65	Kr III	70	3460,09	Kr II	50
3474,578	Cu I	5	3460,08	Xe II	8
3474,55	N IV	3	3459,98	O III	2
3474,23	Xe II	20	3459,52	O III	0
3473,621	F II	2	3459,428	Cu I	25
3473,314	F II	5	3459,38	Ne II	2
3472,964	F II	6	3459,185	Cs II	15
3472,61	Ar III	6	3459,07	O II	0
3472,5706	Ne I	500	3458,50	C I	1
3472,36	Xe I	4	3458,216	Al I	6
3472,141	Cu I	200	3458,020	Ti I	3
3471,818	He I	1	3457,99	O II	1
3471,748	Cu I	2	3457,850	Cu I	750
3471,600	Ar II	3	3457,85	K II	2
3471,3460	Fe I	6	3457,81	Ar I	3
3471,32	Ar III	9	3457,494	Ti I	4
3471,02	Kr III	3	3457,18	Cs	4
3470,92	Cs	4			

λ	Symbol	I	λ	Symbol	I
3457,16	Ne II	4	3444,23	Xe III	60
3456,928	Fe II	5	3444,10	O III	5
3456,87	Kr I	3	3443,88	Cs	4
3456,68	Ne II	4	3443,8775	Fe I	40
3456,661	Ti I	6	3443,83	Xe I	1
3456,390	Ti II	20	3443,70	Ne II	2
3455,48	Cs	4	3443,644	Ti I	5
3455,12	O III	5	3443,640	Al I	9
3454,944	Ar I	20	3443,59	N IV	3
3454,90	Kr I	1	3443,29	Kr II	5
3454,90	O III	2	3442,86	Kr III	6
3454,83	Ne	1	3442,66	Xe I	3
3454,70	N IV	2	3442,58	Ar I	10
3454,686	Cu I	200	3442,5	F II	0
3454,25	Xe III	70	3442,12	Ne II	1
3454,1942	Ne I	100	3440,9899	Fe I	75
3454,098	Ar II	12	3440,80	Ne II	1
3453,8	F II	0	3440,75	Xe II	4
3453,616	Mg I	3	3440,6069	Fe I	150
3453,46	Kr II	3	3440,507	Cu I	250
3453,31	O II	0	3440,39	O III	4
3453,10	Ne II	3	3440,37	Si III	5
3452,657	Al I	2	3440,05	K II	7
3452,657	Ca II	1	3439,46	Kr III	100
3452,470	Ti II	4	3439,347	Al I	6
3452,32	Ar I	3	3439,305	Ti I	8
3452,2760	Fe I	10	3439,242	Si III	3
3451,9166	Fe I	10	3439,094	Ar II	1
3451,33	O III	1	3438,97	Ne II	2
3450,94	O III	4	3438,88	Kr II	3
3450,7641	Ne I	50	3438,04	Ar III	8
3450,36	Cs	6	3438	O VI	—
3450,332	Cu I	750	3437,73	Xe II	3
3450,3304	Fe I	10	3437,147	N II	9
3449,52	Ar I	2	3437,14	N I	4
3448,71	Kr III	10	3436,57	F II	4
3448,281	Ar II	1	3436,57	F III	4
3448,14	Cl II	4	3436,543	Cu I	5
3448,05	O III	0	3436,48	Xe II	1
3447,98	O II	1	3436,112	Fe II	5
3447,7022	Ne I	200	3435,78	Xe III	4
3447,590	Cu I	3	3435,773	Ar II	1
3447,586	He I	15	3435,40	Ti I	3
3447,375	K I	10	3434,1423	Kr I	8
3447,290	Ar II	1	3434	O VI	—
3447,2797	Fe I	8	3433,972	Cu I	3
3447,22	O III	1	3433,69	F II	2
3446,85	Kr III	8	3433,369	Ar II	1
3446,73	O III	2	3433	O VI	—
3446,51	Kr II	50	3432,585	Ar II	3
3446,372	K I	11	3432,49	Xe II	1
3446,34	Xe II	25	3431,737	Ar II	2
3445,20	N IV	2	3431,7217	Kr I	20
3445,1508	Fe I	20	3431,45	Kr I	2
3444,865	Al I	6	3431,03	Kr II	8
3444,409	Mg I	2	3430,990	Ar II	3
3444,403	Ti I	3	3430,60	O III	4
3444,306	Ti II	30			

λ	Symbol	I
3430,417	Ar II	9
3430,4	Cs	4
3430,03	Ar III	2
3429,91	Kr II	3
3429,617	Ar II	7
3429,49	Cs II	3
3428,956	Ti I	4
3428,916	Al II	6
3428,83	Kr III	10
3428,76	Ne II	5
3428,67	O III	3
3428,1948	Fe I	8
3427,71	Kr II	30
3427,42	O III	3
3427,13	K II	2
3427,1213	Fe I	20
3426,862	Na I	6
3426,34	F III	2
3426,27	Kr I	2
3426	O VI	—
3425,57	O IV	0
3424,9433	Kr I	15
3424,43	Ar IV	—
3424,2861	Fe I	10
3424,25	Ar III	9
3423,9120	Ne I	50
3423,73	Kr II	20
3422,6583	Fe I	7
3422,10	Cu I	15
3421,83	K II	4
3421,83	K III	4
3421,615	Ar II	8
3420,82	K III	6
3420,73	Xe II	40
3420,61	O II	3
3420,41	C I	0
3420,166	Cu I	8
3420,00	Xo I	2
3419,87	O II	2
3418,512	Fe I	10
3418,51	Ar I	3
3418,37	Xe I	2
3418,11	Cs	6
3418,007	Ne I	50
3417,9031	Ne I	500
3417,8428	Fe I	12
3417,71	Ne II	5
3417,68	Ar I	4
3417,49	Ar III	7
3417,21	F II	4
3417,04	Xe II	1
3417,02	F II	6
3416,957	Ti II	2
3416,87	Ne II	4
3416,80	Ar I	5
3416,80	F II	1
3416,76	K II	2
3416,58	FII	4
3416,560	Ar II	1
3416,45	F II	4
3416,021	Fe II	5
3415,993	Ti I	5
3415,80	Cu I	200
3415,29	O III	3
3414,82	Ne II	2
3414,80	Kr II	10
3414,663	F II	5
3414,462	Ar II	4
3414,017	Cu I	5
3413,71	O IV	1
3413,53	Ar III	6
3413,343	Cu I	200
3413,20	Xe II	6
3413,1339	Fe I	15
3413,13	Ne II	7
3413,107	Cu I	10
3412,80	Kr I	1
3412,67	Kr II	1
3412,04	F II	2
3411,76	O IV	4
3411,67	Ti I	5
3411,66	F II	3
3411,66	F III	3
3411,38	Ne II	1
3411,313	Cs	10
3410,82	F II	1,5
3409,92	Cl II	5
3409,89	Kr I	2
3409,84	O II	6
3409,809	Ti II	4
3409,75	O IV	2
3409,699	Ar II	2
3409,49	Xe II	8
3409,413	Ar II	1
3409,02	F II	3
3408,97	Kr I	2
3408,68	F II	1,5
3408,612	Ar II	2
3408,13	O III	1
3408,127	N II	5
3407,4611	Fe I	20
3407,38	O II	7
3407,205	Ti II	3
3406,88	Ne II	5
3406,83	F II	1
3406,8021	Fe I	6
3406,626	Cs	10
3406,56	F II	2
3406,361	C II	2
3406,298	Ar II	3
3406,1804	Ar I	30
3405,980	F II	4
3405,97	O IV	—
3405,89	Cl II	3
3405,74	O III	2

λ	Symbol	I	λ	Symbol	I
3405,16	Kr II	80	3392,89	Cl III	8
3405,094	Ti I	5	3392,7812	Ar I	100
3404,77	Ne II	4	3392,78	Ne II	5
3404,66	Cu I	125	3392,713	Ti I	10
3404,3557	Fe I	6	3392,6540	Fe I	15
3404,33	C II	1	3392,63	K II	3
3404,24	K II	6	3392,31	Ar I	3
3403,89	Xe III	8	3392,3058	Fe I	8
3403,66	C II	1	3392,146	C II	2
3403,58	O IV	3	3392,016	Cu I	8
3403,369	Ti I	4	3391,86	Ar IV	—
3403,107	Cu I	5	3391,85	Ar III	15
3402,79	Kr II	2	3390,682	Ti I	10
3402,422	Ti II	8	3390,56	Ne II	2
3402,262	Fe I	5	3390,37	O IV	—
3402,244	Cu I	225	3390,29	Ar I	3
3401,62	F III	2	3390,25	O II	8
3401,53	C II	0	3389,854	Ar I	20
3401,5200	Fe I	6	3389,67	Kr II	5
3401,40	Kr I	5	3389,15	Cs	6
3400,79	Xe I	1	3388,93	Kr III	20
3400,15	Cl III	2	3388,755	Ti II	8
3400,162	Ti I	3	3388,533	Ar II	10
3400,110	Na II	2	3388,46	Ne II	6
3400,07	Xe I	2	3388,365	Ar I	20
3399,983	Cs I	30	3388,134	Fe II	12
3399,71	O III	2	3388,07	Cu I	8
3399,37	Xe II	1	3388,05	Xe II	2
3399,3356	Fe I	15	3387,834	Ti II	50
3399,29	F II	3	3387,600	Ar I	20
3398,78	F II	1	3387,60	Cl III	6
3398,634	Ti I	8	3387,11	Kr II	7
3397,969	Cs I	60	3386,30	Xe II	2
3397,920	Ar I	20	3386,24	Ne II	2
3397,900	Ar II	5	3386,22	Cl III	5
3397,90	Ne II	1	3385,944	Ti I	40
3397,608	Ar II	1	3385,664	Ti I	12
3397,187	Cs	6	3385,55	O IV	6
3397,002	Ar II	1	3385,394	Cu I	2
3396,9774	Fe I	4	3385,23	Kr II	15
3396,83	O IV	2	3384,95	O III	4
3396,70	Fe III	8	3384,948	Cu II	3
3396,63	F II	1	3384,86	K II	6
3396,60	Cs II	2	3384,80	Cu I	15
3396,58	Kr III	15	3384,36	Xe I	1
3396,324	Cu I	10	3384,13	Xe II	40
3395,77	F II	1	3383,9808	Fe I	8
3395,50	Xe II	3	3383,98	Ar I	2
3395,476	Cu I	60	3383,865	Ar II	1
3394,5854	Fe I	5	3383,85	O III	2
3394,574	Ti II	40	3383,761	Ti II	125
3394,26	O III	1	3383,20	Xe I	1
3394,22	F II	2	3382,69	O III	3
3393,946	C II	1	3382,312	Ti I	15
3393,7522	Ar I	250	3382,18	Fe III	6
3393,45	Cl III	8	3382,133	Ar II	3
3393,40	F II	1	3381,49	Ar I	20
3393,35	Ar IV	—	3381,421	Cu I	200
3393,25	Cs	3			

λ	Symbol	I	λ	Symbol	I
3381,34	Xe II	1	3372,88	Ar I	3
3381,33	O IV	—	3372,800	Ti II	100
3381,28	O IV	4	3372,68	Ca III	8
3381,124	Cu I	60	3372,24	F III	1
3381,11	Kr II	20	3372,208	Ti II	10
3381,063	Ar II	1	3371,87	Ne II	4
3380,717	Cu II	10	3371,85	O II	2
3380,62	K II	6	3371,447	Ti I	80
3380,278	Ti II	30	3371,412	Cu II	8
3380,1117	Fe I	8	3371,38	O IV	4
3379,961	Cu II	3	3370,925	Ar II	8
3379,864	Cu I	3	3370,7852	Fe I	10
3379,653	Cu I	5	3370,65	Xe III	4
3379,577	Ar II	4	3370,457	Cu II	30
3379,458	Ar II	4	3370,436	Ti I	40
3379,39	Ne II	1	3370,34	Xe I	1
3379,29	F II	2	3370,23	O II	00
3379,216	Ti I	15	3369,9069	Ne I	700
3379,03	Kr II	15	3369,8076	Ne I	500
3379,0206	Fe I	6	3369,549	Fe I	8
3379,02	Xe III	5	3369,40	O III	00
3379,0	Cs	6	3369,212	Ti II	2
3378,707	Cu I	2	3368,84	Ar I	1
3378,685	Fe I	6	3368,555	Cs II	30
3378,512	Cu II	3	3367,881	Ti I	3
3378,442	Ar II	1	3367,81	Ca III	5
3378,28	Ne II	5	3367,65	F III	1
3378,09	O IV	0	3367,36	N III	7
3377,706	Cu II	5	3367,20	Ne II	6
3377,577	Ti I	30	3367,00	O II	00
3377,485	Ti I	20	3366,72	Xe II	300
3377,44	F II	4	3366,586	Ar II	6
3377,23	Ne II	1	3366,560	Cu II	5
3377,20	O II	7	3366,269	Cu II	10
3377,09	Xe III	2	3366,176	Ti I	5
3376,82	O III	1	3366,176	Ti II	8
3376,66	O III	2	3365,79	N III	3
3376,443	Ar II	12	3365,65	Cu II	15
3376,261	Cs II	2	3365,536	Ar II	8
3375,78	Kr II	3	3365,342	Cu I	750
3375,77	O II	0	3364,52	Cs	4
3375,706	Ti I	3	3364,362	Ar II	2
3375,672	Cu I	30	3364,22	K III	6
3375,6489	Ne I	50	3363,83	O III	1
3375,50	O IV	3	3363,47	Ar I	20
3375,18	Cu I	8	3363,46	O IV	6
3375,16	Xe II	3	3363,300	Ar II	2
3374,96	Kr III	40	3362,89	Ne II	2
3374,953	Cu II	20	3362,81	Xe III	3
3374,77	O II	00	3362,63	O IV	—
3374,352	Ti II	8	3362,38	O III	4
3374,10	Ne II	3	3362,28	Ca I	0
3374,06	N III	6	3362,131	Ca I	35
3373,92	Xe II	2	3362,12	Cu I	2
3373,842	Ar II	3	3362,100	Ti I	3
3373,60	K II	6	3361,918	Ca I	35
3373,594	Cu II	15	3361,90	N III	2
3373,49	F II	5	3361,835	Ti I	10
3373,4823	Ar I	300			

λ	Symbol	I
3361,752	Ar II	6
3361,74	Kr I	2
3361,721	C II	6
3361,28	Ar III	7
3361,263	Ti I	40
3361,213	Ti II	125
3361,051	C II	8
3360,990	Ti I	10
3360,891	C II	3
3360,87	Fe III	6
3360,63	Ne II	5
3360,15	O II	00
3359,48	Ar I	10
3358,96	Xe I	1
3358,8	Cs	4
3358,74	Cu I	2
3358,72	N III	1
3358,49	Ar III	15
3358,479	Ti I	8
3358,32	F III	4
3358,271	Ti I	10
3358,27	Cu I	2
3358,17	Xe I	1
3357,98	Xe III	30
3357,90	Ne II	3
3357,82	F III	1
3357,687	Cs	6
3357,58	Kr II	2
3357,19	C II	0
3356,51	K II	2
3356,35	Ne II	2
3355,98	F III	2
3355,92	O III	3
3355,690	C II	1
3355,47	N III	2
3355,2287	Fe I	6
3355,05	Ne II	7
3354,634	Ti I	60
3354,550	He I	10
3354,474	Cu I	60
3354,34	F III	2
3354,31	O IV	—
3354,29	N III	4
3353,88	Cs	4
3353,78	N III	4
3353,63	Ne II	2
3353,466	Cu I	10
3353,39	Cl II	125
3353,302	C II	2
3352,937	Ti I	6
3352,20	Ar I	1
3352,11	Ar III	4
3352,071	Ti II	5
3352,044	Cu II	8
3351,93	Kr III	100
3351,744	Ne I	25
3351,456	Al II	3
3350,933	Ar II	12
3350,99	O III	4
3350,68	O III	3
3350,44	Xe II	6
3350,361	Ca I	25
3350,209	Ca I	25
3350,07	Cl II	4
3349,76	Xe III	12
3349,463	Cu II	5
3349,445	Cs	10
3349,399	Ti II	125
3349,279	Cu I	450
3349,11	O IV	4
3349,035	Ti II	75
3348,844	Ti II	10
3348,825	Cs I	15
3348,63	Xe I	1
3348,535	Ti I	5
3348,43	F II	0,5
3348,17	Kr III	10
3348,08	O IV	2
3348,05	O III	2
3347,9271	Fe I	6
3347,70	Fe III	8
3347,694	Ar II	1
3347,50	Kr I	2
3347,494	Cs I	30
3347,27	Xe II	3
3347,035	Ca II	1
3346,724	Ti II	15
3346,717	Si III	2
3346,41	F II	0,5
3345,88	Ne II	1
3345,73	Kr I	4
3345,49	Ne II	3
3345,32	K II	6
3345,00	Cs II	2
3344,97	Xe II	4
3344,97	Xe III	4
3344,72	Ar III	20
3344,513	Ca I	8
3344,43	Ne II	5
3344,26	O III	2
3344,004	Cs	10
3343,770	Ti II	10
3343,743	Cu II	20
3342,77	Cu I	5
3342,77	N III	1
3342,48	Kr III	50
3342,454	Cu I	5
3342,151	Ti I	6
3341,905	Fe I	5
3341,875	Ti I	50
3341,875	Ti II	100
3341,746	Ar II	6
3341,507	Ar II	3
3340,74	O III	6
3340,574	Cs	10
3340,5666	Fe I	6
3340,42	Cl III	9
3340,344	Ti II	35
3340,06	Xe III	10
3340,04	Xe I	1
3339,819	Si II	500

λ	Symbol	I
3339,38	Fe III	10
3339,084	Cu II	3
3338,98	Xe III	25
3338,937	Cu II	3
3338,828	Ar II	5
3338,80	Xe II	4
3338,647	Cu II	10
3337,85	Ti II	2
3337,845	Cu I	1500
3337,67	K II	1
3337,6664	Fe I	6
3337,20	Cl II	3
3337,17	Kr I	1
3337,116	Ar II	2
3336,78	O III	3
3336,674	Mg I	20
3336,16	Cl III	5
3336,13	Ar III	25
3336,12	Ne II	2
3335,7699	Fe I	4
3335,215	Cu I	400
3335,192	Ti II	40
3335,16	Kr II	4
3334,87	Ne II	10
3334,47	Kr I	1
3334,26	Xe III	1
3334,2201	Fe I	4
3333,84	Ar I	2
3333,64	Cl II	40
3333,139	Si II	300
3333,00	O III	4
3332,50	Kr III	10
3332,49	O III	1
3332,47	Kr I	1
3332,42	Cl II	15
3332,146	Mg I	19
3332,111	Ti II	30
3331,65	Xe III	40
3331,613	Fe I	4
3331,310	N II	6
3330,78	Ne II	2
3330,76	Kr III	60
3330,40	O III	4
3330,33	Cs	4
3330,314	N II	5
3329,919	Mg I	17
3329,89	Fe III	7
3329,704	N II	5
3329,636	Cu I	225
3329,455	Ti II	70
3329,428	Cs II	10
3329,20	Ne II	4
3329,12	Cl II	150
3329,06	Cl III	8
3328,8667	Fe I	5
3328,730	N II	7
3328,00	Kr I	2
3327,685	Na II	4
3327,46	Xe II	15
3327,34	Ar III	4
3327,16	Ne II	5
3326,762	Ti II	20
3326,16	O III	0
3326,13	Kr II	1
3325,812	Cu I	3
3325,812	Cu II	8
3325,75	Kr III	200
3325,5006	Ar I	100
3325,462	Fe I	4
3325,328	Cu I	3
3325,229	Ti I	3
3325,155	Ti I	3
3324,78	Ar IV	—
3324,754	Ti I	4
3324,573	N II	5
3324,5385	Fe I	4
3324,5	Cs	4
3324,228	Ar II	2
3324,13	F II	0,5
3323,825	Ar I	30
3323,803	Ti I	4
3323,75	Ne II	7
3323,7375	Fe I	7
3323,735	Cu II	5
3323,59	Ar III	9
3323,068	Fe II	8
3322,936	Ti II	75
3322,8	Cs	4
3322,477	Fe I	5
3322,44	Ar I	5
3322,40	K III	6
3321,700	Ti II	25
3321,588	Ti I	8
3321,58	Ar I	5
3321,578	Si III	4
3321,30	F II	2
3321,16	Kr II	8
3320,67	Ar I	2
3320,57	Cl III	7
3320,375	Ar II	2
3320,29	Ne II	2
3320,14	Cl II	30
3320,06	Ar I	3
3319,75	Ne II	3
3319,682	Cu I	150
3319,53	Xe III	2
3319,3446	Ar I	300
3318,362	Ti I	4
3318,098	N II	5
3318,032	Na II	4
3318,024	Ti II	10
3317,825	Ar II	3
3317,54	Ar I	1
3317,44	Xe III	2
3317,218	Cu I	750
3317,140	Cu II	5
3317,121	Fe I	4
3316,86	Cl II	50

λ	Symbol	I	λ	Symbol	I
3316,39	Xe II	6	3305,22	Fe III	10
3316,279	Cu II	20	3305,2	Li II	4
3315,72	Kr II	15	3305,15	O II	6
3315,614	Al II	1	3304,950	Na II	0
3315,498	Cs	10	3304,75	Kr III	30
3315,44	Cl II	100	3303,89	F II	6
3315,324	Ti II	10	3303,72	Cs	4
3314,889	Al II	2	3303,516	Cu II	5
3314,87	Xe III	10	3302,979	Na I	18
3314,7420	Fe I	7	3302,787	Cu I	4
3314,60	Ne II	1	3302,54	Kr I	10
3314,523	Ti I	8	3302,369	Na I	19
3314,49	Ar I	2	3302,28	Kr II	4
3314,422	Ti I	10	3301,88	Ar III	20
3314,30	Xe III	1	3301,75	Kr II	5
3314,059	Cs I	5	3301,71	Ti II	2
3314	O IV	—	3301,60	K II	3
3313,48	Xe II	2	3301,60	Xe III	20
3313,351	Al II	3	3301,56	O II	3
3313,124	Cs I	10	3301,41	F II	3
3312,936	Ar II	3	3301,346	Na II	2
3312,78	Cl II	15	3301,228	Cu II	40
3312,690	Ti I	5	3300,95	Cl III	3
3312,30	O III	5	3300,885	Cu II	20
3311,80	Xe II	2	3300,644	Cu II	10
3311,63	F II	1	3300,444	Cu II	5
3311,52	Cs	4	3300,39	Ar I	20
3311,47	Kr III	50	3300,18	Kr II	4
3311,30	Ne II	3	3299,86	Cs	6
3311,25	Ar III	15	3299,413	Ti I	10
3310,987	Cu I	8	3299,36	O III	3
3310,85	Xe II	1	3299,26	Ar I	2
3310,55	Ne II	1	3299,050	Mg I	4
3310,47	Ar I	3	3299,02	Ar I	1
3310,38	Xe II	3	3298,72	Xe II	6
3310,342	Fe I	4	3298,7	O V	—
3309,78	Ne II	3	3298,418	Ar II	2
3309,730	Ti I	6	3298,1331	Fe I	6
3309,558	Cu I	4	3297,888	Fe II	5
3309,501	Ti I	15	3297,74	Ne II	7
3309,39	Xe II	2	3297,199	Cu II	10
3309,343	Ar II	5	3297,020	Ar II	2
3308,806	Ti II	8	3296,773	He I	7
3308,73	Kr III	1	3296,56	F II	5
3308,391	Ti I	10	3296,19	F II	2
3308,16	Kr III	20	3295,814	Fe II	6
3307,948	Cu I	2500	3295,29	Kr II	3
3307,90	Cl II	50	3295,13	O II	4
3307,229	Ar II	9	3295,103	Cu II	15
3306,879	Ti I	10	3294,903	Ti I	6
3306,80	Xe III	10	3294,37	F II	4
3306,60	O II	6	3294,336	Cu II	3
3306,498	Fe I	6	3294,168	Cu I	5
3306,45	Cl II	40	3293,921	Ar II	9
3306,445	Ar II	5	3293,88	Kr III	4
3306,354	Fe I	20	3293,815	Cu I	2
3306,17	Kr I	7	3293,641	Ar II	10
3305,9719	Fe I	20	3293,334	Cu II	2
3305,77	O III	0	3292,965	Cu I	450
3305,530	Cu I	4			

λ	Symbol	I
3292,827	Cu I	650
3292,5910	Fe I	8
3292,393	Cu I	125
3292,21	Kr III	1
3292,124	Cu II	10
3292,078	Ti I	20
3292,04	Fe III	8
3292,023	Fe I	8
3291,441	Ar II	6
3290,9899	Fe I	5
3290,65	K II	5
3290,541	Cu I	1500
3290,422	Cu II	50
3290,13	O II	5
3289,95	Ar I	3
3289,80	Cl III	7
3289,39	Ar I	3
3289,347	Fe II	7
3289,290	Cs I	2
3289,06	K III	6
3288,81	Fe III	15
3288,605	Cs I	4
3288,575	Ti II	5
3288,428	Ti II	5
3287,92	Xe III	30
3287,69	Kr II	2
3287,64	Ti II	40
3287,59	O II	9
3287,38	Kr II	2
3287,37	Al III	1
3286,7541	Fe I	20
3286,193	Cu I	2
3286,067	Ca I	4
3285,89	Kr III	30
3285,89	Xe III	10
3285,85	Ar III	25
3285,603	Na II	8
3285,25	Kr III	3
3285,10	Ar I	2
3284,70	Xe III	3
3284,5888	Fe I	5
3284,57	O III	4
3283,74	Ar I	1
3283,41	Cl III	6
3283,114	Al III	0,5
3282,716	Cu I	1400
3282,70	Ar I	1
3282,329	Ti II	20
3282,1	Cs	3
3282,08	Kr II	15
3281,94	O III	3
3281,703	Ar II	12
3281,696	Cu II	10
3281,300	Fe II	7
3281,26	Xe II	12
3280,59	Kr I	1
3280,56	Fe III	6
3280,48	Xe II	8
3280,2613	Fe I	8
3279,995	Ti II	4
3279,97	O III	1
3279,937	Ar II	4
3279,815	Cu I	2000
3279,42	Kr III	2
3279,258	Si III	7
3279,25	Ar I	3
3278,93	Ar I	3
3278,922	Ti I	12
3278,922	Ti II	35
3278,79	K III	6
3278,734	Fe I	4
3278,48	Xe III	8
3278,290	Ti II	30
3278,26	Cs	4
3277,69	O II	7
3277,346	Fe II	9
3277,310	Cu I	650
3276,81	Cl II	40
3276,774	Ti II	5
3276,606	Fe II	5
3276,4713	Fe I	4
3276,39	Xe III	8
3276,264	Si III	10
3276,085	Ar II	3
3276,08	Fe III	15
3275,776	Al II	4
3275,72	Ar I	2
3275,68	Cs	4
3275,67	O V	0
3275,639	Ar II	4
3275,293	Ti II	3
3275,20	Ne II	2
3274,94	Xe II	4
3274,661	Ca I	2
3274,220	Na II	5
3274,047	Ti I	5
3273,957	Cu I	10000
3273,53	Fe III	6
3273,52	O II	7
3273,316	Ar II	6
3272,91	Xe II	60
3272,080	Ti II	25
3271,652	Ti II	25
3271,65	Kr III	30
3271,626	Cs II	20
3271,16	Ar I	10
3271,0014	Fe I	15
3270,980	Cs I	1
3270,98	O II	7
3270,79	Ne II	2
3270,562	Ti I	3
3270,477	Cs I	2
3270,474	Ar II	5
3270,456	Si III	6
3269,86	Ne II	3
3269,090	Ca I	1
3268,987	Ar II	5
3268,96	Xe III	80

λ	Symbol	I
3268,48	Kr III	100
3268,314	Cs	10
3268,278	Cu I	650
3268,236	Fe I	5
3268,08	Xe II	1
3267,34	Xe II	3
3267,31	O III	5
3267,22	Ne	1
3267,202	F III	4
3267,135	Cs II	30
3267,05	Xe II	6
3266,88	Fe III	20
3266,34	Ar I	1
3266,08	Xe II	4
3266,023	Cu I	650
3265,924	Cs II	30
3265,6182	Fe I	15
3265,46	O III	10
3265,45	Cl III	0
3265,37	Ne	3
3265,0473	Fe I	8
3264,81	Kr III	150
3264,7	O V	—
3264,33	Kr II	5
3264,29	Ar I	3
3264,164	F III	9
3264,16	F II	7
3263,982	Cs II	5
3263,78	Ar I	3
3263,686	Ti II	4
3263,572	Ar II	12
3263,43	Ne II	3
3263,12	Kr II	1
3263,06	Cs	4
3262,46	Fe III	6
3262,29	Cs	6
3262,272	C III	3
3262,083	Ar II	2
3262,02	Xe II	4
3261,70	Kr IV	3
3261,596	Ti II	60
3261,58	Kr II	8
3260,98	O III	8
3260,87	Ne	3
3260,73	Xe II	2
3260,259	Ti I	3
3260,259	Ti II	3
3260,218	Na II	3
3259,994	Fe I	6
3259,656	Ar II	6
3259,541	C III	2
3259,36	Xe II	12
3259,32	Cl III	6
3259,048	Fe II	10
3258,894	Ar II	2
3258,81	K II	3
3258,773	Fe II	10
3258,664	Si III	12
3258,275	He I	5
3258,00	C III	1
3258,00	Kr I	1
3257,965	Na II	6
3257,5940	Fe I	8
3257,585	Ar I	100
3257,10	Kr I	1
3256,67	Kr II	4
3256,20	Ar I	2
3255,890	Fe II	8
3255,39	Ne II	2
3255,35	Cs	10
3254,800	Si III	7
3254,3628	Fe I	10
3254,250	Ti II	30
3253,98	K II	3
3253,918	Ar II	3
3253,741	Si III	5
3253,602	Fe I	4
3253,43	F III	2,5
3253,401	Si III	7
3253,117	Si III	4
3252,94	O III	2
3252,914	Ti II	40
3252,437	Fe II	5
3252,220	Cu I	650
3251,911	Ti II	30
3251,235	Fe I	8
3250,949	Na II	3
3250,58	Cs	6
3250,56	Xe II	25
3250,469	Cu II	10
3250,34	Ne	1
3250,04	Xe II	2
3249,801	Ar II	15
3249,8	Li II	5
3249,7	O V	—
3249,370	Ti II	2
3249,35	Xe II	1
3248,602	Ti I	15
3248,602	Ti II	50
3248,206	Fe I	10
3248,15	Ne II	3
3248,03	Kr II	6
3247,74	Xe II	6
3247,540	Cu I	10000
3247,5	Cs	4
3247,481	Ar II	3
3247,213	Fe II	9
3247,00	Kr II	12
3246,962	Fe I	6
3246,84	Xe III	10
3246,62	Kr III	5
3246,005	Fe I	8
3245,69	Kr III	300
3245,05	Cl III	2
3244,44	Cl III	5
3244,192	Si IV	1
3244,190	Fe I	5
3244,15	Ne II	5
3243,803	Ti I	4
3243,724	Fe II	8

λ	Symbol	I
3243,689	Ar II	14
3243.6	O V	—
3243,513	Ti I	3
3243,34	Ne II	2
3243,164	Cu I	1500
3242,86	Xe III	100
3242,40	Ar I	2
3242.28	Cs	10
3241,984	Ti II	60
3241,708	Ar II	2
3241,622	Si III	15
3240,44	Kr III	40
3240,20	Kr II	2
3239,91	F II	0,5
3239,664	Ti II	30
3239,52	Kr III	40
3239,436	Fe I	15
3239.3	O V	—
3239,16	Cu I	150
3239,037	Ti II	60
3238,83	Cu II	5
3238,57	O III	5
3238,49	Ar I	1
3238,47	Ne	1
3238,224	Ti I	4
3238,10	F II	0
3237,819	Fe II	8
3237,402	Fe II	5
3237,266	Mg I	3
3236,84	Xe III	25
3236,809	Ar II	6
3236,573	Ti II	70
3236,2231	Fe I	8
3236,122	Ti II	20
3235,713	Cu I	650
3235,57	Ar I	2
3235,21	Kr III	2
3235,175	Ar II	3
3234,026	Na II	4
3234,6138	Fe I	7
3234,517	Ti II	75
3234,491	Ar I	100
3234.16	Cs	6
3233,971	Fe I	12
3233,954	Si III	14
3233,899	Cu I	450
3233,23	Xe II	1
3233,054	Fe I	8
3233,02	Ca III	4
3232,80	Kr I	2
3232,791	Fe II	7
3232,791	Ti I	3
3232.634	Li I	50
3232,38	Ne II	3
3232,280	Ti II	30
3232,15	Kr II	2
3231,97	Ne II	0
3231,75	Cl II	12
3231,702	Fe II	5
3231,315	Ti II	4
3231,266	He I	3
3231,178	Cu I	650
3230,967	Fe I	10
3230,78	Cl III	1
3230,680	Ar II	2
3230,68	Kr I	2
3230,499	Si III	12
3230,211	Fe I	6
3230,16	Ne II	5
3230,021	Ar II	4
3229,91	Ar I	3
3229,50	Ne II	3
3229,397	Ti II	35
3229,193	Ti II	40
3229,03	Xe II	4
3228.8	O V	—
3228,605	Ti II	30
3228,254	Fe I	5
3227,798	Fe I	15
3227,747	Fe II	13
3227.2	Cs	4
3226,771	Ti II	2
3226,602	Cu I	150
3226,57	Kr II	5
3226,541	Cu I	50
3226,129	Ca I	8
3226,128	Ti I	12
3225,976	Na II	4
3225,973	Ar II	6
3225,896	Ca I	8
3225,789	Fe I	20
3225,698	Cu I	5
3225,58	Ar I	20
3225,088	Cu I	2
3225,08	Xe II	15
3224,99	Kr IV	6
3224,85	Kr III	20
3224,82	Ne II	4
3224,664	Cu I	450
3224,241	Ti II	35
3223,74	Kr III	3
3223,52	Kr II	12
3223,519	Ti I	10
3223,435	Cu I	400
3223,01	Si II	20
3223,00	Kr II	6
3222,843	Ti II	35
3222,741	Ti I	3
3222,55	Cl II	7
3222,393	Ar II	6
3222.3	O V	—
3222,24	Kr III	10
3222,069	Fe I	20
3221,625	Ar II	7
3221,381	Ti I	10
3221,35	Cu I	8
3220,65	Cu I	8
3220,62	Kr III	20
3220,60	K II	4

λ	Symbol	I	λ	Symbol	I
3220,44	Si II	10	3210,236	Fe I	8
3220,25	Kr II	6	3210,025	Si II	200
3220,1	O V	—	3209,930	Ca I	2
3219,581	Fe I	12	3209,7	Kr II	7
3219,212	Ti I	8	3209,65	Cs	10
3219,1	Cs	4	3209,64	O IV	3
3218,77	Cu II	3	3209,498	Cu I	4
3218,34	Fe III	6	3209,38	Ne II	3
3218,270	Ti II	25	3209,34	K III	6
3218,21	Ne II	8	3209,297	Fe I	12
3218,204	Cu I	5	3209,030	Ti I	4
3218,10	O II	2	3208,99	Ne II	2
3217,99	Si II	15	3208,475	Fe I	4
3217,942	Ti I	8	3208,28	Kr II	40
3217,669	Ar II	5	3208,231	Cu I	1400
3217,64	Cu I	10	3207,906	Ne	2,5
3217,621	K I	6	3207,897	Ti I	5
3217,380	Fe I	10	3207,655	Ar II	2
3217,155	K I	7	3207,577	Ar II	4
3217,056	Ti II	30	3207,50	Ar I	10
3216,76	O II	1	3207,337	Ti I	5
3216,729	Ar II	8	3207,29	Kr II	1
3216,31	O IV	—	3207,12	O III	1
3216,284	Na II	2	3207,07	Cs	4
3216,25	Kr II	7	3206,825	Ti I	5
3216,249	Si III	7	3206,72	Xe II	1
3216,203	Ti I	3	3206,709	N II	2
3216,08	O II	0	3206,344	Ti I	5
3215,97	O III	1	3205,848	Ti I	5
3215,940	Fe I	12	3205,44	Kr II	2
3215,688	Ar II	3	3205,400	Fe I	15
3215,63	Fe III	8	3205,26	Kr II	4
3215,334	Ca I	5	3204,996	Ar II	8
3215,145	Ca I	5	3204,870	Ti I	6
3215,10	F II	0,5	3204,76	Fe III	6
3214,750	Ti II	4	3204,318	Ar II	9
3214,67	F II	0,5	3204,27	Cs	4
3214,66	Si II	75	3203,872	Si II	100
3214,396	Fe I	8	3203,828	Ti I	15
3214,38	Ne II	5	3203,66	Ar I	10
3214,240	Ti I	12	3203,435	Ti II	3
3214,040	Fe I	20	3203,392	Ar II	3
3213,972	F III	6	3203,104	He II	200
3213,84	Ar I	2	3203,05	Cl II	20
3213,7	Cs	6	3202,85	Ar I	5
3213,70	Ne II	3	3202,740	F II	10
3213,314	Fe II	13	3202,54	Kr II	15
3213,145	Ti I	8	3202,535	Ti II	40
3212,99	Ar I	2	3202,49	Si II	20
3212,516	Ar II	9	3202,12	Cl II	6
3212,29	Xe II	5	3202,04	Xe II	10
3212,186	Na II	6	3201,95	K III	6
3211,992	Fe I	10	3201,796	Mg I	3
3211,99	Ar I	2	3201,68	Xe II	3
3211,568	He I	2	3201,594	Ti I	5
3211,43	Cu I	30	3201,17	F II	1
3210,89	Kr II	7	3201,12	Ar I	3
3210,834	Fe I	10	3201,09	Cs	4
3210,64	Kr II	2	3200,95	O III	1
3210,554	Si III	15	3200,84	Ar I	2
3210,451	Fe II	10	3200,685	N II	2

λ	Symbol	I	λ	Symbol	I
3200,475	Fe I	15	3191,6599	Fe I	7
3200,40	Kr II	50	3191,50	Ar I	2
3200,39	Ar I	100	3191,45	Cl III	9
3199,915	Ti I	100	3191,21	Kr III	80
3199,91	Kr IV	2	3190,874	Ti II	30
3199,53	O IV	1	3190,86	Ne II	2
3199,525	Fe I	15	3190,58	Cl III	4
3199,514	Si II	200	3190,07	K II	5
3199,43	Li II	7	3189,783	Na II	6
3199,22	Xe III	4	3189,52	Ti II	5
3198,920	Ar II	2	3189,28	K II	2
3198,7	Cs	4	3189,20	Cs	4
3198,62	Ne II	5	3189,11	Kr III	100
3197,65	Kr II	4	3189,04	Cl II	20
3197,625	Mg I	2	3188,97	Si II	150
3197,518	Ti II	2	3188,821	Fe I	7
3196,997	Fe I	20	3188,74	Ne II	3
3196,930	Fe I	20	3188,571	Fe I	4
3196,742	He I	2	3188,369	Ar II	1
3196,51	Xe III	25	3187,90	Ar III	6
3196,504	Si III	14	3187,745	He I	200
3196,22	Xe II	25	3187,61	Kr II	4
3196,076	Fe II	10	3187,60	Ne II	2
3195,8	Li II	3	3187,42	Cl II	5
3195,752	Ar II	5	3187,293	Fe II	8
3195,717	Ti II	3	3186,741	Fe II	11
3195,6	Li I	3	3186,63	Ar I	5
3195,5	Cs	4	3186,451	Ti I	60
3195,574	Ar II	2	3186,169	Ar II	5
3195,50	Kr II	2	3186,022	Si III	13
3195,41	Si II	100	3186,017	Cu II	5
3195,12	Ar I	5	3186,01	Kr I	1
3194,93	Ar I	1	3185,99	Si I	10
3194,76	Ti II	6	3185,734	Ar II	3
3194,75	O IV	1	3185,729	Cu II	2
3194,69	Si II	50	3185,72	O IV	0
3194,61	Ne II	4	3185,315	Fe II	5
3194,598	Ar II	4	3185,24	Xe III	40
3194,56	Ti II	8	3185,125	Si III	16
3194,26	Ti II	5	3184,8955	Fe I	7
3194,229	Ar II	9	3184,843	Cu II	15
3194,21	Si II	50	3184,53	Kr I	1
3194,099	Cu I	1500	3184,09	Ti II	2
3193,84	Cl III	0	3183,108	Fe II	8
3193,809	Fe II	11	3182,9798	Fe I	4
3193,75	Xe II	1	3182,57	Ti II	6
3193,6	Cs	4	3182,175	Cu II	10
3193,512	Ar II	1	3181,84	Ti II	8
3193,314	Fe I	8	3181,70	Cl II	7
3193,2268	Fe I	10	3181,39	Xe II	3
3193,09	Si II	150	3181,275	Ca II	15
3192,926	Fe II	9	3181,26	Cl II	5
3192,802	Fe I	8	3181,25	Kr II	5
3192,68	Ti II	4	3181,038	Ar II	12
3192,54	Kr II	2	3180,98	O IV	—
3192,363	Ar II	3	3180,94	Cs II	10
3192,26	Ti II	2	3180,7562	Fe I	5
3192,25	Si II	50	3180,521	Ca I	1
3192,22	Cu I	2			
3192,1	Cs	4			
3191,994	Ti I	80			
3191,72	Ar I	2			

λ	Symbol	I
3180,43	Cl II	7
3180,225	Ti II	2
3180,226	Fe I	20
3180,164	Fe II	7
3179,793	Cu II	5
3179,504	Fe II	8
3179,343	Cu I	2
3179,332	Ca II	18
3179,291	Ti I	3
3179,055	Na II	5
3178,92	Kr II	1
3178,630	Ti II	3
3178,61	Cs	10
3178,015	Fe I	10
3178,01	Fe III	10
3177,965	Cu II	3
3177,80	O IV	0
3177,535	Fe II	10
3177,19	Xe III	5
3176,95	Cl II	5
3176,94	Kr II	15
3176,16	Ne II	3
3176,08	F IV	2
3175,99	Fe III	10
3175,783	Mg II	7
3175,67	Cu I	60
3175,67	Kr II	40
3175,66	Ti II	2
3175,64	Xe II	80
3175,447	Fe I	12
3175,30	Cl II	6
3175,25	Xe II	6
3175,008	Na II	3
3174,80	Ti II	5
3174,725	F III	10
3174,59	Xe II	1
3174,58	C II	0
3174,125	F III	12
3174,09	Fe III	10
3173,71	Ar I	2
3173,66	Cl II	20
3173,58	Ne II	3
3173,355	Cs II	5
3172,97	N III	2
3172,961	Ar I	150
3172,855	Ar II	3
3172,731	Ti I	4
3172,706	Mg II	6
3172,62	C II	1
3172,56	Cl II	6
3172,56	Cs	10
3172,18	Ar I	5
3171,81	K II	2
3171,663	Cu I	5
3171,64	Ar III	2
3171,403	Ar II	3
3171,14	N III	1
3170,93	Kr III	20
3170,925	Ti I	3
3170,63	Kr II	2
3170,346	Fe II	6
3170,23	Cl II	15
3170,03	C II	2
3170,016	C III	4
3169,854	Ca I	1
3169,82	Xe III	5
3169,80	K II	3
3169,73	Cs	4
3169,681	Cu I	500
3169,667	Ar II	15
3169,66	C II	1
3169,45	Cl II	7
3169,30	Ne II	0
3169,2	O II	1
3168,951	Mg II	6
3168,67	Xe II	3
3168,519	Ti II	40
3167,931	C II	8
3167,87	Cl IV	2
3167,859	Fe II	11
3167,74	F IV	1
3167,5762	Ne I	50
3167,487	Na II	2
3167,464	Ar II	3
3166,56	Cu II	5
3166,438	Fe I	6
3165,974	C II	4
3165,878	Mg II	2
3165,861	Fe I	4
3165,710	Si IV	9
3165,70	Ne II	4
3165,467	C II	9
3165,38	Si III	2
3165,288	Ar II	6
3165,27	Xe II	6
3165,1	O II	1
3164,94	Kr II	3
3164,91	Ti II	8
3164,618	Ca I	1
3164,46	Ne II	3
3164,44	Xe II	4
3164,23	Xe II	6
3163,731	Na II	6
3163,535	Ar II	2
3163,281	Si III	2
3163,247	Na II	1
3163,091	Fe II	5
3162,93	Xe II	25
3162,800	Fe II	8
3162,570	Ti II	35
3162,42	F II	1
3162,03	Cu II	3
3161,949	Fe I	8
3161,949	Fe II	5
3161,92	C III	2
3161,755	Ti II	30
3161,68	N V	3
3161,610	Si III	8
3161,456	Ar II	8

λ	Symbol	I	λ	Symbol	I
3161,44	Cl II	20	3150,538	Cu II	3
3161,369	Ar II	7	3150,510	Ar II	4
3161,333	Cs II	2	3150,42	Ar I	1
3161,205	Ti II	25	3149,92	Si II	20
3161,16	Na II	0	3149,561	Si IV	7
3160,70	Xe III	20	3149,508	Cu I	30
3160,658	Fe I	10	3149,36	Cs	10
3160,52	Cl II	10	3149,266	Na II	5
3160,06	Ar I	5	3148,99	Xe II	5
3160,047	Cu I	25	3148,6107	Ne I	100
3159,75	N V	2	3148,57	Cu I	2
3159,75	Xe II	4	3148,333	Cu I	3
3159,55	Ar I	1	3148,202	Ar II	5
3159,53	Na II	0	3148,20	Ar I	1
3158,869	Ca II	17	3148,033	Ti II	12
3158,64	Cu II	5	3147,965	F II	5
3157,887	Fe I	6	3147,86	Cl II	20
3157,60	Ar IV	—	3147,701	Ne I	25
3157,42	Ar III	5	3147,39	Kr II	1
3157,397	Ti II	2	3147,371	Si III	7
3157,15	K II	2	3147,268	Ti I	3
3157,13	C II	0	3146,962	F III	8
3157,040	Fe I	8	3146,821	Cu I	450
3156,63	Kr III	1	3146,422	Ar II	4
3156,629	Cu I	450	3146,260	Ti I	3
3156,274	Fe I	5	3145,900	Ar II	2
3156,11	F III	0	3145,697	Na II	3
3155,670	Ti II	12	3145,63	Ar I	1
3155,4	Li II	2	3145,536	F III	4
3155,09	C III	1	3145,42	Ar I	1
3154,82	Ne II	1	3145,2	Cs	4
3154,75	Cs II	4	3145,02	Xe II	4
3154,387	F III	4	3144,758	Fe II	5
3154,289	Ar II	2	3144,68	O V	1
3154,206	Fe II	12	3144,495	Fe I	6
3154,195	Ti II	12	3144,32	Kr III	9
3153,88	Cs	6	3143,9896	Fe I	8
3153,782	Ar II	4	3143,891	Ar II	3
3153,492	F II	6	3143,756	Ti II	10
3153,4107	Ne I	100	3143,74	Ne II	2
3153,206	Fe I	5	3143,62	Xe II	6
3152,98	Xe III	8	3143,350	Ti I	12
3152,883	Cu II	3	3142,797	Cu I	8
3152,613	Ar II	3	3142,777	F III	3
3152,30	Cs	6	3142,75	K II	2
3152,29	Ar I	3	3142,60	Ar I	3
3152,251	Ti II	15	3142,453	Fe I	6
3151,85	C III	0	3142,444	Cu I	750
3151,82	Xe III	10	3142,04	C II	0
3151,75	Kr III	10	3142,01	Kr IV	3
3151,62	Cu I	8	3141,88	Kr III	20
3151,52	Ar I	3	3141,670	Ti I	10
3151,351	Fe I	10	3141,537	Ti I	15
3151,280	Ca I	4	3141,46	Cs	4
3151,16	Ne II	2	3141,35	Kr III	60
3151,14	Cs	6	3141,35	Ne II	3
3151,049	Cu II	10	3141,164	Ca I	3
3150,93	Kr II	80	3141,16	N IV	3
3150,738	Ca I	4	3140,963	Ar II	2
3150,69	Xe III	20	3140,782	Ca I	3
			3140,44	Kr II	3

λ	Symbol	I
3140,391	Fe I	5
3140,312	Cu I	400
3139,87	Ti I	10
3139,86	Kr II	4
3139,77	O II	4
3139,58	Kr II	20
3139,58	Kr III	15
3139,34	Cl III	8
3139,257	Ar II	4
3139,015	Ar II	12
3138,44	O II	8
3137,92	C II	1
3137,852	Na II	3
3137,72	Cu I	5
3137,629	Ar II	3
3136,481	Ar II	3
3136,43	Fe III	10
3136,20	Kr III	10
3136,003	Ca I	1
3135,906	Si III	3
3135,875	Al II	5
3135,82	Ne II	1
3135,483	Na II	5
3135,364	Fe II	9
3135,10	Kr II	8
3135,069	Ti I	8
3134,90	Ar IV	3
3134,82	O II	10
3134,8	Cs	4
3134,32	O II	3
3134,27	Ar I	2
3134,208	F III	8
3134,1115	Fe I	10
3132,87	Ar I	3
3132,86	O III	6
3132,84	Kr II	4
3132,707	Ti I	6
3132,31	Ar I	2
3132,22	Ne II	2
3131,33	Cu I	5
3131,04	Ar I	2
3130,804	Ti I	15
3130,804	Ti II	15
3130,80	Ar I	20
3130,7	Cs	4
3130,40	Xe II	3
3130,175	Ti I	8
3129,44	O II	6
3129,368	Na II	6
3129,3349	Fe I	5
3129,1	Cs	4
3129,075	Ti I	7
3128,701	Cu I	650
3128,640	Ti I	8
3128,640	Ti II	10
3128,40	Xe II	1
3127,90	Ar III	7
3127,90	Ti I	5
3127,883	Ti II	10
3127,684	Ti I	8
3127,41	N IV	2
3126,40	N II	3
3126,380	Mg I	2
3126,267	Si III	6
3126,1986	Ne I	200
3126,175	Fe I	8
3126,109	Cu I	1400
3126,02	Kr II	6
3125,98	Ar IV	—
3125,96	Cl II	5
3125,73	F II	1
3125,654	Fe I	15
3125,59	F II	0,5
3125,44	Cl II	6
3125,3	Cs	4
3125,208	Na II	1
3125,15	F II	1
3124,762	F III	8
3124,61	Xe III	1
3124,414	Na II	3
3124,39	Kr III	100
3124,28	Cl II	6
3124,268	Ar II	1
3124,19	F II	2
3124,18	F III	3
3124,133	C II	2
3124,02	O II	2
3124,02	Xe II	12
3123,769	Ti I	20
3123,74	Cl III	1
3123,72	Cl II	15
3123,074	Ti I	15
3122,62	O II	6
3122,46	Kr III	20
3122,086	C II	1
3122,065	Ti II	2
3121,87	Xe II	250
3121,71	O III	5
3121,62	Cl II	10
3121,515	F III	12
3120,847	Fe III	20
3120,61	Kr III	30
3120,4364	Fe I	6
3120,435	Cu I	50
3120,212	Ti I	3
3120,12	F II	1
3120,06	Ar I	3
3119,82	Cl II	12
3119,800	Ti II	15
3119,725	Ti I	15
3119,66	Ca III	8
3119,4956	Fe I	6
3118,85	Ti II	2
3118,79	N IV	1
3118,69	F II	0,5
3118,355	Cu I	5
3118,35	Cs	4
3118,130	Ti I	15
3118,02	Ne II	4
3117,899	Ti I	5
3117,85	Ar I	3
3117,75	F II	0

λ	Symbol	I	λ	Symbol	I
3117,669	Ti II	20	3104,396	Na II	4
3117,656	Ca I	1	3104,359	Ar II	5
3117,455	Ti I	6	3103,804	Ti II	50
3116,78	Xe II	2	3103,47	Xe III	3
3116,6337	Fe I	12	3102,975	Ti II	2
3116,63	Ar I	3	3102,953	Ar II	1
3116,590	Fe II	6	3102,73	Xe II	3
3116,348	Cu I	400	3102,585	Ar II	4
3116,22	Ar I	1	3102,517	Ti I	3
3115,73	O III	4	3102,36	Ca I	0
3115,669	F III	10	3102,043	K I	3
3114,96	Ar I	1	3101,790	K I	4
3114,46	Xe III	12	3101,526	Ti I	4
3114,378	Ar II	3	3101,51	Xe II	50
3114,293	Fe II	7	3101,407	Ne	2
3114,10	Ar I	1	3101,004	Ar II	2
3114,092	Ti I	20	3100,6667	Fe I	20
3113,92	Kr II	2	3100,666	Ti I	12
3113,71	O II	1	3100,570	C II	2
3113,579	F III	8	3100,3054	Fe I	20
3113,482	Cu I	50	3100,09	Ar I	5
3112,74	Xe II	20	3099,9695	Fe I	15
3112,482	Ti I	8	3099,928	Cu I	1250
3112,25	Kr III	60	3099,923	Ar II	5
3112,18	Cs	4	3099,91	Xe III	8
3112,050	Ti II	10	3099,8968	Fe I	20
3111,609	Fe III	8	3098,50	Xe II	1
3111,45	Kr II	2	3098,21	Xe II	2
3111,283	Ti I	10	3098,192	Fe I	6
3110,841	Fe III	8	3097,52	O II	0
3110,66	Ar I	3	3097,38	Cs	10
3110,620	Ti II	20	3097,186	Ti II	25
3110,41	Ar III	7	3097,16	Kr III	40
3110,095	Ti II	8	3097,15	Ne II	3
3109,711	Ar II	4	3096,90	Xe II	8
3109,581	Ti I	8	3096,890	Mg I	24
3109,3	Cs	4	3096,826	Si III	16
3108,801	Ar II	2	3096,72	Cl II	25
3108,605	Cu I	2000	3096,52	Kr II	20
3108,58	Ca I	3	3096,424	Ti II	2
3108,452	Cu I	600	3096,296	Fe II	5
3107,978	Fe III	6	3095,86	Cs II	6
3107,82	Xe II	20	3095,81	O III	00
3107,468	Ti I	12	3095,14	Kr II	30
3107,388	Ca I	1	3094,960	Ar II	4
3106,806	Ti I	8	3094,82	Cs	4
3106,33	Xe III	30	3094,53	Xe II	30
3106,234	Ti II	35	3094,08	Ne II	4
3106,16	F II	4	3093,989	Cu I	1500
3106,09	Cl IV	1	3093,813	Ti I	3
3105,68	Kr II	1	3093,65	Si III	5
3105,548	Fe II	5	3093,424	Si III	20
3105,168	Fe II	5	3093,403	Ar II	10
3105,084	Ti II	20	3092,984	Mg I	22
3105,00	K II	6	3092,97	Ar I	1
3104,809	Mg II	8	3092,91	Ne II	2
3104,722	Mg II	9	3092,90	Cl II	8
3104,593	Ti II	3	3092,8386	Al I	20
3104,46	Cl III	6	3092,729	Na II	10
3104,40	Xe II	70			

λ	Symbol	I
3092,7099	Al I	26
3092,41	Xe II	15
3092,31	Cs II	10
3092,22	Cl II	50
3091,6	Cs	4
3091,5786	Fe I	20
3091,32	Ar I	2
3091,065	Mg I	20
3091,06	Xe III	50
3090,47	Xe II	1
3090,18	Ar I	1
3090,137	Ti I	8
3090,051	Ti II	8
3089,401	Ti II	15
3089,17	Ar I	2
3089,053	Cs II	5
3088,92	Xe II	3
3088,910	Ar II	3
3088,9	Cs	4
3088,523	Al II	3
3088,23	Ne II	3
3088,209	Ar II	7
3088,132	Cu I	125
3088,04	O III	2
3088,027	Ti II	75
3087,90	C II	0
3087,81	Ar I	1
3087,34	Xe II	1
3087,31	Ar I	1
3087,047	Na II	2
3086,903	C II	1
3086,78	N II	2
3086,47	Ar I	2
3086,47	Cu I	2
3086,46	Si III	6
3086,236	Si III	25
3085,026	Ar II	5
3084,96	Cu I	2
3084,875	Cs II	5
3084,819	Ti I	4
3084,63	O III	0
3084,155	N II	2
3084,07	Fe III	6
3083,7430	Fe I	20
3083,65	O III	1
3083,64	Ar III	3
3083,54	Xe III	40
3083,363	Si III	2
3083,193	Ar II	1
3083,052	C II	2
3082,979	Ar II	5
3082,87	Xe II	2
3082,62	Xe II	20
3082,381	C II	2
3082,191	N II	4
3082,1529	Al I	24
3081,575	Ti II	5
3081,485	N II	2
3081,46	O II	2
3081,45	Ne	1
3081,222	N II	2
3080,874	Cs II	6
3080,826	Ca I	2
3080,250	Na II	3
3079,175	Ne I	100
3078,875	Ne I	100
3078,733	Na II	1
3078,698	Fe II	8
3078,645	Ti II	50
3078,315	Na II	6
3078,25	N IV	6
3078,15	Ar III	10
3078,07	Cs II	6
3078,018	Fe I	4
3077,523	Si III	4
3077,40	Ar IV	8
3077,168	Fe II	10
3076,971	Ne I	200
3076,68	Cl IV	6
3076,455	Fe II	6
3075,95	O III	0
3075,7214	Fe I	25
3075,225	Ti II	40
3075,19	O III	0
3075,00	K II	3
3074,68	O III	00
3074,665	Al II	6
3074,334	Na II	6
3074,15	Ar I	1
3074,15	O III	0
3073,99	Mg	8
3073,798	Cu I	1400
3073,49	Xe III	10
3073,17	Xe II	2
3072,971	Ti II	40
3072,7	Cs	4
3072,68	Ne II	1
3072,107	Ti II	30
3071,96	Cu I	2
3071,66	O IV	5
3071,39	Xe II	6
3071,36	Cl IV	3
3071,35	Cl II	40
3071,242	Ti II	15
3071,08	Ne II	2
3070,97	Cu I	5
3070,84	Na II	0
3069,73	Cs	4
3069,66	Cl II	5
3068,906	Cu I	15
3068,68	O III	0
3068,238	Si III	7
3068,1749	Fe I	8
3068,06	O III	00
3068	O VI	—
3067,8	Cs	4
3067,30	Xe II	30
3067,2457	Fe I	30
3067,214	Ne	4
3067,8	Cs	4
3067,1196	Fe I	8
3066,889	Ar II	6

λ	Symbol	I
3066,72	Kr II	2
3066,71	F III	2
3066,60	Cs II	10
3066,60	Xe II	1
3066,536	Na II	4
3066,514	Ti II	3
3066,354	Ti II	20
3066,238	Na II	1
3066,220	Ti II	30
3066,145	Al I	5
3066,011	Cu I	3
3065,73	Ar I	1
3065,668	Ne	1,5
3065,315	Fe II	6
3065,120	Ar II	3
3065,11	Ar IV	—
3065,01	O III	00
3064,77	Ar III	10
3064,372	Na II	4
3064,290	Al I	7
3063,7	Cs	4
3063,695	Ne I	200
3063,57	Kr II	3
3063,502	Ti II	4
3063,46	O IV	6
3063,44	Ar I	5
3063,411	Cu I	2500
3063,280	Ti II	2
3063,13	Cl IV	5
3063,13	Kr III	60
3062,82	Ar I	1
3062,7	Cs	4
3062,643	Ar II	3
3062,58	Ne	2
3062,43	Kr III	3
3062,233	Fe II	9
3062,18	K II	5
3062,06	Ar I	3
3061,54	Xe II	12
3061,51	Kr II	6
3061,333	Na II	1
3061,24	Cs	6
3060,976	Cs II	5
3060,909	Ar II	8
3060,84	Cu I	2
3060,84	Kr II	30
3060,64	C II	1
3060,260	Na II	1
3060,12	Cs	6
3059,960	F II	8
3059,924	Al I	4
3059,83	C II	0
3059,741	Ti II	6
3059,30	O III	6
3059,16	Ne II	3
3059,091	C II	3
3059,0871	Fe I	100
3059,029	Al I	4
3058,727	Na II	1
3058,68	O V	0
3058,6	Cs	6
3058,45	C II	2
3058,141	F II	7
3058,090	Ti II	50
3058,00	Cl II	40
3057,4471	Fe I	40
3057,395	Ti II	10
3057,388	Ne I	300
3057,36	Cu I	8
3057,144	Al I	14
3057,083	F II	6
3056,85	C II	1
3056,84	K III	5
3056,802	Fe II	5
3056,740	Ti II	15
3056,72	Kr III	30
3056,49	Xe II	20
3056,28	Ar I	3
3056,157	Na II	6
3056,04	Cs	6
3056,01	Kr II	30
3055,346	Na II	1
3055,31	Kr II	3
3055,2638	Fe I	12
3054,82	Ar III	12
3054,69	Ne II	5
3054,679	Al I	5
3054,56	Cs	4
3054,49	Xe III	15
3054,20	Cs	4
3054,138	Fe III	6
3053,74	Cl II	10
3053,664	Na II	6
3053,5	Cs	4
3053,38	Cu I	10
3053,184	Si II	150
3053,151	Ar II	5
3053,070	Fe I	5
3052,554	Cu I	15
3052,54	O IV	1
3052,07	K III	6
3051,901	Cu I	2
3051,75	Kr II	1
3050,98	Xe II	3
3050,8	Cs	6
3050,57	Ne II	1
3050,211	Na II	1
3050,073	Al I	13
3050,043	Ar II	1
3049,671	C II	3
3049,398	C II	3
3049,23	Kr II	8
3049,139	F III	8
3049,011	Fe II	5
3048,933	C II	2
3048,92	Xe II	3
3048,80	F III	2
3048,784	Ar II	2
3048,766	Ti II	6
3048,69	C II	0
3048,50	Xe II	2

λ	Symbol	I	λ	Symbol	I
3048,30	Si II	50	3039,31	Cs	4
3048,17	Xe II	5	3039,254	F III	7
3048,021	Ar II	2	3039,21	Si II	3
3047,9	O II	0	3038,91	C III	1
3047,76	Xe II	8	3038,706	Ti II	6
3047,6060	Fe I	100	3038,38	Kr II	3
3047,57	Ne II	6	3037,98	Ar IV	6
3047,16	K II	2	3037,98	Cl II	35
3047,13	O III	8	3037,73	Ne II	4
3047,077	F III	4	3037,3901	Fe I	80
3046,93	Kr III	50	3037,35	Xe II	6
3046,685	Ti II	30	3037,287	Si III	8
3046,284	Si III	3	3037,071	Na II	5
3046,27	Xe II	25	3036,986	Fe II	5
3046,079	Ar II	5	3036,96	Ar III	3
3045,949	Ne I	7	3036,887	Ar II	2
3045,9	Cs	4	3036,80	Xe II	30
3045,77	Si II	10	3036,101	Cu I	2500
3045,593	Na II	5	3035,98	Ne II	3
3045,58	Ne II	4	3035,43	O III	4
3045,25	Xe II	30	3034,920	K I	—
3045,085	Ti II	5	3034,761	K I	—
3045,078	Fe I	5	3034,732	Si III	6
3045,076	Si III	5	3034,555	Cu I	3
3045,00	Cl II	10	3034,54	F III	1,5
3044,843	Fe II	5	3034,48	Ne II	5
3044,80	Kr III	6	3034,32	O III	0
3044,75	Xe II	10	3034,16	Kr II	2
3044,16	Ne II	2	3033,71	Xe II	10
3044,028	Cu I	20	3033,510	Ar II	10
3043,932	Si III	7	3033,480	Cu I	2
3043,851	Ti II	5	3033,11	Xe II	6
3043,85	Si II	10	3032,77	Kr II	5
3043,692	Si II	100	3032,66	Si III	4
3043,02	Ne	2	3032,50	O II	1
3043,02	O III	5	3032,41	Cs	4
3042,808	F III	10	3032,08	O II	2
3042,6667	Fe I	15	3031,639	Fe I	15
3042,535	Ti I	3	3031,59	Kr II	5
3042,463	Ar II	1	3031,5	Cs	4
3042,3	Cs	4	3031,215	Fe I	12
3042,191	Si II	30	3030,792	Ne II	2
3042,12	Xe II	12	3030,43	K II	2
3042,0215	Fe I	15	3030,35	Cs	4
3041,7401	Fe I	15	3030,313	Ne I	50
3041,6396	Fe I	10	3030,258	Cu I	10
3041,573	Si II	20	3030,1494	Fe I	15
3041,278	Al II	6	3030,01	Kr II	4
3040,933	Si III	9	3030,000	Si II	100
3040,512	C II	2	3029,730	Ti II	35
3040,4281	Fe I	15	3029,60	Cu I	2
3039,76	O II	1	3029,15	Cs	4
3039,75	Ar IV	—	3029,1	Li II	2,5
3039,746	F III	6	3029,068	Na II	6
3039,714	C II	3	3028,914	Ar II	8
3039,65	Ne II	3	3028,860	Ne II	4
3039,488	Cu I	10	3028,82	O II	1
3039,51	O II	1			

λ	Symbol	I
3028,721	Ar II	3
3028,66	Ca III	6
3028,424	Ne	1
3028,25	Cs	4
3028,04	O IV	0
3027,82	Cu I	5
3027,63	Xe II	2
3027,27	Xe II	3
3027,16	Ar III	5
3027,011	Ne II	4
3027,006	Fe III	6
3026,913	Ne	3
3026,762	Al II	1,5
3026,745	Ar II	5
3026,52	Xe III	8
3026,4637	Fe I	15
3025,8442	Fe I	50
3025,75	O II	1
3025,638	Fe I	15
3024,994	Cu I	100
3024,63	Ne	2
3024,57	O III	4
3024,45	Kr III	80
3024,36	O III	1
3024,074	Al II	1
3024,05	Ar III	12
3024,0337	Fe I	15
3023,86	Ti II	12
3023,83	Fe III	8
3023,80	Xe III	100
3023,668	N II	4
3023,45	O III	5
3022,93	Cl II	30
3022,820	Ti II	15
3022,608	Cu I	300
3022,49	Kr II	5
3022,30	Kr III	50
3022,10	Xo II	2
3021,544	Cu I	300
3021,55	Si II	20
3021,0743	Fe I	150
3020,9	Cs	4
3020,6405	Fe I	200
3020,4918	Fe I	150
3020,37	Cs II	4
3020,29	Xe II	2
3020,0044	Si I	75
3020,001	Fe II	10
3019,78	Xe II	2
3018,9848	Fe I	15
3018,82	Cl II	12
3018,789	Fe III	6
3018,30	Kr II	1
3018,09	Cu I	2
3017,65	Kr II	20
3017,63	O III	5
3017,6288	Fe I	15
3017,43	Xe II	100
3017,348	Ne I	50
3017,348	Ne II	3
3017,187	Ti II	50
3016,185	Fe I	12
3016,15	Ar IV	5
3016,14	O II	1
3015,980	Si II	3
3015,8	Cs	4
3015,52	Xe II	20
3015,400	Na II	6
3015,260	Fe III	7
3014,920	Si II	3
3014,848	Cu I	30
3014,50	O II	1
3014,481	Ar II	6
3014,18	Xe III	6
3013,82	Xe II	2
3013,510	Cu I	2
3013,37	O II	3
3013,167	Fe III	20
3013,091	Si III	5
3012,955	Ne I	50
3012,88	Xe II	1
3012,83	O II	1
3012,129	Ne I	50
3012,041	Cs II	8
3012,005	Cu I	250
3012,00	Kr II	1
3011,482	Fe I	7
3010,838	Cu I	2000
3010,02	Ar III	10
3009,83	O II	1
3009,62	O II	1
3009,5707	Fe I	25
3009,205	Ca I	5
3009,138	Na II	4
3008,83	O II	3
3008,79	O III	3
3008,42	Kr II	8
3008,322	Ti II	2
3008,28	O II	1
3008,1399	Fe I	60
3008,12	Cu I	5
3007,82	Ne	1
3007,793	Fe III	6
3007,74	O II	3
3007,442	Na II	4
3007,275	Fe III	20
3007,1469	Fe I	8
3007,08	O II	3
3006,98	Cl II	20
3006,97	Xe II	2
3006,858	Ca I	6
3006,830	N II	7
3006,82	O II	3
3006,75	Cs II	—
3006,7387	Si I	50

λ	Symbol	I	λ	Symbol	I
3006,05	Cl II	20	2995,94	O II	1
3006,01	O II	2	2995,75	Ti II	5
3005,62	O II	2	2995,524	Al II	1,5
3004,486	Ar II	2	2995,34	Cs	20
3004,39	Cl II	10	2994,958	Ca I	5
3004,35	O III	4	2994,69	Xe III	8
3004,32	Xe III	30	2994,4281	Fe I	100
3003,98	Xe II	40	2994,385	Fe I	5
3003,15	F III	2	2994,280	Al II	1
3003,0323	Fe I	10	2994,273	F III	8
3002,961	Ar II	6	2994,250	Ne I	3
3002,93	O II	1	2994,13	Cu I	5
3002,88	Cs	6	2993,27	Kr III	2
3002,730	Ti I	3	2993,09	Cl II	8
3002,649	Fe II	13	2992,91	Xe III	40
3002,48	Kr II	2	2992,618	C II	18
3002,330	Fe II	5	2992,438	Ne I	200
3002,281	Cu I	10	2992,420	Ne I	200
3002,24	Kr III	6	2992,24	K III	6
3001,920	F III	3	2992,223	K I	1
3001,85	Xe III	10	2992,22	Kr III	60
3001,82	Al II	3	2992,118	K I	1
3001,774	Cu I	2	2992,11	O III	2
3001,663	Ne II	6	2991,82	Cl III	3
3001,617	Fe III	12	2991,780	Cu I	15
3001,271	Cs II	10	2991,73	Xe II	3
3001,24	Cu I	5	2991,637	Fe I	5
3000,9489	Fe I	100	2991,45	Xe III	8
3000,892	Ti I	20	2991,25	Xe III	10
3000,863	Ca I	5	2990,981	Ti I	3
3000,4527	Fe I	8	2990,85	Cs II	2
3000,442	Ar II	9	2990,843	Ar II	2
3000,110	Ar II	5	2990,54	Xe II	12
3000,097	F III	4	2990,488	Ti I	3
3000,059	Fe II	5	2990,3933	Fe I	6
2999,84	Kr II	40	2990,17	Ti II	10
2999,641	Ca I	4	2990,036	Ti I	3
2999,513	Cs	8	2989,30	Ca III	6
2999,5125	Fe I	30	2989,010	Cu I	2
2999,465	F III	6	2988,69	Kr II	3
2999,21	Xe II	15	2988,61	Ca III	7
2999,110	Ar II	2	2988,45	F III	4
2998,43	N V	5	2988,45	F II	3
2998,384	Cu I	150	2987,6453	Si I	150
2998,20	Cs	2	2987,2923	Fe I	10
2998,174	Al II	2	2986,89	Cs	2
2997,74	O II	2	2986,82	Xe II	8
2997,71	O III	2	2986,4569	Fe I	3
2997,513	F III	6	2986,33	Cu II	2
2997,364	Cu I	2000	2986,20	K III	5
2997,309	Ca I	5	2986,18	Xe II	10
2997,301	Fe II	7	2985,926	Cu I	10
2997,2	Cs	2	2985,550	Fe II	13
2997,168	F III	6	2985,464	Ti I	3
2996,63	Cl II	40	2985,33	Kr II	4
2996,60	Kr III	20	2985,3	Cs	2
2996,60	Kr II	20	2985,04	Ar IV	—
2996,51	O III	3			
2996,15	Cs	2			

λ	Symbol	I
2984,830	Fe II	15
2984,785	Fe I	10
2984,76	Ti III	10
2984,63	Xe III	15
2984,479	F III	5
2984,267	Cu I	5
2984,183	Na II	7
2983,94	Kr II	2
2983,91	Cs	2
2983,82	Ne	5
2983,78	O III	9
2983,765	F III	4
2983,66	O III	1
2983,58	N III	6
2983,5714	Fe I	125
2983,290	Ti I	20
2983,22	Kr IV	2
2983,038	Cu I	3
2982,78	Cl II	18
2982,765	Cu I	8
2982,663	Ne I	300
2982,5	Cs	2
2982,34	Kr II	1
2982,23	Xe II	2
2982,123	Cu I	3
2982,106	C III	8
2982,07	N III	1
2982,062	Fe II	8
2982,03	Cs	2
2981,854	Fe I	6
2981,4459	Fe I	20
2981,31	N V	10
2980,922	Ne I	50
2980,90	Cl II	4
2980,78	N V	8
2980,642	Ne I	40
2980,622	Na II	3
2980,538	Fe I	5
2980,510	Si III	5
2980,47	Cl II	2
2979,81	Kr II	20
2979,806	Ne I	50
2979,662	Na II	5
2979,380	Cu I	25
2979,352	Fe II	8
2979,32	Xe II	300
2979,20	Ti II	10
2979,051	Ar II	15
2979,050	Na II	2
2978,87	Kr II	25
2978,87	N III	3
2978,48	Cl II	7
2978,295	Cu I	30
2978,145	F III	4
2977,90	Xe II	5
2977,80	Ti II	7
2977,32	N III	3
2977,258	Cs II	3
2977,222	Fe III	6
2977,132	Na II	3
2976,971	N II	4
2976,81	Cs	2
2976,39	Xe II	8
2976,28	Kr II	3
2976,131	Fe I	5
2975,938	Fe II	5
2975,92	Kr II	3
2975,65	Cs	2
2975,518	Ne I	35
2975,13	Cs	2
2974,991	Na II	6
2974,926	Ti I	4
2974,86	Xe II	20
2974,714	Ne I	300
2974,675	Cu I	10
2974,65	N II	2
2974,527	Ne	1
2974,52	N V	6
2974,236	Na II	2
2974,04	Kr II	25
2973,601	N II	3
2973,46	Cl II	2
2973,2368	Fe I	60
2973,1336	Fe I	60
2973,07	Ne II	1
2972,8	Cs	2
2972,63	Cl II	5
2972,60	N III	4
2972,34	Kr II	2
2972,31	Xe II	8
2972,279	Fe I	3
2971,80	Kr II	4
2971,839	Mg II	1
2971,522	Si IV	1
2971,24	Xe III	8
2970,851	Cs II	5
2970,725	Na II	1
2970,682	Fe II	5
2970,67	Cl III	4
2970,554	Ti I	4
2970,513	Fe II	5
2970,372	Ti I	10
2970,3547	Si I	55
2970,106	Fe I	40
2969,934	Fe II	8
2969,80	Xe II	12
2969,59	C II	0
2969,4759	Fe I	10
2969,45	Xe III	4
2969,3606	Fe I	5
2969,23	Xe II	3
2969,145	Mg II	0
2969,0	Cs	8
2968,836	C II	2
2968,56	Xe III	10
2968,383	Cs II	5
2968,31	Kr III	20
2968,226	Ti I	4

λ	Symbol	I	λ	Symbol	I
2968,020	Mg II	2	2958,35	Kr II	20
2967,87	Mg II	1	2958,30	Ti II	2
2967,868	C II	7	2958,286	Fe III	6
2967,629	C II	3	2957,532	Ar II	3
2967,25	Kr II	80	2957,50	Ti IV	4
2967,244	C I	5	2957,3660	Fe I	30
2967,218	Ti I	25	2957,293	Ne I	8
2967,181	Ne II	3	2956,795	Ti I	25
2966,97	Xe III	10	2956,541	Ar II	4
2966,8997	Fe I	125	2956,30	Kr II	3
2966,89	F III	1	2956,18	Ti I	70
2966,871	C II	5	2955,84	Xe II	2
2966,74	Xe II	1	2955,73	Ne II	7
2966,655	C II	3	2955,388	Ar II	10
2966,187	C II	3	2955,20	Kr III	3
2966,13	Kr II	3	2955,13	F III	2
2965,750	Na II	2	2954,78	Xe II	2
2965,686	Ti I	15	2954,76	Ti II	60
2965,681	Ti I	8	2954,6543	Fe I	5
2965,56	Cl III	6	2954,37	F III	0,5
2965,4	Cs	2	2954,28	Kr II	12
2965,2561	Fe I	20	2953,95	C IV	1
2965,231	Ti I	6	2953,9411	Fe I	50
2965,19	Mg II	0	2953,778	Fe II	11
2965,11	Kr II	2	2953,486	Fe I	5
2965,037	Fe II	10	2953,40	C IV	0
2965,0	Cs	8	2953,10	Ne II	0
2964,98	Xe III	15	2952,7	Li II	0,5
2964,846	C I	2	2952,56	Kr III	50
2964,629	Fe II	9	2952,527	Ne I	5
2964,21	Cl II	2	2952,48	Xe II	2
2964,19	Xe II	12	2952,395	Na II	3
2964,131	Fe II	7	2952,10	Ti II	4
2963,41	Xe II	50	2952,09	Kr III	4
2963,235	Ne II	2	2951,59	Cs	2
2963,230	Fe III	8	2951,58	Xe II	2
2962,953	N II	4	2951,231	Na II	8
2962,8	Cs	2	2951,21	Cu I	5
2962,4	Cs	2	2951,10	Ne II	2
2961,596	F III	5	2950,88	K II	2
2961,272	Fe II	5	2950,35	Cl II	5
2961,165	Cu I	2500	2950,243	Fe I	20
2961,06	Al III	1,5	2950,21	Kr II	30
2961,05	Kr II	4	2949,91	F III	1,5
2960,78	Kr II	5	2949,800	Cs II	5
2960,260	Ar II	5	2949,77	Xe II	4
2960,14	Kr II	40	2949,54	Kr II	15
2960,112	Na II	1	2949,316	Ne I	15
2959,9929	Fe I	10	2949,178	Fe II	10
2959,98	Ti I	5	2949,1	Cl III	1
2959,74	O III	5	2949,043	Ne I	10
2959,71	Ti I	3	2948,94	K III	0
2959,682	Fe I	5	2948,388	Fe III	8
2959,67	Si III	3	2948,38	Ti I	60
2959,666	F III	2	2948,13	Kr III	10
2959,599	Fe II	7	2948,06	Xe III	40
2959,150	Si III	5	2947,8773	Fe I	60
2958,98	Ti II	50			

λ	Symbol	I
2947,85	Cs	2
2947,700	Ti I	3
2947,658	Fe II	13
2947,53	Xe III	40
2947,441	Na II	5
2947,297	Ne I	200
2947,275	Ar II	2
2946,732	Ne I	2
2945,695	Na II	4
2945,47	Ti II	50
2945,368	Cu II	2
2945,25	Xe III	60
2945,23	Cu I	3
2945,106	He I	100
2944,61	Xe II	4
2944,575	Ne I	2
2944,398	Fe II	13
2944,1	Cs	2
2943,495	N II	4
2943,41	Xe II	4
2943,12	Ti II	12
2942,892	Ar II	20
2942,25	Cs II	8
2942,17	N II	3
2942,10	Xe II	20
2941,993	Ti II	50
2941,963	Ti I	60
2941,995	Mg I	13
2941,893	Ar II	1
2941,39	Ti II	8
2941,38	Xe II	8
2841,3438	Fe I	15
2940,953	Cs II	20
2940,22	Xe III	40
2939,91	Kr III	15
2939,72	Xe II	5
2939,70	Kr II	2
2939,55	Fe III	7
2939,506	Fe II	5
2939,453	Cu I	2
2939,13	Xe III	10
2938,868	Cu I	15
2938,69	Ti II	30
2938,56	Kr III	4
2938,5	Cs	20
2938,473	Mg I	12
2938,45	K III	5
2937,811	Fe I	10
2937,766	Cu I	2
2937,725	Na II	5
2937,52	Ti IV	5
2937,293	Ti I	25
2936,9049	Fe I	60
2936,739	Mg I	10
2936,509	Mg II	10
2936,17	Ti II	30
2935,86	Xe II	60
2935,538	Ar II	3
2935,30	Ne II	1
2935,23	Kr III	20
2935,12	C IV	1
2934,80	Xe II	2
2934,60	Cl II	5
2934,065	Na II	2
2934,00	Kr III	10
2933,70	Ne II	2
2933,526	Ti I	25
2933,34	Xe II	1
2933,060	Cu I	20
2932,74	Xe III	25
2932,721	Ne I	100
2932,589	Ar II	8
2932,479	F III	8
2932,06	Kr II	1
2931,699	Cu I	10
2931,483	Ar II	9
2931,27	Ti II	40
2931,09	Cs II	20
2930,883	Na II	1
2930,416	Cu I	5
2930,40	Kr II	2
2930,29	Xe III	20
2930,14	Ti IV	1
2929,312	Ne I	15
2929,121	Fe I	6
2929,0085	Fe I	25
2928,69	Ti II	15
2928,655	N II	3
2928,634	Mg II	9
2928,313	Ti I	30
2927,87	Ti II	2
2927,58	Xe II	2
2926,75	Ti II	10
2926,587	Fe II	12
2926,33	Ar IV	11
2926,33	O III	2
2926,274	Cs II	1
2926,14	O IV	1
2926,057	Cu I	10
2925,623	Ne II	3
2925,439	Cu I	30
2924,882	Cu I	10
2924,642	Ar II	10
2924,52	Al II	3
2924,48	Cs	2
2924,38	Xe II	2
2924,33	Ca III	8
2923,95	Xe II	6
2923,902	Fe III	8
2923,852	Fe I	7
2923,704	Cu I	80
2923,51	Xe III	25
2923,474	Na II	3
2923,212	Cu I	20
2923,050	N II	1

λ	Symbol	I	λ	Symbol	I
2923,03	Xe II	1	2910,729	C II	3
2922,830	Cu I	10	2910,64	Xe II	1
2922,76	N II	1	2910,82	Cs	2
2922,21	Cs	2	2910,44	Ne	2
2922,024	Fe II	5	2910,27	Xe II	3
2921,92	Kr II	4	2910,059	Ne II	5
2921,83	Cs	2	2909,912	Ti II	7
2921,43	O IV	3	2909,77	Al III	2
2921,03	Cs	20	2909,17	Kr III	30
2920,940	Na II	4	2908,957	C II	2
2920,887	F III	4	2908,74	O II	1
2920,6915	Fe I	5	2908,62	Kr II	5
2920,538	F III	6	2908,14	Ti II	4
2920,296	Cu I	10	2907,90	Ca III	2
2919,87	Xe II	40	2907,701	Fe III	12
2919,845	Na II	2	2907,520	Fe I	5
2919,048	Na II	5	2907,497	Fe III	10
2918,77	Ti II	2	2907,18	Xe II	80
2918,28	Ar IV	3	2907,15	Kr II	1
2918,023	Fe I	10	2907,09	C II	1
2917,734	N II	1	2907,05	Al III	10
2917,67	Kr III	10	2906,815	Ne II	3
2917,59	Xe III	20	2906,69	Ti II	20
2917,516	Na II	5	2906,62	O II	3
2916,40	O III	4	2906,56	Xe III	50
2916,335	F III	10	2906,360	Mg I	4
2916,29	O IV	2	2906,34	Al III	3
2916,16	Ne II	1	2906,29	C IV	5
2916,09	Ti II	10	2906,25	Cl II	20
2915,967	Ar II	1	2906,17	Cs	2
2915,78	Kr III	6	2906,011	C II	2
2915,65	O II	1	2905,85	Ne	4
2915,593	Ar II	4	2905,80	Fe III	8
2915,453	Mg I	3	2905,715	C II	2
2915,24	Cs	2	2905,692	Si II	500
2914,932	Ar II	1	2905,662	Cu I	5
2914,89	Ti II	10	2905,655	Ti I	5
2914,652	Cs II	8	2905,301	F III	6
2914,12	Xe III	20	2905,10	Xe II	2
2913,417	Ne I	2	2905,00	O II	2
2913,34	Ti II	10	2904,914	Na II	7
2913,279	F III	8	2904,61	F II	0
2913,267	Al I	3	2904,470	Si IV	2
2913,23	Kr II	4	2904,431	Fe III	12
2913,168	Ne I	200	2904,357	N II	1
2913,08	Ti II	1	2904,29	O II	2
2913,00	Ar IV	12	2904,283	Si II	300
2912,98	O III	2	2904,18	Xe II	3
2912,916	Cu I	2	2903,30	O III	4
2912,1589	Fe I	20	2903,19	Al II	1
2912,082	Ti I	40	2902,923	Mg I	2
2912,06	Cl II	15	2902,68	Xe II	3
2911,90	Xe III	40	2902,47	Fe III	9
2911,85	O II	2	2902,456	Fe II	5
2911,47	Xe III	2	2902,45	Cl II	4
2911,461	Ne I	25	2902,258	Al I	2
2911,215	Cu I	30	2902,08	Al II	2
2911,20	O II	2			

λ	Symbol	I
2901,915	Fe I	5
2901,60	C IV	2
2901,3820	Fe I	5
2901,136	Na II	4
2901,1	Cs	2
2900,04	Kr III	20
2900,02	Ti IV	0
2899,78	Ca III	9
2899,75	Cs II	8
2899,57	Xe III	1
2899,086	N II	1
2898,90	K III	1
2898,48	O III	1
2898,355	Fe I	5
2897,69	Xe III	2
2897,503	N II	4
2897,332	Ar II	6
2897,262	Fe II	8
2897,03	Ne II	2
2896,753	Ar II	10
2896,63	Xe III	30
2896,564	Ar II	2
2895,92	Kr III	1
2895,92	Kr II	1
2895,458	F III	4
2895,45	O III	2
2895,32	Cs	2
2895,22	Xe II	150
2895,215	Fe II	7
2895,131	Si IV	3
2895,076	Fe III	8
2895,05	Ne	1
2895,036	Fe I	8
2894,85	Cs	2
2894,778	Fe II	7
2894,63	Kr II	2
2894,5055	Fe I	10
2894,228	Al I	3
2893,985	Ar II	1
2893,946	Na II	6
2893,889	N II	1
2893,81	Cs	2
2893,70	O III	3
2893,63	Kr III	40
2893,618	Na I	1
2893,11	Ne	1
2892,868	N II	4
2892,47	O II	2
2892,18	Kr III	100
2891,88	O II	1
2891,75	Cs	2
2891,71	Xe III	25
2891,64	Cu I	30
2891,612	Ar II	18
2891,36	Ne II	0,5
2891,050	Ti II	15
2891,046	N II	3
2890,84	Cu I	50
2890,59	Ti II	8
2889,447	F III	8
2889,07	Xe II	10
2888,923	Ti II	15
2888,62	Ti II	10
2888,43	Ne II	1
2888,093	Fe II	5
2887,91	O II	3
2887,807	Fe I	5
2887,559	F III	8
2887,511	Si II	10
2887,456	Ti II	2
2887,41	C III	4
2887,358	Si II	5
2887,12	Xe II	10
2886,67	Cs	20
2886,250	Na II	4
2885,928	Fe II	5
2885,90	O II	1
2885,496	C II	6
2885,408	Cu I	5
2885,36	O III	3
2885,273	N II	6
2884,808	C II	4
2884,77	N IV	4
2884,685	N II	2
2884,55	Kr III	2
2884,42	Cs	8
2884,246	N II	4
2884,21	Kr II	2
2884,20	Al II	4
2884,1955	Cu II	60
2884,12	Ar III	9
2884,099	Ti II	70
2884,01	Cl II′	2
2883,96	O II	4
2883,78	O I	3
2883,745	Cs II	5
2883,71	Xe II	12
2883,702	Fe II	10
2882,99	F IV	0
2882,934	Cu I	1500
2881,852	Ne I	2
2881,80	Ca III	7
2881,70	O III	4
2881,5792	Si I	1000
2881,463	Al II	4
2881,28	O III	2
2881,19	Cs II	15
2881,14	Xe II	1
2881,140	Na II	6
2880,756	Fe II	9
2880,290	Ne I	3
2880,28	Ti II	3
2879,80	O III	0
2879,751	N II	4
2879,743	Cu I	2
2879,327	Ar II	4

λ	Symbol	I	λ	Symbol	I
2879,25	Cs	8	2871,40	F II	5
2879,04	O II	3	2871,399	Ar II	1
2878,95	O I	2	2871,32	Cs	2
2878,86	Cu I	5	2871,270	Na II	5
2878,72	Ar III	5	2871,24	Xe II	50
2878,13	Ne II	0,5	2871,125	Fe II	6
2877,6996	Cu II	40	2871,059	Fe II	6
2877,681	N II	4	2871,022	Ar II	1
2877,418	Ti II	60	2870,61	Kr III	50
2877,3021	Fe I	8	2870,04	Ti II	25
2877,29	Cs	8	2869,993	F III	3
2877,101	Cu I	5	2869,95	Ca III	7
2876,802	Fe II	7	2869,95	Ne II	2
2876,65	Ar III	1	2869,80	Cu I	2
2876,49	F III	3	2869,3083	Fe I	10
2876,49	F II	3	2869,283	Ar II	1
2876,43	Ne II	4	2868,874	Fe II	5
2876,42	Cl II	5	2868,732	Ti II	15
2876,30	O I	1	2868,52	Al II	9
2876,025	Cu I	2	2868,470	Cu I	10
2875,88	F II	2	2868,42	Xe III	1
2875,87	F III	3	2868,41	Cl II	10
2875,79	Ti II	10	2868,33	Cs	8
2875,67	Cu I	10	2867,36	Xe II	2
2875,39	Ti II	15	2867,30	F III	5
2875,346	Fe II	8	2867,30	F II	3
2875,3034	Fe I	5	2866,90	Cs	2
2875,30	Cs	8	2866,76	Xe II	5
2875,240	Cu I	2	2866,73	O III	2
2875,09	Si III	2	2866,65	Ne	5
2874,81	F III	5	2866,6264	Fe I	7
2874,80	F II	4	2866,57	Ca III	7
2874,722	C III	3	2866,37	Cs II	8
2874,626	Si III	4	2865,841	Ar II	4
2874,583	Ar II	3	2865,670	F III	4
2874,560	Cu I	20	2865,45	Cs	2
2874,43	C III	2	2865,21	Cl II	4
2874,40	Ar IV	6	2864,973	Fe II	5
2874,24	Kr III	2	2864,73	Xe II	150
2874,24	C III	0	2863,8644	Fe I	8
2874,22	F II	1	2863,86	Xe III	1
2874,173	Fe I	10	2863,712	C III	4
2874,08	Ti II	2	2863,57	O III	3
2873,82	O III	2	2863,55	Cl II	7
2873,72	Kr II	4	2863,4311	Fe I	8
2873,401	Fe II	10	2862,866	F III	6
2873,13	F II	2	2862,67	Ti IV	1
2873,12	F III	4	2862,52	O III	3
2873,00	Ne II	3	2862,41	Xe III	30
2872,95	Na II	0	2862,40	Cs	8
2872,85	Kr III	5	2862,34	Ti II	30
2872,73	Xe III	2	2862,26	O III	3
2872,663	Ne I	35	2862,26	N III	6
2872,382	Fe II	9	2862,17	Kr II	2
2872,35	Cs	8	2862,07	Cu I	5
2872,3346	Fe I	7	2862,070	Ne I	8
2871,68	Xe III	30	2862,06	Cl II	5
2871,40	F III	8			

λ	Symbol	I	λ	Symbol	I
2861,99	Ti II	20	2851,94	Ar IV	4
2861,90	Xe II	20	2851,7979	Fe I	15
2861,40	F IV	2	2851,743	Cu I	15
2861,38	O III	3	2851,660	Mg I	16
2861,291	Ti II	3	2851,23	Cs	20
2861,060	C II	2	2851,16	Kr III	30
2861,011	Na II	1	2851,087	Ti II	20
2860,85	Cs	8	2850,95	Xe II	3
2860,79	Ti II	4	2850,4	Cs	2
2860,742	Ar II	3	2850,288	Fe III	7
2860,71	Cl II	5	2850,25	Xe III	2
2860,308	F III	9	2849,66	Xe II	8
2859,481	Na II	5	2849,606	Fe II	7
2859,32	Cs	20	2849,050	C III	5
2859,3	Kr IV	3	2848,955	Cs II	3
2859,16	N V	5	2848,91	O III	6
2859,05	Kr III	4	2848,899	Fe II	5
2858,8970	Fe I	5	2848,72	Cu II	2
2858,734	Cu I	200	2848,342	Mg I	14
2858,664	Fe III	7	2848,332	Fe II	7
2858,399	Ti II	8	2848,122	Fe II	7
2858,343	Fe II	11	2848,046	Fe II	8
2858,225	Cu I	50	2847,816	Ar II	3
2858,03	N V	4	2847,66	Xe III	40
2858,01	Ne II	2	2847,655	Cs II	1
2858,00	C II	1	2847,36	Kr II	25
2857,89	O III	3	2847,24	Cs II	3
2857,83	Cs	2	2847,146	Ar II	2
2857,81	Xe III	1	2846,716	Mg I	12
2857,79	Ti II	15	2846,490	Ne I	2
2857,746	Cu II	5	2846,48	Xe II	3
2857,171	Fe II	7	2846,478	Cu I	15
2857,013	C III	1	2846,193	Cs II	10
2856,928	Fe II	8	2846,09	Ti II	15
2856,78	O III	3	2845,92	Xe II	8
2856,660	Cu I	2	2845,84	O III	3
2856,65	Xe II	2	2845,67	Cs	20
2856,616	Ti II	2	2845,5959	Fe I	8
2856,392	Fe II	5	2844,842	Cu I	10
2856,24	Ti II	25	2844,48	Cs	2
2856,2	Kr IV	2	2844,46	Kr II	20
2856,141	Fe II	7	2844,45	Xe II	5
2856,09	Kr III	5	2844,39	O III	2
2855,670	Fe II	9	2844,28	Cl II	4
2855,29	Ar III	8	2844,160	Cu I	15
2854,53	Xe II	60	2844,129	Ar II	4
2854,45	Cs	8	2844,117	C III	2
2854,13	C III	0	2844,09	Ti II	2
2853,922	Ti II	10	2843,9775	Fe I	20
2853,78	O III	1	2843,6314	Fe I	10
2853,23	Ar III	6	2843,485	Fe II	5
2853,22	Kr III	2	2843,369	Ar II	3
2853,13	C III	0	2842,88	Ar III	7
2853,11	Xe II	1	2842,647	Mg I	6
2853,013	Na I	15	2842,632	Ne I	15
2853,0	Kr IV	5	2842,3345	Si I	15
2852,811	Na I	16	2841,914	Ti II	30
2852,415	Cs II	8			
2852,39	Xe II	3			
2852,127	Mg I	50			

λ	Symbol	I	λ	Symbol	I
2841,721	Na II	7	2832,49	Cu I	5
2841,72	F IV	2	2832,46	Xe II	2
2841,7	F III	3	2832,4364	Fe I	25
2841,5	Cs	2	2832,39	Kr II	2
2841,00	Kr III	30	2832,33	Cl II	4
2840,92	Cu I	10	2832,158	Ti II	20
2840,756	Fe II	8	2832,00	Xe II	2
2840,647	Fe II	9	2831,562	Fe II	11
2840,489	Cu II	2	2831,490	Si III	7
4840,4229	Fe I	6	2830,93	Cu II	3
2840,342	Fe II	7	2830,43	Kr II	3
2840,205	Al I	2	2830,36	N II	0
2840,099	Al I	7	2830,31	Cu I	2
2839,819	Fe II	6	2830,25	Ar IV	10
2839,70	Ti II	15	2829,854	Na II	2
2839,622	Si III	5	2829,60	Kr IV	3
2839,57	Xe III	2	2829,423	Cs II	5
2839,57	Xe II	2	2829,42	Cu I	5
2839,555	Na II	4	2829,41	Kr III	6
2839,535	Fe II	7	2829,358	N II	1
2839,20	Kr II	2	2829,18	O IV	2
2838,85	Xe III	3	2829,076	He I	40
2838,85	Xe II	3	2829,045	Cs II	5
2838,79	Kr II	20	2828,87	Ti II	30
2838,1205	Fe I	10	2828,8094	Fe I	7
2838,09	Cs	20	2828,80	Ti II	30
2837,963	Al I	7	2828,634	Fe II	6
2837,856	Al I	2	2828,150	Ti II	60
2837,603	C II	18	2827,98	Ne	0
2837,3685	Cu II	50	2827,91	Cs II	00
2837,300	Fe II	5	2827,90	Xe II	2
2837,28	Cs II	00	2827,8931	Fe I	5
2836,98	Ti IV	0	2827,584	Ne I	3
2836,710	C II	20	2827,45	Xe III	30
2836,60	Ti II	15	2827,431	Fe II	5
2836,35	O II	2	2827,22	Ti II	10
2836,34	O III	4	2826,94	Xe II	5
2836,25	O IV	6	2826,802	Cs II	1
2836,16	Xe II	1	2826,13	F IV	5
2836,08	Kr IV	3	2826,081	F III	5
2835,94	Kr III	6	2826,05	Xe III	20
2835,716	Fe II	9	2825,82	Ne III	5
2835,606	F III	9	2825,687	Fe I	6
2835,59	Cl II	3	2825,609	Ne I	8
2835,4574	Fe I	6	2825,557	Fe I	20
2835,4	Cl III	4	2825,28	Ne III	4
2835,35	Kr II	8	2825,259	Ne I	10
2835,233	Ne I	15	2824,66	Ar III	6
2835,01	Cs	8	2824,47	Ne III	3
2834,472	Si II	5	2824,370	Cu I	1250
2834,30	Cu I	2	2824,12	Cs	8
2834,14	Ti II	10	2823,80	F IV	3
2833,962	F III	8	2823,77	F III	3
2833,18	Xe III	6	2823,635	N II	5
2833,100	Fe II	5	2823,2767	Fe I	20
2833,00	Kr II	100	2823,03	Cs	8
2832,95	Xe III	6	2823,03	Kr II	2
2832,921	Ne I	8			

λ	Symbol	I	λ	Symbol	I
2822,95	Ne III	7	2811,67	Kr III	25
2822,812	C II	2	2811,67	Xe III	8
2822,63	Kr III	6	2811,422	F III	10
2822,63	Kr II	5	2811,112	Mg I	2
2822,05	Cu III	1	2810,82	Cs	20
2821,68	Ne	0	2810,80	Cu II	3
2821,54	C II	1	2810,52	Xe III	1
2821,41	Ti II	8	2810,276	Ti II	50
2820,95	Kr III	4	2809,91	Cs	8
2820,74	F IV	4	2809,806	Fe II	7
2820,70	C II	1	2809,761	Mg I	3
2820,695	F III	4	2809,63	O III	3
2820,632	Al II	1	2809,514	Na II	5
2820,580	Si II	2	2809,50	Ne II	4
2820,44	Ne	0	2809,44	Ar IV	16
2820,36	Ti II	4	2809,35	N IV	2
2820,268	Cs II	5	2809,154	Ti I	5
2820,06	Xe II	4	2809,07	Xe III	8
2820,00	C II	1	2808,99	K II	3
2819,99	Ti II	8	2808,84	O II	2
2819,28	Cs	2	2808,72	Kr II	1
2819,24	C IV	1	2808,685	Na II	1
2819,13	C II	1	2808,56	Xe II	4
2819,02	Xe II	1	2808,07	C III	1
2819,02	Ti III	1	2807,55	Xe II	2
2818,88	Ne	1	2807,46	F IV	0
2818,68	Cu I	4	2807,25	Xe III	10
2818,68	O III	1	2807,02	Ar III	4
2818,624	Fe III	6	2806,985	Fe I	20
2818,302	F III	5	2806,407	Ti II	5
2818,271	Na II	2	2806,39	Xe III	3
2818,26	Ar III	6	2806,31	C III	1
2817,98	Cs	20	2806,168	Ar II	5
2817,838	Ti II	60	2806,07	Kr III	20
2817,53	Kr III	2	2806,00	F IV	1
2817,5047	Fe I	6	2805,990	Ar II	1
2817,37	Ti I	3	2805,791	Fe II	5
2817,110	Si III	9	2805,71	Cu I	5
2816,943	Cs II	20	2805,694	Ti I	6
2816,87	Kr II	30	2805,65	Al II	4
2816,53	O IV	4	2805,17	Cl III	2
2816,46	Kr II	60	2805,13	C III	0
2816,179	Al II	20	2805,08	Xe III	2
2815,94	Xe III	40	2805,00	Ti II	40
2815,57	Ti II	2	2804,5212	Fe I	20
2815,33	Cs	2	2803,97	F III	0,5
2814,685	Ne I	20	2803,686	Cu I	10
2814,48	Kr III	15	2803,60	Kr II	4
2814,47	Xe III	30	2803,60	O IV	2
2813,97	Kr III	15	2803,45	C II	0
2813,88	Ca III	7	2803,441	Fe III	6
2813,613	Fe II	5	2803,20	Kr II	20
2813,558	Cu I	2	2803,11	O II	1
2813,2877	Fe I	30	2803,02	Xe II	5
2813,241	Fe III	10	2802,95	C II	0
2812,96	Cu III	5	2802,704	Mg II	12
2812,74	Cu I	2	2802,556	Cu I	10
2811,781	Mg I	1	2802,50	Xe II	1

λ	Symbol	I	λ	Symbol	I
2802,498	Ti I	15	2793,316	Cs II	5
2802,39	C II	0	2793,18	F III	2
2802,34	Ne III	2	2792,660	Ne I	3
2801,43	C II	5	2792,52	Xe II	1
2801,23	Kr II	2	2792,318	Ne I	20
2801,21	C II	5	2792,16	Cs	8
2800,98	Kr II	2	2792,015	Ne II	5
2800,919	Ar II	1	2791,951	Cu I	5
2800,65	Ti II	30	2791,798	Cu II	10
2800,27	Cl II	4	2791,63	Ca III	6
2800,24	Ne III	3	2790,776	Mg II	9
2800,22	Xe III	20	2790,62	Ti II	3
2799,80	Ne I	2	2790,39	Li II	2
2799,69	Cu II	2	2789,89	O III	3
2799,60	Cl II	4	2789,86	O V	3
2799,536	Cu II	5	2789,83	Kr II	3
2799,47	C III	4	2789,797	Cs II	10
2799,41	Cs II	10	2789,52	Xe II	2
2799,286	Fe II	7	2789,352	F III	3
2799,216	N II	5	2788,96	Ar IV	14
2799,15	C II	1	2788,81	Cs	2
2799,01	O III	2	2788,56	F IV	1
2798,96	Ne	0	2788,258	Fe III	6
2798,95	Ti III	1	2788,24	Cs II	10
2798,73	Ti III	0	2788,105	Fe I	30
2797,998	Mg II	10	2788,093	F III	20
2797,97	O III	1	2788,00	Ti II	8
2797,914	Fe II	5	2787,9331	Fe I	5
2797,7765	Fe I	15	2787,73	Ne III	4
2797,70	C II	1	2787,72	F III	2
2797,65	Xe II	30	2787,38	F III	0,5
2797,44	Cu II	2	2787,04	O IV	8
2797,26	Cu II	2	2787,03	O V	4
2797,11	Ar IV	7	2787,02	Cs	8
2796,80	F IV	2	2786,89	Ne III	3
2796,49	Xe II	2	2786,496	Cu I	10
2796,46	C III	3	2786,17	Ne III	2
2796,37	Cl III	1	2785,99	Ti II	6
2796,26	Kr II	2	2785,96	F IV	3
2795,963	Ne I	8	2785,42	Xe II	3
2795,81	Kr II	30	2785,39	Ar IV	—
2795,613	Ne I	1	2785,29	Ne III	5
2795,528	Mg II	13	2785,26	Kr III	2
2795,500	F III	4	2785,23	Ar III	5
2795,425	Ar II	2	2785,213	Fe II	8
2795,31	Cu II	2	2784,666	Cs II	3
2795,289	Ar II	2	2784,648	Ti II	3
2795,101	Ne I	35	2784,48	O III	2
2794,86	Xe III	20	2784,47	Ar IV	12
2794,592	Ne I	5	2784,10	Cs	8
2794,56	C III	2	2783,696	Fe II	12
2794,50	Cs II	10	2783,65	Ar III	5
2794,26	F IV	3	2783,551	Cu I	20
2794,220	Ne II	3	2783,37	Xe III	12
2794,2	F III	3	2783,30	F III	0
2794,09	O III	5	2783,15	O II	2
2793,888	Fe II	7	2783,03	Ne III	2
2793,485	Cu I	2			

λ	Symbol	I	λ	Symbol	I
2782,972	Mg I	18	2772,1107	Fe I	20
2782,92	Ar IV	3	2772,04	O III	2
2782,73	Xe II	2	2771,27	Ca III	4
2782,592	Cu I	20	2771,184	Fe II	5
2782,47	Cl IV	7	2770,79	Ca I	3
2782,46	O IV	3	2770,64	Cl IV	4
2782,30	Ti II	2	2770,508	Fe II	5
2782,07	Ne I	2	2770,41	Xe II	2
2781,956	F III	4	2770,15	O III	2
2781,68	Ne I	3	2770,06	Ne II	1
2781,42	Ne	1	2769,748	Ar II	4
2781,416	Mg I	18	2769,69	O V	1
2781,350	F III	5	2769,6693	Cu II	50
2781,288	Mg I	8	2769,5	Cs	2
2781,18	F IV	2	2769,354	Fe II	9
2781,05	O IV	7	2769,3	Cl III	3
2781,04	O V	5	2769,2985	Fe I	6
2780,81	Cs	2	2769,153	Fe II	6
2780,55	Ti II	5	2768,934	Fe II	8
2780,065	Cs II	3	2768,878	Cu I	125
2780,023	Ne II	2	2768,54	Kr III	4
2779,97	Kr II	1	2768,349	Mg I	7
2779,9	Cs	8	2767,945	Ar II	2
2779,831	Mg I	20	2767,77	Ne I	2
2779,64	Xe III	5	2767,673	C II	3
2779,51	Kr II	4	2767,5232	Fe I	20
2779,299	Fe II	11	2767,503	Fe II	13
2779,11	Kr II	20	2767,28	Ne I	3
2779,1	Cs	8	2767,00	Xe II	1
2778,99	Kr II	2	2767,00	Ne	7
2778,48	Ti II	2	2767,0	Li II	4
2778,270	Mg I	18	2766,50	O III	2
2778,2214	Fe I	20	2766,371	Cu I	2500
2778,03	F IV	1	2766,364	Ne I	3
2777,96	Kr II	1	2766,20	Xe III	5
2777,892	Fe II	5	2766,13	Ca I	1
2777,89	K II	2	2766,118	C II	2
2777,714	C III	5	2766,095	Cs II	5
2777,65	Ne III	7	2766,07	Ne	6
2776,99	Cs II	15	2765,90	Kr III	2
2776,96	Xe III	10	2765,222	Mg I	5
2776,923	Fe II	5	2765,120	C II	1
2776,690	Mg I	18	2764,821	Ti II	10
2776,26	Ar IV	10	2764,762	Cu I	5
2775,049	Ne I	5	2764,70	Ne	2
2774,86	Xe II	15	2764,648	Ar II	4
2774,70	Kr IV	6	2764,60	F IV	0
2774,691	Fe II	7	2764,60	Ca I	2
2774,59	Kr II	3	2764,42	Cs	20
2774,46	Cs	8	2764,38	Ne	1
2774,099	Ar II	2	2763,88	Cl II	10
2773,73	Ti III	1	2763,809	Cu I	15
2773,55	Xe II	5	2763,804	He I	20
2773,306	Fe III	8	2763,56	Xe II	1
2772,80	Ca I	1	2763,520	Ar II	1
2772,740	Ar II	2	2763,108	Fe I	4
2772,60	Kr II	10	2763,00	Xe III	1
2772,41	Xe III	10			

λ	Symbol	I
2762,922	Ne II	3
2762,815	Al III	9
2762,77	Xe II	2
2762,460	Al II	2
2762,324	Ne I	3
2762,23	Ar III	7
2762,22	Ti II	2
2762,05	Ca I	2
2762,0275	Fe I	15
2761,97	Cs II	8
2761,813	Fe II	9
2761,785	Fe I	18
2761,60	Xe III	12
2761,30	O III	3
2761,291	Ti II	7
2760,852	Al II	1
2760,76	Xe III	6
2760,48	Al III	1
2760,25	Cu I	2
2759,817	Fe I	5
2759,81	F III	5
2759,589	F III	10
2759,323	Ne I	2
2759,02	Kr II	4
2758,69	Cl II	5
2758,64	Ne I	3
2758,36	Xe II	1
2758,35	Ti II	2
2758,066	Ti I	20
2757,92	Ar IV	14
2757,86	Xe II	40
2757,81	Cs II	7
2757,62	Ti II	3
2757,40	Ca I	2
2757,374	Ti I	6
2757,3170	Fe I	10
2757,304	Ar II	3
2757,025	Fe II	5
2756,617	Ne II	3
2756,664	F III	5
2756,53	Kr III	8
2756,512	Fe II	5
2756,3295	Fe I	20
2756,2677	Fe I	20
2756,22	O III	1
2755,82	Ne I	15
2755,737	Fe II	15
2755,69	Cu I	5
2755,556	F III	7
2755,20	Cs	20
2755,307	F III	4
2755,13	O V	2
2754,907	Fe II	6
2754,864	Ar II	2
2754,10	Cl II	25
2753,287	Fe II	12
2752,85	Ti II	4
2752,8	F III	1
2752,47	O III	0
2752,24	O V	0
2751,828	C III	3
2751,810	Cu I	10
2751,8	F III	1
2751,70	Ti II	50
2751,59	Kr II	5
2751,52	Cl II	5
2751,29	Cu I	10
2751,23	Cl IV	5
2751,123	Fe II	6
2751,11	Cs	2
2750,878	Fe I	5
2750,786	Cu I	5
2750,36	Kr III	10
2750,1415	Fe I	25
2749,839	Cs II	8
2749,734	Cu I	2
2749,484	Fe II	12
2749,34	Ca I	1
2749,324	Fe II	14
2749,184	Fe II	13
2749,031	Ti I	5
2748,23	Cs II	15
2748,18	Kr IV	8
2748,065	Al I	3
2747,88	Xe III	8
2747,870	F III	5
2747,46	O II	6
2747,41	Kr II	2
2747,282	C II	12
2746,982	Fe I	20
2746,713	Cu I	20
2746,70	Ti II	30
2746,488	C II	10
2746,483	Fe II	14
2746,31	Kr II	15
2745,49	Ca I	1
2745,452	Cu I	20
2745,2710	Cu II	20
2745,00	O III	2
2744,838	Ti I	5
2744,797	Ar II	6
2744,64	Kr II	1
2744,5287	Fe I	8
2744,51	F IV	1
2744,0691	Fe I	10
2744,05	Kr III	2
2744,04	Xe II	2
2743,89	Ar III	3
2743,62	O V	0
2743,55	K II	4
2743,53	Ne I	15
2743,196	Fe II	14
2743,16	Xe II	2
2743,03	Kr III	3
2742,56	Kr II	40

λ	Symbol	I	λ	Symbol	I
2742,4064	Fe I	30	2733,15	Xe II	25
2742,30	Ti II	8	2733,022	Ar II	4
2742,297	Ti I	15	2732,61	Ne I	1
2742,2554	Fe I	20	2732,504	Ar II	6
2742,13	Kr IV	20	2732,335	Ar II	1
2742,05	Kr III	5	2732,33	Kr II	4
2741,962	Ar II	1	2732,3	Li I	2
2741,84	Kr III	2	2731,993	Mg I	8
2741,397	Fe II	6	2731,93	Cu II	2
2741,186	Li I	10	2731,8	Cs	2
2741,067	Ar II	2	2731,639	Ar II	1
2740,980	Al I	4	2731,583	Ti I	7
2740,912	Ar II	1	2731,528	Ne I	3
2740,80	Xe III	12	2731,46	Xe II	1
2740,73	Cs II	15	2731,44	O V	0
2740,333	Ar II	1	2731,37	N II	1
2740,31	F III	1	2731,358	Ne I	3
2740,11	Kr II	1	2731,141	Ti I	4
2739,808	Ti I	15	2730,95	Ti II	6
2739,7658	Cu II	8	2730,738	Fe II	11
2739,63	F II	0	2730,7	Li II	5
2739,546	Fe II	15	2730,61	C II	1
2739,15	O III	0	2730,55	Kr IV	3
2739,11	F III	1	2730,41	Kr III	5
2738,70	Ti II	3	2730,065	Cs II	5
2738,13	Kr II	1	2729,46	Kr II	30
2737,954	F III	4	2729,35	O V	1
2737,608	Cu I	2	2729,213	C II	2
2737,3422	Cu II	10	2728,707	C II	4
2737,3108	Fe I	20	2728,4	Li II	2
2736,968	Fe II	12	2728,22	Xe III	4
2736,96	K III	0	2728,0212	Fe I	5
2736,91	F V	0	2727,93	F III	2
2736,65	Kr IV	2	2727,80	Cs	2
2736,542	Mg I	12	2727,7	Cl III	2
2736,177	Ne I	5	2727,539	Fe II	13
2735,69	Ne I	8	2727,47	F III	0
2735,614	Fe I	8	2727,420	Ti I	8
2735,613	Ti I	6	2727,382	Fe II	8
2735,4762	Fe I	8	2727,36	C II	2
2735,283	Ti I	10	2727,22	Xe III	4
2735,168	Ne I	3	2726,95	O III	1
2735,14	O III	1	2726,802	Cs II	1
2734,858	Cu I	10	2726,702	Si II	5
2734,85	Cs	2	2726,30	Cs II	0
2734,82	Ca I	2	2726,054	Fe I	6
2734,755	Ne I	2	2725,90	C III	7
2734,702	N II	2	2725,79	Ti II	3
2734,14	Xe II	50	2725,30	C III	7
2733,879	Cs II	5	2725,081	Ti I	10
2733,8	F III	0,5	2724,9542	Fe I	10
2733,5816	Fe I	15	2724,885	Fe II	9
2733,493	Mg I	10	2724,85	C III	6
2733,36	Kr IV	2	2724,84	Ar III	10
2733,34	O II	10	2724,772	Ne I	3
2733,32	He II	100	2724,21	Cs II	10
2733,264	Ti I	30			
2733,26	Kr II	50			

λ	Symbol	I	λ	Symbol	I
2724,03	Cl IV	5	2714,54	Cu I	2
2724,03	Cl III	5	2714,49	Kr II	3
2724,01	O IV	2	2714,412	Fe II	13
2723,953	Cu I	30	2714,38	Cl II	8
2723,95	Cs	8	2714,37	Cl III	2
2723,812	Si IV	3	2714,35	N III	1
2723,5786	Fe I	15	2714,08	N III	3
2723,40	Xe II	1	2714,0	Cs	2
2723,25	F IV	0	2714,00	Cu I	2
2723,25	F III	1	2713,95	N III	5
2723,191	He I	10	2713,76	Ne	1
2723,091	Al II	2	2713,54	F IV	0
2722,737	Fe II	5	2713,5079	Cu II	50
2722,702	Cu I	5	2713,40	O III	2
2722,040	Fe II	5	2712,88	F V	0
2721,6771	Cu II	25	2712,77	Cl II	4
2721,645	Ca I	10	2712,40	Kr II	80
2721,6	Cs	2	2712,388	Fe II	6
2721,28	Xe II	1	2712,32	C II	0
2721,06	F V	0	2711,88	Cu II	5
2720,918	Al II	1	2711,845	Fe II	9
2720,9035	Fe I	40	2711,6560	Fe I	4
2720,62	Cu I	2	2711,6	Cs	2
2720,199	Cu I	15	2711,11	Kr II	2
2720,184	Ar II	2	2710,59	C II	1
2719,90	Kr II	5	2710,5	Cs	2
2719,89	F III	1,5	2710,37	Cl III	7
2719,61	Cl II	4	2710,27	Kr III	2
2719,39	Ti II	2	2710,27	Kr II	3
2719,296	Fe II	5	2709,837	N II	6
2719,097	Cu I	15	2709,82	Cl II	2
2719,025	Fe I	60	2709,60	Cl II	4
2719,0	Cs	2	2709,582	Al II	1,5
2718,90	O II	3	2709,408	F III	3
2718,847	Cu I	2	2709,056	Fe II	7
2718,79	Xe II	1	2709,03	Cl II	10
2718,775	Cu II	35	2709,02	Kr III	1
2718,64	Ti III	0	2709,0	Cs	2
2718,639	Fe II	5	2708,87	O III	1
2718,4365	Fe I	6	2708,4	C II	1
2718,34	F IV	0	2708,34	Kr III	1
2718,14	F III	1	2708,272	Ar II	6
2717,86	Cs II	1	2708,052	Ar II	2
2717,70	Kr II	1	2707,507	Fe I	20
2717,62	Cl III	2	2707,17	F V	2
2717,35	Xe II	30	2707,132	Fe II	6
2717,304	Ti II	3	2706,79	Cs	20
2717,18	Kr II	1	2706,76	Cl II	4
2716,860	Ar II	2	2706,74	Ne	1
2716,218	Fe II	9	2706,66	F IV	1
2716,20	Ti II	4	2706,5829	Fe I	8
2716,16	Kr II	10	2706,566	Fe II	7
2715,8	Cs	2	2705,3	Cs	2
2715,76	Xe II	3	2705,18	Cu I	2
2715,543	Cu I	20	2705,117	Fe III	7
2715,45	O II	5	2704,87	Ca III	6
2715,35	Cu I	5	2704,32	Ne I	2.
2715,19	Kr III	7			

λ	Symbol	I
2704,1	Cs	2
2703,989	Fe II	10
2703,96	F V	1
2703,95	Cs II	3
2703,44	Xe II	10
2703,184	Cu II	30
2702,65	Cu I	10
2702,554	Ne I	3
2702,34	Xe II	2
2702,30	F V	1
2702,22	Xe II	2
2701,95	Ti III	1
2701,766	Ne I	2
2701,719	Ar II	2
2701,653	Ne I	2
2701,36	Cl IV	4
2701,34	Kr II	15
2701,19	Cs II	4
2701,13	Fe III	8
2701,05	O III	3
2700,963	Cu II	30
2700,681	Ne I	2
2700,60	Kr II	3
2700,555	Ne I	8
2700,30	Cs	8
2700,045	Fe III	8
2699,79	Cl III	1
2699,16	Cs	8
2699,1075	Fe I	6
2698,71	Kr III	2
2698,67	C IV	4
2698,56	Cl II	2
2698,52	Ti II	30
2698,46	Cu III	3
2698,414	Fe III	7
2698,145	Mg I	6
2698,07	Kr III	3
2697,75	C IV	4
2697,75	C III	7
2697,462	Fe II	5
2697,42	C III	3
2697,30	Kr III	25
2697,905	Fe III	7
2696,64	N III	1
2696,59	Kr III	25
2696,50	Xe III	8
2696,39	Cu III	6
2696,284	Fe I	5
2696,19	N III	2
2696,119	He I	7
2696,11	O III	2
2696,00	O III	3
2695,70	Kr II	30
2695,49	O III	6
2695,45	F IV	3
2695,45	F III	4
2695,314	Fe III	9
2695,181	Mg I	5
2695,150	Fe III	10
2695,08	F IV	1
2694,81	Kr III	20
2694,63	Cl II	3
2694,538	Fe I	5
2694,080	Cu I	5
2693,98	F V	1
2693,723	Mg I	3
2692,836	Fe II	5
2692,790	F II	5
2692,74	O III	1
2692,597	Fe II	10
2692,596	Ar II	5
2692,45	Mg I	2
2692,15	Ti III	1
2691,86	Kr III	4
2691,83	Cs	2
2691,52	Cl III	5
2691,40	Xe II	1
2691,20	Kr II	2
2690,49	N II	1
2690,23	Kr III	15
2690,033	Ar II	2
2689,90	K III	5
2689,49	Mg I	1
2689,412	Cs II	5
2689,39	Cl II	6
2689,2998	Cu II	50
2689,26	N III	4
2689,2130	Fe I	8
2689,093	Ar II	2
2688,826	Ti I	10
2688,728	Al II	2
2688,37	Kr II	4
2688,11	F IV	2
2688,04	Cl II	150
2687,78	Ca III	8
2687,53	O III	5
2687,395	Ar II	1
2687,03	Xe III	5
2687,03	Xe II	5
2687,01	N III	3
2686,75	Ne	3
2686,73	Ca III	3
2686,60	Cs II	10
2686,322	Ar II	2
2686,14	O III	10
2686,14	Xe II	3
2685,63	Ar III	6
2685,58	Xe III	2
2685,58	Ne	1
2685,40	Cl III	4
2685,33	Ne	1
2685,137	Ti I	3
2684,795	Ti I	5
2684,76	Cl III	5
2684,751	Fe II	10

λ	Symbol	I
2683,65	O III	4
2683,55	Kr II	15
2683,280	Al II	3
2683,094	Ar II	3
2682,63	Ar IV	9
2682,60	F IV	1
2682,41	O IV	2
2682,40	Cl III	3
2682,210	Si II	10
2681,99	Cs	8
2681,42	O III	2
2681,34	Cs	8
2681,19	Kr III	40
2680,88	Cl III	2
2680,72	Kr III	7
2680,685	Ne I	1
2680,433	Na I	7
2680,340	Na I	8
2680,32	Kr III	30
2679,923	Ti I	20
2679,62	Kr III	15
2679,60	N II	1
2679,37	Cl II	5
2679,19	Ne I	2
2679,0626	Fe I	10
2678,92	Cs	20
2678,810	Fe III	6
2678,64	Ne III	25
2678,54	Xe III	1
2678,38	Ar III	9
2678,086	Na II	5
2677,906	Si II	3
2677,90	Ne III	30
2677,87	Ar III	3
2677,87	Ne I	2
2677,81	O III	3
2677,57	Si IV	1
2677,42	F III	0
2677,36	Ne	3
2677,20	Kr II	6
2677,18	Xe II	50
2677,135	He I	5
2677,09	O IV	2
2677,020	Ne I	1
2677,01	Cs	2
2676,95	Cl II	100
2676,46	Ar III	4
2676,428	Cu I	20
2676,08	Cu II	2
2676,00	Kr III	8
2675,78	N II	2
2675,64	Ne I	200
2675,4	Cl III	2
2675,34	O IV	2
2675,31	Kr II	4
2675,249	Si IV	4
2675,24	Ne I	200
2675,120	Si IV	4
2674,62	Cs	2
2674,57	O III	8
2674,54	F III	1
2674,4	Li II	2
2674,170	Ar II	2
2674,02	Ar III	8
2674,0	Cs	2
2673,24	Cs II	6
2673,0	Kr IV	2
2672,959	C III	5
2672,79	Kr III	3
2672,79	Kr II	3
2672,460	Mg I	10
2672,22	Xe II	4
2672,193	Si IV	1
2672,19	Cl II	50
2672,05	Cu I	5
2671,829	Na II	6
2671,43	Cl II	6
2671,318	C III	4
2671,204	Cu I	20
2671,17	Cs II	4
2670,67	Kr III	20
2670,240	C III	3
2670,10	O III	2
2669,960	C II	3
2669,792	Cs II	10
2669,592	Ti I	15
2669,553	Mg I	8
2669,52	Cl III	3
2669,36	Ne	1,5
2669,166	Al II	10
2669,13	Ne I	3
2669,00	Xe III	10
2668,76	Cs	8
2668,25	F III	2
2668,124	Mg I	6
2668,02	Xe II	5
2667,36	Cl II	40
2666,8133	Fe I	8
2666,635	Fe II	10
2666,61	Kr II	6
2666,59	Cu I	2
2666,46	Cl II	20
2666,358	Cs II	1
2666,2910	Cu II	20
2665,69	O III	7
2665,54	Cl III	6
2664,664	Fe II	10
2664,390	F III	3
2664,37	Kr II	4
2664,00	Kr II	8
2663,38	Ne	2
2663,29	Xe II	3
2662,271	He I	4
2663,22	O IV	2

λ	Symbol	I	λ	Symbol	I
2663,20	Cl III	3	2651,550	F III	3
2662,62	Cs II	1	2651,19	Cl III	3
2662,57	Kr II	2	2651,01	Ne	6
2662,29	Cl III	3	2650,96	Kr III	1
2662,29	O IV	2	2650,7	Cs	20
2661,962	Ti I	10	2650,492	Fe II	4
2661,65	Cl III	5	2650,20	Xe III	1
2661,47	Kr II	5	2650,10	Al II	4
2661,22	Kr II	1	2649,840	Cu I	30
2661,00	Xe III	1	2649,67	Kr II	4
2660,996	Na II	7	2649,599	Ar II	2
2660,97	Kr II	8	2649,464	Fe II	4
2660,817	Mg II	8	2649,27	Kr II	20
2660,755	Mg II	8	2649,062	Mg I	4
2660,386	Al I	12	2648,69	Kr III	10
2660,24	Cs II	5	2648,60	Cu II	3
2660,22	Ar III	3	2648,56	Ne I	25
2659,781	Si II	5	2648,43	Kr III	4
2659,60	Kr II	2	2648,21	Ne I	15
2659,28	Xe II	1	2648,19	Cl II	10
2658,74	Cl II	100	2648,18	F IV	0
2658,71	Cs	2	2648,15	Kr II	20
2658,26	Xe III	3	2648,07	Cs II	10
2658,251	Fe II	5	2647,844	Ar II	1
2658,00	Kr III	2	2647,79	Cl II	5
2657,52	Ne I	15	2647,76	Ne I	8
2657,406	Al I	3	2647,5588	Fe I	5
2657,3	Li II	1,5	2647,42	Ne I	200
2657,178	Ti I	10	2647,247	Ar II	6
2657,00	Xe II	5	2646,956	N IV	12
2656,83	Cs	2	2646,88	Cl II	25
2656,475	F III	6	2646,87	N II	0
2656,38	Kr II	15	2646,751	Fe III	6
2656,303	Ar II	2	2646,631	Ti I	40
2656,294	F III	3	2646,206	Mg I	3
2656,17	Ar III	1	2646,20	Cs	8
2655,803	Şi II	3	2646,19	Ne I	15
2655,512	Si III	14	2646,176	N IV	11
2655,39	Xe II	2	2646,08	Ti II	50
2654,921	Ti I	5	2646,02	N II	0
2654,63	Ar III	10	2645,654	N IV	7
2654,056	Ar II	2	2645,645	Ne I	35
2653,95	Kr II	6	2645,539	Si II	5
2653,77	Ar III	4	2645,51	Ne	6
2653,757	F III	5	2645,5	F III	0
2653,66	Kr III	4	2645,47	Ar III	2
2653,491	F III	4	2645,39	Fe III	9
2653,252	F III	2	2645,303	Cu I	20
2652,899	Ar II	1	2644,801	Mg I	2
2652,848	He I	3	2644,802	He I	2
2652,475	Al I	12	2644,69	Cs II	5
2652,065	Cu I	2	2644,253	Ti I	40
2651,958	F III	3	2644,16	Ne I	5
2651,906	Ar II	2	2644,000	Fe I	8
2651,723	F III	3	2643,93	N II	1
2651,71	Cs II	12	2643,92	Cu III	40
2651,693	Cu I	10	2643,427	C II	3
2651,6	Kr IV	2			

λ	Symbol	I
2643,413	N II	2
2643,06	Kr II	20
2642,63	Cs	20
2642,47	Ne I	8
2642,42	Ne III	3
2642,331	C II	3
2642,28	Cl II	4
2642,25	Ne III	2
2642,15	Ti II	20
2642,08	Kr II	4
2641,74	Kr III	2
2641,6468	Fe I	4
2641,550	Cu I	5
2641,54	Cu III	8
2641,53	O III	3
2641,425	C II	8
2641,24	F III	2
2641,12	Xe III	5
2641,089	Ti I	40
2641,07	Ne III	10
2641,0	Cs	2
2641,00	Kr III	4
2640,98	O III	2
2640,92	Cs II	3
2640,894	C II	5
2640,788	Si III	11
2640,74	Kr II	2
2640,68	O III	2
2640,63	F IV	1
2640,56	NeIII	1
2640,560	CII	6
2640,36	Al II	3
2640,34	Ar IV	15
2639,97	Ne I	15
2639,76	Kr III	60
2639,553	Fe II	5
2639,47	F III	3
2639,18	Ne III	5
2639,05	F III	4
2639,04	O III	2
2638,70	Ti II	10
2638,70	Ne III	10
2638,695	Al II	3
2638,32	Kr II	2
2638,263	Al II	4
2637,696	Al II	5
2637,644	Fe II	6
2637,6	Cs	2
2637,54	Xe III	3
2637,14	Cs II	8
2636,906	Ar II	2
2636,725	Al II	6
2636,51	Kr II	3
2636,354	Ar II	2
2636,070	Ne I	25
2635,882	Cs II	1
2635,8100	Fe I	8
2635,60	Ti II	5
2635,37	F IV	3
2635,17	Al II	1
2635,11	K III	5
2635,03	Al II	3
2634,95	Cl II	12
2634,933	Cu I	30
2634,8	F III	0,5
2634,49	F IV	0
2634,41	Kr II	6
2634,20	Xe II	2
2634,17	Cs	2
2634,17	Ca III	6
2634,001	Ar II	2
2633,88	Xe II	2
2633,194	Fe II	5
2633,18	Cl III	5
2632,873	Mg I	8
2632,78	O IV	4
2632,67	Cl III	5
2632,414	Ti I	15
2632,40	Ar III	4
2632,2382	Fe I	4
2631,90	Ar III	7
2631,609	Fe II	8
2631,553	Al II	7
2631,33	Cl II	2
2631,323	Fe II	13
2631,2819	Si I	190
2631,25	Xe II	2
2631,051	Fe II	13
2630,93	F III	1
2630,66	Kr III	15
2630,40	Xe II	6
2630,28	F IV	3
2630,20	Cl II	4
2630,066	Fe II	8
2630,053	Mg I	6
2630,004	Cu I	20
2629,686	F III	8
2629,587	Fe I	5
2629,587	Fe II	8
2629,54	Xe II	5
2628,90	Kr III	25
2628,86	Cs II	2
2628,664	Mg I	3
2628,53	O III	2
2628,46	C II	1
2628,29	Fe II	13
2628,08	Kr III	6
2627,952	Cs II	5
2627,84	Cs	8
2627,75	Kr II	7
2627,74	F IV	2
2627,68	Al II	7
2627,397	Ar II	3
2627,365	Cu I	20
2627,22	Kr II	3

λ	Symbol	I
2626,678	Cu I	10
2626,500	Fe II	6
2626,32	Ar IV	2
2625,711	Ar II	1
2625,666	Fe II	13
2625,64	Kr III	2
2625,51	F IV	0
2625,490	Fe II	9
2625,000	F III	7
2624,92	Ar IV	12
2624,78	Kr II	6
2624,71	Cl III	3
2624,593	Ar II	3
2624,52	Xe III	1
2623,721	Fe II	5
2623,69	O III	3
2623,532	Fe I	5
2623,11	Kr III	1
2623,090	Ar II	1
2622,90	Ne I	15
2622,90	C II	2
2622,875	Cu I	5
2622,85	N III	2
2622,82	Kr II	2
2622,41	O III	3
2622,32	O III	2
2621,879	Ar II	1
2621,87	Cl II	4
2621,668	Fe II	10
2621,39	Xe II	2
2621,36	Ar IV	12
2621,19	N III	1
2621,11	Kr IV	7
2621,1	Cs	2
2621,10	Ne	2
2620,985	Ar II	4
2620,82	Ca III	6
2620,695	Fe II	7
2620,6663	Cu II	5
2620,65	Kr II	6
2620,44	Kr II	40
2620,407	Fe II	6
2620,20	C II	3
2620,05	Cl III	4
2620,04	O IV	3
2619,98	Ar IV	6
2619,940	Ti I	10
2619,80	Cl II	4
2619,77	Ne	1
2619,22	Cs	2
2619,076	Fe II	7
2619,02	Ne	1,5
2618,78	Cl III	4
2618,366	Cu I	2500
2618,0191	Fe I	5
2617,66	Ca I	3
2617,616	Fe II	12
2617,596	Ar II	2
2617,513	Mg I	3
2617,3	F III	1
2617,26	Ar III	1
2617,149	Fe III	8
2616,97	Cl III	4
2616,811	Ar II	3
2616,71	Kr II	10
2616,627	C III	4
2616,62	Ne I	25
2616,27	Cs II	10
2615,87	Ne III	10
2615,68	Ar IV	12
2615,3	Kr IV	8
2615,19	Kr III	3
2615,13	Cl II	10
2614,726	Mg I	2
2614,65	Cl II	5
2614,62	Cs	8
2614,51	Ne III	4
2614,478	C III	5
2614,41	Cu II	8
2614,26	Ne I	5
2613,95	Ar III	3
2613,94	Ne I	2
2613,823	Fe II	13
2613,6	Cs	2
2613,59	Ne I	30
2613,44	Ar III	3
2613,41	Ne III	12
2613,357	Mg I	1
2613,083	F III	5
2612,45	C II	2
2612,1	Cs	2
2611,872	Fe II	13
2611,815	Na II	7
2611,476	Ti I	8
2611,45	Cl III	2
2611,42	Ne III	4
2611,288	Ti I	25
2611,24	Ar IV	3
2611,072	Fe II	6
2610,98	Kr II	10
2610,8	F III	0,5
2610,76	Kr II	1
2610,140	Cs II	1
2610,03	Ne III	15
2610,020	C III	6
2609,83	C III	1
2609,66	Kr III	1
2609,59	O III	4
2609,5	Kr IV	10
2609,50	Cl III	4
2609,44	Cs II	15
2609,31	Cu III	50
2609,122	Fe II	5
2608,90	Xe III	6
2608,72	Cl II	2

λ	Symbol	I
2608,44	Ar IV	7
2608,24	Cl II	2
2608,112	Fe III	7
2608,06	Ar IV	10
2607,52	Xe II	1
2607,087	Fe II	13
2606,93	Xe II	5
2606,8286	Fe I	6
2606,621	Mg I	5
2606,504	Fe II	7
2606,17	Kr IV	5
2606,045	F III	5
2605,67	Cl II	5
2605,6578	Fe I	6
2605,62	C II	1
2605,54	Xe II	50
2605,416	Fe II	6
2605,41	Kr II	1
2605,41	O III	6
2605,40	Cs	20
2605,303	Fe II	6
2605,26	Cu I	3
2605,121	Ti I	25
2605,1	Li II	1,5
2605,040	Fe II	6
2605,04	Cl III	2
2604,883	Ti I	3
2604,863	C II	4
2604,35	Kr III	8
2604,18	Cl II	8
2604,11	Ti II	2
2603,854	Mg I	4
2603,72	Cs	2
2603,72	C II	1
2603,59	Cl III	5
2603,36	Cl II	10
2603,161	C II	3
2602,9	F III	0
2602,495	Mg I	2
2602,39	C II	2
2602,12	Ar III	1
2602,11	Kr II	7
2602,02	C II	2
2601,42	C II	2
2601,16	Cl III	4
2601,05	C II	1
2600,956	Ar II	3
2600,551	F III	4
2600,36	Cs	20
2600,2711	Cu II	20
2599,885	Ti I	25
2599,570	Fe I	6
2599,47	Ar IV	12
2599,396	Fe II	14
2599,230	F III	8
2598,8125	Cu II	20
2598,7	Cs	2
2598,42	Xe II	2
2598,369	Fe II	14
2597,73	Kr II	7
2597,69	O III	8
2597,25	Ar III	3
2597,18	Al II	6
2597,01	Xe II	4
2596,95	Cs	20
2596,86	Xe II	5
2596,79	O III	00
2596,73	Kr II	5
2596,564	Ti I	10
2596,5	F III	1
2595,973	Mg I	2
2595,886	Cs II	3
2595,68	Ne III	20
2595,622	Fe III	8
2595,488	F III	7
2595,36	Kr II	4
2595,295	C IV	3
2595,21	Ne	6
2595,089	C IV	4
2594,965	Na II	1
2594,56	Ne I	2
2594,41	Ar III	1
2594,40	Kr II	4
2594,34	N IV	2
2593,97	Cl III	2
2593,919	Na I	2
2593,869	Na I	3
2593,726	Fe II	7
2593,65	Cu I	2
2593,647	Ti I	3
2593,60	Ne III	30
2593,231	Mg I	1
2593,195	F III	6
2592,804	F III	6
2592,779	Fe II	9
2592,71	C II	1
2592,65	F III	1
2592,627	Cu I	200
2592,48	Kr II	60
2592,45	Cl III	2
2592,43	F III	1
2592,178	Ar II	1
2592,074	Ar II	1
2591,891	Mg I	0
2591,845	C II	4
2591,696	Ar II	1
2591,69	Xe III	4
2591,543	Fe II	10
2591,44	N V	1
2591,410	C II	2
2591,25	Kr II	1
2591,17	Cs	20
2591,15	Ne I	3
2590,938	N II	5
2590,81	N V	2
2590,74	Kr II	2
2590,67	Ne I	10
2590,5290	Cu II	15

λ	Symbol	I	λ	Symbol	I
2590,45	Xe III	2	2577,19	F III	2
2590,34	Ca III	2	2577,1514	Si I	45
2590,247	Ti I	5	2577,13	Cl III	5
2590,09	Cs II	10	2577,12	Cu I	2
2590,04	Ne III	40	2576,97	Xe II	15
2589,48	Ne I	2	2576,865	Fe II	7
2589,47	Kr III	3	2576,74	Cs II	10
2589,3	Cs	2	2576,43	Ti III	5
2589,08	Kr II	30	2575,433	O II	1
2588,80	Cl III	3	2575,397	Al I	8
2588,285	Mg I	3	2575,300	O II	10
2588,23	O III	0	2575,095	Al I	10
2587,999	Fe I	8	2575,07	Cs II	3
2587,948	Fe II	7	2574,945	Mg I	2
2587,09	Ca III	3	2574,838	Fe III	7
2586,95	Al II	6	2574,826	C II	10
2586,9	Kr IV	3	2574,55	Ne I	8
2586,78	Kr III	3	2574,54	Cs II	10
2586,312	Na II	2	2574,368	Fe II	9
2586,274	Ti I	3	2574,13	Cl III	0
2585,876	Fe II	13	2573,596	F III	4
2585,558	Mg I	2	2573,09	Ca II	3
2584,88	Xe II	1	2573,03	Cs II	30
2584,5370	Fe I	8	2572,648	Ti II	5
2584,216	Mg I	1	2572,30	Xe III	1
2584,15	Kr II	3	2572,30	O III	2
2584,038	Fe III	6	2572,248	Mg I	1
2583,760	F III	7	2572,03	Kr II	10
2583,39	Ar III	3	2571,79	Cs II	2
2582,99	O III	3	2571,76	C II	1
2582,901	C I	5	2571,7563	Cu II	10
2582,585	Fe II	10	2571,476	O II	8
2582,5	Cs	2	2571,19	Kr III	6
2582,37	Fe III	8	2571,10	Cl II	8
2582,299	Fe I	6	2571,036	Ti II	20
2581,74	Kr II	5	2570,917	F III	4
2581,05	Cs II	0	2570,908	Mg I	0
2580,803	Ti I	5	2570,841	Fe II	7
2580,67	Cl III	6	2570,800	Cu I	10
2580,587	Mg I	1	2570,57	C II	2
2580,57	Cu I	5	2570,525	Fe II	5
2580,43	Ti III	5	2570,48	Kr III	10
2580,40	Cl II	4	2570,411	Ar II	4
2580,360	Ar II	1	2570,26	Xe III	1
2580,17	Ar III	2	2569,984	Ar II	3
2580,031	F III	6	2569,983	Al I	10
2579,846	F III	3	2569,888	Cu I	10
2579,428	Ar II	2	2569,601	Fe I	6
2579,29	Cu I	20	2569,53	Ar IV	7
2579,0	Kr IV	2	2569,21	O III	2
2578,98	Kr II	2	2569,202	Ar II	4
2578,62	Xe III	2	2569,11	O III	1
2578,36	Xe III	5	2568,69	Cs II	15
2578,27	O III	00	2568,6407	Si I	85
2578,26	Cl III	5	2568,405	Fe II	6
2578,24	O IV	4	2568,25	Cl II	3
2577,923	Fe II	9	2568,17	Cs II	10
2577,888	Mg I	0			

λ	Symbol	I
2568,13	Cl III	4
2568,07	Ar IV	10
2567,983	Al I	10
2567,727	Ar II	1
2567,53	Ti III	8
2567,330	Cu I	2
2567,095	Ar II	1
2566,912	Fe II	9
2566,61	Kr II	1
2566,23	Cl III	1
2566,218	Fe II	5
2566,01	Cl II	5
2565,782	Ar II	3
2565,68	Al II	4
2565,42	Ti III	8
2565,29	Cl II	15
2565,20	Ca I	2
2565,02	Cs II	1
2564,937	Mg I	1
2564,84	Cl II	20
2564,8242	Si I	20
2564,416	Ar II	4
2564,13	Cl II	6
2564,09	Ca I	3
2563,955	Cu I	3
2563,6787	Si I	30
2563,553	Cu I	3
2563,474	Fe II	12
2563,42	Ti III	15
2563,32	Na III	25
2563,319	N II	3
2563,29	Ar III	5
2563,25	Kr III	30
2563,167	Cu I	10
2562,534	Fe II	13
2562,52	Cl III	1
2562,413	F III	4
2562,305	Li I	5
2562,259	Mg I	1
2562,17	Ar IV	12
2562,090	Ar II	6
2562,04	Kr II	1
2561,954	Ar II	1
2561,943	N II	2
2561,94	Kr II	3
2561,79	Ne I	8
2561,545	N II	1
2561,48	Xe II	2
2560,941	Mg I	0
2560,89	Xe II	3
2560,853	Ar II	1
2560,37	Cs	8
2560,272	Fe II	7
2560,243	N II	3
2559,912	Fe II	5
2559,774	Fe II	5
2559,614	Al II	3
2559,50	Cl III	3
2559,281	Ar II	3
2559,210	Si III	14
2559,10	Kr II	8
2558,62	N II	0
2558,60	Ca I	2
2558,20	Ca I	2
2558,08	Kr IV	4
2558,06	O III	8
2558,00	Kr III	5
2557,93	F IV	2
2557,9	Cl III	3
2557,71	Al II	5
2557,55	Kr III	1
2557,4	Li I	—
2557,226	Mg I	0
2557,206	Si II	1
2557,18	Ca I	2
2556,78	Al II	3
2556,586	Ar II	4
2556,58	Ti III	1
2556,36	Kr II	6
2556,12	C II	0
2556,10	F II	4
2556,01	Al II	4
2555,988	Ti II	10
2555,91	Kr II	6
2555,66	C II	1
2555,59	F IV	00
2555,447	Fe II	7
2555,13	Kr III	10
2555,066	Fe II	5
2554,82	Ca I	2
2554,8	Cs	2
2554,530	Si II	10
2554,478	C II	3
2554,47	F IV	1
2554,25	Kr III	8
2554,20	Xe II	1
2553,81	Kr III	1
2553,61	Na III	25
2553,61	O IV	2
2553,422	N II	4
2553,400	Ar II	2
2553,32	Cu II	3
2553,29	Cu I	2
2553,193	Fe I	7
2553,16	Kr III	8
2552,29	F III	0
2552,12	Al II	5
2552,00	Cs II	2
2551,70	Xe II	3
2551,7	Li II	1
2551,64	N II	2
2551,61	F IV	1
2551,571	Ar II	1
2551,49	Kr III	2
2551,17	Cs II	10

λ	Symbol	I	λ	Symbol	I
2551,098	Fe III	6	2544,72	Kr III	3
2551,094	Fe I	8	2544,685	Ar II	6
2550,89	F III	1	2544,63	F III	4
2550,680	Fe II	8	2544,046	Si II	3
2550,65	Cs II	7	2543,98	Cl II	10
2550,23	Al II	3	2543,92	Cs	20
2550,020	Fe II	8	2543,920	Fe I	6
2550,02	K III	6	2543,872	Na I	2
2550,02	K II	6	2543,841	Na I	1
2549,85	Cl II	50	2543,45	C II	2
2549,788	Ar II	3	2543,431	Fe II	5
2549,68	F III	2	2543,4	F III	1
2549,62	O III	2	2543,384	Fe II	9
2549,6142	Fe I	10	2542,89	Na III	10
2549,51	Kr III	2	2542,767	F III	6
2549,453	Fe II	8	2542,733	Fe II	5
2549,399	Fe II	8	2542,68	O III	5
2549,30	Al II	1,5	2542,65	Cl III	2
2549,082	Fe II	7	2542,41	Ti III	1
2548,925	Fe II	5	2542,18	Cs II	1
2548,741	Fe II	7	2542,101	Fe I	6
2548,69	Ti III	1	2541,910	Ti I	20
2548,60	Kr III	4	2541,831	Fe II	7
2548,590	Fe II	6	2541,818	Si III	25
2548,55	Ti III	1	2541,75	Ti IV	8
2548,45	F IV	00	2541,49	Ca III	6
2548,43	Cs II	2	2541,40	Ca I	0
2548,10	F IV	0	2541,096	Fe II	7
2547,98	Ti III	0	2541,03	F III	1
2547,76	Cl II	12	2540,9734	Fe I	10
2547,48	Cu I	10	2540,88	C II	1
2547,45	O III	2	2540,84	Cl III	3
2547,35	C II	1	2540,83	Cs II	3
2547,330	Fe II	5	2540,70	Al II	1
2547,30	Ti IV	3	2540,666	Fe II	6
2547,184	Ar II	2	2540,55	Ar IV	4
2547,0	Kr IV	6	2540,39	C II	3
2546,98	F IV	0	2540,38	Cu I	5
2546,94	Cl II	20	2540,12	Al II	4
2546,866	Ar II	2	2540,037	Ar II	3
2546,85	Ti IV	12	2540,02	Ti III	15
2546,81	C II	2	2539,58	F III	00
2546,67	Kr III	1	2539,50	O III	2
2546,667	Fe II	8	2539,4	Li II	2
2546,43	O III	4	2539,3576	Fe I	7
2546,37	Xe II	3	2539,174	Cs II	5
2546,36	Kr III	1	2539,08	Cs II	10
2546,23	F III	3	2539,003	Fe II	10
2546,1	Cs	2	2538,997	Fe II	8
2546,093	Si III	10	2538,98	C II	2
2546,0	Kr IV	5	2538,94	O IV	3
2545,9795	Fe I	10	2538,809	Fe II	9
2545,642	Ar II	3	2538,67	Cs	2
2545,60	Al II	6	2538,500	Fe II	5
2545,215	Fe II	7	2538,34	Kr II	5
2544,972	Fe II	6	2538,205	Fe II	6
2544,84	Cl II	15	2538,02	Xe II	3
2544,8055	Cu II	*100*			

λ	Symbol	I
2537,873	N II	3
2537,57	Kr III	6
2537,49	N II	0
2537,16	Kr III	1
2537,142	Fe II	5
2536,86	Cu I	2
2536,817	Fe II	9
2536,673	Fe II	7
2536,67	Cu I	2
2536,62	F IV	1
2536,03	Cu I	2
2536,018	Ar II	7
2535,881	Ti II	10
2535,758	Ar II	1
2535,6086	Fe I	8
2535,480	Fe II	7
2535,250	Ar II	3
2534,712	Ar II	7
2534,640	Ti II	20
2534,416	Fe II	9
2534,10	Ar IV	—
2534,08	O III	6
2533,95	Cl III	1
2533,92	Ar III	3
2533,77	C IV	2
2533,644	F III	4
2533,627	Fe II	10
2533,44	Cs	20
2533,31	Xe III	2
2533,16	Al II	1
2532,655	Al II	2
2532,48	Cl III	5
2532,3814	Si I	110
2532,10	Al II	3
2531,76	Cl III	5
2531,76	O IV	2
2531,73	Kr II	1
2531,548	Na II	6
2531,46	Kr III	1
2531,36	Xe II	3
2531,266	Ti II	20
2530,66	F III	2
2530,6	C IV	6
2530,423	Ar II	1
2530,30	O II	8
2530,21	Na III	15
2530,18	Xe II	2
2530,103	Fe II	6
2529,98	C IV	11
2529,97	Mg III	2
2529,92	O IV	3
2529,866	Ti I	4
2529,547	Fe II	10
2529,545	Fe II	10
2529,52	Kr III	1
2529,36	O III	
2529,3048	Cu II	50
2529,221	Fe II	5
2529,1361	Fe I	10
2529,080	Fe II	5
2528,8	Cs	8
2528,679	Ar II	3
2528,5086	Si I	450
2528,49	Xe II	6
2528,318	Ar II	4
2528,08	Cl III	5
2528,08	O IV	2
2527,980	Ti I	5
2527,80	Ti III	15
2527,762	N II	2
2527,7	C IV	1
2527,4358	Fe I	15
2527,16	Kr II	3
2527,16	Fe I	5
2527,102	Fe II	6
2527,03	O II	1
2526,98	Xe II	12
2526,91	O II	7
2526,79	Xe II	12
2526,5929	Cu II	25
2526,477	Al II	1
2526,295	Fe II	9
2526,17	N II	0
2526,076	Ar II	2
2526,071	Fe II	5
2525,69	Ar IV	9
2525,68	Cs	20
2525,619	Ti II	30
2525,51	Kr III	2
2525,48	N II	0
2525,479	Ar II	4
2525,387	Fe II	10
2525,022	Fe I	7
2524,97	Kr III	10
2524,655	Ti II	8
2524,64	Cs II	0
2524,5	Kr IV	5
2524,488	N II	4
2524,48	Ar III	1
2524,46	Xe II	3
2524,41	C IV	9
2524,2939	Fe I	8
2524,27	Kr III	2
2524,1079	Si I	425
2524,09	Xe III	1
2523,7	C IV	4
2523,67	F IV	3
2523,66	Cs II	4
2523,658	Fe I	6
2523,20	O II	1
2523,11	Fe I	5
2523,09	O II	1
2522,8505	Fe I	40
2522,5	F II	0
2522,497	Ar II	4

λ	Symbol	I	λ	Symbol	I
2522,488	Fe I	6	2514,29	Xe II	5
2522,458	N II	4	2514,01	Cl II	3
2522,36	Cu III	25	2513,305	Al I	5
2522,227	N II	7	2513,28	Ar IV	12
2521,9197	Fe I	7	2513,15	Al II	1
2521,814	Fe II	7	2512,92	Kr III	6
2521,590	F III	4	2512,513	Fe II	5
2521,089	Fe II	7	2512,41	Cl II	2
2520,8	Cs	2	2512,363	Fe I	5
2520,791	N II	6	2512,260	Ar II	4
2520,64	Al II	2	2512,210	Na I	4
2520,534	Ti I	10	2512,128	Na I	2
2520,32	Kr III	2	2512,1	Cs	2
2520,222	N II	5	2512,065	C II	12
2519,6305	Fe I	10	2511,92	Kr III	1
2519,514	Al I	1	2511,759	Fe II	10
2519,45	Cl III	5	2511,74	Kr II	3
2519,38	Kr IV	6	2511,734	C II	5
2519,29	Kr III	6	2511,51	Cs	8
2519,222	Al I	4	2511,418	Fe III	6
2519,2023	Si I	350	2511,33	Cl II	3
2519,17	Cs II	00	2511,22	He II	50
2519,17	Xe II	6	2511,16	F III	1
2519,044	Fe II	7	2510,92	Cl III	4
2519,017	Ti I	8	2510,90	Ti II	2
2518,95	Cu II	8	2510,8362	Fe I	15
2518,40	Ar IV	6	2510,624	Ar II	3
2518,26	Ar III	2	2510,56	Kr II	5
2518,15	Cl II	4	2510,52	Xe III	3
2518,1029	Fe I	12	2510,37	Na III	20
2518,02	Kr IV	5	2510,2	Kr IV	2
2517,97	O II	6	2509,23	O IV	8
2517,95	Kr II	8	2509,122	Fe II	5
2517,659	Fe I	8	2509,121	C II	10
2517,506	Si IV	7	2508,91	Ar III	3
2517,448	Ti II	2	2508,,83	Li II	3
2517,40	O IV	7	2508,548	Ar II	1
2517,28	Ar IV	5	2508,31	F IV	0
2517,120	Fe II	6	2507,899	Fe I	6
2517,07	F III	3	2507,84	Kr III	1
2517,0	Kr IV	4	2507,77	O IV	7
2516,791	Ar II	6	2507,333	Ar II	3
2516,27	F IV	00	2507,08	Ne	1
2516,12	Xe II	12	2506,8973	Si I	425
2516,1125	Si I	500	2506,86	Xe II	8
2516,01	Ti III	20	2506,86	Kr III	5
2515,92	Cl II	3	2506,69	Ar III	5
2515,72	Cs II	10	2506,56	Kr II	5
2515,62	F III	4	2506,56	O IV	3
2515,598	Ar II	4	2506,53	Cs II	1
2515,57	F IV	2	2506,295	Na II	2
2515,460	Na II	2	2506,2732	Cu II	30
2515,42	Kr III	10	2506,093	Fe II	7
2515,272	Ar II	3	2505,485	Fe I	5
2515,14	Xe III	1	2504,738	Ar II	3
2515,01	F IV	2	2504,70	O IV	5
2514,383	Fe II	7	2504,60	K II	3
2514,3161	Si I	375			

λ	Symbol	I
2504,42	Ar III	4
2504,25	Al II	1
2504,23	Cl III	5
2504,20	O IV	2
2504,188	N II	4
2503,935	Ar II	4
2503,87	Kr II	7
2503,870	Fe II	7
2503,57	F IV	1
2503,560	Fe II	5
2503,325	Fe II	7
2502,75	Cl II	40
2502,390	Fe II	7
2502,2	Cs	2
2501,970	Si II	5
2501,84	O IV	4
2501,836	Ar II	4
2501,66	F IV	0
2501,1332	Fe I	20
2501,10	F IV	1
2501,04	Xe III	4
2500,928	Si II	3
2500,919	Fe II	5
2500,672	N II	4
2500,64	Kr III	8
2500,397	Ar II	5
2499,96	Ar III	1
2499,825	N II	2
2499,527	Ar II	4
2499,29	O IV	6
2498,95	F IV	0
2498,94	Ti II	2
2498,895	Fe I	10
2498,894	Fe II	10
2498,77	Kr III	3
2498,53	Cl II	30
2497,85	Al II	2
2497,820	Fe II	7
2497,72	F II	2
2497,71	Kr III	15
2497,67	Ca III	5
2497,58	Cu III	20
2497,221	Ar II	3
2497,10	O IV	3
2497,05	Na III	50
2496,97	N II	4
2496,93	Ar IV	3
2496,9	Cs	2
2496,83	N II	5
2496,79	F II	1
2496,5343	Fe I	6
2496,40	Ar III	5
2496,04	Cl II	20
2495,920	Ar II	2
2495,04	Cs	20
2494,92	N II	0
2494,90	Ar III	6
2494,89	Cu I	10

λ	Symbol	I
2494,71	N II	3
2494,66	Kr II	1
2494,114	Ar II	4
2494,01	Kr III	40
2493,999	Fe I	6
2493,940	N II	3
2493,75	O IV	10
2493,40	O IV	7
2493,261	Fe II	12
2493,180	Fe II	12
2493,16	N II	2
2493,153	Na II	5
2492,95	Ar III	3
2492,58	F III	0
2492,2	O III	00
2492,146	Cu I	2000
2492,013	Ar II	3
2491,984	Fe I	8
2491,78	Xe II	5
2491,46	N II	3
2491,392	Fe II	6
2491,37	C II	2
2491,35	Kr III	2
2491,21	N II	3
2491,1562	Fe I	20
2491,036	Ar II	4
2490,87	C II	2
2490,856	Fe II	6
2490,76	Xe II	20
2490,733	Na I	3
2490,6454	Fe I	30
2490,3	Cl III	5
2490,281	N II	4
2489,822	Fe II	8
2489,751	Fe I	15
2489,664	Cu II	5
2489,5	Cs	2
2489,39	Kr II	8
2489,11	Xe II	50
2489,0	O III	00
2488,950	Fe I	6
2488,86	Ar III	12
2488,746	N II	3
2488,74	Cs II	1
2488,3	O III	00
2488,1437	Fe I	40
2488,120	N II	2
2487,62	Kr II	4
2487,50	Kr II	3
2487,356	Fe II	5
2487,064	Fe I	12
2487,03	Kr III	1
2486,91	Cl III	5
2486,906	Ar II	3
2486,693	Fe I	10
2486,69	Xe III	3
2486,49	N III	3

λ	Symbol	I	λ	Symbol	I
2486,371	Fe I	10	2478,568	Fe II	6
2486,345	Fe II	7	2478,556	C I	16
2485,985	Fe I	10	2478,37	Kr III	2
2485,79	F IV	0	2478,05	F IV	1
2485,794	Cu II	20	2477,69	N IV	8
2485,63	Ar III	2	2477,58	Cs	20
2485,42	Cs	20	2477,29	Cl III	2
2485,378	Si IV	1	2477,21	Ti II	2
2485,35	Al II	1	2476,970	Ar II	2
2485,27	O III	00	2476,55	Ar III	6
2485,1	Cl III	3	2476,30	Al II	4
2484,87	Ar III	2	2476,10	Ar III	7
2484,56	N III	4	2476,07	Cs II	10
2484,360	F III	9	2475,89	Xe II	100
2484,27	Cl III	4	2475,73	O III	00
2484,243	Fe II	5	2475,462	Ar II	4
2484,187	Fe I	15	2475,31	F IV	0
2484,11	Ar III	6	2475,260	Al II	4
2484,06	F IV	1	2475,057	Li I	4
2483,62	Kr II	1	2474,90	Kr III	3
2483,531	Fe I	10	2474,818	Cu I	5
2483,43	Xe III	1	2474,8151	Fe I	8
2483,2718	Fe I	60	2474,765	Fe II	6
2483,24	O III	0	2474,69	Na III	40
2483,225	Ar II	2	2474,69	Kr II	2
2483,196	Si III	6	2474,252	Ar II	1
2483,0	Cs	2	2474,22	Ti II	2
2482,99	Kr III	2	2474,06	Kr IV	5
2482,85	N III	1	2473,998	Ar II	4
2482,816	Si IV	2	2473,96	Kr III	4
2482,654	Fe II	8	2473,40	Ne III	10
2482,60	O III	0	2473,3339	Cu II	20
2482,34	Cu III	30	2473,314	Fe II	6
2482,151	Ar II	4	2472,95	Ar III	8
2482,115	Fe II	8	2472,95	Al II	1
2481,77	Cl III	2	2472,909	Fe I	12
2481,508	Si III	3	2472,8962	Fe I	5
2481,478	Ar II	5	2472,69	Cl II	3
2481,04	Kr III	1	2472,52	Ca III	1
2480,861	C III	4	2472,426	Fe II	5
2480,858	Ar II	6	2472,34	Xe III	1
2480,73	O III	1	2471,92	Ar III	6
2480,7	Cs	2	2471,88	Cs II	1
2480,502	C III	4	2471,28	Xe III	3
2480,467	Ar II	3	2471,24	N III	00
2480,41	Cs II	6	2471,07	Cl III	5
2480,158	Fe II	8	2470,987	Ti I	3
2479,7774	Fe I	20	2470,658	Fe II	7
2479,77	F IV	0	2470,48	F III	3
2479,76	Ar III	3	2470,45	Kr II	10
2479,754	Cu I	10	2470,355	Ar II	3
2479,4813	Fe I	6	2470,279	F III	7
2479,055	Ar II	5	2470,18	Xe II	5
2478,85	Kr II	3	2469,876	Ar II	2
2478,82	Xe II	4	2469,712	Fe II	8
2478,79	Ar III	6	2469,58	Cs II	0
2478,709	F III	6	2469,46	Xe II	5
2478,64	Ti II	5	2469,20	Cl III	5

λ	Symbol	I	λ	Symbol	I
2468,86	Na III	30	2459,63	Kr III	5
2468,5006	Cu II	5	2459,45	O IV	3
2468,50	Mg III	3	2459,40	Na III	45
2468,43	Xe II	5	2459,36	Cl II	10
2468,43	Kr III	6	2459,26	N III	0
2468,37	Cl III	3	2459,23	Cs	2
2468,36	N III	0	2458,964	Fe II	5
2468,30	Ar III	2	2458,88	Cu I	5
2468,20	Ne III	4	2458,88	Al II	0,5
2468,12	Cs II	3	2458,782	Fe II	8
2467,732	Fe II	6	2458,05	Al II	2
2467,30	Ar IV	—	2457,954	Ar II	2
2467,10	Ar III	3	2457,8	O III	00
2466,811	Fe II	7	2457,74	Cu I	5
2466,8	Cs	2	2457,72	Kr III	10
2466,72	Cl II	2	2457,5980	Fe I	6
2466,670	Fe II	7	2457,55	Ne III	2
2466,60	Xe II	2	2457,525	Ar II	1
2466,3	Cs	2	2457,32	Cs II	1
2466,24	N III	1	2457,20	Al II	1
2466,162	F III	4	2456,92	F IV	5
2466,162	F II	4	2456,266	Ar II	2
2465,200	Fe II	7	2456,07	Kr II	6
2465,150	Fe I	6	2455,892	Fe II	10
2464,907	Fe II	7	2455,81	F III	0,5
2464,834	F III	8	2455,80	Cs	8
2464,77	Kr II	100	2455,628	Ar II	1
2464,62	Ar III	5	2455,31	Kr II	2
2464,26	Ar III	4	2455,235	Ar II	1
2464,007	Fe II	7	2455,22	Al II	2
2463,900	Fe II	5	2455,080	Ar II	5
2463,79	F IV	2	2455,04	Kr II	2
2463,728	Fe I	6	2454,99	O III	8
2463,38	Ne III	2	2454,98	Ne III	5
2463,280	Fe II	6	2454,706	Fe I	6
2463,27	Kr II	2	2454,63	Ar III	6
2463,04	N III	00	2454,567	Fe II	6
2463,02	Xe III	1	2454,270	Ar II	8
2462,998	Ar II	2	2454,21	O III	0
2462,76	Kr III	3	2454,12	Kr III	3
2462,6483	Fe I	10	2453,935	Fe II	25
2462,56	N III	1	2453,85	N III	4
2462,35	Ne III	6	2453,747	Fe II	15
2462,33	Kr II	2	2453,74	Kr III	1
2462,0	Cs	2	2453,47	Al II	1
2461,857	Fe II	8	2453,28	Kr III	8
2461,83	N II	0	2452,743	Ar II	2
2461,33	F V	1	2452,62	Xe III	3
2461,282	Fe II	8	2452,59	Al II	0,5
2461,270	N II	6	2452,58	Ar IV	4
2461,203	Ar II	1	2452,30	Cl II	10
2460,93	Cu I	5	2452,29	Kr III	10
2460,84	Ne III	1	2452,1180	Si I	70
2460,635	Ar II	2	2452,070	F III	7
2460,453	Fe II	5	2451,7	Kr IV	4
2459,953	Ar II	4	2451,58	F IV	4
2459,82	Al II	4	2451,56	F III	4
2459,74	Kr IV	6			

λ	Symbol	I	λ	Symbol	I
2451,52	Kr III	4	2441,622	F III	8
2450,63	F V	2	2441,60	Xe II	2
2450,541	Ar II	1	2441,288	Ar II	2
2450,44	Ti II	6	2441,06	O III	2
2450,134	Fe II	5	2440,976	Ti I	10
2450,06	O IV	10	2440,89	Kr III	5
2450,062	O III	2	2440,49	Cl II	4
2449,590	Mg II	6	2440,33	Cl II	5
2449,5	F III	5	2440,21	Ti II	5
2449,484	Si III	11	2440,107	Fe I	15
2449,407	Ar II	2	2440,05	Kr III	6
2449,38	O III	2	2440,028	Ar II	4
2449,36	O IV	8	2439,860	Fe II	8
2449,179	Ar II	2	2439,8	Cs	2
2448,58	Cl III	6	2439,78	Kr III	1
2448,21	Cu II	5	2439,744	Fe I	25
2447,92	Ti II	2	2439,69	Cl III	5
2447,747	Fe II	6	2439,34	Ne III	5
2447,743	Ar II	2	2439,300	Fe II	8
2447,71	Ar IV	8	2439,21	Kr III	6
2447,64	Xe III	1	2438,83	O III	5
2447,43	Ar III	1	2438,7674	Si I	65
2447,374	Fe III	7	2438,76	Xe II	1
2447,14	Cl III	6	2438,47	Cu III	25
2446,92	O III	00	2438,174	Fe III	8
2446,462	Fe II	5	2438,09	O II	1
2446,44	Kr II	8	2437,632	Fe II	20
2446,405	Fe II	25	2437,517	Ar II	2
2446,355	Ar II	1	2437,200	Ar II	1
2445,558	Fe II	7	2437,1	Cs	2
2445,55	O II	10	2437,100	Fe II	5
2445,34	Cl II	20	2436,987	Fe II	10
2445,210	Fe I	6	2436,615	Fe II	20
2445,114	Fe II	40	2436,48	Xe III	5
2444,828	Ar II	3	2436,344	Fe I	10
2444,512	Fe II	8	2436,1	Cl III	5
2444,40	Xe II	2	2436,06	O II	5
2444,274	Fo II	10	2435,62	F IV	2
2444,26	O II	5	2435,47	Xe II	6
2444,12	Cl II	7	2435,1545	Si I	300
2443,8728	Fe I	20	2435,1	Cl III	2
2443,842	Fe II	15	2434,988	Fe II	25
2443,69	Ar III	7	2434,96	O III	2
2443,3643	Si I	65	2434,944	Fe II	7
2443,32	Cu II	6	2434,822	Fe II	5
2443,24	Cs II	5	2434,81	C II	1
2443,219	Ar II	2	2434,733	Fe II	7
2443,2	Cs	2	2434,64	Kr III	2
2442,794	Ar II	2	2434,364	Ar II	2
2442,68	Kr IV	5	2434,24	C II	2
2442,67	Cu II	15	2434,229	Fe II	20
2442,67	Ti II	2	2434,13	F III	6
2442,567	Fe I	20	2434,10	Cl II	50
2442,54	Ca III	1	2434,067	Ti I	3
2442,47	Cl III	5	2434,052	Fe II	15
2441,90	Ne III	2	2433,95	F III	3
2441,67	O II	2	2433,62	Ne III	8
2441,637	Cu I	1000			

λ	Symbol	I	λ	Symbol	I
2433,538	O II	9	2424,27	Ar III	6
2433,49	C II	0	2424,247	Ti I	10
2433,26	Cl II	3	2424,143	Fe II	8
2433,211	Ti I	6	2424,01	Cl II	10
2432,90	C II	0	2423,93	Ar III	12
2432,8	Kr IV	1	2423,528	Ar II	5
2432,74	Kr II	8	2423,52	Ar III	12
2432,72	Xe II	12	2423,42	Si II	3
2432,71	Cs II	5	2422,94	Xe II	10
2432,6	Cs	8	2422,91	F III	3
2432,267	Fe II	7	2422,9	Cs	2
2432,12	C II	0	2422,84	O III	5
2431,923	Ar II	1	2422,78	F III	4
2431,69	O III	0	2422,695	Ar II	4
2431,66	O II	0	2422,47	Cl III	4
2431,62	Ar II	2	2422,44	Al III	5
2431.55	N IV	0	2422,12	Xe II	2
2431,025	Fe I	20	2422,089	Ar II	2
2431,08	Ca III	1	2421,81	Ar III	4
2431.07	N IV	2	2421,72	Si II	3
2431,04	Kr III	1	2421.65	N IV	3
2430,876	Fe II	10	2421,502	Ar II	3
2430,78	C II	1	2421,4	Cs	2
2430.41	N IV	3	2421,296	Ti I	10
2430,16	Cl II	30	2421,27	Xe II	20
2430,072	Fe II	7	2421,2	O III	00
2430,032	Ar II	5	2421,00	Na III	8
2430,0	Li II	1	2420,457	Ar II	6
2429,65	O III	0	2420,44	F III	3
2429,446	Ar II	2	2420,19	Si II	3
2429,35	Si III	7	2420,06	Cs	2
2429,35	O III	1	2419,85	Cl II	4
2429,148	Fe II	10	2419,5	Cl III	5
2428,970	Fe II	6	2419,413	Ar II	1
2428,92	Cu II	8	2419,36	F III	0,5
2428,92	Kr III	1	2419,164	Ar II	1
2428,523	Ar II	1	2418,82	Ar III	10
2428,45	Si II	10	2418,704	Ar II	1
2428,367	Fe II	6	2418,60	O II	1
2428,35	Kr II	20	2418,568	Fe III	7
2428,04	Kr IV	3	2418,46	O II	7
2428,02	Cl II	10	2418,41	Kr II	4
2427,79	Cl II	20	2418,362	Ti I	10
2427,70	Al II	3	2417,866	Fe II	6
2427,65	Cs	20	2417,214	Ar II	2
2427,48	Kr III	1	2416,9	Kr IV	4
2427,20	Ar III	4	2416,73	Xe III	4
2426,94	O III	2	2416,605	Cu I	5
2426,70	C II	2	2416,42	Cl III	7
2426.54	N IV	1	2416,05	F III	4
2426,41	Cs	8	2416,00	Ar III	3
2426,36	Kr II	10	2415,84	Ar III	4
2426,35	O III	0	2415,61	Ar III	7
2426,280	F III	4	2415,197	Cu I	5
2425,93	O III	2	2415,13	O II	4
2425,55	O II	2	2415,0	Cs	2
2425.414	Li I	3	2414,94	Kr II	2
2425,15	Cs	20	2414,89	Cs II	8
2425,05	Xe II	40			
2424.73	N IV	2			
2424,659	Ar II	4			
2424,49	Ar III	2			
2424,428	Cu II	50			

λ	Symbol	I
2414,89	Kr II	10
2414,84	Cu II	5
2414,78	Kr III	1
2414,77	Cs	2
2414,52	Xe III	1
2414,224	Ar II	5
2414,18	Cu II	5
2413,97	Ti III	15
2413,81	Kr II	10
2413,78	Ne III	10
2413,69	F III	3
2413,54	Ne III	6
2413,486	Ar II	1
2413,309	Fe II	9
2413,20	Ar III	10
2413,18	Ne III	8
2412,94	Ne III	12
2412,910	Ar II	1
2412,73	Ne III	15
2412,48	Cl II	10
2412,461	Ar II	4
2412,32	Cu III	15
2412,08	Cu III	4
2411,98	Cs	2
2411,60	O II	6
2411,58	Ti I	3
2411,066	Fe II	9
2411,01	Ar III	5
2410,94	Ar II	6
2410,85	Li II	1
2410,80	Ar III	4
2410,72	Xe II	7
2410,517	Fe II	9
2410,34	Ar III	4
2409,74	Xe II	40
2409,702	Ar II	1
2409,503	Ar II	2
2409,06	Kr II	5
2408,96	Cs II	0
2408,943	Ar II	1
2408,52	Kr II	5
2408,207	Ar II	2
2407,862	Ar II	2
2407,49	O II	6
2407,37	O II	1
2407,20	Ar IV	6
2407,10	Kr III	10
2407,10	Cl II	5
2406,90	Cs	2
2406,665	Cu I	1500
2406,658	Fe II	9
2406,647	Ar II	5
2406,58	Na III	18
2406,42	Kr IV	2
2406,41	O II	6
2405,92	Xe II	3
2405,776	Ar II	2
2405,49	Cu III	20
2405,228	Ar II	5
2405,19	Ne IV	1
2405,10	C IV	6
2405,01	F III	4
2404,98	Ar III	6
2404,882	Fe II	9
2404,864	Cu I	2
2404,59	Cl II	5
2404,50	Ar III	3
2404,44	C IV	5
2404,430	Fe II	7
2404,352	Ar II	9
2404,28	Ne IV	0
2403,87	Cl II	3
2403,76	Xe III	1
2403,65	Kr III	3
2403,551	Fe III	6
2403,3378	Cu II	100
2403,32	Cl III	5
2403,29	Kr III	1
2403,237	Ar II	5
2402,96	Kr III	3
2402,450	Fe II	8
2402,402	C II	7
2402,3	Li II	1
2402,10	Kr III	2
2402,05	N IV	5
2401,79	Xe II	2
2401,761	C II	5
2401,7	Cs	2
2401,58	Kr III	1
2400,33	Al III	3
2400,1138	Cu II	20
2400,10	Kr III	4
2399,851	Ar II	3
2399,85	Cl II	3
2399,372	Ar II	2
2399,239	Fe II	9
2399,237	Fe II	9
2399,15	Ar III	12
2398,98	Al III	5
2398,91	Cl II	2
2398,76	Xe II	4
2398,76	Al III	5
2398,559	Ca I	2
2398,372	Ar II	5
2397,548	Ar II	2
2397,29	F III	1
2396,86	Cs	2
2396,04	Mg III	3
2395,63	Ar III	10
2395,625	Fe II	9
2395,408	Fe II	7
2395,150	Mg I	4
2394,92	Cs	2
2394,73	Cl III	5
2394,355	Li I	2
2394,33	O III	4

λ	Symbol	I	λ	Symbol	I
2394,051	Na II	2	2380,80	Ti	4
2393,99	Cu II	2	2380,759	Fe II	7
2393,94	Kr III	40	2380,15	Cs	3
2393,835	Al II	2	2379,863	Ar II	3
2393,6	Cs	2	2379,54	Cs	2
2393,24	Cu II	2	2379,47	Cl III	5
2393,20	Ca III	3	2379,39	Cu II	3
2392,86	Cs II	15	2379,275	Fe II	7
2392,78	Kr II	10	2378,90	O III	4
2392,627	Cu I	2500	2378,83	Cu II	10
2392,33	Xe II	2	2378,395	Al I	7
2392,15	Xe II	2	2378,39	Cu II	2
2392,15	Al II	4	2378,145	Ti I	3
2391,73	Cu III	10	2376,80	Cs II	0
2391,35	Al II	1	2376,69	Kr III	1
2390,878	Ar II	2	2376,435	Fe II	5
2390,866	N II	4	2376,430	Ar II	1
2390,755	Al II	2	2376,29	Cu II	50
2390,50	Kr II	4	2375,82	Cs	2
2390,44	O III	8	2375,73	O II	4
2390,02	Cs	4	2375,52	Kr II	20
2389,9732	Fe I	25	2375,192	Fe II	7
2389,533	Fe III	8	2375,08	C II	4
2389,08	Al II	1	2375,02	Ti III	6
2388,627	Fe II	9	2374,59	Ti I	3
2388,268	Ar II	1	2374,5192	Fe I	10
2388,230	N II	3	2374,496	Al I	4
2388,20	O III	1	2374,255	Si II	5
2388,05	Kr IV	3	2373,68	Kr II	4
2387,933	Ar II	3	2373,6250	Fe II	8
2387,90	Kr III	1	2373,618	Fe I	20
2387,75	Xe II	4	2373,571	Al I	8
2387,3	Cl III	3	2373,548	Li I	1
2387,26	Cs	3	2373,4	Cs	2
2387,026	Na II	2	2373,351	Al I	15
2386,78	N II	1	2373,21	Ne IV	9
2386,14	Xe II	2	2373,122	Al I	7
2385,936	Ar II	1	2372,82	O III	2
2385,85	Xe II	1	2372,7	Cl III	0
2385,42	He II	30	2372,46	N III	2
2384,969	Ar II	4	2372,21	O III	3
2384,95	Ne IV	7	2372,16	Ne IV	7
2384,9441	Cu II	10	2372,070	Al I	10
2384,80	Cu II	5	2371,718	Ar II	4
2384,67	O IV	1	2371,662	Ar II	1
2384,516	Ti I	4	2371,4313	Fe I	15
2384,386	Fe II	7	2370,985	Si IV	3
2384,20	Ne IV	3	2370,7474	Cu II	20
2383,934	Ar II	2	2370,726	Al I	6
2383,92	O III	6	2370,496	Fe II	5
2383,486	Ar II	6	2370,49	N III	3
2383,26	Li II	2	2370,37	Cl III	6
2383,242	Fe II	7	2370,225	Al I	9
2382,955	Ar II	1	2369,960	Fe II	5
2382,32	O III	7	2369,916	Ar II	2
2382,039	Fe II	9	2369,8897	Cu II	100
2381,99	F III	2	2369,62	Xe II	4
2381,138	Ar II	3			

λ	Symbol	I
2369,4567	Fe I	8
2369,304	Al I	10
2369,187	Ar II	2
2368,94	Kr II	3
2368,75	Cs II	0
2368,68	Xe II	5
2368,612	Ar II	1
2368,595	Fe II	7
2368,19	Kr III	4
2368,15	Cu III	20
2368,15	Ar IV	3
2368,112	Al I	8
2367,611	Al I	8
2367,43	N III	4
2367,33	Na III	4
2367,248	Ar II	1
2367,2	F III	1
2367,053	Al I	6
2367,02	Ne	5
2366,972	Si II	5
2366,778	Ar II	2
2366,755	Si IV	2
2366,595	Fe II	5
2366,053	Si II	5
2365,77	Ne	6
2365,52	Kr II	3
2365,49	Al II	1,5
2365,49	Ne IV	4
2365,15	O II	3
2365,03	O II	1
2364,827	Cs	2
2364,826	Fe II	8
2364,81	Cs II	10
2364,70	Kr III	1
2364,33	Si II	3
2364,15	Cu II	3
2364,112	Ar II	5
2364,04	N II	0
2363,51	Fe III	7
2363,28	Ne IV	6
2363,26	Kr III	3
2363,220	Cu I	5
2363,21	Cu III	8
2362,866	Ar II	1
2362,85	Ne	5
2362,74	Kr II	6
2362,68	Cu II	2
2362,68	Ne IV	6
2362,50	Xe II	1
2362,083	Ar II	1
2362,019	Fe II	6
2361,82	Kr III	4
2361,56	Cu III	10
2361,19	Cu II	3
2360,59	Si II	5
2360,4	Kr IV	2
2360,293	Fe II	8
2360,26	Ar III	9
2360,20	Si II	10
2360,14	Kr III	3
2360,058	Ar II	4
2359,997	Fe II	8
2359,67	Cl III	6
2359,51	Ti IV	5
2359,23	Cs	2
2359,111	Fe II	8
2359,11	Ti IV	5
2359,104	Fe II	8
2358,917	Li I	1
2358,70	K II	1
2358,5	Kr IV	3
2358,48	Kr III	3
2358,408	Ar II	2
2357,97	Si II	50
2357,96	Ne IV	10
2357,9	Cs	2
2357,85	Cs II	5
2357,82	Ti II	2
2357,589	Ar II	5
2357,18	Si II	30
2356,95	Cs II	0
2356,90	N II	0
2356,72	Xe II	4
2356,6410	Cu II	10
2356,295	Si II	100
2356,25	Xe II	1
2356,12	Cs	2
2355,17	Ti II	2
2355,0141	Cu II	15
2354,889	Fe II	5
2354,793	Ar II	1
2354,466	Fe II	5
2354,44	Cs II	10
2354,42	Cs	2
2354,135	Ar II	6
2354,12	Ti II	3
2353,96	Cu II	2
2353,68	Kr II	50
2353,52	Xe II	1
2353,426	Ar II	3
2353,09	Si II	20
2352,86	Kr II	2
2352,731	Ar II	2
2352,52	Ne IV	8
2352,33	Ar III	5
2351,911	Cs	2
2351,8	Ar IV	—
2351,74	Cs	8
2351,67	Ar III	7
2351,56	Xe II	4
2351,40	Ca III	1
2351,198	Fe II	5
2351,18	Xe II	4
2350,84	Ne IV	6
2350,67	Ti II	2

λ	Symbol	I
2350,486	Ar II	5
2350,408	Fe I	5
2350,20	Al II	4
2350,174	Si II	20
2349,97	Ti II	3
2349,54	Si II	10
2348,910	Ar II	1
2348,90	O III	4
2348,74	Cu II	15
2348,352	Cu I	2
2348,303	Fe II	8
2348,27	Kr IV	4
2348,099	Fe II	8
2347,7	Cl III	2
2347,54	Al II	2,5
2347,46	Ti II	2
2346,78	Ti III	6
2346,570	Ar II	2
2346,17	Cu III	40
2345,92	Al II	1,5
2345,45	Kr III	6
2345,42	Ar III	5
2345,4	Ar IV	—
2345,327	Fe II	5
2345,17	Ar III	9
2344,69	Al II	1
2344,47	Xe II	12
2344,38	Cs	4
2344,38	Kr II	10
2344,278	Fe II	8
2344,204	Ar II	6
2344,203	Si II	10
2343,959	Fe II	6
2343,56	Ar III	3
2343,492	Fe II	8
2343,13	Cs II	8
2342,31	Ti II	3
2342,30	K II	3
2342,18	Xe II	3
2342,17	Cu II	3
2341,23	Ti II	3
2340,93	Kr IV	2
2340,64	Cl III	6
2340,60	Cl II	2
2340,47	Cs	2
2339,795	Ar II	4
2339,73	Cu II	3
2339,31	O II	3
2339,15	Kr IV	2
2339,01	Ti III	5
2338,961	Fe III	10
2338,005	Fe II	8
2337,88	Cs	4
2337,780	Ar II	6
2336,768	Fe III	10
2336,75	Kr IV	4
2336,45	Cl III	5
2336,17	Cu II	20
2335,90	Cs II	00
2335,90	C IV	2
2335,42	Xe II	2
2334,99	F II	0,5
2334,606	Si II	30
2334,54	Ti II	3
2334,404	Si II	30
2334,33	Ti III	3
2334,12	F II	1
2333,78	F II	2
2333,036	Ar II	2
2332,895	Ar II	1
2332,796	Fe II	8
2332,42	Cs	8
2331,67	Ti III	3
2331,452	Ar II	8
2331,35	Ti III	3
2331,306	Fe II	7
2331,16	O II	0
2330,0	Li II	1
2329,931	Si III	2
2329,93	F II	0,5
2329,905	Fe III	9
2329,578	Mg II	3
2329,357	Ar II	1
2329,3	Kr IV	3
2329,22	Kr III	3
2328,56	Si IV	2
2327,97	O II	2
2327,784	Ar II	2
2327,394	Fe II	7
2327,28	F VI	5
2327,04	Ti III	1
2326,498	Al II	2
2326,340	N II	3
2326,948	Fe III	10
2325,497	Al II	3
2325,16	N II	0
2324,85	Kr IV	1
2324,83	O II	0
2324,427	Ar II	3
2324,358	Fe III	8
2324,20	Al II	4
2323,57	Kr IV	1
2323,50	Cl III	6
2323,35	F VI	7
2323,02	Cl II	4
2323,01	Cu II	8
2322,81	N III	1
2322,32	Kr III	1
2322,23	N III	0
2322,15	O II	3
2322,081	Ar II	2
2322,00	Cl II	1
2321,71	Fe III	10
2321,650	N II	4
2321,562	Al I	9

λ	Symbol	I
2321,28	Cl II	1
2321,1	Cs	6
2321,07	Cs II	10
2320,3585	Fe I	40
2320,33	N III	00
2320,28	Cu III	8
2320,25	Cl II	2
2319,941	N II	4
2319,70	Xe II	7
2319,68	O II	4
2319,561	Cu I	500
2319,52	O III	2
2319,466	Fe III	8
2319,37	Ar III	10
2319,220	Fe III	10
2319,13	Ar III	10
2319,057	Al I	5
2318,83	Mg III	1
2318,35	Ar III	2
2318,09	N IV	6
2318,04	Ar III	12
2317,87	Kr III	1
2317,745	Ar II	5
2317,482	Al I	7
2317,47	Ar III	15
2317,37	O III	3
2317,35	N III	0
2317,046	N II	8
2317,00	Ar III	9
2316,80	Xe II	10
2316,79	O II	3
2316,690	N II	6
2316,493	N II	7
2316,32	Kr II	10
2316,299	Ar II	8
2316,12	O II	3
2315,70	Fe III	10
2315,68	Cs II	6
2315,65	Na II	0
2315,52	Kr II	8
2315,52	O III	4
2315,39	F VI	9
2315,306	Ar II	3
2315,25	N II	0
2315,10	Cu III	4
2314,983	Al I	4
2314,970	Ar II	6
2314,76	O III	2
2314,66	Na III	3
2314,56	N III	1
2314,24	Kr II	6
2313,77	Al II	1
2313,720	Ar II	7
2313,70	Xe II	5
2313,526	Al I	6
2313,1048	Fe I	40
2313,05	O II	3
2312,491	Al I	5
2312,47	Cs II	0
2312,31	Cu III	5
2312,29	Xe III	1
2312,13	N II	0
2312,00	Kr II	6
2311,83	F IV	3
2311,58	O III	2
2311,16	Cs II	0
2311, 035	Al I	4
2310,87	Ca III	0
2309,96	Na III	30
2309,860	Ar II	2
2309,53	N II	1
2309,51	Cu II	6
2309,148	Ar II	6
2308,9999	Fe I	30
2308,70	O III	1
2308,191	Si III	10
2307,72	O II	1
2307,71	Cs II	5
2307,456	Ar II	2
2307,28	Xe II	3
2307,27	Ne	2
2307,266	Ar II	2
2306,61	Ne	6
2306,42	Si III	2
2306,31	Ne V	2
2306,22	He II	20
2305,859	Ar II	2
2305,665	Ti I	12
2305,50	Ne	2
2304,87	Ne	4
2304,627	Fe I	5
2304,60	Xe II	1
2303,94	Ne	3
2303,73	Xe III	1
2303,5815	Fe I	20
2303,422	Fe I	15
2303,116	Cu I	1000
2303,0585	Si I	55
2303,012	Fe III	7
2303,00	Kr III	2
2302,92	Ar III	6
2302,83	O II	5
2302,808	Fe III	8
2302,730	Ti I	10
2302,67	Kr II	3
2302,17	Ar III	15
2302,077	Ar II	4
2301,825	Ar II	3
2301,73	Kr II	6
2301,6849	Fe I	20
2301,173	Fe I	6
2300,930	Si III	8
2300,85	Ar III	10
2300,38	Kr II	6
2300,38	Ne	2
2300,35	O II	8

λ	Symbol	I
2300,179	Ar II	5
2300,139	Fe I	30
2299,98	Xe II	6
2299,852	Ti I	10
2299,72	Ar IV	4
2299,47	Al III	3
2299,47	Cu II	7
2299,36	Xe II	2
2299,2209	Fe I	25
2299,15	Kr III	3
2298,96	Ne	1
2298,662	Fe I	6
2298,51	Cl III	5
2298,36	Al III	0
2298,31	F III	3
2298,29	F IV	5
2298,1699	Fe I	10
2297,879	Ar II	2
2297,82	F IV	2
2297,7877	Fe I	35
2297,15	Ar III	5
2297,14	Na III	25
2296,9279	Fe I	15
2296,873	Si III	10
2296,870	C III	16
2296,79	K II	1
2296,64	Na III	25
2296,52	Xe II	30
2296,24	Ar III	4
2295,859	Fe III	15
2295,476	Si III	6
2295,401	Si I	10
2295,349	Ar II	3
2294,91	Ar III	5
2294,57	Xe II	15
2294,406	Fe I	25
2294,3683	Cu II	40
2294,200	Ti I	3
2294,17	F IV	2
2294,05	Ar III	3
2293,8482	Fe I	25
2293,842	Cu I	2500
2293,745	Ti I	3
2293,49	Ne IV	6
2293,32	O II	6
2293,318	N II	4
2293,14	Ne IV	2
2293,056	Fe III	10
2293,03	Ar III	12
2292,652	N II	3
2292,5249	Fe I	30
2292,40	Xe II	20
2292,39	Ar III	5
2292,25	Ar III	4
2292,130	Ar II	4
2291,850	Fe III	6
2291,81	Cl III	4
2291,652	N II	4
2291,38	Cl II	4
2291,28	Kr III	3
2291,26	Kr IV	6
2291,120	Fe I	15
2291,034	Si I	35
2290,998	Cu II	15
2290,88	O II	6
2290,84	Xe II	2
2290,61	Ar III	6
2290,545	Fe I	9
2290,52	Kr III	1
2290,425	Ar II	3
2290,259	N II	3
2289,84	N II	0
2289,771	Ar II	5
2289,6074	Si I	20
2289,376	Ar II	2
2289,36	Ne VI	1
2289,31	Ar III	4
2289,031	Fe I	10
2288,82	Ar III	2
2288,765	Ar II	4
2288,444	N II	5
2288,36	O III	00
2288,20	O III	2
2288,17	Cl II	7
2288,12	O III	00
2287,79	Kr II	30
2287,630	Fe I	15
2287,2505	Fe I	30
2287,21	O III	1
2287,041	Si IV	5
2286,925	Ar II	4
2286,69	F IV	1
2286,689	N II	6
2286,68	Cs II	5
2286,642	Cu II	15
2286,40	O III	0
2286,0	Cl III	3
2285,94	Xe II	8
2285,801	Ar II	4
2285,79	Ne IV	9
2285,72	Na III	35
2285,66	O III	2
2285,612	Ar II	1
2285,52	Al II	2
2285,24	Xe II	2
2285,22	F IV	2
2285,17	Al II	3
2285,07	O III	00
2284,89	O II	3
2284,60	Cs	2
2284,0864	Fe I	40
2283,994	Ar II	7
2283,93	Cl III	7
2283,753	Ar II	1
2283,6557	Fe I	12

λ	Symbol	I	λ	Symbol	I
2283,652	N II	4	2273,58	Ne	20
2283,42	O II	3	2273,51	N III	1
2283,3045	Fe I	9	2273,280	Ti I	8
2283,243	Ar II	7	2273,24	Kr II	8
2283,079	Fe I	9	2272,816	Fe I	8
2283,07	Kr II	30	2272,8	Cl III	1
2282,621	Ar II	8	2272,79	Cs	2
2282,61	Ne V	1	2272,76	Cs II	0
2282,21	Ar III	7	2272,640	Ar II	2
2281,512	Ar II	1	2272,613	Ti I	8
2281,50	Cs II	00	2272,55	Kr II	1
2281,22	Ar III	7	2272,42	N III	0
2280,85	Ar III	5	2272,0703	Fe I	15
2280,72	F IV	0	2271,97	F IV	3
2280,222	Fe I	8	2271,79	N III	0
2280,02	Cs II	2	2271,778	Fe I	40
2279,964	Ti I	12	2271,69	Cu III	5
2279,96	Cs II	2	2270,91	C V	3
2279,924	Fe I	10	2270,8628	Fe I	18
2279,85	Na III	3	2270,43	N III	2
2279,79	Kr III	4	2270,20	C II	2
2279,68	Ar III	4	2269,70	C II	2
2279,47	Ar III	3	2269,36	C II	0
2279,45	Cu III	2	2269,30	N III	0
2279,36	F II	3	2269,222	Al I	1
2279,10	Ar III	10	2269,14	Ti II	3
2278,98	Ne	10	2269,0990	Fe I	18
2278,48	Na III	40	2269,096	Al I	7
2278,432	Fe III	6	2268,95	Cl III	5
2278,34	Cl III	5	2268,91	C II	1
2278,33	Cu II	7	2268,78	Ti I	4
2278,281	Si I	10	2268,72	Xe II	1
2277,92	C V	2	2267,98	Ti I	4
2277,820	Fe III	8	2267,96	Na III	8
2277,672	Fe I	12	2267,77	C II	0
2277,43	Cu III	4	2267,61	Cs II	20
2277,25	C V	1	2267,466	Fe I	15
2277,096	Fe I	9	2267,42	Fe III	10
2276,870	Fe III	8	2267,29	Cs II	3
2276,703	Ti I	10	2267,28	N III	3
2276,54	Ca III	2	2267,111	Ar II	2
2276,2582	Cu II	35	2267,080	Fe I	9
2276,25	Cl II	4	2266,98	Ne III	5
2276,0263	Fe I	12	2266,903	Fe I	10
2275,471	Ca I	1	2266,80	Xe II	3
2275,358	Ar II	3	2266,441	Ar II	2
2275,187	Fe I	6	2266,16	Ne III	8
2275,054	Ar II	1	2266,08	Cl III	2
2274,923	Ar II	3	2266,014	Al I	3
2274,74	Cu II	3	2265,94	Xe II	2
2274,54	Ne V	0	2265,87	N III	0
2274,12	N III	0	2265,71	Ne V	6
2274,088	Fe I	9	2265,62	Xe II	3
2274,00	Fe III	8	2265,36	Cu II	7
2273,98	Cs II	0	2265,215	Ar II	4
2273,83	Cs II	20	2265,0546	Fe I	20
2273,76	Kr III	3	2265,04	K II	5
2273,65	F IV	2			

λ	Symbol	I
2264,91	Ne III	10
2264,55	Cu II	2
2264,54	Ne IV	4
2264,390	Fe I	45
2264,20	Xe II	2
2264,11	Ne III	3
2264,020	Ti I	5
2263,780	Cu II	35
2263,738	Al I	1
2263,474	Fe I	6
2263,463	Al I	7
2263,39	Ne V	3
2263,212	Cu II	8
2263,21	Ne III	12
2263,079	Cu I	2200
2263,068	Ar II	2
2262,95	Xe II	2
2262,80	O II	1
2262,632	Ar II	2
2262,16	Ne III	2
2262,08	Ne IV	5
2261,592	Fe III	12
2261,23	Ti II	3
2260,547	Fe III	7
2260,528	Cu I	1300
2259,76	Kr III	6
2259,66	O II	2
2259,587	Si I	10
2259,57	Ne V	3
2259,5109	Fe I	15
2259,42	Kr IV	1
2259,22	Xe II	1
2258,35	Cs II	5
2258,342	Ar II	1
2258,02	Ne IV	6
2258,008	Al I	1
2257,965	Ar II	1
2257,82	Cs II	12
2257,788	Fe II	25
2257,406	Fe III	8
2256,897	Fe II	10
2256,79	C II	0
2256,56	Xe II	1
2256,545	Ar II	3
2256,33	Ca III	0
2256,19	C II	2
2256,10	Cs	3
2256,05	Ne V	1
2255,859	Fe I	45
2255,691	Fe II	50
2255,68	C II	1
2255,64	Cl III	2
2255,408	Ar II	3
2255,29	K II	3
2255,23	C II	0
2255,178	Ar II	1
2254,975	Cu II	6
2254,58	Cs II	15
2254,283	Ar II	5
2254,066	Fe II	8
2253,26	Ti II	2
2253,22	Ne VI	3
2253,16	Cl II	30
2253,07	Cl III	7
2253,00	Cu II	2
2252,90	O II	3
2252,72	F V	2
2252,71	He II	10
2252,65	Ca III	2
2252,54	Kr IV	2
2252,248	Ar II	6
2251,8749	Fe I	12
2251,84	Cu II	2
2251,831	Fe II	80
2251,50	Cl II	40
2251,44	Na III	45
2251,403	Ar II	2
2251,17	Na III	20
2250,96	Cl II	20
2250,7911	Fe I	10
2250,32	Kr II	8
2250,09	Ti II	2
2250,00	O II	1
2249,86	Xe II	4
2249,658	Ar II	1
2249,347	Ar II	3
2249,181	Fe II	25
2249,175	Fe II	10
2249,063	Fe II	30
2248,960	Cu II	25
2248,88	N III	25
2248,857	Fe I	25
2248,80	Cs II	0
2248,73	Ar III	7
2247,92	N III	6
2247,76	Ne VI	1
2247,692	Fe II	35
2247,65	N III	2
2247,503	Cu I	2
2247,002	Cu II	75
2246,66	Na III	40
2246,56	Cs	2
2245,975	Ar II	3
2245,6536	Fe I	15
2245,505	Fe II	45
2245,48	Ne V	3
2245,410	Ar II	2
2245,39	Kr II	10
2245,116	Ar II	2
2244,690	Ti I	7
2244,31	Ca III	2
2244,265	Cu I	2300
2244,216	Fe II	8
2244,17	Na III	3
2244,080	Ar II	1

λ	Symbol	I	λ	Symbol	I
2243,662	Ar II	5	2231,571	Cu II	30
2243,405	Fe III	8	2231,512	Fe II	10
2243,10	Cu II	6	2231,423	Ar II	5
2243,05	Al II	4	2231,2138	Fe I	15
2243,0	K II	2	2231,16	Cl III	3
2242,6183	Cu II	50	2231,024	Ar II	1
2242,579	Fe I	15	2230,948	Cu II	30
2242,29	Ar III	6	2230,79	Xe II	1
2242,14	Cu II	6	2230,69	Kr III	1
2242,10	C II	1	2230,492	Ti I	7
2242,05	Cs II	2	2230,40	Cu II	10
2241,86	Xe II	2	2230,317	Ar II	3
2241,858	Ar II	2	2230,30	Na III	50
2241,54	Fe III	12	2230,244	Ti I	4
2241,426	Fe II	20	2230,087	Cu II	30
2241,05	C II	1	2230,084	Cu I	2500
2241,028	Ar II	6	2229,850	Cu II	30
2240,89	K	4	2229,67	Ti I	7
2239,906	Ar II	1	2229,66	O II	0
2239,89	O II	0	2229,648	Ar II	8
2239,43	Na III	45	2229,56	Na III	15
2239,10	O III	2	2229,267	Fe III	10
2239,047	Fe II	25	2229,18	F V	1
2238,974	N II	4	2229,12	Cs	5
2238,750	Ti I	8	2229,0735	Fe I	5
2238,454	Cu I	1100	2228,88	Cs II	10
2238,20	Ti I	6	2228,8700	Cu II	40
2238,155	Fe III	10	2228,761	Fe II	30
2237,82	Ti III	1	2228,1722	Fe I	10
2237,721	Ar II	2	2228,15	O III	3
2237,577	Fe II	20	2227,92	Kr II	30
2237,385	Ar II	1	2227,848	Fe III	7
2237,34	Kr IV	3	2227,775	Cu I	1600
2237,34	Cu I	5	2227,42	Ne V	3
2237,21	N III	—	2227,298	Ar II	5
2237,15	Kr II	4	2227,14	Ti II	2
2236,527	Ar II	3	2227,01	Cs II	00
2236,29	Ne V	2	2226,,798	Ti I	6
2236,278	Cu I	900	2226,773	Cu II	40
2235,908	Fe III	10	2226,19	Na III	8
2235,904	Ar II	2	2225,90	Na III	45
2235,760	Ar II	3	2225,697	Cu I	2100
2235,699	Fe III	6	2225,662	Ar II	6
2235,35	Xe III	1	2225,29	Na III	12
2235,208	N II	4	2225,11	Ti I	8
2234,673	Ar II	6	2224,701	Cu II	15
2234,57	Cs II	0	2224,550	Ar II	1
2233,809	Ti I	8	2224,12	Ne V	1
2233,654	Fe III	6	2223,199	Ti I	7
2233,478	Ar II	4	2222,79	Na III	0
2233,28	Cs	3	2222,763	Fe I	7
2232,690	Fe III	10	2222,066	Ar II	3
2232,548	Fe III	8	2222,01	Si III	2
2232,430	Fe III	10	2221,830	Fe III	10
2232,41	Ne V	4	2221,65	Cu II	2
2232,35	Kr III	1	2221,352	Ar II	1
2232,17	Na III	40	2220,81	Ne IV	1
2231,65	N III	—			

λ	Symbol	I
2220,51	Cs II	9
2220,453	Fe II	6
2220,388	Fe II	25
2220,347	Ar II	2
2219,962	Ar II	6
2219,889	Fe II	20
2219,87	Ca III	1
2219,75	Ti I	5
2219,22	Ca III	2
2219,14	Kr III	1
2218,9148	Si I	50
2218,805	Ar II	4
2218,70	O II	2
2218,504	Cu II	25
2218,41	N II	0
2218,38	Ti I	5
2218,375	Ar II	1
2218,289	Fe II	30
2218,1082	Cu II	50
2218,0569	Si I	120
2217,91	Cs II	3
2217,34	F III	5
2217,34	F II	5
2216,6688	Si I	120
2216,190	Ar II	4
2216,07	Ne III	15
2215,67	O II	1
2215,654	Cu I	1000
2215,60	Kr III	2
2215,100	Cu II	35
2215,094	Fe II	10
2214,77	Ne III	4
2214,67	He II	6
2214,581	Cu I	1600
2214,17	Na III	25
2214,147	Ar II	1
2214,059	Fe II	20
2213,76	Ne III	12
2213,679	Fe II	20
2213,56	Al III	2
2213,15	Cs II	5
2212,96	Kr II	5
2212,741	Cu II	10
2212,63	Ne III	5
2212,40	Cs II	0
2212,29	Kr II	6
2211,85	Ne III	10
2211,7441	Si I	110
2211,71	Kr II	5
2211,30	Cs	3
2211,243	Fe II	12
2211,2364	Fe I	7
2211,16	Na III	1
2211,112	Fe II	5
2211,07	F IV	1
2210,952	Fe II	5
2210,883	Ar II	3
2210,8940	Si I	115
2210,6894	Fe I	9
2210,53	K II	4
2210,321	Ar II	2
2210,2684	Cu II	60
2210,073	Fe III	6
2210,060	Al I	2
2209,795	Cu II	30
2209,66	Al III	1
2209,61	Cs	3
2209,35	Ne III	10
2209,049	Fe II	20
2208,85	Fe III	10
2108,611	Ca II	20
2208,419	Fe II	30
2208,41	Kr II	1
2208,04	Ne III	4
2207,9783	Si I	110
2207,29	Ne III	8
2207,0692	Fe I	6
2206,94	F III	3
2206,153	Fe II	8
2206,088	N II	6
2205,95	Ne III	5
2205,738	Ar II	4
2205,65	Cu I	5
2204,98	Ne III	7
2204,698	Ar II	1
2204,668	Al I	1
2204,619	Al I	1
2204,40	Ca III	3
2204,16	Ne	2
2203,89	Ne	6
2203,88	Ne IV	2
2203,633	N II	3
2202,78	Na III	40
2202,54	C III	1
2202,458	Fe III	8
2202,22	Ne III	7
2202,135	Ar II	1
2201,595	Fe II	5
2201,573	Ar II	1
2201,242	Ar II	2
2201,23	Ne III	4
2200,82	Ne	5
2200,728	Ca I	1
2200,722	Fe I	15
2200,68	Cs	8
2200,498	Cu II	25
2200,37	Na III	2
2200,370	Fe I	10
2199,752	Cu I	1300
2199,583	Cu I	1700
2199,30	Ti III	1
2199,183	Al I	1
2198,01	Na III	2
2197,86	Ne III	7
2197,84	Cu II	1
2197,787	Ca II	2

λ	Symbol	I	λ	Symbol	I
2197,786	Ar II	1	2187,444	Fe II	12
2197,506	N II	4	2187,320	Ar II	6
2197,36	F IV	0	2187,1950	Fe I	40
2197,273	Fe II	5	2186,97	O III	1
2197,15	Cs	6	2186,876	Fe III	6
2197,10	Ne III	3	2186,62	Ne	3
2196,389	Ar II	1	2186,61	He II	4
2196,0428	Fe I	50	2186,485	Fe I	40
2195,674	Cu II	25	2186,31	Cs	3
2195,532	Fe III	6	2185,622	Fe II	8
2195,445	Ar II	5	2185,52	Kr II	5
2195,43	O II	2	2185,489	Ar II	5
2194,92	Ne III	5	2185,13	N III	1
2194,907	Ar II	2	2184,06	Ar III	8
2194,85	Na III	1	2183,980	Fe III	6
2194,251	Al II	1	2183,803	Fe II	10
2193,08	Na III	1	2183,468	Fe II	8
2192,74	Ne	7	2183,301	Fe II	12
2192,674	Fe II	5	2183,30	Ca III	3
2192,607	Al II	1,5	2183,24	Ne III	2
2192,52	N III	1	2182,85	Cu II	6
2192,2678	Cu II	75	2182,74	Na III	15
2192,224	Ar II	2	2182,64	O II	4
2192,06	Ar III	15	2182,28	Ne III	3
2191,935	Fe II	10	2182,14	Cs II	5
2191,838	Fe I	60	2182,13	Cs	5
2191,579	Ar II	4	2182,049	Si III	3
2191,45	Ne	1	2181,720	Cu I	1700
2191,44	O II	2	2181,66	O III	1
2191,39	N III	3	2181,378	Ar II	1
2191,287	Ar II	5	2181,41	Cu II	4
2191,27	Ca III	2	2181,407	Fe II	5
2191,215	Fe III	8	2181,211	Ar II	7
2191,2052	Fe I	10	2181,137	Fe II	8
2191,16	Ne	4	2180,996	Al I	8
2190,511	Ar II	4	2180,89	Ne III	10
2190,42	O II	2	2180,8692	Fe II	12
2190,235	Ar II	2	2180,836	Si III	4
2190,29	Ne III	7	2180,789	Ar II	1
2190,18	Na III	5	2180,74	Cu II	10
2190,00	K II	6	2180,63	Na III	0
2189,6323	Cu II	50	2180,410	Fe III	12
2189,62	C II	1	2180,255	Fe II	12
2189,51	O II	1	2180,247	Ar II	1
2189,47	Cs II	10	2180,089	Ar II	2
2189,36	Cu II	3	2179,60	Cs II	10
2189,20	O II	2	2179,399	Cu II	60
2188,72	C II	1	2178,944	Cu I	1600
2188,52	N III	3	2178,69	Ne III	4
2188,492	Ar II	3	2178,37	Mg III	3
2188,39	C II	2	2178,090	Fe I	35
2188,27	N III	5	2177,79	Kr II	3
2188,22	Ar III	10	2177,73	Ne III	8
2187,88	Cs	3	2177,61	Cs II	3
2187,87	Cs II	3	2177,432	Si I	10
2187,868	Fe II	15	2177,396	Al I	6
2187,678	Fe II	10	2177,22	Ar III	25
2187,48	C II	1			

λ	Symbol	I	λ	Symbol	I
2177,025	Fe II	10	2166,19	Ar III	15
2176,963	C III	4	2165,860	Fe I	20
2176,894	Si III	5	2165,821	Ar II	6
2176,8414	Fe I	6	2165,555	Fe II	10
2176,826	Fe II	20	2165,32	O III	3
2176,67	Ne III	5	2165,24	He II	2
2176,387	Ar II	1	2165,093	Cu I	1300
2175,959	Ar II	2	2164,915	Al I	2
2175,636	Ar II	10	2164,558	Fe II	25
2175,445	Fe II	25	2164,547	Fe I	7
2174,968	Cu II	35	2164,38	Kr II	4
2174,849	Fe II	8	2164,351	Ar II	2
2174,658	Fe III	15	2164,339	Fe II	20
2174,585	Ar II	5	2163,8633	Fe I	6
2174,43	Na III	3	2163,77	Ne III	15
2174,190	Ar II	2	2163,54	Cs	8
2174,168	C II	3	2163,51	Ca III	4
2174,071	Al I	2	2163,370	Fe II	20
2173,848	C II	5	2163,368	Fe I	10
2173,84	Na III	1	2163,32	Na III	1
2173,829	Fe III	7	2162,944	C III	9
2173,720	Fe II	15	2162,88	O III	5
2173,220	Fe II	20	2162,68	Ti II	4
2173,2146	Fe I	8	2162,50	Kr III	3
2173,209	Ar II	1	2162,50	Kr II	3
2172,989	Fe II	15	2162,292	Ar II	2
2172,679	Fe II	8	2162,023	Fe II	20
2172,637	Ar II	2	2161,895	Ar II	3
2172,581	Fe I	6	2161,65	O II	0
2172,341	Ar II	2	2161,582	Fe II	20
2172,25	Kr III	1	2161,5802	Fe I	5
2171,817	Cu I	200	2161,314	Cu II	20
2171,60	Ca III	5	2161,313	Fe II	20
2171,44	F IV	4	2161,270	Fe III	10
2171,418	Ar II	5	2161,22	Ne III	10
2171,312	Ar II	3	2161,161	Fe II	15
2171,2976	Fe I	40	2161,04	Ne III	6
2171,045	Fe III	12	2160,88	Ne III	2
2171,038	Ar II	1	2160,655	Fe III	6
2170,914	Ar II	1	2160,52	O II	0
2170,83	Kr III	2	2160,40	Ca III	2
2170,63	Na III	0	2160,388	Al I	3
2170,23	Ar III	20	2159,927	N II	3
2170,193	Fe II	5	2159,60	Ne III	4
2169,994	Fe II	12	2159,50	Ti II	3
2169,950	Fe II	12	2159,44	Ne III	5
2169,562	Cu I	300	2159,199	Fe II	25
2169,431	Fe II	10	2159,152	Fe II	10
2168,925	Fe II	8	2159,09	Ti II	5
2168,826	Al I	2	2159,046	Ar II	2
2168,26	Ar III	10	2158,883	Ar II	2
2167,880	Fe II	12	2158,755	Ar II	2
2167,70	Cs II	3	2158,518	Fe II	25
2167,401	Fe II	12	2158,49	Fe I	6
2166,952	Fe III	12	2158,472	Fe III	12
2166,773	Fe I	100	2158,43	Kr III	1
2166,67	Na III	3	2158,29	Ti II	2
2166,198	Fe II	20			

λ	Symbol	I	λ	Symbol	I
2157,795	Fe I	5	2146,91	Cu II	8
2157,710	Fe III	12	2146,823	Ar II	2
2157,53	Ar III	3	2146,75	Cs II	10
2157,280	Si III	2	2146,59	N III	00
2156,38	Ar III	3	2146,339	Fe III	6
2156,28	C II	1	2146,062	Fe III	8
2155,839	Fe II	12	2146,058	Fe II	10
2155,588	Ar II	1	2145,74	N III	1
2155,58	Ti II	4	2145,616	Fe III	6
2155,39	C II	0	2145,58	C III	0
2154,70	Ti II	4	2145,555	Al I	3
2154,70	C II	0	2145,48	Cu II	10
2154,635	Al III	1	2145,08	Kr II	10
2153,980	Ar II	3	2144,743	Fe III	7
2153,373	Fe II	12	2144,70	Cu II	2
2153,281	Fe II	5	2144,282	Fe III	8
2153,15	Ne III	2	2143,96	N III	0
2153,068	Ar II	3	2143,827	Fe III	7
2153,06	Cs II	0	2143,81	Ca III	1
2153,0075	Fe I	5	2143,52	Ti I	6
2152,706	Fe III	6	2143,470	Fe III	8
2152,488	Fe II	25	2143,045	Fe III	7
2152,47	Ca III	6	2142,775	N II	6
2152,373	Fe II	12	2142,72	Cu I	5
2151,801	Cu II	25	2142,67	N III	0
2151,78	Ne III	3	2142,49	Kr III	1
2151,776	Fe III	15	2142,49	C III	1
2151,774	Fe II	25	2142,263	Ar II	2
2151,61	N III	0	2142,05	Ti I	5
2151,26	Ne III	5	2141,682	Ar II	2
2151,052	Ar II	6	2141,30	Cs	10
2150,78	Ca I	1	2140,92	C III	1
2150,762	Fe II	10	2140,747	Ar II	2
2150,70	Ne III	8	2140,56	Cu I	2
2150,699	Al I	5	2140,39	Ca III	6
2150,618	Fe II	20	2139,86	C III	1
2150,537	Ar II	2	2139,676	Fe II	25
2149,96	N III	0	2139,48	Cs II	2
2149,92	Ne III	6	2139,41	Ti I	5
2149,47	O II	0	2139,007	N II	4
2149,40	Cu I	10	2138,882	Ar II	3
2149,07	Ar III	3	2138,70	Kr III	1
2148,99	N III	1	2138,59	Ar III	10
2148,9838	Cu II	60	2138,533	Cu I	500
2148,73	Ar III	8	2138,103	Fe II	20
2148,58	Kr III	2	2137,897	C II	5
2148,47	N III	3	2137,735	Fe II	15
2148,38	Ar III	5	2137,417	C II	3
2148,23	O II	0	2137,365	Fe III	8
2148,09	N III	3	2136,73	Ar III	3
2147,95	Ar III	6	2136,560	Si II	50
2147,911	Si I	50	2136,519	Fe II	20
2147,904	Fe III	7	2136,402	Si II	30
2147,79	N III	2	2136,17	Cs II	3
2147,719	Fe II	15	2135,9815	Cu II	75
2147,681	Ar II	2	2134,861	Fe III	9
2147,37	Na III	1	2134,733	Al I	2
2147,27	N III	4			

λ	Symbol	I
2134,72	Mg III	3
2134,3413	Cu II	35
2133,99	Si II	10
2133,990	Fe II	8
2133,87	Ar III	15
2133,85	Kr II	2
2133,77	Cs II	0
2132,304	Ca II	1
2131,99	O II	4
2131,85	Cs	10
2131,76	O II	5
2131,505	Ca II	2
2131,49	Ca III	3
2131,23	Cu II	2
2130,762	Cu I	50
2130,548	Fe II	12
2130,429	Ar II	6
2130,259	Fe II	15
2130,179	N II	5
2130,08	Cu II	8
2129,810	Ar II	3
2129,80	Kr II	1
2129,75	Kr III	1
2129,663	Al I	1
2129,54	Ne	6
2129,427	Ar II	4
2129,20	Ca III	6
2128,750	Ca II	0
2128,22	Ar III	6
2127,967	Fe II	10
2127,89	Ar III	3
2127,69	Cs	6
2127,646	Ar II	3
2127,467	Si IV	4
2127,050	Ar II	2
2126,89	Ti I	5
2126,668	Ar II	4
2126,0449	Cu II	50
2125,706	Ar II	1
2125,50	Ar III	3
2125,272	Ar II	1
2125,24	Cu II	4
2125,16	Ar III	10
2125,098	Cu II	8
2124,80	Cs II	0
2124,35	Cu I	5
2124,27	Ne	7
2124,1225	Si I	100
2123,590	Fe III	8
2123,50	Ti I	7
2123,48	Kr II	3
2123,39	O II	0
2123,362	Al I	1
2122,994	Si I	15
2122,9793	Cu II	50
2122,34	Ar III	8
2122,27	Cs II	1
2121,90	Ti I	6
2121,542	Ar II	1
2121,306	Ar II	2
2121,1945	Si I	10
2120,179	Si IV	3
2119,985	Ar II	3
2119,125	Fe I	5
2119,050	Fe II	12
2118,948	Ar II	1
2118,83	Kr II	12
2118,567	Fe III	6
2118,38	Cu II	2
2118,195	Fe II	8
2117,934	Ar II	1
2117,08	Na III	0
2117,633	Fe II	25
2117,300	Cu II	35
2117,01	Ti I	6
2116,960	Fe II	25
2116,70	Na III	8
2116,687	Ar II	5
2116,588	Fe III	7
2116,00	Kr III	1
2115,55	Cs II	0
2115,1697	Fe I	20
2115,090	Ar II	1
2114,72	C II	0
2114,631	Si I	30
2114,532	Ar II	1
2114,488	Fe I	25
2113,891	Fe III	6
2113,45	Mg III	2
2113,26	Cu I	2
2113,146	Ca II	1
2113,08	Fe I	20
2112,966	Fe I	25
2112,757	Ca II	10
2112,65	Cs II	5
2112,1001	Cu II	30
2111,30	Cu II	6
2111,274	Fe I	20
2110,896	Ar II	2
2110,747	Ar II	2
2110,724	Fe II	15
2110,240	Fe II	25
2110,233	Fe I	30
2109,861	Fe I	25
2109,81	Kr II	5
2109,613	Fe II	25
2109,37	Cl II	2
2109,11	Cs	6
2109,097	Fe II	10
2109,046	Ar II	2
2108,959	Fe I	30
2108,886	Ar II	1
2108,302	Fe I	12
2108,139	Fe II	15
2108,139	Fe I	12

λ	Symbol	I
2108,068	Ar II	2
2107,62	Na III	1
2107,555	Fe II	10
2107,324	Fe III	10
2107,13	O III	5
2106,537	Ar II	1
2106,39	Cu II	2
2106,380	Fe I	25
2106,260	Fe I	20
2106,247	Ar II	1
2106,07	O III	5
2105,935	Ar II	3
2105,45	K II	1
2105,112	Cu I	800
2104,885	Ar II	3
2104,7971	Cu II	40
2104,24	C IV	1
2103,94	C IV	2
2103,799	Fe III	12
2103,353	Ar II	5
2103,235	Ca II	10
2103,213	Si I	3
2103,08	Ti IV	10
2103,048	Fe I	25
2102,99	Cl II	3
2102,910	Fe I	20
2102,3542	Fe I	30
2102,33	Ne	2
2101,49	Cs	8
2101,467	Ar II	1
2101,29	O II	4
2100,961	Fe III	8
2100,7984	Fe I	30
2100,69	O II	1
2100,46	C III	0
2100,144	Fe I	10
2099,91	O II	1
2099,68	Al II	5
2099,60	Na III	1
2099,59	Ne	4
2099,50	Cs II	4
2099,34	Ne	10
2099,332	Fe III	6
2098,97	Kr II	1
2098,953	Fe I	25
2098,72	Cu II	2
2098,62	Mg III	2
2096,56	Ca III	5
2098,386	Cu II	15
2098,181	Fe II	25
2098,123	Ar II	1
2098,081	Fe I	15
2098,00	Ne	1
2097,692	Fe III	12
2097,52	Cs II	0
2097,512	Fe II	25
2097,480	Fe III	15
2097,43	Ne	2
2096,856	N II	5
2096,808	Ar II	1
2096,430	Fe III	6
2096,24	Kr II	15
2096,23	Ne	12
2096,192	N II	4
2095,976	Ar II	1
2095,582	N II	6
2095,54	Ne III	20
2095,20	Al II	5
2095,02	Kr II	1
2094,8	Al II	5,5
2094,77	Cu II	2
2094,30	Al II	6
2094,211	Si I	10
2094,183	N II	3
2094,15	Ne	2
2093,683	Fe II	35
2093,660	Fe I	40
2093,64	Ne	3
2093,606	Cu II	10
2093,37	Kr II	3
2093,13	C II	1
2092,945	Fe III	6
2092,90	O II	1
2092,764	Ar II	3
2092,64	Mg III	4
2092,44	Ne III	12
2092,337	Ar II	3
2091,999	C III	6
2091,97	Cs II	8
2091,90	Ne	4
2091,83	Cs	8
2091,63	C II	2
2091,627	Ar II	5
2091,316	N II	3
2091,312	Fe III	7
2091,17	C II	2
2090,862	Fe I	20
2090,380	Fe I	30
2090,240	Fe III	6
2090,139	Fe III	12
2090,053	Fe III	7
2089,43	Ne III	15
2089,27	O III	2
2089,20	Ne	2
2089,089	Fe III	6
2088,92	Ne III	5
2088,71	Cs II	8
2088,16	Kr II	20
2087,930	Cu II	35
2087,907	Fe III	7
2087,718	Ar II	1
2087,527	Fe II	25
2087,525	Fe I	25
2087,44	Ne III	7
2087,132	Fe III	8

λ	Symbol	I
2087,0	Al II	5
2086,96	Ne III	10
2086,816	Ar II	2
2086,73	Kr II	5
2086,55	Mg III	2
2085,56	Ne	5
2085,295	Cu II	8
2084,54	Kr II	1
2084,4669	Si I	10
2084,349	Fe III	10
2084,33	Cu II	2
2084,117	Fe I	50
2083,87	Kr II	1
2083,75	Cs	8
2083,530	Fe III	6
2083,01	Cs	7
2082,92	Cu II	2
2082,109	Ar II	3
2081,5	Al II	2
2080,912	Fe II	20
2080,53	Kr II	1
2080,357	Ar II	5
2080,34	N IV	6
2080,05	Cs II	8
2080,03	Cu II	2
2079,968	N II	3
2079,86	N III	6
2079,654	Ar II	4
2079,529	Cu I	20
2079,06	Na III	0
2078,989	Fe III	14
2078,95	Ne	15
2078,646	Cu II	40
2078,164	Fe II	8
2077,81	Cu III	2
2077,507	Fe II	12
2077,43	Cs II	8
2077,14	Cs	8
2076,944	N II	4
2076,29	Cs	8
2076,178	Ar II	2
2075,683	Fe II	5
2075,04	Si III	2
2074,89	Ca III	2
2074,74	N III	2
2074,195	Fe II	8
2074,17	O II	1
2074,003	Ar II	1
2073,8	Al II	3
2073,60	Cs II	1
2073,426	Ar II	4
2073,38	Na III	10
2073,147	Fe II	8
2072,86	N III	1
2072,701	Si II	200
2072,016	Si II	200
2071,821	Fe II	10
2071,79	N III	2
2070,63	N III	5
2070,539	Fe III	8
2070,330	Fe II	8
2069,952	Fe II	10
2069,92	Cu II	2
2069,55	Ca III	2
2068,321	Cu I	5
2068,25	N III	6
2068,243	Fe III	12
2067,917	Fe II	20
2067,64	Ca III	2
2067,50	Ti IV	15
2067,41	Na III	18
2067,302	Fe III	6
2066,25	Cu II	8
2066,005	Fe II	15
2065,54	Mg III	5
2065,516	Si I	30
2065,23	Na III	10
2065,18	Ne	20
2064,335	Fe II	25
2064,212	Ar II	5
2063,99	N III	10
2063,761	Ar II	2
2063,672	Fe II	25
2063,50	N III	10
2063,13	Cs II	00
2062,97	Na III	8
2062,62	Ne	2
2062,41	Cu II	10
2062,17	Ca III	3
2061,751	Fe III	9
2061,552	Fe III	10
2061,192	Si I	40
2060,079	Ar II	1
2059,677	Fe III	7
2059,190	Ar II	1
2059,014	Si II	50
2058,72	Na III	8
2058,646	Si II	50
2058,560	Fe III	8
2058,47	Cs	3
2058,136	Si I	15
2058,10	Cs II	3
2058,087	Ar II	3
2058,00	Cs	1
2057,61	Cs	3
2057,514	Ar II	5
2057,332	Fe II	12
2057,058	Fe III	6
2056,64	Ca III	3
2056,30	Cs	4
2056,145	Fe III	7
2056,13	Mg III	3
2055,93	Ne VI	3
2055,855	Fe III	6
2055,270	Fe II	20
2055,17	Na III	10

λ	Symbol	I
2054,9795	Cu II	50
2054,828	Si I	50
2054,54	Ti II	3
2054,43	Cu II	6
2054,27	Cu II	4
2052,53	O III	4
2052,16	C II	2
2051,90	Na III	8
2051,79	C II	2
2051,75	Cs II	0
2051,688	Fe II	25
2050,794	Ar II	5
2050,739	Fe III	7
2050,324	Ar II	2
2049,94	He II	—
2049,913	Si III	2
2049,384	Fe III	7
2048,67	Na III	8
2048,492	Fe II	5
2048,36	Ca III	3
2047,995	Ar II	2
2047,72	Al II	1
2047,65	Cu II	8
2047,14	Ca III	3
2046,65	Ca III	4
2046,492	Ar II	4
2046,25	Cs II	00
2046,00	Cs II	00
2045,62	Cu I	5
2045,41	Na III	18
2045,41	O III	5
2043,8031	Cu II	60
2043,37	Cu III	5
2042,382	Ne VI	3
2042,355	Ar II	3
2041,61	Ca III	4
2041,49	Ti II	3
2041,345	Fe II	25
2041,204	Fe I	25
2041,09	Na III	0
2040,687	Fe II	25
2040,23	Mg III	3
2039,93	Al II	3
2039,507	Fe III	6
2039,490	Ar II	3
2038,35	Ca III	3
2037,70	Na III	10
2037,1269	Cu II	30
2036,435	Fe II	20
2036,42	N IV	1
2036,10	N IV	4
2035,8539	Cu II	30
2035,84	Na III	3
2035,62	N III	2
2035.57	N IV	5
2035,15	Cs II	7
2035,02	N III	3
2034,88	Cl III	3
2034,760	Ar II	2
2033,78	Cs II	0
2033,46	Ca III	4
2033,064	Fe II	25
2032,65	Na III	0
2032,173	Ar II	3
2032,14	Cl III	3
2031,10	Na III	10
2031,023	Cu II	15
2030,01	Ca III	1
2029,182	Fe II	8
2028,558	Ar II	1
2028,55	Na III	25
2028,38	Ca III	2
2027,778	Fe II	5
2027,72	Ca III	2
2026,68	Ca III	2
2026,602	Ar II	2
2025,824	Mg I	9
2025,4902	Cu II	8
2025,44	Na III	8
2025,183	Ar II	3
2025,05	Cs II	5
2024,733	Ar II	2
2024,335	Cu I	200
2024,21	Cl III	3
2023,118	Ar II	2
2022,83	O II	1
2022,73	Ar II	1
2022,29	Cs II	0
2022,27	Na III	10
2022,192	Ne IV	4
2022,14	Al II	2
2021,46	Cl III	3
2021,45	O II	1
2020,83	Ca III	3
2020,739	Fe II	25
2020,44	O II	2
2020,19	Cl III	3
2019,427	Fe II	25
2018,77	Ca III	4
2018,754	Ar II	2
2018,441	Ne IV	3
2018,38	C II	2
2017,94	C II	1
2017,090	Fe II	15
2017,090	Fe I	15
2016,885	Cu II	8
2016,84	C III	1
2016,654	Si II	3
2016,60	O II	2
2016,512	Fe I	5
2016,154	Fe II	10
2016,092	Fe II	10
2016,09	Al II	1
2015,7	C III	0
2015,576	Cu II	5
2015,500	Fe II	20
2015,319	Ar II	2
2014,311	Ar II	1

λ	Symbol	I
2014,10	Ca III	3
2013,268	Fe II	15
2012,96	Cu II	15
2011,88	Na III	30
2011,34	Cl III	1
2010,974	Si I	30
2010,688	Fe II	25
2010,094	C III	5
2009,90	Ca III	1
2009,570	C III	4
2009,327	C III	2
2008,494	Fe III	6
2008,439	Si I	15
2008,43	Na III	8
2007,711	Fe II	12
2007,452	Fe II	15
2007,215	Fe I	15
2007,178	Ar II	1
2007,013	Fe II	12
2006,84	Cl III	4
2006,260	Fe I	15
2005,83	Cs II	00
2005,55	Mg III	0
2005,33	Na III	6
2005,24	Na III	30
2004,914	Ar II	3
2004,80	Na III	0
2004,143	Fe III	8
2003,97	Cl III	0
2003,903	Ar II	2
2003,325	Ar II	1
2003,09	Ca III	3
2002,72	Cl III	3
2000,96	Ca III	4
2000,78	Cu III	3
2000,368	Fe II	30
2000,228	Fe III	9
2000,000	Ar II	2
1999,79	Ca III	4
1999,7000	Cu II	60
1999,588	Fe III	9
1999,430	Fe II	10
1997,345	Ne VII	1
1996,5	Cs	5
1996,420	Fe III	12
1995,62	Na III	3
1995,563	Fe III	12
1995,266	Fe III	7
1994,857	Fe II	20
1994,073	Fe III	13
1993,627	C I	2
1993,289	Fe II	8
1993,262	Fe III	7
1992,858	Fe III	6
1992,196	Fe III	9
1992,060	Ne VII	3
1992,017	Fe III	9
1991,9	Cs	1
1991,848	Si I	50
1991,64	N II	—
1991,613	Fe III	14
1990,53	Al II	7
1990,1	Cs	1
1989,975	Fe III	7
1989,61	Ca III	2
1989,2116	Cu II	30
1988,9950	Si I	15
1988,620	Ar II	3
1988,51	C II	1
1988,09	C II	2
1987,76	C II	3
1987,503	Fe III	15
1987,33	C II	1
1986,3637	Si I	10
1985,58	Na III	30
1985,1	Cs	1
1984,434	Si I	30
1984,288	Fe III	9
1984,027	Fe III	7
1983,831	Ar II	1
1983,7	Cs	1
1983,61	Cl III	5
1983,296	Ar II	1
1983,2341	Si I	20
1982,805	Fe III	8
1982,076	Fe III	6
1981,974	Ne VII	6
1981,74	Ar II	1
1981,5	Cs	1
1981,394	Ar II	2
1980,95	Na III	0
1980,6203	Si I	15
1980,129	Fe I	25
1979,988	Ar II	1
1979,62	C III	1
1979,46	Cl III	3
1979,3124	Cu II	50
1979,31	Mg III	1
1979,2062	Si I	15
1979,16	C III	2
1978,702	Xe III	3
1978,63	Ca III	3
1977,5982	Si I	15
1977,56	Mg III	1
1977,14	Na III	1
1977,02	Cu II	15
1976,8	Cs	1
1976,765	Ar II	3
1976,62	Na III	1
1976,126	Fe III	8
1975,58	Na III	0
1974,5	Cs	1
1974,467	Ar II	3
1973,780	Ar III	4
1973,4837	Ar II	2

λ	Symbol	I	λ	Symbol	I
1972,6	Xe II	5	1956,58	Mg IV	0
1972,270	Ar II	2	1956,48	Na III	0
1972,01	Ca III	1	1956,026	Fe I	30
1971,57	Cs II	3	1955,690	Fe I	20
1971,57	Mg III	0	1955,31	Na III	8
1970,489	Cu II	15	1955,3	Cs	1
1968,21	Cs II	2	1954,975	Fe III	8
1968,03	Ca III	5	1954,966	Si I	100
1967,99	Cu II	2	1954,87	Mg III	0
1966,952	Ar II	1	1954,223	Fe III	10
1966,91	Cs II	2	1953,80	N III	3
1966,740	Fe III	8	1953,66	N III	3
1965,309	Fe III	8	1953,488	Fe III	10
1965,23	Al II	4	1953,322	Fe III	13
1965,04	Na III	18	1953,06	Ca III	4
1964,776	Fe III	8	1952,997	Fe I	20
1964,70	Ca III	5	1952,648	Fe III	11
1964,330	Fe II	12	1952,596	Fe I	30
1964,260	Fe III	7	1952,56	Cu II	5
1964,25	O II	0	1952,262	Fe I	20
1964,169	Fe III	8	1952,20	N III	1
1964,043	Fe I	20	1952,16	Ca III	3
1963,84	O II	2	1951,556	Fe I	25
1963,629	Fe I	15	1951,43	N III	2
1963,110	Fe II	25	1951,21	Na III	40
1963,110	Fe I	25	1951,007	Fe III	12
1962,871	Fe I	20	1950,79	Na III	15
1962,746	Fe I	15	1950,334	Fe III	10
1962,74	Ar III	2	1950,223	Fe I	20
1962,67	Al II	7	1949,81	N III	4
1962,3	Cs	0	1949,564	Si II	100
1962,24	O II	3	1949,331	Si II	10
1962,18	Mg III	0	1949,22	N III	6
1962,164	Ar II	3	1948,79	Ti III	5
1962,100	Fe I	30	1948,372	Fe II	10
1962,031	Fe I	25	1948,31	Ca III	5
1961,4	Cs	2	1946,99	N III	5
1961,3610	Ar II	4	1946,978	Fe I	25
1961,236	Fe I	20	1946,8000	Ar II	2
1961,230	Fe III	6	1946,70	Na III	0
1960,76	Na III	20	1946,49	Cu II	10
1960,70	Al II	3	1946,43	Na III	20
1960,34	O II	1	1946,219	Fe I	10
1960,318	Fe III	13	1946,20	Mg IV	0
1960,129	Fe I	30	1945,504	Si II	3
1959,324	Fe III	8	1945,35	Al II	5
1958,739	Fe I	15	1945,342	Fe III	12
1958,598	Fe I	30	1945,294	Fe I	25
1958,583	Fe III	11	1945,111	Ar II	1
1958,29	Al II	1	1945,1	Cs	1
1958,18	Ca III	2	1945,070	Fe I	20
1958,121	Fe II	5	1944,99	Na III	3
1957,938	Fe III	6	1944,586	Si II	15
1957,831	Fe I	25	1944,586	Cu II	25
1957,83	Ar III	1	1943,481	Fe III	14
1957,51	Cu II	20	1943,40	Na III	6
1957,42	O II	0	1943,12	Ca III	6
1957,02	Ti III	0			

λ	Symbol	I	λ	Symbol	I
1942,3	Cs	4	1926,27	Na III	45
1942,19	Na III	6	1926,18	Ti III	0
1941,77	Na III	0	1926,013	Fe III	10
1941,667	Si II	50	1925,99	Al II	2
1941,61	Na III	0	1925,99	Mg IV	0
1941,50	Mg III	0	1925,987	Fe II	20
1941,40	Ti III	4	1925,0	Cs	0
1941,2	Cs	0	1924,81	Al II	4
1941,0724	Ar II	3	1924,532	Fe III	6
1940,649	Fe I	25	1923,88	Kr III	0
1940,018	Fe III	8	1923,877	Fe III	7
1939,72	Ca III	4	1923,87	Mg III	3
1939,32	Na III	0	1923,86	N III	2
1939,30	Al II	5	1923,4	Cs	2
1938,95	Mg III	0	1923,35	Cl II	4
1938,827	Ne II	8	1923,31	C III	2
1938,901	Fe III	10	1923,14	C III	4
1938,899	Fe II	8	1923,11	N III	2
1938,8	Cs	2	1923,003	Fe III	7
1937,80	Mg III	2	1922,93	C III	5
1937,4	Cs	1	1922,797	Fe II	20
1937,345	Fe III	14	1922,789	Fe III	15
1937,274	Fe I	35	1922,13	Cu II	5
1937,042	Ar II	1	1921,630	Xe III	2
1936,96	Al II	4	1921,49	N III	4
1936,781	Fe II	20	1920,86	N III	8
1936,0	Cs	0	1920,665	Cu II	5
1935,83	Al III	10	1920,32	Cl III	4
1935,79	Ca III	3	1920,12	Na III	6
1935,54	Na III	0	1920,016	Ar II	2
1935,296	Fe II	15	1920,0	Cs	1
1935,2	Cs	8	1919,99	N III	2
1935,18	Ti III	3	1919,71	N III	2
1934,75	Al II	10	1919,515	Ar III	4
1934,54	Al II	10	1919,44	N III	1
1934,528	Fe I	25	1919,3	Cs	1
1933,87	Na III	30	1919,197	Ar II	3
1933,694	Ar II	2	1919,06	N III	0
1933,59	Mg III	0	1918,76	Mg III	1
1932,477	Fe II	15	1918,69	N III	0
1932,43	Al II	5	1918,667	Ar III	4
1932,231	Ar II	2	1918,480	Fe III	7
1931,507	Fe III	14	1918,46	Na III	6
1931,421	Ar II	1	1918,284	Fe III	7
1931,027	C III	4	1918,06	Ar III	1
1930,905	C I	10	1917,960	Fe III	6
1930,9	Cs	1	1917,87	Cl III	4
1930,64	Mg III	3	1917,453	Fe III	9
1930,387	Fe III	15	1917,351	Fe III	8
1930,033	Ne II	8	1917,337	Fe II	15
1930,03	Al II	5	1916,53	Cl III	4
1929,74	Cu II	25	1916,48	O III	2
1929,34	Ti III	1	1916,081	Ne II	10
1928,787	Ne II	1	1915,6	Cs	3
1928,715	Cu III	2	1915,564	Ar III	7
1927,21	Na III	15	1915,083	Fe III	15
1926,99	Al II	1	1914,653	Ar III	3
1926,304	Fe III	18			

λ	Symbol	I	λ	Symbol	I
1914,6	Cs	4	1897,27	Ti III	0
1914,398	Ar III	9	1897,1	Cs	0
1914,09	Cl III	3	1896,904	Xe III	5
1914,086	Kr III	3	1896,803	Fe III	9
1914,056	Fe III	19	1896,8	Cs	0
1913,17	Na III	8	1896,26	Mg III	0
1912,90	Cl III	4	1895,7	K	2
1911,338	Fe III	7	1895,675	Fe II	10
1910,91	Al II	5	1895,456	Fe III	20
1910,669	Fe II	8	1894,49	C III	2
1910,621	Si II	50	1894,17	Ca III	3
1910,401	Fe III	6	1894,006	Fe II	10
1910,2	Cs	0	1893,981	Fe III	11
1910,17	Ca III	4	1893,87	Mg IV	1
1909,74	Ti II	2	1893,245	Si I	200
1909,33	Ti II	2	1892,92	Ca III	1
1908,96	N III	1	1892,7	K	2
1908,46	Mg III	3	1892,030	Si III	3
1908,29	Ti II	3	1890,9	K	2
1908,11	N III	7	1890,75	Na III	12
1908,1	Cs	1	1890,669	Fe III	13
1907,989	Ar II	4	1890,35	Mg III	2
1907,577	Fe III	10	1889,714	Ne II	1
1907,494	Ne II	8	1889,2	Cs	6
1907,46	Ca III	2	1889,06	Cl III	0
1907,28	N III	4	1889,056	N I	2
1907,06	O III	1	1889,029	Ar II	6
1906,89	N III	1	1888,788	Ar II	4
1906,814	Fe III	6	1888,729	Fe II	20
1906,71	Mg IV	0	1888,32	Fe I	12
1906,57	Al II	4	1888,110	Ne II	1
1906,457	Fe III	6	1887,9	K	2
1906,30	Ti II	3	1887,761	Fe I	14
1906,22	N III	1	1887,700	Si I	200
1905,878	Si II	3	1887,48	Na III	15
1904,787	Fe II	15	1887,471	Fe III	8
1904,660	Si I	50	1887,45	N II	4
1904,38	Al II	2	1887,31	Mg III	0
1904,326	Si II	5	1887,197	Fe III	8
1902,89	O III	1	1886,757	Fe III	12
1902,459	Si II	100	1886,387	Ar II	4
1902,402	Fe III	6	1885,75	Na III	4
1901,61	Cl III	5	1885,25	N III	10
1901,55	Mg III	1	1885,125	Fe III	9
1901,331	Si I	1000	1884,596	Fe III	8
1901,31	Ti III	3	1884,0	Cs	6.
1901,096	Fe III	9	1883,799	Ne II	1
1900,7	Cs	1	1883,14	Cl II	3
1900,638	Ar II	4	1882,92	N V	1
1899,834	Ar II	1	1882,36	N V	0
1899,70	Na III	3	1882,250	Cu III	2
1899,271	Ar II	1	1882,047	Fe III	10
1899,17	Al II	4	1881,851	Si I	200
1898,870	Fe III	6	1881,19	Al IV	1
1898,538	Fe II	10	1880,976	Fe II	20
1897,85	Cl III	3	1880,953	Si I	20
1897,7	Cs	0	1880,21	Ne II	3
1897,49	Al II	2	1880,14	Fe I	5
			1880,10	Cl III	3

λ	Symbol	I
1879,788	Ar II	2
1879,46	Mg III	4
1879,419	Ar II	1
1878,60	N II	2
1878,48	Al II	3
1877,989	Fe III	12
1877,6	Cs	1
1877,523	Ar II	4
1877,462	Fe II	20
1877,13	Al II	1
1876,835	Fe II	15
1876,421	Fe I	10
1876,173	Fe II	8
1875,809	Si I	100
1875,536	Fe II	15
1874,907	Xe III	5
1874,838	Si I	500
1874,59	Mg IV	0
1874,22	Na III	0
1873,32	Na III	4
1873,259	Fe I	15
1873,217	N I	1
1873,2	Cs	2
1873,140	Ar II	6
1873,100	Si I	100
1872,582	Ar II	1
1873,052	Fe I	12
1872,45	Na III	1
1872,39	Ca III	5
1872,359	Fe I	15
1872,214	Fe III	6
1871,152	Fe III	9
1870,782	Si II	3
1870,28	Ca III	6
1870,227	Si II	15
1869,828	Fe III	10
1869,43	Na III	1
1869,317	Si II	20
1868,660	Ar II	3
1868,23	Mg III	1
1868,21	N II	0
1867,747	Cu III	50
1866,815	Fe I	10
1866,305	Fe III	9
1866,093	Ar II	1
1866,07	Fe I	12
1865,202	Fe III	7
1864,743	Fe II	20
1862,856	Ar II	1
1862,749	Al III	10
1862,57	N II	2
1862,40	Na III	6
1862,34	Al II	15
1862,318	Fe I	15
1861,3	Cs	0
1861,19	Na III	15
1860,50	Ca III	3
1860,37	N V	6
1860,040	Fe II	20
1859,99	Al II	3
1859,744	Fe II	15
1859,61	Na III	0
1859,3	Cs	2
1859,22	N II	5
1859,20	Na III	0
1858,685	Cu III	1
1858,53	N II	2
1858,19	Mg III	2
1858,05	Al II	10
1857,935	Fe II	12
1857,88	N V	3
1857,83	N II	3
1857,69	N V	3
1857,57	Na III	5
1856,73	Na III	20
1856,690	Fe III	7
1855,95	Al II	8
1855,91	Na III	15
1855,58	Fe I	15
1854,826	Fe III	9
1854,76	Al II	3
1854,72	Ca III	6
1854,715	Al III	10
1854,365	Xe III	6
1854,11	Ne II	1
1853,22	Ne II	1
1853,148	Si I	50
1852,677	Fe III	6
1852,464	Si I	200
1852,11	Cl III	2
1851,791	Si I	30
1851,261	Fe III	6
1850,691	Ca II	2
1850,668	Si I	500
1850,39	Na III	18
1850,24	Na III	20
1849,64	Cl III	0
1849,58	Na III	35
1849,51	Ca III	2
1849,41	N II	1
1849,407	Fe III	7
1848,90	Al II	2
1848,768	Fe II	12
1848,737	Si I	100
1848,74	Cl III	0
1848,231	Fe II	5
1848,144	Si I	200
1848,1	N II	0
1847,54	Na III	10
1847,468	Si I	400
1846,581	Fe II	12
1846,399	N I	6
1846,103	Si I	100
1845,80	N III	4
1845,7	N II	—
1845,64	N III	5
1845,510	Si I	300

λ	Symbol	I
1845,521	Fe III	7
1845,10	Na III	12
1844,590	Fe II	5
1844,547	Fe III	6
1844,4	N II	10
1844,36	Na III	20
1843,9	K	2
1843,765	Si I	200
1843,5	N II	0
1843,43	Na III	2
1843,19	Ar III	2
1843,088	Ca II	1
1842,547	Si III	9
1842,4	N II	1
1841,701	Fe II	10
1841,68	N III	1
1841,440	Si I	200
1841,146	Si I	100
1841,1	N II	—
1840,917	Cu III	200
1840,6	Cs	2
1840,2	K	2
1840,061	Ca II	8
1839,64	Al II	2
1839,59	N III	2
1839,43	Ar III	3
1838,32	Mg III	1
1838,309	Fe III	7
1838,27	Al II	1
1838,11	Na III	6
1838,008	Ca II	7
1838,006	Si I	30
1837,1	K	2
1836,97	Al II	1
1836,739	N I	4
1836,506	Si I	200
1836,42	Ar III	5
1836,36	N II	1
1835,869	Fe II	15
1835,587	N III	6
1835,22	Na III	15
1834,82	Al II	6
1834,254	Xe III	4
1834,039	Ar II	2
1833,31	Cl III	4
1832,87	Al II	8
1832,21	Ti III	0
1832,08	Cl III	4
1831,78	N II	5
1831,525	Ar II	5
1831,31	Ti III	0
1830,771	Ar II	5
1830,458	N II	4
1829,893	Si I	20
1829,42	Ti III	0
1828,61	Al II	10
1828,43	Ca III	1
1828,40	Cl III	5
1828,14	Ti III	1
1827,97	Mg I	8
1826,477	Xe III	7
1826,339	Cu III	10
1825,858	Xe III	3
1825,55	Fe IV	8
1825,44	Na III	10
1825,348	Cu I	100
1825,30	Ti III	0
1824,59	Cl III	3
1824,52	Na III	0
1823,207	Ar II	1
1822,72	Fe IV	2
1822,50	Cl III	6
1822,452	Si I	50
1821,68	Na III	12
1820,47	Mg III	1
1820,339	Cu III	5
1819,845	Xe III	6
1819,29	Fe IV	1
1819,01	Na III	2
1818,55	Al IV	1
1817,73	Cl III	4
1817,445	Si II	10
1817,381	Xe III	2
1817,334	Cu I	20
1817,265	Cu I	20
1816,921	Si II	200
1816,83	Na III	2
1815,61	Fe IV	25
1814,647	Ca II	1
1814,495	Ca II	1
1814,35	Na III	3
1814,068	Si I	500
1813,772	Ar II	1
1813,009	Ar II	1
1812,53	Fe IV	1
1812,17	Ca III	5
1811,70	Na III	5
1811,62	N V	1
1811,09	Ti III	2
1811,08	N V	0
1810,74	Na III	4
1810,26	Cl III	1
1809,316	Fe II	10
1809,092	Si I	50
1808,51	Cl III	4
1808,003	Si II	150
1807,91	Ca III	5
1807,84	Cu II	15
1807,40	Al II	4
1807,337	Ca II	1
1805,5	N III	7
1804,3	N III	6
1804,109	Xe III	2
1803,023	Si III	3
1801,53	Fe IV	5
1801,27	Na III	7
1800,95	Cu II	2

λ	Symbol	I	λ	Symbol	I
1800,75	Mg III	4	1772,01	Cl II	3
1800,24	Ca III	4	1771,829	Ar II	2
1799,122	Si I	100	1770,922	Si I	100
1798,761	Cu III	5	1770,8	K	6
1798,163	Fe II	10	1770,8	Na	6
1797,98	Cl III	2	1770,652	Ar II	1
1797,69	Ti III	1	1770,629	Si I	30
1797,343	Si I	15	1770,554	Fe III	6
1797,10	Ti III	0	1769,762	Si I	15
1794,68	Mg III	3	1769,140	Al I	4
1794,31	Ca III	4	1768,869	Cu III	200
1793,371	Fe II	10	1768,042	Ar II	1
1792,56	Ti III	2	1767,76	Al II	7
1791,91	Cl II	4	1767,24	Cl II	1
1791,80	Na III	8	1767,21	Na III	0
1791,561	Ar II	1	1766,385	Al I	4
1791,50	Mg III	1	1766,346	Si I	20
1791,23	Na III	10	1766,219	Cu III	2
1790,65	Cu II	5	1766,08	N II	1
1790,292	Si I	20	1766,060	Si I	30
1788,86	Ti III	1	1765,82	Al II	4
1788,101	Ar II	3	1765,636	Al I	4
1788,05	Mg III	1	1765,60	Si I	30
1787,997	Fe II	35	1765,4	K	3
1787,902	Cu III	1	1765,366	C I	1
1787,538	Si II	8	1765,13	N II	1
1787,4	K	4	1765,030	Si I	40
1787,4	Na	4	1764,540	Cu I	10
1787,32	Ti III	2	1764,01	Al II	10
1787,10	Cl II	3	1764,0	K II	0
1786,817	Si II	4	1763,93	Mg III	2
1786,738	Fe II	40	1763,909	C I	2
1785,669	Ar II	1	1763,85	Al II	8
1785,262	Fe II	40	1763,84	Na III	3
1785,06	Cl II	1	1763,664	Si I	50
1784,36	Ti III	1	1763,63	N II	2
1783,935	Cu III	5	1762,899	Al I	2
1783,92	Ca III	4	1762,557	Cu III	30
1783,799	Cu III	20	1762,14	Ca III	3
1783,58	Ti III	1	1762,13	Na III	0
1783,36	Mg III	4	1762,00	Al II	5
1783,232	Si I	50	1761,379	Fe II	25
1782,92	Na III	12	1761,155	Cu III	20
1782,587	Ar II	1	1761,05	Na III	1
1780,062	Cu III	5	1760,81	C II	3
1776,826	Si I	100	1760,586	Cu III	10
1776,670	Ar II	1	1760,415	Fe II	20
1776,136	Cu III	20	1760,40	C II	4
1775,983	Fe III	6	1760,15	Al II	7
1775,32	Na III	0	1759,601	Si I	20
1774,820	Cu I	200	1758,01	Mg III	1
1773,697	Cu III	1	1757,223	N I	1
1773,5	K	6	1756,0	Li II	5
1773,29	Ca III	3	1755,810	Ar II	1
1773,09	Mg III	3	1755,012	Cu III	20
1773,00	Na III	0	1754,97	Na III	0
1772,518	Fe II	15	1753,474	Mg II	60
1772,478	Cu III	2	1753,27	Cu II	15
			1753,112	Si I	30

λ	Symbol	I
1752,65	Na III	3
1752,4	K	4
1752,06	Na III	1
1751,827	C I	8
1751,75	N III	10
1751,679	Ar II	2
1751,24	N III	6
1750,749	Xe III	2
1750,664	Mg II	50
1750,56	Al II	6
1750,391	Cu III	500
1750,079	N I	2
1749,3	K	8
1749,3	Na	8
1749,202	Cu I	2
1749,02	Mg III	5
1747,86	N III	9
1747,81	Mg I	5
1747,64	Mg III	4
1747,40	F II	3
1747,404	Si I	50
1746,816	Fe II	20
1746,39	Na III	0
1745,57	F II	2
1745,332	Si I	15
1745,249	N I	30
1744,86	F II	1
1744,61	Ca III	3
1744,50	Cu II	20
1744,08	Mg III	0
1743,884	Si I	30
1743,197	N II	3
1742,724	N I	10
1741,574	Cu I	50
1741,378	Cu III	500
1741,33	Na III	1
1741,2	K	4
1741,135	Cu III	30
1740,378	Si I	20
1740,309	N II	4
1739,64	Al II	5
1739,56	Mg III	0
1739,508	Cu III	300
1739,4	K	2
1738,91	Mg III	6
1738,648	Cu III	10
1738,56	Ca III	3
1738,145	Cu III	30
1737,893	Cu III	30
1737,628	Mg II	10
1737,612	Mg II	10
1736,830	Ar II	1
1736,54	Cu II	10
1734,852	Mg II	10
1734,769	Si I	10
1734,21	Cu II	3
1733,362	Ar II	1
1732,998	Cu III	5
1732,69	Ne II	1
1732,674	Cu I	20
1732,253	Fe II	15
1731,88	Mg III	1
1731,32	Cu I	2
1731,08	Na III	0
1731,038	Fe II	10
1730,81	Mg III	1
1730,576	Cu I	10
1730,04	N III	8
1729,997	Si IV	5
1729,481	N I	1
1729,262	Ar II	1
1729,075	Ar II	1
1728,170	N I	3
1728,139	Cu III	200
1727,377	Si IV	5
1726,88	Ca III	1
1726,394	Fe II	12
1726,275	Cu III	5
1726,006	Si IV	4
1725,664	Cu I	50
1725,402	Fe II	5
1725,138	Ar II	1
1725,01	Al II	15
1725,0	K II	3
1724,963	Fe II	8
1724,847	Fe II	8
1724,810	Cu III	10
1722,534	Si IV	6
1722,379	Cu III	1000
1722,21	C II	0
1722,10	Mg III	2
1721,746	N I	3
1721,66	C II	2
1721,637	Kr III	1
1721,6	K	3
1721,31	Al II	10
1720,99	C II	1
1720,621	Fe II	20
1720,44	C II	0
1720,042	Fe II	10
1719,8	K	3
1719,60	Na III	0
1719,43	Al II	8
1719,346	Ar II	2
1718,680	Ar II	1
1718,551	N IV	20
1718,48	Na III	0
1717,72	Cu II	15
1717,134	Cu III	5
1716,400	Cu III	10
1716,23	Ca III	1
1715,507	Fe II	12
1715,24	Ti III	0
1714,85	Mg III	0
1713,364	Cu I	50

λ	Symbol	I
1713,346	Cu III	5
1713,215	Ar II	2
1713,002	Fe II	20
1711,437	Cu III	30
1711,296	Si II	20
1711,257	Cu III	30
1710,826	Si II	10
1709,678	Fe II	15
1709,4	K	3
1709,396	Cu I	2
1709,036	Cu III	700
1709,018	Si III	3
1708,958	Cu III	200
1708,627	Fe II	8
1707,500	Cu III	5
1707,391	Cu I	5
1707,10	Mg I	3
1705,977	Ar II	1
1705,633	Cu III	400
1705,333	Cu III	300
1704,967	Si II	2
1704,90	F II	0,5
1704,5	K	3
1704,434	Si I	50
1704,072	Cu III	10
1703,843	Cu I	30
1703,78	Mg III	1
1703,5	K	9
1703,43	Mg III	0
1703,218	N V	4
1702,994	Cu III	500
1702,862	Si I	30
1702,349	Cu III	30
1702,25	N V	3
1702,190	Cu III	300
1702,186	Ar II	1
1702,102	Cu III	400
1702,06	N IV	5
1702,045	Fe II	25
1701,358	Ar II	1
1701,292	Cu I	10
1701,023	Cu III	400
1700,626	Si I	30
1700,423	Si I	50
1699,95	N III	4
1699,808	Si III	5
1699,716	Si I	10
1699,32	N III	5
1699,09	Cu II	30
1699,08	N IV	4
1699,00	N III	2
1698,95	Ca III	1
1698,9	K	10
1698,9	Na	10
1698,83	Mg IV	2
1698,183	Ca II	—
1698,16	N III	2
1697,938	Si I	200
1697,32	Mg III	2
1697,19	N III	0
1696,86	N IV	3
1696,800	Fe II	8
1696,54	N III	3
1696,203	Si I	200
1696,202	Cu III	15
1695,507	Si I	50
1694,79	N III	0
1693,461	Si I	20
1693,292	Si I	50
1692,706	Cu III	300
1692,654	Cu I	5
1691,779	Ca II	1
1691,70	Na III	1
1691,289	Fe II	8
1691,076	Cu I	30
1690,786	Si I	30
1690,781	Fe II	8
1689,821	Fe II	10
1689,50	Cl III	1
1689,290	Si I	50
1689,051	Cu III	200
1688,865	Cu I	15
1688,81	Ca III	1
1688,618	Cu III	100
1688,44	Fe IV	2
1688,356	Ne II	4
1688,11	N IV	3
1688,093	Cu I	30
1687,82	N IV	2
1687,60	N IV	1
1687,134	Cu III	600
1687,095	Si I	15
1687,043	Cu I	20
1686,815	Si I	30
1686,475	Fe II	8
1686,214	Cu III	300
1686,19	Al II	5
1685,953	Fe II	5
1685,682	Cu I	25
1685,457	Fe II	8
1684,674	Cu I	20
1684,642	Cu III	500
1683,51	Mg I	1
1683,18	Cl III	0
1683,15	Cu II	40
1683,04	Mg IV	3
1682,695	Cu III	30
1682,675	Si I	30
1682,4	Li II	4
1682,044	Cu III	10
1681,78	Al II	5
1681,683	Ne II	3
1681,481	Cu III	300
1681,45	Fe IV	25
1680,86	Fe IV	1
1680,129	Ca II	1
1680,051	Ca II	—
1680,02	Mg IV	0
1679,56	Mg III	0
1679,388	Fe II	15
1679,151	Cu III	400
1678,74	Na III	1
1677,906	N I	4

λ	Symbol	I	λ	Symbol	I
1677,373	Cu III	200	1662,26	Fe IV	20
1676,818	Si I	10	1662,253	Ar II	1
1676,469	Cu III	15	1661,059	Si II	3
1675,920	N II	1	1660,887	Cu III	30
1675,78	Fe IV	25	1660,484	Si I	10
1675,76	Mg III	0	1660,07	Fe IV	20
1675,744	N II	4	1660,0009	Cu II	20
1675,637	Ar III	4	1659,809	Kr III	2
1675,5	Cs	1	1659,7	K	4
1675,484	Ar III	7	1659,487	Fe II	20
1675,198	Si I	200	1659,28	Mg III	0
1675,1	K	2	1658,92	Mg IV	0
1674,89	Fe IV	5	1658,785	Fe II	15
1674,716	Fe II	10	1658,71	Na III	2
1674,602	Cu III	500	1658,7	Na	40
1673,860	Ca II	1	1658,472	Cu III	200
1673,470	Fe II	15	1658,121	C I	5
1673,440	Cu I	5	1657,907	C I	4
1673,425	Ar III	7	1657,379	C I	2
1673,374	Si IV	150	1657,008	C I	10
1673,315	Si III	7	1656,930	C I	3
1673,241	Ar III	3	1656,61	Fe IV	15
1673,2	Cs	1	1656,3216	Cu II	20
1673,14	Ar III	1	1656,268	C I	5
1672,9	K	3	1656,25	Fe IV	10
1672,77	Cu II	10	1655,922	N V	1
1672,593	Si I	100	1655,318	Cu I	30
1672,18	Fe IV	5	1654,574	Cu III	300
1671,886	Cu III	500	1654,484	Fe II	5
1671,484	Cu I	3	1654,105	Fe II	5
1671,111	Si I	20	1653,9	Li II	8
1671,020	N I	1	1653,399	Cu III	10
1670,81	Al II	15	1653,351	Si I	30
1670,759	Fe II	25	1653,322	Ar II	1
1670,140	Cu III	500	1652,85	Fe IV	20
1669,73	Fe IV	2	1652,26	Mg III	0
1669,671	Ar III	7	1652,010	Cu III	300
1669,52	Na III	3	1651,991	Ca II	1
1669,5	Cs	1	1651,758	Cu III	15
1669,304	Ar III	5	1651,721	Cu I	20
1669,3	K	4	1651,21	Cl IV	1
1669,3	Na	40	1651,013	Si I	20
1669,273	Cu III	10	1650,91	Na III	1
1669,10	Ar III	1	1650,709	Fe II	20
1668,7	K	4	1650,531	Ar II	1
1668,7	Na	40	1650,301	Cu I	5
1668,57	Mg I	0	1650,119	Cu I	5
1668,517	Si I	100	1649,858	Ca II	2
1667,7	Ca	30	1649,583	Fe II	20
1667,618	Si I	100	1649,4573	Cu II	25
1666,369	Si I	50	1649,444	Fe II	15
1664,708	Cu I	10	1649,299	Ar II	1
1664,521	Si I	30	1648,88	Mg III	0
1664,303	Cu I	10	1648,04	Cl IV	0
1663,52	Fe IV	10	1647,493	N I	3
1663,266	Fe II	15	1647,359	Kr III	2
1663,21	Fe IV	10	1647,161	Fe II	25
1663,0017	Cu II	30	1647,05	Fe IV	45

λ	Symbol	I	λ	Symbol	I
1646,187	Fe II	20	1626,411	Cu III	200
1645,06	C III	1	1626,139	Cu III	200
1644,78	Al II	5	1625,919	Fe II	15
1644,441	Ca II	0	1625,707	Si I	30
1644,15	Al II	5	1625,60	Al II	3
1643,770	Ca II	—	1625,534	Si I	15
1643,588	Fe II	15	1625,525	Fe II	20
1643,40	Cl IV	1	1625,500	Cu III	1
1642,86	Mg III	1	1624,07	Na III	12
1642,802	Ca II	—	1623,17	Cu II	30
1642,208	Cu III	2000	1623,102	Fe II	8
1642,187	Fe II	5	1622,867	Si I	100
1642,168	Si IV	4	1622,86	Cl IV	2
1641,761	Fe II	25	1622,44	Cu II	40
1640,496	He II	1	1621,966	N V	1
1640,474	Cu I	5	1621,94	Na III	5
1640,474	He II	10	1621,723	Cu III	3
1640,437	N I	2	1621,685	Fe II	30
1640,335	Ar II	1	1621,4256	Cu II	60
1640,332	He II	5	1621,316	Cu I	20
1640,257	Si I	20	1620,776	Cu III	1
1640,167	Fe II	12	1620,68	C III	1
1640,03	Fe IV	65	1620,389	Si I	20
1639,960	Cu III	10	1620,33	C III	2
1639,403	Fe II	30	1620,05	C III	3
1639,00	Al IV	2	1619.688	N V	12
1638,956	Cu III	300	1619,531	Si I	10
1638,95	Cl IV	0	1618,464	Fe II	25
1638,816	Kr III	3	1618,408	Cu III	5
1638,274	Si I	10	1618,38	Al II	4
1637,400	Fe II	15	1617,9151	Cu II	20
1636,61	Cu II	10	1617,43	Cl IV	1
1636,334	Fe II	30	1616,972	Ar II	1
1635,389	Fe II	35	1616,940	Cu I	20
1634,607	Si IV	1	1616,607	Cu III	300
1634,353	Fe II	20	1616,571	Si I	20
1633,978	Si I	50	1616,41	Al II	4
1633,907	Fe II	15	1616,328	N V	9
1633,64	Na III	4	1616,160	Cu III	15
1633,318	Si I	15	1616,06	N II	1
1633,192	Cu III	1	1615,937	Si I	20
1633,203	Si I	15	1614,650	Si I	10
1632,326	Cu I	5	1614,557	Si I	10
1631,124	Fe II	30	1613,77	Na III	8
1631,134	Si I	75	1612,814	Fe II	20
1630,99	Fe IV	75	1611,849	Al III	8
1630,27	Cu II	25	1611,763	Fe III	7
1629,921	Si I	200	1611,723	Fe III	7
1629,834	Ar II	1	1611,1180	Cu II	10
1629,830	N II	4	1610,97	Na III	4
1629,426	Si I	300	1610,933	Fe II	15
1629,301	Cu III	1	1610,571	Cu III	75
1629,155	Fe II	30	1610,2964	Cu II	15
1629,02	N II	1	1609,757	Cu III	100
1628,825	Ar II	1	1609,599	Cu III	50
1628,295	Cu III	300	1608,900	Si I	10
1628,088	Cu III	50	1608,6396	Cu II	25
1627,42	N II	1	1608,446	Fe II	35
			1607,723	Fe III	9
			1607,542	Cu III	100

λ	Symbol	I
1607,168	Ar II	1
1606,927	Ar II	4
1606,837	Cu III	10
1606,8338	Cu II	40
1606,730	Cu III	300
1606,197	Ar II	3
1605,969	Cu III	300
1605,776	Al III	8
1605,274	Cu II	30
1604,8474	Cu II	20
1604,47	Na III	6
1604,083	Ar II	5
1603,443	Ar II	4
1603,146	Cu III	400
1603,074	Ar II	4
1602,973	N I	1
1602,971	C I	5
1602,91	Na III	5
1602,893	Ar II	2
1602,588	Fe II	12
1602,554	Ar II	2
1602,3882	Cu II	40
1602,250	Cu II	15
1601,289	Fe III	6
1601,211	Fe III	10
1600,694	Ar II	6
1600,194	Cu III	500
1600,133	Ar II	4
1599,597	Ar II	1
1599,44	Al II	3
1599,125	Ar II	1
1598,872	Ar II	1
1598,724	Ar II	2
1598,561	Ar II	1
1598,4024	Cu II	40
1597,950	Si I	30
1597,418	Cu III	10
1597,181	N I	2
1596,141	Ar II	1
1596,02	Al II	3
1595,760	Si I	20
1595,734	Ar II	1
1595,597	Fe III	6
1595,24	Ca III	1
1594,927	Si I	50
1594,548	Si I	150
1594,787	Ar II	1
1593,758	Cu III	1000
1593,581	Ar II	2
1593,5557	Cu II	60
1592,867	N I	3
1592,409	Si I	50
1592,39	Mg III	1
1591,933	Ar II	1
1591,48	C III	2
1590,25	N II	2
1590,229	Ar II	2
1590,1646	Cu II	40
1589,463	Ar II	5
1589,27	Al IV	1
1588,950	Si III	2
1588,551	Cu III	3
1588,295	Fe II	10
1586,783	Si I	20
1586,26	Mg III	3
1586,256	Ar II	2
1586,19	Ca III	4
1586,133	Si I	15
1585,871	Cu I	5
1584,954	Fe II	15
1584,77	Al II	2
1584,45	Al IV	2
1583,83	Ar II	1
1583,799	Cu I	15
1583,683	Cu II	50
1582,849	Cu II	10
1582,04	Al IV	3
1581,991	Cu II	40
1581,243	Fe II	8
1580,960	Ar II	1
1580,93	Al II	2
1580,768	Ar II	2
1580,635	Fe II	25
1580,628	Cu II	30
1580,303	Si I	10
1580,025	Cu II	15
1579,73	Fe IV	3
1579,658	Cu I	5
1579,498	Xe III	5
1579,492	Cu II	30
1579,353	Cu III	15
1578,812	Ar II	3
1577,89	C III	2
1577,32	C III	2
1577,131	N I	1
1576,897	Ar II	3
1576,817	Si I	15
1576,49	C III	3
1575,815	Ar II	3
1575,349	Cu II	5
1575,115	Si I	20
1574,992	Ar II	6
1574,931	Fe II	20
1574,817	Si I	50
1574,68	Fe IV	8
1574,402	Ar II	1
1573,874	Si I	50
1573,831	Fe II	5
1573,650	Si I	20
1573,21	N II	1
1572,97	Al II	1
1572,72	Mg III	4
1571,390	Cu III	1
1571,390	Ar II	1
1571,377	Si I	10

λ	Symbol	I	λ	Symbol	I
1571,31	Ca III	5	1558,543	Fe II	10
1571,21	Fe IV	10	1558,3446	Cu II	30
1571,154	Cu III	3	1558,05	Cl II	1
1570,568	Cu II	3	1557,583	Cu II	20
1570,248	Fe II	20	1557,302	Ar II	1
1570,202	Cu III	30	1557,24	Al IV	5
1570,035	Cu II	2	1556,48	Fe IV	15
1569,886	Kr III	2	1555,698	Cu II	50
1569,670	Fe II	12	1555,48	Ca III	4
1569,426	Cu II	10	1555,134	Cu II	40
1569,35	Al II	1	1555,01	Fe IV	1
1569,322	Si I	10	1554,642	Ca II	4
1569,2123	Cu II	10	1554,438	Xe III	1
1568,655	Cu III	2	1554,17	Fe V	1
1568,564	Cu III	2	1553,893	Cu II	25
1568,172	Si I	15	1553,176	Ca II	4
1568,031	Fe II	8	1553,00	Al IV	1
1567,987	Ar II	4	1552,641	Cu II	50
1567,703	Si I	10	1552,11	Fe IV	15
1566,825	Fe II	20	1551,379	Cu II	30
1566,812	Ar II	1	1551,27	Cl IV	1
1566,54	Fe IV	3	1550,862	Fe III	8
1566,4151	Cu II	40	1550,80	Fe V	2
1565,9240	Cu II	40	1550,771	C IV	19
1565,377	Ar II	1	1550,644	Cu II	30
1565,194	Cu III	5	1550,292	Cu II	3
1565,05	Fe IV	2	1550,196	Fe III	12
1564,589	Si I	10	1549,975	Xe III	2
1564,14	Al IV	1	1549,336	N V	6
1564,066	Si II	5	1549,203	Cu III	10
1563,790	Fe II	25	1549,15	Cl IV	2
1563,765	Si II	10	1548,867	Cu III	300
1563,56	Al II	1	1548,68	Na III	8
1563,30	Fe IV	10	1548,185	C IV	20
1563,189	Cu II	5	1548	N V	—
1563,036	Ar II	1	1547,950	Cu II	10
1562,845	Si II	15	1547,58	Fe IV	15
1562,563	Xe III	4	1547,354	Ar II	1
1562,50	Ca III	6	1546,03	Fe IV	8
1562,451	Si II	10	1545,7	Ca	3
1562,441	Ar II	2	1545,19	Cl IV	2
1561,982	Si I	10	1544,711	Ar II	2
1561,792	Si I	10	1544,674	Cu II	40
1561,790	Cu III	3	1544,50	Fe V	3
1561,435	C I	20	1544,177	Ar II	2
1561,337	C I	5	1544,110	Cu III	2
1560,739	Si I	15	1544,062	Cu III	2
1560,691	C I	15	1543,66	Fe V	2
1560,35	Al II	1	1543,438	Cu III	500
1560,306	C I	8	1543,180	Cu III	2
1560,26	Fe IV	15	1542,562	Cu III	2
1560,184	Ar II	4	1542,499	N I	1
1560,067	Si I	15	1542,177	C I	2
1559,106	Fe II	20	1542,15	Fe IV	15
1559,08	Fe IV	15	1541,970	Cu III	40
1559,072	Ar II	3	1541,7031	Cu II	75
1558,802	Kr III	3	1541,19	Na III	1
1558,706	Fe II	10	1540,77	Fe IV	1

λ	Symbol	I
1540,589	Cu II	30
1540,3889	Cu II	30
1540,231	Cu II	20
1539,74	Al II	10
1539,30	Cl IV	2
1539,128	Fe III	8
1538,67	Fe IV	25
1538,632	Fe III	10
1538,488	Cu II	10
1537,560	Cu II	50
1537,52	Al IV	2
1537,21	Cl IV	3
1535,515	Cu II	15
1535,0024	Cu II	25
1533,976	Cu II	25
1533,445	Si II	1000
1533,27	Fe V	2
1533,25	Cl IV	1
1532,70	Fe V	4
1532,124	Cu II	30
1532,19	Cl IV	1
1531,864	Fe III	7
1531,85	C III	2
1531,8557	Cu II	50
1531,644	Fe III	8
1531,588	Cu III	1
1531,293	Fe III	6
1529,67	Na III	1
1529,28	Cl IV	0
1528,91	Cl II	1
1528,89	Ca III	0
1528,782	Cu II	2
1527,801	Cu II	5
1526,969	Cu II	5
1526,719	Si II	500
1526,15	Al IV	1
1525,794	Cu II	30
1525,653	Cu II	10
1524,857	Cu II	20
1524,67	Fe IV	15
1523,740	Cu II	10
1522,580	Cu III	15
1522,575	Cu II	15
1520,546	Cu III	20
1520,543	Cu II	20
1519,8370	Cu II	60
1519,4917	Cu II	50
1518,221	Si II	5
1517,930	Cu II	10
1517,6312	Cu II	20
1517,162	Cu II	10
1516,910	Si II	60
1516,902	Cu II	5
1514,492	Cu II	50
1514,238	Cu II	10
1513,570	Si II	30
1513,533	Si III	2
1513,360	Cu II	20
1512,457	Cu II	20
1512,174	Cu II	20
1512,072	Si II	50
1511,121	Xe III	3
1510,924	CI	1
1510,502	Cu II	35
1509,454	Xe III	2
1509,101	Si II	100
1508,82	Mg IV	0
1508,741	Si II	3
1508,627	Cu II	30
1508,175	Cu II	25
1506,94	Ca III	3
1506,07	Ti III	10
1506,060	Si III	6
1505,848	Cu II	5
1505,384	Cu II	20
1505,166	Fe III	10
1504,91	Ti III	5
1504,755	Cu II	25
1504,59	Ti III	10
1503,368	Cu II	15
1502,36	Ti III	10
1502,107	Cu III	1
1501,870	Si III	9
1501,333	Cu II	10
1501,3	Cs II	5
1501,191	Si III	10
1500,241	Si III	12
1499,510	Cu II	10
1499,17	Ti III	20
1498,65	Ti III	30
1498,566	Cu II	3
1496,92	Ca III	2
1496,6860	Cu II	35
1496,59	Ti III	1
1496,172	Si III	7
1495,5	N V	2
1495,426	Cu II	25
1495,311	Fe II	15
1495,08	Ti III	1
1494,668	N I	60
1494,658	Cu II	5
1493,7	Li II	6
1493,640	Fe III	9
1493,359	Cu II	25
1492,837	Cu II	30
1492,817	N I	30
1492,684	Cu II	10
1492,624	N I	80
1492,149	Cu II	10
1491,98	Ti III	5
1490,41	Mg IV	0
1488,6373	Cu II	75
1487,35	Fe IV	5
1486,904	Cu III	10
1486,87	Al IV	1
1486,659	Cu III	25
1486,496	N I	2

λ	Symbol	I	λ	Symbol	I
1486,265	Fe III	7	1468,180	Xe III	4
1485,6777	Cu II	40	1468,11	Fe IV	2
1485,513	Si II	100	1468,006	Ar III	2
1485,48	Fe IV	12	1467,841	Ar III	3
1485,318	Cu II	20	1467,450	C I	3
1485,224	Si II	30	1467,384	N I	3
1485,024	Si II	90	1467,25	Ti IV	30
1484,92	Ca III	4	1466,751	Cu II	5
1484,873	Si II	15	1466,723	N I	5
1484,010	Cu III	5	1466,524	Ar II	1
1483,831	Cu III	15	1466,519	Cu II	10
1483,429	Kr III	2	1466,067	Cu II	20
1482,890	Mg II	—	1465,712	Ar III	3
1481,762	Cl	7	1465,542	Cu II	15
1481,750	N I	4	1465,532	Ar III	2
1481,541	Cu II	20	1465,37	Fe V	3
1481,46	Al IV	0	1465,153	Ar II	1
1481,243	Cu III	20	1465,043	Fe II	20
1480,880	Mg II	—	1464,81	Fe IV	40
1480,55	Ca III	2	1464,73	Fe V	6
1479,65	Fe IV	38	1464,176	Ar II	1
1479,49	Fe V	4	1463,771	Cu II	50
1478,30	C III	1	1463,41	Ca III	4
1478,230	Cu II	2	1463,336	C I	6
1478,05	C III	2	1463,25	Fe IV	4
1478,004	Mg II	—	1463,155	Ar II	2
1477,997	Mg II	—	1462,822	N I	2
1477,69	Fe IV	5	1462,67	Fe V	3
1477,68	C III	3	1461,556	Cu II	15
1476,054	Fe II	10	1460,915	Cu III	10
1476,000	Mg II	—	1460,86	Fe V	2
1475,846	Cu II	30	1460,234	Ar III	2
1475,67	Fe IV	28	1460,077	Ar III	4
1475,188	Si II	5	1459,92	Fe IV	40
1474,934	Cu II	20	1459,875	Ar II	1
1474,649	Si II	15	1459,87	Ca III	3
1474,537	Ar II	1	1459,85	Fe V	5
1474,52	Fe IV	2	1459,52	Mg IV	1
1473,9788	Cu II	25	1459,412	Cu II	25
1473,834	Fe II	20	1459,032	C I	2
1473,531	Cu II	15	1458,15	Na III	3
1472,594	Ar II	1	1458,021	Cu III	6
1472,399	Cu II	20	1458,004	Cu II	30
1472,231	C I	0	1457,356	Xe III	3
1472,13	Fe IV	35	1457,253	Si III	5
1471,69	N III	2	1457,175	Cu II	10
1471,06	Cl II	2	1456,23	Fe V	5
1471,02	N III	1	1455,66	Fe IV	25
1470,697	Cu II	40	1455,655	Cu II	5
1470,68	N III	0	1455,59	Fe V	5
1470,54	Fe IV	2	1455,484	Ar II	1
1470,082	C I	1	1455,22	Ti III	40
1469,92	Fe IV	20	1455,200	Cu III	3
1469,691	Cu II	15	1454,71	Fe V	3
1469,621	Xe I	5	1453,67	Fe IV	15
1469,21	Ti IV	15	1452,291	Cu II	20
1469,04	Fe IV	37			
1468,410	C I	3			

λ	Symbol	I	λ	Symbol	I
1451,879	Ar II	1	1428,53	C III	2
1451,75	Ti IV	30	1428,366	Cu II	15
1451,478	Cu III	1	1428,17	C III	2
1450,307	Cu II	25	1428,081	Cu III	5
1450,29	Ti III	2	1427,85	C III	3
1450,165	Cu III	3	1427,835	Cu II	20
1449,70	Al IV	0	1427,589	Cu II	10
1449,056	Cu II	20	1427,27	Na III	0
1448,91	Fe V	6	1426,89	Cl IV	1
1447,47	Al IV	2	1426,78	C III	1
1447,196	Si III	6	1426,45	C III	4
1446,114	N IV	5	1425,282	Cu III	1
1445,982	Cu II	20	1425,079	Cu III	2
1444,692	Cu III	1	1425,00	Al IV	0
1444,1305	Cu II	2	1424,775	Si III	2
1443,541	Cu II	10	1424,747	Fe II	12
1442,1389	Cu II	15	1424,14	Ti III	20
1441,81	Al IV	1	1424,047	Fe II	8
1441,732	Si III	5	1424,020	Cu III	5
1440,95	Cl IV	0	1423,553	Kr III	1
1440,59	Fe V	7	1423,504	Cu III	10
1440,446	Cu III	3	1422,41	Ti III	25
1439,391	Si III	2	1421,97	Cl IV	1
1439,094	Si II	10	1421,760	Cu II	25
1438,983	Cu III	2	1421,69	Ti III	20
1438,931	Si II	4	1421,382	Cu II	5
1438,702	Si III	2	1420,89	Li II	1
1438,37	N IV	3	1420,42	Ti III	15
1438,228	Si III	2	1420,24	Fe V	3
1437,645	Cu III	3	1420,04	Ti III	15
1436,994	Cu III	15	1419,742	Cu II	2
1436,724	Si III	4	1418,811	Cu III	5
1436,376	Cu III	1	1418,423	Cu II	25
1436,233	Cu II	15	1417,781	Si II	5
1436,21	Na III	12	1417,744	Fe II	20
1436,166	Si III	7	1417,58	Al IV	0
1435,776	Si III	8	1417,538	Cu III	10
1435,312	Cu II	10	1417,237	Si III	13
1434,916	Cu II	25	1417,124	Cu III	2
1434,758	Cu II	15	1417,060	Cu III	2
1433,85	Ti III	2	1416,972	Si II	10
1433,837	Cu II	10	1415,478	Cu III	1
1433,749	Ca II	—	1414,897	Cu II	10
1433,690	Si III	6	1414,431	Cu III	3
1432,530	C I	1	1413,707	Fe II	25
1432,503	Ca II	1	1413,39	Cl IV	1
1432,275	Cu III	3	1412,834	Fe II	12
1432,204	Xe III	4	1412,794	Cu III	5
1432,105	C I	2	1412,724	Cu III	5
1431,93	Al IV	2	1412,24	Al IV	0
1431,901	Cu III	3	1411,939	N I	30
1431,671	Cu III	10	1411,69	Cu	30
1431,597	C I	2	1411,510	N I	1
1430,969	Cu III	3	1410,950	Si IV	1
1430,791	Si IV	1	1410,219	Si II	20
1430,61	Fe V	8	1409,52	Al IV	0
1430,373	Cu III	3	1409,51	Fe V	7
1430,243	Cu II	40			
1429,201	Cu III	5			
1428,95	C III	1			

λ	Symbol	I
1409,248	Cu III	1
1409,19	Fe V	6
1409,073	Si II	10
1408,811	Cu II	2
1408,536	Cu III	1
1408,310	Cu III	1
1408,19	Fe V	1
1407,701	Si IV	3
1407,196	Cu III	3
1407,160	Cu II	15
1407,139	Cu III	5
1406,78	Fe V	7
1404,72	Al IV	2
1404,478	Si II	6
1403,783	Si II	5
1403,763	Cu III	1
1403,181	Cu III	10
1402,917	Cu III	1
1402,9	Ca	10
1402,776	Cu II	15
1402,769	Si IV	12
1402,45	Fe V	6
1402,435	Cu III	3
1402,250	Cu III	5
1401,655	Cu III	5
1401,602	Cu III	5
1401,376	Cu III	2
1400,90	Kr III	1
1400,7	Ca	4
1400,30	Fe V	4
1399,355	Cu II	3
1399,190	Cu III	5
1398,636	Cu II	10
1398,379	Cu III	5
1397,99	Fe V	3
1397,581	Fe II	12
1396,527	Cl I	8
1396,417	Cu III	1
1396,231	Ar II	1
1395,274	Cu III	10
1394,77	Fe V	3
1393,755	Si IV	15
1393,496	Xe III	1
1393,4	Ca	8
1393,126	Cu II	10
1390,306	Cu III	10
1389,97	Fe V	0
1389,961	Cl I	6
1389,822	N V	2
1389,688	Cl I	6
1389,528	Cu III	5
1389,514	N V	3
1389,05	Fe V	1
1388,77	Al IV	2
1388,276	Cu III	1
1388,07	Fe V	5
1387,994	Si III	5
1387,31	N III	4
1386,714	Cu III	1
1386,33	Fe V	0
1385,921	Cu III	3
1385,39	Ca III	2
1385,380	Cu III	1
1385,32	Fe V	2
1384,929	Cu III	3
1384,840	Cu III	5
1384,75	Fe V	1
1384,324	Cu III	5
1384,17	Fe V	1
1384,144	Al III	5
1382,765	Ar II	1
1382,561	Cu III	5
1382,228	Ar II	2
1381,250	Fe II	10
1380,723	Ar II	1
1380,18	Fe V	2
1379,884	Ar II	3
1379,670	Al III	3
1379,529	Cl I	11
1379,379	Cu III	1
1379,377	Ar II	1
1378,238	Cu III	1
1377,833	Kr III	2
1377,504	Cu III	30
1377,238	Si III	2
1377,211	Ar II	4
1377,082	Si III	3
1376,956	Ar II	1
1376,807	Cu III	30
1376,45	Fe V	6
1375,688	Si III	2
1375,621	Cu III	5
1375,520	Cu II	3
1375,083	Si III	2
1374,758	Cu III	3
1374,033	Cu III	3
1373,68	Fe V	6
1373,118	Cl I	4
1373,030	Si III	5
1372,899	Cu III	5
1371,840	Cu II	20
1371,652	Si III	3
1371,287	O V	10
1371,26	Al II	2
1371,144	Cu III	10
1371,00	Fe V	4
1370,6	Ca II	3
1370,558	Cu II	2
1369,988	Cu III	1
1369,612	Cu III	5
1369,437	Si III	5
1369,423	Mg II	—
1369,1	Ca II	3
1368,923	Cu III	2
1367,952	Cu II	25
1367,708	Mg II	—
1367,646	Cu III	5

λ	Symbol	I
1367,256	Mg II	—
1367,049	Si III	7
1365,73	Fe V	3
1365,544	Mg II	—
1365,253	Si III	8
1365,14	Fe V	3
1364,90	Fe II	12
1364,165	C I	6
1363,853	Kr III	2
1363,72	Fe V	3
1363,501	Cu II	5
1363,459	Si III	7
1363,449	Cl I	10
1363,031	Ar II	2
1363,00	Fe V	4
1362,771	Fe II	20
1362,598	Cu II	20
1362,366	Si III	5
1361,597	Si III	8
1361,42	Fe V	5
1360,870	Fe II	5
1360,735	Ar II	1
1360,566	N I	1
1359,935	Cu II	5
1359,81	Ca III	1
1359,49	Al IV	0
1359,41	Fe V	1
1359,329	C I	2
1359,010	Cu II	20
1358,764	Cu II	30
1358,524	O I	5
1358,440	Cu III	2
1357,28	Fe V	1
1357,140	C I	3
1356,424	Cu III	5
1356,364	Xe III	3
1355,887	N I	6
1355,825	C I	6
1355,605	O I	2
1355,37	Ca III	1
1355,304	Cu II	15
1354,912	Ar II	2
1354,292	C I	5
1353,964	Cu III	2
1353,73	Al IV	0
1353,718	Si II	100
1352,92	Al III	1
1352,635	Si II	100
1351,837	Cu II	25
1351,657	Cl I	10
1351,330	Ar II	1
1351,271	Cu III	3
1350,658	Si II	20
1350,592	Cu II	15
1350,520	Si II	20
1350,15	Al II	6
1350,057	Si II	150
1349,441	Cu III	5
1348,745	Ar II	1
1348,543	Si II	100
1348,077	Cu III	1
1347,56	N III	0
1347,238	Cl I	12
1347,048	Cu III	3
1346,873	Si II	100
1346,44	N II	0
1346,27	N III	4
1346,062	Cu III	5
1345,69	N III	4
1345,330	N II	1
1343,730	Cu III	5
1343,507	O IV	7
1343,388	Si III	6
1343,338	N II	2
1343,032	Cu III	2
1342,995	O IV	4
1342,678	Kr III	1
1342,535	Ca II	—
1342,392	Si III	7
1342,193	Cu III	3
1341,889	Ca II	1
1341,465	Si III	8
1340,909	Cu II	3
1339,769	Cu II	5
1339,72	Ti III	2
1339,497	Cu III	5
1338,603	O IV	6
1337,572	Cu III	5
1337,39	Na III	6
1335,723	Cl I	9
1335,684	C II	14
1334,94	Ca III	3
1334,556	Cu II	2
1334,515	C II	13
1333,054	Cu II	20
1332,985	Cu III	15
1332,222	Cu II	5
1330,816	N II	3
1330,365	Cu III	2
1329,588	C I	6
1329,103	C I	5
1328,834	C I	3
1328,416	Cu II	5
1327,927	N I	10
1327,60	Ti III	15
1327,178	Cu III	5
1326,964	N IV	0
1326,572	N I	15
1326,394	Cu II	10
1325,685	N IV	1
1325,511	Cu II	3
1324,40	N III	3
1324,033	Cu III	5
1323,98	N IV	2
1323,916	C II	8
1323,811	Cu II	6
1323,187	Cu II	3
1322,627	Cu II	6

λ	Symbol	I	λ	Symbol	I
1321,788	Cu II	5	1301,146	Si III	14
1321,60	N I	4	1299,267	Cu II	10
1320,83	N I	3	1298,95	Ti III	40
1320,687	Cu II	10	1298,960	Si III	18
1319,684	N I	30	1298,891	Si III	15
1319,003	N I	20	1298,67	Ti III	50
1318,582	Cu III	2	1298,394	Cu II	15
1318,13	N I	3	1297,96	Ca III	3
1317,60	Ca III	2	1297,549	Cu II	2
1317,41	N I	2	1296,726	Si III	14
1316,287	N I	1	1296,600	N IV	5
1316,143	Cu III	5	1296,30	C III	2
1315,903	C I	4	1296,088	Fe II	20
1315,484	N I	1	1295,91	Ti III	30
1315,23	N I	2	1295,560	Xe I	8
1314,335	Cu II	30	1294,914	Fe II	12
1314,147	Cu II	15	1294,67	Ti III	50
1313,867	Si III	3	1294,543	Si III	17
1313,471	C I	6	1293,988	Kr III	3
1313,47	N I	3	1293,26	Ti III	30
1313,20	N I	3	1291,969	Si IV	30
1312,86	N I	3	1291,64	Ti III	20
1312,590	Si III	13	1291,594	Fe II	15
1312,44	N I	3	1291,380	C I	1
1312,400	Cu III	10	1290,204	Fe II	15
1312,261	C I	2	1289,983	C I	3
1311,985	C I	2	1289,32	Ti III	30
1311,365	C I	8	1288,633	C I	2
1310,952	N I	25	1288,445	C I	5
1310,646	C I	4	1288,055	C I	1
1310,58	Ca III	1	1287,464	Cu II	15
1310,548	N I	25	1286,38	Ti III	40
1310,057	N I	1	1284,868	Cu II	8
1309,557	N IV	4	1284,793	Ar II	1
1309,463	Cu II	15	1284,218	N IV	3
1309,458	Si II	20	1283,798	Kr III	3
1309,443	Mg II	—	1283,48	Al IV	0
1309,30	N I	3	1283,313	Kr III	3
1309,274	Si II	200	1282,620	Ar II	1
1308,86	N I	3	1282,49	Ti III	3
1308,73	C III	2	1282,450	Cu II	15
1308,296	Cu II	30	1281,50	Ca III	2
1308,280	Mg II	—	1281,458	Cu II	8
1307,875	Mg II	—	1281,098	Cu II	3
1306,714	Mg II	—	1280,852	C I	4
1307,595	Cu III	3	1280,604	C I	2
1306,025	O I	25	1280,43	Fe IV	10
1305,590	Si II	50	1280,362	N I	1
1305,554	Cu II	5	1280,354	Si III	6
1304,866	O I	30	1280,403	C I	2
1304,369	Si II	100	1280,340	C I	6
1303,979	Cu II	2	1280,336	Si IV	20
1303,656	Cu II	2	1280,265	Cu II	5
1303,59	Fe V	1	1280,225	Ar II	1
1303,320	Si III	16	1280,140	C I	2
1302,99	Fe V	1	1279,995	N I	1
1302,586	Kr III	2	1279,898	C I	5
1302,173	O I	30	1279,230	C I	6
1302,13	Al IV	2	1278,943	Kr III	1
			1278,38	Ca III	2

λ	Symbol	I
1277,727	C I	3
1277,551	C I	10
1277,50	Xe I	6
1277,282	C I	9
1277,154	C I	2
1276,800	N II	2
1276,754	C I	4
1276,206	N II	3
1275,801	Fe II	20
1275,570	Cu II	30
1275,247	N II	1
1275,154	Fe II	15
1275,038	N II	4
1275,021	C I	5
1274,880	C I	2
1274,463	Cu II	3
1274,131	C I	5
1274,069	Cu II	3
1273,716	N IV	2
1273,704	Cu II	2
1273,49	Fe IV	2
1273,47	N IV	3
1273,423	Mg II	—
1272,74	N IV	2
1272,720	Mg II	—
1272,70	Al IV	3
1272,638	Fe II	15
1272,160	N IV	4
1272,036	Cu II	8
1272,001	Fe II	25
1271,940	Mg II	—
1271,839	Cu III	2
1271,326	Cu II	2
1271,239	Mg II	—
1271,234	Cu III	5
1271,08	Fe IV	15
1270,54	Ca III	2
1270,28	N IV	5
1270,204	Kr III	5
1268,483	Ar II	1
1268,40	Fe IV	2
1267,633	C I	1
1267,437	Fe II	25
1266,694	Fe II	20
1266,449	C I	3
1266,308	Cu II	10
1265,66	Na III	2
1265,504	Cu II	15
1265,315	Kr III	4
1265,28	Fe IV	15
1265,023	Si II	200
1264,730	Si II	2000
1264,14	Al IV	3
1263,47	Fe IV	15
1262,928	Cu II	3
1262,51	Al IV	1
1261,72	Fe IV	10
1261,560	C I	8
1261,430	C I	5
1261,128	C I	7
1261,000	C I	3
1260,930	C I	4
1260,738	C I	4
1260,670	C I	2
1260,542	Fe II	20
1260,418	Si II	1000
1259,937	Cu III	10
1259,54	Fe IV	30
1259,309	Kr III	3
1258,88	Al II	4
1258,795	Si I	50
1258,75	N II	3
1258,745	Kr III	3
1258,68	Fe IV	2
1257,58	Al IV	3
1257,29	Fe IV	6
1257,190	Ne III	6
1256,68	Na III	1
1256,52	C III	1
1256,490	Si I	40
1255,685	Ne III	5
1255,276	Si I	10
1255,026	Ne III	2
1254,80	Fe IV	10
1254,717	Cu III	3
1254,6	Li II	1
1253,538	C I	0,5
1253,645	Xe III	2
1253,179	Cu II	5
1251,35	Al IV	1
1251,164	Si II	200
1250,433	Si II	150
1250,199	Xe I	2
1250,089	Si II	100
1250,045	Cu II	10
1249,28	Xe IV	—
1248,790	Cu II	5
1248,76	Al IV	2
1248,426	Si II	150
1247,383	C III	3
1246,738	Si II	100
1246,51	N IV	2
1244,92	N IV	1
1244,756	Xe II	5
1244,377	Cu III	10
1243,73	N IV	—
1243,309	N I	15
1243,179	N I	20
1242,804	N V	19
1241,961	Cu II	2
1240,83	Al IV	3
1240,395	Mg II	—
1240,18	Al IV	2
1239,925	Mg II	—
1238,821	N V	20
1238,325	Cu III	1
1237,776	Cu III	3
1237,14	Al IV	4
1235,920	Si II	10
1235,839	Kr I	13
1235,431	Si III	7
1235,40	Na III	4
1233,660	Fe II	8
1233,20	N I	2

λ	Symbol	I
1232,074	Xe III	25
1231,588	N I	1
1231,406	Si II	5
1230,511	C IV	3
1230,288	N I	0
1230,046	C IV	2
1229,94	Al IV	0
1229,40	N I	2
1229,388	Si II	200
1229,172	N I	1
1228,790	N I	10
1228,746	Si II	150
1228,617	Si II	25
1228,437	Si II	10
1228,410	N I	5
1228,30	Al IV	1
1227,788	N I	2
1227,604	Si II	100
1227,226	N I	1
1227,00	N I	3
1226,986	Si II	40
1226,887	Si II	20
1226,831	N I	1
1226,814	Si II	50
1225,85	N I	2
1225,719	N IV	4
1225,372	N I	10
1225,192	N IV	3
1225,089	Xe III	3
1225,028	N I	15
1224,972	Si II	10
1224,960	N IV	1
1224,73	Na III	4
1224,252	Si II	20
1223,907	Si II	20
1223,80	N I	2
1223,71	Cl II	2
1223,44	Na III	4
1223,20	N I	3
1222,635	Si II	5
1221,12	Na III	5
1220,882	Fe II	5
1219,290	Cu III	5
1219,19	Al IV	0
1217,643	O I	2
1217,26	Xe IV	—
1216,896	Kr III	5
1216,78	Al IV	1
1216,117	Si II	10
1215,670	H	3000
1215,340	D	3000
1215,229	T	3000
1215,171	He II	5
1215,088	He II	2
1214,409	Fe II	10
1213,764	Fe II	20
1213,149	Fe II	20
1212,011	Si III	2
1211,93	Al II	3
1211,80	Al IV	0
1210,456	Si III	10
1210,15	Al II	2
1209,19	Al II	1
1208,35	Al II	3
1207,517	Si III	9
1206,533	Si III	30
1206,510	Si III	30
1206,346	Kr III	5
1205,95	Ar III	1
1205,900	Cu II	2
1201,626	Cu II	2
1201,358	Cl I	11
1200,711	N I	30
1200,224	N I	0
1199,718	N I	2
1199,549	N I	50
1198,6	Li II	7
1198,58	C IV	1
1198,47	Al IV	1
1197,84	Ar IV	1
1197,812	C I	0,5
1197,389	Si II	100
1196,29	Xe IV	—
1195,25	Ti IV	5
1194,656	C I	5
1194,496	Si II	250
1194,494	C I	7
1194,291	C I	2
1194,060	C I	5
1194,027	C I	3
1193,674	C I	4
1193,388	C I	3
1193,284	Si II	250
1193,252	C I	10
1193,013	C I	8
1192,923	C I	2
1192,55	N I	2
1192,480	C I	2
1192,261	Cu II	2
1192,04	Xe I	2
1191,99	N I	4
1191,86	Al II	5
1191,855	C I	1
1191,55	Cs II	8
1191,03	N I	5
1190,84	N I	5
1190,52	N I	3
1190,418	Si II	100
1190,354	Ar IV	2
1190,07	Al II	4
1189,628	C I	6
1189,628	N I	5
1189,556	C I	4
1189,244	N I	3
1189,244	C I	4
1189,21	Al II	2
1189,074	C I	3
1188,972	N I	5
1188,935	C I	1
1188,768	Cl I	12
1188,006	N IV	6

λ	Symbol	I
1187,80	Ar IV	1
1185,899	Cu II	2
1184,544	N III	8
1183,998	N I	3
1183,63	Ti IV	5
1183,053	Xe III	8
1183,053	Xe II	7
1183,030	N III	7
1182,018	Si III	3
1180,40	Na III	8
1179,38	Al II	1
1178,65	Cs II	10
1178,004	Si III	8
1177,694	N I	15
1177,48	Al II	4
1176,626	N I	3
1176,508	N I	15
1176,370	C III	3
1175,987	C III	3
1175,711	C III	5
1175,590	C III	2
1175,263	C III	3
1174,933	C III	3
1174,84	N I	3
1174,432	Si III	6
1174,369	Si III	5
1172,55	N I	3
1172,529	Si III	4
1172,02	N I	2
1171,606	Fe II	8
1171,60	N I	2
1171,39	N I	2
1171,067	N I	0
1170,43	Xe	3
1170,276	N I	1
1169,692	N I	1
1169,63	Xe II	2
1169,478	N IV	1
1169,063	N IV	2
1168,990	C IV	4
1168,873	C IV	3
1168,599	N IV	3
1168,537	N I	20
1168,334	N I	8
1167,450	N I	25
1167,35	Al IV	0
1167,0	Li II	2
1165,87	C III	1
1165,566	N I	2
1165,269	Fe II	12
1164,868	Kr I	4
1164,322	N I	8
1163,884	N I	12
1162,66	Al III	0
1162,610	Cu II	3
1161,85	Al IV	0
1161,579	Si III	8
1161,26	N I	2
1160,67	N I	2
1160,45	N I	3
1160,252	Si III	6
1159,858	N I	1
1159,347	Fe II	20
1159,285	N I	1
1159,004	C I	5
1158,729	C I	3
1158,724	Kr III	6
1158,474	Xe II	5
1158,398	C I	2
1158,14	Al II	1
1158,138	C I	8
1158,102	Si III	7
1158,051	N I	2
1158,030	C I	7
1157,871	Cu II	8
1157,825	C I	3
1157,46	Xe IV	—
1157,391	C I	2
1157,333	C I	1
1157,13	Al II	3
1157,021	Cu II	5
1156,782	Si III	4
1156,619	C I	5
1156,502	C I	1
1156,480	Xe III	9
1156,21	Al IV	1
1156,059	C I	2
1155,957	Si III	6
1155,839	C I	1
1154,998	Si III	6
1154,401	Fe II	20
1154,23	N I	3
1153,955	Fe II	15
1153,773	O III	3
1153,52	N I	4
1153,281	Fe II	20
1153,04	Na III	2
1152,882	Fe II	20
1152,75	N I	1
1152,440	Fe II	15
1152,35	N I	4
1152,19	Xe	7
1152,149	O I	2
1152,14	Al II	4
1151,163	Fe II	25
1150,882	O III	2
1150,85	Al IV	0
1150,689	Fe II	20
1150,292	Fe II	20
1149,603	O III	1
1149,47	N I	2
1148,76	N I	4
1148,693	Fe II	8
1148,295	Fe II	30
1147,762	Cu II	8
1147,69	N I	4
1147,413	Fe II	25
1146,963	Fe II	15

λ	Symbol	I	λ	Symbol	I
1145,92	N I	1	1129,626	C I	1
1145,23	N I	2	1129,19	Fe III	7
1145,22	Si III	8	1129,161	C I	6
1145,177	Si III	7	1128,909	Fe II	20
1144,959	Si III	6	1128,748	C I	1
1144,946	Fe II	35	1128,72	Fe III	7
1144,853	Cu II	30	1128,530	Fe II	10
1144,306	Si III	8	1128,340	Si IV	10
1144,24	N I	2	1128,325	Si IV	10
1144,052	Fe II	5	1128,180	Fe II	5
1143,649	N I	5	1128,074	Fe II	25
1143,32	N I	1	1128,02	Fe III	8
1143,235	Fe II	25	1127,907	Si II	40
1142,97	Al II	2	1127,442	Si II	20
1142,73	N I	2	1126,850	Fe II	20
1142,642	Cu II	20	1126,72	Fe III	6
1142,334	Fe II	25	1126,603	Fe II	20
1142,282	Si III	6	1126,425	Fe II	20
1142,03	Al IV	1	1124,883	Fe III	9
1141,745	C II	3	1124,134	Fe II	20
1141,705	C I	1	1123,226	Cu II	5
1141,70	N I	2	1122,858	Fe II	25
1141,630	C II	3	1122,526	Fe III	9
1141,580	Si III	7	1122,485	Si IV	8
1141,20	N I	1	1122,325	C I	4
1140,76	N I	2	1122,30	Na III	0
1140,688	C I	3	1122,179	C I	1
1140,545	Si III	6	1121,987	Fe II	25
1140,391	C I	1	1119,945	Cu II	15
1140,070	C I	1	1118,80	Al IV	4
1139,894	C I	7	1118,42	Xe IV	—
1139,818	N I	1	1117,990	Si IV	4
1139,794	C I	6	1117,706	C I	3
1139,60	F VI	2	1117,686	Si III	4
1139,49	C II	0	1114,414	C I	2
1139,330	C II	3	1113,228	Si III	18
1139,15	N I	1	1112,407	Cu II	5
1139,142	C I	2	1112,086	Fe II	35
1139,037	C I	1	1111,114	Fe II	15
1138,936	C II	2	1110,62	Xe	3
1138,642	Fe II	25	1109,970	Si III	16
1138,625	C I	1	1108,368	Si III	14
1138,545	O III	2	1107,933	C IV	2
1138,039	Fe II	5	1107,600	C IV	1
1136,80	Al IV	3	1106,446	Cu II	3
1136,241	N IV	2	1106,362	Fe II	5
1135,244	N IV	3	1106,215	Fe II	15
1134,981	N I	25	1105,182	Cu II	5
1134,89	Kr I	3	1102,385	Fe II	8
1134,417	N I	25	1101,538	Fe II	20
1134,168	N I	20	1101,293	N I	40
1134,15	Kr I	3	1100,525	Fe II	20
1133,678	Fe II	25	1100,49	Na III	5
1133,413	Fe II	25	1100,46	Xe I	15
1133,117	N IV	4	1100,432	Xe II	10
1132,8	Li II	1	1100,362	N I	30
1131,194	Fe III	7	1100,050	Si IV	1
1130,428	Fe II	25			
1130,344	Xe III	30			
1129,927	C I	1			
1129,777	Fe II	12			

λ	Symbol	I	λ	Symbol	I
1100,026	Fe II	20	1068,66	N I	4
1099,80	N I	2	1068,476	N I	35
1099,153	N I	25	1068,356	Fe II	30
1099,117	Fe II	25	1067,94	Cl II	4
1098,98	N I	2	1067,74	Xe IV	—
1098,78	N I	1	1067,607	N I	35
1098,63	N I	2	1067,37	N I	4
1098,408	Si IV	1	1067,10	Xe I	5
1098,264	N I	40	1066,97	N I	4
1098,103	N I	40	1066,660	Ar I	15
1097,990	N I	25	1066,660	O I	9
1097,245	N I	50	1066,650	Si IV	8
1097,16	Xe IV	—	1066,636	Si IV	8
1097,049	Cu II	25	1066,614	Si IV	8
1096,886	Fe II	30	1066,56	N I	3
1096,793	Fe II	20	1066,391	Xe III	12
1096,749	N I	35	1066,181	Fe III	10
1096,616	Fe II	20	1066,143	Fe III	10
1096,322	N I	35	1066,133	Cu II	20
1095,940	N I	35	1066,121	C II	6
1095,279	N I	4	1066,126	N I	1
1094,401	Cu II	30	1065,883	C II	7
1092,740	C II	2	1065,7822	Cu II	20
1092,422	C II	0	1065,04	Xe IV	—
1092,240	C II	0	1063,982	Fe II	15
1091,930	C II	1	1063,872	Fe III	8
1091,288	Cu II	5	1063,83	Cl II	10
1089,236	Cu II	3	1063,351	N I	1
1088,94	Xe I	10	1063,30	C II	0
1088,41	F V	0	1063,003	Cu II	60
1088,393	Cu II	20	1062,758	Fe II	20
1086,691	N IV	2	1061,708	Fe III	6
1086,269	N IV	1	1060,630	Cu II	60
1086,110	Cu II	5	1059,571	Fe II	20
1086,084	N IV	–	1059,0960	Cu II	60
1085,701	N II	12	1058,796	Cu II	40
1085,542	N II	9	1058,128	Xe III	2
1085,47	Xe	2	1057,503	Si II	15
1084,951	He II	3	1057,050	Si II	30
1084,572	N II	11	1056,9545	Cu II	60
1083,990	N II	10	1056,582	Si IV	12
1083,860	Xe II	5	1055,795	Cu II	40
1083,210	Si III	6	1055,328	Xe III	5
1079,112	Si III	4	1055,269	Fe II	25
1079,08	Cl II	15	1054,6903	Cu II	60
1078,708	N IV	6	1053,90	N I	3
1077,325	Si IV	1	1053,65	N I	3
1076,589	Si III	10	1053,38	N I	5
1076,253	Si III	3	1053,289	Si III	10
1075,24	Cl II	7	1053,03	N I	3
1074,476	Xe II	15	1052,72	N I	2
1073,738	Cu II	30	1052,170	Cu II	20
1071,76	Cl II	10	1052,16	N I	3
1071,656	N I	1	1052,07	N I	3
1071,596	Fe II	30	1051,920	Xe II	10
1071,05	Cl II	20	1051,89	N I	2
1070,821	N I	0	1051,596	Si IV	70
1070,308	Cu II	15			
1069,984	N I	30			
1069,198	N I	2			
1069,193	Cu II	50			
1069,038	Fe II	15			

λ	Symbol	I
1050,399	Cu II	10
1050,153	Cu II	10
1049,93	Al II	2
1049,7556	Cu II	50
1049,65	N V	3
1049,363	Cu II	20
1048,754	Xe III	3
1048,53	Al II	1
1048,272	Xe II	8
1048,218	Ar I	25
1048,218	O I	8
1048,20	N V	2
1047,801	Xe III	10
1044,7434	Cu II	80
1044,69	N I	4
1044,516	Cu II	80
1044,13	N I	5
1043,58	N I	2
1043,12	N I	5
1041,81	Xe IV	—
1041,688	O I	1
1041,306	Xe II	9
1040,941	O I	15
1039,569	Cu II	60
1039,345	Cu II	60
1039,233	O I	20
1038,90	N I	1
1038,76	N I	1
1038,355	Fe III	6
1038,34	N I	3
1037,931	Ar IV	1
1037,680	Xe II	6
1037,64	N I	1
1037,613	O VI	9
1037,38	N I	4
1037,053	Si III	7
1037,017	C II	13
1036,4695	Cu II	60
1036,330	C II	12
1036,16	N IV	8
1035,768	Fe III	6
1035,657	Si III	3
1035,366	Ca III	4
1035,1631	Cu II	8
1034,96	N I	1
1034,848	Ca III	3
1034,39	N I	2
1034,287	Si III	4
1033,920	Si III	8
1033,875	K	2
1033,65	N I	0
1033,5679	Cu II	10
1033,48	N I	3
1032,958	N I	2
1032,851	Si III	3
1032,768	K	2
1032,612	Ca VI	2
1032,438	Xe II	4
1032,19	N I	1
1032,123	Fe III	8
1031,912	O VI	10
1031,7661	Cu II	8
1031,760	Ca III	4
1031,65	N I	2
1031,169	Si III	7
1030,924	Fe III	6
1030,72	N I	2
1030,273	Ca IV	4
1030,261	Cu II	20
1030,020	Kr I	2
1029,747	Cu II	10
1029,566	Ca IV	3
1029,53	N I	1
1028,64	N I	2
1028,560	Ca III	4
1028,3281	Cu II	25
1028,162	O I	8
1027,8312	Cu II	50
1027,433	O I	20
1027,309	Ca IV	5
1027,174	K V	2
1027,04	Xe I	10
1026,790	Fe III	6
1026,28	Xe IV	—
1026,113	Mg II	—
1025,968	Mg II	—
1025,766	O I	9
1025,742	K IV	3
1025,722	H	1000
1025,443	D	1000
1025,350	T	1000
1025,280	He II	2
1024,339	Ca IV	5
1023,820	Ca IV	4
1023,693	Si II	50
1022,1021	Cu II	5
1021,508	Ca VI	4
1021,332	K V	2
1021,139	Ca V	3
1020,699	Si II	25
1020,1075	Cu II	15
1019,789	Fe III	6
1019,6545	Cu II	15
1019,371	Ca III	2
1018,7075	Cu II	50
1018,346	Ca VI	3
1018,286	Fe III	8
1018,0643	Cu II	15
1017,745	Fe III	8
1017,680	Xe III	35
1017,254	Fe III	9
1015,520	Fe II	20
1015,083	Fe II	10
1015,023	Cl III	7
1014,998	Ca III	3
1014,162	Ca V	4
1012,6834	Cu II	3
1012,613	Ca V	3

λ	Symbol	I	λ	Symbol	I
1012,5972	Cu II	25	991,514	N III	14
1012,417	Fe II	25	991,232	Fe III	9
1012,088	Fe II	20	990,88	Al II	1
1011,4362	Cu II	2	990,805	O I	2
1011,037	Fe II	25	990,800	Fe III	6
1010,6395	Cu II	3	990,210	O I	8
1010,453	Cu II	10	990,132	O I	1
1010,376	Xe III	3	989,867	Si II	100
1010,369	C II	10	989,790	N III	16
1010,267	Cu II	30	989,2368	Cu II	8
1010,074	C II	10	988,776	O I	15
1009,854	C II	9	988,66	O I	2
1009,638	Ca V	3	988,581	O I	3
1008,875	N I	1	987,680	Ca V	5
1008,777	Cl III	6	987,656	Cu II	10
1008,7284	Cu II	30	987,336	Ca III	5
1008,5692	Cu II	30	987,281	Kr III	18
1008,08	Ti III	0	985,824	Fe III	8
1007,975	Fe II	25	985,749	Cl IV	4
1007,657	Fe II	20	984,952	Cl IV	7
1007,15	Ti III	1	984,935	Ca III	4
1006,015	N III	6	984,889	Si IV	1
1005,75	Ti III	0	984,530	Cu II	10
1005,365	Si III	7	983,877	Fe III	10
1005,280	Cl III	5	982,115	K VI	2
1004,68	Ti III	2	981,373	Fe III	10
1004,0557	Cu II	30	981,088	Xe III	7
1003,542	Kr I	2	979,919	N III	9
1003,370	Xe III	35	979,842	N III	8
1002,23	Ti III	0	979,418	Cu II	5
1002,095	Ar III	3	978,616	O I	4
1001,544	Ca V	3	977,967	O I	1
1001,048	Kr I	2	977,901	Cl IV	4
1001,0130	Cu II	8	977,745	F I	100
1000,310	Ca V	6	977,567	Cu II	25
1000,056	K VI	2	977,560	Cl IV	6
999,7944	Cu II	5	977,544	Ca III	2
999,493	O I	2	977,4	Al	4
998,397	Ca III	3	977,026	C III	18
998,3063	Cu II	8	976,708	Cu II	10
997,599	Fe III	6	976,678	Xe II	6
997,579	Ca IV	7	976,540	Cu II	10
997,389	Si III	16	976,505	F I	40
997,081	Fe III	7	976,452	O I	5
995,829	Fe II	8	976,217	F I	100
995,50	Xe VII	3	975,825	Ca V	4
995,150	Fe III	6	975,055	Ca VI	3
994,946	Ca V	3	974,759	Cu II	20
994,787	Si III	13	974,124	Xe III	8
994,724	Fe III	6	973,895	F I	350
994,311	Ca IV	6	973,884	O I	4
993,519	Si III	10	973,508	Cu II	2
993,080	Fe III	7	973,437	Ca V	6
992,9533	Cu II	25	973,240	O I	5
992,675	Si II	200	973,212	Cl IV	5
992,370	He II	1	972,769	Xe II	7
991,829	Fe III	6	972,537	H	400
991,579	N III	17			

λ	Symbol	I	λ	Symbol	I
972,401	F I	20	952,46	Xe IV	—
972,272	D	400	952,414	O I	8
972,263	Cu II	2	952,414	N I	—
972,184	T	400	952,304	N I	8
972,118	He II	0,7	951,871	F I	500
971,84	Xe II	3	951,413	Cu II	5
971,818	Xe III	8	951,35	N I	1
971,741	O I	8	951,08	N I	3
969,652	Ca VI	6	951,06	Kr I	0
968,518	K VI	6	950,888	O I	4
968,236	Ca V	3	950,732	O I	—
968,037	Cu II	25	950,344	Fe III	10
967,946	Si III	9	950,114	O I	0
967,197	Fe III	6	949,743	H	200
966,466	Ca V	6	949,485	D	200
966,231	Cu II	3	949,335	He II	0,3
965,540	Xe III	10	948,72	Cl III	1
965,042	N I	10	948,689	O I	4
964,962	Kr II	30	948,540	N IV	5
964,626	N I	5	948,244	N IV	4
963,991	N I	5	948,214	C IV	1
963,34	Kr I	1	948,155	N IV	2
962,896	Ca V	2	948,098	C IV	2
961,901	Fe III	7	947,700	Cu II	2
961,49	Cl II	10	946,97	Cl III	1
960,409	Cu II	20	946,769	Mg II	—
960,325	Xe III	2	946,703	Mg II	—
959,54	N I	3	946,52	Kr I	1
959,22	Xe	2	946,208	C II	2
958,86	Ne I	1	945,981	C II	1
958,705	He II	0,5	945,976	Cu II	50
958,585	Xe III	4	945,860	Cu II	40
958,524	F I	500	945,566	C I	3
958,149	Cu II	40	945,524	Cu II	60
957,70	Xe IV	—	945,45	Kr I	1
956,286	Cu II	25	945,401	T	200
955,99	Al II	1	945,336	C I	2
955,91	N I	3	945,193	C I	1
955,545	F I	750	945,095	Fe II	25
955,438	N I	—	943,910	Fe II	15
955,335	N IV	20	943,328	Cu II	60
955,321	Cu II	5	943,267	Fe II	12
955,265	N I	—	943,22	Cl III	1
954,825	F I	1000	942,52	He II	0,2
954,774	Kr III	4	941,660	Fe II	12
954,378	Cu II	20	939,837	O I	—
954,11	N I	3	939,522	Cu II	10
953,98	N I	6	939,31	Cl III	0
953,975	Xe III	3	939,237	O I	—
953,658	N I	15	939,16	Xe II	12
953,42	Kr I	1	939,159	Fe II	20
953,415	N I	15	939,093	Si III	7
953,40	Cl III	2	938,967	Fe II	10
953,399	N I	6	938,621	O I	—
952,940	O I	4	938,287	K VI	2
952,789	N I	3	938,022	O I	—
952,522	N I	4	937,841	O I	3
952,470	Fe II	10	937,814	Cu II	5
			937,804	H	120

λ	Symbol	I
937,548	D	120
937,464	T	120
937,40	He II	—
936,630	O I	3
936,484	Fe II	8
936,28	Cl III	1
936,060	Si III	3
935,892	Cu II	60
935,405	Xe II	2
935,35	Cu II	20
935,25	Cu II	40
935,24	Xe IV	—
935,20	Al II	1
935,183	O I	4
935,074	Cu II	60
934,703	Fe III	7
933,46	He II	—
933,420	Si IV	5
932,940	Cu II	60
932,687	Fe II	30
932,244	Fe II	30
932,0528	Ar II	10
932,046	O II	10
931,709	Fe II	10
931,667	Si II	5
931,479	O I	—
931,25	Xe II	10
931,200	Si II	5
931,142	Fe II	25
930,94	Cl III	1
930,889	O I	—
930,748	H	80
930,558	Fe II	30
930,495	D	80
930,410	T	80
930,318	K VI	1
930,219	Fe II	30
930,165	Fe II	30
930,030	Fe II	30
929,897	Cu II	5
929,810	Si II	20
929,732	Cu II	2
929,612	Fe II	30
929,538	Fe II	30
929,517	O I	—
929,374	K VI	1
928,470	Fe II	20
928,297	Si II	5
928,107	Fe II	30
927,632	Fe II	8
927,176	Fe II	30
926,900	Fe II	25
926,75	Cs II	20
926,618	Fe II	10
926,24	Xe IV	—
926,226	H	50
926,220	Fe II	60
925,974	D	50
925,890	T	50
925,866	Xe II	5
925,125	Cu II	30
924,970	Fe II	15
924,283	N IV	14
924,239	Cu II	50
923,884	Fe II	30
923,675	N IV	14
923,433	O IV	4
923,353	O IV	6
923,220	N IV	16
923,150	H	40
923,057	N IV	14
922,899	D	40
922,815	T	40
922,519	N IV	14
922,411	Cu II	20
922,017	Cu II	60
922,011	O I	—
921,992	N IV	14
921,364	O IV	5
921,301	O IV	4
920,963	H	30
920,713	D	30
920,629	T	30
920,29	Xe IV	—
919,7815	Ar II	10
919,78	O II	15
919,351	H	20
919,143	Kr III	2
918,581	K VI	2
918,129	H	16
917,498	K V	1
917,434	Kr II	20
917,303	Cu II	20
917,278	Ca III	2
917,257	Xe III	4
917,181	H	12
916,917	Ca III	2
916,700	N II	12
916,682	Ca VI	2
916,429	H	10
916,004	N II	11
915,955	N II	10
915,824	H	8
915,603	N II	10
915,488	Xe III	3
915,329	H	7
915,29	Xe IV	—
914,919	H	6
914,90	Cl	2
914,576	H	5
914,286	H	—
914,209	Cu II	80
913,853	Si II	20
913,264	Si II	3
913,012	Si II	10
912,71	Xe II	8
912,459	Si II	5
912,414	Cu II	3

λ	Symbol	I
912,375	Si II	5
911,753	H	—
911,505	D	—
911,422	T	—
911,384	Kr II	25
911,37	He II	—
910,961	Fe III	6
910,6456	N I	0
910,518	Cu II	15
910,2785	N I	0
909,6976	N I	0
909,209	Si II	3
908,83	N I	2
908,23	N I	3
907,412	Mg II	—
907,375	Mg II	—
907,278	N I	4
906,833	N I	2
906,722	N I	1
906,63	N I	4
906,615	Ca III	2
906,426	N I	15
906,202	N I	10
906,109	Cu II	40
905,829	N I	5
905,53	N I	2
905,338	Fe III	7
905,23	N I	4
904,5	Al	5
904,468	C II	10
904,134	C II	12
903,950	C II	11
903,609	C II	10
901,804	Ar IV	2
901,746	Xe III	7
901,735	Si II	20
901,34	Cs II	20
901,168	Ar IV	9
901,071	Cu II	60
900,362	Ar IV	5
900,360	Fe II	5
899,791	Cu II	50
899,417	Fe III	8
899,405	Si II	10
898,957	O III	8
898,953	K V	3
898,873	Xe III	8
897,801	Kr III	40
897,790	Cu II	15
896,970	Cu II	40
896,753	Cu II	60
896,003	Xe III	20
895,95	Cl II	3
895,406	Xe III	4
894,910	C IV	1
894,351	Ca III	4
894,347	Si IV	3
894,340	Cl V	4
894,310	Ar I	4
894,226	Cu II	40
893,989	Xe III	20
893,905	Al III	5
893,674	Cu II	80
893,56	Cl II	3
892,671	Ca IV	3
892,621	K IV	2
892,417	Fe III	6
892,411	Cu II	50
892,056	Al III	4
891,999	Si II	200
891,833	Xe III	9
891,442	Fe III	8
891,21	Xe IV	—
891,172	Fe III	10
890,982	Kr II	20
890,892	Ca III	2
890,755	Fe III	9
890,567	Cu II	60
889,722	Si II	100
889,276	Xe III	15
888,363	N I	0
888,07	Cl II	4
888,019	N I	0
887,41	N I	4
887,404	Ar III	10
887,404	O III	10
887,24	Xe II	6
886,95	N I	3
886,946	Cu II	60
886,80	N I	3
886,515	Cu II	10
886,33	N I	6
886,302	Kr II	30
885,93	N I	3
885,842	Cu II	25
885,67	N I	5
885,54	Xe II	3
885,36	N I	3
884,824	Cu II	5
884,719	Mg II	—
884,697	Mg II	—
884,516	C III	8
884,430	Cu II	8
884,144	Kr II	30
884,127	Cu II	10
883,837	Cu II	5
883,688	Fe III	6
883,398	Si III	5
883,282	Cu II	5
883,179	Ar III	9
883,159	Ca III	3
883,127	C IV	4
882,88	O I	—
882,184	K VI	2
881,405	K V	3

λ	Symbol	I	λ	Symbol	I
881,088	Fe III	7	869,336	Cu II	25
880,949	Fe III	6	869,062	Cu II	10
880,802	Xe II	5	868,98	N I	5
880,447	Fe III	6	868,869	Kr II	25
880,325	Cu II	5	868,552	K V	1
880,04	Xe VI	2	868,140	K V	1
879,949	Ar I	3	867,921	K V	1
879,906	Cu II	2	867,726	Cu II	8
879,622	Ar III	8	866,805	Ar I	4
879,553	O I	1	866,440	Cu II	5
879,108	O I	1	865,93	N I	3
879,079	O I	1	865,69	Xe IV	—
879,027	O I	1	865,63	N I	5
878,979	O I	1	865,383	Cu II	40
878,790	Xe III	8	864,93	N I	5
878,728	Ar III	12	864,812	Kr II	20
878,728	O III	11	864,695	Ca III	4
878,696	Cu II	50	864,67	Cl II	5
877,97	O I	4	864,199	Cu II	10
877,9	Cs III	7	863,386	Xe III	8
877,885	O I	2	863,15	N I	3
877,839	Cu II	15	862,90	N I	5
877,804	O I	2	862,578	Kr III	35
877,559	Cu II	20	862,15	N I	5
877,11	N I	2	862,011	Cu II	40
877,007	Cu II	25	861,832	Fe III	10
876,79	N I	2	861,761	Fe III	8
876,719	Cu II	20	861,63	O I	—
876,674	Kr III	22	861,15	N I	1
876,32	N I	4	861,071	Xe III	5
876,063	Ar I	4	860,85	N I	4
875,764	N I	0	860,827	Ca VI	2
875,534	Ar III	9	860,205	N II	0
875,534	O III	9	860,15	N I	4
875,25	N I	6	859,838	Fe III	6
875,092	N I	5	859,75	N I	2
874,985	K V	1	859,721	Fe III	8
874,883	K V	1	859,626	Fe III	6
874,045	K III	3	859,35	N I	3
873,865	K III	2	859,040	Kr II	20
873,462	Fe III	8	858,855	Ca IV	3
873,264	Cu II	15	858,80	N I	2
872,313	K III	4	858,602	Fe III	6
872,006	K VI	1	858,59	Xe VIII	3
871,850	N III	0	858,561	C II	9
871,42	Xe II	6	858,482	Cu II	25
871,099	Ar III	10	858,374	N II	1
871,099	O III	10	858,094	C II	8
871,064	Cu II	8	857,76	N I	2
871,01	N I	1	856,80	N I	2
870,825	Kr III	20	856,791	Ca III	4
870,544	Cu II	8	856,768	Al III	5
870,40	N I	3	856,635	Ca III	4
870,346	Xe III	6	856,24	N I	2
870,00	N I	3	855,815	K IV	2
869,965	K V	3	855,701	Cu II	10
869,754	Ar I	2	855,70	N I	2
869,66	N I	5			

λ	Symbol	I	λ	Symbol	I
855,474	Cu II	5	840,46	Xe IV	—
855,040	Al III	4	840,162	Xe III	7
854,923	Ca VI	3	840,029	Ar IV	15
854,789	Si IV	7	839,73	Xe	2
854,771	K V	1	839,63	Cl II	2
854,733	Kr III	25	839,439	K V	1
854,71	Xe II	6	839,30	Cl II	2
854,416	K V	2	838,449	Xe III	3
854,367	Fe III	6	838,048	Fe III	8
852,950	Xe III	25	837,666	Kr III	22
852,898	Cu II	3	837,439	Fe III	7
851,992	Fe III	6	836,837	N II	1
851,842	Fe III	6	836,618	N II	3
851,76	Cu II	2	836,521	Fe III	7
851,70	Cl II	7	836,281	N II	0
851,332	Fe III	7	836,184	N II	3
851,300	Cu II	25	836,126	Si IV	1
851,29	Xe IV	—	836,126	Ar V	2
851,150	Fe III	7	835,792	Ar V	1
851,147	Xe III	8	835,292	O III	16
850,966	Ca III	3	835,096	O III	14
850,76	Cu II	2	835,01	Xe IV	—
850,74	O I	—	835,003	Ar I	6
850,602	Ar IV	25	834,967	Cl IV	5
850,572	Xe III	5	834,944	Fe III	6
850,318	Kr II	6	834,878	Ar V	4
850,142	Si II	10	834,840	Cl IV	5
849,354	Cu II	3	834,67	Cl II	10
849,248	Si IV	1	834,659	Cl IV	3
848,806	Cu II	15	834,462	O II	15
848,074	Si II	5	834,397	Ar I	6
847,924	Fe III	6	833,742	O III	16
847,700	Fe III	6	833,326	O II	15
847,578	Fe III	7	832,927	O III	14
847,425	Fe III	8	832,754	O II	14
846,611	Ca III	3	831,431	Cl IV	4
846,534	Fe III	6	830,785	K V	1
845,925	Fe III	7	830,377	Kr II	18
845,774	Si II	40	829,343	Cu III	5
845,408	Fe III	9	827,85	Cl II	1
844,910	Cu II	5	827,777	Fe III	6
844,616	Cu II	3	827,349	Ar V	3
844,284	Fe III	10	827,055	Ar V	5
844,058	Kr II	25	826,995	Cu II	30
843,772	Ar IV	20	826,432	Kr II	22
843,718	Si II	20	826,395	K V	1
842,950	Ca V	3	826,371	Ar I	2
842,808	Ar I	2	826,1[illegible]	Xe III	4
842,483	Cu II	3	825,559	K V	1
842,35	Xe	5	825,348	Ar I	1
842,035	Kr IV	22	824,881	Xe III	30
842,020	Fe III	6	823,800	Cu II	2
841,41	Cl II	4	823,408	Si III	9
841,105	Cu II	2	823,358	K V	3
840,933	Cl IV	4	823,273	N IV	2
840,921	Ca III	3	823,257	Fe III	6
840,901	Si IV	5	823,210	Xe III	25
840,808	Cl IV	6	823,047	K V	3

λ	Symbol	I	λ	Symbol	I
822,844	Si II	5	806,964	F I	150
822,647	Xe III	4	806,875	Ar I	2
822,159	Ar V	4	806,846	C II	6
821,60	Xe IV	—	806,684	C II	4
821,583	Ca V	1	806,577	K V	2
821,568	K V	1	806,555	C II	7
821,450	Si II	2	806,550	Cu II	3
821,161	Kr II	20	806,384	C II	5
820,918	Si II	3	806,373	K V	2
820,516	Si II	20	805,95	Xe II	5
820,166	Xe III	4	805,763	Kr IV	7
820,15	Xe IV	—	805,101	Si II	10
820,129	Ar I	—	804,45	Xe II	5
818,590	Si II	2	803,826	K V	2
818,147	Kr II	25	803,325	Ca V	1
818,129	Si IV	8	803,066	Xe II	3
817,9	Cs III	1	802,841	Cu III	150
817,058	Ca VI	2	802,8	Cs III	00
817,038	Fe III	7	802,250	O IV	5
816,822	Kr IV	18	802,198	O IV	6
816,805	Ca VI	2	802,122	K V	2
816,466	Ar I	4	801,980	Xe III	15
816,273	Fe III	6	801,913	Ar IV	5
816,233	Ar I	4	801,409	Ar IV	10
816,163	Fe III	6	801,359	Ar I	1
815,049	Si IV	7	801,154	Cu III	200
814,242	Fe III	6	801,086	Ar IV	10
813,882	Cu II	20	800,84	Xe VI	2
813,85	Cs II	20	800,819	Xe III	3
813,382	Fe III	10	800,573	Ar IV	5
812,158	O I	—	800,228	Xe III	2
812,096	O I	3	800,066	Si III	5
811,710	O I	1	799,947	C II	4
811,51	Xe IV	—	799,660	C II	5
811,501	O I	1	799,338	Xe III	8
811,480	Ca VI	1	799,137	Ar I	0
811,284	Fe III	8	799,083	Kr II	9
811,052	O I	4	797,744	Ar I	1
810,997	Cu II	15	797,566	Cu III	100
810,940	Fe III	7	797,452	Cu II	10
810,937	Ca V	4	796,982	F I	3
810,893	K V	1	796,97	Xe IV	—
810,667	O I	1	796,678	Kr II	6
810,215	K V	1	796,661	O II	10
810,119	Xe III	7	796,070	Xe III	12
809,933	Ar I	2	795,774	F I	2
809,770	C II	3	795,36	Cl II	2
809,682	C II	4	795,258	Cu III	2
809,673	K V	2	794,417	F I	10
809,599	F I	125	793,977	Xe III	8
808,860	Xe III	3	793,53	Xe	4
808,840	Fe III	8	793,47	Cl II	3
808,77	Cs II	20	793,34	Cl II	3
808,583	Cu III	20	793,292	Xe III	8
807,855	Fe III	8	793,065	Cu III	100
807,702	Ar I	2	792,971	O I	—
807,547	Fe III	9	792,937	O I	3
807,220	Ar I	2			

λ	Symbol	I	λ	Symbol	I
792,896	Xe III	15	778,172	N V	2
792,536	F I	10	778,059	F I	6
792,510	O I	1	777,712	N V	1
792,237	O I	1	777,55	Cl II	3
792,19	Cl II	2	777,531	F I	4
791,976	O I	3	777,508	Ca VI	1
791,875	F I	12	777,125	Cu III	200
791,516	O I	1	777,010	F I	5
791,371	Cu III	300	776,957	K VI	4
790,203	O IV	16	776,926	F I	4
790,103	O IV	13	776,91	Cl IV	0
790,064	Xe III	5	776,82	Ti IV	10
790,006	F I	7	775,966	Ca VI	3
789,840	Cu III	200	775,957	N II	12
789,01	Cl II	7	775,526	Ca IV	3
788,75	Cl II	4	774,738	K	3
788,462	Cu III	300	774,532	Ca VI	3
788,073	Cu III	400	774,53	Xe	4
787,710	O IV	15	774,522	O V	7
787,62	Cl II	3	774,354	Ca V	3
787,31	Xe II	6	774,088	Ca V	5
787,15	Cl II	1	773,684	Kr II	18
786,464	K V	2	772,975	N III	8
786,141	Ne IV	1	772,891	N III	9
785,968	Kr III	25	772,641	Ca III	2
784,713	K V	2	772,389	Ca VI	1
784,393	C III	3	772,385	N III	12
784,09	Xe	2	771,901	N III	11
783,73	Xe IV	—	771,544	N III	10
783,715	Kr II	20	771,456	K V	3
782,976	F I	5	771,376	K V	3
782,6	Cs III	3	771,103	K VI	5
782,575	F I	2	771,024	Kr II	18
782,378	F I	10	770,928	Ca VI	1
782,084	Kr II	25	770,698	O I	0
781,78	Ti IV	20	770,409	Ne VIII	8
781,654	F I	3	770,379	C IV	0
780,713	F I	5	770,350	O I	—
780,519	F I	10	770,294	O I	1
780,390	F I	15	770,287	K V	3
780,324	Ne VIII	4	770,264	O I	1
780,250	Ne IV	3	770,022	K VI	1
780,030	Xe III	7	769,402	K V	2
779,997	O IV	6	769,355	O I	1
779,972	F I	2	769,152	Ar III	12
779,919	Ca V	4	769,143	Xe III	10
779,910	F I	5	768,411	O I	1
779,905	O IV	10	768,104	Kr III	1
779,824	Ca V	3	767,71	Ar VI	1
779,821	O IV	9	767,06	Ar VI	2
779.781	Xe III	5	766,522	Ca VI	1
779,734	O IV	6	766,202	Kr II	9
779,365	F I	6	765,644	K III	6
779,300	Cu II	8	765,314	K III	4
779,192	F I	2	765,154	Ca VI	6
779,126	Xe III	25	765,148	N IV	15
779,14	Ti IV	20	765,120	Xe III	7
778,718	Ca VI	1	764,358	Ca VI	3
778,603	Cu III	50	764,357	N III	15
778,528	K III	7			

λ	Symbol	I	λ	Symbol	I
763,976	Kr II	11	748,580	F I	4
763,736	Xe III	7	748,409	Ca V	3
763,344	Ca VI	2	748,4	O I	1
763,340	N III	14	748,393	K	2
762,200	Ar II	3	748,364	N II	5
762,001	O V	10	748,338	F I	2
761,861	Ca VI	0	748,291	N V	9
761,790	Xe III	5	748,198	Ar II	4
761,470	Ar IV	5	748,195	N V	8
761,130	O V	10	747,999	F I	2
761,050	Kr II	18	747,848	K VI	2
760,445	O V	12	747,677	K	3
760,439	Ar IV	3	747,553	Cl III	1
760,229	O V	10	747,415	Cl III	1
759,440	O V	10	746,976	N II	8
758,9	Cs III	1	746,864	Cl III	1
758,677	O V	10	746,834	Kr III	5
758,50	Xe IV	—	746,695	Kr III	7
758,559	K V	1	746,627	F I	3
758,317	Ne IV	3	746,350	K IV	8
757,199	K VI	4	745,839	N II	6
757,112	K V	4	745,767	F I	2
757,08	F V	4	745,763	Kr III	3
756,7	O I	1	745,323	Ar II	7
756,563	Cl IV	1	745,264	K IV	10
756,031	Xe III	10	745,205	Cl IV	4
755,8	O I	2	744,925	Ar II	8
755,603	F I	2	744,141	Xe III	6
755,362	Si II	2	743,970	Cu III	30
755,212	Ar IV	3	743,870	Kr III	3
754,93	Ar VI	1	743,721	Ne I	12
754,824	Ar II	3	743,303	Cu III	20
754,673	K IV	8	743,292	K V	2
754,205	Ar IV	4	743,122	Kr II	9
754,194	K IV	3	742,83	Kr VI	8
754,148	F I	2	742,821	Kr II	9
754,144	Xe III	5	742,566	Xe III	15
753,877	K VI	3	741,950	K IV	10
753,303	F I	4	741,293	O II	0
752,884	F I	4	740,85	Xe IV	—
752,762	O III	4	740,838	O II	0
752,051	Kr II	30	740,555	Ca III	3
751,861	F I	4	740,44	Xe VIII	3
751,73	Xe IV	—	740,406	Xe II	3
750,986	Kr III	4	740,270	Ar II	10
750,76	Xe	3	739,949	O II	1
750,610	F I	3	739,177	K VI	1
750,447	Xe III	8	737,979	Xe III	7
750,381	K V	1	737,457	Ar II	1
750,230	K V	3	737,144	K IV	10
750,226	O II	2	736,987	F I	2
750,155	Xe III	8	736,762	Cl VI	3
749,993	K IV	6	736,031	Cu II	25
749,941	Si IV	5	735,892	Ne I	30
749,662	O II	1	735,519	Cu II	20
749,3	O I	1	735,224	Cu III	100
748,946	F I	3	734,5	Cs III	1
748,709	F I	2	733,891	Cl VI	3
			733,314	Xe III	10

λ	Symbol	I	λ	Symbol	I
732,688	Cu III	5	716,09	Xe	2
732,63	Xe IV	—	715,999	K VI	5
732,259	Kr III	4	715,986	Xe III	4
732,026	Cu III	100	715,645	Ar V	3
731,858	K V	2	715,599	Ar V	4
731,030	Xe III	15	715,58	Cl II	3
730,929	Ar II	5	715,530	Cu III	200
730,92	Cl II	3	714,879	C III	1
730,365	Cu III	150	714,772	Kr III	2
730,311	Cl VI	4	714,03	Cl II	2
730,264	Kr III	3	713,999	Kr III	7
730,257	Ca V	5	713,860	N V	8
730,241	Si IV	2	713,81	Ar VIII	5
729,402	Kr II	20	713,518	N V	6
729,39	Cl II	3	713,262	Cu III	10
729,39	Ti IV	0	713,041	K V	1
729,1	Li III	—	712,728	K VI	1
728,94	Cl II	3	712,688	Si IV	1
728,906	Cu III	2	712,668	O V	3
728,810	Fe III	6	712,66	Cl II	3
727,646	Ca V	1	712,473	Cu III	15
727,537	Cl VI	3	712,040	Cu III	5
727,058	Xe III	9	712,036	Kr II	8
726,99	Xe	2	711,834	Cu III	3
726,948	Al III	3	711,190	Xe III	4
726,295	Cu III	10	710,932	K VI	1
725,848	K IV	1	710,677	Xe III	5
725,716	Al III	2	710,576	Xe III	5
725,64	Cl II	2	710,519	K VI	4
725,550	Ar II	4	709,303	Cu II	10
725,309	K VI	2	709,195	Ar V	5
725,27	Cl II	3	709,16	Cl II	2
724,623	Xe III	3	708,85	Kr V	8
724,487	Cu II	15	708,838	K III	4
724,420	K VI	8	708,4	Cs III	1
724,420	K V	8	708,356	Kr III	8
724,129	Cl VI	3	707,43	Cl II	4
723,958	Cu III	20	707,315	O III	4
723,873	Xe III	5	706,298	O III	2
723,71	Xe VII	3	706,224	O III	3
723,361	Ar II	5	705,84	Kr VI	8
722,2	Cs III	2	705,783	Xe III	5
722,036	Kr III	50	705,762	O III	2
722,036	Kr II	50	705,641	K IV	3
721,630	Xe III	4	705,353	Ar V	3
721,199	Xe III	10	705,096	Xe III	12
720,432	K V	6	704,838	Kr III	4
719,85	Kr III	1	704,523	Ar II	4
719,506	Cu III	150	703,906	Xe III	9
719,26	Cl II	1	703,850	O III	18
718,9	Al	3	703,622	Cu III	15
718,89	Xe	3	702,899	O III	17
718,562	O II	16	702,822	O III	16
718,484	O II	17	702,799	Xe III	8
718,171	Cu II	10	702,332	O III	16
718,091	Ar II	4	702,112	Cu III	20
717,15	Cl II	2	701,692	Cu III	15
716,272	K VI	2	700,277	Ar IV	8
			700,271	Cu III	150

λ	Symbol	I
700,24	Ar VIII	10
700,182	Cu III	20
699,72	Ar III	1
699,408	Ar IV	6
699,070	Xe III	5
698,771	Ar II	4
698,541	Xe III	20
698,037	Kr III	20
698,02	Xe VII	10
697,940	Ar II	2
697,930	Cu III	20
697,74	Ar III	2
697,526	Xe III	8
697,489	Ar II	2
696,212	Al III	4
696,206	Ca III	3
696,202	K III	1
695,91	Kr VIII	8
695,824	Ca III	4
695,820	K III	3
695,817	Al III	5
695,604	Kr III	15
695,537	Ar III	6
695,042	K V	3
694,477	K V	2
693,972	Xe III	10
693,510	Cu III	50
693,301	Ar II	2
691,919	Kr III	18
691,557	Cu III	100
691,388	N III	1
691,187	N III	2
691,038	Ar II	1
691,036	Xe III	7
690,689	Si III	2
690,557	Kr II	11
690,526	C III	7
690,40	Xe	1
690,397	Xe III	7
690,250	Cu III	75
690,170	Ar III	8
689,538	Ca VI	3
689,15	Xe IV	—
689,007	Ar IV	12
688,935	Si IV	2
688,933	Cl V	4
688,907	Ca III	3
688,392	Ar IV	7
688,231	Xe III	4
688,085	K	3
687,987	Cu III	100
687,979	Kr III	11
687,55	Cl II	1
687,495	K IV	6
687,355	C II	11
687,059	C II	10
686,903	Cu III	15
686,73	Xe	1

λ	Symbol	I
686,5	Cs III	0
686,489	Ar II	2
686,480	C II	2
686,335	N III	14
686,254	Kr III	20
686,190	Ca III	2
685,816	N III	16
685,812	Kr II	11
685,807	Ca VI	4
685,513	N III	15
685,396	Cu II	2
685,139	Cu II	8
684,996	N III	14
684,490	Cl IV	0
683,97	Xe IV	—
683,666	Kr III	18
683,568	Si IV	1
683,278	Ar IV	10
683,171	Cl V	4
683,135	Si IV	1
682,82	Xe	7
682,791	Kr II	16
682,581	F I	2
682,564	Xe III	7
682,56	Xe V	3
682,171	Cu III	200
681,924	Cl V	4
681,911	Si IV	1
681,908	Ca III	4
681,119	Kr II	16
680,709	F I	2
680,679	Si IV	1
680,58	Xe	2
680,119	Kr III	22
679,400	Ar II	6
679,257	Cl V	3
679,221	Ar II	3
679,217	F IV	16
679,003	F IV	13
678,055	Si III	2
678,0	Al	5
677,951	Ar II	5
677,224	F IV	15
677,154	F IV	13
676,981	Si IV	1
676,785	Cl V	3
676,775	Ca V	1
676,606	Xe III	9
676,564	Cu III	300
676,564	Kr III	25
676,241	Ar III	6
676,241	Ar II	6
676,130	F IV	14
675,601	Cu II	2
674,828	Kr III	8
674,278	Ca VI	2
674,046	Ca VI	1

λ	Symbol	I
673,996	Xe III	9
673,87	Xe	3
673,813	Xe III	9
673,768	O II	7
673,598	Cl III	1
673,477	Si III	5
673,127	Cl III	3
672,948	O II	8
672,941	K IV	5
672,856	Ar II	2
672,826	Kr III	7
672,659	Cu III	50
672,57	Xe IV	—
672,428	Cl IV	0
672,330	Kr III	25
672,293	Si III	4
671,999	N II	6
671,852	Ar II	6
671,770	N II	6
671,718	Si III	2
671,629	N II	6
671,391	N II	8
671,37	Cl VI	4
671,365	Ca V	6
671,354	Si IV	5
671,198	Al III	2
671,175	Kr III	7
671,058	Kr III	7
671,014	N II	6
670,948	Ar II	5
670,881	N II	1
670,813	Kr III	3
670,55	Xe III	2
670,508	N II	1
670,383	Cl III	3
670,300	Kr III	4
670,289	N II	2
670,144	Al III	1
669,949	Cl III	2
669,725	Ca IV	10
668,864	K VI	3
668,827	Kr II	20
668,770	Cl IV	2
668,473	Xe III	4
668,43	Cs II	12
667,49	Cl II	1
666,17	Cl II	2
666,08	Cl II	3
666,010	Ar II	6
665,870	Kr II	9
665,21	Cl II	1
665,09	Xe	4
664,877	Xe III	6
664,844	Kr III	11
664,67	Cl II	2
664,563	Ar II	4
663,67	Cl II	2
663,134	K VI	3
663,08	Cl II	2
663,039	Kr III	20
663,039	Kr II	20
662,454	Cl IV	3
662,15	Cl II	1
661,869	Ar II	5
661,82	Cl II	2
661,402	K VI	3
661,124	Xe III	4
660,280	N II	9
660,124	Xe III	8
660,04	Ne I	2
659,854	Si IV	1
659,852	K VI	3
659,716	Kr III	22
659,538	O III	0
658,758	O III	1
658,637	Kr II	5
658,337	F III	12
657,931	K VI	3
657,828	Xe III	8
657,335	F V	4
657,327	K VI	3
657,327	N II	1
657,320	Cl III	2
657,220	F V	1
657,15	Cs II	5
657,088	Kr II	13
656,878	F III	11
656,772	Cl III	2
656,763	Ca V	6
656,125	F III	10
656,038	Ca IV	15
655,677	Kr II	5
655,09	Cl II	1
654,034	F V	3
653,80	Cl II	1
653,720	Ca III	4
653,696	Cl IV	4
653,332	Si III	8
653,013	Cl III	2
652,223	Si III	6
651,668	Si III	4
651,57	Kr VIII	10
651,550	Ca V	5
651,342	C II	8
651,324	K VI	1
651,269	C II	7
651,216	C II	7
651,198	Kr III	8
651,13	Cl II	1
650,88	Cl II	1
649,4	Cs III	0
648,50	F VI	1
647,879	F V	1
647,876	Ca V	5
647,31	F VI	1

λ	Symbol	I
647,12	Xe IV	—
646,667	Xe III	5
646,570	Ca V	8
646,417	Kr III	20
646,36	F VI	3
646,188	K IV	15
646,10	F VI	1
645,764	Si IV	2
645,167	N II	10
645,02	F VI	1
645,0	Cs III	4
644,963	K V	0
644,825	N II	9
644,621	N II	8
644,521	Kr III	1
644,388	Ar VII	2
644,159	O III	6
644,148	O II	12
643,98	F VI	1
643,404	Kr II	9
643,256	Ar III	9
643,19	Xe	2
643,118	Ca V	6
642,84	Kr III	1
641,875	C II	6
641,883	Ca VI	2
641,808	Ar III	12
641,772	C II	6
641,591	C II	6
641,364	Ar III	5
641,318	Ar VII	2
641,304	Cl III	1
640,928	Cl III	1
640,870	Kr II	5
640,36	Xe	2
639,99	Xe	2
639,982	K V	2
639,981	Kr III	15
639,757	Cl III	1
639,42	Cs II	12
638,42	Cl II	2
639,263	Kr II	5
639,230	Ca III	6
639,226	Cl V	3
639,208	Si IV	1
638,952	Kr II	5
638,668	K V	5
638,23	Cl II	2
638,214	Kr II	4
638,21	Na VI	0
637,928	Ca V	8
637,466	Ar VII	1
637,282	Ar III	20
637,195	K VI	1
637,06	Cl II	1
637,052	Ar VII	4
636,818	Ar III	3
636,62	Cl II	2
636,348	Kr III	1
636,247	C II	4
636,154	Kr II	3
635,988	C II	3
635,87	Cl II	2
635,323	Cl V	4
635,318	Ca IV	8
635,312	Si IV	10
635,180	N II	5
634,265	Kr II	4
634,24	Cl II	1
634,208	Ar VII	2
633,815	Ca VI	2
633,631	Kr III	5
633,375	Kr II	5
633,187	Ca III	5
633,186	Cl V	4
633,183	Si IV	3
633,082	Kr III	7
632,90	Na VI	0
631,550	Kr III	7
631,006	Cl III	1
630,940	K VI	1
630,746	Cl III	1
630,48	Xe IV	—
630,380	Cl III	1
630,306	Ar VII	2
630,302	K VI	2
630,194	F III	7
630,131	F III	6
630,037	Kr III	15
629,729	Ne I	6
629,750	Si IV	2
629,732	O V	15
629,594	Ca VI	2
629,434	N II	2
629,363	Si IV	2
629,355	Ca III	4
629,354	Cl V	3
629,217	Xe III	7
629,161	N II	3
628,874	N V	3
628,744	N V	5
628,581	Kr III	15
627,560	K VI	2
627,393	Xe III	4
626,819	Ne I	6
626,70	Cl II	1
626,40	Xe IV	—
625,852	O IV	14
625,758	Kr III	13
625,404	K V	4
625,130	O IV	14
625,011	Kr III	9
624,617	O IV	13
624,268	Kr III	3
623,768	Cl III	3
623,767	Ar III	5
623,016	K VI	8

λ	Symbol	I
622,795	Kr III	11
622,144	C III	2
621,910	Kr II	5
621,448	Kr III	8
621,280	Cl III	4
621,279	Ca III	3
621,12	Cl II	4
621,071	Kr II	5
621,027	Cl III	3
620,28	Cl II	1
619,548	Kr II	2
619,44	Xe IV	—
619,379	Kr II	2
619,092	Ne I	4
619,025	Cl III	1
618,879	Kr II	3
618,67	Kr VII	1
618,668	Ne I	5
618,515	Kr II	2
618,042	Kr II	4
618,02	Cl II	2
617,750	Kr II	4
617,61	Cl II	1
617,517	Ca VI	4
617,27	Cl II	0
617,068	Kr II	6
617,051	O II	6
617,033	O IV	7
616,933	O IV	3
616,728	Kr III	5
616,485	Kr III	6
616,363	O II	4
616,291	O II	7
616,136	K VI	6
615,623	Ne I	5
615,40	K II	2
615,225	Kr II	4
615,134	Kr II	4
614,015	Ca VI	3
613,336	Kr II	4
612,82	Cs II	7
612,73	Cl II	0
612,621	K II	4
612,371	Ar II	5
612,272	K VI	1
612,082	Si IV	2
612,070	Cl IV	4
612,062	Ca III	5
611,862	K VI	3
611,26	Xe IV	—
611,187	Kr III	8
611,100	Kr III	9
610,850	O III	6
610,746	O III	8
610,740	Ca III	3
610,043	O III	7
609,901	Cl III	0
609,829	O IV	15
609,705	O III	6
609,673	Cl III	4
609,460	Kr III	9
609,275	C III	6
609,168	Ne IV	1
609,025	C III	4
608,904	Si IV	1
608,903	Cl IV	4
608,895	Ca III	3
608,395	O IV	14
608,124	Kr II	5
608,065	F II	7
607,98	Cs II	1
607,931	K II	5
607,472	F II	6
607,31	Cs II	3
607,088	Cl IV	3
607,069	Ca III	3
606,925	F II	5
606,805	F II	8
606,527	Ne IV	5
606,460	Kr III	9
606,345	Cl III	5
606,333	Ca III	2
606,284	F II	6
606,100	Cl III	2
605,908	K IV	1
605,862	Kr III	9
605,855	Cl III	1
605,776	Kr II	5
605,668	F II	7
605,595	Ne IV	2
605,536	Kr II	5
605,316	K IV	1
605,316	Kr II	5
604,590	Cl IV	5
604,355	Kr III	4
604,152	Ar III	10
603,849	Kr III	6
603,666	Kr III	7
603,622	Ca III	3
603,429	K V	8
602,999	Ne IV	2
602,95	Cs II	1
602,858	Ar II	2
602,712	Ne I	4
602,43	Xe IV	—
602,389	Ca VI	0
602,269	K V	5
601,878	N III	0
601,700	Ca VI	5
601,499	Cl IV	5
601,458	N III	1
601,4041	He I	5
601,134	Kr III	7
600,917	Ca VI	6
600,765	K II	6

λ	Symbol	I
600,585	O II	6
600,532	C II	2
600,369	C II	3
600,265	C II	1
600,167	Kr III	5
600,04	Ne I	2
599,944	Kr III	4
599,944	Kr II	4
599,84	Xe VI	3
599,733	Cl IV	2
599,598	O III	18
599,19	Cl II	0
598,968	Kr II	3
598,86	Ne I	1
598,791	Kr II	3
598,698	Ne I	2
598,66	Kr II	0
598,06	Xe IV	—
597,818	O III	15
597,701	Ar II	2
597,194	Kr III	6
596,944	Kr II	4
596,694	Ar VI	4
596,576	Kr III	6
596,401	Kr III	6
596,240	Cl III	4
596,223	Ca III	2
595,990	Cl III	3
595,911	Ne I	3
595,7	Cs III	2
595,530	Kr II	7
595,530	Kr III	7
595,032	C II	7
594,808	C II	6
594,640	Ca III	3
594,636	Cl III	4
594,49	Cl II	0
594,239	Ca V	1
594,096	Ar VI	3
594,090	Kr III	9
593,699	Kr III	7
593,472	Ca V	1
593,404	Ca V	1
592,9	Cs III	0
591,962	Cl III	2
591,82	Ne I	2
591,646	Cl III	4
591,449	Ca III	3
591,434	K IV	1
591,428	Cl III	4
591,4117	He I	20
591,311	K IV	1
591,237	K IV	1
591,118	Cl III	3
591,08	Cs II	3
590,396	Ca VI	3
589,92	Ne I	1
589,82	Cl II	0
589,783	Ar VI	2
589,262	Kr II	5
589,16	Ne I	1
588,921	Ar VI	5
588,77	Cl II	0
587,872	Ca VI	1
587,78	Xe IV	—
587,604	Ca VI	2
587,6	Fe IV	2
587,543	Kr III	4
587,374	Kr III	4
587,311	Ca III	3
587,295	Cl III	4
587,20	Ne I	1
587,078	Cl III	3
587,006	Ar VI	1
586,880	Ca III	2
586,874	Cl III	4
586,54	Xe IV	—
586,322	K V	8
586,30	Ne I	—
586,269	Kr II	1
586,25	Cl II	0
585,950	Kr III	8
585,754	Ar VII	15
585,684	Kr II	2
585,666	C III	6
585,608	C III	6
585,510	K V	5
585,496	C III	5
585,417	C III	8
585,37	Kr VII	8
585,261	C III	6
585,25	Ne I	—
585,140	Kr III	8
584,3340	He I	500
584,10	Cl II	1
583,437	Ar II	2
582,46	Ne I	—
582,150	N II	5
582,140	K III	1
581,496	Kr II	3
581,466	Ca VI	3
581,22	Kr II	1
581,14	Ne I	—
580,967	O II	7
580,64	Ne I	—
580,63	Kr VI	2
580,577	Kr III	6
580,50	Ne I	—
580,444	Cl VI	2
580,400	O II	6
580,342	Kr II	3
580,319	K V	7
580,264	Ar II	3
579,823	Kr III	6
579,775	Ca VI	2

λ	Symbol	I
579,75	Ne I	—
579,40	Ne I	—
579,212	Ar III	3
579,11	Kr II	0
578,82	Ne I	—
578,732	Ca VI	4
578,604	Ar II	2
578,386	Ar III	4
578,220	Kr III	5
578,107	Ar II	2
578,09	Kr III	0
577,737	F IV	4
577,444	Cl VI	1
577,30	Xe IV	—
577,153	Ar III	3
577,108	C II	2
577,01	Kr II	0
576,900	C II	1
576,8	Fe IV	40
576,738	Ar II	2
576,647	Kr II	4
576,419	Cl VI	2
576,349	F IV	2
576,266	F IV	3
576,076	Kr III	4
575,902	Kr II	2
575,716	Kr III	5
575,633	F IV	2
575,582	Cl III	3
575,580	Ca III	6
575,34	Cs II	1
575,30	Cl II	0
574,956	Kr III	5
574,650	N II	6
574,634	K II	1
574,5	Fe IV	50
574,408	Cl III	3
574,398	Ca III	3
574,37	Cl II	3
574,279	C III	12
574,065	O III	00
573,468	Ar III	4
573,364	Ar II	2
573,228	Kr III	13
572,693	Cl III	4
572,691	Ca III	4
572,637	F IV	16
572,336	Ne V	80
572,106	Ne V	25
572,014	Ar II	2
571,983	Kr III	15
571,95	Cl II	1
571,7	Cs III	1
571,435	Cl VI	0
571,384	F IV	15
571,376	Cl VI	1
571,302	F IV	14
570,738	Kr III	4
570,881	Cl VI	0
570,636	F IV	14
570,529	Cl VI	2
570,291	Ca III	2
570,025	Cl VI	4
569,830	Ne V	50
569,759	Ne V	25
569,156	Kr III	7
569,13	Kr VI	5
568,418	Ne V	40
568,04	Xe IV	—
567,794	F III	6
567,737	F III	9
567,676	F III	10
567,629	F III	6
567,479	Cl VI	1
566,77	Cl II	0
566,630	Cl VI	2
566,613	Si III	8
566,490	C III	4
566,04	Xe VII	2
565,879	Kr III	4
565,75	Cl II	0
565,698	Si III	2
565,640	Kr III	5
565,530	C III	7
565,480	Cl VI	6
565,480	Cl III	4
565,463	Ca IV	3
565,289	Si III	1
565,272	Cl III	3
565,262	O IV	16
565,124	Kr III	4
564,645	C II	5
564,529	Ne VII	2
564,514	Cl III	2
564,287	Cl III	4
564,275	Ca VI	2
564,25	Cs II	3
563,58	Cl II	0
562,992	Ne VII	2
562,805	Ne VI	15
562,735	Ne VI	1
562,690	Kr III	5
562,577	C II	3
562,55	Xe VIII	2
562,54	Cl II	0
562,498	C II	3
562,355	C II	3
562,28	Cl II	3
562,250	Ca VI	3
561,738	Cl III	7
561,728	Ne VII	4
561,689	Si IV	1
561,680	Cl III	7
561,670	Ca III	3
561,530	Cl III	7

λ	Symbol	I
561,518	Ca III	3
561,378	Ne VII	2
560,986	Kr III	5
560,788	Kr II	3
560,636	Cl III	1
560,443	C II	5
560,390	Al III	7
560,244	C II	4
560,224	Ar II	2
559,947	Ne VII	3
559,760	N II	0
559,320	Kr II	4
558,66	Xe IV	—
558,634	Kr III	15
558,61	Ne VII	4
558,602	Ca V	10
558,595	Ne VI	5
558,481	Ar V	5
558,385	Cl III	1
558,321	Ar III	5
558,14	Cl II	1
557,118	Cl III	7
557,112	Ca III	3
557,029	K VI	1
556,893	Ar III	6
556,817	Ar II	2
556,605	Cl III	7
556,583	Ca III	3
556,232	Cl III	6
556,215	Ca III	2
555,764	Ar II	1
555,639	Ar VI	4
555,580	Cl VI	3
555,485	Cl VI	20
555,482	Ca V	5
555,262	O IV	16
555,121	O II	5
555,056	O II	5
554,794	Kr III	7
554,655	C III	2
554,619	Cl IV	7
554,615	Ca III	3
554,52	Kr VI	5
554,514	O IV	18
554,275	O III	0
554,210	Cl V	1
554,074	O IV	17
553,470	Ar III	9
553,328	O IV	16
553,297	Cl IV	6
553,123	Ar II	1
552,908	Cl III	2
552,053	Cl VI	2
552,017	Cl IV	7
552,005	Ca	4
551,992	Cl VI	10
551,894	C II	0
551,689	Kr III	4
551,643	Cl V	1
551,371	Ar VI	8
551,117	Cl V	2
551,103	Ca V	2
550,896	Ar II	1
550,706	Cl IV	3
550,481	Ar II	1
550,355	Cl VI	5
550,323	K II	1
550,2	Cs III	2
550,020	Cl IV	4
550,004	Ca III	3
549,568	C II	3
549,507	C II	5
549,375	C II	4
549,317	C II	3
549,219	Cl IV	5
549,201	Ca III	4
549,070	Ca V	3
548,905	Ar VI	5
548,781	Ar II	2
548,652	Kr III	5
548,517	F II	2
548,324	F II	3
547,898	Ca VI	3
547,873	F II	4
547,813	N II	0
547,8	Cs III	2
547,630	Cl V	10
547,456	Ar II	2
547,288	C II	0
547,169	C II	0
547,166	Ar II	2
546,846	F II	6
546,686	Kr III	5
546,547	Kr III	6
546,329	Cl V	6
546,175	Ar II	2
546,123	K II	3
545,114	Cl V	10
545,091	Ca III	2
544,731	Ar VI	4
544,627	K V	1
544,537	K V	1
544,413	Kr III	5
544,03	Kr VI	5
543,973	K IV	1
543,891	Ne IV	150
543,818	Cl V	1
543,730	Ar II	2
543,640	K IV	2
543,475	C II	3
543,420	Kr III	8
543,291	C II	2
543,205	Ar II	2
542,911	Ar II	2
542,868	Cl V	4

λ	Symbol	I	λ	Symbol	I
542,842	Ca	3	535,039	Cl IV	4
542,395	Cl V	3	534,873	K V	2
542,297	Ci V	6	534,727	Cl IV	8
542,296	Si IV	5	534,715	Ca III	4
542,290	Ca V	10	534,26	Ar III	1
542,229	Cl V	8	534,059	K	2
542,073	Ne IV	100	533,809	N II	4
541,284	Cl V	3	533,726	N II	6
541,127	Ne IV	80	533,644	N II	4
540,860	Kr III	4	533,577	N II	5
540,788	Kr III	5	533,504	N II	4
540,0	Li III	—	533,082	Ar II	1
539,853	O II	7	532,716	C II	3
539,731	Ne IV	3	532,413	Ar III	7
539,731	K II	3	531,917	C II	1
539,547	O II	8	531,775	C II	0
539,441	Cl V	0	531,255	Kr III	4
539,086	O II	8	531,18	Xe VII	1
538,977	Cl V	3	530,494	Ar II	1
538,967	Ca IV	2	530,386	C II	4
538,788	Ar III	6	530,306	Kr III	6
538,681	Cl V	4	530,303	Ca V	6
538,648	Ca IV	3	530,290	C II	3
538,595	Cl IV	4	530,268	N III	3
538,544	Kr III	8	530,037	N III	2
538,318	O II	7	529,900	Ar III	9
538,312	C III	13	529,860	N II	5
538,256	O II	10	529,796	K III	8
538,150	C III	12	529,713	N II	3
538,119	Cl IV	6	529,7	Cs III	0
538,075	C III	11	529,627	N II	3
538,032	Cl V	5	529,481	N II	3
537,830	O II	9	529,405	N II	3
537,613	Ca VI	6	529,343	N II	3
537,606	Cl IV	9	528,879	K IV	1
537,461	Cl V	3	528,811	Kr III	4
537,459	Ar III	6	528,746	Ca V	3
537,140	Ar II	1	528,730	Na VI	0
537,0296	He I	200	528,519	K	2
537,006	Cl V	4	528,286	Ca III	8
537,004	Ca IV	2	527,693	Ar V	6
536,965	Ne IV	1	527,617	K IV	3
536,790	Ca IV	2	527,565	K III	1
536,745	Ar III	8	527,064	K IV	2
536,532	Cl V	3	526,870	Ar VIII	1
536,531	Ca IV	2	526,60	Fe IV	60
536,216	K V	2	526,497	Ar II	1
536,150	Cl IV	6	526,457	Ar VIII	5
536,132	Ca III	4	526,448	K IV	4
536,008	Ca VI	0	526,298	F V	3
535,916	Cl V	2	526,28	Fe IV	75
535,666	Cl IV	7	525,795	O III	18
535,647	Ca IV	4	525,687	Kr III	4
535,580	Ar III	7	525,68	Fe IV	100
535,455	Cl V	2	525,292	F V	3
535,288	C III	10	524,683	Ar II	1
535,287	K V	1	524,597	F V	2
535,204	F VI	10			

λ	Symbol	I	λ	Symbol	I
524,189	Ar V	5	509,018	N II	0
524,11	Ti VI	10	508,903	N II	0
523,792	K III	5	508,700	N II	2
523,661	F III	4	508,6431	He I	20
523,001	K IV	5	508,655	Ar III	2
522,791	Ar II	1	508,595	Ar III	4
522,288	F III	3	508,58	Ti VI	12
522,2128	He I	80	508,459	N II	0
522,090	Ar V	3	508,434	Ar III	9
521,813	Ne IV	25	508,384	F III	10
521,742	Ne IV	25	508,182	O III	18
520,611	K III	10	508,079	F V	4
520,493	K	3	507,7178	He I	15
519,723	O VI	2	507,683	O III	17
519,610	O VI	2	507,391	O III	16
519,429	Ar VIII	3	507,0576	He I	10
519,372	K VI	5	506,5702	He I	7
519,329	Ar II	1	506,2000	He I	5
518,910	Ar II	1	506,163	F V	3
518,249	K II	3	506,160	N II	3
518,242	O II	5	506,057	N II	2
517,937	O II	4	506,029	K IV	2
517,250	Ar V	0	505,985	N II	1
517,00	Xe VIII	2	505,9122	He I	4
516,384	Kr III	4	505,761	K IV	1
516,348	Si IV	3	505,6840	He I	3
515,653	K II	3	505,199	Ca VI	8
515,640	O II	4	505,013	Ar II	0,5
515,6165	He I	50	503,649	Ar II	0,5
515,514	K III	4	502,157	Ar II	1
515,320	K V	1	501,649	K VI	2
515,498	O II	5	501,184	Ar II	0,5
515,118	Si IV	2	500,798	Ar II	0,5
514,945	F II	6	500,343	O II	1
514,943	K III	2	500,125	K IV	3
514,350	Na V	0	500,047	K IV	2
514,310	Ar II	1	499,993	K IV	2
514,087	F V	1	499,871	O II	2
513,975	F V	2	499,583	C III	7
513,914	Ar V	1	499,530	C III	9
513,845	N II	2	499,462	C III	8
513,649	F II	4	499,425	C III	7
512,769	Ar III	7	498,911	F IV	4
512,0982	He I	35	498,790	F IV	7
511,886	Ar V	0	498,431	O VI	1
511,565	Ar III	7	498,090	O VI	0
511,527	C III	10	497,910	C III	1
511,497	Ar III	8	497,842	F IV	4
511,215	Al III	4	497,802	F IV	6
511,193	Na V	1	497,363	F IV	5
510,757	N II	3	497,104	K III	15
510,554	Ar II	1	496,650	Ar II	0,5
510,102	Na V	0	495,144	K II	6
509,9979	He I	25	494,686	Ar II	0,5
509,897	N III	4	494,382	Na VI	7
509,601	K V	1	494,160	Na VI	3
509,586	N III	5	493,587	C III	7
509,293	Ca V	2			

λ	Symbol	I
493,519	C III	5
493,464	C III	5
493,396	C III	5
493,364	C III	5
493,341	C III	5
492,649	C III	7
492,228	Ar III	3
491,980	O III	1
491,714	O III	0
491,340	Na VI	6
491,240	Na VI	3
491,121	Ar III	4
491,062	K VI	2
491,050	Ne III	9
490,997	F IV	16
490,698	Ar II	0,5
490,68	Ar III	3
490,566	F IV	13
490,546	Ca III	2
490,423	K VI	2
490,310	Ne III	7
489,641	Ne III	4
489,580	Na VI	5
489,501	Ne III	10
489,196	Ar II	0,5
488,868	Ne III	7
488,782	Ar II	0,5
488,452	Ar III	7
488,120	K VI	10
488,103	Ne III	8
487,988	Ar III	7
487,070	Ne V	3
487,025	Ar III	7
486,95	Al III	1
486,172	Cl IV	8
486,160	Ca	3
485,857	N II	0
485,359	K IV	2
485,636	Ca	3
485,631	O II	4
485,626	K II	2
485,572	O II	1
485,515	Ar III	4
485,515	O II	5
485,513	K II	3
485,465	O II	0
485,359	K IV	2
485,150	Ar III	6
485,086	O II	6
485,084	K II	5
484,600	F II	8
484,445	Ar III	5
484,368	Ca	3
484,200	K III	1
484,116	Ar III	5
484,025	O II	2
483,976	O II	5
483,972	K III	3
483,752	O II	4
483,745	K V	4
483,733	C III	5
483,618	C III	4
483,567	C III	3
482,987	Ne V	50
482,706	K V	4
482,548	Ar III	8
482,408	K III	2
482,107	K III	2
482,1	Li III	—
481,848	Ar III	6
481,755	O II	3
481,704	O II	1
481,635	O II	0
481,587	O III	4
481,587	O II	4
481,381	O III	2
481,361	Ne V	25
481,354	O III	3
481,313	K VI	2
481,281	Ne V	15
480,965	K III	1
480,955	O III	4
480,471	Ca	3
480,406	Ne V	25
480,397	K VI	1
479,485	Ar VII	2
479,379	Ar VII	12
479,185	K III	8
478,305	Ca III	4
477,625	C III	3
476,606	Ca V	2
476,432	Ar III	7
476,201	Cu	20
476,029	K II	2
475,876	N II	1
475,800	N II	3
475,733	Ar VII	2
475,697	N II	2
475,656	Ar VII	8
475,638	N II	1
474,920	K III	9
474,883	N II	2
474,774	N II	0
474,698	N II	0
474,601	N II	0
474,540	N II	0
474,493	N II	0
473,938	Ar VII	4
473,918	Ar III	6
473,207	K VI	2
473,025	Ar III	6
473,021	F II	3
472,710	F II	5
472,392	N III	5
472,347	Cu	20

λ	Symbol	I
472,232	N III	4
472,16	Kr V	3
471,990	F II	6
471,949	F II	3
471,603	O IV	0
471,569	K III	15
471,273	O IV	1
470,408	O II	4
470,089	K III	20
469,968	Ar III	4
469,865	Ne IV	200
469,831	Ar III	4
469,817	Ne IV	200
469,499	K II	—
468,956	Ar III	3
468,766	O II	2
468,467	Ar III	4
468,447	K V	2
467,926	O II	0
467,390	Ar III	6
467,194	Cl IV	3
467,106	Cu	15
466,995	F V	5
466,932	Ar VI	4
466,793	K III	15
466,536	C II	0
466,530	Ar III	5
466,492	C II	2
466,404	C II	1
466,358	C II	0
466,132	Cl IV	3
466,129	Si III	4
465,978	F V	7
465,760	O II	2
465,586	Ar VI	2
465,529	O II	1
465,374	F V	6
465,350	Cl IV	3
465,27	Kr VI	6
465,21	Ne VII	10
465,113	F III	10
465,078	K II	1
464,861	Cl IV	4
464,830	Ca III	3
464,824	Cu	20
464,785	O II	3
464,640	Cu	20
464,370	F V	5
464,310	O II	1
464,292	Cl IV	3
464,284	F III	9
464,270	K VI	10
464,257	Ar VI	4
464,194	O II	2
463,938	Ar V	7
463,737	N IV	3
463,712	Cu	20
463,263	Na V	12
463,011	Cl IV	3
462,596	K V	1
462,415	Ar V	3
462,388	Ne II	14
462,146	Ar VI	4
462,007	Ar VI	25
461,898	Ar VI	1
461,737	K VI	3
461,227	Ar VI	6
461,227	Ar V	6
461,085	Ca	5
461,051	Na V	10
460,725	Ne II	15
460,438	K VI	8
460,202	Ar VI	1
460,058	Ar VI	1
460,050	C III	8
459,897	Na V	7
459,896	O III	1
459,881	Cu	15
459,728	Ar V	1
459,633	C III	15
459,603	Ar VI	3
459,521	C III	14
459,462	C III	13
459,320	Ar VI	10
459,005	K V	3
458,975	Ar V	2
458,422	O II	0
458,155	Si IV	3
458,121	Ar V	3
458,048	K VI	7
458,039	Ar VI	1
457,818	Si IV	4
457,475	Ar VI	20
457,444	Cl III	0
457,323	K VI	1
457,245	Cl III	2
457,177	F II	6
457,169	Cl III	3
457,007	Ar VI	5
456,997	O II	1
456,981	Ca IV	5
456,895	Ne II	5
456,375	Ar VI	3
456,344	Ne II	4
456,328	K IV	8
456,328	K V	4
456,078	N III	1
455,813	Ar VI	2
455,670	K V	1
455,6	Li III	—
455,270	Ne II	7
454,648	Ne II	5
454,112	Si IV	—
454,072	Ne VI	3
453,425	Cu	40

λ	Symbol	I
453,340	N II	1
453,257	N II	0
453,130	Cu	20
452,900	K V	3
452,745	Ne VI	3
452,667	K VI	3
452,654	Cu	30
452,227	K V	2
452,226	N III	11
451,869	N III	10
451,843	Ne VI	2
451,320	K VI	2
451,152	Cu	25
450,732	C III	9
450,565	Ca IV	10
450,20	Kr VI	2
450,08	N V	3
450,079	Ar V	1
450,015	Cu	25
449,708	K V	4
449,493	Ar V	4
449,065	Ar V	18
449,013	K V	3
449,013	K VI	3
448,595	K III	15
448,420	Cu	25
447,840	O VI	0
447,813	Ne II	8
447,712	O VI	0
447,527	Ar V	4
447,085	K V	3
446,995	Cu	25
446,949	Ar V	8
446,926	K	2
446,830	K II	5
446,591	Ne II	7
446,252	Ne II	8
446,036	Ca V	1
446,009	K VI	4
445,997	Ar V	5
445,933	Ca V	1
445,878	K V	1
445,638	O II	4
445,607	K IV	4
445,601	O II	4
445,190	Na V	7
445,046	Na V	6
445,032	Ne II	7
445,018	Ca IV	1
444,999	Cu	15
444,766	Ca IV	3
444,344	K III	15
443,821	Ca IV	15
443,681	O II	0
443,567	K IV	6
442,947	Cl III	2
442,913	K III	3
442,873	O IV	0
442,705	O IV	1
442,518	K IV	2
442,300	K IV	4
442,048	O II	4
442,043	K III	2
442,001	O II	4
441,812	K II	5
441,398	Cl III	3
441,370	K VI	3
441,0	Li III	—
440,905	K IV	4
440,60	Ne VI	0
440,598	O II	2
440,552	O II	3
440,429	K III	15
440,404	Ne VI	1
440,266	Na VI	3
440,245	Cl IV	2
439,700	Ca IV	5
439,255	Ce IV	3
438,930	Ca IV	4
438,910	C II	1
438,869	K III	4
438,647	K V	2
438,023	K V	5
437,825	Cl IV	4
437,773	Ca IV	5
437,683	O II	3
437,332	O II	3
437,271	Ca IV	2
437,216	K III	3
436,85	N V	4
436,649	O II	0
436,563	F II	1
436,510	O II	1
436,279	F II	2
435,676	K III	10
435,649	Ne VI	4
435,634	F II	3
434,975	O III	10
434,840	O III	2
434,722	K III	15
434,646	O III	3
434,570	Ca IV	12
434,280	N III	6
434,256	O III	4
434,246	N III	6
434,129	N III	5
434,066	N III	7
434,014	N III	6
433,911	N III	6
433,774	Cl III	0
433,664	Cl III	0
433,337	C III	8
433,237	Ne IV	50
433,176	Ne VI	4
432,919	Fe V	1
432,340	Fe V	3
431,826	F II	2

λ	Symbol	I
431,545	F II	3
431,541	Fe V	3
431,472	Ne IV	25
430,909	F II	4
430,758	F IV	15
430,624	Fe V	2
430,218	F III	8
430,177	O II	6
430,154	F III	11
430,053	Fe V	1
430,050	Ca III	1
430,041	O II	6
429,923	K II	3
429,918	O II	5
429,716	O II	4
429,656	K II	3
429,647	O II	5
429,557	O II	2
429,511	F III	10
429,438	K VI	2
429,206	Fe V	1
428,909	Fe V	5
428,763	Fe V	5
428,538	K VI	5
428,315	K VI	2
428,292	Fe V	0
428,244	N III	5
428,180	N III	6
428,131	Fe V	3
428,000	Fe V	0
427,918	Fe V	2
427,840	Ne III	3
427,782	Fe V	1
427,442	Fe V	2
427,320	Fe V	1
427,190	Fe V	3
426,969	Fe V	3
426,814	Fe V	4
426,745	Fe V	3
426,609	Fe V	1
426,526	O II	1
426,338	K VI	2
426,097	Fe V	5
426,045	Fe V	5
425,840	Fe V	0
425,589	Fe V	1
425,588	K V	7
425,476	Fe V	1
425,273	O II	0
425,159	K V	5
425,000	Ca V	15
424,75	N V	2
424,733	Fe V	3
424,61	N V	1
424,577	O II	0
424,28	Ti IV	3
423,833	Fe V	2
423,821	Na VI	2
423,58	Ti IV	4
422,713	Cl III	1
422,347	Ne V	5
422,287	Fe V	6
422,214	Ne V	15
422,178	K V	5
422,012	F II	1
421,990	Cl III	3
421,771	Cl III	3
421,765	Fe V	4
421,682	Fe V	2
421,609	Ne IV	150
421,465	Na VI	1
421,446	K V	3
421,045	Fe V	5
420,951	Ne V	15
420,874	Fe V	2
420,807	K VI	4
420,758	N IV	1
420,727	F IV	16
420,546	Fe V	5
420,386	Ne V	10
420,041	F IV	15
419,915	Fe V	3
419,731	K V	1
419,714	C IV	14
419,644	F IV	14
419,525	C IV	13
419,310	K V	2
419,045	K V	2
418,910	N III	6
418,812	O II	0
418,705	N III	7
418,623	K III	6
418,609	C III	2
418,598	O II	1
418,457	Fe V	5
418,160	K VI	2
418,033	Fo V	6
417,874	F II	1
417,595	Na VI	6
417,535	K III	6
417,516	Fe V	0
417,382	Fe V	6
417,280	K IV	3
417,048	Fe V	1
416,910	Fe V	2
416,834	Ne V	25
416,769	C III	5
416,509	K VI	1
416,208	Fe V	5
416,198	Ne V	80
416,001	K III	6
415,972	Fe V	3
415,825	Fe V	1
415,793	K V	4
415,505	Na VI	4
415,465	K V	3

λ	Symbol	I	λ	Symbol	I
415,333	Cl III	1	401,939	Ne VI	25
415,196	Cl III	1	401,639	Fe V	2
415,052	K V	5	401,138	Ne VI	15
415,006	Fe V	4	401,030	Fe V	2
414,870	K III	6	400,951	K VI	5
414,790	Fe V	1	400,824	Ca VI	3
414,465	K V	3	400,722	Na V	10
414,370	Na VI	2	400,676	Mg VI	7
413,797	N III	0	400,625	Fe V	4
413,792	K III	10	400,579	F II	1
413,681	N III	0	400,210	K III	8
412,939	Si IV	—	399,995	Cl VI	8
412,790	K VI	1	399,957	Cl VI	7
412,289	K III	5	399,938	Cl VI	5
412,240	Na IV	8	399,925	Ca VI	0
412,155	Si IV	—	399,85	O IV	4
412,080	K V	6	399,820	Ne VI	5
411,958	C III	3	399,754	K V	4
411,812	Cl III	4	399,71	O IV	2
411,373	Cl III	4	399,688	C III	6
411,333	Na IV	7	399,637	C III	6
411,163	Cl III	3	399,634	Ar IV	3
410,540	Na IV	6	399,62	O IV	2
410,371	Na IV	10	399,50	O IV	3
410,102	K III	8	399,400	K V	3
409,971	Ca III	18	399,289	Mg VI	6
409,737	K III	8	399,084	N III	1
409,615	Na IV	8	399,073	K VI	2
409,325	C III	6	399,045	N III	4
408,959	K III	8	398,885	N III	3
408,682	Na IV	8	398,878	K V	4
408,076	K IV	5	398,86	Ar III	1
407,513	Cl III	0	398,633	K III	3
407,511	F II	4	398,551	C III	1
407,136	Ne II	8	398,546	Ar IV	4
407,053	F II	5	398,363	K V	4
406,484	K III	6	398,168	C III	1
406,274	Cl III	1	398,087	K VI	4
406,102	K VI	2	397,67	Ar III	1
405,852	Ne II	9	397,310	O III	0
405,773	K IV	2	397,231	O III	1
405,675	K VI	2	397,178	Ca	4
405,644	F II	4	397,120	O III	2
405,475	K VI	1	396,917	Ca VI	2
405,333	K VI	2	396,902	Fe V	0
405,178	K VI	2	396,869	Ar IV	4
404,684	K VI	4	396,773	Fe V	3
404,412	K IV	3	396,763	K III	0
403,732	Ca III	20	396,382	Ca III	3
403,372	O II	0	396,38	Ar III	4
403,315	Mg VI	8	396,247	F III	1
403,273	O II	0	396,235	K VI	4
403,262	Ne VI	10	396,055	Ca VI	2
403,087	O II	0	395,968	F III	2
403,035	O II	0	395,92	Ar III	1
402,907	K III	6	395,789	Fe V	2
402,197	Fe V	1	395,558	O III	2
402,104	K III	4			

λ	Symbol	I
395,442	F III	3
395,395	K VI	5
395,155	Fe V	3
394,909	K V	3
394,477	K VI	3
393,911	Fe V	4
393,676	F II	1
393,270	Fe V	5
393,142	K IV	10
392,907	Fe V	6
392,467	K IV	4
392,433	Cl V	5
392,420	Ca III	2
392,322	O II	3
392,274	K IV	2
392,002	O II	3
391,943	O II	2
391,918	K III	4
391,912	O II	1
391,462	K IV	4
390,574	K IV	6
390,415	K IV	5
390,148	Cl V	4
390,137	Ca VI	3
390,114	K III	5
390,055	C III	3
389,750	K VI	2
389,531	K VI	2
389,428	K V	2
389,090	C III	7
389,069	K V	5
389,069	K IV	5
389,005	C III	6
388,965	C III	5
388,920	K IV	3
388,607	Fe V	2
388,500	Fe V	2
388,485	K VI	2
388,233	K VI	4
388,218	Ne IV	100
388,020	Mg VI	3
387,983	Fe V	3
387,800	K V	6
387,787	Mg VI	2
387,775	Fe V	4
387,639	O III	4
387,616	Fe V	4
387,500	Fe V	6
387,482	O III	3
387,398	O III	2
387,372	K III	2
387,353	N IV	4
387,199	Fe V	5
387,141	Ne IV	125
387,080	Ca VI	4
387,077	Ca V	5
386,897	Fe V	4
386,783	Fe V	4
386,737	Fe V	3
386,710	K V	4
386,585	Fe V	1
386,505	K VI	2
386,483	Fe V	1
386,256	Fe V	0
386,254	Ca VI	1
386,203	C III	14
386,156	Fe V	4
386,106	Ca VI	1
385,941	Ca VI	1
385,869	Fe V	5
385,740	Fe V	5
385,689	K V	5
385,091	Ca VI	2
385,023	Fe V	4
385,020	K V	1
384,957	Fe V	6
384,956	K IV	3
384,826	Fe V	1
384,610	Fe V	2
384,516	K V	2
384,400	K V	2
384,212	Fe V	3
384,178	C IV	17
384,172	Ca VI	3
384,095	K IV	5
384,032	C IV	16
384,028	Ca VI	2
383,505	Ca VI	2
383,484	Fe V	3
383,318	K V	2
382,906	K IV	6
382,903	O III	1
382,646	K IV	4
382,487	K IV	3
382,229	K III	6
382,214	O III	1
381,881	Fe V	4
381,849	Ca VI	2
381,702	K IV	4
381,671	Fe V	0
381,606	Ca V	3
381,467	Fe V	0
381,464	Ca VI	2
381,260	Fe V	3
381,152	Fe V	2
380,902	F II	2
380,664	Fe V	3
380,477	K IV	5
380,477	K III	5
380,396	Ca V	5
380,107	Na III	8
380,003	Ca VI	1
379,919	O IV	3
379,877	K III	6
379,775	O IV	4

λ	Symbol	I	λ	Symbol	I
379,765	Ca V	3	373,700	Ca VI	3
379,631	O III	2	373,418	Ca VI	5
379,575	O III	3	373,318	K V	3
379,505	O III	4	373,165	Cl V	2
379,326	Cu	25	373,074	K V	2
379,308	Ne III	7	372,904	Ca V	6
379,294	Fe V	3	372,774	K VI	4
379,279	K IV	2	372,589	Cl V	2
379,138	Ca V	2	372,462	K V	4
379,118	K III	8	372,148	K V	10
379,065	C III	1	372,069	Na II	6
379,032	Fe V	1	371,784	C III	8
378,745	Ca VI	1	371,747	C III	10
378,653	Ca VI	1	371,694	C III	10
378,603	F III	2	371,225	Ca V	6
378,563	F III	1	370,580	K V	3
378,551	Ca VI	1	370,523	K V	3
378,219	K V	3	370,115	K VI	2
378,143	Na III	10	370,022	Ca VI	7
377,763	K V	5	369,743	Na V	3
377,756	Cu	15	369,647	Ca III	5
377,263	K VI	2	369,472	C III	2
377,181	Ca V	5	369,415	C III	5
377,045	O II	0	368,580	K	2
376,745	O II	0	368,303	Ca III	3
376,693	O II	0	368,030	K VI	2
376,686	F II	1	367,557	Na V	2
376,375	Na II	3	367,378	K VI	2
376,279	Ca V	3	367,371	Ca VI	2
376,061	K V	3	367,192	O IV	2
375,955	K IV	6	366,391	F III	6
375,928	F II	1	366,240	Na VI	0
375,793	F II	1	366,169	C III	4
375,718	F II	1	366,110	Na VI	4
375,434	F II	1	366,001	Fe V	3
375,333	Ca V	3	365,874	F III	7
375,300	F II	2	365,858	Fe V	6
375,103	Cl V	2	365,778	C III	1
374,939	K V	5	365,634	Fe V	3
374,864	Fe V	5	365,614	K VI	3
374,744	Ca IV	5	365,594	Ne V	100
374,662	Cl V	1	365,440	Fe V	6
374,464	Fe V	2	365,339	Fe V	3
374,441	N III	12	364,973	Fe V	3
374,436	O III	8	364,940	O III	1
374,331	O III	8	364,867	O III	2
374,31	N III	1	364,795	Fe V	4
374,240	Fe V	4	364,739	O III	3
374,204	N III	11	364,477	Na VI	3
374,165	O III	8	364,292	Fe V	3
374,075	O III	10	363,864	C III	6
374,005	O III	8	363,790	C III	5
374,000	Ca V	4	363,774	Na VI	2
373,997	Ca VI	7	363,761	C III	4
373,911	Cl V	0	363,525	Ca VI	2
373,805	O III	8	363,021	K IV	3
373,776	Cl V	3	362,985	N III	6
373,720	Fe V	5			

λ	Symbol	I	λ	Symbol	I
362,946	N III	8	356,558	O III	0
362,881	N III	8	356,534	Ne II	3
362,833	N III	7	356,436	Ne II	2
362,788	Ca VI	1	356,372	K VI	1
362,612	Ca VI	4	356,260	K IV	3
362,456	Ne II	4	356,246	Ca V	5
362,444	Na VI	4	356,131	Ne II	4
362,154	K IV	3	355,946	Ne II	2
362,085	K IV	5	355,848	Ne II	1
361,838	Cu	15	355,800	K VI	1
361,645	Ca VI	2	355,663	K VI	1
361,427	Ne II	5	355,647	Ne II	3
361,250	Na VI	8	355,469	K VI	1
361,234	Ca VI	2	355,469	O III	5
361,220	Cu	25	355,450	Ne II	2
361,114	Ca VI	4	355,425	Cu	20
360,761	Na IV	6	355,333	O III	5
360,675	C III	5	355,326	Mg V	12
360,635	F IV	1	355,293	O III	3
360,623	C III	7	355,137	O III	6
360,618	Cu	30	355,133	K	2
360,568	K IV	2	355,045	F IV	2
360,367	Na V	8	354,954	Ne II	4
360,557	C III	6	354,927	K V	6
360,319	Na V	8	354,927	K IV	6
359,907	K IV	4	354,627	K V	0
359,873	Cu	50	354,223	Mg V	10
359,730	K IV	6	354,139	K IV	2
359,616	O III	1	353,922	Ne II	2
359,415	O III	2	353,455	K	3
359,385	Ne V	50	353,421	F II	0
359,384	O III	7	353,325	K	3
359,223	O III	8	353,300	Mg V	9
359,016	O III	8	353,206	Ne II	3
359,865	Cu	90	353,094	Mg V	14
358,740	C III	4	353,031	Cu	15
358,721	Ne IV	200	353,000	C III	3
358,578	N III	6	352,946	Ne II	4
358,509	N III	5	352,915	Ca V	9
358,472	Ne V	50	352,750	K V	2
358,469	N III	5	352,463	K V	2
358,401	N III	3	352,237	Ne II	2
358,356	N III	5	352,202	Mg V	10
358,327	N III	5	352,058	N IV	4
358,278	N III	3	351,979	N III	1
358,153	Ca VI	3	351,931	N IV	5
357,973	Ca III	8	351,089	Mg V	12
357,955	Ne V	40	350,878	Ar V	3
357,897	Cu	100	350,703	O III	8
357,831	Ne IV	50	350,394	Ca VI	1
357,685	K VI	3	350,330	C III	2
357,645	K VI	3	350,164	K VI	2
357,534	Ne II	5	349,964	Cu	30
356,795	Ne II	5	349,961	O III	1
356,768	O III	0	349,918	O III	2
356,725	O III	2	349,825	O III	.3
356,625	O III	1	349,793	K V	3
356,615	K VI	3			

λ	Symbol	I
349,504	K V	4
349,494	Ca VI	1
349,155	Mg VI	10
348,927	Ca VI	1
348,800	N III	0
348,690	N III	0
348,650	Ca VI	1
348,413	Cu	15
347,999	K III	3
347,967	Ca VI	3
347,854	Cu	30
347,854	C III	3
347,777	C III	3
347,431	Ca VI	1
347,334	Ca VI	1
347,005	Ca VI	4
346,688	O IV	3
346,372	O IV	4
346,335	Ca VI	2
346,004	Cu	60
345,545	K III	2
345,405	K III	2
345,368	Cu	90
345,309	O III	10
345,201	N IV	3
345,197	K III	2
345,130	Ca IV	4
345,107	N IV	3
345,065	N IV	5
345,023	N IV	3
344,958	Ca IV	4
344,915	N IV	3
344,635	K III	4
344,388	F III	6
344,270	K III	4
344,219	Ca III	2
343,933	Ca IV	5
343,931	F III	4
343,892	F III	7
343,640	Ca V	4
343,468	K IV	3
343,438	Ca IV	4
343,203	Ca IV	6
342,805	K IV	2
342,713	Cu	80
342,703	K IV	2
342,447	Ca IV	5
342,432	Cu	20
342,410	K IV	3
341,924	K III	6
341,924	F III	7
341,483	Cu	25
341,455	Ca IV	4
341,391	O V	0
341,284	Ca IV	4
341,242	C III	7
341,183	Cu	20
341,179	C III	6
341,143	C III	5
340,745	K IV	3
340,528	Ca VI	8
340,462	K IV	6
340,389	Ca III	3
340,286	Ca IV	4
340,037	Ca VI	4
339,940	Ca VI	4
339,887	Cu	30
339,886	Ar V	3
339,800	Ca IV	5
339,773	C III	1
339,463	Ca VI	6
339,436	O IV	0
339,420	Cu	15
339,330	O IV	1
339,009	Ar V	3
338,929	Ca IV	5
338,828	Ca IV	4
338,426	Ar V	2
338,345	N III	2
338,314	Cu	25
338,222	Ar VIII	0
338,161	K VI	3
338,056	Ca V	5
337,998	Ar V	6
337,555	Ar V	3
337,541	Ca V	4
337,257	Ar VIII	2
336,555	Ca IV	15
336,555	Ar V	3
336,554	Ca V	4
336,279	Cu	25
335,919	Cu	25
335,470	Cu	15
335,374	Ca IV	25
335,344	Ca V	5
335,175	K VI	3
335,050	N IV	11
335,016	Cu	20
334,545	Ca V	6
334,204	Cu	30
334,135	Ca V	3
333,910	Na V	9
333,857	Ca V	3
333,570	Ca V	4
333,562	Cu	30
333,438	Ca V	4
333,057	Ca IV	2
332,893	Cu	100
332,808	Ca IV	3
332,550	Na V	8
332,531	Ca IV	5
332,327	N III	3
332,133	N III	2
331,991	Ca IV	5
331,835	Cl IV	2

λ	Symbol	I	λ	Symbol	I
331,50	Ne II	2	324,56	Ne II	2
331,442	Ca IV	4	324,485	Cu	70
331,416	K III	1	324,477	Ca V	5
331,168	K V	1	324,110	Ca V	3
331,06	Ne II	1	323,936	Cl VI	20
330,937	Ca V	6	323,816	Cu	60
330,77	Ne II	3	323,671	N III	4
330,718	Na V	0	323,615	N III	6
330,687	C III	1	323,488	N III	5
330,684	K III	5	323,431	N III	4
330,637	C III	1	323,356	Cl VI	15
330,62	Ne II	2	323,310	Mg IV	18
330,20	Ne II	2	323,223	Ca V	6
329,851	Cu	30	323,175	N IV	7
329,805	Cu	30	322,757	Ca V	5
329,391	Ca IV	3	322,724	N IV	9
329,307	K V	0	322,685	F III	7
329,298	Ca VI	3	322,650	F III	8
329,116	Ca IV	5	322,617	Cu	15
329,053	K III	2	322,575	C III	8
329,047	Cu	100	322,570	N IV	8
328,973	K V	2	322,503	N IV	7
328,933	K III	3	322,166	Ca V	10
328,845	K III	2	321,609	Ca V	6
328,831	Cu	20	321,593	Ca IV	10
328,742	O III	9	321,457	O IV	1
328,737	Cu	40	321,270	N III	1
328,577	Ca IV	1	321,161	N III	2
328,536	Cu	25	321,110	Ca VI	1
328,448	O III	10	321,071	N III	1
328,412	Cu	50	320,999	Mg IV	20
328,08	Ne II	2	320,979	O III	12
327,806	Ca VI	4	320,881	Cl IV	1
327,784	C III	1	320,720	O III	2
327,63	Ne II	2	320,445	Ca VI	2
327,620	Cu	20	320,392	He I	10
327,605	K III	1	320,250	Cl IV	1
327,519	O IV	0	320,192	F IV	1
327,383	Cu	15	320,004	F IV	2
327,376	K V	4	319,996	O III	3
327,33	Ne II	3	319,993	Cl IV	0
327,320	O IV	1	319,695	F IV	3
327,25	Ne II	2	319,638	Na IV	10
327,176	C III	4	319,616	Cl IV	3
327,175	Ca VI	2	319,513	Cl IV	1
327,112	C III	4	319,266	C III	3
327,031	K V	2	318,969	K V	1
326,77	Ne II	3	318,750	Cl IV	1
326,575	Cu	20	318,392	Ca IV	4
326,54	Ne II	5	318,364	Fe VI	3
325,687	Cu	20	318,093	Ca IV	15
325,570	C III	1	317,641	Na VI	6
325,282	Ca V	5	317,319	Fe VI	3
325,278	K III	0	317,265	O III	1
325,161	Cl VI	25	316,998	F III	2
325,038	Cu	20	316,967	O III	3
325,020	Ca V	3	316,947	Ca VI	3
324,607	Cu	50			

λ	Symbol	I	λ	Symbol	I
316 ,823	F III	3	308 ,534	Fe VI	4
316 ,488	F III	4	308 ,383	Fe VI	2
316 ,389	Ca VI	0	308 ,306	O III	2
316 ,115	Ca VI	3	308 ,264	Na V	10
315 ,748	F III	6	308 ,187	Fe VI	2
315 ,539	F III	7	308 ,129	K	2
315 ,537	K V	3	308 ,051	O III	1
315 ,506	Fe VI	3	308 ,007	Fe VI	3
315 ,221	F III	8	307 ,884	Fe VI	0
315 ,181	K V	4	307 ,806	C IV	1
315 ,053	N IV	8	307 ,800	Fe VI	3
315 ,027	Fe VI	4	307 ,404	Fe VI	3
314 ,877	N III	6	307 ,375	Fe VI	4
314 ,850	N III	9	307 ,248	Al VI	7
314 ,814	Fe VI	1	307 ,152	Na V	8
314 ,715	N III	8	307 ,013	Fe VI	2
314 ,676	Mg VI	4	306 ,922	Fe VI	5
314 ,554	Mg VI	3	306 ,882	O IV	7
314 ,395	C III	1	306 ,823	Fe VI	2
314 ,299	Fe VI	3	306 ,621	O IV	8
313 ,92	Ne III	1	306 ,620	K	2
313 ,748	Na VI	5	306 ,460	Fe VI	1
313 ,677	Ne III	3	305 ,918	N III	1
313 ,048	Ne III	4	305 ,879	O III	4
312 ,770	K V	5	305 ,837	Fe VI	1
312 ,608	Na VI	3	305 ,836	O III	8
312 ,505	Cu	20	305 ,769	O III	10
312 ,455	C IV	14	305 ,703	O III	8
312 ,418	C IV	15	305 ,656	O III	9
312 ,311	Mg V	10	305 ,596	O III	8
312 ,263	Fe VI	7	305 ,200	Fe VI	4
312 ,241	Al VI	6	304 ,912	N III	3
311 ,921	Na VI	4	304 ,910	Ca IV	3
311 ,726	O IV	3	304 ,874	N III	4
311 ,702	Fe VI	7	304 ,818	N III	4
311 ,679	O IV	6	304 ,551	Fe VI	7
311 ,628	N III	3	304 ,330	Ca IV	3
311 ,539	N III	2	304 ,221	Fe VI	7
311 ,490	O IV	5	304 ,032	N III	2
311 ,415	F III	4	303 ,981	N III	2
311 ,243	K V	6	303 ,891	N III	2
311 ,236	Fe VI	2	303 ,799	O III	9
311 ,138	Fe VI	1	303 ,783	He II	500
310 ,908	Al VI	6	303 ,693	O III	7
310 ,807	Fe VI	0	303 ,690	K	2
310 ,727	Cu	15	303 ,621	O III	7
310 ,601	Fe VI	4	303 ,558	Fe VI	4
310 ,380	Cu	20	303 ,515	O III	7
310 ,274	Fe VI	5	303 ,468	C III	1
310 ,171	C III	7	303 ,460	O III	7
309 ,852	Al VI	6	303 ,432	C III	4
309 ,627	Fe VI	1	303 ,411	O III	7
309 ,596	Al VI	8	303 ,163	N IV	4
308 ,993	Fe VI	3	303 ,123	N IV	6
308 ,960	Fe VI	3	303 ,079	N IV	4
308 ,664	Fe VI	5	303 ,048	N IV	5
308 ,560	Al VI	6	303 ,023	K VI	1
308 ,559	Ne III	1			

λ	Symbol	I
303,009	N IV	4
302,657	K VI	2
302,28	Na II	0
301,741	Ca III	4
301,432	Na II	1
301,311	Na II	0
301,279	C III	1
301,243	C III	3
301,206	C III	2
301,139	Ca V	0
301,124	Ne III	4
300,997	Fe VI	2
300,503	K V	6
300,455	O III	3
300,316	N IV	3
300,252	K V	7
300,151	Na II	1
299,850	O IV	4
299,820	N III	1
299,803	Fe VI	1
299,710	O IV	2
299,670	N III	0
299,620	O IV	2
299,579	Fe VI	1
299,495	O IV	3
299,315	Ca IV	4
299,275	O III	2
299,217	Cu	15
298,901	Cu	20
297,815	N IV	5
297,768	N IV	3
297,712	N IV	3
297,701	Ar VII	6
297,658	Ar VII	4
297,644	N IV	4
297,621	Ar VII	3
297,595	N IV	3
297,568	Fo VI	8
297,308	Fe VI	7
297,131	Fe VI	2
297,064	K V	5
296,988	Fe VI	6
296,958	Ca IV	6
296,951	C IV	7
296,857	C IV	6
296,808	Fe VI	5
296,723	Fe VI	3
296,554	Ca IV	5
296,317	Fe VI	1
296,270	O III	1
296,169	K V	6
296,012	O III	4
295,944	O III	3
295,886	F III	6
295,874	O IV	2
295,716	O III	6
295,710	F III	5
295,657	O III	6
295,634	Fe VI	4
295,619	O III	5
295,511	O III	3
295,405	F III	1
295,365	F III	2
295,140	O IV	1
295,051	O IV	1
295,042	Fe VI	2
295,014	Fe VI	4
294,960	Fe VI	4
294,853	O IV	1
294,850	Fe VI	4
294,836	K V	6
294,665	Fe VI	4
294,650	O IV	1
294,520	Fe VI	7
294,390	Ne IV	3
294,339	Fe VI	5
294,265	Fe VI	7
294,100	Ne IV	3
294,052	Ar VI	6
294,040	Fe VI	0
293,966	Fe VI	8
293,947	Ne IV	1
293,820	Fe VI	1
293,745	Fe VI	8
293,649	Ne IV	5
293,488	Fe VI	4
293,438	K VI	2
293,429	Ne IV	10
293,384	Fe VI	4
293,332	K VI	3
293,292	Fe VI	4
293,214	Fe VI	1
293,124	Mg VI	4
293,123	Ne IV	15
293,050	K VI	2
293,046	Fe VI	2
293,026	Mg VI	2
293,925	Fe VI	5
292,736	Fe VI	7
292,597	Fe VI	4
292,595	N III	4
292,447	N III	3
292,343	Fe VI	1
292,154	Ar VI	5
292,038	Fe VI	2
291,976	Ca VI	1
291,931	Fe VI	0
291,829	Fe VI	5
291,800	Fe VI	5
291,632	Fe VI	2
291,473	Fe VI	5
291,458	Mg VI	2
291,348	Mg VI	3
291,330	C III	5
291,229	Fe VI	6

λ	Symbol	I
291,203	O IV	1
291,184	Fe VI	6
291,054	O IV	1
291,020	Fe VI	5
290,947	F III	5
290,890	Fe VI	2
290,848	F III	6
290,737	Fe VI	4
290,608	F IV	2
290,577	Fe VI	4
290,499	Fe VI	2
290,461	F IV	2
290,440	F IV	3
290,302	Fe VI	5
290,271	Fe VI	6
290,147	F IV	4
290,146	Fe VI	4
290,089	Fe VI	4
290,038	Fe VI	4
289,933	O IV	1
289,898	O IV	2
289,851	Fe VI	4
289,672	Fe VI	2
289,590	O IV	1
289,520	Fe VI	4
289,469	O IV	2
289,468	Fe VI	3
289,302	Fe VI	4
289,292	O IV	3
289,241	K VI	1
289,143	C IV	9
289,112	Fe VI	5
289,048	C IV	3
288,652	Mg VI	0
288,551	Fe VI	4
288,267	F IV	1
287,657	Ca V	3
287,333	Fe VI	1
287,327	Cl V	3
287,206	Ne IV	10
286,965	Ca V	9
286,947	Ca V	5
286,934	Ne IV	15
286,688	Ne IV	15
286,448	O V	6
286,127	Cl V	2
286,038	O III	0
285,850	K V	2
285,838	O IV	7
285,734	K V	1
285,714	O IV	6
285,563	N IV	5
284,948	Ca V	5
284,794	Ca V	2
284,504	Fe VI	4
284,346	N III	2
284,296	N III	1
283,894	Ne III	3
283,770	Fe VI	5
283,765	K IV	2
283,690	Ne III	5
283,579	N IV	12
283,470	N IV	11
283,420	N IV	10
283,206	Ne III	6
283,178	Ne III	5
283,164	Ar VI	3
282,96	Na II	1
282,827	Na II	0
282,556	Ar VI	1
282,50	Ne III	0
282,440	Cu	20
282,423	Ar VI	6
282,355	K V	3
282,213	O IV	1
282,213	N III	2
282,093	N III	1
281,915	Ar VI	4
281,81	Na II	1
281,744	Cu	40
281,492	Cu	50
281,433	Ar VI	3
281,397	Al V	14
281,390	C III	2
281,350	F III	4
281,207	F III	3
280,992	Ca V	8
280,905	F III	1
280,522	C III	2
280,483	O III	1
280,412	O III	1
280,328	O III	1
280,265	O III	3
280,234	O III	1
280,116	O III	1
280,043	C III	3
280,030	O III	2
280,010	F III	6
279,937	O IV	11
279,834	F IV	3
279,787	O III	3
279,692	F III	7
279,633	O IV	10
279,456	O IV	2
278,699	Al V	16
278,471	Fe VI	3
278,339	Fe VI	5
278,149	Fe VI	5
278,128	Cu	20
277,951	Fe VI	3
277,610	Fe VI	2
277,569	Fe VI	6
277,514	O III	1
277,394	K V	3
277,385	O III	7

λ	Symbol	I
276,947	Fe VI	3
276,895	F III	4
276,786	F III	5
276,581	Mg V	16
275,513	O III	4
275,366	O III	3
275,350	Al VI	6
275,281	O III	2
275,10	Na II	0
274,601	Cu	15
274,552	K IV	3
274,260	F III	6
274,051	C III	2
273,99	Na II	0
273,546	K IV	1
273,417	Cu	20
273,207	F III	2
273,065	K IV	2
272,982	Ca V	4
272,915	F III	3
272,758	F III	3
272,710	F III	4
272,441	Na III	0
272,424	Cu	15
272,336	Ca V	3
272,311	O IV	6
272,270	O IV	6
272,265	Ca V	5
272,174	O IV	7
272,125	O IV	7
272,076	O IV	6
271,989	O IV	6
271,820	K IV	3
271,611	O III	0
271,523	O III	1
271,443	Cu	20
271,440	Ca V	1
271,403	O III	1
271,141	Ca V	4
271,014	C III	1
271,01	Na II	0
270,995	N IV	6
270,982	O V	0
270,740	Cu	20
270,675	F III	4
270,583	C III	1
270,570	Ca V	2
270,494	Ca V	3
270,394	Mg VI	12
270,305	Ca V	6
270,298	Cu	30
270,225	F IV	6
269,98	Na II	0
269,653	Cu	20
269,559	O IV	1
269,225	F IV	2
269,076	F IV	3
269,044	Cu	40

λ	Symbol	I
268,986	Mg VI	10
268,817	F IV	1
268,785	F IV	4
268,773	Cu	50
268,623	Na III	5
268,583	Ca V	2
268,451	O III	1
268,309	Cu	60
267,952	N III	1
267,868	Na III	6
267,851	N III	0
267,772	Ca V	8
267,709	Ne III	2
267,642	Na III	8
267,562	Cu	25
267,516	Ne III	3
267,121	O III	4
267,059	Ne III	3
267,050	O III	3
267,036	K	2
267,030	O III	7
266,985	O III	7
266,967	O IV	5
266,967	O III	6
266,938	K	2
266,932	O IV	6
266,893	Na III	5
266,863	Ca V	3
266,729	O IV	0
266,690	O IV	0
266,378	N V	9
266,344	K	4
266,197	N V	8
266,061	Cu	20
265,641	Cu	50
265,550	O V	4
265,062	O IV	0
264,948	N III	2
264,837	N III	1
264,480	O III	6
264,478	K V	2
264,414	Cu	20
264,339	K V	2
264,338	O III	5
264,257	O III	4
264,029	Cu	15
263,903	O III	0
263,861	O III	3
263,819	K IV	2
263,818	O III	5
263,807	F III	8
263,768	O III	3
263,760	Cu	30
263,728	O III	4
263,716	K IV	2
263,692	O III	3
262,938	Cu	30

λ	Symbol	I	λ	Symbol	I
262,882	O III	1	256,317	He I	150
262,729	O III	0	255,865	F III	7
262,627	C IV	4	255,772	F III	6
262,553	C IV	3	255,725	F III	5
262,442	Cu	20	255,624	F III	1
262,289	O III	0	255,417	Cu	35
262,113	O III	2	255,302	O III	0
261,806	Cu	20	255,252	O IV	5
261,751	F III	6	255,214	Cu	45
261,716	F III	7	255,158	O III	1
261,606	Cu	30	255,044	O III	0
261,200	K II	1	254,772	Cu	70
261,028	K II	1	254,595	F IV	1
261,027	O III	4	254,510	Cu	50
260,967	Cu	25	254,491	F IV	2
260,782	F III	1	254,193	F III	3
260,556	O IV	9	254,162	F III	4
260,498	F III	3	253,786	Cu	15
260,455	N IV	2	253,548	O III	0
260,446	Ca V	3	253,465	Cu	15
260,389	O IV	10	253,106	K VI	1
260,375	F III	3	253,083	Cu	15
260,332	Ar VIII	6	253,082	O IV	7
260,313	F III	4	252,965	K VI	1
260,253	Ar VIII	4	252,948	O IV	6
260,245	Cu	25	252,780	Cu	75
259,978	Ca V	3	252,564	O IV	6
259,871	Cu	25	252,223	Cu	15
259,856	Ca V	3	251,947	Cu	20
259,609	K VI	2	251,816	Ca VI	1
259,576	Ca V	3	251,726	Ne III	2
259,558	Cu	20	251,670	Cu	20
259,542	C IV	7	251,558	Ne III	2
259,471	C IV	6	251,465	Ca VI	4
259,199	Cu	20	251,371	Na III	6
258,927	Cu	80	251,354	Ca IV	3
258,873	K VI	3	251,278	Cu	20
258,411	K VI	1	251,148	O IV	1
258,265	Cu	45	251,145	Ne III	2
258,251	Ca V	3	251,114	O IV	1
258,207	O IV	3	251,026	F IV	10
258,116	O IV	2	250,940	Ar VII	7
258,018	K VI	4	250,515	Na III	8
258,004	Cu	15	250,400	Cu	30
257,976	Ca V	5	250,265	Ca VI	4
257,657	K VI	2	250,153	Ca IV	3
257,626	Cu	20	249,914	Ca VI	3
257,315	Cu	20	249,886	Ar VII	5
256,898	Cu	40	249,415	Cu	15
256,890	F III	2	249,408	Ca IV	3
256,831	K VI	3	249,384	Ar VII	2
256,673	F III	1	249,365	O IV	4
256,525	F III	1	249,228	F IV	1
256,506	O III	3	249,223	O IV	3
256,460	O III	3	249,213	Cu	20
256,425	O III	2	249,125	Si VI	8
256,365	Cu	30	248,744	C V	0
256,360	F III	5			

λ	Symbol	I
248,668	C V	0
248,636	Ca III	4
248,618	O III	2
248,574	O III	1
248,538	O III	1
248,459	O V	6
248,426	Cu	50
248,320	O III	1
248,004	Ne IV	8
247,807	Ne IV	8
247,777	K VI	1
247,742	Cu	25
247,709	N V	7
247,708	K	3
247,564	N V	6
247,561	K	2
247,422	Ne IV	10
247,415	C IV	2
247,357	C IV	1
247,205	N IV	10
247,202	K	2
247,080	O III	1
246,563	O IV	4
246,503	O IV	3
246,465	O IV	2
246,265	O III	3
246,235	K V	1
246,001	Si VI	8
245,860	F III	1
245,830	C IV	5
245,775	C IV	4
245,720	O IV	1
245,002	F III	3
244,907	C IV	10
244,768	F III	4
244,698	F III	2
244,049	O III	2
243,922	F IV	4
243,883	Cl VI	3
243,854	Cl VI	12
243,796	F IV	3
243,760	Al VI	12
243,736	F IV	2
243,364	F III	1
243,208	Cl VI	2
243,194	Cl VI	8
243,027	He II	70
242,883	Cl VI	3
242,631	Ca VI	5
242,592	Ca VI	3
242,439	F V	2
242,384	Ca III	3
242,324	F V	3
242,265	Ca VI	3
242,183	O IV	0
242,140	O IV	3
242,045	O IV	2
241,879	O III	1
241,815	O III	1
241,583	Cu	15
241,037	O III	2
240,979	O III	2
240,758	K III	3
240,730	F III	2
240,721	Ca VI	6
240,550	F III	1
240,371	F IV	7
240,275	F IV	7
240,233	F III	1
240,146	F IV	7
240,079	F IV	9
240,079	O IV	1
240,017	F IV	7
239,935	O IV	0
239,856	F IV	7
239,693	N IV	1
239,618	N IV	4
239,592	O IV	3
239,535	Ca VI	7
239,296	Ca VI	0
239,210	N IV	2
239,196	C IV	1
239,161	N IV	1
239,010	K III	2
238,573	O IV	15
238,361	O IV	14
238,250	C IV	3
238,200	C IV	2
238,099	F IV	1
238,042	F IV	2
238,012	F IV	3
237,983	N IV	3
237,955	F IV	4
237,913	F IV	3
237,903	N IV	2
237,860	N IV	1
237,331	He II	35
237,231	Cl V	2
236,710	O III	1
236,435	Cl V	1
236,071	O IV	1
235,840	F V	1
235,299	Cu	30
234,988	O IV	3
234,701	Ne IV	25
234,347	He II	20
234,316	Ne IV	25
234,258	Mg III	12
234,249	N IV	2
234,195	N IV	4
234,124	N IV	3
233,596	O IV	6
233,561	O IV	8
233,530	C IV	3
233,526	F IV	4

λ	Symbol	I	λ	Symbol	I
233,521	O IV	6	226,608	F V	2
233,495	O IV	7	226,341	F V	2
233,457	O IV	7	226,166	F III	4
233,393	F IV	5	226,091	F III	3
233,297	F IV	2	226,051	F III	2
233,222	F IV	6	226,038	O III	1
233,159	F IV	2	225,628	Ca VI	7
232,673	K V	1	225,497	Cu	25
232,584	He II	13	225,337	Ti V	100
232,531	Ca VI	5	225,299	O IV	5
232,282	Ca VI	6	225,205	N IV	5
231,823	O V	7	225,136	N IV	4
231,730	Mg III	14	225,098	N IV	3
231,454	He II	8	223,841	O IV	0
231,302	O IV	7	223,728	O IV	0
231,240	O IV	6	223,605	Ne IV	25
231,200	O IV	6	223,497	F IV	1
231,144	O IV	4	223,456	F IV	2
231,101	O IV	6	223,394	F IV	3
231,100	F III	3	223,241	Ne IV	25
231,070	O IV	7	222,791	C IV	7
231,031	O IV	3	222,777	O IV	4
231,015	F III	2	222,763	O IV	5
230,875	Ar VIII	7	222,600	Ne IV	40
230,755	O IV	2	222,378	Cu	15
230,686	He II	5	222,235	O V	3
230,682	O IV	1	221,648	O IV	4
230,593	Na III	2	221,515	O IV	0
230,495	Ca VI	5	220,946	Ar VI	5
230,43	C IV	2	220,765	F IV	7
230,139	He II	4	220,352	O V	13
230,117	F III	5	219,896	Ar VI	3
230,040	O IV	0	218,766	Ne IV	5
229,896	O IV	0	218,643	Ne IV	25
229,868	Na III	3	218,483	Ne IV	20
229,736	He II	3	218,343	Ne IV	15
229,734	Ca VI	7	218,184	Ne IV	10
229,437	Ar VIII	5	218,131	Ne IV	20
229,431	He II	2	218,085	N IV	1
229,261	F IV	2	217,830	Ne IV	25
228,988	O III	0	217,777	Ne IV	15
228,898	Ti V	75	217,743	Cu	20
228,893	O III	0	217,640	Ne IV	15
228,645	F IV	1	217,337	Ne IV	15
228,628	Ca VI	7	217,227	N IV	1
228,27	C IV	1	216,960	O IV	0
227,83	He	—	216,454	Cu	25
227,688	O V	5	216,120	Na III	1
227,662	O V	4	216,063	Cu	50
227,636	O V	5	216,018	O V	8
227,549	O V	5	215,843	Ne IV	15
227,510	O V	7	215,711	Ne IV	3
227,468	O V	5	215,671	Na III	4
227,374	O V	5	215,611	Cu	50
227,211	F IV	4	215,396	Ne IV	3
227,101	F IV	5	215,340	Na III	4
227,079	F IV	3	215,245	O V	9
226,944	F IV	6			

λ	Symbol	I
215,230	Na III	4
215,104	O V	8
215,042	Na III	2
215,034	O V	7
214,868	Na III	4
214,865	F III	1
214,804	F III	1
214,596	Na III	2
214,351	K V	2
214,290	O IV	1
214,249	O IV	1
214,235	Na III	4
214,209	O IV	4
214,206	Cu	20
214,155	O IV	6
214,062	F IV	7
214,032	O IV	5
213,978	O IV	4
213,848	F IV	7
213,121	K V	1
213,061	O IV	3
212,974	O IV	3
212,578	O IV	2
212,556	Ne IV	150
212,421	C IV	5
211,808	O IV	0
211,707	Cu	20
211,396	N IV	0
211,109	Cu	50
210,612	Cu	15
210,547	F IV	1
210,480	F IV	1
210,217	Cu	15
209,723	Ca	3
209,648	Cu	30
209,306	N V	1
209,274	N V	2
209,241	Cu	30
208,902	Cu	60
208,899	Ne IV	80
208,734	Ne IV	100
208,549	F IV	2
208,502	Cu	15
208,485	Ne IV	100
208,254	F IV	9
207,925	Cu	35
207,794	O V	10
207,733	Cu	15
207,348	O IV	4
207,282	Cu	20
207,239	O IV	7
207,183	O IV	6
206,842	Cu	35
206,641	C IV	3
206,594	F V	3
206,430	F V	2
206,355	Cu	60
206,155	Na IV	3
206,021	N IV	2
206,002	O IV	1
205,960	N IV	2
205,956	Na IV	4
205,862	K VI	2
205,842	O IV	0
205,778	F V	3
205,772	K VI	2
205,610	Cu	30
205,552	F V	4
205,487	Na IV	6
205,278	Cu	60
205,102	O V	3
204,996	O IV	0
204,908	Ne IV	5
204,905	O IV	0
204,786	Ne IV	15
204,725	Cu	40
204,708	O IV	0
204,531	Ne IV	25
204,270	Ne IV	15
204,056	Cu	15
203,959	Na IV	2
203,935	O V	6
203,890	O V	8
203,890	K III	3
203,851	O V	6
203,823	K III	2
203,821	O V	7
203,783	O V	6
203,432	Cu	60
203,324	Na III	2
203,282	Na III	2
203,152	F IV	1
203,057	C IV	2
203,050	Na III	3
203,048	O IV	5
203,010	Cu	60
202,891	O IV	4
202,760	Na III	3
202,720	Na III	3
202,490	Na III	2
202,393	O V	7
202,335	O V	5
202,282	O V	5
202,226	O V	5
202,191	O V	5
202,184	Na III	4
202,158	O V	5
202,065	Cu	20
201,862	Ti VI	5
201,615	Cu	20
201,465	F IV	4
201,329	Cu	15
201,313	Ti VI	5
201,222	F IV	6
201,160	F IV	8
201,101	F IV	6

λ	Symbol	I	λ	Symbol	I
201,098	O IV	1	194,900	Ti VI	7
201,073	O IV	1	194,840	F VI	1
201,063	F IV	7	194,839	Ne VI	2
201,022	O IV	0	194,796	Cl III	2
201,011	F IV	6	194,623	Ne IV	50
200,995	O IV	2	194,593	O V	8
200,966	O IV	1	194,477	Ne IV	40
200,915	O IV	1	194,306	Na III	1
200,861	F V	1	194,276	Ne IV	100
200,860	Ca V	3	194,166	Na III	0
200,827	O IV	1	194,108	F V	3
200,68	C IV	1	194,032	Na III	1
200,512	Ca V	5	192,906	O V	14
200,341	K VI	1	192,800	O V	13
200,089	F IV	7	192,751	O V	12
200,001	F IV	5	192,747	Ti VI	8
199,934	F IV	5	192,705	Ti VI	1
199,890	Ca V	3	192,635	Ar VII	7
199,849	F IV	5	192,244	O IV	3
199,804	F IV	5	192,206	O IV	5
199,769	Na IV	6	192,169	O IV	4
199,761	F IV	5	192,139	O IV	4
199,759	Ti VI	6	192,041	Ar VII	5
199,607	F IV	1	191, 973	F V	4
199,553	Ca V	6	191,892	F V	3
199,282	Li II	3	191,801	Ca V	2
199,086	F IV	3	191,759	Ar VII	3
199,04	C IV	1	191,752	O IV	3
199,004	F IV	3	191,695	O IV	2
198,974	Ti VI	8	191,640	O IV	0
198,476	F V	1	191,609	O IV	2
198,031	O V	3	191,556	O V	2
197,82	C IV	1	191,480	Ca V	2
197,648	Ca V	2	191,458	O V	1
197,531	Ca V	2	191,439	Ca V	3
197,455	Ti VI	5	191,397	O V	0
197,230	N IV	3	191,000	Na IV	6
197,108	F IV	2	190,839	F V	7
197,007	N IV	1	190,835	Na IV	8
196,978	K VI	1	190,645	Ne IV	15
196,970	Ca V	5	190,571	F V	6
196,968	F IV	1	190,565	Ne IV	25
196,954	N IV	0	190,558	Ca V	3
196,870	F V	1	190,457	Ca V	5
196,713	F V	2	190,440	Na IV	10
196,438	F IV	6	190,363	Ca V	4
196,435	O IV	0	190,250	N V	32
196,390	F IV	5	190,158	N V	20
196,351	F IV	4	190,126	Na IV	6
196,348	O IV	0	189,943	F V	2
196,009	O IV	8	189,346	Na III	1
195,863	O IV	7	188,870	Na III	2
195,621	Ne V	2	188,758	F IV	1
195,553	Ne V	3	188,656	F IV	2
195,538	Na III	0	188,526	Mg III	3
195,368	Ne V	5	188,424	Ne VI	3
195,227	Cl III	3	188,190	O IV	0
194,936	Ne VI	2			

λ	Symbol	I
188,152	O IV	2
187,916	F IV	1
187,240	F IV	3
187,194	Mg III	8
187,105	F IV	2
187,008	F V	4
186,982	O IV	0
186,968	F V	4
186,936	O IV	2
186,915	Ne IV	15
186,879	F V	3
186,872	O IV	1
186,842	F V	5
186,788	F V	4
186,787	Ne IV	5
186,715	F V	4
186,575	Ne IV	150
186,558	F IV	1
186,510	Mg III	9
186,153	N V	62
186,069	N V	52
185,883	K VI	2
185,747	O V	9
185,544	O IV	1
185,540	Ca V	2
185,484	F IV	3
185,479	Ne IV	20
185,455	O V	2
185,384	O IV	0
185,288	Ca V	1
185,102	Ca V	2
184,730	Ne V	10
184,415	Ca V	1
184,315	Ar VIII	5
184,280	Ca V	3
184,273	Ar VIII	3
184,218	Na III	0
184,189	Mg IV	0
184, 117	O VI	9
184,104	Ti VI	4
183,937	O VI	8
183,915	Mg IV	1
183,747	Na III	0
183,575	Na III	0
183,454	O IV	1
183,439	Mg IV	4
183,395	O IV	1
183,353	O IV	0
183,247	Ne IV	12
183,165	Ne IV	15
183,139	O IV	0
183,016	F V	3
182,979	F V	4
182,973	Mg III	2
182,832	O IV	4
182,711	O IV	3
182,282	Na IV	4
182,240	Mg III	3
182,205	O V	2
182,148	Ti VI	5
182,128	Na IV	6
181,995	O IV	4
181,876	O IV	3
181,758	Na IV	8
181,746	N IV	1
181,691	Ne IV	20
181,655	F IV	2
181,614	Ne IV	20
181,571	F IV	4
181,521	F IV	4
181,345	Mg IV	8
181,275	O IV	5
181,265	K VI	1
181,150	O IV	4
180,796	Mg IV	9
180,719	Ar VI	3
180,617	Mg IV	10
180,481	O IV	2
180,402	Ne IV	15
180,351	O IV	1
180,254	Ar VIII	15
180,074	Ar VI	2
180,070	Mg IV	8
179,943	F IV	2
179,827	F IV	1
179,400	Ar VIII	10
178,805	F IV	1
178,724	F IV	1
178,713	O V	2
178,670	F IV	3
178,612	F V	3
178,590	F V	4
178,540	F IV	1
178,434	F V	5
178,126	F IV	1
178,015	Li II	1
177,971	F IV	2
177,808	O IV	2
177,761	O IV	2
177,698	O IV	1
177,659	O IV	0
177,598	O IV	0
177,161	Ne IV	80
176,566	Ar VII	10
176,367	F IV	4
176,007	Ne IV	50
174,920	Ne IV	8
174,880	Ne IV	10
174,698	F V	4
174,558	F V	3
174,558	O V	2
174,513	F V	3
174,490	F V	3
174,303	Ne IV	3
174,220	O IV	3

λ	Symbol	I
174,105	O IV	2
174,008	Na IV	0
173,968	O IV	0
173,932	Ne V	50
173,917	O IV	2
173,851	O IV	1
173,803	O IV	0
173,145	F VI	1
173,082	O VI	13
173,020	F V	1
172,935	O VI	12
172,620	Ne IV	80
172,525	Ne IV	50
172,492	Ne IV	40
172,306	Mg IV	7
172,163	O V	12
171,896	Mg III	0
171,653	Mg IV	8
171,582	Li II	1
171,395	Mg III	4
171,302	F V	2
171,241	F V	1
171,212	Ne VI	2
171,191	O IV	0
171,121	O IV	2
171,114	Ne VI	5
171,076	Na V	1
171,071	O IV	2
171,066	F IV	3
170,988	O IV	0
170,940	O IV	0
170,923	Na V	1
170,802	Mg III	5
170,631	Na V	1
170,227	K IV	1
170,218	O V	5
170,187	F IV	2
169,839	F IV	3
169,790	F IV	3
169,748	F IV	2
169,746	Mg III	1
169,661	F IV	2
169,610	F IV	1
169,586	O V	0
169,502	F IV	2
169,481	F IV	1
169,478	O V	0
169,166	F IV	2
169,150	Mg III	1
168,741	Li II	—
168,590	N V	12
168,544	Na IV	5
168,517	N V	5
168,450	F IV	2
168,409	Na IV	8
168,101	Ne IV	2
168,084	Na IV	10
168,077	O V	4
168,042	O V	4
167,991	O V	8
167,921	Ne IV	5
167,921	Ne V	5
167,858	F V	1
167,837	Ne V	5
167,670	Ne V	25
167,610	Ne V	3
167,510	Na V	1
167,483	Ne V	15
167,145	O IV	0
166,947	N V	52
166,881	N V	44
166,499	F IV	2
166,444	F IV	2
166,246	K IV	1
166,234	O V	5
166,177	F V	10
166,163	K IV	0
166,152	O V	4
166,113	O V	3
165,983	F V	9
165,195	Mg III	0
164,986	O V	2
164,954	Mg III	2
164,841	Na IV	4
164,710	O V	4
164,656	O V	6
164,628	O V	4
164,612	F IV	2
164,578	O V	5
164,450	Ti V	6
164,384	Mg III	2
164,294	Ne V	8
164,178	O V	2
164,159	Mg III	0
164,145	Ne V	10
164,023	Ne V	10
164,015	F VI	1
163,930	Na V	2
163,840	Na IV	4
163,616	Na V	3
163,602	Ne IV	2
163,596	F V	2
163,586	Mg III	0
163,562	Ne IV	12
163,558	F V	5
163,501	F V	4
163,456	F V	3
163,187	Na IV	6
163,140	Ti V	5
163,138	F VI	2
162,565	N V	48
162,494	O V	4
162,445	Na IV	8
162,270	F V	4
162,215	F V	3

λ	Symbol	I
162,172	F V	3
162,121	F V	2
162,082	F V	3
162,053	F V	3
162,013	F V	2
161,686	Al IV	14
161,683	Mg III	0
161,477	F VI	1
161,414	F VI	1
161,341	F VI	1
161,308	F VI	3
161,257	F VI	1
161,174	F VI	1
161,135	Mg III	0
160,804	Mg IV	4
160,471	Ne IV	10
160,230	Mg IV	6
160,141	O IV	0
160,073	Al IV	16
159,755	Mg III	0
159,380	O V	4
159,343	O V	4
159,209	Mg III	0
159,175	Ar VIII	5
158,933	N V	7
158,926	O V	2
158,925	F IV	1
158,923	Ar VIII	8
158,867	N V	4
158,822	Ne IV	15
158,813	O V	1
158,646	Ne IV	15
158,606	O IV	1
158,601	F IV	1
158,553	O IV	0
158,537	F V	4
158,530	Mg III	0
158,105	Ne IV	2
158,090	N V	36
158,063	Ne IV	5
158,030	N V	24
157,862	Ne IV	2
157,782	Na IV	3
157,781	Ne IV	3
157,626	Ne IV	5
157,599	Na IV	1
157,515	F V	1
157,511	Na V	2
157,433	K VI	1
157,209	Na V	3
157,090	Na IV	4
157,036	Na V	2
156,887	Na IV	3
156,873	Ne IV	3
156,780	Na IV	5
156,610	Ne V	2
156,536	Na IV	8
156,480	Ne IV	5
156,247	F VI	6
156,225	O V	3
156,158	O V	2
156,126	O V	1
155,832	Na IV	0
155,781	Na IV	1
155,693	Na IV	2
155,622	Na IV	0
155,515	Na IV	4
155,445	Na IV	3
155,354	Na IV	0
155,248	Na IV	2
155,090	Na IV	1
154,506	F VI	3
154,488	Ne IV	5
153,948	O V	3
153,880	F VI	4
153,741	F VI	3
153,683	N V	6
153,678	F VI	2
153,624	N V	3
153,192	N V	28
153,162	O IV	0
153,136	N V	18
152,591	Mg V	0
152,563	F V	2
152,527	Mg V	1
152,511	F V	4
152,391	F V	3
152,384	Mg V	1
152,355	O IV	0
152,339	F V	2
152,264	O IV	0
152,259	Ar VII	3
152,231	Ne IV	15
152,149	Mg V	3
152,010	Mg V	0
151,876	Ar VII	2
151,817	Ne IV	15
151,698	Ar VII	1
151,548	O V	6
151,481	O V	5
151,449	O V	4
151,424	Ne V	12
151,303	Na V	1
151,303	Na IV	1
151,188	Na V	1
151,127	Na V	4
151,048	Na IV	0
151,005	F IV	1
150,977	F IV	1
150,968	Na V	2
150,968	Na IV	2
150,931	Ne IV	1
150,695	Na IV	2
150,647	Na IV	2
150,545	Na IV	3

λ	Symbol	I	λ	Symbol	I
150,488	N V	5	146,297	Na IV	1
150,429	N V	2	146,262	Ne IV	2
150,297	Na IV	4	146,083	Mg V	6
150,171	N V	14	146,060	Na IV	3
150,124	O VI	9	145,846	Na IV	0
150,116	N V	7	145,742	N V	5
150,088	O VI	10	145,691	F VI	3
149,621	Na VI	0	145,630	F VI	1
149,589	Ne IV	2	145,585	F VI	1
149,442	Na VI	0	145,547	F V	3
149,333	Ar VIII	3	145,489	F VI	1
149,078	O V	2	145,485	Mg V	5
149,034	O V	0	145,462	F VI	1
149,001	Na V	2	145,392	F V	2
148,942	Ne IV	4	145,177	F V	1
148,856	Na V	3	144,979	Na IV	0
148,787	Ne V	3	144,978	N V	4
148,787	Ne IV	3	144,837	O V	1
148,725	Ar VIII	2	144,802	O V	1
148,660	Ne IV	1	144,673	F V	1
148,653	F VI	4	144,661	Na V	1
148,642	Na V	4	144,637	F V	1
148,387	N V	4	144,546	Na V	2
148,328	N V	1	144,392	N V	3
148,168	N V	7	144,330	Na V	2
148,121	Mg IV	2	144,288	Ne IV	1
148,116	N V	4	144,151	Ne IV	2
148,108	F V	1	144,019	Ne IV	2
148,002	F V	5	143,914	N V	2
147,946	F V	4	143,897	F V	1
147,897	Na V	2	143,520	N V	1
147,887	Mg IV	1	143,344	Ne V	15
147,746	Mg IV	4	143,273	Ne V	10
147,632	Mg IV	0	143,241	N V	1
147,535	Mg IV	5	143,219	Ne V	5
147,433	N V	24	142,981	N V	0
147,405	Mg IV	5	142,933	Mg V	6
147,321	Mg IV	4	142,929	Ne IV	3
147,261	O V	3	142,797	N V	0
147,252	Mg IV	3	142,724	Ne V	15
147,132	Ne V	15	142,688	Na IV	0
147,052	Mg IV	4	142,661	Ne V	4
147,006	Mg IV	4	142,593	Ne V	10
146,949	Mg IV	4	142,441	Ne V	10
146,921	N V	3	142,415	Na V	0
146,836	Mg IV	3	142,363	Na IV	1
146,767	N V	6	142,232	Na V	2
146,718	F VI	2	142,232	Na IV	2
146,716	N V	3	142,119	O V	0
146,676	F VI	4	141,154	F VI	2
146,621	Mg V	4	141,040	Na VI	0
146,613	F VI	3	140,966	Mg IV	4
146,576	F VI	2	140,918	Mg IV	2
146,526	Mg IV	4	140,867	Mg IV	4
146,464	Mg V	5	140,833	Na VI	2
146,398	Na VI	0	140,791	Ne V	15
146,397	Na IV	0	140,757	Ne V	15
146,345	O V	1			

λ	Symbol	I	λ	Symbol	I
140,716	Ne V	5	135,953	Mg V	1
140,564	Mg IV	2	135,854	Na V	3
140,523	Mg IV	2	135,791	Na V	3
140,475	Mg IV	2	135,638	Mg V	2
140,425	Mg IV	2	135,523	O V	5
140,414	F V	1	135,397	F VI	3
140,364	N V	16	135,02	Li III	—
140,258	Na V	0	134,539	F V	5
140,176	Mg IV	4	134,532	Na VI	3
140,171	Na V	0	134,407	F V	4
140,127	Ne IV	3	134,272	Na V	2
140,120	Mg IV	4	134,183	Na V	0
140,109	O V	0	134,135	Na VI	0
140,045	O V	0	134,021	Na VI	1
139,995	Mg IV	1	133,994	N V	7
139,900	F VI	7	133,825	Na VI	2
139,869	Na IV	0	133,662	F V	1
139,800	F VI	6	133,521	O V	3
139,758	F VI	5	133,395	O V	0
139,492	O V	0	133,388	Na V	4
139,025	O V	5	133,328	O V	0
138,917	Na V	3	133,242	Al V	2
138,812	Na V	2	133,208	F V	1
138,693	Mg IV	2	133,202	Mg IV	3
138,693	Na VI	2	133,162	Na V	5
138,630	Ne VI	3	133,013	Al V	4
138,628	Na	2	132,885	O V	0
138,440	Ar VIII	5	132,851	O V	2
138,397	Ne VI	3	132,819	F V	2
138,394	Mg IV	2	132,815	Mg IV	3
138,262	Mg IV	3	132,800	O V	1
138,181	F V	1	132,740	Na IV	0
138,108	O V	4	132,699	F V	1
138,054	O V	3	132,630	Al V	10
138,030	O V	2	132,623	Mg V	3
137,966	Mg IV	1	132,511	F V	3
137,945	Na IV	0	132,485	Mg V	5
137,926	Ar VIII	3	132,484	F V	3
137,880	Mg V	6	132,465	Na IV	0
137,814	Mg VI	0	132,453	F V	2
137,748	Mg V	7	132,383	N V	6
137,714	Na IV	0	132,312	O VI	2
137,589	Na VI	0	132,310	F V	1
137,414	Mg V	8	132,219	O VI	1
137,234	Mg V	6	132,211	Na IV	0
137,144	Na IV	0	132,171	Mg V	6
137,062	Na IV	0	131,807	O V	1
136,902	F V	3	131,750	O V	1
136,854	Na IV	1	131,652	Al IV	3
136,748	Na IV	0	131,638	F V	0
136,668	Al V	2	131,635	Na V	3
136,645	Na IV	0	131,516	F V	0
136,550	Na IV	1	131,441	Al V	20
136,435	Na IV	0	131,413	Na V	2
136,429	N V	8	131,345	Na V	3
136,215	Ne V	2	131,254	N V	5
136,128	Mg V	0	131,003	Al V	20
136,089	Ne VI	4	130,848	Al V	20

λ	Symbol	I	λ	Symbol	I
130,723	Na V	1	126,065	Al V	15
130,701	Mg VI	0	125,899	Na V	2
130,680	Na V	2	125,830	Ne V	2
130,630	Mg VI	1	125,811	Mg IV	1
130,630	Mg IV	1	125,600	Mg VI	4
130,431	N V	4	125,600	Mg V	4
130,413	Al V	20	125,525	Al V	15
130,403	Al IV	11	125,461	Na V	3
130,350	Mg IV	3	125,459	Mg VI	3
130,294	Mg VI	2	125,459	Mg IV	0
130,294	Mg IV	2	125,428	Na V	3
130,243	Mg IV	1	125,383	Na VI	0
130,118	Mg IV	2	125,286	Na V	5
130,085	Mg IV	2	125,216	Na V	4
129,969	Mg IV	3	125,206	Mg VI	3
129,942	Na V	1	125,178	Na V	4
129,872	O VI	6	124,990	Mg IV	2
129,855	Mg IV	4	124,870	Mg IV	2
129,811	N V	3	124,850	Na VI	0
129,786	O VI	5	124,759	Mg IV	2
129,710	Mg IV	2	124,649	Mg IV	3
129,729	Al IV	12	124,598	O V	3
129,464	Na IV	0	124,543	Al IV	6
129,337	N V	2	124,538	Mg IV	2
129,040	Na VI	2	124,474	F VI	0
129,034	Ne V	5	124,414	Mg IV	2
128,954	N V	1	124,400	F VI	0
128,793	Ne V	1	124,387	F VI	3
128,662	N V	1	124,153	Na VI	4
128,500	O VII	0	124,059	Na VI	4
128,430	N V	0	124,034	Al IV	8
128,412	O VII	00	123,970	Na VI	2
128,235	O V	0	123,929	Na VI	5
128,229	N V	0	123,868	Na VI	3
128,112	Na	3	123,774	F V	1
128,051	Na V	4	123,744	Na VI	4
128,025	Na V	4	123,712	Ne V	3
127,837	Na VI	4	123,722	Mg IV	0
127,7	Ne VII	2	123,665	F V	0
127,473	Na V	4	123,590	Mg VI	1
127,444	Na V	4	123,588	Mg IV	0
127,036	Na V	0	123,500	Mg IV	3
126,985	Na V	0	123,377	Mg IV	1
126,923	F VI	5	123,273	Mg IV	2
126,920	Na V	0	123,175	F VI	1
126,814	Na V	1	123,134	Na VI	4
126,779	Na V	0	123,091	F VI	0
126,677	Mg V	0	123,051	F VI	0
126,608	Na V	1	123,033	Ar VIII	1
126,557	Na V	2	122,686	Ne VI	10
126,544	Mg V	2	122,624	Ar VIII	0
126,488	Mg VI	1	122,520	Ne VI	20
126,458	Na V	0	122,520	Ne V	20
126,450	Mg VI	1	122,372	O V	0
126,368	Na V	0	122,251	F VI	0
126,280	Mg V	4	122,200	F VI	2
126,210	Na V	1	122,169	F VI	0
126,090	Na V	0			

λ	Symbol	I	λ	Symbol	I
122,128	O V	0	115,399	Mg V	4
122,122	F VI	1	115,093	Mg V	4
122,034	Mg V	4	115,013	Mg V	6
122,018	Na VI	3	114,785	Mg V	6
121,922	Mg V	5	114,759	Al IV	0
121,913	Na VI	3	114,738	Na V	1
121,773	Na VI	4	114,725	Mg VI	0
121,644	Mg V	6	114,700	Na V	1
121,290	Mg VI	3	114,666	Na VI	4
121,263	Na V	0	114,624	Mg VI	0
121,140	Ne VI	5	114,412	Mg VI	0
121,025	Mg VI	5	114,329	Al IV	0
121,004	Na VI	1	114,324	Mg V	3
120,973	Na VI	2	114,285	Mg V	3
120,931	Na VI	3	114,226	Mg V	3
120,331	O VII	00	114,199	Mg V	3
120,283	Mg IV	1	114,059	Mg V	4
120,157	Ar VIII	1	114,029	Mg V	2
120,116	F VI	1	113,990	Mg V	3
120,093	Ar VIII	2	113,934	Mg V	3
120,032	F V	0	113,93	Li III	—
119,986	F V	0	113,870	Ne VI	5
119,684	Na VI	3	113,840	F VI	0
119,447	Mg V	4	113,823	Mg V	1
119,401	Mg V	4	113,756	Al VI	1
119,204	Na VI	0	113,703	Mg V	4
118,984	Al V	6	113,623	Al VI	1
118,968	Si V	20	113,574	Na V	0
118,841	Ne V	1	113,518	Mg V	1
118,810	Mg V	5	113,437	Al VI	3
118,715	Ne V	5	113,414	Mg V	2
118,603	Mg IV	1	113,314	Al VI	1
118,585	Na VI	0	113,217	Mg V	2
118,500	Al V	10	113,189	Mg VI	5
118,500	Na VI	0	113,125	Na VI	4
118,083	Mg V	5	112,950	Na VI	4
117,990	Na V	4	112,448	Na VI	3
117,876	Na V	0	112,347	Na V	0
117,860	Si V	20	112,186	Na V	0
117,699	Na VI	3	112,077	Na V	0
117,609	Na VI	3	112,009	Na VI	3
117,527	Mg VI	1	112,009	Na V	3
117,491	Na VI	4	111,879	Na V	0
117,377	Al IV	0	111,864	Mg VI	4
117,226	Mg VI	3	111,793	Na VI	1
116,968	Mg VI	5	111,780	Al IV	0
116,920	Al IV	5	111,746	Mg VI	4
116,7	Ne VII	5	111,725	Na VI	1
116,459	Al IV	7	111,590	Al IV	1
116,419	O VI	2	111,552	Mg VI	5
116,347	O VI	1	111,552	Na V	0
116,094	F VI	0	111,512	Na V	1
115,824	O VI	4	111,496	Mg V	2
115,780	Na VI	0	111,467	Mg V	2
115,729	Na VI	2	111,419	Mg V	2
115,537	Mg V	4	111,247	Mg V	2
115,5	Ne VII	3	111,200	Al IV	1
115,4	Ne VII	3			

λ	Symbol	I	λ	Symbol	I
111,199	Mg VI	4	107,711	Al V	6
111,199	Mg V	4	107,683	Na VI	5
111,160	Mg VI	3	107,661	Mg V	2
111,142	Ne VI	1	107,620	Al VI	14
111,091	Mg V	3	107,608	Na VI	4
111,031	Mg V	3	107,553	Na VI	3
110,939	Mg V	2	107,535	Na VI	3
110,921	Na V	0	107,288	Na VI	4
110,878	Na V	2	107,227	Na VI	3
110,859	Mg V	4	107,158	Na VI	1
110,824	O VI	1	107,093	Na VI	3
110,817	Na V	2	107,061	Na VI	3
110,809	Mg V	2	107,014	Na VI	2
110,750	Na VI	2	106,580	Na VI	0
110,410	Ne VI	2	106,490	Na V	1
110,220	O VI	0	106,453	Mg	2
110,148	O VI	0	106,399	Na V	1
110,121	Mg V	0	106,302	Na V	1
110,085	Na	2	106,278	Na V	1
110,029	Mg V	1	106,2	Ne VII	7
109,974	Al VI	4	106,125	Na VI	4
109,896	Na VI	5	106,1	Ne VII	7
109,843	Al VI	12	106,077	Na VI	3
109,812	Mg V	2	106,040	Na VI	3
109,763	Na VI	0	105,502	Mg VI	3
109,514	Al VI	20	105,49	Li III	—
109,284	Al VI	7	105,410	Mg VI	2
109,174	Mg V	0	105,066	Mg	5
109,040	F VI	1	104,811	O VI	2
109,021	Al V	3	104,597	Mg VI	5
108,975	F VI	1	104,519	Mg VI	3
108,907	Al IV	0	104,495	Al V	8
108,851	Al V	1	104,466	Al VI	8
108,707	Al V	6	104,447	Al V	10
108,678	Na VI	0	104,432	Mg V	2
108,616	Al V	1	104,363	Al V	10
108,555	Na VI	4	104,344	Al VI	16
108,535	Al IV	0	104,214	Mg V	1
108,462	Al V	10	104,182	Mg V	1
108,446	Al V	3	104,180	Al V	14
108,441	Mg VI	0	104,17	Li III	—
108,404	Al V	5	104,140	Mg V	2
108,385	Al V	10	104,121	Al V	12
108,338	Mg VI	1	104,100	Mg V	2
108,315	Al V	4	104,073	Al V	10
108,148	Mg VI	1	104,047	Al VI	20
108,114	Mg VI	2	103,990	Al V	4
108,112	Al V	12	103,947	Mg V	3
108,057	Al V	12	103,940	Al VI	6
108,017	Na V	2	103,904	Mg V	4
108,015	Mg VI	3	103,881	Al V	14
108,01	Li III	—	103,805	Al V	10
108,004	Al V	5	103,482	Na V	0
107,945	Al V	20	103,40	Li III	—
107,934	Na VI	2	103,333	Mg V	0
107,934	Na V	2	103,210	Na VI	2
107,820	Mg VI	4	103,163	Si VI	2
107,742	Na VI	2			

λ	Symbol	I	λ	Symbol	I
103,1	Ne VIII	6	98,983	Mg VI	4
103,078	Na VI	1	98,983	Mg V	1
103,002	Na VI	0	98,872	Mg V	1
102,906	Mg	3	98,805	Mg V	2
102,9	Ne VIII	5	98,636	Mg V	2
102,86	Li III	—	98,508	Mg VI	3
102,846	Si VI	1	98,444	Mg V	1
102,239	Mg VI	5	98,406	Mg V	1
102,189	Mg VI	5	98,309	Na VI	0
102,079	Mg V	2	98,271	Mg V	2
101,782	Mg V	3	98,235	Mg V	1
101,671	Mg V	3	98,209	Si V	2
101,556	Mg VI	3	98,2	Ne VIII	9
101,508	Mg VI	2	98,1	Ne VIII	9
101,160	Si VI	0	97,686	Mg V	1
101,027	Al VI	3	97,636	Na VI	0
100,963	Si VI	1	97,606	Mg V	2
100,949	Mg V	0	97,563	Mg V	1
100,945	Na V	0	97,5	Ne VII	6
100,919	Al VI	4	97,439	Mg V	2
100,904	Mg VI	4	97,391	Mg V	1
100,894	Al VI	4	97,278	Mg VI	5
100,851	Na V	0	97,251	Mg VI	5
100,702	Mg VI	5	97,143	Si V	10
100,640	Si VI	10	96,973	Mg VI	4
100,639	Al VI	2	96,939	Mg VI	4
100,616	Al VI	12	96,903	Mg VI	0
100,597	Mg	2	96,857	Mg VI	1
100,590	Na VI	1	96,797	Mg VI	1
100,545	Mg	2	96,704	Mg VI	2
100,519	Na VI	3	96,673	Al VI	1
100,469	Na VI	2	96,670	Mg VI	2
99,966	Si VI	10	96,488	Si VI	10
99,788	Mg V	1	96,475	Na VI	3
99,769	Al V	0,5	96,467	Mg VI	0
99,738	Mg VI	3	96,439	Si V	15
99,713	Mg VI	3	96,388	Mg VI	1
99,680	Na VI	1	96,307	Na VI	1
99,617	Na VI	1	96,303	Mg VI	2
99,616	Al V	7	96,256	Mg VI	2
99,610	Mg	2	96,240	Mg VI	1
99,598	Si VI	10	96,196	Na VI	1
99,544	Al V	2	96,159	Mg VI	1
99,500	Na VI	0	96,159	Mg V	0
99,460	Si VI	15	96,150	Al V	1
99,427	Al V	4	96,124	Na VI	0
99,333	Mg VI	4	96,085	Mg VI	1
99,290	Al V	10	96,085	Mg V	1
99,279	Mg VI	4	96,020	Si VI	10
99,203	F VI	0	96,019	Mg V	2
99,200	Al V	1	95,965	Mg V	1
99,105	F VI	0	95,933	Na VI	3
99,095	Si VI	10	95,909	Mg V	1
99,067	Mg V	2	95,835	Al V	2
99,044	F VI	0	95,803	Mg VI	2
99,025	Mg VI	2	95,803	Mg V	2
99,025	Mg V	2	95,720	Al	4
99,004	Na VI	0			

λ	Symbol	I
95,675	Mg VI	3
95,637	Mg VI	3
95,592	Mg V	0
95,556	Mg V	1
95,551	Na	2
95,483	Mg VI	5
95,436	Al VI	2
95,421	Mg VI	4
95,385	Mg VI	4
95,319	Na VI	0
95,263	Na VI	1
95,182	Na VI	1
94,970	Al	3
94,827	Na VI	0
94,793	Mg V	0
94,208	Na VI	1
94,187	Al V	2
94,160	Al V	2
94,117	Al V	2,5
93,981	Al V	2
93,955	Al V	6
93,855	Al V	4
93,755	Al V	7
93,650	Mg	2
93,493	Mg VI	3
93,109	Mg VI	1
93,109	Mg V	1
92,970	Al VI	5
92,964	Mg VI	1
92,875	Al VI	10
92,641	Mg V	0
92,636	Al VI	4
92,626	Al VI	15
92,588	Mg V	0
92,428	Mg V	0
92,409	Mg V	0
91,836	Na VI	0
91,798	Si VI	4
91,750	Al V	1
91,737	Na VI	0
91,475	Na VI	0
91,414	Na VI	0
91,369	Si VI	4
91,332	Al VI	10
91,268	Na VI	1
91,023	Al	10
90,982	Al V	1
90,914	Al V	4
90,897	Mg VI	6
90,858	Al VI	12
90,852	Si V	4
90,746	Na VI	0
90,701	Al V	4
90,646	Al V	2
90,630	Al V	5
90,468	Na VI	3
90,453	Si V	4
90,200	Al VI	20
89,649	Mg VI	0
89,4	Ne VII	3
89,021	Mg VI	0
88,952	Mg VI	2
88,827	Mg VI	2
88,817	Al V	1
88,688	Al V	0
88,688	Al VI	4
88,636	Al V	2
88,539	Al VI	8
88,539	Al V	8
88,469	Al VI	5
88,460	Na VI	1
88,425	Al V	2
88,387	Na VI	0
88,376	Al VI	15
88,340	Na VI	1
88,325	Al VI	2
88,273	Al VI	15
88,270	Na VI	3
88,246	Na VI	2
88,223	Na VI	1
88,170	Al VI	20
88,143	Na VI	2
88,1	Ne VIII	9
88,038	Na VI	1
88,016	Mg	2
87,887	Al VI	5
87,866	Al VI	7
87,802	Al VI	5
87,783	Al VI	5
87,655	Al VI	13
87,629	Al VI	2
87,592	Al VI	10
87,544	Al VI	7
87,524	Na	2
87,406	Mg VI	0
87,334	Al VI	8
87,279	Al V	1
87,211	Na VI	7
87,141	Na VI	1
87,020	Al V	2
86,807	Mg VI	2
86,440	Mg	2
86,417	Mg	2
86,147	Al VI	4
86,097	Al VI	3
86,020	Al VI	3
85,865	Al V	2
85,817	Al VI	7
85,804	Al V	7
85,764	Al VI	8
85,724	Al VI	6
85,662	Al V	1

λ	Symbol	I
85,622	Al VI	6
85,622	Mg VI	3
85,577	Mg VI	2
85,576	Si V	6
85,569	Al VI	4
85,515	Al VI	20
85,423	Al VI	2
85,175	Si V	10
85,153	Mg VI	0
84,745	Mg VI	2
84,082	Si VI	12
83,965	Si VI	0
83,802	Si VI	6
83,729	Si VI	1
83,684	Si VI	1
83,639	Si VI	3
83,639	Na VI	1
83,611	Si VI	8
83,560	Mg VI	2
83,526	Si VI	8
83,519	Mg VI	3
83,403	Mg VI	4
83,358	Si VI	8
83,283	Si VI	1
83,258	Si VI	5
83,128	Si VI	15
83,006	Si VI	4
82,853	Mg VI	1
82,475	Mg VI	1
82,3	Ne VII	5
82,267	Al VI	1
82,238	Mg VI	2
82,082	Al VI	1,5
81,738	Al VI	1
81,584	Na VI	1
81,543	Na VI	1
81,498	Na VI	2
81,113	Si V	2
81,106	Mg VI	3
81,030	Si VI	7
80,930	Mg VI	2
80,908	Si VI	8
80,821	Si VI	8
80,807	Si V	2
80,770	Al VI	1,5
80,725	Si VI	10
80,724	Mg VI	0
80,698	Si VI	10
80,645	Na VI	0
80,577	Si VI	12
80,501	Si VI	10
80,491	Si VI	5
80,449	Si VI	10
80,395	Si VI	5
80,345	Na VI	0
80,075	Mg VI	2
80,032	Mg VI	2
79,880	Mg	2
79,857	Mg VI	4
79,830	Mg VI	4
79,817	Mg VI	2
78,938	Al	6
78,903	Si V	1
78,611	Si V	1
78,508	Al	3
78,459	Al VI	1,5
78,239	Mg VI	0
78,178	Al VI	1
78,112	Al VI	2
77,945	Al VI	10
77,718	Si VI	6
77,511	Mg VI	1
77,429	Si VI	10
77,405	Mg VI	2
76,953	Al VI	1
76,908	Mg VI	0
76,853	Al	4
76,794	Al	4
76,697	Al VI	4
76,618	Al VI	4
75,890	Mg VI	0
75,834	Mg VI	2
75,666	Mg	3
75,587	Si VI	1
75,486	Si VI	1
75,334	Mg VI	1
75,248	Mg VI	1
75,193	Si VI	4
74,892	Al VI	2
74,813	Al VI	1
74,7	Ne VIII	4
74,656	Al VI	5
74,592	Al VI	3
74,574	Mg VI	2
74,504	Al VI	1
74,461	Mg VI	0
74,444	Al VI	6
74,346	Al VI	1
74,319	Mg VI	3
73,6	Ne VIII	8
73,076	Al VI	2
72,926	Al VI	2
72,896	Si VI	1
72,865	Al VI	1
72,810	Al VI	5
72,674	Al	6
72,430	Mg VI	1
71,718	Si VI	0
71,644	Si VI	0
71,561	Si VI	1
71,534	Si VI	1
71,474	Si VI	1
71,384	Si VI	4
71,366	Si VI	3
71,340	Si VI	1

λ	Symbol	I	λ	Symbol	I
71,304	Si VI	0	40,270	C V	—
71,273	Si VI	2	34,973	C V	—
71,181	Si VI	5	33,734	C VI	—
69,631	Al	4	33,426	C V	—
69,448	Si VI	2	32,754	C V	—
69,421	Si VI	1	32,400	C V	—
69,236	Si VI	5	32,188	C V	—
69,204	Si VI	1	32,064	C V	—
69,165	Al	4	29,084	N VI	—
68,223	Al VI	0	28,787	N VI	—
68,167	Al VI	1	28,464	C VI	—
67,3	Ne VIII	8	26,988	C VI	—
66,796	Si VI	0	24,898	N VI	—
66,772	Si VI	0	23,771	N VI	—
65,9	Ne VIII	6	21,804	O VII	—
65,211	Si VI	0	21,602	O VII	—
65,004	Si VI	0	18,627	O VII	—
62,4	Ne VIII	3	17,768	O VII	—
60,7	Ne VIII	3	17,396	O VII	—
40,731	C V	—	17,200	O VII	—

Section V

Auxiliary Tables

FORBIDDEN LINES

λ, Å	I	E_H, eV	E_B, eV	Transition	J
				He I	
4920,35	—	21,22	23,74	$2p\,^1P°–4f\,^1F°$	1—3
4517,43	—	20,96	23,71	$2p\,^3P°–4p\,^3P°$	2—2, 1, 0
4469,92	—	20,96	23,74	$2p\,^3P°–4f\,^3F°$	2—4, 3, 2
4045,16	—	20,96	24,03	$2p\,^3P°–5p\,^3P°$	2—2, 1, 0
4025,49	—	20,96	24,04	$2p\,^3P°–5f\,^3F°$	2—4, 3, 2
4007,81	—	21,22	24,31	$2p\,^1P°–7p\,^1P°$	1—1
3829,47	—	20,96	24,20	$2p\,^3P°–6p\,^3P°$	2—2, 1, 0
3819,25	—	20,96	24,21	$2p\,^3P°–6f\,^3F°$	2—4, 3, 2
3704,79	—	20,96	24,31	$2p\,^3P°–7f\,^3F°$	2—4, 3, 2
3634,10	—	20,96	24,37	$2p\,^3P°–8f\,^3F°$	2—4, 3, 2
3587,16	—	20,96	24,42	$2p\,^3P°–9f\,^3F°$	2—4, 3, 2
601,404	5	0,00	20,61	$1s^2\,^1S–2s\,^1S$	0—0
591,412	20	0,00	20,96	$1s^2\,^1S–2p\,^3P°$	0—1
540,935	—	0,00	22,92	$1s^2\,^1S–3s\,^1S$	0—0
538,896	—	0,00	23,01	$1s^2\,^1S–3p\,^3P°$	0—1
537,33	—	0,00	23,07	$1s^2\,^1S–3d\,^1D$	0—2
523,724	—	0,00	23,67	$1s^2\,^1S–4s\,^1S$	0—0
516,359	—	0,00	24,01	$1s^2\,^1S–5s\,^1S$	0—0
				Li I	
6240,4	2	1,85	3,83	$2p\,^2P°–3p\,^2P°$	$^3/_2$, $^1/_2$—$^3/_2$, $^1/_2$
4636,0	1	1,85	4,52	$2p\,^2P°–4p\,^2P°$	$^3/_2$, $^1/_2$—$^3/_2$, $^1/_2$
4148,4	—	1,85	4,84	$2p\,^2P°–5p\,^2P°$	$^3/_2$, $^1/_2$—$^3/_2$, $^1/_2$
3921,6	—	1,85	5,01	$2p\,^2P°–6p\,^2P°$	$^3/_2$, $^1/_2$—$^3/_2$, $^1/_2$
3195,6	3	0,00	3,88	$2s\,^2S–3d\,^2D$	$^1/_2$—$^3/_2$
2732,3	2	0,00	4,54	$2s\,^2S–4d\,^2D$	$^1/_2$—$^3/_2$
2557,4	—	0,00	4,85	$2s\,^2S–5d\,^2D$	$^1/_2$—$^3/_2$
				C I	
9849,5	—	0,005	1,26	$2p^2\,^3P–2p^2\,^1D$	2—2
9823,4	—	0,002	1,26	$2p^2\,^3P–2p^2\,^1D$	1—2
9808,9	—	0,00	1,26	$2p^2\,^3P–2p^2\,^1D$	0—2
8727,4	—	1,26	2,68	$2p^2\,^1D–2p^2\,^1S$	2—0
4627,3	—	0,005	2,68	$2p^2\,^3P–2p^2\,^1S$	2—0
4621,5	—	0,002	2,68	$2p^2\,^3P–2p^2\,^1S$	1—0
				N I	
10404,1	—	2,38	3,57	$2p^3\,^2D°–2p^3\,^2P°$	$^5/_2$—$^3/_2$, $^1/_2$
10395,4	—	2,38	3,57	$2p^3\,^2D°–2p^3\,^2P°$	$^3/_2$—$^3/_2$, $^1/_2$
5200,41	—	0,00	2,38	$2p^3\,^4S°–2p^3\,^2D°$	$^3/_2$—$^5/_2$
5197,94	—	0,00	2,38	$2p^3\,^4S°–2p^3\,^2D°$	$^3/_2$—$^3/_2$
3466,4	—	0,00	3,57	$2p^3\,^4S°–2p^3\,^2P°$	$^3/_2$—$^3/_2$, $^1/_2$
				N II	
6583,37	—	0,02	1,90	$2p^2\,^3P–2p^2\,^1D$	2—2
6548,06	—	0,01	1,90	$2p^2\,^3P–2p^2\,^1D$	1—2
6527,4	—	0,00	1,90	$2p^2\,^3P–2p^2\,^1D$	0—2
5754,57	—	1,90	4,05	$2p^2\,^1D–2p^2\,^1S$	2—0
3070,8	—	0,02	4,05	$2p^2\,^3P–2p^2\,^1S$	2—0
3063,0	—	0,01	4,05	$2p^2\,^3P–2p^2\,^1S$	1—0
				O I	
6363,82	—	0,02	1,97	$2p^4\,^3P–2p^4\,^1D$	1—2
6300,31	—	0,00	1,97	$2p^4\,^3P–2p^4\,^1D$	2—2
5577,350	—	1,97	4,19	$2p^4\,^1D–2p^4\,^1S$	2—0
2972,3	—	0,02	4,19	$2p^4\,^3P–2p^4\,^1S$	1—0

λ, Å	I	E_H, eV	E_B, eV	Transition	J
				O II	
7330,19	—	3,33	5,02	$2p^3\ ^2D^\circ—2p^3\ ^2P^\circ$	$^3/_2—^3/_2,\ ^1/_2$
7319,92	—	3,32	5,02	$2p^3\ ^2D^\circ—2p^3\ ^2P^\circ$	$^5/_2—^3/_2,\ ^1/_2$
3728,80	—	0,00	3,32	$2p^3\ ^4S^\circ—2p^3\ ^2D^\circ$	$^3/_2—^5/_2$
3726,04	—	0,00	3,33	$2p^3\ ^4S^\circ—2p^3\ ^2D^\circ$	$^3/_2—^3/_2$
				O III	
5006,86	—	0,04	2,51	$2p^2\ ^3P—2p^2\ ^1D$	2—2
4958,93	—	0,01	2,51	$2p^2\ ^3P—2p^2\ ^1D$	1—2
4931,0	—	0,00	2,51	$2p^2\ ^3P—2p^2\ ^1D$	0—2
4363,19	—	2,50	5,35	$2p^2\ ^1D—2p^2\ ^1S$	2—0
				F II	
4869,3	—	0,04	2,59	$2p^4\ ^3P—2p^4\ ^1D$	1—2
4789,5	—	0,00	2,59	$2p^4\ ^3P—2p^4\ ^1D$	2—2
4157,5	—	2,59	5,59	$2p^4\ ^1D—2p^4\ ^1S$	2—0
				F III	
5733,0	—	4,23	6,39	$2p^3\ ^2D^\circ—2p^3\ ^2P^\circ$	$^3/_2—^3/_2,\ ^1/_2$
5721,2	—	4,23	6,39	$2p^3\ ^2D^\circ—2p^3\ ^2P^\circ$	$^5/_2—^3/_2,\ ^1/_2$
				F IV	
4060,23	—	0,08	3,13	$2p^2\ ^3P—2p^2\ ^1D$	2—2
3997,40	—	0,03	3,13	$2p^2\ ^3P—2p^2\ ^1D$	1—2
3532,2	—	3,13	6,64	$2p^2\ ^1D—2p^2\ ^1S$	2—0
				Ne I	
3899,723	2	16,85	20,02	$3s'\ [^1/_2]^\circ—3d\ [^1/_2]^\circ$	1—1
3889,427	5	16,85	20,03	$3s'\ [^1/_2]^\circ—3d\ [3^1/_2]^\circ$	1—3
3887,134	1	16,85	20,04	$3s'\ [^1/_2]^\circ—3d\ [1^1/_2]^\circ$	1—2
3882,698	2	16,85	20,04	$3s'\ [^1/_2]^\circ—3d\ [1^1/_2]^\circ$	1—2
3769,654	5	16,85	20,14	$3s'\ [^1/_2]^\circ—3d'\ [2^1/_2]^\circ$	1—2
3769,449	7	16,85	20,14	$3s'\ [^1/_2]^\circ—3d'\ [2^1/_2]^\circ$	1—3
3768,047	5	16,85	20,14	$3s'\ [^1/_2]^\circ—3d'\ [1^1/_2]^\circ$	1—2
3765,819	5	16,85	20,14	$3s'\ [^1/_2]^\circ—3d'\ [1^1/_2]^\circ$	1—1
				Ne III	
3967,47	—	0,08	3,20	$2p^4\ ^3P—2p^4\ ^1D$	1—2
3868,76	—	0,00	3,20	$2p^4\ ^3P—2p^4\ ^1D$	2—2
3342,9	—	3,20	6,91	$2p^4\ ^1D—2p^4\ ^1S$	2—0
				Na I	
5675,3	3	2,10	4,29	$3p\ ^2P^\circ—4f\ ^2F^\circ$	$^3/_2—^5/_2,\ ^7/_2$
5669,8	3	2,10	4,29	$3p\ ^2P^\circ—4f\ ^2F^\circ$	$^1/_2—^5/_2,\ ^7/_2$
4977,6	1	2,10	4,59	$3p\ ^2P^\circ—5f\ ^2F^\circ$	$^3/_2—^5/_2,\ ^7/_2$
4973,4	1	2,10	4,59	$3p\ ^2P^\circ—5f\ ^2F^\circ$	$^1/_2—^5/_2,\ ^7/_2$
4665,8	—	2,10	4,76	$3p\ ^2P^\circ—6f\ ^2F^\circ$	$^3/_2—^5/_2,\ ^7/_2$
4662,0	—	2,10	4,76	$3p\ ^2P^\circ—6f\ ^2F^\circ$	$^1/_2—^5/_2,\ ^7/_2$
				Na IV	
3445,9	—	0,14	3,86	$2p^4\ ^3P—2p^4\ ^1D$	1—2
3319,3	—	0,00	3,86	$2p^4\ ^3P—2p^4\ ^1D$	2—2
				Na V	
4021,6	—	5,90	8,98	$2p^3\ ^2D^\circ—2p^3\ ^2P^\circ$	$^3/_2—^1/_2$
4017,5	—	5,90	8,98	$2p^3\ ^2D^\circ—2p^3\ ^2P^\circ$	$^5/_2—^1/_2$
4015,3	—	5,90	8,99	$2p^3\ ^2D^\circ—2p^3\ ^2P^\circ$	$^3/_2—^3/_2$
4011,2	—	5,90	8,99	$2p^3\ ^2D^\circ—2p^3\ ^2P^\circ$	$^5/_2—^3/_2$

λ, Å	I	E_H, eV	E_B, eV	Transition	J
				Si I	
10991,52	—	0,78	1,91	$3p^2\,{}^1D$—$3p^3\,{}^1S$	2—0
6589,74	—	0,03	1,91	$3p^2\,{}^3P$—$3p^2\,{}^1S$	2—0
6526,85	—	0,01	1,91	$3p^2\,{}^3P$—$3p^2\,{}^1S$	1—0
				Cl II	
9125,8	—	0,09	1,44	$3p^4\,{}^3P$—$3p^4\,{}^1D$	1—2
8579,5	—	0,00	1,44	$3p^4\,{}^3P$—$3p^4\,{}^1D$	2—2
6152,9	—	1,44	3,46	$3p^4\,{}^1D$—$3p^4\,{}^1S$	2—0
3675,0	—	0,09	3,46	$3p^4\,{}^3P$—$3p^4\,{}^1S$	1—0
3583,2	—	0,00	3,46	$3p^4\,{}^3P$—$3p^4\,{}^1S$	2—0
				Cl III	
8550,5	—	2,25	3,70	$3p^3\,{}^2D^\circ$—$3p^3\,{}^2P^\circ$	$^5/_2$—$^1/_2$
8501,8	—	2,24	3,70	$3p^3\,{}^2D^\circ$—$3p^3\,{}^2P^\circ$	$^3/_2$—$^1/_2$
8481,6	—	2,25	3,71	$3p^3\,{}^2D^\circ$—$3p^3\,{}^2P^\circ$	$^5/_2$—$^3/_2$
8433,7	—	2,24	3,71	$3p^3\,{}^2D^\circ$—$3p^3\,{}^2P^\circ$	$^3/_2$—$^3/_2$
5537,7	—	0,00	2,24	$3p^3\,{}^4S^\circ$—$3p^3\,{}^2D^\circ$	$^3/_2$—$^3/_2$
5517,2	—	0,00	2,25	$3p^3\,{}^4S^\circ$—$3p^3\,{}^2D^\circ$	$^3/_2$—$^5/_2$
3353,4	—	0,00	3,70	$3p^3\,{}^4S^\circ$—$3p^3\,{}^2P^\circ$	$^3/_2$—$^1/_2$
3342,7	—	0,00	3,71	$3p^3\,{}^4S^\circ$—$3p^3\,{}^2P^\circ$	$^3/_2$—$^3/_2$
				Cl IV	
8046,1	—	0,17	1,71	$3p^2\,{}^3P$—$3p^2\,{}^1D$	2—2
7530,9	—	0,06	1,71	$3p^2\,{}^3P$—$3p^2\,{}^1D$	1—2
5322,2	—	1,71	4,04	$3p^2\,{}^1D$—$3p^2\,{}^1S$	2—0
3203,3	—	0,17	4,04	$3p^2\,{}^3P$—$3p^2\,{}^1S$	2—0
3118,3	—	0,06	4,04	$3p^2\,{}^3P$—$3p^2\,{}^1S$	1—0
				Ar I	
5533,20	1	11,62	13,62	$4s\,[1^1/_2]^\circ$—$3d\,[^1/_2]^\circ$	1—1
5396,96	1	11,55	13,84	$4s\,[1^1/_2]^\circ$—$3d\,[^1/_2]^\circ$	2—0
5353,46	20	11,55	13,86	$4s\,[1^1/_2]^\circ$—$3d\,[^1/_2]^\circ$	2—1
5263,02	2	11,55	13,90	$4s\,[1^1/_2]^\circ$—$3d\,[1^1/_2]^\circ$	2—2
5147,34	1	11,83	14,24	$4s'\,[^1/_2]^\circ$—$3d'\,[2^1/_2]^\circ$	1—3
5098,97	20	11,55	13,98	$4s\,[1^1/_2]^\circ$—$3d\,[3^1/_2]^\circ$	2—4
5081,19	1	11,62	14,06	$4s\,[1^1/_2]^\circ$—$3d\,[2^1/_2]^\circ$	1—2
5029,64	5	11,55	14,01	$4s\,[1^1/_2]^\circ$—$3d\,[3^1/_2]^\circ$	2—3
5007,09	2	11,62	14,10	$4s\,[1^1/_2]^\circ$—$3d\,[2^1/_2]^\circ$	1—3
5006,84	2	11,83	14,30	$4s'\,[^1/_2]^\circ$—$3d'\,[1^1/_2]^\circ$	1—1
4976,87	1	11,72	14,21	$4s'\,[^1/_2]^\circ$—$3d'\,[2^1/_2]^\circ$	0—2
4936,50	1	11,72	14,23	$4s'\,[^1/_2]^\circ$—$3d'\,[2^1/_2]^\circ$	0—2
4929,16	2	11,55	14,06	$4s\,[1^1/_2]^\circ$—$3d\,[2^1/_2]^\circ$	2—2
4901,26	2	11,62	14,15	$4s\,[1^1/_2]^\circ$—$3d\,[1^1/_2]^\circ$	1—1
4859,44	5	11,55	14,10	$4s\,[1^1/_2]^\circ$—$3d\,[2^1/_2]^\circ$	2—3
4759,65	1	11,55	14,15	$4s\,[1^1/_2]^\circ$—$3d\,[1^1/_2]^\circ$	2—1
4748,23	5	11,62	14,23	$4s\,[1^1/_2]^\circ$—$3d'\,[1^1/_2]^\circ$	1—2
4744,47	2	11,62	14,24	$4s\,[1^1/_2]^\circ$—$3d'\,[2^1/_2]^\circ$	1—3
4615,15	1	11,55	14,23	$4s\,[1^1/_2]^\circ$—$3d'\,[1^1/_2]^\circ$	2—2
4611,75	1	11,55	14,24	$4s\,[1^1/_2]$— $3d'\,[2^1/_2]$	2—3
				Ar III	
7751,06	—	0,14	1,74	$3p^4\,{}^3P$—$3p^4\,{}^1D$	1—2
7135,80	—	0,00	1,74	$3p^4\,{}^3P$—$3p^4\,{}^1D$	2—2
5191,82	—	1,74	4,12	$3p^4\,{}^1D$—$3p^4\,{}^1S$	2—0
3109,0	—	0,14	4,12	$3p^4\,{}^3P$—$3p^4\,{}^1S$	2—0
3005,1	—	0,00	4,12	$3p^4\,{}^3P$—$3p^4\,{}^1S$	1—0
				Ar IV	
7332,0	—	2,63	4,32	$3p^3\,{}^2D^\circ$—$3p^3\,{}^2P^\circ$	$^5/_2$—$^1/_2$
7262,76	—	2,61	4,32	$3p^3\,{}^2D^\circ$—$3p^3\,{}^2P^\circ$	$^3/_2$—$^1/_2$

λ, Å	I	E_H, eV	E_B, eV	Transition	J
7237,26	—	2,63	4,34	$3p^3\,{}^2D^\circ$—$3p^3\,{}^2P^\circ$	$^5/_2$—$^3/_2$
7170,62	—	2,61	4,34	$3p^3\,{}^2D^\circ$—$3p^3\,{}^2P^\circ$	$^3/_2$—$^3/_2$
4740,20	—	0,00	2,61	$3p^3\,{}^4S^\circ$—$3p^3\,{}^2D^\circ$	$^3/_2$—$^3/_2$
4711,33	—	0,00	2,63	$3p^3\,{}^4S^\circ$—$3p^3\,{}^2D^\circ$	$^3/_2$—$^5/_2$
				Ar V	
7005,67	—	0,25	2,02	$3p^2\,{}^3P$—$3p^2\,{}^1D$	2—2
6435,10	—	0,09	2,02	$3p^2\,{}^3P$—$3p^2\,{}^1D$	1—2
4625,54	—	2,02	4,70	$3p^2\,{}^1D$—$3p^2\,{}^1S$	2—0
				K I	
4642,373	11	0,00	2,67	$4s\,{}^2S$—$3d\,{}^2D$	$^1/_2$—$^5/_2$
4641,876	10	0,00	2,67	$4s\,{}^2S$—$3d\,{}^2D$	$^1/_2$—$^3/_2$
				K IV	
6794,8	—	0,21	2,03	$3p^4\,{}^3P$—$3p^4\,{}^1D$	1—2
6101,83	—	0,00	2,03	$3p^4\,{}^3P$—$3p^4\,{}^1D$	2—2
4510,9	—	2,03	4,78	$3p^4\,{}^1D$—$3p^4\,{}^1S$	2—0
				K V	
6446,5	—	3,00	4,93	$3p^3\,{}^2D^\circ$—$3p^3\,{}^2P^\circ$	$^5/_2$—$^1/_2$
6349,5	—	2,98	4,93	$3p^3\,{}^2D^\circ$—$3p^3\,{}^2P^\circ$	$^3/_2$—$^1/_2$
6316,6	—	3,00	4,97	$3p^3\,{}^2D^\circ$—$3p^3\,{}^2P^\circ$	$^5/_2$—$^3/_2$
6223,4	—	2,98	4,97	$3p^3\,{}^2D^\circ$—$3p^3\,{}^2P^\circ$	$^3/_2$—$^3/_2$
4163,30	—	0,00	2,98	$3p^3\,{}^4S^\circ$—$3p^3\,{}^2D^\circ$	$^3/_2$—$^3/_2$
4122,63	—	0,00	3,00	$3p^3\,{}^4S^\circ$—$3p^3\,{}^2D^\circ$	$^3/_2$—$^5/_2$
				Ca I	
4916,18	—	0,00	2,52	$4s^2\,{}^1S$—$3d\,{}^3D$	0—1
4912,82	—	0,00	2,52	$4s^2\,{}^1S$—$3d\,{}^3D$	0—2
4575,46	—	0,00	2,71	$4s^2\,{}^1S$—$3d\,{}^1D$	0—2
				Ca II	
7323,88	—	0,00	1,69	$4s\,{}^2S$—$3d\,{}^2D$	$^1/_2$—$^3/_2$
7291,46	—	0,00	1,70	$4s\,{}^2S$—$3d\,{}^2D$	$^1/_2$—$^5/_2$
				Ca V	
6086,92	—	0,30	2,33	$3p^4\,{}^3P$—$3p^4\,{}^1D$	1—2
5309,18	—	0,00	2,33	$3p^4\,{}^3P$—$3p^4\,{}^1D$	2—2
3996,3	—	2,33	5,44	$3p^4\,{}^1D$—$3p^4\,{}^1S$	2—0
				Kr I	
5775,56	2	10,03	12,18	$5s\,[1^1/_2]^\circ$—$4d\,[3^1/_2]^\circ$	1—3
5643,04	1	9,91	12,11	$5s\,[1^1/_2]^\circ$—$4d\,[1^1/_2]^\circ$	2—2
5608,37	3	9,91	12,12	$5s\,[1^1/_2]^\circ$—$4d\,[3^1/_2]^\circ$	2—4
5476,58	2	9,91	12,18	$5s\,[1^1/_2]^\circ$—$4d\,[3^1/_2]^\circ$	2—3
5337,72	1	10,03	12,35	$5s\,[1^1/_2]^\circ$—$4d\,[1^1/_2]^\circ$	1—1
5290,76	1	9,91	12,26	$5s\,[1^1/_2]^\circ$—$4d\,[2^1/_2]^\circ$	2—2
5232,06	2	9,91	12,28	$5s\,[1^1/_2]^\circ$—$4d\,[2^1/_2]^\circ$	2—3
				Kr III	
9902,2	—	0,56	1,82	$4p^4\,{}^3P$—$4p^4\,{}^1D$	1—2
6826,9	—	0,00	1,82	$4p^4\,{}^3P$—$4p^4\,{}^1D$	2—2
				Xe I	
6949,76	1	8,44	10,22	$6s\,[1^1/_2]^\circ$—$5d\,[2^1/_2]^\circ$	1—3
6507,50	3	8,31	10,22	$6s\,[1^1/_2]^\circ$—$5d\,[2^1/_2]^\circ$	2—3
4576,60	2	8,31	11,10	$6s\,[1^1/_2]^\circ$—$6d\,[2^1/_2]^\circ$	2—3

λ, Å	I	E_H, eV	E_B, eV	Transition	J
				Xe II	
9487,5	4	0,00	1,31	$5p^5\ ^2P^\circ$—$5p^5\ ^2P^\circ$	$^3/_2$—$^1/_2$
4061,06	3	12,92	15,98	$6s\ ^2P^\circ$—$6d\ ^4F^\circ$	$^1/_2$—$^5/_2$
3978,98	2	15,02	18,14	$6p\ ^2S^\circ$—$7s'\ ^2D$	$^1/_2$—$^5/_2$
2631,25	2	11,27	15,98	$5p^6\ ^2S$—$6p'\ ^2F^\circ$	$^1/_2$—$^5/_2$
				Xe III	
10206,5	1	0,00	1,21	$5p^4\ ^3P$—$5p^4\ ^3P$	2—1
5846,3	6	0,00	2,12	$5p^4\ ^3P$—$5p^4\ ^1D$	2—2
				Cs I	
8053,35	100	1,81	3,35	$5d\ ^2D$—$5g\ ^2G$	$^5/_2$—$^9/_2$, $^7/_2$
7990,68	100	1,80	3,35	$5d\ ^2D$—$5g\ ^2G$	$^3/_2$—$^9/_2$, $^7/_2$
7270,70	15	1,81	3,51	$5d\ ^2D$—$6g\ ^2G$	$^5/_2$—$^9/_2$, $^7/_2$
7219,70	15	1,80	3,51	$5d\ ^2D$—$6g\ ^2G$	$^3/_2$—$^9/_2$, $^7/_2$
6895,005	—	0,00	1,80	$6s\ ^2S$—$5d\ ^2D$	$^1/_2$—$^3/_2$
6848,906	—	0,00	1,81	$6s\ ^2S$—$5d\ ^2D$	$^1/_2$—$^5/_2$

EDGES OF STABLE BANDS OF SOME MOLECULES

(C_2, O_2, N_2, O_2^+, N_2^+, CN, CO, CO^+, NO, NO^+, CO_2, CH, OH, OH^+, NH, SiO, He_2)

λ, Å	I	Molecule	λ, Å	I	Molecule
10420	10	N_2	5958,7	2	C_2
8911,6	10	N_2	5906,0	8	N_2
8722,3	8	N_2	5899,3	8	C_2
8541,8	6	N_2	5861,0	6	CO
7852,5	3	C_2	5858,2	9	CN
7753,2	6	N_2	5854	8	N_2
7626,2	7	N_2	5826,4	9	O_2
7593,7	10	O_2	5804,3	7	N_2
7503,9	7	N_2	5755,2	7	N_2
7386,6	5	N_2	5749,1	6	CO
7210,4	5	CO	5733,0	9	He_2
7083,2	4	C_2	5730,2	8	CN
6927,6	2	CN	5670,5	6	CO
6867,2	8	O_2	5653,1	2	N_2^+
6856,3	8	O_2^+	5647,6	6	CO
6804,0	8	CO	5635,5	9	C_2
6792,5	2	CN	5631,9	9	O_2^+
6788,6	6	N_2	5621,7	6	CO
6704,8	8	N_2	5610,2	10	CO
6685,7	7	CO	5598,3	3	CN
6631,6	9	CN	5585,5	8	C_2
6623,6	9	N_2	5540,7	6	C_2
6620,3	7	CO	5515,6	2	N_2
6544,8	10	N_2	5501,9	4	C_2
6513,5	9	CO	5499,9	6	CO^+
6478,7	10	CN	5478,5	2	N_2
6468,5	10	N_2	5473,3	5	CN
6464,6	10	CO	5442,3	3	N_2
6442,3	6	C_2	5434,9	3	C_2
6433,1	10	CO	5407,1	3	N_2
6418,7	9	O_2^+	5372,8	3	N_2
6399,0	10	CO	5364	4	O_2
6398,7	10	He_2	5354,1	4	CN
6394,7	9	N_2	5351,3	5	CO
6368	10	O_2	5330,5	5	CO
6366,9	5	CO	5307,2	5	CO
6332,2	9	CN	5295,7	8	O_2^+
6322,9	7	N_2	5239,3	3	CN
6252,8	3	N_2	5228,3	4	N_2^+
6244,0	5	CO	5198,2	10	CO
6238,7	7	CO^+	5165,2	10	C_2
6191,7	6	CN	5148,8	5	N_2^+
6191,2	4	C_2	5129,3	6	C_2
6189,4	7	CO^+	5097,7	2	C_2
6122,1	4	C_2	5072,1	5	CO^+
6105,2	5	CO	5070,9	8	CO
6079,9	8	CO	5060,1	5	N_2
6069,7	7	N_2	5053,6	2	N_2
6059,7	3	C_2	5052,7	8	CO
6037,0	8	CO	5039,7	5	CO^+
6026,4	10	O_2^+	5031,7	8	CO
6013,6	7	N_2	5030,8	2	N_2
6010,5	8	CO	4935,8	2	CN
6004,9	3	C_2	4910,9	6	CO^+
5998,9	10	NO^+	4892,1	2	NO
5992,6	6	CN	4879,5	3	CO^+
5980,7	8	CO	4837,1	10	N_2
5959,0	8	N_2	4835,3	10	CO
			4832,6	2	CN

λ, Å	I	Molecule	λ, Å	I	Molecule
4823,5	8	CO	4171,2	5	N_2
4814,7	1	N_2	4141,8	5	N_2
4806,7	8	CO	4137,6	6	CO_2
4802	3	O_2	4124,8	7	CO
4787,3	8	CO	4123,6	7	CO
4737,1	9	C_2	4115,8	9	O_2^+
4723,5	1	N_2	4113,6	4	NO
4715,2	8	C_2	4102,3	9	C_2
4711,2	5	CO^+	4095,4	10	O_2
4709,2	4	N_2^+	4093	1	C_2
4697,6	7	C_2	4082,4	9	O_2^+
4684,8	4	C_2	4068,1	6	C_2
4683,4	5	CO^+	4059,4	8	N_2
4661,3	5	CO	4041,8	3	C_2
4651,8	3	N_2^+	4027,8	6	NO
4649,7	5	N_2	4025,3	3	CH
4606,1	1	CN	4020,6	9	O_2
4602,6	7	CO	4019,7	9	CO^+
4599,7	4	N_2^+	4017,7	9	CO^+
4586,4	7	CO	3999,6	9	CO^+
4574,3	2	N_2	3998,4	9	N_2
4574,0	5	NO	3997,3	9	CO^+
4569,2	7	CO	3989,1	5	He_2
4565,8	8	CO^+	3984,6	1	CN
4545,8	5	He_2	3973,5	9	CO^+
4539,4	8	CO^+	3957,0	7	CO^+
4535,5	7	N_2	3944,7	2	CN
4514,8	1	CN	3943,0	8	N_2
4510,9	10	CO	3940,3	5	N_2
4502,2	1	CN	3914,4	10	N_2^+
4490,2	3	N_2	3912,3	9	O_2
4479,8	3	NO	3909,5	3	CN
4416,7	3	N_2	3893,1	7	CO
4393,1	8	CO	3893	5	OH^+
4382,5	2	C_2	3889,2	5	N_2
4380,3	7	CO	3889,0	4	CH
4372,0	8	O_2	3884,3	3	N_2^+
4371,4	4	C_2	3883,4	10	CN
4368,8	5	C_2	3871,4	9	CN
4365,2	5	C_2	3868,3	6	NO
4355,0	3	N_2	3861,9	8	CN
4343,6	4	N_2	3859,5	9	O_2^+
4315,0	3	CH	3857,9	4	N_2^+
4312,5	10	CH	3854,7	6	CN
4293,7	3	NO	3852,2	10	C_2
4291,8	9	O_2	3840,6	10	O_2
4288,2	3	NO	3830,5	9	O_2^+
4278,1	10	N_2^+	3825,6	6	C_2
4274,3	10	CO^+	3804,9	10	N_2
4272,0	10	CO^+	3795,8	8	CO^+
4269,7	5	N_2	3788,5	10	NO
4252,4	8	CO^+	3777,8	8	CO^+
4248,9	5	CO^+	3755,4	10	N_2
4236,5	9	N_2^+	3741,7	9	O_2
4231,6	8	CO^+	3724,9	8	CO^+
4216,0	9	CN	3711,2	9	CO^+
4214,2	7	O_2	3710,5	8	N_2
4212,9	8	CO^+	3706,6	9	O_2^+
4200,7	4	NO	3672,7	9	O_2
4200,5	6	N_2	3665,0	5	He_2
4199,1	4	N_2^+	3647,2	4	NO
4197,2	8	CN	3629,8	9	O_2^+
4181,0	7	CN	3628,9	1	CN
4172,7	9	O_2	3628	3	O_2

λ, Å	I	Molecule	λ, Å	I	Molecule
3612,4	1	N_2^+	3008,8	4	NO
3607,3	8	C_2	2987,5	9	O_2^+
3603,7	8	O_2^+	2984,2	2	CO^+
3603,0	5	N_2	2977,4	9	CO
3603,0	3	CN	2976,8	6	N_2
3600,8	6	CO^+	2970,0	8	O_2^+
3592,9	7	C_2	2935,7	9	N_2
3590,4	8	CN	2925	2	CO
3587,6	7	C_2	2919,8	9	O_2^+
3585,9	7	CN	2903,9	1	N_2
3584,2	6	CO^+	2901,9	8	O_2^+
3583,9	6	CN	2897,2	3	CO^+
3582,1	6	N_2^+	2890,3	8	O_2
3576,9	10	N_2	2885,2	10	NO
3572,4	10	NO	2875	2	OH
3565	10	OH^+	2859,5	6	NO
3562,2	8	CO_2	2858,1	4	CO^+
3536,7	8	N_2	2839,7	10	O_2^+
3517,7	9	O_2^+	2833,1	10	CO
3516,1	8	O_2	2823,7	9	O_2^+
3511,7	7	CO^+	2820,8	5	CO^+
3500,4	3	CO^+	2819,8	1	N_2
3494,2	8	O_2^+	2811,3	8	OH
3493,3	6	CO	2810,4	2	NO
3428,1	4	OH	2802,6	4	NO
3424,6	4	N_2	2799,7	9	CO
3421,2	9	O_2^+	2785,8	5	CO^+
3399,7	5	C_2	2780,5	7	SiO
3398,1	5	C_2	2777,9	2	N_2
3397,8	9	O_2^+	2776,7	8	O_2^+
3381,3	1	N_2^+	2763,3	2	NO
3376,4	10	NO	2761,9	7	O_2^+
3371,3	10	N_2	2760,6	9	N_2
3370	10	NH	2755,0	6	SiO
3369,6	8	O_2	2752,9	6	CO^+
3360	9	NH	2750	1	CO
3348,0	3	He_2	2747,6	9	NO
3332	5	OH^+	2722,3	6	CO^+
3308,0	2	N_2^+	2722,2	9	NO
3305,7	7	CO	2711,3	3	CO
3298,7	2	N_2^+	2705,3	8	O_2^+
3253,4	5	NH	2680,0	5	NO
3253	3	CO	2672,4	7	CO^+
3242,1	6	CO	2672,2	7	NO
3232,5	7	O_2	2665,3	8	CO
3231,2	9	O_2^+	2660,5	5	N_2
3210,8	9	O_2^+	2638,8	8	CO^+
3198,0	10	NO	2632,7	7	O_2^+
3197,5	5	N_2	2630	2	CO
3159,3	9	N_2	2620,5	6	NO
3143,4	1	CH	2608,5	3	OH
3138	4	CO	2607,2	8	CO^+
3136,0	8	N_2	2603,8	10	N_2
3134,4	8	CO	2602,1	6	NO
3107,5	2	CO^+	2596,9	4	CO
3103,9	6	O_2	2595,7	10	NO
3079,9	5	CO	2591	5	O_2
3064,0	3	CO^+	2587,5	10	NO
3063,6	10	OH	2587,1	5	SiO
3043,6	9	O_2^+	2581,0	9	O_2^+
3035,2	1	NH	2577,7	10	CO^+
3034,9	10	NO	2575,3	8	CO
3028	3	CO	2563,8	5	SiO
3023,0	2	CO^+	2558,6	5	NO

λ, Å	I	Molecule	λ, Å	I	Molecule
2557,3	1	NH	2352,5	6	CO^+
2553	7	O_2	2351,4	6	N_2
2553,3	6	CO	2344,3	5	SiO
2551,8	5	NO	2342,4	4	SiO
2550,7	8	N_2	2332,8	4	N_2
2550,3	7	CO^+	2326,6	3	NO
2545,5	8	O_2^+	2325	10	C_2
2532,8	7	O_2^+	2325,2	9	CO^+
2530,2	3	NH	2317,7	2	NO
2518	9	O_2	2299,6	10	CO^+
2518	2	CO	2298,9	6	SiO
2510,9	6	CO	2295,9	4	CO
2509,9	4	SiO	2269,4	5	NO
2509,8	6	N_2	2268,6	3	CO^+
2504,6	10	CO^+	2261,7	9	CO
2491	4	CO	2260,8	2	N_2
2489,9	5	CO	2257,7	1	CO
2489	10	O_2	2255,9	5	SiO
2488,3	6	O_2	2244,3	2	NO
2487,8	7	NO	2238,3	9	CO
2486,8	10	SiO	2236,3	2	SiO
2478,7	10	NO	2226,8	3	NO
2474,2	10	CO^+	2221,5	10	CO
2471,1	10	NO	2221,3	2	NO
2465	9	O_2	2215,4	2	SiO
2463,2	10	CO	2214,5	5	CO^+
2461,6	9	N_2	2196,8	10	CO
2459,0	4	SiO	2189,8	10	CO^+
2451,8	6	CO	2173,0	9	CO
2448,0	10	N_2	2156	3	NO
2445,8	10	CO^+	2143,9	6	N_2
2440	7	O_2	2141,2	4	NO
2436,3	3	SiO	2137,8	6	CO^+
2433,9	9	CO	2125,9	9	N_2
2427,8	7	NO	2113,1	9	CO
2419,4	8	CO^+	2112,4	8	CO^+
2413,8	8	SiO	2089,9	10	CO
2409,2	6	CO	2067,6	10	CO
2392,6	3	O_2^+	2062	3	NO
2389,7	5	CO	2059,6	6	CO
2388,8	6	CO	2046,3	10	CO
2387,9	5	SiO	2041,2	10	N_2
2377,5	7	N_2	2025,8	9	CO
2370,2	9	NO	2023,5	7	N_2
2365,7	6	SiO	2006,0	4	N_2
2364,5	4	SiO			

MOLECULAR SPECTRUM OF HYDROGEN (H_2)

λ, Å	I	λ, Å	I	λ, Å	I
8724,18	2	6940,421	4	6090,910	7
8663,91	3	6875,272	3	6080,778	10
8546,27	6	6840,886	3	6078,900	3
8528,11	2	6806,438	4	6074,386	5
8520,37	2	6794,083	3	6069,994	8
8486,08	2	6756,262	3	6067,720	7
8444,62	2	6696,755	3	6066,631	7
8443,57	2	6622,590	4	6063,285	8
8398,26	3	6572,044	3	6053,265	5
8381,16	2	6567,047	1	6052,369	6
8366,89	3	6561,736	1	6047,850	5
8349,52	10	6561,064	2	6041,027	3
8330,42	7	6559,131	1	6031,900	10
8273,26	8	6554,036	4	6031,474	5
8240,75	3	6527,355	3	6023,751	10
8222,90	2	6517,695	3	6021,273	9
8164,64	8	6441,502	3	6018,291	10
8130,77	3	6437,822	3	6011,396	3
8054,55	2	6433,490	3	6002,816	9
7997,20	2	6429,314	3	5994,062	10
7970,14	2	6428,113	10	5990,530	4
7812,42	3	6399,475	10	5989,239	5
7789,78	4	6380,110	4	5982,561	6
7732,74	2	6372,209	5	5975,437	10
7685,53	2	6362,483	6	5974,139	4
7661,46	2	6340,574	7	5970,933	3
7650,75	3	6332,486	5	5970,310	5
7606,36	4	6329,814	5	5967,273	4
7603,43	4	6327,063	10	5963,473	4
7597,06	3	6320,379	3	5959,806	6
7544,99	3	6303,485	5	5959,615	4
7541,95	2	6299,420	10	5949,895	10
7538,32	3	6285,388	10	5947,302	6
7524,64	9	6277,103	3	5941,977	3
7506,95	3	6274,841	3	5938,620	10
7459,45	4	6271,313	5	5936,027	4
7449,14	4	6270,536	3	5932,295	3
7396,07	2	6267,966	3	5931,368	10
7395,04	3	6238,388	10	5924,830	9
7374,87	3	6233,014	5	5920,799	4
7350,70	4	6230,256	7	5920,489	5
7350,00	4	6224,809	10	5918,078	5
7328,12	4	6201,178	8	5916,506	5
7309,56	5	6199,387	10	5916,056	5
7295,42	3	6197,113	5	5910,165	3
7288,94	3	6182,989	10	5909,395	3
7269,96	7	6176,234	3	5889,033	5
7254,02	3	6174,888	3	5888,167	10
7253,28	10	6174,089	6	5884,632	8
7244,11	3	6169,637	7	5883,942	6
7240,57	6	6167,736	4	5879,201	3
7231,06	4	6161,605	8	5878,496	10
7230,66	4	6155,628	5	5871,952	6
7210,22	3	6151,470	4	5869,262	4
7195,66	9	6146,192	3	5864,462	4
7184,04	6	6135,395	10	5859,808	5
7179,52	3	6135,145	3	5849,317	8
7176,28	5	6134,313	5	5836,133	8
7168,81	8	6127,242	6	5835,829	3
7112,65	3	6121,787	10	5833,065	4
7095,10	4	6098,218	9	5832,773	5
7049,60	3	6095,956	10	5831,016	4

λ, Å	I	λ, Å	I	λ, Å	I
5822,763	10	5597,636	10	5319,162	3
5822,073	4	5591,420	4	5317,889	5
5819,279	3	5590,094	3	5309,036	5
5817,600	3	5579,591	3	5304,415	5
5816,438	4	5578,712	3	5303,238	4
5814,943	6	5573,920	6	5303,104	9
5812,587	10	5564,506	4	5296,089	3
5811,498	4	5561,741	3	5291,595	9
5806,099	8	5555,102	3	5284,500	9
5791,912	4	5552,525	8	5283,285	3
5788,262	4	5544,771	3	5273,023	3
5785,768	7	5543,496	7	5272,296	10
5785,208	3	5543,112	4	5270,412	4
5778,984	4	5537,466	10	5266,045	10
5775,050	9	5537,288	3	5264,684	3
5774,580	3	5535,980	5	5261,182	7
5773,224	5	5534,054	3	5256,610	7
5766,284	3	5527,346	3	5239,012	4
5762,719	3	5523,965	3	5226,771	6
5760,392	6	5520,881	3	5222,854	3
5759,559	6	5518,472	8	5214,623	6
5757,350	5	5507,853	3	5199,707	4
5755,687	5	5506,341	4	5197,213	4
5741,835	5	5505,522	10	5196,375	8
5740,089	3	5499,581	9	5180,583	7
5736,879	10	5495,964	8	5174,702	6
5735,130	6	5481,083	10	5168,235	4
5731,925	8	5474,856	5	5153,870	6
5728,552	10	5471,577	3	5146,340	5
5727,052	3	5465,192	4	5143,567	3
5723,454	6	5462,990	4	5122,583	4
5716,005	3	5459,598	8	5113,126	7
5713,449	5	5456,983	6	5109,307	4
5709,772	5	5455,316	4	5107,644	3
5703,760	4	5438,777	3	5103,566	4
5703,252	5	5434,822	10	5084,842	9
5700,644	4	5430,871	3	5080,494	7
5696,175	3	5425,887	8	5075,442	4
5694,140	5	5425,200	4	5068,121	7
5692,465	3	5419,893	10	5067,475	5
5691,155	3	5417,797	6	5063,878	6
5689,195	10	5410,219	3	5061,732	4
5684,126	5	5409,692	4	5055,091	9
5683,744	3	5408,789	6	5048,004	5
5683,080	4	5405,328	5	5041,627	8
5682,507	3	5404,746	3	5039,821	9
5670,930	4	5401,053	8	5030,367	9
5662,872	3	5398,970	4	5020,744	4
5661,626	3	5392,285	6	5017,127	5
5655,750	9	5391,144	3	5016,496	6
5652,175	3	5388,166	10	5015,069	8
5642,942	5	5386,195	3	5014,473	4
5642,717	3	5385,508	5	5013,036	10
5634,807	8	5378,391	5	5011,189	8
5634,161	3	5372,445	3	5007,988	8
5630,620	4	5371,896	4	5003,398	9
5627,435	3	5365,902	7	4997,937	5
5624,306	4	5355,912	7	4996,852	3
5623,081	4	5344,792	4	4990,147	4
5620,907	5	5343,166	4	4980,479	6
5612,541	10	5340,820	4	4978,256	4
5604,685	3	5336,581	5	4976,630	3
5601,704	3	5334,264	6	4973,310	8
5600,416	7	5326,775	4	4969,222	5

λ, Å	I	λ, Å	I	λ, Å	I
4966,909	6	4705,260	3	4543,692	5
4956,792	5	4702,562	5	4539,162	5
4955,763	6	4692,040	5	4537,731	4
4952,585	3	4690,184	7	4534,627	6
4949,536	3	4686,772	6	4534,157	7
4942,546	3	4686,142	3	4533,128	4
4939,603	4	4684,654	4	4533,087	5
4939,162	5	4683,824	8	4531,193	3
4935,242	3	4682,341	6	4529,079	5
4934,241	10	4680,432	5	4527,183	3
4933,515	4	4679,092	5	4524,139	7
4932,263	7	4675,312	4	4521,488	3
4928,795	8	4674,958	3	4521,417	3
4928,637	9	4674,530	4	4519,959	3
4928,365	4	4673,097	3	4519,122	3
4925,233	3	4671,305	7	4517,428	3
4924,018	3	4670,652	4	4515,562	3
4919,127	5	4669,278	3	4514,313	3
4908,782	5	4667,791	4	4513,828	3
4908,063	3	4667,083	3	4511,690	4
4906,336	4	4665,585	6	4510,904	5
4891,269	4	4662,811	8	4505,631	6
4878,128	3	4661,402	8	4501,960	6
4875,964	6	4660,395	5	4498,523	4
4874,289	4	4654,056	3	4498,108	10
4873,010	8	4652,999	6	4497,577	3
4869,451	4	4645,344	3	4497,101	3
4867,029	3	4634,622	6	4493,688	5
4866,311	1	4634,032	10	4490,451	9
4863,643	1	4631,849	9	4487,813	4
4861,738	3	4631,460	8	4486,084	8
4860,806	0	4627,986	10	4478,987	3
4860,108	2	4625,584	3	4477,071	5
4858,754	3	4625,311	5	4474,261	6
4856,553	9	4620,756	5	4471,961	3
4849,303	9	4618,304	5	4467,145	8
4842,385	3	4617,528	10	4460,965	10
4838,242	6	4614,588	3	4458,859	3
4832,792	5	4613,116	4	4458,732	3
4831,563	3	4607,403	4	4457,030	3
4824,568	4	4598,493	3	4456,851	4
4822,943	8	4597,206	4	4456,665	3
4817,514	3	4588,678	3	4453,141	3
4813,601	7	4584,499	5	4452,767	3
4801,993	2	4582,592	10	4450,816	5
4797,050	4	4581,538	4	4449,913	6
4793,913	5	4579,994	10	4447,932	6
4789,418	5	4579,454	5	4447,553	9
4780,957	5	4578,015	6	4445,246	7
4777,454	6	4575,878	8	4444,213	3
4770,610	3	4572,710	8	4438,206	3
4763,844	9	4568,130	10	4423,248	3
4756,948	5	4565,544	3	4420,304	4
4743,383	5	4563,728	3	4419,494	4
4742,780	6	4562,222	4	4417,342	5
4742,109	3	4558,606	4	4414,994	3
4740,985	5	4558,318	4	4414,218	4
4724,820	4	4557,393	5	4412,253	8
4723,032	10	4557,125	4	4410,618	3
4721,542	3	4554,158	10	4400,830	3
4719,043	10	4551,724	3	4400,745	4
4713,930	6	4550,983	5	4390,900	4
4711,067	3	4549,896	3	4389,084	3
4709,536	10	4547,967	5	4379,403	4

λ, Å	I	λ, Å	I	λ, Å	I
4354,540	3	4097,433	3	3888,988	3
4343,600	1	4095,533	3	3887,836	2
4341,794	0	4087,755	9	3887,388	1
4339,817	1	4085,243	3	3886,218	1
4339,534	1	4082,383	5	3884,144	3
4337,380	0	4078,843	7	3879,526	4
4336,309	1	4074,100	4	3872,354	4
4335,519	3	4072,961	3	3871,594	5
4332,619	3	4071,235	4	3869,931	3
4327,927	5	4069,631	10	3863,202	4
4306,276	3	4066,877	9	3861,496	4
4303,877	3	4065,617	3	3860,711	4
4303,423	4	4063,631	4	3859,877	3
4289,641	3	4062,457	8	3851,266	3
4253,289	3	4059,254	3	3837,609	0
4233,818	3	4048,451	4	3836,443	2
4233,407	4	4043,567	5	3835,860	0
4224,503	5	4035,567	4	3835,381	1
4223,935	5	4031,757	3	3833,489	0
4222,518	5	4028,333	8	3832,415	0
4222,158	5	4027,377	3	3831,101	1
4212,498	10	4026,605	3	3824,940	2
4210,131	5	4024,734	4	3812,744	2
4209,169	4	4018,899	3	3803,031	5
4205,098	9	4005,943	4	3799,032	3
4200,971	3	4005,492	5	3798,816	1
4199,793	5	4002,770	4	3797,908	2
4198,210	3	4002,074	4	3797,518	5
4195,674	6	4000,840	3	3797,128	2
4182,170	8	3998,254	7	3796,595	6
4180,111	5	3993,848	3	3796,062	1
4179,598	4	3993,255	5	3791,403	2
4177,720	5	3991,145	8	3771,500	3
4177,125	10	3987,363	3	3751,879	2
4175,165	6	3985,702	3	3732,108	2
4171,308	10	3982,633	4	3722,093	2
4165,195	3	3976,835	3	3702,112	3
4163,605	3	3974,772	3	3684,313	5
4161,941	5	3963,144	4	3681,963	2
4159,302	5	3962,328	3	3674,398	7
4156,861	4	3944,272	4	3673,621	3
4156,623	4	3924,409	6	3665,907	3
4133,995	3	3902,620	3	3664,132	3
4106,231	3	3890,700	1	3652,460	2
4101,768	—	3889,299	1	3394,838	2
4101,690	—				

GPSR Compliance

The European Union's (EU) General Product Safety Regulation (GPSR) is a set of rules that requires consumer products to be safe and our obligations to ensure this.

If you have any concerns about our products, you can contact us on ProductSafety@springernature.com

In case Publisher is established outside the EU, the EU authorized representative is:

Springer Nature Customer Service Center GmbH
Europaplatz 3
69115 Heidelberg, Germany

Batch number: 10838292

Printed by Printforce, the Netherlands